1993

Verein Deutscher Zementwerke e.V., Düsseldorf

VERFAHRENSTECHNIK DER ZEMENTHERSTELLUNG

PROCESS TECHNOLOGY OF CEMENT MANUFACTURING

TECHNOLOGIE DES PROCESSUS DE LA PRODUCTION DU CIMENT

BAUVERLAG GMBH · WIESBADEN U'

ISBN 3-7625-3101-3

Die Deutsche Bibliothek – CIP-Einheitsaufnahme

Verfahrenstechnik der Zementherstellung = Process technology
of cement manufacturing/VDZ-Kongress '93;
Verein Deutscher Zementwerke e.V., Düsseldorf. –
Wiesbaden; Berlin: Bauverl., 1994.
ISBN 3-7625-3101-3

NE: VDZ-Kongress <4, 1993, Düsseldorf>;
Process technology of cement manufacturing.

Übersetzungen:
Robin B. C. Baker, Dr. Fritz Feige, Josef Hamacher, J. Müller, W. Rothlauf, G. Spriesterbach, Ute Ulrich.
Redaktion: Verein Deutscher Zementwerke e.V.

Das Werk ist urheberrechtlich geschützt. Jede Verwendung auch von Teilen außerhalb des Urheberrechtsgesetzes ist ohne Zustimmung des Verlags unzulässig und strafbar. Das gilt insbesondere für Vervielfältigungen, Übersetzungen, Mikroverfilmungen sowie die Einspeicherung und Verarbeitung in elektronischen Systemen. Autor(en) bzw. Herausgeber, Verlag und Herstellungsbetrieb(e) haben das Werk nach bestem Wissen und mit größtmöglicher Sorgfalt erstellt. Gleichwohl sind sowohl inhaltliche als auch technische Fehler nicht vollständig auszuschließen.

Bauverlag GmbH, Wiesbaden und Berlin

idler, Wiesbaden-Biebrich
ikentscher, Darmstadt

Geleitwort

Fachleute aus über 50 Nationen sind der Einladung des Vereins Deutscher Zementwerke zum 4. Internationalen Kongreß über die Verfahrenstechnik der Zementherstellung nach Düsseldorf gefolgt. Wir freuen uns, daß der diesjährige Kongreß wie seine Vorgänger wieder auf ein so großes Interesse gestoßen ist, das sich nicht nur in einer Teilnehmerzahl von insgesamt über 1100 Personen, sondern insbesondere in einer großen Anzahl von außerordentlich inhaltsreichen Beiträgen zum Kongreß dokumentiert. Traditionell ist der Kongreß auf die Belange der Praxis und auf einen internationalen Austausch von Erfahrungen unter Fachleuten aus der Industrie ausgerichtet. In 6 Fachbereichen soll aufgezeigt werden, welche Fortschritte seit dem letzten VDZ-Kongreß im Jahr 1985 eingetreten sind und welchen technischen Stand die Verfahrenstechnik der Zementherstellung bis heute erreicht hat. Die zusammenfassende Darstellung des Wissensstands, seine kritische Bewertung wie auch ein Ausblick auf künftige Entwicklungen haben 6 General- und – erstmals bei diesem Kongreß – 12 Fachberichter übernommen. In etwa 130 Einzelbeiträgen wird darüber hinaus über neueste Entwicklungen aus Forschung und Anwendung berichtet. Allen, die hierzu beigetragen haben, danken wir herzlich. Allen Teilnehmern gilt unser Willkommensgruß, den wir mit dem Wunsch verbinden, daß dieser Kongreß die internationalen Bindungen weiter fördern und der Wissenstransfer für jeden einzelnen von uns eine konstruktive Unterstützung bei der Bewältigung künftiger Aufgaben bieten möge.

Düsseldorf, im November 1994

Foreword

Experts from over 50 nations have accepted the invitation from the Association of German Cement Works to attend the 4th International Congress in Düsseldorf on the process technology of cement manufacturing. We are very pleased that this year's congress, like the previous ones, has met with such great interest. This is demonstrated not only by the over 1100 participants, but in particular by the large number of exceptionally substantial contributions to the congress. By tradition the congress is orientated towards practical matters and an international interchange of experience between experts in the industry. The intention is to demonstrate the progress which has been made since the last VDZ congress in 1985 in 6 technical sessions, and to show the technical level now reached by the process technology of cement manufacture. 6 general reports and – for the first time in this congress – 12 technical reports give a summary of the current state of knowledge, evaluate it critically and take a look at future developments. In addition to this there are approximately 130 individual contributions which report on the latest developments from research and applications. We give sincere thanks to all those who have contributed. We welcome all the participants to the conference, and linked to this we hope that this congress will continue to promote international ties and that the transfer of knowledge may offer constructive assistance to every one of us in overcoming future problems.

Düsseldorf, November 1994

Préface

Des spécialistes de plus de 50 Nations ont répondu à l'invitation de l'Association des Cimenteries Allemandes, au 4. Congrès International sur la technique de processus de la fabrication du ciment, à Düsseldorf. Nous nous réjouissons, que le congrès de cette année, comme d'ailleurs ses prédécesseurs, suscite à nouveau un intérêt tellement grand, s'exprimant non seulement par un nombre de participants, de plus de 1100 personnes, mais aussi et surtout par un grand nombre de contributions au congrès, toutes d'un contenu extraordinairement riche. Le congrès est, comme le veut la tradition, dédié aux thèmes de la pratique et à un échange international de l'expérience acquise, entre spécialistes de l'industrie. Dans 6 thèmes de la profession doit être mis en évidence, quels progrès sont apparus depuis le dernier congrès VDZ, en 1985 et quel est l'état technique atteint jusqu'à aujourd'hui, dans la technique de processus de la fabrication du ciment. La présentation condensée de l'état des connaissances, son évaluation critique comme aussi les perspectives des développements futurs, ont été prises en charge par 6 rapporteurs généraux et, pour la première fois à ce congrès, 12 rapporteurs spécialiste de la profession. Dans environ 130 contributions individuelles sont, de plus, traités les développements les plus récents dans la recherche et l'application. Un grand merci à tous ceux qui y ont contribué. Notre salut de bienvenue s'adresse à tous les participants aux journées. S'y ajoute le voeu, que ce congrès favorise encore plus les relations internationales et que le transfert des connaissances puisse offrir une aide solide à chaqun de nous, dans l'accomplissement des tâches futures.

Düsseldorf, en Novembre 1994

Dr. rer. nat. K. Kroboth
Vorsitzender des Vereins Deutscher Zementwerke
President of the Association of German Cement Works
Président de l'Association des Cimenteries Allemandes

Inhaltsverzeichnis

Fachbereich 1: **Einfluß der Verfahrenstechnik auf die Herstellung marktorientierter Zemente**
(Auswahl und Aufbereitung von natürlichen Einsatzstoffen, von Sekundär-Roh- und -Brennstoffen und von Zumahlstoffen, Brennen und Kühlen des Zementklinkers, Mahlen bzw. Mischen und Lagern des Zements, Einbindung von Neben- und Spurenbestandteilen, Anforderungen an die Leistungsfähigkeit von Zementen)

J. Albeck, G. Kirchner	Einfluß der Verfahrenstechnik auf die Herstellung marktorientierter Zemente	2
J. G. M. de Jong	Qualitätssicherung	20
P. Liebl, W. Gerger	Nutzen und Grenzen beim Einsatz von Sekundärstoffen	29
S. N. Banerjee	Optimale Brennbedingungen in Zementöfen bei Verwendung von Kohle geringer und schwankender Qualität	40
L. Ernstbrunner	Klinkerqualität und Betriebserfahrungen beim Einsatz von Reststoffen aus der Papieraufbereitung als Sekundärstoff	47
W. Gerger	Erfahrungen mit spezialisierten Kontrolltechniken beim Einsatz von Sekundärstoffen	51
K. Harr, R. Böing	Bautechnische Leistungsfähigkeit von Zementen mit Zumahlstoffen	58
F. Jung	Anforderungen an Zumahlstoffe für die Zementherstellung	64
Ph. Mathew, M. A. Purohit T. N. Tiwari, A. K. Chatterjee	Marktorientierte Technologie – eine Überlebens- und Wachstumsstrategie	68
K. C. Narang, S. Chaturvedi R. B. Sharma	Wirkung von Zusätzen aus aktiven/inerten Sekundärrohstoffen einschließlich Abfallstoffen auf die Leistungsfähigkeit von Zement	71
I. Odler, Y. Chen	Einfluß des Zerkleinerungsverfahrens auf die Eigenschaften des Zementes	78
K. Raina, L. K. Janakiraman	Einflüsse auf die Qualität von Schachtofen-Klinker	83
G. V. Rao, S. P. Ghosh K. Mohan	Auswirkungen von unterschiedlichen prozeßtechnischen Parametern und des Energieeinsatzes auf die Qualität von Klinker und Zement der gleichen Rohstoffbasis	90
V. Rudert	Gedanken zur Einführung von Qualitätssicherungssystemen für Zement-Herstellwerke	97
H. M. von Seebach, H. Tseng	Herstellung von Zementen unter Verwendung von Sondermüll sowie fossilen Brennstoffen und ihre Eignung bei Anwendung im Trinkwasserbereich	101
S. Sprung, W. Rechenberg G. Bachmann	Umweltverträglichkeit von Zement	112
S. Sprung, W. Rechenberg	Schwermetallgehalte im Klinker und im Zement	121
N. Streit	Beurteilung der Rohmehlreaktivität durch zwei eigenständige und sich ergänzende Verfahren	131
H.-M. Sylla	Einfluß der Klinkerzusammensetzung und Klinkerkühlung auf die Zementeigenschaften	135

Fachbereich 2: **Prozeßführung und Informationsmanagement**
(Messen, Steuern, Regeln, Prozeßführung, Erfassen, Leiten und Verarbeiten von Informationen)

R. Säuberli, U. Herzog H. Rosemann	Prozeßführung und Informationsmanagement	148
A. Scheuer	Erfassen und Verarbeiten von Prozeßdaten	164
M. Riedhammer	Laborautomation	175
F.-J. Barton	Prozeßführung der Zementmahlanlage III im Werk Neubeckum der Dyckerhoff AG	186

H. O. Biggs, N. S. Smith M. Hirayama	Überwachung der Sinterzone durch Messung der Alkaliemissions- und -absorptionsspektren in der Gasphase	191
D. Bonvin, A. Bapst R. Yellepeddi	Ein Röntgenfluoreszenz-Spektrometer mit integrierter Diffraktionsmeßeinrichtung und seine Anwendung in der Zementindustrie	195
D. Edelkott	Regelung des Energiestroms zur Drehofensteuerung	202
D. Espig, V. Reinsch B. Bicher, J. Otto	Computersimulation – wertvolle Entscheidungshilfe für die Optimierung der Zementmahlung	206
A. Fröhlich	Quantitative Klinkerphasenanalyse mittels XRD	212
R. Hasler, E. A. Dekkiche	Erfahrungen der „Holderbank" mit automatischer Prozeßführung mittels „HIGH LEVEL CONTROL"	216
R. Hepper	Vorteile der Prozeßführung mit Expertensystem	220
D. Kaminski	Kostengünstige Produktion eines Klinkers hoher Qualität mit Hilfe eines Expertensystems. Das Beispiel von Ciments Français	225
O. P. Mtschedlow-Petrossian A. S. Koschmai	Die Anwendung der Methoden instrumenteller Elektrochemie zur Qualitätssicherung der Zementherstellung	228
C. de Pierpont, V. Werbrouck V. van Breusegem, L. Chen G. Bastin, V. Wertz	Industrielle Anwendung einer multivariablen linearquadratischen Steuerung für Zementmühlen	230
H. Pisters, M. Becke G. Jäger	PYROEXPERT® – Ein neues Regelungssystem zur Vergleichmäßigung des Klinkerbrennprozesses	234
W. Schlüpmann	Visualisierung von Klassieranlagen im Mörtel- und Baustoffbereich	238
W. Sedlmeir	Ein zukunftssicheres Management-Informationssystem für Zementwerke	241
Z. Senyuva, H. Baerentsen S. Tokkesdal Pedersen	Labor-Informations-Management-System (LIMS) in Zementwerkslaboratorien	247
J. Teutenberg	Höhere Leistungsdichten im Zementlabor mit multimodularer Robotertechnik	252
W. Triebel	Entwicklungsschritte zur Automationstechnik des Werkes Bernburg	255
M. Tschudin	On-line Kontrolle des Mischbettaufbaus mittels PGNAA in Ramos Arizpe (Mexiko)	259
K. Utzinger	Projektablauf der Modernisierung der elektrischen Ausrüstungen in der Cementfabrik Rekingen	267
M. Weichinger	Prozeßdatenerfassung als integrierter Bestandteil des Informationssystems der Perlmooser Zementwerke AG	273

Fachbereich 3: **Brenntechnik und Wärmewirtschaft**
(Vorwärmen, Vorcalcinieren, Brennen, Kühlen, Ofensysteme, Stoffkreisläufe, feuerfestes Futter, Ansätze, Brennstoffe, Brennstoffaufbereitung, Verbrennung, Wärmenutzung)

H. S. Erhard, A. Scheuer	Brenntechnik und Wärmewirtschaft	278
S. Buzzi, G. Sassone	Optimierung des Klinkerkühlerbetriebs	296
T. M. Lowes, L. P. Evans	Auswirkung der Brennerbauart und der Betriebsparameter auf Flammenform, Wärmeübergang, NO_x und SO_3-Kreisläufe	305
I. C. Ahuja, M. N. Rao R. Vasudevan	Umbau vom Naßverfahren zum Halbnaßverfahren im Madukkarai-Zementwerk von ACC	315
P. Bartha, H.-J. Klischat	Klassifikation von Magnesiasteinen nach Spezifikation und Gebrauchswert im Zementdrehrohrofen	320
S. Bayer, J. Zajdlík	Ergebnisse und Erfahrungen mit den Ofenlinien von PSP Prerov	326
A. J. de Beus	Energieeinsparungen bei langen Öfen	330

A. Bhattacharya	Eine einfache Abschätzung der Länge von kohlegefeuerten Vorwärmeröfen auf der Basis von Rohmehl-Thermoanalyse und Wärmeübertragung durch Strahlung auf staubförmige Partikel	338
S. M. Cohen	Anwendungsmöglichkeiten der Wirbelschicht bei der Zementherstellung	342
X. J. Deng	Das CFPC-Verfahren	352
M. Deussner	Die 5000 t/d-Ofenanlage mit PYRORAPID®-Kurzdrehofen und PYROCLON® LowNO$_x$-Calcinator bei ACC, Taiwan	358
J. Duda, B. Werynski	Umbau vom Naß- zum Halbnaßverfahren mit Einsatz von Rückständen der Kohlenaufbereitung	363
H. Fleck	Optimierung eines Satellitenkühlers an einem 3000-t/d-Ofen	367
G. Gaussorgues, R. Houis	HADES – Die thermische Überwachung der Klinkerkühlung in Zementwerken	372
G. Goswami, B. N. Mohapatra J. D. Panda	Ansatzbildung beim Brennen eines fluorhaltigen Zementrohmehls	375
P. Green-Andersen	Umbau von F. L. Smidth/Fuller-Klinkerkühlern	381
H. W. Häfner	Die Kohlenstaubbefeuerung von Zementdrehöfen mit Dosierrotorwaagen	384
J. Hartenstein, U. Bongers R. Prange, J. Stradtmann	Verschleißuntersuchungen an Magnesia-Spinell-Steinen für den Zementdrehrohrofen	389
J. Hartenstein, U. Bongers R. Prange, J. Stradtmann	Verbesserung der Temperaturwechselbeständigkeit von basischen, gebrannten Steinen für den Zementdrehrohrofen durch Zirkondioxid	394
H. Heinrici, A. Burk	Die Kohlenstaubdosierung nach dem Coriolis-Meßprinzip – ein neues Verfahren zur Brennstoffeinsparung	399
G. Kästingschäfer	Technische Neuerungen im Rekuperationsbereich der Klinkerkühler	404
C. G. Manias, G. J. Nathan	Gyro-Therm-Low-NO$_x$-Brenner	408
K. Menzel	Anpassung des Calcinierprozesses an Brennstoffreaktivität und Emissionserfordernisse mit PREPOL® MSC	413
U. Mrowald, R. Hartmann	Auswirkung der Gestaltung der Zyklone für Zyklonvorwärmer auf Abgastemperatur und Druckverlust	416
K. C. Narang, S. Chaturvedi	Einfluß der Verbrennung von Abfallbrennstoffen auf die Prozeßparameter und die feuerfeste Ausmauerung in Zement-Drehrohröfen	420
H. Reckziegel	Betriebserfahrungen mit IKN-Kühlern im Werk Göllheim der Dyckerhoff AG	428
A. Rodrigues, W. S. Oliveira	Magnesia-Spinell-Steine – eine praxisbezogene Betrachtung	431
R. Schneider	PYROSTEP® Rostkühler – Weiterentwicklungen in der Rostkühlertechnik	439
G. Seidel	Zur Einflußnahme der Mahlfeinheit bei Stein- und Braunkohlen auf die Flammenausbildung in Zementdrehöfen	443
E. Steffen, G. Koeberer H. Meyer	Innovation bei Schubrostkühlern mit großen Durchsatzleistungen	449
E. Steinbiß, C. Bauer W. Breidenstein	Entwicklungsstand des PYRO-JET®-Brenners	454
K. von Wedel	Betriebsergebnisse von Pendelrostkühlern mit horizontaler Anströmung des Klinkers	462

Fachbereich 4: **Maßnahmen zum Schutz der Umwelt**
(Emissionen und Immissionen von Staub, Gas, Lärm und Erschütterungen, verfahrenstechnische Minderungsmaßnahmen, Auswahl von Einsatzstoffen und Emissionsprognose, Gasreinigung, Lärmschutz, Rekultivierung, Renaturierung)

J. Kirsch	Maßnahmen zum Schutz der Umwelt	468
H. Xeller	Messen und Verarbeiten von umweltrelevanten Daten	484
K. Kuhlmann	Produktökobilanz – Verfahren für eine ganzheitliche Beurteilung, beispielsweise von Bauprodukten aus Zement und Beton	500

J. Blumbach, L.-P. Nethe	Emissionsminderung von Dioxinen und Furanen	508
K.-H. Boes	Maßnahmen zur Minderung der SO_2-Emission beim Klinkerbrennen im Werk Höver der Nordcement AG	514
R. Bolwerk	Behördliche Umweltschutzanforderungen beim Einsatz alternativer Brenn- und industrieller Reststoffe im Zementdrehrohrofen	519
W. Boßmann	Abbau mit dem Hydraulikbagger als Alternative zur Gewinnungssprengung	525
H. Braig, B. Kirchartz	Maßnahmen zum optimierten Betrieb von Elektrofiltern	529
S. Brandenfels, A. Koszinski V. Rahn	Das Rekultivierungsprogramm der Rüdersdorfer Zement GmbH	533
D. Eckoldt, H. V. Fuchs	Entwicklung von Schalldämpfern für die Zementindustrie	537
H. Fleck	Ersatz eines 3feldrigen Elektrofilters nach 20jährigem Betrieb	542
W. Heine	Staubfreie Klinkerentladung von Schiffen – eine neue Problemlösung	546
Ø. Høidalen, A. Thomassen T. Syverud	NO_x-Reduzierung im norwegischen Zementwerk Brevik – Versuche mit gestufter Brennstoffzufuhr zum Calcinator	550
Ø. Høidalen	Norwegische Umweltanforderungen an den Einsatz von flüssigen Ersatz- brennstoffen in Drehrohröfen	555
S. A. Khadilkar, N. A. Krishnan D. Ghosh, C. H. Page A. K. Chatterjee	Eigenschaften der Ofenstäube aus verschiedenen mit hoch aschehaltigem Kohlenstaub befeuerten Öfen	561
B. Kirchartz	Abscheidung von Spurenelementen beim Klinkerbrennprozeß	570
P. Küllertz, M. Schneider	Der Einfluß von Gewinnungssprengungen auf die Erschütterungsimmission	574
J. Lawton	Leitlinien der europäischen Zementindustrie zum Schutz der Umwelt	578
E. Mikols	Staatliche Umweltschutzregelungen in den USA für die Verbrennung von Sondermüll in Portlandzementöfen	587
J.-J. Østergaard	Optimale Ofenführung zur Einhaltung von Umweltschutzvorschriften	591
B. de Quervain	Weitergehende Rauchgasreinigung für den Einsatz von Trockenklärschlamm als Alternativ-Brennstoff bei der „Holderbank" Zement und Beton, Siggenthal	596
J. A. Riekert, A.D. van Doornum J. M. Gaylard	Programm zur Umweltverbesserung bei einem Zementwerk im Stadtgebiet von Pretoria	601
D. Rose	Technik der Emissionsminderung für Drehofenabgase	606
W. Ruhland, H. Hoppe	Die betriebliche Umsetzung von NO_x-Minderungsmaßnahmen (Bericht des VDZ-Arbeitskreises „NO_x-Minderung")	610
M. Schneider	Die Immissionsverteilung staubförmiger Emissionen aus Zementofenanlagen	614
M. Schneider, P. Küllertz	Gezielte Lärmminderungsprogramme für Zementwerke	617
P. Scur	Entwicklung der Immissionen in der Umgebung der Zementwerke in den neuen Bundesländern am Beispiel Rüdersdorf	621
K. Thomsen	NO_x-Minderung – ein systematisches Konzept	624

Fachbereich 5: Zerkleinerungstechnik und Energiewirtschaft
(Brechen, Mahlen, Trocknen, Klassieren, Kühlen, Energiebedarf, Energie- ausnutzung, Energiemanagement)

H.-G. Ellerbrock, H. Mathiak	Zerkleinerungstechnik und Energiewirtschaft	630
H.-G. Ellerbrock	Gutbett-Walzenmühlen	648
E. Onuma, M. Ito	Sichter in Mahlkreisläufen	660
B. Allenberg	Energiesparpotentiale durch optimierte Regelung der Schüttgutflüsse an Kugelmühlen und Gutbett-Walzenmühlen	673
R. Anantharaman	Energieeinsparung bei Mahlsystemen	678

R. Atzl	Umbau einer Rohmehl-Mahlanlage mit Kugelmühle und Sichter zu einer Kombi-Mahlanlage mit Gutbett-Walzenmühle – Vergleich des Energiebedarfs mit dem einer Walzenschüsselmühle	684
H. Brundiek	Die Loesche-Mühle für die Zerkleinerung von Zementklinker und Zumahlstoffen in der Praxis	689
S. Buzzi	BHG-Mühlen – ein neues Mahlverfahren	697
J. Folsberg, O. S. Rasmussen	Sichtersysteme für Fertigmahlung und Halbfertigmahlung auf der Rollenpresse	701
H. Herchenbach	Einfluß des Antriebes auf die Wirtschaftlichkeit einer Hochdruck-Rollenpresse	706
K. Kumar, A. K. Mullick, J. P. Saxena, A. Pahuja	Analyse des Energieeinsatzes in der indischen Zementindustrie – ein Überblick	710
L. Lohnherr	Steigerung der Mahleffizienz durch verbesserte Kinematik in der Rollenmühle	713
N. Patzelt	Verschleißschutzalternativen für die Beanspruchungsflächen von Gutbett-Walzenmühlen	717
N. Patzelt	Beispiele erfolgreicher Integration der Hüttensandaufbereitung	722
K. Ravi Kumar, Y. V. Satyamurthy, M. A. Purohit, T. N. Tiwari	Verbesserung einer Rohmehlmahlanlage im ACC-Chanda-Werk	726
B. Schiller, H.-G. Ellerbrock	Mahlbarkeit von Zement-Bestandteilen und Energiebedarf von Zementmühlen	730
T. Schmitz	Vermeidung von Vibrationen beim Betrieb der Gutbett-Walzenmühle	736
S. Strasser	Die Fertigmahlung von Rohmaterial unter Druck	740
M. S. Sumner	Moderne Mahlhilfsmittel-Technologie	744
F. Thomart	„AIRFEEL" – eine neue Trennwand zur Regelung der Mahlgutfüllung in Kugelmühlen	747

Fachbereich 6: Betriebstechnik

(Wasser-, Druckluft- und Elektrizitätsversorgung, Gewinnen, Homogenisieren und Vergleichmäßigen, Mischen, Fördern, Lagern, Verpacken)

F. Guilmin	Betriebstechnik	752
J. Patzke, K.-A. Krause	Geplante Instandhaltung	760
U. Jönsson	Mensch im Betrieb	768
G. Adam	Verfahren für den sicheren Ausbruch feuerfester Baustoffe aus Drehrohröfen	773
B. Allenberg	Rationalisierte Instandhaltung von kontinuierlichen Dosiereinrichtungen für Schüttgut unter Berücksichtigung der Qualitätssicherung	776
W. Baumgartner	Computerunterstützte Rohstoffplanung	782
R. Festge	Heutiger Stand der vollautomatischen Verpackungstechnik von Baustoffen aller Art	787
H. Fleck	Einschichtbetrieb eines Steinbruchs mit einem Durchsatz von 1,8 Mio. t/Jahr	793
W. Heine	Transport- und Lagertechnik für Zementklinker – Vorstellung der großen Klinkerlager in Belgien und Thailand	798
R. Hoffmann, H. Steinberg	Einsatz eines Schaufelradbaggers in Kalksteinbrüchen	804
L. P. Maltby, G. G. Enstad, E. Stoltenberg-Hansson	Uniaxial-Tester – ein neues Gerät zur Beurteilung der Fließeigenschaften von Zement	811
B. K. Shrikhande, T. N. Tiwari, M. A. Purohit	Konzept zur Zementwerksmodernisierung und -optimierung	815
U. Spielhagen, H. Karasch	Ein rechnergestütztes Planungs- und Informationssystem für Instandhaltung, Materialwirtschaft und Dokumentation	818
J. Spiess	Strömungsverhältnisse in Turbinenpackern	824

E. Stoltenberg-Hansson **G. Enstad**	Mischen und Dosieren von Klinker – Segregationseffekt und Gegenmaßnahmen .	**830**
B. Thier	Mischtechnik im Werk II der Anneliese Zementwerke AG, Ennigerloh	**835**
W. Wahl	Verbesserung des Verschleißschutzes der Mahlwalzen von Walzenschüsselmühlen	**842**

Contents

Subject 1: Influence of the process technology on the production of market-orientated cements

(Selection and preparation of natural constituents, secondary raw materials and fuels, and of interground additions, burning and cooling of the cement clinker, grinding, blending, and storage of the cement, incorporation of minor and trace constituents, requirements on the performance of cements)

Authors	Title	Page
J. Albeck, G. Kirchner	Influence of process technology on the production of market-orientated cements	2
J. G. M. de Jong	Quality assurance	20
P. Liebl, W. Gerger	Benefits and limitations when using secondary materials	29
S. N. Banerjee	Optimum burning conditions in kilns using inferior and variable quality coal	40
L. Ernstbrunner	Clinker quality and operational experience using residues from paper processing as a secondary material	47
W. Gerger	Experience with specialized control techniques when using secondary materials	51
K. Harr, R. Böing	Structural capabilities of cements with interground additives	58
F. Jung	Specifications for interground additives used in cement production	64
Ph. Mathew, M. A. Purohit, T. N. Tiwari, A. K. Chatterjee	Market driven technology – a strategy for survival and growth	68
K. C. Narang, S. Chaturvedi, R. B. Sharma	Mechanism of addition of active/inert secondary raw materials including wastes on performance of cements	71
I. Odler, Y. Chen	Influence of the method of comminution on the properties of cement	78
K. Raina, L. K. Janakiraman	Parameters influencing vertical shaft kiln clinker quality	83
G. V. Rao, S. P. Ghosh, K. Mohan	Impact of different process-operational parameters and energy inputs on the quality of commercial clinker and cement from the same raw materials	90
V. Rudert	Thoughts on the introduction of quality assurance systems for cement manufacturing plants	97
H. M. von Seebach, H. Tseng	The suitability of cements manufactured with hazardous waste derived fuels and with fossil fuels for drinking water applications	101
S. Sprung, W. Rechenberg, G. Bachmann	Environmental compatibility of cement	112
S. Sprung, W. Rechenberg	Levels of heavy metals in clinker and cement	121
N. Streit	Assessment of raw meal reactivity by two independent and complementary methods	131
H.-M. Sylla	Influence of clinker composition and clinker cooling on cement properties	135

Subject 2: Process control and information management

(Measuring, control, automation, process control, acquisition, transmission and processing of information)

Authors	Title	Page
R. Säuberli, U. Herzog, H. Rosemann	Process control and information management	148
A. Scheuer	Acquisition and processing of process data	164
M. Riedhammer	Laboratory automation	175
F.-J. Barton	Process control of cement mill III in Dyckerhoff AG's Neubeckum works	186
H. O. Biggs, N. S. Smith, M. Hirayama	Kiln sintering zone monitor using alkali spectrum emission and absorption	191
D. Bonvin, A. Bapst, R. Yellepeddi	An integrated diffraction XRF spectrometer and its applications in the cement industry	195
D. Edelkott	Controlling the energy flow to rotary kiln firing systems	202

D. Espig, V. Reinsch B. Bicher, J. Otto	Computer simulation – valuable decision aid for optimizing cement grinding	206
A. Fröhlich	Quantitative clinker phase analysis using XRD	212
R. Hasler, E. A. Dekkiche	The experience of "Holderbank" with automatic process control using "HIGH LEVEL CONTROL"	216
R. Hepper	Advantages of process control with expert systems	220
D. Kaminski	Cost-effective production of high-grade clinker using expert systems. An example from Ciments Français	225
O. P. Mtschedlow-Petrossian A. S. Koschmai	The application of instrumental electrochemical methods for quality assurance in cement production	228
C. de Pierpont, V. Werbrouck V. van Breusegem, L. Chen G. Bastin, V. Wertz	An industrial application of multivariable linear quadratic control to a cement mill	230
H. Pisters, M. Becke G. Jäger	PYROEXPERT® – a new control system for stabilizing the clinker burning process	234
W. Schlüpmann	Visual display of classifying plants in the mortar and building materials sector	238
W. Sedlmeir	A management information system with a secure future for cement works	241
Z. Senyuva, H. Baerentsen S. Tokkesdal Pedersen	LIMS in cement works laboratories	247
J. Teutenberg	Higher performance density in the cement laboratory with multi-modular robot technology	252
W. Triebel	Development steps in the automation system at the Bernburg works	255
M. Tschudin	On-line control of the blending bed composition using PGNAA in Ramos Arizpe (Mexico)	259
K. Utzinger	Progress of the project during modernization of the electrical equipment at the Rekingen cement factory	267
M. Weichinger	Process data acquisition as an integral component of Perlmooser Zementwerke AG's information system	273

Subject 3:	**Burning technology and thermal economy** (Preheating, precalcining, burning, cooling, kiln systems, circulation of volatile substances, refractory linings, coatings, fuels, treatment of fuels, combustion, waste heat utilization)	
H. S. Erhard, A. Scheuer	Burning technology and thermal economy	278
S. Buzzi, G. Sassone	Optimization of clinker cooler operation	296
T. M. Lowes, L. P. Evans	The effect of burner design and operating parameters on flame shape, heat transfer, NO_x and SO_3 cycles	305
I. C. Ahuja, M. N. Rao R. Vasudevan	Wet to semi-wet process conversion at ACC's Madukkarai Cement Works	315
P. Bartha, H.-J. Klischat	Classification of magnesia bricks in rotary cement kilns according to specification and serviceability	320
S. Bayer, J. Zajdlík	Results and experience with the kiln lines from PSP Prerov	326
A. J. de Beus	Energy savings in long kilns	330
A. Bhattacharya	A simple approach to length estimation of coal fired preheater kilns based on raw mix thermal analysis and radiative dust heat transfer	338
S. M. Cohen	Fluid-bed cement clinker applications	342
X. J. Deng	The CFPC Process	352

M. Deussner	5000 t/d kiln plant with PYRORAPID® short rotary kiln and PYROCLON® LowNO$_x$ calciner at ACC, Taiwan	358
J. Duda, B. Werynski	Changeover from the wet to the semi-wet process using residues from coal processing	363
H. Fleck	Optimizing a planetary cooler on a 3000 t/d kiln	367
G. Gaussorgues, R. Houis	HADES – Thermal monitoring of cement plant clinker coolers	372
G. Goswami, B. N. Mohapatra, J. D. Panda	Coating formation during burning of a cement raw mix with inherent fluorine	375
P. Green-Andersen	F. L. Smidth/Fuller clinker cooler retrofit	381
H. W. Häfner	Pulverized coal firing system for rotary cement kilns using rotor weighfeeders	384
J. Hartenstein, U. Bongers, R. Prange, J. Stradtmann	Investigations into the wear on magnesia-spinel bricks for rotary cement kilns	389
J. Hartenstein, U. Bongers, R. Prange, J. Stradtmann	Use of zirconium dioxide to improve the thermal shock resistance of basic burnt bricks for rotary cement kilns	394
H. Heinrici, A. Burk	Pulverized coal metering using the Coriolis measuring principle – a new method of saving fuel	399
G. Kästingschäfer	Technical innovations in recuperation zones of clinker coolers	404
C. G. Manias, G. J. Nathan	Gyro Therm Low NO$_x$ Burner	408
K. Menzel	Using the PREPOL® MSC to adapt the calcining process to fuel reactivity and emission requirements	413
U. Mrowald, R. Hartmann	Effect of the cyclone configuration in cyclone preheaters on the exhaust gas temperature and pressure drop	416
K. C. Narang, S. Chaturvedi	Influence of burning waste-derived fuels on the process parameters and refractory lining in cement rotary kilns	420
H. Reckziegel	Operating experience with IKN coolers in Dyckerhoff AG's Göllheim works	428
A. Rodrigues, W. S. Oliveira	Magnesia spinel bricks – a practical view	431
R. Schneider	PYROSTEP® grate cooler – further developments in grate cooler technology	439
G. Seidel	Influence of the fineness of coal and lignite on the flame formation in rotary cement kilns	443
E. Steffen, G. Koeberer, H. Meyer	Innovations with reciprocating grate coolers with large throughput capacities	449
E. Steinbiß, C. Bauer, W. Breidenstein	Current state of development of the PYRO-JET® burner	454
K. von Wedel	Operational experience with pendulum clinker coolers with horizontal air jets	462

Subject 4: Measures to protect the environment

(Emission and immission of dust, gas, noise and vibrations, process-technological measures to reduce annoyances from industrial processes, selection of constituents and emission prognosis, gas cleaning, noise protection, recultivation, renaturalization)

J. Kirsch	Measures to protect the environment	468
H. Xeller	Measuring and processing environmentally relevant data	484
K. Kuhlmann	Product ecobalance – a method for complete assessment of, for example, building products made of cement and concrete	500
J. Blumbach, L.-P. Nethe	Reduction in the emission of dioxins and furans	508
K.-H. Boes	Measures to reduce the SO$_2$ emission during clinker burning at Nordcement AG's Höver works	514
R. Bolwerk	Official environmental protection requirements when using alternative fuels and industrial residues in cement rotary tube kilns	519

W. Boßmann	Quarrying with a hydraulic excavator as an alternative to primary blasting	525
H. Braig, B. Kirchartz	Measures for optimized operation of electrostatic precipitators	529
S. Brandenfels, A. Koszinski, V. Rahn	The recultivation programme at Rüdersdorfer Zement GmbH	533
D. Eckoldt, H. V. Fuchs	Development of sound absorbers for the cement industry	537
H. Fleck	Replacement of a 3-compartment electrostatic precipitator after 20 years' operation	542
W. Heine	Dust-free clinker unloading from ships – a new solution to the problem	546
Ø. Høidalen, A. Thomassen, T. Syverud	Reducing NO_x at the Brevik cement works in Norway – Trials with stepped fuel supply to the calciner	550
Ø. Høidalen	Norwegian environmental requirements when using liquid substitute fuels in rotary tube kilns	555
S. A. Khadilkar, N. A. Krishnan, D. Ghosh, C. H. Page, A. K. Chatterjee	Kiln dust characteristics from high ash coal fired kilns of different processes	561
B. Kirchartz	Removal of trace elements in the clinker burning process	570
P. Küllertz, M. Schneider	The influence of primary blasting on vibration immission	574
J. Lawton	The European cement industry's approach to the environment	578
E. Mikols	Federal environmental regulations on portland cement kilns that burn hazardous wastes in the United States	587
J.-J. Østergaard	Optimal kiln control under environmental regulations	591
B. de Quervain	Extensive flue gas cleaning when using dry sewage sludge as an alternative fuel at the "Holderbank" Cement und Beton, Siggenthal	596
J. A. Riekert, A. D. van Doornum, J. M. Gaylard	Environmental improvement programme at a cement plant in the Pretoria urban area	601
D. Rose	Technology of emission reduction for rotary kiln exhaust gases	606
W. Ruhland, H. Hoppe	Practical application of measures for reducing NO_x (Report from the VDZ Working Group "NO_x reduction")	610
M. Schneider	The immission distribution of dust emissions from cement kiln plants	614
M. Schneider, P. Küllertz	Specific noise reduction programme for cement works	617
P. Scur	Development of the immissions in the vicinity of the cement works in the new Federal Länder using Rüdersdorf as an example	621
K. Thomsen	NO_x reduction – a systematic approach	624

Subject 5: Comminution technology and energy management
(Crushing, grinding, drying, classifying, cooling, energy consumption, energy utilization and management)

H.-G. Ellerbrock, H. Mathiak	Comminution technology and energy management	630
H.-G. Ellerbrock	High-pressure grinding rolls	648
E. Onuma, M. Ito	Separators in grinding circuits	660
B. Allenberg	Potential for saving energy through optimized control of bulk material flows at ball mills and high-pressure grinding rolls	673
R. Anantharaman	Energy economics in grinding systems	678
R. Atzl	Conversion of a raw material grinding plant with ball mill and classifier to a combination grinding plant with high-pressure grinding rolls – comparison of the energy consumption with that of a roller grinding mill	684
H. Brundiek	The Loesche mill for comminution of cement clinker and interground additives in practical operation	689

S. Buzzi	BHG mill – a new grinding system	697
J. Folsberg, O. S. Rasmussen	Separator system for roller press finish and semi-finish grinding	701
H. Herchenbach	Influence of the drive on the cost-effectiveness of high-pressure grinding rolls	706
K. Kumar, A. K. Mullick, J. P. Saxena, A. Pahuja	Energy use analysis in the Indian cement industry – an overview	710
L. Lohnherr	Increasing the grinding efficiency through improved kinematics in the roller grinding mill	713
N. Patzelt	Alternative means of protecting the working surfaces of high-pressure rolls against wear	717
N. Patzelt	Examples of successful integration of the granulated blast furnace slag processing system	722
K. Ravi Kumar, Y. V. Satyamurthy, M. A. Purohit, T. N. Tiwari	Raw mill upgrading at ACC-Chanda works	726
B. Schiller, H.-G. Ellerbrock	Grindability of cement components and power consumption of cement mills	730
T. Schmitz	Avoiding vibrations when operating high-pressure grinding rolls	736
S. Strasser	Finish grinding of raw material under pressure	740
M. S. Sumner	Modern grinding additive technology	744
F. Thomart	The „AIRFEEL" – a new diaphragm for controlling the material level in a ball mill	747

Subject 6: Plant engineering

(Supply of water, compressed air, and electricity, winning, homogenizing and equalizing, blending, materials handling, storage, packaging)

F. Guilmin	Plant engineering	752
J. Patzke, K.-A. Krause	Planned maintenance	760
U. Jönsson	People at work	768
G. Adam	Safe methods of breaking refractory materials out of rotary tube kilns	773
B. Allenberg	Rationalized maintenance of continuous metering equipment for bulk materials with due regard to quality assurance	776
W. Baumgartner	Computer-aided raw material planning	782
R. Festge	Present state of fully automatic packaging technology for all types of building materials	787
H. Fleck	Single-shift operation in a quarry with a throughput of 1.8 million t/year	793
W. Heine	Transport and storage technology for cement clinker – Description of the large clinker stores in Belgium and Thailand	798
R. Hoffmann, H. Steinberg	The use of bucket-wheel excavators in limestone quarries	804
L. P. Maltby, G. G. Enstad, E. Stoltenberg-Hansson	The uniaxial tester – a new apparatus for assessment of flow properties of cement	811
B. K. Shrikhande, T. N. Tiwari, M. A. Purohit	Approach to cement plant upgrading and optimization	815
U. Spielhagen, H. Karasch	A computer-aided planning and information system for maintenance, material management and documentation	818
J. Spiess	Flow conditions in rotary impeller packers	824
E. Stoltenberg-Hansson, G. Enstad	Blending and dosage of clinker – resulting effect of segregation and ways to counteract it	830
B. Thier	Blending technology in Works II of Anneliese Zementwerke AG, Ennigerloh	835
W. Wahl	Improving the wear protection on the grinding rollers in roller grinding mills	842

Sommaire

Séance Technique 1: **Influence de la technologie des procédés sur la fabrication de ciments axés sur le marché**

(Sélection et préparation de matières naturelles, de matières secondaires, de combustibles secondaires et d'ajouts, cuisson et refroidissement du clinker de ciment, mouture, mélange et stockage du ciment, incorporation de constituants secondaires et d'éléments de trace, exigences à la performance des ciments)

J. Albeck, G. Kirchner	Influence de la technologie des procédés sur la fabrication de ciments axés sur le marché	2
J. G. M. de Jong	Assurance qualité	20
P. Liebl, W. Gerger	Exploitation et limites de l'utilisation de matières secondaires	29
S. N. Banerjee	Conditions de cuisson optimales à l'intérieur des fours à ciment, en utilisant du charbon de qualité inférieure et variable	40
L. Ernstbrunner	Qualité du clinker et expériences acquises au cours de service lors de l'emploi de matières résiduaires provenant de la préparation du papier comme matière première secondaire	47
W. Gerger	Expériences acquises avec des techniques de contrôle spéciales lors de l'utilisation de matières premières secondaires	51
K. Harr, R. Böing	L'aptitude technique des ciments à additifs	58
F. Jung	Exigences relatives aux additifs destinés à la fabrication du ciment	64
Ph. Mathew, M. A. Purohit T. N. Tiwari, A. K. Chatterjee	Technologie axée sur le marché – stratégie de survie et de croissance	68
K. C. Narang, S. Chaturvedi R. B. Sharma	Influences des additifs de matières premières secondaires actives/intertes y compris déchets sur les performances du ciment	71
I. Odler, Y. Chen	L'influence du procédé de broyage sur les propriétés du ciment	78
K. Raina, L. K. Janakiraman	Paramètres influenant la qualité du clinker provenant d'un four vertical	83
G. V. Rao, S. P. Ghosh K. Mohan	Impact de divers paramètres de process et de l'utilisation d'énergie sur la qualité du clinker de même matière première	90
V. Rudert	Réflexions sur l'introduction de systèmes d'assurance de la qualité dans les usines de fabrication de ciment	97
H. M. von Seebach, H. Tseng	Fabrication de ciments avec utilisation de déchets spéciaux et combustibles fossiles et leur conformité pour emploi dans le domaine de l'eau potable	101
S. Sprung, W. Rechenberg G. Bachmann	Les ciments et l'environnement	112
S. Sprung, W. Rechenberg	Teneurs en métaux lourds du clinker et du ciment	121
N. Streit	L'évaluation de la réactivité de la farine crue moyennant deux procédés indépendants et complémentaires	131
H.-M. Sylla	Influence de la composition et du refroidissement du clinker sur les propriétés du ciment	135

Séance Technique 2: **Gestion du procédé et management d'informations**

(Mesurage, contrôle, réglage, conduite du processus, collection, transmission et traitement d'informations)

R. Säuberli, U. Herzog H. Rosemann	Gestion du procédé et management d'informations	148
A. Scheuer	Saisie et traitement de données de process	164
M. Riedhammer	Automatisation de laboratoire	175
F.-J. Barton	Conduite du process de l'atelier de broyage de ciment III à l'usine Neubeckum de la Dyckerhoff AG	186

Sommaire

H. O. Biggs, N. S. Smith M. Hirayama	Surveillance de la zone de cuisson par relevé de mesures des spectres d'absorption et d'émissions d'alcalis durant la phase gazeuse	191
D. Bonvin, A. Bapst R. Yellepeddi	Spectre de fluorescence et de diffraction des rayons X intégrée et ses possibilités d'application dans l'industrie cimentière	195
D. Edelkott	Régulation de flux d'énergie de la chauffe du four rotatif	202
D. Espig, V. Reinsch B. Bicher, J. Otto	Simulation sur ordinateur — aide à la décision appréciable pour l'optimisation du broyage du ciment	206
A. Fröhlich	Analyse quantitative des phases du clinker par DRX	212
R. Hasler, E. A. Dekkiche	Expérience acquise à la firme „Holderbank" avec une conduite de process automatique au moyen de „HIGH LEVEL CONTROL"	216
R. Hepper	Avantages de la conduite du process au moyen d'un système expert	220
D. Kaminski	Comment produire un clinker de qualité à un côut minimum grâce au Système Expert. L'exemple des Ciments Français	225
O. P. Mtschedlow-Petrossian A. S. Koschmai	Emploi des méthodes d'electrochimie instrumentale pour l'assurance de la qualité dans la fabrication du ciment	228
C. de Pierpont, V. Werbrouck V. van Breusegem, L. Chen G. Bastin, V. Wertz	Application industrielle d'un système de conduite linéaire-quadratique multivariable pour broyeur à ciment	230
H. Pisters, M. Becke G. Jäger	PYROEXPERT® — Un nouveau système de régulation pour la régularité du process de cuisson du clinker	234
W. Schlüpmann	Visualisation d'installations de classement dans le secteur des mortiers et matériaux du bâtiment	238
W. Sedlmeir	Un système de gestion de l'information sûr dans l'avenir pour les cimenteries	241
Z. Senyuva, H. Baerentsen S. Tokkesdal Pedersen	Système Management Information laboratoire (LIMS) dans des laboratoires de cimenterie	247
J. Teutenberg	Plus grande capacité de travail au laboratoire du ciment avec technique de robotisation multimodulaire	252
W. Triebel	Etapes de développement pour la technique d'automation à l'usine Bernburg	255
M. Tschudin	Contrôle en temps réel d'une constitution de lits de mélange au moyen de PGNAA à Ramos Arizpe (Mexique)	259
K. Utzinger	Déroulement du projet de modernisation des équipements électriques à la cimenterie Rekingen	267
M. Weichinger	Acquisition des données du process comme partie intégrante du système d'information de la Perlmooser Zementwerke AG	273

Séance Technique 3: Technique de cuisson et économie de chaleur

(Préchauffage, précalcination, cuisson, refroidissement, types de fours, circulation des matières, revêtements réfractaires, formation de dépôts, combustibles, préparation de combustibles, combustion, utilisation de l'énergie thermique)

H. S. Erhard, A. Scheuer	Technique de cuisson et économie de chaleur	278
S. Buzzi, G. Sassone	Optimisation de l'exploitation de refroidisseur de clinker	296
T. M. Lowes, L. P. Evans	Influence du type de brûleur et des paramètres d'exploitation sur la forme de la flamme, le transfert de chaleur, l'émission de NO_x et les circuits de SO_3	305
I. C. Ahuja, M. N. Rao R. Vasudevan	Modification du procédé humide en procédé semi-humide à la cimenterie de Madukkarai de ACC	315
P. Bartha, H.-J. Klischat	Classification de briques de magnésie selon spécification et utilisation appropriée dans le four à ciment rotatif	320
S. Bayer, J. Zajdlík	Les lignes de four de PSP Prerov, résultats et expérience acquise	326
A. J. de Beus	Economie d'énergie en présence de fours longs	330

A. Bhattacharya	Simple évaluation de la longueur des fours préchauffeurs chauffés au charbon basée sur l'analyse thermique de farine crue et la transmission de chaleur par rayonnement de particules de poussière	338
S. M. Cohen	Possibilités d'utilisation de lit fluidisé pour la fabrication de ciment	342
X. J. Deng	Procédé CFPC	352
M. Deussner	La ligne de four de 5000 t/j avec four rotatif court PYRORAPID® et calcinateur PYROCLON®-LowNO$_x$ chez ACC, Taïwan	358
J. Duda, B. Werynski	Reconversion de voie humide en voie semihumide avec emploi de résidus du traitement du charbon	363
H. Fleck	Optimisation d'un refroidisseur planétaire sur un four de 3000 t/j	367
G. Gaussorgues, R. Houis	HADES – surveillance thermique du refroidissement du clinker dans les cimenteries	372
G. Goswami, B. N. Mohapatra J. D. Panda	Phénomène de croûtage au cours de la cuisson d'une farine crue contenant du fluor	375
P. Green-Andersen	Transformation de refroidisseurs de clinker F. L. Smidth/Fuller	381
H. W. Häfner	La chauffe au poussier de charbon des fours à ciment rotatifs avec des bascules doseuses à rotor	384
J. Hartenstein, U. Bongers R. Prange, J. Stradtmann	Essais d'usure sur briques magnésie-spinelle pour le four à ciment rotatif	389
J. Hartenstein, U. Bongers R. Prange, J. Stradtmann	Amélioration de la résistance aux changements de température de briques basiques, cuites, pour le four à ciment rotatif, à l'aide d'oxyde de zirkone	394
H. Heinrici, A. Burk	Le dosage du poussier de charbon selon le principe de mesure Coriolis – Un nouveau procédé pour économiser le combustible	399
G. Kästingschäfer	Nouveautés techniques dans la zone de récupération des refroidisseurs de clinker	404
C. G. Manias, G. J. Nathan	Brûleur Gyro-Therm-Low-NO$_x$	408
K. Menzel	Adaption du process de calcination à la réactivité du combustible et aux exigences posées par l'émission, avec PREPOL® MSC	413
U. Mrowald, R. Hartmann	Effet de la forme des cyclones de préchauffeurs à cyclones sur la température des gaz d'exhaure et la perte de charge	416
K. C. Narang, S. Chaturvedi	Influence de la combustion de combustibles de déchets sur les paramètres du process et le garnissage réfractaire de fours rotatifs de cimenterie	420
H. Reckziegel	Expérience acquise en exploitation avec les refroidisseurs IKN à l'usine Göllheim de la Dyckerhoff AG	428
A. Rodrigues, W. S. Oliveira	Briques de spinelle-magnésie – aspect pratique	431
R. Schneider	Refroidisseurs à grille PYROSTEP® – Nouveaux développements de la technique des refroidisseurs à grille	439
G. Seidel	Sur l'influence de la finesse de mouture de charbons et lignites sur la formation des flammes dans les fours rotatifs à ciment	443
E. Steffen, G. Koeberer, H. Meyer	Innovation dans les refroidisseurs à grille poussée de grande capacité de débit	449
E. Steinbiß, C. Bauer W. Breidenstein	Etat actuel du développement du brûleur PYRO-JET®	454
K. von Wedel	Résultats obtenus en exploitation avec des refroidisseurs à grille pendulaire avec soufflage horizontal du clinker	462

Séance Technique 4: Mesures de protection de l'environnement

(Emissions et immissions de poussière, gaz, bruit et vibrations, réduction par des mesures de la technologie des procédés, sélection des matières et pronostic d'émission, épuration des gaz, protection contre les bruits, replantation, restauration de la végétation)

J. Kirsch	Mesures de protection de l'environnement	468

Sommaire

H. Xeller	Mesure et traitement des données concernant l'environnement	484
K. Kuhlmann	Bilan écologique du produit – Procédé pour une évaluation globale, par l'exemple de produits de construction à base de ciment et de béton	500
J. Blumbach, L.-P. Nethe	Réduction des émissions de dioxines et de furanes	508
K.-H. Boes	Contremesures pour réduire l'émission SO_2 dans la cuisson du clinker à l'usine Höver de la Nordcement AG	514
R. Bolwerk	Exigences de protection de l'environnement posées par le Ministère Public pour l'utilisation de combustibles alternatifs et de résidus industriels dans le four rotatif à ciment	519
W. Boßmann	Extraction avec pelle hydraulique comme alternative au tir d'abattage	525
H. Braig, B. Kirchartz	Règles pour l'exploitation optimale des électrofiltres	529
S. Brandenfels, A. Koszinski, V. Rahn	Le programme de recultivation de la Rüdersdorfer Zement GmbH	533
D. Eckoldt, H. V. Fuchs	Développement de silencieux pour l'industrie cimentière	537
H. Fleck	Remplacement d'un électrofiltre à trois champs après 20 ans de service	542
W. Heine	Déchargement sans poussière de bâteaux transportant du clinker – une nouvelle solution du problème	546
Ø. Høidalen, A. Thomassen, T. Syverud	Réduction NO_x à la cimenterie norvégienne Brevik – Essais avec alimentation étagée en combustible au calcinateur	550
Ø. Høidalen	Exigences norvégiennes côté environnement pour l'utilisation de combustibles alternatifs liquides dans les fours rotatifs	555
S. A. Khadilkar, N. A. Krishnan, D. Ghosh, C. H. Page, A. K. Chatterjee	Propriétés des poussières de four de différents fours chauffés au poussier de charbon riche en cendres	561
B. Kirchartz	Séparation d'éléments traces dans le processus de cuisson du clinker	570
P. Küllertz, M. Schneider	L'influence des tirs d'abattage sur les immissions d'ébranlement	574
J. Lawton	L'approche de l'industrie cimentière européenne pour la protection de l'environnement	578
E. Mikols	Réglementations d'Etat en matière de protection de l'environnement appliquées aux USA pour l'incinération de déchets spéciaux dans les fours à ciment de Portland	587
J.-J. Østergaard	Conduite de four optimale pour respect des réglementations sur la protection de l'environnement	591
B. de Quervain	Epuration des gaz de fumée encore améliorée pour l'utilisation de boue de décantation séchée comme combustible alternatif chez „Holderbank" Cement und Beton, Siggenthal	596
J. A. Riekert, A. D. van Doornum, J. M. Gaylard	Programme visant à l'amélioration des normes antipollution dans une cimenterie dans la banlieue de Prétoria	601
D. Rose	Technique de réduction des émissions pour gaz d'exhaure des fours rotatifs	606
W. Ruhland, H. Hoppe	La réalisation industrielle d'actions de réduction de NO_x (Contribution du groupe de travail VDZ „Réduction de NO_x")	610
M. Schneider	La dispersion en immission des émissions sous forme de poussière des installations de fours à ciment	614
M. Schneider, P. Küllertz	Programmes ciblés pour la réduction du bruit des cimenteries	617
P. Scur	Evolution des immissions dans le voisinage des cimenteries dans les nouveaux Länder Fédéraux, à l'exemple de Rüdersdorf	621
K. Thomsen	Diminution de NO_x – un concept systématique	624

Séance Technique 5: Technique de broyage et gestion d'énergie
(Broyage, mouture, séchage, classification, refroidissement, consommation d'énergie, utilisation et gestion d'énergie)

H.-G. Ellerbrock, H. Mathiak	Technique de broyage et gestion d'énergie	630
H.-G. Ellerbrock	Broyeurs à cylindres à lit de matière	648
E. Onuma, M. Ito	Séparateurs dans les circuits de broyage	660
B. Allenberg	Potentiels d'économie d'énergie dans la régulation optimale des courants de matière sur broyeurs à boulets et broyeurs à rouleaux	673
R. Anantharaman	Economie d'énergie dans les systèmes de broyage	678
R. Atzl	Modification d'un atelier de broyage du cru avec broyeur à boulets et séparateur en un atelier de broyage tandem avec broyeur à rouleaux — comparaison de la consommation d'énergie avec celle d'un broyeur à galets	684
H. Brundiek	Le broyeur Loesche pour la comminution de clinker à ciment et de matières cobroyées dans la pratique	689
S. Buzzi	Broyeurs BHG — un nouveau procédé de broyage	697
J. Folsberg, O. S. Rasmussen	Systèmes de séparateurs pour broyage de produit fini et de produit semi-fini pour presse à rouleaux	701
H. Herchenbach	Influence de la motorisation sur la rentabilité d'une presse à rouleaux haute pression	706
K. Kumar, A. K. Mullick, J. P. Saxena, A. Pahuja	Analyse de l'utilisation d'énergie dans l'industrie cimentière indienne — un aperçu	710
L. Lohnherr	Augmentation de l'efficacité de broyage par amélioration de la cinématique dans le broyeur à galets	713
N. Patzelt	Moyens de protection des surfaces des broyeurs à rouleaux, soumises à l'usage	717
N. Patzelt	Exemples d'intégration réussie de la préparation du laitier granulé	722
K. Ravi Kumar, Y. V. Satyamurthy, M. A. Purohit, T. N. Tiwari	Amélioration d'un atelier de broyage cru à l'usine ACC Chanda	726
B. Schiller, H.-G. Ellerbrock	Broyabilité des composants du ciment et besoin d'énergie des broyeurs à ciment	730
T. Schmitz	Comment éviter les vibrations dans la marche du broyeur à rouleaux	736
S. Strasser	Le broyage fini sous pression de la matière première	740
M. S. Sumner	Technologie moderne avec mise en oeuvre d'additifs de broyage	744
F. Thomart	L'„AIRFEEL" — une nouvelle cloison destinée au contrôle du niveau de matière dans les broyeurs à boulets	747

Séance Technique 6: Technique d'exploitation
(Alimentation en eau, en air comprimé et en électricité, extraction, homogénéisation et régulation, mélange, manutention, stockage, emballage)

F. Guilmin	Technique d'exploitation	752
J. Patzke, K.-A. Krause	Programmation de la maintenance	760
U. Jönsson	L'homme au travail	768
G. Adam	Méthodes pour le dégarnissage en toute sécurité des matériaux réfractaires dans les fours rotatifs	773
B. Allenberg	Maintenance rationalisée d'équipements de dosage en continu pour matières en vrac, en tenant compte de l'assurance de la qualité	776
W. Baumgartner	Planification des matières premières assistée par ordinateur	782
R. Festge	Etat actuel de la technique d'ensachage totalement automatisée de matériaux de construction de toute sorte	787

H. Fleck	Exploitation non postée d'une carrière avec un débit de 1,8 Mt/an	793
W. Heine	Technique de transport et de stockage du clinker à ciment — Présentation des grands stockages de clinker en Belgique et en Thaïlande	798
R. Hoffmann, H. Steinberg	Utilisation d'un excavateur à roue-pelle dans des carrières de calcaire	804
L. P. Maltby, G. G. Enstad E. Stoltenberg-Hansson	Testeur uniaxial — un appareil nouveau pour tester les propriétés d'écoulement du ciment	811
B. K. Shrikhande, T. N. Tiwari M. A. Purohit	Concept de modernisation et d'optimisation d'une cimenterie	815
U. Spielhagen H. Karasch	Un système de planification et d'information, assisté par ordinateur, pour la maintenance, la gestion du matériel et la documentation	818
J. Spiess	Conditions d'écoulement dans les ensacheuses à turbine	824
E. Stoltenberg-Hansson G. Enstad	Mélange et dosage de clinker — effet de ségrégation et remèdes	830
B. Thier	Technique de mélange à l'usine II de la Anneliese Zementwerke AG, Ennigerloh	835
W. Wahl	Amélioration de la protection contre l'usure des galets de broyeurs à piste	842

Tabla de materias

Tema de ramo 1: **Influjo de la tecnología del proceso sobre la fabricación de los cementos que el mercado demanda**
(Selección y preparación de materias primas y naturales, de materias primas de sustitución, de combustibles de sustitución y de materias apropiadas para molienda, cocción y enfriamiento del clínker, molienda, mezcla y almacenaje de cemento, incorporación de elementos secundarios y elementos de traza, exigencias a los cementos de gran rendimiento)

J. Albeck, G. Kirchner	Influjo de la tecnología del proceso sobre la fabricación de los cementos que el mercado demanda	2
J. G. M. de Jong	Aseguramiento de la calidad	20
P. Liebl, W. Gerger	Utilidad y límites del empleo de materias secundarias	29
S. N. Banerjee	Condiciones de cocción óptimas en los hornos de cemento, utilizando carbón de calidad inferior y variable	40
L. Ernstbrunner	Calidad del clínker y experiencias adquiridas con el empleo de materias residuales procedentes de la preparación del papel como material secundario	47
W. Gerger	Experiencias adquiridas con técnicas de control especiales, utilizando materias primas secundarias	51
K. Harr, R. Böing	La aptitud técnica de los cementos con aditivos	58
F. Jung	Condiciones que tienen que reunir los aditivos destinados a la fabricación del cemento	64
Ph. Mathew, M. A. Purohit T. N. Tiwari, A. K. Chatterjee	Tecnología orientada hacia el mercado – una estrategia para sobrevivir y crecer	68
K. C. Narang, S. Chaturvedi R. B. Sharma	Influjo de los aditivos de materias primas secundarias activas/inertes, incluyendo los desechos, sobre la capacidad del cemento	71
I. Odler, Y. Chen	Influencia del proceso de molienda sobre las propiedades del cemento	78
K. Raina, L. K. Janakiraman	Factores que influyen en la calidad del clínker producido en el horno vertical	83
G. V. Rao, S. P. Ghosh K. Mohan	Repercusiones de diferentes parámetros del proceso y el empleo de energía sobre la calidad del clínker y del cemento fabricados con la misma materia prima	90
V. Rudert	Reflexiones sobre la introducción de sistemas de aseguramiento de la calidad en las fábricas cementeras	97
H. M. von Seebach, H. Tseng	Fabricación de cementos, utilizando basuras especiales y combustibles fósiles, y su aptitud para ser empleados en el ámbito del agua potable	101
S. Sprung, W. Rechenberg G. Bachmann	Los cementos y el medio ambiente	112
S. Sprung, W. Rechenberg	Metales pesados contenidos en el clínker y el cemento	121
N. Streit	Evaluación de la reactividad de la harina cruda mediante dos procedimientos independientes y complementarios	131
H.-M. Sylla	Influjo de la composición y del enfriamiento del clínker sobre las propiedades del cemento	135

Tema de ramo 2: **Control del proceso y gestión de informaciones**
(Medición, control, regulación, control del proceso, comprender, dirigir y elaborar informaciones)

R. Säuberli, U. Herzog H. Rosemann	Control del proceso y gestión de informaciones	148
A. Scheuer	Registro y tratamiento de los datos de proceso	164
M. Riedhammer	Automatización de los laboratorios	175

F.-J. Barton	Control del proceso de la planta de molienda de cemento III de la factoría de Neubeckum, de Dyckerhoff AG	186
H. O. Biggs, N. S. Smith, M. Hirayama	Control de la zona de sinterización midiendo los espectros de emisión y de absorción de álcalis durante la fase gaseosa	191
D. Bonvin, A. Bapst, R. Yellepeddi	Un espectrómetro integrado por difracción y por fluorescencia de rayos X y sus aplicaciones en la industria del cemento	195
D. Edelkott	Regulación del flujo de energía hacia el sistema de combustión del horno rotatorio	202
D. Espig, V. Reinsch, B. Bicher, J. Otto	Simulación por ordenador – una ayuda valiosa para la optimización de la molienda de cemento	206
A. Fröhlich	Análisis cuantitativo de fases de clínker mediante XRD	212
R. Hasler, E. A. Dekkiche	Experiencias adquiridas por „Holderbank" con el control automático de procesos, mediante el empleo del „HIGH LEVEL CONTROL"	216
R. Hepper	Ventajas del control de procesos mediante el sistema experto	220
D. Kaminski	Producción a bajo coste de un clínker de alta calidad, con ayuda del sistema experto. El ejemplo de Ciments Français	225
O. P. Mtschedlow-Petrossian, A. S. Koschmai	La aplicación de los métodos de la electroquímica instrumental para el aseguramiento de la calidad en la fabricación del cemento	228
C. de Pierpont, V. Werbrouck, V. van Breusegem, L. Chen, G. Bastin, V. Wertz	Aplicación industrial de un sistema de mando multivariable, lineal-cuadrático, para molinos de cemento	230
H. Pisters, M. Becke, G. Jäger	PYROEXPERT® – Un nuevo sistema de regulación destinado a estabilizar el proceso de cocción del clínker	234
W. Schlüpmann	La visualización de las instalaciones de clasificación en el campo de los morteros y materiales para la construcción	238
W. Sedlmeir	Un sistema de gestión de la información que ofrece un futuro seguro a las fábricas de cemento	241
Z. Senyuva, H. Baerentsen, S. Tokkesdal Pedersen	Sistema de gestión de informaciones (LIMS) para laboratorios	247
J. Teutenberg	Mayor rendimiento en los laboratorios de cemento mediante la robótica multimodular	252
W. Triebel	Etapas de desarrollo de la técnica de automatización de la fábrica de Bernburg	255
M. Tschudin	Control on-line de la formación del lecho de mezcla por medio del PGNAA en Ramos Arizpe (México)	259
K. Utzinger	Desarrollo del proyecto de modernización de los equipos eléctricos en la fábrica de cemento de Rekingen	267
M. Weichinger	Registro de datos del proceso como parte integrante del sistema de información de Perlmooser Zementwerke AG	273

Tema de ramo 3: Técnica de cocción y energía térmica

(Precalentamiento, precalcinación, cocción, enfriamiento, tipos de hornos, circulación de materiales, revestimientos refractarios, formación de costras, combustibles, preparación de combustibles, combustión, utilización de energía térmica)

H. S. Erhard, A. Scheuer	Técnica de cocción y economía térmica	278
S. Buzzi, G. Sassone	Optimización del servicio de los enfriadores de clínker	296
T. M. Lowes, L. P. Evans	Influjo del tipo de quemador y de los parámetros de servicio sobre la forma de la llama, la transmisión del calor, la emisión de NO_x y los circuitos de SO_3	305

I. C. Ahuja, M. N. Rao R. Vasudevan	Transformación de la vía húmeda en vía semihúmeda en la fábrica de cemento de Madukkarai de ACC	315
P. Bartha, H.-J. Klischat	Clasificación de los ladrillos de magnesia, según su especificación y utilidad en el horno rotatorio de cemento	320
S. Bayer, J. Zajdlík	Las líneas de horno PSP Prerov — resultados obtenidos y experiencias adquiridas	326
A. J. de Beus	Ahorro de energía en los hornos largos	330
A. Bhattacharya	Evaluación sencilla de la longitud de los hornos con precalentador y calefacción de carbón, sobre la base del análisis térmico del crudo y de la transmisión del calor por radiación a las partículas en forma de polvo	338
S. M. Cohen	Posibilidades de aplicación del lecho fluidizado a la fabricación del cemento	342
X. J. Deng	El procedimiento CFPC	352
M. Deussner	La instalación de horno para 5000 t/d con horno corto PYRORAPID® y calcinador PYROCLON®-LowNO$_x$ en ACC, Taiwán	358
J. Duda, B. Werynski	Transformación de la vía húmeda en vía semihúmeda, utilizando residuos del tratamiento de carbón	363
H. Fleck	Optimización de un enfriador de satélites en un horno para 3000 t/d	367
G. Gaussorgues, R. Houis	HADES — Control térmico del enfriamiento de clínker en las fábricas de cemento	372
G. Goswami, B. N. Mohapatra J. D. Panda	Formación de adherencias durante la cocción de un crudo de cemento que contiene fluoruro	375
P. Green-Andersen	Transformación de enfriadores de clínker F. L. Smidth/Fuller	381
H. W. Häfner	La calefacción por carbón pulverizado de los hornos rotatorios de cemento, empleando básculas dosificadoras de rotor	384
J. Hartenstein, U. Bongers R. Prange, J. Stradtmann	Estudios del desgaste en los ladrillos de espinela magnésica para hornos rotatorios de cemento	389
J. Hartenstein, U. Bongers R. Prange, J. Stradtmann	Mejora de la resistencia a los cambios de temperatura de los ladrillos básicos, cocidos, destinados al horno rotatorio de cemento, por medio de dióxido de zirconio	394
H. Heinrici, A. Burk	La dosificación del carbón pulverizado según el principio de medición de Coriolis — un nuevo procedimiento para ahorrar combustible	399
G. Kästingschäfer	Novedades técnicas en la zona de recuperación de los enfriadores de clínker	404
C. G. Manias, G. J. Nathan	Quemador Gyro-Therm-Low-NO$_x$	408
K. Menzel	Adaptación del proceso de calcinación a la reactividad del combustible y a las exigencias relacionadas con las emisiones, mediante el PREPOL® MSC	413
U. Mrowald, R. Hartmann	Repercusiones de la forma de los ciclones de los precalentadores sobre la temperatura de los gases de escape y la pérdida de presión	416
K. C. Narang, S. Chaturvedi	Influjo de la combustión de desechos sobre los parámetros del proceso y el revestimiento refractario de los hornos rotatorios de cemento	420
H. Reckziegel	Experiencias adquiridas durante el servicio con enfriadores IKN en la factoría de Göllheim, de Dyckerhoff AG	428
A. Rodrigues, W. S. Oliveira	Ladrillos de espinela magnésica — aspectos prácticos	431
R. Schneider	Enfriador de parrilla PYROSTEP® — Nuevos desarrollos en la técnica de los enfriadores de parrilla	439
G. Seidel	El influjo de la finura de molido de hulla y de lignito sobre la formación de la llama en los hornos rotatorios de cemento	443
E. Steffen, G. Koeberer H. Meyer	Innovaciones en el campo de los enfriadores de parrilla de vaivén, de gran capacidad de rendimiento	449
E. Steinbiß, C. Bauer W. Breidenstein	Estado de desarrollo del quemador PYRO-JET®	454

K. von Wedel	Resultados obtenidos durante el servicio con enfriadores de parrilla pendulares con chorros de aire horizontales	462

Tema de ramo 4: Medidas de protección del medio ambiente

(Emisiones e inmisiones de polvo, gas, ruido y trepidaciones, medidas de reducción usando la tecnología del proceso, elección de material y prognóstico de emisión, instalaciones de tratamiento de gases, control de niveles ruidosos, replantación, restauración de paisajes)

J. Kirsch	Medidas de protección del medio ambiente	468
H. Xeller	Medición y tratamiento de datos relevantes para el medio ambiente	484
K. Kuhlmann	Balance ecológico del producto – Procedimiento para una evaluación integral, por ejemplo de productos de cemento y hormigón	500
J. Blumbach, L.-P. Nethe	Reducción de emisiones de dioxinas y de furanos	508
K.-H. Boes	Medidas para reducir las emisiones de SO_2 durante la cocción del clínker en la fábrica de Höver, de Nordcement AG	514
R. Bolwerk	Prescripciones oficiales en materia de protección del medio ambiente, en relación con la utilización de combustibles alternativos y de residuos industriales en los hornos rotatorios de cemento	519
W. Boßmann	Extracción por medio de la excavadora hidráulica, como alternativa de la explotación por voladura	525
H. Braig, B. Kirchartz	Medidas para optimizar el servicio con filtros eléctricos	529
S. Brandenfels, A. Koszinski, V. Rahn	El programa de restauración de paisajes de la Rüdersdorfer Zement GmbH	533
D. Eckoldt, H. V. Fuchs	Desarrollo de silenciadores para la industria del cemento	537
H. Fleck	Sustitución de un electrofiltro de 3 compartimientos, tras 20 años de servicio	542
W. Heine	Descarga del clínker de los barcos, sin producir polvo – una nueva solución del problema	546
Ø. Høidalen, A. Thomassen, T. Syverud	Reducción de NO_x en la fábrica de cemento noruega de Brevik – Ensayos efectuados con la alimentación escalonada de combustible al calcinador	550
Ø. Høidalen	Exigencias medioambientales noruegas en cuanto al empleo de combustibles secundarios líquidos en los hornos rotatorios	555
S. A. Khadilkar, N. A. Krishnan, D. Ghosh, C. H. Page, A. K. Chatterjee	Características de los polvos procedentes de hornos con calefacción por combustión de carbón pulverizado, de elevado contenido de cenizas	561
B. Kirchartz	Separación de elementos-traza durante el proceso de cocción del clínker	570
P. Küllertz, M. Schneider	El influjo de la explotación por voladuras sobre las inmisiones debidas a las vibraciones del suelo	574
J. Lawton	La industria del cemento y la protección del medio ambiente	578
E. Mikols	Normas estatales en materia de protección ambiental, establecidas en EE.UU. para la incineración de basuras especiales en los hornos de cemento Portland	587
J.-J. Østergaard	Conducción óptima del horno para cumplir las normas de protección ambiental	591
B. de Quervain	Depuración mejorada de los gases de humo, en relación con el empleo de los lodos secos como combustible alternativo en „Holderbank" Cement und Beton, Siggenthal	596
J.A. Riekert, A.D. van Doornum, J. M. Gaylard	Programa medioambiental de una fábrica de cemento situada en el término municipal de Pretoria	601
D. Rose	Técnica de reducción de las emisiones de los gases de escape de los hornos rotativos	606
W. Ruhland, H. Hoppe	Aplicación práctica de medidas para la reducción de NO_x (Informe del grupo de trabajo VDZ „Reducción NO_x")	610

M. Schneider	La distribución de las inmisiones en forma de polvo, procedente de plantas de hornos de cemento	614
M. Schneider, P. Küllertz	Programas bien enfocados para la reducción de los ruidos producidos en las fábricas de cemento	617
P. Scur	Evolución de las inmisiones a proximidad de las fábricas de cemento en los 5 nuevos Estados federados, citando el ejemplo de Rüdersdorf	621
K. Thomsen	La reducción de NO_x – un concepto sistemático	624

Tema de ramo 5: Técnica de trituración y economía energética
(Trituración, molienda, secado, tamizado, enfriamiento, consumo energético, utilización de energía, management de energía)

H.-G. Ellerbrock, H. Mathiak	Técnica de trituración y economía energética	630
H.-G. Ellerbrock	Molinos de cilindros y lecho de material	648
E. Onuma, M. Ito	Separadores en los circuitos de molienda	660
B. Allenberg	Potenciales de ahorro de energía por medio de la regulación optimizada de los flujos de material a granel en los molinos de bolas y los molinos de cilindros y de lecho de material	673
R. Anantharaman	Ahorro de energía en sistemas de molienda	678
R. Atzl	Transformación de una planta de molienda de crudo, con molino de bolas y separador, en planta de molienda combinada, con molino de cilindros y lecho de material – Comparación del consumo de energía con el de un molino de cubeta y rodillos	684
H. Brundiek	El molino Loesche para la molienda de clínker de cemento y de adiciones, durante el servicio industrial	689
S. Buzzi	Molinos BHG – un nuevo procedimiento de molienda	697
J. Folsberg, O. S. Rasmussen	Sistemas de separadores para la molienda de acabado y de semiacabado mediante la prensa de rodillos	701
H. Herchenbach	Influjo del mando sobre la rentabilidad de una prensa de rodillos, de alta presión	706
K. Kumar, A. K. Mullick J. P. Saxena, A. Pahuja	Análisis del empleo de energía en la industria del cemento india – un resumen	710
L. Lohnherr	Aumento de la eficiencia de molienda mejorando la cinemática dentro del molino de rodillos	713
N. Patzelt	Alternativas de protección de las superficies de los molinos de cilindros y lecho de material, sometidas al desgaste	717
N. Patzelt	Ejemplos de integración bien lograda de la preparación de escoria siderúrgica	722
K. Ravi Kumar, Y. V. Satyamurthy M. A. Purohit, T. N. Tiwari	Mejoras introducidas en una instalación de molienda de crudo de la factoría ACC-Chanda	726
B. Schiller, H.-G. Ellerbrock	Molturabilidad de los componentes del cemento y consumo de energía de los molinos de cemento	730
T. Schmitz	Evitando vibraciones durante la marcha del molino de cilindros y lecho de material	736
S. Strasser	La molienda de acabado de la materia prima bajo presión	740
M. S. Sumner	Tecnología moderna de empleo de coadyuvantes de molienda	744
F. Thomart	„AIRFEEL" – un nuevo tabique destinado al control del nivel del material dentro de los molinos de bolas	747

Tema de ramo 6: Técnica operativa

(Abastecimiento de agua, de aire a presión y de electricidad, explotación, homogeneización, mezcla, transporte, almacenaje, embalaje)

F. Guilmin	Técnica operativa	752
J. Patzke, K.-A. Krause	Planificación del mantenimiento	760
U. Jönsson	El hombre y su trabajo	768
G. Adam	Métodos para la demolición segura de materiales refractarios empleados en los hornos rotativos	773
B. Allenberg	Mantenimiento racionalizado de equipos de dosificación continua de materiales a granel, teniendo en cuenta el aseguramiento de la calidad	776
W. Baumgartner	Planificación de materias primas con ayuda del ordenador	782
R. Festge	Estado actual de la técnica de envasado totalmente automático de todo tipo de materiales para la construcción	787
H. Fleck	Explotación de una cantera en un solo turno, con un caudal de 1,8 millones de t/año	793
W. Heine	Técnica de transporte y de almacenamiento de clínker de cemento – Presentación de los grandes almacenes de clínker existentes en Bélgica y Tailandia	798
R. Hoffmann, H. Steinberg	Empleo de una excavadora de rueda de palas en canteras	804
L. P. Maltby, G. G. Enstad, E. Stoltenberg-Hansson	Aparato de ensayo uniaxial – un nuevo aparato para el ensayo de las propiedades reológicas de los cementos	811
B. K. Shrikhande, T. N. Tiwari, M. A. Purohit	Concepto de modernización y optimización de una fábrica de cemento	815
U. Spielhagen, H. Karasch	Un sistema de planificación y de información computerizado para mantenimiento, gestión de materiales y documentación	818
J. Spiess	Condiciones de flujo dentro de las turboensacadoras	824
E. Stoltenberg-Hansson, G. Enstad	Mezcla y dosificación de clínker – efecto de segregación y remedios	830
B. Thier	Técnica de mezcla en la factoría II de Anneliese Zementwerke AG, de Ennigerloh	835
W. Wahl	Mejorando la protección contra el desgaste de los rodillos de molienda de los molinos de cubeta y rodillos	842

Eröffnungsansprache
Aufgaben und Ziele des 4. Internationalen VDZ-Kongresses
Von **Dr.-Ing. G. Mälzig**, Ennigerloh/Deutschland
Obmann des Tagungsbeirats

Zum vierten Mal hat der Verein Deutscher Zementwerke die Zementproduzenten und die Fachleute aus der Anlagen- und Zulieferindustrie zu einem internationalen Kongreß eingeladen. Nach einer Zeitspanne von 8 Jahren hat es eine Vielzahl von Entwicklungen gegeben, die es rechtfertigen, wie schon bei den vorangegangenen Tagungen 1971, 1977 und 1985 erneut einen weltweiten Erfahrungsaustausch über die Verfahrenstechnik der Zementherstellung zu führen. Ein besonderes Gewicht erhält bei dieser Tagung die Umsetzung der verfahrenstechnischen Erkenntnisse bei der Durchführung umweltschonender Herstellungsprozesse sowie bei der Herstellung marktorientierter und umweltverträglicher Produkte. Hierbei kommt es in erster Linie darauf an, weniger die wissenschaftlichen Erkenntnisse zu vertiefen, als vielmehr die damit in der Praxis gesammelten Erfahrungen und künftige Entwicklungen zu erörtern.

Wenn auch nicht zu übersehen ist, daß der Zement in der Grundzusammensetzung und Grundkonzeption seiner Herstellung schon fast 150 Jahre alt wird, so ist er doch in der Verfeinerung und Spezialisierung, auch für neue Anwendungsgebiete, letzten Endes jung geblieben. Gemessen an der Schnellebigkeit anderer industrieller Güter handelt es sich hier um ein traditionelles und bewährtes Produkt, das allen zuverlässigen Prognosen zufolge auch in der Zukunft ein unverzichtbarer Baustoff für jede Volkswirtschaft sein wird. Gerade in der jüngsten Vergangenheit, unter dem Eindruck großer politischer Veränderungen in Osteuropa und in anderen Teilen der Welt, ist deutlich zu erkennen, daß der Bauindustrie und damit auch der Zementproduktion eine neue enorme Bedeutung zukommt. Der Verein Deutscher Zementwerke ist daher davon überzeugt, daß nach einer Pause von 8 Jahren wiederum ein entsprechend großes Interesse für eine solche Veranstaltung besteht. Die Zahl der Teilnehmer aus über 50 Staaten wie auch die Zahl der schriftlich eingereichten Kurzbeiträge beweist, daß die Annahme richtig war.

Ähnlich wie bei den früheren internationalen Kongressen über die Verfahrenstechnik der Zementherstellung wurde auch in diesem Jahr der gesamte Bereich in 6 Schwerpunktthemen unterteilt, die mit jeweils einem Generalbericht und – das ist neu bei der diesjährigen Veranstaltung – mit je zwei zusätzlichen Fachberichten in den Plenarsitzungen vorgestellt werden. Außerdem wurde die Möglichkeit geschaffen, in Parallelsitzungen die zu jedem der 6 Komplexe ausgewählten Kurzbeiträge vorzutragen. Die insgesamt etwa 130 Beiträge stammen aus über 20 Ländern. Der Tagungsbeirat hatte daher große Mühe, aus den sehr qualifizierten Beiträgen diejenigen auszuwählen, die im Verlauf der Veranstaltung auch mündlich vorgetragen werden können. Hierbei kam es in erster Linie darauf an, ein abgerundetes Bild über den jeweiligen Fachbereich zu vermitteln. Alle Kurzbeiträge werden jedoch zusammen mit den General- und Fachberichten in deutscher und englischer Sprache im Tagungsband veröffentlicht.

Während der Parallelsitzungen wird diesmal auch Gelegenheit zu einer Diskussion der General-, Fach- und Kurzberichte gegeben, um den Erfahrungsaustausch weiter zu intensivieren. Diese Veränderung gegenüber den früheren Kongressen nimmt Rücksicht auf den Wunsch und die Anregung von Teilnehmern früherer Kongresse.

Im Fachbereich 1 werden mit dem Generalbericht „Einfluß der Verfahrenstechnik auf die Herstellung marktorientierter Zemente" vor allem die Anforderungen der Verbraucher an die Leistungsfähigkeit der Zemente, zusätzlich aber auch die Gesichtspunkte Qualitätssicherung sowie Nutzen und Grenzen des Einsatzes von Sekundärstoffen erörtert. Kennzeichnend für die Entwicklung der letzten Jahre war es, daß in den Staaten, in denen ein Zementmarkt mit freiem Wettbewerb existiert, der Zementanwender zunehmend Produkte verlangt, die innerhalb der Grenzen einer Norm ein hohes Qualitätsniveau und eine hohe Gleichmäßigkeit aufweisen, in ihren Eigenschaften auf das jeweilige Anwendungsgebiet ausgerichtet und zudem insgesamt umweltverträglich sind.

Der Fachbereich 2 stellt die rasante Entwicklung vor, die in den letzten Jahren beim Erfassen und Verarbeiten von Prozeßdaten, bei der Prozeßführung, beim Informationsmanagement sowie bei der Laborautomation stattgefunden hat. Hier werden Themen der detaillierten Überwachung des Brennprozesses, aber auch der Prozeßführung in Zementmahlanlagen, der Fortschritte in der Meßtechnik und bei der Modernisierung der elektrischen Ausrüstung einer Zementfabrik behandelt, neue Regelsysteme zur Vergleichmäßigung des Klinkerbrennprozesses vorgestellt sowie Vorteile der Prozeßführung mit den in letzter Zeit vermehrt diskutierten Expertensystemen erörtert.

Im Fachbereich 3 wird mit dem Generalbericht „Brenntechnik und Wärmewirtschaft" der ganze Bereich um den Ofen behandelt, beginnend mit den Vorwärmersystemen bis hin zu den modernen Entwicklungen z. B. bei den Klinkerkühlern, den feuerfesten Zustellungen oder den Brennerbauarten. Von Bedeutung ist sicherlich die Feststellung, daß ein wesentlicher Fortschritt mit dem Einsatz der Vorcalciniertechnik erzielt wurde, daß aber Modernisierungsmaßnahmen bei hohem Kapitaleinsatz nur durch Stillegung veralteter Produktionskapazitäten und Einsatz kostengünstiger Brennstoffe rentabel waren oder künftig sein werden.

Die Maßnahmen zum Schutz der Umwelt, die im Fachbereich 4 behandelt werden, bilden einen weiteren Schwerpunkt der Tagung. Hier hat sich gezeigt, daß in der letzten Zeit neben die Fortentwicklung der Meßtechnik und der Datenverarbeitung sowie von Minderungsmaßnahmen eine übergeordnete Betrachtung der Umweltverträglichkeit des gesamten Prozesses getreten ist, die im Rahmen einer Ökobilanz die Umweltbelastung wie auch die durch technische Maßnahmen oder den Gebrauch des Produkts erzielte Umweltbelastung gegeneinander abwägt. Galten noch vor wenigen Jahren Umweltschutz und Wirtschaftlichkeit als Gegensatz, so wissen wir heute, daß es durchaus möglich und auch nötig ist, Ökologie und Ökonomie in Einklang zu bringen. Ziel der Bemühungen sollte daher sein, für die Produktionsprozesse weniger Energie zu verbrauchen und die Emissionen soweit wie sinnvoll zu mindern. Hierbei zeichnet sich aber inzwischen ab, daß die Erfüllung extremer Anforderungen nicht nur die Kosten erheblich steigert, sondern manche Schutz- und Vorsorgemaßnahmen kaum noch zu merkbaren Verbesserungen z. B. bei der Immission führen und inzwischen häufig mehr Energie benötigen als durch die Verbesserungen technischer Anlagen an anderer Stelle eingespart werden kann.

Der Fachbereich 5 „Zerkleinerungstechnik und Energiewirtschaft" umfaßt primär den Bereich Brechen, Mahlen und Klassieren und natürlich auch wieder schwerpunktmäßig den hierfür nötigen Energiebedarf sowie die Fortschritte bei der Energieausnutzung und im Energiemanagement. Einen breiten Raum in diesem Themenkomplex nimmt neben der Entwicklung von Hochleistungssichtern die Technik der Gutbett-Walzenmühlen ein, die noch beim 3. Kongreß 1985 als eine ganz neue Technik vorgestellt wurde. Inzwischen sind mehrere hundert Anlagen dieser Art in Betrieb, und der Bedarf, hierüber Erfahrungen auszutauschen, ist mit Sicherheit groß. Bemerkenswert ist, daß das Problem der Verfügbarkeit noch nicht zufriedenstellend gelöst wurde und insbesondere das theoretisch diesem Zerkleinerungsverfahren innewohnende Energiesparpotential technisch noch nicht voll ausgenutzt werden konnte, da die Gutbett-Walzenmühle aus Gründen der Produktqualität bisher nur im zweistufigen Zerkleinerungsprozeß oder bei der Halbproduktmahlung eingesetzt werden konnte.

Der Fachbereich 6, mit dem Generalbericht „Betriebstechnik" und den Fachberichten „Geplante Instandhaltung" sowie „Mensch im Betrieb" behandelt in erster Linie den Bereich der Hilfsbetriebe, der Wasser-, Druckluft- und Elektroversorgung, aber auch die Gewinnung, die Homogenisierung, das Fördern und Lagern sowie das Verpacken der Produkte und stellt neue Entwicklungen vor. Wie selbstverständlich stehen auch hier die Themen Energiesparen und Umweltschutz, dazu aber auch die Arbeitssicherheit im Vordergrund des Interesses.

Am letzten Tag des Kongresses soll es den Teilnehmern ermöglicht werden, sich noch eingehender und im Detail durch Besichtigung von Produktionsstätten zu informieren und dabei den Erfahrungsaustausch zu vertiefen. Die Exkursionen führen in Zementwerke der Umgebung von Düsseldorf und ermöglichen den Besuch von Forschungs- und Entwicklungszentren der Anlagenhersteller und der Zuliefererindustrie bis hin zum Besuch von Kraftwerken oder des Forschungsinstitutes der Zementindustrie. Der Tagungsbeirat hofft, daß damit die Anregungen und Informationen der Vortragsveranstaltungen abgerundet werden können und die Tagung insgesamt dazu beiträgt, durch neue bzw. erneuerte Kontakte den Erfahrungsaustausch weltweit zu fördern.

Opening speech

Tasks and objectives of the 4th International VDZ Congress

By **Dr.-Ing. G. Mälzig,** Ennigerloh/Germany
Chairman of the Congress Advisory Board

For the fourth time the Association of German Cement Works has invited cement producers and experts from the plant manufacturers and supporting industries to an international congress. After an interval of 8 years there have been a great many developments which, as with the previous conferences in 1971, 1977 and 1985, justify another world-wide interchange of information on the process technology of cement manufacture. At this conference there is particular emphasis on implementing the process engineering findings while carrying out production processes which conserve the environment and while producing market-orientated and environmentally compatible products. This is essentially less a matter of extending scientific knowledge than of discussing accumulated practical experience and future developments.

Although it should not be forgotten that the basic composition of cement and the basic concept of its production are now almost 150 years old, it has nevertheless been kept young through improvement and specialization, including new applications. Measured against the short lives of other industrial goods this is a traditional and well tried product which, according to all reliable predictions, is going to remain an indispensable construction material for every national economy. In the very recent past, under the influence of great political changes is Eastern Europe and in other parts of the world, it is clear that enormous importance is being attached to the construction industry, and with it also to the cement industry. The Association of German Cement Works is therefore convinced that after an interval of 8 years such an event will again raise a corresponding high level of interest. That this assumption is correct is proved by the number of participants from over 50 countries and by the number of short written contributions submitted.

As in the previous international congresses on the process technology of cement manufacturing the whole field is again subdivided into 6 main topics, each of which is introduced in the plenary sessions by a general report and – this is new for this year's event – two additional special reports.

Opportunities have also been made for selected short contributions on each of the 6 complex topics to be delivered in parallel sessions. The total of approximately 130 contributions come from more than 20 countries. The congress advisory board therefore had great difficulty in selecting from the very competent contributions those which could also be delivered orally during the event. In this case it was primarily a matter of conveying a rounded picture of the particular technical field. However, all the short contributions are being published in German and English together with the general and special reports in the conference proceedings.

This time there will also be an opportunity during the parallel sessions for a discussion on the general, special and short reports to extend the exchange of information still further. This departure from previous congresses is in accordance with the wishes and suggestions of the participants of earlier congresses.

Technical Session 1 with a general report on the "Influence of process technology on the production of market-orientated cements" deals principally with the consumers' demands on the capabilities of cements, but also with the aspects of quality assurance as well as the benefits and limitations of the use of secondary materials. A feature of the development in recent years has been that in countries where there is a cement market with free competition the cement user has increasingly required products which, within the limits of a standard, exhibit high levels of quality and uniformity, have properties aimed at a particular application, and are also environmentally compatible.

Technical Session 2 introduces the meteoric development which has taken place in the last few years in the acquisition and processing of process data, in process control, in information management, and in laboratory information. Topics covering detailed monitoring of the burning process are dealt with, as are process control in cement grinding plants, progress in methods of measurement and in the modernization of electrical equipment in a cement factory. New control systems for smoothing the clinker burning process are presented and there is a discussion of the advantages of process control using the expert systems which have been much in the limelight recently.

The whole field relating to kilns is dealt with in Technical Session 3 with the general report on "Burning technology and thermal economy", starting with the preheater systems and going on to modern developments, e.g. in clinker coolers, refractory linings, as well as fuels and types of burners. Definitely important is the observation that an important step forward was achieved with the use of precalciner technology, but that modernization measures with high capital input were, or in the future will be, profitable only by closing down obsolete production capacity and using low-cost fuels.

The measures taken to protect the environment which are dealt with in Technical Session 4, form another focal point in the conference. Here it has become apparent that an overall examination of the environmental compatibility of the entire process is making its appearance alongside the onward development of measuring technology, data processing and abatement measures. Within the framework of an ecobalance this examination weighs the environmental burden against the relief to the environmental load achieved by technical measures or through the use of the product. It was taken as true only a few years ago that environmental protection and cost-effectiveness were opposites, but we now know that it is entirely possible, and also essential, to reconcile ecology and economy. Efforts should therefore be directed at using less energy for the production process and reducing emissions as far as is sensible. However, it is now becoming apparent that not only does fulfilling extreme demands add greatly to the costs but that some protective and precautionary measures lead to hardly any appreciable improvements, e.g. in the immission, and in the meanwhile often require more energy than can be saved elsewhere by improving industrial plants.

Technical Session 5 "Comminution technology and energy management" is primarily concerned with the field of crushing, grinding and classifying, and naturally also focusses on the associated energy consumption and the progress in energy utilization and in energy management. In this complex topic a great deal of space is taken not only by the development of high-efficiency classifiers but also by the technology of high-pressure grinding rolls, which was introduced as a completely new technology at the 3rd Congress in 1985. Since then several hundred of these systems have come into operation, and there is certainly a great need for exchange of information on the subject. It is noteworthy that the problem of availability has not yet been satisfactorily solved and that, in particular, the energy-saving potential theoretically inherent in this comminution process cannot yet be fully utilized because, for reasons of product quality, it has so far only been possible to use high-pressure grinding rolls in two-stage comminution processes or for grinding intermediate products.

Technical Session 6, with a general report on "Plant engineering" and the special reports "Planned maintenance" and "People at work" deals primarily with the area of auxiliary operations, the supply of water, compressed air and electricity, but also quarrying, blending, conveying and storing, as well as packing the products, and also presents new developments. Here again the topic of energy saving and environmental protection is naturally present, but interest is also focussed on safety at work.

On the last day of the congress it will be possible for participants to gather information even more comprehensively and in detail by visiting production sites and extending the interchange of information. The excursions will go to cement works in the neighbourhood of Düsseldorf and make it possible to visit the research and development centres of the plant manufacturers and of the supporting industries as well as power stations or the Cement Industry's Research Institute. The congress advisory board hopes that this will round off the ideas and information from the lecture sessions, and, as a whole, contribute to the conference by promoting world-wide interchange of information through new or renewed contacts.

Discours d'inauguration
Missions et objectifs du 4ème Congrès International du VDZ
De **Dr.-Ing. G. Mälzig,** Enningerloh/Allemagne
Président de la Commission Consultative du Congrès

Pour la 4ème fois, l'Association des Cimenteries Allemandes a invité les producteurs de ciment et les hommes du métier des industries d'équipements et de fournitures, à un congrès international. Après huit années écoulées, bon nombre de développements ont été réalisés, qui justifient comme déjà aux sessions précédentes de 1971, 1977 et 1985, d'effectuer à nouveau un échange des expériences au niveau mondial, sur la technique de processus de la fabrication du ciment. Un accent particulier est mis, au cours de cette session, sur la transposition des connaissances acquises dans la technique de processus lors de l'application de procédés de fabrication respectant l'environnement, ainsi que lors de la fabrication de produits orientés sur le marché et compatibles avec l'environnement. Dans ce contexte, il s'agit en premier lieu, moins d'approfondir les connaissances scientifiques, mais plutôt de citer l'expérience collectée dans la pratique et les développements à venir.

Bien qu'il ne faut pas oublier, que le ciment dans ses compositions et les conceptions fondamentales de sa fabrication, a presqu'un siècle et demi d'âge, un affinement et une spécialisation, aussi pour des domaines d'utilisation nouveaux, sont finalement restés d'actualité. Mesuré à la courte vie d'autres produits industriels, il s'agit là d'un produit traditionnel et éprouvé qui sera, d'après tous les pronostics fiables, un matériau de construction indispensable aussi dans le futur, quelle que soit l'économie politique. Justement dans le passé le plus récent, sous la pression de grands changements politiques en Europe de l'Est et dans d'autres parties du monde, il apparaît clairement que l'industrie du bâtiment et aussi la production de ciment revêtent une énorme importance nouvelle. L'Association des Cimenteries Allemandes est de ce fait convaincue, qu'il existe, après une pause de 8 années, à nouveau un relativement grand intérêt pour une telle manifestation. Le nombre des participants de plus de 50 Etats, aussi bien que le nombre des contributions courtes proposées par écrit, prouvent que cette supposition était juste.

Un peu comme lors des congrès internationaux antérieurs, sur la technique de processus de la fabrication du ciment, l'ensemble du domaine a été divisé en 6 thèmes majeurs, présentés respectivement par un exposé général et, ceci est nouveau pour la manifestation de cette année, par respectivement deux exposés spécifiques présentés lors des sessions plénières. De plus, a été aménagée la possibilité d'exposer, dans des sessions parallèles, les contributions condensées choisies pour chacun des 6 domaines. Les environ 130 contributions au total proviennent de plus de 20 pays. La Commission consultative du Congrès avait donc grande peine, de sélectionner parmi les contributions très qualifiées celles, qui peuvent aussi être présentées oralement au cours de la manifestation. Dans ce contexte, il était important en premier lieu, de donner un aperçu global de chaque domaine spécifique. Néanmoins, toutes les contributions condensées seront publiées conjointement aux exposées généraux et spécifiques en langue allemande et anglaise dans le volume des compte-rendus des journées.

Pendant les sessions parallèles est aménagée cette fois-ci aussi la possibilité d'une discussion des contributions générales, spécifiques et condensées, afin de rendre l'échange d'expérience acquise encore plus intensif. Ce changement, par rapport aux congrès antérieurs, tient compte des désirs et suggestions des participants à des congrès antérieurs.

Au thème 1 sont abordés, avec l'exposé général „Influence de la technology des procédés sur la fabrication de ciments axés sur le marché", avant tout les besoins des utilisateurs quant à la performance des ciments, mais aussi les points de vue assurance de la qualité ainsi que justification et limites de l'utilisation de matières secondaires. Il est caractéristique, pour l'évolution au cours des dernières années, que dans les pays où existent des marchés de ciment à libre concurrence, l'utilisateur du ciment demande de plus en plus des produits, qui offrent un haut niveau de qualité et une grande régularité dans les limites d'une norme et qui sont orientés, dans leurs propriétés, sur le champ d'utilisation respectif tout en respectant globalement l'environnement.

Le thème 2 présente le développement fulgurant réalisé, au cours des dernières années, dans la saisie et le traitement de données du process, dans la gestion de l'information ainsi que dans l'automatisation des laboratoires. Ici sont traités des sujets de la surveillance détaillée du processus de cuisson, mais aussi de la conduite du process dans les ateliers de mouture, des progrès dans la technique de mesure et dans la modernisation des équipements électriques d'une cimenterie.

Sont présentés aussi des nouveaux systèmes de régulation pour le déroulement régulier de processus de cuisson de clinker, ainsi que des avantages apportées dans la conduite du process par les systèmes expert, de plus en plus évoqués ces derniers temps.

Au thème 3 est traité, avec l'exposé général „Technique de cuisson et économiè de chaleur", tout le secteur de four, en commençant par les systèmes de préchauffage et allant jusqu'aux développements récents, p. ex. sur les refroidisseurs de clinker, les garnissages réfractaires ou les types de construction des brûleurs et les combustibles. Significantif y est certainement le progrès essentiel réalisé avec mise en œuvre de la technique de précalcination, mais aussi, que l'action de modernisation entraînant un gros effort en capitaux ne peut pas être, ou ne sera rentable dans le futur, que par la mise à l'arrêt de capacités de production obsolètes et par l'utilisation de combustibles peu coûteux.

Les actions pour la protection de l'environnement, traitées au thème 4, forment un autre point fort des journées. Ici s'est révélée, qu'outre le développement continu des techniques de mesure et du traitement des données, ainsi que des actions de réduction, une considération à un niveau plus élevé, de la compatibilité avec l'environnement de l'ensemble du processus s'est imposée, jugeant comparativement, dans le cadre d'un écobilan, la charge sur l'environnement comme aussi les actions positives pour l'environnement réalisées par des mesures techniques ou l'emploi du produit. Alors qu'il y a peu d'années, la protection de l'environnement et la rentabilité étaient encore perçues comme contradictoires, nous savons aujourd'hui, qu'il est tout à fait possible et aussi nécessaire, de concilier écologie et économie. Le but des efforts devrait donc être, de consommer moins d'énergie pour les processus de fabrication et de réduire, autant que raisonnable, les émissions. Mais là, si se révèle déjà que la satisfaction d'exigences extrêmes accroît, non seulement de manière considérable les coûts, mais qu'au contraire, certaines actions de protection et de prévention ne peuvent pratiquement plus apporter des améliorations perceptibles, p. ex. lors de l'immission et nécessitent maintenant, fréquemment, plus d'énergie qu'il est possible d'économiser, en autre lieu, par l'amélioration des installations techniques.

Le thème 5 „Technique de broyage et gestion d'énergie" concerne, en premier lieu, le secteur concassage, mouture et classement et naturellement, là aussi comme point fort, ses besoins d'énergie ainsi que les progrès réalisés dans l'utilisation correcte et la gestion de l'énergie. Une large place occupe dans ce thème, outre le développement de séparateurs à haut rendement, la technique des broyeurs à rouleaux et lit de matière, présentée lors du 3ème congrès 1985 encore comme une technique tout à fait nouvelle. Depuis, plusieurs centaines d'installations sont en service, et le besoin d'échanger l'expérience y acquise est certainement très grand. Il faut signaler, que le problème de la disponibilité n'a pas encore été résolu de manière satisfaisante et que, surtout, le potentiel d'économie d'énergie propre à ce procédé de fragmentation n'a pas encore pu être pleinement exploité techniquement, parce que le broyeur à rouleaux n'a pu être utilisé jusqu'à présent, en raison de la qualité du produit, que dans des processus de fragmentation à deux étages ou pour la mouture de produits semi-finis.

Le thème 6, avec l'exposé général „Technique d'exploitation" et les exposés spécifiques „Maintenance planifiée" ainsi que „L'homme au travail", traite en premier lieu le secteur des ateliers annexes, de l'approvisionnement en eau, air comprimé et électricité, mais aussi de l'extraction et homogénéisation, du transport et du stockage, ainsi que de l'emballage des produits, et présente des développements nouveaux. Comme de bien entendu, ici aussi les sujets économie d'energie, protection de l'environnement et, en plus, sécurité du travail, sont au premier plan de l'intérêt.

Le dernier jour du congrès, doit être donnée aux participants la possibilité de s'informer encore plus à fond et en détail, par des visites de sites de production et d'approfondir en même temps l'échange des expériences acquises. Les excursions conduisent dans des cimenteries des environs de Düsseldorf, avec la possibilité de visiter les centres de recherche et de développement de constructeurs d'installations et de l'industrie de fournitures et même, de centrales électriques ou de l'Institut de Recherche de l'Industrie Cimentière. Le conseil du congrès espère, que cela permette de compléter les motivations et informations reçues au cours des sessions d'exposés et que les journées contribuent à encourager, grâce aux contacts nouveaux ou renouvelés, l'échange, au niveau mondial, des expériences acquises.

Fachbereich 1

Einfluß der Verfahrenstechnik auf die Herstellung marktorientierter Zemente

(Auswahl und Aufbereitung von natürlichen Einsatzstoffen, von Sekundär-Roh- und -Brennstoffen und von Zumahlstoffen, Brennen und Kühlen des Zementklinkers, Mahlen bzw. Mischen und Lagern des Zements, Einbindung von Neben- und Spurenbestandteilen, Anforderungen an die Leistungsfähigkeit von Zementen)

Subject 1

Influence of the process technology on the production of market-orientated cements

(Selection and preparation of natural constituents, secondary raw materials and fuels, and of interground additions, burning and cooling of the cement clinker, grinding, blending, and storage of the cement, incorporation of minor and trace constituents, requirements on the performance of cements)

Séance Technique 1

Influence de la technologie des procédés sur la fabrication de ciments axés sur le marché

(Sélection et préparation de matières naturelles, de matières secondaires, de combustibles secondaires et d'ajouts, cuisson et refroidissement du clinker de ciment, mouture, mélange et stockage du ciment, incorporation de constituants secondaires et d'éléments de trace, exigences à la performance des ciments)

Tema de ramo 1

Influjo de la tecnología del proceso sobre la fabricación de los cementos que el mercado demanda

(Selección y preparación de materias primas y naturales, de materias primas de sustitución, de combustibles de sustitución y de materias apropiadas para molienda, cocción y enfriamiento del clínker, molienda, mezcla y almacenaje de cemento, incorporación de elementos secundarios y elementos de traza, exigencias a los cementos de gran rendimiento)

Einfluß der Verfahrenstechnik auf die Herstellung marktorientierter Zemente

Influence of process technology on the production of market-orientated cements

Influence de la technologie des procédés sur la fabrication de ciments axés sur le marché

Influjo de la tecnología del proceso sobre la fabricación de los cementos que el mercado demanda

Von **J. Albeck** und **G. Kirchner**, Ulm/Deutschland

Generalbericht 1 · Zusammenfassung – Die Qualitätsanforderungen des Zementverarbeiters und der Wunsch des Zementherstellers nach einer wirtschaftlichen Produktion von Zementen auf hohem Qualitätsniveau unter Berücksichtigung der Umweltverträglichkeit von Produktion und Produkt sind eng miteinander verflochten und werden auch künftig die gesamte Verfahrenstechnik der Zementherstellung beeinflussen. Die Anforderungen an marktorientierte Zemente mit hoher Leistungsfähigkeit führen zu einer größeren Anzahl von Zementen mit definierten Eigenschaften, die auf das jeweilige Anwendungsgebiet ausgerichtet sind. Diese Anforderungen werden je nach Anwendungsgebiet von reinen Portlandzementen mit unterschiedlicher Mahlfeinheit und Korngrößenverteilung oder von Zementen mit unterschiedlichen Zumahlstoffen und Zumahlstoffanteilen erfüllt. Die Auswahl der für die Herstellung des Portlandzementklinkers einzusetzenden Roh- und Brennstoffe sowie der zu verwendenden Zumahlstoffe muß sich vor diesem Hintergrund an den bautechnischen Eigenschaften des Zements sowie an der Umweltverträglichkeit der Produktion und des Zements orientieren. Beim gemeinsamen Mahlen von Zement mit mehreren Hauptbestandteilen kommt der unterschiedlichen Mahlbarkeit der einzelnen Komponenten eine entscheidende Bedeutung zu. Durch Variation der Zumahlstoffe, der Zumahlstoffanteile und/oder der Zementfeinheit können die gewünschten Zementeigenschaften zielsicher eingestellt werden. Die Herstellung von Zement mit mehreren Hauptbestandteilen durch getrenntes Mahlen der Einzelkomponenten und anschließendes Mischen stellt hohe Anforderungen an die Homogenität und die Gleichmäßigkeit der Vorprodukte. Beim Mischen mehlfeiner Vorprodukte mit festgelegter und auch erzeugter Korngrößenverteilung können die Zementeigenschaften nur noch durch Verändern der prozentualen Anteile der Einzelkomponenten beeinflußt werden. Deshalb müssen die Korngrößenverteilungen der einzelnen Vorprodukte sehr sorgfältig aufeinander abgestimmt werden. Unter diesen Voraussetzungen kann das getrennte Mahlen und Mischen von Zementen mit Zumahlstoffen wirtschaftlich und technologisch sinnvoll sein.*

General report 1 · Summary – The quality demands of those working with cement and the desire of the cement producer to produce cement cost-effectively at a high quality level taking into account the environmental compatibility of the production process and of the product are closely interlinked, and will continue interlinked in the future to influence the entire process technology of cement manufacture. The demand for market-orientated high-performance cements is leading to quite a large number of cements with specific properties which are aimed at a particular area of application. Depending on the application, these demands are met by pure Portland cements with different finenesses and particle size distributions or by cements with different, and differing proportions of interground additives. Against this background the selection of the raw materials and fuels for producing the Portland cement clinker and of the interground additives used must be guided by the structural properties of the cement and by the environmental compatibility of the production process and of the cement. When intergrinding a cement with several main constituents the differing grindabilities of the individual components are of critical importance. The required cement properties can be obtained accurately by varying the additives, the proportion of additives and/or the cement fineness. The production of cement with several main constituents by separate grinding of the individual components followed by mixing places high demands on the homogeneity and uniformity of the intermediate products. When mixing intermediate products of flour fineness with defined and generated particle size distributions the cement properties can only be influenced further by changes in the percentages of the individual components. The particle size distributions of the individual intermediate products must therefore be matched very carefully to one another. Under these conditions it is possible to produce cements of higer quality with a lower expenditure of grinding energy than for intergrinding.*

Rapport général 1 · Résumé – Les impératifs de qualité du consommateur de ciment et le souhait du fabricant visant à une production rentable et d'un haut niveau de qualité, compte tenu des problèmes de compatibilité avec l'environnement de la production et du produit, sont étroitement liés, et influeront, à l'avenir, sur l'ensemble de la

technologie des procédés de fabrication des ciments. Les exigences en matière de ciments très performants, axés sur le marché, conduisent, en effet, à un nombre croissant de ciments ayant des propriétés définies et orientées sur un domaine d'application. Ces exigences sont remplies, selon le domaine d'application, par les ciments de Portland proprement dits de différentes finesse de broyage et répartition granulométrique, ou par des ciments avec additifs de broyage et proportions d'additifs les plus variés. Le choix des combustibles et matières premières utilisés pour la fabrication du ciment de Portland ainsi que des additifs de broyage utilisés, doit être orienté, dans cette perspective, sur les propriétés constructives du ciment et sur la compatibilité de la production et du ciment vis-à-vis de l'environnement. Durant l'opération de broyage commune du ciment avec plusieurs composants principaux, la différente broyabilité des divers composants est un point qui revêt une grande importance. En variant les additifs de broyage, leur proportion et/ou la finesse du ciment, on parvient davantage à régler les propriétés du ciment demandées. La fabrication de ciment à plusieurs composés principaux par broyage séparé des divers composants et ensuite mélange, pose des exigences élevées en matière d'homogénéité et d'uniformité des pré-produits. Lors du mélange de produits de prébroyage à farine fine ayant une répartition granulométrique définie mais aussi produite, on ne peut influer sur les propriétés du ciment qu'en modifiant le pourcentage de ses différents composants. C'est la raison pour laquelle il faut que les répartitions granulométriques des divers pré-produits soient harmonisées entre elles avec beaucoup de soin. Dans ces conditions on parvient à fabriquer des ciments de qualité supérieure avec des dépenses énergétiques inférieures, comparativement, au broyage commun.

Informe general 1 · Resumen – *Las exigencias de calidad del usuario del cemento así como el deseo del fabricante de conseguir una producción rentable de cementos de alta calidad, teniendo en cuenta la inocuidad para el medio ambiente, tanto de la producción como del producto, van íntimamente relacionados e influirán, también en el futuro, en la tecnología de los procesos de fabricación del cemento. Las exigencias respecto de los cementos de alta calidad, que el mercado demanda, conducen a un mayor número de cementos, de características bien definidas, cada uno de los cuales está orientado hacia un campo de aplicación determinado. Estas exigencias se ven cumplidas, según el campo de aplicación de que se trate, por los cementos Portland puros, de diferente finura de molienda y distribución granulométrica, o bien por cementos fabricados con diferentes adiciones y proporciones de las mismas. La elección de las materias primas y combustibles empleados en la fabricación del clinker de cemento Portland así como de las adiciones utilizadas para ello, debe regirse, en estas condiciones, por las características técnicas del cemento y por el impacto medioambiental del proceso de producción y del cemento. Durante la molienda de cementos de varios componentes principales, tiene una importancia decisiva la molturabilidad de los distintos componentes. Variando las adiciones, las proporciones de las mismas y/o la finura del cemento, es posible regular, con precisión, las características deseadas del cemento. La fabricación de cementos de diferentes componentes principales, mediante la molienda separada y la subsiguiente mezcla de los mismos, supone el cumplimiento de elevadas exigencias en cuanto a la homogeneidad y uniformidad de los productos previos. Al mezclar productos previos, en forma de harinas y con distribuciones granulométricas preestablecidas y realizadas, las características de los cementos sólo pueden ser influidas, variando los porcentajes de los componentes individuales. Por esta razón, las distribuciones granulométricas de los diferentes productos previos tienen que ser armonizados cuidadosamente entre sí. En estas condiciones, se pueden fabricar cementos de alta calidad, con un consumo de energía de molienda inferior, en comparación con la molienda conjunta.*

Influjo de la tecnología del proceso sobre la fabricación de los cementos que el mercado demanda

1. Einleitung

Zement ist ein homogenes Massengut. Für die Herstellung von Zement sind neben wirtschaftlichen Gesichtspunkten die Umweltverträglichkeit von Produktion und Produkt sowie in zunehmendem Maße die Sicherung einer marktorientierten Zementqualität von großer Bedeutung. Dadurch spielen Maßnahmen, die der Sicherung der gewünschten Qualität dienen und die eine entsprechende Qualitätssteuerung erfordern, eine immer größere Rolle [1]. Das ist vor allem darauf zurückzuführen, daß der Markt als Qualitätsanforderung nicht nur den Nachweis der Leistungsfähigkeit auf hohem Niveau, sondern vor allem eine Gleichmäßigkeit der Zementeigenschaften fordert. Diese Entwicklung wird sich fortsetzen, weil Zement zunehmend in die Qualitätssicherungssysteme der zementverarbeitenden Industrie eingebunden wird. Die daraus resultierenden Fragestellungen im Hinblick auf die Produktionssteuerung und Produktionsüberwachung bei der Zementherstellung sind in [2] umfassend zusammengestellt.

Die derzeit gültigen nationalen Zementnormen sowie der Entwurf der europäischen Zementnorm ENV 197 legen im wesentlichen nur die Mindestanforderungen an die Zement-

1. Introduction

Cement is a homogeneous bulk material. Not only the economic aspects, but also environmental compatibility of production and product and, to an increasing extent, assurance of a market-orientated grade of cement are of great importance in the production of cement. This means that an ever greater part is being played by measures which are used to assure the required quality and which require appropriate quality control [1]. This is primarily because as a quality requirement the market not only demands evidence of the high level of performance but also, and most importantly, uniformity of the cement properties. This trend will continue because cement is being increasingly incorporated into the quality assurance systems of the industries which use cement. The resulting problems relating to production control and product monitoring during cement manufacture are comprehensively summarized in [2].

Current national cement standards and the draft of the European Cement Standard ENV 197 essentially only establish the minimum requirements for the cement properties with relatively wide ranges of fluctuation [3]. On the other hand the cement purchaser, and hence the market, increasingly

eigenschaften bei relativ großen Schwankungsbreiten fest [3]. Demgegenüber verlangen die Zementabnehmer und damit der Markt zunehmend eine Anpassung der Zementeigenschaften an den jeweiligen Anwendungsbereich. Deshalb ist die Ausrichtung des homogenen Massengutes „Zement" auf marktorientierte Zemente mit hoher Leistungsfähigkeit für den jeweiligen Anwendungsbereich eine wesentliche Voraussetzung für einen dauerhaften Markterfolg [4–7].

In den letzten Jahrzehnten hat es im Bereich der Zementverfahrenstechnik nicht zuletzt aus wirtschaftlichen Gründen eine Reihe von bedeutenden Entwicklungen gegeben, die die Eigenschaften der Zemente und ihre Qualität mehr oder weniger stark beeinflußt haben. Beispielhaft erwähnt seien an dieser Stelle die Weiterentwicklung des Klinkerbrennprozesses hin zu Drehrohröfen mit mehrstufigen Zyklonvorwärmern, die Mahlung von Rohmaterialien und Zement auf Umlaufmahlanlagen mit trennscharfen Sichtern sowie der zunehmende Einsatz von Gutbett-Walzenmühlen auf der Rohstoff- und Zementseite. Aber nicht nur wirtschaftliche Gesichtspunkte sind die treibende Kraft für die Weiterentwicklung verfahrenstechnischer Prozesse. Eine zunehmende Rolle spielen in den letzten Jahren die Umweltverträglichkeit der Zementproduktion und der Zemente in den verschiedenen Anwendungsbereichen.

Die aufgezeigte Verflechtung zwischen den Qualitätsanforderungen des Zementverarbeiters und dem Wunsch des Zementherstellers nach einer wirtschaftlichen Produktion von Zementen auf hohem Qualitätsniveau unter Berücksichtigung der Umweltverträglichkeit von Produktion und Produkt bestimmt in entscheidendem Maße die gesamte Verfahrenstechnik der Zementherstellung. Welche Entwicklungen sich unter diesem Aspekt in den letzten Jahren vollzogen haben und zukünftig zu erwarten sind, ist Inhalt der nachfolgenden Ausführungen.

2. Leistungsfähigkeit marktorientierter Zemente

Bei der Beantwortung der Frage, was Zemente heute leisten müssen, ist ein Blick in die Anwendungsstatistik von Zement nach DIN 1164 sehr hilfreich. **Bild 1** gibt die Mengenaufteilung des in Westdeutschland in den Jahren 1980 bis 1991 hergestellten Zements auf die Anwendungsbereiche Transportbeton, Betonfertigteile/Betonwaren, Sackware und Sonstige wieder. Die Darstellung läßt erkennen,

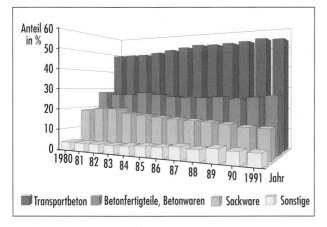

BILD 1: Prozentuale Aufteilung des in Westdeutschland von 1980 bis 1991 hergestellten Zements auf die Anwendungsbereiche Transportbeton, Betonfertigteile/Betonwaren, Sackware und Sonstige (nach Statistiken des Bundesverbandes der Deutschen Zementindustrie)

FIGURE 1: Percentage distribution of the cement manufactured in West Germany from 1980 to 1991 between the applications of ready-mix concrete, precast concrete parts/concrete products, bagged products and others (according to statistics from the Federation of the German Cement Industry)

Transportbeton	= ready-mix concrete
Betonfertigteile, Betonwaren	= precast concrete parts, concrete products
Sackware	= bagged products
Sonstige	= others

require the cement properties to be suited to particular areas of application. It is therefore essential for long-term market success that the homogeneous bulk material "cement" is directed towards high performance market-orientated cements for specific areas of application [4–7].

In recent decades there have been a number of important developments in cement process technology, not least on economic grounds, which have had a greater or lesser effect on the properties and quality of the cements. Examples which could be mentioned here are the continued development of the clinker burning process resulting in rotary tube kilns with multi-stage cyclone preheaters, the grinding of raw materials and cements in closed-circuit grinding plants with highly selective classifiers, and the increasing use of high-pressure grinding rolls on the raw material and cement sides. However, economic aspects are not the only driving forces for the continued development of process technology. The environmental compatibility of cement production and of cements in their various applications has played an increasing part in recent years.

The overall process technology of cement manufacture is being determined to a critical extent by this close connection between the quality demands of cement users and the cement producer's wish to produce cement cost-effectively at a high quality level while taking the environmental compatibility of the production process and of the product into account. The following comments will deal with the relevant developments which have taken place in recent years or are to be expected in the future.

2. Performance of market-orientated cements

When answering questions about what cements have to achieve nowadays it is very helpful to take a look at the application statistics for cements conforming to DIN 1164. **Fig. 1** shows how the quantity of cement produced in West Germany in the years 1980 to 1991 is divided between the applications of ready-mixed concrete, precast concrete parts/concrete products, bagged products and others. It can be seen from the diagram that the importance of ready-mix concrete as the main consumer group has increased continuously over the years shown. The cement sales in the precast concrete parts and concrete products industry have remained virtually unaltered although the production quantities in these two areas have increased during the period under consideration. The reason for this apparent stagnation is that more highly developed manufacturing processes have significantly decreased the cement content per m^3 of fresh concrete, especially in the production of concrete products. There is a continuous drop in the proportion of bagged products. They are increasingly replaced by finished products such as ready-mix mortar. The use of cements for other applications is increasing slowly but at a low level.

Fig. 2 shows the percentages of the cement strength classes used in West Germany. The proportion of cements with high and very high strengths increases continuously with the increase in precast cement parts and there is a corresponding decrease in the proportion of cements in the normal strength class. Added to this is the increase in proportion of cements in the high strength classes at the expense of the normal strength class due to the increased use of concrete additives like coal fly-ash and stone dust – including use in ready-mix concrete. **Fig. 3** shows a comparable trend for all Western European countries. Information from the Cembureau shows that the proportion of cements in the normal strength class has fallen between 1980 and 1991 by the same amount as the increase in the proportion of cements with high and very high strengths.

Among the cements in the bottom "normal" strength class, there has been a considerable increase in the proportion of cements containing additives (**Fig. 4**) in Western Europe during the last 10 years. In the EC the proportion of Portland cements in this strength class in 1991 was only 29.3 %. The reason for this trend lies partly in the fact that cements containing additives such as granulated blast furnace slags, pozzolanas, fly ash or limestone have proved to be very good

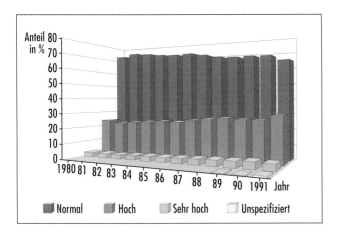

BILD 2: Prozentuale Aufteilung des in Westdeutschland von 1980 bis 1991 hergestellten Zements auf die Zementfestigkeitsklassen Normal, Hoch, Sehr Hoch und Unspezifiziert (nach Statistiken des Bundesverbandes der Deutschen Zementindustrie)
FIGURE 2: Percentage distribution of the cement manufactured in West Germany from 1980 to 1991 between the normal, high, very high and unspecified cement strength classes (according to statistics from the Federation of the German Cement Industry)

Normal = normal
Hoch = high
Sehr hoch = very high
Unspezifiziert = unspecified

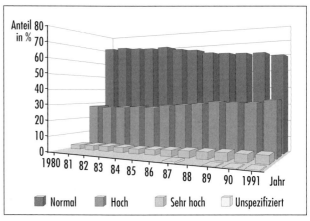

BILD 3: Prozentuale Aufteilung des in Westeuropa von 1980 bis 1991 hergestellten Zements auf die Festigkeitsklassen Normal, Hoch, Sehr Hoch und Unspezifiziert. Länderbasis: B, D, DK, E, F, GB, GR, I, IRE, L, NL, P (nach Statistiken des Cembureau)
FIGURE 3: Percentage distribution of the cement manufactured in Western Europe from 1980 to 1991 between the normal, high, very high and unspecified cement strength classes. Countries involved: B, D, DK, E, F, GB, GR, I, IRE, L, NL, P (according to statistics from the Cembureau)

Normal = normal
Hoch = high
Sehr hoch = very high
Unspezifiziert = unspecified

daß die Bedeutung des Transportbetons als Hauptabnehmergruppe über die aufgezeigten Jahre stetig zugenommen hat. Der Zementabsatz in die Betonfertigteil- und Betonwarenindustrie ist nahezu unverändert geblieben, obwohl die Produktionsmenge in diesen beiden Bereichen in dem betrachteten Zeitraum angestiegen ist. Der Grund für diese scheinbare Stagnation liegt darin, daß insbesondere bei der Herstellung von Betonwaren weiterentwickelte Fertigungsverfahren den Zementgehalt je m^3 Frischbeton deutlich abgesenkt haben. Der Anteil der Sackware geht kontinuierlich zurück. Sie wird zunehmend durch Fertigprodukte wie z.B. Fertigmörtel ersetzt. Die Verwendung von Zementen für sonstige Anwendungsgebiete steigt auf niedrigem Niveau langsam an.

Bild 2 zeigt den prozentualen Anteil der Zementfestigkeitsklassen in Westdeutschland. Mit dem Anstieg vorgefertigter Betonteile nimmt der Anteil von Zementen mit hoher bzw. sehr hoher Festigkeit kontinuierlich zu und der Anteil an Zementen der Festigkeitsklasse Normal entsprechend ab. Hinzu kommt, daß auch bei Transportbeton durch den verstärkten Einsatz von Betonzusatzstoffen wie Steinkohlenflugaschen und Gesteinsmehlen der Anteil an Zementen hoher Festigkeitsklassen zu Lasten der Festigkeitsklasse Normal zunimmt. Eine vergleichbare Entwicklung läßt **Bild 3** für alle Länder Westeuropas erkennen. Nach Angaben des Cembureau ist der Anteil der Zemente der Festigkeitsklasse Normal zwischen 1980 und 1991 im gleichen Maße gefallen, wie der Anteil der Zemente mit hoher und sehr hoher Festigkeit zugenommen hat.

Bei den Zementen der unteren Festigkeitsklasse Normal zeigte sich in den letzten 10 Jahren in Westeuropa eine beachtliche Zunahme der Anteile der Zemente mit Zumahlstoffen (**Bild 4**). Portlandzemente dieser Festigkeitsklasse hatten in der EG im Jahr 1991 nur noch einen Anteil von 29,3 %. Der Grund für diese Entwicklung liegt zum einen darin, daß sich Zemente mit Zumahlstoffen wie Hüttensand, Puzzolan, Flugasche oder Kalkstein durch ihre sehr guten Verarbeitungseigenschaften zur Herstellung von Transportbeton, Estrichen u.a. sehr gut bewährt haben. Zum anderen kann es für den Zementhersteller unter bestimmten Voraussetzungen vorteilhaft sein, steigende Brennstoffkosten zur Herstellung von Portlandzementklinker durch die Verwendung geeigneter Zumahlstoffe zu kompensieren. Um jedoch die Leistungsfähigkeit solcher

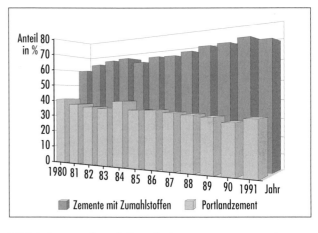

BILD 4: Prozentuale Aufteilung des in Westeuropa von 1980 bis 1991 hergestellten Zements der Festigkeitsklasse Normal auf Portlandzement und Zemente mit Zumahlstoffen. Länderbasis: B, D, DK, E, F, GB, GR, I, IRE, L, NL, P (nach Statistiken des Cembureau)
FIGURE 4: Percentage distribution of cement of the normal strength class manufactured in Western Europe from 1980 to 1991 between Portland cement and blended cements. Countries involved: B, D, DK, E, F, GB, GR, I, IRE, L, NL, P (according to statistics from the Cembureau)

Zemente mit Zumahlstoffen = blended cements
Portlandzement = Portland cement

for producing ready-mix concrete, flow screeds, etc., because of their good workability properties. Under some circumstances it can also be an advantage for the cement producer to offset increasing fuel costs for producing Portland cement clinker by the use of suitable additives. However, it usually requires finer grinding to keep the performance of such cements at the same high level, and the energy-saving effect is diminished by the increased grinding energy required [8].

The specifications for the performance of cements in concrete arise from the specific requirements of the users of the concrete and of the buildings (**Fig. 5**). While the concrete user tends to be more interested in specific properties of the unset concrete, the attention of the building user is concentrated mainly on the properties of the hardened concrete.

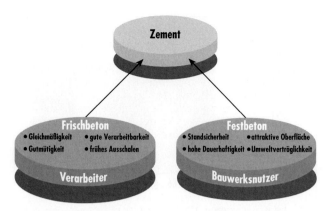

BILD 5: Anforderungen des Verarbeiters und Bauwerknutzers an die Leistungsfähigkeit von Zement im Frisch- und Festbeton
FIGURE 5: Requirements of cement and building users for the performance of the cement in unset and hardened concrete

Zement	= cement
Frischbeton	= unset concrete
Gleichmäßigkeit	= uniformity
gute Verarbeitbarkeit	= good workability
Gutmütigkeit	= tolerance
frühes Ausschalen	= early formwork stripping
Verarbeiter	= concrete user
Festbeton	= hardened concrete
Standsicherheit	= stability
attraktive Oberfläche	= attractive surface
hohe Dauerhaftigkeit	= high durability
Umweltverträglichkeit	= environmental compatibility
Bauwerksnutzer	= building user

Zemente auf gleich hohem Niveau zu halten, ist in der Regel eine stärkere Feinmahlung erforderlich. Der Mehrbedarf an Mahlenergie läßt dabei den Energiespareffekt insgesamt schrumpfen [8].

Die Anforderungen an die Leistungsfähigkeit von Zementen im Beton ergeben sich aus den speziellen Anforderungen des Betonverarbeiters und des Bauwerknutzers (**Bild 5**). Während der Verarbeiter eher an bestimmten Frischbetoneigenschaften interessiert ist, gilt das Hauptaugenmerk des Bauwerknutzers vorrangig den Festbetoneigenschaften. Zement wird heute im Frischbeton vor allem an seiner Gleichmäßigkeit und an seinen wesentlichen Verarbeitungseigenschaften beurteilt. Der Festbeton soll zu jedem Zeitpunkt die der Nutzung des Bauwerks entsprechende ausreichende Standsicherheit und Dauerhaftigkeit aufweisen. Darüber hinaus spielen die Umweltverträglichkeit und die Oberflächengestaltung von Beton eine zunehmende Rolle.

Tabelle 1 soll verdeutlichen, daß aus den unterschiedlichen Einsatzgebieten von Beton sehr unterschiedliche Anforderungen an den Frisch- und Festbeton und damit auch an den zu verwendenden Zement gestellt werden. Ausgewählt

TABELLE 1: Anforderungen an den Frisch- und Festbeton sowie an den eingesetzten Zement in der Transportbeton-, Betonwaren- und Betonfertigteilindustrie und Auswahl der in diesen Anwendungsbereichen zu bevorzugenden Zemente

	Anforderungen an den Beton	Anforderungen an den Zement im Beton	Bevorzugter Zement
Transportbeton	○ gute Verarbeitbarkeit ○ Gutmütigkeit ○ frühes Ausschalen ○ geringe Reißneigung ○ hohe Dauerhaftigkeit	○ Abstimmung der Feinteile im Zement auf die Feinteile im Betonzuschlag ○ geringes Ansteifen ○ genügend hohe Frühfestigkeit ○ moderate Wärmeentwicklung ○ hoher Frostwiderstand, geringe Carbonatisierung	○ Zement mit Zumahlstoff
Betonwaren	○ hohe Grünstandsfestigkeit ○ hohe Dauerhaftigkeit ○ gute Verarbeitbarkeit	○ hohe Frühfestigkeit ○ hoher Frost- und Tausalzwiderstand	○ Portlandzement hoher Festigkeit
Betonfertigteile	○ frühes Ausschalen, schnelles Vorspannen ○ hohe Dauerhaftigkeit ○ gute Verarbeitbarkeit, gute Sichtbetonqualität	○ sehr hohe Frühfestigkeit ○ hoher Frostwiderstand, hohe Endfestigkeit	○ Portlandzement hoher und sehr hoher Festigkeit

Nowadays, cement in the unset concrete is assessed mainly on its uniformity and its essential workability properties. At any given time the hardened concrete should have sufficient stability and durability to match the use of the structure. The environmental compatibility and the surface finish of the concrete are also playing an increasing part.

TABLE 1: Requirements for the unset and hardened concrete and for the cement used in the ready-mix concrete, concrete products and precast concrete parts industries, and selection of the cements preferred for these applications

	Requirements for the concrete	Requirements for the cement in the concrete	Preferred cement
Ready-mix concrete	– good workability – tolerance – early formwork stripping – low cracking tendency – high durability	– matching the fines in the cement to the fines in the aggregate – little stiffening – adequate early strength – moderate heat development – high freeze-thaw resistance, low carbonation	– blended cement
Concrete products	– high green strength – high durability – good workability	– high early strength – high freeze-thaw and de-icing salt resistance	– high strength Portland cement
Precast concrete parts	– early formwork stripping, rapid prestressing – high durability – good workability, good fair-faced concrete quality	– very high early strength – high freeze-thaw resistance, high final strength	– high and very high strength Portland cement

Table 1 is intended to illustrate the fact that the different areas of use of concrete place very different demands on the unset and hardened concrete and therefore also on the cement to be used. The selection covers cements to be used for ready-mix concrete, for concrete products and for precast concrete parts, and these are listed in the table. The first column in the table gives the requirements for the particular concrete. The second column gives the requirements for the cement in these concretes, and the third column lists the cement or cements which are particularly suitable for these applications.

With ready-mix concrete the properties of the unset concrete such as workability and compactability, and the ability to maintain consistency over fairly long periods, even at fairly high temperatures, are very important. The associated cement properties are a cement grading curve which is suited to the fine fractions of the concrete aggregate and a suitable setting behaviour. Early stripping of the formwork without excessive expenditure on curing, little tendency to crack, and high durability are very important properties of the hardened cement. The cement properties required to achieve this are a sufficiently high early strength, moderate heat evolution, high freeze-thaw resistance, and low carbonation. Cements containing additives match this requirement profile for cement for producing ready-mix concrete particularly well. In the production of concrete products and precast concrete parts the properties – besides workability – which are especially important are a good green strength, high early strength for early formwork stripping and prestressing, a sufficiently high final strength, and high durability. Portland cements of high and very high early strength which are characterized by high early strength and good final strength and by high freeze-thaw resistance with low carbonation are particularly suitable for this purpose.

wurde der Einsatz von Zement für Transportbeton, für Betonwaren und für Betonfertigteile, die in der Tafel untereinander angegeben sind. In der Tabelle sind in der ersten Spalte die Anforderungen an den jeweiligen Beton, in der zweiten Spalte die Anforderungen an den Zement in diesen Betonen und in der dritten Spalte der oder die für diese Aufgaben besonders geeigneten Zemente angegeben.

Bei Transportbeton sind die Frischbetoneigenschaften mit der Verarbeitbarkeit und Verdichtungswilligkeit sowie das Halten der Konsistenz über einen längeren Zeitraum auch bei höheren Temperaturen von großer Bedeutung. Die zugehörigen Zementeigenschaften sind eine auf die Feinanteile des Betonzuschlages abgestimmte Sieblinie des Zements und ein günstiges Erstarrungsverhalten. Bei den Festbetoneigenschaften sind ein frühes Ausschalen ohne übermäßig großen Aufwand zur Nachbehandlung, eine geringe Reißneigung und eine hohe Dauerhaftigkeit sehr wichtig. Die dafür notwendigen Zementeigenschaften sind eine genügend hohe Frühfestigkeit, eine moderate Wärmeentwicklung, ein hoher Frostwiderstand und eine geringe Carbonatisierung. Diesem Anforderungsprofil an Zement für die Herstellung von Transportbeton entsprechen insbesondere Zemente mit Zumahlstoffen. Bei der Produktion von Betonwaren und Betonfertigteilen sind neben der Verarbeitbarkeit vor allem eine gute Grünstandsfestigkeit, eine hohe Frühfestigkeit für ein frühes Ausschalen und Vorspannen, eine genügend hohe Endfestigkeit sowie eine hohe Dauerhaftigkeit von großer Bedeutung. Gut geeignet hierfür sind Portlandzemente hoher und sehr hoher Festigkeit, die sich durch eine hohe Früh- und eine gute Endfestigkeit sowie durch einen hohen Frostwiderstand bei niedriger Carbonatisierung auszeichnen.

Die **Bilder 6** bis **9** zeigen zwei Beispiele für marktorientierte Zemente und ihre Leistungsfähigkeit für den speziellen Anwendungsbereich. Bild 6 zeigt den Einsatz von Spritzbeton im Tunnelbau. Zur Herstellung von Spritzbeton werden in der Regel Zemente verwendet, deren Erstarrungsablauf durch die Zugabe von Betonzusatzmitteln auf Basis Alkalialuminat sehr stark beschleunigt wird. Diese Spritzbetontechnologie kann während der Nutzung des Bauwerkes zu Schwierigkeiten führen, wie sie auf Bild 7 ersichtlich sind. Das auf die Spritzbetonschale auftretende Gebirgswasser wird in der Regel durch ein Drainagesystem abgeleitet. Das Wasser kann jedoch während des Kontaktes mit dem Spritzbeton Calciumhydroxid aus dem hydratisierten Zement

Figs. 6 to 9 show two examples of market-orientated cements and their performance in specific areas of application. Fig. 6 shows the use of gunned concrete in tunnel construction. As a rule, concretes with setting characteristics which are very sharply accelerated by the addition of alkali-aluminate-based concrete additives are used for producing gunned concrete. This gunned concrete technology can lead to problems during the use of the structure, as can be seen in Fig. 7. The rock water appearing on the gunned concrete shell is normally removed through a drainage system. However, during contact with the gunned concrete the water can leach calcium hydroxide out of the hydrated cement and alkalis out of the accelerator. The calcium hydroxide is precipitated through reaction with carbonic acid to form calcium carbonate and can lead to the encrustation of the drainage system shown in Fig. 7 with correspondingly high operating costs, while the constituents of the accelerator can pass into the surface and ground water where they are detrimental to the water quality. Rapid setting, low-calcium-sulphate Portland and Portland slag cements have therefore been developed very recently which do not need the addi-

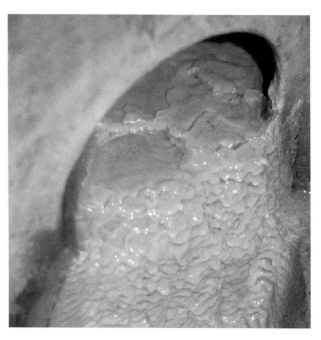

BILD 7: Ausfällung von Calciumcarbonat im Drainagesystem eines Tunnelbauwerkes
FIGURE 7: Precipitation of calcium carbonate in the drainage system of a tunnel structure

BILD 6: Sicherung des Streckenvortriebes im Tunnelbau durch Einsatz von Spritzbeton
FIGURE 6: Securing the heading in tunnel construction using gunned concrete

BILD 8: Bau des massiven Beton-Stabbogens der Maintalbrücke Veitshöchheim auf der Schnellbahnstrecke Hannover-Würzburg der Deutschen Bundesbahn
FIGURE 8: Construction of the solid arched girder of the Maintal bridge at Veitshöchheim on the Hannover-Würzburg section of the suburban rail system of the federal railway

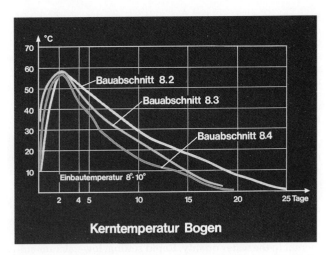

BILD 9: Temperaturverlauf im erhärteten Kernbeton des Stabbogens nach Bild 8
FIGURE 9: Temperature profile in the hardened mass concrete of the arched girder shown in Fig. 8

Bauabschnitt 8.2	= construction section 8.2
Bauabschnitt 8.3	= construction section 8.3
Bauabschnitt 8.4	= construction section 8.4
Einbautemperatur 8–10	= laying temperature 8–10
Kerntemperatur Bogen	= Mass concrete temperature of the arched girder
Tage	= days

sowie Alkalien aus dem zugesetzten Beschleuniger herauslösen. Während das Calciumhydroxid durch Reaktion mit Kohlensäure zu Calciumcarbonat ausgefällt wird und zu der im Bild 7 dargestellten Versinterung der Drainage mit entsprechend hohen Betriebskosten führen kann, können die Bestandteile des Beschleunigers in das Oberflächen- und Grundwasser gelangen und dort die Wasserqualität beeinträchtigen. Deshalb wurden in jüngster Zeit schnell erstarrende, calciumsulfatarme Portland- und Eisenportland-Zemente entwickelt, die ohne Beschleunigerzugabe auskommen. Diese Zemente besitzen sowohl die notwendigen Frühfestigkeiten als auch die Endfestigkeiten der geforderten Betonfestigkeitsklasse B 35 und ergeben eine deutlich verminderte Eluation umweltrelevanter Stoffe aus dem Spritzbeton [9, 10].

Ein weiteres Beispiel für die Herstellung und Anwendung eines leistungsfähigen Zements stellt der Bau der Maintalbrücke Veitshöchheim auf der Bundesbahn-Schnellbahnstrecke Hannover-Würzburg dar (Bild 8). Das Brückenbauwerk überspannt den Main mit einem massiven Stabbogen von 162 m Spannweite und einer Querschnittsfläche von ca. 11 m². Der Beton für den Bogen mußte die Anforderungen an die Festigkeitsklasse B 55 erfüllen. Weiterhin sollte die Temperaturerhöhung des erhärteten Betons in den ersten beiden Tagen 50 °C nicht überschreiten. Zur Herstellung des Betons für den Bogen wurde ein langsam erhärtender, relativ grob gemahlener Portlandzement der Festigkeitsklasse Z 45 F mit guter Nacherhärtung eingesetzt. Der Zementgehalt im Beton betrug 300 kg/m³, zusätzlich wurden 80 kg Flugasche/m³ Beton und 1 Gew.-% Fließmittel bei einem Wasser-Zement-Wert von 0,53 verwendet. Der Frischbeton wurde mit Flüssig-Stickstoff auf eine Temperatur von maximal 10 °C gekühlt, um eine Rißbildung infolge zu hoher Temperaturunterschiede zwischen dem Bauteilinneren und der Betonoberfläche zu verhindern. Bild 9 zeigt, daß die Temperaturentwicklung in allen Betonierabschnitten innerhalb des angestrebten Bereiches lag.

3. Einsatzstoffe und Brennprozeß

Ziel des Klinkerbrennprozesses ist die Herstellung eines Portlandzementklinkers mit gleichbleibend hoher Qualität. Der Weg zu diesem Ziel wird durch eine Vielzahl materialspezifischer und verfahrenstechnischer Einflußgrößen bestimmt. Auf einige von ihnen wird nachfolgend näher eingegangen.

tion of accelerators. These cements have both the necessary early strength and also the final strength of the required B 35 concrete strength class, and produce significantly reduced elution of environmentally relevant substances from the gunned concrete [9, 10].

Another example of the manufacture and use of an efficient cement is represented by the construction of the Maintal bridge at Veitshöchheim on the Hannover-Würzburg section of the suburban rail system of the federal railway (Fig. 8). The bridge structure spans the Main with a solid arched girder with a span of 162 m and a cross-sectional area of approximately 11 m². The concrete for the arch had to fulfil the specifications for strength class B 55. In addition to this the temperature rise of the hardened concrete was not to exceed 50 °C in the first two days. The concrete for the arch was produced using a slow-hardening relatively coarsely ground, Portland cement of the Z 45 F strength class with good secondary hardening characteristics. The concrete had a cement content of 300 kg/m³, and 80 kg fly ash per m³ concrete and 1 wt.% plasticizer were also used with a water/cement ratio of 0.53. The unset concrete was cooled with liquid nitrogen to a maximum temperature of 10 °C in order to prevent crack formation as the result of excessively high temperature differences between the inside of the component and the concrete surface. Fig. 9 shows that the temperature development in all the concreting sections lay within the required range.

3. Raw materials and burning process

The aim of the clinker burning process is to produce a Portland cement clinker of uniformly high quality. The route to this objective is determined by a number of material-specific process engineering factors. Some of these will be discussed in detail below.

Table 2 shows the most important characteristic variables for the quality of the raw material mix, of the fuels, and of the clinker produced. The raw material mix and the fuels used should have high levels of homogeneity and uniformity. The industrial equipment needed for this, such as blending beds and homogenizing silos, reached a high technical level years ago. Inhomogeneous distribution of raw material constituents of varying reactivity in the raw meal with respect to chemical composition and/or particle size distribution can result not only in unstable kiln operation, but also in fluctuations in the grindability of the Portland cement clinker and in the strength development of the cement. This also applies to the compositions and particle size distributions of the fuels used, where fairly large fluctuations can severely disrupt the kiln system [12]. The Portland cement clinker should have a high level of homogeneity and uniformity so that the required cement properties can be achieved reliably with an industrially appropriate safety margin. With a high

TABLE 2: Parameters for characterizing the quality of the raw materials and fuels used and of the Portland cement clinker produced

	Quality parameters
Raw material mix	– homogeneity, uniformity – chemical composition – particle size distribution
Fuel	– homogeneity, uniformity – calorific value – chemical composition (sulphur, chlorine, ash) – particle size distribution
Clinker	– homogeneity, uniformity – chemical composition – mineral phases, content of C_3S and C_3A – reactivity of C_3S and C_3A – degree of sulphatization of the alkalis – CaO_{fr} – particle size distribution – grindability

TABELLE 2: Kenngrößen zur Charakterisierung der Qualität der eingesetzten Roh- und Brennstoffe und des erzeugten Portlandzementklinkers

	Qualitätskenngröße
Rohmaterialgemisch	○ Homogenität, Gleichmäßigkeit ○ Chemische Zusammensetzung ○ Korngrößenverteilung
Brennstoffe	○ Homogenität, Gleichmäßigkeit ○ Heizwert ○ Chemische Zusammensetzung (Schwefel, Chlor, Asche) ○ Korngrößenverteilung
Klinker	○ Homogenität, Gleichmäßigkeit ○ Chemische Zusammensetzung ○ Mineralphasen, Gehalt an C_3S und C_3A ○ Reaktivität von C_3S und C_3A ○ Sulfatisierungsgrad der Alkalien ○ CaO_{fr} ○ Korngrößenverteilung ○ Mahlbarkeit

Tabelle 2 zeigt die wichtigsten Kenngrößen für die Qualität der Rohmaterialmischung, der Brennstoffe und des erzeugten Klinkers. Die Rohmaterialmischung und die eingesetzten Brennstoffe sollen sich durch eine hohe Homogenität und Gleichmäßigkeit auszeichnen. Die dafür notwendigen technischen Einrichtungen wie Mischbetten und Homogenisiersilos haben bereits seit Jahren einen hohen technischen Stand erreicht. Eine inhomogene Verteilung unterschiedlich reaktiver Rohmaterialbestandteile im Rohmehl hinsichtlich ihrer chemischen Zusammensetzung und/oder Korngrößenverteilung kann nicht nur einen instabilen Ofengang nach sich ziehen, sondern auch zu Schwankungen in der Mahlbarkeit des Portlandzementklinkers und der Festigkeitsentwicklung des Zementes führen. Das gilt auch für die Zusammensetzung und Korngrößenverteilung der eingesetzten Brennstoffe, bei denen größere Schwankungen das Ofensystem empfindlich stören können [12]. Der Portlandzementklinker sollte eine hohe Homogenität und Gleichmäßigkeit aufweisen, damit die angestrebten Zementeigenschaften mit einem technisch sinnvollen Vorhaltemaß zielsicher erreicht werden. Mit einem hohen C_3S-Gehalt und einer hohen Reaktivität der C_3S-Phase lassen sich im Regelfall Portlandzemente mit einer genügend hohen Frühfestigkeit, einer stetigen Festigkeitsentwicklung und einer hohen Endfestigkeit herstellen. Der C_3A-Gehalt und die C_3A-Reaktivität sollten im mittleren Bereich liegen. Die Alkalien sollten durch eine gezielte Sulfatisierung über den Rohstoff oder über den Brennstoff als Alkalisulfat vorliegen, damit der Zement ein gemäßigtes Erstarrungsverhalten und damit ein geringes Ansteifen im Frischbeton aufweist. Eine gute Mahlbarkeit des Klinkers ist ein wesentlicher Faktor für eine wirtschaftliche Zementherstellung.

Die in Tabelle 2 vorgestellten Qualitätskenngrößen des Klinkers werden zum Teil durch die Bedingungen beim Brennen und Kühlen des Klinkers stark beeinflußt (**Tabelle 3**). Die Zusammensetzung des Brenngutes, das im Ofensystem in Abhängigkeit von der Zeit zu durchlaufende Temperaturprofil, eine oxidierende Ofenatmosphäre, eine ausreichend hohe Sintertemperatur und das beim Kühlen des Klinkers auftretende Temperatur-Zeitprofil haben entscheidenden Einfluß auf die chemisch-mineralogische Zusammensetzung der Klinkerphasen und das sich ausbildende Klinkergefüge. Darüber hinaus ist die Bedeutung von Kreislaufvorgängen innerhalb des Ofensystems und der Sulfatisierungsgrad der Alkalien auf Ofenbetrieb, Klinkerbildung und Reaktivität der Klinkerphasen zu beachten [13–18].

C_3S content and a highly reactive C_3S phase it is usually possible to produce Portland cements with sufficiently high early strength, continuous strength development, and high final strength. The C_3A content and the C_3A reactivity, should lie in the middle range. The alkalis should be present as alkali sulphates as a result of carefully controlled sulphatization through the raw material or through the fuel, so that the cement has moderated setting characteristics and therefore slight stiffening in the unset concrete. Good clinker grindability is an important factor for cost-effective cement manufacture.

TABLE 3: Variables from rotary kiln operation which influence the formation of Portland cement clinker

— Rate of heating of the raw material mix and its residence time in the individual temperature zones
— Kiln atmosphere
— Sintering temperature
— Rate of clinker cooling
— Cyclic phenomena within the kiln system

The characteristic quality variables of the clinker listed in Table 2 are in some cases heavily influenced by the clinker burning and cooling conditions (**Table 3**). The composition of the kiln feed, the temperature profile in the kiln system which the material passes through as a function of time, an oxidizing kiln atmosphere, a sufficiently high sintering temperature, and the temperature-time profile occurring during the cooling of the clinker all have a marked influence on the chemical-mineralogical composition of the clinker phases and on the internal clinker structure which is formed. It is also necessary to bear in mind the importance of cyclic processes within the kiln system and of the degree of sulphatization of the alkalis for kiln operation, clinker formation and reactivity of the clinker phases [13–18].

TABLE 4: Requirements for the acceptability of secondary raw materials and fuels

— Structural properties of the cement
— Environmental compatibility of the cement and concrete
— Environmental pollution caused by the production process
— Uniformity of the production process
— Cost-effectiveness

Secondary raw materials and secondary fuels have been used increasingly in the cement manufacturing process in recent years. The specifications listed in **Table 4** have proved necessary for industrial acceptability of these materials. In particular, the use of secondary materials must not impair the structural properties of the cement, the environmental compatibility of the cement and concrete, or the production process and the measures taken for environmental protection. The secondary materials should also have the most uniform possible composition and already be highly homogenized. Finally, the use of secondary materials must be cost effective, bearing in mind the increased capital and operating costs for quality and environmental precautions [19–22].

4. Characterization of the different cement components

The previous comments have indicated that in the future there will be an even greater market requirement for cements of a generally high quality level which are intended to fulfil the user's varying different requirements in specific instances. Under these circumstances there will obviously be a trend towards a larger number of cements optimized for particular applications.

4.1 Portland cement clinker

Portland cement clinker is the component of Portland cements, and of blended cements, which determines their properties. The properties listed in **Table 5** represent the

TABELLE 3: Einflußgrößen des Drehofenbetriebes auf die Portlandzementklinkerbildung

- Aufheizgeschwindigkeit des Rohmaterialgemisches und seine Verweildauer in den einzelnen Temperaturbereichen
- Ofenatmosphäre
- Sintertemperatur
- Kühlgeschwindigkeit des Klinkers
- Kreislaufvorgänge innerhalb des Ofensystems

TABELLE 4: Anforderungen an die Verwertbarkeit von Sekundärroh- und -brennstoffen

- Bautechnische Eigenschaften von Zement
- Umweltverträglichkeit von Zement und Beton
- Umweltbelastung durch den Herstellungsprozeß
- Gleichmäßigkeit des Herstellungsprozesses
- Wirtschaftlichkeit

In den letzten Jahren haben Sekundärrohstoffe und Sekundärbrennstoffe vermehrten Eingang in den Zementherstellungsprozeß gefunden. Für eine technische Verwertung dieser Stoffe haben sich die in **Tabelle 4** genannten Anforderungen als notwendig erwiesen. Der Einsatz von Sekundärstoffen darf vor allem die bautechnischen Eigenschaften des Zements, die Umweltverträglichkeit von Zement und Beton sowie den Herstellungsprozeß und die dabei getroffenen Maßnahmen zur Umweltvorsorge nicht beeinträchtigen. Darüber hinaus sollten die Sekundärstoffe eine möglichst einheitliche Zusammensetzung aufweisen und bereits in einer hohen Gleichmäßigkeit vorliegen. Schließlich muß der Einsatz von Sekundärstoffen unter Berücksichtigung der erhöhten Investitions- und Betriebskosten für Qualitäts- und Umweltvorsorge wirtschaftlich sein [19—22].

4. Charakterisierung der verschiedenen Zementbestandteile

Die vorangegangenen Ausführungen haben gezeigt, daß der Markt künftig in noch stärkerem Maße Zemente fordert, die auf generell hohem Qualitätsniveau die im Einzelfall unterschiedlichen Anforderungen des Verarbeiters erfüllen sollen. Bei dieser Sachlage ist die Tendenz zu einer größeren Anzahl auf bestimmte Anwendungsbereiche optimierter Zemente unübersehbar.

4.1 Portlandzementklinker

Der Portlandzementklinker ist der eigenschaftsbestimmende Bestandteil von Portlandzementen und von Zementen mit Zumahlstoffen. Die in **Tabelle 5** angegebenen Eigenschaften stellen die Wunschanforderungen an einen optima-

TABLE 5: Requirements for the properties of Portland cement clinker

— Homogeneity, uniformity
— High C_3S content
— High C_3S reactivity
— Medium C_3A reactivity
— Good grindability

ideal requirements for an optimum Portland cement clinker. It is only in extremely rare cases that the properties are all united in one industrial clinker but, through carefully controlled combination of Portland cement clinker with other main cement components, the cement manufacturer has the opportunity to offer market-orientated cements at a high quality level [23, 24].

TABLE 6: Other possible main cement components in addition to Portland cement clinker (as specified in ENV 197, Part 1)

— Granulated blast furnace slag
— Pozzolanas (natural, industrial)
— Fly ash (Si-rich, Ca-rich)
— Limestone
— Burnt shale
— Silica dust
— Filler
— Sulphate agent

4.2 Additives

Table 6 gives a summary of the main cement components other than Portland cement clinker which are described in the European Pre-standard for cement ENV 197, Part 1. The spectrum ranges from latent-hydraulic granulated blast furnace slag, through the pozzolanic materials of natural or industrial origin, to the inert materials. The sulphate agents, which are very important for the hydration behaviour, are also listed [25]. The important requirements for suitable additives are listed in **Table 7**. In principle, the additives should have comparable requirement profiles to that of the clinker. For example, additives must be homogeneous and have a high level of uniformity so that the cement produced also has uniform properties. Where there are strong fluctuations in the chemical compositions or other properties, it may be necessary to homogenize the additive in a blending bed or homogenizing silo. There must be no detrimental effect to the soundness or freeze-thaw resistance of the cement produced. The latent-hydraulic and pozzolanic additives should have the highest possible hydraulic reactivity and contribute substantially to strength formation. If the additives already have a high level fineness as delivered, then this can have a beneficial effect on the hydraulic properties and on the economic acceptability. This also applies to additives of good grindability. In many countries the requirements for suitable additives are controlled by standards or licences. In view of the increased use of additives in cement it is absolutely essential that the requirement profiles for additives are set at a high level [4, 26—36].

TABLE 7: Requirements for the properties of additives

— Homogeneity, uniformity
— Soundness
— Freeze-thaw resistance
— High reactivity
— Suitable particle size distribution
— Good grindability

4.3 Selecting suitable additives

From the available additives it is now necessary to select an additive or additives which suit an existing Portland cement clinker. With the knowledge that the materials listed take part in very different ways in the hydration of the cement,

TABELLE 5: Anforderungen an die Eigenschaften von Portlandzementklinker

- Homogenität, Gleichmäßigkeit
- hoher C₃S-Gehalt
- hohe C₃S-Reaktivität
- mittlere C₃A-Reaktivität
- gute Mahlbarkeit

len Portlandzementklinker dar. Diese Eigenschaften sind aber in den seltensten Fällen alle in einem technischen Klinker vereint. Der Zementhersteller hat jedoch die Möglichkeit, durch die gezielte Kombination von Portlandzementklinker mit anderen Zementhauptbestandteilen marktorientierte Zemente auf hohem Qualitätsniveau anzubieten [23, 24].

4.2 Zumahlstoffe

Tabelle 6 gibt einen Überblick, welche weiteren Zementhauptbestandteile neben Portlandzementklinker die europäische Vornorm für Zement ENV 197, Teil 1, beschreibt. Das Spektrum reicht von dem latent-hydraulischen Hüttensand über die puzzolanisch reagierenden Stoffe natürlichen oder industriellen Ursprungs bis hin zu den inerten Stoffen. Auch die für den Hydratationsablauf sehr wichtigen Sulfatträger sind aufgelistet [25]. Die wesentlichen Anforderungen an geeignete Zumahlstoffe sind in **Tabelle 7** zusammengestellt. Grundsätzlich gilt, daß die Zumahlstoffe ein vergleichbares Anforderungsprofil wie der Klinker aufweisen sollen. So müssen Zumahlstoffe homogen und in hoher Gleichmäßigkeit vorliegen, damit der hergestellte Zement ebenfalls gleichmäßige Eigenschaften erreicht. Bei stärkeren Schwankungen in der chemischen Zusammensetzung oder in anderen Eigenschaften kann eine Homogenisierung des Zumahlstoffes in einem Mischbett oder Homogenisiersilo erforderlich sein. Die Raumbeständigkeit und der Frost-

TABELLE 6: Weitere mögliche Zementhauptbestandteile neben Portlandzementklinker (nach ENV 197, Teil 1)

- granulierte Hochofenschlacke
- Puzzolan (natürlich, industriell)
- Flugasche (Si-reich, Ca-reich)
- Kalkstein
- gebrannter Schiefer
- Silikastaub
- Füller
- Sulfatträger

but all have lower reactivities than the Portland cement clinker or are inert, it is possible to make the following general statements [37, 38]:

- With the exception of very slag-rich blast furnace cements the proportion of Portland cement clinker is predominant in all multi-component cements.
- If cements of the lowest "normal" strength class are to be made from a Portland cement clinker with a relatively low rate of hydration and a high strength potential, then they can only be coarsely ground. They then have low early strengths and a tendency to bleed in the concrete. As market-orientated cements, they should therefore be combined with a finely ground additive of low reactivity. The required coarse grinding of the clinker fraction which is combined with finely ground additives makes it possible to produce cements with excellent unset concrete properties and very good early and final strengths.
- On the other hand, Portland cement clinker with a high rate of hydration which, because of its phase composition, has to be very finely ground for the required 28 day strength and therefore produces a very reactive Portland cement which tends to stiffen early, should be combined with an additive which hydrates slowly and only contributes to strength formation at a later time. This requires sufficiently long curing.
- For reasons of economy or application technology it may be appropriate to use several additives simultaneously.

The particular importance of the sulphate agent or sulphate agent mix on the setting behaviour of the cement must be mentioned here. Due to the change in reactivity of the cement when other cement components are used it is normally necessary to re-optimize the type or quantity of the sulphate agent [17, 39, 40].

5. Grinding operation and cement properties

The required cement properties in the concrete are heavily dependent on the properties of the main components and of their proportions in the cement. However, they can be varied within wide limits by changing the fineness or particle size distribution of the cement. Adequate grinding of the cement is therefore an effective process measure for reliable and cost-effective production of market-orientated cements.

Electrical power consumption has been lowered significantly in recent years by the introduction of new grinding systems for producing cement [41–44]. This was achieved primarily through the use of high-pressure grinding rolls for preliminary comminution of the mill feed [45–47]. In contrast to ball mills with their frictional and impact stressing, the mill feed in high-pressure grinding rolls is comminuted by pressure stressing. The following three process variants have proved successful for cement grinding:

- Preliminary comminution in high-pressure grinding rolls followed by fine grinding in a ball mill which is operated in close circuit with a classifier.
- So-called hybrid grinding, i.e. preliminary comminution in high-pressure grinding rolls followed by fine grinding in a ball mill – classifier circuit with partial return of the classifier tailings to the high-pressure grinding rolls.
- Combined grinding, i.e. preliminary grinding in a circuit containing high-pressure grinding rolls and a classifier, followed by fine grinding in a ball mill which may possibly be operated with another classifier.

5.1 Grinding Portland cement

Fig. 10 shows the power consumption of different grinding systems, using the same clinker from one production works, as a function of the generated specific surface area of the ground Portland cement [18, 53]. This does not take account of the power consumption of the classifier and conveyors. The upper curve gives the average power consumption of closed-circuit grinding plants with ball mills which are used in the described works for grinding Portland cement. The lower curve shows the power consumption of high-pressure

TABELLE 7: Anforderungen an die Eigenschaften von Zumahlstoffen

- Homogenität, Gleichmäßigkeit
- Raumbeständigkeit
- Frostwiderstand
- hohe Reaktivität
- günstige Korngrößenverteilung
- gute Mahlbarkeit

widerstand des hergestellten Zementes dürfen nicht nachteilig beeinflußt werden. Die latent-hydraulischen und puzzolanischen Zumahlstoffe sollten eine möglichst hohe hydraulische Reaktivität besitzen und maßgeblich zur Festigkeitsbildung beitragen. Liegen Zumahlstoffe bereits im Anlieferungszustand in hoher Feinheit vor, so kann das die hydraulischen Eigenschaften sowie die wirtschaftliche Verwertung günstig beeinflussen. Das gilt auch für Zumahlstoffe mit einer guten Mahlbarkeit. Die Anforderungen an geeignete Zumahlstoffe sind in vielen Ländern über Normen oder Zulassungen geregelt. Im Hinblick auf den verstärkten Einsatz von Zumahlstoffen im Zement ist es dringend erforderlich, das Anforderungsprofil an Zumahlstoffe auf hohem Niveau festzulegen [4, 26–36].

4.3 Auswahl geeigneter Zumahlstoffe

Aus den verfügbaren Zumahlstoffen gilt es nun, den oder die passenden Zumahlstoffe zu einem vorhandenen Portlandzementklinker auszuwählen. Vor dem Hintergrund, daß die genannten Stoffe in sehr unterschiedlicher Weise an der Hydratation des Zementes teilnehmen, aber alle eine geringere Reaktivität als der Portlandzementklinker aufweisen oder inert sind, lassen sich dazu folgende allgemeine Aussagen treffen [37, 38]:

— Sieht man von dem sehr hüttensandreichen Hochofenzement ab, so überwiegt in allen Mehrkomponentenzementen der Portlandzementklinkeranteil.

— Wenn man aus einem Portlandzementklinker mit einer relativ niedrigen Hydratationsgeschwindigkeit und einem hohem Festigkeitspotential Zemente der unteren Festigkeitsklasse „Normal" herstellen will, dann können diese nur grob aufgemahlen werden. Sie weisen dann niedrige Frühfestigkeiten und die Tendenz zum Wasserabsondern im Beton auf. Als marktorientierter Zement sollten sie daher mit einem fein aufgemahlenen, wenig reaktiven Zumahlstoff kombiniert werden. Auf diese Weise lassen sich wegen der notwendigen groben Aufmahlung des Klinkeranteils und seiner gleichzeitigen Kombination mit fein aufgemahlenen Zumahlstoffen Zemente mit hervorragenden Frischbetoneigenschaften und sehr guten Früh- und Endfestigkeiten herstellen.

— Demgegenüber sollten Portlandzementklinker mit einer hohen Hydratationsgeschwindigkeit, die aufgrund ihrer Phasenzusammensetzung für die angestrebte 28-Tage-Festigkeit sehr fein aufgemahlen werden müssen und deshalb einen sehr reaktiven und zum frühen Ansteifen neigenden Portlandzement ergeben, mit einem Zumahlstoff kombiniert werden, der langsam hydratisiert und erst zu einem späteren Zeitpunkt zur Festigkeitsbildung beiträgt. Voraussetzung ist eine ausreichend lange Nachbehandlung.

grinding rolls followed by a hammer mill as the disagglomerator and operated in closed circuit with a cyclone recirculating air classifier for fine grinding Portland cement from the same clinker. The middle curve gives the power consumption of hybrid grinding plants based on information from the cement machinery manufacturers. The dotted lines indicate the power consumption for combined grinding. In this process variant the greater the proportion of fines which are produced in the high-pressure grinding rolls (lower dotted line) the better is the power utilization. Efforts are therefore being made to produce Portland cements in high-pressure grinding rolls without subsequent grinding in a ball mill. However, the operational trials carried out so far indicate that Portland cements produced in this way have steeper particle size distributions than Portland cements from ball mills. **Fig. 11** shows the different particle size distributions of Portland cements of the same specific surface area after grinding in high-pressure grinding rolls and in a ball mill.

The RRSB granulometric diagram has become established in the cement industry as the way of representing the particle size distribution of cement. In this type of representation the particle size distribution of a material is uniquely characterized by two parameters, the slope n and the position parameter x'. The slope of the particle size distribution is a measure of the ratio between the coarse and the fine components in the cement, while the position parameter describes the level of fineness of the cement. For the same specific surface area, a cement with a steep particle size distribution must, because of its lower proportion of fines, be more finely ground than a cement with a flatter distribution which has a higher proportion of fines. However, it is particularly important in relation to the cement properties that, for a given specific surface area, a cement with a steep particle size distribution always has a higher water demand for a certain consistency than a cement of the same specific surface area with a flatter particle size distribution. The water demand of cement is very important for the concrete manufacturer because increased water demand for the same cement strength generally reduces the concrete strength and also the durability of the concrete [51]. The steep particle size distribution generated when Portland cement is finish ground in high-pressure grinding rolls can therefore be a disadvantage for the quality of the unset and the hardened concrete.

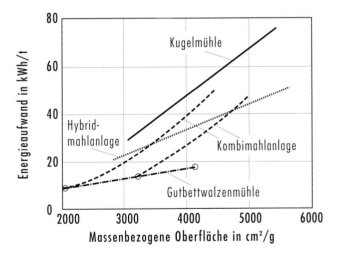

BILD 10: Energieaufwand verschiedener Mahlsysteme in Abhängigkeit von der erzeugten massebezogenen Oberfläche des ermahlenen Portlandzements bei Verwendung des gleichen Portlandzementklinkers

FIGURE 10: Power consumptions of different grinding systems in relation to the generated specific surface area of the ground Portland cement when using the same Portland cement clinker

Energieaufwand in kWh/t	= Power consumption in kWh/t
Kugelmühle	= ball mill
Hybridmahlanlage	= hybrid grinding plant
Kombimahlanlage	= combined grinding plant
Gutbettwalzenmühle	= high-pressure grinding rolls
Massenbezogene Oberfläche in cm²/g	= Specific surface area in cm²/g

— Anwendungstechnische oder wirtschaftliche Aspekte können den gleichzeitigen Einsatz mehrerer Zumahlstoffe sinnvoll machen.

Nicht unerwähnt bleiben soll an dieser Stelle die besondere Bedeutung des Sulfatträgers oder Sulfatträgergemisches auf das Erstarrungsverhalten des Zements. Durch die Veränderung der Reaktivität des Zements beim Einsatz weiterer Zementbestandteile müssen üblicherweise die Art und/oder die Menge des Sulfatträgers neu optimiert werden [17, 39, 40].

5. Mahlbetrieb und Zementeigenschaften

Die angestrebten Zementeigenschaften im Beton hängen in starkem Maße von den Eigenschaften der Hauptbestandteile und deren mengenmäßigen Anteilen im Zement ab. Sie lassen sich jedoch in weiten Grenzen durch Mahlfeinheit und Korngrößenverteilung des Zements variieren. Ein adäquates Mahlen des Zements ist daher eine wirksame verfahrenstechnische Maßnahme, um marktorientierte Zemente zuverlässig und wirtschaftlich herzustellen.

Durch die Einführung neuer Mahlsysteme für die Zementherstellung konnte der elektrische Energiebedarf in den letzten Jahren deutlich gesenkt werden [41–44]. Das gelang vor allem durch den Einsatz von Gutbett-Walzenmühlen zur Vorzerkleinerung des Mahlgutes [45–47]. Im Gegensatz zu Kugelmühlen mit einer Reib- und Schlagbeanspruchung wird in Gutbett-Walzenmühlen das Mahlgut durch Druckbeanspruchung zerkleinert. Für die Zementmahlung haben sich folgende drei Verfahrensvarianten bewährt:

— Die Vorzerkleinerung in einer Gutbett-Walzenmühle mit anschließender Feinmahlung in einer Kugelmühle, die im Kreislauf mit einem Sichter betrieben wird.

— Die sogenannte Hybridmahlung, d.h. die Vorzerkleinerung in einer Gutbett-Walzenmühle mit anschließender Feinmahlung in einem Kugelmühle-Sichter-Kreislauf bei teilweiser Rückführung des Sichtergrobgutes zur Gutbett-Walzenmühle.

— Die Kombimahlung, d.h. die Vormahlung in einem Gutbett-Walzenmühle-Sichter-Kreislauf mit anschließender Feinmahlung in einer Kugelmühle, die evtl. mit einem weiteren Sichter betrieben wird.

5.1 Mahlen von Portlandzement

Im **Bild 10** ist der Energieaufwand verschiedener Mahlsysteme in Abhängigkeit von der erzeugten massebezogenen Oberfläche des ermahlenen Portlandzements unter Verwendung des gleichen Klinkers eines Herstellerwerkes dargestellt [18, 53]. Der Energiebedarf von Sichter und Fördereinrichtungen ist dabei nicht berücksichtigt. Die obere Kurve gibt den mittleren Energiebedarf von Kreislaufmahlanlagen mit Kugelmühlen an, die in dem beschriebenen Werk zur Portlandzementmahlung betrieben werden. Die untere Kurve zeigt den Energiebedarf einer Gutbett-Walzenmühle mit nachgeschalteter Hammermühle als Desagglomerator bei der Feinmahlung von Portlandzement aus dem gleichen Klinker im Kreislauf mit einem Zyklon-Umluftsichter. Die mittlere Kurve gibt den Energiebedarf von Hybridmahlanlagen nach Angaben eines Zementmaschinen-Herstellers an. Die gestrichelten Linien beschreiben den Energiebedarf für die Kombimahlung. Je größer bei dieser Verfahrensvariante der Feingutanteil ist, der in der Gutbett-Walzenmühle erzeugt wird (untere gestrichelte Linie), umso größer ist die Energieausnutzung. Deshalb sind Bestrebungen vorhanden, Portlandzemente in der Gutbett-Walzenmühle ohne Nachmahlung in einer Kugelmühle herzustellen. Die bisher durchgeführten Betriebsversuche zeigen allerdings, daß so hergestellte Portlandzemente gegenüber Portlandzementen aus Kugelmühlen eine steilere Korngrößenverteilung aufweisen. **Bild 11** zeigt die unterschiedliche Korngrößenverteilung von Portlandzement mit gleicher spezifischer Oberfläche nach der Mahlung in einer Gutbett-Walzenmühle und in einer Kugelmühle.

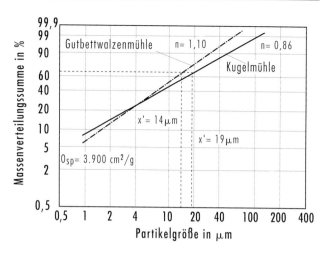

BILD 11: Korngrößenverteilung von Portlandzement, der im Technikumsmaßstab aus dem gleichen Portlandzementklinker auf einer Kreislaufmahlanlage mit Kugelmühle bzw. auf einer Gutbettwalzenmühle mit nachgeschaltetem Sichter ermahlen wurde

FIGURE 11: Particle size distributions of Portland cements which were ground on pilot-plant scale from the same Portland cement clinker in a closed-circuit grinding plant with ball mill and in high-pressure grinding rolls with downstream classifier

Massenverteilungssumme in %	= Cumulative mass distribution in %
Gutbettwalzenmühle	= high-pressure grinding rolls
Kugelmühle	= ball mill
O_{sp}	= A_{sp}
Partikelgröße in μm	= Particle size in μm

There is great interest in developments in which two highly selective classifiers with cuts at different particle sizes are positioned downstream of the high-pressure grinding rolls, and the mixture of these particle sizes provides a significantly flatter particle size distribution. This development is currently being tested for the first time on an industrial scale [48].

When finish grinding Portland cement with high-pressure grinding rolls particular attention must be paid to the sulphate agent added to retard the setting. Because of the very short residence time in the roller gap and the resulting low temperature of the material being ground, there is practically no dewatering of the gypsum setting retarder. A mixture of β-hemihydrate and anhydrite must therefore be added to the classifier circuit. This mixture must have sufficient fineness for it to be uniformly distributed in the clinker meal mass flow. If this is not ensured, it can result in a significant deterioration of the unset concrete properties [17, 39, 52–55].

5.2 Production of blended cements by intergrinding

There is a significant decrease in grinding resistance from granulated blast furnace slag, through clinker and a trass, to limestone. **Fig. 12** shows the particle size distributions of these materials after comminution in a laboratory mill. For the same position parameter x' of 16 μm the slope for the less easily ground granulated blast furnace slag and clinker is 0.9, for the trass it is 0.7, and for the easily ground limestone it is 0.5 [1, 56, 57]. Such differences in grindability must be taken into account during intergrinding; the fine grinding of the various cement components does not take place individually or independently of one another, but is constrained by the settings of the mill and classifier and, in particular, by the grindability of the cement components. These differences in grindability also affect the overall particle size distribution of the cement.

Fig. 13 compares the particle size distributions of a clinker meal, and of clinker/slag and clinker/limestone materials of the same specific surface area after intergrinding in a laboratory ball mill [58]. It can be seen that addition of only 17 wt. % of the easily ground limestone is sufficient to widen the particle size distribution of the combined ground material sig-

Im Bereich der Zementindustrie hat es sich eingebürgert, die Korngrößenverteilung des Zements im RRSB-Diagramm darzustellen. Bei dieser Art der Darstellung wird die Korngrößenverteilung eines Stoffes durch zwei Parameter eindeutig gekennzeichnet, durch die Steigung n und den Lageparameter x'. Während die Steigung der Korngrößenverteilung ein Maß für das Verhältnis zwischen den groben und den feinen Bestandteilen des Zements ist, beschreibt der Lageparameter das Feinheitsniveau des Zements. Für die gleiche spezifische Oberfläche muß ein Zement mit einer steilen Korngrößenverteilung wegen seines geringeren Feinkornanteils feiner aufgemahlen werden als ein Zement mit einer flacheren Verteilung, der einen höheren Feinkornanteil aufweist. Hinsichtlich der Zementeigenschaften ist jedoch vor allem von Bedeutung, daß Zement mit einer vorgegebenen spezifischen Oberfläche bei einer steilen Korngrößenverteilung stets einen höheren Wasseranspruch für eine bestimmte Konsistenz aufweist als Zement gleicher spezifischer Oberfläche mit einer flacheren Korngrößenverteilung. Der Wasseranspruch von Zement ist für die Betonherstellung von großer Bedeutung, da ein steigender Wasseranspruch bei gleicher Zementfestigkeit in der Regel die Betonfestigkeit und auch die Dauerhaftigkeit des Betons mindert [51]. Die bei der Fertigmahlung von Portlandzement auf der Gutbett-Walzenmühle erzeugte steile Korngrößenverteilung kann somit von Nachteil für die Frisch- und Festbetonqualität sein. Von großem Interesse sind deshalb Entwicklungen, bei denen der Gutbett-Walzenmühle zwei trennscharfe Sichter mit Trennschnitten bei unterschiedlichen Korngrößen nachgeschaltet werden, deren Gemisch eine deutlich flachere Korngrößenverteilung liefert. Diese Entwicklung wird zur Zeit erstmals großtechnisch erprobt [48].

Ein besonderes Augenmerk muß bei der Fertigmahlung von Portlandzement auf der Gutbett-Walzenmühle dem als Erstarrungsverzögerer zuzusetzenden Sulfatträger gewidmet werden. Wegen der sehr geringen Verweilzeit im Walzenspalt und der daraus resultierenden geringen Mahlguttemperatur wird der Gips als Erstarrungsverzögerer praktisch nicht entwässert. In Abhängigkeit von der Reaktivität des Klinkers muß deshalb ein Gemisch aus β-Halbhydrat und Anhydrit in den Sichterkreislauf zugegeben werden. Dieses Gemisch muß eine ausreichende Feinheit aufweisen, damit es sich gleichmäßig im Klinkermehlmassestrom verteilt. Ist das nicht gewährleistet, kann daraus eine deutliche Verschlechterung der Frischbetoneigenschaften resultieren [17, 39, 52–55].

5.2 Herstellung von Zementen mit Zumahlstoffen durch gemeinsames Mahlen

Der Mahlwiderstand nimmt von Hüttensand über Klinker und Traß zum Kalkstein signifikant ab. In **Bild 12** sind die Korngrößenverteilungen dieser Stoffe nach ihrer Zerkleinerung in einer Labormühle dargestellt. Bei einem gleichen Lageparameter x' von 16 μm liegt das Steigungsmaß für den schwerer mahlbaren Hüttensand und den Klinker bei 0,9, für den Traß bei 0,7 und für den leicht mahlbaren Kalkstein bei 0,5 [1, 56, 57]. Solche Unterschiede in der Mahlbarkeit müssen bei der gemeinsamen Vermahlung beachtet werden, da die Feinmahlung der verschiedenen Zementbestandteile nicht einzeln oder unabhängig voneinander betrieben wird, sondern von den Einstellungen von Mühle und Sichter sowie vor allem von der Mahlbarkeit der Zementbestandteile erzwungen wird. Darüber hinaus wirken sich diese Unterschiede in der Mahlbarkeit auf die Gesamtkorngrößenverteilung des Zements aus.

Bild 13 zeigt im Vergleich die Korngrößenverteilungen eines Klinkermehls, eines Klinker/Hüttensand- und eines Klinker/Kalkstein-Mahlguts gleicher massebezogener Oberfläche nach dem gemeinsamen Mahlen in einer Laborkugelmühle [58]. Man erkennt, daß schon Zusätze von 17 Gew.-% des leicht mahlbaren Kalksteins genügen, um die Korngrößenverteilung des gesamten Mahlguts deutlich zu verbreitern. Demgegenüber hat das Klinker/Hüttensand-Mahlgut im Vergleich zum reinen Klinkermehl eine etwas

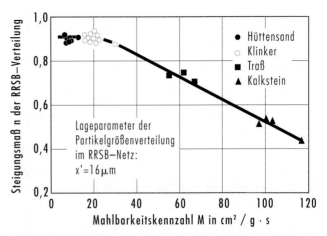

BILD 12: Steigungsmaß der RRSB-Verteilung von Hüttensand-, Klinker-, Traß- und Kalksteinmehlen mit gleichem Lageparameter x' in Abhängigkeit von der Mahlbarkeitskennzahl nach Labormahlungen

FIGURE 12: Slope of the RRSB granulometric distribution of slag, clinker, trass and limestone meals with the same position parameter x', as a function of the grindability parameter from laboratory grinding trials

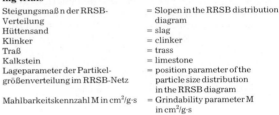

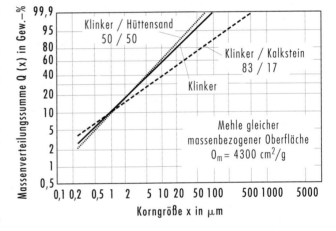

BILD 13: Korngrößenverteilungen eines Klinkermehls sowie eines gemeinsam gemahlenen Klinker/Hüttensand- und Klinker/Kalkstein-Gemischs mit gleicher massebezogener Oberfläche

FIGURE 13: Particle size distributions of a clinker meal and of interground clinker/slag and clinker/limestone mixes of the same specific surface area

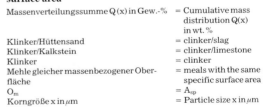

nificantly. On the other hand, the clinker/slag material has a somewhat steeper particle size distribution than the pure clinker meal. These differing particle size distributions occur because during intergrinding the less easily ground material builds up in the coarser fractions and the more easily ground material builds up in the finer fractions. The particle size distribution of the more easily ground material therefore becomes flatter and of the less easily ground material becomes steeper [37, 59–62].

steilere Korngrößenverteilung. Diese unterschiedlichen Korngrößenverteilungen sind darauf zurückzuführen, daß sich beim gemeinsamen Mahlen der schwerer mahlbare Stoff in den gröberen und der leichter mahlbare Stoff in den feineren Fraktionen anreichert. Die Korngrößenverteilung des leichter mahlbaren Stoffes wird dadurch flacher und die des schwerer mahlbaren Stoffes steiler [37, 59–62].

Die Auswirkung dieses Sachverhaltes bei gemeinsamer Vermahlung der Ausgangsstoffe soll am Beispiel der Herstellung von Portlandkalksteinzementen und von hüttensandhaltigen Zementen dargestellt werden. **Bild 14** zeigt den Wasseranspruch dieser beiden Zementarten bei gleicher 28-Tage-Normdruckfestigkeit von 52 N/mm² in Abhängigkeit vom Zumahlstoffanteil für einen Zementleim mit Normkonsistenz nach EN 196, Teil 3 [61]. Die untere Kurve zeigt, daß der Wasseranspruch mit steigendem Kalksteinanteil stetig abnimmt, obwohl der Klinkeranteil mit steigender Menge des inerten Kalksteins immer feiner aufgemahlen werden muß. Dieser Effekt ist darauf zurückzuführen, daß zum einen der wasserzehrende Einfluß des feiner gemahlenen Klinkeranteils durch den teilweisen Ersatz des Klinkers mit Kalkstein kompensiert wird. Zum anderen wird mit steigendem Kalksteinanteil die gesamte Korngrößenverteilung des Portlandkalksteinzements flacher, wodurch sich das mit Wasser gefüllte Lückenvolumen zwischen den gemahlenen Klinkerpartikeln im Zementleim durch Ausfüllen mit feinen Kalksteinpartikeln verringert [37, 38, 49, 63, 64].

Bei hüttensandhaltigen Zementen zeigen die Zemente gleicher Festigkeit mit steigenden Hüttensandanteilen im Gegensatz zum Portlandkalksteinzement keinen Rückgang des Wasseranspruchs. Zwar halten sich wie beim Portlandkalksteinzement mit abnehmendem Klinkeranteil die gegenläufigen Einflüsse einer feineren Aufmahlung der reaktiveren Klinkerkomponente und der Ersatz des Klinkers durch die weniger reaktive Hüttensandkomponente in etwa die Waage, eine zusätzliche Verringerung des Wasseranspruchs tritt jedoch nicht ein, weil die hüttensandhaltigen Zemente wegen ihrer steileren Korngrößenverteilung einen geringeren Feinanteil zum Füllen des Lückenvolumens im Zementleim besitzen [59, 61, 62, 65].

Die flache Korngrößenverteilung des Portlandkalksteinzementes wirkt sich auf einige wichtige Frischbetoneigenschaften besonders vorteilhaft aus. **Bild 15** zeigt das Wasserabsondern von Frischbetonen, die mit Portlandzement und

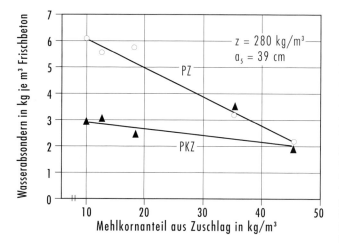

BILD 15: Wasserabsondern vergleichbarer Frischbetone, die unter Verwendung von Portlandzement und Portlandkalksteinzement desselben Lieferwerkes hergestellt wurden, in Abhängigkeit vom Mehlkorngehalt des Betonzuschlags
FIGURE 15: Bleeding from comparable unset concretes, produced using Portland cement and Portland limestone cement from the same production works, as a function of the fines content of the concrete aggregate

Wasserabsondern in kg je m³ Frischbeton	= Bleeding in kg per m³ unset concrete
z	= c
a₅	= spread
Mehlkornanteil aus Zuschlag in kg/m³	= Proportion of fines in the aggregate in kg/m³

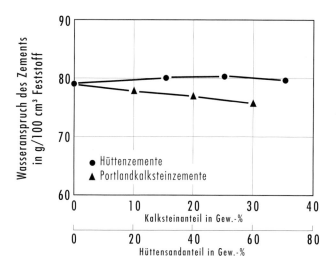

BILD 14: Wasseranspruch von Portlandkalksteinzementen und Hüttenzementen gleicher Normdruckfestigkeit, die durch gemeinsames Mahlen hergestellt wurden, in Abhängigkeit vom Zumahlstoffanteil
FIGURE 14: Water demand of Portland limestone cements and slag cements of the same standard strength produced by intergrinding, as a function of the proportion of additive

Wasseranspruch des Zements in g/100 cm³ Feststoff	= Water demand of the cement in g/100 cm³ solids
Hüttenzemente	= slag cements
Portlandkalksteinzemente	= Portland limestone cements
Kalksteinanteil in Gew.-%	= Proportion of limestone in wt. %
Hüttensandanteil in Gew.-%	= Proportion of slag in wt. %

The effect that this has when the original materials are interground will be illustrated using the example of the production of Portland limestone cements and of cements containing granulated blast furnace slag. **Fig. 14** shows the water demands of these two types of cement for the same 28 day standard compressive strength of 52 N/mm² as a function of the proportion of additive for a cement paste of standard consistency conforming to EN 196, Part 3 [61]. The lower curve shows that the water demand decreases continuously with increasing proportion of limestone although the clinker fraction has to be ground ever finer with the increasing quantity of inert limestone. This is partly because the water-consuming effect of the more finely ground clinker fraction is offset by partial replacement of the clinker with limestone. With increasing proportion of limestone the overall particle size distribution of the Portland limestone cement also becomes flatter; this reduces the water-filled voids between the ground clinker particles in the cement paste by filling them with fine limestone particles [37, 38, 49, 63, 64].

Slag-containing cements of the same strength show no reduction in the water demand with increasing proportion of slag, unlike the Portland limestone cement. In fact, as with Portland limestone cement, the opposed effects of finer grinding of the more reactive clinker component and replacement of the clinker by the less reactive slag component are approximately balanced with decreasing proportion of clinker, but there is no additional reduction of the water as the cements containing slag, because of their steeper particle size distributions, have smaller proportions of fines for filling the voids in the cement paste [59, 61, 62, 65].

The flat particle size distribution of Portland limestone cement has a particularly favourable effect on some important properties of the unset concrete. **Fig. 15** shows the bleeding from unset concretes which were made with Portland cement and Portland limestone cement with 18 wt. % limestone using concrete aggregates containing differing quantities of fines. Both cements came from the same manufacturing works and had comparable 28 day standard com-

Portlandkalksteinzement mit 18 Gew.-% Kalkstein unter Verwendung von Betonzuschlägen mit unterschiedlichem Mehlkorngehalt hergestellt wurden. Beide Zemente stammen aus demselben Herstellwerk und hatten eine vergleichbare 28-Tage-Normdruckfestigkeit. Bei einem Zementgehalt von 280 kg/m³ und einem Ausbreitmaß von 39 cm zeigten die Betone mit mehlkornarmen Zuschlägen unter Verwendung von Portlandkalksteinzement ein besseres Zusammenhaltevermögen des Frischbetons und ein deutlich geringeres Wasserabsondern mit gleichmäßigeren Sichtbetonflächen im Vergleich zu den Betonen mit Portlandzement [37, 66, 67].

5.3 Herstellung von Zementen mit Zumahlstoffen durch getrenntes Mahlen und Mischen

Das getrennte Mahlen und anschließende Mischen der mehlfeinen Komponenten kommt ausschließlich für Zemente mit Zumahlstoffen in Betracht. Die Sulfatträger sollten dabei gemeinsam mit einem der Zementbestandteile der Mühle aufgegeben werden, da ein getrenntes Mahlen und nachträgliches Zumischen des Calciumsulfates zu Agglomerationen und damit zu einer nicht optimalen Verteilung im Zement führen kann [52, 53].

Für ein getrenntes Mahlen und anschließendes Mischen werden häufig zwei Vorteile genannt. Während sich beim gemeinsamen Vermahlen je nach Mahlbarkeit der Bestandteile erzwungenermaßen Anreicherungen im groben oder feinen Bereich ergeben, ist es beim getrennten Mahlen möglich, die Korngrößenverteilung der einzelnen Zementbestandteile auf ihre Wirksamkeit im Zement und Beton einzustellen. Die Annahme, man könnte nun bei der Zementherstellung den Anteil dieser getrennt gemahlenen Bestandteile in weiten Grenzen variieren, hat sich als nur bedingt richtig erwiesen. Es ist zum Beispiel nicht ohne weiteres möglich, aus den gleichen Klinker- und Hüttensandmehlen marktorientierte Zemente mit niedrigen und hohen Hüttensandgehalten – also Portlandhüttenzement und Hochofenzement – herzustellen. Das liegt daran, daß bei gegebenen Korngrößenverteilungen der einzelnen Bestandteile nur ein einziges Mischungsverhältnis eine optimale Korngrößenverteilung des fertigen Gemisches liefert. Je weiter man sich von diesem optimalen Mischungsverhältnis entfernt, desto ungünstiger werden die Zementeigenschaften. Im konkreten Einzelfall kann das zum Beispiel bedeuten, daß man aus den gleichen Bestandteilen Mehle mit unterschiedlichen Korngrößenverteilungen mahlen muß oder sich in der Bandbreite der Mischungsverhältnisse beschränkt. Im Hinblick auf die Zementeigenschaften kann daher die gemeinsame Vermahlung mehr Freiheitsgrade in der Verfahrenstechnik aufweisen als eine getrennte Mahlung mit anschließendem Mischen. Diese Nachteile müssen durch eine entsprechend intensive Eigenüberwachung bei der Herstellung der Vorprodukte und beim Mischen des Zementes kompensiert werden [1].

6. Schlußfolgerungen

Aus den vorangegangenen Ausführungen lassen sich folgende Schlußfolgerungen ziehen:

— Die engen Verflechtungen zwischen den Qualitätsanforderungen des Zementverarbeiters und dem Wunsch des Zementherstellers nach einer wirtschaftlichen Produktion von Zementen auf hohem Qualitätsniveau unter Berücksichtigung der Umweltverträglichkeit von Produktion und Produkt werden künftig in noch stärkerem Maße als bisher die gesamte Verfahrenstechnik der Zementherstellung beeinflussen. Die Zementverarbeiter verlangen zunehmend Zemente, die in ihren Eigenschaften möglichst weitgehend auf die jeweiligen Anwendungsgebiete ausgerichtet sind. Andererseits sind der Zementhersteller und der Zementverarbeiter daran interessiert, für Zement und Beton neue Anwendungsfelder zu erschließen. Beide Richtungen erfordern einen intensiven Informationsaustausch zwischen Zementhersteller und Zementverarbeiter. Die verstärkten Anforde-

pressive strengths. For a cement content of 280 kg/m³ and a flow table spread of 39 cm, the concretes made with low-fines aggregates using Portland limestone cement showed a better cohesive ability of the unset concrete and significantly lower bleeding with more uniform exposed concrete surfaces than the concretes made with Portland cement [37, 66, 67].

5.3 Manufacture of cements with additives by separate grinding and mixing

Separate grinding followed by mixing of the meal-fine components only comes into consideration for cements with additives. The sulphate agents should be fed to the mill together with one of the cement constituents, as separate grinding and subsequent addition of the calcium sulphate can lead to agglomeration and less than optimum distribution in the cement [52, 53].

Two advantages are frequently quoted for separate grinding followed by mixing. During intergrinding the components inevitably build up in the coarse or fine range depending on grindability, but with separate grinding it is possible to adjust the particle size distribution of the individual cement components for their effectiveness in the cement and concrete. The assumption that it would then be possible to vary the proportions of these separately ground components over a wide range during cement production has only proved correct within limits. For example, it is not directly possible to manufacture market-orientated cements with low and high content of slag – i.e. Portland slag cement and blast furnace cement – from the same clinker and slag meals. This is because with given particle size distributions of the individual components only a single mixing ratio gives an optimum particle size distribution for the finished mix. The further the system moves away from this optimum mix ratio, the less favourable are the cement properties. In a specific case, this can mean it is necessary to grind meals with different particle size distributions from the same components, or else the mixing ratios must be limited in spread. With respect to the cement properties, intergrinding can therefore exhibit more degrees of freedom in process technology than separate grinding followed by mixing. These disadvantages have to be offset by correspondingly intensive internal monitoring during production of the intermediate products and while the cement is being mixed [1].

6. Conclusions

The following conclusions can be drawn from the preceding remarks:

— In future the entire process technology of cement production will be influenced to an even greater extent than before by the close interdependence between the quality requirements of the cement users and the desire of the cement producer for cost-effective production of cements at a high quality level while taking account of the environmental compatibility of production and product. The cement users are to an increasing extent requiring cements with properties which are adapted as extensively as possible to specific areas of application. The cement manufacturer and the cement users are also interested in opening up new areas of application for cement and concrete. Both directions require intensive exchange of information between cement manufacturer and cement user. The increased demands made on high-performance market-orientated cements is leading to a greater number of cements with defined properties. Depending on the application, these requirements are met by pure Portland cements with differing levels of fineness and particle size distributions, or by cements with different additives and proportions of additives. The selection of suitable additive or additives is governed by the desired properties of the unset and hardened concrete.

— Portland cement clinker is the component in Portland cements and blended cements which determines their

rungen an marktorientierte Zemente mit hoher Leistungsfähigkeit führen zu einer größeren Anzahl von Zementen mit definierten Eigenschaften. Je nach Anwendungsgebiet werden diese Anforderungen von reinen Portlandzementen mit unterschiedlicher Mahlfeinheit und Korngrößenverteilung oder von Zementen mit unterschiedlichen Zumahlstoffen und Zumahlstoffanteilen erfüllt. Die geforderten Frisch- und Festbetoneigenschaften bestimmen die Auswahl des oder der geeigneten Zumahlstoffe.

– Der Portlandzementklinker ist der eigenschaftsbestimmende Teil von Portlandzementen und von Zementen mit Zumahlstoffen. Deshalb müssen hohe Anforderungen an eine möglichst gleichmäßige Roh- und Brennstoffqualität für die angestrebte Klinkerzusammensetzung gestellt werden. Die Auswahl der Roh- und Brennstoffe zur Klinkerherstellung muß sich an den bautechnischen Eigenschaften des Zements sowie an der Umweltverträglichkeit der Produktion und des Zements orientieren. Die Brenn- und Kühlbedingungen beeinflussen die Klinkereigenschaften in starkem Maße. Die Zusammensetzung des Brenngutes, das im Ofensystem durchlaufene Temperaturprofil, die Ofenatmosphäre, die Sintertemperatur und die Bedingungen beim Kühlen des Klinkers haben entscheidenden Einfluß auf die Zusammensetzung der Klinkerphasen und das Klinkergefüge. Darüber hinaus ist die Bedeutung von Kreislaufvorgängen innerhalb des Ofensystems sowie des Sulfatisierungsgrades der Alkalien auf Ofenbetrieb, Klinkerbildung und Reaktivität der Klinkerphasen zu beachten.

– Die Zumahlstoffe müssen ein dem Portlandzementklinker vergleichbares Anforderungsprofil aufweisen, um die gewünschten Zement- und Betoneigenschaften zu erreichen. Bei den latent-hydraulischen und puzzolanischen Stoffen sollen der Anteil und die Reaktivität der zur Festigkeitsbildung beitragenden Stoffe möglichst hoch sein. Inerte Stoffe sollen eine möglichst gute Mahlbarkeit aufweisen. Portlandzementklinker mit niedriger Hydratationsgeschwindigkeit, die wegen ihrer hohen 28-Tage-Festigkeit für bestimmte Zementfestigkeitsklassen nur relativ grob aufgemahlen werden können und deshalb niedrige Frühfestigkeiten und die Tendenz zum Wasserabsondern im Beton aufweisen, sollten mit einem fein aufgemahlenen, wenig oder nicht reaktiven Zumahlstoff wie z. B. Kalkstein kombiniert werden. Portlandzementklinker mit hoher Hydratationsgeschwindigkeit, die für die angestrebte 28-Tage-Festigkeit sehr fein aufgemahlen werden müssen, sollten mit einem gröber aufgemahlenen, langsam hydratisierenden Zumahlstoff wie z. B. Hüttensand kombiniert werden.

– Beim gemeinsamen Mahlen von Zement mit mehreren Hauptbestandteilen kommt der unterschiedlichen Mahlbarkeit der einzelnen Komponenten eine entscheidende Bedeutung zu. Durch Variation der Zumahlstoffe, der Zumahlstoffanteile und/oder der Zementfeinheit können die gewünschten Zementeigenschaften zielsicher eingestellt werden. Bei der Herstellung marktorientierter Zemente durch getrenntes Mahlen und Mischen müssen die Vorprodukte für die zu mischenden Zemente die jeweils passenden Korngrößenverteilungen aufweisen. Unterschiedliche Zementarten mit den gleichen Ausgangsstoffen erfordern deshalb unterschiedlich gemahlene Vorprodukte. Diese Vorprodukte müssen getrennt hergestellt und gelagert werden. Eine intensive Eigenüberwachung der Vorprodukte und der gemischten Zemente ist deshalb für die angestrebte Produktqualität zwingend erforderlich. Unter diesen Bedingungen kann das getrennte Mahlen und Mischen von Zementen mit Zumahlstoffen wirtschaftlich und technologisch sinnvoll sein.

properties. High standards must therefore be fixed for the most uniform possible quality of the raw material and fuel to achieve the required clinker composition. The selection of the raw materials and fuels for clinker production must be governed by the structural properties of the cement and the environmental compatibility of the production and of the cement. Clinker properties are greatly affected by the burning and cooling conditions. The composition of the kiln feed, the temperature profile along the kiln system, the kiln atmosphere, the sintering temperature and the conditions during the cooling of the clinker, have a decisive influence on the composition of the clinker phases and the internal clinker structure. The importance of cyclic phenomena within the kiln system and of the degree of sulphatization of the alkalis on kiln operation, clinker formation and reactivity of the clinker phases also has to be taken into account.

– The additives must have a specification profile which is comparable with that of the Portland cement clinker in order to achieve the required cement and concrete properties. With the latent-hydraulic and pozzolanic materials, the proportion and the reactivity of the materials contributing to the strength formation should be as high as possible. Inert materials should have the best possible grindability. Portland cement clinker with a low rate of hydration which, because of its high 28 day strength, can only be relatively coarsely ground for certain cement strength classes and therefore has low early strengths and a tendency to bleed in the concrete, should be combined with a finely ground additive which has little or no reactivity, e. g. limestone. Portland cement clinker with a high rate of hydration which has to be very finely ground for the required 28 day strength should be combined with a more coarsely ground, slowly hydrating additive, e. g. granulated blast furnace slag.

– When intergrinding cement with several main components, the different grindabilities of the individual components are critically important. The required cement properties can be accurately adjusted by varying the additives, the proportions of additives, and/or the cement fineness. When manufacturing market-orientated cements by separate grinding and mixing, the intermediate products for the cements which are to be blended must in each case have matching particle size distributions. Different types of cement made from the same original materials therefore require intermediate products with different levels of grinding. These intermediate products must be produced and stored separately. Intensive internal monitoring of the intermediate products and of the blended cements is therefore absolutely essential to achieve the required product quality. Under these conditions the separate grinding and mixing of blended cements can be cost-effective and technologically appropriate.

Literaturverzeichnis

[1] Verein Deutscher Zementwerke e.V.: Verbesserung von Homogenität und Gleichmäßigkeit der Zementeigenschaften beim Zementherstellungsprozeß und der dafür erforderliche Energieaufwand.
Antrag auf Forschungsförderung bei der Arbeitsgemeinschaft Industrieller Forschungsvereinigungen e. V., Köln 1993.

[2] de Jong, J.G.M.: Quality-Assurance-European Developments. Fachbericht 1.1 zum VDZ-Kongreß '93 „Verfahrenstechnik der Zementherstellung" vom 27. 9. bis 1. 10. 1993 in Düsseldorf.

[3] Wischers, G.: „Leistungsfähigkeit" als Kriterium für die Normung von Zement und Beton. Betonwerk + Fertigteil-Technik 56 (1990) Nr. 3, S. 51–60.

[4] Schmidt, M.: Zement mit Zumahlstoffen – Leistungsfähigkeit und Umweltentlastung. Zement-Kalk-Gips 45 (1992) Nr. 2, S. 64–69; Nr. 6, S. 296–301.

[5] Schmidt, M.: Baustoffe für die Bauaufgaben von morgen – Perspektiven der Baustoff-Forschung. Betonwerk + Fertigteil-Technik 59 (1993) Nr. 4, S. 64–75.

[6] Schuhmacher, P., und Schmidt, M.: Der Qualitätsfahrstuhl ist nie besetzt. Betonwerk + Fertigteil-Technik 59 (1993) Nr. 3, S. 52–64.

[7] Shah, S.P.: Betontechnologie der Zukunft. Betonwerk + Fertigteil-Technik 59 (1993) Nr. 2, S. 39–45.

[8] Scheuer, A., und Ellerbrock, H.-G.: Möglichkeiten der Energieeinsparung bei der Zementherstellung. Zement-Kalk-Gips 45 (1992) Nr. 5, S. 222–230.

[9] Mayer, L.: Beton – keine Gefahr für Boden und Grundwasser. Vortrag auf dem Deutschen Betontag 1993 am 14. 5. 1993 in Berlin.

[10] Gebauer, B.: Spritzbeton ohne Beschleuniger. Hoch- und Tiefbau (1992) Nr. 3, S. 105–108.

[11] Toepsch, N.: Storage and stockpile-homogenization of raw materials for the cement industry. World Cement (1991) Nr. 6, S. 9–20.

[12] Sprung, S.: Maßnahmen und Möglichkeiten zur Qualitätssteuerung im Zementwerk. Zement-Kalk-Gips 43 (1990) Nr. 7, S. 340–346.

[13] Kreft, W., Scheubel, B., und Schütte, R.: Klinkerqualität, Energiewirtschaft und Umweltbelastung. Zement-Kalk-Gips 40 (1987) Nr. 3, S. 127–133.

[14] Kreft, W.: Die Unterbrechung von Stoffkreisläufen mit Berücksichtigung der integrierten Weiterverwertung im Zementwerk. Zement-Kalk-Gips 40 (1987) Nr. 9, S. 447–450.

[15] Scheuer, A.: Beurteilung der Betriebsweise von Klinkerkühlern und ihr Einfluß auf die Klinkereigenschaften. Zement-Kalk-Gips 41 (1988) Nr. 3, S. 113–118.

[16] Schürmann, W., Scheuer, A., und Sylla, H.-M.: Optimierung des Satellitenkühlerbetriebs durch gezielte Eindüsung von Wasser. Zement-Kalk-Gips 44 (1991) Nr. 8, S. 393–397.

[17] Thormann, P., und Schmitz, Th.: Die Beeinflussung der Zementeigenschaften durch Klinkermahlung mit der Gutbettwalzenmühle. Zement-Kalk-Gips 45 (1992) Nr. 4, S. 188–193.

[18] Verein Deutscher Zementwerke e.V., Forschungsinstitut der Zementindustrie: Tätigkeitsbericht 1990–1993. Beton-Verlag GmbH, Düsseldorf 1993.

[19] Kirsch, J.: Umweltentlastung durch Verwertung von Sekundärbrennstoffen. Zement-Kalk-Gips 44 (1991) Nr. 12, S. 605–610.

[20] Liebl, P., und Gerger, W.: Nutzen und Grenzen des Einsatzes von Sekundärstoffen. Fachbericht 1.2 zum VDZ-Kongreß '93 „Verfahrenstechnik der Zementherstellung" vom 27. 9. bis 1. 10. 1993 in Düsseldorf.

[21] Sprung, S.: Umweltentlastung durch Verwertung von Sekundärrohstoffen. Zement-Kalk-Gips 45 (1992) Nr. 5, S. 213–221.

[22] Terry, M.S.: Bypass dust handling. World Cement (1992) Nr. 4, S. 30–33.

[23] Harrison, T.A., und Spooner, D.C.: The properties and use of concretes made with composite cements. Cement and Concrete Association (1986), S. 23–26.

[24] Massazza, F.: The role of the additions to cement in the concrete durability. il cemento (1987) Nr. 4, S. 359–382.

[25] ENV 197: Zement. Zusammensetzung, Anforderungen und Konformitätskriterien. Teil 1: Allgemein gebräuchlicher Zement. Vornorm, Dezember 1992.

[26] Asim, M.E.: Die Verarbeitung von Hochofenschlacken zu Zumahlstoffen. Zement-Kalk-Gips 45 (1992) Nr. 10, S. 519–528.

[27] Baron, J., und Douvre, C.: Technical and economical aspects of the use of limestone filler additions in cement. World Cement (1987) Nr. 4, S. 100–104.

[28] Cohen, M.D., und Bentur, A.: Durability of portland cement-silica fume pastes in magnesium sulfate and sodium sulfate solutions. ACI Materials Journal (1988) Nr. 5/6, S. 148–157.

[29] Efes, Y., und Lühr, H.-P.: Bauaufsichtliche Gesichtspunkte für die Verwendung von Steinkohlenflugaschen. Beton- und Stahlbetonbau (1987) Nr. 8, S. 219–224.

[30] Gutteridge, W.A., und Dalziel, J.A.: Filler Cement: The effect of the secondary component on the hydratation of portland cement. Part I: Fine non-hydraulic filler. Cement and Concrete Research 20 (1990), S. 778–782.

[31] Gutteridge, W.A., und Dalziel, J.A.: Filler Cement: The Effect of the secondary component on the hydratation of portland cement. Part II: Fine hydraulic binders. Cement and Concrete Research 20 (1990), S. 853–861.

[32] Narang, K.C.: Portland and Blended Cement. 9th International Congress on the Chemistry of Cement, New Delhi, India, 1992, Volume I, Congress Reports, S. 213–257.

[33] Sersale, R.: Advances in Portland and Blended Cements. 9th International Congress on the Chemistry of Cement, New Delhi, India, 1992, Volume I, Congress Reports, S. 261–302.

[34] Siebel, E., und Sprung, S.: Einfluß des Kalksteins im Portlandkalksteinzement auf die Dauerhaftigkeit von Beton. Beton 41 (1991) Nr. 3, S. 113–117 und Nr. 4, S. 185–188.

[35] Sprung, S., und Siebel, E.: Beurteilung der Eignung von Kalkstein zur Herstellung von Portlandkalksteinzement (PKZ). Zement-Kalk-Gips 44 (1991) Nr. 1, S. 1–11.

[36] Wierig, H.-J., und Restorff, B.: Konsistenz und Ansteifen des Frischbetons. Beton 41 (1991) Nr. 6, S. 282–287.

[37] Albeck, J., und Sutej, B.: Eigenschaften von Betonen aus Portlandkalksteinzement. Beton 41 (1991) Nr. 5, S. 240–244 und Nr. 6, S. 288–291.

[38] Locher, C.H.: Zum Einfluß verschiedener Zumahlstoffe auf das Gefüge von erhärtendem Zementstein in Mörteln und Betonen. Dissertation Rheinisch-Westfälische Technische Hochschule Aachen, 1988.

[39] Basile, F., Biagini, S., Ferrari, G., und Collepardi, M.: Effect of the gypsum state in industrial cements on the action of superplasticizers. Cement and Concrete Research 17 (1987) Nr. 17, S. 715–722.

[40] Wolter, H.: Einfluß der Calciumsulfatformen und der Mischdauer auf das Ansteifen und Erstarren des Zementes. Zement-Kalk-Gips 42 (1989) Nr. 7, S. 372–375.

[41] Patzelt, N.: Entwicklungstendenzen in der Mahltechnik. Zement-Kalk-Gips 42 (1989) Nr. 5, S. 264–268.

[42] Hirayama, M., und Obana, H.: Operational results of roller mill on grinding normal and white cement. World Cement (1986) Nr. 1/2, S. 22–24.

[43] Furukawa. T., Obana, H., Misaka, T., Tamashige, T., und Miyabe, Y.: New Technology in Grinding Process. J. of Research of the Onoda Cement Company 42 (1990) H. 123, S. 24–45.

[44] Schneider, L.T.: Energy saving clinker grinding systems. World Cement (1985) Nr. 3, S. 49–58.

[45] Patzelt, N.: Umlaufmahlung mit der Gutbett-Walzenmühle. TIZ International Powder Magazine 113 (1989) Nr. 5, S. 383–386.

[46] Patzelt, N.: Hybrid-, Kombi- und Fertigmahlung mit der Gutbett-Walzenmühle. Zement-Kalk-Gips 43 (1990) Nr. 7, S. 347–351.

[47] Schneider, G., Gudat, G., und Schneider, V.: Betriebserfahrungen mit Gutbett-Walzenmühlen bei der Zementmahlung. Zement-Kalk-Gips 42 (1989) Nr. 4, S. 175–178.

[48] Krupp Polysius AG: persönliche Mitteilung.

[49] Wolter, A., und Dreizler, I.: Einfluß der Rollenpresse auf die Zementeigenschaften. Zement-Kalk-Gips 41 (1988) Nr. 2, S. 64–70.

[50] Wüstner, H., Dreizler, I., und Oberheuser, G.: Einsatz von Rollenpressen in Mahlanlagen für Kohle, Zementrohstoffe und Zement. Zement-Kalk-Gips 40 (1987) Nr. 7, S. 345–353.

[51] Gebauer, J.: Einfluß der Zementmahlung auf die Eigenschaften des frischen Betons. Holderbank News (1988) Nr. 5, S. 14–17.

[52] Kupper, D., und Knobloch, O.: Fertigmahlung von Zement mit der Gutbett-Walzenmühle POLYCOM. Teil I: Untersuchungen an Mischungen von Klinkermehlen und Sulfatträgern. Zement-Kalk-Gips 44 (1991) Nr. 1, S. 21–27.

[53] Rosemann, H., Hochdahl, O., Ellerbrock, H.-G., und Richartz, W.: Untersuchungen zum Einsatz einer Gutbett-Walzenmühle zur Feinmahlung von Zement. Zement-Kalk-Gips 42 (1989) Nr. 4, S. 165–169.

[54] Schmitz, T.: Die Beeinflussung des C_3A-Umsatzes bei Portlandzementen, gemahlen mit der Gutbettwalzenmühle. Diplomarbeit Technische Universität Clausthal, 1989.

[55] Tang, F. J., und Gartner, E. M.: Influence of sulphate source on portland cement hydration. Advances in Cement Research 1 (1988) Nr. 2, S. 67–74.

[56] Ellerbrock, H.-G., und Schiller, B.: Energieaufwand zum Mahlen von Zement. Zement-Kalk-Gips 41 (1988) Nr. 2, S. 57–63.

[57] Schiller, B.: Mahlbarkeit der Hauptbestandteile des Zements und ihr Einfluß auf den Energieaufwand beim Mahlen und die Zementeigenschaften. Schriftenreihe der Zementindustrie Heft 54, Beton-Verlag Düsseldorf, 1992.

[58] Schiller, B., und Ellerbrock, H.-G.: Mahlbarkeit von Zement-Bestandteilen und Energiebedarf von Zementmühlen. Zement-Kalk-Gips 42 (1989) Nr. 11, S. 553–557.

[59] Blunk, G., Brand, J., Kollo, H., und Ludwid, U.: Zum Einfluß der Korngrößenverteilung von Hüttensand und Klinker auf die Eigenschaften von Hochofenzementen. Zement-Kalk-Gips 41 (1988) Nr. 12, S. 616–623.

[60] Opoczky, L.: Mahltechnische und Qualitätsfragen bei der Herstellung von Kompositzementen. Zement-Kalk-Gips 46 (1993) Nr. 3, S. 136–140.

[61] Schiller, B., und Ellerbrock, H.-G.: Mahlung und Eigenschaften von Zementen mit mehreren Hauptbestandteilen. Zement-Kalk-Gips 45 (1992) Nr. 7, S. 325–334.

[62] Ellerbrock, H.-G., Sprung, S., und Kuhlmann, K.: Korngrößenverteilung und Eigenschaften von Zement. Teil III: Einflüsse des Mahlprozesses. Zement-Kalk-Gips 43 (1990) Nr. 1, S. 13–19.

[63] Tsivilis, S., Tsimas, S., und Moutsatsou, A.: Contribution to the problems arising from the grinding of multicomponent cements. Cement and Concrete Research 22 (1992), S. 95–102.

[64] Opoczky, L.: Verlauf der Korngrößenverteilung bei der gemeinsamen Mahlung einer Klinker-Kalkstein-Mischung. Zement-Kalk-Gips 45 (1992) Nr. 12, S. 648–651.

[65] Opoczky, L. O., Verdes, S., und Török, M.: Grinding technology for producing high-strength cement of high slag content. Powder Technology 48 (1986), S. 91–98.

[66] Krell, J., und Wischers, G.: Einfluß der Feinststoffe im Beton auf Konsistenz, Festigkeit und Dauerhaftigkeit. Beton 38 (1988) Nr. 9, S. 401–404.

[67] Uchikawa, H., Uchida, S., und Okamura, T.: Influence of fineness and particle size distribution of cement on fluidity of fresh cement paste, mortar and concrete. CAJ Proceedings of Cement and Concrete 43 (1989), S. 42–47.

Qualitätssicherung*)
Quality assurance*)
Assurance qualité

Aseguramiento de la calidad

Von **J. G. M. de Jong,** Maastricht/Niederlande

Fachbericht 1.1 · Zusammenfassung – Die 1988 verabschiedete Europäische „Bauproduktenrichtlinie" basiert auf einer Reihe von wesentlichen Anforderungen, die bei der Erstellung der Europäischen Normen für Baustoffe berücksichtigt werden müssen. Der Europäische Normenausschuß für Zement, CEN/TC 51, hat sowohl die Europäische Norm EN 196 („Prüfverfahren") als auch die Europäische Vornorm für „Allgemein gebräuchliche Zemente" ENV-197, Teil 1, ausgearbeitet und mittlerweile angenommen. Um zu bescheinigen, daß Zement den Anforderungen der technischen Norm entspricht, wurde ein Verfahren entwickelt, das zur Konformitätsbescheinigung durch eine zugelassene Zertifizierungsstelle führt. Das Verfahren basiert auf der Prüfung von Zement sowohl durch den Hersteller als auch durch die zugelassene Stelle, einer werkseigenen Produktionskontrolle durch den Zementhersteller und der Bewertung durch die Überwachungsstelle. In zahlreichen Ländern ist dieses „Europäische Zement-Zertifizierungsverfahren" nun auch national angenommen worden. Es wird erwartet, daß diesem Verfahren 1994 der Status einer Europäischen Vornorm (ENV) zuerkannt wird. Dann ist der Weg zum Zement mit EG-Zeichen geschaffen.

Qualitätssicherung

Special report 1.1 · Summary – The European "Construction Products Directive" (CPD) adopted in 1988 is based on a number of essential requirements which must be taken into account when European Standards for Construction Materials are formulated. The European Standard Committee for Cement, CEN/TC 51, has drawn up, and then approved, the European Standard EN 196 ("Test methods") and the European pre-standard for "Cements in Common Use", ENV 197, Part 1. For attestation that a cement conforms to the requirements of the technical standard a scheme has been developed which leads to certification of conformity by an approved certification body. The scheme is based on testing of cement both by the manufacturer and by the approved body – factory production control operated by the cement producer and its assessment by the inspection body. In a number of countries this "European Cement Certification Scheme" has now also become adopted as the national system. It is expected that this scheme will be awarded the status of a European pre-standard (ENV) in.1994. The way is then open for cement carrying the EC mark.

Quality assurance

Rapport spécial 1.1 · Résumé – La „Directive" Européenne sur les „matériaux de contruction" adoptée en 1988 repose sur toute une série d'exigences fondamentales, qui ont dû être prises en compte pour l'établissement de normes Européennes pour matériaux de construction. Le Comité Européen des Normes pour le ciment, CEN/TC 51, a élaboré et adopté entre-temps aussi bien la norme Européenne EN 196 („méthode d'essai") que la prénorme Européenne pour „les ciments à usage courant" ENV-197, partie 1. Pour certifier que le ciment répond aux exigences de la norme technique, il a été mis au point une méthode conduisant à l'attestation de conformité par un organisme d'attestation agréé. Cette méthode est basée sur le contrôle du ciment tant par le fabricant que par l'organisme agréé, par un contrôle de production propre à l'usine du fabricant de ciment et par l'appréciation du service de surveillance. Dans de nombreux pays cette „méthode d'attestation de certificat Européen sur le ciment" a été maintenant adoptée au plan national. On s'attend à ce qu'en 1994 le statut de prénorme Européenne (ENV) soit attribué à cette procédure. La voie du ciment frappé du sceau CE sera alors ouverte.

Assurance qualité

Informe de ramo 1.1 · Resumen – La Directiva europea relativa a los „materiales para la construcción", aprobada en 1988, se basa en una serie de requerimientos esenciales que hay que tener en cuenta al establecer las Normas europeas de materiales para la construcción. El Comité Europeo de Normalización para el cemento, CEN/TC 51, ha elaborado y aprobado, entretanto, la Norma europea EN 196 („Métodos de ensayo") así como la Norma previa europea para „cementos de uso general", ENV-197, parte 1ª. Con el fin de poder certificar que un cemento cumple los requerimientos de la norma técnica, se ha desarrollado un procedimiento que conduce a la extensión de un Certificado de Conformidad por parte de una Entidad certificadora autorizada. Esta procedimiento se basa en el control del cemento, tanto por parte del fabricante como por parte de la Entidad autorizada, un control de la producción por el fabricante del cemento y la evaluación por la Entidad de control. En numerosos países, este „Procedimiento europeo de certificación de cementos" ha sido aprobado también a nivel nacional. Se espera que a este procedimiento se le conceda, en 1994, la categoría de Norma previa europea" (ENV). Después querará abierto el camino hacia los cementos con simbolo CE.

Aseguramiento de la calidad

---
*) Überarbeitete Fassung eines Vortrages zum VDZ-Kongreß '93, Düsseldorf (27.9.–1.10.1993)
 Revised text of a lecture to the VDZ Congress '93, Düsseldorf (27.9.–10.1993)

1. Bauproduktenrichtlinie (CPD)

Das CE-Zeichen ist ein Zertifizierungszeichen mit magischer Wirkung, gleichsam ein Passierschein für den europäischen Markt, dem größten Verbrauchermarkt in der Welt. Um es genauer zu illustrieren: das gesamte Bruttosozialprodukt der europäischen Länder und die Absatzzahlen in der Bauindustrie übertreffen beispielsweise die Zahlen der USA oder Japans. Dieser Tatbestand unterstreicht die große Bedeutung der Entwicklungen in Europa, welche die Schaffung eines offenen Marktes für Bauprodukte zum Ziele haben (**Bild 1**).

BILD 1: Zertifizierungszeichen des Europäischen Zementmarktes
FIGURE 1: Certification mark of the European cement market

Allerdings existiert das CE-Zeichen bisher noch nicht in der europäischen Bauwelt. Der Weg dahin ist lang und beschwerlich. Wie ist die gegenwärtige Sachlage in der Bauwelt, speziell mit Blick auf den Zement, einem Produkt, das uns alle angeht?

Seit 1985 hat die Europäische Kommission (EC) Richtlinien für eine Vielzahl von Produkten formuliert. Die Harmonisierungsrichtlinien, welche die Grundlage dafür bilden, Produkte aus unterschiedlichen Herkunftsländern zu bewerten und unzweideutig untereinander zu vergleichen, enthalten allgemein formulierte Spezifikationen, sogenannte wesentliche Anforderungen, für den Bereich der Sicherheit, für die Gesundheit, die Umwelt und den Verbraucherschutz. Die europäischen Richtlinien sind Gesetze auf europäischer Ebene, und die Mitgliedsstaaten der EEC sind verpflichtet, diese Richtlinien in ihre nationale Gesetzgebung aufzunehmen.

Die Bauproduktenrichtlinie 89/106/EEC erschien bereits am 21.12.1988 [1]. Diese sollte bis 1991 durchgesetzt bzw. in die nationale Gesetzgebung aufgenommen werden. Allerdings haben die letzten EEC-Länder diese Richtlinie bis 1993 noch nicht in Kraft gesetzt. Außerdem erwies sich ihre Interpretation als sehr uneinheitlich, und schließlich sind auch eine Reihe von Problemen im Zusammenhang mit ihrer Durchführung aufgetreten. Diese Europäische Bauproduktenrichtlinie (CPD), offiziell als „Ratsdirektive zur Annäherung von Gesetzen, Verordnungen und administrativen Bestimmungen der Mitgliedsstaaten" bezeichnet, basiert auf einer Reihe wesentlicher Anforderungen, welche die technischen Eigenschaften eines Bauproduktes beeinflussen können (**Tabelle 1**).

TABELLE 1: Grundlegende Anforderungen der CPD

Mechanische Festigkeit und Standsicherheit
Brandschutz
Hygiene, Gesundheit und Umweltschutz
Nutzungssicherheit
Schallschutz
Energieeinsparung und Wärmeschutz

Die Grundanforderungen sind in zahlreichen Dokumenten enthalten. Sie müssen berücksichtigt werden, wenn die europäischen Normen für Baustoffe formuliert werden (**Tabelle 2**). Bedauerlicherweise sind diese Dokumente noch nicht fertiggestellt. Ihre Veröffentlichung wird jedoch bis Ende 1993 erwartet. Produkte, die unter die Europäische Bauproduktenrichtlinie fallen, müssen mit dem CE-Zeichen versehen werden. Das CE-Zeichen ist eine Bescheinigung des Herstellers oder Importeurs gegenüber der Regie-

1. Construction products directive (CPD)

The CE Mark: a certification mark with a magic name, a passport to the European market, the largest consumer market in the world. By way of illustration: the total gross national product of the European countries and the sales in the construction industry exceed those of, for example, the United States or Japan. This fact demonstrates the great importance of the developments in Europe aimed at achieving an open market for construction products (**Fig. 1**).

However, the CE Mark does not yet exist in the European construction world. The road is long and difficult. What is the current state of affairs in the construction world, especially with regard to the product that concerns us all: cement?

Since 1985 the European Commission (EC) has formulated directives for a variety of products. The harmonization directives, which enable products from different countries of origin to be evaluated and compared unambiguously, contain generally formulated specifications (so called Essential Requirements) in the fields of safety, health, the environment and consumer protection. European directives are "laws at the European level", and the member states of the EEC are obliged to incorporate these directives in their national legislation.

A directive concerning construction products appeared in 1988: Construction Products Directive 89/106/EEC (adopted 21. 12. 1988) [1]. This should have been implemented or incorporated into national legislation in 1991. However, the last EEC countries did not put it into force until 1993. Furthermore, the interpretation of this directive is far from uniform, and as a consequence there are a number of problems associated with its implementation.

This European Construction Products Directive (CPD), officially entitled the "Council Directive on the approximation of laws, regulations and administrative provisions of the Member States relating to construction products" is based on a number of essential requirements which may influence the technical characteristics of a product (**Table 1**).

TABLE 1: Fundamental requirements of the CPD

Mechanical resistance and stability
Safety in case of fire
Hygiene, health and the environment
Safety in use
Protection against noise
Energy and heat retention

These fundamental requirements are elaborated in a number of interpretative documents which must be taken into account when the European standards for construction materials are formulated (**Table 2**). Unfortunately these documents are not yet ready. It is expected that they will be published in late 1993.

TABLE 2: The steps from essential requirements of the CPD to standards

Essential requirements
↓
Interpretative documents
↓
Technical specifications
↓
Attestation of conformity (and use of the CE mark)

Products that fall under the CPD must be provided with the CE Mark. The CE Mark is an attestation by the producer or importer to the government that the product conforms to the

TABELLE 2: Die Stufen der wesentlichen Anforderungen der CPD bis hin zu den Normen

wesentliche Anforderungen
↓
Grundlagendokumente
↓
technische Spezifikationen
↓
Konformitätsbescheinigung
(und Anwendung des CE-Zeichens)

rung, daß sein Produkt den erwähnten Anforderungen entspricht. Das CE-Zeichen steht für „Conformité Européenne". Die Europäische Kommission hat einen ständigen Ausschuß für das Bauwesen (SCC) eingesetzt, der sich mit der Durchführung der Richtlinie, den Prüfvorschriften, Qualitätsnachweisen, Normen usw. befaßt. Jedes Mitgliedsland hat zwei Vertreter in diesem Ausschuß. Um die Einführung der Richtlinie zu unterstützen, hat der Ausschuß eine Anzahl von Empfehlungspapieren herausgegeben. Diese bilden eine Art Leitfaden für die Hauptabschnitte, welche die Qualitätsüberwachung in den europäischen Normen betreffen. Einen Überblick über diese „Leitpapiere" mit besonderer Bedeutung für die Durchführung und Anwendung der Richtlinie wird in [2] gegeben.

2. Europäische Zementnorm

Die grundlegenden Anforderungen der Bauproduktenrichtlinie beziehen sich nicht nur auf Bauprodukte, sondern auch auf Bauwerke und sind deshalb nur sehr allgemein formuliert. Es ist wichtig, daß diese allgemeinen Anforderungen durch genaue und nachweisbare Beschreibungen auf Einzelprodukte umgesetzt werden. Zu diesem Zweck hat die Europäische Kommission in Übereinstimmung mit Artikel 4.1. der Bauproduktenrichtlinie dem Europäischen Normenausschuß CEN ein Mandat zur Ausarbeitung der Europäischen Normen erteilt. Diese Normen enthalten die technischen Spezifikationen zur Herstellung eines Produktes, das unter die Bauproduktenrichtlinie fällt.

Für die Produkte Zement und Baukalk ist diese Aufgabe dem Normenausschuß CEN/TC 51 übertragen worden. Die verschiedenen Stufen sind in **Bild 2** dargestellt.

Eine harmonisierte Norm
— definiert die wesentlichen Eigenschaften des Produktes,
— entwickelt die geeigneten Methoden zur Messung dieser Eigenschaften,
— legt die Normungsebenen oder Klassen fest und
— zeigt das Verfahren für die Bescheinigung der Konformität des Produktes auf.

Für Zement sind die Prüfmethoden in der Europäischen Norm EN-196 beschrieben.

Die Europäische Vornorm für allgemein gebräuchliche Zemente ENV 197, Teil 1, war im Frühjahr 1991 verabschiedet worden. Diese Vornorm beschreibt detailliert die Zusammensetzung, die Eigenschaften sowie die Konformitätskriterien für traditionelle und bewährte, allgemein gebräuchliche Zemente, die in Europa produziert bzw. verwendet werden. Die meisten Länder sind gegenwärtig damit befaßt, ihre nationalen Normen zu modifizieren oder neue Normen auf der Grundlage der Europäischen Vornorm vorzubereiten. Die durch das CEMBUREAU (European Cement Association) erstellte **Tabelle 3** gibt eine zusammenfassende Darstellung, wie in den einzelnen Ländern die ENV 197, Teil 1, zur Anwendung gelangen soll.

In der Zwischenzeit wird in den Arbeitsgruppen auf der Ebene des Europäischen Normenausschusses die Arbeit unvermindert fortgeführt, um eine umfassende Europäische Norm vorzubereiten. Die Vornorm wird dabei mehr auf Leistungsfähigkeit orientiert umgeschrieben werden.

above mentioned requirements of the directive. CE stands for "Conformité Européenne".

The European Commission (EC) has set up a Standing Committee for Construction (S.C.C.) to deal with the implementation of the directive, including the rules for testing, quality attestation, marks, etc. There are two representatives of each member state on the Committee.

For the purpose of introducing the directive the committee has published a number of documents containing recommendations. These form a sort of guideline for the sections concerning quality monitoring in the European standards. An overview of these "guidance papers" dealing with specific matters related to the implementation and application of the directive is given in [2].

2. European cement standard

The fundamental requirements in the CPD do not apply to construction products but to construction works and are formulated in very general terms. It is therefore essential that these general requirements are translated into concrete, verifiable specifications for individual products. For this purpose the EC, in conformity with Article 4.1 of the CPD, has given the European standardization organization CEN a mandate to formulate European standards. These European standards contain the technical specification for producing a product that falls under the CPD.

In the case of "Cement and Building Limes" this task has been given to CEN/TC 51. The various steps are given in **Fig. 2**.

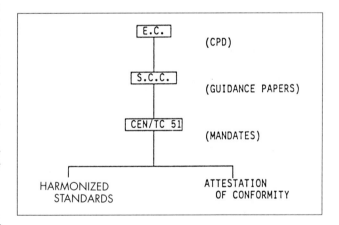

BILD 2: Die verschiedenen Stufen von der „Europäischen Bauproduktenrichtlinie (CPD) 1988" bis hin zu harmonisierten Normen
FIGURE 2: The different steps from European "Construction Products Directive (CPD) — 1988" to harmonized standards

Bauproduktenrichtlinie = CPD
Leitpapiere = Guidance Papers
Mandate = Mandates
harmonisierte Normen = harmonized Standards
Bescheinigung der Konformität = attestation of conformity

A harmonized standard:

— defines the relevant characteristics of the product,

— develops the relevant methods to measure these characteristics,

— sets up the levels or classes,

— indicates the procedure for the attestation of conformity.

Regarding Cement the test methods are described in the European standard EN-196.

The European pre-standard for common cements (ENV 197, part 1), was adopted in the spring of 1992. This pre-standard details composition, specifications and conformity criteria for the traditional and well tried common cements used in Europe.

TABELLE 3: Nationale Anwendung/Ziel ENV 197-1
TABLE 3: National use / intention of ENV 197-1

Land / Country	ENV 197-1 verfügbar machen / Make ENV 197-1 available	Nationale Normen überarbeiten / Modify national standard(s)	Neue nationale Normen vorbereiten / Prepare new national standard(s)	ENV 197-1 einführen und nationale Normen zurückziehen / Introduce ENV 197-1 / withdraw national standard(s)
A	[]	×		
B	[]		0	
DK	[]			
F	[]		0	
D	[]		0	
GR	[]			
IRL	[]	×		
I	[]			*
L			0	
NL	[]		0	
N	[]		0	
PL			0	
P	[]		0	
E	[]	×		
S	[]		0	
CH				*
TR		×		
GB	[]	×		

3. Konformitätsbescheinigung

3.1 CE-Zeichen

Das CE-Zeichen, eine Konformitätsbescheinigung der Europäischen Kommission, beweist, daß ein Produkt den in der harmonisierten Europäischen Norm spezifizierten Anforderungen entspricht und daß die zur Erbringung des Nachweises erforderlichen formalen Verfahren befolgt worden sind, d.h. das Produkt ordnungsgemäß geprüft und zertifiziert wurde. Das von der Zertifizierungsstelle erteilte CE-Zeichen ist und soll auch kein Qualitätszeichen sein. Das CE-Zeichen soll nur die Übereinstimmung mit minimalen gesetzlichen Anforderungen ausweisen. Zemente, die diese Kennzeichnung tragen, können in allen EEC-Ländern frei gehandelt und frei ihrer beabsichtigten Verwendung zugeführt werden. Die in Gang gekommenen Verhandlungen zwischen den Ländern der EEC und der EFTA lassen eine Erweiterung der bisherigen Übereinkommen durch Einbeziehung der EFTA-Länder erwarten. In diesem Zusammenhang wird auf V. Rudert [3] verwiesen, der sich ebenfalls mit diesem Thema beschäftigt hat.

3.2 CEN/TC 51/WG 13

Das Europäische Komitee CEN hat seinen Technischen Gremien die Aufgabe gestellt, die Anforderungen und Vorschriften als einen Teil der Normen zur Bewertung der Konformität vorzubereiten (Beschluß BT 129/1991). In einem späteren Stadium wurden die entsprechenden CEN-Richtlinien veröffentlicht [4]. Die Zementindustrie wie auch CEN/TC 51 begrüßten diese Entwicklung. Es erwies sich als wichtig, alle interessierten Parteien (Normung und Zertifizierung) zusammen zu bringen, um zu einem vollständigen und einheitlichen Verfahren zu gelangen.

Auf der Basis des vorstehend erwähnten Beschlusses BT 129/1991 wurde Mitte 1991 durch CEN/TC 51 die Arbeitsgruppe 13 berufen (Beurteilung der Konformität).

3.3 Auswahl des Verfahrens

Eine der Bestimmungen der Bauproduktenrichtlinie besteht darin, daß der Produkthersteller für die Bescheinigung verantwortlich ist, d.h. seine Produkte den Anforderungen der technischen Norm entsprechen müssen, die von CEN unter dem Mandat der Kommission verabschiedet worden ist. Die Konformität ist durch ein Prüfverfahren auf

Following the adoption of this ENV 197-1 most countries are now in the process of modifying their National Standards or preparing new standards based on relevant sections of the European pre-standard. **Table 3**, which was set up by CEMBUREAU (the European Cement Association), summarizes the way in which it is intended to use ENV 197-1 in each country.

Meanwhile, work continues unabated at the level of the relevant CEN/TC 51 working groups to prepare the full European standard. The pre-standard will be rewritten in a more performance-oriented way.

3. Attestation of conformity

3.1 CE Mark

A CE Mark – an EC Certification of Conformity – demonstrates that the relevant product conforms to the requirements specified in the harmonized European Standard, and that the necessary formal procedures to demonstrate compliance have been followed (i.e. the product has been properly tested and certified). A CE Mark – issued by the Certification Body – will not be, nor is it intended to be, a Quality Mark. The CE Mark will only demonstrate compliance with the minimum legal requirements. Cements bearing the CE Mark will be allowed free movement and free use for their intended purpose throughout the EEC. The on-going negotiations between the EEC and the EFTA are expected to lead to an extension of the above arrangements to include the EFTA countries.

In this context the paper of V. Rudert [3] for this congress is interesting.

3.2 CEN/TC 51/WG 13

CEN gave its Technical Committees the task of preparing, as part of the standards, the requirements and rules for the evaluation of conformity (Resolution BT 129/1991). At a later stage corresponding CEN guidelines were published [4]. The cement industry was happy with this development, as was CEN/TC 51. It is important to produce a complete and coherent scheme and to bring all interested parties (standardization and certification) together.

der Basis der technischen Spezifikationen in Übereinstimmung mit Anhang III der Bauproduktenrichtlinie nachzuweisen.

In Abhängigkeit von der Bedeutung des Produktes innerhalb eines Bauwerkes gibt es verschiedene Bescheinigungsstufen. In dem oben erwähnten Anhang III wird unterschieden zwischen der

— Bescheinigung der Konformität durch eine zugelassene Zertifizierungsstelle und der

— Konformitätserklärung durch den Produkthersteller.

Zu einem späteren Zeitpunkt wurde Anhang III detailliert als Baublatt 92/097 [5] der Europäischen Kommission ausgearbeitet.

Bezüglich der Bestimmungen von Artikel 13 und Anhang III schreibt die Bauproduktenrichtlinie die durchzuführenden Verfahren sowie die bei der Wahl des Verfahrens zu berücksichtigenden Kriterien vor. Die zu wählende Bescheinigungsstufe hängt von der gegenwärtigen Situation (gängige Praxis in den Mitgliedsländern) und der Bedeutung des Produktes (Risikograd) in bezug auf die Erfüllung einer oder mehrerer der sechs wesentlichen Anforderungen ab. Auf Ersuchen des ständigen Ausschusses für das Bauwesen wurde durch CEN/TC 51 auf der Basis von Leitpapier 8 die Frage untersucht, welches System der Konformitätsbescheinigung für Zement bevorzugt werden sollte.

Da Zement im Hinblick auf die Erfüllung wesentlicher Anforderungen für ein Bauwerk eine große Rolle spielt, kam der Ausschuß zu dem Schluß, daß ein bestimmtes Verfahren zur Konformitätsbescheinigung führen müßte und daß eine Konformitätserklärung allein unzweckmäßig und unzulänglich wäre. Diese Schlußfolgerung basierte auf folgenden Überlegungen:

— Zement spielt eine wichtige Rolle im Hinblick auf die Anforderungen an die mechanische Festigkeit und Standsicherheit (was jedoch nicht bedeuten soll, daß andere Anforderungen vernachlässigt werden sollten).

— Zement ist ein kostengünstiges Massenprodukt; er wird an Kunden verkauft, die ihn für eine große Spannweite von mehr oder weniger anspruchsvollen Anwendungen einsetzen, wobei in den meisten Fällen der beabsichtigte Einsatz zu dem Zeitpunkt nicht bekannt ist, wenn der Zement produziert wird.

— Obgleich Zement nur ein Bestandteil des Betons ist, ist er der wichtigste, weshalb es notwendig erscheint, daß seine Konformitätsüberwachung auf einem sehr strengen Verfahren basieren sollte, um seine Qualität garantieren zu können.

— Eine Untersuchung der Prüfverfahren, wie sie in den CEN-Mitgliedsländern angewendet werden, hat deutlich gezeigt, daß die Produktzertifizierung durch eine zugelassene Stelle der am meisten verbreiteten Praxis entspricht.

Diese Ansicht wurde auch von den beiden Europäischen Vereinigungen der Transportbeton- (ERMCO) und der Betonfertigteilindustrie (BIBM) unterstützt, welche die größten Zementverbraucher vertreten. CEN/TC 104 „Beton" bezog ein ähnlichen Standpunkt. Als die größten Zementverbraucher drängen sie darauf, daß die Konformitätsbescheinigung des Produktes, die ihnen die Qualität ihres Betons garantiert, auf dem strengsten Verfahren basieren sollte.

Dieser Standpunkt macht deutlich, daß nicht nur der Beton selbst, sondern auch seine wesentlichen Bestandteile nach den gleichen Zertifizierungsmaßstäben behandelt werden sollten, d.h. Fremdzertifizierung. Der ständige Ausschuß für das Bauwesen hat in der Zwischenzeit empfohlen, daß die Zertifizierung für allgemein gebräuchliche Zemente erforderlich ist. Diese Empfehlung wird gegenwärtig von der Kommission geprüft. Wegen des Fehlens von Grundlagendokumenten hat die EC bisher noch keinen Standpunkt mit Blick auf die Ausarbeitung von Verfahren zur Konformitätsbescheinigung von Bauprodukten, insbesondere von allgemein gebräuchlichen Zementen, bezogen. Eine Entschei-

On the basis of the above mentioned resolution BT 129/1991, in mid 1991 CEN/TC 51 set up working group 13 (Assessment of Conformity).

3.3 Choice of the procedure

One of the provisions of the CPD is that the manufacturer shall be responsable for the attestation that products conform to the requirements of a technical standard adopted by CEN under mandates given by the commission. Conformity is to be established by means of testing on the basis of the technical specifications in accordance with Annex III of the CPD.

There are various levels of attestation, depending upon the product's importance in construction works. In the above mentioned Annex III a distinction is drawn between the following systems of conformity attestation:

— Certification of conformity by an approved certification body,

— Declaration of conformity by the manufacturer.

In a later stage this Annex was elaborated in more detail in the European Commission Paper Construct 92/097 [5].

According to the provisions of Article 13 the CPD, by referring to Annex III, indicates the procedures to be followed and the criteria to be taken into account with regard to the choice of the procedure. The level of attestation of conformity to be chosen will depend on the actual situation (current practice in the member states) and the importance of the product (degree of risk) in relation to the achievement of one or more of the six essential requirements. At the request of SCC CEN/TC 51 took the guidance paper 8 as the basis for considering the question of which system of conformity attestation should be preferred for cement.

Because cement plays such an important role with respect to the essential requirements for construction work, CEN/TC 51 considered that a scheme leading to certification of conformity was required and that declaration of conformity would be unsuitable and even inadequate. This conclusion was based on the following considerations:

— cement plays an important role with respect to the requirements for "Mechanical Resistance and Stability". (This does not mean, however, that its effects on other requirements should be neglected);

— cement is a bulky and low-cost product, sold to customers who use it for a large variety of more or less demanding applications; in most cases its intended use is not known when it is produced;

— although cement is only one of the constituents of concrete, it is the most active and it is therefore important that control of its conformity should be based on a very strict procedure in order to guarantee its quality;

— an examination of the procedures used in the CEN Member countries clearly shows that the product certification procedure by an approved body is by far the most prevalent.

This view was endorsed by the two European associations representing the largest cement consumers, the ready-mixed concrete (ERMCO) and the precast concrete industries (BIBM). CEN/TC 104 "Concrete" took a similar view. As the major consumers, they strongly urge that the attestation of conformity of the product that guarantees the quality of their concrete should be based on the strictest procedure.

This position implies that not only concrete itself but also its active constituents should have the same level of attestation, i.e. third-party product certification. The SCC has recommended in the meantime that product certification is required for common cements and this recommendation is presently being considered by the Commission. Unfortunately, owing to the absence of the interpretative documents, the EC has not yet adopted a position with regard to the specification of procedures for the attestation of conformity of building products, especially common cements. A decision should be taken soon as a matter of some urgency!

dung sollte aus Dringlichkeitsgründen bald herbeigeführt werden.

4. Europäische Zementzertifizierung

4.1 Historische Entwicklung

Neben der kürzlich veröffentlichten Bauproduktenrichtlinie ergriffen im Jahre 1989 die europäische Zementindustrie und ihre Vereinigung, das CEMBUREAU, die Initiative, ein einheitliches und brauchbares europäisches Verfahren zur Konformitätsbescheinigung von Zement vorzubereiten. Ein erster Entwurf wurde 1990 der Europäischen Kommission übergeben. Wie im vorstehenden Kapitel 3.3 beschrieben, erteilten mehrere interessierte Stellen auf Drängen des ständigen Ausschusses für das Bauwesen ihre Ratschläge in bezug auf das erforderliche Niveau der Konformitätsbescheinigung von Zement.

In der Zwischenzeit haben das CEMBUREAU und die Arbeitsgruppe 13 von CEN/TC 51 (3.2) das „Europäische Zertifizierungsverfahren für Zement" ausgearbeitet. Ein Entwurf wurde im vergangenen Jahr den Mitgliedern von CEN/TC 51 zur nationalen Diskussion und Kommentierung zugeschickt. Daraus entstand ein überarbeitetes Dokument, das „Europäische Zement-Zertifizierungsverfahren" (zweiter Entwurf, überarbeitet im Mai 1993), in welches die Kommentare der Mitgliedsländer weitgehend eingearbeitet wurden. Dieses Dokument ist einerseits streng genug, um Anwendern das notwendige Vertrauen in das Verfahren zu geben, andererseits auch wieder so flexibel, um es an unterschiedliche Situationen und Praktiken anpassen zu können.

4.2 Grundgedanke

In Übereinstimmung mit den im Abschnitt 3 formulierten Ausgangspunkten basiert das System auf dem Verfahren (i) von Anhang III der Bauproduktenrichtlinie und ähnelt der Mehrheit der Verfahren, die gegenwärtig in den CEN-Mitgliedsländern angewandt werden.

Desweiteren berücksichtigt dieses System die Grundsätze des Globalen Konzeptes für Zertifizierung und Prüfwesen [6], die durch die SCC herausgegebenen Leitpapiere zur Konformitätsbescheinigung [2] sowie die entsprechenden Abschnitte europäischer Normen in bezug auf Qualitätssicherung und Produktzertifizierung [7]. In diesem Zusammenhang müssen die EN 29002 (Qualitätssicherungs- bzw. Qualitäts-Managementsysteme und das Leitpapier Nr. 7) besonders erwähnt werden. Es muß nachdrücklich betont werden, daß das Zertifizierungsverfahren die Zertifizierung des Produktes „Zement" betrifft und kein Zertifizierungsverfahren des Qualitätssystems darstellt.

Das Verfahren basiert auf zwei prinzipiellen Elementen:
— auf der Prüfung des fertigen Zementes (Produktkontrolle) und
— auf der werkseigenen Produktionskontrolle und ihrer Beurteilung/Bewertung.

Das Verfahren spezifiziert die Maßnahmen, die durch den Hersteller und die Zulassungsstelle zu ergreifen sind, um die geforderte Produktqualität zu erzielen. **Bild 3** gibt einen groben Überblick über das Europäische Zementzertifizierungsverfahren.

4.3 Produktkontrolle

Die Produktkontrolle, worunter die Kontrolle der Qualität auf Einhaltung der Konformitätskriterien zu verstehen ist, gilt als ausgeführt, wenn das Fertigprodukt geprüft worden ist (beim Zement an den Versandstellen im Werk oder Depot). Eine derartige Qualitätsprüfung ist sowohl durch den Hersteller (Eigenüberwachungsprüfung) als auch durch die Zulassungsstelle (Fremdüberwachungsprüfung) durchzuführen. Das Verfahren der Eigenüberwachung schließt auch eine Beschreibung des statistischen Verfahrens für die vom Zementhersteller durchzuführende kontinuierliche Überwachung ein.

Die Überwachungsstelle hat zu prüfen, ob die bei der Eigenüberwachung durch den Zementhersteller erzielten Ergeb-

4. European cement certification

4.1 Historical development

Further to the recently published CPD, in 1989 the European Cement Industries and their Association, Cembureau, took the initiative to prepare a possible single European scheme for the attestation of conformity of cement. A first draft was sent to the EC in 1990. As it was described in Chapter 3.3 above several concerned parties gave an advice upon the required level of attestation for cement upon request of SCC.

Meanwhile Cembureau and Working Group 13 of CEN/TC 51 (3.2) have elaborated the "European Cement Certification Scheme". A draft was sent to the Members of CEN/TC 51 last year and circulated for national discussion and submission of comments. A revised document – "The European Cement Certification Scheme" (second draft, revised May 1993) – was then formulated, taking account of as many as possible of the comments received from the member countries.

This document meets the condition that it must be rigid enough to give users confidence in the scheme but flexible enough to accommodate different situations and practices.

4.2 Principle

In accordance with the points of departure formulated in section 3.3, the system is based on procedure (i) of Annex III of the CPD and closely resembles the majority of schemes currently applied in the CEN member countries.

Furthermore the scheme takes into account the principles of the Global Approach to Certification and Testing [6], the SCC's Guidance Papers on the Attestation of Conformity [2], Relevant clauses of European Standards related to quality assurance and product certification [7].

In this context special mention must be made of EN 29002 (Quality systems) and Guidance Paper No. 7. It must be emphasized that the Certification Scheme concerns certification of the product "cement" and is not a scheme for certification of the quality system.

CEMENT	MANUFACTURING OF CEMENT	
TESTING	WORKS QUALITY MANUAL	
Audit Control / Auto Control	ASSESSMENT by the inspection body	OPERATION by the cement producer
PRODUCT CONTROL	FACTORY PRODUCTION CONTROL	
EUROPEAN CEMENT CERTIFICATION		

BILD 3: Grober Überblick über das Europäische Zertifizierungsverfahren
FIGURE 3: Global overview of the European Certification Scheme

Zement	= cement
Zementherstellung	= Manufacturing of cement
Prüfung	= Testing
Werksqualitätshandbuch	= Works quality manual
Fremdüberwachung	= Audit Control
Eigenüberwachung	= Auto control
Bewertung durch die Überwachungsstelle	= Assessment by the inspection body
Anwendung durch den Zementhersteller	= Operation by the cement producer
Produktkontrolle	= Product control
Werkseigene Produktionskontrolle	= Factory production control

nisse mit den Anforderungen der Produktnorm übereinstimmen. Die Fremdüberwachungsprüfung von Proben wird grundsätzlich als eine Genauigkeitsüberprüfung der durch den Hersteller vorgelegten Prüfergebnisse durchgeführt. Die dafür anzuwendenden statistischen Verfahren sind im Zertifizierungsdokument beschrieben.

4.4 Werkseigene Produktionskontrolle

Die vom Hersteller durchzuführende werkseigene Produktionskontrolle basiert auf dem Leitpapier Nr. 7 der SCC und der EN 29002. In das Verfahren der werkseigenen Produktionskontrolle wurden dabei allerdings nur die Abschnitte solcher Dokumente einbezogen, die für die Produktion und Prozeßkontrolle von Zement wichtig sind. In diesem Zusammenhang muß ausdrücklich bemerkt werden, daß die Produktionskontrolle ausschließlich in der Verantwortung des Herstellers liegt. Bei dieser Kontrolle geht es darum, die geforderten Produkteigenschaften zu gewährleisten. Im Hinblick auf die Tatsache, daß die Qualität des Endprodukts von vorrangiger Bedeutung ist, kommt sowohl der Zementmahlung als auch dem Zementversand die höchste Priorität innerhalb des Produktionsprozesses zu.

Die Aufgaben der zugelassenen Zertifizierungsstelle betreffen die Überwachung, die Bewertung und Annahme der werkseigenen Produktionskontrolle. Die Dokumentation und die Verfahren des Herstellers sind im Werksqualitätshandbuch beschrieben.

Die Überwachungsstelle

— überprüft, ob das Qualitätshandbuch des Werkes mit den Anforderungen des Zertifizierungsverfahrens übereinstimmt,

— prüft, ob die Produktionskontrolle nach dem Werksqualitätshandbuch erfolgt,

— kontrolliert die vom Hersteller durchgeführten Management-Reviews des Qualitätslenkungssystems.

Es muß hervorgehoben werden, daß die oben erwähnten Aufgaben von Hersteller und zugelassener Überwachungsstelle sich ausschließlich auf bereits vorhandene Zementwerke beziehen. Wenn ein neues Zementwerk seine Produktion aufnimmt oder eine neue Zementart bzw. eine neue Zementklasse eingeführt wird, dann muß sowohl durch den Hersteller als auch durch die Zulassungsstelle eine Erstprüfung durchgeführt werden. Eine entsprechende Erstüberwachung durch die zugelassene Stelle muß in diesem Falle auch für das betreffende Werk und sein System der werkseigenen Produktionskontrolle vorgenommen werden.

4.5 Nichtkonformität

Das Europäische Verfahren enthält auch Verfahren, die von der zugelassenen Stelle für den Fall der Nichtkonformität anzuwenden sind. Die Verfahren beziehen sich in erster Linie auf die Einleitung von Maßnahmen durch die Zertifizierungsstelle für den Fall der Nichtkonformität der Ergebnisse aus der Eigenüberwachung bzw. Fremdüberwachung mit den Anforderungen der zugehörigen Produktnorm (d.h. ENV 197-1). Diese Maßnahmen sind klar spezifiziert, um ihre einheitliche Anwendung durch alle Zertifizierungsstellen sicherzustellen.

4.6 Anforderungen an Überwachungsstellen und Prüflabors

Um eine zufriedenstellende Anwendung des vorgeschlagenen Verfahrens für die Europäische Zementzertifizierung zu gewährleisten, müssen die Anforderungen für die zugelassenen Überwachungsstellen spezifiziert und die Prüflabors in die Zementzertifizierung mit einbezogen werden. Das CEMBUREAU hat dazu eine Reihe von Empfehlungen herausgegeben, welche die übereinstimmende Ansicht von Experten aus der Zementindustrie darstellen. Im Moment ist nicht beabsichtigt, diese Empfehlungen in das Zertifizierungsdokument aufzunehmen. Es ist jedoch unverkennbar, daß diese Empfehlungen im Zusammenhang mit der Anwendung des Zertifizierungsverfahrens von großem Nutzen sein können.

The scheme is based on two principal elements:
— testing of the finished cement (product control),
— factory production control and its assessment.

It specifies the measures to be taken by the manufacturer and the approved body in order to achieve the required product quality. **Fig. 3** gives a global overview of the European Cement Certification Scheme.

4.3 Product Control

Product Control, which means the control of quality to meet the conformity criteria as specified in the relevant product standards, is achieved by testing the finished product (i.e. cement at the point(s) of release from the factory or depot). Such testing is carried out by both the manufacturer (Auto Control Testing) and the approved body (Audit Testing). The auto control scheme includes a description of the statistical procedure for continuous inspection operated by the cement manufacturer.

The inspection body checks that the results of the manufacturer's auto control conform with the requirements of the product standard. Audit testing of samples is performed principally as a check on the accuracy of the manufacturer's test results.

Statistical procedures are described in the certification document for this purpose.

4.4 Factory Production Control

The manufacturer operates Factory Production Control, based on the S.C.C. Guidance Paper 7 and EN 29002. Of course only the clauses from the relevant documents have been included in the Factory Production Control Scheme, that are relevant to the production and process control of cement. It must be emphasized that factory production control is the manufacturer's responsibility. Factory production control is the control of production by the manufacturer to enable the required product characteristics to be achieved. In view of the fact that the quality of the final product is of primary importance, the cement grinding and dispatch of the cement are given greatest priority in the production process.

The tasks of the approved certification body concern the surveillance, assessment and acceptance of factory production control. The manufacturer's documentation and procedures are described in the Works Quality Manual. The Inspection Body:

— verifies that the Works Quality Manual complies with the requirements of the certification scheme;

— verifies that the factory production control has been carried out according to the Works Quality Manual;

— inspects the manufacturer's management reviews of the quality control system.

It must be pointed out that the above mentioned tasks for the manufacturer and the approved body apply exclusively to already established cement production. If a new plant is started up, or a new type or class of cement is introduced, initial testing of the cement must be carried out by both the manufacturer and the approved body, and the approved body must carry out an initial inspection of the factory and the factory production control system.

4.5 Non-Conformity

The European Scheme includes procedures to be adopted by the Approved Body in the event of non-conformity. These procedures give priority to measures to be taken by the Certification Body in the case of non-conformity of the results of auto control or audit testing with the requirements of the relevant product standard (i.e. ENV 197-1). These measures are clearly specified to achieve uniform implementation by all Certification Bodies.

5. Zukünftige Entwicklungen

5.1 Umwandlung von der ENV in EN 197-1

Im Prinzip besteht die Europäische Zementnorm ENV 197-1 aus drei Teilen:

— den Zementarten (Zusammensetzung, Bestandteile),

— den Qualitätsanforderungen (mechanische, physikalische und chemische Anforderungen),

— den Konformitätskriterien.

Die Diskussionen in WG 6 von CEN/TC 51 – beauftragt mit der Umwandlung der ENV 197-1 in die EN 197-1 – waren hauptsächlich auf den ersten Teil, d. h. auf die Zementarten konzentriert. In diesem Zusammenhang wird auch auf eine Arbeit von F. Jung [8] verwiesen.

Von Teil 2 – „Qualitätsanforderungen" – wird erwartet, daß er ohne irgendwelche großen Probleme in allen CEN-Ländern angenommen wird.

Der dritte Teil – Konformitätskriterien; Abschnitt 9 der ENV 197, Teil 1 – ist unverkennbar eng mit dem vorgeschlagenen Zementzertifizierungsverfahren verknüpft. In jedem Falle wird der Zementhersteller ein System der Eigenüberwachung anwenden, um die Konformität in Übereinstimmung mit den Anforderungen von Abschnitt 9 zu gewährleisten.

Aus diesem Grunde hat CEN/TC 51 entschieden, daß die WG 13 – das ist die Arbeitsgruppe, welche das Europäische Zementzertifizierungsverfahren formuliert hat – für die Durchsicht des in Abschnitt 9 beschriebenen Qualitätslenkungsverfahrens verantwortlich zeichnen und diesen Abschnitt so überarbeiten soll, daß er in die Europäische Norm EN 197-1 aufgenommen werden kann. Diese Verfahrensweise wird dazu führen, daß es zu einer guten Übereinstimmung zwischen der Europäischen Zementnorm und dem Europäischen Zementzertifizierungsverfahren kommen wird.

In diesem Zusammenhang soll erwähnt werden, daß CEN/TC 51, Arbeitsgruppe 10 (Putz- und Mauerbinder) und Arbeitsgruppe 11 (Baukalk) auf der Basis der Dokumente von WG 13 ebenfalls Zertifizierungsverfahren ausarbeiten werden, um die Europäische Vornorm ENV 413 (Putz- und Mauerbinder) und die ENV 459 (Baukalk) zu vervollständigen.

5.2 Europäisches Zementzertifizierungsverfahren

5.2.1 Annahme als nationales Verfahren

Wie bereits erwähnt, ähnelt das vorgeschlagene Zertifizierungsverfahren den nationalen Verfahren, wie sie in einer Reihe von Ländern schon angewendet werden. In vielen CEN-Ländern wurden die nationalen Zementnormen in Übereinstimmung mit der Europäischen Vornorm für allgemein gebräuchliche Zemente, ENV 197, Teil 1 (**Tabelle 3**) teilweise oder sogar gänzlich modifiziert.

Die Modifizierung bietet auch die Möglichkeit, gleichzeitig die nationalen Zementzertifizierungsverfahren in geeigneter Weise zu ändern. In diesem Zusammenhang verdient es erwähnt zu werden, daß in England und Belgien auf der Grundlage des Europäischen Verfahrens neue Zertifizierungsverfahren zum 1.5.1992 und 1.1.1993 bereits eingeführt wurden. In vielen anderen Ländern, so in Norwegen, Spanien, Deutschland und den Niederlanden, ist man dabei, die national angewandten Verfahren dem Europäischen Verfahren anzugleichen. In diesen Ländern baut das Zertifizierungsverfahren auf den nationalen Normen auf, die zur Übereinstimmung mit der ENV 197-1 entweder modifiziert oder nicht modifiziert werden können.

5.2.2 Annahme als ENV

Das vorgeschlagene Europäische Zementzertifizierungsverfahren (doc. no. CEN/TC 51/WG 13/N 13) wurde auf der letzten (19.) Sitzung von CEN/TC 51 in Oslo verabschiedet. Das Dokument wird von WG 13 im Zeitraum 1993/1994 fertiggestellt. Es wird als Abrundung den Abschnitt „Verwen-

4.6 Requirements for Inspection Bodies and Testing Laboratories

In order to ensure the satisfactory application of the proposed scheme for the European Certification of Cement, requirements need to be specified for approved inspection bodies and testing laboratories involved in the certification of cement. Cembureau has issued a set of recommendations representing the consensus view of cement industry experts. At the moment it is not intended that these should be included in the Certification Document.

Obviously these papers can be of considerable use in the context of the application of the Certification Scheme.

5. Future Developments

5.1 Change from ENV to EN 197-1

In principle the European cement standard ENV 197-1 consists of three parts:

— cement types (composition; constituents),

— quality requirements (mechanical, physical, chemical),

— conformity criteria.

The discussions in WG6 of CEN/TC 51 – charged with transforming ENV 197-1 into an EN 197-1 – have mainly been concentrated on the first part: cement types. In this context the paper by F. Jung [8] is interesting, published on the occasion of the VDZ Congress.

Part two – "Quality Requirements" – has been or is expected to be accepted without any great problems in all CEN countries.

The third part – Conformity Criteria; clause 9 of ENV 197 part 1 – is obviously closely connected with the proposed Cement Certification Scheme. In any case the manufacturer will operate a system of auto control to ensure conformity in accordance with the requirements of this clause 9.

For these reasons CEN/TC 51 has decided that WG 13 – i.e. the working group that formulated the European Cement Certification Scheme – should be responsible for reviewing the statistical quality control scheme, described in clause 9 and for making the changes needed to allow the clause to be considered for inclusion in a full European Standard EN 197-1. This will ensure that there is a good agreement between the European Cement Standard and the European Cement Certification Scheme.

In this context it must be mentioned that the CEN/TC 51 Working Groups 10 (Masonry Cement) and 11 (Building Lime) will also formulate certification procedures, on the basis of the documents drawn up by WG 13, to complement the European pre-standard ENV 413 ("Masonry Cement"), and ENV 459 ("Building Lime") respectively.

5.2 European Cement Certification Scheme

5.2.1 Adoption as national scheme

As already mentioned, the proposed certification scheme is similar to the national schemes already used in a number of countries. In many CEN countries the national cement standards have been partly or wholly modified in line with the European pre-standard for common cements, ENV 197 part 1 (Table 3).

The modification also offers the option of carrying out appropriate modifications to the national cement certification schemes at the same time. In this connection it is worth noting that, in the UK and Belgium, new certification schemes based on the European Scheme were introduced on 1. 5. 1992 and 1. 1. 1993 respectively. In many other countries (including Norway, Spain, Germany and the Netherlands) the National scheme is being modified in line with the European Scheme. In these countries the certification scheme refers to national standards, which may or may not be modified in line with ENV 197-1.

5.2.2 Adoption as ENV

The proposed "European Cement Certification Scheme" (doc.no. CEN/TC51/WG13/N13) was approved during the

dung des CE-Zeichens durch Versandstellen" enthalten (die Möglichkeit der Zuerkennung des CE-Zeichens für Zement aus Versandstellen einschließlich Transport). Im Laufe des Jahres 1994 wird CEN/TC 51 das Zertifizierungsverfahren zur Verabschiedung als ENV-Norm vorlegen[1]. Sobald die Europäische Kommission über die Stufe der Konformitätsbescheinigung entschieden hat, wird der Status der Europa-Norm (EN) gegeben sein.

5.3 Zukünftige Anwendung des CE-Zeichens

Wie bereits erwähnt, sind die sechs „Grundlagendokumente" bisher noch nicht fertiggestellt; sie sollen bis Ende 1993 herauskommen. Ohne Grundsatzdokumente kann es in der nahen Zukunft keine harmonisierten Europäischen Normen und deshalb auch kein CE-Zeichen geben. Auf der anderen Seite muß abgewartet werden, bis die Europäische Kommission den Zusammenhang zwischen Normung und Zertifizierung auf europäischer Ebene deutlicher herausgearbeitet hat.

Zu diesem Zeitpunkt kann erwartet werden, daß das Europäische Zementzertifizierungsverfahren in Europa nach und nach an Boden gewinnt: ein ausgereiftes und einheitliches Verfahren, das alle Aspekte der Zertifizierung behandelt – wirklich einmalig auf der Welt!

---

[1] Der Entwurf des Europäischen Zementzertifizierungsverfahrens „Zement – Beurteilung der Konformität" wurde bei der 20. Sitzung von CEN/TC 51 am 9./10. 6. 1994 in Luxemburg als Europäische Vornorm angenommen (Anmerkung des Herausgebers).

latest (19th) meeting of CEN/TC 51 in Oslo. The document will be completed by WG13 in the period 1993/1994. This will involve rounding off the section "Use of the EC Mark by intermediaries" (concerning the possibility of awarding a CE Mark to cement dispatched from dispatching centres, including the transport of cement). In the course of 1994 CEN/TC 51 will present the certification scheme for approval as an ENV standard[1]. When the CE will decide on the level of the attestation of conformity, the status of EN will be given.

5.3 Future use of the CE Mark

As already mentioned, the six "interpretative documents" are not yet ready and are expected to appear in late 1993. Without policy documents there can be no harmonized European standards and therefore no CE Mark in the near future. On the other hand it must be waited until the European Commission has defined the relationship between standardization and certification at European level more clearly.

Until that moment the European Cement Certification Scheme can be expected to gain ground gradually in Europe: a complete and coherent scheme, dealing with all aspects of certification, unique in the world!

---

[1] The draft of the European Cement Certification Scheme "Cement – Conformity evaluation" was approved as european prestandard at the 20th meeting of CEN/TC 51 in Luxemburg on 9./10.6.1994 (Note of the editor).

Literature

[1] Construction Products Directive: European Council Directive of 21. 12. 88 on the approximation of laws, regulations and administrative provisions of the Member States relating to construction products (89/106/EEC).

[2] Guidance Papers prepared by the European Commission's Standing Committee for Construction

Guidance Paper No. 5: Information to accompany the EC mark for construction products.

Guidance Paper No. 6: Guidelines for the designation of approved bodies in the field of the Council Directive 89/106/EEC on construction products.

Guidance Paper No. 7: Guidelines for the performance of the factory production control for construction products.

Guidance Paper No. 8: Guidelines for the choice of conformity attestation procedure.

Guidance Paper No. 9: Guidelines for the certification of construction products by an approved certification body.

Guidance Paper No. 10: Guidelines for the assessment and certification of the factory production control by an approved body.

[3] Rudert, V.: Gedanken zur Einführung von Qualitätssicherungssystemen für Zement-Herstellwerke. Vortrag zum VDZ-Kongreß 1993, Fachbereich 1.

[4] CEN/CS, TC Guidelines on requirements in European standards concerning the evaluation of conformity (dd. 5. 2. 1993).

[5] European Commission Paper Construct 92/097 (rev. 1), Note for Reflexion (dealing with a global approach for the choice of procedures for attestation of conformity within the framework of putting the CPD into operation).

[6] Global Approach to Certification and Testing Communication from the European Commission to the Council (COM (89) 209 final – SYN 208). Approved in principle by the Council on 21. 12. 89 and Council decision of 13. 12. 90 concerning the modules for the various phases of the conformity assessment procedures which are intended to be used in the technical harmonization Directives (90/683/EEC).

[7] European standard EN 29002 "Quality systems – Model for quality assurance in production and installation".

[8] Jung, F.: Anforderungen an Zumahlstoffe für die Zementherstellung. Vortrag zum VDZ-Kongreß 1993, Fachbereich 1.

Nutzen und Grenzen beim Einsatz von Sekundärstoffen

Benefits and limitations when using secondary materials

Exploitation et limites de l'utilisation de matières secondaires

Utilidad y limites del empleo de materias secundarias

Von **P. Liebl** und **W. Gerger**, Gmunden/Österreich

Zusammenfassung – Reststoffe aus industriellen Prozessen bzw. aus der qualitätsgesteuerten Abfallsammlung können als Sekundärroh- bzw. Sekundärbrennstoffe bei der Portlandzementerzeugung verwendet werden. Der Einsatz ist an Voraussetzungen gebunden und dadurch fallweise eingeschränkt.
- Genehmigungsrechtliche und emissionsrelevante Auflagen müssen zielsicher eingehalten werden.
- Die Qualität des Zements darf hinsichtlich seiner bautechnischen Leistungsfähigkeit und seiner Umweltverträglichkeit nicht beeinträchtigt sein.
- Der Erzeugungsprozeß darf nicht beeinträchtigt werden, und die Sicherheit am Arbeitsplatz muß gewährleistet sein.
- Es darf keine zusätzliche Umweltbelastung entstehen.

Aus diesen Anforderungen ergeben sich Kriterien, die den Einsatz von Sekundärstoffen in ihrer Menge begrenzen oder gänzlich ausschließen. Wenn unter Beachtung dieser Kriterien, unter Berücksichtigung des Aufwands für erforderliche Investitionen und unter Inkaufnahme von erhöhtem Kontroll- und Qualitätssicherungsaufwand die Wirtschaftlichkeit des Einsatzes gegeben ist, kann ein wertvoller Beitrag zur Umweltentlastung geleistet werden. Beim Einsatz von Sekundärstoffen kann unter Nutzung vorhandener Standorte sowie der verfahrenstechnischen Möglichkeiten des Zementerzeugungsprozesses eine umweltfreundliche, reststofffreie und natürliche Ressourcen schonende Entsorgungstechnologie genutzt und zugleich eine Kostenminderung für Roh- und Brennstoffe bei der Zementerzeugung erzielt werden.

Nutzen und Grenzen beim Einsatz von Sekundärstoffen

Summary – Residues from industrial processes and from quality-controlled refuse collection can be used as secondary raw materials or secondary fuels in the production of Portland cement. This usage is linked to conditions and is restricted by them in each individual case.
- Legal licensing and emission-relevant conditions must be met dependably.
- The quality of the cement must not be impaired with respect to its structural capabilities or its environmental compatibility.
- The production process must not be impaired and the safety of the workplace must be ensured.
- There must be no additional pollution of the environment.

These requirements give rise to criteria which limit the quantity of secondary materials used or even exclude them entirely. If, taking these criteria into account and bearing in mind the necessary capital expenditure and allowing for the increased expenditure on control and quality assurance, it is cost-effective to use these materials then a valuable contribution is made to relieving the load on the environment. When using secondary materials it is possible to take advantage of an environmentally friendly, residue-free disposal technology which preserves natural resources by using existing locations and the process engineering capabilities of cement production system; at the same time it is possible to reduce the raw material and fuel costs for cement production.

Benefits and limitations when using secondary materials

Résumé – Les matières résiduelles provenant de procédés industriels ou de système de collecte avec gestion de la qualité, peuvent être utilisées comme matières premières secondaires ou combustibles secondaires pour la fabrication du ciment de Portland. Leur utilisation est néanmoins liée à certaines conditions, limitatives dans certains cas.
- Les obligations en matière d'émissions et de droit d'autorisation doivent être respectées pour chaque cas.
- La qualité du ciment ne doit nullement être affectée tant du point de vue de ses performances techniques que de sa compatibilité avec l'environnement.
- Le procédé de fabrication ne doit pas être affecté et la sécurité au poste de travail doit être garantie.
- Aucune charge supplémentaire ne doit peser sur l'environnement.

De ces exigences précitées résultent des critères qui limitent l'utilisation de matières secondaires au point de vue quantité ou même les excluent entièrement. Dans le respect de ces critères avec prise en compte des moyens engagés pour les investissements et contrôle et assurance qualité élevés, si la rentabilité de l'utilisation est assurée, une contribution appréciable pourra être apportée à l'environnement. Si l'on utilise des matières secondaires on peut, tout en se servant de lieux d'implantation existants et de possibilités de procédé de fabrication de ciment, mettre à profit une technologie propre

Exploitation et limites de l'utilisation de matières secondaires

ménageant les ressources naturelles, sans matières résiduelles, et, en même temps, assurer une diminution des coûts de matières premières et de combustibles pour la fabrication du ciment.

Resumen – *Las materias residuales procedentes de los procesos industriales o de la recogida de desechos, según determinados criterios de calidad, pueden ser aprovechadas como materias secundarias o combustibles secundarios en la fabricación de cementos Portland. Dicho aprovechamiento queda limitado por diferentes requisitos:*
- *Hay que observar estrictamente determinadas exigencias legales en materia de permisos y emisiones.*
- *La calidad del cemento no debe sufrir ningún menoscabo, en cuanto a su capacidad técnica y de preservación del medio ambiente.*
- *El proceso de fabricación no debe ser afectada de forma negativa y ha de quedar garantizada la seguridad en el trabajo.*

De estos requisitos resultan unos criterios que limitan de forma cuantitativa el empleo de materias secundarias o que lo excluyen totalmente. A condición de que el citado empleo sea rentable, teniendo en cuenta los referidos criterios y las inversiones necesarias, y si se aceptan los gastos adicionales en concepto de controles y de aseguramiento de la calidad, se podrá contribuir, de forma considerable, a la protección del medio ambiente. Al emplear materias secundarias, y aprovechando los emplazamientos existentes así como las posibilidades tecnológicas del proceso de fabricación del cemento, se podrá emplear una tecnología de eliminación de residuos que no afecta al medio ambiente, que está libre de materias residuales y que permite preservar los recursos naturales. Y al mismo tiempo se puede conseguir una reducción del coste de materias primas y de combustibles necesarios para la fabricación del cemento.

Utilidad y límites del empleo de materias secundarias

1. Einleitung und Definitionen

Eine Möglichkeit der bewußten Gestaltung und Optimierung des Produktionsprozesses, die dem Zementhersteller zur Verfügung steht, ist der gezielte Einsatz von Sekundärstoffen im Zementwerk. Dieser kann mit nicht unbeträchtlichem Nutzen für den Anwender verbunden sein und damit gerade in wirtschaftlich schwierigen Zeiten eine Chance des Gegensteuerns bieten. Andererseits sind dem Einsatz von Sekundärstoffen aber auch vielfältige Grenzen gesetzt, deren Nichtbeachtung und Überschreitung mit einem beachtlichen Gefahrenpotential verbunden ist. So erfordert es den verantwortungsbewußten Einsatz der material- und verfahrenstechnischen Kenntnisse des erfahrenen Zementherstellers, z.B. hinsichtlich der Bilanzierung von Stoffströmen, um den vollen Nutzen aus dem Einsatz von Sekundärstoffen bei gleichzeitiger Vermeidung von Gefahren ziehen zu können.

Im klassischen Zementherstellungsprozeß werden Roh- und Brennstoffe überwiegend natürlichen Ursprungs wie Kalkstein, Mergel und Tonkomponeten bzw. Kohle, Heizöl oder Erdgas eingesetzt. Diese können unter der Beachtung bestimmter Voraussetzungen und Grenzen sowie in gezielter Nutzung der bekannten und bewährten Verfahrenstech-

1. Introduction and definitions

Carefully controlled use of secondary materials in a cement works constitutes one option available to the cement manufacturer for organizing and optimizing the production process. This can be linked with not inconsiderable benefits for the user, and therefore offers a chance of improvement during these economically difficult times. On the other hand, the use of secondary materials is also subject to a great many limitations which are linked with considerable potential risks if disregarded or exceeded. The experienced cement manufacturer's knowledge of material technology and process engineering, e.g. with regard to balancing material flows, has to be applied responsibly in order to extract the full benefit from the use of secondary materials while at the same time avoiding the dangers.

In the classical cement manufacturing process the raw materials and fuels used, such as limestone, marl and clay components and coal, fuel oil, or natural gas, are of predominantly natural origin. By bearing in mind certain conditions and limits and making carefully controlled use of the known and proven process technology of the cement manufacturing process, these can be replaced by so-called secondary materials – which are also designated as residues from other

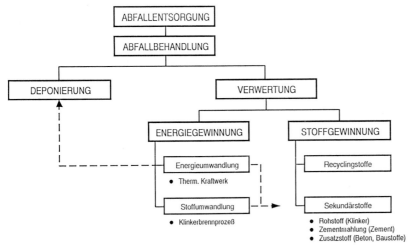

BILD 1: Schematischer Ablauf der Abfallentsorgung [1]

FIGURE 1: Diagrammatic representation of the disposal of waste materials [1]

Abfallentsorgung	= Waste disposal
Abfallbehandlung	= Waste treatment
Deponierung	= Landfill site
Verwertung	= Utilization
Energiegewinnung	= Energy recovery
Stoffgewinnung	= Material recovery
Energieumwandlung	= energy conversion
Recyclingstoffe	= recycled materials
Therm. Kraftwerk	= thermal power station
Stoffumwandlung	= material conversion
Klinkerbrennprozeß	= clinker burning process
Sekundärstoffe	= secundary materials
Rohstoff	= raw material
Klinker	= clinker
Zementmahlung	= cement grinding
Zement	= cement
Zusatzstoff	= additives
Beton	= concrete
Baustoffe	= construction materials

nik des Zementherstellungsprozesses durch sogenannte Sekundärstoffe ersetzt werden, die auch als Reststoffe aus anderen industriellen Produktionen oder aus der Abfallentsorgung bezeichnet werden. Diese Reststoffe sind entsprechend dem Ablaufschema der Abfallentsorgung (**Bild 1**) zunächst Abfälle, bis sie oder die aus ihnen gewonnenen Stoffe oder erzeugte Energie dem Wirtschaftskreislauf zugeführt werden [1].

Für die Zementherstellung eignen sich Reststoffe dann als Sekundärstoffe, wenn nach Prüfung und gegebenenfalls erforderlicher Aufbereitung die Verwendbarkeit gesichert ist.

Sekundärrohstoffe sind somit Reststoffe, die alternativ zu Primärrohstoffen eingesetzt werden können. Sekundärbrennstoffe sind brennbare Reststoffe, die Alternativen zu Primärbrennstoffen, d. h. zu fossilen Brennstoffen, darstellen. Eine weitergehende Unterteilung von Sekundärrohstoffen kann erfolgen für solche, die bei der Zementmahlung zugesetzt werden können, indem man sie als Sekundärzumahlstoffe charakterisiert. Diese Definitionen grenzen die Begriffe nicht scharf gegeneinander ab, wenn beispielsweise Sekundärbrennstoffe nennenswerte Anteile an unbrennbaren Bestandteilen aufweisen, so daß neben der energetischen Nutzung auch noch eine stoffliche Verwertung gegeben ist und der Begriff Sekundärrohstoff nicht ausgeschlossen werden kann. Dieser Gedankengang ist dem Zementerzeuger seit jeher vertraut durch die Verwendung aschereicher Kohlen als Primärbrennstoff, die bei der Mischungsrechnung zur Berücksichtigung des Ascheanteils als Rohstoffkomponente (= Sekundärrohstoff) gezwungen hat.

Vorrangiges Ziel der Verwertung muß die Stoffgewinnung im Sinne der ursprünglichen Verwendung im Recycling sein. Gelingt diese Zielsetzung nicht oder nur teilweise, dann ist die Verwertung als Sekundärstoff in jedem Fall der Deponierung vorzuziehen.

Da die Hinzunahme von bislang nicht verwendeten Sekundärstoffen grundsätzlich geeignet erscheint, schädliche Umwelteinwirkungen hervorzurufen oder in anderer Weise die Allgemeinheit oder die Nachbarschaft zu gefährden, erheblich zu benachteiligen oder zu belästigen, muß der Zementhersteller entsprechend der nationalen Gesetzgebung die genehmigungsrechtlichen Voraussetzungen dafür schaffen. Die bestehenden Verflechtungen zwischen Abfallgesetzgebung, Genehmigungsrecht sowie wirtschaftlichen Gesichtspunkten haben dazu geführt, daß sowohl auf nationaler als auch auf internationaler Ebene Listen und Richtlinien im Entstehen begriffen sind, die Klarheit für Behörden und Betreiber schaffen sollen. Ein detailliertes Eingehen auf dieses umfangreiche Sachgebiet, das sich überdies im

industrial production processes or from waste disposal. According to the waste disposal flow sheet (**Fig. 1**) these residues are waste materials until they, or the materials extracted from them, or the energy generated, are fed into the economic cycle [1].

Residues are suitable as secondary materials for cement manufacture when their acceptability has been established after they have been tested and, if necessary, processed.

Secondary raw materials are therefore residues which can be used as alternatives to primary raw materials. Secondary fuels are combustible residues which represent alternatives to primary fuels, i.e. to fossil fuels. A further subdivision of secondary raw materials can be made for those which can be added during the cement grinding process in that they are designated secondary additives. These definitions do not draw sharp boundaries between the terms. If, for example, secondary fuels contain appreciable proportions of incombustible components then use is made of the material as well as of the energy, and the term secondary raw material could also be applied. This train of thought has been familiar to the cement manufacturer for a long time through the use of high-ash coal as primary fuel, where the ash content has to be taken into account as a raw material component (= secondary raw material) in the mix calculation.

The primary object of the utilization must be to use the recycling system to recover the material for its original use. If this objective fails or is only partially successful, then utilization as a secondary material is always preferable to dumping on a landfill site.

Basically, the addition of secondary raw materials which have not previously been used appears to be a suitable way of causing harmful environment effects or in other ways threatening, disadvantaging, or annoying the public or the neighbourhood, so the cement manufacturer must comply with the legal licensing conditions issued by the national legislature. The existing interrelationships between waste legislation, licensing law and economic aspects have led to the situation where lists and guidelines are being drawn up both at the national and international level; these are intended to clarify the situation for the authorities and the operators. Any detailed examination of this extensive area, which is in a state of flux, would be beyond the scope of this article.

2. Types of secondary materials

In principle, the cement manufacturing process provides four opportunities for the introduction of materials (**Fig. 2**)

— as a raw material,
— in the secondary firing system for precalcination,

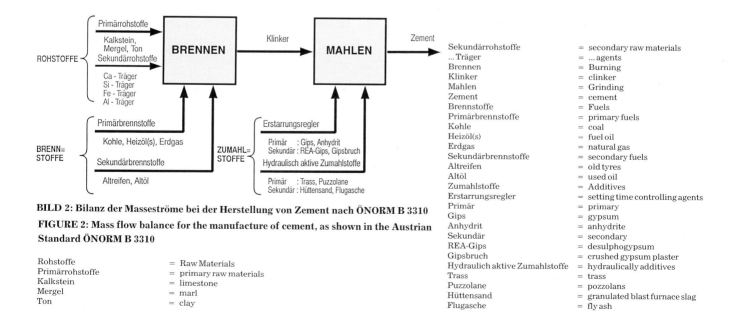

BILD 2: Bilanz der Massestrome bei der Herstellung von Zement nach ÖNORM B 3310

FIGURE 2: Mass flow balance for the manufacture of cement, as shown in the Austrian Standard ÖNORM B 3310

Rohstoffe	= Raw Materials
Primärrohstoffe	= primary raw materials
Kalkstein	= limestone
Mergel	= marl
Ton	= clay
Sekundärrohstoffe	= secondary raw materials
... Träger	= ... agents
Brennen	= Burning
Klinker	= clinker
Mahlen	= Grinding
Zement	= cement
Brennstoffe	= Fuels
Primärbrennstoffe	= primary fuels
Kohle	= coal
Heizöl(s)	= fuel oil
Erdgas	= natural gas
Sekundärbrennstoffe	= secondary fuels
Altreifen	= old tyres
Altöl	= used oil
Zumahlstoffe	= Additives
Erstarrungsregler	= setting time controlling agents
Primär	= primary
Gips	= gypsum
Anhydrit	= anhydrite
Sekundär	= secondary
REA-Gips	= desulphogypsum
Gipsbruch	= crushed gypsum plaster
Hydraulich aktive Zumahlstoffe	= hydraulically additives
Trass	= trass
Puzzolane	= pozzolans
Hüttensand	= granulated blast furnace slag
Flugasche	= fly ash

raschen Wandel befindet, würde über den Rahmen dieses Beitrages hinausgehen.

2. Arten von Sekundärstoffen

Grundsätzlich erlaubt der Zementerzeugungsprozeß vier Möglichkeiten der Materialzufuhr (**Bild 2**)

— als Rohmaterial,
— in der Sekundärfeuerung zur Vorcalcinierung,
— in der Hauptfeuerung des Drehrohrofens,
— als Zumahlstoff (bzw. auch als Hilfsstoff) bei der Zementmahlung.

Sekundärrohstoffe ersetzen je nach chemischer Zusammensetzung Rohmaterialkomponenten bzw. eignen sich als Korrekturstoffe für die Rohmischung. Nach ihrem Hauptbestandteil bzw. der für die Mischungsberechnung wirksamen Elemente kann man sie in Calcium-, Silicium-, Eisen-, Aluminium-, Schwefelträger und andere unterteilen.

TABELLE 1: Gruppen von Sekundärrohstoffen für die Herstellung von Portlandzement-Klinker [1]

Ca-Träger	z. B.	Industriekalk Kalkschlamm Karbidschlamm Trinkwasserschlamm
Si-Träger	z. B.	Gießereialtsand Mikrosilika
Fe-Träger	z. B.	Kiesabbrand Synth. Hämatit Rotschlamm Zinnschlacke Konverterstäube Walzzunder
Si-, Al-, Ca-Träger	z.B.	Steinkohle – Braunkohle Flugasche Wirbelschichtasche Schlacken Reststoffe aus Natursteinaufbereitung
S-Träger	z. B.	REA-Gips Chemiegips
F-Träger	z. B.	CaF_2-Filterschlamm

Die **Tabelle 1** enthält eine Zusammenstellung von Gruppen von Sekundärrohstoffen [1].

Die Auflistung in Tabelle 1 will und kann nicht vollständig sein. Die Eignung der verschiedenen Reststoffe unterschiedlichster Herkunft muß in jedem Einzelfall geprüft werden.

Sekundärbrennstoffe können sowohl bei der primärseitigen Hauptfeuerung als auch in der Sekundärfeuerung zur Calcinierung des Rohmehles eingesetzt werden [2]. Wegen der aus Qualitätsgründen erforderlichen hohen Brennguttemperaturen von ca. 1450 °C und des erforderlichen Sauerstoffüberschusses stehen in der Hauptfeuerung ideale Verbrennungsbedingungen für Sekundärbrennstoffe zur Verfügung. In der Sekundärfeuerung können, weil die zur Calci-

— in the main firing system for the rotary kiln,
— as an additive (or as a grinding aid) in the cement grinding process.

Depending on their chemical compositions, secondary raw materials can replace raw material components or are suitable as correction materials for the raw mix. They can be subdivided into calcium, silicon, iron, aluminium, and sulphur agents and others depending on their main components or on the elements which affect the mix calculation.

TABLE 1: Groups of secondary raw materials for the manufacture of Portland cement clinker [1]

Ca agents	e. g.	industrial lime lime slurry carbide slurry drinking water slurry
Si agents	e. g.	used foundry sand microsilica
Fe agents	e. g.	roasted pyrites synthetic haematite red mud tin slag converter dust mill scale
Si, Al, Ca agents	e. g.	coal – lignite fly ash fluidized bed ash slag residues from natural stone working
S agents	e. g.	desulphogypsum chemical gypsum
F agents	e. g.	CaF_2 filter slurry

Table 1 contains a list of groups of secondary raw materials [1].

The listing in Table 1 will not be, and cannot be, exhaustive. The suitability of the different residues of very varied origin has to be tested in each individual case.

Secondary fuels can be used both in the main firing system on the primary side and also in the secondary firing system for calcining the raw meal [2]. Ideal combustion conditions for secondary fuels are available in the main firing system because of the high material temperature of approximately 1450 °C required on quality grounds and because of the oxygen excess needed. Low grade and lump fuels can be used in the secondary firing system because the temperatures required for calcining do not have to be so high [3, 4].

Table 2 contains a summary of solid, liquid and gaseous secondary fuels, but here again there is no claim to completeness [5].

3. Limitations on the use of secondary materials

The use of secondary raw materials and fuels is dependent on compliance with various constraints. Limitations arise through:

TABELLE 2: Sekundärbrennstoffe für die Herstellung von Portlandzement-Klinker [5]

Feste Sekundärbrennstoffe		Flüssige Sekundärbrennstoffe	Gasförmige Sekundärbrennstoffe
z. B.		z. B.	z. B.
Papierabfälle	Holzabfälle	Teer	Deponiegas
Abfälle der Papierindustrie	(Rinde, Holzspäne, Sägespäne)	Säureharz	Pyrolysegas
Petrolkoks	Reisspreu	Altöl	
Graphitstaub	Olivenkerne	Petrochemische Abfälle	
Holzkohle	Kokosnußschalen	Abfälle der Farbenindustrie (Lackrückstände)	
Kunststoffabfälle	Hausmüll	Chemieabfälle	
Gummiabfälle	BRAM	Lösungsmittelabfälle	
Altreifen	Shredder	Destillationsrückstände	
Batteriekästen	Ölhaltige Erde	Wachssuspensionen	
Bleicherde	Klärschlamm	Asphaltschlamm	
aktivierter Bentonit		Ölschlamm	

TABLE 2: Secondary fuels for the manufacture of Portland cement clinker [5]

Solid secondary fuels		Liquid secondary fuels	Gaseous secondary fuels
e. g.		e. g.	e. g.
paper wastes	wood wastes	tar	landfill gas
wastes from the paper industry	(bark, shavings, sawdust)	acid sludge	pyrolysis gas
petroleum coke	rice chaff	used oil	
graphite dust	olive kernels	petrochemical wastes	
charcoal	coconut shells	wastes from the paint industry (varnish residues)	
plastics residues	household refuse	chemical wastes	
rubber residues	refuse-derived fuel	solvent wastes	
old tyres	shreddings	distillation residues	
battery cases	oil-bearing earths	wax suspensions	
podsol	sewage sludge	asphalt slurry	
activated bentonite		oil sludge	

nierung erforderlichen Temperaturen nicht so hoch sein müssen, auch minderwertige bzw. auch stückige Brennstoffe eingesetzt werden [3, 4].

Tabelle 2 enthält eine Zusammenstellung von festen, flüssigen und gasförmigen Sekundärbrennstoffen, wobei auch hier kein Anspruch auf Vollständigkeit erhoben werden kann [5].

3. Grenzen des Einsatzes von Sekundärstoffen

Der Einsatz von Sekundärroh- bzw. Brennstoffen ist von der Erfüllung von Randbedingungen abhängig. Grenzen ergeben sich durch:

— Genehmigungsrechtliche Auflagen,
— Produktqualität,
— Verfahrensführung,
— zusätzliche Umweltbelastung,
— Sicherheit am Arbeitsplatz,
— mangelnde Akzeptanz bei Behörden, Anrainern, Kunden,
— Aufwand durch Investitionen und Betriebskosten.

Grundsätzlich können diese Voraussetzungen für jede Art von Einsatzmaterial gelten. Eine Bewertung kann sich nur auf die Beurteilung der Eigenschaften, unabhängig von Herkunft bzw. Entstehung, beziehen.

In der Regel begrenzen genehmigungsrechtliche Einschränkungen den Einsatz von Sekundärroh- und Brennstoffen nach Art und fallweise auch nach Menge. So sieht beispielsweise die gewerberechtliche und abfallrechtliche Genehmigung zur thermischen Verwertung brennbarer Flüssigabfälle bei den Gmundner Zementwerken eine Kategorisierung von Sekundärbrennstoffen nach begrifflicher Klassifikation in Form von Schlüsselnummerhauptgruppen des Abfallkataloges vor (**Tabelle 3**) [6].

Eine allgemeine verbale Charakteristik klassifiziert am ehesten nach Herkunft mit wenig konkreten Grenzen. Die Praxis der Sammlung von unterschiedlichen Mengen unterschiedlicher Herkunftsart sowie die notwendigerweise vorzunehmende Mischung bei der Zwischenlagerung eröffnet

TABELLE 3: Gewerbe- und abfallrechtlich genehmigte Abfälle gemäß Abfallkatalog ÖNORM S 2100 bzw. S 2101 zur thermischen Entsorgung brennbarer Flüssigabfälle bei den Gmundner Zementwerken H. Hatschek AG

541	Abfälle von Mineralölen
544	Abfälle von Emulsionen und Gemischen von Mineralölprodukten
547	Mineralölschlämme
548	Rückstände aus Mineralölraffinerien
553	Abfälle von halogenfreien, organischen Lösungsmitteln und Lösungsmittelgemischen
555	Abfälle von Farb- und Anstrichmitteln

— legal licensing regulations,
— product quality,
— process management,
— additional environmental pollution,
— health and safety at work
— unacceptability with authorities, neighbours, clients,
— expenditure through investment and operating costs.

In principle, these conditions can apply to any type of material used. Any evaluation can only relate to assessment of the properties, regardless of origin.

As a rule, legal licensing restrictions place limits on the use of secondary raw materials and fuels according to type and also sometimes according to quantity. For example, licensing in accordance with the industrial code and waste regulations for thermal exploitation of combustible liquid waste at the Gmunden cement works provides for a categorization of secondary fuels based on a conceptual classification in the form of main groups of key numbers from the waste catalogue (**Table 3**) [6].

TABLE 3: Waste materials licensed in accordance with industrial and waste legislation as listed in waste catalogue ÖNORM S 2100 and S 2101 for thermal disposal of liquid wastes at the Gmunden cement works of H. Hatschek AG

541	Wastes from mineral oils
544	Wastes from emulsions and mixtures of mineral oil products
547	Mineral oil sludges
548	Residues from mineral oil refineries
553	Wastes from halogen-free organic solvents and solvent mixtures
555	Wastes from paints and varnishes

Classification is first made by a general verbal characteristic based on origin with few definite limits. The practice of collecting different quantities of varying origin and the inevitable mixing which takes place during intermediate storage opens up possibilities for a different classification. More appropriate is the supplementary method of specifying limits for measurable properties and concentrations of constituents. Compliance with these limits has to be proved specifically for each individual delivery by chemical analysis or physical test methods (**Table 4**).

Provision can be made for quantity restrictions on the basis of the concentration of certain constituents such as, for example, restricting the mass flow of combustible liquid waste as a function of PCB content.

Proof that the use of secondary materials does not lead to any additional environmental pollution normally has to be carried out within the framework of a trial operation with accompanying emission measurements. If used oils containing PCBs are to be used then, in addition to regulations relating to safety and input checks on the secondary materials, limits are usually allocated to other parameters such as, for

Möglichkeiten unterschiedlicher Einordnung. Zweckmäßiger ist das zusätzliche Verfahren der Vorgabe von Grenzwerten für meßbare Eigenschaften und Konzentrationen von Inhaltsstoffen. Die Einhaltung dieser Grenzwerte ist im konkreten Fall an jeder Einzelanlieferung durch chemischanalytische bzw. physikalische Prüfverfahren nachzuweisen (**Tabelle 4**).

TABELLE 4: Grenzwerte für Flüssigabfälle, Auflagen zur thermischen Entsorgung brennbarer Flüssigabfälle bei den Gmundner Zementwerken H. Hatschek AG

● Einzelanlieferung

– PCB	$\leq$	80 ppm
– Pb	$\leq$	5000 ppm
– Hg	$\leq$	2 ppm
– Tl	$\leq$	10 ppm
– Cd	$\leq$	60 ppm
– F	$\leq$	600 ppm
– S	$\leq$	5 %
– Cl	$\leq$	1 %
– N	$\leq$	5 %

● Tankinhalt

– H$_2$O	$\leq$	20 %
– H$_o$	$\geq$	15000 kJ/kg
– Fördermenge:		
PCB > 10 ppm		max 3000 kg/h
PCB < 10 ppm		max 4000 kg/h

● **Jahresfracht PCB** $\leq$ **200 kg**

In Abhängigkeit von der Konzentration von bestimmten Inhaltsstoffen können Mengenbegrenzungen vorgesehen werden, wie beispielsweise eine Begrenzung des Masseflusses an brennbarem Flüssigabfall in Abhängigkeit vom PCB-Gehalt.

Der Nachweis, daß es beim Einsatz von Sekundärstoffen zu keiner zusätzlichen Umweltbelastung kommt, muß in der Regel im Rahmen eines Probebetriebes mit begleitenden Emissionsmessungen geführt werden. Neben Auflagen, die die Sicherheitstechnik und die Eingangskontrolle der Sekundärstoffe betreffen, werden meist zusätzliche Parameter mit Grenzwerten belegt, wie z.B. das 2,3,7,8 TCDD-Äquivalent mit 0,1 ng/m^3_N, wenn PCB-haltige Altöle eingesetzt werden sollen. Problematisch ist die Übertragung von Grenzwerten für „Verbrennungsanlagen für Abfälle und ähnliche brennbare Stoffe" auf Anlagen zur Zementerzeugung, ohne deren spezifischer prozeßabhängiger Emissionssituation Rechnung zu tragen. Grenzwerte, wie sie die 17. Verordnung zum Deutschen BImSchG für NO$_x$, CO und Org. C festlegt, sind aus Gründen, die von der Art des eingesetzten Sekundärstoffes unabhängig sind, nach dem Stand der Technik der Emissionsminderung von Zementöfen zielsicher nicht einzuhalten.

Dem ökologisch und ökonomisch sinnvollen Einsatz von Sekundärstoffen werden aus formalen Gründen Grenzen gesetzt, die vom Zementerzeuger beim erklärten Willen zur Emissionsminderung unter Beachtung des Prinzips der Verhältnismäßigkeit nicht überwunden werden können [7].

Aus der Forderung nach erstklassiger Produktqualität und uneingeschränkten Gebrauchseigenschaften sowie aus der Forderung, daß es zu keiner zusätzlichen Umweltbelastung und keiner Beeinträchtigung der Prozeßführung kommen darf, ergeben sich ausschließende bzw. mengenbegrenzende Kriterien.

Diese Kriterien gelten selbstverständlich auch für natürliche Rohmaterialien bzw. für fossile Brennstoffe. Die erhöhten Anforderungen des Umweltschutzes, aber auch die Diskussion um Sekundärstoffe, bringen durch die Anwendung einer verfeinerten Analytik auch bei natürlichen Roh- und Brennstoffen mitunter Erkenntnisse, die ihren Einsatz in Frage stellen bzw. begrenzen.

Zur Gewährleistung einer kontinuierlichen Zementqualität sollten Sekundärstoffe während der Produktionsperioden

TABLE 4: Limiting values for liquid wastes; regulations for thermal disposal of combustible liquid wastes at the Gmunden cement works of H. Hatschek AG

● Single delivery

– PCB	$\leq$	80 ppm
– Pb	$\leq$	5000 ppm
– Hg	$\leq$	2 ppm
– Tl	$\leq$	10 ppm
– Cd	$\leq$	60 ppm
– F	$\leq$	600 ppm
– S	$\leq$	5 %
– Cl	$\leq$	1 %
– N	$\leq$	5 %

● Tank contents

– H$_2$O	$\leq$	20 %
– c.v. (gross)	$\geq$	15000 kJ/kg
– transport quantity:		
PCB > 10 ppm		max 3000 kg/h
PCB < 10 ppm		max 4000 kg/h

● **Annual PCB loading** $\leq$ **200 kg**

example, 0.1 ng/m^3 (standard state) for the 2,3,7,8 TCDD-equivalent. There is a problem with transferring limiting values for "incineration plants for wastes and similar combustible substances" to plants for manufacturing cement without taking into account the specific process-dependent emission situation. Limiting values, such as those laid down in the 17th Ordinance of the BImSchG (German Federal Immission Protection Act) for NO$_x$, CO and organic C cannot, for reasons which are independent of the type of secondary material used, be met reliably using state of the art of emission reduction systems for cement kilns.

The ecologically and economically appropriate use of secondary substances is beset on formal grounds by restrictions which cannot be overcome by the cement producer for the purposes of emission while keeping a reasonable balance between cost and effect [7].

The requirement for first class product quality and unrestricted service characteristics, as well as the requirement that there must be no additional environmental pollution and no impairment of the process operation give rise to exclusive or quantity-restricting criteria.

These criteria obviously also apply to natural raw materials and to fossil fuels. The heightened requirements of environmental protection and also the debate about secondary materials sometimes, because of the application of improved methods of analysis, give rise to findings, even with natural raw materials and fuels, which place their use in question or restrict them.

To ensure continuous cement quality there should be unrestricted availability of secondary materials during the production period. This often necessitates the creation of additional special storage capacity which can jeopardize the cost effectiveness of utilizing the material. The consistency of the secondary materials may also have to be adapted to suit the existing storage, conveying and metering equipment. If, for a good quality clinker, the use of a certain secondary raw material produces changes in the cement properties, such as colour or hardening characteristics, then changes cannot be carried out at will and would be unreasonable for the market. Fluctuations in the composition of the secondary materials must be detected by appropriate quality controls, which requires considerable expenditure on analytical equipment, and also on appropriately qualified personnel.

If the fluctuations cannot be overcome with the existing blending equipment, then the plant itself imposes limits on the use of certain secondary materials.

In principle, it has to be possible to integrate secondary raw materials into the raw mix scheme in order to satisfy the quality requirements for the cement clinker. The lime standard must not fall below a relatively high minimum level in

uneingeschränkt verfügbar sein. Häufig erfordert das die Schaffung von zusätzlicher spezieller Lagerkapazität, was die Wirtschaftlichkeit der Verwertung in Frage stellen kann. Auch die Konsistenz der Sekundärstoffe erfordert gegebenenfalls Anpassungen der vorhandenen Lager-, Förder- und Dosiereinrichtungen. Bewirkt die Verwendung eines bestimmten Sekundärrohstoffes bei einwandfreier Klinkerqualität Veränderungen der Zementeigenschaften, wie z. B. Farbe bzw. Erhärtungscharakteristik, werden Umstellungen nicht beliebig oft durchführbar und dem Markt nicht zumutbar sein. Schwankungen der Zusammensetzung der Sekundärstoffe müssen durch entsprechende Qualitätskontrolle erfaßt werden, wodurch ein beträchtlicher Aufwand an analytischer Ausrüstung aber auch an Personal mit entsprechender Qualifikation notwendig wird.

Wenn die Schwankungen mit den vorhandenen Homogenisierungseinrichtungen nicht beherrscht werden können, sind dem Einsatz bestimmter Sekundärstoffe anlagenspezifische Grenzen gesetzt.

Sekundärrohstoffe müssen grundsätzlich in das Rohmischungskonzept integrierbar sein, um den Qualitätsanforderungen an den Zementklinker zu genügen. Für die Ansprüche der modernen Bauindustrie an die Leistungsfähigkeit des Zements wird ein relativ hoher Mindestkalkstandard nicht unterschritten werden dürfen. Es wird daher jeweils zu prüfen sein, ob beispielsweise bei Verwendung von Flugasche als Rohmaterial [8] eine Kompensation mit einer calciumhaltigen Komponente in ausreichendem Maße zustandegebracht werden kann. Vorteilhaft ist, wenn über eine Reinkalksteinkomponente verfügt wird. Selbst dann kann ein erhöhter Aluminiumgehalt zu negativen Auswirkungen bei der Klinkerqualität führen und daher zu einem mengenbegrenzenden Faktor für den Sekundärrohstoff werden.

Siliciumträger wie Gießereialtsande eignen sich zur Steuerung des Silikatmoduls, wenn die Altsande frei von Phenol oder Furanharzen bzw. unter den Bedingungen ihrer primären Verwendung entstandenen Zersetzungs- und Reaktionsprodukten sind. Sind sie das nicht, können diese flüchtigen Verbindungen im Zuge der Mahltrocknung emittiert werden. Überhaupt ist bei Rohmaterialien die Frage des Anteils an verflüchtigbaren organischen Bestandteilen zu klären, um daraus resultierende Umweltbelastungen abzuschätzen und gegebenenfalls durch Mengenbegrenzung einzuschränken.

Die Grenzen des Einsatzes von Sekundärrohstoffen mit flüchtigen organischen Verbindungen werden erweitert, wenn diese Stoffe im Bereich höherer Temperaturen des Ofensystems eingebracht werden. Die möglicherweise auftretende Inhomogenität der Klinkerzusammensetzung kann dieser Verfahrenstechnik allerdings auch Grenzen setzen [8].

Wie in natürlichen Rohmaterialien ist auch in Sekundärrohstoffen der Anteil an Magnesium ein mengenbegrenzendes Element, zumal zur Verhinderung des Magnesiatreibens die europäische Norm 5% als Grenzwert im Klinker festsetzt.

Eisenträger dienen vorwiegend als Korrekturmaterial zur Erhöhung der Schmelzphase und zur Verbesserung der Brennbarkeit sowie zur Absenkung des C_3A-Gehaltes bei der Herstellung von Klinker für erhöht sulfatbeständige Zemente.

Mit eisenhaltigen Sekundärrohstoffen, aber auch mit Sekundärbrennstoffen, werden in der Regel Schwermetalle in das Ofensystem eingetragen. Bei Schwermetallen mit geringer Flüchtigkeit resultieren hohe Einbindungsraten, so daß kaum mit einem Beitrag zur Erhöhung der Emissionen gerechnet werden muß [9–13]. Werden erhöhte Mengen an Schwermetallen im Klinker gebunden – wie das beispielsweise bei Zink der Fall sein kann, wenn Autoreifen als Sekundärbrennstoff verwendet werden – ist zu klären, ob das Prinzip der uneingeschränkten Verwendung der Zemente dadurch nicht beeinträchtigt wird. Während bei Zinkkonzentrationen bis ca. 500 mg/kg noch keine Einflüsse auf Zementeigenschaften festgestellt werden [6], geben verschiedene Untersuchungen Anhaltspunkte für Veränderun-

order to satisfy the modern construction industry's demands for the performance of the cement. It is therefore always necessary to check whether, for example, when using fly-ash as a raw material [8] sufficient compensation can be achieved with a calcium-containing component. It is an advantage if a pure limestone component is available. Even then an increased aluminium content can have detrimental effects in the clinker quality and therefore become a quantity-limiting factor for the secondary raw material.

Silicon agents, such as used foundry sand, are suitable for controlling the silica modulus provided that the used sand is free from phenol or furan resin, and from decomposition and reaction products produced under the conditions of primary use. If this is not the case these volatile compounds may be emitted during the drying and grinding process. The first question to be answered with raw materials concerns the fraction of volatilizable organic constituents, so that the resulting environmental pollution can be established, and if necessary limited by restricting the quantity.

The limits on the use of secondary raw materials containing volatile organic compounds are widened if these materials are introduced into the higher temperature regions of the kiln system. Any inhomogeneity of the clinker composition which may occur can, however, also set limits to this procedure [8].

The proportion of magnesium in secondary raw materials is a quantity-limiting factor in the same way as for natural raw materials, especially as the European Standard stipulates 5% as the limiting value in the clinker in order to prevent expansion due to magnesia.

Iron agents are used predominantly as correction materials for increasing the liquid phase and for improving the burnability, as well as for lowering the C_3A content when producing clinker for cements with increased sulphate resistance.

As a rule heavy metals are introduced into the kiln system with secondary raw materials which contain iron, and also with secondary fuels. High fixation rates are achieved with heavy metals of low volatility, so it is hardly necessary to take account of any contribution towards increased emissions [9–13]. If increased quantities of heavy metals are combined in the clinker – as, for example, can be the case with zinc when car tyres are used as a secondary fuel – it is necessary to resolve whether or not this prejudices the principle of unrestricted use of the cements. Although no influence on cement properties can be detected at zinc concentrations up to approximately 500 mg/kg [6], various investigations indicate that there are changes in the workability characteristics at higher concentrations [14–18]. In doubtful cases it is recommended not to go substantially beyond the levels of concentrations dictated by experience [19].

In contrast to the situation with elements of low volatility the cement production process offers no security for readily volatile elements such as mercury, with the result that residues which contain relevant quantities of mercury are excluded from use [20]. The same applies to thallium which builds up in the recirculating dust unless suitable measures are provided for relieving these recirculating systems [21–22].

Internal recirculation of alkalis and chlorides within the kiln can cause process problems at critical concentrations through coating and ring formation. If there is no process engineering equipment available for controlling the alkali and chloride concentrations, then it is necessary to restrict the quantities of any components which contain these compounds [23].

Raw materials which contain sulphur can have a beneficial effect on the clinker properties if there is an excess of alkalis under oxidizing combustion conditions [24]. Levels of sulphurization above 100% or the presence of the sulphur as sulphide in the raw material can lead to SO_2 emission and force measures to be taken where emission limits are exceeded [25, 26]. Phosphates and fluorides affect the clinker properties if they enter the kiln system with secondary materials.

gen der Verarbeitungseigenschaften bei höheren Konzentrationen [14–18]. Im Zweifelsfall wird es angeraten sein, über das Band des Konzentrationsbereiches, wie es den jeweiligen Erfahrungen entspricht, nicht wesentlich hinauszugehen [19].

Im Gegensatz zu schwer flüchtigen Elementen bietet der Zementerzeugungsprozeß bei leicht flüchtigen Elementen wie Quecksilber keine Sicherheiten, so daß Reststoffe, die relevante Mengen an Quecksilber enthalten, vom Einsatz auszuschließen sind [20]. Ähnliches gilt für Thallium, das in Kreislaufstäuben angereichert wird, wenn keine geeigneten Maßnahmen zur Entlastung dieser Kreisläufe vorgesehen sind [21–22].

Innere Ofenkreisläufe von Alkalien und Chloriden können bei kritischen Konzentrationen Verfahrensstörungen durch Ansatz und Ringbildung verursachen. Sollten verfahrenstechnische Einrichtungen zur Steuerung der Alkali- und Chloridkonzentrationen nicht vorhanden sein, sind Mengenbegrenzungen bei jenen Komponenten vorzunehmen, die diese Verbindungen enthalten [23].

Schwefelhaltige Rohmaterialien können sich dann positiv auf die Klinkereigenschaften auswirken, wenn bei oxidierenden Brennbedingungen Alkaliüberschuß gegeben ist [24]. Sulfatisierungsgrade über 100 % bzw. das Vorliegen des Schwefels als Sulfid im Rohmaterial können zu SO_2-Emission führen und zwingen zu Maßnahmen, wenn Emissionsgrenzwerte überschritten werden [25, 26]. Einflüsse auf die Klinkereigenschaft werden von Phosphaten und Fluoriden ausgeübt, wenn diese mit Sekundärstoffen in das Ofensystem gelangen.

Die regelmäßige analytische Erfassung von Elementen, die produktqualitäts-, umwelt- und verfahrenstechnisch relevant sind, in Einsatzstoffen, im Klinker, in Stäuben und Zement sowie ihre Bilanzierung ist ein notwendiges Instrument zur Überwachung der Produktion [27].

Grundsätzlich gelten analoge Kriterien für den Einsatz von Sekundärbrennstoffen, wobei je nach Ort des Brennstoffeintrages unterschiedliche Grenzen zu setzen sind. So erfordern zum Beispiel toxische organische Brennstoffbestandteile zu ihrer weitestgehenden Zersetzung hohe Temperaturen und ausreichend lange Verweilzeiten. Ihr Einsatz muß daher auf die Hauptfeuerung beschränkt bleiben. So wurde beispielsweise der Nachweis erbracht, daß Altöle mit erhöhten PCB-Gehalten in der Hauptfeuerung zufriedenstellend zersetzt werden, so daß die Bildung von Dioxinen und Furanen aus diesen Verbindungen in relevanten Mengen nicht zustandekommt und ein Grenzwert von 0,1 ng/m^3_N für das 2,3,7,8 TCDD-Äquivalent mit Sicherheit unterschritten werden kann [28]. Sowohl die Emission von Dioxinen und Furanen als auch von polykondensierten Aromaten und anderen organischen Verbindungen unterscheidet sich in der Regel nicht von der Emission dieser Verbindungen beim Einsatz von fossilen, konventionellen Brennstoffen [6].

Der Einsatz von Sekundärbrennstoffen in der Sekundärfeuerung stößt dort auf Grenzen, wo die Menge des eingesetzten Brennstoffes zu übermäßiger Erwärmung des Abgases führt bzw. wo durch Ansatz und Krustenbildung der Materialfluß so behindert wird, daß eine gleichmäßige Verfahrensführung nicht mehr möglich ist.

Wegen ihres unmittelbaren Einflusses auf die Gebrauchseigenschaften von Zement müssen für den Einsatz von Sekundärstoffen als Zumahlstoffe besondere Anforderungen gestellt werden. Hier kommen ausschließende und mengenbegrenzende Kriterien wie Feuchtigkeitsgehalt, Korngrößenzusammensetzung, Mahlbarkeit, Farbe, aber auch Geruch in Betracht. Kriterien, die einen Einfluß auf Frischmörteleigenschaften ausüben, wie die Beeinflussung des Wasseranspruchs, der Rheologie und der Ausbildung von Luftporen, sind zu berücksichtigen. In gleicher Weise sind bei hydraulisch wirksamen Zumahlstoffen Eignungskriterien hinsichtlich des Beitrages zur Festigkeitsentwicklung nachzuweisen [29–31].

Aus Rauchgasentschwefelungsanlagen, z.B. von thermischen Kraftwerken, fällt mitunter hochwertiger REA-Gips an, der sich als Zumahlstoff zur Regelung des Erstarrungs-

To monitor the production process it is essential to make regular analytical measurements of those elements in the feed materials, in the clinker, in the dusts, and in the cement which are relevant to product quality, environment and process technology, and to carry out balances on them [27].

In principle, analogous criteria apply to the use of secondary fuels, but differing limits should be set depending on the fuel input location. For example, toxic organic fuel constituents require high temperatures and adequate residence times to decompose them as fully as possible. Their use must therefore be confined to the main firing system. It has been shown, for example, that used oils containing high levels of PCBs are satisfactorily decomposed in the main firing system, with the result that dioxins and furans are not formed in relevant quantities from these compounds, and their levels can definitely be kept below a limit of 0.1 ng/m^3 (standard state) for the 2,3,7,8 TCDD equivalent [28]. The emission of dioxins and furans, and also of poly-condensed aromatics and other organic compounds, do not normally differ from the emissions of these compounds when using conventional fossil fuels [6].

The use of secondary fuels in the secondary firing system comes up against limits where the quantity of fuel used leads to excessive heating of the exhaust gas or where the material flow is so obstructed by coating and encrustation that uniform process operation is no longer possible. Separate specifications must be set for the use of secondary materials as additives because of their direct effect on the service properties of cements. In this case, exclusive and quantity-restricting criteria, such as moisture content, particle size composition, grindability, colour, and also odour, come into consideration. Criteria which affect the properties of the unset mortar, such as the effects on the water demand, the rheology and the formation of air voids, have to be taken into account. Similarly, it is necessary to establish suitability criteria for hydraulically active additives with regard to their contribution to strength development [29–31].

Flue gas desulphurization plants, e.g. in thermal power stations, sometimes provide high-grade desulphogypsum which is suitable as an additive for controlling the setting behaviour of cement [32–34]. A fairly high proportion of water-soluble phosphates, e.g. in chemical gypsums, can be detrimental to the workability characteristics of cements [35].

Residues should be excluded if they have characteristics which cannot be controlled or can only be controlled at unacceptably high cost. This is the case if these materials have high levels of radioactivity, or if they have characteristics which can place the safety of the employees at risk, such as, for example, infectious materials. Limits also have to be imposed where the capabilities of the personnel are not sufficient to ensure that they can deal safely with dangerous substances as a daily routine.

One criterion, which can sometimes set up barriers against the use of secondary materials which are very hard to overcome, is the acceptability of this mode of operation with neighbours, inhabitants in the vicinity, and possibly even with clients. In a process lasting more than 3 years for licensing the use of combustible liquid wastes, with voluntary participation of a citizens committee which was formed from members of a citizens action group, a doctors action group, regional politicians and representatives of the authorities, experience has shown that with intensive information, the highest possible transparency, and great patience it is possible to achieve a certain willingness to compromise [6].

Continuous emission measurements, technical safety process monitoring and specialized checking of the deliveries as well as the capability of the personnel can create an important basis for trust, which ultimately benefits the company's image [27].

4. Benefits when using secondary materials

The use of secondary materials has to be beneficial for the cement producer. The benefits can lie in the improved cost

verhaltens des Zements eignet [32–34]. Ein höherer Anteil an wasserlöslichen Phosphaten, z.B. in Chemiegipsen, kann zu negativer Beeinflussung der Verarbeitungseigenschaften der Zemente führen [35].

Auszuschließen sind Reststoffe, wenn sie Eigenschaften haben, die nicht oder nur mit unvertretbar hohem Aufwand beherrscht werden können. Das ist dann der Fall, wenn diese Stoffe eine erhöhte Radioaktivität aufweisen bzw. wenn sie Eigenschaften haben, welche die Sicherheit der Arbeitnehmer gefährden können, wie z.B. infektiöses Material. Grenzen sind auch dort zu setzen, wo die Qualifikation des Personals nicht ausreicht, um in der täglichen Praxis den sicheren Umgang mit gefährlichen Stoffen zu gewährleisten.

Ein Kriterium, das dem Einsatz von Sekundärstoffen mitunter schwer überwindliche Grenzen entgegensetzen kann, ist die Akzeptanz dieser Vorgangsweise bei Anrainern, Bewohnern in der näheren Umgebung sowie eventuell auch bei Kunden. In einem mehr als drei Jahre dauernden Verfahren zur Genehmigung des Einsatzes von brennbaren Flüssigabfällen unter freiwilliger Beteiligung eines Bürgerbeirates, der aus Mitgliedern einer Bürgerinitiative, einer Ärzteinitiative, Regionalpolitikern und Behördenvertretern gebildet wurde, konnte die Erfahrung gemacht werden, daß es mit intensiver Information, höchster Transparenz und großer Geduld gelingt, eine gewisse Konsenzbereitschaft zu erreichen [6].

Kontinuierliche Emissionsmessungen, sicherheitstechnische Prozeßüberwachung und spezialisierte Kontrollen der Anlieferungen sowie die Kompetenz des Personals können eine wichtige Vertrauensbasis schaffen, die letztlich dem Image des Unternehmens zugute kommt [27].

4. Nutzen beim Einsatz von Sekundärstoffen

Der Einsatz von Sekundärstoffen muß für den Zementerzeuger von Nutzen sein. Der Nutzen kann in der verbesserten Wirtschaftlichkeit des Zementerzeugungsprozesses durch Senkung von Brennstoff- und Rohstoffkosten liegen. In gleichem Sinne kann eine Senkung der Herstellungskosten durch verringerten Brennstoffeinsatz aus der Verbesserung der Brennbarkeit resultieren. Liegen Sekundärstoffe in mehlfeiner Form vor, können Aufwendungen für Rohsteingewinnung, Brechen und Mahlenergie eingespart werden.

Je nach Art des Reststoffes und der Alternativen, die es für dessen Entsorgung gibt, läßt sich Entsorgungsentgelt in unterschiedlicher Höhe erzielen. In der Regel allerdings muß die Funktion der qualitätskontrollierten Sammlung, der Zwischenlagerung zur Überbrückung von Ofenstillstandszeiten sowie eine entsprechende Aufbereitung zur Anpassung an vorhandene Lager-, Förder- und Dosiereinrichtungen oder zur Qualitätsverbesserung in Anspruch genommen und abgegolten werden.

Nutzen kann auch darin liegen, daß Qualitäten erzeugt werden können, die mit der gegebenen Rohstoffbasis allein nicht erzielbar wären. Gegebenenfalls kann eine Anpassung an marktkonforme Qualitäten erzielt werden, die ohne Nutzung von Sekundärstoffen nicht möglich wäre. Die damit verbundene Schonung des Abbaues natürlicher Rohstoffe ist sowohl für den Werksbetreiber als auch für die Allgemeinheit von Bedeutung, insbesondere wenn fossile Brennstoffe für solche Prozesse, bei denen Sekundärstoffe nicht eingesetzt werden können, dadurch auf längere Sicht zur Verfügung stehen. Der Einsatz von Sekundärstoffen im Zementerzeugungsprozeß bringt für die Allgemeinheit beträchtlichen Nutzen. Die richtige, umweltgerechte und rückstandsfreie Entsorgung unter begleitender Kontrolle, wie es sich durch die Eigenschaften des Zementerzeugungsprozesses ergibt, verhindert vielfältigen Mißbrauch aus Unwissenheit aber auch aus gewinnorientierter Fahrlässigkeit und minimiert die Gefahren, die bei unzweckmäßiger Entsorgung von Abfallstoffen entstehen können.

So ist die Altölverbrennung im Zementwerk der Entsorgung in Kleinanlagen nicht nur wegen der besseren Kontrol-

effectiveness of the cement production process as a result of reduced fuel and raw materials costs. The production costs can also be lowered through reduced use of fuel resulting from improved burnability. If secondary materials are available in a very fine form, then it is possible to save expenditure on power for raw stone quarrying, crushing and grinding.

It may be possible to obtain disposal fees of varying levels depending on the type of residue and the available alternatives for disposing of it. As a rule, however, it is necessary to enlist and pay for the functions of quality controlled collection, intermediate storage to bridge kiln stoppage times, and appropriate processing to adapt the materials to existing storage, conveying and metering equipment, or to improve the quality.

There can also be benefits from the fact that it is possible to produce grades which could not be achieved with the given basic raw materials alone. It may perhaps be possible to adjust the output to make grades which conform to market requirements in a way that could not have been achieved without using secondary materials. The associated conservation in the excavation of natural raw materials is important both for the works operator and also for the general public, especially if it means that fossil fuels will be available for longer for those processes in which secondary materials cannot be used. The use of secondary materials in the cement production process brings considerable benefits for the general public. Correct residue-free disposal in a manner appropriate to the demands of the environment with accompanying checks, such as is provided through the characteristics of the cement production process, prevents a great deal of misuse through ignorance and also through profit-orientated negligence; it also minimizes the dangers which can occur through inappropriate disposal of wastes.

For example, the combustion of used oil in a cement works is to be preferred to disposal in small plants, not only because of the better controllability. The problem of disposal of old tyres would not be solved yet if the cement industry had not guaranteed the disposal. Although processes such as pyrolysis and recycling of the materials crop up again and again in discussion, they have not yet proved to be alternatives

The reduction of nitrogen oxides, such as normally occurs when using old tyres in the secondary firing system due to the carbon monoxide concentrations produced locally, should be regarded as a benefit in the sense of a lessening of emission [36]. The use of secondary fuels results in a reduction of emissions as a whole, because the same exhaust gas volume which is needed for producing cement also solves a disposal problem which otherwise would have entailed separate emissions during separate disposal in incineration plants. This applies in particular to the reduction of the total emission of CO_2.

There the CO_2 emission is also lowered if additives reduce the proportion of clinker in the end product; in addition to this cement grades can be produced which have become indispensable for specific applications in modern construction technology.

If residues cannot be used by recycling or as secondary materials, they have to be dumped in landfill sites. Their use as secondary materials leads to considerable conservation of landfill volume.

5. Opportunities and dangers when using secondary materials

Increased environmental awareness, combined with the desire for an intact and viable environment, is causing enormous difficulties in the search for new locations for landfill sites and refuse incineration plants.

The use of existing cement industry locations provides an opportunity for integrating the industrial capacities needed for incineration and material utilization into regional and national wastes schemes.

lierbarkeit vorzuziehen. Das Problem der Entsorgung von Altreifen wäre bis heute nicht gelöst, wenn die Zementindustrie die Entsorgung nicht sichergestellt hätte. Verfahren wie Pyrolyse und stoffliches Recycling, obwohl immer wieder in Diskussion, haben sich bis heute nicht als Alternativen erweisen können.

Die Reduktion von Stickoxiden, wie sie in der Regel beim Einsatz von Altreifen in der Sekundärfeuerung aufgrund der örtlich entstehenden Kohlenmonoxidkonzentrationen zustande kommt, ist als Nutzen im Sinne einer Emissionsminderung zu sehen [36]. Die Verwendung von Sekundärbrennstoffen erbringt insgesamt eine Minderung der Emission, da mit demselben Abgasvolumen, das zur Zementerzeugung erforderlich ist, gleichzeitig eine Entsorgungsaufgabe gelöst wird, die sonst bei separater Entsorgung in Verbrennungsanlagen separate Emissionen mit sich gebracht hätte. Hier ist insbesondere die Verringerung der Gesamtemission an CO_2 hervorzuheben.

Ebenfalls gelingt eine Reduktion der CO_2-Emission, wenn Zumahlstoffe den Anteil an Klinker im Endprodukt verringern, wobei darüber hinaus noch Zementqualitäten erzeugt werden, die für spezielle Anwendungen in der modernen Bautechnik unverzichtbar geworden sind.

Können Reststoffe nicht durch Recycling oder als Sekundärstoffe genutzt werden, müssen sie deponiert werden. Ihr Einsatz als Sekundärstoff führt zu einer beträchtlichen Schonung von Deponievolumen.

5. Chancen und Gefahren beim Einsatz von Sekundärstoffen

Im Zusammenhang mit dem gesteigerten Umweltbewußtsein, verbunden mit dem Wunsch nach einer intakten und lebenswerten Umwelt, ist die Suche nach neuen Standorten für Deponien und Müllverbrennungsanlagen auf enorme Schwierigkeiten gestoßen.

Mit der Nutzung vorhandener Standorte der Zementindustrie ist eine Chance gegeben, die Verbrennungs- bzw. Materialnutzungskapazitäten in regionale und nationale Abfallkonzepte zu integrieren.

Neben dem Nutzen von Standorten und Anlagen wird auch das Know-how der Zementerzeuger im Umgang mit großen Masseströmen an Rohmaterial, großen Masseströmen an gasförmigen, flüssigen und festen Brennstoffen und mit allen Anforderungen an die Sicherheitstechnik sowie die Erfahrung im Homogenisieren von Stoffströmen, die Erfahrung im Berechenbarmachen von Materialien durch zweckmäßige Probenahme, Analyse und daraus abzuleitende Maßnahmen genutzt werden.

Wie überall, wo die Chance auf Nutzen besteht, liegt im Einsatz von Sekundärstoffen auch ein beträchtliches Gefahrenpotential. Dieses Gefahrenpotential resultiert daraus, daß durch unvorhergesehene Störfälle, durch Unwissenheit, aber auch durch Schlamperei und Leichtsinn, Grenzen des Einsatzes von Sekundärstoffen mißachtet werden können. Daher soll nicht in der Hoffnung auf schnellen Gewinn mit untauglichen Mitteln und halbherzigen Konzepten vorgegangen werden, sondern mit hohem Verantwortungsbewußtsein eine der Tragweite des Vorhabens und seinen Auswirkungen angepaßte Projektierung und Realisierung vorgenommen werden, selbst wenn die eine oder andere Maßnahme zunächst als übertrieben oder in Anbetracht der aufzuwendenden Kosten als unnötig erachtet werden sollte.

In addition to the benefit of sites and plants, it would also be possible to make use of the cement manufacturers' know-how in dealing with large bulk flows of raw materials, large mass flows of gaseous, liquid and solid fuels, and with all the requirements of safety technology, and of their experience in homogenizing material flows and in turning materials into calculable quantities by appropriate sampling and analyses and measures derived from them.

Wherever there is a chance of benefits in the use of secondary materials there is also considerable potential danger. This potential danger results from the fact that limitations on the use of secondary materials can be disregarded through unforcseen malfunctions, through ignorance, but also through slackness and negligence. It is no good proceeding with unsuitable means and half-hearted schemes in the hope of rapid profit; the planning and execution work suited to the importance of the project and its consequence must be undertaken with great awareness of the responsibility, even if some of the measures are initially considered excessive or unnecessary in view of the costs involved.

Literaturverzeichnis

[1] Sprung, S.: Umweltentlastung durch Verwertung von Sekundärrohstoffen. Zement-Kalk-Gips 45 (1992) Nr. 5, S. 213–221.

[2] Kirsch, J.: Umweltentlastung durch Verwertung von Sekundärbrennstoffen. Zement-Kalk-Gips 44 (1991) Nr. 12, S. 605–610.

[3] Steinbiß, E.: Erfahrungen mit der Vorcalcinierung unter Berücksichtigung von Ersatzstoffen. Zement-Kalk-Gips 32 (1979) Nr. 5, S. 211–221.

[4] Hochdahl, O.: Erfahrungen und Gesichtspunkte beim Einsatz von Ersatzbrennstoffen. Zement-Kalk-Gips 31 (1978) Nr. 9, S. 421–424.

[5] Kreft, W.: Einsatz von Abfallstoffen als Energie- und Rohstoffersatz bei der Zementherstellung. TIZ, Vol. 112 (1988) No. 2, S. 123–127.

[6] Gerger, W., und Liebl, P.: Thermische Verwertung von Sekundärbrennstoffen bei den Gmundner Zementwerken. Zement-Kalk-Gips 44 (1991) Nr. 9, S. 457–462.

[7] Kroboth, K., Kuhlmann, K., und Xeller, H.: Stand der Technik der Emissionsminderung in Europa. Zement-Kalk-Gips 43 (1990) Nr. 3, S. 121–131.

[8] Borgholm, H. E.: Umweltentlastung durch Verwertung von Flugasche als Rohmehlkomponente. Zement-Kalk-Gips 45 (1992) Nr. 4, S. 163–170.

[9] Maury, H.-D., und Pavenstedt, R. G.: Primär- und Sekundär-Brennstoffe beim Klinkerbrennen. Zement-Kalk-Gips 42 (1989) Nr. 2, S. 90–93.

[10] Sprung, S., Kirchner, G., und Rechenberg, W.: Reaktionen schwer verdampfbarer Spurenelemente beim Brennen von Zementklinker. Zement-Kalk-Gips 37 (1984) Nr. 10, S. 513–518.

[11] Kirchner, G.: Reaktionen des Cadmiums beim Klinkerbrennprozeß. Zement-Kalk-Gips 38 (1985) Nr. 9, S. 535–539.

[12] Kirchner, G.: Verhalten der Schwermetalle beim Brennen von Zementklinker. Zement-Kalk-Gips 39 (1986) Nr. 10, S. 555–557.

[13] Sprung, S.: Spurenelemente – Anreicherung und Minderungsmaßnahmen. Zement-Kalk-Gips 41 (1988) Nr. 5, S. 251–257.

[14] Sprung, S., und Rechenberg, W.: Reaktionen von Blei und Zink bei der Zementherstellung. Zement-Kalk-Gips 36 (1983) Nr. 10, S. 539–548.

[15] Sprung, S., und Rechenberg, W.: Die Reaktionen von Blei und Zink beim Brennen von Zementklinker. Zement-Kalk-Gips 31 (1978) Nr. 7, S. 327–329.

[16] Knöfel, D.: Beeinflussung einiger Eigenschaften des Portlandzementklinkers und des Portlandzementes durch ZnO und ZnS. Zement-Kalk-Gips 31 (1978) Nr. 3, S. 157–161.

[17] Lieber, W., und Gebauer, J.: Einbau von Zink in Calciumsilikate. Zement-Kalk-Gips 22 (1949) Nr. 4, S. 161–164.

[18] Lieber, W.: Einfluß von Zinkoxid auf das Erstarren und Erhärten von Portlandzementen. Zement-Kalk-Gips 20 (1967) Nr. 3, S. 91–95.

[19] Sprung, S., und Rechenberg, W.: Schwermetallgehalte im Klinker und Zement. VDZ-Kongreß '93 (in Vorbereitung).

[20] Weisweiler, W., und Keller, A.: Zur Problematik gasförmiger Quecksilberemissionen aus Zementwerken. Zement-Kalk-Gips 45 (1992) Nr. 10, S. 529–532.

[21] Kirchner, G.: Das Verhalten des Thalliums beim Brennen von Zementklinker. Schriftenreihe der Zementindustrie, Heft 47, Beton-Verlag, Düsseldorf 1986.

[22] Kirchner, G.: Thalliumkreisläufe und Thalliumemissionen beim Brennen von Zementklinker. Zement-Kalk-Gips 40 (1987) Nr. 3, S. 134–144.

[23] Sprung, S.: Technologische Probleme beim Brennen des Zementklinkers, Ursache und Lösung. Schriftenreihe der Zementindustrie, Heft 43, Beton-Verlag, Düsseldorf 1992.

[24] Sprung, S.: Emissionsprognosen beim Einsatz von Abfallbrennstoffen. Zement-Kalk-Gips 37 (1984) Nr. 10, S. 519–522.

[25] Bonn, W., und Hasler, R.: Verfahren und Erfahrung einer rohstoffbedingten SO_2-Emission im Werk Untervaz der Bündner Cementwerke. Zement-Kalk-Gips 43 (1990) Nr. 3, S. 139–143.

[26] Schütte, R.: Möglichkeiten der Entstehung und Minderung von SO_2-Emissionen in Zementwerken. Zement-Kalk-Gips 42 (1989) Nr. 3, S. 128–133.

[27] Gerger, W.: Erfahrungen mit spezialisierten Kontrolltechniken beim Einsatz von Sekundärstoffen. Vortrag zum VDZ-Kongreß 1993, Fachbereich 1.

[28] Krogbeumker, G.: Sicherheitseinrichtungen bei zusätzlicher Verbrennung PCB-haltiger Altöle im Zementdrehofen. Zement-Kalk-Gips 41 (1988) Nr. 4, S. 188–192.

[29] Jung, F.: Anforderungen an Zumahlstoffe für die Zementherstellung. Vortrag zum VDZ-Kongreß 1993, Fachbereich 1.

[30] Schmidt, M.: Zement mit Zumahlstoffen – Leistungsfähigkeit und Umweltentlastung, Teil 1. Zement-Kalk-Gips 45 (1992) Nr. 2, S. 64–69.

[31] Braun, H., und Gebauer, J.: Möglichkeiten und Grenzen der Verwendung von Flugaschen im Zement. Zement-Kalk-Gips 36 (1983) Nr. 5, S. 254–258.

[32] Krogbeumker, H.: Lagerung und Dosierung von feuchtem REA-Gips im Zementwerk Phönix. Zement-Kalk-Gips 45 (1992) Nr. 3, S. 105–109.

[33] Mosch, H.: Die Eignung von Anhydrit aus Rauchgasschwefeldioxid als Sulfatträger. Zement-Kalk-Gips 39 (1986) Nr. 1, S. 33–35.

[34] Weiler, H., Hamm, H., und Hüller, R.: Aufbereitung von feuchten feinteiligen Rauchgasgipsen. Zement-Kalk-Gips 36 (1983) Nr. 11, S. 608–614.

[35] Lieber, W.: Wirkung anorganischer Zusätze auf das Erstarren und Erhärten von Portlandzement. Zement-Kalk-Gips 26 (1973) Nr. 2, S. 75–79.

[36] Scheuer, A.: Minderung der NO_x-Emission beim Brennen von Zementklinker. Zement-Kalk-Gips 41 (1988) Nr. 1, S. 37–42.

Optimale Brennbedingungen in Zementöfen bei Verwendung von Kohle geringer und schwankender Qualität

Optimum burning conditions in kilns using inferior and variable quality coal

Conditions de cuisson optimales à l'intérieur des fours à ciment, en utilisant du charbon de qualité inférieure et variable

Condiciones de cocción óptimas en los hornos de cemento, utilizando carbón de calidad inferior y variable

Von **S. N. Banerjee,** Satna/Indien

Zusammenfassung – Hohe Energiekosten zwingen die Zementhersteller, Energieeinsparmöglichkeiten zu prüfen. Eine Möglichkeit zur Kosteneinsparung bietet der Einsatz aschereicher Kohle. Zur Beurteilung des Ascheeinflusses auf die Eigenschaften des Zements kann die Brennbarkeit des Ofenmehls, gekennzeichnet durch den Freikalkgehalt, herangezogen werden. Dazu wurden Rohmehlmischungen unterschiedlicher Feinheiten mit verschiedenen Ascheanteilen bei 1000 bis 1500 °C zwischen 1 und 4 h im Laborofen gebrannt. An den aus diesen Klinkern hergestellten Zementen wurde die Normdruckfestigkeit nach 3, 7 und 28 Tagen ermittelt. Eines der wesentlichen Ergebnisse war, daß offenbar eine optimale Kombination zwischen Brenndauer und Brenntemperatur besteht, bei der die höchste Druckfestigkeit für eine bestimmte Ofenmehlzusammensetzung gegeben ist. Schärferes oder schwächeres Brennen führte zu geringeren Festigkeiten. Bei optimalen Brennbedingungen werden Einsparungen an thermischerEnergie für den Brennprozeß und an elektrischer Energie für die Zementmahlung möglich.

Summary – High-energy costs are forcing cement producers to test possible ways of saving energy. One option for cost saving is offered by the use of high-ash coal. The burnability of kiln meal, characterized by free lime content, can be used for assessing the effect of ash on the properties of the cement. Raw meals of different finenesses containing varying percentages of ash were burnt in the laboratory furnace at 1000 to 1500°C between 1 and 4 hours. The standard compressive strengths of the cements produced from these clinkers were measured after 3, 7 and 28 days. One important finding was that there is clearly an optimum combination of burning time and burning temperature at which the highest compressive strength is obtained for a given kiln meal composition. Harder or softer burning led to lower strengths. Under optimum burning conditions it is possible to make savings in thermal energy for the burning process and in electrical energy for the cement grinding process.

Résumé – Les coûts d'énergie très élevés obligent les fabricants de ciment à étudier des possibilités d'économiser de l'énergie. C'est l'emploi de charbon riche en cendres qui offre la possibilité de réduire les frais. Pour évaluer l'influence des cendres sur les propriétés du ciment, on peut se servir de l'aptitude à la cuisson de la farine du four, caractérisée par la teneur en chaux libre. A ce sujet, on a cuit, dans un four de laboratoire, des mélanges de farine crue de différentes finesses et de différentes teneurs en cendres, à des températures de 1000 à 1500 °C et pendant une durée de 1 à 4 h. Puis, on a déterminé, à l'aide des ciments fabriqués avec ces clinkers, la résistance à la compression après 3, 7 et 28 jours, conformément aux normes. Un des resultats essentiels obtenu a permis d'établir qu'il existe apparemment une combinaison optimale entre la durée et la température de cuisson, pour laquelle on peut atteindre une résistance à la compression maximale, avec une composition déterminée de la farine du four. Une cuisson plus forte ou plus faible a conduit à des résistances inférieures. Si les conditions de cuisson sont optimales, il est possible de faire des économies d'énergie thermique, au cours du processus de cuisson, et des économies d'énergie électrique au cours du broyage du ciment.

Resumen – Los elevados costes de energía obligan a los fabricantes de cementos a estudiar las posibilidades de ahorrar energía. El empleo de carbones ricos en volátiles ofrece la posibilidad de reducir costes. Para evaluar el influjo de las cenizas sobre las propiedades del cemento, se puede recurrir a la aptitud a la cocción de la harina, caracterizada por el contenido de cal libre. A este respecto, se han cocido, en un horno de laboratorio, unas mezclas de harina cruda de diferentes finuras y contenidos de cenizas, a temperaturas comprendidas entre 1000 y 1500 °C, durante 1 a 4 horas. Con los cementos fabricados a partir de estos clínkeres, se ha determinado luego la resistencia

a la compresión a los 3, 7 y 28 días, según las normas. Uno de los resultados esenciales obtenidos ha sido el hecho de que existe, por lo visto, una combinación óptima entre la duración y la temperatura de cocción, para la cual se establece una resistencia a la compresión máxima con una determinada composición de la harina del horno. Una cocción más fuerte o más débil conduce a resistencias menores. Si las condiciones de cocción son óptimas, se pueden conseguir ahorros de energía térmica, durante el proceso de cocción, y ahorros de energía eléctrica durante la molienda del cemento.

1. Einführung

Ziel dieser Untersuchungen war die Beurteilung der Auswirkungen verschiedener Parameter auf die Brennbedingungen. Dazu gehören

— verschiedene Rohstoffarten,

— die physikalischen, chemischen und mineralogischen Eigenschaften der Rohstoffe,

— das Konzept der Brennbarkeit und seine Rolle beim Klinkerbrennen und

— die Energiekosten.

Die hohen Energiekosten zwingen die Zementhersteller heute zu einer Überprüfung des Verfahrens, um gerade angesichts täglich schwankender Kohlequalität Energiesparmöglichkeiten zu ermitteln.

Zementöfen dienen der Herstellung von Klinker oder, genau genommen, von Alit. Der wichtigste und praktisch langsamste Prozeß bei der Klinkerbildung ist die Bildung von Alit aus CaO und C_2S. C_3S ist der wirksamste hydraulische Bestandteil des Zements. Bei den meisten Bauarbeiten muß recht schnell eine hohe Betonfestigkeit erreicht werden. Allgemein ist die Erfahrung zu machen, daß die Qualität und die Preise der Brennstoffe und der elektrischen Energie sich nur in einer Richtung verändern — im Sinne einer zunehmenden Verschlechterung. Deshalb muß angesichts der Vorgaben der Prozeßführung und der Qualitätskontrolle Klinker mit hohem Alitgehalt unter minimalem Energieeinsatz hergestellt werden. Dazu gehört gegebenenfalls auch eine Verbesserung der Brennbarkeit der Rohmischung. Ferner ist für die Bildung eines gut granulierten Klinkers zu sorgen. Die Herstellung eines granulierten statt eines staubigen Klinkers trägt nicht nur zu einer reibungslosen und wirtschaftlichen Ofenführung bei, sondern ermöglicht darüber hinaus Energieeinsparungen in der Zementmühle.

2. Art der Rohstoffe und Auswirkungen auf den theoretischen Wärmebedarf

2.1 Tonmineralien

Es wird von folgenden Stoffen im Rohgemisch ausgegangen:

TABELLE 1: Chemische Analyse von Tonmineralien und deren Einfluß auf den theoretischen Wärmebedarf

	Kaolinit	Montmorillonit	Illit
		Gew.-%	
Glühverlust	12,93	6,18	7,05
SiO_2	46,65	60,80	56,17
Al_2O_3	38,83	22,19	19,75
Fe_2O_3	0,59	4,10	5,22
TiO_2	0,10	0,22	0,97
CaO	0,11	0,28	0,28
MgO	0,39	2,78	1,87
SO_3	0,17	0,74	1,37
K_2O	0,29	0,11	4,46
Na_2O	0,10	2,17	0,61
Silikatmodul	1,18	2,31	2,25
Tonerdemodul	55,80	5,41	3,78
Theoretischer Wärmebedarf (kcal/kg Klinker)	419,5	403,1	401,6

1. Introduction

The objective of these investigations was the assessment of the effects of different parameters on the burning conditions. Among these are
— different types of raw materials, the
— physical, chemical and mineralogical properties of raw materials, the
— concept of burnability and its role in clinker burning and the
— cost of energy.

The high cost of energy today is forcing cement manufacturers to re-examine the process to see where energy savings are possible, especially with day to day varying coal quality.

Cement kilns exist to make clinker, or more specifically, to make alite. The most important, and virtually the slowest, process in clinker formation is the formation of alite from CaO and C_2S. C_3S is the most powerful hydraulic constituent of the cement. The most construction activities require the attainment of rather rapid high concrete strength. The general experience is that quality and prices of kiln fuel and electric power change to only one direction, from bad to worse. Therefore, due to the targets of process operation and quality control, it is necessary to make clinker with high alite contents by using the minimum amount of energy for all types. That includes improving the raw mix burnability, if necessary. It also means to ensure the formation of a well nodulized clinker. The production of a nodular, rather than dusty clinker is favourable not only for smooth, economical kiln operation but also for energy saving in the cement mill as well.

2. Type of raw materials and their effect on theoretical heat demand

2.1 Clay minerals

The substances assumed to be present in the raw mix are:
— Calcium carbonate ($CaCO_3$),
— Quartz (SiO_2),
— Clay minerals (Kaolinite, Montmorillonite, Illite),
— Iron oxides, -hydroxides (e.g. $Fe(OH)_3$),
— Magnesium carbonate ($MgCO_3$),

TABLE 1: Chemical analysis of clay minerals and its influence on the theoretical heat demand

	Kaolinite	Montmorillonite	Illite
		wt.-%	
Loss on ignition	12.93	6.18	7.05
SiO_2	46.65	60.80	56.17
Al_2O_3	38.83	22.19	19.75
Fe_2O_3	0.59	4.10	5.22
TiO_2	0.10	0.22	0.97
CaO	0.11	0.28	0.28
MgO	0.39	2.78	1.87
SO_3	0.17	0.74	1.37
K_2O	0.29	0.11	4.46
Na_2O	0.10	2.17	0.61
Silica ratio	1.18	2.31	2.25
Alumina ratio	55.80	5.41	3.78
Theoretical heat demand (kcal/kgcl)	419.5	403.1	401.6

TABLE 2: Zusammensetzung und Eigenschaften in der Zementindustrie eingesetzter indischer Kohle (Gew.-%)

Kohle Nr.	1	2	3	4	5	6	7	8	9	10
Feuchtigkeit	5,74	5,76	6,62	3,65	8,23	5,70	5,66	8,39	6,73	6,53
Asche	22,45	24,64	35,39	28,49	28,87	30,06	31,03	26,09	39,21	41,12
Flüchtige Bestandteile	24,30	23,41	27,30	27,89	24,90	27,87	25,63	30,45	24,74	25,03
Gebundener Kohlenstoff (Differenzverfahren)	47,51	46,19	30,69	39,97	38,00	36,37	37,68	35,07	29,32	27,32
Oberer Heizwert (kcal/kg)	5650	5350	4270	5390	4730	5070	4790	4920	4000	3680
Elementaranalyse										
Kohlenstoff	56,99	56,29	44,62	53,13	48,31	51,05	48,96	50,18	37,63	38,07
Wasserstoff	1,14	2,97	3,41	3,25	3,33	3,47	3,05	3,36	2,36	3,74
Schwefel	0,44	0,46	0,51	1,96	0,47	0,87	0,50	0,50	0,53	0,60
Stickstoff	1,27	1,26	1,04	1,24	1,18	1,20	1,22	1,24	1,14	0,98
Feuchtigkeit	5,74	5,76	6,62	3,65	8,23	5,70	5,66	8,39	6,73	6,53
Mineralbestandteile (1,1 x % Aschegehalt)	24,70	27,10	38,93	31,34	31,76	33,07	34,13	28,70	43,13	45,23
Kohleaschenanalyse										
Glühverlust	0,28	0,33	0,56	0,28	0,23	0,20	0,99	1,18	0,35	0,40
SiO_2	63,11	62,19	61,01	53,61	60,78	62,86	63,11	58,30	61,76	66,41
Fe_2O_3	6,88	6,90	5,62	12,89	5,65	5,55	5,17	6,26	4,98	4,89
$R_2O_3 - Fe_2O_3$	25,76	26,29	28,71	25,02	28,07	27,65	26,81	28,64	29,11	24,99
CaO	2,05	2,06	2,08	4,79	2,63	1,20	1,47	2,33	1,36	1,34
MgO	0,69	0,70	0,60	0,62	0,81	0,73	0,95	0,70	0,73	0,43
SO_3	0,51	0,52	0,71	1,55	1,37	0,45	0,35	1,32	0,12	0,25
Unlöslicher Rückstand	–	–	–	–	–	–	–	–	–	–
Na_2O	0,15	0,17	0,22	0,32	0,32	0,25	0,18	0,15	0,22	0,42
K_2O	0,95	1,19	0,87	1,22	0,67	1,17	0,75	0,65	1,07	0,67

TABLE 2: Composition and properties of Indian coal used in the cement industry [wt.-%]

Coal number	1	2	3	4	5	6	7	8	9	10
Moisture	5.74	5.76	6.62	3.65	8.23	5.70	5.66	8.39	6.73	6.53
Ash	22.45	24.64	35.39	28.49	28.87	30.06	31.03	26.09	39.21	41.12
Volatile matter	24.30	23.41	27.30	27.89	24.90	27.87	25.63	30.45	24.74	25.03
Fixed carbon (by difference)	47.51	46.19	30.69	39.97	38.00	36.37	37.68	35.07	29.32	27.32
Calorific Value (Gross) (kcal/kg)	5650	5350	4270	5390	4730	5070	4790	4920	4000	3680
Ultimate Analysis										
Carbon	56.99	56.29	44.62	53.13	48.31	51.05	48.96	50.18	37.63	38.07
Hydrogen	1.14	2.97	3.41	3.25	3.33	3.47	3.05	3.36	2.36	3.74
Sulphur	0.44	0.46	0.51	1.96	0.47	0.87	0.50	0.50	0.53	0.60
Nitrogen	1.27	1.26	1.04	1.24	1.18	1.20	1.22	1.24	1.14	0.98
Moisture	5.74	5.76	6.62	3.65	8.23	5.70	5.66	8.39	6.73	6.53
Mineral matter (1.1 x % ash cont.)	24.70	27.10	38.93	31.34	31.76	33.07	34.13	28.70	43.13	45.23
Coal ash analysis										
Loss on ignition	0.28	0.33	0.56	0.28	0.23	0.20	0.99	1.18	0.35	0.40
SiO_2	63.11	62.19	61.01	53.61	60.78	62.86	63.11	58.30	61.76	66.41
Fe_2O_3	6.88	6.90	5.62	12.89	5.65	5.55	5.17	6.26	4.98	4.89
$R_2O_3 - Fe_2O_3$	25.76	26.29	28.71	25.02	28.07	27.65	26.81	28.64	29.11	24.99
CaO	2.05	2.06	2.08	4.79	2.63	1.20	1.47	2.33	1.36	1.34
MgO	0.69	0.70	0.60	0.62	0.81	0.73	0.95	0.70	0.73	0.43
SO_3	0.51	0.52	0.71	1.55	1.37	0.45	0.35	1.32	0.12	0.25
Insoluble residue	–	–	–	–	–	–	–	–	–	–
Na_2O	0.15	0.17	0.22	0.32	0.32	0.25	0.18	0.15	0.22	0.42
K_2O	0.95	1.19	0.87	1.22	0.67	1.17	0.75	0.65	1.07	0.67

- Calciumcarbonat (CaCO$_3$),
- Quarz (SiO$_2$),
- Tonmineralien (Kaolinit, Montmorillonit, Illit),
- Eisenoxide, -hydroxide (z. B. Fe(OH)$_3$),
- Magnesiumcarbonat (MgCO$_3$),
- Calciumsulfat (CaSO$_4$),
- Alkalien,
- Verunreinigungen durch Kohlenasche.

Die verschiedenen Arten und Mengenanteile mineralischer Inhaltsstoffe beeinflussen den theoretischen Wärmebedarf aufgrund unterschiedlicher Reaktionsenthalpien innerhalb eines Bereichs von ca. 10%. Wie die **Tabelle 1** zeigt, schwankt der theoretische Wärmebedarf eines Rohmehlgemischs mit konstanter chemischer Zusammensetzung bei Verwendung der bekannten Rechenmethoden je nach Art und Zusammensetzung der Tonmineralien zwischen 402 und 420 kcal/kg Klinker (1683 bzw. 1759 kJ/kg Klinker).

2.2 Freikalk

Ein Anstieg des Freikalkgehalts führt zu einer deutlichen Abnahme der Alitbildung, wie aus einer Berechnung der Phasenzusammensetzung des Klinkers nach Bogue hervorgeht. Dementsprechend geht die exotherme Enthalpie der C$_3$S-Bildung innerhalb der gesamten Reaktionsenthalpie der Klinkerbildung bei einem Anstieg des Freikalkgehalts von 1 auf 3 Gew.-% von rund 118 auf 105 kcal/kg Klinker (von 494 auf 440 kJ/kg Klinker) zurück. Im Hinblick auf die Klinkereigenschaften ist die Wirkung von freiem Kalk auf die Raumbeständigkeit nicht sehr stark, soweit der Freikalkgehalt nicht deutlich über 3 Gew.-% liegt. Eine Beeinträchtigung der Klinkerphasenbildung ist nicht zu erwarten, wenn auf dem 210-μm-Sieb nur ein geringer Prozentsatz an Kalkrückständen oder Quarz zurückbleibt.

2.3 Kohlenasche

Die Menge und Zusammensetzung der Kohlenasche beeinflussen bei einem vorgegebenen Gesamtverbrauch an thermischer Energie die Zusammensetzung und die Eigenschaften des Klinkers sehr nachhaltig. Die Eigenschaften der Schwankungen der indischen Kohle gehen aus der **Tabelle 2** hervor. Entsprechend dieser Vielfalt wurden die Auswirkungen des Anteils an Kohlenasche und ihrer Zusammensetzung auf den Kalksättigungsgrad (KSG), den Silikatmodul und den Tonerdemodul, die Phasenzusammensetzung des Klinkers und den theoretischen Wärmebedarf (TWB) der Klinkerbildung beurteilt. Das Ergebnis dieser für ver-

- Calcium sulphate (CaSO$_4$),
- Alkalies,
- Adulterations caused by coal ash.

The different type and quantity of mineral components influence within a range of approximately 10% the theoretical heat demand, due to different reaction enthalpies. Utilizing the well known methods of calculation, the different kind and composition of clay minerals change the theoretical heat demand of a raw meal mix with constant chemical composition between about 402 and 420 kcal/kgcl (1683 and 1759 kJ/kgcl), as can be seen from **Table 1**.

2.2 Free lime

An increasing free lime content leads to a drastic decrease in alite formation, as a calculation of the phase composition of the clinker, according to Bogue, shows. Accordingly, the exothermic enthalpy of formation of C$_3$S within the total of reaction enthalpy of the clinker formation decreases from about 118 to 105 kcal/kgcl (494 to 440 kJ/kgcl) with growing free lime content from 1 to 3 wt.-%. With view to clinker properties, the effect of free lime on expansion is not very high, provided the free lime content does not exceed too much a figure of 3 wt.-%. No harmful effect on clinker phase formation is to be expected if calcareous residue or quartz are not present as residue in high percentage on the 210 micron sieve.

2.3 Coal ash

Quantity and composition of coal ash influence, at given total thermal energy comsumption, to a great extent the composition and the properties of clinker. The property range of Indian coal follows from **Table 2**. Following this variety, the effect of the portion of coal ash and its composition on the lime saturation factor (LSF), the silica and alumina module, the phase composition of the clinker and the theoretical heat demand (HRA) of clinker formation have been evaluated. The result of this calculation, obtained for different raw meal mixes, is summarized in **Table 3**. It can be concluded that the negative effect on clinker properties, which has to be expected under the given conditions, obviously indicates to be somewhat more severe than that on the theoretical heat demand.

3. Burnability

After the general raw material assessment different raw mixes are composed for burnability tests after adjustment

TABLE 3: Reaktionswärme von Klinker bei verschiedenen Anteilen an Kohlenasche und unterschiedlicher Brenngutzusammensetzung

Brenngut KSG	Kohlenasche	KSG	SM	TM	C$_3$S	C$_2$S	C$_3$A	C$_4$AF	CV	Cl	LV	TWB
Gew.-%					Gew.-%				Gew.-%			kcal/kgcl
105,71	0,00	105,60	2,47	0,93	89,43	−9,73	3,24	12,78	22,47	31,01	24,75	428,67
105,71	30,06	91,55	2,44	1,11	56,91	20,23	5,34	12,96	30,87	40,30	27,33	398,31
105,71	35,39	87,29	2,39	1,20	45,29	30,69	6,36	13,02	34,08	43,86	28,55	387,85
105,71	41,12	81,34	2,51	1,23	29,64	46,02	6,66	12,96	37,34	47,44	28,80	371,10
107,88	0,00	107,77	2,55	1,04	92,34	−13,1	4,03	11,44	20,37	28,48	23,84	431,33
107,88	30,06	93,16	2,51	1,24	59,70	17,02	6,11	11,65	28,82	37,84	26,44	399,94
107,88	35,39	88,76	2,45	1,33	48,06	27,50	7,11	11,78	32,13	41,50	27,71	389,46
107,88	41,12	82,61	2,57	1,36	32,40	42,86	7,38	11,72	35,37	45,06	27,93	372,58
109,28	0,00	109,12	2,49	1,06	94,55	−15,4	4,21	11,50	20,19	28,29	24,12	434,45
109,28	30,06	94,22	2,46	1,25	61,84	14,78	6,25	11,75	28,68	37,68	26,71	402,91
109,28	35,39	89,71	2,41	1,34	50,10	25,36	7,26	11,84	31,95	41,32	27,96	392,31
109,28	41,12	83,43	2,52	1,37	34,30	40,86	7,56	11,78	35,25	44,94	28,21	375,40

schiedene Rohmehlgemische angestellten Berechnung wird in der **Tabelle 3** zusammenfassend dargestellt. Es läßt sich der Schluß ziehen, daß die negative Beeinflussung der Klinkereigenschaften, die unter den gegebenen Umständen zu erwarten ist, etwas stärker als bei dem theoretischen Wärmebedarf ausfallen dürfte.

3. Brennbarkeit

Entsprechend der Rohstoffzusammensetzung wurden zur Überprüfung der Brennbarkeit verschiedene Rohmehlmischungen nach entsprechender Einstellung mit Kohlenasche mit einem C_3S-Gehalt von 55 Gew.-% im Klinker zusammengestellt. Bei einer Brennbarkeitsprüfung an Rohmehlmischungen wurden die in der **Tabelle 4** dargestellten Ergebnisse erzielt.

Die Ergebnisse zeigen, daß sowohl die Feinheit als auch die Brenntemperatur im Hinblick auf die richtige Sinterung des Klinkers maßgebend sind. Eine ganze Reihe von Variablen beeinflussen die Qualität des Zementklinkers. Dazu gehö- with coal ash to give a C_3S content of 55 wt.-% in the clinker. A burnability test was carried out on raw mixes with the following results in **Table 4**.

The results show that both the fineness and burning temperature are decisive for proper sintering of the clinker. Quite a lot of variables effect the quality of cement clinker. These include the burning and cooling regime and the physical and chemical properties of the raw materials. The cement properties depend on the clinker quality and the addition of gypsum. The results of investigations (**Table 5**) suggest that there is an optimum combination of burning temperature and time for each raw mix.

The effect of decreased burning temperatures and increased fineness of the raw mix is indicated in **Table 6**.

These figures indicate that reducing the burning temperature by means of fine grinding leads to an improved quality.

The influence of different temperatures and sintering time on the standard compressive strength of the cements is obvious from **Table 7**. The results show a positive correlation between free lime and strength.

TABLE 3: Heat of reaction of clinker with different coal ash portions and different kiln feed composition

kiln feed LSF	coal ash	LSF	SM	AM	C_3S	C_2S	C_3A	C_4AF	CV	CI	LV	HRA
	wt.-%				wt.-%					wt.-%		kcal/kgcl
105.71	0.00	105.60	2.47	0.93	89.43	−9.73	3.24	12.78	22.47	31.01	24.75	428.67
105.71	30.06	91.55	2.44	1.11	56.91	20.23	5.34	12.96	30.87	40.30	27.33	398.31
105.71	35.39	87.29	2.39	1.20	45.29	30.69	6.36	13.02	34.08	43.86	28.55	387.85
105.71	41.12	81.34	2.51	1.23	29.64	46.02	6.66	12.96	37.34	47.44	28.80	371.10
107.88	0.00	107.77	2.55	1.04	92.34	−13.1	4.03	11.44	20.37	28.48	23.84	431.33
107.88	30.06	93.16	2.51	1.24	59.70	17.02	6.11	11.65	28.82	37.84	26.44	399.94
107.88	35.39	88.76	2.45	1.33	48.06	27.50	7.11	11.78	32.13	41.50	27.71	389.46
107.88	41.12	82.61	2.57	1.36	32.40	42.86	7.38	11.72	35.37	45.06	27.93	372.58
109.28	0.00	109.12	2.49	1.06	94.55	−15.4	4.21	11.50	20.19	28.29	24.12	434.45
109.28	30.06	94.22	2.46	1.25	61.84	14.78	6.25	11.75	28.68	37.68	26.71	402.91
109.28	35.39	89.71	2.41	1.34	50.10	25.36	7.26	11.84	31.95	41.32	27.96	392.31
109.28	41.12	83.43	2.52	1.37	34.30	40.86	7.56	11.78	35.25	44.94	28.21	375.40

TABELLE 4: Brennbarkeitsprüfung an Kohlenasche enthaltenden Rohmischungen unterschiedliche Feinheit

Feinheit der Rohmischung (Rückstand + 90 μm in Gew.-%)	19,4	14,3	10,3
Freikalkgehalt (Gew.-%) Brenntemperatur (°C)			
Bei 1000 °C	38,0	38,0	37,0
1400 °C	3,1	2,5	1,7
1450 °C	1,3	0,9	0,6
1500 °C	0,5	0,4	0,3
Brennbarkeit	Leicht	Leicht	Leicht

TABLE 4: Burnability test on coal ash containing raw mixes of different fineness

Fineness of raw mix (residue + 90 μm in wt.-%)	19.4	14.3	10.3
Free lime content (wt.-%) Burning temperature (°C)			
At 1000 °C	38.0	38.0	37.0
1400 °C	3.1	2.5	1.7
1450 °C	1.3	0.9	0.6
1500 °C	0.5	0.4	0.3
Burnability	Easy	Easy	Easy

TABELLE 5: Druckfestigkeit von Zementen aus Klinkern (KSG = 93, SM = 2,25), die unter verschiedenen Brennbedingungen im Hinblick auf einen konstanten Freikalkgehalt gebrannt wurden

Temperatur °C	Zeit h	Freies CaO Gew.-%	Druckfestigkeit (%)		
			3 d	7 d	28 d
1350	4	0,7	100	100	100
1400	3	0,7	117	106	109
1450	2	0,8	103	106	110
1500	1	0,4	94	97	99

TABLE 5: Compressive strength of cements from clinkers (LSF = 93, SM = 2.25) burnt to constant free lime under different burning regimes

Temperature °C	Time h	Free CaO wt.-%	Compressive strength (%)		
			3 d	7 d	28 d
1350	4	0.7	100	100	100
1400	3	0.7	117	106	109
1450	2	0.8	103	106	110
1500	1	0.4	94	97	99

Finally, the impact of the raw mix composition, characterized by the LSF, is shown by the test results in **Table 8**.

When the lime combination was maintained at 88%, the 3 d and 7 d strength increased significantly with increasing lime saturation and free lime. However, the 28 d strength of a 94 and 97 LSF raw mix decreased. By burning a mix of 97% LSF at a higher temperature and for a longer time the 3 d and 7 d strength decreased but the 28 d strength increased. An LSF of 101, burnt at a moderate temperature of 1400 °C, gave the highest relative strength.

The potential C_3S content according to Bogue showed only a poor correlation to strength and it follows that some parameters other than compound composition are important.

From the above mentioned data the importance of the burning temperature is apparent. Factors affected by the burning regime are the quantity, viscosity and composition of the liquid phase, the lattice stability and crystal size of the silicates and finally the quantity of oxides in solid solution in the silicates. Microscopically, flux contents showed little variation but silicate crystal size increased when burning temperature or time increased. Low temperature burning produces small crystals, possibly in a greater state of strain, which may result in enhanced reactivity.

4. Conclusion

1. The outstanding indication is that there is an optimum combination of time and temperature giving the best strength for a stated raw mix.
2. Harder or softer burning, whether by changing temperature or time, leads to less strength.
3. The optimum burning conditions vary from raw mix to raw mix. For a given raw mix the conditions appear somewhat harder to achieve with view to 28 d strength than for 3 d strength.
4. Increasing lime saturation leads to higher strength and higher free lime contents.

TABLE 6: Effect on quality of reduced burning temperatures permitted by fine grinding

Raw mix	LSF: 95					
Residue + 90 µm	14 wt.-%			8 wt.-%		
Burning time (h)	2	2	2	2	2	2
Temperature (°C)	1450	1500	1550	1350	1400	1450
Free lime (wt.-%)	3.6	2.8	2.2	3.0	1.9	1.4
Lime Comb. %	94.6	95.8	96.7	95.5	97.1	97.9
Compressive strength of cement (%)						
3 d	100	92.6	81.5	103.7	101.9	98.1
7 d	100	97.5	96.3	102.5	107.4	102.5
28 d	100	95.2	88.7	100.8	104.8	105.6

TABLE 7: Effect of increasing temperature and time on free lime, lime combination and quality

Raw mix	Lime saturation 95%					
Temp./Time (°C/h)	1350/1	1350/2	1400/1	1450/1	1500/1	1500/2
Free CaO (wt.-%)	6.1	3.1	3.0	2.8	1.6	1.0
Lime Comb. (%)	87.8	92.3	92.5	92.8	94.7	95.7
Compressive strength (%)						
3 d	100	103.5	96.4	89.3	85.7	83.9
7 d	100	106.5	111.7	110.4	101.3	96.1
28 d	100	115.4	113.5	124.0	113.5	118.3

flußte Faktoren sind die Menge, die Viskosität und die Zusammensetzung der flüssigen Phase, die Gitterstabilität und Kristallgröße der Silikate sowie schließlich die Menge der in den Silikatmischkristallen enthaltenen Oxide. Unter dem Mikroskop waren die Schmelzanteile nicht sehr verschieden, doch stieg die Kristallgröße bei einer Zunahme der Brenntemperatur oder der Brenndauer an. Ein Brennen bei niedrigen Temperaturen erzeugt kleine Kristalle, die sich möglicherweise in einem ausgeprägteren Spannungszustand befinden, was zu erhöhter Reaktionsfreudigkeit führen kann.

4. Schlußfolgerung

1. Eine der wesentlichen Feststellungen lautet, daß eine optimale Kombination zwischen Zeit und Temperatur besteht, die bei einer bestimmten Rohmehlmischung zu der höchsten Festigkeit führt.
2. Schärferes oder schwächeres Brennen durch Veränderung der Temperatur oder der Brenndauer verringert die Festigkeit.
3. Die optimalen Brennbedingungen sind von einer Rohmehlmischung zur anderen verschieden. Bei einer bestimmten Rohmehlmischung sind diese Bedingungen im Hinblick auf die Festigkeit nach 28 Tagen anscheinend leichter zu erreichen als bei der Festigkeit nach 3 Tagen.
4. Eine Erhöhung der Kalksättigung führt zu größerer Festigkeit und höheren Freikalkgehalten.
5. Die auf der Raumbeständigkeit beruhenden herkömmlichen Anforderungen an den Freikalkgehalt führen in der Praxis oft zu unnötig scharfem Brennen und damit zu einem potentiellen Festigkeitsverlust. Einige Rohmehlmischungen ergeben deshalb bei einer höheren Kalksättigung weniger Festigkeit als bei niedrigeren Sättigungswerten. In solchen Fällen dürfte die Herstellung eines groben, rückstandshaltigen Materials, das gewöhnlich für eine hohe Verbrennungstemperatur verantwortlich ist, die Verwendung einer Mischung mit höherem Kalksättigungsgrad erlauben, die sich unter optimalen Bedingungen brennen läßt, so daß eine potentiell größere Festigkeit mit einem zufriedenstellenden Freikalkgehalt erzielt werden kann.

TABELLE 8: Festigkeit von Zementen aus Rohmischungen unterschiedlicher Kalksättigungsgrade, die im Hinblick auf eine ähnliche Kalkkombination gebrannt wurden
TABLE 8: Strength of cements from raw mixes of different lime saturations, burnt to give similar combination

KSG	89	91	94	97	97	101
Temperatur/Zeit (°C/h)	1450/1	1400/1	1350/1	1350/1	1500/2	1400/2
Freies CaO (Gew.-%)	0,7	1,8	4,2	6,1	1,0	3,4
Kalkkomb. (%)	87,7	88,2	87,8	87,8	95,7	95,4
C_3S (Bogue) (Gew.-%)	36,7	37,6	35,4	33,5	54,4	53,1
Druckfestigkeit (%)						
3 d	100	114,3	148,6	160,0	134,3	154,3
7 d	100	107,0	126,3	135,0	129,0	159,6
28 d	100	107,3	106,4	94,5	111,8	121,8

KSG	= LSF
Temperatur/Zeit	= Temperature/Time
Freies CaO	= Free CaO
Gew.-%	= wt.-%
Kalkkomb.	= Lime Comb.
Druckfestigkeit	= Compressive strength

5. The conventional requirements for free lime content, based on soundness, often lead in practice to harder than optimum burning and correspondingly to a loss of potential strength. Therefore, some raw mixes give less strength at higher than at lower lime saturation for this reason. In such cases, the production of a coarse residue containing material, which is usually responsible for high combustion temperature, should permit to use a mix of higher lime saturation to be burnt under optimum conditions thus realising a potentially increased strength with satisfactory free lime content.

Klinkerqualität und Betriebserfahrungen beim Einsatz von Reststoffen aus der Papieraufbereitung als Sekundärstoff

Clinker quality and operational experience using residues from paper processing as a secondary material

Qualité du clinker et expériences acquises au cours de service lors de l'emploi de matières résiduaires provenant de la préparation du papier comme matière première secondaire

Calidad del clinker y experiencias adquiridas con el empleo de materias residuales procedentes de la preparación de papel como material secundario

Von **L. Ernstbrunner**, Waldeck-Wopfing/Österreich

Zusammenfassung – Die stoffliche Verwertung von Altpapier führt zu einem drastischen Ansteigen des bei der Altpapieraufbereitung anfallenden Papierreststoffes, der bisher deponiert wurde. Weiter ansteigende Deponierungskosten zwingen jedoch zur Verwertung dieses Papierschlammes. In einem zum Patent angemeldeten Verfahren wird im Zementwerk Wopfing/Österreich seit Anfang 1990 Papierschlamm als Sekundärstoff eingesetzt. Dabei werden die Drehofenabgase für die Trocknung genutzt. Der Schlamm enthält etwa 50 Gew.-% Wasser. Der Trocknungsrückstand besteht zur Hälfte aus für die Papiererzeugung nicht mehr geeigneten brennbaren Zellulosefasern, der Rest aus natürlichen mineralischen Füllstoffen, wie Kalkstein und Kaolin. Dieser im Bereich des Ofeneinlaufes eingesetzte Rückstand vermindert den Einsatz fossiler Brennstoffe, wobei gleichzeitig der im Papierreststoff enthaltene Füllstoff als Rohstoff beim Klinkerbrennprozeß zur Verfügung steht. Der Beitrag befaßt sich mit den Kosten und der Umweltverträglichkeit des Verfahrens im Vergleich zu anderen Verbrennungsverfahren sowie mit den Erfahrungen bei der behördlichen Genehmigung und dem Betrieb der Anlage. Seit Oktober 1992 läuft das Genehmigungsverfahren für ein Erweiterungsprojekt, das eine Verdoppelung der eingesetzten Papierreststoffmenge vorsieht. Im Zuge dieser Investition ist die Errichtung eines neuen Wärmetauschers mit MSC-Calcinator zur Minderung der NO_x-Emission geplant.

Klinkerqualität und Betriebserfahrungen beim Einsatz von Reststoffen aus der Papieraufbereitung als Sekundärstoff

Summary – Recycling of waste paper leads to a drastic increase in the paper residues produced during waste paper processing which up until now have been placed in landfill sites. However, increasing landfill costs are making it necessary to utilize this paper slurry. Paper slurry has been used at the Wopfinger cement works in Austria since the beginning of 1990 as a secondary material in a process for which a patent has been applied. The rotary kiln exhaust gases are used for the drying. The slurry contains about 50 wt.% water. The residue after drying consists roughly of half combustible cellulose fibres which are not suitable for making paper and the rest of natural mineral fillers, such as limestone and kaoline. This residue introduced near the kiln inlet of the cement rotary kiln reduces the use of fossil fuels; at the same time the fillers contained in the paper residue are available as raw material for the clinker burning process. This lecture deals with the costs and environmental compatibility of the process when compared with other methods of combustion, as well as with experience with official licensing and operation of the plant. Since October 1992 the licensing procedure has been under way for an extension project intended to double the quantity of paper residues used. During this capital investment it is planned to build a new preheater with MSC calciner to reduce the NO_x emission.

Clinker quality and operational experience using residues from paper processing as a secondary material

Résumé – L'utilisation de vieux papiers conduit à une augmentation massive de matières résiduaires résultant de la préparation de vieux papiers, matières qui ont été mises en dépôt jusqu'à présent. Cependant, l'augmentation des frais de mise en dépôt obligent à utiliser ces boues de papier. Selon un procédé, pour lequel un brevet a été demandé, les boues de papier sont utilisées, depuis début janvier 1990, comme matière secondaire dans la cimenterie de Wopfing/Autriche. Pour en assurer le séchage, on utilise les gaz de fumée du four rotatif. Les boues contiennent environ 50% en poids d'eau. Les résidus du séchage sont composés, pour la moitié environ, de fibres de cellulose inflammables, qui ne sont plus aptes à la fabrication de papier, le reste étant composé de filler minéral naturel, tel que le calcaire et le caolin. Ces résidus employés dans la zone d'entrée du four à ciment, réduisent l'emploi de combustibles fossiles et, en même temps, la matière de remplissage contenue dans les résidus de papier est disponible comme matière première dans le processus de cuisson de clinker. Le présent article traite des coûts et de l'effet sur l'environnement du procédé par rapport à d'au-

Qualité du clinker et expériences acquises au cours de service lors de l'emploi de matières résiduaires provenant de la préparation du papier comme matière première secondaire

tres procédés de combustion ainsi que des expériences acquises en relation avec l'autorisation officielle et l'exploitation de l'installation. Depuis octobre 1992, une procédure d'autorisation est en cours concernant un projet d'extension prévoyant un doublement de la quantité de résidus utilisés. Dans le cadre de cet investissement, il est prévu de monter un nouvel échangeur de chaleur avec calcinateur MSC, destiné à réduire les émissions de NO_x.

Resumen – *El aprovechamiento de papel viejo conduce a un aumento masivo de materias residuales resultantes de la preparación de dicho papel, las cuales se llevaban hasta ahora a un depósito. Sin embargo, los crecientes costes de los depósitos hacen necesario un aprovechamiento de esta pasta de papel. Desde primeros de enero de 1990, se está utilizando, en la fábrica de cemento de Wopfing/Austria, pasta de papel como materia secundaria, según un procedimiento, para el cual se ha solicitado una patente. Para el secado de dicha pasta se aprovechan los gases de escape del horno rotatorio. La pasta contiene aproximadamente un 50% en peso de agua. La mitad de los residuos del secado se componen de fibras de celulosa inflamables, que ya no son aptas para la fabricación de papel, y el resto se compone de filler minerales, naturales, tales como caliza y caolín. Estos residuos, empleados en la zona de entrada del horno de cemento, reducen el empleo de combustiles fósiles, en tanto que el filler contenido en los residuos de papel se puede aprovechar como materia prima en el proceso de cocción del clínker. El presente artículo trata del coste así como del efecto de este procedimiento sobre el medio ambiente, en comparación con otros procedimientos de combustión, y de las experiencias adquiridas en relación con la autorización oficial y la explotación de la instalación. Desde octubre de 1992, está pendiente el procedimiento de autorización referente a un proyecto de ampliación de la instalación, el cual comprende la duplicación de la cantidad de residuos de papel utilizados. Dentro del marco de esta inversión, está previsto el montaje de un nuevo intercambiador de calor, equipado con un calcinador MSC y destinado a reducir las emisiones de NO_x.*

Calidad del clinker y experiencias adquiridas con el empleo de materias residuales procedentes de la preparación de papel como material secundario

In den dicht besiedelten Industriestaaten kam es in den letzten Jahren durch gesetzliche Auflagen zum Schutz der Umwelt und einer möglichen Gefährdung des Grundwassers zu einer dramatischen Verknappung an Deponieraum und zu einem enormen Anstieg der Deponierungskosten für Abfälle. Durch gesetzliche Regelungen, wie zum Beispiel die Verpackungsverordnung, wird die Vermeidung und die stoffliche Verwertung von Abfällen gefordert. Der zum Schutz der Umwelt wünschenswerte Einsatz von bis zu 100% Altpapier, wie zum Beispiel bei der Tissue-Papiererzeugung, führt jedoch zu einem höheren Anteil an Papierreststoff als bei dem herkömmlichen Zellstoffverfahren.

Im **Bild 1** sind die Mengen von Reststoffen dargestellt, die in der österreichischen Papierindustrie im Jahr 1992 anfielen. Die Säule rechts außen zeigt zum Vergleich den Papierreststoffanfall in der BRD. Die wirtschaftliche Grundlage für die Realisierung des Projektes war die Entwicklung der Deponiepreise für Papierreststoff (**Bild 2**).

Bei der Aufbereitung von Altpapier fallen zum Beispiel Faserreststoffe und Sortierrückstände an. Zur Zeit erfolgt die Entsorgung des Papierreststoffes hauptsächlich durch Deponieren, Ausbringen in der Landwirtschaft, Kompostieren und Verwendung in der Ziegelindustrie. In den letzten Jahren versucht die Papierindustrie außerdem, den Papierreststoff in eigenen, sehr teuren Verbrennungsanlagen thermisch zu entsorgen. Das größte Problem stellt dabei die Entsorgung der anfallenden Asche dar, die häufig 25% der Papierreststoffmenge ausmacht.

In einem zum Patent angemeldeten Verfahren im Zementwerk Wopfing in Österreich wird seit 1990 nach einem mehrjährigen Genehmigungsverfahren mit einer Serie von Umweltmessungen, die noch während des Pilotbetriebes durchgeführt wurden, Papierreststoff eingesetzt. **Bild 3** zeigt das Verfahrensschema. Das Verfahren arbeitet denkbar einfach. Nach dem Wärmetauscherturm (12) wird ein Teilstrom des Ofenabgases (15) abgezogen, über einen Trockner (5) geleitet und abgekühlt in das System zurückgeführt (17). Damit steht die gesamte Ofenabgasmenge als Trägergas für die nachgeschaltete Rohmühle voll zur Verfügung. Der auf ca. 10–15% Restfeuchte getrocknete Papierreststoff wird über eine Schleuse (8) am Ofeneinlauf (9) aufgegeben, ähnlich wie zum Beispiel bei der Verbrennung von Autoreifen. Für die Trocknung des Papierreststoffes mit

BILD 1: Reststoffe aus der Papierindustrie in Österreich
FIGURE 1: Residues from the paper industry in Austria

Tonnen/Jahr = tonnes/year
Holzabfälle = wood wastes
Sortierungsrückstände = sorting wastes
Rinde = bark
Aschen = ash
Schlacken = slag
Papierreststoff in Österreich = paper residues in Austria
Papierreststoff in der BDR = paper residues in Germany

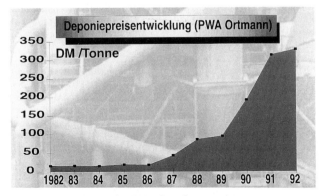

BILD 2: Entwicklung der Deponiepreise
FIGURE 2: Growth of landfilling charges

DM/Tonne = DM/tonne

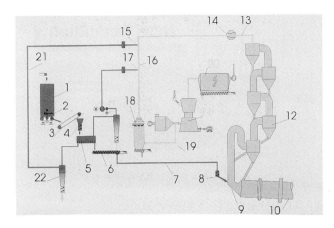

BILD 3: Verfahrensschema
FIGURE 3: Process flowsheet

Drehofenabgasen kann somit ein Teil der Abgaswärme genutzt werden. Weiterhin wird die anfallende Asche zur Gänze als Sekundärrohstoff für die Klinkererzeugung verwendet. Wie umfangreiche Berechnungen ergeben haben, ist das Verfahren auch für solche Anlagen geeignet, bei denen im Verbundbetrieb keine überschüssige Wärmemenge mehr verfügbar ist. In solchen Fällen wird ein Teil der Energie aus dem Papierreststoff für die Erhöhung der Abgastemperatur nach dem Wärmetauscherturm verwendet. Dies geschieht durch verfahrenstechnische Maßnahmen am Wärmetauscher.

Eine wesentliche Voraussetzung für den Einsatz von Papierreststoffen ist die Forderung, daß hierdurch kein negativer Einfluß auf die Klinkerqualität entstehen darf. Der Papierreststoff hat im Anlieferungszustand einen Wassergehalt von mehr als 50 %. Der Feststoffanteil besteht zu je etwa 25 % aus brennbaren Papierfasern und Asche. Die Analyse der Asche geht aus **Bild 4** hervor. Danach weist die Asche keine ungewöhnliche Zusammensetzung auf. Das ist darauf zurückzuführen, daß die in der Papierindustrie verwendeten Füllstoffe aus hochreinen natürlichen Kalkstein- und Kaolinvorkommen stammen. Die Sollzusammensetzung des Klinkers wird durch eine Anpassung der Rohmehl- an die jeweilige Aschezusammensetzung erreicht. Dementsprechend sind je nach der ursprünglich gegebenen Rohmehlzusammensetzung mitunter Korrekturen durch Eisenerz oder Kalkstein vorzunehmen.

Es stellt sich die Frage, ob ein solcher Aufwand für ein Zementwerk bzw. eine Papierfabrik überhaupt lohnend ist. Geht man jedoch davon aus, daß das Deponieren von Papierreststoff immer schwieriger und teurer wird und zieht man die verhältnismäßig hohen Investitionskosten für eine Verbrennungsanlage in der Papierfabrik in Höhe von 20 bis 30 Mio. DM, die zusätzlichen Deponiekosten für die Asche und die Betriebskosten einer solchen Anlage in Höhe von 1,5 Mio. DM pro Jahr in Betracht, so wird klar, daß die Verbrennung von Papierreststoffen sowohl für die Papier- als auch für die Zementindustrie von Interesse sein kann. Dementsprechend ist vorgesehen, die Entsorgungsmöglichkeiten der zu einem großen deutschen Papierunternehmen gehörenden Firma PWA Ortmann im Werk Wopfing weiter auszubauen. Danach soll die durch eine Erweiterungsinvestition in der Papierfabrik anfallende zusätzliche Menge von 30.000 t Papierreststoff je Jahr ab 1995 in Wopfing verarbeitet werden. **Bild 5** zeigt den Umfang der Investition. Es wird ein komplett neuer Wärmetauscherturm mit Tertiärluftleitung nach dem MSC-Verfahren errichtet. Zum Vergleich ist die bisherige Bauhöhe der Anlage durch die hellen Linien angedeutet. Der Neubau bietet darüber hinaus den Vorteil, mit Hilfe verfahrenstechnischer Maßnahmen den NO_x-Gehalt des Ofenabgases zu mindern (**Bild 6**). Ähnlich wie in Deutschland wird auch in Österreich ab 1996 eine drastische Reduktion der NO_x-Emissionen vorgeschrieben.

In the densely populated industrial countries, legislation to protect the environment and the potential risk to groundwater quality have led in recent years to a dramatic reduction in the amount of available landfilling space and a huge rise in the cost of land disposal of waste. Prevention and recycling of wastes is being encouraged by statutory provisions such as the Packaging Directive. However, the environmentally desirable use of up to 100 % wastepaper, for instance in tissue paper production, gives rise to a larger proportion of paper residue than is the case with the conventional chemical pulp process.

Figure 1 depicts the quantities of residues generated in the Austrian paper industry in 1992. The bar at the far right shows, by way of comparison, the volume of paper residue generated in Germany. The economic basis for carrying out the project was the growth of landfilling charges for paper residue (**Figure 2**).

Processing of wastepaper gives rise to, for instance, fibre residues and sorting wastes. At present this paper residue is mainly disposed of by landfilling, discharge for agricultural purposes, composting, and utilization in the brick-making industry. In recent years the paper industry has also been trying to dispose of the paper residue by thermal means in its own, very expensive, incinerators. Here, the biggest problem is disposal of the resulting ash, which often makes up 25 % of the volume of the paper residue.

In a process used at the Wopfing Cement Works in Austria, paper residues have been in use since 1990 following a licensing procedure lasting several years and involving an environmental monitoring programme with a series of environmental measurements which were carried out during pilot-scale operation. A patent for the process has been applied for. **Figure 3** shows the process flowsheet. The operating principle is extremely simple. Downstream of the preheater tower (12), a side stream of the kiln exhaust gas (15) is drawn off, passed through a dryer (5) and recirculated to the system after cooling (17). In this way the entire kiln exhaust gas is fully available for use as carrier gas for the downstream raw-grinding mill. The paper residue, which is dried to a residual moisture content of approx. 10–15 %, is fed in via a lock (8) at the kiln inlet (9) as in burning of car tyres, for example. In this way the paper residues can be dried with rotary kiln exhaust gases using some of the heat content of the exhaust gas. In addition, all the resulting ash is used as a secondary raw material for clinker production. As extensive calculations have shown, the process is also suitable for plants with interconnected operation, where no surplus heat is available. In such cases part of the paper residue's energy content is used to increase the exhaust gas temperature downstream of the preheater tower. This is achieved through the engineering design of the preheater.

An important condition for the use of paper residues is the requirement that the quality of the clinker should not be adversely affected. In as-delivered condition, the paper residue has a water content of over 50 %. The solids content comprises combustible paper fibres and ash in the proportion of about 25 % each. The ash analysis is given in **Figure 4**. It shows that the composition of the ash is not abnormal. This is due to the fact that the fillers used in the paper indus-

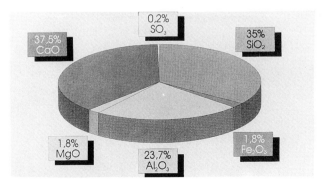

BILD 4: Zusammensetzung der Asche
FIGURE 4: Ash composition

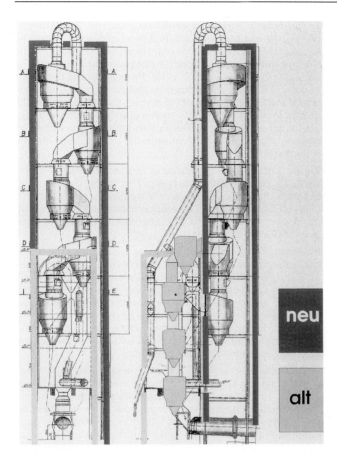

BILD 5: Maßstäbliche Darstellung der Alt- und Neuanlage
FIGURE 5: Scale drawing of the existing and new plants

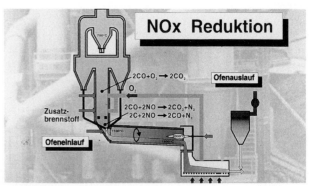

BILD 6: Prinzip der NO_x-Reduktion
FIGURE 6: Principle of NO_x reduction

Zusatzbrennstoff	= auxiliary fuel
Ofeneinlauf	= kiln inlet
Ofenauslauf	= kiln outlet

Es ist beabsichtigt, den NO_x-Ausstoß durch die im Bild 6 dargestellte mehrstufige Verbrennung ohne den Einsatz der Ammoniaktechnologie zu halbieren.

Der Einsatz des Reststoffes aus der Papierindustrie ermöglicht es demnach, auf ökonomisch wie ökologisch sinnvolle Weise den Einsatz fossiler Brennstoffe und natürlicher Rohstoffe unter Ausnutzung der im Papierreststoff enthaltenen Rohstoffe zur vermindern, Deponieraum zu sparen und den NO_x-Ausstoß des Zementwerkes zu senken.

try originate from highly pure natural limestone and kaolin deposits. The desired clinker composition is achieved by matching the composition of the raw meal to the respective ash composition. Accordingly, depending on the originally specified raw meal composition adjustments have to be made from time to time by adding iron ore or limestone.

The question arises of whether such expense is at all worthwhile for a cement works or a paper mill. On the assumption, however, that landfilling of paper residues will become more and more difficult and expensive and considering the relatively high capital cost of an incinerator at the paper mill, amounting to DM 20–30 million, the additional cost of landfilling the ash and the operating costs of this incinerator amounting to DM 1.5 million p.a., it becomes clear that combustion of paper residues may be of interest to both the paper industry and the cement industry. It is therefore planned to further extend the disposal facilities of PWA Ortmann, which is owned by a large German paper company, at the Wopfing works. The additional 30 000 t p.a. of paper residue resulting from an extension project at the paper mill will then be processed at Wopfing beginning in 1995. **Figure 5** illustrates the extent of the project. A complete new preheater tower with tertiary air duct based on the MSC process is being built. By way of comparison, the present height of the plant is indicated by the light lines. The new unit also has the advantage of reducing the NO_x content of the kiln exhaust gas by process engineering means (**Figure 6**). In similar fashion to Germany, a drastic cut in NO_x emissions is also prescribed in Austria from 1996. It is planned to halve the level of NO_x emissions by means of the multi-stage combustion system illustrated in **Figure 6** without using ammonia-based technology.

Thus the use of this residue from the paper industry enables the use of fossil fuels and natural raw materials to be reduced by making use of the raw materials contained in the paper residue, and also to economise on landfilling space and to cut the cement works' NO_x emissions, in an economically and ecologically worthwhile manner.

Erfahrungen mit spezialisierten Kontrolltechniken beim Einsatz von Sekundärstoffen

Experience with specialized control techniques when using secondary materials

Expériences acquises avec des techniques de contrôle spéciales lors de l'utilisation de matières premières secondaires

Experiencias adquiridas con técnicas de control especiales, utilizando materias primas secundarias

Von **W. Gerger,** Gmunden/Österreich

Zusammenfassung – *Der Einsatz neuer und unbekannter Roh- und Brennstoffe, insbesondere der Einsatz von Sekundärstoffen, erfordert die*
– *Gewährleistung einer erstklassigen Produktqualität, den*
– *Ausschluß von zusätzlichen Umweltbelastungen, eine vom*
– *Sekundärstoffeinsatz unbeeinträchtigte Prozeßtechnik, die*
– *Wahrung der Sicherheit der Arbeitnehmer sowie die*
– *Akzeptanz bei Kunden, Anrainern, Behörden und Mitarbeitern.*
Der Nachweis der Erfüllung dieser Voraussetzungen und häufig darüber hinausgehende behördliche Auflagen machen mitunter den Einsatz bislang für ein Zementwerk neuartiger Prüftechniken nötig. Organische Analytik, insbesondere der Einsatz von Gaschromatographen, ist für ein Zementlabor eine neuartige Aufgabe. Ausgehend von der auflagengemäßen regelmäßigen Kontrolle des PCB-Gehaltes in Sekundärbrennstoffen als Bestandteil der Betriebsgenehmigung zur thermischen Entsorgung brennbarer Flüssigabfälle eignet sich die GC-Analytik auch zur qualitativen und quantitativen Erfassung von Extrakten aus Feststoffproben, Abgasbestandteilen, organischem Gesamtkohlenstoff und Mahlhilfsmitteln in Zementen. Die regelmäßige Spurenelementbestimmung in Sekundärbrennstoffen, ebenfalls als behördliche Auflage der Betriebsgenehmigung, erfordert die Etablierung einer erweiterten Analysentechnik mit Röntgenfluoreszenz. Damit besteht die Möglichkeit, umfangreiches Datenmaterial für Spurenelemente in Rohmaterialien, Brennstoffen, Klinker und Stäuben zu sammeln, wodurch Bilanzierungen von Schwermetallen über längere Zeiträume ermöglicht werden. Die routinemäßige Anwendung weitergehender Kontrolltechniken schafft ein Instrumentarium zum tieferen Einblick und besserem Verständnis des Zementerzeugungsprozesses.

Summary – *The use of new and unknown raw materials and fuels, especially the use of secondary materials, requires*
– *a guarantee of a first class product quality,*
– *exclusion of any additional environmental pollution,*
– *a process technology which is not impaired by the use of secondary materials,*
– *preservation of the safety of the employees, and*
– *acceptability with clients, neighbours, authorities and staff.*
Evidence that these requirements and the official conditions which frequently go beyond them have been fulfilled is, among other things, making it necessary to use test methods which are new to cement works. Organic analysis, especially the use of gas chromatography, is a new type of problem for a cement laboratory. Used initially for regular checking of the PCB content in secondary fuels required as part of the plant licensing for thermal disposal of combustible liquid wastes, GC analysis is also suitable for qualitative and quantitative measurement of extracts from solid samples, exhaust gas constituents, organic total carbon, and grinding aids in cements. Regular measurement of trace elements in secondary fuels, also an official condition in plant licensing, requires the establishment of an extended analytical technique with X-ray fluorescence. This makes it possible to collect extensive data for trace elements in raw materials, fuels, clinker and dusts, which facilitate balances of heavy metals over fairly long time periods. Routine application of more extensive test methods creates an instrument which provides a deeper insight and better understanding of the cement production process.

Résumé – *L'emploi de matières premières et de combustibles nouveaux et inconnus, notamment de matières premières secondaires, requiert la garantie d'une qualité parfaite du produit, l'exclusion de répercussions négatives sur l'environnement, une technique de processus indépendante de l'emploi des matières secondaires, la sauvegarde de la sécurité du personnel ainsi que l'acceptation de la part des clients, des voisins, des Autorités compétentes et des collaborateurs. La preuve de l'accomplissement de ces conditions et, très souvent, les exigences des Autorités entraînent parfois l'emploi de techniques de contrôle encore inconnues dans une cimenterie. L'analyse organique, notamment l'emploi de la chromatographie gazeuse, constitue une tâche nouvelle pour un laboratoire à ciment. En partant du contrôle périodique, exigé par les Autorités, de*

la teneur en PCB des combustibles secondaires, en tant qu'élément de l'autorisation officielle relative à l'élimination technique de résidus liquides inflammables, l'analyse GC est apte également à l'enregistrement qualitatif et quantitatif d'extraits d'échantillons de solides, de constituants de gaz de fumée, de carbone organique total et d'adjuvants de broyage, contenus dans les ciments. La détermination périodique des éléments de trace dans les combustibles secondaires, exigée également par les Autorités dans le cadre de la procédure d'autorisation, requiert l'établissement d'une technique d'analyse plus large, utilisant la méthode par fluorescence X. De cette manière, il existe la possibilité de recueillir des dates copieuses concernant les éléments de trace contenus dans les matières premières, les combustibles, les clinkers et les poussières, ce qui permet d'établir des bilans de métaux lourds pendant des périodes prolongées. L'application par routine de techniques de contrôle encore plus poussées, offre des instruments permettant d'obtenir des connaissances plus approfondies et de mieux comprendre le processus de fabrication du ciment.

Experiencias adquiridas con técnicas de control especiales, utilizando materias primas secundarias

Resumen – El empleo de materias primas y combustibles nuevos y desconocidos, sobre todo de materias primas secundarias, requiere
– la garantía de una calidad perfecta del producto,
– la exclusión de repercusiones negativas adicionales sobre el medio ambiente,
– una técnica de procesos independiente del empleo de materias primas secundarias,
– la salvaguardia de la seguridad del personal y
– la aceptación de estas materias por parte de los clientes, vecinos, Autoridades competentes y colaboradores.

La prueba del cumplimiento de estos requisitos y, a veces, otras exigencias aún más rigurosas decretadas por las Autoridades competentes en esta materia, hacen necesaria, en ciertos casos, la aplicación de nuevas técnicas de control, desconocidas en las fábricas de cemento. La analítica orgánica y, especialmente, el empleo de la cromatografía de gases (CG), constituyen una tarea nueva para un laboratorio de cemento. Partiendo del control periódico, exigido por las Autoridades, del contenido de PCB de los combustibles secundarios, como elemento de la autorización oficial del método de eliminación térmica de residuos líquidos inflamables, el análisis CG es apto también para la medición cualitativa y cuantitativa de extractos de muestras de sólidos, de constituyentes de gases de escape, del carbono orgánico total y de los coadyuvantes de la molienda contenidos en el cemento. La determinación periódica de los elementos-traza en los combustibles secundarios, exigida también por las Autoridades para el procedimiento de autorización, require el establecimiento de una técnica analítica más amplia, incluyendo el análisis por fluorescencia de rayos X. De esta forma, existe la posibilidad de recoger amplios datos referentes a los elementos-traza contenidos en las materias primas, combustibles, clínkeres y polvos, lo cual permite establecer balances de metales pesados durante largos períodos de tiempo. La aplicación por rutina de técnicas de control más avanzadas es un instrumento que permite obtener conocimientos más profundos y comprender mejor el proceso de fabricación del cemento.

1. Routinemäßige Eingangskontrolle von Sekundärbrennstoffen

Die Betriebsgenehmigung zur „Thermischen Entsorgung brennbarer Flüssigabfälle" bei den Gmundner Zementwerken Hans Hatschek AG legt für Inhaltsstoffe und Heizwert der angelieferten Sekundärbrennstoffe Grenzwerte fest [1]. Diese sind an jeder einzelnen Anlieferung zu prüfen. Dazu wird eine repräsentative Probe mittels Röntgenfluoreszenz auf den Gehalt der Spurenelemente Quecksilber, Thallium, Cadmium, Blei, Zink sowie Schwefel und Chlor geprüft. Die Analyse erfolgt mit einem Sequenzröntgenfluoreszenzgerät SRS 300 in Einweggefäßen mit Hostaphanfolie unter Heliumatmosphäre. Zur Berücksichtigung von Matrixeffekten wird für die Analyse der Schwermetalle die geräteeigene Rhodium-Compton-Linie, für die Elemente Schwefel und Chlor die Rhodium-Lα_1-Linie mitvermessen. Jeweils wird eine Untergrundkorrektur durchgeführt. Darüber hinaus werden die Sekundärbrennstoffe auf Fluor (mit einer ionenselektiven Elektrode), Stickstoff (nach Kjeldahl), Wasser (nach Karl-Fischer) sowie auf polychlorierte Biphenyle (nach DIN 51527), Heizwert, Sedimente und Viskosität geprüft.

In der **Tabelle 1** sind Ergebnisse dieser regelmäßigen Eingangskontrolle von brennbaren Flüssigabfällen aus den Jahren 1989 bis 1992 zusammengestellt.

Seit 1989 wurden jährlich steigende Mengen Sekundärbrennstoff übernommen, geprüft und verfeuert. Die Anzahl der Analysen – jede repräsentiert ca. 22 bis 25 Tonnen angelieferten Materials – veranschaulicht den steigenden Substitutionsgrad von Primärbrennstoffen.

1. Routine testing of incoming secondary fuels

The operating licence for the "thermal disposal of combustible liquid wastes" at Gmundner Zementwerke Hans Hatschek AG stipulates limit values for the constituents and the calorific value of the secondary fuels delivered [1]. Each individual delivery has to be tested for compliance with these limits. This is done by testing a representative sample for its content of the trace elements mercury, thallium, cadmium, lead, zinc, sulphur and chlorine by means of X-ray fluorescence. The analysis is carried out with an SRS 300 sequential X-ray fluorescence unit in expendable containers with Hostaphan film in a helium atmosphere. For the heavy metal analysis the equipment's intrinsic rhodium-Compton line and for the elements sulphur and chlorine the rhodium-Lα_1 line are measured at the same time to allow for matrix effects. An adjustment for background levels is made in each case. The secondary fuels are also tested for fluorine (with an ion sensitive electrode), nitrogen (by the Kjeldahl method), water (by the Karl Fischer method) and polychlorinated biphenyls (according to DIN 51527), and with regard to their calorific value, sediments and viscosity.

Table 1 lists results of these regular tests of incoming combustible liquid wastes from 1989 to 1992.

Since 1989, the quantity of secondary fuels accepted, tested and burnt has risen year by year. The number of analyses performed, each of which represents approx. 22–25 tonnes of as-delivered material, illustrates the growing degree of substitution of primary fuels.

Besides the mean values with minima and maxima, Table 1 contains the range of variation of selected parameters used in the tests on the incoming fuels.

TABELLE 1: Entsorgung brennbarer Flüssigkeitsabfälle, Ergebnisse der Eingangskontrollen

	1989	1990	1991	1992
Anzahl der Analysen	101	228	470	638
PCB (DIN x 5) (mg/kg)	19,4 3,0–99,5	12,7 3,5–67,0	6,4 n.n.–27,0	1,4 n.n.–54,5
Blei (mg/kg)	521 196–3943	373 198–593	334 194–391	298 5–770
Zink (mg/kg)	–	–	515 289–586	503 2–1311
Chlorid (%)	0,117 0,047–0,511	0,072 0,033–0,123	0,097 0,051–0,50	0,098 0,006–1,00
Wasser (%)	1,4 0,3–7,0	5,5 0,5–15,4	7,2 0,9–26,8	8,3 0,0–26,4
Heizwert (Hu) (kJ/kg)	41583 38724–42119	39533 34035–42981	38714 32991–41635	38175 27210–43656

TABLE 1: Disposal of combustible liquid wastes, results of tests of incoming goods

	1989	1990	1991	1992
Number of analyses	101	228	470	638
PCBs (DIN x 5) (mg/kg)	19.4 3.0–99.5	12.7 3.5–67.0	6.4 n.d.–27.0	1.4 n.d.–54.5
Lead (mg/kg)	521 196–3943	373 198–593	334 194–391	298 5–770
Zinc (mg/kg)	–	–	515 289–586	503 2–1311
Chloride (%)	0.117 0.047–0.511	0.072 0.033–0.123	0.097 0.051–0.50	0.098 0.006–1.00
Water (%)	1.4 0.3–7.0	5.5 0.5–15.4	7.2 0.9–26.8	8.3 0.0–26.4
Calorific value (net) (kJ/kg)	41583 38724–42119	39533 34035–42981	38714 32991–41635	38175 27210–43656

Die Tabelle 1 enthält neben den Mittelwerten mit Minima und Maxima die Spannweite der Variation ausgewählter Parameter der Eingangskontrolle.

Während im Zuge des Probebetriebes 1989 Einsatzmaterial mit künstlich erhöhtem Anteil an PCB bzw. Blei zum Einsatz gekommen ist, repräsentieren die Maxima in den folgenden Jahren extreme Qualitäten der eingesetzten Sekundärbrennstoffe. Die Mittelwerte an polychlorierten Biphenylen nehmen infolge des Verbotes der Produktion von PCBs von Jahr zu Jahr ab. In einem Großteil der angelieferten Altöle sind PCBs nicht mehr nachweisbar. Die Nachweisgrenze der Routineanalyse liegt bei 1 mg/kg. Werden fallweise doch höhere Konzentrationen nachgewiesen, sind in den Proben Chlophene identifizierbar.

Ebenfalls fallend ist die Konzentration an Blei in Altölen als Folge des vermehrten Einsatzes bleifreier Treibstoffe. Höhere Konzentrationen treten jedoch in Destillationsrückständen bzw. auch in Lackschlämmen auf. Während die Zinkkonzentration in Altölen bei ca. 500 mg/kg liegt, treten ebenfalls höhere Zinkkonzentrationen in Lack- und Destillationsrückständen auf. Selektives qualitätskontrolliertes Sammeln von Altöl erzielt Chloridkonzentrationen unter 0,1 %. Mit höheren Chloridgehalten muß bei Lösungsmittelgemischen gerechnet werden.

Die Wassergehalte in den angelieferten Sekundärbrennstoffen zeigen eine von Jahr zu Jahr steigende Tendenz, dem bei der Bewertung der Materialien hinsichtlich Vergütung bzw. Entsorgungsentgelt Rechnung getragen wird. Überhaupt dient neben dem Heizwert jeder der routinemäßig geprüften Parameter als Kriterium für die preisliche Einstufung der Sekundärbrennstoffe.

Als positiver Effekt der erhöhten Wassergehalte in den Sekundärbrennstoffen resultiert eine geringere Stickstoffoxidentwicklung in der Primärfeuerung.

2. Organische Analyse mittels Gaschromatographie

Zur Analyse von PCB nach DIN 51527 werden nach flüssigchromatographischer Vorreinigung sechs ausgewählte Biphenyle mit einem Gaschromatographen, der mit einem Elektroneneinfangdetektor ausgerüstet ist, aufgetrennt, identifiziert und qualifiziert. Die Multiplikation mit dem Faktor 5 erbringt eine Abschätzung des Gesamt-PCB-Gehaltes und steht repräsentativ für 209 PCB-Kongenere. Das Peakmuster des Gaschromatogrammes ermöglicht dem erfahrenen Laboranten eine weitergehende Beurteilung des angelieferten Sekundärbrennstoffes.

Zur Durchführung von gaschromatographischen Untersuchungen ist das Laboratorium mit zwei Gaschromatographen (HP 5890A) ausgerüstet. Einer ist für die routinemäßige Analyse von PCBs in brennbaren Flüssigabfällen spezialisiert und mit einem automatischen Probengeber und einer Chemstation (HP-Vectra) zur Ablaufsteuerung halbautomatisiert.

Der zweite Gaschromatograph ist mit einem massenselektiven Detektor, mit einem Flammenionisationsdetektor und

Whereas in the course of the trial running in 1989 feed materials with artificially increased PCB and lead contents were used, the maxima in the subsequent years represent extreme qualities of the secondary fuels used. The average polychlorinated biphenyl content decreases from year to year due to the ban on PCB production. In a large proportion of the waste oils delivered, PCBs are no longer detectable. The detectability limit in the routine analysis is 1 mg/kg. Where higher concentrations are still occasionally detected, chlophenes can be identified in the samples.

Also on a downward trend is the concentration of lead in waste oils due to the increased use of unleaded motor fuels. However, higher concentrations occur in distillation residues and paint sludge. While the zinc concentration in waste oils is approx. 500 mg/kg, higher concentrations are again found in paint and distillation residues. Selective quality-controlled collection of waste oil achieves chloride concentrations of less than 0.1 %. Elevated chloride contents are likely in mixed solvents.

The water contents of the secondary fuels delivered show a rising trend from year to year; this is taken into account when valuing the materials with regard to compensation or the disposal fee. In any case, not only the calorific value but each of the routinely tested parameters serves as a criterion for pricing the secondary fuels.

A beneficial effect of the increased water content of the secondary fuels is reduced nitrogen oxide formation in the primary firing system.

2. Organic analysis by gas chromatography

For the analysis of PCBs according to DIN 51527, after pre-purification by liquid chromatography six selected biphenyls are separated, identified and quantified using a gas chromatograph equipped with an electron capture detector. Multiplication by a factor of 5 gives an estimate of the total PCB content and is representative of 209 members of the PCB family. The peak pattern of the gas chromatogram enables an experienced laboratory chemist to make a more extensive analysis of the secondary fuel delivered.

For performing gas chromatography testing, the laboratory is equipped with two gas chromatographs (HP 5890A). One of them is reserved for routine analysis of PCBs in combustible liquid wastes and is semi-automated with an automatic sample injector and a "Chemstation" (HP-Vectra) for sequence control.

The second gas chromatograph is equipped with a mass-sensitive detector, a flame ionisation detector and an electron capture detector and is a universal instrument for the qualitative and quantitative analysis of organic compounds in mixtures of substances.

The gas chromatograph with mass-sensitive detection is used among other things for testing solids for organic constituents. This relates to the testing of raw materials (especially secondary raw materials) for volatile organic con-

einem Elektroneneinfangdetektor ausgestattet und ist insgesamt ein universelles Gerät für die qualitative und quantitative Analyse organischer Verbindungen in Stoffgemischen.

Der Gaschromatograph mit massenselektiver Detektion wird unter anderem zur Untersuchung von Feststoffen auf organische Inhaltsstoffe verwendet. Das betrifft die Untersuchung von Rohmaterialien – insbesondere von Sekundärrohstoffen – auf flüchtige organische Bestandteile, die im Zuge des Mahltrocknungsprozesses emittiert werden können. Diese Stoffe werden mit geeigneten Lösungsmitteln extrahiert, im Gaschromatographen aufgetrennt und mit dem Massenspektrometer identifiziert. Zum gaschromatographischen Nachweis von organischen Bestandteilen in Rohmaterialien müssen diese bei Temperaturen unterhalb 400° verflüchtigbar sein. Beim Vorliegen von schwer flüchtigen organischen Verbindungen (z. B. Kerogenen), wie sie häufig in Primärrohstoffen vorkommen, müssen diese durch thermische Vorbehandlung in leichter flüchtige Bruchstücke zerlegt werden, die dann mit dem Gaschromatographen nachgewiesen werden können. In entsprechenden Extrakten von Zementproben ist es möglich, mit dem Gaschromatographen organische Mahlhilfsmittel zu identifizieren und quantifizieren. Die Bestimmung des Anteils von Mahlhilfsmitteln in Zement erlaubt die Durchführung von Massenbilanzierungen. Optimierungen hinsichtlich der Art und Menge von Mahlhilfsmitteln werden ermöglicht.

Hinweise auf die Art allfällig verwendeter Betonzusatzmittel können durch Eluation von Betonproben und gaschromatographische Analyse erhalten werden.

Die gaschromatographische Analyse von organischen Abgasbestandteilen gelingt bei ausreichend hohen Konzentrationen im Abgas durch direkte Injektion repräsentativer Gasproben. Für Abgasbestandteile, die in geringer Konzentration vorliegen, ist eine Konzentrationserhöhung durch Adsorption auf geeignete Adsorbentien bzw. durch Anreicherung in Absorptionslösungen erforderlich.

3. Massenbilanzen für Blei und Zink

Spurenelemente in Rohmaterialien, Brennstoffen, Klinker und Stäuben werden regelmäßig mittels Röntgenfluoreszenz analysiert. Das Probenmaterial wird in Achat- bzw. Zirkonoxidmahlwerkzeugen auf Analysenfeinheit gebracht, mit Wismutoxid als internem Standard versehen und unter Berücksichtigung des Untergrundes analysiert. Es werden Nachweisgrenzen von 2–6 mg/kg erreicht [2–4]. Die regelmäßige Analyse von Spurenelementen erbringt eine breite Datenbasis, die die Durchführung von Massenbilanzen über größere Zeiträume ermöglicht.

In der Regel ergeben sich bei Einzeldatensätzen je nach Ofenbetrieb und Aufbau von inneren und äußeren Staubkreisläufen Bilanzfehlbeträge, die auch von Analysenfehlern bzw. von ungenauer Bestimmung der Massenflüsse mitverursacht werden.

Die Mittelwertbildung über größere Zeiträume gleicht Analysenfehler statistisch aus, gestattet Massenflüsse genauer zu erfassen und erbringt Verbesserungen der Ergebnisse, die verläßlichere Aussagen über Einbindegrade von Schwermetallen ermöglichen (**Bild 1**).

Im Bild 1 sind Monatsdurchschnitte für Massenbilanzen des Jahres 1992 für Blei dargestellt. Bei nur geringfügig schwankenden Konzentrationen der Einnahmen aus dem Rohmaterial sind vorwiegend die Einnahmen aus den Flüssigabfällen für das Bleikonzentrationsniveau im System bestimmend.

Während sich nach Inbetriebgehen des Ofens nach längerem Stillstand durch den Aufbau innerer Kreisläufe zunächst niedrigere Einbindungsgrade ergeben, gleichen sich nach längerem Ofenbetrieb Einnahmen und Ausgaben an. Nach Erreichen eines dynamischen Gleichgewichtes resultiert eine Erhöhung der Bleikonzentration im Klinker, und es werden fallweise höhere Ausgaben als Einnahmen registriert. Insgesamt kann im Jahresmittel ein hoher Einbindegrad nachgewiesen werden. Die Monatsmittel der

stituents which might be emitted during the drying and grinding process. These substances are extracted with suitable solvents, separated in the gas chromatograph and identified with the mass spectrometer.

For gas-chromatographic detection of organic constituents in raw materials, the latter must be capable of being volatilized at temperatures below 400 °C. If low-volatile organic compounds (e. g. kerogens) such as often occur in primary raw materials are present, they must be broken down by thermal pretreatment into higher-volatile fractions which can then be detected by the gas chromatograph.

Using suitable extracts of cement samples, it is possible with the gas chromatograph to identify and quantify organic grinding aids . Measurement of the grinding aid content of cement enables material balances to be drawn up. Optimization with regard to the type and quantity of grinding aids becomes possible.

Indications of the nature of concrete additives, if used, may be obtained by elution of concrete samples and gas-chromatographic analysis.

Gas-chromatographic analysis of organic exhaust gas constituents is achieved, if the concentrations in the exhaust gas are high enough, by direct injection of representative gas samples. For exhaust gas constituents present in low concentration, it is necessary to increase the concentration by adsorption on suitable adsorbents or by concentration in absorption solutions.

3. Material balances for lead and zinc

Trace elements in raw materials, fuels, clinker and dust are regularly analysed by X-ray fluorescence. The sample material is brought to analytical fineness in agate or zirconium oxide grinding implements, after which bismuth oxide is added as an internal standard and the material is analysed, allowance being made for background levels. Detectability limits of 2–6 mg/kg are achieved [2–4].

Regular analysis of trace elements provides a broad stock of data which enables material balances to be drawn up for extended periods.

As a rule, depending on kiln operation and the formation of internal and external dust cycles, deficits arise in individual sets of data which are caused by, among other things, analytical errors or inaccurate measurement of the material flows.

Averaging over extended periods compensates for analytical errors, allows material flows to be measured more accurately and brings improvements in the results which enable more reliable information to be obtained on the levels of retention of heavy metals.

Figure 1 shows the monthly averages of the material balances of 1992 for lead. As the concentrations of the intakes from the raw materials vary only slightly, it is mainly the intakes from the liquid wastes that determine the lead concentration level in the system.

Whereas after starting up the kiln after a prolonged shutdown the retention levels are initially lower due to the formation of internal cycles, after a prolonged period of kiln operation the intakes and outputs balance each other out. When a dynamic equilibrium has been attained, an increase in the lead concentration in the clinker results and higher outputs than intakes are occasionally recorded. In overall terms, a high retention level can be demonstrated on average for the year. The average monthly outputs as a percentage of the lead intakes ranged between 83 and 109 %. The average lead retention level for the year was 97 %.

A similar picture is evident for zinc (**figure 2**). The total intake stems mainly from intakes of zinc from old tyres. In normal service, some 20 % of the energy demand for burning is met from old tyres. During the first few months of kiln operation, higher intakes than outputs were generally recorded, as with lead. After a sufficient kiln operating time, establishment of the equilibria, and conversion of the zinc to retainable compounds, the monthly balances came into balance. Any accumulated surpluses were worked off in the course of a prolonged kiln operating period.

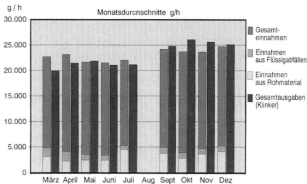

BILD 2: Massenbilanzen für das Element Zink im Jahr 1992
FIGURE 2: Material balances for zinc, 1992

Monatsdurchschnitte = Monthly averages
Gesamteinnahmen = Total intakes
Einnahmen aus Flüssigabfällen = Intakes from liquid wastes
Einnahmen aus Rohmaterial = Intakes from raw materials
Gesamtausgaben (Klinker) = Total outputs (clinker)
März = March
Mai = May
Juni = June
Juli = July
Okt = Oct
Dez = Dec
(All other names of months same as in German)

TABLE 2: Annual averages from the material balances for lead and zinc

Material balance, 1992 (n = 45 sets of data)		Pb	Zn
Intakes	(Pb/Zn)		
Raw materials	(approx. 5/33 ppm)	451 g/h	3036 g/h
Fuels			
Heavy fuel oil	(approx. 1/2 ppm)	2 g/h	4 g/h
Old tyres	(approx. 40 ppm/1.4%)	52 g/h	18564 g/h
Waste oil	(298/503 ppm)	715 g/h	1211 g/h
Total intakes		1220 g/h	22815 g/h
Outputs			
Clinker	(approx. 20/380 ppm)	1180 g/h	22411 g/h
Emissions	(0.5/1 µg/m³ (stp))	<0.1 g/h	<0.2 g/h
Outputs as a % of intakes		approx. 97	approx. 98

The outputs of zinc as a percentage of the intakes (monthly averages) were between 88 and 112%. The average zinc retention level for the year was 98%. The material balances for zinc and lead on average for 1992 based on 45 sets of data are reproduced in **Table 2**.

The dust from the external dust cycle of the Dopol rotary preheater kiln with built-in drying and grinding exhibits a significant accumulation of lead. In the case of zinc, on the other hand, only low accumulations were found. In the course of a year, however, no significant increase in this concentration level in relation to the kiln operating time was apparent for either zinc or lead [5, 6].

4. Continuous measurement of emissions, especially of total organic carbon

As stipulated by the licensing conditions, a device for continuous measurement of emissions of the exhaust gas constituents NO_x, dust and especially organic carbon was installed.

NO_x is measured as NO with an Ultramat 5 unit (Siemens) after converting the NO_2 present with a CGO-K NO_2/NO converter (Hartmann & Braun). Total organic carbon is measured with a flame ionisation detector (JUM VE 7). The instantaneous values measured are logged by a data computer (Talas) and converted to standard conditions. NO is converted to NO_2, while organic carbon is converted to C_{12} and related to an O_2 content of 10%. The data are transmitted

4. Kontinuierliche Emissionsmessung — insbesondere organischer Gesamtkohlenstoff

Auflagengemäß wurde eine Einrichtung zur kontinuierlichen Emissionsmessung der Abgasbestandteile NO_x, Staub und insbesondere für organischen Kohlenstoff installiert.

NO_x wird nach Umwandlung des vorhandenen NO_2 mit einem NO_2/NO-Konverter CGO-K (Hartmann & Braun) als NO mit einem Ultramat 5 (Siemens) gemessen. Gesamtorganischer Kohlenstoff wird mit einem Flammenionisationsdetektor (JUM VE 7) bestimmt. Die jeweils gemessenen Momentanwerte werden von einem Meßwertrechner (Talas) erfaßt und auf Normbedingungen umgerechnet. NO wird zu NO_2, organischer Kohlenstoff zu C_{12} umgerechnet und auf einen O_2-Gehalt von 10 % bezogen. Die Daten werden über einen Lichtleiter zu einer Auswerteeinheit im Laboratorium gesendet. Die Auswerteeinheit ist eine UNIX Workstation HP 330/9000, mit der auf Basis einer angepaßten Software die Datenaufbereitung erfolgt. Es können Minutenmittel für einen Zeitraum von 24 Stunden, Halbstundenmittel, Mittel über beliebige Zeiträume, Qualifizierung von Meßwerten über Betriebsparameter und Grenzwerte, Fraktilenberechnung, Klasseneinteilung sowie grafische Darstellungen der Emissionsdaten durchgeführt werden.

Mit den **Bildern 3** und **4** sind Beispiele einer Monatsauswertung für organischen Kohlenstoff in Form der Halbstundenmittelwerte und der Summenhäufigkeitskurve dargestellt.

Aus der Summenhäufigkeitsverteilung in Bild 4 sind zwei Grundgesamtheiten zu erkennen. Eine repräsentiert die normalerweise aus Rohmaterial und Feuerung stammenden Anteile an organischem Kohlenstoff. Die zweite resultiert aus Betriebszuständen mit hohem Anteil an Sekundärfeuerung (Altreifen), die durch ein höheres Niveau an Kohlenmonoxid — registriert von einer kontinuierlich arbeitenden Meßstelle im Wärmetauscher — und eine dementsprechend niedrige O_2-Konzentration gekennzeichnet sind.

In der **Tabelle 3** sind Ergebnisse der Auftrennung des Anteils an gesamtorganischem Kohlenstoff aufgrund gaschromatographischer Untersuchungen des Abgases dargestellt:

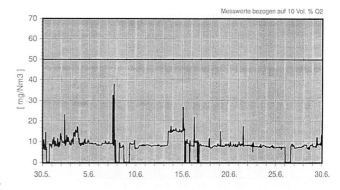

BILD 3: Kontinuierliche Messung der Emission organischer Kohlenstoffverbindungen (Halbstundenmittel) Juni 1992
FIGURE 3: Continuous measurement of emissions of organic carbon compounds (Half-hourly averages) June 1992

Meßwerte bezogen auf 10 Vol.% O_2 = Data referred to 10 % vol.% O_2
mg/Nm³ = mg/m³ (stp)

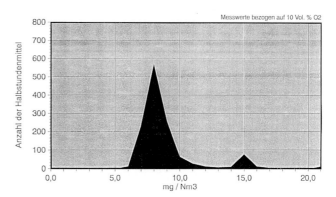

BILD 4: Summenhäufigkeitsverteilung von Halbstundenmittelwerten der Emission organischer Kohlenstoffverbindungen Juni 1992
FIGURE 4: Cumulative frequency distribution of half-hourly averages of emissions of organic carbon compounds June 1992

mg/Nm³ = mg/m³ (stp)
Meßwerte bezogen auf 10 Vol.% O_2 = Data referred to 10 % vol.% O_2
Anzahl der Halbstundenmittel = No. of half-hourly averages

TABELLE 3: Organische Bestandteile im Ofenabgas

	Organischer Kohlenstoff Gaschromatographisch		Summe organischer Kohlenstoff
	[mg/Nm³]	% der Summe organisch C FID	Kontinuierliche Emissionsmessung FID [mg/Nm³]
Methan	2,3 –3,2	19–33	
Äthen	2,4 –4,3	21–45	
Acetylen	0,7 –1,7	6–14	8–12
Propan/Propen	1,4 –1,8	12–16	
Äthan	0,07–0,6	1– 5	

Von 8–12 mg organischem Gesamtkohlenstoff/Nm³, gemessen mit dem Flammenionisationsdetektor der kontinuierlichen Emissionsmessung, werden 19–33 % als Methan identifiziert, 21–45 % als Äthen, 6–14 % als Acetylen, 12–16 % als Propan/Propengemisch und 1–5 % als Äthan.

Die Verwendung von Gaschromatographen zur Erfüllung gewerbebehördlicher Auflagen erweitert das Instrumentarium eines klassischen Werkslaboratoriums. Neben der Sicherstellung der Einhaltung von Grenzwerten zum Schutze der Umwelt und der Produktqualität lassen sich zusätzliche wertvolle Informationen zur verbesserten Beschreibung des Zementerzeugungsprozesses gewinnen.

via an optical fibre cable to an evaluation unit in the laboratory. The evaluation unit is a UNIX HP 330/9000 workstation on which the data editing is carried out on the basis of a customized software package. Averages per minute for a 24-hour period, half-hourly averages or averages over any desired period can be calculated; qualification of data relating to operating parameters and limit values, fractile calculations and classifications may be performed, and charts of the emissions data may be plotted.

Figures 3 and **4** show examples of a monthly evaluation for organic carbon in the form of the half-hourly averages and the cumulative frequency curve.

The cumulative frequency distribution in Figure 4 shows two populations. One represents the organic carbon contents which normally originate from the raw materials and the firing system. The other results from operating states with a high proportion of secondary firing (old tyres); these are characterized by a higher level of carbon monoxide (recorded by a continuously on-line sensor in the preheater) and a correspondingly lower O_2 concentration.

Table 3 shows results from the separation of the proportion of total organic carbon due to gas-chromatographic testing of the exhaust gas.

Of 8–12 mg total organic carbon/m(stp) measured with the flame ionisation detector used for continuous emission measurement, 19–33 % is identified as methane, 21–45 % as ethene, 6–14 % as acetylene, 12–16 % as a mixture of propane and propene and 1–5 % as ethane.

The use of gas chromatographs to meet conditions imposed by the industrial regulatory authorities extends the range of

Literaturverzeichnis

[1] Gerger, W., und Liebl, P.: Thermische Verwertung von Sekundärbrennstoffen bei den Gmundner Zementwerken. Zement-Kalk-Gips 44 (1991) Nr. 9, S. 457–462.

[2] Kraeft, U., und Ruch, C.: Quantitative röntgenfluoreszenzanalytische Bestimmung von Spurenelementen in technischen Produkten. Zement-Kalk-Gips 35 (1982) Nr. 3, S. 136–139.

[3] Van Eenbergen, A.F.P.: Röntgenfluoreszenzanalyse zur Spurenelementbestimmung im Zementwerk. Zement-Kalk-Gips 44 (1991) Nr. 4, S. 238–241.

[4] Uhlig, S., et al.: Contaminated Soil Mapping (Heavy Metals). A Comparison of Ation Absorption Spectrometry (AAS) and X-ray Fluorescence Analysis (XRF). Fresenius Envir Bull 1: 741–747 (1992). Birkhäuser Verlag, Basel/Switzerland.

[5] Sprung, S., und Rechenberg, W.: Die Reaktionen von Blei und Zink beim Brennen von Zementklinker. Zement-Kalk-Gips 31 (1978) Nr. 7, S. 327–329.

[6] Sprung, S., und Rechenberg, W.: Reaktionen von Blei und Zink bei der Zementherstellung. Zement-Kalk-Gips 36 (1983) Nr. 10, S. 539–548.

TABLE 3: Organic Constituents of kiln exhaust gas

	Gas-chromatographic analysis of organic carbon		Total organic carbon
	$[mg/m^3]$ (stp)	% of total organic C FID	Continuous emission measurement FID $[mg/m^3 (stp)]$
Methane	2.3 –3.2	19–33	
Ethene	2.4 –4.3	21–45	
Acetylene	0.7 –1.7	6–14	8–12
Propane/Propene	1.4 –1.8	12–16	
Ethane	0.07–0.6	1– 5	

tools available to a conventional works laboratory. Besides ensuring compliance with limit values for protecting the environment and maintaining product quality, additional valuable information leading to an improved description of the cement production process can be obtained.

Bautechnische Leistungsfähigkeit von Zementen mit Zumahlstoffen

Structural capabilities of cements with interground additives

L'aptitude technique des ciments à additifs

La aptitud técnica de los cementos con aditivos

Von **K. Harr** und **R. Böing,** Leimen/Deutschland

Zusammenfassung – Angeregt durch die Ausarbeitung einer europäischen Zementnorm wurde auch in Deutschland die in DIN 1164 beschriebene Palette unterschiedlicher Zemente durch neue, in anderen Ländern bereits seit Jahrzehnten eingeführte Zemente mit Zumahlstoffen erweitert. Hierzu gehören in erster Linie Portlandkalksteinzement (PKZ) und Zemente mit Steinkohlenflugaschen (FAZ und FAHZ). Sie ermöglichen es, den je nach Verwendungsgebiet teilweise sehr unterschiedlichen Anforderungen der Zementverbraucher und Bauwerksnutzer an die Leistungsfähigkeit von Zement im Beton individueller zu entsprechen. Zemente mit Zumahlstoffen sind nach umfassenden Prüfungen – u.a. hinsichtlich ihrer Festigkeit und ihrer Dauerhaftigkeitseigenschaften im Beton einschließlich des Frost- und Frost-Tausalzwiderstands – den regional hergestellten Normzementen bautechnisch wenigstens gleichwertig, wenn die Zumahlstoffe sorgfältig ausgewählt und qualitätsüberwacht sind und wenn geeignete herstelltechnische Maßnahmen ergriffen werden. Sie dürfen entsprechend der jeweiligen Zulassung des Instituts für Bautechnik wie Normzemente verwendet werden und sind für nahezu alle Anwendungsgebiete einsetzbar. Portlandkalksteinzemente ergeben vielfach Betone mit besonders günstigen Verarbeitungseigenschaften.

Summary – Stimulated by the preparation of a European Cement Standard, the range of different cements described in DIN 1164 has also been extended in Germany by new cements with interground additives which have already been used for decades in other countries. These are primarily Portland limestone cements and cement containing coal fly-ash. They make it possible for the capabilities of the cement in the concrete to be matched more individually to the requirements of the cement consumer and building user, which sometimes vary very considerably depending on the area of application. If the interground additives are carefully selected and monitored for quality, and if suitable production measures are taken, then comprehensive tests have shown that cements with interground additives are at least equivalent structurally to the locally produced standard cements with respect, among other things, to strength and durability characteristics in concrete, including resistance to frost and the combination of frost and de-icing salt. According to the relevant approval from the Institute of Structural Engineering, they can be used like standard cements and for virtually all areas of application. Portland limestone cements often result in concretes with particularly favourable workability properties.

Résumé – Sous l'influence de l'élaboration d'une norme européenne relative aux ciments, on a élargi également en Allemagne la gamme des différents ciments décrits dans DIN 1164, tout en y ajoutant des ciments nouveaux à additifs, utilisés dans d'autres pays depuis plusieurs décennies. En font partie, notamment, le ciment Portland (PKZ) et les ciments contenant des cendres volantes de houille (FAZ et FAHZ). Selon les différents champs d'application, ces ciments permettent de répondre, d'une façon plus individuelle, aux exigences très variables des consommateurs et des utilisateurs de bâtiments, en ce qui concerne la capacité du ciment à l'intérieur du béton. Les ciments contenant des additifs sont, d'après des études détaillées, – entre autres du point de vue de leur résistance et de leurs caractéristiques de durabilité à l'intérieur du béton, y compris la résistance au gel et au sel antigel – au moins équivalents aux ciments normalisés, de fabrication régionale, si les additifs ont été soigneusement choisis et surveillés du point de vue qualité et, si on a pris des mesures techniques appropriées lors de leur fabrication. Ils peuvent être employés, après avoir été homologués par l'Institut technique du bâtiment et des travaux publics, comme des ciments normalisés et, ils sont utilisables dans presque tous les domaines d'application. Les ciments Portland donnent souvent des bétons ayant des caractéristiques d'ouvrabilité particulièrement avantageuses.

Resumen – Bajo el influjo de la elaboración de una norma europea para cementos, se ha ampliado también en Alemania la gama de cementos descritos en la norma DIN 1164, mediante la adición de cementos nuevos, con aditivos, introducidos en otros países ya desde hace varios decenios. Entre estos cementos cuentan, en primer lugar, los Port-

land con filler calizo y los cementos con cenizas volantes de hulla. Según los diferentes campos de aplicación, estos cementos permiten responder, de forma individual, a las distintas exigencias de los consumidores de cemento y de los utilizadores de edificios, en cuanto a la capacidad de los cementos dentro del hormigón. Los cementos con aditivos son, según estudios detallados, realizados al efecto, – p.ej. respecto de su resistencia y sus características de durabilidad dentro del hormigón, incluyendo su resistencia a las heladas y a las sales de descongelación, – al menos equivalentes a los cementos normalizados, de fabricación regional, a condición de que los aditivos se elijan cuidadosamente, que se controle su calidad y que se tomen las medidas técnicas apropiadas para su fabricación. Después de haber sido homologados por el Instituto técnico de la construcción, estos cementos pueden ser empleados, igual que los cementos normalizados, en casi todos los campos de aplicación. Los cementos Portland con filler calizo dan, muchas veces, hormigones, cuyas características de trabajabilidad son particularmente ventajosas.

1. Einleitung

In Anlehnung an die ENV 197 [1] wird die DIN 1164 [2] überarbeitet, so daß in Zukunft auch die Zumahlstoffzemente Portlandkalksteinzement (PKZ) und Flugaschehüttenzement (FAHZ) darin ihren Platz haben werden (**Tabelle 1**). Die folgenden Ausführungen beschränken sich auf diese beiden Zemente.

Die Erstzulassung eines FAHZ 35 F in Deutschland erfolgte 1982, die eines PKZ 35 F im Jahr 1985, so daß inzwischen von diesen beiden Zementen genügend langfristige Erfahrungen vorliegen. Da nach den jetzigen Zulassungen bei der Betonherstellung unter Verwendung von FAHZ 35 F keine weitere Steinkohlenflugasche zugesetzt werden darf, liegen für FAHZ im Bereich Transportbeton weniger Erfahrungen vor. Der FAHZ 35 F wird daher überwiegend als Sackzement, z.B. im Estrichbereich, eingesetzt. Für den PKZ 35 F gilt diese Einschränkung jedoch nicht.

2. Herstellung der Zumahlstoffzemente PKZ 35 F und FAHZ 35 F

Bei der Herstellung von PKZ 35 F und FAHZ 35 F werden die jeweiligen Zumahlstoffe mit dem Klinker gemeinsam vermahlen. Hiermit wird eine optimale Kornverteilung erreicht.

Der günstig abgestimmte Feinstanteil der Flugasche bzw. die sich bei der PKZ-Herstellung ergebende Anreicherung des Kalksteinmehles im Feinstbereich des Zementes ermöglichen es, Zwickelräume im Zementleim durch dieses Material zu füllen und damit das darin befindliche Wasser zu verdrängen. Dadurch wird bei beiden Zementen der Wasseranspruch sowohl im Mörtel als auch im Beton verringert. Die günstige Kornverteilung und die höhere Feinheit dieser beiden Zemente führen im Vergleich zum PZ 35 F zu einem besseren Wasserrückhaltevermögen und einem geringeren

1. Introduction

DIN 1164 [2] is being revised in conformity with ENV 197 [1] so that in future it will also include the cements with interground additives – Portland limestone cement and pfa cement (**Table 1**). The following remarks are confined to these two cements.

An FAHZ 35 F cement (hereinafter pfa cement) was first approved in Germany in 1982 and a PKZ 35 F cement (hereinafter Portland limestone cement) in 1985, which means that by now sufficient long-term experience has been obtained with both these cements. Since according to the current approvals the addition of further coal fly ash when manufacturing concrete using a pfa cement is prohibited, less experience has been gained with pfa cement in the ready-mixed concrete sector. Therefore pfa cement is mainly used as bagged cement, e.g. for floor screeds. However, Portland limestone cement is not subject to this restriction.

2. Production of the cements with interground additives – Portland limestone cement and pfa cement

In the production of Portland limestone cement and pfa cement, the respective interground additives are ground together with the clinker, which achieves an optimum particle size distribution.

The well-coordinated ultrafine particle content of the fly ash and the concentration of the pulverized limestone in the ultrafine portion of the cement particle size range which occurs in the manufacture of Portland limestone cement enable voids in the cement paste to be filled with this material, displacing the water there. In this way, with both cements the water demand is reduced, both in mortar and in concrete. The favourable particle size distribution and the

TABELLE 1: Zementarten und Zusammensetzungen (Auszug aus [1] Tafel 1)

Zementart	Bezeichnung	Kennzeichnung	Hauptbestandteile Massenanteil in %			
			Portlandzementklinker K	Hüttensand S	Kieselsäurereiche Flugasche V	Kalkstein L
I	Portlandzement	I	95–100	–	–	–
II	Portlandhüttenzement	II/A-S	80–94	6–20	–	–
		II/B-S	65–79	21–35	–	–
	Portlandpuzzolanzement	II/A-P	80–94	–	–	–
		II/B-P	65–79	–	–	–
	Portlandflugaschezement	II/A-V	80–94	–	6–20	–
	Portlandschieferzement	II/A-T	80–94	–	–	–
		II/B-T	65–79	–	–	–
	Portlandkalksteinzement	II/A-L	80–94	–	–	6–20
	Portlandflugaschehüttenzement	II/A-SV	65–79	10–20	10–20	–
III	Hochofenzement	III/A	35–64	36–65	–	–
		III/B	20–34	66–80	–	–

Nebenbestandteile aller Zementarten: 0–5 Massenprozent

TABLE 1: Cement types and compositions (extract from [1], Table 1)

Cement type	Designation	Code	Main constituents — content in wt.%			
			Portland cement clinker K	Granulated blast-furnace slag S	Silica-rich fly ash V	Limestone L
I	Portland cement	I	95–100	—	—	—
II	Portland slag cement	II/A-S	80–94	6–20	—	—
		II/B-S	65–79	21–35	—	—
	Portland pozzolanic cement	II/A-P	80–94	—	—	—
		II/B-P	65–79	—	—	—
	Portland fly ash cement	II/A-V	80–94	—	6–20	—
	Portland slate cement	II/A-T	80–94	—	—	—
		II/B-T	65–79	—	—	—
	Portland limestone cement	II/A-L	80–94	—	—	6–20
	Portland pulverized fuel ash (pfa) cement	II/A-SV	65–79	10–20	10–20	—
III	Blast furnace cement	III/A	35–64	36–65	—	—
		III/B	20–34	66–80	—	—

Minor constituents of all cement types: 0–5 wt.%

Bluten des Mörtels bzw. Betons. Als Zumahlstoffe kommen nur qualitativ hochwertige Kalksteine sowie prüfzeichenfähige Steinkohlenflugaschen zum Einsatz.

3. Zementtechnische Leistungsfähigkeit

Bei den Prüfungen an Frischmörtel liegen die Rohdichten der Normmörtel mit FAHZ 35 F und PKZ 35 F höher als diejenigen mit PZ 35 F, EPZ 35 F sowie HOZ 35 L (**Bild 1**). Der Grund dafür liegt in der zuvor erwähnten Zwickelfüllung durch die Feinstanteile der Flugasche bzw. des Kalksteinmehles und die hierdurch verbesserte Verarbeitbarkeit und Verdichtungswilligkeit des Mörtels. Dadurch wird auch gleichzeitig der Grobporenanteil im Mörtel verringert, so daß sich ein dichteres Gefüge ergibt.

Bei gleichem w/z-Wert zeigt z.B. der Frischmörtel mit PKZ 35 F und FAHZ 35 F ein deutlich höheres Ausbreitmaß (**Bild 2**). Dies weist auf den geringen Wasseranspruch dieser Zemente hin.

4. Betontechnische Leistungsfähigkeit

4.1 Verarbeitungseigenschaften im Frischbeton

Die mit PKZ 35 F und FAHZ 35 F hergestellten Betone haben im Vergleich zu Betonen mit PZ 35 F, EPZ 35 F und HOZ 35 L bei gleicher Konsistenz einen deutlich geringeren Wasserbedarf. Dies kommt besonders bei feinststoffarmen Sanden zum Ausdruck. Bei 20°C Frischbeton-Temperatur

higher fineness of these two cements result in a better water retaining capability and less bleeding of the mortar or concrete compared with PZ 35 F Portland cement. Only high-quality limestones and officially certifiable coal fly ashes are used as interground additives.

3. Cement-specific capabilities

In the tests with freshly-mixed mortar, the bulk densities of the standard mortars containing pfa cement and Portland limestone cement are higher than those containing PZ 35 F Portland cement, EPZ 35 F Portland slag cement and HOZ 35 L blast furnace cement (**Figure 1**). The reason lies in the previously mentioned filling of voids by the ultrafine particles of the fly ash or the pulverized limestone and the resulting improved workability and compressibility of the mortar. This also reduces the macropore content of the mortar, leading to a denser structure.

At the same w/c ratio, the freshly-mixed mortar containing Portland limestone cement and pfa cement has a markedly larger flow table spread (**Figure 2**). This points to the low water demand of these cements.

4. Concrete-specific capabilities

4.1 Processing qualities in green concrete

Concretes made with Portland limestone cement and pfa cement have a significantly lower water demand for the

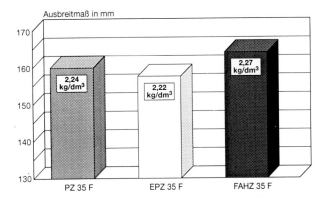

BILD 1: Ausbreitmaß und Frischrohdichte von Normmörtel nach EN 196 [3]
FIGURE 1: Flow table spread and bulk density in freshly-mixed condition of standard mortar in accordance with EN 196 [3]

Ausbreitmaß in mm	= Spread in mm
PZ 35 F	= PZ 35 F Portland cement
EPZ 35 F	= EPZ 35 F Portland slag cement
FAHZ 35 F	= FAHZ 35 F pfa cement

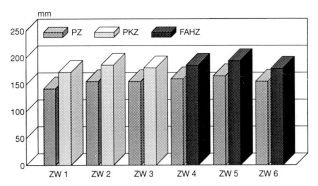

BILD 2: Ausbreitmaß nach EN 196 [3] an Normmörtel, 15 Minuten nach der Wasserzugabe
FIGURE 2: Flow table spread according to EN 196 [3], measured with standard mortar 15 minutes after addition of water

PZ	= Portland cement
PKZ	= Portland limestone cement
FAHZ	= pfa cement
ZW	= CW

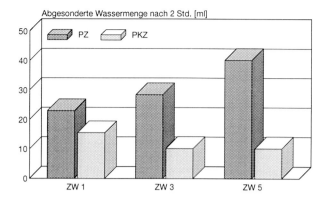
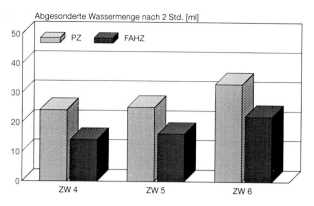

BILD 3: Blutneigung von PKZ-Feinmörteln im Vergleich mit PZ-Feinmörteln
FIGURE 3: Tendency to bleeding of fine mortars made with Portland limestone cement compared with fine mortars made with Portland cement

Abgesonderte Wassermenge nach 2 Std. [ml]	= Volume of water segregated after 2 hours (ml)
PZ	= Portland cement
PKZ	= Portland limestone cement
FAHZ	= pfa cement
ZW	= CW

BILD 4: Blutneigung von FAHZ-Feinmörteln im Vergleich mit PZ-Feinmörteln
FIGURE 4: Tendency to bleeding of fine mortars made with pfa cement compared with fine mortars made with Portland cement

Abgesonderte Wassermenge nach 2 Std. [ml]	= Volume of water segregated after 2 hours (ml)
PZ	= Portland cement
PKZ	= Portland limestone cement
FAHZ	= pfa cement
ZW	= CW

kann diese Wassereinsparung bis zu 5 % betragen. Auch bei höheren Frischbeton-Temperaturen geht diese Eigenschaft nicht verloren.

Beim Verarbeiten zeigen die mit PKZ 35 F bzw. FAHZ 35 F hergestellten Betone ein gutes Zusammenhaltevermögen. Die Betone sind homogen und plastisch. Ein leistungsfähiger Frischbeton hat nur eine geringe Neigung zum Bluten. Zur Einschätzung der Zumahlstoffzemente PKZ und FAHZ 35 F wurden vergleichende Prüfungen mit PZ 35 F durchgeführt. Die Prüfung des Blutens erfolgte an einem Feinmörtel, welcher mit dem jeweiligen Zement mit einem w/z-Wert von 0,60 hergestellt und in einen Standzylinder eingefüllt wurde.

Die nach zwei Stunden abgesetzte Wassermenge wurde volumetrisch festgestellt. Das Wasserabsondern war in allen Fällen bei den PKZ- und FAHZ-Mörteln deutlich reduziert, wie aus den nächsten beiden **Bildern 3** und **4** zu ersehen ist.

4.2 Festbeton

Zur Beurteilung der Festbetoneigenschaften wurden alle Prüfungen an Betonen mit gleichem w/z-Wert durchgeführt. Der geringere Wasseranspruch des PKZ 35 F bzw. FAHZ 35 F wurde also nicht berücksichtigt. Werden die Betone jedoch auf gleiche Konsistenz eingestellt, so ist dies in den meisten Fällen bei PKZ 35 F und FAHZ 35 F mit geringerem w/z-Wert möglich. Damit ergeben sich noch dichtere und dauerhaftere Betone.

Ein wesentliches Leistungskriterium für die Dauerhaftigkeit eines Betons ist der Frostwiderstand. Sowohl an der TU Hannover als auch in der Forschung, Entwicklung und Beratung von Heidelberger Zement wurde der Frostwiderstand der mit PKZ 35 F und FAHZ 35 F hergestellten Betone im Vergleich zu PZ 35 F geprüft. Dies geschah in beiden Instituten nach dem Würfelfrostverfahren des VDZ [4, 5]. Hiernach werden Betone, die nach 100 Frosttauwechseln nicht mehr als 10 M.-% Abwitterungen aufweisen, als frostwiderstandsfähig eingestuft. Wie die **Bilder 5** und **6** zeigen, können mit beiden Zumahlstoffzementen Betone mit ausreichendem Frostwiderstand hergestellt werden.

Die Dauerhaftigkeit eines Betons z. B. für Fahrbahndecken zeigt sich in seinem Frost-Tausalz-Widerstand. Diese Eigenschaft wird in Anlehnung an die österreichische Betonnorm ÖNORM B 3303 [6] geprüft. Die hiernach in der Zulassungsprüfung [5] unter Verwendung von PKZ 35 F und FAHZ 35 F hergestellten Betone ergaben nach der Beurteilung des zuständigen Sachverständigenausschusses des Deutschen Institutes für Bautechnik, Berlin, einen ausreichenden Frost-Tausalz-Widerstand. Die Ergebnisse sind in den

same consistency than concretes containing 35 F Portland cement, 35 F Portland slag cement and 35 L blast furnace cement. This is particularly apparent in the case of sands with a low ultrafine particle content. At a green concrete temperature of 20 °C, this water saving may be as much as 5 %. Even at higher green concrete temperatures, this property is not lost.

During working, concretes made with Portland limestone cement or pfa cement exhibit good cohesion. The concretes are homogeneous and plastic. A high-quality green concrete has only a slight tendency to bleed. Comparative tests were performed with PKZ 35 F Portland cement to evaluate the cements with interground additives – Portland limestone cement and pfa cement. The bleeding tests were carried out with a fine mortar, which was made with the respective cement using a w/c ratio of 0.60 and put into a glass jar.

The volume of water which had settled after two hours was measured. With the Portland limestone cement and pfa cement mortars, water segregation was significantly reduced, as can be seen from the following **Figures 3** and **4**.

4.2 Hardened concrete

In order to assess the qualities of the cement with regard to hardened concrete, all the tests were conducted on concretes made with the same w/c ratio, so the lower water demands of the Portland limestone cement and pfa cement were not taken into account. However, if concretes are made with the same consistency then for Portland limestone cement and pfa cement this is usually possible with a lower w/c ratio. This produces even denser and more durable concretes.

An important measure of a concrete's durability is its frost resistance. Both at the Technical University of Hannover and in the research, development and technical support department of Heidelberger Zement, the frost resistance of concretes made with Portland limestone cement and pfa cement was tested in comparison with PZ 35 F Portland cement. In both places the cube frost resistance method developed by the German Cement Works Association (VDZ) was used for this purpose [4, 5]. According to this method, concretes which after 100 freeze-thaw cycles exhibit weathering losses of not more than 10 wt. % are classed as frost-resistant. As **Figures 5** and **6** show, concretes with adequate frost resistance could be produced with both of the cements with interground additives.

The durability of a concrete for use as, for instance, carriageway surfacing is shown by its combined resistance to frost and de-icing salt. This property is tested on the basis of the

Bildern 7 und **8** dargestellt. Auch die etwas größeren Masseverluste der mit FAHZ 35 F hergestellten Betone im Vergleich zu einem unter Verwendung von PZ 35 F hergestellten Beton lassen nach Expertenmeinung keine baupraktischen Nachteile bzgl. des Frost-Tausalz-Widerstandes erwarten.

Der Frost-Tausalz-Widerstand wurde nicht nur durch Messungen im Labor nachgewiesen. Bereits im November 1982 waren im Werksbereich eines Zementwerkes unter Verwendung von PKZ 35 F Verkehrsflächen hergestellt worden. Diese befinden sich im Verladebereich von Silofahrzeugen und unterliegen dadurch einer sehr starken Verkehrsbeanspruchung. Die Straßenflächen wurden jedes Jahr im Winter intensiv mit Salz beaufschlagt. Bis heute sind keine Schäden durch Frost, Tausalzeinwirkung sowie mechanischen Abrieb durch Schwerlastverkehr zu erkennen.

Die bautechnische Leistungsfähigkeit eines Betons zeigt sich darüber hinaus in seinem Carbonatisierungsverhalten. Die **Bilder 9** und **10** zeigen das Carbonatisierungsverhalten von mit PKZ 35 F bzw. FAHZ 35 F hergestellten Betonen im Vergleich zu Beton aus PZ 35 F und HOZ 35 L (Hüttensand-

Austrian concrete standard NORM B 3303 [6]. The concretes produced according to this standard in the approval test [5] using Portland limestone cement and pfa cement had, in the opinion of the relevant expert committee of the German Institute of Structural Engineering (Berlin), an adequate resistance to frost and de-icing salt. The results are shown in **Figures 7** and **8**. Even the somewhat higher losses in weight of concretes made with pfa cement compared with a concrete made with PZ 35 F Portland cement are unlikely, according to the experts, to give rise to any practical drawbacks with regard to their resistance to frost and de-icing salt.

The concretes' resistance to frost and de-icing salt was not only demonstrated by laboratory measurements. As long ago as November 1982, traffic surfaces had been laid in the production area of a cement works using PKZ 35 F Portland limestone cement. These surfaces are located in the bulk transport vehicle loading area and are thus subject to a very high traffic load. Every year in winter, the road surfaces are heavily salted. No damage due to frost, the effects of de-icing salt or mechanical wear from the passage of heavy vehicles has been evident so far.

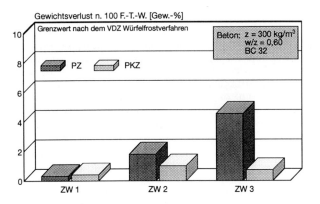

BILD 5: Ergebnisse der Frost-Prüfung (VDZ-Würfelfrostverfahren [4]) an Betonen mit PZ und PKZ
FIGURE 5: Results of frost resistance tests (VDZ cube frost resistance method [4]) on concretes made with Portland cement and Portland limestone cement

Gewichtsverlust n. 100 F.-T.-W. [Gew.-%]	= Loss in weight after 100 freeze-thaw cycles (wt. %)
Grenzwert nach dem VDZ Würfelfrostverfahren	= Limit value according to the VDZ cube frost resistance method
Beton	= Concrete
z = 300 kg/m³	= Cement = 300 kg/m³

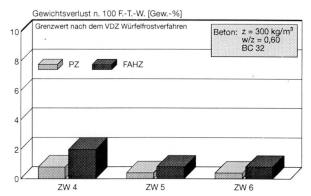

BILD 6: Ergebnisse der Frost-Prüfung (VDZ-Würfelfrostverfahren [4]) an Betonen mit PZ und FAHZ
FIGURE 6: Results of frost resistance tests (VDZ cube frost resistance method [4]) on concretes made with Portland cement and pfa cement

w/z	= w/c
PZ	= Portland cement
PKZ	= Portland limestone cement
FAHZ	= pfa cement
ZW	= CW

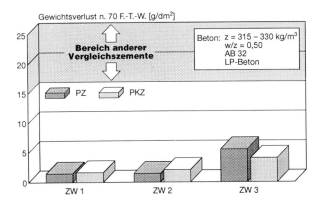

BILD 7: Ergebnisse der Frost-Tausalz-Prüfung nach [5] an Betonen mit PZ und PKZ
FIGURE 7: Results of tests of combined resistance to frost and de-icing salt according to [5] on concretes made with Portland cement and Portland limestone cement

Gewichtsverlust n. 70 F.-T.-W. [g/dm²]	= Loss in weight after 70 freeze-thaw cycles [g/dm²]
Bereich anderer Vergleichszemente	= Range of other cements used for comparison
Beton	= Concrete
z =	= Cement

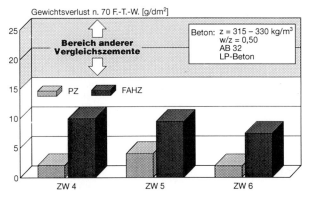

BILD 8: Ergebnisse der Frost-Tausalz-Prüfung nach [5] an Betonen mit PZ und FAHZ
FIGURE 8: Results of tests of combined resistance to frost and de-icing salt according to [5] on concretes made with Portland cement and pfa cement

w/z = 0,50	= w/c = 0.50
LP-Beton	= Air-entrained concrete
PZ	= Portland cement
PKZ	= Portland limestone cement
FAHZ	= pfa cement
ZW	= CW

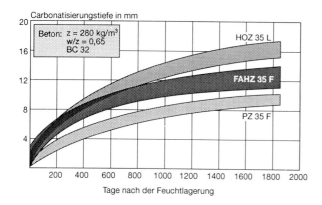

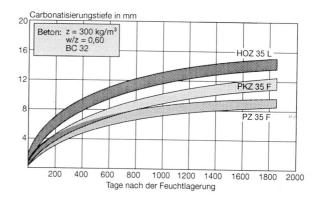

BILD 9: Carbonatisierung von Beton aus PZ, HOZ und FAHZ unter Laborbedingungen
FIGURE 9: Carbonation of concrete made from Portland cement, blast furnace cement and pfa cement under laboratory conditions

BILD 10: Carbonatisierung von Beton aus PZ, HOZ und PKZ unter Laborbedingungen
FIGURE 10: Carbonation of concrete made from Portland cement, blast furnace cement and Portland limestone cement under laboratory conditions

Carbonatisierungstiefe in mm	= Carbonation depth in mm
Beton	= Concrete
z =	= Cement =
w/z = 0,65	= w/c = 0.65
HOZ	= Blast furnace cement

FAHZ	= pfa cement
PZ	= Portland cement
PKZ	= Portland limestone cement
Tage nach der Feuchtlagerung	= Days after storage in humid atmosphere

anteil ca. 50 %). Näheres zur Versuchsdurchführung geht aus [5] hervor. Der PKZ- bzw. FAHZ-Beton zeigt eine geringfügig höhere Carbonatisierungsgeschwindigkeit als der Portlandzement-Beton, liegt aber niedriger als der Hochofenzement-Beton. Die jahrzehntelangen Erfahrungen mit Beton, der unter Verwendung von HOZ 35 L hergestellt wurde, zeigen aber, daß dieses Ergebnis bei sachgerechtem Vorgehen in der Verarbeitungsweise einschließlich guter Nachbehandlung nicht auf Nachteile für die Dauerhaftigkeit der Bauwerke schließen läßt.

6. Zusammenfassung

Die bisherigen Erfahrungen mit der Verwendung von PKZ 35 F und FAHZ 35 F zeigen, daß sich insbesondere bei den Frischbetoneigenschaften eine Reihe von Vorteilen ergibt. Hierzu zählen:

— Niedriger Wasseranspruch
— Gute Verarbeitbarkeit
— Gute Verdichtungswilligkeit
— Gutes Zusammenhaltevermögen
— Verbesserung der Kornverteilung im Mörtel und Beton bei Verwendung von Sanden mit geringem Feinkornanteil
— Geringes Wasserabsondern (Bluten)
— Günstiges Erstarrungsverhalten auch bei höheren Temperaturen.

Die Verbesserungen der Frischbetoneigenschaften wirken sich auch günstig auf den Festbeton aus. Die Herstellung von dichteren und dauerhafteren Betonen ist damit möglich. Die Leistungsfähigkeit des PKZ 35 F wurde z. B. beim Bau des Flughafens München II an verschiedenen Bauwerken unter Beweis gestellt.

Literaturverzeichnis

[1] pr ENV 197-1 „Zement-Zusammensetzung, Anforderungen und Konformitätskriterien — Teil 1: Allgemein gebräuchlicher Zement", Januar 1992.

[2] DIN 1164 Teil 1 „Zement-Zusammensetzung, Anforderungen", Entwurf Juli 1993.

[3] EN 196 „Prüfverfahren für Zement", Teil 1, „Bestimmung der Festigkeit", März 1990.

[4] Siebel, E.: „Würfelfrostverfahren", RILEM Seminar 1984, Hannover, S. 151–159.

[5] Prüfplan für die Zulassungsprüfungen von Portlandkalksteinzement PKZ 35 F — Fassung November 1988, Institut für Bautechnik, Berlin.

[6] Ö-Norm 3303, Betonprüfung, Ausgabe 3/83.

A concrete's structural capabilities are also reflected in its carbonation behaviour. **Figures 9** and **10** illustrate the carbonation behaviour of concretes made with Portland limestone cement and pfa cement compared with concrete made from PZ 35 F Portland cement and HOZ 35 L blast furnace cement (granulated blast-furnace sand content approx. 50 %). Details of the test procedure are given in [5]. The concretes made with Portland limestone cement and pfa cement exhibit a carbonation rate which is slightly higher than that of the Portland cement concrete but lower than that of the blast furnace cement concrete. However, the decades of experience gained with concrete made using HOZ 35 L blast furnace cement show that with correct application of the method of working, including correct curing, this result does not imply a loss of durability of the structures.

6. Summary

Experience to date with the use of Portland limestone cement and pfa cement reveals a number of advantages, especially with regard to the green concrete properties. These include:

— A low water demand
— Good workability
— Good compressibility
— Good cohesion
— Improvement of the particle size distribution in mortar and concrete when using sands with a low fines content
— Low water segregation (bleeding)
— Favourable setting properties, even at elevated temperatures.

The improvements in the green concrete properties have a beneficial effect on the hardened concrete as well — denser and more durable concretes can be produced. The capabilities of Portland limestone cement have been put to the test in, for example, various structures during the building of Munich's second airport.

Anforderungen an Zumahlstoffe für die Zementherstellung

Specifications for interground additives used in cement production

Exigences relatives aux additifs destinés à la fabrication du ciment

Condiciones que tienen que reunir los aditivos destinados a la fabricación del cemento

Von **F. Jung,** Wien/Österreich

Zusammenfassung – Die europäische Zementvornorm ENV 197 zählt 24 zumahlstoffhaltige Zementarten auf. Die dafür verwendeten zugelassenen Zumahlstoffe fallen bei industriellen Prozessen an oder stammen aus natürlichen Vorkommen. Sie unterscheiden sich wesentlich in Art und Wirkungsweise. Die Eignungsprüfung von Zumahlstoffen umfaßt die Erstprüfung und die laufende Überwachung. Die Erstprüfung muß die grundsätzliche Eignung nachweisen. Dazu sind Prüfungen am Zumahlstoff selbst, aber auch am damit hergestellten Zement und Beton notwendig. Insbesondere für die Prüfung der Gebrauchstauglichkeit und der Dauerbeständigkeit des Betons stehen jedoch noch keine einheitlichen europäischen Prüfmethoden zur Verfügung. Daher müssen derzeitig bei der Erstprüfung noch jeweils Prüfmethoden und Anforderungen unter Nutzung der bei den zu erwartenden Beanspruchungen vorliegenden Erfahrungen und nationalen Vorschriften festgelegt werden. Die laufende Überwachung, aufgezeigt am Beispiel der österreichischen Zumahlstoffe, muß sicherstellen, daß keine die Eignung des Zumahlstoffs beeinträchtigende Änderungen eingetreten sind. Auf diese Weise kann auch mit zumahlstoffhaltigen Zementen das Image des Betons als eines wirklich beständigen Werkstoffs bewahrt werden.*

Summary – *The European draft cement standard ENV 197 lists 24 types of cement containing interground additives. The approved interground additives used for this purpose are produced in industrial processes or come from natural deposits. They differ substantially in type and mode of operation. Suitability testing for interground additives covers initial testing and continuous monitoring. The initial testing has to prove the basic suitability. This requires tests on the interground additive itself, but also on the cement and concrete produced with it. However, as yet there are no consistent European test methods available specifically for testing the performance capabilities and durability of the concrete. For the initial testing it is therefore still necessary to stipulate the test methods and requirements, making use of the national regulations and the available experience concerning the expected stresses. Continuous monitoring, demonstrated using the example of Austrian interground additives, has to ensure that no changes have occurred which might impair the suitability of the interground additive. In this way it is possible to protect the image of concrete as a truly resistant material, even with cements containing interground additives.*

Résumé– *La norme préliminaire européenne de ciment ENV 197 énumère 24 types de ciments contenant des additifs. Les additifs admissibles à ce sujet sont obtenus au cours de processus industriels ou bien ils proviennent de gisements naturels. Ils se distinguent essentiellement selon leur nature et leur effet. L'examen d'aptitude des additifs comprend le premier examen ainsi que la surveillance continue. Le premier examen doit prouver l'aptitude fondamentale de l'additif. A ce sujet, il faut effectuer des essais sur les additifs eux-mêmes, mais également sur le ciment et le béton fabriqués à partir de ces additifs. En ce qui concerne notamment l'examen d'aptitude à l'usage et la durabilité du béton, on ne dispose pas encore de méthodes de contrôle européennes uniformes. Pour cette raison, il est actuellement nécessaire, lors du premier examen, de déterminer dans chaque cas particulier les méthodes de contrôle et les exigences, tout en profitant des expériences acquises ainsi que des prescriptions nationales existantes. Le contrôle continu, qui est expliqué en citant l'exemple des additifs utilisés en Autriche, doit assurer qu'il n'y ait aucun changement susceptible de porter préjudice à l'aptitude des additifs. De cette manière, il est possible, même à l'aide de ciments contenant des additifs, de conserver l'image du béton en tant que matériau vraiment résistant et durable.*

Resumen – *La norma previa europea para cemento ENV 197 menciona 24 tipos de cemento que contienen aditivos. Los aditivos admitidos para ello se obtienen en los procesos industriales o provienen de yacimientos naturales. Se distinguen claramente, los unos de los otros, por su naturaleza y efecto. El examen de aptitud de los aditivos comprende un primer examen y luego un control continuo. En el primer examen hay que comprobar la aptitud fundamental de los mismos. Para ello, se deben llevar a cabo en-*

sayos con los aditivos mismos y también con el cemento y el hormigón fabricados con dichos aditivos. En lo que se refiere, especialmente, al examen de aptitud para el uso y de durabilidad del hormigón, no se dispone aún a nivel europeo de métodos de control uniformes. Por esta razón, es necesario actualmente, en cuanto al primer examen, determinar en cada caso, según las solicitaciones previstas, los métodos de control y las exigencias correspondientes, de acuerdo con las experiencias adquiridas y las prescripciones nacionales existentes. El control continuo, que se explica citando el ejemplo de los aditivos empleados en Austria, debe asegurar el que no se haya producido ningún cambio susceptible de afectar la aptitud de los aditivos. De esta forma, se puede mantener la imagen del hormigón como material realmente resistente y duradero, aun utilizando cemento con aditivos.

Zemente mit Zumahlstoffen gibt es schon seit langem, etwa die Hüttenzemente oder die Zemente mit Puzzolanen wie den Traßzement. Sie besaßen häufig regionale Bedeutung und hatten sich meist jahrzehntelang in ihrem Verkaufsgebiet unter den dort gegebenen Bedingungen bewährt.

Das hat sich geändert. Die europäische Zementvornorm ENV 197, geltend nur für allgemein gebräuchlichen Zement, nicht für sulfatbeständige oder andere Spezialzemente, nennt 24 Zementarten mit Zumahlstoffen, basierend auf sehr unterschiedlichen Zumahlstofftypen, wobei alle derzeit in Europa am Markt befindlichen Zemente aufgenommen wurden, auch wenn Praxiserfahrungen nur regional vorlagen.

Dieser stärkere Einsatz der Zumahlstoffe ist darauf zurückzuführen, daß viele dieser Stoffe wirtschaftlich sind, häufig ansonsten sogar deponiert werden müßten, aber auch darauf, daß die Forderungen vieler moderner Betonbauweisen sich mit zumahlstoffhaltigen Zementen besonders gut erfüllen lassen. Beispiele dafür sind Beton für massige Bauteile oder Spritzbeton. In vielen Ländern sind zumahlstoffhaltige Zemente bereits die Regel.

Wenn nun heute für den Klinker schon detaillierte Qualitätsvorschreibungen existieren, so stellt sich auch bei den Zumahlstoffen die Frage nach ihrer Qualitätssicherung, übersteigt doch bei einigen Zementarten der europäischen Vornorm der Anteil an Zumahlstoff bereits denjenigen an Klinker. Die Vorgangsweise sei anhand der Verhältnisse in Österreich geschildert, werden doch dort Zemente sowohl mit Hüttensand als auch mit Flugasche seit Jahrzehnten in großem Umfang mit Erfolg eingesetzt, auch oder insbesondere für sensible Bauwerke wie Talsperren, Fernsehtürme oder große Brücken.

Die Qualitätsanforderungen sind bei Zumahlstoffen schwieriger festzulegen als bei Portlandzementklinker, ist doch der Rahmen viel größer. Das wirkt sich bereits beim festigkeitsbildenden Wert aus. Nur ein Teil der Zumahlstoffe trägt zur Festigkeit, nämlich durch Bildung von hydraulischen Verbindungen, bei, wie etwa Hüttensand oder Traß. Glasig erstarrte Schmelzkammerflugaschen sind im Vergleich dazu hydraulisch nahezu inert, sie senken aber unter gewissen Bedingungen durch ihre Kugelform und Kornverteilung den Wasserbedarf bei unveränderter Betonverarbeitbarkeit so stark ab, daß sich daraus in der Praxis ebenfalls ein Festigkeitsbeitrag ergibt. Weiters ist bei der Prüfung zu berücksichtigen, daß etwa Silikastaub bereits bei Zusätzen von wenigen Prozent des Zementgewichts die Eigenschaften des Betons deutlich verändert, währenddessen bei Hüttensand oder Flugasche dafür viel höhere Zusätze nötig sind, ja bis etwa 20 bis 25 % der Portlandzementcharakter des Betons erhalten bleibt. Bei Hüttensand und Flugasche wird daher der festigkeitsbildende Wert an einem Gemisch Zement zu Zumahlstoff von 70:30 im Vergleich zum zumahlstoff-freien Zement geprüft. Österreichischer Traß ist wegen seiner ansteifenden Wirkung bei einem Verhältnis von 75:25 statt 70:30 zu prüfen. Für andere Zumahlstoffe ist das Verhältnis von Fall zu Fall festzulegen. Der Einfluß der Zumahlstoffe auf Erstarrungszeiten und Raumbeständigkeit ist in sinngemäß gleicher Weise zu prüfen.

Die Abstimmung der Prüfmethode auf den zu untersuchenden Zumahlstoff ist auch bei der chemischen Prüfung zu beachten. Beim Chlorid — wichtig im Hinblick auf die Ver-

Cements containing interground additives, such as slag cements or pozzolanic cements like trass cement, have been around for a long time. They have often been of local significance and have usually proved their worth over many decades in their area of sale under the conditions prevailing there.

This has changed. The European draft cement standard ENV 197, which applies only to universal cements and not to sulphate-resistant cements or other speciality types, lists 24 types of cement containing interground additives, based on a wide variety of types of interground additive. All the cements currently on the market in Europe have been included, even where practical experience was only available on a local basis.

This increased use of interground additives is attributable to the fact that many of these substances are cost-effective and would otherwise even have to be landfilled in many cases, but also to the fact that the specifications of many modern methods of concrete construction can be met particularly well with cements containing interground additives. Examples are concrete for massive structural elements or shotcrete. In many countries, cements containing interground additives are already the rule.

Since detailed quality specifications already exist for clinker, the question of quality assurance arises for interground additives as well, especially since in some of the types of cement listed in the draft European standard the proportion of interground additives actually exceeds that of clinker. The approach will be described with reference to the situation prevailing in Austria where cements containing both granulated blast-furnace slag and fly ash have been successfully used on a large scale for several decades, even or especially for sensitive structures such as dams, TV towers or large bridges.

Quality specifications for interground additives are more difficult to lay down than for Portland cement clinker; after all, the scale of the problem is much larger. This applies even to the strength-developing value. Only certain interground additives, such as granulated blast-furnace slag or trass, contribute to cement strength through formation of hydraulic bonds. In comparison, vitrified fly ash from wet-bottom boilers is almost hydraulically inert; under certain conditions, however, owing to its spherical particle shape and size distribution it reduces the water demand, while leaving the workability of the concrete unchanged, to such an extent that in practice it also contributes to strength. Furthermore, it should also be borne in mind in testing that silica dust, for instance, significantly changes the properties of the concrete even with additions amounting to only a few percent of the cement weight, whereas with granulated blast-furnace slag or fly ash much larger quantities of additive are needed to do so, and even up to 20–25 % the Portland cement character of the concrete is preserved. The strength-developing value of granulated blast-furnace slag and fly ash is therefore tested using a ratio of cement to interground additive of 70:30 in comparison with cement without interground additives. Austrian trass is required to be tested at a ratio of 75:25 instead of 70:30 due to its rigidity-developing effect. For other interground additives, the ratio has to be determined on a case-by-case basis. The influence of the interground additives on setting times and soundness should be tested in an analogous manner.

meidung der Korrosion der Stahlbewehrung – gilt max. 0,10 % wie bei Zement. Auch für Sulfat werden ähnliche Grenzwerte wie bei Zement mit 3 bis 5 % genannt. Dabei ist jedoch zu prüfen, in welcher Form die Schwefelverbindungen vorliegen. Ein Beispiel für unschädliche Schwefelverbindungen ist das viele Jahre umstrittene, heute aber eindeutig als ungefährlich anerkannte Calciumsulfid des Hüttensandes sowie das im österreichischen Traß als unlösliches Alunit gebundene Sulfat. Beim chemisch bestimmten MgO ist zu prüfen, ob es tatsächlich als langsam reagierender, gefährlicher Periklas vorliegt oder in ungefährlichen ternären Verbindungen gebunden ist. Beim freien Calciumoxid (Freikalk) ist wiederum die Brenntemperatur zu berücksichtigen, die z. B. bei Flugasche in einem viel größeren Bereich schwanken kann als beim Klinker, was dazu führt, daß der Freikalk in der Flugasche sogar gefährlicher sein kann als im Klinker. Dies zeigt sich dann bei der Prüfung der Raumbeständigkeit.

Die chemische Prüfung ist zu ergänzen durch physikalische Prüfungen wie z. B. der Feinheit und der Porosität. Insbesondere die Porosität kann die Wirkung von Zusatzmitteln oder die Frostbeständigkeit des Betons verändern. Diesbezüglich sei auf die grundlegenden Untersuchungen des VDZ über Kalkstein als Zumahlstoffe verwiesen. Weiter ist zu untersuchen, ob Stoffe, welche die Erhärtung beeinflussen, inhomogen verteilt vorliegen. So kann sich bei der Rauchgasentschwefelung Sulfat an der Oberfläche der Flugaschekörner durch Kondensation anreichern und zu Erstarrungsstörungen führen, sofern nicht die Reaktionsgeschwindigkeit des Klinkers und des als Erstarrungszeitregler zugesetzten Gipses darauf abgestimmt wird.

Die Aufzählung ist nicht vollständig. Einflüsse insbesondere auf die Beständigkeit und Gebrauchstauglichkeit des Betons können häufig nicht am Zumahlstoff oder Zement, sondern nur am Beton geprüft werden, z. B. Einflüsse auf die Karbonatisierung bzw. Korrosion der Stahlbewehrung oder auf die Frost- und Frosttausalzbeständigkeit. Dafür sind europäische einheitliche Prüfmethoden und Beurteilungskriterien zwar in Ausarbeitung, stehen aber noch nicht zur Verfügung. Bereits jetzt läßt sich jedoch sagen, daß die Prüfungen hohe Kosten verursachen und das Ergebnis erst nach Monaten oder Jahren vorliegt. Sie können daher kaum in der laufenden Überwachung durchgeführt werden, sondern konzentrieren sich auf die Erst- oder Zulassungsprüfung.

Auch für diese gibt es leider noch kein allgemein gültiges Schema. Angesichts der großen Unterschiede in Aufbau und Wirkungsweise der Zumahlstoffe wird es schwierig, wahrscheinlich aber auch nicht sinnvoll und notwendig sein, alle Zumahlstoffe nach völlig gleichem Schema und in gleichem Umfang zu prüfen. Bis auf weiteres müssen Art und Umfang der Prüfungen in der Regel noch jeweils von der fremdüberwachenden Prüfanstalt festgelegt werden, die dafür zweifellos umfangreiche chemische und betontechnische Kenntnisse und langjährige Erfahrungen auf dem Gebiet der Zementherstellung und Zementanwendung benötigt.

Ist die grundsätzliche Eignung des Zumahlstoffs in der Erstprüfung nachgewiesen, kann darauf die laufende Überwachung aufgebaut werden. Dabei ist zu berücksichtigen, daß viele Zumahlstoffe industriellen Prozessen entstammen, bei denen die Qualitätssicherung primär nicht auf den Zumahlstoff, sondern auf Eisengewinnung, Stromerzeugung u. dgl. ausgerichtet ist. Die laufende Überwachung hat daher sicherzustellen, daß keine die Zumahlstoffqualität beeinträchtigende Änderungen eingetreten sind, wie z. etwa bei Flugasche bei einem Wechsel der Kohle oder einer deutlichen Änderung der Verbrennungstemperatur möglich sein kann. Für Zumahlstoffe aus natürlichen Vorkommen mit ihren unvermeidbaren Schwankungen gelten sinngemäß gleiche Bedingungen. Die Details der laufenden Überwachung sind immer mit der fremdüberwachenden Prüfanstalt festzulegen, die auch die Erstprüfung durchgeführt hat. Nur dann kann sie die Änderungen beurteilen.

Matching of the test method to the interground additive under investigation should also be practised in chemical testing. For chloride, which is important with regard to prevention of corrosion of the steel reinforcement, a maximum of 0.10 % applies as for cement. For sulphate, too, limits similar to those for cement of 3–5 % are stated. Here, however, the form in which the sulphur compounds are present should be checked. Examples of harmless sulphur compounds are the calcium sulphide in granulated blast-furnace slag, which was a subject of controversy for many years but today is definitely acknowledged to be innocuous, and the sulphate which is bound in Austrian trass as insoluble alunite. In the case of chemically determined MgO, it should be checked whether it is really present in the form of slow-reacting, harmful periclase or is bound in harmless ternary compounds. In the case of free calcium oxide (free lime), the burning temperature should be taken into account; with fly ash, for instance, this temperature may vary over a much wider range than with clinker, which means that free lime may actually be more harmful in the fly ash than in the clinker. This becomes apparent when testing the soundness.

Chemical testing should be supplemented by physical testing, e.g. of fineness and porosity. The porosity in particular may alter the effect of additives or the frost resistance of the concrete. On this subject, the fundamental research conducted by the VDZ into the use of limestone as an interground additive should be consulted. It should also be investigated whether substances which influence setting are unevenly distributed. In flue gas desulphurisation, for instance, sulphate may build up on the surface of the fly ash particles due to condensation and interfere with setting unless the rates of reaction of the clinker and the gypsum added as a setting time regulator are adjusted accordingly.

The list is not exhaustive. Factors influencing, in particular, the stability and fitness for use of the concrete often cannot be examined by testing the interground additive or the cement but only by testing the concrete; for example, factors influencing carbonation or corrosion of the steel reinforcement, or the frost resistance or the combined resistance to frost and de-icing salt. Standard European test methods and evaluation criteria are in preparation, but are not yet available. However, it can already be stated that the tests are expensive to carry out and the results only become available after several months or years. Therefore they can scarcely be performed as part of continuous monitoring, but are restricted to initial or approval testing.

Even for these, a universal system is unfortunately not yet available. Given the wide differences in the structure and action of the interground additives, however, it will be difficult to test every interground additive according to exactly the same system and to the same extent; but it will probably not be worthwhile or necessary to do so. For the time being, the nature and scope of the tests generally have to be specified by the testing agency acting as the external monitor, which for this purpose undoubtedly requires extensive chemical and concrete-specific knowledge and long experience in the field of cement manufacture and use.

If the basic suitability of the interground additive has been demonstrated in the initial testing, continuous monitoring may be based on this. Here, it should be borne in mind that many interground additives originate from industrial processes in which quality assurance is primarily directed not at the interground additive but at iron production, power generation, etc. The continuous monitoring therefore has to ensure that no changes have occurred which might impair the quality of the interground additive, such as might occur with fly ash if the coal is changed or the combustion temperature is substantially varied. Similar conditions apply to interground additives from natural deposits with their unavoidable variations. The details of the regular monitoring should always be fixed in consultation with the testing agency acting as external monitor, which also performed the initial testing. Only then can it evaluate the changes.

Der Einsatz der Zumahlstoffe im Zement sichert im Vergleich zum Zusatz auf der Baustelle neben der Einhaltung der Zusatzmenge auch homogene Verteilung und Abstimmung in Feinheit und chemischer Reaktion auf den verwendeten Klinker. Durch die beschriebenen QS-Maßnahmen können so mit zumahlstoffhaltigen Zementen die in Mitteleuropa vergleichsweise besonders hoch entwickelten Bautechnologien beibehalten und sogar besonders gut genützt werden.

Compared with on-site addition of additives, the use of interground additives in cement ensures not only that the required quantity of additive is adhered to but also that the additive is distributed evenly and matched to the clinker used with regard to fineness and chemical reaction. By using the quality assurance measures described, cements containing interground additives can help to sustain the construction technologies which, in comparative terms, are extremely highly developed in Central Europe and to make especially good use of them.

Marktorientierte Technologie – eine Überlebens- und Wachstumsstrategie

Market driven technology – a strategy for survival and growth

Technologie axée sur le marché – stratégie de survie et de croissance

Tecnología orientada hacia el mercado – una estrategia para sobrevivir y crecer

Von **Ph. Mathew, M. A. Purohit, T. N. Tiwari** und **A. K. Chatterjee**, Bombay/Indien

Zusammenfassung – Viele ältere Zementwerke in Indien müssen die Zementproduktion einstellen oder nach Alternativen suchen, um ihr Überleben zu sichern. Die Zementfabrik Mancherial ist ein solches Werk. Um in diesem Umfeld überleben zu können, hat das Werk eine einzigartige, marktorientierte Strategie entwickelt und sich dadurch eine Nische auf dem durch hohe Wertschöpfung gekennzeichneten Markt für Spezialzemente geschaffen. Trotz der geringen Qualität des Kalksteins mit extrem schlechter Brennbarkeit und Granulierfähigkeit und trotz veralteter Anlagen und einer Kohle mit hohem Aschegehalt ist Mancherial die erste indische Zementfabrik, die Tiefbohrzement, der die API G-Spezifikationen erfüllt, herstellt und auch eines der führenden Werke für die Erzeugung von sulfatwiderstandsfähigem Zement in Indien. Der Vortrag erläutert die praktizierte marktorientierte Strategie im einzelnen, ebenso die technischen Verbesserungen bei der Rohmehlherstellung und beim Betriebsablauf, die Umwandlung von Schwächen in Stärken und die Vorteile solcher Strategien für kleine, rohstoffarme Werke, die mit größeren und technologisch fortschrittlicheren Werken erfolgreich konkurrieren wollen.

Summary – Many older Indian cement plants are forced to stop cement production or look for alternative possibilities to ensure survival. The Mancherial Cement Works is one such plant. To survive in this environment the plant has adopted a unique market-driven strategy by creating a niche for itself in the high value-added speciality cements market. Despite poor quality limestone with extremely low burnability and nodulization characteristics, older generation equipment and coal with a high ash content, Mancherial has become the first Indian plant to produce oil well cement conforming to API G specifications and also one of the leading producers of sulphate resistant cement in India. This paper details the market-driven strategy adopted, the technical improvements carried out in raw mix design and plant operation, the transformation of inherent weaknesses into strengths and the advantages of such strategies for small, resource-poor plants to compete successfully with larger and more technologically advanced rivals.

Résumé – En Inde, nombreuses sont les cimenteries qui doivent arrêter leur production de ciment ou rechercher des alternatives pour assurer leur survie, comme par exemple la cimenterie de Mancherial. Pour pouvoir survivre dans un pareil contexte, cette usine a développé une stratégie unique, axée sur le marché, se créant ainsi un créneau dans un segment de ciments spéciaux caractérisé par une valeur ajoutée élevée. Malgré la faible qualité du calcaire d'une combustibilité et granulation très mauvaises, et en dépit d'installations très anciennes et d'un charbon à teneur élevée en cendres, Mancherial est la première cimenterie indienne à fabriquer du ciment en puits profonds, répondant à la spécification API G, mais aussi une des usines de tout premier plan en Inde dans la fabrication de ciment résistant au sulfate. Le présent exposé explique en détail aussi bien la stratégie appliquée axée sur le marché, de même que les améliorations techniques pour la fabrication de farine crue et durant les phases du process, la transformation des points faibles en points forts et, enfin, les avantages de telles stratégies pour de petites usines, pauvres en matières premières, qui souhaitent concourir avec succès face à des usines plus importantes, aux moyens technologiques plus avancés.

Resumen – En la India, numerosas fábricas de cemento viejas tienen que dejar de producir o buscar otras alternativas para asegurar su supervivencia, como por ejemplo la fábrica de cemento Mancherial. Con el fin de poder sobrevivir en semejante entorno, esta fábrica ha desarrollado una estrategia única, orientada hacia el mercado, abriéndose un segmento en un mercado de cementos especiales, caracterizado por elevadas plusvalías. A pesar de la mala calidad de la piedra caliza y de su reducida aptitud para la cocción y la granulación, y a pesar de que las instalaciones están anticuadas y que el carbón disponible contiene una elevada proporción de cenizas, Mancherial es la primera fábrica india productora de cemento para perforaciones profundas, que cumple las especificaciones API G, y también una de las principales fábricas

indias productoras de cemento resistente a los sulfatos. En el presente artículo se explica detalladamente la estrategia practicada, orientada hacia el mercado, así como las mejoras técnicas introducidas en la molienda y en el desarrollo del proceso de producción, la transformación de los puntos débiles en puntos fuertes así como las ventajas que presentan tales estrategias para fábricas pequeñas, carentes de materias primas, que tienen que competir con éxito con empresas más grandes y técnicamente más avanzadas.*

1. Einleitung

In Indien werden neben alten Anlagen mit geringer Kapazität und Naß- bzw. Halbnaßöfen auch moderne Anlagen mit Trockenöfen mit Vorwärmer sowie Vorcalcinierung betrieben. Für diese Situation waren drei Hauptfaktoren maßgebend:

1. Während mehrerer Jahrzehnte unterlag die Zementindustrie der staatlichen Aufsicht. Die festgelegten Preise waren unwirtschaftlich, und die Branche war wirtschaftlich ungesund.
2. Der staatliche Einfluß verminderte sich allmählich in den Jahren 1982 bis 1989. Dadurch setzte eine Hochkonjunktur ein, und viele moderne Werke wurden gebaut. Allerdings fehlte den meisten eingesessenen Herstellern das Geld für den Neubau solcher Anlagen.
3. Die Verteilung der Zementwerke ist uneinheitlich. So sind 70 % der installierten Kapazität in nur 5 der 24 indischen Bundesstaaten konzentriert. Eine Überschußproduktion führte daher rasch zu einem Verdrängungswettbewerb.

Die in diesen Gegenden gelegenen älteren Werke fanden sich in einem Teufelskreis gefangen. Mit ineffektiver Technik und unwirtschaftlicher Betriebsgröße waren ihre Herstellkosten deutlich höher als die ihrer neuen Mitbewerber. Außerdem verfielen die Preise, so daß die finanziellen Mittel für eine Modernisierung der Anlagen nicht erwirtschaftet werden konnten.

2. Das ACC-Werk Mancherial

Das Zementwerk Mancherial der Associated Cement Companies Ltd. (ACC) ist typisch für die aufgezeigte Situation. Der vorhandene Lepolrostofen wird mit einer installierten Leistung von 0,3 Mio. t/a betrieben. Das Werk liegt im Bundesstaat Andhra Pradesh, mit einer Überschußproduktion von rd. 286 % des regionalen Bedarfs. Die überwiegende Produktion stammt aus modernen Vorwärmeröfen mit Vorcalcinierung. Der Ertrag geriet daher unter starken Druck, die Rentabilität war gering bzw. nicht vorhanden und das Überleben fraglich. Nach dem Abwägen verschiedener Möglichkeiten entschloß man sich, am Markt orientiert, zur Herstellung von Spezialzementen, wie Tiefbohrzement und Zement mit hohem Sulfatwiderstand, die eine höhere Wertschöpfung versprachen. Dazu waren verschiedene technologische Neuerungen und technische Verbesserungen erforderlich. Mit geschickten Lösungen konnten bisherige Schwächen in Stärken verwandelt werden. Dadurch hat sich das Werk zu einem bedeutenden Hersteller von Tiefbohrzementen und Zementen mit hohem Sulfatwiderstand nach den API G-Normen entwickelt. Die Orientierung am Markt führte nicht nur zum Überleben, sondern sogar zu einer gewissen Prosperität trotz widriger Umstände.

3. Technische Strategie

Die Hauptanlagenteile sind in **Tabelle 1** aufgeführt. Die wesentlichen Unterschiede in der Herstellung von Portlandzement und den Spezialzementen sind in **Tabelle 2** zusammengestellt. Wegen der begrenzten Möglichkeiten, ältere Anlagenteile unter Betrieb anzupassen, wurde die grundlegende Feineinstellung bei einer Portlandzementherstellung vorgenommen, um die Betriebsparameter für die Spezialzemente festzustellen. Die Rohmaterialmischung wurde intern in der Forschungs- und Beratungsabteilung von ACC eingestellt. Die Überwachung der chemischen Zusammensetzung ist besonders wichtig, da nur eine Kohle mit schwankendem Aschegehalt als Brennstoff zur

1. Introduction

The Indian cement industry has old plants with small capacity wet and semi-wet process kilns as well as large modern plants based on dry process technology with preheater and precalciner kilns. This situation has arisen due to three main factors.

1. For several decades, the Indian cement industry was regulated by the Government. The price fixed was unrenumerative and the industry was in poor financial health.
2. From 1982 to 1989, governmental controls were gradually withdrawn. This set off a boom and many large modern plants were installed. However most existing manufacturers lacked the financial resources to install such modern plants.
3. The distribution of cement plants in the country is non-uniform and 70 % of the installed capacity is concentrated in just five of the twenty four states in India. A surplus situation and cut-throat competition rapidly developed in these areas.

The older plants sited in these areas found themselves trapped in a vicious cycle. With inherently inefficient technology and uneconomic scale of operation, their costs of manufacture were substantially higher than those of their newer competitors. On the other hand, the price slashing in the market meant that they could not generate the financial resources required for modernization of their plants.

2. ACC's Mancherial Works

The Mancherial Cement Works of The Associated Cement Companies Ltd (ACC) is typical of such plants. It operates a Lepol grate kiln with an installed capacity of only 0.3 MTPA. It is located in the state of Andhra Pradesh which has a massive surplus with an installed capacity of 286 % of the local demand. The bulk of the production is from modern preheater precalciner kilns.

In this situation the margins were under severe pressure, the profitability was poor or non-existent, and the very survival of the plant under question. After considering various options, the plant decided to adopt a market driven approach and target the high value-added speciality cements market with Oil Well Cement (OWC) and Sulphate Resistant Cement (SRC) which provide substantially higher realisations. The successful realization of this strategy necessitated several technological innovations and technical improvements. The plant also found ingenious solutions to turn inherent weaknesses into strengths. Today the plant has become a major Indian manufacturer of Oil Well Cement and High Sulphate Resistant Cement conforming to API G-specifications. The market driven strategy has resulted not only in survival but even prosperity in a hostile environment.

3. Technical Stratety

The major equipments are given in **Table 1**. The major process engineering differences between OPC production and special cement production were identified and are listed in **Table 2**.

Due to the limited possibilities for adjustments during operation in the older generation of equipment, substantial fine-tuning was carried out during an OPC production run to determine the operating parameters for special cement production. The raw mix design was carried out in-house at ACC's Research & Consultancy Directorate. Control of the chemical composition is critical especially since coal with

TABELLE 1: Aufstellung der Hauptanlagenteile	
Rohmühlen	Doppelrotor 3,6 m · 8,0 m 2 · 3,7 m Turboabscheider
Öfen	Lepolrost 3,6 m · 58 m Rostkühler
Zementmühlen	offener Kreislauf — 3 Kammer 2,2 m · 12 m

TABLE 1: List of major equipment	
Raw mills	Double rotator 3.6 m · 8.0 m 2 · 3.7 m Turbo-Separators
Kilns	Lepol grate based 3.6 m · 58 m Grate cooler
Cement mills	open circuit — 3 chamber 2.2 m · 12 m

TABELLE 2: Hauptunterschiede in der Prozeßtechnik zwischen der Herstellung von Portlandzement und Spezialzementen (Tiefbohrzement und Zement mit hohem Sulfatwiderstand)

- Rohmaterialmischung (C_3S, C_3A)
- Feinheit des Brennguts
- Homogenität des Brennguts
- Sinterzonentemperatur
- F/CaO-Gehalt
- Zementzusammensetzung
- Zementfeinheit

TABLE 2: Major process engineering differences between manufacture of OPC and special cements (oil well and sulphate resistant cements)

- Raw mix design (C_3S, C_3A)
- Kiln feed fineness
- Homogeneity of kiln feed
- Burning zone temperature
- F/CaO content
- Cement composition
- Cement fineness

Verfügung steht. Der Gipsgehalt und die erforderliche spezifische Oberfläche wurden in ausführlichen Labor- und Betriebsuntersuchungen festgelegt.

Um die Mahlfeinheit des Rohmehls festzulegen, war es erforderlich, den Auslaufkonus und die Gegenflügel des Sichters abzustimmen sowie die Mahlkörpergattierung zu verändern. Die Zusammensetzung des Rohmehls wurde dadurch optimiert, daß verschiedene Chargen aus mehreren Silos abgezogen und unter statistischen Gesichtspunkten gemischt wurden.

Die Temperaturverteilung in der Sinterzone war eine besonders schwierige Aufgabe, da nur eine minderwertige Kohle mit einem Aschegehalt von 30 Gew.-% und einem Heizwert von 4200 kcal/kg (17550 kJ/kg) als Brennstoff zur Verfügung steht. Dazu wurde das Brennerrohr für den Betrieb mit sauerstoffangereicherter Luft verändert. In Indien ist dies übrigens der bisher einzige Versuch dieser Art. Wegen der Empfindlichkeit des Lepolrostes gegen Ofenstaub, die zu häufigen Betriebsstörungen führte, mußte auch der Ofenbetrieb optimiert werden. Dadurch wurde auch ein Überbrennen oder Schwachbrand des Klinkers verhindert.

In den Zementmahlanlagen wurden die Mahlkörpergattierung, der Mühlenfüllgrad und der Mühlenluftstrom optimiert, um die erforderliche Mahlfeinheit und Korngrößenverteilung zu erzielen.

4. Der Wandel von der Schwäche zur Stärke

Zum Schutz seiner Marktnische ist es dem Werk gelungen, seine größte Schwäche in Stärke zu verwandeln. Im Gegensatz zu seinen Mitbewerbern mit ihren großen Produktionseinheiten verfügt Mancherial über viele kleine Einheiten, wie aus der Aufstellung in Tabelle 1 hervorgeht. Dies ergab einen entscheidenden Vorteil, da das Werk rasch von einer Zementqualität zu einer anderen wechseln oder, falls erforderlich, auch zwei Zementarten gleichzeitig herstellen kann. Verluste, die beim Wechsel der Qualitäten auftreten können, sind dadurch ebenfalls deutlich vermindert.

5. Schlußfolgerung

Die Erfahrung des Zementwerks Mancherial ist ein gutes Beispiel für angewandte Marktorientierung in Situationen, in denen konventionelle Überlebensstrategien (Expansion, Optimierung, Steigerung der Leistungsfähigkeit usw.) nicht möglich sind. Eine derartige Strategie bietet kleinen, rohstoffarmen Werken die Chance zum erfolgreichen Wettbewerb mit großen und technologisch besser ausgerüsteten Anbietern.

varying ash content is used as a fuel. The gypsum percentage and the specific surface requirements for cement grinding were determined through detailed laboratory and plant scale trials.

The raw meal fineness control strategy evolved combined adjustment of the separator bottom cone along with changes in the grinding media pattern and the separator auxiliary fan. A two pronged strategy was evolved for the blending silos involving firstly an optimisation of the blending parameters and secondly a statistical mixing procedure consisting of simultaneous withdrawal from two or more silos.

Increasing the burning zone temperatures was a particularly challenging task as the only fuel available is a low grade coal with ash content as high as 30 wt.-% and a calorific value of 4200 kcal/kg coal (17550 kJ/kg). Increases of burning zone temperatures were achieved by modification of the burner pipe. Additionally, in the only attempt of its kind in India, an oxygen enrichment device was developed in-house in order to supply oxygen-enriched air for combustion and to boost the flame temperature. Kiln operating strategies were optimised to minimise kiln upsets in view of the sensitivity of the Lepol grate to dust generation in the kiln and the necessity to prevent both under- and over-burning of clinker.

In the cement grinding section, changes in the grinding media pattern and optimisation of operating parameters such as ventilation and mill filling levels were carried out in order to obtain the required fineness and particle size distribution.

4. Turning Weakness into Strength

In order to protect its niche market, the plant has successfully turned its main weakness into its major strength. Unlike its competitors who have large size production units, Mancherial has multiple small units as can be seen from the list of equipment in Table 1. However this has been turned into a critical advantage as it enables the plant to switch quickly from one type of cement to another or even to produce two grades of cement and clinker simultaneously if required. The loss due to flushing of the system when switching from one grade to another is also lower due to the lower volumes involved.

5. Conclusion

The Mancherial Cement Works experience is a good illustration of the use of a market driven approach in situations where conventional survival strategies (expansion, optimisation, improving efficiencies etc.) are not possible. Such a strategy holds out the promise for small, resource poor plants to compete successfully with large, more technologically advanced competitors.

Wirkung von Zusätzen aus aktiven/inerten Sekundärrohstoffen einschließlich Abfallstoffen auf die Leistungsfähigkeit von Zement

Mechanism of addition of active/inert secondary raw materials including wastes on performance of cements

Influences des additifs de matières premières secondaires actives/inertes y compris déchets sur les performances du ciment

Influjo de los aditivos de materias primas secundarias activas/inertes, incluyendo los desechos, sobre la capacidad del cemento

Von **K. C. Narang, S. Chaturvedi** und **R. B. Sharma,** New Delhi/Indien

Zusammenfassung – Verschiedene hydraulisch aktive oder inerte anorganische Stoffe werden heute bei der Entwicklung von Zementen mit der Absicht verwendet, Energie, Kapital und Zeit zu sparen und die Umwelt zu schonen. Energiesparende Rezepturen, die auf dem Einsatz minderwertiger Stoffe aus natürlichen und industriellen Quellen basieren, sind sowohl beim Brenn- als auch beim Mahlprozeß erfolgreich erprobt worden. Was die Verwendungsmöglichkeiten betrifft, unterscheiden sich solche Materialien zwar kaum von anderen, aber die Wirkung derartiger Zusätze auf die Leistungsfähigkeit des Zements ist unterschiedlich und hängt von den physikochemischen und mineralogischen Eigenschaften sowie von der Menge ab, in der solche Zusätze zumahlstoffhaltigen Zementen beigegeben werden. Die Zusatzmenge hat einen erheblichen und variablen Einfluß, d.h., bei geringerer Dosierung ist ihre Wirkung hauptsächlich auf den Füllereffekt und die Keimbildung zurückzuführen, bei mittlerer Dosierung auf Aktivierung und chemische Reaktionen und bei hoher Dosierung auf die Kombination aller genannten Faktoren und den Verdünnungseffekt. Im Vortrag werden die Wirkungen und Einflüsse solcher Zusätze auf die Leistung des Endprodukts sowie das Energiesparpotential und die wirtschaftlichen Gegebenheiten dargelegt.

Summary – Various hydraulically active/inert inorganic materials are now being used in formulating various cement compositions with a view to conserving energy, capital, the environment and time. Low energy formulations based on the use of low grade materials from natural and industrial sources have been successfully tried both in pyroprocessing and blending stages. These materials do not differ much in utilization potential but the role and mechanism of these additions in the performance characteristics of cement differ depending on the physicochemical and mineralogical characteristics and quantity of addition of such materials to blended cements. The quantity of addition has an important and varying influence, i.e. at lower dosage the effect is mainly due to fine powder effect and nucleation, at moderate dosage the effect is mainly due to activation and chemical reactions, and at higher dosage it is due to a combination of the above two and the dilution effect. The possible role and mechanism of such additions in the performance of the end product is analysed in the paper together with energy saving potential and economics.

Résumé – Diverses matières minérales hydrauliques actives et inertes sont, de nos jours, utilisées pour mettre au point des ciments destinés à économiser l'énergie, les capitaux, le temps et, en même temps, ménager l'environnement. Des recettes économisant l'énergie, basées sur l'application de produits de faible qualité de sources naturelle et industrielle, ont été testées avec succès aussi bien au point de vue cuisson que broyage. Pour ce qui est de leur application possible, ces produits ne présentent que peu de différence par rapport aux autres, mais l'effet des additifs sur les performances du ciment est variable et dépend des propriétés physico-chimiques et minéralogiques mais aussi de la quantité dans laquelle sont ajoutés aux ciments contenant du produit de broyage ces additifs. L'ajout a une influence considérable mais variable: l'effet, en cas de faible dosage, porte essentiellement sur l'influence du filler et la formation de germes, en cas de dosage moyen il porte sur l'activation et les réactions chimiques et enfin si le dosage est élevé il s'agit d'une réaction combinée de tous les facteurs précités et d'un effet de dilution. L'exposé traite de l'impact et des effets de ces additifs sur le rendement du produit fini ainsi que l'économie d'énergie potentielle et les données économiques.

Resumen – Hoy en día se utilizan diversas materias inorgánicas, hidráulicamente activas o inertes, en la composición de los cementos, con el fin de ahorrar capital, energía y tiempo y de preservar el medio ambiente. Tanto en el proceso de cocción como en el de molienda, se han probado con éxito recetas que permiten ahorrar energía y que están basadas en el empleo de materias de poco valor, procedentes de fuentes naturales e industriales. En cuanto a las posibilidades de aplicación, estas materias apenas se distinguen las unas de las otras, pero su efecto sobre la capacidad de los cementos es diferente y depende de las características físico-químicas y mineralógicas así como de las cantidades en que las mismas se mezclen a los cementos con aditivos. Dichas cantidades tienen un influjo considerable y variable, es decir al añadirse poca cantidad, el efecto se deberá principalmente al influjo del filler y a la formación de gérmenes. Con una dosis media, la adición repercutirá sobre la activación y las reacciones químicas y, con dosis elevadas, sobre el conjunto de factores mencionados y sobre el efecto de dilución. En el presente articulo se explican las repercusiones e influjos de tales adiciones sobre la capacidad del producto final así como sobre el potencial de ahorro de energía y las condiciones económicas.

Influjo de los aditivos de materias primas secundarias activas/inertes, incluyendo los desechos, sobre la capacidad del cemento

1. Einführung

Die Zugabe aktiver/inerter Rohstoffe in verschiedenen Stadien der Zementherstellung stellt, insbesondere angesichts der mit diesen Materialien verbundenen Kosteneinsparungen und Umweltschutzaspekten, eine sehr verbreitete Praxis dar. Der Einsatz solcher Rohstoffe im Mahlprozeß hat sich gut bewährt. Zu den verwendeten Sekundärrohstoffen gehören gebrannter Ton, Schiefer, Flugasche, Schlacken, Reishülsenasche, Bimsstaub sowie kolloidales/amorphes Siliziumdioxid, und die inerten Sekundärrohstoffe sind Kalksteinmehl, Ofenstaub, halbkalzinierte Mineralien usw. Im Hinblick auf die Zugabe von Zusatzstoffen schwanken die Dosierungen zwischen 0 und 65 Gew.-%, doch hängt ihr jeweiliger Beitrag zu der Produktleistung nicht nur von der zugegebenen Menge, sondern auch von den physikalisch-chemischen und mineralogischen Eigenschaften dieser Stoffe und ihrer Ursprungsgeschichte ab. Diese Zusätze besitzen an sich nur geringe oder gar keine zementartigen Eigenschaften, doch wirken einige von ihnen in fein verteilter Form und durch Reaktion mit Calciumhydroxid in Gegenwart von Feuchtigkeit bei Raumtemperatur oder hohen Temperaturen oder durch Verminderung der Gelporosität wie Zement [1–4].

Zum besseren Verständnis des Reaktionsmechanismus verschiedener puzzolanartiger/inerter Materialien wurden schon früher Untersuchungen dazu vorgenommen. Allerdings läßt sich noch nicht genau vorhersagen, welcher Faktor eine Prognose der Leistungsfähigkeit puzzolanartiger oder inerter Materialien und ihres Beitrags zu den Leistungsdaten des Endprodukts ermöglicht [5–8].

Nach den in dem letzten Jahrzehnt veröffentlichten Berichten zu urteilen, beeinflussen wahrscheinlich folgende Faktoren das Leistungsverhalten des Endprodukts:

— Zunahme der Geloberfläche,

— Feinmehleffekt,

— chemische Reaktionen,

— Verbesserung der Gelstruktur,

— Verdünnungseffekt.

Diese Faktoren können sich bei einem bestimmten Sekundärrohstoff je nach der zugegebenen Menge einzeln oder in Kombination auswirken. Zur Beurteilung der Wirkung unterschiedlicher Beimischungsprozentsätze wurden für die vorliegende Studie gebrannter Ton, Reishülsenasche, Flugasche, Ofenstaub und Kalksteinmehl ausgewählt.

2. Wirkung puzzolanartiger Zusätze

Zur Untersuchung des Einflusses der Zugabe verschiedener puzzolanartiger Materialien wurden mit einem breiten Spektrum von Zusatzstoffen Prüfungen durchgeführt. Auf der Grundlage der Herkunft der Puzzolane wurden Klinker aus 2 verschiedenen Quellen und Puzzolane dreierlei Herkunft ausgewählt und in der vorliegenden Studie eingesetzt. Die Kalkreaktivität sämtlicher Puzzolane war mit 58–61 kg/cm² fast gleich. Allerdings unterscheiden sich die Puzzolane in ihren physikalischen Eigenschaften, vor allem der Blaine-Feinheit, deutlich. Reishülsenasche (RHA) ist dabei mit

1. Introduction

The addition of active/inert raw materials at various stages of cement manufacturing has been a very common practice, particularly in view of the cost savings and environmental protecting factors associated with these materials. The use of such materials at the grinding stages has been well accepted. The various active secondary raw materials identified are burnt clay, shale, fly ash, slags, rice husk ash, pumicite, collodial/amorphous silica and the inert secondary raw materials are limestone, kiln dust, semi-calcined minerals, etc. From the point of view of additive addition the dosages of such materials have varied in the range of 0–65 wt-%, however, their individual role to the product performance depends not only on the quantum of addition but also on the physicochemical and mineralogical characteristics of these materials and their origin history. These additives by themselves possess little or no cementitious properties but some of them provide cementitious properties in finely divided form and through reaction with calcium hydroxide in presence of moisture at ambient or elevated temperatures or through reducing gel porosity [1–4].

To understand the mechanism of reaction of different pozzolanic/inert materials, investigations on the mechanism of reactions were attempted in past. However, it is not yet possible to predict with certainty which factor can be used to predict the performance of either pozzolanic materials or inert materials and their role in the performance of the end product [5–8].

Based on the reports, published during the last decade, the various factors which are probably playing an influencing role in the performance of the end product are:

— Increase in gel surface area,

— Fine powder effect,

— Chemical reactions,

— Improving the gel structure,

— Dilution effect.

These factors can act singularly or in combination for a given secondary raw material depending on the quantum of the addition. Therefore in order to see the effect of the varying percentages of additon, burnt clay, rice husk ash, fly ash, kiln dust and limestone have been selected for the present study.

2. Effect of pozzolanic additions

In order to study the varying effects of addition of different pozzolanic materials a broad range of additions have been considered for analysis. Clinkers from 2 different sources and pozzolanic materials of 3 different origins have been selected on the basis of origin of pozzolanic materials and are used in the present study. The lime reactivity of all the pozzolanic materials was nearly the same, i.e. 58–61 kg/cm². However, the pozzolanic materials differ significantly in their physical characteristics, particularly in Blaine fineness. Rice husk ash (RHA) is the finest of all with a Blaine surface of 11000 cm²/g. The other two materials have the same but lower fineness [9]. Chemical compositions of the

TABELLE 1: Typische chemische Zusammensetzung puzzolanartiger/inerter Sekundärrohstoffe

Stoff		Charge %	SiO_2 %	Fe_2O_3 %	Al_2O_3 %	CaO %	MgO %	Na_2O %	K_2O %	SO_3 %
Gebrannter Ton	(GT)	1,12	75,62	4,65	11,56	2,58	1,40	0,86	1,54	—
Reishülsenasche	(RHA)	2,9	90,06	0,7	0,55	2,67	1,14	1,18	0,35	—
Flugasche	(FA)	8,22	58,91	4,96	20,73	1,23	3,23	0,35	1,17	—
Ofenstaub	(OS)	28,13	18,89	2,46	5,84	41,42	1,55	0,28	0,47	0,71
Kalkstein	(KS)	41,96	2,72	0,96	0,71	53,65	0,15	—	—	—
Klinker	(KL1)	2,24	23,36	3,30	4,80	63,81	1,50	0,35	0,48	—
für GT, RHA, FA,	(KL2)	1,45	21,36	3,00	5,45	66,34	1,49	0,21	0,47	—
Zement für OS, KS		2,6	20,96	2,59	4,11	61,46	4,86	0,79	0,14	2,24

TABLE 1: Typical chemical compositions of the pozzolanic/inert secondary raw materials

Materials		LOI %	SiO_2 %	Fe_2O_3 %	Al_2O_3 %	CaO %	MgO %	Na_2O %	K_2O %	SO_3 %
Burnt clay	(BC)	1.12	75.62	4.65	11.56	2.58	1.40	0.86	1.54	—
Rice husk ash	(RHA)	2.9	90.06	0.7	0.55	2.67	1.14	1.18	0.35	—
Fly ash	(FA)	8.22	58.91	4.96	20.73	1.23	3.23	0.35	1.17	—
Kiln dust	(KD)	28.13	18.89	2.46	5.84	41.42	1.55	0.28	0.47	0.71
Limestone	(LS)	41.96	2.72	0.96	0.71	53.65	0.15	—	—	—
Clinker used	(CL1)	2.24	23.36	3.30	4.80	63.81	1.50	0.35	0.48	—
for BC, RHA, FA.	(CL2)	1.45	21.36	3.00	5.45	66.34	1.49	0.21	0.47	—
Cement used for KD, LS.		2.6	20.96	2.59	4.11	61.46	4.86	0.79	0.14	2.24

TABELLE 2: Typische physikalische Kennwerte puzzolanartiger/inerter Sekundärrohstoffe

Stoffe		Raumgewicht g/cm³	Bond-Index	Kalkreaktivität kg/cm²	Mineralogische Kennwerte
Gebrannter Ton	(GT)	1,3	10,5	58	Quarz, Mullit, sehr niedriger Glasgehalt
Reishülsenasche	(RHA)	0,96	—	58	kieselsäurehaltiges Glas
Flugasche	(FA)	1,2	9,6	61	Quarz, Mullit, unverbrannter Kohlenstoff, Glas (25–30 Gew.-%)
Ofenstaub	(OS)				Kalkspat, Quarz, Glimmer, freier Kalk
Klinker 1		1,25	16,9	—	C_3S − 45,30 Gew.-% C_2S − 37,53 Gew.-% C_3A − 7,14 Gew.-% C_4AF − 10,03 Gew.-%
Klinker 2		1,30	17,6	—	C_3S − 66,85 Gew.-% C_2S − 14,66 Gew.-% C_3A − 9,37 Gew.-% C_4AF − 9,12 Gew.-%

TABLE 2: Typical physical characteristics of the pozzolanic/inert secondary raw materials

Materials		Bulk density g/cm³	Bond index	Lime reactivity kg/cm²	Mineralogical characteristics
Burnt clay	(BC)	1.3	10.5	58	quartz, mullite, very little glass content
Rice husk ash	(RHA)	0.96	—	58	silicious glass
Fly ash	(FA)	1.2	9.6	61	quartz, mullite, unburnt carbon, glass (25–30 wt-%)
Kiln dust	(KD)				calcite, quartz, mica, free lime
Clinker –1		1.25	16.9	—	C_3S − 45.30 wt-% C_2S − 37.53 wt-% C_3A − 7.14 wt-% C_4AF − 10.03 wt-%
Clinker –2		1.30	17.6	—	C_3S − 66.85 wt-% C_4S − 14.66 wt-% C_3A − 9.37 wt-% C_4AF − 9.12 wt-%

TABELLE 3: Leistungsfähigkeit von unter Verwendung von Flugasche, Reishülsenasche, gebranntem Ton, Ofenstaub und Kalkstein hergestellten Zementen

Zement	Druckfestigkeit (kg/cm²)				Abbindezeit			Abbindewärme (cal/g)		
	1d	7d	28d	90d	K	AAZ	EAZ (Minuten)	1d	7d	28d
NPZ (A)	132	313	422	525	27,4	152	262	38	59	79
*	145	355	435	590	28,0	81	236	54	70	90
NPZ (I)	170	485	640	720	37,0	160	300	49	68	79
RHA 5	158	316	450	495	29,2	55	315	—	—	—
*	180	365	464	507	30,6	35	175	—	—	—
RHA 10	120	325	450	495	32	105	270	33	52	61
*	135	318	459	525	34,0	170	285	51	62	66
RHA 20	101	338	438	497	35	75	365	—	—	—
*	120	325	460	505	36,4	95	275	—	—	—
RHA 30	97	296	420	455	39,6	135	305	26	41	49
*	98	295	402	467	40,0	190	345	42	54	60
GT 5	140	288	430	530	28,4	107	287	—	—	—
*	152	313	395	505	28,0	65	265	—	—	—
GT 10	110	245	375	500	29,2	165	280	27	40	60
*	130	285	352	480	28,4	128	223	40	56	71
GT 20	94	228	330	420	30	97	262	—	—	—
*	106	255	395	495	29,2	74	284	—	—	—
GT 30	77	194	295	395	31,4	195	355	20	31	48
*	87	218	325	435	30,0	150	245	31	47	63
FA 5	146	273	465	555	31,2	160	290	—	—	—
*	160	317	425	500	28,6	50	265	—	—	—
FA 10	114	247	450	515	31,6	170	360	29	47	66
*	132	307	410	515	29,8	125	250	42	59	78
FA 20	96	235	417	545	32,4	140	310	—	—	—
*	114	247	395	540	32,0	70	275	—	—	—
FA 30	76	195	373	488	35,2	192	347	23	39	52
*	94	198	315	478	35,0	140	350	33	48	67
OS 2	310	520	690	760	37,0	180	350	60	73	96
OS 3	282	480	680	740	37,0	183	355	—	72	87
OS 5	202	448	580	720	37,0	210	370	45	54	102
OS 8	80	230	306	425	38,0	260	400	30	31	47
KS 3	270	466	590	710	37,0	145	235	—	—	—
KS 5	273	450	600	710	37,0	150	240	—	—	—

NPZ (A) – Für RHA, FA und GT verwendeter normaler Portlandzement.
NPZ (I) – Für OS, KS verwendeter normaler Portlandzement.
* – Werte für Klinker Nr. 2.
K – Konsistenzwasser (%)

TABLE 3: Performance of cements made by using fly ash, rice husk ash, burnt clay, kiln dust and limestone

Cement	Compressive strength (kg/cm²)				Setting time			Heat of hydration (cal/g)		
	1d	7d	28d	90d	c	IST	FST (minutes)	1d	7d	28d
OPC (A)	132	313	422	525	27.4	152	262	38	59	79
*	145	355	435	590	28.0	81	236	54	70	90
OPC (I)	170	485	640	720	37.0	160	300	49	68	79
RHA 5	158	316	450	495	29.2	55	315	—	—	—
*	180	365	464	507	30.6	35	175	—	—	—
RHA 10	120	325	450	495	32	105	270	33	52	61
*	135	318	459	525	34.0	170	285	51	62	66
RHA 20	101	338	438	497	35	75	365	—	—	—
*	120	325	460	505	36.4	95	275	—	—	—
RHA 30	97	296	420	455	39.6	135	305	26	41	49
*	98	295	402	467	40.0	190	345	42	54	60
BC 5	140	288	430	530	28.4	107	287	—	—	—
*	152	313	395	505	28.0	65	265	—	—	—
BC 10	110	245	375	500	29.2	165	280	27	40	60
*	130	285	352	480	28.4	128	223	40	56	71
BC 20	94	228	330	420	30	97	262	—	—	—
*	106	255	395	495	29.2	74	284	—	—	—
BC 30	77	194	295	395	31.4	195	355	20	31	48
*	87	218	325	435	30.0	150	245	31	47	63
FA 5	146	273	465	555	31.2	160	290	—	—	—
*	160	317	425	500	28.6	50	265	—	—	—
FA 10	114	247	450	515	31.6	170	360	29	47	66
*	132	307	410	515	29.8	125	250	42	59	78
FA 20	96	235	417	545	32.4	140	310	—	—	—
*	114	247	395	540	32.0	70	275	—	—	—
FA 30	76	195	373	488	35.2	192	347	23	39	52
*	94	198	315	478	35.0	140	350	33	48	67
KD 2	310	520	690	760	37.0	180	350	60	73	96
KD 3	282	480	680	740	37.0	183	355	—	72	87
KD 5	202	448	580	720	37.0	210	370	45	54	102
KD 8	80	230	306	425	38.0	260	400	30	31	47
LS 3	270	466	590	710	37.0	145	235	—	—	—
LS 5	273	450	600	710	37.0	150	240	—	—	—

OPC (A) – Ordinary portland cement used for RHA, FA, BC.
OPC (I) – Ordinary portland cement used for KD, LS.
* – Values for Clinker no. 2.
c – Water of consistency (%)

einer Blaine-Oberfläche von 11000 cm²/g am feinsten. Die anderen beiden Rohstoffe sind weniger fein [9]. Die chemische Zusammensetzung der Materialien ist sehr unterschiedlich, wobei RHA am kieselsäurereichsten ist. Für die Studie wurde Gips verwendet. Die Eigenschaften der eingesetzten Materialien sind in den **Tabellen 1** und **2** aufgeführt.

Insgesamt wurden mit 2 Klinkern und 3 Puzzolanen 24 Portlandpuzzolan- (PPZ) und 2 Normalportlandzemente (NPZ) hergestellt. Die Wirkung der Zusatzmenge wurde durch Zugabe von 5, 10, 20 und 30 Gew.-% dieser Puzzolane zum Basiszement überprüft. Zwei Kontroll-NPZ wurden außerdem mit zwei Klinkern für eine Paralleluntersuchung hergestellt. In jedem Fall wurde der Gipsgehalt konstant gehalten. Alle Zemente einschließlich des NPZ wurden durch gemeinsames Vermahlen hergestellt, wobei alle Komponenten in den erforderlichen Anteilen vorgemischt, in eine Gutbettwalzenmühle gefüllt und bis auf eine Blaine-Feinheit von 3300 ± 50 cm²/g vermahlen wurden. Die hergestellten Proben wurden auf Festigkeitsentwicklung, Abbindezeit, Hydratationswärme und Hydratationsprodukte untersucht. Die Ergebnisse der Studie sind in der **Tabelle 3** aufgeführt, und die Beobachtungen lassen sich wie folgt zusammenfassen:

Die Festigkeitsentwicklung der beiden aus zwei Klinkern hergestellten Basiszemente verläuft in allen Altersstufen fast in dem gleichen Bereich, außer nach 90 Tagen, wobei der aus Klinker Nr. 2 hergestellte Zement eine höhere

materials are significantly different with RHA being most silicious. The same mineral gypsum was used throughout the study. The characteristics of the materials used are given in **Table 1** and **2**.

In all 24 portland pozzolana cement (PPC) and 2 ordinary portland cement (OPC) samples were prepared using two clinkers and three pozzolanic materials. The effect of the quantum of additon was analysed by adding 5, 10, 20 and 30 wt-% of these pozzolanic materials to the base cement. Two control OPC were also made with the two clinkers for parallel study. In all cases the level of gypsum was kept constant. All the cement including OPC have been made by the process of intergrinding in which all the components were mixed in the required proportions and introduced in a synchronised twin roll mill and ground to the Blaine fineness of 3300 ± 50 cm²/g. Prepared samples were analysed for strength development, setting time, heat of hydration and hydration products. The results of the study are given in **Table 3** and the observations are summarized as follows:

The strength development of the two base cements made from two clinkers is nearly in the same range at all ages, except at 90 days, where the cement made from clinker no. 2 has a higher strength (590 kg/cm²) as compared to cement made from clinker no. 1 (525 kg/cm²). For the cement, made from clinker no. 1, the strength development with addition of 5 wt-% pozzolanic material indicates an increase in

Festigkeit (590 kg/cm²) als der Zement aus Klinker Nr.1 (525 kg/cm²) besitzt. Bei dem Zement aus Klinker Nr. 1 weist die Festigkeitsentwicklung nach Zugabe von 5 Gew.-% Puzzolan gegenüber gewöhnlichem Portlandzement (NPZ) nach 1, 28 und 90 Tagen eine Festigkeitszunahme auf, außer nach 7 Tagen; dabei nimmt bei allen Puzzolanzugaben die Festigkeit ab. Die Geschwindigkeit der Festigkeitszunahme erreicht bei Zementen auf RHA-Basis nach einem Tag ein Maximum (20%) und sinkt dann nach 7 Tagen auf 1%, worauf nach 28 bzw. 90 Tagen erneut ein Anstieg um 7 bzw. 6% erfolgt. Die Geschwindigkeit der Festigkeitszunahme in frühen Altersstufen verlangsamt sich mit zunehmendem Puzzolangehalt (10–30 Gew.-%), außer bei RHA, deutlich. Die Geschwindigkeit der Festigkeitszunahme bei PPZ nach 1, 7 und 28 Tagen ist bei Zement auf RHA-Basis besser als bei PPZ aus gebranntem Ton und Flugasche. Nach 90 Tagen ist die Festigkeit jedoch in allen Fällen stets niedriger als bei den Basiszementen, und die Abnahmerate liegt zwischen 6 und 13%.

Bei den aus Klinker Nr. 2 hergestellten Zementen ist die Geschwindigkeit der Festigkeitsentwicklung bei den Zementen auf RHA-Basis nach Zugabe von 5 Gew.-% Puzzolan am höchsten und folgt dem gleichen Trend wie bei Klinker Nr. 1, außer daß die Wirkung stärker in Erscheinung tritt. Die geringste Festigkeitsentwicklung wird in jedem Prüfalter bei gebranntem Ton festgestellt. Wenn die Puzzolanzugabe erhöht wird, ist diese Tendenz auch bei Klinker Nr. 1 festzustellen, sieht man von dem Ergebnis nach 90 Tagen ab, wo die Abnahmegeschwindigkeit höher als bei Klinker Nr. 1 ist.

Der Unterschied in den Merkmalen der Festigkeitsentwicklung bei den drei Puzzolanzusätzen weist- vor allem angesichts der ähnlichen Kalkreaktivität dieser Puzzolane- darauf hin, daß die Kalkreaktivität nicht das einzige Leistungskriterium darstellt, sondern daß auch die chemischen und mineralogischen Merkmale der Klinker viel zur Festigkeitsentwicklung beitragen.

Die Hydratationswärme der aus drei Puzzolanen hergestellten PPZ ist in jedem Alter stets geringer als die der Basis-NPZ. Dies gilt für aus beiden Klinkern hergestellte Zemente. Bei Zement auf RHA-Basis ist die Abnahme nach 1 Tag am geringsten und nach 28 Tagen am höchsten, während es in den beiden anderen Fällen umgekehrt ist. Bei Flugaschezementen ist die Hydratationswärme in jeder Altersstufe niedriger. Dies spricht für die Auffassung, daß die Reaktion nicht in jedem Fall nach dem gleichen Mechanismus abläuft und daß der Mechanismus, vor allem bei Zementen auf RHA-Basis, gerade in frühen Altersstufen anders aussieht.

Die Abbindezeit der aus drei Puzzolanen hergestellten Zemente weist deutliche Unterschiede auf. Bei Zementen auf RHA-Basis nimmt die Abbindezeit zuerst ab und dann wieder gegenüber der des Basis-NPZ zu. Bei gebranntem Ton und Flugasche nimmt die Abbindezeit bei allen Dosierungen zu. Diese Erscheinung spiegelt die unterschiedlichen Gelcharakteristika dieser hydratisierten Systeme wider. Bei Zement auf RHA-Basis liefert RHA die für eine schnellere Reaktion benötigte größere Oberfläche, was im Vergleich mit gebranntem Ton und Flugasche in den frühen Stadien zu einer größeren Geloberfläche führt. Dies liegt an dem Umstand, daß gebrannter Ton und Flugasche zum Teil grobkörniger sind und inerte kristalline Anteile enthalten, die nicht in diesem Maße zu Hydratationsprodukten bzw. Geloberflächen beitragen.

Die Röntgenanalyse hydratisierter NPZ-Pasten und verschiedener Puzzolanzemente zeigt Ettringit, unhydratisierte Zementbestandteile, $Ca(OH)_2$, CSH, Gehlenit, Quarz usw. Das Vorhandensein von Aluminiumhydroxid nach Zugabe von Puzzolanen wird ebenfalls deutlich. Die vorhandenen Mengen an $Ca(OH)_2$ und Gehlenit werden durch die Zugabe von Puzzolan und mit dem Abbindealter ebenfalls stark verändert. Die Röntgenanalyse zeigt außerdem, daß die Hydratationsprodukte fast die gleichen, aber in der Menge verschieden sind.

strength over OPC at 1, 28 and 90 days, except that at 7 days, there is a decrease in strength with all the pozzolanic additions. The rate of increase in strength is maximum in the case of RHA based cements at 1 day (20%) and then it is reduced to 1% at 7 days which is again increased by 7 and 6% at 28 and 90 days, respectively. The rate of development of strength at early ages slows down considerably with increase in the pozzolana content (10–30 wt-%), except in case of RHA. The rate of strength development of PPC at 1, 7 and 28 days is better for RHA based cement compared to burnt clay and fly ash based PPC. However, at 90 days the strength is always lower than that of the base cements in all cases and the rate of decrease is between 6–13%.

For the cements made from clinker no. 2 the rate of strength development with the addition of 5 wt-% pozzolanic materials is the highest (24%) in case of RHA based cements and follows the same trend as for clinker no. 1 with the exception that the effect is more prominent. The lowest strength development is observed at all ages in case of burnt clay. As the pozzolanic addition is increased, the same trend is observed with clinker no. 1 excepting the result at 90 days when the rate of decrease is higher than compared to clinker no. 1.

The difference in strength development characteristics of the three pozzolanic additions, more particularly in view of the similar lime reactivity of these pozzolanic materials, indicates that the lime reactivity is not the only criterion contributing to the performance but it is the chemical and mineralogical characteristic of the clinkers which also play an important role in strength development.

Heat of hydration of the PPC's made from three pozzolanas is always lower than that of the base OPC's at all ages. This is true for cements made from both clinkers. In case of RHA based cement the decrease is minimum at 1 day and maximum at 28 days while in the other two cases it is reverse. In the case of cements made with fly ash the heat of hydration is lower at all ages. This supports the view that the reaction in all cases is not following the same mechanism and that the mechanism of RHA based cements is different, particularly at early ages.

Setting time of the cements made from three pozzolanic materials differs significantly. In case of RHA based cements the setting time first drops and then increases when compared to the base OPC. In the case of burnt clay and fly ash, the setting time increases with all doses. This phenomenon reflects the difference in gel characteristics of these hydrated systems. In the case of RHA based cement RHA provides the higher surface area necessary for a faster reaction, resulting in higher gel area as compared to burnt clay and fly ash at early ages. This is due to the fact that burnt clay and fly ash partly contain coarser grains and inert crystalline substances which do not contribute in such extent to hydration products and gel areas, respectively.

The XRD study of hydrated pastes of OPC and different pozzolanic cements indicates the presence of ettringite, unhydrated cement minerals, $Ca(OH)_2$, CSH, gehlenite, quartz, etc. The presence of aluminate hydrate is also indicated with the addition of pozzolanic materials. The quantity of $Ca(OH)_2$ and gehlenite gets altered to a great extent with the addition of pozzolana and also with the age of hydration. XRD study also reveals that the hydration products are nearly the same but their quantity differs.

Thermal analysis of the hydrated cement pastes indicates that in all the pozzolanic cements with up to 10 wt-% pozzolanic additions the weigth loss is very small up to 100°C and then it becomes steep. While in case of PPC's made with smaller pozzolanic doses the drop is lower due to an increase in gel surface area caused by fine additions which enable the gel to hold higher quantities of water. At higher levels of additions the gel area is small and therefore can hold less water. The fact that later age phenomena are pozzolanic in nature is confirmed by peaks at 480–520°C for $Ca(OH)_2$ and between 750–800°C for CSH.

Bei einer thermischen Analyse der hydratisierten Zementpasten ergibt sich für alle Puzzolanzemente mit bis zu 10 Gew.-% Puzzolanzugabe bis 100°C ein sehr niedriger Gewichtsverlust, der darüber jedoch steil ansteigt. Bei mit weniger Puzzolanzusatz hergestellten PPZ ist der Rückgang dagegen wegen der Vergrößerung der Geloberfläche geringer, da feine Beimengungen dem Gel das Halten größerer Wassermengen ermöglichen. Bei größeren Zusatzmengen hat das Gel eine kleine Oberfläche und kann darum weniger Wasser halten. Der Puzzolancharakter der zu einem späteren Zeitpunkt auftretenden Reaktionen wird durch Maxima von Ca(OH)$_2$ bei 480–520°C und von CSH bei 750–800°C bestätigt.

Aufgrund der obigen Untersuchungen ist davon auszugehen, daß an dem Mechanismus der Puzzolanreaktionen bei mit Flugasche, gebranntem Ton und Reishülsenasche hergestelltem PPZ zwei Aspekte beteiligt sind. Die anfängliche Wirkung der Puzzolanzugabe wird dem als Feinmehleffekt bezeichneten physikalischen Phänomen zugeschrieben. Diese Erscheinung verursacht katalytische Reaktionen durch Erhöhung der Hydratationsgeschwindigkeiten von C_3S und C_3A, wobei eine größere Geloberfläche entsteht, sowie durch Ausfüllung von Hohlräumen im Gel, was die Porosität verringert und eine höhere Frühfestigkeit ergibt. Mit steigender Puzzolanzugabe wird die Kontinuität der Reaktion durch die Anwesenheit weiterer Materialien beeinträchtigt, so daß die Porosität trotz der verfügbaren größeren Oberfläche nicht vermindert wird und die Festigkeit der Systeme in den Anfangsstadien bei höheren Dosierungen deshalb niedrig ist. Mit zunehmender Hydratationsdauer neigen Puzzolane zu erhöhter Kalkabsorption, woraus sich eine Verstärkung der Puzzolanreaktionen ergibt. In dem vorliegenden Fall läßt sich die Wirkung niedriger Dosierungen, d. h. einer Zugabe von 5 Gew.-%, nicht unterscheiden. Bei Zugaben von 10–20 Gew.-% ist dies deutlicher zu erkennen. Oberhalb von 20 Gew.-% beginnt der Verdünnungseffekt den Vorgang zu bestimmen. Diese Erscheinung ist aus den Festigkeitswerten, den Röntgenanalysen und den thermischen Untersuchungen deutlich abzulesen [10–12].

3. Wirkung inerter Zusätze

Die Wirkung inerter Zusätze unterscheidet sich von der von Puzzolanzusätzen. Als inerte Stoffe werden Kalksteinmehl und Ofenstaub verwendet. In der vorliegenden Studie wurde die Zugabe von bis zu 8 Gew.-% Ofenstaub und bis zu 5 Gew.-% Kalksteinmehl untersucht [13, 14]. Insgesamt wurden für die Studie 10 Proben von NPZ und zumahlstoffhaltigen Zementen hergestellt. Die Ergebnisse sind in der **Tabelle 3** aufgeführt, und die Beobachtungen lassen sich wie folgt zusammenfassen:

Die Zugabe von bis zu 5 Gew.-% Kalksteinmehl bewirkt nach 1 Tag eine deutliche Zunahme der Geschwindigkeit der Festigkeitsentwicklung (60%), doch erfolgt der Festigkeitsverlauf in späteren Stadien nicht nur langsam, sondern auch in geringerem Maße als bei dem Basiszement, da die Wirkung in den Anfängen auf den Feinmehleffekt und in späteren Phasen auf das Verdünnungsphänomen zurückzuführen ist [15].

Die Festigkeitsentwicklung bei Zugabe von Ofenstaub ergibt nach 1 Tag für eine Zugabe von 2 Gew.-% Ofenstaub den höchsten Wert, der nach der Zugabe von bis zu 8 Gew.-% Ofenstaub auf einen beträchtlichen Wert zurückgeht. Die Erscheinung in späteren Stadien zeigt eine Festigkeitsentwicklung der gleichen Größenordnung wie bei dem Basiszement, doch sinkt die Festigkeit bei Zugabe von 8 Gew.-% Ofenstaub nach 7, 28 und 90 Tagen sehr stark auf nur 53, 51 und 41%. Ganz offensichtlich erhöht die Zugabe von bis zu 5 Gew.-% Ofenstaub die Geschwindigkeit der Festigkeitsentwicklung aufgrund des Feinmehleffekts in Verbindung mit der Hydratation von Zementbestandteilen im Ofenstaub. Demgegenüber führt die Zugabe von mehr als 5 Gew.-% dieser Stoffe eine deutliche Abnahme der Festigkeitsentwicklung herbei, was auf das Vorliegen eines Verdünnungseffekts hindeutet.

Based on the above studies, it can be suggested that there are two aspects involved in the mechanism of pozzolanic reactions in PPC made with fly ash, burnt clay and rice husk ash. The initial action of pozzolanic addition is attributed to the physical phenomena termed as fine powder effect. The phenomenon causes catalytic reactions by way of increasing the hydration rates of C_3S and C_3A, providing a higher gel surface area and also by filling gel voids, which reduce the porosity and result in higher early strength. As the addition of pozzolana is increased, the continuity of reaction is disturbed by the presence of extra materials and therefore, in spite of providing higher surface area, the porosity does not get reduced and therefore strength of systems in the initial stages with higher doses is low. As the age of hydration increases, the pozzolanic materials have a tendency to absorb more lime which leads to intensified pozzolanic reactions. In this particular case the effect of lower doses, i.e. of a 5 wt-% addition, can not be distinguished. This can be better seen for additions of 10–20 wt-%. Beyond 20 wt-% it is the dilution effect which takes over as the controlling mechanism. This phenomena is clearly seen from the data on the strength, XRD and thermal studies [10–12].

3. Effect of inert additions

The effect of inert additions is different from that of pozzolanic additions. The used inert materials are limestone and kiln dust. In the present study the addition of kiln dust up to 8 wt-% and limestone addition up to 5 wt-% were investigated [13, 14]. All in all 10 samples of OPC and blended cements were prepared in the study. The results are given in **Table 3** and the observations are summarized as follows:

The addition of limestone up to 5 wt-% causes marked increase in the rate of strength development (60%) at 1 day, however, the later age strength development is not only slow but is in fact less than the base cement because at early ages the effect is due to fine powder effect and at later ages it is due to dilution phenomena [15].

The strength developments with the addition of kiln dust indicate that the rate of strength development with a 2 wt-% addition of kiln dust is the highest (82%) after 1 day which decreases to a considerable rate with the addition of kiln dust up to 8 wt-%. The phenomenon at later ages displays that strength development is in the same range as that one of the base cement, but with a 8 wt-% addition of kiln dust the strength is reduced to a drastic rate of only 53, 51 and 41% at 7, 28 and 90 days. It is therefore obvious that kiln dust addition up to 5 wt-% increases the rate of strength development which is due to fine powder effect, coupled with hydration of cement minerals in the kiln dust. However, the addition of these materials beyond 5 wt-% causes a drastic reduction of strength development indicating the presence of dilution phenomena.

In the case of cement made with the addition of kiln dust the heat of hydration after 1 day is increased at 2 wt-% addition and then it is reduced at higher additions. This trend is also similar at later ages. With the addition of kiln dust from 2 to 8 wt-% the initial (IST) and final setting time (FST) increase at a constant rate from 160 of the base cement to 260 and from 300 to 400 minutes, respectively. The degree of calcination of kiln dust has a significant effect on the strength development and a calcination level of 10% is found ideal for the maximum additive contribution.

Literaturverzeichnis

[1] Lea, F. M.: Pozzolanas and pozzolanic cement. The chemistry of cement and concrete, III Edition, Edward Arnold (pub) Ltd. 1983.

[2] Massazza, F.: Chemistry of pozzolanic additions and mixed cements. VIth Int. Symp. on Chemistry of Cement. Moskau, 1974.

[3] Sersale, R.: Structure and characterisation of pozzolanas amd fly ash. VIIth Int. Symp. on Chemistry of Cement. Paris, (4) 1980.

Bei unter Zugabe von Ofenstaub hergestelltem Zement nimmt die Abbindewärme bei 2 Gew.-% nach 1 Tag zu, um dann bei höheren Zugabemengen wieder zurückzugehen. Eine ähnliche Tendenz ist auch in späteren Stadien zu verzeichnen. Bei Zugabe von 2–8 Gew.-% Ofenstaub steigen die anfängliche Abbindezeit (AAZ) und die Endabbindezeit (EAZ) konstant von 160 Minuten des Basiszements auf 260 bzw. von 300 auf 400 Minuten. Der Entsäuerungsgrad des Ofenstaubs wirkt sich nachhaltig auf die Festigkeitsentwicklung aus, und ein Entsäuerungsgrad von 10 % hat sich im Hinblick auf eine maximale Beimischung als ideal erwiesen.

[4] Chopra, S.K., und Narang, K.C.: Industrial waste in cement manufacture — status and prospects. Chemical age of India, 29, (7) 1978, S. 543–552.

[5] Narang, K.C.: Portland and blended cements. IXth Int. Congress on Chemistry of Cement. New Delhi, Indien, (I) 1992, S. 213–260.

[6] Sersale, R.S.: Advances in portland and blended cements. IXth Int. Congress on Chemistry of Cement. New Delhi, Indien, (I) 1992, S. 261.

[7] Chatterji, S.: Pozzolanic property of natural and synthetic pozzolanas: A comparative study. Ist Int. Symp. on the Use of Fly ash, Silica fume, Slag and other mineral Byproducts in Concrete. SP-79, Kanada, 1983.

[8] Kokubu, M.: Fly ash and fly ash cements. Vth Int. Symp. on Chemistry of Cements. Tokyo, Japan, (IV) 1968, S. 75–113.

[9] Sharma, R.B.: Investigations of cementitious binders containing artificial pozzolanic materials. A Ph.D. Thesis, Agra University, Indien, April 1986.

[10] Hara, N. et al.: Suitability of rice husk ash obtained by fluidized bed combustion for blended cements. IXth Int. Congress on Chemistry of Cement. New Delhi, Indien, (III) 1992, S. 72–78.

[11] Sersale, R. et al.: Chemical and physical properties of fly ashes and influence on the mechanical performance of the result cement mortars. Cemento, 4, 1986, S. 565.

[12] Subba, Rao und James, J.: Hydration of husk ash-lime paste. Cemento, 4, 1987, S. 383.

[13] Gosh, S. P., Padmanathan, P., und Mohan, K.: Study of effects of kiln dust addition to portland and blended cements. IXth Int. Congress on Chemistry of Cement. New Delhi, Indien, (III) 1992, S. 159–165.

[14] Vernet, C.: Mechanism of limestone filler reactions in the system. IXth Int. Congress on Chemistry of Cement. New Delhi, Indien, (II) 1992, S. 430.

[15] CRI: Establishing alternative avenues of using kiln dust from the indian cement plants, August 1983.

Einfluß des Zerkleinerungsverfahrens auf die Eigenschaften des Zementes

Influence of the method of comminution on the properties of cement

L'influence du procédé de broyage sur les propriétés du ciment

Influencia del proceso de molienda sobre las propiedades del cemento

Von **I. Odler** und **Y. Chen,** Clausthal/Deutschland

Zusammenfassung – Untersucht wurde der Einfluß der Zementfeinheit, der Art und Menge des verwendeten Sulfatträgers sowie der Klinkerzusammensetzung auf die Eigenschaften von Portlandzementen nach Feinmahlung in einer Kugelmühle sowie nach einer dem Gutbettverfahren ähnlichen Hochdruckzerkleinerung. Die Feinmahlung im Gutbett führte generell zu einer Verschlechterung des Fließverhaltens von Zementleim (Ausbreitmaß), zu einer Verkürzung der Erstarrungszeit und zu einer erhöhten Anfangsreaktivität des Zements, erkennbar an der Intensität des ersten Peaks bei kalorischen Messungen. Entsprechende Untersuchungen an reinen Klinkermineralen ergaben, daß durch die Gutbettzerkleinerung die Reaktivität des C_3S erhöht und die Ettringit-Monosulfatumwandlung deutlich beschleunigt wurden.

Summary – The influence of cement fineness, the type and quantity of the sulphate agent used, and the clinker composition on the properties of Portland cements was investigated after fine grinding in a ball mill and after a high-pressure comminution process resembling interparticulate comminution. In general terms, interparticulate fine grinding led to a deterioration in the flow behaviour of the cement paste (flow table spread), to a shortening of the setting time, and to an increased initial reactivity of the cement, recognizable by the intensity of the first peak during calorific measurement. Corresponding investigations with pure clinker minerals showed that the reactivity of the C_3S was increased and the ettringite-monosulphate conversion was significantly accelerated by the interparticulate comminution.

Résumé – On a étudié l'influence de la finesse du ciment, de la nature et de la quantité du porteur de sulfates utilisé ainsi que de la composition du clinker sur les propriétés des ciments Portland après le broyage fin dans un broyeur à boulets ainsi qu'après un broyage à haute pression, comparable au procédé par lit de matière. Le broyage fin à l'aide du lit de matière a conduit, en général, à une détérioration de la réologie de la pâte de ciment (indice d'extension), à une diminution du temps de prise et à une augmentation de la réactivité initiale du ciment, que l'on reconnaît par le premier peak lors des essais calorimétriques. Des études correspondantes, effectuées sur des minéraux de clinker purs, ont donné pour résultat que, dû au broyage par lit de matière, la réactivité du C_3S a été augmentée et en même temps la transformation d'ettringite-monosulfate a été nettement accélérée.

Resumen – Se ha estudiado la influencia de la finura del cemento, de la naturaleza y cantidad del portador de sulfatos utilizado así como de la composición del clínker, sobre las propiedades de los cementos Portland, después de la molienda fina del clínker en un molino de bolas, y después de la molienda del mismo a alta presión, comparable al sistema de lecho de material. La molienda fina con ayuda del lecho de material ha conducido, en general, a un empeoramiento de la reología de la pasta de cemento (índice de extensión), a una reducción del tiempo de fraguado y a un aumento de la reactividad inicial del cemento, que se reconoce por la intensidad del primer peak durante las mediciones con el calorímetro. De los estudios correspondientes, realizados con minerales de clínker puros, se desprende que, debido a la molienda en el lecho de material, aumentó la reactividad del C_3S, en tanto que se aceleró notablemente la transformación de ettringita-monosulfato.

1. Einleitung

Es ist bekannt, daß durch den Einsatz einer Gutbettwalzenmühle (Rollenpresse) der Energiebedarf bei der Zementvermahlung wesentlich gesenkt werden kann. Gleichzeitig zeigt sich aber auch, daß der so hergestellte Zement einen höheren Wasserbedarf und ein schnelleres Erstarren ausweist [1–5]. Um eine derartige Verschlechterung der Quali-

1. Introduction

It is well known that the use of high-pressure grinding rolls can substantially reduce the consumption of energy in cement grinding. However, the resulting cement has proved to have a higher water demand and a faster rate of setting [1–5]. To prevent such an impairment of the quality of the cement, grinding by means of high-pressure grinding rolls

2. Materials and procedure

The compositions of the clinkers used were as shown in Table 1. A white clinker with an increased C_3A content and a highly sulphate-resistant clinker with a reduced C_3A content were included in the investigation in addition to a standard Portland clinker.

TABLE 1: Composition of the cement clinkers used in wt.%

	PC	PC-white	PC-SR
CaO	65.0	68.3	66.1
Si_2O	21.4	23.7	24.1
Al_2O_3	5.6	4.7	2.6
Fe_2O_3	2.4	0.2	3.9
MgO	1.8	0.7	1.2
K_2O	1.0	0.9	0.7
Na_2O	1.3	0.2	0.2
SO_3	1.3	0.8	0.5
TiO_2	0.0	0.1	0.2

In order to simulate the process of comminution by means of high-pressure grinding rolls, precrushed material was comminuted by pressure in a hydraulic press and then disagglomerated. At the same time another portion of material was ground in a laboratory ball mill. By subsequent classification in an air classifier, samples with approximately the same Blaine specific surface and a similar particle size distribution (measured with a laser granulometer) were obtained from the resulting comminuted clinkers. In the case of the two clinker phases, classification in the air classifier was not possible due to the smaller original quantities, so the material passing a 70 μm screen was used in this case. To assess the influence of the method of comminution on the particle shape, the shape factor was determined for the samples by the method described in [5].

To control setting, calcium sulphate in dihydrate form (CSH_2) and, in one case, also in hemihydrate form ($CSH_{0.5}$) was mixed with the ground clinkers. To determine the influence of the comminution process on the water demand of the cement, the spread was measured using pastes. For this purpose a method was developed which in principle resembles that described in DIN 1060. The paste is put into a plastic ring (d = 50 mm, h = 25 mm) which is placed on a level sheet of glass. After removing the ring, the sheet is lifted 10 mm and dropped fifteen times, causing the paste to spread out on the sheet. The diameter of the paste is measured and termed the "spread". At a constant water-cement ratio, this value increases as the water demand of the cement falls. In the investigations, the spread of the material commi-

gemessen und als „Ausbreitmaß" bezeichnet. Bei konstantem Wasser-Zement-Verhältnis steigt dieser Wert mit abnehmendem Wasserbedarf des Zementes. Bei den Untersuchungen wurde das Ausbreitmaß des in der Kugelmühle zerkleinerten Materials zunächst durch entsprechende Wasserzugabe auf etwa 10 cm eingestellt. Das Fließverhalten der im Gutbett-Verfahren hergestellten Parallelprobe wurde daraufhin mit demselben w/z-Wert ermittelt und damit verglichen. Außerdem wurde zur Beurteilung der Reaktivität die in den ersten zwei Stunden freigegebene Hydratationswärme ermittelt.

3. Resultate

In **Tabelle 2** sind die Zusammensetzung und Eigenschaften der untersuchten Zemente zusammengefaßt. Verglichen wurden jeweils Zementpaare, die eine gleiche spezifische Oberfläche aufwiesen und die sich lediglich durch das angewendete Zerkleinerungsverfahren unterschieden. Die bestehenden Unterschiede in der spezifischen Oberfläche bei den ebenfalls untersuchten Klinkerphasen sind der unterschiedlichen Aufbereitungsweise zuzuschreiben.

Neben Resultaten an Zementen, die aus einem normalen Portlandklinker durch Vermahlung auf 3000 cm^2/g mit 3 Gew.-% SO_3 erhalten wurden, enthält die Tabelle auch Resultate an Zementen, die besonders fein (PZ-f) oder grob (PZ-g), mit erhöhtem Gipsgehalt (PZ-5) oder mit Halbhydrat anstelle von Gips (PZ-H) hergestellt wurden. Ebenfalls enthalten sind Angaben über Zemente, hergestellt aus Spezialklinkern (PZ-weiß oder PZ-HS), und reine Klinkerphasen. Dabei wurde das C_3A allein und zusammen mit Gips untersucht.

Ein Vergleich der gefundenen Formfaktoren ergab, daß dieser Wert jeweils bei den im Gutbettverfahren zerkleinerten Proben höher war, was auf eine mehr von der Kugelform abweichende Geometrie dieser Zementpartikel hindeutet.

nuted in the ball mill was first adjusted to approx. 10 cm by adding an appropriate quantity of water. Then the flow behaviour of the parallel sample produced by high compression grinding was determined using the same w-c ratio and compared with the first value. The heat of hydration evolved in the first two hours was also measured to assess the reactivity.

3. Results

The composition and properties of the cements investigated are summarized in **Table 2**. In each case, pairs of cements having the same specific surface and differing only in the method of comminution were compared. The differences in the specific surface of the clinker phases which were also investigated are attributable to the different methods of preparation.

Besides results for cements obtained from a standard Portland clinker by grinding to 3000 cm^2/g with 3 wt.% SO_3, the table also contains results for cements produced with a particularly fine (PC-f) or particularly coarse (PC-c) size range, with an increased gypsum content (PC-5) or with hemihydrate instead of gypsum (PC-H). It also includes data for cements made from special clinkers (PC-white or PC-SR) and pure clinker phases. The C_3A was studied on its own and together with gypsum.

A comparison of the shape factors determined showed that this value was always higher for the samples comminuted by the high compression method. This suggests that the geometry of these cement particles deviates more from a spherical shape.

A comparison of the spreads showed, in all cases, a significant deterioration in the flow behaviour of cements comminuted by the high compression method. This effect could not be remedied by varying the type or quantity of sulphate or by means of the clinker composition. The setting time of

TABELLE 2: Ergebnisse der Untersuchungen an verschiedenen Zementen

Zem.	Mahl-weise	Form-faktor	sp. Oberf. (cm^2/g)	SO_3 (%)	Ausbreitmaß W/Z	(cm)	W/Z	Erstarrungszeit (h) Anfang	Ende	W* (J/g)
PZ	KM#	1,78	3000	3	0,35	9,9	0,35	3:50	5:00	15,0
	GB	2,10	3100	3	0,35	5,6	0,35	2:40	3:55	23,3
PZ-5	KM	1,78	3000	5	0,35	9,8	0,35	4:05	5:20	15,6
	GB	2,10	3100	5	0,35	6,5	0,35	2:55	4:15	18,9
PZ-H (HH)	KM	1,78	3000	3	0,35	8,9	0,35	2:30	4:15	20,4
	GB	2,10	3100	3	0,35	6,3	0,35	3:00	5:00	21,5
PZ-f	KM	1,82	4500	3	0,56	10,5	0,56	4:40	5:55	21,8
	GB	2,23	5000	3	0,56	8,0	0,56	4:00	5:20	24,0
PZ-g	KM	1,78	900	3	0,32	10,2	0,32	6:30	7:45	
	GB	2,30	900	3	0,32	8,3	0,32	5:20	6:30	
PZ-Weiß	KM	1,51	2550	3	0,50	10,5	0,50	4:15	5:35	20,1
	GB	1,56	2500	3	0,50	9,2	0,50	3:50	5:00	27,4
PZ-HS	KM	1,34	1950	3	0,33	9,7	0,35	1:30	3:10	3,8
	GB	1,56	2000	3	0,33	8,5	0,35	1:00	3:00	6,3
C_3S	KM	1,23	3500	–	0,40	9,1	0,40	13:15	16:05	2,7
	GB	1,39	2600		0,40	7,8	0,40	10:25	12:55	3,6
C_3A	KM	1,26	4250	–						295
	GB	1,39	2950							289
C_3A	KM	1,26	4250	15						124
	GB	1,39	2950	15						235

\# KM: Kugelmühle; GB: Gutbettzerkleinerung
* Die gesamte Wärmefreigabe (J/g) in den ersten drei Stunden (erster Wärmepeak)

TABLE 2: Results of the investigations with various cements

Cem.	Grinding method	Shape factor	Sp. surf. (cm^2/g)	SO$_3$ (%)	Spread w-c	(cm)	Setting time (h) w-c	Initial	Final	W* (J/g)
PC	BM#	1.78	3000	3	0.35	9.9	0.35	3:50	5:00	15.0
	HP	2.10	3100	3	0.35	5.6	0.35	2:40	3:55	23.3
PC-5	BM	1.78	3000	5	0.35	9.8	0.35	4:05	5:20	15.6
	HP	2.10	3100	5	0.35	6.5	0.35	2:55	4:15	18.9
PC-H (HH)	BM	1.78	3000	3	0.35	8.9	0.35	2:30	4:15	20.4
	HP	2.10	3100	3	0.35	6.3	0.35	3:00	5:00	21.5
PC-f	BM	1.82	4500	3	0.56	10.5	0.56	4:40	5:55	21.8
	HP	2.23	5000	3	0.56	8.0	0.56	4:00	5:20	24.0
PC-c	BM	1.78	900	3	0.32	10.2	0.32	6:30	7:45	
	HP	2.30	900	3	0.32	8.3	0.32	5:20	6:30	
PC-white	BM	1.51	2550	3	0.50	10.5	0.50	4:15	5:35	20.1
	HP	1.56	2500	3	0.50	9.2	0.50	3:50	5:00	27.4
PC-SR	BM	1.34	1950	3	0.33	9.7	0.35	1:30	3:10	3.8
	HP	1.56	2000	3	0.33	8.5	0.35	1:00	3:00	6.3
C$_3$S	BM	1.23	3500	–	0.40	9.1	0.40	13:15	16:05	2.7
	HP	1.39	2600	–	0.40	7.8	0.40	10:25	12:55	3.6
C$_3$A	BM	1.26	4250	–						295
	HP	1.39	2950	–						289
C$_3$A	BM	1.26	4250	15						124
	HP	1.39	2950	15						235

\# BM: ball mill; HP: high compression comminution
* Total heat evolution (J/g) in the first three hours (first heat peak)

Ein Vergleich der Ausbreitmaße ergab in allen Fällen eine merkliche Verschlechterung des Fließverhaltens bei Zementen aus der Gutbettzerkleinerung. Dieser Effekt konnte durch Variieren der Sulfatart oder -menge oder durch die Klinkerzusammensetzung nicht behoben werden. Die Erstarrungszeit aller Zemente und des Tricalciumsilicats wurde durch die Gutbettzerkleinerung generell verkürzt. Die einzige Ausnahme war der Zement mit Halbhydrat als Sulfatträger.

Die Reaktivität der Zemente, beurteilt nach der Wärmefreigabe zu Beginn der Hydratation, wurde durch das Vermahlen im Gutbettverfahren generell erhöht. Dieser Effekt der Gutbettzerkleinerung wurde durch eine Erhöhung der Sulfatzugabe oder einen Einsatz von Halbhydrat gemindert, aber nicht völlig eliminiert. Ein Vergleich der einzelnen Klinkerarten, die zu Zementen in der Kugelmühle vermahlen wurden, ergab, daß die anfänglich freigegebene Hydratationswärme beim Weißzement mäßig, beim HS-Zement deutlich gesenkt wurde. Erwartungsgemäß waren die mit dem C$_3$A-reichen Klinker gefundenen Werte höher und jene mit C$_3$A-freiem Klinker deutlich niedriger als die des normalen Portlandklinkers.

Kalorimetrische Messungen an den Klinkerphasen ergaben, daß im Falle des C$_3$S die erste Wärmefreigabe der im Gutbett-Verfahren zerkleinerten Probe trotz kleinerer spezifischer Oberfläche um etwa 50 % erhöht war. Im Falle des C$_3$A ohne Sulfatzugabe war die erste Wärmefreigabe durch die unterschiedliche Zerkleinerungsweise kaum beeinflußt (**Bild 1a**). In den Proben mit zugegebenem Calciumsulfat war die Wärmefreigabe durch zwei exotherme Peaks charakterisiert (**Bild 1b**). Der erste entsprach der anfänglichen C$_3$A-Hydratation und Ettringitbildung, der andere der Ettringit-Monosulfatumwandlung, nachdem die zur Verfügung stehende Sulfatmenge verbraucht war. Ein Vergleich der beiden kalorimetrischen Kurven ergab, daß durch die Gutbettzerkleinerung die Wärmefreigabe früher einsetzte.

all the cements and the tricalcium silicate was generally shortened by high compression comminution. The sole exception was the cement containing hemihydrate as sulphate agent.

The reactivity of the cements, assessed according to the heat evolution at the start of setting, was generally increased by grinding by the high compression method. This effect of high compression comminution was lessened, but not completely eliminated, by increasing the quantity of added sulphate or through the use of hemihydrate. A comparison of the individual types of clinker ground to cements in the ball mill showed that the heat of setting initially evolved was moderately reduced in the case of the white cement and significantly reduced in the case of the highly sulphate-resistant cement. As expected, the values measured with the C$_3$A-rich clinker were higher, and those measured with C$_3$A-free clinker were markedly lower, than those of the standard Portland clinker.

Calorific measurements for the clinker phases showed that in the case of the C$_3$S the initial heat evolution of the sample comminuted by the high compression method was increased by about 50 % despite the smaller specific surface. In the case of the C$_3$A without added sulphate, the initial heat evolution was scarcely influenced by the different methods of comminution (**Fig. 1a**). For the samples with added calcium sulphate, the heat evolution was characterised by two exothermic peaks (**Fig. 1b**). The first corresponded to the initial C$_3$A hydration and ettringite formation, while the second corresponded to the ettringite-monosulphate conversion after the available quantity of sulphate had been used up. A comparison of the two calorimetric curves showed that comminution by the high compression method caused the heat evolution to begin earlier.

4. Diskussion und Schlußfolgerung

Die an Zementen verschiedener Zusammensetzung und Kornfeinheit durchgeführten Untersuchungen bestätigten frühere Berichte über den Einfluß des Zerkleinerungsverfahrens auf die Eigenschaften vergleichbar hergestellter Zemente sowie über die Beeinflussung der Kornform durch die Wahl des Zerkleinerungsverfahrens. Es kann angenommen werden, daß diese unterschiedliche Korngeometrie einer der Gründe für den erhöhten Wasserbedarf des im Gutbett-Verfahren zerkleinerten Zementes ist.

Die Untersuchungen bestätigen weiterhin die erhöhte Reaktivität der C_3A-Phase in Klinkern, die im Gutbettverfahren zerkleinert wurden. Neu ist dagegen der Befund, daß durch die Gutbettzerkleinerung ebenfalls die Reaktivität der C_3S-Phase erhöht wird. Es darf angenommen werden, daß eine beschleunigte Reaktion dieser Phase im Anfangsstadium der Hydratation ebenfalls zu einer Verkürzung der Erstarrungszeit und möglicherweise auch zur Erhöhung des Wasseranspruches beiträgt.

Die bisherigen Resultate deuten darauf hin, daß die Erhöhung des Wasseranspruches und das ebenfalls geänderte Erstarrungsverhalten der Zemente, die in der Gutbettmühle vermahlen waren, offensichtlich von mehreren Faktoren abhängt. Dabei erscheint es denkbar, daß einzelne Faktoren bei verschiedenen Zementen eine unterschiedliche Rolle spielen können.

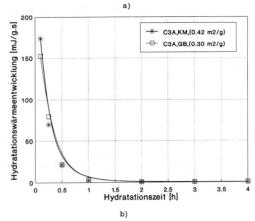

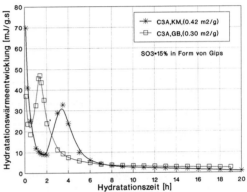

BILD 1: Hydratationswärmefreigabe von C_3A, vermahlen in einer Kugelmühle und im Gutbettverfahren a) reines b) C_3A mit Gipszugabe

FIGURE 1: Heat of hydration evolved by C_3A ground in a ball mill and by the high compression method a) Pure C_3A, b) C_3A with added gypsum

KM	= ball mill
GB	= high compression
Hydratationswärmeentwicklung	= Heat of hydration evolved
Hydratationszeit	= Hydration time
in Form von Gips	= in the form of gypsum

4. Discussion and conclusion

The investigations conducted with cements of varying composition and fineness confirmed previous reports concerning the influence of the method of comminution on the properties of cements produced in similar ways and the ability of the method of comminution to influence the particle shape. This different particle geometry is assumed to be one of the reasons for the increased water demand of the cement comminuted by the high compression method.

The experiments also confirm the increased reactivity of the C_3A phase in clinkers comminuted by the high compression method. A new finding, on the other hand, is that high compression comminution also increases the reactivity of the C_3A phase. An accelerated reaction of this phase in the initial stages of setting presumably also contributes to a reduction of the setting time and possibly also to the increase in the water demand.

The results to date indicate that the increase in the water demand and the accompanying change in the setting behaviour of the cements ground in highpressure rollers obviously depend on several factors. The individual factors may conceivably play differing roles in different cements.

Literaturverzeichnis

[1] Wüstner, H., Dreizler, I., und Oberhauser, C.: Einsatz von Rollenpressen in Mahlanlagen für Kohle, Zement-Rohstoffe und Zement. Zement-Kalk-Gips 40 (1987) Nr. 7, S. 501–505.

[2] Rosemann, H., Hochdahl, O., Ellerbrock, H.-G., und Richartz, W.: Untersuchungen zum Einsatz einer Gutbett-Walzenmühle zur Feinmahlung von Zement. Zement-Kalk-Gips 42 (1989) Nr. 4, S. 165–169.

[3] Ellerbrock, H.-G., Richartz, W., und Sprung, S.: Auswirkungen des Zerkleinerungsablaufs in Gutbett-Walzenmühlen auf die Eigenschaften von Portlandzement. Technisch-wissenschaftliche Zementtagung Düsseldorf 31. Januar – 1. Februar 1990.

[4] Odler, I., und Chen, Y.: Einfluß des Mahlens in der Gutbett-Walzenmühle auf die Eigenschaften des Portlandzements. Zement-Kalk-Gips 43 (1990) Nr. 4, S. 188–191.

[5] Chen, Y., and Odler, I.: Effect of the grinding technique on the shape of cement particles. Proc. 14th Internat. Conf. on Cement Microscopy, Cosa Mesa 1992, S. 22–28.

[6] Ellerbrock, H.-G., Sprung, S., und Kuhlmann, K.: Korngrößenverteilung und Eigenschaften von Zement, Teil III: Einfluß des Mahlprozesses. Zement-Kalk-Gips 43 (1990) Nr. 1, S. 13–19.

[7] Wolter, A., und Dreizler, I.: Einfluß der Rollenpresse auf die Zementeigenschaften. Zement-Kalk-Gips 41 (1988) Nr. 2, S. 64–70.

[8] Thormann, P., und Schmitz, Th.: Die Beeinflussung der Zementeigenschaften durch Klinkermahlung in der Gutbettwalzenmühle. Zement-Kalk-Gips 45 (1991) Nr. 4, S. 188–193.

Einflüsse auf die Qualität von Schachtofen-Klinker

Parameters influencing vertical shaft kiln clinker quality

Paramètres influenant la qualité du clinker provenant d'un four vertical

Factores que influyen en la calidad del clínker producido en el horno vertical

Von **K. Raina** und **L. K. Janakiraman,** New Delhi/Indien

Zusammenfassung – In den Entwicklungsländern, z. B. in Indien, findet man häufig kleinere Zementwerke, die mit dem Schwarzmehl-Verfahren und der Schachtofen-Technologie arbeiten. Das einem solchen Werk zur Verfügung stehende Rohmaterial ist normalerweise von unterschiedlicher Zusammensetzung, so daß die Rohmehlzusammensetzung entsprechend variiert und angepaßt werden muß. Zu beachten ist, daß das Rohmehlverhalten während des Sintervorgangs im Schachtofen sehr stark von seiner chemischen und mineralogischen Zusammensetzung einerseits und von der Mahlfeinheit, Temperatur und Verweilzeit im Ofen andererseits beeinflußt wird. Für die Untersuchung wurden drei industrielle Rohmehle ausgewählt. Sie wurden im Schachtofen gebrannt, wobei Kalkgehalt, Mahlfeinheit und Verweilzeit im Ofen variiert wurden. Um die Qualität der so erzeugten Klinker zu beurteilen, wurden das Klinkerlitergewicht und der freie Kalk bestimmt und die Röntgenbeugungsverfahren angewandt. Morphologische Untersuchungen wurden mittels optischer Mikroskopie und Rasterelektronenmikroskopie durchgeführt. Die Brennbarkeitseigenschaften des Rohmehls beeinflußten das Temperaturprofil des Materials und spielten somit eine entscheidende Rolle für die Klinkerqualität. Klinker, die aus der Verarbeitung von Rohmehl im Schachtofen gewonnen werden, unterscheiden sich bezüglich ihrer Morphologie und Phasenverteilung ganz erheblich von Klinkern, die im Drehofen hergestellt werden. Der Kurzbeitrag faßt die Ergebnisse einer Untersuchung des Einflusses verschiedener Parameter auf die Klinkerqualität zusammen.

Summary – Small scale cement plants in developing countries like India are more popular with small entrepreneurs and are mostly based on the black meal process using vertical shaft kiln (VSK) technology. The raw material available to a plant is normally of variable composition and as a consequence, the raw mix design needs variation and adjustment. It must be taken into consideration that the raw meal behaviour during the sintering process in VSK is greatly influenced by its chemical and mineralogical compositions on the one hand and by fineness, temperature and retention time in the kiln on the other hand. Three industrial raw meals have been chosen for the study. They were burnt in the VSK with varying lime content, fineness and retention time. Clinkers thus obtained have been examined for their quality by clinker litre weight, free lime determination and X-ray diffraction methods. Morphological studies were carried out using optical microscopy (OM) and SEM techniques. The burnability characteristics of raw meal influenced the temperature profile of the material and hence played a decisive part in clinker quality. Clinkers obtained from processing a raw meal in VSK are very different from those obtained in rotary kiln with respect to morphology and phase distribution. The present paper highlights the results of investigations undertaken to establish the effect of different parameters on the quality of clinker.

Résumé – Dans les pays en voie de développement comme en Inde, par exemple, on trouve souvent de petites cimenteries travaillant selon le procédé de la farine noire et technologie du four vertical. La matière première disponible dans une telle usine est généralement de composition très variée de sorte que la composition de la farine crue doit être adaptée en conséquence. A noter que la tenue de la farine crue durant la phase de clinkérisation dans le four vertical dépend très largement de sa composition chimique et minéralogique, d'une part, mais aussi de la finesse de mouture, de la température et de la durée de séjour au four. Trois types de farine crue industrielle ont été choisis pour l'examen. La cuisson a été réalisée dans le four vertical, la teneur de chaux, la finesse de mouture et le temps de séjour au four ayant été modulés. Pour apprécier la qualité du clinker produit, le poids au litres de clinker et la chaux libre ont été déterminés et le procédé de diffraction des rayons X a été appliqué. Des analyses morphologiques ont été effectuées à l'aide de la microscopie optique et d'un microscope électronique à balayage de surface. Les propriétés de combustibilité de la farine crue influent sur le profil thermique de la matière et jouent ainsi un rôle déterminant sur la qualité du clinker. Les clinkers produits à partir de la transformation de farine crue dans un four vertical, présentent tant au point de vue morphologie que répartition des phases des différences considérables par rapport aux clinkers fabriqués dans un four rotatif. L'exposé succinct récapitule les résultats de l'analyse de l'influence de divers paramètres sur la qualité du clinker.

Factores que influyen en la calidad del clínker producido en el horno vertical

Resumen – *En los países en vías de desarrollo, por ejemplo en la India, se encuentran a menudo pequeñas fábricas que emplean el procedimiento de la harina negra y la tecnología del horno vertical. La materia prima de que disponen estas fábricas suele ser de composición heterogénea, de modo que la composición de la harina cruda varía también y debe ser adaptada. Hay que tener en cuenta que el comportamiento del crudo durante el proceso de sinterización en el horno vertical está muy sujeto, por una parte, al influjo de su composición química y mineralógica y, por otra parte, a la finura de molienda, la temperatura y el tiempo de permanencia en el horno. Para llevar a cabo un estudio correspondiente, se han elegido tres crudos industriales, que han sido cocidos en el horno vertical, variando el contenido de cal, la finura de molienda y el tiempo de permanencia dentro del horno. Para poder evaluar la calidad del clínker producido en estas condiciones, se ha determinado el peso del litro de clínker y la cal libre, aplicándose los procedimientos por difracción de rayos X. Se han llevado a cabo estudios morfológicos mediante el microscopio óptico y el microscopio electrónico reticulado. La aptitud a la cocción del crudo ha influido en el perfil de temperaturas del material, repercutiendo así de forma decisiva en la calidad del clínker. Los clínkeres obtenidos a partir del tratamiento del crudo en el horno vertical se distinguen, respecto de su morfología y distribución de fases, notablemente de los clínkeres fabricados en el horno rotatorio. Este breve artículo da un resumen de los estudios referentes al influjo de diferentes parámetros sobre la calidad del clínker.*

1. Einführung

In Indien findet man immer häufiger kleine Zementwerke, die mit der Schachtofen-Technologie und dem Schwarzmehlverfahren arbeiten [1]. Das erweist sich als sehr vorteilhaft, da auf diese Weise auch kleinere Kalklagerstätten genutzt und Verbrauch und Wirtschaftlichkeit auf lokaler Ebene gesteigert werden können. Zur Zeit sind in ganz Indien mehr als 200 solcher Anlagen in Betrieb.

Hinsichtlich des Sinterprozesses und Kühlvorgangs unterscheidet sich die Wärmebehandlung der Rohmehlgranalien im Schachtofen von derjenigen im konventionellen Drehofen. Das Rohmehlverhalten wird während des Sintervorgangs im Schachtofen sehr stark von der chemischen und mineralogischen Zusammensetzung einerseits und von der Mahlfeinheit, Temperatur und Verweilzeit im Ofen andererseits beeinflußt. Somit erfolgt die Klinkerbildung und die Bildung der Klinkerphasen wie C_3S, C_2S, C_3A und C_4AF in äußerst komplexen physikalisch-chemischen Prozessen.

Im Schachtofen werden bei Anwendung des Schwarzmehlverfahrens [2] die Roh- und Brennstoffe (normalerweise Kohle mit geringer Flüchtigkeit) gemeinsam zu einem feinen Pulver trocken vermahlen und granuliert. Diese Granalien werden dann über einen Drehteller dem Ofen aufgegeben. Die Wärmebehandlung im Schachtofen erfolgt in verschiedenen Reaktionszonen, d.h. von oben gesehen in der Trocken-, Calcinier- und Sinterzone, die nicht völlig getrennt sind, sondern ineinander übergehen. Die Kühlzone erstreckt sich über zwei Drittel des Ofens, so daß der Klinker langsam abkühlt. Die von einem Roots-Gebläse gelieferte und von unten in die Kühlzone aufsteigende Verbrennungsluft entzieht der sich abwärts bewegenden Klinkersäule die Wärme. Im Schachtofen wird der Klinker innerhalb von 7 bis 8 Stunden von 1450 °C auf 60 °C abgekühlt [3].

Da die mit der Schachtofen-Technologie arbeitenden Miniwerke normalerweise nicht die Mittel für den Einsatz moderner Abbautechniken im Steinbruch haben, müssen sie Rohmaterial mit unterschiedlicher Zusammensetzung verarbeiten, so daß die Rohmehlzusammensetzung entsprechend variiert. Die Qualität des Klinkers wird von der Rohmischung, der Mahlfeinheit des Rohmehls und der Verweilzeit in den verschiedenen Zonen des Ofens entscheidend beeinflußt. Dieser Beitrag beschreibt die Untersuchungen, die an drei industriellen Rohmehlen mit dem Ziel vorgenommen wurden, die Auswirkung dieser Parameter auf die Klinkerqualität zu ermitteln.

2. Versuche

Um festzustellen, welche Auswirkung kalkreiche und -arme Rohmischungen haben, wurden drei industrielle Rohmischungen ausgewählt, nämlich RM-1 mit sehr hohem Kalkgehalt, RM-2 mit sehr wenig Kalk und RM-3 mit optimalem Kalkgehalt. Diese Rohmischungen wurden in einem

1. Introduction

The popularity of small scale cement plants based on vertical shaft kiln (VSK) technology using black meal process is increasing in India [1]. It has proved highly useful in exploiting small deposits of limestone and helpful in boosting local consumption and economy. At present more than 200 of such plants are in operation throughout India.

Thermal treatment of raw mix nodules both in terms of clinkerisation and cooling is different in VSK from conventional rotary kiln. Raw mix behaviour during its sintering process is greatly influenced by the chemical and mineralogical compositions on one hand and by fineness, temperature and retention time in the kiln on the other hand. This explains the formation of clinker through extremely complex physico-chemical processes and its transformation into phases such as C_3S, C_2S, C_3A and C_4AF.

In VSK using black meal process [2] the raw materials and fuels (usually low volatile coal), are interground dry to a fine powder and nodulized. These nodules are then fed to kiln through a revolving feeder. Thermal treatment in VSK is carried out through various zones of reactions (starting from top of the kiln) — drying zone, calcination zone and sintering zone — which have a very thin dividing line and as such merge together. Two-third portion of the kiln is for cooling zone resulting in slow cooling of clinker. The combustion air supplied by roots blower ascending from below in cooling zone absorbs heat from the descending clinker. The clinker is cooled in vertical shaft kiln from 1450 °C to 60 °C in 7–8 hours [3].

Since VSK mini cement plants cannot normally apply modern techniques for query operations, due to investment limitations, variable raw materials are to be used in the plant and hence raw mix may be variable. The quality of clinker is decisively influenced by (a) raw mix design, (b) fineness of raw meal and (c) retention time in various zones. The present communication deals with the investigations conducted on three industrial raw mixes to study the impact of these parameters on clinker quality.

2. Experimental

To study the effect of underlime and overlime raw mixes, three industrial raw mixes, RM-1 an overlime mix, RM-2 an underlime mix and RM-3 an optimum lime mix, were chosen. These raw mixes were burnt in a VSK with varying fineness and retention time. Clinkers thus obtained, CL-1, CL-2 and CL-3, were studied for their quality by (a) clinker litre weight, (b) free lime determination using Frank extraction method, (c) optical microscopy (on polished clinker section), using Carl Zeiss polarising microscope, (d) SEM (on fractured clinker sample) using Philips CM 301 scanning electron microscope, (e) x-ray diffraction (on powdered clinker sample), using Philips x-ray diffractometer model PW 1120.

3. Results and discussion

3.1 Effect of overlime and underlime raw mix

Raw mix design parameters represented by moduli values for RM-1, RM-2 and RM-3 along with the corresponding clinker quality of the clinkers CL-1, CL-2 and CL-3, represented by the value of free lime and clinker litre weight, are shown in **Table 1** and **2**.

The various mineral phases present in clinker and their relative percentages, as obtained by Bogue's equation and by optical microscopy, are given in **Table 3**. The granulometric results indicate crystal size of alite and belite phases.

Optical micrographs (**Figs. 1, 2** and **3**) give information on the distribution of crystals and porosity created in the clinker due to the presence of intergrated fuels in the raw mix. This is the main reason why clinker obtained from VSK is lighter than that of rotary clinker.

SEM studies on all the three clinkers indicate the porous nature both at macro and micro levels. Alite and belite grains are uniformly distributed and well formed (**Fig. 4, 5** and **6**).

X-ray diffraction studies on clinker indicate the presence of all major clinker phases i.e. C_3S, C_2S, C_4AF and C_3A (**Fig. 7**).

CL-1 and CL-2 are showing the presence of free lime which is characteristic of overlime or underlime clinkers.

Performance characteristics of the corresponding cements, OPC-1, OPC-2 and OPC-3 (**Table 4**) indicate that the physical performance of cement prepared from over- and underlime raw mix (OPC-1 and 2) gives high expansion whereas optimum lime raw mix (OPC-3) is passing all physical tests.

TABLE 1: Modul values of industrial raw mix

Raw mix	Silica modulus	Alumina modulus	Lime saturation factor
RM-1	1.65	1.3	1.01
RM-2	2.2	1.3	0.70
RM-3	2.0	1.3	0.90

TABLE 2: Quality assessment by clinker litre weight and free lime

Clinker	Clinker litre weight, (g/l)	Free lime, wt-%
CL-1	950	6.8
CL-2	1000	4.5
CL-3	1150	1.9

TABLE 3: Mineral phase composition by Bogue's equation and optical microscopy (OM)

	Phases present	Quantity (%) acc. to Bogue	OM	Granulometry (μm) Min.	Max.	Av.
CL-1	C_3S	59.5	55	5	20	15
	C_2S	16.08	20	5	20	18
	C_3A	9.52	20			
	C_4AF	13.00				
CL-2	C_3S	40.76	40	8	30	39
	C_2S	33.22	42	7	42	25
	C_3A	12.31	18			
	C_4AF	10.95				
CL-3	C_3S	52.7	50	5	35	25
	C_2S	18.60	30	5	30	21
	C_3A	9.60	20			
	C_4AF	14.33				

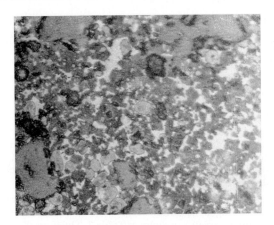

BILD 1: Lichtmikroskopisches Bild eines kalkreichen Klinkers (256fach)
FIGURE 1: Optical micrographs of overlime mix clinker (× 256)

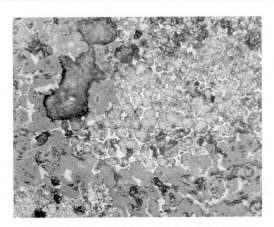

BILD 2: Lichtmikroskopisches Bild eines kalkarmen Klinkers (256fach)
FIGURE 2: Optical micrographs of underlime mix clinker (× 256)

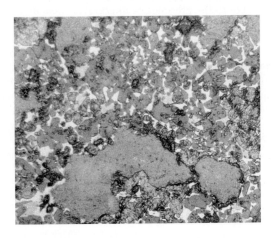

BILD 3: Lichtmikroskopisches Bild eines Klinkers mit optimalem Kalkgehalt (256fach)
FIGURE 3: Optical micrographs of optimum lime mix clinker (× 256)

BILD 4: REM-Aufnahme eines kalkreichen Klinkers
FIGURE 4: SEM of overlime mix clinker

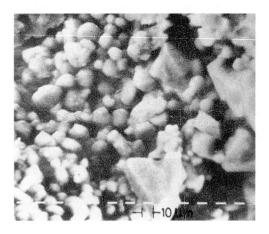

BILD 5: REM-Aufnahme eines kalkarmen Klinkers
FIGURE 5: SEM of underlime mix clinker

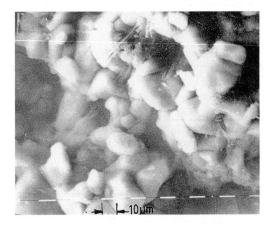

BILD 6: REM-Aufnahme eines Klinkers mit optimalem Kalkgehalt
FIGURE 6: SEM of optimum lime mix clinker

Die rasterelektronenmikroskopische Untersuchung aller drei Klinker zeigt den porösen Gefügeaufbau im Makro- und Mikrobereich. Die Alit- und Belitpartikel sind gleichmäßig verteilt und gut ausgebildet (**Bild 4, 5 und 6**).

Die Röntgendiffraktometrie des Klinkers zeigt, daß alle Hauptklinkerphasen, d.h. C_3S, C_2S, C_4AF und C_3A, vorhanden sind (**Bild 7**).

Die Klinker CL-1 und CL-2 enthalten freien Kalk, wie er für Klinker mit sehr hohem oder sehr niedrigem Kalkgehalt typisch ist.

Overlime mix is characterised with high LSF (RM-1) which promotes the formation of small alite crystals (Fig. 1). An increase in the proportion of lime renders complete combustion more difficult as evident by clinker litre weight and free lime values (Table 2). The cement produced from this clinker showed increase in early strength but failed in autoclave expansion test (Table 4) and had a tendency of slow setting. The size and amount of alite crystals could be increased with high burning temperature or longer retention time in the burning zone due to increased mobility of the reacting ions in liquid phase.

Die Leistungsmerkmale der aus solchem Klinker hergestellten Zemente OPC-1, OPC-2 und OPC-3 (**Tabelle 4**) zeigen, daß Zemente aus sehr kalkreichem oder sehr kalkarmem Klinker (OPC-1 bzw. OPC-2) ein starkes Kalktreiben aufweisen. Der Zement aus dem Klinker mit optimalem Kalkgehalt (OPC-3) besteht demgegenüber alle physikalischen Prüfungen.

TABELLE 4: Physikalische Eigenschaften des Zements

Zement	Festigkeit, kg/cm² Tage			Raumbeständigkeit		Erstarrungszeit	
				Le Chatelier	Autoklav	Beginn	Ende
	3	7	28	(mm)	(%)	(min)	
OPC-1	300	500	520	8	nicht bestanden	90	480
OPC-2	180	235	310	15	nicht bestanden	40	100
OPC-3	230	380	500	1	0,2	75	130

Eine sehr kalkreiche Rohmischung (RM-1) weist einen hohen Kalksättigungsgrad auf, der zur Bildung kleiner Alitkristalle führt (Bild 1). Eine Erhöhung des Kalkgehalts erschwert die komplette Sinterung, was sich im Klinkerlitergewicht und Freikalkgehalt (Tabelle 2) ausdrückt. Der aus diesem Klinker hergestellte Zement zeigte zwar eine höhere Anfangsfestigkeit, bestand aber dem Autoklavversuch nicht (Tabelle 4) und neigte dazu, nur langsam zu erstarren. Die Größe und Menge der Alitkristalle könnte aufgrund einer erhöhten Mobilität der Bestandteile in der Klinkerschmelze nur bei einer hohen Brenntemperatur oder einer längeren Verweilzeit in der Sinterzone gesteigert werden.

Eine sehr kalkarme Mischung zeichnet sich durch einen hohen SiO_2-Gehalt aus, der normalerweise auf tonhaltige Bestandteile zurückzuführen ist. Diese Mischungen mit viel SiO_2, das meist als Quarz vorliegt, erfordern höhere Brenntemperaturen. In solchen Fällen erfolgte trotz Erhöhung des Schmelzphasenanteils keine Granulation, und der im Ofen erzeugte Klinker blieb weitgehend staubförmig. Der Klinker hatte dementsprechend ein niedriges Litergewicht und einen hohen Gehalt an freiem Kalk (Tabelle 2). Die optische Mikroskopie zeigte, daß Belit in großer Menge vorhanden war (**Tabelle 3**). Das Auftreten von Belitanhäufungen war in den optischen Schliffbildern deutlich zu erkennen (Bild 2).

Eine optimal zusammengesetzte Rohmischung enthält CaO, SiO_2, Al_2O_3 und Fe_2O_3 in einem ausgewogenen Verhältnis. Der Kalkgehalt wird vom Tonerdemodul, dem Kalksättigungsgrad und dem Silikatmodul bestimmt. Diese Parameter, die die Klinkerqualität entscheidend beeinflussen, sollten folgende Werte aufweisen:

    Tonerdemodul: 2,0
    Kalksättigungsgrad: 0,85–0,95
    Silikatmodul: 2,0

Wenn diese Werte eingehalten wurden, konnte ein Zement hergestellt werden, der die erforderlichen Phasen enthielt. Unter dem optischen Mikroskop zeigte dieser Klinker eine gleichmäßige Verteilung gut ausgebildeter Alit- und Belitphasen. Der Alit war prismatisch und hatte eine polygonale Form (Bild 3). Die Poren waren entweder von Alit- oder Belitkristallen oder von beiden umgeben.

Im Rasterelektronenmikroskop (**Bild 8**) war gut ausgebildeter Alit mit scharfen Kanten zu erkennen, die durchschnittliche Alitkorngröße lag zwischen 10 und 40 μm. Auch ideal geformte Belitpartikel mit der charakteristischen Streifung konnten nachgewiesen werden. Die durchschnittliche Belitkorngröße war 10–35 μm.

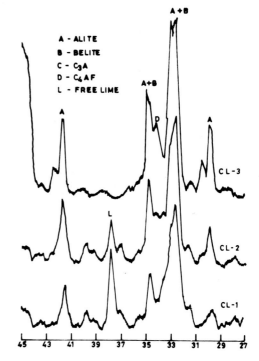

BILD 7: Röntgenbeugungsdiagramme eines Schachtofenklinkers mit den Hauptklinkerphasen
FIGURE 7: X-ray diffractograms of VSK clinkers showing major cement phases

TABLE 4: Physical data of cement

Cement	Strength, kg/cm² (days)			Soundness		Setting time	
				Le Chatelier	Autoclave	I	F
	3	7	28	(mm)	(%)	(min)	
OPC-1	300	500	520	8	failed	90	480
OPC-2	180	235	310	15	failed	40	100
OPC-3	230	380	500	1	0.2	75	130

The underlime mix is characterized with high silica content which is usually coming from clayey portions. These mixes having high content of SiO_2, mostly in the form of quartz, require higher burning temperature. In such cases even on increasing the liquid phase agglomeration did not take place and the kiln produced dusty clinker. Thus the clinker had low clinker litre weight and high free lime (Table 2). Optical microscopy data showed the presence of belite in a large quantity (Table 3). The presence of belite in clusters was clear by optical micrograph (Fig. 2).

For optimum raw mix, the lime content is to be balanced with silica, alumina and iron. The hydraulic modulus, lime saturation factor and silica modulus control the lime content. These values for high clinker quality should be of the following order:

— Hydraulic modulus: 2.0
— Lime saturation factor: 0.85–0.95
— Silica modulus: 2.0

Working in this range a cement was produced with the required amount of phases. Optical microscopy study of this clinker showed even distribution of well developed alite and belite phases. Alite are prismatic and polygonal in shape (Fig. 3). Pores are surrounded by either alite or belite crystals or both.

SEM data (Fig. 8) showed well formed alite with sharp edges, average alite grain size varied between 10–40 μm. Idealy formed belite grains with characteristic striction were also seen. Average belite grain range between 10–35 μm.

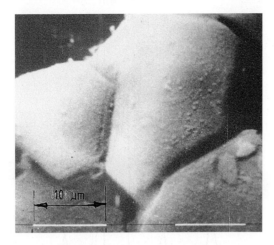

BILD 8: REM-Aufnahmen des optimal zusammengesetzten Klinkers mit gut ausgebildeten Alit- und Belitkristallen

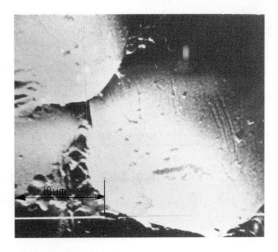

FIGURE 8: SEM showing well developed alite and belite of optimum mix clinker

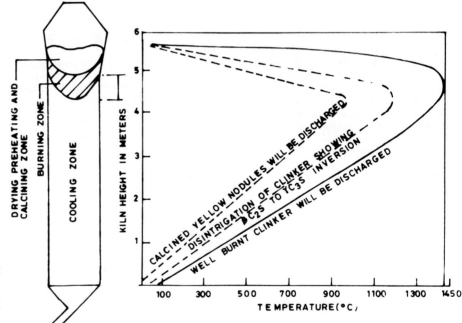

BILD 9: Einfluß des Temperaturprofils auf die Klinkerphasenbildung im Schachtofen
FIGURE 9: Effect of temperature profile on formation of clinker in VSK

3.2 Auswirkung der Temperatur

Das Temperaturprofil des Ofens spielt eine wichtige Rolle bei der Reaktionskinetik der Rohmischung, ebenso wie die Verweilzeit der Granalien in jeder Ofenzone (**Bild 9**).

3.3 Auswirkung der Rohstoff-Feinheit

Bei der Untersuchung der Morphologie und anderer Eigenschaften des Klinkers wurde festgestellt, daß die Feinheit des Rohmehls insgesamt und die Feinheit der einzelnen Rohmehlkomponenten und Brennstoffe eine wichtige Rolle beim Brand spielten. Die Kokszündung und die Kalkentsäuerung im Schwarzmehl traten fast gleichzeitig ein [4, 5]. Wenn der Koks zu fein gemahlen wird, zündet er, ehe die Kalkentsäuerung beginnt. Dadurch treten höhere CO-Verluste auf, die sich in einer erhöhten Abgastemperatur, ungebranntem Material und mehr primärem Freikalk im ausgetragenen Klinker bemerkbar machen.

Wenn hingegen der Brennstoff gröber gemahlen wird, tritt die Kalkentsäuerung zu früh ein, was zu einer verstärkten Verlagerung der Sinterzone nach unten führt. Unverbrannter Restkoks, der die Brennzone durchläuft, verursacht Reduktionen im Klinker. Der Koks der Granalien kann nicht mit dem aus dem Inneren der Granalien entweichen-

3.2 Effect of temperature

The temperature profile of the kiln plays an important role for the reaction kinetics of the raw mix, along with retention time of noduls in each zone (Fig. 9).

3.3 Effect of raw materials fineness

During the course of the studies on the morphology and other properties of clinker, it was observed that the fineness of raw meal as a whole and the individual fineness of raw materials and fuel played an important role in burning. The ignition of coke and decarbonization of limestone in the black meal occured almost simultaneously [4, 5]. If coke is ground too fine, its ignition takes place early followed by decarbonization of limestone. This results in more CO_2 losses, which is indicated by higher waste gas temperature, presence of unburnt material and more primary free lime in the discharged clinker.

On the other hand, if fuel is ground coarser decarbonation of limestone takes place early resulting in increasing sintering zone downwards. A portion of residual coke passing the burning zone results in reducing the clinker. The coke of the nodules cannot react with the CO_2 escaping from the core of the nodules to form CO and again react with kiln air (oxygen)

den CO_2 reagieren, um CO zu bilden und dann wieder mit der Verbrennungsluft (Sauerstoff) zu CO_2 zu reagieren. Im Inneren der Granalien steht kein CO_2 mehr zur Verfügung, da die Entsäuerung bereits stattgefunden hat. Der Koks reagiert daher mit Sauerstoff aus dem Eisenoxid und reduziert es zu einer niedrigeren Oxidationsstufe oder sogar zu metallischem Eisen. Reduziertes Eisenoxid bildet nicht die gewünschten Klinkerphasen, sondern Mischkristalle mit einem tieferen Schmelzpunkt, der die Tendenz zur Bildung großer Schollen hat. Das wird durch einen braunen Kern im Innern erkennbar, der auf einen niedrigen Oxidationsgrad bzw. auf metallisches Eisen hinweist [6]. Unter solchen Bedingungen steigt auch der primäre Freikalkgehalt im Klinker, und die Phasenbildung verläuft nicht normal. Die erwünschte Feinheit von Koks und Rohstoffen ist erreicht, wenn die Mahlfeinheit des Schwarzmehls einen Rückstand von 6–7 Gew.-% auf dem 90-μm-Maschensieb aufweist.

4. Schlußfolgerungen

Die Faktoren, die zur Produktion von Zement mit einer bestmöglichen Qualität führen, sind ein optimaler Kalkgehalt und Feinheitsgrad der einzelnen Rohstoffkomponenten. Diese Parameter lassen sich leicht kontrollieren. Wenn der Kalkgehalt in der Rohmischung vor dem Brennen eingestellt und das richtige Temperaturprofil im Ofen während des Brennens aufrechterhalten wird, entsteht ein Klinker mit gut ausgebildeten Alit- und Belitphasen.

Danksagung

Die Autoren danken dem Generaldirektor des National Council for Cement and Building Materials, New Delhi, für die Genehmigung zur Veröffentlichung dieses Beitrags.

to form CO_2. In the granule centre CO_2 will not be available any more as decarbonization has already taken place. Coke takes up oxygen from iron oxide and reduces it to lower oxidation state or even to metallic iron. Reduced iron oxide does not form desired clinker minerals but forms solid solution of lower melting point which has the tendency to form big lumps. This is indicated by the presence of brown core in the centre and low degree of oxidation, as an indication of metallic iron [6]. Under such conditions also primary free lime content in clinker will increase and phase formation will not be proper. The individual fineness of coke and raw materials could be achieved by maintaining a black meal fineness within the range of 6–7 wt-% retaining on 90 μm sieve.

4. Conclusion

Facts influencing the production of cement having the most desirable qualities are the optimum lime content and fineness of various raw materials. These parameters can easily be controlled. Adjusting the lime content in raw mix before burning, maintaining the proper temperature profile in the kiln during burning, a clinker having well developed alite and belite phases is formed.

Acknowledgement

Authors are grateful to the Director General of National Council for Cement and Building Materials, New Delhi, for permission to publish this paper.

Literature

[1] S p e n c e, R. J. S.: Small scale production of cementitious materials, Published by Intermediate Technology Publication Ltd, 1980.

[2] H a i s a s h i   K o n o: An analysis of the reaction zone of shaft kilns, Kakagu Koguku, 29 (2), 101–106, 1965.

[3] J a i n, V. K., R a o, D. B. N., G u p t a, S. K., and S i n g h, D. K.: Technology of mini cement plant, National Seminar of Mini Cement Plants, New Delhi, 41–53, 1982.

[4] S p o h n, E.: Cement Shaft Kiln of Today, Zement-Kalk-Gips 11 (7), 285–290, 1958.

[5] N a r j e s, A.: What is the present position of Shaft Kiln, Zement-Kalk-Gips (9), 409–418, 1960.

[6] G e o r g e   S a m u e l, S u b b a   R a o, V. V., and C h o p r a, S. K.: Influences of the burning process on the morphology and microstructure of Indian Cement Clinkers, Proceedings of the Sixth International Conference on Cement Microscopy, Texas, USA, (20) 1984.

Auswirkungen von unterschiedlichen prozeßtechnischen Parametern und des Energieeinsatzes auf die Qualität von Klinker und Zement der gleichen Rohstoffbasis

Impact of different process-operational parameters and energy inputs on the quality of commercial clinker and cement from the same raw materials

Impact de divers paramètres de process et de l'utilisation d'énergie sur la qualité du clinker de même matière première

Repercusiones de diferentes parámetros del proceso y el empleo de energía sobre la calidad del clínker y del cemento fabricados con la misma materia prima

Von **G. V. Rao, S. P. Ghosh** und **K. Mohan,** New Delhi/IND

Zusammenfassung – *Die Qualität marktüblicher Klinker und Zemente, die aus den gleichen Rohstoffen, Brennstoffen und Gipsen hergestellt, in Naß- und Trockenöfen mit Zyklonvorwärmer und Vorcalcinieranlagen unterschiedlicher Leistungsfähigkeit gebrannt und mittels verschiedener Systeme gemahlen und homogenisiert wurden, wurde in Abhängigkeit von der Kapazitätsauslastung und vom Energieverbrauch der jeweiligen Produktionseinheit verglichen. Der Vergleich der Rohstoff-Mahlung bei einer Naßofenanlage von 600 t/d mit einer nach dem Trockenverfahren arbeitenden und mit Vorcalcinator ausgerüsteten Anlage mit einem Tagesdurchsatz von 3000 t zeigt, daß die Vertikal-Rollenmühle ein viel gröberes Ausgangsmaterial bei gleichem Energieverbauch, aber 10% weniger Mühlenkapazitätsauslastung zu einem bedeutend feineren Produkt vermahlt als die Naßmühle. Obgleich der nach dem Naßverfahren hergestellte Klinker mit 30% höherem Brennstoffenergieverbrauch eine bessere Klinkergranulometrie und einen geringeren freien Kalkgehalt trotz einer um 50 bis 100°C niedrigeren Sinterzonentemperatur aufweist, hatte der normale Portlandzement (OPC), der aus einem nach dem Trockenverfahren erzeugten Klinker mit 7,5% niedrigerem Mahlenergieverbrauch und 10% geringerer Mahlfeinheit hergestellt wurde, eine um 5 – 10% höhere Druckfestigkeit nach unterschiedlich langer Erhärtung. Ein ähnlicher Vergleich zwischen einem Zyklonvorwärmer-Ofen mit 1200 t/d und einer Vorcalcinieranlage mit 3000 t/d zeigte, daß die Rohmehlmahlanlage mit geschlossenem Kreislauf der ersten Anlage 5 – 10% mehr Energie benötigte als die Vertikal-Rollenmühle der zweiten Anlage, um die gleiche Produktfeinheit bei gleicher Mühlenkapazitätsauslastung zu erreichen. Ein besseres Druck-Temperaturprofil in den Zyklonen und ein höherer Calciniergrad führten zu der vermuteten besseren Klinkerqualität der zweiten Anlage mit einem 10% niedrigeren Brennstoffenergieverbrauch. Die Produktion eines besseren Portlandzements aus diesem Klinker, der in Kugelmühlen mit einer Kapazität, die fast derjenigen der Anlage 1 entsprach, hergestellt worden war, führte bei vergleichbar hoher Mörtelfestigkeit nicht nur zu einer 20%igen Verringerung des Erstarrungsbeginns und -endes und zu einer um 40 – 50% stärkeren Dehnung, sondern erforderte auch eine 10 – 15% feinere Mahlung mit 5% höherem Energieverbrauch. Hintergrund dieser Wechselwirkungen sind Unterschiede in den chemischen, physikalischen und mineralogischen Eigenschaften und in der Brennbarkeit der Rohstoffe sowie in den prozeßtechnischen Parametern.*

Summary – *The quality of commercial clinkers and cement, manufactured from identical raw materials, fuel and gypsum, burnt in wet and dry kilns with SP and precalciner of different capacities, using different systems for size reduction and homogenization has been compared with respect to capacity utilization and energy consumption under each unit operation. The comparison of the raw feed grinding process between a 600 t/d wet process and a 3000 t/d dry process precalciner plant shows that the vertical roller mill (VRM) grinds a much coarser feed to a significantly finer product with the same energy consumption but 10% less mill capacity utilization than the wet raw mill. Although the clinker from a wet process with a 30% higher fuel energy consumption shows better clinker granulometry and smaller free lime content despite a 50 to 100°C lower burning zone temperature, the OPC produced from the dry process clinker with 7.5% lower*

grinding energy consumption and 10% lower fineness shows 5 to 10% higher compressive strength at different ages. Similar analysis between a 1200 t/d SP and 3000 t/d-precalciner plant showed that the closed circuit raw mill of the first plant has a 5 to 10% higher energy consumption than the VRM in the second plant for attaining the same product fineness with the same mill capacity utilization. A better P-T profile and degree of calcination in the cyclones ensured a supposed higher quality of the clinker in the second plant with 10% less fuel energy consumption. Better production of OPC from this clinker made in ball mills of nearly identical capacity like in plant 1 led, at comparable high mortar strength, not only to a 20% loss of initial and final setting time and a 40 to 50% higher expansion but also required a 10 to 15% finer grinding with 5% more energy consumption. The background for these interactions are differences in chemical, physical and mineralogical properties and in the burnability of raw materials as well as in the different process-operational parameters.

Impact de divers paramètres de process et de l'utilisation d'énergie sur la qualité du clinker de même matière première

Résumé – La qualité de clinkers et ciments courants sur le marché – fabriqués à partir de matières premières, combustibles et gypses semblables, cuits dans des fours par voie humide et par voie sèche avec préchauffeur à cyclones et installations de précalcination de différente capacité et broyés et homogénéisés selon divers systèmes – a été comparée en fonction du degré d'utilisation de la capacité et de la consommation d'énergie de l'unité de production concernée. D'une étude comparative du broyage de matière première dans un four à voie humide de 600 t/j avec une installation – équipée d'un précalcinateur et travaillant selon le procédé par voie sèche – d'une production journalière de 3000 t, il ressort que le broyeur à galets vertical broie une matière de départ beaucoup plus grossière que le broyeur à voie humide avec la même consommation d'énergie, mais 10% en moins d'utilisation de capacité du broyeur avec un produit sensiblement plus fin. Et bien que le clinker fabriqué selon le procédé humide avec 30% de plus d'énergie de combustible, accuse une meilleure granulométrie et une teneur en chaux libre plus faible malgré une température de zone de cuisson de 50 à 100°C en moins, le ciment de Portland courant (OPC), fabriqué à partir d'un clinker fabriqué selon le procédé sec, avec une consommation d'énergie de broyage diminuée de 7,5% et une réduction de la finesse de mouture de 10%, présentait une résistance à la pression plus élevée de 5 à 10% selon une prise de durée variable. Une comparaison semblable entre un four avec préchauffeur à cyclones de 1200 t/j et une installation de précalcination de 3000 t/j, a prouvé que l'installation de broyage de cru en circuit fermé de la première installation, a dépensé 5 à 10% d'énergie en plus que le broyeur à galets vertical de la deuxième installation, pour obtenir la même finesse de produit avec un degré d'utilisation du broyer identique. L'amélioration du profil thermique de pression dans les cyclones et le degré de calcination plus poussé ont entraîné une meilleure qualité du clinker supposée de la deuxième installation avec une diminution de 10% de la consommation d'énergie. La production de ciment de Portland à partir de ce clinker fabriqué dans des broyeurs à boulets d'une capacité correspondant presque à celles de l'installation 1, a, avec une résistance de mortier élevée comparable, non seulement entraîné une diminution de 20% du début et de la fin de prise et une augmentation de la dilatation de 40 à 50% mais a nécessité également un broyage plus fin de 10 à 15% avec consommation d'énergie accrue de 5%. A noter qu'en toile de fond de ces changements d'effets il faut voir les différences de propriétés chimiques, physiques et minéralogiques, la combustibilité des matières premières et, enfin, les paramètres du process.

Repercusiones de diferentes parámetros del proceso y el empleo de energía sobre la calidad del clínker y del cemento fabricados con la misma materia prima

Resumen – Se ha comparado la calidad de cementos y clínkeres corrientes en el mercado, fabricados con la misma materia prima, combustibles y yesos, cocidos en hornos de vía seca y de vía húmeda, con precalentador de ciclones e instalaciones de precalcinación, de diferente capacidad de producción y molidos y homogeneizados por medio de diferentes sistemas. Esta comparación se ha hecho en función del grado de utilización de la capacidad y del consumo de energía de cada unidad de producción. La comparación de la molienda de materias primas en el caso de una planta de horno de vía seca, de 600 t/d, con una instalación de vía seca, equipada con precalcinador, de 3000 t/d, muestra que el molino de rodillos vertical trabaja con un material de alimentación mucho más grueso, con el mismo consumo de energía, pero con un grado de utilización de la capacidad del molino un 10 % menor, consiguiendo un producto mucho más fino que el molino de vía húmeda. Aunque el clínker fabricado por vía húmeda, con un consumo de combustible que resulta un 30 % mayor, presenta una mejor granulometría y un menor contenido de cal libre, a pesar de que la temperatura de la zona de sinterización es un 50–100 °C más baja, el cemento Portland normal (OPC), fabricado con un clínker obtenido por vía seca, con un consumo de energía de molienda más bajo (–7,5 %) y una menor finura de molienda (–10 %), tenía una reistencia a la compresión un 5 a 10 % mayor, según el tiempo de endurecimiento. Una comparación similar entre un honro con precalentador de ciclones, de 1200 t/d, y una planta de precalcinación, de 3000 t/d, dio como resultado que la planta de molienda de crudo, de circuito cerrado, de la primera instalación necesitaba más energía (+ 5–10 %) que el molino de rodillos vertical de la segunda instalación, para lograr la misma finura del producto, con el mismo aprovechamiento de la capacidad del molino. Un mejor perfil de presión y temperatura dentro de los ciclones y un mayor grado de calcinación han permitido, tal como se esperaba, obtener un clínker de mejor calidad en la segunta instala-

ción, con un menor consumo de energía (–10 % de combustible). La producción de un cemento Portland de mejor calidad a partir de este clínker, obtenido en molinos de bolas, cuya capacidad equivalía casi a la de la primera instalación, dio por resultado no solamente la consecución de un mortero de elevada resistencia en ambos casos y una disminución del comienzo y del fin del fraguado (–20 %) y una mayor dilatación (+ 40–50 %), sino que exigía también una molienda un 10–15 % más fina, con un consumo de energía mayor (+ 5 %). Las causas de estas acciones recíprocas residen en las diferencias existentes en cuanto a las propiedades químicas, físicas y mineralógicas así como en la aptitud a la cocción de las materias primas y en los diferentes parámetros tecnológicos.

1. Einführung

Die Wechselwirkung zwischen den wesentlichen Eigenschaften von Rohstoffen und Brennstoff einerseits und den verfahrens- und betriebstechnischen Parametern andererseits hat in jedem Fall eine typische Auswirkung auf die Qualität der Endprodukte Klinker und Zement [1–6]. Außerdem führen die jeweiligen spezifischen Merkmale der Betriebsweise und Verfahrenssteuerung sehr häufig zu unterschiedlicher Qualität von Klinkern, die aus grundsätzlich ähnlichen Rohstoffen hergestellt wurden und umgekehrt [7–10]. Die vorliegende Untersuchung analysiert den Einfluß von Kalkstein identischer Provenienz, der chemischen Zusammensetzung und ähnlicher Zusatz- und Korrekturstoffe bei unterschiedlichen prozeß- und betriebstechnischen Parametern und versucht, diese mit der Klinker- und Zementqualität zu korrelieren.

Die über mehrere Jahre gewonnenen Betriebsergebnisse von Naß- und Halbtrockenöfen, Trockenöfen mit Zyklonvorwärmer und Vorcalcinieranlagen in verschiedenen Teilen Indiens liefern eine Fülle von Daten für die Untersuchung der „Ursache-Wirkungs"-Beziehung zwischen den Klinkereigenschaften und den prozeß- und betriebstechnischen Maßnahmen, die mit dem Ziel durchgeführt wurde, die beobachteten deutlichen Unterschiede in der Klinker- und Zementqualität zu erklären. Von den möglichen Kombinationen vergleichbarer Werke beschränkt sich die vorliegende Arbeit auf den Vergleich einer Naßofenanlage von 600 t/d (A 1) und einer in der Nähe gelegenen, mit Vorcalcinator ausgerüsteten Anlage mit einem Tagesdurchsatz von 3 000 t (A 2) in Südindien, die beide Rohstoffe derselben Provenienz verwenden, und zwischen einer Trockenofenanlage mit Zyklon-Vorwärmern mit einem Tagesdurchsatz von 1200 t (B 1) und einer mit Vorcalcinator ausgerüsteten Anlage, die einen Tagesdurchsatz von 3 000 t (B 2) hat und in derselben Region (Nordindien) liegt; beide Anlagen arbeiten mit dem gleichen Roh- und Brennstoff. Den Standort der Werke zeigt **Bild 1**.

2. Rohstoff und Ausrüstung der Anlagen

Außer Werk A 2, das kalkhaltigen Schieferton als Zusatzstoff verwendet, arbeiten alle Werke mit mittelhartem Kalkstein als Hauptkomponente bei einem Anteil von etwa 98,0 bis 98,5 Gew.-% sowie eisenhaltigen Korrekturstoffen, wie Eisenerz oder Hämatit, und Tonen unterschiedlicher Zusammensetzung. Das nach dem Naßverfahren arbeitende Werk mit 600 t/d (A 1) verwendet Kalkstein mit 97,0 bis 97,5 Gew.-% $CaCO_3$ und Laterit. Das Werk A 2 mit einem Tagesdurchsatz von 3 000 t verwendet Kalkstein mit 88 bis 90 Gew.-% $CaCO_3$, kalkhaltigen Schieferton (8–10 Gew.-%) und Laterit. Sowohl die Anlage mit einem Tagesdurchsatz von 1200 t (B 1) als auch diejenige mit einem Tagesdurchsatz von 3 000 t (B 2) im nördlichen Teil Indiens verwendet neben Laterit Kalkstein mit 97,0 bis 97,5 Gew.-% $CaCO_3$ aus demselben Steinbruch. Laterit ist ein in tropischen Ländern vorkommendes verwittertes Gestein und enthält hauptsächlich SiO_2, Al_2O_3 und Fe_2O_3 in unterschiedlichen Mengen. Der Kalkstein ist ein verhältnismäßig kieselsäurereiches Material mit etwa 9 bis 13 Gew.-% SiO_2, das hauptsächlich als Quarz (Alpha-Quarz) vorliegt. Die eingesetzte bituminöse Kohle hat einen Heizwert von 4 000 ± 500 kcal/kg (16 750 ± 2 100 kJ/kg).

1. Introduction

The interaction between the intrinsic properties of raw materials and fuel on the one hand and the process and operational control parameters on the other hand leaves in each case a characteristical impact on the quality of the end product clinker and cement [1–6]. Furthermore, the specific characteristics of operation and process control in individual cases very often lead to different qualities of clinker obtained from basically similar raw material ingredients and vice versa [7–10]. The present study analyses the behaviourial pattern of limestones of same sources, the chemical composition, similar types of additives and correctives under different processes and plant operational parameters, and attempts to link them with the clinker and cement quality.

Operational results of wet and semi-dry kilns, dry kilns with suspension preheaters and precalcinator plants in different parts of India for several consecutive years provide considerable information for the study of "Cause-Effect" relationships between the clinker properties and the process technological and operational measures in order to explain the observed major differences in clinker and cement quality. Among several combinations of comparable plants this article will be confined to the comparison between a 600 t/d wet process (A1) and a 3 000 t/d precalcinator plant (A2) nearby in South India, both using raw materials from same sources, and between a 1200 t/d dry process plant with suspension preheaters (B1) and a 3 000 t/d precalcinator plant (B2) in the same premises (North India), both using the same raw material and fuel. The location of the plants is shown in **Fig. 1**.

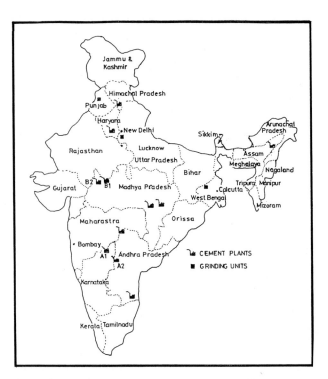

BILD 1: Landkarte Indiens mit den Standorten der Werke
FIGURE 1: Index map of India showing plant locations

Zementwerke = cement plants
Mahlanlagen = grinding units

2. Raw material and plant equipment

All the plants, except A2 using calcareous shale as additive, use medium hard limestone as the main component in a portion of about 98.0 to 98.5 wt.-% with ferruginous correctives like iron ore or hematite besides clays of different composition. The 600 t/d wet process plant (A1) uses limestone with 97 to 97.5 wt.-% $CaCO_3$ and laterite. The 3000 t/d plant (A2) uses limestone with 88 to 90 wt.-% $CaCO_3$, calcareous shale (8–10 wt.-%) and laterite. Both the 1200 t/d (B1) and 3000 t/d (B2) plants in the northern parts of India use limestone with 97 to 97.5 wt.-% $CaCO_3$ from the same quarry, and laterite. Laterite is a typical weathered rock found in tropical countries and mainly contains silica, alumina and iron oxide in varying amounts. The limestone is a fairly siliceous material of about 9 to 13 wt.-% SiO_2 with predominating free quartz (α-quartz). The employed bituminous coal has a calorific value of 4000 ± 500 kcal/kg (16750 ± 2100 kJ/kg).

The plant equipments for the 3000 t/d units A2 and B2 comprise impact hammer crushers, vertical roller mills for the raw material grinding, a twin steam 4-stage suspension preheater (SP) with fluidized bed calciner (A2), a single stream 4-stage suspension preheater with a new flash furnace calcinator (B2), a ball mill (A2) and a vertical roller mill (B2) for coal grinding and open circuit ball mills for cement grinding.

The 600 t/d wet (A1) and the 1200 t/d (B1) plant comprise jaw crushers, wet (A1) and closed circuit (B1) ball mills for raw meal grinding, a 4-stage suspension preheater kiln (B1), air swept mills for coal grinding, open circuit (A1) and closed circuit (B1) ball mills for cement grinding. The cement produced from all the plants conforms to the Indian standard IS 8112-1989 for 43-grade OPC with a 28 d compressive strength exceeding 43 MPa (N/mm²).

3. Comparative assessment

Some of the most important data of the four plants are presented in the **Tables 1** and **2**. Despite the lowest silica and free silica content, the limestone of plant A2 gives the most coarse crusher output, presumably due to its

— more compact structure on account of cementing the major mineral grains in contrast to matrix filling in other limestones and due to the

— coarser grain size of the α-quartz.

This limestone therefore has the highest grindability index according to Bond. The vertical roller mill of plant A2 produces a much coarser feed compared with the wet mill grinding of plant A1. With nearly the same fineness of the raw feed the vertical roller mill needs about 15% more electric power (kWh/t) and simultaneously shows a 15% less capacity utilization.

During kiln operation the raw feed of the wet kiln (A1) needs about 50 K less sintering temperature (1375°C ± 25) but the clinker has a longer residence time as compared to the material flow in the 4-stage suspensive preheater and precalciner kiln (1425°C ± 25) of plant A2. Additionally the A1 kiln feed leads to a more homogeneous granulometry of the clinker containing the same CaO and SiO_2 but a higher flux portion than the A2 feed [11–13]. Smaller crystal sizes of alite and belite in the clinker of A1 plant explain its better grindability compared with A2 plant clinker, both plants using open circuit mills (OC). With almost the same mineral phase composition the A1 clinker is ground to a 10% finer surface with 1 kWh/t more electric energy. A more homogeneous microstructure and higher fineness of the cement from A1 clinker accounts for its 5 to 10% less setting time and a 10 to 15% higher compressive strength at all ages.

The comparison between the plants B1 and B2, using the same raw materials and fuel, shows that the operational results are the same with respect to the raw meal grinding stage, although the closed circuit ball mill of plant B1 consumes about 5% more energy than the vertical roller mill of plant B2. Modern longitudinal blending facilities at the B2 plant ensure more uniform quality of raw meal and coal. In kiln operation a better pressure and temperature profile and

TABLE 1: Physico-mineralogic, granulometric, communition & dissociation characteristics of raw materials and fuel characteristics

SN	Parameters studied		600 TPD – WET (A1)	3000 TPD – PC (A2)	1200 TPD – SP (B1)	3000 TPD – PC (B2)
1	2		3	4	5	6
1	Limestone					
a.	Physical Nature		Medium Hard, Slabby	Hard to Medium, Slabby	Medium Hard, Compact	
b.	Compact Strength kg/sq cm		435 ± 50	550 ± 25	500 ± 25	
c.	Mineralogy Calcite:Quartz		10 ± 3	12.5 ± 5	5 ± 2	
d.	Texture-Major Minerals Bonded by		Matrix of Iron-Ore and Chlorite	Cemented by Clay Minerals	Matrix of Iron-Ore, Illite, Muscovite	
e.	Grain Size of Major Minerals (mm)		Calcite Quartz Max 0.09 0.24 Min 0.03 0.04 Avg 0.06 0.08	Calcite Quartz Max 0.08 0.24 Min 0.01 0.05 Avg 0.06 0.18	Calcite Quartz Max 0.30 0.25 Min 0.10 0.08 Avg 0.15 0.12	
f.	Minor (Accessory) Minerals and Contents (%)		Iron-Ore & Chlorite (5–12)	Clay Minerals (5–12)	Hematite, Illite & Muscovite (8–10)	
g.	Chem. Comp (%) Avg. and Range		SiO_2 10.5 ± 1.5 R_2O_3 3.0 ± 1.0 CaO 47.5 ± 1.0	SiO_2 9.0 ± 1.5 R_2O_3 3.3 ± 0.5 CaO 48.0 ± 1.0	SiO_2 12.5 ± 1.0 R_2O_3 3.8 ± 1.0 CaO 44.8 ± 1.0	
h.	Free Silica (% of Total Silica)		70.0 ± 5.0	82.0 ± 5.0	70.0 ± 5.0	
i.	Bond's Work Index Range (kWh/sh. t)		10.6–12.0	11.0–13.5	9.0–10.5	
2	Limestone crushing					
a.	Crusher Type and Capacity		Jaw (Primary)] Double Roll] 200 Hammer (Sec)] TPH	Impact Hammer Mill 600 TPH	Double Toggle Jaw 400 TPH Rev. Imp. Hammer-250 TPH	Impactor (Blow Bar) Rotor-1000 TPH
b.	Crusher Feed Size Maximum (mm)		800 x 750 x 750 (Primary) 200 (Secondary)	1400 x 1400 x 1600	840 x 840 x 1050 (Primary) 260 (Secondary)	1600 x 1200 x 1000
c.	Crusher Output (final) Size %		1.8 ± 0.5 of + 25 mm	16.0 ± 4.0 of + 25 mm	12.0 ± 3.0 of + 25 mm	12.0 ± 3.0 of + 25 mm
3	Raw mill					
a.	Type and Capacity		Ball Mill (Open Circuit) 55 TPH	Vertical Roller Mill 265 TPH	Ball Mill (Closed Circuit) 125 TPH	Vertical Roller Mill 310 TPH
b.	Product Size (% on + 72 & 170 mesh)		+72 +170 3.7 ± 0.5 19.0 ± 1.0	+72 +170 0.7 ± 0.25 20.0 ± 1.0	+72 +170 1.8 ± 0.2 17.5 ± 1.0	+72 +170 1.6 ± 0.5 18.0 ± 1.0
c.	Grinding Energy Consumption (kWh/t)		20.25 ± 0.5	23.50 ± 0.5	22.80 ± 0.5	22.00 ± 0.5
d.	Mill Capacity Utilisation (%)		105–110	92–95	95–98	95–100
4.	Minor Constituents in Limestone and Additives (%)	Na_2O K_2O SO_3	Limestone Laterite 0.06–0.10 0.07–0.11 0.15–0.25 0.10–0.13 — —	Limestone Shale Laterite 0.06–0.08 0.25–0.33 0.25–0.39 0.19–0.21 1.50–1.70 0.13–0.17 0.12–0.14 0.06–0.10 —		Limestone Laterite 0.03–0.06 — 0.18–0.25 — — —
5.	Limestone Thermal Dissociation, Range, Degree cent.		780–1000 (Peak 870)	740–915 (Peak 880)	685–870 (Peak 825)	
6.	Kiln Feed Composition (%)	SiO_2 R_2O_3 CaO	10.5 ± 0.5 6.5 ± 0.5 45.5 ± 0.5	11.5 ± 0.5 5.0 ± 0.5 44.5 ± 0.5	13.0 ± 0.5 6.0 ± 0.5 44.5 ± 0.5	12.0 ± 0.5 5.0 ± 0.5 44.5 ± 0.5
7.	Coal					
a.	Proximate Analysis (%)	VM FC Ash	30 ± 5 30 ± 5 37 ± 5	28 ± 3 30 ± 3 37 ± 2	22 ± 2 41 ± 2 32 ± 2	22 ± 2 41 ± 2 30 ± 2
b.	Calorific Value kcal/kg		3600 ± 250	3650 ± 250	4400 ± 150	4500 ± 100
c.	Pulv. Coal Fineness Residue (%) on	+72 +170	2.5 ± 0.5 24 ± 2	Precalcinator Kiln 5.0–10.0 2.0– 5.0 30.0–40.0 20.0–30.0	1.6– 2.0 16.0–19.0	1.0– 2.0 16.0–20.0
d.	Ash Absorption (%)		9–10	8.5– 9 8– 9	8–8.5	
e.	Coal Ash Composition (%)	SiO_2 Al_2O_3 Fe_2O_3 CaO	59.0 ± 1.5 25.5 ± 3.0 8.0 ± 0.4 4.0 ± 0.2	61.0 ± 2.0 56.0 ± 3.0 25.0 ± 3.0 28.5 ± 3.0 6.0 ± 2.0 4.5 ± 2.0 4.5 ± 1.0 Traces	55.0 ± 3.0 28.0 ± 3.0 4.0 ± 2.0 Traces	
f.	Minor Constituents		\multicolumn{4}{c}{Ash from this category (Grade – D) Bituminous Coal contains: Na_2O 0.3–0.5 %, K_2O 0.5–1.0 %, SO_3 3.2–4.1 % & Cl 0.009–0.013 %}			

TABLE 2: Process and operational parameters in clinkerization, clinker quality and cement properties

SN	Parameters Studied			Plant and Process							
				600 TPD WET (A1)		3000 TPD-PC (A2)		1200 TPD-SP (B1)		3000 TPD-PC (B2)	
1	2			3		4		5		6	
1	Draft (mmWg) &			Pressure	Temp	Pressure	Temp	Pressure	Temp	Pressure	Temp
	temp (deg. C)		Cyclone I	–	–	450–500	390–420	650	350–400	450–460	500–600
	in suspension		Cyclone II	–	–	275–320	650–690	400–450	540–580	250–280	600–710
	preheater,		Cyclone III	–	–	200–260	760–800	250–280	680–720	200–210	700–800
	precalcinator		Cyclone IV	–	–	105–145	800–880	120–180	850–880	80–100	840–850
	& kiln		Precalcinator	–	–						
			Riser Duct	–	–	35–40	880–900			880–950	
			Kiln Inlet	60	185–200	15–25	960–1000	20–30	960–1050	40–60	975–1000
2	Burning Zone Temp (deg. C)			1375 ± 25		1425 ± 25		1450 ± 25		1425 ± 25	
3	Degree of Calcination (After SP/PC)			–		90 ± 5		35 ± 3		90 ± 5	
4	Heat Cons. K. cal/kg Clinker			1250 ± 25		875 ± 25		950 ± 25		925 ± 25	
5	Kiln (A) L:D			36.8		14.8		14.7		14.6	
	(B) Opt. RPM			0.9–1.0		2.8–3.0		1.5–1.6		2.8–3.0	
6	Cooler (A) Type			Recupol		Reciprocating Grate		Grate (Folax)		Reciprocating Grate	
	(B) Clinker Exit. Deg C			100		80–100		80–100		80–100	
7	Clinker (A) Liter Wt (gm)			1200		1125 = 25		1175 = 25		1300 = 25	
	(B) Clinker Granulometry %]		+25.0 mm	3–5		9–15		10–15		13–15	
			+12.5 mm	10–15		20–50		20–25		30–45	
	(C) Chem Comptn %		SiO_2	22.0 ± 1.0		22.0 ± 1.0		22.75 ± 0.75		22.50 ± 0.5	
			Al_2O_3	6.8 ± 0.5		6.6 ± 0.5		5.50 ± 0.50		5.50 ± 0.5	
			Fe_2O_3	4.5 ± 0.5		3.5 ± 0.5		4.50 ± 0.5		4.00 ± 0.5	
			CaO	65.5 ± 0.5		65.5 ± 0.5		64.00 ± 0.5		64.50 ± 0.5	
	(D) Mineral Phases (Microscopy & XRD)		C_3S	40 ± 10		42 ± 5		40 ± 5		37 ± 3	
			C_2S	30 ± 10		30 ± 5		27 ± 5		35 ± 5	
			C_3A	10 ± 2		10 ± 2		8 ± 2		8 ± 2	
			C_4AF	14 ± 2		12 ± 2		13 ± 2		12 ± 2	
8	Gypsum (A) Type			Chemical		Chemical		Mineral		Mineral	
	(B) Purity			85 %		86–90 %		70–74 %		70–74 %	
	(C) % ADN			5		4		5–6		5–6	
9	Cement Grinding (A) Mill Type Cap.			Ball Mill 35 TPH (OC)		Ball Mill 100 TPH (OC)		Ball Mill 80 TPH (OC)		Ball Mill 100 TPH (OC)	
	(B) Energy Cons. (kWh/t)			33.7 ± 1		34.8 ± 1		32.2 ± 1		32.2 ± 1	
	(C) Fineness sq cm/gm			3100 ± 300		2850 ± 200		2600 ± 200		3000 ± 100	
	(D) Expansion LC (mm)			1.0		1.4–2.5		0.5–2.5		1.0–3.5	
	Autocl. (%)			0.04–0.08		0.1–0.3		0.05–0.26		0.2–0.4	
	(E) Setting Time Initial (min)			50–100		70–90		90–100		50–75	
	Final			120–200		150–180		155–200		105–135	
	(F) Compressive Strength (kg/sq cm)		3 Days	290 ± 25		270 ± 50		230 ± 25		250 ± 25	
			7 Days	400 ± 25		350 ± 50		330 ± 25		330 ± 50	
			28 Days	520 ± 50		500 ± 25		450 ± 50		470 ± 50	

Jedoch erleichtern ein um 10 % höherer Alitgehalt und eine bessere Mikrohomogenität die Mahlbarkeit des Klinkers aus dem Werk B 2. Das Werk B 1 verfügt über Durchlaufmahlanlagen mit einer Leistung von 80 t/h und das Werk B 2 über Durchlaufmahlanlagen mit einer Leistung von 100 t/h, bezogen auf einen Zement mit einer durchschnittlichen Mahlfeinheit (Blaine) von 2600 cm²/g bzw. 3000 cm²/g. Die feinere Mahlung des Klinkers der Anlage B 2, der den gleichen C_3A-Gehalt aufweist und die gleiche Menge an Naturgips enthält wie der Klinker der Anlage B 1, erklärt die kürzere Erstarrungszeit des Zements und eine um 10 bzw. 5 % höhere Normdruckfestigkeit nach 3 und 7 Tagen.

Bei einem Vergleich der zwei Calcinieranlagen mit einem Tagesdurchsatz von je 3000 t muß zunächst die schlechtere Mahlbarkeit des Kalksteins aus dem Werk A 2 berücksichtigt werden. Die höhere Mahlfeinheit des Rohmehls aus der Wälzmühle des Werks B 2 führt zu einem etwas höheren Rückstand auf dem 72-mesh-Sieb und zu einem geringeren Rückstand auf dem 170-mesh-Sieb bei einer nur wenig

a higher degree of precalcination in the kiln feed of the B2 kiln equipped with a flash furnace precalciner provide a 25 kcal/kg cl (105 kJ/kg cl) less thermal energy consumption for the clinkerization process. This clinker has a higher than normal liter weight and coarser granulometry. However, a 10 % higher alite content and a better microhomogeneity enable easier grinding of B2 plant clinker. The B1 plant uses open circuit ball mills (OC) with a capacity of 80 t/h and the B2 plant open circuit ball mills (OC) with a capacity of 100 t/h with reference to an average cement fineness (Blaine) of 2600 and 3000 cm²/g respectively. A finer grinding of B2 plant clinker with the same C_3A content and the same quantity of natural (mineral) gypsum addition as for B1 plant clinker explains a considerably less setting time for B2 plant cement and its 10 to 5 % higher standard compressive strength on 3 and 7 d age respectively.

The comparison between the two 3000 t/d calcinator plants has at first to consider a more difficult grindability of the limestone from the A2 plant. The finer feed output of the ver-

höheren Kapazitätsauslastung (310 t/h) von bis zu 5 % gegenüber der Wälzmühle des Werks A 2 mit einer insgesamt niedrigeren Leistung (265 t/h). Der kieselsäurereichere Kalkstein des Werks B 2 führt zu etwa 50 K geringeren Temperaturen bei der Dissoziation, da er einen höheren Gehalt an Flußmitteln und alkalischen Verunreinigungen aufweist. Im Werk A 2 wird eine Kohle mit einem im Vergleich zu Werk B 2 20 % höheren Aschegehalt und einem 20 % niedrigeren Heizwert verwendet.

Das Druck-Temperaturprofil im Vorwärmer und Vorcalcinator des Ofensystems liegt im Bereich des Optimums. Das höhere Litergewicht und die gröbere Granulometrie des Klinkers aus Werk B 2 sind auf die leichtere Brennbarkeit, die zur Bildung gröberer Klinkergranalien führt, zurückzuführen [14]. Beachtenswert ist, daß das Werk A 2 Kalkstein mit höherem Kalk- und niedrigerem SiO_2-Gehalt verwendet als das Werk B 2, jedoch den Ofen mit einer aschereicheren Kohle, die zu höherer Ascheaufnahme führt, befeuert. Dennoch weist der Klinker aus der Anlage A 2 einen höheren CaO- und einen geringeren SiO_2-Gehalt auf als der Klinker aus Anlage B 2 [15]. Bei identischer Art und Kapazität der Zementmühlen kann der Klinker des Werks B 2 aufgrund seiner Phasenzusammensetzung und höheren Mikrohomogenität in der Klinkerphasenverteilung auf eine um 5 % größere spezifische Oberfläche bei einem um 5 % geringeren Energieverbrauch vermahlen werden. Das homogenere Feingefüge des Klinkers der Anlage A 2 führt zu einer gegenüber der Anlage B 2 etwa 10 % höheren Festigkeit bei allen Prüfterminen. Das schnellere Erstarren des Zements aus Werk B 2 könnte auf den niedrigeren C_3A-Gehalt und den Einsatz eines besser kristallin ausgebildeten Naturgipses im Vergleich zum nur teilweise kristallinen chemischen Gips des Werkes A 2 zurückzuführen sein.

4. Schlußfolgerungen

Die Modernisierung des Verfahrens und die betriebliche Kontrolle der unterschiedlichen Parameter sowie die gründliche Analyse der Grundeigenschaften der Einsatzstoffe führen zusammen zu einem besseren Verständnis des Qualitätsgewinns oder -verlustes des Endprodukts Zement.

tical roller mill in the B2 plant leads to a somewhat higher residue on the 72 mesh sieve and to a smaller residue on the 170 mesh sieve with a marginal higher capacity (310 t/h) utilization of up to 5 % in comparison with the vertical roller mill of plant A2 with an altogether lower capacity (265 t/h). The more siliceous limestone of the B2 plant shows an about 50 K lower temperature range for thermal dissociation because of its higher content of fluxes and alkali impurities. In the A2 plant a coal with a comparatively 20 % higher ash content and a 20 % lower calorific value as in plant B2 is employed.

The pressure and temperature profile in the preheating and precalcining part of the kiln system is in the optimum. A higher liter weight and a coarser granulometry of the B2 plant clinker is due to its easier burning giving rise to larger clinker nodules [14]. It is noteworthy that the A2 plant uses limestone with higher lime and lower silica content than the B2 plant but a coal with higher ash content and ash absorption then the B2 kiln. Despite this the A2 plant clinker has a higher lime and a lower silica content than the B2 plant clinker [15]. With the same type and capacity of cement mills the B2 plant clinker can be ground to 5 % more specific surface with 5 % less electric energy because of its phase composition and a higher microinhomogeneity in its mineral phase distribution. The more homogeneous microstructure of the A2 plant clinker accounts for its 10 % higher strength at all ages compared with B2 plant cements. Earlier setting of B2 plant cement may be attributed to its lower C_3A content and the use of more crystalline natural (mineral) gypsum in comparison with partially crystalline chemical gypsum used by the A2 plant.

4. Conclusion

Modernization of the process and operational control of its different parameters as well as an in-depth analysis of the basic and intrinsic properties of the inputs are complementary to each other in understanding the quality gains or the deficits of the final product cement.

Literature

[1] Heilman, T.: 3rd ICCC, London (1952), p. 711.
[2] Heilman, T.: 6th ICCC, Moscow (1974), p. I-207.
[3] Sinha, S.K., Handoo, S.K., and Chatterjee, A.K.: 7th ICCC, Paris, Vol. 2 (1980), p. I. 108–113.
[4] Luding, U., and Ebrahim, S.E.: 7th ICCC, Paris, Vol. II (1980), p. I-115.
[5] Christensen, N.H.: Cement and Concrete Research (1979), 9, p. 285.
[6] Fundal, E.: World Cement Technology (1979), 10 (6), p. 195.
[7] Chromy, S.: 7th ICCC, Paris (1980), p. I-56.
[8] Abdul Maula, S., and Odler, I.: World Cement Technology (1980), 11 (7), p. 330.
[9] Ghosh, S.P.: Advances in Cement Technology, Pergamon Press (1981), p. 26.
[10] Chatterjee, A.K.: Advances in Cement Technology, Pergamon Press (1981), p. 56–57.
[11] Chatterjee, T.K.: Advances in Cement Technology, Pergamon Press (1981), p. 74–79.
[12] Viswanathan, V.N., and Ghosh, S.N.: Advances in Cement Technology, Pergamon Press (1981), p. 193–199.
[13] Chatterjee, T.K.: Cement & Concrete Science & Technology (1991), Vol. I, Part I-Process in Cement & Concrete, ABI Books, New Delhi, p. 14.
[14] Kurdowskii, W.: Advances in Cement Technology, Pergamon Press (1981), p. 59.
[15] Werynski, B., and Werynska, A.: 7th ICCC, Paris (1980), Vol. II, p. I-182.

Gedanken zur Einführung von Qualitätssicherungssystemen für Zement-Herstellwerke

Thoughts on the introduction of quality assurance systems for cement manufacturing plants

Réflexions sur l'introduction de systèmes d'assurance de la qualité dans les usines de fabrication de ciment

Reflexiones sobre la introducción de sistemas de aseguramiento de la calidad en las fábricas cementeras

Von **V. Rudert,** Wiesbaden/Deutschland

Zusammenfassung – Zemente, die innerhalb des gemeinsamen europäischen Wirtschaftsraumes verwendet werden sollen, bedürfen zukünftig einer Zertifizierung, die die Übereinstimmung des Produktes mit harmonisierten europäischen Normen oder Zulassungen feststellt. Kernstück des Konformitätsnachweises ist ein dokumentiertes werkseigenes Produktionskontrollsystem (Qualitätssicherungssystem). Für die Zertifizierung der Zementherstellung wird vorgeschlagen, die umfassende Betrachtung eines Zementwerkes (Integralmodell) zugunsten einer Aufgliederung in die einzelnen Betriebsteile (Differentialmodell) entsprechend dem Verfahrensablauf der Herstellung aufzugeben. Das Differentialmodell ermöglicht, daß sich die Audits der Überwachungsstelle auf den Verfahrensschritt beschränken, in dem Zement aus Komponenten hergestellt wird. Der Verein Deutscher Zementwerke hat für seine Mitgliedswerke ein dreistufiges Ebenenmodell für ein Qualitätssicherungssystem mit Rahmenhandbuch und Unterhandbüchern erarbeitet, in denen das Qualitätssicherungskonzept und die produktionsbezogenen Maßnahmen im Rahmen der werkseigenen Produktionskontrolle für die Überwachungsstelle beschrieben werden. Qualitätssicherungssysteme sollen unternehmensweit, bereichsübergreifend erarbeitet und eingeführt werden. Der Unternehmensleitung wächst mit der Durchdringung des Qualitätsgedankens in alle Bereiche und Ebenen eine anspruchsvolle Führungsaufgabe zu. Die Motivation kann durch Workshops und Gruppenarbeit gefördert werden und erhöht die Akzeptanz für die anfänglich bürokratisch erscheinende Dokumentation.

Summary – Cements which are to be used in the EEC will in future require a certificate which establishes that the product conforms to co-ordinated European standards or licences. Proof of conformity is centred on a documented production control system within the works (quality assurance system). For certification of cement production it is proposed to abandon the overall consideration of a cement plant (integral model) in favour of a break-down into the individual operating sections (differential model) corresponding to the process sequence during manufacture. The differential model means that the audits by the monitoring agency are confined to the process step in which the cement is produced from components. The German Cement Works' Association has drawn up for its member works a three-stage level model for a quality assurance system, with an outline manual and subsidiary manuals in which the quality assurance concept and the production-related measures are described for the monitoring agency within the framework of the works' internal production control system. Quality assurance systems should be drawn up and introduced to cover the entire company and embrace all sections. With the penetration of the quality concept into all areas and levels, the company management is taking on a demanding leadership function. Motivation can be promoted through workshops and group work, and increases the acceptability of the documentation which appears at first to be bureaucratic.

Résumé – Les ciments destinés à être utilisés à l'intérieur du marché commun européen, nécessiteront à l'avenir un certificat prouvant la conformité du produit avec les normes harmonisées européennes ou avec des agrégations. La partie essentielle de la preuve de conformité est un système de contrôle documenté, propre à chaque usine (système d'assurance de la qualité). Quant aux certificats relatifs à la fabrication du ciment, on propose d'abandonner l'idée globale d'une cimenterie (modèle intégral) en faveur d'une séparation en plusieurs sections (modèle différentiel), conformément au déroulement du processus de fabrication. Le modèle différentiel permet au personnel du poste de contrôle de se limiter à la section, dans laquelle le ciment est fabriqué à partir de différents composants. La Verein Deutscher Zementwerke a élaboré, pour ses usines membres, un modèle à trois étapes, destiné à un système d'assurance de la qualité, y compris un manuel cadre ainsi que des sous-manuels, dans lesquels sont décrits le concept d'assurance de la qualité et les mesures à prendre en fonction de la production, dans le cadre du contrôle interne de la qualité. Les systèmes d'assurance de la qualité doivent être élaborés et introduits, d'une manière globale, dans l'entreprise entière. La

direction de l'entreprise acquiert une tâche exigeante, au fur et à mesure que l'idée de la qualité pénètre dans tous les domaines, à tous les niveaux. La motivation peut être favorisée à ce sujet, en organisant des workshops et des travaux en équipe, ce qui paraîtra au début assez bureaucratique.

Resumen – *Para los cementos destinados a ser empleados dentro del mercado común europeo se necesitará en el futuro un certificado, en el que se establezca la conformidad del producto con las normas o admisiones armonizadas a nivel europeo. La pieza clave de la prueba de conformidad es un sistema de control documentado de la producción, propio de cada fábrica (sistema de aseguramiento de la calidad). Para la certificación de la fabricación del cemento se propone abandonar la visión global de una fábrica de cemento (modelo integral) a favor de una división de la instalación en sus distintas secciones (modelo diferencial), de acuerdo con el desarrollo del proceso de producción. El modelo diferencial permite que las auditorías de los servicios de vigilancia se limiten a la etapa del proceso, en la que se produce el cemento a partir de los diferentes componentes. La Verein Deutscher Zementwerke ha elaborado para sus empresas miembros un modelo de tres etapas, destinado a un sistema de aseguramiento de la calidad, con manual general y submanuales, en los cuales se describe el concepto de aseguramiento de la calidad así como las medidas a tomar en función de la producción, dentro del marco del control interno de la producción, llevado a cabo por el servicio de vigilancia. Los sistemas de aseguramiento de la calidad deben ser elaborados e introducidos de forma global, en toda la empresa. A los directivos de las empresas les espera una tarea exigente, a media que la idea de la calidad penetre en todos los sectores, a todos los niveles. La motivación para ello puede ser alentada mediante workshops y trabajos en grupos, con lo cual se aceptará mejor el trabajo de documentación, que se considera, al principio, bastante burocrático.*

Reflexiones sobre la introducción de sistemas de aseguramiento de la calidad en las fábricas cementeras

Die Bescheinigung der Konformität von Zement mit den Anforderungen der zukünftigen Europäischen Norm EN 197 setzt aufgrund der EG-Richtlinie über Bauprodukte vom 21. 12. 1988 [1] voraus, daß der Hersteller von Zement über ein werkseigenes Produktionskontrollsystem verfügt und daß eine Zertifizierungsstelle (z.B. eine Güteüberwachungsgemeinschaft) zusammen mit einer Überwachungsstelle in die Beurteilung und Überwachung der Produktionskontrolle und der Eigenschaften des Zements eingeschaltet ist. Die Güteüberwachungsgemeinschaft und die kooperierende Überwachungsstelle sollten hierfür besonders geeignet sein.

Die EG-Richtlinie über Bauprodukte wurde mit Datum vom 10. August 1992 als deutsches Bauproduktengesetz (Bau PG) in nationales Recht überführt. Das Bauordnungsrecht liegt in Deutschland im Zuständigkeitsbereich der Länder; diese setzen über Verordnungen in den Landesbauordnungen die Vorgaben des Bauproduktengesetzes in die Praxis um.

Die Bescheinigung der Konformität wird in der Richtlinie des Rates im Artikel 13 geregelt; dieser enthält auch die Empfehlung, daß — vermutlich mit Blick auf die wirtschaftliche Zumutbarkeit — dem jeweils am wenigsten aufwendigen Verfahren, das mit den Sicherheitsanforderungen vereinbar ist, der Vorzug zu geben ist. Nun ist für Zement zwar das aufwendigste Konformitätsbescheinigungsverfahren (Verfahren i) entsprechend der Richtlinie für Bauprodukte gewählt worden, jedoch ist sowohl europäisch als auch national noch ein verborgener Gestaltungsspielraum verfügbar.

WAS durch die Überwachung zukünftig erfaßt werden muß, ist also weitgehend festgelegt. Dem Hersteller obliegt es, eine werkseigene Produktionskontrolle einzurichten, durchzuführen und gegenüber einer Überwachungsstelle nachzuweisen; zusätzlich ist wie bisher eine Eigenüberwachung des Produktes durchzuführen. Die für die Überwachung zugelassene Stelle (Fremdüberwacher) nimmt wie in der Vergangenheit eine Erstprüfung des Produktes vor, führt eine Erstinspektion des Werkes und der werkseigenen Produktionskontrolle durch und prüft das Produkt anhand von Einzelzugriffen auf Konformität bezüglich der Anforderungsnorm und der Daten der Eigenüberwachung. Die laufende Überwachung, Beurteilung und gegebenenfalls Anerkennung der Organisation und Durchführung der werkseigenen Produktionskontrolle ist dagegen eine neue zusätzliche Aufgabe für die überwachende Stelle. Aber noch können sich die Regelwerke schaffenden Gremien auf weniger

Under the EC Building Products Directive of 21.12.88 [1], certification of the conformity of cement with the specifications of the future European Standard EN 197 is conditional upon the cement manufacturer having an internal production control system and a certification agency (e.g. a quality monitoring association) working together with a monitoring agency being brought in to assess and monitor the production control system and the properties of the cement. The quality monitoring association and the monitoring agency with which it works should be specifically suited to this purpose.

The EC Building Products Directive was enshrined in national law as the German Building Products Act with effect from 10 August 1992. In Germany, building regulations are the responsibility of the states, which implement the provisions of the Building Products Act by means of regulations in the building ordinances of each state.

The certification of conformity is regulated in Article 13 of the Council Directive. This article also contains the recommendation that (presumably with a view to what can reasonably be expected financially) preference should be given in each case to the least expensive method that is compatible with the safety requirements. In fact, for cement the most expensive conformity certification method (Method i) according to the Building Products Directive has been chosen; however, latent scope still exists for shaping the course of events at both the European and the national level.

WHAT the monitoring process has to cover in future is therefore largely fixed. It is the manufacturer's responsibility to set up and implement an internal production control system and furnish proof thereof to a monitoring agency; in addition, internal product monitoring is required as at present. As in the past, the agency licensed to carry out the monitoring (external monitor) carries out initial testing of the product, conducts an initial inspection of the works and the internal production control system, and tests the product for conformity with regard to the specification standard and the data from internal monitoring on the basis of spot checks. The continuous monitoring, assessment and, where applicable, approval of the organisation and implementation of the internal production control is, on the other hand, a new and additional task for the monitoring agency. But it is still possible for the bodies drawing up the regulations to agree on less expensive or more streamlined forms of internal production control (**Figure 1**). The certificate of conformity will apply to cement; thus (also in the spirit of the future standard EN 197) only the areas of the works where cement is actu-

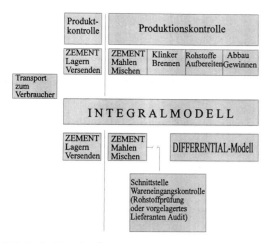

BILD 1: Definition des Überwachungsbereiches Zement-Herstellwerk

FIGURE 1: Definition of the area to be monitored in a cement manufacturing plant

Produktkontrolle	= Product control
Produktionskontrolle	= Production control
Zement	= Cement
Lagern	= Storage
Versenden	= Shipment
Mahlen	= Grinding
Mischen	= Blending
Klinker	= Clinker
Brennen	= Burning
Rohstoffe	= Raw materials
Aufbereiten	= Preparation
Abbau	= Extraction
Gewinnen	= Recovery
Transport zum Verbraucher	= Transport to user
Integralmodell	= Integral model
Differential-Modell	= Differential model
Schnittstelle	= Dividing line
Wareneingangskontrolle	= Incoming goods inspection
(Rohstoffprüfung oder vorgelagertes Lieferanten Audit)	= (Inspection of raw materials or supplier's audit in advance)

aufwendige oder schlankere Formen der werkseigenen Produktionskontrolle einigen (**Bild 1**). Die Konformitätsbescheinigung soll für Zement gelten, somit sind – auch im Sinne der zukünftigen EN 197 – nur diejenigen Betriebsteile durch die Überwachungsstelle auf ein Kontrollsystem zu überprüfen, in denen Zement tatsächlich hergestellt wird. Ein INTEGRALMODELL von der Rohstoffgewinnung bis zum Versand ist von den Regelwerken nicht gefordert und würde durch die Benachteiligung der Werke mit eigener Gewinnung und eigenem Brennaggregat gegenüber Mahl- und Mischwerken zu einer Wettbewerbsverzerrung führen. Somit ist das selektivere DIFFERENTIALMODELL eines Qualitätsicherungssystems (QSS) für ein Zementherstellwerk mit der wesentlichen Schnittstelle der Wareneingangskontrolle vor der eigentlichen Zementherstellung besser auf die zukünftige europäische Zertifizierung zugeschnitten. Damit hätte man gestaltend darauf Einfluß genommen, WO die Überwachungsstelle Aufgaben bei der Überwachung der werkseigenen Produktionskontrolle zugewiesen bekommt, oder anders ausgedrückt, man hätte damit das zu überwachende Zementherstellwerk eindeutig definiert.

Es bleibt nun noch zu überlegen, WIE ein Zementwerk unter den Prämissen des Wettbewerbs im Markt die Art der Zertifizierung nach außen dokumentieren kann und sollte (**Bild 2**). Hierfür läßt sich das EBENENMODELL heranziehen. Es gliedert sich in zwei Ebenen, die der Überwachung und Einsichtnahme durch die Überwachungsstelle zugänglich sein sollten, das Rahmenhandbuch und die Unterhandbücher. Dabei folgen die Unterhandbücher in ihrer Gliederung der Ablauforganisation des Herstellprozesses und den zugeordneten Dienstleistungseinheiten. Die Ebene der zugeordneten Betriebshandbücher enthält alle quantitativen, prozeßspezifischen Größen und das zugehörige firmenspezifische verfahrenstechnische Know-how und sollte somit einer externen Überwachung nicht zugänglich gemacht werden. Im Gegensatz zum Integralmodell sind neben dem Rahmenhandbuch nur diejenigen Unterhandbücher für ein Audit heranzuziehen, die sich mit der Herstellung, der Lagerung, Verpackung und dem Versand sowie mit den funktional zugeordneten Betriebseinheiten der internen Qualitätssicherung befassen.

Jedes Ebenenelement, ob Rahmenhandbuch oder einzelne Unterhandbücher, kann im Sinne eines Qualitätsicherungssystems alle Qualitätselemente (z.B. der Reihe DIN/ISO 9000) enthalten oder nur diejenigen, die aus der Sicht der Bauaufsicht und der Zertifizierungsstelle für die Konformitätsbescheinigung für erforderlich gehalten werden (**Bild 3**). Im ersteren Fall würde es sich wiederum um ein INTEGRALMODELL handeln. Nur ein zertifiziertes Qualitätssicherungssystem würde ein Integralmodell bezogen auf die QS-Elemente voraussetzen; jedoch ist ein zertifiziertes QSS im Sinne einer integralen Anwendung der DIN/ISO 9000 weder bei der Zementherstellung noch bei den Kunden gefordert, geschweige denn erforderlich. Die Broschüre [2] der Bundesverbände Kies- und Sandindustrie e.V., Mörtel-

ally produced need to be checked by the monitoring agency for the operation of a control system. An INTEGRAL MODEL extending from extraction of the raw materials to shipment of the finished product is not required by the regulations and, by putting works with their own extraction operations and burning unit at a disadvantage compared with grinding and blending works, would lead to a distortion of competition. Thus the more selective DIFFERENTIAL MODEL, where the essential dividing line is the incoming goods inspection prior to the actual cement production, is better suited to the future European certification process. In this way a formative influence would have been exerted on WHERE the monitoring agency has tasks assigned to it with regard to the monitoring of the internal production control system; in other words, the cement manufacturing plant to be monitored would thus have been clearly defined.

It remains to be considered HOW a cement works can and should outwardly document the nature of the certification under the conditions of a competitive market (**Figure 2**). For this purpose, the LEVEL MODEL of a quality assurance system (QAS) may be applied to a cement works. It is divided into two levels, which should be accessible for monitoring and inspection by the monitoring agency: the outline manual and the subsidiary manuals. The structure of the subsidiary manuals follows the operational organisation of the production process and the related service units. The level of the related operating manuals includes all the quantitative process-specific data and the associated proprietary process know-how, and should therefore not be made accessible to an external monitoring body. In contrast to the integral model, an audit should cover, besides the outline manual, only those subsidiary manuals which deal with production,

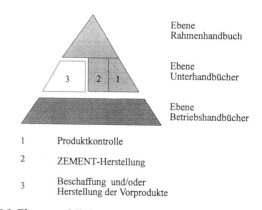

1 Produktkontrolle
2 ZEMENT-Herstellung
3 Beschaffung und/oder Herstellung der Vorprodukte

BILD 2: Ebenenmodell eines Qualitätssicherungs-Handbuches
FIGURE 2: Level model of a quality assurance manual

Ebene Rahmenhandbuch	= Outline manual level
Ebene Unterhandbücher	= Subsidiary manuals level
Ebene Betriebshandbücher	= Operating manuals level
Produktkontrolle	= Product control
Zement-Herstellung	= Cement production
Beschaffung und/oder Herstellung der Vorprodukte	= Procurement and/or manufacture of feed materials

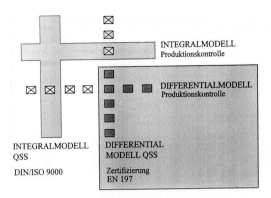

BILD 3: Differentialmodell der Produkt-Zertifizierung
FIGURE 3: Differential model for product certification

Integralmodell	= Integral model
Differentialmodell	= Differential model
Produktionskontrolle	= Production control
QSS	= QAS
Zertifizierung	= Certification

industrie e.V. und Transportbetonindustrie „Qualitätsmanagement und Überwachung bei der Herstellung von Baustoffen" macht aus der Sicht der Kunden der Zementindustrie deutlich, daß die Produktzertifizierung zur Erlangung des CE-Zeichens als Konformitätsnachweis mit einer harmonisierten europäischen Norm in keinem Zusammenhang zur Zertifizierung von Qualitätssicherungs- oder -managementsystemen steht. In gleicher Weise äußerte sich der Bundesverband Deutsche Beton- und Fertigteilindustrie e.V., eine weitere wesentliche Kundengruppierung der Zementindustrie [3]. Somit sollte als Umfang für die Produkt-Zertifizierung jeweils das Differentialmodell für die Definition des Zementherstellwerkes und für die Definition des Qualitätssicherungssystems angestrebt werden. Das sind die verbliebenen verborgenen Möglichkeiten für die Gestaltung der Europäischen Zementzertifizierung.

Für die formale äußere Darstellung der Konformität ergibt sich für den Zementmarkt der Bundesrepublik Deutschland somit zukünftig das in **Bild 4** gezeigte System. Danach wird lediglich das bisherige Überwachungszeichen durch das CE-Zeichen ersetzt. Daneben steht weiterhin das Zeichen des Vereins Deutscher Zementwerke als Fremdüberwacher. Damit ist eine zwar differenzierte, aber mit hoher fachlicher Kompetenz ausgestattete Fremdüberwachung auf Lieferscheinen, Geschäftspapieren und Gebinden dokumentiert. Die formale Bestätigung eines integralen QS-Systems durch einen fachfremden Auditor ist zwar möglich, sollte jedoch wegen der zu erwartenden inhaltlichen Schwächen vermieden und weiterhin der damit zu beauftragenden Güteüberwachungsgemeinschaft überlassen bleiben.

Qualitätssicherungssysteme sollten aber wesentlicher integraler Grundpfeiler der Unternehmenskultur sein [4] und sind ein eher nach innen, denn nach außen gerichtetes Instrument im Wettbewerb. Schon heute kann ein Qualitäts-Sicherungssystem bei einem guten Dutzend akkreditierter Zertifizierungsstellen zertifiziert werden, jedoch läuft eine zu sehr nach außen getragene Darstellung auch Gefahr, daß das Qualitätssicherungssystem zum Werbeprospekt verkommt.

Literaturverzeichnis

[1] Richtlinie des Rates vom 21. Dezember 1988 zur Angleichung der Rechts- und Verwaltungsvorschriften der Mitgliedsstaaten über Bauprodukte (89/106/EWG). Amtsblatt der Europäischen Gemeinschaften (1989) Nr. L 40, S. 12–26.

[2] Bundesverband der Deutschen Kies- und Sandindustrie e.V., Bundesverband der Deutschen Mörtelindustrie e.V. und Bundesverband der Deutschen Transportbetonindustrie e.V.: Qualitätsmanagement und Überwachung bei der Herstellung von Baustoffen. (1993), S. 1–31.

[3] Pesch, L.: Qualitätssicherungs-Handbücher − nachgewiesene Qualitätssicherung. BETONWERK + FERTIGTEIL-TECHNIK BFT (1993) Nr. 1, S. 65–73.

[4] von Engelhardt, A.: Qualität als Differenzierungsmerkmal. BETONWERK + FERTIGTEIL-TECHNIK BFT (1991) Nr. 9, S. 41–44.

storage, packaging and shipment and with the functionally related company units responsible for the internal quality assurance.

Each level element, whether it is the outline manual or individual subsidiary manuals, may, in the spirit of a quality assurance system, contain either all the quality elements (e.g. of the DIN/ISO 9000 series) or only those elements which the construction supervision authority and the certification agency consider necessary for the certification of conformity (**Figure 3**). With the former, it would again be a case of an INTEGRAL MODEL. Only a certified quality assurance system would necessitate an integral model based on the QA elements; however, a certified QAS in the sense of a full application of DIN/ISO 9000 is not stipulated (let alone necessary) either for cement production or for customers. The brochure [2] "Qualitätsmanagement und Überwachung bei der Herstellung von Baustoffen"[Quality management and monitoring during the production of construction materials] published by the Federal Associations of the Sand and Gravel Industry, the Mortar Industry and the Ready-Mixed Concrete Industry makes it clear that from the point of view of the cement industry's customers, product certification in order to obtain the CE symbol as proof of conformity with a harmonised European standard has nothing to do with the certification of quality assurance or management systems. The Federal Association of the German Concrete and Prefabricated Materials Industry, another important alliance of customers of the cement industry, has said much the same thing [3]. Thus it is the differential model for defining the cement manufacturing plant and the quality assurance system which should always be aimed at as the framework for product certification. These are the remaining latent possibilities for shaping the European cement certification system.

In future, the formal outward representation of conformity for the German cement market will therefore be the system shown in **Figure 4**. Here, the existing monitoring symbol is simply replaced by the CE symbol. The symbol of the German Cement Works' Association, as the external monitor, will continue to stand alongside it. In this way, a complex yet highly professional external monitoring system is documented on delivery notes, commercial documents and packing drums. While formal certification of an integral QA system by a non-specialist auditor is possible, this should be avoided due to the likely weaknesses with regard to content and should remain the preserve of the quality monitoring association to be entrusted with the task.

BILD 4: Äußere Dokumentation der Konformität
FIGURE 4: Outward documentation of conformity

bisher	= current
zukünftig	= future
Produktüberwachung und Überwachung der werkseigenen Produktionskontrolle	= Product monitoring and monitoring of the internal production control system

However, quality assurance systems should be a fundamental cornerstone of management philosophy [4] and are an instrument of competition which is directed inward rather than outward. Certification of a quality assurance system can already be obtained from over a dozen accredited certification agencies; however, with a representation where there is too much emphasis on the outward aspect, there is a risk of the quality assurance system degenerating to the level of an advertising brochure.

Herstellung von Zementen unter Verwendung von Sondermüll sowie fossilen Brennstoffen und ihre Eignung bei Anwendung im Trinkwasserbereich

The suitability of cements manufactured with hazardous waste derived fuels and with fossil fuels for drinking water applications

Fabrication de ciments avec utilisation de déchets spéciaux et combustibles fossiles et leur conformité pour emploi dans le domaine de l'eau potable

Fabricación de cementos, utilizando basuras especiales y combustibles fósiles, y su aptitud para ser empleados en el ámbito del agua potable

Von **H. M. Seebach**, Atlanta, und **H. Tseng**, Southdown/USA

Zusammenfassung – Als die amerikanische Zementindustrie in großem Umfang begann, zur Ergänzung fossiler Brennstoffe solche zu verwenden, die aus gefährlichen Abfallstoffen gewonnen wurden, behaupteten einige Umweltschützer, daß „alle Giftstoffe des Sondermülls im Zement sein werden" und daß infolgedessen Bauten und Trinkwasserversorgungsanlagen im Lauf der Jahre „die Öffentlichkeit vergiften werden". Ehe das Programm zur Gewinnung von Brennstoffen aus Sondermüll umgesetzt wurde, hatte Southdown dieses Thema untersucht und war aufgrund der von Sprung et al. veröffentlichten wissenschaftlichen Daten zu der Feststellung gelangt, daß aus Zement, der unter Verwendung von aus Sondermüll gewonnenen Brennstoffen und Kohle hergestellt wird, Schwermetalle nicht auf andere Weise ausgelaugt werden als bei normalen Zementen. Außerdem wurde festgestellt, daß die Menge an organischen Substanzen, die unter den Bedingungen in der Sinterzone möglicherweise unzersetzt bleiben, vernachlässigbar sein würde. Aufgrund fortgesetzter Befürchtungen in der Öffentlichkeit leitete Southdown ein umfangreiches von NSF International durchgeführtes Versuchsprogramm ein. Dieses Versuchsprogramm war so angelegt, daß das Auslaugen von Schwermetallen aus Mörtelwürfeln nach den in der ANSI-Norm 61 beschriebenen Bedingungen für Bauteile, die mit Trinkwasser in Berührung kommen, geprüft werden konnte. Das Versuchsprogramm umfaßte auch die intensive Überprüfung des Gehalts organischer Rückstände im Klinker, die auf die Verbrennung von Müll zurückzuführen sind. Der Beitrag beschreibt im einzelnen die Ergebnisse der vorangegangenen Untersuchungen und die Ergebnisse des Versuchsprogramms. Alle Ergebnisse führten zu der Schlußfolgerung, daß Zemente oder daraus hergestellte Betone keinen wesentlichen Unterschied, der auf die bei der Herstellung verwendeten Brennstoffe zurückzuführen wäre, aufweisen. Die Versuchsergebnisse zeigen auch sehr deutlich, daß das Auslaugen von Schwermetallen und der Gehalt an organischen Substanzen erheblich unter der von der EPA veröffentlichten maximalen Schadstoffwerten (MCL – Maximum Contaminant Levels) für Trinkwasser liegen und deutlich niedriger sind, als die in der NSF-Norm 61 festgelegten maximal zulässigen Werte, (MAL – Maximum Allowable Levels), die 10% der MCL-Werte betragen.

Summary – As the U.S. cement industry started to extensively use hazardous waste derived fuels to supplement fossil fuels, some environmental activists alleged that "all the toxins of the waste will be in the cement" and that, as a consequence, our structures and drinking water supplies will, over the years to come, "poison the public". Prior to implementing its waste fuels program, Southdown had investigated this subject and determined, based upon scientific data published by Sprung et al., that cement manufactured with hazardous waste derived fuels and with coal only would not leach out metals in any different way. It was also determined that the amount of organics which could possibly survive the sintering zone condition would be negligible. Based upon continued public fears, Southdown entered into an extensive testing program to be performed by NSF International. This testing program was designed to investigate the leaching of metals from mortar cubes under conditions described in ANSI Standard 61 for components that come in contact with drinking water. The testing program also included the extensive search for organic residues from waste combustion in the clinker. The paper describes in detail the results of the previous investigations and the results of the testing program. All results lead to the conclusion that cement or concrete made thereof does not show any material difference between the fuels used to manufacture it. The test results also show very clearly that the leaching of metals and the content of organics are considerably lower than the Maximum Contaminant Levels (MCL) published by EPA for drinking water, and considerably below the Maximum Allowable Levels (MAL) as defined in NSF Standard 61 as 10% of the MCL.

Fabrication de ciments avec utilisation de déchets spéciaux et combustibles fossiles et leur conformité pour emploi dans le domaine de l'eau potable

Résumé – Quand l'industrie cimentière américaine commença, en complément des combustibles fossiles, à utiliser sur une vaste échelle ceux qui provenaient de déchets dangereux, certains défenseurs de la nature affirmèrent que „toutes les substances toxiques des déchets spéciaux allaient se retrouver dans le ciment" et que, de ce fait, les édifices et installations d'approvisionnement d'eau allaient „contaminer la population" au fil des ans. Avant que le programme de fabrication de combustibles à partir de déchets ne se concrétise, Southdown s'était penché sur ce sujet et était parvenu, à la lumière des données scientifiques publiées par Sprung et al., à la conclusion que le ciment fabriqué à partir de combustibles et de charbon obtenus avec l'utilisation de déchets spéciaux, ne présente pas plus de métaux lourds que les ciments courants. Il a, en outre, été constaté que la quantité de substances organiques, qui dans ces conditions restent probablement à l'état non décomposé durant la phase de cuisson, sont négligeables. Vu les craintes réitérées formulées par le grand public Southdown a lancé un vaste programme expérimental effectué par NSF International. Ce programme a été conçu de manière à pouvoir tester le lessivage de métaux lourds de cubes de mortier dans les conditions, décrites sous la norme ANSI 61, pour composants entrant en contact avec l'eau potable. Ce progamme expérimental comprenait aussi l'examen intensif de la teneur en résidus organiques du clinker, imputables à l'incinération des déchets. L'exposé décrit en détail aussi bien les résultats des tests précédents que ceux du programme expérimental. Tous les résultats permettent de conclure que les ciments ou bétons fabriqués à partir de là ne présentent pas de différence notable, décelable due à la présence de combustibles utilisés pour leur fabrication. Et les résultats des tests montrent aussi très clairement que la présence de métaux lourds et la teneur en substances organiques se situent sensiblement en-dessous des valeurs de matières toxiques maximales publiées par l'EPA (MCL – Maximum Contaminant Levels = niveau de contamination maximum) pour eau potable, et sont nettement inférieures aux valeurs maximales admises (MAL – Maximum Allowable Levels) définies par la norme NSF 61, stipulant 10% des valeurs MCL.

Fabricación de cementos, utilizando basuras especiales y combustibles fósiles, y su aptitud para ser empleados en el ámbito del agua potable

Resumen – Cuando la industria del cemento norteamericana empezó a emplear, aparte de los combustibles fósiles, combustibles obtenidos a base de desechos peligrosos, ciertos defensores del medio ambiente manifestaron que „todos los tóxicos contenidos en las basuras especiales estarán presentes en el cemento" y que, por consiguiente, los edificios e instalaciones para agua potable con los años „contaminarán a la población". Antes de llevarse a la práctica el programa encaminado a obtener combustibles a partir de basuras especiales, Southdown había estudiado este problema, sobre la base de los datos científicos publicados por Sprung et al., llegando a la conclusión que en los cementos que se fabriquen utilizando combustibles obtenidos a partir de basuras especiales y carbón, los metales pesados no son lixiviados de otra forma que en los cementos normales. Además, se ha comprobado que la cantidad de sustancias orgánicas que puedan no ser descompuestas en la zona de sinterización, puede ser descuidada. Debido a las constantes preocupaciones del público, Southdown inició un amplio programa de ensayos, llevado a cabo por NSF International. Dicho programa estaba concebido de forma tal que permitía comprobar la lixiviación de metales pesados en los cubos de mortero, de acuerdo con las condiciones descritas en la norma ANSI 61 para piezas o elementos en contacto con el agua potable. El programa de ensayos abarca, asimismo, el detenido control del contenido de residuos orgánicos del clínker, debidos a la incineración de basuras. En el presente artículo se describen detalladamente los resultados de los ensayos previos así como los del citado programa de ensayos. Todos los resultados obtenidos han llevado a la conclusión que los cementos, o los hormigones fabricados con los mismos, no presentan diferencias esenciales que puedan achacarse a los combustibles utilizados. Los resultados de los ensayos muestran también de forma muy clara que la lixiviación de metales pesados así como el contenido de sustancias orgánicas quedan muy por debajo de los índices máximos de contaminación del agua potable, publicados por EPA (MCL – Maximum Contaminant Levels) y son también sensiblemente más bajos que los valores máximos admisibles (MAL – Maximum Allowable Levels), fijados en la norma NSF 61, los cuales llegan al 10% de los valores MCL.

1. Einleitung

Die moderne Gesellschaft sieht sich vielen Umweltproblemen gegenüber. Eines der bedrohlichsten ist die Behandlung der enormen Mengen von Industrie- und Haushaltsabfällen, die Jahr für Jahr entstehen. Eine Lösung für einen Teil des Problems ergibt sich durch die Nutzung bestimmter Arten von Sondermüll als Zusatzbrennstoff in Zementöfen. Die Technologie dazu befindet sich nicht mehr im experimentellen Stadium, sie wird seit vielen Jahren in Europa und Kanada bereits weitgehend angewendet. Begrenzte Anwendungen in den USA haben ebenfalls zu zufriedenstellenden Ergebnissen geführt [1], jedoch wurde hier diese Technik bis heute zu wenig eingesetzt, da die Öffentlichkeit verständlicherweise besorgt darüber ist, ob die Verbrennung von Sondermüll aller Art wirklich ungefährlich für die Umwelt ist [2]. Während es zu dieser Technologie noch eine Reihe von Fragen gibt, die zur Zufriedenheit der Öffentlichkeit beantwortet werden müssen, beschäftigt sich dieser Beitrag mit einem der Hauptprobleme, nämlich

1. Introduction

Modern society faces many environmental challenges. One of the most threatening is the management of the vast amounts of industrial and household wastes which are generated each year. A solution to part of this problem is available through the utilization of certain hazardous wastes as supplemental fuels in cement kilns. The technology involved in this process is not experimental; it has been used extensively in Europe and Canada for many years. Limited applications in the U.S. have also produced satisfactory results [1], but this technology has been under-utilized to date because the public is understandably concerned whether any type of hazardous waste incineration is environmentally safe [2]. While there are many questions which must be answered to the public's satisfaction regarding this technology, this paper deals with one of the major concerns, the effect that burning hazardous waste fuels in cement kilns has on the final product of the process, cement, which serves as the "glue" for concrete. Concrete is used for

der Auswirkung, die die Verbrennung von Sondermüll in Zementöfen auf das Endprodukt des Prozesses, den Zement, hat, der als Bindemittel für Beton dient. Beton wird praktisch für alles eingesetzt, von Autobahnen, Gebäuden und Kläranlagen bis hin zu Staudämmen und Druckrohren für die Trinkwasserversorgung. Es ist daher entscheidend, daß die Industrie absolut sicherstellt, daß die Verwendung bestimmter Arten von Sondermüll beim Herstellungsprozeß von Zement sicher ist und für die Gesellschaft in keiner Weise zusätzliche Umweltprobleme mit sich bringt.

2. Sondermüll als Zusatzbrennstoff

Beim Einsatz von Brennstoffen aus Sondermüll als teilweiser Ersatz von herkömmlichen fossilen Brennstoffen beim Brennen von Zementklinker werden vollkommen neue Bestandteile oder Elemente in das Ofensystem eingebracht. Brennstoffe aus Sondermüll enthalten vorwiegend Kohlenstoff und Wasserstoff aus den organischen Bestandteilen. Diese Bestandteile sind grundsätzlich die gleichen wie in fossilen Brennstoffen. Kohlenstoff und Wasserstoff liegen jedoch in anderen Verbindungen vor. Deren Verbrennung kann sich auch schwieriger gestalten, wobei höhere Temperaturen und längere Verweilzeiten im Ofen erforderlich sind. Die für die Verbrennung und den vollständigen Zerfall notwendigen Mindesttemperaturen und -verweilzeiten liegen jedoch deutlich unter den Temperaturen und Verweilzeiten, wie sie in Brennanlagen für Zementklinker vorherrschen. Darüber hinaus sind die Mengen und die Zusammensetzung der als Zusatzbrennstoff verwendbaren Brennstoffe aus Sondermüll klar geregelt [3]. Die Regelungen schreiben praktisch die nahezu vollständige Zerstörung der organischen Verbindungen (über 99,99 %) und eine ständige Kontrolle geeigneter Verbrennungsbedingungen vor.

Die mit dem Sondermüll eingetragenen anorganischen Bestandteile, vorwiegend SiO_2 und verschiedene Metalloxide, sind im Ofensystem ebenfalls nicht unbekannt. Diese Bestandteile werden durch die großen Feststoffströme in der Ofenanlage verdünnt und aufgrund der hohen Temperaturen und langen Verweilzeiten im Ofen in der kristallinen Matrix des Zementklinkers chemisch gebunden. Auch hier ist als wirksamstes Mittel zur Überwachung der Emissionen der Gesamteintrag an emissionsrelevanten Metallen gesetzlich geregelt. Werden beispielsweise, bezogen auf den Gesamtwärmebedarf, 35 % des fossilen Brennstoffs durch Abfallbrennstoffe ersetzt, so können in einer 2000-t/d-Ofenanlage 5 t/h Sondermüll und 8,5 t/h Kohlenstaub verfeuert werden. Ersatzbrennstoffe können einen erhöhten Metalleintrag in die Ofenanlage mit sich bringen, da der Gehalt einzelner Metalle in Brennstoffen aus Sondermüll häufig höher sein kann als in fossilen Brennstoffen. Um diese Zu- oder Abnahme zu bewerten, müssen die Massenströme im Ofensystem wie auch die Zu- und Abnahme der jeweiligen Metallkonzentrationen berechnet werden.

Als Beispiel für einen erhöhten Metalleintrag soll Cadmium als eines der 11 Metalle mit gesetzlich festgelegten Grenzwerten dienen [3]. Auf der Basis der Grenzwerte für Metallemissionen aus der betrachteten Ofenanlage darf der gesamte Cadmiumeintrag, unabhängig davon, ob er aus Sondermüllbrennstoffen, fossilen Brennstoffen oder Rohstoffen stammt, einen Grenzwert von 2,0 lb/h (0,909 kg/h) nicht überschreiten. Aufgrund der in den Rohstoffen und den Brennstoffkombinationen enthaltenen Cadmiummengen wird dieser Höchstwert für den gesamten Cadmiumeintrag nicht immer erreicht, aber er wird hier dazu verwendet, um die maximale Zunahme des Cadmiumgehalts im Produkt aufzuzeigen. Für die hier gewählte Ofenanlage mit einer Tageskapazität von 2000 t/d (75 t/h) werden 123,750 kg/h Rohmaterial benötigt. Der Ofenanlage werden außerdem die oben angegebenen Brennstoffströme zugeführt, nämlich 8500 kg/h Kohlenstaub und 4545 kg/h Brennstoffe auf Sondermüllbasis. Der Cadmiumeintrag aus Rohmaterial und Kohle beträgt 0,136 kg/h. Bei Verwendung von Brennstoff aus Sondermüll dürften dem Ofen höchstens noch etwa 0,773 kg Cd/h zugeführt werden. Dies entspricht dem 5,7fachen der im Brenngut und der Kohle enthaltenen Menge. Da der überwiegende Teil des Cadmiums, nämlich

everything from highways, buildings and sewer systems, to dams and pressure pipes for drinking water systems. It is, therefore, crucial that industry has made absolutely sure that the utilization of certain hazardous wastes in the cement making process is safe and will not, in any way, add to the environmental problems of society.

2. Hazardous wastes as supplemental fuel

When hazardous waste fuels are used in the pyroprocessing of cement clinker to supplement traditional fossil fuels, entirely new substances or elements are not introduced into the kiln system. The organic elements present in the hazardous waste fuels consisting primarily of C and H, are basically the same as those present in fossil fuels; however, they will be of different configurations. They may also be more difficult to combust, requiring higher temperatures and longer retention times, buth both of these parameters required for combustion and destruction are considerably below the condition prevailing in a pyroprocessing system for cement clinker. Furthermore, the amounts and composition of the hazardous waste fuels used as supplement fuel are carefully regulated [3]. These regulations require that the organic compounds are virtually destroyed (over 99.99 %) and that suitable combustion conditions are continuously monitored.

The inorganic compounds introduced with the hazardous waste, primarily SiO_2 and various metals are also not new to the kiln system. These compounds are diluted in the massive streams of solids of the kiln system and are chemically bonded by the high temperatures and during the long retention times into the crystalline matrix of the cement clinker. Again, the total input of the critical metals is regulated as the most effective means to control metals emissions. For example, when replacing, on a total heat input basis, 35 % of fossil fuel a typical 2000 t/d kiln system might use 5 t/h of hazardous waste fuel and 8.5 t/h of pulverized coal. This fuel replacement may result in a higher metals input into the kiln system because, generally, some metal contents of the hazardous waste fuel will be higher than the metals content of the fossil fuel. To put this increase or decrease in perspective, however, the mass flows in the kiln system have to be considered and the increase or decrease in metals concentration has to be calculated.

As an example for the metals increase, Cadmium is being used as one of the 11 regulated metals [3]. Due to metal emissions limitations of the kiln system considered the total Cd input, regardless whether from hazardous waste fuels, fossil fuel or from raw materials, is limited to 2.0 lbs/hr (0.909 kg/h). This maximum for the total Cd-input will not always be reached due to the Cd-content of the incoming feed and fuel composition, but it is used to demonstrate the maximum Cd-increases of the product. For the specific pyroprocessing system with a capacity of 2000 std/d (75 mt/h) the total raw feed is 123,750 kg/h. Also fed into the kiln system are the fuel streams outlined above: 8500 kg/h of pulverized coal and 4545 kg/h of hazardous waste fuels. The typical Cd-input into the kiln system with raw feed and the coal is 0.136 kg/h. With the hazardous waste fuel a maximum of 0.773 kg/h could be fed to the kiln, which is 5.7 times the amount contained in the raw materials and the coal. Since most of the Cadmium, typically > 99.55 % [1] remains in the solids of the kiln system, the Cd resulting from the raw materials and coal will result in a Cd concentration of 1.8 ppm in the clinker. The Cd from the hazardous waste fuel will result in a maximum increase of 10 ppm, increasing the Cd content of the product from 1.8 to 11.7 ppm.

This total maximum Cd concentration needs to be compared to the concentration of other metals in the clinker, which typically show 50 000 to 60 000 ppm of Al_2O (and TiO_2) and 30 000 to 40 000 ppm of Fe_2O_3 (and Mn_2O_3) and 5 000 ppm of $Na_2O + K_2O$. This ratio of concentration shows how easy the trace metals, including Cd, are being incorporated into the crystalline matrix of the clinker. However, to complete the calculation of metals increase one also needs to consider that clinker is ground with 5 % gypsum to form cement and that concrete, which forms the water storage and transporta-

über 99,55 % [1], in den Feststoffen der Ofenanlage verbleibt, kann sich aus der Cadmiumzufuhr durch Rohstoffe und Kohle eine Cadmiumkonzentration von 1,8 ppm im Klinker einstellen. Das Cadmium aus dem Abfallbrennstoff führt zu einer maximalen Steigerung von 10 ppm. Damit erhöht sich der Cadmiumgehalt des Brennprodukts von 1,8 auf 11,7 ppm.

Die sich insgesamt einstellende Höchstkonzentration an Cadmium muß mit den Konzentrationen anderer Metalle im Klinker verglichen werden, die normalerweise zwischen 50 000 und 60 000 ppm für Al_2O_3 (und TiO_2), 30 000 bis 40 000 ppm für Fe_2O_3 (und Mn_2O_3) sowie 5 000 ppm für Na_2O+K_2O liegen. Dieses Konzentrationsverhältnis zeigt, wie leicht die Spurenmetalle einschließlich Cadmium in die kristalline Matrix des Klinkers eingebunden werden. Um jedoch die Berechnung des Anstiegs im Schwermetallgehalt zu vervollständigen, muß außerdem berücksichtigt werden, daß Klinker mit einem Zusatz von 5 % Gips zu Zement vermahlen wird und daß Beton, der unter anderem für den Bau von Wasserspeichern und Wasserleitungssystemen verwendet wird, nur 10–15 % Zement enthält. Somit vermindert sich die durch den Einsatz von Abfallbrennstoffen hervorgerufene Veränderung des Cadmiumgehalts im Beton auf 1 bis 1,5 ppm. Ähnliche Berechnungen für andere Spurenmetalle untermauern die Schlußfolgerung, daß der durch den Einsatz von Abfallbrennstoffen bei der Klinkerherstellung bedingte Anstieg der Konzentrationen von Spurenmetallen kaum zu messen ist und häufig innerhalb derselben Bandbreite liegt, wie sie bei den verschiedenen Rohmaterialien und Brennstoffen sowie bei den Zuschlägen anzutreffen ist, die zur Betonherstellung verwendet werden. Dennoch ist es wichtig zu wissen, ob Metalle, die bei der Verwendung von Abfallbrennstoffen zusätzlich eingetragen werden, eine Bedrohung für die menschliche Gesundheit oder die Umwelt darstellen.

3. Ergebnisse der TCLP- und Durchlässigkeitsprüfung

Eine umfassende Studie über die Auslaugbarkeit von Metallen aus Zement und Mörtel wurde von Hansen und Miller [4] durchgeführt. Sie unterzogen nach ASTM-C-109 hergestellte Mörtelprüfkörper (Würfel) dem TCLP-Auslaugverfahren (Toxicity Characteristic Leaching Procedure). Mit diesem Verfahren wird bestimmt, ob ein Stoff deponiefähig ist. Zur Herstellung der Mörtelprüfkörper wurde Zement verwendet, der zum einen mit Abfallbrennstoff und zum anderen ausschließlich unter Verwendung von Kohle hergestellt wurde. Für diesen Versuch müssen die Mörtelprüfkörper gebrochen und auf eine Größe von unter 3 mm gemahlen werden. Die Fraktionen ≤ 3 mm werden zur Auslaugung in Säurelösung verwendet. Die Ergebnisse dieser Versuche sind in **Tabelle 1** dargestellt.

Die Tabelle zeigt die Metallmengen im TCLP-Eluat in mg/l. Diese Konzentrationen werden mit den RCA-Grenzwerten (Resource Recovery and Conservation Act) verglichen, die in Zeile 1 der Tabelle 1 aufgeführt sind. So schwankt zum Beispiel die Bleikonzentration des Eluats der Mörtelprüfkörper zwischen 0,045 und 0,70 mg/l. Obwohl der unterste Bereich für einen nur mit Kohle und der obere Bereich für einen mit Abfallbrennstoff und Kohle hergestellten Zement gilt, konnte bei den durchgeführten Versuchen kein signifikanter Unterschied zwischen den Zementen nachgewiesen werden. Diese Schlußfolgerung läßt sich auf alle Schwermetalle übertragen. Es konnte kein signifikanter Unterschied festgestellt werden zwischen Zementen, die mit Abfallbrennstoffen, kombiniert mit fossilen Brennstoffen, oder ausschließlich mit fossilen Brennstoffen hergestellt wurden. Dies läßt die Schlußfolgerung zu, daß die Auslaugung von Metallen aus Mörtel unabhängig davon ist, ob der Zement mit Brennstoffen aus Sondermüll hergestellt wurde oder nicht.

Ein Vergleich der Ergebnisse aus Tabelle 1 mit den EPA-Grenzwerten (Environmental Protection Agency) zeigt, daß alle Mörtelprüfkörper die Grenzwerte einhielten, d.h. sie können mit dem Boden in Berührung kommen. Ebenso gilt

tion systems, contains 10–15 % cement. This decreases the change of the Cd-content in the concrete resulting from the use of hazardous waste fuels to 1 to 1.5 ppm. Similar calculations for other trace metals will support the conclusion that the increases of trace metals concentrations resulting from the use of hazardous fuel during clinker production, can hardly be measured and many frequently even be within the range seen with different raw materials, fuels and aggregates used for concrete production. Nevertheless, it is important to know whether any additional metals resulting from the utilization of hazardous waste fuels could pose a threat to human health or the environment.

3. TCLP and permeation test results

An extensive study on the leachability of metals from cement and mortar was performed by Hansen and Miller [4]. They subjected mortar cubes which were prepared in accordance with ASTM-C-109 to the Toxicity Characteristic Leaching Procedure (TCLP). This procedure is used to determine whether a material may be "put on the ground" (land filled). The mortar cubes were made using cement produced with hazardous waste fuels and coal and cements produced with coal only. For this testing procedure the mortar cubes have to be destroyed and ground to sizes below 3 mm. The ≤ 3 mm fractions are then subjected to the acid for leaching. The results of these tests are given on **Table 1**.

This table shows the quantities of metals in the TCLP leachate in mg/l. These concentrations are compared to the threshold limits as given in RCA (Resource Recovery and Conservation Act), which are shown in line 1 of Table 1. For instance, the lead concentration in the leachate from the mortar cubes varies from 0.045 to 0.70 mg/l. While the lower end of the range represents a cement manufactured with coal only and the upper end of the range represents a cement manufactured with coal and hazardous waste, no significant difference between the cements can be found in these tests. This conclusion can be generalized for all the metals. No significant difference could be established between the cement produced with hazardous waste fuel and fossil fuel and the cement produced with fossil fuel only. This allows the conclusion that the leaching of metals from mortar or cement is independent of whether or not the cement was manufatured with hazardous waste fuels.

A comparison of the results given in Table 1 with the EPA limits shows that all mortar cubes meet the restrictions, i.e., they can be placed on the ground. Again, this conclusion applies for all cement samples regardless of whether the cement was made with fossil fuel and hazardous waste fuel or with fossil fuel only.

Another series of tests were made to investigate leaching of metals from porous, concrete-like samples [5]. This test set-up is much more realistic than the TCLP procedure since it allows water to permeate through the sample while the mortar sample is left in tact.

The methodology was to produce mortar cylinders by mixing cement, sand and water, along with discrete amounts of two metals, thallium and chromium. Thallium was used since it is extremely soluble in water and it could be expected that all of the thallium would be leachable if it was not firmly bonded into the crystalline structure of the hydrated hardened cement. Chromium was used because of the relatively high quantities in the ppm ranges which are contained in all cements whether they were made with conventional fuels only or also with hazardous waste fuels. To make the experiment a conservative one, the two metals thallium and chromium were added to the water for sample preparation as soluble salts. This puts both metals completely into solution at the beginning of the experiment, whereas in the manufacturing process utilizing waste fuels the metals are completely incorporated in the clinker structure and consequently are not in the solution when the cement is mixed with water to form concrete. Sprung [5] prepared his samples in such a way that the results would reflect how much of each metal was leached while water permeated through the sample. He therefore, prepared very

TABELLE 1: Ergebnisse der TCLP-Auslaugversuche (nach Hansen und Miller)
TABLE 1: Comprehensive multi-plant TCLP extraction results (after Hansen and Miller)

Werk	Typ	pH	Ag	Cr	Se	Pb	Ba	As	Cd
RCA-Grenzwerte	NA	NA	5	5	1	5	100	5	1
Grenzwerte	NA	NA	0,002	0,002	0,043	0,016	0,001	0,033	0,001
Blank	NA	2.85	ND	ND	ND	ND	ND	ND	ND
Ott. Sand	NA	2,89	ND	ND	ND	ND	0,003	ND	ND
A*	Würfel	11,73	0,039	0,098	0,044	0,053	0,474	ND	0,001
A*	Zement	12,72	0,052	0,622	0,032	0,097	0,938	0,01	0,003
B	Würfel	11,46	0,039	0,103	ND	0,035	0,448	ND	0,001
C*	Würfel	11,14	0,042	0,155	0,049	0,066	0,448	ND	0,002
D*	Würfel	11,29	0,037	0,061	ND	0,065	0,302	0,05	0,003
E*	Würfel	11,37	0,032	0,044	ND	0,052	0,288	ND	0,002
F	Würfel	11,13	0,039	0,122	nD	0,053	0,352	0,036	0,002
G	Würfel	11,25	0,036	0,078	ND	0,055	0,528	ND	0,001
H	Würfel	11,15	0,039	0,132	ND	0,054	1,29	ND	0,001
I	Würfel	11,09	0,043	0,174	ND	0,06	0,598	ND	0,001
J	Würfel	10,49	0,035	0,139	ND	0,054	0,404	ND	0,001
K*	Würfel	11,23	0,036	0,244	ND	0,065	1,07	ND	0,002
L*	Würfel	11,12	0,04	0,288	ND	0,062	1,09	0,054	0,002
M	Würfel	11,7	0,039	0,189	ND	0,048	0,366	ND	0,002
N	Würfel	11,22	0,04	0,173	ND	0,065	1,24	ND	0,001
O	Würfel	11.29	0,046	0,131	ND	0,068	0,28	ND	0,002
P*	Würfel	11,32	0,036	0,1	ND	0,064	0,404	0,038	0,002
Q*	Würfel	11,3	0,037	0,103	ND	0,07	0,405	0,049	0,002
R	Würfel	11,42	0,036	0,061	ND	0,066	0,389	0,034	0,002
S	Würfel	11,99	0,038	0,044	ND	0,064	0,668	ND	0,002
T	Würfel	11,23	0,039	0,103	ND	0,058	0,144	0,033	0,002
U	Würfel	10,99	0,036	0,088	ND	0,051	0,439	ND	0,001
V*	Würfel	11,34	0,038	0,148	ND	0,058	0,635	ND	0,001
W	Würfel	11,16	0,038	0,162	ND	0,045	0,27	ND	0,001
X	Würfel	11,15	0,036	0,018	ND	0,049	0,791	ND	0,001
Y	Würfel	11,25	0,037	0,139	ND	0,051	0,417	ND	0,001
Z	Würfel	11,32	0,037	0,01	ND	0,063	0,848	ND	0,002
AA	Würfel	11,36	0,035	0,067	ND	0,05	0,843	ND	0,001
BB*	Würfel	11,14	0,037	0,165	ND	0,048	0,42	ND	0,002
CC	Würfel	10,95	0,033	0,139	ND	0,057	0,147	0,041	0,002
DD*	Würfel	11,1	0,037	0,25	ND	0,62	0,456	ND	0,002
EE	Würfel	11,37	0,035	0,22	ND	0,044	0,55	ND	0,001
FF	Würfel	11,01	0,034	0,092	ND	0,059	0,341	0,034	0,002
K*	Beton	6,6	0,034	0,027	0,057	0,062	0,695	0,069	0,002

k.N.: kein Nachweis, unterhalb der Nachweisgrenze
Alle Werte außer dem pH-Wert in ppm
* Diese Proben wurden mit Abfallbrennstoff hergestellt

ND – Not detected, below detection limit
All values except pH are in ppm
* These samples represent the use of waste fuel

Werk = Plant   RCA-Grenzwerte = RCA Limits   Würfel = Cube   Beton = Concrete
Typ = Type   Grenzwerte = Det'n Limits   Zement = Cement

dies für alle Zementproben unabhängig davon, ob der Zement mit Brennstoffen aus Sondermüll und fossilen Brennstoffen oder nur mit fossilen Brennstoffen hergestellt wurde.

Eine weitere Versuchsreihe wurde durchgeführt, um das Auslaugen von Schwermetallen aus porösen, betonähnlichen Proben zu untersuchen [5]. Diese Untersuchungsmethode ist wesentlich realistischer als das TCLP-Verfahren, da das Wasser die Probe durchdringen kann, wobei das Mörtelgefüge aber intakt bleibt.

Zunächst wurden durch Mischen von Zement, Sand und Wasser Mörtelzylinder unter Zugabe von gezielten Mengen der Schwermetalle Thallium und Chrom hergestellt. Thallium wurde deshalb verwendet, weil es extrem wasserlös-

porous mortar. In fact, the porosity was so great, or the permeability of the samples was so great, that they were several orders of magnitude greater than normal concrete. The porosity of the test sample, however, was necessary to get water to permeate the sample rather than just expose the surface. After preparation of the porous mortar cylinders they were allowed to hydrate for 28 days. Thereafter, they were placed in the testing apparatus as shown in **Fig. 1** and water was added and the pressure was increased and maintained at 14.7 lbs/sq.in. The permeated water which was collected on the top of the sample was used for determination of chromium and thallium concentrations. Since it can be expected that metals concentration in the permeated water would decrease with increasing time and quantities of permeation, the first 100 ml of permeated water were used

lich ist und demzufolge ein Auslaugen des gesamten Thalliums erwartet werden konnte, falls es nicht fest im kristallinen Gefüge des hydratisierten und erhärteten Zements eingebunden wird. Chrom wurde verwendet, weil dieses Element in allen Zementen im ppm-Bereich in relativ hohen Mengen enthalten ist, und zwar unabhängig davon, ob die Zemente ausschließlich mit herkömmlichen Brennstoffen oder mit Abfallbrennstoffen hergestellt wurden. Für den Modellversuch wurden die beiden Schwermetalle Thallium und Chrom dem zur Probenherstellung benötigten Wasser als lösliche Salze zugegeben. Dadurch werden beide Metalle gleich zu Beginn des Versuchs vollständig gelöst, wohingegen sie im Herstellungsprozeß mit Einsatz von Abfallbrennstoffen vollständig im Klinkergefüge eingebunden werden und sich demzufolge nicht in Lösung befinden, wenn dem Zement zur Betonherstellung Wasser zugegeben wird. Sprung [5] stellte die Prüfkörper auf diese Weise her, um zu prüfen, welcher Anteil des jeweiligen Schwermetalls bei einem Durchfluß von Wasser ausgelaugt wird. Es wurde demnach ein sehr poröser Mörtel hergestellt. In der Tat war die Porosität so groß bzw. die Durchlässigkeit der Prüfkörper so hoch, daß der Durchlässigkeitsbeiwert um einige Größenordnungen über der eines normalen Betons lag. Diese Porosität der Prüfkörper war jedoch erforderlich, damit das Wasser die Probe durchdringen konnte und nicht nur die Oberfläche der Prüfkörper dem Wasser ausgesetzt war. Nach der Herstellung erhärteten die porösen Mörtelzylinder 28 d. Danach wurden sie in eine Versuchsapparatur (**Bild 1**) eingesetzt, die mit Wasser gefüllt wurde. Der Wasserdruck wurde auf 14,7 psi (1 bar) erhöht. Das durch den Prüfkörper gedrückte Wasser wurde an der Oberfläche aufgefangen und zur Bestimmung der Chrom- und Thalliumkonzentration verwendet. Da davon auszugehen ist, daß die Metallkonzentration im austretenden Wasser mit zunehmender Zeit und Menge abnimmt, wurden zur Analyse nur die ersten 100 ml Wasser verwendet. Die darin festgestellten Konzentrationen stellen demnach den ungünstigsten Fall dar.

Die Ergebnisse dieser Durchlässigkeitsprüfungen sind in **Bild 2** zusammengefaßt. Das Bild zeigt, daß die von Sprung [5] verwendeten Modellmörtel Durchlässigkeitsbeiwerte von 10^{-9} bis 10^{-7} m/s aufwiesen. Bei dieser Porosität wurden 10^{-2} bis 10^{-4}% oder 10^{-4} bis 10^{-7} Mengenanteile der ursprünglichen Thallium- und Chromkonzentration mit dem durchtretenden Wasser ausgelaugt. Das zeigt, daß nur sehr kleine Anteile ausgelaugt werden, wenn Wasser die Mörtelproben durchströmen kann. Die Auslaugraten zeigen deutlich, daß mindestens 99,99% des leichtlöslichen Thalliums und Chroms in den Hydratationsprodukten, d.h. im erhärteten Zement, fest eingebunden sind. Eine ähnliche Metalleinbindung kann für weitere Metalle wie Cadmium, Blei usw. angenommen werden, da ihre Einbindung in die Hydratationsprodukte auf gleiche Weise erfolgt. Sie werden in das kristalline Gefüge des erhärteten Zements eingebunden.

Mit den von Sprung [5] für porösen Mörtel angegebenen Auslaugraten können die aus Beton für Druckrohre auslaugbaren Schwermetallmengen durch Extrapolation auf die Porosität dieser Betonqualität abgeschätzt werden. Die Extrapolation für Durchlässigkeitsbeiwerte von Beton zwischen 10^{-14} und 10^{-16} m/s ist in Bild 2 dargestellt. Bei diesen Beiwerten würden die Auslaugraten einen Bereich von 10^{-9} bis 10^{-12}% erreichen. Dies sind Bereiche, die deutlich machen, daß theoretisch nur ein extrem geringer Teil des Thalliums und Chroms oder generell von Schwermetallen aus Beton ausgelaugt werden kann. Es ist jedoch anzumerken, daß bei der für Beton typischen Porosität weder Wasser noch andere Flüssigkeiten den Beton durchdringen können (bei Durchlässigkeitsbeiwerten von 10^{-14} m/s würde Wasser hierfür einen unendlich langen Zeitraum benötigen). Demzufolge ist der Oberflächenkontakt zwischen Beton und Wasser der einzig entscheidende Faktor für das Auslaugen von Schwermetallen aus Beton, wie er für Druckleitungen oder andere Betonprodukte, die mit Wasser in Berührung kommen, verwendet wird. Da Beton wasserundurchlässig ist, kommen nur die Oberflächenauslaugung und Löslichkeitseffekte für weitere Untersuchungen in Frage.

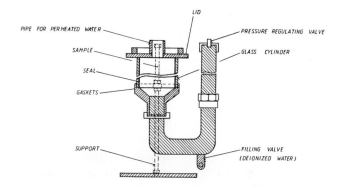

BILD 1: Versuchseinrichtung zur Prüfung der Durchlässigkeit und Auslaugbarkeit von Schwermetallen aus porösem Mörtel
FIGURE 1: Apparatus for Testing Permeability and Heavy Metal Leachability from Porous Mortar

Deckel	= lid
Überlaufrohr	= pipe for permeated water
Druckregelventil	= pressure regulating valve
Probe	= sample
Glaszylinder	= glass cylinder
Dichtung	= seal
Dichtringe	= gaskets
Haltestange	= support
Füllventil	= filling valve

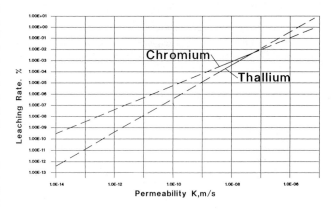

BILD 2: Auslaugrate von Schwermetallen in Abhängigkeit vom k-Wert (m/s)
FIGURE 2: Heavy Metal Leaching Rate vs. Concrete Permeability

Auslaugrate (%) = leaching rate (%)
Durchlässigkeitsbeiwert k (m/s) = permeability

for the analysis of metals. This would give the most conservative results.

The results of these permeation tests are given in **Fig. 2**. This figure shows that the mortar samples used by Sprung [5] have a permeability factor of 10^{-9} to 10^{-7} m/s. At this porosity, 10^{-2} to 10^{-5} percent or 10^{-4} to 10^{-7} parts of the original total thallium and chromium concentration had leached into the permeating water. This indicates that very small fractions are leached when water permeates through mortar samples. The leaching rates show clearly that at least 99.99% of the highly soluble thallium and chromium had been tied up in the products of hydration, i.e., in the hardened cement. A similar assumption for metals tie-up can be made for other metals like cadmium, lead, etc. since their tie-up in the products of hydration are of the same pattern: they get incorporated into the crystalline structure of the hardened cement.

With the leaching rates given by Sprung [5] for every porous mortar, the amounts of metals leaching out from pressure pipe concrete may be estimated by extrapolation to porosities of this type of concrete. This extrapolation is shown in Fig. 2 with porosity parameters of concrete being 10^{-14} to 10^{-16} m/s. At these porosity parameters, the leaching rates would get into the 10^{-9} to 10^{-12}% range. These are ranges which clearly indicate that only an extremely minute portion of thallium and chromium, or metals in general

4. Surface leaching test results

Since surface effects and leaching are the determining factors when concrete or concrete products come in contact with drinking water, the testing as performed in accordance with ANSI/NSF Standard 61 [6] was determined to be appropriate. This standard was developed to test the suitability of products which come in direct contact with drinking water. The standard sets extremely tight limits as it prescribes as maximum allowable levels (MAL) only 1/10 of the maximum contaminate level (MCL) defined by the U.S.E.P.A. as drinking water standard [7]. While the most current MCL's are health based, and consequently on the conservative side, it seems extremely conservative and totally protective to human health to only allow 1/10 of these maximum contaminate levels to originate from one specific product which has been in contact with the water.

Extensive testing has been performed on mortar cubes from a Southdown cement plant at the laboratories of NSF International in accordance with Standard 61 [8]. Prior to analyzing the extraction water for 14 inorganic constituents the mortar cubes were prepared and hardened in accordance with ASTM-Standard C-109. Then the cubes were stored for the leaching and extraction reaction in water which for one series was buffered at pH 5 and for the other series buffered at pH 10. At the two pH values the cubes were stored for 72 ± 4 hours as prescribed by ANSI/NSF Standard 61. After the exposure the extraction waters were collected and analyzed for dissolved arsenic, barium, cadmium, mercury, lead, selenium, silver, beryllium, nickel, antimony, thallium, as well as for gross alpha and beta nuclide. The results of the analyses were normalized using a surface area-to-volume ratio of 65.62 cm^2/l, which reflect the surface area of a 2-foot diameter (60.96 cm) concrete pressure pipe filled with water for the 72 ± 4 hour period.

The results of the surface leaching tests are given in **Tables 2, 3** and **4**. The tables also state, in the right hand columns, the maximum concentration levels (MCL) as specified by U.S.E.P.A. [7], generally referred to as the "Drinking Water Standard", and the maximum allowable levels (MAL) according to Standard 61 [6]. Table 2 shows the results for mortar exposure at pH 5 and Table 3 the respective results for exposure at pH 10. It becomes obvious from the 2 Tables, that the majority of metals was not present at detectable levels in most of the extracts. The exceptions are listed in Table 4. Of all the metals analyzed Sb, Cd and Cr were only found in some of the extract waters of the pH = 5 exposures and Cr and Ni were found in some of the extract waters of the pH = 10 exposures. It should be noted, however, that only Cr was found in a considerable number of extracts. The extracts from mortar cubes showing Cr were 19 from the cement made with fossil fuel and 5 from the cement made with hazardous waste derived fuel supplementing the coal. After exposure to pH = 10 Sb was found above detection limit in 3 extracts from mortar cubes made with cement manufactured with fossil fuel. Sb was also found above detection limit after exposure at pH = 5 in 6 extracts from cubes made with cement manufactured with waste derived fuel. Cd was found above detection limit at pH = 5 in one extract. In all other cases, Cd, Cr and Ni were only found at detectable levels in 1 out of 20 extracts analyzed. Only Sb was found to be statistically higher in the extracts from the 6 cubes with waste derived fuel than from 3 cubes (out of 20) with fossil fuel. This finding applies for exposure of the mortar to pH = 5 and to pH = 10, i.e. over a

TABELLE 2: Schwermetallgehalte unterhalb der Nachweisgrenze in Eluaten mit pH 5 und die daraus berechneten normierten Konzentrationen (MCL und MAL)
TABLE 2: Metals present at non-detectable levels in pH = 5 extract waters and the corresponding normalized concentrations, MCLs, and MALs

Metall	Brennstoffart	Zahl der Analysen	Nachweise[1]	Nachweisgrenze (mg/l)	Normierte Konzentration[2] (mg/l)	MCL[3]/ MAL[4] (mg/l)
Arsen	fossiler Brennstoff	20	0	0,001	$4,80 \times 10^{-5}$	0,05/ 0,005
	Abfallbrennstoff	20	0	0,001	$4,80 \times 10^{-5}$	
Barium	fossiler Brennstoff	20	0	0,01	$4,80 \times 10^{-4}$	2,0/ 0,2
	Abfallbrennstoff	20	0	0,01	$4,80 \times 10^{-4}$	
Beryllium	fossiler Brennstoff	20	0	0,001	$4,80 \times 10^{-5}$	0,001/ 0,0001
	Abfallbrennstoff	20	0	0,0005	$2,40 \times 10^{-5}$	
Blei	fossiler Brennstoff	20	0	0,001	$4,80 \times 10^{-5}$	0,015/ 0,0015
	Abfallbrennstoff	20	0	0,001	$4,80 \times 10^{-5}$	
Nickel	fossiler Brennstoff	20	0	0,01	$4,80 \times 10^{-4}$	0,1/ 0,01
	Abfallbrennstoff	20	0	0,01	$4,80 \times 10^{-4}$	
Quecksilber	fossiler Brennstoff	20	0	0,0002	$9,60 \times 10^{-6}$	0,002/ 0,0002
	Abfallbrennstoff	20	0	0,0002	$9,60 \times 10^{-6}$	
Selen	fossiler Brennstoff	20	0	0,001	$4,80 \times 10^{-5}$	0,05/ 0,005
	Abfallbrennstoff	20	0	0,001	$4,80 \times 10^{-5}$	
Silber	fossiler Brennstoff	20	0	0,0001	$4,80 \times 10^{-6}$	0,10/ 0,010
	Abfallbrennstoff	20	0	0,0001	$4,80 \times 10^{-6}$	
Thallium	fossiler Brennstoff	20	0	0,001	$4,80 \times 10^{-5}$	(0,002 oder 0,001)/ (0,0002 oder 0,0001)
	Abfallbrennstoff	20	0	0,001	$4,80 \times 10^{-5}$	

[1] Die Anzahl der Nachweise bezieht sich auf die Anzahl von Proben mit Werten oberhalb der Nachweisgrenze von insgesamt 20 analysierten Proben
[2] Aus der Division der Nachweisgrenze durch 3 und Multiplikation des Ergebnisses mit 0,144 erhalten, was dem NF aus Gleichung 7 entspricht
[3] MCL (Maximum Contaminant Level) ist die höchstzulässige Konzentration von Verunreinigungen im Wasser, das an Verbraucher in einem öffentlichen Netz geliefert wird. Die Werte wurden direkt den Empfehlungen für Trinkwasser der US-Umweltschutzbehörde ("Drinking Water Regulations and Health Advisories" des Office of Water, U.S.E.P.A., Washington, D.C.) entnommen, veröffentlicht im November 1991
[4] MAL (Maximum Allowable Levels) Höchstzulässige Konzentration nach Definition der Norm ANSI/NSF 61, MAL entspricht 10 % MCL

[1] detected refers to the number of samples, out of the twenty analyzed, that had levels above the detection limit
[2] obtained by dividing the detection limit by 3 and multiplying the resulting number by 0.144 which corresponds to NF obtained in Equation 7
[3] MCL-maximum contaminant level-maximum permissible level of contaminant in water which is delivered to any user of a public water system. Specific values were taken directly from "Drinking Water Regulations and Health Advisories" published November 1991 by the Office of Water, U.S. EPA., Washington, DC
[4] MAL-maximum allowable levels-defined in ANSI/NSF Standard 61 as 10 % of the MCL

Metall	= Metal
Brennstoffart	= Fuel Type
Zahl der Analysen	= Analyzed
Nachweise[1]	= detected[1]
Nachweisgrenze	= detection limit
Normierte Konzentration[2]	= Normalized concentration[2]
Arsen	= Arsenic
Blei	= Lead
Quecksilber	= Mercury
Selen	= Selenium
Silber	= Silver
fossiler Brennstoff	= fossil fuel
Abfallbrennstoff	= waste derived fuel
oder	= or

Bereich. Insgesamt zeigen diese Ergebnisse jedoch deutlich, daß kein erkennbarer Unterschied im Auslaugverhalten der Oberfläche zwischen den mit unterschiedlichen Brennstoffkombinationen hergestellten Zementen besteht.

Wenn man die in Tabelle 4 angegebenen Eluat-Konzentrationen auf die Bedingungen bei der Lagerung eines Betonrohres mit einem Durchmesser von etwa 60 cm bezieht, d. h. die gemessenen Konzentrationen von Sb, Cd, Cr und Ni auf das Betonrohr normiert, dann liegen die Konzentrationen beträchtlich unter den MCL-Werten, die U.S.E.P.A. für Trinkwasser vorschreibt. Nach der Normierung liegen die gefundenen Konzentrationen auch mindestens 2 Größenordnungen unter den entsprechenden maximal zulässigen Mengen (MAL) nach ANSI/NSF 61. Demnach ist festzustellen, daß die Mörtelprüfkörper aus Zement, der mit fossilen Brennstoffen und Abfallbrennstoffen hergestellt wurde, ohne Risiko für Trinkwasser verwendet werden kann.

Tabelle 5 zeigt die normierten Ergebnisse der in den Eluaten gemessenen gesamten β- und α-Aktivität. Im allgemeinen ist die in den Eluaten gemessene normierte Aktivität sehr

wide pH range. Overall, these findings, however, clearly show that there is no discernable difference in the surface leaching behavior between cement made with fossil fuel and cement made with waste derived fuel.

Furthermore, when relating the extract concentration shown in Table 4 to the exposure of a 2 foot diameter pipe, i. e., when multiplying these concentrations of Sb, Cd, Cr and Ni by the normalization factor, the resulting concentrations are considerably lower than the maximum contaminant levels (MCL) as prescribed for drinking water by the U.S.E.P.A. After normalization all resulting concentrations are also at least 2 orders of magnitude below the corresponding maximum allowable levels (MAL) as defined in the ANSI/NSF Standard 61. Consequently, it can be stated that the mortar cubes made with cement manufactured with fossil fuel and the cubes made with cement manufactured with waste derived fuel can safely be exposed to drinking water.

Table 5 shows the normalized results of the gross beta and gross alpha activities as measured in the extracts. In general, the normalized activity in the extracts is very low, i.e. below

TABELLE 3: Schwermetallgehalte unterhalb der Nachweisgrenze in Eluaten mit pH 10 und die daraus berechneten normierten Konzentrationen (MCL und MAL)
TABLE 3: Metals present at non-detectable levels in pH = 10 extract waters and the corresponding normalized concentrations, MCLs, and MALs

Metall	Brennstoffart	Zahl der Analysen	Nachweise[1]	Nachweisgrenze (mg/l)	Normierte Konzentration[2] (mg/l)	MCL[3]/ MAL[4] (mg/l)
Antimon	fossiler Brennstoff	20	0	0,001	$4,80 \times 10^{-5}$	(0,01 oder 0,005)/
	Abfallbrennstoff	20	0	0,001	$4,80 \times 10^{-5}$	(o,001 oder 0,0005)
Arsen	fossiler Brennstoff	20	0	0,001	$4,80 \times 10^{-5}$	0,05/
	Abfallbrennstoff	20	0	0,001	$4,80 \times 10^{-5}$	0,005
Barium	fossiler Brennstoff	20	0	0,01	$4,80 \times 10^{-4}$	2,0/
	Abfallbrennstoff	20	0	0,02	$4,80 \times 10^{-4}$	0,2
Beryllium	fossiler Brennstoff	20	0	0,001	$4,80 \times 10^{-5}$	0,001/
	Abfallbrennstoff	20	0	0,0005	$2,40 \times 10^{-5}$	0,0001
Cadmium	fossiler Brennstoff	20	0	0,0001	$4,80 \times 10^{-6}$	0,005
	Abfallbrennstoff	20	0	0,0001	$4,80 \times 10^{-6}$	0,005
Blei	fossiler Brennstoff	20	0	0,001	$9,60 \times 10^{-5}$	0,015/
	Abfallbrennstoff	20	0	0,001	$9,60 \times 10^{-5}$	0,0015
Quecksilber	fossiler Brennstoff	20	0	0,0002	$9,60 \times 10^{-6}$	0,002/
	Abfallbrennstoff	20	0	0,0002	$9,60 \times 10^{-6}$	0.0002
Selen	fossiler Brennstoff	20	0	0,001	$4,80 \times 10^{-5}$	0,05/
	Abfallbrennstoff	20	0	0,001	$4,80 \times 10^{-5}$	0,005
Silber	fossiler Brennstoff	20	0	0,0001	$4,80 \times 10^{-6}$	0,10/
	Abfallbrennstoff	20	0	0,0001	$4,80 \times 10^{-6}$	0,010
Thallium	fossiler Brennstoff	20	0	0,001	$4,80 \times 10^{-5}$	(0,002 oder 0,001)/
	Abfallbrennstoff	20	0	0,001	$4,80 \times 10^{-5}$	(0,0002 oder 0,0001)

[1] Die Anzahl der Nachweise bezieht sich auf die Anzahl von Proben mit Werten oberhalb der Nachweisgrenze von insgesamt 20 analysierten Proben
[2] Aus der Division der Nachweisgrenze durch 3 und Multiplikation des Ergebnisses mit 0,144 erhalten, was dem NF aus Gleichung 7 entspricht
[3] MCL (Maximum Contaminant Level) ist die höchstzulässige Konzentration von Verunreinigungen im Wasser, das an Verbraucher in einem öffentlichen Netz geliefert wird. Die Werte wurden direkt den Empfehlungen für Trinkwasser der US-Umweltschutzbehörde ("Drinking Water Regulations and Health Advisories") des Office of Water, U.S.E.P.A., Washington, D.C.) entnommen, veröffentlicht im November 1991
[4] MAL (Maximum Allowable Levels) Höchstzulässige Konzentration nach Definition der Norm ANSI/NSF 61, MAL entspricht 10 % MCL

[1] detected refers to the number of samples, out of the twenty analyzed, that had levels above the detection limit
[2] obtained by dividing the detection limit by 3 and multiplying the resulting number by 0.144 which corresponds to NF obtained in Equation 7
[3] MCL-maximum contaminant level-maximum permissible level of contaminant in water which is delivered to any user of a public water system. Specific values were taken directly from "Drinking Water Regulations and Health Advisories" published November 1991 by the Office of Water, U.S. EPA, Washington, DC
[4] MAL-maximum allowable levels-defined in ANSI/NSF Standard 61 as 10 % of the MCL

Metall	= Metal
Brennstoffart	= Fuel Type
Zahl der Analysen	= Analyzed
Nachweise[1]	= detected[1]
Nachweisgrenze	= detection limit
Normierte Konzentration[2]	= Normalized concentration[2]
Antimon	= Antimony
Arsen	= Arsenic
Blei	= Lead
Quecksilber	= Mercury
Selen	= Selenium
Silber	= Silver
fossiler Brennstoff	= fossil fuel
Abfallbrennstoff	= waste derived fuel
oder	= or

gering, d.h. sie liegt unterhalb eines Niveaus, das im Trinkwasser bedenklich wäre. In allen Fällen sind die α- und β-Aktivitäten wesentlich geringer als die Grenzwerte für Trinkwasser nach MCL und MAL, und zwar unabhängig davon, welcher Brennstoff bei der Zementherstellung verwendet wurde.

Diese umfangreichen Versuche liefern einen klaren Hinweis darauf, daß speziell der Zement aus dem Southdown-Werk, sei er nun mit fossilen Brennstoffen oder mit Brennstoffen auf Abfallbasis und fossilen Brennstoffen hergestellt, nach Lagerung der damit hergestellten Mörtelprüfkörper in Wasser mit pH 5 bzw. pH 10 extrem niedrige Schwermetall- und Nuklidkonzentrationen aufweist. Aufgrund dieser Ergebnisse wurde mit NSF International Übereinstimmung darüber erzielt, daß weitere Versuche mit Zement aus anderen Southdown-Werken, die auch Abfallbrennstoffe einsetzen, zu keinem anderen Ergebnis führen

a level that would raise concern in drinking water. In all cases the alpha and beta activities are much lower than acceptable in drinking water in accordance with the MCL and MAL levels regardless of which fuel was used to manufacture the cement.

These extensive tests provide a clear indication that cement from this specific Southdown plant, whether manufactured with fossil fuel or with waste derived fuel and fossil fuel, shows extremely low metal concentrations and nuclide concentrations in extracts after respective mortar cubes had been exposed to pH = 5 and pH = 10 water. Based upon these results, a consensus was reached with NSF International, that any testing of cements from other Southdown plants using waste derived fuel will not show different results. Consequently, no additional exposure and surface leaching tests were performed using cements from 2 other Southdown facilities.

TABELLE 4: Schwermetallgehalte oberhalb der Nachweisgrenze in Eluaten mit pH 5 oder pH 10 und die daraus berechneten normierten Konzentrationen (MCL und MAL)
TABLE 4: Metals present at detectable levels in at least one of the pH = 5 or pH = 10 extract waters and the corresponding normalized concentrations, MCLs, and MALs

Metall	Brennstoffart	Zahl der Analysen	Nachweise[1]	Nachweisgrenze (mg/l)	Mittlere Konzentration[1] (mg/l)	Normierte Konzentration[2] (mg/l)	MCL[3]/ MAL[4] (mg/l)
Antimon pH = 5	fossiler Brennstoff	20	3	0,001	0,001	$4{,}80 \times 10^{-5}$	(0,01 oder 0,005)/
	Abfallbrennstoff	20	6	0,001	0,001	$4{,}80 \times 10^{-5}$	(0,001 oder 0,0005)
Cadmium pH = 5	fossiler Brennstoff	20	0	0,0001	0,0001	$4{,}80 \times 10^{-6}$	0,005/
	Abfallbrennstoff	20	1	0,0001	0,00011	$5{,}28 \times 10^{-6}$	0,0005
Chrom pH = 5	fossiler Brennstoff	20	1	0,001	0,001	$4{,}80 \times 10^{-5}$	0,1/
	Abfallbrennstoff	20	0	0,001	0,001	$4{,}80 \times 10^{-5}$	0,01
Chrom pH = 10	fossiler Brennstoff	20	19	0,001	0,003	$1{,}44 \times 10^{-4}$	0,1/
	Abfallbrennstoff	20	5	0,001	0,002	$9{,}60 \times 10^{-5}$	0,01
Nickel pH = 10	fossiler Brennstoff	20	0	0,01	0,01	$4{,}80 \times 10^{-4}$	0,1/
	Abfallbrennstoff	20	1	0,01	0,011	$5{,}28 \times 10^{-4}$	0,01

[1] Zur Vereinfachung der Berechnung des Mittelwerts wurde die Nachweisgrenze benutzt als die Konzentration für Proben, deren Konzentrationsniveau als nicht nachweisbar gekennzeichnet ist
[2] Aus der Division der Nachweisgrenze durch 3 und Multiplikation des Ergebnisses mit 0,144 erhalten, was dem NF aus Gleichung 7 entspricht
[3] MCL (Maximum Contaminant Level) ist die höchstzulässige Konzentration von Verunreinigungen im Wasser, das an Verbraucher in einem öffentlichen Netz geliefert wird. Die Werte wurden direkt den Empfehlungen für Trinkwasser der US-Umweltschutzbehörde („Drinking Water Regulations and Health Advisories" des Office of Water, U.S.E.P.A., Washington, D.C.) entnommen, veröffentlicht im November 1991
[4] MAL (Maximum Allowable Levels) Höchstzulässige Konzentration nach Definition der Norm ANSI/NSF 61, MAL entspricht 10% MCL

[1] to facilitate the calculation of the mean, the detection limit was used as the concentration for samples with analyte levels reported as non-detectable
[2] obtained by dividing the detection limit by 3 and multiplying the resulting number by 0.144 which corresponds to NF obtained in Equation 7
[3] MCL-maximum contaminant level-maximum permissible level of contaminant in water which is delivered to any user of a public water system. Specific values were taken directly from "Drinking Water Regulations and Health Advisories" published November 1991 by the Office of Water, U.S. EPA, Washington, DC
[4] MAL-maximum allowable levels-defined in ANSI/NSF Standard 61 as 10% of the MCL

Metall = Metal
Brennstoffart = Fuel Type
Zahl der Analysen = Analyzed
Nachweise[1] = detected[1]
Nachweisgrenze = detection limit

Mittlere Konzentration[1] = mean concentration[1]
Normierte Konzentration[2] = Normalized concentration[2]
Antimon = Antimony
Chrom = Chromium
oder = or

TABELLE 5: Normierte α- und β-Aktivität, Jahresdosisfaktoren (Erwachsene) und normierte potentielle Dosis
TABLE 5: Normalized Alpha and Beta Activities, Adult Annual Dose Factors, and Normalized Potential Doses

Nuklid	Brennstoffart	Normierte Konzentration (pCi/l)	Jahresdosis-Faktor (Erwachsene, Ganzkörper)	Jahresdosis-Faktor (Erwachsene, organspezifisch)	Potentielle Ganzkörper-Dosis (mrem/a)	Potentielle organspezifische Dosis (mrem/a)
β-Aktivität	fossiler Brennstoff	0,056	NA	NA	NA	NA
	Abfallbrennstoff	0,055				
Kalium[40]	fossiler Brennstoff	0,56	$2{,}0 \times 10^{-2}$	NA	0,0112	NA
	Abfallbrennstoff	0,55			0,011	
Strontium[90]	fossiler Brennstoff	0,017	$2{,}3 \times 10^{-3}$	4,6	$3{,}91 \times 10^{-5}$	0,0782
	Abfallbrennstoff	0,017			$3{,}91 \times 10^{-5}$	0,0782

Nuklid = Nuclide
Brennstoffart = Fuel type
Normierte Konzentration = Normalized Concentration
Jahresdosis-Faktor (Erwachsene, Ganzkörper) = Adult Annual Dose Factor, Total Body
Jahresdosis-Faktor (Erwachsene, organspezifisch) = Adult Annual Dose Factor, Organ Specific

Potentielle Ganzkörper-Dosis (mrem/a) = Potential Dose, Total Body mrem/yr
Potentielle organspezifische Dosis (mrem/a) = Potential Dose, Organ Specific mrem/yr
β-Aktivität = β-Activity
Kalium[40] = [40]potassium
Strontium[90] = [90]strontium

würden. Infolgedessen wurden keine zusätzlichen Lagerungs- und Oberflächenauslaugversuche mit Zementen aus zwei anderen Southdown-Werken vorgenommen.

5. Untersuchung des Gehalts organischer Bestandteile in Klinkerproben

Falls organische Bestandteile des Abfallbrennstoffs die im Ofensystem herrschenden Bedingungen überstehen sollten, müßten sie in dem aus dem Klinkerkühler kommenden Klinker erkennbar sein. Um diese Frage zu klären, wurden

5. Organic scans of clinker samples

If organic constituents from the hazardous waste derived fuel should survive the conditions in the kiln system they should show up in the clinker as discharged from the clinker cooler. Therefore, the extracts from clinker samples were scanned for organics using gas chromatography. On some clinker samples testing for polychlorinated dibenzodioxins and dibenzofurans (PCDD and PCDF) was also performed. For all these tests clinker samples from all Southdown facilities using waste derived fuel were used. Also, during a

Auszüge aus Klinkerproben gaschromatographisch auf organische Bestandteile untersucht. Einige Klinkerproben wurden auch auf ihren Gehalt an polychlorierten Dibenzodioxinen und Dibenzofuranen (PCDD und PCDF) untersucht. Für alle Versuche wurden Klinkerproben aus sämtlichen Southdown-Werken herangezogen, die Abfallbrennstoffe einsetzen. Es wurden auch während eines Zeitraums, in dem in diesen Werken ausschließlich fossiler Brennstoff verfeuert wurde, Klinkerproben für organische Untersuchungen sowie PCDD- und PCDF-Analysen entnommen [8, 9, 10].

Bei den gaschromatographischen Messungen wurde mit Ausnahme von Phthalaten keiner der gesuchten 8 250 organischen Stoffe in einer Konzentration gefunden, die oberhalb der Nachweisgrenze lag. Es ist äußerst unwahrscheinlich, daß diese Verbindungen die Temperaturen im Ofen oder im Klinkerkühler überstehen können. Daraus war zu schließen, daß die Phthalate aus einer Kontamination der Proben stammen müssen. Zudem war die gemessene Phthalat-Konzentration in den Klinkerextrakten so gering, daß sie nach der Normierung um mehrere Größenordnungen unter dem zulässigen Höchstwert für Verunreinigungen (MCL) nach U.S.E.P.A. und unter den zulässigen Höchstwerten (MAL) nach ANSI/NSF 61 lag.

Die Dioxin- und Furan-Analysen wurden bei extrem niedrigen Nachweisgrenzen durchgeführt. In sehr wenigen Klinkerproben wurden Spuren im Bereich von 10^{-10} bis 10^{-11} mg/l für Gesamtdioxin und Gesamtfuran gefunden. Im Vergleich zu den vorgeschlagenen zulässigen MCL-Höchstwerten für Gesamtdioxin von 10^{-8} mg/l und zulässigen MAL-Höchstwerten von 10^{-9} mg/l lagen die Meßwerte wiederum um ein bis zwei Größenordnungen unterhalb der relevanten Grenzwerte. Darüber hinaus konnte kein erkennbarer Unterschied zwischen Klinker, der mit Abfallbrennstoffen, und solchem, der ausschließlich mit fossilen Brennstoffen hergestellt wurde, festgestellt werden.

Die umfangreiche analytische Arbeit zur Ermittlung von Art und Anteil auslaugbarer Schwermetalle und organischer Stoffe zeigte deutlich, daß Zement zur Verwendung bei Produkten geeignet ist, die mit Trinkwasser in Berührung kommen, und zwar unabhängig von dem bei seiner Herstellung eingesetzten Brennstoff. Dies wurde auch sehr ausführlich von Calucci et al. berichtet [1].

Danksagung

E. J. Marston und J. B. Tompkins von der Firma Southdown trugen durch viele anregende Diskussionen zu diesem Artikel bei. Geschäftsleitung und Mitarbeiter der Southdown-Werke in Knoxville – TN, Fairborn – OH und Louisville – KY, haben ganz erheblich dazu beigetragen, die schwierige Probenentnahme und Mörtelherstellung gemäß Übereinkunft mit NSF International durchzuführen.

period when only fossil fuel was used in these facilities clinker was sampled for organic scans and for PCDD and PCDF analyses [8–10].

During the GC scans none of the 8250 organics on the target list were found at detection limits in the clinker extracts with the exception of phthalates. It is extremely unlikely that these compounds survived the kiln system or even the clinker cooler temperatures. It was, therefore, concluded that the phthalates must originate from sample contamination. Furthermore, the phthalate concentration in the clinker extracts was so low that it was, after normalization, orders of magnitude below the maximum contaminant level (MCL) as specified by U.S.E.P.A. and below the maximum allowable levels (MAL) as defined in the ANSI/NSF Standard 61.

The analyses for dioxin and furans were performed with extremely low detection limits. In very few clinker samples traces in the 10^{-10} to 10^{-11} mg/l range for total dioxins and total furans were analyzed. With proposed total dioxin maximum concentration levels of 10^{-8} mg/l and maximum allowable levels of 10^{-9} mg/l these results are again one to two orders of magnitude below levels of concern. Furthermore, no discernible difference between clinker manufactured with waste derived fuels and clinker manufactured with fossil fuel only were found.

The extensive analytical work in respect to leachable metals and organic showed clearly that cement is suitable to be used for products which come in contact with drinking water, regardless of the type of fuel used during the manufacturing process. This was also reported in great detail by Calucci et al. [11].

Acknowledgement

E. J. Marston and J. B. Tompkins at Southdown contributed to this paper through many inspiring discussions. Management and staff of Southdown's Knoxville, TN, Fairborn, OH, and Louisville, KY facilities went through many efforts to implement and perform the delicate sampling and mortar preparation protocol as agreed with NSF International.

Literature

[1] von Seebach, H.M., and Tompkins, J.B.: Metals Emissions Are Predictable; Rock Products, April 1991.

[2] Kleppinger, E.W., and Carnes, R.A.: Cement Kiln Incineration of Hazardous Waste, A critique; EWK & Associates, Washington, DC.

[3] Burning of Hazardous Waste in Boilers and Industrial Furnaces, 40CFR, Parts 260, 261, 264, 254; Federal Register 42504–42519, August 27, 1991.

[4] Hansen, Eric R., and MacGregor Miller, F.: Leaching Study of Portland Cement Using the TCLP Procedure. Paper presented at the PCA Conference on Emerging Technologies in Resource Recovery and Emission Reduction in the Cement Industry, Sept. 19–20, 1990, Dallas, TX.

[5] Sprung, S., and Rechenberg, W.: The Bonding of Heavy Metals in Secondary Material by Solidifying with Cement, 1988, Beton 38.

[6] ANSI/NSF Standard 61. Testing of Drinking Water Products issued by NSF International, Ann Arbor, MI, 48105.

[7] USEPA; Office of Water: Drinking Water Regulations and Health Advisory, Federal Register, Nov. 1991.

[8] NSF International: A Comparison of Metal and Organic Concentration in Cement and Clinker Made with Fossil Fuels to Cement and Clinker Made with Waste Derivatives as Fuels. Final Report of Nov. 13, 92, Submitted to Southdown, Inc. Houston, TX.

[9] NSF International: Organic Concentrations in Clinker Made with Waste Derived Fuels at the Kosmos Cement Plant. Final Report of Nov. 13, 1992, Submitted to Southdown, Inc., Houston, TX.

[10] NSF International: Organic Concentrations in Clinker made with Waste Derived Fuels of the Southwestern Cement Plant. Final Report of Dec. 21, 1992, Submitted to Southdown, Inc., Houston, TX.

[11] Calucci, M., Epstein, P., and Bartley, B.: A Comparison of Metal and Organic Concentrations in Cement and Clinker Made with Fossil Fuels to Cement and Clinker Made with Waste Derived Fuels. Paper presented at the "Emerging Technologies Symposium on Cement and Concrete in the Global Environment", March 10–11, 1993, Chicago, IL.

Umweltverträglichkeit von Zement
Environmental compatibility of cement

Les ciments et l'environnement

Los cementos y el medio ambiente

Von **S. Sprung, W. Rechenberg** und **G. Bachmann,** Düsseldorf/Deutschland

Zusammenfassung – Jeder Zement enthält geringe Konzentrationen an Schwermetallen. Sie stammen aus den überwiegend verwendeten natürlichen Einsatzstoffen sowie auch Sekundärstoffen und gelangen über den Klinkerbrennprozeß in das Produkt. Erste Untersuchungen hatten zum Ziel, die Art und Langzeitstabilität der Einbindung von Elementen wie Cr, Hg und Tl im Zementstein des Betons zu prüfen. Die hierfür verwendeten Betonprüfkörper lagerten in normalem Leitungswasser sowie in Wasser, das kalklösende Kohlensäure enthielt. Für die Lagerung wurde u.a. das Trogverfahren angewendet. Das Wasser/Feststoff-Verhältnis betrug 10:1. Aus den Untersuchungen an unzerkleinerten Betonprismen (4 × 4 × 16 cm) ging hervor, daß die Auslaugrate der leicht wasserlöslichen Elemente insgesamt sehr gering war. Sie nahm mit zunehmender Betondichtigkeit (W/Z-Wert) und mit zunehmendem Betonalter (Nachbehandlung) ab, stieg jedoch bei einem lösenden chemischen Angriff gegenüber den in Leitungswasser gelagerten Prüfkörpern etwas an. Die ausgelaugte Menge fiel nach kurzer Zeit auf einen Wert nahe Null ab, sobald der Vorgang durch Diffusion gesteuert wird. Für die Elemente Hg und Tl lagen die Werte der ersten Auslaugungen unterhalb der analytischen Bestimmungsgrenze. Die derzeit vorliegenden Ergebnisse lassen erwarten, daß die Umweltverträglichkeit von Zement durch Schwermetalle aus den überwiegend natürlichen Einsatzstoffen nicht beeinträchtigt wird.

Summary – Every cement contains small concentrations of heavy metals. They come from the natural materials which are used predominantly, as well as from secondary materials, and pass via the clinker-burning process into the product. The objective of initial investigations was to test the type and long-term stability of the fixation of elements such as Cr, Hg and Tl in the hardened cement paste of the concrete. The concrete test pieces used for this purpose were stored in normal tap water and in water which contained lime-dissolving carbonic acid. The trough method, among others, was used for the storage. The water/solids ratio was 10:1. Investigations on uncrushed concrete prisms (4 × 4 × 16 cm) showed that the overall leaching rate of the readily water-soluble elements was very low. It decreased with increasing concrete impermeability (W/C value) and with increasing concrete age (curing), but increased slightly during a dissolving chemical attack when compared with test pieces stored in tap water. Provided the process is controlled by diffusion the quantity leached out drops after a short time to a value close to zero. For the elements Hg and Tl the values from the first leachings lay below the limits of analytical determination. From the results which are available at present it can be expected that the environmental compatibility of cement will not be impaired by heavy metals from the predominantly natural materials used.

Résumé – Tout ciment contient de faibles concentrations de métaux lourds. Ces concentrations proviennent surtout des matières naturelles ainsi que des matières secondaires utilisées et, elles passent au produit au cours des processus de cuisson du clinker. Les premières études qui ont été effectuées avaient pour objectif d'examiner la nature et la stabilité à long terme des inclusions d'éléments tels que Cr, Hg et Tl dans la pâte de ciment après prise dans le béton. Les éprouvettes de béton utilisées à ce sujet étaient placées dans l'eau de distribution normale ainsi que dans l'eau contenant du dioxyde de carbone dissolvant la chaux. Quant au placement des éprouvettes, on a employé, entre autres, le procédé à auge. Le rapport eau-solide était de 10:1. Les études effectuées sur des prismes en béton non concassés (4 × 4 × 16 cm) ont démontré que l'indice de lessivage des éléments facilement solubles dans l'eau était, dans son ensemble, très réduit. Il descendait au fur et à mesure qu'augmentait la densité (valeur W/Z) et l'âge du béton (traitement ultérieur); cependant, il augmentait légèrement dans le cas d'une attaque chimique à effet de dissolution, par rapport aux éprouvettes placées dans l'eau de distribution. La quantité lessivée descendait, après peu de temps, jusqu'à une valeur d'à peu près zéro, dès que le processus de diffusion était contrôlé par la diffusion. Quant aux éléments Hg et Tl, les valeurs correspondant au premier lessivage se situaient au-dessous de la limite de détermination analytique. Selon les résultats disponibles actuellement, on peut s'attendre à ce que le comportement à l'environnement des ciments ne sera pas influencé de façon négative par les métaux lourds provenant des matières utilisées, qui sont principalement naturelles.

Resumen – Todo cemento contiene pequeñas concentraciones de metales pesados. Proceden de las materias primas naturales, mayormente utilizadas, así como de las materias primas secundarias y entran en el producto a través del proceso de cocción del clínker. Los primeros estudios llevados a cabo tuvieron por objetivo examinar el modo

y la estabilidad a largo plazo de la inclusión de elementos tales como Cr, Hg y Tl en la pasta de cemento después de su endurecimiento en el hormigón. Las probetas de hormigón empleadas para ello estaban metidas en agua corriente del grifo y en agua que contenía ácido carbónico disolvente de la cal. Para la colocación de dichas probetas se empleó, entre otros métodos, el de la artesa. La relación agua/sólido era de 10:1. De los ensayos efectuados con prismas de hormigón sin desmenuzar (4 x 4 x 16 cm) se desprende que el índice de lixiviación de los elementos fácilmente solubles en agua era, en su conjunto, bastante bajo. Disminuía a medida que aumentaba la densidad (valor W/Z) y la edad del hormigón (tratamiento posterior). Sin embargo, descendía ligeramente en el caso del ataque químico disolvente, en comparación con las probetas metidas en agua del grifo. La cantidad lixiviada bajaba, al cabo de poco tiempo, a un valor próximo a cero, en cuanto el proceso se controlaba por difusión. Respecto de los elementos Hg y Tl, los valores de las primeras lixiviaciones quedaban por debajo del límite de determinación analítica. Los resultados obtenidos hasta ahora permiten esperar que la inocuidad del cemento para el medio ambiente no sea afectada por los metales pesados provenientes de las materias primas utilizadas, que son mayormente naturales.

1. Einleitung

Die Umweltverträglichkeit von Zement wird meist als Summe aller Aufwendungen, wie z.B. Landschaftsverbrauch bei der Rohstoffgewinnung sowie Transport und Energieverbrauch einschließlich der Auswirkungen des Herstellprozesses auf die Umwelt, beurteilt [1]. Der Begriff „Umweltverträglichkeit" kann aber auch im engeren Sinn als eine Stoffeigenschaft aufgefaßt werden. In diesem Fall ist ausschließlich die Wirkung von Zement bzw. des daraus hergestellten Betons und seiner Inhaltsstoffe auf die menschliche Gesundheit, den Boden oder das Wasser zu bewerten.

Jeder Zement enthält geringe Konzentrationen an Schwermetallen [2–4]. Sie stammen überwiegend aus den natürlichen Einsatzstoffen, ggf. auch aus Sekundärstoffen, und gelangen über den Klinkerbrennprozeß und durch die Zumahlstoffe in den Zement und dementsprechend in den Beton. Die Spannweite der Spurenelementkonzentration von Portlandzement entspricht dabei in erster Näherung einem Konzentrationsbereich, wie er in natürlichen Böden anzutreffen ist [4]. Für eine ökologische Bewertung ist allerdings nicht der absolute Gehalt an Schwermetallen, sondern vielmehr die Mobilität bzw. die Bindungsstabilität der betrachteten Elemente im erhärteten Beton von Bedeutung. Die Stabilität einer solchen Bindung kann pauschal über die zeitabhängige Auslaugbarkeit beurteilt werden.

Untersuchungen an Reststoffen, die mit Zement verfestigt wurden, haben gezeigt, daß die Eluierbarkeit vieler Schwermetallverbindungen hierdurch drastisch herabgesetzt werden kann [5–8]. Maßgebend für die Größenordnung einer noch auslaugbar bleibenden Menge ist der Aufbau eines ausreichend dichten Gefüges der verfestigten Masse. Je dichter ein solches Gefüge ist, desto geringer ist die Schwermetallabgabe. Einige Elemente, wie z.B. Arsen, Blei und Zink, werden durch Zement offenbar weitgehend chemisch in den Hydratphasen gebunden oder bei pH-Werten von über 12 in der Porenlösung des Betons als Hydroxid ausgefällt [5–8].

Für die Beurteilung der Schwermetallbindung im Beton ist davon auszugehen, daß ein nach DIN 1045 [9] hergestellter Beton im Gegensatz zu verfestigten Massen immer wasserdicht ist. Verfahren, mit denen das Auslaugverhalten praxisgerecht geprüft werden soll, müssen demnach die Stabilität einer Einbindung von Schwermetallen in den oberflächennahen Zonen des Betons berücksichtigen. Dabei spielt die Diffusionsgeschwindigkeit der Schwermetalle aus dem inneren Bereich in die Randzone eine wesentliche Rolle [10]. Die Untersuchungen beschränkten sich zunächst auf die Art und Langzeitstabilität der Bindung leicht löslicher Verbindungen der Elemente Chrom (CrVI), Quecksilber (Hg) und Thallium (Tl) im Zementstein des Betons.

2. Prüfkörperherstellung

2.1 Zement

Als Bindemittel wurde ein Portlandzement 35 F verwendet, dessen natürliche Schwermetallgehalte an Cr, Hg und Tl in **Tabelle 1** wiedergegeben sind.

1. Introduction

The environmental compatibility of cement is usually assessed as the sum of all factors, such as depletion of the countryside during the excavation of raw materials, as well as transport and energy consumption and the effects of the manufacturing process on the environment [1]. However, the term "environmental compatibility" can also be taken in the narrower sense as a material property. In this case it is only the effect of cement, and of the concrete manufactured from it, and its constituents on human health and on soil and water which has to be evaluated.

Every cement contains small concentrations of heavy metals [2–4]. They come predominantly from the natural feed materials, and sometimes also from secondary materials, and pass via the clinker burning process and the interground additives into the cement, and therefore into the concrete. To a first approximation the range of trace element concentrations in Portland cement corresponds to the concentration range found in natural soils [4]. For an ecological evaluation, however, it is not the absolute content of heavy metals, but rather the mobility or the stability of bonding of the elements under consideration in the hardened concrete which are of importance. The stability of such bonding can be assessed as a whole by the amount which can be leached out in a given time.

Investigations on residual materials which have been stabilized with cement have shown that the elutability of many heavy metals can be drastically reduced by this method [5–8]. The formation of a sufficiently dense microstructure of the solidified material governs the order of magnitude of the quantity of material which is still leachable. The denser such a microstructure the lower is the elution of heavy metal. It appears that some elements, such as arsenic, lead and zinc, are substantially chemically bound in the hydrate phases by cement, or are precipitated as hydroxides at pH values above 12 in the pore solution of the concrete [5–8].

For assessing the heavy metal bonding in concrete it is to be assumed that, unlike consolidated materials, a concrete manufactured in accordance with DIN 1045 [9] is always water-tight. Methods which are intended to provide a test of leaching behaviour which resembles practical conditions must therefore take account of the stability of bonding of the heavy metals in the zones of the concrete close to the surface. In this situation the rate of diffusion of the heavy metals from the inner region into the boundary zone plays an important part [10]. The investigations were initially confined to the type and long-term stability of the bonding of readily soluble compounds of the elements chromium (CrVI), mercury (Hg) and thallium (Tl) in the hardened cement paste in concrete.

2. Producing the test pieces

2.1 Cement

A 35 F Portland cement was used as the binder, with the natural levels of the heavy metals Cr, Hg and Tl as shown in **Table 1**.

TABELLE 1: Gehalte einiger Spurenelemente im PZ 35 F; Angaben in g/t
TABLE 1: Levels of some trace elements in PZ 35 F Portland cement; data in g/t

Element	Gehalt
Chrom	79
Quecksilber	< 0,02
Thallium	< 0,2

Element	=	element
Gehalt	=	content
Chrom	=	chromium
Quecksilber	=	mercury
Thallium	=	thallium

Der Gesamtchromgehalt des PZ 35 F liegt mit 79 mg/kg in der oberen Hälfte des aus dem Schrifttum bekannten Bereichs [11], während die Gehalte an Quecksilber und Thallium mit < 0,02 mg/kg bzw. < 0,2 mg/kg kleiner als die Bestimmungsgrenze des atomabsorptionsspektrometrischen Verfahrens sind [12].

2.2 Zuschlag

Als Zuschlag bis zu Korngrößen von 2 mm wurde Quarzmehl 0/0,2 mm und Quarzsand 0/2 mm und für den Anteil 2/8 mm Rheinkies verwendet. Das Zuschlaggemisch wurde aus den 4 Korngruppen 0/0,2 mm, 0/2 mm, 1/2 mm und 2/8 mm so zusammengesetzt, daß seine Sieblinie im oberen Bereich zwischen den Sieblinien A und B des **Bildes 1** der DIN 1045 verlief.

Im oberen Teil der **Tabelle 2** sind die Gesamtgehalte der Schwermetalle in den einzelnen Korngruppen und im unteren Teil ihr wasserlöslicher Anteil aufgeführt [11, 12]. Der Gesamtgehalt wurde nach dem Verfahren der Feststoffatomisierung [12], der wasserlösliche Anteil aus einem wäßrigen Auszug ermittelt [11, 13]. Aus Tabelle 2 geht hervor, daß die Zuschlagfraktionen teilweise ähnlich hohe Gesamtchromgehalte aufweisen wie der Zement. Der wasserlösliche Anteil liegt im Bereich von 0,01 g/t. Das läßt darauf schließen, daß das Cr praktisch ausschließlich in mineralischer Bindung vorliegt. Die Elemente Hg und Tl waren meist nicht nachzuweisen.

TABELLE 2: Schwermetallgehalt im Zuschlag
TABLE 2: Heavy metal levels in the aggregate

Korngrößengruppe in mm	Chrom in g/t	Quecksilber in g/t	Thallium in g/t
Gesamtgehalt			
0/0,2	2,19	< 0,01	< 0,2
0/2	15,6	< 0,01	< 0,2
1/2	5,9	< 0,01	< 0,2
2/8	62,8	0,03	< 0,2
Wasserlöslicher Anteil			
0/0,2	< 0,01	< 0,01	< 0,01
0/2	0,01	< 0,01	< 0,01
1/2	< 0,01	< 0,01	< 0,01
2/8	0,01	< 0,01	< 0,01

Korngrößengruppen	=	particle size group
Chrom	=	chromium
Quecksilber	=	mercury
Thallium	=	thallium
Gesamtgehalt	=	total content
Wasserlöslicher Anteil	=	water-soluble fraction

The total chromium content of the PZ 35 F Portland cement of 79 mg/kg lies in the upper half of the range given in the literature [11], while the levels of mercury and thallium of < 0.02 mg/kg and < 0.2 mg/kg respectively are lower than the limit of determination of the atomic absorption spectrometry method [12].

2.2 Aggregate

0/0.2 mm quartz meal and 0/2 mm quartz sand were used for the aggregate of particle size up to 2 mm, and Rhine gravel for the 2/8 mm fraction. The aggregate mix was made up from the 4 particle size groups (0/0.2 mm, 0/2 mm, 1/2 mm and 2/8 mm) in such a way that its grading curve lay in the upper region between the grading curves A and B in **Fig. 1** of DIN 1045.

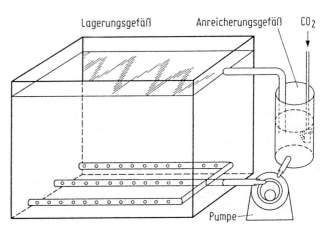

BILD 1: Versuchsanordnung zur Elution von Betonprüfkörpern mit CO_2-haltigem Wasser
FIGURE 1: Test rig for the elution of concrete test pieces with water containing CO_2

Lagerungsgefäß	=	storage vessel
Anreicherungsgefäß	=	enrichment vessel
Pumpe	=	pump

The total levels of the heavy metals in the individual particle size groups are listed in the upper part of **Table 2**, and their water-soluble fractions are listed in the lower part [11, 12]. The total levels were determined using the solids atomization method [12], and the water-soluble fractions were determined from aqueous extracts [11, 13]. It can be seen from Table 2 that some of the aggregate fractions contain similar total chromium levels to the cement. The water-soluble fractions lie around 0.01 g/t. From this it can be concluded that practically all the Cr present is mineralogically combined. In most cases the elements Hg and Tl could not be detected.

2.3 Fresh concrete

The composition of the fresh concrete can be seen from **Table 3**.

The fresh concretes were made with a cement content of 300 kg/m³ concrete and varying water/cement (w/c) ratios from 0.50 to 0.70. All the concretes were practically fully compacted. Aqueous solutions containing 100 mg of the respective heavy metal per litre were used for the mixing water. A reference concrete ("blank concrete") was also produced with a mixing water which contained no heavy metals. In this case the quantity of heavy metal in the concrete consisted exclusively of the sum of the heavy metals in the cement and in the aggregates.

2.4 Test pieces

4 × 4 × 16 cm test pieces with projections were produced from the fresh concrete. They were removed from the moulds after 1 d moist storage and then stored until 7 days old at 20°C and 100 % relative humidity in a cloud chamber from which drips were excluded.

2.3 Frischbeton

Die Zusammensetzung des Frischbetons geht aus **Tabelle 3** hervor.

Die Frischbetone wurden mit einem Zementgehalt von 300 kg/m³ Beton und unterschiedlichen Wasserzementwerten (w/z) von 0,50 bis 0,70 hergestellt. Alle Betone ließen sich praktisch vollständig verdichten. Als Anmachwasser dienten wäßrige Lösungen mit 100 mg des jeweiligen Schwermetalls je Liter. Außerdem wurde ein Vergleichsbeton (Nullbeton) hergestellt, dessen Anmachwasser keine Schwermetalle enthielt. In diesem Falle setzte sich die Schwermetallmenge im Beton ausschließlich aus der Summe des Schwermetallgehalts im Zement und in den Zuschlagstoffen zusammen.

2.4 Prüfkörper

Aus dem Frischbeton wurden 4 × 4 × 16 cm-Prüfkörper mit Meßzapfen hergestellt. Sie wurden nach 1 d Feuchtlagerung entformt und danach bis zum Alter von 7 d bei 20 °C und 100 % r.F. in einer Nebelkammer unter Ausschluß von Tropfwasser gelagert.

3. Untersuchung

Die Prüfkörper wurden nach Ablauf ihrer jeweiligen Nachbehandlung in einem Trog der Einwirkung von Düsseldorfer Leitungswasser (Trinkwasser) ausgesetzt [14–16]. Das Leitungswasser wurde in einer weiteren Versuchsreihe mit CO_2 angereichert. Hiermit sollte zusätzlich die Auswirkung eines sehr starken chemischen Angriffs und eine dadurch mögliche Erhöhung der Schwermetallfreisetzung geprüft werden [14–17].

3.1 Schwermetallelution

Im Bild 1 ist ein Lagerungsgefäß zur Elution von Betonprüfkörpern – in diesem Beispiel mit CO_2-angereichertem Wasser – schematisch dargestellt [14–16]. In das Wasser, das sich in einem Anreicherungsgefäß befindet, wird gasförmiges CO_2 eingeleitet. Aus dem Anreicherungsgefäß wird das Wasser mit einer Pumpe durch eine Lochgabel in das eigentliche Lagerungsgefäß gedrückt. Das Wasser steigt in dem Lagerungsgefäß gleichförmig mit einer Strömungsgeschwindigkeit von rd. $4 \cdot 10^{-5}$ cm/s auf und fließt durch einen Überlauf in das Anreicherungsgefäß zurück. Das Wasser kann daher als schwach strömend nach DIN 4030 [18] gekennzeichnet werden.

Messungen haben gezeigt, daß der Gehalt des Wassers an kalklösender Kohlensäure in dem Lagerungsbecken mit den Prüfkörpern nach dem Neubefüllen bzw. nach einem Wasserwechsel auf rd. 140 mg/l ansteigt und danach im Verlauf einer Woche auf rd. 100 mg/l abfällt. Daraus ergibt sich im Mittel ein ständiger Gehalt an kalklösender Kohlensäure von rd. 120 mg/l, der als sehr stark betonangreifend nach DIN 4030 [18] zu kennzeichnen ist. Die in ein Lagerungsgefäß eingesetzten 3 Prüfkörper wogen rd. 1,8 kg und wiesen eine Oberfläche von 864 cm² auf. Sie waren der Einwirkung von 18 l Wasser ausgesetzt. Das entspricht einem Masse/Volumen-Verhältnis von 1:10 nach DIN 38 414, Teil 4 (DEV S4) [13]. Das Oberflächen/Volumen-Verhältnis betrug rd. 0,05 cm^{-1} und entsprach damit annähernd den Vorgaben der KTW-Empfehlungen [19].

Die Meßzapfen der Prüfkörper ruhten in Kerben eines Kunststoffgestells, das in das Lagerungsgefäß nach Bild 1 eingesetzt wird. Dadurch war sichergestellt, daß das Wasser die Prüfkörper allseitig umströmte, jedoch nicht unmittelbar anströmen kann, wodurch sich ein zusätzlicher mechanischer Abtrag praktisch ausschließen ließ [20]. **Bild 2** zeigt ein Kunststoffgestell mit einem eingelegten Prüfkörper.

Für die Lagerung in unverändertem Leitungswasser entfiel das Anreicherungsgefäß, so daß das aus dem Überlauf austretende Wasser unmittelbar zur Pumpe floß. Das Wasser in dem Lagerungsbecken ohne CO_2-Anreicherung wurde ebenfalls wöchentlich gewechselt.

TABELLE 3: Herstellung und Nachbehandlung von Prüfkörpern
TABLE 3: Production and curing of test pieces

Frischbeton, undotiert	
Zement	PZ 35 F
Zementgehalt	300 kg/m³ Beton
Zuschlag, quarzitisch	A/B 8
Wasserzementwert	0,50 bis 0,70
Frischbeton, dotiert	
Dotiert mit	Cr, Hg, Tl
Zugesetzte Menge	100 mg/l Anmachwasser
Festbeton	
Prüfkörper	4 cm × 4 cm × 16 cm
1 d	In der Form
6 d	20 °C, 100 % rel. Feuchtigkeit

Frischbeton, undotiert	=	fresh concrete, undoped
Zement PZ 35 F	=	cement PZ 35 F Portland cement
Zementgehalt 300 kg/m³ Beton	=	cement content 300 kg/m³ concrete
Zuschlag, quarzitisch A/B 8	=	aggregate, quarzitic A/B 8
Wasserzementwert 0,50 bis 0,70	=	water/cement ratio 0.50 to 0.70
Frischbeton, dotiert	=	fresh concrete, doped
Dotiert mit Cr, Hg, Tl	=	doped with Cr, Hg, Tl
Zugesetzte Menge	=	quantity added
100 mg/l Anmachwasser	=	100 mg/l mixing water
Festbeton	=	hardened concrete
Prüfkörper 4 cm × 4 cm × 16 cm	=	test piece 4 cm × 4 cm × 16 cm
1 d In der Form	=	1 d in the mould
6 d 20 °C, 100 % rel. Feuchtigkeit	=	6 d 20 °C, 100 % rel. humidity

3. Investigation

When the curing was completed the test pieces were exposed to the action of Düsseldorf tap water (drinking water) in a trough [14–16]. In another test series the tap water was enriched with CO_2. This was intended to test the effect of a very strong chemical attack, possibly resulting in an increase in the liberation of heavy metals [14–17].

3.1 Heavy metal elution

The storage vessel for elution of the concrete test pieces – in this case with CO_2-enriched water – is shown diagrammatically in Fig. 1 [14–16]. Gaseous CO_2 is introduced into the water in an enrichment vessel from which the water is forced by a pump into the actual storage vessel through perforated branched tubes. The water rises uniformly in the storage vessel with a flow velocity of approximately $4 \cdot 10^{-5}$ cm/s and flows through an overflow back into the enrichment vessel. The water can therefore be described as flowing gently as defined in DIN 4030 [18].

Measurements have shown that the level of lime-dissolving carbondioxide in the water in the storage basin containing the test pieces after being newly filled or after a water change rises to approximately 140 mg/l and then falls to approximately 100 mg/l over the period of a week. On average this gives a continuous level of lime-dissolving carbondioxide of approximately 120 mg/l which, according to DIN 4030 [18], attacks concrete strongly. The 3 test pieces placed in the storage vessel weighed about 1.8 kg and had a surface area of 864 cm². They were exposed to the action of 18 l water. This corresponds to a mass/volume ratio of 1:10 as defined in DIN 38 414, Part 4 (DEV S4) [13]. The surface area to volume ratio is approximately 0.05 cm^{-1} and therefore corresponds approximately to the targets set by the KTW recommendations [19].

The projections on the test pieces rest in notches in a plastic pedestal which is placed in the storage vessel shown in Fig. 1. This ensures that the water flows around all sides of

TABELLE 4: Mittelwerte und Extremwerte von Schwermetallen im Trinkwasser für die Elution von Beton; Angaben in mg/l
TABLE 4: Average values and extreme values of heavy metals in the drinking water used for elution of concrete; data in mg/l

Element	Kleinstwert	Mittelwert	Höchstwert
Chrom	$< 0{,}1 \cdot 10^{-3}$	$0{,}5 \cdot 10^{-3}$	$2{,}1 \cdot 10^{-3}$
Quecksilber	$< 0{,}1 \cdot 10^{-3}$	$0{,}25 \cdot 10^{-3}$	$1{,}9 \cdot 10^{-3}$
Thallium	$< 0{,}5 \cdot 10^{-3}$	$0{,}5 \cdot 10^{-3}$	$< 0{,}5 \cdot 10^{-3}$

Element = element
Kleinswert = lowest measured value
Mittelwert = mean value
Höchstwert = highest measured value
Chrom = chromium
Quecksilber = mercury
Thallium = thallium

3.2 Probenahme und Analyse

Aus dem Zulauf des Leitungswassers wurden während des Befüllens der Lagerungsgefäße verschiedene Teilvolumen von etwa 250 ml entnommen, zu einer Mischprobe vereinigt und analysiert. **Tabelle 4** gibt die jeweiligen Mittelwerte und die zugehörigen Extremwerte des Schwermetallgehalts in den Mischproben des Trinkwassers für die bisherige Versuchsdauer wieder.

Nach Ablauf einer Woche wurde unmittelbar vor dem Entleeren der Lagerungsbecken eine Probe von etwa 1 l entnommen. In dieser Einzelprobe und in den Zulaufmischproben wurden neben den Elementen Cr, Hg und Tl auch das Chlorid, das Sulfat, die Alkalien, das Calcium und der pH-Wert bestimmt. Die wöchentlichen Mischproben aus dem Leitungswasserzulauf waren zur Korrektur der im Leitungswasser nach einer Woche gefundenen Bestandteile erforderlich. Dabei zeigte sich unter anderem, daß der Sulfatgehalt des Lagerungswassers stets geringfügig niedriger war als der des Leitungswasserzulaufs. Daraus geht hervor, daß Beton Bestandteile eines anstehenden Wassers aufnehmen und binden kann.

BILD 2: Kunststoffgestell zur Aufnahme von 4 cm × 4 cm × 16 cm-Prismen mit Meßzapfen
FIGURE 2: Plastic pedestal for holding 4 cm × 4 cm × 16 cm prisms with projections

the test piece, but cannot flow directly at it, and practically eliminates any additional mechanical abrasion [20]. **Fig. 2** shows a plastic pedestal with a test piece on it.

The enrichment vessel is dispensed for storage in untreated tap water and the water leaving the overflow flows directly to the pump. The water in the storage basin without CO_2-enrichment is also changed weekly.

3.2 Sampling and analysis

Various subsamples of about 250 ml were taken from the tap water feed during the filling of the storage vessel; these were combined to form a composite sample and then analyzed. **Table 4** shows the respective mean values and the associated extreme values of the levels of heavy metal in the composite samples of the drinking water for the duration of the test so far.

After a week had expired a sample of approximately 1 litre was taken immediately before emptying the storage basin. The levels of chloride, sulphate, alkalis and calcium and the pH value were measured in these individual samples and in the composite feed samples as well as the elements Cr, Hg and Tl. The weekly composite sample from the tap water feed was needed for correcting the constituents found in the storage water after a week. This showed, among other things, that the sulphate content in the storage water was always slightly lower than in the tap water supply. This is because the concrete is able to absorb and bind some constituents from the adjacent water.

4. Results

4.1 Chromium

The individual chromium values measured during the investigation of the eluates are shown in **Fig. 3** as the amount of chromium eluted during the leaching period. In all cases the leaching was carried out on three 4 × 4 × 16 cm test pieces with a total mass of approximately 1.8 kg. In each case the concretes contained 300 kg cement per m³. The chromium content of the concrete was consistently 11 mg/kg. However, the concretes differed in their water/cement ratios which lay between 0.5 and 0.7. A further test parameter was the differing leaching medium.

The shapes of the curves in Fig. 3 show that over about the first 3 weeks (21 days) $5.0 \cdot 10^{-5}$ to $2.1 \cdot 10^{-5}$ g Cr (depending on the water/cement ratio) were eluted from the concrete into the drinking water (hollow symbols) for a total quantity in the concrete of 11 mg. After that the quantity eluted de-

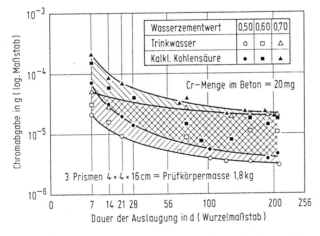

BILD 3: Über die Elutionszeit kumulierte Chromabgabe von Beton in Abhängigkeit vom Wasserzementwert und der Art des Auslaugmediums. Gesamt-Cr: 18 mg
FIGURE 3: Cumulative chromium elution from concrete during the elution period as a function of the water/cement ratio and the type of leaching medium. Total Cr = 18 mg.

Chromabgabe in g (log. Maßstab) = chromium elution in g (log scale)
Dauer der Auslaugung in d (Wurzelmaßstab) = leaching time in d (quare root scale)
3 Prismen 4 × 4 × 16 cm = Prüfkörpermasse 1,8 kg = three 4 × 4 × 16 cm prisms = 1.8 kg test piece mass
Wasserzementwert = water/cement ratio
Trinkwasser = drinking water
Kalkl. Kohlensäure = lime-dissolving carbondioxide
Cr-menge im Beton = 20 mg = Cr quantity in the concrete = 20 mg

4. Ergebnisse

4.1 Chrom

Die bei der Untersuchung der Eluate ermittelten Chromeinzelwerte wurden im **Bild 3** als Chromabgabe über der Auslaugdauer wiedergegeben. Ausgelaugt wurden stets 3 Prüfkörper 4 cm x 4 cm x 16 cm mit einer Gesamtmasse von ca. 1,8 kg. Die Betone enthielten jeweils eine Zementmenge von 300 kg/m^3. Der Chromgehalt des Betons betrug einheitlich 11 mg/kg. Die Betone unterschieden sich jedoch im Wasserzementwert, der zwischen 0,5 und 0,7 lag. Ein weiterer Versuchsparameter war das unterschiedliche Auslaugmedium.

Der Verlauf der Kurven im Bild 3 zeigt, daß die Betone etwa in den ersten 3 Wochen (21 Tage) in Abhängigkeit vom Wasserzementwert $5{,}0 \cdot 10^{-5}$ g bis $2{,}1 \cdot 10^{-5}$ g Cr an das Trinkwasser (offene Symbole) bei einer Gesamtmenge im Beton von 11 mg abgaben. Danach nahm die eluierte Menge zunächst rasch, danach bis 200 d langsam auf $1{,}9 \cdot 10^{-5}$ g bis $2{,}6 \cdot 10^{-6}$ g Cr ab. Die Kurven näherten sich dabei asymptotisch einem Endwert. Der Verlauf deutet auf diffusionskontrollierte Vorgänge hin.

Die ausgelaugte Chrommenge nahm zu, wenn statt des im Kalk-Kohlensäure-Gleichgewicht befindlichen Trinkwassers ein chemisch sehr stark angreifendes saures Wasser mit einem Gehalt von rd. 120 mg kalklösende Kohlensäure je Liter auf den Beton einwirkte. Aus einem dichten Beton (w/z = 0,50; geschlossene Kreise) wurde dann soviel Chrom ausgelaugt wie aus einem weniger dichten Beton (w/z = 0,60; offene Vierecke) durch Trinkwasser. Die durch das saure Wasser eluierte Menge nahm außerdem mit abnehmender Dichtigkeit des Betons (w/z = 0,70; geschlossene Dreiecke) deutlich zu. Diese erhöhte Elution ließe sich jedoch durch bautechnische Maßnahmen verhindern, da ein nach DIN 4030 [18] chemisch sehr stark angreifendes Wasser nicht unmittelbar auf Beton einwirken darf [9, 17].

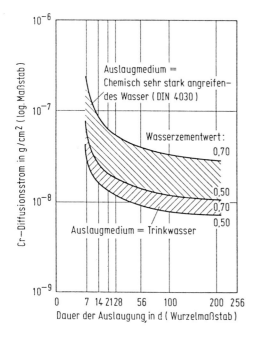

BILD 4: Über die Elutionszeit aus der Betonoberfläche austretender kumulierter Chrom-Diffusionsstrom in Abhängigkeit vom Wasserzementwert und der Art des Auslaugmediums.
FIGURE 4: Cumulative chromium diffusion flow leaving flow leaving the concrete surface during the elution period as a function of the water/cement ratio and the type of leaching medium

Cr-Diffusionsstrom in g/cm^2 (log. Maßstab)	= Cr diffusion flow in g/cm^2 (log scale)
Dauer der Auslaugung in d (Wurzelmaßstab)	= leaching time in d (square root scale)
Auslaugmedium = Chemisch sehr stark angreifendes Wasser (DIN 4030)	= leaching medium = chemically very strongly corrosive water (DIN 4030)
Wasserzementwert	= water/cement ratio
Auslaugmedium = Trinkwasser	= leaching medium = drinking water

creased, rapidly at first and then slowly, to $1{.}9 \cdot 10^{-5}$ to $2{.}6 \cdot 10^{-6}$ g Cr at 200 days. The curves approached their final values asymptotically. The shape points to diffusion-controlled processes.

The quantity of chromium leached out increases if the concrete is attacked by a chemically very strongly corrosive acidic water containing approximately 120 mg lime-dissolving carbondioxide per litre, instead of by drinking water with its equilibrium between lime and carbondioxide. As much chromium was then leached from a dense concrete (w/c = 0.50; solid circles) as from a less dense concrete (w/c = 0.60; hollow squares) by drinking water. The quantity eluted by the acid water also increased with decreasing density of the concrete (w/c = 0.70; solid triangles). This increased elution can, however, be prevented by structural engineering measures as a chemically very strongly corrosive water as defined in DIN 4030 [18] must not be allowed to act directly on concrete [9, 17].

As already mentioned, the curves shown in Fig. 3 point to a diffusion-controlled process. To investigate this further the measured values were expressed in terms of the surface areas of the test pieces. The individual points have been omitted in **Fig. 4**.

In the two regions shaded in opposite directions the respective diffusion flows per unit surface area fall as a function of the leaching medium. The lower narrow region characterizes the elution possible by a chemically non-corrosive drinking or ground water. The upper wide region shows the leaching of chromium during "very strong" chemical attack on concrete, but this can only occur to the extent shown if the recognized rules of construction technology [9] are not observed. The results therefore also show how a diffusion-controlled leaching process can be accelerated if a "very strong" chemical attack as defined in DIN 4030 is superimposed on it. The dissolving chemical attack is itself diffusion-dependent and is greatly accelerated if, for example, the concentration of the lime-dissolving carbondioxide exceeds a value of 100 mg/l [9, 15–18]. Test methods in which, to test its leaching behaviour, the hardened cement paste in the concrete is exposed to an attack by strong mineral acids, such as nitric or sulphuric acids [21–25], should therefore be designated, at least in some cases, as being as equally remote from practical conditions as those test methods in which the protective silica gel layer produced during a chemical attack [20, 26] is, in addition to the action of the acid, removed mechanically from the concrete surface [27].

4.2 Comparison with other heavy metals

The heavy metals mercury and thallium showed difXusion-controlled leaching characteristics similar to those of chromium. However, the quantieties leached out lay at such a low level that it was not possible to carry out a detailed investigation. Use was therfore only made of the results of those trials in which the concretes were doped through the mixing water by adding an extra 100 mg heavy metal per litre of mixing water.

Fig. 5 shows a bar chart of the total quantities of elements in mg leached out from the concretes with water/cement ratios of 0.50 during the test duration so far of approximately 200 days. The rear row shows the results of the doped concretes and the front row the results of the unchanged concretes with natural levels of the elements. The left-hand group of columns show the values obtained with drinking water as the leaching medium and the right-hand group the values obtained with lime-dissolving carbondioxide as the leaching medium. The elemental content in the doped concrete was approximately 12 mg higher than in the undoped concrete (for 3 test pieces in each case).

It can be seen from the diagram that generally, regardless of the extra addition, the quantities of heavy metal leached out by drinking water after about 200 days were extremely small. Comparatively more was eluted from the doped concretes, but there is no relationship which can be detected at present between the increase and the elevated heavy metal

Wie bereits ausgeführt, deuten die im Bild 3 wiedergegebenen Kurven auf einen diffusionskontrollierten Prozeß hin. Um dem weiter nachzugehen, wurden die Meßwerte auf die Oberfläche der Prüfkörper bezogen. In der Darstellung im **Bild 4** wurde auf eine Wiedergabe der einzelnen Punkte verzichtet.

In die beiden gegenläufig schraffierten Felder fallen die jeweiligen oberflächenbezogenen Diffusionsströme in Abhängigkeit vom Auslaugmedium. Der untere schmale Bereich kennzeichnet eine durch chemisch nicht angreifendes Trink- bzw. Grundwasser mögliche Elution. Der obere breite Bereich gibt das Auslaugen von Chrom bei „sehr starkem chemischen" Angriff auf Beton wieder, der sich jedoch nur dann in dem dargestellten Ausmaß auswirken kann, wenn die anerkannten Regeln der Bautechnik [9] nicht beachtet werden. Das Ergebnis zeigt demnach auch, wie ein diffusionskontrollierter Auslaugprozeß beschleunigt werden kann, wenn er von seinem „sehr starken" chemischen Angriff nach der Definition der DIN 4030 überlagert wird. Der lösende chemische Angriff ist seinerseits diffusionsabhängig und beschleunigt sich erheblich, wenn z. B. die Konzentration der einwirkenden kalklösenden Kohlensäure einen Wert von 100 mg/l übersteigt [9, 15–18]. Prüfverfahren, bei denen der Zementstein im Beton zur Überprüfung des Auslaugverhaltens einem Angriff starker Mineralsäuren, wie z.B. Salpeter- oder Schwefelsäure ausgesetzt wird [21–25], sind demnach — zumindest teilweise — als ebenso praxisfremd zu bezeichnen wie solche Prüfverfahren, bei denen die bei einem chemischen Angriff entstehende Kieselgelschutzschicht [20, 26] zusätzlich zur Säurewirkung mechanisch von der Betonoberfläche entfernt wird [27].

4.2 Vergleich mit anderen Schwermetallen

Die Schwermetalle Quecksilber und Thallium zeigten im Vergleich zu Chrom ein ähnliches, diffusionskontrolliertes Auslaugverhalten. Die ausgelaugten Mengen lagen jedoch auf einem so niedrigen Niveau, daß eine weitergehende Auswertung nicht vorgenommen werden konnte. Daher wurden hierfür nur die Ergebnisse der Versuche herangezogen, bei denen die Betone über das Anmachwasser zusätzlich mit jeweils 100 mg Schwermetall je Liter Anmachwasser dotiert wurden.

Bild 5 zeigt in einem Säulendiagramm die insgesamt während der bisher abgelaufenen Prüfzeit von rd. 200 Tagen aus den Betonen mit einem Wasserzementwert von 0,50 ausgelaugten Elementmengen in mg. Die hintere Reihe gibt die Ergebnisse der dotierten Betone, die vordere Reihe die Ergebnisse der unveränderten Betone mit den natürlichen Elementgehalten wieder. Die linke Säulengruppe zeigt die mit dem Auslaugmedium Trinkwasser, die rechte Säulengruppe die mit dem Auslaugmedium kalklösende Kohlensäure erhaltenen Werte. Der Elementgehalt war in jeweils 3 Prüfkörpern aus dotiertem Beton um rd. 12 mg höher als in dem undotierten Beton.

Generell geht aus der Darstellung hervor, daß die durch Trinkwasser nach etwa 200 d ausgelaugten Schwermetallmengen, unabhängig von der zusätzlichen Dotierung, außerordentlich gering waren. Aus den dotierten Betonen wurde zwar vergleichsweise mehr eluiert, jedoch steht der Zuwachs in keinem derzeitig erkennbaren Zusammenhang mit dem erhöhten Schwermetallgehalt dieser Betone. Daraus ist zu schließen, daß die Elemente im Zementstein offenbar in Verbindung mit nur geringer Löslichkeit vorliegen und dementsprechend der Diffusionsgradient zum auslaugenden Wasser klein ist. Durch die kalklösende Kohlensäure nahm die ausgelaugte Menge nur bei den dotierten Betonen erkennbar zu. Insgesamt waren jedoch die ausgelaugten Mengen mit höchstens 0,50 mg auch unter diesen extremen Verhältnissen nach 200 Tagen gering. Bei den Elementen Tl und Hg waren praktisch keine Veränderungen festzustellen.

4.3 Schlußfolgerungen

Aus den Ergebnissen im Bild 5 geht hervor, daß selbst aus den Betonen, die mit Schwermetallen künstlich angerei-

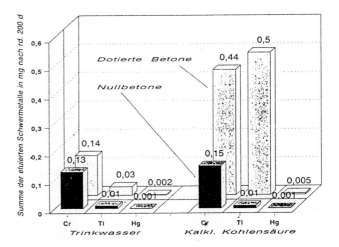

BILD 5: Vergleich der nach ca. 200 Tagen insgesamt aus Betonen mit einem Wasserzementwert von 0,50 ausgelaugten Schwermetallmengen in Abhängigkeit vom Auslaugmedium.
FIGURE 5: comparison of the total quantities of heavy metals leached from concretes with water/cement ratios of 0.50 after approximately 200 days as a function of the leaching medium

Summe der eluierten Schwermetalle in mg nach rd. 200 d = total eluted heavy metals in mg after approx. 200 d
Trinkwasser = drinking water
Kalkl. Kohlensäure = lime-dissolving carbondioxide
Dotierte Betone = doped concretes
Nullbetone = "blank" concretes

content of these concretes. From this it can be concluded that the elements are clearly present in the hardened cement paste in compounds which are only slightly soluble so that the diffusion gradient towards the leaching water is small. It was only with the doped concretes that the quantities leached out were increased to a detectable extent by the lime-dissolving carbondioxide. As a whole, however, the quantities leached out after 200 days, with a maximum of 0.50 mg, were low even under these extreme conditions. Practically no changes could be detected with the elements Tl and Hg.

4.3 Conclusions

It follows from the results in Fig. 5 that even with the concretes which were artificially enriched with heavy metals and exposed to a very severe attack by lime-dissolving carbondioxide, the total quantities leached out are very small. Relating these quantities of heavy metals to the total volume of water involved during the test period as a result of the weekly water change gives the concentrations listed in **Table 5**. The limits in the German Drinking Water Regulations [28] are also listed for comparison. Line 1 of Table 5 shows the leaching medium used and Line 2 the different water-cement ratios of the concrete. Lines 4 to 6 contain the heavy metal concentrations for the water in which the undoped concretes with natural levels of heavy metals were stored, and Lines 8 to 10 contain the corresponding data for the concretes with artificially elevated levels of heavy metal.

The relatively highest concentrations of Cr, Hg and Tl are found in Lines 8 to 10 on the right-hand side (lime-dissolving carbondioxide, w/c = 0.70). However, they would have to be increased by factors of approximately 42 and 25 respectively to reach the corresponding limits of the German Drinking Water Regulations (Line 12) for Cr and Hg. The Regulations [28] do not specify a limit for thallium, so the limit for lead has been used for comparison. The thallium concentrataon would have to be increased by a factor of 20 in order to reach this limit.

The heavy metal concentrations in water generally decrease with decreasing water/cement ratio. During the action of drinking water on concrete the concentrations of the heavymetals were substantially lower than with water containing CO_2. they lie below the limit in the Drinking Water Regulations by a factor of at least 100.

chert und einem sehr starken Angriff durch kalklösende Kohlensäure ausgesetzt waren, die insgesamt ausgelaugten Mengen sehr klein sind. Werden diese Schwermetallmengen auf das sich innerhalb des Versuchszeitraums durch den wöchentlichen Wasserwechsel ergebende gesamte Wasservolumen bezogen, ergeben sich die in **Tabelle 5** zusammengestellten Konzentrationen. Zum Vergleich sind die Grenzwerte der Trinkwasserverordnung (TrinkwV) [28] mit aufgeführt. Zeile 1 der Tabelle 5 gibt die verwendeten Auslaugmedien und Zeile 2 die unterschiedlichen Wasserzementwerte der Betone wieder. Die Zeilen 4 bis 6 enthalten die Schwermetallkonzentrationen des Wassers, in dem die undotierten Betone mit natürlichem Schwermetallgehalt lagerten und die Zeilen 8 bis 10 die entsprechenden Angaben zu den Betonen mit künstlich erhöhtem Schwermetallgehalt.

Die relativ höchsten Konzentrationen an Cr, Hg und Tl finden sich in den Zeilen 8 bis 10, rechts (kalklösende Kohlensäure, w/z = 0,70). Sie müßten jedoch mit einem Faktor von rd. 42 bzw. 25 erhöht werden, um die entsprechenden Grenzwerte der TrinkwV (Zeile 12) für Cr und Hg zu erreichen. Für Thallium nennt die TrinkwV [28] keinen Grenzwert. Deshalb wurde der Bleigrenzwert zum Vergleich herangezogen. Die Thalliumkonzentration müßte demnach um den Faktor 20 erhöht werden, um diesen Grenzwert zu erreichen.

Die Schwermetallkonzentrationen im Wasser nahmen generell mit kleiner werdendem Wasserzementwert ab. Bei der Einwirkung von Trinkwasser auf Beton waren die Konzentrationen der Schwermetalle im Vergleich zu CO_2-haltigem Wasser wesentlich geringer. Sie liegen wenigstens um den Faktor 100 unter dem Grenzwert der TrinkwV.

Aus diesem Vergleich geht deutlich hervor, daß Beton selbst dann uneingeschränkt im Trinkwasserbereich einsetzbar bleibt, wenn der Schwermetallgehalt künstlich erhöht würde.

5. Schlußbemerkung

Die Stoffeigenschaft „Umweltverträglichkeit" gibt Auskunft über die Frage, ob ein Stoff wie Zement bzw. der daraus hergestellte Beton Substanzen an die Umwelt abgeben und damit zu einer Belastung beitragen kann oder ob er sich umweltneutral verhält. Um die Umweltverträglichkeit zu prüfen, wurden Betone mit unterschiedlicher Gefügedichtigkeit hergestellt und mit Trinkwasser bzw. mit chemisch sehr stark angreifendem, kalklösende Kohlensäure enthaltendes Wasser in Trögen ausgelaugt. Die Betone wurden mit 300 kg PZ 35 F je m³ Beton und Wasserzementwerten von 0,50 bis 0,70 mit einem Zuschlaggemisch entsprechend der Sieblinie A/B 8 hergestellt. Teilweise wurden die Schwermetallgehalte des Betons künstlich durch Zusatz von je 100 mg Cr, Hg bzw. Tl/l Anmachwasser erhöht. Aus dem Frischbeton wurden 4 × 4 × 16 cm-Prüfkörper hergestellt. Die Prüfkörper lagerten 1 d in der Form und danach, vor Tropfwasser geschützt, 7 d in einer Nebelkammer bei 20 °C und rd. 100 % r. F.

Aus den Analysen der Eluate ging hervor, daß die Abgabe von Schwermetallen aus Beton an ein anstehendes Wasser den Diffusionsgesetzen gehorcht. Die ausgelaugten Mengen hingen deutlich von der Gefügedichtigkeit des Betons ab. Aus dichtem Beton (w/z = 0,50) wurden deutlich weniger Schwermetalle ausgelaugt als aus Beton mit erhöhtem Kapillarporenvolumen (w/z = 0,70). Die eluierten Mengen nahmen außerdem zu, wenn weniger dichter Beton mit chemisch sehr stark angreifendem Wasser statt mit Trinkwasser ausgelaugt wurde. Die sich dabei im Gesamteluat einstellenden Konzentrationen lagen jedoch mindestens um den Faktor 20 unter dem Grenzwert der Trinkwasserverordnung. Bei gefügedichten Betonen stellten sich Konzentrationen ein, die wenigstens um den Faktor 100 den Grenzwert der Trinkwasserverordnung unterschritten. Daraus geht hervor, daß sachgerecht hergestellter Beton uneingeschränkt in Trinkwasseranlagen eingesetzt werden kann.

TABELLE 5: Konzentration aus der insgesamt eluierten Schwermetallmenge und dem Totalvolumen nach 200 d; Angaben in mg/l
TABLE 5: Concentration made up from the total quntity of heavvy metal wluted and the total volume after 200 days; data in mg/l

1	Auslaug- medium	\multicolumn{3}{c}{Trinkwasser}			\multicolumn{3}{c}{Kalklösende Kohlensäure}	
2	Wasser- zementwert	0,50	0,60	0,70	0,50	0,60	0,70
3	\multicolumn{7}{c}{Undotierte Betone mit natürlichem Schwermetallgehalt}						
4	Chrom	$3 \cdot 10^{-4}$	$3 \cdot 10^{-4}$	$3 \cdot 10^{-4}$	$3 \cdot 10^{-4}$	$7 \cdot 10^{-4}$	$8 \cdot 10^{-4}$
5	Quecksilber	$2 \cdot 10^{-6}$	$2 \cdot 10^{-6}$	$2 \cdot 10^{-6}$	$2 \cdot 10^{-6}$	$2 \cdot 10^{-6}$	$2 \cdot 10^{-6}$
6	Thallium	$2 \cdot 10^{-5}$	$2 \cdot 10^{-5}$	$2 \cdot 10^{-5}$	$2 \cdot 10^{-5}$	$2 \cdot 10^{-5}$	$2 \cdot 10^{-5}$
7	\multicolumn{7}{c}{Dotierte Betone (100 mg Schwermetall/l Anmachwasser)}						
8	Chrom	$3 \cdot 10^{-4}$	$3 \cdot 10^{-4}$	$3 \cdot 10^{-4}$	$9 \cdot 10^{-4}$	$11 \cdot 10^{-4}$	$12 \cdot 10^{-4}$
9	Quecksilber	$4 \cdot 10^{-6}$	$4 \cdot 10^{-6}$	$5 \cdot 10^{-5}$	$1 \cdot 10^{-5}$	$2 \cdot 10^{-5}$	$4 \cdot 10^{-5}$
10	Thallium	$6 \cdot 10^{-5}$	$6 \cdot 10^{-5}$	$1 \cdot 10^{-4}$	$1 \cdot 10^{-3}$	$1 \cdot 10^{-3}$	$2 \cdot 10^{-3}$
11	\multicolumn{7}{c}{Grenzwerte der Trinkwasserverordnung (TrinkwV)}						
12	Chrom	\multicolumn{6}{c}{$5 \cdot 10^{-2}$}						
13	Quecksilber	\multicolumn{6}{c}{$1 \cdot 10^{-3}$}						
14	Thallium	\multicolumn{6}{c}{n. b. $(4 \cdot 10^{-2})*)$}						

*) n. b.: Nicht begrenzt (ersatzweise Bleigrenzwert)
*) n. b.: unlimited (as an alternative limit value for lead)

Auslaugmedium	=	leaching medium
Trinkwasser	=	drinking water
Kalklösende Kohlensäure	=	lime-dissolving carbondioxde
Wasserzementwert	=	water/cement ratio
Undotierte Betone mit natürlichem Schwermetallgehalt	=	undoped concretes with natural heavy metal content
Chrom	=	chromium
Quecksilber	=	mercury
Thallium	=	thallium
Dotierte Betone (100 mg Schwermetall/l Anmachwasser)	=	doped concretes (100 mg heavy metal/l mixing water)
Grenzwerte der Trinkwasser- verordnung (TrinkwV)	=	limit values of the Drinking Water Regulations (TrinkwV)

It is clear from this comparison taht conctrete can still be used without reservation in the dringking water sector even if the heavy metal content has been artificially raised.

5. Final comment

The material property "environmental compatibility" provides information about whether a material such as cement, or the concrete manufactured from it, gives off substances into the environment and therefore contributes to pollution or whether it exhibits environmentally neutral behaviour. In order to test their environmental compatibility concretes were manufactured with microstructures of differing densities and were leached in troughs with drinking water and with water containing chemically very strongly corrosive, lime-dissolving, carbondioxide. The concretes were manufactured with 300 kg PZ 35 F Portland cement per m³ concrete and water/cement ratios of 0.50 to 0.70 with an aggregate mix corresponding to the grading curve A/B 8. In some cases the levels of heavy metals in the concrete were raised artificially by the addition of 100 mg each of Cr, Hg and Tl/l to the mixing water. 4 × 4 × 16 cm test pieces were stored for one day in the mould and then for 7 days in a cloud chamber (protected from drips) at 20°C and approx. 100% relative humidity.

From the analyses of the eluate it can be seen that the elution of heavy metals from concrete into the adjacent water obeys the diffusion laws. The quantities leached out depended to a significant extent on the density of the concrete's microstructure. Substantially smaller quantities of heavy metals were leached from dense concrete (w/c = 0.50) than from concretes with increased capillary void volumes (w/c = 0.70).

Literaturverzeichnis

[1] Wischers, G., und Kuhlmann, K.: Ökobilanz von Zement und Beton – Abwägende Gegenüberstellung von ökologisch entlastenden und belastenden Einwirkungen auf die Umwelt. Zement-Kalk-Gips 44 (1991) Nr. 11, S. 545–553.

[2] Sprung, S.: Spurenelemente – Anreicherung und Minderungsmaßnahmen. Zement-Kalk-Gips 41 (1988) Nr. 5, S. 251–256.

[3] Forschungsinstitut der Zementindustrie: Tätigkeitsbericht 1990–1993, S. 37, Chemisch-mineralogische Prüfungen. Beton-Verlag, Düsseldorf 1993.

[4] Sprung, S., und Rechenberg, W.: Schwermetallgehalte im Klinker und im Zement. Zement-Kalk-Gips 47 (1994) Nr. 5, S. 258–263.

[5] Sprung, S., und Rechenberg, W.: Einbindung von Schwermetallen in Sekundärstoffen durch Verfestigen mit Zement. Beton 38 (1988) Nr. 5, S. 193–198.

[6] Sprung, S., und Rechenberg, W.: Bindung umweltrelevanter Sekundärstoffe durch Verfestigen mit Zement. Zement und Beton 34 (1989), Nr. 2, S. 54–61.

[7] Rechenberg, W., und Sprung, S.: Probenvorbereitung zur Beurteilung der Auslaugung umweltrelevanter Spurenelemente aus zementverfestigten Stoffen. Abwassertechnik 41 (1990) Nr. 3, S. 24–27 u. Nr. 4, S. 33–35.

[8] Rechenberg, W., und Sprung, S.: Umweltsicheres Deponieren von Abfallstoffen durch Verfestigen mit Zement. In: B. Böhnke (Hrsg.): Gewässerschutz, Wasser, Abwasser, Bd. 118, S. 178–203. Gesellschaft zur Förderung der Siedlungswasserwirtschaft, Aachen 1990.

[9] DIN 1045 (07.88): Beton- und Stahlbeton. Bemessung und Ausführung. Beuth-Verlag, Berlin-Köln.

[10] Rechenberg, W., und Spanka, G.: Verfahren zur Prüfung des Auslaugverhaltens zementverfestigter Stoffe. In: Österreich. Bundesmin. Umwelt, Jugend und Familie (Hrsg.): RILEM-Workshop Auslaugverhalten von Beton und zementgebundenem Material, Wien, Juni 1992. Mittlg. Forschinst. Vereinig. Österreich. Zementind., Heft 42, S. 21–26. Wien 1992.

[11] Pisters, H.: Chrom im Zement und Chromatekzem. Zement-Kalk-Gips 19 (1966) Nr. 10, S. 467–472.

[12] VDZ-Arbeitskreis „Analytische Chemie": Bestimmung von Spurenelementen in Stoffen der Zementherstellung. Schriftenreihe der Zementindustrie, Heft 55. Beton-Verlag, Düsseldorf 1993.

[13] DIN 38 414, Teil 4 (10.84): Deutsche Einheitsverfahren zur Wasser-, Abwasser- und Schlammuntersuchung. Schlamm und Sedimente (Gruppe S): Bestimmung der Eluierbarkeit mit Wasser (S4). Beuth-Verlag, Berlin-Köln.

[14] Locher, F. W.: Chemischer Angriff auf Beton. beton 17 (1967), H. 1, S. 17–19 und H. 2, S. 47–50 sowie Betontechnische Berichte 1967, S. 19-34. Beton-Verlag, Düsseldorf 1968.

[15] Locher, F. W., und Sprung, S.: Die Beständigkeit von Beton gegenüber kalklösender Kohlensäure. beton 25 (1975), H. 7, S. 241–245 sowie Betontechnische Berichte 1975, S. 91–104. Beton-Verlag, Düsseldorf 1976.

[16] Locher, F. W., Rechenberg, W. und Sprung, S.: Beton nach 20jähriger Einwirkung von kalklösender Kohlensäure. beton 34 (1984) H. 5, S. 193–198 sowie Betontechnische Berichte 1984/85, S. 41–56. Beton-Verlag, Düsseldorf 1986.

[17] Rechenberg, W., und Siebel, E.: Chemischer Angriff auf Beton. Hinweise zur Anwendung der DIN 4030. Schriftenreihe der Zementindustrie, Heft 53. Beton-Verlag, Düsseldorf 1992.

The quantities eluted out also increased when less dense concrete was leached with chemically very strongly corrosive water instead of with drinking water. However, the concentrations found in the total eluate lay below the limits set by the German Drinking Water Regulations by a factor of at least 20. Concretes with dense microstructures gave concentrations which were lower than the limits in the Drinking Water Regulations by a factor of at least 100. From this it can be seen that correctly manufactured concrete can be used without any reservations in drinking water systems.

[18] DIN 4030, Teil 1 (06.91): Beurteilung betonangreifender Wässer, Böden und Gase. Grundlagen und Grenzwerte. Beuth-Verlag, Berlin-Köln.

[19] N.N.: Mitteilungen aus dem Bundesgesundheitsamt: Gesundheitliche Beurteilung von Kunststoffen und anderen nichtmetallischen Werkstoffen im Rahmen des Lebensmittel- und Bedarfsgegenständegesetzes für den Trinkwasserbereich. Bundesgesundh. Bl. 20 (1977) H. 9, S. 124–129.

[20] Grube, H., und Rechenberg, W.: Betonabtrag durch chemisch angreifende saure Wässer. beton 37 (1987) H. 11, S. 446–451 und H. 12, S. 495–498 sowie Betontechnische Berichte 1986–88, S. 117–141. Beton-Verlag, Düsseldorf 1989.

[21] Gruber, H.: Elutionsverhalten von Blei- und Cadmiumverbindungen in Feststoffrückständen aus Rauchgasreinigungen. GIT Fachz. Lab 28 (1984) Nr. 7, S. 603–605.

[22] Straub, H., Hösel, G. und Schenkel, W. (Hrsg.): Handbuch über die Sammlung, Beseitigung und Verwertung von Abfällen aus Haushaltungen, Gemeinden und Wirtschaft: Richtlinie für das Vorgehen bei physikalischen und chemischen Untersuchungen im Zusammenhang mit der Beseitigung von Abfällen. EW/77. 46. Lieferung 1977. Erich Schmidt Verlag, Berlin.

[23] Sloot, van der, H. A., de Groot, G. J., und Wijkstra, J.: Leaching Characteristics of Construction Materials and Stabilization of Construction Materials and Stabilization Products Containing Waste Materials. S. 125–149. Amer. Soc. Test. Mat., Stand. Techn. Publ. 1033, Philadelphia, Pa, 1989.

[24] Groot, de, G.J. Wijkstra, J., Hoede, D. und van der Sloot, H. A.: Leaching Characteristics of Selected Elements from Coal Fly Ash as a Function of the Acidity of the Contact Solution and the Liquid/Solid Ratio. S. 170–183. Amer. Soc. Test. Mat., Stand. Techn. Publ. 1033, Philadelphia, Pa, 1989.

[25] NVN 5432 (03.90): Waste Materials, Construction Materials and Stabilized Waste Products. Nederlands Normalisatie-Instituut, Rijswijk.

[26] Koelliker, E.: Zur hydrolytischen Zersetzung von Zementstein und zum Verhalten von Kalkzuschlag bei der Korrosion von Beton durch Wasser. Betonwerk + Fertigteil-Technik 52 (1986) Nr. 4, S. 234–239.

[27] Schweizer Bundesamt für Abfälle, Eidgenössisches Departement des Inneren (Hrsg.): Bericht zum Entwurf für eine technische Verordnung über Abfälle (TVA). Richtlinie zur Durchführung des Eluats-Tests für Inertstoffe und endlagerfähige Reststoffe. Bern 1988.

[28] Verordnung über Trinkwasser und über Brauchwasser für Lebensmittelbetriebe (Trinkwasserverordnung – TrinkwV). BGBl., Teil 1, 1975, Nr. 16, Bonn 15.02.1975, geändert: BGBl., Teil 1, 1977, Nr. 89, Bonn 24.12.1977, BGBl., Teil 1, 1980, Nr. 32, Bonn 05.07.1980. In der Fassung BGBl., Teil 1, 1986, Nr. 22, S. 760/773, Bonn 28.05.1986, zuletzt geändert BGBl., Teil 1, Nr. 66 S. 2613 2629, Bonn 12.12.1990.

Schwermetallgehalte im Klinker und im Zement

Levels of heavy metals in clinker and cement

Teneurs en métaux lourds du clinker et du ciment

Metales pesados contenidos en el clínker y el cemento

Von **S. Sprung** und **W. Rechenberg**, Düsseldorf/Deutschland

Zusammenfassung – Der Spurenelementgehalt des Klinkers hängt im wesentlichen von den jeweiligen Gehalten der Elemente in den Roh- und Brennstoffen ab. Die Spurenelementgehalte von Rohstoffen unterschiedlicher geologischer Herkunft weisen eine verhältnismäßig geringe Konzentration und Schwankungsbreite auf. Das trifft auch für die Brennstoffe Steinkohle und Braunkohle zu. Sekundärstoffe können dagegen einzelne Schwermetalle in höherer Konzentration enthalten. Beim Brennprozeß werden nicht oder nur wenig flüchtige Elemente, wie z.B. Cr, Ni, V und Zn, vorwiegend im Klinker gebunden. Der Gehalt dieser Schwermetalle in einem Klinker, der unter teilweiser Verwertung von Sekundärstoffen gebrannt wird, unterscheidet sich innerhalb der üblichen Schwankungsbreite in der Regel nur wenig von dem solcher Klinker, für deren Herstellung ausschließlich natürliche Einsatzstoffe verwendet wurden. Das ist immer dann der Fall, wenn die in das Ofensystem eingebrachten Spurenelementmengen hierdurch insgesamt nicht wesentlich ansteigen. Eine auf diese Zusammenhänge abgestimmte Verwendung von Sekundärstoffen erlaubt demnach die Herstellung von Zementen, deren Schwermetallgehalt praktisch mit dem in normalen Böden übereinstimmt.

Summary – The trace element content of clinker depends essentially on the relevant levels of the elements in the raw materials and fuels. The trace elements in raw materials with different geological origins are present in relatively low concentrations and ranges of variation. This also applies to coal and lignite fuels. On the other hand secondary materials can contain individual heavy metals in fairly high concentrations. During the clinker burning process, most of the elements which are not volatile or are only slightly so, such as Cr, Ni, V and Zn, become fixed in the clinker. As a rule, the content of these heavy metals in a clinker which is burnt with partial utilization of secondary substances differs only slightly, within the usual range of fluctuation, from that of those clinkers manufactured exclusively from natural materials. This is always the case when there is no substantial overall increase in the quantities of trace elements introduced into the kiln system. Use of secondary materials adapted to these relationships therefore makes it possible to produce cements with levels of heavy metals which practically coincide with those in normal soils.

Résumé – La teneur en éléments de trace du clinker dépend essentiellement des différents éléments contenus dans les matières premières et les combustibles. La teneur en éléments de trace des matières premières de différentes origines géologiques présente une concentration et une variation relativement réduites. Cela est le cas également de la houille et du lignite servant de combustibles. Les matières secondaires, par contre, peuvent contenir des concentrations plus élevées de certains métaux lourds. Au cours du processus de cuisson, des éléments volatiles ou légèrement volatiles, tels que Cr, Ni, V et Zn, sont fixés surtout dans le clinker. La teneur en métaux lourds d'un clinker, cuit en utilisant partiellement des matières secondaires, se distingue normalement, dans les limites habituelles, très peu du clinker pour la fabrication duquel on utilise exclusivement des matières naturelles. Ceci est toujours le cas lorsque les quantités d'éléments de trace introduites dans le système de four n'augmentent pas, de ce fait, d'une manière importante. L'utilisation de matières secondaires accordées à cette situation permet donc de fabriquer des ciments, dont la teneur en métaux lourds correspond pratiquement à celle des sols naturels.

Resumen – El contenido de elementos-traza del clínker depende esencialmente del contenido de estos elementos en las materias primas y en los combustibles. Los contenidos de elementos-traza de las materias primas de distinta procedencia geológica presentan una concentración y variación relativamente pequeñas. Lo mismo se puede decir de los combustibles hulla y lignito. Las materias secundarias, sin embargo, pueden contener algún metal pesado en concentración más altas. Durante el proceso de cocción son combinados elementos no volátiles o poco volátiles, como por ejemplo Cr, Ni, V y Zn, sobre todo dentro del clínker. El contenido de estos metales pesados en un clínker, para cuya cocción se aprovechan en parte materias secundarias, por regla general, se distingue muy poco – dentro de los márgenes de variación normales – de aquellos clínkeres que se fabrican exclusivamente de materias primas naturales. Esto siempre es el caso, cuando las cantidades de elementos-traza introducidos en el sistema de horno, en su conjunto, no aumentan mucho por este hecho. Una utilización de materias secundarias que tenga en cuenta estas relaciones permite, por lo tanto, la fabricación de cementos, cuyo contenido de metales pesados coincide prácticamente con el de los suelos normales.

---

*) Überarbeitete Fassung eines Vortrages zum VDZ-Kongreß '93, Düsseldorf (27.9.–1.10.1993)
Revised text of a lecture to the VDZ Congress '93, Düsseldorf (27.9.–10.1993)

1. Einleitung

Zur industriellen Herstellung von Zementen nach DIN 1164 werden als Hauptbestandteile Portlandzementklinker, Hüttensand, Traß und gebrannter Ölschiefer sowie Steinkohlenflugasche mit Prüfzeichen und Kalksteinmehl eingesetzt. Die Zumahlung von Calciumsulfaten dient der gezielten Steuerung von Verarbeitungseigenschaften und der Festigkeitsentwicklung. Bei den Rohstoffen zum Brennen des Zementklinkers handelt es sich vorwiegend um Stoffe natürlichen Ursprungs, wie z.B. Kalkstein, Kalkmergel, Ton, Sand sowie gelegentlich auch um Calciumsulfat und andere Korrekturstoffe. Als Energieträger zum Brennen des Rohstoffgemisches werden in erster Linie Stein- und Braunkohle sowie derzeit in untergeordnetem Maße auch Heizöl S verwendet.

Die natürlichen Einsatzstoffe können in begrenztem Umfang durch Sekundärstoffe ersetzt werden. Darunter sind Reststoffe aus anderen industriellen Produktionen oder aus der Abfallentsorgung zu verstehen. Ihre Verwendung als Wirtschaftsgut unterliegt dabei einer Reihe von einschränkenden Kriterien. So dürfen Sekundärstoffe weder die bautechnischen Eigenschaften und die Umweltverträglichkeit von Zement und Beton beeinträchtigen, noch den Herstellungsprozeß stören und die Umweltbelastung steigern [1, 2]. Das gilt insbesondere für den Klinkerbrennprozeß.

Die Eigenschaften des Portlandzementklinkers werden in erster Linie von der Konzentration der Haupt- und Nebenelemente bestimmt, die mit den Einsatzstoffen eingebracht werden. Ihre Summe übersteigt in der Regel 99 Gew.-% [2, 3]. Darüber hinaus enthalten die Ausgangsstoffe zur Klinkerherstellung auch Spurenelemente. Art und Konzentration werden bei Stoffen natürlichen Ursprungs von deren geochemischer Verteilung in den Lagerstätten bestimmt. Die Konzentration kann daher in verhältnismäßig weiten Grenzen schwanken. Bei der Verwertung von Sekundärstoffen hängt die damit verbundene Zufuhr von Spurenelementen sehr wesentlich von der Herkunft und Vorgeschichte dieser Stoffe ab. Maßgebendes Beurteilungskriterium für die Umweltbelastung durch den Herstellungsprozeß und die Umweltverträglichkeit des Produkts sind demnach der Gesamteintrag in das Ofensystem und die Art der durch chemische Reaktion entstehenden Verbindungen.

Die Zufuhr von Elementen, die – wie z.B. Thallium und Quecksilber – leicht flüchtig sind oder leicht flüchtige Verbindungen bilden, muß begrenzt werden, um eine Anreicherung durch Kreisläufe innerhalb des Ofensystems zu vermeiden. Wachsende äußere Kreisläufe können zu erhöhten Emissionen führen, wenn sie nicht unterbrochen werden. Hierdurch würden dann Reststoffe entstehen, die wiederum auf geeignete Weise zu entsorgen sind [3-5].

Nicht oder schwer flüchtige Elemente werden demgegenüber in erster Linie in das Brennprodukt Klinker eingebunden [6-9]. Über die Höhe der sich einstellenden Spurenelementkonzentration entscheidet die Gesamtzufuhr durch alle im Einzelfall verwendeten natürlichen Roh- und Brennstoffe sowie Sekundärstoffe. Auf der Basis der bis heute aus Bilanzuntersuchungen an Ofenanlagen gewonnenen Erkenntnisse ist es demnach möglich, die Auswirkungen eines teilweisen Ersatzes von natürlichen Einsatzstoffen durch Sekundärstoffe auf den Spurenelementgehalt des Klinkers und die Emission über eine Prognoserechnung abzuschätzen.

1. Introduction

The main constituents used in the industrial manufacture of cements conforming to DIN 1164 are Portland cement clinker, granulated blast furnace slag, trass and burnt oil shale, certified coal fly ash, and limestone. Calcium sulphates are interground to achieve accurate and specific control of workability characteristics and strength development. The raw materials for burning the cement clinker are predominantly materials of natural origin, such as limestone, lime marl, clay, sand and possibly also calcium sulphate and other correction materials. The fuels used for burning the raw material mix are primarily coals and lignites and also, to a lesser extent, heavy fuel oil.

To a limited extent the natural feed materials can be replaced by secondary materials. These are understood to include residual materials from other industrial production processes and from waste disposal. Their use as commercial goods is subject to a series of restrictive criteria. Secondary materials must, for example, neither impair the structural properties and environmental compatibility of cement and concrete nor upset the production process or increase its environmental impact [1, 2]. This is particularly applicable to the clinker burning process.

The properties of Portland cement clinker are determined primarily by the concentration of the principal and secondary elements which are introduced with the feed materials. As a rule their total exceeds 99 wt. % [2, 3]. The starting materials for clinker production also contain trace elements. The type and nature are determined in substances of natural origin by their geochemical distribution in the deposits. The concentrations can therefore fluctuate within relatively wide margins. In the assessment of secondary materials the associated input of trace elements depends very substantially on the origin and previous history of the materials. The total input into the kiln system and the types of compounds produced by chemical reactions are therefore important assessment criteria for the environmental impact caused by the production process and for the environmental compatibility of the product.

The input of elements which – like thallium and mercury – are readily volatile or form readily volatile compounds has to be limited in order to avoid any build-up of concentration caused by recirculation within the kiln system. Growing external recirculating systems can lead to increased emissions unless they are interrupted. This would result in residual materials which then have to be suitably disposed of [3-5].

On the other hand, elements which are only slightly or not at all volatile are mainly combined in the product of the burning process, namely clinker [6-9]. The level of the trace element concentration which occurs is determined by the total input from all the natural raw materials and fuels as well as the secondary materials used in the particular instance. The findings obtained by taking balances at kiln plants show that it is possible to use a predictive calculation to estimate the effects on the trace element content of the clinker and on the emission caused by partial replacement of natural feed materials by secondary materials.

The following sections deal with the advance calculation of the levels of trace elements in clinker with varied use of secondary fuels. The calculated concentrations are also compared with those obtained from analytical investigations of

TABELLE 1: Kalkstandard (KSt), Silicatmodul (SM) und Tonerdemodul (TM) von Betriebsklinkern aus deutscher Produktion

1988	H	M	N
KSt	101	96	90
SM	4,2	2,5	1,4
TM	4,2	2,3	0,6

H: Höchstwert, M: Mittelwert, N: Niedrigstwert

TABLE 1: Lime standard (LSt), silica ratio (SR) and alumina ratio (AR) of industrial clinkers produced in Germany

1988	H	M	L
LSt	101	96	90
SR	4.2	2.5	1.4
AR	4.2	2.3	0.6

H: highest value, M: mean value, L: lowest value

TABELLE 2: Haupt- und Nebenbestandteile von Roh- und Brennstoffen; Angaben in Gew.-%
TABLE 2: Main and secondary constituents of raw material and fuels; data in wt.%

Bestandteil	Kalkstein		Ton		Braunkohle		Steinkohle		Altreifen	
	Glv.-hlt.*)	Glv.-fr.**)	Glv.-hlt.	Glv.-fr.	Glv.-hlt.	Glv.-fr.	Glv.-hlt.	Glv.-fr.	Glv.-hlt.	Glv.-fr.
Brennbares	–	–	–	–	95,0	–	80,0	–	84,4	–
CO_2	37,4	–	6,6	–	–	–	–	–	–	–
CaO	47,6	76,0	8,4	9,0	1,8	36,0	0,6	3,0	–	–
SiO_2	10,0	16,0	57,0	61,0	0,7	14,0	7,0	35,0	–	–
Al_2O_3	1,0	1,6	18,0	19,3	0,4	8,0	3,6	18,0	–	–
Fe_2O_3	0,8	1,3	5,0	5,4	0,2	4,0	5,6	28,0	15,6	100,0
Summe	96,8	94,9	95,0	97,7	98,1	62,0	96,8	84,0	100,0	100,0

*) Glv.-hlt: Glühverlusthaltig = inc. l.o.i.: including loss on ignition
**) Glv.-fr.: Glühverlustfrei = l.o.i.-free: loss-on-ignition-free
Bestandteil = constituent
Kalkstein = limestone
Ton = clay
Braunkohle = lignite
Steinkohle = coal
Altreifen = used tyres
Glv.-hlt. = inc. l.o.i.
Glv.-fr. = l.o.i.-free
Brennbares = combustibles
Summe = total

TABELLE 3: Heizwert und Aschegehalt von Brennstoffen zur Deckung des Brennstoffenergiebedarfs

Brennstoffe	Brennstoffenergiebedarf 3350 kJ/kg Kl (800 kcal/kg Kl)		Aschegehalt Gew.-%
	Heizwert H_u		
	kJ/kg	kcal/kg	
Steinkohle	28050	6700	20,0
Braunkohle	23030	5500	5,0
Altreifen	33500	8000	15,6

TABLE 3: Calorific value and ash content of fuels for covering the fuel energy requirement

Fuels	Fuel energy requirement 3350 kJ/kg clinker (800 kcal/kg clinker)		ash content wt.%
	net calorific value		
	kJ/kg	kcal/kg	
Coal	28050	6700	20.0
Lignite	23030	5500	5.0
Used tyres	33500	8000	15.6

Die nachfolgenden Abschnitte befassen sich mit der Vorausberechnung des Gehaltes von Spurenelementen im Klinker bei variablem Einsatz von Sekundärbrennstoffen. Darüber hinaus werden die berechneten Konzentrationen mit denjenigen verglichen, die sich aus analytischen Untersuchungen an praktisch allen in Deutschland industriell hergestellten Klinkern und Portlandzementen ergaben. Als Beurteilungs- sowie auch Vergleichsmaßstab dienten die Schwermetallgehalte natürlicher Gesteine und Böden.

2. Spurenelemente im Klinker und Zement

2.1 Ausgangssituation

Der Prognoserechnung lagen die Ergebnisse analytischer Untersuchungen an Klinker-, Zement- und Sulfatträgerproben aus allen deutschen Zementwerken zugrunde. Aus den Klinkeranalysen wurden die Spannweiten und arithmetischen Mittelwerte des Kalkstandards (KSt), des Silicatmoduls (SM) und des Tonerdemoduls (TM) berechnet. Die Werte sind in **Tabelle 1** zusammengestellt. Die Mittelwerte der Zusammensetzung dienten dazu, mit Hilfe der hydraulischen, silicatischen und aluminatischen Abweichungen [10] den spezifischen, auf 1 kg dieses Klinkers bezogenen Bedarf an Kalkstein und Ton zu berechnen. Die für die Rechnung verwendete Zusammensetzung der beiden Rohstoffkomponenten geht aus **Tabelle 2** hervor. Berücksichtigt wurden außerdem der jeweilige Aschegehalt (**Tabelle 3**) und die Aschezusammensetzung (Tabelle 2) der Brennstoffe Stein- und Braunkohle sowie der für die Prognoserechnung beispielhaft verwendeten Altreifen. Der Gehalt nicht verbrennbarer Bestandteile in Altreifen betrug rd. 16 Gew.-% und besteht praktisch vollständig aus metallischem Eisen, das während des Brennprozesses bei ausreichend hohem Sauerstoffüberschuß zu Eisen(III)oxid oxidiert und in den Klinkerphasen gebunden wird. Die Brennstoffmenge ergab sich aus einem angenommenem mittleren spezifischen Energiebedarf für den Klinkerbrand von rd. 3350 kJ/kg Kl

practically all the clinkers and Portland cements manufactured industrially in Germany. The levels of heavy metal in natural rocks and soils are used as standards both for assessment and for comparison.

2. Trace elements in clinker and cement

2.1 Initial situation

The predictive calculation was based on the results of analytical investigations on samples of clinker, cement and sulphate agents from all German cement works. The ranges and arithmetic means of the lime standard (LSt), silica ratio (SR) and alumina ratio (AR) were calculated from the clinker analyses. The values are summarized in **Table 1**. The mean values are used to calculate the specific consumption of limestone and clay relative to 1 kg of this clinker [10]. The compositions of the two raw material components used for the calculation can be found in **Table 2**. Not only are the relevant ash content (**Table 3**) and ash composition (Table 2) of the coal and lignite fuels taken into account but also those of the used tyres used by way of example for the predictive calculation. The content of incombustible constituents in used tyres amounts to about 16 wt.% and consists almost entirely of metallic iron; with a sufficiently high excess of oxygen this is oxidized during the burning process to iron(III) oxide and is fixed in the clinker phase. The quantity of fuel is found from an assumed average specific energy consumption for clinker burning of about 3350 kJ/kg clinker (800 kcal/kg clinker). In the predictive calculation the respective natural fuels used were replaced by used tyres in steps up to a level corresponding to 25% of the fuel energy requirement (**Table 4**).

Heavy metals are introduced into the burning process in varying quantities with all raw materials and fuels. The highest (H), mean (M) and lowest (L) values of 10 heavy metals in the starting materials limestone and clay as well as in the fuels

TABELLE 4: Bezogene Roh- und Brennstoffmassen; Angaben in kg/kg Klinker
TABLE 4: Specific masses of raw material and fuel; data in kg/kg clinker

Reifenanteil am Brennstoff-energiebedarf in %	0	5	10	15	20	25
	\multicolumn{6}{c}{Braunkohle}					
Kalkstein	1,3865	1,3843	1,3932	1,3801	1,3779	1,3755
Ton	0,1735	0,1757	0,1768	0,1799	0,1821	0,1845
Kohle	0,1455	0,1382	0,1309	0,1236	0,1164	0,1091
Reifen	–	0,005	0,01	0,015	0,02	0,025
	\multicolumn{6}{c}{Steinkohle}					
Kalkstein	1,3868	1,3848	1,3823	1,3806	1,3783	1,3759
Ton	0,1732	0,1752	0,1777	0,1794	0,1817	0,1841
Kohle	0,1194	0,1134	0,1075	0,1015	0,0955	0,0896
Reifen	–	0,005	0,01	0,015	0,02	0,025

Reifenanteil am Brennstoffenergiebedarf in % = Proportion of fuel energy requirement covered by tyres in %
Braunkohle = lignites
Kalkstein = limestone
Ton = clay
Kohle = coal
Reifen = tyres
Steinkohle = coals

TABELLE 5: Schwermetallgehalte in Roh- und Brennstoffen; Angaben in g/t [4–9, 11–14]
TABLE 5: Heavy metal levels in raw materials and fuels; data in g/t [4–9, 11–14]

Element		Kalkstein	Ton	Steinkohle	Braunkohle	Altreifen
As	H	12	23	13	0,4	–
	M	6	18	7	0,3	20
	N	0,2	13	1	0,2	–
Be	H	0,4	–	1,5	–	–
	M	0,2	3	0,9	0,04	0,05
	N	< 0,01	–	0,2	–	–
Pb	H	21	40	27	1,5	760
	M	7	17	16	1,1	410
	N	0,3	10	5	0,7	60
Cd	H	0,5	0,2	0,71	0,10	10
	M	0,07	0,16	0,39	0,08	8
	N	0,02	0,05	0,07	0,06	5
Cr	H	12	90	50	6,1	–
	M	9	60	25	4,2	97
	N	0,7	20	1	2,3	–
Hg	H	0,1	0,15	0,61	0,14	0,43
	M	0,03	0,03	0,33	0,07	0,17
	N	0,005	0,02	0,05	< 0,01	0,10
Ni	H	13	70	37	4,6	–
	M	4,5	69	19	2,8	77
	N	1,4	11	1	1,0	–
Tl	H	0,8	0,9	1,2	0,3	0,3
	M	0,27	0,6	0,7	0,2	0,25
	N	0,06	0,2	0,2	0,1	0,2
V	H	80	170	50	25	–
	M	45	134	30	13	5,3
	N	10	98	10	1	–
Zn	H	57	110	150	–	20500
	M	23	87	85	22	15000
	N	1,0	55	20	–	9300

Element = element
Kalkstein = limestone
Ton = clay
Steinkohle = coals
Braunkohle = lignites
Altreifen = used tyres
H:Höchstwert = H:highest value
M:Mittelwert = M:mean value
N:Niedrigstwert = N:lowest value

(800 kcal/kg Kl). Bei der Prognoserechnung wurde der jeweilige natürliche Brennstoff schrittweise durch Altreifen bis zu 25 % des Brennstoffenergiebedarfs ersetzt (**Tabelle 4**).

Mit allen Roh- und Brennstoffen werden Schwermetalle in unterschiedlicher Menge in den Brennprozeß eingebracht. In **Tabelle 5** sind die Höchst- (H), Mittel- (M) und Niedrigstwerte (N) von 10 Schwermetallen in den Ausgangsstoffen Kalkstein und Ton sowie in den Brennstoffen zusammengestellt [4–9, 11–14]. **Tabelle 6** enthält die Spurenelementkonzentrationen von natürlichem Gipsstein [13, 14]. Vereinfachend wurde dabei angenommen, daß bei der Mahlung von Portlandzement nur Dihydrat als Calciumsulfatträger verwendet wird.

TABELLE 6: Schwermetallgehalt im Gips in g/t [13]
TABLE 6: Heavy metal levels in gypsum in g/t [13]

Element	Höchstwert	Mittelwert	Niedrigstwert
As	2,1	1,3	0,5
Be	0,9	0,2	<0,01
Pb	20	11	8
Cd	0,6	0,17	0,08
Cr	33	10	3,7
Hg	0,08	0,05	<0,005
Ni	13,5	4,4	0,7
Tl	–	<0,2	–
V	27	10	<1
Zn	61	17	1

Element	= element
Höchstwert	= highest value
Mittelwert	= mean value
Niedrigstwert	= lowest value

2.2 Spurenelementkonzentration im Klinker und Zement

Mit den spezifischen Masseströmen aus Tabelle 4 und den Konzentrationsangaben in Tabelle 5 wurden die Schwermetallgehalte eines Klinkers mittlerer Zusammensetzung (Tabelle 1) berechnet. Der Rechnung lagen durch Bilanzmessung an Zyklonvorwärmeranlagen ermittelte mittlere Einbindungsgrade zu Grunde. Sie betragen bei den schwer flüchtigen Elementen As, Be, Cr, Ni, V und Zn 90 %, bei Pb 80 % und bei Cd 75 % [4–9, 11, 12, 15]. Die Einbindungsgrade leicht flüchtiger Elemente, wie z. B. Tl und Hg, wurden mit weniger als 1 %, bezogen auf die zugeführte Menge, hierfür nicht weiter berücksichtigt [15].

Ein Vergleich der prognostizierten Schwermetallkonzentrationen, die in einem mit Stein- oder Braunkohle bzw. Altreifen gebrannten Portlandzementklinker zu erwarten sind, mit den durch Untersuchung sämtlicher in Deutschland produzierter Betriebsklinker gemessenen Spannweiten [14] geht aus **Tabelle 7** hervor. Danach ist davon auszugehen, daß die prognostizierte Konzentrationsspanne nicht oder schwer flüchtiger Elemente etwa im Bereich der an Betriebsklinkern ermittelten Konzentrationen liegt, obwohl die Betriebsklinker eine wesentlich größere Streubreite in den Hauptbestandteilen und damit auch in Art und Zusammensetzung der Ausgangsstoffe gegenüber dem Durchschnittsklinker aufweisen. Der Vergleich zeigt aber, daß es – ausgehend von den Analysendaten der jeweiligen Einsatzstoffe – möglich ist, den im Klinker zu erwartenden Schwermetallgehalt im Einzelfall mit hinreichender Treffsicherheit zu prognostizieren und hieran eine Bewertung der Einsatzmöglichkeiten von Sekundärstoffen zu knüpfen.

Für eine Abschätzung der zu erwartenden Konzentrationen nicht oder schwer flüchtiger Elemente in Portlandzement ist zusätzlich nur noch der Sulfatträger zu berücksichtigen. Vereinfachend kann dabei angenommen werden, daß der Anteil des Sulfatträgers in allen Zementen einheitlich 5 Gew.-% beträgt und der Sulfatträger nur aus Gips (Tabelle 6) besteht. Bei dem hier gewählten Beispiel mit Alt-

are listed in **Table 5** [4–9, 11–14]. **Table 6** gives the trace element concentrations in natural gypsum rock [13, 14]. As a simplification it was assumed that dihydrate was the only calcium sulphate agent used when grinding Portland cement.

2.2 Trace element concentrations in clinker and cement

The levels of heavy metal in a clinker of average composition (Table 1) were calculated with the specific mass flows from Table 4 and the concentration data in Table 5. The calculation was based on average degrees of fixation determined by balance measurements on cyclone preheater plants. These were 90 % for the elements of low volatility As, Be, Cr, Ni, V and Zn, 80 % for Pb, and 75 % for Cd [4–9, 11, 12, 15]. No further consideration was given to the degrees of fixation of readily volatile elements such as Tl and Hg of less than 1 % relative to the input quantities [15].

The predicted heavy metal concentrations which can be expected in Portland cement clinker burnt with coal or lignite and used tyres are compared in **Table 7** with the ranges measured by investigating all the industrial clinkers produced in Germany [14]. From this it is clear that the predicted concentration range of elements which are not volatile or only slightly so lies approximately in the range of concentrations measured on industrial clinkers, although the industrial clinkers exhibit a substantially larger range of scatter in the main constituents, and therefore also in the type and composition of the starting materials, than the average clinker. However, the comparison shows that – starting from the analysis data for the particular feed material – it is possible to predict the heavy metal content to be expected in the clinker in the individual instance with sufficient precision and from this to form a judgement about the feasibility of using secondary substances.

For estimating the expected concentrations of elements which are not volatile or are only slightly so in Portland cement the only further factor to be taken into consideration is the sulphate agent. For this purpose it can be assumed as a simplification that all cements contain a uniform proportion 5 wt. % sulphate agent and that the sulphate agent only consists of gypsum (Table 6). In the example selected here with used tyres as a secondary fuel it turned out that the levels of arsenic (As), beryllium (Be), chromium (Cr), nickel

TABELLE 7: Berechnete Schwermetallgehalte im Klinker ohne und mit 25 % Altreifen-Einsatz im Vergleich mit den an Betriebsklinkern bestimmten Spannweiten [14]
TABLE 7: Calculated heavy metal levels in clinker with and without the use of 25 % used tyres as compared with the ranges measured in industrial clinkers [14]

Element		Schwermetallgehalte im Klinker	
		Prognose	Meßwerte
		g/t	
Arsen	(As)	3 – 21	2 – 15
Beryllium	(Be)	0,5 – 1	<0,2 – 1,1
Blei	(Pb)	2 – 46	5 –105
Cadmium	(Cd)	0,03– 0,8	0,01– 1,5
Chrom	(Cr)	4 – 35	10 – 90
Quecksilber	(Hg)	–	<0,01
Nickel	(Ni)	4 – 32	10 – 50
Thallium	(Tl)	–	<0,01
Vanadium	(V)	29 –131	20 –100
Zink	(Zn)	12 –562	40 –350

Element	= element
Schwermetallgehalte im Klinker	= heavy metal levels in the clinker
Prognose	= predicted
Meßwerte	= measured
Arsen	= arsenic
Blei	= lead
Chrom	= chromium
Quecksilber	= mercury
Zink	= zinc

reifen als Sekundärbrennstoff stellte sich heraus, daß die Gehalte an Arsen (As), Beryllium (Be), Chrom (Cr), Nickel (Ni) und Vanadium (V) vom Anteil der Altreifen am Brennstoffenergiebedarf praktisch unabhängig waren. Die Werte lagen bei Einsatz von Steinkohle geringfügig über denen bei Braunkohle als Hauptenergieträger.

Demgegenüber stieg die Konzentration der Elemente Zn, Pb und Cd im Portlandzement rechnerisch an, da sie bei einem teilweisen Ersatz von Stein- oder Braunkohle durch Altreifen etwas stärker angereichert werden. Die Veränderung des Gehalts dieser Elemente im Zement bei Verfeuerung steigender Altreifenmengen ist im **Bild 1** dargestellt. Aufgetragen wurde der Spurenelementgehalt in Portlandzement in Abhängigkeit vom Altreifenanteil beim Klinkerbrand. Die untere Begrenzung eines schraffierten Feldes gibt die berechneten Niedrigstwerte bei Einsatz von Braunkohle und die obere Begrenzung die Höchstwerte bei Einsatz von Steinkohle als Hauptbrennstoff wieder. Die Felder stellen demnach die bei gegebenem KSt, SM und TM des Klinkers mögliche Schwankungsbreite des Elementgehalts in einem Portlandzement mittlerer Zusammensetzung dar.

Aus dem Bild geht hervor, daß der Cadmiumgehalt bei teilweisem Ersatz der Braunkohle durch Altreifen nur geringfügig von 0,03 g/t auf 0,13 g/t ansteigt. Steinkohle als Hauptbrennstoff führt insgesamt zu höheren Cd-Gehalten im Zement. Ein etwas höheres Konzentrationsniveau weist der Pb-Gehalt auf. Er wird jedoch in Abhängigkeit von der verfeuerten Altreifenmenge nur wenig verändert. Etwas deutlicher ausgeprägt ist demgegenüber der Zuwachs des Zinkgehalts, der aufgrund der Zn-Mengen im Altreifengummi bei Braunkohlefeuerung erwartungsgemäß von 11 g/t auf etwa 210 g/t im Zement ansteigen kann. Auch hier liegen die Werte bei Steinkohlefeuerung höher. Beeinträchtigungen der Zementeigenschaften ließen sich in diesen Konzentrationsbereichen nicht feststellen.

Ein Vergleich der berechneten Schwermetallkonzentrationen mit denen an Portlandzementen einer Güteüberwachungsperiode gemessenen Konzentrationen [14] geht aus **Tabelle 8** hervor. Danach ist generell festzustellen, daß ähnlich wie beim Klinker die prognostizierten Konzentrationen der Elemente As, Be, Cd, Ni, V und Zn praktisch der Spannweite der gemessenen Konzentrationen entsprechen. Die größere Abweichung des Höchstwerts der gemessenen Spannweite zum berechneten Gehalt der Elemente Pb, Cd, Cr, Ni und Zn beruht darauf, daß innerhalb des untersuchten Kollektivs von etwa 70 Stichproben meistens nur ein

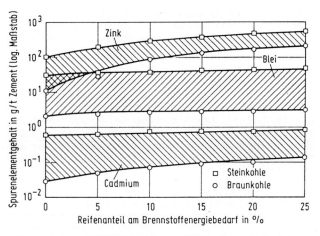

BILD 1: Abhängigkeit des Gehalts von Spurenelementen im Zement vom Reifenanteil am Brennstoffenergiebedarf

FIGURE 1: Dependence of the levels of trace elements in the cement on the proportion of the fuel energy requirement covered by tyres

Spurenelementgehalt in g/t Zement (log. Maßstab)	= trace element content in g/t cement (log scale)
Zink	= zinc
Blei	= lead
Cadmium	= cadmium
Steinkohle	= coal
Braunkohle	= lignite
Reifenanteil am Brennstoffenergiebedarf in %	= proportion of fuel energy requirement covered by tyres in %

(Ni) and vanadium (V) were practically independent of the proportion of the fuel energy requirement covered by used tyres. The values when using coal as the main fuel were slightly higher than those for lignite.

On the other hand there is an increase in the calculated concentrations of the elements Zn, Pb and Cd in the Portland cement as they become enriched to a somewhat greater extent when coal or lignite is partially replaced by used tyres. The change in the levels of these elements in the cement when burning increasing quantities of used tyres is shown in **Fig. 1**. The trace element content in the Portland cement is plotted against the proportion of used tyres when burning the clinker. The lower boundary of a shaded area shows the calculated minimum value when using lignite and the upper boundary shows the maximum value when using coal as the main fuel. The areas between them therefore represent the possible range of fluctuation of the levels of the elements in a Portland cement of average composition for a given LSt, SR and AR in the clinker.

It can be seen from the diagram that on partial replacement of the lignite by used tyres the cadmium content increases slightly from 0.03 to 0.13 g/t. Using coal as the main fuel leads to higher overall levels of Cd in the cement. The Pb content shows a somewhat higher concentration level. However, it changes only slightly as a function of the quantity of used tyres burnt. Somewhat more pronounced is the increase in zinc content which, because of the quantities of Zn in the used tyre rubber, can be expected to increase from 11 g/t with lignite firing to about 210 g/t in the cement. Here again the values are higher for coal firing. No impairment of the cement properties could be found in these concentration ranges.

A comparison of the calculated heavy metal concentrations with the concentrations measured in Portland cements during a quality monitoring period [14] is shown in **Table 8**. From this it can be seen in general terms that, as with the clinker, the predicted concentrations of the elements As, Be, Cd, Ni, V and Zn practically correspond to the ranges of the measured concentrations. The greater deviation of the maximum value of the measured ranges from the calculated levels of the elements Pb, Cd, Cr, Ni, and Zn are due to the fact that within the population of about 70 spot samples investigated in most cases only one cement exhibited this relatively high concentration value.

TABELLE 8: Schwermetallgehalte im Portlandzement
TABLE 8: Heavy metal levels in Portland cement

Element		Schwermetallgehalte im Portlandzement	
		Prognose	Meßwerte
		g/t	
Arsen	(As)	2,9 – 20	2 – 14*)
Beryllium	(Be)	0,5 – 1	< 0,2 – 1*)
Blei	(Pb)	2,1 – 45	5 – 254
Cadmium	(Cd)	0,03 – 0,8	0,03 – 5,5
Chrom	(Cr)	4,1 – 35	25 – 124
Quecksilber	(Hg)	–	< 0,02 – 0,12
Nickel	(Ni)	3,5 – 31	17 – 97
Thallium	(Tl)	–	< 0,02 – 4,1
Vanadium	(V)	28 – 126	19 – 96*)
Zink	(Zn)	11 – 534	21 – 679

*) Aus Klinker- (Tabelle 7) und Gipsanalyse (Tabelle 6) berechnet
*) Calculated from clinker (Table 7) and gypsum (Table 6) analyses

Element	=	element
Schwermetallgehalte im Portlandzement	=	heavy metal levels in the Portland cement
Prognose	=	predicted
Meßwerte	=	measured
Arsen	=	arsenic
Blei	=	lead
Chrom	=	chromium
Quecksilber	=	mercury
Zink	=	zinc

TABELLE 9: Schwermetallgehalte in Bodenausgangsgesteinen und Böden [13, 20]; Angaben in g/t
TABLE 9: Heavy metal levels in rocks from which the soils originated and soils [13, 20]; data in g/t

Spalte		1	2	3	4	5
Element		Kalkstein	Ton	Boden, weltweit	Boden, USA	Boden, Schottland
Arsen	(As)	0,1– 2,8	3 – 25	5	–	–
Beryllium	(Be)	<0,2– 12	2 – 18	–	1	5
Blei	(Pb)	0,4– 17	1 – 219	10	20	20– 80
Cadmium	(Cd)	~0,04	<0,02– 500	0,5	–	–
Chrom	(Cr)	1,2– 16	33 –1500	200	53	5–3000
Nickel	(Ni)	–	–	40	20	10– 800
Quecksilber	(Hg)	~0,04	~0,5	0,08	–	–
Thallium	(Tl)	–	0,2 – 0,7	–	–	–
Vanadium	(V)	5 –3000	30 –3600	100	76	20– 250
Zink	(Zn)	<0,1–1900	2 –1300	50	54	–

Spalte	= column		Arsen	= arsenic
Kalkstein	= limestone		Blei	= lead
Ton	= clay		Chrom	= chromium
Boden, weltweit	= soil, worldwide		Quecksilber	= mercury
Boden, USA	= soil, USA		Zink	= zinc
Boden, Schottland	= soil, Scotland			

Zement diesen verhältnismäßig hohen Konzentrationswert aufwies.

Die Gehalte der flüchtigen Elemente wie Tl und Hg sind in den Zementen lagerstättenbedingt in der Regel außerordentlich gering und unterschreiten in der überwiegenden Mehrzahl Werte von 1g/t. Abweichungen hiervon können jedoch dann auftreten, wenn Stäube zur Unterbrechung von Kriesläufen aus dem Ofensystem abgezogen und bei der Zementmahlung in Anteilen zugesetzt werden, die nicht der im entsprechenden Zeitraum erzeugten Klinkermenge entsprechen [2]. Eine Prognose des im Zement zu erwartenden insgesamt geringen Tl- oder Hg-Gehalts ist daher nicht generell, sondern nur im Einzelfall möglich.

3. Bewertung des Spurenelementgehalts im Zement

Da der Gehalt umweltrelevanter Schwermetalle in erster Linie von den natürlichen Einsatzstoffen sowie aber auch von ersatzweise eingesetzten Sekundärstoffen bestimmt wird, liegt es nahe, vergleichend zu prüfen, ob und gegebenenfalls wie weit die Konzentrationen von denen in natürlichen Gesteinen oder Böden abweichen. Bei diesem Vergleich könnte der Zement als technisches, der Boden als natürliches Umwandlungsprodukt von Gesteinen angesehen werden [13]. Die Art und Langzeitbeständigkeit einer chemischen oder adsorptiven Bindung von Schwermetallen im erhärteten Zementstein des Betons wird an anderer Stelle erörtert [16–19].

In **Tabelle 9** sind beispielhaft Schwermetallgehalte aus geochemischen Untersuchungen an Kalkstein, Ton und verschiedenen Böden zusammengestellt [13, 20]. Daraus geht hervor, daß die Konzentration der Schwermetalle teilweise eine erhebliche Schwankungsbreite aufweisen kann. Ähnlich große Schwankungsbreiten ergeben sich für Böden auf silicatischen, tonigen und carbonatischen Sedimenten sowie auf magmatischen Gesteinen in Deutschland, die hier nicht aufgeführt sind [21]. **Tabelle 10** enthält Zahlenwerte, die die häufig in deutschen Kulturböden gegebene Konzentration kennzeichnen sollen [22]. Sie gelten als Orientierungsdaten oder bei Anwendung der Klärschlammverordnung [23] auch als Richtwerte [22, 24]. Die angeführten Werte geben den Gesamtgehalt und nicht den pflanzenverfügbaren oder resorbierbaren Anteil wieder [z.B. 25–28]. Sie liegen damit offenbar weit auf der „sicheren Seite". Es ist nicht auszuschließen, daß ähnliche Richtwerte auch bei der künftigen Bodenschutzgesetzgebung eingeführt werden [28, 29]. Weiterhin sind in Tabelle 10 „tolerierbare" Gesamtgehalte in Böden sowie zum Vergleich Schwellenwerte der Landesanstalt für Ökologie, Landschaftsentwicklung und Forstplanung (LÖLF) [29] und Grenzwerte für kontaminierte Böden aufgeführt.

The levels of the volatile elements such as Tl and Hg in the cements are, depending on the deposit, normally exceptionally low, and the great majority lie below 1 g/t. However, departures from this can occur if dust is withdrawn from the kiln system to interrupt recirculating systems and is then added during the cement grinding in proportions which do not correspond to the quantity of clinker produced in the corresponding period of time [2]. The low overall levels of Tl and Hg to be expected in the cement therefore cannot be predicted in general terms, but only in individual cases.

3. Evaluation of the trace element content in cement

The levels of environmentally relevant heavy metals are determined primarily by the natural feed materials, but also by the secondary materials used as replacements, so it is important to make comparative checks as to whether and, if necessary, to what extent the concentrations deviate from those in natural rocks or soils. In these comparisons the cement could be regarded as an industrial conversion product of the rocks, and the soil as a natural conversion product [13]. The type and long term durability of chemical or adsorptive fixation of heavy metals in the hardened cement paste in the concrete is discussed elsewhere [16–19].

Table 9 lists examples of levels of heavy metals from geochemical investigations on limestone, clay and various soils [13, 20]. From this it can be seen that the concentrations of the heavy metals can in some cases exhibit a considerable range of fluctuation. Similarly large ranges of fluctuation occur in Germany in soils on siliceous, argillaceous and carbonatic sediments and on magmatic rocks which are not listed here [21]. **Table 10** contains numerical values for characterizing the concentrations frequently specified in German arable soils [22]. They are used as guide values and also, when applying the Sewage Sludge Regulation [23], as standard values [22, 24]. The listed values give the total content and not percentage which is available to a plant or can be resorbed [25–28]. They obviously therefore lie well on the "safe side". It cannot be ruled out that similar standard values will also be introduced in future soil protection legislation [28, 29]. Table 10 also lists "tolerable" total levels in soils as well as, for comparison, threshold values from the state institute for ecology, agriculture and forest planning (LÖLF) [29], and limiting values for contaminated soils.

The heavy metal concentration ranges predicted for Portland cement under defined conditions from Table 8 (white columns) are contrasted in **Fig. 2** with the range of standard values (shaded columns) and the total heavy metal levels which are still tolerable (base lines under the triangles) in air-dry arable soils taken from Table 10. For the comparison it

TABLE 10: Standard values '80. Guidelines for tolerable total levels of some elements in arable soils. Extract as given in [27] and [26]; data in g/t

Element		Total content in air-dry soil			
		standard values, commonly used	tolerable values	threshold value (LÖLF)	special or contaminated soils
Arsenic	(As)	2 – 20	20	40	< 8000
Beryllium	(Be)	1 – 5	10	–	< 2300
Lead	(Pb)	0.1– 20	100	300	< 4000
Cadmium	(Cd)	0.1– 50	3	2 (pH < 6.5)	< 200
Chromium	(Cr)	2 – 50	100	100	< 20 000
Nickel	(Ni)	2 – 50	50	100	< 10 000
Mercury	(Hg)	0.1– 1	2	2	< 500
Thallium	(Tl)	0.1– 0.5	1	1	< 40
Vanadium	(V)	10 –100	50	–	< 1000
Zinc	(Zn)	3 – 50	300	500	< 20 000

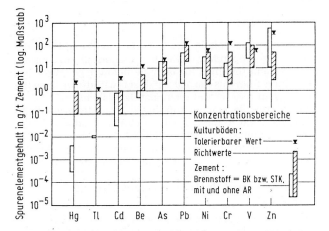

FIGURE 2: Levels of trace elements in the cement when burning lignite and coal with and without used tyres when compared with the standard values and tolerable levels in arable soils

Spurenelementgehalt in g/t Zement (log. Maßstab)	=	trace element content in g/t cement (log scale)
Konzentrationsbereiche	=	concentration ranges
Kulturböden	=	arable soils
Tolerierbarer Wert	=	tolerable value
Richtwerte	=	standard value
Zement	=	cement
Brennstoff = BK bzw. STK.	=	fuel = lignite or coal
mit und ohne AR	=	with or without used tyres

should be noted that the data for levels of heavy metals relate to materials with very large differences in loss on ignition. While cements have a loss on ignition of about 1 wt.%, the loss on ignition of soils can reach values of over 30 wt.%. This means that the range of the soil standard values and the tolerable total level would have to be moved to significantly higher concentrations in a comparison based on loss-on-ignition-free states.

Overall, the diagram shows that most of the heavy metal concentrations in cements lie within, and in some cases significantly below, the ranges of values which apply as standard values for normal air-dry soils. In the case of the elements V and Zn the standard value ranges may lie somewhat below the ranges for cement. The tolerable total levels are not exceeded with the exception of V and Zn which, on the basis of the LÖLF values (Table 10), are assessed as being of significantly lower environmental relevance.

Also important is the discovery that the heavy metal content in Portland cement is not, or not greatly, increased when part of the natural feed materials is replaced by secondary materials. A precondition for this is, however, that the secondary materials considered are selected in such a way and used in such quantities that they neither increase the emission, disrupt the industrial manufacturing process, impair the service properties of the cement, nor jeopardize the environmental compatibility of the concrete. Based on these criteria the feasibility of utilizing a secondary material can therefore be assessed after a prediction of the expected emission and of the heavy metal content in the cement. It also has to be taken into account that cement does not come directly into contact with the environment, e.g. with the soil. Cement is an intermediate product for manufacturing concrete. In concretes of normal composition the cement content is matched by an approximately 5- to 6-times greater quantity of aggregate which therefore reduces the quantity of trace elements introduced with the cement by the same factor. This means that the quantities of heavy metal in the concrete originating from the cement are correspondingly lower. The levels of heavy metal in the concrete originating from the cement therefore as a rule lie in or below the ranges of the corresponding elements in arable soils, and in particular below the currently specified tolerance thresholds [28].

4. Conclusion

Extensive investigations on feed materials for the burning process, on clinkers, and on cements have shown that the levels in cement of heavy metals which are not, or are only slightly, volatile can be predicted by taking account of the conditions prevailing in the individual instance. The ranges of the calculated levels correspond in order of magnitude to the values measured by analysis.

ten. Im Fall der Elemente V und Zn kann die Richtwertspanne etwas unterhalb des Bereichs für Zement liegen. Die tolerierbaren Gesamtgehalte werden mit der Ausnahme von V und Zn, deren Umweltrelevanz jedoch z. B. aufgrund der LÖLF-Werte (Tabelle 10) als deutlich geringer beurteilt wird, nicht überschritten.

Wesentlich ist darüber hinaus die Feststellung, daß sich der Schwermetallgehalt im Portlandzement nicht oder nicht sonderlich erhöht, wenn ein Teil der natürlichen Einsatzstoffe gegen Sekundärstoffe ausgetauscht wird. Voraussetzung hierfür ist jedoch, daß die in Frage kommenden Sekundärstoffe so ausgewählt und in solchen Mengen eingesetzt werden, daß hierdurch weder die Emission ansteigt, der technische Herstellungsprozeß gestört, die Gebrauchseigenschaften des Zements beeinträchtigt noch die Umweltverträglichkeit des Betons in Frage gestellt werden. Die Verwendungsmöglichkeit eines Sekundärstoffs kann demnach anhand dieser Kriterien nach einer Prognose der zu erwartenden Emission und des Schwermetallgehalts im Zement beurteilt werden. Hierbei ist darüber hinaus zu berücksichtigen, daß Zement nicht direkt mit der Umwelt, wie z. B. mit Boden, in Berührung kommt. Zement ist ein Zwischenprodukt für die Betonherstellung. Auf den Zementgehalt in Betonen üblicher Zusammensetzung entfällt dabei ein etwa 5- bis 6fach größerer Zuschlaganteil, der demnach die mit dem Zement eingebrachte Spurenelementmenge um den gleichen Faktor herabsetzt. Das bedeutet, daß die vom Zement herrührende Schwermetallmenge im Beton dementsprechend geringer ausfällt. Der vom Zement herrührende Schwermetallgehalt des Betons liegt daher in der Regel im oder unterhalb des Bereichs der entsprechenden Elemente in Kulturböden, insbesondere aber unterhalb der derzeit angegebenen Toleranzschwelle [28].

4. Schlußbetrachtung

Umfangreiche Untersuchungen an Einsatzstoffen für den Brennprozeß, an Klinker sowie an Zementen haben gezeigt, daß der Gehalt nicht oder schwer flüchtiger Schwermetalle im Zement unter Berücksichtigung der im Einzelfall vorliegenden Bedingungen prognostiziert werden kann. Die Spanne der rechnerisch ermittelten Gehalte stimmt in der Größenordnung mit den analytisch gemessenen Gehalten überein.

Ein teilweiser Austausch natürlicher Einsatzstoffe gegen Sekundärstoffe ist möglich, ohne daß hierdurch der Schwermetallgehalt merklich über die üblichen, von der geochemischen Verteilung der Elemente in den Lagerstätten abhängigen Spannweiten der Schwermetallkonzentration im Portlandzement ansteigt. Beurteilungskriterien für die Einsatzfähigkeit von Sekundärstoffen sind in diesem Zusammenhang neben der Einsatzmenge die Art und Anteile der darin enthaltenen umweltrelevanten Schwermetalle.

Die Spannweite der relevanten Schwermetallkonzentration in Portlandzementen entspricht weitgehend den entsprechenden Konzentrationsbereichen dieser Elemente in Kulturböden, wobei zum Teil erhebliche Unterschiede in den Glühverlusten unberücksichtigt bleiben.

Literaturverzeichnis

[1] Wischers, G.: Möglichkeiten und Grenzen des Recycling in der Zementindustrie. Arbeitsgemeinschaft industrieller Forschungsvereinigungen (AiF) – Herbsttagung des Wissenschaftlichen Rats 1988, S. 7–14.

[2] Sprung, S.: Umweltentlastung durch Verwertung von Sekundärrohstoffen. Zement-Kalk-Gips 45 (1992) Nr. 5, S. 213–221.

[3] Sprung, S.: Spurenelemente – Anreicherung und Minderungsmaßnahmen. Zement-Kalk-Gips 41 (1988) Nr. 5, S. 251–256.

Partial replacement of natural feed materials by secondary materials is possible without causing the heavy metal levels to rise appreciably above the usual ranges of heavy metal concentrations in Portland cement which are dependent on the geochemical distributions of the elements in the deposits. In this context the assessment criteria for the usability of secondary materials are, in addition to the quantity used, the type and proportions of the environmentally relevant heavy metals contained in them.

The ranges of the relevant heavy metal concentrations in Portland cements correspond substantially to the concentration ranges of these elements in arable soils, which does not take account of the sometimes considerable differences in loss on ignition.

[4] Kirchner, G.: Das Verhalten des Thalliums beim Brennen von Zementklinker. Schriftenreihe der Zementindustrie, Heft 47. Beton-Verlag, Düsseldorf 1986.

[5] Kirchner, G.: Thalliumkreisläufe und Thalliumemissionen beim Brennen von Zementklinker. Zement-Kalk-Gips 40 (1987) Nr. 3, S. 134–144.

[6] Sprung, S., und Rechenberg, W.: Die Reaktionen von Blei und Zink beim Brennen von Zementklinker. Zement-Kalk-Gips 31 (1978) Nr. 7, S. 327–329.

[7] Sprung, S.: Technologische Probleme beim Brennen des Zementklinkers, Ursache und Lösung. Schriftenreihe der Zementindustrie, Heft 43. Beton-Verlag, Düsseldorf 1982.

[8] Sprung, S., und Rechenberg, W.: Reaktionen von Blei und Zink bei der Zementherstellung. Zement-Kalk-Gips 36 (1983) Nr. 10, S. 539–548.

[9] Sprung, S., Kirchner, G., und Rechenberg, W.: Reaktionen schwer verdampfbarer Spurenelemente beim Brennen von Zementklinker. Zement-Kalk-Gips 37 (1984) Nr. 10, S. 513–518.

[10] Kühl, H.: Zement-Chemie, Bd. II, 3. Auflg. VEB Verlag Technik, Berlin 1958.

[11] Kirchner, G.: Reaktionen des Cadmiums beim Klinkerbrennprozeß. Zement-Kalk-Gips 38 (1985) Nr. 9, S. 535–539.

[12] Kirchner, G., und Rechenberg, W.: Spurenelementbilanzen von Zementdrehöfen. In: B. Welz (Hrsg.): Fortschritte der atomspektrometrischen Spurenanalytik. VCH Verlagsgesellschaft, Weinheim 1986.

[13] Wedepohl, K. H.: Handbook of Geochemistry. Springer-Verlag, Berlin–Heidelberg–New York.

[14] Verein Deutscher Zementwerke – Forschungsinstitut der Zementindustrie: Tätigkeitsbericht 1990–93.

[15] Verein Deutscher Zementwerke – Forschungsinstitut der Zementindustrie: Tatigkeitsbericht 1987–90.

[16] Sprung, S., und Rechenberg, W.: Einbindung von Schwermetallen in Sekundärstoffen durch Verfestigen mit Zement. beton 38 (1988) Nr. 5, S. 193–198.

[17] Schmidt, M.: Verwertung von Müllverbrennungsrückständen zur Herstellung zementgebundener Baustoffe. beton 38 (1988) Nr. 6, S. 238–245.

[18] Sprung, S., und Rechenberg, W.: Bindung umweltrelevanter Sekundärstoffe durch Verfestigen mit Zement. Zement Beton 34 (1989) Nr. 2, S. 54–61.

[19] Rechenberg, W.: Auslaugverhalten und Bewertung der Prüfergebnisse, Generalbericht. RILEM-Workshop: Auslaugverhalten von Beton und zementgebundenem Material. Juni 1992, S. 26/35. Mittlg. Forschungsinst. Vereinig. Österreich. Zementind., Nr. 42, Wien 1992.

[20] Rösler, H. J., und Lange, H.: Geochemische Tabellen. Ferdinand Enke Verlag, Stuttgart 1976.

[21] Golwer, A.: Geogene Schwermetallgehalte in mineralischen Böden von Hessen. Dechema Fachgespräche Umweltschutz 1988, S. 137–141.

[22] Kloke, A.: Orientierungsdaten für tolerierbare Gesamtgehalte einiger Elemente in Kulturböden. Mittlg. VDLUVA 1980, Nr. 1/3, S. 9–10.

[23] Klärschlammverordnung (AbfKlärV) v. 25. 7. 1982. BGBl 1982, Teil 1, S. 734–739.

[24] Kloke, A.: Grundlagen zur Ermittlung von nutzungsbezogenen, höchsten akzeptierbaren Schadstoffgehalten in innerstädtischen und stadtnahen Böden. 2. Intern. TNO/BMFT-Kongr. Altlastsanierung, S. 291–303. BMFT und UBA (Hrsg.), Bonn–Berlin 1988.

[25] Binder, S., Sokal, D., und Maugham, D.: Estimating Soil Ingestion: The Use of Trace Elements in Estimating the Amount of Soil Ingested by Young Children. Arch. Environm. Health 41 (1986) Nr. 6, S. 341–345.

[26] Clausing, P., Brunekreef, B., und van Wijnen, J. H.: A Method of Estimating Soil Ingestion by Children. Int. Arch. Occup. Environm. Health 59 (1987) S. 73–82.

[27] Barnes, R. M.: Childhood Soil Ingestion: How much Dirt do Kids Eat? Anal. Chem. 62 (1990) Nr. 19, S. 1023A–1033A.

[28] Kloke, A.: Zur Problematik und Begründung von Schwellenwerten für Schwermetalle in Böden. In: Beurteilung von Schwermetallkontaminationen im Boden. S. 77–86. DECHEMA, Frankfurt/Main, 1989.

[29] Landesanstalt für Ökologie, Landschaftsentwicklung und Forstplanung (LÖLF): Mindestuntersuchungsprogramm Kulturböden. Recklinghausen, Jan. 1988.

Beurteilung der Rohmehlreaktivität durch zwei eigenständige und sich ergänzende Verfahren

Assessment of raw meal reactivity by two independent and complementary methods

L'évaluation de la réactivité de la farine crue moyennant deux procédés indépendants et complémentaires

Evaluación de la reactividad de la harina cruda mediante dos procedimientos independientes y complementarios

Von **N. Streit**, Köln/Deutschland

Zusammenfassung – Die Rohmehlreaktivität beeinflußt die Zementqualität und die Wirtschaftlichkeit des Klinkerbrandes. Zur Bewertung der Rohmehlreaktivität wurden zwei Methoden verglichen. Sie weisen völlig andersartige Lösungsansätze auf und haben sich beide in der Praxis bewährt. Bei der einen Methode wird mit einer dynamischen Laborbrennapparatur der Brennprozeß so praxisnah wie möglich simuliert. Die zweite Methode bewertet die Rohmehlreaktivität indirekt über die chemische Heterogenität der Kornklassen. Im Rahmen eines Forschungsprojektes wurde eine Vielzahl von Rohmehlen nach beiden Methoden untersucht. Trotz der völlig andersartigen Lösungswege zeigte sich eine von der Tendenz her ähnliche Bewertung der Rohmehle. Beide Methoden wiesen jedoch Schwachpunkte auf. Bei den Versuchen konnte vor allem die Auswertung der Fraktionsanalyse optimiert werden. Insgesamt erlaubten beide Verfahren für sich eine verhältnismäßig verläßliche Beurteilung. Durch Kombination beider Methoden wurde nicht nur die Aussagesicherheit erheblich gesteigert, sondern es wurden auch die Ursachen sichtbar, die zu Abweichungen bei Einzelergebnissen geführt haben. Daraus ließen sich verfahrenstechnische Hinweise zur Verbesserung der Brennbarkeit von Rohmehlen ableiten.

Beurteilung der Rohmehlreaktivität durch zwei eigenständige und sich ergänzende Verfahren

Summary – Raw meal reactivity affects cement quality and the cost-effectiveness of the clinker burning process. Two methods for evaluating raw meal reactivity are compared. They show entirely different approaches to the problem and have both proved successful in practice. In one method a dynamic laboratory burning apparatus simulates the burning process as closely as possible to practical conditions. The second method evaluates the raw meal reactivity indirectly through the chemical heterogeneity of the particle size classes. A large number of raw meals were investigated by both methods as part of a research project. In spite of the completely different approaches the evaluations of the raw meals showed similar trends. However, both methods have weak points. In particular, it was possible to optimize the evaluation of the fraction analysis during the tests. On the whole both methods permit relatively dependable assessments to be made. Combining the two methods not only considerably increased the reliability of the information, but also clarified the reasons for deviations in individual results. From this it was possible to derive process engineering information for improving the burnability of raw meals.

Assessment of raw meal reactivity by two independent and complementary methods

Résumé – La réactivité de la farine crue exerce une influence sur la qualité du ciment et sur la rentabilité de la cuisson du ciment. Afin d'évaluer la réactivité de la farine crue, deux méthodes ont été comparées, qui offrent des solutions tout à fait différentes et qui ont fait leurs preuves dans la pratique. Quant à la première de ces méthodes, le processus de cuisson est simulé, en s'orientant le plus possible à la pratique, à l'aide d'un appareil de cuisson, de laboratoire. La deuxième méthode permet d'évaluer la réactivité de la farine crue de façon indirecte, à l'aide de la hétérogénéité chimique des fractions granulométriques. Dans le cadre d'un projet de recherche, on a étudié un grand nombre de farines crues en employant les deux méthodes susmentionnées. Malgré que les manières de procéder soient tout à fait différentes, on a pu constater que l'évaluation des farines crues accusait une tendance similaire. Cependant, les deux méthodes ont des points faibles. Lors des essais, on a pu optimiser surtout l'évaluation de l'analyse granulométrique. Chacun de ces procédés a permis de faire une évaluation relativement sûre. En combinant les deux méthodes, on a pu augmenter non seulement de façon considérable la fiabilité des dates obtenues, mais on a découvert également les causes ayant entraîné des divergences dans les résultats individuels. Ceci a permis de dériver certains renseignements technologiques permettant d'améliorer la cuisson des farines crues.

L'évaluation de la réactivité de la farine crue moyennant deux procédés indépendants et complémentaires

Resumen – La reactividad de la harina cruda influye en la calidad del cemento y en la rentabilidad de la cocción del clínker. Para evaluar la reactividad de la harina cruda, se han comparado dos métodos, que ofrecen soluciones totalmente diferentes y que han probado su eficacia en la práctica. En cuanto al primero de estos métodos, el proceso de cocción es simulado con ayuda de un aparato de cocción, de laboratorio, orientándose lo más posible hacia la práctica. El segundo método permite evaluar la reactividad de la harina cruda de forma indirecta, con ayuda de la heterogeneidad química de las fracciones granulométricas. Dentro del marco de un proyecto de investigación, se ha estudiado un gran número de harinas crudas, empleando los dos métodos arriba mencionados. A pesar de las soluciones totalmente diferentes, resulta una tendencia similar en la evaluación de las harinas crudas. No obstante, ambos métodos tienen sus puntos débiles. Durante los ensayos, se ha podido optimizar, sobre todo, la evaluación del análisis granulométrico. Se puede decir que cada uno de estos métodos, por sí solo, permite una evaluación relativamente segura. Combinando estos métodos, se ha podido aumentar no solamente de forma considerable la fiabilidad de los datos obtenidos, sino que se han puesto de manifiesto también las causas de las divergencias registradas en los resultados individuales. Todo ello ha permitido derivar informaciones tecnológicas susceptibles de mejorar la aptitud a la cocción de las harinas crudas.

Evaluación de la reactividad de la harina cruda mediante dos procedimientos independientes y complementarios

Die Rohmehlreaktivität beeinflußt in starkem Maße die Zementqualität und die Wirtschaftlichkeit des Klinkerbrandes. Im Forschungszentrum der KHD Humboldt Wedag AG wird die Reaktivität eines Rohmehls mit einem indirekten Prüfverfahren über dessen chemisch-granulometrische Heterogenität ermittelt. Nach I. Dreizler und H.-U. Schäfer [1] wird das Rohmehl zunächst mittels Handsiebung in sechs Kornklassen zerlegt (**Tabelle 1**). Von den Fraktionen werden die Massenanteile und die chemischen Hauptbestandteile bestimmt. Die Auswertung erfolgt über die Abweichungen der Konzentrationen bzw. der Moduli in den einzelnen Fraktionen gegenüber der berechneten Gesamtprobe. Als Kennwert für die Rohmehlaktivität wurde die „gewichtete mittlere Abweichung des Kalkstandards" (ΔKST_{gew}) gewählt. Berechnet wird dieser Wert, indem die Abweichungen im Kalkstandard einer jeden Fraktion gewichtet und dann addiert werden.

Durch den Vergleich der ermittelten Kennwerte mit Betriebserfahrungen wurde ferner eine Bewertungsskala erarbeitet. Um die Bewertung zu optimieren, wurde zusätzlich eine Labormethode herangezogen, aus der sich ein quantifizierbarer Kennwert für die Brennbarkeit ableiten läßt. Die Versuchsbedingungen dieser Methode müssen hierfür so praxisnah wie möglich gewählt werden.

Raw meal reactivity greatly influences cement quality and the economic efficiency of clinker burning. At the Research Centre of KHD Humboldt Wedag AG, the reactivity of a raw meal is ascertained by means of an indirect testing method based on its chemical-granulometric heterogeneity. Following the method described by I. Dreizler and H.-U. Schäfer [1], the raw meal is first divided into six particle size classes by manual screening. The proportions by weight and the main chemical constituents of the fractions are determined. The evaluation is carried out on the basis of the deviations of the concentrations or the moduli in the individual fractions compared with the calculated complete sample. The "weighted mean deviation of the lime standard" (ΔLST_{wt}) was chosen as the coefficient of raw meal reactivity. This value is calculated by weighting and then adding together the deviations in the lime standard of each fraction.

In addition, an assessment scale was drawn up by comparing the coefficients thus determined with operational experience. To optimise the assessment, use was also made of a laboratory method from which a quantifiable coefficient of burnability can be derived. The experimental conditions chosen for this method should approximate as closely as possible to practical conditions.

TABELLE 1: Heterogenitätsuntersuchung eines Rohmehls (Rohmehl 14) und die daraus abgeleitete gewichtete mittlere Abweichung des Kalkstandards (ΔKST_{gew})
TABLE 1: Heterogeneity analysis of a raw meal (raw meal 14) and the weighted mean deviation of the lime standard (ΔLST_{wt}) derived from it

Kornklassen Particle size classes	>200 μ	200–125 μ	125–90 μ	90–63 μ	63–32 μ	<32 μ	gesamt Total
Anteil % Proportion in %	1,50	3,10	6,40	10,60	25,20	53,20	100,00
SiO_2	36,08	26,28	22,31	21,26	17,07	9,79	14,55
Al_2O_3	7,46	4,78	3,18	2,48	2,05	2,58	2,62
Fe_2O_3	7,80	4,82	2,97	2,00	1,55	1,57	1,89
CaO	23,97	33,64	38,48	40,22	42,93	45,58	43,20
KST II/LST II	20,86	40,85	56,46	63,09	83,81	144,80	95,88
SM/SR	2,36	2,74	3,63	4,75	4,74	2,36	3,23
KM/LM	4,84	5,50	7,02	8,57	8,33	3,79	5,56
TM/AR	0,96	0,99	1,07	1,24	1,32	1,64	1,38
ΔSiO_2	21,53	11,73	7,76	6,71	2,52	−4,76	
ΔAl_2O_3	4,84	2,16	0,56	−0,14	−0,57	−0,04	
ΔFe_2O_3	5,91	2,93	1,08	0,11	−0,34	−0,32	
ΔCaO	−19,23	−9,56	−4,72	−2,98	−0,27	2,38	
ΔKST II/ΔLST II	−75,02	−55,04	−39,42	−32,80	−12,07	48,92	
$\Delta SM/\Delta SR$	−0,86	−0,49	0,40	1,52	1,52	−0,87	
$\Delta KM/\Delta LM$	−0,73	−0,06	1,45	3,01	2,77	−1,77	
$\Delta TM/\Delta AR$	−0,42	−0,39	−0,31	−0,14	−0,06	0,26	
$\Delta KST_{gew}/\Delta LST_{wt}$							14,15

$\Delta KST_{gew} = \Sigma((KST_{Fraktion} - KST_{gesamt}) \cdot Anteil_{Fraktion})$ $\Delta LST_{wt} = \Sigma((LST_{fraction} - LST_{total}) \cdot Proportion_{fraction})$

TABELLE 2: Bewertungsskalen für die Heterogenität (ΔKST_{gew}) und die Garbrandreaktivität ($\Sigma K' 2$)
TABLE 2: Assessment scales for heterogeneity (ΔLST_{wt}) and finishing burn reactivity ($\Sigma K'2$)

Sinterverhalten Sintering behaviour	Heterogenität (ΔKST_{gew}) Heterogeneity (ΔLST_{wt})
gut/good	< 0,5
normal/normal	0,5–3,0
erschwert/difficult	3,0–6,0
sehr schwer/very difficult	> 6,0

Sinterverhalten Sintering behaviour	Garbrandreaktivität ($\Sigma K'2$) Finishing burn reactivity ($\Sigma K'2$)
sehr gut/very good	> 2,9
gut/good	2,9–2,7
mittel/average	2,7–2,5
mäßig/mediocre	2,5–2,3
schlecht/poor	< 2,3

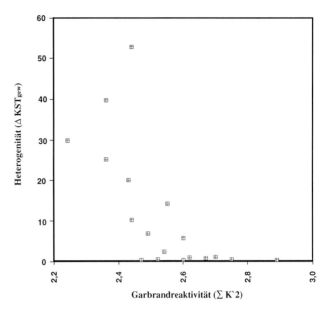

BILD 1: Korrelation von Heterogenität (ΔKST_{gew}) und Garbrandreaktivität ($\Sigma K' 2$)
FIGURE 1: Correlation between heterogeneity (ΔLST_{wt}) and finishing burn reactivity ($\Sigma K'2$)

Heterogenität (ΔKST_{gew}) = Heterogeneity (ΔLST_{wt})
Garbrandreaktivität = Finishing burn reactivity

Hierfür bot sich der Reaktivitätstest mit dynamischer Laborbrennapparatur (DLA) des Instituts Weimar der ZAB Dessau an [2]. Innerhalb eines bestimmten Aufheiz- und Abkühlprogramms wird der Kalkbindungsgrad K' eines Brenngutes auf unterschiedlichen Temperaturstufen bestimmt. Durch Addition der ermittelten Werte für den Kalkbindungsgrad wird im Temperaturintervall von 800 bis 1200 °C der Kennwert für den Niedertemperaturbereich ($\Sigma K' 1$) und im Intervall von 1200 bis 1450 °C der Kennwert für den Garbrandbereich ($\Sigma K' 2$) ermittelt.

Bisher wurden 18 Rohmehle nach beiden Methoden vergleichend bearbeitet. Bei Gegenüberstellung von Heterogenität (ΔKST_{gew}) und Garbrandreaktivität ($\Sigma E' 2$) (**Bild 1**) zeichnet sich zwar eine tendenzielle Abhängigkeit zwischen den beiden Kennwerten ab, die Korrelation ist aber nicht befriedigend. Nach den Bewertungsskalen für die beiden Methoden in **Tabelle 2** werden z. B. die aufgrund der „Heterogenität" als „gut" sinterbar eingestuften Rohmehle in ihrer auf der „Garbrandreaktivität" basierenden Bewertung nur von „gut bis mäßig" eingestuft. Dies entspricht immerhin drei Bewertungsklassen.

Außerdem liefert die „Heterogenität" grundsätzlich schlechtere Bewertungen als die „Garbrandreaktivität". So wird nur bei einer Probe die „Garbrandreaktivität" mit „schlecht" beurteilt, wogegen die Hälfte der Proben aufgrund ihrer „Heterogenität" als schlecht sinterbar eingestuft werden. Daraus wird ersichtlich, daß die Korrelation dieser beiden Kennwerte nicht befriedigend ist. Außerdem stimmen die Bewertungsskalen der beiden Methoden nicht gut überein.

Die Tatsache, daß diese beiden völlig unterschiedlichen Methoden zumindest tendenziell ähnliche Ergebnisse lieferten, ermutigte jedoch zur Fortsetzung der Untersuchungen. Dabei zeigte sich, daß die Abweichung im SiO$_2$-Gehalt wesentlich besser mit der „Garbrandreaktivität" ($\Sigma E' 2$) korreliert ist. Außerdem wurde bereits 1990 [1] auf die Zusammenhänge zwischen Heterogenität, Rohmehlfeinheit und Sinterverhalten hingewiesen. Deshalb war es naheliegend, die Rohmehlfeinheit als einen entscheidenden Faktor in einen aus der Heterogenität abgeleiteten Kennwert für das Sinterverhalten einfließen zu lassen.

Für den neu entwickelten Kennwert (H) werden die gewichteten Abweichungen im SiO$_2$-Gehalt als Absolutbeträge addiert (**Tabelle 3**). Dadurch wird verhindert, daß sich positive und negative Abweichungen aufheben. Als Maß für die Rohmehlfeinheiten geht der Rückstand auf dem 90-µm-Sieb als Faktor in die Berechnung ein. Werden die neuen, aus der „Heterogenität" abgeleiteten Kennwerte (H) und die Werte für die „Garbrandreaktivität" gegenübergestellt, so ist eine klare Abhängigkeit zu erkennen (**Bild 2**). Dies ist besonders bemerkenswert, wenn die völlig andersartigen Lösungsansätze beider Methoden berücksichtigt werden. Eine Regres-

The reactivity test using a dynamic laboratory burning apparatus from the Weimar institute of ZAB Dessau was suitable for this purpose. Within a specific heating and cooling regime, the lime combination value K' of a kiln feed is determined at different temperature levels. Adding together the lime combination values thus determined gives the coefficient for the low-temperature range ($\Sigma K'1$) in the temperature range 800 – 1200 °C, and the coefficient for the finishing burn range ($\Sigma K'2$) in the range 1200 – 1450 °C.

So far, 18 raw meals have been compared using both methods. If the heterogeneity (ΔLST_{wt}) and the finishing burn reactivity ($\Sigma E'2$) are plotted against each other (**Fig. 1**), a broad relationship between the two coefficients is apparent, but the correlation is not satisfactory. According to the assessment scales shown in **Table 2** for the two methods, the raw meals which, for instance, are classed as having "good" sintering properties on the basis of their "heterogeneity" are classed as "good to mediocre" when assessed on the basis of their "finishing burn reactivity". This does at least correspond to three assessment classes.

Moreover, the "heterogeneity" generally supplies poorer assessments than the "finishing burn reactivity". Thus, only in the case of one sample was the "finishing burn reactivity" classed as "poor", whereas on the basis of their "heterogeneity", half the samples were classed as having poor sintering behaviour. It is clear from this that the correlation between these two coefficients is unsatisfactory, nor do the assessment scales of the two methods correspond well.

The fact that these two completely different methods produce at least broadly similar results nevertheless provided an incentive for continued research. There turned out to be a significantly better correlation between the deviation in the SiO$_2$ content and the "finishing burn reactivity" ($\Sigma E'2$). Moreover, the relationships between heterogeneity, raw meal fineness and sintering behaviour had been referred to back in 1990 [1]. Therefore it seemed reasonable that, as a deciding factor, the raw meal fineness should have some influence on a coefficient of sintering behaviour which is derived from the heterogeneity of the raw meal.

For the newly developed coefficient (H), the weighted deviations in the SiO$_2$ content are added together as absolute amounts (**Table 3**). This prevents positive and negative deviations from cancelling each other out. The oversize on the 90 µm screen is taken as a measure of the raw meal fine-

TABELLE 3: Berechnung des neuen aus der Heterogenität abgeleiteten Kennwerts (H) sowie neue Bewertungsskalen für diesen Kennwert und die Garbrandreaktivität ($\Sigma K' 2$)
TABLE 3: Calculation of the new coefficient (H) derived from the heterogeneity and new assessment scales for this coefficient and for the finishing burn reactivity ($\Sigma K'2$)

Sinterverhalten Sintering behaviour	Kennwert (H) Coefficient (H)	Garbrandreaktivität ($\Sigma K'2$) Finishing burn reactivity ($\Sigma K'2$)
sehr gut/very good	<5	>2,8
gut/good	5–17	2,6–2,8
normal/normal	17–64	2,4–2,6
schlecht/poor	>64	<2,4

Kennwert (H) = $\Sigma (| SiO_{2\,Fraktion} - SiO_{2\,gesamt} | \cdot Anteil_{Fraktion}) \cdot R.90 \mu m$

Coefficient (H) = $\Sigma (| SiO_{2\,fraction} - SiO_{2\,total} | \cdot proportion_{fraction}) \cdot R.90 \mu m$

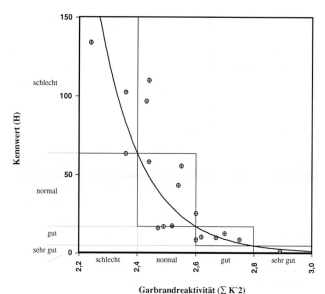

BILD 2: Korrelation von dem aus der Heterogenität abgeleiteten Kennwert (H) und der Garbrandreaktivität ($\Sigma K' 2$)
FIGURE 2: Correlation between the coefficient (H) (derived from the heterogeneity) and the finishing burn reactivity ($\Sigma K'2$)

Kennwert (H) = Coefficient (H)
schlecht = poor
normal = normal
gut = good
sehr gut = very good
Garbrandreaktivität = Finishing burn reactivity

sion nach einer Exponentialfunktion entspricht dabei den vorhandenen Daten besser als eine lineare Regression.

Mit Hilfe dieser Regression lassen sich Bewertungsskalen für beide Methoden festlegen, die in ca. 85 % aller Fälle zu einer gleichen Beurteilung führen (unterer Teil in Tabelle 3). Bei den restlichen 15 % unterscheiden sich die Beurteilungen um maximal eine Klasse, wobei es sich meist um Grenzfälle handelt. Durch eine Kombination beider Methoden läßt sich nicht nur die Verläßlichkeit der Bewertung steigern, sondern es wird neben dem Sinterverhalten auch ein breites Feld an zusätzlichen Informationen über das Rohmehl erschlossen. Dadurch sind Hinweise für die verfahrenstechnische Durchführung möglich, die sowohl die Wirtschaftlichkeit als auch die Zementqualität verbessern können. So liefert z. B. der Reaktivitätstest mit dynamischer Brennapparatur zusätzliche Informationen über die Kalzination und die Festkörperreaktionen im Niedertemperaturbereich. Ferner erlaubt die Heterogenitätsuntersuchung, aus dem Kennwert (H) näherungsweise eine nach wirtschaftlichen und verfahrenstechnischen Gesichtspunkten optimale Rohmehlfeinheit abzuleiten. Schließlich lassen sich bei Rohmehlen mit hohen MgO-Gehalten Aussagen hinsichtlich zu erwartender Raumbeständigkeitsprobleme machen, wenn MgO-Gehalt mit in die Heterogenitätsuntersuchung einbezogen wird. Zeigt sich hierbei eine Anreicherung des MgO in den groben Fraktionen, so läßt sich durch feineres Aufmahlen die Raumbeständigkeit verbessern [3, 4].

Insgesamt haben die vergleichenden Untersuchungen gezeigt, daß die chemisch-granulometrische Heterogenität verläßliche Aussagen über das Sinterverhalten von Rohmehlen erlaubt. Die Auswertung wurde optimiert und dieses von der KHD Humboldt Wedag AG entwickelte Prüfverfahren als eigenständige Methode bestätigt.

Literaturverzeichnis

[1] Dreizler, I., und Schäfer, H.-U.: Fraktionsanalysen zur Bewertung der Heterogenität von Rohmehlen. Zement-Kalk-Gips 43 (1990) Nr. 9, S. 445–451.

[2] Kieser, J., Krähner, A., und Gathemann, B.: Modell zur Bestimmung der Rohmehlreaktivität unter praxisähnlichen Bedingungen. Zement-Kalk-Gips 32 (1979) Nr. 9, S. 442–447.

[3] Dreizler, I., und Knöfel, D.: Der Einfluß von Magnesiumoxid auf die Zementeigenschaften. Zement-Kalk-Gips 35 (1982) Nr. 10, S. 537–590.

[4] Dreizler, I.: Herstellung von Portlandzement aus magnesiumreichen Rohstoffen. Zement-Kalk-Gips 41 (1988) Nr. 5, S. 243–250.

ness and is included as a factor in the calculation. If the new coefficients (H) derived from the "heterogeneity" and the "finishing burn reactivity" values are plotted against each other, a clear relationship is evident (**Fig. 2**). This is especially remarkable considering the completely different approaches of the two methods. Here, a regression based on an exponential function corresponds better to the available data than a linear regression.

With the aid of this regression, assessment scales can be established for both methods which produce the same assessment in about 85 % of cases (lower part of Table 3). In the case of the remaining 15 %, the assessments differ by at most one category; most of these are borderline cases. Combining the two methods not only enables the reliability of the assessment to be increased, but a broad range of additional information on the raw meal besides its sintering behaviour is also gained. This may produce indications with regard to implementation in terms of engineering practice which can improve both efficiency and cement quality. Thus, for instance, the reactivity test using a dynamic burning apparatus supplies additional information on calcination and the solid state reactions in the low-temperature range. Furthermore, the heterogeneity analysis enables the best possible raw meal fineness from the commercial and process engineering aspects to be estimated from the coefficient (H). Lastly, for raw meals with a high MgO content, it is possible to make an assessment regarding probable soundness problems if the MgO content is included in the heterogeneity analysis. If this reveals that the MgO is concentrated in the coarse fractions, the soundness can be improved by grinding the meal more finely [3–4].

In all, the comparative studies have shown that reliable judgments as to the sintering behaviour of raw meals can be made on the basis of their chemical-granulometric heterogeneity. The evaluation was optimised and this testing method developed by KHD Humboldt Wedag AG was confirmed as being an independent method.

Einfluß der Klinkerzusammensetzung und Klinkerkühlung auf die Zementeigenschaften

Influence of clinker composition and clinker cooling on cement properties

Influence de la composition et du refroidissement du clinker sur les propriétés du ciment

Influjo de la composición y del enfriamiento del clinker sobre las propiedades del cemento

Von **H.-M. Sylla,** Düsseldorf/Deutschland

Zusammenfassung – Untersuchungen über den Einfluß unterschiedlicher Kühlung auf Erstarren und Festigkeitsentwicklung von Zementen gleicher Mahlfeinheit aus im Labor gebrannten Klinkern mit konstantem Kalkstandard und Schmelzphasenanteil, aber mit verschiedenen Silicat- und Tonerdemodulen zeigten, daß die Festigkeit und das Erstarren durch die Ausbildung und Zusammensetzung der Grundmasse des Klinkers beeinflußt wird. Die Normdruckfestigkeit des Zements nach 28 Tagen nimmt mit steigendem Tricalciumaluminatgehalt (C_3A) bei langsamer Klinkerkühlung deutlich ab. Bei schneller Kühlung sind dagegen die Festigkeitsunterschiede weniger deutlich ausgeprägt. Zemente aus Klinker mit niedrigem C_3A-Gehalt reagieren auf Kühleinflüsse hinsichtlich Erstarrungsverhalten und Festigkeitsentwicklung weniger empfindlich als Zemente aus Klinker mit höherem C_3A-Gehalt. C_3A-reiche Zemente erstarren generell schneller. Die Verarbeitbarkeit läßt sich jedoch durch darauf abgestimmte Sulfatträgergemische aus Anhydrit und Halbhydrat verbessern. Durch unterschiedliche Kühlbedingungen werden das Gefüge und die Zusammensetzung der Klinkerphasen verändert, wie mikroskopische, rasterelektronenmikroskopische und röntgenfluoreszenzanalytische Untersuchungen zeigten.

Einfluß der Klinkerzusammensetzung und Klinkerkühlung auf die Zementeigenschaften

Summary – Investigations were carried out on the influence of different types of cooling on the setting and strength development of cements of the same fineness made from clinkers burnt in the laboratory with constant lime standard and proportion of melt phase, but with differing silica and alumina ratios. These showed that the strength and setting are influenced by the form and composition of the clinker matrix. With fairly slow clinker cooling the standard compressive strength of cement after 28 days decreases significantly with increasing tricalcium aluminate content (C_3A). But with more rapid cooling the strength differences are less strongly marked. Cements made from clinker of low C_3A content react less sensitively to cooling effects with regard to setting behaviour and strength development than do cements made from clinker of higher C_3A content. C_3A-rich cements generally set more rapidly. However, the workability can be improved by suitable sulphate agent mixtures of anhydrite and hemihydrate. Investigations with microscopes, scanning electron microscopes and X-ray fluorescence analysis have shown that the structures and compositions of the clinker phases are changed by different cooling conditions.

Influence of clinker composition and clinker cooling on cement properties

Résumé – Des études relatives à l'influence de différentes manières de refroidissement sur la prise et le développement de la résistance de ciments de la même finesse de broyage, fabriqués à l'aide de clinkers cuits au laboratoire et présentant un degré de saturation en chaux et une proportion de phase liquide constants, tandis que les modules siliciques et alumineux étaient différents, ont démontré que la résistance et la prise des ciments sont influencées par la formation et la composition de la masse fondamentale du clinker. La résistance à la compression normalisée du ciment après 28 jours diminue nettement, au fur et à mesure qu'augmente la teneur en aluminate tricalcique (C_3A), si le refroidissement du clinker est assez lent. Par contre, si le refroidissement du clinker est rapide, les différences en ce qui concerne la résistance sont moins nettes. Les ciments fabriqués sur la base de clinkers à teneur en C_3A réduite réagissent de manière moins sensible aux influences du refroidissement, quant au comportement à la prise et au développement de la résistance, que les ciments fabriqués de clinkers à teneur en C_3A plus élevée. La prise des ciments riches en C_3A est, en général, plus rapide. Cependant, il est possible d'améliorer l'ouvrabilité des ciments moyennant des mélanges appropriés de porteurs de sulfates, c'est-à-dire d'anhydrite et de semi-hydrate. Dû aux différentes conditions de refroidissement, la structure et la composition des phases de clinker sont modifiées, conformément aux études microscopiques ainsi qu'aux études effectuées à l'aide du microscope électronique à balayage et aux études par fluorescense X.

Influence de la composition et du refroidissement du clinker sur les propriétés du ciment

Resumen – Estudios referentes al influjo de diferentes modos de enfriamiento sobre el fraguado y el desarrollo de las resistencias de cementos de la misma finura de molienda, fabricados a partir de clínkeres cocidos en el laboratorio, de standard de cal y proporción de fase líquida constantes, pero de módulos silícicos y alumínicos diferen-

Influjo de la composición y del enfriamiento del clinker sobre las propiedades del cemento

tes, han mostrado que la resistencia y el fraguado son influenciados por la formación y la composición de la masa básica del clínker. La resistencia a la compresión del clínker a los 28 días, según las normas, disminuye notablemente a medida que aumenta el contenido de aluminato tricálcico (C_3A), si el enfriamiento del clinker es lento. Por el contrario, si el enfriamiento del clínker es rápido, las diferencias en cuanto a resistencia son menos marcadas. Los cementos fabricados con clínker de reducido contenido de C_3A reaccionan de forma menos sensible frente a los influjos del enfriamiento, en lo que se refiere al comportamiento al fraguado y al desarrollo de la resistencia, que los cementos fabricados con clínkeres de contenido de C_3A más alto. El fraguado de cementos ricos en C_3A suele ser más rápido. Sin embargo, es posible mejorar la trabajabilidad de los cementos por medio de mezclas apropiadas de portadores de sulfatos, a base de anhidrita y semi-hidrato. Debido a diferentes condiciones de enfriamiento cambian la estructura y la composición de las fases de clínker, tal como lo han demostrado unos ensayos efectuados con ayuda del microscopio, del microscopio electrónico reticulado y del análisis por fluorescencia de rayos X.

1. Einleitung

Untersuchungen an Klinkerkühlern in verschiedenen Werken haben gezeigt, daß ein Vergleich der von der Kühlerbauart ausgehenden Einflüsse auf die Klinkereigenschaften nicht ohne weiteres möglich ist [1–8]. Darüber hinaus ging aus Untersuchungen über die Reaktivität von technischen Klinkern hervor, daß bei gleichartigen Kühlbedingungen eine unterschiedliche Ofenmehlzusammensetzung zwangsläufig zu einer veränderten Phasenzusammensetzung sowie zusätzlich auch zu Gefügeänderungen führte [9]. Bei gleicher chemischer Klinkerzusammensetzung, aber unterschiedlicher Kühlung, traten neben der Kristallvergröberung der Grundmassephasen Calciumaluminat C_3A und Aluminatferrit $C_2(A,F)$ Veränderungen in der Kristallausbildung der Calciumsilicate Alit (C_3S) und Belit (C_2S) auf [10]. Im Schrifttum liegen zum Einfluß der Kühlung und der Klinkerzusammensetzung auf die Zementeigenschaften sehr unterschiedliche und zum Teil widersprüchliche Aussagen vor [11–41].

Um die Auswirkung unterschiedlich großer Kühlraten bei veränderter Klinkerzusammensetzung auf die Eigenschaften von Zement deutlicher erkennen zu können, war es erforderlich, ergänzend zu Untersuchungen an technischen Klinkern unter definierten Bedingungen Klinker im Labor zu brennen und zu kühlen.

2. Versuchsdurchführung

2.1 Klinkerzusammensetzung

Zur Prüfung des Einflusses der Zusammensetzung und der Kühlung des Klinkers auf das Erstarrungsverhalten und die Festigkeitsentwicklung der daraus hergestellten Zemente wurde die Rohmehlzusammensetzung so gewählt, daß der Kalkstandard und der Schmelzphasenanteil konstant blieben. Die Gehalte an Aluminat C_3A und Aluminatferrit $C_2(A,F)$ wurden zwischen jeweils 7 und 13 Gew.-% in zwei Gew.-%-Schritten variiert, wobei die Summe beider Phasen stets 20 Gew.-% betrug. Darüber hinaus wurde eine extrem zusammengesetzte Mischung mit 15 Gew.-% C_3A und 5 Gew.-% $C_2(A,F)$ in die Untersuchung einbezogen. Zur Herstellung der Rohmehlgemische wurden chemisch reine Ausgangsstoffe wie $CaCO_3$, $Al(OH)_3$, Fe_2O_3, Quarz (SiO_2), MgO und K_2CO_3 verwendet. Die theoretische Zusammensetzung, die Moduln und die nach R. H. Bogue berechnete Phasenzusammensetzung der daraus hergestellten Klinker sind in **Tabelle 1** wiedergegeben.

2.2 Brenn- und Kühlbedingungen

Aus dem Rohstoffgemisch wurden unter Zusatz von Wasser mit einem Laborgranulierteller Pellets von etwa 0,5 bis 1 cm Durchmesser hergestellt. Die Pellets wurden in Platintiegeln zwei Stunden bei 110 °C getrocknet, anschließend in einem auf 600 °C vorgeheizten Laborkammerofen in 45 min auf 1450 °C erhitzt und danach 30 min bei dieser Temperatur gesintert. Zur Veränderung der Vorkühlgeschwindigkeiten wurden die Klinkerproben im abgeschalteten Ofen bis zu einer Temperatur von 1400, 1250 oder 1150 °C unterschiedlich langsam und anschließend außerhalb des Ofens an Luft in Platintiegeln gleich schnell abgekühlt.

1. Introduction

Tests with clinker coolers at various plants have shown that comparing the effects of the cooler design on the properties of clinker is not a straightforward task [1–8]. Moreover, investigations of the reactivity of commercially-produced clinkers revealed that under similar cooling conditions, a different kiln meal composition inevitably led to a change in the phase composition and also to changes in the structure [9]. With the same chemical composition of the clinker but different cooling, not only crystal enlargement of the matrix phases calcium aluminate (C_3A) and aluminoferrite ($C_2(A,F)$) but also changes in the crystal shape of the calcium silicates alite (C_3S) and belite (C_2S) occurred [10]. The literature contains widely differing and to a certain extent contradictory statements regarding the influence of cooling and clinker composition on cement properties [11–41].

To enable the effects of different cooling rates with changes in clinker composition on the properties of cement to be discerned more clearly, it was necessary, in addition to tests on commercially-produced clinkers, to burn and cool clinker in the laboratory under defined conditions.

2. Testing procedure

2.1 Clinker composition

To test the influence of the composition and cooling of the clinker on the setting and strength development of the cements made from it, a raw meal composition was chosen in which the lime standard and the proportion of liquid phase remained constant. The contents of aluminate (C_3A) and aluminoferrite ($C_2(A,F)$) were each varied between 7 and 13 wt.% in two wt.% steps; the sum of the two phases was always 20 wt.%. In addition, a blend of abnormal composition containing 15 wt.% C_3A and 5 wt.% $C_2(A,F)$ was included in the investigation. The raw meal mixes were prepared with chemically pure raw materials such as $CaCO_3$, $Al(OH)_3$, Fe_2O_3, quartz (SiO_2), MgO and K_2CO_3. The theoretical composition, the moduli and the Bogue phase composition of the resulting clinkers are reproduced in **Table 1**.

2.2 Burning and cooling conditions

The raw material mix was formed into pellets approx. 0.5 to 1 cm in diameter on a laboratory granulating table with addition of water. The pellets were dried in platinum crucibles for two hours at 110 °C, then heated to 1450 °C in 45 min. in a laboratory chamber furnace preheated to 600 °C and then sintered at 1450 °C for 30 min. In order to vary the precooling rates, the clinker specimens were cooled at different rates with the furnace switched off to a temperature of 1400, 1250 or 1150 °C and then cooled at the same rate outside the furnace in air in platinum crucibles.

2.3 Structure and phase composition of the clinker

The structure of the laboratory-produced clinker was microscopically examined by means of polished sections after etching with a solution of dimethylammonium citrate (DAC) in alcohol and 10 % caustic potash solution. Etched and gold-sputtered polished sections were prepared for examination

2.3 Gefüge und Phasenzusammensetzung des Klinkers

Das Gefüge der Laborklinker wurde mikroskopisch an Anschliffen nach Ätzung mit alkoholischer Dimethylammoniumcitratlösung (DAC) und mit 10%iger Kalilauge untersucht. Um den Kühleinfluß, insbesondere bei den Calciumsilicaten Alit und Belit, besser erkennen zu können, wurden parallel hierzu geätzte und mit Gold besputterte Anschliffe für rasterelektronenmikroskopische Untersuchungen hergestellt.

Da sich auch bei gleicher chemischer Zusammensetzung des Rohstoffgemisches die Kühlbedingungen auf die Zusammensetzung und die Gitterabstände der einzelnen Phasen auswirken können, wurden die Klinkerproben weiterhin röntgendiffraktometrisch an Pulverpräparaten untersucht. Für die Beurteilung von Gitteränderungen wurden beim Alit dessen koinzidenzfreie Interferenz bei d = 17,61 nm, beim Aluminat und Aluminatferrit die Hauptinterferenzen bei d = 26,98 bzw. 26,36 nm herangezogen. Um den Einfluß der Kühlung auf die Schmelzphasen und deren Gitterveränderungen deutlicher erkennen zu können, wurden die Silicate mit einem Methanol/Salicylsäure-Gemisch aus dem Klinkergefüge herausgelöst [42]. Die chemische Zusammensetzung der Schmelzphasen wurde am Rückstand röntgenfluoreszenzanalytisch untersucht.

2.4 Herstellen und Prüfen der Zemente

Die Laborklinker wurden in einer Kugelmühle auf eine einheitliche spezifische Oberfläche von 3 500 ± 100 cm²/g nach Blaine gemahlen. Untersuchungen mit einem Laser-Granulometer zeigten, daß die Kornverteilung in allen Proben etwa gleich war. Aus den Klinkermehlen wurden durch Zumischen von Anhydrit ($CaSO_4$) oder Halbhydrat ($CaSO_4 \cdot 0.5 H_2O$) oder Gemischen aus Anhydrit und Halbhydrat in Abstufungen von 20 Gew.-% jeweils 6 Zemente mit einem Sulfatgehalt von 3,0 Gew.-% SO_3 hergestellt. Zur Herstellung von Halbhydrat wurde Gips 16 Stunden bei 140°C getrocknet. Die optimale Verzögerung des Erstarrens der Zemente ergab sich dann, wenn das Angebot an Sulfat- und Aluminationen je nach Reaktionsfähigkeit des Klinkers zu Beginn der Hydratation gerade so groß war, daß sich ausschließlich Ettringit bilden konnte [43].

Das Erstarren der unterschiedlich zusammengesetzten Zemente wurde mit einem Penetrometer untersucht [44]. Der Zementleim wies bei den Zementen aus dem gleichen Klinker, der jedoch jeweils unterschiedlich schnell gekühlt worden war, einen konstanten Wasserzementwert auf. Es war allerdings erforderlich, je nach Klinkerzusammensetzung den Wasserzementwert von 0,28 bis 0,32 zu verändern. Daher ist es nicht möglich, die Ergebnisse der Erstarrungsprüfung für Zemente aus Klinkern unterschiedlicher Zusammensetzung unmittelbar miteinander zu vergleichen.

Die Druckfestigkeiten nach 2 und 28 Tagen wurden an Kleinprismen 1,5 × 1,5 × 6 cm aus Mörtel nach DIN 1164, Teil 7, geprüft.

3. Untersuchungsergebnisse

3.1 Chemische Zusammensetzung des Klinkers

Als Ausgangsgemische für die Klinkerherstellung wurden die Rohmehlmischungen nach **Tabelle 1** verwendet. Die daraus gebrannten Klinker wurden mittels Röntgenfluoreszenz analysiert. Mit Ausnahme des K_2O-Gehalts, der sich durch teilweises Verdampfen während des Brands um ca. 25 bis 40% verminderte, stimmten die Klinkeranalysen mit denen der Rohmehle im Rahmen der Analysengenauigkeit überein. Bei langsamer Klinkervorkühlung zeigte sich in der Tendenz eine geringfügig erhöhte Verdampfbarkeit des Kaliums.

3.2 Gefüge und Phasenzusammensetzung

Die mikroskopische Untersuchung zeigte, daß Gefüge und Phasenzusammensetzung des Klinkers in besonderem Maß von der Rohmehlzusammensetzung und den Bedingungen

TABELLE 1: Berechnete Chemische Zusammensetzung, Moduln und nach R. H. Bogue berechnete Phasenzusammensetzung der Rohmehlmischungen; Angaben in Gew.-%
TABLE 1: Calculated chemical composition, moduli and Bogue phase composition of the raw meal mixes; figures in wt.%

Bestandteil/ constituent	1	2	3	4	5
SiO_2	21,68	21,81	21,94	22,06	22,19
Al_2O_3	5,37	5,71	6,04	6,38	6,71
Fe_2O_3	4,27	3,61	2,96	2,30	1,64
CaO	66,18	66,37	66,56	66,76	66,96
MgO	1,50	1,50	1,50	1,50	1,50
K_2O	1,00	1,00	1,00	1,00	1,00
KSt	96,3	96,1	95,9	95,8	95,7
SM	2,2	2,3	2,4	2,5	2,7
TM	1,3	1,6	2,0	2,8	4,1
C_3S	62,5	60,9	59,4	58,0	56,5
C_2S	15,1	16,6	18,1	19,5	21,0
C_3A	7,0	9,0	11,0	13,0	15,0
C_4AF	13,0	11,0	9,0	7,0	5,0

by scanning electron microscope at the same time to enable the influence of cooling, especially on the calcium silicates alite and belite, to be more readily discerned.

Since even with raw material mixes of identical chemical composition the cooling conditions may affect the composition and lattice spacing of the individual phases, the clinker specimens were also examined with an X-ray diffractometer using powdered specimens. For the evaluation of lattice changes, in the case of alite its coincidence-free interference at d = 17.61 nm and in the case of aluminate and aluminoferrite the main interferences at d = 26.98 and 26.36 nm respectively were used. To discern more clearly the influence of cooling on the liquid phases and their lattice changes, the silicates were dissolved out of the clinker structure with a mixture of methanol and salicylic acid [42]. The chemical composition of the liquid phases was investigated by X-ray fluorescence analysis using the residue.

2.4 Production and testing of cements

The laboratory-produced clinker was ground in a ball mill to a uniform specific surface of 3,500 ± 100 cm²/g (Blaine). Examination with a laser granulometer showed that the particle size distribution was approximately the same in all the specimens. From each of the ground clinkers, six cements with a sulphate content of 3.0 wt.% SO_3 were produced by adding anhydrite ($CaSO_4$) or hemihydrate ($CaSO_4 \cdot 0.5 H_2O$) or mixtures of anhydrite and hemihydrate in 20-wt.% steps. Hemihydrate was produced by drying gypsum for 16 hours at 140°C. Optimum retardation of cement setting occurred when, depending on the reactivity of the clinker at the start of hydration, the supply of sulphate and aluminate ions was just large enough to allow only ettringite to form [43].

The setting of the cements of different composition was examined using a penetrometer [44]. With the cements made from the same clinker but cooled at different rates, the cement paste exhibited a constant water-cement ratio. However, it was necessary to vary the water-cement ratio between 0.28 and 0.32 depending on the clinker composition. Therefore it is not possible to make a direct comparison of the results of the setting tests for cements made from clinkers of differing composition.

The compressive strengths after 2 and 28 days were tested using small prism shapes measuring 1.5 x 1.5 x 6 cm, made from mortar conforming to DIN 1164, Part 7.

beim Kühlen des Klinkers abhängen. Das geht aus den mikroskopischen und rasterelektronenmikroskopischen **Bildern 1** bis **6** hervor. Die Bilder 1 bis 4 stammen von einem Klinker mit 13 Gew.-% C_3A und 7 Gew.-% $C_2(A,F)$ und die Bilder 5 und 6 von einem Klinker mit 7 Gew.-% C_3A und 13 Gew.-% $C_2(A,F)$, wobei jeweils eine Klinkerprobe nach kurzer Vorkühlung dem Ofen bei 1400 °C bzw. nach längerer Vorkühlung bei 1150 °C entnommen worden waren.

Bild 1 gibt das Gefüge des schneller vorgekühlten Klinkers mit 13 Gew.-% C_3A wieder. In den Zwickeln zwischen den Alitkristallen sind hauptsächlich grobkörnige und zahlreiche kleine, mittelgraue Aluminat- und deutlich weniger und kleine, hell reflektierende Aluminatferritkristalle zu erkennen. Der größte Teil der Alitkristalle ist während des Kristallisationsprozesses von der kalkärmeren Schmelze am Rand korrodiert worden. Als Folge davon hat sich ein hierfür typischer Belitsaum (sekundäre Belitbildung) um die Alite gebildet.

Bild 2 zeigt im Vergleich dazu das Gefüge des gleichen, aber langsamer vorgekühlten, C_3A-reichen Klinkers. Das dunkelgraue C_3A ist grobkörnig und zum Teil stengelig ausgebildet – ein Zeichen für K_2O-haltiges Aluminat – und gut von dem hell reflektierenden, ebenfalls grobkörnig ausgebildeten $C_2(A,F)$ getrennt. Die Belitsäume an den Korngrenzen des Alits sind zum größten Teil verschwunden. Statt dessen sind die vielen kleinen Belitkristalle zu größeren Kristallen mit unregelmäßigen Umrissen zusammengewachsen. Die Alite sind zum Teil rekristallisiert und häufig als idiomorphe Kristalle mit scharfkantigen Umrissen ausgebildet.

Besonders deutlich läßt sich der unterschiedliche Kühleinfluß an den geätzten, rasterelektronenmikroskopisch untersuchten Proben erkennen (Bilder 3 und 4). Im Bild 3 des aluminatreichen und bei 1400 °C dem Ofen entnommenen Klinkers ist deutlich der durch die Strukturätzung reliefartig erhöhte Saum der kleinen Belitkristalle an den Alitkorngrenzen zu erkennen. Da C_3A durch die Ätzmittel DAC und KOH nur angefärbt wird, werden die Grundmassephasen im erhabenen Bereich des Bildes durch die Sekundärelektronen nicht differenziert. Bei dem gleichen, aber langsam bis 1150 °C vorgekühlten Klinker (Bild 4) fehlt dieser Saum vollständig. Durch Sammelkristallisation haben sich aus den vielen kleinen Kristallen einzelne größere, unregelmäßig geformte Belite, zum Teil mit Zwillingslamellierung, gebildet (z. B. links, im oberen Drittel des Bildes 4). Bei den gerundeten Einschlüssen der scharfkantig ausgebildeten Alitkristalle handelt es sich um Belitkristalle, die während des Aufheizprozesses als primäre Bildung von den kristallisierenden Alitkristallen eingeschlossen wurden.

Ein wesentlich anderes Gefüge wies der Klinker mit dem geringeren C_3A-Gehalt von 7 Gew.-% auf (Bilder 5 und 6). Das C_3A (Bild 5) ist kleinkörnig im $C_2(A,F)$ verteilt. Die Alit-

3. Test results

3.1 Chemical composition of clinker

The raw meal mixes shown in Table 1 were used as the initial mixes for clinker production. The clinkers burnt from them were analysed by X-ray fluorescence. With the exception of the K_2O content, which decreased during burning by approx. 25–40% due to partial evaporation, the clinker analyses corresponded to those of the raw meals within the tolerance range of analytical accuracy. With slower precooling of the clinker, there was a tendency towards a slightly increased evaporation of the potassium.

3.2 Structure and phase composition

The microscopic examination showed that the structure and phase composition of the clinker depend to a very large extent on the composition of the raw meal and the conditions prevailing during clinker cooling. This is evident from **Figures 1–6**, (photomicrographs and scanning electron microscope photographs). Figures 1–4 originate from a clinker containing 13 wt.% C_3A and 7 wt.% $C_2(A,F)$ and Figures 5 and 6 from a clinker containing 7 wt.% C_3A and 13 wt.% $C_2(A,F)$; in each case, one clinker specimen was removed from the kiln at 1400 °C after brief precooling and one at 1150 °C after longer precooling.

Figure 1 depicts the structure of the more rapidly precooled clinker containing 13 wt.% C_3A. In the interstices between the alite crystals, mainly large but numerous small medium-grey aluminate crystals and markedly fewer small, brightly reflecting aluminoferrite crystals can be seen. Most of the alite crystals have been corroded at the edges during the crystallization process by the melt, which is lower in lime. As a result, a typical belite border (secondary belite formation) has formed around the alites.

Figure 2 shows, in comparison, the structure of the same C_3A-rich clinker with slower precooling. The dark grey C_3A is coarse-grained and partly spiky in shape – indicative of K_2O-bearing aluminate – and is well separated from the brightly reflecting, also coarse-grained $C_2(A,F)$. The belite borders at the grain boundaries of the alite have largely disappeared. Instead, the many small belite crystals have united to form larger crystals with irregular contours. The alites have partly recrystallized and frequently take the form of idiomorphic crystals with sharp-edged contours.

The varying influence of the cooling can be seen particularly clearly in the case of the etched specimens examined with the scanning electron microscope (Figures 3 and 4). In Figure 3, showing the aluminate-rich clinker removed from the kiln at 1400 °C, the border of small belite crystals, raised relief-like by the etching of the structure, is clearly discerna-

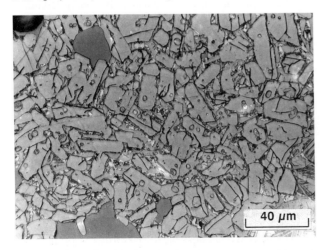

BILD 1: Mikroskopisches Gefüge von Klinker, der bei 1400 °C dem Ofen entnommen wurde und 13 Gew.-% C_3A und 7 Gew.-% $C_2(A,F)$ enthält
FIGURE 1: Microstructure of clinker removed from the kiln at 1400 °C and containing 13 wt.% C_3A and 7 wt.% $C_2(A,F)$

BILD 2: Mikroskopisches Gefüge von Klinker, der bei 1150 °C dem Ofen entnommen wurde und 13 Gew.-% C_3A und 7 Gew.-% $C_2(A,F)$ enthält
FIGURE 2: Microstructure of clinker removed from the kiln at 1150 °C and containing 13 wt.% C_3A and 7 wt.% $C_2(A,F)$

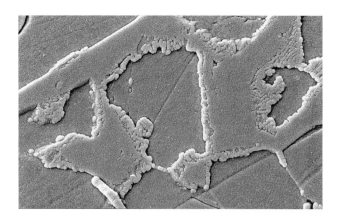

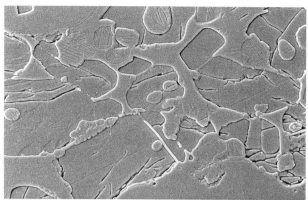

BILD 3: Alitkristalle mit starken Resorptionserscheinungen und deutlichem Belitsaum im Klinker, der bei 1400 °C dem Ofen entnommen wurde und 13 Gew.-% C_3A und 7 Gew.-% $C_2(A,F)$ enthält. Rasterelektronenmikroskopische Aufnahme
FIGURE 3: Alite crystals with marked resorption phenomena and a clear belite border in clinker removed from the kiln at 1400 °C and containing 13 wt. % C_3A and 7 wt. % $C_2(A,F)$. Scanning electron microscope photograph

BILD 4: Alitkristalle ohne Resorptionserscheinungen und sekundär gebildete, gestreifte Belitkristalle in der Grundmasse von Klinker, der bei 1150 °C dem Ofen entnommen wurde und 13 Gew.-% C_3A und 7 Gew.-% $C_2(A,F)$ enthält. Rasterelektronenmikroskopische Aufnahme
FIGURE 4: Alite crystals with no resorption phenomena and striped belite crystals of secondary formation in the matrix of clinker removed from the kiln at 1150 °C and containing 13 wt. % C_3A and 7 wt. % $C_2(A,F)$. Scanning electron microscope photograph

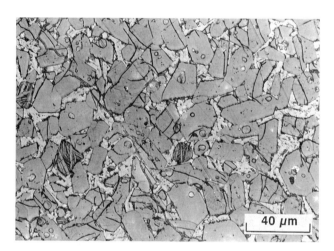

BILD 5: Mikroskopisches Gefüge von Klinker, der bei 1400 °C dem Ofen entnommen wurde und 7 Gew.-% C_3A und 13 Gew.-% $C_2(A,F)$ enthält
FIGURE 5: Microstructure of clinker removed from the kiln at 1400 °C and containing 7 wt. % C_3A and 13 wt. % $C_2(A,F)$

BILD 6: Mikroskopisches Gefüge von Klinker, der bei 1400 °C dem Ofen entnommen wurde und 7 Gew.-% C_3A und 13 Gew.-% $C_2(A,F)$ enthält
FIGURE 6: Microstructure of clinker removed from the kiln at 1150 °C and containing 7 wt. % C_3A and 13 wt. % $C_2(A,F)$

kristalle sind nahezu idiomorph ausgebildet und weisen auf Grund des geringeren CaO-Bedarfs der Schmelze weniger stark korrodierte Randzonen auf. Bei den kreuzgestreiften Kristallen handelt es sich um primär gebildete Belitkristalle. Bei langsamer Vorkühlung des gleichen Klinkers (Bild 6) sind noch deutlicher die idiomorphen Alitkristalle und die großflächige Ausbildung der hell reflektierenden $C_2(A,F)$-Kristalle zu erkennen, die die Zwickel zum Teil vollständig ausfüllen. Das C_3A ist nicht mehr fein verteilt, sondern vorrangig in kleineren Zwickeln angereichert. Bei den punktförmigen Kristallen handelt es sich um kleine Belitkristalle, die sich als tertiäre Ausscheidung aus der Schmelze während des Abkühlens gebildet haben. Rasterelektronenmikroskopische Untersuchungen bestätigten den mikroskopischen Befund (**Bild 7**).

Obwohl bei den langsam gekühlten Proben die Stabilitätsgrenze des Alits unterschritten wird, ist entgegen anderen Untersuchungsergebnissen [45, 46] kein Zerfall des Alits in Belit und freies CaO eingetreten. Das hat die naßchemische Bestimmung des freien CaO bestätigt. Alle unter 1250 °C gekühlten Klinker weisen Gehalte an freiem CaO von unter 0,2 Gew.-% auf.

ble at the alite grain boundaries. As C_3A is only stained by the etchants DAC and KOH, the secondary electrons do not differentiate between the matrix phases in the raised area of the photo. With the same clinker, but this time slowly precooled to 1150 °C (Figure 4), this border is completely absent. From the many small crystals, individual larger, irregularly shaped belites, in some cases with twin lamination, have formed by accretive crystallization (e. g. at the left in the top third of Figure 4). The rounded inclusions of the sharp-edged alite crystals are belite crystals which were captured, as primary formations, by the crystallizing alite crystals during preheating.

A substantially different structure was exhibited by the clinker with the lower C_3A content of 7 wt. % (Figures 5 and 6). The C_3A (Figure 5) is distributed in the $C_2(A,F)$ as small particles. The alite crystals are almost idiomorphic in shape and have less severely corroded edges due to the lower CaO demand of the melt. The cross-striped crystals are belite crystals of primary formation. With slower precooling of the same clinker (Figure 6), the idiomorphic alite crystals and the extensive development of the brightly reflecting $C_2(A,F)$ crystals, which in some cases completely fill the interstices,

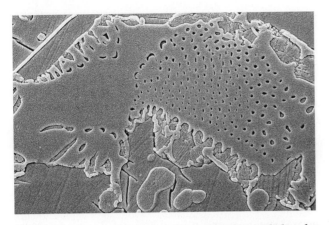

BILD 7: Sekundäre und tertiäre Bildung von Belit im Klinker, der bei 1150°C dem Ofen entnommen wurde und 13 Gew.-% C_3A und 7 Gew.-% $C_2(A,F)$ enthält. Rasterelektronenmikroskopische Aufnahme
FIGURE 7: Secondary and tertiary formation of belite in clinker removed from the kiln at 1150°C and containing 13 wt.% C_3A and 7 wt.% $C_2(A,F)$. Scanning electron microscope photograph

3.3 Chemische Zusammensetzung der Grundmasse

Aus den Gefügeuntersuchungen ging hervor, daß während der Kühlung des Klinkers Resorptionen an den Aliten, Sammelkristallisationen bei Belit und Alit, Kristallvergröberungen von Aluminat und Aluminatferrit und Ausheilungsprozesse an korrodierten Aliten eintreten. Das Ausmaß der Gefügeänderungen war je nach Rohmehlzusammensetzung unterschiedlich groß. Frühere Untersuchungen [10] hatten gezeigt, daß bei langsamer Klinkerkühlung und trotz ausbleibender Alitresorption die Festigkeitswerte der daraus hergestellten Zemente abfielen. Diese Beobachtungen lassen darauf schließen, daß während des Kühlprozesses – vor allem in der Grundmasse – Stofftransporte stattfinden, die die Klinker- und Zementeigenschaften verändern können. Zur Klärung dieser Frage wurden daher die Calciumsilicate der unterschiedlich zusammengesetzten und gekühlten Klinker mit in Methanol gelöster Salicylsäure weggelöst und der Gehalt der chemischen Hauptbestandteile im Rückstand röntgenfluoreszenzanalytisch untersucht. Die Vollständigkeit der Abtrennung der Silicate wurde zur Kontrolle röntgendiffraktometrisch überprüft.

Die auf diese Weise ermittelte Zusammensetzung der Grundmasse ist in **Tabelle 2** zusammengestellt. Daraus geht hervor, daß während des Kühlprozesses ein Stoffaustausch zwischen den Calciumsilicaten und der Grundmasse stattfindet. In dem C_3A-reicheren Klinker (13 Gew.-% C_3A) wandern Fe_2O_3, Al_2O_3 und MgO bei langsamerer Vorkühlung in die Calciumsilicate (Verarmung der Schmelze) und umge-

are even more clearly discernable. The C_3A is no longer finely distributed but is principally concentrated in smaller interstices. The dot-like crystals are small belite crystals formed as tertiary precipitation from the melt during cooling. Scanning electron microscope examinations confirmed the findings obtained with the microscope (**Figure 7**).

In contrast to other research results [45, 46], no decomposition of the alite into belite and free CaO occurred even though the slowly cooled specimens fell below the limit of alite stability. This was confirmed by wet-chemical determination of the free CaO content. All the clinkers cooled to below 1250°C had free CaO contents of less than 0.2 wt.%.

3.3 Chemical composition of the matrix

The structure examinations revealed that during clinker cooling, resorption at the alites, accretive crystallization of belite and alite, enlargement of aluminate and aluminoferrite crystals, and repair of corroded alite crystals take place. The extent of the structural changes varied according to the composition of the raw meal. Previous studies [10] had shown that with slow clinker cooling, the strengths of the cements made from the clinker decreased despite the absence of alite resorption. These observations imply that during cooling, mass transfers which may change the properties of the clinker and cement occur, especially in the matrix. In order to clarify this question, the calcium silicate content of the clinkers of different composition and cooling rate was therefore dissolved away with a solution of salicylic acid in methanol, and the contents of the main chemical constituents in the residue were analysed by X-ray fluorescence. For control purposes, the completeness of silicate removal was checked with an X-ray diffractometer.

The matrix compositions determined in this way are listed in **Table 2**. It is evident from this that mass transfer takes place between the calcium silicates and the matrix during cooling. In the C_3A-rich clinker (13 wt.% C_3A), with slower precooling Fe_2O_3, Al_2O_3 and MgO migrate into the calcium silicates (depletion of the melt) and CaO migrates into the matrix (enrichment). In contrast, in the case of clinkers containing less C_3A (7 wt.% C_3A), the slower cooling causes Fe_2O_3, Al_2O_3 and CaO to build up in the matrix and the silicates have correspondingly lower contents. Some of the MgO passes, as in the case of the clinkers with a higher C_3A content, from the matrix into the silicates.

The chemically determined results explain the findings of the structure examinations. As was evident from Figures 1 and 3, an Al_2O_3-rich matrix partly resorbs the alite crystals during cooling. The CaO is needed for crystallization of the matrix phases C_3A and $C_2(A,F)$. The withdrawal of the CaO from the boundary zone of the alite leaves a border of belite, which with even slower cooling becomes detached from the edges of the alite and forms larger belite crystals. The corroded surface of the alites is made smooth by absorption of

TABELLE 2: Chemische Zusammensetzung der Grundmasse aus unterschiedlich gekühlten Klinkern und Differenz der Bestandteile zwischen schneller und langsamer Vorkühlung in Abhängigkeit von der Zusammensetzung; Angaben in Gew.-%
TABLE 2: Chemical composition of the matrices of clinkers cooled at different rates and differences in constituent contents between rapid and slow precooling in relation to the composition; figures in wt.%

$C_3A/C_2(A,F)$	Temperat. °C	SiO_2	ΔSiO_2	Al_2O_3	ΔAl_2O_3	Fe_2O_3	ΔFe_2O_3	CaO	ΔCaO	MgO	ΔMgO
7/13	1400 1150	4,49 3,16	−1,33	23,51 24,43	+1,02	18,93 19,26	+0,33	47,01 47,70	+0,69	3,61 3,18	−0,43
9/11	1400 1150	4,87 3,82	−1,05	24,40 25,02	+0,62	15,95 16,39	+0,44	48,41 49,51	+1,10	3,70 3,12	−0,58
13/7	1400 1150	4,20 3,43	−0,77	28,93 28,27	−0,66	10,21 9,46	−0,75	49,34 52,00	+2,64	3,81 3,22	−0,59
15/5	1400 1150	3,88 3,22	−0,66	31,58 30,59	−0,99	7,32 6,29	−1,03	50,65 53,67	+3,02	3,86 3,51	−0,35

Anmerkung: + = Anreicherung,
− = Verarmung chemischer Bestandteile in der Grundmasse

Notes: + = concentration
− = depletion of chemical constituents in matrix

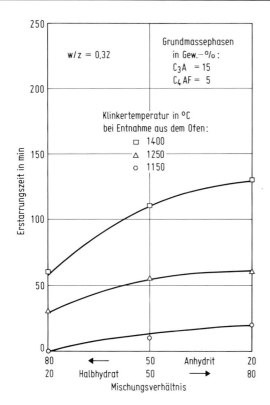

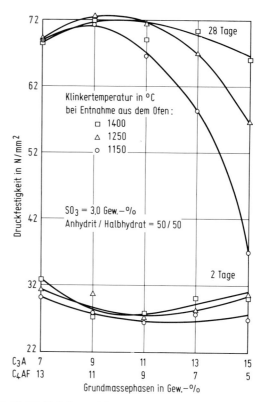

BILD 11: Einfluß der Kühlung von Klinker mit 15 Gew.-% C₃A und 5 Gew.-% C₂(A,F) auf das Erstarren von Zement mit verschiedenen Gemischen aus Anhydrit und Halbhydrat
FIGURE 11: Influence of the cooling of clinker containing 15 wt. % C_3A and 5 wt. % $C_2(A,F)$ on setting of cement containing various mixtures of anhydrite and hemihydrate

Erstarrungszeit in min	= setting time in min
w/z = 0,32	= w/c = 0.32
Grundmassephasen in Gew.-%	= Matrix phases in wt.%
Klinkertemperatur in °C bei Entnahme aus dem Ofen	= clinker temperature in °C on removal from the kiln
Halbhydrat	= hemihydrate
Anhydrit	= anhydrite
Mischungsverhältnis	= Mix ratio

BILD 12: Einfluß der Kühlung von Klinker in Abhängigkeit vom C₃A- und C₂(A,F)-Gehalt auf die Druckfestigkeit nach 2 und 28 Tagen
FIGURE 12: Influence of clinker cooling in relation to the C_3A and $C_2(A,F)$ contents on compressive strength after 2 and 28 days

Druckfestigkeit in N/mm²	= compressive strength in N/mm²
Klinkertemperatur in °C bei Entnahme aus dem Ofen	= clinker temperature in °C on removal from the kiln
Anhydrit/Halbhydrat = 50/50	= anhydrite/hemihydrate = 50/50
Grundmassephasen in Gew.-%	= Matrix phases in wt.%

Aluminatgehalt im Klinker zu. Zur Erstarrungsregelung müssen demnach mit steigendem C₃A-Gehalt und geringerer Vorkühlungsrate halbhydratreichere Sulfatträgergemische bei der Zementmahlung verwendet werden.

Das wesentliche Ergebnis der Festigkeitsuntersuchungen bestand darin, daß sich die Vorkühlgeschwindigkeit des Klinkers insbesondere bei Zementen mit höherem C₃A-Gehalt deutlich auf die 28d-, jedoch weniger ausgeprägt auf die 2d-Festigkeit von Normmörtel auswirkt. Dem Schrifttum sind zum Teil einander widersprechende Ergebnisse zum Einfluß des C₃A-Gehalts auf die Festigkeit zu entnehmen [28–41]. Diese Untersuchungen deuten jedoch darauf hin, daß optimale 28d-Normfestigkeiten bei vergleichbarer Klinkerzusammensetzung offenbar mit einem bestimmten C₃A-Gehalt des Klinkers erzielt werden können, der bei diesen Versuchen im Bereich von etwa 9 bis 11 Gew.-% C₃A lag.

Die deutlich ausgeprägte Abnahme der 28d-Normfestigkeit mit steigendem C₃A-Gehalt und langsamer Vorkühlung des Klinkers ist sicherlich nicht allein auf die Abnahme des rechnerischen Alitgehalts (Tabelle 1) zurückzuführen. Im Schrifttum wird festgestellt, daß Belitsäume um Alitkristalle durch langsame Vorkühlung des Klinkers entstehen [46]. Aus den hier vorliegenden Untersuchungsergebnissen geht jedoch hervor, daß die Saumbildung weniger auf die Vorkühlung des Klinkers zurückzuführen ist, sondern in erster Linie nur von der Zusammensetzung des Klinkers, speziell der Zusammensetzung der Grundmasse, abhängt. Mit steigendem C₃A-Gehalt in der Grundmasse wurden die Belitsäume an den Alitkorngrenzen, insbesondere bei schneller Vorkühlung des Klinkers, deutlich breiter. Sie vergrößerten sich demgegenüber bei langsamer Vorkühlung nicht, sondern lösten sich vielmehr von den Alitkorngrenzen und wuchsen zu unregelmäßig ausgebilde-

hand, with a further rise in the C_3A content of the clinker, the 28-day strength falls significantly. This occurs particularly dramatically where very slow clinker precooling has taken place.

In contrast, the precooling rate and the C_3A content of the clinker have only a minor effect on the 2-day strength. At 9–11 wt.%, a minimum level of strength development is apparent. The higher initial strengths, albeit only slightly marked, are found with rapidly precooled clinkers. The composition of the sulphate agent has an insignificant effect on the strength development; however, the sulphate agent blend consisting of 50 wt.% each of anhydrite and hemihydrate proved beneficial.

4. Conclusion

The tests carried out on 15 cements made from four clinkers with C_3A contents ranging from 7 to 15 wt.% showed that for the same grinding fineness and sulphate content, setting depends not only on the C_3A content but also to a great extent on the clinker cooling and the composition of the sulphate agent mixture. Slow precooling increases the reactivity of the aluminate, and its reaction with the sulphate agent added to control setting therefore changes, so the influence of precooling rises as the aluminate content of the clinker increases. With increasing C_3A contents and falling precooling rates, it is therefore necessary to use sulphate agent mixtures richer in hemihydrate in cement grinding to control the setting.

The main outcome of the strength tests was that the clinker precooling rate had a significant effect on the 28-day strength of standard mortar, especially in the case of cements with a higher C_3A content, but the effect on the 2-

ten Belitkristallen zusammen. Bisher wurde vermutet, daß der Saum aus dem weniger hydraulisch wirkenden Belit gegenüber dem hydraulisch aktiveren Alit eine Festigkeitsminderung verursachen kann [46]. Da sich jedoch bei langsamerer Vorkühlung die Belitsäume auflösten und dementsprechend keine Blockierung der Alitoberfläche mehr eintreten kann, muß für die Festigkeitsabnahme bei langsamer Kühlung eine andere Ursache maßgebend sein. Erhärtet durch röntgendiffraktometrisch nachweisbare Gitteränderungen, besonders bei den Grundmassephasen, und die Tatsache, daß es auf Grund des Stoffaustausches zwischen den Calciumsilicaten und der Grundmasse während der Vorkühlung zu einer Calciumanreicherung in den Grundmassephasen bei gleichzeitiger Abnahme der Alitmenge kommt, ist davon auszugehen, daß hierdurch die Reaktivität des Klinkers insgesamt beeinträchtigt wird. Sie tritt bei C_3A-reichen Klinkern besonders deutlich hervor und führt zu merklichen Verlusten der 28d-Festigkeit bei Normmörteln. Diesen Zusammenhängen soll noch durch weitere Versuche nachgegangen werden.

5. Zusammenfassung

An 15 Zementen aus 4 Klinkern mit C_3A-Gehalten zwischen 7 und 15 Gew.-% wurde das Erstarrungsverhalten, die Festigkeitsentwicklung, das Klinkergefüge und die Zusammensetzung der Grundmasse bei unterschiedlicher Klinkerkühlung untersucht. Die Rohmehlzusammensetzung wurde so gewählt, daß der Kalkstandard stets einen Wert von rd. 96 aufwies und der Schmelzphasenanteil mit 20 Gew.-% konstant blieb. Aus den unterschiedlich zusammengesetzten und gekühlten Klinkern wurden Zemente mit einem SO_3-Gehalt von 3,0 Gew.-% hergestellt. Das Sulfatträgergemisch bestand aus unterschiedlichen Anteilen an Halbhydrat und Anhydrit.

Die Normdruckfestigkeit der Zemente nach 28 Tagen durchläuft mit steigendem Aluminatgehalt (C_3A) ein Optimum, das bei etwa 9 Gew.-% C_3A liegt. Vor allem bei langsamer Vorkühlung des Klinkers und C_3A-Gehalten über 9 bis 15 Gew.-% nimmt die Druckfestigkeit merklich, bei C_3A-Gehalten unter 9 bis 5 Gew.-% weniger deutlich ab. Bei schneller Vorkühlung sind die Festigkeitsunterschiede in Abhängigkeit vom C_3A-Gehalt geringer.

C_3A-reiche Zemente erstarren generell schneller oder weisen einen höheren Wasseranspruch auf. Die Verarbeitbarkeit von Zementen mit C_3A-Gehalten bis 13 Gew.-% läßt sich im allgemeinen durch darauf abgestimmte Sulfatträgergemische mit höheren Halbhydratanteilen steuern. Durch unterschiedliche Vorkühlgeschwindigkeiten des Klinkers werden das Gefüge und die Zusammensetzung der Klinkerphasen verändert. Die CaO-Anreicherung in den Schmelzphasen, die besonders deutlich bei C_3A-reichen Klinkern und langsamer Vorkühlung in Erscheinung tritt, führt zu einer Abnahme des Alitgehalts und damit zu einer Verminderung der hydraulischen Aktivität.

Schrifttum

[1] Scheubel, B.: Mineralogische Untersuchungen auf Struktur und Zusammensetzung der Schmelzphase in Portland-Zement-Klinkern. Dissertation 1985, Erlangen-Nürnberg.

[2] Locher, F.W.: Einfluß der Klinkerherstellung auf die Eigenschaften des Zements. Zement-Kalk-Gips 28 (1975), S. 265–272.

[3] Kreft, W., Scheubel, B., und Schütte, R.: Klinkerqualität, Energiewirtschaft und Umweltbelastung – Einflußnahme und Anpassung des Brennprozesses. Teil I: Basisbetrachtungen. Zement-Kalk-Gips 40 (1987), S. 127–133.

[4] Kreft, W., Scheubel, B., und Schütte, R.: Klinkerqualität, Energiewirtschaft und Umweltbelastung – Einflußnahme und Anpassung des Brennprozesses. Teil II: Erfahrungen aus der Praxis. Zement-Kalk-Gips 40 (1987), S. 243–258.

day strength was less marked. The results given in the literature regarding the influence of the C_3A content on strength are to a certain extent contradictory [28–41]. These investigations indicate, however, that with similar clinker compositions, optimum 28-day standard strengths can evidently be achieved with a specific C_3A content of the clinker, which in these tests was in the range of approx. 9–11 wt.% C_3A.

The marked decrease in the 28-day standard strength with increasing C_3A content and slow clinker precooling is undoubtedly not solely attributable to the decrease in the calculated alite content (Table 1). The literature states that belite borders form around alite crystals due to slow clinker precooling [46]. The test results under discussion show, however, that the formation of borders is attributable not so much to the precooling of the clinker, but primarily depends on the clinker composition and specifically on the composition of the matrix. As the C_3A content of the matrix increased, the belite borders at the alite grain boundaries became markedly wider, especially with rapid clinker precooling. With slower precooling, on the other hand, they did not grow larger, but instead became detached from the alite grain boundaries and united to form irregularly-shaped belite crystals.

It has been presumed until now that the border consisting of the less hydraulically active belite may cause a reduction in strength compared with the more hydraulically active alite [46]. However, since with slower precooling the belite borders became detached and consequently the surface of the alite can no longer be obstructed, the decrease in strength with slow cooling must have a different cause. Substantiated by lattice changes which are verifiable with an X-ray diffractometer, especially in the case of the matrix phases, and the fact that due to the mass transfer between the calcium silicates and the matrix, a build-up of calcium in the matrix phases accompanied by a decrease in the quantity of alite takes place during precooling, it may be assumed that the reactivity of the clinker as a whole is impaired as a result. This is particularly evident in the case of C_3A-rich clinkers and leads to significant reductions in 28-day strength with standard mortars. These relationships will be investigated by means of further tests.

5. Summary

The setting behaviour, strength development, clinker structure and matrix composition at various rates of clinker cooling were studied using 15 cements made from four clinkers with C_3A contents ranging from 7 to 15 wt.%. The composition of the raw meal was such that the value of the lime standard was always approx. 96 and the liquid percentage was constant at 20 wt.%. Cements with an SO_3 content of 3.0 wt.% were produced from the clinkers of different composition and cooling rate. The sulphate agent mixture consisted of hemihydrate and anhydrite in various proportions.

As the aluminate (C_3A) content increases, the standard compressive strength of the cements after 28 days traverses an optimum level at approx. 9 wt.% C_3A. Especially with slow precooling of the clinker and C_3A contents of over 9 to 15 wt.%, the compressive strength decreases appreciably, while at C_3A contents of less than 9 down to 5 wt.% it falls less markedly. With rapid precooling, the differences in strength in relation to the C_3A content are smaller.

C_3A-rich cements generally set faster or have a higher water demand. The workability of cements with C_3A contents up to 13 wt.% can generally be controlled by means of tailored sulphate agent mixtures with fairly high hemihydrate contents. Different clinker precooling rates change the structure and composition of the clinker phases. The build-up of CaO in the liquid phases, which is particularly marked with C_3A-rich clinkers and slow precooling, leads to a decrease in the alite content and thus to a reduction in hydraulic activity.

kehrt CaO in die Grundmasse (Anreicherung). Demgegenüber reichern sich bei C₃A-ärmeren Klinkern (7 Gew.-% C₃A) als Folge der langsameren Kühlung Fe₂O₃, Al₂O₃ und CaO in der Grundmasse an, und die Silicate weisen entsprechend geringere Gehalte auf. Ein Teil des MgO geht – wie bei den C₃A-reicheren Klinkern – aus der Grundmasse in die Silicate über.

Die chemisch ermittelten Ergebnisse erklären den Befund der Gefügeuntersuchungen. Wie aus Bild 1 und 3 hervorgeht, resorbiert eine Al₂O₃-reiche Grundmasse während des Abkühlens teilweise die Alitkristalle. Das CaO wird zur Kristallisation der Grundmassephasen C₃A und C₂(A,F) benötigt. Der Entzug des CaO aus der Randzone des Alits hinterläßt einen Belitsaum, der sich bei noch langsamerer Kühlung vom Alitrand löst und größere Belitkristalle bildet. Durch Aufnahme von Fe₂O₃, Al₂O₃ und MgO aus der Grundmasse wird die korrodierte Oberfläche der Alite geglättet. Aus **Tabelle 2** geht weiter hervor, daß C₃A-ärmerer Klinker deutlich weniger CaO austauscht. Diesem Ergebnis entspricht der im Bild 5 sichtbar kleinere Korrosionssaum an den Alitkorngrenzen. In einem C₃A-ärmeren Klinker dient das aus dem Alit resorbierte CaO hauptsächlich zur Kristallisation des CaO-ärmeren C₂(A,F).

Aus der Tabelle geht außerdem hervor, daß sich bei langsamerer Klinkerkühlung der SiO₂-Gehalt in der Grundmasse vermindert. Das ist darauf zurückzuführen, daß hierbei als tertiäre Bildung Belit entsteht, der sich in kleinen, gerundeten Kristallen im Aluminatferrit ausscheidet.

3.4 Erstarren

Um eine möglichst lange Verarbeitbarkeit von Zement zu erzielen, muß das Sulfatangebot in der Porenlösung gerade so hoch bemessen sein, daß der vor Beginn der Ruheperiode hydratisierende C₃A-Anteil ausschließlich als Ettringit gebunden wird (Optimierung) [43]. Das Erstarrungsverhalten wurde an den Zementen aus den unterschiedlich zusammengesetzten und gekühlten Klinkern ermittelt. Die in Abhängigkeit vom Anhydrit- und Halbhydratanteil in den Zementen gemessene Erstarrungszeit ist in den **Bildern 8 bis 11** dargestellt.

Bild 8 zeigt das Erstarrungsverhalten der Zemente aus Klinker mit 7 Gew.-% C₃A. Bei allen Kühlraten und Sulfatträgergemischen erstarren die Zemente normgerecht. Die Erstarrungszeit wird jedoch bei halbhydratreichen Sulfatträgern erwartungsgemäß deutlich gegenüber anhydritreichen Sulfatträgern durch sekundäre Gipsbildung verkürzt. Offenbar weisen Zemente aus schneller vorgekühltem Klinker (Entnahmetemperatur 1400 °C) einen etwas erhöhten Halbhydratbedarf auf.

Nach Bild 9 erstarren Zemente aus Klinker mit 9 Gew.-% C₃A ebenfalls normgerecht. Kühlgeschwindigkeit und Sulfatträgerzusammensetzung wirken sich jedoch nur geringfügig auf die Erstarrungszeit aus. Die Zemente aus diesem Klinker erstarren allerdings mit der halbhydratreichen Sulfatträgermischung auf Grund sekundärer Gipsbildung etwas schneller.

Demgegenüber reicht das Sulfatangebot bei den Zementen aus C₃A-reichem Klinker nicht in allen Fällen für eine ausreichend lange Verzögerung des Erstarrens aus (Bild 10). Das ist besonders ausgeprägt bei den Zementen aus sehr langsam gekühltem Klinker mit 20 bis 50 Gew.-% Halbhydratanteil. Das schnellere Erstarren ist auf Syngenit- und teilweise auf Monosulfatbildung zurückzuführen. Eine Erhöhung des Halbhydratanteils verbesserte generell das Erstarrungsverhalten.

Aus **Bild 11** geht hervor, daß Zemente aus besonders C₃A-reichem Klinker nur dann genügend langsam erstarren, wenn der Klinker schnell vorgekühlt wurde und der Halbhydratanteil möglichst hoch ist. Die kurze Erstarrungszeit wird in erster Linie durch Monosulfatbildung verursacht, wie ergänzende thermoanalytische und rastermikroskopische Untersuchungen zeigten. Die erhöhte Klinkerreaktivität kann in Übereinstimmung mit Ergebnissen früherer Untersuchungen nur mit größeren Anteilen schnell löslichen Halbhydrats als Sulfatträger aufgefangen werden [43].

Fe₂O₃, Al₂O₃ and MgO from the matrix. It is also evident from Table 2 that in clinker containing less C₃A, significantly less CaO transfer takes place. The smaller margin of corrosion at the alite grain boundaries that is visible in Figure 5 corresponds to this result. In a clinker containing less C₃A, the CaO resorbed from the alite mainly serves to crystallize the C₂(A,F), which contains less CaO.

It is also evident from the table that with slower clinker cooling, the SiO₂ content of the matrix falls. This is attributable to tertiary formation of belite, which is precipitated in small, rounded crystals in the aluminoferrite.

3.4 Setting

To achieve the longest possible workability of the cement, the supply of sulphate in the pore solution must be just high enough for the proportion of C₃A which hydrates before the start of the dormant period to be bonded as ettringite alone (optimisation) [43]. The setting behaviour was measured using the cements made from the clinkers of different composition and cooling rate. The setting times measured in relation to the anhydrite and hemihydrate content of the cements are shown in **Figures 8 to 11**.

Figure 8 illustrates the setting behaviour of the cements made from clinker containing 7 wt.% C₃A. The cements set in conformity with the standard at all cooling rates and with all sulphate agent mixtures. With hemihydrate-rich sulphate agents, however, the setting time is, as expected, significantly reduced compared with anhydrite-rich sulphate agents due to secondary gypsum formation. Cements made from more rapidly precooled clinker (removal temperature 1400 °C) clearly have a somewhat higher demand for hemihydrate.

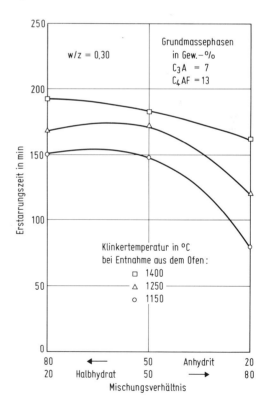

BILD 8: Einfluß der Kühlung von Klinker mit 7 Gew.-% C₃A und 13 Gew.-% C₂(A,F) auf das Erstarren von Zement mit verschiedenen Gemischen aus Anhydrit und Halbhydrat
FIGURE 8: Influence of the cooling of clinker containing 7 wt.% C₃A and 13 wt.% C₂(A,F) on setting of cement containing various mixtures of anhydrite and hemihydrate

Erstarrungszeit in min	= setting time in min
w/z =	= w/c =
Grundmassephasen in Gew.-%	= Matrix phases in wt.%
Klinkertemperatur in °C bei Entnahme aus dem Ofen	= clinker temperature in °C on removal from the kiln
Halbhydrat	= hemihydrate
Anhydrit	= anhydrite
Mischungsverhältnis	= Mix ratio

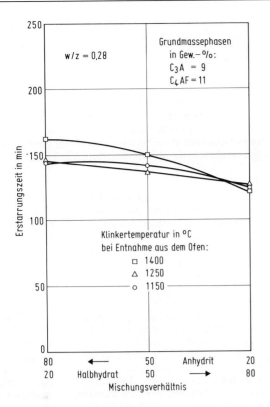
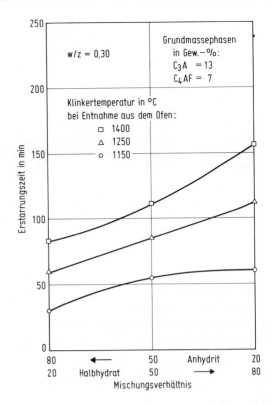

BILD 9: Einfluß der Kühlung von Klinker mit 9 Gew.-% C₃A und 11 Gew.-% C₂(A,F) auf das Erstarren von Zement mit verschiedenen Gemischen aus Anhydrit und Halbhydrat
FIGURE 9: Influence of the cooling of clinker containing 9 wt. % C₃A and 11 wt. % C₂(A,F) on setting of cement containing various mixtures of anhydrite and hemihydrate

Erstarrungszeit in min	= setting time in min
w/z =	= w/c =
Grundmassephasen in Gew.-%	= Matrix phases in wt.%
Klinkertemperatur in °C bei Entnahme aus dem Ofen	= clinker temperature in °C on removal from the kiln

BILD 10: Einfluß der Kühlung von Klinker mit 13 Gew.-% C₃A und 7 Gew.-% C₂(A,F) auf das Erstarren von Zement mit verschiedenen Gemischen aus Anhydrit und Halbhydrat
FIGURE 10: Influence of the cooling of clinker containing 13 wt. % C₃A and 7 wt. % C₂(A,F) on setting of cement containing various mixtures of anhydrite and hemihydrate

Halbhydrat	= hemihydrate
Anhydrit	= anhydrite
Mischungsverhältnis	= Mix ratio

3.5 Festigkeit

Die Ergebnisse der Festigkeitsprüfung nach 2 und 28 Tagen sind für Zemente aus den unterschiedlich zusammengesetzten und vorgekühlten Klinkern und einem Sulfatträgergemisch aus je 50 Gew.-% Anhydrit und Halbhydrat im **Bild 12** dargestellt. Danach deutet sich an, daß die Druckfestigkeit nach 28 Tagen mit steigendem C_3A-Gehalt ein Optimum bei etwa 9 bis 11 Gew.-% C_3A durchläuft. Die unterschiedliche Vorkühlgeschwindigkeit hat hierauf praktisch keinen Einfluß. Bei weiter steigendem C_3A-Gehalt des Klinkers nimmt die 28d-Festigkeit demgegenüber deutlich ab. Das ist besonders drastisch der Fall bei sehr langsamer Vorkühlung des Klinkers.

Auf die 2d-Festigkeit wirken sich die Vorkühlgeschwindigkeit und der C_3A-Gehalt des Klinkers dagegen nur geringfügig aus. Zwischen 9 und 11 Gew.-% deutet sich ein Minimum in der Festigkeitsentwicklung an. Wenn auch nur wenig ausgeprägt, ergeben sich die höheren Anfangsfestigkeiten bei schnell vorgekühlten Klinkern. Die Zusammensetzung des Sulfatträgers wirkte sich auf die Festigkeitsentwicklung nur unwesentlich aus, jedoch erwies sich die Sulfatträgermischung mit je 50 Gew.-% Anhydrit und Halbhydrat als vorteilhaft.

4. Schlußfolgerung

Die an 15 Zementen aus 4 Klinkern mit C_3A-Gehalten zwischen 7 und 15 Gew.-% durchgeführten Untersuchungen zeigten, daß das Erstarren bei gleicher Mahlfeinheit und gleichem Sulfatgehalt nicht nur vom C_3A-Gehalt, sondern auch in starkem Maß von der Kühlung des Klinkers und von der Zusammensetzung des Sulfatträgergemisches abhängt. Da eine langsame Vorkühlung die Reaktionsfähigkeit des Aluminats erhöht und sich daher dessen Reaktion mit dem zum Regeln des Erstarrens zugesetzten Sulfatträger verändert, nimmt der Einfluß der Vorkühlung mit steigendem

According to Figure 9, cements made from clinker containing 9 wt. % C_3A also set in conformity with the standard. However, the cooling rate and the composition of the sulphate agent have only a minor effect on the setting time. Cements made from this clinker set slightly faster with the hemihydrate-rich sulphate agent mix due to secondary gypsum formation.

In contrast, with cements made from C_3A-rich clinker the sulphate supply is not sufficient in every case for adequate retardation of setting (Figure 10). This is particularly evident with the cements made from very slowly cooled clinker with a hemihydrate proportion of 20–50 wt. %. The faster setting is due to syngenite formation and partly to monosulphate formation. An increase in the proportion of hemihydrate generally improved the setting behaviour.

Figure 11 shows that cements made from exceptionally C_3A-rich clinker only set slowly enough if the clinker is rapidly precooled and the hemihydrate content is as high as possible. The short setting time is primarily caused by monosulphate formation, as has been shown by additional thermoanalytical and scanning electron microscope investigations. In conformity with results from previous studies, the increased clinker reactivity can only be compensated with larger proportions of rapidly soluble hemihydrate as sulphate agent [43].

3.5 Strength

The results of the strength tests after 2 and 28 days are shown in **Figure 12** for cements made from the clinkers of different composition and precooling rate and a sulphate agent blend consisting of 50 wt. % each of anhydrite and hemihydrate. Based on these results, it appears that with an increasing C_3A content the compressive strength after 28 days traverses an optimum at approx. 9–11 wt. % C_3A. The different rate of precooling has practically no influence on this. On the other

[5] Scheuer, A.: Beurteilung der Betriebsweise von Klinkerkühlern und ihr Einfluß auf die Klinkereigenschaften. Zement-Kalk-Gips 41 (1988), S. 113–118.

[6] Schürmann, W., Scheuer, A., und Sylla, H.-M.: Optimierung des Satellitenkühlerbetriebs durch gezielte Eindüsung von Wasser. Zement-Kalk-Gips 44 (1991), S. 393–397.

[7] Sylla, H.-M., und Steinbach, V.: Einfluß der Klinkerkühlung auf die Zementeigenschaften. Zement-Kalk-Gips 41 (1988), S. 13–20.

[8] Schürmann, W.: Untersuchung zur thermischen Beurteilung von Gegenstrom-Klinkerkühlern der Zementindustrie und deren Einfluß auf die Klinkerqualität. Dissertation 1993. Clausthal-Zellerfeld.

[9] Sylla, H.-M.: Einfluß der Klinkerkühlung auf Erstarren und Festigkeit von Zement. Zement-Kalk-Gips 28 (1975), S. 357–362.

[10] Sylla, H.-M.: Einfluß reduzierenden Brennens auf die Eigenschaften des Zementklinkers. Zement-Kalk-Gips 34 (1981), S. 618–630.

[11] Lerch, W., und Taylor, W.C.: Some effects of heat treatment of portland cement clinker. Concrete, Cem. Mill Sect. 45 (1937), S. 199–217.

[12] Schwachheim, O.: Versuche mit schnell und langsam abgekühltem Klinker. Zement 25 (1936), S. 291.

[13] Chatterji, S., und Jeffery, J.W.: The effect of various heat treatments of the clinker on the early hydration of cement pastes. Mag. Concr. Res. 16 (1964) No. 46, S. 3–10.

[14] Lieber, W.: Einfluß der Klinkerkühlung auf die Zementeigenschaften. Unveröffentlicht.

[15] Chandler, W.R.: Effect of different cooling conditions on the quality of cement clinker. Rock Products 37 (1934) No. 4, S. 46–48.

[16] Narjes, A.: Über den Einfluß der Dampfbehandlung auf Zementklinker verschiedener Zusammensetzung. Dissertation TH Aachen (1958), Schriftenreihe der Zementindustrie, H. 21 (1958). Zement-Kalk-Gips 12 (1959), S. 129–136.

[17] Akatsu, K., und Higuchi, K.: On the strength and color of the portland cement clinker burned under reduction atmosphere. Cem. Assoc. Japan, Rev. 24. Gen. Meeting (1970), S. 23–25.

[18] Tomita, K., Ogawa, T., Abe, S., und Sagiya, I.: The effect of cooling rate of cement clinker on the strength of cement.
Cem. Assoc. Japan, Rev. 23. Gen. Meeting (1969), S. 81–86.

[19] Ono, Y., Kawamura, S., und Soda, Y.: Microscopic observations of alite and belite and the hyraulic strength of cement. Proc. V. Intern. Symp. Chem. Cem., Tokyo (1968), Part 1, S. 275–284.

[20] Wolter, A.: Einfluß des Ofensystems auf die Klinkereigenschaften. Zement-Kalk-Gips 38 (1985), S. 612–614.

[21] Billhardt, H.-W.: Erfahrungen mit einem Rohrkühler an einer 2500-t/d-Kurzdrehofenanlage mit Vorcalcination. Zement-Kalk-Gips 39 (1986), S. 122–124.

[22] Sprung, S.: Einflüsse der Verfahrenstechnik auf die Zementeigenschaften. Zement-Kalk-Gips 38 (1985), S. 577–585.

[23] Jepsen, O.L.: Zementfestigkeit und ihre Beziehung zur Kühlgeschwindigkeit und Kühlertyp. Zement-Kalk-Gips 29 (1976), S. 62–64.

[24] Herchenbach, H.: Verfahren der Zementkühlung und Auswahlkritrien für die gebräuchlichsten Kühlsysteme. Zement-Kalk-Gips 31 (1978), S. 62–64.

[25] Chatterjee, T.K., und Gosh, S.N.: The effect of cooling rate on cement properties – a review. World Cement Technology 11 (1980), S. 252–257.

[26] Dreizler, I.: Microscopic examination of high magnesium type clinkers as a contribution to clarify the causes of expansion due to magnesia. Proc. 3rd Int. Conf. Cem. Microscopy (1981), S. 33–35.

[27] Dreizler, I., und Knöfel, D.: Der Einfluß von Magnesiumoxid auf die Zementeigenschaften. Zement-Kalk-Gips 39 (1982), S. 537–550.

[28] Knöfel, D.: Beziehungen zwischen Chemismus, Phasengehalt und Festigkeit bei Portlandzementen. Zement-Kalk-Gips 32 (1979), S. 448–454.

[29] Schmitt-Henco, C.: Einfluß der Zusammensetzung des Klinkers auf Erstarren und Anfangsfestigkeit von Zement. Zement-Kalk-Gips 26 (1973), S. 63–66.

[30] Locher, F.W.: Die Festigkeit des Zements. Beton 26 (1976) Nr. 7, S. 247–249 u. Nr. 8, S. 283–286.

[31] Locher, F.W.: Erstarren und Anfangsfestigkeit von Zement. Zement-Kalk-Gips 26 (1973), S. 53–62.

[32] Sutej, C., und Vrgoc, K.: Zur Abhängigkeit der Zementfestigkeit von der chemischen Zusammensetzung des Klinkers. Zement-Kalk-Gips 26 (1973), S. 497–500.

[33] Alexander, K.M.: The relationship between strength and the composition and fineness of cement. Cem. Concr. Res. 2 (1972), S. 663–680.

[34] Bogue, R.H., und Lerch, W.: Hydration of portland cement compounds. Ind. Eng. Chem. 26 (1934), S. 837–841.

[35] Ish-Shalom, M., und Bentur, A.: Effects of aluminate and sulphate contents on the hydration and strength of portland cement pastes and mortars. Cem. Concr. Res. 2 (1972), S. 653–662.

[36] Alexander, K.M., Taplin, H., und Wardlaw, J.: Correlation of strength and hydration with composition of portland cement. Proc. 5th Intern. Symp. Chem. Cem., Tokyo 1968, Paper III-72, Part III, S. 152–166.

[37] Celani, A., Moggi, P.A., und Rio, A.: The effect of tricalcium aluminate on the hydration of tricalciumsilicate and portland cement. Proc. 5th Intern. Symp. Chem. Cem., Tokyo 1968, Paper II-134, Part II, S. 492–502.

[38] Popovics, S.: Comparison of various measurements concerning the kinetics of hydration of portland cements. Proc. 5th Intern. Symp. Chem. Cem., Tokyo 1968, Paper III-1. Part III, S. 129–137.

[39] Popovics, S.: Phenomenological approach to the role of C_3A in the hardening of portland cement pastes. Cem. Concr. Res. 6 (1976), S. 343–350.

[40] Stürmer, S., und Seidel, G.: Einfluß des Tricalciumaluminat-(C_3A)Gehaltes im Portlandzement auf Brennverhalten und Phasenbestand der Klinker sowie Festigkeitsentwicklung der Zemente. Baustoffind. 33 (1990), S. 144–147.

[41] Müller, A., Stürmer, S., und Stark, J.: Zum Einfluß des C_3A-Gehalts auf die Beständigkeit von Zementmörtel. Zement-Kalk-Gips 44 (1991), S. 190–193.

[42] Takashima, S., und Amano, F.: On the content of tricalcium aluminate in portland cement. Cem. Assoc. Jap., Rev. 13th Gen. Meeting 1959, S. 7–9.

[43] Locher, F.W., Richartz, W., und Sprung, S.: Erstarren von Zement. Teil II: Einfluß des Calciumsulfatzusatzes. Zement-Kalk-Gips 33 (1980), S. 271–277.

[44] Sprung, S.: Einfluß der Mühlenatmosphäre auf das Erstarren und die Festigkeit von Zement. Zement-Kalk-Gips 27 (1974), S. 259–267.

[45] Mohan, K., und Glasser, F.P.: The thermal decomposition of Ca_3SiO_5 at temperatures below $1250°C$. I. Pure C_3S and the influence of excess CaO or Ca_2SiO_4. Cem. Concr. Res. 7 (1977), S. 1–7.

[46] Hofmänner, F.: Portlandzement-Klinker. Kleine Gefügekunde. Holderbank Management und Beratung AG, 1973.

ZEMENT · KALK · GIPS
CEMENT · LIME · GYPSUM

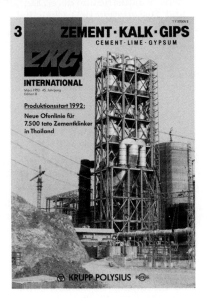

Fachzeitschrift für die Zement-, Kalk- und Gips-Industrie
Journal for the Cement, Lime and Gypsum Industries

ZKG-International gilt weltweit als die führende Fachzeitschrift für die Bindemittel Zement, Kalk und Gips und deren Herstellung. Wichtige Fachprobleme werden wissenschaftlich behandelt und die neuesten Forschungsergebnisse übermittelt.

Besondere Aktivität entwickelt ZKG-International bei der Behandlung der gesamten Verfahrenstechnik der Zement-, Kalk- und Gips-Herstellung von der Rohstoffgewinnung bis zum Versand. Durch die sowohl technische wie mathematische Behandlung neuer Verfahren, die Beschreibung ihrer Wirksamkeit in der Praxis, durch eingehende Systemschilderungen und Reportagen aus neuen Werken in aller Welt, trägt diese Zeitschrift dazu bei, die Innovationen in den Technologien der Bindemittel allen Fachleuten in den Zement-, Kalk- und Gips-Werken des In- und Auslandes bekanntzumachen.

ZKG-INTERNATIONAL erscheint monatlich.
Ausgabe B Deutsch/Englisch.
Bitte fordern Sie ein kostenloses Probeheft an.

ZKG International is acknowledged world-wide as the leading trade journal on cement, lime and gypsum and the manufacture of these materials. Important technical issues are studied scientifically and the latest research findings are revealed.

ZKG is particularly active in covering the whole range of process engineering involved in the manufacture of cement, lime and gypsum, from the extraction of the raw materials to the finished product. Through the technical and mathematical treatment of new processes with descriptions of their effectiveness in actual practice and through detailed accounts of systems and reports on new plants all over the world, ZKG brings the innovations of cement, lime and gypsum technology to the notice of all experts specialized in the manufacture of these materials at home and abroad.

ZKG INTERNATIONAL is published monthly.
Edition B German/English.
Please ask for sample copies.

BAUVERLAG GMBH
D-65173 Wiesbaden · Germany
Tel. (49) 6123 / 7 00-0
Fax (49) 6123 / 7 00-122

Bureck-Erlenkötter Fördersysteme

Unsere Erfahrung - Ihre Sicherheit
Our experience - your quarantee

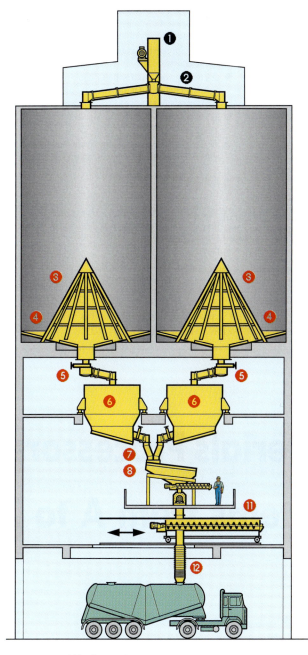

Wir planen und liefern **einzelne Förderanlagen** und **komplette Systeme.**

We design and deliver **individual conveyors** and **complete systems.**

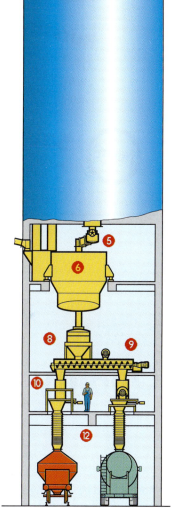

Loseverladung Zementsilo
bulk loading cement silos

1 Becherwerk
2 Materialaufgabe mit BE-Weiche
3 Silodruckentlastung
4 sektionsgesteuerte Silobelüftungsböden
5 Abzugsrinne mit Drehdosierschieber
6 Wiegeeinrichtung mit belüfteten Böden
7 Abzugsrinne mit Drehdosierschieber
8 Siebmaschine
9 Verteilerschnecke
10 Trichtermobil - längs- und quer verfahrbar
11 Schneckenmobil - längs- und quer verfahrbar
12 Verladegarnitur

1 bucket elevator
2 material feed with BE-switching
3 silo pressure relief
4 sectionally controlled silo aeration floors
5 outlet chute with rotary metering gate
6 weighing device with aerated floors
7 outlet chute with rotary metering gate
8 screening machine
9 spreader screw
10 mobile hopper - longitudinally and transversely traversable
11 mobile screw - longitudinally and transversely traversable
12 loading set

Siemensstraße 26 · 59269 Beckum · Postfach 1265 · 59242 Beckum · ☎ (0 25 21) 93 66-0 · Fax (0 25 21) 93 66-20

Safe with Pressure and Explosion

Bursting Discs
- DN 3 – 3000 · 0.01 – 7000 bar ·
- leak tight · maintenance free ·
- corrosionresistant ·

Safety Experts

Rembe® GmbH · D-59918 Brilon
Germany · Fax (02961) 507 14
Phone (02961) 7405-10
Teletex 296134 = REMBE
Telex 17296134

e116

Im Vorteil sein . . .

mit Sprengstoffen, Zündmitteln und Sprengzubehör aus Schönebeck und Gnaschwitz

- Gelatinöse Gesteinssprengstoffe (Gelamone) in abgestuften Leistungsparametern für alle Über- und Untertageanwendungsfälle
- ANC - Sprengstoffe und patronierte Emulsionssprengstoffe
- Lieferung von elektrischen Zündern, Sprengschnüren, Zündleitungen und sonstigem Sprengzubehör
- Expansivmittel zur Zerstörung von Mauerwerk, Fels, Beton
- Sevice und Beratung durch unsere Fachleute einschließlich bei Sprengerschütterungen

Vertrieb:
Anhaltinische Chemische Fabriken (ACF) GmbH
Magdeburger Str. 241, 39218 Schönebeck (Elbe)
Tel. (03928) 78 58 70 und 78 58 19
Fax (03928) 78 58 07

sowie die Auslieferungslager des Unternehmens:
AL Bad Berka, 99438 Tannroda · Tel. (036458) 21158, Fax 22044
AL Tarthun, 39435 Egeln · Telefon/Fax (039268) 2253
AL Lugau, 09385 Lugau (Erzgebirge) · Telefon/Fax (037295) 2569
AL Goes, 01769 Pirna · Telefon/Fax (03501) 762278

For Raw Materials Processors:
Quality Control Systems from A to Z

Market leaders in many sectors use them; Pfaff's automatic quality control systems.
They withdraw samples from a continuous stream of material and send them to be laboratory by tube transport. There they receive, prepare and evaluate the samples. They then use a computer to feed back data influencing the production process.

A complete range of systems to reduce waste and optimize quality. Just ask us about it.

PFAFF AQS GMBH
P.O. Box 240228 · D-42232 Wuppertal
Tel. 0202/26602-0 · Fax 0202/2660212

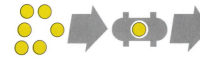

Sampling   Transport   Preparation   Analysis

Tradition und Innovation: Investitionen für die Zukunft

HUMBOLDT WEDAG ZAB GMBH

Anlagen zur Herstellung von

- Zement nach dem Trockenverfahren
- Zement nach dem Halbnaßverfahren
  Umstellung von Naßanlagen auf Halbnaßanlagen, einschließlich Schlammentwässerung
- Zement nach dem Gips-Schwefelsäure-Verfahren
  Ofenanlagen mit Drehrohrofen als langen Trockenofen bzw. kurzer Drehrohrofen mit Vorwärmer
- Kalk
  Ofenanlagen mit Drehrohrofen nach dem Halbnaßverfahren
- Sinterdolomit
  Ofenanlagen mit Drehrohrofen und Rostvorwärmer
- Kalkmilch
  Mahlanlagen mit Kugelrohrmühle als Naßmühle
- Kalkschlammregenerierung
  Ofenanlage mit Drehrohrofen
- Blähton
  Ofenanlage-Vorwärmer
  Kühler in Kaskadenanordnung

Anlagen zur

- Klärschlammverwertung
  Thermische Klärschlammbehandlung mit Pulsationsreaktor
- Rückstandsverbrennung

sonstige Anlagen

- Brech- und Klassieranlagen
- Pneumatische Fördersysteme
- Wasserschleusenbau

Unsere weltweite Erfahrung und die breite Angebotspalette sind Ihre Erfolgsgaranten

Brauereistraße 13 · D-06847 Dessau · Telefon: (03 40) 73 10 -0 · Telefax: (03 40) 7 31 06 76 · Telex: 319 350 zab d

Praxis im Zementlabor?

Practice in the Cement Lab?

Die bessere Lösung

The improved solution

Innovative Systeme für die Baustoffprüfung

TONICOMP III
Für die Festigkeitsprüfung von Baustoffen könnten wir Ihnen auch ein Universalgenie anbieten. Aber für die rationelle Druck- und Biegezugprüfung an Prismen 40 x 40 x 160 mm ist TONICOMP III der überlegene Spezialist.

Jetzt mit TONITROL: Testprogramm aufrufen, starten und ab geht die Post: vollautomantisch und mit normgerechten Prüfparametern.

Innovative Systems for Building Material Testing

TONICOMP III
We could also offer a universal genius for compressive strength tests on building materials. But TONICOMP III is the convincing specialist for rational compression and bend tests on prisms 40 x 40 x 160 mm.

Now with TONITROL: Call up the program and away you go: fully automatic and with test parameters according to standards.

Baustoffprüfsysteme

Toni Technik

Toni Technik Baustoffprüfsysteme GmbH
Gustav-Meyer-Allee 25 · D-13355 Berlin
Tel. (0 30) 4 63 30 96 · Fax (0 30) 4 63 30 99

 **BAUVERLAG**

Adreßbuch Umweltschutz

Handbuch für Presse, Behörden, Wissenschaft, Wirtschaft, Verbände, Bürgerinitiativen

Hrsg. von der Deutschen Umweltstiftung. 3., neubearbeitete und erweiterte Auflage 1993.
522 Seiten DIN A 5.
Kart. DM 38,50
öS 300,– / sFr 38.50
ISBN 3-7625-2984-1

Mit diesem Handbuch liegt eine umfassende Sammlung überwiegend nichtkommerzieller Umweltschutzadressen vor.
Es enthält die Anschriften und Telefonnummern der einzelnen Institutionen, zusätzlich die Namen und Durchwahlnummern von Behördenleitern, Vorsitzenden und Pressesprechern der Verbände, von umweltpolitischen Sprechern der Parteien und Beauftragten der Kirchen, von Ökologie-Redakteuren bei Presse, Rundfunk und Fernsehen.

Umweltfreundliche Beschaffung

Handbuch zur Berücksichtigung des Umweltschutzes in der öffentlichen Verwaltung und im Einkauf

Hrsg. vom Umweltbundesamt. 3., neubearbeitete und erweiterte Auflage 1993.
600 Seiten DIN A 5 mit ca. 15 Abbildungen und 70 Tabellen.
Kart. DM 32,–
öS 250,– / sFr 32.–
ISBN 3-7625-2882-9

In welchen Bereichen heute neue, die Umwelt weniger belastende Materialien, Produkte und Verfahren angeboten werden, läßt sich in „Umweltfreundliche Beschaffung" nachlesen. Insgesamt neun Produktgruppen werden hinsichtlich ihrer Umweltverträglichkeit untersucht und beschrieben.

BAUVERLAG GMBH
D-65173 Wiesbaden
Tel. (0 61 23) 700-0 · Fax (0 61 23) 700-122

LAGERN   DOSIEREN   FÖRDERN

Kohlenstaubanlage zur Kesselbeschickung, sowie Entaschungs- und Ascheverladeanlage.

Gravimetrische Dosieranlage AIRDOS® zum direkten Produkteintrag in pneumatische Fördersysteme.

Siloaustragssystem ROTEX für schwerfließende Schüttgüter mit anschließendem pneumatischem Transport.

Anspruchsvolle Aufgaben erfordern sichere Lösungen.

Gleich, welches Schüttgut Sie lagern, dosieren, pneumatisch fördern, homogenisieren, wir liefern dafür eine Komplettlösung.
Denn die Optimierung des Schüttguthandlings haben wir uns zur Aufgabe gemacht. Zentrale Schwerpunkte dieser Arbeit sind das Lagern, Fördern und Dosieren der unterschiedlichsten Schüttgüter. Basierend auf langjähriger Erfahrung verfügen wir über ein umfassendes Produktprogramm und liefern maßgeschneiderte, innovative Lösungen.

Fragen Sie uns.

ANLAGENTECHNIK

Südstraße 14
66780 Rehlingen
Telefon 0 68 35 / 21 95- 21 97
Telefax 0 68 35 / 38 73

*Umwelt-
und energiebewußt.*

ROHRBACH
ZEMENT

Portlandzementwerk
Dotternhausen
Rudolf Rohrbach
Kommanditgesellschaft
D-72359 Dotternhausen
Telefon (0 74 27) 79-0
Telefax (0 74 27) 79-300
Telex 7 62 896 zemdo d

Fachbereich 2

Prozeßführung und Informationsmanagement

(Messen, Steuern, Regeln, Prozeßführung, Erfassen, Leiten und Verarbeiten von Informationen)

Subject 2

Process control and information management

(Measuring, control, automation, process control, acquisition, transmission and processing of information)

Séance Technique 2

Gestion du procédé et management d'informations

(Mesurage, contrôle, réglage, conduite du processus, collection, transmission et traitement d'informacions)

Tema de ramo 2

Control del proceso y gestión de informaciones

(Medición, control, regulación, control del proceso, comprender, dirigir y elaborar informaciones)

Prozeßführung und Informationsmanagement*)
Process control and information management*)

Gestion du procédé et management d'informations

Control del proceso y gestión de informaciones

Von **R. Säuberli, U. Herzog**, Holderbank/Schweiz, und **H. Rosemann**, Lägerdorf/Deutschland

Prozeßführung und Informationsmanagement

Generalbericht 2 · Zusammenfassung – Die Aufgaben der Prozeßführung wurden sowohl für das Steuer-, Instrumentierungs- und Überwachungssystem als auch für das Betriebspersonal in den letzten Jahren immer komplexer. Produktqualität und -vielfalt, Umweltschutz, Sekundärbrennstoffe und Energiewirtschaft erfordern eine sensiblere Betriebsführung. Die steigende Informationsfülle wird durch Informationsmanagement-Systeme aufgefangen. Die Informationsdichte wächst laufend und betrifft sämtliche Ebenen der CIM-Pyramide. Auf den verschiedenen Ebenen werden funktionsgerechte Hardware und Software eingesetzt und untereinander durch Kommunikationsbusse zu einem integrierten Prozeßführungs- und Informationssystem verbunden. Um das effizient und wirtschaftlich zu realisieren, sind internationale Standards erforderlich, durch welche die Kommunikationsschnittstellen sowie die Hard- und die Software der verschiedenen Sensoren, speicherprogrammierbaren Steuergeräte, Personal-Computer, Workstations und Server vereinheitlicht werden. Die Standardisierungsbemühungen zeigen weltweit erste Erfolge, lassen aber für die Zukunft noch ein breites Feld für Aktivitäten offen. Die Ergonomie der Systeme spielt für die Akzeptanz der Prozeßsteuerer eine entscheidende Rolle. Die Modularität, die Flexibilität und die Dokumentation sind für den Instandhalter sehr wichtig. Ohne Dokumentation sind die heutigen Systeme nicht mehr wartbar. Die Entwicklung geht demzufolge in Richtung selbstdokumentierender Systeme. Auf der Prozeßleitebene und höher werden heute immer mehr Geräte eingesetzt, die sich in der Büroautomation weltweit bewährt haben. Der Daten- und Informationsverbund der technischen mit der administrativen Seite ist noch nicht in vielen Werken realisiert. Die Wartbarkeit der installierten Systeme muß periodisch überprüft werden. Die Strategie des Ersatzes, resp. der Modernisierung sowie die zu wählende Konzeption sind Sache des höheren Managements. Sie dürfen nicht mehr nur dem „Elektriker" oder dem „Administrator" überlassen werden. Demgegenüber muß die technische Konzeption und Planung nach dem Prinzip „Bottom-up" gemacht und das Werkspersonal frühzeitig mit einbezogen werden. In modernen Anlagen hat sich das Anforderungsprofil vieler Arbeitsplätze stark verändert. Der Personalbestand wird immer kleiner und die Aufgabenvielfalt immer größer. In Zukunft verlangen die Software-Entwicklung und die Personalschulung die größte Aufmerksamkeit. Ein oft etwas vernachlässigter Aspekt ist das Engineering, wobei ein großer Teil der Software-Erstellung und Parametrierung dazugehört. Auch dieser Aspekt wird in Zukunft noch wichtiger.

Process control and information management

General report 2 · Summary – Process control tasks have become increasingly complex in recent years both for the control, instrumentation and monitoring systems and for the operating personnel. Product quality and diversity, environmental protection, secondary fuels, and energy management require more sensitive plant management. The increasing abundance of information is picked up by information management systems. There is a constant increase in information density which affects every level in the CIM (Computer Integrated Manufacture) pyramid. Hardware and software with appropriate functions are installed at the different levels and linked with one another through communication buses to an integral process control and information system. To make this work efficiently and cost-effectively there is a need for international standards which standardize the communication interfaces and the hardware and software, the various sensors, programmable controllers, PCs, workstations and servers. The efforts at standardization are showing some initial world successes but are still leaving a wide field open for future activities. The ergonomics of the systems play a critical part in their acceptance by process controllers. Modularity, flexibility, and documentation are very important for those responsible for maintenance. It is not possible to maintain modern systems without documentation, so development is moving in the direction of self-documentation. Increasing numbers of machines, which have proved their value worldwide in office automation, are being used at the process control level and above. Many works have not yet achieved the data and information link between the technical and administrative sides. The maintainability of the installed system must be checked periodically. The strategy of replacement as opposed to modernization, and the scheme to be selected, are matters for higher management. They can no longer just be left to the "electrician" or the "manager". However, the technical concept and the planning must be carried out on the "bottom-up" principle, and the works personnel

---

*) Überarbeitete Fassung eines Vortrages zum VDZ Kongreß '93, Düsseldorf (27. 9.–1. 10. 1993)
Revised text of a lecture to the VDZ Congress '93, Düsseldorf (27. 9.–1. 10. 1993)

must be included at an early stage. In modern plants there have been great changes in
the requirement profiles for many of the jobs. The staffing level becomes ever lower and
the job diversity ever greater. Software development and personnel training will call
for the greatest care in the future. An aspect which is often somewhat neglected is the
engineering side, which includes a large part of the generation and parameterization
of the software. This is another aspect which will become more important in the
future.

Gestion du procédé et management d'informations

Rapport général 2 · Résumé – Les tâches de gestion du process sont devenues, au cours
des dernières années, de plus en plus complexes aussi bien pour les systèmes de commande, instrumentation et surveillance que pour le personnel d'exploitation. La qualité et grande variété des produits, la protection de l'environnement, les combustibles
secondaires et l'économie d'énergie exigent une grande sensibilité de gestion. La multiplicité des informations croissantes est absorbée par les systèmes de management d'informations. La densité des informations ne cesse de croître et concerne tous les niveaux
de la pyramide CIM. Des matériels et logiciels fonctionnels utilisés aux niveaux les plus
variés sont reliés entre eux par des bus de communication pour former un système d'informations et de gestion des opérations du process intégré. Pour parvenir à une réalisation efficace et rentable il faut disposer de normes internationales permettant d'harmoniser les interfaces de communication ainsi que les matériels et logiciels des différents
capteurs, des appareils de commande programmable, des ordinateurs personnels, des
postes de travail et des serveurs. Les efforts de standardisation enregistrent à l'échelon mondial leurs premiers succès, mais ils constituent pour l'avenir un vaste champ
d'activités. L'ergonomie des systèmes joue un rôle décisif dans l'acceptabilité des systèmes de commande de process. La modularité, la flexibilité et la documentation sont
de la plus grande importance pour le spécialiste chargé de la maintenance. Sans documentation il n'est plus possible d'entretenir les systèmes actuels. Et l'évolution va,
de ce fait, dans le sens de systèmes ayant leur propre documentation. Au niveau de la
direction du process et à un échelon plus élevé, on fait appel de nos jours de plus en
plus à des appareils qui ont largement fait leurs preuves en bureautique. L'interconnexion données-informations entre les parties technique et administrative n'est pas encore réalisée dans nombre d'usines. Il faut contrôler périodiquement les possibilités de
maintenance des systèmes installés. La stratégie du remplacement ou de la modernisation de ces systèmes, ainsi que la conception à retenir, entrent dans les compétences
du haut management. Ces décisions ne doivent plus être prises au niveau des „électriciens" ou de l'„administrateur". La conception technique et la planification, par contre, doivent être effectuées selon le principe du „bottom-up" et, dès le début, avec la
participation du personnel de l'installation. Dans les installations modernes, le profil
des postes de travail a changé énormément. Les affectifs diminuent, tandis que les
tâchent deviennent chaque fois plus multiples. A l'avenir, la mise au point du software
ainsi que la formation du personnel exigeront une attention primordiale. Un aspect souvent un peu négligé est l'engineering, qui comprend une grande partie de la mise au
point du software et de la fixation des différents paramètres. Cet aspect deviendra également plus important à l'avenir.

Control del proceso y gestión de informaciones

Informe general 2 · Resumen – Las tareas relacionadas con el control del proceso se
han vuelto, en los últimos años, cada vez más complejas, en cuanto se refiere a los sistemas de mando, instrumentación y vigilancia y también al personal de servicio. La
calidad y variedad del producto, la protección del medio ambiente, los combustibles secundarios y el ahorro de energía requieren un control más sensible del proceso. El creciente volumen de informaciones se abarca mediante los sistemas de gestión de informaciones. La densidad de las informaciones aumenta constantemente y afecta a todos
los niveles de la pirámide CIM. En los diferentes niveles se emplean el hardware y el
software apropiados, conectados entre sí por medio de buses de comunicación, de modo
que se obtiene un sistema integrado de control del proceso y de gestión de informaciones. Para conseguir este objetivo de forma eficiente y económico, son necesarias unas
normas internacionales, mediante las cuales se llegue a unificar los inferface de comunicación así como el hardware y el software de los diferentes sensores, controladores programables, ordenadores personales, workstations y servers. Los esfuerzos de normalización dan, a escala mundial, los primeros resultados, pero dejan todavía para el
futuro un amplio campo de actividades. La ergonomía de los sistemas desempeña un
papel decisivo en cuanto a la aceptación de los controladores del proceso. La modularidad, la flexibilidad y la documentación son de gran importancia para el encargado
del mantenimiento. Sin la citada documentación, el mantenimiento de los sistemas actuales ya no es posible. Por lo tanto, el desarrollo técnico va en dirección de los sistemas de autodocumentación. A nivel de control de procesos se emplean, hoy en día, cada
vez más los aparatos que han probado su eficacia, a nivel mundial, en la automatización de las oficinas. En muchas fábricas no se ha llegado aún a realizar el necesario
intercambio de datos e informaciones entre la parte técnica y la parte administrativa.
Hay que comprobar periódicamente las posibilidades de mantenimiento de los sistemas
instalados. La estretegia de su sustitución o de su modernización así como el concepto a elegir son cosas que incumben al alto management. No se deben ya dejar al criterio del „electricista" o del „administrador". El concepto técnico y la planificación, en
cambio, han de llevarse a cabo según el principio del „bottom-up", incorporando a tiempo al personal de la fábrica. En las plantas modernas, el perfil de muchos puestos de
trabajo ha cambiado sustancialmente. La plantilla queda cada vez más reducida, en
tanto que aumenta la diversidad de las tareas planteadas. En el futuro, el desarrollo

del software y el entrenamiento del personal exigirán la máxima atención. Un aspecto a veces algo descuidado es la ingeniería, que comprende gran parte de la confección del software y de la fijación de los diferentes parámetros. También este aspecto va a ser más importante en el futuro.

1. Einleitung

Die Prozeßtechnik und die maschinelle Ausrüstung für die Zementproduktion wurden in den vergangenen Jahren zunehmend komplexer. Aber nicht nur die Verfahren, sondern auch die Infrastrukturen und die Anforderungen an die Umweltverträglichkeit wurden anspruchsvoller.

Durch die laufende Optimierung auf praktisch allen Gebieten werden die eingebauten Reserven aufgebraucht und die Sicherheitsgrenzen immer enger gesetzt. Somit steigen auch die Anforderungen an die Anlagenbetreiber. Diese Tendenz wird sich in Zukunft noch durch die vermehrte Öffnung der Märkte, steigende Energiekosten, erhöhtes Umweltbewußtsein, erhöhte Qualitätsanforderungen, Produkthaftpflicht und Personalminimierung verstärken.

Die folgenden Ausführungen sollen das weite, multi-disziplinäre Gebiet der Prozeßführung und des Informationsmanagements aufdecken, strukturieren, heutige Systeme und Techniken aufzeigen und auf zukünftige Anforderungen und Technologien verweisen. Dabei kann kein Anspruch auf Vollständigkeit erhoben werden.

2. Anforderungen an Prozeßführungs- und Informations-Systeme

Die Prozeßführungs- und Informations-Systeme sind heute nicht nur ein Mittel, um die Produktionsanlagen überhaupt betreiben zu können, sondern viel mehr integrierter Anlagen-Bestandteil und absolute Notwendigkeit für die effiziente, kostengünstige Herstellung von Qualitätsprodukten sowie für die wirkungsvolle Betriebsführung und -überwachung. Von der Anlagengröße und vom Aufgabenbereich her gesehen, verlangt die Zementindustrie sehr umfangreiche Systeme. In einer typischen Produktionslinie, Brecherei bis Zementsilos, von 3 000 t Zement pro Tag sind heute ca. 900 Motore und Ventile installiert (zukünftig ca. 1100).

Das heißt die Systeme müssen technisch für die in folgender **Tabelle 1** aufgeführten Kapazitäten ausgelegt werden:

TABELLE 1: Anforderungen an Prozeßführungssysteme für eine Zementproduktionslinie, z.B. 3 000 t/d mit ca. 900 Antrieben, zukünftig ca. 1100

Bezeichnung	heute	zukünftig
Digitale Eingänge	6 000	8 000
Digitale Ausgänge	1 300	2 000
Analog-Eingänge	350	1 000
Regelkreise (PID)	40	70
Alarmpunkte	10 000	15 000
Gerechnete Werte	150	500
Archivierte Werte pro Minute	1 000	4 000

Die Anforderungen an die Qualität, die Zuverlässigkeit und die Reproduzierbarkeit der Daten und Informationen nimmt mit zunehmender Verfeinerung der Anlagenoptimierung zu. Die Forderung nach mehr Intelligenz für automatische Entscheidungen an vorderster Front verlangt schnelle und zuverlässige Kommunikationssysteme zwischen allen Funktionen.

Der heute noch klassische Aufbau der Prozeßführungs- und Informationssysteme zeichnet sich durch eine relativ klare, pyramidenförmige Hierarchie aus (**Bild 1**).

Was die Ebenen besonders charakterisiert, sind die Anforderungen bezüglich Abfrage- und Verarbeitungszeiten (Bild 1). Auf den prozeßnahen Ebenen werden unbedingt Echtzeit- und Multitasking-Fähigkeiten verlangt.

1. Introduction

The process technology and mechanical equipment for cement production have become increasingly complex in past years. Not only the processes but also the infrastructure and the demand for environmental compatibility have become more exacting.

Due to continuous optimization in practically all areas the built-in reserves are being used up and the safety limits set ever tighter. This also increases the demands on the plant operators. These trends will be intensified even further in the future through the increased openness of the markets, increasing energy costs, increased environmental awareness, higher quality specifications, product liability and minimization of personnel.

The following comments are intended to reveal and classify the wide, multi-disciplinary field of process control and information management, give an indication of present systems and procedures, and touch on future requirements and technologies. No claim can be made for completeness.

2. The demands made on process control and information systems

Nowadays, process control and information systems are not just a means of operating production plants; they are integrated parts of the plant and are essential for efficient, low-cost manufacture of quality products and for effective plant control and monitoring. From the point of view of plant size and range of tasks, the cement industry calls for very extensive systems. Nowadays, a typical production line, from crushing plant to cement silos, for 3 000 t cement per day contains approximately 900 motors and valves (increasing to about 1100 in the future).

This means that the systems must be designed technically for the capacities listed below in **Table 1**:

TABLE 1: Requirements for process control systems for a cement production line, e.g. 3 000 t/d with approximately 900 drives, increasing to about 1 100 in the future

Designation	Present	Future
Digital inputs	6 000	8 000
Digital outputs	1 300	2 000
Analogue inputs	350	1 000
Control loops (PID)	40	70
Alarm points	10 000	15 000
Calculated values	150	500
Values filed per minute	1 000	4 000

The demands on the quality, reliability and reproducibility of the data and information is increasing with increased refinement of plant optimization. The requirement for more intelligence for automatic decisions at the very front line requires rapid and reliable communication systems between all functions.

The classical structure of the process control and information systems which still exists is distinguished by a relatively clear, pyramidal hierarchy (**Fig. 1**).

What particularly characterizes the levels are the requirements with regard to scanning and processing times (Fig. 1). Real-time and multi-tasking capabilities are essential at the levels close to the process.

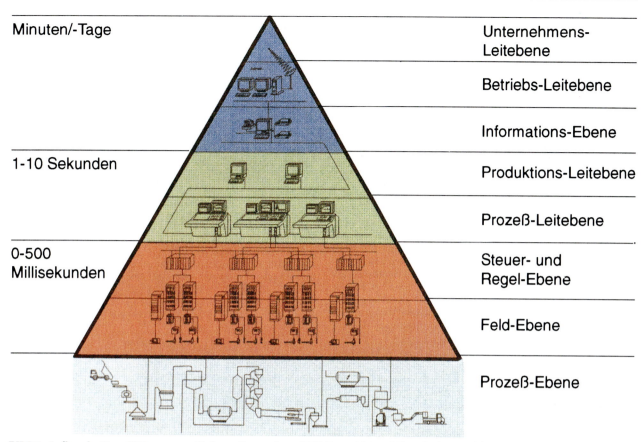

BILD 1: Aufbau der Prozeßführungs- und Informationssysteme (CIM-Pyramide) und die geforderten Abfrage- und Verarbeitungszeiten
FIGURE 1: Structure of the process control and information systems (CIM pyramid) and the required scanning and processing times

Minuten	= minutes	Prozeß-Leitebene	= process control level
Tage	= days	Steuer-und Regel-Ebene	= open- and closed-loop control level
Unternehmens-Leitebene	= company control level	Feld-Ebene	= field level
Betriebs-Leitebene	= plant control level	Prozeß-Ebene	= process level
Informations-Ebene	= information level	Sekunden	= seconds
Produktions-Leitebene	= production control level	Millisekunden	= milliseconds

Die alten Systeme mit wändefüllenden Anzeigen und Schaltelementen und mit Handprotokollierung genügen nicht mehr. Gefordert werden Informationen, die

— zielgerichtet,
— schnell,
— detailliert,
— plausibel,
— nach dem Empfänger sortiert,
— bereichsübergreifend,
— am Ort sind, wo man sie braucht,
— archivierbar und rückgewinnbar sind.

In bezug auf die Instandhaltung müssen die Systeme trotz der Komplexität ihrer Aufgaben einfach, durchschaubar, systematisch, standardisiert, modular und flexibel aufgebaut sein.

Die Investitionskosten für Prozeßführungs- und Informations-Systeme der Zementindustrie von beispielsweise 4 bis 6 Mio. DM pro Produktionslinie teilen sich auf in

— ca. 30–40 % für Hardware und
— ca. 60–70 % für Software und Engineering.

Für den Instandhalter ist es darum vor allem wichtig, die Software und den Engineeringteil komplett dokumentiert zu bekommen und zu halten. Ohne saubere, umfängliche Dokumentation sind diese Systeme nicht mehr wartbar. Der Ausbildungsbedarf für den Einstieg und das weitergehende Training aller Systemanwender und vor allem auch der Instandhalter soll möglichst gering sein.

Der Anforderungskatalog für die Prozeßführungs- und Informations-Systeme ist, heute und in Zukunft, sehr mannigfaltig. Dennoch kann die technische Machbarkeit für die

The old systems with indicating instruments and switch elements covering the walls, and with manual logging, can no longer cope. Information is required which

— is specific,
— is rapid,
— is detailed,
— is plausible,
— is sorted according to addressee,
— covers all areas,
— is located where it is needed,
— can be stored and recovered.

In spite of the complexity of their tasks the systems must, for the sake of maintenance, be of simple, clear, systematic, modular and flexible design.

The investment costs for process control and information systems for the cement industry of, for example, 4 to 6 million DM per production line are divided into

— approximately 30–40 % for hardware, and
— approximately 60–70 % for software and engineering.

The most important thing for the maintenance engineer is to obtain, and maintain, complete documentation for the software and the engineering section. It is no longer possible to maintain these systems without clear complete documentation. The training requirement for the initiation and on-going training of all system users, especially those responsible for the maintenance, should be as small as possible.

The list of requirements for the process control and information systems is, now and for the future, very varied. However, technical feasibility can be regarded as already substantially established as far as the cement industry is con-

Zementindustrie schon weitgehend als gegeben angesehen werden. Die sehr hohen Erwartungen gegenüber der Betriebssicherheit und der Verfügbarkeit der Systeme werden heute im allgemeinen erfüllt, z.B. maximal ein Ofenstop pro Jahr, ausgelöst durch den Ausfall des Prozeßleitsystems.

3. Systemstrukturen

Die vorgestellte allgemeine CIM-Pyramide ordnet die verschiedenen Automationsaufgaben unterschiedlicher Ebenen zu. Die verschiedenen Funktionen lassen sich aber nicht immer so klar trennen.

Ein modularer, dezentraler Aufbau ist heute bei allen Leitsystemen vorhanden. Bei den meisten installierten Systemen fehlt die Integration der Pyramidenspitze, d.h. der Unternehmens- und Betriebsleitebene. Auf der unteren Ebene, dem eigentlichen Prozeßleitsystem, ist eine Abkehr von herstellerspezifischer Hardware und Basissoftware feststellbar. Außer auf der Steuer- und Regelebene, wo „Echtzeit-Verarbeitung" verlangt wird, werden überall Systeme eingesetzt, die sich in der Büroautomation bewährt haben (**Bild 2**). Die technische Software, die für Prozeßrechner geschrieben wurde, nimmt ab, die für PCs und Workstations nimmt zu. Das Ziel, die Realisierung von offenen Systemen (open systems), rückt damit näher.

Als weltweit anerkannte Standards gelten heute PCs und Workstations, die mit den Betriebssystemen Windows, VMS oder UNIX zu betreiben sind und über selbsterklärende graphische Benutzeroberflächen bedient werden. Diese dezentralen Einzelkomponenten werden über Netzwerk wie z.B. Ethernet mit TCP/IP-Protokoll miteinander vernetzt, wodurch von allen Arbeitsplätzen auf die Standard-Software wie z.B. relationale Datenbanken mit SQL zugegriffen werden kann. Bei den Datenbus-Systemen (oder Netzwerken) gibt es zwei unterschiedliche Konzepte. Entweder verbindet ein einziger leistungsfähiger Bus, der meist redundant ausgeführt ist, alle Computersysteme untereinander oder es wird ein prozeßnaher Bus für den schnellen Echtzeit-Datenaustausch und ein zusätzlicher Informationsbus für den Austausch großer Datenmengen eingesetzt.

Prozeßleitsysteme für das gesamte Zementwerk mit allen Betriebsabteilungen werden sowohl von den namhaften Unternehmen der Elektroindustrie, Zementanlagen-Herstellern oder auch von Zementherstellern selbst entwickelt.

Dabei sind zwei vor allem historisch bedingte Strömungen festzustellen:

- Die Vertreter der DCS-Systeme (Distributed Control System) realisieren die Visualisierung, Optimierung und Informationsdarstellung auf Workstations und die Prozeßsteuerung und -regelung auf speicher-programmierbaren Steuergeräten (SPS) oder auf Prozeßrechnern.
- Die Vertreter der PC/SPS-Systeme hingegen verwenden ausschließlich Personal-Computer und speicher-programmierbare Steuergeräte.

Die beiden System-Typen kommen einander immer näher.

4. Feldebene

Die unterste Ebene in einem Prozeßführungssystem ist die Feldebene (Bild 1). Sie beinhaltet:

- die Sensorik,
- die Schützenschränke und
- die Verkabelung

und bildet somit die eigentliche Schnittstelle zum Prozeß, zur Mechanik und zur Umwelt.

Was hier nicht richtig erfaßt wird, kann weiter oben im System nur schwerlich, wenn überhaupt, korrigiert werden. Mit wachsendem Automatisierungsgrad, durch neue Anlagen- und Maschinenkonzepte sowie durch belastungsoptimierte Maschinen und Installationen stieg die Zahl der Sensoren und Aktoren überdurchschnittlich an, wie **Tabelle 1** und **Bild 3** erkennen lassen. Es wird aber auch eine transpa-

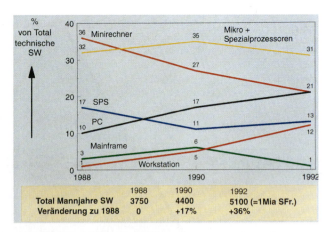

BILD 2: Statistik über die Erstellung von technischer Software pro Computer-Technologie und Zeit
FIGURE 2: Statistics relating to the generation of technical software for each type of computer technology and with time

von Total technische SW	= of total technical software
Minirechner	= mini computers
Mikro + Spezialprozessoren	= micro and specialist processors
SPS	= PLC systems
Total Mannjahre SW	= Total software man-years
Veränderung zu 1988	= Change from 1988

cerned. The very high expectations in relation to operational reliability and availability of the systems are now generally fulfilled, e.g. a maximum of one kiln stop per year caused by failure of the process control system.

3. System structures

The universal CIM (Computer Integrated Manufacture) pyramid assigns the various automation tasks to different levels. However, the various functions cannot always be separated so clearly.

All control systems now have modular, decentralized structures. In the majority of installed systems the integration of the tip of the pyramid, i.e. the company and plant control levels, is missing. Manufacturer-specific hardware and basic software is being rejected at the lowest control level, the actual process control system. Systems which have proved successful in office automation (**Fig. 2**) are used universally except in the open- and closed-loop control level where "real-time processing" is required. There is a decrease in the technical software written for process computers (minicomputers) and an increase in that for the PCs and workstations. The objective of producing "open" systems is therefore drawing closer.

PCs and workstations which are operated with Windows, VMS or UNIX operating systems through self-explanatory graphics user interfaces are now recognized as standard throughout the world. These decentralized individual components are linked to one another through networks such as Ethernet with TCP/IP protocol, whereby all workstations can access standard software such as relational databases with SQL. There are two different schemes with the data bus systems (or networks). Either a single high-capacity bus, usually with design redundancy, links all the computer systems to one another, or else a bus close to the process is used for rapid real-time data exchange, with an additional information bus for the exchange of large quantities of data.

Process control systems for the entire cement works with all plant departments are being developed both by the well-known companies in the electrical industry and the cement plant manufacturers, and also by the cement manufacturers themselves.

Two trends, of mainly historical origin, can be found here:

- The representatives of the DCS systems (Distributed Control Systems) execute the visual display, optimization and information display on workstations, and the open- and closed-loop process control on PLC equipment or process computers.

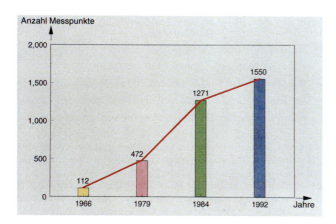

BILD 3: Anzahl der Meßpunkte für 4 charakteristische Zementwerksinstallationen (Angaben Krupp Polysius)
FIGURE 3: Number of measuring points for 4 typical cement works installations (data from Krupp Polysius)

Anzahl Messpunkte	= Number of measuring points
Jahre	= Year

rente, analytische Prozeßführung verlangt, was wiederum eine wesentliche Erhöhung des Informationsbedarfes, auch auf der untersten Automatisierungsebene, erfordert.

4.1 Sensorik

Hinter dem Begriff „Sensorik" verbergen sich die Verfahren zur Erfassung und Verarbeitung von Prozeßdaten. Auf die Vielfalt und die Bedeutung einer zuverlässigen Meßdatenerfassung hat Scheuer [1] hingewiesen.

Generell ist dazu festzustellen, daß sich in der Zementindustrie in den letzten Jahren verschiedene Meßsysteme zur Überwachung der Ofenmanteltemperatur etabliert haben. Über die Erfahrungen mit einem On-line-System für die Rohmaterial-Analyse hat Tschudin [2] näher berichtet.

Andere Wünsche an die Erfassung von Prozeßdaten werden demgegenüber noch nicht befriedigend erfüllt. So werden vor allem On-line-Systeme in Zukunft noch mehr gefragt sein, z. B. für die Messung

— der Abgasmenge,
— der relativen Feuchte von Gasströmen,
— der Sekundärluft-Temperatur und -Menge,
— der Feinheit bzw. Korngrößenverteilung von Zement,
— von geeigneten Zustandsgrößen zur Überwachung der Sinterzone.

Über ein Regelkonzept zur kontinuierlichen Erfassung der Sekundärluft-Temperatur und des Brennstoffenergiestroms hat Edelkott berichtet [3]. Auch für die Überwachung der Sinterzone wurde bereits von Biggs et al. [4] ein neues Meßverfahren vorgestellt. Ob sich videobasierte Bilderkennungs- und Auswertesysteme, z.B. im Bereich Ofenauslauf, durchsetzen werden, bleibt abzuwarten. Laufend an Bedeutung gewinnen die Messungen im Zusammenhang mit dem Umweltschutz.

Verbesserungen finden sich in erster Linie bei der Signalverarbeitung. Schwerpunkte bei den Entwicklungen sind die Kommunikationsschnittstellen der Sensor- und Aktorsysteme.

Bild 4 zeigt eine grobe Übersicht über die Migration von der konventionellen analogen Technik mit Schraubendreher-Parametrierung über die teildigitalisierte Smart-Technik zur digitalen Feldbus-Technik. Aus **Bild 5** geht hervor, daß einer international einheitlichen Feldbus-Technik heute die größten Zukunftschancen eingeräumt werden können.

4.2 Schützenschränke

In den MCC-Schränken sind die Schütze zum leistungsmäßigen Ein- und Ausschalten sowie die Schutz- und Überwa-

— The representatives of the PC/PLC systems, on the other hand, only use personal computers and PLC equipment.

The two types of system are becoming ever closer to one another.

4. Field level

The lowest level in a process control system is the field level (Fig. 1). It contains:

— the sensory technology,
— the contactor cubicles, and
— the cabling

and therefore forms the actual interface to the process, to the machinery and to the environment.

Anything which is not dealt with correctly here can only be corrected with difficulty, if at all, further up the system. As can be seen from **Table 1** and **Fig. 3**, the number of sensors and actuators is increasing out of all proportion with the increasing level of automation due to new designs of plant and machinery and to load-optimized machines and installations. However, there is also a need for a transparent, analytical process control system, which in turn requires a substantial increase in the information requirement, even at the lowest level of automation.

4.1 Sensory technology

The term "sensory technology" covers the processes for acquiring and processing process data. Scheuer [1] has given an indication of the diversity and importance of a reliable system for acquisition of measurement data.

It is also generally found that various measuring systems for monitoring kiln shell temperature have become established in the cement industry in recent years. Tschudin [2] has given a detailed report of the results with an on-line system for raw material analysis.

However, other requirements for the acquisition of process data have not yet been satisfactorily fulfilled. In particular,

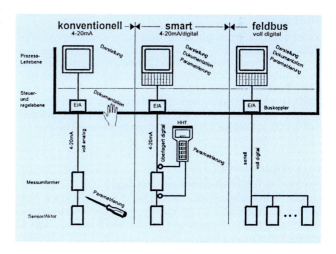

BILD 4: Migration der Sensortechnik von Konventionell über Smart zu Feldbus
FIGURE 4: Migration of sensor technology from conventional, via smart, to field bus

konventionell	= conventional
Feldbus	= field bus
Prozess-Leitebene	= process control level
Darstellung	= display
Dokumentation	= documentation
Parametierung	= parameterization
Steuer- und Regelebene	= open- and closed-loop control level
E/A	= I/O
Buskoppler	= bus coupler
Messumformer	= measuring transducer
voll analog	= fully analogue
Sensor / Aktor	= sensor/actuator
überlagert digital	= digital overlay
seriell	= serial
voll digital	= fully digital

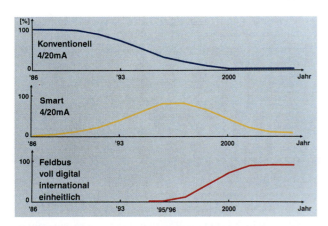

BILD 5: Geschätzter zukünftiger Einsatz der drei Sensor-Techniken
FIGURE 5: Estimated future use of the three sensory technologies

Konventionell	= conventional
Feldbus	= field bus
voll digital	= fully digital
international	= internationally
einheitlich	= standardized
Jahr	= Year

there is going to be an increasing demand for on-line systems, for example, for measuring

— the quantity of exhaust gas,
— the relative moisture content of gas flows,
— the temperature and volume of the secondary air,
— the fineness and particle size distribution of cement,
— suitable variables for monitoring the state of the sintering zone.

Edelkott [3] has described a control scheme for continuous measurement of the secondary air temperature and the fuel energy flow. Biggs et al. [4] have also already suggested a new measuring system for monitoring the sintering zone. It remains to be seen whether video-based image recognition and evaluation systems, e.g. at the kiln outlet region, will become generally accepted. Measurements associated with environmental protection are becoming increasingly important.

Most of the improvements are in signal processing. The developments have focused on the communication interfaces of the sensor and actuator systems.

Fig. 4 outlines the general move from conventional analogue technology with screwdriver parameterization via partially digitalized smart technology to digital field bus technology. From **Fig. 5** it can be seen that it is now conceded that the greatest future prospects lie in an internationally standardized field bus technology.

4.2 Contactor cubicles

The contactors for switching the power circuits on and off, and the protection and monitoring elements for the drive motors, are mounted in the MCC cubicles. The drive-related, modular, standardized structure has become generally accepted here. A change can be detected in the cement industry from the fully compartmentalized drawer technology to the lighter, more economical tray technology (**Fig. 6**).

chungselemente der Antriebsmotoren montiert. Der antriebsbezogene, modulare und standardisierte Aufbau hat sich hier durchgesetzt. Dabei ist in der Zementindustrie ein Wandel von der voll geschotteten Schubladentechnik zur leichteren, wirtschaftlicheren Platinentechnik festzustellen (**Bild 6**).

Die sicherungslose Technik ist heute Standard, und die praktischen Erfahrungen mit der integrierten Bauweise sind sehr positiv. Bei dieser sind Stotz, Sicherheitstrenner, Schütz, Thermik, Kurzschluß-Schutz und Phasenüberwachung in einem Gerät vereint.

Ganz neu auf dem Markt werden solche integrierte Geräte mit eigener Buskopplung angeboten. Durch Schlaufen dieses Busses von Antrieb zu Antrieb wird es in Zukunft möglich sein, alle Signale seriell zum und vom Leitsystem zu übertragen.

Es ist auch denkbar, die heute übliche zentrale Struktur in Zukunft aufzureißen und alle MCC-Funktionen mit den heutigen Vorort-Bedien- und Überwachungsfunktionen beim Antrieb zu installieren, was dann Smart Motor Control, SMC genannt werden könnte. Dazu ist es zur Zeit noch zu früh. Die Übertragungsgeschwindigkeit auf dem Bus ist noch ungenügend, und die Wirtschaftlichkeit in unserer Anlagentopographie kann noch nicht bewiesen werden.

4.3 Verkabelung

Die Zementfabrik ist charakterisiert durch ihre dreidimensionale Weitläufigkeit einerseits und durch ihre klare Aufteilung in Prozeßabschnitte andererseits. Die Verkabelung, insbesondere die Signalverkabelung, muß beiden Eigenschaften Rechnung tragen. Die bis heute bewährte parallele Verkabelung weist in den meisten Fällen eine Struktur auf, wie sie in **Bild 7** dargestellt ist. Für eine typische Produktionslinie muß heute mit ca. 100000–150000 Anschlußpunkten gerechnet werden. Diesem Aspekt wird oft zu wenig Aufmerksamkeit geschenkt.

Bei den digitalen Feldsignalen kann man über die letzten 15 Jahre einen fortschreitenden Wechsel von 220 oder 110 Volt Wechselspannung zu 48 Volt Gleichspannung feststellen.

Mit zunehmender Kapselung und Elektronifizierung der Feldgeräte geht der Trend weiter zu 24 Volt Gleichspannung.

Der Feldbus wird sich generell erst durchsetzen, wenn er einheitlich und international genormt zur Verfügung steht. Das wird noch einige Jahre dauern, denn die Feldbusentwicklungen und die Feldbusstandardisierungen der unterschiedlichen internationalen Interessenvereinigungen ver-

BILD 6: Schützenschrank (MCC) in Platinentechnik
FIGURE 6: Contactor cubicle (MCC) using printed circuit board technology

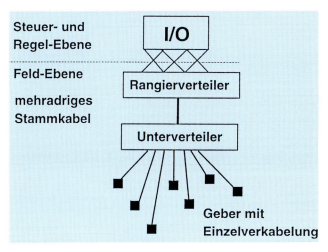

BILD 7: Typische Parallel-Verkabelung
FIGURE 7: Typical parallel cabling

Steuer- und Regel-Ebene	= open- and closed-loop control level
Feld-Ebene	= field level
mehradriges Stammkabel	= multi-core main cable
Rangierverteiler	= marshalling rack
Unterverteiler	= sub-distribution board
Geber mit Einzelverkabelung	= sensors with individual cabling

laufen gegenwärtig noch sehr turbulent und vielschichtig. Es ist abzuwarten, welcher Bus sich im Markt als meistangewendeter zum „De-facto-Standard" entwickeln wird. Die weitergehende Dezentralisierung der Intelligenz und deren Ausnutzung und Lenkung setzt geeignete Kommunikationswege, d. h. Feldbusse, voraus.

Aufgrund der oben angesprochenen Topographie von Zementanlagen sollte das Bussystem die im **Bild 8** dargestellte Baumstruktur zulassen. Bei konsequenter Anwendung der Bussysteme werden die diskreten Ein/Ausgabe-Einheiten der Automatisierungsgeräte (z. B. der SPS) verschwinden. Punkt/Punkt-Verbindungen auf der Feldebene für viele Daten und Informationen über große Distanzen oder durch mit elektromagnetischen Feldern verseuchte Umgebung (z. B. hohe Parallelströme) wurden in letzter Zeit immer mehr durch Glasfaserleitungen realisiert.

5. Steuer- und Regelebene

Auf der sehr prozeßnahen Steuer- und Regelebene werden die grundlegenden Aufgaben der Maschinensteuerung und

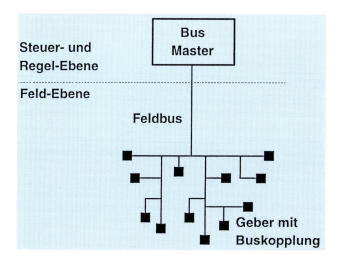

BILD 8: Feldbus mit Baumstruktur
FIGURE 8: Field bus with tree structure

Steuer- und Regel-Ebene	= open- and closed-loop control level
Feld-Ebene	= field level
Bus Master	= master bus
Feldbus	= field bus
Geber mit Buskopplung	= sensors with bus coupling

Fuseless technology is now standard, and very good practical results have been achieved with the integrated designs. These combine safety switches, contactors, thermal protection, short-circuit protection, and open-phase protection in a single piece of equipment.

Integrated equipment of this type with its own bus coupler has recently been offered on the market. By looping this bus from drive to drive it will in the future be possible to transmit all signals serially to and from the control system.

It is also possible that the present normal centralized structure could be eliminated in the future and all MCC functions installed at the drive with the local operating and monitoring functions, which could then be called Smart Motor Control (SMC), but it is still too soon to do this. The transmission speed along the bus is still insufficient, and it is not yet possible to prove its cost-effectiveness in our plant topography.

4.3 Cable installation

A cement factory is characterized both by its three-dimensional extent, and by its clear division into process sections. The cabling, especially the signal cabling, has to take both characteristics into account. In the majority of cases the parallel cabling system which has proved successful so far has the type of structure shown in **Fig. 7**. Nowadays it is necessary to reckon on about 100000–150000 connection points for a typical production line. This aspect often receives insufficient attention.

Over the last 15 years it has been possible to discern a progressive change from 220 or 110 volt a.c. to 48 volt d.c. for the digital field signals.

With increasing encapsulation and eletronification the trend is going further to 24 volt d.c.

The field bus will only become generally accepted if it is available in a consistent and internationally standardized form. This will still take several years as the development and standardization of field buses by the various international associations involved is at present still following a very turbulent and complex path. It still has to be seen which bus will develop to be the most-used in the market and become the "de-facto" standard. Continuing decentralization of intelligence, and its utilization and control, require suitable communication routes, i.e. field buses.

Due to the above-mentioned topography of cement plants the bus system should make the tree structure shown in **Fig. 8** feasible. With consistent application of the bus systems the discrete input/output units of the automation equipment (e. g. the PLC system) will disappear. Increasing use has been made recently of fibre optical lines for point-to-point connections at the field level to deal with large quantities of data and information over great distances and through surroundings affected by electromagnetic fields (e. g. high parallel currents).

5. Open- and closed-loop control level

The basic task of the machinery control system and instrumentation are carried out at the open- and closed-loop control level, which lies very close to the process. Important elements in this are the protection of people and machinery by an appropriately designed machine and motor control system with interlocks, local controls and alarm signals, process automation by sequential controls and path selection, and detailed alarm processing. This level is also used for the acquisition and processing of measurement data, closed control circuits and mix formulation control systems, as well as for preparing the data and information for the next highest level.

The computer-based PLC equipment which has been in use for a number of years makes it possible to program and run the control functions and the instrumentation in a single piece of equipment. The programming can be looked after by the works electrician. This equipment can be used anywhere, has a good price/performance ratio, and works very reliably; the latest generations of the equipment are pro-

der Instrumentierung ausgeführt. Wesentliche Elemente sind hierbei der Schutz von Mensch und Maschine über eine entsprechend ausgelegte Maschinen- und Motorensteuerung mit Verriegelungen, Vorortsteuerungen und Gefahrmeldungen, aber auch die Prozeßautomatisierung durch Sequenzsteuerungen und Weganwahl sowie die detaillierte Alarmaufbereitung. Darüber hinaus werden in dieser Ebene die Meßdatenerfassung und Meßdatenverarbeitung, geschlossene Regelkreise und Rezeptursteuerungen aufgebaut und die Daten und Informationen für die nächst höhere Ebene aufbereitet.

Die seit mehreren Jahren eingesetzten computerbasierten speicherprogrammierbaren Steuergeräte ermöglichen das Programmieren und Abarbeiten der Steuerung und Instrumentierung in einem einzigen Gerätetyp. Die Programmierung kann vom Betriebselektriker gewartet werden. Diese Geräte sind universell einsetzbar, weisen ein günstiges Preis/Leistungsverhältnis auf, sind in den neuesten Gerätegenerationen mit komplexen Reglerfunktionen ausgestattet und arbeiten sehr zuverlässig. Sie verfügen heute auch über sehr schnelle und leistungsfähige Datenbusse für die Kommunikation zwischen den Geräten sowie zu den lokalen Ein-/Ausgangsmodulen (Remote I/O). Für die Automatisierung kann somit ein verteiltes modulares Steuer- und Regelsystem aufgebaut werden, das der Werksstruktur entspricht und die gewünschte hohe Betriebssicherheit und Flexibilität gewährleistet. Der Einsatz von lokalen Ein-/Ausgabemodulen reduziert den Verkabelungsaufwand.

Bild 9 enthält das Beispiel eines dezentralisierten Prozeßleitsystems. Kennzeichnend ist, daß jede Produktionsanlage über ihr eigenes Automatisierungsgerät verfügt. Das einzige zentrale Element des Systems, der zentrale Kommunikationsbus, kann redundant ausgelegt werden. Die lokalen E/A-Module sind in den jeweiligen Schützengerüst-Räumen untergebracht. **Bild 10** zeigt die fünf Automatisierungsgeräte einer ganzen Produktionslinie. Sie sind direkt über dem SPS-Programmierplatz im Elektroraum neben dem zentralen Kommandoraum installiert. **Bild 11** zeigt die lokalen E/A-Module mit der dazugehörigen standardisierten, modularen Rangierverkabelung.

6. Prozeßleitebene

Das oberste Ziel eines Zementwerks ist die kostengünstige Herstellung von Produkten konstanter, definierter Qualität. Dazu ist eine sensible Steuerung notwendig, die trotz Automatismen die genaue Beobachtung und den zeitweiligen Eingriff durch den Prozeßsteuerer, den Prozeßingenieur oder den Produktionsleiter erfordert.

Es ist die Aufgabe der Prozeßleitebene, auf eine möglichst effiziente und ergonomische Art und Weise dieses Anliegen zu ermöglichen. Dieses Fenster im Prozeß wird auch Mensch-Maschinen-Interface (MMI) genannt.

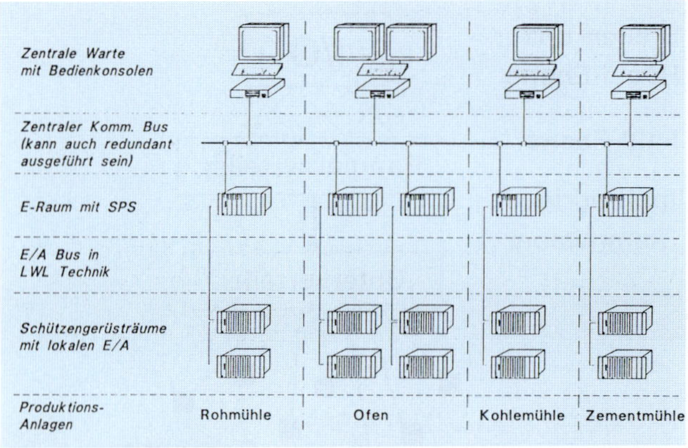

BILD 9: Modulares dezentrales Prozeßleitsystem für die Zementproduktion
FIGURE 9: Modular decentralized process control system for cement production

Zentrale Warte mit Bedienkonsolen	= central control room with operating consoles
Zentraler Komm. Bus (kann auch redundant ausgeführt sein)	= central communication bus (can also have redundant design)
E-Raum mit SPS	= electrical equipment room with PLC system
E/A Bus in LWL Technik	= I/O bus using optical waveguide technology
Schützengerüsträume mit lokalem E/A	= contactor frame rooms with local I/O
Produktions-Anlagen	= production plants
Rohmühle	= raw mill
Ofen	= kiln
Kohlemühle	= coal mill
Zementmühle	= cement mill

vided with complex controller functions. They now also have very rapid and efficient data buses for communication between the pieces of equipment and with the local input/output modules (remote I/O). It this therefore possible to build a distributed modular open- and closed-loop control system for the automation which corresponds to the works structure and ensures the required high level of operational reliability and flexibility. The use of local input/output modules reduces the expenditure on cabling.

Fig. 9 shows an example of a decentralized process control system. A characteristic feature is that each production department has its own automation equipment. The only central element of the system, the central communication bus, can have design redundancy. The local I/O modules are housed in the respective MCC rooms. **Fig. 10** shows the five PLC's for an entire production line. They are installed directly above the PLC programming PC in the electrical equipment room adjacent to the central control room. **Fig. 11** shows the local I/O modules with their associated, standardized, modular marshalling rack.

6. Process control level

The main objective of a cement works is to manufacture products of constant, defined, quality at a favourable price. This needs a sensitive control system which, in spite of automation, requires accurate monitoring and occasional intervention by the operator, the process engineer or the production manager.

It is the task of the process control level to facilitate these matters in the most efficient and ergonomic way possible. This window in the process is also called the man-machine interface (MMI).

It fulfils the functions

— of displaying the process, (nowadays almost exclusively with colour-graphics screens),

— operating the process by switch, keyboard, light pen, touch-screen and mouse or trackball, as well as

— issuing alarms and logging messages.

It also contains the short-term data recording system (up to one week) for displaying historical data on recorders or

BILD 10: SPS Automatisierungsgeräte einer Produktionslinie mit Programmierplatz
FIGURE 10: PLC automation equipment for a production line with programming position

Es erfüllt die Funktionen

- der Visualisierung des Prozesses, heute fast ausschließlich durch Farbgraphik-Bildschirme;
- der Bedienung des Prozesses über Schalter, Tastatur, Lichtgriffel, Touch-Screen und Maus oder Trackball sowie
- der Alarmausgabe und Meldeprotokollierung.

Darüber hinaus beinhaltet es die Kurzzeit-Datenaufzeichnung (bis zu einer Woche) für das Darstellen der historischen Daten auf Schreibern oder Graphikbildschirmen und Druckern und die Datenvorverarbeitung für die oberen Ebenen.

6.1 Ausrüstung der Leitwarte

Im Leitstand laufen alle für die Prozeßführung notwendigen Informationen zusammen. Die immer großräumiger werdenden Regelkreise werden hier entweder von Hand oder automatisch geschlossen. Alle Maschinen werden von hier aus gruppenweise gestartet oder gestoppt. Durch die Zentralisierung der Prozeßführung und die generell erhöhten Anforderungen muß die Flut der Prozeßinformation vorverarbeitet und dem Produktionssteuerer auf Wunsch oder bei Alarmsituationen gezielt angezeigt werden. Dieses „Management by Exception"-Konzept hat sich seit Jahren bewährt. Weil große Flexibilität gefordert wird, werden heute nur noch computergestützte Bildschirmgeräte verwendet, wie beispielhaft im **Bild 12** dargestellt.

Die Computer-Revolution hat auch das Erscheinungsbild der Leitstände radikal verändert. Kennzeichnend ist, daß heute meist nur noch Standard-Hardware eingesetzt wird, die sich in vielen verschiedenen Anwendungen weltweit bewährt hat. Das heißt, der spezialisierte Prozeßrechner wird durch den „General Purpose Computer" (PC oder Workstation) ersetzt. Dieselbe Entwicklung ist auch im Softwarebereich und bei den Netzwerk-Technologien feststellbar. Mehr und mehr werden nur noch im Markt bewährte „Standards" eingesetzt.

Heute befinden sich viele Leitstände noch in einer Übergangsphase, in der eine Kombination von Fließbild- und Bildschirm-Technologie genutzt wird (**Bild 13**). **Bild 14** zeigt eine Leitwarte nach neuestem Stand. Dank des hohen Automatisierungsgrads wird eine ganze Produktionslinie nur noch von einem Produktionssteuerer gefahren und ist ausschließlich mit Bildschirmgeräten ausgerüstet, ohne die Notwendigkeit von Schreibern. Diese Entwicklung wird sich fortsetzen, so daß davon auszugehen ist, daß manche Leitstände nur noch tagsüber „bemannt" sein werden, wie es in der Kraftwerksindustrie, z.B. bei Wasserkraftwerken, heute schon üblich ist.

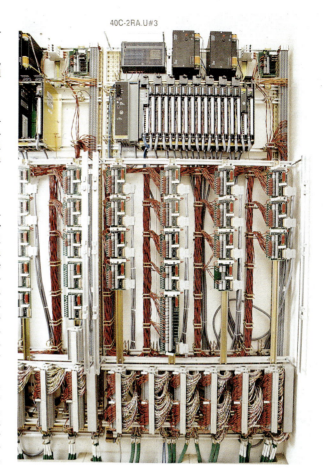

BILD 11: Lokale SPS-E/A-Module mit dem dazugehörenden Rangierverteiler (standardisiert, modularisiert)
FIGURE 11: Local PLC I/O modules with associated marshalling rack (standardized, modular)

graphics screens and printers, and the data preprocessing system for the upper levels.

6.1 Equipment in the control room

All the information needed for guidance and surveillance of the process is brought together at the control centre. This is where the ever more extensive control loops are closed, either manually or automatically. All machines are started and stopped from here in groups. Because of the centralization of the process control and the generally increased demands, the flood of process information has to be preprocessed and displayed selectively to the production controller on request or predefined automatically in alarm situations. This concept of "management by exception" has proved successful for years. Because great flexibility is required, only computer-aided display terminals of the type shown in **Fig. 12** are now used.

The computer revolution has also radically changed the appearance of the control centres. It is a characteristic feature nowadays that standard hardware which has proved its value worldwide in many different applications is normally the only type used. This means that the specialized process computer is being replaced by the "General Purpose Computer" (PC or workstation). The same trend can also be seen in the software sector and in network technology. To an ever increasing extent, market-proven "standards" are the only ones being used.

At present, many control centres are still in a transitional phase, using a combination of mimic panel and screen technologies (**Fig. 13**). **Fig. 14** shows a state-of-the-art control room. Thanks to the high level of automation an entire production line is run by only one operator and is provided exclusively with display terminals, without the need for recorders. This development will progress, so it can be

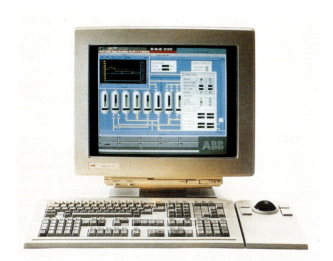

BILD 12: Prozeß-Visualisierung und -Bedienung mittels Bildschirmgerät, Tastatur und Trackball
FIGURE 12: Process display and operation using display terminal, keyboard and trackball

BILD 13: Leitstandwarte in gemischter Technologie (Übergangsphase, PCO Olten)
FIGURE 13: Control room using mixed technology (transition phase, PCO Olten)

BILD 14: Moderner Kommandoraum, ausgerüstet ausschließlich mit Bildschirmgeräten und Tastenfeldern (HCB Rekingen)
FIGURE 14: Modern control room, fitted entirely with display terminals and keyboards (HCB Rekingen)

BILD 15: Kommandoraum mit Großbildschirmen
FIGURE 15: Control room with large screens

Die Wartentechnologie muß auch an diese Anforderungen angepaßt werden, was zu einer weiteren Reduzierung der Informationsflut durch Datenvorverarbeitung sowie der Anzahl der Bildschirme durch „Multiwindow-Technik" und Großbildschirme führen wird (**Bild 15**).

Der fortschreitende Ausbau des Automatisierungsgrades und die weitere Erhöhung der Zuverlässigkeit ermöglichen auch Rationalisierungen im Personalbereich. Die Aufgabenvielfalt für Produktionssteuerer hingegen hat in den letzten Jahren stetig zugenommen. Diesem Umstand muß mit vermehrter und fortlaufender Schulung Rechnung getragen werden.

6.2 Laborautomation

Die Laborautomation hat eine sehr große Bedeutung erlangt. Ihre besondere Aufgabe besteht darin, die qualitätsrelevanten Daten der Eingangsstoffe und der Produkte für eine qualitätssichernde Prozeßführung kontinuierlich zur Verfügung zu stellen. Über den Stand der Laborautomation hat Riedhammer [6] berichtet. Über spezielle Lösungen auf diesem Gebiet gibt es Beiträge von Bonvin et al. [7], Senyuva et al. [8], Fröhlich [9], Teutenberg [10], Triebel [11] und Mtschedlow-Petrossian et al. [12].

7. Produktionsleitebene

Die Funktionen der Produktionsführung haben sich noch nicht überall klar ausgebildet. Sie werden demzufolge noch von vielen Herstellern integrierter Prozeßleitsysteme auf den Computern der Prozeßleitebene realisiert. Heutzutage, mit der Etablierung von schnellen und zuverlässigen Com-

assumed that many control centres will only be "manned" during day time, as is already usual in the power station industry, e.g. with hydroelectric power stations.

Control room technology must also be adapted to these requirements; this will lead to further reduction in the flood of information by data preprocessing, and in the number of the screens by "multiwindow technology" and large screens (**Fig. 15**).

The progressive expansion in the level of automation and the further increase in reliability are also making it possible to rationalize the staffing. However, the diversity of tasks for the operator has increased continuously over the last few years. This must be dealt with by increased and continuous training.

6.2 Laboratory automation

Laboratory automation has become extremely important. Its particular function is to ensure that the data from the raw materials and the products relevant to quality is continuously available to ensure a quality driven process control. Riedhammer [6] has given a report on the present state of laboratory automation, and there are contributions from Bonvin et al. [7], Senyuva et al. [8], Fröhlich [9], Teutenberg [10], Triebel [11] and Mtschedlow-Petrossian et al. [12] on special solutions in this field.

7. Production control level

The functions of production control have not yet developed clearly in all areas. As a consequence many manufacturers of integrated process control systems still execute them on the computers of the process control level. Nowadays, with the establishment of rapid and reliable computer networks, increasing numbers of dedicated computer systems are being used for this range of tasks. Fully automatic running of the rotary kiln, including cooler and mills, has now become completely accepted.

7.1 High level control (HLC)

For more than 20 years attempts have been made to carry out the clinker burning process fully automatically with computer systems. However, with a few exceptions, the attempts to describe and control the process with multi-variable, linear, mathematical models came to nothing. One exception was a scheme which has been described by de Pierpont et al. [13].

The actual breakthrough in the automation of rotary kilns and mills only came with the introduction of new heuristic control methods, such as Fuzzy Logic, rule-based logic and on-line expert systems, coupled with the availability of powerful computers. In Europe alone there are now more than six suppliers of systems for "running" cement kilns and ball

puternetzwerken, werden immer mehr eigenständige Rechnersysteme für diesen Aufgabenbereich eingesetzt. Das vollautomatisierte Fahren des Drehofens inklusive Kühler sowie der Mühlen hat sich heute voll durchgesetzt.

7.1 Höhere Prozeßautomatisierung

Seit über 20 Jahren wird mit Computersystemen versucht, den Klinkerbrennprozeß vollautomatisch durchzuführen. Allerdings schlugen die Versuche, den Prozeß mit multivariablen, linearen mathematischen Modellen zu beschreiben und zu regeln, von wenigen Ausnahmen abgesehen, fehl. Eine Ausnahme stellt ein Konzept dar, über das von de Pierpont et al. [13] berichtet wurde.

Erst die Einführung von neuen heuristischen Regelungsmethoden, wie Fuzzy-Logik, regelbasierender Logik und Online Expertensystemen, gepaart mit der Verfügbarkeit von leistungsfähigen Computern, hat zum eigentlichen Durchbruch bei der Automatisierung von Drehöfen und Mühlen geführt. Es gibt heute allein in Europa mehr als sechs Anbieter von Systemen für das vollautomatische „Fahren" von Zementöfen und Kugelmühlen, über deren praktische Anwendungen sowie Erfahrungen im Detail von Hasler et al. [14], Hepper [15], Kaminski [16] und Pisters et al. [17] berichtet wurde. Die neueste Generation dieser Systeme hat eine vollgraphische Bediener- und Programmieroberfläche und ist somit viel anwenderfreundlicher geworden (**Bild 16**). Die Systeme können daher direkt vom Verfahrensingenieur programmiert und angepaßt werden. Die neuesten Erfahrungen in der Praxis sind sehr positiv, so daß sich z.B. die Holderbank entschieden hat, diese Technik gruppenweit einzusetzen. Selbstlernende (autoadaptive) Prozeßregelsysteme oder Simulatoren scheinen auf der Basis sogenannter neuronaler Netze in weiterer Zukunft möglich zu werden.

7.2 Produktionsplanung

Neben dem manuellen oder automatischen Fahren der verschiedenen Produktionsprozesse wird in der Leitwarte auch bestimmt, wann, was und wieviel produziert werden kann und soll. Die dazu erforderliche Produktionsplanung wird heute noch fast ausschließlich von Hand ausgeführt, wird jedoch computergestützt in naher Zukunft erfolgen.

Die dabei zu berücksichtigenden Parameter sind:

— der Zement(Klinker)verkauf,

— die Lagerbestände von Roh-, Zwischen- und Endprodukten,

— notwendige Instandhaltungsarbeiten,

— Tarif und Verfügbarkeit der Elektro-Energie,

— die Kapazitäten der Produktionsanlagen sowie deren Laufzeiten.

Moderne Zementwerke werden heute mit immer weniger Personal betrieben, die Lagerkapazität ist kleiner (Umlaufvermögen) geworden, und es müssen immer mehr Zementsorten produziert werden. Die Tarifstruktur für elektrische Energie und Leistung wird immer komplizierter. In England gibt es heute Leistungstarifkurven mit Tarifintervallen von 30 Minuten, die sich alle 24 Stunden je nach Nachfrage und Produktionskapazität ändern. Berücksichtigt man zusätzlich die Forderungen nach reduziertem Umlaufvermögen, den Einsatz von Alternativbrennstoffen und denkbare Auflagen im Bereich der Emissionen in Form von Kontingenten, so kann eine künftige Produktionsplanung nur noch mit Hilfe von Computern erstellt werden. Dazu ist der On-line- und Echtzeit-Zugriff zu den Produktionsdaten erforderlich.

8. Informationsebene

Auf der Informationsebene werden die Daten und Informationen weiter verdichtet und entsprechend der verschiedenen Bedürfnisse der Hierarchiestufen aufbereitet. Ein wichtiger Aspekt für das Führen und Optimieren einer Produktionslinie ist das Sammeln und Auswerten aller Daten und Informationen aus dem Prozeß sowie aus allen Hilfsbetrie-

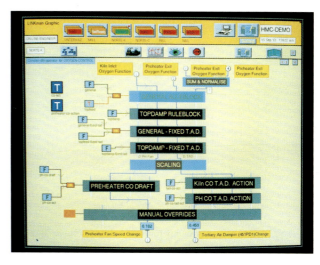

BILD 16: Vollgraphische, objektorientierte Bediener- und „Programmieroberfläche" eines Prozeßregelungssystems der neuesten Generation

FIGURE 16: Full graphics, object-orientated, operator and programming interface for the latest generation of process control system

mills fully automatically; Hasler et al. [14], Hepper [15], Kaminski [16] and Pisters et al. [17] have given detailed descriptions of their practical applications and results. The latest generation of these systems has full graphics operator and programming interfaces and has therefore become much more user friendly (**Fig. 16**). The systems can therefore be programmed and adapted directly by the process engineer. The latest practical results have been excellent, with the result that Holderbank, for example, has decided to use this technology throughout the group. Self-learning (autoadaptive) process control systems or simulators based on so-called neural networks appear to be a possibility in the more distant future.

7.2 Production planning

In addition to the manual or automatic operation of the various production processes, decisions are also made in the control room over when, what, and how much is to be produced. The production planning needed for this is at present still almost entirely carried out by hand, but will become computer-aided in the near future.

The parameters which have to be taken into account are:

— the cement (clinker) sales,

— the stocks of raw, intermediate and end products,

— necessary maintenance work,

— tariff and availability of electrical power,

— the capacities of the production plants and their running times.

Modern cement works are now being operated with fewer and fewer personnel, the storage capacity (current assets) has become smaller, and more and more grades of cement have to be produced. The tariff structure for electrical power is becoming increasingly complex. In England there are now power tariff curves with tariff intervals of 30 minutes which change every 24 hours depending on demand and production capacity. When the requirements for reduced current assets, the use of alternative fuels, and possible instructions in the sphere of emissions in the form quotas, are also taken into account then it will only be possible to approach any future planning with the aid of computers. This requires on-line and real-time access to the production data.

8. Information level

The data and information are further compressed at the information level and processed to suit the various requirements of the hierarchical stages. Another aspect for the operation and optimization of a production line is the collection

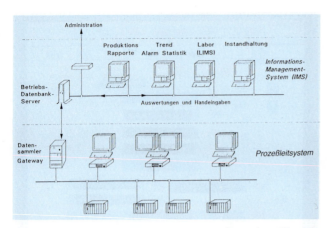

BILD 17: Struktur und Ankoppelung eines Informations-Management-Systems
FIGURE 17: Structure and interfacing of an information management system

Administration	= administration
Betriebs-Datenbank-Server	= plant database server
Datensammler	= data channeller
Produktions Rapporte	= production reports
Trend Alarm Statistik	= trends alarms statistics
Labor	= laboratory
Instandhaltung	= maintenance
Informations-Management-System (IMS)	= Information Management System (IMS)
Auswertung und Handeingaben	= evaluations and manual inputs
Prozeßleitsystem	= process control system

ben. Prozeßleitsysteme messen und zeigen die momentanen Werte und die der näheren Vergangenheit. Mittels Meldungsprotokoll-Listen und Trenddarstellungen läßt sich der Prozeßverlauf um Stunden oder Tage zurückverfolgen. Die Datenspeichermedien reichen aber bei den meisten Prozeßleitsystemen nicht weiter als eine Woche zurück. Um Aussagen und Vergleiche über den historischen Prozeßzustand machen zu können, müssen die Rohdaten in komprimierter und vorverarbeiteter Form bis zu einem Jahr gespeichert und dargestellt werden können. Gerade der Einsatz vollautomatischer Prozeßregel- und Produktionssteuerungssysteme verlangt nach einer verbesserten Analyse der jeweiligen und der vergangenen Prozeßzustände. Aus diesem Grund werden Informations-Management-Systeme (IMS) entwickelt, deren Vorläufer heute bereits in Teilbereichen eingesetzt werden.

Bei modernen IMS-Systemen wird ausschließlich handelsübliche „Standard"-Hard- und Software eingesetzt. **Bild 17** zeigt die mögliche Struktur eines IMS-Systems mit zentralem Datenbank-Server und verteilten Auswertestationen.

Eine zentrale Datenbank muß die automatisch erfaßten Prozeßwerte, die gerechneten Werte und Handeingabewerte bis zu einem Jahr abspeichern. Von mehreren Auswertestationen aus müssen alle Daten vorkonfigurierbar ausgewählt und dargestellt werden können. Die verschiedenen Abteilungen eines Werkes, wie Produktion, Instandhaltung, Labor sowie Umweltschutz und Energie, verlangen dabei nach unterschiedlich ausgewählten Daten. Rapporte und Histogramme bilden die Grundlage für die Überwachung und Optimierung der Produktion. Die Auswerte-Software muß aber auch über gewisse Standardfunktionen verfügen:

— Korrelationsdarstellungen, wie beispielhaft in **Bild 18** dargestellt, und Langzeittrends helfen bei der Prozeßanalyse für die automatische Prozeßführung.

— Statistical Process Control (SPC) erlaubt eine enge und zeitrichtige Überwachung von qualitätsrelevanten Daten.

— Fehlermeldungs-Statistiken, sogenannte Pareto-Charts, unterstützen das Instandhaltungspersonal beim Suchen der Schwachstellen.

Über Details und erste Erfahrungen mit IMS haben Weichinger [18] und Sedlmeir [19] berichtet.

and evaluation of all data and information from the process and from all auxiliary operations. Process control systems measure and indicate the instantaneous values and those from the recent past. By using message protocol lists and trend displays it is possible to backtrack over the course of the process for hours or days. However, in the majority of process control systems the data storage media do not extend back for more than one week. In order to provide information and draw comparisons over historical process conditions it must be possible to store and display the raw data in compressed and preprocessed form for up to a year. The use of fully automatic process control and production control systems requires improved analysis of the present and past process conditions. For this reason information management systems (IMS) are being developed, and their forerunners are already in use in limited areas.

Modern IMS only use normal commercial "standard" hardware and software. **Fig. 17** shows the possible structure of an IMS with central database server and distributed evaluator stations.

A central database has to store the automatically acquired process values, the calculated values and manual input values for up to a year. It must be possible to select and display all data in preconfigured form from several evaluator stations. The various departments in a works, such as production, maintenance, laboratory, as well as environmental protection and power, require data selected in different ways. Reports and histograms form the basis for monitoring and optimizing the production. However, the evaluation software must have certain standard functions:

— correlation displays, like the example shown in **Fig. 18**, and long-term trends help in the process analysis for the automatic process control system.

— Statistical process control (SPC) allows data which is relevant to quality to be monitored closely and at the correct times.

— Fault-message statistics, so-called Pareto charts, assist the maintenance personnel when looking for weak points.

Weichinger [18] and Sedlmeir [19] have described details and initial results with IMS's.

9. Plant control level

As a rule the management functions are now concentrated in a small number of staff, so information must be available rapidly from various plant sections which may also be in separate places or locations.

The information systems for the plant management should, as far as possible, cover all sections and, among other things, should

— provide complete information about the process, the product quality and all environmentally relevant data,

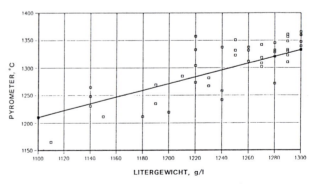

BILD 18: Korrelation zwischen Klinkerqualität und Brennzonentemperatur für die Optimierung eines automatischen Ofenregelsystems
FIGURE 18: Correlation between clinker quality and burning zone temperature for optimizing an automatic kiln control system

Litergewicht = Litre weight

9. Betriebsleitebene

Da sich die Leitungsfunktionen heute in der Regel auf wenige Mitarbeiter konzentrieren, müssen Informationen aus verschiedenen, möglicherweise auch räumlich und örtlich getrennten Betriebsteilen schnell verfügbar sein.

Die Informationssysteme für die Betriebsleitung sollen möglichst abteilungsübergreifend arbeiten und dabei unter anderem

— eine vollständige Information über den Prozeß, die Produktqualität und alle umweltrelevanten Daten liefern,

— die Planung der Instandhaltung und Materialwirtschaft unterstützen,

— die Pflege und Verwaltung von technischen CAD-Dokumenten erleichtern,

— alle kostenrelevanten Uraufschreibungen wie Lohnstunden, kWh-Verbräuche etc. automatisch erfassen, verdichten und an die betriebswirtschaftlichen Kostenrechnungsstellen übergeben,

— beim Versand von Zement die auftrags- und kundenbezogenen Daten erfassen und zur Rechnungsstellung an die übergeordneten Systeme weiterleiten.

Für diese Aufgabenstellung werden Hard- und Softwarelösungen benötigt, die eine durchgängige Datenverarbeitung vom Prozeß bis zur Administration zulassen.

10. Unternehmensleitebene

Wenn an dieser Stelle auch originär über die Zementerzeugung und zementherstellenden Unternehmen gesprochen wird, so muß doch zur Kenntnis genommen werden, daß viele Unternehmen ihre Tätigkeitsfelder diversifiziert haben. So findet man heute in vielen Unternehmen neben der Sparte Zement noch Sparten für Frischbeton, Betonfertigteile, Zuschlagstoffe, Baustoffhandel usw. Diese Situation stellt natürlich auch erhöhte Anforderungen an die Flexibilität und Aktualität der Informationssysteme der Unternehmensleitebene.

Infolge der rasanten Entwicklung im Hardwarebereich gibt es heute im Gegensatz zu den früher propagierten Großrechnern sehr leistungsfähige, preisgünstige Microcomputer (PCs und Workstations) in lokalen und standortübergreifenden Netzwerken (LAN und WAN). Damit eröffnen sich vielfältige neue Einsatzmöglichkeiten: Zum Beispiel durch die Einbeziehung und sinnvolle Verdichtung der von Produktionsanlagen gewonnenen Daten in die administrative Ebene und durch Zugriff auf zentralverwaltete Datenbanken und Register, entweder im Unternehmen oder bei Drittanbietern, z.B. über Umweltbestimmungen, Kredit-Auskunfteien usw. Die Unternehmensleitebene kann mit der ganzen Palette von Systemen und Funktionen des Bereichs Büroautomation kombiniert werden. Allerdings ist die Entwicklung wirklich offener Systeme immer noch viel langsamer als gewünscht.

11. Ausblick

Auch in Zukunft soll Zement und nicht die Information produziert werden. Automation sowie bessere und rechtzeitige Information werden dazu beitragen, daß mehr Qualitätsprodukte kostengünstiger auf den Markt gelangen.

Den geschilderten Forderungen an das Prozeßführungs- und Informationssystem kann durch alte Techniken früher oder später nicht mehr entsprochen werden. Zudem haben moderne Systeme sowohl auf der Hardware- wie auch auf der Softwareseite die unangenehme Eigenschaft der Kurzlebigkeit, da mit einer durchschnittlichen Einsatzdauer von ca. 7–8 Jahren gerechnet werden muß. Die Wartbarkeit muß periodisch überprüft werden. Obsolete Systeme, die niemand mehr kennt und warten kann und für die es keine Ersatzteile mehr gibt, werden immer kostspieliger. Ob sie laufend oder periodisch aufzudatieren sind, ist eine Frage der Strategie. Das bedeutet, daß die Modernisierung von Steuerungs-, Überwachungs- und Informationssystemen in die langfristige Planung einbezogen werden muß und somit

— assist in planning the maintenance and management of materials,

— make it easier to look after and administer technical CAD documents,

— collect automatically all cost-relevant original records such as hours of paid work, kWh consumption, etc., compress them, and transfer them to the works administration cost-accounting offices,

— collect the order- and client-related data when despatching cement, and route it on to the accounting centre in the high-ranking systems.

The nature of this task requires hardware and software solutions which permit continual data processing from the process right up to the administration.

10. Company control level

Although this article deals primarily with cement production and cement manufacturing companies it must be noted that many companies have diversified their activities. In many companies, for example, it is found that there is not only a cement division but also divisions for fresh concrete, precast concrete parts, aggregates, trade in construction materials, etc. This situation naturally also places greater demands on the flexibility and updating capacity of the information systems at the company control level.

In contrast to the large computers which used to be favoured, the vigorous development in the hardware sector has now resulted in very powerful, low-cost microcomputers (PCs and workstations) in local and translocational networks (LAN and WAN). This opens up a variety of possible new applications. These include, for example, the incorporation and appropriate compression of the data obtained from the production plants in the administrative level, or access to centrally administered databases and registers, either in the company or at third-party information providers, e.g. through environmental provisions, credit reference agencies, etc. The company control level can be combined with the whole range of systems and functions from the sphere of office automation. However, the growth of truly open systems is, as ever, still much slower than desired.

11. Prospects

In the future it will still be cement, and not information, that is to be produced. Automation, as well as better and correctly timed information, will contribute to more quality products reaching the market at lower cost.

Sooner or later the old technologies will no longer be able to meet the requirements outlined for the process control and information systems. In addition to this, modern systems have the annoying characteristic of short life, both on the hardware and the software sides; it is necessary to reckon on an average working life of about 7–8 years. Maintainability must be checked periodically; obsolete systems which no-one can recognize or maintain any more and for which there are no longer any spare parts become increasingly costly. Whether they should be updated continuously or periodically is a question of strategy. This means that the modernization of control, monitoring and information systems must be incorporated in the long-term planning, and is therefore a matter for higher management. The technical design and planning must be carried out on the "bottom-up" principle.

During the technical planning phase of a modernization project the works personnel of all levels and disciplines must be included in the decision-making process; experience has shown that this has a lasting beneficial effect on the motivation of the personnel as well as on the acceptability, and hence the availability and degree of utilization, of the new system. However, against this there is often a certain blinkered approach in the plant which ought to be countered by including outside specialists in the project team. Important elements of a modernization project, based on a completed project, can be found in Utzinger [5].

eine Sache des höheren Managements ist. Die technische Konzeption und Planung muß nach dem Prinzip „Bottom-up" gemacht werden.

Während der Ausführungsphase von Modernisierungsprojekten muß das Werkspersonal aller Stufen und Disziplinen in die Entscheidungsfindung einbezogen werden, was sich erfahrungsgemäß nachhaltig positiv auf die Motivation des Personals sowie auf die Akzeptanz und damit auf die Verfügbarkeit und den Nutzungsgrad des neuen Systems auswirkt. Demgegenüber steht jedoch oft eine gewisse Betriebsblindheit, welcher mit der Integration von externen Fachleuten ins Projektteam begegnet werden sollte. Wesentliche Elemente eines Modernisierungsprojektes anhand eines realisierten Projektes finden sich bei Utzinger [5].

Die Dokumentation der Systeme spielt im Zyklus der Modernisierungsprojekte eine immer wichtigere Rolle. Die laufenden Entwicklungen gehen in Richtung der automatischen Selbstdokumentation mit Rückdokumentation und des „systemneutralen Engineering- und Dokumentations-Systems". Dieses sichert den Datenbestand von einer zur nächsten Generation.

Viel Potential für die Zukunft liegt in der „vielbesungenen Öffnung" der Systeme in hersteller-unabhängige, voll integrierbare Systemeinheiten. Auch wird die Zukunft mehr Standards bringen, vor allem auch De-facto-Standards, die durch die Akzeptanz auf dem Markt ihre Auswahl finden. Diese teils erwünschten, teils von außen aufgezwungenen technischen Standards werden dem Unternehmen nur dienlich sein, wenn es gelingt, den vollen Nutzen daraus zu ziehen. Das wiederum wird nur gelingen, wenn die Anforderungen an das Prozeßführungs- und Informationssystem klar definiert sind. Dabei geht es um die mittel- und langfristigen Strategien und die Konzeptionen, die Sache des Managements sind. Die Definition der Anforderungen darf in Zukunft nicht mehr nur dem „Elektriker" oder dem „Administrator" überlassen werden.

Optimale Lösungen und Resultate können nur durch interdisziplinäres Denken und Teamwork erreicht werden. Berücksichtigt werden müssen dabei – wie das im **Bild 19** schematisch dargestellt wird – insbesondere die sich im zeitlichen Wandel der Technologie abzeichnenden Trends in der Aufteilung der Investitions- und Betriebskosten bei steigenden Anforderungen an die Leistungsfähigkeit der Systeme. Ohne auf Einzelheiten einzugehen ist aus der Darstellung zu entnehmen, daß hierbei der Software-Entwicklung und der Personalschulung die größte Aufmerksamkeit zu schenken ist, vor allem auch dann, wenn die von den meisten Zementunternehmen geforderte Unabhängigkeit von Lieferanten gewährt bleiben soll. Das Anforderungsprofil wird sich in modernen Anlagen für viele Arbeitsplätze stark verändern. Der Personalbestand wird immer kleiner und die Aufgabenvielfalt immer größer.

Ein sehr wichtiger, oft vernachlässigter, in Zukunft noch an Bedeutung gewinnender Aspekt ist dabei das Engineering. Mit einer umsichtigen Planung können die zukünftigen Anforderungen an das Anwender- sowie an das Instandhaltungspersonal stark reduziert werden. Der österreichische Schriftsteller Johannes Mario Simmel hat einmal gesagt:

„Je leichter ein Buch zu lesen ist, desto schwerer wurde es geschrieben!"

In Analogie dazu könnte man für ein Zementwerk sagen:

„Je leichter eine Anlage zu fahren und instandzuhalten ist, desto schwerer wurde sie geplant!"

Zeichenerklärung

Symbol	Bezeichnung
Dezentrale E/A	SPS Ein- und Ausgänge in lokalem, vom Prozessor unabhängigen Rackmodul
HLC	(High Level Control) = Automatische Ofen- und Mühlenregelung
IMS	Information Management System

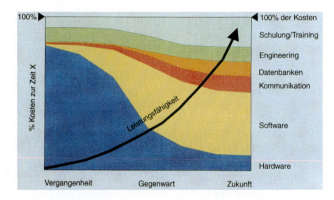

BILD 19: Leistungsfähigkeit der Systeme und Aufteilung der Investitions- und Betriebskosten im zeitlichen Wandel der Technologie (Allen Bradley Co.)
FIGURE 19: System capabilities and breakdown of the investment and operating costs with changing technology (Allen Bradley Co.)

Leistungsfähigkeit	= efficiency
Vergangenheit	= past
Gegenwart	= present
Zukunft	= future
100% der Kosten	= 100% of the cost
Schulung/Training	= instruction/training
Datenbanken	= databases
Kommunikation	= communication

In the cycle of modernization projects an increasingly important part is played by the documentation of the systems. Current developments are moving in the direction of automatic self-documentation – with automatic documentation of changes – and of the "system-neutral engineering and documentation system". This ensures continuity of data from one generation to the next.

Great future potential lies in the "much vaunted opening up" of systems into fully integrated system units which are independent of the manufacturer. The future will also bring more standards, especially de-facto standards which are selected through their acceptability to the market. These industrial standards, which in part are welcome and in part are imposed from outside, will only be useful to the company if it can extract the full benefit from them. This will, in turn, only be successful if the requirements for the process control and information systems are clearly defined. This involves medium- and long-term strategies and schemes, which are matters for higher management. In the future it will no longer be possible to leave the definition of the requirements to the "electrician" or the "manager".

Optimum solutions and results can be achieved only through interdisciplinary thinking and team work. This requires particular attention – as shown diagrammatically in **Fig. 19** – to the emerging trend in the change of technology with time in the breakdown of investment and operation costs with the increasing demands on the efficiencies of the systems. Without going into details it can be seen from the diagram that very great care must be taken with software development and personnel training, especially if the independence from suppliers required by the majority of cement companies is to be maintained. In modern plants there will be sharp changes in requirement profiles for many of the jobs. The staffing level becomes ever lower and the job diversity ever greater.

A very important, often neglected, aspect which will gain in significance in the future, is the engineering. The future demands on the users and maintenance personnel can be greatly reduced with prudent planning. The Austrian writer Johannes Mario Simmel once said:

"The easier a book is to read, the harder it was to write!"

In the same way, it could be said of a cement works:

"The easier a plant is to run and maintain, the harder it was to design!"

MMI	Mensch-Maschinen-Interface (Beobachten und Bedienen)
PC	Personal Computer ISA/EISA Norm
QCS	(Quality Control System) = Labor-System
SPS	Speicher-Programmierbare-Steuerung
WS	Workstation

Explanation of symbols

Symbol	Designation
Decentralized I/O	PLC system inputs and outputs in the local processor-independent rack module
HLC	High Level Control = automatic kiln and mill control system
IMS	Information Management System
MMI	Man-Machine Interface (monitoring and operating)
PC	Personal Computer, ISA/EISA standard
QCS	Quality Control System = laboratory system
PLC	Programmable Logic Controller
WS	Workstation

Literaturverzeichnis

[1] Scheuer, A.: Erfassen und Verarbeiten von Prozeßdaten. Zement-Kalk-Gips 46 (1993) Nr. 11, S. 689–694.

[2] Tschudin, M.: On-line Kontrolle des Mischbettaufbaus mittels PGNAA in Ramos Arizpe (Mexiko). Vortrag zum VDZ-Kongreß '93, Fachbereich 2.

[3] Edelkott, D.: Regelung des Energiestroms zur Drehofenfeuerung. Vortrag zum VDZ-Kongreß '93, Fachbereich 2.

[4] Biggs, H. O., Smith, S., und Hirayama, M.: Überwachung der Sinterzone durch Messung der Alkaliemissions- und -absorptionsspektren in der Gasphase. Vortrag zum VDZ-Kongreß '93, Fachbereich 2.

[5] Utzinger, K.: Projektablauf Modernisierung der elektrischen Ausrüstungen in der Zementfabrik Rekingen. Vortrag zum VDZ-Kongreß '93, Fachbereich 2.

[6] Riedhammer, M.: Laborautomation. Zement-Kalk-Gips 46 (1993) Nr. 11, S. 696–702.

[7] Bonvin, D., Juchli, K., und Yellepeddi, R.: Ein integriertes Röntgen-Diffraktions- und -Fluoreszenzspektrometer und seine Anwendungsmöglichkeiten in der Zementindustrie. Vortrag zum VDZ-Kongreß '93, Fachbereich 2.

[8] Senyuva, Z., Baerentsen, H., und Tokkesdal Pedersen, S.: Labor-Informations-Management-System (LIMS) in Zementwerkslaboratorien. Vortrag zum VDZ-Kongreß '93, Fachbereich 2.

[9] Fröhlich, A.: Quantitative Klinkerphasenanalyse mittels XRD. Vortrag zum VDZ-Kongreß '93, Fachbereich 2.

[10] Teutenberg, J.: Höhere Leistungsdichten im Zementlabor mit multimodularer Robotertechnik. Vortrag zum VDZ-Kongreß '93, Fachbereich 2.

[11] Triebel, W.: Entwicklungsschritte zur Automationstechnik des Werkes Bernburg. Vortrag zum VDZ-Kongreß '93, Fachbereich 2.

[12] Mtschedlow-Petrossian, O. P., und Koschmai, A. S.: Die Anwendung der Methoden instrumenteller Elektrochemie zur Qualitätssicherung der Zementherstellung. Vortrag zum VDZ-Kongreß '93, Fachbereich 2.

[13] de Pierpont, C., Werbrouck, V., van Breusegem, V., Chen, L., Bastin, G., und Wertz, V.: Industrielle Anwendung einer multivariablen linear-quadratischen Steuerung für Zementmühlen. Vortrag zum VDZ-Kongreß '93, Fachbereich 2.

[14] Hasler, R., und Dekkiche, E. A.: Erfahrungen der Firma „Holderbank" mit automatischer Prozeßführung mittels „HIGH LEVEL CONTROL". Vortrag zum VDZ-Kongreß '93, Fachbereich 2.

[15] Hepper, R.: Vorteile der Prozeßführung mit Expertensystem(en). Vortrag zum VDZ-Kongreß '93, Fachbereich 2.

[16] Kaminski, D.: Kostengünstige Produktion eines Klinkers hoher Qualität mit Hilfe eines Expertensystems. Das Beispiel von Ciments Français. Vortrag zum VDZ-Kongreß '93, Fachbereich 2.

[17] Pisters, H., Becke, M., und Jäger, G.: PYROEXPERT — Ein neues Regelungssystem zur Vergleichmäßigung des Klinkerbrennprozesses. Vortrag zum VDZ-Kongreß '93, Fachbereich 2.

[18] Weichinger, M.: Prozeßdatenerfassung als integrierter Bestandteil des Informationssystems der Perlmooser Zementwerke AG. Vortrag zum VDZ-Kongreß '93, Fachbereich 2.

[19] Sedlmeir, W.: Ein zukunftssicheres Management-Informationssystem für Zementwerke. Vortrag zum VDZ-Kongreß '93, Fachbereich 2.

Erfassen und Verarbeiten von Prozeßdaten*)

Acquisition and processing of process data*)

Saisie et traitement de données de process

Registro y tratamiento de los datos de proceso

Von **A. Scheuer**, Leimen/Deutschland

Fachbericht 2.1 · Zusammenfassung – Aus der Sicht der Verfahrenstechnik gehören zum Erfassen und Verarbeiten von Prozeßdaten die Automatisierung des Prozesses sowie seine verfahrenstechnische Optimierung durch betriebliche und bauliche Maßnahmen. Neben dem kontinuierlichen Einsatz von Betriebsmeßgeräten sind hierfür einmalig durchzuführende Untersuchungen erforderlich, die deutlich mehr Information über verfahrenstechnische Abläufe liefern, als das mit einer noch so aufwendigen Routine möglich wäre. Erst wenn eine Anlage verfahrenstechnisch optimiert worden ist und die dafür erforderliche Betriebsweise festliegt, kann und sollte sie automatisiert werden. Letzteres erfordert die Formulierung von Prozeßzielen, zugehörigen Prioritäten und Prozeßstrategien. Die Zielgrößen sind dabei häufig nicht nur von der verarbeitenden Verfahrensstufe selbst abhängig, es können vielmehr auch vorgelagerte oder nachgeordnete Stufen von entscheidendem Einfluß sein. Um das herausfinden zu können, benötigt der Verfahrensingenieur ein Informationssystem, bei dem die Umwelt- und Labordaten schon auf der Informations- bzw. Betriebs-Leitebene zusammengeführt werden. Durch stetigen Abgleich von Planungsvorgaben mit dem aktuellen Produktionsgeschehen werden damit Zusammenhänge transparent gemacht, und es können Prozeßziele und -strategien einzelner Anlagenteile im Sinne übergeordneter Ziele angepaßt werden. Durch seine breite Informationsbasis zur Historie der eigenen Produktion ist ein derartiges Konzept ein wirkungsvolles Instrument der Produktionsplanung, -lenkung und -dokumentation.

Erfassen und Verarbeiten von Prozeßdaten

Special report 2.1 · Summary – From the process engineering point of view the automation of a process and its process engineering optimization by operation and constructional measures are part of the acquisition and processing of process data. In addition to the continuous use of operational measuring equipment this also requires one-off investigations which supply significantly more information about process engineering processes than would be possible with a routine, however sophisticated it was. Only when a plant has been optimized from the process engineering point of view and the mode of operation for this has been established can, and should, it be automated. This last requires the formulation of process targets, associated priorities and process strategies. The target variables are frequently dependent not just on the particular process stage being processed – upstream or downstream stages often also have a critical influence. In order to discover this the process engineer requires an information system in which the environmental and laboratory data are brought together at the information and plant management level. Relationships are made transparent by continuous comparison of the design information with what is currently happening in production; process targets and strategies for individual plant sections can then be adjusted in the direction of higher level objectives. Such a concept is an effective instrument for production planning, steering and documentation through its wide information base on the plant's own production history.

Acquisition and processing of process data

Rapport spécial 2.1 · Résumé – Selon la technologie des procédés, l'automatisation du process et son optimisation grâce à des moyens d'exploitation et mesures constructives, font partie de la saisie et du traitement des données du process. Outre l'utilisation en continu d'appareils de mesure d'exploitation, il s'avère nécessaire de procéder ici, à titre exceptionnel, à des recherches permettant de fournir beaucoup plus d'informations sur les phases d'exploitation du process que ce ne serait le cas avec des opérations de routine, aussi sophistiquées soient-elles. Ce n'est, en effet, que lorsqu'une installation a été optimisée au point de vue process et que le mode de fonctionnement requis a été définitivement fixé qu'elle peut et doit être optimisée. Ce dernier point exige la définition des objectifs du process, les priorités à fixer et les stratégies du process à appliquer. Il arrive alors souvent que les valeurs visées ne dépendent pas seulement de la phase d'exploitation à traiter, il se peut, au contraire, que des phases situées en amont ou en aval jouent un rôle décisif. Pour trouver cela, l'ingénieur de process a besoin d'un système d'informations avec lequel les données de laboratoire et d'environnement sont regroupées à un niveau central d'exploitation ou d'information. Une remise à niveau constante entre les données de planification et les valeurs de la production réelle permet de rendre transparent les rapports, et les objectifs et stratégies du process de

Saisie et traitement de données de process

---

*) Überarbeitete Fassung eines Vortrages zum VDZ-Kongreß '93, Düsseldorf (27. 9.–1. 10. 1993).
Revised text of a lecture to the VDZ Congress '93, Düsseldorf (27. 9.–1. 10. 1993)

différentes parties d'installation isolées peuvent ainsi être adaptés pour parvenir aux objectifs supérieurs. Grâce à sa large base d'informations visant à l'historique de la propre production, ce concept est un instrument efficace pour la planification, la direction et la documentation de la production.

Registro y tratamiento de los datos de proceso

Informe de ramo 2.1 · Resumen – Desde el punto de vista de la tecnología de procesos, forman parte del registro y tratamiento de los datos de proceso la automatización del proceso así como su optimización tecnológica a través de medidas industriales y constructivas. Aparte del empleo continuo de aparatos de medición fiables, se requieren a este respecto unas investigaciones únicas, que suministran bastante más informaciones sobre los procesos tecnológicos que las conseguidas mediante operaciones de rutina, por costosas que sean. Una instalación no puede ni debe ser automatizada hasta que no se haya optimizado su funcionamiento tecnológico y determinado su modo de operación. Esto requiere la determinación de los objetivos, prioridades y estrategias del proceso. Por eso, muchas veces las magnitudes prefijadas no dependen solamente de la etapa tecnológica de tratamiento propiamente dicha, sino puede haber también otras etapas preconectadas o postconectadas, de influencia decisiva. Para averiguarlo, el ingeniero de procesos necesita un sistema informativo, en el que los datos referentes al medio ambiente y los de laboratorio se reúnan ya a nivel de información o de dirección de la explotación. Mediante la comparación continua de los datos de planificación con la situación actual de la explotación, estas relaciones llegan a ser transparentes. De esta forma, se pueden adaptar los objetivos y estrategias de procesos de las distintas secciones de una instalación, teniendo en cuenta los objetivos prioritarios. Gracias a su amplia base de información sobre la historia de la producción propia, semejante concepto representa un instrumento eficaz de la planificación, la dirección y la documentación de la producción.

1. Einleitung

Zahlreiche Forschungsarbeiten auf den Gebieten der Verfahrenstechnik und der Zementchemie haben die Zusammenhänge im Zementherstellungsprozeß weitgehend transparent werden lassen [1]. Vorausschauend und vor allem kontinuierlich angewendet wurden sie bisher jedoch nur wenig, da es sowohl an geeigneten Meßsystemen als auch an geeigneten Informationsmanagement-Systemen fehlte. Mit der Einführung der Digitaltechnik konnte dieser Zustand zumindest teilweise behoben werden [2]. Nunmehr gilt es, das oben genannte Wissen und die Zusammenhänge im Zementherstellungsprozeß im Sinne übergeordneter Zielgrößen umzusetzen. Nach heutigem Erkenntnisstand gehören hierzu als Schwerpunkte die verfahrenstechnische Optimierung des Prozesses durch betriebliche und bauliche Maßnahmen, die Schaffung geeigneter Meßsysteme, die Festlegung der Verfahrensführung in allen Betriebszuständen sowie deren Überwachung im Dauerbetrieb.

Produktqualität und -vielfalt, Umweltschutz sowie steigender Kostendruck erfordern zunehmend eine sensible Betriebsführung bei maximaler Ausnutzung der Stellbereiche. In einem modernen Leitstand, in dem die Produktion und die Qualität aller Produktionsanlagen eines Werks zentral überwacht und gesteuert werden, ist das nicht immer im gewünschten Umfang möglich. Zweckmäßigerweise werden hier deshalb sogenannte Expertensysteme eingesetzt, die den Produktionssteuerer entlasten und den Prozeß im oben genannten Sinne führen helfen.

Die Einführung von Expertensystemen ist nicht unumstritten. Das liegt nicht zuletzt daran, daß beim Kauf der für die Integration des Systems notwendige Aufwand bagatellisiert und damit notorisch unterschätzt wird. Kritisch gestaltet sich insbesondere die Synchronisation der Bedienungsführung von Produktionssteuerern und Rechner. Hierfür ist erheblicher Aufwand bei der Ausbildung und Programmierung erforderlich. Nachfolgend wird ein Konzept beschrieben, das auf bereits angewendeten Verfahrensanweisungen aufbaut und damit diesem Umstand Rechnung trägt. In Verbindung mit einem werksübergreifenden Datenerfassungs- und -verarbeitungssystem ist ein derartiges Konzept auch ein wesentlicher Teil eines modernen Qualitätssicherungssystems.

1. Introduction

A great deal of research in the fields of process technology and cement chemistry has largely explained the interrelationships existing in the cement manufacturing process [1]. However, until now little use has been made of them on a forward-looking or continuous basis due to the lack of suitable measuring systems and suitable management information systems. This situation was at least partially remedied with the introduction of digital technology [2]. This knowledge and these interrelationships must now be implemented in the cement manufacturing process in the context of higher-ranking target variables. According to current perception the emphasis here is on process engineering optimization of the process through operational and constructional measures, the creation of suitable measuring systems, the designation of the process control for all operating conditions, and its monitoring during continuous operation.

Product quality and diversity, environmental protection and rising cost pressures are producing an increasing demand for sensitive plant control systems with maximum utilization of the operating range. In a modern control centre where the production and quality of all the production systems in a works are monitored and controlled centrally this cannot always be achieved to the desired extent. In this case it is more appropriate to use so-called expert systems which relieve the load on the production controller and help him run the process in line with the above-mentioned concepts.

The introduction of expert systems is not without its opponents. This is not least because when the system is purchased the effort needed to integrate it is played down and therefore notoriously underestimated. Synchronization of the operating control of the production controllers and the computers proves to be particularly critical, and requires considerable effort during training and programming. A scheme is described below which builds on process instructions which are already in use and therefore takes this circumstance into account. A scheme of this type, coupled with a data acquisition and processing system covering the entire works, is also an important part of a modern quality assurance system.

2. Ziele einer Teil- oder Vollautomatisierung

Die Automatisierung der Prozesse und Abläufe im Zementwerk und die damit verbundene Informationsbeschaffung und -verarbeitung sind kein Selbstzweck; sie sind ergebnisorientierte Bestandteile investiver Maßnahmen. Folgende Ziele stehen dabei im Vordergrund:

- Die *produktionsintegrierte Qualitätssicherung;* Schlagwörter sind hier die Steigerung der Produktqualität im Hinblick auf homogene und reproduzierbare Produkteigenschaften sowie die Steuerung einer markt- und kundengerechten Produktion.

- Der *produktionsintegrierte Umweltschutz;* hierzu zählen unter anderem die Vermeidung von Emissionen und Abfällen schon am Ort der Entstehung sowie die Vermeidung von Störzuständen.

- Der *wirtschaftliche Nutzen;* er ist Vorbedingung für die Verfügbarkeit der notwendigen Investitionsmittel und gehört selbstverständlich auch zu dieser Auflistung. Erreicht wird er durch Reduktion der Wartungs-, Energie- und Personalkosten sowie durch einen maximalen Zeitanteil des ungestörten Normalbetriebes und damit eine optimale Ausbeute der eingesetzten Rohstoffe und Zwischenprodukte sowie der Anlage selbst.

- Die Realisierung von Automatisierungsaufgaben führt zu einer schrittweisen Umschichtung der Belegschaft mit gleichzeitiger Reduzierung von Arbeitsplätzen. Zwangsläufig entstehen dadurch personelle und soziale Probleme im Betrieb. Ziel einer Automatisierungsmaßnahme ist es also nicht nur, Rationalisierungspotentiale zu schaffen, sondern notwendigerweise auch die damit gegebenen personellen und sozialen Probleme im Umfeld zu lösen. Hierzu gehört insbesondere eine umfangreiche Aus- und Weiterbildung der Mitarbeiter. Automatisierung führt nicht zur menschenleeren Fabrik, sondern zu einer notwendigen Kooperation von vordenkenden Menschen mit den ihre Gedanken fortführenden Anwendern. Das geht nur gut, wenn beide ausreichend voneinander wissen und sich respektieren, ohne daß sie sich mitunter jemals gesehen haben. Aufgrund dieser Abhängigkeit müssen Automatisierungsmaßnahmen zwangsläufig auch zur *Verbesserung der Arbeitsplatzqualität* führen.

3. Voraussetzungen

3.1 Verfahrenstechnische Optimierung

Um Automatisierungsmaßnahmen erfolgreich einführen zu können, sollte der Prozeß verfahrenstechnisch optimiert werden. Neben dem kontinuierlichen Einsatz von Betriebsmeßgeräten sind hierfür insbesondere diskontinuierliche Untersuchungen erforderlich, die deutlich mehr Informationen über verfahrenstechnische Abläufe liefern, als das mit einer noch so aufwendigen Routine möglich ist.

Die Tradition des Vereins Deutscher Zementwerke, hierfür geeignete Meß- und Analyseverfahren zu entwickeln und zu erproben sowie die damit gemachten Erfahrungen in Merkblättern festzuhalten, kann daher nicht hoch genug bewertet werden. Ein typisches Beispiel ist das vom Ausschuß Verfahrenstechnik in 1992 herausgebrachte Merkblatt „Durchführung und Auswertung von Drehofenversuchen" [3]. Gemeinsam mit den Anlagenbauern wurden darin neben der Auswertung insbesondere auch Angaben zur Durchführung solcher Untersuchungen einschließlich wesentlicher Hinweise zur Meß- und Analysentechnik aufgenommen.

Solche ergänzenden Untersuchungen führen dazu, daß der Betreiber seine Anlage im Sinne der vorgenannten Zielgrößen hundertprozentig versteht. Auf dieser Basis können dann im Laufe der Zeit überflüssig gewordene Anlagenteile, Antriebe, Meßstellen und Regelkreise eliminiert, fehlerhaft arbeitende kontinuierliche Meß- und Regelgeräte repariert und der Prozeß im Sinne der Zielgrößen optimiert werden.

2. Objectives of a partial or full automation system

The automation of processes and sequences in a cement works and the associated acquisition and processing of information are not objectives in themselves; they are result-orientated components of an investment. Emphasis is placed on the following objectives:

- *Production-integrated quality assurance;* the key concepts here are increased product quality in respect of homogeneous and reproducible product properties, and the controls for a production system which conforms to market and client requirements.

- *Production-integrated environmental protection;* this includes, among other things, avoiding emissions and waste materials right at the point of origin, and avoiding fault conditions.

- *Cost-effective yield;* this is a requirement if the necessary investment resources are to be available, and clearly also belongs in this list. It is achieved by the reduction of maintenance, energy and personnel costs and through a maximum percentage of time spent in undisrupted normal operation – and hence optimum yield from the raw materials used, from the intermediate products and from the plant itself.

- Putting the automation functions into effect leads to a gradual rearrangement of the staff with a simultaneous reduction in the number of jobs. This inevitably gives rise to personnel and social problems in the plant. The objective of any automation measure is therefore not only to create opportunities for rationalization but also, of necessity, to solve the accompanying personnel and social problems. In particular, this includes extensive education and onward training of the employees. Automation does not lead to deserted factories but to essential cooperation between those who develop the concepts and those who apply them. This is only successful when both parties know enough about each other and respect one another, sometimes without having actually seen each other. Because of this dependency, any automation measures must inevitably also lead to *improvements in job quality.*

3. Requirements

3.1 Process engineering optimization

If automation measures are to be introduced successfully then it is necessary to optimize the process engineering side of the process. In addition to the continuous use of operational measuring equipment there is also a particular need for discontinuous investigations which supply significantly more information about process engineering processes than would be possible with a routine, however sophisticated.

The tradition of the German Cement Works' Association of developing and testing methods of measuring and analysis which are suitable for this purpose and of recording the results in leaflets cannot therefore be too highly praised. A typical example of this is the leaflet "Procedure and evaluation for rotary kiln trials" issued in 1992 by the process engineering committee [3]. This was compiled in cooperation with the plant manufacturers, and in addition to the evaluation, it also contained information on carrying out such investigations, including important advice on methods of measuring and analysis.

These supplementary investigations lead to the situation where the operator has a complete understanding of his plant in the context of the above-mentioned target variables. On this basis it is then possible, with time, to eliminate plant sections, drives, measuring points and control loops which have become superfluous, to repair continuous measuring and control equipment which is operating incorrectly, and to optimize the process in line with the target variables, It is also possible to recognize disturbance variables and limit their effects on the target variables by operational or constructional measures.

Außerdem können Störgrößen erkannt und in ihren Auswirkungen auf die Zielgrößen bereits durch betriebliche oder bauliche Maßnahmen begrenzt werden.

3.2 Zuverlässigkeit der Meßsysteme und Stellorgane

Als weitere Voraussetzung für eine gewinnbringende Automatisierung, ist die Zuverlässigkeit der verwendeten Meßsysteme und Stellorgane sicherzustellen. Im Sinne übergeordneter Prozeßziele kommt es dabei weniger auf absolute Genauigkeit, sondern mehr auf Plausibilität und Reproduzierbarkeit an. Da das auch heute noch eine sehr große Herausforderung ist, sollte für die Prozeßführung vorzugsweise eine minimale Anzahl zuverlässiger Meßgrößen verwendet werden.

Die auf dem Markt angebotene Anzahl der Meßgeräte ist durch die Einführung der Mikroelektronik unüberschaubar groß geworden. In der Regel stehen mehrere Meßprinzipien für die gleichen Meßgrößen zur Verfügung. Die Auswahl des richtigen Sensors fällt daher häufig schon sehr schwer. Zusätzlich müssen betriebliche und verfahrenstechnische Randbedingungen berücksichtigt werden. Einen Leitfaden für die Auswahl von Sensoren kann es daher nicht geben.

Anders als bei den Sensoren besteht bei der Entwicklung von zementwerkstauglichen Meßsystemen noch dringender Handlungsbedarf. Die Möglichkeiten der heutigen Sensortechnik werden bei weitem noch nicht voll ausgeschöpft. Wie aufwendig andererseits die Entwicklung eines solchen Meßsystems sein kann, zeigt Edelkott am Beispiel gemessener Energieströme [4]. Die Entwicklung eines Meßsystems ist in der Regel sehr teuer. Anderseits führt der Kauf eines Systems von der Stange, verbunden mit einer Standardinstallation, fast immer zum Mißerfolg, weil er nicht den Zementwerksbedingungen angepaßt wurde. Mißerfolge auf diesem Gebiet führen aber zwangsläufig zu Akzeptanzverlusten beim Werkspersonal. Das ist als weitere Voraussetzung für eine erfolgreiche Einführung einer Automatisierungsmaßnahme unbedingt zu vermeiden. Bei der Einführung neuer Meßsysteme ist also auch Vorsicht geboten.

4. Prozeßanalyse

Wenn eine Anlage verfahrenstechnisch optimiert worden ist, die Betriebsweise eindeutig festliegt und die dafür erforderlichen Meßtechniken und Stellorgane zuverlässig funktionieren, kann und sollte sie automatisiert werden. Vom Verfahrensingenieur sind hierfür folgende Unterlagen bereitzustellen:

— Ein vollständiges Verfahrensfließbild, aus dem außer dem grundsätzlichen Aufbau der Anlage auch Art und Wirkungsweise der Maschinenelemente ersichtlich sind,

— Eine vollständige Meßstellenliste,

— Eine vollständige Beschreibung der erforderlichen Basis-Regelkreise,

— Die Grobfunktionspläne für die Steuerung und Verriegelung,

— Vollständige Verfahrensanweisungen,

— Prozeßziele mit Vorgaben und Prioritäten,

— Prozeßstrategien.

Die ersten 5 Punkte können in aller Regel auf der Basis bereits vorhandener Unterlagen ausgearbeitet werden.

Prozeßziele, deren Prioritäten sowie Prozeßstrategien sollten demgegenüber aus einer systematischen Betrachtung des Prozesses und unter Berücksichtigung der vorgenannten Zielgrößen folgen. Zu dieser Prozeßanalyse gehört zunächst eine Auflistung und anschließend eine Bewertung der beeinflussenden und der beeinflußten Größen. Nachfolgend wird das am Beispiel des Klinkerbrennprozesses gezeigt.

3.2 Reliability of the measuring systems and control units

Another precondition for profitable automation is ensuring that the measuring systems and control units used are reliable. In the context of the higher-ranking process objectives this is less a matter of absolute accuracy, and more of plausibility and reproducibility. Even now this is still a great challenge, so it is best if the process control system uses the minimum number of reliable measured variables.

With the introduction of microelectronics there are now an enormous number of measuring devices available on the market. As a rule there are several measuring principles available for the same measured variables, so it is often very difficult to select the correct sensor. It is also necessary to take plant and process engineering constraints into account, so no manual can be provided for selecting sensors.

In contrast to the situation with sensors, there is still an urgent need for measuring systems to be developed which are suitable for cement works. The potential of modern sensor technology is still far from being exhausted, but Edelkott has shown how much effort can be required to develop such a measuring system, using the example of measured energy flows [4]. As a rule it is very expensive to develop a measuring system. On the other hand the purchase of an off-the-shelf system, linked with a standard installation, almost always leads to failure because it has not been adapted to cement works conditions, and failure in this area leads inevitably to loss of acceptability by the works personnel. It is essential that this situation is avoided as acceptability is another requirement for successful introduction of automation. Caution is therefore advised when introducing new measuring systems.

4. Process analysis

A plant can, and should, be automated when its process engineering side has been optimized, the mode of operation is clearly established, and the necessary measuring technology and control units are functioning reliably. The process engineer needs to provide the following documents for this purpose:

— a complete process flow diagram which shows not only the basic structure of the plant but also the nature and mode of operation of the mechanical elements,

— a complete list of measuring points,

— a complete description of the basic control loops needed,

— outline logic diagrams for the control and interlock systems,

— complete process instructions,

— process objectives with targets and priorities,

— process strategies.

The first 5 points can normally be prepared from existing documents.

Process objectives, their priorities and process strategies should, however, follow from systematic observation of the process, bearing in mind the above-mentioned target variables. This process analysis includes first a listing and then an evaluation of the influencing and influenced variables. This is illustrated below using the example of the clinker burning process.

4.1 Clinker burning process

4.1.1 Steady-state operation

Fig. 1 summarizes the kiln feed variables which influence the clinker process, while **Fig. 2** gives the influenced variables in the clinker burning process. Seen from the point of view of control technology Fig. 1 contains manipulated and disturbance variables and Fig. 2 contains controlled variables. Each variable in Fig. 1 influences one or more of the variables in Fig. 2. The size of the effects varies from plant to plant and depends on the process technology. In a Lepol kiln, for example, the H_2O content of the nodules, the parti-

4.1 Klinkerbrennprozeß

4.1.1 Stationärer Betrieb

Im **Bild 1** sind die den Klinkerprozeß beeinflussenden Größen des Brennguts und im **Bild 2** die beinflußten Größen des Klinkerbrennprozesses zusammengestellt. Regelungstechnisch gesehen enthält Bild 1 Stör- oder Stellgrößen und Bild 2 Regelgrößen. Jede Größe in Bild 1 beeinflußt eine oder mehrere Größen in Bild 2. Die Auswirkungen sind dabei je nach Verfahrenstechnik und von Anlage zu Anlage unterschiedlich groß. In einem Lepolofen spielen z.B. der H_2O- Gehalt der Granalien, die Korngrößenverteilung und der mineralogische Aufbau des Ofenmehls eine ausschlaggebende Rolle für die Abgaszusammensetzung und den Garbrand des Klinkers. In einem Vorcalcinierofen bewirken demgegenüber Schwankungen des Massenstroms und/ oder des $CaCO_3$-Gehalts besonders starke Veränderungen der Abgaszusammensetzung und des Ofendurchsatzes.

A) Brenngut
- Massenstrom
- $CaCO_3$ -Gehalt
- H_2O-Gehalt
- Kalkstandard
- Silikatmodul
- Tonerdemodul
- Nebenbestandteile (K_2O, SO_3, Cl)
- Korngrößenverteilung
- Mineralogische Herkunft

BILD 1: Beeinflussende Größen des Brennguts

A) Kiln feed
- mass flow
- $CaCO_3$ content
- H_2O content
- lime standard
- silica ratio
- alumina ratio
- secondary constituents (K_2O, SO_3, Cl)
- particle size distribution
- mineralogical origins

FIGURE 1: Influencing variables associated with the kiln feed

- Abgasmassenstrom
- Abgaszusammensetzung
- Abgastemperatur
- Vorentsäuerungsgrad
- Kreisläufe
- Ansätze
- Garbrand des Klinkers
- Verbrennungsluftenergiestrom
- Energieveruststrom
- Klinkerdurchsatz
- Spez. Arbeitsbedarf

BILD 2: Beeinflußte Größen des Klinkerbrennprozesses

- exhaust gas mass flow
- exhaust gas composition
- exhaust gas temperature
- degree of precalcination
- recirculating systems
- coating
- finish burning of the clinker
- combustion air energy flow
- energy flow loss
- clinker throughput
- specific power consumption

FIGURE 2: Influenced variables associated with the clinker burning process

Bei einem Lepolofen kommt es daher insbesondere darauf an, den H_2O-Gehalt der Granalien auf möglichst niedrigem Niveau zu vergleichmäßigen. Gleichzeitig sollte das Brenngut ausreichend fein und möglichst konstant aufgemahlen sein. Die mineralogische Zusammensetzung sollte ebenfalls so wenig wie möglich schwanken.

Bei einem Vorcalcinierofen kommt es demgegenüber auf eine besonders gute Dosiergenauigkeit des Ofenmehls auch im Sekundenbereich und einen konstanten Calciumcarbonatmassenstrom an. Geringe Schwankungsbreiten dieser Größen bewirken einen niedrigen spezifischen Abgasmassenstrom und damit maximale Klinkerdurchsätze.

cle size distribution and the mineralogical structure of the kiln meal play a critical part in the exhaust gas composition and the finishing burn of the clinker. In a precalciner kiln, on the other hand, fluctuations in the mass flow and/or in the $CaCO_3$ content cause particularly severe changes in the exhaust gas composition and kiln throughput.

With a Lepol kiln it is therefore particularly important to keep the H_2O content of the nodules constant at the lowest possible level. At the same time the kiln feed should be ground sufficiently finely and as consistently as possible. The mineralogical composition should also fluctuate as little as possible.

With a precalciner kiln, on the other hand, the important factors are good accuracy of the metered feeding of the kiln feed on a time scale of seconds, and a constant calcium carbonate mass flow. When the range of fluctuation of these variables are low the specific exhaust gas flow is also low, and therefore the clinker throughput is maximized.

Fig. 3 contains the influencing variables associated with the fuel. In this case the uniformity of the fuel mass flows plays a dominant role. Even very slight short-term fluctuations sometimes have a considerable effect on the exhaust gas composition. In such cases the plant has to be operated with an excessively high excess air ratio to avoid CO peaks, which increases the fuel energy consumption due to the increasing exhaust gas losses. Under- and over-control of the supply of energy to the kiln system and continuous changes in the flame shape also cause fluctuations in the levels of the solid phases, the proportions of liquid phase, and the levels of secondary constituents such as K_2O and SO_3 in the clinker. The particle size distributions of the C_3S and C_3A also vary.

The mass and energy flows within the kiln system can also be subject to significant fluctuations (**Fig. 4**). The entry of false air, for example, has a considerable effect on the gas

Bild 3 enthält die beeinflussenden Größen des Brennstoffs. Hier spielt insbesondere die Gleichmäßigkeit der Brennstoffmassenströme eine herausragende Rolle. Schon geringfügige Kurzzeitschwankungen beeinflussen die Abgaszusammensetzung z.T. beträchtlich. In solchen Fällen muß die Anlage zur Vermeidung von CO-Spitzen mit überhöhter Luftzahl betrieben werden, wodurch der Brennstoffenergiebedarf wegen zunehmender Abgasverluste ansteigt. Durch Unter- und Übersteuerung der Energiezufuhr zur Ofenanlage sowie ständiger Veränderung der Flammenform schwanken ferner die Gehalte der festen Phasen, die Schmelzphasenanteile und die Gehalte an Nebenbestandteilen, wie z.B. K_2O und SO_3, im Klinker. Außerdem verändert sich die Korngrößenverteilung des C_3S und des C_3A.

Die Massen- und Energieströme innerhalb der Ofenanlage können ebenfalls deutlichen Schwankungen unterliegen (**Bild 4**). Eintretende Falschluft beeinflußt z.B. den Gasvolumenstrom in der Ofenanlage beträchtlich. Mangelnde Recuperation im Klinkerkühler und/oder reduzierendes Brennen sind dann die Folge.

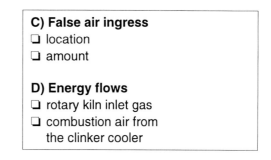

FIGURE 3: Influencing variables associated with the fuel

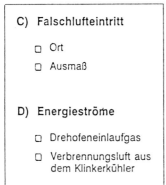

BILD 3: Beeinflussende Größen des Brennstoffs

FIGURE 4: Influencing variables associated with false air ingress and energy flows

BILD 4: Beeinflussende Größen des Falschlufteintritts und der Energieströme

Zwischen Brenngutvorwärmer und Drehofen ist es der Energiestrom des Drehofeneinlaufgases, der sich aufgrund von Veränderungen sowohl der Gastemperatur als auch der Enthalpie verdampfter und/oder dissoziierter Verbindungen, insbesondere KCl und $CaSO_4$, im Gas in starkem Maß verändern kann. Da die übrigen Energieströme des Brenngutvorwärmers bei unveränderter Brenngut- und Brennstoffdosierung praktisch konstant sind, wirken sich solche Schwankungen ausschließlich auf den Grad der Entsäuerung aus, mit dem das Brenngut in den Drehofen eintritt. Veränderungen des Vorentsäuerungsgrads beeinflussen unmittelbar den Energiebedarf und das Fließverhalten des Brennguts im Drehofen. Weitere Folgen sind ungleichmäßig gebrannter Klinker und sich verändernde Ansatzverhältnisse im Drehofen mit negativen Auswirkungen auf den Betrieb der Ofenanlage.

Der Energiestrom der vorgewärmten Verbrennungsluft aus dem Klinkerkühler ist ebenfalls von entscheidender Bedeu-

volume flow in the kiln system and results in inadequate recuperation in the clinker cooler and/or reducing burning conditions.

The energy flow of the rotary kiln inlet gas between the kiln feed preheater and the rotary kiln can vary sharply because of changes not only in the gas temperature but also in the enthalpy of vaporized and/or dissociated compounds, especially KCl and $CaSO_4$, in the gas. With unchanged metered feeding of the kiln feed and fuel the other energy flows in the kiln feed preheater are practically constant, so such fluctuations act exclusively on the degree of calcination with which the kiln feed enters the rotary kiln. Changes in the degree of precalcination have a direct effect on the energy consumption and flow behaviour of the kiln feed in the rotary kiln. Other consequences are unevenly burnt clinker and changing coating conditions in the rotary kiln, with detrimental effects on the operation of the kiln system.

The energy flow of the preheated combustion air from the clinker cooler is also critically important to the kiln operation. Fluctuations in this energy flow also produce changes in the high temperature zone of the rotary kiln in a way that cannot be neglected. The combustion air energy flow depends on the mass flow of the hot clinker, its temperature and the degree of energy recuperation in the clinker cooler. With constant particle size distribution of the clinker the latter depends in turn on the capacity flow ratio of combustion air to hot clinker.

Changes in the combustion air energy flow cause fluctuations in the clinker phase composition and in the particle size distributions of the C_3S and C_3A. The levels of secondary constituents such as K_2O and SO_3 in the clinker also fluctuate. With uncontrolled supply of fuel energy the fluctuations also result in an increase in the fuel energy required for stable sintering which is as complete as possible. There is sometimes also a considerable increase in the NO_x emissions.

In the context of the objectives mentioned in Section 2 any changes to the influencing variables should be avoided as far as possible. Blending beds and homogenizing equipment for the raw materials and by-pass plants for relieving undesirable recirculating systems are suitable for this purpose.

tung für den Ofenbetrieb. Schwankungen dieses Energiestroms bewirken in nicht zu vernachlässigender Weise auch Veränderungen in der Hochtemperaturzone des Drehofens. Der Verbrennungsluftenergiestrom hängt vom Heißklinkermassenstrom, dessen Temperatur und dem Energierückführungsgrad des Klinkerkühlers ab. Letzterer ist bei konstanter Korngrößenverteilung des Klinkers wiederum vom Kapazitätsstromverhältnis Verbrennungsluft/Heißklinker abhängig.

Veränderungen des Verbrennungsluftenergiestroms haben Schwankungen der Phasenzusammensetzung im Klinker und der Korngrößenverteilung des C_3S sowie des C_3A zur Folge. Desweiteren schwanken auch die Gehalte an Nebenbestandteilen, wie z. B. K_2O und SO_3, im Klinker. Bei ungeregelter Brennstoffenergiezufuhr haben die Schwankungen außerdem einen Anstieg des Brennstoffenergiebedarfs zur Folge, der für eine dauerhafte und möglichst vollständige Sinterung bereitgestellt werden muß. Darüber hinaus steigt die NO_x-Emission z. T. beträchtlich an.

Im Sinne der im Abschnitt 2 genannten Ziele sollten Veränderungen der beeinflussenden Größen soweit wie möglich vermieden werden. Dazu sind Mischbetten und Homogenisierungseinrichtungen für die Einsatzstoffe sowie Bypaßanlagen zur Entlastung von unerwünschten Stoffkreisläufen geeignet. Da diese vorbeugenden Maßnahmen häufig nicht ausreichen, sollten verbleibende Veränderungen durch geeignete Stelleingriffe ausgeglichen werden. Je nach Prozeßstrategie und Verfahrenstechnik ergeben sich hieraus jedoch sehr unterschiedliche Prozeßziele.

These preventive measures are often inadequate so any remaining changes have to be offset by appropriate control actions. However, this results in widely differing process objectives depending on the process strategy and process technology.

Process strategy: Low-alkali clinker

Fig. 5 lists those essential process objectives for a precalciner kiln with by-pass which could be relevant for the manufacture of, for example, low-alkali clinker. These include the NO content in the exhaust gas, the hot meal temperature, the alkali content of the clinker, and the energy flow in the rotary kiln firing system.

The NO concentration in the exhaust gas is a rapid and reliable indicator of the combustion temperature in the rotary kiln firing system, so it can be used for controlling the evaporation of alkalis from the kiln feed. **Fig. 6** shows the relationship measured by Hansen [5] between the NO concentration and the K_2O content of the clinker, using the example of a precalciner plant with by-pass. The upper line applies to a normal raw material and the lower line to a raw material to which 0.05% chloride in the form of calcium chloride has been added to lower the K_2O in the clinker. It can be seen from this diagram that in both cases there is an almost linear relationship between NO concentration and K_2O content. In this case the NO concentration lay between 800 and 2200 ppm. Even at about 300 ppm NO the free lime was less than one percent, but the free lime was of secondary importance here.

Wesentliche Prozeßziele

❏ NO-Gehalt im Abgas größer x_1 ppm
❏ Heißmehltemperatur größer x_2 °C
❏ Alkali-Gehalt im Klinker kleiner x_3 M.-%
❏ Energiestrom Drehofenfeuerung gleich x_4 kJ/kg Kli

BILD 5: Wesentliche Prozeßziele für die Herstellung von Niedrig-Alkali-Klinker in einem Vorcalcinierofen mit Bypaß

Prozeßstrategie: Niedrig-Alkali-Klinker

Bild 5 enthält für einen Vorcalcinierofen mit Bypaß die wesentlichen Prozeßziele, die z. B. für die Herstellung von Niedrig-Alkali-Klinker relevant sein können. Hierzu zählen der NO-Gehalt im Abgas, die Heißmehltemperatur, der Alkali-Gehalt im Klinker und der Energiestrom der Drehofenfeuerung.

Da die NO-Konzentration im Abgas ein schneller und zuverlässiger Indikator für die Verbrennungstemperatur in der Drehofenfeuerung ist, kann sie zur Steuerung der Alkaliausdampfung aus dem Brenngut genutzt werden. **Bild 6** zeigt dazu einen von Hansen [5] gemessenen Zusammenhang zwischen der NO-Konzentration und dem K_2O-Gehalt des Klinkers an einer Vorcalcinieranlage mit Bypaß als Beispiel. Die obere Linie gilt für ein normales Rohmaterial und die untere Linie für ein Rohmaterial, dem zur Absenkung von K_2O im Klinker 0,05 % Chlorid als Calciumchlorid zugegeben wurde. Aus dem Bild geht hervor, daß in beiden Fällen ein nahezu linearer Zusammenhang zwischen der NO-Konzentration und dem K_2O-Gehalt bestand. Die NO-Konzentration lag dabei zwischen 800 und 2200 ppm. Ein Prozent Freikalk wurde schon bei ca. 300 ppm NO unterschritten. Der Freikalkgehalt war hier jedoch von untergeordneter Priorität.

Eine hohe Heißmehltemperatur bedeutet einen hohen Zweitbrennstoffenergieanteil und damit geringe spezifische Drehofenabgasmengen bei gleichzeitig hohen Alkaligehalten im Drehofenabgas. Für einen niedrigen Energieverlust durch die fühlbare Wärme des Bypaßgases ist des-

---

Important process objectives

❏ NO content in the exhaust gas greater than x_1 ppm
❏ hot meal temperature greater than x_2 °C
❏ alkali content in the clinker less than x_3 wt.%
❏ energy flow, rotary kiln firing system, equal to x_4 kJ/kg clinker

FIGURE 5: Important process objectives for the production of low-alkali clinker in a precalciner kiln with by-pass

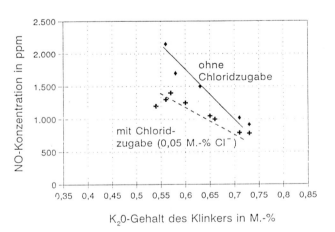

BILD 6: NO-Konzentration als Funktion des K_2O-Gehaltes im Klinker bei der Herstellung von Niedrig-Alkali-Klinker mit und ohne Chloridzugabe [5]

FIGURE 6: NO concentration as a function of the K_2O content of the clinker when producing low-alkali clinker with and without the addition of chloride [5]

NO-Konzentration in ppm = NO concentration in ppm
K_2O-Gehalt des Klinkers in M.-% = K_2O content in the clinker in wt.%
ohne Chloridzugabe = without chloride addition
mit Chloridzugabe (0,05 M.-% Cl⁻) = with chloride addition (0.05 wt.% Cl⁻)

halb eine möglichst hohe Heißmehltemperatur wünschenswert. Ihre Kontrolle dient somit der Begrenzung der Energieverluste.

Der Alkaligehalt im Klinker ist bei der vorgegebenen Prozeßstrategie das Prozeßziel mit der höchsten Priorität. Er folgt aus Laboruntersuchungen an Stichproben und steht erst mit erheblicher Zeitverzögerung zur Verfügung. Die NO-Konzentration im Drehofenabgas zeigt anstehende Änderungen demgegenüber schon schneller und vor allem kontinuierlich an. Allerdings liefert sie noch keine eindeutige Aussage über den Zustand der Sinterzone. Außerdem korreliert sie mit der Regelgröße „Alkali-Gehalt im Klinker" häufig nur ungenügend. Eine vollständige Überwachung folgt dagegen mit der Ersatzgröße „Energiestrom Drehofenfeuerung" [4]. Sie beinhaltet den Brennstoffenergiestrom der Drehofenfeuerung und den Verbrennungsluftenergiestrom aus dem Klinkerkühler und damit auch die Einflüsse durch den Massenstrom, die Temperatur und die Korngrößenverteilung des Klinkers aus der Sinterzone. Bei konstantem Energiestrom zur Drehofenfeuerung auf möglichst hohem Niveau bei gleichzeitig hohem NO_x-Gehalt lassen sich somit in einem gewissen Rahmen gezielt Alkaligehalte im Klinker einstellen. Zu hoher Energiestrom führt demgegenüber zu einer Überhitzung der Sinterzone, die sich häufig erst zu spät durch einen Anstieg der NO-Konzentration bemerkbar macht.

Prozeßstrategie: Maximaler Klinkerdurchsatz

Bild 7 enthält die Prozeßziele eines konventionellen Zyklonvorwärmerofens, der mit maximalem Klinkerdurchsatz betrieben werden soll. Das Abgasgebläse wird in solch einem Fall ständig mit maximaler Drehzahl gefahren. Die Brennstoffmenge wird in Abhängigkeit vom CO-Gehalt eingestellt. Hilfsmittel ist somit wiederum die Gasanalyse im Abgas. Allerdings wird jetzt der CO-Gehalt verwandt.

Wesentliche Prozeßziele

- Abgasmassenstrom maximal
- CO-Gehalt im Abgas zwischen x_1 und x_2 ppm
- CaO_{frei}-Gehalt im Klinker zwischen x_3 und x_4 M.-%

BILD 7: Wesentliche Prozeßziele für die Einhaltung einer maximalen Produktionsleistung in einem konventionellen Zyklonvorwärmerofen ohne Bypaß

Der Zusammenhang geht aus **Bild 8** hervor, das beispielhaft die CO-Gehalte und den Klinkerdurchsatz bei unterschiedlichen O_2-Gehalten im Drehofeneinlaufgas zeigt. Aus dem Verlauf der Kurven geht hervor, daß die CO-Konzentration mit abnehmendem O_2-Gehalt stark zunimmt. Diese gegenüber der O_2-Konzentration ausgeprägte Abhängigkeit der CO-Konzentration kann dazu genutzt werden, dem Ofen ständig die maximale Brennstoffenergiemenge zuzuführen. Der Durchsatz der Ofenanlage wird dann in Abhängigkeit vom Freikalkgehalt hinterhergefahren. Im vorgestellten Beispiel konnte der Klinkerdurchsatz so durch einen um ca. 500 ppm erhöhten CO-Gehalt immerhin von ca. 1650 auf ca. 1780 t/d gesteigert werden. Letzteres ist insbesondere auf die deutlich verminderten Abgasverluste zurückzuführen.

Die im Bild gezeigte Abhängigkeit der CO-Konzentration vom O_2-Gehalt ist für jede Ofenanlage neu festzulegen. Vor ihrer Anwendung sollten allerdings Brennstoffdosierung und -förderung, Einmischung des Brennstoffs in die Verbrennungsluft und Abdichtung gegen Falschlufteinbruch auf der Basis von verfahrenstechnischen Untersuchungen schon so optimiert worden sein, daß die CO-Kurve möglichst weit links liegt. Dadurch verbessert sich die Anwendbarkeit der Prozeßstrategie maßgeblich, und es kann der Anlage bei gleichem Abgasmassenstrom deutlich mehr Brennstoff zugeführt werden.

A high hot meal temperature signifies a high proportion of secondary fuel energy and hence low specific quantities of rotary kiln exhaust gas containing high levels of alkali. The highest possible hot meal temperature is therefore desirable for a low energy loss through the sensible heat in the by-pass gas, so it is checked to limit the energy losses.

For the given process strategy the alkali content in the clinker is the process objective with the highest priority. It is obtained from laboratory tests on spot samples and is only available after a considerable time delay. However, incipient changes are shown much sooner and, above all, continuously, by the NO concentration in the rotary kiln exhaust gas. However, this still does not provide any definite information about the conditions in the sintering zone. Apart from this, it often has inadequate correlation with the controlled variable "alkalic content in the clinker". On the other hand full monitoring is achieved using the substitute variable "energy flow, rotary kiln firing system" [4]. This covers the fuel energy flow from the rotary kiln firing system and the combustion air energy flow from the clinker cooler, and therefore also includes the effects of the mass flow, the temperature and the particle size distribution of the clinker from the sintering zone. With constant energy flow to the rotary kiln firing system at the highest possible level coupled with a high NO_x content it is then possible to set specific alkali levels in the clinker within certain limits. On the other hand, an excessively high energy flow leads to overheating of the sintering zone, which is often only shown too late by an increase in the NO concentration.

Process strategy: Maximum clinker throughput

Fig. 7 gives the process objectives for a conventional cyclone preheater kiln which is to be operated at maximum clinker throughput. In such a situation the exhaust gas fan is run at maximum speed at all times. The quantity of fuel is adjusted on the basis of the CO level. The gas analysis in the exhaust gas is therefore again an aid, but it is now the CO content which is used.

Important process objectives

- maximum exhaust gas mass flow
- CO content in the exhaust gas between x_1 and x_2 ppm
- $CaO_{(free)}$ content in the clinker between x_3 and x_4 wt.%

FIGURE 7: Important production objectives for maintaining maximum production capacity in a conventional cyclone preheater kiln without by-pass

The relationship can be seen from **Fig. 8** which shows an example of CO content and clinker throughput at different levels of O_2 in the rotary kiln inlet gas. The shapes of the curves show that the CO concentration increases sharply with decreasing O_2 content. This pronounced dependence of the CO concentration on the O_2 can be used for supplying the kiln with the maximum quantity of fuel energy at all times. The throughput of the kiln system is then controlled as a function of the free lime content using a feed-back system. By increasing the CO content by approximately 500 ppm in the example described, the clinker throughput was raised from about 1650 to about 1780 t/d. This is mainly attributable to the significantly reduced exhaust gas losses.

The dependence of the CO concentration on the O_2 content shown in the diagram has to be determined separately for each kiln system. However, before it is put to use the fuel metering and conveying system, the blending of the fuel into the combustion air, and the seal against ingress of false air should have been optimized on the basis of process engineering investigations so that the CO curve lies as far to

Die Abhängigkeit der CO-Konzentration vom O_2-Gehalt ist ein typisches Beispiel dafür, daß der Anwendung einer automatisierten Prozeßführung zunächst eine verfahrenstechnische Optimierung vorausgehen sollte. Letzteres gilt im übrigen auch für die zuvor erläuterte Steuerung der Heißmehltemperatur, die bei Mehldurchschuß in der untersten Zyklonstufe nur eine Vergleichmäßigung eines schlechten Betriebspunktes bedeutet. Unter Berücksichtigung solcher Einschränkungen können jedoch einfache Vorgaben von Prozeßzielen den Zeitanteil des ungestörten Normalbetriebs durch rechtzeitiges Eingreifen in den Prozeß deutlich erhöhen.

4.1.2 Instationärer Betrieb und Notfälle

Für den instationären Betrieb und für Notfälle ist der Aufwand sehr viel größer. **Bild 9** veranschaulicht das am Beispiel eines Lepolofens. Die Vorgaben für den stationären Betrieb beschränken sich hier auf die Granulation, die Abgaszusammensetzung, den Betrieb des Rostvorwärmers, den Garbrand des Klinkers und den Betrieb des Klinkerkühlers. Im instationären Betrieb sind demgegenüber Vorgaben für „Aufheizen", „An- und Abfahren", „Kasten-Putzen", „Einlauf/Gewölbe-Putzen", „Kühler voll", „Sinterzone überhitzt", „Schwachbrand", „Trockenkammer zu heiß" usw. erforderlich. Notfälle können entstehen durch Stromausfall, Maschinendefekt, „Feuerfest"-Schaden oder Notstop. Beim Aufheizen kann desweiteren unterschieden werden zwischen Aufheizen aus dem kalten Zustand mit neuer Feuerfest-Ausmauerung, Aufheizen aus dem kalten Zustand ohne neue Feuerfest-Ausmauerung oder Aufheizen aus dem warmen Zustand.

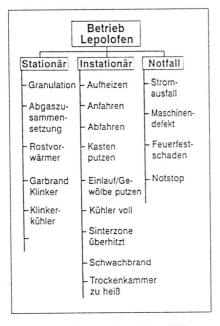

BILD 9: Konzept eines übergeordneten Prozeßführungssystems für den stationären und instationären Betrieb sowie für den Notfall am Beispiel eines Lepolofens

Diese Liste ließe sich beliebig fortsetzen. Trotzdem oder gerade deshalb ist eine systematische Integration der dazugehörigen Prozeßziele und -strategien in eine begleitende oder automatisierte Prozeßführung als Unterstützung der Produktionssteuerer immer lohnenswert. Im Sinne der anfangs aufgelisteten Zielgrößen führt sie zu einem in allen Betriebszuständen beherrschten Prozeß.

4.2 Weitere Anwendungsgebiete

Weitere klassische Anwendungsgebiete der automatisierten Prozeßführung sind:

— die Rohmehlmahlung,

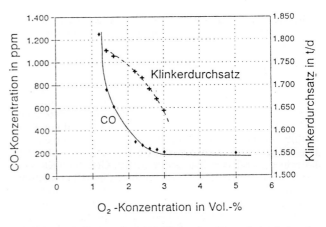

BILD 8: CO-Konzentration und Klinkerdurchsatz als Funktion der O_2-Konzentration im Drehofeneinlaufgas

FIGURE 8: CO concentration and clinker throughput as functions of the O_2 concentration in the rotary kiln inlet gas

CO-Konzentration in ppm	= CO concentration in ppm
O_2-Konzentration in Vol.-%	= O_2 concentration in vol.%
Klinkerdurchsatz	= clinker throughput
Klinkerdurchsatz in t/d	= clinker throughput in t/d

the left as possible. This make a considerable improvement in the usability of the process strategy, and significantly more fuel can be fed to the system for the same exhaust gas mass flow.

The dependence of the CO concentration on the O_2 content is a typical example of the fact that the application of automated process control should be preceded by process engineering optimization. The latter is in fact also true of the control system for the hot meal temperature explained above which, in the case of meal flushing through into the bottom cyclone stage, is just a question of sorting out a weak point in the plant. However, by taking such limitations into account and making timely interventions in the process, simple targets for process objectives can make a significant increase in the proportion of time spent in undisturbed normal operation.

4.1.2 Unsteady-state operation and emergencies

The expenditure for unsteady-state operation and emergencies is very much greater. This is illustrated in **Fig. 9** using

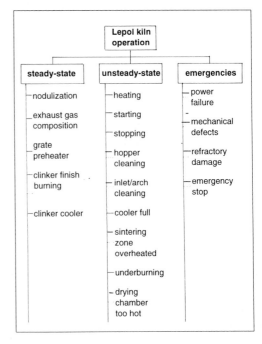

FIGURE 9: Scheme for a higher-ranking process control system for steady-state and unsteady-state operation and for emergencies, using the example of a Lepol kiln

— die Klinkerlagerung (Stichwort ist hier das Sortieren von Klinker),
— die Zementmahlung und
— das Ansteuern der Zementsilos, d. h. die Zementlagerung.

Die Prozeßanalyse erfolgt wie beim Klinkerprozeß in Anlehnung an die verfahrenstechnischen und betrieblichen Randbedingungen und Belange des jeweiligen Werks.

4.3 Übergeordnete Prozeßstrategien

Die Verknüpfung der in Abschnitt 4.1 und 4.2 aufgelisteten Verfahrensstufen im Sinne übergeordneter Prozeßstrategien bietet weitere langfristige Optimierungsmöglichkeiten.

Die NO_x-Emission hängt z.B. von der Rohmehlmahlung, dem Klinkerbrennprozeß und der Qualitätssteuerung ab. Bei der Rohmehlerzeugung sind es insbesondere die Gleichmäßigkeit des Kalkstandards und die mineralogische Herkunft des Rohmehls, die einen deutlichen Einfluß auf die NO_x-Emission ausüben können. In der Drehofenfeuerung zählen dazu der Gasgehalt und die Korngrößenverteilung des Brennstoffs, die Kurzzeitkonstanz des Brennstoffmassenstroms, die Brennstoffvermischung und damit die Luftzahl, der Energiestrom aus Verbrennungsluft und Brennstoff sowie die Kreisläufe an Nebenbestandteilen. In der Zweitfeuerung sind es die Art und Zugabestelle des Brennstoffs. Indirekt beinflussen auch Qualitätserfordernisse, wie C_3S- und C_3A-Gehalt des Klinkers, die NO_x-Emission. Durch langfristiges Erfassen und Verarbeiten dieser Prozeßdaten können die wesentlichen Einflußgrößen herausgearbeitet und die Abhängigkeiten im Sinne niedriger NO_x-Emissionen angewendet werden.

Die Zementfestigkeit kann als weiteres Beispiel in praktisch allen Verfahrensstufen beeinflußt werden. Sie hängt u. a. ab vom Kalkstandard und der mineralogischen Herkunft des Rohmehls, der Gleichmäßigkeit des Energiestroms zur Drehofenfeuerung, der Luftzahl und den Kreisläufen an Nebenbestandteilen im Drehofen, der qualitätsorientierten Sortierung des Klinkers, der Rezeptur des Zements, dem Aufgabegutmassenstrom zur Mühle, der Korngrößenverteilung des Zements und der Temperatur in der Mühle.

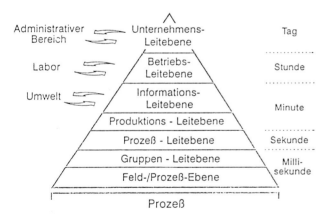

BILD 10: Datenverbund von Prozeß, Umwelt, Labor und Administration als Instrument der Produktionsplanung, -lenkung und -dokumentation

Einzelziele sind somit nicht nur von der verarbeitenden Verfahrensstufe selbst abhängig. Es können vielmehr auch vorgelagerte oder nachgeordnete Stufen von entscheidendem Einfluß sein. Um das herauszufinden und die Einzeldaten sinnvoll verarbeiten zu können, wird ein Informationssystem benötigt, bei dem die Umwelt- und Labordaten schon auf der Informations- bzw. Betriebs-Leitebene zusammengeführt werden. **Bild 10** zeigt den von Säuberli u. a. in [2] vorgestellten Datenverbund für diese Art von Produktionssteuerung mit den einzelnen Ebenen als Beispiel.

the example of a Lepol grate. The targets for steady-state operation are confined here to nodulization, exhaust gas composition, operation of the grate preheater, finish burn of the clinker, and operation of the clinker cooler. For unsteady-state operation, on the other hand, it is necessary to have targets for "heating", "starting and stopping", "hopper cleaning", "inlet/arch cleaning", "cooler full", "sintering zone overheated", "underburning", "drying chamber too hot", etc. Emergencies can arise through power failure, mechanical defects, refractory damage, or emergency stops. When heating up it is also necessary to differentiate between heating up from cold with new refractory brickwork, heating up from cold without new refractory brickwork, and heating up from warm.

This list could be continued indefinitely. In spite of this, or precisely because of it, the systematic integration of the associated process objectives and strategies into an accompanying or automated process control system is always a worthwhile aid to production control. In the context of the target variables listed at the beginning it leads to a process which is under control in all operating conditions.

4.2 Other areas of application

Other classical applications of automated process control are:

— the raw meal grinding system,
— the clinker storage system (the keyword here is the grading of the clinker),
— the cement grinding system, and
— the selection of the cement silos, i.e. the cement storage system.

The process analysis is carried out in the same way as for the clinker process bearing in mind the process engineering and plant constraints and the requirements of the particular works.

4.3 Higher-level process strategies

Linking the process steps listed in Sections 4.1 and 4.2 in the context of higher-level process strategies offers further long-term opportunities for optimization.

The NO_x emission depends, for example, on the raw meal grinding, the clinker burning process, and the quality control system. During the production of raw meal it is the uniformity of the lime standard and the mineralogical origins of the raw meal which have a significant influence on the NO_x emission. In addition to these, the kiln firing system also includes the gas content and particle size distribution of the fuel, the short-term constancy of the fuel mass flow, the fuel blending and hence the excess air ratio, the energy flow from combustion air and fuel, and the recirculating systems of secondary constituents. The secondary firing system includes the type of fuel and its feed points. Quality requirements, such as C_3S and C_3A content in the clinker and the NO_x emission, have an indirect influence. By long-term collection and processing of this process data it is possible to work out the essential influencing variables and apply the relationships in order to obtain lower NO_x emissions.

Another example is the cement strength, which can be influenced at practically every process stage. It depends on, among other things, the lime standard and the mineralogical origins of the raw meal, the uniformity of the energy flow to the rotary kiln firing system, the excess air ratio and the recirculating systems of secondary constituents in the rotary kiln, the quality-orientated grading of the clinker, the cement mix formulation, the mass flow of feed material to the mill, the particle size distribution of the cement, and the temperature in the mill.

Individual objectives are therefore not just dependent on the actual process stage being processed; in fact the preceding or subsequent stages can also have a decisive influence. To be able to find this out and process the individual data

Durch stetigen Abgleich von Planungsvorgaben mit dem aktuellen Produktionsgeschehen werden damit Zusammenhänge transparent gemacht, und es können Prozeßziele und -strategien einzelner Anlagenteile im Sinne übergeordneter Ziele angepaßt werden. Durch seine breite Informationsbasis zur Historie der eigenen Produktion ist ein derartiges Konzept ein wirkungsvolles Instrument der Produktionsplanung, -lenkung und -dokumentation.

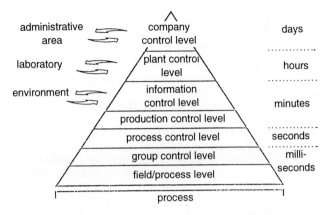

FIGURE 10: Data combination from process, environment, laboratory and administration as an instrument for planning, control and documentation of the production

Literaturverzeichnis

[1] 1. bis 4. Internat. Kongreß über die Verfahrenstechnik der Zementherstellung. Kongreßbände, Verein Deutscher Zementwerke e.V., Düsseldorf, 1971, 1977, 1985 und 1993.

[2] Säuberli, R., Herzog, U., und Rosemann, H.: Prozeßführung und Informationsmanagement. Zement-Kalk-Gips 46 (1993) Nr. 11, S. 679–688.

[3] VDZ-Merkblatt: Durchführung und Auswertung von Drehofenversuchen. Verein Deutscher Zementwerke, Düsseldorf, 1992.

[4] Edelkott, D.: Regelung des Energiestroms zur Drehofenfeuerung. Vortrag zum VDZ-Kongreß 1993, Fachbereich 2.

[5] Hansen, E. R.: Reduction of clinker and SO_2, NO_x emissions from preheater kilns. Portland Cement Association GTC fall meeting, September 1986.

appropriately requires an information system in which the environmental and laboratory data are brought together right at the information or plant control level. By way of example, **Fig. 10** shows the data combination presented by Säuberli et al. in [2] for this type of production control system with the individual levels. Interrelationships are made transparent by constant comparison of design targets with actual production events, and process objectives and strategies of individual plant sections can be adjusted in line with higher-level objectives. Through its wide information base founded on its own production system a scheme of this type is an effective instrument in production planning, control and documentation.

Laborautomation*)
Laboratory automation*)

Automatisation de laboratoire

Automatización de los laboratorios

Von **M. Riedhammer,** Dotternhausen/Deutschland

Laborautomation

Fachbericht 2.2 · Zusammenfassung – Der Markt erwartet beim Einsatz von Zement eine Qualität auf hohem Niveau bei gleichzeitig hoher Gleichmäßigkeit in den Eigenschaften. Die Forderung nach Vergleichmäßigung löst wiederum prozeßsteuernde Maßnahmen bei der Rohmehlherstellung, beim Brennprozeß und bei der Zementmahlung aus. Die Laborautomation trägt entscheidend dazu bei, die Maßnahmen in die Praxis umzusetzen. Die Laborautomation umfaßt nach heutigem Erkenntnisstand als Schwerpunkte die Probenahme, -aufbereitung und -analytik sowie die Datenerfassung, -verarbeitung und -bereitstellung. In Sonderfällen hat sich herausgestellt, daß das System der Laborautomation zweckmäßigerweise durch Expertensysteme unterstützt werden sollte. Alle heute gebräuchlichen Automationssysteme sind dazu geeignet, betriebliche Abläufe zu überwachen, diese mit Hilfe entsprechender Konzepte zu steuern oder zu regeln und damit das jeweils angestrebte Arbeitsziel zu erreichen. In dieser Kombination stellt die Laborautomation einen wesentlichen Teil eines modernen Qualitätssicherungssystems dar. Der möglichen Einsparung von Schichtpersonal stehen allerdings erhebliche Investitionskosten gegenüber. Der Anspruch auf ein hohes Qualitätsniveau bei hoher Gleichmäßigkeit der Produkteigenschaften fordert also auch seinen Preis.

Laboratory automation

Special report 2.2 · Summary – The cement market expects a high level of quality coupled with high uniformity of properties when using this building material. The demand for uniformity is in turn initiating process control measures in the production of raw meal, in the burning process, and during cement grinding. Laboratory automation makes an important contribution towards implementing these measures. According to current understanding the main functions covered by laboratory automation are sampling, sample processing and sample analysis, as well as the acquisition, processing and retrieval of data. In special cases it has proved advisable to support the laboratory automation system with expert systems. All the present normal automation systems are suitable for monitoring plant processes and for applying open- and closed-loop controls to them with the aid of suitable schemes, thereby achieving the particular work objective required. In this combination the laboratory automation system represents an important part of a modern quality assurance system. However, the possible saving in shift personnel have to be set against substantial investment costs. The demand for high quality coupled with high uniformity of the product properties exacts its price.

Automatisation de laboratoire

Rapport spécial 2.2 · Résumé – Les attentes du marché du ciment utilisant ce matériau sont une qualité de haut niveau associée à des caractéristiques élevées et constantes. L'exigence en matière d'homogénéisation déclenche, à son tour, des opérations de gestion de process pour la fabrication de la farine crue, pour la cuisson et pour le broyage du ciment. L'automatisation du laboratoire contribue largement à la concrétisation de ces opérations. L'automatisation du laboratoire comprend, d'après le niveau de connaissance actuel, comme points forts la prise d'échantillons, leur traitement et analyse ainsi que la saisie, le traitement et la mise à disposition des données. Dans certains cas spéciaux, il a été constaté que le système d'automatisation du laboratoire peut de manière efficace être assisté par des systèmes experts. Tous les systèmes d'automatisation actuellement usuels sont en mesure de surveiller les phases d'exploitation, de les piloter et régler à l'aide de concepts appropriés et de parvenir ainsi à l'objectif visé. Dans cet ensemble l'automatisation du laboratoire représente un maillon essentiel dans un système moderne de sécurité qualité. Face à une possible réduction de personnel travaillant en équipe les frais d'investissement à engager sont considérables. Mais n'est-ce pas là le prix à payer d'une qualité élevée associée à des caractéristiques élevées et constantes?

Automatización de los laboratorios

Informe de ramo 2.2 · Resumen – Los consumidores de cemento esperan, hoy en día, de este material un elevado nivel de calidad y, al mismo tiempo, una gran uniformidad de características. La exigencia de uniformidad, a su vez, da origen a las medidas de mando del proceso de fabricación de la harina cruda, del proceso de cocción y de la molienda del cemento. La automatización de los laboratorios contribuye, de forma decisiva, a llevar estas medidas a la práctica. Según el estado actual de los conocimientos, la automatización de los laboratorios abarca, esencialmente, la toma, preparación

---

*) Überarbeitete Fassung eines Vortrages zum VDZ-Kongreß '93, Düsseldorf (27. 9.–1. 10. 1993).
Revised text of a lecture to the VDZ Congress '93, Düsseldorf (27. 9.–1. 10. 1993)

y análisis de muestras así como el registro, tratamiento y suministro de datos. Se ha visto que, en algunos casos especiales, convendría apoyar al sistema de automatización de laboratorios mediante sistemas expertos. Todos los sistemas de automatización empleados actualmente son apropiados para controlar las operaciones de servicio, asegurar el mando o la regulación de las mismas con ayuda de los conceptos correspondientes, alcanzando así el objetivo fijado en cada caso. En esta combinación, la automatización de los laboratorios constituye una parte esencial de un sistema moderno de aseguramiento de la calidad. Sin embargo, frente al posible ahorro de mano de obra de los turnos, hay notables gastos de inversión. La exigencia de un alto nivel de calidad, junto con una gran uniformidad de las características de producción, tienen, desde luego, su precio.

1. Einleitung

Der Verarbeiter von Zement erwartet heute von diesem Baustoff neben der selbstverständlichen Einstellung der Qualität auf hohem Niveau auch deren Einhaltung in engen Grenzen. Die Forderung nach Vergleichmäßigung der Qualität bedingt prozeßsteuernde Maßnahmen bei Herstellung des Rohmehls, Ofenführung und Zementmahlung, deren Realisierung ohne Laborautomation schwierig ist.

Da Laborautomationssysteme in größerem Umfang in der Zementindustrie genutzt werden [1-6], ist es Aufgabe dieses Berichtes, den heutigen Stand der Laborautomation mit den Schwerpunkten Aufbereitung, Klinkerprobenahme und -analytik sowie Datenerfassung und -bereitstellung zu beleuchten.

2. Struktur der Laborautomation

Bild 1 zeigt einen Überblick über ein Laborautomationssystem im Werkszusammenhang. In Bild 1 links in der Senkrechten ist der Produktionsablauf schematisiert dargestellt.

Die Proben aus den verschiedenen Betriebsabteilungen werden mit Hilfe eines Rohrpostsystems der automatisierten Probenvorbereitung im Labor zugeführt. Je nach vorgesehener Analyse werden die Proben unterschiedlich aufbereitet. Die in der analytischen Meßtechnik ermittelten Daten fließen dem Prozeßrechner zu. Außerdem verarbeitet der Prozeßrechner alle Betriebsdaten des Laborsystems, die der internen Systemsteuerung dienen. Die analytischen Daten sind, nach einem Soll-Ist-Vergleich, Grundlage der Prozeßregelung. Zur weiteren Nutzung werden die im Laborsystem ermittelten Daten einem zentralen Rechner zugeführt, der auch Daten anderer Betriebs- und Verwaltungssysteme verarbeitet. Aufbereitete kompaktierte Daten dienen der Geschäftsleitung sowie den Verwaltungs- und Betriebsabteilungen als Grundlage strategischer und operativer Maßnahmen im Unternehmen.

Mit Bild 1 sind die wesentlichen Komponenten eines automatisierten Laborsystems in ihrer inneren Zuordnung sowie in ihrer Zuordnung zum Produktionsbetrieb definiert.

3. Komponenten des Laborautomationssystems

Probenehmer können weitgehend automatisch repräsentativ über den Gesamtförderstrom oder gezielt für Teilfraktionen die für die Untersuchung erforderlichen Probeninkremente entnehmen. Nach Materialeigenschaften und geografischer Lage der Probeentnahmestelle innerhalb der Betriebsanlage werden unterschiedliche Probenahmesysteme verwendet. Nicht automatisierte Probenahmen, zum Beispiel von Rohstoffen, Bohrmehlen aus den Steinbrüchen, Versandzementen, ergänzen die automatisiert entnommenen Proben. Die automatisch entnommenen Proben werden in der Versandstation gesammelt, gemischt, geteilt und in das Rohrpostsystem eingegeben. Der Probenüberschuß wird in die Produktion zurückgeführt.

Das im Arbeitsablauf nächste Teilsystem des automatisierten Labors ist die Probenaufbereitung. Der integrierte Aufbereitungsautomat [1, 2] beinhaltet Probenempfänger,

1. Introduction

Nowadays, cement users not only expect as a matter of course that the quality of this construction material will be set at a high level, but also that it will be kept within narrow limits. The demand for uniformity of quality means that process control measures are needed during raw meal production, kiln operation and cement grinding, which would be difficult to achieve without laboratory automation.

As laboratory automation systems are used fairly widely in the cement industry [1-6] the purpose of this report is to examine the present state of laboratory automation, with emphasis on the preparation system, clinker sampling and analysis, and data acquisition and retrieval.

2. Structure of laboratory automation

Fig. 1 outlines a laboratory automation system in the context of a works. The production sequence is shown diagrammatically in the left-hand column of the diagram. The samples from the different plant sections are fed to the automated sample preparation system in the laboratory with the air of a pneumatic tube conveyor system. The samples are prepared in different ways depending on the intended analysis. The data obtained during the analysis flow to the process computer. The process computer also processes all operating data from the laboratory system which is used for the internal system control. After a comparison of setpoint and actual values the analytical data form the basis for the process control. The data obtained in the laboratory system are fed for further use to a central computer which also processes data from other operating and administrative systems. Prepared and compressed data serve the management and the administrative and plant departments as the basis for strategic and operational measures in the company.

Fig. 1 defines the essential components of an automated laboratory system in their relationship to one another and to the production plant.

3. Components of the laboratory automation system

Samplers working largely automatically are able to take the sample increments needed for the investigation either representatively over the entire flow or specifically for partial fractions. Different sampling systems are used depending on the properties of the material and the geographical position of the sampling point within the plant. The samples taken automatically are augmented by non-automated sampling of, for example, raw materials, drillings from the quarries, and cement for despatch. The samples taken automatically are collected at the despatch station, mixed, split, and fed into the pneumatic tube conveyor system. The surplus is returned to the production system.

The automated laboratory's next subsystem in the operating cycle is the sample preparation. The integral automated preparation system [1, 2] contains sample receiver, metering equipment, grinding aid addition, fine mills, tablet press, tablet cleaning, and sufficient space for retention samples. The internal transport is carried out by slewing arms and movable beakers. The prepared tablets are transferred to the

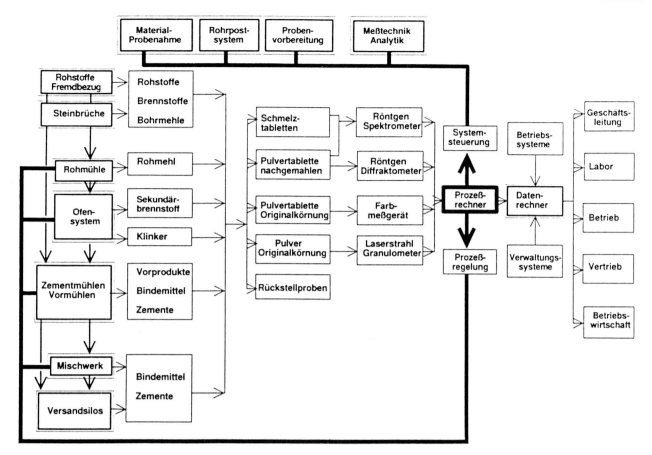

BILD 1: Laborautomationssystem im Werkszusammenhang
FIGURE 1: Laboratory automation system in the context of a works

Material-Probenahme	= material sampling
Rohrpostsystem	= pneumatic tube conveyor system
Probenvorbereitung	= sample preparation
Meßtechnik Analytik	= analysis
Rohstoffe Fremdbezug	= raw materials external supply
Steinbrüche	= quarries
Rohmühle	= raw mill
Ofensystem	= kiln system
Zementmühlen Vormühlen	= cement mills primary mills
Mischwerk	= blender
Versandsilos	= despatch silos
Rohstoffe	= raw materials
Brennstoffe	= fuels
Bohrmehle	= drillings
Rohmehl	= raw meal
Sekundärbrennstoff	= secondary fuel
Klinker	= clinker
Vorprodukte	= intermediate products
Bindemittel	= binders
Zemente	= cements
Schmelztabletten	= fused tablets
Pulvertablette nachgemahlen	= powder tablets reground
Pulvertablette Originalkörnung	= powder tablets orig. particle size
Pulver Originalkörnung	= powder orig. particle size
Rückstellproben	= retention samples
Röntgen Spektrometer	= X-ray spectrometer
Röntgen Diffraktometer	= X-ray diffractometer
Farbmeßgerät	= colorimeter
Laserstrahl Granulometer	= laser beam granulometer
Systemsteuerung	= system control
Prozeßrechner	= process computer
Prozeßregelung	= process control
Betriebssysteme	= operating systems
Datenrechner	= data computer
Verwaltungssysteme	= administrative systems
Geschäftsleitung	= management
Labor	= laboratory
Betrieb	= plant
Vertrieb	= sales
Betriebswirtschaft	= industrial administration

Dosierungen, Mahlhilfezugabe, Feinmühlen, Tablettenpresse, Tablettenreinigung sowie ausreichend Platz für Rückstellproben. Die inneren Transporte werden mittels Schwenkarmen und verfahrbarer Becher durchgeführt. Die vorbereiteten Tabletten werden über ein Förderbandsystem den Analysengeräten übergeben. Die Originalmehlproben zum Laserstrahlgranulometer werden z.B. mittels einer Minirohrpostanlage transportiert.

Bei Steigerung der Probenanzahl und Ausweitung der vom Laborautomationssystem durchzuführenden Aufgaben wurde gelegentlich der Weg der Installation eines zweiten Aufbereitungsautomaten gegangen. Flexibler ist es, bei steigendem Probenaufkommen auf die Modultechnik überzugehen [7-9]. Ausgehend von einem Basismodul sind Erweiterungsmodule zu installieren und die zu erfüllenden Aufbereitungs- und Dosieraufgaben unterschiedlichen Modulen zuzuordnen.

Bild 2 zeigt ein Basismodul mit den wesentlichen Funktionen. Mögliche Erweiterungsmodule (**Bild 3**) beinhalten neben Probenempfänger und Dosierung zum Beispiel Lasergranulometer mit Alkoholaufbereitung, Farbmeßge-

analysis equipment by a conveyor belt system. The samples of original meal for the laser beam granulometer can be transported by, for example, a mini pneumatic tube conveyor system.

In some instances a second automatic preparation has been installed where the number of samples increased and the tasks to be carried out by the laboratory automation system were broadened. It is more flexible, however, to change over to modular technology when the number of samples increases [7-9]. Starting from a base module it is possible to install extension modules and assign the preparation and metering tasks to be performed to different modules.

Fig. 2 shows a base module with the essential functions. Possible extension modules (**Fig. 3**) could contain, in addition to the sample receiver and metering system, laser granulometers with alcohol preparation, colorimeter, loss on ignition determination, fusion digestion equipment and appropriate parking areas. This modular technology operates with small robots integrated into the particular modules. The sample preparation system operates in a similar way with a central robot, but with a considerably larger design of robot because

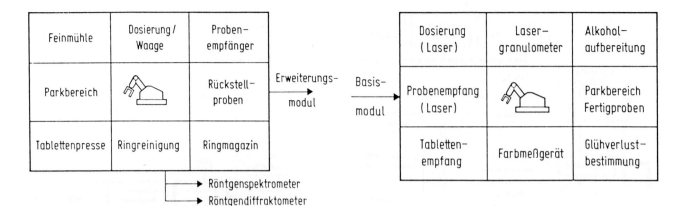

BILD 2: Probenaufbereitung mittels Modultechnik: Basismodul
FIGURE 2: Sample preparation using modular technology: base module

Feinmühle	= fine mill
Parkbereich	= parking area
Tablettenpresse	= tablet press
Dosierung/Waage	= metering/weighing
Ringreinigung	= ringcleaning
Probenempfänger	= sample receiver
Rückstellproben	= retention samples
Ringmagazin	= ringmagazine
Erweiterungsmodul	= extension module
Röntgenspektrometer	= X-ray spectrometer
Röntgendiffraktometer	= X-ray diffractometer

BILD 3: Probenaufbereitung mittels Modultechnik: Erweiterungsmodul
FIGURE 3: Sample preparation using modular technology: extension module

Basismodul	= base module
Dosierung (Laser)	= metering (laser)
Probenempfang (Laser)	= sample receiver (laser)
Tablettenempfang	= tablet receiver
Lasergranulometer	= laser granulometer
Farbmeßgerät	= colorimeter
Alkoholaufbereitung	= alcohol preparation
Parkbereich Fertigproben	= parking area finished samples
Glühverlustbestimmung	= loss on ignition determination

rät, Glühverlustbestimmung, Schmelzaufschlußgerät und entsprechende Parkbereiche. Diese Modultechnik arbeitet mit kleinen, in die jeweiligen Module integrierten Robotern. Auf ähnliche Weise, jedoch wegen der erforderlichen Reichweite in erheblich größerer Robotorauslegung, arbeitet das System der Probenaufbereitung mit Zentralroboter [6]. In einem abgeschlossenen Raum wird dabei die gleiche Aufgabenstruktur bearbeitet wie in den modular unterteilten Aufbereitungssystemen. Die analytische Meßtechnik liegt außerhalb und wird mit Förderbändern angefahren.

Eine Alternative stellt der verfahrbare Zentralroboter dar [10], der durch eine weitere Bewegungsachse in der Lage ist, einen kreisförmigen Arbeitsbereich auf ein Oval auszuweiten (**Bild 4**). Damit ist es möglich, eine größere Anzahl von Geräten zu bedienen. Nach Herstellerangaben ist die Zentralrobotertechnik nach entsprechender Schnittstellenschaffung in der Lage, auch ältere Geräte in ein automatisiertes Labor einzubeziehen.

Die analytische Meßtechnik besteht üblicherweise aus Röntgenspektrometern und -diffraktometern, Laserstrahlgranulometern sowie fotometrischen Farbmeßgeräten. Die Entwicklung der Röntgenanalytik hat in den letzten Jahren weitere Fortschritte gemacht [11–14]. Der Prozeßrechner organisiert die Systemsteuerung der Laborautomation mit ihren Teilsystemen, wie das Schema in Bild 1 zeigt. Darüber hinaus werden vom Prozeßrechner aus die Dosierbandwaagen der Roh- und Zementmühlen und des Ofens sowie die Drehzahl des Zementmühlensichters geregelt.

4. Aufgaben des Laborautomationssystems

Mit der strukturellen Einordnung der Laborautomation in den Ablauf der Zementherstellung liegen deren Aufgaben fest. Hierzu zählen die Eingangskontrolle zugefahrener Rohstoffe, wie zum Beispiel Ton und Sand, und die Abbauplanung durch Bohrmehluntersuchungen, auf die nicht näher eingegangen wird.

Eine wesentliche Teilaufgabe stellt die Regelung der Rohmehlherstellung aus den Komponenten Kalkstein, Ton und Sand in Kombination mit der Regelung der Ofenaufgabe dar. Das ist insbesondere dann von Bedeutung, wenn Anteil und Zusammensetzung des mineralischen Rückstandes eines Brennstoffes, wie im Beispiel des Ölschiefers (**Bild 5**), bei der Einstellung der gewünschten Sollzusammensetzung des Brenngutes berücksichtigt werden müssen. Der Prozeß-

of the required reach [6]. It works in a closed space to the same task structure as in the preparation systems divided into modules. The analysis equipment lies outside and is reached by conveyor belts.

An alternative is the travelling robot [10] which has an additional axis of movement and is capable of extending a circular operating area into an oval (**Fig. 4**). This makes it possible to serve a greater number of items of equipment. According to the manufacturer's literature the central robot technology is also capable, after appropriate interfaces have been created, of incorporating older items of equipment into an automated laboratory.

The methods of analysis usually consist of X-ray spectrometers and diffractometers, laser beam granulometers and photometric colour-measuring equipment. There has been further progress in the development of X-ray analysis in recent years [11–14]. The process computer organizes the system control for the automated laboratory with its subsystems as shown in the diagram in Fig. 1. The process com-

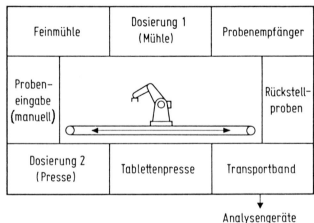

BILD 4: Probenaufbereitung mittels Zentralroboter (verfahrbar)
FIGURE 4: Sample processing using central (mobile) robot

Feinmühle	= fine mill
Dosierung 1 (Mühle)	= metering 1 (mill)
Probenempfänger	= sample receiver
Probeneingabe (manuell)	= sample input (manual)
Rückstellproben	= retention samples
Dosierung 2 (Presse)	= metering 2 (press)
Tablettenpresse	= tablet press
Transportband	= transport belt
Analysengeräte	= analysis equipment

rechner regelt die Rohmehlherstellung auf der Basis der Analysen der Rohstoffe und der mineralischen Bestandteile des Ölschiefers an Mühlenproben. Die Anteile von Rohmehl und Ölschiefermehl bei der Ofenaufgabe werden mittels gleichzeitig durchgeführter Analysen an Proben aus den Ofenaufgabematerialien geregelt. Beide Regelkreise sind miteinander vernetzt, um eine schnelle Anpassung der Sollwerte bei der Rohmehlherstellung zu ermöglichen [1]. Prinzipiell läßt sich dieses Regelkonzept auch bei Einsatz anderer ballastreicher Brennstoffe anwenden.

Die automatisierte Klinkerprobenahme und -analytik hat neben der kontrollierenden Funktion für die im Vorfeld des Ofens ablaufenden regeltechnischen Vorgänge im wesentlichen die Aufgabe, die chemische Zusammensetzung und den Freikalkgehalt des Klinkers festzustellen sowie Probemengen für das zement- und mörteltechnische Laboratorium bereitzustellen. Dabei ist die Probenahmetechnik für die bei der Klinkeranalytik erzielten Ergebnisse von wesentlicher Bedeutung.

Es deuten sich Entwicklungen an, die Phasenzusammensetzung quantitativ zu ermitteln und die Informationen der Prozeßregelung des Drehofens zuzuführen [12]. Einblicke in den Stand des Wissens vermittelt auch der Beitrag von Fröhlich [13].

Die stündliche Ermittlung der Klinkeranalyse, speziell des Freikalkes, an Stichproben gibt dem Betriebsmann eine nahezu lückenlose Information über den zeitlichen Verlauf des Brennprozesses insbesondere bei gestörtem Ofenbetrieb durch Ansatzbildung und -austrag aus dem Ofen. Diese Effekte werden schon durch eine Schichtdurch-

puter also controls the belt weighfeeders for the raw mills and cement mills and for the kiln, as well as the rotational speed of the cement mill classifier.

4. Duties of the laboratory automation system

The duties of the laboratory automation are fixed when it is structurally integrated into the cement manufacturing process. These duties include the checking of incoming raw materials, such as clay and sand, and planning the excavations through investigating the drillings; these will not be described in any greater detail.

An important subsidiary duty is to control the production of raw meal from the limestone, clay and sand components, combined with controlling the kiln feed. This is especially important if the proportion and composition of the mineral residue of a fuel, e.g. of the oil shale (**Fig. 5**), has to be taken into account when setting the required target composition for the kiln feed. The process computer controls the raw meal production on the basis of the analyses of the raw materials and the mineral constituents of the oil shale from mill samples. The proportions of raw meal and oil shale meal in the kiln feed are controlled by means of analyses carried out simultaneously on samples of the kiln feed materials. The two control loops are linked with one another to enable the target values to be adjusted rapidly during raw meal production [1]. In principle, this control concept can also be applied when using other fuels containing large proportions of inerts.

In addition to the controlling functions for the control procedures leading up to the kiln, the main task of the automated

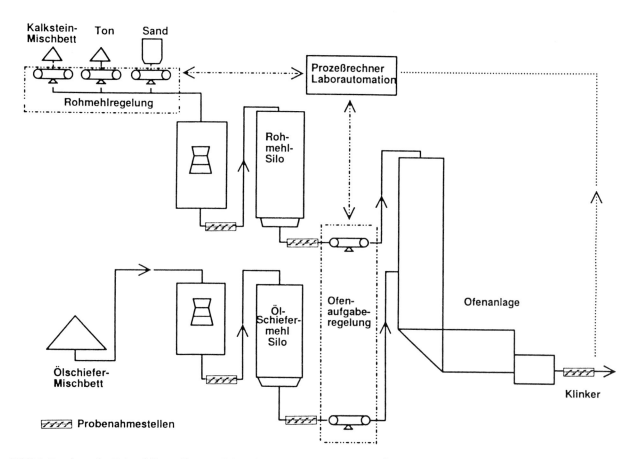

BILD 5: Regelung der Rohmehlherstellung und der Ofenaufgabe bei Einsatz von Ölschiefer als Sekundärbrennstoff im Schwebegaswärmetauscher
FIGURE 5: Control system for raw meal production and kiln feed using oil shale as a secondary fuel in the suspension preheater

Kalkstein-Mischbett	= limestone blending bed		
Ton	= clay	Ölschiefer-Mischbett	= oil shale blending bed
Sand	= sand	Öl-Schiefermehl-Silo	= oil shale meal silo
Rohmehlregelung	= raw meal control	Ofenaufgaberegelung	= kiln feed control
Prozeßrechner Laborautomation	= process computer automated laboratory	Ofenanlage	= kiln system
Rohmehl Silo	= raw meal silo	Klinker	= clinker

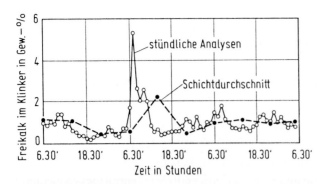

BILD 6: Freikalkgehalt des Klinkers: Vergleich von Stichproben mit Schichtdurchschnittsproben
FIGURE 6: Free lime content in the clinker: comparison of spot samples with shift average samples

Freikalk im Klinker in Gew.-%	= Free lime in the clinker in wt.%
stündliche Analysen	= hourly analyses
Schichtdurchschnitt	= shift average sample
Zeit in Stunden	= Time in hours

schnittsprobe, wie die gestrichelte Kurve in **Bild 6** zeigt, verwischt. Tagesdurchschnittsproben sind in diesem Zusammenhang indiskutabel.

Die Ermittlung der Klinkeranalyse und des Freikalkgehaltes an Teilfraktionen, die entweder durch gezieltes Heraussieben aus dem Klinkergesamtmassestrom oder durch eine Begrenzung des offenen Probenahmequerschnittes erreicht werden, sind nach Untersuchungen im eigenen Hause problematisch, so daß in der Regel die Analysen an der Gesamtfraktion vorzuziehen sind.

Die Zementmühlenregelung kann, angepaßt an qualitative und wirtschaftliche Grundforderungen, auf verschiedene Zielgrößen hin erfolgen. Zu den Zielgrößen zählen zum Beispiel Granulometrie, Sulfat- oder CO_2-Gehalt des Zements, volle Kapazitätsnutzung, Minimierung der elektrischen Arbeit, Farbe des Zements. Die Zementmühlenregelung auf der Basis der im Laserstrahlgranulometer festgestellten Korngrößenverteilung des Zements und weiterer Parameter ist als sogenanntes Expertensystem erfolgreich realisiert [15–17].

Die Komponentenregelung auf der Basis eines vorgegebenen Sulfatgehaltes wird unter anderem bei Rohrbach Zement in Dotternhausen betrieben. Bei vorgegebenem Sulfatgehalt wird die bereits den Anhydrit als Erstarrungsregler beinhaltende Menge an gebranntem Ölschiefer so geregelt, daß Überschreitungen im Sulfatgehalt ausgeschlossen werden. Dabei werden zusätzliche qualitative Vorgaben mit einbezogen.

Last but not least ist die selbstverständliche Aufgabe eines automatisierten Laborsystemes dessen Integration in das Qualitätssicherungssystem des Unternehmens. Sprung [3] geht in einer Veröffentlichung von 1990 ausführlicher auf QS-Systeme bei der Zementherstellung ein. Alle Zahlen, die im Rahmen der Laborautomation, aber auch in den zement- und mörteltechnischen Prüfbereichen anfallen, stehen für die Qualitätssicherungshandbücher zur Verfügung. Die Einführung von zertifizierungsfähigen QS-Systemen wird dadurch erleichtert, daß bei Ermittlung und Aufarbeitung aller Labor- und Betriebsdaten bereits ein Standard existiert, der an vielen Stellen über das Geforderte hinausgeht. Insofern ist die Erstellung von QS-Handbüchern vielfach nur noch ein formeller Akt zur normgerechten Bereitstellung der ohnehin vorhandenen Informationen.

5. Datenbereitstellung und -bearbeitung

Wesentlicher Teil der Laborautomation ist die Datenbereitstellung und -bearbeitung. Dabei beruht der Labordatenanfall, wie **Bild 7** zeigt, nur zum Teil auf der eigentlichen Laborautomation. **Bild 8** enthält den aus der Laborautomation gezogenen Teildatenfluß. Beginnend im Kalksteinbruch bis hin zu den Versandzementen hat diese Laborautomation täglich etwa 130 Untersuchungsvorgänge durchzuführen.

clinker sampling and analysis system is to determine the chemical composition and the free lime content of the clinker as well as to provide quantities of samples for the cement and mortar laboratory. The sampling method used here is of considerable importance for the results obtained during the clinker analysis.

There are indications that developments are being made in determining phase compositions quantitatively and conveying the information to the rotary kiln's process control system [12]. The contribution from Fröhlich [13] also provides insights into the current state of knowledge.

Hourly determination of the clinker analysis, especially the free lime, on spot samples provides the operator with virtually uninterrupted information about the behaviour of the burning process with time, especially during upset kiln operation caused by the formation and discharge of coating from the kiln. These effects are smoothed out by a shift average sample, as shown by the broken line in **Fig. 6**. Daily average samples do not come into consideration here.

Our own investigations have indicated that problems are caused by determining the clinker analysis and the free lime content on part fractions which are obtained either by carefully controlled screening from the total clinker mass flow, or by restricting the open sampling cross-section; analyses on the entire fraction are therefore normally to be preferred.

When it has been adapted to basic qualitative and economic requirements the cement mill control system can be based on a variety of target variables. These include such parameters as granulometry, sulphate or CO_2 content in the cement, full utilization of capacity, minimizing the electrical power, cement colour. Cement mill control systems based on the cement particle size distribution measured with the laser beam granulometer and on other parameters are being put into practice successfully as so-called expert systems [15–17].

The component control system based on a predetermined sulphate content is used, among other places, at Rohrbach Zement in Dotternhausen. The quantity of burnt oil shale, which already contains anhydrite as a setting regulator, is controlled so that the sulphate content cannot exceed the predetermined level. Other qualitative targets are also included.

Last but not least is the obvious task of an automated laboratory system of fitting into the firm's quality assurance system. In a publication in 1990, Sprung [3] gave details of a quality assurance system used in cement manufacturing. All figures which are produced within the automated laboratory and also in the cement and mortar test areas are available for the quality assurance manuals. The introduction of certified quality assurance systems is made easier by the fact that a standard, which in many places goes beyond what is required, already exists for determining and developing all laboratory and operating data. In this respect the creation of quality assurance manuals is in many cases only a formal act for standardized presentation of information which is available anyway.

5. Provision and processing of data

The provision and processing of data forms an important part of the automated laboratory. As shown in **Fig. 7**, the supply of laboratory data is only partially based on the actual automated laboratory. **Fig. 8** shows the subsidiary flow of data drawn from the automated laboratory. This automated laboratory has to carry out about 130 investigation procedures daily, starting in the limestone quarry and ending with the cement for despatch.

Of these, the analyses shown by dots are used for the on-line control system. All the others are used for production control and monitoring the standards. As shown by **Fig. 9**, all data flows to the central computer where it forms the basis for the laboratory information management system (LIMS). Supplemented by the data from other operating and administrative systems, they serve as a management information system for running the company.

```
┌─────────────────────────────────┐   ┌─────────────────────────────────┐
│    - Laborautomation            │   │    - automated laboratory       │
│                                 │   │                                 │
│    - Periphere Geräte           │   │    - peripheral equipment       │
│      -- C, S-Analysator         │   │      - C, S analyzer            │
│      -- Oberflächenmessung      │   │      - surface area measurement │
│      -- Atomabsorption          │   │      - atomic absorption        │
│      -- Titroprozessor          │   │      - titration processor      │
│                                 │   │                                 │
│    - Prüfungen nach DIN 1164    │   │    - testing to DIN 1164        │
│                                 │   │                                 │
│    - Betontechnik, Anwendungs-  │   │    - concrete technology,       │
│      technik                    │   │      application technology     │
│    - Lagerungs- und             │   │                                 │
│      Raumklimadaten             │   │    - storage and air conditioning data │
└─────────────────────────────────┘   └─────────────────────────────────┘
```

BILD 7: Gesamtdatenanfall im Werkslabor

FIGURE 7: Total data produced in the works laboratory

Stoff Ort	Kalkstein Bruch	Rohmehl Mühle	Rohmehl Ofen	Ölschiefer Mühle	Ölschiefer Ofen	Klinker Ofen	Ölschiefer gebrannt Kraftwerk	Zemente Mühlen	Zemente Versand
Proben pro Tag	6	24	12	24	12	24	3	18	8
Element-Analyse	▨	▨	▨	▨	▨	▨	▨	▨	▨
CaO_fr						▨			
Granulometrie								▨	
Farbe								▨	
Regelung		⋯	⋯	⋯	⋯	⋯			
Steuerung	▨	▨	▨	▨	▨	▨		▨	
Normen						▨		▨	

BILD 8: Teildatenanfall aus der Laborautomation am Beispiel von Rohrbach-Zement

FIGURE 8: Data produced from the automated laboratory based on the example of Rohrbach Zement

Stoff	= material
Ort	= location
Proben pro Tag	= samples per day
Element-Analyse	= elemental analysis
CaO_{fr}	= CaO_{fr}
Granulometrie	= granulometry
Farbe	= colour
Regelung	= closed-loop control
Steuerung	= open-loop control
Normen	= standards
Kalkstein	= limestone
Bruch	= quarry
Rohmehl	= raw meal
Mühle	= mill
Ofen	= kiln
Ölschiefer	= oil shale
Klinker	= clinker
Ölschiefer gebrannt	= burnt oil shale
Kraftwerk	= power station
Zemente	= cements
Mühlen	= mills
Versand	= despatch

Für die On-line-Regelung werden hiervon die gepunktet gekennzeichneten Analysen verwendet. Alle anderen dienen der Produktionssteuerung und Normüberwachung. Alle Daten fließen, wie **Bild 9** zeigt, in den zentralen Rechner ein und bilden dort die Basis für das Labor-Informations-Management-System (LIMS). Ergänzt durch die Daten weiterer Betriebs- und Verwaltungssysteme dienen sie als Management-Informations-System der Unternehmensführung.

Der Prozeßrechner Laborautomation organisiert die gesamte Systemtechnik, wobei alle peripheren Probenahme-, Transport- und Aufbereitungssysteme wie auch alle analytischen Geräte über frei programmierbare Steuerungen verfügen. Der Prozeßrechner gibt die Daten zu festgelegten Zeiten an den Datenrechner der zentralen EDV im rechten Teilbild ab. Der Bediener des LIMS holt sich für weitergehende Kontroll- und Auswertungsarbeiten auch Daten aus dem bereits abgelegten Datenpool der zentralen EDV zurück. Alle Bereiche des Labors sind an den EDV-BUS der zentralen EDV mittels Terminal angeschlossen, wobei einzelne Bereiche über einen zwischengeschalteten PC weitergehende Auswertungsarbeiten durchführen können. Standardstatistikprogramme, wie Korrelationen, Trends und ähnliche sind auch ohne PC direkt über den EDV-BUS im zentralen Datenrechner zu bearbeiten.

Weitere Nutzer dieses Systems sind, beginnend bei der Geschäftsführung, alle Mitarbeiter des Vertriebes, der Bauberatung und der Betriebsleitung.

Die **Bilder 10** und **11** zeigen, in welcher Form aktuelle Daten direkt nach Messung auf dem Bildschirm im Bedienerraum des LIMS auflaufen. Bild 10 zeigt eine On-line-Grafik der Klinkeranalytik einschließlich der Standards und der nach Bogue errechneten Phasen. In Bild 11 ist eine On-line-Grafik einer Zementmühle mit den wesentlichen Daten und den

The automated laboratory process computer organizes the entire system technology, in which all peripheral sampling, transport and preparation systems, as well as all the analysis equipment, have access to PLC systems. The process computer outputs the data at fixed times to the data computer of the central electronic data processing (EDP) system in the right-hand part of the diagram. For more extensive checking and evaluation work the LIMS operator also retrieves data from the pool of data already filed in the central EDP system.

All sections of the laboratory are linked by terminals to the EDP bus of the central EDP system, and individual sections are able to carry out further evaluation work via an intermediate PC. Standard statistics programs, such as correlations, trends and the like, can also be run directly on the central data computer via the EDP bus without using a PC.

Other users of this system are the management, all the sales staff, the construction advisory service, and the plant management.

Figs. 10 and **11** show the form in which current data are displayed on the screens in the LIMS operator room directly after measurement. Fig. 10 shows an on-line graphics display of the clinker analysis including the lime standard and the calculated Bogue phases. Fig. 11 is an on-line graphics display for a cement mill showing the essential data and the proportions of the components. For simplicity the graphics are shown on the screen without dimensions. Seconds after each measurement in the automated laboratory the newly measured values appear in this graphics display alongside the values from the last 2 to 3 days. Baerentsen et al. [18] and Sedlmeir [19] give detailed descriptions of special management systems.

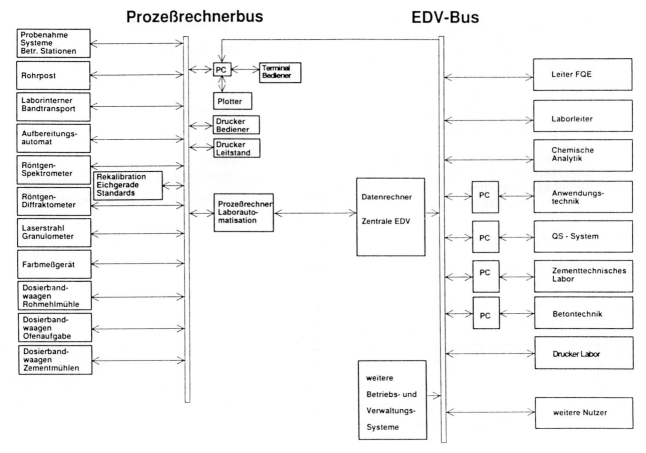

BILD 9: Labor-Informations-Management-System

FIGURE 9: Laboratory information management system

Probenahme Systeme Betr. Stationen	= sampling systems plant stations	Drucker-Bediener	= printer operator
Rohrpost	= pneumatic tube conveying system	Drucker-Leitstand	= printer control centre
Laborinterner Bandtransport	= laboratory internal belt transport	Prozeßrechner Laborautomation	= process computer laboratory automation
Aufbereitungsautomat	= automatic preparation	Datenrechner Zentrale EDV	= data computer central EDP
Röntgen-Spektrometer	= X-ray spectrometer	Weitere Betriebs- und Verwaltungs-Systeme	= other operational and administrative systems
Röntgen-Diffraktometer	= X-ray diffractometer	Leiter FQE	= head of FQE
Laserstrahl-Granulometer	= laser beam granulometer	Laborleiter	= head of laboratory
Farbmeßgerät	= colorimeter	Chemische Analytik	= chemical analysis
Dosierbandwaagen Rohmehlmühle	= belt weigh-feeder raw mill	Anwendungstechnik	= application technology
Dosierbandwaagen Ofenaufgabe	= belt weigh-feeder kiln feed	QS-System	= quality assurance system
Dosierbandwaagen Zementmühlen	= belt weigh-feeder cement mills	Zementtechnisches Labor	= cement laboratory
Rekalibration Eichgerade Standards	= recalibration calibration lines standards	Betontechnik	= concrete technology
PC	= PC	Drucker Labor	= laboratory printer
Terminal-Bediener	= terminal operator	Weitere Nutzer	= other use
Plotter	= plotter		

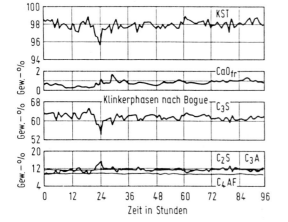

BILD 10: On-line-Grafik aus LIMS-Bildschirm für die Klinkerqualität

FIGURE 10: On-line graphics from the LIMS display screen for clinker quality

Gew.-%	= Wt.%
Zeit in Stunden	= Time in hours
KST	= LSt
Klinkerphasen nach Bogue	= Bogue clinker phases

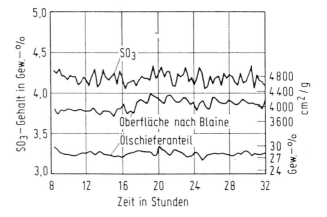

BILD 11: On-line-Grafik auf dem LIMS-Bildschirm für eine Zementmühle

FIGURE 11: On-line graphics from the LIMS display screen for a cement mill

SO_3-Gehalt in Gew.-%	= SO_3 content in wt.%
Oberfläche nach Blaine	= Blaine surface area
Ölschieferanteil	= proportion of oil shale
Zeit in Stunden	= Time in hours

Komponentenanteilen dargestellt. Die Grafiken erscheinen auf dem Bildschirm aus Gründen der Vereinfachung dimensionslos. Nach jeder Messung in der Laborautomation laufen Sekunden später die neu gemessenen Werte im Vergleich mit den Werten der letzten 2 bis 3 Tage in diese Grafik ein. Baerentsen et al. [18] sowie Sedlmeir [19] gehen auf spezielle Managementsysteme näher ein.

6. Personalaspekte und Bereitschaftsdienst

Die Personalausstattung kann bei Einführung der Laborautomation durch den Verzicht auf die Schichtlaboranten deutlich verringert werden. Es genügt nach der Erfahrung bei Rohrbach Zement [1] in Dotternhausen ein qualifizierter Mitarbeiter, um das System zu bedienen und weiter zu optimieren.

Daneben benötigt man 2 bis 3 Mitarbeiter für den unumgänglichen Bereitschaftsdienst. Diese Mitarbeiter müssen einfache Rechnerprobleme und mechanische Störungen in den Systemen notfalls mit Hilfe der Schichtelektriker oder -schlosser beheben können.

In Dotternhausen wurde der Bereitschaftsdienst durch ein Rechnerprogramm vereinfacht, das über das Cityruf-System in einzustellenden festen Zeitabschnitten, zum Beispiel jede Stunde, automatisch auf dem Cityruf-Empfänger die für die Prozeßregelung wichtigsten Meßwerte und Analysen anzeigt. Begrenzt ist dieses System nur durch die maximale Anzahl von 4×80 Zeichen im System des Cityrufes.

7. Erfolg der Laborautomation

Die Vergleichmäßigung der Qualität bei Betrieb einer Laborautomation steht außer Frage, jedoch ist der Grad der Vergleichmäßigung nicht in Zahlen auszudrücken. Das liegt daran, daß eine Informationsverdichtung über eine wesentlich häufigere Analytik bei Einführung des automatisierten Labors gegenüber dem klassischen chemischen Labor eintritt. Da die Anzahl der Werte um mindestens eine Zehnerpotenz nach oben schnellt, ist ein Vergleich, nach der Standardabweichung, nicht zulässig.

Um aber trotzdem Hinweise auf die Gleichmäßigkeit der Qualität von Zwischenprodukten und Zementen zu geben, werden im **Bild 12** die Variationskoeffizienten der wesentlichen Oxide und Standards des Rohmehls an der Ofenaufgabe beispielhaft für ein Werk dargestellt. Die Variationskoeffizienten als Quotient von Standardabweichung zu arithmetischem Mittelwert liegen danach bei 1 bis 4 %. Der Auswertung liegt ein halbes Produktionsjahr mit 2120 Einzelwerten zugrunde. In gleichen Größenordnungen bewegen sich die entsprechenden Variationskoeffizienten für den Klinker.

Variationskoeffizienten [n = 2120]	
SiO2	1,8
Al2O3	2,7
Fe2O3	2,2
CaO	0,9
Kstd.	1,5
Si.-M	1,0
T.-M	3,9

BILD 12: Gleichmäßigkeit von Rohmehl

Wie die Werte der Rohmehlqualität stammen auch die in **Bild 13** angegebenen Normfestigkeiten der Zemente aus dem Datenpark von Rohrbach Zement. Für sieben vom

6. Personnel aspects and support service

The introduction of laboratory automation can cause a significant decrease in the staffing level by dispensing with shift laboratory staff. Experience at Rohrbach Zement [1] in Dotternhausen indicates that one competent employee is sufficient for operation and further optimization of the system.

In addition to this, 2 to 3 employees are needed for the inevitable support service. These employees must be able to rectify simple computer problems and mechanical faults in the systems, if necessary with help from the shift electrician or fitter.

In Dotternhausen the support service is simplified by a computer program which, at given adjustable time intervals, e.g. every hour, uses the Cityruf system (a computer network system) to give an automatic indication on the Cityruf receiver of the measured values and analyses which are most important for the process control system. This system is restricted only by the maximum number of 4×80 characters in the Cityruf system.

7. Success of laboratory automation

There is no question that the quality is smoothed when a laboratory automation system is in operation; however, the degree of smoothing cannot be expressed numerically. This is because the information is compressed because of the substantially more frequent analyses – as compared with the classical chemical laboratory – when the automated laboratory is introduced. The number of values shoots up by at least a power of ten, so any comparison based on standard deviations is not admissible.

In spite of this, it is still possible to provide information about the uniformity of quality of intermediate products and cements; by way of example, **Fig. 12** shows the coefficients of variation of the important oxides and of the lime standard of the raw meal in the kiln feed for one works. This shows that the coefficients of variation (standard deviation divided by arithmetic mean) range from 1 to 4 %. The evaluation is based on half a production year with 2120 individual values. The corresponding coefficients of variation for the clinker are of the same order of magnitude.

Coefficients of variation [n = 2120]			
SiO$_2$	1,8	LSt	1,5
Al$_2$O$_3$	2,7	Si ratio	1,0
Fe$_2$O$_3$	2,2	AL ratio	3,9
CaO	0,9		

FIGURE 12: Uniformity of raw meal

Like the raw meal quality values, the standard strengths of the cements in **Fig. 13** also come from Rohrbach Zement's database. For seven of the cement grades produced by the company the standard compressive strengths after 2 and 28 days were evaluated for those periods during which no systematic changes were made in the grinding conditions as a result of changed quality targets. The values therefore come from operating periods in which the control of the cement mills is based on constant quality targets without external intervention.

With individual values from n = 26 to 116 the standard deviations of the compressive strength after 2 days ranges between 1.2 and 1.7 N/mm^2, and the standard deviations of compressive strengths after 28 days range between 1.5 and 1.9 N/mm^2.

If the statistically determined standard deviations of the compressive strength test in accordance with DIN 1164, including all work procedures, are set at 0.7 N/mm^2 for the 2-day strength and 0.5 N/mm^2 for the 28-day strength – these values for the standard deviation of the test itself come from

Zementart	D 2		D 28	
	[N/mm2]		[N/mm2]	
	X	± S	X	± S
PÖZ 35 F	25,4	1,4	51,8	1,5
PÖZ 35 F Terrament	25,1	1,7	50,7	1,6
PÖZ 45 F	34,9	1,5	59,9	1,7
PÖZ 45 F Terrament	34,5	1,4	59,5	1,6
PÖZ 55	42,6	1,3	63,7	1,8
PZ 35 F	26,5	1,4	48,2	1,9
PZ 45 F	37,1	1,2	56,8	1,5

BILD 13: Gleichmäßigkeit der Normdruckfestigkeit von verschiedenen Zementen

Cement type	D 2		D 28	
	[N/mm2]		[N/mm2]	
	X	± S	X	± S
PÖZ 35 F	25,4	1,4	51,8	1,5
PÖZ 35 F Terrament	25,1	1,7	50,7	1,6
PÖZ 45 F	34,9	1,5	59,9	1,7
PÖZ 45 F Terrament	34,5	1,4	59,5	1,6
PÖZ 55	42,6	1,3	63,7	1,8
PZ 35 F	26,5	1,4	48,2	1,9
PZ 45 F	37,1	1,2	56,8	1,5

FIGURE 13: Uniformity of the standard compressive strengths of different cements

Unternehmen erzeugte Zementqualitäten wurden die Normdruckfestigkeiten nach 2 und 28 Tagen für Zeiträume ausgewertet, bei denen eine systematische Veränderung der Mahlbedingungen durch veränderte Qualitätsvorgaben nicht vorgenommen wurde. Deshalb stammen die Werte aus Betriebszeiten, in denen die Regelung der Zementmühlen ohne Eingriffe von außen auf konstanten Qualitätsvorgaben basierte.

Bei Einzelwerten von n = 26 bis 116 bewegen sich die Standardabweichungen der Druckfestigkeit nach 2 Tagen zwischen 1,2 und 1,7 N/mm^2, die Standardabweichungen der Druckfestigkeit nach 28 Tagen zwischen 1,5 und 1,9 N/mm^2.

Werden die Standardabweichung der Druckfestigkeitsprüfung nach DIN 1164 einschließlich aller Arbeitsvorgänge mit 0,7 N/mm^2 für die 2-Tage-Festigkeit und 0,5 N/mm^2 für die 28-Tage-Festigkeit – diese Werte für die Standardabweichung der Prüfung selbst stammen aus dem Forschungsinstitut der Zementindustrie – statistisch ermittelt, so liegen die Standardabweichungen der reinen Zementqualität zwischen 1,0 und 1,2 N/mm^2.

Die vorgestellten Werte für die Gleichmäßigkeit von Rohmehl und Zementen lassen den Schluß zu, daß bei Betrieb einer Laborautomation die bisher übliche Fremdüberwachung der Versandzemente ausreicht, um die Qualität des Produktes selbst, jedoch auch des gesamten Produktionsablaufes im Sinne einer Performanceprüfung sicher zu beschreiben.

8. Ausblick

Bei extrem schwankenden Qualitäten des Kalksteins aus dem Steinbruch ist das System der Laborautomation zur Rohmehlregelung nur bedingt verwendbar. Weitere Verbesserungen der Regelung durch Einbeziehen der heute von Hand vorgenommenen Korrekturen in die Software könnte, ähnlich wie bei der Zementmühlenregelung als sogenanntes Expertensystem realisiert, für Abhilfe sorgen.

Ebensowenig gehen heute in die Rohmehlregelung betriebswirtschaftliche Grundlagen und Voraussetzungen der Lagerstättenverfügbarkeit ein. Im Rahmen der Ressourcenschonung und Kostenminimierung besteht deshalb über die reine Technik hinausgehender Handlungsbedarf.

Neue on-line-analytische Verfahren könnten zukünftig eine Vereinfachung der Laborautomation bewirken. Ein bereits genutztes Verfahren zur chemischen Analytik basiert auf der Anregung der Atomkerne der zu messenden Elemente durch Neutronenbeschuß. Die Sekundäremissionen werden gemessen und mit Eichproben verglichen [20, 14].

Zur Ermittlung der Korngrößenverteilung durch On-line-Analyse von Rohmehl und Zement gibt es erste betriebliche Erfahrungen [21].

the cement industry's Research Institute – then the standard deviations of the cement quality on its own lie between 1.0 and 1.2 N/mm^2.

The values given for the uniformity of raw meal and cements lead to the conclusion that when a laboratory automation system is in operation the external monitoring of the cement for despatch, which is already general practice, is sufficient to give a reliable description not only of the quality of the product itself but also of the entire production sequence in the context of performance testing.

8. Outlook

If there are extreme fluctuations in the quality of the limestone from the quarry then the laboratory automation system is only of limited use for controlling the raw meal. This could be remedied by further improvements to the control system by using software – in the form of a so-called expert system in the same way as for the cement mill control – for the corrections which at present are carried out by hand.

Equally few fundamental industrial management principles and preconditions for availability of deposits are input into the raw meal control system. In the context of preservation of resources and cost minimization there is therefore a need for action which goes beyond pure technology.

In the future, laboratory automation could be simplified by new on-line methods of analysis. One method of chemical analysis already in use is based on excitation of the atomic nuclei of the elements to be measured by neutron bombardment. The secondary emissions are measured and compared with calibration samples [20, 14].

The first operational results are now available for the measurement of particle size distribution by on-line analysis of raw meal and cement [21].

9. Resumé

To summarize, it can be stated that the laboratory automation systems currently in existence are suitable for monitoring, and for providing open- and closed-loop control for operational sequences. With their help and in combination with these open- and closed-loop control systems in the works it is possible to improve the uniformity of product quality to a high level. The great increase in the quantity of measured data and information requires management systems for processing the data for all areas of production, administration and management of a company to suit the natures of the different tasks.

Laboratory automation is a suitable way of making a saving in shift laboratory personnel. Against this must be set the not inconsiderable investment costs of 3 to 5 million DM per outfit. The demand for high quality exacts its price.

9. Resümee

Zusammenfassend ist festzustellen, daß die heute existierenden Laborautomationssysteme geeignet sind, betriebliche Abläufe zu überwachen, zu steuern und zu regeln. Mit ihrer Hilfe und in Kombination mit diesen Steuerungs- und Regelsystemen im Werk kann die Gleichmäßigkeit der Produktqualität auf hohem Niveau verbessert werden. Die Vervielfachung von Meßdaten und Informationen erfordert Managementsysteme, die die Daten für alle Produktions-, Verwaltungs- und Führungsbereiche eines Unternehmens den differenzierten Aufgabenstellungen entsprechend aufbereiten.

Die Laborautomation ist geeignet, Schichtlaborpersonal einzusparen. Dem stehen nicht unerhebliche Investitionskosten von 3 bis 5 Mio. DM je nach Ausstattung gegenüber. Der Anspruch auf hohe Qualität fordert also auch seinen Preis.

Literaturverzeichnis

[1] Riedhammer, M.: Prozeßsteuerungsmaßnahmen bei Rohrbach Zement in Dotternhausen. Zement-Kalk-Gips 38 (1985) Nr. 7, S. 359–364.

[2] Triebel, W.: Zentrale, zukunftsorientierte Qualitäts- und Prozeßsteuerung im Zementwerk Karlstadt. Zement-Kalk-Gips 39 (1986) Nr. 9, S. 482–487.

[3] Sprung, S.: Maßnahmen und Möglichkeiten zur Qualitätssteuerung im Zementwerk. Zement-Kalk-Gips 43 (1990) Nr. 7, S. 340–346.

[4] Galvez, J. A., und Teutenberg, J.: Moderne Laborautomation mit Robotertechnik im Zementwerk Monjos der Uniland Cementera S.A., Spanien. Zement-Kalk-Gips 44 (1991) Nr. 6, S. 299–306.

[5] Martinez Ynzenga, J. I., Bergenfelt, C., und Pedersen, S. T.: Laboratory automation by robotics. Zement-Kalk-Gips 43 (1990) Nr. 6, S. 277–281.

[6] Gecks, W., und Pedersen, S. T.: Robotics – an efficient tool for laboratory automation. Zement-Kalk-Gips 44 (1991) Nr. 6, S. 275–280.

[7] Teutenberg, J.: Höhere Leistungsdichten im Zementlabor mit multimodularer Robotertechnik. Vortrag zum VDZ-Kongreß 1993, Fachbereich 2.

[8] Eggert, A., und Teutenberg, J.: Qualitätssicherung und -steuerung durch modulare und flexible Systemtechnik – POLAB. Zement-Kalk-Gips 45 (1992) Nr. 2, S. 70–78.

[9] Triebel, W.: Entwicklungsschritte zur Automationstechnik des Werkes Bernburg. Vortrag zum VDZ-Kongreß 1993, Fachbereich 2.

[10] Jäger, G.: Werksprospekt Humboldt Prozeßautomation GmbH, 1992.

[11] Juchli, K., Bonvin, D., und Yellepeddi, R.: Ein integriertes Röntgen-Diffraktions- und Fluoreszenzspektrometer und seine Anwendungsmöglichkeiten in der Zementindustrie. Vortrag zum VDZ-Kongreß 1993, Fachbereich 2.

[12] Beilmann, R., und Brüggemann, H.: Quantitative XRD clinker phase analysis, a tool for process optimization and quality control. Ciments, Bétons, Plâtres, Chaux, No. 791–4/91, S. 247–251.

[13] Fröhlich, A.: Quantitative Klinkerphasenanalyse mittels XRD. Vortrag zum VDZ-Kongreß 1993, Fachbereich 2.

[14] Tschudin, M.: On-line Kontrolle des Mischbettaufbaus mittels PGNAA in Ramos Arizpe (Mexiko). Vortrag zum VDZ-Kongreß 1993, Fachbereich 2.

[15] Espig, D., Reinsch, V., Bicher, B., und Otto, J.: Computersimulation – Wertvolle Entscheidungshilfe für die Optimierung der Zementmahlung. Vortrag zum VDZ-Kongreß 1993, Fachbereich 2.

[16] Hepper, R.: Vorteile der Prozeßführung mit Expertensystemen. Vortrag zum VDZ-Kongreß 1993, Fachbereich 2.

[17] de Pierpont, C., van Breusegem, V., Chen, L., Bastin, G., und Wertz, V.: Industrielle Anwendung einer multivariablen linear-quadratischen Steuerung für Zementmühlen. Vortrag zum VDZ-Kongreß 1993, Fachbereich 2.

[18] Baerentsen, H., Petersen, T. S., und Senyuva, Z.: Labor-Informations-Management-System (LIMS) in Zementwerkslaboratorien. Vortrag zum VDZ-Kongreß 1993, Fachbereich 2.

[19] Sedlmeir, W.: Ein zukunftssicheres Management-Informationssystem für Zementwerke. Vortrag zum VDZ-Kongreß 1993, Fachbereich 2.

[20] Glorieux, G.: Operating experience with on-line analyzer for automatic raw mix control in Belgian Cement Plant. Ciments, Bétons, Plâtres, Chaux, No. 789–2/91, S. 77–83.

[21] Höffl, K., und Folgner, T.: Granulometrische On-line-Bewertung von Partikelschüttungen unter Anwendung des radiometrischen Transmissions-Meßprinzips. Zement-Kalk-Gips 44 (1991) Nr. 7, S. 371–375, und Zement-Kalk-Gips 45 (1992) Nr. 8, S. 403–418.

Prozeßführung der Zementmahlanlage III im Werk Neubeckum der Dyckerhoff AG

Process control of cement mill III in Dyckerhoff AG's Neubeckum works

Conduite du process de l'atelier de broyage de ciment III à l'usine Neubeckum de la Dyckerhoff AG

Control del proceso de la planta de molienda de cemento III de la factoría de Neubeckum, de Dyckerhoff AG

Von **F.-J. Barton,** Beckum/Deutschland

Zusammenfassung – Im Werk Neubeckum der Dyckerhoff AG wurde eine neue Zementmahlanlage mit Gutbett-Walzenmühle und Kugelmühle in Hybridschaltung und integrierter Hüttensandtrocknung errichtet. Bei diesem Neubau wurde erstmals ein Prozeßleitsystem eingeführt. Schon während der Projektierung wurden die Möglichkeiten der Automatisierungs- und Rechnertechnik genutzt, um damit ein Optimum an Betriebsverfügbarkeit und Bedienungsfreundlichkeit bei der gegebenen Komplexität der Anlage zu realisieren. Mittels umfangreicher Programme im Prozeßleitsystem für eine komfortable Rezepturverwaltung mit Sortenanwahl, Verfahrens- und Laborparameteranwahl wurden die Aufgabenstellungen umgesetzt. Hierbei wird durch minimalen Bedienaufwand ein Maximum an Parametern zum Betreiben der Mahlanlage in den Prozeß eingebracht. Eine vorausschauende Steuerungslogik mit aufwendigen Aggregatschwerpunkten unterstützt dabei das Bestreben nach maximaler Verfügbarkeit der Gesamtanlage. Eine weitere Aufgabe bestand darin, die neue Technik innerhalb der Werksstruktur anzusiedeln und das Prozeßleitsystem mit Fremdsystemen zu koordinieren.

Prozeßführung der Zementmahlanlage III im Werk Neubeckum der Dyckerhoff AG

Summary – A new cement grinding plant was built at Dyckerhoff AG's Neubeckum works, with high-pressure grinding rolls and ball mill in a hybrid circuit and an integral system for drying granulated blast furnace slag. A process control system was introduced for the first time in this new plant. Even during the planning stage use was made of the potential of the automation and computation technology to achieve optimum availability and operator friendliness with the given complexity of the plant. Extensive programs were used to convert the functions to be performed into a process control system for convenient mix management with product type selection and selection of process and laboratory parameters. This means that the maximum number of parameters for operating the grinding plant are introduced into the process with minimum operator involvement. The efforts to achieve maximum availability of the entire plant are supported by a forward-looking control logic with sophisticated aggregate focal points. Another task was to locate the new technology within the works structure and to co-ordinate the process control system with outside systems.

Process control of cement mill III in Dyckerhoff AG's Neubeckum works

Résumé – A l'usine Neubeckum de la Dyckerhoff AG a été construite une nouvelle installation de broyage de ciment, avec presse à rouleaux et broyeur à boulets en marche hybride et installation de séchage du laitier intégrée. Avec cet atelier nouveau a été introduit, pour la première fois, un système de conduite du process. Déjà pendant l'étude ont été utilisées les possibilités de la technique d'automatisation et de calcul, afin de réaliser ainsi un optimum de disponibilité de service et de convivialité compte tenu de la complexité donnée de l'atelier. Les tâches imposées ont été traduites au moyen de programmes exhaustives dans le système de conduite du process, pour une gestion confortable des compositions avec choix préliminaire des sortes, appel des paramètres du process et du laboratoire. Ainsi est introduit un maximum de paramètres pour l'exploitation de l'atelier de broyage, avec un effort minimal de conduite. Une logique prévoyante, avec des points forts importants des équipements, assiste alors l'effort pour obtenir la disponibilité maximale de l'ensemble de l'atelier. Une autre tâche était d'assimiler la nouvelle technique dans la structure interne de l'usine et de coordonner le système de conduite du process avec des systèmes étrangers.

Conduite du process de l'atelier de broyage de ciment III à l'usine Neubeckum de la Dyckerhoff AG

Resumen - En la factoría de Neubeckum de Dyckerhoff AG se ha instalado una nueva planta de molienda de cemento con molino de cilindros y lecho de material y molino de bolas, en conexión híbrida y con secado integrado de escorias de horno alto. En esta nueva construcción se ha introducido, por primera vez, un sistema de control de proceso. Se han aprovechado ya durante la fase de planificación las posibilidades que ofrecen las técnicas de automatización e informáticas, con el fin de conseguir un máximo de disponibilidad y de facilidad de servicio, teniendo en cuenta la complejidad de la instalación. Las diferentes tareas planteadas han sido realizadas mediante amplios programas, dentro del sistema de control de proceso, para conseguir una gestión cómoda de recetas, con selección de tipos de productos y selección de parámetros tecnológicos y de laboratorio. Con ello, se introduce en el proceso, con un mínimo de es-

Control del proceso de la planta de molienda de cemento III de la factoría de Neubeckum, de Dyckerhoff AG

Fachbereich 2 · Subject 2 · Séance Technique 2 · Tema de ramo 2

fuerzos, un máximo de parámetros para el funcionamiento de la planta de molienda. Una lógica de control precavida, que incluye sofisticados componentes, ayuda a conseguir una máxima disponibilidad del conjunto de la instalación. Otra tarea ha sido la de implantar la nueva técnica dentro de la estructura de la fábrica y de coordinar el sistema de control de proceso con sistemas extraños.

1. Einleitung

Bei dem vorgestellten Automatisierungsprojekt handelt es sich um eine Zementmahlanlage mit Gutbett-Walzenmühle und Kugelmühle in Hybridschaltung und integrierter Hüttensandtrocknungsanlage. Die Aufgabenstellung bestand darin, 5 verschiedene Portlandzemente, 3 Hochofenzemente, Flugaschezement, Hochhydraulischen Kalk und Hüttensandmehl mit einer Mahlanlage herstellen zu können. Bedingt durch die Vielzahl der Produkte und die damit verbundenen unterschiedlichen Zielsilos wurde von der Steuerungstechnik ein Höchstmaß an Optimierung im Bedienungsablauf gefordert.

2. Die Prozeßanlagenstruktur

Das Automatisierungsprojekt wurde mit dem Prozeßleitsystem „Contronic P" von Hartmann & Braun realisiert. Für die Zementmühle III ergibt sich eine Anlagenstruktur, die im **Bild 1** hierarchisch gegliedert dargestellt ist. An die Betriebsleitebene als oberste Hierarchieebene wurde auch Hardware der Firma Digital Equipment mit entsprechender Software angebunden. Die zweite Hierarchiestufe stellt die Prozeßleitebene dar, die das Führen, Protokollieren, Optimieren und die Kopplung zum Laborleitsystem übernimmt. In der Feldebene ist die Kopplung zu Fremdsystemen wie beispielsweise speicherprogrammierbaren Steuerungen, zum Waagendatenkonzentrator sowie zu den Prozeßstationen angeordnet.

3. Bedienen der Zementmühle III mit dem Prozeßleitsystem

Ein besonderes Augenmerk wurde auf die Bedienung der Zementmühle III gelegt. So sollten nicht nur die alten Taster gegen eine neue Folientastatur getauscht werden. Vielmehr

1. Introduction

The automation project described deal with a cement grinding plant with high-pressure grinding rolls and ball mill in a hybrid circuit and an integral system for drying granulated blast furnace slag. The problem was how to produce 5 different Portland cements, 3 blast furnace cements, fly ash cement, hydraulic lime and powdered blast furnace slag with one grinding plant. Owing to the large number of products and the different target silos involved, the control system was required to provide maximum optimisation of the operating sequences.

2. The process plant structure

The automation project was carried out with Hartmann & Braun's "Contronic P" process control system. Cement mill III has a plant structure as shown, subdivided hierarchically, in **Fig. 1**. The plant management level, being the highest rank in the hierarchy, also has Digital Equipment hardware and appropriate software connected to it. The second rank in the hierarchy is the process management level, responsible for control, logging, optimisation and coupling with the laboratory management system. The field level contains the coupling with outside systems such as PLC systems, the weighing data concentrator and the processing stations.

3. Operation of cement mill III with the process control system

Special attention was paid to the operation of cement mill III. It was not just to be a case of exchanging the old control switches for a new sealed keyboard. With the large number of products and the frequent changes involved, the whole operating philosophy was to be altered and the control room

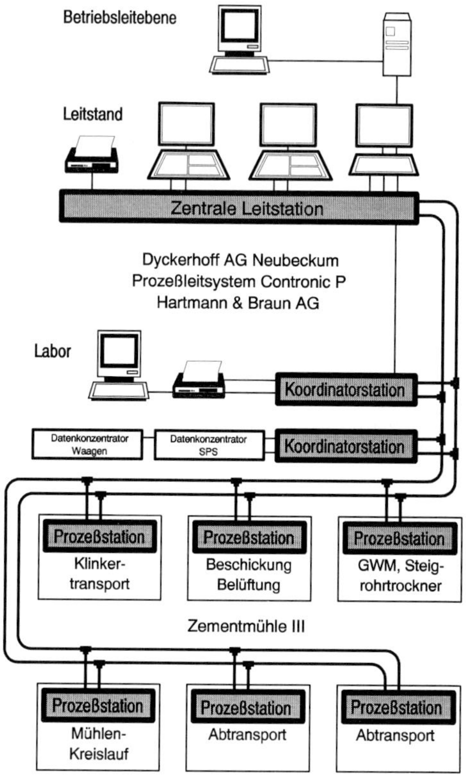

BILD 1: Hierarchischer Aufbau des Prozeßleitsystems
FIGURE 1: Hierarchic structure of process control system

Betriebsleitebene	= plant management level
Leitstand	= control room
Zentrale Leitstation	= central control station
Labor	= laboratory
Koordinatorstation	= coordinating station
Datenkonzentrator	= data concentrator
Waagen	= weighing
SPS	= PLC system
Prozeßstation	= process station
Klinkertransport	= clinker transport
Beschickung	= feed
Belüftung	= aeration
GWM	= high-pressure grinding rolls
Steigrohrtrockner	= pneumatic conveyor dryer
Zementmühle III	= cement mill III
Mühlen-Kreislauf	= mill circuit
Abtransport	= dispatch

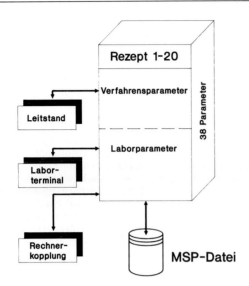
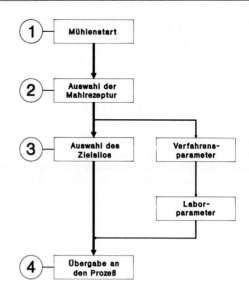

BILD 2: Aufbau und Abruf der Rezepturdateien
FIGURE 2: Creation and poling of mix files

Rezept	= mix
Leitstand	= control room
Verfahrensparameter	= process parameters
Laborterminal	= laboratory terminal
Laborparameter	= laboratory parameters
38 Parameter	= 38 parameters
Rechnerkopplung	= computer connection
MSP-Datei	= mass storage file

BILD 3: Einschaltschema für Mühlenstart
FIGURE 3: Start-up sequence for starting mill

Mühlenstart	= start mill
Auswahl der Mahlrezeptur	= select grinding mix
Auswahl des Zielsilos	= select target silo
Verfahrensparameter	= process parameters
Laborparameter	= laboratory parameters
Übergabe an den Prozeß	= transfer to the process

sollte sich die Bedienphilosophie dahingehend verändern, daß der Leitstandsfahrer wegen der Vielzahl der Produkte und den damit verbundenen häufigen Sortenwechseln mit so wenig Befehlen wie möglich die Mühle bedient.

3.1 Aufbau der Rezepturdatei

Ausgehend von dem Gedanken, durch einen Tastendruck alle prozeßrelevanten Größen für die Herstellung einer bestimmten Produktqualität an den Prozeß zu übergeben, wurden die Parameter in Form von Rezeptdateien zusammengefaßt, wie es in **Bild 2** veranschaulicht ist. Die Anzahl dieser Dateien richtet sich nach der Anzahl der Mahlrezepturen. Die Rezeptdatei ist für jedes Mahlprodukt auf dem Massenspeicher hinterlegt und wird bei Bedarf von diesem geladen und beinhaltet je Datei 38 Parameter.

3.2 Ablauf des Mühlenstarts

Für den Start der Mühle wird ein Bedienvariablenbild aufgerufen. In ihm sind alle verfahrenstechnischen Anwahlen zum Betreiben der Mühle hinterlegt, z.B. der Mühlenstart, der Betrieb mit Gutbett-Walzenmühle und auch mit dem Steigrohrtrockner. Die einzelnen Schritte des Mühlenstarts sind in **Bild 3** aufgeführt. Bei der Anwahl des Rezepturbildes wird zwischen aktuellem und zukünftigem Mahlungsprozeß unterschieden und direkt eine Sorte, z.B. PZ 45 F, angewählt. Das in **Bild 4** dargestellte Rezepturbild ist beispielhaft für die komfortable Bedienerführung.

Nach Abfrage der Rezepturdatei bezüglich der Zielsilos erscheint das Siloverteilungsbild mit der aktuellen Rezeptur und den möglichen Zielsilos. Daraufhin erfolgt die Auswahl des Zielsilos und danach eine Mitteilung des Systems über die Verfügbarkeit des Weges mit seinen Antrieben.

Mit der Übernahme der Werte und deren Übergabe an den Prozeß hat die zuvor gestartete Mühle ihre aktuelle Rezeptur und Verfahrensparameter übernommen. Somit wurden also nur maximal vier Anwahlen getroffen, um die komplette Mühle mit Komponentendosierung und Silobeschickung in Betrieb zu nehmen. Eine Änderung der Beteiligungen und Anteile der Komponenten für die aktuelle und künftige Mahlung erfolgt durch ein separates Laborbild vom Leitstand oder mittels eines externen Terminals aus dem Labor und in Zukunft direkt über den Rechner der Laborautomation.

operator was to work the mill with the fewest commands possible.

3.1 Creation of mix files

Based on the idea that all relevant values for making a certain grade of product should be introduced into the process by pressing a button, the parameters were combined in the form of mix files as illustrated in **Fig. 2**. The number of files depends on the number of grinding mixes. The mix file for each ground product is deposited in the mass storage and loaded from it as; each file contains 38 parameters.

3.2 Mill start-up process

To start the mill an operating variable display is first called up. It contains all the process options for operating the mill, e.g. start-up, operation with high-pressure grinding rolls and operation with the pneumatic conveyor dryer. The various steps in starting the mill are listed in **Fig. 3**. The mix display option distinguishes between a current and a future grinding process and directly selects one type, e.g. PZ 45 F. The mix display in **Fig. 4** is an example of operator-friendly control.

When the mix file is interrogated about target silos the silo distribution display appears with the current mix and the possible target silos. The target silo is then selected, after which a system message is received concerning the availability of the path and its drives.

With the transfer of the values and their introduction into the process the previously started mill has accepted the current mix and process parameters. Thus only a maximum of four selections have to be made to set the whole mill in operation with constituent metering and silo feeding. Any change in the inclusion and proportions of constituents for current and future grinding is produced by a separate laboratory display from the control room or by means of an external terminal from the laboratory; in the future it will be produced directly by the laboratory automation computer.

The process parameters are adapted to the process values by means of the option display; this is mainly a repository for mass flow presettings. When the changes have been entered

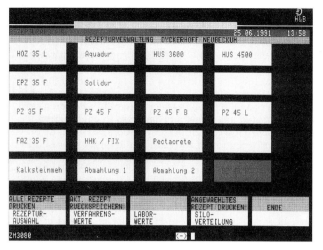

BILD 4: Menü-Darstellung zur Auswahl der gewünschten Rezeptur
FIGURE 4: Menu for selecting required mix

Rezepturauswahl	= mix selection
Rezepturverwaltung	= mix management
Kalksteinmehl	= powdered limestone
Abmahlung	= changeover grinding
Alle Rezepte drucken	= print all mixes
Akt. Rezept rückspeichern	= re-save current mix
Angewähltes Rezept drucken	= print selected mix
Ende	= end
Rezeptur-Auswahl	= mix selection
Verfahrens-Werte	= process values
Labor-Werte	= laboratory values
Silo-Verteilung	= silo distribution

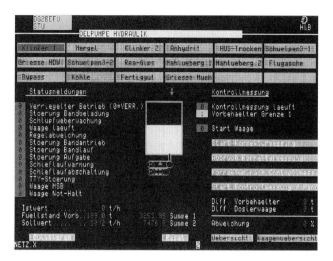

BILD 5: Darstellung der Waagensteuerung im Prozeßleitsystem
FIGURE 5: Weighing control in process control system

Oelpumpenhydraulik	= oil pump hydraulics
Klinker	= clinker
Anhydrit	= anhydrite
HUS-Trocken	= dry granulate blast furnace slag
Schuelpen	= scab
Griesse HDW	= tailings
Rea-Gips	= desulphogypsum
Mahlueberg.	= grinding changeover
Flugasche	= fly ash
Kohle	= carbon
Fertiggut	= finished product
Griesse Mueh	= tailings mill

Statusmeldungen	= Status Messages
Verriegelter Betrieb	= interlocked operation
0=VERR.	= 0=INTERLOCKED
Stoerung Bandbeladung	= belt loading malfunction
Schlupfueberwachung	= slip monitoring
Waage laeuft	= weigher working
Regelabweichung	= control deviation
Stoerung Bandantrieb	= belt drive malfunction
Stoerung Bandlauf	= belt movement malfunction
Stoerung Aufgabe	= feed malfunction
Schieflaufwarnung	= off-track running warning
Schieflaufabschaltung	= stop off-track running
TTY-Stoerung	= TTY fault
Waage MSB	= weigher MSB
Waage Not-Halt	= weigher emergency stop
Istwert	= actual value
Fuellstand Vorb.	= filling level, feed container
Sollwert	= desired value
Summe	= total
Quittieren	= acknowledge
Netz	= network
Kontrollmessung	= check measurement
Kontrollmessung laeuft	= check measurement running
Vorbehaelter Grenze 1	= feed container limit 1
Start Waage	= start weigher
Start Korrekturmessung	= start corrective measurement
Abbruch Korrekturmessung	= stop corrective measurement
Korrektur nach Kontrollmess.	= correction after check measurement
Start Kontrollmessung/Hand	= start check measurement/manual
Diff. Vorbehaelter	= difference feed container
Diff. Dosierwaage	= difference metering weigher
Abweichung	= deviation
Uebersicht	= summary
Waagenuebersicht	= weighing summary

Die verfahrenstechnischen Parameter werden über das Anwahlbild der Verfahrenswerte angepaßt. Hier sind in erster Linie Massenstromvoreinstellungen hinterlegt. Nach erfolgreicher Eintragung werden die Änderungen auf der Festplatte archiviert und stehen beim nächsten Aufruf automatisch zur Verfügung.

4. Datenübertragung zwischen Prozeßleitsystem und Waagensteuerung

Bei der Anbindung der Waagen an das Prozeßleitsystem wird eine Variante praktiziert, bei der die Regelung der Waagen vom Waagenhersteller realisiert wurde und die Komponentenvorgabe von dem Prozeßleitsystem erfolgt. Dazu ist jede Waage mit einem eigenen Steuergerät ausgerüstet, das wiederum entsprechend dem Schema in **Bild 5** mit dem zentralen Prozeßrechner in Verbindung steht. Zur Optimierung der Waagenjustage und Fehlerdiagnose ist im Prozeßleitsystem pro Waage ein Anwahlbild mit sämtlichen der successfully they are archived on the hard disk and are automatically available next time they are called up.

4. Data transmission between the process control system and weighing control

There is a different arrangement for the connection of the weighers to the process control system, with control systems for the weighers being provided by the weigher manufacturer and the constituents being preset by the process control system. Each weigher is fitted with its own control device, which is again connected to the central processing computer as shown in **Fig. 5**. As a means of optimising weigher adjustment and error diagnosis an option display for each weigher is included in the process control system, with all the status messages connected with the weigher and container measurements. The control room operator can initiate a measurement check.

Waage und Behältermessung angeschlossenen Status-Meldungen hinterlegt. Eine Kontrollmessung kann vom Operator im Leitstand eingeleitet werden.

5. Betriebssicherheit durch Silobeschickungslogik

In der Mahlanlage befindet sich eine pneumatische Förderanlage zum Abtransport des Fertiggutes mit einer Vielzahl von Antrieben. Aus den Erfahrungen der vorhandenen Zementmühle II ist bekannt, daß gerade beim Zielsilowechsel Störungen an den Rohrweichen auftreten, die einen Stillstand der Mühle zur Folge haben. Eine Erhöhung der Betriebssicherheit bei der Zementmühle III wird durch eine überarbeitete Aggregateanordnung in Verbindung mit einer umfangreichen Steuerungslogik erreicht. Dabei wird der Wegaufbau zum Zielsilo, im Gegensatz zur bisherigen Steuerung, parallel zur Förderung in die Mahlübergangsbehälter ausgeführt und dem Leitsystem rückgemeldet, ob der Weg zum Silo erfolgreich aufgebaut wurde. Ist dies nicht der Fall, bleibt genügend Zeit, den für die Förderung offenen Weg zu wählen bzw. den Fehler bei den Rohrweichen zu beseitigen. Nach erfolgreichem Positionieren der Rohrweichen erfolgt bei der Umstellung lediglich das Schalten von 2 Kugelhähnen, d.h. die Anzahl der Fehlerquellen ist auf ein Minimum reduziert.

6. Betriebserfahrungen

Seit Juli 1991 ist die neue automatisierte Zementmahlanlage erfolgreich in Betrieb. Aus elektrotechnischer Sicht sind die Erwartungen an den Einsatz des Prozeßleitsystems voll erfüllt worden. Der Aufwand während der Projektierungsphase hat sich durch die konsequente Nutzung der Möglichkeiten der Automatisierungstechnik für die Rezepturverwaltung, die Silobeschickungslogik sowie durch die komfortable Darstellung von Zusatzinformationen bei der täglichen Arbeit des Bedienpersonals und der Elektromitarbeiter bezahlt gemacht.

5. Operational reliability provided by the silo feed logic system

The grinding plant contains pneumatic conveying equipment for carrying away the finished product, with several drives. It is known from experience with existing cement mill II that malfunctions occur at the diverters in the pipes when the target silo is changed, leading to stoppage of the mill. Operational reliability at cement mill III is increased by a sophisticated aggregate arrangement in conjunction with an extensive control logic system. In contrast with the previous control, creation of the path to the target silo is carried out in parallel with conveying to the transitional grinding container, and the control system is informed whether the path to the silo has been created successfully. If not, there is still time to select a path which is open for conveying or to deal with the error at the pipe diverters. Once the diverters have been positioned successfully the change is effected simply by operating two ball valves, so sources of error are minimised.

6. Operating experience

The new automated cement grinding plant has been operating successfully since July 1991. The process control system has fulfilled all expectations from the electrotechnical point of view. The outlay during the planning phase has paid off in the daily work of operators and electrical engineers, through the appropriate use of the capabilities of the automation technology for mix management, the silo feed logic system and the user-friendly presentation of additional information.

Überwachung der Sinterzone durch Messung der Alkaliemissions- und -absorptionsspektren in der Gasphase

Kiln sintering zone monitor using alkali spectrum emission and absorption

Surveillance de la zone de cuisson par relevé de mesures des spectres d'absorption et d'émissions d'alcalis durant la phase gazeuse

Control de la zona de sinterización midiendo los espectros de emisión y de absorción de álcalis durante la fase gaseosa

Von **H. O. Biggs, N. S. Smith** und **M. Hirayama,** Lucerne Valley/USA

Zusammenfassung – Aufgrund der weitverbreiteten Anwendung der automatischen Ofensteuerung ist die Überwachung der Sinterzone in Zementdrehöfen besonders wichtig geworden. Zur Zeit werden Zweifarben-Pyrometer, NO_x-Analysegeräte, Ofendrehmomentmesser usw. zur Beurteilung der Bedingungen in der Sinterzone eingesetzt. Es ist bekannt, daß die Atmosphäre der Ofensinterzone aufgrund des Phänomens der Alkalikreisläufe einen hohen Gehalt an Alkalidämpfen hat und daß diese Dämpfe stets in metastabilem Zustand sind. Wenn sich die Bedingungen in der Sinterzone ändern, beobachtet man Alkaliemissions- und Alkaliabsorptionsspektren. Bei einem Anstieg der Temperatur in der Sinterzone geht der Atomzustand der Alkalien vom Normalzustand in einen angeregten Zustand über, der durch ein Absorptionsspektrum gekennzeichnet ist und umgekehrt. Die Form der Überwachung ist zwar gleich, nur daß die verfügbaren Zweifarben-Pyrometer innen drei Filter haben, um die optische Intensität bei drei verschiedenen Wellenlängen zu messen; die Emission bzw. Absorption des jeweiligen Spektrums wird dann durch Berechnung ermittelt. Das Überwachungssystem wurde 1990 im Werk Cushenbury, Kalifornien, der Mitsubishi Cement Corporation installiert und ist seit mehr als 2 Jahren in Betrieb. Der Vorteil der Messungen ist darin zu sehen, daß diese nicht durch die Flammenausbildung oder andere Betriebsfaktoren, wie die spezifische Strahlung beim Klinkeraustrag, beeinträchtigt werden. Die gemessenen Werte werden je nach Emissions- oder Absorptionsverhalten des K-Spektrums in der Gasatmosphäre der Sinterzone mit der Temperatur korreliert.

Summary – With the wide application of automatic kiln control it has become more important to monitor the sintering zone condition in cement rotary kilns. Currently, two-colour pyrometers, NO_x-analyzers, kiln torque meters, etc. are used in practice for judging the state of the kiln. It is well known that the atmosphere in the kiln sintering zone has a high alkali vapour content due to the alkali circulation phenomenon, and that the alkali vapour is always in a meta-stable condition. Changes in the sintering zone alkali spectrum emission and absorption are observed as functions of the state of the kiln. When the sintering zone temperature increases the atomic state of the alkalis shifts from the ground state to an excited state, which appears in the absorption of the alkali spectrum and vice versa. Although the form of the monitoring is identical, three sets of filters are placed inside existing two colour pyrometers to detect the optical intensity at three different wave lengths, and the emission or absorption of the specific spectrum is obtained by calculation. The monitor was installed in the Cushenbury plant of Mitsubishi Cement Corporation, California, in 1990 and has been working for more than 2 years. The advantage of the measurement is that it is not affected by flame ondition or other operation factors like the emissivity of the discharging clinker. The detection only correlates with the temperature as a function of the emission or absorption behaviour of the K-spectrum in the sintering zone gas atmosphere.

Résumé – Etant donné que l'automatisation de la conduite des fours s'est largement répandue, la surveillance de la zone de cuisson dans les fours rotatifs de cimenterie revêt une importance toute particulière. Actuellement, pour juger des conditions régnant durant la zone de cuisson, on fait appel à des pyromètres à deux couleurs, des appareils d'analyse de NO_x, des couplemètres de four etc. On sait par exemple que l'atmosphère de la zone de cuisson du four présente, en raison du phénomène des circuits alcalins, une teneur élevée en vapeurs alcalines, lesquelles se trouvent toujours à l'état métastable. Quand les conditions de la zone de cuisson changent, on observe des spectres d'absorption et d'émission d'alcalis. Dès accroissement de la température dans la zone de cuisson, l'état atomique des alcalis passe de l'état normal à l'état excité, et il est caractérisé par un spectre d'absorption et réciproquement. La forme de supervision est la même, seulement les pyromètres disponibles à deux couleurs ont trois filtres

à l'intérieur pour mesurer l'intensité optique sur trois longueurs d'onde différentes; l'émission ou l'absorption du spectre correspondant est évaluée par calcul. Ce système de surveillance qui a été installé en 1990 à l'usine de Cushenbury, Californie, de la Mitsubishi Cement Corporation, est en service depuis plus de deux ans. L'avantage de ce relevé des mesures est que celles-ci ne sont pas affectées par la formation de la flamme ou d'autres facteurs d'exploitation comme le rayonnement spécifique lors du déchargement du clinker. Les valeurs mesurées sont, en fonction de la courbe d'emission ou d'absorption du spectre K dans l'atmosphère gazeuse de la zone de cuisson, mises en correlation avec la température.

Control de la zona de sinterización midiendo los espectros de emisión y de absorción de álcalis durante la fase gaseosa

Resumen – Debido a la aplicación generalizada del mando automático de los hornos, ha adquirido especial importancia la vigilancia de la zona de sinterización en los hornos rotatorios de cemento. Actualmente, se están empleando pirómetros a dos colores, aparatos de análisis de NO_x, dispositivos de medición del momento de giro del horno, etc., para evaluar las condiciones reinantes en la zona de sinterización. Es sabido que la atmósfera de la zona de sinterización del horno, debido al fenómeno de los circuitos de álcalis, tiene un elevado contenido de vapores alcalinos y que estos vapores se encuentran siempre en un estado metaestable. Al variar las condiciones de la zona de sinterización, se pueden observar espectros de emisión y de absorción de álcalis. Si aumentan las temperaturas en la zona de sinterización, el estado atómico de los álcalis pasa del normal al de excitación, el cual está caracterizado por un espectro de absorción, y viceversa. El modo de vigilancia es el mismo, pero los pirómetros disponibles, a dos colores, llevan por dentro tres filtros para medir la intensidad óptica de tres longitudes de ondas diferentes; la emisión o la absorción de cada espectro se determina entonces por medio de cálculos. El sistema de vigilancia ha sido instalado en 1990 en la factoría de Cushenbury, California, de la Mitsubishi Cement Corporation y se encuentra en servicio desde hace más de 2 años. La ventaja de las mediciones reside en el hecho de que éstas no son afectadas por la forma de la llama u otros factores de servicio, tales como las radiaciones específicas, producidas durante la descarga del clínker. Los valores de medición son puestos en correlación con la temperatura, en función del comportamiento a la emisión o a la absorción del espectro K en la atmósfera gaseosa de la zona de sinterización.

1. Vorwort

Das Werk Cushenbury der Mitsubishi Cement Corporation wurde 1957 gebaut und ist seither mehrmals erweitert und modernisiert worden.

Die letzte Maßnahme dieser Art wurde 1982 beendet, als ein neuer Ofen mit Vorwärmer/Vorcalcinieranlage mit einer Nennleistung von 5000 t/d installiert wurde, der die damals betriebenen drei Naßöfen ersetzte.

Als diese Anlagen 1988 von der Mitsubishi Cement Corporation gekauft wurden, plante man kleinere Änderungen und Verbesserungen im Ofenbetrieb, um die Klinkerproduktion zu erhöhen. Nach Überprüfung verschiedener Betriebsfaktoren, wie Brennstoff und Luftdurchsatz, erkannte man, daß es besonders wichtig war, die Sinterzone des Ofens präzise zu überwachen. Damals hatte der Ofen ein NO_x-Analysegerät zur Überwachung der Brennbedingungen. Zunächst dachte man an den Einbau eines Zweifarbenpyrometers. Der Ofen in Cushenbury hatte eine sogenannte lange Flamme, die hauptsächlich auf eine hohe Brennstoffrate in der Vorcalcinieranlage zurückzuführen war, und eine Kühlzone, die sich am Ofenausgang bildete. Dadurch war es schwierig, mit einem Zweifarbenpyrometer die genaue Brennzonentemperatur zu ermitteln.

Das Alkalispektrometer wird dank seines Prinzips von den Betriebsbedingungen des Ofens (lange oder kurze Flamme) nicht beeinflußt. Deshalb wurde ein solches Instrument in die Anlage eingebaut und arbeitet seither problemlos.

2. Meßprinzip

Es ist bekannt, daß wegen des Phänomens der Alkalikreisläufe die Atmosphäre der Ofenbrennzone einen hohen Gehalt an Alkalidämpfen hat.

Da die Gastemperatur im Inneren des Ofens sehr hoch ist, liegt der Alkalidampf im metastabilen Zustand vor, d.h. sein Atomzustand ändert sich bei geringer Anregung vom Normal- in den Anregungszustand und umgekehrt.

Bei jedem Übergang wird die Absorption oder Emission des Alkalispektrums je nach Energieunterschied beobachtet.

1. Preface

Mitsubishi Cement Corporation's Cushenbury Plant was originally constructed in 1957 and has gone through several stages of expansion and modernization in its history. The most recent of these was completed in 1982, when a new preheater/ precalciner kiln rated at 5000 tons per day was installed, which replaced the three wet-process kilns which were in use at the time.

In 1988, when these facilities were purchased by Mitsubishi Cement Corporation, some minor modification and improvement in the kiln operation were planned to increase clinker production. As several operation factors such as fuel and air ratio were reconsidered, it became more important to be able to monitor the kiln sintering zone accurately. The kiln had at that time a NO_x analyzer to monitor the burning condition.

The application of a two color pyrometer was first considered. But kiln condition at Cushenbury were a so-called long flame, caused mainly by a high fuel ratio in the precalciner, and a cooling zone that was formed at the discharge end of the kiln. This made it difficult for a two color pyrometer to indicate the exact burning zone temperature.

The alkali spectrometer, because of its principle, is not affected by kiln operating condition (long flame/short flame). In 1990, the instrument was installed in the plant and has been working without any problems since then.

2. Principle of Measurement

It is well known that a high content of alkali vapor exists in the atmosphere of the kiln burning zone due to the alkali circulation phenomenon.

Because the gas temperature inside the kiln is quite high, the alkali vapor is in a metastable condition, namely, its atomic state shifts easily from a ground state to an excited state and vice-versa.

In each transition state, the absorption or the emission of the spectrum corresponding to the difference of the energy level is observed.

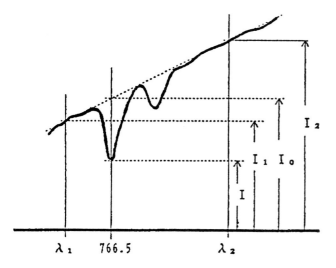

BILD 1: Meßprinzip
FIGURE 1: Principle of measurement

Das Spektrum von Kalium (K) bei 766,5 nm ist daher ein sehr guter Indikator der Bedingungen in der Sinterzone des Zementdrehofens.

Die Beobachtungen während des Betriebs eines Zementdrehofens zeigen, daß ein Absorptionsspektrum eintritt, wenn der Ofen stabil ist und die Sinterzonentemperatur auf einem relativ hohen Niveau gehalten wird, und daß bei einem gestörten Ofenbetrieb mit fallender Sinterzonentemperatur ein Emissionsspektrum auftritt.

Daraus kann geschlossen werden, daß das aufgrund der optischen Intensität des K-Spektrums bei der objektiven Wellenlänge (766,5 nm) ermittelte Intensitätsverhältnis im Vergleich zu dem bei zwei anderen Wellenlängen errechneten Intensitätsverhältnis eine zuverlässige Größe für die Beurteilung und Diagnose der Bedingungen in der Ofensinterzone darstellt.

Ein schematisches Modell zeigt **Bild 1**.

Der Detektor soll die optische Intensität bei drei verschiedenen Wellenlängen messen.

Das Absorptionsverhältnis (η) des objektiven K-Spektrums, das herangezogen wird, um die Bedingungen in der Ofensinterzone zu ermitteln, wird dann mit der folgenden Formel errechnet:

$$\eta = \frac{I}{I_o} \frac{I(\lambda_2 - \lambda_1)}{I_1(\lambda_2 - \lambda_o) + I_2(\lambda_o - \lambda_1)}.$$

3. Meßgeräte

Um die optische Intensität bei drei verschiedenen Wellenlängen zu messen, wurden, wie beim Zweifarbenpyrometer, optische Filter verwendet.

Wie **Bild 2** zeigt, sieht das Instrument einem Zweifarbenpyrometer ähnlich, nur daß es drei optische Filter enthält, die die optische Intensität bei drei verschiedenen Wellenlängen ermitteln und das Verhältnis I_1/I und I_2/I als zwei analoge Signale angeben.

Diese Signale werden dann in den Rechner eingegeben und auf das Absorptionsverhältnis (η) umgerechnet, das dem Ofenbetriebspersonal als Informationswert mitgeteilt wird. **Bild 3** gibt die effektiven Betriebsdaten wieder.

4. Schlußfolgerungen

Die Vorteile der Alkalispektrometer-Methode zur Überwachung der Ofensinterzone können folgendermaßen zusammengefaßt werden: Die gemessenen Daten werden nicht durch die Flammenausbildung beeinträchtigt, da bei dieser

The spectrum of Potassium (K) at 766.5 nm gives a quite remarkable indication of the sintering zone condition of the cement rotary kiln.

According to observation during the actual operation of a cement rotary kiln, spectrum absorption occurs when the kiln is stable and the sintering zone temperature is being kept relatively high, while spectrum emission occurs when the sintering zone temperature comes down because of an upset condition in the kiln.

Hence, it can be concluded that the index obtained in such a way at the optical intensity of the K-spectrum observed at the objective wave length (766.5 nm) and compared with the index which is calculated from the intensity at two other wave lengths will be useful for judging and diagnosing the kiln sintering zone condition.

A schematic model is shown in **Fig. 1**.

The detector should measure the optical intensity at three different wave lengths.

The absorption ratio (η) of the objective K-spectrum, which is used to determine the kiln sintering zone condition, is then calculated by the following formula.

$$\eta = \frac{I}{I_o} \frac{I(\lambda_2 - \lambda_1)}{I_1(\lambda_2 - \lambda_o) + I_2(\lambda_o - \lambda_1)}$$

3. Measuring Equipment

In order to detect the optical intensity in three different wave lengths optical filter technology, which is also used in the two color pyrometer, was applied.

As shown in **Fig. 2**, the physical appearance of the instrument is similar to the two color pyrometer but it contains three sets of optical filters which detect the optical intensity in three different wave lengths and outputs the ratio I_1/I and I_2/I as two analog signals.

The signals then are input to the computing unit and converted to the absorption ratio (η), which is given to the kiln operator as an information value. **Fig. 3** shows the actual operation data.

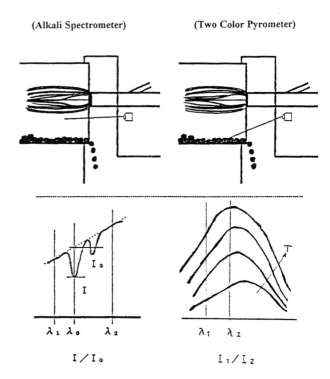

BILD 2: Vergleich mit Zweifarbenpyrometer
FIGURE 2: Comparison with two-color pyrometer
Zweifarbenpyrometer = Two Color Pyrometer

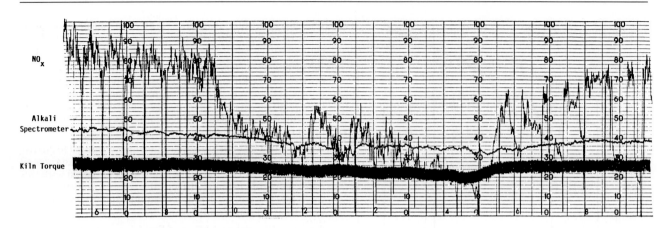

BILD 3: Grafische Darstellung von Betriebsergebnissen
FIGURE 3: Chart of the operational results
Ofendrehmoment = Kiln Torque

Methode die Messungen in der Sinterzonenatmosphäre gemacht werden, und die gemessenen Daten werden auch nicht durch die spezifische Strahlung beim Klinkeraustrag beeinträchtigt, da das Spektrometer nicht die Oberflächentemperatur des Klinkers sondern die Emission bzw. Absorption des K-Spektrums im Gas mißt.

Es gibt keine Teile, die eine aufwendige Wartung und Instandhaltung erfordern, wie die Gasentnahmesonde eines NO_x-Analysegeräts. Das Betriebspersonal erhält einen festen Meßwertebereich und erkennt somit ohne Schwierigkeiten, wie die Betriebsparameter zur Steuerung der Sinterkonditionen des Ofens eingestellt werden müssen.

4. Conclusion

The advantages of the alkali spectrometer method for monitoring the kiln sintering zone are as follows: Measured data are not affected by flame condition as it is looking at the sintering zone atmosphere, and measured data are not affected by the emissivity of the clinker because the spectrometer does not measure the clinker surface temperature but measures the emission/absorption of the K-spectrum in the gas.

There are no parts which require complicated maintenance, such as the gas sampler in a NO_x analyzer. A fixed range of measured data is given to the operator which makes it easy to understand where the operating parameters must be kept to control the kiln sintering zone.

# Ein Röntgenfluoreszenz-Spektrometer mit integrierter Diffraktionsmeßeinrichtung und seine Anwendung in der Zementindustrie

An integrated diffraction XRF spectrometer and its applications in the cement industry

Spectre de fluorescence et de diffraction des rayons X intégrée et ses possibilitiés d'application dans l'industrie cimentière

Un espectrómetro integrado por difracción y por fluorescencia de rayos X y sus aplicaciones en la industria del cemento

Von **D. Bonvin, A. Bapst** und **R. Yellepeddi,** Ecublens/Schweiz

Zusammenfassung – Die analytische Qualitätskontrolle in der Zementindustrie umfaßt heute die genaue Analyse verschiedener Bestandteile im Klinker und Zement. Neben den wichtigsten Oxiden werden der freie Kalk (CaO), Kohlenstoff (aus $CaCO_3$), Sulfid und Sulfat usw. analytisch bestimmt. Es ist deshalb von großem Vorteil, diese Substanzen mit einem einzigen Instrument analysieren zu können, um alle Angaben für die Prozeßsteuerung der Zementproduktion zu erhalten. Es wurde ein neues Instrument entwickelt, das Röntgenfluoreszenz (XRF) mit Röntgen-Diffraktometrie (XRD) kombiniert und somit diese Forderung erfüllt. Es arbeitet mit einer einzigen Erregerquelle, einem Probenhalter und einer Röntgenstrahlenoptik mit enger Kopplung, so daß es einerseits eine Elementanalyse und andererseits eine Phasenanalyse in einem spezifischen Bereich durchführen kann. Der freie Kalkgehalt wird mittels XRD bestimmt, und der Kohlenstoff kann mit einem XRF-Monochromator, der eigens zu diesem Zweck entwickelt wurde, ermittelt werden. Zur Differenzierung von Sulfid und Schwefel verwendet man die Satelliten-Peaks im XRF-Diagramm. Der XRF-Teil selbst kann je nach Anwendungsfall konfiguriert werden. Es können mehrere Monochromatoren (feststehende Kanäle) für die wichtigsten Oxide verwendet werden, während ein Goniometer für die fortlaufende und flexible Analyse gewährleistet, daß die Anforderungen der Zukunft im Bereich des Umweltschutzes erfüllt werden. Zweifellos hat dieser Ansatz große Vorteile gegenüber den herkömmlichen Systemen, die z.B. in der Zementindustrie angewandt werden, wo sowohl chemische- als auch Phasenanalysen erforderlich sind.

Summary – Analytical quality control in the cement industry now involves a more complete analysis dealing with various constituents in clinker and cement. These include, apart from the key oxides, the analysis of free lime (CaO), carbon (from $CaCO_3$), sulphide and sulphate etc. It is therefore highly desirable to analyze these components in a single instrument to obtain the complete information for the process control of cement production. A new instrument has been developed which combines the X-ray Fluorescence (XRF) and X-ray Diffractometry (XRD) in order to meet these challenges. It uses a single excitation source, sample support and closely coupled X-ray optics to perform an elemental analysis on the one hand and phase analysis in a specific range on the other hand. The free lime is analysed using the XRD section while carbon can be analysed by an XRF monochromator specially designed for this purpose. The sulphide and sulphur differentiation can be achieved using the satellite peaks in XRF. The XRF section itself can be configured according to the application. Several monochromators (fixed channels) may be used for the major oxides while a goniometer for a sequential and flexible analysis will meet the changing future requirements and environmental control. Clearly, this approach has significant advantages over the conventional systems for applications, e.g. in a cement plant, where both chemical and phase analysis are required.

Résumé – Le contrôle de qualité analytique dans l'industrie cimentière comporte de nos jours l'analyse exacte de divers composants contenus dans le clinker et le ciment. Outre les oxydes les plus importants, la chaux libre (CaO), le carbone (provenant du $CaCO_3$), le sulfure et le sulfate etc. sont déterminées par analyse. Il est donc extrêmement important d'analyser ces substances avec un seul instrument pour obtenir toutes les données pour le pilotage du procédé de la production de ciment. Un nouvel instrument mis au point combinant la fluorescence X (XRF) et la diffraction des rayons X (XRD),

permet de répondre à cette exigene. Il travaille avec une seule source d'excitation, un porte-échantillon et un objectif de rayons X avec couplage étroit de manière à pouvoir réaliser en un domaine spécifique une analyse d'éléments, d'un côté, et une analyse de phase, de l'autre. La teneur en chaux libre est déterminée par le XRD, et le carbone peut être évalué par un monochromateur XRF, ayant été développé exprès dans ce but. Pour différencier le sulfure du soufre on utilise les pointes de satellite sur le diagramme XRF. La partie XRF peut, selon son application, être configurée. On peut utiliser plusieurs monochromateurs (canaux fixes) pour les oxydes les plus importants alors qu'un goniomètre pour analyse continue et flexible assure que les exigences futures en matière de protection de l'environnement seront respectées. Il ne fait pas de doute que cette solution présente de grands avantages par rapport aux systèmes conventionnels utilisés, par exemple, dans l'industrie cimentière, où des analyses tant chimiques que de phase sont indispensables.

Un espectrómetro integrado por difracción y por fluorescencia de rayos X y sus aplicaciones en la industria del cemento

__Resumen__ – El control analítico de la calidad abarca, hoy en día, en la industria del cemento, el análisis exacto de diferentes componentes del clínker así como del cemento. Aparte de los óxidos más importantes, se determinan analíticamente la cal libre (CaO), el carbono (de $CaCO_3$) el sufuro y el sulfato, etc. Por esta razón, es muy ventajoso el poder analizar estas sustancias con un solo aparato, con el fin de obtener todos los datos necesarios para el control del proceso de producción de cemento. Ha sido desarrollado un nuevo instrumento, que combina la fluorescencia por rayos X (XRF) con la difractometría de rayos X (XRD), cumpliendo así dicha exigencia. Funciona con una sola fuente de excitación, un portamuestras y una óptica de rayos X, de acoplamiento estrecho, de modo que se puede efectuar, por un lado, un análisis de elementos y, por otro lado, un análisis de fases, dentro de un ámbito específico. El contenido de cal libre se determina mediante XRD, y el carbono se puede determinar por medio de un monocromador XRF, especialmente desarrollado para ello. Para la diferenciación de sulfuro y de azufre, se utilizan los „peaks" de satélites en el diagrama XRF. La parte XRF puede configurarse, según los casos de aplicación. Se pueden emplear varios monocromadores (canales fijos) para los óxidos más importantes, mientras que un goniómetro garantiza, en el análisis continuo y flexible, el que se cumplan las exigencias futuras en materia de protección del medio ambiente. Este método tiene, sin duda, grandes ventajas en comparación con los sistemas tradicionales, empleados por ejemplo en la industria del cemento, donde se requieren tanto los análisis químicos como los de fases.

1. Einleitung

Traditionell dient die Röntgenfluoreszenz-Technik (XRF) zur chemischen Elementaranalyse von Klinker und Zement [1], während die Röntgen-Diffraktometrie (XRD) und die sogenannte Titration als naßchemische Methode zur Phasenanalyse von Klinker und Zement angewendet werden.

Zur Qualitätssicherung des Zements muß auch der freie Kalkgehalt (CaO) im Klinker ständig überwacht werden. Ein übermäßig hoher Gehalt an freiem Kalk führt zu unerwünschten Auswirkungen wie z. B. Volumenausdehnung, erhöhten Erstarrungszeiten oder verminderten Festigkeiten. Außerdem erlaubt es eine ständige Kontrolle des Freikalkgehaltes, die optimale Fahrweise der Ofenanlage zu bestimmen, wodurch sowohl eine maximale Reaktivität als auch eine Reduzierung des spezifischen Wärmeverbrauches gewährleistet werden können.

In der Zementindustrie besteht darüber hinaus die wachsende Forderung sowie das Interesse, den Calciumcarbonatgehalt ($CaCO_3$) als Hauptbestandteil des Kalksteins im Fertigprodukt zu überwachen. Neuere Vorschriften [2] in Europa erlauben die Zugabe von Kalkstein als Füller bei der Zementmahlung bis zu Anteilen von 30 % in Abhängigkeit von der geforderten Zementsorte. Deshalb ist es sehr wichtig und auch ökonomisch interessant geworden, den Gehalt an Calciumcarbonat im Zement auf einem relativ schnellen Weg zu bestimmen, um die Qualität und Konformität des Endproduktes [3] zu gewährleisten.

Da eine mineralogische Information durch das Röntgenfluoreszenz-Spektrum nicht zur Verfügung gestellt werden kann – bei der XRF-Analyse wird nur der gesamte Ca-Gehalt in der Probe einschließlich des freien Kalkgehaltes bestimmt – ist normalerweise ein separates Röntgendiffraktometer erforderlich, um sowohl qualitative als auch quantitative Strukturdaten ermitteln zu können [4]. Technisch bedingt kann durch das Röntgen-Diffraktometer eine effektive Kalksteinbestimmung nicht erfolgen. Deshalb

1. Introduction

Traditionally, the X-ray fluorescence technique (XRF) is used to perform chemical elemental analysis on clinker and cement [1] while X-ray diffractometry (XRD) or wet chemical methods like titration are used to determine the phase content in clinker or cement.

Free lime (CaO) in clinkers has to be closely monitored for the quality control of cement. Excess free lime results in undesirable effects such as volume expansion, increased setting time or reduced strength. In addition, constant monitoring of free lime allows to determine and maintain the optimum operating point of the kiln in order to obtain maximum reactivity and reduce thermal consumption. There is also a growing demand and interest in the cement industry to monitor the concentration of limestone, of which calcium carbonate ($CaCO_3$) is the main constituent, in the final product. Recent European regulations [2] permit the addition of limestone as filler up to concentrations of 30 % depending on the type of cement required. Therefore, it becomes very important and economically interesting to control the concentrations of calcium carbonate in cement in a relatively fast way in order to guarantee the quality and conformity of the final product [3].

Since mineralogical information is not available from XRF spectra (for instance XRF gives only the total Ca concentration in the sample including free CaO), separate XRD equipment is normally required to obtain qualitative and quantitative structural data [4]. Due to technical restrictions diffractometers do not achieve effective limestone determinations and therefore they are used only for free lime analysis in cement plants [5]. This means that two X-ray instruments are to be controlled and maintained with possibly different sample transport mechanisms resulting in significant costs for the user.

For the first time, ARL has developed an Integrated Diffraction XRF Spectrometer, the ARL 8600 S Total Cement Analyzer [6–8], in which a diffraction channel involving

werden diese Geräte in der Zementindustrie [5] auch nur für die Freikalkanalyse benutzt. Das bedeutet, daß zwei Röntgen-Geräte installiert und mit den unterschiedlichen Probentransport-Einrichtungen ausgerüstet werden müssen, womit beträchtliche Kostenaufwendungen für den Nutzer verbunden sind.

Erstmalig hat ARL ein Röntgenfluoreszenz-Spektrometer mit integrierter Diffraktionsmeßeinrichtung, den Total Cement Analyzer ARL 8600 S [6–8] entwickelt, d.h. innerhalb eines XRF-Spektrometers einen variablen Diffraktionskanal installiert. Die Vorteile eines in dieser Weise kombinierten Gerätesystems sind unverkennbar:

— Die XRF- und XRD-Analyse kann an der gleichen Probe und unter völlig gleichen Bedingungen durchgeführt werden, wodurch die Kosten für zusätzliche Hardware gesenkt werden und die Gesamtanalyse zuverlässiger wird.
— Die Röntgenfluoreszenzanalyse bleibt unbeeinflußt von der im Spektrometer installierten Diffraktionsmeßeinrichtung.
— Software und Datenverarbeitungsverfahren werden gemeinsam für die Messungen mit dem kombinierten Analysengerät genutzt.

Als schnelles und zugleich flexibles System kann der XRF-Teil selbst je nach Anwendungsfall konfiguriert werden auf der Basis

— feststehender XRF-Kanäle für die simultane Analyse von allen Zementoxiden,
— der Diffraktionsmeßeinrichtung, die in der Lage ist, den Freikalk- und den Calcitgehalt zu messen,
— des ARL Moiré-Streifen-Goniometers zur Durchführung einer sequentiellen XRF-Analyse für jedes der 83 Elemente des periodischen Systems.

Der Diffraktionskanal ist in der Lage, sowohl qualitative Scans als auch quantitative Analysen zu liefern. Das wird ermöglicht durch Anwendung des Moiré-Streifenmechanismus [9], einer durch ARL entwickelten und inzwischen erprobten Technologie. Da die Lage eines Peaks und der Untergrund in einem Röntgen-Diffraktometer von unterschiedlichen Parametern (wie z.B. Korngröße oder Matrixeffekte) abhängig sind, führt eine Peaksuche und eine Peakintegration prinzipiell zu einer genauen Analyse. In diesem Beitrag werden nur Peakintensitäten verwendet, da signifikante Peakverschiebungen nicht beobachtet werden konnten.

2. Freikalkanalyse

Untersucht wurden eine Serie von Klinkerstandards, die durch die Holderbank naßchemisch auf freien Kalkgehalt analysiert worden sind. Die als Pulver angelieferten Klinker wurden 10 s mit zwei 200-mg-Mahlkugeln gemahlen und dann mit einem Druck von 15 t 40 s lang zu Pellets gepreßt. **Bild 1** zeigt ein Diffraktogramm, aufgenommen mit der Diffraktionsmeßeinrichtung an drei verschiedenen Klinkerpellets. Die beiden Peaks, welche die C_3S- und CaO-Phasen kennzeichnen, sind gut aufgelöst und nicht gestört durch Probenfluoreszenz. Wie aus der Größe des Peaks, der einen Freikalkgehalt von 0,5 % repräsentiert, entnommen werden kann, ist die Empfindlichkeit mehr als ausreichend, um den Freikalkgehalt zu kontrollieren. **Bild 2** zeigt Ergebnisse und **Tabelle 1** die relevanten Parameter eines Eichprogramms, das die gemessenen Peakintensitäten von 6 Klinkerstandards nutzend, durchgeführt wurde. Die Standardabweichung dieser Bewertung entspricht im Durchschnitt den Abweichungen aus den chemischen Untersuchungen und den über die Röntgenfluoreszenzanalyse gefundenen Konzentrationen. Die Ergebnisse unterstreichen die Genauigkeit der Analyse. Der erzielte Wert von 0,08 % ist als gut zu bezeichnen, er liegt im Wertebereich der naßchemischen Analyse.

Die Kurzzeitstabilität wurde an einer Klinkerprobe gemessen, die 11mal 100 s lang auf den Freikalkgehalt mit der Diffraktionsmeßeinrichtung und 40 s lang mit den feststehen-

innovative technology is incorporated into the existing XRF system. The advantages of this combined system are obvious:

— XRF and XRD can be performed on the same sample and under identical conditions thus cutting down all the extra costs for additional hardware and ensuring more reliable and stable total analysis.
— XRF performance is not affected by the addition of this channel.
— Software and data treatment methods are common to XRF and XRD measurements.

A rapid and flexible system can be configured based on

— fixed XRF channels for simultaneous analysis of all cement oxides,
— the diffraction channel which has the capability of measuring free lime (CaO) and calcite ($CaCO_3$) contents.
— ARL's Moiré fringe goniometer for a sequential XRF analysis on any of 83 elements of the periodic table and back-up of the fixed channels.

The diffraction channel is capable of making qualitative scans and also quantitative analysis. This is made possible by using ARL's proven technology, namely the Moiré fringe positioning mechanism [9]. Since the peak positions and backgrounds in XRD are sensitive to different parameters (e.g. grain size, matrix effects), peak search and peak integration can be done for an accurate analysis. However in this study were only used peak intensities since no significant peak shifts have been observed.

2. Free lime analysis

A series of clinker standards have been investigated, which were analysed for free lime by wet chemistry at Holderbank. The powders have been ground for 10 sec with two 200 mg grinding pills. They were then pressed into pellets at 15 tons for 40 sec. **Fig. 1** shows the diffractogram recorded with the diffraction channel on three different clinker pellets. The two peaks, assigned to C_3S and CaO phases, are well resolved and not interfered by any fluorescence from the sample. As can be seen from the size of the peak representing 0.5 % of free lime, the sensitivity is more than adequate to monitor the free lime content. There was carried out a calibration program using the peak intensities measured on 6 clinker standards. The results are shown in **Fig. 2** along with

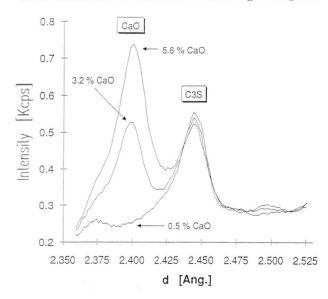

BILD 1: Diffraktionsmuster, aufgenommen mit dem Total Cement Analyzer ARL 8600 S im Freikalkbereich von Klinkerproben mit verschiedenen CaO-Konzentrationen
FIGURE 1: Diffraction pattern obtained with ARL 8600 S Total Cement Analyzer in the free lime region of clinker samples with varying CaO concentrations

Intensität = Intensity
Abstand = Spacing

TABELLE 1: Ergebnisse einer Regressionsanalyse zum Freikalkgehalt
TABLE 1: Regression results on free lime

Proben-bezeichnung	Intensität	Konzentration Concentration		
Sample name	Intensity	Chemisch Chemical	Gemessen Found	Differenz Difference
	[Kcps]	[%]	[%]	[%]
FLS 1	0,453	2,47	2,48	0,01
FLS 2	0,552	3,17	3,45	0,28
FLS 3	0,256	0,46	0,56	0,10
FLS 4	0,380	1,79	1,77	−0,02
FLS 5	0,411	2,02	2,08	0,06
FLS 6	0,285	0,91	0,85	−0,06
FLS 7	0,285	0,98	0,84	−0,14
FLS 8	0,292	0,98	0,91	−0,07
FLS 9	0,294	1,05	0,93	−0,12
FLS 10	0,333	1,19	1,32	0,13
FLS 11	0,344	1,22	1,42	0,20
FLS 13	0,357	1,45	1,55	0,10
FLS 14	0,309	0,89	1,08	0,19
FLS 15	0,277	0,82	0,76	−0,06
FLS 16	0,338	1,40	1,37	−0,03
FLS 17	0,357	1,76	1,54	−0,22
FLS 18	0,547	3,74	3,40	−0,34
Standardabweichung Standard error of estimate				0,164
Empfindlichkeit Sensitivity				102 cps/%
Nachweisgrenze (100 s) Limit of detection (100 s)				413 ppm

den XRF-Kanälen gemessen wurde. Die Ergebnisse enthält **Tabelle 2**, in der zu Vergleichszwecken auch die XRF-Ergebnisse der wichtigsten Oxide zu finden sind. Die Werte zeigen, daß das neue Analysensystem eine bemerkenswerte Kurzzeitstabilität besitzt: Die relative Standardabweichung von < 1 % liegt fast um den Faktor 10 besser als vergleichsweise bei einer naßchemischen Analyse. Schließlich wurde auch die Langzeitstabilität geprüft. Dabei konnten hervorragende Ergebnisse mit σ-Werten innerhalb der statisch zulässigen Werte ohne eine monotone Trift über eine Zeitdauer von 50 h erreicht werden.

TABELLE 2: Typische Reproduzierbarkeit von 11 Freikalkbestimmungen im Klinker. Eingeschlossen sind auch die Ergebnisse aus der üblichen XRF-Analyse
TABLE 2: Typical reproducibility (11 runs) of free lime analysis in clinker. Included are also the results of usual XRF analysis

Versuch Run	Freikalk Kanal Free lime channel	Konzentrationen/Concentrations [%]					
		CaO XRF	SiO$_2$ XRF	Al$_2$O$_3$ XRF	Fe$_2$O$_3$ XRF	MgO XRF	K$_2$O XRF
1	2,62	64,78	20,46	4,87	3,80	2,02	1,05
2	2,60	64,79	20,44	4,84	3,79	2,02	1,06
3	2,58	64,76	20,42	4,85	3,81	2,00	1,05
4	2,62	64,75	20,44	4,87	3,80	2,01	1,05
5	2,58	64,78	20,44	4,86	3,80	2,02	1,05
6	2,63	64,75	20,45	4,86	3,79	2,01	1,05
7	2,59	64,78	20,44	4,85	3,79	2,03	1,05
8	2,59	64,77	20,42	4,86	3,80	2,01	1,06
9	2,58	64,76	20,44	4,85	3,80	2,00	1,05
10	2,62	64,75	20,44	4,85	3,81	2,02	1,05
11	2,58	64,77	20,44	4,86	3,80	2,00	1,05
Durchschn. Average	2,599	64,767	20,439	4,856	3,799	2,013	1,051
Standard-abweichung St. dev.	0,024	0,014	0,011	0,009	0,007	0,010	0,004
RSD	0,91 %	0,02 %	0,06 %	0,19 %	0,18 %	0,50 %	0,42 %

the relevant parameters in **Table 1**. The Standard error of estimate is an average of the differences between chemical and found concentrations. It is an estimation of the accuracy of analysis. The value of 0.08 % obtained is well within the capability of wet chemistry.

The short term stability has been tested by measuring a clinker sample 11 times for 100 sec on free lime with the XRD channel and 40 sec on the fixed XRF channels. The results are summarised in **Table 2** where XRF results on major oxides are also included for comparison. They show the remarkable short term stability of the system: < 1 % relative standard deviation, which is close to 10 times better than that which can be obtained by wet chemistry. Finally, a long term stability test has been performed. Excellent results are obtained with sigma values well within statistics and no monotonic drift over 50 hours.

3. Limestone additions in cement

Carbonate content can be measured by X-Ray Fluorescence (XRF) using the carbon line (CKα). However, XRF analysis is not directly correlated to a phase (here CaCO$_3$): it only gives the total carbon concentration in the sample. XRF analysis of carbon is also subject to some difficulties:

— The fluorescence yield of light elements like carbon is rather poor. Therefore, the sensitivity of the technique is low, even when using 3 kW systems and does not allow to get high reproducibility of results. Of course, this is even worse when using only a 200 W X-ray tube.

— Due to matrix absorption effects, the carbon fluorescence only escapes from a very thin layer at the surface of the sample. This means that the volume of sample effectively measured for carbon analysis by XRF is extremely small.

— Surface contamination and addition of binding/grinding agents (usually organic materials, e.g. stearic acid) can produce inconsistent XRF results due to their carbon content. Binding agents are used to improve the stability of the pellet under vacuum.

— When carbon is measured by XRF, all errors are multiplied by a factor of 8 when converting to limestone concentrations.

On the other hand, the innovative X-Ray Diffraction (XRD) channel integrated in the ARL 8600 S Total Cement Analyzer is capable of analysing the specific phase of CaCO$_3$. In addition, XRD intensities are not affected by the factors mentioned above, because

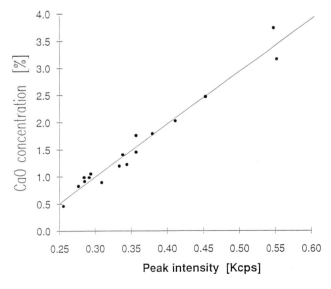

BILD 2: Eichkurve von Freikalk im Klinker bei Anwendung des Total Cement Analyzer ARL 8600 S
FIGURE 2: Calibration curve on free lime in clinker using the ARL 8600 S Total

Freikalk-Konzentration = CaO Concentration
Peak-Intensität = Peak Intensity

3. Kalkstein als Zumahlstoff im Zement

Der Carbonatgehalt kann mit Hilfe der Röntgenfluoreszenzanalyse (XRF) gemessen werden, indem die Kohlenstoff-Linie benutzt wird (CKα). Allerdings liefert die XRF-Analyse nicht ein Ergebnis, das direkt mit der interessierenden Phase (CaCO₃) korreliert. Die Analyse liefert nur den gesamten Kohlenstoffgehalt in der Probe. Die XRF-Analyse von Kohlenstoff bringt einige Schwierigkeiten mit sich. Diese Schwierigkeiten sind:

- Die Fluoreszenzausbeute bei leichten Elementen wie z. B. Kohlenstoff ist ziemlich gering. Deshalb ist auch die Empfindlichkeit der verwendeten Technik entsprechend niedrig, die eine hohe Reproduzierbarkeit von Ergebnissen nicht erlaubt, auch nicht, wenn 3-kW-Systeme benutzt werden. Die Ergebnisse fallen notwendigerweise noch schlechter aus, wenn nur eine 200-W-Röntgenröhre verwendet wird.

- Wegen der sogenannten Matrix-Absorptionseffekte stammt die Kohlenstoff-Fluoreszenzstrahlung nur von einer sehr dünnen Schicht der Probenoberfläche. Das bedeutet, daß das effektiv gemessene Probenvolumen bei einer Kohlenstoffanalyse extrem niedrig liegt.

- Die Oberflächenkontamination und Zugabe von Bindern bzw. Mahlhilfsmitteln (gewöhnlich organische Substanzen wie z. B. Stearinsäure) können wegen ihres Kohlenstoffgehaltes zu inkonsistenten Ergebnissen führen. Bindemittel werden bekanntlich dann verwendet, wenn die Stabilität von Pellets unter Vakuum verbessert werden soll.

- Schließlich multiplizieren sich alle Fehler um den Faktor 8, wenn der Kohlenstoffgehalt mittels der Röntgenfluoreszenzanalyse gemessen und daraus die Kalksteinkonzentration ermittelt wird.

Auf der anderen Seite ist die im Gesamtzementanalysator ARL 8600 S integrierte Diffraktionsmeßeinrichtung in der Lage, das Phasenbild von CaCO₃ aufzunehmen. Außerdem sind die dabei gemessenen Intensitäten nicht belastet durch die oben erwähnten Faktoren, da

- die hohe Energie der gegebenen Strahlung es erlaubt, ein im Vergleich zum XRF-Gerät viel größeres Probenvolumen zu analysieren (um ca. 10 × größer), was diese Analysenmethode wesentlich repräsentativer macht,

- die Bestimmung der CaCO₃-Phase durch Oberflächenkontamination, organische Binde- oder Mahlhilfsmittel nicht verfälscht werden kann und

- schließlich die Calcitphase, die von Interesse ist, direkt gemessen werden kann.

Ein traditionelles Diffraktometer ist nicht in der Lage, Calcit zu analysieren, weil die Empfindlichkeit des Gerätes dazu nicht ausreicht und Überlappungen durch Linien der Röhrenanode (Cu-Anode) bestehen. Im Total Cement Analyzer wird eine Rh-Anode benutzt.

4. Ergebnisse und Diskussion

Es wurden Serien von industriellen Zementproben als Pulver untersucht. Dabei wurden alle Proben ohne den Zusatz von Bindemitteln 40 s mit 15 t gepreßt. **Bild 3** zeigt die XRD-Scans von 3 Zementproben mit unterschiedlichen CaCO₃-Konzentrationen. Zwei unterschiedliche Peaks, die dem Calcit und den C₃S-Phasen zuzuordnen sind, können auf jedem der Scans identifiziert werden. Die beiden Peaks, die sich gut voneinander unterscheiden, ermöglichen eine quantitative Analyse ohne Korrektur von Überlappungen.

Bild 4 gibt die Eichkurve wieder, die erhalten wurde durch die Peakintensitäten von CaCO₃ aus einer Serie von 8 Zementstandards. Die Ergebnisse der Regressionsanalyse sind in **Tabelle 3** zusammengestellt. Die Standardabweichung der Bestimmung von 0,08 bescheinigt die ausgezeichnete Korrelation zwischen den nominellen Konzentrationen (ausgedrückt als CO₂) und den aus der Röntgendiffraktometrie erhaltenen Intensitäten. Auch die Tests zur Kurz- und Langzeitstabilität wurden durchgeführt. Über einen

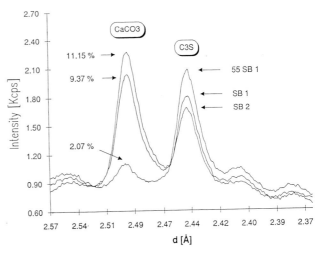

BILD 3: XRD-Scans an 3 Zementpellets mit unterschiedlichem CaCO₃-Gehalt
FIGURE 3: XRD scans on three cement pellets containing different concentrations of CaCO₃

Intensität — Intensity
Abstand — Spacing

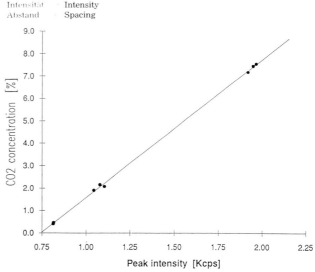

BILD 4: Eichkurve von 8 Zementstandards. Die gemessene Intensität des CaCO₃-Peaks wurde ohne Korrekturen von Background und Überschneidungen benutzt
FIGURE 4: Calibration curve obtained using 8 cement standards. CaCO₃ peak intensity is used as measured (no background or overlap corrections)

CO₂-Konzentration = CO₂ concentration
Peak-Intensität = Peak intensity

- the high energy of the diffracted radiation allows to measure a much larger volume of sample (about 10 times larger) than in XRF; this makes the XRD analysis more representative.

- surface contamination, organic binders or grinding aids do not alter the CaCO₃ phase content and will therefore not alter the limestone analysis.

- the phase of interest, calcite, is measured directly.

A traditional diffractometer cannot achieve the calcite analysis because of lack of sensitivity and overlap by X-ray tube lines (from the Cu anode of the tube). A Rh anode is used in the Total Cement Analyzer.

4. Results and Discussion

A series of industrial cement samples were used as powders. All samples were pressed at 15 tons for 40 sec without binder. **Fig. 3** shows the XRD scans on three cement samples containing different concentrations of CaCO₃. Two distinct peaks assigned to calcite and C₃S phases can be identified in each of the scans. The two peaks are well separated, thus enabling quantitative analysis without correction for overlap. **Fig. 4** presents the calibration curve obtained using the

Durchschnitt von 21 Analysen (jede über 100 s) wurde die hervorragende Standardabweichung von 0,024 % bei einem CO_2-Gehalt von 7,17 % ($CaCO_3$ als CO_2 ausgedrückt) erhalten. Meßergebnisse aus dem kontinuierlichen Betrieb des Gerätes über einen Zeitraum von 60 h zeigt **Bild 5**. Beide Tests liefern deutlich den Nachweis, daß das Röntgenfluoreszenzspektrometer mit integrierter Diffraktionsmeßeinrichtung eine ausgezeichnete Meßstabilität besitzt.

TABELLE 3: Ergebnisse einer Regressionsanalyse eines Zementes mit Kalkstein als Zumahlstoff
TABLE 3: Regression results for limestone addition in cement

Probenbezeichnung Sample name	Intensität Intensity [Kcps]	Konzentrationen Concentrations		
		Gegeben Given [%]	Gemessen Found [%]	Differenz Difference [%]
Zement CPJ 1 Ciment CPJ 1	1,917	7,17	7,23	0,06
Zement CPJ 2 Ciment CPJ 2	1,964	7,55	7,52	−0,03
Zement CPJ 3 Ciment CPJ 3	1,946	7,45	7,41	−0,04
Zement HPR Ciment HPR	1,044	1,90	1,85	−0,05
Zement HP 1 Ciment HP 1	1,103	2,07	2,21	0,14
Zement HP 2 Ciment HP 2	1,077	2,15	2,05	−0,10
Zement HTS 1 Ciment HTS 1	0,815	0,45	0,44	−0,01
Zement HTS 2 Ciment HTS 2	0,813	0,40	0,42	0,02
Standardabweichung Standard error of estimate			0,08	
Empfindlichkeit Sensitivity			162 cps/%	
Nachweisgrenze (100 s) Limit of detection (100 s)			505 ppm	

5. Schlußfolgerungen

Mit dem Total Cement Analyzer ARL 8600 S mit integrierter Diffraktionsmeßeinrichtung kann der Freikalkgehalt im Klinker und der Calciumcarbonatgehalt im Zement mit hoher Empfindlichkeit, Zuverlässigkeit und ausgezeichneter Meßstabilität bestimmt werden. Zwei Gerätekonfigurationen stehen dafür zur Verfügung:

Der Total Cement Analyzer ARL 8660 S

— für die routinemäßige Prozeßkontrolle von Rohmaterialmischung, Klinker und Zement mit 8 oder 9 feststehenden Monochromatoren für die oxidischen Bestandteile wie Na_2O, MgO, Al_2O_3, SiO_2, SO_3, K_2O, CaO und Fe_2O_3.

— für die Freikalkbestimmung und Kontrolle der Kalksteinzugabe mit Hilfe der integrierten Diffraktionsmeßeinrichtung.

— mit der XRF-386-Software und/oder der anwendungsbezogenen Kontrollsoftware für das Datenhandling.

Der Total Cement Analyzer ARL 8680 S

— für die routinemäßige Prozeßkontrolle von Rohmaterialmischung, Klinker und Zement mit beispielsweise 5 oder mehr feststehenden Monochromatoren für die oxidischen Hauptbestandteile wie Al_2O_3, SiO_2, CaO, SO_3, Fe_2O_3 und 3 oder weitere oxidische Bestandteile wie Na_2O, MgO und K_2O, die durch das automatische Goniometer analysiert werden.

— für die flexible Analyse jedes anderen Elements vom Bor bis zum Uran unter Nutzung des vollautomatischen Goniometers, welches auch eine zusätzliche Analysenkapazität für die mit den feststehenden Kanälen gemessenen oxidischen Hauptkomponenten bereithält.

$CaCO_3$ peak intensity in a set of 8 cement standards. The regression results are summarised in **Table 3**. The standard error of estimate (SEE) of 0.08 % attests the very good correlation obtained between the nominal concentrations (expressed as CO_2) and the XRD intensities.

Short term and long term stability tests were carried out. An average of 21 analyses (each for 100 sec) gave the excellent standard deviation of 0.024 % at a level of 7.17 % CO_2 ($CaCO_3$ expressed as CO_2). Data over 60 hours of continuous operation are presented in **Fig. 5**. Both tests clearly show the exceptional stability of measurement of the integrated XRD channel.

5. Conclusions

Using the diffraction channel integrated into the ARL 8600 S Total Cement Analyzer, free lime in clinker and limestone ($CaCO_3$) in cement can be quantified with good sensitivity, reliability and excellent stability. Two configurations of instrument are available:

ARL 8660 S Total Cement Analyzer

— Routine raw mix, clinker and cement process control with 8 or 9 fixed monochromators for oxide components (e.g. Na_2O, MgO, Al_2O_3, SiO_2, SO_3, K_2O, CaO, Fe_2O_3).

— Free lime and limestone addition monitoring with the integrated diffraction channel.

— Data handling through the XRF 386 software and/or engineering company plant control software.

ARL 8680 S Total Cement Analyzer

— Routine raw mix, clinker and cement process control with for example 5 or more fixed monochromators for key oxide components (e.g. Al_2O_3, SiO_2, SO_3, CaO, Fe_2O_3) and 3 or more oxide components to be analysed by the automatic goniometer (e.g. Na_2O, MgO, K_2O).

— Flexible analysis of any other elements in the range of boron to uranium using the fully automatic goniometer which provides also back-up analysis capability to the key oxide components measured on fixed channels.

— Free lime and limestone addition monitoring with the integrated diffraction channel.

— Data handling through the XRF 386 software and/or engineering company plant control software.

The combination of XRF and XRD in the same instrument can now provide complete quality control of clinker and cement. Separate instruments or methods are no longer required resulting in significant savings from increased operator efficiency and lower running costs.

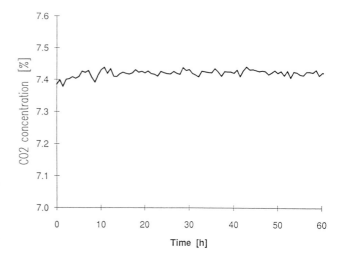

BILD 5: Langzeitstabilität der $CaCO_3$-Analyse über eine Zeitdauer von 60 h
FIGURE 5: Long term stability for $CaCO_3$ analysis over 60 hrs

CO_2-Konzentration = CO_2 concentration
Zeit = Time

- für die Freikalkbestimmung und Kontrolle der Kalksteinzugabe mit Hilfe der integrierten Diffraktionsmeßeinrichtung.
- mit der XRF-386-Software und/oder der anwendungsbezogenen Kontrollsoftware für das Datenhandling.

Mit der Gerätekombination eines Röntgenfluoreszenzspektrometers mit den Meßeinrichtungen eines Diffraktometers kann nunmehr die komplette Qualitätskontrolle von Klinker und Zement durchgeführt werden. Gesonderte Einrichtungen bzw. Methoden sind dafür nicht mehr länger erforderlich. Mit dieser Neuentwicklung sind signifikante Einsparungen sowohl durch die höhere Effektivität der Analysendurchführung als auch der Analysendauer verbunden.

Danksagung

Die Autoren bedanken sich bei Frau G. Khouri von Ciments de Sibline für die Klinkerproben und bei Dr. P. Bürki sowie Dr. R. Stenger von der Holderbank Management and Consulting Ltd für die Übernahme der Freikalkbestimmung. Außerdem bedanken sich die Autoren bei den Herren X. Dupont-Wavrin und J.Y. Clément von Ciments Lafarge für die übrigen Proben sowie die nützlichen Diskussionen über die Kalksteinbestimmung im Zement.

Acknowledgement

The authors thank Mrs. G. Khouri of Ciments de Sibline (LBN) for the clinker samples and Dr. P. Bürki and Dr. R. Stenger of Holderbank Management and Consulting (CH) for their free lime analysis. They also thank Mr. X. Dupont-Wavrin and Mr. J.Y. Clément of Ciments Lafarge for the samples and useful discussions on limestone analysis in cement.

Literaturverzeichnis

[1] Anzelmo, J. A., and Boyer, B. W.: ASTM C-114 rapid method qualification of pressed powders and fused beads by X-ray fluorescence spectrometry. World Cement 18 (1987).

[2] Cement Composition, specifications and conformity criteria. European prestandard ENV 197-1, 1992.

[3] Carbonate additions to cement. ASTM publication, STP 1064, Edited by Klieger/Hooton (1990).

[4] Taylor, H. F. W.: Cement Chemistry. Academic Press (1990).

[5] Price, B. J.: Process control elemental analysis of cement-making materials using automated XRF spectrometers. World Cement 17 (1986).

[6] Adamson, B. W., and Yellepeddi, R. S.: Chemical analysis of cement — a more complete solution. World Cement, August 1992.

[7] Bonvin, D., Bapst, A., Yellepeddi, R. S., and Larsen, O. R.: Free lime determination in clinker using the ARL 8600 S Total Cement Analyzer. X-ray Spectrometry, Oktober 1992.

[8] Yellepeddi, R. S., Bapst, A., and Bonvin, D.: Determination of limestone addition in cement manufacture using the ARL 8600 S Total Cement Analyzer. World Cement, August 1993.

[9] Bonvin, D., Yellepeddi, R. S., and Matula, G.: XRF analysis of light elements in cement — new developments. World Cement, September 1990.

Regelung des Energiestroms zur Drehofensteuerung

Controlling the energy flow to rotary kiln firing systems

Régulation de flux d'énergie de la chauffe du four rotatif

Regulación del flujo de energía hacia el sistema de combustión del horno rotatorio

Von **D. Edelkott**, Düsseldorf/Deutschland

Zusammenfassung – Das Ziel durchgeführter Untersuchungen bestand darin, ein Konzept für die Regelung der Brennstoffzufuhr in Drehofenfeuerungen zu entwickeln und im praktischen Betrieb zu erproben. Als Grundlage des Reglerentwurfs dient ein einfacher Modellansatz, mit welchem die wesentlichen Charakteristiken des dynamischen Verhaltens einer Zementdrehofenanlage beschrieben werden. Dieses Modell liefert eine anschauliche Erklärung für das häufig zu beobachtende oszillierende Verhalten eines Drehofen/Klinkerkühler-Systems, welches sich in langanhaltenden zyklischen Schwankungen wesentlicher Prozeßparameter äußert. Das Verständnis dieser Vorgänge erlaubt eine qualitative Beschreibung des Einflusses von Kühlerbetrieb, Brennerposition, Flammeneinstellung und Ofenzustand auf die Schwingungsfähigkeit des Systems Drehofen/Klinkerkühler. Damit wird es möglich, einerseits durch gezielte Variation der verfahrenstechnischen Parameter die Schwingungsneigung des Systems zu dämpfen und andererseits Restschwankungen durch eine gegensteuernde Regelung der Brennstoffzufuhr zu kompensieren. Als wesentliche Störgröße des Brennprozesses ergibt sich aus der Modellbetrachtung der Sekundärluftenergiestrom, über den Austragsschwankungen des Drehofens in die Sinterzone rückgekoppelt werden. Als Regelgröße ist der Brennstoffenergiestrom zu betrachten, der insbesondere bei inhomogenen Brennstoffen durch eine Mengenregelung nicht ausreichend zu vergleichmäßigen ist. Daher wurde für die Erfassung dieser beiden Energieströme eine neue Meßtechnik entwickelt, erprobt und in eine sehr robust parametrierte Brennstoffregelung eingebunden.

Regelung des Energiestroms zur Drehofensteuerung

Summary – The object of the investigations carried out was to develop a concept for controlling the feed of fuel to rotary kiln firing systems and to test it in practical operation. The design of the controller was based on a simple model approach which describes the essential characteristics of the dynamic behaviour of a rotary cement kiln system. This model provides a clear explanation for the oscillating behaviour of a rotary-kiln/clinker-cooler system, which is frequently observed and is expressed in long-term cyclic fluctuations of important process parameters. Understanding these processes facilitates qualitative description of the influence of the cooler operation, burner position, flame adjustment and kiln state on the capacity of the rotary-kiln/clinker-cooler system to oscillate. This makes it possible to damp the tendency of the system to oscillate by carefully controlled variation of the process parameters, and also to compensate for residual oscillations through counter-control of the fuel feed. Examination of the model showed that the secondary air energy flow, by which the discharge fluctuations from the rotary kiln are fed back into the sintering zone, is an important disrupting variable in the clinker burning process. The fuel energy flow should be regarded as a controlled variable which cannot be evened out sufficiently with a volumetric control system, especially with inhomogeneous fuels. A new measuring technique for measuring the two energy flows was therefore developed, tested, and incorporated into a very robust parameterized fuel control system.

Controlling the energy flow to rotary kiln firing systems

Résumé – Le but des recherches menées était de développer un concept de régulation de l'alimentation en combustible des chauffes des fours rotatifs et de le vérifier en exploitation pratique. Comme base du dessin du régulateur sert une formule de modèle simple, avec laquelle sont décrites les caractéristiques essentielles du comportement dynamique d'une installation de four à ciment rotatif. Ce modèle fournit une explication claire pour le comportement oscillant, souvent observé, d'un système four rotatif/refroidisseur de clinker, se traduisant par des variations cycliques de longue durée, de paramètres essentiels du process. La compréhension de ces phénomènes permet une description qualitative de l'influence du fonctionnement du refroidisseur, de la position de la lance, du réglage de la flamme et de l'état du four, sur la possibilité d'oscillation du système four rotatif/refroidisseur de clinker. Ainsi, il devient possible d'une part, d'amortir la tendance à l'oscillation du système, par variation ciblée des paramètres de technologie du process et, d'autre part, de compenser les variations résiduelles par un réglage opposé de l'alimentation en combustible. Comme grandeur perturbante essentielle, se révèle, lors de l'étude du modèle, le flux d'énergie de l'air secondaire, par

Régulation de flux d'énergie de la chauffe du four rotatif

---

*) Überarbeitete Fassung eines Vortrages zum VDZ-Kongreß '93, Düsseldorf (27.9.–1.10.1993)
 Revised text of a lecture to the VDZ-Congress '93, Düsseldorf (27.9.–1.10.1993)

lequel sont répercutées sur la zone de clinkérisation les variations du débit de sortie du four rotatif. Comme grandeur de réglage, est à considérer le flux d'énergie du combustible, qui ne peut pas être lissé suffisamment, surtout avec des combustibles inhomogènes, au moyen d'une régulation quantitative. Pour cette raison a été élaborée et experimentée une nouvelle technique de mesure pour la saisie de ces deux flux d'énergie; elle a été insérée dans une régulation du combustible très robustement paramétrée.

Resumen – *El objetivo de los ensayos realizados ha sido el de desarrollar un concepto para la regulación de la alimentación de combustible en los sistemas de combustión de los hornos rotatorios y de probar el mismo en la práctica. El diseño del regulador se basa en un modelo sencillo, que permite describir las características esenciales del comportamiento dinámico de una instalación de horno rotatorio para cemento. Este modelo ofrece una explicación clara del comportamiento oscilante, muchas veces observado, de un sistema de horno rotatorio y enfriador de clínker, que se traduce por variaciones cíclicas de larga duración, de parámetros esenciales del proceso. La comprensión de estos fenómenos permite una descripción cualitativa del influjo que tienen el servicio del enfriador, la posición del quemador, la regulación de la llama y el estado del horno, sobre la facultad de oscilación del sistema de horno rotatorio y enfriador de clínker. Con ello es posible, por un lado, mediante la variación de los parámetros tecnológicos, amortiguar la tendencia a la oscilación del sistema y, por otro lado, compensar oscilaciones residuales, mediante una regulación contraria de la alimentación del combustible. Como magnitud de perturbación esencial del proceso de cocción resulta, según este modelo, el flujo de energía del aire secundario, mediante el cual las variaciones de la descarga del horno rotatorio repercuten en la zona de sinterización. Se debe considerar como magnitud de regulación el flujo de energía del combustible, el cual no se puede homogeneizar lo suficiente, sobre todo en el caso de los combustibles no homogéneos. Por esta razón, se ha desarrollado una nueva técnica para la medición de estos dos flujos de energía, la cual ha sido sometida a ensayo y luego incorporada a una regulación parametrada, muy robusta, del combustible.*

Regulación del flujo de energía hacia el sistema de combustión del horno rotatorio

1. Einleitung

Ziel der durchgeführten Untersuchungen war es, ein Konzept für die Regelung der Brennstoffzufuhr zur Drehofenfeuerung zu entwerfen. Die einzelnen Schritte dieser Entwicklung umfaßten dabei sowohl den theoretischen Entwurf der Reglerstruktur und die Schaffung der meßtechnischen Voraussetzungen als auch die praktische Erprobung.

Kennzeichnend für eine Regelung ist ein geschlossener Wirkungskreis. Ein solcher Regelkreis besteht einerseits aus der durch Störgrößen beeinflußten Regelstrecke, also der zu regelnden Drehofenanlage, und andererseits aus dem Regler. Dessen Aufgabe ist es, den aktuellen Wert der Regelgröße laufend mit der von außen vorzugebenden Führungsgröße zu vergleichen und bei auftretenden Abweichungen über Stellorgane, z.B. über die Brennstoffdosiereinheit, ausgleichend auf die Regelstrecke einzuwirken. Sofern diese Stelleingriffe jedoch z.B. zu heftig oder aber zum falschen Zeitpunkt erfolgen, kann dies statt zu der gewünschten Prozeßberuhigung zu einer Anregung von Prozeßstörungen führen. Daher muß vor dem Auslegen und Parametrieren einer Regeleinrichtung das zeitliche Verhalten der vorgegebenen Regelstrecke untersucht und beschrieben werden.

2. Modellierung der Systemdynamik

Zur Beschreibung der Systemdynamik einer Drehofenanlage wurde ein einfaches Modell entwickelt, das als Grundlage für den Reglerentwurf diente und darüber hinaus auch eine plausible Erklärung für häufig zu beobachtendes oszillierendes Verhalten von Drehofenanlagen liefert. Bei solchem instationären Anlagenverhalten tritt eine nur durch systeminterne Rückkoppelungseffekte zu erklärende Schwingung auf, die mehr oder weniger stark gedämpft sein kann und eine für die betreffende Ofenanlage typische Periode aufweist. In extremen Fällen konnte sogar eine trotz konstanter äußerer Einflußgrößen über mehrere Tage stabile Oszillation eines Drehofensystems beobachtet werden. Dieses Schwingungsverhalten äußert sich nicht nur durch Schwankungen des Materialaustrags aus dem Drehofen, sondern vielmehr durch entsprechende zyklische Schwankungen einer ganzen Reihe von Prozeßparametern.

Eine Erklärung für dieses Verhalten ist durch eine Wirkungskette möglich, die die Wechselwirkung zwischen

1. Introduction

The object of the investigations which were carried out was to develop a scheme for controlling the feed of fuel to rotary kiln firing systems. The individual stages in this development work included both the theoretical development of the controller structure and the creation of the required measurement technology, as well as the practical testing.

A control system is characterized by a closed action loop. Such a control loop consists on the one hand of the controlled system affected by disruptive variables, i.e. the rotary kiln system which has to be controlled, and on the other hand of the controller. Its task is to make continuous comparisons between the current value of the controlled variable and the externally specified reference input and, if there is a difference, to make adjustments to the controlled system through control units, e.g. through the fuel metering unit. If these control interventions are, for example, too violent or are made at the wrong time then instead of the intended calming effect they can have a disruptive effect on the process. It is therefore necessary to investigate and describe the time characteristics of the given controlled system before designing and parameterizing any control unit.

2. Modelling the system dynamics

To describe the system dynamics of a rotary kiln plant a simple model was developed which served as the basis for developing the controller and also provided a plausible explanation for the oscillating behaviour often observed in rotary kiln plants. During such unsteady-state plant behaviour oscillations occur which can only be explained by feedback effects within the system; they can be damped to a certain extent and have a period which is typical of the kiln plant concerned. In extreme cases stable oscillations have been observed in a rotary kiln system in spite of the fact that the external influencing variables had been constant for several days. This oscillating behaviour is apparent not just in fluctuations in the material discharge from the rotary kiln but also through corresponding cyclic fluctuations in a whole series of process parameters.

This behaviour can be explained by the following chain of actions which describe the interaction between rotary kiln

Drehrohrofen und Klinkerkühler beschreibt: Eine Erhöhung der Sekundärlufttemperatur führt zu einer heißeren Ofenflamme und damit zu einem höheren Schmelzphasenanteil in der Sinterzone. Dadurch erhöht sich der Transportwiderstand des Brennguts, was einen vorübergehend verringerten Klinkeraustrag des Drehofens zur Folge hat. Dies wiederum führt zu einer Abnahme der Sekundärlufttemperatur und in der Folge auch des Schmelzphasenanteils in der Sinterzone. Das Resultat ist ein erhöhter Materialaustrag des Drehofens, somit also auch ein Anstieg der Sekundärlufttemperatur, und der beschriebene Zyklus, der je nach Ofenanlage eine Periode zwischen ca. 15 Minuten und mehreren Stunden aufweisen kann, beginnt von neuem. Nach diesem Ansatz ist das oszillierende Verhalten eines Drehofen-Klinkerkühler-Systems geprägt von der Wechselwirkung zwischen dem Austragsverhalten des Drehofens und der aus dem Kühler rückgeführten Sekundärluftenergie (**Bild 1**).

Um für ein solches System das Zeitverhalten auch quantitativ diskutieren zu können, muß es in einzelne in sich rückwirkungsfreie Elemente unterteilt werden.

Im vorliegenden Fall handelt es sich dabei zum einen um den Klinkerkühler, dessen Eingangsgröße der Heißklinkerenergiestrom und dessen Ausgangsgröße der Sekundärluftenergiestrom ist. Das zweite Element dieses Wirkungskreises stellt der Drehrohrofen dar, als dessen Eingangsgröße der sich als Summe von Brennstoff- und Sekundärluftenergiestrom ergebende zugeführte Energiestrom anzusehen ist, und dessen Ausgangsgröße der Heißklinkerenergiestrom ist.

Die Darstellung des Zusammenhangs zwischen einer Ausgangsgröße und einer zeitvariablen Eingangsgröße ist für solche rückwirkungsfreie Übertragungsglieder über die Ermittlung einer Sprungantwortkurve möglich. Für den Klinkerkühler ist eine solche Antwortfunktion relativ einfach zu beschreiben, da Schwankungen des Materialaustrags aus dem Drehofen und somit des Heißklinkerenergiestroms zu einer annähernd proportionalen und nur geringfügig zeitversetzten Änderung der Temperatur der Sekundärluft führen.

Komplizierter ist dagegen die Reaktion des Drehofens auf eine sprunghafte Veränderung der zugeführten Energieströme. Entsprechend der Darstellung in **Bild 2** bewirkt eine plötzliche Erhöhung der Sekundärluft-Energiezufuhr zum Drehofen nach einer gewissen Verzugszeit einen vorübergehend rückläufigen Energieaustrag durch Heißklinker, da aufgrund einer erhöhten Schmelzphasenbildung mehr Material im Ofen zurückgehalten wird. Diese Vergrößerung des Materialspeichers ist in der Praxis an einem Anstieg des Drehofenantriebsstromes zu erkennen. Nach dem Durchlaufen eines Minimums steigt der Materialaustrag aus Bilanzgründen letztlich wieder auf den ursprünglichen Wert, der Energiestrom des Heißklinkers steigt aufgrund der erhöhten Temperatur sogar über das Ausgangsniveau hinaus an.

Dieses Übertragungsverhalten von Drehofen und Klinkerkühler läßt sich mathematisch beschreiben, indem die zunächst für Klinkerkühler und Drehofen getrennt aufgestellten Übertragungsfunktionen als gekoppeltes System behandelt werden. Damit wird es durch Variation der die Übertragungsfunktionen beschreibenden Parameter möglich, einerseits die Einflüsse von Ansatzverhältnissen, Flammenform und Brennerpositionierung auf die Schwingungsneigung eines Drehofensystems zu diskutieren, andererseits stellt das Modell auch die Grundlage für den Entwurf einer Energiestromregelung dar. Aus diesen Modellbetrachtungen ergibt sich als Hauptanforderung an eine Prozeßregelung, daß sie Schwankungen der Energiezufuhr zum Drehofen ausgleichen muß, die aus veränderlichen Heizwerten der eingesetzten Brennstoffe und/oder aus kurzfristigen Veränderungen des Sekundärluftenergiestroms resultieren. Voraussetzung zur Umsetzung eines solchen Regelkonzeptes ist eine zuverlässige meßtechnische Erfassung der genannten Regelgrößen „Sekundärluftenergiestrom" und „Brennstoffenergiestrom".

and clinker cooler: An increase in the secondary air temperature leads to a hotter kiln flame and hence to a higher proportion of liquid phase in the sintering zone. This increases the transport resistance of the material in the kiln, resulting in a temporary reduction in the clinker discharge from the kiln. This in turn leads to a fall in the secondary air temperature and, as a consequence, also in the proportion of liquid phase in the sintering zone. The result is an increased discharge of material from the rotary kiln, and therefore also a rise in the secondary air temperature. The cycle, which can have a period of between about 15 minutes and several hours depending on the kiln system, then starts again. According to this approach the oscillating behaviour of a system consisting of rotary kiln and clinker cooler is characterized by the interaction between the discharge behaviour of the rotary kiln and the secondary air energy transported back from the cooler (**Fig.1**).

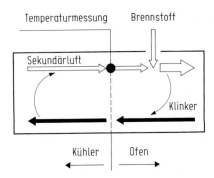

BILD 1: Schema der Wechselwirkungen zwischen Klinkerkühler und Drehofen
FIGURE 1: Diagram showing the interactions between clinker cooler and rotary kiln

Temperaturmessung	=	temperature measurement
Brennstoff	=	fuel
Sekundärluft	=	secondary air
Kühler	=	cooler
Ofen	=	kiln
Klinker	=	clinker

If the time behaviour of such a system is also to be examined quantitatively it must be subdivided into individual elements which are in themselves free from feedback.

In the present case this involves, firstly, the clinker cooler where the input variable is the hot clinker energy flow and the output variable is the secondary air energy flow. The second element of this action loop is represented by the rotary kiln where the input variable can be regarded as the incoming energy flow given by the sum of the fuel and secondary air energy flows, and the output variable is the hot clinker energy flow.

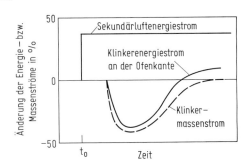

BILD 2: Sprungantwortkurve zur Beschreibung des Übertragungsverhaltens eines Drehofens
FIGURE 2: Step-response curve for describing the transfer behaviour of a rotary kiln

Änderung der Energie- bzw. Massenströme in %	=	changes in the energy and mass flows in %
Zeit	=	time
Sekundärluftenergiestrom	=	secondary air energy flow
Klinkerenergiestrom an der Ofenkante	=	clinker energy flow at the kiln outlet
Klinkermassenstrom	=	clinker mass flow

BILD 3: Schema zur Ermittlung des Brennstoffenergiestroms aus einer kontinuierlichen Sauerstoffbilanz
FIGURE 3: Scheme for determining the fuel energy flow from a continuous oxygen balance

Brennstoffenergie	= fuel energy
Auswerterechner für O_2-Bilanzierung	= evaluation computer for O_2 balance
Gasanalyse	= gas analysis
Abgasmenge	= exhaust gas volume
Kontinuierliche Messungen im Rohgas	= continuous measurements in the raw gas

3. Meßtechnik

Bei der Messung des Brennstoffenergiestroms müssen neben den zugeführten Brennstoffmengen auch relativ kurzfristige Änderungen der Heizwerte mit erfaßt werden. Daher wird zur Zeit ein Meßverfahren erprobt, das den Brennstoffenergiestrom nach dem in **Bild 3** dargestellten Verfahrensschema indirekt über eine kontinuierliche Sauerstoff-Bilanz ermittelt. Die Hauptschwierigkeit bestand dabei in einer hinreichend genauen Erfassung der Abgasmenge.

Zur Messung der Sekundärlufttemperatur wird ein akustisches Meßverfahren eingesetzt, wie es in **Bild 4** schematisch dargestellt ist. Dieses Verfahren nutzt die Temperaturabhängigkeit der Schallgeschwindigkeit, um über die Laufzeit, die ein Schallimpuls zum Durchqueren der Sekundärluft benötigt, direkt auf deren Temperatur zu schließen. Der Vorteil gegenüber anderen Meßverfahren liegt im wesentlichen darin, daß diese Messung nicht durch Strahlungseinflüsse beeinflußt wird und annähernd trägheitslos auf Temperaturänderungen reagiert.

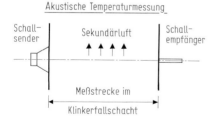

BILD 4: Prinzip der akustischen Temperaturmessung
FIGURE 4: Principle of the acoustic temperature measuring system

Akustische Temperaturmessung	= acoustic temperature measurement
Schallsender	= sound emitter
Sekundärluft	= secondary air
Schallempfänger	= sound pick-up
Meßstrecke im Klinkerfallschacht	= measurement section in the clinker fall chute

Die genannten Meßverfahren liefern direkte und eindeutige Informationen über die wesentlichen Prozeßparameter und erlauben daher ein gezielteres Eingreifen, als das aus den bisher verfügbaren indirekten Prozeßindikatoren ableitbar ist. Ihre betriebliche Verfügbarkeit stellt somit nicht nur die Voraussetzung für die Installation der beschriebenen Regeleinrichtung dar, sondern ist auch für die manuelle Prozeßführung eine wertvolle Hilfe.

The relationship between an output variable and a time-variable input variable can be represented for such feedback-free transfer elements by determining a step-response curve. Such a response function is relatively easy to describe for the clinker cooler because fluctuations in the material discharged from the rotary kiln, and hence in the hot clinker energy flow, lead to an almost proportional, and only slightly time-delayed, change in the temperature of the secondary air.

The reaction of the rotary kiln to a step change in the incoming energy flows is, however, more complicated. According to the representation in **Fig. 2** a sudden rise in the energy supply from the secondary air to the rotary kiln produces, after a certain delay time, a temporary drop in the energy discharged through the hot clinker as more material is held back in the kiln because of increased formation of liquid phase. This enlargement in the quantity of material retained can in practice be detected by a rise in the kiln drive current. After passing through a minimum the material discharge then rises again to the original value to satisfy the mass balance but, because of the increased temperature, the energy flow in the hot clinker rises above the original level.

This transfer behaviour of the rotary kiln and clinker cooler can be described mathematically, and the transfer functions initially set up separately for the clinker cooler and rotary kiln are treated as a coupled system. In this way it is possible, by varying the parameters which describe the transfer function, to examine the influence of coating conditions, flame shape and burner positioning on the oscillating tendencies of a rotary kiln system. In addition to this the model also represents the basis for developing an energy flow control system. This modelling shows that the main requirement of a process control system is that it must smooth the fluctuations in the energy supply to the rotary kiln resulting from varying calorific values of the fuels used and/or from short-term changes in the secondary air energy flow. A precondition for implementation of such a control concept is a reliable method of measuring the controlled variables mentioned, namely "secondary air energy flow" and "fuel energy flow".

3. Measurement technology

When the fuel energy flow is measured it is necessary to measure not only the quantities of fuel supplied but also any relatively short-term changes in the calorific values. A method of measurement is therefore being tested at present which determines the fuel energy flow indirectly through a continuous oxygen balance based on the process scheme shown in **Fig. 3**. The main problem lies in measuring the quantity of exhaust gas with sufficient accuracy.

An acoustic method of measurement, shown diagrammatically in **Fig. 4**, is used for measuring the secondary air temperature. This method makes use of the temperature dependence of the speed of sound to draw a direct inference about the secondary air temperature from the transit time required for a sound pulse to pass through the secondary air. The main advantage over other methods of measurement is that this measurement is not affected by radiation and reacts almost instantaneously to temperature changes.

The methods of measurement referred to supply direct and unambiguous information about the important process parameters and therefore permit more accurately controlled intervention than can be deduced from the indirect process indicators previously available. It is therefore essential for the installation of the described control equipment that these methods of measurement are operationally available, and what is more they represent a valuable aid for manual process operation.

Computersimulation – wertvolle Entscheidungshilfe für die Optimierung der Zementmahlung

Computer simulation – valuable decision aid for optimizing cement grinding

Simulation sur ordinateur – aide à la décision appréciable pour l'optimisation du broyage du ciment

Simulación por ordenador – una ayuda valiosa para la optimización de la molienda de cemento

Von **D. Espig, V. Reinsch, B. Bicher** und **J. Otto,** Freiberg/Deutschland

Zusammenfassung – Neben der Standardtechnik zur Prozeßüberwachung und -steuerung wird die Wirtschaftlichkeit von Zementmahlanlagen durch qualifizierte ingenieurtechnische Betreuung gewährleistet. Diese Betreuung erfordert eine permanente Prozeß- und Systemanalyse, um zum richtigen Zeitpunkt die günstigsten Maßnahmen zur Optimierung der Zementmahlung durchzuführen. Diese Maßnahmen dienen letztlich der Qualitätssicherung und der Kostensenkung durch verbesserte Energieausnutzung. Eine gründliche und ständige Systemanalyse ist jedoch mit äußerst hohem Zeitaufwand verbunden. Außerdem muß eine wachsende Informations- und Datenflut verarbeitet werden. Bewährtes methodisches Instrumentarium zur off-line Bilanzierung und Informationsverdichtung sowie neue Möglichkeiten, das Verhalten von Mahlkreisläufen zu analysieren und zu simulieren, wurden deshalb in einem Softwarepaket mit der Bezeichnung „CGC – Calculation of Grinding Circuits" zusammengefaßt. CGC – vorerst für den PC-Einsatz verfügbar – besitzt eine anwenderfreundliche Bedieneroberfläche, leistungsfähige Datenbankkomponenten und ermöglicht die Darstellung von Ergebnissen der Mahlkreislaufuntersuchung oder von Simulationen in Diagrammen und Tabellen.

Summary – The cost-effectiveness of cement grinding plants is ensured not only by standard methods of process monitoring and control, but by competent process engineering care. This care requires permanent process and system analysis so that the measures for optimizing the cement grinding process can be carried out at the correct time. All this serves ultimately to ensure quality and to lower costs through improved energy utilization. However, any fundamental and continuous system analysis is linked with exceptionally high time expenditure, and it is also necessary to process an increasing flood of information and data. Well tried systematic instruments for off-line balances and information compaction as well as new ways of analyzing and simulating the behaviour of grinding circuits were therefore combined in a software package known as „CGC – calculation of grinding circuits". CGC – available at present for use with PCs – has a user-friendly operator interface, powerful database components, and allows the results of the grinding circuit investigation or of simulations to be displayed as diagrams and tables.

Résumé – Mise à part la technique standard de surveillance et conduite du process, la rentabilité des ateliers de broyage de ciment est garantie par une supervision qualifiée relevant de la technique de l'ingénieur. Cette supervision exige une analyse constante du process et du système, afin de prendre au moment correct les mesures les plus favorables pour l'optimisation du broyage du ciment. Tout cela sert, en dernier lieu, à l'assurance de la qualité et à l'abaissement des coûts, par une utilisation améliorée de l'énergie. Une analyse du système, approfondie et permanente, est par contre liée à une très forte dépense de temps. De plus, il faut traiter un afflux croissant d'informations et de données. Une instrumentation méthodique, éprouvée, pour l'établissement de bilans en temps différé et une densification de l'information, ainsi que des possibilités nouvelles d'analyser et de simuler le comportement de circuits de broyage, ont pour cette raison été groupées dans un paquet de programmes intitulé „CGC – Calculation of Grinding Circuits" (Calcul de circuits de broyage). CGC – pour l'instant disponible pour PC – possède une surface de manipulation conviviale, des composants puissants de banque de données et permet une représentation des résultats de l'étude du circuit de broyage ou de simulations, en diagrammes et tableaux.

Resumen – Aparte de la técnica standard de control y mando del proceso, queda garantizada la rentabilidad de las plantas cementeras mediante una ingeniería cualificada. Este asesoramiento requiere un análisis permanente de procesos y sistemas, con el fin de poder efectuar, en el momento más oportuno, las medidas adecuadas para la optimización de la molienda del cemento. Estas medidas sirven, al fin y al cabo, para el aseguramiento de la calidad y la reducción de los costes, gracias a un mejor aprovechamiento de la energía. Sin embargo, un análisis detallado y constante requiere muchísimo tiempo. Además, hay que tratar un creciente flujo de informaciones y de da-

tos. Por esta razón, se han reunido en un paquete de software llamado „CGC - Calculation of Grinding Circuits" (Cálculo de circuitos de molienda) unos instrumentos sistemáticos y de probada eficacia, destinados al establecimiento off-line de balances y la densificación de la información, así como nuevas posibilidades de analizar y simular el comportamiento de los circuitos de molienda. El CGC – disponible de momento para el empleo mediante ordenador personal – posee una superficie de usuario fácil de manejar y unos componentes de banco de datos muy potentes y permite la representación de resultados de estudios del circuito de molienda o de simulaciones en forma de diagramas y tablas.

1. Einleitung

In zunehmendem Maße werden beim Betreiben von Zementmahlanlagen Steuerstrategien eingesetzt, die wissensbasiert sind und in denen das Fuzzy-Konzept zum Tragen kommt. Die Verarbeitung der on-line-Meßwerte und das Informationsmanagement hierfür sind Komponenten eines modernen Prozeßleitsystems.

Zur verfahrenstechnischen Bewertung der Effektivität von Mahlanlagen müssen jedoch auch Prozeßgrößen herangezogen werden, die noch nicht on-line zur Verfügung stehen und einer separaten Informationsaufbereitung bedürfen. Hierzu gehören granulometrische Informationen zu den Stoffströmen sowie Kenngrößen zur Charakterisierung von Ausgangsmaterial und Produkteigenschaften. Insbesondere kann mit Hilfe der granulometrischen Informationen eine detaillierte Bilanzierung der Masseströme im Mahlkreislauf vorgenommen werden. Auf dieser Basis ist die Bewertung der energetischen Effektivität der Mahlung möglich. Eine solche Iststandsaufnahme ist stets mit dem Ziel verbunden, Schwachstellen im System zu lokalisieren, deren Wirkung zu quantifizieren und die kostengünstigste Variante zur Verbesserung der Arbeitsweise der Mahlanlage abzuleiten. Dies erfordert es, den Zusammenhang zwischen verfahrenstechnischer Bilanz und dem jeweiligen maschinentechnischen Zustand der Mahlanlage herzustellen.

Mit dem Computereinsatz bietet sich die Möglichkeit, den Gesamtaufwand einer Prozeßanalyse wesentlich zu reduzieren und die Aussagesicherheit zu erhöhen. Voraussetzung ist eine Softwarelösung, die sowohl die erforderlichen Komponenten zur Systembewertung als auch zur modellgestützten Systemsimulation besitzt. Die notwendige Funktionalität einer Software, die diesen Ansprüchen genügt, geht aus dem im **Bild 1** dargestellten Schema hervor. Das Schema verdeutlicht das enge Wechselspiel zwischen Zielstellung, experimentellen Untersuchungen, ausgewählten Modell-Identifikationsmodulen und den Komponenten zur verfahrenstechnischen Simulation der Kreislaufmahlung. Über entsprechende Datenbanken werden die konkreten Angaben zur Mühle, zum Sichter und zu deren Zusammenschaltung im Fließschema gestellt. Analog hierzu ist problemorientiert die Verbindung zu den einzusetzenden Berechnungsmethoden und Simulationsmodellen sowie zur benötigten verfahrenstechnischen Informationsmenge herzustellen.

1. Introduction

Increasing use is being made in the operation of cement grinding plants of control strategies which are knowledge based and in which the fuzzy concept is applied. The processing system for on-line measured values and the information management system required for this purpose form part of a modern process control system.

However, for process engineering evaluation of the effectiveness of grinding plants it is also necessary to use process variables which are not available on-line and which require a separate information editing system. This includes granulometric information about the mass flows, and parameters for characterizing the starting materials and product characteristics. The granulometric information in particular can be used to carry out a detailed balance of the mass flows in the grinding circuit. This makes it possible to evaluate the energy efficiency of the grinding system. Such an actual-value recording system is always linked with the objective of locating weak points in the system, quantifying their effects, and deriving the most cost-favourable options for improving the mode of operation of the grinding system. For this it is necessary to establish the relationship between the process engineering balance and the corresponding mechanical state of the grinding system.

The use of computers offers the chance of making a substantial reduction in the total expenditure on a process analysis, and of increasing the reliability of the information. This requires a software solution which possesses the components necessary both for system evaluation and for the model-aided system simulation. The block diagram in **Fig. 1** shows the functionality required of software which satisfies these demands. The diagram illustrates the close interplay between setting the objective, experimental investigations, selected model-identification modules, and the components for process engineering simulation of the grinding circuit. The concrete information about the mill, the classifier, and their interconnection in the flow diagram is made available through appropriate databases. In analogy with this, it is necessary from the problem point of view to establish the connection between the methods of calculation and the simulation models to be used and the quantity of process engineering information required.

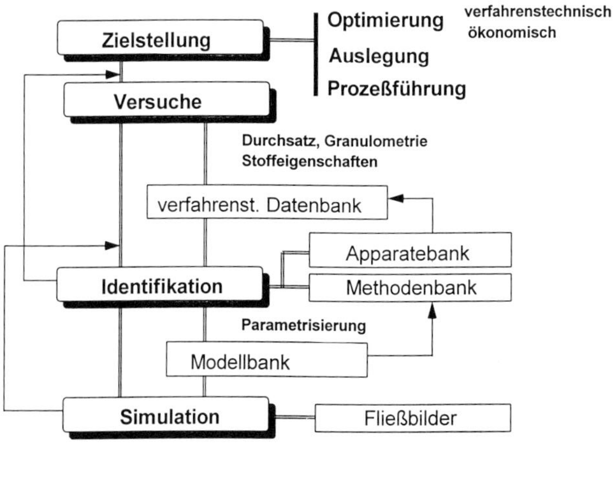

BILD 1: Software im Entscheidungsprozeß
FIGURE 1: Software in the decision-making process

verfahrenstechnisch	= process engineering
ökonomisch	= costs
Zielstellung	= setting the objective
Versuche	= trials
Identifikation	= identification
Simulation	= simulation
Optimierung	= optimization
Auslegung	= design
Prozeßführung	= process control
Durchsatz, Granulometrie Stoffeigenschaften	= throughput, granulometry, physical properties
verfahrenst. Datenbank	= process engineering database
Apparatebank	= equipment database
Methodenbank	= method database
Parametrisierung	= parameterization
Modellbank	= model database
Fließbilder	= flow diagrams

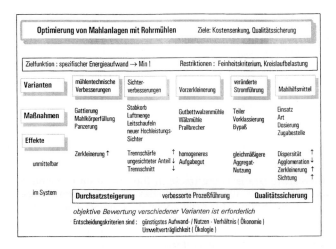

BILD 2: Optimierung von Mahlanlagen
FIGURE 2: Optimizing grinding plants

Optimierung von Mahlanlagen mit Rohrmühlen	= Optimization of grinding plants with tube mills
Ziele: Kostensenkung, Qualitätssicherung	= Objectives: reduced costs, quality assurance
Zielfunktion: spezifischer Energieaufwand → Min!	= target function: specific energy consumption → min!
Restriktionen: Feinheitskriterium, Kreislaufbelastung	= restrictions: fineness criteria, recirculating load
Varianten	= options
Maßnahmen	= measures
Effekte	= effects
unmittelbar	= direct
im System	= in the system
mühlentechnische Verbesserungen	= improvements in mill technology
Gattierung Mahlkörperfüllung	= charge grading grinding media filling
Panzerung	= protective lining
Zerkleinerung	= comminution
Durchsatzsteigerung	= increased throughput
verbesserte Prozeßführung	= improved process control
Qualitätssicherung	= quality assurance
objektive Bewertung verschiedener Varianten ist erforderlich	= objective evaluation of the different options is needed
Entscheidungskriterien sind:	= decision criteria are:
günstigstes Aufwand-/Nutzen-Verhältnis (Ökonomie)	favourable cost/benefit ratio (economy)
Umweltverträglichkeit (Ökologie)	environmental compatibility (ecology)
Sichterverbesserungen	= classifier improvements
Stabkorb, Luftmenge, Leitschaufeln, neuer Hochleistungs-Sichter	= bladed rotor, quantity of air, guide vanes, new high-efficiency classifier
Trennschärfe ungesichteter Anteil	= sharpness of separation unclassified fraction
Trennschnitt	= cut
Vorzerkleinerung	= pre-comminution
Gutbettwalzmühle, Wälzmühle, Prallbrecher	= high-pressure grinding rolls, roller grinding mill, impact crusher
homogeneres Aufgabegut	= more homogeneous feed material
veränderte Stromführung	= changed flow regime
Teiler Vorklassierung Bypaß	= splitter preliminary classification bypass
gleichmäßigere Aggregat-Nutzung	= more uniform utilization of plant units
Mahlhilfsmittel	= grinding aids
Einsatz, Art, Dosierung, Zugabestelle	= use, type, metered addition, feed point
Dispersität, Agglomeration, Zerkleinerung, Sichtung	= dispersity, agglomeration, comminution, classification

Das Spektrum verfahrenstechnischer Möglichkeiten zur Verbesserung der Arbeitsweise einer Mahlanlage und der damit erreichbaren Effekte wird im **Bild 2** veranschaulicht.

2. Anwendungsbeispiele für die Computersimulation

Leistungsfähigkeit und Möglichkeiten der Computersimulation mit Hilfe der speziell für die genannte Fragestellung entwickelten Softwarelösung CGC — Calculation of Grinding Circuits [1] sollen an den Anwendungsbeispielen der Prozeßanalyse eines Zementmahlkreislaufs mit Mittenaustragsmühle, der Sichteraustausch-Effekte im Mahlkreislauf und der Variantenrechnungen für vorzerkleinertes Aufgabegut aufgezeigt werden.

2.1 Prozeßanalyse eines Zementmahlkreislaufs mit Mittenaustragsmühle

Die Ziele der Untersuchung waren die Bilanzierung und Bewertung der Mahlanlage für einen Referenzzustand, die Ermittlung spezieller Kennfelder zum Sichter und die Aufstellung eines Arbeitsdiagramms der Mahlanlage, das die Relation zwischen Anlagendurchsatz, Streutellerdrehzahl und Fertiggutfeinheit wiedergibt und so die optimale Fahrweise für jede Produktfeinheit verdeutlicht.

Die korrekte verfahrenstechnische Bilanzierung der Mahlgutströme stellt, wie aus **Bild 3** ersichtlich, im Falle der Mittenaustragsmühle eine recht anspruchsvolle Teilaufgabe dar. Aufgrund der unvollständigen Information führt nur eine iterative Vorgehensweise zum Ziel, die einen ständigen Wechsel zwischen Bilanzierung und Simulation erfordert und ohne rechnertechnische Hilfsmittel nicht zu bewältigen

The range of process engineering options for improving the mode of operation of a grinding system and the effects which this can achieve are illustrated in **Fig. 2**.

2. Application examples for computer simulation

The effectiveness and capabilities of computer simulation will be illustrated with the aid of the CGC (Calculation of Grinding Circuits) software solution [1] developed specifically for the problem referred to, using the application examples of process analysis of a cement grinding circuit with central discharge mill, classifier replacement effect on the grinding circuit, and variant calculations for pre-comminuted feed material.

2.1 Process analysis of a cement grinding circuit with central discharge mill

The objectives of the investigation were to balance and evaluate the grinding system for a reference state, to determine specific performance characteristics of the classifier, and to prepare a working diagram for the grinding system which reflects the relationship between plant throughput, distributor plate speed and finished product fineness, and therefore illustrates the optimum mode of operation for each product fineness.

As can be seen from **Fig. 3**, correct process engineering balancing of the material flows is a really demanding task in the case of the central discharge mill. Because of the incomplete information, the objective can only be reached by an iterative procedure, which requires continuous interchange between balancing and simulation and could not be accomplished without calculation aids. For this purpose CGC falls back on a special low-parameter calculation model [3] which provides suitable approximations and essential relationships for the three basic processes of comminution, particle transport and classification.

The Tromp curve, described by a special 4-parameter function equation (**Fig. 4**), is used in CGC for characterizing the

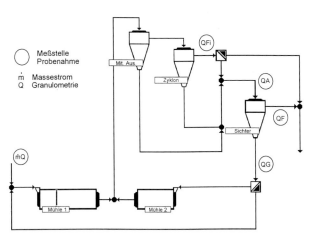

BILD 3: Mahlkreislauf mit Mittenaustragsmühle
FIGURE 3: Grinding circuit with central discharge mill

Meßstelle	= measuring point	Mühle 2	= mill 2
Probenahme	= sampling	Mit. Aus.	= central discharge
Massestrom	= mass flow	Zyklon	= cyclone
Granulometrie	= granulometry	Sichter	= classifier
Mühle 1	= mill 1		

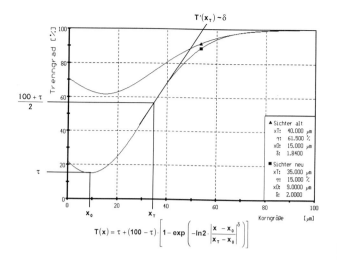

BILD 4: Trennkurvenvergleich
FIGURE 4: Comparison of separation curves

Trenngrad = Degree of separation
Korngröße = Particle size
Sichter alt = old classifier
Sichter neu = new classifier

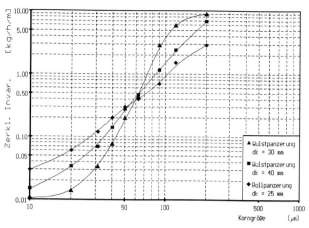

BILD 5: Durchsatzcharakteristik einer Trommelmühle
Variable: Korngröße
Parameter: Panzerung/Gattierung
FIGURE 5: Throughput characteristic of a drum mill
variable: particle size
parameters: lining/charge grading

Zerkl. Invar. = Comminution invariance
Korngröße = Particle size
Wulstpanzerung = bead armouring
Rollpanzerung = grooved armouring

wäre. CGC greift hierfür auf ein spezielles parameterarmes Berechnungsmodell zurück [3], mit dem geeignete Approximationen und wesentliche Zusammenhänge für die drei Grundprozesse Zerkleinern, Teilchentransport und Klassieren zusammengefaßt zur Verfügung stehen.

Zur Charakterisierung der Arbeitsweise des Sichters dient in CGC die Trompkurve, beschrieben durch einen speziellen vierparametrischen Funktionsansatz (**Bild 4**). Damit gelingt es, die „Angelhakenform" im Bereich kleiner Korngrößen nachzubilden [4]. Dieser Funktionsansatz kann in gleicher Weise auch für andere Klassierstellen im Mahlkreislauf (hier: Mittenaustrag und Staubzyklon) verwendet werden. Wird der Sichter mit unterschiedlichen Massenströmen beaufschlagt und mit unterschiedlichen Streutellerdrehzahlen betrieben, so wirkt sich dies auf die Trennkurve aus [5]. Sichteraufgabemassestrom als Prozeßvariable und Streutellerdrehzahl als Stellgröße beeinflussen die Hauptparameter Trennkorngröße x_T und Teilungsmenge (ungesichteter Anteil) τ in charakteristischer Weise, was sich über Kennfelder quantitativ erfassen läßt. CGC bietet für die nötigen Auswerteschritte vom Bilanzausgleich über die interaktive Trennkurvenanpassung bis hin zur Aufstellung der Parameter-Kennfelder eine optimale Unterstützung.

Eine ähnlich aussagekräftige Kennlinie, wie dies die Trompkurve zur Klassierung darstellt, war bislang für die Mühle noch nicht in Gebrauch. Ergänzend zur Energieausnutzung als bewährtem integralen Kennwert wird deshalb vorgeschlagen, die im **Bild 5** dargestellte „Durchsatzcharakteristik" zu verwenden, die aus dem erwähnten Berechnungsmodell abgeleitet wurde. Auf die Interpretation und Bedeutung dieser Kennlinie wird in [2] näher eingegangen.

Diese neue Kennlinie bietet auf Grund ihrer spezifischen Invarianzeigenschaften vorteilhafte Möglichkeiten zur differenzierten Bewertung der korngrößenabhängigen Zerkleinerungseffektivität, zum Vergleich verschiedener Mühlenzustände (im Bild 5 z. B. für verschiedene Gattierungen und Panzerungen ausgeführt) und zur Vorhersage der Produktfeinheit des Mühlenaustrags bei Änderung von Aufgabemassestrom, Korngrößenverteilung des Mühlenaufgabegutes, Transportverhalten des Mahlgutes und der Mahlraumlänge.

Das hier im Überblick dargelegte methodische Vorgehen ermöglicht detaillierte Simulationsrechnungen, in deren Ergebnis das im **Bild 6** wiedergegebene Arbeitsdiagramm aufgestellt wurde. Deutlich erkennbar sind am Verlauf der eingetragenen Isolinien der zu jeder Feinheitsforderung (R 63) maximale Anlagendurchsatz und die zugehörige Sichtereinstellung.

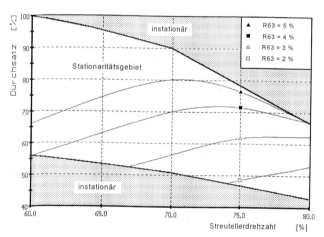

BILD 6: Arbeitsdiagramm mit Feinheits-Isolinien
Variable: Streutellerdrehzahl
Durchsatz
FIGURE 6: Work diagram with fineness isolines
variable: distributor plate speed
throughput

Durchsatz = Throughput
Streutellerdrehzahl = Distributor plate speed
instationär = unsteady state
Stationaritätsgebiet = steady state region

mode of operation of the classifier. In this way it is possible to model the „fish-hook shape" in the range of small particle sizes [4]. This function equation can also be used in a similar manner for other classifier positions in the grinding circuit (here: central discharge and dust cyclone). If the classifier is supplied with different mass flows and operated with different distributor plate speeds, then this affects the separation curve [5]. Classifier feed mass flow as the process variable and distributor plate speed as the manipulated variable have characteristic effects on the main parameters of cut size x_T and partition quantity (unclassified fraction) τ, which can be detected quantitatively by performance characteristics. CGC offers optimum support for the necessary evaluation stages from establishing the balance, through the interactive matching of the separation curve, right up to establishing the parameter performance characteristics.

A similarly informative characteristic curve, as represented by the Tromp curve for classification, has not so far been used for the mill. In addition to energy utilization as a proven integral parameter it is therefore recommended that the „throughput characteristic" shown in **Fig. 5** should be used;

2.2 Sichteraustausch — Effekte im Mahlkreislauf

Eine beträchtliche Anzahl von Mahlanlagen wird noch mit wenig trennscharfen Sichtern einer veralteten technischen Generation betrieben. Mit der Entscheidung, einen solchen Sichter durch einen neuen, trennscharfen Sichter zu ersetzen, ist für den Mahlanlagenbetreiber die Frage nach Aufwand (Investition) und Nutzen (Effekt) verbunden. Mittels Computersimulation können die zu erwartenden Effekte hinsichtlich Durchsatzsteigerung, Veränderung des Kornbandes und Veränderung der Kreislaufbelastung im Detail vorhergesagt werden. Dies setzt die Kenntnis über den Istzustand (Referenzzustand) der Mahlanlage mit dem bisherigen Sichter voraus, auf dessen Basis die Wirkung einer veränderten Sichtertrennkurve (s. Bild 4) ausgewiesen werden kann. Die konkreten Effekte zur Durchsatzsteigerung, zur Kreislaufbelastung und die mit der veränderten Stromführung erreichbare Qualitätssicherung werden in der nachstehenden **Tabelle 1** angegeben. **Bild 7** veranschaulicht die Fließbildsimulation.

TABELLE 1: Vergleich der Meßwerte im Referenzzustand mit den Simulationsergebnissen
(Hochleistungssichter: Ziel — 3500 cm^2/g nach Blaine)

	Dimension	Referenzzustand	Simulation
Durchsatz, relativ	%	100	119
Umlaufzahl		3,11	2,61
Parameter der Sichtertrennkurve			
x_T	μm	40,4	35,0
τ	%	61,5	15,0
Sichteraufgabegut			
$D_{45\mu m}$	%	59,5	46,8
Fertiggut			
$D_{45\mu m}$	%	84,2	92,5
d'_{RRSB}	μm	21,4	19,4
n_{RRSB}		0,83	0,95
spez. Oberfläche nach Blaine	cm^2/g	3180	3520

2.3 Variantenrechnungen für vorzerkleinertes Aufgabegut

Die Ausrüstung von Roh- und Zementmahlanlagen mit einer separaten Vorzerkleinerungsstufe ist zentraler Punkt bei Rekonstruktionsmaßnahmen und bei der Neuerrichtung von Mahlanlagen der Zementindustrie. Mittels Computersimulation kann die Frage geklärt werden, welche Durchsatzsteigerungen erwartet werden können. Entscheidend hierfür sind die durch Vorzerkleinerung erreichte Aufgabegutfeinheit und die Feinheitsforderung an das Fertig-

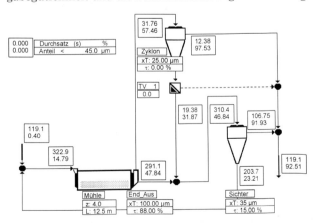

BILD 7: Fließbildvisualisierung einer Mahlkreislauf-Simulation mit CGC
FIGURE 7: Flow chart display of a grinding circuit simulation with CGC

Durchsatz	= throughput	End Aus	= end discharge
Anteil	= fraction	Sichter	= classifier
Mühle	= mill	Zyklon	= cyclone

this is derived from the calculation model referred to. The interpretation and significance of this characteristic curve is described in detail in [2].

Because of its specific invariance characteristics, this new characteristic curve offers benefits for refined evaluation of comminution effectiveness as a function of particle size, for comparison of different mill conditions (listed in Fig. 5, for example, for different charge gradings and protective linings), and for predicting the product fineness of the mill output on changing the feed mass flow, the particle size distribution of the mill feed, the transport behaviour of the mill feed, and the length of the grinding chamber.

The methodical procedure outlined here makes it possible to carry out detailed simulation calculations, the results of which were used to draw up the working diagram shown in **Fig. 6**. The shapes of the plotted isolines clearly show the maximum plant throughput and the associated classifier setting for each fineness requirement (R63).

2.2 Classifier replacement — effect on the grinding circuit

A considerable number of grinding systems are still being operated with classifiers of an obsolete technical generation with poor sharpness of separation. The decision to replace such a classifier by a new more selective classifier is, for the grinding plant operator, linked with the question of expenditure (investment) and benefit (effect). By using computer simulation it is possible to make a detailed prediction of the expected effects with respect to increase in throughput, change in particle size range, and change in the load of recycled material. This presupposes knowledge of the actual state (reference state) of the grinding system with the existing classifier which can then be used as the basis for identifying the effect of a changed classifier separation curve (see Fig. 4). The actual effects on the increase in throughput, the quantity of material recycled, and the quality assurance which can be achieved with the changed flow regime are given in **Table 1** below. **Fig. 7** illustrates the flow sheet simulation.

TABLE 1: Comparison of the values measured in the reference state with the simulation results
(high-efficiency classifier: target — 3500 cm^2/g Blaine)

	Dimensions	Reference state	Simulation
Throughput, relative	%	100	119
Rotational speed		3.11	2.61
Parameters of the classifier separation curve			
x_T	μm	40.4	35.0
τ	%	61.5	15.0
Classifier feed material			
$D_{45\mu m}$	%	59.5	46.8
Finished product			
$D_{45\mu m}$	%	84.2	92.5
d'_{RRSB}	μm	21.4	19.4
n_{RRSB}		0.83	0.95
Blaine specific surface area	cm^2/g	3180	3520

2.3 Variance calculations for pre-comminuted feed material

Equipping raw material and cement grinding systems with a separate pre-comminution stage is a central point in reconstruction measures and when building new grinding systems in the cement industry. By using computer simulation it is possible to answer the question as to what increases in throughput can be expected. Critical factors here are the feed material fineness achieved by pre-comminution and the fineness requirement for the finished product. The working diagram shown in **Fig. 8** was established for a specific example using CGC. The relationship between the d80 particle size of the pre-comminuted feed material and the

gut. Mittels CGC wurde für ein konkretes Beispiel das in **Bild 8** wiedergegebene Arbeitsdiagramm aufgestellt. Aus dem Kennfeld ist der Zusammenhang zwischen d80-Korngröße des vorzerkleinerten Aufgabegutes und möglichem Anlagendurchsatz für eine vorgegebene — hier konstante — Fertiggutfeinheit ersichtlich. Berücksichtigt wurde außerdem die technische Belastungsgrenze für die Mahlgutförderung im System. Der Sichtereinfluß wird durch die Trennkorngröße als Variable wiedergegeben. Außerdem wurde in dieses Diagramm der Grenzfall der Durchlaufmahlung eingetragen. Auf diese Weise lassen sich die resultierenden komplexen Zusammenhänge anschaulich verdeutlichen. Anzustreben sind auf Grund der hierfür einbezogenen Größen ein möglichst hoher Grad der Vorzerkleinerung und eine Sichtereinstellung, die zur jeweiligen Trennkorngröße im Knickpunkt der Isolinien führt. Diese Fahrweise ermöglicht den höchsten Durchsatz für den konventionellen Mühle-Sichter-Kreislauf.

In ähnlicher Weise wie für die beschriebenen Beispiele kann CGC auch für weitere Probleme als Entscheidungshilfe eingesetzt werden, z.B. für die Quantifizierung von Panzerungs- und Gattierungseffekten und für den Entwurf und das scale up von Mahlanlagen.

3. Zusammenfassung und Ausblick

Computersimulation erweist sich auch bei der Optimierung von Mahlanlagen der Zementindustrie zunehmend als unentbehrliche Entscheidungshilfe, um die energetische Effektivität von Mahlprozessen zu verbessern und die Qualität der Mahlprodukte zu sichern.

Der Nutzen der vorgestellten Methodik und der dafür entwickelten Softwarelösung für zukünftige Anwendungen ist vielfältig. An dieser Stelle sollen die Aspekte der drastischen Reduzierung kostenaufwendiger experimenteller Untersuchungen, der Erhöhung der Aussagesicherheit über optimale Betriebszustände, der Minimierung des Entscheidungsrisikos bei Rekonstruktionsmaßnahmen und des besseren Informationsmanagement hervorgehoben werden. Simulationsmethoden können zukünftig auch direkt in vorhandene Prozeßsteuerungssysteme eingebunden werden, womit eine qualifizierte, modellgestützte Optimalsteuerung realisierbar wird.

Literaturverzeichnis

[1] Reinsch, V., Espig, D., Bicher, B., und Otto, J.: CGC-Computersimulation als Entscheidungsgrundlage für die Optimierung von Mahlkreisläufen. Kurzfassung des Beitrags zur 1. Tagung „Aufbereitungstechnik und Recycling"; Freiberg, 10./11. 12. 1992.

[2] Espig, D., und Reinsch, V.: Computer aided grinding circuit optimization utilizing a new mill efficiency curve. Paper to be presented at the 8th Europan Symposium on Comminution, May 17–19, 1994 Stockholm, Sweden.

[3] Espig, D.: Berechnungsgleichungen und Gesetzmäßigkeiten für Mahlanlagen mit Trommelmühlen. Diss. 1985 an der TH Leuna-Merseburg, 116 S.

[4] Lippek, E., und Espig, D.: Forschungsarbeiten zur mathematischen Modellierung von Trockenmahlanlagen. Freib. Forsch.-H. A 602 (1978), S. 77–87.

[5] Lippek, E., Espig, D., und Fiala, G.: Mathematisches Sichtermodell für die Simulation von Mahlanlagen. Freib. Forsch.-H. A 774 (1988), S. 37–46.

[6] Aßmus, R., Reichelt, D., und Espig, D.: Simulationsrechnungen zur Zementmahlung in einer Rohrmühle mit vorgeschalteter Walzmühle. Zement-Kalk-Gips 46 (1993) Nr. 10, S. 639–642.

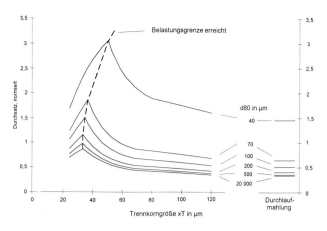

BILD 8: Durchsatzkennfeld unter Berücksichtigung von Zielfeinheit, Vorzerkleinerung und Kreislaufbedingungen
FIGURE 8: Throughput performance data taking account of target fineness, preliminary comminution and recirculation conditions

Durchsatz, normiert	= Throughput, normalized
Belastungsgrenze erreicht	= loading limit achieved
Trennkorngröße xT in μm	= Cut size xT in μm
Durchlaufmahlung	= open-circuit grinding

possible plant throughput for a given — in this case constant — finished product fineness can be seen from the performance characteristics. The engineering loading limit for conveying the material in the system was also taken into account. The influence of the classifier is described by using the cut size as a variable. The limiting case of open-circuit grinding is also plotted in this diagram. In this way it is possible to provide a visual illustration of the resulting complex relationships. The variables used for this purpose should be used to aim for the highest possible degree of pre-comminution and for a classifier setting which leads to the particular cut size at the salient points of the isolines. This mode of operation enables the maximum throughput to be achieved with the conventional mill classifier circuit.

In a similar way to the example described, the CGC can also be used as a decision aid for other problems, e.g. for quantification of the effects of mill lining and charge grading, and for the design and scale-up of grinding systems.

3. Summary and outlook

Computer simulation is also proving increasingly to be an essential decision aid for optimizing grinding systems in the cement industry to improve the energy efficiency of grinding processes and to ensure the quality of the ground products.

The methodology described and the software solution developed for it have many benefits for future applications. The drastic reduction in expensive experimental investigations, the increase in the reliability of information about optimum operating conditions, minimization of decision risk during reconstruction work, and better information management are aspects which should be stressed here.

In the future it will also be possible to incorporate simulation methods directly into existing process control systems, so that a refined model-assisted optimized control system can be put into effect.

Quantitative Klinkerphasenanalyse mittels XRD
Quantitative clinker phase analysis using XRD

Analyse quantitative des phases du clinker par DRX

Análisis cuantitativo de fases de clínker mediante XRD

Von **A. Fröhlich**, Sötenich/Deutschland

Zusammenfassung – *Auf der Basis der synthetisierten Klinkerphasen C_3S, C_2S, C_3A und C_4AF wurden Eichproben definierter Zusammensetzung für die röntgenographischen Untersuchungen erstellt. Durch die Zugabe eines internen Standards wurden die unterschiedlichen Matrixeffekte von Eich- und Klinkerprobe berücksichtigt. Zementklinkerproben aus dem Produktionsprozeß wurden auf den Gehalt von C_3S, C_2S, C_3A und C_4AF hin untersucht und die röntgenographisch ermittelten Werte den rechnerisch nach Bogue ermittelten Werten gegenübergestellt. Es zeigte sich, daß die röntgenographisch bestimmte Menge an gut kristallisiertem C_3S größenordnungsmäßig der nach Bogue errechneten Menge entspricht. Große Differenzen zwischen Experiment und Berechnung ergaben sich jedoch bei den Phasen C_3S, C_3A und C_4AF. Das ist auf die mehr oder weniger gut ausgebildete Kristallinität dieser Klinkerphasen zurückzuführen, die durch Schwankungen der Prozeßführung von Ofen und Kühler bedingt ist. Die Summe des röntgenographisch ermittelten Phasenbestands korrelierte mit dem gemessen NO-Gehalt der Ofenabgase. Sie kann demnach als Indiz für unterschiedliche Produktionsbedingungen im Brenn- und Kühlprozeß dienen.*

Quantitative Klinkerphasenanalyse mittels XRD

Summary – *The synthesized clinker phases C_3S, C_2S, C_3A and C_4AF were used to make up calibration samples of defined composition for radiographic investigations. The differing matrix effects of calibration and clinker samples were taken into account by the addition of an internal standard. Cement clinker samples from the production process were tested for their levels of C_3S, C_2S, C_3A and C_4AF; the values determined radiographically were compared with the values calculated according to Bogue. It was found that for well crystallized C_3S the quantities determined radiographically were of the same order of magnitude as the quantities calculated according to Bogue. However, large differences between experiment and calculation were found with the C_3S, C_3A and C_4AF phases. This is attributed to the variable quality of the crystallinity of the clinker phases caused by fluctuations in the process control of kiln and cooler. The sum of the constituent phases determined radiographically correlated with the measured NO content of the kiln exhaust gas. It can therefore serve as an index for differing production conditions in the burning and cooling process.*

Quantitative clinker phase analysis using XRD

Résumé – *Partant des phases synthétiques du clinker C_3S, C_2S, C_3A et C_4AF ont été fabriqué des étalons de composition définie pour les investigations par rayons X. Par addition d'un étalon interne a été tenu compte des effets de matrice différents des échantillons d'étalonnage et de clinker. Des échantillons de clinker à ciment issus de process de production ont été étudiés en vue de leur teneur en C_3S, C_2S, C_3A et C_4AF et les valeurs obtenus par analyse DRX ont été comparées aux valeurs calculées d'après Bogue. Il s'est révélé, que la quantité trouvée par analyse DRX du C_3S bien cristallisé correspond, du point de vue ordre de grandeur, à la quantité calculée d'après Bogue. Des différences importantes, entre expérimentation et calcul, sont apparues par contre pour les phases C_3S, C_3A et C_4AF. Ceci est à mettre sur le compte de la cristallinité plus ou moins bien établie de ces phases du clinker occasionnée par les variations de conduite du four et du refroidisseur. La somme de la teneur en phases déterminée par DRX corrèle avec la teneur en NO mesurée des gaz d'exhaure du four. Elle peut alors servir d'indice pour des conditions de marche différentes dans le process de cuisson et de refroidissement.*

Analyse quantitative des phases du clinker par DRX

Resumen – *Sobre la base de las fases sintetizadas de clínker C_3S, C_2S, C_3A y C_4AF se han fabricado unas muestras calibradas de composición definida para llevar a cabo los estudios por rayos X. Mediante la adición de un standard interno, se han tenido en cuenta los diferentes efectos de matriz de la muestra calibrada y de la de clínker. Se han estudiado unas muestras de clínker de cemento, procedentes del proceso de producción, para determinar el contenido de C_3S, C_2S, C_3A y C_4AF, comparándose los valores obtenidos por el análisis XRD con los calculados según Bogue. Resulta que la cantidad de C_3S bien cristalizado, determinada por rayos X, corresponde, en cuanto a su magnitud, a la calculada según Bogue. Sin embargo, ha habido grandes diferencias, entre experimento y cálculo, en las fases C_3S, C_3A y C_4AF. La causa de ello es la mejor o peor cristalinidad de estas fases de clínker, debido a las variaciones producidas en los procesos del horno y del enfriador. La suma de fases constituyentes, determinadas por XRD, guardan correlación con el contenido de NO medido en los gases de escape. Sirve, por lo tanto, de indicio de la existencia de diferentes condiciones de marcha en los procesos de cocción y de enfriamiento.*

Análisis cuantitativo de fases de clínker mediante XRD

1. Einleitung

Die quantitative Klinkerphasenanalyse mittels Röntgenbeugung beinhaltet als zentrales Problem die Eichung des Röntgendiffraktometers. Hierbei besteht die Schwierigkeit in der Erfassung der absoluten Phasengehalte der gewählten Eichproben aus den Drehrohrofenklinkern. Die herkömmliche Methode basiert auf lichtmikroskopischen Untersuchungen, bei denen die Phasengehalte durch lineare Analyse bestimmt werden.

Beilmann und Brüggemann [1] stellten 1991 fest, daß die gemessenen Röntgenintensitäten nur schlecht mit den optisch ermittelten Phasenkonzentrationen korrelieren, aus Gründen der fehlerbehafteten optischen Bestimmungsmethode.

Diese Untersuchung basiert auf Eichkurven, die mittels synthetischer Klinkerphasengemische erstellt wurden. Die Synthese der Klinkerphasen erfolgte nach der Gelmethode, wie sie für siliziumhaltige Phasen von Hamilton und Henderson beschrieben wurde [2]. Die intensitätsverfälschende Wirkung der unterschiedlichen Probenmatrizen, Laborklinker – Produktionsklinker, wurde durch die Bestimmung des Matrixfaktors durch einmalige Verwendung eines internen Standards korrigiert. Dem untersuchten Material wurden Stichproben der 3/7 mm Produktionsklinkerfraktion entnommen.

Deutliche Unterschiede zwischen Labor- und Produktionsklinker ergaben sich vor allem im Kristallinitätsgrad der Phasen, wie die **Bilder 1a, b** am Beispiel des Reflexes 2,699 Å der Phase C_3A zeigen.

Die gut ausgebildeten Kristalle des synthetischen Klinkers sind eindeutig anhand der JCPDS-Datei [3] zu identifizieren. Die gute Kristallinität hat zur Folge, daß bei der Bestimmung von C_3S und C_4AF in dieser Arbeit auf Reflexe zurückgegriffen wurde, die bisher bei ähnlichen Untersuchungen unberücksichtigt blieben (**Tabelle 1**).

Die aus Reflexmaxima bestimmten Steigungsfaktoren der Eichkurven für C_3S, C_3A, und C_4AF bewegen sich in relativ engen Grenzen, für C_2S sind aufgrund der geringen Intensität des verwendeten Reflexes hohe Steigungswerte der Eichkurve ermittelt worden.

Es zeigt sich, daß die Klinkerphasensumme, die auf diese Weise ermittelt wurde (**Tabelle 2**), zwischen 70 % und 75 %

1. Introduction

A central problem in quantitative clinker phase analysis by X-ray diffraction is the calibration of the X-ray diffractometer. The difficulty here lies in measuring the absolute phase contents of the calibration samples selected from rotary kiln clinker. The traditional method is based on light-optical microscopy in which the phase contents are determined by linear analysis.

In 1991 Beilmann and Brüggemann [1] established that the measured X-ray intensities show poor correlation with the optically determined phase concentrations because of the deficiencies in the optical method of measurement.

This investigation is based on calibration curves which were drawn up using synthetic clinker phase mixes. The clinker phases were synthesized by the gel method as described by Hamilton and Henderson [2] for phases containing silicon. The intensity-distorting effects of the differing sample matrices in the laboratory clinker and production clinker were corrected by determining the matrix factor by once-only use of an internal standard. Spot samples from the 3/7 mm production clinker fraction were taken from the material under investigation.

Significant differences were found between the laboratory and production clinker, especially in the degree of crystallinity of the phases, as is shown in **Figs. 1a** and **b** for the example of the 2.699 Å peak of the C_3A phase.

The well formed crystals of the synthetic clinker can be clearly identified using the JCPDS data [3]. As a consequence of the good crystallinity, this work has measured C_3S and C_4AF using peaks which have previously been disregarded in similar investigations (**Table 1**).

The gradient factors of the calibration curves determined from the peak maxima for C_3S, C_3A and C_4AF fluctuate within relatively narrow limits, but high gradient factors were measured for C_2S because of the low intensity of the peak used.

It can be seen that the clinker phase total which has been determined by this method (**Table 2**) varies between 70 % and 75 %. This apparently unsatisfactory result is improved if the cryptocrystalline matrix is taken into account through the proportion of melt phase calculated in accordance with Lea [4]. In this case the totals of the clinker phase analysis

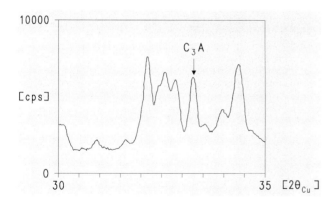

BILD 1a: Klinkerbeugungsdiagramm Laborklinker
FIGURE 1a: Clinker diffraction diagram laboratory clinker

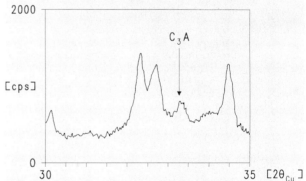

BILD 1b: Klinkerbeugungsdiagramm Produktionsklinker
FIGURE 1b: Clinker diffraction diagram production clinker

TABELLE 1: Charakteristische Daten der verwendeten Röntgenreflexe

Phase	d[Å]	$2\Theta_{Cu}$	Indizierung	rel. Intensität	JCPDS
C_3S	2,613	34,35	445, 405	90	31– 301
C_2S_β	2,877	30,95	021	21	33– 302
C_3A	2,699	33,26	044	100	32– 149
C_4AF	1,9237	47,21	202	35	30– 226
C	2,41	37,44	200	100	37–1497
Quarz-Standard	3,384	26,74	101	100	12– 708

TABLE 1: Characteristic data for the X-ray peaks used

Phase	d[Å]	$2\Theta_{Cu}$	indexing	rel. Intensity	JCPDS
C_3S	2.613	34.35	445, 405	90	31– 301
C_2S_β	2.877	30.95	021	21	33– 302
C_3A	2.699	33.26	044	100	32– 149
C_4AF	1.9237	47.21	202	35	30– 226
C	2.41	37.44	200	100	37–1497
quartz standard	3.384	26.74	101	100	12– 708

TABELLE 2: Klinkerphasendaten zweier Produktionsklinker aus gleichmäßigem Drehrohrofenbetrieb

TABLE 2: Ckinker phase data for two production clinkers from uniform rotary kiln operation

Nr.	Phase	XRD	Bogue	KST	Nr.	Phase	XRD	Bogue	KST
1	C_3S	60,8	57,47	94,7	2	C_3S	59,5	56,07	94,6
	C_2S	7	20,60			C_2S	4	21,31	
	C_3A	4,8	12,95			C_3A	5,3	13,74	
	C_4AF	3	5,51			C_4AF	3	5,66	
	C	0,6				C	0,7		
	LEA		24,97			LEA		25,99	
$\Sigma_{XRD+LEA}$		101			$\Sigma_{XRD+LEA}$		98		

No.	Phase	XRD	Bogue	LSt	No.	Phase	XRD	Bogue	LSt
1	C_3S	60.8	57.47	94.7	2	C_3S	59.5	56.07	94.6
	C_2S	7	20.60			C_2S	4	21.31	
	C_3A	4.8	12.95			C_3A	5.3	13.74	
	C_4AF	3	5.51			C_4AF	3	5.66	
	C	0.6				C	0.7		
	LEA		24.97			LEA		25.99	
$\Sigma_{XRD+LEA}$		101			$\Sigma_{XRD+LEA}$		98		

schwankt. Dieses scheinbar unbefriedigende Ergebnis verbessert sich, wenn die kryptokristalline Matrix durch den berechneten Schmelzanteil nach Lea [4] berücksichtigt wird. In diesem Fall bewegen sich die Summen aus Klinkerphasenanalyse und berechneter Schmelzphase zwischen 98% und 102%. Diese Verfahrensweise begründet sich durch die Meßmethode.

2. Messung der Intensität im Reflexmaximum

Der Produktionsprozeß mit schnellem Brennen und schnellem Kühlen erzeugt Klinkerphasen, die schlecht kristallisiert sind, d.h. deren Röntgenreflexintensitätsverteilung von denen der gut kristallisierten Laborklinkerphasen abweicht. Als Folge davon und wegen der Nichtberücksichtigung der akzessorischen Anteile ist die ermittelte Klinkerphasenmenge zu niedrig, was durch die Addition des nach Lea berechneten Schmelzphasenanteils bedingt korrigiert werden kann, da der nun doppelt addierte Anteil von C_3A und C_4AF in etwa der Menge der Akzessorien entspricht. Somit stellt die vorgestellte Methode für schlecht- bzw. teilkristallisierte Klinkerphasen eine semi-quantitative Bestimmungsmethode dar.

Im Fall von C_3S, dessen Mengenanteil als gut kristallisierte Phase sehr genau bestimmt werden kann, zeigt sich ein gemessener Anteil, der größenordnungsmäßig dem nach Bogue [5] errechneten Wert nahekommmt. Dies trifft jedoch nur bei gleichmäßigem Ofengang zu. Die Calciumaluminat- und Calciumaluminatferritphase ist schlecht kristallisiert bzw. liegt als Glasphase vor und kann infolgedessen quantitativ nicht in ihrer Gesamtheit erfaßt werden.

Auffällig ist auch der geringe Anteil an C_2S. Diese Besonderheit ist nicht allein auf den bei dieser Phase relativ großen Meßfehler zurückzuführen, da sich tendenziell die bekannten C_2S-Anteile unter besonderen Brennbedingungen bestimmen lassen. Die Reaktion $C_2S + C \rightarrow C_3S$ ist eine Festkörperreaktion. Solche Reaktionen verändern in drastischem Maße den inneren Aufbau der Kristalle. Veränderungen dieser Art sind in den Röntgenbeugungsdiagrammen durch die Intensitätsverteilung einiger Röntgenreflexe festzustellen. Im Fall des verwendeten Röntgenbeugungsreflexes von C_2S bedeutet dies eine „Intensitätverschmierung" über einen größeren Beugungswinkelbereich. Bei unveränderter Meßmethode, nämlich Bestimmung der Lage des Reflexmaximums und Messung der Intensität im Maximum, wird bei gleichem C_2S-Gehalt und mangelhaftem Kristallisationsgrad der Probe eine zu geringe Menge bestimmt. So ist die Intensitätsverteilung des gewählten Röntgenreflexes verantwortlich für zu niedrig bestimmte C_2S-Gehalte. Die Intensitätsverteilung, d.h. der Grad der Kristallinität hängt jedoch von den Brennbedingungen ab, so daß in Abhängigkeit vom Ofengang sehr unterschiedliche C_2S-Mengen ermittelt wurden. Bei gleichmäßigem Ofengang pendelt die Summe aus Klinker- und Glasphasen zwischen 98% und 102% und der Anteil an C_2S zwischen 4% – 8%. Störungen im Ofengang können größere Schwankungen hervorrufen, wie **Bild 2** darstellt. Auf diesem Bild ist das Stickstoffmonoxid (NO)-Stundenprofil eines Produktionstages dargestellt mit den Schwankungen, die durch den Eingriff des Drehofenbrenners bei schwierigen Rohstoffverhältnissen hervorgerufen werden. Nach Scheuer [6]

and calculated melt phase vary between 98% and 102%. The reason for this procedure lies in the method of measurement.

2. Measuring the intensity at the peak maximum

The production process with rapid burning and rapid cooling produces clinker phases which are poorly crystallized, i.e. their X-ray peak intensity distributions deviate from those of the well crystallized laboratory clinker phases. As a consequence, and because the accessory phases are not taken into account, the measured quantities of clinker phase are too low. This can be corrected to a certain extent by adding the proportion of melt phase calculated according to Lea, as the now doubled proportions of C_3A and C_4AF correspond approximately to the quantities in the accessory phases. For poorly or partially crystallized clinker phases the method described therefore represents a semi-quantitative method of determination.

In the case of C_3S, which is a well crystallized phase, the proportion can be determined very accurately, and the measured quantities have orders of magnitude which approximate to the calculated Bogue values [5]. However, this is only true for uniform kiln operation. The calcium aluminate and calcium aluminoferrite phases are poorly crystallized or are present as glassy phase, and therefore cannot be measured quantitatively in their entirety.

The small proportion of C_2S is also very noticeable. This peculiarity cannot be attributed solely to the relative large errors in measurement for this phase, as it is possible to measure the trend in the known proportions of C_2S under particular burning conditions. The reaction $C_2S + C \rightarrow C_3S$ is a solid-state reaction. Such reactions cause drastic changes to the internal crystal structure. Changes of this type can be seen in the X-ray diffraction diagrams through the intensity distribution of some X-ray peaks. In the case of the X-ray diffraction peak used for C_2S this means a "blurring" of the intensity over a fairly wide range of diffraction angle. If the method of measurement is unchanged, namely determination of the position of the peak maximum and measurement

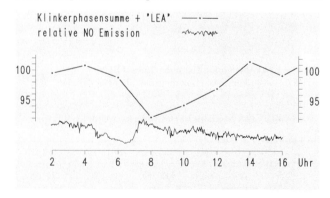

BILD 2: NO-Stundenprofil und Klinkerphasensumme bei der Produktion von Drehrohrofenklinker

FIGURE 2: NO time profile and clinker phase total during the production of rotary kiln clinker

Klinkerphasensumme + „LEA" = clinker phase total + "LEA"
relative NO-Emission = relative NO emission

besteht zwischen den NO-Werten und dem Betriebszustand des Drehrohrofens ein enger Zusammenhang. Diese Korrelation findet sich in den ermittelten Klinkerphasenwerten wieder.

3. Korrelation mit Prozeßparametern

Als besonderer Betriebszustand, bei dem die Homogenisierung große Abweichungen in der Rohmaterialzusammensetzung nicht ausgleichen konnte, ist folgendes Beispiel anzusehen (Bild 2): Ab 4.30 Uhr ist ein Schwachwerden des Ofens zu beobachten, das seinen Ausdruck im Absinken der NO-Werte findet. Die Klinkerphasensumme um 4.00 Uhr beträgt 101 %. Selbst um 6.00 Uhr ist ein Summenwert von 99 % zu messen. Die Reaktion des Drehofenbrenners, die Rohmehlaufgabenmenge zu reduzieren, hat starke Auswirkungen auf den NO-Wert und die Klinkerphasensumme. Die Verminderung der Rohmehlaufgabenmenge führt zu einem Temperaturanstieg im Ofen, der NO-Wert steigt an. Gleichzeitig fällt die Summe der Klinkerphasen auf Kosten von C_2S von 99 % um 6.00 Uhr auf 92 % um 8.00 Uhr. Die Temperaturerhöhung führt zur Bildung von C_3S aus C_2S und C. Die Reaktionsbereitschaft ist durch einen gestörten kristallinen Aufbau gekennzeichnet und wird bei schnellem Abkühlen erkennbar. Röntgenographisch ergibt sich ein zu niedriger C_2S-Gehalt. Im Laufe des Tages werden dann wieder Werte um 100 % erreicht. Die weiteren Tageswerte pendeln bei gleichmäßigem Ofengang um 100 %. Es zeigt sich, daß die Menge an C_2S, die auf diese Weise ermittelt wird, ein Indiz für die Gleichmäßigkeit der Ofenführung ist. Dies setzt jedoch eine gleichmäßige Rohmehlzusammensetzung voraus.

C_3A und C_4AF liegen im Ofen als Schmelzphasen vor, sind also weitgehend von den Schwankungen des Ofenbetriebs unberührt. Diese Phasen kristallisieren je nach Kühlerbetrieb mehr oder weniger grobkörnig. In Analogie zum Kristallinitätsgrad bzw. der röntgenographisch ermittelten Quantität von C_2S und der damit verbundenen Aussage über den Ofengang besteht somit die Möglichkeit, im Fall der semi-quantitativen Bestimmung von C_3A und C_4AF Rückschlüsse auf die Betriebsbedingungen im Klinkerkühler zu ziehen. Inwieweit dies zutrifft, ist Gegenstand weiterer Untersuchungen.

4. Zusammenfassung

Zusammenfassend kann bemerkt werden, daß diese Art der Eichung des Röntgendiffraktometers dem realen Klinkerphasengehalt des Produktionsklinkers für gut kristallisierte Klinkerphasen nahekommt. Augenscheinlich besteht zwischen Klinkerphasensumme bzw. der bestimmten C_2S-Menge und dem NO_{therm} eine Korrelation. Der Grad der Kristallinität der Klinkerphasen C_2S, C_3A und C_4AF kann als Maß für Brenn- und Kühlbedingungen herangezogen werden, dient somit als zusätzliches Instrument der Ofensteuerung und kann Hinweise auf wechselnde Zementfestigkeiten liefern.

Literaturverzeichnis

[1] Beilmann, R., und Brüggemann, H.: Quantitative XRD clinkerphase analysis, a tool for process optimization and quality control. Ciments, betons, platres, chaux, Nr. 791, (1991) 4, S. 247–251.

[2] Hamilton, D. L., und Henderson, C. M. B.: The preparation of silicate compositions by a gelling method. Mineralogical Magazine 36 (1968), S. 832.

[3] Joint Comitee on Powder Diffraction Standards

[4] Lea, F. M.: The Chemistry of cement and concrete. Edward Arnold (Publishers) Ltd (1970), S. 128–129.

[5] Bogue, R. H.: Calculation of the compoundds in portland cement. Industrial and Engineering Chemistry 1 (1929), S. 192–197.

[6] Scheuer, A.: Theoretische und betriebliche Untersuchungen zur Bildung und Abbau von Stickstoffmonoxid in Zementdrehofenanlagen. Schriftenreihe der Zementindustrie, Heft 49, S. 76–77, Verein Deutscher Zementwerke, Düsseldorf, 1987.

of the intensity at the maximum, the same C_2S content and a deficient degree of crystallization in the sample gives too low a reading. The intensity distribution of the chosen X-ray peak is therefore responsible for the excessively low levels of C_2S measured. However, the intensity distribution, i.e. the degree of crystallinity, depends on the burning conditions, so very different quantities of C_2S are measured depending on the kiln operation. For consistent kiln operation the sum of clinker and glassy phases fluctuates between 98 % and 102 % and the proportion of C_2S between 4 % and 8 %. Disruptions in the kiln process can cause quite large fluctuations, as shown in **Fig. 2**. This diagram shows the time profile for nitrogen monoxide (NO) for one production day, with the fluctuations which were caused by the intervention of the kiln burner during difficult raw material conditions. According to Scheuer [6] there is a close connection between NO values and the operating condition of the rotary kiln. This correlation emerges again in the measured clinker phase values.

3. Correlation with process parameters

The following example can be regarded as a particular operating condition where the blending system was not able to smooth out large deviations in the raw material composition (Fig. 2): At 4.30 hours the kiln is observed to be underburning, expressed as a drop in the NO values. At 4.00 hours the clinker phase total is 101 %. A total value of 99 % can still be measured at 6.00 hours. The reaction of the kiln burner to reduce the raw meal feed has a sharp effect on the NO value and the clinker phase total. The reduction in raw meal feed leads to a temperature rise in the kiln, the NO value rises. At the same time the clinker phase total drops at the expense of the C_2S from 99 % at 6.00 hours to 92 % at 8.00 hours. The temperature rise leads to the formation of C_3S from C_2S and C. The reactiveness is characterized by a distorted crystalline structure and becomes recognizable on rapid cooling. Radiography shows too low a C_2S content. Values around 100 % are then reached again during the course of the day. The rest of the day's values fluctuate around 100 % with uniform kiln operation. It is apparent that the quantity of C_2S determined in this way provides an indication of the uniformity of the kiln operation. However, this does presuppose consistent raw meal composition.

C_3A and C_4AF are present in the kiln as melt phases and are therefore largely unaffected by fluctuations in kiln operation. These phases crystallize into coarse particles of a size which depends on the cooler operation. In analogy with the degree of crystallinity or the quantity of C_2S determined radiographically and the associated information about the kiln process, semi-quantitative determination of C_3A and C_4AF therefore provides an opportunity for drawing conclusions about the operating conditions in the clinker cooler. The extent to which this holds true is the subject of further investigations.

4. Summary

To summarize, it can be said that this type of calibration of the X-ray diffractometer approximates the real clinker phase content of production clinker for well crystallized clinker phases. Evidently there is a correlation between the clinker phase total, or the measured quantity of C_2S, and NO_{therm}. The degree of crystallinity of the C_2S, C_3A and C_4AF clinker phases can be used as a measure of the burning and cooling conditions, and therefore serves as an additional instrument for kiln control, and can provide pointers to varying cement strengths.

Erfahrungen der „Holderbank" mit automatischer Prozeßführung mittels „HIGH LEVEL CONTROL"

The experience of "Holderbank" with automatic process control using "HIGH LEVEL CONTROL"

Expérience acquise à la firme „Holderbank" avec une conduite de process automatique au moyen de „HIGH LEVEL CONTROL"

Experiencias adquiridas por „Holderbank" con el control automático de procesos, mediante el empleo del „HIGH LEVEL CONTROL"

Von **R. Hasler** und **E. A. Dekkiche**, Holderbank/Schweiz

Zusammenfassung – Die „Holderbank" hat sich entschlossen, High Level Control zur automatischen Prozeßführung in ihren Werken einzusetzen. Dazu wird das Produkt LINKman Graphic von der Firma ABB verwendet, welches einerseits auf dem bereits 8 Jahre eingeführten LINKman Classic basiert, andererseits neu auf einem äußerst benutzerfreundlichen Expertenprogramm mit objektorientierter, vollgraphischer Programmierung und Einbezug von Fuzzy Logic Techniken aufbaut. LINKman Graphic kommt vorkonfiguriert ins Werk und wird dort in zwei Schritten feinangepaßt. Die ersten Erfahrungen haben gezeigt, daß eine sorgfältige Vorbereitung im Werk sowie eine intensive Auseinandersetzung mit diesem neuen Werkzeug Schlüsselpunkte für den Erfolg von High Level Control sind. Nach relativ kurzer Zeit kann mit einer guten Verfügbarkeit des Systems gerechnet werden. Damit werden zu einem frühen Zeitpunkt bereits sehr ansprechende Ergebnisse, u. a. bezüglich erzeugter Klinkerqualität, erhöhter Produktivität und verminderter Emission erreicht. Allerdings ist die Optimierungszeit erst nach ca. einem Jahr nach der Inbetriebnahme abgeschlossen. Mit dem Entschluß, das HLC einzuführen, mußte für die Leitstandfahrer ein neues Trainingskonzept entwickelt werden, um neben einem hohen Wissensstand auch die Fähigkeit für eine optimale manuelle Ofenführung zu erhalten. Unter dieser Vorgabe setzt „Holderbank" zusätzlich den selbst entwickelten Simulator ein.

Summary – "Holderbank" has decided to use High Level Control for automatic process control in its works. The product LINKman Graphic from ABB is being used for this purpose; it is based partly on the LINKman Classic introduced 8 years ago, and is also recently building on an exceptionally user-friendly expert program with object-orientated, fully graphic, programming using fuzzy logic techniques. LINKman Graphic comes to the works preconfigured and is then finely tuned in two stages. Initial experience has shown that careful preparation in the works and intensive dissection with this new tool are key points for the success of High Level Control. The system can be counted on to provide a high level of availability after a relatively short time. It has already achieved some very attractive results at an early stage relating, among other things, to the quality of the clinker produced, increased productivity, and reduced emissions. However, the optimization period has only ended about a year after commissioning. When it was decided to introduce the HLC it became necessary to develop a new training scheme for the control room operators so that they retained not only a high level of understanding, but also the capacity for optimum manual kiln control. "Holderbank" uses a self-developed simulator for this purpose.

Résumé – La firme „Holderbank" a décidé d'utiliser High Level Control pour la conduite automatique du process dans ses usines. Pour cela est utilisé le produit LINKman Graphic de la firme ABB, basé d'une part sur le LINKman Classic déjà introduit depuis 8 ans et qui, d'autre part, se complète de manière nouvelle par un système expert extrêmement convivial, avec une programmation orientée sur l'objet, totalement graphique et incluant des techniques de Fuzzi Logic. LINKman Graphic arrive préconfiguré à l'usine et est sur place adapté finement en deux étapes. Les premières expériences ont montré, qu'une préparation soigneuse à l'usine ainsi qu'une explication intensive sur cet outil nouveau sont les points clé du succès de High Level Control. Après relativement peu de temps, on peut compter sur une bonne disponibilité du système. Ainsi sont obtenus, très tôt déjà, des résultats très prometteurs, entre autre en ce qui concerne la qualité du clinker produit, une productivité accrue et une moindre émission. Toutefois, la période d'optimisation n'est terminée qu'environ une année après la mise en service. La décision de l'introduction du HLC a impliqué le développement d'un nouveau concept de formation pour les conducteurs de la salle de commande, afin de sauvegarder aussi, à côté d'une grande connaissance du métier, la capacité d'une conduite manuelle optimale du four. Dans ces conditions, Holderbank utilise en complément le simulateur élaboré par ses soins.

Resumen – *La „Holderbank" se ha decidido a emplear en sus fábricas High Level Control para el control automático de procesos. Para ello, utiliza el producto LINKman Graphic de la firma ABB, basado, por una parte, en el LINKman Classic, introducido ya desde hace 8 años, y, por otra parte, en un nuevo programa experto, muy fácil de manejar, de programación totalmente gráfica y orientada hacia el objeto, icluyendo técnicas de Fuzzy Logic. El LINKman Graphic llega a la fábrica previamente configurado, donde tiene lugar su adaptación fina, en dos etapas. Las primeras experiencias adquiridas han demostrado que una cuidadosa preparación dentro de la fábrica así como un estudio intensivo de este nuevo útil de trabajo constituye la clave del éxito de High Level Control. Tras un período relativamente corto, se puede contar con una buena disponibilidad del sistema. De esta forma, se consiguen pronto unos resultados muy prometedores, entre ellos los referentes a la calidad del clínker producido, el aumento de productividad y la reducción de emisiones. No obstante, el tiempo de optimización queda concluido solamente un año, más o menos, después de la puesta en servicio. Junto con la decisión de introducir el HLC, ha surgido la necesidad de desarrollar un nuevo programa de entrenamiento para los conductores del centro de control, con el fin de que éstos adquieran no solamente un alto nivel de conocimientos, sino también la capacidad de conducir el horno manualmente, de forma óptima. Para este fin, la „Holderbank" utiliza, adicionalmente, un simulador desarrollado por ella.*

Experiencias adquiridas por „Holderbank" con el control automático de procesos, mediante el empleo del „HIGH LEVEL CONTROL"

1. Einleitung

Auf Grund der positiven Ergebnisse von „High Level Control"-(HLC)-Systemen hat sich „Holderbank" für den ABB LINKman Graphic zur ständigen Prozeßführungsoptimierung in ihren Zementwerken entschieden.

„Holderbank" hat ein möglichst benützerfreundliches HLC-System gesucht, welches von den Werksingenieuren und bis zu einem gewissen Grade auch von den Leitstandfahrern beherrscht werden kann. Dabei wurde bewußt kein eigenes System entwickelt, damit auch auf lange Sicht eine dynamische Produktweiterentwicklung durch einen möglichst großen, über „Holderbank" hinausgehenden Anwenderkreis gewährleistet ist. Parallel zu diesem Automatisierungsschritt wurde auch die Notwendigkeit erkannt, eine Trainingsstation für Leitstandfahrer zu schaffen, um die manuelle Ofenführung auf hohem Niveau zu halten.

2. „Holderbank's" HLC-Konzept

Wie in **Bild 1** dargestellt, ist im Leitstand eine Bedienungsstation für den Leitstandfahrer mit einer speziellen, verein-

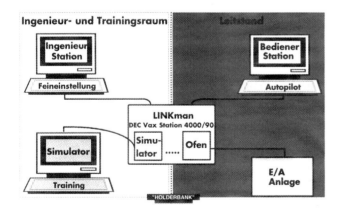

BILD 1: Prinzipieller Aufbau des HLC-Konzeptes, bei dem neben einer weitgehend automatischen Prozeßführung auch eine periodische Leitstandsfahrer-Schulung vorzusehen ist
FIGURE 1: Basic structure of the HLC concept, in which provision is made not only for substantially automatic process control but also for regular training of the control room operator

Ingenieur- und Trainingsraum	= engineering and training room
Ingenieur-Station	= engineer station
Feineinstellung	= fine tuning
Simulator	= simulator
Training	= training
LINKman	= LINKman
DEC Vax Station 4000/90	DEC Vax station 4000/90
Simulator	= simulator
Ofen	= kiln
Leitstand	= control room
Bediener-Station	= operator station
Autopilot	= autopilot
E/A-Anlage	= I/O system

1. Introduction

Because of the good results obtained with "High Level Control" – (HLC) – systems "HOLDERBANK" has decided to use the ABB LINKman Graphic in its cement works for continuous process control optimization.

"Holderbank" had been looking for an HLC system which was as user-friendly as possible and which can be mastered by the works engineers and, up to a certain level, also by the control room operators. In order to ensure continued dynamic development of the product in the long-term through the largest possible user circle extending beyond "Holderbank", a deliberate decision was made not to develop an in-house system. Parallel to this automation step, it was also recognized that there was a need to create a training station for control room operators in order to maintain manual kiln operation at a high level.

2. "Holderbank's" HLC concept

As is shown in **Fig. 1**, the control room contains an operating station for the control room operator with a special, simplified, keyboard with which he can retrieve the main functions of LINKman as well as trend displays. The engineering station is located in a separate room from which the process engineer carries out fine adjustments to the LINKman, or modifies it, or instals new applications. It is also here that the PC-based "Holderbank" kiln simulator is connected for periodic control room operator training.

The most important feature of LINKman Graphic is its user-friendliness. Because object-orientated, fully graphic, programming is used it is not necessary for the user to have any special programming knowledge. The user-friendliness is also supported through the use of leading expert system technology.

LINKman Graphic uses Gensym's G2, one of the most widely used real-time expert systems, with which the expert knowledge specific to the works is acquired and displayed. Expert systems are relatively complex programs which cannot be mastered directly by the untrained process engineer at the works. A so-called tool kit was therefore necessary with which it is simple to create, change, or fine-tune the user program (**Fig. 2**).

3. Experience with various HLC systems

The potential for savings through the use of HLC is heavily dependent on the state of the plant. In practically all cases an improved clinker quality is obtained with a smaller variability of the free lime value. A significant increase in production is achieved in plants which are not already producing far above nominal capacity and which do not suffer frequent interruptions in operation due to irregular operation (e.g. blockages). Lower lining consumption is found in all cases in the long-term. From this it can be concluded that the ther-

fachten Tastatur vorgesehen, mit welcher er die Hauptfunktionen von LINKman sowie Trenddarstellungen abrufen kann. In einem separaten Raum befindet sich die Ingenieurstation, von wo aus der Prozeßingenieur LINKman feineinstellt, modifiziert oder neue Applikationen erstellt. Ferner ist hier der PC-basierte „Holderbank"-Ofensimulator für die periodische Leitstandfahrerschulung angeschlossen.

Das wesentliche Merkmal von LINKman Graphic ist die Benützerfreundlichkeit. Durch Verwendung von objektorientierter, vollgraphischer Programmierung muß der Anwender keine besonderen Programmierkenntnisse mitbringen. Die Benützerfreundlichkeit wird auch durch die Verwendung von führender Experten-System-Technologie unterstützt.

LINKman Graphic benützt Gensym's G2, eines der am weitest verbreiteten Realtime-Expertensysteme, mit welchem das werksspezifische Expertenwissen erfaßt und dargestellt wird. Experten-Systeme sind relativ komplexe Programme, die der ungeübte Prozeßingenieur im Werk nicht ohne weiteres beherrscht. Deshalb war ein sogenanntes Toolkit (Werkzeugkasten) erforderlich, mit dem Anwenderprogramme auf einfache Weise erstellt, abgeändert oder feineingestellt werden können (**Bild 2**).

3. Erfahrungen mit verschiedenen HLC-Systemen

Das Einsparungspotential durch Einsatz von HLC hängt maßgeblich vom Zustand der Anlage ab. In praktisch allen Fällen wird eine verbesserte Klinkerqualität mit einer kleineren Variabilität des Freikalkwertes erzielt. In Anlagen, die nicht bereits weit über der Nominalkapazität produzieren oder die häufige Betriebsunterbrechungen wegen unregelmäßiger Fahrweise (z. B. Verstopfungen) erleiden, wird eine deutliche Produktionssteigerung erreicht. Langfristig wird in allen Fällen ein geringerer Futterverbrauch beobachtet. Daraus läßt sich schließen, daß die thermische Belastung der ganzen Anlage gleichmäßiger wird, was wiederum eine positive Auswirkung auf die Instandhaltungskosten hat. Signifikante Energieeinsparungen (thermische Energie bei Öfen und elektrische Energie bei Mühlen) werden dann erzielt, wenn die Anlage weit vom optimalen Punkt entfernt läuft, der durch das Verfahren und den Anlagentyp gegeben ist. Bei allen Anlagen wird eine Reduktion der Emissionen (NO_x, SO_2, CO) registriert, sofern nicht bereits vorher sehr tiefe Werte erreicht wurden. In **Bild 3** sind die erzielten Resultate zusammengefaßt.

Mittlerweile sind in der „Holderbank"-Gruppe eine ganze Anzahl von LINKman Graphic-Anlagen in Betrieb gegangen, bis Ende 1993 waren es 15. Die Inbetriebsetzung erfolgt schrittweise, wie in **Bild 4** dargestellt. Nach ca. einem Jahr kann mit der gewünschten Verfügbarkeit von 90–95 % und damit mit nachhaltigen Verbesserungen gerechnet werden.

Vor der eigentlichen Inbetriebnahme wird das Werk in einem Vorprojekt auf LINKman vorbereitet, und zwar in bezug auf Instrumentierung und Überprüfung der Stellglieder sowie auf Integrierung ins bestehende Leitsystem. Der Inbetriebsetzungsingenieur kommt dann für die erste Inbetriebsetzung mit einem Standardprogramm, welches an die spezifischen Bedürfnisse angepaßt wird, ins Werk. Am Ende dieser Phase von ca. 5 Wochen wird eine typische Verfügbarkeit von 70–80 % erreicht. Dann übernimmt der für LINKman verantwortliche Werksingenieur bereits das

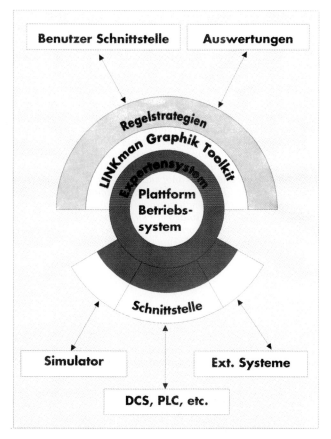

BILD 2: Einbindung des Expertensystems in die Prozeßführung
FIGURE 2: Incorporating the expert system into the process control

Benutzer-Schnittstelle	= user interface
Auswertungen	= evaluations
Regelstrategien	= control strategies
LINKman Graphik Toolkit	= LINKman Graphic toolkit
Expertensystem	= expert system
Plattform Betriebssystem	= platform operating system
Schnittstelle	= interface
Simulator	= simulator
Ext. Systeme	= external systems

mal loading of the entire plant is more uniform, which in turn has a positive effect on maintenance costs.

Significant savings in energy (thermal energy for kilns and electrical energy for mills) are achieved whenever the plant is operating at some distance from the optimum point – which is specific to the process and the plant type. A reduction in the emissions (NO_x, SO_2, CO) is recorded in all plants unless very low values have already been achieved. The results obtained are summarized in **Fig. 3**.

The "Holderbank" group has now brought quite a number of LINKman Graphic systems into operation – 15 by the end of 1993. Commissioning is carried out in stages, as shown in **Fig. 4**. The required availability of 90–95 %, accompanied by lasting improvements, can be expected after approximately one year.

Before the actual commissioning, the works is prepared for the LINKman in a preliminary study; this relates to the

BILD 3: Typische Resultate der optimierten Prozeßführung

Automatischer Betrieb	80–95 %
Einsparung therm. Energieverbrauch	0– 3 %
Einsparung elektr. Energie (Mühlen)	0–10 %
Produktionssteigerung	0– 5 %
Einsparung Futterverbrauch	0–30 %
Erhöhung des mittleren Freikalkwertes im Klinker	0–25 %
Reduzierung der Klinkerqualitätsschwankungen	0–50 %
Reduzierung der NO_x-Emissionen	0–20 %

FIGURE 3: Typical results of optimized process control

Automatic operation	80–95 %
Saving in thermal energy consumption	0– 3 %
Saving in electrical energy (mills)	0–10 %
Increase in production	0– 5 %
Saving in lining consumption	0–30 %
Increase in the average clinker free lime content	0–25 %
Reduction in clinker quality fluctuations	0–50 %
Reduction in NO_x emissions	0–20 %

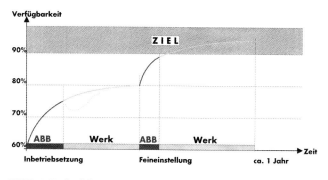

BILD 4: Verlauf des Automatisierungsfortschritts bei der Inbetriebnahme und anschließender Feinanpassung des LINKman-Systems
FIGURE 4: Progress of the automation during commissioning and subsequent fine adjustment of the LINKman system

Verfügbarkeit	= Availability
Inbetriebsetzung	= commissioning
Feineinstellung	= fine tuning
ca. 1 Jahr	= approx. 1 year
Zeit	= Time
Werk	= works
Ziel	= target

System und beobachtet und verfeinert die Strategien während mehrerer Monate. Während dieser Zeit steigt der LINKman-Betrieb auf typische 80 % der Gesamtzeit.

Darauf erfolgt schließlich die Feineinstellung. Basierend auf den Beobachtungen durch das Werk wird die Regelstrategie nochmals grundsätzlich angepaßt und verfeinert, so daß der LINKman-Betrieb gegen 90 % steigt. Nun ist es wiederum das Werk, das in einer weiteren Beobachtungs- und Feinanpassungsphase die gewünschten 95 % Verfügbarkeit erreicht.

Der LINKman Graphic wird wegen seiner großen Transparenz von den Betrieben sehr gut akzeptiert. Die ersten Erfahrungen haben aber auch gezeigt, daß eine sorgfältige Vorbereitung im Werk bezüglich zuverlässiger, repräsentativer Meßgeber und präziser Stellglieder sowie eine intensive Auseinandersetzung mit diesem neuen Werkzeug die wichtigsten Erfolgsfaktoren sind.

4. Schlußfolgerungen

Die vorliegenden Erfahrungen belegen, daß ein vollautomatischer Ofen- und Mühlenbetrieb durch Einsatz von HLC-Systemen möglich ist. HLC ist aber keine Anlage, die man kauft und nach einem Abnahmetest einfach laufen läßt wie die übrigen Installationen. HLC ist vielmehr ein Werkzeug zur ständigen Verbesserung der Produktivität, ein Werkzeug der Zukunft, das es möglich macht, bezüglich Prozeßführung an der Spitze zu bleiben.

instrumentation and checking the actuators as well as to integration into the existing control system. The commissioning engineer then comes to the works for the initial commissioning with a standard program which is adapted to the specific requirements. At the end of this phase of approximately 5 weeks, a typical availability of 70–80 % is achieved. At this stage the works engineer responsible for the LINKman takes over the system and observes and refines the strategies over a period of several months. During this time the LINKman operation increases to a typical value of 80 % of the total time. This is then followed by the fine adjustment. Based on observations by the works, the control strategy is again fundamentally tuned and refined so that the LINKman operation increases towards 90 %. It is now the turn of the works again to achieve the required 95 % availability in a further observation and fine-tuning phase.

Because of its great transparency the LINKman Graphic is very well accepted by the operators. However, initial experience has also shown that the most important success factors are careful preliminary preparation in the works with regard to reliable, representative, sensors and precise actuators as well as intensive explanation about the new tool.

4. Conclusions

The available experience proves that fully automatic kiln and mill operation can be achieved by using HLC systems. HLC is not, however, a plant which is just bought and then simply left to run after an acceptance test like some other installations. HLC is much more a tool for continuous improvement of productivity, a tool for the future which makes it possible to keep process control at peak performance.

Vorteile der Prozeßführung mit Expertensystem
Advantages of process control with expert systems

Avantages de la conduite du process au moyen d'un système expert

Ventajas del control de procesos mediante el sistema experto

Von **R. Hepper**, Neubeckum/Deutschland

Zusammenfassung – Expertensysteme dienen in der automatisierten Prozeßführung dazu, um aus der intelligenten Verknüpfung von aktuellen Prozeßwerten mit den Erfahrungswerten in Wissensdatenbanken das jeweilige Optimum für den Prozeß zu ermitteln und die erforderlichen Stellgrößen für den Prozeß zu liefern. In der Zementindustrie werden zunehmend Expertensysteme für die Steuerung des Brenn- und Mahlprozesses installiert. Die von Polysius für den Ofen- und Mühlenbetrieb entwickelten Expertensysteme KCE und MCE gehen weit über die übliche Abarbeitung konventioneller Regelstrategien hinaus. Vielmehr suchen diese Systeme – analog einem erfahrenen Bedienungsmann – stets diejenigen Prozeßführungsstrategien, mit denen die Anlagen an ihrer oberen Leistungsgrenze betrieben werden können. Auf diese Weise werden das betriebliche und wirtschaftliche Optimum hinsichtlich Produktmenge, Energieeinsatz, Standzeiten und Reaktion auf Störungen erreicht. Das effektiv zu erreichende Verbesserungspotential hängt von der jeweils vorhandenen Anlagenkonfiguration und ihrem Zustand ab und muß daher am konkreten Anwendungsfall überprüft werden.

Vorteile der Prozeßführung mit Expertensystem

Summary – Expert systems are used in automated process control for determining the current optimum for the process by intelligent combination of current process values with the empirical values in the knowledge database, and for supplying the necessary correcting variables for the process. Expert systems are being increasingly installed in the cement industry for controlling the burning and grinding processes. The KCE and MCE expert systems developed by Polysius for kiln and mill operation go far beyond the usual processing of conventional control strategies. These systems – like an experienced operator – are always looking for those process control strategies with which the plants can be operated at their upper performance limits. This achieves the operational and economic optimum with respect to product quantity, energy input, service life and reaction to faults. The improvement potential which can be achieved in practice depends on the actual plant configuration in each case and its condition, and must therefore be checked for the specific application.

Advantages of process control with expert systems

Résumé – Dans la conduite automatisée du process, les systèmes expert servent à définir, à partir d'un mariage intelligent des valeurs du process momentanées et du capital d'expérience collecté dans des banques de données „cognitives", l'optimum instantané du process et à fournir les grandeurs de réglage du process. Les systèmes expert KCE et MCE, développés par Polysius pour l'exploitation du four et du broyeur, dépassent de loin les limites connues des stratégies de régulation conventionnelles. Bien plus, ces systèmes – de même qu'un conducteur expérimenté – cherchent à chaque instant les stratégies de conduite de process appropriées permettant d'exploiter l'atelier à la limite supérieure de sa capacité. De cette façon est atteint l'optimum d'exploitation, des points de vue quantité du produit, consommation d'énergie, durées de service et réaction aux perturbations. Le potentiel d'amélioration effectivement à atteindre dépend bien sûr de la configuration propre de l'unité de production et de son état et doit donc être évalué dans des conditions concrètes d'exploitation.

Avantages de la conduite du process au moyen d'un système expert

Resumen – Los sistemas expertos sirven, en el control automático de los procesos, para determinar las condiciones óptimas de los procesos, a partir de la combinación inteligente de los valores momentáneos, obtenidos en los mismos, con los datos basados en la experiencia y almacenados en los bancos de datos, así como para suministrar los valores de regulación necesarios para cada proceso. En la industria del cemento se están instalando, cada vez más, estos sistemas expertos para el control de los procesos de cocción y de molienda. Los sistemas expertos KCE y MCE, desarrollados por Polysius para el servicio del horno y del molino, pasan con creces de los límites normales de las estrategias de regulación convencionales. Estos sistemas buscan más bien, – de forma análoga a un experimentado operador – aquellas estrategias de control del proceso, con las que las instalaciones puedan trabajar lo más cerca posible de su límite superior de capacidad. De esta forma, se consigue un óptimo, tanto técnico como económico, en cuanto se refiere a cantidad del producto, consumo de energía, tiempo de duración y reacción frente a las perturbaciones. El potencial efecto de mejora que se pueda lograr depende en cada caso de la configuración y del estado de la instalación, por lo que debe estudiarse en cada caso concreto.

Ventajas del control de procesos mediante el sistema experto

1. Einleitung

Die Fortschritte auf den Gebieten der Soft- und Hardwareentwicklung eröffnen der Automatisierungstechnik heute neue, teils revolutionäre Wege zur Sicherung, Steuerung und Optimierung von verfahrenstechnischen Prozessen. Neue Softwareprodukte aus dem Gebiet der „Künstlichen Intelligenz" sind wesentliche Bestandteile dieser Entwicklung. Mit ihrer Hilfe verfügt man über die Möglichkeit, schwierige und komplizierte Prozesse so zu beeinflussen, daß hohe Ansprüche an die Effizienz der Produktion, die Gewährleistung der erforderlichen Produktqualität, die Auslastung vorhandener Ressourcen und die Entlastung der Anlagenfahrer erfüllt werden können.

2. Kennzeichen von Expertensystemen

Expertensysteme dienen in der automatisierten Prozeßführung dazu, aus der Verknüpfung von Prozeßwerten und dem Wissen (Expertenwissen) von Anlagenherstellern und -betreibern das Optimum für einen Prozeß zu ermitteln und die erforderlichen Stellgrößen für den Prozeß zu liefern. Speziell für mathematisch schwer beschreibbare und nichtlineare Prozesse, wie sie in der Zementherstellung beim Brenn- und Mahlprozeß vorkommen, haben sich Expertensysteme in letzter Zeit zunehmend bewährt.

Der strukturelle Aufbau ist bei Expertensystemen, unabhängig vom verwendeten Softwareprodukt, annähernd gleich. Sie bestehen einerseits aus einem fest strukturierten Programm, das wichtige Funktionen, wie die Kommunikation mit dem Benutzer, die automatische Erfassung von Daten (z.B. Prozeßwerten) sowie die Verarbeitung des neuen bzw. vorhandenen Wissens durchführt, und andererseits dem wichtigsten Bestandteil des gesamten Systems, der Wissensbasis. In ihr ist das Wissen über den zu führenden Prozeß enthalten und kann durch den Anlagenbetreiber bei Veränderungen ergänzt und der Anlagentechnik angepaßt werden.

Expertensysteme zur Regelung verfahrenstechnischer Prozesse unterscheiden sich maßgeblich durch die in der Wissensbasis hinterlegten Strategien zur Beeinflussung der Prozesse. Dieser Sachverhalt trifft verstärkt bei der Anwendung solcher Regelungsalgorithmen zu, die über eine einfache Sollwertvorgabe an vorhandene Regelkreise hinausgehen und direkt menschliches Verhalten, mit Hilfe eines Expertensystems, zur Optimierung der Prozeßführung einsetzen.

3. Anforderungen

Expertensysteme, die zur Steuerung verfahrenstechnisch komplizierter Prozesse genutzt werden, müssen wichtige Anforderungen erfüllen. Besonders zu nennen sind in diesem Zusammenhang die Gewährleistung verfahrenstechnisch richtiger Vorgehensweisen bei der Beeinflussung von Stellgrößen, die Realisierung geforderter Kriterien gemessen an der Qualität des herzustellenden Produktes, die Diagnostizierung auftretender Fehlfunktionen von Sensoren oder Aggregaten, die Bereitstellung von Handlungsvorschlägen, der Betrieb des zu führenden Prozesses an der Leistungsgrenze und die Ausnutzung aller vorhandenen Ressourcen.

Betrachtet man diesen umfangreichen Forderungskatalog als realisierbaren Stand der Technik und die immer kürzer werdenden Produktlebenszyklen von Maschinen, verbunden mit der Einführung neuer Verfahren, so läßt dies den Schluß zu, daß die Anlagenbauer mit dem größten Know-how und der engsten Zusammenarbeit mit Anlagenbetreibern einen signifikanten Wettbewerbsvorsprung bei der Entwicklung von intelligenten Systemen zur Prozeßführung und -optimierung besitzen müssen.

Die Vorteile der Prozeßführung mit einem Expertensystem sollen anhand der Beschreibung des Polysius Expertensystems MCE (Mill Control Expertsystem), das zur Führung eines Mahlsystems (Kugelmühle mit Sichterkreislauf) eingesetzt wird, aufgezeigt werden.

1. Introduction

Progress in the areas of software and hardware development is now opening up new and, in some cases, revolutionary ways for automation technology to protect, control and optimize chemical engineering processes. New software products from the field of "artificial intelligence" are important constituents in this development. They make it possible to govern difficult and complicated processes in such a way as to fulfil the heavy demands on the efficiency of the production system, on ensuring the required product quality, on relieving existing resources, and on reducing the load on plant operators.

2. Characteristic features of expert systems

Expert systems are used in automated process control systems to determine the optimum conditions for a process by linking process values and the expert knowledge obtained from plant manufacturers and operators, and to provide the necessary manipulated variables for the process. Expert systems have proved increasingly valuable in recent times, especially for non-linear processes which are difficult to describe mathematically, like the burning and grinding processes in cement manufacture.

Regardless of the software product used, expert systems have virtually the same structure. It consists, on the one hand, of a rigidly structured program which carries out important functions, such as communication with the user, automatic acquisition of data (e.g. process values) as well as the processing of new or existing knowledge, and on the other hand, of the most important part of the entire system, the knowledge base. This contains the knowledge about the process to be controlled and, if there are changes, can be expanded by the plant operator and adapted to the plant technology.

Expert systems for controlling chemical engineering processes differ considerably in the strategies contained in the knowledge base for governing the process. This is particularly true when applying those control algorithms which go beyond simply specifying setpoint values in existing control circuits and, with the aid of an expert system, use human behaviour for optimizing the process control.

3. Requirements

Expert systems which are used for controlling complicated chemical engineering processes must fulfil important requirements. Particular mention should be made in this connection of ensuring the correct process engineering procedure when influencing manipulated variables, of realizing the required criteria with regard to the quality of the product to be produced, of diagnosing any incorrect functioning of sensors or units, of providing recommended courses of action, of operating the process to be controlled at the limit of performance, and of making full use of all available resources.

If this extensive list of requirements is regarded as achievable state of the art, and when the ever-shorter product life cycles of machines linked with the introduction of new processes are taken into consideration, then this leads to the conclusion that the plant manufacturer with the greatest know-how and the closest cooperation with plant operators must have a significant competitive edge in the development of intelligent systems for process control and optimization.

The advantages of process control with an expert system will be demonstrated by describing the Polysius expert system MCE (Mill Control Expert system) which is used for controlling a grinding system (ball mill with classifier circuit).

Fig. 1 illustrates very clearly that in contrast to the expert systems used previously, MCE is not a system with the sole task of controlling individual sub-processes. In fact the development of this system achieved a step forward in the field of higher-ranking process control aimed at an optimum

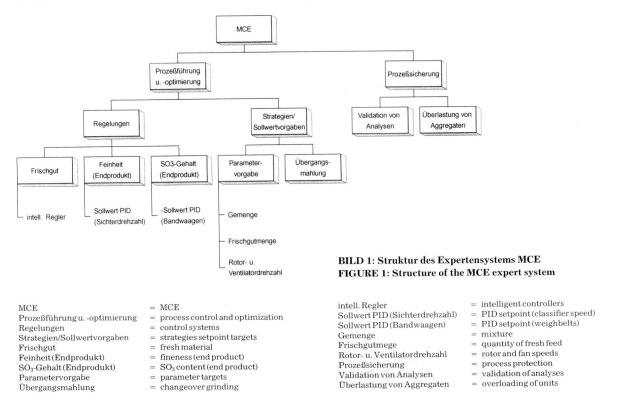

BILD 1: Struktur des Expertensystems MCE
FIGURE 1: Structure of the MCE expert system

MCE	= MCE
Prozeßführung u. -optimierung	= process control and optimization
Regelungen	= control systems
Strategien/Sollwertvorgaben	= strategies setpoint targets
Frischgut	= fresh material
Feinheit (Endprodukt)	= fineness (end product)
SO$_3$-Gehalt (Endprodukt)	= SO$_3$ content (end product)
Parametervorgabe	= parameter targets
Übergangsmahlung	= changeover grinding
intell. Regler	= intelligent controllers
Sollwert PID (Sichterdrehzahl)	= PID setpoint (classifier speed)
Sollwert PID (Bandwaagen)	= PID setpoint (weighbelts)
Gemenge	= mixture
Frischgutmege	= quantity of fresh feed
Rotor- u. Ventilatordrehzahl	= rotor and fan speeds
Prozeßsicherung	= process protection
Validation von Analysen	= validation of analyses
Überlastung von Aggregaten	= overloading of units

Wie **Bild 1** sehr deutlich veranschaulicht, stellt MCE im Gegensatz zu den bisher verbreiteten Expertensystemen kein System dar, welches nur die Regelung einzelner Teilprozesse zur Aufgabe hat. Vielmehr wurde mit der Entwicklung dieses Systems ein Fortschritt auf dem Gebiet der übergeordneten Prozeßführung zum Zwecke einer optimalen Betriebsweise des Gesamtsystems, verbunden mit einer Entlastung des Anlagenpersonals erreicht. Bisher installierte Anwendungen in den Zementwerken zeigen, daß die realisierten neuen Konzepte eine sehr hohe Akzeptanz finden. Maßgebliche Ursache für diesen Erfolg ist sicherlich die Tatsache, daß MCE direkt vor Ort, zusammen mit erfahrenen Betriebsleuten, entwickelt wurde.

4. Prozeßführung und -optimierung

Bild 2 zeigt am Beispiel einer Mahlanlage die wichtigsten Ein- und Ausgangssignale, die MCE benutzt, um die nachfolgend beschriebenen Leistungsfunktionen erfüllen zu können. Hauptschwerpunkte von MCE sind die Realisierung und Koordinierung der für den Ablauf des Mahlprozesses wichtigen Regelungen wie die Frischgutaufgabe, die Produktfeinheit sowie des SO$_3$-Gehaltes im Endprodukt.

Da die Anforderungen an eine verfahrenstechnisch optimale Frischgutaufgabe mit konventioneller Regelungstechnik, auf Grund der möglichen Situationsvielfalt, nur beschränkt erfüllt werden können, kommt im MCE zur Erfüllung dieser Teilaufgabe kein PID-Regler zum Einsatz. Vielmehr wurde ein Regelungsalgorithmus entwickelt, der direkt dem Vorgehen eines motivierten, erfahrenen Anlagenfahrers entspricht und so, neben einer wesentlich verbesserten Dosierung der Frischgutaufgabe, auch erstmals eine Online-Beurteilung des Auslastungsgrades der Mahlanlage ermöglicht.

Das Grundkonzept der vom Expertensystem verwirklichten Regelungsstrategie entspricht im wesentlichen der von Inbetriebnehmern und erfahrenen Anlagenfahrern angewandten Vorgehensweise zur Optimierung eines Mahlprozesses. Diese erstmals in einem Expertensystem hinterlegten menschlichen Vorgehensweisen gestatten es, jederzeit die Mahlanlage in den tatsächlich optimalen Betriebszustand zu bringen.

Die Feinheit des Endproduktes wird mittels Veränderungen der Sollwerte von vorhandenen PID-Reglern für die Rotor- bzw. Ventilatordrehzahl ständig so eingestellt, daß

mode of operation of the entire system, linked with relieving the load on the plant personnel. Applications installed so far in cement works show that the new concepts being put into practice are finding a very high level of acceptance. Important grounds for this success are certainly the fact that MCE was developed directly on site together with experienced plant personnel.

4. Process control and optimization

Using the example of a grinding plant, **Fig. 2** shows the most important input and output signals which MCE uses to be able to fulfil the performance functions described below. The main emphasis in MCE is placed on the implementation and coordination of the important control systems for the grinding process, such as fresh material feed, product fineness and SO$_3$ content in the end product.

Because of the multiplicity of possible situations, conventional control technology can only fulfil the process engineering requirements for optimum fresh material feed to a limited extent, so no PID controllers are used in the MCE for fulfilling this sub-function. Instead, a control algorithm was developed, which corresponds directly to the procedure of a motivated, experienced, plant operator; this not only enables substantially improved metered feeding of the fresh material feed to be achieved, but also for the first time makes it possible to evaluate the level of capacity utilization of the plant system on-line.

The basic concept of the control strategies implemented by the expert system corresponds essentially to the procedure used by the commissioning engineer and experienced plant operators for optimizing a grinding process. This human procedure, installed for the first time in an expert system, makes it possible to bring the grinding plant to the actual optimum operating state at any time.

The fineness of the end product is continuously adjusted by changing the setpoint value of existing PID controllers for the rotor and fan speeds, so that the required quality criteria lie in defined, permissible, tolerance ranges. On arrival of the latest analysis results, the MCE imitates the procedure of a plant operator, and not only changes the classifier parameters but also, when necessary, makes corrections to the fresh material feed to match the changed grinding situation. When new analysis values are received the expert system also reacts to deviations from the permissible tolerance band by

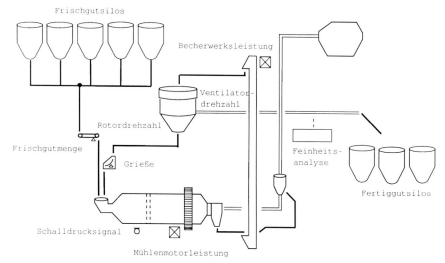

BILD 2: Ein- und Ausgangsgrößen des Expertensystems MCE
FIFURE 2: Input and output variables of the MCE expert system

Frischgutsilos	= fresh material silos
Becherwerksleistung	= bucket elevator power
Ventilatordrehzahl	= fan speed
Rotordrehzahl	= rotor speed
Frischgutmenge	= quantity of fresh material
Grieße	= tailings
Schalldrucksignal	= noise signal
Mühlenmotorleistung	= mill motor power
Feinheitsanalyse	= fineness analysis
Fertiggutsilos	= finished product silos

die geforderten Qualitätskriterien in definierten, zulässigen Toleranzbereichen liegen. In Anlehnung an die Vorgehensweise eines Anlagenfahrers verändert MCE beim Eintreffen der aktuellen Analysenergebnisse nicht nur die Sichterparameter, sondern nimmt, wenn erforderlich, zur Anpassung an die veränderte Mahlsituation auch Korrekturen der Frischgutaufgabe vor. Durch eine Veränderung der Aufgabe von Komponenten, die maßgeblich den SO₃-Gehalt im Endprodukt beeinflussen, wie Gips oder Anhydrit, reagiert das Expertensystem beim Eintreffen neuer Analysenwerte auf Abweichungen vom zulässigen Toleranzband.

5. Strategien

Durch das ständige parallele Übernehmen der aktuellen Betriebsparameter ist MCE in der Lage, die während einer Inbetriebnahme einmal eingestellten Parameter selbständig auf die aktuellen Mahlbedingungen zu adaptieren. Damit wird eine fortlaufende Anpassung wichtiger Parameter durch einen Systemingenieur vermieden. MCE ist jederzeit durch eine automatische Erkennung des selektierten Silos in der Lage, die aktuell produzierte Zementsorte festzustellen. Sollwertveränderungen bzw. -vorgaben sind deshalb auch sortenabhängig im Expertensystem integriert.

Nach einem Stop der Mahlanlage, ohne Zementsortenwechsel, stellt das Expertensystem die letzten Betriebsparameter (Gemenge, Sichter, Frischgut) als Startwerte ein. Damit wird zum einen die Startphase der Mahlanlage signifikant verkürzt und zum anderen die Anlage sofort an den alten Betriebspunkt gebracht.

Mit Hilfe anwenderspezifisch hinterlegter Strategien für den definierten, anlagenfahrerunabhängigen Prozeß der Übergangsmahlung zwischen verschiedenen Zementsorten realisiert das Expertensystem einen oft geäußerten Wunsch der Anlagenbetreiber. In Abhängigkeit der momentanen und der nachfolgend zu mahlenden Zementsorte führt MCE ereignisgesteuert Handlungsvorgänge, wie z. B. die Veränderung der Bandwaagenbeteiligungen oder der Sichterparameter, automatisch durch.

Neben den allgemeinen Regelungsfunktionen erfüllt MCE auch wichtige Funktionen auf dem Gebiet der Prozeßsicherung. Zielstellung dieser realisierten Automatisierungsfunktion ist es, den Betrieb des Mahlsystems zu verbessern bzw. auch in schwierigen Situationen aufrecht zu erhalten.

Alle dem System übermittelten aktuellen Analysenergebnisse werden auf ihre Sinnfälligkeit überprüft. Dies geschieht unter anderem durch eine Betrachtung der abgelaufenen Prozeßsituation und der bisher eingegangenen Analysenwerte. Stellt das System nicht plausible Analysen fest, so kann neben der Aufforderung zu einer neuen Probennahme auch eine Korrektur der eingetroffenen Werte vorgenommen werden. Diese Vorgehensweise hat sich gut

changing the feed of components which have a controlling effect on the SO₃ content in the end product, such as gypsum or anhydrite.

5. Strategies

By continuous parallel acceptance of the current operating parameters MCE is capable of making automatic adjustments to parameters, which were first set during commissioning, to suit the current grinding conditions. This avoids continuous adjustment of important parameters by a systems engineer. MCE is able at any time to establish the type of cement currently being produced through automatic recognition of the selected silo. Setpoint changes or preset values are therefore also integrated into the expert system as a function of the product type.

After a stoppage of the grinding plant without changing cement type the expert system installs the last operating parameters (quantity, classifier, fresh material) as starting values. For one thing, this significantly shortens the starting phase of the grinding plant and, for another, the plant is immediately brought to the old operating point.

With the aid of installed user-specific strategies the expert system puts into effect a wish, often voiced by the plant operator, for a defined process for a changeover grinding system between different cement types which is independent of the plant operator. Depending on the types of cement being ground and to be ground, the MCE automatically carries out event-driven action procedures such as, for example, changing the weighbelt proportions or the classifier parameters.

In addition to the general control functions MCE also fulfils important functions in the field of process protection. The object of this automation function is to improve the operation of the grinding system and to maintain it even in difficult situations.

All the current analysis results transferred to the system are checked for plausibility. This is carried out, among other things, by considering the previous process situation and the analysis values input so far. If the system does not discover plausible analyses then not only can a request be put in for a fresh sample to be taken, but also a correction can be made to the values received. This procedure has proved very successful as it has resulted in a substantial reduction in incorrect interventions in the grinding process.

MCE observes critical units (where there is a possible risk of breakdown) and if, for example, the danger of a drive breaking down due to overloading is recognized as being very probable, then it undertakes corrective actions which override the control system in the interests of maintaining the overall grinding process.

bewährt, da somit fehlerhafte Eingriffe in den Mahlprozeß wesentlich verringert werden können.

MCE beobachtet kritische Aggregate (eventuell ausfallgefährdet) und nimmt, wenn z. B. die Gefahr des Ausfalls eines Antriebs auf Grund von Überlastung als sehr wahrscheinlich erkannt wird, regelungsübergreifende Stellhandlungen vor, die im Interesse der Aufrechterhaltung des gesamten Mahlprozesses stehen.

6. Ergebnisse

Die mit dem Einsatz des Polysius-Expertensystems MCE erzielten Ergebnisse lassen sich in prozeßspezifische, anwenderspezifische und ökonomische Resultate unterteilen.

Die prozeßtechnischen Vorteile resultieren daraus, daß der Mahlprozeß in allen Situationen, also auch beim An- und Abfahren und bei Übergangsmahlungen, sicher beherrscht wird. Daher wird die Anlage tatsächlich in ihrem Zerkleinerungsmaximum betrieben, und ein Vollaufen der Grobmahlkammer wird sicher vermieden. Die Regelung der Frischgutaufgabe bewirkt einen gleichmäßigen Lauf der Mahlanlage und somit ein homogenes Zerkleinerungsverhalten.

Die anwenderspezifischen Vorteile ergeben sich aus der hohen Akzeptanz und der daraus resultierenden permanenten Nutzung des Systems, da alle Eingriffe verstanden und nachvollzogen werden können. Das breite Leistungsspektrum bewirkt eine Entlastung des Bedienpersonals. Das in der Wissensbasis gespeicherte Wissen kann vom Anwender leicht erweitert oder modifiziert werden, und MCE adaptiert alle wichtigen Parameter parallel zum laufenden Prozeß.

Die ökonomischen Vorteile werden durch Energieeinsparungen und Durchsatzsteigerungen erzielt. Darüber hinaus wird durch die schonende Fahrweise der Verschleiß abgesenkt. Die Integration in eine bestehende Anlage findet ohne Beeinträchtigung des normalen Produktionsablaufs statt.

7. Ausblick

Expertensysteme werden sich immer mehr in der Prozeßsteuerung durchsetzen und im Laufe der Zeit normaler Bestandteil eines Automationssystems sein. Doch nicht nur von den Expertensystemen kann ein Innovationsschub erwartet werden. Auch andere Teilgebiete der „Künstlichen Intelligenz", wie künstliche neuronale Netzwerke oder genetische Algorithmen werden für interessante Neuerungen sorgen.

6. Results

The results achieved with the use of the Polysius MCE expert system can be divided into process-specific, user-specific, and economic results.

The process engineering advantages result from the fact that the grinding process is managed reliably in all situations, including starting and stopping and changeover grinding. The plant is therefore actually operated at its comminution maximum and reliably avoids any overfilling of the coarse grinding chamber. Control of the fresh material feed produces a uniform process in the grinding plant and therefore homogeneous comminution behaviour.

The user-specific advantages result from the high acceptance and the resulting permanent utilization of the system, as all interventions can be understood and reconstructed. The wide performance spectrum relieves the load on the operating personnel. The knowledge stored in the knowledge base is easily expanded or modified by the user, and the MCE adapts all the important parameters in parallel to the current process.

The economic advantages are achieved through energy savings and increases in output. The careful mode of operation also lowers the wear. Integration into an existing plant takes place without detriment to the normal course of production.

7. Outlook

Expert systems are being used more and more extensively in process control and in the course of time will become normal components in automation systems. However, it is not only from expert systems that an impetus for innovation can be expected. Other areas of „artificial intelligence" such as artificial neuronal networks or genetic algorithms will also supply interesting innovations.

Kostengünstige Produktion eines Klinkers hoher Qualität mit Hilfe eines Expertensystems. Das Beispiel von Ciments Français

Cost-effective production of high-grade clinker using expert systems. An example from Ciments Français

Comment produire un clinker de qualité à un côunt minimum grâce au Système Expert. L'exemple des Ciments Français

Producción a bajo coste de un clínker de alta calidad, con ayuda del sistema experto. El ejemplo de Ciments Français

Von **D. Kaminski,** Guerville/Frankreich

Zusammenfassung – 1989 installierte Ciments Français das erste Experten-System für die automatische Steuerung des Zementwerkes in Bussac, Frankreich. Die Probleme dieses Werkes waren häufige Instabilität des Brennprozesses aufgrund des schwierigen Rohmaterials, ein hoher spezifischer Energieverbrauch, ein überhöhter Verschleiß des Feuerfestmaterials sowie als Folge dieser Probleme überforderte Prozeßsteuerer. Das Werk Bussac war daher sehr an einer Prozeßsteuerung interessiert, die eine kostengünstige Produktion eines qualitativ hochwertigen Klinkers mit hoher Gleichmäßigkeit und ohne größere Eingriffe in den Prozeß sicherstellte. Von TECHNODES wurde die Installation des Experten-Systems TOP-EXPERT vorgeschlagen, das sich durch eine leichte Handhabung und vielfältige Anwendungsmöglichkeiten auszeichnet. Diese Computersteuerung ist in der Sprache der Prozeßsteuerer zu bedienen und erlaubt es dem Betriebspersonal, die Funktion des Systems nachzuvollziehen und zu verstehen. Die Erfolge dieser ersten Installation waren Einsparungen an thermischer und elektrischer Energie sowie die Steigerung der Produktion bei geringerem Verschleiß des Feuerfestmaterials und höhere Gleichmäßigkeit der Zementqualität. Heute verfügt TECHNODES, das Forschungszentrum von Ciments Français, mit weltweit mehr als 60 installierten Systemen über große Erfahrungen beim Einsatz von Zementwerkssteuerungen.*

Summary – In 1989 Ciments Français installed the first expert system for automatic control of the cement works in Bussac, France. The problems in this works were frequent instabilities in the burning process caused by the difficult raw material, a high specific energy consumption, and excessive wear of the refractory material, as well as over-taxed process controllers as a result of these problems. The Bussac works was therefore very interested in a process control system which guaranteed cost-effective production of a qualitatively high-grade clinker of high uniformity and without violent interventions in the process. TECHNODES recommended installation of the TOP-EXPERT expert system which is characterized by its ease of operation and versatility. This computer control system is operated in the process controller's language and permits the operating personnel to reconstruct and understand the functioning of the system. This first installation successfully achieved savings in thermal and electrical energy as well as an increase in production with lower wear of the refractory material and greater uniformity of the cement quality. TECHNODES, the research centre of Ciments Français, with more than 60 systems installed throughout the world, now has extensive experience with the use of control systems for cement works.*

Résumé – En 1989, les Ciments Français ont installé le premier Système Expert de conduite automatique de cimenterie à l'usine de Bussac, France. Les problèmes de cette usine étaient un procédé souvent instable dû à une matière difficile, une consommation énergétique spécifique élevée, une consommation de réfractaires importante et en conséquence, des opérateurs de conduite très chargés. L'usine de Bussac était donc très intéressée par une conduite de process permettant de produire, à un coût minimum, un clinker de qualité constante grâce à une conduite plus douce sans interventions sévères. TECHNODES a donc proposé l'installation du Système Expert TOPEXPERT, facile à l'emploi et qui présente maintes possibilités d'applications. Cette application informatique utilise le language des opérateurs et permet au personnel de l'usine d'analyser et de comprendre son fonctionnement. Les résultats de cette première installation ont été des économies en énergie thermique et électrique ainsi qu'une augmentation de la production avec simultanément une consommation de réfractaires réduite et une qualité plus constante du ciment. Aujourd'hui, TECHNODES, Centre de Recherche des Ciments Français, a acquis une grande expérience dans la conduite des cimenteries, avec plus de 60 installations dans le monde.*

Resumen – En 1989, Ciments Français instaló el primer sistema experto destinado al control automático de la fábrica de cemento de Bussac, Francia. Los problemas de esta fábrica eran la frecuente inestabilidad del proceso de cocción, debido a un material difícil, un elevado consumo específico de energía, un excesivo desgaste del material refractario y, como consecuencia de todos estos problemas, una carga excesiva para los operarios. Por esta razón, la fábrica de Bussac estaba muy interesada en un control del proceso que garantizara la producción de un clínker de alta calidad y de gran uniformidad, a costes ventajosos, sin necesidad de intervenir demasiado en el proceso. TECHNODES propuso la instalación del sistema experto TOP-EXPERT, caracterizado por un fácil manejo y múltiples aplicaciones. Este sistema de control por ordenador utiliza el lenguaje de los operadores y permite al personal de servicio analizar y comprender el funcionamiento del mismo. Los éxitos de esta primera instalación han sido el ahorro de energía térmica y eléctrica así como el aumento de la producción, con un reducido desgaste del material refractario y una mayor uniformidad de la calidad del cemento. Actualmente, TECHNODES, el Instituto de Investigación de Ciments Français, que tiene instalados más de 60 sistemas en todo el mundo, cuenta con mucha experiencia en sistemas de control para fábricas de cemento.

Producción a bajo coste de un clínker de alta calidad, con ayuda del sistema experto. El ejemplo de Ciments Français

1. Einführung

Seit 1989 arbeitet die Gruppe Ciments Français in ihren Werken mit Expertensystemen, um eine bessere Steuerung der Zementöfen zu gewährleisten. Heute hat das industrielle Forschungszentrum Technodes S.A. auf der Grundlage von über 60 Installationen innerhalb der Gruppe und bei anderen Zementwerken ein Know-how erworben, mit dem es weltweit führend in diesem speziellen Bereich geworden ist.

Um ein einfaches und zugleich leistungsfähiges Instrument voll nutzen zu können, hat Technodes S.A. ein eigenes Expertensystem entwickelt. TOPEXPERT ist in der Sprache der Prozeßsteuerung zu bedienen und erlaubt es dem Betriebspersonal, die Funktion des Systems nachzuvollziehen und zu verstehen (**Bild 1**). Deshalb ist es mit diesem System möglich, Anwendungen zu realisieren, die den spezifischen Gegebenheiten und Anforderungen eines jeden Werks entsprechen. Solche Anwendungen werden in enger Zusammenarbeit mit dem Kunden entwickelt, so daß dieser das System versteht und in der Praxis voll beherrscht.

2. Steuerungsstrategien für NO_x

Auf der Grundlage seiner Erfahrung hat das Zentrum Technodes S.A. verschiedene Steuerungsstrategien für die Anwendung von Expertensystemen entwickelt, wie z.B. Strategien für die Senkung der Produktionskosten und die Steigerung der Produktivität. Wegen der erhöhten Anforderungen im Bereich des Umweltschutzes wurden in den letzten Jahren z.B. zusätzliche Meßgeräte insbesondere zur Messung der NO_x-Emissionen in den Industrieanlagen eingebaut.

Die Meßergebnisse und anschließenden Untersuchungen, die gezeigt haben, daß die Schwankung der NO_x-Werte ein wesentlicher Parameter der Ofenstabilität ist, veranlaßten Technodes S.A., Steuerungsstrategien für diesen Parameter zu entwickeln. Tatsächlich zeigt die Erfahrung, daß die Korrelation zwischen NO_x, freiem Kalk und der Brennzonentemperatur stabil ist, wenn die CO-Werte unter einer Grenze von etwa 0,3 % bleiben. Gegenüber den oben erwähnten Parametern hat NO_x den Vorteil, daß es schon sehr früh bei Überbrand oder plötzlichem Einbruch der Sinterbedingungen reagiert. Die Reaktion auf die Brennzonentemperatur kann durch die Analyse des freien Kalks nach einigen Minuten oder aber mit einer Verzögerung von mehr als einer halben Stunde erkannt werden.

Deshalb ist es sehr interessant, die NO_x-Schwankungen in die Expertensysteme zu integrieren, um möglichst schnell reagieren zu können und dadurch Abweichungen zu verhindern und einen stabilen Ofenbetrieb zu gewährleisten. Stabilität ist der unmittelbar zu erzielende Vorteil nach Einführung einer Ofensteuerung durch ein Expertensystem. **Bild 2** zeigt ein Beispiel dafür, wie der NO_x-Wert als Parameter genutzt werden kann.

In den **Bildern 3** und **4** werden zur Information typische Ergebnisse beispielhaft dargestellt. Bild 3 zeigt, wie sich die Integration kontinuierlicher NO_x-Messungen in das Exper-

1. Introduction

With the main objective of improving the control of cement kilns, Ciments Franaçis Group have generalized, since 1989, the use of expert systems in their plants. Today, with more than 60 applications in operation within the Group as well as among other cement manufacturers, Technodes S.A., the Industrial Research Center, have built up a know-how which puts them as one of the world leader in this specific domain.

In order to have the full control of a simple yet powerful tool, Technodes S.A. have developed their own expert system. TOPEXPERT puts in line "schemas and rules" which permit the reproduction of the operator's reasoning, as shown in **Fig. 1**. Thanks to this structure, it is possible to put in operation applications responding to the needs and requirements of each plant. These applications are developed in close relationship with customers, thus permitting a throughout understanding and a quite total hand-on control.

2. Control strategies based on NO_x

By experience, Technodes S.A. have developed several different control strategies in the applications of expert systems. Among those strategies can be quoted the reduction of production cost and the increase in productivity. For example, these last years, with the stress on the requirements related to environment, additional measuring devices have

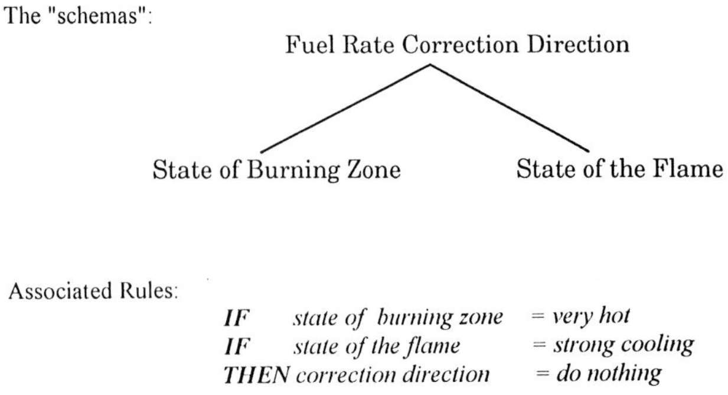

The "schemas":

```
           Fuel Rate Correction Direction
                     /\
                    /  \
       State of Burning Zone    State of the Flame
```

Associated Rules:

IF	state of burning zone	= very hot
IF	state of the flame	= strong cooling
THEN	correction direction	= do nothing

BILD 1: Die „Schemata und Regeln" des Expertensystems
FIGURE 1: Structure of "schemes and rules" of the expert system

Korrekturanweisung Brennstoffzugabe/Fuel Rate Correction Direction

Zustand der Brennzone/State of Burning Zone

Vereinbarte Regeln: Wenn Zustand der Brennzone „sehr heiß",
　　　　　　　　　dann Korrekturanweisung „Tue nichts!"
Associated Rules: If state of burning zone "very hot",
　　　　　　　　then correction direction "do nothing"

Zustand der Flamme/State of the Flame

Vereinbarte Regeln: Wenn Zustand der Flamme „stark kühlend",
　　　　　　　　　dann Korrekturanweisung „Tue nichts!"
Associated Rules: If state of the flame "strong cooling",
　　　　　　　　then correction direction "do nothing"

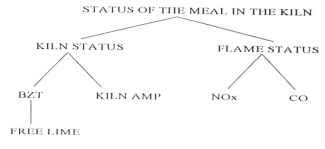

BILD 2: Schema für die Beurteilung der Sinterbedingungen auf Basis des NO$_x$-Werts und weiterer Signale
FIGURE 2: Scheme for judging the sintering conditions by the NO$_x$ rate and additional signals

Zustand des Ofenmehls im Drehofen	= Status of the meal in the kiln
Ofenzustand	= Kiln status
BZT	= BZT
Ampereaufnahme Ofenantrieb	= Kiln Amp
Freier CaO-Gehalt	= Free lime
Flammenzustand	= Flame status
NO$_x$-Gehalt	= NO$_x$
CO-Gehalt	= CO

The Kiln expert system

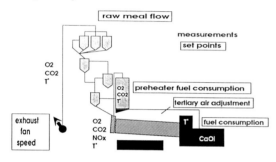

TYPICAL RESULTS

- gains: −3 % thermal energy
  −2 % electrical energy
  −10 % refractory

- + 5 % production

BILD 3: Variable und typische Ergebnisse des Ofen-Expertensystems
FIGURE 3: Variables and typical results of the kiln expert system

Rohmehlaufgabe	= raw meal flow
Messungen	= measurements
Sollwerte	= set points
Brennstoffverbrauch Vorwärmer	= preheater fuel consumption
Einstellung der Tertiärluftmenge	= tertiary air adjustment
Drehzahl Abgaslüfter	= exhaust fan speed
Brennstoffverbrauch	= fuel consumption
Typische Ergebnisse	= Typical results
Zielstellungen	= gains
Senkung des therm. Energieverbrauchs um 3 %	= −3 % thermal energy
Senkung des elektr. Energieverbrauchs um 2 %	= −2 % electrical energy
Senkung des Verbrauchs an Feuerfestmaterial um 10 %	= −10 % refractory
Steigerung des Durchsatzes um 5 %	= +5 % production

tensystem auf den Verbrauch an thermischer und elektrischer Energie des Ofensystems auswirkt und welchen vorteilhaften Einfluß sie auf die Klinkerproduktion und den Verbrauch an Feuerfestmaterial hat. Bild 4 schließlich vermittelt einen Eindruck vom Wirkungsgrad des Zementmahlprozesses, der beim Einsatz von Expertensystemen erzielt werden kann.

Dennoch müssen in jedem einzelnen Fall die besonderen Gegebenheiten der Anlage berücksichtigt werden. Um eine optimale Dienstleistung anbieten zu können, bedarf es einer engen Zusammenarbeit mit dem Kunden.

The ball mill

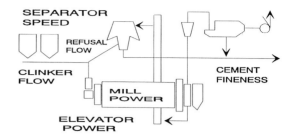

THE RESULTS:

- production increase of up to 10 %
- decrease of 7 % Kwh per Ton.
- removal of quality disturbances

BILD 4: Variable und typische Ergebnisse des Kugelmühlen-Expertensystems
FIGURE 4: Variables and typical results of the ball mill expert system

Sichterdrehzahl	= Separator speed
Grießrücklauf	= Refusal flow
Zementfeinheit	= Cement fineness
Klinkeraufgabe	= Clinker flow
Energieaufnahme Mühle	= Mill power
Energieaufnahme Becherwerk	= Elevator power
Die Ergebnisse	= The results
Durchsatzsteigerung bis zu 10 %	= production increase of up to 10 %
Senkung des spez. Energieaufwands um 7 % pro Tonne	= decrease of 7 % Kwh per ton
Beseitigung von Qualitätsstörungen	= removal of quality disturbances

been implemented in industrial plants and particularly in the field of NO$_x$ emission measurements.

Using available results of measurements and following research studies showing that NO$_x$ variation is a significant parameter of the kiln stability state, Technodes S.A. have implemented control strategies to use this parameter. Effectively, experience shows that the correlation between NO$_x$, free lime and the burning zone temperature is stable if the CO rate remains acceptable (less than 0.3 % usually). Compared to the above mentioned parameters, the NO$_x$ rate has the advantage of reacting early in time as soon as the situation of overburning or of a sudden drop of the sintering status occurs. In the case of the burning zone temperature, the time for respond may amount e. g. to some minutes and to more than half an hour of delay for the result of free lime analysis.

Therefore, it is very interesting to integrate the variation of NO$_x$ in the expert system to react as soon as possible and thus avoiding important deviations and consequently getting a stable behaviour of the kiln. Stability is, moreover, the most immediate benefit after an implementation of a kiln control by expert system. An example for a possible use of the NO$_x$ rate is shown in **Fig. 2**.

In **Figs. 3** and **4**, typical results are given as an example for information purposes. Fig. 3 shows the effect of integrating the result of continuous NO$_x$ measurements into the expert system on the consumption of thermal and electrical energy of the kiln system as well as the advantageous development of the clinker production and refractory consumption figures. Fig. 4, finally, gives an impression of the efficiency in the use of the expert system for a cement grinding process.

Nevertheless, in each single case, the special situation of a plant must be taken into consideration. That needs a very close cooperation with the customer to optimize the offered services.

Die Anwendung der Methoden instrumenteller Elektrochemie zur Qualitätssicherung der Zementherstellung*)

The application of instrumental electrochemical methods for quality assurance in cement production*)

Emploi des méthodes d'electrochimie instrumentale pour l'assurance de la qualité dans la fabrication du ciment

La aplicación de los métodos de la electroquímica instrumental para el aseguramiento de la calidad en la fabricación del cemento

Von **O. P. Mtschedlow-Petrossian** und **A. S. Koschmai**, Charkow/Ukraine

Zusammenfassung – In der Praxis ist es bisher nicht möglich, die bautechnischen Eigenschaften des Zementsteins anhand der physikalisch-chemischen Parameter des Systems Zement-Wasser zu prognostizieren und damit die mechanische Prüfung adäquat zu ersetzen. Im erhärtenden System Zement-Wasser spielen sich eine Reihe elektro-chemischer Vorgänge ab, die in unmittelbarem Zusammenhang mit den für die bautechnischen Eigenschaften wesentlichen Bildungsmechanismen stehen. Die Untersuchung solcher Vorgänge im Erhärtungsprozeß mit Methoden der instrumentellen Elektrochemie ermöglicht es daher, ein Verfahren zur frühen Prognostizierung bautechnischer Eigenschaften zu entwickeln. Bei der neuen Methode wird eine in einen Stromkreis eingebundene Zementleimprobe über einen Hochfrequenzrelaxationsgenerator angeregt und die elektrochemischen Vorgänge innerhalb der Zementpaste über eine spezielle Transmittersubstanz gemessen. Die praktische Anwendung dieser Methode erlaubt es, die bautechnischen Eigenschaften des Zementsteins, wie mechanische Festigkeit, Wasserundurchlässigkeit, Abbindezeit, Korrosionsbeständigkeit etc., bereits drei Stunden nach dem Anmischen der Probe mit hoher Zuverlässigkeit zu bestimmen. Der Fehler des Schnellverfahrens beträgt bei der Bestimmung der Druckfestigkeit maximal ± 1,95 MPa und ist somit kleiner als 4,5%.

Die Anwendung der Methoden instrumenteller Elektrochemie zur Qualitätssicherung der Zementherstellung

Summary – So far it has not been possible in practice to predict the structural properties of hardened cement paste on the basis of the physical and chemical parameters of the cement-water system, and hence to provide an adequate replacement for mechanical testing. A series of electrochemical processes occur in the hardening cement-water system which are directly related to the formation mechanisms which are essential for the structural properties. Investigation of such processes in the hardening process using instrumental electrochemical methods therefore makes it possible to develop a method for early prediction of structural properties. In the new method a cement paste sample incorporated in an electrical circuit is energized by a high-frequency relaxation generator and the electrochemical processes within the cement paste are measured through a special transmitter substance. The practical application of this method means that the structural properties of the hardened cement paste, such as mechanical strength, water impermeability, setting time, corrosion resistance, etc., can be determined with a high level of reliability only three hours after mixing the sample. When measuring the compressive strength the rapid method has a maximum error of ± 1.95 MPa which is therefore less than 4.5%.

The application of instrumental electrochemical methods for quality assurance in cement production

Résumé – Jusqu'à présent, il est impossible dans la pratique de pronostiquer les propriétés dans l'oeuvre du ciment durci et de remplacer ainsi convenablement l'essai mécanique, au seul moyen des paramètres physico-chimiques du système ciment-eau. Dans le système ciment- eau se produisent une série de phénomènes électrochimiques, qui sont en relation immédiate avec les mécanismes de formation essentiels pour les propriétés dans l'oeuvre. L'étude de tels phénomènes dans le processus de durcissement, à l'aide des méthodes de l'électrochimie instrumentale, rend donc possible d'élaborer un procédé pour le pronostic précoce des propriétés dans l'oeuvre. Avec la nouvelle méthode, une éprouvette de pâte de ciment insérée dans un circuit électrique est excitée au moyen d'un générateur de relaxation de haute fréquence et les phénomènes électrochimiques au sein de la pâte de ciment sont mesurés au moyen d'une substance de transmission spéciale. L'application pratique de cette méthode permet de déterminer avec une haute fiabilité, déjà trois heures après le gâchage de l'éprouvette, les propriétés du ciment durci dans l'oeuvre telles que résistance mécanique, imperméabilité à l'eau, temps de prise, résistance à la corrosion etc. La précision de la méthode rapide est, pour la résistance en compression, de max. ± 1,95 MPa et l'erreur est ainsi inférieur à 4,5%.

Emploi des méthodes d'electrochimie instrumentale pour l'assurance de la qualité dans la fabrication du ciment

*) Gekürzte Fassung / Abbreviated version.

La aplicación de los métodos de la electroquímica instrumental para el aseguramiento de la calidad en la fabricación del cemento

Resumen – En la práctica no es posible, hasta ahora, diagnosticar las características técnicas de la pasta de cemento después de su endurecimiento en el hormigón mediante los parámetros físico-químicos del sistema cemento-agua y de sustituir así adecuadamente el examen mecánico. En el sistema cemento-agua en fase de endurecimiento tienen lugar una serie de procesos electroquímicos, directamente relacionados con los mecanismos de formación, que son esenciales para las características técnicas. El estudio de tales procesos, originados durante la fase de endurecimiento del hormigón, aplicando métodos de la electroquímica instrumental, permite desarrollar un procedimiento de pronóstico precoz de las características técnicas. Con este nuevo método una muestra de pasta de cemento, incorporada a un circuito eléctrico, es excitada a través de un generador de relajación, de alta frecuencia, midiéndose los procesos electroquímicos que se desarrollan dentro de la pasta de cemento, con ayuda de una sustancia de transmisión especial. La aplicación práctica de este método permite determinar, con un alto grado de fiabilidad, las características técnicas del cemento después de su endurecimiento en el hormigón, como por ejemplo la resistencia mecánica, la impermeabilidad al agua, el tiempo de fraguado, la resistencia a la corrosión, etc., y eso sólo tres horas después de preparar la mezcla. La precisión del procedimiento rápido para la determinación de la resistencia a la compresión es de ± 1,95 MPa, de modo que el error es inferior al 4,5 %.

Für die Qualitätssicherung der Zementherstellung wäre es wünschenswert, mittels eines geeigneten Prüfverfahrens eine frühzeitige Prognose der bautechnischen Eigenschaften des produzierten Zementes stellen zu können. Daher fehlte es in der Vergangenheit nicht an Versuchen, aus der Beobachtung des zeitlichen Verlaufs einiger physikalisch-chemischer Parameter des erhärtenden Zement-Wasser-Systems quantitative Schlußfolgerungen bezüglich der Festigkeitseigenschaften des Zementsteins abzuleiten. Die Praxis hat jedoch gezeigt, daß ein solcher Weg bisher nicht zum Erfolg geführt hat.

Um dieses Problem einer Schnellprüfung mathematisch formulieren zu können, wurde aus der Gesamtheit der bautechnischen Eigenschaften zunächst nur die Druckfestigkeit R des Zementsteins betrachtet. Bei der Beschreibung der Festigkeitsentwicklung müssen zum einen die physikalisch-chemischen Eigenschaften des erhärtenden Zement-Wasser-Systems über einen geeigneten festzulegenden Parameter A berücksichtigt werden und zum anderen auch der zeitliche Verlauf dieser Größen. Dieser funktionale Zusammenhang ist in der folgenden Gleichung wiedergegeben:

$$R = F(A, t).$$

Das Prinzip des entwickelten Schnellprüfverfahrens zur Prognose der Endfestigkeit beruht darauf, nicht etwa unmittelbar die Festigkeitsentwicklung zu messen und daraus auf spätere Termine zu extrapolieren, sondern vielmehr einen geeigneten Indikator K zu bestimmen, der eindeutig mit der Festigkeit korreliert. Wenn dies gelingt, so kann nach entsprechender Kalibrierung aus der zu messenden Indikatorgröße K auf die Endfestigkeit des zu untersuchenden Zementleims geschlossen werden.

Um eine solche „Expreßprognose" zu realisieren, wurde eine elektrochemische Analyseapparatur entwickelt, das „σ-Chemotron". Als theoretische Grundlage dieses Verfahrens wurde eine Modellvorstellung entwickelt, die die elektrochemischen Vorgänge bei der Erhärtung des Zement-Wasser-Systems beschreibt [1]. Der instrumentelle Aufbau besteht aus einem Hochfrequenz-Relaxationsgenerator, in dessen Stromkreis eine elektrochemische Zelle eingebunden ist. Diese Zelle ist mit Zementleim gefüllt, der einerseits als elektrischer Widerstand fungiert und gleichzeitig die elektrische Kapazität der Zelle beeinflußt. Die Eigenfrequenz dieser als R-C-Glied zu beschreibenden Zelle hängt daher vom Erhärtungszustand der Zementpaste ab und dient als die gesuchte Indikatorgröße K.

Die Erprobung dieses Schnellverfahrens im Labor und in der Betriebspraxis hat gezeigt, daß bereits 3 Stunden nach dem Anmischen der Probe eine Prognose der Druckfestigkeit des ausgehärteten Zementsteins mit hoher Zuverlässigkeit möglich ist. Die Absolutabweichung von den mechanischen Prüfverfahren wurde zu ± 1,2 MPa bestimmt, was einer relativen Abweichung von nur 3 % entspricht.

For the purposes of quality assurance in cement production it would be helpful if the structural properties of the cement produced could be predicted at an early stage by a suitable test method. Countless attempts have been made to draw quantitative conclusions about the strength properties of hardened cement paste by observing the behaviour with time of some physico-chemical parameters of the hardening cement-water system. However, practical experience has shown that so far none of these attempts has been successful.

To enable the problem of rapid testing to be formulated mathematically, the compressive strength R of the hardened cement paste was first considered separately from all the other structural properties. In describing the development of strength it is necessary to consider firstly the physico-chemical properties of the hardening cement-water system in terms of a suitable defined parameter A and secondly the behaviour with time of these variables. The functional relationship is expressed by the following equation:

$$R = F(A, t).$$

The rapid test method developed for predicting the final strength is based on the principle that, rather than measuring the development of strength directly and extrapolating from it to later times, a suitable indicator K which has a clear correlation with strength should be determined. If this is possible then information about the final strength of the cement paste under investigation can be derived from the measured indicator variable K after suitable calibration.

An apparatus for electrochemical analysis, the "σ-Chemotron", has been developed to achieve this "express prediction". A model describing the electrochemical processes involved in the setting of the cement-water system was produced as the theoretical basis for the method [1]. The instrumentation consists of a high-frequency relaxation generator with an electrochemical cell incorporated in its electrical circuit. The cell is full of cement paste, which both acts as an electrical resistance and affects the electrical capacitance of the cell. The natural frequency of the cell, which may be described as an RC element, thus depends on the state of hardening of the cement paste and serves as the required indicator variable K.

Laboratory and industrial trials of the rapid method have shown that the compressive strength of the hardened cement paste can be predicted very reliably only three hours after mixing the sample. The absolute deviation from the mechanical test method was found to be ± 1.2 MPa, corresponding to a relative deviation of only 3 %.

Literaturverzeichnis

[1] Koschmai, A. S., und Mtschedlow-Petrossian, O. P.: Theory and practice of cement strength characteristics prediction. 9th Internat. Congr. Chem. Cem. India 1992, vol. IV, p. 532–533.

Industrielle Anwendung einer multivariablen linearquadratischen Steuerung für Zementmühlen

An industrial application of multivariable linear quadratic control to a cement mill

Application industrielle d'un système de conduite linéaire-quadratique multivariable pour broyeur à ciment

Aplicación industrial de un sistema de mando multivariable, lineal-cuadrático, para molinos de cemento

Von **C. de Pierpont, V. Werbrouck, V. van Breusegem, L. Chen, G. Bastin** und **V. Wertz**, Louvain-la-Neuve/Belgien

Zusammenfassung – Es war das Ziel der Forschungsarbeit, die Zementmühlenkreisläufe zu optimieren. Das Forscherteam bestand aus Slegten, einem Mühlenspezialisten, „Ciments d'Obourg", einem Zementhersteller, und „CESAME", einem Spezialisten für Steuerungssysteme an der Universität Louvain (Leuven). Das Steuerungssystem basiert auf einem neuen Ansatz und einer neuen Methodologie. Die in der Industrie erzielten Ergebnisse lassen den Schluß zu, daß die Mühlensteuerung am besten durch die Steuerung des Mahlgutes und des Luftstroms bewerkstelligt werden kann. Mit einem Hochleistungssteuerungssystem können ein stabiler Mühlenumlauf, kürzere Übergangszeiten (z.B. Wechsel des Zementtyps, Umlaufschwankungen), optimale Betriebsbedingungen und höhere Feinheit und Qualität erzielt werden. Das vorgestellte Steuerungssystem hält zwei Materialströme, d.h. Grieße und Fertigprodukt, stabil, da es gleichzeitig die Umfangsgeschwindigkeit des Sichterkorbs und die Aufgabemenge steuert. Dieses System verwirklicht die derzeit modernste Steuerungstechnologie, und das Herzstück dieser Steuerung ist eine multivariable linear-quadratische Gauss'sche Steuereinheit, die auf Modellen beruht.

Industrielle Anwendung einer multivariablen linear-quadratischen Steuerung für Zementmühlen

Summary – The object of the research was to optimize cement grinding circuits. The team involved included Slegten, a grinding expert, "Ciments d'Obourg", a cement producer, and "CESAME", a control system specialist located at the University of Louvain. The control system uses a new approach and new methodology. From industrial results it can be concluded that the best way to operate a grinding circuit is to control the material and air flow. A stable circuit, reduced transition periods (such as cement type change, circuit fluctuations), optimized operating conditions and increased fineness and quality can be achieved when using a high performance control system. The perfected controller maintains two flows – reject material and finished product – in a stable state by acting simultaneously on the separator cage speed and the feed rate. This system applies the most up-to-date control technology, and the heart of this control is based on a Linear Quadratic Gaussian Multivariable Controller based on models.

An industrial application of multivariable linear quadratic control to a cement mill

Résumé – Le but du présent travail de recherche visait l'optimisation des circuits de broyeurs à ciment. L'équipe des chercheurs regroupait Slegten, spécialiste de broyeurs, „Ciments d'Obourg", fabricant de ciment, et „CESAME", spécialiste de systèmes de pilotage à l'Université de Louvain (Leuven). Le système de conduite est basé à la fois sur la nouvelle hypothèse et la nouvelle méthodologie. Les résultats obtenus dans l'industrie permettent de conclure que la conduite du broyeur est préférable par la conduite de la matière à broyer et du flux d'air. Avec un système de conduite très performant on obtient un cycle de broyage stable, des temps de transition plus courts (par exemple changement du type de ciment, fluctuations de cycle), des conditions d'exploitation optimales ainsi qu'une finesse de mouture et une qualité très élevées. Le système de conduite présenté permet de maintenir stables deux flux de matière, c'est-à-dire refus et produit fini, du fait qu'il commande en même temps la vitesse périphérique de la rouecage du séparateur et la quantité d'alimentation. Ce système concrétise la technologie de conduite la plus moderne actuellement, et le coeur de ce système est l'unité de commande Gauss linéaire-quadratique multivariable, reposant sur des modèles.

Application industrielle d'un système de conduite linéaire-quadratique multivariable pour broyeur à ciment

Resumen – El presente trabajo de investigación ha tenido por objetivo la optimización de los circuitos de molienda de los molinos de cemento. El grupo de investigación estaba compuesto por Slegten, un especialista en molinos, „Ciments d'Obourg", un fabricante de cemento y „CESAME", un especialista en sistemas de mando, de la Universidad de Lovaina (Louvain). El citado sistema de mando se basa en una hipótesis y una metodología nuevas. Los resultados obtenidos a nivel industrial permiten concluir que

Aplicación industrial de un sistema de mando multivariable, lineal-cuadrático, para molinos de cemento

el mando del molino se efectúa mejor mediante el control del material a moler y del flujo de aire. Con un sistema de mando de alto rendimiento se puede conseguir un ciclo de molienda estable, tiempos de transición más cortos (por ejemplo cambio del tipo de cemento, fluctuaciones de ciclo), condiciones de servicio óptimas, una mayor finura y una mejor calidad. El sistema de mando que se presenta mantiene estables dos flujos de material, o sea el de gruesos y el de producto acabado, ya que controla al mismo tiempo la velocidad circunferencial de la cesta del separador y el caudal de alimentación. Este sistema aplica la tecnología de mando más moderna actualmente, y la pieza clave del mismo es una unidad de mando de Gauss, multivariable y lineal-cuadrático, basada en modelos.

1. Vorwort

Das Ziel dieser Forschungsarbeit war die Optimierung von Zementmühlenkreisläufen. Um erfolgreich arbeiten zu können, wurde ein Team zusammengestellt, das aus Slegten, einem Mühlenspezialisten, „Ciments d'Obourg", einem Zementhersteller, und „CESAME", einem Spezialisten für Steuerungssysteme an der Universität Louvain (Leuven), bestand. Das Steuerungssystem basiert auf einem neuen Ansatz und einer neuen Methodologie. Das Prinzip, nach dem das System arbeitet, und einige in der Industrie erzielten Ergebnisse werden in diesem Beitrag vorgestellt.

Der Mahlkreislauf läßt sich am besten durch die Steuerung des Mahlguts und des Luftstroms steuern. Mit einem Hochleistungssteuerungssystem können ein stabiler Mühlenumlauf, kürzere Übergangszeiten, optimale Betriebsbedingungen und höhere Feinheit und Qualität erzielt werden.

Bild 1 zeigt die schematische Darstellung einer Kugelmühle mit Sichter; die Mühle arbeitet im geschlossenen Kreislauf. Der Materialfluß muß an mehreren Stellen des Kreislaufs konstant gehalten werden. Das bezieht sich auf das Mühlenaufgabegut, den Mühlenaustrag bzw. das Sichteraufgabegut, den Sichterrückstand und das Fertigprodukt oder den Sichterdurchlauf. Wenn Mühle und Sichter mit konstanter Aufgabemenge beschickt werden, kann der Mahlkreislauf optimiert werden. Diese Parameter sind noch wichtiger, wenn die Anlage mit einem Hochleistungssichter ausgerüstet ist, da solche Sichter ein konstantes Verhältnis zwischen Luftmenge und durchlaufendem Mahlgut erfordern.

2. Steuerungsstrategie

Nach eingehender Untersuchung aller Kreislaufparameter und bestehender Steuerungssysteme wurde eine neue Strategie entwickelt, um zwei Materialströme, d.h. Grieße und Fertigprodukt, konstant zu halten. Im Gleichgewichtszu-

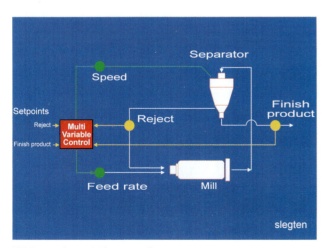

BILD 1: Schematische Darstellung eines Zementmahlkreislaufs
FIGURE 1: Scheme of cement grinding circuit

Sichter	= separator
Geschwindigkeit	= speed
Sollwerte	= setpoints
Grieß	= reject
Fertigprodukt	= finish product
Multivariable Steuerung	= multivariable control
Aufgabemenge	= feed rate
Mühle	= mill
t/h	= tons/h

1. Preface

The objective of our research was to optimize cement grinding circuits. To conduct a successful program, a team was created. It included Slegten, a grinding expert, "Ciments d'Obourg", a cement producer, and "CESAME", a control system specialist, located at the University of Louvain. The control system uses a new approach and new methodology. The principle of this system and some industrial results are presented in this paper.

The best way to operate a grinding circuit is to control the material and air flow. When using a high performance control system, as results can be achieved a stable circuit, reduced transition periods, optimized operating conditions and increased fineness quality.

Fig. 1 shows a sketch of a ball mill operating in closed circuit with a separator. The material flow in several locations of the circuit must be kept constant. This includes fresh feed to the mill, mill discharge or separator feed, separator rejects and the finished product or fines. This means that the mill and the separator are fed constantly which allows to optimize the circuit. These parameters are even more important if the installation is equipped with a high efficiency separator, since these separators need to have a constant ratio between the quantity of air and material going through them.

2. Control Strategy

After an in depth study of all the circuit parameters and existing control systems, a new strategy was developed to keep the flow of rejects from the separator and the flow of finished product constant. At equilibrium, the flow of finished product equals that of the fresh feed. To be able to control these two flow rates and keep them constant (rejects and finished product), actions are taken on the speed of the separator cage and the feed rate of fresh material to the mill.

Numerous tests were carried out to analyze the circuit's reactions. Two key results were obtained. First, it was possible to make stochastic models of a grinding circuit. Using these models, specific circuit reactions can be simulated on a personal computer. This provides numerous advantages such as on-line optimization of setpoints. Secondly, strong interrelations can be observed between the actions (speed of the separator and feed rate of fresh material) and the reactions (rejects and finished product).

To control these two flows in spite of the interrelations, the heart of the control system uses a Linear Quadratic Gaussian Multivariable controller based on the models. This controller maintains the two flows stable by acting simultaneously on the separator speed and the feed rate. This system applies the most up-to-date control technology.

The second level of the control system, not described in this presentation, uses the air flow through the mill and separator, the power absorbed by the mill and the noise level in the first mill chamber as additional parameters.

3. Operational results

Fig. 2 shows the control system applied to a circuit producing two different cement types over a period of fifty hours. The yellow lines on the top show how the controlled flows (rejects and finished product) are following the red setpoint lines. The setpoint for the rejects is kept constant at 400 t/h and the setpoint for the finished product is changing according to the fineness. The green lines are the actions taken on the separator speed and the feed rate.

stand entspricht der Fertigproduktstrom dem Strom des Aufgabeguts. Um diese beiden Fließgeschwindigkeiten (Grieße und Fertigprodukt) zu steuern und konstant zu halten, werden Maßnahmen zur Steuerung der Umfangsgeschwindigkeit des Sichterkorbs und der Mühlenaufgabemenge ergriffen.

Es wurden zahlreiche Versuche unternommen, um die Reaktionen im Kreislauf zu analysieren. Zwei wichtige Feststellungen wurden dabei gemacht. Erstens war es möglich, stochastische Modelle des Mahlkreislaufs zu erstellen und mit ihrer Hilfe bestimmte Kreislaufreaktionen auf einem PC zu simulieren. Einer der damit erzielten Vorteile ist die Online-Optimierung der Sollwerte. Zweitens wurde eine starke Wechselwirkung zwischen den ergriffenen Maßnahmen (Umfangsgeschwindigkeit des Sichterkorbs und Mühlenaufgabemenge) und den darauf erfolgenden Reaktionen (Grieße und Fertigprodukt) beobachtet.

Um diese beiden Materialströme trotz der genannten Wechselwirkungen zu steuern, besteht das Herzstück des Steuerungssystems aus einer linear-quadratischen multivariablen Steuerungseinheit, die auf der Grundlage der Modelle entwickelt wurde. Diese Einheit wirkt gleichzeitig auf die Sichterdrehzahl und die Mühlenaufgabe und hält so die beiden Materialströme stabil. Dieses System entspricht modernster Steuerungstechnologie.

Die zweite Ebene des Steuerungssystems, auf die in diesem Beitrag nicht eingegangen wird, arbeitet mit dem durch die Mühle und den Sichter gehenden Luftstrom, der von der Mühle aufgenommenen Energie und dem Geräuschpegel in der ersten Mahlkammer als zusätzlichen Parametern.

3. Betriebsergebnisse

Bild 2 zeigt das in einem 50stündigen Mahlkreislauf für zwei verschiedene Zemente eingesetzte Steuerungssystem. Die gelben Kurven oben stellen dar, wie die gesteuerten Materialströme (Grieße und Fertigprodukt) den roten Sollwertkurven folgen. Der Sollwert für die Grieße wird konstant auf 400 t/h gehalten, und der Sollwert für das Fertigprodukt variiert je nach Mahlfeinheit. Die grünen Kurven sind die hinsichtlich der Sichterdrehzahl und Aufgaberate ergriffenen Maßnahmen.

Wenn sich die Drehzahl des Sichters ändert, ist mit Abweichungen in der Mahlfeinheit zu rechnen. Wie jedoch auf der untersten Darstellung zu sehen ist, bleibt die weiße Kurve (effektive Mahlfeinheit) sehr nahe an der roten Sollwertkurve, auch bei Änderung des Zementtyps und einer daraus folgenden Sollwerterhöhung der Mahlfeinheit von 4300 auf 4500 cm²/g (Blaine). Da es sich um ein multivariables Steuerungssystem handelt, sind die Schwankungen der Sichterdrehzahl nur gering, so daß die Mahlfeinheit des Fertigprodukts nicht beeinträchtigt wird.

Bei Kreislaufschwankungen reagiert das Steuerungssystem sehr elastisch. Die multivariable Steuerungseinheit veranlaßt sofort die Auffüllung des Kreislaufs mit der optimalen Mahlgutmenge, um den Kreislauf so schnell wie möglich zu stabilisieren. **Bild 3** zeigt die Situation bei Mangel an Aufgabegut aufgrund eines Defekts. Wenn sich der Vorbunker entleert, wird der Materialstrom zur Mühle unterbrochen, aber der Mahlkreislauf geht weiter. Die Mühle arbeitet weiter, und das Material verläßt den Kreislauf als Fertigprodukt. Das bedeutet, daß sich immer weniger Zement im Mahlkreislauf befindet. In einem solchen Fall sinkt die Fertigproduktmenge um 60 t/h und die Menge der Grieße von 400 auf 200 t/h. Das stellt eine gewaltige Schwankung dar, die von den meisten Steuerungssystemen nicht bewältigt werden kann. Das multivariable Steuerungssystem kann nicht nur diese Situation meistern, sondern man kann auch beobachten, wie schnell es den Mahlkreislauf wieder in den Gleichgewichtszustand bringt.

Der Einbau eines solchen Systems in einem Zementwerk erfordert einen Sichter mit regelbarer Drehzahl und Mengenmesser für Grieße und Fertigprodukt. Das System läßt sich leicht mit einem industriellen PC installieren und in jede vorhandene Rechnerarchitektur integrieren.

As the separator speed changes, fineness variations could be expected. But, as can be noticed on the bottom graph, even with a change in the cement type which resulted in a setpoint increase of the fineness from 4300 to 4500 cm²/g (Blaine), the white line (actual fineness) stays very close to the red setpoint line. Thanks to the multivariable control system, the fluctuations of the separator speed are slight, thus the fineness of the finished product is not affected.

In case of circuit fluctuations, the control system reacts accordingly and very smoothly. The multivariable controller immediately fills the circuit with the optimum levels of material in order to stabilize the circuit as quickly as possible.

Fig. 3 shows a lack of fresh feed due to a breakdown. As the feed bin empties, the fresh feed to the mill is interrupted, but the rest of the circuit continues to operate. The mill continues to grind and allows the material to leave the circuit as finished product. This means there is less and less cement in

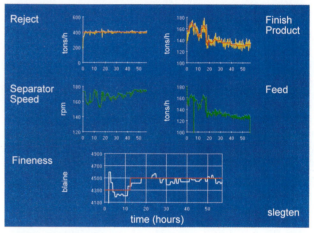

BILD 2: Steuerungssystem eines Mahlkreislaufs für zwei verschiedene Arten von Zement
FIGURE 2: Control system of a grinding circuit producing two different cement types

Grieß	= reject
Fertigprodukt	= finish product
Sichterdrehzahl	= separator speed
t/h	= tons/h
Zeit (Stunden)	= time (hours)
Mahlfeinheit	= fineness
Aufgabe	= feed

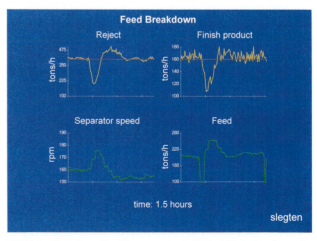

BILD 3: Verhalten des Steuerungssystems bei Unterbrechung des Materialzuflusses
FIGURE 3: Behaviour of the control system at a feed breakdown

Grieß	= reject
Fertigprodukt	= finish product
Unterbrechung in der Materialaufgabe	= feed breakdown
U/Min.	= rpm
Sichterdrehzahl	= separator speed
Zeit: 1,5 Stunden	= time: 1.5 hours
t/h	= tons/h
Aufgabe	= feed

4. Schlußfolgerungen

Das Forschungsprogramm wurde aufgelegt, um den Mahlkreislauf in Zementwerken zu verbessern. Die Zusammenarbeit zwischen einem Mühlenforschungszentrum, einem Zementhersteller und einem Experten für Steuerungssysteme führte zu einer umfassenden Lösung für den gesamten Mahlkreislauf. Das so entwickelte System basiert auf einer linear-quadratischen Gauss'schen multivariablen Steuerungseinheit, die für die spezifischen Kreislaufmodelle ausgelegt wurde. Das bedeutet, daß das System den einzelnen Mahlkreisläufen genauestens angepaßt werden kann.

Es entspricht modernster Technologie, ist dennoch einfach anzuwenden und reduziert den Überwachungsbedarf durch das Bedienungspersonal. Das System gewährleistet einen stetigen Materialfluß in der Mühle und im Sichter. Außerdem verbessert es die Gleichmäßigkeit der Mahlfeinheit des Fertigprodukts und verkürzt die Übergangszeiten aufgrund von Störungen im Mahlkreislauf. Es stellt auch in schwierigen Situationen ein elastisches, sicheres und zuverlässiges Steuerungssystem dar.

the circuit. In such a case, the finished product decreases by 60 t/h and the rejects from 400 to 200 t/h. This is a huge fluctuation that most control systems can not handle. The multivariable control system is not only capable of mastering this type of situation, but one can notice how fast the control system brings the circuit back to equilibrium.

The installation of such a system in a cement plant requires a variable speed separator, a reject flow meter and a finished product flow meter. The system is easily installed using an industrial PC and can be integrated into any existing computer architecture.

4. Conclusion

The research program was initiated to inprove the grinding operation of cement producers. The collaboration between a grinding research center, a cement manufacturer and a control system expert allowed to develop a complete solution for the entire grinding circuit. This system is based on a linear quadratic Gaussian multivariable controller which is designed considering the specific circuit models. This means that the system is tailor made to individual circuits.

It applies the latest technologies, but remains easy to use and reduces operator supervision. It ensures a steady flow of material in the mill and the separator. It also improves the stability of the finished product fineness and has the capability to reduce the transition periods due to any disturbances in the circuit. It provides a smooth, secure, reliable control system, even under difficult conditions.

PYROEXPERT® – Ein neues Regelungssystem zur Vergleichmäßigung des Klinkerbrennprozesses

PYROEXPERT® – a new control system for stabilizing the clinker burning process

PYROEXPERT® – Un nouveau système de régulation pour la régularité du process de cuisson du clinker

PYROEXPERT® – Un nuevo sistema de regulación destinado a estabilizar el proceso de cocción del clínker

Von **H. Pisters, M. Becke,** Beckum, und **G. Jäger,** Köln/Deutschland

Zusammenfassung – Die stetig steigenden Anforderungen an die Qualität und Gleichmäßigkeit des Zements sowie die Minimierung des Energieeinsatzes, der Verschleißbeanspruchung und der Emissionen bei dessen Herstellung stellen an die Produktionsanlagen und deren verfahrenstechnische Führung erhöhte Ansprüche. Zur Realisierung eines Ofenregelsystems, das diesen Anforderungen genügt, wurde ein Regelungskonzept entwickelt, das die Stärken eines thermodynamischen Prozeßmodells mit denen einer heuristischen Kontrollkomponente kombiniert. In das Prozeßmodell werden aktuelle Entwicklungen im Bereich der Sensorik sowie der aus den Meßsignalen abgeleiteten virtuellen Größen Brennstoff- und Sekundärluftenergiestrom integriert. Die Heuristik-Komponente bildet das menschliche Prozeßführungsverhalten nach und ist durch die Verwendung eines Expertensystems mit Fuzzy-Regelung anwenderfreundlich und flexibel aufgebaut. Besonderer Wert wurde auf eine detaillierte Erklärungskomponente gelegt, die es dem Produktionssteuerer ermöglicht, die Handlungsweise des Expertensystems und die daraus resultierenden Stelleingriffe nachzuvollziehen. Da sich das zeitliche Verhalten der Regelstrecke ändern kann, ist dem System ein neuronales Netzwerk übergeordnet. Das Netzwerk nimmt eine automatische Anpassung der Regelung durch Einstufung von Prozeß-Situationen über einen Lernmodus vor.

Summary – The constantly increasing requirements for quality and uniformity of the cement and for minimizing the energy input, the wear stresses and the emission during its manufacture, are placing increased demands on the production plants and their process engineering control. To achieve a kiln control system which satisfies these requirements, a control concept was developed which combines the strengths of a thermodynamic process model with those of a heuristic control component. Current developments in the field of sensory analysis as well as the virtual variables consisting of the fuel and secondary air energy flows derived from the measurement signals are integrated in the process model. The heuristic component imitates the human process control behaviour and is built up in a user-friendly and flexible manner by using an expert system with fuzzy control. Particular emphasis was placed on a detailed explanation component, which makes it possible for the production controller to reconstruct the expert system's mode of operation and the resulting control actions. The temporal behaviour of the controlled member can change, so the system is provided with a higher-ranking neural network. The network undertakes automatic adaptation of the control system by rating process situations through a learning mode.

Résumé – Les exigences toujours croissantes de qualité et de régularité du ciment, ainsi que la minimisation de la consommation d'énergie, de l'exposition à l'usure et des émissions au cours de sa fabrication demandent des efforts plus grands aux unités de production et à leur conduite du process. Afin de réaliser un système de régulation du four, susceptible de satisfaire ces exigences, a été développé un concept de régulation combinant la puissance d'un modèle de process thermodynamique avec celle d'une composante de contrôle heuristique. Dans le modèle de process sont intégrés des acquis récents dans le domaine des capteurs, ainsi que les grandeurs virtuelles flux d'énergie du combustible et de l'air secondaire, dérivées des signaux de mesure. La composante heuristique simule le comportement humain de conduite de process et se présente conviviale et flexible par l'emploi d'un système expert avec régulation Fuzzy. Une importance particulière a été accordée à une composante d'explication détaillée, permettant au conducteur de la production de répéter la manière d'agir du système expert et les interventions de réglage, qui en résultent. Comme le comportement dans le temps de la ligne de régulation peut changer, le système est supervisé par un réseau neuronal. Le réseau effectue une adaptation automatique, par insertion de situations du process au moyen d'un module d'apprentissage.

Resumen – *Los crecientes requerimientos en cuanto a la calidad y uniformidad del cemento así como la reducción del consumo de energía, del desgaste y de las emisiones durante el proceso de fabricación del cemento van unidos a mayores exigencias que tienen que cumplir las instalaciones de producción así como el personal de control. Para crear un sistema de regulación del horno capaz de satisfacer estos requerimientos, se ha desarrollado un concepto de regulación que combina las ventajas de un modelo termodinámico de proceso con las de un componente de control heurístico. Este modelo de proceso integra los avances actuales logrados en el campo de los sensores y los de las magnitudes virtuales, como son el combustible y el flujo de aire secundario, derivadas de las señales de medición. El componente heurístico simula el comportamiento humano frente a la conducción del proceso, siendo flexible y de fácil aplicación, debido al empleo de un sistema experto con regulación Fuzzy. Se ha hecho hincapié, especialmente, en un detallado componente explicativo, el cual permite al responsable de la producción analizar y reconstruir el modo de operación del sistema experto y los ajustes de ahí resultantes. Puesto que el comportamiento de la línea de regulación puede cambiar con el tiempo, el sistema es supervisado por una red neuronal de orden superior. Esta red procede a una adaptación automática de la regulación, mediante la inserción de situaciones del proceso a través de un módulo de aprendizaje.*

PYROEXPERT® – Un nuevo sistema de regulación destinado a estabilizar el proceso de cocción del clínker

1. Einleitung

Vor dem Hintergrund steigender Anforderungen hinsichtlich Produktqualität und Umweltverträglichkeit gilt es, insbesondere den Klinkerbrennprozeß durch Anwendung neuer Regelkonzepte weiter zu optimieren. Daher wurde unter Einbeziehung neuer Entwicklungen auf dem Gebiet der Sensorik und Regelungsstrategien [1–3] das rechnergestützte Regelungssystem PYROEXPERT® entwickelt und bei den Readymix Zementwerken mit dem Ziel eingesetzt, den Klinkerbrennprozeß zu vergleichmäßigen.

2. Regelungskonzept

Das Regelungskonzept nutzt sowohl die Stärken unscharfer heuristischer als auch diejenigen exakter prozeßmodellbasierter Regelungsstrategien zur Harmonisierung der Fahrweise und Optimierung auf vorgegebene Zielparameter. Expertensystem- und Fuzzy-Technik erlauben durch umgangssprachliche Formulierung die einfache Einbeziehung von Erfahrungswissen (Heuristiken) in die Regelung. Damit wird das Bedienverhalten modelliert (Bedienermodell), im Gegensatz zum Prozeßmodell, welches das Prozeßverhalten beschreibt.

Die Funktionsweise eines PYROEXPERT®-Fuzzy-Reglers ist in **Bild 1** dargestellt. Die Prozeßmeßwerte werden fuzzyfiziert, d. h., unscharf gemacht. Es wird festgestellt, ob z. B. die Brennzonen-Temperatur TBZ eher noch dem Term „normal" oder schon mehr dem Term „zu hoch" zuzuweisen ist und in welchem Maße dies der Fall ist. Eine Bewertung könnte lauten: TBZ gehört mit dem Maß 0,7 dem Term „normal" und mit 0,3 dem Term „zu hoch" an.

Die unscharfe Bewertung wird mit anderen unscharfen Daten durch umgangssprachlich formulierte Regeln verknüpft. In diesen Regeln steckt das Erfahrungswissen, die Ofenanlage in verschiedenen Prozeßsituationen führen zu können, ohne daß ein vollständiges Prozeßmodell vorliegt.

1. Introduction

With the increasing requirements for product quality and environmental compatibility it is particularly important that the clinker burning process should be further improved by applying new control concepts. The computer-aided control system PYROEXPERT®, which incorporates new advances in the field of sensory analysis and control strategies [1–3], has therefore been developed. It has been installed at the Readymix cement works with a view to stabilising the clinker burning process.

2. Control concept

The control concept uses the strengths both of indefinite heuristic strategies and of exact strategies based on a process model to harmonise operation and achieve predetermined target parameters. Empirical knowledge (heuristics) can easily be input in the control by expert-system and fuzzy technology with the use of natural language. This models the operating behaviour (operator model) in contrast to the process model, which describes the process behaviour.

The operation of a PYROEXPERT® fuzzy control is illustrated in **Fig. 1**. The process measurements are fuzzified, i.e. made indefinite. The control finds out e. g. whether the burning zone temperature BZT should be described as "normal" or perhaps as "too high", and to what extent this applies. A possible assessment might be: the temperature is 0.7 "normal" and 0.3 "too high".

The indefinite evaluation is related to other indefinite data by rules formulated in natural language. These rules contain empirical knowledge of how to operate the kiln plant in various processing situations without having a complete process model. The rules are put in the form: "IF temperature BZT too high AND O_2 content too low AND ..... THEN reduce fuel". The results of thus linking the data are also indefinite, e. g. "Reduce fuel slightly". In a final step they are

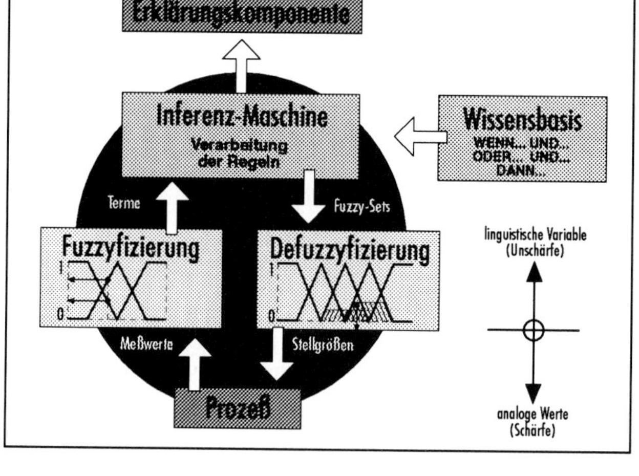

BILD 1: PYROEXPERT®-Fuzzy-Regler
FIGURE 1: PYROEXPERT® fuzzy control

Erklärungskomponente	= explanation program
Inferenz-Maschine	= inference machine
Verarbeitung der Regeln	= processing the rules
Wissensbasis	= knowledge base
WENN ... UND	= IF ... AND
ODER ... UND ... DANN ...	= OR ... AND ... THEN ...
Terme	= terms
Fuzzy-Sets	= fuzzy sets
Fuzzyfizierung	= fuzzifying
Defuzzyfizierung	= defuzzifying
linguistische Variable	= linguistic variable
(Unschärfe)	= (indefiniteness)
Meßwerte	= readings
Stellgrößen	= manipulated variables
Prozeß	= process
analoge Werte	= analog values
(Schärfe)	= (definiteness)

Die Regeln sind in der Form „WENN Temperatur TBZ zu hoch UND O₂-Gehalt zu niedrig UND ... DANN Brennstoff reduzieren" formuliert. Die ebenfalls unscharfen Resultate dieser Verknüpfung, z.B. „Brennstoff geringfügig reduzieren", werden in einem letzten Schritt wieder defuzzifiziert, also wieder in einen scharfen Wert umgewandelt, z.B. „40 kg/h Brennstoff abziehen". Durch die unscharfe Bewertung der Meßdaten werden Entscheidungen auf eine dem menschlichen Denken nahe Art gefällt.

3. Ergebnisse

Die mit dem Regelungssystem PYROEXPERT® ausgestattete Drehofenanlage ist mit einem 4stufigen Wärmetauscher und einem Rohrkühler ausgerüstet. Die Leistung der Anlage beträgt 3 000 t/d.

In den nachfolgenden Bildern werden die Vorteile einer rechnergestützten Betriebsweise der einer manuellen Fahrweise des Ofenbetriebes gegenübergestellt. Hierzu werden typische Situationen herausgegriffen, wobei der dargestellte Zeitraum von 20 Stunden aus Gründen der Übersichtlichkeit und zur Verdeutlichung der Regelungseffekte gewählt wurde.

Bilder 2 und **3** zeigen in der Gegenüberstellung den zeitlichen Verlauf des Ofenankerstromsignals und des relativen Klinkermassenstroms. Der relative Klinkermassenstrom ist dabei definiert als Verhältniszahl zwischen dem Klinkermassenstrom, abgeleitet aus der elektrischen Leistungsaufnahme des Klinkertransportes und dem theoretischen Massenstrom aus der Brenngutzufuhr.

Diese Darstellungsweise ermöglicht die unmittelbare Beurteilung der Massespeicherungs- und Freisetzungsvorgänge im Drehofen, die aus maschinen-, verschleiß- und verfahrenstechnischer Sicht von ausschlaggebender Bedeutung sind. Repräsentiert durch die Bilder 2 und 3, zeigt die Erfahrung, daß der PYROEXPERT®-Betrieb eine deutlich schonendere Fahrweise der Drehofenanlage ermöglicht, da sowohl die Häufigkeit der Ansatzfälle als auch deren Intensität wesentlich reduziert werden konnte. Daraus abgeleitet, liefert die rechnergestützte Ofenführung einen wichtigen Beitrag zur Erreichung der Zielsetzung, die Verfügbarkeit der Drehofenanlage weiter zu erhöhen.

Als weiterer positiver Effekt des PYROEXPERT®-Betriebes wurde eine wesentliche Stabilisierung der Brennzonenkonditionen festgestellt. Zur Veranschaulichung dieses Ergebnisses sind in den **Bildern 4** und **5** die jeweils typischen Verläufe der Meßsignale Sinterzonentemperatur, NO-Konzentration im Abgas und des Schallpegelsignals am Rohrkühlerauslauf dargestellt.

Die Messung des Schallpegels am Rohrkühlerauslauf ermöglicht bei entsprechender Aufbereitung dieses Signals

defuzzified, i.e. converted back to a definite value, e.g. "Deduct 40 kg/h of fuel". Indefinite evaluation of the measurement data arrives at decisions in a manner similar to human thought.

3. Results

The rotary kiln plant equipped with the PYROEXPERT® control system is fitted with a 4-stage preheater and a rotary cooler. The output of the plant is 3 000 t/d.

The following figures compare the advantages of computer-aided and manual operation of the kiln. Typical situations are used, the 20-hour period being selected to give an overall view and clarify the effects of the control.

Figs. 2 and **3** compare the timing of the kiln armature current signal and the relative clinker mass flow. The relative clinker mass flow is defined as the ratio of the clinker mass flow, derived from the electrical power consumption for the clinker transport, to the theoretical mass flow from the supply of kiln feed.

This type of representation allows direct assessment of the mass hold-up and release processes in the kiln, which are of prime importance where mechanical, wear and process technology are concerned. With reference to Figs. 2 and 3, experience shows that PYROEXPERT® enables the rotary kiln plant to be run far more smoothly, since coating falls have been considerably reduced in both frequency and intensity. Computer-aided operation derived from experience makes an important contribution towards achieving the goal of further increasing plant availability.

Another positive effect of operating with PYROEXPERT® is found to be that burning zone conditions are considerably stabilised. To illustrate this result **Figs. 4** and **5** show typical shapes of the measuring signals for sintering zone temperature and NO concentration in the waste gas, and the sound level signal at the rotary cooler outlet. The degree to which the clinker has been burnt can be continuously assessed by measuring the sound level at the cooler outlet, with appropriate treatment of the signal. Referring to Figs. 4 and 5, computer-aided kiln operation can be shown to have the following advantages in a continuous process:

— Automatic operation guarantees optimum development of burning zone conditions by avoiding quality-reducing temperature drops in the sintering zone, and thus leads to considerable improvement in product quality.

— Avoidance of energy flow peaks in the burning zone and over-burning of the clinker keeps NO emissions to a low level while avoiding marked peak values.

— Stabilisation of the burning process and the consequent running of the rotary kiln plant within an economically

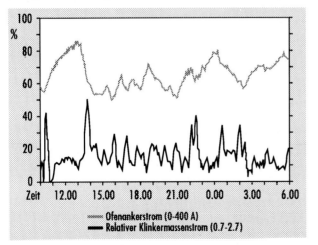

BILD 2: Zeitlicher Verlauf des Ofenankerstromsignals und des relativen Klinkermassenstroms für den manuellen Ofenbetrieb
FIGURE 2: Time behaviour of kiln armature current signal and of relative clinker mass flow for manual kiln operation

Ofenankerstrom = kiln armature current
Relativer Klinkermassenstrom = relative clinker mass flow

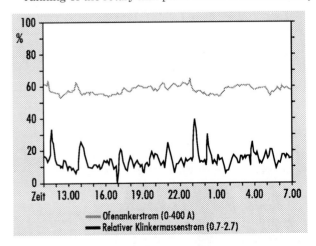

BILD 3: Zeitlicher Verlauf des Ofenankerstromsignals und des relativen Klinkermassenstroms für den PYROEXPERT®-Betrieb
FIGURE 3: Time behaviour of kiln armature current signal and of relative clinker mass flow for PYROEXPERT® operation

Ofenankerstrom = kiln armature current
Relativer Klinkermassenstrom = relative clinker mass flow

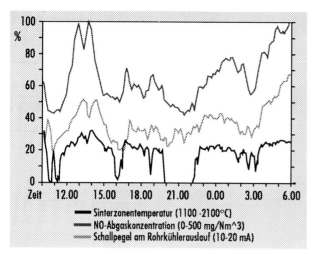

BILD 4: Zeitlicher Verlauf der Sinterzonentemperatur, der NO-Abgaskonzentration und des Schallpegels am Rohrkühlerauslauf für den manuellen Ofenbetrieb
FIGURE 4: Time behaviour of sintering zone temperature, NO waste gas concentration and sound level at rotary cooler outlet for manual kiln operation

Sinterzonentemperatur	= sintering zone temperature
NO-Abgaskonzentration	= NO waste gas concentration
Schallpegel am Rohrkühlerauslauf	= Sound level at kiln outlet

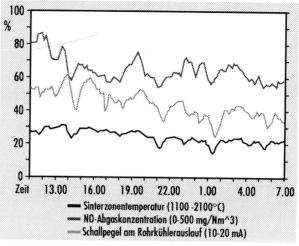

BILD 5: Zeitlicher Verlauf der Sinterzonentemperatur, der NO-Abgaskonzentration und des Schallpegels am Rohrkühlerauslauf für den PYROEXPERT®-Betrieb
FIGURE 5: Time behaviour of sintering zone temperature, NO waste gas concentration and sound level at rotary cooler outlet for PYRO-EXPERT® operation

Sinterzonentemperatur	= sintering zone temperature
NO-Abgaskonzentration	= NO waste gas concentration
Schallpegel am Rohrkühlerauslauf	= Sound level at kiln outlet

eine kontinuierliche Beurteilung des Klinkerbrenngrades. Unter Bezugnahme auf die Bilder 4 und 5 konnten im Dauerbetrieb folgende Vorteile der rechnergestützten Ofenfahrweise nachgewiesen werden:

— Der Automatikbetrieb gewährleistet einen optimalen Verlauf der Brennzonenkonditionen durch Vermeidung qualitätsmindernder Temperatureinbrüche in der Sinterzone und führt damit zu einer erheblichen Verbesserung der Produktqualität.

— Durch die Vermeidung von Energiestromspitzen in der Brennzone bzw. dem Überbrennen des Klinkers liegen die NO-Emissionen bei gleichzeitiger Vermeidung ausgeprägter Spitzenwerte auf einem niedrigen Niveau.

— Die Vergleichmäßigung des Brennprozesses und der daraus resultierende Betrieb der Anlage in einem ökonomisch günstigen Arbeitsbereich führte dazu, daß der spezifische Brennstoffenergiebedarf um gut 2 % und der elektrische Energiebedarf der Drehofenanlage um ca. 3 % gesenkt werden konnte. Nicht eingerechnet sind dabei die Einsparungen infolge höherer Verfügbarkeit der Ofenanlage.

— Die beobachtete Stabilisierung der Klinkergranulometrie führte zu einer deutlichen Verbesserung des Rohrkühlerbetriebes und dessen thermischen Wirkungsgrades durch eine Reduzierung der Klinkerstaubkreisläufe.

Gleichzeitig haben die Erfahrungen gezeigt, daß das Regelungssystem aufgrund seines transparenten Regelwerkes und seiner plausiblen, weichen Stelleingriffe eine hohe Akzeptanz beim Leitstandspersonals erfährt, die in der Einschaltquote des Reglers von über 95 % zum Ausdruck kommt. Die sehr komfortable Benutzeroberfläche des Systems erlaubt eine einfache Parametrierung des Reglers. Das im Expertensystem abgebildete Fuzzy-Regelwerk kann auf der Basis einer grafischen Bedieneroberfläche auf sehr einfache Weise modifiziert und ergänzt werden.

4. Ausblick

In einem weiteren Optimierungsschritt ist die Integration eines Brennstoff-Energiestromreglers in das PYROEXPERT®-System geplant. Die zur Messung dieses Energiestroms notwendige Sensorik befindet sich zur Zeit in der Erprobung.

Um bei Änderungen des zeitlichen Verhaltens der Regelstrecke optimale Regelergebnisse zu erzielen, soll die Lernfähigkeit „Künstlicher Neuronaler Netze" genutzt werden, um das Regelwerk automatisch anzupassen. Das Netz „lernt" dazu in einem Trainingsmodus, die Prozeßsituation an Hand von Beispielen einzustufen.

favourable operating range has led to a lowering of the specific fuel energy requirement of the plant by a good 2 % and of the electrical energy requirement by about 3 %. This does not include savings resulting from greater availability of the plant.

— The stabilisation of clinker granulometry which has been observed has led to a clear improvement in rotary cooler operation and in its thermal efficiency through a reduction in the recirculation of clinker dust.

At the same time, experience has shown that the control system, with its transparent control mechanism and its smooth, self-explanatory intervention, is well accepted by the control room staff, as evidenced by the controller's switch-on rate of over 95 %. The system's very convenient user interface makes it easy to parameterise the control. The fuzzy control mechanism shown in the expert system can very easily be modified and enhanced on the basis of a graphic operator interface.

4. Prospects

As a further optimising step it is planned to incorporate a fuel energy flow control in the PYROEXPERT® system. The sensory analysis required to measure the energy flow is at present undergoing trials.

As a means of obtaining optimum control results when there are changes in the time characteristics of the controlled member, "artificial neuronal network" machine learning is to be used to adapt the control system automatically. In a learning mode, the network "learns" how to categorize the process situation from examples.

Literaturverzeichnis

[1] Edelkott, D., und Scheuer, A.: Messung des Sekundärluft-Energiestromes zur Steuerung des Klinkerbrennprozesses. Vortrag zur Fachtagung „Zement Verfahrenstechnik" — 16. Fortsetzung des VDZ-Ausschusses — Verfahrenstechnik, Düsseldorf (27. Februar 1992).

[2] Schmidt, K.D., Scheuer, A., und Gardeik, H.O.: Signalgrößen zur Steuerung und Regelung des Klinkerbrennprozesses und der Klinkerqualität. Zement-Kalk-Gips 41 (1988), Nr. 3, S. 105–112.

[3] Bauer, C., Jäger, G., Kaufmann, M., Patzer, J., und Walen, K.H.: Optimierungssystem PYROEXPERT sichert Wirtschaftlichkeit der Klinkerproduktion. Zement-Kalk-Gips 46 (1993), Nr. 4, S. 182–192.

Visualisierung von Klassieranlagen im Mörtel- und Baustoffbereich

Visual display of classifying plants in the mortar and building materials sector

Visualisation d'installations de classement dans le secteur des mortiers et matériaux du bâtiment

La visualización de las instalaciones de clasificación en el campo de los morteros y materiales para la construcción

Von **W. Schlüpmann,** Oelde/Deutschland

Zusammenfassung – Die Verfügbarkeit von Maschinen und Anlagen sowie die Qualität der Produkte läßt sich durch eine kontinuierliche Überwachung während des Betriebes und eine sichere Bedienung erheblich verbessern. Hierzu werden in zunehmendem Maße auf allen Bedienungs- und Beobachtungsebenen Visualisierungssysteme eingesetzt, von einfachen Textdisplays bis zu kompletten PC-Einheiten. Am Beispiel eines Zementterminals wird deutlich gemacht, welche Vorteile ein Visualisierungssystem für den Betrieb dieser Anlage mit sich bringt. Die Prozeßdaten werden dem Bedienungspersonal in anschaulicher Weise in Form von Bildern und Tabellen auf dem Prozeßmonitor angezeigt. Durch Einblenden einer Störmeldezeile wird auf die Art, die Ursache sowie auf den Ort der Störung hingewiesen. Die Prozeßdaten und die Störmeldungen werden auf der Festplatte archiviert und können über einen Drucker als Schicht- oder Monatsprotokoll ausgedruckt werden. Somit wird durch ein Visualisierungssystem ein direkter und schneller Zugriff auf alle Systemkomponenten in allen Verfahrensebenen möglich.

Summary – The availability of machines and systems and the quality of the products can be greatly improved by continuous monitoring during operation and by more reliable control. Visual display systems are being used increasingly for this purpose on all operating and monitoring levels, from simple text displays to complete PC units. The advantages associated with a visual display system for operating a plant are illustrated clearly using the example of a cement terminal. The process data are shown descriptively to the operating personnel on the process monitor in the form of diagrams and tables. The type, cause and location of a fault are indicated by superimposing a fault message line. The process data and the fault messages are stored on the hard disc and can be printed out as shift or monthly logs. A visual display system therefore facilitates direct and rapid access to all system components at every process level.

Résumé – La disponibilité de machines et d'installations peut être nettement améliorée par une surveillance en continu pendant la marche et une conduite sûre. Pour ce faire sont, de plus en plus, mis en oeuvre des systèmes de visualisation, à tous les niveaux de conduite et d'observation, allant de simples impressions de texte jusqu'aux unités complètes de PC. Avec l'exemple d'un terminal à ciment sont expliqués les avantages apportés au fonctionnement de cette installation, par un système de visualisation. Les données du process sont présentées au personnel de conduite sur l'écran du process, de manière claire sous forme d'images et de tableaux. Par l'apparition d'une ligne de message de perturbation, sont indiqués la nature, la cause et le lieu de la perturbation. Les données du process et les messages d'alerte sont archivés sur un disque dur et peuvent être repris comme protocole de poste ou du mois au moyen d'une imprimante. Le système de visualisation rend ainsi possible une intervention directe et rapide sur tous les composants du système, à tous les niveaux du process.

Resumen – La disponibilidad de las máquinas e instalaciones así como la calidad de los productos se puede mejorar considerablemente mediante una vigilancia continua durante el servicio y una manipulación segura. Para ello, se emplean cada vez más, y a todos los niveles de servicio y de observación, los sistemas de visualización, desde los simples displays de textos hasta las unidades completas de PC. Citando el ejemplo de una terminal para cemento, el autor del presente artículo nos explica cuáles son las ventajas de un sistema de visualización para el funcionamiento de esta instalación. Los datos del proceso se le muestran al personal de servicio de manera muy clara, en forma de imágenes y tablas que aparecen en la pantalla. Mediante la aparición de una línea de aviso de perturbaciones, se señala el tipo, causa y localización de la avería. Los datos del proceso así como los avisos de perturbación se archivan en el disco duro, pudiéndose establecer con ellos un protocolo de turno o mensual, con ayuda de una impresora. De esta forma, el sistema de visualización permite obtener un acceso directo y rápido a todos los componentes del sistema, a todos los niveles del proceso.

Die Verfügbarkeit von Maschinen und Anlagen sowie die Qualität der Produkte läßt sich durch eine kontinuierliche Überwachung während des Betriebes, rechtzeitiges Erkennen von Störungen, Diagnosefunktionen und sichere Bedienung erheblich verbessern. Hierzu werden in zunehmendem Maße auf allen Bedienungs- und Beobachtungsebenen Visualisierungssysteme eingesetzt, die von einfachen Textdisplays bis zu kompletten PC-Systemen reichen.

Am Beispiel eines Zementterminals in der Türkei wird deutlich, welch hoher Automatisierungsgrad erreicht werden kann, wenn konsequent Anlagenausrüstung und Steuerungstechnik aufeinander abgestimmt werden.

Das neue Zementterminal besteht aus vier Silos. Unter jedem Silo ist eine Losebeladung installiert. Im Packereigebäude sind zwei Rotopackeranlagen aufgebaut. Luftrinnen sind so angeordnet, daß von jedem Silo jede Packanlage über Becherwerke beschickt werden kann. Der Zement wird in 50-kg-Säcke gefüllt, die von Aufsteckern auf die Rotopacker geschossen werden. Die Verladung der Säcke auf die LKWs erfolgt z. Z. durch zwei automatische Verlademaschinen (Autopacs).

Die Datenübertragung wurde so ausgeführt, daß sie durchgängig von den Maschinensteuerungen über die Anlagensteuerungen bis zum Visualisierungssystem im Leitstand reicht. Jede Maschine sowie die Anlagenkomponenten für den Siloabzug, Material- und Sacktransport werden durch eigene SIMATIC S5-115U gesteuert. Über den Siemens H1-Bus wird der Datenverkehr zwischen den Anlagensteuerungen und dem Visualisierungssystem abgewickelt. Über einen Kommunikationsprozessor, einen PC und einen Kartenleser ist das Visualisierungssystem mit der Versandautomatisierung verbunden. Entscheidend ist, daß die Veradedaten für die Autopacs, die auf Magnetkarten gespeichert sind, in den an den Verladestellen installierten Kartenlesern eingelesen und über den zusätzlichen Siemens L1-Bus an die einzelnen Autopacs übertragen werden. Die Autopacs werden über ein Kamerasystem von dem Bediener im Leitstand auf die LKWs positioniert und die Verladung gestartet. Im Leitstand wurde zusätzlich zum Visualisierungssystem ein Leuchtschaltbild mit Gruppenanzeigen installiert. Die Visualisierung besteht im wesentlichen aus IBM-Komponenten in Industrieausführung, wie z. B. einem PC-386 mit 4-MB-RAM und 80-MB-Festplatte zur Datenarchivierung. Dieser ist mit einer Schnittstellenkarte zur Busankopplung, einer Prozeßbedientastatur sowie einem Protokolldrucker ausgestattet. Als Visualisierungssoftware wurde das „Control View" von Allen Bradley auf dem Rechner installiert. Die Software besteht aus einzelnen Softwaremodulen, so daß eine optimale Anpassung an die jeweilige Automatisierungsaufgabe gewährleistet ist. Das vollgraphische System vermindert den Projektierungsaufwand bei der Erstellung der Anlagenfließbilder und Prozeßvariablen sowie bei späteren Anpassungen und Erweiterungen. Die Bild- und Prozeßbedienung wird über eine Prozeßbedientastatur realisiert. Im einzelnen sind als Bedienmöglichkeiten die Fließbildanwahl über die Prozeßbedientastatur, die manuelle Aufschaltung des Störmeldebildes durch Betätigen einer Funktionstaste, die Vorgabe der Sackzahl und des Lagenmusters für die Autopacs sowie die Vorwahl, aus welchem Silo abgezogen wird und zu welcher Packanlage der Zement transportiert werden soll, gegeben.

Das gesamte Terminal wird durch Fließbilder, in denen alle wichtigen Prozeßinformationen mit aktuellem Zustand enthalten sind, dargestellt. Die Bilder sind hierarchisch aufgebaut, so daß der Bediener über Funktionstasten von der Anlagenübersicht (**Bild 1**) bis zum Detailbild (**Bild 2**) blättern kann. Im einzelnen sind in den Bildern die Betriebszustände von Maschinen- und Anlagenkomponenten, die durch Farbumschlag deutlich gemacht werden, sowie die Füllstände der Ringsilos dargestellt. Durch Betätigen einer Funktionstaste wird das Störmeldebild aufgeschaltet. Die einzelnen Störungen werden mit Uhrzeit, Melde-Nummer, Klartext und Zustand angezeigt. Weiterhin wird farblich dargestellt, wann die Meldung gekommen ist und gegangen ist.

The availability of machines and plants and the quality of the products can be greatly improved by continuous monitoring during operation, quick observation of trouble, diagnostic functions and reliable operation. Visual display systems ranging from simple text displays to complete PC systems are being used increasingly at all operating and observation levels.

The high degree of automation which can be achieved by consistently coordinating the plant equipment with the control technology is demonstrated by a cement terminal in Turkey.

The new cement terminal consists of four silos, each with a bulk loading unit installed below it. Two rotary packer plants have been set up in the packing plant building. Pneumatic trough conveyors are arranged so that each packing plant can be supplied by bucket elevators from each silo. The cement is filled into 50 kg bags which are positioned on the rotary packers by bag applicators. The bags are at present being put onto lorries by two automatic loading machines (Autopacs).

The data transmission system is designed so that it extends universally from the machine control system via the plant control system to the visual display unit in the control room. Each machine and each plant component for silo discharge, material transport and bag transport is controlled by its own SIMATIC S5-115U. Data traffic between the plant control system and the visual display system is handled by the Siemens H1 bus. The visual display system is connected to the dispatch automation system by a communications processor, a PC and a card reader. An important feature is that the loading data for the Autopacs, which are stored on magnetic cards, are read into the card readers installed at the loading points and transmitted to the individual Autopacs by the additional Siemens L1 bus. The Autopacs are positioned on the lorries and the loading process started by the operator in the control room by means of a camera system. An illuminated circuit diagram with group displays is installed in the control room in addition to the visual display system. The visual display system essentially consists of IBM industrial components, e.g. a PC 386 with 4 MB RAM and an 80 MB hard disc for archiving the data. The PC is equipped with an interface card for connection to the bus, a keyboard for operating the process and a log printer. The visual display software installed in the computer is Allen Bradley's "Control View". It consists of individual modules to ensure optimum adaptation to each particular automation job. The fully graphic system reduces project expenditure on the creation of plant flow sheets and process variables and on subsequent adaptations and extensions. The display and the process are operated with a keyboard. The specific operations which can be carried out are flow sheet selection by keyboard, manual superimposition of the fault message display by pressing a function key, stipulation of the number of bags and the layer pattern for the Autopacs, and preselection of the silo from which the material is to be taken and the packing plant to which the cement is to be transported.

The whole terminal is represented by flow charts containing all the important processing information with the current status. The displays are in a hierarchic arrangement so that the operator can use function keys to page through from a general view of the plant (**Fig. 1**) to the required detail (**Fig. 2**). They show the operating state of machine and plant components, emphasised by a change of colour, and the filling levels of the ring silos. The fault message display can be superimposed by pressing a function key. Individual faults are displayed with the time, message number, some plaintext and their status. Colours also indicate when the message came and went.

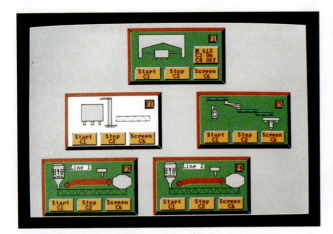

BILD 1: Anlagenübersicht
FIGURE 1: General view of plant

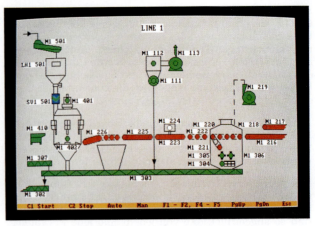

BILD 2: Packanlageübersicht
FIGURE 2: General view of packing plant

Die Störmeldedaten werden auf der Festplatte gespeichert und können später als Schichtprotokoll auf dem vorhandenen Drucker ausgedruckt werden. Zusätzlich können Produktionsdaten, wie z.B. Betriebsstunden einzelner Anlagenteile, Sackzahl usw., als Schichtprotokoll oder Monatsprotokoll auf der Festplatte gespeichert und auf dem Drukker ausgedruckt werden. Auf einer speziell eingerichteten Bildschirmseite werden Transportwege angezeigt. Der Bediener kann über Funktionstasten den Transportweg auswählen und freigeben. Das gleiche Verfahren kann auch für eine Sortenvorwahl eingesetzt werden. Ein Sortenwechsel ist daher in kürzester Zeit durchführbar. Auf einer zusätzlichen Bildschirmseite werden Datenfelder für die Beschickung der Autopacs angezeigt. Über die Prozeßbedientastatur können in die Datenfelder Sackzahl und Lagenmuster, mit dem die Autopacs die LKWs beladen sollen, eingegeben werden.

Das Visualisierungssystem im Leitstand gewährt dem Bedienpersonal jederzeit einen umfassenden Einblick in den Zustand der Anlage und in den Prozeß. Der Prozeß und die Anlage werden soweit wie möglich im ungestörten Betrieb automatisch innerhalb der festgelegten Grenzen und Toleranzen gefahren. Bei Störungen wird eine schnelle und eindeutige Meldung, Fehlererkennung und Fehlerortung angezeigt und im Gefahrenfall das Stillsetzen der Anlage bzw. des betroffenen Anlagenteils gewährleistet.

The fault message data are stored on hard disc and may be printed out later as a shift log. Production data, such as hours of operation in various parts of the plant, numbers of bags etc, may also be stored on hard disc as shift or monthly logs and printed out. Conveying routes are displayed on a specially created screen page. The operator can select and release a route by means of function keys, and the same procedure may be used for selecting different product types. A change of type can therefore be carried out in a very short time. Data fields for feeding the Autopacs are displayed on an additional screen page. The number of bags and the layer pattern with which the Autopacs to load the lorries are entered in the data fields through the process operating keyboard.

The visual display system in the control room gives the operators a comprehensive view of the state of the plant and the process at all times. As far as possible the plant and process are run automatically within the set limits and tolerances during undisturbed operation. If malfunctions occur a quick, clear message with fault recognition and fault location is displayed, and the plant or affected plant section is shut down if there is any danger.

Ein zukunftssicheres Management-Informationssystem für Zementwerke

A management information system with a secure future for cement works

Un système de gestion de l'information sûr dans l'avenir pour les cimenteries

Un sistema de gestión de la información que ofrece un futuro seguro a las fábricas de cemento

Von **W. Sedlmeir,** Erlangen/Deutschland

Zusammenfassung – *In den meisten Zementwerken ist ein Informationssystem nur fragmentarisch vorhanden, z. B. als Protokollierfunktion am Leitstandssystem oder Laborrechner. Diese „Insellösungen" sind entwicklungstechnische „Sackgassen", denn sie entziehen sich einer werksweiten Integration und verbauen somit eine ganzheitliche Sicht auf die relationalen Prozeß- und Produktionsdaten. Rationalisierung und Optimierung des Prozesses erfordern aber die Auswertung aller relevanten Daten und Informationen. Ein ganzheitliches, umfassendes Informationssystem kann außerdem in Zukunft für den Nachweis der Qualität bestimmter Produktions-Chargen zweckmäßig sein. Das CEMAT-MIS Management-Informationssystem ermöglicht eine lückenlose Erfassung und Archivierung aller Prozeß-, Produktions- und Anlagendaten. Ein Datenprozessor-PC empfängt die Daten von den S5-Geräten der untersten Ebene, bereitet sie auf und speichert sie in einen zentralen Fileserver. Jeder Benutzer hat Zugriff auf die aktuellen und historischen Daten und kann alle Daten und Informationen auf seinem PC analysieren. Die PCs können beliebig im Werk verteilt sein. Für die Datenanalyse und die Protokollgestaltung werden Standard- PC-Programme verwendet. So kann jeder Benutzer seine Protokolle und Grafiken leicht selbst erstellen. CEMAT-MIS läuft auf Standard-PCs (386 oder höher) und benutzt MS-Windows und NOVELL-Netware. Praktische Erfahrungen aus mehreren Werken liegen vor.*

Ein zukunftssicheres Management-Informationssystem für Zementwerke

Summary – *In the majority of cement works an information system only exists on a fragmentary basis, e.g. as a logging function in the control room system or laboratory computer. These "island solutions" are "dead ends" as far as development technology is concerned as they elude any works-wide integration and therefore obstruct a complete view of the relational process and production data. Rationalization and optimization of the process requires evaluation of all relevant data and information. Apart from this, a complete comprehensive information system may be appropriate in the future for proving the quality of certain production batches. The CEMAT-MIS management information system facilitates uninterrupted acquisition and filing of all process, production, and plant data. A data processor PC receives the data from the lowest level S5 machines, processes it, and files it in a central file server. Every user has access to the current and historical data and can analyze all data and information on his PC. The PCs can be distributed anywhere in the works. Standard PC programs are used for the data analysis and log layout so it is easy for users to draw up their logs and graphics themselves. CEMAT-MIS runs on standard PCs (386 or higher) and uses MS-Windows and NOVELL netware. Practical experience has been obtained from several works.*

A management information system with a secure future for cement works

Résumé – *Dans la plupart des cimenteries, un système de gestion de l'information n'existe que sous forme fragmentaire, p. ex. comme fonction de récapitulation au système de la salle de conduite ou au niveau de l'ordinateur du laboratoire. Ces solutions „insulaires" constituent, pour le développement technique, des „voies sans issue", parce qu'elles se soustraient à une intégration à l'ensemble de l'usine et obstruent ainsi une vue globale sur les données relationnelles du process et de la production. La rationalisation et l'optimisation du process exigent pourtant l'exploitation de toutes les données et informations relevantes. De plus, un système d'information complet, exhaustif peut être opportun dans l'avenir pour la preuve de certaines charges de production. Le système de gestion de l'information CEMAT-MIS rend possible une saisie et un archivage sans faille, de toutes les données du process, de la production et de l'installation. Un PC à processeur de données reçoit toutes les données des équipements S5 du niveau le plus bas, les traite et les stocke dans un serveur central de fichiers. Chaque utilisateur a accès aux données actuelles ou antérieures. Il peut analyser toutes les données et informations sur son PC propre. Les PC peuvent être disséminés à volonté dans l'usine. Pour l'analyse des données et la configuration des protocoles sont utilisés des programmes de PC standards. Ainsi, chaque utilisateur peut configurer lui-même aisément ses protocoles et graphiques. CEMAT-MIS tourne sur PC standard (386 ou plus) et utilise MS-Windows et NOVELL-Netware. L'expérience pratique faite dans plusieurs usines est disponible.*

Un système de gestion de l'information sûr dans l'avenir pour les cimenteries

Resumen – En la mayoría de las fábricas de cemento un sistema de gestión de la información sólo existe de forma muy incompleta, por ejemplo como elemento para establecer protocolos en el centro de control o a nivel del ordenador de laboratorio. Estas soluciones „aisladas" constituyen un „callejón sin salida", desde el punto de vista del desarrollo tecnológico, ya que se oponen a una integración en toda la fábrica, impidiendo de esta forma una visión global de los datos del proceso tecnológico y de la producción. Sin embargo, la racionalización y la optimización del proceso requieren la evaluación de todos los datos e informaciones relevantes. Además, un sistema de gestión de la información, integral y completo, puede ser útil en el futuro para comprobar la calidad de determinados lotes del producto. El sistema de gestión de la información CEMAT-MIS permite el registro completo y el archivo de los datos del proceso, de la producción y de la instalación. Un PC con procesador de datos recibe estos últimos de los aparatos S5 del nivel más bajo, y tras su tratamiento los almacena en un „fileserver" central. Todo usuario tiene acceso a los datos, tanto actuales como antiguos, pudiendo analizar todos ellos mediante su ordenador. Los PC pueden distribuirse a voluntad por toda la fábrica. Para el análisis de datos y la configuración de los protocolos se utilizan programas standard para PC. Así, los usuarios pueden establecer fácilmente sus protocolos y gráficos por sí mismos. CEMAT-MIS funciona con PC standard (tipo 386 o más alto) y utiliza MS-Windows y NOVELL-Netware. Se cuenta con experiencias prácticas adquiridas en varias fábricas.

Un sistema de gestión de la información que ofrece un futuro seguro a las fábricas de cemento

1. Anforderungen

Ein Management-Informationssystem ist dann zukunftssicher, wenn es folgende Forderungen erfüllt: Es muß modular ausgebaut und innoviert werden können, es muß offen sein für funktionelle Erweiterungen, und es muß vom Benutzer selbst, dessen individuellen Bedürfnissen entsprechend, leicht angepaßt werden können. Es muß also flexibel sein.

Ein Management-Informationssystem soll eine umfassende Auswertung aller Daten und Informationen aus Prozeß, Produktion und Anlage ermöglichen. In **Tabelle 1** ist aufgeführt, welche Daten und Informationen ein Informationssystem erfassen, verwalten und auswerten muß. Daten, die nicht elektrisch erfaßt werden können, werden über Eingabemasken eingegeben und wie automatisch erfaßte Daten archiviert, weiterverarbeitet und ausgewertet. Auch solche Eingabemasken können vom Betreiber selbst erstellt werden und über eine Schnittstelle zwischen CEMAT-MIS und einer fremden Datenbank ausgetauscht werden.

Für die Betriebsberichte braucht man Verbrauchs- und Produktionsmengen, Stromverbräuche, Silofüllstände und Materialbewegungen, Meßwerte und die Laufzeiten, eventuell auch die Reparaturzeiten der Aggregate. Warnmeldungen, Alarmmeldungen und alle Schalthandlungen des Produktionssteuerers werden für die Betriebs-Ablaufprotokolle benötigt. Auch Sollwertverstellungen, Grenzwertänderungen sowie Silo- und Wegeumschaltungen müssen lückenlos erfaßt werden. Für Qualitätsüberwachung und Produktionsnachweis müssen Labordaten, Vorgabesollwerte, Silozuordnungen, aber auch z. B. die eingestellten Grenzwerte, in das Informationssystem eingespeist werden.

Die Aufgaben eines Informationssystems kann man kurzgefaßt so formulieren:

> Es muß Daten und Informationen aus allen Anlagenteilen und Systemen einheitlich erfassen und archivieren sowie gemeinsam jedem Entscheidungsträger auf seinem Schreibtisch-PC für individuelle Auswertungen zur Verfügung stellen.

2. „Individuelle" Auswertungen

Daten und Informationen müssen zueinander in Beziehung gesetzt werden. Sie müssen gemeinsam analysiert werden können, damit sie eine Aussagekraft bekommen. So sagt z. B. der spezifische Energieverbrauch einer Produktionsanlage in kJ/kg nichts aus, wenn man nicht die Labordaten für Rohmehl und Klinker, die Produktionsmenge, die Abgasverwertung, aber auch Produktionsunterbrechungen und Prozeßstörungen mit berücksichtigt. Ein Informationssystem muß also Daten und Informationen individuell selektieren und miteinander verknüpfen können. Auch mathematische Berechnungen oder statistische Auswertungen müssen frei wählbar sein, damit die Daten zu aussagekräftigen Ergebnissen und vergleichbaren Kennwerten verdich-

1. Requirements

To have a secure future a management information system (MIS) must satisfy the following requirements: it must be of modular construction and capable of being updated, it must be capable of functional extension, and it must be easy for the user himself to adapt to meet his needs. It must therefore be flexible.

A management information system must enable comprehensive analysis of all the data and information from the process, from production and from the plant itself. **Table 1** lists the data and information which an information system has to capture, manage and analyse. Data which cannot be captured electronically are input via input masks and archived, processed and analysed in the same way as data captured automatically. The operator can produce his own input masks, and exchange them between CEMAT-MIS and an outside database via an interface.

The information needed to produce management reports includes consumption and production figures, power consumption, level of material in the silo and material movements, readings and running times, and possibly also plant downtime.

Warning and alarm signals and all switching operations by the production controller are needed for the process operating reports. It is also essential to capture all set-point adjustments, limit-value changes and all silo and path switching. Laboratory data, predetermined setpoint values, silo allocation and also such data as limit values must be fed into the information system for quality control and production monitoring.

The functions of an information system can be briefly summarised as follows:

> It must capture and file data and information from all plant sections and systems in an integrated manner, and present all decision-makers jointly with this information on their desktop PCs for individual analysis.

2. "Individual" analysis

Data and information have to be related to each other. It must be possible to analyse them jointly so that they can be meaningfully interpreted. For example, the specific energy consumption of a production plant in kJ/kg is only meaningful when considered together with laboratory data for raw meal and clinker, production quantity, exhaust gas utilization, and also production stoppages and process malfunctions. An information system must therefore be capable of selecting individual items of data and information and interrelating them. Mathematical calculations and statistical analyses must also be freely selectable, so that meaningful results and comparable characteristic values can be abstracted from the data. It must also be possible to present the results as a report, a graphic or a list, as appropriate. The kiln operator needs detailed information, e.g. tables of meas-

TABLE 1: Summary of the various types of data and information to be captured by an information system.

Measured values
Levels in silos
Quantities consumed
Production quantities
Power consumption
Material movements

Running times
Repair times

Alarm messages
Control room
switching operations Changes to: setpoint values, limit values

Predetermined setpoint
values Silo allocations

Laboratory data
Emission data

---

ured values over 24 hours. The production manager only needs a summary of the key production figures, and the Works Manager may need only a graphic display of the most important characteristic values. **Fig. 1** shows an example of this for one kiln.

CEMAT-MIS condenses data to provide meaningful information. The network display, for example, enables a rapid assessment to be made of how the current kiln production characteristic values compare with the nominal values. The kiln actual values can be compared with various benchmark values, e. g. to the guaranteed characteristic values, the best figures achieved, or the last month's averages.

3. Design for a Management Information System with a secure future

An information system is an infrastructure system and must, like an energy supply system, be carefully planned for the long term, and systematically implemented. It is also important that standard hardware and software products are used, and that an open communication network is selected so that external systems can also feed their data into the information system. Data interfaces and data definitions must be standardised so that data from all systems can be uniformly managed and analysed.

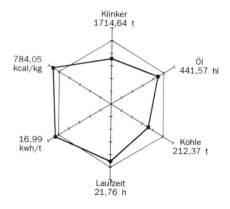

FIGURE 1: Network diagram of variances between current actual production figures and nominal values.

Kennwerte Ofen 1, Woche 12 / Prod. C2 = Characteristic values for kiln 1 week 12 / prod. C2
Klinker = clinker
Öl = oil
Kohle = coal
Laufzeit = running time

TABELLE 2: Speicherkapazitäten und Ausbauwerte des CEMAT-MIS

Meßwerte	4 000
Mengenwerte	10 000
Laufzeiten	2 000
Alarm-/Betriebsmeldungen	100 000
Wartungsmeldungen	10 000
Archivkapazität	1 Jahr
Benutzer-PCs	10–50

TABLE 2: CEMAT-MIS memory capacities and expansion values

Measured values	4 000
Quantity values	10 000
Running times	2 000
Alarm/operating messages	100 000
Maintenance messages	10 000
Archive capacity 1 year	1 Jahr
User PCs	10–50

4. Struktur des CEMAT-MIS

Ein CEMAT-MIS-System besteht aus vernetzten PCs, die koordiniert ihre Aufgaben parallel abwickeln. Mit diesem Multiprocessing-Konzept können mehr Daten in kürzerer Zeit bearbeitet werden als mit einem Prozeß-Rechner, in dem alle Funktionen zentral von nur einer CPU bearbeitet werden müssen. In **Tabelle 2** sind die wichtigsten Ausbaudaten des CEMAT-MIS-Systems aufgeführt.

CEMAT-MIS bietet genügend Speicherplatz (im GigaByte-Bereich), der auch für sehr große Zementwerke ausreicht. Die Datenbanken des CEMAT-MIS erweitern sich selbst ohne Umprogrammierung und Rekonfiguration und ohne Betriebsunterbrechung, wenn eine neue Meßstelle oder ein zusätzlicher Zähleingang angeschlossen werden. Auch neue Benutzer-PCs können während des Betriebs an das Netz angeschlossen werden.

Im **Bild 2** ist der prinzipielle Aufbau des Systems dargestellt. CEMAT-MIS basiert auf einem PC LAN. Über den SINEC H1 Werksbus (Ethernet Bus nach IEEE 802.3) empfängt das CEMAT-MIS zyklisch alle Daten von den Simatic-S5-Geräten auf der Prozeßebene. So brauchen keine Signale doppelt verdrahtet zu werden. Alle Benutzer haben Zugriff auf die gleichen Daten im Fileserver, auch gleichzeitig. Die Zahl der Benutzer-PCs ist nur durch die NOVELL-Lizenz begrenzt.

Alle Daten werden auf der untersten Ebene im Prozeßleitsystem von den Simatic-S5-Geräten erfaßt, überwacht und auf Plausibilität überprüft. Nur geprüfte Daten werden über den SINEC H1 Werksbus an das Informationssystem und parallel dazu an das Leitstandssystem geschickt. Es gibt also keine mehrfache Signalverkabelung und Signalerfassung sowie keine unterschiedlichen Anzeigen im Leitstand und im Informationssystem.

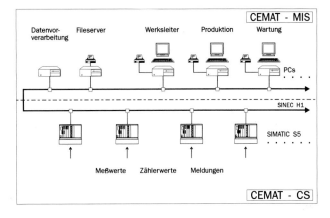

BILD 2: Bus-Struktur des CEMAT-MIS-Systems
FIGURE 2: The CEMAT-MIS system bus structure

Datenvorverarbeitung	= data preprocessing
Fileserver	= file server
Werksleiter	= Works Manager
Produktion	= production
Wartung	= maintenance
Meßwerte	= measured values
Zählerwerte	= counter inputs
Meldungen	= messages

Systematic processing of raw data is crucial for consistency of data and of results, and thus ultimately for the quality of an MIS. This is achieved by using "plausibility checks", and will be illustrated using the example of the treatment of pulse signals for quantity values. For example, resources consumed during no-load operation and during repairs must be counted separately, and must not be allocated to production. It must also be possible at the raw data stage to allocate consumption and production quantities separately to products, product changeover phases and tariff rate periods. It is also necessary to ensure that input pulses are not converted several times in the system, otherwise rounding errors and inconsistent results will arise. On the other hand, pulses from the weigher during calibration must be suppressed completely. Quantity values that are missing or incorrect must be detected and identified, and this information permanently stored with the data. It must also be possible to input information and correct the results manually. Manual inputs must be identified as such and stored in the database.

Plausibility checks of this type are also necessary for measured values and for messages. It is for example necessary to ensure that the process control system produces only one fault message in the event of a plant malfunction, and not two or more.

4. Structure of the CEMAT-MIS

A CEMAT-MIS system consists of networked PCs carrying out their function in parallel as a coordinated system. This multi-processing system enables more data to be processed in less time than with a process control computer where all the functions have to be processed centrally by just one CPU. The key system parameters of the CEMAT-MIS system are listed in **Table 2**.

The CEMAT-MIS offers sufficient memory (measured in gigabytes) to accommodate even very large cement works. The CEMAT-MIS databases are self-expanding without reprogramming or reconfiguration; new measuring points or additional counter inputs can be connected without interrupting the process. New PC users can also be added to the network while the system is running.

Fig. 2 shows the basic structure of the system. The CEMAT-MIS is based on a PC LAN. The CEMAT-MIS receives all data cyclically from the Simatic S5 devices on the process level via the SINEC H1 works bus (Ethernet bus to IEEE 802.3), so no signals need to be double-wired. All users have access to the same data in the file server, and at the same time. The number of user PCs is limited only by the NOVELL licence.

All data are captured, monitored and checked for plausibility at the lowest level in the process control system by the Simatic S5 devices. Only data that have been verified are sent via the SINEC H1 works bus to the information system, and in parallel to the control room system. There is therefore no multiple signal cabling and signal acquisition, and no different displays in the control room and in the information system.

A data processor PC prepares the data and saves them in a central file server PC. All authorized users can now access data and information on the file server, and display, analyse

WOCHENPRODUKTIONS-PROTOKOLL:		ZEMENT-BEREICH							
Woche: 12		29.03.1992 06:00	MO	DI	MI	DO	FR	SA	SO	WOCHE

				MO	DI	MI	DO	FR	SA	SO	WOCHE
ZEMENTMÜHLE Nr. 6		Produkt: P3									
556.WF1.UZ1	Klinker		t	2951,00	3362,00	2597,00	2784,00	2426,00	1732,00	2446,00	18298,00
556.WF2.UZ1	Schlacke		t	125,00	144,00	111,00	111,00	103,00	74,00	105,00	775,00
556.WF3.UZ1	Gips		t	120,00	157,00	117,00	112,00	97,00	82,00	129,00	814,00
556.WFX.UZ1	Gesamtaufgabe		t	**3196,00**	**3662,00**	**2825,00**	**3008,00**	**2627,00**	**1889,00**	**2680,00**	**19886,00**
BETRIEBSZEIT											
556.ML1.UZ1	Mühlen-Laufzeit		h	23,76	21,65	22,80	23,51	21,70	15,16	22,10	157,41
556.GEN.UZ3	Mühlen-Standzeit		h	0,24	2,35	1,20	0,49	2,30	8,84	1,90	10,59
	Produktions-Kennzahl		t/h	134,52	129,04	123,92	127,93	121,04	124,59	121,26	126,34
ELEKTRISCHER VERBRAUCH											
556.GEN.UZ1	Mahlanlage		kWh	61805,00	73755,00	58387,00	60220,00	56472,00	39844,00	58339,00	408821,00
556.GEN.UZ4	Mühlen-Effizienz		kWh/t	19,34	20,14	20,67	20,02	21,50	21,09	21,77	20,56
ZEMENT SILOS											
586.SB4.UZ1	Silo 4		t								
586.SB5.UZ1	Silo 5		t								
586.SB6.UZ1	Silo 6		t								
586.SB7.UZ1	Silo 7		t								
Anmerkungen:								geprüft:			

BILD 3: Wochenprotokoll einer Zementmahlanlage

Wochenproduktions-Protokoll = Weekly production report:
Zement-Bereich = Cement sector
Week: 12   MO = MON
            DI = TUE
            MI = WED
            DO = THU
            FR = FRI
            SA = SAT
            SO = SUN
            Woche = Week
Zementmühle Nr. 6 = cement mill No. 6
Produkt: P3 = product: P3
Klinker = clinker

FIGURE 3: Weekly report of a cement grinding plant

Schlacke = slag
Gips = gypsum
Gesamtaufgabe = total feed
Betriebszeit = running time
Mühlen-Laufzeit = mill running time
Mühlen-Standzeit = mill downtime
Produktions-Kennzahl = production rate
Elektrischer Verbrauch = electrical power consumption
Mahlanlage = grinding plant
Mühlen-Effizienz = mill efficiency
Zement Silos = cement silos
Anmerkungen = Remarks
geprüft = checked

Ein Datenprozessor-PC bereitet die Daten auf und speichert sie in einem zentralen Fileserver-PC. Jeder berechtigte Benutzer kann nun Daten und Informationen vom Fileserver abrufen und auf seinem PC anzeigen, analysieren und ausdrucken lassen. Alle Daten werden nur mit ihrem Anlagenkennzeichen angesprochen. Interne Dateinamen, Adressen etc. braucht der Anwender nicht zu kennen. Die PCs können beliebig in den Büros und Werkstätten im Werk verteilt sein.

5. Datenanalyse und Protokollgestaltung

Für die Datenanalyse und die Protokollgestaltung sind im CEMAT-MIS Standard-PC-Programme wie Lotus 1-2-3 oder Excel integriert. So kann jeder Benutzer alle Arten von Protokollen, z.B. Monatsberichte, Betriebsdatenprotokolle, Alarm- und Betriebsablaufprotokolle, Meßwertetabellen etc., selbst erstellen und auf seine individuellen Anforderungen zuschneiden. Programmierkenntnisse sind dazu nicht erforderlich.

Alle Meß- und Zählerwerte, die in einem Protokoll verarbeitet werden sollen, wählt man in einem Fenster per Mausklick an und übernimmt sie per Tastendruck in den Datensatz. Beispielsweise wählt man für das Protokoll „ZM6-Produktionsdaten für Betriebsleitung" in einem Definitionsfenster die Zeitbasis „Wochenwerte" und kombiniert in einem weiteren Definitionsfenster den Datensatz, die Zeitbasis und ein vorbereitetes Formular zu einem neuen Wochenprotokoll für die Produktionsdaten der Zementmahlanlage 6. Das Ergebnis ist im **Bild 3** dargestellt.

CEMAT-MIS nutzt alle Darstellungsformen, die Excel und Lotus 1-2-3 bieten, so z.B. die im **Bild 4** zu erkennende Bal-

and print them out on their PCs. All data are addressed by their plant code. The user does not need to know internal file names, addresses, etc. The PCs can be located in offices and workshops as required.

5. Data analysis and report formats

Standard PC programs such as Lotus 1-2-3 or Excel are integrated in the CEMAT-MIS for data analysis and report generation. All users can thus produce their own reports of all types (e.g. monthly reports, operating data reports, alarm and process reports, tables of measured values, etc.), and adapt them to their own individual requirements. No knowledge of programming is required.

All the measured values and counter inputs to be incorporated in a report are selected in a window by clicking with a mouse, and added to the data set with a keystroke. For the "CM6 Works Management Production Data" report for example, you select the time base "Weekly figures" in a definition window, and then in another definition window combine the data set, the time base and a prepared form to produce a new weekly report for Cement Mill 6 production data. The result is shown in **Fig. 3**.

CEMAT-MIS uses all the types of display provided by Excel and Lotus 1-2-3, such as the bar chart shown in **Fig. 4**, as well as column charts, line graphs and pie charts, presented in 2D or 3D, as line or area diagrams. The user can display the information in different ways if required, and select the most suitable for the particular data analysis required.

CEMAT-MIS offers various graph window displays for sets of readings, which can be extrapolated on-line (**Fig. 5**). Cursor lines facilitate the plotting of data. An overview window

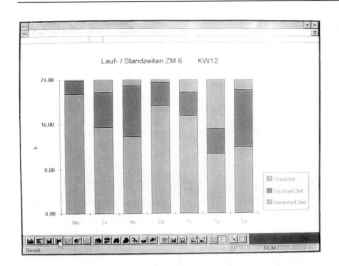

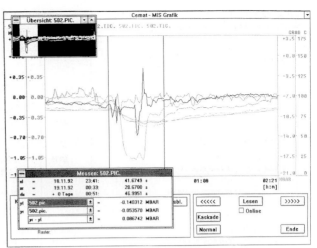

BILD 4: Säulendarstellung der Produktionszeiten einer Zementmahlanlage
FIGURE 4: Column chart of production times for a cement grinding plant

Lauf-/Standzeiten ZM6 = running/downtimes cement mill No. 6
Standzeit = downtime
Hochtarifzeit = high tariff time
Niedrigtarifzeit = low tariff time

BILD 5: Kurvenfenster-Darstellung verschiedener Meßwertreihen
FIGURE 5: Graph window display of various data sets

kendarstellung, aber auch Säulen-, Kurven- und Kreisdiagramme, in 2D- oder 3D-Darstellung, Linien- und Oberflächendiagramme. Für jede Datenauswertung kann sich der Benutzer die geeignete Darstellung auswählen und auch probeweise anzeigen lassen.

Für Meßwertereihen bietet CEMAT-MIS verschiedene Kurvenfenster-Darstellungen an, die auch online weitergeschrieben werden können (**Bild 5**). Cursorlineale erleichtern das Ausmessen von Meßpunkten. In einem Übersichtsfenster wird angezeigt, welcher Ausschnitt gerade vergrößert dargestellt ist. Kurven können überlagernd oder separiert dargestellt werden. Statistische und mathematische Operationen werden durch Anklicken in einem Auswahlfenster aktiviert. Dazu zählen Korrelationen, Kreuzkorrelationen, Regressionsanalysen bis hin zu FFT-Frequenzanalysen (**Bild 6**).

6. Schlußbetrachtung

Mit einem schnell entwickelten PC-Programm sind die an das Informations-Management zu stellenden Anforderungen nicht zu erfüllen. Ein solches System ist vielmehr immer werksspezifisch und es lebt, d.h. es muß immer wieder neuen Bedürfnissen angepaßt werden können. Mit CEMAT-MIS bietet Siemens ein flexibles, preiswertes und standardisiertes Werkzeug, das unter MS-Windows und NOVELL-Netware zu betreiben ist. Praktische Erfahrungen aus mehreren Werken liegen vor.

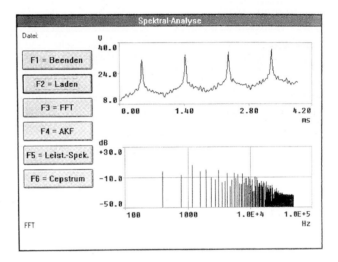

BILD 6: CEMAT-MIS-Darstellung einer FFT-Spektralanalyse
FIGURE 6: CEMAT-MIS display of an FFT spectrum analysis

shows the position of the detail that is currently displayed in enlarged form. Graphs can be displayed separately or overlaid. Statistical and mathematical operations can be initiated by clicking a selection window. This includes correlations, cross-correlations, regression analysis through to FFT frequency analyses (**Fig. 6**).

6. Conclusion

The demands made of information management cannot be met with a hastily developed PC program. Such a system always has to be works-specific, and is alive, i.e. it always has to be capable of adapting to new needs. The Siemens CEMAT-MIS system provides a flexible, cost-effective and standardized tool which runs under MS Windows and NOVELL netware. There are case histories from many plants.

Labor-Informations-Management-System (LIMS) in Zementwerkslaboratorien

LIMS in cement works laboratories

Système Management Information laboratoire (LIMS) dans des laboratoires de cimenterie

Sistema de gestión de informaciones (LIMS) para laboratorios

Von **Z. Senyuva,** Mersin/Türkei, **H. Baerentsen** und **S. Tokkesdal Pedersen,** Valby/Dänemark

Labor-Informations-Management-System (LIMS) in Zementwerkslaboratorien

Zusammenfassung – Wenn man sich außerhalb der Zementindustrie umsieht, stellt man fest, daß „Labor-Informations-Management-Systeme (LIMS)" schon seit einigen Jahren von Laboratorien, die hohe Anforderungen in bezug auf Qualitätssicherung und Datensicherung erfüllen müssen, eingesetzt werden. Wenn auch die Laboratorien eines Zementwerks wahrscheinlich nicht alle Aspekte eines typischen LIMS-Systems nutzen können, bieten ihnen die Funktionen eines solchen Systems doch große Vorteile. FLS Automation A/S, ein Unternehmen mit 20 Jahren Erfahrung als Hersteller von Labor-Automationssystemen für die Zementindustrie, bietet seit kurzem ein generell nutzbares Konzept an. Dabei wurden die Datenverarbeitungs- und -darstellungseinheiten des FLS-QCX-Systems so weiterentwickelt, daß daraus ein LIMS-System entstanden ist, das als zementindustrie-spezifisch gelten kann. Das System unterstützt nicht nur die Laborautomation und spezielle chemische Überwachungsaufgaben, sondern ist auch so konzipiert, daß es die unterschiedlichen Informationsanforderungen aller Mitarbeiterbereiche erfüllt. Die 1993 im CIMSA-Werk Mercin in der Osttürkei installierte Anlage arbeitet mit den LIMS-Einrichtungen eines QCX-Systems.

LIMS in cement works laboratories

Summary – Looking outside the cement industry, the "Laboratory Information Management Systems (LIMS)" have been used for quite some years for laboratories with high quality assurance and data security requirements. Although the laboratories of a cement plant will probably not need all aspects of a typical LIMS system, there are substantial benefits to be gained by applying LIMS functionality. With its 20 year experience as a supplier of laboratory automation systems to the cement industry, FLS Automation A/S has recently marketed a more general system concept where the data processing and reporting facilities of the FLS-QCX System have been developed into what could be described as a cement industry specific LIMS system. Besides supporting laboratory automation and specialised chemical control tasks, the system is designed to meet the various data information requirements for all staff levels. The system installed in 1993 at the Cimsa Mercin plant in Eastern Turkey uses the LIMS facilities of a QCX System.

Système Management Information laboratoire (LIMS) dans des laboratoires de cimenterie

Résumé – Il suffit de regarder hors de l'industrie cimentière pour constater que „des systèmes de Management Information laboratoire" (LIMS) sont utilisés depuis de nombreuses années par des laboratoires devant assurer des exigences élevées en matière d'assurance qualité et de sauvegarde des données. Même si les laboratoires d'une cimenterie ne peuvent probablement pas utiliser tous les aspects d'un système LIMS typique, il n'empêche que les fonctions d'un tel système sont, pour eux, du plus grand intérêt. FLS Automation A/S, entreprise ayant une expérience de 20 ans comme fabricant de systèmes d'automatisation de laboratoires pour l'industrie cimentière, propose depuis peu un concept d'application générale. Pour ce faire, les unités de représentation et de traitement de données du système FLS-QCX ont été améliorées de manière à créer un système LIMS capable de répondre aux besoins spécifiques de l'industrie cimentière. Et ce système assiste non seulement l'automatisation des laboratoires et les fonctions de surveillance proprement chimiques mais il est conçu également pour fournir les renseignements les plus divers au personnel. L'unité installée en 1993 à l'usine CIMSA Mercin en Turquie Orientale travaille avec les équipments LIMS du système QCX.

Sistema de gestión de informaciones (LIMS) para laboratorios

Resumen – Mirando un poco fuera de la industria del cemento, se ve que los „sistemas de gestión de informaciones (LIMS) para laboratorios" ya se vienen empleando, desde hace varios años, en laboratorios muy exigentes en cuanto a aseguramiento de la calidad y de los datos. Aunque los laboratorios de las fábricas de cemento probablemente no podrán aprovechar todos los aspectos de un típico sistema LIMS, las funciones de semejante sistema les ofrecen grandes ventajas. FLS Automation A/S, una empresa que cuenta con 20 años de experiencias como fabricante de sistemas de automatización de laboratorios de la industria cementera, ofrece, desde hace poco, un concepto de empleo universal. A este respecto, las unidades de tratamiento y representa-

ción de datos del sistema FLS-QCX han sido perfeccionadas de tal forma que se ha conseguido un sistema LIMS, el cual se puede considerar como sistema específico para la industria del cemento. Este sistema apoya no solamente a la automatización de los laboratorios y determinadas tareas de vigilancia, sino que está concebido también de forma tal que cumple los distintos requerimientos de información del personal. La planta instalada en 1993 en la factoría CIMSA Mercin, en Turquía Oriental, trabaja con los equipos LIMS de un sistema QCX.

1. Das QCX-LIMS-System im Zementwerk CIMSA

Das Zementwerk CIMSA Cimento Sanayi ve Ticaret AS (nahe der Stadt Mersin in der Südost-Türkei) hat ein Labor-Automationsprojekt mit FLS Automation als Hauptlieferant in Angriff genommen. Das Projekt begann 1992, und Phase 1 (**Bild 1**) wurde im Frühjahr 1993 abgeschlossen.

Die Computer-Hardware besteht aus einer VAX/VMS-Zentraleinheit mit drei Arbeitsplätzen, die mit Ethernet verknüpft sind. Zwei der drei Arbeitsplätze liegen im neuen Produktionslabor und der dritte eine Etage unter dem Zentrallabor. Der Rechner ist mit folgenden Geräten verbunden: zwei Röntgenfluoreszenzspektrometern und einem Röntgendiffraktometer, einem pneumatischen Probentransportsystem mit drei Probenahmepunkten, drei schon vorhandenen allgemeinen Steuerungssystemen für verschiedene Produktionslinien und einem Listen-PC, der die Daten automatisch vom QCX-System und vom allgemeinen Steuerungssystem überträgt und sie zu Managementlisten verarbeitet.

Die Software besteht aus der ersten Installation einer neuen Generation des FLS-QCX-Systems, das zwei wesentliche neue Entwicklungen darstellt: Erstens die Übertragung auf eine neue Hardware-Plattform mit einer modernen Window-Benutzerschnittstelle (entsprechend dem MOTIF-Standard) und hochauflösender Grafik. Die zweite neue Entwicklung ist die Neugestaltung der früheren QCX-Anwendungsmodule, die jetzt in eine allgemein anwendbare LIMS (**L**aboratory **I**nformation **M**anagement **S**ystem)-Architektur passen, welche sämtliche Labor-, chemische Qualitätskontroll- und Qualitätssicherungsfunktionen eines Zementwerks unterstützt. Die im Werk CIMSA installierten Anwendungsmodule sind QCX/Laboratory, QCX/Auto Sampling, QCX/Proportioner und QCX/Pile. In den folgenden Phasen des Labor-Automatisierungsprojekts werden die QCX/Quarry und QCX/RoboLab-Module installiert und die Zahl der Probenahmepunkte erhöht.

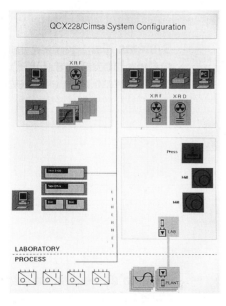

BILD 1: Konfiguration des LIMS-Systems im Zementwerk CIMSA
FIGURE 1: Configuration of the LIMS system at CIMSA cement works

Laborverfahren = Laboratory process
Werkslabor = Plant Lab
drücken = Press
Mühle = Mill

1. The QCX-LIMS system at CIMSA cement works

CIMSA Cimento Sanayi ve Ticaret AS cement works (near the town of Mersin in south-east Turkey) has introduced a laboratory automation project with FLS Automation as the selected main supplier. The project started in 1992 and Phase 1, outlined in **Fig. 1**, was completed in the spring of 1993.

The computer hardware comprises a VAX/VMS cpu with three workstations linked to an Ethernet. Two of the workstations are located in the new production laboratory and the third is located one floor below in the general laboratory. The computer is connected to two X-ray fluorescence spectrometers and one X-ray diffractometer, a pneumatic sample transport system with three sampling points, three existing general control systems covering various production lines, a report PC where data automatically transferred from the QCX system and the general control systems are digested into management reports.

The software is the first installation of a new generation FLS-QCX system which represents two major developments: One is transfer to a new hardware platform introducing modern window type user interface (according to the MOTIF standard) and high resolution graphics. The other main development is the re-design of the previous QCX application modules to fit into a generally applicable LIMS (**L**aboratory **I**nformation **M**anagement **S**ystem) architecture supporting all laboratory, chemical quality control and quality assurance functions at a cement plant. The application modules installed at CIMSA are QCX/Laboratory, QCX/AutoSampling, QCX/Proportioner and QCX/Pile. In the subsequent phases of the laboratory automation project the QCX/Quarry and QCX/RoboLab modules will be installed, and also the number of automatic sampling points are to be increased.

2. Design criteria for QCX-LIMS functionality

The new generation of the QCX/Laboratory module supports any number of analytical instruments as well as the entering of data for all manual analyses. Originally the QCX system was a dedicated X-ray based chemical quality control system for a production laboratory operating round the clock. The new generation has facilities for all laboratory activities based on the LIMS technology. The main design criteria for the system's development are listed in **Table 1**.

The system must be able to handle fast analysis procedures when necessary. It must support any degree of automation in terms of sampling, sample transport, sample preparation and sample analysis, including robotics and bar code technology.

TABLE 1: Design criteria for the system's development

QCX/LIMS Design Criteria

— Flexible laboratory configuration

— Support for automatic sampling/transport/preparation/analysis equipment, including robotics

— Compliant with international quality assurance standards

— Simple and intuitive routine function interface

— Versatile engineering functions

— Reliable and robust

— Support for bar code technology

— Possible to downgrade and to disable tell-tale functions

2. Auslegungskriterien für die QCX-LIMS-Funktionsvielfalt

Die neue Generation des QCX/Laboratory-Moduls unterstützt jede Art von analytischen Instrumenten sowie die Dateneingabe für alle manuellen Analysen. Ursprünglich war das QCX-System ein speziell auf der Röntgentechnik beruhendes chemisches Qualitätskontrollsystem für ein Produktionslabor mit 24-Stunden-Betrieb. Die neue Generation hat Einrichtungen für alle Laborfunktionen auf Basis der LIMS-Technologie. Die wichtigsten Auslegungskriterien bei der Entwicklung des Systems werden in **Tabelle 1** aufgezählt.

TABELLE 1: Auslegungskriterien für die Systementwicklung

QCX/LIMS-Auslegungskriterien
- flexible Laborkonfiguration
- Unterstützung für Analysegeräte mit automatischer Probenahme/Probentransport/Probenaufbereitung einschließlich Robotik
- Übereinstimmung mit internationalen Qualitätssicherungsnormen
- Schnittstelle für einfache und intuitive Routinefunktionen
- vielseitige technische Funktionen
- zuverlässig und robust
- Unterstützung für Strichcode-Technologie
- Möglichkeit der Reduzierung und Ausschaltung von Anzeigen, die dem Datenschutz unterliegen

Das System muß in der Lage sein, gegebenenfalls Schnellanalysen durchzuführen. Es muß jeden Automatisierungsgrad in den Bereichen Probenahme, Probentransport, Probenvorbereitung und Probenanalyse einschließlich Roboter- und Strichcode-Technologie unterstützen.

Das System muß flexibel sein, damit es im selben Maße wie die Forderungen der Kunden wachsen kann. Deshalb sollte es sich leicht mit vorhandenen und neu hinzukommenden Laborgeräten, dem Betriebsüberwachungssystem und dem administrativen EDV-System verknüpfen lassen.

Das System muß den internationalen de-facto-Normen für Qualitätssicherung, wie GLP, EN 45001, ISO 9002, entsprechen.

Die routinemäßigen Laborarbeiten müssen einfach und intuitiv durchzuführen sein. Die Benutzerschnittstellen der Analysegeräte sollten einheitlich, d.h. unabhängig vom Fabrikat, und vorzugsweise in der Landessprache gestaltet sein.

Das System muß umfassende technische Funktionen bieten einschließlich eines leicht anzuwendenden Listenprogrammgenerators und eines vielseitig verwendbaren Grafikpakets. Außerdem sollte das System so konfiguriert werden können, daß es eine allgemeine Schnittstelle mit SQL-relationalen Datenbanken hat.

Das System muß zuverlässig und robust sein − eine unabdingbare Voraussetzung in Verbindung mit der Überwachung kontinuierlich betriebener extrem kapitalintensiver Geräte.

Das LIMS-Konzept wurde ursprünglich für Laboratorien entwickelt, die ein besonders hohes Maß an Datenschutz erfordern. Im Produktionslabor jedoch liegt das Schwergewicht auf der Optimierung des Datendurchsatzes. Deshalb können die Anforderungen des betreffenden Zementwerks auch mit weniger System-„Bürokratie" und Anzeigevorrichtungen erfüllt werden, ohne die Qualitätssicherung zu beeinträchtigen. Neben den allgemeinen LIMS-Auslegungskriterien erfüllt der QCX/Laboratory-Modul alle spezifischen Bedingungen für eine Zementanalyse.

The system must be flexible, allowing it to grow in step with increasing customer requirements. It should therefore be easy to link up with existing and new laboratory equipment, the plant control system and the administrative EDP system.

The system must comply with international de facto standards for quality assurance, such as GLP, EN 45001, ISO 9002.

Routine laboratory operations must be simple to perform and based on intuition. The user interfaces of analytical equipment should be consistent, independent of equipment make, and preferably in the local language.

The system must provide comprehensive engineering functions, including an easy-to-use report generator and a versatile graphics package. In addition, the system should be configurable with a general interface to SQL relational databases.

The system must be reliable and robust − an absolute necessity in connection with the control of continuously operating, highly capital intensive equipment.

The LIMS concept was originally developed for laboratories which require an extremely high degree of data security. In the production laboratory, however, focus is on optimising the throughput. It is therefore possible to downgrade the "bureaucracy" and "tell tale" facilities of the system to meet the needs of the plant in question without compromising the quality assurance. In addition to the general LIMS design criteria, all the specific cement analytical requirements are maintained in the QCX/Laboratory module.

3. Examples of QCX-LIMS functionality

In a brief survey it is not possible to describe all the LIMS functionality of the new generation QCX System. Consequently, only a few typical operation procedures and system facilities are highlighted.

Fig. 3 shows the lay-out of a workstation where routine analysis jobs take place. To the left on the screen is shown the "Sample Log-In" window. Sample log-in can be done

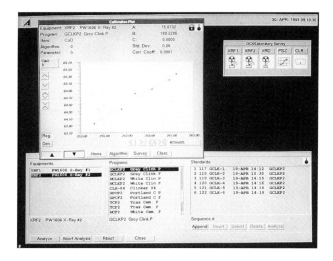

BILD 2: Bildschirmaufbau für die Kalibrierung
FIGURE 2: Screen layout for calibration

Kalibrierkurve	= Calibration plot
Geräte	= Equipment
Programm	= Program
Posten	= Item
Algorithmus	= Algorithm
Parameter	= Parameter
Standardabweichung	= Std. Dev.
Korrelationskoeffizient	= Corr. Coeff.
Regelabweichung	= Reg. Dev.
Überblick	= Survey
schließen	= Close
K-Zahl	= KCount
Analyse beginnen	= Analyze
Analyse abbrechen	= Abort Analysis
rückstellen	= Reset

3. Beispiele für die QCX-LIMS-Funktionsvielfalt

In einem kurzen Überblick ist es nicht möglich, die ganze Funktionsvielfalt der neuen Generation des QCX-Systems zu beschreiben. Folglich sollen nur einige typische Arbeitsverfahren und Systemfunktionen herausgestellt werden.

Bild 3 zeigt die Anordnung eines Arbeitsplatzes für Routine-Analysen. Auf der linken Seite befindet sich das „Sample Log-in"- (Probenprotokolleingabe-)Fenster. Die Probenprotokolleingabe kann manuell durch Anklicken der Zeile der entsprechenden Probengruppe erfolgen, normalerweise werden die Proben aber automatisch nach dem routinemäßigen Probenahmeverfahren protokolliert. Wenn, wie im Werk CIMSA, automatische Probenahmesysteme vorhanden sind, werden die Daten bei der Probenahme automatisch protokolliert. Während der Protokollierung werden die vordefinierten Analysen und die damit verbundenen Parameter den Proben zugeordnet. Deshalb braucht das Bedienungspersonal bei Beginn der Analyse keine Vorgaben zu machen, es sei denn, die vorgegebenen Standards sollen geändert werden. Die Probenanalyse beginnt beim rechten „Analysis"-Fenster. Um die Analysenfolge in Gang zu setzen, wird einfach der entsprechende Knopf am Gerät gedrückt. Mit den Analyseergebnissen wird entsprechend den vorgegebenen Spezifikationen verfahren, einschließlich automatischer Annahme, Verwerfung und Validierung auf der Grundlage verschiedener Grenzwerte und der vom Laboranten, der die Analyse angefordert hat, gesetzten Prioritäten. Oben links befindet sich das „Status"-Fenster, das alle zur Konfiguration gehörenden Laborinstrumente überwacht. Jedes Instrument wird durch ein Ikon dargestellt, das je nach Instrumentenstatus die Farbe ändert und die Probennummer während der Analyse anzeigt. Eine laufende Analyse kann durch Anklicken des entsprechenden Instrumenten-Ikons unterbrochen werden.

Bild 2 zeigt den Bildschirmaufbau bei der Kalibrierung der Instrumente. Die Funktion wird in zwei Masken unterteilt, d.h. in „Measure Standards" und „Analysis Program Calibration". Die Benutzerschnittstelle ist für alle Instrumente gleich, auch wenn die Kalibrierungsverfahren der einzelnen Instrumente sehr unterschiedlich sind. Dieses Prinzip ist eines der wichtigsten Auslegungskriterien und vereinfacht den Umgang mit dem System. Wenn später einmal automatische Probenladeeinrichtungen bzw. Probenmagazine im Werk CIMSA eingebaut werden, wird das „Measure Standards"-Verfahren voll automatisiert. Zur Zeit jedoch ist kei-

manually by simply clicking the line of the relevant sample group, but normally samples are automatically logged according to the routine sampling procedure. When automatic sampling systems are installed — as is the case at CIMSA — automatic log-in takes place at the time of sampling. At the time of logging-in, the predefined analytical tests and the associated parameters are assigned to the samples. When the analysis begins the operator therefore needs not to specify anything unless he wants to modify the default analysis conditions. Sample analysis is performed from the right hand "Analysis" window. To start an analysis sequence, it is necessary only to click the relevant instrument button. The analysis results are treated according to predefined specifications including automatic accept/reject/validation procedures based on various check limits and the priorities given by the operator who ordered the analysis. In the upper left-hand corner is shown the "Status" window monitoring all the configured laboratory instruments. Each instrument is represented by an icon that changes colour according to the instrument status and displays the sample number during analysis. A current analysis may be interrupted by clicking the specific instrument icon.

Fig. 2 shows the screen layout when instrument calibration is performed. The function is split into two tasks, namely "Measure Standards" and "Analysis Program Calibration". Note that the user interface is identical for all instruments, even if the calibration procedures may vary greatly from instrument to instrument. This principle, one of the key design criteria, simplifies system operation. When at a later stage, automatic sample loaders/sample magazines are installed at CIMSA the "Measure Standards" procedure becomes fully automatic, but at present none of the connected analysis instruments are equipped with automatic sample loaders, so the operator has to confirm each analysis during the sequence. The upper window shows the resulting calibration in the form of calibration curves which can be edited interactively by graphic editing techniques. Any standard can be excluded from the current calibration — and subsequently inserted again — merely by clicking in the plot area. The system automatically re-calculates after each modification.

Fig. 3 shows an example of a screen used for supervising product quality control. In the left-hand window is shown one of the graphical presentations from the comprehensive reporting package of the QCX/Laboratory module, namely a trend plot of selected analysis data. It is possible to scroll through old data by clicking the bar below the plot. In the

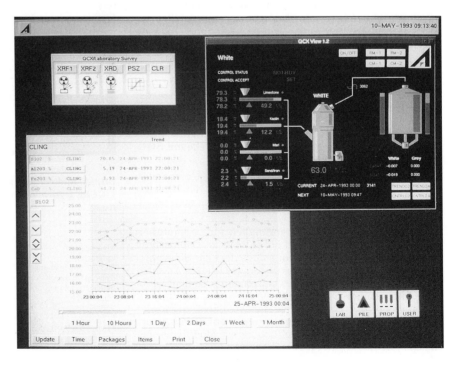

BILD 3: Bildschirm für die Überwachung der Qualitätskontrolle eines Rohmehls
FIGURE 3: Screen for supervising product quality control with the raw mill as an example

Folgenummer = Sequence
anhängen = Append
einfügen = Insert
auswählen = Select
löschen = Delete

nes der mit dem System verbundenen Analyseinstrumente mit automatischen Probenladeeinrichtungen ausgerüstet, so daß das Bedienungspersonal jede einzelne Analyse der Analysenfolge einzeln bestätigen muß. Das obere Fenster zeigt die erfolgte Kalibrierung in Form von Kalibrierungskurven, die interaktiv mit dem Grafikeditor redigiert werden können. Jeder Standard kann aus der laufenden Kalibrierung ausgeschlossen und anschließend wieder eingefügt werden; dazu wird nur der entsprechende Kurventeil angeklickt. Nach jeder Änderung führt das System eine neue Berechnung durch.

Bild 3 zeigt einen Bildschirm für die Überwachung der Produktqualitätskontrolle. Im Fenster links ist eine grafische Darstellung aus dem umfangreichen Listenpaket des QCX/Laboratory-Moduls zu sehen, d.h. eine Trendkurve ausgewählter Analysedaten. Es ist möglich, durch Anklicken des Balkens unter der Grafik den Bildschirm zu alten Daten zurückzurollen. Das Fenster oben rechts zeigt die Überwachung der Rohmehlproduktion. Das Bild zeigt live-Daten von den speicherprogrammierten Steuerungen und andere Daten vom QCX/Proportioner-Modul.

An allen Arbeitsplätzen können alle Fenster in jeder Kombination gemischt und ikonisiert werden. Die Fenster können so konfiguriert werden, daß an einem bestimmten Arbeitsplatz nur die gewünschten Anwendungen angezeigt werden.

4. Schlußbemerkungen

Als dieser Beitrag geschrieben wurde, war das System im Werk CIMSA erst wenige Wochen in Betrieb, aber die bei Inbetriebnahme und in den ersten Wochen gemachten Erfahrungen erfüllen die Erwartungen, die an die LIMS-Technologie in einem zementindustrie-spezifischen Umfeld gestellt werden. Die rationelle und leicht bedienbare Benutzerschnittstelle bietet eine solide Grundlage für die Verwaltung aller Analysedaten, die in einem modernen Zementwerk mit einem zeitgemäßen Qualitätssicherungskonzept erfaßt werden, das bei CIMSA Teil einer Strategie in Richtung auf die ISO-9000-Zertifizierung der Zementproduktion ist.

upper right-hand corner window raw mill production is supervised. The picture is fed by live data from the control system PLCs and other data from the QCX/Proportioner module.

All windows can be mixed and iconised in any combination on all workstations. They may be configured so that only desired applications appear on a given workstation.

4. Concluding remarks

At the time of writing, the system at CIMSA has been in operation for a few weeks only, but the experiences gained during commissioning and the first operation weeks confirm the expectations to the LIMS technology applied in a cement specific setup. The efficient and easy-to-use operator interface provides a solid basis for administration of all analysis data generated at a modern cement plant in accordance with an up-to-date quality assurance concept, which at CIMSA is part of a part of a strategy to achieve ISO 9000 certification of the cement production.

Höhere Leistungsdichten im Zementlabor mit multimodularer Robotertechnik

Higher performance density in the cement laboratory with multi-modular robot technology

Plus grande capacité de travail au laboratoire du ciment avec technique de robotisation multimodulaire

Mayor rendimiento en los laboratorios de cemento mediante la robótica multimodular

Von **J. Teutenberg,** Neubeckum/Deutschland

Zusammenfassung – Durch den Einzug der multimodularen Robotertechnik in die Zementlabore konnten die Qualitätsüberwachung noch verbessert und Probendurchsatz und Verfügbarkeit wesentlich gesteigert werden. Mit dem in der Roboterversion von POLAB realisierten Modulkonzept bietet Polysius der Zementindustrie ein auf ihre spezifischen Betriebsbedürfnisse abgestimmtes Laborsystem an. Zum Gesamtsystem gehören generell eine repräsentative Probenahme, eine Anlage für den schnellen Probentransport, eine auf Besonderheiten von Stoffproben abgestimmte Probenaufbereitung und Analysentechnik sowie zuverlässige Steuerungs- und Regeleinrichtungen zur Sicherung der Qualität und deren Überwachung.

Summary – By introducing multi-modular robot technology into cement laboratories it has been possible to improve the quality monitoring still further and make substantial increases in the sample throughput and availability. With its modular concept put into practice with the robot version of POLAB, Polysius is offering the cement industry a laboratory system adapted to suit their specific plant requirements. The entire system normally comprises a representative sampling system, a plant for rapid sample transport, a sample processing system and analysis technology suited to the special features of the material samples, and reliable open- and closed-loop control equipment for ensuring and monitoring the quality.

Résumé – Par l'introduction de la technique de robotisation multimodulaire dans les laboratoires du ciment, la surveillance de la qualité a pu encore être améliorée et la fréquence des éprouvettes et la disponibilité ont pu être nettement accrues. Avec le concept modulaire réalisé dans la version robotisée de POLAB, Polysius offre à l'industrie cimentière un système de laboratoire taillé sur mesure pour ses besoins d'exploitation spécifiques. Le système en entier comprend généralement un échantillonnage représentatif, un équipement pour l'acheminement rapide des échantillons, une préparation des éprouvettes accordée avec les particularités des échantillons de matière et une technique d'analyse ainsi que des équipements de conduite et de réglage pour l'assurance de la qualité et de sa surveillance.

Resumen – Con el empleo de la robótica multimodular en los laboratorios de cemento ha sido posible mejorar el control de la calidad y aumentar notablemente el caudal de muestras y el índice de disponibilidad. Polysius ofrece a la industria cementera, con el concepto modular realizado mediante la versión robotizada de POLAB, un sistema de laboratorio adaptado a las necesidades específicas del servicio. Forman siempre parte del sistema en su conjunto, una toma de muestras representativas, una instalación para el transporte rápido de las muestras, una técnica de preparación y análisis de muestras, teniendo en cuenta las características de las muestras de material, así como unos equipos de mando y regulación fiables, para el aseguramiento de la calidad y la vigilancia correspondiente.

1. Einleitung

Die Erweiterung der Produktpalette und ein verschärfter Wettbewerb zwingen die Zementindustrie zu ständig steigenden Anforderungen an Qualitätssicherungsmaßnahmen. Es ist die Aufgabe des Labors, die Eignung der Produkte durch Einhaltung der Normen mit sehr engen Toleranzen sicherzustellen. Das bedingt eine ständige Überwachung der Ausgangs-, Zwischen- und Endprodukte mit direkten und automatischen Regel- und Steuereingriffen in den Verfahrensablauf.

Somit wird eine flexible und bewährte Systemtechnik erforderlich, die die anspruchsvollen Aufgaben der repräsentati-

1. Introduction

Widening product ranges and heightened competition are forcing the cement industry to make ever increasing demands on quality assurance systems. It is the task of the laboratory to ensure the suitability of products by complying with standards within very tight tolerances. This requires continuous monitoring of the input, intermediate and end products with direct and automatic open- and closed-loop control interventions in the process.

Flexible and proven system technology is needed which can provide a cost-effective solution for the demanding tasks of representative sampling, rapid sample transport, individual

ven Probenentnahme, des schnellen Probentransports, der individuellen Probenvorbereitung, der auf die Produktion abgestimmten Analysentechnik und der Schaffung zuverlässiger Steuer- und Regelmechanismen für die Qualitätssicherung und -kontrolle wirtschaftlich löst.

2. Kriterien für die Auswahl der Robotertechnik

2.1 Aufgabenstellung

Für die Probenvorbereitung im Zementlabor müssen relativ kleine Gewichte und geringe Volumina mit schnellen und sicheren Bewegungsabläufen über kurze Transportwege präzise in Verbindung mit der Geräteausrüstung gehandhabt werden. Von wesentlicher Bedeutung ist dabei eine hohe Positionier- und Greifgenauigkeit mit guter Langzeitreproduzierbarkeit. Bezüglich Personenschutz sind die DIN- und die VDI-Richtlinien einzuhalten. Abhängig vom Probenaufkommen und der Analysentechnik müssen relativ einfache Erweiterungen möglich sein, wie z.B. eine Parallelschaltung der Systeme mit Back-up-Funktionen.

2.2 Lösung

Zur Lösung der Aufgabe ist die Auswahl eines Kleinroboters mit intelligenter Software und hoher Verfügbarkeit erforderlich. Die Baugruppen und der Roboter werden zusammen in Modulbautechnik eingehaust (**Bild 1**). Damit sind nur kurze Wege erforderlich. So werden schnelle Bewegungsabläufe erreicht und eine problemlose und sichere Wartung der Baugruppen und Geräte im On-Line-Betrieb ermöglicht. Daher werden störende und unnötige Abschaltungen bei Pflege- und Wartungsarbeiten vermieden. Bei dieser Gerätetechnik ist jederzeit ein freier Personenzugang im On-Line-Betrieb gegeben.

3. Vom Baustein zur individuellen Modultechnik

3.1 Modulare Systemtechnik

Abhängig vom Probenaufkommen, dem gewünschten Automatisierungsgrad für das Produktionsverfahren, der gewählten Ausbaustufen und der erforderlichen Art der Probenvorbereitung und -analyse werden für jede individuelle Anwendung aus Bausteinen die erforderlichen Module mit integrierten Kleinrobotern konfiguriert. Durch diese flexible Bauweise ist einmal die optimale Systemtechnik und jederzeit ein problemloser und schneller Ausbau von zu Beginn kleinen Anwendungen über mittlere bis zu komplexen Automationssystemen gegeben.

Bild 2 zeigt als Beispiel die Konfiguration eines Basismoduls für einen Durchsatz von bis zu 10 Proben je Stunde. Ein derartiges Modul löst im wesentlichen folgende Aufgaben:

- Gleichzeitige Eingabe von bis zu 34 manuellen Proben (1),
- Handhabung der Rohrpostbüchsen für das im Verfahrensablauf automatisch entnommene Probengut (2),

sample preparation, analytical techniques suited to the type of production, and the creation of reliable open- and closed-loop control mechanisms for quality assurance and quality control.

2. Criteria for the selection of robot technology

2.1 Problem definition

Sample preparation in a cement laboratory requires relatively small weights and low volumes to be handled accurately with rapid and reliable movement cycles over the short transport distances associated with the apparatus. Of particular importance is the high precision of positioning and gripping with good long-term reproducibility. Personnel safety must comply with the DIN and VDI guidelines. Relatively simple extensions must be possible to suit the sample intake and the analytical technology, e.g. parallel configuration of the system with back-up functions.

2.2 Solution

To solve the problem it is necessary to choose a small robot with intelligent software and a high level of availability. The structural sub-assemblies and robots are housed together in a modular system (**Fig. 1**) and only short distances are involved. This means that rapid movement cycles can be achieved, and reliable, trouble-free, maintenance can be carried out to the structural sub-assemblies during on-line operation. Disruptive and unnecessary switching off during cleaning and maintenance work is therefore avoided. This equipment technology allows unrestricted personnel access at any time during on-line operation.

3. From the sub-module to individual module technology

3.1 Modular system technology

The necessary modules with small integral robots are configured from sub-modules for each individual application to suit the sample intake, the required degree of automation for the production process, the selected extension stages and the type of sample processing and analysis required. This flexible design provides the best possible system technology and allows rapid, trouble-free, extension at any time from small initial applications, via medium-sized ones, right up to complex automation systems.

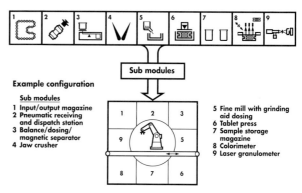

BILD 2: Beispielkonfiguration eines Basismoduls
FIGURE 2: Example of the configuration of a basic module

Beispielh. Konfiguration	= Example configuration
Submodule	= Sub modules
1 Input/Output-Magazin	1 Input/output magazine
2 Pneumatische Empfangs- und Sendestation	1 Pneumatic receiving and dispatch station
3 Wägen, Dosieren, Magnetabscheiden	3 Balance/dosing/magnetic separator
4 Backenbrecher	4 Jaw crusher
5 Feinmühle mit Mahlhilfsmitteldosierung	5 Fine mill with grinding aid dosing
6 Tablettenpresse	6 Tablet press
7 Probenlagermagazin	7 Sample storage magazine
8 Kalorimeter	8 Calorimeter
9 Laser-Granulometer	9 Laser granulometer

BILD 1: Außenansicht der eingehausten Baugruppen und Roboter
FIGURE 1: External view of the enclosed sub-assemblies and robots

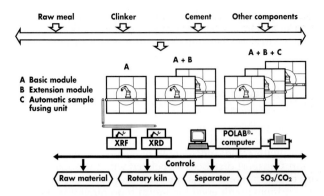

BILD 3: Erweiterungsmöglichkeiten der Modultechnik
FIGURE 3: Extension options using module technology

Rohmehl	= Raw meal
Klinker	= Clinker
Zement	= Cement
andere Komponenten	= Other components
A Basis-Modul	= A Basic module
B Erweiterungsmodul	= B Extension module
C Automatische Probenschmelz-Einrichtung	= C Automatic sample fusing unit
Rohmaterial	= Raw material
Drehofen	= Rotary kiln
Sichter	= Separator
SO_3/CO_2	= SO_3/CO_2

— gravimetrische Dosierung für die individuellen Probenvorbereitungs- und Analyseverfahren (3),

— Feinvermahlung mit variablen, der anschließenden Analyse angepaßten Verfahrensschritten (z. B. Vor- und Hauptprobe) (5),

— Pressen des Probegutes (6),

— Rückstellung dosierter Teilmengen für physikalische Untersuchungen (7),

— Farbmessung (8),

— Lasergranulometer (9).

3.2 Mögliche Ausbaustufen

Wie bereits ausgeführt, definiert der Betreiber den Leistungsumfang bezüglich der Ausbaustufen, der auf sein Produktionsverfahren optimal abgestimmt ist. Dazu zählen u. a. die Anzahl der automatischen Probenahmestellen, das erforderliche tägliche Probenaufkommen für die eindeutige Qualitätssicherung sowie die notwendigen Analysegeräte, die Auswertung der Meßergebnisse und die erforderlichen Regel- und Steuereingriffe.

Die im **Bild 3** als Beispiel dargestellten Systeme A, A + B, A + B + C unterscheiden sich durch die Bestückung der Automaten nach Anzahl der Roboter, der Analysengeräte und der Aufbereitungsaggregate (Mühlen, Presse, Aufschluß und Zusatzmodule wie z. B. Rohrpost, ELA-Magazin und Rückstellprobenbehälter).

Reicht z. B. eine Modulbaugruppe für den Probendurchsatz nicht mehr aus, so werden diese Modulbaugruppen, z. B. das Basis-Modul, sofort oder später einfach doppelt ausgeführt und parallel geschaltet. Erreicht wird in diesem Fall ein erhöhter Probendurchsatz mit zusätzlichen Back-up-Funktionen.

TABELLE 1: Mittelwerte MW und Standardabweichung s als Beispiel der Reproduzierbarkeit bei gleichzeitiger Eingabe von 20 identischen Klinkerproben

	KST	SM	TM	FrCaO
MW	100,18	2,57	1,599	1,79
S	0,12	0,01	0,0064	0,05

3.3 Ergebnisbetrachtung

Die zuvor beschriebene Systemtechnik zeichnet sich zum einem durch einen hohen Probendurchsatz mit guter Systemverfügbarkeit und zum anderen durch sehr gute Reproduzierbarkeitswerte in der Bearbeitung eines breiten Probenspektrums aus. Bei der gleichzeitigen Eingabe von 20 Klinkerproben werden die im Beispiel aufgeführten Reproduzierbarkeitswerte erreicht (**Tabelle 1**).

Fig. 2 shows an example of the configuration of a basic module for a throughput of up to 10 samples per hour. In essence, a module of this type performs the following tasks:

— simultaneous input of up to 34 manual samples (1),

— handling the pneumatic despatch capsules for the sample materials sampled automatically during the course of the process (2),

— gravimetric metered feeding for the individual methods of sample preparation and analysis (3),

— fine grinding with varying process stages to suit the following analysis (e. g. preliminary and main samples) (5),

— pressing the sample material (6),

— retention of metered sub-samples for physical investigations (7),

— colour measurement (8),

— laser granulometer (9).

3.2 Possible extension stages

As already mentioned, the operator defines the range of performance regarding the expansion stages which is best suited to his production process. This includes, among other things, the number of automatic sampling points, the daily sample intake required for definite quality assurance, as well as the necessary analytical equipment, the evaluation system for the measurements, and the required open- and closed-loop control intervention systems.

The systems A, A+B, A+B+C shown as examples in **Fig. 3** differ in the components within the automated systems based on the number of robots, the analytical equipment and the processing units (mills, presses, decomposition and auxiliary modules, such as pneumatic tube conveyor, ELA magazine and sample storage magazines).

If, for example, one module assembly is no longer sufficient for the sample throughput then this module assembly, e. g. the basic module, is either immediately or later simply doubled and arranged in parallel. This provides increased sample throughput with additional back-up functions.

TABLE 1: Mean values MV and standard deviation s as an example of the reproducibility on simultaneous input of 20 identical clinker samples

	LSt	SM	AR	FrCaO
MW	100.18	2.57	1.599	1.79
S	0.12	0.01	0.0064	0.05

3.3 Examination of the results

The system technology described above is characterized not only by high sample throughput with good system availability but also by very good reproducibility values when processing a wide sample spectrum. The reproducibility values listed by way of example in **Table 1** were obtained from the simultaneous input of 20 clinker samples.

Entwicklungsschritte zur Automationstechnik des Werkes Bernburg

Development steps in the automation system at the Bernburg works

Etapes de développement pour la technique d'automation à l'usine Bernburg

Etapas de desarrollo de la técnica de automatización de la fábrica de Bernburg

Von **W. Triebel,** Ulm/Deutschland

Zusammenfassung – Anfang der 80er Jahre entschied sich Schwenk, den auf den Gebieten der Meß- und Regeltechnik erreichten Fortschritt in einem Zuge zur Verbesserung der Steuerung des Zementherstellungsprozesses in den eigenen Werken zu nutzen. In einem mehrjährigen Entwicklungsprozeß wurden die Überlegungen zusammen mit der Firma Krupp Polysius konkretisiert und realisiert. Mitte der 80er Jahre waren alle Schwenk-Zementwerke mit modernster Leit- und Labortechnik ausgerüstet, die in den Folgejahren laufend erweitert und optimiert wurde. Die dabei gewonnenen Erfahrungen konnten für das Konzept der Automationstechnik des Werkes Bernburg genutzt werden. Die gesamte Qualitätskontrolle von Eingangsstoffen und Zementen wird dort über das modular aufgebaute POLAB 3-System abgewickelt, das mit zwei automatischen Probenaufbereitungslinien ausgestattet ist und neben chemisch-mineralogischen Analysen mittels Röntgenfluoreszenz- und Röntgendiffraktometrie auch eine automatische Feinheitskontrolle und Farbmessung von Zementen ermöglicht. Um die anfallenden Datenmengen trotz ihrer Vielfalt und Komplexität optimal nutzen zu können, werden die Labordaten in einer überlagerten Datenbank abgelegt und über ein komplexes Statistikprogramm miteinander korreliert. Die dabei herauszufilternden signifikanten Zusammenhänge stehen dann dem Betriebspersonal für weitere Schlußfolgerungen zur Verfügung.

Summary – At the start of the 80s Schwenk decided to make use of the progress achieved in the fields of measuring and control engineering in a move to improve the control of the cement production process in their own works. In a development process covering several years these thoughts were defined and put into practice in cooperation with Krupp Polysius. By the mid 80s all the Schwenk cement works were equipped with very modern control and laboratory technology which were continuously extended and optimized in the following years. The experience obtained was utilized for designing the automation system at the Bernburg works. The entire quality control of the input materials and the cements are handled there by the modular POLAB 3 system which has two automatic sample processing lines. This means that, in addition to chemical and mineralogical analyses by X-ray fluorescence and X-ray diffraction, it is also possible to carry out automatic fineness checking and colour measurement of cement. So that optimum use can be made of the quantity of data obtained in spite of its diversity and complexity, the laboratory data are filed in an overlay database and correlated with one another by a complex statistics program. The significant relationships sifted from this are then available to the operating personnel to draw further conclusions.

Résumé – Au début des années 80 Schwenk s'est décidé à profiter en une seule opération du progrès réalisé dans les secteurs de la technique de mesure et régulation, pour améliorer la conduite du process de fabrication dans ses propres usines. Au cours d'un processus de développement de plusieurs années, les réflexions ont été concrétisées et réalisées en collaboration avec la firme Krupp Polysius. Au milieu des années 80, toutes les cimenteries Schwenk étaient équipées des techniques les plus modernes de conduite et de laboratoire; celles-ci ont été, dans les années suivantes, continuellement élargies et optimisées. L'expérience acquise ainsi a pu être mise à profit pour la conception de la technique d'automation de l'usine Bernburg. L'ensemble des contrôles de qualité des matières de départ et des ciments s'y déroule au moyen du système POLAB 3 à architecture modulaire, équipé de deux lignes automatiques de préparation des échantillons et qui permet, en plus des analyses chimiques-minéralogiques, par fluorescence et diffractométrie X aussi un contrôle automatique de la finesse et la colorimétrie des ciments. Afin de pouvoir exploiter de façon optimale les quantités de données recueillies, malgré leur multitude et complexité, les données du laboratoire sont stockées dans une banque de données de niveau plus haut et corrélées entre elles par un programme complexe de statistique. Les relations significatives à filtrer ainsi sont disponibles pour le personnel de conduite, pour des conclusions ultérieures.

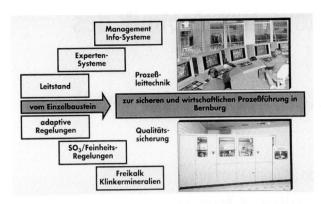

BILD 2: Konzept der Prozeßleittechnik und Laborautomation des Werkes Bernburg
FIGURE 2: Scheme for process control technology and laboratory automation at the Bernburg works

Management Info-Systeme	management information systems
Experten-Systeme	expert systems
Leitstand	control station
Prozeß-Leittechnik	process control technology
vom Einzelbaustein	from single modules
zur sicheren und wirtschaftlichen Prozeßführung in Bernburg	to Bernburg's reliable and cost-effective process control system
adaptive Regelungen	adaptive control systems
SO₃/Feinheits-Regelungen	SO₃/fineness controls
Freikalk Klinkermineralien	free lime clinker minerals
Qualitätssicherung	quality assurance
Automatisationskonzept Bernburg	Bernburg's automation scheme

sowohl für den Drehofenprozeß als auch für den Rollenmühlenbetrieb und die Zementmahlung.

Durch die politischen Ereignisse der Wiedervereinigung Deutschlands ergab sich die Herausforderung, die Automationstechnik des neuen Werkes in Bernburg, aufbauend auf obigen Erfahrungen, dem letzten Stand der Technik Rechnung tragend und in die Zukunft weisend, zu konzipieren (**Bild 2**). Was dabei erreicht wurde und in Zukunft weitergeführt werden soll, wird im folgenden dargestellt: Die gesamte Qualitätskontrolle des Zementwerkes von der Rohmaterial- bis zur Zementseite wird über das neueste POLYSIUS-System POLAB 3 in Modultechnik über zwei vollkommen getrennte Probenaufbereitungslinien mit insgesamt 5 Robotern einschließlich automatischem Schmelzgerät abgewickelt. Wie in **Bild 3** dargestellt, sind dabei alle Möglichkeiten der Röntgenfluoreszenzanalyse, der Röntgendiffraktometrie mit Erfassung von Freikalk und Klinkerphasen, der automatischen Feinheitskontrolle von Zementen, der Farbmessung von Zementen etc. einbezogen.

Da die Ofenanlage Bernburg wegen des hohen Chlorgehaltes mit einem Bypass ausgerüstet werden mußte, dessen Minimierung oberstes Ziel ist, wurde erstmals eine Heißmehlprobenahme am Ofeneinlauf konzipiert, die es ermöglichen soll, kontinuierliche Heißmehlproben aus diesem kritischen Bereich zu entnehmen und ebenfalls über das automatische Probensystem zu analysieren. Steuerungsmäßig kommt das Leitsystem POLYSIUS POLCID DC zum Einsatz, in welches die Steuerungssysteme KCE und MCE entsprechend dem oben vorgestellten Expertensystem integriert und implementiert werden. Bei dem Expertensystem KCE für die Ofenanlage werden erstmalig die automatisch ermittelten Werte für den Freikalkgehalt und für die Klinkermineralien in die Regelstrategie einbezogen.

Wie in **Bild 4** zu erkennen ist, sollen alle Prozeß- und Qualitätssteuerungssysteme durch ein überlagertes Informations-Management-System (IMS) miteinander verbunden und vernetzt werden, so daß die Möglichkeiten der statistischen Auswertung mit entsprechender nachgeschalteter Optimierung voll genutzt werden können. Es hat sich gezeigt, daß die aus den modernen Systemen fließenden Informationen und Signale in so großer Zahl anfallen, daß ihre Vielfalt und Komplexität vom Betriebspersonal nicht mehr beherrscht werden kann. Die Zielsetzung lautet des-

plex integrated system without the use of robot technology and modular design, and it placed corresponding demands on servicing and maintenance.

For the first time in the cement industry the process control system incorporated the ability to carry out automatic fineness analysis and control of cement fineness, as well as automatic determination and continuous control of the SO₃ content.

At the Karlstadt works, for example, these methods made it possible to combine the previously decentralized control stations for controlling raw meal composition, homogenization, and cement quality centrally in the laboratory. It became apparent that this not only had a rationalizing effect and produced a significant increase in the quality standard, but also relieved the respective production controllers of a great deal of tiring routine work. Because of these successes it then seemed appropriate that the capabilities of computer-controlled process control should also be included. The "Kiln Control" system was installed at a rotary kiln and the "Mill Control" systems was installed at one of the large cement mills jointly with Krupp-Polysius.

After great efforts to optimize the systems over a fairly long period it became clear that in the long term these systems would be too complex, too maintenance-intensive, and not sufficiently flexible. The solution could only lie in software which contained self-learning, adaptive components based on methods of artificial intelligence. The "MCE" (Mill Control Expert) control system, which is structured like an expert system, was therefore developed in collaboration with Krupp-Polysius. This system is capable of fully automatic management of every operational condition of a cement mill from start to finish, including all changes of cement type.

After these successes it was clear that intelligent control systems are the only ones able to meet the high demands of cement technology. Further development work was therefore concentrated on these systems, both for the rotary kiln process and for roller mill operation and cement grinding.

The political events of the re-unification of Germany presented the challenge of devising forward-looking automation technology for the new works in Bernburg, building on the experience described above and taking the latest state of technology into account (**Fig. 2**). A description of what was achieved and will be done in the future is given below:

The entire quality control of the cement works from the raw materials to the cement is regulated by the latest POLYSIUS system – POLAB 3 – using modular technology, and by two completely separate sample preparation lines with a total of 5 robots including automatic fusion equipment. As is shown in **Fig. 3**, this incorporates all the

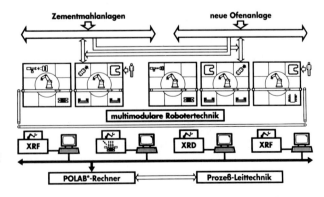

BILD 3: Modularer Aufbau des automatischen Produktionslabors
FIGURE 3: Modular structure of the automated production laboratory

Zementmahlanlagen	cement grinding plants
neu Ofenanlage	new kiln plant
multimodulare Robotertechnik	multimodular robot technology
XRF	XRF
XRD	XRD
POLAB®-Rechner	POLAB® computer
Prozeß-Leittechnik	process control technology
Qualitätssicherung Bernburg	Bernburg quality assurance system

Resumen – A principios de los años 80, Schwenk se decidió a aprovechar de una vez, en sus propias fábricas, los adelantos logrados en los campos de la regulación y la medición, con el fin de mejorar el control del proceso de fabricación del cemento. En un proceso de desarrollo, de varios años de duración, se concretizaron y se realizaron estas ideas, junto con la firma Krupp Polysius. A mediados de los años 80, todas las fábricas de cemento de Schwenk estaban equipadas con la técnica de control y de laboratorio más moderna, la cual fue ampliada y optimizada continuamente en los años siguientes. Las experiencias adquiridas a este respecto han podido aprovecharse para el concepto de la técnica de automatización de la fábrica de Bernburg. El control total de calidad de los materiales de partida y de los cementos se realiza allí por medio del sistema modular POLAB 3, equipado con dos líneas automáticas de preparación de muestras, que permite llevar a cabo no solamente los análisis químico-mineralógicos con ayuda de la fluorescencia de rayos X y de la difracción de rayos X, sino también un control automático de la finura y una medición del color de los cementos. Con el fin de poder utilizar la gran cantidad de datos de laboratorio, a pesar de su variedad y complejidad, se archivan los mismos en un banco de datos de nivel superior y se ponen en relación los unos con los otros mediante un programa estadístico muy complejo. Las relaciones significativas que hay que filtrar en esta operación son puestas entonces a disposición del personal de servicio, que podrá sacar de ellas las conclusiones pertinentes.

Etapas de desarrollo de la técnica de automatización de la fábrica de Bernburg

Anfang der 80er Jahre stellte sich im Rahmen des Programms zur Modernisierung und Neustrukturierung der Schwenk'schen Zementwerke auch die Aufgabe, die mittlerweile verfügbaren, modernen Möglichkeiten der Laborautomation und der Prozeß-Leittechnik mit einzubeziehen (**Bild 1**).

Erstes Ziel dabei war, zunächst die vielen vorhandenen, lokalen Leitstände zwar funktionsmäßig beizubehalten, ihre Bedienung aber mit den Methoden des modernen Datentransfers und des Signalaustausches örtlich zu verlagern und damit eine Zentralisierung der Prozeßführung möglich zu machen, ohne die gesamte Steuerung vom Grunde her erneuern zu müssen. Parallel dazu wurde zusammen mit der Fa. Krupp-Polysius die Laborautomation Polab installiert, die, damals noch als integriertes System ohne den Einsatz von Robotertechnik und Modulbauweise, sehr komplex war und dementsprechende Anforderungen an Wartung und Instandhaltung stellte.

Zum ersten Mal in der Zementindustrie wurden dabei die Möglichkeit der automatischen Feinheitsanalyse und Feinheitssteuerung von Zementen sowie der automatischen Bestimmung des SO_3-Gehaltes und dessen kontinuierliche Steuerung in die Prozeßführung miteinbezogen.

Mit diesen Methoden gelang es z.B. im Werk Karlstadt, die vorher dezentralen Leitstände für die Steuerung der Rohmehlzusammensetzung, der Homogenisierung und der Zementqualität im Labor zentral zusammenzufassen. Es zeigte sich, daß damit nicht nur ein Rationalisierungseffekt und eine signifikante Erhöhung des Qualitätsstandards verbunden waren, sondern auch eine starke Entlastung der jeweiligen Produktionssteuerer von ermüdender Routine. Aufgrund dieser Erfolge erschien es vielversprechend, nunmehr auch die Möglichkeiten der rechnergesteuerten Prozeßführung miteinzubeziehen. In Zusammenarbeit mit Krupp-Polysius wurden die Systeme „Kiln Control" an einem Drehofen und „Mill Control" an einer der großen Zementmühlen installiert.

Nach erheblichen Optimierungsbemühungen über einen längeren Zeitraum wurde deutlich, daß diese Systeme auf Dauer zu komplex, zu wartungsaufwendig und zu wenig flexibel sein würden. Die Lösung konnte nur in einer Software liegen, die nach den Methoden der künstlichen Intelligenz selbstlernende, adaptive Komponenten enthält. Deshalb wurde zusammen mit Krupp-Polysius das Steuerungssystem „MCE" (Mill-Control-Expert) entwickelt, das wie ein Expertensystem aufgebaut ist. Dieses System ist in der Lage, jeden Betriebszustand einer Zementmühle, angefangen von deren Start bis hin zum Abstellen inklusive aller Zementsortenwechsel, vollautomatisch zu beherrschen.

Nach diesen Erfolgen wurde deutlich, daß nur intelligente Regelsysteme den hohen Anforderungen der Zementtechnologie gerecht werden können. Die weiteren Entwicklungen konzentrieren sich daher auf diese Systeme, und zwar

The programme for modernizing and restructuring the Schwenk cement works at the start of the 80s included the task of incorporating the modern capabilities of laboratory automation and process control technology which had become available by then (**Fig. 1**).

The first objective was to retain the function of the many existing local control stations for the time being, but to use methods of modern data transfer and signal exchange to relocate their operation, thereby making it possible to centralize the process control without having to replace the entire control system from the bottom up. At the same time the Polab laboratory automation system was installed jointly with Krupp-Polysius. This was then still a very com-

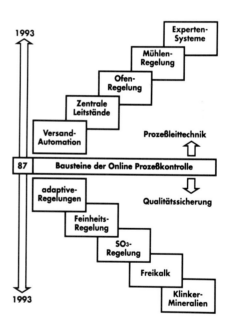

BILD 1: Komponenten der Prozeß- und Laborautomation
FIGURE 1: Components of the process and laboratory automation

Experten-Systeme	expert systems
Mühlen-Regelung	mill control
Ofen-Regelung	kiln control
Zentrale Leitstände	central control stations
Versand-Automation	despatch automation
Prozeß-Leittechnik	process control technology
Baust...der Online Prozeßk.	on-line process control modules
Qualitätssicherung	quality assurance
adaptive Regelungen	adaptive control systems
Feinheits-Regelung	fineness control
SO_3-Regelung	SO_3 control
Freikalk	free lime
Klinker-Mineralien	clinker minerals
Von der Idee zur Bausteinerprobung	From concept to module testing

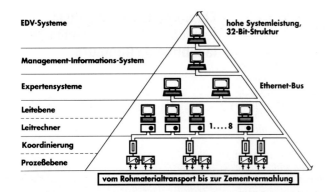

BILD 4: Hierarchische Struktur der Prozeßleittechnik
FIGURE 4: The hierarchical structure of process control technology

EDV-Systeme	EDP systems
Management-Informations-Sys.	management information system
Expertensysteme	expert systems
Leitebene	control level
Leitrechner	control computers
Koordinierung	coordination
Prozeßebene	process level
hohe Systemleistung,	high system capacity,
32-Bit-Struktur	32 bit structure
Ethernet-Bus	Ethernet bus
vom Rohmaterialtr.	from raw material transport to
Zementverm.	cement grinding

halb, alle im Prozeß anfallenden Daten inklusive der Qualitätsdaten des Labors zunächst in einem Datenbanksystem abzuspeichern. Anschließend werden diese Daten mit einem überlagerten, komplexen Statistikprogramm teils automatisch, teils nach vorgegebenem Algorithmus beliebig miteinander kombiniert und über Korrelations- und Regressionsanalysen auf ihren funktionalen Zusammenhang hin untersucht.

Das Endziel soll sein, die verschiedensten Daten aus dem Prozeß- und Qualitätsbereich ständig nach statistischen Methoden mit einem automatisch arbeitenden Programm miteinander in Beziehung zu setzen und diejenigen Größen auszufiltern, bei denen der Korrelationskoeffizient auf einen genügend signifikanten Zusammenhang hindeutet. Diese Daten stehen dann dem Betriebspersonal für weitere Überlegungen und Schlußfolgerungen zur Verfügung.

Mit der gesamten Automationstechnik des Werkes Bernburg wurde ein Instrument geschaffen, mit dem nicht nur den zukünftigen Anforderungen durch ein Zertifizierungssystem, sondern auch der ständigen Herausforderung nach Minimierung des Energieeinsatzes und Maximierung der Qualitätskriterien voll entsprochen werden kann.

BILD 5: Architektonische Gestaltung des neuen zentralen Leitstand-Gebäudes
FIGURE 5: Architectonic form of the new central control room building

capabilities of X-ray fluorescence analysis and X-ray diffractometry with the measurement of free lime and the clinker minerals, automatic checking of cement fineness, cement colour measurement, etc.

One prime objective is to minimize the bypass, which is needed at the Bernburg cement plant because of the high chloride content. For the first time a hot meal sampling system was therefore designed for the kiln inlet so that hot meal samples could be taken continuously from the critical region and analyzed using the automatic sample system. This was controlled by the POLYSIUS POLCID DC control system, within which the KCE and MCE control systems corresponding to the expert systems described above are integrated and implemented. The KCE expert system for the kiln plant made it possible for the first time for the automatically determined values for free lime and for the clinker minerals to be incorporated in the control strategy.

As can be seen in **Fig. 4**, all process and quality control systems should be linked and networked with one another by an overlay information management system (IMS) so that full use can be made of the potential of statistical evaluation followed by appropriate optimization. It has been found that information and signals flow from modern systems in such great quantities that the operating personnel can no longer cope with their diversity and complexity. The aim is therefore to store all the process data, including the quality data from the laboratory, in a database system for the time being. These data are combined with one another by a complex overlay statistical program, partially automatically and partially in accordance with predetermined algorithms, and then investigated for functional relationships using correlation and regressional analyses.

The final aim should be to relate the very different data from the process and quality sectors to one another continuously by statistical methods with an automatic program, and to filter out those variables where the correlation coefficient points to a sufficiently significant relationship. These data are then available to the operating personnel for further examination and conclusions.

The overall automation technology at the Bernburg works has created an instrument which is fully capable of meeting not only the future demands caused by a certification system, but also the continuous requirement to minimize the use of energy and maximize the quality criteria.

On-line Kontrolle des Mischbettaufbaus mittels PGNNA in Ramos Arizpe (Mexiko)

On-line control of the blending bed composition using PGNAA in Ramos Arizpe (Mexico)

Contrôle en temps réel d'une constitution de lits de mélange au moyen de PGNAA à Ramos Arizpe (Mexique)

Control on-line de la formación del lecho de mezcla por medio del PGNAA en Ramos Arizpe (México)

Von **M. Tschudin**, Schweiz

Zusammenfassung – Erstmals in der „Holderbank"-Gruppe wurde im neuen Werk Ramos Arizpe von Cementos Apasco (Mexiko) ein auf „Prompt-Gamma-Neutronen-Aktivierungs-Analyse" (PGNAA) basierender Analysator zur Kontrolle des Aufbaus zweier Mischbetten installiert. Ausschlaggebend für die Anschaffung des PGNAA-Gerätes waren, neben anderen Vorteilen wie Erfassung des integralen Materialstromes und Echtzeitanalyse, insbesondere das Vermeiden eines konventionellen Probenahme- und Aufbereitungsturms. Die vertraglichen Garantiewerte wurden bei der Inbetriebnahme und während des Routinebetriebes einer intensiven Überprüfung unterzogen. Dabei zeigte sich, daß die Präzision der Elementanalysen – ausgedrückt als Standardabweichungen von wiederholten Messungen an einem Kalibrationsblock – zu keinen Beanstandungen Anlaß gab. Die Garantiewerte für die Genauigkeit hingegen – definiert als Maß der Abweichung zwischen PGNAA und Röntgenspektrometer – konnten noch nicht für alle Elemente erreicht werden. Die weitgehend systematischen und reproduzierbaren Abweichungen erlauben aber trotzdem einen kontrollierten und nahe dem Zielwert liegenden Aufbau der Mischbetten. Anfängliche Betriebsprobleme ergaben sich in erster Linie durch Staubeintritt in das Analysatorgehäuse und durch Verklebung bei feuchten Rohstoffen. Die Wartung beschränkt sich im Normalbetrieb auf das periodische Ersetzen der Auskleidung des Analysatorschachts, das – wie in diesem Fall – bei einem Durchsatz von maximal 900 t/h zu nicht unerheblichen Kosten führen kann. Die prinzipiell positiven Erfahrungen mit PGNAA stellen eine zukunftweisende Alternative zur traditionellen Mischbettkontrolle dar.

Summary – An analyzer based on "Prompt Gamma Neutron Activation Analysis" (PGNAA) was installed for the first time in the "Holderbank" group in the new Ramos Arizpe works of Cementos Apasco (Mexico) for controlling the composition of two blending beds. In addition to other advantages, such as measurement of the integral material flow and real-time analysis, a decisive factor in the purchase of the PGNAA equipment was the avoidance of a conventional sampling and processing tower. The contractual guarantee values were submitted to intensive checking at commissioning and during routine operation. This showed that the precision of the elemental analysis – expressed as standard deviations of repeated measurements on a calibration block – gave no cause for complaint. On the other hand the guarantee values for the accuracy – defined as a measure of the deviation between PGNAA and X-ray spectrometer – could not be achieved for all elements. In spite of this the largely systematic and reproducible deviations permit controlled formation of the blending beds with compositions close to the target value. Initial operating problems arose primarily through dust ingress into the analyzer housing and through sticking with moist raw materials. In normal operation the maintenance is confined to periodic replacement of the lining of the analyzer shaft which, at a maximum throughput of 900 t/h as is the case here, can lead to not inconsiderable costs. In principle, the results with PGNAA were positive and the system represents a forward-looking alternative to traditional blending bed control.

Résumé – Pour la première fois dans le groupe „Holderbank" a été installé, à la nouvelle usine Ramos Arizpe de Cementos Apasco (Mexique) un analyseur basé sur „l'analyse par activation neutronique Prompt Gamma" (PGNAA) pour le contrôle de la constitution de deux lits de mélange. Décisif pour l'acquisition de l'instrument PGNAA a été, mis à part d'autres avantages comme saisie de l'intégralité du courant matière et analyse en temps réel, d'éviter la construction d'une tour conventionelle d'échantillonnage et de préparation. Au moment de la mise en service et pendant la marche de routine, les valeurs garanties par contrat ont été soumises à une vérification intensive. Il s'y est montré, que la précision des analyses élémentaires – exprimée en déviations standard de mesures répétées sur un bloc de calibrage – n'a donnée lieu à aucune contestation. Les valeurs garanties pour l'exactitude – définie comme mesure de l'écart entre PGNAA et spectromètre X – n'ont par contre pas encore pu être atteintes pour tous les éléments. Les écarts en grande partie systématiques et reproductibles permettent, néanmoins, la constitution contrôlée et située près de la valeur

de consigne, des lits de mélange. Des problèmes de fonctionnement au début ont résulté en premier lieu de pénétration de poussières dans le logement de l'analyseur et de colmatage avec des matières premières humides. En exploitation normale, la maintenance se limite au remplacement périodique du revêtement du puits de l'analyseur, ce qui – comme dans le cas présent – peut conduire à des dépenses non négligeables avec un débit maximum de 900 t/h. L'expérience principalement positive désigne PGNAA comme alternative pleine d'avenir au contrôle traditionnel des lits de mélange.

Control on-line de la formación del lecho de mezcla por medio del PGNAA en Ramos Arizpe (México)

***Resumen** – El grupo „Holderbank" ha instalado por primera vez, en la factoría Ramos Arizpe de Cementos Apasco (México), un analizador basado en el „análisis por activación de neutrones Prompt Gamma", para el control de la formación de dos lechos de mezcla. Ha sido decisivo, para la adquisición del aparato PGNAA, aparte de otras ventajas, tales como el registro del flujo integral de material y el análisis en tiempo real, el hecho de que se pueda prescindir de una torre convencional de toma y preparación de muestras. Los valores de garantía, fijados por contrato, han sido sometidos a un control intensivo en el momento de la puesta en servicio y también durante el servicio normal. Resulta que la precisión de los análisis de elementos – expresada como desviaciones patrón de repetidas mediciones en un bloque calibrador – no ha dado lugar a ninguna clase de reclamaciones. Sin embargo, los valores de garantía referentes a la precisión – definida como medida de la desviación entre el PGNAA y el espectrómetro de rayos X – no se han alcanzado aún para todos los elementos. Las desviaciones mayormente sistemáticas y reproducibles permiten, no obstante, una formación del lecho de mezcla controlada y próxima al valor pretendido. Los problemas iniciales eran debidos, sobre todo, a la entrada de polvo en el interior de la carcasa del analizador y a las pegaduras producidas por las materias húmedas. El mantenimiento se limita, durante el servicio normal, al cambio periódico del revestimiento del pozo del analizador, lo que puede conducir – como en el caso presente – para un caudal máx. de aprox. 900 t/h, a unos gastos bastante elevados. Las experiencias adquiridas con el PGNAA, que son en principio positivas, representan una alternativa prometedora del control tradicional de los lechos de mezcla.*

1. Einleitung

Im Zuge der Projektierung eines neuen Zementwerkes für Cementos Apasco in Ramos Arizpe/Mexiko wurden für die Rohmaterialkontrolle verschiedene Alternativen studiert, nachdem die Entscheidung zugunsten des integrierten Mischbettverfahrens gefällt worden war. Als mögliche Alternativen für die Kontrolle des Mischbettaufbaus verblieben der traditionelle Probenahmeturm und die seit wenigen Jahren verfügbare Technologie der „Prompt-Gamma-Neutronen-Aktivierungs-Analyse" (PGNAA). Kontinuierliche Fortschritte bei der praktischen Anwendung dieser Technologie sowie vielversprechende analytische Tests beim zur Zeit einzigen Hersteller von industriell einsetzbaren Geräten haben dazu geführt, daß im Jahre 1991 erstmalig in der „Holderbank"-Gruppe ein solcher Analysator (Bulk Material Analyzer) von Gamma-Metrics installiert wurde.

Ausschlaggebend für die Anschaffung des Gerätes waren die für diese Technologie charakteristischen Vorteile der Erfassung des integralen Materialstromes und der Echtzeitanalyse. Die Alternative eines konventionellen mechanischen Probenahmeturmes fiel letztlich bei ähnlichen Investitions-, aber erwarteten höheren Unterhaltskosten aus der Entscheidung.

Der vorliegende Bericht stellt die Installation in Ramos Arizpe vor und gibt einen Überblick über die beobachtete analytische Leistungsfähigkeit des Gerätes. Außerdem wird über allgemeine Betriebserfahrungen in den ersten zweieinhalb Jahren berichtet.

2. Installation

Bild 1 zeigt schematisch die Rohmaterialaufbereitung in Ramos Arizpe. Der Analysator selbst ist, wie aus **Bild 2** ersichtlich, in einem mehrgeschossigen Turm untergebracht, der zwischen Brecher und Mischbett steht. Über ein Aufgabesilo mit einer Kapazität von 30 t, das die Mengenschwankungen des Materials aus dem Brecher ausgleicht, gelangt das gebrochene Material in den Analysatorschacht. Die Durchflußrate wird mit Hilfe eines unterhalb des Analysators befindlichen Gurtförderers mit variabler Geschwindigkeit geregelt. Der in Ramos Arizpe installierte Analysatortyp (3612L) weist einen Schachtquerschnitt von 30 cm × 90 cm auf und erlaubt eine maximale Durchflußrate von 900 t/h. Mit Hilfe einer Ultraschallsonde und eines Regelkreises wird

1. Introduction

During the planning of a new cement works for Cementos Apasco at Ramos Arizpe (Mexico) various alternatives for raw material control were studied after a basic decision in favour of integrated blending beds had been taken. The two alternatives finally arrived at for controlling blending bed composition were the traditional sampling tower and the "prompt gamma neutron activation analysis" technology (PGNAA) which had only been available for a few years. The continual advances in practical application of this technology and the promising analytical tests carried out by the then sole manufacturer of industrial-scale apparatus led to the first installation of such a bulk material analyzer (BMA) produced by Gamma-Metrics in the Holderbank group in 1991.

The decisive factors in the purchase of the PGNAA apparatus were the method's characteristic advantages of measurement of the integral material flow and real-time analysis. The alternative of a conventional mechanical sampling tower was eventually discarded since capital costs were similar but costs for its maintenance were expected to be higher.

This report describes the installation at Ramos Arizpe and its observed analytical performance. It also covers general operating experience during the first two and a half years.

2. Installation

Fig. 1 is a diagram showing the raw material processing at Ramos Arizpe. The analyzer itself is, as can be seen from **Fig. 2**, housed in a multi-storey tower between the crusher and the blending bed. The crushed material passes through a 30 t feed bin which compensates for fluctuations in the quantity of material from the crusher, and into the analyzer chute. Throughput is controlled by a variable-speed conveyor belt below the analyzer. The type of analyzer installed at Ramos Arizpe (3612L) has a chute section of 30 × 90 cm, allowing a maximum throughput of 900 t/h. The filling level of the feed bin is kept constant by an ultrasonic probe and a control loop. This ensures that the analyzer chute is always full of material. Hence measuring stability is guaranteed and there is no neutron leakage. The analyzer is controlled and

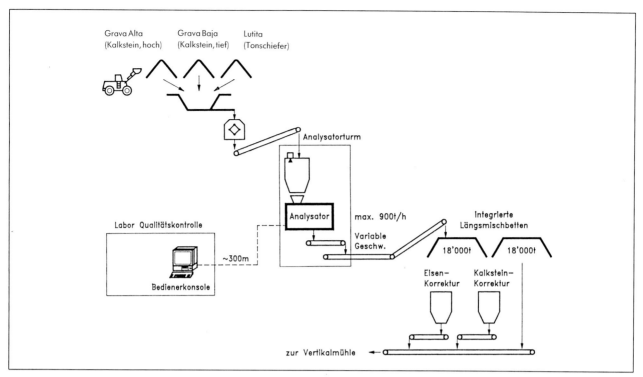

BILD 1: Schematische Darstellung der Rohmaterialaufbereitung bei Cementos Apasco mit eingezeichneter Position des PGNA-Analysators
FIGURE 1: Diagrammatic representation of raw material preparation system at Cementos Apasco, showing the position of the PGNA analyzer

Kalkstein, hoch	= limestone, high	Bedienerkonsole	= operator console
Kalkstein, tief	= limestone, low	Variable Geschw.	= variable speed
Tonschiefer	= clay shale	integrierte Längsmischbetten	= integrated linear blending beds
Analysatorturm	= analyzer tower	Eisen-Korrektur	= iron correction
Analysator	= analyzer	Kalkstein-Korrektur	= limestone correction
Labor	= laboratory	zur Vertikalmühle	= to the vertical mill
Qualitätskontrolle	= quality control		

BILD 2: Blick auf den Analysatorturm mit dem PGNA-Analysator auf der zweiten Etage
FIGURE 2: View of the analyzer tower with the PGNA analyzer on the second level

monitored by an operator console located in the quality control laboratory about 300 m away. All operations and controls can be carried out from this computer, so no access to the analyzer is necessary apart from regular maintenance.

Ramos Arizpe uses three raw material constituents of very different chemical composition, a high-grade limestone of relatively constant composition, a medium- to low-grade limestone of greater variability, and a shale (**Table 1**). The three materials are put into heaps in front of the crusher and fed to it with a front-end loader. Experience has shown that a far more homogeneous blending bed is obtained if the constituents are fed in so that the bed is always near the target value during its formation. The integral composition of the blending bed displayed at the operator console enables the feed ratio to be adjusted at any time. The beds are finalized slightly below the target value for lime saturation. Fine correction of the raw mix is effected by adding high-grade limestone and iron corrective material before the mill.

TABLE 1: Chemical composition of raw material components used (average values for 1992)

Constituent %	Grava Alta (limestone, low)	Grava Baja (limestone, high)	Lutita (clay shale)
loss on ignition	41.0	32.8	11.3
SiO_2	4.7	19.3	53.3
Al_2O_3	0.9	3.8	12.4
Fe_2O_3	0.7	1.3	4.7
CaO	51.0	37.9	10.3
MgO	0.8	2.4	2.0
SO_3			1.0
K_2O	0.20	0.91	2.20
Na_2O	0.06	0.21	1.16
lime saturation	348	64	6
silica modulus	2.9	3.8	3.1
alumina ratio	1.3	2.9	2.6

der Füllungsgrad des Aufgabesilos konstant gehalten. Dadurch wird sichergestellt, daß der Analysatorschacht jederzeit mit Material gefüllt ist. Damit ist die Meßstabilität gewährleistet, und es tritt keine Neutronenstrahlung aus. Gesteuert und überwacht wird der Analysator mittels einer Bedienerkonsole, die sich im Labor der Qualitätskontrolle, ungefähr 300 m vom Analysatorstandort entfernt, befindet. Sämtliche Operationen und Kontrollen können von diesem Rechner aus durchgeführt werden, so daß sich der Zugang zum Analysator, von regelmäßigen Wartungsarbeiten abgesehen, erübrigt.

Ramos Arizpe verwendet 3 Rohmaterialkomponenten mit deutlich unterschiedlichen chemischen Zusammensetzungen: einen hochprozentigen Kalkstein mit relativ konstanter Zusammensetzung, einen mittel- bis tiefprozentigen Kalkstein mit erhöhter Variabilität und einen Tonschiefer (**Tabelle 1**). Vorratshaufen der 3 Komponenten werden vor dem Brecher bereitgestellt und mit einem Schaufellader diesem aufgegeben. Die Erfahrung hat gezeigt, daß die Homogenität des Mischbettes deutlich besser ausfällt, wenn die Komponenten so aufgegeben werden, daß sich das Mischbett während des Aufbaus immer nahe dem Zielwert befindet. Die auf der Bedienerkonsole ablesbare integrierte Zusammensetzung des Mischbettes ermöglicht jederzeit die Anpassung des Aufgabeverhältnisses. Abgeschlossen wird der Aufbau der Mischbetten hinsichtlich der Kalksättigung jeweils leicht unter dem Zielwert. Die Feinabstimmung der Rohmischung wird durch Zugabe von hochprozentigem Kalkstein und einer Eisenkorrektur vor der Mühle vorgenommen.

TABELLE 1: Chemische Zusammensetzung der verwendeten Rohmaterialkomponenten
(Mittelwerte des Jahres 1992)

Bestandteil %	Grava Alta (Kalkstein, hoch)	Grava Baja (Kalkstein, tief)	Lutita (Tonschiefer)
Glühverlust	41,0	32,8	11,3
SiO_2	4,7	19,3	53,3
Al_2O_3	0,9	3,8	12,4
Fe_2O_3	0,7	1,3	4,7
CaO	51,0	37,9	10,3
MgO	0,8	2,4	2,0
SO_3			1,0
K_2O	0,20	0,91	2,20
Na_2O	0,06	0,21	1,16
Kalksättigung	348	64	6
Silikatmodul	2,9	3,8	3,1
Tonerdemodul	1,3	2,9	2,6

3. Kalibration und analytische Leistungstests

Um die analytische Leistungsfähigkeit des Analysators zu spezifizieren, wurden aufgrund von Vorversuchen Garantiewerte für die Präzision (Wiederholbarkeit) und die Genauigkeit vertraglich festgehalten und während der Inbetriebnahme sowie im Verlaufe der ersten Monate überprüft. Der Analysator wurde beim Hersteller vorkalibriert, die endgültige Kalibration und deren Überprüfung jedoch erst während der Installation vorgenommen. Der ganze Prozeß im Werk selbst dauerte ungefähr eine Woche. Neben der Kalibration der 7 chemischen Elemente Silicium, Aluminium, Eisen, Calcium, Magnesium, Kalium und Natrium wurden weitere Parameter wie Dichte, Feuchtigkeit und Durchflußrate des Analysators kalibriert.

Zur Kalibration der chemischen Zusammensetzung dienten 6 mit werkspezifischen Rohmaterialien gefüllte Kalibrationsblöcke. Die Materialien wurden so ausgesucht, daß der ganze chemische Bereich abgedeckt war. Eine sorgfältige Aufbereitung, Homogenisierung und Beprobung der Materialien lieferte die Proben für die Referenzanalysen mittels Röntgenfluoreszenz, die als Grundlage zur Kalibration dienten. Um analytische Konsistenz zu gewährleisten, wurde das Röntgenspektrometer des Zementwerkes, mit dem die Überprüfung der Mischbetten durchgeführt wurde, mit denselben Materialien kalibriert wie der PGNA-Analysator.

3. Calibration and analytical performance tests

As a means of determining the analytic performance of the analyzer guarantee values for precision (repeatability) and accuracy, based on preliminary tests, were laid down in the contract and checked at commissioning and during the first few months of operation. The analyzer was pre-calibrated at the manufacturer's works but final calibration and checking were carried out during installation. The whole process on the actual site took about a week. Apart from the seven chemical elements silicon, aluminium, iron, calcium, magnesium, potassium and sodium other parameters were calibrated such as density, humidity and the analyzer throughput.

Six calibration blocks filled with raw materials specific to the plant were used for calibrating the chemical composition. The materials were selected so as to cover the full chemical range. Careful preparation, homogenisation and sampling of the materials provided the specimens for reference analysis by X-ray fluorescence which served as a basis for calibration. To ensure analytic consistency the plant X-ray spectrometer used to carry out the blending bed tests described below was calibrated with the same materials as the PGNA analyzer.

3.1 Precision

Analytical precision – expressed as standard deviations of repeated measurements – was determined by measuring one of the calibration blocks a total of 18 times at 10-minute intervals over a period of 3 hours. Analytical repeatability was found to be good for all the elements covered, and the results obtained were sometimes even far better than required (**Table 2**).

TABLE 2: Results of precision measurements (statistical repeatability) on one of the six calibration blocks. The statistical values relate to eighteen readings at 10-minute intervals

Constituent	Measured value standard deviation (%, abs.)	Guarantee standard deviation (%, abs.)	Concentration (wt. %)
SiO_2	0.16	0.27	17.71
Al_2O_3	0.07	0.21	2.19
Fe_2O_3	0.02	0.03	0.48
CaO	0.13	0.19	41.35
MgO	0.04	0.31	1.71
K_2O	0.04	0.04	1.03
Na_2O	0.04	0.08	0.16

3.2 Accuracy determined on the basis of the calibration blocks

The accuracy of the apparatus – defined as a measure of the deviation between PGNAA and X-ray spectrometer – was first checked using the six calibration blocks. The strict guarantee values for the statistical accuracy thus measured could not be achieved for some elements such as iron, magnesium and potassium (**Table 3**).

TABLE 3: Accuracy of the analyzer determined with the six calibration blocks. The reference values are X-ray fluorescence analyses of the calibration materials

Constituent	Measured value RMSD*) (%, abs.)	Guarantee RMSD (%, abs.)	Concentration range (wt. %)
SiO_2	0.93	1.44	0.7–55.7
Al_2O_3	0.28	0.71	0.1–14.7
Fe_2O_3	0.17	0.09	0.1–5.3
CaO	1.38	1.87	7.6–54.5
MgO	0.64	0.43	0.8–2.8
K_2O	0.17	0.12	0.02–2.25
Na_2O	0.14	0.16	0.03–1.52

*) RMSD = root mean square deviation

3.1 Präzision

Die analytische Präzision – ausgedrückt als Standardabweichung von wiederholten Messungen – wurde bestimmt, indem einer der Kalibrationsblöcke über einen Zeitraum von 3 Stunden insgesamt 18 mal in 10-Minuten-Intervallen gemessen wurde. Es zeigte sich, daß die analytische Wiederholbarkeit aller erfaßten Elemente gut ist und teilweise sogar deutlich bessere Werte lieferte als gefordert (**Tabelle 2**).

TABELLE 2: Ergebnisse der Präzisionsmessungen (statistische Wiederholbarkeit) an einem der 6 Kalibrationsblöcke. Die statistischen Werte beziehen sich auf 18malige 10-Minuten-Messungen

Bestandteil	gemessene Werte, Standardabweichung (%, abs.)	Garantie, Standardabweichung (%, abs.)	Konzentration (Gew.%)
SiO_2	0,16	0,27	17,71
Al_2O_3	0,07	0,21	2,19
Fe_2O_3	0,02	0,03	0,48
CaO	0,13	0,19	41,35
MgO	0,04	0,31	1,71
K_2O	0,04	0,04	1,03
Na_2O	0,04	0,08	0,16

3.2 Genauigkeit bestimmt aufgrund der Kalibrationsblöcke

Die Genauigkeit des Gerätes – ausgedrückt als Abweichung zwischen PGNAA und Röntgenspektrometer – wurde zuerst anhand der 6 Kalibrationsblöcke überprüft. Die strengen Garantiewerte für die auf diese Weise gemessene statische Genauigkeit konnten für einige Elemente wie Eisen, Magnesium und Kalium nicht eingehalten werden (**Tabelle 3**).

TABELLE 3: Genauigkeit des Analysators, bestimmt aufgrund der 6 Kalibrationsblöcke. Als Referenzwerte dienten Röntgenfluoreszensanalysen der Kalibrationsmaterialien

Bestandteile	gemessene Werte, RMSD*) (%, abs.)	Garantie, RMSD (%, abs.)	Konzentrationsbereich (Gew.%)
SiO_2	0,93	1,44	0,7–55,7
Al_2O_3	0,28	0,71	0,1–14,7
Fe_2O_3	0,17	0,09	0,1–5,3
CaO	1,38	1,87	7,6–54,5
MgO	0,64	0,43	0,8–2,8
K_2O	0,17	0,12	0,02–2,25
Na_2O	0,14	0,16	0,03–1,52

*) RMSD = Root Mean Standard Deviation

3.3 Genauigkeit bestimmt aufgrund der Mischbettüberprüfung

Wichtiger jedoch als die an den Kalibrationsblöcken bestimmte Genauigkeit ist die praxisrelevante Prüfung der Übereinstimmung von tatsächlicher Mischbettzusammensetzung mit der vom Analysator ermittelten Zusammensetzung. Zu diesem Zweck wurden dazu geeignete, d.h. sich nahe oder direkt auf Zielwert befindliche Mischbetten während des Abbaus überprüft. Auf eine direkte Beprobung während des Mischbettabbaus mußte verzichtet werden, da sich die notwendigen Probenmengen als nicht prozessierbar erwiesen. Aus diesem Grund wurde das Rohmehl nach der Mühle stündlich analysiert und aus dem gewichteten Mittel und unter Berücksichtigung der Korrekturstoffe der Vergleichswert zur PGNAA bestimmt. Zu erwähnen ist, daß sich die für die Tests vereinbarten Garantiewerte für Silicium, Aluminium, Eisen und Calcium nicht direkt auf die Elemente beziehen, sondern auf die daraus resultierenden Moduli Kalksättigung, Silikatmodul und Tonerdemodul.

In einer ersten Serie wurden 4 Mischbetten kontrolliert. Weitere Mischbetten konnten aufgrund eines technischen

3.3 Accuracy determined on the basis of blending bed tests

However, it is of more practical importance to establish how closely the actual blending bed composition corresponds to that determined by the analyzer than to establish accuracy on the basis of the calibration blocks. Suitable blending beds, i.e. beds at or near the target value, were tested for this purpose during reclaiming. There could be no direct sampling while the bed was being reclaimed as it proved impossible to process the necessary quantities of the specimens. For this reason the raw meal leaving the mill was analyzed hourly and the value compared with PGNAA was determined from the weighted mean, allowing for the corrective substances. It should be mentioned that the guarantee values for silicon, aluminium, iron and calcium agreed for the tests relate not to the actual elements but to the resultant moduli – lime saturation, the silica modulus and the alumina ratio.

Four blending beds were checked in a first test series. Others could only be examined after recalibration owing to a technical defect in one of the detectors. This explains the grouping of the results in the following diagrams.

Examples for moduli, and for two elements, are given in **Figs. 3** and **4**, and the full results are set out in **Table 4**. The most important results are:

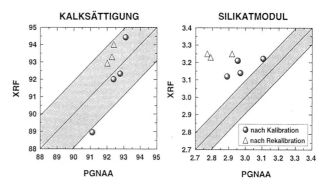

BILD 3: Genauigkeit des PGNA-Analysators aufgrund der Überprüfung von Mischbetten. Als Beispiel wiedergegeben sind die Moduli Kalksättigung und Silikatmodul. Die schattierten Balken geben die vertraglich garantierten Bereiche an.
FIGURE 3: Accuracy of the PGNA analyzer based on testing blending beds. The lime saturation factor and the silica modulus are taken as examples. The shaded bars indicate the contractually guaranteed ranges.

Kalksättigung	= lime saturation
Silikatmodul	= silica modulus
nach Kalibration	= after calibration
nach Rekalibration	= after recalibration

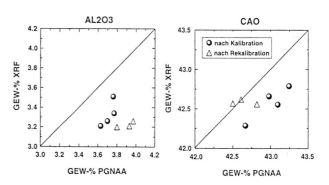

BILD 4: Genauigkeit aufgrund der Überprüfung von Mischbetten. Beispiel für ein Element mit ungenügender Übereinstimmung (Aluminium) und zufriedenstellender Übereinstimmung (Calcium)
FIGURE 4: Accuracy based on testing blending beds. An example of an element where results do not correspond closely enough (aluminium) and one where they correspond satisfactorily (calcium)

Gew.	= wt
nach Kalibration	= after calibration
nach Rekalibration	= after recalibration

TABELLE 4: Genauigkeit des Analysators, bestimmt durch die Überprüfung der Mischbetten
(PGNAA-Resultat minus Röntgenspektrometer-Resultat)

Bestandteil	nach Kalibration				nach Rekalibration				
%	#1	#2	#3	#4	#5	#6	#7	Garantie	Konzentration
SiO_2	−0,36	−0,12	0,00	0,27	−0,20	−0,03	0,05		1,46
Al_2O_3	0,25	0,43	0,44	0,42	0,71	0,72	0,60		3,2
Fe_2O_3	0,00	−0,07	−0,17	−0,19	0,01	0,04	−0,08		1,3
CaO	0,39	0,32	0,55	0,46	−0,17	−0,01	0,26		42,5
MgO	0,07	0,04	−	−0,01	−0,02	−0,06	−0,08	±0.25	1,2
K_2O	0,25	0,24	−	0,30	−0,13	−0,08	−0,08	±0.05	0,7
Na_2O	−0,03	0,12	−	0,08	0,34	0,33	0,28	±0,06	0,3
Kalksättigung	2,17	0,44	0,37	−1,33	−0,92	−1,62	−1,02	±1,8	93
Silikatmodul	−0,23	−0,26	−0,17	−0,11	−0,48	−0,47	−0,33	±0,093	3,2
Tonerdemodul	0,19	0,52	0,64	0,78	0,56	0,47	0,64	±0,065	2,5

TABLE 4: Accuracy of the analyzer determined by testing the blending beds
(PGNAA result minus X-ray spectrometer result)

Constituent	after calibration				after recalibration				
%	#1	#2	#3	#4	#5	#6	#7	guarantee	concentration
SiO_2	−0.36	−0,12	0.00	0.27	−0.20	−0.03	0.05		1.46
Al_2O_3	0.25	0.43	0.44	0.42	0.71	0.72	0.60		3.2
Fe_2O_3	0.00	−0.07	−0.17	−0.19	0.01	0.04	−0.08		1.3
CaO	0.39	0.32	0.55	0.46	−0.17	−0.01	0.26		42.5
MgO	0.07	0.04	−	−0.01	−0.02	−0.06	−0.08	±0.25	1.2
K_2O	0.25	0.24	−	0.30	−0.13	−0.08	−0.08	±0.05	0.7
Na_2O	−0.03	0.12	−	0.08	0.34	0.33	0.28	±0.06	0.3
lime saturation	2.17	0.44	0.37	−1.33	−0.92	−1.62	−1.02	±1.8	93
silica modulus	−0.23	−0.26	−0.17	−0.11	−0.48	−0.47	−0.33	±0.093	3.2
alumina modulus	0.19	0.52	0.64	0.78	0.56	0.47	0.64	±0.065	2.5

Defektes eines der Detektoren erst nach einer Rekalibration überprüft werden. Das erklärt die Gruppierung der Resultate in den nachfolgenden Bildern.

Beispiele für Moduli, aber auch für zwei Elemente sind in **Bild 3** und **4** gegeben, vollständige Resultate finden sich in **Tabelle 4**. Die wichtigsten Ergebnisse waren:

− Die Kalksättigungen der geprüften Mischbetten liegen mit Ausnahme eines einzigen Wertes innerhalb der garantierten 1-Sigma-Bandbreite von 1.8 Einheiten.

− Für die Silikat- und Tonerdemoduli sind signifikante Abweichungen von den geforderten Werten offensichtlich. Alle Werte liegen außerhalb der Spezifikationen, jedoch in systematischer Art und Weise.

− Die Hauptursache für die Abweichungen der Moduli liegt darin begründet, daß Aluminium, gemessen mit PGNAA, signifikant höhere Werte ergibt. Die anderen Hauptelemente Silicium, Eisen und Calcium zeigen dagegen eine zufriedenstellende Übereinstimmung.

− Elemente, die für das Zementwerk im praktischen Betrieb nicht von vorrangiger Bedeutung sind (Magnesium, Alkalien), zeigen unterschiedliche und verglichen mit den Kalibrationsblöcken teilweise widersprüchliche Ergebnisse. Das hängt einerseits mit der Unempfindlichkeit der Analysenmethode für einige Elemente zusammen (v. a. Magnesium), aber auch mit den in den Mischbetten gemessenen tiefen Konzentrationsniveaus für Kalium und Natrium (0.7 bzw. 0.3 %).

Für einige der beschriebenen Phänomene konnten bis zum jetzigen Zeitpunkt keine befriedigenden Erklärungen gefunden werden. Wichtig bei der Beurteilung der Leistungsdaten und für die praktische Anwendung ist aber die nochmalige Feststellung, daß die Abweichungen systematischer Natur sind.

4. Betriebserfahrungen

Ob der Analysator die in ihn gesetzten Erwartungen erfüllen kann, wird nicht nur durch die Leistungsdaten bestimmt, sondern auch durch die praktische Erfahrung im Routine-

− Except for one value, lime saturation in the beds tested comes within the guaranteed 1-sigma bandwidth of 1.8 units.

− There are clearly significant deviations from the required values for the silica modulus and alumina ratio. All the values are outside the specifications, but the deviations are systematic.

− The main cause of the deviations of the moduli is that aluminium gives significantly higher values when measured by PGNAA, whereas results for the other main elements, silicon, iron and calcium, have a satisfactory similarity.

− Elements which are not of prime importance to the actual operations at Ramos Arizpe (magnesium and alkalis) produce different results, sometimes inconsistent with the calibration blocks. This has to do with the insensitivity of the analytical methods for some elements (especially magnesium) and also with the low concentration of potassium and sodium measured in the blending beds (0.7 and 0.3 % respectively).

So far no satisfactory explanations have been found for some of the findings described. In judging performance data and for practical purposes however, it is important to repeat that the deviations are of a systematic nature.

4. Operating experience

Whether the analyzer can meet expectations will be determined not by performance data but by practical experience in routine operation. The following points have emerged clearly from the first two and a half years' operating experience:

− The systematic deviations observed allow the composition of the blending beds to be controlled close to the target value. The addition of limestone corrective before grinding only involves small quantities (0−3 %) and in favourable cases is quite unnecessary.

betrieb. Die Betriebserfahrungen der ersten zweieinhalb Jahre haben folgendes klar gezeigt:

– Die beobachteten systematischen Abweichungen erlauben einen kontrollierten Aufbau der Mischbetten nahe dem Zielwert. Die Zugabe von Korrekturkalkstein vor der Mühle ist in günstigen Fällen überhaupt nicht notwendig oder nur in Mengen von 0–3 %.

– Das Ofenmehl ist sowohl kurzfristig als auch langfristig von hoher Gleichmäßigkeit, trotz eines nur kleinen und ineffizienten Homogenisierungssilos. Als Beispiel für kurzzeitige Schwankungen sind die Resultate eines Monates (Februar 1993) in **Bild 5** wiedergegeben. Trotz einiger „Ausreißer" infolge Mischbettwechsels liegt die Standardabweichung der Einzelanalysen der Kalksättigung unter 1 %. Über das erste ganze Betriebsjahr 1992 gesehen lag die Standardabweichung der Kalksättigung, bezogen auf Tagesmittelwerte, bei 1,4 %. Für 1993 deuten die Resultate darauf hin, daß die Gleichmäßigkeit der Ofenaufgabe im Vergleich zum Vorjahr nochmals verbessert werden kann.

Die guten Resultate rühren nicht nur von einer allgemein zufriedenstellenden Leistung des Analysators her, sondern sind auch auf die Aufmerksamkeit und die Erfahrung des Bedienungspersonals zurückzuführen. Der Analysator ist zwar ein entscheidendes Glied in der Rohmaterialaufbereitung, andere Faktoren wie eine fachgerechte Bedienung, kurz- bis langfristige Abbauplanung etc., spielen jedoch eine ebenso wichtige Rolle.

5. Betriebsprobleme

Die Phase der Inbetriebnahme war durch verschiedene Betriebsprobleme gekennzeichnet, die einerseits auf technische Ursachen zurückzuführen waren und andererseits mit Software-Problemen zu tun hatten. Einige der nachstehenden Probleme konnten in der Zwischenzeit gelöst oder zumindest vermindert werden, andere bedürfen nach wie vor einer Lösung:

– Staubeintritt in das Analysatorgehäuse aufgrund von Dichtungsproblemen am Übergang Aufgabesilo – Analysatorschacht: Dieses Problem konnte zwischenzeitlich durch technische Modifikationen gelöst werden.

– Teilweises oder vollständiges Blockieren des Materialflusses durch Anhaften von feuchtem und klebrigem Material im untersten Teil des Aufgabesilos am Über-

– The kiln feed is highly uniform both in the short and long term in spite of the fact that the homogenising silo is small and inefficient. Results for one month (February 1993) are given in **Fig. 5** as an example of the short-term fluctuations. In spite of some freak results due to changes of blending bed the standard deviation of individual lime saturation analyses is less than 1 %. For the first complete year of operation, 1992, the standard deviation of lime saturation relative to daily averages was approximately 1.4 %. For 1993 the results indicate that uniformity of kiln feed can be further improved as compared with the previous year.

The good results are due not only to the generally satisfactory performance of the analyzer but also and to a considerable degree to the attentiveness and experience of the operating personnel. Although the analyzer is a significant element in raw material processing, other factors such as professional operation, short and long-term planning of quarrying, etc., also play an important part.

5. Operating problems

The commissioning phase was characterised by various operating problems, some with technical causes, others involving software difficulties. Some of the problems described below have by now been solved or at least eased while others still have to be dealt with:

– Dust ingress into the analyzer housing owing to sealing problems at the transition from feed bin to analyzer chute. This has now been dealt with by technical modifications.

– Partial or complete blockage of material flow owing to moist and adhesive material sticking in the bottom of the feed bin at the transition to the analyzer chute. The same effect was initially produced by foreign bodies such as pieces of wood or roots.

– Strong vibration of the analyzer housing at maximum material flow. As an immediate measure this was greatly reduced by slightly diminishing throughput and lowering the filling level in the bin. Releasing the attachment between the analyzer housing and the following chute extension, in conjunction with other modifications, also helped to reduce vibration.

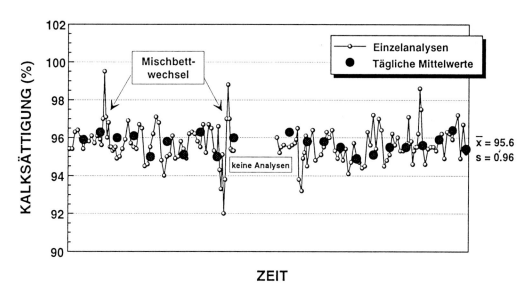

BILD 5: Kalksättigung des Ofenmehls als Funktion der Zeit während des Monats Februar 1993
(Mittelwert und Standardabweichung der Kalksättigung aus 1- bis 2stündigen Einzelanalysen)
FIGURE 5: Lime saturation factor of the kiln feed as a function of time during the month of February 1993 (The average and standard deviation of the lime saturation factor relate to individual analyses taken every 1 to 2 hours)

Kalksättigung	= lime saturation		Einzelanalysen	= individual analyses
Mischbettwechsel	= blending bed changeover		Tägliche Mittelwerte	= daily average values
keine Analysen	= no analyses			

gang zum Analysatorschacht: Denselben Effekt bewirkten zu Beginn auch Fremdkörper wie Holzteile oder Wurzeln.

— Starke Vibrationen des Analysatorgehäuses bei maximalem Materialfluß: Als Sofortmaßnahme konnten diese durch eine leicht reduzierte Durchflußmenge und durch einen geringeren Füllungsgrad im Silo deutlich vermindert werden. Ebenfalls zur Verminderung beigetragen hat, daß in Zusammenhang mit anderen Modifikationen die Befestigung zwischen Analysatorgehäuse und nachfolgender Schachtverlängerung gelöst wurde.

— Übermäßiger Verschleiß des untersten Teils der Analysatorschachtauskleidung: Der Verschleiß stellt nach wie vor ein Hauptproblem dar. Verschiedene Modifikationen haben aber Verbesserungen gebracht.

— Korrektur und Anpassung der Software: Verschiedene Änderungen an der Bedienersoftware wurden verwirklicht, die sowohl offensichtliche Fehler und Mängel betrafen, aber auch die Bedienungsfreundlichkeit verbesserten. So mußten u.a. Formeln für die Berechnung von Qualitätsparametern (Kalksättigung, Silikatmodul etc.) angepaßt werden.

6. Schlußfolgerungen

Der Neutronen-Aktivierungs-Analysator zur Kontrolle des Mischbettaufbaus bei Cementos Apasco ist für die hohe Qualität der Produkte von entscheidender Bedeutung. Die hohe Gleichmäßigkeit des Ofenmehls ist dafür ein deutlicher Beweis. Mit Hilfe des Gerätes ist es möglich, die Mischbetten nahe am Zielwert zu fahren. Die Hauptvorteile liegen in der Echtzeitanalyse des gesamten Materialstromes und im Wegfall konventioneller Probennahme, -aufbereitung und -analysentechnik.

Investitionskosten von ungefähr 900 000 US$ für den Analysator und die notwendigen Installationen sind vergleichbar mit einem traditionellen Probenahmeturm oder liegen sogar etwas tiefer. Betriebs- und Unterhaltskosten dagegen sind erheblich und übersteigen die anfänglichen Erwartungen.

Dennoch kann nach zweieinhalbjährigen Betriebserfahrungen und trotz der teilweise noch anhängigen Probleme festgestellt werden, daß die wesentlichen Anforderungen an das Gerät erfüllt werden konnten. Die Prompt-Gamma-Neutronen-Aktivierungs-Analyse als analytische Methode in der Zementindustrie verspricht nicht nur bei der Rohmaterialkontrolle, sondern auch in weiteren Bereichen der Qualitätskontrolle zukünftig interessant zu werden.

Aufgrund der mit dem Analysator gewonnenen Erfahrungen wurde für ein weiteres Projekt in der mexikanischen Apasco Gruppe ein Analysator von Gamma-Metrics bestellt. Es handelt sich dabei um den Prototypen eines sogenannten Bandanalysators (Cross-Belt Analyzer), der direkt über einem bestehenden Transportband installiert wird. Viele der Nachteile des Schachtanalysators (Notwendigkeit eines speziellen Turmes, Betriebsschwierigkeiten mit klebrigem Material, hoher Verschleiß der Schachtauskleidung etc.) können bei diesem Analysatortyp umgangen werden. Die Inbetriebnahme des Bandanalysators ist Ende 1993 erfolgt.

— Excessive wear occurred in the bottom part of the analyzer chute lining. This is still the main problem at Ramos Arizpe, but is known to be a problem at other installations. Various modifications have however brought improvements.

— Correction and adaptation of the software have been carried out through various changes to the operator software, both dealing with obvious errors and defects and making it more operator-friendly. Among other changes formulae for calculating quality parameters (lime saturation, silica modulus, etc.) had to be adapted.

6. Conclusions

The neutron activation analyzer for control of the blending bed composition at Ramos Arizpe is of decisive importance to the high quality of the products, as evidenced by the uniformity of the kiln feed. The apparatus enables the blending beds to be run at or near the target value. The main advantages reside in the real-time analysis of the whole material flow and the elimination of conventional sampling, preparation and analysis.

Capital costs of about 900 000 US$ for the analyzer and the necessary installations are comparable with a traditional sampling tower or even somewhat lower. On the other hand operating and maintenance costs are considerable and higher than originally expected.

After two and a half years' operating experience and in spite of the as yet unsolved problems it can nevertheless be said that the expectations for PGNAA have been fulfilled. The potential of prompt gamma neutron activation analysis as an analytical method in the cement industry is not restricted to raw material control; it is a technology with a promising future which may eventually be applied to other quality control areas.

On the basis of the experience gained with the Ramos Arizpe analyzer an analyzer has been ordered from Gamma Metrics for another project in the Mexican Apasco Group. It is the prototype of a so-called cross belt analyzer, which is installed directly above an existing conveyor belt. Many of the disadvantages of the chute analyzer (the need for a special tower, operating difficulties with sticky material, heavy wear on the chute lining etc) can be avoided with this type of apparatus. The cross belt analyzer has been commissioned at the end of 1993.

Projektablauf der Modernisierung der elektrischen Ausrüstungen in der Cementfabrik Rekingen

Progress of the project during modernization of the electrical equipment at the Rekingen cement factory

Déroulement du projet de modernisation des équipements électriques à la cimenterie Rekingen

Desarrollo del proyecto de modernización de los equipos eléctricos en la fábrica de cemento de Rekingen

Von **K. Utzinger**, Rekingen/Schweiz

Zusammenfassung – Nach zehn erfolgreichen Betriebsjahren des Werkes Rekingen-Mellikon der „Holderbank" Cement und Beton (HCB), Schweiz, zeigte es sich, daß ein Teil der elektrischen Ausrüstungen ersetzt werden müßte. Ausgeschöpfte Erweiterungsmöglichkeiten, die Liefer- und Wartungssituation einiger elektronischer Ausrüstungskomponenten und weitere Gründe, bewogen die HCB, ein Konzept für die Modernisierung unter der Mithilfe der „Holderbank" Management und Beratungs AG (HMB) zu erarbeiten. Von der Signalerfassung im Feld, den Schützengerüsten, den speicherprogrammierbaren Steuerungen und dem Leitsystem bis hin zur Neugestaltung des Leitstandes wurde alles in die Planung einbezogen und den besonderen Bedingungen einer Zementfabrik angepaßt. In der anschließenden Evaluationsphase wurden die Automationskomponenten nach Pflichtheften festgelegt. Es wurde großes Gewicht auf ein übersichtliches und lieferantenunabhängiges Gesamtkonzept gelegt. Die Planung der Modernisierung erfolgte zum größten Teil in der HCB selbst. Das war nur unter konsequenter Ausnützung der PC-gestützten Hilfswerkzeuge CAE und CAD möglich. Aber auch die Standardisierung von Prozeßabläufen, Hardware und Software hat zu einer Vereinfachung des Planungsaufwandes beigetragen. Die Ausführung der Modernisierung erfolgte in drei Teilschritten, wobei geplante Revisionsstillstände der Produktionsanlagen genutzt wurden. Bereits bei der Wiederinbetriebnahme und beim späteren Produktionsbetrieb mit den modernisierten elektrischen Ausrüstungen zeigten sich die Vorteile des neuen Konzeptes. Der problemlose Betrieb und eine geringe Fehlerrate führten zu einer hohen Akzeptanz beim Betriebspersonal.

Summary – After the Rekingen-Mellikon works of "Holderbank" Cement und Beton (HCB), Switzerland, had been operating successfully for 10 years it became apparent that some of the electrical equipment needed to be replaced. Exhausted options for expansion, the supply and maintenance situation of some electrical equipment components, and other reasons decided the HCB to work out a scheme for modernization with the assistance of "Holderbank" Management und Beratung AG (HMB). Everything was included in the planning from in-situ signal measurement, protective framework, the programmable control system, and the management system up to new configuration of the control room, were adapted to suit the particular conditions of a cement factory. In the subsequent evaluation phase the automation components were defined on the basis of performance specifications. Great emphasis was placed on a total concept which was clearly laid out and independent of the supplier. To a great extent the modernization was planned by HCB itself. This was only possible with consistent utilization of the PC-assisted auxiliary tools, CAE and CAD. Standardization of process cycles, hardware and software has also contributed to simplifying the complexity of the planning. The modernization took place in three stages, making use of planned inspection stoppages of the production plants. The advantages of the new scheme became apparent immediately on restarting and during the later production operation with the modernized electrical equipment. The trouble-free operation and low error rate led to a high level of acceptability with the operating personnel.

Résumé – Après dix années d'exploitation satisfaisante de l'usine Rekingen-Mellikon de la Holderbank Cement et Beton (HCB), Suisse, il est apparu qu'il était opportun de remplacer une partie des équipements électriques. L'épuisement des possibilités d'extension, les conditions de fourniture et de maintenance de quelques composants d'équipement électronique et d'autres raisons, ont incité HBC d'élaborer un concept pour la modernisation avec l'assistance de „Holderbank" Management et Conseil AG (HMB). De l'acquisition des signaux sur site, en passant par les armoires à relais, les conduites à mémoire programmable et le système de commande, jusqu'à la nouvelle configuration de la salle de commande, tout a été pris en compte dans la planification et adapté aux conditions particulières d'une cimenterie. Dans la phase suivante ont été définis les composants d'automation d'après des cahiers de charge. Une grande importance a été accordée à un concept global transparent et indépendant des fournisseurs. La planification de la modernisation a été effectuée, en plus grande partie, à HCB même. Cela n'a été possible, que par l'utilisation conséquente des outils sur PC,

CAE et CAD. Mais la standardisation des déroulements du process, de l'instrumentation et des logiciels a aussi contribuée à la simplification des travaux de planification. La réalisation de la modernisation s'est effectuée en trois étapes partielles, utilisant pour ce faire des arrêts pour révision des installations de production. Déjà à la remise en exploitation et au cours de l'exercise de production avec les équipements électriques modernisés, se sont révélés les avantages du nouveau concept. Le fonctionnement sans problème et un faible taux d'erreurs ont suscité une très bonne acceptation de la part du personnel d'exploitation.

Resumen – Tras diez años de servicio satisfactorio de la factoría de Rekingen-Mellikon, perteneciente a „Holderbank" Cement und Beton (HCB), Suiza, resultó que parte de los equipos eléctricos tenían que ser sustituidos. Las escasas posibilidades de ampliación, la situación de suministro y mantenimiento de ciertos componentes electrónicos y otras razones más, llevaron a HCB a elaborar un concepto de modernización, con la ayuda de „Holderbank" Management und Beratung AG (HMB). Desde el registro de señales in situ, los portarrelés, los controladores programables y el sistema de control hasta la reforma del centro de control, todo fue incorporado al proyecto y adaptado a las condiciones específicas de una planta cementera. En la subsiguiente fase de evaluación, los componentes de la automatización se fijaron de acuerdo con el pliego de condiciones. Se dio mucha importancia a que el concepto en su conjunto fuese claro e independiente de los suministradores. La planificación de la modernización se llevó a cabo mayormente en la misma HCB, y sólo fue posible aprovechando los útiles auxiliares CAE y CAD, con soporte por PC. Pero también la estandardización de las etapas del proceso, del Hardware y el Software, ha contribuido a simplificar los trabajos de planificación. La modernización se efectuó en tres etapas, aprovechando las paradas por revisión de las instalaciones de producción. En el momento de la nueva puesta en servicio y, más tarde, durante la producción con ayuda de los equipos eléctricos modernizados, aparecieron las ventajas del nuevo concepto. El buen funcionamiento así como el bajo índice de errores contribuyeron a que este concepto fuera muy bien acogido por el personal de servicio.

Desarrollo del proyecto de modernización de los equipos eléctricos en la fábrica de cemento de Rekingen

1. Konzeptwahl

Im Jahr 1985 wurde erkannt, daß große Teile der elektrischen Ausrüstungen im Zeitraum der folgenden zehn Jahre ersetzt werden mußten. Die Hauptgründe für diese Maßnahmen lagen im Erreichen der technischen Lebensdauer, dem Fehlen von Erweiterungsmöglichkeiten und der unzweckmäßigen Funktionalität der damaligen Einrichtungen.

Die technische Lebensdauer einer elektrischen Ausrüstungskomponente ist erschöpft, wenn Ersatzteile nicht mehr beschafft werden können und der Hersteller keine Reparaturen mehr ausführen kann. Mit einer RbM-Studie (Risk based Maintenance) konnte dem Management gezeigt werden, daß ab einem gewissen Zeitpunkt, auf Grund der abgelaufenen technischen Lebensdauer von wichtigen Ausrüstungskomponenten, eine planmäßige und durchlaufende Produktion nicht mehr sichergestellt werden kann.

Die Erweiterbarkeit von elektrischen Ausrüstungen kann als erschöpft bezeichnet werden, wenn auf Grund von Änderungen und Erweiterungen sämtliche Reserven aufgebraucht worden sind, Erweiterungen nicht mehr möglich oder nicht mehr mit dem gleichen Konzept oder dem erforderlichen Standard ausgeführt werden können. In Rekingen wurden in den vergangenen Betriebsjahren äußerst viele und teilweise umfangreiche Änderungen und Modifikationen ausgeführt. Die Überlegungen zu zukünftigen Projekten zeigten, daß die Erweiterbarkeit der bestehenden Ausrüstungen an Grenzen stieß.

Die Funktionalität einer bestehenden Ausrüstung ist nicht mehr zweckmäßig, wenn durch neue Aufgaben die Anforderungen höher gestellt werden, die bestehende Ausrüstung diese jedoch nicht oder aber nur noch teilweise erfüllen kann. Die langjährige Betriebserfahrung in Rekingen zeigte, daß die Rahmenbedingungen für die Zementproduktion, und damit auch die Anforderungen an die elektrischen Ausrüstungen, dauernd verändert bzw. enger gesteckt wurden (Energieverbrauchs-Optimierung, Einsatz von Alternativbrennstoffen, Umweltauflagen, Personalreduktionen und die Zement-Qualität).

Zusammen mit dem Produktions- und Instandhaltungspersonal sowie durch die Auswertung der Berichte der elektrischen Instandhaltung wurden daraufhin die Schwachstellen der bestehenden Ausrüstungen aufgedeckt und die Anforderungen an ein neues Konzept erarbeitet.

1. Choice of scheme

In 1985 it was recognized that large sections of the electrical equipment would have to be replaced during the next ten years. This was mainly because the equipment had reached the end of its technical lifetime and did not function appropriately, and because of the lack of options for expansion.

The technical lifetime of an electrical component is exhausted when replacement parts can no longer be obtained and the manufacturer cannot carry out any more repairs. Using an RbM (Risk based Maintenance) study the management were shown that it is no longer possible to guarantee continuous planned production beyond a certain point because the technical lifetimes of important equipment components have expired.

The potential of electrical equipment for expansion can be taken as exhausted when all the reserves have been used up by alterations and extensions, or when extensions are no longer possible or can no longer be carried out to the same scheme or at the required standard. A great many changes and modifications, some of them extensive, had been carried out at Rekingen during the previous years. Consideration of future projects indicated that the potential of the existing equipment for expansion had come up against limiting factors.

An existing item of equipment no longer counts as functional when, due to the greater demands of new duties, the existing equipment can fulfil them only partially or not at all. The many years of operational experience at Rekingen have shown that the prevailing conditions for cement production, and therefore also the demands on the electrical equipment, were constantly changing and being tightened (Optimization of energy consumption, use of alternative fuels, environmental regulations, reductions in personnel, and cement quality).

The weak points of the existing equipment were therefore uncovered in consultation with the production and maintenance personnel and by evaluating the reports from the electrical maintenance department, and the specifications were drawn up for a new scheme.

2. Entwicklung und Auswahl der neuen Ausrüstungen

Auf Grund dieser Anforderungen konnte dann ohne Zeitdruck die Entwicklung und die Auswahl der neuen Ausrüstungskomponenten eingeleitet werden. Die Signalinstallationen führen die Prozeßsignale aus dem Feld auf die SPS (Speicher Programmierbare Steuerung). Die Cementfabrik entwickelte selbst ein neuzeitliches Konzept, welches den Installationsaufwand, die Fehlerquellen und die Installationskosten stark senken konnte (**Bild 1 + 2**).

Über die Schützen-Gerüste (SG) werden alle Antriebe mit Elektrizität versorgt. In Zusammenarbeit mit der Firma ABB wurden sicherungslose SG, basierend auf dem MNS-System (Modulare Niederspannungs-Schaltanlagen) entwickelt, welche preisgünstiger als herkömmliche Einschubsysteme und wesentlich einfacher änder- und erweiterbar waren (**Bild 3**).

Mit der SPS werden im Werk Rekingen sämtliche MSR-Aufgaben (Messen Steuern Regeln) abgedeckt. Das Ziel war es, für diese wichtige Komponente ein neuzeitliches und leistungsfähiges Gerät zu finden, das gleichzeitig auch eine Sicherheit für eine lange technische Lebensdauer bietet. Über 15 verschiedene Fabrikate und Typen wurden auf alle wichtigen Kriterien geprüft und nach einem klaren und neutralen Punktesystem bewertet. Der Entscheid fiel auf das Fabrikat SIEMENS mit dem Einsatz eines Zentralgerätes S5-155 in Verbindung mit einer dezentralen Peripherie S5-115.

Ein Leitsystem ist das Fenster zum Prozeß und das Herzstück jeder Automation. Es muß in der Lage sein, bestehende konventionelle Leitsysteme und Instrumentierungen vollständig zu ersetzen. Die Auswahl des geeigneten Leitsystemes gestaltete sich sehr zeitintensiv, denn viele unterschiedliche Systeme waren einer genauen Beurteilung zu unterziehen. Die Definition und die Überprüfung der Leistungsfähigkeiten als Voraussetzung für die vergleichende Bewertung erschwerten die Auswahl. Zudem ist die technische Entwicklung im Bereich der Leitsystem-Technik so rasant, daß alle sechs bis zwölf Monate eine sorgfältig getroffene Auswahl, infolge von Neuerungen, wieder hinfällig

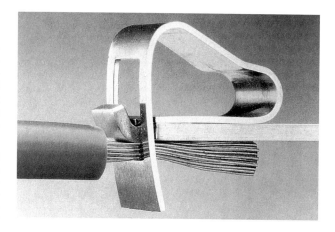

BILD 1: Die schraubenlose Klemme gewährt bei Einzeldrahtverbindungen einen sicheren und schnellen Anschluß
FIGURE 1: The screwless terminal ensures secure and rapid connection for single-wire connections

2. Development and selection of the new equipment

These specifications then made it possible to start developing and selecting the new equipment components without any time pressure. The signal installations carry the process signals from the field to the PLC (programmable logic control) system. The cement works itself developed a modern scheme which sharply reduced installation work, sources of error and installation costs (**Figs. 1 + 2**).

All drives are supplied with power through the contactor racks. Switchless contactors, based on modular low-voltage switchgear systems, were developed in collaboration with the firm of ABB; these were less expensive that conventional plug-in unit systems, and also substantially simpler to change and extend (**Fig. 3**).

The PLC system covers all the measurement and control functions at the Rekingen works. The aim was to find modern and efficient equipment for these important components which also offers a long technical lifetime. Over 15 dif-

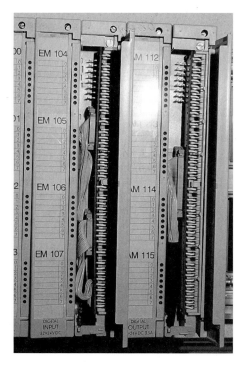

BILD 2: Die Schneidklemmtechnik bietet bei mehradrigen Standardverbindungen mit minimalstem Zeitaufwand größte Sicherheit gegen Falschanschlüsse
FIGURE 2: For standard multi-core connections the insulation piercing connecting system offers maximum security against incorrect connections with minimum time expenditure.

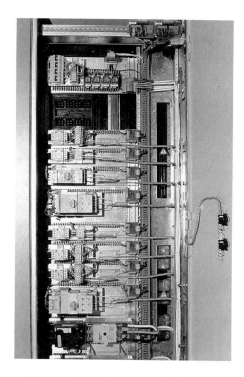

BILD 3: MNS-Schützengerüst mit standardisierten Motorenabgängen verschiedener Größe und Funktion
FIGURE 3: Modular low-voltage switchgear contactor rack with standardized motor feeders with different sizes and functions

BILD 4: Leitstand mit optimaler Raumorganisation und Raumgestaltung (Wettbewerbsgewinner)
FIGURE 4: Control room with optimum space organization and spatial design (competition winner)

BILD 5: Wettbewerbsteilnehmer mit Leitstandspulten aus einem handelsüblichen Büro-Normmöbel-Programm
FIGURE 5: Competition participant with control room desks from a normal commercial product range of standard office furniture

werden kann. In Zusammenarbeit mit der Holderbank Management und Beratung (HMB) wurde ein Leitsystem, basierend auf einer handelsüblichen PC-Hardware und der Leitsystem-Software der HMB, ausgewählt.

Der Leitstand ist ein 24-Stunden-Arbeitsplatz, und an das Bedienpersonal werden hohe Anforderungen gestellt. Deshalb muß die Neukonzeption eines Leitstandes die Funktionalität und die Arbeits-Ergonomie besonders berücksichtigen. Alle wichtigen Anforderungen wurden in einem Pflichtenheft festgehalten und drei Spezialfirmen zu einem Architektur-Wettbewerb eingeladen. Damit konnte einerseits die optimale Lösung gewählt werden, und andererseits konnten wertvolle Ideen von anderen Lösungen übernommen werden (**Bild 4 + 5**).

3. Planung der Modernisierung

Nach Erstellen des Konzeptes der Automation konnten die Investitionskosten und der zeitliche Ablauf des gesamten Vorhabens festgelegt werden. Die Modernisierung mußte zeitlich gestaffelt in drei Teilprojekte aufgeteilt werden:

- 1990/1991 Umbau Zementmahlanlage und Zuschlagtransporte, Umbauzeit 6 Wochen, Kosten SFr. 1,6 Mio.
- 1991/1992 Umbau Ofen und Leitstand 1. Etappe, Umbauzeit 6 Wochen, Kosten SFr. 2,4 Mio.
- 1992/1993 Umbau Rohmaterialtransport, Rohmehlmahlanlage, Kohlemahlanlage und Leitstand 2. Etappe, Umbauzeit 10 Wochen, Kosten SFr. 1,8 Mio.

Die Planung der Modernisierung wurde zum größten Teil mit eigenem Personal durchgeführt. Durch den konsequenten Einsatz von CAE-Hilfsmitteln (Computer Aided Engineering) konnten nicht nur der Arbeitsaufwand, sondern auch die Fehlerquellen drastisch gesenkt werden.

In einem ersten Schritt mußten sämtliche Fließschemata der Anlage neu erfaßt und auf CAD (Computer Aided Design) gezeichnet und sämtliche Ausrüstungskomponenten mit dem HAC-Code (Holderbank Anlagen Code) eindeutig bezeichnet werden. Sämtliche Ausrüstungs-Komponenten wurden in eine HDRS-Datenbank (Holderbank Data Retrieval System) eingegeben, welche zum wichtigsten Hilfsmittel für die Planung, die Inbetriebnahme und die künftige Instandhaltung wurde.

Ein weiterer wichtiger Planungs-Schritt betraf die Standardisierung der eingesetzten Ausrüstungs-Komponenten. Damit läßt sich nicht nur der Bestell-Aufwand vereinfachen, sondern es kann auch Geld und Lagerplatz eingespart werden.

Die gesamte SPS-Programmierung erfolgte im Werk selbst, wobei eine bewährte Standardsoftware der HMB eingesetzt wurde. Immer wiederkehrende Funktionen wurden mittels Standardbausteinen programmiert und die verfahrenstechnischen Abläufe der Produktions-Maschinen mit einer

ferent makes and types were checked for all important criteria, and evaluated on a clear, neutral, points system. The decision was made in favour of the SIEMENS equipment, using a central S5-155 device in conjunction with a decentralized S5-115 peripheral.

A control system is the window on the process and the heart of any automation system. It must be capable of completely replacing existing conventional control systems and instrumentation. Selection of a suitable control system turned out to be very time-intensive because many different systems had to be submitted to a precise assessment process. Defining and checking the capabilities as a precondition for the comparative evaluation made the selection procedure more difficult. In addition to this, technical development in the field of control system technology has been so meteoric that a carefully made selection can be invalidated after six to twelve months as a result of innovations. A control system was selected in collaboration with Holderbank Management und Beratung (HMB) which was based on normal commercial PC hardware and the HMB control system software.

The control room is a 24-hour work station, and high demands are made on the personnel, so the new control room scheme had to pay particular attention to functionality and working ergonomics. All the important requirements were laid down in a performance specification, and three specialist companies were invited to an architecture competition. This meant that it was possible to select the optimum solution, and at the same time useful ideas could be adopted from other solutions (**Figs. 4 + 5**).

3. Planning the modernization

After the scheme for the new automation system had been drawn up it was possible to establish the investment costs and the time schedule for the entire project. The modernization had to be divided into three sub-projects with staggered timing:

- 1990/1991 Conversion of cement grinding plant and additive transport systems, conversion time 6 weeks, cost SFr. 1.6 million.
- 1991/1992 Conversion of kiln and control room stage 1, conversion time 6 weeks, cost SFr. 2.4 million.
- 1992/1993 Conversion of raw material transport system, raw meal grinding plant, coal grinding plant, and control room stage 2, conversion time 10 weeks, cost Sfr. 1.8 million.

The majority of the modernization planning was carried out by the works personnel themselves. By logical use of CAE (computer aided engineering) tools it was possible to make a drastic reduction not only in the amount of work but also in the sources of error.

Schrittkette festgehalten. Mit dieser Art der Programmierung wird es möglich, daß sich auch außenstehende Personen in den Programmen sehr schnell zurechtfinden und eine klare Schnittstelle zwischen dem Verfahrens- und dem Automationsingenieur vorliegt.

Das Konzept der Signalinstallationen, der SPS-Programme und des Leitsystemes ist so ausgelegt, daß der Elektriker kein SPS-Programmiergerät braucht, um sich Klarheit bei der Störungssuche zu schaffen. Die Planung des neuen Leitstandes wurde mit einem kleinen Team sehr effizient abgewickelt. Die übrigen Beteiligten der Zementfabrik wurden regelmäßig über den Stand der Arbeiten orientiert.

Die ergonomisch richtige Auslegung eines Bedienungspultes legt dessen wichtigste Hauptabmessungen fest, welche in jedem Fall einzuhalten sind. Aus diesem Grund wurden alle technischen Ausrüstungen, welche nicht unbedingt in die Pulte zu integrieren waren, an einem anderen Ort eingebaut. Auch die Wahl der verwendeten Konstruktions-Materialien für das Pult wurde sorgfältig getroffen, denn sämtliche mit dem menschlichen Körper in Berührung stehenden Konstruktionsteile dürfen die Wärme nicht ableiten. Die Pulte wurden im Raum so angeordnet, daß die Übersichtlichkeit der gesamten Anlage gewährleistet wird.

Die Farbgebung des Raumes wurde bewußt in hellen Tönen gehalten, denn dadurch erscheint dieser größer, als er tatsächlich ist. Die Beleuchtung wurde ebenfalls sehr sorgfältig geplant, damit keine störenden Reflexionen auf den Bildschirmen entstehen konnten. Die Kommunikations-Anlagen (Funk und Telefon) und akustische Alarmsignale wurden sinnvoll in die Akustik-Anlage (Radio, Tonband) integriert (**Bild 6**).

4. Umbau der Ausrüstungen und Inbetriebnahme

Der Umbau der Ausrüstungen war die kritischste Phase des gesamten Projektes. Durch gezielte Vorarbeiten konnte sie aber stark vereinfacht werden, und viele unvorhergesehene Schwierigkeiten, wie Platzprobleme in den Elektroräumen sowie die Auslastung und die Linienführungen von Kabeltrassen, ließen sich vorab entschärfen.

Die Installationsarbeiten wurden durchwegs mit eigenem Personal realisiert, wobei während der eigentlichen Stillstände noch Fremdpersonal zugemietet wurde. Ein genauer Terminplan garantierte den rechtzeitigen Abschluß der einzelnen Arbeitsschritte.

Der Umbau des Leitstandes selbst wurde zum größten Teil von Fremdfirmen ausgeführt. Pünktlich wurden die Bedienungspulte angeliefert und innerhalb eines Tages aufgestellt. Nur einen weiteren Tag nahm die Montage der Leitsystem-Ausrüstungen in die Pulte und die Wiederinbetriebnahme der Zementmahlanlage in Anspruch. Diese wohl einmalige Rekordzeit ist eindeutig auf das gewählte Konzept des Leitsystems, die einfache Verkabelung und die gut durchdachte Pultkonstruktion zurückzuführen.

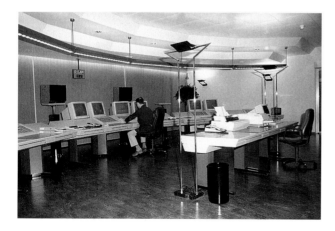

BILD 6: Umgebauter Leitstand nach der ersten Etappe
FIGURE 6: Converted control room after the first stage

The first stage was to draw up all the plant flow sheets again, transfer them to CAD (computer aided design), and mark all the equipment components clearly with the Holderbank plant code. All the equipment components were input into a HRDS (Holderbank data retrieval system) data base, which became a very important tool for planning, commissioning and future maintenance.

Another import planning step concerned the standardization of the equipment components used. This not only simplified the ordering procedure, but also saved money and storage space.

All the programming of the PLC system took place in the works itself, using well-tried standard software from HMB. Functions which kept recurring were programmed by standard modules, and the sequential process engineering operations of the production machines were recorded in step sequences. With this type of programming it is possible for outsiders to find their way around the program very quickly, and there is a clear interface between the process engineer and the automation engineer.

The scheme for the signal installations, the PLC system programmes and the control system is designed so that the electrician does not need any PLC programming equipment to clarify the situation during fault-finding. The planning of the new control room was managed very efficiently by a small team. The others in the cement works who were affected were kept regularly informed about the state of the work.

The ergonomically correct design of an operating desk defines its most important dimensions which must always be adhered to. For this reason all technical equipment which did not have to be integrated in the desk was installed somewhere else. The materials of construction used for the desk were also selected carefully, as any parts of the structure coming into contact with the human body must not conduct heat away. The desks were positioned in the room to ensure that the entire system was easy to survey.

The colouring of the room was deliberately kept to light tones, because this makes it appear larger than it actually is. The lighting was also very carefully planned so that there were no annoying reflections on the screens. The communication systems (radio communication and telephone) and acoustic alarm signals were suitably integrated into the acoustic system (radio, tape recorder) (**Fig. 6**).

4. Conversion of the equipment, and commissioning

The conversion of the equipment was the most critical phase of the entire project. However, it was greatly simplified by carefully controlled preliminary work, so that many unforseen difficulties, such as space problems in the electrical equipment rooms and the utilization and layout of the cable routes, were ironed out in advance.

All the installation work was carried out by the works personnel, although additional outside personnel were engaged during the actual shut-down periods. An accurate schedule guaranteed that the individual stages of the work were completed on time.

Most of the conversion of the control room itself was carried out by outside firms. The operators' desks were delivered punctually and erected within a day. The assembly of the control system equipment in the desks and the re-commissioning of the cement grinding plant only required one further day. This really exceptional record timing is clearly attributable to the scheme chosen for the control system, the simple cabling system and the well-thought-out desk design.

The control room operators were included when the modernized sections of the process were re-commissioned; they themselves initiated all the switching operations and interpreted the incoming messages and faults independently.

Bei der Wiederinbetriebnahme der modernisierten Prozeßabschnitte wurden die Leitstandfahrer mit einbezogen,
indem diese alle Schaltoperationen selbst auslösten und die
ankommenden Meldungen und Störungen selbständig
interpretierten.

5. Schlußbetrachtung

Es hat sich als wichtig erwiesen, daß für die Entwicklungs-
und Auswahlphase genügend Zeit zur Verfügung steht.
Damit können nicht nur zweckmäßige, sondern auch
kostengünstige Lösungen erarbeitet werden. Leistungsfähige Hilfsmittel, wie PC-Anlagen mit der zugehörigen Software und einer optimalen Standardisierung, erleichtern die
umfangreiche Planung einer Modernisierung.

Die Veranstaltung eines Architektur-Wettbewerbes für die
Gestaltung des Leitstandes hat sich als positiv erwiesen. Es
ist aber wichtig, daß die Teilnehmer ganz klare Vorgaben
und Richtlinien für ihre Aufgabe bekommen.

Durch die gezielte Einbeziehung des Betriebs- und Instand-
haltungspersonales während des gesamten Projektes iden-
tifizierten sich die Mitarbeiter mit dem neuen System, was
die Übergabe der modernisierten Anlage wesentlich vereinfachte.

Insgesamt läßt sich sagen, daß ein gut durchdachtes, sorgfältig geplantes und präzise ausgeführtes Projekt einen
guten Abschluß gefunden hat.

5. Final comments

It has proved important that sufficient time was available for
the development and selection phases. This meant that it
was possible to work out solutions which were not only
appropriate but also economical. Powerful tools such as PC
systems with their associated software, and an optimum
level of standardization are a help in the extensive planning
work for a modernization project.

Organizing an architectural competition for the design of the
control room proved beneficial. However, it is important
that the participants are given very clear information and
guidelines for their task.

Through carefully controlled involvement of the operating
and maintenance personnel during the entire project the
staff identified with the new system, which greatly
simplified the handover of the modernized plant.

To sum up, it can be said that a well-thought-out, carefully
planned, and precisely executed project has come to a successful conclusion.

Prozeßdatenerfassung als integrierter Bestandteil des Informationssystems der Perlmooser Zementwerke AG

Process data acquisition as an integral component of Perlmooser Zementwerke AG's information system

Acquisition des données du process comme partie intégrante du système d'information de la Perlmooser Zementwerke AG

Registro de datos del proceso como parte integrante del sistema de información de Perlmooser Zementwerke AG

Von **M. Weichinger**, Wien/Österreich

Zusammenfassung – In den Werken der Perlmooser Zementwerke werden wichtige Prozeßdaten wie Produktionsmengen, Wärme- und Energieverbräuche, Temperaturen, Drücke usw. erfaßt und zur Auswertung allen zuständigen Personen auf verschiedenen Ebenen und in unterschiedlichem Verdichtungsmaß zur Verfügung gestellt. Die Prozeßdatenerfassung dient dabei im wesentlichen dazu, die Darstellung des Anlagenzustandes auf jedem PC im Werk in graphischer Form zu ermöglichen und statistische Auswertungen von langfristig gespeicherten Daten (mehrere Jahre mit einer Auflösung von 10 Minuten) mit Hilfe einer Statistiksoftware vorzunehmen. Darüber hinaus erlaubt sie die Aufbereitung umweltrelevanter Meßdaten für die Vorlage bei den Behörden. Derzeitig wird daran gearbeitet, den neuen Normenentwurf (ÖNORM M9412) vollständig zu implementieren. Die Wandlung der Analogwerte wird entweder durch eine A/D-Karte direkt im PC oder durch eine vorgelagerte frei programmierbare Steuerung durchgeführt. Die Erfassung der Daten erfolgt durch einen PC, der im Token-Ring-Netzwerk des jeweiligen Werkes betrieben wird. Die Datenübertragung in die Zentrale erfolgt mittels Standleitung, von wo eine Datenweiterleitung in die operativen Systeme erfolgt. Die Installation des BDE-Systems benötigt nur einen sehr geringen Aufwand an Hardware. Eigene Softwareentwicklung konnte in engen Grenzen gehalten werden. Der Einsatz des „Information Delivery Systems" SAS hat wesentlich zur Vereinfachung der Auswertearbeiten geführt.

Summary – In Perlmooser Zementwerke's works important process data such as production quantities, heat and power consumption, temperatures, pressures, etc. are collected and made available at different levels and in different degrees of condensation to all the appropriate personnel for evaluation. The process data acquisition system serves primarily for facilitating the display of the plant status on any PC on the works in graphic form and for undertaking statistical evaluations of data stored for long periods (several years with a resolution of 10 minutes) with the aid of statistical software. It also facilitates the preparation of environmentally relevant test data for submission to the authorities. At present, work is in hand for complete implementation of the new draft standard (ÖNORM M9412). Conversion of the analogue values is carried out either by an A/D card directly in the PC or by a preliminary programmable control system. Data acquisition takes place through a PC which is operated in the token ring network of the works concerned. The data is transmitted to the control centre by dedicated line from where it is routed onward into the operative systems. Installation of the industrial data acquisition system required very little expenditure on hardware. It was possible to keep in-house software development within tight limits. The use of the SAS "Information Delivery System" has substantially simplified the evaluation work.

Résumé – Dans les usines de la Perlmooser Zementwerke sont acquises dés données de process importantes, comme quantités de production, consommations de chaleur et d'énergie, températures, pressions etc. et mises à disposition, pour l'exploitation, à toutes les personnes concernées, à différents niveaux et sous densité différente. L'acquisition des données du process sert alors essentiellement à permettre la représentation de l'état de l'installation, sous forme graphique, sur chaque PC de l'usine et à effectuer l'exploitation statistique de données conservées sur longue durée (plusieurs années, avec une résolution de 10 minutes), à l'aide d'un logiciel de statistique. En plus, elle permet le traitement de données de mesure concernant l'environnement, pour les présenter aux instances officielles. Actuellement, est entrepris d'intégrer complètement le nouveau projet de Norme (ÖNORM M9412). La transformation des valeurs analogues est effectuée, soit au moyen d'une carte A/D directement dans le PC, soit par une commande librement programmable en amont. L'acquisition des données est réalisée sur un PC

fonctionnant dans le réseau Token-Ring de l'usine concernée. La transmission vers la centrale se fait par bus permanent, d'où a lieu un acheminement des données vers les systèmes opérationnels. L'implantation du système BDE ne nécessite qu'une très faible dépense en matériel. Le développement interne de logiciels a pu être réduit au strict minimum. L'utilisation de „Information Delivery Systems", SAS, a contribué essentiellement à la simplification des travaux d'exploitation.

Registro de datos del proceso como parte integrante del sistema de información de Perlmooser Zementwerke AG

Resumen – En las fábricas de cemento de Perlmooser Zementwerke se registran importantes datos del proceso, tales como cantidades de producción, consumo de calor y de energía, temperaturas, presiones, etc., y se ponen a la disposición de las personas competentes, en distintos niveles y con diferentes grados de densidad. A este respecto, el registro de los datos del proceso sirve, esencialmente, para hacer posible la representación gráfica del estado de la instalación en cada PC disponible en la fábrica y proceder a la evaluación estadística de los datos almacenados a largo plazo (varios años, con una resolución de 10 minutos), con ayuda de un software de estadística. Además, permite la preparación de datos de medición relevantes para el medio ambiente y aptos para ser presentados a las Autoridades. Actualmente, se está trabajando en la implementación completa del nuevo proyecto de norma (ÖNORM M9412). La conversión de los valores análogos se efectúa o bien mediante una tarjeta A/D directamente en el PC o bien por medio de un mando preconectado, libremente programable. El registro de los datos se realiza mediante un PC que funciona en la red Token-Ring de la fábrica en cuestión. La transmisión de datos a la central se lleva a cabo mediante una línea permanente, y desde allí los datos son retransmitidos hacia los sistemas operativos. El montaje del sistema BDE requiere muy poco hardware. El desarrollo propio de software ha quedado muy limitado. El empleo del „Information Delivery System" SAS ha contribuido esencialmente a simplificar los trabajos de evaluación.

1. Einführung

Im Jahr 1988 wurde im Werk Mannersdorf der Perlmooser Zementwerke AG ein Expertensystem zur Regelung der Drehofenanlage in Betrieb genommen. Bei diesem Programm ist als Nebenprodukt ein Datenerfassungssystem implementiert. Es wurde vor allem von den Betriebsingenieuren für unterschiedliche verfahrenstechnische Auswertungen und Optimierungen verwendet und entwickelte sich in kurzer Zeit zu einem vielbenutzten eigenständigen System. Aus diesem Grund wurde 1990 damit begonnen, das System für diesen Einsatzzweck zu optimieren und auch in den anderen Werken der Perlmooser einzuführen. Die wohl wesentlichste Änderung war, daß das ursprünglich auf einer VAX unter VMS realisierte System für die Verwendung auf PC's umgeschrieben wurde.

2. Hardware-Aufbau

In **Bild 1** ist die Hardware-Übersicht des Systems vereinfacht dargestellt. Generell werden PC-Netze eingesetzt, die auf einem Token-Ring basieren. Als Betriebssystem kommt OS/2 zum Einsatz. In jedem Werk existiert ein PC für die Prozeßwerterfassung (PWE) der jeweiligen Anlagenbereiche. In **Bild 2** sind die charakteristischen Daten eines Prozeßwerterfassungsrechners zusammengestellt. Die Wandlung der Analogwerte erfolgt zum Teil auf einer 12-Bit Analog-Digitalwandlerkarte im PC. Durch vorgelagerte Multiplexer können 128 verschiedene Meßwerte erfaßt werden. In SPS-Systemen bereits gewandelte Werte können direkt übernommen werden, wobei der PWE-Rechner jedoch die Synchronisation übernehmen muß. Nur dann ist gewährleistet, daß die Meßwerte zeitlich vergleichbar sind. Digitale Eingänge werden in der gleichen Art erfaßt wie die Analogwerte.

Derzeit werden rund 100 bis 200 verschiedene Meßgrößen pro Werk erfaßt. Die Messung erfolgt alle 2 Sekunden. Im PWE-Rechner werden alle 10 Minuten Mittelwerte gebildet, über den Token-Ring zum File-Server des Werkes übertragen und dort gespeichert. Nur bei Netzausfällen erfolgt eine lokale Zwischenspeicherung am PWE-Rechner. Auf dem File-Server stehen die Daten mehrerer Jahre on-line zur Verfügung. Pro Jahr und Werk muß mit einem Speicherbedarf von rund 25 Mbyte gerechnet werden. In der Firmenzentrale sind alle Daten zusätzlich auf einer Optical Disk gespeichert.

1. Introduction

In 1988 an expert system for controlling the rotary kiln plant was brought into operation at Perlmooser Zementwerke AG's Mannersdorf works. A data acquisition system is implemented in this program as a secondary product. It was used chiefly by the plant engineers for various process engineering evaluation and optimization tasks; in a short time it developed into a much used independent system. A start was therefore made in 1990 to optimize the system for this application and also to introduce it into the other Perlmooser works. The most important alteration was that the system originally implemented on a VAX under VMS was transcribed for use on PCs.

2. Hardware configuration

Fig. 1 gives a simplified overview of the system's hardware. PC networks are generally used, based on a token ring. OS/2 is used as the operating system. There is one PC in each

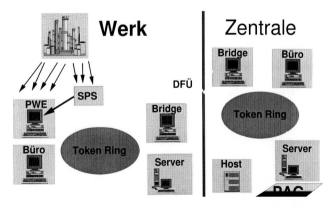

BILD 1: Schematischer Aufbau der Prozeßwerterfassung (PWE) der Perlmooser Zementwerke AG
FIGURE 1: Diagrammatic configuration of the process data acquisition system

Werk	= works
Zentrale	= headquarters
DFÜ	= remote data transmission
SPS	= PLC
PWE	= process data acquisition
Büro	= office

BILD 2: Schnittstellen und Kapazitätsdaten der Prozeßwerterfassung

- Analog/Digitalwandlung im PC und in SPS
- Digitaleingänge im PC und in SPS
- Derzeit ca. 100 bis 200 Werte pro Anlage
- Meßfrequenz 0,5 HZ
- Mittelwertbildung über 10 Minuten
- Speichern der Mittelwerte am PC-Server
- Mehrere Jahre im direkten Zugriff

3. Nutzungsmöglichkeiten

Das System hat den Vorteil, daß es unabhängig von den eingesetzten SPS-Systemen und gleichwertig auch bei Relais-Steuerungen arbeitet. Dadurch sind anlagenübergreifende und sogar werksübergreifende Auswertungen möglich. Der Zugriff auf die Daten kann über das Netzwerk von jedem PC im gesamten Werk aus erfolgen. Die Zugriffsrechte sind über einen Passwordschutz geregelt. Über eine Bridge werden die Daten der einzelnen Werke in die Zentrale überspielt. Die Übertragung erfolgt nur einmal täglich. Auf einen on-line Zugriff wurde verzichtet, um die Leitungsbelastung zu minimieren.

In **Bild 3** ist eine Übersicht über die Nutzungsmöglichkeiten der Prozeßwerterfassung gegeben. Mit dem bei uns eingesetzten Informationssystem SAS lassen sich alle erforderlichen statistischen Auswertungen durchführen. So werden in den Werken z.B. automatisch Tagesberichte erstellt, die der Werksleitung detailliert und übersichtlich die wichtigsten Prozeßparameter des vergangenen Tages darstellen. Die Daten der Prozeßwerterfassung sind auch Grundlage für das unternehmensweite Berichtswesen. In diesem sind Produktions- und Verkaufsmengen sowie spezifische Energieverbräuche täglich erfaßt. Daraus werden dann verdichtete Monats- und Jahresberichte erzeugt. Weitere wichtige Standardauswertungen sind z.B. die nach TA-Luft vorgeschriebenen Emissionsgrenzwerte. Derzeit wird daran gearbeitet, mit diesem System alle Anforderungen des diesbezüglichen Normenentwurfes der ÖNORM M9412 zu erfüllen.

BILD 3: Einsatzmöglichkeiten der Prozeßwerterfassung

- Auswertungen
  - Tages- und Monatsberichte
  - Emissionswerte (Halbstundenmittelwerte nach TA-Luft)
  - Umfangreiche Statistik mittels Standardsoftware SAS
- Laufende Anlagenbeobachtung
  - „Lebende" Diagramme der Anlage
  - Auf jedem PC abrufbar
- Integration mit
  - Kommerzieller Software SAP (vor allem RM)
  - Berichtswesen
  - Chemischen Datenbanken

Die besondere Stärke des Systems liegt aber darin, daß nicht nur Standardauswertungen möglich sind. Ein relativ großer Personenkreis, der vor allem Betriebsingenieure, Werkschemiker usw. umfaßt, hat vollständig wahlfreien Zugriff auf alle Daten. Mit SAS steht ein Werkzeug für komplexe statistische Verfahren, Grafik, Forecasting usw. zur Verfügung. Als Beispiel ist in **Bild 4** eine Grafik dargestellt, die die SO_2-Emissionen einer Ofenanlage wiedergibt. Darin bleiben alle Werte unberücksichtigt, bei denen aus anderen Meßwerten auf Anlagenstörungen geschlossen werden kann. Dann wurden aus den Einzelwerten Halbstundenmittelwerte gebildet. Nach den Aufgabemengen auf die Roh-

works for the process data acquisition from the plant sectors in question. The characteristic data from a process data acquisition computer are summarized in **Fig. 2**. Some of the conversion of the analogue values takes place in a 12-bit analog-digital converter card in the PC. 128 different measured values can be accommodated through interfaced multiplexers. Values which have already been converted in PLC systems can be received directly, but the process data acquisition computer must take over the synchronization. Only in this way is it possible to ensure that the measured values are comparable on a time basis. Digital inputs are collected in the same way as the analogue values.

FIGURE 2: Characterization of the process data acquisition system

- analog/digital conversion in the PC and PLC
- digital inputs into the PC and PLC
- currently approx. 100 to 200 values per plant
- measurement frequency 0.5 Hz
- mean values calculated over 10 minute periods
- mean values filed at the PC server
- several years directly accessible

At present about 100 to 200 different measured variables are collected at each works. The measurements take place every two seconds. Mean values are calculated in the process data acquisition computer every 10 minutes, and transmitted via the token ring to the works' file server where they are stored. Only in cases of network failure is local intermediate storage carried out at the process data acquisition computer. The data are available on-line for several years at the file server. A storage requirement of about 25 Mbyte must be expected per year and per works. All the data are also stored on an optical disc at the company headquarters.

3. Possible applications

The system has the advantage that it works independently of the PLC systems used and also works equally well with relay control systems. This means that evaluations covering all the plants, or even all the works, are possible. The data can be accessed via the network from every PC in the entire works. Right of access is controlled by a password security system. The data from the individual works are re-recorded in the headquarters via a bridge. Transmission only takes place once per day. On-line access is dispensed with in order to minimise the line loading. **Fig. 3** summarises the possible applications of the process data acquisition system. All necessary statistical evaluations can be carried out with the memory and sequence request control information system used. For example, daily reports which provide the works management with the most important process parameters

FIGURE 3: Applications of the process data acquisition system

- Evaluations
  - daily and monthly reports
  - emission values (half-hour values as specified in German Clean Air Regulations)
  - extensive statistics using the memory and sequence request control software SAS
- Continuous plant monitoring
  - "living" diagrams of the plant
  - accessible from every PC
- Integration with
  - commercial SAP software (especially RM)
  - reporting
  - chemical data bases

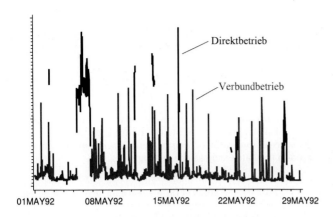

BILD 4: Vergleich der SO$_2$-Emissionen bei Verbund- und Direktbetrieb als Auswertungsbeispiel der Prozeßwerterfassung
FIGURE 4: SO$_2$ emissions in direct and interconnected operation

Direktbetrieb = direct operation
Verbundbetrieb = interconnected operation

mühle wurde in Direktbetrieb und Verbundbetrieb unterschieden. Auf diese Art kann die Einbindung von SO$_2$ sehr gut quantifiziert werden.

Eine weitere Anwendung, die vom Personal sehr positiv aufgenommen wird, ist ein Programm, das auf PC's alle erfaßten Prozeßdaten on-line darstellt. Da ein multi-tasking-fähiges System eingesetzt wird, kann somit jeder Befugte vom Anlagenfahrer über den Betriebsingenieur bis zum Werksdirektor den momentanen Anlagenzustand laufend mitverfolgen. Durch das eingesetzte Netzwerk ist dies auf jedem PC des Werkes möglich. Das Programm erlaubt die Darstellung der Meßwerte in Kurvenform, wobei die Zeitachse zwischen einer Stunde und einem Tag skalierbar ist. Die Zusammenstellung der Kurven am Schirm ist zu jeder Zeit frei wählbar. Ein verschiebbares Lineal erlaubt die Ablesung der Werte in digitaler Form. Die Kurven sind relativ, die digitalen Werte absolut skaliert. In einem eigenen Fenster ist die Anzeige von gleitenden Mittelwerten in digitaler Form möglich.

Eine wichtige Voraussetzung für die statistischen Auswertungen ist die Verbindung der PWE mit verwandten Systemen. So sind gemeinsame Auswertungen zwischen der PWE, dem Berichtswesen und den Laborsystemen, die im wesentlichen eine Chemiedatenbank und eine Datenbank mit den Ergebnissen der Festigkeitsprüfungen enthalten, möglich.

4. Zusammenfassung

Zusammenfassend kann festgestellt werden, daß die Einführung eines Systemes zur Prozeßwerterfassung als Bestandteil des Informationssystems der Perlmooser Zementwerke AG wesentliche Vorteile zur langfristigen Analyse verfahrenstechnischer Zusammenhänge der Zementanlagen gebracht hat. Darüber hinaus konnte damit die Überleitung der in der Anlage erfaßten Daten in die entsprechenden Systeme weitgehend automatisiert werden.

from the previous day in a detailed and clearly arranged manner, are prepared automatically in the works. The data from the process data acquisition system also form the basis for the companywide reporting. This gives daily coverage of production and sales quantities and specific energy consumptions, from which summarized monthly and yearly reports are then compiled. Other important standard evaluations are, for example, the emission limits prescribed by TA Luft (German Clean Air Regulations). At present work is being carried out so that this system will fulfil all the requirements of the related standard draft of the ÖNORM M9412.

However, the particular strengths of the system lie in the fact that standard evaluations are not the only ones possible. A relatively large circle of personnel consisting mainly of plant engineers, works chemists, etc., has complete and unrestricted access to all data. The memory and sequence request control provides a tool for complex statistical methods, graphics, forecasting, etc. As an example, **Fig. 4** shows a graphics display which reproduces the SO$_2$ emissions from a kiln plant. This omits all values which, on the basis other measured values, can be attributed to plant malfunctions. Half-hour mean values were then calculated from the individual values. Direct operation and interconnected operation were differentiated on the basis of the quantity of feed to the raw mill. This method is also very effective for quantifying the fixation of SO$_2$.

Another application which is very positively accepted by the personnel, is a program which gives an on-line display on PCs of all the process data collected. As a system is used which is capable of multi-tasking, this means that every authorized person, from the plant operator through the plant engineer to the works director, can follow the instantaneous state of the plant on a continuous basis. Because of the network used, this is possible on any PC in the works. The program allows the measured values to be displayed in the form of curves, in which the time axis can be scaled between one hour and one day. The combination of the curves on the screen can be freely selected at any time. A movable ruler allows the values to be read off in digital form. The curves have relative scales and the digital values have absolute scales. It is possible to display the floating mean values in digital form in a special window.

One important precondition for the statistical evaluations is the connection of the process data acquisition system with related systems. This makes joint evaluations possible between the process data acquisition system, the reporting system, and the laboratory systems which essentially contain a chemical database and a database with the results of the strength tests.

4. Summary

To summarize, it can be said that the introduction of a system for process data acquisition as a component of the information system at Perlmooser Zementwerke AG has brought substantial advantages for the long-term analysis of process engineering correlations in the cement plants. It has also meant that the transfer of the data collected in the plant to the appropriate systems has been largely automated.

Elpro
Individual Solutions of Electrical Systems and Projects

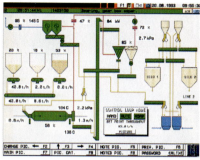

- **Cement plants**
- **Crushing and classifying plants**
- **Grinding mills**

We offer you the complete electrical equipment inclusive of integrated drive and automation technology as individual solution. For more than four decades we have equipped cement plants and related industrial plants of the building materials industry.

Our many years' experience in

- Engineering
- Development
- Production
- Installation
- Commissioning

will satisfy your requirements.

Good reasons to get in touch with us

Leit- und Energietechnik GmbH

Basic Industries Division

Rhinstraße 100 · D-12673 Berlin
Phone +49-30/54 607 516 · Fax +49-30/54 607 345

A COMPANY OF THE ELPRO GROUP

INTELLIGENCE
STRAIGHT AHEAD!

PRODUX® -PLUS — trend-setting control system for custom-fit process management combines transparent process operations and high system availability to ensure economic plant utilization.

WOKURS®
PYROEXPERT — the new generation of kiln automation on the basis of expert knowledge and neuro-fuzzy. Savings up to 1 Mio. DM/year are reached by optimized kiln operation, increased productivity and uniform product quality at minimum fuel demand.

MILLEXPERT — the expert system for automated process management of grinding systems helps to reach optimal and trouble-free plant operation, to meet high quality requirements, to reduce operating costs and to increase production capacity.

ROMIX® — gets raw meal quality safely under control by advanced control system and automated laboratory operation. Achieving uniform raw meal quality and economic use of raw material components becomes reality - with ROMIX®!

SCANEX® — the rotary kiln diagnostic system shows the kiln shell temperature and refractory thickness, thereby improves operational reliability, reduces production costs and increases productivity. Operates also in data-network environments. Test systems are available for You!

All arguments speak in favour of systems offered by KHD Humboldt Wedag Processautomation GmbH. Ask us for additional details.

KHD Humboldt Wedag Processautomation GmbH
Aachener Straße 340-346 · D-50933 Köln (Braunsfeld) · Germany
Telefon 02 21/54 05 05 · Telefax 02 21/54 05-302 · Telex 8 882 786

HUMBOLDT WEDAG Processautomation

SIEMENS

Wir bieten die Formel für konkurrenzfähige Produktion

"Man nehme ein bißchen von diesem und etwas von jenem…" – wenn die "Rezepturen" für die technische Ausrüstung der Zementindustrie so einfach wären … Nicht jede Technik paßt hier so einfach zusammen. Und es bedarf schon besonderer Fähigkeiten, alle Möglichkeiten der Technik für Effizienzsteigerungen auszuschöpfen.

Für diese Aufgaben haben Sie in Siemens einen erfahrenen und kompetenten Partner. Einen, der das Thema Zement in allen technischen und logistischen Facetten kennt und Werke der Zementindustrie seit Jahrzehnten ausrüstet.

Einen, der bekannt ist für viele Innovationen und einen, der die Gesamtverantwortung für den Bau einer ganzen Anlage übernimmt – oder auch "nur" für die Schaltanlage. Oder die Leittechnik. Oder die Automatisierung. Oder die Motoren. Oder … oder …

Was immer Sie brauchen, um Ihr Projekt auf Zukunftskurs zu bringen: Wir von Siemens bieten Ihnen die technischen und logistischen Voraussetzungen dafür. Außerdem das Know-how und die finanzielle Standfestigkeit, die solche Projekte erfordern.

Dabei ist Augenmaß schon seit jeher unsere Stärke. Ob Sie also nur eine Teilleistung vergeben möchten oder ein Projekt im Ganzen: Bei Siemens ist alles möglich. Wir binden lokale Ressourcen in unsere Projekte mit ein. Wir haben die Spezialisten, die Erfahrung und die weltweite Infrastruktur, um selbst extrem große und komplexe Projekte zeit- und budgetgerecht abwickeln zu können. Anlagen in aller Welt beweisen dies.

Sprechen Sie uns an, wenn Sie Fragen haben:
Siemens AG
ANL A 71, Germany
Tel. (049) 9131/7-2 20 30
Fax (049) 9131/7-2 70 08

INNOVATIONS

From Siemens.
For Cement.

PC-BLAINE-STAR

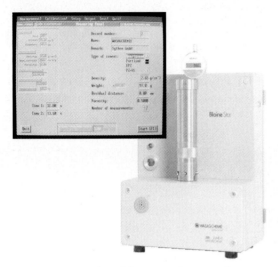

The most comfortable way of measuring specific surfaces.
Highest precision together with easy operation is gurananteed by our new Blaine-Instrument working with any PC.

For further information, please contact:

Werkstraße 111
D-45721 Haltern   Telefon (+49) (0 23 64) 6 89-2 57
Postfach 104   Telefax (+49) (0 23 64) 6 89-2 97
D-45713 Haltern   Telex   8 29 873 wsgvk d

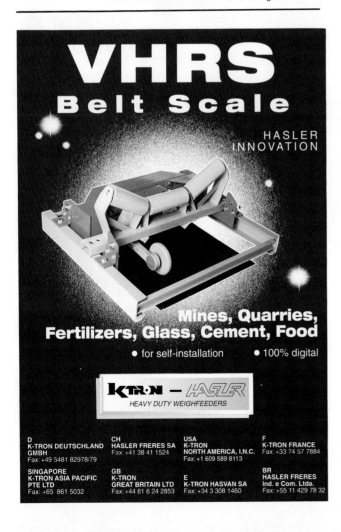

 BAUVERLAG

Schalungstechnik mit System

Bewährte Methoden – Neue Entwicklungen

Von F. H. Hoffmann.
1993. 232 Seiten mit
rd. 115 Abbildungen
und rd. 15 Tabellen.
Format 17 x 24 cm.
Gebunden DM
138,–/öS 1.076,–/
sFr 138,–
ISBN 3-7625-2609-5

Die Bedeutung, die dem sinnvollen Einsatz von Schalungen bezüglich Qualitätssicherung und Wirtschaftlichkeit im Betonbau zukommt, wird meist unterschätzt. Dabei sind Schalungsarbeiten inzwischen ein wichtiger, wenn nicht der wichtigste Kostenfaktor im Betonbau, zumal sie etwa ein Drittel der Rohbaukosten und zwei Drittel des Arbeitskräfteaufwands ausmachen.

Eine wirtschaftliche Rohbauausführung ist nur dann gewährleistet, wenn Betonkonstruktionsformen schon bei der Planung auf vorhandene und bewährte Schalungsverfahren abgestimmt werden. Schalungsarbeiten bestimmen aufgrund ihres hohen Arbeitsanteils den Bauablauf und haben somit Priorität im Bauablaufplan. Deshalb müssen diese Bereiche in der Planung und Arbeitsvorbereitung sowie bei der Aus- und Weiterbildung von Ingenieuren stärker als bisher berücksichtigt werden.

Diese Forderungen formulierte der Autor in vielen Fachzeitschriftenartikeln, die hier gesammelt veröffentlicht werden. Anhand der Aufsätze läßt sich der technische Wandel im Schalungsbau verfolgen. Dieser Rückblick auf 25 Jahre Schalung im Betonbau zeigt die Kontinuität einiger Problemstellungen, aber auch die Möglichkeiten, die die Schalungs- und Gerüsttechnik bei optimaler Nutzung heute bietet.

BAUVERLAG GMBH
D-65173 Wiesbaden
Tel. (0 61 23) 700-0 · Fax (0 61 23) 700-122

A U T O M A T I O N

Everything from a single source - because complex cement facilities are more than just the sum of their parts.

■ As a universal supplier, AEG Daimler-Benz Industrie has the electrical engineering expertise and specific know-how required to set up and equip cement factories. Whether it's direct business or orders from consortiums, we can supply individual systems or complete turnkey facilities. If desired, AEG Daimler-Benz Industrie can also help arrange the appropriate financing.

■ Our advantageous scope of products and services ranges from power supply to drive technology; from automation to auxiliary systems as lighting, signalling and monitoring systems.

■ We at AEG Daimler-Benz Industrie undertake and coordinate all tasks of assembly and installation of complete electrical systems:

■ Consulting, Planning
■ Production
■ Assembley, Start-up
■ After sales service
■ Training.

Ge△matics

■ With the automation system Geamatics, AEG Daimler-Benz Industrie is one of the few companies in the world that can offer the cement industry comprehensive, single source solutions.

AEG Aktiengesellschaft
Projects and Drive Systems Division
Geamatics Building
Lyoner Strasse 9
D - 60528 Frankfurt/Main
Germany
Phone: (+49 69) 66 49 - 0
Fax: (+49 69) 66 49 - 30 00

Daimler-Benz Industrie

AEG

Computereinsatz in Umwelt- und Verfahrenstechnik

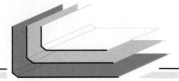

GRAINsoft

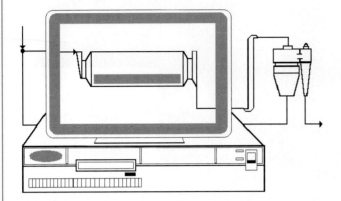

Ihr kompetenter Partner für

Engineering

Informationsmanagement

in Betrieb, Labor, R & D

Innovative und praxisbewährte Lösungen

☐ Optimierung von Mahlanlagen

☐ Planung und Qualitätssicherung

☐ Qualifizierung von Routine - Auswertungen

GRAINsoft GmbH   Chemnitzer Straße 40 D - 09599 Freiberg   Tel.+49 3731 797 300   Fax+49 3731 797 221

Belt Scale Uniband 802
- High performance at low price
- Easy to install
- Maintenance-free

Instrument - Technology of the Future

Rembe® GmbH · D-59918 Brilon
Germany · Fax (02961) 507 14
Phone (02961) 7405-20
Teletex 296134 = REMBE
Telex 17296134

e227

BAUVERLAG

DICTIONARY OF CEMENT

Manufacture and Technology

By Dipl.-Ing. C. van Amerongen. German-English/English-German in one volume.
2nd revised and enlarged edition 1986, 335 pp. Size 13,5 x 20,5 cm.
Hardcover DM 130,- (plus postage)

Besides numerous terms relating to mechanical engineering, electrical engineering, chemical engineering, physics etc., this dictionary contains above all terminology of the rock and associated products industry, quarrying, excavating machinery, haulage vehicles, crushing and grinding machinery, materials handling, automation, fuel and heat engineering, refractory materials, chemical and physical testing and many other subjects connected with cement and its manufacture.

This dictionary is a great assistance to those, who value perfect linguistic communication and exact translations of the corresponding technical terms.

BAUVERLAG GMBH · D-65173 Wiesbaden

KRUPP POLYSIUS

Machines, Plants and Services

As a leading supplier of plants and systems for the cement industry, Krupp Polysius is used to dealing with challenges. Demands for shorter delivery times and for the development of innovative, energy-saving and non-polluting technologies are just two examples of modern challenges which we have successfully overcome.

In Krupp Polysius you have an eminently competent partner, adviser and supplier, with a comprehensive and technologically advanced range of products and services, centered on pyro-processing, comminution and automation.

And this not limited to new plants. We also specialize in modernising existing facilities to reduce operating costs and ensure compliance with the increasingly strict environmental protection regulations.

Krupp Polysius sets the standards - increasing the quality and efficiency of modern industrial plants and cutting the implementation time between order award and commencement of production. Plants constructed by Krupp Polysius rank among the most productive in the world.

Krupp Polysius AG
Graf-Galen-Straße 17
D-59269 Beckum
Tel.: 02525/99-0
Telefax: 02525/992100

A company of the Krupp Anlagenbau group.

PILLARD

Die bewährte Brennergeneration...

220 x in Auftrag
175 x in Betrieb

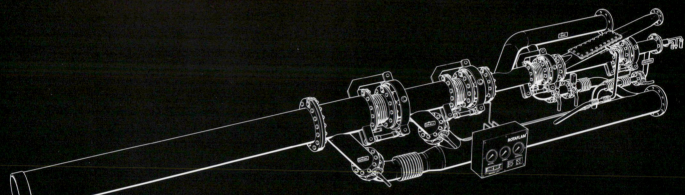

Beispiel: Brenner für Kohle, Gas, Öl

... mit den vielen Vorteilen

ROTAFLAM®

Umweltentlastung	Verringerung der NO_x-Emmission bis über 50%
Kosteneinsparung	Senkung des spezifischen Wärmeverbrauchs durch drastische Reduzierung des Primärlufteinsatzes
Verfahrentechnische Verbesserungen	Außerordentlich stabile Flamme innerhalb eines sehr großen Regelbereiches.
Wirtschaftlichkeit	sehr kurze Amortisationszeit.

EGCI PILLARD	13, Rue R.-Teissère	F-13272 MARSEILLE Cédex 8	Tel. 91.80.90.21	Tlx 430330	Fax 91.25.72.71
PILLARD FEUERUNGEN GMBH	Aarstraße 168	D-65232 TAUNUSSTEIN	Tel. 06128/242.0	Tlx 4182728	Fax 06128/242112
PILLARD ESPAÑA	Esteban Terradas 9	E-28036 MADRID	Tel. (91) 733 12 50		Fax (91) 7 33 01 95

Fachbereich 3

Brenntechnik und Wärmewirtschaft

(Vorwärmen, Vorcalcinieren, Brennen, Kühlen, Ofensysteme, Stoffkreisläufe, feuerfestes Futter, Ansätze, Brennstoffe, Brennstoffaufbereitung, Verbrennung, Wärmenutzung)

Subject 3

Burning technology and thermal economy

(Preheating, precalcining, burning, cooling, kiln systems, circulation of volatile substances, refractory linings, coatings, fuels, treatment of fuels, combustion, waste heat utilization)

Séance Technique 3

Technique de cuisson et économie de chaleur

(Préchauffage, précalcination, cuisson, refroidissement, types de fours, circulation des matières, revêtements réfractaires, formation de dépôts, combustibles, préparation des combustibles, combustion, utilisation de l'énergie thermique)

Tema de ramo 3

Técnica de cocción y energía térmica

(Precalentamiento, precalcinación, cocción, enfriamiento, tipos de hornos, circulación de materiales, revestimientos refractarios, formación de costras, combustibles, preparación de combustibles, combustión, utilización de energía térmica)

Brenntechnik und Wärmewirtschaft*)
Burning technology and thermal economy*)
Technique de cuisson et économie de chaleur

Técnica de cocción y economía térmica

Von **H. S. Erhard,** Heidelberg, und **A. Scheuer,** Leimen/Deutschland

Generalbericht 3 · Zusammenfassung – Durch den Einsatz der Vorcalciniertechnik sind beachtliche Fortschritte erzielt worden. In Ländern mit steigendem Zementabsatz etablierte sich diese Technik durch den Neubau von kompletten Ofenlinien. In den meisten Industrieländern wurden vermehrt bestehende Anlagen modernisiert und in ihrer Produktionskapazität erweitert. Rentabel waren diese Modernisierungen nur in Verbindung mit der Stillegung alter Anlagen oder mit dem vermehrten Einsatz von Sekundärstoffen. Den niedrigsten Wärmeverbrauch und die höchste Produktionsleistung erreichen Ofensysteme, die mit trockenem Rohmehl über einen mehrstufigen Zyklonvorwärmer beschickt werden. Soweit keine zusätzliche Rohmaterialtrocknungsleistung erforderlich ist, werden diese Ofensysteme mit wenigstens 4 und höchstens 6 Vorwärmerstufen ausgerüstet. Moderne Ofensysteme sind desweiteren mit einem Calcinator ausgestattet. Brennstoffwärme kann man mit einer Vorcalcinieranlage allerdings nicht einsparen. Die Vorteile liegen vielmehr bei den niedrigen Investitions- und Betriebskosten und bei einer verbesserten Prozeßführung. Die Drehöfen weisen heute L/D-Verhältnisse zwischen 11 : 1 und 17 : 1 auf. Kurze Öfen können auf zwei Laufrollenstationen gelagert werden. Dadurch sinken die Investitionskosten, und die Sicherheit gegen Überlastungszustände nimmt zu. Für die Herstellung von Sonderprodukten ist der Wirbelschichtofen eine interessante Alternative. Zuverlässige anlagentechnische Lösungen für den industriellen Dauerbetrieb sind jedoch noch nicht auf dem Markt. Außerdem sind die spezifischen Herstellkosten deutlich höher als bei konventionellen Verfahren. Die Klinkerkühlung wurde in den letzten Jahren erheblich verbessert. Heute sind Rostkühler der „3.Generation" erfolgreich eingeführt, für Ofenkapazitäten unterhalb 4000 t/d auch noch Satelliten- und Rohrkühler. Den Stoffkreisläufen im Ofen ist besondere Bedeutung beizumessen. Chloridkreisläufe können mit niedrigen Bypaßraten beherrscht werden. Zu hohe Schwefelkreisläufe resultieren aus Reduktionsreaktionen mit dem Brennstoff. Ihnen kann durch feuerungstechnische Maßnahmen begegnet werden. Eine Absenkung des Alkaligehalts erfordert hohe Bypaßraten. Hierbei bietet der Vorcalcinierofen Vorteile. Bei der feuerfesten Ausmauerung zeichnet sich ein Trend weg von Magnesia-Chromsteinen hin zu Magnesia-Spinellstein ab. Im Vorwärmer und Kühler werden demgegenüber zunehmend feuerfeste Massen eingesetzt. Die Stromerzeugung aus Abwärme weist einen geringen Wirkungsgrad von ca. 15 % auf. Sie sollte daher nur dann angewendet werden, wenn keine verfahrenstechnischen Alternativen zur Brennstoffenergieeinsparung zur Verfügung stehen.

General report 3 · Summary – Substantial progress has been achieved through the use of precalcining technology. In countries with increasing cement sales it has become established by the construction of complete new kiln lines. In the majority of industrial countries there was more a tendency for existing plants to be modernized and their production capacities extended. This modernization was only profitable when linked with the closure of old plants or with increased use of secondary materials. The lowest heat consumptions and the highest production capacities are achieved by kiln systems which are fed with dry raw meal through a multi-stage cyclone preheater system. Provided no additional raw material drying capacity is needed these kiln systems are equipped with a minimum of 4 and a maximum of 6 preheater stages. This cannot be used for saving fuel heat, the advantages lie more in the low capital and operating costs and in improved process control. Rotary kilns now have length/diameter ratios between 11:1 and 17:1. Short kilns for readily burnable raw meals can be supported on two supporting roller stations. This lowers the capital costs and gives increased protection against overload conditions. The fluidized bed kiln is an interesting alternative for producing special products, but there are as yet no reliable plant solutions for continuous industrial operation on the market. The specific production costs are also significantly higher than with conventional processes. Clinker cooling has been greatly improved in recent years. Grate coolers of the "3rd generation" have now been successfully introduced; planetary and rotary coolers are still used for kiln capacities below 4000 tpd. Particular importance is attributed to recirculating systems within the kiln. Chloride cycles can be overcome with low bypass rates. Excessive sulphur cycles result from reduction reactions with the fuel. They can be countered by combustion engineering measures. High bypass rates are needed to lower the alkali content. Precalcining kilns offer

---

*) Überarbeitete Fassung eines Vortrages zum VDZ Kongreß '93, Düsseldorf (27. 9.–1. 10. 1993)
Revised text of lecture to the VDZ Congress '93, Düsseldorf (27. 9.–1. 10. 1993)

advantages here. Refractory linings are showing a trend from magnesia-chrome towards magnesia-spinel bricks, while monolithic refractories are being used increasingly in preheaters and coolers. Power generation from waste heat has a low efficiency of about 15 %, so it should only be used when there is no other available process alternative for saving fuel energy.

Technique de cuisson et économie de chaleur

Rapport général 3 · Résumé – L'utilisation de la technique de précalcination a permis de réaliser des progrès considérables. Dans les pays où la vente de ciment croît, elle s'est établies sous forme de projet de construction de lignes de four complètes. La plupart des pays industriels ont modernisé en nombre croissant leurs installations existantes, et élargi leur capacité de production. Cette politique de modernisation s'est avérée rentable uniquement en liaison avec l'arrêt d'anciennes installations ou avec l'utilisation accrue de matières secondaires. Les systèmes de four alimentés en farine crue sèche par un préchauffeur à cyclones à plusieurs étages, obtiennent une consommation calorifique basse et une capacité de production élevée. Dans la mesure où aucune puissance de séchage de matière brute ne s'avère nécessaire, ces systèmes de four sont équipés d'au moins 4 et au plus 6 étages de préchauffeur. Une telle solution ne permet pas d'économiser la chaleur du combustible. Les avantages sont intéressants plutôt au niveau des frais d'investissement et d'exploitation qui sont bas et dans une gestion du process améliorée. Les fours rotatifs accusent de nos jours un rapport L/D compris entre 11:1 et 17:1. Les fours courts à farine facilement combustible peuvent être montés sur deux postes de roulement à galets. Il en résulte une diminution des frais d'investissement et une sécurité accrue vis-à-vis des phénomènes de surcharge. Pour la fabrication de produits spéciaux le four à lit fluidisé constitue une alternative intéressante. Il n'existe, sur le marché, pas encore de solutions fiables pour une exploitation industrielle en continu. De plus, les frais de fabrication spécifiques sont nettement plus élevés que pour les procédés conventionnels. Le refroidissement de clinker a subi des améliorations notables durant les dernières années. A l'heure actuelle, les refroidisseurs à grille de la „3ème génération" sont introduits avec succès, et pour des capacités de four inférieures à 4000 t/j on trouve encore des refroidisseurs à satellites et des refroidisseurs tubulaires. Les circuits de matière du four revêtent une importance particulière. Les circuits de chlorure peuvent être maîtrisés avec de faibles taux de dilution. Des circuits de soufre trop élevés résultent de réactions de réduction avec le combustible. Des mesures au niveau du dispositif de chauffe permettent de remédier à cette situation. Une diminution de la teneur en alcali exige des taux de dilution élevés. Le four de précalcination offre des avantages en pareil cas. Pour ce qui est du garnissage réfractaire on note une tendance abandonnant la brique en magnésie-chrome et allant en faveur de la brique en magnésie-spinelle. Dans le préchauffeur et le refroidisseur, on utilise, en revanche, de plus en plus de masses réfractaires. La production de courant à partir de chaleur perdue représente un faible rendement de l'ordre de 15 %. Il est donc préférable de ne l'utiliser que si l'on ne dispose pas de procédé alternatif pour économiser l'énergie du combustible.

Técnica de cocción y economía térmica

Informe general 3 · Resumen – Gracias al empleo de la técnica de precalcinación, se han conseguido importantes adelantos tecnológicos y operativos. En los paises con creciente venta de cemento, dicha técnica se ha establecido, debido al montaje de nuevas líneas completas de hornos. En la mayoría de los paises industrializados, se han modernizado mayormente las instalaciones existentes, aumentando su capacidad de producción. Estas modernizaciones han sido rentables solamente en combinación con la parada de instalaciones viejas o el empleo, en mayor grado, de materias primas secundarias. El menor consumo térmico y el mayor rendimiento de paso lo consiguen los sistemas de horno alimentados con crudo seco, a través de un precalentador de ciclones de varias etapas. Si no hace falta un secado adicional del crudo, estos sistemas de horno van equipados con 4 etapas de precalentamiento, como mínimo, y 6, como máximo. Los sistemas de horno modernos van provistos, además, de un calcinador. Sin embargo, con la precalcinación no se puede ahorrar energía térmica procedente de combustibles. Las ventajas residen, más bien, en los reducidos costes de inversión y de explotación así como en un mejor control del proceso. Los hornos rotatorios presentan, hoy en día, unas relaciones de longitud/diámetro de 11:1 y 17:1. Los crudos difíciles de cocer requieren unos recintos de horno más grandes que los crudos fáciles de cocer. Los hornos cortos pueden apoyarse en dos estaciones de rodillos de rodadura. Debido a ello, se reducen los gastos de inversión y aumenta la seguridad contra sobrecargas, gracias el apoyo estáticamente determinado. Para la fabricación de productos especiales, el horno de recinto turbulento representa una alternativa interesante. En este horno, la temperatura del material sometido a cocción y la atmósfera gaseosa se pueden ajustar de forma muy precisa, con lo cual se consigue una gama muy variada de productos. Sin embargo, no se encuentran aún en el mercado soluciones tecnológicas fiables para el servicio industrial continuo. Además, los gastos específicos de fabricación son notablemente más elevados que en el caso de los procedimientos convencionales. Los usuarios así como los constructores de las instalaciones han hecho grandes esfuerzos, con el fin de mejorar el enfriamiento del clínker. Actualmente, están funcionando con éxito enfriadores de parrilla de la „3ª generación". Para capacidades de hornos inferiores a 4000 t/día, bajo determinadas condiciones marginales, también los enfriadores de tambor rotatorio o de satélites pueden ser soluciones económicas y de reducidas emisiones. Hay que prestar especial atención a los circuitos de material dentro del horno. Los circuitos de cloruros se pueden controlar con reducidos índices de by-pass. Los cir-

cuitos de azufre demasiado altos resultan de las reacciones reductoras con el combustible. Se puede hacer frente a los mismos, tomando medidas respecto de la técnica de combustión. Una reducción del contenido de álcalis requiere elevados índices de by-pass. A este respecto, el horno de precalcinación presenta ventajas. En los revestimientos refractarios se divisa una tendencia hacia el abandono de los ladrillos de cromomagnesita, a favor de los ladrillos de espinela magnésica. En el precalentador y en el enfriador se utilizan cada vez más masas refractarias. La producción de corriente eléctrica, aprovechando el calor residual, presenta un rendimiento eléctrico del 15 % aprox. Solo se puede aplicar de forma económica si no existen otras alternativas tecnológicas de ahorro energético en los combustibles y si se cumplen, además, otros requisitos esenciales.

1. Einleitung

Das Herzstück eines Zementwerkes ist der Ofen, in dem der Portlandzementklinker gebrannt wird. Traditionell werden hierfür verschiedene Verfahren angewendet. Weltweit verbreitet sind Schachtöfen, lange Naß- oder Trockenöfen, Öfen mit Rost- und Zyklonvorwärmer sowie Vorcalcinieröfen.

In den letzten Jahren wurden durch den Einsatz der Vorcalciniertechnik beachtliche verfahrens- und produktionstechnische Fortschritte erzielt. In Ländern mit steigendem Zementabsatz etablierte sich die Vorcalciniertechnik durch den Neubau von kompletten Ofenlinien. Dabei standen insbesondere Kapazitätserweiterungen im Vordergrund. In den meisten traditionellen Industrieländern stagnierte demgegenüber der Zementabsatz. Der Bau von Neuanlagen und damit auch die Einführung der Vorcalciniertechnik verliefen daher nur zögerlich.

Bild 1 zeigt die Anzahl der in den letzten 8 Jahren von namhaften europäischen und nordamerikanischen Anlagenbauern*) in den verschiedenen Regionen gebauten Ofenlinien. Demnach haben insbesondere Zementhersteller in Asien und Afrika umfangreiche Investitionsprogramme abgewickelt. In Europa und Amerika sind überwiegend bestehende Anlagen modernisiert worden. Die dabei realisierten Neuerungen auf dem Gebiet der Brenntechnik sowie einige Ergebnisse von Entwicklungsarbeiten sind Gegenstand dieser Arbeit.

2. Naßverfahren

Weltweit wird immer noch ein erheblicher Teil des Zementklinkers auf der Basis naß aufbereiteter Rohmaterialien hergestellt. Das ist eine Folge der geschichtlichen Entwicklung

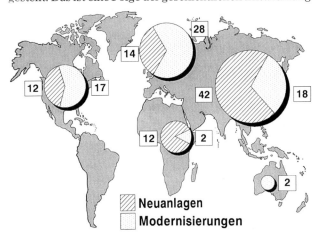

BILD 1: Anzahl der in den letzten 8 Jahren in den verschiedenen Regionen der Welt gebauten Ofenlinien (Ergebnis einer Umfrage bei namhaften europäischen und nordamerikanischen Anlagenbauern)
FIGURE 1: Number of kiln lines built in the last 8 years in the different regions of the world (result of a questionnaire to well-known European and North American plant manufacturers)

Neuanlagen = new plants
Modernisierungen = modernization projects

---

*) Den Unternehmen FCB, F.L. Smidth & Co A/S, Fuller International Inc., KHD Humboldt Wedag und Krupp Polysius sei an dieser Stelle für ihre Unterstützung gedankt.

1. Introduction

The heart of a cement works is the kiln in which the Portland cement clinker is burnt. Traditionally, a variety of processes have been used for this purpose. Shaft kilns, long wet or dry kilns, kilns with grate and cyclone preheaters, and precalciner kilns are spread throughout the world.

Considerable progress in process and production technology has been achieved in recent years through the use of precalciner technology. In countries with increasing cement sales, precalciner technology has become established through the construction of complete new kiln lines. The particular emphasis in these cases has been on increased capacity. However, cement sales have stagnated in the majority of traditional industrialized countries, so the construction of new plants, and hence also the introduction of precalciner technology, has only progressed very slowly.

Fig. 1 shows the number of kiln lines constructed in the last 8 years by well known European and North American plant manufacturers*) in the various regions. This shows that the cement producers in Asia and Africa have conducted particularly extensive investment programmes. In Europe and America the modernization of existing plants has predominated. This article describes the innovations achieved in the field of burning technology and some results from development work.

2. Wet Process

A considerable proportion of the cement clinker throughout the world is still produced using wet-processed raw materials. This is a consequence of the historical development of energy costs and of the process technology of cement production. However, the moisture of the available raw materials also plays an important part.

The fuel energy consumption of approximately 5500 to 6000 kJ/kg clinker for the wet process is comparatively high when compared to the dry process (**Table 1**). The figure for long dry kilns with internal fittings, for example, is approximately 4500 kJ/kg clinker, and for short dry kilns with 4-stage cyclone preheaters it is approximately 3300 kJ/kg clinker. Modern precalciner kilns with 6-stage preheaters only need approximately 3000 kJ/kg clinker. However, the sensible heat in the rotary kiln exhaust gas from such kilns is only sufficient for drying approximately 6% moisture in the raw material.

TABLE 1: Fuel energy consumption of different burning systems

Burning process	Fuel energy consumption kJ/kg clinker
– Wet process	5500–6000
– Long dry kiln with internal fittings	4500
– Short dry kiln with 4-stage cyclone preheater	3300
– Precalciner kiln with 6-stage cyclone preheater	3000

---

*) Thanks are due to the firms of FCB, F.L. Smidth & Co. A/S, Fuller International Inc., KHD Humboldt Wedag, and Krupp Polysius for their assistance.

der Energiepreise und der Verfahrenstechnik der Zementherstellung. Darüber hinaus spielt aber auch die Feuchte der verfügbaren Rohmaterialien eine wesentliche Rolle.

Der Brennstoffenergiebedarf des Naßverfahrens ist mit ca. 5500 bis 6000 kJ/kg Kli gegenüber dem Trockenverfahren vergleichsweise hoch (**Tabelle 1**). Lange Trockenöfen mit Einbauten liegen z. B. bei ca. 4500 kJ/kg Klinker und kurze Trockenöfen mit 4stufigem Zyklonvorwärmer bei ca. 3300 kJ/kg Klinker. Moderne Vorcalcinieröfen mit 6stufigem Vorwärmer verbrauchen demgegenüber nur ca. 3000 kJ/kg Kli. Die fühlbare Wärme im Drehofenabgas solcher Öfen reicht allerdings nur noch zum Trocknen von ca. 6% Feuchte im Rohmaterial.

TABELLE 1: Brennstoffenergiebedarf unterschiedlicher Brennverfahren

Brennverfahren	Brennstoffenergiebedarf kJ/kg Klinker
– Naßverfahren	5500–6000
– Langer Trockenofen mit Einbauten	4500
– Kurzer Trockenofen mit 4stufigem Zyklonvorwärmer	3300
– Vorcalcinierofen mit 6stufigem Zyklonvorwärmer	3000

Theoretisch ist es möglich, bestehende Naßanlagen auf moderne Vorcalciniertechnik mit trockener Aufbereitung umzustellen. Allerdings kostet diese Umrüstung im Bereich der Rohmaterialaufbereitung mehr als im Bereich des Ofens (**Tabelle 2**). Angesichts dieses Aufwandes wird man versuchen, die bestehende Naßaufbereitung weiter zu nutzen und den Ofen als zusätzlichen Heißgaserzeuger zu verwenden.

TABELLE 2: Relative Investitionskosten für den Bau eines Klinkerwerks [1]

Bezeichnung	Anteil der Investitionskosten in %	
– Rohmateriallagerung	33	
– Rohmühle	17	60
– Rohmehllagerung	10	
– Ofenanlage	40	

Darüber haben z. B. J. C. Ahuja et al. [2] berichtet, indem sie den Einsatz von 5 Naßöfen durch einen Vorcalcinierofen bei gleichzeitiger Steigerung des Klinkerdurchsatzes um ca. 50% beschrieben haben. J. Duda und B. Werynski [3] haben über einen ähnlichen Umbau berichtet. Zur weiteren Kostenminderung setzen sie Kohleschiefer als Sekundärroh- und Sekundärbrennstoff im Calcinator ein. Neuanlagen werden demgegenüber auch bei sehr feuchtem Rohmaterial noch mit trocken arbeitender Aufbereitung ausgerüstet, die, wenn nötig, mit einem zusätzlichen Heißgaserzeuger versorgt wird [4]. In diesem Fall spielt der hohe Anteil der Investitionskosten für die Rohmaterialaufbereitung keine Rolle.

Grundsätzlich ist die Modernisierung einer Ofenlinie oder die komplette Umstellung eines Werkes vom Naß- auf das Trockenverfahren unrentabel, wenn damit nicht gleichzeitig andere Kostenvorteile erzielt werden. Das kann eine deutliche Produktionssteigerung in Verbindung mit steigendem Absatz, die Stillegung von Altanlagen und/oder der vermehrte Einsatz von Sekundärroh- und Sekundärbrennstoffen sein. **Bild 2** zeigt dazu als Beispiel das Verfahrensschema einer von F. L. Smidth modernisierten Anlage eines dänischen Zementwerkes [5]. Die Anlage besteht aus einem Schubrostkühler, dem Drehofen mit der Tertiärluftleitung, dem Calcinator, zweistufigem Zyklonvorwärmer, Schlagpralltrockner und Zyklonabscheider. Im Bild ist diese Anlage einsträngig dargestellt. In die Schlagpralltrockner werden zwei Schlammkomponenten eingedüst, eine Kreide-Sand-Komponente mit ca. 30% Wasser und eine aufgeschlämmte Kiesabbrand-Komponente mit ca. 50% Wasser. Als Tonkomponente und Brennstoffträger wird ferner

Theoretically, it is possible to convert existing wet plants to modern precalciner technology with dry materials processing. However, this conversion costs more in the raw materials processing sector than in the kiln sector (**Table 2**). In view of this expenditure, attempts are being made to make further use of the existing wet processing system and to use the kiln as an additional hot gas generator.

TABLE 2: Relative investment costs for the construction of a clinker works [1]

Designation	Proportion of investment costs in %	
– Raw material storage	33	
– Raw mills	17	60
– Raw meal storage	10	
– Kiln plant	40	

J. C. Ahuja et al. [2] have given a report on this subject, in which they describe the replacement of five wet kilns by one precalciner kiln with a simultaneous increase in the clinker throughput by about 50%. J. Duda and B. Werynski [3] have reported on a similar conversion. To reduce the cost still further they use coal shale as a secondary raw material and secondary fuel in the calciner. However, new plants are equipped with dry processing systems even where there are very moist raw materials and, if necessary, are supplied with additional hot gas generators [4]. In this case the high proportion of the investment costs for the raw materials processing plant plays no part.

Basically, the modernization of a kiln line or the complete conversion of a works from wet to dry process, is not profitable unless other cost advantages can be achieved at the same time. These may be a significant increase in production in conjunction with increasing sales, the closure of old plants, and/or increased use of secondary raw materials and secondary fuels. As an example, **Fig. 2** shows the process flow diagram for a plant in a Danish cement works modernized by F. L. Smidth [5]. The plant consists of a reciprocating grate cooler, the rotary kiln with tertiary air duct, the calciner, two-stage cyclone preheater, impact crusher-dryer and cyclone separator. In the diagram, the plant is shown with a

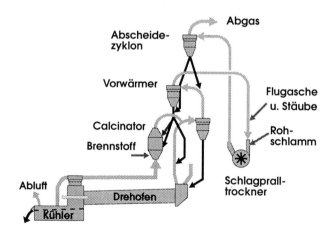

BILD 2: Schematische Darstellung einer Vorcalcinieranlage mit Naßaufbereitung bei der Aalborg Portland A/S [5]
FIGURE 2: Diagrammatic representation of a precalciner plant with wet processing system at Aalborg Portland A/S [5]

Abscheidezyklon	= separating cyclone
Vorwärmer	= preheater
Calcinator	= calciner
Brennstoff	= fuel
Abluft	= exhaust air
Kühler	= cooler
Abgas	= exhaust gas
Drehofen	= rotary kiln
Flugasche u. Stäube	= fly ash and dust
Rohschlamm	= raw slurry
Schlagpralltrockner	= impact crusher-dryer

ca. 0,165 kg/kg Klinker Kraftwerksflugasche aufgegeben, die noch etwa 7% Restkohlenstoff enthält [6]. Im Trockner wird das Brenngutgemisch vollständig getrocknet und auf etwa 150 °C vorgewärmt. Von dort wird es pneumatisch zu den oben gelegenen Abscheidezyklonen gefördert, abgeschieden und der ersten Vorwärmstufe aufgegeben, wo sich das Brenngut auf ca. 700 °C erwärmt. Anschließend wird das Brenngut aufgeteilt und in Teilströmen in die Steigleitung bzw. in die Calcinatoren zugegeben. Das Brenngut gelangt dann mit ca. 860 °C in den Drehofen, in dem die für die Klinkerbildung maßgebenden Reaktionen ablaufen.

Nach dem Vorwärmer beträgt die Abgastemperatur ca. 700 °C und nach dem Schlagpralltrockner ca. 150 °C. Die Gastemperaturen werden über Brenngut-Bypässe, die Aufteilung der Gasmassenströme über Schieber in den Gassteigeschächten und die Anteile der Aufgabegutmassenströme mit Hilfe stündlicher Klinkeranalysen gesteuert.

Der Drehofen besitzt die Abmessungen von ⌀ 4,75 m × 74 m. Die Produktionskapazität beträgt heute ca. 5 500 t/d, der spezifische elektrische Energiebedarf ca. 36 kWh/t Kli und der spezifische Brennstoffenergiebedarf ca. 4290 kJ/kg Kli. Auf der Basis von 7 % Restkohlenstoff in der Flugasche entspricht das einem Primärenergiebedarf von 3 840 kJ/kg Kli. Neben den günstigen Brennstoffenergiebedarf und dem gegenüber vergleichbaren Naßöfen vervielfachten Durchsatz kann das Werk heute vollständig auf die Zugabe von Ton verzichten. Können solche Vorteile nicht genutzt werden, treten Maßnahmen zur Prozeßoptimierung in den Vordergrund, wie sie z. B. A. J. de Beus [7] beschreibt.

3. Trockenverfahren

3.1 Ofen mit Rostvorwärmer

Bild 3 zeigt einen Lepolofen, der nach dem Halbtrockenverfahren arbeitet. Für dieses Verfahren stehen schon seit Jahrzehnten ausgereifte und bewährte Aggregate zur Verfügung. Charakteristisch ist die Brenngutaufgabe in Form von Granalien. Lepolöfen verbrauchen nicht mehr Brennstoffenergie, jedoch deutlich weniger elektrische Energie als Zyklonvorwärmeröfen gleicher Produktionskapazität. Aufgrund ihrer systembedingten Nachteile, wie

- relativ hohe spezifische Anschaffungs- und Betriebskosten,
- hohe Anforderungen an Qualität und Gleichmäßigkeit der Rohmaterialien,
- geringes Abwärmepotential für die Mahltrocknung und
- Einschränkungen bei der Verwertung minderwertiger Brennstoffe,

haben sie ihre frühere Bedeutung nicht halten können. Mittelfristig ist daher zu erwarten, daß die Anzahl der Lepolöfen weltweit weiter zurückgehen wird, vor allem im Zusammenhang mit Rationalisierungsmaßnahmen, also der Konzentration von Produktionskapazität auf große Einheiten [8]. Andererseits lassen sich durch betriebliche Verbesserungsmaßnahmen, wie

- Betrieb mit Vorcalcinierung,
- Einsatz von Sinterhilfen und Mineralisatoren sowie

single string. Two slurry components are injected into the impact crusher-dryer – a chalk/sand component with approximately 30% water and a slurried roasted pyrites component with approximately 50% water. Power station fly-ash which still contains about 7% residual carbon is also added as a clay component and a source of fuel at a rate of approximately 0.165 kg/kg clinker [6]. The kiln feed mix is completely dried in the dryer and heated to about 150°C. From there it is conveyed pneumatically to the top separating cyclones, separated, and fed to the first preheater stage where the kiln feed is heated to approximately 700°C. The kiln feed is then divided and fed in sub-streams into the riser duct or into the calciners. The kiln feed with a temperature of approximately 860°C then passes into the rotary kiln where the reactions essential for clinker formation take place.

The exhaust gas temperature after the preheater is approximately 700°C and after the impact crusher-dryer is approximately 150°C. The gas temperatures are controlled by kiln feed bypasses, the distribution of the gas mass flows using dampers in the gas riser shafts, and the proportions of the mass flows of the feed material with the aid of hourly clinker analyses.

The rotary kiln has dimensions of 4.75 m diameter × 74 m, and its production capacity is now approximately 5 500 t/d. The specific electrical power consumption is approximately 36 kWh/t clinker and the specific fuel consumption is approximately 4 290 kJ/kg clinker. Assuming 7 % residual carbon in the fly ash, this corresponds to a primary energy consumption of 3 840 kJ/kg clinker. In addition to the favourable fuel energy consumption and the throughput, which is many times higher than for comparable wet kilns, the works is now able to dispense entirely with the addition of clay. If it is not possible to make use of such advantages, then the emphasis should be placed on measures for process optimization, such as those described by A. J. de Beus [7].

3. Dry process

3.1 Kiln with grate preheater

Fig. 3 shows a Lepol kiln using the semi-dry process. There are units available for this process which have been perfected and proven over decades. The characteristic feature is the kiln feed in the form of nodules. Lepol kilns require no more fuel energy, and significantly less electrical energy, than cyclone preheater kilns of the same production capacity. They have not been able to maintain their earlier importance because of disadvantages inherent in the system such as

- relatively high purchase and operating costs,
- high demands on the quality and uniformity of the raw materials,
- low waste heat potential for drying and grinding, and
- limitations in the use of low grade fuels.

In the medium term it is therefore to be expected that the number of Lepol kilns throughout the world will continue to decrease, chiefly in conjunction with rationalization measures, i.e. the concentration of production capacity in large units [8]. However, in many cases it is possible to achieve appreciable energy savings and increases in throughput through operational improvements, such as

- operation with precalcination,
- use of sintering aids and mineralizers, and
- by making the nodulizing and sintering conditions more consistent.

3.2 Kiln with cyclone preheater

The lowest heat consumption and the highest clinker throughput is achieved by kiln systems which are fed with dry raw meal through a multi-stage cyclone preheater.

For information, **Table 3** shows statistically evaluated operating data for 4-, 5-, and 6-stage cyclone preheater kilns,

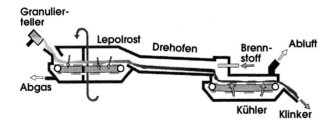

BILD 3: Schematische Darstellung eines Lepolofens
FIGURE 3: Diagrammatic representation of a Lepol kiln

Granulierteller	= nodulizing table	Abluft	= exhaust air
Lepolrost	= Lepol grate	Abgas	= exhaust gas
Drehofen	= rotary kiln	Kühler	= cooler
Brennstoff	= fuel	Klinker	= clinker

TABELLE 3: Betriebsdaten von 4-, 5- und 6stufigen Zyklonvorwärmeröfen
(Ergebnis einer Umfrage bei namhaften europäischen und nordamerikanischen Anlagenbauern)
TABLE 3: Operating data from 4-, 5-, and 6-stage cyclone preheater kilns
(results of a questionnaire to well-known European and North American plant manufacturers)

Bezeichnung	Einheit	Zyklonwärmetauscher		
		4stufig	5stufig	6stufig
Anzahl der Anlagen		34	34	8
Anlagen mit Tertiärluftleitung (% der Gesamtzahl)	%	79	88	100
Anlagen mit Ausbrandregelung im Calcinator	%	9	6	25
Kühlerbauart – Rostkühler – Satellitenkühler – Rohrkühler	%	 85 12 3	 91 9 	 100
L/D-Verhältnis min/max		14,9 ± 1,7*) 10,8/17,5	14,8 ± 1,8*) 9/17,5	12,9 ± 0,9*)
Klinkerdurchsatz	t/d	2805 ± 1060*)	3531 ± 1914*)	4790 ± 2076*)
Spez. Ofenraumbelastung min/max	t/m^{3}d	3,14 ± 0,94*) 1,74/5,25	3,9 ± 0,6*) 2,3/4,8	4,7 ± 0,6*) 4/5,4
Brennstoffenergiebedarf	kJ/kg Kli	3380 ± 365*)	3097 ± 103*)	2991 ± 67*)
Druckverlust nach Vorwärmer	mbar	49,7 ± 10,5*)	50,9 ± 9,1*)	54,9 ± 10,6*)
Spez. Abgasmenge	m$_N^3$/kg Kli	1,5 ± 0,2*)	1,4 ± 0,2*)	1,3–1,4
Abgastemperatur	°C	363 ± 29*)	325 ± 25*)	276 ± 16*)
Klinkeraustrittstemperatur	°C	105 ± 22*)	113 ± 21*)	k.A.

*) Standardabweichung

Bezeichnung	= designation	Rohrkühler	= rotary cooler
Einheit	= units	L/D-Verhältnis min/max	= L/D ratio min/max
Zyklonwärmetauscher	= cyclone preheater	Klinkerdurchsatz	= kiln throughput
4stufig	= 4-stage	Spez. Ofenraumbelastung min/max	= specific kiln volume loading min/max
5stufig	= 5-stage	Brennstoffenergiebedarf	= fuel energy consumption
6stufig	= 6-stage	Druckverlust nach Vorwärmer	= pressure drop after preheater
Anzahl der Anlagen	= number of plants	Spez. Abgasmenge	= specific exhaust gas volume
Anlagen mit Tertiärluftleitung (% der Gesamtzahl)	= plants with tertiary air ducts (% of total number)	Abgastemperatur	= exhaust gas temperature
Anlagen mit Ausbrandregelung im Calcinator	= plants with burn-out control in the calciner	Klinkeraustrittstemperatur	= clinker discharge temperature
Kühlerbauart	= cooler type	Standardabweichung	= standard deviation
Rostkühler	= grate cooler	kJ/kg Kli	= kJ/kg clinker
Satellitenkühler	= planetary cooler	m$_N^3$/kg Kli	= m^3(stp)/kg clinker
		k.A.	= no data

– Vergleichmäßigung der Granulier- und Sinterbedingungen

in vielen Fällen nennenswerte Energieeinsparungen und Durchsatzsteigerungen realisieren.

3.2 Ofen mit Zyklonvorwärmer

Den niedrigsten Wärmeverbrauch und den höchsten Klinkerdurchsatz erreichen Ofensysteme, die mit trockenem Rohmehl über einen mehrstufigen Zyklonvorwärmer beschickt werden.

Tabelle 3 zeigt zur Orientierung statistisch ausgewertete Betriebsdaten von 4-, 5- und 6stufigen Zyklonvorwärmeröfen, die im Rahmen einer Umfrage ermittelt wurden. Danach ist der Wärmeverbrauch des 5stufigen Systems um ca. 270 kJ/kg Kli, der des 6stufigen Systems nochmals um ca. 120 kJ/kg Kli geringer als beim vergleichbaren 4stufigen Vorwärmer. Ähnlich verhalten sich die Abgastemperaturen nach dem Vorwärmer. Überlagert sind bei diesen statistischen Angaben jedoch noch die Einflüsse von Rohmaterial, Brennstoffart, Ofendurchsatz und Kühlerbauart. Die Druckverluste unterscheiden sich demgegenüber nur wenig.

which were obtained from a questionnaire. These show that the heat consumption of the 5-stage system is approximately 270 kJ/kg clinker lower than that of the comparable 4-stage preheater, and that that of the 6-stage system is a further 120 kJ/kg clinker lower. The exhaust gas temperatures after the preheater show a similar behaviour pattern. The effects of raw material, type of fuel, kiln throughput and cooler design are, however, superimposed on this statistical information. On the other hand, there is little difference between the pressure drops.

Preheaters are designed not only for the lowest heat consumption of the kiln line and for the power consumption resulting from the pressure drop and specific exhaust gas mass flow, but also for the heat consumption of the drying and grinding plants.

Fig. 4 shows the raw material drying capacity of kiln systems as a function of heat consumption of the kiln plant. According to this, it is possible to dry raw materials with an initial moisture content of 5 to 6% with the exhaust gas from a 6-stage cyclone preheater, of up to approximately 7% with the exhaust gas from a 5-stage preheater, and of up to approximately 9% with the exhaust gas from a 4-stage pre-

Vorwärmer werden nicht nur nach dem niedrigsten Wärmeverbrauch der Ofenlinie und dem aus Druckverlust und spezifischem Abgasmassenstrom resultierenden Arbeitsbedarf ausgelegt, sondern auch nach dem Wärmebedarf der Mahltrocknungsanlagen.

Bild 4 zeigt die Rohmaterialtrocknungsleistung von Ofensystemen in Abhängigkeit vom Wärmeverbrauch der Ofenanlage. Danach kann mit dem Abgas eines 6stufigen Zyklonvorwärmers Rohmaterial mit einer Eingangsfeuchte von 5 bis 6 %, mit einem 5stufigen bis ca. 7 % und mit einem 4stufigen bis ca. 9 % getrocknet werden, jeweils gerechnet mit Mühlenlaufzeiten von 20 bis 22 h/d. Mit Verwertung der Kühlerabluft können nochmals ca. 4 % mehr Feuchte getrocknet werden.

Neben der reinen Wärmewirtschaft müssen natürlich auch die Investitions- und sonstigen Betriebskosten betrachtet werden. In der Regel sind moderne Ofensysteme deshalb mit wenigstens 4 und höchstens 6 Vorwärmerstufen ausgerüstet. Erhöhter Mahltrocknungsbedarf bzw. Bedarfsspitzen werden in solchen Anlagen mit einem zusätzlichen Heißgaserzeuger oder mit einem Brenngutbypaß unter Umgehung der obersten Zyklonstufe abgedeckt.

3.2.1 Wärmetauscherstufe

In den vergangenen Jahren ist die Zyklonwärmetauscherstufe weiter optimiert worden. Zielgrößen waren dabei die Abscheidung des Brennguts, der Druckverlust, die Flexibilität des Durchsatzes und der Platzbedarf. Gesucht wurde eine Wärmetauscherstufe, die alle Anforderungen optimal erfüllt.

Bild 5 zeigt als Beispiel ein modernes Systemdesign von Krupp Polysius. Der rechteckige Gaseintrittsquerschnitt ist nach außen gelegt, und störende Ecken sind vermieden worden. Die Rotationsachsen von Gas und Brenngut sind nicht mehr identisch, und das Gas tritt durch ein kurzes Tauchrohr aus. Die Mehlleitungen sind mit Pendelklappen abgedichtet und münden in einem Streukasten. So ausgerüstet bilden Zyklon, Gas- und Brenngutleitung für den Strömungsverlauf jeweils eine in sich geschlossene Einheit. Die Stoßverluste der Gase werden dadurch reduziert, und das Brenngut wird besser abgeschieden.

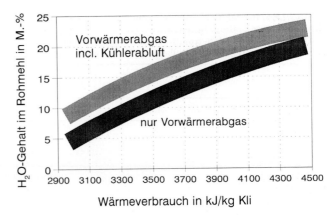

BILD 4: Rohmaterialtrocknungsleistung von Ofensystemen nach [9]
FIGURE 4: Raw material drying capacity of kiln systems according to [9]

H₂O-Gehalt im Rohmehl in M.-%	= H₂O content in the raw meal in wt. %
Vorwärmerabgas incl. Kühlerabluft	= preheater exhaust gas incl. cooler exhaust air
nur Vorwärmerabgas	= preheater exhaust gas only
Wärmeverbrauch in kJ/kg Kli	= Heat consumption in kJ/kg clinker

heater, each based on mill running times of 20 to 22 h/d. Approximately 4 % more moisture can be dried if the cooler exhaust air is utilized.

The investment and other operating costs must also naturally be taken into account as well as the pure thermal economy. As a rule, modern kiln systems are therefore equipped with a minimum of 4 and a maximum of 6 preheater stages. Such plants cover increased drying and grinding requirements or peaks in demand with an additional hot gas generator, or with a kiln feed bypass which avoids the top cyclone stage.

3.2.1 Preheater stage

The cyclone preheater stage has been further optimized in recent years. The target variables were: separation of the kiln feed, pressure drop, flexibility of throughput, and space requirements. The object was to provide a preheater stage which gave optimum fulfilment of all the requirements.

As an example, **Fig. 5** shows a modern system design from Krupp Polysius. The axis of rotation of gas and kiln feed are no longer identical and the gas emerges through a short outlet duct. The meal lines are closed off with flap valves and discharge into a distributor box. When designed in this way, the cyclone, gas and kiln feed line each form an inherently closed unit for the flow process. This reduces the gas impact losses and gives better separation of the kiln feed.

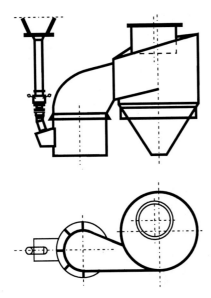

Bild 5: Moderne Zyklonwärmetauscherstufe, Bauart Polysius
FIGURE 5: Modern cyclone preheater stage, designed by Polysius

U. Mrowald und R. Hartmann [10] haben über Modellversuche zur Optimierung der Zyklonbauformen berichtet. Danach sollen moderne Zyklonvorwärmer geringere Abgastemperaturen von 40 bis 50 K und geringere Druckverluste von ca. 15 % aufweisen.

In bestehenden Vorwärmern kann der Druckverlust ebenfalls abgesenkt werden [11, 12]. In den oberen Zyklonen

U. Mrowald and R. Hartmann [10] have described model trials for optimizing cyclone designs. According to this, modern cyclone preheaters should have exhaust gas temperatures which are 40 to 50 K lower, and pressure drops which are approximately 15 % lower.

It is also possible to reduce the pressure drop in existing preheaters [11, 12]. For example, the outlet ducts in the upper cyclones can be retrofitted with cross baffles. Other well-tried measures are flow guide plates and compressed air cannons for removing coating at the horizontal gas inlet to the cyclone. It is possible to retrofit outlet ducts in the bottom cyclone using suspended elements. They have the effect of increasing the degree of separation from, for example, 60 to 70 %. This reduces the exhaust gas losses in a 4-stage cyclone preheater by approximately 30 kJ/kg clinker, but does also increase the pressure drop in the preheater by approximately 1.5 mbar.

The kiln feed acceleration in the gas also affects the pressure loss in the cyclone preheater and, in particular, the flexibility of clinker throughput. For example, a high kiln feed momentum against the gas flow increases the pressure drop as a result of the additional effort needed to accelerate the kiln feed. If the kiln feed momentum is too high locally, then meal flushing also occurs, which causes considerable

können z.B. Leitkreuze in den Tauchrohren nachgerüstet werden. Strömungsleitbleche und Druckluftstoßgeräte zur Ansatzbeseitigung am waagerechten Gaseintritt des Zyklons sind weitere bewährte Maßnahmen. Im untersten Zyklon können Tauchrohre in Elementbauweise nachgerüstet werden. Sie bewirken, daß der Abscheidegrad von z.B. 60 auf 70 % ansteigt. Dadurch sinken die Abgasverluste einer 4stufigen Zyklonvorwärmeranlage zwar um ca. 30 kJ/kg Kli, der Druckverlust im Vorwärmer steigt aber gleichzeitig um ca. 1,5 mbar an.

Die Brenngutbeschleunigung im Gas beeinflußt ebenfalls den Druckverlust des Zyklonvorwärmers und besonders die Flexibilität des Klinkerdurchsatzes. Ein hoher Brenngutimpuls gegen die Gasströmung erhöht z.B. den Druckverlust infolge zusätzlichen Aufwands für die Brenngutbeschleunigung. Ist der Brenngutimpuls lokal zu hoch, tritt ferner Mehldurchschuß auf. Dadurch verschlechtert sich der Wärmeübergang im Vorwärmer merklich. Konstruktiv vermeiden läßt sich ein Mehldurchschuß durch eine Querschnittsverengung in der Gasleitung vor dem Brennguteitritt, durch Streukästen oder Prallschieber zur Brenngutauffächerung, durch Brenngutumlenkung in Gasrichtung und durch Verlängerung des Beschleunigungsweges über dem Gasquerschnitt. Nur als letzte Notlösung bis zur Durchführung notwendiger Korrekturen besteht noch die Möglichkeit, den spezifischen Gasdurchsatz der Anlage zu erhöhen.

Bild 6 zeigt maßstäblich einen mit Hilfe von Stoff- und Energiebilanzen ermittelten Mehldurchschuß in einem 4stufigen Zyklonvorwärmer als Beispiel. Die Massenströme wurden im Rahmen eines Betriebsversuches bei ca. 10 % reduziertem Ofendurchsatz ermittelt. Aus dem Bild geht zunächst hervor, daß sich die Brenngutmassenströme im Vorwärmer gegenüber dem Aufgabegutmassenstrom aufgrund unvollständiger Abscheidung in den Zyklonen deutlich vergrößerten. Zwischen Zyklon 3 und 4 wurde das Brenngut desweiteren nicht vollständig vom entgegenströmenden Gas mitgerissen und fiel teilweise direkt in den Drehofeneinlauf. Dadurch stieg die Gastemperatur im Zyklon 4 unvertretbar hoch von 830 auf 980 °C an.

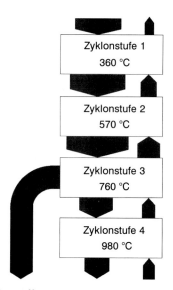

BILD 6: Feststoffmassenströme und -temperaturen in einem Zyklonvorwärmer mit Mehldurchschuß aus der Stufe 3 in den Drehofen [13]
FIGURE 6: Solid mass flows and temperatures in a cyclone preheater with meal flushing from Stage 3 into the rotary kiln [13]
Zyklonstufe = cyclone stage

Das Beispiel zeigt, daß bestehende Ofenanlagen aufgrund ungenügend funktionierender Zyklonwärmetauscherstufen häufig nur bei Nenndurchsatz problemlos betrieben werden können. Leider gilt das auch für Ofenanlagen neueren Datums. Moderne Vorcalcinieranlagen mit 5stufigem Wärmetauscher, die einen Druckverlust von ca. 40 mbar aufweisen, sollten jedoch auch noch mit 70 bis 80 % ihres Nenn-

deterioration in the heat transfer in the preheater. Meal flushing can be avoided structurally by reducing the cross-section in the gas duct before the kiln feed inlet, by using distribution boxes or impact deflectors to spread out the kiln feed, by diverting the kiln feed into the direction of gas flow, and by extending the acceleration path over the gas cross-section. The option of raising the specific gas throughput through the system is only a last emergency solution until necessary corrections can be carried out.

Fig. 6 shows a true-to-scale example of meal flushing in a 4-stage cyclone preheater determined with the aid of mass and energy balances. The mass flows were measured during a plant trial at approximately 10 % reduced kiln throughput. The first thing to be seen from the diagram is that, due to incomplete separation in the cyclones, the mass flows of kiln feed in the preheater were considerably higher than the feed mass flow. Furthermore, the kiln feed was not fully entrained by the opposing gas flow between cyclones 3 and 4, and some of it dropped directly into the kiln inlet. This caused an unacceptable increase in the gas temperature in cyclone 4 from 830 to 980°C.

This example shows that trouble-free operation of existing kiln plants can often only be achieved at their rated throughputs because of inadequate functioning of the cyclone preheater stages. Unfortunately this also applies to kiln plants of fairly recent date. However, it should be possible to operate a modern precalciner plant with a 5-stage preheater, and a pressure drop of approximately 40 mbar, at 70 to 80 % of its rated throughput without also having to make a drastic increase in the excess air ratio.

3.2.2 Calciner

In modern kiln systems the calciner is positioned between the cyclone preheater and the rotary kiln. This is intended to calcine the preheated raw meal to such an extent with the aid of a secondary firing system that the following rotary kiln is essentially only required for the actual clinker mineral formation. Conventional kiln plants are also being increasingly operated with second firing systems. It is here in particular that secondary fuels, such as used oils, old tyres, shreds of rubber, oil shale, and coal shale, are used [14].

The combustion air needed for this type of secondary firing system can either be passed through the rotary kiln or through a separate gas duct, the so-called tertiary air duct. The precalciner processes are therefore divided into processes with and without tertiary air ducting, known as AS (air separate) and AT (air through) processes respectively.

The process without tertiary air duct is shown in **Fig. 7**. It is used both for kilns with grate preheaters and for conventional kilns with cyclone preheaters. However, in Variant A the proportion of energy in the secondary firing system is restricted to approximately 15 to a maximum of 20 % in order to ensure complete combustion of the fuel in the gas duct or in the hot chamber. Higher proportions of energy up to about 30 % are only possible in kilns with cyclone preheaters where the combustion chamber between the rotary kiln inlet and the lowest cyclone stage has been enlarged. Variant B shows an example of this type of plant. The enlarged burner chamber shown is known as the "calciner".

The process with tertiary air duct, shown in **Fig. 8**, is confined to kilns with cyclone preheaters and calciners which are able to draw tertiary air from the cooler or kiln hood. Because of the separate supply of combustion air, the proportion of fuel energy used in the calciner can be as high as approximately 60 %, or even over 70 % for kiln feed prepared by the wet process.

Finally, there is yet one further way of differentiating between the systems which relates to the combustion air in the calciner. The so-called pure air calciners are operated without the addition of kiln gas (Variant D), while the others burn the fuel in a mixture of air and kiln gas (Variant C). Any combination of the systems described can be used for 2- or multi-string cyclone preheaters [15].

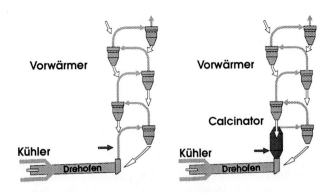

Variante A   **Variante B**

BILD 7: Vorcalcinierverfahren ohne Tertiärluftführung
— Zyklonvorwärmer mit Zweitfeuerung (Variante A)
— Zyklonvorwärmer mit Calcinator (Variante B)

FIGURE 7: Precalciner processes without tertiary air duct
— cyclone preheater with secondary firing system (Variant A)
— cyclone preheater with calciner (Variant B)

Vorwärmer	= preheater
Calcinator	= calciner
Kühler	= cooler
Drehofen	= rotary kiln

durchsatzes betrieben werden können, ohne daß gleich die Luftzahl drastisch erhöht werden muß.

3.2.2 Calcinator

Zwischen Zyklonvorwärmer und Drehofen ist bei modernen Ofensystemen der Calcinator angeordnet. In diesem soll mit Hilfe einer Zweitfeuerung das vorgewärmte Rohmehl so weit entsäuern, daß der nachgeschaltete Drehofen im wesentlichen nur noch für die eigentliche Klinkermineralbildung benötigt wird. Aber auch konventionelle Ofenanlagen werden zunehmend mit einer zweiten Feuerung betrieben. Dabei kommen insbesondere sekundäre Brennstoffe zum Einsatz, z. B. Altöl, Altreifen, Gummischnitzel, Ölschiefer und Kohlenschiefer [14].

Die für eine derartige Zweitfeuerung erforderliche Verbrennungsluft kann entweder durch den Drehofen geführt werden oder in einer gesonderten Gasleitung, der sogenannten Tertiärluftleitung. Die Vorcalcinierverfahren werden deshalb in Verfahren mit und ohne Tertiärluftführung in AS-(„Air Separate") und AT-Verfahren („Air Through") unterteilt.

Das Verfahren ohne Tertiärluftführung ist im **Bild 7** dargestellt. Es wird sowohl bei Öfen mit Rostvorwärmer als auch bei konventionellen Öfen mit Zyklonvorwärmer praktiziert. Bei der Variante A ist der Energieanteil in der Zweitfeuerung jedoch auf ca. 15 bis max. 20 % begrenzt, um noch einen vollständigen Ausbrand des Brennstoffs im Gaskanal oder in der Heißkammer zu gewährleisten. Höhere Energieanteile bis etwa 30 % sind nur bei Öfen mit Zyklonvorwärmer möglich, deren Brennraum zwischen Drehofeneinlauf und unterster Zyklonstufe vergrößert wurde. Variante B zeigt eine derartige Anlage als Beispiel. Der dargestellte vergrößerte Brennraum wird „Calcinator" genannt.

Das Verfahren mit Tertiärluftführung, im **Bild 8** dargestellt, ist auf Öfen mit Zyklonvorwärmer und Calcinator mit der Möglichkeit der Tertiärluftentnahme am Kühler oder Ofenkopf beschränkt. Aufgrund der gesonderten Verbrennungsluftführung sind Brennstoffenergieanteile im Calcinator von bis ca. 60 % möglich, bei naß aufbereitetem Brenngut sogar über 70 %.

Letztlich gibt es noch eine weitere Unterscheidungsmöglichkeit der Systeme, die sich auf die Verbrennungsluft im Calcinator bezieht. Die sogenannten Reinluftcalcinatoren werden ohne Zumischen von Ofengas betrieben (Variante D), während in den anderen der Brennstoff in einer Mischung aus Luft und Ofengas verbrennt (Variante C). Bei zwei- oder mehrsträngigen Zyklonvorwärmern können die vorgestellten Systeme beliebig kombiniert werden [15].

The introduction of precalcining which, from its nature, would be better described as rapid calcining, caused considerable changes in the process technology of clinker manufacture. However, precalcining does not save fuel heat. On the contrary, many plants would have a higher specific heat consumption if the throughput had not also been raised at the same time. The advantages of modern precalciner plants lie more in the lower specific investment and operating costs, higher production capacity, improved process control, and additional options for reducing emissions.

Continued development work in recent years has placed particular emphasis on the flexibility of the calciners. Criteria which play a part in this are

— suitability for different primary and secondary fuels,
— burn-out behaviour of the secondary fuel with respect to residual carbon and CO,
— control of the excess air ratio and the temperature with respect to NO_x,
— coating behaviour.

K. Menzel [16] has described a calciner operated in several stages which makes it possible to reduce the NO_x by up to 50 %, even when using inactive petroleum coke. By using dust recirculation it is also possible to improve the coating-forming effects of the recirculating sulphur system at the kiln inlet.

One result of development work which could give fresh impetus to burning technology in the medium term is the so-called 2-stage calciner. The intention is to achieve virtually 100 % precalcination in the calciner, as far as possible independently of the raw materials, by using a short rotary kiln and reducing the exhaust gas losses with an alkali bypass, and/or by using a fluidized-bed kiln instead of a rotary kiln. This ought not to produce any excessive temperatures. Kobe Steel and Nihon Cement have developed the 2-stage calciner for this purpose; the second stage with its lower CO_2 partial pressure makes considerably higher levels of precalcination possible than could be achieved in one stage [17]. **Fig. 9** shows the process flow chart for the test plant with a production capacity of 60 t/d. About 50 % of the total fuel is burnt in the 1st calciner stage, approximately 20 % in the second calciner stage, and only approximately 30 % in the rotary

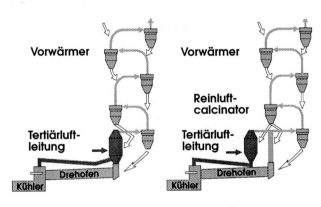

Variante C   **Variante D**

BILD 8: Vorcalcinierverfahren mit Tertiärluftführung
— Zyklonvorwärmer mit Mischgascalcinator und Tertiärluftleitung (Variante C)
— Zyklonvorwärmer mit Reinluftcalcinator und Tertiärluftleitung (Variante D)

FIGURE 8: Precalciner processes with tertiary air duct
— cyclone preheater with mixed gas calciner and tertiary air duct (Variant C)
— cyclone preheater with pure air calciner and tertiary air duct (Variant D)

Vorwärmer	= preheater
Mischgascalcinator	= mixed gas calciner
Reinluftcalcinator	= pure air calciner
Tertiärluftleitung	= tertiary air duct
Kühler	= cooler
Drehofen	= rotary kiln

Die Einführung der Vorcalcinierung, die ihrem Wesen nach besser als Schnellcalcinierung zu bezeichnen wäre, veränderte die Verfahrenstechnik der Klinkerherstellung maßgeblich. Brennstoffwärme wird jedoch mit der Vorcalcinierung nicht eingespart. Im Gegenteil, manche Anlage hätte einen höheren spezifischen Wärmebedarf, wäre nicht gleichzeitig auch der Durchsatz gestiegen. Die Vorteile moderner Vorcalcinieranlagen liegen vielmehr in niedrigeren spezifischen Investitions- und Betriebskosten, hoher Produktionskapazität, verbesserter Prozeßführung und zusätzlichen Möglichkeiten zur Emissionsminderung.

In den vergangenen Jahren wurde insbesondere die Flexibilität der Calcinatoren weiterentwickelt. Kriterien, die dabei eine Rolle spielen, sind

— Tauglichkeit für unterschiedliche Primär- und Sekundärbrennstoffe,

— Ausbrandverhalten des Zweitbrennstoffs im Hinblick auf Restkohlenstoff und CO,

— Steuerung der Luftzahl und der Temperatur im Hinblick auf NO_x,

— Ansatzverhalten.

K. Menzel [16] hat einen mehrstufig betriebenen Calcinator beschrieben, der eine NO_x-Minderung von bis zu 50 % auch mit reaktionsträgem Petrolkoks ermöglicht. Durch Staubrezirkulation kann darüber hinaus der Schwefelkreislauf im Ofeneinlauf hinsichtlich seiner ansatzbildenden Wirkungen positiv beeinflußt werden.

Ein Ergebnis von Entwicklungsarbeiten, die der Brenntechnik mittelfristig neue Impulse verleihen können, ist der sogenannte 2-Stufen-Calcinator. Mit dem Einsatz eines Kurzdrehofens, der Reduzierung der Abgasverluste bei einem Alkalibypaß und/oder dem Einsatz eines Wirbelschichtofens anstelle eines Drehofens sollte möglichst unabhängig von den Rohmaterialien ein Vorcalciniergrad von annähernd 100 % im Calcinator erreicht werden. Andererseits sollten dabei keine übermäßigen Temperaturen auftreten. Kobe Steel hat dazu gemeinsam mit Nihon Cement den 2-Stufen-Calcinator entwickelt, der in der 2. Stufe bei niedrigerem CO_2-Partialdruck deutlich höhere Vorcalciniergrade ermöglicht, als dies in einer Stufe möglich wäre [17]. **Bild 9** zeigt das Verfahrensfließbild der Versuchsanlage mit einer Produktionskapazität von 60 t/d. In der 1. Calcinierstufe werden etwa 50 % des gesamten Brennstoffs verfeuert, in der 2. Calcinierstufe ca. 20 % und im Drehofen nur noch ca. 30 %. Im 1. Calcinator betrug der CO_2-Gehalt im Abgas ca. 33 % und in der 2. Stufe ca. 16 %. Dadurch konnten Vorcalcinierraten von ca. 98 % bei Temperaturen unterhalb 850 °C erreicht werden.

Folgende Vorteile hätte ein 2-Stufen-Calcinator:

— Im Falle eines Alkalibypasses sind die Energieverluste geringer.

— Die Bypaßstaubmenge wird minimal bei gleichzeitig hoher Alkalikonzentration.

— Die Ansatzneigung wird geringer.

— Der Drehofen kann kürzer gebaut oder durch einen Wirbelschichtofen ersetzt werden.

Der 2-Stufen-Calcinator könnte somit eine Vorstufe auf dem Weg zum Wirbelschicht-Zementofen sein. Gerade dazu hat es in jüngster Vergangenheit wieder verstärkte Bemühungen zur Entwicklung eines industrietauglichen Systems gegeben.

3.3 Wirbelschichtverfahren

Bild 10 zeigt das Verfahrensschema eines Wirbelschichtofens für einen Durchsatz von 300 t/d, wie ihn IHI gemeinsam mit Chichibu Cement aufgrund von Untersuchungen an einer 40-t/d-Versuchsanlage bauen würde [17, 18]. Das neue System besteht im wesentlichen aus einem 3stufigen Vorwärmer, einem Calcinator, einem Sinterofen und 2 hintereinandergeschalteten Kühlern. Der Calcinator arbeitet ebenfalls 2stufig. Die untere Kammer dient gleichzeitig als

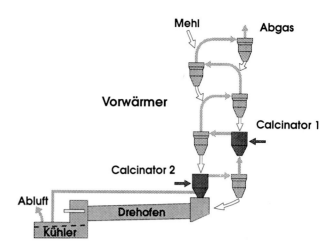

BILD 9: 2-Stufen-Calcinator von Kobe Steel und Nihon Cement [7]
FIGURE 9: 2-stage calciner, by Kobe Steel and Nihon Cement [7]

Mehl	= meal
Abgas	= exhaust gas
Vorwärmer	= preheater
Calcinator	= calciner
Abluft	= exhaust air
Kühler	= cooler
Drehofen	= rotary kiln

kiln. The CO_2 content in the exhaust gas in the 1st calciner was approximately 33 % and approximately 16 % in the 2nd stage. This made it possible to achieve precalcining levels of approximately 98 % at temperatures below 850°C.

A 2-stage calciner would have the following advantages:

— The energy losses are lower if there is an alkali bypass.

— The bypass dust is reduced in quantity and has a high alkali concentration.

— There is a reduced tendency to form coating.

— The rotary kiln can be made shorter or replaced by a fluidized-bed kiln.

The 2-stage calciner could therefore be a preliminary stage on the route to fluidized-bed cement kilns. Very recently there have been increased efforts in precisely this direction to develop an industrially acceptable system.

3.3 Fluidized-bed process

Fig. 10 shows the process flow sheet which IHI, working jointly with Chichibu Cement, would use for building a fluidized-bed kiln for a throughput of 300 t/d on the basis of investigations with a 40 t/d test plant [17, 18]. The new system consists essentially of a 3-stage preheater, a calciner, a sintering kiln, and 2 coolers arranged in series. The calciner operates in two stages, and the lower chamber also acts as a dust separator for the fluidized-bed kiln. The kiln itself is a reactor with Venturi nozzles. It consists of a free zone and zones containing fluidized bed in states of high and low expansion. The fuel is fed both to the calciner and to the fluidized-bed kiln. On start-up the system is fed exclusively with crushed clinker, so-called "seed clinker", and then above 1300°C with precalcined raw meal. **Table 4** gives the operating results for the 40 t/d test plant over the period of a week. From the table it can be seen that, as before, the heat and power consumption of 4450 kJ/kg clinker and 70 kWh/t clinker respectively are very high. The target values for a 300 t/d plant of 3200 kJ/kg clinker and 45 kWh/t clinker are still very high. It also seems that problems of coating behaviour of the kiln feed in the reactor have not yet been

TABLE 4: Operating results from a 40 t/d fluidized bed kiln test plant [17]

– Trial duration:	1 week
– Clinker throughput:	40 t/d
– Heat consumption:	4450 kJ/kg clinker
– Power consumption:	70 kWh/t clinker

TABELLE 4: Betriebsergebnisse einer 40-t/d-Wirbelschichtofen-Versuchsanlage [17]

– Versuchsdauer:	1 Woche
– Klinkerdurchsatz:	40 t/d
– Wärmeverbrauch:	4450 kJ/kg Kli
– Stromverbrauch:	70 kWh/t Kli

Staubabscheider für den Wirbelschichtofen. Der Wirbelschichtofen selbst ist ein Reaktor mit Venturidüse. Er besteht aus einer Freiraumzone sowie Zonen mit hoch und niedrig expandierender Wirbelschicht. Der Brennstoff wird sowohl dem Calcinator, als auch dem Wirbelschichtofen aufgegeben. Zum Anfahren des Systems wird ausschließlich gebrochener Klinker, sogenannter „Saatklinker", aufgegeben, oberhalb 1300 °C dann vorentsäuertes Rohmehl. **Tabelle 4** enthält Betriebsergebnisse der 40-t/d-Versuchsanlage über die Dauer einer Woche. Aus der Tabelle geht hervor, daß Wärme- und Strombedarf mit 4450 kJ/kg Kli bzw. 70 kWh/t Kli nach wie vor sehr hoch ausfallen. Auch die Zielwerte einer 300-t/d-Anlage liegen mit 3200 kJ/kg Kli bzw. 45 kWh/t Kli noch sehr hoch. Darüber hinaus scheinen Fragen des Ansatzverhaltens des Brenngutes im Reaktor noch nicht zuverlässig geklärt zu sein. Es bleibt somit festzuhalten, daß der Wirbelschichtofen mit seinen Betriebsdaten sich derzeitig nicht mit großen Vorcalcinieröfen messen kann. Bei Bedarf kleiner Ofendurchsätze von z. B. 300 t/d hat dagegen der Schachtofen auf absehbare Zeit noch deutliche Vorteile, sowohl im Energiebedarf, als auch hinsichtlich der Betriebssicherheit. Das ist auch der Grund für weitere Entwicklungsarbeiten in der Schachtofentechnologie, vor allem in China und Indien [17].

Bei der Herstellung von Sonderprodukten ist der Wirbelschichtofen allerdings eine interessante Alternative. Das Brenngut wird praktisch ohne Zeitverzug auf Sintertemperatur aufgeheizt, dort eine definierte Zeit gehalten und dann von Sintertemperatur sehr schnell auf unter 800 °C abgekühlt. Dadurch kann mit gleichem Brenngut die Klinkerqualität positiv beeinflußt werden. Außerdem können Temperatur und Gasatmosphäre in einem gestuften System örtlich gezielt und sehr gleichmäßig eingestellt werden. Das ermöglicht eine breite Produktpalette.

S.M. Cohen [19] beschreibt z. B. ein Wirbelschichtverfahren, das speziell zur Verwertung von Bypaß- und Filterstäuben entwickelt wurde. Mit dem Verfahren soll sich der Anfall von alkalihaltigem Staub auf ca. 10 % der Ausgangsmasse verringern lassen. Ein Anteil von 90 % des Staubs wird zu Klinker mit weniger als 0,6 M.-% Na₂O-Äquivalent verarbeitet.

4. Drehofen

Trotz aller Möglichkeiten, die der Wirbelschichtofen bietet, wird der Drehofen wegen der günstigen spezifischen Betriebs- und Investitionskosten noch auf lange Sicht ohne Konkurrenz bleiben. Moderne Vorcalciniertechnologien in Verbindung mit moderner Prozeßsteuerung werfen jedoch die Frage auf, wie lang der Drehofen sein muß. Hier gehen die Anlagenbauer unterschiedliche Wege. Nach KHD ist ein Länge/Durchmesser-Verhältnis von 11:1 für Vorcalcinieröfen ausreichend [20]. Derartige Öfen werden ausnahmslos auf zwei Laufringstationen gelagert. Die übrigen Anlagenbauer bevorzugen demgegenüber höhere L/D-Verhältnisse bis 17:1, wobei diese den Einfluß der Sinterbarkeit der Rohmehle stärker betonen.

Der Wegfall der dritten Laufrollenstation und die geringere Ofenmasse erlauben eine Investitionskostenreduzierung von ca. 17 %, bezogen auf Drehofen, Tertiärluftleitung, Feuerfest-Auskleidung, Bauteil und Montage [20]. Ein weiterer, ganz wesentlicher Vorteil des Kurzdrehofens ist seine statisch bestimmte Lagerung. Überlastungszustände, wie sie beim 3- oder 4fach gelagerten Ofen z.B. durch Fundamentsenkungen und/oder Ofenrohrverkrümmungen auftreten können, scheiden beim Kurzdrehofen aus.

Den Vorteilen des Kurzdrehofens steht allerdings eine gewisse Unsicherheit bei schwer sinterbaren Rohmehlen

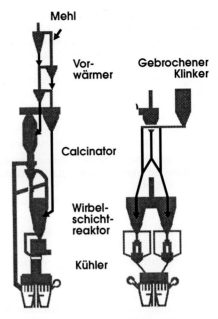

BILD 10: Entwurf eines 300-t/d-Wirbelschichtofens von Ishikawajima Harima Heavy Industries Company Ltd (IHI) und Chichibu Cement [17, 18]
FIGURE 10: Design for a 300 t/d fluidized bed kiln, by Ishikawajima Harima Heavy Industries Company Ltd (IHI) and Chichibu Cement [17, 18]

Mehl	= meal
Vorwärmer	= preheater
Calcinator	= calciner
Wirbelschichtreaktor	= fluidized bed calciner
Kühler	= cooler
Gebrochener Klinker	= crushed clinker

solved reliably. It is therefore still the case that the fluidized-bed kiln cannot at present compete with large precalciner kilns on the basis of its operating data. However, where there is a need for smaller kiln throughputs of, for example, 300 t/d the shaft kiln still has significant advantages for the foreseeable future, both in energy consumption and in operational reliability. This is the reason for further development work in shaft kiln technology, chiefly in China and India [17].

The fluidized-bed kiln is, however, an interesting alternative for the manufacture of special products. The kiln feed is heated to sintering temperature with practically no time delay, held there for a definite time, and then cooled very rapidly from the sintering temperature to below 800°C. For the same kiln feed, this can have a beneficial effect on clinker quality. In a stepped system, the local temperature and gas atmosphere can also be adjusted accurately and very consistently, allowing a wide product range to be produced.

S. M. Cohen [19], for example, describes a fluidized-bed process which was developed specifically for utilizing bypass and filter dusts. The process is intended to reduce the output of alkali-containing dust to approximately 10 % of the initial mass. Ninety percent of the dust is processed to form clinker containing less than 0.6 wt. % Na₂O.

4. Rotary kiln

In spite of all the options offered by the fluidized-bed kiln, the rotary kiln will still remain without competition for a long time because of the favourable specific operating and investment costs. Modern precalciner technology used in conjunction with modern process control does, however, throw up the question of the correct length for the rotary kiln. The plant constructors have gone their own ways here. According to KHD a length/diameter ratio of 11:1 is sufficient for precalciner kilns [20]. Such kilns are, without exception, supported on two tyre stations. However, the other plant manufacturers prefer higher L/D ratios up to 17:1, and place greater stress on the influence of the sinterability of the raw meals.

gegenüber. Der Garbrand des Klinkers soll in solchen Fällen mit Hilfe eines größeren Ofendurchmessers, einer niedrigeren Ofendrehzahl und vor allem einer kürzeren Flamme beliebig verändert werden können. Letzteres hängt aber insbesondere von der Flexibilität des Drehofenbrenners ab. Ein moderner Brenner allein ist noch kein Garant für eine optimale Drehofenflamme. Vielmehr kommt es darauf an, den Kühlerbetrieb sowie Brennstoffaufbereitung, -dosierung, -transport und -vermischung im Hinblick auf Energiebedarf, Emission, Ofendurchsatz, Qualität des Klinkers und Ansatzverhältnisse im Drehofen für jeden betrieblichen Einzelfall erneut zu optimieren [21].

5. Klinkerkühler

Klinkerkühler spielen eine herausragende Rolle für die Wärmewirtschaft der Ofenanlage. Sie sind deshalb in den letzten Jahren ein Schwerpunkt der konstruktiven und betrieblichen Optimierungsbemühungen gewesen [22–28]. Eine dauerhafte Verbesserung des Kühlbereichswirkungsgrades bedeutet nennenswerte Einsparungen beim Wärmeverbrauch des Drehofens.

Gebräuchliche Bauarten sind Rostkühler und für kleinere und mittlere Ofenkapazitäten auch noch Satelliten- und Rohrkühler. Unterhalb 3000–4000 t/d sind konventionelle Wärmetauscheröfen mit Satellitenkühler nach wie vor eine Alternative zu Vorcalcinieröfen mit Rost- oder Rohrkühler. Die **Tabellen 5** und **6** zeigen typische Energiebilanzen für einen 5stufigen Zyklonvorwärmerofen mit einer Kapazität von 2500 t/d, der in einem Fall mit Satellitenkühler ausgestattet ist und im anderen Fall mit Rostkühler, Tertiärluftleitung und Calcinator. Demnach verbraucht der Ofen mit Satellitenkühler zwar ca. 180 kJ/kg Kli mehr Brennstoffenergie, jedoch ist der Arbeitsbedarf ca. 7 kWh/t Kli niedriger.

Ein Drehofen mit Satellitenkühler ist einfacher im Betrieb und kostengünstiger in der Anschaffung. Bei der Dimensionierung des Kühlers ist die zu erwartende Korngrößenverteilung des Klinkers jedoch besonders zu beachten [29].

Zunehmende Emissionsbegrenzungen erfordern künftig stark steigende Investitions- und Betriebskosten für die Behandlung ungenutzter Kühlerabluft. Abluftfreie Satelliten- oder Rohrkühler, die wegen ihrer niedrigeren Sekundärlufttemperatur zu niedrigeren NO_x-Werten beitragen können, bieten deshalb durchaus bei bestimmten, standortspezifischen Randbedingungen Vorteile.

Elimination of the third support roller station and the lower kiln weight allow investment costs to be reduced by approximately 17 % based on the rotary kiln, tertiary air duct, refractory lining, structural components, and erection [20]. Another very important advantage of the short rotary kiln is its statically determined support system. Short rotary kilns eliminate any overload situations of the type which can occur with kilns with 3- or 4-point support systems as a result, for example, of foundation settlement and/or bends in the kiln tube.

However, the advantages of the short rotary kiln have to be set against some uncertainty with raw meals which are difficult to sinter. It is said that in such cases the finish burn of the clinker can be varied as required with the aid of a larger kiln diameter, a lower kiln speed and, above all, a shorter flame. However, this last is particularly dependent on the flexibility of the rotary kiln burner. A modern burner alone is still no guarantee of an optimum rotary kiln flame. It is more a question of optimizing the cooler operation and the fuel preparation, metering, transport, and blending systems for each individual operational situation with respect to energy consumption, emission, kiln throughput, quality of clinker, and coating conditions in the rotary kiln [21].

5. Clinker cooler

Clinker coolers play a prominent part in the thermal economy of the kiln plant. In recent years they have therefore formed a focus for efforts aimed at constructional and operational optimization [22–28]. Permanent improvement of the efficiency of the cooling sector means appreciable savings in the heat consumption of the rotary kiln.

The usual types are grate coolers, but planetary and rotary coolers are still also used for smaller and medium sized kiln capacities. Below 3000–4000 t/d conventional preheater kilns with planetary coolers are still alternatives to precalciner kilns with grate or rotary coolers. **Tables 5** and **6** show typical energy balances for a 5-stage cyclone preheater kiln with a capacity of 2500 t/d which, in one case, has planetary coolers and in the other a grate cooler, tertiary air duct and calciner. This shows that although the kiln with planetary coolers needs approximately 180 kJ/kg clinker more fuel energy, the power consumption is approximately 7 kWh/t clinker lower.

TABELLE 5: Brennstoffenergieverbrauch 5stufiger Zyklonvorwärmeröfen

Energieverluste in kJ/kg Kli	5stufiger Zyklonvorwärmerofen	
	mit Satellitenkühler	mit Rostkühler, Tertiärluftltg. und Calcinator
– Vorwärmer	744	778
– Ofen	246	149
– Kühler	610	489
– Reaktionswärme	1721	1721
– Fühlbare Wärme	–121	–118
Brennstoffenergieverbrauch	3200	3019

TABLE 5: Fuel energy consumption of 5-stage cyclone preheater kilns

Energy losses in kJ/kg clinker	5-stage cyclone preheater kiln	
	with planetary cooler	with grate cooler, tertiary air duct and calciner
– Preheater	744	778
– Kiln	246	149
– Cooler	610	489
– Heat of reaction	1721	1721
– Sensible heat	–121	–118
Fuel energy consumption	3200	3019

TABELLE 6: Stromverbrauch 5stufiger Zyklonvorwärmeröfen

Energieverbrauch in kWh/t Kli	5stufiger Zyklonvorwärmerofen	
	mit Satellitenkühler	mit Rostkühler, Tertiärluftltg. und Calcinator
– Abgasgebläse	4,8	6,0
– Ofenantrieb	3,0	1,5
– Kühler incl. Entstaubung	0	7,0
– Sonstiges	1,5	1,5
Summe	9,3	16,0

TABLE 6: Power consumption of 5-stage cyclone preheater kilns

Power consumption in kWh/t clinker	5-stage cyclone preheater kiln	
	with planetary cooler	with grate cooler, tertiary air duct and calciner
– Exhaust gas fan	4.8	6.0
– Kiln drive	3.0	1.5
– Cooler including dedusting system	0	7.0
– Other	1.5	1.5
Total	9.3	16.0

6. Stoffkreisläufe

Stoffkreisläufe im Ofensystem können den Ofenbetrieb und die Qualität des Klinkers maßgeblich beeinflussen. Sie entstehen aus Nebenbestandteilen im Rohmaterial und Brennstoff, wie z.B. Schwefel und Chlor. Im Verlauf des Brennprozesses reagieren sie mit den Alkalien der Rohstoffe zu Alkalisulfaten und -chloriden, aber auch zu Calciumsulfat. Im Bereich hoher Temperaturen verdampfen oder dissoziieren diese Verbindungen dann ganz oder teilweise, werden mit den Gasen in kältere Bereiche transportiert, kondensieren dort und gelangen mit dem Brenngut wieder in Bereiche hoher Temperaturen, wo sie erneut ganz oder teilweise verdampfen oder dissoziieren. Dadurch können sich starke innere Kreisläufe im System Ofen – Vorwärmer ergeben [30].

Neben dem Abscheidegrad der einzelnen Anlagenbereiche ist für das Kreislaufverhalten der flüchtigen Bestandteile vorwiegend deren Schmelz- und Siedepunkt sowie die Temperaturabhängigkeit ihres Dampfdrucks von Bedeutung [31]. **Bild 11** zeigt als Beispiel die Dampfdrücke einiger Alkaliverbindungen in Abhängigkeit von der Temperatur. Daraus geht hervor, daß der Dampfdruck der Alkalichloride bei Sintertemperaturen von 1450 °C den Umgebungsdruck von 1 bar annimmt, so daß die Alkalichloride in der Sinterzone nahezu vollständig verdampfen können. Da der Schmelzpunkt der Alkalichloride bei 800 °C liegt, bildet sich zwischen Sinterzone und Vorwärmer ein ständig anwachsender Kreislauf aus, der leicht zu einer 50fachen Anreicherung des Alkalichlorids im Brenngut an der Schnittstelle Drehofen/Vorwärmer führen kann. Demgegenüber verdampfen Alkalisulfate in der Sinterzone aufgrund ihres niedrigeren Dampfdrucks nur teilweise und reichern sich nicht so stark im Ofensystem an.

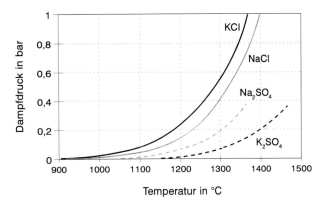

BILD 11: Dampfdruck einiger Alkaliverbindungen in Abhängigkeit von der Temperatur [31]
FIGURE 11: Vapour pressures of some alkali compounds as functions of temperature [31]

Dampfdruck in bar = vapour pressure in bar
Temperatur in °C = temperature in °C

Betriebstechnisch sind solche Kreisläufe nicht erwünscht, weil die dabei anfallenden Salze mit dem Heißmehl verstärkt Ansätze im Drehofeneinlauf und im unteren Bereich des Vorwärmers bilden sowie die Fließeigenschaften des Brennguts verschlechtern. Deshalb führt man häufig einen geringen Teilgasstrom des Drehofenabgases, der hohe Anteile an flüchtigen Bestandteilen enthält, über einen Bypaß ab.

Bild 12 zeigt schematisch die Bypaßentnahme von F. L. Smidth als Beispiel, die vorzugsweise stirnseitig oberhalb des Drehofens im Gaskanal angeordnet wird. Durch den Bypaß sollen die Schadstoffe möglichst noch gasförmig abgesaugt und durch eine schnelle Kühlung mit Kaltluft kondensiert und abgeführt werden. Für einen effektiven Bypaßbetrieb, d.h. maximale Schadstoffreduzierung bei minimalen Wärme- und Materialverlusten, kommt es jedoch sehr auf die richtige Gestaltung des Bypaßabzuges an. Ein zu hoher Staubgehalt des Bypaßgases erhöht den Wärmebedarf, zu niedriger Staubgehalt führt zu Ansätzen im Abzug und ggf.

A rotary kiln with planetary coolers is simpler to operate and less expensive to purchase. However, particular attention must be paid to the expected particle size distribution of the clinker when dimensioning the cooler [29].

Increasing emission restrictions will in the future require sharply increasing investment and operating costs for dealing with unused cooler exhaust air. Exhaust-air-free planetary or rotary coolers, which can contribute to lower NO_x values because of their lower secondary air temperatures, therefore definitely offer advantages under certain specific local conditions.

6. Recirculating systems

Recirculating materials within the kiln system can have a considerable effect on the kiln operation and the quality of the clinker. They come from secondary constituents, such as sulphur and chlorine, in the raw material and fuel. During the burning process they react with the alkalis in the raw materials to form alkali sulphates and chlorides, and also calcium sulphate. In the high temperature regions these compounds vaporize or dissociate, either entirely or partially, are transported with the gases into the colder areas where they condense, and then pass with the kiln feed back into the areas of higher temperature again, where they are once more entirely or partially vaporized or dissociated. This can result in heavy internal recirculation within the kiln-preheater system [30].

In addition to the level of deposition in the individual plant areas, the most important factors for the recirculating behaviour of the volatile constituents are their melting and boiling points and the temperature dependence of their vapour pressures [31]. As an example, **Fig. 11** shows the vapour pressures of some alkali compounds as functions of temperature. From this it can be seen that the vapour pressures of the alkali chlorides take on the ambient pressure of 1 bar at sintering temperatures of 1450°C, with the result that alkali chlorides can be virtually fully vaporized in the sintering zone. As the melting point of the alkali chlorides is 800°C there is a constantly increasing recirculating system formed between the sintering zone and the preheater, which can easily lead to a 50-times enrichment of the alkali chloride in the kiln feed at the rotary kiln/preheater interface. Alkali sulphates, on the other hand, are only partially vaporized in the sintering zone because of their lower vapour pressures and so do not build up so severely in the kiln system.

Such recirculating systems are undesirable from the operational point of view because the resulting salts form increased amounts of coating with the hot meal at the rotary kiln inlet and in the lower part of the preheater, and also have a detrimental effect on the flow properties of the kiln feed. Part of the rotary kiln exhaust gas flow, which contains high proportions of volatile constituents, is therefore often removed through a bypass.

As an example, **Fig. 12** shows a diagram of the bypass off-take used by F.L. Smidth which is normally located in the gas duct above the end of the kiln. The harmful materials should be drawn off through the bypass while, as far as possible, they are still gaseous, and then condensed by rapid cooling with cold air, and removed. Correct configuration of the bypass off-take is very important for effective bypass operation, i.e. maximum reduction in harmful materials with minimum heat and material loss. Excessive dust content in the bypass gas increases the heat requirement and too low a dust content leads to build up at the off-take and possibly to SO_2 emissions in the bypass exhaust gas. **Table 7** lists guide values for the intakes of K_2O, Na_2O, SO_3 and Cl which will still be admissible for the various kiln systems without using a bypass. However, in practice it is often necessary to accept higher values.

In precalciner plants with tertiary air ducts the concentration of recirculating materials in the kiln exhaust gas gradually increases because of the lower specific gas mass flow. Precalciner plants therefore require bypasses at lower intakes of chlorine, and also of SO_3.

zu SO₂-Emissionen im Bypaßabgas. Richtwerte für die Einnahmen an K₂O, Na₂O, SO₃ und Cl, die ohne Bypaß für die verschiedenen Ofensysteme noch verträglich sein sollen, sind in der **Tabelle 7** zusammengestellt. In der Praxis müssen jedoch häufig auch noch höhere Werte in Kauf genommen werden.

In Vorcalcinieranlagen mit Tertiärluftleitung ist die Konzentration an Kreislaufstoffen im Ofenabgas aufgrund des geringeren spezifischen Gasmassenstroms graduell höher. Infolgedessen benötigen Vorcalcinieranlagen schon bei niedrigeren Chlor-, aber auch bei niedrigeren SO₃-Einnahmen einen Bypaß.

TABELLE 7: Zulässige Einnahmen an K₂O, Na₂O, SO₃ und Cl aus Rohmehl und Brennstoff [32] (Ofensysteme ohne Bypaß in mg/kg Kli)

Komponente	Konventionelle Öfen	Vorcalcinieröfen mit Tertiärluftltg.
K₂O + Na₂O	max. 15	max. 15
SO₃	max. 16	max. 12
Cl	max. 0,23	max. 0,2

Der Anteil der Alkali- und Calciumsulfate im Kreislauf hängt von der Temperatur und der Verweilzeit des Brennguts im Drehofen ab. Daneben können aber auch Zweitbrennstoffe, z.B. Altreifen oder Restkoks aus dem Calcinator, die in das Heißmehl des Drehofens gelangen, die Sulfate des Brennguts unter Bildung von SO₂ reduzieren. Dadurch kann die SO₂-Konzentration im Drehofeneinlauf auf Werte von bis zu 2 Vol.-% ansteigen, wodurch die Ansatzbildung im Bereich des Ofeneinlaufs, des Calcinators und der untersten Zyklonstufen zunimmt [33]. Sulfatreiche Ansätze bilden sich dann bevorzugt an den Stellen, an denen Falschluft in die Ofenanlage eintritt und/oder an Orten hoher Turbulenz. **Bild 13** zeigt als Beispiel den Schwefelkreislauf in einer Zyklonvorwärmeranlage bei Betrieb einer Zweitfeuerung mit stückigen Zweitbrennstoffen im Drehofeneinlauf. Aus dem Bild geht hervor, daß durch die Reaktion des Zweitbrennstoffs mit dem Brenngut im Drehofeneinlauf SO₂ freigesetzt wird und daß sich der Schwefelkreislauf dadurch verstärkt. Bei Inbetriebnahme oder bei Ausfall der Zweitfeuerung wird sich deshalb der Schwefelkreislauf erhöhen bzw. vermindern. Kurzfristig ergeben sich daraus erhebliche Schwankungen der SO₃-Konzentration im Klinker um

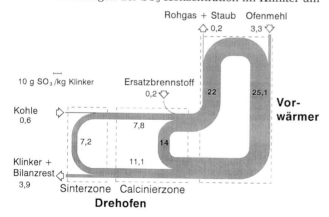

BILD 13: Einfluß der Verbrennung von grobstückigen Zweitbrennstoffen in der Calcinationszone einer Drehofenanlage mit Zyklonvorwärmer auf die Schwefelbilanz [34]
FIGURE 13: Effect on the sulphur balance of burning secondary fuels in coarse lump form in the calciner zone of a rotary kiln system with cyclone preheater [34]

Rohgas + Staub	= raw gas + dust
Ofenmehl	= kiln meal
10 g SO₃/kg Klinker	= 10 g SO₃/kg clinker
Ersatzbrennstoff	= substitute fuel
Kohle	= coal
Vorwärmer	= preheater
Klinker + Bilanzrest	= clinker + balance figure
Sinterzone	= sintering zone
Calcinierzone	= calcining zone
Drehofen	= rotary kiln

TABLE 7: Admissible intakes of K₂O, Na₂O, SO₃ and Cl from raw meal and fuel [32] (kiln systems without bypass, in mg/kg clinker)

Components	Conventional kilns	Precalciner kilns with tertiary air ducts
K₂O + Na₂O	max. 15	max. 15
SO₃	max. 16	max. 12
Cl	max. 0.23	max. 0.2

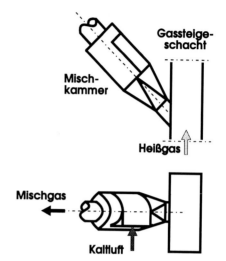

BILD 12: Gasbypaß am Steiggaskanal mit Kaltluftabkühlung, Bauart F. L. Smidth
FIGURE 12: Gas bypass at the gas riser duct with cold air cooling, designed by F.L. Smidth

Mischkammer	= mixing chamber
Mischgas	= mixed gas
Gassteigeschacht	= gas riser duct
Heißgas	= hot gas
Kaltluft	= cold air

The proportions of alkali sulphates and calcium sulphates in the recirculating system depend on the temperature and residence time of the kiln feed in the rotary kiln. However, secondary fuels, e.g. old tyres or coke residue from the calciner, which pass into the hot meal in the rotary kiln, can reduce the sulphates in the kiln feed with the formation of SO₂. This can increase the SO₂ concentration at the rotary kiln inlet to values of up to 2 vol.%, which increases the formation of coating at the kiln inlet, the calciner and the lowest cyclone stages [33]. Sulphate-rich coatings tend to form at points where false air enters the kiln system and/or at locations of high turbulence. As an example, **Fig. 13** shows the recirculating sulphur system in a cyclone preheater plant operating with a secondary firing system using secondary fuel in lump form at the rotary kiln inlet. It can be seen from the diagram that SO₂ is liberated by the reaction of the secondary fuel with the kiln feed at the rotary kiln inlet, and that this increases the recirculation of sulphur. On start up, or if the secondary firing system fails, the sulphur recirculation will therefore increase or decrease. In the short-term this results in considerable fluctuations in the SO₃ concentration in the clinker around the theoretical target mean value. One condition for uniform clinker quality is therefore also that secondary fuel in lump form is fed as uniformly as possible.

Such recirculating systems are undesirable from the process engineering point of view because, as a result of vaporization and condensation or chemical reaction of these materials with the kiln feed, they transport energy from a high temperature level into a colder region of the kiln. This can raise the level of precalcination in the preheater by up to 20%, corresponding to an energy release of about 400 kJ/kg clinker. In precalciner plants it is then necessary to reduce the proportion of secondary fuel. Fluctuations in the intensity of recirculation can also cause significant changes in the degree of calcination, which inevitably lead to operating problems.

den angestrebten rechnerischen Mittelwert. Voraussetzung für gleichbleibende Klinkerqualität ist daher auch eine möglichst gleichmäßige Zufuhr der stückigen Zweitbrennstoffe.

Verfahrenstechnisch sind solche Kreisläufe unerwünscht, weil sie infolge Verdampfung und Kondensation oder chemischer Reaktion dieser Stoffe mit dem Brenngut Energie von einem hohen Temperaturniveau in kältere Bereiche des Ofens verschleppen. Dadurch kann der Vorentsäuerungsgrad im Vorwärmer um bis zu 20 % ansteigen. Das entspricht einer Energiefreisetzung von etwa 400 kJ/kg Klinker, die bei Vorcalcinieranlagen dazu zwingt, den Anteil des Zweitbrennstoffs zu vermindern. Darüber hinaus können Schwankungen der Kreislaufintensität deutliche Änderungen des Entsäuerungsgrads hervorrufen, die unweigerlich zu Betriebsstörungen führen.

Das Chloridproblem kann mit vergleichsweise niedrigen Bypaßraten beherrscht werden. Zu hohe Schwefelkreisläufe resultieren hauptsächlich aus Reduktionsreaktionen mit dem Brennstoff und sollten deshalb verhindert werden. Brennstoffauswahl, Flammen- und Ofenführung stellen dabei wesentliche Optimierungsfelder dar.

Kritisch wird es, wenn mit Hilfe eines Bypasses der Alkaligehalt des Klinkers gesenkt werden soll. Hier bietet der Vorcalcinierofen mit seinem niedrigen spezifischen Abgasmassenstrom Vorteile. Durch Zugabe von $CaCl_2$ kann der Bypaßbetrieb in solchen Fällen noch wirksamer gestaltet werden [35].

7. Feuerfeste Ausmauerung

Eine optimale feuerfeste Zustellung der Ofenanlage ist eine wichtige Voraussetzung für eine kostengünstige Klinkerproduktion. Die Bestrebungen der Feuerfestindustrie, durch neu entwickelte Stein- und Massenqualitäten die Widerstandsfähigkeit der Ausmauerung gegen thermischen, chemischen und mechanischen Verschleiß zu erhöhen, sind anerkennenswert. Andererseits haben aber der Einsatz von Kohle und einer breiten Palette von Sekundärbrennstoffen auch zu schwierigeren Betriebsbedingungen für das Feuerfestmaterial geführt. Deshalb konnte eine deutliche Reduzierung des Feuerfestverbrauchs in den letzten 8 Jahren nicht nachgewiesen werden. Der Verbrauch liegt derzeitig in Deutschland bei ca. 0,6 bis 0,8 kg/t Kli. Die Einführung der Calcinator-Technik und das Bemühen um einen gleichmäßigen Ofenbetrieb lassen erwarten, daß Verbesserungen noch bevorstehen. Beispielsweise sind bei optimal betriebenen Vorcalcinier-Anlagen auch Verbrauchswerte unter 0,5 kg/t Klinker über längere Zeiträume nachgewiesen.

Wesentliche Voraussetzungen für niedrige Verschleißraten des Feuerfestmaterials sind optimale Bedingungen im stationären und instationären Betrieb [36].

Erforderlich sind dafür:
- kontinuierlicher Betrieb,
- konstante Flammenform,
- konstanter Energiestrom zur Drehofenfeuerung (Brennstoff + Sekundärluft),
- konstanter Ansatz in der Sinterzone,
- konstante Brenngutzusammensetzung,
- konstante SO_3- und Cl-Kreisläufe im Ofen sowie
- eindeutige An- und Abfahrtechniken für jeden Betriebszustand.

Zu den mechanischen Voraussetzungen für einen niedrigen Feuerfestverschleiß gehören des weiteren geringere Ovalitäten der Laufringschüsse, geringe Manteldeformationen, abgestimmte Steifigkeiten der Ofenschüsse, ein ausreichendes Abfangen von Axialschüben der Feuerfestausmauerung im Ofen sowie der sachgemäße Einbau, möglichst mit modernen Ausmauerungsgeräten.

Diese Voraussetzungen müssen während des Betriebes durch geeignete Überwachungsmaßnahmen sichergestellt

The chloride problem can be controlled with comparatively low bypass rates. Excessively high sulphur recirculation results mainly from reducing reactions with the fuel and should therefore be prevented. Choice of fuel and control of flame and kiln are therefore important fields for optimization.

It becomes critical when the alkali content of the clinker is to be lowered with the aid of a bypass. In this situation the precalciner kiln with its low specific exhaust gas mass flow offers advantages. In such cases the bypass operation can be made more effective by addition of $CaCl_2$ [35].

7. Refractory lining

Optimum refractory lining for the kiln plant is an important pre-condition for cost-effective clinker production. The refractories industry has made commendable efforts to increase the resistance of the brickwork to thermal, chemical and mechanical wear through newly developed grades of brick and monolithic material. However, the use of coal and a wide range of secondary fuels has also led to more difficult operating conditions for the refractories. For this reason, there has been no significant reduction in the consumption of refractories in the last 8 years. At present the consumption in Germany is approximately 0.6 to 0.8 kg/t clinker. The introduction of calciner technology and efforts to achieve uniform kiln operation mean that further improvements can be expected. For example, consumption values of less than 0.5 kg/t clinker have been obtained over fairly long periods with correctly operated precalciner plants.

Optimum conditions in steady-state and unsteady-state operation are essential for achieving low wear refractory rates [36].

This requires:
- continuous operation,
- constant flame shape,
- constant flow of energy to the rotary kiln firing system (fuel + secondary air),
- constant coating in the sintering zone,
- constant kiln feed composition,
- constant SO_3 and Cl recirculating systems in the kiln, and
- clear starting and stopping techniques for every operating situation.

The mechanical conditions for low refractory wear also include fairly low ovality of the kiln shell at the tyres, low shell deformation, matched stiffnesses of the cylindrical kiln shell sections, adequate support of the axial thrust of the refractory brickwork in the kiln, and correct installation – if possible with modern bricking rigs.

These preconditions have to be ensured during operation by suitable monitoring methods. These include, not only a reliable and well documented method of measuring shell temperature, but also a slip monitoring system for loose kiln tyres, and suitable cooling fans for local cooling of the kiln shell at the tyres. Experience shows that the installation of tyres with fixed splines in critical regions with varying coating conditions can lead to longer lining lives.

Short service life of the refractory lining can also be caused by deficient quality and uniformity of the refractories. In this connection the VDZ (German Cement Works' Association) has demonstrated the informative value of acceptance tests, using the example of five types of brick [37]. According to this, the accuracy of the dimensions and shape, and the external condition were very useful for assessing the uniformity of the production batch. Considerable defect levels were found in the brick symmetry. The task of the manufacturers of refractory products is therefore to take effective measures to minimize such quality defects, for example through the introduction of total quality management in the production system.

In the burning zone there is a trend away from magnesiachrome bricks towards dolomite and magnesia-spinel

werden. Das sind neben einer zuverlässig arbeitenden und gut dokumentierenden Manteltemperaturmeßtechnik bei losen Laufringen eine Schlupfüberwachung sowie geeignete Kühlgebläse zur lokalen Abkühlung der Laufringschüsse. Der Einbau fest verzahnter Laufringe in kritischen Bereichen mit wechselnden Ansatzverhältnissen kann erfahrungsgemäß zu höheren Ausmauerungsstandzeiten führen.

Eine geringe Lebensdauer der feuerfesten Zustellung kann aber auch in mangelnder Qualität und Gleichmäßikeit der Feuerfesterzeugnisse begründet sein. In diesem Zusammenhang hat der Verein Deutscher Zementwerke am Beispiel von 5 Steinsorten die Aussagekraft von Abnahmeprüfungen dargestellt [37]. Danach konnte die Maß- und Formhaltigkeit sowie äußere Beschaffenheit sehr gut zur Beurteilung der Gleichmäßigkeit der Produktionscharge herangezogen werden. Beträchtliche Fehlerquoten ergaben sich bei der Steinsymmetrie. Aufgabe der Hersteller von feuerfesten Produkten ist deshalb, wirksame Maßnahmen zur Minimierung solcher Qualitätsmängel zu ergreifen, beispielsweise durch die Einführung eines Total Quality Management (TQM) in der Produktion.

In der Brennzone geht der Trend weg von den Magnesia-Chromsteinen hin zu Dolomit- und Magnesia-Spinellsteinen [38–40]. Zirkondioxid soll dabei die Temperaturwechselbeständigkeit deutlich verbessern [41]. Außerhalb des Drehofens hat sich der Trend zur Anwendung feuerfester Massen weiter verstärkt. In Deutschland beträgt der Anteil am gesamten Verbrauch ca. 10%. Im Drehrohrofen werden feuerfeste Massen allerdings auch zukünftig nur in der Auslaufzone und im Einlaufkonus eingebaut werden. Brennerdüsen werden ausschließlich mit Feuerbeton umkleidet, der Einbau von Massen in Ofenköpfen, Kühlern und Vorwärmern nimmt zu.

Die monolithische Auskleidung wird nicht mehr nur als Ersatz für Formsteine angesehen. Auch großflächige Wandpartien, die bis vor kurzem ausschließlich mit Steinen gemauert wurden, sind neuerdings auf Gieß- und Spritzmassen umgestellt worden. Neben der Verschleiß- und Feuerfestigkeit kommt es auch bei monolithischen Bauteilen besonders auf Alkalibeständigkeit und gute Aufheizbarkeit an. Probleme gibt es immer wieder mit der Verankerung monolithischer Ausmauerungen. Weder keramische, noch Stahl- oder Gußanker bieten bei thermisch, mechanisch und chemisch hoch belasteten Bereichen, wie in Steiggaskanälen und Calcinatorschächten, zuverlässig befriedigende Ergebnisse. Vor allem an der Übergangszone von Arbeits- zu Isolier-Ausmauerung treten bei reduzierender Gasatmosphäre und Salzkondensation deutliche Korrosionsprobleme auf, welche die Stabilität der monolithischen Ausmauerung gefährden. Pragmatische Lösungen, wie z.B. dickere Anker, können durchaus dazu beitragen, die Lebensdauer zu verlängern. Hochgezüchtete Legierungen scheiden demgegenüber aus Kostengründen aus.

8. Wärmenutzung

Ofen- und Kühlerabgase lassen sich oft nicht vollständig in Trocknungsanlagen verwerten. In diesem Fall kann ein Teil des Wärmeinhalts zur Aufwärmung von Heißwasser, z.B. für Sozial- oder Bürogebäude oder für ein Fernwärmenetz, genutzt werden [42, 43]. Dafür muß es jedoch einen Abnehmer in unmittelbarer Nähe geben. Erschwerend wirkt sich ferner aus, daß die Zementöfen häufig in der Zeit großen Heißwasserbedarfes, z.B. in den Wintermonaten, zu längeren Reparaturen abgestellt werden.

Eine andere Möglichkeit besteht darin, einen Teil der Abgasenthalpie in elektrische Energie umzuwandeln [44]. Diese Technik ist in Asien, insbesondere in Japan, weit verbreitet. Fünf Gründe waren und sind dafür ausschlaggebend:

— Die Feuchtigkeit des Rohmaterials liegt unter 5%, erhebliche Abgasenthalpien können nicht genutzt werden.

— Die Abgastemperaturen der Öfen sind mit 350 bis 450°C außergewöhnlich hoch.

bricks [38–40]. Zirconium dioxide is said to produce a substantial improvement in spalling resistance [41]. Outside the rotary kiln there has been a further increase in the trend towards the use of monolithic refractories, and in Germany they account for approximately 10% of the total consumption. Even in the future, however, monolithic refractories in rotary tube kilns will only be installed in the outlet zones and in the inlet cones. Burner nozzles are always clad with refractory concrete, and the installation of monolithic refractories in kiln hoods, coolers and preheaters is on the increase.

Monolithic lining is no longer regarded solely as a replacement for shaped bricks. Even large wall areas which, until recently, were exclusively lined with bricks, have recently been changed to casting and gunning materials. The important properties of monolithic components are not only their wear resistance and refractoriness but also their resistance to alkalis and their heating-up characteristics. There are always problems with anchoring monolithic linings. Neither ceramic nor steel or cast anchors can provide reliably satisfactory results in areas subjected to high thermal, mechanical and chemical loads, such as in gas riser ducts and calciner shafts. Significant corrosion problems which threaten the stability of the monolithic lining occur with reducing gas atmospheres and salt condensation, especially at the transition zone from working to insulating lining. Pragmatic solutions, such as thicker anchors, can definitely contribute to increasing the lining life. Specially selected alloys are ruled out on cost grounds.

8. Utilization of Heat

Drying plants often cannot make full use of the kiln and cooler exhaust gases. In this case some of the heat content can be used for heating hot water, e.g. for council or office buildings, or for a piped district heating system [42, 43]. However, for this there must be a consumer in the immediate neighbourhood. There is also the difficulty that cement kilns are frequently shut down for long repairs at the time of greatest hot water demand, e.g. in the winter months.

Another option consists of converting part of the exhaust gas enthalpy into electrical power [44]. This technology is widely used in Asia, especially in Japan. This was and is governed by five factors:

— The moisture content of the raw material is under 5%, so a considerable amount of the exhaust gas enthalpy cannot be utilized.

— The exhaust gas temperatures from the kilns of 350° to 450°C are unusually high.

— The average clinker capacity of the cement works is approximately 9000 t/d.

— The production plan does not provide an absolute guarantee of continuous power supply to an acceptable level.

— The cost of power is very high when compared with the rest of the world.

In Japan alone, 27 precalciner and 4 cyclone preheater kilns have each been fitted with one or two waste heat boilers in the last decade, and supply a total of 16 turbines with steam. With an average exhaust gas temperature of 395°C and an average cooler exhaust air temperature of 282°C, these plants generate about 32 kWh/t clinker – about 1/3 of the total electrical power needed for manufacturing Portland cement.

These figures are based on a published questionnaire to the Japanese cement industry [45]. Given the constraint of the five criteria listed the plants have an acceptable return on investment. If even one of the conditions does not apply – if, for example, the clinker capacity is only 3000–4000 t/d or less – then power generation is no longer profitable. The efficiency of the power generation system of only approximately 15% is also very low. Power generation from the waste heat from a rotary kiln is therefore not an alternative for the majority of cement works, especially if there are other possible process engineering ways of saving fuel.

- Die durchschnittliche Klinkerkapazität der Zementwerke liegt bei ca. 9 000 t/d.
- Die kontinuierliche Stromversorgung gemäß Produktionsplan ist zu akzeptablen Bedingungen nicht restlos gesichert.
- Die Strompreise liegen im Weltvergleich sehr hoch.

Allein in Japan sind im letzten Jahrzehnt 27 Vorcalcinier- und 4 Zyklonvorwärmeröfen mit einem oder zwei Abhitzekesseln ausgerüstet worden, die insgesamt 16 Turbinen mit Dampf versorgen. Bei einer mittleren Abgastemperatur von 395 °C und einer mittleren Kühlerablufttemperatur von 282 °C erzeugen diese Anlagen etwa 32 kWh/t Klinker, etwa 1/3 des insgesamt benötigten elektrischen Energiebedarfs zur Herstellung von Portlandzement.

Die vorgenannten Zahlen basieren auf einer veröffentlichten Umfrage der japanischen Zementindustrie [45]. Unter der Einschränkung der vorgenannten 5 Kriterien haben die Anlagen einen akzeptablen „return on investment". Schon wenn eine der Voraussetzungen nicht gegeben ist, wenn z. B. die Klinkerkapazitäten bei nur 3 000–4 000 t/d oder darunter liegen, ist die Stromerzeugung nicht mehr rentabel. Außerdem liegt der Wirkungsgrad der Stromerzeugung mit ca. 15 % sehr niedrig. Die Stromerzeugung aus der Abwärme eines Drehofens ist somit für die meisten Zementwerke keine Alternative, zumal dann, wenn es noch verfahrenstechnische Möglichkeiten zur Brennstoffenergieeinsparung gibt.

9. Ausblick

Die Entwicklung neuer Brenntechniken ist gegenwärtig nur in Randbereichen bei der Wirbelschichttechnik zu sehen. Mittelfristig wird die bewährte Drehofentechnik mit modernen Vorwärmer- und Vorcalciniersystemen nicht zu ersetzen sein.

Moderne Ofenanlagen werden immer mehr zum Knotenpunkt eines vernetzten Systems, in dem Anlagen- und Verfahrenstechnik, Roh- und Brennstoffvielfalt, Umweltschutzmaßnahmen und Produktqualität wesentliche Randbedingungen unterschiedlicher Wichtung darstellen. Daraus leitet sich ab, daß eine Ofenanlage immer mehr zu einem „Maßanzug" für die spezielle Werkssituation wird und selbst Modernisierungsmaßnahmen an vorhandenen Anlagen nicht mehr nach allgemein gültigen Auslegungsregeln geplant werden können.

Das Streben nach niedrigen Kosten gebietet jedoch, daß diese Randbedingungen nicht zu komplizierten und reparaturaufwendigen Anlagen führen. Gefordert wird seitens der Betreiber eine hohe Betriebssicherheit und Verfügbarkeit, aber gleichzeitig auch Produktionsflexibilität.

Ziel ist der gleichmäßige Betrieb, der nicht nur der Produktion und den Kosten dient, sondern vor allem auch dem Umweltschutz und der Produktqualität. Dazu werden zunehmend modernste Steuerungs-, Leit- und Führungssysteme notwendig, die unter Einsatz thermodynamischer Prozeßmodelle, heuristischer Kontrollkomponenten und künstlicher neuronaler, selbstadaptiver Netzwerke die Prozeßoptimierung ermöglichen.

Brenntechnik und Wärmewirtschaft werden somit auch künftig im Mittelpunkt der konstruktiven und verfahrenstechnischen Optimierungsbemühungen von Anlagenbauern und Zementwerksbetreibern stehen.

9. Outlook

The development of new burning technologies can at present only be seen in the marginal areas with the fluidized-bed technology. The well-tried rotary kiln technology with modern preheater and precalciner systems is not going to be displaced in the medium term.

The modern kiln plant is increasingly becoming the node for a network system in which plant and process technology, diversity of raw materials and fuel, environmental protective measures and product quality, represent fundamental constraints of varying importance. From this it can be deduced that kiln plants are being increasingly "tailor-made" for the specific works situation, and even modernization work on existing plants can no longer be planned by following generalized design rules.

However, the efforts to achieve low costs mean that these constraints must not lead to complicated plants which are expensive to repair. The operator requires high operational reliability and availability but also production flexibility.

The objective is to achieve uniform operation which serves the interests not only of production and costs, but also, and in particular, of environmental protection and product quality. This increasingly requires the use of very modern control and management systems which make it possible to optimize the process using thermodynamic process models, heuristic control components, and synthetic neuronal self-adaptive networks.

Burning technology and thermal economy will therefore continue to remain at the centre of the efforts by plant manufacturers and cement works operators to optimize designs and process technology.

Literaturverzeichnis

[1] Grydgaard, P. E.: Conversion of wet process plants to semi-dry. European Seminar on Improved Technologies for the Rational Use of Energy in the cement Industry. Proceedings, Commission of the European Communities (DG XVII), Brüssel, 1993, pp. 147–161.

[2] Ahuja, J.C., Rao, M.N., und Vasudevan, R.: Umbau vom Naßverfahren zum Halbnaßverfahren im Madukkarai-Zementwerk von ACC. Vortrag zum VDZ-Kongreß '93, Fachbereich 3.

[3] Duda, J., und Werynski, B.: Umbau vom Naß- zum Halbnaßverfahren mit Einsatz von Rückständen der Kohleaufbereitung. Vortrag zum VDZ-Kongreß '93, Fachbereich 3.

[4] Vandelli, M., und Denti, E.: Two dry lines replacing two wet lines. European Seminar on Improved Technologies for the Rational Use of Energy in the cement Industry. Proceedings, Commission of the European Communities (DG XVII), Brüssel, 1993, pp. 137–144.

[5] Borgholm, H. E., und Nielsen, P. B.: Commissioning the world's largest semi-dry process kiln system. World Cement (1989) No. 3, pp. 74–79.

[6] Borgholm, H. E.: Umweltentlastung durch Verwertung von Flugasche als Rohmehlkomponente. Zement-Kalk-Gips 45 (1992) Nr. 4, S. 163–170.

[7] de Beus, A. J.: Einsparungen bei langen Öfen. Vortrag zum VDZ-Kongreß '93, Fachbereich 3.

[8] Dumas, J.: Engineering and energy savings. European Seminar on Improved Technologies for the Rational Use of Energy in the cement Industry. Proceedings, Commission of the European Communities (DG XVII), Brüssel, 1990, pp. 109–117.

[9] Krupp Polysius AG, Mitteilung.

[10] Mrowald, U., und Hartmann, R.: Auswirkungen der Gestaltung der Zyklone für Zyklonvorwärmer auf Abgastemperatur und Druckverlust. Vortrag zum VDZ-Kongreß '93, Fachbereich 3.

[11] von Seebach, H.M., Karrasch, G., und Bauer, K.: Optimierung bestehender Drehofenanlagen. TIZ international Vol. 114 (1990) No. 1, pp. 21–25.

[12] Kreft, W., und Günnewig, L.: Einsparung von Wärmeenergie durch Sanierungs- und Umbaumaßnahmen in Rohmehlwärmetauschern. Zement-Kalk-Gips 41 (1988) Nr. 7, S. 322–327.

[13] Scheuer, A., und Ellerbrock, H.-G.: Möglichkeiten der Energieeinsparung bei der Zementherstellung. Zement-Kalk-Gips 45 (1992) Nr. 5, S. 222–230.

[14] Liebl, P., und Gerger, W.: Nutzen und Grenzen beim Einsatz von Sekundärstoffen. Zement-Kalk-Gips 46 (1993) Nr. 10, S. 632–638.

[15] Garrett, H.M.: Precalciners today – a review. Rock Products/July 1985, pp. 39–61.

[16] Menzel, K.: Anpassung des Calcinierprozesses an Brennstoffreaktivität und Emissionserfordernisse mit PREPOL MSC. Vortrag zum VDZ-Kongreß '93, Fachbereich 3.

[17] Trends in Japanese cement technology. World Cement (1990) No. 6, pp. 234–244.

[18] Uchikawa, H.: The future for Cement/Concrete in Japan. Onoda Pacific Conference 1990, Sydney, 1990, pp. 1–62.

[19] Cohen, S.M.: Anwendungsmöglichkeiten der Wirbelschicht bei der Zementherstellung. Vortrag zum VDZ-Kongreß '93, Fachbereich 3.

[20] Wolter, A.: Pyrorapid-Kurzdrehofen – Vorteile für alle Rohmaterialien. Zement-Kalk-Gips 43 (1990) Nr. 9, S. 429–432.

[21] Lowes, T.M., und Evans, L.P.: Auswirkungen der Brennerbauart und der Betriebsparameter auf die Flammenform, auf Wärmeübergang, NO_x und SO_3-Kreisläufe. Zement-Kalk-Gips 46 (1993) Nr. 12, S. 761–768.

[22] Buzzi, S., und Sassone, G.-J.: Optimierung des Klinkerkühlerbetriebes. Zement-Kalk-Gips 46 (1993) Nr. 12, S. 755–760.

[23] Green-Andersen, P.: Umbau von F.L. Smidth/Fuller-Klinkerkühlern. Vortrag zum VDZ-Kongreß '93, Fachbereich 3.

[24] Kästingschäfer, G.: Technische Neuerungen im Rekuperationsbereich der Klinkerkühler. Vortrag zum VDZ-Kongreß '93, Fachbereich 3.

[25] Reckziegel, H.: Betriebserfahrungen mit IKN-Kühlern im Werk Göllheim der Dyckerhoff AG. Vortrag zum VDZ-Kongreß '93, Fachbereich 3.

[26] Schneider, R.: PYROSTEP Rostkühler – Weiterentwicklungen in der Rostkühlertechnik. Vortrag zum VDZ-Kongreß '93, Fachbereich 3.

[27] Steffen, E., Koeberer, G., und Meyer, H.: Innovation bei Schubrostkühlern mit großen Durchsatzleistungen. Vortrag zum VDZ-Kongreß '93, Fachbereich 3.

[28] von Wedel, K.: Betriebsergebnisse von Pendelrostkühlern mit horizontaler Anströmung des Klinkers. Vortrag zum VDZ-Kongreß '93, Fachbereich 3.

[29] Scheuer, A.: Beurteilung der Betriebsweise von Klinkerkühlern und ihr Einfluß auf die Klinkereigenschaften. Zement-Kalk-Gips 41 (1988) Nr. 3, S. 113–118.

[30] Rosemann, H.: Einfluß der Brennstoffart und des Einsatzortes auf Energieverbrauch, Betriebsverhalten und Emission von Drehofenanlagen. Zement-Kalk-Gips 41 (1988) Nr. 3, S. 231–236.

[31] Goes, C.: Über das Verhalten der Alkalien beim Zementbrennen. Schriftenreihe der Zementindustrie, Heft 24, 1960.

[32] F.L. Smidth & Co A/S, Mitteilung.

[33] Scheuer, A., Schmidt, K.D., Gardeik, H.O., und Rosemann, H.: Einflüsse auf die Entstehung von gasförmigen Schadstoffen bei Drehofenanlagen der Zementindustrie und Primärmaßnahmen zu ihrer Minderung. IF-Die Industriefeuerung, Heft 38, S. 65–78.

[34] Tätigkeitsbericht 1984–87. Verein Deutscher Zementwerke, Düsseldorf, April 1987, S. 55.

[35] Hansen, E.R.: Reduction of clinker alkali and SO_2, NO_x emissions from preheater kilns. Portland cement association GTC fall meeting, September 1986.

[36] Xeller, H.: Entwicklung der Brenntechnik in der Zementindustrie und Anforderungen an die feuerfeste Auskleidung. XXVI. Internationales Feuerfest-Kolloquium. Kongreßband, Forschungsinstitut der Feuerfest-Industrie, Bonn 1983, S. 1–26.

[37] Scheuer, A., Sylla, H.-M., Kühle, W., und Rosemann, H.: Annahmeprüfung und Anforderungen an die Gleichmäßigkeit von Drehofensteinen. Zement-Kalk-Gips 42 (1989) Nr. 2, S. 57–67.

[38] Bartha, P., und Nachtwey, W.: Klassifikation von Magnesiasteinen nach Spezifikation und Gebrauchswert im Zementdrehofen. Vortrag zum VDZ-Kongreß '93, Fachbereich 3.

[39] Hartenstein, J., Bongers, U., Prange, R., und Stradtmann, J.: Verschleißuntersuchungen an Magnesia-Spinell-Steinen für den Zementdrehrohrofen. Vortrag zum VDZ-Kongreß '93, Fachbereich 3.

[40] Rodrigues, A., und Oliveira, W.S.: Magnesia-Spinell-Steine – eine praxisbezogene Betrachtung. Vortrag zum VDZ-Kongreß '93, Fachbereich 3.

[41] Hartenstein, J., Bongers, U., Prange, R., und Stradtmann, J.: Verbesserung der Temperaturwechselbeständigkeit von basischen, gebrannten Steinen für den Zementdrehrohofen durch Zirkondioxid. Vortrag zum VDZ-Kongreß '93, Fachbereich 3.

[42] Bouquelle, J.-F.: Heat recovery on the smoke of the cement kilns and utilization of the recovered energy. European seminar on energy efficiency in the cement industry. Proceedings, Commission of the European Communities (DG XVII), Brüssel, 1990, pp. 78–87.

[43] Ahlkvist, B.: District heating based on waste heat from clinker cooler. European seminar on energy efficiency in the cement industry. Proceedings, Commission of the European Communities (DG XVII), Brüssel, 1990, pp. 73–77.

[44] Steinbiß, E.: Traditional and advanced concepts of waste heat recovery in cement plants. European seminar on energy efficiency in the cement industry. Proceedings, Commission of the European Communities (DG XVII), Brüssel, 1990, pp. 57–72.

[45] Onissi, T.R., und Munakata, N.: Elektrische Energieerzeugung aus dem Abgas von Drehofenanlagen. Zement-Kalk-Gips 45 (1992) Nr. 11, S. 564–570.

Optimierung des Klinkerkühlerbetriebs*)

Optimization of clinker cooler operation*)

Optimisation de l'exploitation de refroidisseur de clinker

Optimización del servicio de los enfriadores de clínker

Von **S. Buzzi** und **G. Sassone**, Casale Monferrato/Italien

Fachbericht 3.1 · Zusammenfassung – Das Rekuperationsverhalten des Klinkerkühlers beeinflußt maßgeblich die Wirtschaftlichkeit des Klinkerbrennprozesses. Neben der Aufwärmung der Verbrennungsluft durch die bestmögliche Wiedergewinnung der Klinkerabwärme des Ofens hat der Kühler jedoch auch weitere betriebliche sowie qualitäts- und umweltrelevante Aufgaben zu erfüllen. Heute werden zur Klinkerkühlung weltweit hauptsächlich Rost-, Rohr- und Satellitenkühler eingesetzt, die sich vor allem in der Art der Wärmeübertragung und der Verteilung ihrer Wärmeverluste unterscheiden. Rohr- und Satellitenkühler werden heute kaum noch gebaut, u. a. weil sie bei sehr großen Anlagen kaum einsetzbar sind, die Kaltklinkertemperatur deutlich höher als bei Rostkühlern ist und weil der größte Wärmeverlust, die Strahlungsabwärme, kaum sinnvoll genutzt werden kann. Der Rostkühler ist die weltweit verbreitetste Kühlerbauart und wurde in Einheiten bis zu 10 000 t/d realisiert. Bei diesen Kühlertypen ist die für die Calciniertechnik erforderliche Tertiärluftentnahme am leichtesten möglich. Aus diesen Gründen ist verständlich, daß die Weiterentwicklung dieses Kühlers in den vergangenen Jahren am stärksten betrieben wurde. Die Firma IKN entwickelte 1984 die Idee des „Widerstandsrostes", der durch verbesserte Wärmeübertragung zu niedrigeren Kaltklinkertemperaturen, höheren Sekundärlufttemperaturen bei gleichzeitig niedrigerem Kühlluftbedarf führte. Die inzwischen auch von anderen Firmen nachvollzogene Entwicklung neuer Rostplattenkonstruktionen, die Steuerung der Klinkerbetthöhe durch eine Stauwand und die gezielte regelbare Verteilung der Kühlluft ermöglichen heute einen Kühlbereichswirkungsgrad von bis zu 75 %.

Optimierung des Klinkerkühlerbetriebs

Special report 3.1 · Summary – The recuperation characteristics of the clinker cooler have a great influence on the cost-effectiveness of the clinker burning process. In addition to heating the combustion air by optimum recovery of the waste heat in the clinker from the kiln the cooler also has to fulfil other tasks affecting not only the operation but also the quality and the environment. Nowadays grate, rotary and planetary coolers are the main types used for clinker cooling; they differ mainly in the type of heat transfer and the distribution of their heat losses. Hardly any rotary or planetary coolers are being built now, principally because they cannot really be used with very large plants, the cold clinker temperatures are significantly higher than with grate coolers, and because it is hardly possible to make practical use of the largest heat loss, the radiant waste heat. The grate cooler is the world's most common type of cooler and has been built in units up to 10 000 t/d. The withdrawal of tertiary air needed for calcining systems is easiest with this type of cooler. For these reasons it is understandable that continued development work in recent years has been concentrated on this cooler. In 1984 IKN developed the idea of the "resistance grate" which, through improved heat transfer, led to lower cold clinker temperatures and higher secondary air temperatures combined with a lower requirement for cooling air. The development of new grate plate designs which has taken place in the meantime, by other firms too, the control of the clinker bed depth by a dam wall, and the accurately controlled distribution of cooling air over the grate has now made it possible to achieve efficiencies of up to 75% in the cooling sector.

Optimization of clinker cooler operation

Rapport spécial 3.1 · Résumé – Le pouvoir de récupération du refroidisseur de clinker a une influence considérable sur la rentabilité du processus de cuisson du clinker. Outre le réchauffement de l'air de combustion par une récupération de chaleur perdue du clinker du four, le refroidisseur doit assurer d'autres fonctions d'exploitation ainsi que des paramètres de qualité et de lutte pour la protection de l'environnement. De nos jours le refroidissement de clinker utilise, de par le monde, essentiellement des refroidisseurs à grille, des refroidisseurs tubulaires et des refroidisseurs à satellites, qui se diférencient surtout au point de vue mode de transfert de chaleur et répartition des pertes de chaleur. Les refroidisseurs tubulaires comme les refroidisseurs à satellites ne sont guère plus utilisés à l'heure actuelle, notamment du fait qu'ils ne sont pratiquement pas utilisables dans de très grandes installations, du fait que la température du clinker froid est sensiblement plus élevée que dans le cas des refroidisseurs à

Optimisation de l'exploitation de refroidisseur de clinker

---

*) Überarbeitete Fassung eines Vortrages zum VDZ-Kongreß '93, Düsseldorf (27. 9.–1. 10. 1993).
  Revised text of a lecture to the VDZ Congress '93, Düsseldorf (27. 9.–1. 10. 1993).

grille et, enfin, parce que la plus grande partie de la perte de chaleur, à savoir la chaleur perdue de rayonnement, peut être judicieusement réutilisée. Le refroidisseur à grille est celui qui est le plus répandu à travers le monde, il a été réalisé sous forme d'unités atteignant 10 000 t/j. Il est également le type de refroidisseur facilitant le plus l'évacuation d'air tertiaire nécessaire à la technique de calcination. Pour ces raisons, on comprend que des perfectionnements notables aient été apportés à ce refroidisseur au cours des dernières années. La société IKN a, en 1984, été à l'origine de l'idée de la „grille de résistance", qui a conduit à une amélioration du transfert de chaleur avec des températures de clinker froid plus basses et des températures d'air secondaire élevées et, simultanément, un besoin d'air de refroidissement moins grand. Le développement effectué entre temps par d'autres sociétés sur de nouvelles constructions à grilles, le pilotage de la hauteur du lit de clinker par l'entremise d'une paroi de retenue et, enfin, la répartition réglable de l'air de refroidissement par la grille sont autant d'éléments qui, de nos jours, facilitent des taux de rendement de la zone de refroidissement pouvant atteindre 75%.

Optimización del servicio de los enfriadores de clínker

Informe de ramo 3.1 · Resumen – El comportamiento a la recuperación del enfriador de clínker influye de forma notable en la rentabilidad del proceso de cocción del clínker. Aparte del calentamiento del aire de combustión mediante la mejor recuperación posible del calor de escape del clínker, el enfriador tiene que cumplir otras tareas relacionadas con el servicio, la calidad del producto y el medio ambiente. Hoy en día se emplean, a nivel mundial, para el enfriamiento del clínker sobre todo enfriadores de parrilla, de tambor rotatorio y de satélites, los cuales se distinguen entre sí esencialmente por su forma de transmisión del calor y de distribución de las pérdidas de calor. Pero actualmente, los enfriadores de tambor rotatorio y de satélites apenas se construyen ya, entre otras cosas porque practicamente no se pueden emplear en las instalaciones muy grandes. Ademas, la temperatura del clínker enfriado es notablemente más alta en comparación con los enfriadores de parrilla, y la mayor pérdida de calor, o sea la pérdida por radiación, apenas puede aprovecharse razonablemente. El enfriador de parrilla es el tipo más extendido en el mundo y ha sido realizado en unidades con capacidad de producción de hasta 10 000 t/d. En estos tipos de enfriadores, resulta más fácil la recogida del aire terciario necesario para la técnica de calcinación. Por estas razones, se comprende fácilmente que en los últimos años el interés de los fabricantes se haya centrado en el perfeccionamiento de este tipo de enfriadores. La empresa IKN desarrolló en 1984 la idea de la „parrilla resistente", mediante la cual se consiguió, una mejor transmisión del calor, que condujo a temperaturas más bajas del clínker enfriado a temperaturas más altas del aire secundario y, al mismo tiempo, a un menor consumo de aire frío. El desarrollo de nuevos tipos de placas de parrilla, realizado entretanto también por otras empresas, el control de la altura del lecho de clínker mediante una pared de retención y la distribución controlada del aire de enfriamiento permiten, hoy en día, alfanzar un rendimiento de enfriamiento de hasta un 75%.

1. Einleitung

Der erste Klinkerkühler für einen Drehrohrofen in einem Zementwerk wurde in den Jahren 1890 bis 1900 geplant und gebaut. Es handelte sich dabei um ein getrenntes Drehrohr, das eine Kühlung des Zementklinkers auf eine Temperatur ermöglichte, die eine weitere mechanische Behandlung im nächsten Prozeßschritt erlaubte. Es ist anzunehmen, daß damals die Kaltklinkertemperatur unter 350 °C gelegen hat. Der Planeten-Kühler wurde etwa 30 Jahre später erfunden, zuletzt der Rost-Kühler in den vierziger Jahren dieses Jahrhunderts.

Der Bericht befaßt sich zunächst generell mit den Aufgaben der Klinkerkühlung und den Anforderungen an den Kühlerbetrieb, die Klinkerqualität und den Umweltschutz. Anschließend werden die Klinkerkühlerbauarten behandelt und vergleichende Beurteilungen unter den genannten Gesichtspunkten vorgenommen. Von Bedeutung ist auch die Erörterung der Frage, welche Maßnahmen zur Optimierung des Kühlerbetriebs in letzter Zeit durchgeführt wurden oder in Zukunft erwartet werden können. Neben der allgemeinen Entwicklung soll an Beispielen aufgezeigt werden, welche vielfältigen Erfahrungen ganz speziell auch im Unternehmen Fratelli Buzzi in den zurückliegenden Jahren gesammelt wurden.

2. Aufgaben und Anforderungen

Im modernen Klinkerbrennprozeß werden zur Kühlung Rohr-, Satelliten- und Rostkühler eingesetzt. Die Kühler haben folgende Hauptaufgaben zu erfüllen:

— Bestmögliche Rückgewinnung der fühlbaren Klinkerwärme (1200 bis 1500 kJ/kg Klinker),

1. Introduction

The first clinker cooler for a rotary tube kiln in a cement works was designed and built in the years 1890–1900. This was a separate rotating tube which made it possible to cool the cement clinker to a temperature which allowed further mechanical treatment to be carried out in the next process step. It can be assumed that at that time the cold clinker temperature was less than 350°C. The planetary cooler was invented about 30 years later, followed finally by the grate cooler in the forties of this century.

This report starts by dealing in general terms with the functions of clinker cooling and the demands placed on cooler operation, clinker quality, and environmental protection. It then goes on to deal with the types of clinker coolers, and undertakes a comparative assessment under the aspects mentioned. It is also important to discuss which measures for optimizing cooler operation have been carried out recently, or can be expected in the future. In addition to the general development, examples will be used to indicate the varied experience which has been accumulated over the past years specifically in the firm of Fratelli Buzzi.

2. Functions and Requirements

Modern clinker burning processes use rotary, planetary and grate coolers for cooling. The cooler has to fulfil the following main functions:

— optimum recovery of the sensible clinker heat (1200 to 1500 kJ/kg clinker),

— heating the secondary, and possibly also tertiary, air required as combustion air to the highest possible temperature to minimize the consumption of fuel energy,

— Aufheizung der als Verbrennungsluft benötigten Sekundär- und gegebenenfalls auch Tertiärluft auf möglichst hohe Temperatur zur Minimierung des Brennstoffenergieverbrauchs,

— Anpassung der Kühlgeschwindigkeit des Klinkers an die Qualitätsanforderungen,

— Klinker-Endkühlung auf möglichst niedrige Temperaturen mit Rücksicht auf den Transport, die Lagerung und die Zementmahlung.

Neben diesen Hauptaufgaben müssen moderne Kühler eine Reihe von Anforderungen an den Betrieb erfüllen. Hierzu zählen im wesentlichen die

— Verfügbarkeit, der
— Schutz der Umwelt und die
— Erzielung der gewünschten Klinkerqualität.

Die Forderung nach einer hohen Verfügbarkeit bedeutet, daß der Kühler über eine möglichst lange Zeitdauer fortdauernd und regelmäßig betrieben werden kann, was vor allem eine hohe mechanische Haltbarkeit der Maschine voraussetzt und zu einem gleichmäßigen und wirtschaftlichen Ofenbetrieb beiträgt.

Der Forderung nach dem Schutz der Umwelt kann dadurch Rechnung getragen werden, indem der Kühler mit den notwendigen Einrichtungen ausgerüstet ist, mit deren Hilfe die Emission von Staub je nach Bauart ganz unterbunden oder minimiert wird. Ähnliches gilt für die Emission von Lärm. Hierauf soll im folgenden nicht weiter eingegangen werden.

Die Kühlgeschwindigkeit kann zu einem erheblichen Teil die Klinkerqualität und damit insbesondere die Mahlbarkeit sowie die Verarbeitungseigenschaften und die Festigkeitsentwicklung der daraus hergestellten Zemente beeinflussen. Dabei ist zwischen der Kühlgeschwindigkeit in der Vorkühlzone des Drehofens und derjenigen im Klinkerkühler zu unterscheiden [1–4]. Die Einflüsse der Kühlgeschwindigkeit auf die Klinkereigenschaften sollen hier nicht näher behandelt werden.

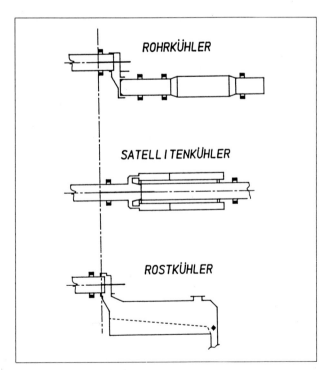

BILD 1: Gebräuchliche Kühlerbauarten
FIGURE 1: Types of cooler in normal use

Rohrkühler = rotary cooler
Satellitenkühler = planetary cooler
Rostkühler = grate cooler

— matching the rate of cooling of the clinker to the quality requirements,

— final clinker cooling to the lowest possible temperature, taking transport, storage and cement grinding into consideration.

In addition to these main functions, modern coolers must also fulfil a series of operational demands. These are essentially

— availability,
— protection of the environment, and the
— achievement of the desired clinker quality.

The demand for high availability means that the cooler should be able to operate continuously and consistently for as long as possible; this contributes to even and cost-effective kiln operation; the main requirement being high mechanical durability of the machinery.

The demand for protection of the environment can be dealt with by fitting the cooler with the necessary equipment which, depending on the design, will either minimize the emission of dust or suppress it entirely. The same applies to the emission of noise. This will not be discussed in any greater detail below.

The rate of cooling can have a very considerable influence on the clinker quality and therefore, in particular, on its grindability, as well as on the workability characteristics and strength development of the cements produced from it. It is necessary to differentiate between the rate of cooling in the precooling zone in the rotary kiln and that in the clinker cooler [1–4]. The effects of rate of cooling on clinker properties will not be dealt with in detail here.

3. Types of cooler

The types of cooler used nowadays for cooling clinker are essentially the three different types shown diagrammatically in **Fig. 1** [2]. These are the rotary, planetary and grate coolers. The coolers differ principally in the

— method of heat transfer,
— length and design of the pre-cooling zone,
— level of the hot clinker inlet temperature, and in their
— controllability.

The start of the pre-cooling zone is marked by the dot-dash line.

The influencing factors mentioned can affect the fuel energy consumption of the kiln plant in different ways. Above all, investigations have shown that a reduction in the energy loss flow, and hence an increase in the efficiency, generally leads to a disproportionately large saving (by a factor of 1.4) in the fuel energy for the entire plant [4].

3.1 Rotary cooler

The rotary cooler is one of the oldest designs of clinker cooler. It is built either as a continuation of the rotary kiln or, for space reasons, in the opposite direction. It can be assumed to account for about 5% of the clinker coolers in use throughout the world [1].

The heat transfer from hot clinker to cooling air in a rotary cooler takes place by a combination of counter- and crossflow. **Table 1** contains a summary of important technical and technological data relating to rotary coolers. The throughput range of rotary coolers lies between ≤ 2000 t/d for older kiln plants and about 4500 t/d for newer ones. A characteristic feature is the rotational speed which can be adjusted independently of the rotary kiln. With a somewhat longer pre-cooling zone than for grate coolers, the clinker inlet temperature for rotary coolers often lies between 1200 and 1400°C [3]. As with planetary coolers, only as much cooling air can be used as is needed for complete combustion with the lowest possible surplus of oxygen, so the quantity of air is frequently insufficient to achieve cold clinker tempera-

3. Kühlerbauarten

Zur Kühlung des Klinkers werden heute im wesentlichen drei unterschiedliche Bauarten eingesetzt, die schematisch im **Bild 1** dargestellt sind [2]. Es handelt sich hierbei um Rohr-, Satelliten- und Rostkühler. Die Kühler unterscheiden sich vor allem in der

— Art der Wärmeübertragung, der
— Länge und Ausbildung der Vorkühlzone, der
— Höhe der Heißklinker-Eintrittstemperatur sowie in ihrer
— Regelbarkeit.

Der Beginn der Vorkühlzone ist durch die strichpunktierte Linie markiert.

Die genannten Einflüsse können sich in unterschiedlicher Weise auf den Brennstoffenergiebedarf der Ofenanlage auswirken. Untersuchungen haben gezeigt, daß vor allem eine Verringerung des Energieverluststroms und damit eine Steigerung des Wirkungsgrads generell zu einer überproportionalen, mit einem Faktor von 1,4 eingehenden Einsparung an Brennstoffenergie für die Gesamtanlage führt [4].

3.1 Rohrkühler

Die Rohrkühler zählen zu den ältesten Klinkerkühler-Bauarten. Sie werden entweder fortlaufend oder aus Platzgründen auch gegenläufig zum Drehrohrofen gebaut. Ihr Anteil an den weltweit im Einsatz befindlichen Klinkerkühlern kann mit ca. 5 % angenommen werden [1].

Im Rohrkühler erfolgt der Wärmeübergang vom Heißklinker auf die Kühlluft in einer Kombination von Gegen- und Kreuzstrom. Eine Zusammenstellung wesentlicher technischer und technologischer Daten von Rohrkühlern enthält **Tabelle 1**. Der Durchsatzbereich von Rohrkühlern liegt zwischen ≤ 2000 t/d bei älteren und etwa 4500 t/d bei neueren Ofenanlagen. Kennzeichnend ist die unabhängig vom Drehofen einstellbare Drehzahl. Bei einer gegenüber Rostkühlern etwas verlängerten Vorkühlzone liegt die Klinkereintrittstemperatur von Rohrkühlern häufig zwischen 1200 und 1400 °C [3]. Da ähnlich wie bei Satellitenkühlern nur so viel Kühlluft eingesetzt werden kann, wie zur vollständigen Verbrennung bei möglichst niedrigem Sauerstoffüberschuß benötigt wird, reicht die Luftmenge häufig nicht aus, um damit Kaltklinkertemperaturen von unter 300 °C zu erzielen. Versuche zur Optimierung des Rohrkühlerbetriebs haben daher schon immer das Ziel verfolgt, den Wärmeübergang durch die geometrische Gestaltung und Anordnung von Einbauten zu verbessern oder die Kaltklinkertemperatur über die Wärmeverluste und durch Eindüsen von Wasser zu steuern. Zur optimalen Anpassung von Einbauten nach Art, Anordnung und Zonenlänge bei gegebener Korngrößenverteilung des Klinkers ist für Rohr- und Satellitenkühler eine Modellrechnung [3] entwickelt worden.

TABELLE 1: Technische und technologische Daten von Rohrkühlern

Bezeichnung	Maßeinheit	Zahlenwerte
Durchsatz	t/d	< 2000–4500
L/D-Verhältnis	—	ca. 10:1
Drehzahlbereich	min⁻¹	1–3
Neigung	%	3–5
Spezifische Kühlluftmenge	m³$_N$/kg Kli	0,8–1,1
Klinkereintrittstemperatur	°C	1200–1400
Kaltklinkertemperatur	°C	200–400
Kühlbereichswirkungsgrad	%	56–70

Aufgrund von Untersuchungen an technischen Anlagen hat sich herausgestellt, daß die Klinkerkühlung bei Rohr-, aber auch bei Satellitenkühlern, in starkem Maße von der Klinkerkorngröße abhängt [2, 4]. Danach nimmt — wie in **Bild 2** dargestellt — die Klinkerkühlung als Differenz zwischen Eintritts- und Kaltklinkertemperatur mit gröber werdender

TABLE 1: Technical and technological data for rotary coolers

Designation	Units	Numerical values
Throughput	t/d	< 2000–4500
L/D ratio	—	approx. 10:1
Speed range	rpm	1–3
Slope	%	3–5
Specific cooling air volume	m³(stp)/kg clinker	0.8–1.1
Clinker inlet temperature	°C	1200–1400
Cold clinker temperature	°C	200–400
Cooling sector efficiency	%	56–70

tures of under 300 °C. Attempts to optimize rotary cooler operation have therefore always been aimed at improving the heat transfer through geometric configuration and arrangement of internal fittings, or controlling the cold clinker temperature through the heat losses and by injecting water. A mathematical model [3] has been developed for rotary and planetary coolers to match the type, arrangement and zone lengths of the internal fittings to given particle size distributions of the clinker.

It has emerged from investigations on industrial plants that the clinker cooling in rotary, and also planetary, coolers is heavily dependent on the clinker particle size [2, 4]. This shows that — as demonstrated in **Fig. 2** — the clinker cooling (difference between inlet and cold clinker temperatures)

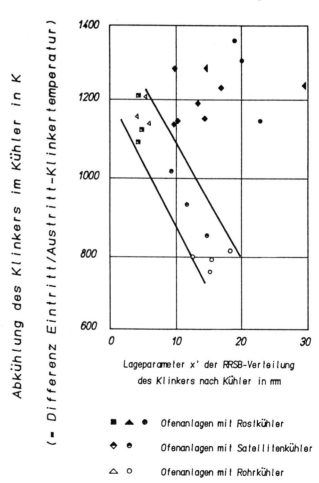

BILD 2: Klinkerkühlung in Abhängigkeit von der Klinkerkorngröße
FIGURE 2: Clinker cooling as a function of clinker particle size

Abkühlung des Klinkers im Kühler in K (= Differenz Eintritt/Austritt-Klinkertemperatur)	= clinker cooling in the cooler in K (= difference between clinker inlet and outlet temperatures)
Lageparameter x' der RRSB-Verteilung des Klinkers nach Kühler in mm	= positional parameter x' of the RRSB distribution of the clinker after the cooler in mm
Ofenanlagen mit Rostkühler	= kiln plant with grate cooler
Ofenanlagen mit Satellitenkühler	= kiln plant with planetary cooler
Ofenanlagen mit Rohrkühler	= kiln plant with rotary cooler

3.2 Planetary coolers

During the period 1980 to 1986 planetary coolers accounted for about 10% of new installations. A planetary cooler consists of 9 to 11 individual cooling tubes which are attached rigidly and rotationally symmetrically to the kiln. As with the rotary cooler, the heat transfer takes place by counterflow as well as − predominantly in the region of the internal fittings − by cross-flow. **Table 2** contains the essential data for this type of cooler.

TABLE 2: Technical and technological data for planetary coolers

Designation	Units	Numerical values
Throughput	t/d	< 3 000 − 4 000
L/D ratio	−	9 − 12
Specific cooling air volume	m^3(stp)/kg clinker	0.8 − 1.0
Clinker inlet temperature	°C	1 100 − 1 250
Cold clinker temperature	°C	200 − 300
Cooling sector efficiency	%	60 − 68

The throughput of this cooler is at present restricted to approximately 4 000 t/d. Other characteristic features are that the cooler cannot be controlled, the specific quantity of cooling air corresponds to the quantity of combustion air required, and with extended pre-cooling zone, the clinker inlet temperature is lower than with rotary coolers. Against the advantage of an operation which is free from exhaust air and therefore from dust, is the disadvantage that for a rotary kiln with calciner there is at present no practically proven way of drawing off tertiary air. In other respects, the operation of planetary coolers is governed by similar relationships to those for rotary coolers.

H. Fleck has given a report in [5] on particular findings when operating a rotary kiln with a planetary cooler designed for a throughput of 3 000 t/d.

3.3 Grate Cooler

The grate cooler is the most widely used type of cooler. In the years 1980 to 1986 the grate cooler held approximately 85% of the market. In the last 10 to 15 years the development of the grate cooler has been determined predominantly by the trend towards production units with clinker throughputs larger than 4 000 t/d with precalcination and tertiary air offtake, as well as by the desire for greater cost-effectiveness and environmental friendliness.

With these coolers, it is necessary to differentiate in principle between reciprocating and travelling grate coolers. The clinker in reciprocating grate coolers is transported by the forwards and backwards movement of the rows of grate plates, while in travelling grate coolers it is transported by a circulating continuous grate. The moving rows of grate plates in reciprocating grate coolers are linked by an oscillating or pendulum frame. The grate coolers normally used today include both the travelling grate cooler and the reciprocating grate cooler, which can be built with inclined, horizontal, combined, or stepped grates.

Short grate coolers, which are primarily operated only as recuperators and work in combination with a roller crusher and a gravitation cooler represent special designs. With a cooler concept of this type it is possible to fulfil special requirements for withdrawal of secondary and tertiary air, for clinker quality and for environmental protection. The roller crusher positioned between individual grate sections to improve the heat transfer by secondary size reduction is now state of the art. G. Koeberer et al. [6] have also given a report on their recent experience.

Development of the roll crusher for use in grate coolers was fostered for a number of years starting in 1978 at the Robilante cement works by Fratelli Buzzi in conjunction with Claudius-Peters. Until 1990 this crusher was the only one in the world incorporated in a large kiln system.

Grate coolers are normally operated by the cold air or the recirculating air methods, and are generally supplied with larger quantities of fresh air than are required for combustion in the burning process. This makes it possible to achieve low cold clinker temperatures of 80 to 100°C. From the summary in **Table 3**, which is partially based on information provided by R. Schneider, it can be seen that the specific volume of cooling air for grate coolers of small and large capacity normally lies between 1.6 and 2.6 m³(stp) per kg clinker [2, 3, 7]. More recent developments, such as in grate plate design and control of the clinker bed depth by a dam wall, are said to have reduced the specific volume of cooling air in continuous operation to only slightly more than 1.4 m³(stp) per kg clinker; this has been the subject of a report by K. von Wedel [8]. Altogether, these developments have contributed to the situation where the efficiency of the cooler and the cooling sector in day-to-day operation has been increased to 75 %, not only in new plants but also in existing clinker coolers by modernization and conversion work [9].

TABLE 3: Technical and technological data for reciprocating grate coolers

Designation	Units	Numerical values
Throughput	t/d	700 – >10 000
Grate area loading	t/m²d	26–55 (100)
Grate slope	degree	up to 10
Specific cooling air volume	m³(stp)/kg clinker	(1,4) 1,6–2,6
Clinker inlet temperature	°C	1300–1400
Cold clinker temperature	°C	70–120
Cooling sector efficiency	%	60–75

Comparison of the three types of cooler shows that the energy losses in the cooling sector lie between 400 and 600 kJ/kg clinker regardless of the type. For about the same total energy consumption of the kiln system and the same total loss there are characteristic differences in the distribution of the losses:

— With grate coolers, approximately 75 % of the energy loss occurs with the exhaust air, approximately 20 % with the clinker and the remaining 5 % as wall heat and other losses.

— With rotary and planetary coolers, the greatest energy loss of about 65 % occurs as wall heat loss, and approximately 35 % through incompletely cooled clinker.

The enthalpy of the cooler exhaust air can be used for drying raw materials.

A development project, which started in 1984 and even now is not complete, has led to an appreciable improvement in the uniformity of the cooler operation, its availability and its recuperation characteristics; the project has concentrated on a carefully controlled cooling air system and new grate plate designs, [10]. The concept of the "resistance grate" developed by IKN has now been assimilated by all plant manufacturers into a variety of versions. This also emerges from investigations by G. Kästingschäfer [11] and P. Green-Andersen [12].

In principle, and without going into details, the different types of plates, of which three designs are shown in **Fig. 3**, are able to supply more tightly restricted quantities of air to specific individual plates or aeration sectors. The new types of grate plates designed for carefully controlled air supply have narrow air outlet slots. This produces a sharp air jet at about 40 m/s inclined in the direction of conveying or horizontally below a cushion of clinker.

Schlitz-Plattentypen

Bauart IKN

Bauart Cl. Peters   Bauart Polysius

BILD 3: Rostplatten verschiedener Anlagenlieferanten
FIGURE 3: Grate plates from different plant suppliers

Schlitz-Plattentypen	= types of slotted plate
Bauart IKN	= IKN plate
Bauart Cl. Peters	= Cl. Peters plate
Bauart Polysius	= Polysius plate

etwa 40 m/s in Förderrichtung geneigt oder horizontal unter einem Klinkerpolster erzeugt.

Die Roste mit horizontaler Anströmung des Klinkers haben den Begriff „Widerstandsrost" geprägt. Sie weisen nur noch eine offene Rostfläche von 2,5 gegenüber 4 % bei älteren Rostkühlern auf. Damit wird eine gleichmäßige Luftverteilung erzielt, die allein vom Rostwiderstand und nicht vom Widerstand des Klinkerbetts abhängt. Das führt zu einer über die Rostfläche gleichmäßig verteilten, schnellen Kühlung des Klinkers und zu einem verbesserten Wärmeaustausch.

Über eine Messung der Temperaturverteilung auf der Oberfläche des Klinkerbetts mit Hilfe von Infrarot-Scannern erscheint es heute außerdem möglich, die Verteilung der Kühlluft zu steuern [11–13], wie unter anderem auch G. Gaussorgues und Mitautoren gezeigt haben [13].

Die Verfügbarkeit von Rostkühlern – gemessen an der Zahl der Stillstände und deren Dauer – soll sich durch die neuen Rostplattenkonstruktionen wesentlich verbessert haben. Nach eigener Erfahrung trifft diese Feststellung für die festen, nicht immer für die beweglichen Rostplatten zu. Vergleichende Angaben für die unterschiedlichen Kühlerbauarten enthält das VDZ-Merkblatt „Klinkerkühler" [1].

4. Betriebserfahrungen

Zur Optimierung des Kühlerbetriebs stehen beispielhaft vielfältige Erfahrungen aus dem eigenen Unternehmen zur Verfügung, die seit 1976 gesammelt werden konnten.

— 1976 wurde ein kurzer Rostkühler (Rekuperator) in Kombination mit einem g-Kühler für große Durchsätze in Betrieb genommen;

— 1978 ging ein Rollen-Brecher in Betrieb bei Klinkertemperaturen von 700–800 °C;

— 1980 wurde ein abluftfreier Rostkühler in Betrieb gesetzt, der mit 100 % Umluft betrieben wurde;

— 1986 erfolgte die Erprobung des damals neuen IKN-Systems mit dem besonderen Ziel, die spezifische Belastung des Rekuperators auf ein extremes Niveau von mehr als 100 t/m² × d zu erhöhen.

Beispielhaft soll hier die Entwicklung der Anlage mit Rekuperator und g-Kühler zur abluftfreien Klinkerkühlung bei Presacementi Italien näher beschrieben werden. 1976 wurde ein 4stufiger Dopol AT-Ofen mit einem Durchsatz von 2500 t/d errichtet, der mit einem kurzen Rostkühler (Rekuperator) in Kombination mit einem von Claudius Peters neu entwickelten, aus 5 Abschnitten bestehenden g-Kühler ausgerüstet war (**Bild 4**). Diese Lösung wurde unter anderem auch aus Umweltschutzgründen gewählt, wodurch der Erwerb der Bau- und Betriebsgenehmigung von den zuständigen Behörden erleichtert werden konnte.

The term "resistance grate" has been coined for grates with horizontal clinker aeration. They have an open grate area of only 2.5 % as against the 4 % in the older grate coolers. This achieves uniform air distribution which depends solely on the grate resistance and not on the resistance of the clinker bed. It leads to rapid cooling of the clinker distributed uniformly over the grate surface, and to improved heat transfer.

It also now appears possible to control the distribution of the cooling air through measurement of the temperature distribution over the surface of the clinker bed using infrared scanners [11–13], as has been demonstrated by, among others, G. Gaussorgues and co-authors [13].

The availability of grate coolers – based on the number and duration of stoppages – is said to have been substantially improved by the new grate plate designs. From our own experience this is true for the fixed grate plates, but not always for the moving ones. The VDZ leaflet "Clinker Coolers" [1] contains comparative information for the different types of coolers.

4. Operational experience

Our own company has had a great deal of experience with the optimization of cooler operation which has been gathered since 1976; for example:

— In 1976 a short grate cooler (recuperator) combined with a gravitation cooler was commissioned for large throughputs;

— in 1978 a roll crusher was brought into operation for clinker temperatures of 700–800 °C;

— in 1980 an exhaust-air-free grate cooler was brought into operation, working with 100 % air circulation;

— in 1986 the then new IKN system was tested with the particular objective of raising the specific loading of the recuperator to the extremely high level of more than 100 t/m² × d.

The development of the system with recuperator and gravitation cooler for exhaust-air-free clinker cooling at Presacementi, Italy, will be described in detail here by way of example. A 4-stage Dopol AT kiln with a throughput of 2500 t/d was built in 1976. This was equipped with a short grate cooler (recuperator) combined with a gravitation cooler newly developed by Claudius Peters and consisting of 5 sections (**Fig. 4**). This solution was selected for, among other things, environmental protection reasons, which

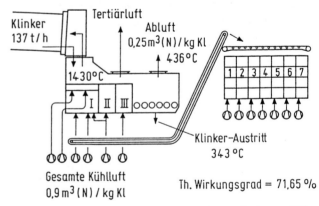

BILD 4: Kurzer Rostkühler in Verbundschaltung mit einem g-Kühler

FIGURE 4: Short grate cooler used in combination with a gravitation cooler

Klinker	= clinker
Tertiärluft	= tertiary air
Abluft	= exhaust air
0,25 m³ (N)/kg Kl	= 0.25 m³ (stp)/kg clinker
Klinkeraustritt	= clinker outlet
Gesamte Kühlluft	= total cooling air
Th. Wirkungsgrad	= thermal efficiency
0,9 m³ (N)/kg Kl	= 0.9 m³ (stp)/kg clinker

Der Transport des heißen Klinkers zum g-Kühler wurde durch ein Tiefbeckenband realisiert. Nach nur kurzer Betriebszeit stellte sich heraus, daß die vorgesehene Kühlluftmenge für die Rostkühler mit nur 0,8 m_N^3/kg Kli zu gering war. Der Klinker erreichte am Rostaustritt anstelle der theoretisch berechneten Temperatur von 400 bis 500°C die wesentlich höhere Temperatur von 700°C, was vor allem zu einer thermischen Überlastung des g-Kühlers und der Fördereinrichtungen führte. Daher wurde zunächst die Kühlluftmenge am Rekuperator um 0,4 m_N^3/kg Kli erhöht. Die entstehende Abluft wurde entweder in die Rohrmühle oder direkt in das Elektrofilter zur Entstaubung der Vorwärmeabgase geleitet. Weiterhin wurde der Hammerbrecher durch einen gezahnten 6-Walzen-Brecher mit hydraulischen, langsam drehenden (0 bis 5 min^{-1}) Haegglunds-Motoren ersetzt und der g-Kühler schließlich mit einer verstärkten Schleppkette für die Klinkerverteilung ausgerüstet.

In einer zweiten Ausbaustufe nach weiteren drei Betriebsjahren wurden zusätzliche Änderungen am Rekuperator durchgeführt:

— Das Walzengleitlager wurde durch Rollenlager ersetzt.
— Die Walzenwelle wurde verlängert, um die Wärme- und Staubbeanspruchung an den Lagern zu mindern.
— Der Rostschwingrahmen wurde mit hydraulischem Antrieb versehen.
— Der g-Kühler wurde entsprechend der ursprünglichen Planung von 5 auf 7 Abschnitte erweitert, um der Erhöhung der Ofenkapazität auf 2900 t/d durch eine AT-Zweitfeuerung Rechnung zu tragen.

1986 wurde geplant, den Ofendurchsatz auf 3600 t/d durch eine AS-Vorcalcination mit 50% Brennstoffanteil und Tertiärluftleitung zu erhöhen. Dabei sollte die Größe und Geometrie des Kühlers möglichst nicht verändert werden, was einer Kühlerrostbelastung von mehr als 100 t/m² × d entsprochen hätte. Der Kühler wurde zu diesem Zweck von IKN in mehreren Schritten umgerüstet:

— Von den 39 Plattenreihen wurden in zwei Schritten 16 Reihen mit IKN-Platten und einem gesteuerten Belüftungssystem ausgerüstet (**Bild 5**), beginnend am Rosteinlauf.
— Im Jahre 1989 wurde der Rost durch ein selbstentwickeltes seitliches Abdichtungssystem, durch verstärkte Längsträger im Schwingrahmen und durch Einsatz von Stahlschläuchen zur Belüftung der Rostträger verbessert.
— Die Luftverteilung in den Unterrostkammern wurde verändert.
— Die hydraulischen Haegglunds-Motoren wurden durch Elektro-Getriebemotoren mit Geschwindigkeiten bis 4 min^{-1} an den 6 Brechwalzen ersetzt.

Seit 1990 weist der Rostkühler einen regelmäßigen Betrieb auf mit verhältnismäßig langen Zeitabschnitten von bis zu ca. 4000 h zwischen betriebsbedingten Stillständen. Der durchschnittliche Klinkerdurchsatz betrug dabei ca. 3400 t/d.

5. Schlußbemerkung

Nach allen Optimierungsversuchen im Unternehmen Fratelli Buzzi wie auch in anderen Zementwerken läßt sich heute feststellen, daß die Entwicklung im Kühlerbau noch nicht abgeschlossen ist. Der Trend zur Verminderung der Kühlluftmenge bei Rostkühlern ist unverkennbar, wobei der thermische Kreuzstrom auch bei den zukünftigen Hochleistungskühlsystemen zur Anwendung gelangen wird.

Mit zuverlässigen und auch im hohen Temperaturbereich einsetzbaren Walzenbrechern dürfte das Stufenrost-Verfahren mit Zwischenbrecher die besten energetischen Aussichten besitzen. Bei einem sehr gleichmäßigen Kühlerbetrieb könnte auch ein getrenntes Elektrofilter als Einrichtung zur Entstaubung der Kühlerabluft an Bedeutung gewinnen.

made it easier to obtain building and operating approval from the relevant authorities. The hot clinker was transported to the gravitation cooler by a deep pan conveyor. After a short operating period it was found that the volume of cooling air of only 0.8 m^3(stp) per kg clinker provided for the grate cooler was too low. Instead of the theoretical calculated temperature of 400 to 500°C the clinker at the grate outlet reached the substantially higher temperature of 700°C; the main result of this was thermal overloading of the gravitation cooler and the conveying equipment. The volume of cooling air to the recuperator was therefore first increased by 0.4 m^3(stp) per kg clinker. The exhaust air produced was ducted either into the tube mill or directly into the electrostatic precipitator for dedusting the preheater exhaust gases. The hammer crusher was replaced by a toothed 6-roll crusher with slow speed (0 to 5 rpm) hydraulic Haegglunds motors, and finally the gravitation cooler was fitted with a strengthened drag chain for the clinker distribution.

Further changes to the recuperator were carried out in a second expansion stage after a further three years in operation:

— The plain bearing for the roller was replaced by roller bearings.
— The roller shaft was extended in order to reduce the action of heat and dust on the bearings.
— The grate oscillating frame was provided with a hydraulic drive.
— Following the original plan, the gravitation cooler was extended from 5 to 7 sections to allow for raising the kiln capacity to 2900 t/d by an AT secondary firing system.

In 1986 it was planned to raise the kiln throughput to 3600 t/d with an AS precalciner taking 50% of the fuel and a tertiary air duct. The size and geometry of the cooler was as far as possible to remain unchanged, which would have corresponded to a cooler grate loading of more than 100 t/m² × d. The cooler was converted by IKN in several stages:

— Of the 39 plate rows, 16 rows were equipped with IKN plates and a controlled aeration system (**Fig. 5**) in two stages, starting at the grate inlet.
— In 1989 the grate was improved by a side sealing system developed in-house, by reinforced longitudinal beams in the oscillating frame, and by the use of steel tubing for aerating the grate support beams.
— The air distribution in the undergrate chambers was changed.
— The hydraulic Haegglunds motors were replaced by electric geared motors with speeds up to 4 rpm on the 6 crushing rolls.

Since 1990 the grate cooler has provided regular operation with relatively long time intervals of up to about 4000 h between process stoppages. The average clinker throughput has been 3400 t/d.

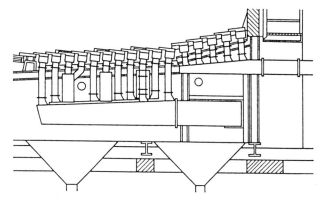

BILD 5: Beplattung des kurzen Rostkühlers im Zementwerk Presacementi/Italien
FIGURE 5: Plating system for the short grate cooler at the Presacementi cement works in Italy

Literaturverzeichnis

[1] Verein Deutscher Zementwerke: Rost-, Satelliten- und Rohrkühler in der Zementindustrie. Merkblatt Vt 8 des Ausschusses Verfahrenstechnik 1989.

[2] Scheuer, A.: Beurteilung der Betriebsweise von Klinkerkühlern und ihr Einfluß auf die Klinkereigenschaften. Zement-Kalk-Gips 42 (1988) Nr. 3, S. 113—118.

[3] Schürmann, W.: Untersuchungen zur thermischen Beurteilung von Gegenstrom-Klinkerkühlern der Zementindustrie und deren Einfluß auf die Klinkerqualität. Dissertation TU Clausthal 1993.

[4] Rosemann, H.: Theoretische und betriebliche Untersuchungen zum Brennstoffenergieverbrauch von Zementdrehofenanlagen mit Vorcalcinierung. Schriftenreihe der Zementindustrie H. 48, Beton-Verlag, Düsseldorf 1988.

[5] Fleck, H.: Optimierung eines Satellitenkühlers an einem 3000-t/d-Ofen. Vortrag zum VDZ-Kongreß 1993, Fachbereich 3.

[6] Koeberer, G., Meyer, H., und Steffen, E.: Innovation bei Schubrostkühlern mit großen Durchsatzleistungen. Vortrag zum VDZ-Kongreß 1993, Fachbereich 3.

[7] Schneider, R.: PYROSTEP-Rostkühler — Weiterentwicklungen in der Rostkühlertechnik. Vortrag zum VDZ-Kongreß 1993, Fachbereich 3.

[8] Wedel v., K.: Betriebsergebnisse von Pendelrostkühlern mit horizontaler Anströmung des Klinkers. Vortrag zum VDZ-Kongreß 1993, Fachbereich 3.

[9] Reckziegel, H.: Betriebserfahrungen mit IKN-Kühlern in Göllheim. Vortrag zum VDZ-Kongreß 1993, Fachbereich 3.

5. Final comments

After all the optimization trials at Fratelli Buzzi and in other cement works it is clear that development in cooler construction is not yet over. There is an unmistakable trend towards reducing the volume of cooling air for grate coolers, and future high efficiency cooling systems will continue to use thermal cross flow.

The use of reliable roll crushers which can also be used in the high temperature sector means that the stepped grate system with intermediate crusher probably has the best energy prospects. With very uniform cooler operation, increasing use could be made of separate electrostatic precipitators for dedusting the cooler exhaust air.

[10] Wedel v., K., und Wagner, R.: Sind Kühlroste Klinkerkühler oder Wärmerekuperatoren? Zement-Kalk-Gips 37 (1984) Nr. 5, S. 244—247.

[11] Kästingschäfer, G.: Brenntechnik und Wärmewirtschaft — Technische Neuerungen im Rekuperationsbereich der Klinkerkühler. Vortrag zum VDZ-Kongreß 1993, Fachbereich 3.

[12] Green-Andersen, P.: F. L. Smidth/Fuller Clinker Cooler Retrofit. Vortrag zum VDZ-Kongreß 1993, Fachbereich 3.

[13] Gaussorgues, G., und Houis, R.: Thermal monitoring of cement plant clinker coolers. Vortrag zum VDZ-Kongreß 1993, Fachbereich 3.

Auswirkung der Brennerbauart und der Betriebsparameter auf Flammenform, Wärmeübergang, NO$_X$ und SO$_3$-Kreisläufe*)

The effect of burner design and operating parameters on flame shape, heat transfer, NO$_X$ and SO$_3$ cycles*)

Influence du type de brûleur et des paramètres d'exploitation sur la forme de la flamme, le transfert de chaleur, l'émission de NO$_X$ et les circuits de SO$_3$

Influjo del tipo de quemador y de los parámetros de servicio sobre la forma de la llama, la transmisión del calor, la emisión de NO$_X$ y los circuitos de SO$_3$

Von **T. M. Lowes** und **L. P. Evans**, Greenhithe, Großbritannien

Fachbericht 3.2 · Zusammenfassung – Bei den im Auftrag des CEMFLAME-Konsortiums bei der International Flame Research Foundation (IFRF) durchgeführten Untersuchungen wurden die Grundlagen der Vermischung, der Verbrennung und des Wärmeübergangs in Zementöfen beim Einsatz von Ein- und Mehrkanalbrennern für feste Brennstoffe erarbeitet. Die wesentlichen Vorgänge der NO$_x$-Bildung und -Minderung werden zusammen mit einer Bewertung der wichtigsten Maßnahmen zur Vermeidung von SO$_3$-Kreisläufen dargestellt. Die wichtigsten quantitativen und qualitativen Kriterien für die Brennerkonstruktion werden in bezug auf Drall, Impuls, Luftzahl, Korngröße und Anordnung der Kanäle mit dem Ziel beschrieben, den Wärmeübergang, die Flammenform, die NO$_x$-Emissionen und die SO$_3$-Kreisläufe zu optimieren. Außerdem wird gezeigt, welchen Einfluß diese Parameter auf die Ausbildung reduzierender Bedingungen im Ofen und damit auf die Klinkerqualität haben. Für die Vorhersage und Optimierung der Betriebsparameter von Brennern werden ein vorhandenes mathematisches Modell angewandt und zukünftige Möglichkeiten aufgezeigt. Es werden Beispiele beschrieben und auf die weltweiten Erfahrungen beim Betrieb und der Verbesserung der Konstruktion von Ein- und Mehrkanalbrennern sowie ihre Auswirkungen auf NO$_x$-Emissionen, Klinkerqualität und SO$_3$-Kreisläufe Bezug genommen.

Auswirkung der Brennerbauart und der Betriebsparameter auf Flammenform, den Wärmeübergang, NO$_x$ und SO$_3$-Kreisläufe

Special report 3.2 · Summary – The fundamentals of mixing, combustion and heat transfer relative to both monotube and multichannel solid fuel burners in cement kilns are reviewed with reference to the work done by the CEMFLAME Consortium at the International Flame Research Foundation (IFRF). The essential kinetics of NO$_x$ formation and reduction are outlined together with an assessment of the essential features which can be used to minimise SO$_3$ cycles. The essential quantitative and qualitative design criteria are identified for both types of burner with reference to swirl level, channel position, momentum, particle size and oxygen level, in order to optimise heat transfer, flame shape, NO$_x$ emission and SO$_3$ cycles. The impact of these design criteria on clinker quality is also identified with reference to the avoidance of reducing conditions. An existing mathematical model for the prediction and optimisation of burner operating characteristics is used and its future role identified. Examples are drawn and reference made to world industrial experience in operating and modifying monotube and multichannel burners and their impact on NO$_x$ emission, clinker quality and SO$_3$ cycles.

The effect of burner design and operating parameters on flame shape, heat transfer, NO$_X$ and SO$_3$ cycles

Rapport spécial 3.2 · Résumé – Au cours de recherches effectuées pour le compte du consortium CEMFLAME auprès de „International Flame Research Foundation (IFRF)", il a été procédé à l'élaboration des bases du mélange, de la combustion et du transfert de chaleur dans les fours à ciment lors de l'utilisation de brûleurs à un ou plusieurs conduits à combustibles solides. Les opérations de formation et de réduction de NO$_x$ sont représentées conjointement à une évaluation des plus importantes mesures permettant d'éviter les circuits de SO$_3$. Les critères quantitatifs et qualitatifs essentiels pour l'étude constructive du brûleur sont décrits aux points de vue mise en rotation, impulsion, coefficient d'air, granulométrie et disposition des conduits dans le but d'optimiser le transfert de chaleur, la forme de la flamme, les émissions de NO$_x$ et les circuits de SO$_3$. En plus, on montre quelle influence ces paramètres exercent sur la formation de conditions réductrices dans le four et, par voie de conséquence, sur la qualité du clinker. Pour faire des prévisions et optimiser les paramètres d'exploitation des brûleurs, on fait appel à un modèle mathématique existant, ce qui permet d'illustrer les possibilités futures. Des exemples sont fournis avec référence à l'expérience faite dans le monde entier dans l'exploitation et l'amélioration de la construction de brûleurs à un et plusieurs conduits ainsi qu'aux effets correspondants sur les émissions de NO$_x$, la qualité du clinker et les circuits de SO$_3$.

Influence du type de brûleur et des paramètres d'exploitation sur la forme de la flamme, le transfert de chaleur, l'émission de NO$_x$ et les circuits de SO$_3$

---

*) Überarbeitete Fassung eines Vortrages zum VDZ-Kongreß '93, Düsseldorf (27. 9.–1. 10. 1993)
Revised text of a lecture to the VDZ Congress '93, Düsseldorf (27. 9.–1. 10. 1993)

Informe de ramo 3.2 · Resumen – Durante las investigaciones realizadas en la International Flame Research Foundation (IFRF), por encargo del grupo CEMFLAME, han sido elaboradas las bases del mezclado, la combustión y la transmisión del calor en los hornos de cemento, que emplean quemadores de combustibles sólidos, de uno y de varios canales. Se describen los fenómenos esenciales relacionados con la formación y la reducción de NO_x, junto con una evaluación de las principales medidas destinadas a evitar los circuitos de SO_3. Los criterios de cantidad y calidad más importantes, en cuanto a la construcción de los quemadores, se describen con respecto al rayado, el impulso, el indice de aire, la granulometría y la disposición de los canales, con el fin de optimizar la transmision del calor, la forma de la llama, las emisiones de NO_x y los circuitos de SO_3. Además, se muestra cuál es el influjo de estos parámetros sobre la creación de condiciones reductoras dentro del horno y, con ello, sobre la calidad del clinker. Para la predicción y la optimizacion de los parámetros de servicio de los quemadores, se emplea un modelo matemático existente y se mencionan las posibilidades futuras. Se describen varios ejemplos y se hace referencia a las experiencias adquiridas a nivel mundial en el servicio y perfeccionamiento constructivo de quemadores, de uno y de varios canales, así como a sus repercusiones en las emisiones de NO_x, la calidad del clinker y los circuitos de SO_3.

Influjo del tipo de quemador y de los parámetros de servicio sobre la forma de la llama, la transmisión del calor, la emisión de NO_x y los circuitos de SO_3

1. Einleitung

Die Flamme spielt beim Klinkerbrennprozeß im Drehofen eine wichtige Rolle. Daß die Klinkerproduktion unmittelbar von der Drehofenflamme abhängt, ist aber nicht die Hauptursache für ihre Bedeutung. Eine ungünstige Flammeneinstellung kann zu einem reduzierend gebrannten Zementklinker mit niedriger Qualität, niedrigen Gehalten an C_3S und wasserlöslichen Alkalien, zu erhöhten SO_2-Emissionen und/oder Verstopfungen im Vorwärmer, zu erhöhten NO_x-, CO- und Kohlenwasserstoffemissionen, erhöhtem Brennstoff- und elektrischem Energieverbrauch, sinkender Standzeit der Feuerfestauskleidung und vermindertem Klinkerdurchsatz führen.

Während einige dieser Probleme erst in den letzten 10 Jahren in Erscheinung traten, sind andere bereits seit Jahren bekannt und waren bisher Anlaß zu zahlreichen Mythen und Spekulationen um die Flamme. Die Hauptursache dafür, der Mangel an echten Kenntnissen etwa über die Frage, was für eine gut ausgebildete Flamme wesentlich ist, besteht darin, daß es im allgemeinen schwierig ist, die Drehofenflamme zu sehen, da sie meßtechnisch nicht zugänglich ist, und da sie bisher nicht in dem Maße Gegenstand wissenschaftlicher Untersuchungen war, wie etwa Kraftwerksfeuerungen, notwendige Finanzierungen durch den Staat oder die Europäische Kommission nicht zur Verfügung standen, zumal die Zementwerke im Normalfall nicht dem Staat gehören.

Traustel und Ruhland [1] unternahmen die ersten Untersuchungen, um die Vorgänge im Zementdrehofen zu klären. Lowes [2] gab in seinem Vortrag, „Kohlenflamme im Zementofen" zum Internationalen VDZ-Kongreß '85 einen Überblick über den Stand des Wissens. Das war eine für damalige Zeit nützliche Zusammenfassung, allerdings noch zu einem Zeitpunkt, da beispielsweise Petrolkoks nicht verfügbar war, über NO_x- und SO_2-Emissionen niemand sprach und eine entsprechende Auswahl von Ofenbrennern zur Reduzierung von NO_x nicht zur Verfügung stand.

Wegen der Bedeutung, die einer Drehofenflamme zukommt, und wegen der Verwendung von Billigbrennstoffen zur ökonomischen Produktion von Zementklinker war es in der Folge für viele Unternehmen notwendig geworden, Werksuntersuchungen sowie Forschungsarbeiten durchzuführen. Während die meisten der Forschungsarbeiten den Universitäten übertragen wurden und auf mathematische Modelle begründet waren, formierte sich 1990 das CEM-FLAME Konsortium mit dem Ziel, bei der International Flame Research Foundation (IFRF) in den Niederlanden praxisorientierte Forschungsarbeiten über die Verbrennung von Kohle und Petrolkoks in Zementdrehöfen zu finanzieren. Das Konsortium umfaßt 16 Organisationen, die entweder selbst Zement herstellen, Zementanlagen bauen oder mit Forschungs- und Entwicklungsaufgaben befaßt sind. Auf diesem Wege konnten Gelder in Höhe von 1500000 Hfl zum Aufbau einer Flammenversuchsanlage und zur Durchführung von Versuchen zusammengebracht werden.

1. Introduction

The flame is an essential part of clinker production. While without it there would be no production, this is not the main reason for it's importance. An inadequate flame can produce reduced clinker with poor workability, low C_3S and less water soluble alkalis, increased SO_2 emission and/or increased preheater blockages, high NO_x emissions, increased CO and hydrocarbon emissions, increased fuel and electricity consumption, decreased refractory life, lower production rates.

While some of these problem areas have only arisen in the last decade, many have been recognised for years and have given rise to much myth and speculation surrounding the flame.

The main reasons, the lack of real knowledge of what is essential for a "good flame", is that it is difficult to see, impossible to measure and has not been subjected to the same degree of research as flames from boilers, due to little Government or EC finance being available and the fact that cement manufacture is not normally Government owned.

Traustel and Ruhland [1] made the first real steps to develop an understanding of what was happening in the cement kiln. Lowes [2] summarised the state of knowledge with a paper on "Cement Kiln Coal Flames" at the last VDZ congress. This was a reasonable summary of what was required in 1985, however that was before petroleum coke, NO_x and SO_2 emissions and a wider range of international coals became available, as well as a range of burners aimed at reducing NO_x while maintaining good kiln operation.

Due to the importance of the flame and the use of low cost fuels to economic clinker production it has been necessary for many companies to carry out plant investigations and fund some research. While most of the research funded has been at universities and is mathematical modelling based, a CEMFLAME consortium was formed in 1990 to finance and direct research work on the combustion of coal and petroleum coke in cement kilns, at the International Flame Research Foundation (IFRF). The consortium comprises 16 organisations who either produce cement, build cement plants or carry out research and development. It raised 1500000 Hfl to construct a pilot furnace and carry out trials.

This paper will seek to update fundamentals described in 1985 paper and extend them to NO_x and SO_2, outline the major findings of the first CEMFLAME trials and indicate the future requirements.

2. Fundamentals of combustion and burner design

The combustion and heat transfer of coal flames in cement kilns are essentially controlled by fundamental jet mixing laws [3]. **Fig. 1** shows a schematic diagram of the aerodynamic mixing and combustion that takes place at the beginning of the burning zone in a cement kiln. For a specific coal it's ignition distance and combustion pattern is primarily dictated by the temperature and rate of entrain-

Dieser Beitrag will versuchen, die schon 1985 angerissenen Grundlagen zu aktualisieren, auf die NO$_x$- und SO$_2$-Problematik zu erweitern, die wichtigsten Erkenntnisse aus den ersten Versuchen von CEMFLAME mitteilen und zukünftige Anforderungen formulieren.

2. Grundlagen zur Verbrennung Brennerauslegung

Die Verbrennung und der Wärmeübergang bei Kohlenstaubflammen in Zementdrehöfen wird im wesentlichen bestimmt durch die grundlegenden Gesetze der Mischung von Freistrahlen [3]. **Bild 1** zeigt schematisch die Vorgänge der aerodynamischen Vermischung und Verbrennung, die im ersten Teil der Brennzone eines Drehofens stattfinden. Für eine bestimmte Kohle sind der Zündweg sowie das Verbrennungsbild in erster Linie durch die Temperatur und Geschwindigkeit der zugeführten Sekundärluft festgelegt. Die Geschwindigkeit, mit der die Sekundärluft in die Flamme eingemischt wird, ist eine Funktion des durch den Ofenbrenner erzeugten Impulses. Z.B. hat eine unverdrallte, mit einer Primärluftgeschwindigkeit von 90 m/s stabilisierte Petrolkoksflamme eine sehr kurze Zündstrecke, wenn die Sekundärlufttemperatur 900 °C beträgt. Der Zündweg wächst auf 1,5 bis 2,0 m an, wenn die Sekundärlufttemperatur nur 600 °C beträgt. Das ist im wesentlichen darauf zurückzuführen, daß die flüchtigen Bestandteile des Petrolkokses unterhalb von 900 °C nicht freigesetzt werden, weil der Gehalt an leichtflüchtigen Bestandteilen während des Herstellungsprozesses bereits extrahiert worden ist.

Sobald die Zündung erfolgt ist, laufen im Drehofen einige kritische Reaktionen ab, von denen wesentlich die Qualität der Verbrennung abhängt:

$$H + O_2 \rightarrow OH + O \quad (1)$$
$$C_nH_m + O \rightarrow C_{n-1}H_m + CO \quad (2)$$
$$CO + OH \rightarrow CO_2 + H \quad (3)$$
$$2CO + O_2 + M \rightarrow 2CO_2 + M \quad (4)$$
$$H_2O + O \rightarrow 2OH \quad (5)$$
$$2C + O_2 \rightarrow 2CO \quad (6)$$

Eine besonders kritische Reaktion wird durch die Gleichung (1) wiedergegeben. Wenn nicht genügend Sauerstoff vorhanden ist, der mit dem Wasserstoff und den Kohlenwasserstoffen reagiert, die bei der Entgasung unter nichtstöchiometrischen Verhältnissen [2] entstanden sind, werden die OH-Radikale nicht in ausreichendem Maße erzeugt, um mit dem nach Gleichung (2) und (6) gebildeten CO zu reagieren. In der Brennzone entstehen reduzierende Bedingungen. Die Reaktion nach Gleichung (4) verläuft um ca. 100mal langsamer als die Reaktion nach Gleichung (3) und ist deshalb nicht signifikant für die Vermeidung von reduzierenden Bedingungen. Es wird oftmals angenommen, daß eine Erhöhung des Sauerstoffgehaltes am Ofeneinlauf zu einer Absenkung des CO-Gehaltes führt. Das ist nur dann der Fall, wenn der gesamte Sauerstoffgehalt im Flammenbereich nicht ausreicht. Die Erhöhung des Brennerimpulses ist eine wirkungsvolle Maßnahme zur Verminderung der CO-Bildung [2], wenn eine Erhöhung des Sauerstoffgehaltes im Ofeneinlaufbereich nicht möglich ist.

Brennerkonstruktionen können eingeteilt werden in solche mit oder ohne Drall. Brenner ohne Drall besitzen normalerweise nur einen einzigen Kanal und sind dann meistens mit Direktfeuerungen verbunden. Brenner mit Drall sind gewöhnlich Mehrkanalkonstruktionen und in indirekten Feuerungssystemen integriert. Brennerkonstruktionen ohne Drall haben manchmal zwei Primärluftkanäle und/oder einen sog. Bluff Body, um die Vermischung zu verbessern, die Zünddistanz zu verkürzen und über einen zusätzlichen Brennerimpuls eine innere Rezirkulationszone zu schaffen. Die Flammen drallfreier Brenner benötigen einen Primärluftanteil zwischen 10 bis 30 %, der unter einem Halbwinkel von 10° eingeblasen wird, so daß die Flammengase nach ca. drei Ofendurchmessern auf die Brennzone an der Ofenwand stoßen. Es ist wichtig, daß die Gase kein CO bzw. keinen Kohlenstoff wenigstens einen halben Ofendurchmesser vor diesem Punkt mehr enthalten, da andernfalls

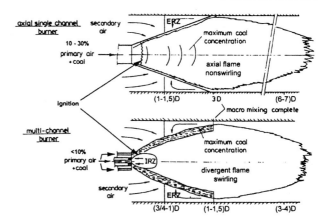

BILD 1: Kohlenstaubflamme im Zementdrehofen
FIGURE 1: Cement kiln coal flame

Einkanalbrenner	= axial single channel burner
Mehrkanalbrenner	= multi channel burner
Primärluft	= primary air
Sekundärluft	= secondary air
maximale Kohlenstaubkonzentration	= maximum coal concentration
axiale Flamme	= axial flame
unverdrallt	= nonswirling
divergierende Flamme	= divergent flame
verdrallt	= swirling
vollständige Makrovermischung	= macro mixing complete

ment of the secondary air. The rate of entrainment of secondary air is a function of burner momentum.

For example a pet coke flame from a non swirling burner with a primary air velocity of 90 m/s will have a very short ignition distance with a secondary air temperature of 900 °C and an ignition distance of 1.5 to 2.0 metres with a secondary air temperature of 600 °C. This is essentially due to the fact that pet coke volatiles are not released until around 900 °C, because the low temperature volatiles have been extracted in the manufacturing process.

Once ignition has occurred the critical reactions as far as a good combustion in cement kilns is concerned are:

$$H + O_2 \rightarrow OH + O \quad (1)$$
$$C_nH_m + O \rightarrow C_{n-1}H_m + CO \quad (2)$$
$$CO + OH \rightarrow CO_2 + H \quad (3)$$
$$2CO + O_2 + M \rightarrow 2CO_2 + M \quad (4)$$
$$H_2O + O \rightarrow 2OH \quad (5)$$
$$2C + O_2 \rightarrow 2CO \quad (6)$$

The most critical reaction is represented by formula (1). Unless sufficient oxygen is available to react with the hydrogen/hydrocarbons produced during devolatilisation in near stoichiometric proportions [2], then OH radicals will not be produced in sufficient quantities to react with the CO produced from reactions (2) and (6) and reducing conditions will occur in the burning zone. Reaction (4) is around 100 times slower than (3) and is not significant in avoiding reduction.

It is often thought that increasing the back end oxygen will decrease the CO level. This is only true if the overall oxygen level in the flame region is not adequate. Increasing the burner momentum has proved to be effective in eliminating CO [2] when increasing back end oxygen has failed.

Burners can be categorised as either with or without swirl. Burners without swirl normally have only one channel and are associated with direct firing systems, while burners with swirl are normally multi-channel and are associated with indirect firing systems. Non swirling burners can sometimes have two primary air channels and/or a bluff body to enhance mixing and reduce ignition distance, via extra burner momentum and the creation of an internal reverse flow zone of hot gases. The flames from non swirling burners have between 10 and 30 % primary air and expand at around

reduzierende Bedingungen in der Kernströmung und in den äußeren Rezirkulationszonen herrschen.

Drallbrenner werden gewöhnlich mit einem Primärluftanteil bis zu 10 % betrieben, der über insgesamt drei Kanäle zugeführt wird. Dabei handelt es sich um einen Drall-, einen Axial- und einen Kohleförderluftkanal. Die dabei realisierten Luftgeschwindigkeiten können Schallgeschwindigkeit haben. Sie verfügen über eine drallinduzierte innere Rückströmzone, welche die Verbrennungsgase in die Nähe des Kohleeintrags zurückfördert und eine frühe Entgasung und Verbrennung verursacht. In Abhängigkeit von der Drallzahl, der Kanalanordnung und der Sekundärlufttemperatur können allerdings in der Brennzone reduzierende Bedingungen entstehen, weil der Kohlestrahl sich schneller ausdehnt als er verbrennen kann, so daß Ausbrandprobleme von CO und Kohlenstoff auftreten. Die drallinduzierte Rückströmzone ist jedoch von keinerlei Bedeutung, wenn die Flammenfront nicht nahe genug ist, um die heißen Gase zurückzuführen. Unter diesen Umständen kann unverbrannter Kohlenstoff mit dem Brenngut in Kontakt kommen. Die Kriterien für den Brennerimpuls, die wichtig sind, um sowohl bei verdrallten als auch nichtverdrallten Flammen reduzierende Bedingungen zu vermeiden, wurden bereits im Beitrag zum VDZ-Kongresses '85 [2] mitgeteilt. **Bild 2** zeigt eine typische Auswahl empfohlener Betriebsbedingungen für einen Drallbrenner als Funktion von Brennstoffverbrauch, Klinkerausstoß, Ofendurchmesser und Primärluftanteil.

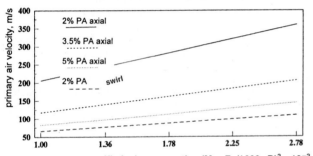

BILD 2: Primärluftgeschwindigkeit in Abhängigkeit vom spezifischen Brennstoffverbrauch
FIGURE 2: Primary air velocity/specific fuel consumption

Primärluftgeschwindigkeit	= primary air velocity
spezifischer Brennstoffverbrauch	= specific fuel consumption
axial	axial
Drall	= swirl
Klinkerdurchsatz in t/h	= Mc – tph
spez. Brennstoffverbrauch in kcal/kg	= Fc – kcals/kg
Ofendurchmesser in m	= D (m)
Förderluft	= transport air
Drallwinkel 45°	= swirl vanes 45 deg

Außer dem Brenner mit seiner erforderlichen Impulsaufteilung ist eine gleichmäßige Anströmung der heißen Sekundärluft um den Brenner herum sowie eine geeignete Korngrößenverteilung des Brennstoffes zu realisieren. Es wurde auch gefunden, daß ein zentrisch in der Ofenachse angeordneter Brenner reduzierende Bedingungen in der Brennzone vermeiden kann, da die Flamme an ihrer Unterseite mit mehr Sauerstoff versorgt wird.

Eine Daumenregel für die Festlegung der Feinheit von Kohlenstaub besagt, daß der R90-Rückstand nicht mehr als 50 % des Gehaltes an flüchtigen Bestandteilen betragen sollte. Obgleich diese Regel allgemein für gut gehalten wird, enthält sie keine Aussage über die Aufheizgeschwindigkeit eines Partikels und seine Reaktivität, nach der man beurteilen könnte, ob es sich bei dem Brennstoff z. B. um Petrolkoks oder eine bituminöse Kohle handelt. Die Freisetzung der flüchtigen Bestandteile und ihre Verbrennung ist der bestimmende Faktor bei der Kohleverbrennung im Zementdrehofen. Die Reduzierung des R90-Rückstandes unter 10 % ist ein weniger wirkungsvoller Weg zur Verbesse-

a 10° half angle, with the flame gases impinging on the burning zone at around three kiln diameters. It is essential that the gases contain no CO or carbon at least half a kiln diameter before this point, otherwise reduction will occur in the plug flow and external recirculation zones.

Flames from swirling burners have up to 10 % primary air from three channels one with swirl and two others, including the coal transport line. Air velocities may be up to sonic. They have a swirl induced internal reverse flow zone which brings back combustion gases to near the coal injection point and causes early devolatisation and combustion.

However depending on the level of swirl, the location of the channels and the secondary air temperature, reducing conditions can be produced in the burning zone due to the coal jet expanding faster than it can be combusted, due to both carbon and CO burnout problems.

Neither swirl nor bluff body induced internal reverse flow zone will be of any value if the flame front is not near enough for it to bring back hot gases. All that will occur under these circumstances is unburnt carbon impinging on the burning zone.

Burner momentum criteria to avoid reducing conditions for both swirling and non swirling flames have been specified in the previous VDZ paper [2]. **Fig. 2** shows a typical set of recommended operating conditions for a swirl burner as a function of fuel consumption, clinker output, kiln diameter and primary air percentage.

In addition to the burner having the necessary momentum, a uniform flow of high temperature secondary air has to be developed around the burner and the coal has to have the appropriate particle size distribution. It has been found that ensuring that the burner is centralised along the kiln axis has eliminated reducing conditions in the burning zone due to the provision of more oxygen to the underside of the flame.

The rule of thumb which generally exists for particle size is that the R90-residue should be not more than 50 % of the volatile matter. While this generally holds good it makes no allowance for particle heating rates and the fact that the reactivity of char is similar whether it comes from pet coke or a bituminous coal. The evolution of volatiles and their combustion is the controlling factor in coal burnout in cement kilns.

Decreasing R90-residue below 10 % is not as an effective way of improving burnout as increasing secondary air temperature and/or the rate of particle heating [4]. In some situations with high secondary air preheat pet coke R90-residues of 25 % can be used without the onset of reducing conditions.

3. NO_x emissions

Much has been written about NO_x formation and destruction. It has been often claimed that emissions from power station are all fuel NO_x and emissions from cement kilns are all thermal NO_x. Neither of these statements are correct. As far as the cement kiln is concerned there is an underlying fuel NO_x and a thermal NO_x which is a function of burning zone temperature. The principal reactions for thermal NO formation are:

$$N_2 + O \rightarrow NO + N \quad E = 75\,kcal/mole \quad (7)$$
$$N + O_2 \rightarrow NO + O \quad E = 6\,kcal/mole \quad (8)$$
$$H + O_2 \rightarrow OH + O \quad E = 16.5\,kcal/mole \quad (9)$$
$$O_2 \rightarrow 2O \quad E = 59\,kcal/mole \quad (10)$$

The formation of thermal NO is post flame front and is dictated by the temperature and oxygen concentration. Maximum formation rate of NO is with around 10 % excess air for the combustion of the volatiles. In these circumstances the oxygen atoms for reaction (7) are provided from the oxidation of hydrogen/hydrocarbons via reaction (9) and super equilibrium O and OH concentrations occur. If there is insufficient oxygen mixed into the flame to combust the volatiles near to stoichiometric then the oxygen atoms

rung des Ausbrandes als etwa die Erhöhung der Sekundärlufttemperatur und/oder der Partikelaufheizgeschwindigkeit [4]. Unter bestimmten Bedingungen, wie z.B. hohe Sekundärlufttemperaturen, kann Petrolkoks mit einem R90-Rückstand von 25% verwendet werden, ohne daß dabei reduzierende Bedingungen auftreten.

3. NO_x-Emissionen

Über die Bildung und den Zerfall von NO_x ist viel geschrieben worden. Es ist dabei häufig behauptet worden, daß es sich bei den Emissionen in Kraftwerken um Brennstoff-NO_x, bei den Emissionen von Zementdrehöfen hingegen immer nur um thermisches NO_x handelt. Keine dieser Erklärungen ist korrekt. Die Bildung von Brennstoff-NO_x und thermischem NO_x ist beim Klinkerbrennprozeß eine Funktion der Brennzonentemperatur. Die grundlegenden Reaktionen bei der Bildung von thermischem NO sind:

$$N_2 + O \rightarrow NO + N \quad E = 75\,kcal/mol \quad (7)$$
$$N + O_2 \rightarrow NO + O \quad E = 6\,kcal/mol \quad (8)$$
$$H + O_2 \rightarrow OH + O \quad E = 16,5\,kcal/mol \quad (9)$$
$$O_2 \rightarrow 2O \quad E = 59\,kcal/mol \quad (10)$$

Die Bildung von thermischem NO erfolgt hinter der Flammenfront und wird bestimmt durch die Temperatur und den Sauerstoffgehalt. Die maximale Bildungsrate von NO erfolgt bei einem Luftüberschuß von ca. 10% für die Verbrennung der flüchtigen Bestandteile. Unter diesen Bedingungen werden die Sauerstoffatome für Reaktion (7) durch die Oxidation von Wasserstoff bzw. Kohlenwasserstoffen über Reaktion (9) bereitgestellt und es kommt zu O- und OH-Konzentrationen, die weit über dem chemischen Gleichgewicht liegen (Faktor 100). Wenn die Flamme nicht genügend Sauerstoff zur nahstöchiometrischen Verbrennung der flüchtigen Bestandteile enthält, werden die Sauerstoffatome später im Ofen über die höhere Aktivierungsenergie nach Reaktionsgleichung (10) gebildet. Die Reaktionen nach Gleichung (7) und (9) verlaufen im Verbund und ergeben eine Gesamtaktivierungsenergie von 91,5 kcal/mol. Das bedeutet im wesentlichen, daß kein thermisches NO unterhalb einer Temperatur von 1600°C gebildet wird und das produzierte NO unter gleichen Mischbedingungen des Brenners in starkem Maße von der Sintertemperatur und dem Gehalt an freiem Kalk abhängt.

Bild 3 zeigt den Zusammenhang zwischen den NO-Gehalten am Ofeneinlauf und dem Freikalkgehalt als Funktion der Sintertemperatur.

Bild 4 zeigt den Zusammenhang zwischen dem NO-Gehalt am Ofeneinlauf, dem Primärluftanteil, dem Luftüberschuß sowie dem Brennerimpuls für einen typischen drallfreien Brenner. Es ist zu sehen, daß bei optimalem Brennerimpuls [2] die NO-Emission ein Maximum annimmt. Das ist deshalb so, weil die Reaktionen nach den Gleichungen (1) und (9) tatsächlich die gleichen sind und für die Bildung von NO sowie die Oxidation von CO bestimmend sind.

Der Betrieb mit einer „High Level Control" ist am wirkungsvollsten, wenn Brenner in der Nähe ihres optimalen Impulses arbeiten. Diese Betriebsweise kann aufgrund der niedrigeren Brenntemperaturen tatsächlich zu einem niedrigeren NO-Niveau führen. Die Bildung von Brennstoff-NO_x vollzieht sich bei niedrigeren Temperaturen als die von thermischem NO_x und wird bestimmt durch den verfügbaren Sauerstoff während der Verbrennung der flüchtigen Bestandteile, die den Stickstoff enthalten. Diese Vorgänge vollziehen sich nach dem folgenden Reaktionsschema:

$$HNC + O \rightarrow NH + CO \quad (11)$$
$$NH + O_2 \rightarrow NO + OH \quad (12)$$

Dabei führt die begrenzte Bildung von Sauerstoffatomen zu einer Reduzierung der Bildung sowohl des Brennstoff-NO (11) als auch des thermischen NO (7). Auch eine reduzierte Temperatur in der Brennzone wird das NO, welches auf dem thermischen Wege gebildet wird, absenken. Für Ofenbrenner, die mit einem hohen Primärluftanteil betrieben werden müssen, ist eine maximale Umwandlung von flüchtigem

will be provided later in the kiln via the higher activation energy reaction (10), less NO will therefore be formed. Reaction (7) and (9) combine to give an overall activation energy of 91.5 kcal/mole. This in effect means that no thermal NO is formed below 1600°C and the amount produced for the same burner mixing conditions is very dependent on clinker combinability temperature and free lime.

Fig. 3 shows a typical relationship between kiln back end NO levels and free lime as a function of combinability temperature.

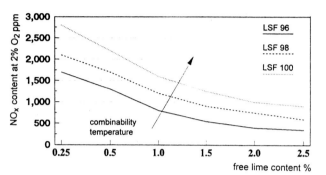

BILD 3: NO_x-Konzentration als Funktion des Freikalkgehaltes
FIGURE 3: Typical NO_x content vs free lime

NO_x-Konzentration bei 2% O_2 in ppm = NO_x content at 2% O_2 ppm
Freikalkgehalt in % = free lime content
Sintertemperatur = combinability temperature

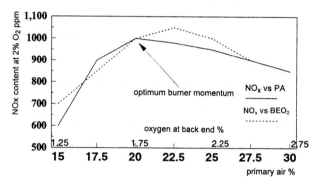

BILD 4: NO_x-Konzentration als Funktion des Primärluftanteils und des Luftüberschusses
FIGURE 4: Typical NO_x content vs primary and excess air

Sauerstoffkonzentration im Ofeneinlauf % = Oxygen at back end %
Optimaler Brennerimpuls = Optimum burner momentum

Fig. 4 shows a relationship between NO at the kiln back end, primary and excess air and burner momentum for a typical non swirling burner. It can be seen that at the optimum burner momentum [2] for the elimination of reducing conditions the NO emission peaks, this is because reactions (1) and (9) are in fact the same and are critical for the formation of NO and the combustion of CO.

The operation of High Level Control is at it's most effective when burners are operated near their optimum momentum and can in fact lead to lower NO levels, due to lower burning temperatures via better control. Fuel NO formation is formed at a lower temperature than thermal NO and is controlled by the availability of oxygen during the combustion of the volatiles which contain the nitrogen.

A simple schematic reaction scheme is:

$$HNC + O \rightarrow NH + CO \quad (11)$$
$$NH + O_2 \rightarrow NO + OH \quad (12)$$

Clearly limiting the formation of oxygen atoms will reduce the formation of NO from both the fuel (11) and thermal NO (7) routes. Reducing burning zone temperature will also reduce the NO formed via the thermal route. For burners with a high primary air, a maximum conversion of the volatile fuel nitrogen into NO is possible as there could be enough oxygen available to combust the volatiles near to stoichiometric without a significant amount of mixing.

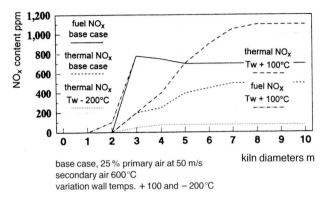
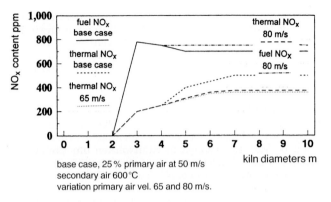

BILD 5: Bildung von Brennstoff- und thermischem NO$_x$ in Abhängigkeit der Wandtemperatur
FIGURE 5: Fuel and thermal NO$_x$ content vs wall temperature

NO$_x$-Konzentration in ppm	= NO$_x$ content ppm
Ofendurchmesser	= kiln diameters
Brennstoff-NO$_x$	= fuel NO$_x$
thermisches NO$_x$	= thermal NO$_x$
Basisfall, 25% Primärluft bei 50 m/s	= base case, 25% primary air at 50 m/s
Sekundärluft 600°C	= secondary air 600°C
Veränderung der Wandtemperatur +100 und 200°C	= Variation wall temps. +100 and 200°C

BILD 6: Bildung von Brennstoff- und thermischem NO$_x$ in Abhängigkeit der Primärluftgeschwindigkeit
FIGURE 6: Fuel and thermal NO$_x$ content vs primary air velocity

| Primärluftgeschwindigkeit 65 und 80 m/s | = Variation primary air vel. 65 and 80 m/s |

Brennstoffstickstoff in NO möglich, da in diesem Falle genügend Sauerstoff zur Verbrennung der flüchtigen Bestandteile in der Nähe stöchiometrischer Verhältnisse verfügbar ist, ohne daß dabei eine wesentliche Vermischung stattfindet.

Die **Bilder 5** und **6** enthalten Voraussagen [4] zur NO-Bildung in Abhängigkeit von Wandtemperatur und Brenneraustrittsgeschwindigkeit für einen Drehofen nach dem Halbnaß-Verfahren, der mit 100% Petrolkoks und einem Primärluftanteil von 25% betrieben wird. Die Abhängigkeit des thermischen NO von der Wandtemperatur und der Geschwindigkeit der Verbrennungsluft bzw. die Unabhängigkeit von Brennstoff-NO von diesen beiden Parametern kann deutlich aus den beiden Diagrammen entnommen werden. Die Unabhängigkeit des Brennstoff-NO von der Austrittsgeschwindigkeit der Verbrennungsluft aus dem Brenner ist möglicherweise auf den Betrieb des Ofens mit einem Primärluftanteil von 25% zurückzuführen, der damit höher liegt als der Anteil der Flüchtigen.

Es ist unverkennbar, daß eine NO-Reduzierung durch Brennermodifikationen besonders bei Öfen mit großen Durchmessern erreicht werden kann. Die Entgasung in einem kleinen Primärluftstrahl scheint heute der Schlüssel zu sein, um sowohl die Bildung von Brennstoff-NO als auch von thermischem NO zu reduzieren. Die Vermeidung von reduzierenden Bedingungen im Prozeß führt über eine optimale Brennerkonstruktion und einen optimalen Brennerbetrieb.

4. SO$_2$-Emissionen und SO$_3$-Kreisläufe

Die Vermeidung von reduzierenden Bedingungen ist eine Voraussetzung für den Betrieb eines jeden nach dem Naß- oder Halbnaßverfahren arbeitenden Ofens, um die Verordnungen zur SO$_2$-Emission einzuhalten. Während beim Trokkenverfahren reduzierende Bedingungen die SO$_2$-Emissionen nicht notwendigerweise erhöhen, treten beträchtliche betriebliche Probleme durch sulfatische Ablagerungen im Bereich des Ofeneinlaufs, im Steigschacht und in den Zyklonen auf.

Bild 7 zeigt den Einfluß reduzierender Bedingungen auf die SO$_2$-Emission einer nach dem Halbnaß-Verfahren arbeitenden Ofenanlage, ausgelöst durch die Zurücknahme des Sauerstoffgehaltes im Ofeneinlauf. Unter normalen Betriebsbedingungen lag bei einer Sauerstoffkonzentration von 2% der SO$_2$-Gehalt um 50 ppm. Bei einer Senkung des Sauerstoffniveaus auf eine Größenordnung von 1,0 bis 1,5% wurden SO$_2$-Emissionen bis zu 800 ppm gemessen. Der in dieser Ofenanlage erzeugte Klinker besaß einen Alkaligehalt von 0,6% Na$_2$O-Äquivalent.

Bild 8 zeigt den Einfluß der Sinterzonentemperatur (NO$_x$-Konzentration) und der Ofeneinlaufkonzentrationen von O$_2$

Predictions [4] of the sensitivity of both fuel and thermal NO to wall temperature and burner injection velocity for a semi-wet kiln operating at 100% pet coke with a 25% primary air are shown in **Figs. 5** and **6**. The dependence of thermal NO on wall temperature and firing velocity and the independence of fuel NO on both parameters can be seen. The independence of fuel NO on firing velocity is probably due to the 25% primary air being higher than the percentage volatiles.

NO reduction by burner modifications can obviously be achieved, particularily in large diameter kilns. Devolatilisation into a minimum of primary air appears to be the key to reduce the formation of both fuel and thermal NO. The skill being to avoid reducing conditions in the process via an optimum burner design and operation.

4. SO$_2$ emission/SO$_3$ cycles

Avoiding reducing conditions is a must for any wet/semi wet kiln to meet SO$_2$ emission regulations. While for a dry process reducing conditions do not necessarily increase SO$_2$ emissions, considerable process problems occur due to sulphate deposits at the kiln back end, in the gas riser and in the cyclones.

Fig. 7 shows the impact of reducing conditions created via lowering the back end oxygen, on the SO$_2$ emission from a semi wet process. Normal operation was at 2% oxygen with around 50 ppm of SO$_2$, decreasing the oxygen level to the range 1.0–1.5%, produced SO$_2$ emissions up to 800 ppm. This effect was for a clinker with alkalis of 0.6% Na$_2$O equivalent with the SO$_3$ at the molal equivalent.

Fig. 8 shows the impact on stage IV SO$_3$ of burning zone temperature (NO$_x$) and kiln back end oxygen and CO, for a four stage dry process kiln with a ratio of SO$_3$ to Na$_2$O equivalent of 2. Generally it was found that a stage IV SO$_3$ of less than 3.5% enabled stable operation to be maintained. The impact of higher NO$_x$ levels on stage IV SO$_3$ was significantly less, than the reducing conditions which were produced at 1.4% kiln back end oxygen and as evidenced by an increase in CO from 500 to 2500 ppm.

CaSO$_4$ is stable thermodynamically up to around 1450°C in the presence of oxygen. Even above this temperature it's dissociation is kinetically controlled. If 2000 ppm [5] of CO is in contact with the CaSO$_4$ at temperatures above 1000°C, it will begin to be reduced to CaO and SO$_3$, resulting in severely increased SO$_3$ concentrations in stage IV. This effect can be clearly seen in Fig. 8. CO has a similar effect on the dissociation of alkali sulphates, but at higher temperatures.

Similar observations have been made by Flament [6] using between 50 and 100% pet coke, with clinker SO$_3$ to alkali ratios of up to 3. CO levels of 2000 ppm can be the result of

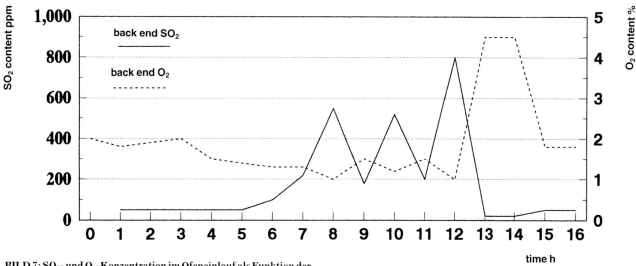

BILD 7: SO$_2$- und O$_2$-Konzentration im Ofeneinlauf als Funktion der Zeit bei einem Ofen nach dem Halbnaßverfahren
FIGURE 7: Typical SO$_2$ vs back end O$_2$ for semi wet kiln

SO$_2$-Konzentration in ppm = SO$_2$ content ppm
O$_2$-Konzentration in % = O$_2$ content %
Zeit in h = time h
Ofeneinlauf = back end

und CO auf den SO$_3$-Gehalt in Stufe IV für eine Ofenanlage mit 4stufigem Zyklonvorwärmer. In der Ofenanlage wurde ein Klinker mit einem Sulfatisierungsgrad von 2,0 produziert. Generell wurde gefunden, daß bei einem gemessenen SO$_3$-Gehalt kleiner 3,5 % an Zyklonstufe IV ein stabiler Ofenbetrieb gewährleistet werden kann. Der Einfluß des NO$_x$-Niveaus auf den SO$_3$-Gehalt in Stufe IV ist deutlich geringer als der reduzierender Bedingungen, die bei einer Sauerstoff-Konzentration von 1,4 % am Ofeneinlauf durch einen Anstieg des CO-Gehalts von 500 auf 2500 ppm gekennzeichnet war.

In Anwesenheit von Sauerstoff ist CaSO$_4$ bis zu Temperaturen um 1450 °C thermodynamisch stabil. Selbst oberhalb dieser Temperatur ist die Dissoziation von CaSO$_4$ noch kinetisch kontrolliert. Wenn CO mit Konzentrationen von 2000 ppm [5] bei Temperaturen oberhalb von 1000 °C mit dem CaSO$_4$ in Kontakt tritt, beginnt dieses in CaO und SO$_3$ zu zerfallen, begleitet von stark ansteigenden SO$_3$-Konzentrationen in Zyklonstufe IV. Dieser Effekt kann deutlich in Bild 8 verfolgt werden. CO übt eine ähnliche Dissoziationswirkung auf Alkalisulfate aus, allerdings erst bei höheren Temperaturen.

Ähnliche Beobachtungen wurden auch von Flament [6] gemacht, der Petrolkoks mit einem Anteil zwischen 50 und 100 % bei Alkali-/Schwefel-Verhältnissen bis zu 3,0 einsetzte. CO-Konzentrationen von 2000 ppm können das Ergebnis einer schlechten Brennerkonstruktion, einer unzulänglichen Betriebsweise oder auch eines zu niedrigen O$_2$-Gehaltes im Ofeneinlauf sein. Petrolkoks ist gegenwärtig zu Preisen erhältlich, die bei 50 % internationaler Kohlepreise liegen. Es gibt deshalb auch einen ganz klaren kommerziellen Beweggrund, den Einfluß der Brennerkonstruktion auf den CO-Level am Flammenende und in der Zone der Rohrströmung zu ermitteln, damit insbesonders die nach dem Trockenverfahren arbeitenden Ofenanlagen mit 100 % Petrolkoks betrieben werden können.

5. CEMFLAME-Konsortium

Vorliegender Beitrag, der Probleme und neue Erkenntnisse auf der Grundlage betrieblicher Erfahrungen mit Ofenbrennern zur Reduzierung der NO$_x$-Emissionen aufzeichnet, soll zugleich eine Lücke schließen helfen bei der Beantwortung der Frage nach der Gewährleistung der Zementqualität, der Verwendung minderwertiger Kohle und hoch schwefelhaltigen Petrolkokses. Diese Problemstellungen waren auch der Hintergrund für die Gründung des CEMFLAME-Konsortiums. Die ersten Versuchsreihen wurden bereits im Jahre 1992 durchgeführt, die nächste Versuchsreihe wird im

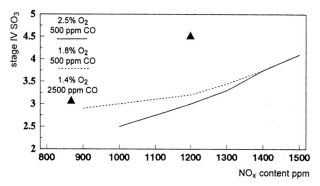

BILD 8: SO$_3$-Gehalt in Zyklonstufe IV als Funktion der NO$_x$- und O$_2$-Konzentration im Ofeneinlauf
FIGURE 8: Stage IV SO$_3$ vs back end NO$_x$ and O$_2$

SO$_3$-Gehalt in Zyklonstufe IV in % = stage IV SO$_3$ %
NO$_x$-Konzentration in ppm = NO$_x$ content ppm

poor burner design and operation as well as low back end oxygen. Pet coke is currently available at around 50 % the cost of international coal. There is therefore a strong commercial reason to identify the impact of burner design on CO level at the edge of the flame and in the plug flow zone, to enable dry processes in particular, to operate at 100 % pet coke.

5. CEMFLAME consortium

The paper so far has outlined the problems and knowledge gaps associated with operating burners to reduce NO$_x$, maintaining cement quality and using lower grade coal and high sulphur pet coke. It was this situation that was the reason for formation of the CEMFLAME Consortium.

The first series of trials were carried out last year, another campaign is due in 1994, with some possible funds from the ECSC.

Members of the Consortium for the first campaign were, Blue Circle Cement, Castle Cement, ENCI, CBR, Ciments Français, FCT, Fincem, Heracles, Holderbank, IFRF, Italcementi, KHD, Lafarge Coppée, Pillard, Unicem, VDZ.

Fig. 9 shows the experimental kiln used for the investigation. It has an internal diameter of 0.78 m and a length to diameter ratio of 15:1. The fuel thermal input was designed to be 250 kW, with secondary air temperatures of 600–1000 °C available. Measurements could be made of all flame gases and temperatures at the exit of the kiln and at 15 different axial locations. The experimental programme was developed by the Consortium.

The following objectives of the experimental campaign were developed:

— to improve the understanding of cement kiln flames;

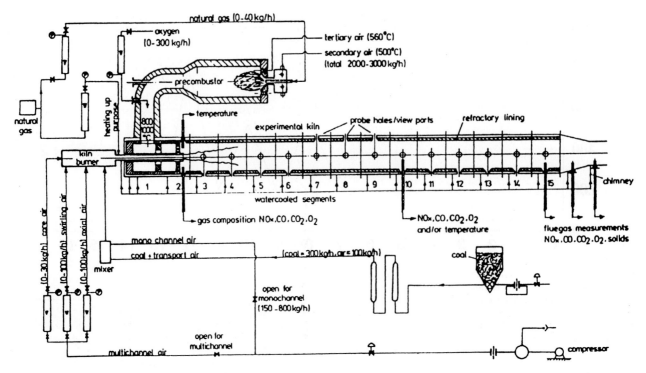

BILD 9: CEMFLAME – Versuchsofen im IFRF
FIGURE 9: CEMFLAME kiln at IFRF

Erdgas	= natural gas
Sauerstoff	= oxygen
Axialluft	= swirling air
zum Zwecke der Vorwärmung	= heating up purpose
Tertiärluft	= tertiary air
Sekundärluft	= secondary air
Vorbrennkammer	= precombustor
Temperatur	= temperature
Ofenbrenner	= kiln burner
Versuchsofen	= experimental kiln
Sondenöffnungen	= probe holes
Feuerfestausmauerung	= refractory lining
Schornstein	= chimney
Kompressor	= compressor
Kohle	= coal
wassergekühlte Segmente	= water cooled segments
Abgasmessungen	= flue gas measurements
Feststoffe	= solids
Gaszusammensetzung	= gas composition
Luft für Einkanalbrenner	= mono channel air
Kohleförderluft	= coal and transport air
offen bei Einkanalbrenner	= open for monochannel
offen bei Mehrkanalbrenner	= open for multichannel
Mischer	= mixer

Jahre 1994, möglicherweise mit Forschungsmitteln der EU, stattfinden.

Die Mitglieder des Konsortiums während der ersten Versuchsetappe waren Unternehmen wie: Blue Circle Cement, Castle Cement, ENCI, CBR, Ciments Français, FCT, Fincem, Heracles, Holderbank, IFRF, Italcementi, KHD, Lafarge Coppée, Pillard, Unicem und der VDZ.

Bild 9 zeigt den Versuchsofen, der für die Untersuchungen benutzt wurde. Die Brennkammer dieses Ofens hat einen lichten Durchmesser von 0,78 m und ein Längen/Durchmesser-Verhältnis von 15:1. Der Brennstoffeinsatz war auf einen Leistungswert von 250 kW festgelegt worden bei verfügbaren Sekundärlufttemperaturen von 600 bis 1000 °C. Die Flammentemperaturen konnten am Ausgang sowie auch an 15 verschiedenen Stellen entlang der Achse des Versuchsofens gemessen werden. Das Versuchsprogramm war von den Mitgliedern des Konsortiums aufgestellt worden. Für die Versuche waren die folgenden Zielstellungen vorgegeben:

– Verbesserung des Erkenntnisstandes über Drehofenflammen,

– Untersuchung des Einflusses von Brennerdesign, Brennerbetrieb und Brennstofftyp auf die Energieausnutzung und die Bildung von NO_x,

– Schaffung von ingenieurmäßigen Grundlagen über die Auswirkung von Brennstoffänderungen, über die Brennstoffausnutzung, die Wärmeübertragung und die NO_x-Bildung,

– Vergleich des Betriebsverhaltens von verkleinerten Versionen industrieller Brenner.

Die durchgeführten Experimente über zwei Perioden dauerten 4 Wochen. Bei den Experimenten wurden hoch- und mittelflüchtige Kohlen sowie Petrolkoks eingesetzt. Der Einfluß von Brennstoffart, Mahlfeinheit, Brennerkonstruk-

– to determine the influence of burner design, operation and fuel type on thermal efficiency and NO_x formation;

– to create engineering guidelines for the effect of fuel variation, fuel efficiency, heat transfer and NO_x characteristics;

– to compare the performance of scaled down versions of industrial burners.

The experiments were carried out over 2 periods totalling 4 weeks using high and medium volatile coals and pet coke. The effect of fuel type, the R90-residue, burner type and settings and secondary air temperature on NO_x emission, burnout and heat transfer were investigated. In all over 150 flames were investigated, the majority input/output but with around 25 detailed involving flame probing. Some of the main conclusions are as follows: the major parameter effecting NO_x emissions was the amount of oxygen in the jet at the point of ignition. Therefore the initial mixing rate between the fuel and the combustion air is extremely important. The higher the jet momentum the higher the mixing rate and thus the higher the NO_x. The longer the ignition distance, the higher is the potential to entrain oxygen and therefore to produce high NO_x. **Fig. 10** demonstrates this influence, with a calculation of the amount of air mixed with the fuel at ignition, for high and medium volatile coals for multichannel burners. The scatter at any particular mixing ratio is due to the local micro mixing between the air and coal channels. Fuel type can have a major effect on NO_x emission.

The experiments indicated that the higher the coals volatile content, the lower the NO_x. This relates to the impact of fuel type on ignition distance, and of finer grinding of the coal on peak flame temperature. In general it was concluded that R90-residues of less than 10 % were to be avoided.

Heat transfer distribution in the burning zone was affected by fuel type, burner type, tangential and axial momentum.

tion, Brennereinstellung und Sekundärlufttemperatur auf die NO_x-Emissionen, das Ausbrandverhalten sowie den Wärmeübergang wurden untersucht. Mehr als 150 verschiedene Flammen wurden getestet, in der Mehrzahl durch input/output-Messungen, aber auch durch ca. 25 detaillierte Flammenprofilmessungen. Eine der wichtigsten Schlußfolgerungen ist, daß der Sauerstoffgehalt in der Zündfront der Flamme die entscheidende Einflußgröße auf die NO_x-Entstehung ist. Deshalb ist die anfängliche Vermischung zwischen Brennstoff und Verbrennungsluft von großer Wichtigkeit. Je höher der Strahlimpuls ist, desto stärker ist die Vermischung und desto höher die NO_x-Konzentration. Je größer die Zünddistanz, desto größer die Wahrscheinlichkeit, Sauerstoff einzumischen und NO_x zu erzeugen. **Bild 10** zeigt diesen Einfluß für hoch- und mittelflüchtige Kohlen an Mehrkanalbrennern. Die Streubreite um jedes einzelne Mischungsverhältnis ist auf lokale Vermischungen zwischen der aus den einzelnen Kanälen austretenden Luft und Kohle zurückzuführen.

Die Brennstoffart kann einen größeren Einfluß auf die NO_x-Emission haben. Die Experimente haben gezeigt, daß die NO_x-Emission niedriger ausfällt, je höher die flüchtigen Bestandteile einer Kohle sind. Dieses Ergebnis steht in Zusammenhang mit dem Einfluß der Brennstoffart auf den Zündweg und dem Einfluß feiner gemahlener Kohle auf die Spitzentemperatur der Flamme. Allgemein wurde festgestellt, daß ein R90-Rückstand von kleiner 10 % zu vermeiden ist.

Die Wärmestromverteilung wird durch die Brennstoffart, die Brennerkonstruktion sowie den tangentialen und axialen Impuls beeinflußt. Die Art des Brennstoffs erwies sich als entscheidend für die Länge des Zündweges. Der beste Wärmeübergang wurde bei der Verfeuerung von hoch- und mittelflüchtigen Kohlen erreicht.

Für die Brennerkonstruktion ist das Vorhandensein und die Lage des Drallkanals sowie die Drallzahl von Bedeutung. Im allgemeinen führt eine Vergrößerung des tangentialen Impulses zu einem schnelleren und verstärkten Wärmeübergang im Drehofen. Allerdings kann das mit einer wesentlich breiteren Flamme verbunden sein, was sowohl zu reduzierend gebranntem Klinker als auch zu kürzeren Standzeiten der Feuerfestauskleidung führen kann. Unter den Bedingungen eines Einkanalbrenners konnte oberhalb eines Brennerimpulses von 7N/MW in der Brennzone keine Verbesserung des Wärmeübergangs mehr festgestellt werden. Ein Mehrkanalbrenner arbeitet vergleichsweise bei 3N/MW. Diese beiden Zahlenangaben befinden sich in Übereinstimmung mit den praktischen Erfahrungen der Konsortiums-

The effect of fuel type was dominated by the ignition distance, the highest rate of heat transfer coming from high and medium volatile coal.

The effect of burner type was dependent on the existence and location of the swirl channel, and the swirl level employed. Generally increasing the tangential momentum produced earlier and higher rates of heat transfer. However this could be associated with a rapid spreading of the flame, which could cause reduction of clinker or shorten refractory life. Generally for a burner momentum above 7 N/MW there was no increase in heat transfer in the burning zone for monotube burners. 3 N/MW is the comparable momentum for a multichannel burner. Both these values are in general agreement with the industrial experience of the Consortium. The information has allowed the design criteria for low NO_x burners for both swirling and non swirling burners to be specified.

The main criteria are:
— a low specific momentum in the range of 3 to 7 N/MW;
— a R90-residue as high as possible, consistent with avoiding reducing conditions;
— flame ignition as close to the burner as possible;
— the minimum of air mixed with the coal volatiles prior to ignition, achieved by low primary air inputs and velocities, and the optimum use of an internal recirculation zone generated by bluff body and/or swirl.
— the coal channel positioned at the inside of the swirling air channel to create a fuel rich internal reverse flow zone;
— a quarl which will allow flame stabilisation and limit the radial spread of swirling flames.

Several Consortium members have already applied some of the information gaining benefits at least equivalent to the total programme cost. This success has encouraged the Consortium to seek to fund another campaign in 1994. The main aim will be 100 % pet coke without reducing conditions.

6. Conclusions

— Optimum burner design and operation ensures that there are sufficient OH radicals to combust CO in order to avoid reducing conditions in the burning zone, which minimises SO_2 emission and SO_3 cycles.
— The principle reaction for the formation of OH radicals also produces O atoms, which is the controlling mechanism for both thermal and fuel NO_x formation.

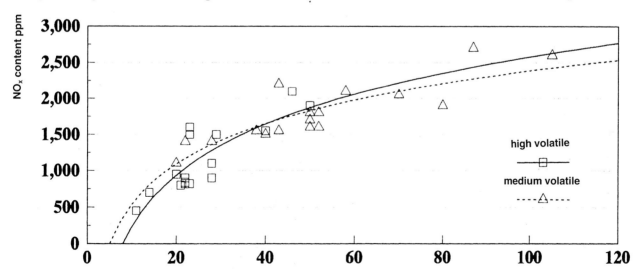

BILD 10: NO_x-Konzentration als Funktion der vor der Zündung eingemischten Verbrennungsluft
FIGURE 10: NO_x content vs combustion air mixed prior to ignition

Anteil der vor Zündung eingemischten Verbrennungsluft in % = total air mixed with coal prior to ignition %
hochflüchtig = high volatile
mittelflüchtig = medium volatile

mitglieder. Die Ergebnisinformationen haben es ermöglicht, Auslegungskriterien für Low-NO$_x$-Brenner mit und ohne Drall zu formulieren.

Die wichtigsten Kriterien sind:

— ein niedriger spezifischer Impuls im Bereich von 3 bis 7 N/MW,

— ein R90-Rückstand so hoch als möglich, vereinbar mit dem Ziel, reduzierende Bedingungen zu vermeiden,

— die Zündung der Flamme so dicht als möglich am Brennermund,

— die minimale Vermischung von Luft und den flüchtigen Bestandteilen der Kohle vor der Zündung, was erreicht wird durch einen niedrigen Primärluftanteil und durch niedrige Austrittsgeschwindigkeiten sowie die Einstellung einer inneren Rezirkulationszone, erzeugt durch Einbauten und/oder Drall,

— die Positionierung des Kohlekanals innerhalb des Drallkanals, um eine brennstoffreiche innere Rückströmzone zu erzeugen,

— eine Brennerkonstruktion, welche die Flamme stabilisiert und die radiale Ausbreitung der verdrallten Flamme begrenzt.

Mehrere Mitglieder des Konsortiums haben bereits durch die Verwertung der Ergebnisse Einsparungen erzielt, die mindestens in der Größenordnung der Versuchskosten liegen. Dieser Erfolg hat das Konsortium ermutigt, nach Finanzierungsmöglichkeiten für eine weitere Versuchsetappe im Jahre 1994 zu suchen. Die Hauptzielstellung für die nächste Versuchsetappe wird sein, bis zu 100% Petrolkoks einzusetzen.

6. Schlußfolgerungen

— Der optimale Entwurf und Betrieb eines Brenners erfordern, daß genügend OH-Radikale zur Verbrennung von CO vorhanden sind, um reduzierende Bedingungen in der Brennzone zu vermeiden, wodurch schließlich auch SO$_2$-Emissionen und SO$_3$-Kreisläufe minimiert werden.

— Die grundlegende Reaktion zur Bildung der OH-Radikalen führt auch zur Produktion von Sauerstoffatomen. Dies ist der bestimmende Mechanismus für die Bildung von thermischem und Brennstoff-NO$_x$.

— CEMFLAME hat gezeigt, daß die Reduzierung von NO$_x$ möglich ist, ohne daß die CO-Emissionen und SO$_3$-Freisetzung ansteigen, wenn sichergestellt wird, daß die flüchtigen Bestandteile der Kohle mit Hilfe des Brenners in einer sauerstoffarmen Atmosphäre freigesetzt werden und dadurch auch die frühe Bildung von Sauerstoffatomen minimiert wird.

— Low-NO$_x$-Brenner, die zugleich in der Brennzone reduzierende Bedingungen vermeiden sollen, werden deshalb den Kohlestrahl ohne oder nur mit wenig Primärluft in den Feuerungsraum einbringen müssen.

— Die vorstehend beschriebenen Bedingungen werden am besten durch einen Brenner erreicht, der mit einem Primärluftanteil von 5 bis 7% arbeitet, dessen Kohlestrahl vom Drall und von den anderen axialen Kanälen umschlossen wird, die Bildung einer drallinduzierten inneren Rückströmzone begünstigt, ohne dabei eine zu starke Expansion des Brennstoffstrahls zu verursachen.

— CEMFLAME hat sich darauf eingestellt, einen langen Weg zu beschreiten, um die Auslegungsgrundlagen von Brennern zu optimieren, wobei die Fortschritte bei der mathematischen Modellierung seit Mitte der 80er Jahre einen Stand erreicht haben, der es gestattet, unter Anwendung der heutzutage verfügbaren technischen Möglichkeiten [4] die Brennerauslegung für aktuelle Ofenprojekte zu verfeinern.

— CEMFLAME has shown that it may be possible to reduce NO$_x$ without increasing CO emission and therefore SO$_3$ volatilisation, by ensuring that the coal's volatiles are released at the burner into a oxygen lean atmosphere and hence minimising the early formation of O atoms.

— Burners to avoid reducing conditions in the burning zone and have at the same time a low NO$_x$ will therfore have the coal jet with either zero or low primary air, being heated to devolatilisation via it's partial penetration of a swirl induced internal reverse flow zone.

— These conditions will be most readily achieved via a burner with 5 to 7% primary air with a coal jet inside the swirl and axial channels, surrounded by a tulip quarl which enhances the formation of a swirl induced internal reverse flow zone without causing the coal jet to expand too rapidly.

— CEMFLAME will go a long way to optimising the design principles, however the development of capabilities of mathematical modelling since the last VDZ Congress, means that burner designs for actual kilns could be refined using the techniques that are now available [4].

Literaturverzeichnis

[1] Ruhland, W.: Investigation of flames in a rotary cement kiln. IFRF Doc No H 10/a/1.

[2] Lowes, T.M., and Lorimer, A.D.J.: Cement Kiln Coal Flames. Zement-Kalk-Gips 39 (1986) Nr. 2, pp. 69–71.

[3] Thring, M.W., and Newby, M.P.: Fourth International Symposium on Combustion, pp. 789–796.

[4] Lockwood, F.C., and Shen, B.: Numerical study of the design and operation of coal flames in a Blue Circle cement kiln. Preliminary Report ICSTM July 1993.

[5] Haspel, D.W., and Taylor, R.A.: High Level Kiln Control. Zement-Kalk-Gips 39 (1986) Nr. 4, pp. 183–185.

[6] Hayhurst, A.N., and Tucker, R.F.: The reductive regeneration of sulphated limestone for flue gas desulphurisation: Thermodynamic considerations of converting calcium sulphate to calcium oxide. Jnl Inst Energy, December 1991, 64, pp. 212–229.

[7] Flament, G.: Burning 100% pet coke with high sulphur — low alkali raw meal in a 5200 t/day precalciner kiln. European Seminar on "Improved Technologies for the Rational Use of Energy". Berlin October 1992.

Umbau vom Naßverfahren zum Halbnaßverfahren im Madukkarai-Zementwerk von ACC

Wet to semi-wet process conversion at ACC's Madukkarai Cement Works

Modification du procédé humide en procédé semi-humide à la cimenterie de Madukkarai de ACC

Transformación de la vía húmeda en vía semihúmeda en la fábrica de cemento de Madukkarai de ACC

Von **I. C. Ahuja, M. N. Rao** und **R. Vasudevan**, Bombay/Indien

Zusammenfassung – Die Geschäftsleitung von ACC suchte nach Möglichkeiten zur Verbesserung der Wirtschaftlichkeit des Werks Madukkarai, das 5 Öfen betrieb, die nach dem Naßverfahren auf der Basis vorbehandelter Kalksteinaufschlämmung arbeiteten. Nach ausgedehnten Schlamm-Filtrationsversuchen im Pilotmaßstab wurde beschlossen, auf das Halbnaßverfahren überzugehen und die Naßmühlen, die Flotationsanlage usw. beizubehalten. Das Konzept bestand darin, den nach dem Flotationsverfahren aufbereiteten Schlamm in 4 Vakuum-Drehfiltern zu filtern und den Kuchen in einem Brecher mit Abgasen aus einem 2-Stufen-Vorwärmer zu trocknen und zu desagglomerieren. Um den thermischen Energiegehalt des Vorwärmer-Abgases je nach Energiebedarf für die Kuchentrocknung variieren zu können, konnten 1 oder 2 wirksame Zyklonstufen eingeschaltet werden. Es wurde eine Vorcalcinieranlage, die für 72 % des gesamten Brennstoffverbrauchs ausgelegt war, eingebaut, um die Größe des neuen Ofens zu verringern. Ein Rostkühler mit Duotherm-Schaltung sollte die Wärmebilanz durch Nutzung der Abwärme auch in der Kohlemahlanlage verbessern. Das neue System kostete 470 Millionen Rs und wurde 1989 in Betrieb genommen. Die erzielten Egebnisse sind besser als die errechneten Produktionszahlen von 1500 t/d und 950 kcal/kg Klinker. Zur Zeit wird ein Plan zur weiteren Kapazitätserhöhung auf 2100 t/d und 900 kcal/kg Klinker in die Tat umgesetzt.

Umbau vom Naßverfahren zum Halbnaßverfahren im Madukkarai-Zementwerk von ACC

Summary – ACC's management explored alternatives for improving the economic viability of Madukkarai works which was operating 5 wet kilns based on pretreated limestone slurry. After extensive pilot scale slurry filtration trials, it was decided to change over to a semi-wet process while retaining the wet raw mills, the flotation plant, etc. The scheme consisted of filtering flotated slurry in 4 rotary vacuum filters and of drying and disagglomerating the cake in a drying crusher with waste gases from a two stage preheater. The number of effective cyclone stages could be varied between 1 and 2 for balancing the thermal energy content of the preheater exhaust gas with heat demand for cake drying. A precalciner designed for 72 % of the total fuel consumption was incorporated with a view to reducing the new kiln size. A grate cooler with Duotherm circuit was provided to improve the heat economy of the kiln by waste heat utilization also in the coal mill systems. The project was commissioned in 1989 at a cost of Rs. 470 million. The results achieved are better than the rated production figures of 1500 t/d and 950 kcal/kg of clinker. A scheme to further enhance the capacity to 2100 t/d and 900 kcal/kg of clinker is under implementation.

Wet to semi-wet process conversion at ACC's Madukkarai Cement Works

Résumé – La Direction de ACC a recherché des possibilités d'amélioration de la rentabilité de l'usine de Madukkarai comportant 5 fours, qui travaillent selon le procédé humide sur la base de la mise en suspension de calcaire prétraité. Après des tests de filtration de boue poussés à l'échelle expérimentale il a été décidé d'adopter le procédé semi-humide tout en conservant les broyeurs par voie humide, l'installation de flottation etc. Le concept consistait à filtrer la boue traitée selon le procédé de flottation dans 4 filtres rotatifs sous vide et, ensuite, à sécher et désagglomérer le gâteau dans un broyeur aux gaz effluents provenant du préchauffeur à deux étages. Afin de pouvoir varier le contenu énergétique thermique des fumées du préchauffeur selon les besoins énergétiques nécessaires au séchage du gâteau, une ou deux phases de cyclone performantes ont été intercalées. Une installation de précalcination, prévue pour 72 % de la consommation globale de combustible, a été installée pour diminuer la taille du nouveau four. Un refroidisseur à grille avec dispositif de commande Duotherm était destiné à améliorer le bilan thermique par récupération de la chaleur perdue utilisée également dans l'installation de broyage de charbon. Ce nouveau système qui a coûté 470 millions de Rs, a été mis en service en 1989. Les résultats obtenus sont meilleurs que les chiffres de production calculés de 1500 t/j et 950 kcal/kg de clinker. A l'heure actuelle un plan d'augmentation de capacité portée à 2100 t/j et 900 kcal/kg de clinker est en voie de réalisation.

Modification du procédé humide en procédé semi-humide à la cimenterie de Madukkarai de ACC

Resumen – La Dirección de ACC buscaba posibilidades de mejorar la rentabilidad de la fábrica de Madukkarai, en la que funcionaban 5 hornos de vía húmeda, sobre la base de la suspensión de piedra caliza previamente tratada. Después de extensos ensayos de filtración de lodos, a escala experimental, se decidió adoptar el procedimiento semihúmedo, conservando los molinos de vía húmeda, la instalación de flotación, etc. El concepto consistía en filtrar el lodo, preparado según el procedimiento de flotación, mediante 4 filtros rotativos al vacío, y luego secar y desaglomerar la torta en una trituradora, aprovechando los gases de escape de un precalentador de dos etapas. Con el fin de poder variar el contenido de energía térmica de los gases de escape del precalentador para el secado de la torta, según las necesidades de energía, se podían conectar una o dos etapas de ciclones eficaces. Se incorporó una instalación de precalcinación, dimensionado para un 72 % del consumo total de combustible, con vistas a una reducción del tamaño del nuevo horno. Un enfriador de parrilla, con circuito Duotherm, tenía que mejorar el balance térmico, aprovechando el calor residual también en la planta de molienda de carbón. El nuevo sistema costó 470 millones de Rs y fue puesto en servicio en 1989. Los resultados obtenidos son mejores que las cifras de producción calculadas, de 1500 t/d y 950 kcal/kg de clínker. Actualmente, se está realizando un proyecto de ampliación de la capacidad a 2100 t/d y 900 kcal/kg de clínker.

Transformación de la vía húmeda en vía semihúmeda en la fábrica de cemento de Madukkarai de ACC

Die Madukkarai-Zementwerke von ACC (Associated Cement Companies) arbeiteten bisher im Naßverfahren. Sie verfügen über 5 Brennöfen mit einem Klinkerausstoß von insgesamt 1180 t pro Tag bei einem Brennstoffverbrauch von 1450 kcal/kg Klinker. Der Kalkstein weist einen relativ niedrigen $CaCO_3$-Gehalt von 76 bis 77 % auf. Etwa 60 % des Kalksteins müssen naß aufbereitet werden. Die steigenden Preise für Kohle und deren sinkende Qualität beeinträchtigten die Rentabilität und den Betrieb des Werks erheblich. ACC erarbeitete mehrere Varianten zur Kapazitätserhöhung und Modernisierung der Anlage, um die Wirtschaftlichkeit des Werks zu verbessern.

Eine vollständige Umstellung auf das Trockenverfahren wurde ausgeschlossen, da auf die Naßaufbereitung des Kalksteins nicht verzichtet werden kann. Filterversuche im Labormaßstab ergaben, daß der Schlamm vakuumfiltrierbar ist. Weitere Versuche mit einem Filter mit Abzug durch Kratzer im Pilotmaßstab zeigten ein schnelles Verstopfen des Gewebes, wodurch sich der erzeugte Filterkuchen innerhalb von 100 Betriebsstunden auf weniger als die Hälfte reduzierte. Nachfolgende Versuche mit einem Filter mit Gurtabzug im Pilotmaßstab zeigten bei kontinuierlicher Wasserwäsche kein Verstopfen des Gewebes. 190 Betriebsstunden bestätigten, daß man aus Schlamm mit 35 bis 36 % Feuchtigkeit bei einem Durchsatz von 485 bis 540 kg/h pro m^2 Filterfläche und bei 665 bis 730 mbar Unterdruck einen Filterkuchen mit 15 bis 17 % Feuchtigkeit gewinnen kann. Bei der Konstruktion des Heißaufbereitungssystems wurde eine Kuchenfeuchtigkeit von bis zu 18 % zugrunde gelegt.

Ermutigt durch die Ergebnisse bei der Filtrierung wurden verschiedene Möglichkeiten der Verfahrensumstellung in Betracht gezogen. Schließlich wurde ein Halbnaß-Verfahren mit einer Tagesleistung von 1500 t und mit Filterkuchen als Eingangsstoff gewählt. Die vorhandenen Anlagen zur Rohmaterialaufbereitung und die Zementmühlen wurden mit kleineren Verbesserungen beibehalten, so daß sie dem erhöhten Durchsatz gerecht werden konnten. Eine neue Ofenlinie wurde in Betracht gezogen, da die vorhandenen Brennöfen zu klein waren, und weil die alte Anlage weitergefahren werden kann bis die neue Anlage stabil läuft.

Für Schlammfilter und Heißaufbereitung kam man zu folgendem Konzept:

Eine Gruppe von vier Drehvakuumfiltern (einschließlich eines Reservefilters) mit Gurtabzug wurde zur Herstellung von 130 t/h Filterkuchen auf Trockenbasis gewählt. Eine Zwischenlagerung des Kuchens wurde nicht vorgesehen.

Zum Zerkleinern und Trocknen des Filterkuchens wurde ein kombinierter Brecher/Trockner gewählt. Dieser verwendet die Abgase des Vorwärmers. Zur Abscheidung des getrockneten Filterkuchens werden zwei Zyklone eingesetzt. Als Zwischenspeicher zwischen der Filteranlage und dem Vorwärmer wurde ein 60-t-Aufgabebunker installiert, um Schwankungen der Filterkuchenzufuhr zu kompensie-

ACC's (Associated Cement Companies) Madukkarai Cement Works was a wet process plant, having five kilns with a total clinker capacity of 1180 t/d at a fuel consumption of 1450 kcal/kg of clinker. The limestone is of low grade with a $CaCO_3$ content of 76 to 77 % requiring wet beneficiation of about 60 % of the limestone. The increasing costs of coal and the deteriorating coal quality were seriously affecting the profitability and the operation of the plant. Various possibilities for plant modernisation and capacity enhancement were worked out by ACC to improve the financial viability of the plant.

The complete conversion to the dry process was ruled out as the limestone needs wet beneficiation. Lab scale filtration tests indicated that the slurry could be vacuum filtered. Further tests with a pilot scale scraper discharge filter showed a rapid blinding of the cloth reducing the cake production rate to less than half within 100 hours of operation. Subsequent trials with a pilot scale belt discharge filter did not show cloth blinding on continuous water washing. 190 hours of operation confirmed that a filter cake of 15 to 17 % moisture could be produced from a slurry containing 35 to 36 % moisture, at a rate of 485 to 540 kg/hr m^2 filtration area, under a vacuum of 665 to 730 mbar. A cake moisture of up to 18 % was considered for the design of the pyroprocessing system.

Encouraged by the filtration results, various options for the process conversion were considered and finally, a 1500 t/d semi-wet process plant using filter cake as feed material was chosen. The existing raw material preparation and the cement grinding systems were retained with minor upgradation to meet the increased capacity requirement. A new kiln line was considered as the existing kilns were small and

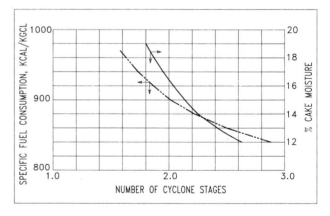

BILD 1: Brennstoffverbrauch und Kuchenfeuchtigkeit in Abhängigkeit der Zyklonstufen
FIGURE 1: Fuel consumption and cake moisture as a function of the number of cyclone stages

spezifischer Brennstoffverbrauch = Specific fuel consumption
Feuchtigkeit des Filterkuchens = Cake moisture
Anzahl der Zyklonstufen = Number of cyclone stages

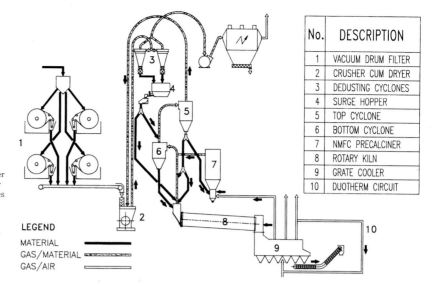

BILD 2: Fließschema der Drehofenanlage
FIGURE 2: Flowsheet of the kiln system

Vakuum-Trommelfilter	= Vacuum drum filter
kombinierter Brecher/Trockner	= Crusher cum dryer
Entstaubungszyklone	= Dedusting cyclones
Ausgleichsbunker	= Surge hopper
oberer Zyklon	= Top cyclone
unterer Zyklon	= Bottom cyclone
NMFC-Vorcalcinierer	= NMFC precalciner
Drehofen	= Rotary kiln
Rostkühler	= Grate cooler
Duotherm-Kreislauf	= Duotherm circuit

No.	DESCRIPTION
1	VACUUM DRUM FILTER
2	CRUSHER CUM DRYER
3	DEDUSTING CYCLONES
4	SURGE HOPPER
5	TOP CYCLONE
6	BOTTOM CYCLONE
7	NMFC PRECALCINER
8	ROTARY KILN
9	GRATE COOLER
10	DUOTHERM CIRCUIT

LEGEND
MATERIAL
GAS/MATERIAL
GAS/AIR

ren. Zur Überwachung der Aufgabemenge zum Vorwärmer wurde eine Massenstrom-Meßeinrichtung vorgesehen.

Die Anzahl der Zyklonstufen, die für verschiedene Feuchtigkeitsstufen des Kuchens und für verschiedene Betriebszustände erforderlich wurde, wurde durch detaillierte Computersimulationen ermittelt. **Bild 1** zeigt den erwarteten Brennstoffverbrauch und die mögliche Kuchenfeuchtigkeit für eine gegebene Anzahl von Zyklonstufen bei einer angenommenen Abgastemperatur von 160 °C. Man sieht, daß für etwa 16,5 % Kuchenfeuchtigkeit 2 Stufen angemessen sind, während bei 18 % Feuchtigkeit um die 1,85 Stufen erforderlich sind. Im Hinblick auf den erwarteten Feuchtigkeitsgehalt des Kuchens wurde ein 2stufiger Vorwärmer mit Einstellmöglichkeit auf Zwischenstufen zwischen 1 und 2 gewählt. Obwohl auch ein 3stufiger Vorwärmer mit Kuchen mit geringerem Feuchtigkeitsgehalt aufnehmen könnte, müßte er in der Nähe von 2 Stufen betrieben werden, wodurch sich lediglich eine Einsparung von etwa 8 bis 10 kcal/kg Klinker ergibt, während sich der Energieverbrauch für das Abgasgebläse um 10 % erhöht. Daher wurde aufgrund der Gesamtenergiebilanz ein 2stufiger Vorwärmer gewählt.

Eine Abschätzung ergab, daß ein Fließbett-Vorcalcinierer bis zu 72 % des gesamten Brennstoffbedarfs benötigt. Dabei wurde eine Temperatur des zugeführten Materials von 540 bis 565 °C vorausgesetzt. Die Kühlerkonstruktion erhielt einen Duotherm-Kreislauf in der 2. und 3. Kammer zur Erhöhung des Wirkungsgrads bei der Wärmerückgewinnung ebenso wie zur Erhöhung der Ablufttemperatur zum Trocknen der Kohle.

Das Konzept wurde schließlich vervollständigt durch einen Drehofen mit 2stufigem Vorwärmer/Vorcalcinierer. **Bild 2** zeigt das Fließbild für die in dieser Form einmalige Drehofenanlage. Die genauen Daten der verschiedenen Verfahrenseinrichtungen zeigt **Tabelle 1**.

also that the old plant can be operated till the new line is stabilised.

The following process concept was arrived at for the slurry filtering and pyroprocessing system: A set of four (including one stand-by) rotary vacuum filters with belt discharge was selected for producing the filter cake at a rate of 130 t/h on dry basis. No intermediate storage of cake was envisaged.

A crusher dryer was selected to disagglomerate and dry the filter cake using the waste gases from the preheater. A pair of dedusting cyclones were used for collecting the dried cake. A 60 tons surge hopper was provided as a buffer between filtration plant and preheater to even out the fluctuations in the cake feed rate. A solid flow meter was provided for regulating the feed to the preheater.

The number of cyclone stages required for various cake moistures and different operating conditions were determined by detailed computer simulation studies. **Fig. 1** shows the expected fuel consumption and cake moisture that can be dried for a given number of cyclone stages for a fixed waste gas temperature of 160 °C. It can be seen that for about 16.5 % of cake moisture, 2 stages are adequate, while around 1.85 stages are needed for 18 % of moisture. In view of the expected moisture content of the cake, a 2 stage preheater with the flexibility to operate between 1 and 2 stages was chosen. Although a 3 stage preheater could tolerate a cake with less moisture, it needs to be operated close to 2 stages resulting in a saving of only about 8 to 10 kcal/kg of clinker, while the power consumption of waste gas fan increases by about 10 %. Thus on total energy basis, a 2 stage preheater was selected.

A fluidised bed precalciner was considered to fire upto 72 % of the total fuel. A feed material temperature of 540 to 565 °C was anticipated. A duotherm circuit was incorporated in the cooler design for the 2nd and 3rd compartment for a better recuperation efficiency as well as to enhance the waste air temperature for drying the coal.

The concept thus finalised of a 2 stage preheater/precalciner kiln system was unique. **Fig. 2** gives the flow sheet of the kiln system. The details of the various process equipment selected are given in **Table 1**.

TABELLE 1: Wichtige Anlagedaten

Vorrichtung	Größe/Bezeichnung
Vakuumfilter	$4 \times \varnothing\,3{,}66 \times 6{,}1$ m
kombin. Brecher/Trockner	ET 355×250
Entstaubungszyklone	$2 \times \varnothing\,4{,}15$ m
Vorwärmer	2stufige Zyklone $\varnothing\,7{,}2$ m
Vorcalcinierer (NMFC)	$\varnothing\,6$ m $\times 25{,}5$ m hoch
Drehofen	$\varnothing\,3{,}75$ m $\times 54$ m lang
Klinkerkühler	$49{,}288$ m² X-sec

TABLE 1: Details of major equipment

Equipment	Size/Designation
vacuum filters	4 Nos. $\times 3{,}66\,\varnothing \times 6{,}1$ m
crusher cum dryer	ET 355×250
dedusting cyclones	$2 \times 4{,}15$ m $\varnothing$
preheater	2 stage – $7{,}2$ m $\varnothing$ cyclones
precalciner (NMFC)	6 m $\varnothing \times 25{,}5$ m height
rotary kiln	$3{,}75$ m $\varnothing \times 54$ m long
clinker cooler	$49{,}288$ m² q. X-sec

Der Naßkuchenausstoß der Filter wird durch kombinierte Regelung von Vakuum und Trommelgeschwindigkeit geregelt. Zwischen dem Materialgewicht im Ausgleichsbunker und dem Ausgang der Filteranlage wurde eine Kaskadenregelung integriert, um die Durchsätze von Drehofen und Filteranlage abzugleichen.

Die Anlage wurde im Januar 1989 zu einem Preis von 470 Mio. Rupien errichtet. Die Leistung der Filter war besser als vorher angenommen. Sie erzeugten Filterkuchen mit 13 bis 15 % Feuchtigkeit. Der niedrigere Feuchtigkeitsgehalt des Kuchens und der geringere Wirkungsgrad des Kühlers ergaben eine Abgastemperatur des Vorwärmers von 620 bis 630 °C und am elektrischen Abscheider eine Abgastemperatur von 200 °C. Schon bald nach Übergabe der Anlage wurde ein Brennstoffverbrauch von 985 kcal/kg Klinker erreicht. Etwa 10 % des dem Drehofen zugeführten Schlamms wurde daher am Brecher/Trockner zugegeben zur Verringerung der Gastemperatur am elektrostatischen Abscheider auf 160 °C und zur Verbesserung von dessen Staubfangleistung. Auch die Entstaubungszyklone wurden modifiziert, um den Wirkungsgrad von 80 auf 93 % zu verbessern, was die Staublast am Eingang des Abscheiders verringerte.

Probleme mit Ansatzbildung in der Einlaufkammer des Drehofens wurden durch Begrenzung der Gastemperatur im Ofeneinlauf auf unter 1050 °C und durch Zufuhr von 10 bis 15 % des Materials aus dem oberen Zyklon in den Drehofeneinlauf zur Kühlung der Gase gelöst.

Nach dem Einfahren arbeitete der Drehofen gleichmäßig, und der Nenndurchsatz von 1500 t pro Tag wurde erreicht. Eine nachfolgende Optimierung der Anlage verbesserte den mittleren Durchsatz auf 1800 t pro Tag, gelegentlich wurden 2000 t pro Tag erreicht. Der Brennstoffverbrauch wurde auf etwa 910 bis 920 kcal/kg Klinker reduziert. **Tabelle 2** zeigt die Wärmebilanz der Drehofenanlage. **Bild 3** zeigt das Schema der Anlage vor bzw. nach Umbau. Der verwendete Brennstoff ist ein Gemisch aus aschereicher unterbituminöser Kohle, Lignit und landwirtschaftlichen Abfällen wie Reishülsen, Erdnußschalen usw. mit einem Heizwert von etwa 3700 kcal/kg.

TABELLE 2: Wärmebilanz

Wärmeeintrag:	kcal/kg Kl.
meßbare Wärme von Luft, Brennstoff, Kuchen usw.	50
Verbrennungswärme des Brennstoffs	912
	962
Wärmeabgabe:	
Klinkerisierungswärme	408
fühlbare Wärme der Abgase	140
fühlbare Wärme der Kühlluft	126
Feuchtigkeitsverdampfung aus dem Kuchen	173
Strahlungsverluste	83
fühlbare Wärme des Klinkers	22
Verschiedene	10
	962

Gegenwärtig wird ein Projekt erwogen, das die Kapazität des Drehofens bei einem Brennstoffverbrauch von etwa 845 kcal/kg Klinker auf 2400 t pro Tag erhöhen soll. Es ist beabsichtigt, eine parallele Vorwärmlinie zu installieren, um zum einen die Kapazität der Anlage und zum anderen die äquivalente Anzahl der Vorwärmstufen auf mehr als 2 zu erhöhen. Damit sollen die Anforderungen an die Trocknung eines Filterkuchens von 13 bis 14 % Feuchtigkeit erfüllt werden. Diese Anlage hat dann den Vorteil, den Durchsatz des Brennofens zu erhöhen, während der Druckabfall im Vorwärmer vermindert wird. Der erhöhte Druckabfall im Bre-

The wet cake production from the filters is regulated by controlling the vacuum and the drum speed together. A cascade control between the weight of the material in the surge bin and the output of the filtration plant was incorporated for balancing the kiln and filter production rates.

The project was implemented at a cost of Rs. 470 million in January 1989. The performance of the filters were better than anticipated and produced a cake with 13 to 15 % of moisture. The lower cake moisture and the lower cooler efficiency resulted in a preheater exit gas temperature of 620 to 630 °C and an exit gas temperature of 200 °C at the electric precipitator. A fuel consumption of 985 kcal/kg of clinker was achieved soon after commissioning. About 10 % of the kiln feed slurry was therefore injected at the crusher dryer inlet to bring down the gas temperature at the electrostatic precipitator to 160 °C and improve its dust collection efficiency. The dedusting cyclones were also modified to improve the efficiency from 80 to 93 %, to reduce the dust load at the precipitator inlet.

Problems with coating formation at the kiln inlet chamber were solved by limiting the kiln outlet temperature to less than 1050 °C and feeding 10 to 15 % of the top cyclone material into the kiln riser duct for cooling the gases.

On stabilisation, the kiln operated smoothly and the rated output of 1500 t/d was achieved. A subsequent optimisation of the system improved the average output to 1800 t/d and on several occasions, 2000 t/d were achieved. The fuel consumption has been reduced to about 910 to 920 kcal/kg of clinker. **Table 2** shows the heat balance of the kiln system. **Fig. 3** shows the flowsheet of the kiln before and after conversion. The fuel being used is a mixture of high ash sub-

TABLE 2: Heat balance

Heat inputs:	kcal/kgcl
sensible heat of air, fuel, cake etc.	50
combustion heat of fuel	912
	962
Heat outputs:	
heat of clinkerisation	408
sensible heat of exhaust gases	140
sensible heat of cooler vent air	126
cake moisture evaporation	173
radiation loss	83
sensible heat of clinker	22
miscellaneous	10
	962

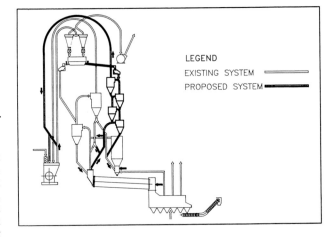

BILD 3: Schema der Anlage vor bzw. nach Umbau
FIGURE 3: Flowsheet of the kiln before and after conversion

bestehende Anlage = Existing system
vorgeschlagene Anlage = Proposed system

cher/Trockner würde dadurch begrenzt, daß ein Teil der Gase direkt zum Ausgang des Brechers/Trockners geleitet würde. Der erste Rost des vorhandenen Klinkerkühlers soll dann auch durch einen Hochleistungskühlrost ersetzt werden, um den Durchsatz und den Wirkungsgrad des Rostkühlers zu steigern. Die Anlage in Madukkarai würde auch nach Realisierung dieser Vorschläge weiterhin in ihrer Verfahrensführung einzigartig bleiben.

bituminous coal, lignite and agricultural wastes like rice husks, groundnut shells, etc., having a net calorific value of about 3700 kcal/kg.

Presently a project is under consideration to increase the kiln capacity to 2400 t/d at a fuel consumption of about 845 kcal/kg of clinker. It is proposed to install a parallel preheater string to increase the capacity of the system as well as the eqiuvalent number of preheater stages to more than 2, so as to match the drying requirement of 13 to 14% of cake moisture. This system has the advantage of increasing the kiln output while reducing the preheater pressure drop. The increased pressure drop in the crusher dryer would be limited through the bypassing part of the gases directly to the outlet of the crusher dryer. The first grate of the existing clinker cooler is also to be replaced by a high efficiency cooling grate for increasing the capacity and efficiency of the grate cooler. The Madukkarai Plant would retain its uniqueness after the implementation of these proposals.

Klassifikation von Magnesiasteinen nach Spezifikation und Gebrauchswert im Zementdrehrohrofen*)

Classification of magnesia bricks in rotary cement kilns according to specification and serviceability*)

Classification de briques de magnésie selon spécification et utilisation appropriée dans le four à ciment rotatif

Clasificación de los ladrillos de magnesia, según su especificación y utilidad en el horno rotatorio de cemento

Von **P. Bartha** und **H.-J. Klischat**, Göttingen/Deutschland

Zusammenfassung – Die historische Entwicklung der Ausmauerung von Zementdrehöfen ist durch den Wandel von einer Schamottezustellung hin zur basischen Zustellung gekennzeichnet. Diese Umstellung hat speziell in den hochbelasteten Bereichen (Übergangszone und zentrale Sinterzone) zu einer erheblichen Verbesserung der Standzeiten geführt. Zur Zeit steht eine Vielzahl von basischen Steinsorten mit unterschiedlichen Werkstoffeigenschaften zur Verfügung. Bei feuerfesten Steinen steht die Standfestigkeit unter Beanspruchung im praktischen Einsatz im Vordergrund des Interesses. Der Beitrag enthält eine detaillierte Klassifikation von Magnesiasteinen nach Spezifikation und Gebrauchswert. Die wichtigsten Korrosionsbeanspruchungen der Steinkomponenten sowie die sich daraus ergebenden Anforderungen an die Eigenschaften werden dargestellt. Von entscheidender Bedeutung sind dabei die angestrebten thermomechanischen und thermochemischen Eigenschaften der Steine und deren Prüfwerte. Anhand dieser Kriterien wird ein Überblick über die zur Verfügung stehenden basischen Steinsorten sowie über zukünftige Entwicklungen gegeben.

Summary – The historical development of the brickwork in rotary cement kilns is characterized by the change from fire clay lining to basic lining. This changeover led to a great improvement in service life, especially in the highly stressed regions (transition zone and central sintering zone). There are now a large number of types of basic brick available with different material properties. Interest in refractory bricks is centred on their stability under load in practical use. This paper contains a detailed classification of magnesia bricks according to specification and serviceability. An explanation is given of the most important corrosion stresses on the brick components and the resulting requirements for the properties. The desired thermomechanical and thermochemical properties of the bricks and their test values are of critical importance in this context. These criteria are used to give a review of the types of basic brick available and of future developments.

Résumé – L'histoire du développement de la maçonnerie des fours à ciment rotatifs est caractérisée par une évolution du garnissage chamotte vers le garnissage basique. Ce changement a conduit à une nette amélioration des durées de service, spécialement dans les sections fortement sollicitées (zone de transition et centre de la zone de clinkérisation). Actuellement, sont disponibles une multitude de sortes de briques de propriétés variées. Pour les briques réfractaires, la durée de service sous charge dans l'utilisation pratique est à l'avant-plan de l'intérêt. La contribution comporte une classification détaillée de briques de magnésie, d'après spécification et opportunité d'utilisation. Les contraintes de corrosion des composants des briques ainsi que les exigences pour les propriétés, qui en résultent, sont présentées. D'une signification décisive sont, dans ce contexte, les propriétés thermomécaniques et thermochimiques visées des briques et leurs valeurs prouvées par essai. A l'aide de ces critères, est donnée une vue d'ensemble des sortes de briques disponibles ainsi que des développements futurs.

Resumen – La historia del revestimiento refractario de los hornos rotatorios de cemento se caracteriza por la sustitución del revestimiento refractario de chamota por el de ladrillos básicos. Este cambio ha conducido a una considerable mejora de la duración del revestimiento refractario, especialmente en las zonas sometidas a elevadas solicitaciones (zona de transición y zona central de sinterización). Actualmente, se dispone de una gran variedad de ladrillos básicos, de diferentes propiedades. En el caso de los ladrillos refractarios, lo que más interesa es su duración de servicio bajo carga. El presente artículo contiene una clasificación detallada de ladrillos de magnesia, según su especificación y utilidad. Se describen las principales solicitaciones por corrosión de

---

*) Überarbeitete Fassung zum VDZ-Kongreß '93, Düsseldorf (27.9.–1.10.1993)
Revised text of lecture to the VDZ Congress '93, Düsseldorf (27.9.–1.10.1993)

los componentes de los ladrillos así como las exigencias de ahí resultantes en cuanto a las características de los mismos. A este respecto, son de importancia decisiva las propiedades termomecánicas y termoquímicas pretendidas de los ladrillos así como los valores de ensayo de los mismos. Con ayuda de estos criterios, se pasa revista a los tipos de ladrillos básicos disponibles y a las posibilidades de desarrollo futuro.

Die feuerfeste Auskleidung von Zementdrehöfen ist seit über 50 Jahren als Folge der Entwicklung in der Brenntechnologie durch den Wechsel von Schamotte- und hochtonerdehaltigen Steinsorten zu basischen Steinsorten gekennzeichnet (**Bild 1**) [1, 2]. Bei einer Diskussion über Feuerfestzustellungen in der Zementindustrie gilt deshalb heute das Hauptaugenmerk dem Magnesiastein. Er dominiert im Drehrohr durch die Länge seiner Einbaustrecke. Bei Vorcalcinieröfen sind das über 70 % der Gesamtofenlänge im Vergleich zu klassischen Wärmetauscher-Öfen mit ≈ 48 %, Lepolöfen mit ≈ 40 % und langen Naß- und Trockenöfen mit ≈ 23 %.

Wenn Stillstände durch Futterschäden verursacht werden, ist in den meisten Fällen eine Reparatur in den basisch zugestellten Zonen notwendig. Die Auswahl der Steinsorten mit direktem Einfluß auf die Stillstandshäufigkeit ist entscheidend für die Verfügbarkeit des Ofens und damit für die Herstellungskosten des Zementes.

Magnesia ist deshalb die erste Wahl für die kritischen Zonen eines Zementdrehofens aufgrund folgender positiver Eigenschaften unter den anwendbaren feuerfesten Oxiden:

1. Höchster Widerstand gegenüber Temperatur,
2. Höchster Widerstand gegenüber chemischer Beanspruchung.

Allerdings haben reine Magnesia-Steine für den Einsatz im Zementdrehrohr einen entscheidenden Werkstoffnachteil: Das Material hat einen zu hohen Elastizitätsmodul und daher eine zu geringe Bruchzähigkeit, um dem im Zementdrehrohr herrschenden mechanischen Beanspruchungen einen ausreichenden Widerstand entgegenzusetzen. (Der Elastizitäts-Modul von Magnesia-Steinen liegt bei über 80 kN/mm² im Vergleich zu Magnesiachromit-Steinen mit 25 kN/mm² und Magnesiaspinell-Steinen in der gleichen

For over 50 years, the refractory lining of rotary cement kilns has been characterised by the changeover from fireclay and high-alumina bricks to basic bricks as a result of developments in burning technology (**Fig. 1**) [1, 2]. When discussing refractory linings in the cement industry, the main focus of attention today is therefore the magnesia brick. It dominates in rotary kilns due to the length over which it is installed. In the case of precalcining kilns, this amounts to over 70% of the total kiln length compared with ≈ 48% in classical preheater kilns, ≈ 40% in Lepol kilns and ≈ 23% in long wet- and dry-process kilns.

If shutdowns are caused by lining damage, in most cases a repair in the zones lined with basic bricks is necessary. The selection of the types of brick which directly influence the frequency of shutdowns is a crucial factor with regard to kiln availability and thus to the cost of cement production.

Magnesia is therefore the first choice for the critical zones of a rotary cement kiln as it has the following advantages among the usable refractory oxides:

1. Optimum resistance to heat,
2. Optimum resistance to chemical attack.

However, pure magnesia bricks have a decisive material drawback with regard to their use in rotary cement kilns: the material has too high a modulus of elasticity and therefore too low a fracture toughness to offer sufficient resistance to the mechanical stresses prevailing in a rotary cement kiln. (The modulus of elasticity of magnesia bricks is over 80 kN/mm² compared with magnesia-chromite bricks with 25 kN/mm² and magnesia-spinel bricks with around the same figure.) Pure magnesia bricks have therefore not found favour since trial applications were carried out in the 1950s.

Subsequently, magnesia bricks were appropriately modified by means of additives so as to obtain optimum properties in use. The classification of modified magnesia bricks is essentially determined by the type and quantity of the modifiers used (**Table 1**). They all essentially have the purpose of optimizing the modulus of elasticity in a manner relevant to the application without adversely affecting the bricks' chemical and thermal resistance.

TABELLE 1: Klassifikation MgO-haltiger feuerfester Steine
TABLE 1: Classification of refractory bricks containing MgO

Steintyp Brick type	Kurzbezeichnung Short form	MgO-Gruppe MgO-Group
Magnesia²/Magnesit Magnesia²/Magnesite	M	98-95-90-85
Magnesia-Kohlenstoff Magnesia Carbon	MG	90-85-80-70-60-50-40-30
Magnesia-Dolomit-Kohlenstoff Dolomit-Kohlenstoff Magnesia-Dolomite-Carbon Dolomite-Carbon	MDG	C³ nach Verkokung >7% C³ after coking >7%
Magnesia-Chromit (Magnesit-Chrom u. Chrom-Magnesit) Magnesia Chromite (Magnesite-Chrome and Chrome-Magnesite)	MC	80-70-60-50-40-30
Chromit (Chromerz) Chromite (Chrome ore)	C	<30 (Cr_2O_3 >30)
Magnesia-Dolomit Magnesia-Dolomite	MD	80-70-60-50
Dolomit Dolomite	D	40-30
Forsterit Forsterite	F	50-40
Magnesia-Spinell Magnesia-Spinel ($MgO\text{-}MgO\cdot Al_2O_3$, $MgO\text{-}MgO\cdot Cr_2O_3$)	S	80-70-60
Kalk Lime	CA	CaO >70

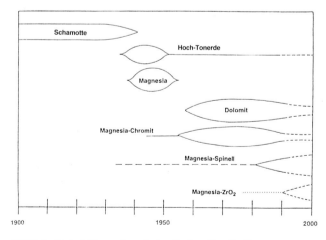

BILD 1: Entwicklung feuerfester Steine für die Brennzone von Zementdrehöfen
FIGURE 1: Development of refractory bricks for the burning zone of rotary cement kilns

Schamotte = fireclay
Hoch-Tonerde = high-alumina
Magnesia = magnesia
Dolomit = dolomite
Magnesia-Chromit = magnesia-chromite
Magnesia-Spinell = magnesia-spinel
Magnesia-ZrO_2 = magnesia-ZrO_2

Größenordnung.) Reine Magnesia-Steine haben sich deshalb nach Versuchseinsätzen in den 50iger Jahren nicht durchgesetzt.

In der folgenden Zeit wurden Magnesia-Steine durch Zusätze entsprechend modifiziert, um optimale Gebrauchseigenschaften einzustellen. Die Klassifikation modifizierter Magnesia-Steine richtet sich dabei im wesentlichen nach Art und Menge der verwendeten Modifikatoren (**Tabelle 1**). Alle dienen im wesentlichen dazu, den Elastizitätsmodul anwendungsrelevant zu optimieren, ohne gleichzeitig die chemische und thermische Widerstandsfähigkeit negativ zu beeinflussen.

Dieses Ziel wurde seit den 50iger Jahren durch Verwendung geeigneter Magnesiachromit-Steinsorten realisiert.

Magnesiachromit-Steine mit hohem silikatischen Bindungsanteil (vor allem Forsterit, Monticellit, Merwinit und hohe Anteile an Belit) haben als erste magnesitische Steinsorten die Schamotte- und hochtonerdehaltigen Steinsorten im Zementdrehrohr abgelöst. Diese Steinsorten hatten entscheidende Vorteile in ihren Gebrauchseigenschaften; sie sind aber der thermischen und chemischen Belastung moderner Drehofensysteme nicht gewachsen. Sie werden heute nur noch in kleineren Öfen verwendet.

„Direktgebundene Magnesiachromit-Steinsorten" haben zwar eine erhöhte mechanische Festigkeit bei hohen Temperaturen, der Chromanteil hat aber u.a. einen negativen Einfluß auf die erforderliche Bruchzähigkeit für Drehofensteine. Hauptsächlich in Japan in großen Öfen werden jedoch immer noch sogenannte „super direct bonded bricks" eingesetzt.

Für die Standardauskleidung in der Sinterzone von Zementdrehöfen wurden die Magnesiachrom-Steinsorten mit hohem Silikatanteil durch Steine, die eine sogenannte „Mischbindung" aufweisen, abgelöst. Ein typisches Beispiel dieser Steinsorte ist Perilex 80 [3]. Auch hier sind Aluminium, Chrom und zum großen Teil Eisen Bestandteile der „Steinbindung", sie sind aber auch Bestandteile der verwendeten Modifikatoren, wie Chromerz.

In Gegenwart von Alkalien in der Ofenatmosphäre ist aber auch diese Steinsorte nicht ausreichend widerstandsfähig; außerdem ist sie empfindlich gegen reduzierende Atmosphäre aufgrund ihres hohen Eisengehaltes (Reduktion von Magnesioferrit zu Magnesiowüstit; Auflösung der Chromerzkomponente; Bourdouard'sches Gleichgewicht!).

Eine Steinsorte, die nach sorgfältiger Entwicklung fast allen Anforderungen eines modernen Drehofenbetriebes gerecht wird, bilden Magnesiaspinell-Steine, bestehend aus MgO (Periklas) und MgO · Al_2O_3 (Spinell) sowie minimierten Anteilen von Calzium, Silizium und Eisen – frei von Chromerz [4–9]. Als chromerzfreie Alternative sind sie grundsätz-

Since the fifties this goal has been achieved by using suitable magnesia-chromite bricks.

Magnesia-chromite bricks containing a high proportion of silicate compounds (especially forsterite, monticellite, merwinite and high proportions of belite) were the first magnesitic bricks to replace fireclay bricks and high-alumina bricks in rotary cement kilns. These types of brick had decisive advantages with regard to their properties in use; however, they cannot cope with the thermal stresses and chemical attack encountered in modern rotary kiln systems. Nowadays they are only used in smaller kilns.

While "direct-bonded magnesia-chromite bricks" have high mechanical strength at high temperatures, the chromium content has, among other things, an adverse effect on the required fracture toughness for rotary kiln bricks. However, so-called "super direct-bonded bricks" are still used in large kilns, mainly in Japan.

For the standard refractory lining in the sintering zone of rotary cement kilns, magnesia-chromite bricks with a high silicate content are replaced by bricks with so-called "mixed bonding". A typical example of this type of brick is Perilex 80 [3]. Here again, aluminium, chromium and to a large extent iron are constituents of the "brick bond", but are also constituents of the modifiers used, e. g. chrome ore.

In the presence of alkalis in the kiln atmosphere, however, even this type of brick is not sufficiently resistant; in addition, it is sensitive to a reducing atmosphere due to its high iron content (reduction of magnesioferrite to magnesiowuestite; dissolution of the chrome ore constituent; Boudouard equilibrium!).

One type of brick which, after painstaking development work, meets almost all the requirements of modern rotary kiln operation is the magnesia-spinel brick, consisting of MgO (periclase) and MgO · Al_2O_3 (spinel) together with minimal percentages of calcium, silicon and iron; it is free of chrome ore [4–9]. As a chrome-ore-free alternative, they are suitable in principle for all parts of the burning zone. In some cases, in the central sintering zone they are unable to bring their beneficial properties to bear in full if a stable, protective coating has formed. In this case, despite their longer life the price/campaign life ratio may at the moment be poorer than for magnesia-chromite bricks. However, this ratio changes immediately if there are legal requirements for the disposal of used magnesia-chromite bricks [5].

Dolomite bricks, which are listed here for the sake of completeness, are another chrome-ore-free type of brick for parts of the burning zone where a stable coating forms.

The properties of this type of brick are adequately known, so they will not be discussed further [10, 11].

Magnesia-zirconium oxide bricks are the latest innovation in the development of chrome-ore-free magnesia bricks. The

TABELLE 2: Charakterisierung der untersuchten Steinsorten
TABLE 2: Characterisation of bricks investigated

	Magnesia-ZrO_2 / Magnesia-ZrO_2	Magnesia-Spinell / Magnesia-Spinel	Magnesia-Chromit / Magnesia-Chromite	Dolomit (D 40) / Dolomite (D 40)
MgO	94–96	94–96	80–85	36,8
Al_2O_3		3–5	2–4	1
ZrO_2	3–5			
Cr_2O_3			3–5	
CaO	1	1	2,5	60
SiO_2	<0,5	<0,5	1,5	<1
Fe_2O_3	<0,5	<0,5	6–8	<1
Schüttdichte (g/cm³) / bulk density (g/cm³)	2,85–3,00	2,85–2,95	2,90–3,00	2,75
Porosität s (%) / porosity s (%)	17–19	16–18	18–20	4**
KDF*)/N/mm²	50	45	50	500

*) KDF ≙ Kaltdruckfestigkeit
**) Teer gesättigt
**) tar impregnated

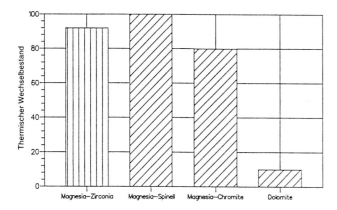

BILD 2: Temperaturwechselbeständigkeit nach DIN 51068
FIGURE 2: Thermal shock resistance to DIN 51068

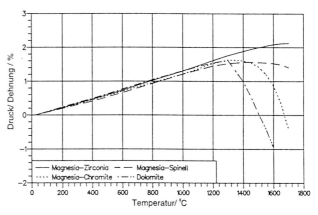

BILD 3: Druckerweichen nach DIN 51053
FIGURE 3: Softening under load to DIN 51053

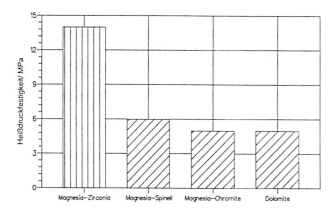

BILD 4: Heißdruckfestigkeit bei 1500 °C
FIGURE 4: Hot compressive strength at 1500 °C

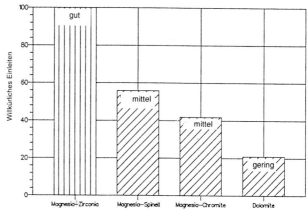

BILD 5: Widerstand gegen Druckfließen nach DIN 51053 (dimensionslos normiert)
FIGURE 5: Resistance to creep under pressure to DIN 51053 (standardised on a dimensionless basis)

lich für alle Bereiche der Brennzone geeignet. In einigen Fällen können sie in der zentralen Sinterzone bei stabilem, schützendem Ansatz ihre vorteilhaften Eigenschaften nicht voll zur Geltung bringen. Hier kann trotz längerer Lebensdauer das Verhältnis von Preis zu Laufzeit z.T. ungünstiger als bei Magnesiachromit-Steinen sein. Dieses Verhältnis ändert sich aber sofort, wenn die Entsorgung gebrauchter Magnesiachromit-Steine durch den Gesetzgeber vorgeschrieben wird [5].

Dolomitsteine, die hier der Vollständigkeit halber aufgelistet sind, repräsentieren ebenfalls eine chromerzfreie Steinsorte für ansatzstabile Bereiche der Brennzone.

Die Eigenschaften dieser Steinsorte sind hinlänglich bekannt, so daß hierauf nicht näher eingegangen werden soll [10, 11].

Magnesia-Zirkonoxid-Steine stellen die jüngste Innovation bei der Entwicklung chromerzfreier Magnesia-Steine dar. Die Wirkungsweise von Zirkonoxid in feuerfesten basischen Steinen ist an sich seit längerem bekannt [12–14]. Der hohe Preis dieses Oxides, welches etwa 7mal so teuer wie Spinell und 30mal so teuer wie Chromerz ist, hat die Einführung dieser Steinsorte in der Zementindustrie u.a. sicher zeitlich verzögert.

In den folgenden Ausführungen werden einige signifikante Werkstoffeigenschaften von realen Steinsorten einander gegenüber gestellt. Die untersuchten Steinsorten sind in **Tabelle 2** charakterisiert. Der Vollständigkeit halber wurde eine typische Dolomitstein-Sorte gegenübergestellt.

Die untersuchte Magnesia-Zirkonoxidsteinsorte weist demnach folgende signifikante Merkmale auf:

1. Die Temperaturwechselbeständigkeit nach DIN 51068 liegt auf einem vergleichbar hohen Niveau wie die der untersuchten Magnesiaspinell-Steinsorte (**Bild 2**).

action of zirconium oxide in basic refractory bricks has in fact been known for some time [12–14]. The high price of this oxide, which is about seven times as expensive as spinel and 30 times as expensive as chrome ore, has undoubtedly delayed the introduction of this type of brick into the cement industry.

In the following remarks, some significant material properties of commercial types of brick are compared. The types of brick investigated are characterized in **Table 2**. For the sake of completeness, a typical dolomite brick was included in the comparison.

The magnesia-zirconium oxide brick investigated therefore exhibits the following significant features:

1. The thermal shock resistance to DIN 51068 is of a similarly high level to that of the magnesia-spinel brick tested (**Fig. 2**).

2. The start of softening under load to DIN 51053 is significantly higher than for the magnesia-chromite and magnesia-spinel bricks tested (**Fig. 3**).

3. The hot compressive strength at 1500 °C (cylinder 50 mm in diameter and 50 mm tall) is significantly higher for the magnesia-zirconium oxide brick than for all the other types of brick tested (**Fig. 4**).

4. Resistance to creep under pressure to DIN 51053 is significantly higher than for all the other types of brick tested (**Fig. 5**).

Based on investigations of the chemical behaviour of the three magnesitic types of brick tested, the following statements may be made:

5. Resistance to clinker melts is much higher than with the magnesia-spinel brick and the magnesia-chromite brick tested, especially with regard to the zirconium oxide used as the modifier.

2. Der Beginn des Druckerweichen nach DIN 51053 ist signifikant höher als für die untersuchten Magnesiazirchromit- und Magnesiaspinell-Steinsorten (**Bild 3**).

3. Die Heißdruckfestigkeit bei 1500°C (Zylinder 50 mm Durchmesser, 50 mm Höhe) ist für den Magnesiazirkonoxid-Stein signifikant höher als bei allen anderen untersuchten Steinsorten (**Bild 4**).

4. Der Widerstand gegen Druckfließen nach DIN 51053 ist signifikant höher als bei allen anderen untersuchten Steinsorten (**Bild 5**).

Aus Untersuchungen zum chemischen Verhalten der drei untersuchten magnesitischen Steinsorten lassen sich folgende Aussagen machen:

5. Der Widerstand gegenüber Klinkerschmelzen ist vor allem in bezug auf den eingesetzten Modifikator Zirkonoxid ganz wesentlich höher als bei dem Magnesiaspinell- und dem untersuchten Magnesiachromit-Stein.

6. Der Widerstand gegenüber Alkalioxiden, Alkalisalzen, Schwefeldioxid, Kohlendioxid und Kohlenmonoxid ist mit dem des untersuchten Magnesiaspinell-Steines vergleichbar und höher als bei der untersuchten Magnesiachromit-Steinsorte.

Aufgrund des besonderen Eigenschaftsprofils von Magnesiazirkonoxid-Steinen ist zunächst ein Einsatz in den ansatzfreien Bereichen der Brennzone von Drehöfen vorgesehen, bei denen tendenziell ein erhöhter Angriff durch Flüssigphase (Klinkerschmelze) bzw. eine erhöhte Temperaturbelastung zu erwarten ist.

Die Entwicklung chromerzfreier Magnesiasteine hat notwendigerweise zu einer tendenziellen Erhöhung des MgO-Gehaltes geführt. Die damit verbundene Erhöhung der Wärmeleistung durch das Mauerwerk wird deshalb in der Praxis immer wieder kritisch hinterfragt. In Zusammenhang damit steht auch der Wunsch nach „Ansatzbildung" als Steineigenschaft.

Aus den verschiedenen theoretischen Modellen [15–17] und der empirischen Erkenntnis geht jedoch hervor, daß die Ansatzbildung – genau so wie die Aufbaugranulation von Rohmehl beim Lepolprozeß bzw. die Klinkergranulation im Drehofen – eine Funktion des Anteils und der Zusammensetzung der Flüssigphase sowie der damit im Gleichgewicht stehenden Betriebsbedingungen ist. Deshalb kann die Bildung eines Ansatzes in den ansatzfreien Zonen der Brennzone auch nicht durch eine feuerfeste Steinsorte „erzwungen" werden.

In den Ofenzonen, in denen sich ein stabiler Ansatz, Ringe oder Anbackungen ausbilden, ist lediglich eine „haftverstärkende" oder „haftvermindernde" Wirkung zu erwarten, die abhängig ist von den chemisch-physikalischen Phasengrenzflächenreaktionen, die zwischen feuerfester Auskleidung und Brenngut stattfinden.

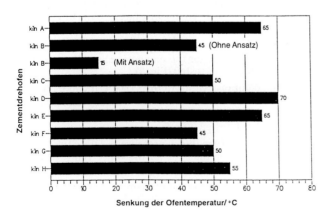

BILD 7: Manteltemperaturabsenkung durch Verwendung von „Hohlfußsteinen" in 5 realen Zementdrehöfen
FIGURE 7: Lowering of kiln shell temperature by using "clog bricks" in five commercial rotary cement kilns

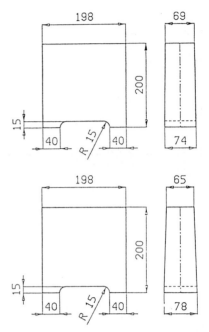

BILD 6: „Hohlfußstein" mit mantelseitiger Aussparung zur Wärmedämmung
FIGURE 6: "Clog brick" with shell-side hollow for thermal insulation

6. Resistance to alkali oxides, alkali salts, sulphur dioxide, carbon dioxide and carbon monoxide is comparable to the magnesia-spinel brick tested and higher than that of the magnesia-chromite brick tested.

Due to the exceptional characteristics of magnesia-zirconium oxide bricks, it is planned to install them first of all in the coating-free parts of the burning zone of rotary kilns, where there is a tendency towards severe attack by the liquid phase (clinker melt) and/or severe temperature stresses.

The development of chrome-ore-free magnesia bricks has inevitably led to a tendency towards an increased MgO content. The associated increase in thermal conductivity is therefore repeatedly subjected in practice to critical analysis. Also connected with this is the desire for "coating formation" as a brick property.

However, it is evident from the various theoretical models [15–17] and empirical knowledge that coating formation, just like the nodulizing of raw meal in the Lepol process and clinker granulation in the rotary kiln, is a function of the proportion and composition of the liquid phase and of the operating conditions which are in equilibrium with it. Therefore the formation of a coating in the coating-free parts of the burning zone cannot be "forced" by using a particular type of refractory brick.

In the parts of the kiln where a stable coating, rings or accretions form, only an "adhesion-increasing" or "adhesion-reducing" effect can be expected; this effect depends on the chemico-physical reactions at the phase boundary surfaces which take place between the refractory lining and the kiln feed.

The question of thermal insulation of rotary cement kilns has been examined several times; the reader is therefore referred to the corresponding literature [18–20]. To reduce the shell temperature in coating-free zones of the rotary kiln which are lined with basic brick, our firm recommends a brickwork structure using so-called "clog bricks" (**Fig. 6**). The results so far from five practical applications show a reduction of the shell temperatures in the coating-free zones of between 45 and 70°C (**Fig. 7**).

The campaign life results so far and the achievable temperature reductions give reason to expect that, when compared with conventional magnesia-chromite bricks, the thermal drawbacks of chrome-ore-free magnesia bricks with a high MgO content can be offset by using so-called "clog bricks" with hollows towards the shell side of the kiln.

Die Frage der Wärmedämmung von Zementdrehöfen ist mehrfach untersucht worden, so daß hier auf die entsprechende Literatur verwiesen wird [18—20]. Um die Manteltemperatur in ansatzfreien basisch ausgekleideten Zonen des Drehrohrs zu senken, wird von unserem Unternehmen eine Mauerwerkskonstruktion mit sogenannten „Hohlfußsteinen" empfohlen (**Bild 6**). Die bisherigen Ergebnisse aus 5 Praxiseinsätzen zeigen für die ansatzfreien Zonen eine Absenkung der Manteltemperaturen zwischen 45 und 70 °C (**Bild 7**).

Die bisherigen Laufzeitergebnisse und die erzielbaren Temperaturabsenkungen versprechen, daß die kalorischen Nachteile chromerzfreier Magnesiasteine mit hohem MgO-Gehalt gegenüber den klassischen Magnesiachromit-Steinen durch den Einsatz von sogenannten „Hohlfußsteinen" mit Aussparungen zur Mantelseite des Ofens ausgeglichen werden können.

Die Entwicklung hochreiner chromerzfreier Magnesiaspinell-Steine ist in den vergangenen 10 Jahren weit fortgeschritten. Gemeinsam mit den klassischen Magnesiachromit-Steinen — und den neu entwickelten Magnesiazirkonoxid-Steinen — ermöglichen sie einen differenzierten Einsatz in Abhängigkeit vom Anforderungsprofil verschiedener Ofensysteme und Ofenzonen.

The development of highly pure chrome-ore-free magnesia-spinel bricks has come a long way in the past ten years. Together with the classical magnesia-chromite bricks — and the newly developed magnesia-zirconium oxide bricks — they enable a diversified pattern of use to suit the requirements of various kiln systems and kiln zones.

Literaturverzeichnis

[1] Odanaka, S., Hara, K., Kusunose, H., und Tokunaga, K.: Refractories for Cement Production, History and Current Status. Taikabutsu Overseas 6 (1986) Nr. 2, S. 62—71.

[2] Bartha, P., und Hotz, G.: Stand der feuerfesten Zustellung von Drehofenanlagen. Refratechnik-Bericht Nr. 19.

[3] Bartha, P.: Perilex 80 — Standardstein für die Brennzone von Zementdrehöfen. Refratechnik Bericht Nr. 34, 1989.

[4] Bartha, P.: Direktgebundene Periklasspinellsteine und ihr Einsatz in der Zementindustrie. Zement-Kalk-Gips 35 (1982) Nr. 10, S. 530—536.

[5] Bartha, P., und Schultes, H.: Feuerfeste Zustellung von Zementdrehöfen — Periklasspinellsteine und Magnesiachromsteine im Vergleich. TIZ-Fachberichte 108 (1984) Nr. 8, S. 507—514.

[6] Bartha, P.: Chemisch-mineralogische Charakterisierung ungebrauchter und gebrauchter feuerfester Steine auf der Basis $MgO \cdot Al_2O_3$. Zement-Kalk-Gips 38 (1985) Nr. 2, S. 96—99).

[7] Bartha, P.: Magnesia-Spinellsteine — Eigenschaften, Herstellung und Anwendung. Refratechnik Nr. 35, 1987.

[8] Weibel, G.: Chromerzfreie Feuerfeststeine für die Zustellung der Übergangs- und Sinterzonen von Zementdrehöfen. Refratechnik-Kolloquium Malmö 1989, Vortragsband, Göttingen 1989, S. 45—58.

[9] Klischat, H.-J., und Bartha, P.: Further development of magnesia-spinel bricks with their own specific properties for lining the transition- and sintering zones of rotary cement kilns. Refratechnik-Bericht Nr. 37, Göttingen 1982.

[10] Bartha, P.: Mineralogisch-chemische Veränderungen an Dolomitsteinen während ihres Einsatzes in Zementdrehöfen. Refratechnik-Bericht Nr. 13, Göttingen 1972.

[11] Münchberg, W., und de Jong, J.G.M.: Verhalten von keramisch gebundenen Dolomitsteinen in Zementdrehrohröfen bei Infiltration von Ofengasen. XVII. Internationales Feuerfest-Kolloquium Aachen, 1974.

[12] Patentschrift DE 22 49 814 C3
Gebrannter feuerfester Formkörper.

[13] Patentschrift DE 26 46 430 C2
Verfahren zur Herstellung von gebrannten feuerfesten Magnesiasteinen.

[14] Patentschrift DE 36 14 604 C2
Feuerfestzusammensetzung.

[15] Konopicky, K.: Beitrag zur Frage der Ansatzbildung in Drehöfen. Zement-Kalk-Gips 4 (1951), S. 240—245.

[16] Weibel, G.: Die Bedeutung der physikalischen Werkstoffeigenschaften für die Weiterentwicklung basischer feuerfester Steine. Refratechnik-Kolloquium Göttingen 1986, Vortragsband, S. 46—72.

[17] Mussgnug, G.: Ansatzbildung in Zementdrehöfen und Futterhaltbarkeit. Zement-Kalk-Gips 41 (1948), S. 41—46.

[18] Opitz, D.: Die Wandwärmeverluste bei kurzen Trockenöfen mit Vorwärmern. Zement-Kalk-Gips 56 (1967), S. 177.

[19] Weibel, G.: Die Bedeutung der physikalischen Werkstoffeigenschaften für die Weiterentwicklung basischer feuerfester Steine. Refratechnik-Kolloquium Göttingen 1986, Vortragsband, S. 46—72.

[20] Nachtwey, W., und Weibel, G.: Theorie und Praxis der Wärmeisolation basisch zugestellter Zonen in Drehöfen. Zement-Kalk-Gips 3 (1988), S. 125—128.

Ergebnisse und Erfahrungen mit den Ofenlinien von PSP Prerov

Results and experience with the kiln lines from PSP Prerov

Les lignes de four de PSP Prerov, résultats et expérience acquise

Las líneas de horno PSP Prerov – resultados obtenidos y experiencias adquiridas

Von **S. Bayer** und **J. Zajdlík,** Prerov/Tschechische Republik

Zusammenfassung – Prerovské strojírny Prerov – PSP, ein Lieferant von technologischen Ausrüstungen für Zementwerke, nahm in der letzten Zeit rekonstruierte Ofenlinien in der CSFR, Brasilien, Italien und Slowenien in Betrieb. Die Ofenlinie in Hranice, CSFR, ist für einen Durchsatz von 2500 t/d ausgelegt und besteht aus einem doppelsträngigen Schacht-Zyklon-Wärmetauscher mit Calcinierkammer, einem Drehofen und einem Rostkühler, aus dem die Tertiärluft entnommen wird. Als Brennstoff dient Steinkohle mit einem Heizwert von 25000 kJ/kg. Eine weitere Ofenlinie mit einem Durchsatz von 2800 t/d wurde in Italien in Betrieb genommen. Sie besteht aus einem einsträngigen Zyklonvorwärmer mit einem speziell für die Verbrennung von Petrolkoks gestalteten Calcinator, dem Drehofen (4,6 × 67 m), der Tertiärluftleitung und einem Rostkühler. Als Brennstoff dient eine Mischung aus Petrolkoks und Kohle (70 : 30). Der neueste Typ des Vorwärmers wurde im Zementwerk Anhovo-Slowenien eingesetzt. Es handelt sich um einen fünfstufigen Zyklonvorwärmer mit gegenüber herkömmlichen Bauarten um 25% vermindertem Druckverlust. Als Brennstoff wird an diesem Ofen (5 × 82 m) Gas eingesetzt. Für die Rekonstruktion der Ofenlinie in Brasilien wurden ebenfalls neue druckverlustarme Zyklone eingesetzt und in den bestehenden Wärmetauscherturm eingepaßt. Die vollständige Rekonstruktion der kompletten Ofenlinie dauerte nur 95 Tage.

Summary – Prerovské strojírny Prerov – PSP – a supplier of industrial equipment for cement works, has recently commissioned some reconstructed kiln lines in the Czech Republic, Brazil, Italy and Slovenia. The kiln line at Hranice in the Czech Republic is designed for an output of 2500 t/d and consists of a double-string shaft cyclone preheater with calcining chamber, a rotary kiln and a grate cooler from which the tertiary air is drawn. Coal with a calorific value of 25000 kJ/kg is used as the fuel. Another kiln line with a capacity of 2800 t/d was commissioned in Italy. It consists of a single-string cyclone preheater with calciner specifically designed for burning petroleum coke, the rotary kiln (4.6 × 67 m), tertiary air duct and grate cooler. The fuel is a mixture of petroleum coke and coal (70:30). The latest type of preheater was used in the Anhovo-Slowenia cement works. This is a five-stage cyclone preheater with a pressure drop which is 25% lower than with traditional designs. Gas is used as the fuel in this kiln (5 × 82 m). New low-pressure-drop cyclones were also used for rebuilding the kiln line in Brazil and were adapted to suit the existing preheater tower. The entire reconstruction of the complete kiln line took only 95 days.

Résumé – Prerovské strojírny Prerov – PSP, un fournisseur d'équipements technologiques pour cimenteries, a récemment mis en service des lignes de four reconstruites dans la CSFR, au Brésil, en Italie et en Slovénie. La ligne de four à Hranice, CSFR, est conçue pour une capacité de 2500 t/j et se compose d'un préchauffeur double voie à tour et cyclones avec une chambre de précalcination, d'un four rotatif et d'un refroidisseur à grille, d'où est prélevé l'air secondaire. Comme combustible sert du charbon, d'un pouvoir calorifique de 25000 kJ/kg. Une autre ligne de four, d'une capacité de 2800 t/j, a été mise en service en Italie. Elle possède un préchauffeur simple à cyclones, avec une chambre de précalcination spécialement conçue pour la chauffe au coke de pétrole, un four rotatif (4,6 × 67 m), un conduit d'air tertiaire et un refroidisseur à grille. Le combustible est un mélange de coke de pétrole et de charbon (70:30). Le type le plus nouveau de préchauffeur a été mis en oeuvre à la cimenterie Anhovo – Slovénie. Il s'agit d'un préchauffeur à cinq étages de cyclones avec, par rapport aux types de construction conventionnels, une perte de charge réduite de 25%. Ce four (5 × 82 m) est chauffé au gaz. Des cyclones nouveaux à faible perte de pression ont également été mis en place lors de la reconstruction de la ligne de four au Brésil et adaptés à la tour existante. L'entière reconstruction de la ligne de four complète n'a duré que 95 jours.

Resumen – Prerovské strojírny Prerov – PSP, un suministrador de equipos tecnológicos para fábricas de cemento, ha puesto en servicio últimamente unas líneas de horno reconstruidas en CSFR, Brasil, Italia y Eslovenia. La línea de horno de Hranice, CSFR, está dimensionado para un rendimiento de paso de 2500 t/d y consta de un intercambiador de calor de dos ramales, de pozo y ciclones, con cámara de calcinación, un

horno rotatorio y un enfriador de parrilla, del cual se extrae el aire terciario. Como combustible se emplea hulla con un valor calorífico de 25 000 kJ/kg. Otra línea de horno con un rendimiento de paso de 2 800 t/d se puso en servicio en Italia. Consiste en un precalentador de ciclones de un solo ramal y un calcinador concebido especialmente para quemar coque de petróleo, el horno rotatorio (4,6 × 67 m), la tubería de aire terciario y un enfriador de parrilla. Como combustible se emplea una mezcla de coque de petróleo y carbón (70:30). El tipo de precalentador más reciente fue utilizado en la fábrica de cemento de Anhovo, Eslovenia. Se trata de un precalentador de ciclones, de 5 etapas, que presenta una pérdida de presión un 25 % menor en comparación con los tipos tradicionales. Como combustible se emplea gas para este horno (5 × 82 m). Para la reconstrucción de la línea de horno en el Brasil se utilizaron asimismo nuevos ciclones, de reducida pérdida de presión, los cuales se adaptaron a la torre de intercambiador existente. La reconstrucción completa de toda la línea de horno sólo duró 95 días.

1. Fabrikationsprogramm der Maschinenbaufabrik Prerov für die Zementindustrie

Die PSP Prerov AG ist im Jahre 1952 als ein auf die Herstellung von Anlagen für Zementwerke, Kalkhütten, Ziegelfabriken, Keramikwerke und Gesteinsaufbereitungsanlagen orientierter Betrieb entstanden. Das Unternehmen hat seinen Sitz in Prerov, einer Stadt mit 50 000 Einwohnern in Mähren, im östlichen Teil der Tschechischen Republik.

Neben Brechern, Mahlanlagen, Dreh-, Schacht- und Tunnelöfen, Wärmetauschern, Kühlern, Vorhomogenisierungs- und Förderanlagen und Packmaschinen produziert die PSP AG Prerov Anlagen zur Entschwefelung von Abgasen, für Zuckerfabriken und für die chemische Industrie, weiterhin leichtere Schaltgetriebe, Kupplungen, einige Baustahlkonstruktionen und eine ganze Reihe von weiteren Erzeugnissen.

Im Bereich der Brenntechnik hat die PSP AG – außer anderen Ofentypen – fast 200 Drehrohröfen, davon 51 Naßöfen und 39 Öfen für die Zementherstellung im Trockenverfahren, vorwiegend mit Wärmetauschern, geliefert.

Zu den traditionellen ausländischen Abnehmern der Zementanlagen gehören Brasilien und auf dem asiatischen Kontinent der Mittlere und der Nahe Osten. In den letzten Jahren hat sich die Ausfuhr von Mahl- und Brecheranlagen sowie von Ofenstraßen nach Italien, dem Iran und nach Mittelamerika erfolgreich entwickelt. Dort schätzt man die Anpassungsfähigkeit der Firma und ihrer Anlagen an die oft spezifischen Forderungen der Kunden sowie in den weniger industriell entwickelten Gebieten den erheblichen Umfang an hochqualifizierter technischer Hilfe bei der Inbetriebnahme der von uns gelieferten, aber auch fremder Anlagen.

So wurde zum Beispiel vor einigen Jahren in Brasilien ein parallellaufender Wärmetauscher mit Calcinator von KHD mit einer Kapazität von ca. 500 t/Tag in den Betonturm des klassischen Fuller-Wärmetauschers mit einer Kapazität von 1 700 t Klinker/Tag eingebaut. In demselben Turm nahm PSP Prerov den Umbau des ursprünglichen Wärmetauschers zu einem Wärmetauscher mit teilweiser Vorcalcinierung vor, der die Gesamtkapazität der Linie auf 2 700 bis 3 000 t Klinker/Tag erhöht. Die umgebaute Ofenlinie wird mit schwerem Heizöl betrieben, das auf 240 bis 260 °C vorgewärmt wird und im Durchschnitt 5 % Schwefel enthält. Als Calcinierbrennstoff wird minderwertige Kohle verwendet. Der rechte Teil der Abbildung stellt den Umbau der ursprünglichen Zyklone auf Zyklone mit reduziertem Druckverlust dar.

Trotz der hohen Brenntemperaturen, die durch den umgebauten, sehr wirksamen Wärmetauscher hervorgerufen werden, muß im Ofen Brenngut mit einem Schmelzphasenanteil von ca. 27 %, der vorwiegend Eisen enthält, gebrannt werden. Die Montage und Inbetriebnahme dieser Ofenlinie brachte allen beteiligten Lieferanten eine Reihe von unschätzbaren Erfahrungen.

2. Erfahrungen mit dispersen Wärmetauschern von PSP Prerov

Der für die Firma typische Gegenstrom-Schachtwärmetauscher charakterisierte die Ofenlinien für das trockene Herstellungsverfahren von PSP Prerov bis in die Hälfte der 80er

1. Maschinenbaufabrik Prerov's manufacturing range for the cement industry

PSP Prerov AG was founded in 1952 for the production of plants for cement, lime, brick, ceramic and rock-processing works. The business has its head office in Prerov, a town of 50 000 inhabitants in Mähren in the eastern part of the Czech Republic.

Apart from crushers, grinding plant, rotary, shaft and tunnel kilns, preheaters, coolers, pre-homogenising and conveying plant and packing machines, PSP AG Prerov also manufactures plants for desulphurising exhaust gases, for sugar factories and the chemical industry, as well as lighter control gear, couplings, some steel constructions and a whole series of other products.

In the combustion field PSP AG have supplied almost 200 rotary kilns – apart from other types of kiln – of which 51 were wet kilns and 39 kilns for dry-process cement production, most supplied with preheaters.

Traditional foreign purchasers of the cement plant include Brazil and the Middle and Near East. Exportation of grinding and crushing plant and kiln lines to Italy, Iran and Central America has developed successfully in recent years. The adaptability of the company and its plant to the customers' often very specific requirements is appreciated in these countries, as is the amount of highly qualified technical assistance provided in the less industrially developed areas for commissioning our own plant as well as that from other companies.

A few years ago in Brazil, for example, a preheater with a KHD calciner with a capacity of approx. 500 t/day was installed in parallel in the concrete tower of the classic Fuller preheater which had a capacity of 1 700 t clinker/day. In the same tower PSP Prerov carried out conversion of the original preheater to one with partial pre-calcining, which raised the total capacity of the line to 2 700–3 000 t clinker/day. The converted kiln line is run with heavy fuel oil which is preheated to 240 to 260 °C and contains an average of 5 % sulphur. Low-grade coal is used as the calcining fuel. Conversion of the original cyclones to cyclones with reduced pressure drop is shown at the right-hand side of the illustration.

In spite of the high burning temperatures caused by the very effective converted preheaters, kiln feed with approximately 27 % liquid phase, consisting largely of iron, has to be burnt in the kiln. The erection and commissioning of the kiln line provided some invaluable experience for all the suppliers involved.

2. Experience with PSP Prerov's dispersion preheaters

The company's typical counter-current shaft preheaters were characteristic of PSP Prerov's kiln lines for the dry process until mid-way through the eighties. The simplicity, self-supporting construction and low pressure drop obtained with this preheater offered clear advantages until the world oil crisis and the subsequent rise in fuel prices.

The preheating of the raw material along the shaft walls as it rotates and falls counter-current to the preheating gases was theoretically the optimum thermodynamic solution. In practice however the energy efficiency of the long shaft was

Jahre. Die Einfachheit, die selbsttragende Konstruktion und der niedrige Druckverlust dieses Vorwärmertyps boten bis zur weltweiten Ölkrise und der folgenden Verteuerung des Brennstoffs klare Vorteile.

Die Vorwärmung des entlang der Schachtwände unter Rotation und gegenläufig zu den vorwärmenden Gasen sinkenden Rohstoffes war eine theoretisch optimale thermodynamische Lösung. Der energetische Wirkungsgrad eines langen Schachtes wurde in der Praxis jedoch durch die Rezirkulation des Rohstoffes von kälteren in heißere Bereiche des Wämetauschers reduziert. Nach Einführung der Vorcalciniertechnik erwies sich außerdem der Anschluß von Gleichstromcalcinatoren an diesen Schacht als problematisch.

Diese Schwierigkeiten wurden durch die Entwicklung des kombinierten Schacht- und Zyklonwärmetauschers COMBI überwunden, bei dem eine weitere Zyklonstufe oberhalb des Schachtes und in einigen Fällen auch unterhalb des Schachtes eingefügt wurde. Für Ofenlinien mit einem höheren Zweitfeuerungsanteil und vergleichsweise trockenem Rohstoff bildet der Calcinierkanal mit dem eigenen „heißen" Zyklon die unterste – fünfte Wärmetauscherstufe. Es wurden bereits 12 solcher Wärmetauscher in drei- bis fünfstufiger Ausführung, in Ein- sowie Zweistrangschaltung und mit unterschiedlichen Stufen der Vorcalcinierung geliefert.

Ein vierstufiger Wärmetauscher des Typs COMBI ohne Zweitbrennstoff hat den thermischen Wirkungsgrad eines klassischen fünfstufigen Zyklonvorwärmers, aber einen um ca. 15 mbar niedrigeren Druckverlust. Dies gilt auch für den Fall, daß dieser nicht mit druckverlustarmen Zyklonen ausgerüstet ist. Für diesen Wärmetauscher erteilte PSP Prerov bereits ausländische Lizenzen.

Die Abneigung einiger Zementhersteller gegenüber Schachtwärmetauschern beliebigen Typs führte bei PSP Prerov zur Entwicklung einer Reihe von Zyklonvorwärmern, die vorwiegend fünfstufig und mit unterschiedlicher Kapazität gebaut wurden.

Für diese Wärmetauscher ist eine neue Form der Zyklone charakteristisch, die eine Reduzierung der Druckverluste um ca. 25 % im Vergleich mit klassischen Zyklonen ermöglicht, und dies auch bei einer Verkleinerung ihres Innendurchmessers von z. B. 7,0 auf 6,5 m bei Anlagen mit einer Kapazität von 2000 bis 2400 t Klinker/Tag (je nach Brennstoffart). Diese Wärmetauscher mit der Gruppenbezeichnung LUCE – Low Underpressure Cyclone Exchanger – sind mit Zyklonen mit niedrigem Druckverlust in der zweiten bis fünften Stufe von oben ausgerüstet und werden mit dem ursprünglichen Calcinierkanal von PSP Prerov geliefert, der je nach den Umständen mit einer Tertiärluftleitung oder mit einer Calcinierkammer ergänzt wird. Im Mai 1993 wurde im Zementwerk Anhovo in Slowenien der Umbau des ursprünglichen Schachtwärmetauschers von PSP Prerov auf den Typ LUCE 5ST abgeschlossen.

Der erste Einsatz druckverlustarmer Zyklone ermöglichte in einem brasilianischen Werk eine Erhöhung der Rohmehlaufgabemenge von 100 auf 124 t/h. Dabei wurden die alten Zyklone an Ort und Stelle umgebaut.

3. Anlagen zur Vorcalcinierung des Rohstoffes

PSP Prerov entwickelte und liefert die Vorcalciniertechnik zur teilweisen sowie vollständigen Vorcalcinierung.

Für die italienische Gesellschaft COLACEM nahm PSP Prerov im Jahre 1985 in der Toscana den ersten Calcinierkanal für eine Ofenlinie, die nach dem AT-Verfahren (Air Through) arbeitet, in Betrieb. Obwohl als Brennstoff ein Gemisch mit 20–40 % des schwierig brennenden Petrolkokses verfeuert wird, wurden nach dem anschließenden Zyklon 0,01 % CO bei 0,5 % O_2 nicht überschritten. Die Rohstoff- und Gastemperaturen unterschieden sich um weniger als 10 K. Bei Verwendung dieses Kanaltyps in Anlagen mit einem Zweitfeuerungsanteil von nur 5–10 % können der Luftüberschuß aus der Hauptfeuerung genutzt und dadurch die Abgasverluste vermindert werden. Überdies

reduced by recirculation of the raw material from colder to hotter parts of the preheater. When precalcining technology was introduced problems were also experienced in connecting co-current calciners to this shaft.

These difficulties were overcome by the development of the COMBI combined shaft and cyclone preheater, in which a further cyclone stage was inserted above the shaft and in some cases also below it. For kiln lines with a larger secondary firing component and comparatively dry raw material the calcining duct with its own "hot" cyclone forms the fifth and lowermost preheating stage. Twelve such preheaters have already been supplied in the three- to five-stage model, with a single and double-string arrangement and with different precalcining stages.

A four-stage preheater of the COMBI type without secondary fuel has the thermal efficiency of a classic five-stage cyclone preheater but the pressure drop is about 15 mbar lower. This also applies in cases where no low-pressure-drop cyclones are fitted. PSP Prerov has already granted foreign licences for such preheaters.

The fact that some cement producers disliked shaft preheaters of any kind led PSP Prerov to develop a series of cyclone preheaters, largely of the five-stage type with different capacities.

The main characteristic of these preheaters is a new form of cyclone which can have a pressure drop about 25 % lower than with traditional cyclones, although the inside diameter is decreased from e. g. 7.0 to 6.5 m in plants with a capacity of 2000 to 2400 t clinker/day (according to the type of fuel). The group name for these preheaters is LUCE – low underpressure cyclone exchangers. They are equipped with low-pressure-drop cyclones at the second to fifth stage down and are supplied with PSP Prerov's original calcining duct, completed by a tertiary air duct or a calcining chamber according to circumstances. Conversion of PSP Prerov's original shaft preheater at the Anhovo cement works in Slovenia to the LUCE 5ST type was completed in May 1993.

When low-pressure-drop cyclones were used for the first time in a Brazilian works the raw meal feed rate was increased from 100 to 124 t/h. The old cyclones were converted in situ.

3. Plant for pre-calcining the raw material

PSP Prerov has developed and can provide the technology for partial or complete pre-calcining.

In 1985 in Tuscany PSP Prerov commissioned the first calcining duct for a kiln line operated by the AT (air through) process, for the Italian company COLACEM. Although the firing fuel is a blend containing 20 to 40 % of petroleum coke which is difficult to burn, the CO level downstream of the adjoining cyclone did not exceed 0.01 % CO at 0.5 % O_2. The raw material and gas temperatures differed by less than 10 K. The excess air from the main firing unit may be utilised when this type of duct is used in plant with a secondary firing component of only 5 to 10 %, thereby diminishing waste gas losses. The air excess also evens out the burn-out and limits the formation of harmful deposits.

The calcining duct was developed by using model trials so that stable operation is possible even with varying gas speeds. It enables the tertiary air duct to be connected to the lower part so that the proportion of secondary fuel can be raised to 40 to 50 %. Furthermore, the calcining chamber may be incorporated in the tertiary air duct if very intensive precalcination is required and the secondary fuel is unsuitable for the calcining process.

The first, swirl duct section forms the lower, rectangular part of the duct to which the raw material separated in the last cyclone or in the shaft, and possibly also the tertiary air, are fed.

vergleichmäßigt er den Ausbrand und beschränkt die Bildung von schädlichen Ansätzen.

Der Calcinierkanal wurde anhand von Modellversuchen so entwickelt, daß auch bei wechselnden Gasgeschwindigkeiten ein stabiler Betrieb möglich ist. Er ermöglicht einerseits den Anschluß der Tertiärluftleitung an den unteren Teil und damit eine Erhöhung des Anteils des Zweitbrennstoffs auf 40 – 50%. Andererseits kann die Calcinierkammer in die Tertiärluftleitung eingebaut werden, falls eine sehr intensive Vorcalcination verlangt wird und der Zweitbrennstoff für den Calcinierprozeß wenig geeignet ist.

Den ersten, wirbelnden Kanalteil bildet der untere, rechteckige Teil des Kanals, dem der im letzten Zyklon oder im Schacht abgeschiedene Rohstoff und gegebenenfalls die Tertiärluft zugeführt werden.

Der mittlere, zylinderförmige Teil ermöglicht beliebige Verdrehungen des Wärmetauschers gegenüber dem Ofen (was bei Umbauten von Bedeutung ist) und leitet die Dispersion in den Wirbelkopf des Kanals. Dort laufen gleichzeitig der Ausbrand des Brennstoffs und die restliche Vorcalcinierung ab.

Ist ein ausreichender Unterdruck durch eine Verengung des Ofeneinlaufbereichs entstanden, kann man den Austritt des Calcinatorgases aus der gleichströmigen Calcinierkammer in den Quadratquerschnitt einmünden lassen. Diese Ausführung ermöglicht eine zweistufige Vorcalcination und die Beschränkung des NO_x-Gehaltes in den Abgasen. Sie eignet sich für Kapazitäten von 2000 bis 3500 t Klinker täglich und wird in diesem Jahr in zwei Zementfabriken in Europa in Betrieb genommen. Eine von ihnen ist mit dem fünfstufigen Zweistrang-Wärmetauscher COMBI, die andere mit dem vierstufigen Zyklonvorwärmer ausgerüstet. Die Betriebsergebnisse dieser zwei Linien werden nach ausführlichen Messungen in der Fachpresse veröffentlicht.

The central, cylindrical part allows for any required rotation of the preheater relative to the kiln (an important feature in conversions) and guides the dispersion into the top swirl section of the duct. Here the fuel burns out and the rest of the precalcination takes place.

If an adequate negative pressure has been created by narrowing the kiln inlet section the outlet for the calciner gas from the co-current calcining chamber may discharge into the square cross-section. With this design there can be two-stage precalcination and the NO_x content of the waste gases can be restricted. The design is suitable for daily capacities of 2000 to 3500 t clinker and is being put into operation in two European cement works this year, one equipped with the COMBI five-stage two-string preheater and the other with the four-stage cyclone preheater. Operating results from these two lines will be published in the technical press when detailed measurements have been taken.

Energieeinsparungen bei langen Öfen
Energy savings in long kilns

Economie d'énergie en présence de fours longs

Ahorro de energía en los hornos largos

Von **A. J. de Beus,** Costa Mesa/USA

Zusammenfassung – In diesen High-Tech-Zeiten mit Zementanlagen, die nach dem Trockenverfahren mit 5-stufigen Zyklonvorwärmern, Calcinieranlagen, Vollautomation und Computern arbeiten, gibt es immer noch eine ganze Reihe von Zementwerken in der Welt, die lange Öfen nach dem Trocken- und Naßverfahren betreiben. Viele dieser Werke mit langen Öfen können nicht zum Trockenverfahren mit Vorwärmer umgebaut werden, weil entweder kein Kapital zur Verfügung steht oder keine Aussicht auf eine Rendite besteht, die die Investition rechtfertigen würde. In diesem Beitrag werden die verschiedenen Möglichkeiten zur Einsparung und Wiedergewinnung von Energie in langen Zementöfen kurz beschrieben. Der Beitrag befaßt sich hauptsächlich mit dem Wärmeaustausch innerhalb des Ofens, d.h. mit dem System der Ketten- und Staueinbauten und der Wender. Ofenketten und Wender sind nichts Neues. In den meisten Fällen wurden sie in kurzen Ofenabschnitten angebracht. Neu ist die Tatsache, daß man heute bessere Produkte zur Verfügung hat, die es den Ingenieuren gestatten, tiefer in diese langen Öfen hineinzugehen und Ketten und Wender aus einer rostfreien Speziallegierung einzubauen, die Temperaturwechsel-Beanspruchungen und hohen Temperaturen widerstehen können. Die in der Vergangenheit gemachten Erfahrungen haben zu einem sehr wirksamen Wärmeaustauschsystem für lange Öfen geführt.

Summary – In this stage of high technology, dry process cement plants with 5 stage vertical preheaters, with calciners, full automation, and computers, there are still quite a number of cement plants around the world that operate long dry and wet process kilns. Many of these long kiln plants cannot change over to preheater dry process kilns because they either do not have the capital or there is no economic return justifying the investment. In this paper the various methods of saving and recovering energy in long cement kilns are briefly described. The main subject will deal with the heat exchange system inside the kiln which means the system of chains, dams and tumblers. Kiln chains and tumblers are not a new idea. In most instances, they were installed in short sections of the kiln. What is new today is the availability of better products so that engineers can go deeper into these long kilns and install special alloy stainless chains and tumblers that can withstand thermal shock and high temperatures. The experience of the past has evolved into a really effective heat exchange system for long rotary kilns.

Résumé – Malgré notre époque high tech marquée par des cimenteries fonctionnant selon le procédé humide avec des préchauffeurs à cyclones à 5 étages, des installations de calcination, une automatisation intégrale et des ordinateurs, il existe toujours, de par le monde, toute une série de cimenteries utilisant des fours longs selon le procédé sec et le procédé humide. Parmi celles-ci, nombre d'usines à fours longs ne peuvent pas être converties au procédé sec avec préchauffeur du fait d'un manque de capital disponible ou parce que les perspectives de rendement ne justifient pas un tel investissement. Le présent exposé décrit brièvement les différentes possibilités d'économie et de récupération d'énergie dans des fours de cimenterie longs. Il se penche essentiellement sur le problème de l'échange thermique à l'intérieur du four, c'est-à-dire le système des équipements internes à chaîne, de retenue et d'inverseur. Les chaînes de four et inverseurs ne sont pas nouveau. Dans la plupart des cas ils ont été installés dans la portion de four courte. Ce qui est nouveau c'est le fait que l'on dispose, de nos jours, de meilleurs produits permettant aux ingénieurs d'aller plus loin dans ces fours longs et d'installer des chaînes et inverseurs en alliage inoxydable spécial, capables de résister aux températures élevées et contraintes de changement de température. L'expérience recueillie par le passé a conduit à un système d'échange thermique très efficace pour fours longs.

Resumen – En esta época de alta tecnología, en la que se emplean instalaciones de fabricación de cemento que funcionan por vía seca, con precalentadores de ciclones de 5 etapas, instalaciones de calcinación, con automatización total de los equipos, y ordenadores, siguen existiendo aún muchas plantas cementeras por todo el mundo que utilizan hornos rotatorios largos de vía seca y de vía húmeda. Muchas de estas fábricas que emplean hornos largos no pueden ser transformadas en instalaciones de vía seca con precalentador, porque no se dispone del capital necesario o bien porque las perspectivas de obtener beneficios no justifican la inversión. En el presente artículo se

describen, de forma resumida, las distintas posibilidades de ahorrar y de recuperar energía en los hornos largos para cemento. Se trata, sobre todo, el tema del intercambio de calor dentro del horno, es decir el sistema de dispositivos internos, tales como cadenas, dispositivos de retención y de inversión. Las cadenas y dispositivos de inversión para hornos no son nada nuevo. En la mayoría de los casos se instalaban en tramos de horno cortos. Lo nuevo es que hoy en día se dispone de mejores productos que permiten a los ingenieros penetrar más en los hornos largos y montar cadenas y dispositivos de inversión fabricados de una aleación especial inoxidable y resistente a elevadas temperaturas así como a las solicitaciones debidas al cambio de temperaturas. Las experiencias adquiridas en el pasado han permitido lograr un sistema de intercambio de calor muy eficaz los hornos largos.

1. Problemstellung

Im Jahr 1973 wurde vorhergesagt, daß die weltweit verfügbaren Erdölvorkommen bis 1990 erschöpft wären und der dann noch verbliebene kärgliche Rest über 100,00 US $/Barrel kosten würde. Damals wurde auch behauptet, daß es bis 1990 keine Zementöfen mit Naßverfahren mehr gäbe. Heute jedoch ist Öl in beliebigen Mengen zum Preis von 17,00 US $/Barrel verfügbar, und weltweit sind in der Zementindustrie immer noch etwa 650 Naßöfen in Betrieb. Zusätzlich existiert noch eine Vielzahl langer Trockenöfen zur Herstellung von Kalk, Magnesium, Aluminium, Eisen und Kieselgur, ganz zu schweigen von Trommeltrocknern.

Dieser Artikel beschäftigt sich in erster Linie mit langen Zementöfen in den Staaten der ehemaligen Sowjetunion und Osteuropas. Aber auch in Amerika sind in der Zementindustrie noch viele Naßöfen im Einsatz, die jedoch durch den Einbau der unten beschriebenen Einrichtungen bereits modernisiert wurden.

Warum wurden die Werke mit langen, im Naßverfahren betriebenen Öfen im ehemaligen Ostblock nicht modernisiert? Seit mehr als 25 Jahren sind Werke mit dem vierstufigen Trockenverfahren bekannt. Einige wichtige Gründe hierfür sind die immer noch relativ günstigen Brennstoffpreise, der in der ehemaligen Sowjetunion relativ niedrige Preis für Zement, der Kapitalmangel oder der Mangel an harter Währung. Letztlich stellt auch die enorm schwankende Zementnachfrage Investitionen immer wieder in Frage. In den USA gab es in den letzten 12 Jahren zwei starke Rezessionen auf dem Bausektor. In der ehemaligen Sowjetunion liegt die Produktion bei nur 40% der installierten Leistung. Darüber hinaus ist für Werke, die über Rohstoffvorkommen mit 25–35% Feuchtigkeitsgehalt verfügen, eine Umstellung auf das Trockenverfahren nicht sinnvoll.

Eine Möglichkeit zur Verminderung der Brennstoffkosten ist der Einsatz von Abfallbrennstoffen. Lange Öfen sind ideal für die Verbrennung von Abfallstoffen geeignet, die in Vorwärmeröfen Probleme verursachen können. Bestimmte chemische Elemente und Metalloxide können zu Anbackungen und Verstopfungen in Zyklonvorwärmer- oder Alkali-Bypass-Systemen führen.

Es ist davon auszugehen, daß auch in 20 Jahren weltweit immer noch mehr als 800 lange Öfen in Betrieb sein werden, davon mindestens 500 Zementöfen. Deshalb ist es wirtschaftlich sinnvoll, Geld in eine vorhandene, nach dem Naßverfahren betriebene Anlage zu investieren, um Öfen und Mühlen leistungsfähiger zu machen. Die meisten dieser Investitionen amortisieren sich innerhalb von zwei Jahren, so daß eine Anlage mit Naßverfahren während der restlichen Lebensdauer noch sehr wirtschaftlich betrieben werden kann.

Derzeit beträgt in vielen der früheren Sowjetrepubliken der Anteil der Energiekosten nahezu 70% der Herstellungskosten, die Personalkosten liegen dagegen bei nur 1–2%. Oberstes Anliegen ist daher die Senkung der Kosten für thermische und elektrische Energie. Der mittlere Brennstoffenergieverbrauch liegt bei 1800 kcal/kg Klinker. Fast alle Mahlanlagen, Rohmühlen und Fertigmahlanlagen, arbeiten im offenen Kreislauf. Die wenigsten verfügen über geeignete Meßsysteme, um günstige Verbrauchszahlen für thermische und elektrische Energie erzielen zu können.

1. The problem and what can be done

In 1973 there were predictions that there would be no oil left in 1990, that whatever there was left would cost more than $ 100,— a barrel. At that same time some people said that by 1990 there would not be any wet process cement kilns left. Well, today, there is plenty of oil at $ 17.00 a barrel and there are about 650 wet process cement kilns left. In addition, there are still many other long kilns left, processing lime, carbon, magnesia, alumina, iron, diatomaceous earth. Not to speak about all the rotary dryers.

This paper deals primarily with long cement kilns located in the ex Soviet republics and Eastern European countries. There are also many wet process kilns in America but those kilns have already been modernized with the systems described below.

Why have not these long wet kiln plants been modernized? After all, the 4 stage dry process type of plant already exists more than 25 years. There are various reasons, old reasons and new reasons. Some of the old reasons are: Still the relatively low cost of fuel, and secondly, the still relatively low price of cement in the ex Soviet Union, thirdly the lack of capital or, in the case of the ex Soviet Union, the lack of hard currency and the lack of capital, fourthly terrible cycling cement markets. Just when one feels comfortable to make an investment, the cement market collapses again. In the U.S., in the last 12 years there were 2 very severe recessions in construction. In the ex Soviet Union production is 40% of the installed capacity. Fifthly, for some plants it just does not make sense to convert to dry process because the raw materials are quarried with 25–35% moisture content.

Some new reasons are the use of waste fuels. Long kilns are ideal to burn certain waste fuels which might cause operating problems in preheater kilns. Certain chemical elements and metal oxides can clog up the cyclones of a preheater system or clog up the alkalies bypass system.

It can be safely assumed that, 20 years from now, there will still be more than 800 long kilns in operation, of which at least 500 cement kilns. Therefore, it makes economic sense to invest money in the existing wet plant process to make the kilns and mills much more efficient. Most of the investments can be recovered within 2 years. Thus, still much money can be saved over the rest of the life of the wet process plant.

Today, the total energy costs in many former Soviet republics approach 70% of the manufacturing costs. The labor costs are only 1–2%. Thus, the main thrust right now is to reduce the thermal and electrical energy costs. The average fuel consumption is 1800 kcal/kg clinker. Almost all the grinding mills, raw and finish, are open circuit mills. Few people have measuring devices to really come up with good numbers on thermal and electrical consumption.

This paper deals primarily with the kilns and coolers with some side references to raw materials preparation, moisture reduction in slurry and grinding. A long kiln can be equipped with chains, lifters, tumblers, dams, insulating refractory. Half of the kiln length can be filled up with some sort of heat exchanger. If it is done properly, the thermal energy reduction can be lowered to 1350 kcal/kg clinker the exact value depending upon the moisture content in the slurry. Obviously, a slurry with 40% moisture will cost more energy than a slurry with 33% moisture.

Dieser Beitrag beschäftigt sich vorwiegend mit der Optimierung von Öfen und Kühlern mit einigen Anmerkungen zur Problematik der Rohmaterialaufbereitung, der Verminderung der Rohschlammfeuchte und der Mahlung. Ein langer Ofen kann mit Einbauten wie Ketten, Hebern, Wendern, Stauringen und einer isolierenden Ausmauerung ausgerüstet werden. In mehr als der Hälfte des Ofens können verschiedene Arten von Wärmetauschern eingebaut werden. Bei richtiger Ausführung kann der thermische Energieverbrauch auf 1350 kcal/kg Klinker gesenkt werden, wobei der genaue Wert vom Feuchtigkeitsgehalt des Rohschlamms abhängt. Selbstverständlich erfordert die Trocknung von Rohschlamm mit 40 % Feuchte mehr Energie als mit nur 33 %.

2. Vergleich mit modernen Zyklonvorwärmeranlagen

Die meisten Rohmaterialvorkommen der ehemaligen Sowjetunion haben einen hohen Feuchtigkeitsgehalt. Demnach läge der Verbrauch an thermischer Energie, einschließlich Trocknung, für eine 4- bis 5stufige Zyklonvorwärmeranlage in Rußland bei etwa 850 kcal/kg Klinker. Das sind 500 kcal/kg Klinker weniger als bei einem leistungsfähigen langen Ofen mit Naßverfahren. Bei den heutigen Brennstoffpreisen entspricht das etwa 5,00 US $/t Klinker. Der Umbau zum Trockenverfahren einer typischen Anlage mit 1 Mio. Tonnen Jahresproduktion würde 100 Mio. US $ kosten. Allein die Zinslast für 100 Mio. US $ läge bei 7 Mio. US $ pro Jahr, entsprechend 7,00 US $/t Kl. Gegenwärtig könnten die 100 Mio. US $ bei einer Amortisationszeit von 20 Jahren zum Preis von 6,72 US $ pro Tonne Zement finanziert werden.

Werden 50 % des Brennstoffs durch „kostenlose" Sekundärbrennstoffe ersetzt, so können die Brennstoffkosten für einen Naßofen tatsächlich geringfügig unter denen eines Ofens mit Trockenverfahren liegen. In Europa gibt es mehrere Werke, die nach dem Naß- bzw. Trockenverfahren mit gleichen Brennstoffkosten arbeiten, da sie „kostenlose" Abfälle und Sondermüll mit einem Anteil von 50 % am thermischen Energieverbrauch verbrennen.

Auf der Basis dieser Überlegungen besteht in den ehemaligen Sowjetrepubliken keine Notwendigkeit für eine übereilte Konversion vom Naß- zum Trockenverfahren. Es ist wirtschaftlich günstiger, beim Naßverfahren zu bleiben. Bei der Neuplanung eines Werkes sollte natürlich in den meisten Fällen eine Anlage mit der neuesten Technologie des Trockenverfahrens gebaut werden.

Es ist also finanziell sicherer, Geld in die Verbesserung des Wirkungsgrades langer Öfen zu investieren. Diese Investition kann sich innerhalb von 2 Jahren amortisieren, da die langen Öfen in den meisten Werken noch 10 bis 15 Jahre in Betrieb sein werden.

3. Möglichkeiten der Energieeinsparung bei Naßöfen

Folgende Maßnahmen kommen hierfür in Frage:

— Einbau von Ketten, Rippen und Dämmen;
— Verbesserung des Klinkerkühlerbetriebs;
— Verminderung der Rohschlammfeuchte durch im geschlossenen Kreislauf betriebene Mühlen;
— Einsatz von Abfallbrennstoffen;
— Verwendung von Brennern mit hohem Wirkungsgrad;
— Installation von Steuer- und Regelsystemen.

Mit der Durchführung von nur einigen dieser Maßnahmen könnte der Gesamtverbrauch an thermischer und elektrischer Energie um etwa 25 % gesenkt und die Ofen- und Mühlenleistung um 25 % gesteigert werden.

Notwendige Investitionen

Die Gesamtkosten der oben genannten Punkte belaufen sich auf weniger als 4 Mio. US $. Eine Anlage mit Halbnaßverfahren mit einer Jahresproduktion von 1 Mio. t kostet

2. Comparison with the modern dry process plant

Most ex Soviet raw materials have high moisture contents in the quarried materials. When including the drying of such raw materials the typical thermal energy consumption of a 4–5 stage preheater plant in Russia would be about 850 kcal/kg clinker. That is 500 kcal/kg clinker lower than a very efficient long wet process kiln. That is worth about $ 5,00 per ton of clinker, on the basis of today's fuel prices. To convert a typical 1 million ton/year plant to dry process might cost 100 million dollars. The interest cost alone on 100 million is 7 million/year = $ 7,00/ton. Actually, on a 20 year recovery of investment basis, the 100 million can be recovered at a cost of $ 6.72/ton of cement.

If 50 % of the fuel can be "no cost" waste fuels then, a long wet process kiln's fuel cost can actually be slightly lower than a dry process kiln. There are several plants in Europe who operate wet process kilns at the same fuel cost as dry process kilns because they burn "no cost" hazardous and non hazardous waste fuels at a rate of 50 % of total fuel.

On the basis of the above considerations there is no need to rush with conversion wet to dry in the ex Soviet republics. It is cheaper to stay wet. Of course, if a new plant is going to be built anyway, one should build, in most cases, the latest technology dry process plant.

Thus, it is financially safe to invest money in making long kilns more efficient. The investment can be fully recovered within 2 years while those long kilns will still be in operation, in most plants, for more than 10–15 years.

3. How to save energy in wet process plants?

1. Install effective chain systems, tumblers and dams
2. Improve clinker cooler operation
3. Reduce moisture in slurry by close circuiting the raw grinding mills
4. Use waste fuels
5. Use effective kiln burners
6. Control and monitoring systems

Implementation of some of these measures could reduce total energy consumption, thermal and electrical, by 25 %. Kiln and raw mill capacities would be increased by 25 %.

Cost of wet plant modifications

The total costs of items 1. through 6. is less than 4 million dollars. A semi wet process plant cost about 20 million U.S. dollars, for a one million ton/year plant. The thermal energy consumption for a semi wet process plant varies between 1100–1200 kcal/kg clinker, the exact amount depending upon the final moisture content in the filter cake. Conversion to a semi wet process is economically viable however, one would still need about 20 million U.S. dollars.

4. Chains, tumblers and dams

Quality of chains

Figs. 1 and 2 show the arrangement of kiln chains, tumblers and dams. Of course we all know about kiln chains for a long time. What is new is that there are chains available now that can handle gas temperatures up to 1200 degrees C, on a sustained basis. Chains in the hottest zone are subject to thermal cycling, corrosion by alkalisulphates and mechanical stress at high temperatures. Commercially available stainless materials such as 18–8 (18 % chrome, 8 % nickel) and 25–12 can, of course, be used. Chains made from such materials have welded links. Other types of chains are cast chains. Complete strands of chains are cast to order. Such strands will have no welded links. Then, there are other types of cast chains where every 28th link is carefully welded in an inert atmosphere.

etwa 20 Mio. US $. Der thermische Energieverbrauch einer Halbnaßanlage schwankt zwischen 1100 und 1200 kcal/kg Klinker, wobei die genauen Verbrauchszahlen von der Endfeuchtigkeit des Filterkuchens abhängen. Ein Umbau zum Halbnaßverfahren ist wirtschaftlich machbar, jedoch würden dazu auch rund 20 Mio. US $ benötigt.

4. Ketten, Rippen und Dämme

Kettenqualität

Die **Bilder 1** und **2** zeigen die Anordnung der Ketten, Rippen und Dämme im Ofen. Natürlich sind Ofenketten seit langem bekannt. Neu dabei ist, daß heute Ketten verfügbar sind, die für Gastemperaturen bis 1200°C auf Dauer geeignet sind. In der Hochtemperaturzone sind Ketten Temperaturschwankungen, Korrosionsangriffen durch Alkalisulfate und mechanischer Belastung bei hohen Temperaturen ausgesetzt. Marktübliche rostfreie Werkstoffe wie 18–8 (18% Chrom, 8% Nickel) und 25–12 können verwendet werden. Die Glieder dieser Ketten sind geschweißt. Eine andere, bessere Möglichkeit besteht darin, komplette Kettenstränge nach Kundenauftrag zu gießen. Diese Arten von Gußketten haben entweder keine geschweißten Glieder oder jedes 28. Verbindungsstück ist sorgfältig in inerter Atmosphäre geschweißt.

Die geschweißten Ketten sind am billigsten, aber ihre Qualität ist schlechter als die von Gußketten. In der heißesten Zone des Kettensystems öffnen sich häufig die Schweißnähte schon kurz nach dem Einbau im Ofen. Sind die Schweißnähte einmal offen, zieht sich das Kettenglied bei hoher Temperatur und mechanischer Zugspannung auf, und die Kette geht verloren. Bei gegossenen Ketten können diese Schäden nicht auftreten.

Ein weiterer Nachteil der marktüblichen schmiedeeisernen Ketten ist, daß im Gegensatz zu gegossenen Ketten die Legierung nicht so spezifisch gewählt werden kann, daß sie Temperaturschwankungen, Korrosion und mechanischen Spannungen bei hohen Temperaturen standhält. Eine gute Faustregel wäre etwa:

Wärmebehandelte Ketten aus Kohlenstoffstahl	0–300°C
Ketten aus Kohlenstoffstahl	300–500°C
Geschweißte ferritische rostfreie Ketten	500–600°C
Geschweißte austenitische (oder gegossene) Ketten	600–850°C
Gegossene austenitische rostfreie Ketten	850–1200°C

Die genannten Temperaturen sind Gastemperaturen. In der Praxis sollten die letzten 25 bis 30% der Kettenlänge austenitische rostfreie gegossene Ketten sein. Die hochwertigen gegossenen Ketten schützen das restliche Kettensystem. Fehlen sie, entstehen in der Mitte des Ketteneinbaus klaffende Löcher.

Letztlich ist es wirtschaftlich günstiger, 200 000 US $ in Ketten mit einer Lebensdauer von 5 Jahren zu investieren, als im gleichen Zeitraum 2 bis 3 Mal 150 000 US $ für schlechtere Qualitäten auszugeben.

Qualität von Rippen und Dämmen

Auch Rippen sind nichts Neues. Das technische Problem mit Rippen oder allen vorstehenden Steinen in Öfen besteht darin, daß die Isothermen den Umrissen der vorstehenden Steine oder Rippen folgen. Das bedeutet, daß die Isotherme durch die Rippe länger ist als durch den darunterliegenden Stein. Dies verursacht äußere Spannungen, was dazu führt, daß die Rippe an der Kontaktlinie mit dem Stein abbricht. Um dies zu verhindern, müssen die Rippen von hoher Qualität, vorgegossen und thermisch vorbehandelt sein und mit einem speziellen inneren Gerüst aus rostfreiem Draht und Ankern versehen sein. Das gleiche gilt für Dämme aus Feuerfestmaterial. **Bild 2** zeigt den kleinen Eintrittswinkel des Damms an der Ausmauerung. Das erleichtert den Rohmaterialfluß und begrenzt Spannungen im Damm.

Die Verwendung von Produkten hoher Qualität erlaubt dem Betreiber, 30% der Ofenlänge mit Ketten und weitere 20% mit Rippen und Dämmen zu versehen.

The welded chains are the cheapest and, the cheaper they are the poorer the quality. In the hottest zone of the chain system, the welds of many welded stainless chains will open, shortly after they are installed in the kiln. Once the welds are open the link will elongate under high temperature and the mechanical tensile stress and the chain will be lost. This cannot happen to cast chains.

Another disadvantage of commercially wrought chains is that one must accept the alloy that is commercially available. With cast chains one can tailor the alloy i.e. specially design the alloy to withstand the thermal cycling, the corrosion and the high mechanical stresses at elevated temperatures. A good general practice would be:

Heat treated carbon steel chains	0– 300 degrees C
Carbon steel chains	300– 500 degrees C
Welded ferritic stainless chains	500– 600 degrees C
Welded austenitic (or cast) chains	600– 850 degrees C
Cast austenitic stainless chains	850–1200 degrees C

The above temperatures are gas temperatures. In actual practice, the last 25–30% of the chain system length should be austenitic stainless cast chains. The high grade cast chains protect the rest of the chain system. If they are not there, gaping holes appear in the middle of the chain system.

It is better to install $ 200 000 of chains lasting 5 years then to install 2–3 times $ 150 000 worth of cheaper chains in that same period of time.

Quality of tumblers and dams

Tumblers are not new either. The technical problems with tumblers or any raised bricks in a kiln is that the isotherms

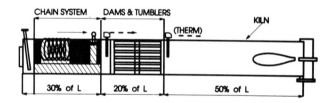

HEAT EXCHANGE SYSTEM LONG CEMENT KILN

BILD 1: Wärmetauschersystem im langen Zementofen
FIGURE 1: Heat exchange system in long cement kiln

Ketten	= Chain System
Dämme & Rippen	= Dams & Tumblers
Thermoelement	= Therm
Ofen	= Kiln
Ofenstaub	= Kiln dust
Thermoelement	= Thermo couple
Ofenlänge	= Kiln length

CROSSECTION OAM

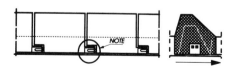

BILD 2: Rippen an der Ausmauerung im Detail
FIGURE 2: Details of refractory tumblers

Ofenlängsschnitt	= Crossection cam
Ausschnitt	= note

Verbesserung der Wärmeübertragung im Ofen

Der Zweck dieser Einbauten besteht darin, das Rohmaterial in einem Naßofen mit gegebener Ofenlänge besser als bisher vorzuwärmen. Tatsächlich kann ein optimiertes Ketten- und Rippensystem die erforderliche Ofenlänge verringern.

Das Brenngut wird nach Verlassen der Ketteneinbauten durch die verschiedenen Abschnitte, die von den Rippen und Dämmen gebildet werden, transportiert. Auf diese Weise wird das Brenngut umgewälzt und die Wärme von den heißen Gasen, dem Ofenfutter, den Dämmen und Rippen an der Ofenwand auf das Material übertragen.

Bei der Anordnung von Ketteneinbauten sind die Details sehr wichtig. Ketten von 200 t Gewicht können bei unsachgemäßem Einbau Probleme bei Staubentwicklung, Materialfluß und Ofenbetrieb hervorrufen, während der Betrieb bei richtigem Einbau problemlos funktioniert.

Die wichtigen Details sind Form und Größe der Kettenglieder, die Dichte jedes Kettenabschnitts, die Kettenlänge, die Konstruktion der Befestigungen sowie die Anzahl der Startspiralen über dem Durchmesser.

Die Entscheidung für Kettengirlanden oder -vorhänge ist oft eine Frage der Wartung. Theoretisch ist der Wirkungsgrad von Girlanden bei entsprechender Wartung geringfügig höher. Die meisten Zementwerke sind jedoch zu diesem Aufwand häufig nicht in der Lage und haben sich deshalb für Vorhang-Systeme entschieden. Dennoch ist gelegentlich bei extrem plastischem (klebrigem) Material und bei hoher Staubbeladung im Ofen ein Girlanden-System erforderlich. Für Öfen, in denen Filterkuchen gebrannt werden (Halbnaßverfahren), sind Girlanden besser geeignet als Kettenvorhänge.

Das Gewicht der Ketten liegt normalerweise bei etwa 15% der täglichen Ofenleistung. Produziert ein Ofen also 1800 t/d, sollte das Gewicht der Ketten bei ca. 270 t liegen.

5. Verbesserung des Klinkerkühlerbetriebs

Bild 3 zeigt das Schema eines Rostkühlers. Die meisten Klinkerkühler in der ehemaligen Sowjetunion sind zu groß und sehr schlecht ausgerüstet. Ist ein Kühler zu breit, verteilt sich der Klinker nur schlecht über die ganze Fläche, und für gewöhnlich ist das Klinkerbett deshalb flach. Dies bewirkt einen schlechten Wärmeaustausch und kann im Extremfall zur Überhitzung und Schädigung des Rostes führen. Ein störanfälliger Klinkerkühlerbetrieb legt oft die gesamte Ofenanlage still. Es sind auch keine Vorrichtungen zum Schutz des Klinkerkühlers vor starker Überhitzung infolge eines Hitzestaus vorhanden. Wird ein Rostkühler auch nur ein einziges Mal stark überhitzt, kann sich der Hauptrahmen verwerfen, der Kühler ist dann nicht mehr ausgerichtet und kann bis zur nächsten Überholung nicht mehr richtig arbeiten.

Die meisten Klinkerkühler in der ehemaligen Sowjetunion sollten mit mehr Lüftern zur Kühlung, besseren Dichtungen unter dem Rost, einem feststehenden luftgekühlten Einlauf, Luftkanonen, modernen Rostplatten, einem durch feuerfeste Ausmauerung verengten Einlauf, Betthöhen bis zu 50 cm und einer Regeleinrichtung ausgerüstet werden. Bei sehr grobstückigem Klinker ist ein zwischengeschalteter Walzenbrecher günstiger als ein nachgeschalteter Klinkerbrecher. Viele Klinkerbrecher in sowjetischen Kühlern sind nicht über die volle Breite eingebaut.

Die grundlegenden Steuereinrichtungen müßten aus

- einer Regelung des Druckes unter dem Rost zur Steuerung der Rostgeschwindigkeit,
- einem gekoppelten Master-Slave-System für die Mehrfachantriebe und
- einer Regelung des Unterdruckes am Ofenkopf zur Steuerung des Kühler- und des Kühlerabluftgebläses

bestehen. Die Drosselklappen arbeiten gegenläufig, d.h. wenn die Klappe des Kühlergebläses öffnet, schließt die des Abgasgebläses. Das Kühlerabluftgebläse ist außerdem oft zu klein ausgelegt und kann bei Betriebsstörungen einen Überdruck im Ofenkopf nicht vermeiden.

follow the contour of the raised bricks or the tumblers. That means that the isotherm through the tumbler is longer than through the underlying brick. The difference in length causes external stresses i.e. the tumbler wants to break off at the brick line. To prevent that, one needs high quality tumblers, precast and kiln fired and equipped with a specially designed internal stainless wire and anchor structure. The same applies to the refractory dams. Notice in Fig. 2 the small entry angle of the refractory dam. That is to facilitate flow of the raw materials and to limit stresses in the dam.

The high quality of chains, dams and tumblers permits the engineer to have a chain system length of 30% of kiln length, followed by tumblers and dams for another 20% of kiln length.

How the process works

The whole purpose of these systems is to preheat the raw materials much better than they are heated now in a typical wet process kiln over a given length of the kiln. In fact, a much improved chain and tumbler system replaces kiln length.

As one can see, after the raw materials discharge from the chain system they cascade from one refractory section into the next, the refractory sections being formed by the tumblers and dams. Thus the raw materials are rolled over, the core of the load exposed, everything being surrounded by the hot gases, hot kiln lining, hot refractory dams and hot tumblers.

The details of a chain system design are very important. One engineer could install a 200 ton system that causes a lot of problems in dust, material flow and kiln operation. Another engineer could install a same weight system without any problems.

The details are the shape and size of the links, the density of each chain section, the lengths of the chains, the design of the attachments, the number of starting spirals on the diameter.

Whether to install garlands or curtain chains is often a matter of maintenance skills. Theoretically, garlands are slightly more efficient than curtain chains but only if they are maintained properly. Most cement plants are unable to maintain garlands properly thus, they have opted for all curtain chain systems. Nevertheless, there is a need for an all garland system, from time to time, if the materials are extremely plastic (sticky) and if there is a high dust loss from the kiln. Filter cake kilns (semi wet process) may have a higher need for gar-

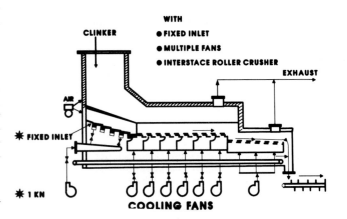

BILD 3: Schematische Darstellung des Klinkerkühlers
FIGURE 3: Schematic drawing clinker cooler

mit	= with
festem Einlauf	= fixed inlet
mehreren Lüftern	= multiple fans
zwischengeschalteter Walzenbrecher	= interstage roller crusher
Klinker	= Clinker
Abgas	= Exhaust
Luft	= Air
fester Einlaß	= fixed inlet
Kühlergebläse	= cooling fans

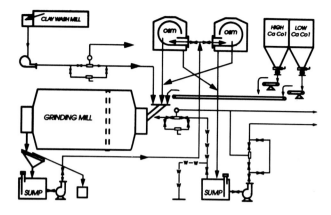

BILD 4: Mahlwerk mit geschlossenem Kreislauf
FIGURE 4: Closed circuit raw grinding

Tonschlämmühle = Clay wash mill
Viel = high
Wenig = low
Mühle = Grinding mill
Sumpf = Sump

6. Optimierung der Rohmaterialaufbereitung

Mahlung des Rohmaterials im geschlossenen Kreislauf

Bild 4 zeigt eine charakteristische Mahlanlage für Rohmehl mit geschlossenem Kreislauf. Das Mahlgut wird auf spezielle feststehende, gekrümmte Siebe gepumpt. Eine Strahlpumpe entlädt den Schlamm am Boden des Siebes unter Druck. Die Strahlpumpe kann auf beide Seiten des um 270° gewölbten Siebbodens geschwenkt werden. Zur beidseitigen Beladung des Siebes können auch 2 Strahlpumpen eingesetzt werden. Das Sieb besteht normalerweise aus rostfreien Keilstäben mit einem Abstand von 0,5 mm. Ist das Rohmaterial jedoch auf Grund von Quarzanteilen sehr abrasiv, kann auch Synthetikmaterial erfolgreich eingesetzt werden. Für einen Ofen mit einer Kapazität von 500 000 t/a sind vier Siebe erforderlich. Jedes Sieb benötigt eine eigene Pumpe, um den Druck wirkungsvoll zu regeln.

Das Mahlen im geschlossenen Kreislauf spart 4 kWh/t an elektrischer Energie und reduziert die Feuchtigkeit des Brennguts um 3–4 %. Beim Mahlen im offenen Kreislauf werden höhere Feinanteile erzeugt, was mit einem Mehrverbrauch von 4 kWh/t an elektrischer Energie verbunden ist. Auch wird für einen Rohschlamm mit hohem Anteil an sehr feinen Partikeln mehr Feuchtigkeit in der Mühle benötigt, um deren Verstopfen zu verhindern. Eine größere Feinfraktion macht den Rohschlamm zäher und erfordert mehr Wasser, um ihn pumpfähig zu machen.

Die gewölbten Siebe haben einen entwässernden Effekt, d. h. ein Teil des Wassers läuft mit dem Mahlrückstand in die Mühle zurück. Die Mühle selbst braucht weniger Wasser, da das Mahlgut gröber ist. Das Mahlgut durchläuft die Mühle schneller, und Feingut wird durch die Siebe abgezogen. Das umlaufende Mahlgut beträgt ca. 200 %. Deshalb sind die Abzugspumpen der Mühle und die Siebe auf 200 % Leistung auszulegen.

Es ist offensichtlich, daß mit der Naßmahlung im geschlossenen Kreislauf sehr viel thermische und elektrische Energie eingespart werden kann. Deshalb sollte das Mahlen im offenen Kreislauf der Vergangenheit angehören.

Umstellung auf das Halbnaßverfahren

Die 3 Hauptsysteme zur Rohmaterialtrocknung beim Halbnaßverfahren sind Sprühtrockner, Filterpressen und Zentrifugen. In einem Sprühtrockner wird der Rohschlamm sehr schnell getrocknet. Mit Filterpressen und Zentrifugen kann der Feuchtigkeitsgehalt auf 15–20 % verringert werden. Zentrifugen sind in Kalkrückgewinnungsanlagen von Papiermühlen weit verbreitet. Dabei sind die Haupteinsatzstoffe Kalziumkarbonate. In der Zementherstellung werden sie jedoch aufgrund der niedrigen Sedimentationsge-

land chains although there are some filter cake kiln in operation that have 100 % curtain chains.

The total weight of a chain system is usually not higher than 15 % of the daily kiln capacity. Thus, if a kiln produces 1800 tons per day, such kilns should have a chain weight of approximately 270 tons.

5. Improved clinker cooler operation

Fig. 3 shows a schematic of a grate clinker cooler. Most ex Soviet clinker coolers are too large and are very poorly equipped. When a clinker cooler is too wide there is poor clinker distribution over the full width of the cooler and, usually a low bed depth. This causes bare grates, loss of grates and poor heat exchange. Poor clinker cooler operation often shuts down the kiln. Also, there are no provisions to protect the clinker cooler against a severe case of overheating usually as a result of a hot push. When grate clinker cooler has been severely overheated just once, the main frame could warp, the cooler will be out of alignment and will never run right again until the next maintenance shutdown.

Most ex Soviet clinker coolers should be equipped with more cooling air fans, better undergrate seals, a fixed air-cooled inlet end, air cannons, latest technology grate with slanted holes, a narrowed inlet end (with refractory walls), bed depths up to 50 cm, a control system. In some cases, where large clinker is produced, an interstage roller crusher would be better than the after cooler clinker breakers. Many clinker breakers on the Soviet coolers are not full width.

The basic control systems would consist of:

— an undergrate pressure control regulating the speed of the cooler,
— a master-slave system for the multiple drives, and
— a hood draft control system that manipulates the dampers on certain cooling fans and on the cooler exhaust fan.

These dampers work opposite i.e. when the cooling fan damper opens, the exhaust fan damper closes. Also, the cooling exhaust fan is often too small and cannot evacuate the air under upset conditions which then causes a positive hood pressure.

6. Moisture reduction in the kiln feed

Closed circuit raw grinding

Fig. 4 shows a typical closed circuit raw grinding system. The mill discharge is pumped to special screens. These are stationary curved screens. An injector discharges the slurry under pressure at the bottom of the curve. The injector can swivel to either the left or right bottom of the 270 degree curved screen. Also 2 injectors can be used loading the screen on both ends. The screen is usually made of stainless wedge bars spaced 0.5 mm apart. However, when the raw materials are very abrasive due to the presence of free silica, a synthetic material can be successfully used. For a 500 000 tons/year kiln 4 screens are needed. Each screen must have its own individual pump for effective control of pressures.

Closed circuit grinding saves 4 kWh/ton in electric power and reduces slurry moisture in the kiln feed by 3–4 %. In open circuit grinding the mill produces unnecessary superfines. This over grinding demands 4 kWh/ton more power. Also, a slurry with many superfines requires a higher moisture content in the mill in order to prevent plug up of the mill and, the superfines make the viscosity of the slurry higher requiring more water to have pumpability.

The curved screens have a dewatering effect i.e. some of the water is returned with the rejects to the mill. But the mill itself requires less water because the load is coarser. The load is pushed through the mill faster and fines are removed with the screens. The circulating load is about 200 %. Thus, the mill discharge pumps and screens us must be designed for a 200 % load.

schwindigkeit, die durch die anderen Bestandteile einer für Zement charakteristischen Rohmischung verursacht wird, kaum verwendet.

Filterpressen sind in verschiedenen Zementwerken im Einsatz und haben sich sehr gut bewährt. Für Wartungszwecke sind Ersatzfilter erforderlich. Die Filteranlagen bestehen aus Tanks, Pumpen, Kompressoren, Mehrfachfiltern und Filterkuchenwaagen. Ein nachteiliger Nebeneffekt beim Einsatz von Brenngut aus Filterkuchen ist die hohe Staubentwicklung, da in einem solchen Ofen nasse Ketteneinbauten zum Auffangen des Staubs fehlen. Um den Vorteil der Brennstoffersparnis beim Halbnaßverfahren voll auszunutzen, muß der gesamte Staub in den Ofen zurückgeführt werden.

7. Einsatz von Sekundärbrennstoffen

Einige europäische Zementwerke verwenden bis zu 6 verschiedene Arten von Sekundärbrennstoffen. Dabei kann im Ofen je ein Brenner für den Hauptbrennstoff und einer für den Abfallbrennstoff installiert werden. Um wirtschaftlich arbeiten zu können, müssen Sekundärbrennstoffe kostenlos oder zu sehr niedrigen Kosten zur Verfügung stehen. Laboranalysen und ständige chemische Überwachung sind notwendig. Die Bevölkerung am Ort muß überzeugt werden, daß der Einsatz von Sekundärbrennstoffen keine gesundheitlichen Probleme mit sich bringt. Als Sekundärbrennstoffe können beispielsweise Abfallstoffe aus dem Kohlebergbau, Lösungsmittel, Altöle, Holzabfälle, Pkw- und Lkw-Altreifen eingesetzt werden. Beim Naßverfahren müssen Altreifen in der Ofenmitte zugeführt werden. In den USA und auch in Rußland wurden entsprechende Systeme entwickelt und sind erprobt.

8. Moderne Drehofenbrenner

Viele Öfen in den Staaten der ehemaligen Sowjetunion werden mit Erdgas gefeuert. Neuartige Gasbrenner erzielen heute sehr hohe Wirkungsgrade bei gleichzeitig niedriger NO_x-Emission. Bei einem in Australien entwickelten Brenner wird die Nachverbrennung durch eine besonders konstruierte Düse in den Flammenkern gezogen, wodurch Berichten zufolge ein sehr hoher Wirkungsgrad erreicht wird. Brenner für Kohle, Öl und Sondermüll sind normalerweise Dreikanalbrenner mit mehreren Brennstoff- und Luftkanälen, deren Düsen für eine bessere Durchmischung des Brennstoffs mit der Verbrennungsluft sorgen. Erfahrungsgemäß amortisiert sich ein Brenner mit der neuesten Technik in weniger als einem Jahr.

9. Steuer- und Regelsysteme

Ein stabiler und sicherer Betrieb des Ofen-Kühler-Systems mit einem Minimum an Instrumentierung ist von grundlegender Bedeutung. Ein Versuch, den Ofenbetrieb per Computer zu steuern, sollte erst unternommen werden, nachdem die Betriebsdaten eines modernisierten Ofen-Kühler-Systems über einen längeren Zeitraum erfaßt wurden.

Der Ofen sollte mit einer Analysen- und Regeleinrichtung für Sauerstoff und Kohlenmonoxid zum Schutz der Elektrofilter und zur Überwachung und Regelung des Anteils von überschüssigem Sauerstoff in den Ofenabgasen ausgerüstet werden. Teure Ketteneinbauten aus rostfreiem Stahl müssen ebenfalls geschützt werden. Deshalb darf die Brennstoffzufuhr bei vollem Ofenbetrieb nie abgeschaltet werden, während das Ofengebläse weiterläuft. Dadurch würde die Luft mit 21 % Sauerstoff durch die rotglühenden Ketten geblasen, was sie zerstören würde.

Wenn also die Brennstoffzufuhr plötzlich unterbrochen wird, sollten die Gebläseklappen sofort automatisch geschlossen werden, um den Sauerstoffanteil unter 10 % zu halten. Natürlich muß ein minimaler Luftdurchzug bestehen bleiben, um Überdruck im Ofenkopf zu vermeiden.

Wenn die Brenngutzufuhr unterbrochen wird, müssen Bildschirm- und akustische Alarmmeldungen erfolgen. Andernfalls würde sich der Ofeneinlauf schnell aufheizen und zur Zerstörung der Ketteneinbauten führen.

Obviously, closed circuit wet raw grinding saves a lot of thermal and electrical power. No one should continue with open circuit grinding!

Semi wet process

The 3 main systems are: Spray drying, filter plate presses, centrifuges. In a spray dryer the slurry is flash dried. Filters squeeze out the moisture, down to 15–20 %. Centrifuges get the moisture down to 15–20 %. Centrifuges are widely used in lime recovery plants in paper mills. The main ingredients there are calcium carbonate. However, in cement plants they are hardly ever used because of the poor sedimentation rate. The poor sedimentation rate is caused by the other components of a typical cement raw mix.

The filter plate presses are used in various cement plants. They work very well. Back up filters are necessary for maintenance purposes. It is a large installation with tanks, pumps, compressors, multiple filters, cake weighing equipment. A bad side effect of slurry cake kiln feed is the high dust loss because in a filter cake kiln there are no wet chains to catch the dust. To take advantage of the fuel savings of a semi wet process all the dust must be returned to the kiln.

7. Waste fuels

Some European cement plants use as many as 6 different kinds of waste fuels. They may have 2 burners on the same kiln, one for the main fuel, one for the waste fuels. Of course, waste fuels must be free or have very low cost. Laboratory analysis and constant chemical control is required. The local population must be convinced that the use of any hazardous fuels will not cause a health problem. Some waste fuels are: Coal mine rejects, solvents, waste oils, wood products, automobile and truck tires. In a wet process kiln any rubber tires must be introduced in the middle of the kiln. Various systems exists, in the U.S. and also in Russia.

8. Kiln burners

Many ex Soviet kilns burn gas. There are new gas burners on the market that are very efficient and also reduce NO_x. One burner used in Australia draws the secondary combustion into the heart of the flame with a specially designed nozzle. It is, reportedly, the most efficient gas burner in operation. Burners for coal, oil, hazardous fuels are usually 3 channel burners with multiple fuel and airflows through the nozzle all designed for a better mixing of fuel and air. Experience shows that the purchase of a latest technology kiln burner pays for itself in less than a year.

9. Control and monitoring systems

Finally, control and monitoring systems. A stable and safe operation of the kiln-cooler is essential which requires a minimum of instrumentation. A computer operated kiln should not be attempted before one has collected, over a period of time, the operating data of a modernized kiln-cooler system.

The kiln should be equipped with an oxygen/CO analyzing and control system to protect the electrostatic precipitator and to monitor and control the percent excess oxygen in the kiln off gases. Expensive stainless kiln chains must also be protected i.e. in full kiln operating condition, the fuel should never be shut off while a full draft remains on the kiln. That would put 21 % of oxygen through a red hot chain system and burn the chains.

Thus, when for some reason the fuel suddenly is interrupted, the induced draft fan dampers should start closing immediately to keep oxygen below 10 %. Of course a minimum draft must be maintained to keep the hood from going positive.

There must be monitoring and audible alarms when the kiln feed is interrupted. Otherwise the back end will heat up rapidly and chains will be lost.

Each kiln should have a kiln drive amps vertical strip recorder. This helps the operator in controlling his kiln.

Die folgenden Betriebsparameter sollten kontinuierlich gemessen und auf einem Schreiber im Leitstand protokolliert werden:

- der Ofenantriebsstrom, um dem Leitstandsfahrer die Ofensteuerung zu erleichtern;
- die Brenngut- und Gastemperaturen direkt unterhalb der Ketteneinbauten sowie
- die Gastemperaturen im Ofeneinlauf und am Eingang des Staubabscheiders.

Darüber hinaus sollten die Öfen für einen sicheren und wirtschaftlichen Betrieb mit

- einem Meßtopf für den Rohschlammdurchsatz mit Sekundenzähler auf der Kontrolltafel im Leitstand (bei Naßöfen),
- einem System zur Regelung und Aufzeichnung des Brennstoffmassenstroms,
- einer Wägeeinrichtung für den in den Ofen zurückgeführten Staub sowie
- einer Infrarot-Scanner-Vorrichtung zur Aufzeichnung der Ofenmanteltemperaturen

ausgestattet sein.

Each kiln should have a material and gas temperature recorder i.e. a vertical strip recorder on which are simultaneously recorded the gas and material temperatures just below the chain system.

Each kiln must have a feed end gas temperature recorder with 2 temperatures recorded i.e. the temperature of the gas in the feed end housing and the temperature of the gas in the precipitator inlet.

Each wet kiln should have a slurry flow measuring pot with seconds counter on the control panel.

Each kiln should have a good fuel control system and recording of fuel flows.

Each kiln should preferably have a weighing system for the dust returned to the kiln.

Finally, each kiln should have a infrared shell temperature scanning system with recording of temperatures.

Eine einfache Abschätzung der Länge von kohlegefeuerten Vorwärmeröfen auf der Basis von Rohmehl-Thermoanalyse und Wärmeübertragung durch Strahlung auf staubförmige Partikel

A simple approach to length estimation of coal fired preheater kilns based on raw mix thermal analysis and radiative dust heat transfer

Simple évaluation de la longueur des fours préchauffeurs chauffés au charbon basée sur l'analyse thermique de farine crue et la transmission de chaleur par rayonnement de particules de poussière

Evaluación sencilla de la longitud de los hornos con precalentador y calefacción de carbón, sobre la base del análisis térmico del crudo y de la transmisión del calor por radiación a las partículas en forma de polvo

Von **A. Bhattacharya**, Bombay/Indien

Zusammenfassung – Es wurde eine einfache Methode zur Abschätzung der Länge von kohlegefeuerten Vorwärmeröfen für ein bestimmtes Rohmehl, eine bestimmte Kohle und Durchsatzmenge entwickelt. Der Ofen wird in die drei Abschnitte „flammfreie Zone", „Flammenzone" und „Brennereinbautiefe" eingeteilt. Das mit TG (Thermogravimetrie), DTG (Differentialthermogravimetrie) und DTA (Differentialthermoanalyse) ermittelte Enthalpie/Temperatur-Diagramm des Rohmehls einschließlich der Kohlenasche zwischen Umgebungs- und Sintertemperatur wird dazu verwendet, um den gesamten Energiebedarf pro 100°C Anstieg der Materialtemperatur zu errechnen, der die Länge eines Segments in der „flammenfreien Zone" festlegt. Die Material- und Gastemperaturen am Ofeneinlauf und an der Schnittstelle von der „flammenfreien Zone" zur „Flammenzone" sind angenommene Werte. Die in jedem Segment vom Ofengas durch Strahlung einschließlich der Strahlung von Staubpartikeln übertragene Wärme wird dem Energieinhalt des Segments gleichgesetzt. Damit werden die Segmentlängen in der „flammenfreien Zone" errechnet. Die Länge der Kohlenstaubflamme wird auf Basis der Ruhland'schen Gleichung mit den erforderlichen Parameterdaten errechnet. Die Länge der Brennereinbautiefe ergibt sich aus den Daten von Öfen mit ähnlichen Abmessungen. Die Summe der drei Längen entspricht etwa der bei der Ofenauslegung zu berücksichtigenden Gesamtlänge.

Summary – A simple method is developed to estimate the length of suspension preheater cement rotary kilns for a given raw mix, coal and throughput. The kiln is divided into the three sections "non-flame zone", "flame zone" and "burner pipe projection". The enthalpy-temperature relationship of the raw mix with coal ash from ambient to sintering temperature, determined by TG, DTG and DTA, is used to compute the total segmental energy demand for every 100°C rise in material temperature which constitutes the length of a segment in the non-flame zone. Material and gas temperatures are assumed at the kiln inlet as also at the interface "non-flame zone/flame zone". Heat transferred from the kiln gas in each segment by radiation, incorporating dust influence, is equated with segmental heat to compute the segment lengths in the non-flame zone. The length for the pulverized coal flame is calculated using Ruhland's equation with data of relevant parameters. Burner pipe projection length is available from data of kilns with similar dimensions. The total sum of the three lengths approximates the length to be designed.

Résumé – Une méthode simple d'estimation de la longueur des fours préchauffeurs à chauffe au charbon a été mise au point pour un certain type de farine crue, de charbon et de débit. Le four est subdivisé en trois secteurs: „la zone sans flamme", „la zone avec flamme" et „profondeur de montage du brûleur". L'enthalpie ou le diagramme de température – calculé par TG (thermogravimétrie), DTG (thermogravimétrie différentielle) et DTA (thermoanalyse différentielle) – de la farine crue et des cendres de charbon entre la température ambiante et la température de cuisson a servi à calculer l'ensemble des besoins énergétiques par tranche d'accroissement de température de matière de 100°C, lesquels, à leur tour, définissent la longueur d'un segment dans la „zone sans flamme". Les températures de la matière et du gaz à l'entrée du four et au point d'intersection de la „zone sans flamme" et de „la zone avec flamme" sont des va-

leurs supputées. La chaleur transmise dans chaque segment par le gaz du four du fait du rayonnement y compris le rayonnement des particules de poussière, est considérée comme la valeur énergétique du segment, ce qui permet de calculer les longueurs de segment compris dans la „zone sans flamme". La longueur de la flamme de poussière de charbon est calculée sur la base de l'équation de Ruhland avec les données des paramètres nécessaires. La longueur de la profondeur d'installation du brûleur résulte des données des fours de dimensions similaires. La somme des trois longeurs correspond en gros à la longueur totale prise en compte lors du dimensionnement du four.

Resumen – *Se ha desarrollado un método sencillo para la evaluación de la longitud de los hornos con precalentador y calefacción de carbón, teniendo en cuenta una harina cruda, un carbón y un caudal de paso determinados. El horno se divide en tres tramos, es decir en „zona sin llama", „zona con llama" y „profundidad de montaje del quemador". El diagrama de entalpía/temperatura del crudo, determinado por medio de la TG (termogravimetría), la DTG (termogravimetría diferencial) y la DTA (termoanálisis diferencial), incluyendo la ceniza de carbón entre la temperatura ambiente y la de sinterización, se emplea para calcular la demanda total de energía por cada 100 °C de aumento de temperatura del material, que determina la longitud de un segmento de la „zona sin llama". Las temperaturas de material y de gas a la entrada del horno y en el punto de intersección entre la „zona sin llama" y la „zona con llama", son valores supuestos. El calor que el gas del horno transmite en cada segmento por radiación, incluyendo la radiación de las partículas de polvo, se equipara al contenido de energía del segmento. Con ello se calculan las longitudes de los segmentos en la „zona sin llama". La longitud de la llama de polvo de carbón se calcula sobre la base de la ecuación de Ruhland, con los datos de los parámetros relevantes. La longitud de la profundidad de montaje del quemador resulta de los datos de los hornos de dimensiones similares. La suma de las tres longitudes corresponde, más o menos, a la longitud total a tener en cuenta para el dimensionamiento del horno.*

Evaluación sencilla de la longitud de los hornos con precalentador y calefacción de carbón, sobre la base del análisis térmico del crudo y de la transmisión del calor por radiación a las partículas en forma de polvo

Der vorliegende Artikel entwirft ein neuartiges Konzept zur Abschätzung der Länge von kohlegefeuerten Drehöfen mit Zyklonvorwärmer und Planetenkühler für ein bestimmtes Rohmehl, eine bestimmte Kohle und Durchsatzmenge. Die Methode besteht aus

1. einer experimentellen Voruntersuchung, nämlich der Rohmehl-Thermoanalyse,
2. der Berechnung der Wärmestrahlung von Gas und Staubpartikeln und
3. der Berechnung der Flammenlänge mit Hilfe bewährter Vorgehensweisen aus der Literatur.

Ausgehend vom kalten Ende läßt sich der Ofen der Länge nach in drei Abschnitte einteilen, nämlich die flammenfreie Zone, die Flammenzone und die Vorkühlzone innerhalb des Ofens.

Der theoretische Wärmebedarf zur Klinkerbildung für ein Rohmehl läßt sich unter Berücksichtigung der eingebundenen Kohlenaschen durch die Techniken der Thermoanalyse (TG, DTG und DTA) experimentell bestimmen. Für jedes gewählte Intervall der Materialtemperatur läßt sich die benötigte Gesamtwärme abschätzen, sofern das Verhältnis Enthalpie/Temperatur zwischen Umgebungs- und Sintertemperatur bekannt ist.

Die flammenfreie Zone wird beispielsweise in 5 Segmente unterteilt. Beginnend am Ofeneinlauf mit t_m = 800 °C und einer angenommen Gastemperatur von 1100 °C wird die Segmentwärme für jeweils 100 °C Anstieg von t_m aus der Enthalpie/Temperatur-Relation abgeschätzt. Die Gastemperatur am Ende des Segmentes 1 berechnet sich aus der Gleichsetzung der erforderlichen Segmentwärme mit der fühlbaren Wärme, die vom Gasstrom übertragen wird. Dessen spezifische Wärme wird berechnet als Funktion der Gaszusammensetzung und -temperatur. Die Wärmeübertragung in den Segmenten der flammenfreien Zone wird modelliert als Wärmeübergang zwischen Feststoff und staubhaltigem Gasstrom, wobei die Staubpartikel im Ofengas als unabhängige Strahler betrachtet werden. Aus Gründen der Vereinfachung wird angenommen, daß die Zusammensetzung des Staubes und die Verteilung der Partikelgrößen über die gesamte Zone konstant bleiben.

Die von den CO_2- und H_2O-Gehalten des Ofengases bei den jeweils vorliegenden Partialdrücken ausgehenden konvek-

This paper projects a novel concept developed to estimate the length of coal fired rotary kilns with suspension preheater and planetary cooler for given raw materials, coal and throughput. The method consists of 1) an experimental input, namely the thermal analysis of the raw mix, 2) the computation of the heat transfer by radiation from the gas and the dust particles and 3) the calculation of the flame length with the help of well established procedures given in the literature.

Starting from the cold end, the kiln may be divided into three independent zones of length, namely the non-flame zone, the flame zone and the cooling zone inside the kiln.

The practical heat of clinkerization of the raw mix with coal ash as a component can be experimentally determined by the thermal analysis techniques (TG, DTG and DTA). For any chosen interval of the material temperature the total heat can be estimated, if the enthalpy-temperature relationship from ambient to sintering temperature is known.

The non-flame zone is subdivided into a number of segments, say 5 based on the rise in material temperature t_m. Starting at the kiln inlet with t_m of 800 °C and an assumed gas temperature of 1100 °C, the segmental heat is estimated for every 100 °C rise in t_m from the enthalpy temperature relationship. The gas temperature at the end of segment 1 is calculated by equating segmental heat requirement with the sensible heat transferred by the gas stream. Its specific heat is computed as a function of the gas composition and its temperature. The heat transfer in the segments of the non-flame zone is considered between the solids and the gas stream containing dust particles which are regarded as independant radiators. For the sake of simplicity the dust composition and the particle size distribution are assumed to remain unchanged throughout the zone.

The convective and radiative heat fluxes from CO_2 and H_2O in the gas stream at their respective partial pressures as well as the radiative heat flux from the dust particles are calculated with the help of appropriate equations taking into account the emissivities of the gas and the particles. The dust radiates both to the solids and the refractory wall and the net dust heat transfer is the sum of these two radiations. The effective heat flux in a segment is thus the sum of the convective and the radiative heat fluxes from the gas and the

TABELLE 1: Enthalpiewerte für ein typisches Rohmehl

TABLE 1: Enthalpy values for a typical raw mix with increasing temperature

Temperaturintervall $t_0 = 25°C$ Temperature Interval $25°C\,t_o$	Enthalpie kJ/kg Klinker Enthalpy kJ/kg clinker	Temperaturintervall $t_0 = 25°C$ Temperature Interval $25°C\,t_o$	Enthalpie kJ/kg Klinker Enthalpy kJ/kg clinker
800 °C	1324	1200 °C	3211
900 °C	1923	1300 °C	3384
1000 °C	3015	1400 °C	3744
1100 °C	3104	1450 °C	3939

TABELLE 2: Längen der Segmente in der flammenfreien Zone
Durchmesser der Ofenmantels: 3,95 m, Ausstoß: 900 t/d
TABLE 2: Segmentwise length in the non-flame zone
Kiln shell diameter: 3.95 m, Capacity: 900 TPD

Segment Segment Number	Temperatur Feststoff (°C) Solids Temp. (°C)	Segmentwärme (kJ/kg Klinker) Segment Heat kJ/kg clinker	Segmentlänge (m) Segment Length (m)
1	800– 900	598	15,1
2	900–1000	1092	14,7
3	1000–1100	89	0,9
4	1100–1200	107	1,1
5	1200–1250	56	0,6
		Gesamt:	32,4
		Total:	32.4

tiv und durch Strahlung übertragenen Wärmeströme werden ebenso wie die Strahlungswärmen der Staubpartikel in geeigneten Berechnungsgleichungen berücksichtigt, wobei auch die unterschiedlichen Emissionsverhältnisse der Gase und der Feststoffpartikel in die Berechnung eingehen. Der Staub strahlt sowohl in Richtung des Brennguts als auch der Ausmauerung. Die Nettowärmeübertragung des Staubes entspricht somit der Summe dieser beiden Strahlungsanteile. Der effektive Wärmefluß in einem Segment ergibt sich daher als Summe der durch Strahlung und Konvektion vom Ofengas auf das Brenngut übertragenen Wärmeströme abzüglich des Wärmeverlustes an die Umgebung. Für die Länge eines Segmentes in der flammenfreien Zone gilt daher

$$\frac{\text{Erforderliche Wärme im Segment} \times \text{Klinkerausstoß}}{\text{Effektiver Wärmefluß im Segment}}$$

Dieses Vorgehen läßt sich für jedes Segment wiederholen bis zu einer Materialtemperatur von 1250 °C, die einer berechneten Gastemperatur von 1900 °C im Übergang von der flammenfreien zur Flammenzone entspricht. Diese wird hier definiert als Gastemperatur an der Flammenspitze. Diese Temperatur entspricht auch ungefähr der tatsächlichen Flammentemperatur, die sich aus dem Heizwert der Kohle und der Wärmebilanz von Dissoziations- und Verbrennungsprodukten berechnet.

Tabelle 1 zeigt die experimentell bestimmten Enthalpiewerte für ein typisches Rohmehl im Temperaturbereich zwischen 800 und 1450 °C bezogen auf eine Temperatur von 25 °C. **Tabelle 2** zeigt die berechneten Segmentlängen in der flammenlosen Zone eines Ofens bei einem Durchsatz von 900 t pro Tag. Man erkennt, daß rund 90 % der Zonenlänge zur vollständigen Calcinierung erforderlich sind. Die Vernachlässigung der Wärmestrahlung des Staubes würde zu einem deutlich niedrigeren effektiven Wärmestrom und infolgedessen zu einer um 25 % erhöhten Länge dieser Zone führen.

Die Länge der Flammenzone wird mit Hilfe der halbempirischen Gleichungen abgeschätzt, die von Ruhland für spezielle Anwendungen bei Kohlestaubflammen ausschließlich für Zementdrehöfen entwickelt wurden. Er definiert die Flammenlänge als Länge zwischen der Brennerspitze und demjenigen Ofenquerschnitt, in dem die CO-Konzentration fast Null ist. Ruhland hat das dimensionslose Verhältnis von Flammenlänge zu Durchmesser der Brennerdüse mit Parametern wie Kohleaufgabe, primären und sekundären Luftströmen in Beziehung gesetzt, wobei sowohl deren Dichten als auch die Dichte der Verbrennungsgase als konstant angenommen werden. Die Flammenlänge für konstante Dichte wurde mit dem Dichtekorrekturfaktor multipliziert, um die tatsächliche Länge der Flamme zu bestimmen.

Die Daten, die zur Lösung der Gleichungen von Ruhland erforderlich sind, wie Analysedaten der Kohle und deren

dust particles minus the heat loss to the ambient. The length of a segment in the non-flame zone is thus

$$\frac{\text{Segmental heat requirement} \times \text{clinker production}}{\text{Effective heat flux in the segment}}$$

This procedure can be repeated for each segment up to a material temperature of 1250 °C corresponding to a calculated gas temperature of 1900 °C at the non-flame/flame zone interface which in this work is defined in terms of gas temperature at the flame tip. This temperature is also approximately the actual flame temperature which is calculated from the calorific value of coal and accounting for the heat of dissociation of the combustion products.

Table 1 shows the experimentally determined enthalpy values for a typical raw mix in the temperature range between 800 °C and 1450 °C with respect to a base temperature of 25 °C. **Table 2** gives the calculated segmentwise length of the non-flame zone of a kiln with a capacity of 900 t/d. It can be seen, that about 90 % of the zone length is required for the completion of the calcination. Neglecting the dust radiation would lead to a substantial lower effective heat flux and thus to about 25 % more length of this zone.

The flame zone length is estimated with the help of semi-empirical equations developed by Ruhland for specific application to pulverised coal flames exclusively for cement rotary kilns. He defines the flame length as the length between the burner tip and the kiln cross section where the CO concentration is almost zero. Ruhland correlated the dimensionless ratio of flame length to burner nozzle diameter with parameters such as coal rate, primary and secondary air flow rates, their densities as well as that of combustion gases being considered constant. The flame length for constant density has to be multiplied by the density correction factor in order to determine the actual length of the flame.

Data required for solving Ruhland's equations such as analysis of coal and its calorific value, heat consumption, coal rate, combustion air quantity, ratio and temperatures of primary and secondary air, excess air, effective diameters of the kiln and the burner nozzle etc., are readily available. The velocity of the primary air-coal dust mixture at the burner nozzle can be checked with its correlation with the kiln internal diameter for optimum flame characteristics. The calculated flame length and the data assumed for the case study are presented in **Table 3**.

The cooling zone in a rotary kiln with planetary coolers consists of the dam ring and its length is assumed to be approximately equal to the projected length of the burner pipe. The latter is introduced inside the kiln up to a distance of 3 m to 3.5 m from the outlet, as may be found from data on kilns of similar design. The burner is assumed to be a straight pipe positioned in the centre of the kiln's cross section.

TABLE 3: Flame length calculation	
Assumptions	
Coal Calorific Value	24 600 kJ/kg
Heat Consumption	3 550 kJ/kg clinker
Kiln Diameter, Effective	3.55 m
Burner Nozzle Diameter, Effective	0.27 m
Primary Air : Secondary Air	20 : 80
Excess Air	10 %
Burner Nozzle Gas Velocity	69.5 m/s
Density Correction Factor	1.5
$(l/d)_p$ = const	55
$(l/d)_{flame}$	83
Flame Length (l)	22.5 m

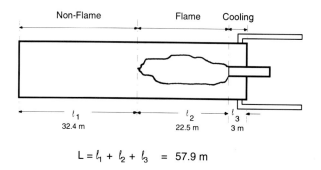

FIGURE 1: Calculated kiln length

Kiln Shell Diameter: 3.95 m
Design Length: 58 m
Non-Flame
Flame
Cooling

Fig. 1 shows the comparison of the calculated and the design length of a 900 t/d rotary kiln with planetary coolers. The sum of the three zonal lengths almost matches the design length, the non-flame zone constitutes about 56% of the total kiln length. The author is fully aware of the limitations in the practice in estimating the segmental heat requirement from the thermal analysis experiments. The concept may however be used to estimate the kiln length with fair accuracy.

Literature

Bhattacharya, A.: Simulating Minerals Pyroprocessing by Thermal Analysis, J. Thermal Analysis, 40 (1993) under publication.

Biermann, L.: Wärmestrahlung staubhaltiger Gase in Dampfkesseln, Doctoral Thesis (1968), Universität Stuttgart.

Roals, J.:.Mathematische Formulierung des Wärmeüberganges im Drehofen, Zement-Kalk-Gips 23 (1970), No. 8, S. 368–377.

Ruhland, W.: Über die Länge von Kohlenstaubflammen im Drehofen, VDZ-Schriftenreihe, No. 32 (1965).

Weber, B.: Heat Transfer in Cement Rotary Kilns, Zement-Kalk-Gips publication No. 9 (1960).

Anwendungsmöglichkeiten der Wirbelschicht bei der Zementherstellung

Fluid-bed cement clinker applications

Possibilités d'utilisation de lit fluidisé pour la fabrication de ciment

Posibilidades de aplicación del lecho fluidizado a la fabricación del cemento

Von **S. M. Cohen,** Bethlehem/USA

Zusammenfassung – Es werden zwei große Entwicklungsprogramme zur Anwendung der Wirbelschichttechnologie bei der Verarbeitung von Zementklinker beschrieben. Das erste ist ein System, das entwickelt wurde, um staubförmige Abfallstoffe aus dem Zementwerk aufzubereiten und dadurch die Menge solcher Abfälle auf 10% oder zu einem potentiell alkalireichen Nebenprodukt zu reduzieren. Die wiederverwendeten 90% entsprechen in den meisten Fällen der Zusammensetzung eines alkaliarmen Klinkers. Das aus dem Abgasstrom abgeschiedene potentiell alkalireiche Nebenprodukt besteht zu 85 bis 95% aus Alkalisalzen, die als Kaliumquelle oder direkt als Düngemittel verwendet werden können. Das zweite System, das mit finanzieller Förderung des Gasforschungsinstituts entwickelt wurde, ist ein System zur Verarbeitung von Rohmehlen für Standardzemente und Spezialzemente bei gutem Brennstoffnutzungsgrad und geringen Wartungskosten. Dieses System kombiniert einen Schachtvorwärmer mit einer Wirbelschicht-Reaktionszone und einem Kontaktkühler in Einzelschachtausführung. Es wurden Einheiten bis zu einem Durchsatz von 50 t/h erprobt. In beiden Fällen werden Ergebnisse aus Pilotanlagen vorgestellt.

Summary – Two major development programmes will be described which use fluid-bed technology in processing cement clinker. The first is a system developed to reprocess cement plant waste dust materials to reduce waste dust products to 10% or to a potential high alkali by-product. The recovered 90% will be in the composition of a low alkali clinker in most cases. The waste gas potential by product will be a 85–95% alkali salt for use as a source of potassium or directly as a fertilizer. The second system, developed with funding supplied by the gas research institute, is a system for processing standard cement raw mixes and special cements in a system with good fuel efficiency and low maintenance costs. This system incorporates a free fall preheater to a fluid-bed reaction zone and a contact cooler in a single shaft design. Units up to 50 t/h production have been evaluated. In both cases actual pilot plant results are given.

Résumé – Deux grands programmes de développement portant sur l'utilisation de la technologie de lit fluidisé ont été décrits pour la transformation du clinker de cimenterie. Le premier de ces programmes concerne un système mis au point pour traiter des déchets pulvérulents provenant de la cimenterie et réduire ainsi la quantité de ces déchets à 10%, ou obtenir un produit secondaire potentiellement riche en composés alcalins. Les 90% réutilisables répondent, dans la plupart des cas, à la composition d'un clinker pauvre en composé alcalin. Quant au produit secondaire potentiellement riche en composés alcalins, extrait du flux des gaz de fumées, il est constitué pour 85 à 95% de sels alcalins, qui peuvent être utilisés comme source de potassium ou directement comme engrais. Le deuxième système, qui a été développé avec le concours financier de l'Institut de recherche du gaz, est un système de transformation de farines crues pour ciments standards et ciments spéciaux avec bon taux d'utilisation du combustible et faibles frais d'entretien. Ce système combine un préchauffeur à cuve à une zone de réaction à lit fluidisé et refroidisseur de contact de type à cuve individuelle. Des unités allant jusqu'à une capacité de production de 50 t/h ont été testées. Dans les deux cas précités, les résultats obtenus présentés proviennent d'installations expérimentales.

Resumen – Se describen dos grandes programas de desarrollo referentes a la tecnología del lecho fluidizado, utilizada para el tratamiento del clínker de cemento. El primero es un sistema desarrollado con el fin de preparar materias de desecho en forma de polvo, procedentes de la fábrica de cemento, y de reducir así la cantidad de tales desechos a un 10 % o de obtener un producto secundario potencialmente rico en álcalis. El 90 % recuperado corresponde, en la mayoría de los casos, a la composición de un clínker pobre en álcalis. El producto secundario potencialmente rico en álcalis, separado de la corriente de gases de escape, se compone en un 85 – 95 % de sales alcalinas, que pueden aprovecharse como fuente de potasio o directamente como abono. El segundo

sistema, desarollado con el apoyo financiero del Instituto de Investigación del Gas, es un sistema de tratamiento de la harina cruda para cementos standard y cementos especiales, con un buen grado de utilización del combustible y reducidos costes de mantenimiento. Este sistema combina el precalentador de cuba vertical con una zona de reacción, de lecho fluidizado, y un enfriador de contacto, de cuba individual. Se sometieron a ensayo unidades con un caudal de paso de hasta 50 t/h. En ambos casos se presentan los resultados obtenidos con plantas piloto.

1. Einleitung

Seit der Entwicklung des sogenannten Pyzel-Prozesses in den späten 50er Jahren beschäftigt sich die Fuller Company mit der Herstellung von Zementklinker in der Wirbelschicht. Das Grundkonzept des Pyzel-Prozesses bestand bekanntlich darin, daß ein chemisch gut abgestimmtes Zement-Rohmehl, das in das heiße Wirbelbett eines fertig gebrannten Zementklinkers eingeleitet wird, reagiert und auf den heißen Oberflächen neue Klinkerphasen bildet. Das Partikelwachstum erfolgt dabei in sehr dünnen Schichten, so daß es bedingt durch das Erweichen des reagierenden Materials anscheinend nicht genügend große Adhäsionskräfte gibt, um die großen, sich schnell bewegenden Partikel zu agglomerieren. **Bild 1** zeigt ein vereinfachtes Fließschema des Pyzel-Prozesses [1]. Dieser Prozeß war drei Jahre lang an einer Pilotanlage mit einem Durchsatz von 110 t/d getestet worden.

Obgleich der Pyzel-Prozeß bisher noch nicht vermarktet wurde, hat die Fuller Company zu keiner Zeit das Konzept aufgegeben, Zementklinker in der Wirbelschicht herzustellen.

Während dieser Zeit wurde ein neues Konzept zur Staubaufbereitung unter Anwendung der Wirbelschichttechnologie entwickelt, das zu einem neuen Prozeß führte, der in den Jahren 1986 und 1987 [2–4] patentiert wurde und nunmehr kommerziell angeboten wird.

Wegen der wachsenden Verwendung von alternativen Brennstoffen beim Zementbrennprozeß, der zunehmenden Nutzung von Bypass-Systemen zur Herstellung eines Zementklinkers mit niedrigem Alkaligehalt und auch aus umwelttechnischen Erwägungen stellt heute die Verwertung von Abfallstäuben ein größeres Problem dar als je zuvor. Diese Entwicklung hat schließlich dazu geführt, daß sich das neue System zur Staubaufbereitung eines allgemeinen Interesses erfreut.

2. Aufbereitung des Prozeßstaubes

Um den Bypassstaub in diesem Prozeß verarbeiten zu können, muß er pelletiert werden, wobei die völlig trockenen Pellets, welche die Abmessungen von 4 Maschen × 20 Maschen haben, unter Zugabe von bis zu 90 % des für den Prozeß erforderlichen Brennstoffes (Kohle oder Koks) hergestellt werden. Neben der Wärmefreisetzung bewirkt die Zugabe von Kohle bzw. Koks während der Verbrennung reduzierende Bedingungen innerhalb der Granalien, was zu einer verstärkten Verdampfung der Alkalien führt. Das entwickelte System zur Staubpelletierung ist dem Fließschema in **Bild 2** zu entnehmen. Die Verwendung eines Bindemittels bzw. eines die Pelletierung unterstützenden Mittels ist grundsätzlich materialabhängig. Einige Materialien erwiesen sich ohne ein entsprechendes Bindemittel als nicht zu pelletieren. In solchen Fällen hat sich die Zugabe von Portlandzement in einer Größenordnung von ca. 2 bis 5 % bewährt.

Der Feststoff wird mit einem Wasseranteil von 8–15 % in einer Mischtrommel gemischt, damit der freie Kalk im Staub reagieren kann. Anschließend wird die Mischung dann zu Pellets verarbeitet. Die Pelletierung selbst kann auf einem Granulierteller oder in einer Presse erfolgen. Dann

1. Background

Fuller Company has been involved in the fluid bed production of cement clinker since the late 1950's with the development of the Pyzel Process. The basic concept of the Pyzel Process was that when a fine, properly designed (chemically) raw mix is introduced into a hot fluidized bed of finished clinker, it will react to form new cement compounds on the hot surfaces. The particle growth is in such thin layer increments that there is apparently not sufficient adhesive force, caused by the softening of the reacting material, to cement together with the large rapidly moving particles. **Fig. 1** presents a simplified basic flow sheet for the Pyzel Process [1]. This process was tested in a 110 TPD pilot plant for three years.

Although the Pyzel Process was never commercialized, Fuller Company has never given up on the fluid bed approach to produce cement clinker.

It was during this development that a new approach using fluid bed technology was applied to dust recovery, which led to the new process which was patented in 1986 and 1987 [2–4] and is now being offered commercially.

Due to the growing use of alternate fuel burning in the cement processing operations, the increased use of bypass systems to lower clinker alkali levels and environmental concerns, the disposal of waste dust has become a greater problem than ever before. This has created the current new interest in the dust recovery system.

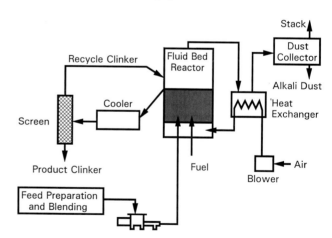

BILD 1: Ofenstaubverwertung im Fuller-Pyzel-Wirbelschichtverfahren
FIGURE 1: Kiln dust beneficiation Fuller-Pyzel fluid bed process

Zurückgeführter Klinker	= recycle clinker
Kühler	= cooler
Siebeinrichtung	= screen
Klinker	= product clinker
Rohmaterialaufbereitung und -mischung	= feed preparation and blending
Wirbelschichtreaktor	= fluid bed reactor
Brennstoff	= fuel
zum Schornstein	= stack
Staubabscheider	= dust collector
Alkalistaub	= alkali dust
Wärmetauscher	= heat exchanger
Luft	= air
Gebläse	= blower

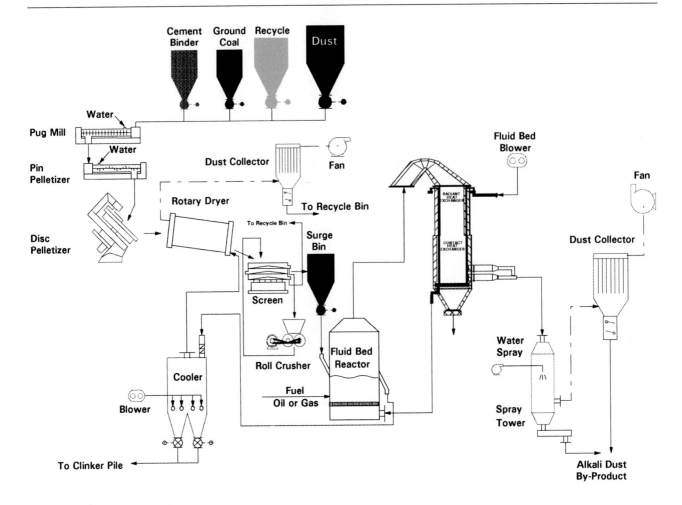

BILD 2: Fließbild einer Anlage zur Staubrückgewinnung mittels Wirbelschicht
FIGURE 2: Fluid bed dust recovery system

Zementbindemittel	= cement binder
Kohlenstaub	= ground coal
Ofenstaub	= dust
Mischer	= pug mill
Wasser	= water
Presse	= pin pelletizer
Gebläse	= fan
Drehkolbengebläse für Wirbelschicht	= fluid bed blower
Drehtrockner	= rotary dryer
zum Kreislaufgutbunker	= to recycle bin
Staubabscheider	= dust collector
Granulierteller	= disc pelletizer
Aufgabebunker	= surge bin
Siebeinrichtung	= screen
Kühler	= cooler
Walzenbrecher	= roll crusher
Wirbelschichtreaktor	= fluid bed reactor
Wassereinsprühung	= water spray
Gebläse	= blower
Heizöl od. Gas	= fuel oil or gas
Sprühturm	= spray tower
zum Klinkerlager	= to clinker pile
Alkali-Staub (Nebenprodukt)	= alkali dust by-product

2. Process dust feed preparation

This process does require the klin dust, as produced, to be pelletized into totally dry 4 mesh x 20 mesh sized pellets with up to 90% of the process coal or coke fuel added to the pellets. In addition to adding heat to the system, the coke or coal addition increases the alkali volatilization levels by producing a reducing condition within each pellet during firing. The pelletization system developed is as shown in the overall process flow sheet in **Fig. 2**. The use of a binder and pellet curing is material related. It has been found that some dust catch material will not pelletize without a binder. In these cases we use approximately 2% to 5% Portland Cement as a binder.

Solid materials with 8–15% are mixed with water in a pug mill to react any free lime in the collected dust. The mixture is then formed into pellets. This is done using pin type and disc pelletizer. The pelletized material is then sent to a dryer. The dryer can be either a fluid bed or rotary unit and is supplied with drying gas from the processes clinker cooler system. Dust generated from dryer is collected and recycled back to the pelletizing system along with undersized pellets after screening. Oversized pellets from the screening are crushed and rescreened.

3. Fluid bed system

The properly sized and dry material is metered into a fluidized bed reactor for thermal processing. The fluid bed reactor system is divided into an upper thermal processing chamber and a lower plenum by means of a gas permeable refractory grid or a restrictive opening. Air under pressure is supplied to the plenum for passage upward through the grid at a velocity in the range of 1.8 – 3 m/s. This maintains the bed material in a highly active fluidized state above the grid area. The final fuel control above that contained in the feed pellets is supplied by direct injection into the bed using gas or oil. This fuel would be the final 10% required by the process after pellet fuel burnout to maintain a 2380 to 2400 °F (1300– 1320°C) bed temperature.

werden die Pellets in einer Wirbelschicht oder einem Trommeltrockner mit Hilfe der Klinkerkühlerabluft getrocknet. Der während des Trocknungsprozesses anfallende Staub wird gesammelt und dem nach der Pelletierung abgesiebten Unterkorn zugeführt, während das Überkorn zerkleinert und anschließend wieder abgesiebt wird.

3. Wirbelschichtsystem

Das sorgfältig aufbereitete und auch getrocknete Material wird dann zur thermischen Behandlung in den Wirbelschichtreaktor dosiert. Der Wirbelschichtreaktor ist unterteilt in einen oberen thermischen Reaktionsraum und in eine darunter befindliche Luftkammer, die durch einen gasdurchlässigen feuerfesten Wirbelboden abgetrennt ist. Die

Luftkammer wird mit Druckluft beaufschlagt, die den Wirbelboden nach oben mit einer Geschwindigkeit von 1,8 bis 3,0 m/s durchströmt. Dadurch wird das Materialbett oberhalb des Wirbelbodens in einen hochaktiven Fluidisierungszustand versetzt. Die Feinabstimmung des über den in den Pellets hinaus enthaltenen Brennstoffes erfolgt durch die Direktzuführung von Gas oder Heizöl in das Fließbett. Der Anteil dieses direkt zugeführten Brennstoffs beträgt etwa 10% und ist notwendig, um die Temperatur im Fließbett bei 1300 bis 1320°C zu halten.

Die Luftversorgung des Prozesses erfolgt mit einem Drehkolbengebläse. Die Abgase des Reaktors werden durch einen Wärmetauscher geleitet, der die eintretende Fluidisierungsluft indirekt vorwärmt.

Die Ausschleusung des fertig gebrannten Produkts über eine Überlaufleitung zu einem Schachtkühler erfolgt entsprechend der Dosierung des Aufgabegutes. Die Menge des austretenden Klinkers hängt somit unmittelbar von der Aufgabemenge ab, mit der der Reaktor beaufschlagt wird. Das in den Schachtkühler gelangte Produkt wird mit Luft von einem Ventilator oder aus einer anderen Quelle gekühlt. Die heißen Abgase aus dem Kühler werden in der Trocknungsstufe bei der Aufbereitung ausgenutzt. Die Alkalien und Schwefelverbindungen im Aufgabematerial werden während des Prozesses verdampft und mit dem Abgas aus dem Wirbelschichtreaktor ausgetragen.

Der dem Reaktor nachgeschaltete Wärmetauscher ist im oberen Bereich mit einem keramisch ausgekleideten Rohrleitungssystem zum Schutz gegen Alkalienangriff und Ansatzbildung ausgestattet. Im unteren Bereich hat das Rohrleitungssystem direkten Kontakt mit dem heißen Abgas. Das Rohrleitungssystem ist im Wärmetauscher parallel zur Gasströmung angeordnet. Im Wärmetauscher wird die durch den Hauptventilator bereitgestellte Fluidisierungsluft auf eine Temperatur von ca. 500°C vorgewärmt. Die mit Feinstaub beladenen Reaktorabgase werden durch einen hocheffektiven Staubabscheider gereinigt.

4. Chemische Analysenergebnisse aus dem Prozeß

Tabelle 1 enthält beispielhaft die chemischen Analysenergebnisse aus einem aktuellen Test an der Pilotanlage, verglichen mit berechneten Ergebnissen für den Stand einer technischen Ofenanlage. Die Zusammensetzung des Originalofenstaubes wurde dabei nicht zur Optimierung der endgültigen Klinkerzusammensetzung verändert. **Tabelle 2** zeigt

The air supply system for the process includes a positive displacement blower. The reactor exhaust gas goes to the heat exchanger which is used to preheat incoming fluidization air indirectly.

As material is supplied to the fluidized bed, product is discharged through an overflow pipe to a shaft cooler by displacement. The clinker discharge rate depends upon the rate of feed material to the vessel. Product discharged to the shaft cooler is cooled with cooling air supplied from a fan or another source. Hot off gases from the cooler are used in the drying step during feed preparation. Alkalies and sulfur compounds within the feed material are volatilized from material and carried out of the reactor by the fluidizing gases.

The heat exchanger used has a first compartment consisting of ceramic buried pipes to protect against alkali attack and build-ups. The lower section has direct hot gas contact with heat exchange pipes, both are parallel to the gas flow. The heat exchange system is used to preheat fluidizing gases from the main blower to a temperature of approximately 932°F (500°C). The final dust laden gases go to a high-efficiency dust collector.

4. Process chemical results

Example of actual test chemical results taken from pilot plant operations are compared to our predicated method for a given plant dust in **Table 1**. There was no attempt made to adjust the chemistry of the original plant dust in order to alter final clinker analyses. **Table 2** shows the results after chemical adjustments (5% limestone). The actual clinker chemistry obtained in the pilot unit is shown in both tables.

The chemical analysis of the final potential by-product dust catch materials produced during the pilot programs was found to contain 35 to 50% K_2O as sulfates and chlorides. It has already been determined that this final dust product may have some commercial value either as a source of potassium or for direct fertilizer use. This must be determined on a case by case basis, however.

Based on the results obtained, the actual removal percentages of alkalies and sulfur were calculated and shown in **Table 3**.

It has also been determined as part of our development program that the disposal of the process's final dust catch can be reduced even further by using a small leaching system. This could also increase the potential value for this end-product

TABELLE 1: Mögliche Zementzusammensetzung
Ausgangsmaterial: 88% Staub, 12% Kohle
TABLE 1: Potential cement compounds
Material: 88% dust, 12% coal

Chem. Bestandteile chemical compounds	Staub dust	Kohle coal	88% Staub dust	12% Kohle coal	Pellet-Aufgabe pellet feed	Glühverlustfrei loss free	Vol. Verl. vol. loss	Klinker berechnet calcul. volatil.	Klinker Pilotanlage pilot plant
SiO_2	14,66	4,16	12,90	0,50	13,40	20,62		22,88	22,49
Al_2O_3	3,51	2,24	3,09	0,27	3,36	5,17		5,74	5,84
Fe_2O_3	2,36	2,73	2,08	0,33	2,41	3,71		4,12	3,99
CaO	42,30	0,39	37,22	0,05	37,27	57,36		63,64	63,98
MgO	1,25	0,06	1,10	0,01	1,11	1,71		1,90	1,87
K_2O	2,82	0,16	2,48	0,02	2,50	3,85	3,47	0,38	0,23
Na_2O	0,34	0,04	0,30	0,00	0,30	0,46	0,32	0,14	0,13
SO_3	4,68	0,15	4,12	0,02	4,14	6,37	5,73	0,64	0,79
Cl	0,27	0,00	0,24	0,00	0,24	0,37	0,35	0,02	—
Verlust @900°C loss	27,55	89,90	24,24	10,79	35,03	0,00		0,00	0,00
Total	99,74	99,93			99,62	9,87		99,46	99,32
							S/R	2,32	
Kühlerphasen cement compounds							C_3S	40,66	41,40
							C_2S	34,92	33,30
							C_3A	8,24	8,70
							C_4AF	12,54	12,10
Gesamtalkalien als Na_2O total alkali as Na_2O						3,00		0,39	0,28

TABELLE 2: Zementzusammensetzung
Ausgangsmaterial: 83 % Staub, 12 % Kohle, 5 % Kalkstein
TABLE 2: Potential cement compounds
Material: 83 % dust, 12 % coal, 5 % limestone

Chem. Bestandteile chemical compounds	Staub dust	Kohle coal	Kalkstein limestone	88 % Staub dust	12 % Kohle coal	5 % Kalkstein limestone	Pellet-Aufgabe pellet feed	Glüh-verlustfrei loss free	Vol. Verl. vol. loss	Klinker berechnet calcul. volatil.	Pilot-anlage pilot plant
SiO$_2$	14,66	4,16	1,25	12,17	0,50	0,06	12,73	19,83		21,90	21,44
Al$_2$O$_3$	3,51	2,24	0,48	2,91	0,27	0,02	3,20	4,99		5,51	6,20
Fe$_2$O$_3$	2,36	2,73	0,21	1,96	0,33	0,01	2,30	3,58		3,95	2,72
CaO	42,30	0,39	52,76	35,11	0,05	2,64	37,80	58,90		65,05	65,51
MgO	1,25	0,06	1,78	1,04	0,01	0,09	1,14	1,78		1,97	1,95
K$_2$O	2,82	0,16	0,21	2,34	0,02	0,01	2,37	3,69	3,32	0,37	0,32
Na$_2$O	0,34	0,04	0,06	0,28	0,00	0,00	0,28	0,44	0,31	0,13	0,12
SO$_3$	4,68	0,15	0,20	3,88	0,02	0,01	3,91	6,09	5,48	0,61	1,01
Cl	0,27	0,00	0,00	0,23	0,00	0,00	0,23	0,36	0,34	0,02	—
Verlust @900°C loss	27,55	89,90	43,11	22,87	10,79	2,16	35,82	0,00		0,00	0,05
Total	99,74	99,93	100,10				99,66		9,45	99,51	

S/R	2,32	2,30
Klinkerphasen cement compounds C$_3$S	55,64	55,30
C$_2$S	20,81	19,80
C$_3$A	7,92	8,30
C$_4$AF	12,01	11,80
Gesamtalkalien als Na$_2$O total alkali as Na$_2$O 2,87	0,37	0,33

dagegen die Ergebnisse nach einer chemischen Korrektur über den Kalksteingehalt. In beiden Tabellen ist die tatsächliche Klinkerzusammensetzung, wie sie an der Pilotanlage erzielt wurde, dargestellt.

Die chemische Analyse des während der Pilotversuche als Nebenprodukt abgeschiedenen Materials ergab einen K$_2$O-Gehalt von 35 bis 50 % in Form von Sulfaten und Chloriden. Wie bereits erwähnt, besitzt dieses Staubprodukt einen gewissen kommerziellen Wert, da es entweder zur Kaliumherstellung oder direkt als Dünger verwendet werden kann. Das muß jedoch von Fall zu Fall entschieden werden.

Die aus diesen Ergebnissen ermittelten Minderungsraten für Alkalien und Schwefel sind in **Tabelle 3** zusammengestellt.

Ein weiteres Ergebnis der Untersuchung ist, daß die Reststaubmenge mit Hilfe eines Auslaugungsverfahrens noch weiter reduziert werden kann. Dadurch wird der Kaliumgehalt im Staub erhöht, was zu einer Steigerung des Marktwertes dieses Nebenproduktes beiträgt. Der anfallende Staub ist ein bis zu 60 % wasserlösliches Produkt und könnte bis zu 98 % der Natriumsalze bzw. 97 % der Kaliumsalze enthalten. Durch Eindampfen könnte ein Alkalisalz von 96 % gewonnen werden.

Die festen Rückstände, die bei der Auslaugung des Staubes anfallen, können zur Staubaufgabe zurückgeführt werden. Dieses feuchte Feststoffprodukt kann bei der Pelletierung mit dem Frischgut zusammengeführt werden, ohne daß dadurch nennenswert die Zusammensetzung des Aufgabegutes verändert wird. Die Feuchtigkeit im Feststoff kann bei der Materialaufbereitung in der Mischtrommel vorteilhaft ausgenutzt werden. Bei Anwendung dieses Konzeptes kann die Reststaubmenge auf nur 5 % des zu Prozeßbeginn aufgegebenen Staubes reduziert werden (**Bild 3**).

5. Schwermetallgehalte der Produkte

Aufgrund allgemeinen Interesses wurden auch die Schwermetallkonzentrationen der verschiedenen Produkte des Aufbereitungsverfahrens bestimmt. Vorläufige Ergebnisse deuten darauf hin, daß die im Ausgangsmaterial enthaltenen Schwermetalle in dem während des Prozesses anfallenden Feststoff verbleiben (**Tabelle 4**). Es ist allerdings zu erkennen, daß die Schwermetalle Pb, Se und Ti im Reststaub aufkonzentriert sind. Die Ergebnisse der auslaugbaren Metallgehalte sind in **Tabelle 5** dargestellt und zeigen,

TABELLE 3: Abbau der Alkali- und Schwefelkonzentrationen
(Massen-%, trocken)
TABLE 3: Alkali and sulfur removal
(Weight %, dry basis)

Verbindungen compounds	Aufgabe feed wie erhalten as received	Glüh-verlustfrei loss free	Klinker clinker	Reduzierung in % % reduction
K$_2$O	2,50	3,30	0,23	92,3
Na$_2$O	0,30	0,46	0,13	71,7
SO$_3$	4,14	6,37	0,79	87,6
Cl	0,24	0,37	—	99,0

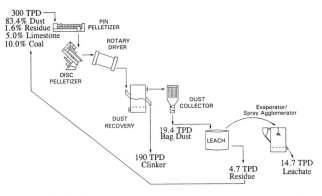

BILD 3: Fließbild einer Anlage zur Staubrückgewinnung mit Rückführung ausgelaugter Feststoffe
FIGURE 3: Dust recovery flow sheet with leachate solid recycle system

t/d	= tpd
Staub	= dust
Rückstände	= residue
Kalkstein	= limestone
Kohle	= coal
Mischer	= pug mill
Drehtrockner	= rotary dryer
Granulierteller	= disc pelletizer
Staubabscheider	= dust collector
Verdampfer/Sprühagglomerator	= evaporator/spray agglomerator
Staubrückgewinnung	= dust recovery
t/d Filterstaub	= tpd bag dust
t/d Klinker	= tpd clinker
t/d Rückstände	= tpd residue
t/d Auslaugung	= tpd leachate

TABELLE 4: Prozeßprodukte, Beispiel 2 —
Neue Analyse des Metallgehalts im Staub
TABLE 4: Process products, example 2 —
new dust metal analysis

Element element	Maß- einheit unit	Aufgabegut composite feed	Reaktor- austrittsgut reactor discharge	Filterstaub baghouse dust
Ag	ppm	<1,2	<1,2	5
As	ppm	23	20	16
Ba	ppm	460	917	205
Be	ppm	<1,2	<1,2	<1,2
Cd	ppm	<1,2	<1,2	14
Cr	ppm	31	70	81
Hg	ppm	0,2	<0,1	0,7
Ni	ppm	6	15	3
Pb	ppm	51	<1,5	842
Sb	ppm	<1,2	<1,2	<1,2
Se	ppm	<1,2	<1,2	8
Ti	ppm	2	<1,2	35

TABELLE 5: Neue TCLP-Analyse des Metallgehalts im Staub
TABLE 5: Example 2 — new dust TCLP metal analysis

Element element	Maß- einheit unit	Aufgabe- gut composite feed	Kühler- Produkt cooler product	Filter- staub baghouse dust	TCLP Grenzwert Conc. limit
Ag	mg/l	<0,003	<0,003	<0,003	5
As	mg/l	0,018	0,02	0,217	5
Ba	mg/l	0,8	3,1	0,4	100
Be	mg/l	<0,003	<0,003	<0,003	0,007
Cd	mg/l	<0,003	<0,003	0,2	1
Cr	mg/l	0,023	0,017	0,389	5
Hg	mg/l	<0,003	<0,003	<0,003	0,2
Ni	mg/l	0,007	0,004	<0,003	70
Pb	mg/l	<0,003	0,005	0,029	5
Se	mg/l	0,005	0,008	0,12	1
Ti	mg/l	<0,003	<0,003	0,565	7

daß kein Konzentrationswert irgendeines Metalls innerhalb des Rückgewinnungssystems die Grenzwerte überschreitet, die durch die Environmental Protection Agency [5] gesetzt worden sind.

6. Bewertung des Klinkers

Der Klinker, der für die vorstehenden Untersuchungen produziert worden ist, besaß einen Mahlbarkeitsindex nach Bond von 15,1. Der Klinker wurde labormäßig zu Zement aufgemahlen und erreichte die in **Tabelle 6** aufgelisteten Druckfestigkeiten. Daraus wird auch der Einfluß der chemischen Regulierung deutlich.

7. Kommerzielles Anlagensystem

Der berechnete spezifische Brennstoffenergieverbrauch für eine kleine Produktionsanlage mit einem Durchsatz von 80 t/d Klinker liegt bei 1143 kcal/kg Klinker. Eine neue Pilotanlage, wie in **Bild 4** dargestellt, wurde inzwischen für Untersuchungen mit technischen Ofenstäuben gebaut. Dabei wurden im Vergleich zum konventionellen Zementprozeß sehr niedrige NO_x-Emissionen gemessen. Bei einem Testprogramm lagen die typischen Emissionen bei 0,68 kg NO_x/t Klinker und 1,32 kg SO_2/t Klinker.

8. Entwicklung eines neuen Zementofens (CAF)

Mit der Unterstützung und finanziellen Mitteln des Gas Research Institute (GRI) und der Southern California Gas Company wurde im Jahre 1985 mit den damaligen AVCO Research Laboratories (heute Textron Defense Systems) in Everett, MA, ein Vertrag über die Entwicklung eines neuen Konzeptes zum Klinkerbrennen in einem Schacht abge-

which can be sold as a by-product by increasing potassium level in the final recovered product. Up to 60% of the dust produced is a water soluble product that would contain 98% of sodium salts, 97% potassium salts present, and if recovered by evaporation could produce a 96% alkali salt product.

The solid residues from leaching of final dust can be recycled back into the process dust feed system. This wet solid product can be recycled to pelletizing system with the new incoming dust without changing process input chemistry greatly. Moisture in solids would be used in pug mill of preparation system. Using this approach, final desirable end-product can be reduced to 5% of starting dust feed as shown in **Fig. 3**.

5. Heavy metals evaluation – solid products

Because of concerns expressed by some, the heavy metal concentrations of various products from the dust recovery system program were determined. Preliminary results indicate that heavy metals in the starting dust material are being maintained in the solids produced from the system in the study conducted (**Table 4**). It is shown that Pb, Se, and Ti were concentrated in the end process dust catch materials. The TCLP results (leachable levels of metals) determined and shown in **Table 5** show that none of the levels of any solid products from recovery system exceeds the concentration limits set by the Environmental Protection Agency [5].

6. Clinker evaluation

The actual clinker produced by the example given had Bond Grindability Work Index of 15.1. The clinkers produced were ground into laboratory cements with the resultant strength listed in **Table 6**. The effect of chemical adjustment is shown.

TABELLE 6: Physikalische Analysenergebnisse an laborgemahlenem Zement
TABLE 6: Physical analysis lab ground cement

Eigenschaft property	Staub wie erhalten as received dust 5% Gips gypsum	eingestellte Zusammensetzung adjusted chemistry 5% Gips gypsum
— Autokl. Ausdehnung (%) Autoclave expansion	0,27	0,27
— Erstarrungszeit: time of set:		
Erstarrungsbeginn n. Gilmore Gilmore initial set (min)	90	105
Erstarrungsende n. Gilmore Gilmore final set (min)	150	180
Erstarrungsbeginn n. Vicat Vicat initial set (min)	98	135
Erstarrungsende n. Vicat Vicat final set (min)	150	180
— Luftgehalt im Mörtel (%) air content of mortar	9,4	4,0
— Druckfestigkeit compressive strength (N/mm^2)		
1 Tag one day	7,2	7,7
3 Tage three day	11,8	16,9
7 Tage seven day	16,2*	21,0
28 Tage twenty-eight day	30,7	35,8

Die untersuchten Proben entsprechen den physikalischen Anforderungen der ASTM C150 für einen Portlandzement Typ I aus der 7- und 28-Tagefestigkeit mit Ausnahme der Festigkeitswerte, die in der Tabelle gekennzeichnet sind.

The above samples meet the physical requirements of ASTM C150 for Type I Portland Cement at the 7 day and 28 day test period except where marked with an asterisk.

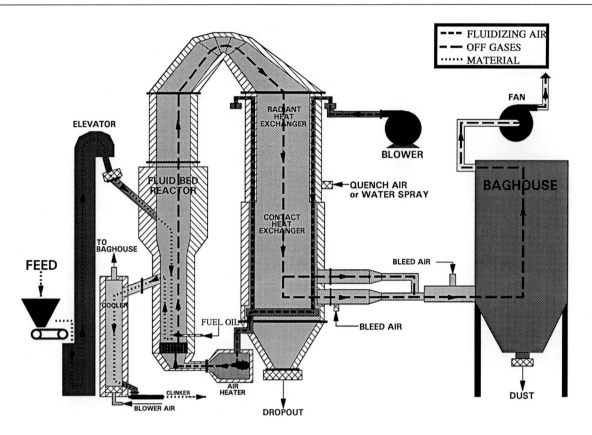

BILD 4: Pilotanlage zur Rückgewinnung von Zementofenstaub mit schematischer Darstellung der Material- und Gasströme
FIGURE 4: Pilot cement plant dust recovery system material and gas flow schematic

Fluidisierungsluft	= fluidizing air
Abgase	= off gases
Material	= material
Becherwerk	= elevator
Strahlungs-Wärmetauscher	= radiant heat exchanger
Gebläse	= fan
Drehkolbengebläse	= blower
Wirbelschicht-Reaktor	= fluid bed reactor
Kühlluft oder Wassereindüsung	= quench air or water spray
Schlauchfilter	= baghouse
Materialaufgabe	= feed
zum Schlauchfilter	= to baghouse
Konvektiver Wärmetauscher	= contact heat exchanger
Verdünnungsluft	= bleed air
Kühler	= cooler
Heizöl	= fuel oil
Klinker	= clinker
Gebläseluft	= blower air
Luft-Vorwärmer	= air heater
Staub	= dust

TABELLE 7: Vorteile des Verfahrens
TABLE 7: Process Benefits

Vorteil / benefit	Bemerkung / remark
— Kapitalkosten	Werden niedriger erwartet als bei Drehofensystemen
capital cost	expected lower on an installed basis than rotary kiln system
— Instandhaltung	Wesentlich niedriger wegen des Fehlens bewegter Teile
maintenance	much lower due to no moving parts
— Umweltschutz	Niedrigere NO_x- und SO_2-Emissionen als bei konv. Ofensystemen
environment	lower NO_x and SO_2 emissions than conventional system
— Flächenbedarf	Niedriger als bei Drehofensystemen
land usage	less than rotary kiln system
— Flexibilität	Geeignet für Mini-Zementwerke mit einem geringen Anstieg der Betriebskosten
flexibility	suitable for mini plants with little increase in operating costs
— Brennstoffverbrauch	Wird niedriger erwartet als bei irgendeinem konventionellen Ofensystem
fuel consumption	expected to be lower than any operating conventional kiln system
— Alkaligehalt	Bei vergleichbarer Rohmaterialbasis im Vergleich zu konventionellen Ofensystemen ein Fertigprodukt mit einem niedrigeren Gesamtalkaligehalt erwartet
low alkali	product expected to have lower total alkali than conventional systems on same feed stock

schlossen. 1989 trat die Fuller Company dem Entwicklungsteam bei, welches zu dieser Zeit durch die CTL Laboratories der Portland Cement Association und durch zwei technische Berater unterstützt wurde.

Zu dieser Zeit wurde entschieden, daß die gewünschten Ergebnisse am besten durch eine Kombination des nach dem freien Fall arbeitenden Vorwärmersystems gemäß einem Patent von GRI/AVCO [6] mit der Wirbelschichttechnologie der Fuller Company [2–4] erreicht werden könnte, um die in **Tabelle 7** zusammengestellten Vorteile zu erreichen. Das daraus entwickelte Konzept erfordert ein pelletiertes Aufgabematerial ähnlich dem System zur Staubrückgewinnung, allerdings mit dem Unterschied, daß ein Zement als Bindemittel benutzt und keine Kohle oder Koks im Pellet enthalten sein darf. Die Zugabe von Kohle bzw. Koks zum Rohmehl konnte wegen des im System verwendeten Vorwärmers nicht aufrecht erhalten werden. Das neue CAF-System ist in erster Linie auf die Befeuerung mit Erdgas oder Flüssigbrennstoffen ausgerichtet.

7. Commercial system

The calculated fuel consumption for a small 80 mtpd clinker production was calculated at 1143 kcal/kg clinker. A new pilot plant installation, as depicted in **Fig. 4**, has been built for plant dust evaluations. It has been determined that very low NO_x emission levels compared to conventional cement processes, will result from this system. Typical NO_x and SO_2 emissions are 0.68 kg/ton clinker NO_x and 1.32 kg/ton clinker SO_2 for one particular test program.

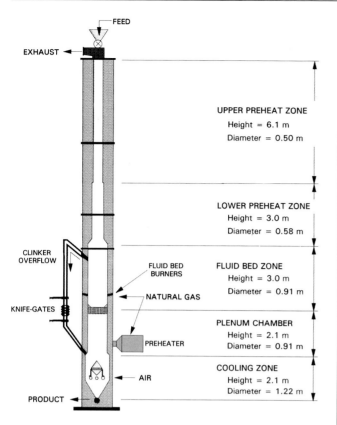

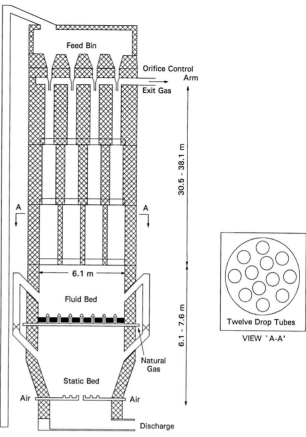

BILD 5: Schema der Pilotanlage eines neuen Zementbrennofens
FIGURE 5: Cement advanced furnace pilot system

Materialaufgabe	= feed
Abgas	= exhaust
Obere Vorwärmzone	= upper preheat zone
Höhe	= height
Durchmesser	= diameter
Untere Vorwärmzone	= lower preheat zone
Klinkerüberlauf	= clinker overflow
Brenner für Wirbelschicht	= fluid bed burners
Wirbelschichtzone	= fluid bed zone
Erdgas	= natural gas
Schieber	= knife-gates
Luftkammer	= plenum chamber
Vorwärmer	= preheater
Kühlzone	= cooling zone
Luft	= air
Produkt	= product

BILD 6: Schema des neuen Zementbrennofens (kommerzielle Anlage: 50 t/h)
FIGURE 6: Advanced cement furnace schematic commercial plant (50 tons/hour)

Aufgabetrichter	= feed bin
Kontrolle der Austrittsöffnung	= orifice control arm
Gasaustritt	= exit gas
Fließbett	= fluid bed
12 Fallschächte	= twelve drop tubes
Schnitt	= view
Erdgas	= natural gas
Statisches Bett	= static bed
Luft	= air
Entleerung	= discharge

Die in **Bild 5** schematisch dargestellte Pilotanlage wurde durch die Fuller Company errichtet. Die ersten Untersuchungen wurden an dieser Anlage Mitte 1990 durchgeführt, wobei Rohmehle aus bestehenden Zementwerken verarbeitet wurden.

Annähernd 300 t Pellets wurden bis zum heutigen Tage verbraucht. Die chemische Zusammensetzung der Rohmehlpellets und des daraus gebrannten Klinkers enthält **Tabelle 8**. Die physikalischen Untersuchungsergebnisse eines CAF-Klinkers, verglichen mit einem Drehofenklinker, befinden sich in **Tabelle 9**. Die Mahlbarkeitsuntersuchungen lieferten einen Arbeitsindex nach Bond von 10,1.

Typische Betriebsparameter sind in **Tabelle 10** zusammengestellt. Die NO_x- und SO_2-Emissionen der Klinkerbrennanlage nach dem CAF-Verfahren, verglichen mit einem heutigen Anlagensystem, zeigt **Tabelle 11**.

Die kommerzielle Wirbelschichtanlage, so wie sie in **Bild 6** dargestellt ist, wird eine dreimal so lange Vorwärmzone wie die Pilotanlage haben, wodurch ein spezifischer Brennstoffverbrauch im Bereich von 700 bis 800 kcal/kg Klinker erreicht werden soll. Das Ofensystem, das Gegenstand eines erst zuletzt herausgegebenen Patentes ist [7], soll für Produktionseinheiten mit einer Jahresproduktion bis zu 360 000 t Klinker angeboten werden.

8. New advanced cement furnace development (CAF)

In 1985 under sponsorship and funding of the Gas Research Institute (GRI) and Southern California Gas Company, a contract was awarded to the then AVCO Research Laboratories (now Textron Defence Systems), Everett, MA, to develop a new concept shaft clinkering device. In 1989, Fuller Company joined the development team, which at that time included review and advice from CTL Laboratories of the Portland Cement Association and two cement plant technical consultants.

It was decided at that time it would be best to achieve the desired end results by putting together the preheater free-fall system of GRI/AVCO patent [6] and the fluid bed technology of Fuller Company patents [2–4], with the resulting system with the projected benefits of such a system as listed in **Table 7**. The concept required a pellet feed system similar to that of the dust recovery system. In this case, the use of a cement binder and elimination of the coal or coke in the pellet is used. The coal or coke addition could not work within the preheat portion of the system. The CAF system is primarily a natural gas or liquid fuel fired system.

A designed pilot plant system as shown in **Fig. 5** was constructed at Fuller Company. First tests were conducted in mid 1990 with the final development now completed.

TABELLE 8: Prozeßchemie der Produkte aus der CAF-Pilot-Zementbrennanlage
TABLE 8: CAF pilot plant products process chemistry

Chem. Bestandteile chemical compounds	Rohmaterial-Aufgabe raw feed	Glüh-verlustfrei calc. loss free	Klinker aktuell actual clinker	Staub-produkt dust product
anal. no.	37689	—	40113	40560
Verlust 105 °C loss @ 105 °C	0,39	0	0	0
SiO_2	14,23	21,82	22,25	14,80
Al_2O_3	3,53	5,41	5,49	3,93
Fe_2O_3	1,16	1,78	2,03	1,93
CaO	41,84	64,15	65,12	47,32
MgO	2,32	3,56	4,05	3,06
K_2O	0,73	1,12	0,08	6,14
Na_2O	0,17	0,26	0,13	0,88
SO_3 (total)	0,91	1,40	0,12	6,90
P_2O_5	0,09	0,14	0,20	0,13
TiO_2	0,15	0,22	0,23	0,16
Mn_2O_3	0,04	0,06	0,07	0,05
Glüh-verlust 900 °C loss @ 900 °C	34,78	0,00	0,17	14,39
total	99,95		99,94	99,69
free lime			1,37	
Gesamtalkali als Na_2O total alkali as Na_2O		1,02	0,18	

Berechnete Klinkerphasengehalte
calculated cement compounds

C_3S			56,4	56,1
C_2S			20,2	21,6
C_3A			11,3	11,1
C_4AF			5,4	7,9

TABELLE 10: Betriebsparameter der Pilotanlage nach dem GRI/CAF-Verfahren (ausgewählte Untersuchung)
TABLE 10: Operating data GRI/CAF pilot system (selected test)

Betriebsparameter operating data	Maßeinheit unit	Zahlenwert value
Bettmasse bed inventory	kg	1090
Betriebstemperatur des Fließbettes operating bed temperature	°C	1320
Vorwärmer Eingangstemperatur preheater inlet temperature	°C	1290
Vorwärmer Austrittstemperatur preheater outlet temperature (exit)	°C	950
Kühleraustrittstemperatur produced cooler temperature	°C	30
Fluidisierungsluft fluidizing air		
Vorwärmertemperatur unter dem Rost preheater temperature under grid	°C	500
Volumenstrom volume rate	m^3_N/min	28
Aufgabemassenstrom feed rate	kg/h	1410
Klinkermassenstrom clinker production rate	kg/h	787
Staubverlust dust loss	kg/h	100
Gasverbrauch (Erdgas) fuel rate (natural gas)	m^3_N/min	142
O_2-Gehalt im Abgas exit gas O_2 content	%	3,4
Freikalkgehalt im Klinker product free lime	%	1,0–1,35

Literature

[1] Cohen, S.M.: Pyzel Process. Portland Cement Association Technical Meeting – September 1970.

[2] Cohen, S.M. (assigned to Fuller Company): U.S. Patent No. 4584022 – Cement Plant Dust Recovery System. Issued April 22, 1986.

TABELLE 9: Zementeigenschaften nach Versuchen der GRI/TEXTRON DEFENSE vom 15.–18. 4. 1991
TABLE 9: Product evaluation cement tests GRI/TEXTRON DEFENSE 15–18 April 1991

Laborgemahlene Zemente laboratory ground cements	Drehofen-verfahren rotary kiln	CAF-Verfahren CAF
— Luftgehalt nach ASTM C 185 air content		
Volumen % volume	10,04	9,4
— Normalkonsistenz nach ASTM C 187 normal consistency		
Wasser/Zement-Verhältnis water/cement-ratio	24,0	24,5
— Erstarrungszeit nach ASTM C 191 setting time		
Erstarrungsbeginn n. Vicat initial Vicat set	150	105
Erstarrungsende n. Vicat final Vicat set	195	180
— Erstarrungszeit nach ASTM C 266 setting time		
Erstarrungsbeginn n. Gilmore initial Gilmore set	105	90
Erstarrungsende n. Gilmore final Gilmore set	210	195
— Druckfestigkeit nach ASTM C 109 in N/mm^2 compressive strength		
3 Tage (Einzelproben) 3 day specimens	20,4 / 21,5 / 21,7	15,3 / 15,8 / 16,1
3 Tage (Durchschnitt) 3 day average	21,2	15,8
7 Tage (Einzelproben) 7 day specimens	26,7 / 25,8 / 26,9	23,9 / 22,4 / 23,2
7 Tage (Durchschnitt) 7 day average	26,4	23,2
28 Tage (Einzelproben) 28 day specimens	32,1 / 33,0 / 32,8	31,2 / 35,7 / 32,6
28 Tage (Durchschnitt) 28 day average	32,6	33,2

TABELLE 11: NO_x/SO_2-Emissions-Werte
TABLE 11: Emissions NO_x/SO_2

A. Kohlebefeuerte Zementofenanlagen (PCA Erhebungen von 1982)
   coal fired plants (PCA survey 1982)

— Konvent. Anlage nach dem Naß-Verfahren
   conventional wet plant
   NO_x 2,28 kg/t Klinker
   SO_2 5,70 kg/t Klinker
— Konvent. Anlage nach dem Trockenverfahren
   conventional dry plant
   NO_x 2,59 kg/t Klinker
   SO_2 4,08 kg/t Klinker
— Anlage mit Vorcalcinierung
   preheater/precalciner plant
   NO_x 2,92 kg/t Klinker
   SO_2 1,41 kg/t Klinker

B. Untersuchungsergebnisse von der GRI/AVCO Pilotanlage
   GRI/AVCO pilot plant test results

— Fuller 46 cm FBR
   Testprogramm (Mai 1990)
   test program (May 1990)
   NO_x 0,86 kg/t Klinker
   SO_2 < 0,09 kg/t Klinker

Tests were conducted on local raw mixes from operating cement plants. Approximately 300 tons of pellets have been tested to date. The chemistry of the raw pellets and resulting clinker products are shown in **Table 8**. The physical test results comparing the actual plant rotary kiln clinker product and the CAF furnace are shown in **Table 9**. Clinker grindability results gave a 10.1 Bond Work Index.

[3] Cohen, S.M. (assigned to Fuller Company): U.S. Patent No. 4595416 – Methods and Apparatus for Producing Cement Clinker Including White Cement. Issued June 17, 1986.

[4] Cohen, S.M. (assigned to Fuller Company): U.S. Patent No. 4682948 – Method and Apparatus For Producing Cement Clinker Including White Cement. Issued July 28, 1987.

[5] An Analyses of Selected Trace Metals in Cement and Kiln Dusts. Portland Cement Association (1992).

[6] AVCO Research Laboratory Textron (assigned to GRI): U.S. Patent No. 4975046 – Cement Shaft Suspension Process. Issued 4 December, 1990.

[7] Cohen, S.M., and Litka, A.F. (assigned to GRI): Cement Advanced Furnace and Process.

Typical operating conditions are shown in **Table 10**. Emissions of NO_x and SO_2 from a CAF raw mix clinker operation compared to the current systems is shown in **Table 11**.

The commercial system as depicted in **Fig. 6** will have three times higher preheat length than the pilot unit which will bring the system projected fuel consumption to 700–800 kcal/kg clinker range. The system will be offered in units producing up to 400000 STPY (360000 MTPY) and is currently covered in the latest patent issued [7].

Das CFPC-Verfahren
The CFPC Process

Procédé CFPC

El procedimiento CFPC

Von **X. J. Deng,** Iowa/USA

Zusammenfassung – Der zirkulierende Vorwärmer/Calcinator-Prozeß (CFPC – Circulating Fluidized Preheater-Calciner Process) kombiniert die Zementherstellung mit der Energieerzeugung. Er verwendet sowohl eine zirkulierende Wirbelschicht als Vorwärmer- und Vorcalciniereinheit für die Klinkerproduktion als auch einen Brennraum für die Energieerzeugung. Das CFPC-Verfahren kann eine hohe Schlupfgeschwindigkeit und eine hervorragende Durchmischung erreichen, die zur Optimierung des Wärmeaustauschs und zu optimalen Reaktionsbedingungen zwischen den feinkörnigen Feststoffen und dem Gas führt. Eine Verweilzeit von optimaler Dauer gewährleistet die vollständige Verbrennung auch geringwertiger Brennstoffe. Die Bildung von NO_x kann durch eine gestufte Verbrennung bei relativ niedrigen Temperaturen kontrolliert werden. Die Absorption der Schwefelverbindungen durch frischen CaO-haltigen Staub führt zu abnehmenden SO_x-Werten innerhalb des Ofens. Die Mischung von Asche und Rauchgas-Entschwefelungsgips aus der Energieerzeugung kann als Rohstoffkomponente zusammen mit entsäuertem Rohmehl aus dem CFPC-Verfahren ohne Kühlung verwendet werden. Die erste CFPC-Anlage wird zur Zeit im Zementwerk Huhhot in China gebaut.

Summary – The Circulating Fluidized Preheater-Calciner (CFPC) process is a combination of cement production and power generation. It uses a circulating fluidized bed both as a preheater and precalciner for the clinker production and as a combustor for power generation. CFPC can obtain high slip velocity and an excellent mixing which leads to the optimization of heat exchange and to optimum reaction conditions between fine-grained solids and gas. Optimal long residence time allows the complete combustion also of low-grade fuel. The formation of NO_x can be controlled by using staged combustion at relatively low temperatures. The SO_x level is decreased inside the furnace by absorption of sulfur compounds on fresh CaO dust. The mixture of ash material and desulfurization gypsum from power generation can be used as a raw material component together with decarbonated raw meal from the CFPC without cooling. The first CFPC plant is now being built at the Huhhot Cement Plant in China.

Résumé – Le process de circulation préchauffeur/calcinateur (CFPC – Circulating Fluidized Preheater-Calciner Process) combine la fabrication de ciment à la production d'énergie. En effet, il utilise aussi bien le lit fluidisé circulant comme unité de préchauffeur et de précalcination pour la production de clinker qu'une chambre de combustion pour la production d'énergie. Le procédé CFPC permet d'obtenir une vitesse de glissement élevée et un excellent brassage, avec pour effet une optimisation de l'échange thermique et de très bonnes conditions de réaction entre les solides à grains fins et le gaz. Un temps de séjour d'une durée optimale assure la cuisson intégrale y compris les combustibles de moindre qualité. La formation de NO_x est contrôlée grâce à une cuisson progressive à températures relativement basses. L'absorption de composés sulfurés due à la poussière fraîche contenant du CaO entraîne une diminution des valeurs SO_x dans le four. Le mélange de cendres et de gypse désulfuré des gaz de fumées provenant de la production d'énergie peut, en tant que composants de matière première, être utilisé sans refroidissement conjointement avec la farine crue décarbonatée issue du procédé CFPC. La première installation CFPC est, à l'heure actuelle, construite à la cimenterie Huhhot en Chine.

Resumen – El procedimiento de circulación precalentador/calcinador (CFPC – Circulating Fluidized Preheater-Calciner Process) combina la fabricación de cemento con la producción de energía. Emplea tanto un lecho fluidizado circulante como unidad de precalentamiento y de precalcinación, para la fabricación del clínker, como también un recinto de combustión para la producción de energía. El procedimiento CFPC puede alcanzar una alta velocidad de deslizamiento y una excelente mezcla, que permite la optimización del intercambio de calor y la consecución de condiciones de reacción óptimas entre los sólidos de grano fino y el gas. Un tiempo de permanencia óptimo garantiza la combustión total, incluso de combustibles de peor calidad. La formación de NO_x se puede controlar mediante una combustión escalonada, a temperaturas relativamente bajas. La absorción de los compuestos de azufre por medio de polvo fresco que contiene CaO conduce a la reducción de los valores de SO_x dentro del horno. La mezcla de ceniza y de yeso de la desulfuración de los gases de humo, procedente de la producción de energía, puede emplearse como componente de la materia prima, junto con el crudo descarbonatado, producto del procedimiento CFPC, sin calentamiento previo. La primera planta CFPC se está construyendo actualmente en la fábrica de cemento Huhhot, en China.

1. Introduction

The People's Republic of China had become the number one of the cement producers among the world cement manufacturers by 1985. The cement industry has been one of China's biggest consumers of electricity because of its higly energy intensive process. In 1989 this industry consumed about 20 billion kilowatt-hours of electricity. Because of the insufficient electricity supply many cement plants operated at half capacity or stopped operations at that time [1].

Cogeneration is one practical solution to remedy this grave situation. Cogeneration in cement plants is not new, however, China's only cogeneration practice had been the Dry Kiln with Boiler (DB) process. The First Fluidized Preheater-Calciner (FPC) process was introduced to the Fushun Cement Plant in 1986. As a modified process the CFPC process was developed by the author to improve the insufficient decarbonation rate and thermal stability of the FPC process [2]. The CFPC process was listed in The 8th Five-Year Plan (1990 to 1995) as a research & development programm.

2. Cogeneration in Cement Plants

Three types of cogeneration systems are known for cement plants. The DB process, the (New) Suspension Preheater (SP/NSP) process and the FPC/CFPC process.

1. The DB process was popular in the 1930's. This process has a heat consumption of 5850 to 6270 kJ/kg of clinker and supplies 70 to 80% of the plant's electricity demand. This process has a low thermal efficiency and some interacting problems between cement production and electricity production. The DB process is replaced by the emerging preheating technology (SP) in the 1960's and the precalcining technology (NSP) in the 1970's.

2. The SP technology was popular due to its higher thermal efficiency and the NSP technology had the advantage of capacity enhancement. However, these two processes require more electricity. Cogeneration within SP/NSP process was developed in Japan to convert the thermal energy of the exhaust gas at temperatures of 200 to 400 °C to electricity because of the high electricity prices. The heat consumption of these processes is 3130 to 3550 kJ/kg of clinker and supplies about 20% of demanded electricity. The cogeneration within the SP/NSP process is economic only when the cement capacity of the plant is larger than 3000 to 4000 t/d [3].

3. FPC/CFPC processes: China has more than 5000 cement plants. However, only four plants have a capacity of 3000 to 4000 t/d, about 60 plants are medium-sized and the rest are all on a smaller scale. The FPC/CFPC process uses a circulating fluidized bed furnace both as a preheater and precalciner for cement production and as a combustion chamber for power production. These processes are economic for both large and medium-sized cement plants since exhaust gas with high temperature is available for power generation.

Table 1 shows a comparison of these three Cogeneration Processes.

3. FPC/CFPC Process Outline

In 1986 the first FPC process was constructed in the 2 kiln system between kiln and waste heat boiler. This caused an

TABLE 1: Comparison of Three Cogeneration Processes

	DB	SP/NSP	FPC/CFPC
gas temperature (°C)	700–900	200–400	850
heat consumption	high	low	low
electrical efficiency	high	low	high
investment	low	high	medium
NO_x, SO_x	high	low	low

TABELLE 1: Vergleich der drei Arten der Kraft-Wärme-Kopplung

	DB	SP/NSP	FPC/CFPC
Gastemperatur (°C)	700–900	200–400	850
Wärmeverbrauch	hoch	niedrig	niedrig
elektrischer Wirkungsgrad	hoch	niedrig	hoch
Kosten	niedrig	hoch	mittel
NO_x, SO_x	hoch	niedrig	niedrig

3. Übersicht über das FPC/CFPC-Verfahren

Im Jahre 1986 wurde die erste FPC-Anlage in einer Ofenlinie zwischen Ofen und Abhitzekessel installiert. Damit konnte die Klinkerproduktion um 4 t Klinker pro Stunde und die Stromerzeugung um 1.300 kW gesteigert werden. Der Nettogewinn lag bei etwa 5 Mio. Yuan (ca. 0,9 Mio. US $) jährlich bei einer Anfangsinvestition von rund 1,8 Mio. Yuan (ca. 0,33 Mio. US $). Die ökonomischen Vorteile liegen auf der Hand. Prozeßtechnisch sind noch einige Verbesserungen notwendig, wie z. B. die Erhöhung des Calcinationsgrades, die Verbesserung der thermischen Stabilität zum leichteren Anfahren der Anlage und der Einsatz von Sekundärbrennstoffen.

Die grundlegende Idee zur Verbesserung des FPC-Verfahrens beruht auf der Zirkulation des Rohmehls in der Wirbelschicht. Das CFPC-Verfahren wurde mit Hilfe der Forschungsarbeiten von Reh [4] und Warshawsky [5] entwickelt. Nach Angaben der Fuller-Company [5] kann der Vorentsäuerungsgrad bei 100prozentiger Zirkulation des Mehls und einer Temperatur von 900°C um 4 bis 5% erhöht werden. Zementrohmehl gehört zum Mehltyp A nach Geldart, ist jedoch dem Typ C sehr ähnlich. In einer zirkulierenden Wirbelschicht können hohe Relativgeschwindigkeiten erreicht werden, so daß eine optimale Durchmischung erreicht wird. Dies führt zu einer Optimierung des Wärmeaustausches und zu optimalen Reaktionsbedingungen zwischen feinkörnigen Feststoffen und dem Gas [4]. Zementrohmehl ist ein heterogenes Pulver aus mehreren Komponenten. Daher ist eine gute Durchmischung eine wichtige Voraussetzung für die Produktion von qualitativ hochwertigem Zementklinker. Große Turbulenz und lange Verweildauer in der Wirbelschicht ermöglichen einen hohen Wirkungsgrad bei der Verbrennung sowie die Verwendung heizwertarmer Brennstoffe. Das CFPC-Verfahren ist schematisch in **Bild 2** dargestellt.

Das Rohmehl wird direkt in die Wirbelschicht aufgegeben, wo es vorgewärmt und entsäuert wird. Das entsäuerte Mehl strömt zusammen mit dem Gas in die Entstaubungszyklone.

increase in clinker production of 4 t/h of clinker and in power production of 1300 kilowatt. The net benefit was about 5 million Yuan (about 0.9 million US Dollars) annually. The initial investment was about 1.8 million Yuan (about 0.33 million US Dollars). The economic benefits are obvious. The process still leaves some areas for improvement such as the decarbonation rate, the thermal stability for an easy start-up and the utilization of waste fuel.

Circulation of the material is the key idea for improving the FPC process. The CFPC process was proposed under the influence of research works by Reh [4] and Warshawsky [5]. The Fuller Company reported [5] that the decarbonation rate increases 4 – 5 % with 100 % of circulating rate at 900 °C. The cement raw meal powder belongs to Geldart's powder type A but close to type C. The circulating fluid bed can obtain high slip velocity and an excellent mixing which leads to the optimization of the heat exchange and to optimum reaction conditions between fine-grained solids and gas [4]. Cement raw meal is a multi-component heterogeneous powder. Therefore good mixing is one important manufacturing factor for producing high quality cement. High turbulence and long residence time in the CFPC allows a high combustion efficiency and the utilization of low-grade fuels. The CFPC process is shown schematically in **Fig. 2**.

The raw meal is introduced directly into the CFPC where the meal is preheated and precalcined. Then the decarbonated meal streams with the gas flow to the de-dusting cyclones where the meal is separated. One part of the collected material returns to the CFPC to increase the decarbonation rate, the other part enters the kiln to increase the production rate. The hot waste gas flows with a temperature of 800 to 900 °C to the boiler to produce electricity. The waste air from the cooler is introduced as staging air to the CFPC at the middle part.

The CFPC program: The first CFPC plant is now being built in the No. 2 kiln system in Huhhot Cement Plant, Inner Mongolia. By the CFPC process the capacity of the kiln (3.0 meters in diameter and 54 meters in length) will be increased and the electricity production goes up to more than 160 kWh/t of clinker. The plant will be self-sustaining and will save about 17700 tons of equivalent standard coal per year compared with the old process at the same cement output.

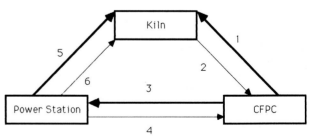

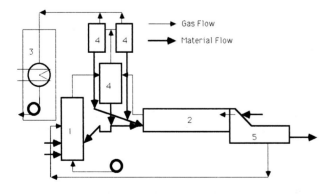

BILD 1: Das CFPC-Verfahren
FIGURE 1: The CFPC Process

| Gas | = Gas Flow |
| Feststoffe | = Material Flow |

1. CFPC — 1. CFPC
2. Drehofen — 2. Cement Kiln
3. Kraftwerk — 3. Power Station
4. Zyklone — 4. Cyclones
5. Kühler — 5. Cooler

BILD 2: Das goldene Dreieck
FIGURE 2: The golden triangle

Drehofen = Kiln
Kraftwerk = Power Station

1. Entsäuertes Rohmehl + Aschen + REA-Gips für Zementproduktion
1. Decarbonated raw meal + Ash materials + FGD Gypsum for cement production
2. Kühlerabluft als sekundäre Verbrennungsluft für die Wirbelschicht
2. Hot gas from the cooler as secondary air for CFPC combustion
3. Abgas vom CFPC für die Stromerzeugung
3. Exhaust gas from CFPC for power generation
4. Elektrizität aus dem Kraftwerk für den CFPC-Betrieb
4. Electricity from power station for CFPC operation
5. Elektrizität aus dem Kraftwerk für die Zementproduktion
5. Electricity from power station for cement production
6. Flugasche und Gips aus der Entschwefelung aus dem Kraftwerk für die Zementproduktion
6. Fly ash and FGD gypsum from power station for cement production

Ein Teil des dort abgeschiedenen Materials wird zur Wirbelschicht zurückgeleitet, um die Entsäuerungsrate zu erhöhen, der andere Teil wird in den Ofen geführt. Das heiße Abgas strömt mit einer Temperatur von 800 bis 900°C zu einem Abhitzekessel, in dem Dampf zur Produktion von elektrischer Energie erzeugt wird. Die Kühlerabluft wird als Zwischenluft (Verbrennungsluft) im mittleren Teil der Wirbelschicht zugeführt.

Das CFPC-Programm: Die erste CFPC-Anlage wird derzeit in einer Ofenlinie im Zementwerk Huhhot in der Inneren Mongolei installiert. Durch das CFPC-Verfahren wird sich die Kapazität des Ofens (3 m Durchmesser und 54 m lang) erhöhen und die Stromerzeugung auf über 160 kWh/t Klinker ansteigen. Die Anlage wird bezüglich des Stromverbrauchs autark sein und im Vergleich zum alten Prozeß mit dem gleichen Zementausstoß etwa 17 700 t SKE pro Jahr sparen.

4. Emissionsüberwachung

Umwelttechnisch sind in der Zementindustrie unter anderem Partikel-, NO_x- und SO_2-Emissionen von Bedeutung.

1. Zementstaub ist nicht giftig, nicht korrosiv, nicht brennbar und nicht explosiv. Die Partikelemissionen aus der CFPC-Anlage werden durch Verwendung hocheffizienter Zyklonabscheider und Elektrofilter unter Kontrolle gehalten.

2. Die Stickstoffoxidemissionen von Zementwerken bestehen in der Hauptsache aus NO (über 90%) und NO_2 (unter 10%). Die NO-Emissionen werden in der Atmosphäre weitgehend in NO_2 umgewandelt. Das NO_2 kann dann zum Bestandteil sauren Regens oder trockener Deposition werden. NO kann bei der Verbrennung auf zwei Arten entstehen. Das sogenannte thermische NO resultiert aus der thermischen Zersetzung und Oxidation von Stickstoff aus der Verbrennungsluft. Das Brennstoff-NO resultiert aus der Oxidation von Stickstoffverbindungen des Brennstoffs. Der Prozeß der Zementklinkerherstellung erfordert hohe Temperaturen und eine oxidierende Atmosphäre, was unvermeidlich zur Bildung von thermischem NO führt. Je höher die Flammentemperatur und je größer der Luftüberschuß, desto mehr thermisches NO wird gebildet. Je höher die Stickstoffkonzentration im Brennstoff und je höher die Luftzahl, desto mehr Brennstoff-NO entsteht.

Beim CFPC-Verfahren wird mehr als die Hälfte des Brennstoffs in der Sekundärfeuerung verbrannt, bei der die Temperatur vergleichsweise niedriger als diejenige im Ofen (über 1450°C) ist, so daß die Gesamtmenge des thermischen NO reduziert wird. Die Menge an Brennstoff-NO, das sich in der Wirbelschicht bildet, wird durch einen gestuften Verbrennungsprozeß kontrolliert. Deshalb wird die heiße Kühlerabluft in die mittlere Stufe der Wirbelschicht zugegeben. Die erhöhte Zirkulationsrate der Feststoffe führt auch zu einer Verminderung der NO_x-Emissionen, da die Verweildauer des Brennstoffs in der reduzierenden Zone vergrößert wird.

3. Schwefeloxide und deren Reaktionsprodukte sind ebenfalls Bestandteile des sauren Regens und trockener Deposition. SO_2 ist die einzige Verbindung, deren Emissionen aus technischen Prozessen höher sind als die aus natürlichen. Schwefeldioxid-Emissionen resultieren direkt aus der Verbrennung des Brennstoffs oder der Umsetzung des pyrithaltigen Rohmehls im Vorwärmer. Zusätzlich können SO_2-Emissionen aus der Zersetzung von Alkali- oder Calciumsulfaten entstehen. Beim Brennen von Zementklinker findet die Entschwefelung während des Brenn- und Mahlprozesses statt. Diese Prozesse können mehr als 90% des SO_2 absorbieren. Deshalb kann die Zementindustrie Brennstoffe mit hohem Schwefelgehalt einsetzen, ohne Probleme mit SO_2-Emissionen hervorzurufen. Bei der Verbrennung bzw. Entsäuerung in der Wirbelschicht wird keine zusätzliche Rauchgasentschwefelung benötigt. Der Kalkstein im Zementrohmehl wirkt als reaktives Absorbens und entfernt SO_2 durch Bildung eines stabilen Reaktionsproduktes,

4. Emission Control

Emission problems in the cement industry are particulate, NO_x and SO_x emissions.

1. Cement dust is non-toxic, non-corrosive, non-flammable and non-explosive. The particulate emission from the CFPC process is kept under control by using high efficiency cyclone collectors and electrostatic precipitators.

2. Nitrogen oxides are mainly NO (more than 90%) and NO_2 (less than 10%) in cement plants. The NO emissions are largely converted to NO_2 in the atmosphere. The NO_2 can then become a component of acid rain or dry deposition. There are two kinds of NO formation. The thermal NO results from the thermal decomposition and oxidation of nitrogen in the combustion air. The fuel NO results from the oxidation of nitrogenous compounds in the fuel. The cement clinkering process requires high temperatures and an oxidizing atmosphere which leads to an inevitable amount of thermal NO. The higher the flame temperature and the more the excess air the more thermal NO is formed. The higher the nitrogen concentration in the fuel and the more the excess air the more fuel NO is formed.

In the CFPC process more than half of the fuel is burned in this secondary firing where the temperature is relatively lower than that in the kiln (larger than 1450°C), so that the total amount of thermal NO is greatly reduced. The amount of fuel NO, formed in the fluidized bed, is controlled by a staged combustion process. Therefore the hot waste air from the cooler is injected to the intermediate stage of the CFPC. The higher circulating rate of the solid material also leads to a lower NO_x emission by increasing the residence time of the fuel in the reducing area.

3. The SO_x and their reaction products are also components of acid rain and dry deposition. SO_2 is the only compound of which the emissions from artificial processes exceed those from natural processes. Sulfur dioxide emissions result directly from the combustion of fuel or the roasting of pyrite-containing raw materials in cement plants. In addition, SO_2 emissions can result from the decompositon of alkali or calcium sulfates. The cement production process, such as the burning and grinding process, is the ideal means for desulfurization. These processes can absorb more than 90% of the SO_2. Thus the cement industry can use high-sulfur fuel without the problem of SO_x emissions. The CFPC process has an inherent emission control and needs no additional flue gas desulfurization. The limestone in the cement raw meal acts as reactive sorbents to remove SO_2 by forming a stable product, namely calcium sulfate, and thus eliminates the need for expensive flue gas scrubbers. The reaction between sulfur dioxide and limestone has been described in two consecutive steps, the endothermic calcination and the exothermic sulfation.

The temperature for coal combustion in a fluidized bed, for carbonate decomposition and for precalcining of cement raw meal is almost the same about 850°C. This is why the CFPC can be used both as a precalciner for cement production and as a low-pollution combustor for power generation. To obtain a good result usually a Ca/S ratio of greater than two is used in the FBC. The ratio of Ca/S in the CFPC is much higher than that in the FBC and the SO_2 emissions are lower. The SO_2 emissions are decreased by increasing the circulating ratio.

5. Waste Management and Optimization

In its environmental protection aspect the advantages of the CFPC process are not only low emissions but also the possibility of using waste fuels and recycling waste. This eliminates the costly problems of ash disposal that normally represents a major project expenditure.

1. In the CFPC process waste fuel can be used both as an additive fuel and as a part of raw material for the cement production. Kühle reported [7] that oil shale with a calorific

nämlich Calciumsulfat. Deshalb sind keine teuren Naßreinigungsverfahren für das Rauchgas erforderlich. Die Reaktion zwischen Schwefeldioxid und Kalkstein wurde in zwei aufeinanderfolgenden Schritten beschrieben: der endothermen Calcinierung und der exothermen Sulfatisierung.

Die Temperaturen zur Verbrennung von Kohle in einer Wirbelschicht, zur Karbonatzersetzung und zur Vorcalcinierung von Zementrohmehl liegen bei etwa 850°C fast gleich. Deshalb kann der CFPC-Prozeß sowohl als Vorcalcinator für die Zementherstellung als auch als emissionsarme Brennkammer zur Stromerzeugung genutzt werden. Um eine gute Entschwefelungsrate zu erzielen, wird üblicherweise in der stationären Wirbelschicht ein Ca/S-Verhältnis von über 2 eingestellt. Das Ca/S-Verhältnis in der zirkulierenden Wirbelschicht ist wesentlich höher und die SO_2-Emission dadurch niedriger. Die SO_2-Emission sinkt mit Erhöhung der Zirkulationsrate.

5. Abfallwirtschaft und Optimierung

Im Hinblick auf den Umweltschutz liegen die Vorteile des CFPC-Verfahrens nicht nur in den geringeren Emissionen, sondern auch in der Möglichkeit der Verwendung von Brennstoffen aus Abfällen und dem Abfallrecycling. Dadurch wird die kostenintensive Ascheentsorgung vermieden.

1. Im CFPC-Verfahren läßt sich Brennstoff aus Abfällen sowohl als Zusatzbrennstoff als auch als Rohmehlkomponente nutzen. Kühle berichtet, daß Ölschiefer mit einem Heizwert zwischen 7 500 und 8 400 kJ/kg in einem Wirbelschichtcalcinator bei etwa 850°C verwendet wurde [7]. Der Wirkungsgrad der Verbrennung und der Vorentsäuerungsgrad lagen bei 99 bzw. 95 %. Anderen Berichten zufolge wurden Abfallbrennstoffe mit hohem Aschegehalt jedenfalls problemlos verfeuert. Dies ist im Hinblick auf das CFPC-Verfahren besonders wichtig, da 15 bis 20 % der Grubenkohle in China Abraum sind. Die Chinesen haben langjährige Erfahrungen mit der Behandlung von Minenabraum in Wirbelschichtfeuerungen und als Brennmaterialien [8].

2. Ein besonderes Merkmal des CFPC-Verfahrens ist, daß Asche und Gips aus der Entschwefelung zusammen mit entsäuertem Rohmehl zur Zementherstellung eingesetzt werden können. Die Asche kann als Al_2O_3 und SiO_2-Komponente des Rohmehls verwendet werden. Der REA-Gips läßt sich anstelle von Naturgips zur Regulierung der Erstarrungszeit des Zementes verwenden, da der Calciumsulfatanteil beider Gipse ähnlich ist. Die Verwendung von Asche als Zuschlagstoff zum Rohmehl ist aus zwei Gründen besser als deren Verwendung als Zuschlagstoff zum Zement [9]: Einerseits entstehen bei der Verbrennung bei niedrigen Temperaturen aus der Asche keine Komponenten mit hydraulischen Eigenschaften. Andererseits kann die heiße Asche direkt ohne Kühlung in den Ofen gegeben werden. Einer der wichtigsten Punkte bei der Zementklinkerproduktion unter Verwendung von Abfall ist, daß die Zusammensetzung so konstant wie möglich sein sollte.

3. Im CFPC-Verfahren bilden Energiefluß und Materialfluß ein Dreieck.

Das Abgas aus der Klinkerproduktion wird zur Stromerzeugung eingesetzt, und die im Kraftwerk erzeugte elektrische Energie wird zur Zementherstellung und zum Betrieb der Wirbelschicht verwendet. Die Asche und der Gips aus der Rauchgasentschwefelung in der Wirbelschicht und dem Kraftwerk werden zusammen mit entsäuertem Rohmehl für die Klinkerherstellung verwendet. Das entsäuerte Rohmehl enthält feines frisches CaO, das als Absorptionsmittel bei den Entschwefelungsreaktionen wirkt. Aufgrund der Zugabe von Asche und Gips aus der Rauchgasentschwefelung sollte der Anteil von Rohstoffen im Rohmehl und die Gipsmenge im Zement entsprechend den unterschiedlichen Bedingungen angepaßt werden. Durch gemeinsame Nutzung der Kohlelager sowie der Transport- und Aufbereitungseinrichtungen für die Zementproduktion und die Stromerzeugung lassen sich Kosten einsparen [9, 10].

value of 7500 to 8400 kJ/kg was used in a circulating fluidized bed calciner at about 850°C and the combustion efficiency and the decarbonation rate were 99 % and 95 % respectively. High ash waste fuel was also reported to be used in that plant without problems. This is particularly important for the CFPC since about 15 – 20 % of China's mine coal are waste. The Chinese have an extensive experience in dealing with mine waste and refuses for fluid bed boiler and building materials [8].

2. One of the CFPC process' unique features is that the ash material and desulfurization gypsum can be used with decarbonated raw meal to produce cement. The ash material can be used as the component of Al_2O_3 and SiO_2 of the raw meal. The desulfurization gypsum can be used instead of natural gypsum to regulate the cement setting time, since the calcium sulfate component of both gypsums have a similar effect. Using ash as a raw meal additive is better than using it as a cement additive for two reasons [9]: 1. If this kind of ash material is fired at low temperatures no compound with puzzolanic or hydraulic properties is produced. 2. The hot ash material can directly enter the kiln to produce cement without cooling. One of the most important points in producing cement by using waste is that the composition should be as constant as possible.

3. In the CFPC process the energy flow and materials flow form a triangle.

The exhaust gas from the cement production is used for power production and the power generated from the power station is used for cement production and CFPC operation. The ash material and the flue gas desulfurization (FGD) gypsum from the CFPC and the power station are used with decarbonated raw meal for the cement production. The decarbonated raw meal contains fine fresh CaO which acts as an absorber for the desulfurization reactions. Because of the addition of ash material and FGD gypsum the proportion of raw materials to raw meal and the amount of gypsum to cement should be changed according to different conditions. Costs can be saved by sharing a coal storage as well as the transportation and preparation system by the cement production and power production [9, 10].

6. Conclusions

The CFPC process represents the hybridization of cement production and power generation. The advantages of using circulating fluidized bed technology are energy saving, low emissions, utilization of low-grade fuels and no disposal problems for ash and FGD gypsum. As a profitable venture this precalcining/power generation system ensures an inexpensive, reliable and self-sustaining electrical power supply. It is welcome in China. The CFPC process is an example of applying the circulating fluidized bed technology in the cement industry to a comprehensive resource utilization in a cost effective yet environmental sound way, especially the consumption of the waste within the process itself.

Literature

[1] Wang, Y., et al.: Electricity Conservation and Electricity Production are Important Programs in Technical Advancement of Cement Industry. Cement Tech. No. 5, 1989 (in Chinese).

[2] Deng, X.J.: An Investigation on The Process and Its Heat Economics of Fluidized Preheat-calciner Plant Incorporating Cogeneration. Master's Thesis, June 1990 (in Chinese).

[3] Lang, T.A.: The Production of Electrical and Thermal Energy from the Exhaust Gas Heat of Preheat Kilns. 26th IEEE Conference, 1984.

[4] Reh, L.: The Circulating Fluid Bed Reactor – A Key to Efficient Gas/Solid Processing. Circulating Fluidized Bed Technology, 1986.

6. Schlußfolgerungen

Das CFPC-Verfahren stellt die Zusammenführung von Zementproduktion und Stromerzeugung dar. Die Vorteile aus der Anwendung der zirkulierenden Wirbelschicht sind Energieeinsparung, niedrige Emissionen, die Möglichkeit des Einsatzes heizwertarmer Brennstoffe sowie keine Entsorgungsprobleme für Asche und Gips aus der Rauchgasentschwefelung. Als gewinnbringendes Unternehmen stellt diese kombinierte Anlage zum Vorcalcinieren und zur Stromerzeugung eine kostengünstige, verläßliche und autarke Energieversorgung sicher. Sie wird in China gut angenommen. Das CFPC-Verfahren ist ein Beispiel für die Anwendung der zirkulierenden Wirbelschichttechnologie in der Zementindustrie für eine umfassende Nutzung der Ressourcen in einer kostengünstigen, jedoch umweltschonenden Art und Weise, insbesondere aufgrund der Verwendung der anfallenden Reststoffe im Prozeß selbst.

[5] Warshawsky, J., et al.: Verbesserung der Vorcalcinierung durch Rückführung des Brenngutes. Zement-Kalk-Gips, No. 3, 1986.

[6] Hansen, P.F.B., et al.: Sulfur Capture on Limestone under Periodically Changing Oxidizing and Reducing Conditions. 3rd Int'l Conf. on Circulating Fluidized Beds, 1990.

[7] Kühle, K.,: Calcinieren von Zementrohmehl in der „Zirkulierenden Wirbelschicht" im Werk Port la Nouvelle/Frankreich. Zement-Kalk-Gips, No. 5, 1984.

[8] Xu, M.: Utilization and Application of Washery Refuse and Mine Waste from Hard Coal Mines of the P.R. China. Aufbereitungs-Technik, No. 9, 1986.

[9] Campbell, H.W.: Cogeneration in the Cement Industry. Calmat's Colton California Project. 28th IEEE Conference, 1986.

[10] Hochdahl, O.: Brennstoff und Wärmewirtschaft. Zement-Kalk-Gips, No. 12, 1986.

Die 5000 t/d-Ofenanlage mit PYRORAPID®-Kurzdrehofen und PYROCLON®-LowNO$_x$-Calcinator bei ACC, Taiwan

5000 t/d kiln plant with PYRORAPID® short rotary kiln and PYROCLON® LowNO$_x$ calciner at ACC, Taiwan

La ligne de four de 5000 t/j avec four rotatif court PYRORAPID® et calcinateur PYROCLON®-LowNO$_x$ chez ACC, Taïwan

La instalación de horno para 5000 t/d con horno corto PYRORAPID® y calcinador PYROCLON®-LowNO$_x$ en ACC, Taiwán

Von **M. Deussner**, Köln/Deutschland

Zusammenfassung – Im Frühjahr 1992 wurde die neue Ofenanlage für 5000 t/d bei der Asia Cement Corporation/Taiwan in Betrieb genommen. Besonders hervorzuheben sind der PYRORAPID®-Kurzdrehofen und der PYROCLON®-LowNO$_x$-Calcinator, die erstmals in diesem Leistungsbereich eingesetzt wurden. Der PYRORAPID®-Drehofen mit 5,0 m Durchmesser und 55 m Länge (L/D = 11:1) ist auf 2 Rollenstationen gelagert und daher statisch eindeutig definiert. Neben den Vorteilen bei Investitionen und mechanischer Verfügbarkeit bietet das Kurzofen-Konzept insbesondere verfahrenstechnische Vorteile. Zur Senkung der Stickoxid-Emission wurde der Calcinator in PYROCLON®-LowNO$_x$-Schaltung ausgeführt. Innerhalb des reduzierenden Bereiches werden die in der Ofenflamme gebildeten Stickoxide reduziert. In der oberen Zone des Calcinators wird das Ofenabgas mit Tertiärluft vermischt und eine oxidierende Atmosphäre eingestellt. Die Tertiärluft wird am Ofenkopf abgezogen und über 2 Tertiärluftleitungen dem 2strängigen PYROCLON®-LowNO$_x$-Calcinator zugeführt. Die streng getrennte Ausführung der beiden Vorwärmer/Calcinatorstränge ab Ofeneinlaufkammer führt zu einem sehr einfachen und flexiblen Betriebs- und Regelverhalten. Obwohl ein Rohmaterial mit relativ geringen Feuchtigkeitsgehalten zur Verfügung steht, wurde der Vorwärmer nur 4stufig ausgeführt, um den Wirkungsgrad einer nachgeschalteten Stromerzeugung mit einer Dampfturbine zu erhöhen. Bereits 6 Wochen nach dem Start lief die Ofenanlage in kontinuierlichem Betrieb mit einem Durchsatz, der 10% oberhalb der Garantie lag. Im Abnahmetest nach weiteren 6 Wochen wurden alle Garantiewerte erreicht.

Die 5000 t/d-Ofenanlage mit PYRORAPID®-Kurzdrehofen und PYROCLON®-LowNO$_x$-Calcinator bei ACC, Taiwan

Summary – The new kiln plant for 5000 t/d at the Asia Cement Corporation, Taiwan, came into operation early in 1992. The plant is centred round the PYRORAPID® short rotary kiln and the PYROCLON® LowNO$_x$ calciner which were used for the first time in this output range. The PYRORAPID® rotary kiln with a diameter of 5.0 m and a length of 55 m (L/D = 11:1) is supported on two roller stations and is therefore uniquely statically defined. In addition to the advantages of capital costs and mechanical availability the short kiln concept also offers particular process engineering advantages. The calciner was built in the PYROCLON® LowNO$_x$ configuration to lower the emission of nitrogen oxides. The nitrogen oxides formed in the kiln flame are reduced in the reducing region. The kiln exhaust gas is mixed with tertiary air in the upper zone of the calciner producing an oxidizing atmosphere. The tertiary air is drawn from the kiln hood and fed to the 2-string PYROCLON® LowNO$_x$ calciner through 2 tertiary air ducts. The strictly separated design of the two preheater/calciner strings from the kiln inlet chamber leads to a very simple and flexible operating and control behaviour. Although a raw material with a relatively low moisture content is available the preheater was designed with only four stages in order to raise the efficiency of a down-stream power generating system using a steam turbine. Only six weeks after start-up the kiln plant was running in continuous operation at an output which was 10% above the guarantee value. All the guarantee values were achieved in the acceptance test after a further six weeks.

5000 t/d kiln plant with PYRORAPID® short rotary kiln and PYROCLON® LowNO$_x$ calciner at ACC, Taiwan

Résumé – Au printemps 1992 a été mise en service la nouvelle ligne de four pour 5000 t/j chez Asia Cement Corporation/Taïwan. Les noyaux de l'installation sont le four rotatif court PYRORAPID® et le calcinateur PYROCLON®-LowNO$_x$, utilisés pour la première fois dans cette plage de capacité. Le four rotatif PYRORAPID®, avec un diamètre de 5,0 m et une longueur de 55 m (L/D = 11:1) est supporté par deux stations de galets et est, ainsi, défini sans équivoque sur la plan statique. À côté des avantages du point de vue investissement et disponibilité mécanique, le concept du four court offre surtout des avantages de technologie de process. Pour abaisser les émissions d'oxydes d'azote, le calcinateur a été réalisé en couplage PYROCLON®-LowNO$_x$. Au sein de la section réductrice, sont réduits les oxydes d'azote formés dans la flamme du four. Dans la zone supérieure du calcinateur, le gaz de sortie du four est mélangé avec l'air tertiaire et une atmosphère oxydante est réalisée. L'air tertiaire est prélevé à la tête du four et est amené au calcinateur à double voie PYROCLON®-LowNO$_x$ par deux conduits

La ligne de four de 5000 t/j avec four rotatif court PYRORAPID® et calcinateur PYROCLON®-LowNO$_x$ chez ACC, Taïwan

d'air tertiaire. La construction strictement séparée des deux voies préchauffeur/ calcinateur, à partir de la boîte à fumée du four, conduit à un comportement d'exploitation et de régulation très simple et flexible. Bien que la matière première disponible possède une humidité relativement faible, le préchauffeur n'a été conçu qu'à 4 étages, afin d'améliorer le rendement d'une génération d'électricité avec turbine à vapeur, montée en aval. Déjà 6 semaines après la mise en marche, la ligne de four tournait en fonctionnement continu avec un débit situé 10% au-dessus de la garantie. Aux essais de réception, 6 semaines plus tard, toutes les valeurs garanties ont été atteintes.

Resumen – *En la primavera de 1992 se puso en servicio, en la compañía Asia Cement Corporation/Taiwán, la nueva instalación de horno para 5000 t/d. Cabe destacar, especialmente, el horno corto PYRORAPID® y el calcinador PYROCLON®-LowNO$_x$, empleados por primera vez para esta capacidad de producción. El horno corto PYRORAPID®, de 5,0 m de diámetro y 55 m de longitud (L/D = 11:1), se apoya en dos estaciones de rodillos, con lo que queda claramente definido desde el punto de vista estático. Aparte de las ventajas respecto de las inversiones y de la disponibilidad mecánica, el concepto de horno corto ofrece, sobre todo, otras ventajas tecnológicas. Para reducir las emisiones de óxidos de nitrógeno, se ejecutó el calcinador en conexión PYROCLON®-LowNO$_x$. Dentro de la zona reductora se reducen los óxidos de nitrógeno formados en la llama del horno. En la zona superior del calcinador se mezclan los gases de escape del horno con aire terciario, creándose una atmósfera oxidante. El aire terciario se elimina a través de la cabeza del horno, siendo conducido al calcinador PYROCLON®-LowNO$_x$ de dos ramales, mediante dos tuberías para aire terciario. La ejecución estrictamente separada de los dos ramales de precalentador/calcinador, a partir de la cámara de entrada del horno, conduce a un comportamiento muy sencillo y flexible en cuanto al servicio y a la regulación. Aunque se dispone de crudo con un contenido de humedad relativamente reducido, el precalentador se ejecutó sólo con 4 etapas, con el fin de aumentar el rendimiento de un generador de electricidad postconectado, con turbina de vapor. 6 semanas después de la puesta en servicio, la instalación de horno ya funcionaba en servicio continuo, alcanzando un rendimiento de paso un 10 % superior al límite de garantía. Durante el test de recepción, efectuado después de otras 6 semanas, se alcanzaron todos los valores garantizados.*

La instalación de horno para 5000 t/d con horno corto PYRORAPID® y calcinador PYROCLON®-LowNO$_x$ en ACC, Taiwán

Die ASIA CEMENT CORPORATION (ACC) betreibt heute in ihren Werken Hsin Chu und Hualien fünf Ofenanlagen, die alle von KHD Humboldt Wedag geliefert wurden. Mit einer Produktionskapazität von 6,4 Mio. Jahrestonnen ist ACC der größte Zementproduzent in Taiwan. Die Ofenanlage III im Werk Hualien wurde Ende Mai 1992 in Betrieb genommen und ist so ausgelegt, daß die Garantieleistung von 4800 t/d um bis zu 10 % überschritten werden kann. Kernstücke der Anlage sind der HUMBOLDT-Wärmetauscher mit PYROCLON-LowNO$_x$-Calcinator und der PYRORAPID-Kurzdrehofen.

1. HUMBOLDT-Wärmetauscher

Obwohl das Rohmaterial relativ trocken ist, wurden die beiden Wärmetauscherstränge nur 4stufig ausgeführt, damit die Abgastemperaturen den Wirkungsgrad einer nachgeschalteten Stromerzeugung erhöhen (**Tabelle 1, Bild 1**).

In Hualien III wurde unsere Zyklongeneration mit optimierter Zyklongeometrie eingesetzt, die gute Abscheideleistungen bei niedrigem Druckabfall ermöglicht. Als Kompromiß zwischen niedrigem elektrischen Energiebedarf einerseits und geringem Bauvolumen andererseits erfolgte die Zyklondimensionierung für einen Druckverlust von 47 mbar. Die Durchmesser der Zyklone II bis IV betragen 7,0 m, die der Stufe I 4,4 m.

In die Abgasführung der Wärmetauscher und des Kühlers wurden Dampfkessel integriert, die eine Wasserdampfturbine mit einer installierten elektrischen Leistung von 11,8 MW speisen. Selbst bei in Taiwan häufigen Ausfällen des öffentlichen Netzes, die sonst zu Stillständen des gesam-

The ASIA CEMENT CORPORATION (ACC) is today operating five kiln installations at its Hsin Chu and Hualien works, all supplied by KHD Humboldt Wedag. ACC is the biggest cement producer in Taiwan with a production capacity of 6.4 million tonnes p.a. Kiln plant III at the Hualien works came into operation at the end of May 1992 and is designed to allow the guaranteed output of 4800 t/d to be exceeded by up to 10 %. The plant is centred round the HUMBOLDT preheater, the PYROCLON LowNO$_x$ calciner and the PYRORAPID short rotary kiln.

1. The HUMBOLDT preheater

Although the raw material is relatively dry the two preheater strings have been designed with only four stages so that the exit gas temperatures can raise the efficiency of a downstream power generating system (**Table 1, Fig. 1**).

Our cyclone generation at Hualien III benefits from optimised cyclone geometry, allowing good separation with only a small drop in pressure. The cyclones have been dimensioned for a pressure loss of 47 mbar as a compromise between a low power requirement and a small overall volume. The diameters of cyclones II to IV are 7.0 m and those at stage I 4.4 m.

Steam generators have been incorporated in the exit gas ducts from the preheater and cooler, feeding a steam turbine with an installed capacity of 11.8 MW. As well as reducing electricity costs they enable the essential parts of the plant to be kept in operation even during power cuts, which are frequent in Taiwan and would otherwise stop the whole cement works.

TABELLE 1: Spezifikation Wärmetauscher und Calcinator

HUMBOLDT-Wärmetauscher mit PYROCLON-LowNO$_x$-Calcinator	
Stufen	4
Stränge	2
Zyklondurchmesser	7,0/4,4 m
Druckverlust	47 mbar
Stromerzeugung aus Abgaswärme	
installierte Leistung	11,8 MW$_{el.}$

TABLE 1: Preheater and calciner specifications

HUMBOLDT preheater with PYROCLON LowNO$_x$ calciner	
stages	4
strings	2
cyclone diameter	7.0/4.4 m
pressure drop	47 mbar
power generation from exhaust gas heat	
installed capacity	11.8 MW$_{el.}$

BILD 1: PYROCLON®-Wärmetauscher
FIGURE 1: PYROCLON® preheater

2. PYROCLON-LowNO$_x$-Calcinator

Die beiden Calcinatoren wurden zur Senkung der Stickoxidemission als PYROCLON-Low-NO$_x$-Calcinatoren ausgeführt, die gleichzeitig zwei wesentliche Aufgaben erfüllen: zum einen die hochgradige Vorentsäuerung des Rohmehls, wie ein „normaler" PYROCLON-Calcinator. Zum anderen wird das NO$_x$ abgebaut, das in der Ofenfeuerung aufgrund der hohen Flammentemperaturen entsteht (**Bild 2**).

Der NO$_x$-Abbau wird in dem konstruktiv einfachen LowNO$_x$-Calcinator durch die Schaffung einer reduzierenden Gasatmosphäre mit verstärkter Kohlenmonoxid-Bildung erreicht. Es werden hier vier funktionelle Bereiche unterschieden:

Im Reinluftbereich wird der größte Teil des Calcinatorbrennstoffes in der Tertiärluft verbrannt und damit der größte Teil der Energie zur Ensäuerung des Rohmehls bereitgestellt.

Im LowNO$_x$-Bereich wird der kleinere Teil des Calcinatorbrennstoffes im Ofenabgas verbrannt und eine reduzierende Atmosphäre unter Bildung von Kohlenmonoxid erzeugt. Vereinfacht dargestellt reagiert dort das CO mit NO zu CO$_2$ und N$_2$. Die Anwesenheit von Rohmehl und Kohlenstaub hat dabei eine katalytische Wirkung. Durch die Aufteilung des Heißmehls aus Stufe III kann die Temperatur im LowNO$_x$-Bereich gezielt beeinflußt werden.

In dem Strähnenbereich tritt durch die nahezu parallele Zusammenführung beider Gasströme mit annähernd gleicher Geschwindigkeit bis zur 180°-Umlenkung eine nur langsame Vermischung ein, so daß eine ausreichend lange Reaktionszeit gewährleistet ist.

Hieran schließt sich bis zum Eintritt in den untersten Zyklon der Durchmischungsbereich an, in dem die Strähnen aufgelöst werden und der Restausbrand des CO zu CO$_2$ stattfindet. Bis zu 50 % des im Ofen gebildeten NO$_x$ kann so reduziert werden.

2. The PYROCLON LowNO$_x$ calciner

As a means of lowering nitrogen oxide emissions the two calciners are of the PYROCLON LowNO$_x$ type, which have two essential functions: firstly, a high-degree precalcination of the raw meal, like a "normal" PYROCLON calciner, and secondly, decomposition of the NO$_x$ which is formed in the kiln firing system by the high flame temperatures (**Fig. 2**). NO$_x$ decomposition is achieved in the structurally simple LowNO$_x$ calciner by providing a reducing gas atmosphere with enhanced carbon monoxide formation. There are four functional regions:

In the pure air region a major proportion of the calciner fuel is burnt in the tertiary air, thereby producing most of the energy for precalcining the raw meal.

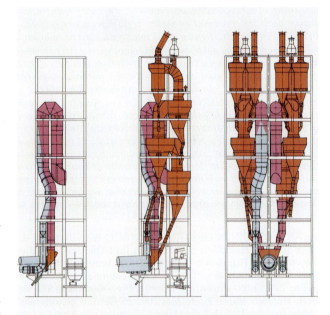

BILD 2: Zyklonvorwärmer mit PYROCLON-LowNO$_x$®-Calcinator
FIGURE 2: Cyclone preheater with PYROCLON LowNO$_x$® calciner

In the LowNO$_x$ region a minor proportion of the calciner fuel is burnt in the exhaust gas from the kiln and a reducing atmosphere is produced with the formation of carbon monoxide. In simplified terms the CO reacts with NO to form CO$_2$ and N$_2$. The presence of raw meal and coal dust has a catalytic effect. Division of the hot meal from stage III enables the temperature to be specifically controlled in the LowNO$_x$ region.

In the strand region the almost parallel convergence of the two gas streams at approximately the same speed as far as the 180° bend causes only gradual mixing, which ensures an adequate reaction time.

From here to the inlet of the lowest cyclone is the thorough mixing region in which the strands are broken up and the residual CO burnt out to form CO$_2$. Up to 50 % of the NO$_x$ formed in the kiln can be reduced in this way.

3. The PYRORAPID short rotary kiln

Another centrepiece of the new Hualien III kiln plant is the PYRORAPID short rotary kiln (**Table 2, Fig. 3**). Its mechanical and processing advantages have already been discussed at length [1–3].

The main point about the process is that the length of the sintering zone is independent of the length of the kiln; the sintering zone forms below the flame and ensures optimum clinker formation. Genuine short kilns improve the kinetics

TABELLE 2: Spezifikation Ofen und Kühler	
PYRORAPID-Kurzdrehofen und Klinkerkühler	
Abmessungen	5,0 × 55 m
Ofenantrieb	3,0 Upm
Produktion	4800–5300 t/d
Wärmebedarf	3050 kJ/kg Kli.
Kühler-Rostfläche	150 m^2
Rekuperationsgrad	71 %

TABLE 2: Kiln and cooler specifications	
PYRORAPID short rotary kiln and clinker cooler	
dimensions	5.0 × 55 m
kiln drive	3.0 rpm
production	4800–5300 t/d
heat consumption	3050 kJ/kg Kli.
grate area of cooler	150 m^2
recuperation rate	71 %

3. PYRORAPID-Kurzdrehofen

Das nächste Kernstück der neuen Ofenanlage Hualien III ist der PYRORAPID-Kurzdrehofen (**Tabelle 2, Bild 3**). Seine mechanischen und verfahrenstechnischen Vorteile wurden bereits mehrfach diskutiert [1–3].

Die verfahrenstechnische Kernaussage lautet: Die Länge der Sinterzone bildet sich unabhängig von der Ofenlänge unter der Flamme aus und gewährleistet eine optimale Klinkerbildung. Echte Kurzöfen verbessern die Reaktionskinetik und sind einfacher zu bedienen als normal lange Öfen. Kurzöfen definieren sich über das L/D-Verhältnis, nicht über die Anzahl der Laufringe.

In Hualien wurde ein echter Kurzofen mit 5,0 m Durchmesser und 55 m Länge installiert, also L/D = 11. Der Ofenantrieb ist für 3 Umdrehungen/min ausgelegt.

Bei dem Klinkerkühler hat sich die ACC für einen 3-Rost-Kühler mit Zwischenbrecher der Firma Claudius Peters entschieden, mit einer gesamten Rostfläche von 150 m^2 und einem garantierten Rekuperationsgrad von 71 %.

Für diese Ofenlinie wurde eine Wärmeverbrauchsgarantie von 3050 kJ/kg Klinker abgegeben, entsprechend 728 kcal/kg Klinker. Für die Kohlenstaubfeuerung im Ofen ist der vielfach bewährte PYRO-JET-Brenner eingesetzt, der auch einen Beitrag zur Minderung der NO$_x$-Emission liefert.

Die Tertiärluft wird vorteilhafterweise am Ofenkopf abgezogen, in zwei Absetzkammern entstaubt und über zwei parallele Tertiärluftleitungen den beiden Wärmetauscher-Strängen zugeführt. Die streng getrennte Ausführung der beiden Vorwärmer-Calcinatorstränge ab Ofeneinlaufkammer bzw. Ofenkopf führt zu einem sehr einfachen und flexiblen Betriebs- und Regelverhalten.

4. Rohstoffe

Das Rohmehl wird aus Kalkstein mit stark wechselnden Kieselsäuregehalten hergestellt, der mit zwei Tonsorten und einem Eisenträger verschnitten wird (**Tabelle 3**). Es wurde zu Beginn der Betriebszeit relativ fein gemahlen, um den Anteil grober Bestandteile größer 200 µm zu minimieren. Aus dem Rohmehl wird ein Klinker mit Kalkstandard 93 gebrannt, aus dem überwiegend ein Zement der Qualität PZ45F hergestellt wird.

5. Betriebsergebnisse

Der Leistungsnachweis, an dem nur 2 KHD-Ingenieure zur Überwachung beteiligt waren, fand während einer ganz normalen Betriebsphase statt und wurde von dem Personal der ACC geleitet (**Tabelle 4**). Bei einer Leistung von 4853 t/d lag der elektrische Energiebedarf für Ofenantrieb und Abgasventilatoren innerhalb der Meßtoleranz. Der Ofenantrieb verbrauchte bei 2,4 – 2,6 Upm nur 1,3 kWh/t Klinker. Der erreichte Wärmeverbrauch lag deutlich unter dem Garantiewert, obwohl der Anlagenbetrieb zu diesem Zeitpunkt noch nicht optimiert war.

Der Betrieb des neuesten und größten Kurzofens hat bestätigt, daß aufgrund der im Kurzofen besseren Reaktionskinetik die mikroskopische Struktur des Klinkers ausgezeichnet ist. Vergleichende Untersuchungen zeigen im Kurzofenklinker feinste Alitkristalle, die innig mit Belit verwachsen sind. Im Gegensatz dazu steht ein chemisch gleichartiger

of the reaction and are simpler to operate than normal long kilns. Short kilns are defined by the L/D ratio, not by the number of tyres.

A genuine short kiln with a diameter of 5.0 m and a length of 55 m, i.e. L/D = 11, is installed at Hualien. The kiln drive is designed for 3 revolutions per minute.

With regard to the clinker cooler the ACC decided on a 3-grate cooler with an intermediate clinker breaker from Claudius Peters, with a total grate area of 150 m^2 and a guaranteed recovery rate of 71 %.

The heat consumption guarantee given for this kiln line was 3.050 kJ/kg clinker, corresponding to 728 kcal/kg clinker. The well-tried PYRO-JET burner is used for pulverized coal firing in the kiln and also helps to reduce NO$_x$ emissions. The tertiary air is advantageously drawn from the kiln hood, cleaned in two settling chambers and fed to the two-string preheater through two parallel tertiary air ducts. The strictly separated design of the two preheater/calciner strings from the kiln inlet chamber leads to a very simple and flexible operating and control behaviour.

4. The raw materials

The raw meal is made from limestone with a widely varying silica content, blended with two types of clay and an iron agent (**Table 3**). It is ground relatively finely at the beginning of the operating period to minimise the proportion of coarse constituents of over 200 µm. A clinker of lime standard 93 is burnt from the raw meal, most of it being used to produce a PZ45F grade cement.

5. Operating results

The assessment of performance, with only two KHD engineers involved in monitoring, took place during an

BILD 3: PYRORAPID®-Kurzdrehofen mit Tertiärluftleitungen
FIGURE 3: PYRORAPID® short rotary kiln with tertiary air ducts

TABLE 3: Raw meal and clinker

	Raw meal
limestone	88.0 wt.-%
clay	10.9 wt.-%
iron ore	1.1 wt.-%
fineness	8.5 % R 90 µm

	Clinker
LSt II	93.0
SM	2.96
AR	1.49

TABLE 4: Guarantee and performance assessment

start-up: 23.5.92 *performance assessment: 25.8.92		guaranteed	achieved*
clinker output	t/d	4800	4853
fuel energy demand clinker	kJ/kg clinker	3050	2915
power requirement	kWh/t clinker	8.3	8.42
free lime content	wt.%	<1.0	0.85

entirely normal operating phase and was managed by the ACC staff (**Table 4**). With an output of 4853 t/d the power requirement for the kiln drive and exit gas fans came within the measured tolerance. The kiln drive only used 1.3 kWh/t of clinker at 2.4 to 2.6 rpm. The heat consumption achieved was well within the guarantee value although plant operation had not been optimised at that stage.

Operation of the newest and largest short kiln has confirmed that the microscopic structure of the clinker is excellent thanks to the improved kinetics of the reaction in that kiln. Comparative tests show that the short kiln clinker contains extremely fine alite crystals coalescing intimately with belite. This is contrasted with a chemically similar clinker from the long pre-calcining kiln, which has a relatively coarse structure. The quality and grindability of the short kiln clinker are better in practice.

We shall end this report with a short description of the commissioning schedule. A mere four weeks after the kiln flame had been lit a preliminary kiln test was carried out at a production rate of 4850 t/d. Only two weeks later production was far above 5200 t/d.

Exactly three months after commissioning started the official acceptance test was carried out with the results given above. Today the kiln plant is running stably with much coarser raw meal (14% residue on 90 µm), with output ranging from 5200 to 5300 t/d.

Umbau vom Naß- zum Halbnaßverfahren mit Einsatz von Rückständen der Kohlenaufbereitung

Changeover from the wet to the semi-wet process using residues from coal processing

Reconversion de voie humide en voie semihumide avec emploi de résidus du traitement du charbon

Transformación de la vía húmeda en vía semihúmeda, utilizando residuos del tratamiento de carbón

Von **J. Duda** und **B. Werynski,** Opole/Polen

Zusammenfassung – Die in Polen noch überwiegende Anwendung des Naßverfahrens mit hohem Aufwand an thermischer Energie gab Anlaß zu umfangreichen Modernisierungen. Dabei mußte neben der Energieeinsparung auch nach Einsatzmöglichkeiten für industrielle Reststoffe gesucht werden, um die begrenzten Vorräte an natürlichen Rohstoffen zu schonen. Beim Umbau großer Öfen von 5 × 185 m im Zementwerk WARTA vom Naß- zum Trockenverfahren sollte der Mangel an kalkarmen Rohstoffen durch den Einsatz von Rückständen der Kohlenaufbereitung kompensiert werden. Hierfür erwiesen sich kohlehaltige Reststoffe als geeignet. Sie konnten auf diese Weise entsorgt werden und führten zu einer Einsparung an Brennstoffenergie von ca. 1000 kJ/kg Kl. Da die Kosten eines Umbaus zum Trockenverfahren hauptsächlich aufgrund der erforderlichen Investitionen für die Rohmehlherstellung zu hoch waren, wurde ein spezielles Halbnaßverfahren entwickelt, das die Möglichkeit eines späteren Umbaus zum Trockenverfahren bietet. Hierfür wird ein kalkreicher Rohstoff, aufbereitet als Schlamm mit einem Wassergehalt von ca. 30 – 32%, zunächst in einem Trommeltrockner vorgetrocknet. Das Heißgas wird der Stufe des Vorwärmers entnommen, in der der kohlehaltige Reststoff verbrannt wird. Das Gemisch aus Asche und vorgewärmtem Brenngut wird teilweise entsäuert dem auf 92 m verkürzten Ofen zugeführt. Mit diesem Verfahren soll der Energiebedarf verringert und der Durchsatz um ca. 20% erhöht werden. Gegenwärtig wird die umgebaute Ofenanlage in Betrieb genommen.

Summary – Use of the wet process, which is still predominant in Poland, with its high expenditure of thermal energy was the reason for extensive modernization. As well as the saving in energy it was also necessary to look for possible applications for industrial residues in order to preserve the limited reserves of natural raw materials. When converting the large 5 × 185 m kiln at the WARTA cement works from the wet to the dry process the intention was to compensate for the lack of low-lime raw materials by using residues from coal processing. Carbonaceous residues proved suitable for this purpose. It offered a means of disposing of them and led to a saving in fuel energy of approximately 1000 kJ/kg clinker. The costs of conversion to the dry process were too high, mainly because of the capital expenditure required for producing raw meal, so a special semi-wet process was developed which offers the opportunity for later conversion to the dry process. A lime-rich raw material prepared as slurry with a water content of approximately 30 – 32% is first predried in a drum drier. The hot gas is taken from the stage of the preheater in which the carbonaceous residue is burnt. The mixture of ash and preheated kiln feed is fed in a partially calcined state to the kiln which has been shortened to 92 m. This process is intended to reduce energy consumption and raise the output by about 20%. The converted kiln plant is now being commissioned.

Résumé – L'emploi de la voie humide, encore prédominant en Pologne et lié à une forte dépense d'énergie thermique, a été le motif de modernisations étendues. Dans ce contexte et outre l'économie d'énergie il a fallu chercher aussi des possibilités d'utilisation de résidus industriels, afin de préserver les réserves de matières premières naturelles. Lors de la conversion des grands fours de 5 × 185 m à la cimenterie WARTA, de la voie humide en voie sèche, la pénurie de matières premières pauvres en calcaire devait être compensée par l'utilisation de résidus du traitement du charbon. Ceux-ci ont pu être, de cette façon, éliminés avec, en contrepartie, une économie d'énergie du combustible d'environ 1000 kJ/kg Kl. Les coûts d'une reconstruction pour voie sèche étant trop élevés, surtout à cause des investissements nécessaires pour la préparation de la farine crue, on a développé un procédé semihumide spécial, offrant les possibilités d'une conversion ultérieure en voie sèche. Pour ce faire, une matière première riche en calcaire, préparé en pâte avec une teneur en eau d'environ 30 – 32%, est d'abord préséchée dans un tambour-sécheur. Les gaz chauds sont prélevés à l'étage du préchauffeur où est brûlé le résidu contenant du charbon. Le mélange de cendres et de matière à cuire préchauffée est partiellement décarbonaté et amené au four raccourci à 92 m. Avec ce procédé, il est prévu de diminuer la consommation d'énergie et d'augmenter le débit d'environ 20%. La ligne de four modifiée est actuellement mise en service.

Resumen – *El empleo de la vía húmeda, que predomina aún en Polonia y que exige un elevado consumo de energía térmica, dio lugar a amplias medidas de modernización. Aparte del ahorro de energía, hubo que buscar también posibilidades de utilización de residuos industriales, con el fin de preservar las escasas reservas de materias primas naturales. Con la transformación de los grandes hornos de vía húmeda, de 5×185 m, existentes en la fábrica de cemento WARTA, en hornos de vía seca se quiso compensar la falta de materias primas pobres en cal con el empleo de residuos procedentes del proceso de preparación de carbón. Para ello resultaron aptas las materias residuales que contenían carbón. De esta forma, dichos residuos podían ser eliminados, lográndose un ahorro de combustibles del orden de 1 000 kJ/kg de clínker. Puesto que los gastos de la transformación en vía seca resultaban demasiado altos, sobre todo por las inversiones necesarias para la fabricación del crudo, se desarrolló un procedimiento especial, semihúmedo, que ofrece la posibilidad de una futura transformación en vía seca. Después de preparar un material rico en cal, en forma de lodo y con un contenido de agua del 30 – 32 %, éste se seca previamente por medio de un tambor secador. Los gases calientes empleados para ello proceden de aquella etapa del precalentador en la que se queman los residuos que contienen carbón. La mezcla de cenizas y de material a cocer, precalentado, es parcialmente descarbonatada y conducida al horno, el cual ha quedado reducido a 92 m. Con este procedimiento se pretende reducir la demanda de energía y aumentar el rendimiento de paso en un 20 % aprox. Actualmente, la instalación de horno, debidamente transformada, se está poniendo en servicio.*

Transformación de la vía húmeda en vía semihúmeda, utilizando residuos del tratamiento de carbón

1. Allgemeines

In der polnischen Zementindustrie wird der Zement zu über 50 % nach dem Naßverfahren hergestellt. Der damit verbundene hohe Energieverbrauch zwang die Werke zu Modernisierungsmaßnahmen zur Verringerung des thermischen Energieverbrauchs. Bei den geplanten Maßnahmen, die zudem mit dem Ziel der Produktionssteigerung und des Einsatzes minderwertiger Brennstoffe durchgeführt wurden, sollten so weit wie möglich die vorhandenen Maschinen und Anlagenteile genutzt werden. Den Mangel an natürlichen Rohstoffen mit niedrigem Kalkgehalt versuchte man durch den Einsatz von Rückständen aus der Kohlenaufbereitung (Kohlenschiefer) zu kompensieren, der in mineralischer und chemischer Hinsicht den Rohstoffanforderungen für die Klinkerherstellung entspricht. Es war vorgesehen, zunächst große ölbefeuerte Naßöfen ($\emptyset$ 5 × 185 m) umzurüsten. Als im Jahr 1980 das Zementwerk WARTA II wegen seines hohen Energieverbrauchs und des allgemeinen Heizölmangels stillgelegt werden mußte, wurde diese Gelegenheit zur Modernisierung der Anlage wahrgenommen. Dabei standen verschiedene technische Varianten zur Diskussion: der Umbau zum Trockenverfahren, der sich technisch und hinsichtlich der späteren Betriebskosten als am günstigsten erwies, konnte jedoch wegen der damit verbundenen hohen Investitionskosten nicht realisiert werden. Die hohen Kosten wären vor allem durch den dann notwendigen vollständigen Umbau der Anlagen zur Rohstoffaufbereitung entstanden. Daher entschied man sich zur Umrüstung zum Halbnaßverfahren, welches dank der weiteren Nutzung der bestehenden Rohschlammaufbereitung mit einem um die Hälfte geringeren Investitionsaufwand durchgeführt werden konnte.

2. Umbau zum Halbnaßverfahren

Die Kombination von Naß- und Trockenverfahren zum sogenannten Halbnaßverfahren ist eine häufig angewandte Technologie zur Modernisierung von Naßöfen. Dabei wird das Rohmaterial als Schlamm aufbereitet, getrocknet, gefiltert und dann wie beim Trockenverfahren dem Vorwärmer zugeführt. Durch die Verwertung von Kohlenschiefer als Rohstoffkomponente und als Sekundärbrennstoff wurde jedoch eine neue Technologie erforderlich, die im Institut für Mineralische Baumaterialien in Oppeln (Opole) entwickelt wurde.

Dabei wird die hochkalkhaltige Rohstoffkomponente durch Naßvermahlung zu Schlamm mit einem Wassergehalt von 30 – 32 % aufbereitet. Der Kohlenschiefer wird dagegen trocken zu Mehl vermahlen. **Bild 1** zeigt das technologische Schema der modernisierten Anlage. Zwischen der III. und IV. Stütze des Ofens wurden etwa 20 m des Ofenrohres entfernt. Der untere Teil mit den Maßen $\emptyset$ 5 × 92 m wird jetzt als

1. General

In the Polish cement industry over 50% of cement is produced by the wet process. The high energy consumption involved has forced the works to modernise in order to reduce thermal energy expenditure. The measures planned were also aimed at increasing production and using low-grade fuels, and the intention was to use existing machinery and plant components as far as possible. There was an attempt to compensate for the lack of natural low-lime raw materials by using residues from coal processing (bituminous shale) which corresponded mineralogically and chemically to the raw material requirements for clinker production. The plan was to convert large oil-fired wet kilns first (diameter 5 × 185 m). When the WARTA II cement works had to be shut down in 1980 owing to its high energy consumption and the general shortage of fuel oil, the opportunity was taken to modernise the plant. Various possible processes were discussed; conversion to the dry process, which was considered the most favourable both technically and in respect of the subsequent operating costs, was not feasible because of the high capital expenditure involved. The high costs would have arisen mainly in the complete conversion of the raw material processing plant which would have been necessary. It was therefore decided to convert to the semi-

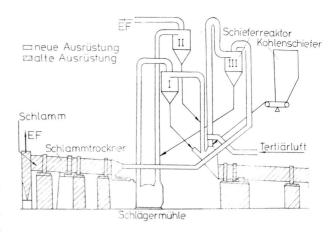

BILD 1: Schema der umgebauten Anlage
FIGURE 1: Layout of the plant after conversion

Schieferreaktor	= shale reactor
Kohlenschiefer	= bituminous shale
neue Ausrüstung	= new equipment
alte Ausrüstung	= old equipment
Schlamm	= slurry
EF	= EP
Schlammtrockner	= slurry dryer
Tertiärluft	= tertiary air
Schlägermühle	= impact pulveriser

BILDER 2: Innenkonstruktion der Mühle mit dem Schlagrad
FIGURES 2: Internal construction of the pulveriser, with the impact wheel

Ofen genutzt. Der obere Teil wurde mit einem Antrieb und Einbauten versehen und dient jetzt als Drehtrommeltrockner für den Kalkschlamm. Zwischen Trockner und Ofen wurde ein zweistufiger Zyklonwärmetauscher und ein Reaktor für die Verbrennung des Kohlenschiefers installiert. Der Kalkschlamm wird vor Eintritt in den Ofen im Trockner mit einem Gemisch von Ofenabgasen und den Abgasen der Kohlenschieferverbrennung getrocknet. Die auf etwa 10 % Wassergehalt vorgetrockneten Granalien werden dann einer Schlägermühle zugeführt, in der die Restfeuchte ausgetrieben und die Granalien zerkleinert werden. **Bild 2** zeigt die Innenkonstruktion der eingesetzten Mühle mit dem Schlagrad.

Der Kohlenstoffanteil des in den Reaktor dosierten Schiefermehles wird mit einem Gemisch aus Tertiärluft und einem Teil der Ofenabgase verbrannt. Die verbleibenden Feststoffe (Schieferasche) werden in die Ausgangsleitung der Schlägermühle dosiert und dort mit dem Kalkmehl vermischt. In den zwei Zyklonstufen wird das Rohmehl homogenisiert und vorgewärmt und dann dem Ofen zugeführt. Die Ofen- und Trocknerabgase werden in bestehenden Elektrofiltern gereinigt. Die abgeschiedenen Stäube werden dem Ofen entweder über die Schlägermühle oder über den Schieferreaktor wieder zugeführt. Im Rahmen der Modernisierungsmaßnahmen wurde weiterhin der störanfällige und unzureichend arbeitende Rostkühler vom Typ

BILD 3: Zentraler Leitstand
FIGURE 3: Central control room

wet process, which could be carried out at half the capital cost by making use of the existing raw slurry processing unit.

2. Conversion to the semi-wet process

The combination of wet and dry processes in a so-called semi-wet process is a technology which is frequently used for modernising wet kilns. The raw material is prepared as a slurry, dried, filtered, then fed to the preheater as in the dry process. However the use of bituminous shale as a raw material component and a secondary fuel made it necessary to develop a new technology, and this was done at the Mineral Building Materials Institute in Opole.

The lime-rich raw material component is treated by wet grinding to give a slurry with a water content of 30 – 32%, whereas the bituminous shale is ground dry to a powder. **Fig. 1** shows the layout of the modernised plant. About 20 m of the kiln shell has been removed between kiln supports III and IV, and the lower part measuring 5 m in diameter by 92 m is now used as the kiln. The upper part has been provided with a drive and fittings and now acts as a rotary drum-type dryer for the lime slurry. A two-stage cyclone preheater and a reactor for burning the bituminous shale have been installed between the dryer and the kiln. Before the lime slurry enters the kiln it is dried in the dryer by a mixture of exhaust gases from the kiln and from burning the bituminous shale. The nodules are pre-dried to a water content of about 10% then fed to an impact pulveriser in which the residual moisture is removed and the nodules reduced in size. **Fig. 2** shows the internal construction of the pulveriser used, with the impact wheel.

The carbon component of the powdered shale-type residue metered into the reactor is burnt with a mixture of tertiary air and some of the exhaust gases from the kiln. The remaining solids (shale-type ash) are metered into the outlet pipe from the impact pulveriser, where they are mixed with the powdered lime. The raw meal is homogenised and preheated at the two cyclone stages then fed to the kiln. The exhaust gases from the kiln and dryer are cleaned in existing electrostatic precipitators. The dusts separated are recycled to the kiln either via the impact pulveriser or via the shale-type residue reactor.

During the modernisation work the VOLGA 75 grate cooler, which was prone to trouble and inadequate, was replaced by a new cooler constructed by WARTA. The process control is provided by a modern computer-aided control system. An

VOLGA 75 durch einen neuen selbstkonstruierten Kühler ersetzt. Die Prozeßführung erfolgt über ein modernes computergestütztes Regel- und Steuersystem. Für die Antriebssteuerung wird ein freiprogrammierbares Mikroprozessor-System ELWRO 80 eingesetzt. **Bild 3** zeigt den zentralen Leitstand.

3. Schlußfolgerungen

Die Verwertung von Kohlenschiefer als Rohstoffkomponente mit niedrigem Kalkgehalt und gleichzeitig als Wärmequelle für den Schlammtrocknungsprozeß führt zu einer Einsparung an Brennstoffenergie von etwa 1000 kJ/kg Kl. Der gesamte Wärmeverbrauch der Ofenanlage, inklusive dem Anteil des Kohlenschiefers, beträgt ca. 5000 kJ/kg. Bei der Modernisierung wurde praktisch die gesamte vorhandene maschinelle Ausrüstung genutzt. Die Anlage wurde zunächst ohne Einsatz des Kohlenschiefers in Betrieb genommen. In der Zwischenzeit wurden die Aufbereitungsanlagen und Dosiervorrichtungen für das Schiefermehl errichtet. Die Ergebnisse der Inbetriebnahme und des Versuchsbetriebs lassen erwarten, daß mit der modernisierten Anlage höhere Klinkerqualitäten produziert werden können als zuvor.

ELWRO 80 programmable microprocessor system is used for the drive control. **Fig. 3** shows the central control room.

3. Conclusions

The use of bituminous shale as a low-lime raw material component and also as a heat source for the slurry drying process provides a saving in fuel energy of about 1000 kJ/kg clinker. The entire heat consumption of the kiln plant, including that of the bituminous shale, is approx. 5000 kJ/kg. Virtually all the existing machinery has been utilised in the modernisation. The plant was first started up without using the bituminous shale. Since then the preparation plant and metering equipment for the powdered shale have been installed. The results of commissioning and trial operation suggest that higher quality clinker can be produced by the modernised plant than was previously possible.

Optimierung eines Satellitenkühlers an einem 3000-t/d-Ofen

Optimizing a planetary cooler on a 3000 t/d kiln

Optimisation d'un refroidisseur planétaire sur un four de 3000 t/j

Optimización de un enfriador de satélites en un horno para 3000 t/d

Von **H. Fleck,** Harburg/Deutschland

Zusammenfassung – Zum Zeitpunkt der Inbetriebnahme des Ofens (5,5 × 89 m) im Jahre 1974 gab es wenig Erfahrungen mit großen Satellitenkühlern (10 Rohre, 2,25 × 20,4 m). Dementsprechend wurden Art und Anordnung der Einbauten ähnlich wie bei einem vorhandenen 1200-t/d-Ofen ausgelegt. Sowohl die Form als auch die Werkstoffe der Einbauten erwiesen sich nicht als günstig im Hinblick auf Streuverhalten, Wärmerückgewinnung und Standzeit. Sie wurden jedoch bis zur Funktionsunfähigkeit durch Korrosion und Bruch beibehalten. 1988/89 wurden die Kühlrohre komplett erneuert und neue Einbauten installiert. Für die Gestaltung und die Werkstoffauswahl bei den Einbauten wurden die in der Zwischenzeit vom Werk und von den Lieferanten gesammelten Erfahrungen berücksichtigt. Der Vergleich des Betriebs mit den alten und neuen Kühlrohren zeigte, daß mit dem Umbau die Kaltklinkertemperatur bei gleichem Ofendurchsatz (2600 t/d) auf Werte von unter 120 °C gesenkt und gleichzeitig die Sekundärlufttemperatur deutlich erhöht werden konnte.

Summary – At the time when the kiln (5.5 × 89 m) was commissioned in 1974 there had been little experience with large planetary coolers (10 tubes, 2.25 × 20.4 m). The type and arrangement of the internal fittings were therefore designed like those in an existing 1200 t/d kiln. Both the shape and the materials of the internal fittings proved unsuitable in respect of dispersion effect, heat recovery and service life. However, they were retained until they became unserviceable through corrosion and breakage. In 1988/89 the cooler tubes were completely renewed and new internal fittings installed. The experience gathered in the meantime by the works and by the suppliers was taken into account when designing the internal fittings and choosing the materials. Comparison of the operation with the old and new cooler tubes showed that with the modification it was possible to lower the cold clinker temperature for the same kiln output (2600 t/d) to values of less than 120°C and at the same time to make a significant increase in the secondary air temperature.

Résumé – Au moment de la mise en service du four (5,5 × 89 m), en 1974, peu d'expérience était disponible sur les grands refroidisseurs planétaires (10 tubes, 2,25 × 20,4 m). Par conséquent, la nature et la diposition des aménagements internes avaient été conçues à peu près comme celles d'un four existant de 1200 t/j. Aussi bien la forme que les matériaux des aménagements se sont révélés défavorables, du point de vue comportement de dispersion, récupération d'énergie et durée de vie. Ils ont pourtant été conservés jusqu'à leur incapacité de fonctionnement due à la corrosion et à la casse. En 1988/89, les tubes de refroidissement ont été renovés complètement et des aménagements nouveaux ont été installés. Pour la configuration et le choix des matériaux des aménagements, ont été mises à profit les expériences acquises, entre temps, par l'usine et les fournisseurs. La comparaison entre l'exploitation des vieux et des nouveaux tubes a montré, que la modification a permis, pour un même débit du four (2600 t/j), d'abaisser la température du clinker refroidi à des valeurs au-dessous de 120°C et d'augmenter, en même temps, nettement la température de l'air secondaire.

Resumen – Cuando se puso en servicio el horno de 5,5 × 89 m, en el año 1974, se contaba con pocas experiencias respecto de los grandes enfriadores de satélites (10 tubos, 2,25 × 20,4 m). Así, el tipo y la disposición de los dispositivos internos se realizaron más o menos de la misma forma que los de un horno existente, para 1200 t/d. Pero tanto la forma como los materiales de los dispositivos internos resultaron poco ventajosos con respecto al comportamiento a la dispersión, recuperación del calor y tiempo de duración. Sin embargo, se conservaron hasta que quedaron inservibles por corrosión y rotura. En 1988/89, se renovaron los tubos enfriadores por completo, instalándose también dispositivos interiores nuevos. En cuanto al diseño y a los materiales empleados para la fabricación de los mencionados dispositivos internos, se tuvieron en cuenta las experiencias adquiridas, entretanto, por la fábrica misma y por los suministradores. De la comparación entre el servicio con los tubos enfriadores antiguos y los nuevos resultó que la transformación permitió reducir la temperatura del clínker enfriado a valores inferiores a 120 °C, con el mismo rendimiento del horno (2600 t/d), a la vez que aumentó notablemente la temperatura del aire secundario.

1974 ging die 3 000-t/d-Ofenanlage, über die berichtet wird, in Betrieb (**Bild 1**). Der Ofen mit 5,5 m ∅ und 89 m Länge hat 10 Kühlrohre der Abmessung 2,25 × 20,4 m. 1973, als der Ofen bestellt und die Kühler ausgelegt wurden, gab es wenig Erfahrungen mit großen Satellitenkühlern. Die Einbauten wurden daher entsprechend den Einbauten eines vorhandenen 1200-t/d-Ofens ausgelegt, der eine Betriebszeit von 3 Jahren hatte. Sowohl die Form als auch die Werkstoffe der Einbauten erwiesen sich im großen Ofen nicht als günstig im Hinblick auf Streuverhalten, Wärmerückgewinnung und Standzeit. Trotzdem wurde aber die Form und Anzahl der Einbauten auch bei Reparaturen beibehalten. Ab 1985 wurden Reparaturen nicht mehr durchgeführt, da schon feststand, daß die Kühlrohre ausgetauscht werden sollten. Durch Bruch und Korrosion waren die Einbauten Ende 1988 nur noch zum Teil funktionsfähig, wie die **Bilder 2-6** zeigen.

Im Winter 1988/89 wurden die Kühlrohre komplett erneuert und neue Einbauten installiert.

Für die Gestaltung und die Werkstoffauswahl der Einbauten wurden die in der Zwischenzeit im Werk und von den Lieferanten gemachten Erfahrungen berücksichtigt, wobei beschlossen wurde, die komplette Auskleidung an einen Hersteller zu vergeben. Der Austausch erfolgte innerhalb einer Arbeitswoche, da die neuen Rohre am Boden bereits komplett ausgerüstet wurden. Die Ausmauerung wurde ebenfalls am Boden durchgeführt. Auch die alten Loslager- und Festlagerschalen wurden ausgetauscht und die Krümmer komplett erneuert. Die Gießmasse wurde bereits im Herstellerwerk eingebracht, die Ofenausläufe wurden mit Stahlgußtöpfen ausgekleidet.

In den neuen Rohren wurden sehr unterschiedliche Hubelemente und Schaufeln in insgesamt 7 Zonen eingebaut, die in **Bild 7** dargestellt sind. **Bild 8** zeigt den Übergang von Kammfutter zu Zone A1 und Zone A2 mit Umwälztöpfen und spiralförmigen Hubelementen, die aufgrund ihrer großen Oberfläche für einen guten Wärmeaustausch sorgen sollen.

The 3 000 t/d kiln plant which is the subject of this report went into operation in 1974. It is 5.5 m in diameter, 89 m long and has 10 cooler tubes measuring 2.25 × 20.4 m. In 1973, when the kiln was ordered and the coolers designed, there had been little experience with large planetary coolers. The internal fittings were therefore designed like those in an existing 1200 t/d kiln which had been in operation for 3 years. Both the shape and the materials of the fittings proved unsuitable in respect of dispersion effect, heat recovery and

BILD 3: Blick durch ein altes Kühlrohr vom Kammfutter aus in Richtung Auslauf mit teilweise verschlissenen Einbauten der Zonen B, C und D
FIGURE 3: View through an old cooler tube from the refractory lifters towards the outlet with partly worn internal fittings in zones B, C and D

BILD 4: Schaufeln im Übergang von Zone B zu Zone C im alten Kühlrohr
FIGURE 4: Scoops at the transition from zone B to zone C in the old cooler tube

BILD 1: Ofenanlage mit einer Leistung von 3 000 t/d
FIGURE 1: Kiln plant with a 3 000 t/d output

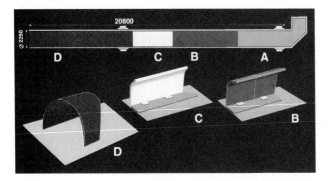

BILD 2: Schnitt durch ein altes Kühlrohr mit den Einbauten
FIGURE 2: Section through an old cooler tube with the internal fittings

BILD 5: Detail der Zone C mit abgebrochenen Schaufeln im alten Kühlrohr
FIGURE 5: Detail of zone C with scoops broken off in the old cooler tube

Auf **Bild 9** ist der Bereich des Mauerwerks mit Umwälztöpfen und Hubelementen, der Bereich der glatten Hubschaufeln, Zone B, mit den anschließenden, becherförmigen geschlitzten und geschlossenen Schaufeln der Zone C und C1 zu sehen. Auf der Nahaufnahme der Zone C und C1, **Bild 10**, sind das Waffelmuster in den Wandplatten, das den Verschleiß minimieren soll, und die schräggesetzten Leisten, die den Materialtransport im Rohr verbessern sollen, gut zu erkennen.

Die Platten sind hohl, der Hohlraum ist zur Isolierung mit AlSi-Matten ausgefüllt. Sie werden nicht verschraubt, sondern mit einem Nut/Feder-System unter die verschraubten Hubelemente gesteckt. Dadurch können sich die Platten ausdehnen, und die Bruchgefahr durch Verspannung wird vermieden.

Bild 11 zeigt einen Blick vom Kühlerauslauf her. Es zeigt die Hubwinkel der Zone E vor dem Stauring und die schräggestellten, geschlossenen Becher der Zone D.

service life. However the shape and number of these fittings were retained although they required frequent repair. From 1985 no more repairs were carried out as it had been decided to replace the cooler tubes. By the end of 1988 the internal fittings had become unserviceable through corrosion and breakage, as shown in **Figs. 2–6**.

In the winter of 1988/89 the cooler tubes were completely renewed and new internal fittings installed.

The experience gathered in the meantime by the works and by the suppliers was taken into account when designing the internal fittings and choosing the materials, and it was decided to have the whole lining made by one manufacturer. The exchange was carried out within one working week as the new tubes had already been completely fitted out on the ground. The lining work also took place on the ground. The old floating bearing and fixed bearing shells were exchanged and the elbows completely renewed. The castable refractory had already been installed in the manufacturer's works, and the kiln outlets were lined with cast steel pots.

BILD 6: Auslaufseite des Kühlrohres mit verschlissenen und abgebrochenen Bogenblechen der Zone D im alten Kühlrohr
FIGURE 6: Outlet side of old cooler tube with curved metal sheets worn and broken off in zone D

BILD 8: Zone A 1 mit den Umwälztöpfen und Zone A 2 mit den spiralförmigen Hubelementen im neuen Kühlrohr
FIGURE 8: Zone A 1 with the circulating pots and zone A 2 with the spiral lifting devices in the new tube

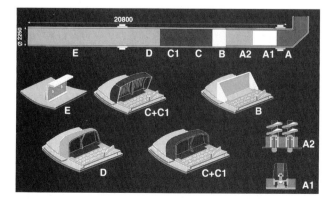

BILD 7: Schnitt durch ein neues Kühlrohr mit den verschiedenen Einbauten
FIGURE 7: Section through a new cooler tube with the various internal fittings

BILD 9: Blick von Zone C zum Kühlrohreinlauf im neuen Kühlrohr
FIGURE 9: View from zone C to the tube inlet in the new tube

BILD 10: Detail der Zone C und C 1 mit Hub- und Kielschaufeln und den Grundplatten mit Waffelmuster und Förderspirale im neuen Kühlrohr
FIGURE 10: Detail of zones C and C 1 with lifting and keel scoops and base plates with waffle pattern and spiral conveyor in the new tube

BILD 12: Blick vom Wärmetauscher auf die Kühlermontage
FIGURE 12: View from preheater to cooler assembly

BILD 11: Blick vom Kühlerauslauf her auf die Zone E mit den kleinen, auf Winkeln verschraubten Hubwinkeln
FIGURE 11: View from cooler outlet to zone E with the small lifting angles screwed onto angles

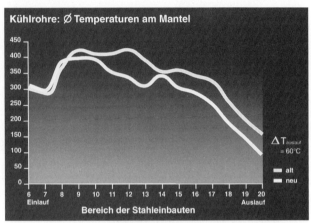

BILD 13: Mittleres Temperaturprofil an der Außenwand der alten und neuen Rohre
FIGURE 13: Average temperature profile at the outside wall of the old and new tubes

Kühlrohre: ∅ Temperaturen am Mantel	= Cooler tubes: shell temperatures
$\Delta T_{Auslauf}$	= ΔT_{outlet}
alt	= old
neu	= new
Einlauf	= inlet
Auslauf	= outlet
Bereich der Stahleinbauten	= region with steel internal fittings

Alt				Neu			
Zone	Länge m	Gegenstand	Gewicht kg	Zone	Länge m	Gegenstand	Gewicht kg
A	4,5	Kammfutter	4530	A	1,5	Kammfutter	1590
B	4,5	Schaufel Typ A + Grundpl.	5150	A1	1,2	Rührkegel + FF - Steine	1290
C	2,1	Schaufel Typ B + Grundpl.	2405	A2	1,6	Brechkegel + FF - Steine	1120
D	8,9	Stahlblechschaufel	7490	B	1,2	Umrührer + Grundpl.	1892
				C+ C1	5,4	Hubschaufel + Grundpl.	9583
				D	2,4	Streuschaufel + Grundpl.	3139
				E	6,7	Streuwinkel	1658
Summe:			19575				20272

TABELLE 1: Gewichtsvergleich der Einbauten in den alten und den neuen Rohren
TABLE 1: Comparison of weights of internal fittings in the old and new tubes

alt	= old
neu	= new
Kühlrohr: Einbauten	= Cooler tube: internal fittings
Zone	= zone
Länge	= length
Gegenstand	= item
Gewicht	= weight
Kammfutter	= cam lining
Schaufel Type A + Grundpl.	= type A scoop + base plate
Stahlblechschaufel	= steel plate scoop
Rührkegel + FF-Steine	= agitator cone + refractory bricks
Brechkegel + FF-Steine	= breaker cone + refractory bricks
Umrührer + Grundpl.	= agitator + base plate
Hubschaufel + Grundpl.	= lifter scoop + base plate
Streuschaufel + Grundpl.	= spreader scoop + base plate
Streuwinkel	= spreader angle
Summe	= total

A large variety of lifting devices and scoops are fitted in the total of 7 zones of the new tubes which can be seen in the **Fig. 7**. **Fig. 8** shows the transition from the refractory lifters to zone A1 and zone A2 with circulating pots and spiral lifting devices. These latter are designed to provide good heat exchange through their large surface area.

The brickwork area with circulating pots and lifting devices and the smooth lifting scoop area, zone B, with the adjoining, bucket-shaped, slotted and closed scoops of zones C and C1 can be seen in **Fig. 9**. The waffle pattern on the wall plates to minimise wear, and the inclined bars to improve material feed in the tube can easily be recognised from the close-up photograph of zone C and C1 in **Fig. 10**.

The plates are hollow and the cavity is filled with AlSi mats for insulation. Instead of being bolted on they are inserted below the bolted lifting devices with a tongue and groove system. In this way the plates can expand and the risk of breakage through distortion is avoided.

Fig. 11 is a view from the cooler outlet. It shows the lifting angles in zone E before the dam ring, and the inclined, closed buckets in zone D.

Erection was carried out with two 120 t mobile cranes (**Fig. 12**). The tubes were lined on the ground and the scoops

Die Montage wurde mit zwei 120-t-Mobilkränen durchgeführt (**Bild 12**). Die Rohre wurden am Boden ausgemauert, und die Schaufeln und Panzerbleche wurden eingebaut. Die gesamte Montage dauerte 1 Woche, wobei täglich 2–3 Rohre gewechselt wurden.

Die Gewichte der Einbauten sind in **Tabelle 1** zusammengestellt. Die Einbauten in den neuen Kühlern sind um 700 kg schwerer als in den alten Kühlern, was zu einer zusätzlichen dynamischen Belastung der Aufhängung führt. Die Fest- und Loslager haben aber genügend Reserven, um die zusätzliche Belastung aufzunehmen. Nach einem Durchsatz von ca. 250 000 t Klinker/Kühlrohr ist ein nennenswerter Verschleiß der Gußteile nicht zu erkennen. Das Kammfutter ist inzwischen weitgehend glatt geworden. Dieses mußte im Winter 1993/94 erneuert werden. Zwischen den Umrührertöpfen und den spiralförmigen Einbauten wurde versuchsweise in einigen Rohren Stampfmasse mit Korund und in einigen Rohren Formsteine mit 35 % Al_2O_3 eingebaut.

Die Stampfmasse zeigt inzwischen an den Gußteilen tiefe Auswaschungen. Bei den nächsten Reparaturen werden diese Bereiche mit Feuerfest-Steinen ausgekleidet.

Nicht gelöst ist die Feuerfest-Auskleidung der Rückhaltebrücken in den Krümmern. Die eingebrachte Masse platzt häufig ab. Bei jedem Ofenstillstand sind mehrere Brücken zu reparieren.

Im Winter 1994/95 soll ein Versuch mit Fertigelementen aus FF-Masse sowie mit 2 verschiedenen Gußelementen an je 3 Brücken zu einer optimierten Lösung dieser Problemstelle führen.

Einen Vergleich der Wandtemperaturen der alten und neuen Kühlrohre zeigt **Bild 13**. Die Ofenleistung betrug in beiden Fällen ca. 2 800 t/d. Auf dem Bild grün dargestellt ist die Temperatur der neuen Rohre, rot die Temperatur der alten Rohre. Die Wandtemperatur am Auslauf ist um ca. 60 °C niedriger als bei den alten Rohren. Dies entspricht einer Temperaturabsenkung des Klinkers nach Kühler von ca. 40 °C auf eine Temperatur von unter 120 °C. Die Sekundärlufttemperatur stieg um 80–100 °C an, die Flamme wurde heißer und kürzer. Um die alte Flammenform wieder herzustellen, wurde der Axialluftanteil der Primärluft erhöht. Die NO_x-Werte, die auf ca. 750 mg/Nm^3 gestiegen waren, gingen wieder auf das ursprüngliche Niveau von 350–450 mg/Nm^3 zurück. Der Wärmebedarf konnte durch die Kühlererneuerung mit den geänderten Einbauten um fast 40 kJ/kg abgesenkt werden.

and liners were installed. The whole erection process took one week with 2 to 3 tubes being changed per day.

The weights of the internal fittings are set out in the table. The fittings in the new coolers are 700 kg heavier than those in the old ones, creating an additional dynamic load on the suspension system. The fixed and floating bearings have enough reserves to take up the additional load. No appreciable wear on the cast parts can be seen after a throughput of about 250 000 t clinker per cooler tube. But the refractory lifters have largely worn smooth and have been renewed during winter 1993/94. In some tubes monolithic refractory with corundum and in others shaped bricks with 35 % Al_2O_3 have been fitted experimentally between the agitator pots and the spiral internal fittings. The monolithic refractory now shows deep erosion at the cast parts. These areas will be lined with refractory bricks next time repairs are carried out.

The question of the refractory lining for the retaining bridges in the elbows has not been resolved. The material used often chips off, and several bridges have to be repaired each time the kiln is stopped.

In winter 1994/95 pre-fabricated elements made of refractory material with 2 different castings are to be tried on 3 bridges each, to provide an optimised solution to the problem.

Fig. 13 compares the wall temperatures of the old and new cooler tubes. The kiln output was about 2 800 t/d in both cases. The temperature of the new tubes is shown in green and that of the old ones in red. The wall temperature at the outlet is about 60 °C lower than in the old tubes. This corresponds to a drop in clinker temperature after the cooler of about 40 °C to a temperature of below 120 °C. The secondary air temperature rose by 80–100 °C and the flame grew hotter and shorter. The axial component of the primary air was increased to restore the old flame shape. The NO_x values, which had risen to approx. 750 mg/m^3(stp), returned to the original level of 350–450 mg/m^3(stp). The renewal of the cooler and changes in the internal fittings reduced the heat requirement by almost 40 kJ/kg.

HADES – Die thermische Überwachung der Klinkerkühlung in Zementwerken

HADES – Thermal monitoring of cement plant clinker coolers

HADES – Surveillance thermique du refroidissement du clinker dans les cimenteries

HADES – Control térmico del enfriamiento de clínker en las fábricas de cemento

Von **G. Gaussorgues** und **R. Houis**, Massy/Frankreich

Zusammenfassung – Das Überwachungssystem HADES wurde von HGH, einem Optronik-Spezialisten, in Zusammenarbeit mit Ciments Lafarge in Frankreich entwickelt, um die Klinkertemperatur in Klinkerkühlern genau messen zu können. HADES ist ein System zur Darstellung der Temperaturverteilung innerhalb eines breiten Gesichtsfelds (180° × 70°), das fortlaufend die „Wärmekarte" des Klinkerbetts in Echtzeit liefert. Es besteht aus einem Infrarot-Endoskop, das mit einem für den kontinuierlichen Betrieb bei sehr hohen Temperaturen (bis zu 1000°C) ausgelegten Analysatorkopf ausgerüstet ist. Das Wärmebild hat 270 Zeilen zu je 1000 Pixel und einen Dynamikbereich von 100°C bis 1500°C. Alle 40 Sekunden werden die thermischen Daten der gesamten Kühleroberfläche abgefragt. Das Optikelement innerhalb des Kühlers wird mittels Luftkühlung auf niedrigen Temperaturen gehalten. Ein Sicherheitssystem kann im Falle der Überhitzung bei Ausfall des Kühlventilators den optischen Meßwertgeber automatisch herausziehen, und der optische Zugang durch die Wand des Klinkerkühlers wird mechanisch verschlossen. Das Meßwertgeber-Signal wird durch einen Lichtleiter zu einem Computer übertragen, der das Klinkerbett-Thermogramm in Falschfarbendarstellung temperaturgeeicht anzeigt. Die Temperaturhöhe eines jeden Bereichs ermöglicht die Überwachung und Steuerung des Kühlers, und örtliche Änderungen der Belüftung verursachen einen stetigen Wärmegradienten entlang des Klinkerbetts. Die Steuerung verhindert die örtliche Überhitzung auf dem Rost und erlaubt gleichzeitig die größtmöglichen Energieeinsparungen. Die reduzierten Wärmedaten werden zum Prozeßüberwachungs-Computer geleitet.

Summary – The HADES monitoring system was developed by HGH, specialists in optronic systems, in cooperation with Ciments Lafarge, France, to obtain an accurate temperature measurement of the clinker in clinker coolers. HADES is a thermal imaging system with a wide field of view (180° × 70°) which continuously delivers in real time the thermal map of the clinker bed. It is a real infrared endoscope equipped with an analyzing optical head designed for continuous operation in very high temperature environments up to 1000°C. The thermal image has 270 lines of 1000 pixels per line and a dynamic range of 100°C to 1500°C. A complete thermal scanning of the clinker cooler surface is made every 40 seconds. The optical head inside the cooler is maintained at low temperature by air cooling. A security system can automatically extract the optical sensor in case of overheating due to failure of the air cooling fan, and optical access through the wall of the clinker cooler is locked by a mechanical shutter. The signal from the sensor is transmitted through optical fiber to a computer which displays the clinker bed thermogram in a false colour mode, calibrated in temperatures. The thermal level of each area allows a control monitoring of the cooler, and local changes of the ventilation cause a steady thermal gradient along the clinker bed. The control prevents local overheating on the grate while saving as much energy as possible. The reduced thermal data are transmitted to the process monitoring computer.

Résumé – Le système de surveillance HADES a été mis au point par HGH, un spécialiste d'optique électronique, en coopération avec Ciments Lafarge, France, pour mesurer exactement la température du clinker dans les refroidisseurs de clinker. HADES est un système de représentation de la répartition de la température à l'intérieur d'un vaste champ visuel (180° × 70°), capable de fournir en permanence et en temps réel, la „carte de la température" du lit du clinker. Il comprend un endoscope infrarouge, équipé d'une tête d'analyseur dimensionnée pour une exploitation en continu à des températures très élevées allant jusqu'à 1000°C. L'image par rayonnement thermique a 270 lignes ayant chacune 1000 pixels et une zone dynamique de 100° à 1500°C. Les données thermiques de toute la surface du refroidisseur sont consultées toutes les 40 secondes. L'élément optique dans le refroidisseur est maintenu à basses températures par refroidissement par air. Un système de sécurité peut, en cas de surchauffe – le ventilateur de refroidissement étant en panne, retirer automatiquement le capteur de mesure optique, l'accès optique à travers la paroi du refroidisseur de clinker est alors fermée mécaniquement. Le signal du capteur de mesure est transmis par un conducteur d'éclairage à un ordinateur qui affiche en température étalonnée le thermogramme du

lit de clinker représenté en couleurs fausses. La hauteur de la température de chaque zone permet de surveiller et de commander le refroidisseur, et les changements de ventilation entraînent un gradient de température stable le long du lit du clinker. Le système de commande prévient une surchauffe locale sur la grille et permet, en même temps, les économies d'énergie les plus grandes possible. Les données thermiques réduites sont transmises à l'ordinateur de surveillance du process.

Resumen – El sistema de control HADES ha sido desarrollado en Francia por HGH, especialista en óptica y electrónica, en colaboración con Ciments Lafarge, con el fin de poder medir con precisión la temperatura del clínker dentro de los enfriadores de clínker. HADES es un sistema para la representación de la distribución de la temperatura dentro de un amplio campo visual (180° × 70°), capaz de suministrar, de forma continua y en tiempo real, el „mapa de temperaturas" del lecho de clínker. Se compone de un endoscopio infrarrojo, equipado con una cabeza analizadora, dimensionada para servicio continuo a temperaturas muy altas (hasta 1000 °C). La imagen térmica consta de 270 renglones de 1000 pixel c/a y una zona dinámica de 100 °C a 1500 °C. Cada 40 segundos se registran los datos térmicos de toda la superficie del enfriador. El elemento óptico dentro del enfriador se mantiene a bajas temperaturas por enfriamiento de aire. En caso de sobrecalentamiento por avería del ventilador de enfriamiento, hay un sistema de seguridad que puede sacar automáticamente el indicador óptico del valor de medición, cerrándose mecánicamente el acceso óptico a través de la pared del enfriador de clínker. La señal del indicador del valor de medición se transmite a un ordenador, por medio de un conductor óptico. Dicho ordenador indica el termograma del lecho de clínker en colores falsos y temperaturas contrastadas. El valor de temperatura de cada zona permite el control y mando del enfriador. Los cambios locales de la ventilación dan lugar a un gradiente de calor constante a lo largo del lecho de clínker. El sistema de mando impide el sobrecalentamiento local sobre la parrilla, permitiendo al mismo tiempo los mayores ahorros posibles de energía. Los datos térmicos reducidos son transmitidos al ordenador de control del proceso.

HADES – Control térmico del enfriamiento de clínker en las fábricas de cemento

Die thermischen Verfahrensschritte des Zementherstellungsprozesses erfordern einen enormen Energieaufwand. Eine Möglichkeit der Energieeinsparung ist die effiziente Steuerung der Wärmeübertragung vom Kühl- bis zum Vorwärmersystem.

Das HADES-System (Heat Area Detection and Evaluation System) wurde von der Firma HGH in Zusammenarbeit mit Ciments Lafarge in Frankreich entwickelt, um die Klinkertemperatur in Klinkerkühlern genau messen zu können.

HGH, eine französische Firma und Spezialist für Optronik-Systeme für militärische und industrielle Anwendungen, verfügt über beachtliche Erfahrungen mit Infrarotsensoren zur thermischen Überwachung von Drehrohröfen.

Das Unternehmen bietet seit vielen Jahren schlüsselfertige Lösungen mit dem selbst entwickelten thermischen Linearscanner ATL 020 einschließlich einem kompletten Softwarepaket an.

Eine der ersten Methoden zur thermischen Überwachung des Klinkerkühlers war die Verwendung des vorhandenen Infrarot-Linearscanners ATL 020, dem eine zweite Scanachse hinzugefügt wurde. Dadurch wurde jedoch die Öffnung in der Wand des Kühlers zu groß, so daß ein kontinuierlicher Betrieb des Sensors unmöglich war.

HADES arbeitet mit einer ganz neuen, für die Wehrtechnik entwickelten Technologie zur Darstellung der Temperaturverteilung innerhalb eines breiten Gesichtsfelds (180° × 70°), das fortlaufend die Temperaturverteilung des Klinkerbetts in Echtzeit liefert.

Es ist ein periskopisches System mit 180 mm Durchmesser, das mit einem Sensorkopf ausgerüstet und darauf ausgelegt ist, 24 Stunden am Tag bei sehr hohen Temperaturen betrieben zu werden. Die Temperatur kann aufgrund der hohen Luftvorwärmung im Klinkerkühler und der Wärmestrahlung des Klinkerbettes, das sich mit einem sehr großen schwarzen Strahler mit 1400 °C vergleichen läßt, auf bis zu 1000 °C ansteigen.

Das Wärmebild hat 270 Zeilen zu je 1000 Pixel. Dies ergibt eine Auflösung von 3 cm × 3 cm für die Klinkerbettoberfläche. Der Meßbereich der Temperaturmessung geht von

The thermal phases of the cement production process is achieved with a huge energy consumption.

One way to save energy is to have an efficient control of heat transfers from the cooling system to the preheating system.

HADES (Heat Area Detection and Evaluation System) was developed by HGH in cooperation with Ciments LAFARGE France, to obtain an accurate temperature measurement of clinker in the clinker cooler.

HGH, a French company specialist in optronic systems for military and industrial applications has a strong experience with infrared sensors for thermal monitoring of rotary kilns.

This company is proposing since many years turn key solutions including its own thermal line scanner ATL 020, with a complete software package.

One of the first technical approaches of the clinker cooler thermal surveillance was to use the existing ATL 020 infrared line scanner by giving it a second axis scanning. In that case, the hole in the wall of the cooler had to be too large and it was not possible to have a continuous running of the sensor.

Using a new very original technology coming from military development, HADES is a thermal imaging system with a wide field of view (180° x 70°) which continuously delivers in real time the thermal map of the clinker bed.

It is a periscopic system of 180 mm diameter equipped with an analysing optical head designed for twenty four hours a day operation in very high temperature environment, up to 1000 °C due to air temperature in the clinker cooler and to the radiant thermal emission of the clinker bed which can be compared to a very large black body at 1400 °C.

The thermal image has 270 lines of 1000 pixels per line giving a resolution of 3 cm x 3 cm on the clinker surface. The thermal measurement dynamic range is going from 100 °C to 1500 °C for each pixel with a thermal sensitivity better than 5 °C. A complete thermal scanning of the clinker cooler surface is made every 40 seconds.

The optical head inside the cooler is maintained at low temperature by air cooling. The infrared spectral response of the

100°C bis 1500°C für jeden Bildpunkt bei einer Temperaturempfindlichkeit von weniger als 5°C. Alle 40 Sekunden werden die thermischen Daten der gesamten Kühleroberfläche abgefragt.

Das in den Kühler eingebaute Optikelement wird mit Luft gekühlt. Das Infrarot-Signal des Sensors ist durch Verwendung eines Quecksilber-Cadmium-Tellur-Detektors mit thermoelektrischer Kühlung über einen großen Temperaturbereich mit dem maximalen Strahlungskontrast des Klinkers verknüpft. Das Infrarotsignal wird mit 12 Bit digitalisiert.

Ein Sicherheitssystem kann im Falle der Überhitzung, z.B. bei Ausfall des Kühlventilators, den optischen Sensor automatisch herausziehen, und die Öffnung in der Wand des Klinkerkühlers wird mechanisch verschlossen.

Das Meßsignal wird mittels Lichtwellenleiter über einen seriell-parallelen Wandler zu einem 386er Personalcomputer übertragen. Mit Hilfe einer speziellen Software werden die Signale in Echtzeit erfaßt und verarbeitet und in Form eines Thermogramms in Falschfarben-Darstellung, die entsprechend der Temperatur geeicht ist, abgestellt.

Die verschiedenen Temperaturbereiche, die entsprechend der Kühlergeometrie auf dem Thermogramm aufgezeigt werden, ermöglichen die Überwachung des Kühlerbetriebs, so daß durch örtliche Veränderung der Belüftung eine gleichmäßige Temperaturverteilung im Klinkerbett erreicht wird.

Diese Steuerung verhindert die örtliche Überhitzung auf dem Rost und erlaubt gleichzeitig eine größtmögliche Energieeinsparung. Der überwachte Bereich wird in 30 Zonen eingeteilt, wobei für jede Zone ein Alarmgrenzwert gewählt werden kann.

Die Software umfaßt Module zur Echtzeitüberwachung, zur Erfassung und Verarbeitung aufgenommener Bilder auf Festplatte, zur Anzeige der Thermogramme, Zoomfunktionen für die Temperaturverteilungen, statistische Berechnungen und 3D-Darstellungen.

Die reduzierten thermischen Daten werden an das Prozeßleitsystem übertragen.

sensor is fitted to the maximum radiant contrast of the clinker along a wide temperature range by using a mercury-cadmium telluride detector thermo-electric cooled. The infrared signal is 12 bits digitized.

A security system can automatically extract the optical sensor in case of overheating due to failure of the air cooling fan, and optical access through the wall of the clinker cooler is locked by a mechanical shutter.

The signal from the sensor is transmitted through optical fiber to a 386 PC computer including an acquisition serial parallel conversion board. The real time acquisition and processing software displays the clinker bed thermogram in false colour mode, calibrated in temperature.

The thermal level of each area defined on the thermogram according to the clinker cooler geometry allows a control monitoring of the cooler and local changes of the ventilation cause a steady gradient along the clinker bed.

This control prevents local overheating on the grate while saving as much energy as possible. The surveillance is made on 30 zones with the choice of alarm thresholding for each zone.

The software includes: realtime surveillance, acquisition and processing of recorded images on hard disk, display of thermal images, thermal profile zooming, statistical calculation, 3D.

The reduced thermal data are transmitted to the process monitoring computer.

Ansatzbildung beim Brennen eines fluorhaltigen Zementrohmehls

Coating formation during burning of a cement raw mix with inherent fluorine

Phénomène de croûtage au cours de la cuisson d'une farine crue contenant du fluor

Formación de adherencias durante la cocción de un crudo de cemento que contiene fluoruro

Von **G. Goswami, B. N. Mohapatra** und **J. D. Panda,** Rajgangpur/Indien

Zusammenfassung – Es wurde die Ansatzbildung in einer kohlebefeuerten Vorcalcinieranlage durch flüchtige Bestandteile, besonders Fluor, untersucht. Gasförmige Fluoride bilden sich während der Verarbeitung von Zementrohmehlen, die Kalkstein mit etwa 0,11 % F enthalten. Die flüchtigen Bestandteile K_2O, Na_2O, F, Cl und SO_3 wurden in einer Reihe von Ansatzproben, die an verschiedenen Stellen des Ofens entnommen wurden, untersucht, um ihre Schwankungsbreite und Auswirkung festzustellen. Die Ansatzproben wurden mineralogisch durch Röntgendiffraktometrie (XRD) und Thermoanalyse (Differentialthermoanalyse, DTA, Thermogravimetrie, TG) untersucht. Dabei wurde festgestellt, daß Spurrit ($2C_2S \cdot CaCO_3$) und Anhydrit ($CaSO_4$) die mit der Ansatzbildung am häufigsten einhergehenden Phasen sind. Außer den normalen Klinkerphasen wurde in einigen Ansatzproben ein Spurritgehalt von bis zu 50 % gefunden. Bei Ölfeuerung und infolge der gelegentlich hohen Absorption von Kohleasche wiesen besonders die an Steigleitungen entnommenen Ansatzproben bis zu 17 % SO_3 auf, was einem hohen Gehalt an $CaSO_4$ entspricht, welches harte, schieferähnliche Ablagerungen verursacht. Bei normaler Kohlefeuerung führt der entstehende Spurrit bei Anwesenheit von Fluor im allgemeinen zu Ansatzbildung und folglich zu einem Materialstau.

Summary – Coating formation in a coal-fired precalciner kiln system with respect to volatile components, particularly fluorine, has been examined. The gaseous fluorine is generated during processing of a raw mix containing limestone with around 0.11% inherent F. The volatiles, K_2O, Na_2O, F, Cl and SO_3 in a number of coating samples, collected from different points of the kiln have been estimated to examine their variation and effect. Mineralogy of the coating samples has been examined by X-Ray Diffractometry (XRD) and Thermal Analysis (DTA/TG). Spurrite ($2C_2S \cdot CaCO_3$) and anhydride ($CaSO_4$) are found to be the main two phases associated with the process of coating formation. In addition to the normal clinker phases, a spurrite content of up to 50% has been recorded in some of the coating samples. In the case of oil firing, and also due to occasional high absorption of coal-ash, coating samples particularly from the riser ducts, recorded up to 17% SO_3 corresponding to a high amount of $CaSO_4$, causing hard slaty deposits. In general, during normal coal firing, the formation of spurrite in the presence of F causes coating and subsequently jamming of the material.

Résumé – L'étude a porté sur le phénomène de croûtage dû à des composants volatils, notamment le fluor, dans une installation de précalcination chauffée au charbon. Les fluorures gazeux se forment durant la transformation de la farine crue de ciment qui contient du calcaire avec environ 0,11 % F. Les composants volatils K_2O, Na_2O, F, Cl et SO_3 ont été analysés sur toute une série d'échantillons de croûtage, qui ont été prélevés en différents points du four pour en définir leur marge de fluctuation et effet. Les échantillons de croûtage ont été analysés du point de vue minéralogique par diffraction des rayons X (XRD) et analyse thermique (thermoanalyse différentielle, DTA, thermogravimétrie, TG). Lors de ces analyses il a été constaté que le spurrit ($2C_2S \cdot CaCO_3$) et l'anhydrite ($CaSO_4$) sont les phases accompagnant le plus souvent le croûtage. Outre les phases de clinkérisation normales, on a trouvé dans certains échantillons de croûtage une teneur de spurrit allant jusqu'à 50%. En présence d'une chauffe au fuel et suite à une absorption élevée par moments de cendres de charbon, les échantillons de croûtage prélevés sur les conduites montantes accusaient jusqu'à 17% de SO_3, ce qui correspond à une teneur élevée en $CaSO_4$, qui entraîne des dépôts durs, de type schisteux. Avec une chauffe au charbon normale, la spurrit se formant en présence de fluor entraîne généralement le phénomène de croûtage et, par voie de conséquence, un bourrage de matière.

Resumen – Se ha estudiado la formación de adherencias en una instalación de precalcinación con calefacción de carbón, debido a los componentes volátiles, sobre todo el flúor. Se forman fluoruros gaseosos durante el tratamiento del crudo de cemento, que contienen caliza con 0,11 % aprox. de F. Los componentes volátiles K_2O, Na_2O, F, Cl y SO_3 han sido estudiados en toda una serie de muestras de adherencias obtenidas en diferentes puntos del horno, con el fin de averiguar su margen de variación así como

sus repercusiones. Las muestras de adherencias han sido estudiadas desde el punto de vista mineralógico, por medio de la difracción de rayos X (XRD), el análisis térmico (análisis térmico diferencial, DTA, y la termogravimetría TG). A este respecto, se ha podido comprobar que la espurrita ($2C_2S \cdot CaCO_3$) a la anhidrita ($CaSO_4$) son las fases más asociadas a la formación de adherencias. Aparte de las fases normales de clínker se ha encontrado en algunas muestras de adherencias un contenido de espurrita de hasta un 50 %. En el caso de la calefacción por petróleo, y debido a la absorción a veces muy alta de cenizas de carbón, han sido sobre todo las muestras de adherencias tomadas en los tubos ascendentes las que tenían hasta un 17 % de SO_3, lo que corresponde a un elevado contenido de $CaCO_4$, que causa sedimentaciones duras, similares a los esquistos. En el caso de las calefacciones normales de carbón, la espurrita que se forma en presencia de flúor, conduce en general a la formación de adherencias y, por lo tanto, a la retención del material.

1. Einleitung

Der Einfluß flüchtiger Bestandteile auf die Ansatzbildung und den Materialfluß insbesondere in modernen im Gegenstrom betriebenen Ofenanlagen ist zum Gegenstand intensiver Forschung geworden. Von den verschiedenen flüchtigen Bestandteilen wurden in Verbindung mit der Ansatzringbildung insbesondere Fluor (F^-) und Chlor (Cl^-) untersucht. In den meisten Fällen wird berichtet, daß das Mineral Spurrit ($2C_2S$, C^-)[1] mit der Materialanhäufung und Ringbildung in Zusammenhang steht [1–3]. Andererseits berichtet Fundal vom Auftreten von Fluorellestadit ($3C_2S$, $3CS$, CaF_2) insbesondere in Gegenwart hoher SO_3-Gehalte [4]. Kürzlich überprüften Bolio-Arceo und Glasser [5] mehrere vorangegangene Untersuchungen und zeigten, daß die Spurritbildung aus Zementrohmehl mindestens 0,2 % an Mineralisator wie z. B. CaF_2 oder $CaCl_2$ benötigt. Dabei kamen sie zu dem Schluß, daß die Bedingungen, die zur Spurritbildung führten, häufig mit denjenigen zusammenfielen, die zur Ringbildung führten.

In Weiterführung unserer Untersuchung über im Zementrohmehl natürlich vorkommendes Fluor [6, 7, 8] untersucht die vorliegende Studie dessen Auswirkung auf Materialfluß und Ansatzbildung in einer kohlebefeuerten Vorcalcinationsanlage. Der Kalkstein im Rohmehl enthält durchschnittlich 0,11 % Fluor. Dies ist viel mehr als die mittlere Konzentration von F^- im Zementrohmehl, die nach der Literatur 0,05 % beträgt [9]. Die Konzentrationen der anderen im Zementrohmehl vorhandenen flüchtigen Bestandteile sind durchschnittlich: $K_2O = 0,7$ %, $Na_2O = 0,08$ %, $SO_3 = 0,2$ % und $Cl = 0,01$ %. Im Ofen kommt es gelegentlich zu Verstopfungen in den Zyklonstufen II bis IV sowie zur Bildung dikker Ansätze in Zyklon V, im Ofeneinlauf und im Bereich der Sekundärfeuerung. Eine mäßige Ansatzbildung findet auch auf der Oberfläche der Ausmauerung mit Mg-Chrom-Steinen in der Sinterzone statt.

2. Untersuchte Materialien

2.1 Zyklonverstopfungen

Die Zusammensetzungen der flüchtigen Bestandteile von Materialien, die den Zyklonen bei Verstopfungen entnommen wurden, sind in **Tabelle 1** aufgeführt. Es wurde beobachtet, daß sowohl F^- als auch Alkalien, insbesondere K_2O, von Zyklon II nach V hin vermehrt auftreten, wohingegen für Cl^- und SO_3 kein derartiger Trend beobachtet wurde. Die mineralogischen Untersuchungen der Proben mit Rönt-

1. Introduction

The effect of volatiles on the coating formation and material flow particularly in the modern closed kiln system has become a subject of intensive investigation. Out of the different volatiles studied in connection with ring formation fluorine (F^-) and chlorine (Cl^-) have been examined very extensively. In most of the cases the mineral spurrite ($2C_2S$, C^-)[*] has been reported to be associated with the material build-up and ring formation [1–3]. On the other hand, Fundal [4] reported fluor-ellestadite ($3C_2S$, $3CS$, CaF_2) particularly in presence of high amount of SO_3. Recently Bolio-Arceo and Glasser [5], while demonstrating that the formation of spurrite from cement raw mix required at least 0.2 % mineralizer as either CaF_2 or $CaCl_2$, reviewed a number of previous investigations and concluded that the conditions leading to spurrite formations frequently coincided with those which led to ring formations.

In continuation of our investigation of fluorine naturally present in the cement raw mix [6–8] the present study examines its effect on material flow and coating formation in a coal-fired precalciner kiln system. The limestone used in the raw mix contains, in average, 0.11 % F. This is much higher than the mean concentration of F^- in cement raw mix which is reported to be 0.05 % [9]. The other volatiles present in the raw mix used in the particular plant in general are: $K_2O = 0.7$ %, $Na_2O = 0.08$ %, $SO_3 = 0.2$ % and $Cl = 0.01$ %. The kiln experiences occasional jamming of the feed material in the cyclones II to IV and thick coating formations in the cyclone V, kiln inlet and in the pre-burning zone. Moderate coating formation also takes place on the surface of the Mg-chrome brick lining in the burning zone.

2. Materials examined

2.1 Jamming in pre-heater cyclones

Volatile contents in the jamming materials collected from the cyclones in different occasions are recorded in **Table 1**. It is observed that both F^- and alkalis, particularly K_2O increase from the cyclone II to V, whereas, no such trend is marked in case of Cl^- and SO_3. Mineralogical analyses of the samples by X-Ray Diffractometry (XRD) and thermal analysis (DTA/TG) record the mineral spurrite normally both in cyclone IV and V materials and occasionally even in cyclone III. In addition anhydrite ($CaSO_4$) is found in cyclone IV material. Another significant feature is the almost complete dissociation of dolomite even in the

TABELLE 1: Anteil der flüchtigen Bestandteile in den Ansätzen (M.-%)

Probe	Zyklon	F^-	K_2O	Na_2O	Cl^-	SO_3
1	II	0,10	0,95	0,15	0,03	0,85
2	III	0,22	1,14	0,13	0,04	0,70
3	IV	0,20	1,07	0,16	0,05	0,60
4	V	0,29	3,56	0,15	0,02	0,65

TABLE 1: Volatile contents in the jamming materials (wt.%)

Sl. No.	Cyclone No.	F^-	K_2O	Na_2O	Cl^-	SO_3
1.	II	0.10	0.95	0.15	0.03	0.85
2.	III	0.22	1.14	0.13	0.04	0.70
3.	IV	0.20	1.07	0.16	0.05	0.60
4.	V	0.29	3.56	0.15	0.02	0.65

[1]) Es werden die in der Zementchemie üblichen Bezeichnungen angewandt.

*) Common cement chemistry notations are used.

TABELLE 2: Anteil der flüchtigen Bestandteile (M.-%) im Material aus Zyklon V (Normalbetrieb) und Zyklon III (verstopft)					
Probe	F$^-$	Cl$^-$	SO$_3$	Na$_2$O	K$_2$O
Zyklon III	0,19	0,02	1,9	0,16	0,55
Zyklon V	0,21	0,01	3,3	0,16	1,12

TABLE 2: Volatile contents (wt.%) in normal (Cyclone V) and jamming material — (Cyclone III)					
Sample	F$^-$	Cl$^-$	SO$_3$	Na$_2$O	K$_2$O
Cyclone III	0.19	0.02	1.9	0.16	0.55
Cyclone V	0.21	0.01	3.3	0.16	1.12

gendiffraktometrie (XRD) und Thermoanalyse (DTA/TG) weisen das Mineral Spurrit normalerweise sowohl im Mehl aus den Zyklonen IV und V und gelegentlich sogar aus Zyklon III auf. Zusätzlich tritt Anhydrit (CaSO$_4$) in Proben aus Zyklon IV auf. Ein weiteres bedeutsames Merkmal ist die fast vollständige Dissoziation von Dolomit sogar im gestauten Material des Zyklons II. Gelegentlich zeigen Mehle mit identischem F$^-$-Gehalt (siehe **Tabelle 2**) einen unterschiedlichen mineralogischen Aufbau und wirken sich folglich unterschiedlich auf den Ofenbetrieb aus. Die mineralogische Untersuchung (**Bild 1**) der beiden in Tabelle 2 aufgeführten Proben zeigt, daß Dolomit sowohl im Material aus Zyklon III als auch aus Zyklon V fehlt. Im Vergleich mit dem Rohmehl (**Bild 1a**) enthält das Material aus Zyklon III (**Bild 1b**) weniger Calcit, Quarz und Phlogopit, zeigt jedoch zusätzlich die Bildung von Spurrit, Periklas und freiem CaO. Im unter normalen Betriebsbedingungen entnommenem Material aus Zyklon V (**Bild 1c**) ist bei fast vollständiger Calcinierung nur sehr wenig Calcit vorhanden, wobei der größte Teil des später entstehenden CaO zu Ca(OH)$_2$ hydratisiert. Die Quarz- und Phlogopitmengen vermindern sich nicht. Während Periklas und CaSO$_4$ neu gebildet werden, fehlt Spurrit.

2.2 Ansätze

Die Zusammensetzung der flüchtigen Bestandteile in den Ansatzproben, die bei verschiedenen Betriebszuständen aus unterschiedlichen Bereichen des Drehofens entnommen wurden, ist in **Tabelle 3** aufgeführt. Es zeigt sich, daß sowohl F$^-$ als auch K$_2$O sich ganz besonders am Ofeneinlauf und im mittleren Bereich der Sinterzone akkumulieren. Alle Proben mit Ausnahme derer vom Ofeneinlauf und der, die dem Ofenauslauf am nächsten lag, enthalten Spurrit. Die Ansatzdicke in einer Entfernung von 33 m vom Ofenauslauf betrug etwa 43 cm. Aus der kombinierten DTA/TG-Analyse läßt sich ein Spurritgehalt dieser Probe von etwa 50 % abschätzen (siehe **Bild 2**). Das Mineral Sylvin (KCl) wurde in den Proben aus dem Ansatz der Sinterzone zwischen 11,5 m und 14,5 m nachgewiesen. Das Röntgen-Diffraktometriemuster (Cu-K$_\alpha$) der Probe an der Stelle 14,5 m zeigt sowohl Spurrit als auch KCl (**Bild 3**). Die Bildung harter, schieferartiger Ansätze in den Steigschächten deutet auf hohe Konzentrationen sowohl von SO$_3$ als auch von Cl$^-$ hin (Tabelle 3). In diesen schieferartigen Proben wurde sowohl CaSO$_4$ als auch Ca$_3$SiO$_4$Cl$_2$ nachgewiesen.

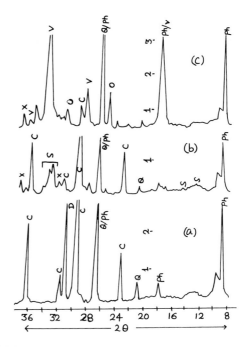

BILD 1: Röntgenmuster (Cu-K$_\alpha$)
FIGURE 1: XRD patterns (Cu-K$_\alpha$)

(a) Rohmehlmischung (a) the raw mix
(b) Material aus Zyklon III (b) Cyclone III materials
(c) Material aus Zyklon V (c) Cyclone V materials

Calcit/Calcite (C), Quarz/Quartz (Q), Phlogopit/Phlogopite (Ph), Dolomit/Dolomite (D), Spurrit/Spurrite (S), CaO (X), Ca(OH)$_2$(V), CaSO$_4$(O)

cyclone II jamming material. Occasionally materials having identical F$^-$ contents (**Table 2**) are found to show different mineralogical assemblages and consequently different effect on kiln operation. The mineralogical examination (**Fig. 1**) of the two samples reported in Table 2 shows that dolomite is absent both in the cyclone III and V materials. In comparison to the raw mix (**Fig. 1a**) the cyclone III material (**Fig. 1b**) contains reduced amount of calcite, quartz and phlogopite but shows additional formations of spurrite, periclase and free CaO. In the normal cyclone V material (**Fig. 1c**) calcite is very much reduced, showing almost complete calcination, wherein most of the free CaO generated later gets hydrated to Ca(OH)$_2$. Quartz and phlogopite do not show any reduction in quantity. While periclase and CaSO$_4$ are newly formed, spurrite is absent.

2.2 Coatings

Volatile contents in the coating samples collected from different parts of the rotary kiln in different occasions are shown in **Table 3**. It is found that both F$^-$ and K$_2$O show heavy accumulation in the kiln-inlet and in the middle part of the burning zone. Spurrite is recorded in all the samples except in the kiln inlet and in the sample nearest to the discharge end. The ring formation at 33 M from the discharge end was about 43 cm thick. From combined DTA/TG, the spurrite content of this sample is estimated to be about 50 % (**Fig. 2**). The mineral sylvite (KCl) was found to be present in the coating samples from 11.5 M to 14.5 M in the burning zone. The XRD pattern (Cu-Kα) of the sample from 14.5 M shows the presence of both spurrite and KCl (**Fig. 3**). Hard slaty coating formations in the riser ducts, show heavy concentration of both SO$_3$ and Cl$^-$ (Table 3). Both CaSO$_4$ and Ca$_3$SiO$_4$Cl$_2$ are found to be present in these slaty samples.

TABELLE 3: Anteil der flüchtigen Bestandteile in Ansatzproben (M.-%)					
Probe	Ort*	F$^-$	K$_2$O	Na$_2$O	Cl$^-$
1	(Ofeneinlauf)	0,36	6,1	0,47	0,03
2	36,6 m	0,17	2,32	0,16	0,04
3	33,0 m	0,22	1,62	0,16	0,02
4	14,5 m	0,38	3,90	0,27	0,25
5	8,0 m	0,26	3,30	0,23	0,25
6	3,8 m	0,21	1,76	0,19	0,02

* Abstand zum Ofenauslauf in m

3. Diskussion der Ergebnisse

Unter den mineralischen Übergangsphasen herrscht in allen Proben Spurrit vor. Dies beginnt normalerweise beim Zyklon IV und geht weiter bis zum Ansatz in der Sinterzone

TABELLE 4: Anteil der flüchtigen Bestandteile in Proben aus den Steigleitungen (M.-%)

Probe	Zyklon	F⁻	K₂O	Na₂O	Cl⁻	SO₃
1	II	0,15	1,56	0,38	0,60	16,8
2	III	0,25	0,43	0,41	1,10	9,3
3	IV	0,14	0,78	0,56	0,66	3,8

mit Ausnahme des Ofeneinlaufs. Gelegentlich wird auch über Spurrit in Zyklon III berichtet. Interessanterweise findet man, daß Materialien mit identischen flüchtigen Bestandteilen (siehe Tabelle 2) eine Ansatzbildung im Zyklon III, nicht jedoch im Zyklon V hervorrufen. Die mineralogische Untersuchung zeigt, daß bei Spurritbildung im Zyklon II im Zyklon V kein Spurrit zu finden war. Bekanntlich wird Spurrit bei Abwesenheit der Klinkerphasen nach der Reaktion

$$2\,SiO_2 + 5\,CaCO_3 \rightarrow Ca_5\,(SiO_4)_2\,CO_3 + 4\,CO_2$$

gebildet [5]. Aufgrund der verminderten Anteile silikathaltiger Phasen, wie Quarz und Phlogopit, war offensichtlich, daß diese Reaktion im untersuchten Zyklon III tatsächlich stattfand (Bild 1c). Andererseits legt das Vorhandensein hoher Anteile von Quarz, Phlogopit und freiem CaO, umgewandelt in Ca(OH)₂, im Zyklon V (Bild 1c) das Fehlen der Reaktion von CaO mit SiO₂ nahe. Daraus kann geschlossen werden, daß in dieser besonderen Situation die Temperatur im Zyklon III zumindest geringfügig höher war als diejenige, die normalerweise im Zyklon V herrscht, und daß die Temperaturdifferenz ausreichte, um in den beiden Zyklonen völlig unterschiedliche Verhältnisse zu schaffen. Dies wiederum legt den Schluß nahe, daß in Gegenwart von F⁻ die Ansätze im Vorwärmer sehr empfindlich gegenüber Temperaturschwankungen reagieren. Verdampfungsversuche bei 800–1000 °C an Zementrohmehlen, die natürliches bzw. zugegebenes F⁻ enthielten, haben gezeigt, daß die Verflüchtigung des natürlichen F⁻ deutlich langsamer abläuft als die des zugegebenen (**Tabelle 5**). Dies läßt vermuten, daß bei diesem Temperaturniveau eine größere Menge an F⁻ zur Verfügung steht, wenn es sich um Rohmehl mit natürlich vorhandenem F⁻ handelt, als wenn es sich um Rohmehl mit der gleichen Menge an zugegebenem F⁻ handelt. Daher ist es denkbar, daß bei gleicher F⁻-Menge und identischen Temperaturbedingungen ein Mehl mit natürlichem F⁻ eher zu Ansätzen neigt, als wenn das F⁻ der Rohmischung zugegeben wurde. Im untersuchten Rohmehl liegt das natürliche F⁻ in Form von Phlogopit (K₂O, 6MgO · Al₂O₃, 6 SiO₂ · 2H₂O) vor [7], und das zugegebene F⁻ rührt von CaF₂ her. Es ist ebenfalls von Bedeutung, daß Fluorsilikat ein effizienteres Flußmittel und Mineralisator ist als CaF₂ [10].

Die mineralogische Analyse einer Anzahl von Ansatzproben bestätigt, daß, obwohl Spurrit vorwiegend in der Aufwärm- und Übergangszone auftritt und dort die Bildung von Rin-

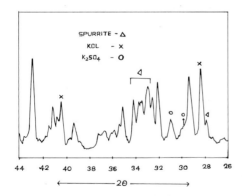

BILD 3: Röntgenmuster (Cu-Kα) von Ansätzen aus der Sinterzone (Probe 4 in Tabelle 3). Man erkennt Spurrit (S), Sylvit (K), K₂SO₄(H) zusammen mit weiteren Mg-Chrom-Stein- und -Zementphasen
FIGURE 3: XRD pattern (Cu-Kα) of coatings in the burning zone (Sample No. 4, Table 3), showing spurrite (S), Sylvite (K), K₂SO₄ (H) with other Mag-chrome brick and cement phases

TABLE 3: Volatile contents in coating samples (wt.%)

Sl. No.	Locations*	F⁻	K₂O	Na₂O	Cl⁻
1.	(Kiln inlet)	0.36	6.1	0.47	0.03
2.	36.6 M	0.17	2.32	0.16	0.04
3.	33.0 M	0.22	1.62	0.16	0.02
4.	14.5 M	0.38	3.90	0.27	0.25
5.	8.0 M	0.26	3.30	0.23	0.25
6.	3.8 M	0.21	1.76	0.19	0.02

* distance from the discharge end of the kiln in metres

TABLE 4: Volatile contents in the samples from riser ducts (wt.%)

Sl. No.	Cyclone No.	F⁻	K₂O	Na₂O	Cl⁻	SO₃
1.	II	0.15	1.56	0.38	0.60	16.8
2.	III	0.25	0.43	0.41	1.10	9.3
3.	IV	0.14	0.78	0.56	0.66	3.8

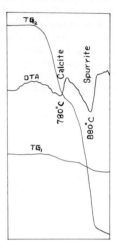

BILD 2: DTA/TG-Analyse von Ansätzen der Übergangszone (Probe 3 aus Tabelle 3), Spurrit-Gehalt 50%
FIGURE 2: DTA/TG of coatings in the pre-burning zone (Sample No. 3, Table 3), spurrite content 50%

3. Discussion

Amongst the transitional mineral phases, spurrite is found to be the most prevalent one in all the samples, normally starting from cyclone IV to the burning zone coating, except in the kiln inlet. Occasionally it is reported from the cyclone III material also. Interestingly it is found that materials with identical volatiles (Table 2) cause build-up in the cyclone III but not in cyclone V. Mineralogical examinations show that, while there was spurrite formation in cyclone III material, there was no spurrite in cyclone V material. In absence of clinker phases, spurrite is known to form through the reaction

$$2\,SiO_2 + 5\,CaCO_3 \rightarrow Ca_5\,(SiO_4)_2\,CO_3 + 4\,CO_2 \qquad [5]$$

That this type reaction took place in the examined cyclone III sample was evident from the reduced quantity of silica bearing phases, quartz and phlogopite (Fig. 1c). On the other hand presence of high amount of quartz, phlogopite and free CaO (converted to Ca (OH)₂) in the cyclone V material (Fig. 1c) suggests a lack of the reaction between CaO and SiO₂. From this it can be inferred that in the particular occasion, temperature in the cyclone III was at least slightly higher than that normally prevailing in the cyclone V and that temperature difference was enough to cause entirely different situation

TABLE 5: Volatilization of F^- from cement raw mixes, with inherent (In) and additional fluorine (Ad)
(Laboratory furnace, soaking time 1 hour at each temperature)

Raw Mix	Volatilization %		
	800 °C	900 °C	1000 °C
In	19	35	36
Ad	23	37	43

in the two cyclones. This suggests that in presence of F^-, material build-up in the pre-heater becomes very sensitive to temperature fluctuation. Examination of volatilization from cement raw mix in respect of inherent F^- and added F^- at the temperature range 800–1000 °C has shown that the volatilization of inherent F^- is noticeably lower than that of the added F^- (**Table 5**). It suggests that at this temperature level, a higher quantity of F^- is available in the raw mix with inherent F^- than in the raw mix with identical amount of added F^-. Thus it is conceivable that at the same F level and in identical temperature condition, raw mix with inherent F^- is more prone to build-up than the raw mix with added F^-. In the investigated raw mix inherent F^- occurs in the form of phlogopite (K_2O, $6MgO \cdot Al_2O_3$, $6SiO_2 \cdot 2H_2O$) [7] and the source of added F^- is CaF_2. It is also quite relevant that fluosilicate is reported to be a more effective fluxing and mineralizing agant than CaF_2 [10].

Mineralogical analysis of a number of coating samples establishes that although maximum occurrence of spurrite is confined to the pre-burning zone, causing ring formation occasionally as thick as 43 mm, it also occurs in the coating of the burning zone. Chlorine in the form of sylvite (KCl) is also associated with spurrite in coating formation in the burning zone. Both the minerals, with relatively low melting points might have been formed during cooling down of the kiln or in the inner layer of the coating.

In the cyclone II and sometimes in the cyclone III, where neither spurrite nor anhydrite is detectable, the presence of F^- and Cl^- might have caused the inception of some melt formation which is regarded essential for both spurrite and ring formation [5]. A similar case of pre-heater build-up in the 3rd stage cyclone was reported to be associated with as low as 0.034 % Cl^- [11]. SO_3, at least up to around 3 % may not cause any serious build-up problem as evident from the fact that cyclone V material (Table 2) with 3.3 % SO_3 did not cause such problem. However excessive SO_3 and Cl^- (**Table 4**) generated from the coal in the pyroclone may cause the slaty formation in the riser duct, wherein they occur as $CaSO_4$ and $Ca_3SiO_4Cl_2$ respectively. That the SO_3 and Cl^- in the riser ducts were caused due to high absorption of coal-ash originated from the pyroclone firing system was evident from accompanying high Al_2O_3 (12–20 %) and low CaO contents (32–51 %).

Alkali cycles are found to be not so significant in the kiln examined as evident from absence of regular coating formation in the cyclones. In presence of F^-, volatile cycles of alkalis and sulphates and consequently tendency for alkali/sulphate based deposits are known to be reduced [11]. Alkali may be only associated with F^- and Cl^- in moderate coating formation in the burning zone, where it occurs in the formation of sylvite (KCl).

4. Conclusion

During pyroprocessing of a cement raw mix with inherent F^-, jamming in the pre-heater and coating formation, both in the pre-burning and burning zones of the rotary kiln are mainly caused by F^- in association with Cl^-. In presence of F^-, the process of material build-up becomes very sensitive to temperature fluctuation.

[6] Goswami, G., Mohapatra, B. N., and Panda, J. D.: Phase formation during sintering of a magnesium and fluorine containing raw mix in a cement rotary kiln. Zement-Kalk-Gips 43 (1990) No. 5, pp. 253–256.

[7] Goswami, G., Mohapatra, B. N., and Panda, J. D.: Effect of fluosilicate on cement raw mix burnability and kiln build-up, Zement-Kalk-Gips 44 (1991) No. 12, pp. 634–637.

[8] Goswami, G., Mohapatra, B. N., and Panda, J. D.: Effect of fluorine-bearing limestone on clinker quality. 9th Int. Cong. Chem. of Cement, New Delhi, 1992, Vol. II, pp. 365–371.

[9] Bucchi, R.: Influence of the nature and preparation of raw materials on the reactivity of the raw mix. 9th Int. Cong. Chem. of Cement, Paris (1980), Vol. I P-1/33.

[10] Lea, F.M.: The chemistry of Cement and Concrete, Edward Arnold (Publishers) Ltd, London, Third edition (1970), p. 157.

[11] Kumar, K., and Irani, D. B.: Prevention of pre-heater build-up by optimization of operation – A case study, NCB Quest – 1988, May, P.

[12] Moir, G. K., and Glasser, F. P.: Mineralisers, Modifiers and activators in the clinkering process. 9th. Int. Cong. Chem. of Cem., New Delhi (1992) Vol. 1, pp. 125–152.

Umbau von F. L. Smidth/Fuller-Klinkerkühlern
F. L. Smidth/Fuller clinker cooler retrofit

Transformation de refroidisseurs de clinker F. L. Smidth/Fuller

Transformación de enfriadores de clínker F. L. Smidth/Fuller

Von **P. Green-Andersen**, Valby/Dänemark

Zusammenfassung – Die jüngsten Fortschritte bei Rostkühlern haben zu einer erheblich verbesserten Wärmerückgewinnung mit entsprechender Senkung des Brennstoffverbrauchs beim Klinkerbrennen geführt. Außerdem wurden die Außenabmessungen des Rostkühlers und somit die Baukosten verringert. Diese Verbesserungen wurden an der Konstruktion der Rostplatten und der Luftzufuhrsysteme vorgenommen, die alle Teil des neuen FLS-Hochleistungsrostkühlers sind. Der Kühlerrost hat zwei verschiedene Konstruktionsteile. Der erste Teil des Kühlers, die Rückgewinnungszone, ist mit Rosten für einen kontrollierten Mengenfluß (CFG – Controlled Flow Grates) ausgestattet. Der zweite Teil des Kühlers, die Nachkühlzone, enthält das Rostsystem mit reduzierter Durchlaßmenge (RFT – Reduced Fall Through Grate System). Es wird gezeigt, wie im Vergleich zu einem konventionellen Rostkühler der Kühler mit modifiziertem Rostsystem den Gesamtenergieverbrauch und die Kapitalkosten des gesamten Ofensystems senken kann.

Umbau von F. L. Smidth/Fuller-Klinkerkühlern

Summary – The recent improvements on grate coolers have led to a substantially better heat recuperation with a corresponding reduction of the fuel consumption of the clinker burning process. Furthermore the outer dimensions of the grate cooler are decreased and consequently the cost of civil work reduced. These improvements, which are all part of the new high efficiency FLS grate cooler, are in the design of the grate plates as well as the air supply system. The cooler grate has two different layouts. The first part of the cooler, the recuperation zone, has the Controlled Flow Grates (CFG). The last part of the cooler, the after cooling zone, has the Reduced Fall Through grate system (RFT). It is shown how a conventional grate cooler, compared to a cooler with a modified grate system, can reduce the total energy consumption and the capital cost for the entire kiln system.

F. L. Smidth/Fuller clinker cooler retrofit

Résumé – Les récentes améliorations opérées sur les refroidisseurs à grille ont entraîné une sensible amélioration de la récupération de chaleur avec, en même temps, diminution de la consommation de combustible durant la cuisson du clinker. De plus, les dimensions extérieures du refroidisseur à grille et, par là même, les frais de construction ont pu être réduits. Ces améliorations ont été effectuées au niveau de la conception des plaques de grille et des systèmes d'alimentation d'air, qui font tous partie du refroidisseur à grille à haut rendement FLS. La grille du refroidisseur présente deux éléments de construction différent l'un de l'autre. La première partie du refroidisseur, la zone de récupération de chaleur, est équipée de grilles en vue d'un flux de matière contrôlée (CFG – Controlled Flow Grates). Quant à la deuxième partie, la zone de refroidissement secondaire, elle contient le système des grilles à débit réduit (RFT – Reduced Fall Through Grate system). L'étude montre comment, par rapport comparativement à un refroidisseur conventionel, le refroidisseur avec système à grilles modifié permet de réduire la consommation globale d'énergie et les frais d'investissement de tout le système du four.

Transformation de refroidisseurs de clinker F. L. Smidth/Fuller

Resumen – Los recientes progresos logrados en los enfriadores de parrilla han permitido mejorar, de forma considerable, la recuperación de calor, con una correspondiente reducción del consumo de combustibles durante la cocción del clínker. Además, se han reducido las dimensiones exteriores del enfriador de parrilla y con ello los gastos de construcción. Las mencionadas mejoras han sido llevadas a cabo en las placas de la parrilla y en el sistema de conducción de aire, que son, todos ellos, elementos del nuevo enfriador de parrilla de alto rendimiento FLS. La parrilla del enfriador lleva en su construcción dos elementos diferentes. La primera parte del enfriador, o sea la zona de recuperación, va equipada con parrillas que permiten un flujo de material controlado (CFG – Controlled Flow Grates). La segunda parte del enfriador, o sea la zona de enfriamiento posterior, consta del sistema de parrillas con caudal reducido (RFT – Reduced Fall Through Grate System). Se desprende de este estudio cómo este enfriador, equipado con un sistema de parrillas modificado, puede reducir el consumo total de energía así como los gastos de inversión de todo el sistema de horno, en comparación con un enfriador de parrilla convencional.

Transformación de enfriadores de clínker F. L. Smidth/Fuller

1. Einleitung

In nur drei Wochen wurden drei Klinkerkühler mit Coolax-CFG-Kühlanlagen (Roste für einen kontrollierten Mengenfluß mit Luftregelung) nachgerüstet, einer Neuentwicklung der Firmen F. L. Smidth & Co. und Fuller. Jeder Kühler hat einen Durchsatz von 5000 t Klinker pro Tag.

Mechanisch ebenso wie betriebstechnisch gilt der Klinkerkühler als die komplexeste Einheit im Klinkerbrennprozeß. Demzufolge wurden in letzter Zeit verschiedene Typen von Klinkerkühlern entwickelt, in denen die Wärmerückgewinnung verbessert und der Kühlluftstrom und damit auch die Überschußluft deutlich reduziert wurden. Gleichzeitig wurde die Rostbelastung von etwa 40 auf etwa 50 Tonnen pro Quadratmeter und 24 Stunden bei gleicher Klinkertemperatur erhöht.

Nach mehrmonatigen Laborversuchen und Testläufen wurde die neue F. L. Smidth/Fuller-Coolax-CFG-Anlage innerhalb von nur drei Wochen an drei Werke ausgeliefert. Es wurden mit Gesamtenergieeinsparungen (Ofen und Kühler) von über 35 kcal pro kg Klinker äußerst zufriedenstellende Ergebnisse erzielt.

Die Nachrüstung mit dem Coolax-CFG-Rostkühler bietet die folgenden Verbesserungen:

1. Die Sekundärlufttemperatur wird deutlich erhöht.
2. Die Staubzirkulation zwischen Ofen, Calcinierer und Kühler wird deutlich reduziert.
3. Die Tertiärlufttemperatur bleibt infolge der fast vollständigen Elimination von Staub in der Absetzkammer unverändert.
4. Die Kühlerabluft wird vermindert.
5. Die Temperatur der Kühlerabluft bleibt gleich oder verringert sich.
6. Größere Verbesserungen beim Betrieb führen zu weniger Anpassungen des Kühlluftstroms und der Rostgeschwindigkeit, wodurch die Ausbildung von pilzförmigen Ansätzen und eines „Red River" vermieden werden.

2. Luftverteilungssystem

In den Umbauprojekten wurde der jeweils gesamte erste Rost durch das Coolax-CFG-System ersetzt. Es besteht aus 23 aktiven Reihen von Rostplatten, von denen 12 beweglich sind. Der Coolax-CFG-Umbau ist in fünf Sektoren unterteilt, die jeweils aus separaten Gebläsen mit Luft versorgt werden. Die ersten dreizehn Reihen Rostplatten sind in drei Sektoren und die nachfolgenden zehn Reihen Rostplatten in zwei Sektoren eingeteilt.

1. Introduction

Within only three weeks, three clinker coolers have been successfully retrofitted with the Coolax-CFG (Controlled Flow Grate) cooling system recently developed by F. L. Smidth & Co. and Fuller Company, each cooler handling 5000 tonnes of clinker per day.

Mechanically as well as operationally, the clinker cooler may be considered the most complex unit in the clinker burning process. As a result, different types of clinker coolers have recently been developed in which the heat recuperation has been improved and the cooling air flow and consequently the amount of excess air have been considerably reduced. At the same time, the grate load has typically been increased from approximately 40 to approximately 50 tonnes per square metre per 24 hours based on the same clinker temperature.

Following several months of laboratory testing and trial runs, the new F. L. Smidth / Fuller Coolax-CFG system was commissioned at three plants in only three weeks. Extremely satisfactory results have been achieved with total heat savings (kiln and cooler) of more than 35 kcal per kg clinker.

The Coolax-CFG grate cooler retrofit offers the following improvements:

1. Secondary air temperature is considerably increased.
2. Circulation of dust between kiln, calciner and cooler is considerably reduced.
3. Tertiary air temperature is unchanged due to almost complete elimination of dust in the settling chamber.
4. The amount of excess air is reduced.
5. The excess air temperature is unchanged or reduced.
6. Major improvements in operation result in fewer adjustments of the cooling air flow and grate speed, thereby avoiding "snowman formations" or "red rivers"

2. Air distribution system

In the retrofit projects the entire first grate has been replaced with the Coolax-CFG system, and there are 23 active rows of grate plates of which 12 are movable. The Coolax-CFG retrofit consists of five sectors each supplied with air from a separate fan. The first thirteen rows of grate plates are divided into three sectors, and the subsequent ten rows of grate plates are divided into two sectors.

All segments (thirteen in all) are supplied with cooling air via a separate duct with an adjustable damper. The first three fans in the Coolax-CFG system (1R, 1C and 1L) each supply cooling air to three segments. The remaining two fans (2R and 2L) each supply air to two segments (**Fig. 1.**). The air ducts continue to the bottom of the cooler from which air is

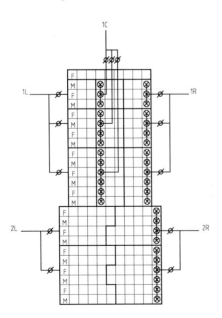

BILD 1: Erster Rost. Luftverteilung
FIGURE 1: Grate one. Air distribution system

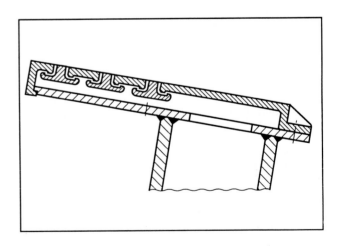

BILD 2: Rost mit Luftregelung (CFG)
FIGURE 2: Control flow Grate (CFG)

Alle Segmente (dreizehn insgesamt) werden über einen separaten Luftkanal mit einstellbarem Schieber mit Kühlluft versorgt. Die ersten drei Lüfter im Coolax-CFG-System (1R, 1C und 1L) liefern Kühlluft für jeweils drei Segmente. Die übrigen beiden Lüfter (2R und 2L) versorgen zwei Segmente (**Bild 1**). Die Luftkanäle setzen sich am Boden des Kühlers fort, von wo aus die Luft über vertikale, feste Verbindungsstücke auf die stationären Roste und über vertikale, bewegliche Verbindungen auf die beweglichen Rostplatten verteilt wird. Die von FLS verwendeten Kupplungen sind aus mechanisch beweglichen Verbindungen aufgebaut, die daher keiner schädigenden Biegebeanspruchung ausgesetzt sind. Die Verbindungselemente bestehen aus Stahlrohr mit Kugelkalotten an den Enden, die wie Kugelgelenke in die zylindrischen Außenstücke passen.

3. Rostplatten

Die CFG-Platte ist für höheren Strömungswiderstand ausgelegt. Zusammen mit einer effizienteren Steuerung der Kühlluft verbessert dies die Luftverteilung, eliminiert „Red River" und Nebenluft und gewährleistet eine bessere Wärmerückgewinnung (**Bild 2**). Das interne Strömungsmuster der Rostplatte stellt eine optimale Kühlung der Platte und eine ausgezeichnete Wärmebeständigkeit sicher. Die Hauptvorteile der CFG-Rostplatte sind:

— verbesserte Wärmerückgewinnung,
— ausgezeichnete Wärmebeständigkeit infolge besserer innerer Kühlung,
— verbesserte Verschleißfestigkeit und niedrige thermische Belastung,
— vermindertes Durchfallen von Klinker unter den Rost, bessere Verteilung der Kühlluft auf der Rostfläche infolge der Strömungsführung,
— niedriger Verstopfungsfaktor, keine Taschenbildung, selbstreinigend.

4. Regelstrategie

Die angewendete Regelstrategie ist mehr oder weniger die gleiche wie für herkömmliche Rostkühler. Eine Ausnahme bildet das Portlandwerk in Aalborg (Dänemark), in dem auch der neue CoolScanner installiert wurde.

5. Beschreibung

Diese Scannereinheit entspricht dem bekannten CemScanner auf PC-Basis, der den Kühler parallel zur Mittelachse abtastet. Eine vollständige Erfassung der Temperaturen auf der Oberseite der Klinkerschicht dauert 15 Sekunden. Der Scanner blickt durch eine Öffnung an der Oberseite des Kühlers. Außer während der Messung ist diese Öffnung durch eine Schwenkklappe verschlossen. Die Meßergebnisse des Scanners stehen auf einem PC im Kontrollraum zur Verfügung.

6. Betriebserfahrung

Die Scanneranlage liefert dem Bedienpersonal in Echtzeit Informationen über die Temperaturverteilung im Kühler. Dies ist ein genaueres Verfahren als die Verwendung der herkömmlichen Kühlerkamera und hilft bei der Optimierung des Kühlerbetriebes und der Vermeidung eines „Red River". **Bild 3** zeigt das Temperaturbild des ersten Klinkerbettrostes in Grad C.

7. Zeitplan des Umbaus

Unter der Voraussetzung einer sorgfältig geplanten Aufstellung läßt sich der Umbau des gesamten ersten Rostes während eines Ofenstillstandes von weniger als drei Wochen durchführen.

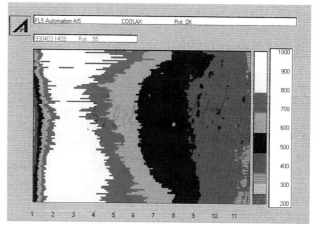

BILD 3: CoolScanner. Temperaturbild des ersten Klinkerbettrostes
FIGURE 3: CoolScanner. Image picture of clinker bed grate one

distributed to the stationary grates via vertical, fixed joints and to the movable grate plates via vertical, flexible joints. The flexible joint used by FLS is a mechanical flexible connection that has no parts which are exposed to destructive bending. It consists of a steel tube with spherical surfaces in the ends, which fits like a ball joint with cylindric outer parts.

3. Grate plates

The CFG plate is designed for larger flow resistance. This, combined with more efficient control of the cooling air, improves the air distribution, eliminates red rivers and blowing-through and ensures better heat recuperation (**Fig. 2**). The internal flow pattern in the grate plate ensures optimal cooling of the grate plate and excellent heat resistance. The main advantages of the CFG grate plate are:

— Improved heat recuperation.
— Excellent heat resistance due to better internal cooling.
— Improved wear resistance and low thermal load.
— Reduced falling through of clinker to the undergrate compartments.
— Better surface distribution of the cooling air due to the internal flow characteristics.
— Low clogging factor, no pockets, and self-cleaning.

4. Control strategy

The control strategy applied is more or less the same as for a traditional grate cooler, except the Aalborg Portland Plant in Denmark, where also the new CoolScanner was installed.

5. Description

The Scanner unit itself is the same scanner used in our well known PC based CemScanner, which in the cooler scans along a line parallel to the center line. A complete scan of the temperatures on top of the clinker layer takes 15 sec. The scanner looks through an opening in the roof of the cooler. Except for the measuring period this opening is closed by a swivel damper. The results from the scanner are available on a PC in the control room.

6. Operation experience

The Scanner system provides the operator with information on real-time temperature distribution in the cooler. This is a more precise method than using the traditional cooler camera and helps to optimise cooler operation and eliminate red rivers. **Fig. 3** (in degree Celsius), image a picture of the clinker bed grate one.

7. Retrofit time schedule

Provided the erection is carefully planned, retrofit of the entire first grate can be carried out in less than three weeks reckoned from kiln shutdown to kiln start-up.

Die Kohlenstaubbefeuerung von Zementdrehöfen mit Dosierrotorwaagen

Pulverized coal firing system for rotary cement kilns using rotor weighfeeders

La chauffe au poussier de charbon des fours à ciment rotatifs avec des bascules doseuses à rotor

La calefacción por carbón pulverizado de los hornos rotatorios de cemento, empleando básculas dosificadoras de rotor

Von **H. W. Häfner**, Augsburg/Deutschland

Zusammenfassung – Mehr als 300 Dosierrotorwaagen wurden in den letzten Jahren weltweit an die Zementindustrie ausgeliefert und erfolgreich in Betrieb genommen. Von den gelieferten Dosierrotorwaagen wurde die überwiegende Anzahl im Austausch gegen vorhandene Kohlenstaub-Dosiereinrichtungen eingesetzt. Die Dosierrotorwaage stellt meßtechnisch durch ihre hohe Kurzzeit-Dosiergenauigkeit und apparativ wegen ihres kompakten Aufbaus eine optimale Lösung zur gravimetrisch kontinuierlichen Dosierung von Kohlenstäuben entsprechend den Anforderungen des Zementbrennprozesses dar. Die Zusammenfassung aller für die Kohlenstaubfeuerung erforderlichen Funktionselemente wie Kohlenstaubaustrag aus dem Silo, Wägestrecke, Stellantrieb und Direktaufgabe in die Brennerblasleitung führen zu dem besonders einfachen Aufbau der Dosierrotorwaage. Das Meß-, Steuer- und Regelsystem, ergänzt durch eine Silokontrollmeßeinrichtung und eine automatische Kalibrierung, machen die Dosierrotorwaage zu einem wichtigen Stellglied des Zementbrennprozesses. Das international patentierte Dosierrotorwaagensystem entspricht den Anforderungen nach Wirtschaftlichkeit, Qualitätssicherung, Verfügbarkeit und Umweltschutz bei der Dosierung von Kohlenstaub, von Ersatzbrennstoffen und Rohmehl.

Die Kohlenstaubbefeuerung von Zementdrehöfen mit Dosierrotorwaagen

Summary – In the last few years more than 300 rotor weighfeeders have been supplied to the cement industry all over the world and brought successfully into operation. The great majority of the rotor weighfeeders supplied have been used as replacements for existing pulverized coal metering equipment. Both from the point of view of measurement technology (thanks to its short-term metering accuracy) and from the structural point of view (because of its compact design) the rotor weighfeeder represents the best possible solution for continuous gravimetric metering of pulverized coal to meet the requirements of the cement burning process. Combination of all the functional elements needed for a pulverized coal firing system, such as pulverized coal extraction from the hopper, weighing section, actuator, and direct feed into the burner air pipe, results in the particularly simple design of the rotor weighfeeder. The instrumentation and control system supplemented by hopper checking equipment and automatic calibration makes the rotor weighfeeder one of the most important controlling elements in the cement burning process. The internationally patented rotor weighfeeder meets the specifications for cost-effectiveness, quality assurance, availability and environmental protection for metering pulverized coal, as well as replacement fuels and raw meal.

Pulverized coal firing system for rotary cement kilns using rotor weighfeeders

Résumé – Plus de 300 bascules doseuses à rotor ont été livrées à l'industrie du ciment partout dans le monde ces dernières années et mises en service avec succès. Les bascules doseuses à rotor fournies ont été prépondéremment mises en place en substitution d'équipements existants de dosage de poussier de charbon. Du point de vue technique de mesure, la bascule doseuse à rotor constitue, grâce à sa haute précision de dosage à court temps et, du point de vue instrumental, par sa construction compacte, une solution optimale pour le dosage gravimétrique en continu des poussiers de charbon, correspondant aux exigences du process de cuisson du ciment. L'intégration de tous les éléments fonctionnels nécessaires à la chauffe au poussier de charbon, tels reprise du poussier de charbon au silo, circuit de pesage, entraînement de réglage et introduction directe dans le conduit de soufflage du brûleur conduit à la construction particulièrement simple de la bascule doseuse à rotor. Le système de mesure, de commande et de régulation, complété par un dispositif de mesure et contrôle du silo et un calibrage automatique, font de la bascule doseuse à rotor un élément important de réglage du process de cuisson du ciment. Le système de la bascule doseuse à rotor, breveté sur le plan international, correspond aux exigences d'économie, d'assurance de la qualité, de disponibilité et de protection de l'environnement dans le dosage de poussier de charbon ainsi que de combustibles de remplacement ou de farine crue.

La chauffe au poussier de charbon des fours à ciment rotatifs avec des bascules doseuses à rotor

Resumen – En los últimos años, se han suministrado más de 300 básculas dosificadoras de rotor a la industria cementera de todo el mundo, las cuales han sido puestas en servicio con éxito. La mayor parte de las básculas dosificadoras de rotor suministradas han sustituido a otras instalaciones dosificadoras de carbón pulverizado, ya existentes. Desde el punto de vista de la técnica de medición, la báscula dosificadora de rotor constituye una solución óptima para la dosificación gravimétrica continua de carbón pulverizado, debido a su alta precisión de dosificación a corto plazo, y también por su forma compacta, de acuerdo con las exigencias del proceso de cocción del cemento. La integración de todos los elementos necesarios para la preparación del carbón pulverizado, tales como la extracción del carbón pulverizado del silo, el tramo de pesaje, el accionamiento regulador y la alimentación directa al conducto de soplado del quemador, hacen que la forma constructiva de la báscula dosificadora de rotor sea particularmente sencilla. El sistema de medición, mando y regulación, completado con una instalación de control del silo, y un calibrado automático convierten la báscula dosificadora de rotor en un elemento regulador importante para el proceso de cocción del cemento. El sistema de báscula dosificadora de rotor, patentado a nivel internacional, cumple los requerimientos de rentabilidad, aseguramiento de la calidad, disponibilidad y protección del medio ambiente, en cuanto a la dosificación de carbón pulverizado, combustibles sustitutivos y harina cruda.

La calefacción por carbón pulverizado de los hornos rotatorios de cemento, empleando básculas dosificadoras de rotor

Einleitung

Forderungen nach Wirtschaftlichkeit, Qualitätssicherung, Verfügbarkeit und Umweltschutz geben der gravimetrischen kontinuierlichen Dosierung von Kohlenstaub, Ersatzbrennstoffen und Rohmehl einen besonderen Stellenwert beim Zementbrennprozeß (**Bild 1**).

Der moderne Zementbrennprozeß verlangt gravimetrisch arbeitende Stellglieder, bei denen Schüttgutaustrag, Stellantrieb und direkte Schüttgutübergabe in den Brennprozeß in einem baulich einfachen geschlossenen Dosiergerät zusammengefaßt sind. Hohe Kurzzeitgenauigkeit, permanent gravimetrische Betriebsweise, verzögerungsfreies Folgen der Prozeßführungsgröße, Funktionsüberwachung und hohe Verfügbarkeit sind Forderungen, die die nachfolgend dargestellte Dosierrotorwaage erfüllt.

Differentialdosierwaagen, Prallplatten oder Coriolis-Schüttgutdosiereinrichtungen haben im Gegensatz zu Dosierrotorwaagen eine nachführende Regelung und benötigen u. a. zusätzliche Dosiergeräte und Schneckenpumpen.

Die durch die nachführende Regelung bedingte unzureichende Kurzzeitgenauigkeit zusammen mit der durch die nachgeschaltete Schneckenpumpe entstehende Totzeit kann zu zufälligen oder pulsierenden Schüttstromschwankungen und damit zu Betriebsstörungen führen. Die gestiegenen Anforderungen an Wirtschaftlichkeit, Verfügbarkeit, Qualitätssicherung und Umweltschutz aus dem Zementbrennprozeß haben in den letzten Jahren weltweit zum erfolgreichen Einsatz von mehr als 300 Dosierrotorwaagen geführt.

Die Weiterentwicklung der Dosierrotorwaage zu einem System gravimetrischer Rotorstellglieder für den Zement-

Introduction

Requirements for cost-effectiveness, quality assurance, availability and environmental protection give the continuous gravimetric metering of pulverised coal, replacement fuels and raw meal a special status in the cement burning process (**Fig. 1**).

The modern cement burning process requires gravimetric control elements in which extraction of bulk material, actuation and direct transfer of bulk material to the combustion process are combined in a structurally simple, closed metering unit. Short-term metering accuracy, a permanent gravimetric function, lag-free tracking of the process control variable, monitoring and availability are requirements which are met by the rotor weighfeeder described below.

In contrast with rotor weighfeeders, differential weighfeeders, deflectors or Coriolis metering equipment have a follow-up control system and, among other things, require additional metering devices and screw pumps.

Inadequate short-term accuracy caused by the follow-up control, together with the dead-time caused by the downstream screw pump, may lead to random or pulsating fluctuation in the bulk flow and thus to malfunctioning. The increased specifications for cost-effectiveness, availability, quality assurance and environmental protection in the cement burning process have led to the successful use of more than 300 rotor weighfeeders worldwide in recent years.

Further development of the rotor weighfeeder into a system of gravimetric rotor control elements for the cement burning process has been carried out consistently, so that the advantages of this method can be extended to rotor weighfeeders for bulk materials such as fly ash, sewage sludge, additives and raw meal (**Fig. 2**).

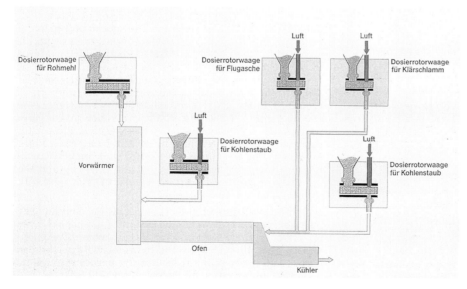

BILD 1: Dosierrotorwaagen als gravimetrische Stellglieder am Zementdrehrohrofen
FIGURE 1: Rotor weighfeeders as gravimetric controlling elements in rotary cement kilns

Luft	= air
Dosierrotorwaage	= rotor weighfeeder
für Rohmehl	= for raw meal
für Flugasche	= for fly ash
für Klärschlamm	= for sewage sludge
für Kohlenstaub	= for pulverised coal
Vorwärmer	= preheater
Ofen	= kiln
Kühler	= cooler

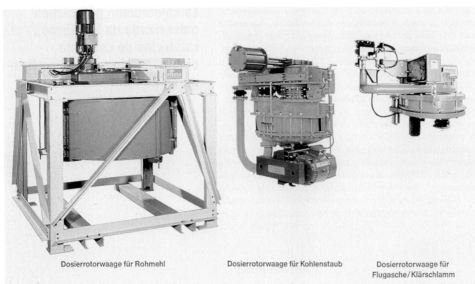

BILD 2: Dosierrotorwaagenbaureihe für den Zementdrehrohrofen
FIGURE 2: Rotor weighfeeder series for rotary cement kilns

Dosierrotorwaage	= rotor weighfeeder
für Rohmehl	= for raw meal
für Flugasche/	= for fly ash/
Klärschlamm	= sewage sludge
für Kohlenstaub	for pulverised coal

brennprozeß wurde konsequent betrieben, um die Vorteile dieser Technik auch auf Dosierrotorwaagen für Schüttgüter wie Flugasche, Klärschlamm, Additive und Rohmehl zu übertragen (**Bild 2**).

Funktion

Die Dosierrotorwaage benutzt das Wirkprinzip einer gravimetrisch arbeitenden horizontal liegenden Durchblasschleuse (**Bild 3**). Das Schüttgut wird durch den Rotor (Zellenrad) direkt aus dem Silo abgezogen, über die Wägestrecke geleitet, in die pneumatische Förderleitung dosiert und mit der vom Gebläse gelieferten Trägerluft direkt in den Prozeß gefördert.

Die Wägeachse A-A läuft quer durch die Verbindungskompensatoren zwischen Schüttgutaustrag, pneumatischer Förderleitung und Rotor. In der Wägeeinrichtung B wird die momentan in der Rotorwägestrecke wirkende Schüttgutmasse erfaßt. Die Wägeelektronik speichert die Schüttgutmasse bis kurz vor die Übergabe des Schüttgutes in die pneumatische Förderleitung. Entspechend dem vorgegebenen Förderstärken-Sollwert und der gespeicherten Schüttgutmasse wird die erforderliche Rotorwinkelgeschwindigkeit errechnet und durch den Rotorantrieb eingestellt. Durch dieses Prinzip wird eine Störgrößenkompensation der Dosierrotorwaage mit dem Ergebnis hoher Kurzzeitgenauigkeit realisiert. Das in **Bild 4** links dargestellte Diagramm stellt die in der Rotorwägestrecke wirkende Schüttgutmomentanlast (m/a) und zeitversetzt die umgekehrt proportional eingestellte Rotorwinkelgeschwindigkeit (a/t) neben der Ist-Dosierleistung (t/h) dar.

Signifikant für Kohlenstaub ist das sich fortlaufend ändernde Schüttgewicht, das in der Rotorwägestrecke als Schüttgutmomentanlast erfaßt wird.

Die Dosierrotorwaage kann durch die Wirkung ihrer Störgrößenkompensation auch bei größeren Schüttgewichtsschwankungen des Kohlenstaubes eine äußerst gleichmäßige Beschickung des Brenners gewährleisten.

Die Beurteilung der Dosiergüte einer Dosierrotorwaage kann letztlich nur über die Gasanalysemessung am Drehrohrofen zuverlässig erfolgen.

Das in **Bild 4** rechts dargestellte Diagramm zeigt die Entwicklung des CO-Gehaltes in Abhängigkeit vom O_2-Gehalt des Drehrohrofenabgases bei Einsatz

a) eines Schüttstrommessers mit geregelter Aufgabezellenradschleuse und nachführender Regelung,

b) einer Dosierrotorwaage mit Störgrößenkompensation.

Dosierrotorwaage

Durch die Zusammenfassung von Schüttgutaustrag, Meßstrecke, Stellantrieb, pneumatische Förderung in einem

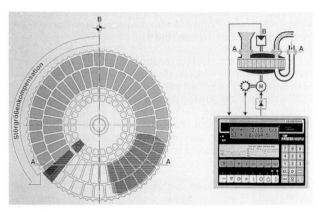

BILD 3: Regelschema der Dosierrotorwaage mit Störgrößenkompensation
FIGURE 3: Diagram of rotor weighfeeder control with compensation for disturbance variables

Störgrößenkompensation = compensation for disturbance variables

Operation

The rotor weighfeeder works on the principle of a horizontal pneumatic air-lock valve with a gravimetric action (**Fig. 3**). The bulk material is extracted directly from the silo by the rotor (star feeder), passed over the weighing section, fed into the pneumatic delivery line and transported directly to the process by the carrier air supplied by the fan.

The weighing axis A-A extends transversely through the flexible joints between the bulk material extraction, pneumatic delivery line and rotor. The mass of bulk material instantaneously operative in the rotor weighing section is recorded in the weigher B. The electronic weighing system stores the mass until shortly before the material is transferred to the pneumatic delivery line. The necessary angular speed of the rotor is calculated and set by the rotor drive, according to the predetermined setpoint conveying value and the stored mass. Disturbance variables in the rotor weighfeeder are compensated on this principle and greater short-term accuracy results. The left-hand diagram in **Fig. 4** represents the instantaneous load of bulk material (m/a) active in the rotor weighing section and, offset in time, the inversely proportional set rotor angular velocity (a/t), with the actual metered output (t/h).

A significant feature of pulverised coal is the continuously changing bulk density recorded as the instantaneous load in the rotor weighing section. Even if the bulk density of pulverised coal fluctuates greatly, the rotor weighfeeder can ensure extremely uniform feed to the burner through its action in compensating for disturbance variables.

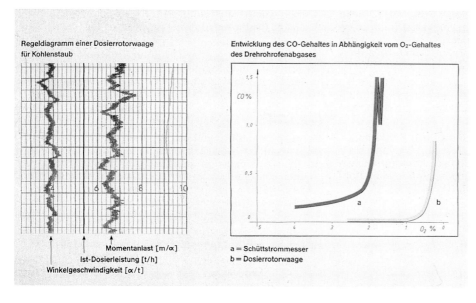

BILD 4: Diagramme einer Dosierrotorwaage für Kohlenstaub
FIGURE 4: Diagrams for a rotor weighfeeder for pulverised coal

Regeldiagramm einer Dosierrotorwaage für Kohlenstaub	= Control diagram for a rotor weighfeeder for pulverised coal
Momentanlast	= instantaneous load
Ist-Dosierleistung	= actual metered output
Winkelgeschwindigkeit	= angular velocity
Entwicklung des CO-Gehaltes in Abhängigkeit vom O₂-Gehaltes des Drehrohrofenabgases	= Development of CO content versus O₂ content of exhaust gas from rotary kiln
Schüttstrommesser	= bulk flow meter
Dosierrotorwaage	= rotor weighfeeder

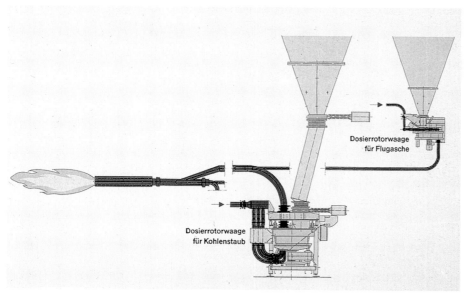

BILD 5: Kohlenstaubfeuerung mit gravimetrisch dosierter Beigabe von Flugasche
FIGURE 5: Pulverised coal firing with gravimetrically metered addition of fly ash

Dosierrotorwaage	= rotor weighfeeder
für Flugasche	= for fly ash
für Kohlenstaub	= for pulverised coal

Gerät und den Direktanschluß an das Schüttgutsilo ergibt sich der klare, einfache Aufbau dieses geschlossenen gravimetrischen Dosiergerätes (**Bild 5**). Modularer Aufbau und einfacher Service sind weitere Kennzeichen des weltweit patentierten Dosierrotorwaagensystems. Besonderer Wert wurde auf die mehrjährige Standzeit der austauschbaren Verschleißteile gelegt. Für Kohlenstaub ist die Dosierrotorwaage druckstoß- und flammendurchschlagsicher ausgeführt.

Das Schüttgut wird durch die Trägerluft mit Strömungsgeschwindigkeiten zwischen 20 – 40 m/sek. aus dem feinteiligen Rotor in die Brennerblasleitung ausgetragen, wobei sowohl sehr niedrige Dosierleistungen pulsationsfrei dosiert als auch Förderstrecken von über 200 m zum Brenner realisierbar sind. Dosierrotorwaagen für die Kohlenstaubfeuerung bilden zusammen mit dem Kohlenstaubtransport zum Brenner eine sorgfältig aufeinander abgestimmte Einheit.

Bei der gravimetrischen Dosierung von Flugaschen, getrockneten Klärschlämmen u.a. zusammen mit Kohlenstaub in den Drehrohrofenbrenner zeichnet sich die Dosierrotorwaage durch ihre Multifunktionen aus. Die kontinuierliche Brennstoffmischung und die Mehrfacheinspeisung in die gemeinsame Brennerblasleitung stehen hierbei im Vordergrund.

Schlußbetrachtung

Der klare und einfache Aufbau, die Störgrößenkompensation mit dem Ergebnis hoher Kurzzeitgenauigkeit der

Ultimately, the quality of the metering by a rotor weighfeeder can only be assessed reliably by gas analysis at the rotary kiln.

The right-hand diagram in **Fig. 4** shows the development of the CO content versus the O₂ content of the exhaust gas from the kiln

a) with the use of a bulk flow meter with a controlled rotary vane feed lock and a follow-up control and

b) with the use of a rotor weighfeeder with compensation for disturbance variables.

The rotor weighfeeder

The combination of bulk material extraction, a measuring section, an actuator, pneumatic conveying in one unit and the direct connection to the bulk material silo give this closed, gravimetric equipment its clear, simple design (**Fig. 5**). Modular construction and simple servicing are further features of the rotor weighfeeder system, which is patented worldwide. Another important feature is that parts which are subject to wear and exchangeable have a service life of several years. For applications with pulverised coal the weighfeeder is made proof to pressure shock and flame penetration. The bulk material is transported by the carrier air from the finely divided rotor into the burner air pipe at flow speeds between 20 and 40 m/sec; very small quantities can be metered without pulsation, and conveying distances of over 200 m to the burner can be obtained. Rotor weighfeeders for firing pulverised coal are combined into

Dosierung, Unempfindlichkeit gegen wechselnde Drücke und Temperaturen, einfache Wartung und eine Kalibrierautomatik mit Funktionsüberwachung runden die Dosierrotorwaagen zu wichtigen Stellgliedern im Zementbrennprozeß ab.

Neben der Kohlenstaubdosierung sind auch für Flugasche, Klärschlamm und Rohmehl Dosierrotorwaagen als gravimetrisch arbeitende Stellglieder am Zementdrehrohrofen verfügbar (Bild 1 und 2).

Mit der Dosierrotorwaage wird den heutigen Forderungen nach Wirtschaftlichkeit, Qualitätssicherung, Verfügbarkeit und Umweltschutz bei der gravimetrisch kontinuierlichen Aufgabe von Kohlenstaub, Ersatzbrennstoffen und Rohmehl in den Zementbrennprozeß voll entsprochen.

carefully coordinated units with the means for conveying the coal to the burner.

The rotor weighfeeder is distinctive in its versatility in gravimetric metering of fly ash, dried sewage sludge and the like into the kiln burner together with pulverised coal. Here the most important features are continuous blending of the fuel and multiple feeding into the common burner air pipe.

Conclusions

The clear, simple design, compensation for disturbance variables, resultant high short-term metering accuracy, insensitivity to changes in pressure and temperature, easy maintenance and automatic calibration with monitoring make the rotor weighfeeder one of the most important control elements in the cement burning process.

Appropriate models for fly ash, sewage sludge and raw meal as well as for pulverised coal are available as gravimetric control elements at rotary cement kilns (Figs. 1 and 2).

The rotor weighfeeder meets all present-day specifications for cost-effectiveness, quality assurance, availability and environmental protection in continuous gravimetric feeding of pulverised coal, replacement fuels and raw meal to the cement burning process.

Verschleißuntersuchungen an Magnesia-Spinell-Steinen für den Zementdrehrohrofen

Investigations into the wear on magnesia-spinel bricks for rotary cement kilns

Essais d'usure sur briques magnésie-spinelle pour le four à ciment rotatif

Estudios del desgaste en los ladrillos de espinela magnésica para hornos rotatorios de cemento

Von **J. Hartenstein, U. Bongers, R. Prange** und **J. Stradtmann,** Wülfrath/Deutschland

Zusammenfassung – Um die Standzeit feuerfester Werkstoffe auch bei stärkerer Thermoschockbeanspruchung zu verlängern, wurde die Entwicklung von spinellhaltigen Steinen ohne Zusatz chromhaltiger Rohstoffe weiter fortgeführt. In der Zwischenzeit liegen vielfältige Erfahrungen aus dem industriellen Einsatz von Spinellsteinsorten der zweiten und dritten Generation in der Übergangs- und Sinterzone von Zementöfen vor. Die Betriebsbedingungen durch Alkali- und Sulfatbelastung unterscheiden sich teilweise beträchtlich. Die Veränderungen des Ofenfutters während der Ofenreise wurden mit Hilfe von chemischen, mineralogischen und mikroanalytischen Verfahren an gebrauchten Steinen untersucht. Daraus wurden Schlußfolgerungen über den Verschleißmechanismus gezogen. Die gewonnenen Erfahrungen führten zur Entwicklung und Erprobung eines neuartigen reinen Magnesiasteines mit Zirkondioxidzusatz, der eine verbesserte Beständigkeit gegen Thermoschock und Alkaliangriff aufweist und für den thermisch hochbeanspruchten Bereich des Zementofens vorgesehen ist.

Verschleißuntersuchungen an Magnesia-Spinell-Steinen für den Zementdrehrohrofen

Summary – The development of bricks containing spinel without the addition of raw materials containing chromium has been continued so that the service life of refractory materials can be extended, even under fairly severe thermal shock conditions. There is now a great deal of experience available from the industrial use of spinel bricks of the second and third generation in the transition and sintering zones of cement kilns. In some cases there are considerable differences in the operating conditions caused by alkali and sulphate attack. The changes in the kiln lining during the kiln campaign were investigated using chemical, mineralogical and micro-analytical methods on used bricks, and conclusions were drawn about the wear mechanism. The experience gained led to the development and testing of a new type of pure magnesia brick with added zirconium dioxide. This has an improved resistance to thermal shock and alkali attack and is intended for the zones of high thermal stress in cement kilns.

Investigations into the wear on magnesia-spinel bricks for rotary cement kilns

Résumé – Le développement de briques spinelle sans addition de matières premières contenant du chrome a été poursuivi, afin de prolonger la durée de vie des matériaux réfractaires, aussi lors d'une plus forte exposition aux chocs thermiques. Entre temps, ont été accumulées des expériences nombreuses dans l'utilisation industrielle de sortes de briques spinelle de la seconde et troisième génération, dans les zones de transition et de clinkérisation des fours à ciment. Les conditions de fonctionnement sous attaque alcali et sulfate diffèrent parfois énormément. Les altérations de la maçonnerie du four, au cours de la campagne du four, ont été étudiées à l'aide de procédés chimiques, minéralogiques et microanalytiques sur des briques usées. Ceci a permis de tirer des conclusions quant au mécanisme d'usure. L'expérience ainsi acquise a conduit au développement et à la mise à l'épreuve d'une brique pure de type nouveau, avec addition d'oxyde de zirkone, qui présente une résistance améliorée contre les chocs thermiques et l'attaque alcali et qui est destiné aux zones hautement sollicités thermiquement du four rotatif.

Essais d'usure sur briques magnésie-spinelle pour le four à ciment rotatif

Resumen – Con el fin de prolongar la duración de los materiales refractarios sometidos a grandes esfuerzos por choques térmicos, se han seguido desarrollando los ladrillos de espinela, sin adición de materias primas que contengan cromo. Entretanto, se cuenta con múltiples experiencias adquiridas con el empleo industrial de diferentes tipos de ladrillos de espinela, de la segunda y tercera generación, en las zonas de transición y de sinterización de hornos para cemento. Las condiciones de servicio son, en parte, muy dinstintas, debido al ataque por álcalis y sulfatos. Los cambios producidos en el forro del horno durante el tiempo de marcha del mismo han sido estudiados con ayuda de procedimientos químicos, mineralógicos y microanalíticos, efectuados en ladrillos usados. Luego se han sacado las correspondientes conclusiones respecto del mecanismo de desgaste. Las experiencias adquiridas han conducido al desarrollo y ensayo de un nuevo ladrillo de magnesia pura, con adición de dióxido de zirconio, el cual presenta una mejor resistencia a los choques térmicos y a los ataques por álcalis y que está destinado a la zona del horno rotatorio sometida a grandes esfuerzos térmicos.

Estudios del desgaste en los ladrillos de espinela magnésica para hornos rotatorios de cemento

1. Einleitung

Seit ca. 60 Jahren werden Magnesia- und Magnesia-Chrom-Steine in der Brennzone des Zementdrehrohrofens eingesetzt. Aufgrund von Verschleißuntersuchungen wurden die verwendeten Steinsorten weiterentwickelt und in ihren Eigenschaften optimiert [1–7]. In der zweiten Hälfte der 70er Jahre kam es mit erneuter Verwendung von Kohle als Brennstoff vor allem in Zementdrehrohröfen hoher spezifischer Leistung zu unbefriedigenden Haltbarkeiten des feuerfesten Futters in den ansatzfreien und ansatzinstabilen Bereichen der Sinterzone. Außerdem entstehen bei der Verwendung von Magnesia-Chrom-Steinen durch Reaktion mit Alkalien umweltschädliche Chromatverbindungen, deshalb wurden chromfreie feuerfeste Produkte entwickelt [8–12].

In den 70er Jahren wurden in Japan Magnesia-Spinell-Steine entwickelt und in Zementdrehrohröfen eingesetzt, wobei eine 1,5- bis 2fache Haltbarkeitssteigerung im Vergleich zu direktgebundenen Magnesia-Chrom-Steinen erzielt wurde. Magnesia-Spinell-Steine haben sich seitdem im Zementdrehrohrofen bewährt und werden von verschiedenen Herstellern weltweit angeboten. Sie haben die Verfügbarkeit der Ofenanlagen erhöht und den spezifischen Feuerfestverbrauch beachtlich verringert. Vorzugsweise werden sie in den Randbereichen der Sinterzone, in einigen Öfen auch in der Sinterzone eingesetzt.

2. Produkte

Die Produktpalette der Spinellsteine ist umfangreich, es werden Steine mit Spinellgehalten von 5 bis 25 % in Kombination mit Magnesia-Sintern unterschiedlicher Fe_2O_3-Gehalte angeboten. Der Spinell-Anteil kann entweder aus reinem synthetischen MA-Spinell-Sinter bestehen oder durch Reaktion einer Tonerde-Zugabe beim Steinbrand erzeugt werden. „Low-Cost-Produkte" mit geringem Spinellgehalt enthalten eisenreiche Magnesia-Sinter, „High-Performance-Produkte" dagegen hohe Anteile synthetischen MA-Spinells und sehr reine Magnesia-Sinter. Bisher wurden die Produkte ausschließlich nach Entwicklungsgenerationen unterschieden, wobei Steine mit „Reaktions-Spinell" zur 1. Generation gehören, solche mit synthetischen

1. Introduction

Magnesia and magnesia-chrome bricks have been used in the burning zone of rotary cement kilns for about 60 years. Based on investigations into wear, the bricks used were further developed and their characteristics were optimised [1–7]. In the second half of the seventies, with the renewed use of coal as fuel the durability of the refractory lining in the coating-free areas and areas with an unstable coating of the sintering zone became unsatisfactory, especially in rotary cement kilns of high specific output. Moreover, when using magnesia-chrome bricks reaction with alkalis gives rise to ecologically harmful chromate compounds; therefore chromium-free refractory products were developed [8–12].

In the seventies, magnesia-spinel bricks were developed in Japan and used in rotary cement kilns, thereby achieving a 1.5- to 2-fold increase in durability compared with direct-bonded magnesia-chrome bricks. Since then magnesia-spinel bricks have proved their worth in rotary cement kilns and are marketed all over the world by various manufacturers. They have increased kiln availability and reduced the specific consumption of refractories appreciably. They are chiefly used in the marginal areas of the sintering zone, and in some kilns in the sintering zone itself.

2. Products

The product range of spinel bricks is extensive; bricks with spinel contents ranging from 5 to 25% in combination with magnesia sinters with various Fe_2O_3 contents are available. The spinel content may consist either of pure synthetic MA-spinel sinter or be produced during brick-burning by the reaction of an alumina additive. "Low-cost" products with a low spinel content contain iron-rich magnesia sinter; "high-performance" products, on the other hand, contain high proportions of synthetic MA-spinel and highly pure magnesia sinter. Hitherto, the products have been distinguished solely

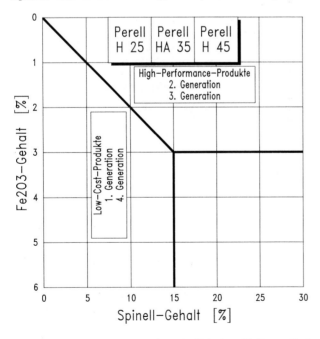

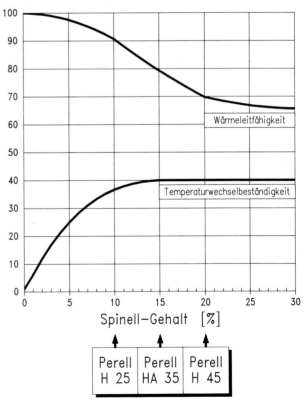

BILD 1: Einteilung von Magnesia-Spinell-Steinen für Zementdrehrohröfen
FIGURE 1: Classification of magnesia-spinel bricks for rotary cement kilns

Fe_2O_3-Gehalt = Fe_2O_3 content
Spinell-Gehalt = Spinel content
Low-Cost-Produkte = Low-cost products
High-Performance-Produkte = High-performance products
1./2./3./4. Generation = 1st/2nd/3rd/4th generation

BILD 2: Eigenschaften Magnesia-Spinell-Steine Wülfrather Qualitäten
FIGURE 2: Properties of magnesia-spinel bricks products from Wülfrath

Wärmeleitfähigkeit = Thermal conductivity
Temperaturwechselbeständigkeit = Thermal shock resistance
Spinell-Gehalt = Spinel content

schem Spinell-Sinter zur 2. Generation. Durch gezielte Veränderungen des Mikrogefüges entstanden hieraus die 3. und 4. Generation (**Bild 1**) [13].

3. Eigenschaften

Die thermo-mechanischen Eigenschaften von Magnesia-Spinell-Steinen werden wesentlich durch den Spinellgehalt bestimmt. Die Temperaturwechselbeständigkeit wird bis zu 15% Spinellanteil verbessert (**Bild 2**). Die Wärmeleitfähigkeit verringert sich und erreicht bei 30% Spinellanteil fast das Niveau des reinen Spinells. Die Reaktionsbeständigkeit wird mit steigendem Spinellgehalt verschlechtert, außerdem nehmen die Rohstoffkosten proportional zum Spinellgehalt zu. Die beste Kombination der genannten Eigenschaften erfordert einen Spinellgehalt von 13 bis 17%. Der Perell HA 35 wurde darauf abgestimmt, zusätzlich wurde das Mikrogefüge modifiziert (3. Generation), um mit möglichst geringem Spinellgehalt eine hohe Reaktions- und Temperaturwechselbeständigkeit sowie eine geringe Wärmeleitfähigkeit zu erhalten.

4. Verschleiß

Die Ursache für den Verschleiß feuerfester Steine im Zementdrehrohrofen läßt sich auf mechanische, thermische und chemische Belastungen zurückführen. Die in der Praxis festgestellte Häufigkeitsverteilung der verschiedenen Verschleißarten zeigt, daß Magnesia-Spinell-Steine oft bei der Kombination thermo-chemischer Beanspruchung und reaktiven Infiltrationen überfordert werden (**Bild 3**).

Die thermo-chemischen Verschleißfälle können auftreten, wenn Magnesia-Spinell-Steine in der Sinterzone eingesetzt werden. Hier kommt es dann bei hohen Temperaturen zur direkten Reaktion des Spinells mit freiem CaO. Neben Festkörperreaktionen erfolgen auch Schmelzbildungen, die eine Korrosion des Spinells und eine Rekristallisation der Magnesiamatrix verursachen. Die Haltbarkeit der Magnesia-Spinell-Steine ist in diesem Fall gering (**Bild 4**) [14].

Die Mehrzahl der Verschleißfälle werden durch Salzinfiltrationen verursacht. Aufgrund des Temperaturgefälles in der feuerfesten Ausmauerung und der porösen Struktur sind Kondensationen von Salzen in bestimmten Steinzonen unvermeidlich. Solange das molare Alkali-/Sulfat-Verhältnis nahe 1 liegt, finden keine chemischen Reaktionen mit Steinmaterial statt (inerte Infiltration). Es kommt hauptsächlich zu einer Füllung des Porenvolumens und damit zur Veränderung von Wärmeleitfähigkeit und Temperaturwechselbeständigkeit. Die Haltbarkeit der Magnesia-Spinell-Steine wird zwar verringert, ist aber besser als bei thermo-chemischer Belastung (**Bilder 4 und 5**).

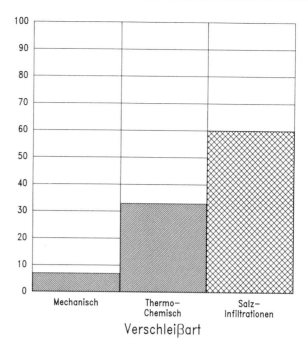

BILD 3: Verteilung der Verschleißarten von Magnesia-Spinell-Steinen in Zementdrehrohröfen
FIGURE 3: Distribution of the types of wear of magnesia-spinel bricks in rotary cement kilns

Verschleißart	= Type of wear
Mechanisch	= Mechanical
Thermo-Chemisch	= Thermochemical
Salz-Infiltrationen	= Salt infiltration

according to their development generation, bricks with "reaction-formed spinel" belonging to the first generation and bricks containing synthetic spinel sinter to the second generation. These were used to produce the third and fourth generations by deliberate modifications to the microstructure (**Fig. 1**) [13].

3. Properties

The thermomechanical properties of magnesia-spinel bricks are essentially a function of their spinel content. Thermal shock resistance is improved up to a spinel content of 15% (**Fig. 2**). Thermal conductivity falls and at a spinel content of 30% almost reaches the level of pure spinel. The resistance to reaction deteriorates as the spinel content increases; in addition, the raw material costs increase in proportion to the spinel content. The best combination of the stated properties requires a spinel content of 13–17%. The Perell HA 35 brick was geared to this figure; the microstructure was also modified (third generation) so as to achieve high resistance to reaction and thermal shock as well as low thermal conductivity with the lowest possible spinel content.

4. Wear

The cause of wear of refractory bricks in rotary cement kilns can be attributed to mechanical and thermal stresses and chemical attack. The frequency distribution of the various kinds of wear which is determined in practice shows that magnesia-spinel bricks are often overtaxed by the combination of thermochemical attack and reactive infiltration (**Fig. 3**).

Instances of thermochemical wear may occur when magnesia-spinel bricks are used in the sintering zone. Here, at high temperatures the spinel reacts directly with free CaO. Besides solid-state reactions, melts form which cause corrosion of the spinel and recrystallisation of the magnesia matrix. In such a case the durability of magnesia-spinel bricks is poor (**Fig. 4**) [14].

The majority of the instances of wear are caused by salt infiltration. Condensation of salts in certain brick zones is unavoidable due to the temperature gradient in the refractory lining and the porous structure. No chemical reactions

Thermo-Chemischer Verschleiß	Verschleiß durch Infiltrationen aus dem Ofenraum	
Temperaturen > 1400 °C	Temperaturen < 1400 °C	Temperaturen < 1400 °C
Infiltrationen durch Klinker, freies CaO, kaum Einfluß von Alkalien	Infiltrationen durch Na, K, Cl, SO$_3$, CO$_2$, Verhältnis Alk/SO$_3$ = 1	Infiltrationen durch Na, K, Cl, SO$_3$, CO$_2$, Verhältnis Alk/SO$_3$ > 1
keine Bildung von Salzen	Bildung von Salzen NaCl, KCl, Na$_2$SO$_4$, K$_2$SO$_4$	Bildung von Salzen NaCl, KCl, Na$_2$SO$_4$, K$_2$SO$_4$
Reaktionsablauf 7 MA + 12 C = C$_{12}$A$_7$ + 7 M	keine Reaktion MA + C	Reaktionsablauf 7 MA + 11 C + (Alk) = C$_{11}$(Alk)A$_7$ + 7 M
Schmelzenbildung, Spinell-Korrosion, Matrix-Rekristallisation	Salzkondensation	Salzkondensation, Schmelzenbildung, Spinell-Korrosion
Starke Veränderung aller Eigenschaften	Erhöhung der WLF, Verminderung der TWB	Starke Veränderung aller Eigenschaften
Schneller Verschleiß	Langsamer Verschleiß	Schneller Verschleiß

BILD 4: Verschiedene Verschleißmechanismen an Magnesia-Spinell-Steinen aus Zementdrehrohröfen

Treten Alkaliüberschüsse auf, d. h. das molare Alkali-/Sulfat-Verhältnis ist größer 1, werden neben den Salzkondensationen in den Poren auch die MA-Spinelle durch Reaktionen mit den Alkalien abgebaut (reaktive Infiltration). Dabei bilden sich Alkali-Calcium-Aluminate, deren Schmelzpunkte deutlich unter den feuerseitigen Steintemperaturen liegen. Diese Korrosion verändert alle Steineigenschaften beträchtlich. Die Haltbarkeit der Magnesia-Spinell-Steine wird damit erheblich verkürzt (**Bilder 4 und 6**).

5. Schlußfolgerungen

Befriedigende Laufzeiten von Magnesia-Spinell-Steinen in der Haupt-Sinterzone können nur mit stabilem Ansatz erreicht werden. Unter diesen Bedingungen ist aber der Einsatz von Dolomitsteinen weitaus wirtschaftlicher. Um lange Standzeiten von Magnesia-Spinell-Steinen in den Übergangszonen zu erreichen, sollte im Zementklinker ein molares Alkali-/Sulfat-Vehältnis nahe 1 angestrebt werden. Ist dies aus betrieblichen Gründen nicht möglich, so kann der Magnesia-Spinell-Stein keine lange Laufzeit erbringen.

6. Zusammenfassung

Magnesia-Spinell-Steine haben die Futterhaltbarkeiten in den Übergangszonen allgemein verbessert. Unter hohen Alkali-Belastungen wird jedoch die Laufzeit stark vermindert. Der Einsatz in der Haupt-Sinterzone ist in vielen Fällen nicht wirtschaftlich.

Literaturverzeichnis

[1] Trojer, F.: Über die Degeneration feuerfester basischer Steine während ihrer Verwendung. Radex-Rundschau Nr. 6 (1954), S. 214–224.

[2] Trojer, F.: Einige mikroskopische Betrachtungen über das Verhalten von Chromit in feuerfestem Chrommagnesit. Radex-Rundschau Nr. 3/4 (1958), S. 123–130.

with the brick material take place (inert infiltration) provided that the molar alkali/sulphate ratio is close to 1. The main result is to fill the pore volume, producing a change in thermal conductivity and thermal shock resistance. Although the durability of the magnesia-spinel bricks is reduced, it is better than in the case of thermochemical attack (**Figs. 4 and 5**).

Thermochemical wear	Wear due to infiltration from the kiln interior	
Temperatures > 1400 °C	Temperatures < 1400 °C	Temperatures < 1400 °C
Infiltration by clinker, free CaO, little influence by alkalis	Infiltration by Na, K, Cl, SO_3, CO_2, Alkali/SO_3 ratio = 1	Infiltration by Na, K, Cl, SO_3, CO_2, Alkali/SO_3 ratio > 1
No formation of salts	Formation of salts NaCl, KCl, Na_2SO_4, K_2SO_4	Formation of salts NaCl, KCl, Na_2SO_4, K_2SO_4
Reaction sequence 7 MA + 12 C = $C_{12}A_7$ + 7 M	No reaction MA + C	Reaction sequence 7 MA + 11 C + (Alk) = $C_{11}(Alk)A_7$ + 7 M
Melt formation, Corrosion of spinel, Recrystallisation of matrix	Salt condensation	Salt condensation, Melt formation, Corrosion of spinel
Major changes in all properties	Increase in thermal conductivity, Decrease in thermal shock	Major changes in all properties
Rapid wear	Slow wear	Rapid wear

FIGURE 4: Various wear mechanisms acting on magnesia-spinel bricks from rotary cement kilns

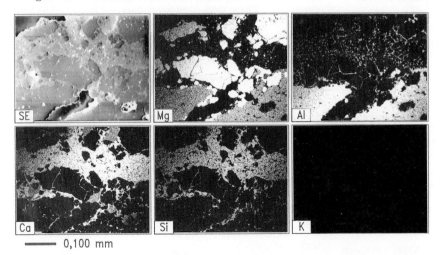

BILD 5: Beispiel einer inerten Infiltration aus dem Ofenraum an der Feuerseite von Magnesia-Spinell-Steinen
SEM-Bild und Element-Verteilungs-Bilder
FIGURE 5: Example of inert infiltration from inside the kiln on the hot face of magnesia-spinel bricks
SEM photos an element distribution photos

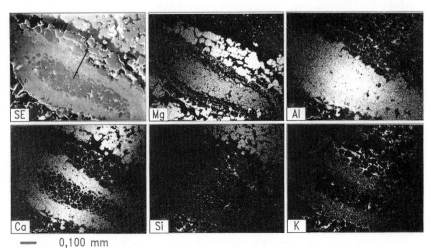

BILD 6: Beispiel einer reaktiven Infiltration aus dem Ofenraum an der Feuerseite von Magnesia-Spinell-Steinen
SEM-Bild und Element-Verteilungs-Bilder
FIGURE 6: Example of reactive infiltration from inside the kiln on the hot face of magnesia-spinel bricks
SEM photo and element distribution photos

[3] Trojer, F.: Futterzerstörungen durch Alkalisulfid- und sulfatphasen im Zement-Rotierofen. Radex-Rundschau Nr. 2 (1961), S. 546–552.

[4] Zednicek, W.: Mineralogische Untersuchungen an gebrauchten basischen feuerfesten Steinen zur Beurteilung von Verschleißvorgängen. Radex-Rundschau Nr. 4/5 (1966), S. 323–341.

[5] Kienow, S., und Seeger, M.: Die Zerstörung basischer Steine in der Sinterzone in Zementdrehöfen durch Alkalien, Sulfate und Vanadinverbindungen. Zement-Kalk-Gips Nr. 12 (1966). S. 555–560.

[6] Hums D., et al.: Zur Zerstörung von Magnesit-Chromsteinen in der Sinterzone von Zementdrehöfen. Tonindustrie-Zeitung Bd. 92, Nr. 6 (1968), S. 212–216.

[7] Opitz, D.: Verschleiß feuerfester Drehofenausmauerungen. Zement-Kalk-Gips Nr. 6 (1969), S. 262–264.

[8] Derie, R.: Änderungen an einem Chrom-Magnesit-Stein nach 60 Tagen Drehofenbetrieb. Zement-Kalk-Gips Nr. 6 (1969), S. 265–270.

[9] Kleinevoss A., et al.: Verschleißursachen bei magnesitisch zugestellten Brennzonen von Zementdrehöfen. Ber. Dt. Keram. Ges. Bd. 52, Nr. 5 (1975), S. 130–135.

[10] Barthel, H.: Beanspruchung und Verschleiß von magnesitischen Zustellungsmaterialien in Zementdrehöfen. Zement-Kalk-Gips Nr. 7 (1976), S. 308–312.

[11] Bray, D. J.: Toxicity of Chromium Compounds Formed in Refractories. Am. Ceram. Soc. Bull. Bd. 64, Nr. 7 (1985), S. 1012–1016.

[12] Fleck, H. K.: Erfahrungen mit Magnesium-Aluminium-Spinellsteinen in einem 3000-t/d-Drehofen. Zement-Kalk-Gips Nr. 2 (1992), S. 86–91.

[13] Routschka, G. und Majdic, A.: Feuerfeste Erzeugnisse für die Zementindustrie im Spiegel der Literatur. Zement-Kalk-Gips Nr. 9 (1983), S. 475–485.

[14] Hideaki N., et al.: Reaction between Synthetic Spinel and CaO. Taikabutsu Overseas Bd. 12, Nr. 3 (1992), S. 30–35.

If there is excess alkali, that is, if the molar alkali/sulphate ratio is greater than 1, besides salt condensation in the pores, the MA-spinels are decomposed by reaction with the alkalis (reactive infiltration). This results in formation of alkali-calcium aluminates, the melting points of which are well below the temperatures at the hot faces of the bricks. This corrosion changes all the properties of the brick considerably. The durability of the magnesia-spinel bricks is thus seriously reduced (**Figs. 4** and **6**).

5. Conclusions

Satisfactory campaign lives of magnesia-spinel bricks in the main sintering zone can only be achieved with a stable coating. Under these conditions, however, it is far more economical to use dolomite bricks. To achieve long service lives for magnesia-spinel bricks in the transition zone, a molar alkali/sulphate ratio close to 1 should be aimed at in the cement clinker. If this is not possible for operational reasons, then the magnesia-spinel brick cannot have a long campaign life.

6. Summary

Magnesia-spinel bricks have generally improved the durability of refractory linings in the transition zones. However, severe alkali attack seriously shortens their campaign life. Their use in the main sintering zone is in many cases uneconomic.

Verbesserung der Temperaturwechselbeständigkeit von basischen, gebrannten Steinen für den Zementdrehrohrofen durch Zirkondioxid

Use of zirconium dioxide to improve the thermal shock resistance of basic burnt bricks for rotary cement kilns

Amélioration de la résistance aux changements de température de briques basiques, cuites, pour le four à ciment rotatif, à l'aide d'oxyde de zirkone

Mejora de la resistencia a los cambios de temperatura de los ladrillos básicos, cocidos, desinados al horno rotatorio de cemento, por medio de dióxido de zirconio

Von **J. Hartenstein, U. Bongers, R. Prange** und **J. Stradtmann**, Wülfrath/Deutschland

Zusammenfassung – Der Verschleiß der feuerfesten Ausmauerung von Zementdrehrohröfen wird u. a. durch kurzzeitige Temperaturwechsel hervorgerufen. Magnesiachromsteine zeichnen sich zwar durch hohe Thermoschockbeständigkeit aus, sind jedoch schwierig zu entsorgen. Auf der Suche nach einem Ersatzwerkstoff sollte die hohe Temperaturwechselbeständigkeit dieses Steintyps erhalten bleiben. In der Sinterzone des Zementofens haben sich Dolomitsteine wegen ihrer hohen Beständigkeit gegen chemische Reaktionen mit Zementklinker bewährt. Es lag deshalb nahe, sowohl auf Basis von Magnesia als auch von Dolomit nach Möglichkeiten der Verbesserung der Temperaturwechselbeständigkeit und weiterer Eigenschaften, die sich aus den Anforderungen im Zementofen ergeben, zu suchen. In die Entwicklungen wurden auch zirkonhaltige Steinsorten für den Einsatz in der Hochtemperaturzone des Zementofens mit einbezogen. Die bereits vorliegenden Ergebnisse aus der praktischen Erprobung von Magnesiasteinen mit Zirkondioxidzusatz zeigen, daß sich dieser Werkstoff sowohl durch hohe Temperaturwechselbeständigkeit als auch durch eine verbesserte Reaktionsbeständigkeit und erhöhte Festigkeit auszeichnet.

Verbesserung der Temperaturwechselbeständigkeit von basischen, gebrannten Steinen für den Zementdrehrohrofen

Summary – The wear to the refractory lining of a rotary cement kiln is caused, among other things, by short-term temperature fluctuations. Magnesia-chrome bricks are characterized by their high thermal shock resistance, but are difficult to dispose of. The high thermal shock resistance of this type of brick should be retained when looking for a replacement material. Dolomite bricks have proved successful in the cement kiln sintering zones due to their high resistance to chemical reaction with the cement clinker. It was therefore obvious that the search for ways of improving thermal shock resistance and other properties needed in cement kilns should be based on magnesia and dolomite. Bricks containing zirconium for use in the high temperature zones in cement kilns were also included in the development work. Results already available from the practical testing of magnesia bricks containing added zirconium dioxide show that this material has a high thermal shock resistance as well as improved reaction resistance and increased strength.

Use of zirconium dioxide to improve the thermal shock resistance of basic burnt bricks for rotary cement kilns

Résumé – L'usure de la maçonnerie réfractaire de fours à ciment rotatifs est provoquée, entre autre, par des changements rapides de température. Les briques magnésie-chrome excellent, bien sûr, par leur haute résistance aux chocs thermiques, mais leur élimination est difficile. Lors de la recherche d'un matériau de remplacement, la haute résistance aux variations de température de ce type de brique devait être conservée. Dans la zone de clinkérisation du four à ciment, les briques dolomie ont fait leur preuve à cause de leur haute résistance contre les réactions chimiques avec le clinker à ciment. Il était donc tentant de chercher, à partir aussi bien de magnésie, que de dolomie, des possibilités d'amélioration de la résistance aux variations de température et, aussi, d'autres propriétés, convenant aux exigences du four à ciment. Les développements ont aussi pris en compte des sortes de briques contenant du zirkone, pour l'utilisation dans la partie haute température du four à ciment. Les résultats déjà disponibles des essais pratiques de briques magnésie avec addition d'oxyde de zirkone montrent, que ce matériau se distingue à la fois par une haute résistance aux variations de température et par une résistance améliorée aux attaques et une résistance mécanique accrue.

Amélioration de la résistance aux changements de température de briques basiques, cuites, pour le four à ciment rotatif, à l'aide d'oxyde de zirkone

Resumen – El desgaste del revestimiento refractario de los hornos rotatorios de cemento lo producen, en parte, los cambios rápidos de temperatura. Es cierto que los ladrillos de cromomagnesita se caracterizan por su elevada resistencia a los choques térmicos, pero su eliminación posterior resulta difícil. Al buscar un material alternativo, había que procurar mantener la alta resistencia de este tipo de ladrillos a los cambios térmicos. En la zona de sinterización del horno de cemento han probado su eficacia los ladrillos de dolomita, debido a su alta resistencia a las reacciones químicas con el clínker de cemento. Por esta razón, ha sido lógico buscar posibilidades de mejorar, sobre la base de magnesia y de dolomita, la resistencia a los cambios de temperatura de los ladrillos y otras características requeridas en el horno de cemento. En los trabajos de desarrollo han sido incluidos también los ladrillos a base de zirconio, destinados a ser empleados en la zona de altas temperaturas del horno de cemento. Los resultados del ensayo práctico de los ladrillos de magnesita con adición de dióxido de zirconio, disponibles entretanto, muestran que este material se distingue tanto por su alta resistencia a los cambios de temperatura como por una mejor resistencia a las reacciones y una mejor resistencia mecánica.

Mejora de la resistencia a los cambios de temperatura de los ladrillos básicos, cocidos, desinados al horno rotatorio de cemento, por medio de dióxido de zirconio

1. Einleitung

Basische Steine werden seit ca. 60 Jahren erfolgreich im Zementdrehrohrofen verwendet. Seit über 25 Jahren werden Dolomitsteine in der Sinterzone eingesetzt. Die Umstellung von Öl auf Kohle als Brennstoff hat seit der zweiten Hälfte der 70er Jahre zu einer Verkürzung der Laufzeiten der Magnesia-Chrom-Steine geführt. In diesen Steinen entstehen durch die Reaktion mit Alkalien umweltschädliche Chromatverbindungen.

In den letzten Jahren hat sich aber gezeigt, daß unter bestimmten Betriebsbedingungen sowohl Magnesia-Spinell-Steine als auch reine Dolomitsteine überfordert sind. Insofern war es notwendig, einen Weg zu finden, um das unzureichende Temperatur-Wechsel-Verhalten von Dolomit-, Magnesia- und Dolomit-Magnesia-Steinen zu verbessern. Seit etwa 1970 ist bereits bekannt, daß der Zusatz geringer Mengen von körnigem Zirkondioxid das Temperatur-Wechsel-Verhalten von Magnesia-Steinen deutlich verbessern kann [1]. Vermutlich aus Kostengründen wurde diese Entwicklung zunächst nicht weiter verfolgt.

Der ZrO_2-Zusatz zu Dolomit-Steinen wurde Anfang der 80er Jahre erstmalig in den USA erprobt [2, 3]. Der Einsatz solcher Dolomit-Zirkonia-Steine in der Sinterzone hat sich inzwischen auch in europäischen Zementdrehrohröfen bewährt. Durch die Notwendigkeit Magnesia-Chromatsteine zu ersetzen, lag es nahe, die Verwendung von Zirkondioxid in Magnesia-Steinen erneut zu untersuchen.

2. Wirkungsweise von Zirkondioxid

Die als Ausgangsstoffe für feuerfeste Produkte vielfach verwendeten Stoffe Calciumoxid und Magnesiumoxid haben die höchste lineare Wärmedehnung aller feuerfesten Oxide. Dies führt bei starken Temperaturänderungen eines feuerfesten Steins zwangsläufig zu Spannungen, welche die Eigenfestigkeiten des Körpers überschreiten können. Infolge dessen kommt es zum Bruch. Dies bedeutet, daß Steine, die vorwiegend aus diesen beiden Oxiden bestehen, nur eine geringe Temperaturwechselbeständigkeit besitzen.

Bei Oxiden mit geringer linearer Wärmedehnung wie ZrO_2 oder Al_2O_3 induzieren Temperaturänderungen lediglich Spannungen unterhalb der Eigenfestigkeit. Daher verfügen feuerfeste Produkte aus diesen Materialien über eine bessere Temperaturwechselbeständigkeit (**Bild 1**).

Werden zwei Oxide mit hoher und niedriger linearer Wärmedehnung in körniger Form gemischt und keramisch gebunden, so tritt eine deutliche Verbesserung des Temperatur-Wechsel-Verhaltens auf. Diese Tatsache kann mit verschiedenen Hypothesen gedeutet werden, wie z.B. der Erzeugung von Eigenspannungen oder von Mikrorissen. In jedem Fall halten heterogene Mischungen thermisch induzierten Spannungen besser stand, als reines Magnesiumoxid oder Calciumoxid für sich allein. Der notwendige Anteil des zuzusetzenden Oxids mit geringer Wärmedehnung kann umso niedriger werden, je höher die Differenz der Wärmedehnung zum anderen Oxid ist. Als Beispiel können hier Magnesia-Chrom, Magnesia-Spinell und Magnesia-Zirkonia

1. Introduction

Basic bricks have been successfully used in rotary cement kilns for some 60 years. For more than 25 years, dolomite bricks have been used in the sintering zone. Since the second half of the seventies the change from oil to coal as fuel has led to a decrease in the campaign lives of magnesia-chrome bricks. Ecologically harmful chromate compounds form in these bricks as a result of the reaction with alkalis.

However, it has turned out in recent years that under certain operating conditions, both magnesia-spinel bricks and pure dolomite bricks are overtaxed. It was therefore necessary to find a way to improve the inadequate thermal shock behaviour of dolomite, magnesia and dolomite-magnesia bricks. It has been known since as far back as 1970 or so that the addition of small amounts of granular zirconium dioxide is able to improve the thermal shock behaviour of magnesia bricks significantly [1]. At the time this development was not pursued further, presumably for reasons of cost.

Trials with the addition of ZrO_2 to dolomite bricks were first carried out in the USA in the early eighties [2, 3]. The use of such dolomite-zirconia bricks has since proved its worth in European rotary cement kilns as well. It became logical to re-examine the use of zirconium dioxide in magnesia bricks because of the need to replace magnesia-chromate bricks.

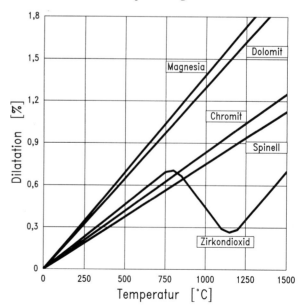

BILD 1: Vergleich linearer Wärmedehnungen basischer Feuerfestmaterialien
FIGURE 1: Comparison of linear thermal expansion of basic refractory materials

Dilatation	= Dilatation
Dolomit	= Dolomite
Magnesia	= Magnesia
Chromit	= Chromite
Spinell	= Spinel
Zirkondioxid	= Zirconium dioxide
Temperatur	= Temperature

angeführt werden. Wie im ersten Bild zu erkennen ist, dehnen sich Chromit und Spinell in deutlich geringerem Maß mit steigender Temperatur aus als Dolomit oder Magnesia. Zirkondioxid zeigt ein davon verschiedenes Dehnungsverhalten: Die anfängliche Ausdehnung bis 750 °C geht im Bereich zwischen 750 °C und 1100 °C wieder zurück.

Bei Dolomit-Steinen wird ebenfalls eine Verbesserung der Temperaturwechselbeständigkeit durch ZrO_2 erreicht. Infolge der Reaktion von CaO und ZrO_2 zu Ca-Zirkonat bilden sich Vorspannungen im Gefüge.

Aus den oben aufgeführten Erkenntnissen wurden der Dolomit-Stein Sindoform K 11 105, der Magnesia-Stein Permag HZ 96 sowie ein neuer Dolomit-Magnesia-Stein VST 2014 entwickelt. Der Sindoform K 11 105 ist heute bei vielen Zementdrehrohröfen in der Sinterzone erfolgreich im Einsatz. Der Permag HZ 96 ist seit Frühjahr 1993 in praktischer Erprobung (**Bild 2**).

3. Eigenschaften

Der Permag HZ 96 weist trotz niedriger Porosität und Gasdurchlässigkeit eine hohe Temperaturwechselbeständigkeit auf. Der Infiltrationswiderstand ist dadurch hoch. Es wird erwartet, daß diese Steinsorte bei großen Alkalibelastungen (molares Alkali-/Sulfat-Verhältnis ≫ 1) im Gegensatz zum Magnesia-Spinell-Stein reaktionsbeständig und stabil ist. Dadurch ist er für den Einsatz in Übergangszonen prädestiniert.

Der Sindoform K 11 105 hat sich mittlerweile in den Sinterzonen vieler Zementdrehrohröfen bewährt. In Bereichen

2. Action of zirconium dioxide

Calcium oxide and magnesium oxide, which are often used as raw materials for refractory products, have the highest linear thermal expansion of all refractory oxides. In the event of severe variations in the temperature of a refractory brick, this inevitably leads to stresses which may exceed the inherent strengths of the brick. As a result, fracture occurs. This means that bricks which mainly consist of these two oxides have low resistance to thermal shock.

In the case of oxides with low linear thermal expansion such as ZrO_2 or Al_2O_3, temperature variations only induce stresses which are below the inherent strength of the material. Refractory products made from these materials are therefore more resistant to thermal shock (**Fig. 1**).

If two oxides with high and low rates of linear thermal expansion are mixed in granular form and ceramic-bonded, a marked improvement in thermal shock behaviour occurs. This fact may be interpreted by means of various hypotheses, for instance the generation of internal stresses or the formation of microcracks. In any case, heterogeneous mixtures withstand thermally induced stresses better than pure magnesium oxide or calcium oxide alone. The higher the difference in the rate of thermal expansion relative to the other oxide, the lower is the percentage of the low-thermal-expansion oxide which needs to be added. Examples are magnesia-chrome, magnesia-spinel and magnesia-zirconia. As can be seen from Fig. 1, as the temperature rises chromite and spinel expand to a significantly lesser extent than dolomite or magnesia. Zirconium dioxide exhibits a different

Wülfrather Feuerfest-/Keramik-Gruppe Forschung & Entwicklung		Sindoform K 11 105 Gebrannter, direkt gebundener Dolomit-Stein	Permag HZ 96 Gebrannter, direkt gebundener Magnesia-Stein	VST 2014 Versuchsprodukt Gebrannter, direkt gebundener Dolomit-Magnesia-Stein
MgO	%	37,5	94,0	49,5
CaO	%	58,0	1,0	46,0
SiO_2	%	1,0	0,5	0,8
Al_2O_3	%	0,6	0,2	0,5
Fe_2O_3	%	0,8	0,5	0,6
ZrO_2	%	2,0	2,0	2,0
Rohdichte	g/cm³	2,80	2,93	2,87
Porosität, offen	Vol.%	15,0	18,0	14,5
Druckfestigkeit	MPa	45	45	45
Anwendung		Zement-Drehrohröfen, Sinterzone	Zement-Drehrohröfen, Übergangszonen	Zement-Drehrohröfen, Sinterzone
besondere Eigenschaften		gute Ansatzbildung, gute TW-Beständigkeit	keine Ansatzbildung, gute TW-Beständigkeit, alkalibeständig, chromfrei	gute Ansatzbildung, gute TW-Beständigkeit, geringe $CO_2(SO_3)$-Aufnahme

BILD 2: Vergleich der Eigenschaften von Dolomit-, Magnesia- und Dolomit-Magnesia-Steinen

Wülfrath refractories/ceramics group Research and development		Sindoform K 11 105 Burnt, direct-bonded dolomite brick	Permag HZ 96 Burnt, direct-bonded magnesia brick	VST 2014 Test product Burnt, direct-bonded dolomite-magnesia brick
MgO	%	37.5	94.0	49.5
CaO	%	58.0	1.0	46.0
SiO_2	%	1.0	0.5	0.8
Al_2O_3	%	0.6	0.2	0.5
Fe_2O_3	%	0.8	0.5	0.6
ZrO_2	%	2.0	2.0	2.0
Bulk density	g/cm³	2.80	2.93	2.87
Porosity, open	Vol.%	15.0	18.0	14.5
Compressive strength	MPa	45	45	45
Application		Rotary cement kilns, Sintering zone	Rotary cement kilns, Transition zones	Rotary cement kilns, Sintering zone
Special properties		Good coating formation, Good thermal shock resistance	No coating formation, Good thermal shock resistance, alkali-resistant, chromium-free	Good coating formation, Good thermal shock resistance, Low $CO_2(SO_3)$ absorption

FIGURE 2: Comparison of the properties of dolomite-, magnesia- and dolomite-magnesia bricks

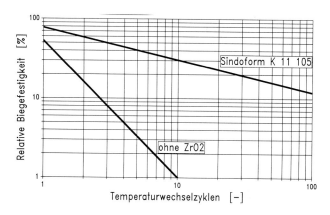

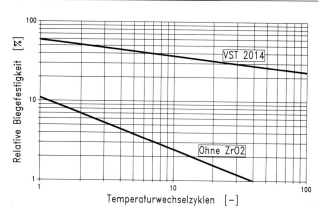

BILD 3: Restbiegefestigkeit nach Temperaturwechsel Dolomit-Steine
FIGURE 3: Residual flexural strength after thermal shock Dolomite bricks

Relative Biegefestigkeit = Relative flexural strength
Temperaturwechselzyklen = Thermal shock cycles
ohne ZrO₂ = without ZrO₂

BILD 4: Restbiegefestigkeit nach Temperaturwechsel Magnesia-Steine
FIGURE 4: Residual flexural strength after thermal shock Magnesia bricks

Relative Biegefestigkeit = Relative flexural strength
Temperaturwechselzyklen = Thermal shock cycles
ohne ZrO₂ = without ZrO₂

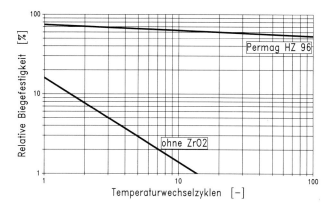

BILD 5: Restbiegefestigkeit nach Temperaturwechsel Dolomit-Magnesia-Steine
FIGURE 5: Residual flexural strength after thermal shock Dolomite-magnesia bricks

Relative Biegefestigkeit = Relative flexural strength
Temperaturwechselzyklen = Thermal shock cycles
ohne ZrO₂ = without ZrO₂

mit stabilem Ansatz weist er gegenüber der normalen, zirkonoxidfreien Sorte deutliche Verlängerungen der Standzeit auf. Voraussetzung hierfür ist jedoch, daß Alkalien nicht im Überschuß vorhanden sind, was ein gleichbleibendes molares Alkali-/Sulfat-Verhältnis nahe 1 voraussetzt.

Untersuchungen an gebrauchten Dolomit-Zirkonia-Steinen haben inzwischen gezeigt, daß das Alkali-/Sulfat-Verhältnis oft außerhalb des für Dolomit-Steine günstigen Bereichs liegen kann. Deshalb wurden die Versuchssteine VST 2014 auf der Basis von Dolomit-Magnesia-Zirkonia mit sehr geringer Porosität und Permeabilität entwickelt. Die Verwendung von Magnesia in Verbindung mit einem besonders dichten Gefüge bewirkt bei guter Temperaturwechselbeständigkeit eine verbesserte Reaktionsbeständigkeit gegenüber gasförmigen Infiltraten, wie Alkalien, Cl, SO₃, CO₂ und nicht zuletzt auch H₂O.

4. Temperaturwechselbeständigkeit

Die Temperaturwechselbeständigkeit wird nach bestehender Prüfpraxis durch die Anzahl der Temperaturzyklen zwischen 1000 °C und Raumtemperatur gekennzeichnet, die eine Probe übersteht. Die Anzahl der erreichten Zyklen bis zum Bruch unter 0,3 MPa Last wird als Maß für das Temperatur-Wechsel-Verhalten verwendet, wobei nach 30 Zyklen die Prüfung beendet wird. Für die Beschreibung der Effekte, die durch den ZrO₂-Zusatz eintreten, ist dieses Verfahren nicht mehr ausreichend, selbst eine Steigerung auf 100 Zyklen führt nicht mehr zum Bruch der Proben.

expansion behaviour: the initial expansion up to 750 °C is reversed in the range 750–1100 °C.

ZrO₂ also improves in thermal shock resistance with dolomite bricks. The reaction of CaO and ZrO₂ to form calcium zirconate leads to the formation of initial stresses in the structure.

Based on these findings, the dolomite brick Sindoform K 11 105, the magnesia brick Permag HZ 96 and a new dolomite-magnesia brick VST 2014 were developed. Sindoform K 11 105 is now successfully in service in the sintering zones of many rotary cement kilns. Permag HZ 96 has been undergoing commercial trials since spring 1993 (**Fig. 2**).

3. Properties

Permag HZ 96 has a high thermal shock resistance despite its low porosity and gas permeability. Its resistance to infiltration is therefore high. In contrast to magnesia-spinel bricks, this brick is expected to prove resistant to reaction and stable under severe alkali attack (molar alkali/sulphate ratio $\gg$ 1). This makes it an ideal product for use in transition zones.

Sindoform K 11 105 has now proved its value in the sintering zones of many rotary cement kilns. In zones with a stable coating it has a markedly longer service life than the normal zirconium-oxide-free type. A condition for this, however, is that alkalis should not be present in excess; this necessitates a constant molar alkali/sulphate ratio close to 1.

Tests on used dolomite-zirconia bricks have now shown that the alkali/sulphate ratio may often lie outside the favourable range for dolomite bricks. The experimental bricks VST 2014 were therefore developed on the basis of dolomite-magnesia-zirconia with very low porosity and permeability. The use of magnesia in conjunction with a particularly dense structure produces, along with good thermal shock resistance, improved resistance to reaction vis-à-vis gaseous infiltrating substances such as alkalis, Cl, SO₃, CO₂ and not least H₂O.

4. Thermal shock resistance

According to current testing practice, thermal shock resistance is indicated by the number of temperature cycles between 1000 °C and room temperature that a specimen withstands. The number of cycles achieved to fracture under a load of 0.3 MPa is used as a measure of thermal shock behaviour, the test being terminated after 30 cycles. This method is no longer adequate to describe the effects of the addition of ZrO₂; even an increase to 100 cycles does not result in fracture of the specimens.

According to a modified test method, the flexural strength of various sets of specimens is determined after 1, 3, 10 and 30 temperature cycles; testing under a load of 0.3 MPa is again

Nach einem geänderten Prüfverfahren wird die Biegefestigkeit an verschiedenen Probensätzen nach 1, 3, 10 und 30 Temperaturzyklen bestimmt, wobei die Prüfung unter 0,3 MPa Last weiterhin nach jedem Zyklus stattfindet. Die so erhaltenen „Rest-Biegefestigkeits-Werte" werden in Relation zur ursprünglichen Biegefestigkeit gesetzt. Dieses Verfahren führt zu einer exponentiell verlaufenden Kurve der Festigkeitsminderung, welche das Verhalten der Steine deutlich besser beschreibt als das bisherige Verfahren. Die **Bilder 3, 4** und **5** zeigen jeweils den Einfluß des ZrO_2-Zusatzes.

5. Zusammenfassung

Die Verwendung von ZrO_2 in basischen feuerfesten Steinen hat die Entwicklung dichter temperaturwechselbeständiger Steinsorten ermöglicht, die eine hohe Reaktionsbeständigkeit gegenüber gasförmigen Infiltraten aufweisen.

Literaturverzeichnis

[1] Deutsches Bundespatent DE 22 49 814 C3

[2] Griffin, D.J., et al.: The Development and Use of Dolomite-Zirconia-Brick in Cement Rotary Kilns. Proceedings of the 2nd International Conference on Refractories (1987), S. 609 – 622.

[3] Stradtmann, J., et al.: Recent Development of Dolomitic Refractories for Application in the Steelmaking Process. Proceedings of the 2nd International Conference on Refractories (1987), S. 349 – 362.

carried out after each cycle. The "residual flexural stress values" thus obtained are plotted against the original flexural strength. This method produces an exponentially-shaped strength decrease curve which describes the bricks' behaviour much better than the previous method. **Figs. 3, 4** and **5** show the influence of adding ZrO_2 for each type of brick.

5. Summary

The use of ZrO_2 in basic refractory bricks has enabled dense, thermal-shock-resistant bricks to be developed with a high resistance to reaction vis-à-vis gaseous infiltrating substances.

Die Kohlenstaubdosierung nach dem Coriolis-Meßprinzip – ein neues Verfahren zur Brennstoffeinsparung

Pulverized coal metering using the Coriolis measuring principle – a new method of saving fuel

Le dosage du poussier de charbon selon le principe de mesure Coriolis – Un nouveau procédé pour économiser le combustible

La dosificación del carbón pulverizado según el principio de medición de Coriolis – un nuevo procedimiento para ahorrar combustible

Von **H. Heinrici** und **A. Burk,** Darmstadt/Deutschland

Zusammenfassung – Die Wirtschaftlichkeit der Klinkerproduktion wird maßgeblich von der benötigten Energiemenge bestimmt. Mit Hilfe einer gravimetrischen Dosierung können die an Kurzzeitdosierkonstanz und Langzeitgenauigkeit zu stellenden Anforderungen am besten erfüllt werden. Außerdem muß ein Dosiersystem einfach in eine Anlage zu integrieren sein und eine hohe Betriebssicherheit aufweisen. Die Anforderungen lassen sich mit einem Dosiersystem kostensparend erfüllen, das aus einem Massendurchflußmeßgerät nach dem Coriolis-Prinzip und einem geregelten Zuteiler besteht. Das Meßverfahren wird seit einigen Jahren in der industriellen Praxis zur Messung von Massenströmen, zum Beispiel Kalkstein oder Flugasche, eingesetzt. Seit kurzem steht eine Ausführung zur Verfügung, die den Vorschriften zur Verhütung von Staubexplosionen in Kohlenstaubanlagen entspricht. Die Verwirklichung des Meßprinzips erfordert ein mit konstanter Winkelgeschwindigkeit rotierendes Meßrad. Wird ein Schüttgutstrom über das Meßrad geleitet, so ist ein zusätzliches Antriebsdrehmoment nötig, um die Winkelgeschwindigkeit konstant zu halten. Dieses meßbare Drehmoment ist dem Massenstrom streng proportional. Das Dosiersystem zeichnet sich dadurch aus, daß Massenstromschwankungen sehr schnell erfaßt und ausgeregelt werden können. Der Einbau in eine Feuerungsanlage wird durch geringen Einbauraum und verminderte Anforderungen an die Entstaubung erleichtert.

Summary – The cost-effectiveness of clinker production is determined to a great extent by the quantity of energy required. The requirements for short-term metering constancy and long-term accuracy are best fulfilled with the aid of a gravimetric metering system. A metering system must also be simple to integrate into a plant and show a high level of operational reliability. These requirements can be met at a low cost using a metering system consisting of a mass flow meter based on the Coriolis principle and a controlled feeding device. This method of measurement has been used industrially for some years for measuring the mass flows of, for example, limestone and fly-ash. Recently a design has become available which meets the specifications for prevention of dust explosions in pulverized coal plants. The measuring principle requires a measuring wheel rotating at constant speed. If a flow of bulk material is passed over the measuring wheel then an additional drive torque is needed to keep the angular velocity constant. This measurable torque is exactly proportional to the mass flow. A feature of the metering system is that fluctuations in mass flow can be detected and stabilized very rapidly. Installation in a firing system is made easier by the small installation space and reduced dedusting requirements.

Résumé – La rentabilité de la production du clinker est essentiellement déterminée par la quantité d'énergie requise. Les exigences à poser au sujet de la régularité du dosage en temps court et l'exactitude à long terme peuvent être satisfaites au mieux à l'aide d'un dosage gravimétrique. De plus, un système de dosage doit pouvoir être aisément intégrable dans l'installation et montrer une haute sécurité de fonctionnement. Ces exigences peuvent être satisfaites à peu de coût avec un système de dosage se composant d'un appareil de mesure de débit massique selon le principe Coriolis et d'un alimenteur régulé. Ce procédé de mesure est utilisé depuis quelques années dans l'industrie, pour mesurer des débits massiques, comme p. ex. du calcaire ou des cendres volantes. Depuis peu, est disponible une version correspondant aux prescriptions pour éviter les explosions de poussières dans les ateliers à poussier de charbon. La réalisation du principe de mesure nécessite une roue de mesure tournant à vitesse angulaire constante. Quand un courant de matière en vrac est conduit sur la roue de mesure, ceci requiert un couple d'entraînement supplémentaire, afin de maintenir constant la vitesse angulaire. Ce couple facile à mesurer est strictement proportionnel au débit massique. Ce système de mesure se caractérise par le fait, que les variations du débit massique

peuvent être rapidement détectées et lissées. Le montage dans une installation de chauffe est facilité par un faible encombrement et peu d'exigences côté dépoussiérage.

Resumen *– La rentabilidad de la producción de clínker queda determinada, en gran parte, por la cantidad de energía requerida. Con ayuda de una dosificación gravimétrica se pueden cumplir mejor las exigencias respecto de la regularidad a corto plazo y la precisión a largo plazo de la dosificación. Además, el sistema de dosificación debe ser fácil de incorporar a una instalación y presentar una elevada seguridad de servicio. Estas exigencias se pueden cumplir, a bajo coste, mediante un sistema de dosificación compuesto de un medidor del flujo de masas, que funciona según el principio de Coriolis, y un alimentador regulado. Este procedimiento de medición se viene aplicando desde hace algunos años en la industria para medir los flujos de masas, por ejemplo de caliza o de cenizas volantes. Desde hace poco, se dispone de un sistema que cumple las prescripciones en materia de prevención de explosiones producidas en las instalaciones de carbón pulverizado. La realización del principio de medición requiere una rueda de medición que gire a una velocidad angular constante. Si un flujo de material a granel pasa por esta rueda, hace falta un momento de accionamiento adicional, con el fin de mantener constante la velocidad angular. Este momento de giro, que se puede medir, es rigurosamente proporcional al flujo de masas. El sistema de dosificación se distingue por el hecho de que las variaciones del flujo de masas se pueden detectar y estabilizar rápidamente. Su montaje en una instalación de combustión queda facilitado por un reducido espacio necesario y reducidas exigencias en cuanto al desempolvado.*

La dosificación del carbón pulverizado según el principio de medición de Coriolis – un nuevo procedimiento para ahorrar combustible

Die Wirtschaftlichkeit der Klinkerproduktion wird zu einem großen Teil von der benötigten Energiemenge bestimmt. Sie ist gekennzeichnet durch einen geringen Verbrauch an Primärenergie (Brennstoff) und Sekundärenergie (Antriebe), sowie geringe Kosten für Investition und Instandhaltung.

Die hohe Wirtschaftlichkeit, die engen Grenzwerte der TA Luft hinsichtlich Stickoxid- und Kohlenmonoxidemissionen, sowie eine sichere Prozeßführung verlangen die zuverlässige Bildung des Brennstoff/Luft-Gemischs für die Befeuerung des Drehrohrofens. Hierzu ist eine der wesentlichen Voraussetzungen eine kurzzeitkonstante und langzeitstabile Brennstoffzuteilung.

Diese Aufgaben lassen sich nur mit Hilfe einer gravimetrischen Dosierung der Brennstoffe erfüllen. Auerdem muß ein mögliches Dosiersystem einfach in eine Anlage zu intregrieren sein und eine hohe Betriebssicherheit aufweisen.

Die heute verfügbaren Dosiersysteme erfüllen diese Anforderungen nur zum Teil. Dies wird bei den Fragen nach Kurzzeit- und Langzeitgenauigkeit deutlich. Ein Differentialdosierwaagensystem bietet zwar eine sehr gute Langzeitgenauigkeit, hat aber infolge der notwendigen Mittelung der Meßwerte beim Erreichen der Kurzzeitgenauigkeit Grenzen. Andere Systeme zeigen prinzipbedingt noch größere Probleme. Die Langzeitgenauigkeit läßt sich hier nur mit Kontrollmeßeinrichtungen (KME), d.h. einem verwogenen Vorbehälter, erreichen (**Tabelle 1**).

Ein Dosiersystem, das aus einem Massendurchflußmeßgerät nach dem Coriolis-Prinzip und einem geregelten Zuteiler besteht, bietet nun die Möglichkeit, die Anforderungen an Kurzzeitdosierkonstanz und Langzeitgenauigkeit zu erfüllen. Weiterhin können mit diesem System Kosteneinsparung im peripheren Bereich der Dosierung realisiert werden.

The economics of clinker production is determined largely by the amount of energy required. Cost-effective production entails low consumption of primary energy (fuel) and secondary energy (drives) and low capital and maintenance costs.

Economic production, safe operation and the narrow limits laid down by German clean air legislation in respect of nitrogen oxide and carbon monoxide emissions all require reliability in the formation of the fuel/air mix for firing rotary kilns. One of the essential prerequisites for this is that the fuel feed should be constant in the short term and stable in the long term.

This is only possible with gravimetric metering of the fuels. Any potential metering system must also be simple to integrate into a plant and show a high level of operational reliability.

The metering systems now available only partly meet these requirements. This becomes clear when considering questions of short-term and long-term accuracy. Although a differential gravimetric system provides very good long-term accuracy it has limits in achieving short-term accuracy owing to the need to average the measurements. Other systems have still greater inherent problems. Long-term accuracy can only be obtained here with controlling and measuring equipment (CME), i.e. with a weighed feed container.

A metering system consisting of a mass flow meter based on the Coriolis principle and a controlled feeding device can fulfil the requirements for short-term metering constancy and long-term accuracy. The system can also save costs in the peripheral metering area.

The metering principle

The metering principle makes use of the Coriolis force (force of inertia) which acts on a particle of material moving out-

TABELLE 1: Dosiersysteme
Erfüllung der Anforderungen

Dosier-systeme	Kurzzeit-konstanz	Genauigkeit	Kosten	Sekundär-energie
Differential-dosierwaage	teilweise erfüllt	erfüllt	hoch	hoch
Dosier-rotorwaage	nicht erfüllt	erfüllt mit KME	mittel	niedrig
Prallplatte mit Dosierer	nicht erfüllt	erfüllt mit KME	hoch	hoch

TABLE 1: Metering systems
Meeting the requirements

Metering systems	Short-term constancy	Accuracy	Cost	Secondary energy
Differential weighing	partly met	met	high	high
Weighing with rotor	not met	met with CME	moderate	low
Baffle with metering equipment	not met	met with CME	high	high

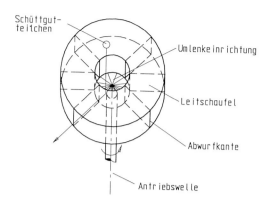

BILD 1: Coriolis-Meßprinzip
FIGURE 1: The Coriolis measuring principle

Schüttgutteilchen	= particles of bulk material
Umlenkeinrichtung	= deflector
Leitschaufel	= guide vane
Abwurfkante	= discharge edge
Antriebswelle	= drive shaft

Das Meßprinzip

Das Meßprinzip nutzt die Coriolis-Kraft (Trägheitskraft), die auf ein Masseteilchen wirkt, das sich in einem drehenden System von innen nach außen bewegt. Die Verwirklichung des Meßprinzips erfordert daher ein rotierendes Meßrad, das sich mit einer konstanten Winkelgeschwindigkeit dreht. Es wird durch einen Elektromotor angetrieben. **Bild 1** zeigt eine schematische Darstellung des Meßrades.

Wird ein Schüttgutstrom über das Meßrad geleitet, so wird er in radiale Richtung umgelenkt. Nach der Umlenkung wird das Schüttgut von den Leitschaufeln des Meßrades erfaßt. Hierbei werden die Partikel durch die Zentrifugalkraft in radialer Richtung und die Corioliskraft in Umfangsrichtung beschleunigt. Bei der Bewegung der Schüttgutpartikel über die Leitschaufeln wirken verschiedene Kräfte ein (**Bild 2**).

In radialer Richtung wirken die Zentrifugalkraft Fz und die entgegengesetzt wirkende Reibkraft Fr. Die Corioliskraft Fc dagegen wirkt in tangentialer Richtung und bewirkt ein Reaktionsmoment M, das vom Antrieb des Meßrades ausgeglichen werden muß und mit einer Drehmomentmessung erfaßt wird.

Der Schüttgutmassenstrom ist dem aufzuwendenden Drehmoment direkt proportional. Reibungskräfte zwischen Schüttgutpartikeln und Meßrad oder auch zwischen Schüttgutschichten haben keinen Einfluß auf das Meßergebnis. Das Meßverfahren ist demzufolge geeignet, Schüttgutmassenströme mit einer hohen Genauigkeit von weniger als ± 0,5 % Abweichung bezogen auf den eingestellten Massenstrom zu messen. Infolge einer extrem kurzen Verweilzeit des Schüttgutes im Meßgerät (siehe **Bild 3**) im Bereich von Sekundenbruchteilen werden Schwankungen im Kohlenstaubmassenstrom sehr schnell erfaßt.

Die Anwendung

Zur Ausregelung von Massenstromschwankungen ist ein geeigneter, geregelter Zuteiler erforderlich. Abhängig von der Dosiertechnik können verschiedene Möglichkeiten realisiert werden. **Bild 4** zeigt die Lösung der Pilotanlage, bei der eine Dosierschnecke als geregelter Zuteiler eingesetzt wird.

Für diese Pilotanwendung wurde ein Coriolis-Meßgerät unterhalb des Abwurfs der Dosierschnecke einer Differentialdosierwaage nachgerüstet. Da das Coriolis-Meßgerät direkt in die Förderleitung einspeist, wurde die Schneckenpumpe umgangen und stillgesetzt. Zwischen dem Abwurf der Dosierschnecke und dem Coriolis-Meßgerät ist eine vertikal fördernde Zellenschleuse zur Entkoppelung des Förderleitungsdrucks eingebaut. Anstelle der Differentialdosierwaage übernimmt nun das Coriolis-Meßgerät die Erfas-

wards in a rotating system. It therefore requires a measuring wheel rotating at constant angular velocity. The wheel is driven by an electric motor and is shown diagrammatically in **Fig. 1**.

If a flow of bulk material is passed over the measuring wheel it is deflected in a radial direction. After being deflected it is picked up by the guide vanes of the wheel, with the particles being accelerated in a radial direction by centrifugal force and in a circumferential direction by the Coriolis force. Various forces are at work in the movement of the particles over the guide vanes.

The centrifugal force Fz and opposing frictional force Fr are exerted in a radial direction, whereas the Coriolis force Fc acts in a tangential direction and produces a reaction torque M. The torque has to be compensated for by the measuring wheel drive and is recorded by a torque measuring device.

The mass flow of bulk material is directly proportional to the torque to be applied. Frictional forces between particles of bulk material and the measuring wheel or between layers of bulk material have no effect on the measured result. The process can accordingly measure mass flows of bulk material with a high degree of accuracy, with less than ± 0.5 % deviation from the set flow. Owing to the extremely short dwell time of the bulk material in the measuring equipment (see **Fig. 3**), only fractions of a second, fluctuations in the mass flow of pulverized coal can be detected very quickly.

The application

A suitable controlled feeding device is necessary to stabilise fluctuations in the mass flow. There are various options

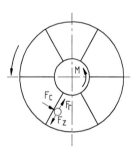

BILD 2: Kräfte bei der Bewegung der Partikel
FIGURE 2: Forces involved in moving the particles

BILD 3: Coriolis-Massenstrommeßgerät
FIGURE 3: The Coriolis mass flow measuring device

sung des Schüttgutstroms. Der ermittelte Istwert wird in der Auswerteelektronik mit dem Sollwert verglichen und die Dosierschnecke entsprechend nachgeregelt. Die zu einer Kontrollwaage umfunktionierte Differentialdosierwaage hat die Aufgabe, die Kurzzeitkonstanz und die Langzeitstabilität des Coriolis-Dosiersystems nachzuweisen.

Im Einlaufbereich der Dosierschnecke sind Rührwerke eingebaut, die das Schüttgut homogenisieren und einen gleichmäßigen Füllgrad der Dosierschnecke sicherstellen. Das Coriolis-Massendurchflußmeßgerät läßt sich in ein solches System einfach einbauen, da es von dem pneumatischen Fördersystem ausgeblasen wird und das Schüttgut direkt in die Förderleitung eingetragen wird. Um die Dosierschnecke vor dem Überdruck der Förderleitung zu schützen, wird eine Zellenschleuse als Dichtschleuse eingesetzt. Sie arbeitet bei konstanter Drehzahl mit teilgefüllten Kammern, sodaß Pulsationen vermieden werden. Infolge der Teilfüllung und einer neuentwickelten, verschleißgeschützten Zellenradabdichtung werden hohe Standzeiten der Zellenradschleuse erreicht.

Die in **Bild 5** dargestellte Lösung stellt eine Variante dar, durch die der bauliche Aufwand für ein Kohlenstaubdosiersystem erheblich reduziert werden kann.

Hier dient eine Dosierschleuse mit vertikaler Drehachse, die das Schüttgut in horizontaler Richtung bewegt, als geregelter Zuteiler. Sie sperrt zugleich das darüber liegende Silo gegen die Förderleitung ab. Ein großer Einlaufdurchmesser verhindert zusammen mit einem Rührwerk eine Brückenbildung des Schüttgutes und bewirkt eine gute und gleichmäßige Füllung der Kammern.

Im Spätsommer 1993 wurde die Pilotanlage in Betrieb genommen. Seitdem wird der Drehrohrofen mit wenigen geplanten kurzzeitigen Unterbrechungen im 24-Stunden-Betrieb von dem Coriolis-Dosiersystem mit Brennstoff versorgt. Die Ergebnisse bestätigen und übertreffen teilweise die Erwartungen (**Tabelle 2**).

Die über den Vergleich mit der Kontrollwaage nachgewiesene Kurzzeitkonstanz, auch als Dosierkonstanz bezeichnet, erreicht überwiegend Werte besser als 1 % bei 10 Sekunden

depending on the metering method. **Fig. 4** shows a pilot plant where a metering screw is used as the controlled feeding device.

For this pilot application a Coriolis measuring device was retrofitted below the discharge point of the metering screw of a differential metering weigher. Since the Coriolis device feeds directly into the delivery line the spiral pump has been bypassed and stopped. A vertically conveying rotary lock for isolating the delivery line pressure is fitted between the discharge from the metering screw and the Coriolis device. The Coriolis device takes over the recording of the stream of bulk material from the differential metering weigher. The actual value determined is compared with the desired value in the electronic evaluating system, and the metering screw is readjusted accordingly. The differential metering weigher, now operating as a check weigher, is responsible for indicating the short-term constancy and long-term stability of the Coriolis metering system.

Agitators are fitted in the inlet section of the metering screw to homogenise the bulk material and ensure uniform filling of the screw. The Coriolis mass flow measuring device can easily be included in such a system, since the pneumatic conveying system blows through it and the bulk material is carried directly into the delivery line. A rotary lock is provided as a seal to protect the metering screw from the excess pressure in the delivery line. It works at a constant speed with partly filled chambers, so that pulsation is avoided. The rotary lock has a long service life owing to the partial filling and a newly developed, wear-protected lock seal.

The arrangement shown in **Fig. 5** is another version which can considerably reduce the constructional outlay on a pulverized coal metering system.

Here a metering lock with a vertical axis of rotation, which moves the bulk material in a horizontal direction, is used as the controlled feeding device. It also isolates the hopper above it from the delivery line. A large inlet diameter together with an agitator prevents the bulk material from bridging and ensures good, uniform filling of the chambers.

The pilot plant came into operation in late summer 1993. Since then the rotary tube kiln has been supplied with fuel

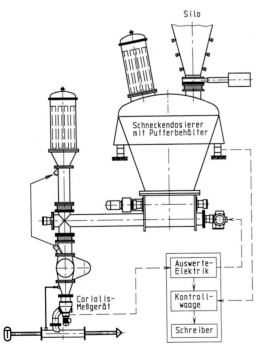

BILD 4: Pilotanwendung Coriolis-Dosiersystem
FIGURE 4: A pilot application of the Coriolis metering system

Silo = hopper
Schneckendosierer mit Pufferbehälter = screw metering unit with buffer container
Auswerte-Elektrik = electric evaluating system
Coriolis-Meßgerät = Coriolis measuring device
Kontrollwaage = check weigher
Schreiber = recording instrument

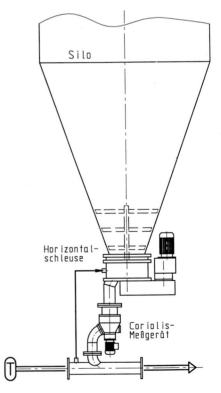

BILD 5: Coriolis-Dosiersystem mit Horizontalschleuse
FIGURE 5: Coriolis metering system with horizontal lock

Horizontalschleuse = horizontal lock
Coriolis-Meßgerät = Coriolis measuring device

TABELLE 2: Coriolis-Meßgerät mit Dosierer
Erfüllung der Anforderungen

Dosier-systeme	Kurzzeit-konstanz	Genauigkeit	Kosten	Sekundär-energie
Differential-dosierwaage	teilweise erfüllt	erfüllt	hoch	hoch
Dosier-rotorwaage	nicht erfüllt	erfüllt mit KME	mittel	niedrig
Prallplatte mit Dosierer	nicht erfüllt	erfüllt mit KME	hoch	hoch
Coriolis-Meßgerät mit Dosierer	erfüllt	erfüllt mit KME	niedrig	niedrig

TABLE 2: Coriolis measuring device with metering equipment
Meeting the requirements

Metering systems	Short-term constancy	Accuracy	Cost	Secondary energy
Differential weighing	partly met	met	high	high
Weighing with rotor	not met	met with CME	moderate	low
Baffle with metering equipment	not met	met with CME	high	high
Coriolis measuring device with metering equipment	met	met with CME	low	low

TABELLE 3: Kurzzeitdosierkonstanz

Sollförderstärke	:	10 933 kg/h
Versuchs-Nr.	:	537

Nr.	Wägeergebnis kg	Meßwert kg
0	836,11	0,00
1	805,98	−30,13
2	776,17	−29,81
3	745,70	−30,46
4	715,50	−30,20
5	685,28	−30,23
6	654,89	−30,39
7	624,36	−30,53
8	593,81	−30,54
9	563,66	−30,16
10	533,19	−30,47

Meßintervall	:	9,953 sec
Meßwert soll	:	−30,23 kg
Meßwert gemittelt	:	−30,29 kg
Dosiergenauigkeit	:	+ 0,21 %
Standardabweichung	:	0,23 kg
Dosierkonstanz	:	0,76 %

TABLE 3: Short-term metering constancy

Desired conveying strength	:	10 933 kg/h
Test no.	:	537

Nr.	Weighed result kg	Measured value kg
0	836.11	0.00
1	805.98	−30.13
2	776.17	−29.81
3	745.70	−30.46
4	715.50	−30.20
5	685.28	−30.23
6	654.89	−30.39
7	624.36	−30.53
8	593.81	−30.54
9	563.66	−30.16
10	533.19	−30.47

Measuring interval	:	9.953 sec
Measured value − desired	:	−30.23 kg
Measured value − mean	:	−30.29 kg
Metering accuracy	:	+ 0.21 %
Standard deviation	:	0.23 kg
Metering constancy	:	0.76 %

Meßintervallen, ermittelt als Standardabweichung nach DIN 1319 (**Tabelle 3**).

Durch die direkte Einspeisung in die Förderleitung unter Umgehung der Schneckenpumpe wurde der Förderleitungsdruck von 0,9 auf 0,6 bar reduziert. Es kann zukünftig nicht nur die Investition der Schneckenpumpe eingespart, sondern auch die Gebläseleistung reduziert werden. Dadurch verringert sich der Sekundärenergiebedarf der Kohlenstaubdosierung um ein Drittel. Dies ist eine erste reale Verbesserung durch Einsparung an Energiekosten von ca. DM 40 000/Jahr.

Das Istleistungssignal des Coriolis-Dosiersystems ist gegenüber dem anderer Dosierprinzipien frei von äußeren Einflüssen wie z. B. Wind oder Vibrationen und daher ausgesprochen stabil. Für die Ofenführung resultiert aus dieser Störunanfälligkeit eine Verbesserung.

Der Betrieb der Pilotanlage zeigt, daß die Wirtschaftlichkeit der Klinkerproduktion durch ein konstanteres und genaueres Brennstoff/Luft-Gemisch sowie durch Reduzierung von installierter Antriebsleistung verbessert wird.

Die Gefahr der Grenzwertüberschreitung bei den Abgasemissionen reduziert sich aufgrund der guten Dosierkonstanz drastisch.

Durch die von äußeren Einflüssen freie und schnelle Istwerterfassung des Coriolis-Meßgerätes sowie der daraus resultierenden schnellen Regelung ist eine flexible Ofenführung auf der Basis stabiler Parameter möglich.

by the Coriolis metering system in a 24-hour operation with only a few brief, planned interruptions. The results confirm and in some cases exceed expectations (**Table 2**).

The short-term constancy, also termed metering constancy, evidenced by comparison with the check weigher, normally reaches values better than 1 % at 10-second measuring intervals, determined as a standard deviation to DIN 1319.

Direct feeding into the delivery line, bypassing the screw pump, reduced the pressure in the delivery line from 0.9 to 0.6 bar. In future not only can the capital cost of the screw pump be saved, but the fan capacity can also be reduced, thereby decreasing the secondary energy requirement for pulverized coal metering by one-third. This is a very real improvement obtained by saving energy costs of about DM 40 000 p.a.

Unlike other metering methods the Coriolis system has an actual-value signal which is free from external influences such as wind or vibrations and is therefore extremely stable. This stability leads to an improvement in kiln operation.

It has been found in operating the pilot plant that clinker production becomes more economic by having a more constant and accurate fuel/air mix and reducing the installed drive capacity.

The danger of exceeding the limits for exhaust gas emission is greatly reduced by the very constant metering action.

The Coriolis measuring device recording the actual values quickly and unaffected by external influences, and the resultant quick control action, facilitate flexible kiln operation based on stable parameters.

Technische Neuerungen im Rekuperationsbereich der Klinkerkühler

Technical innovations in recuperation zones of clinker coolers

Nouveautés techniques dans la zone de récupération des refroidisseurs de clinker

Novedades técnicas en la zona de recuperación de los enfriadores de clínker

Von **G. Kästingschäfer,** Köln/Deutschland

Zusammenfassung – Für die Wirtschaftlichkeit des Klinkerbrennprozesses ist die Verbesserung des Rekuperationsverhaltens der Klinkerkühler von besonderer Bedeutung. In modernen Schubrostkühlern wird das durch eine insgesamt gleichmäßigere Klinkerverteilung auf dem Kühlerrost und eine darauf abgestimmte Kühlluftverteilung erreicht. Die Kühlroste werden dazu in einzelne Belüftungsfelder unterteilt, die individuell belüftet werden. Der ofenfallende Klinker wird außerdem durch Schubbewegungen des ersten Rostes schnell querverteilt. Nach Abschluß der Anfangskühlung und Querverteilung wird die Klinkerschicht auf die maximal belüftbare Schichtdicke (ca. 800 mm) aufgebaut. Das geschieht durch unterschiedliche Transportgeschwindigkeiten der geneigten und der horizontalen Jet-Ring-Platten. Zur Beurteilung der Gleichmäßigkeit der Klinkerkühlung kann die Oberflächentemperatur des Klinkers mit einem Infrarotscanner erfaßt werden. Diese Technologie bietet in der Zukunft weitere Möglichkeiten zur Optimierung der Regelungs- bzw. Steuerungsstrategie von Klinkerkühlern.

Technische Neuerungen im Rekuperationsbereich der Klinkerkühler

Summary – Improving the recuperating behaviour of the clinker cooler is very important for the cost-effectiveness of a clinker burning process. In modern reciprocating grate coolers this is achieved by a much more uniform clinker distribution on the cooler grate and a cooling air distribution system adapted to suit this. The cooler grates are subdivided into individual aeration fields which are aerated independently. In addition to this there is rapid lateral distribution of the clinker falling from the kiln through reciprocating movements of the first grate. After the initial cooling and lateral distribution is completed the clinker layer is set to the maximum bed depth which can be aerated (approx. 800 mm). This is carried out by differing transport speeds of the inclined and horizontal jet-ring plates. The surface temperature of the clinker can be measured with an infrared scanner to assess the uniformity of clinker cooling. This technology offers further possibilities in the future for optimizing the open- and closed-loop control strategies for clinker coolers.

Technical innovations in recuperation zones of clinker coolers

Résumé – Pour la rentabilité du process de cuisson du clinker, l'amélioration du pouvoir de récupération du refroidisseur de clinker est d'une importance particulière. Dans les refroidisseurs modernes à grille poussée, cela est obtenu par une répartition globalement plus régulière sur la grille de refroidissement et par une distribution correspondante de l'air de refroidissement. A cette fin, les grilles de refroidissement sont divisées en champs de ventilation distincts. De plus, le clinker tombant du four est rapidement réparti latéralement par les mouvements de poussée de la première grille. Le refroidissement primaire et la répartition terminés, la couche de clinker est portée à l'épaisseur maximale ventilable (environ 800 mm). Cela est obtenu par des vitesses de transport différentes des plaques „Jet-Ring" inclinées et horizontales. Pour évaluer la régularité du refroidissement du clinker, il est possible de prendre la température superficielle du clinker à l'aide d'un scanner infrarouge. Cette technologie offre, dans le futur, d'autres possibilités pour l'amélioration des stratégies de régulation et de conduite des refroidisseurs de clinker.

Nouveautés techniques dans la zone de récupération des refroidisseurs de clinker

Resumen – Para la rentabilidad del proceso de cocción del clínker es de especial importancia la mejora del comportamiento a la recuperación de los enfriadores de clínker. En los modernos enfriadores de parrilla de vaivén se consigue esto mediante una distribución generalmente más uniforme del clínker sobre la parrilla del enfriador y una correspondiente distribución del aire de enfriamiento. Para ello, las parrillas de

Novedades técnicas en la zona de recuperación de los enfriadores de clínker

enfriamiento se dividen en varios sectores de aireación individual. Además, el clínker que sale del horno se reparte rápidamente en sentido transversal, debido a los movimientos de vaivén de la primera parrilla. Después de terminar el enfriamiento inicial y la repartición transversal, se va formando la capa de clínker hasta alcanzar el espesor máximo admisible para la aireación (unos 800 mm). Esto se logra mediante diferentes velocidades de transporte de las placas „jet-ring", inclinadas y horizontales. Para evaluar la uniformidad del enfriamiento de clínker se puede medir la temperatura en la superficie del clínker por medio de un scanner infrarrojo. Esta tecnología ofrecerá en el futuro más posibilidades de optimización de la estrategia de regulación y mando de los enfriadores de clínker.

1. Einleitung

In modernen Schubrostkühlern sind die Kühlroste in einzelne Belüftungsfelder unterteilt, die individuell belüftet werden. Die gezielte Einstellung der Luftmengen auf die einzelnen Belüftungsfelder wird durch die permanente Erfassung der Oberflächentemperatur des Klinkers, z.B. mit einem Infrarot Line Scanner, wesentlich verbessert.

2. Belüftungssystem mit Einzel- und Mehrplattenbelüftung

Die Kühlroste in Schubrostkühlern sind in individuell belüftbare Felder eingeteilt, die mit speziellen Platten bestückt sind (**Bild 1**). Die Platten sind so konstruiert, daß der Klinkerstaub nicht in das Belüftungssystem gelangen kann. Einige Anbieter versehen den thermisch hoch belasteten Einlaufbereich mit besonders geschützten Rostplatten.

1. Introduction

In modern reciprocating-grate coolers the cooler grates are subdivided into individual aeration fields which are aerated independently. Accurately controlled adjustment of the volumes of air to the individual aeration fields is greatly improved by permanently recording the surface temperature of the clinker e.g. with an infrared line scanner.

2. Aeration system with single and multiple plate aeration

The cooler grates in reciprocating grate coolers are divided into fields which can be aerated individually and which are fitted with special plates (**Fig. 1**). The plates are designed not to let the clinker dust into the aerating system. Some suppliers provide specially protected grate plates in the inlet zone, where the thermal stress is high. In the REPOL-RS jet ring plates with clinker boxes are used. After the inclined inlet zone the cooler grate is horizontal, and the horizontal cooling surfaces are fitted with jet ring plates without clinker boxes. The aeration fields grow larger in the direction of flow, corresponding to the wider clinker distribution. A seal air fan is used in addition to the fans supplying the individual aeration fields.

The jet ring plate with a clinker box which is inserted in the very hot inlet zone is illustrated in **Fig. 2**. The box, which is permanently filled with clinker, covers over 70% of the sur-

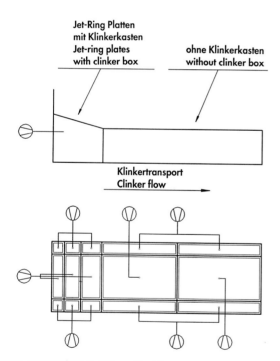

BILD 1: REPOL®RS-Belüftung Rost 1
FIGURE 1: REPOL®RS-aeration of grate 1

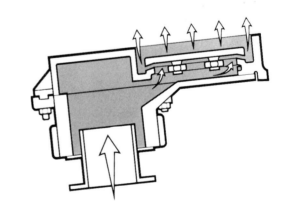

Im REPOL-RS werden Jet-Ring-Platten mit Klinkerkasten eingesetzt. Nach dem geneigten Einlaufbereich verläuft der Kühlrost horizontal. Die horizontalen Kühlflächen sind mit Jet-Ring-Platten ohne Klinkerkasten bestückt. In Transportrichtung werden die Belüftungsfelder entsprechend der zunehmenden Klinkerverteilung größer. Zusätzlich zu den Ventilatoren, die die einzelnen Belüftungsfelder versorgen, wird ein Sperrluftventilator eingesetzt.

Die im thermisch hoch belasteten Einlaufbereich eingesetzte Jet-Ring-Platte mit Klinkerkasten ist im **Bild 2** dargestellt. Der permanent mit Klinker gefüllte Klinkerkasten bedeckt mehr als 70 % der Plattenoberfläche und schützt so die Platten auch bei extremer thermischer Beanspruchung

BILD 2: Jet-Ring-Platte mit Träger
FIGURE 2: Jet-ring-plate with support

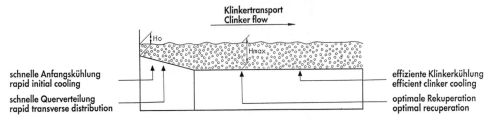

BILD 3: Stufenkonzept REPOL®RS
FIGURE 3: Step concept REPOL®RS

vor Überhitzung und Verschleiß. Die durch die zwei ringförmigen Jet-Ringe geführte Luft kühlt direkt die schmalen verbleibenden Plattenbereiche, die mit dem heißen Klinker in Berührung kommen. Der engste Luftdurchtritt in der Platte ist so bemessen, daß die Platte einen Vorwiderstand von 25 mbar bei einer Plattenbeaufschlagung von 100 Nm³/m² min hat.

3. Stufenkonzept

Stufenkonzept heißt: zunehmende Klinkerschicht in Transportrichtung des Kühlers, wie auf **Bild 3** dargestellt.

Der ofenfallende Klinker wird separiert und haufenförmig auf den Kühlrost abgeworfen. Dieser Klinkerhaufen muß möglichst schnell belüftet werden, um das Verfritten des glühenden Klinkers zu verhindern. Dafür ist im Einlaufbereich des Kühlers eine relativ dünne Klinkerschicht (ca. 500 mm) erforderlich. Anschließend kann das Klinkerbett kontinuierlich auf ca. 800 mm erhöht werden, um die Rekuperation und die Klinkerkühlung zu optimieren.

Anfangskühlung und Querverteilung

Die schnelle Querverteilung des Klinkers, schon im Einlaufbereich des Kühlers, wird durch die Schubbewegung jeder 2. Plattenreihe erreicht. Unterstützt wird die Querverteilung durch die Zwangsbelüftung des Klinkerbettes mit dem feinmaschigen Belüftungssystem. Die intensive, gleichmäßige Belüftung und die mechanische Bewegung ermöglichen eine schnelle Anfangskühlung und Querverteilung des Klinkers. Praktische Erfahrungen und Betriebsmessungen am REPOL-RS zeigen, daß die Querverteilung des Klinkers nach dem Einlaufbereich (ca. 12 Plattenreihen) vollständig abgeschlossen ist.

Erhöhung der Klinkerschicht

Nach der erfolgten Anfangskühlung und Querverteilung des Klinkers wird die Klinkerschicht auf die max. belüftbare Schichtdicke (ca. 800 mm) erhöht. Die Erhöhung des Klinkerbettes geschieht durch das unterschiedliche Transportverhalten der geneigten und der horizontalen Jet-Ring-Platten. Die um 4° geneigten Platten im Einlaufbereich des Kühlers transportieren den Klinker schneller als die folgenden horizontalen Kühlflächen.

4. Beurteilung und Einstellung der Luftverteilung

Die optimale Einstellung der Luftversorgung der einzelnen Felder ist für die Effizienz von Schubrostkühlern außerordentlich wichtig. Zum einen muß die Luftmenge im Einlaufbereich minimiert werden, um optimal zu rekuperieren, zum anderen muß für eine ausreichende Plattenkühlung und gleichmäßige Klinkerbelüftung gesorgt werden, um das Verfritten des Klinkers zu verhindern.

Einstellung der Luftverteilung

Die Luftversorgung der einzelnen Belüftungsfelder wird mit manuell betätigten Drosselklappen in den Zuführungsleitungen außerhalb des Kühlers eingestellt (Bild 1). Diese manuelle Einstellung ist ausreichend, da die Luftverteilung nur während der Inbetriebnahme des Kühlers und abschließend bei der Kühler- und Anlagenoptimierung eingestellt wird.

face of the plates and thus protects them from overheating and wear even under extreme thermal conditions. The air flowing through the two circular jet rings directly cools the remaining narrow plate areas which come into contact with the hot clinker. The narrowest air passage in the plate is of a size which gives the plate an external resistance of 25 mbar when impinged on by a force of 100 Nm³/m² min.

3. The step concept

The „step concept" means that the clinker layer becomes deeper in the flow direction of the cooler as illustrated in **Fig. 3**.

The clinker falling from the kiln is separated and thrown onto the cooling grate in heaps. The glowing clinker in the heaps must be aerated as quickly as possibly to prevent it from sintering. So the layer in the inlet zone of the cooler must be relatively shallow (approx. 500 mm). The clinker bed can then be raised continuously to approx. 800 mm to optimise recuperation and cooling.

Initial cooling and transverse distribution

Rapid transverse distribution of the clinker, even in the inlet zone of the cooler, is achieved by reciprocating movements of alternate rows of plates and assisted by forced aeration of the clinker bed with the fine-mesh aerating system. The intensive uniform aeration and the mechanical movement allow rapid initial cooling and transverse distribution of the clinker. Practical experience and measurements taken when the REPOL-RS is in service show that the transverse distribution of the clinker is completed by the end of the inlet zone (about 12 rows of plates).

Raising the clinker layer

After initial cooling and lateral distribution of the clinker the layer is raised to the maximum thickness which can be aerated (approx. 800 mm). The bed is raised by the different conveying action of the inclined and horizontal jet ring plates. The plates in the inlet zone of the cooler, which are inclined at 4°, convey the clinker faster than the horizontal cooling surfaces which follow.

4. Assessment and adjustment of air supply

Optimum setting of the air supply to the individual fields is extremely important to the efficiency of reciprocating grate coolers. On the one hand the volume of air in the inlet zone must be minimised to obtain optimum recovery, while on the other hand adequate plate cooling and uniform clinker aeration must be ensured to prevent the clinker from sintering.

Adjustment of air distribution

The air supply to the individual aeration fields is set with manually operated throttle valves in the feed pipes outside the cooler (Fig. 1). This manual setting is sufficient, since air distribution is only adjusted during the commissioning of the cooler and finally when the cooler and plant are optimised.

Recording clinker aeration

With the differing clinker aeration in the inlet zone the manometer-read air pressure in the individual feed pipes is not

Erfassung der Klinkerbelüftung

Bei der unterschiedlichen Klinkerbelüftung im Einlaufbereich ist der mit Manometern gemessene Luftdruck in den einzelnen Zuführungsleitungen keine ausreichende Kenngröße zur Beurteilung einer gleichmäßigen Klinkerbelüftung. Eine direkte Kenngröße für die Qualität der Klinkerbelüftung bzw. der Luftverteilung ist die Klinkertemperatur auf der Oberfläche des Klinkerbettes unmittelbar nach dem Einlaufbereich des Kühlers. Hier muß sichergestellt sein, daß der vollständig querverteilte Klinker eine gleichmäßige Oberflächentemperatur über die gesamte Kühlerbreite hat.

Beim REPOL-RS wird für die Temperaturerfassung ein Infrarotscanner eingesetzt. Der Scanner tastet eine Linie quer zur Transportrichtung des Klinkers ab (**Bild 4**). Die Temperatur des Klinkers wird so permanent erfaßt und graphisch als Thermovision dargestellt. Der Temperaturverlauf ist ein direktes Abbild der Klinkerbelüftung. Damit kann anhand der Thermovision die Luftverteilung optimiert werden.

5. Meßergebnisse (Thermovision)

Die Aneinanderreihung von zeitlich folgenden Temperaturscans (Thermovision, **Bild 5**) verdeutlicht die Unterschiede im Temperaturprofil des Klinkerbettes, die bei einer unzureichenden Klinkerbelüftung entstehen.

Die in diesem Bild dargestellte Thermovision wurde während der Testphase des Temperaturscanners in einem herkömmlichen Schubrostkühler mit Kammerbelüftung erzeugt. Auf der linken Seite ist deutlich ein Red River mit Temperaturen von ca. 800 °C zu erkennen.

Diese extremen Zustände können durch die Einzel- und Mehrplattenbelüftung in modernen Schubrostkühlern kaum auftreten. Die Erfassung der Klinkertemperatur mit der graphischen Darstellung als Thermovision dient zur Optimierung der Kühler. Die Kühlluftmenge wird minimiert, die Rekuperation und Klinkerkühlung verbessert.

6. Perspektive

Die permanente Erfassung der Klinkertemperatur über die gesamte Kühlerbreite bietet weitere Möglichkeiten bei der Optimierung des Klinkerkühlers, z.B. bei der Regelungs- und Steuerungsstrategie, die den gesamten Brennprozeß beeinflussen.

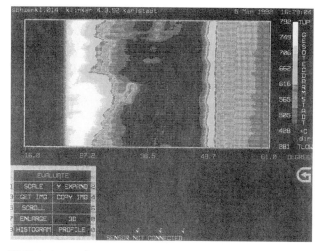

BILD 4: REPOL®RS-Thermovision
FIGURE 4: REPOL®RS-thermovision

an adequate indication for judging the uniformity of clinker aeration. A direct indication of the quality of clinker aeration and air distribution is the clinker temperature on the surface of the bed immediately after the inlet zone of the cooler. It is here that we must ensure that the fully distributed clinker has a uniform temperature over the whole width of the cooler.

In the REPOL-RS an infrared scanner is used to record temperature. It scans a line transversely to the direction of flow of the clinker. The clinker temperature is thus permanently recorded and displayed graphically as a thermovision. The temperature behaviour is a direct image of the clinker aeration, and thus air distribution can be optimised by the thermovision.

5. Measured results (thermovision)

Alignment of a time sequence of temperature scans (thermovision, **Fig. 5**) brings out the differences in the temperature profile of the bed which arise when the clinker is inadequately aerated.

The thermovision shown in this figure was produced during the test phase of the temperature scanner in a conventional reciprocating grate cooler with compartmentalised aeration. A red river with temperatures of about 800 °C can be seen clearly on the left hand side.

These extreme conditions can hardly occur with single and multiple plate aeration in modern reciprocating grate coolers. The purpose of recording the clinker temperature and displaying it graphically as a thermovision is to optimise the cooler. The volume of cooling air is minimised and recovery and clinker cooling improved.

6. Prospects

Permanent recording of clinker temperature over the whole width of the cooler provides further possibilities for optimising the cooler, e.g. in respect of its control strategy, which will affect the whole burning process.

Gyro-Therm-Low-NO$_x$-Brenner
Gyro Therm Low NO$_x$ Burner

Brûleur Gyro-Therm-Low-NO$_x$

Quemador Gyro-Therm-Low-NO$_x$

Von **C. G. Manias** und **G. J. Nathan,** Adelaide/Australien

Zusammenfassung – Ein von Forschern der Universität Adelaide entdecktes neues Gasstrahl-Strömungsphänomen hat zur Entwicklung eines neuartigen und dennoch einfachen Gasbrenners (Präzessions-Strahlbrenner) geführt, der enorme Vorteile bietet. Die Entwicklung erfolgte in den letzten beiden Jahren und ist eine Gemeinschaftsarbeit der Adelaide Brighton Cement Ltd. und einer Forschergruppe an der Universität Adelaide. Sie basiert auf Versuchen im industriellen Maßstab und einer noch andauernden Grundlagenforschung. Der Präzessions-Strahlbrenner hat die Fähigkeit bewiesen, eine extrem kurze, scharfe und helle Gasflamme ohne Verwendung von Primärluft und bei extrem geringer NO$_x$-Bildung im Vergleich zu anderen Gasbrennern zu erzeugen. Während einer Reihe von Betriebsversuchen in einem 300 t/d-Naßofen mit Gasfeuerung wurde der Erdgasbrenner sowohl mit Präzessions-Strahldüse, die eine kurze, scharfe Flamme erzeugt, als auch mit einfacher Rohrdüse, die eine lange, träge Flamme erzeugt, in Tandemanordnung ausgerüstet. Die anteilige Gasmenge der beiden Düsen kann variiert werden, um die Eigenschaften der Flamme zu verändern. Die angegebenen Daten bestätigen die Vorteile des Präzessions-Strahlbrenners in Form niedriger NO$_x$-Emissionen, einer kurzen, scharfen, helleren und stabileren Flamme auch bei einem weiten Regelbereich.

Summary – A new flow phenomenon of gas jets discovered by Adelaide University researchers has led to the development of a novel, yet simple gas burner (the Precessing Jet Burner), with enormous benefits. The development has taken place over the past two years and has been a collaborative effort between Adelaide Brighton Cement Ltd. and Adelaide University Researchers. There has been a combination of plant scale trials and continuing fundamental research applied to the development. The Precessing Jet Burner (P.J. Burner) has demonstrated a capability to produce an extremely short, sharp and luminous gas flame, without the use of any primary air and with extremely low NO$_x$ generation in comparison to other gas burners. During series of plant trials on a 300 t/d gas fired wet kiln, the natural gas burner was used with both a Precessing Jet (P.J.) nozzle, which by itself produces a short sharp flame, and plain pipe nozzle, which by itself produces a long, lazy flame, in tandem. Gas proportions between the two can be varied to adjust flame characteristics. The data given confirm the attributes of the P.J. Burner as low NO$_x$ emission, a short, sharp and more luminous and stable flame even over a wide range of turndown.

Résumé – Un phénomène d'écoulement à jet de gaz nouvellement découvert par l'Université d'Adelaide a conduit au développement d'un brûleur à gaz de conception nouvelle mais simple (brûleur à jet de gaz de précession), présentant d'énormes avantages. Cette mise au point a été réalisée durant les deux dernières années avec la coopération de Adelaide Brighton Cement Ltd. et d'une groupe de chercheurs de l'Université d'Adelaide. Elle repose sur des essais à échelle industrielle et un travail de recherche fondamentale encore en cours actuellement. Le brûleur à jet de gaz de précession a prouvé qu'il était en mesure de produire une flamme gazeuse claire, vive et extrêmement courte sans utilisation d'air primaire et avec très faible formation de NO$_x$ en comparaison avec d'autres brûleurs. Durant toute une série de tests de marche dans un four à voie humide de 300 t/j avec système de chauffe à gaz, le brûleur à gaz naturel a été équipé en tandem aussi bien d'un éjecteur de précession, produisant une flamme courte et vive qu'une buse simple produisant une flamme longue et lente. La quantité de gaz respective des deux buses peut varier pour changer les propriétés de la flamme. Les données indiquées confirment les avantages du brûleur à jet de précession sous forme de faibles émissions de NO$_x$ d'une flamme courte, vive, plus claire et plus stable, et ce dans une large zone de régulation.

Resumen – Un nuevo fenómeno de flujo del chorro de gas, descubierto por los investigadores de la Universidad de Adelaide, ha dado lugar al desarrollo de un quemador de gas nuevo, pero francamente sencillo (quemador de chorro de gas, de precesión), el cual ofrece unas ventajas enormes. Los trabajos de desarrollo se ha llevado a cabo en los dos últimos años; se trata de una cooperación entre Adelaide Brigthon Cement Ltd. y un grupo de investigadores de la Universidad de Adelaide. Se basa en unos ensayos efectuados a escala industrial y una investigación básica aún sin concluir. El quemador de chorro de gas, de precesión, ha demostrado que es capaz de formar una llama de gas extremadamente corta, viva y clara, sin emplear aire primario y con una formación muy reducida de NO$_x$ en comparación con otros quemadores de gas. Durante una serie ensayos prácticos, realizados en un horno de vía húmeda, de 300 t/d, con combustión de gas, el quemador de gas natural ha sido equipado en tandem,

tanto con una tobera a chorro, de precesión, que produce una llama corta y viva, como también con una tobera sencilla que deja una llama larga y lenta. El caudal proporcional de gas de ambas toberas se puede variar, con el fin de modificar las características de la llama. Los datos indicados confirman las ventajas del quemador de chorro de gas, de precesión, que consisten en una reducida emisión de NO_x, una llama corta, viva, más clara y estable, incluso para un amplio margen de regulación.

Einleitung

Der vorliegende Artikel beschreibt einen Erdgasbrenner für Drehöfen der Zementindustrie, der in Zusammenarbeit der Universität von Adelaide mit der Firma Adelaide Brighton Cement entwickelt wurde.

Adelaide Brighton Cement (ABC) hat rund 25 Jahre Erfahrung mit dem Betrieb von Zementöfen mit Erdgas. Gegenwärtig werden alle 12 Ofenlinien der Gruppe mit Erdgas befeuert. Die Probleme im Zusammenhang mit der langen Flamme konventioneller Brenner, die strahlungsarm ist und hohe NO_x-Konzentrationen im Drehofenabgas erzeugt, sind allgemein bekannt. Der GYRO-THERM-LOW-NO_x-Erdgasbrenner wurde unter Zugrundelegung der neuartigen und patentierten Präzessions-Strahltechnik (Precessing Jet Technology) entwickelt.

ABC hat Betriebsversuche in technischem Maßstab durchgeführt, die die folgenden Merkmale des GYRO-THERM-LOW-NO_x-Brenners bewiesen:

— einfache Konstruktion ohne Primärluft,
— kurze Flamme und direkte Regelung der Flammenform für optimale Produktqualität,
— gute Wärmeübertragung durch verstärkte Strahlung und verbesserten Wirkungsgrad,
— exzellente Flammenstabilität in einem hohen Stellbereich (100:1),
— 50%ige Reduzierung der Stickoxidemissionen.

Funktionsweise

Die in der Prozeßtechnik gegenwärtig eingesetzten Erdgasbrenner verfügen über Hochgeschwindigkeitsdüsen für Gas und Primärluft, die eine gute Vermischung von Brennstoff und Luft sowie eine stabile Flamme gewährleisten. Der GYRO-THERM-LOW-NO_x-Brenner nutzt ein natürlich auftretendes Strömungsphänomen zur Erzeugung von Turbulenz, die um ein Mehrfaches größer ist als die eines einfachen turbulenten Strahls. Die Düse besteht, wie in **Bild 1** dargestellt, aus einer axialsymmetrischen Kammer, die einen großen, sich plötzlich öffnenden Expansionsraum am Eingang und eine kleine Lippe am Ausgang hat. Der Brennstoffstrahl, der in die Kammer eintritt, läuft asymmetrisch an der Innenseite der Wandung des Expansionsraumes entlang und wird am Ausgang der Düse durch starke lokale Druckunterschiede in einem großen Winkel (typischerweise 45°) von der Düsenachse abgelenkt. Bild 1 zeigt ein vereinfachtes Schemabild der Strömung. Große azimutale Druckgefälle rufen eine Kreiselbewegung (Präzession) des

Introduction

This paper describes a natural gas kiln burner developed jointly by the University of Adelaide and Adelaide Brighton Cement.

Adelaide Brighton Cement has some 25 years' experience firing cement kilns with natural gas. At present all 12 kiln lines in the group are natural gas fired. The problems associated with the long flame of conventional burners, which produce poor luminosity and high NO_x generation are well known. The GYRO-THERM LOW NO_x gas burner was developed using the new and patented Precessing Jet technology.

Full scale plant trials conducted by ABC have shown the GYRO-THERM LOW NO_x gas burner has the following features:

— simple in construction, using no primary air,
— short flame length and on-line flame shape control for optimum product quality,
— high luminosity giving good radiant heat transfer and improved efficiency,
— excellent flame stability over a wide turndown ratio (100:1),
— 50% reduction of NO_x emissions.

Principle of operation

Natural gas burners currently available to the process industries rely on high momentum gas or primary air jets to entrain air into the fuel for combustion to occur.

The GYRO-THERM LOW NO_x burner utilises a naturally occurring flow phenomena to generate turbulence in a jet which is of scale several times larger than that of a simple turbulent jet. The nozzle, shown in **Fig. 1**, consists of an axi-

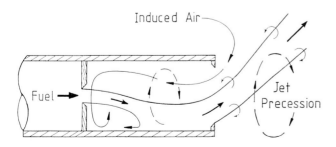

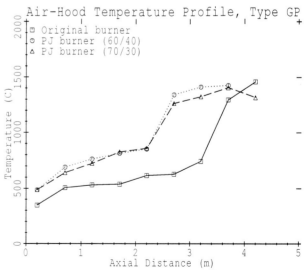

BILD 1: Schema des Brenners mit Präzessions-Strahl und eine vereinfachte Darstellung des Strömungsphänomens
FIGURE 1: A schematic diagram of the precessing jet burner and a simplified representation of the flow phenomenon

Brennstoff = Fuel
Luftzufuhr = Induced Air
Strahlpräzession = Jet precession

BILD 2: Temperaturprofil im Ofenkopf und in der Flammenfront mit unterschiedlichen Brennstoffanteilen durch den GYRO-THERM
FIGURE 2: Temperature profile through the air hood and in the flame front with different flow rates

Originalbrenner = Original burner
Präzessions-Strahlbrenner = PJ burner
Axialer Abstand = Axial Distance

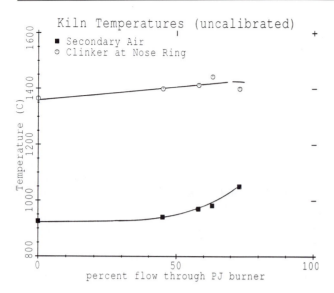

BILD 3: Temperaturen der Sekundärluft und des Klinkers an der Ofenkante
FIGURE 3: Temperatures of the secondary air and the clinker at the nose ring

Sekundärluft	= Secondary air
Klinker am Auslaufring	= Clinker at nose ring
Prozentualer Anteil des Brennstoffstroms durch den Präzessions-Strahlbrenner	= Percent flow through PJ burner

Strahls und des gesamten Strömungsfeldes in der Kammer um die Düsenachse hervor. Die Präzession des Gasstrahls hat eine weitgehende Durchmischung des Brennstoffs mit der Luft zur Folge, was eine geringe Zünddistanz, frühe Wärmeabgabe und gleichmäßige Flammentemperatur bewirkt. Die Flammenform scheint eine gewisse Stufung des Verbrennungsprozesses innerhalb einer großflächigen Turbulenz hervorzurufen, was mit der beobachteten Erhöhung der Flammenhelligkeit übereinstimmt.

Betriebserfahrungen

Bei den ersten Betriebsversuchen wurde eine GYRO-THERM-Düse parallel zu den normalen Ofenbrennern installiert und die Brennstoffanteile zwischen den beiden aufgeteilt, um die resultierende Flamme und den Effekt der Strahlpräzession auf den Verbrennungsprozeß beurteilen zu können.

symmetric chamber, which has a large sudden expansion at its inlet and small lip at its exit. The fuel jet which enters the chamber reattaches assymetrically to the inside of the cavity wall and, at the nozzle exit, is deflected at a large angle (typically 45°) from the nozzle axis by strong local pressure gradients. A simplified schematic of the flow is also shown in Fig. 1. Strong azimuthal pressure gradients cause the jet and the entire flow field within the chamber to precess about the nozzle axis. The frequency of precession is proportional to the jet velocity and inversely proportional to the scale of the burner and is typically 50 cycles per second. The gas jet precession results in large scale mixing of fuel and air giving reduced stand-off distance, early heat release and uniform flame temperature. The flame also appears to give some "self-staging" of the combustion process within the large-scale turbulence which is consistent with the observed increase in flame luminosity.

Operating experience

In the early plant trials, a GYRO-THERM nozzle was installed in parallel with the normal kiln burners and fuel proportions adjusted between the two to assess the resulting flame an PJ effect on the process.

1. Reduced flame length

Increasing fuel percentages through the GYRO-THERM nozzle produced an enormous visual impact. With 70% of the total fuel through the GYRO-THERM burner, the burning zone was reduced to about one half of its original length and its location moved from around 11 m to about 3 m in from the nose ring. The plant data measurements confirmed the dramatic visual evidence

— As shown in **Fig. 2**, suction pyrometer temperature readings in the firing hood and axially into the kiln, clearly confirm the early heat release of the GYRO-THERM.
— **Fig. 3** shows the increasing secondary air temperature and clinker temperature at the kiln discharge.
— **Figs. 4** and **5** show the clinker microscopy using the original burner and the GYRO-THERM respectively. The smaller well-formed crystals in Fig. 5 are clear evidence of better product quality associated with the GYRO-THERM.

2. Flame luminosity

The radiant heat transfer of energy from the flame to the kiln charge is highly dependent on the flame luminosity. The thick yellow coloured flame of the GYRO-THERM is in stark contrast to the relatively clear flame of the plain pipe burner. Laboratory measurements have shown that the luminosity

BILD 4: Schlecht ausgebildete Alitkristalle eines Klinkers, der mit dem vorhandenen Brenner hergestellt wurde
FIGURE 4: Clinker produced with the existing burner showing poorly formed alite

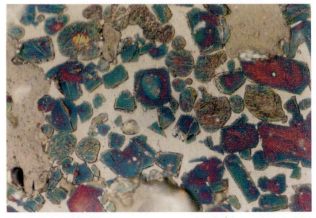

BILD 5: Gut geformte, kleine Alit-Kristalle eines Klinkers, der mit einem Brennstoffanteil von 60% durch den GYRO-THERM hergestellt wurde
FIGURE 5: Well formed clear small alite crystals, highly coloured showing good reactivity. This clinker formed using approximately 60% of the fuel fixed through the GYRO-THERM

1. Verminderte Flammenlänge

Zunehmende Brennstoffanteile in der GYRO-THERM-Düse hatten enorme sichtbare Auswirkungen auf die Flamme. Bei einem Anteil von 70 % des GYRO-THERM-Brenners am Gesamtbrennstoffdurchsatz reduzierte sich die Verbrennungszone auf die Hälfte ihrer ursprünglichen Länge, und ihr Ort verschob sich von Ofenmeter 11 auf ca. 3 (Entfernung vom Ofenauslauf). Die Messungen der Anlagedaten bestätigten die auffällige optische Erscheinung:

— Wie in **Bild 2** dargestellt, bestätigen die mit einem Absaugpyrometer im Ofenkopf und entlang der Ofenachse gemessenen Temperaturverläufe die frühe Wärmeabgabe des GYRO-THERM-Brenners.

— **Bild 3** zeigt die Zunahme der Sekundärlufttemperatur und der Klinkertemperatur am Ofenausgang.

— Die **Bilder 4** und **5** zeigen das mikroskopische Klinkergefüge jeweils bei Verwendung des ursprünglichen und des GYRO-THERM-Brenners. Die kleineren gutgeformten Kristalle in Bild 5 sind ein klarer Beweis höherer Qualität des Endproduktes bei Verwendung des GYRO-THERM-Brenners.

2. Flammenhelligkeit

Der Strahlungswärmeübertrag von der Flamme auf das Brenngut hängt in starkem Maße von der Flammenhelligkeit ab. Die straffe gelbe Flamme des GYRO-THERM steht in starkem Kontrast zur relativ klaren Flamme des Brenners mit einfachem Einblasrohr. Messungen im Labor haben gezeigt, daß die Helligkeit des GYRO-THERM-Brenners etwa 20 Mal größer ist als die eines handelsüblichen Drallbrenners, dessen Strahlungscharakteristik der eines einfachen Einblasbrenners ähneln dürfte.

Der verbesserte Wärmeübergang führt zu einem erhöhten Wirkungsgrad des Ofens, was im Betrieb durch einen erhöhten Klinkerausstoß und einen reduzierten Brennstoffverbrauch bestätigt wird. **Bild 6** zeigt die verminderten Ofeneinlauftemperaturen, **Bild 7** den verminderten Brennstoffverbrauch und den erhöhten Klinkermassenstrom.

3. NO$_x$-Emission

Die Entstehung von Stickstoffoxiden (NO, NO$_2$, N$_2$O) ist eine unvermeidliche Folge der Hochtemperaturverbrennung, wie sie in Zementöfen die Regel ist. Es existiert ein zunehmender ökologischer Druck zur Minimierung der NO$_x$-Emissionen, und viele Technologien wie etwa Rauchgasrückführung, Stufenverbrennung, katalytische und nicht katalytische Reduktion sind in verschiedenem Maße im Einsatz. Jedoch sind diese oft teuer in der Installation und/oder haben nachteilige Auswirkungen auf den Prozeß. Im Fall der Zementindustrie kann auch die Produktqualität dadurch ungünstig beeinflußt werden.

Der GYRO-THERM-Brenner vereinigt geringe NO$_x$-Bildung mit verbesserter Qualität und höherem Wirkungsgrad. **Bild 8** zeigt die Verminderung der NO$_x$-Emissionen bei zunehmenden Anteil des GYRO-THERM-Einsatzes. Ein in der Größenordnung von 50–75 % verminderter Stickstoffoxidausstoß wurde zuverlässig nachgewiesen.

4. Ofenbetrieb

Ein stabiler Ofenbetrieb und stabile Ansatzprofile sind der Beweis für die stabile Flamme des GYRO-THERM-Brenners mit einem Anteil von 70 % am gesamten Brennstoffverbrauch. Bei höherem Anteil wurde festgestellt, daß die Flamme für die derzeit vorhandene Ausmauerung zu kurz wird und ein regelmäßiges Abbrechen des Ansatzes und Störungen des Ofenbetriebs hervorruft. Möglicherweise kann mit einem anderen Ofenfutter ein Brennstoffanteil des GYRO-THERM-Brenners von 100 % ermöglicht werden. Für unterschiedliche Brennstoffanteile sind die Betriebsdaten der Anlage in **Tabelle 1** zusammengestellt.

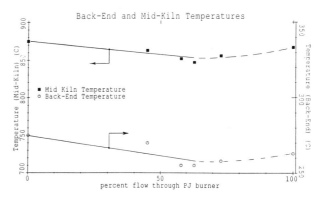

BILD 6: Temperaturen im Ofeneinlauf und in der Ofenmitte in Abhängigkeit des Brennstoffanteils durch den GYRO-THERM
FIGURE 6: Back-end and mid-kiln temperature, related to the increased percentage of fuel through the GYRO-THERM

Temperatur in der Ofenmitte (°C) = Temperature Mid-Kiln
Temperatur im Ofeneinlauf = Back-End Temperature
Anteil des Brennstoffs durch = percent flow through PJ burner
den Präzessions-Strahlbrenner (%)

of the GYRO-TERM burner is 20 times that of a commercial swirl burner, whose radiation characteristics appear to resemble those of a plain pipe burner.

The improved heat transfer leads to increased kiln efficiency, with plant operation confirming increased kiln outputs and reduced fuel consumption. **Fig. 6** shows the reduced kiln backend temperatures and **Fig. 7** shows reduced fuel usage and increased outputs.

3. NO$_x$ Emissions

Generation of NO$_x$ (NO, NO$_2$, N$_2$O) is an unavoidable consequence of high temperature combustion, as exists in cement kilns. There are increasing environmental pressures on minimising NO$_x$ emissions and many techniques such as flue gas recirculation, staged combustion, catalytic and non-catalytic reduction are in various stages of use. However, these are often expensive to install and/or produce a detrimental effect on the process. In the case of the cement industry, product quality may also suffer.

The GYRO-THERM combines low NO$_x$ generation with improved quality and efficiency. **Fig. 8** shows the reduction in NO$_x$ with increasing fuel percentage through the GYRO-THERM. Reductions in the NO$_x$ emissions, in the order of 50–75% have been consistently demonstrated.

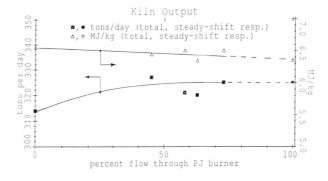

BILD 7: Abhängigkeit der Klinkerleistung und des Brennstoffverbrauchs vom Brennstoffanteil durch den GYRO-THERM
FIGURE 7: The increase in kiln output and reduction in fuel consumption, associated with increased percentage of fuel through the GYRO-THERM

Klinkerleistung = Kiln Output
t/d = tons/day
Brennstoffanteil durch = percent flow through PJ burner
Präzessions-Strahlbrenner (%)

TABELLE 1: Betriebsdaten der Ofenanlage bei unterschiedlicher Brennstoffaufteilung auf die Brenner (Stundenmittelwerte)

Anteil des Präzessions-strahlbrenners (%)	0	45	58	63	73	100 [1]
Klinkerleistung (t/d)	314	327	321	320	325	304 (325)[2]
Brennstoffenergie-verbrauch (MJ/kg)	6,55	6,44	6,50	6,34	6,50	6,35 (5,95)[2]
NO_x-Emissionen (ppm)	1879	1365	1154	1093	1062	873
CO-Emissionen (ppm)	63	77	137	27	12	307
O_2 (%)	3,4	2,3	2,2	2,5	2,8	3,3
Sekundärlufttemperatur (°C)	926	941	970	982	1051	1055
Ofeneinlauftemperatur (°C)	275	270	255	255	258	263
Temperatur in der Kalzinierzone (°C)	875	863	852	848	856	867
Frei-Kalkgehalt (%)	0,8	1,2	1,0	1,4	1,1	1,3
Ono-Microscopy-Rating	410	420	425	435	425	440
Position der Rohmischung im Ofen (m vom Auslaufring)	10	7	5,5	5	4	3
Kaltklinkertemperatur am Ofenausgang (°C) [3]	1365	1400	1413	1443	1400	1370

[1] Instabiler Ofenbetrieb/Unstable kiln operation
[2] Berechnet auf 8h-Versuchszeitraum/Based on a test period of 8 hrs.
[3] Temperaturen nicht repräsentativ/Temperatures are not representative

Anteil des Präzessionsstrahlbrenners (%)	= % P. J. burner
Klinkerleistung (t/d)	= Kiln output, t/d
Brennstoffenergieverbrauch (MJ/kg)	= Fuel consumption, MJ/kg
NO_x-Emissionen (ppm)	= NO_x emissions, ppm
CO-Emissionen (ppm)	= CO emissions, ppm
O_2 (%)	= % O_2
Sekundärlufttemperatur (°C)	= Secondary air, °C
Ofeneinlauftemperatur (°C)	= Back end, °C
Temperatur in der Kalzinierzone (°C)	= Mid kiln, °C (calcining zone)
Frei-Kalkgehalt (%)	= Clinker free lime, %
Ono-Microscopy-Rating	= Ono Microscopy Rating
Position der Rohmischung im Ofen (m vom Auslaufrinmg)	= Raw mix position in kiln – m from nose ring
Kaltklinkertemperatur am Ofenausgang (°C) [3]	= Kiln discharge clinker temperature, °C [3]

Schlußbemerkung

Der Artikel stellt Ergebnisse vor, die mit einer industriellen Anwendung eines Gasbrenners erzielt wurden. Der Brenner wurde durch die gemeinsame Anstrengung einer Universität und einer Zementfabrik konstruiert und entwickelt. Er basiert auf der neu entdeckten und patentierten Präzessions-Strahltechnik. Der Brenner vereinigt Eigenschaften wie einfache Bauweise, hohen Prozeßwirkungsgrad und bessere Produktqualität ebenso mit einer deutlichen Reduktion der NO_x-Emissionen. Mit dem GYRO-THERM-Brenner ist es möglich, während des Betriebs des jeweiligen Prozesses die Flamme auf die Anforderungen einzustellen und den Prozeßbedingungen anzupassen. Die Brennereigenschaften bieten große Vorteile für Zementwerke und andere Branchen, die Erdgas und möglicherweise in naher Zukunft auch andere Brennstoffe verfeuern.

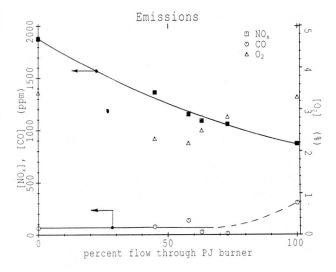

BILD 8: NO_x-Emission in Abhängigkeit des Brennstoffanteils durch den GYRO-THERM
FIGURE 8: NO_x emissions, as a function of fuel fired through the GYRO-THERM

Brennstoffanteil durch den Präzessions-Strahlbrenner (%) = percent flow through PJ burner

4. Kiln operation

Stable kiln operation and stable coating patterns are evidence of the stable flame established by the GYRO-THERM, with up to 70% of fuel through the GYRO-THERM. At levels above this, the flame has been found to be too short for the present refractories installed, causing regular coating breakaways and kiln upsets. It is possible that a new bricking pattern in the kiln would allow 100% fuel flow through a GYRO-THERM nozzle. The operational data for different fuel flow rates are summarized in **Table 1**.

Conclusion

The paper has presented results achieved in an industrial application of a gas burner designed and developed by the joint efforts of a university and a cement company and is based on newly discovered and patented jet flow technology. The burner combines attributes of simplicity, increased process efficiency and improved product quality, as well as substantial reductions in NO_x emissions. The GYRO-THERM burner incorporates a feature for adjustable flame characteristics during operation to suit the process needs. The burner attributes indicate major advantages to cement plants and other industries operating with natural gas fuel and possibly other fuels in the near future.

Anpassung des Calcinierprozesses an Brennstoffreaktivität und Emissionserfordernisse mit PREPOL® MSC

Using the PREPOL® MSC to adapt the calcining process to fuel reactivity and emission requirements

Adaptation du process de calcination à la réactivité du combustible et aux exigences posées par l'émission, avec PREPOL® MSC

Adaptación del proceso de calcinación a la reactividad del combustible y a las exigencias relacionadas con las emisiones, mediante el PREPOL® MSC

Von **K. Menzel**, Köln/Deutschland

Zusammenfassung – Für den Zementhersteller bietet der Einsatz von Calcinatoren mit mehrstufiger Verbrennung die Vorteile, Umweltanforderungen besser gerecht werden zu können und in der Brennstoffwahl eine höhere Flexibilität zu erreichen. Der von Polysius zu diesem Zweck entwickelte und bereits mehrfach in Neuanlagen und zur Nachrüstung vorhandener Anlagen eingesetzte Calcinator mit Mehrstufenverbrennung PREPOL® MSC erfüllt diese Bedürfnisse der Zementindustrie sowohl bezüglich der Umstellung auf verschiedene Brennstoffe als auch bezüglich der Schadstoffemissionen. Der Hauptvorteil der neuen Calcinator-Generation liegt in einer nennenswerten Reduzierung der NO_x-Emission, wie die bisher vorliegenden Betriebsergebnisse zeigen.

Summary – The use of calciners with multi-stage combustion offers the cement producer the advantages of being able to comply better with environmental requirements and of having greater flexibility in the choice of fuel. The calciner with PREPOL® MSC multi-stage combustion system has been developed by Polysius for this purpose and is already installed in many new plants and retrofitted in existing plants. It fulfils the needs of the cement industry both with respect of switching to different fuels and with respect to emission of harmful materials. The main advantage of the new generation of calciners lies in a notable reduction in the NO_x emission, as is shown by the operating results already achieved.

Résumé – L'utilisation de précalcinateurs à combustion étagée offre au producteur de ciment la possibilité de mieux pouvoir respecter l'environnement et d'atteindre une meilleure flexibilité dans le choix du combustible. La chambre de précalcination à combustion étagée PREPOL® MSC, développée à cette fin par Polysius et déjà utilisée à plusieurs reprises dans des installations neuves ou pour moderniser des lignes existantes, satisfait ces besoins de l'industrie cimentière, aussi bien pour le passage à des combustibles différents, que pour la réduction de l'émission de substances nocives. L'avantage principal de la nouvelle génération de la chambre de précalcination réside en diminution remarquable des émissions NO_x, démontrée par les résultats de fonctionnement obtenus à ce jour.

Resumen – Para los fabricantes de cemento, el empleo de calcinadores con combustión en varias etapas ofrece las ventajas de poder cumplir mejor las exigencias del medio ambiente y de conseguir una mayor flexibilidad en cuanto a la elección de los combustibles. El calcinador PREPOL® MSC, de varias etapas de combustión, desarrollado por Polysius para este fin y que ha sido utilizado ya varias veces en instalaciones nuevas así como para equipar instalaciones ya existentes, cumple estos requerimientos de la industria del cemento, tanto con respecto a la elección de diferentes combustibles como a la emisión de substancias nocivas. La principal ventaja de la nueva generación de calcinadores reside en una reducción notable de las emisiones de NO_x, tal como lo demuestran los resultados obtenidos hasta ahora durante el servicio.

Moderne Drehofenanlagen zur Herstellung von Zementklinker arbeiten nach dem Vorcalcinationsverfahren mit Tertiärluft. Während sich die Entwicklung der Calcinatoren in den vergangenen Jahren auf die Anpassung der Brennstoffe konzentrierte, ist nun im Zuge zunehmenden Umweltbewußtseins auch die Reduzierung der NO_x-Emission durch eine Mehrstufenverbrennung im Calcinator hinzugekommen. Die POLYSIUS-Lösung dazu heißt PREPOL MSC.

Modern rotary kiln plants for producing cement clinker operate by the precalcining process with tertiary air. Whereas in previous development of calciners the emphasis has been on fuel adaptation, increasing concern about the environment has brought in the additional factor of reduction of NO_x emissions through multi-stage combustion in the calciner. The POLYSIUS solution to the problem is called PREPOL MSC.

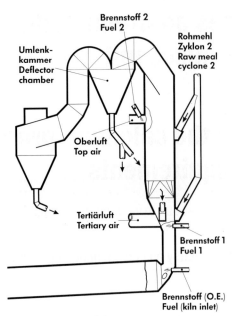

BILD 1: PREPOL®-MSC-Calcinator
FIGURE 1: PREPOL®-MSC-Calciner

Der Reaktionsraum des PREPOL MSC besteht aus dem Ofeneinlaufbereich und der Calcinierschleife, wie in **Bild 1** dargestellt. Brennstoff wird in den Ofeneinlauf entgegen der Gasströmungsrichtung und auf ein oder zwei Niveaus der Calcinierschleife eingegeben. Verbrennungsluft und Rohmehl werden auf den zwei Niveaus der Calcinierschleife zugegeben.

In der ersten Stufe wird mit Hilfe des Ofeneinlaufbrenners eine geringe Brennstoffmenge entgegen der Strömungsrichtung in die Drehofenabgase geblasen. Hierbei wird der Brennstoff pyrolisiert und verbraucht die im Abgas noch vorhandene geringe Sauerstoffmenge. So wird eine reduzierende Zone geschaffen, in der vorhandene Stickstoffoxide zu Stickstoff und Wasserdampf umgewandelt werden. Der aufgrund des Sauerstoffmangels unvollständig umgesetzte Brennstoff wird anschließend im Calcinator nachverbrannt.

Neben den auf diese Weise zum Teil im Ofeneinlaufbereich reduzierten Stickstoffoxiden aus der Drehofenfeuerung ist auch eine NO_x-Neubildung aus dem Brennstoffstickstoff im Calcinator bei den hier vorherrschenden relativ niedrigen Temperaturen möglich. Daher ist es erforderlich, auch den größten Teil des Calcinierbrennstoffes unter reduzierenden Bedingungen zu verbrennen. Dazu wird in der untersten Feuerungsebene des Calcinators die Luftzahl so eingestellt, daß der Brennstoff unterstöchiometrisch verbrennt. Neben der Unterdrückung der NO_x-Bildung aus dem Calcinatorbrennstoff wird durch diese Maßnahme zusätzlich ein weiterer Abbau des aus dem Drehrohr stammenden NO_x erreicht.

Durch entsprechende Aufteilung der Rohmehlaufgabe auf die beiden Calcinatorniveaus kann die Temperatur in der Reduktionszone günstig beeinflußt werden.

Nach der entsprechend bemessenen Reduktionszone wird in der oberen Feuerungsebene des Calcinators ausreichend Verbrennungsluft, auch Oberluft genannt, hinzugegeben und eine weitere kleine Brennstoffmenge für die Nachverbrennung zugeführt, falls dies notwendig erscheint.

Durch die Integration einer Umlenkkammer in die Calcinierschleife wird für eine gute Einmischung der Oberluft und für eine hohe Turbulenz gesorgt. Dies gewährleistet den vollständigen CO-Umsatz. Dieser ist besonders wichtig, wenn sehr reaktionsträge Brennstoffe mit geringen Anteilen an flüchtigen Bestandteilen zum Einsatz kommen, die auch nach Oberluftzugabe nur langsam nachverbrennen. Darüber hinaus werden durch die Umlenkung 30–50 % des im Gasstrom vorhandenen festen Materials abgeschieden. Dieses Material kann nun entweder zur Erhöhung der Verweilzeit von Mehl und nicht vollständig ausgebranntem Brennstoff in die Reduktionszone zurückgeführt werden oder aber als fertig calciniertes Material in den Drehofen

The reaction chamber of the PREPOL MSC consists of the kiln inlet zone and the calcining loop as shown in **Fig. 1**. Fuel is brought into the kiln inlet in the opposite direction to the gas flow and fed to one or two levels of the calcining loop. Combustion air and raw meal are supplied to both levels of the loop.

At the first stage the burner at the kiln inlet injects a small quantity of fuel into the exhaust gases from the kiln countercurrently to the direction of flow. The fuel is pyrolysed and uses up the small amount of oxygen still present in the exhaust gas. A reducing zone is thus created, in which existing nitrogen oxides are converted to nitrogen and steam. The fuel which is incompletely reacted owing to the lack of oxygen is then post-combusted in the calciner.

Apart from the nitrogen oxides from the kiln firing system which are partly reduced in this way in the inlet zone, fresh NO_x may also form from the fuel nitrogen in the calciner at the relatively low temperatures prevailing here. Most of the calcining fuel must therefore also be burnt under reducing conditions. For this purpose the air ratio at the lowest firing level of the calciner is set so that the fuel burns in sub-stoichiometric proportions. As well as suppressing NO_x formation from the calciner fuel this measure also results in further decomposition of the NO_x from the kiln. The temperature in the reduction zone can be favourably controlled by appropriate division of the raw meal feed to the two calciner levels.

Downstream of the reduction zone of appropriate dimensions an adequate supply of combustion air, also described as top air, is passed into the upper firing level of the calciner, and a further small quantity of fuel is fed in for secondary combustion if this appears necessary.

Incorporation of a deflector chamber in the calcining loop results in good mixing of the top air and high turbulence, thus ensuring complete CO conversion. This is particularly important if less active fuels with small proportions of volatile constituents are used, undergoing only gradual secondary combustion even after the addition of top air. Furthermore, the deflection separates 30–50% of the solid material in the gas stream. This material can either be recycled to the reduction zone to lengthen the dwell time of the meal and incompletely burnt fuel or passed into the kiln as a ready-calcined material. Lengthening the dwell time of unreacted fuel particles in the calciner thus diminishes the proportion of fuel discharged from it, an effect which is also important for less active fuels such as petroleum coke.

After a series of industrial trials with the burner at the kiln inlet and with combustion air supplied in stages, the first plant with the complete PREPOL MSC design came into operation at the beginning of 1993. The NO_x readings are

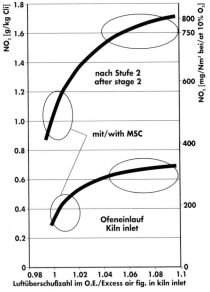

BILD 2: Mehrstufenverbrennung MSC
FIGURE 2: Multi-stage combustion MSC

gegeben werden. Eine Erhöhung der Verweilzeit von nicht umgesetzten Brennstoffpartikeln im Calcinator verringert somit den Brennstoffanteil, der aus dem Calcinator ausgetragen wird. Dies ist ebenfalls für reaktionsträge Brennstoffe, wie z. B. Petrolkoks, von Bedeutung.

Nach einer Reihe von Industrieversuchen mit dem Ofeneinlaufbrenner und gestufter Verbrennungsluftzufuhr konnte Anfang 1993 die erste Anlage mit dem vollständigen Konzept des PREPOL MSC in Betrieb gesetzt werden. Die Ergebnisse der NO_x-Messungen zeigt **Bild 2**. Dargestellt sind die NO_x-Werte über der Luftüberschußzahl am Ofeneinlauf, wobei auf der linken Achse die Skalierung in g NO_2/kg Kli und auf der rechten Achse in mg NO_2/Nm³ gewählt wurde. Bei Betrieb ohne gestufte Verbrennung liegt die Basisemission nach dem Calcinierzyklon im Bereich von 1,6 g NO_2/kg Kli bzw. 0,6 g NO_2/kg Kli am Ofeneinlauf bei deutlich oxidierenden Bedingungen. Mit MSC-Betrieb sinken diese Werte auf 1,1 bzw. 0,35 g NO_2/kg Kli nach Calcinator bzw. am Ofeneinlauf. Dies bedeutet, daß trotz der schon sehr niedrigen Basisemission mit dem MSC-Calcinator eine Gesamtemissionsminderung von durchschnittlich 30 % erreicht wird. Bezogen auf den Ofeneinlaufzustand werden sogar Minderungen von über 40 % erzielt.

Die sehr niedrige Grundemission dieser Anlage läßt sich durch den geringen Luftüberschuß auch ohne MSC-Betrieb am Ofeneinlauf und durch die Art der tangentialen Tertiärluftzuführung sowie einen ständigen Tertiärluftstrom durch die Oberluftleitung erklären. Durch diesen ständigen Tertiärluftstrom in der Oberluftleitung wird der Calcinator bereits im Normalbetrieb partiell reduzierend betrieben.

Die in diesem Anwendungsfall eingesetzte Brennstoffmischung besteht zu 80 % aus Petrolkoks und zu 20 % aus Steinkohle. Der Brennstoff hat einen Stickstoffgehalt von 1,89 % und nur 15 % flüchtige Bestandteile. Typische NO_x-Emissionen von Anlagen ohne PREPOL MSC, die im wesentlichen mit Petrolkoks befeuert werden, liegen im Durchschnitt bei 2,4–2,8 g NO_2/kg Kli. Die hier beschriebene Anlage weist also mit Emissionen von nur ca. 1,1 g NO_2/kg Kli erheblich geringere Werte auf. Ein reaktionsfreudigerer und weniger Stickstoff enthaltender Brennstoff läßt noch höhere Entstickungsraten und weniger Brennstoff-NO_x erwarten. **Bild 3** zeigt die CO-Konzentration an den Punkten Ofeneinlauf, vor und nach Umlenkkammer sowie nach Calcinatorzyklon. Mit dem reaktionsträgen Brennstoffgemisch werden trotz deutlichen Sauerstoffmangels in den Reduktionszonen nur relativ geringe CO-Werte erzeugt. Bis zu 90 % des in der Reduktionszone erzeugten CO's werden in der Umlenkkammer nachverbrannt. Verbleibendes Rest-CO wird problemlos im nachfolgenden Calcinatorzyklon umgesetzt. Bei den hier eingesetzten reaktionsträgen Brennstoffen wird die Möglichkeit der Rezirkulation von Material aus der Umlenkkammer zurück in den Calcinator ständig genutzt. Mit Beginn der Inbetriebnahme wurde das Material aus der Umlenkkammer direkt zum Drehofen geführt. Die Anlage konnte aufgrund der hohen Schwefeleinnahme aus dem Petrolkoks nur mit erhöhtem Reinigungsaufwand betrieben werden. Nach Umstellung auf vollständige Rückführung des Materials stellte sich ein deutlich besseres Betriebsverhalten ein. Der Reinigungsaufwand beschränkt sich nun weitgehend auf Kontrollen. Durch die Rezirkulation haben sich bezüglich der Höhe der Schadstoffkreisläufe keine Änderungen ergeben. Die Gleichmäßigkeit und die Höhe des Glühverlustes der Heißmehlproben zeigen, daß die Rezirkulation den Calcinatorbetrieb vergleichmäßigt und so die gesamte Ofenlinie stabilisiert.

Im Laufe des Jahres 1993 werden drei weitere Anlagen in Betrieb genommen, in denen oben beschriebenes Verfahren in unterschiedlichen Konfigurationen bei Einsatz verschiedener Brennstoffe zur Anwendung kommt. Hiermit wird ein wesentlicher Schritt zur Einführung des Systems in den Markt realisiert werden, das bei hoher Wirtschaftlichkeit eine effektive Möglichkeit zur Senkung von Emissionen und somit zu einer umweltfreundlicheren Zementproduktion darstellt.

given in **Fig. 2**. The figure shows the NO_x values versus the air excess ratio at the kiln inlet, with a scale in g NO_2/kg clinker on the left axis and in mg NO_2/m³(stp) on the right axis. In operation without multi-stage combustion the basic emission after the calcining cyclone is within the range from 1.6 g NO_2/kg to 0.6 g NO_2/kg clinker at the kiln inlet under distinctly oxidising conditions. With MSC operation these values drop to 1.1 and 0.35 g NO_2/kg clinker respectively after the calciner and at the kiln inlet. This means that a total average drop in emissions of 30 % is achieved despite the already very low basic emission with the MSC calciner. Drops of even over 40 % are obtained relative to the state at

The very low basic emission from this plant can be explained by the low air excess at the kiln inlet even without MSC operation, by the nature of the tangential tertiary air supply and by a constant tertiary airstream through the top air pipe. The constant tertiary airstream in the top air pipe gives the calciner a partially reducing action even in normal operation.

The fuel mix used in this application consists of 80 % petroleum coke and 20 % pit coal. The fuel has a nitrogen content of 1.89 % and only 15 % volatile constituents. Typical NO_x emissions from plant without PREPOL MSC, fired essentially with petroleum coke, average about 2.4 to 2.8 g NO_2/kg clinker. The plant described here has considerably lower values with emissions of only about 1.1 g NO_2/kg clinker. With a more reactive fuel containing less nitrogen still higher denitrification rates and less fuel NO_x could be expected. **Fig. 3** shows the CO concentration at the kiln inlet, before and after the deflector chamber and after the calcining cyclone. Only relatively low CO values are obtained with the less active fuel mix in spite of the definite lack of oxygen in the reduction zones. Up to 90 % of the CO produced in the reduction zone is post-combusted in the deflector chamber. The remaining CO is converted without any problems in the downstream calcining cyclone. When the plant first came into operation the material was taken from the deflector chamber straight to the kiln. The plant could only be operated with an increased amount of cleaning owing to the high intake of sulphur from the petroleum coke. A far better mode of operation has been obtained after conversion to complete recycling of the material. The cleaning work is now largely restricted to checks. Recirculation has not changed in level of the recirculating pollutant systems. The uniformity and level of loss on ignition of the hot meal samples show that recirculation smooths calciner operation and thus stabilises the entire kiln line.

Three more installations are coming into operation in 1993, in which the process described above will be applied in different configurations and with different fuels. This will be an important step in launching the system which, being very cost-effective, is an effective means of lowering emissions and thus making cement production more environmentally friendly.

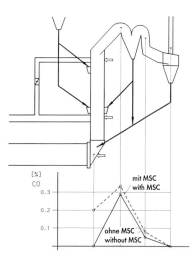

BILD 3: CO-Ausbrand im Calcinator
FIGURE 3: CO-combustion rate in calciner

Auswirkung der Gestaltung der Zyklone für Zyklonvorwärmer auf Abgastemperatur und Druckverlust

Effect of the cyclone configuration in cyclone preheaters on the exhaust gas temperature and pressure drop

Effet de la forme des cyclones de préchauffeurs à cyclones sur la température des gaz d'exhaure et la perte de charge

Repercusiones de la forma de los ciclones de los precalentadores sobre la temperatura de los gases de escape y la pérdida de presión

Von **U. Mrowald** und **R. Hartmann**, Köln/Deutschland

Zusammenfassung – Energieeinsparungen beim Zementbrennen sind möglich durch eine verbesserte Wärmerückgewinnung aus den Ofen- bzw. Calcinatorabgasen. Deshalb ist eine möglichst weitgehende Mehlabscheidung in allen Zyklonstufen sehr wichtig. Zyklone, die auf besonders niedrige Druckverluste ausgelegt sind, können die genannte Aufgabe nur bedingt erfüllen. Theoretische Berechnungen und praktische Versuche hatten daher zum Ziel, eine Zyklongeometrie zu finden, mit der neben einer guten Abscheideleistung ein möglichst niedriger Druckabfall erreicht werden kann. Als Parameter wurden in erster Linie die Tauchrohrlänge, das Verhältnis Tauchrohrdurchmesser zu Zyklondurchmesser, die Gasgeschwindigkeit am Eintritt und im Tauchrohr sowie die Querschnittsform des Eintritts variiert. Daneben wurde untersucht, welchen Einfluß die Neigung des Konus haben kann. Die Untersuchungen an technischen Anlagen mit einer Produktion von 900-5000 t/d haben gezeigt, daß sich die Druckverluste des neugestalteten Vorwärmers um etwa 15% vermindern ließen und die Rohgastemperaturen in Abhängigkeit von der Stufenzahl und dem Anteil der PYROCLON-Feuerung merklich gegenüber den Zyklonvorwärmern alter Bauart abnehmen.

Auswirkung der Gestaltung der Zyklone für Zyklonvorwärmer auf Abgastemperatur und Druckverlust

Summary – Energy savings can be made in the cement burning process by improved heat recovery from the kiln or calciner exhaust gases. It is therefore very important to have most complete possible meal separation in all cyclone stages. Cyclones which are designed for particularly low pressure drops can only fulfil this task to a limited extent. Theoretical calculations and practical trials were therefore used to find a cyclone geometry which achieves not only a good separating performance but also the lowest possible pressure drop. The parameters varied in the first instance were the dip tube length, the ratio of dip tube diameter to cyclone diameter, the gas velocity at the inlet and in the dip tube, and the cross-sectional shape of the inlet. The influence of the slope of the cone was also investigated. Investigations on industrial plants with production capacities of 900 – 5000 t/d show that the pressure drops in the newly designed preheater have been lowered by about 15% and, when compared with cyclone preheaters of other designs, the raw gas temperatures are reduced to an appreciable extent which depends on the number of stages and the proportion of PYROCLON firing.

Effect of the cyclone configuration in cyclone preheaters on the exhaust gas temperature and pressure drop

Résumé – Les économies d'énergie dans la cuisson du ciment sont possibles au moyen d'une meilleure récupération de chaleur sur les gaz d'exhaure du four et de la chambre de précalcination. Pour cela, une séparation aussi complète que possible de la farine, à tous les étages de cyclones, est très importante. Des cyclones conçus pour des pertes de charge particulièrement faibles ne peuvent y satisfaire que partiellement. Des calculs théoriques et des études pratiques conduites avaient donc comme objectif, de trouver une géométrie de cyclone avec laquelle pouvaient être réalisés à la fois un bon rendement de séparation et une perte de charge aussi faible que possible. En premier lieu ont été variés, comme paramètres, la longuer du tube plongeur, le rapport diamètre tube plongeur: diamètre cyclone, la vitesse des gaz à l'entrée et dans le tube plongeur et la forme de la section transversale de l'entrée. Parallèlement, a été étudiée l'influence de la pente du cône. Les études effectuées dans des installations industrielles d'une production de 900–5000 t/j ont révélé, que les pertes de charge du préchauffeur à configuration nouvelle ont pu être réduites d'environ 15% et que les températures des gaz d'exhaure bruts ont, compte tenu du nombre d'étages et de la part de chauffe PYROCLON, diminuées nettement par rapport au préchauffeurs à cyclones du type ancien.

Effet de la forme des cyclones de préchauffeurs à cyclones sur la température des gaz d'exhaure et la perte de charge

Fachbereich 3 · Subject 3 · Séance Technique 3 · Tema de ramo 3

Repercusiones de la forma de los ciclones de los precalentadores sobre la temperatura de los gases de escape y la pérdida de presión

Resumen – *Durante la cocción del cemento es posible lograr un ahorro de energía mediante una mejor recuperación del calor contenido en los gases de escape del horno y del calcinador. Por esta razón, es muy importante una máxima separación de harina cruda en todas las etapas de ciclones. Los ciclones concebidos para pérdidas de presión particularmente bajas, sólo pueden cumplir en parte esta tarea. Los cálculos teóricos y los ensayos prácticos llevados a cabo han tenido, pues, por objetivo encontrar una geometría para los ciclones que permita conseguir un buen rendimiento de separación y también una caída de presión lo más reducida posible. Los parámetros variados han sido, en primer lugar, la longitud de los tubos de inmersión, la relación entre el diámetro de los tubos de inmersión y el diámetro de los ciclones, la velocidad de los gases a la entrada y en el tubo de inmersión así como la forma de la sección de entrada. Además, se ha estudiado el influjo que puede tener la inclinación del cono. Los estudios realizados en instalaciones técnicas con una producción de 900 – 5000 t/d han demostrado que existe la posibilidad de reducir las pérdidas de presión del precalentador de nuevo diseño en un 15% aprox., disminuyendo notablemente las temperaturas de los gases brutos, en función del número de etapas y la proporción de la instalación de combustión PYROCLON, en comparación con los precalentadores de tipo antiguo.*

Zyklonvorwärmer von KHD Humboldt Wedag sind seit 40 Jahren markantes Zeichen moderner Zementfabriken. Das Brennverfahren mit Zyklonvorwärmer hat sich mittlerweile weltweit durchgesetzt, weil Zyklonvorwärmer einfach zu bedienen sind und einen sicheren und gleichmäßigen Brennprozeß gewährleisten. Sie bestehen aus mehreren übereinander angeordneten Zyklonen, die mit Gasleitungen verbunden sind. In diesen Gasleitungen vollzieht sich der Wärmeaustausch zwischen dem heißen Ofenabgas und dem kälteren Rohmehl fast vollständig.

Der Wärmeübergang wird in erster Linie bestimmt durch die Temperaturdifferenz zwischen Gas und Feststoff sowie von der Verteilung der Brenngutpartikel im Gasstrom. Je höher die Temperaturdifferenz, desto schneller der Wärmeaustausch. Je homogener die Verteilung des Mehls im Gasstrom, desto gleichmäßiger der Erwärmungsgrad.

Das Rohmehl durchläuft nacheinander mehrere Vorwärmstufen bis zum Einlauf in den Drehrohrofen. Die notwendige Trennung des Rohmehls aus der Gasphase geschieht in den Zyklonen. Für die Zyklone wird ein maximaler Abscheidegrad verlangt, um Mehlkreisläufe innerhalb des Vorwärmersystems gering zu halten. Diese Mehlkreisläufe erhöhen den spezifischen Wärmebedarf und den Druckverlust des Vorwärmers. Bei hohen Mehlkreisläufen ist es durchaus möglich, daß sich die zweifache Mehlmenge im Vorwärmersystem befindet.

Dadurch wird die Mehldispergierung im Gasstrom schlechter, und das Temperaturniveau verschiebt sich zu höheren Werten. Darüber hinaus erfordert die übermäßige Hebearbeit zusätzliche elektrische Energie für den Ventilator.

Zyklone mit hohem Abscheidegrad senken also den Druckverlust und verbessern den Wärmeübergang. Das Rohmehl nimmt mehr Wärmeenergie auf, die Abgastemperatur sinkt entsprechend.

Auch bei sehr guten Abscheidern lassen sich jedoch die Mehlkreisläufe nicht vollkommen vermeiden. Aber wie hoch sind diese Mehlkreisläufe bzw. der Abscheidegrad? Wie kann man den Abscheidegrad der Zyklone beeinflussen? Der Vorwärmerbetrieb und die komplizierten Strömungsverhältnisse im Zyklon lassen es nicht zu, den Abscheidegrad während der Klinkerproduktion sicher zu bestimmen. Infolgedessen weicht man auf Modellversuche aus, um diese Fragen zu beantworten.

Zwischen Abscheidegrad und Druckverlust eines Zyklons besteht ein enger Zusammenhang, der durch die Tauchrohrgeometrie wesentlich beeinflußt wird.

Da jeder Zyklon mit einem Tauchrohr ausgestattet sein sollte, haben wir uns gefragt, wie kurz dieses Tauchrohr sein darf. **Bild 1** verdeutlicht die Ergebnisse unserer Untersuchung zur Auswirkung der Tauchrohrlänge auf den Abscheidegrad und den Druckverlust eines Modellzyklons. Auf der Abszisse ist die Tauchrohrlänge im Verhältnis zur Höhe des Gaseintritts dargestellt. Die linke Ordinate zeigt den Abscheidegrad in %, die rechte Ordinate gibt den

For 40 years KHD Humboldt Wedag cyclone preheaters have been the mark of a modern cement works. During that time the combustion process with a cyclone preheater has become established worldwide, since cyclone preheaters are easy to operate and ensure a safe, even burning process. They consist of several superimposed cyclones linked by gas ducts. Heat exchange between the hot exhaust gas from the kiln and the colder raw meal takes place almost entirely in the gas ducts.

Heat transfer is determined primarily by the difference in temperature between gas and solid and the distribution of fuel particles in the gas flow. The greater the difference in temperature the faster the heat exchange. The more homogeneously the meal is distributed in the gas flow the more uniform is the heating.

The raw meal passes through several pre-heating stages in succession before entering the rotary kiln. The necessary separation of the raw meal from the gas phase takes place in the cyclones. Maximum separating efficiency is required at the cyclones in order to minimise recirculation of the meal within the preheater system. This recirculation increases the specific heat requirement and the pressure drop of the preheater. When large quantities of meal are recirculating the preheater system may quite possibly contain twice the normal amount.

Meal dispersion in the gas flow becomes poorer as a result and the temperature moves up to higher levels. The excess lifting work also requires additional power for the fan.

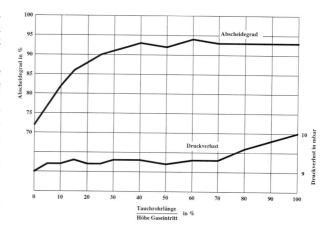

BILD 1: Einfluß der Tauchrohrlänge auf den Abscheidegrad und den Druckverlust eines Modellzyklons

FIGURE 1: Effect of top outlet duct length on the separation efficiency and pressure drop of a model cyclone

Abscheidegrad	= separation efficiency
Druckverlust	= pressure drop
Tauchrohrlänge	= top outlet duct length
Höhe Gaseintritt	= height of gas inlet

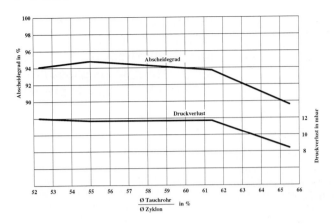

BILD 2: Einfluß des Tauchrohrdurchmessers auf den Abscheidegrad und den Druckverlust eines Modellzyklons
FIGURE 2: Effect of top outlet duct diameter on the separation efficiency and pressure drop of a model cyclone

Abscheidegrad	= separation efficiency
Druckverlust	= pressure drop
Tauchrohr	= top outlet duct
Zyklon	= cyclone

Druckverlust in mbar an. Man erkennt, daß der Abscheidegrad mit zunehmender Tauchrohrlänge zunächst rasch ansteigt.

Ab einer Tauchrohrlänge, die 40–50 % der Gaseintrittshöhe überdeckt, ist bereits der maximal mögliche Abscheidegrad erreicht. Der Druckverlust ändert sich bis zu dieser Tauchrohrlänge fast nicht. Zur weiteren Erforschung des Zusammenhangs zwischen Abscheidegrad und Druckverlust werden unterschiedliche Tauchrohrdurchmesser untersucht. Im **Bild 2** ist auf der Abszisse das Verhältnis von Tauchrohrdurchmesser und Zyklondurchmesser in % aufgetragen. Die linke Ordinate zeigt den Abscheidegrad in %. Die rechte Ordinate gibt den Druckverlust in mbar wieder. Diese Darstellung zeigt, daß Tauchrohre mit Durchmessern, die 55 bis 60 % des Zyklondurchmessers betragen, hohe Abscheidegrade bei akzeptablen Druckverlusten bewirken.

Weitere Untersuchungen betrafen die Geometrie des Gaseintritts in den Zyklon. Hier wurde erkannt, daß die Einlaufspirale weit um den Zyklon geführt werden sollte, und daß die abgeschrägte Bodenfläche unerwünschte Mehlablagerungen vermeidet. Der Eintrittsquerschnitt sollte länglich, d. h. hoch und schmal ausgeführt werden.

Der Zyklonkonus dient als Austragshilfe für das abgeschiedene Rohmehl. Je steiler dieser verläuft, desto ungestörter fließt das Mehl aus dem Zyklon. Die drei zuletzt genannten Merkmale verbessern deutlich den Abscheidegrad des Zyklons, haben aber kaum Einfluß auf seinen Druckverlust.

Unter Anwendung von Ähnlichkeitskriterien wurden die aus den Modellversuchen gewonnenen Erkenntnisse auf den Vorwärmerzyklon übertragen. **Bild 3** zeigt als Resultat die neue Zyklonkonstruktion.

Der Zyklon besitzt ein Tauchrohr mit einer Länge, die ca. 50 % der Gaseintrittshöhe abdeckt, und einen Tauchrohrdurchmesser, der ca. 55 % des Zyklondurchmessers aufweist. Der neue Zyklon hat außerdem eine 270° Einlaufspirale mit abgeschrägter Bodenfläche und einen steilen Konus.

Zyklonvorwärmer dieser neuen Bauart sind von KHD Humboldt Wedag mehrfach in Betrieb gesetzt worden, und wir konnten ihre positiven Kennwerte nachweisen. Der Vergleich zur Vorgeneration weist zwei wesentliche Fortschritte aus: Erstens wurde der Abscheidegrad deutlich verbessert, denn die Staubmengen im Vorwärmerabgas liegen um mehr als 25 % niedriger. Zweitens hat sich wie erwartet der verbesserte Abscheidegrad auch auf die Abgastemperaturen ausgewirkt. Diese liegen jetzt um ca. 30 °C niedriger als zuvor.

Ein Vergleich dieser Abgastemperaturen in Abhängigkeit von Stufenzahl und PYROCLON-Feuerungsanteil zeigt **Bild 4**. Meßwerte von Anlagen der Vorgeneration sind durch

Cyclones with good separating efficiency thus lower the pressure drop and improve heat transfer. The raw meal absorbs more heat energy and the exhaust gas temperature drops accordingly.

Even with very good separators meal recirculation cannot be avoided altogether. But how much meal is circulating and how high is separating efficiency? How can the separating efficiency of the cyclones be controlled? With the preheater in operation and the complex flow conditions in the cyclone it is not possible to determine separating efficiency reliably during clinker production. Consequently model trials are used to answer these questions.

There is a close correlation between the separating efficiency and the pressure drop of a cyclone, and this is influenced considerably by the geometry of the top outlet duct.

As every cyclone needs to have a top outlet duct we wondered how short this duct could be. **Fig. 1** shows the results of our study of the effect of the length of the top outlet duct on the separation efficiency and pressure drop of a model cyclone. The length of the top outlet duct relative to the height of the gas inlet is shown on the x-axis. The left ordinate gives the separation efficiency as a percentage and the right ordinate the pressure drop in mbar. It will be seen that separation efficiency at first rises rapidly with an increase in the length of the top outlet duct.

The maximum possible separation efficiency is reached when the top outlet duct has a length covering 40 to 50 % of the height of the gas inlet. There is hardly any change in the pressure drop up to this length. We investigated the correlation between separation efficiency and pressure drop further by studying different top outlet duct diameters.

In **Fig. 2** the ratio of top outlet duct diameter to cyclone diameter is entered as a percentage on the x-axis. The left ordinate shows separation efficiency as a percentage. The right ordinate shows the pressure drop in mbar. These curves show that top outlet ducts with diameters of 55 to 60 % of the cyclone diameter give high levels of separation efficiency with acceptable pressure drops.

Other investigations concerned the geometry of the gas inlet into the cyclone. Here it was realised that the inlet spiral should be carried a long way round the cyclone and that an inclined bottom surface avoids undesirable meal deposits. The cross-section of entry should be elongated, i.e. tall and narrow.

The cone of the cyclone aids in discharging the separated raw meal. The steeper it is, the more trouble-free the flow of meal from the cyclone. The last three features mentioned greatly improve the separation efficiency of the cyclone but have no effect on its pressure drop.

The discoveries made in the model trials were transferred to the preheater cyclone using similarity criteria. **Fig. 3** shows the resultant new cyclone design.

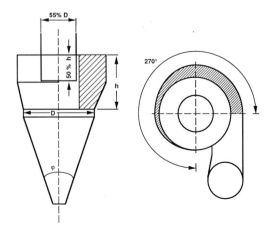

BILD 3: Neue Zyklonkonstruktion
FIGURE 3: New cyclone design

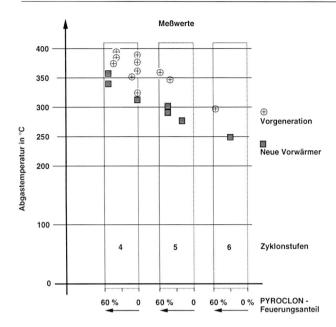

BILD 4: Vergleich der Abgastemperaturen zwischen neuen Vorwärmern und Vorwärmern der Vorgeneration
FIGURE 4: Comparison of exhaust gas temperatures with new preheaters and previous generation preheaters

Abgastemperatur	= exhaust gas temperature
Meßwerte	= measured values
Vorgeneration	= previous generation
Neue Vorwärmer	= new preheaters
Zyklonstufen	= cyclone stages
PYROCLON Feuerungsanteil	= proportion of PYROCLON firing

The cyclone has a top outlet duct of a length covering approx. 50% of the height of the gas inlet and a top outlet duct diameter equal to approx. 55% of the cyclone diameter. It also has a 270° inlet spiral with an inclined bottom surface and a steep cone.

Several cyclone preheaters of this new design have been commissioned by KHD Humboldt Wedag, and we have evidence of their positive characteristics. They provide two important advances from the previous generation: Firstly, separation efficiency has been greatly improved, and the quantities of dust in the exhaust gas from the preheater are more than 25% lower. Secondly, the improved separation efficiency has, as expected, also affected exhaust gas temperatures. These are now about 30 °C lower than before.

Fig. 4 compares these exhaust gas temperatures versus the number of stages and the proportion of PYROCLON firing. Measurements from plants of the previous generation are indicated by circles and those for the new generation by squares. Regardless of the number of cyclone stages and the proportion of PYROCLON firing, exhaust gas temperatures are far lower with the new preheaters.

In addition to this advantageous effect on fuel costs the second advantage is that the pressure drop of the preheater is reduced by about 15%, so electrical power can be saved.

Kreise symbolisiert, die der neuen Vorwärmergeneration durch Quadrate. Unabhängig von der Anzahl der Zyklonstufen und vom PYROCLON-Feuerungsanteil sind die Abgastemperaturen deutlich niedriger mit den neuen Vorwärmern.

Neben diesem für die Brennstoffkosten vorteilhaften Einfluß liegt der zweite Vorteil darin, daß der Druckverlust des Vorwärmers um ca. 15% vermindert wird und sich somit elektrische Energie einsparen läßt.

Einfluß der Verbrennung von Abfallbrennstoffen auf die Prozeßparameter und die feuerfeste Ausmauerung in Zement-Drehrohröfen

Influence of burning waste-derived fuels on the process parameters and refractory lining in cement rotary kilns

Influence de la combustion de combustibles de déchets sur les paramètres du process et le garnissage réfractaire de fours rotatifs de cimenterie

Influjo de la combustión de desechos sobre los parámetros del proceso y el revestimiento refractario de los hornos rotatorios de cemento

Von **K. C. Narang** und **S. Chaturvedi,** New Delhi/Indien

Zusammenfassung – Die Verwendung von Brennstoffen aus Abfall in Zement-Drehrohröfen ist in den letzten Jahren häufiger und interessanter geworden. Der Drehrohrofen stellt ein ideales Medium für die Rückgewinnung von Ressourcen dar, weil viele gefährliche und ungefährliche Abfallstoffe dank der hohen Temperatur und hohen Absorptionsfähigkeit des Feststoffs im Inneren des Drehofens zerstört werden können. Der bisher in Indien für Zement-Drehrohröfen eingesetzte Abfallbrennstoff besteht aus Autoreifen, Altöl, verbrauchten organischen Lösemitteln, Hausmüll, Reisschalen, Holzabfällen usw. Die Untersuchungen haben sich bisher auf die Umwandlung von Abfall in nutzbare Brennstoffe und die Einsatzmenge konzentriert, die ohne schädliche Auswirkungen auf die Umwelt fossile Brennstoffe ersetzen kann. Die Abfallbrennstoffe können den Prozeß der Zementherstellung beeinflussen. Von besonderem Interesse ist ihre langfristige Wirkung auf die verschiedenen Arten von feuerfesten Ausmauerungen von Zement-Drehrohröfen. Deshalb erscheint eine Prognose bezüglich ihres Einflusses auf den Brennprozeß wünschenswert. Von besonderem Interesse sind die Auswirkungen der Nebenbestandteile in Abfallbrennstoffen auf die Klinker- und Zementqualität.

Summary – The use of waste-derived fuels in cement rotary kilns has become more common and attractive in recent years. The rotary kiln provides an ideal medium for resource recovery because many hazardous and non-hazardous wastes can be easily destroyed inside the rotary kiln due to high temperature and great absorbility of the kiln solid mass. The various waste fuels for use in cement rotary kilns so far established in India are tyres, liquid waste oil, spent organic solvents, municipal refuse, rice husks, wood wastes, etc. The studies so far were concentrated on conversion of waste fuels into useable fuels and the quantity of these fuels which could replace fossil fuels without impact on the environment. The waste fuels may influence the cement manufacturing process. Of special interest is their long term effect on the different types of refractory lining of cement rotary kilns. Therefore a prediction of possible influence on the burning process seems to be necessary. Of more particular interest is the influence of minor constituents in burnt waste fuels on the quality of clinker and cement.

Résumé – L'utilisation de combustibles provenant de déchets dans des fours rotatifs de cimenterie est devenue de plus en plus fréquente et présente un intérêt accru. Le four tubulaire rotatif constitue un moyen idéal pour récupérer des ressources, du fait que de nombreux déchets dangereux et non-dangereux sont détruits à l'intérieur du four grâce à la température élevée y régnant et à la capacité d'absorption du solide. Le combustible utilisé jusqu'ici en Inde dans les fours tubulaires rotatifs de cimenterie est tiré de déchets tels que les pneus de voiture, l'huile usagée, les solvants organiques usagés, les déchets domestiques, les restes de riz et de bois, etc. Les études se sont jusqu'à présent concentrées sur la transformation des déchets en combustibles utiles et quantité d'utilisation, capables de remplacer les combustibles fossiles sans effet préjudiciable sur l'environnement. Les combustibles tirés des déchets peuvent influer sur le processus de fabrication du ciment. Leur effet à long terme sur les différents types de garnissages réfractaires de fours tubulaires rotatifs de cimenteries revêt un intérêt particulier. Voilà pourquoi un pronostic quant à leur effet sur le processus de cuisson paraît souhaitable. Les effets des composants secondaires dans les combustibles provenant de déchets sur la qualité du clinker et du ciment sont particulièrement intéressants.

Resumen – *La utilización de combustibles, procedentes de desechos, en los hornos rotatorios de cemento se ha vuelto más frecuente y más interesante en los últimos años. El horno rotatorio representa un medio ideal para la recuperación de recursos, ya que permite destruir muchos desechos peligrosos o no peligrosos, gracias a las elevadas temperaturas y la gran capacidad de absorción del sólido dentro del horno rotatorio. Los desechos utilizados hasta ahora como combustible en la India para hornos rotatorios de cemento consisten en neumáticos, aceite usado, disolventes orgánicos usados, basura doméstica, cáscaras de arroz, desperdicios de madera, etc. Los estudios se han centrado hasta ahora en la transformación de los desechos en combustibles útiles y en aquellas cantidades capaces de sustituir los combustibles fósiles, sin repercutir de forma negativa en el medio ambiente. Los combustibles procedentes de desechos pueden influir en el proceso de fabricación del cemento. Es de especial interés su influjo a largo plazo sobre los diferentes tipos de revestimiento refractario de los hornos rotatorios de cemento. Por esta razón, parece deseable un pronóstico respecto de su influjo sobre el proceso de cocción. Interesa, sobre todo, conocer el efecto que producen los elementos secundarios contenidos en los desechos sobre la calidad del clínker y del cemento.*

Influjo de la combustión de desechos sobre los parámetros del proceso y el revestimiento refractario de los hornos rotatorios de cemento

1. Einleitung

Durch die laufend fortschreitende Industrialisierung wachsen die Mengen von gefährlichem und nicht gefährlichem Müll sowie von Nebenprodukten beängstigend schnell an. Jedoch müssen diese Abfallprodukte gleich welcher Art so entsorgt werden, daß die Umwelt nicht geschädigt wird. Hohe Entsorgungskosten, die infolge der strengeren Umweltgesetzgebung entstehen, setzen die verursachende Industrie zunehmend unter Druck. Im Gegensatz dazu stehen andere Branchen dem Problem sich erschöpfender Quellen an konventionellen Brennstoffen gegenüber. Dabei besteht die Möglichkeit, die Abfallprodukte als teilweisen Ersatz für natürliche Brennstoffe einzusetzen. Umweltfreundliche, sparsame und leistungsfähige Technologien sind erforderlich, um das Potential dieser Abfallprodukte zu nutzen und beide Seiten, Lieferanten und Entsorger, zusammenzubringen. Solange auch wirtschaftliche Aspekte berücksichtigt werden, gewinnen alle Parteien und tragen so zu einer lebenswerten Umwelt bei [1–3].

Dieser Idee folgend wurden in den vergangenen Jahren die Drehrohröfen der Zementindustrie als eine der Möglichkeiten zur Wiederverwertung solcher Abfallprodukte erkannt. Früher wurde brennbarer Sondermüll als eine Bedrohung für Boden, Luft und Wasser angesehen, jetzt jedoch kann er gefahrlos bei der Herstellung von Portlandzement als teilweiser Ersatz für konventionelle Brennstoffe wiederverwertet werden. Die grundlegenden Vorteile beim Einsatz von Abfällen einschließlich Sondermüll als Brennstoff im Drehrohrofen resultieren aus der Tatsache, daß die hohen Temperaturen im Ofen, die lange Verweildauer und die Ofendynamik den vollständigen Abbau organischer Stoffe sicherstellen. Nichtbrennbare Spurenmetalle und anorganische Bestandteile werden in den Klinker eingebunden, wodurch Entsorgungsprobleme entfallen. Zudem ist die Ofenatmosphäre naturgemäß alkalisch, was bei der Neutralisierung der verschiedenen sauren Verbindungen, die bei der Verbrennung von Müll und insbesondere Sondermüll entstehen, hilfreich ist [4].

Während Studien zur Verwendung von Abfallstoffen in Zementdrehrohröfen unter Umwelt- und ökonomischen Gesichtspunkten durchgeführt wurden, sind andere Gesichtspunkte, wie der Einfluß auf die chemische Zusammensetzung des Zements, auf die Lebensdauer der Ausmauerung, auf den Ofenbetrieb infolge der Bildung von Kreisläufen, auf die Qualität des Endproduktes infolge der Einbindung von Verunreinigungen und auf die Lebensdauer der Produktionseinrichtungen und deren Wartung noch kritisch zu untersuchen. Die verfügbaren Daten lassen sich nicht verallgemeinern, da in jeder Ofenanlage unterschiedliche Brennstoffe verwendet und andere Kombinationen von Einsatzstoffen und Prozeßparametern auftreten und daher unabhängig voneinander untersucht werden müssen.

1. Introduction

With the ever increasing industrialisation the amount of hazardous and non-hazardous byproducts/wastes are piling up at an alarming rate. However, irrespective of the nature of byproducts/wastes, these are needed to be disposed of such that the disposal itself does not degrade the environment further. High cost of disposing such materials due to strict pollution laws puts an economic pressure on the producer industry. In contrast, there are industries which are facing the problem of depleting sources of conventional fuels, and there exists a possibility of diverting wastes/byproducts as a part replacement of fuel. Environmentally friendly, economical and efficient systems are needed to exploit the potential of these byproducts/wastes by bringing together industries, and as long as the producer pays the user, everyone gains and creates an environment in which we want to live [1–3].

In this pursuit, the cement rotary kiln has been identified as one major source for recycling such wastes in recent years. Wastes that have fuel value and are hazardous were previously considered to threaten soil, air and water but now can be safely reused as part substitute for conventional fuel being used in the production of Portland cement. The basic advantages of using waste fuels including hazardous materials in the cement rotary kiln are due to the fact that the high temperature profile in the kiln, the long residence time and the dynamic situation in the kiln ensure the complete destruction of organic materials. The trace metals and inorganic components which are not combustible get included in the clinker eliminating subsequent disposal problems. Further the environment in the kiln is naturally alkaline which helps in neutralising the various acidic compounds generated while burning wastes, particularly hazardous wastes [4].

While studies have been carried out on the utilisation of these materials in cement rotary kilns considering environmental and economic aspects yet the other aspects of the issue, i.e. effect on the chemistry of cement, on refractory lining life, on process due to recirculation phenomenon, quality of final product due to incorporation of impurities and life of plant equipment and maintenance need critical examination. The data available cannot be generalised as each kiln system is using different fuels and has a different set of raw materials and process parameters and needs to be evaluated independently.

2. Sources of waste fuels

Various sources of waste-derived fuels are the chemical and the petroleum manufacturing industries, metal producing, processing and metal finishing industries, petroleum refineries, solvent reclamation, pharmaceutical manufacturing industries, agriculture based industries, municipal

2. Herkunft von Abfallbrennstoffen

Die verschiedenen Quellen der Abfallbrennstoffe sind die chemische und Erdölindustrie, die Metallherstellung, -verarbeitung und -bearbeitung, Erdölraffinerien, Lösungsmittelrückgewinnung, Pharmaindustrie, Agrarindustrie, städtische Entsorgungsbetriebe und Hausmüll [5]. Im allgemeinen lassen sich Abfallbrennstoffe in zwei Kategorien einteilen:

— feste Abfallbrennstoffe,
— flüssige Abfallbrennstoffe.

Feste Abfallbrennstoffe

Die unterschiedlichen festen Abfallbrennstoffe umfassen Altreifen, verbrauchte Tiegelauskleidungen, landwirtschaftliche Abfälle, feste Erdölderivate, kommunale und medizinische Abfälle sowie festen Sondermüll usw. [5]. Aus der oben genannten Liste werden Altreifen am weitestgehenden akzeptiert und in Zementdrehöfen eingesetzt. Allein in den Vereinigten Staaten liegen gegenwärtig mehr als 2 Mrd. Altreifen auf Halde. Der Energieinhalt dieser Reifen entspricht 20 Mio. t Kohle [6]. Mehr als 88% der Masse von Altreifen besteht aus Kohlenstoff, Wasserstoff, Stickstoff und Sauerstoff. Reifen haben einen geringen Feuchtigkeitsgehalt (< 1%) und enthalten im Mittel 1,5% Schwefel. Der mittlere Heizwert beträgt 32 500 kJ/kg. Um den Schwefelgehalt aus den Altreifen aufzunehmen, sind geeignete Anpassungen bei der Mischung der Rohstoffe erforderlich [7].

Ein weiteres Potential an festen Abfallbrennstoffen sind verbrauchte Tiegelauskleidungen aus der Aluminiumverarbeitung. Dies sind verschlissene Auskleidungen aus Kohlenstoffsteinen, kontaminiert mit Natrium, Aluminium und Fluor. Der Heizwert liegt bei etwa 18 500 kJ/kg. Der hohe Alkali- (11%) und Fluorgehalt (10%) der Tiegelauskleidungen erfordert eine besondere Berücksichtigung im Hinblick auf die Prozeßregelung und die herzustellende Zementart. Aufgrund des erhöhten Gehalts an $Al_2O_3 + Fe_2O_3$ (> 21%) der Tiegelauskleidungen muß die Zusammensetzung des Rohmehls in geeigneter Weise geändert werden [8].

Brennstoffe aus landwirtschaftlichen Abfällen (ADF – Agricultural Derived Waste Fuels) umfassen Gräser und Stroh, Abfälle aus der Baumwollherstellung und -verarbeitung, Raffinerieabfälle aus der Rohrzuckerverarbeitung, Trester aus der Weinherstellung, Hölzer aus Obst- und Weinbau, Reisschalen und viele andere Stoffe. ADF weisen einen erhöhten Halozellulosegehalt auf, dafür einen geringeren Lignin- und deutlich höheren Eiweißgehalt. ADF lassen sich in die beiden Kategorien Biobrennstoffe und Biomasse einteilen. Landwirtschaftliche Biobrennstoffe weisen höhere Konzentrationen anorganischer Bestandteile (15–21%) auf mit hohen Konzentrationen von SiO_2 und Alkalien, die potentiell die Schmelztemperatur der Asche verringern und so Verschlackungsprobleme hervorrufen können [5].

Festbrennstoffe auf Erdölbasis umfassen Ölschiefer, Teersande, extraschweres Rohöl usw. Von diesen wurde besonders Ölschiefer in verschiedenen Studien untersucht. Der Feuchtigkeitsgehalt von Ölschiefer ist normalerweise gering (0,5–1,5%). Die Elementaranalyse von Ölschiefer weist Kohlenstoff in einer Größenordnung von 4–25%, Schwefel von 0,6–2,0% und Wasserstoff von 0,5–3,0% auf. Der Gehalt an mineralischen Bestandteilen liegt über 68%. Der charakteristische Heizwert beträgt zwischen 2 000 und 15 600 kJ/kg. Die geschätzten Vorräte an Ölschiefer betragen 2×10^{16} TJ, das sind etwa 38% der Energie der bekannten Weltkohlevorräte. Vor der Verwendung von Ölschiefer als Ersatz für konventionellen Brennstoff sollte der höhere Schwefelgehalt im Hinblick auf die Prozeßregelung kritisch untersucht werden.

Brennstoffe auf der Basis von kommunalen Abfällen bestehen aus Papier, Zellstoffen, Plastik, Holz, Textilien, Gummi, Leder, Nahrungsmittelabfällen, Glas und vielen anderen Mischstoffen. Der Chlorgehalt liegt zwischen 1 und 5%. Der Heizwert kann zwischen 22 670 und 45 750 kJ/kg liegen. Auf-

and household refuse, etc. [5]. In general waste-derived fuels can be classified into two categories:

— solid waste-derived fuels,
— liquid waste-derived fuels.

Solid waste-derived fuels

The various solid waste-derived fuels include tyre-derived fuels, spent pot lining, agriculture waste-derived fuels, petroleum related solid fossil fuels, municipal and medical waste-derived fuels and solid hazardous waste fuels, etc. [5]. Out of all the listed solid waste fuels, tyres are most widely accepted and used in cement rotary kilns. In the United States alone more than 2 billion waste tyres are currently stockpiled. The energy value of these tyres is equivalent to 20 million tonnes of coal [6]. Tyre-derived fuels are composed of C, H, N, O, which is more than 88% of the tyre mass. Tyres are low in moisture (< 1%) and averaging 1.5% of sulfur. These contain an average calorific value of 32,500 kJ/kg. Therefore suitable adjustments are required in the raw mix to accommodate the sulfur coming from tyre-derived fuels [7].

Another potential solid waste fuel is spent pot lining which is worn out carbon block lining contaminated with Na, Al, F in aluminium industry. It has a calorific value of 18,500 kJ/kg. The high alkali (11%) and fluoride (10%) contents of spent pot lining call for attention from the point of view of process control and type of cement to be made. The higher content of $Al_2O_3 + Fe_2O_3$ (> 21%) in the spent pot lining necessitates raw mix to be altered suitably [8].

Agricultural-derived waste fuels (ADF) include grasses and straws, cotton growing and processing wastes, cane sugar refining waste, residuals from manufacturing wine from grapes, orchard and vineyard pruning, rice husks and host of other materials. ADF are higher in halocellulose content, lower in lignin content and substantially higher in protein content. ADF can be divided into two categories, i.e. biofuels and biomass. Agricultural biofuels have higher ash content than biomass. Biofuels have higher concentrations of inorganic components (15 – 21%) which include high concentrations of SiO_2 and alkali which has a potential for lowering the ash fusion temperature and creating slagging problems [5].

Petroleum related solid fossil fuels include oil shale, tar sands, oil sands, super heavy crude oil, etc. Out of the above oil shale is prominent and subjected to various studies. Typical moisture content of oil shale is low (0.5 – 1.5%). The ultimate analysis of oil shale indicates carbon in the range of 4 – 25% with sulfur as 0.6 – 2.0%. Hydrogen is 0.5 – 3.0%. Mineral matter content is more than 68% by weight. Characteristic heating values range between 2,000 – 15,600 kJ/kg. Estimated oil shale reserves are about $2 \cdot 10^{16}$ TJ which is about 38% of the energy of the world measured reserves of coal. Hence before going for using oil shale as replacement of conventional fuel, the higher sulfur content should be critically examined with respect to process control parameters [5].

Municipal waste based fuels are composed of paper, pulp related products, plastics, wood, textile, rubber and leather, food waste, glass and host of other such miscellaneous materials. The chlorine content ranges from 1 – 5.0%, and the higher heating value can range from 22,670 – 45,750 kJ/kg. Therefore, in view of the very high chlorine content and the low volatile trace metal contents, a very careful evaluation is needed to use this waste in the kiln [5, 9].

Liquid waste-derived fuels

The various liquid waste-derived fuels include waste from mineral oils and synthetic oils, emulsions and mixtures of mineral oil products, solvent-derived fuels, mineral oil sludge, residues from mineral oil refineries, waste automobile oils, colouring agents, paints, inks, varnishes, etc. Solvent-derived fuels are hazardous waste materials as they are characteristically ignitable. They generally contain halogenated organics and trace metals. They possess relatively high heat values and will sustain the combustion without addition of auxiliary fuels.

grund des hohen Chlorgehalts und des niedrigen Gehalts an flüchtigen Spurenmetallen ist vor dem Einsatz dieser Abfälle in Zementdrehöfen eine sehr sorgfältige Bewertung vorzunehmen [5, 9].

Flüssige Abfallbrennstoffe

Die verschiedenen Abfallbrennstoffe umfassen mineralische und synthetische Öle, Emulsionen und Gemische von Mineralölprodukten, Brennstoffe auf Lösungsmittelbasis, Mineralölschlamm, Reststoffe aus Erdölraffinerien, Altöle aus Kraftfahrzeugen, Farbzusatzstoffe, Farben, Lacke, Tinten usw. Brennstoffe auf Lösungsmittelbasis sind wegen deren leichten Entflammbarkeit grundsätzlich Sondermüll. Im allgemeinen enthalten sie organische Halogenverbindungen und Spurenmetalle. Sie weisen relativ hohe Heizwerte auf und können ohne Zugabe weiterer Brennstoffe eingesetzt werden.

Der Ausdruck „Sondermüll" ist ein juristischer Ausdruck, der sich auf Stoffe mit bestimmter Entflammbarkeit, Korrosivität oder Toxizität bezieht. Diese Stoffe sind insbesondere polychlorierte Biphenyle, belastete Abfälle und Hexachlorbenzol, Pentachlorphenol, Schlämme aus der Ölraffinierung, Abfälle aus der Produktion von Pestiziden und Herbiziden usw. Die Arten von flüssigem Sondermüll sind so unterschiedlich wie die Quellen, aus denen sie stammen [4].

Herstellungsverfahren

Zementklinker kann im Naß-, Halbnaß- oder Trockenverfahren hergestellt werden. Von diesen drei Verfahren bietet das Naßverfahren aufgrund der folgenden Vorteile die besten Möglichkeiten zur Verwendung von Abfallbrennstoffen:

— höchstes Potential an Energieeinsparung,
— geringere Kreislaufbildung,
— bessere Überwachung der Stoffbilanz und der Prozeßregelung.

Das Trockenverfahren ist wegen des Auftretens von Kreisläufen am wenigsten für den Einsatz von Abfallbrennstoffen geeignet. Wenn der Brennstoff nicht frei von Chlor ist, ist die Einrichtung eines Bypasses unumgänglich.

3. Einfluß auf die Prozeßparameter

Bei Zementdrehöfen mit Zyklonvorwärmer und Calcinator können beim Einsatz von Abfallbrennstoffen Kreisläufe von einigen chemischen Elementen auftreten. Beim Einsatz von Abfallbrennstoffen ist zu erwarten, daß sich die Konzentrationen von Chloriden, Sulfaten und Alkalien drastisch und von Kadmium, Arsen, Blei und Thallium geringfügig erhöhen. Bei modernen Zyklonvorwärmeröfen beträgt die Grenzkonzentration für Alkalien und Sulfate 1 % und für Chloride 0,1 % im Prozeß. Für diese kreislaufbildenden Elemente, d.h. Alkalien, Sulfate, Chloride und deren Verbindungen, sind kritische Bilanzen aufzustellen. In den meisten Fällen wird es unumgänglich sein, aus Qualitätsgründen Zusatzstoffe zur Steuerung des Prozesses zuzugeben. Jedoch hat sich bei den meisten Zementöfen, die mit Brennstoffen aus Abfällen betrieben werden, gezeigt, daß Abfallbrennstoffe mit höheren Alkaligehalten eine Kompensation durch die Zugabe von Sulfat erfordern. Während in Naßöfen alle Arten von Brennstoffen verwendet werden können, reagieren Trockenöfen, insbesondere solche mit Calcinator, sehr empfindlich auf solche Brennstoffe. Überall dort, wo ein Bypass vorhanden war, hat sich Berichten zufolge der Wirkungsgrad des Ofens durch die Verwendung von Brennstoffen auf der Basis von Abfällen erhöht [10, 11].

Blei und andere wenig flüchtige Spurenelemente werden teilweise in den Klinker eingebunden. Dabei wird im Ofen viel schneller ein Gleichgewichtszustand erreicht. Beim Erreichen dieses Zustandes ist die Aufnahme von Blei mit den unbehandelten Einsatzstoffen im Gleichgewicht mit der Ausschleusung von Blei durch den Klinker und den emittierten Staub [12].

The term "hazardous wastes" is a legal indication referring to materials with specific ignitability, corrosivity, toxicity. These materials are specifically polychlorinated bifinyles, laden wastes and hexachloro benzene, pentachloro phenol, oil refinery sludge, pesticides and herbiside manufacturing residuals, etc. These liquid hazardous wastes are as diverse as the sites on which they are found [4].

Type of processes

Cement may be produced by one of the three processes, i.e. wet, semidry and dry. Of the three processes the wet process offers the best solutions to the use of waste-derived fuels due to the following advantages:

— highest potential for energy savings,
— recirculation phenomenon not so critical,
— better control of material balance and process control.

The dry process offers a maximum resistance to the use of waste-derived fuels because of the recirculation phenomenon and, unless the fuel is chloride free, a bypass system would be indispensable.

3. Effect on process parameters

Cyclic phenomena associated with chemical elements due to the use of waste fuels are expected to occur in kilns with cyclone preheaters and precalciners. With the use of waste fuels it is expected that the concentration of chlorides, sulfates and alkalies will increase substantially and that of Cd, As, Pb, Th will increase marginally. The modern dry process rotary kilns operating with cyclone preheaters have limit concentrations for alkalies and sulfate of 1 % and for chlorides of 0.1 % in the system. Critical balances have to be established between these recirculating elements, i.e. alkalies, sulfates, chlorides and their ratios. In most cases it would be indispensable to add additional components for controlling the process and to compensate for quality. However some of the kilns using waste-derived fuels have confirmed that waste fuels having higher alkali content need additional sulfates to compensate for high alkali. While wet process kilns can accept any kind of fuel, the dry process kiln, particularly the precalciner kiln, will be very sensitive to the use of such fuels, and wherever a bypass system is in use, the efficiency of the same is reported to have improved by the use of waste-derived fuels [10, 11].

Lead and other low volatile trace elements get partially incorporated in the clinker, and the state of equilibrium in the kiln is reached much sooner. On attaining that condition, the intake of lead with the untreated input materials is balanced by equal output of lead with the clinker and the emitted dust [12].

It is reported that at cyclone preheater kilns the preheater acts as an effective barrier to thallium so that most of the thallium stored up in the kiln system is accumulated in the kiln/preheater cycle. In case of grate preheater kilns no internal thallium cycle is developed. Flame characteristics should be critically examined while using waste-derived fuels as flame length and flame shape change significantly with quality and quantity of waste-derived fuels (particularly the volatile content). It is observed that, with high volatile waste-derived fuels, petroleum coke and other low volatile coals can also be made compatible. A very close control of the stack emissions and the process is essential. From the operating kilns different experiences have been reported in terms of changes in the operating parameters and the quality of clinker with different fuels. While some have experienced that the use of waste fuels has resulted in an increased heat consumption, lower output and required higher degree of skills and alertness at the plant, others have not felt the same. This clearly indicates that each kiln is to be optimised to get best results. In any case the liquid fuels contain moisture, and the heat required to evaporate the same has to be accounted for.

Es wird berichtet, daß bei Zyklonvorwärmeröfen der Vorwärmer als eine effektive Barriere für Thallium wirkt, so daß sich der überwiegende Teil dieses Metalls, der sich in der Ofenanlage anreichert, im Ofen-Vorwärmerkreislauf ansammelt. Bei Rostvorwärmeröfen entsteht kein interner Thalliumkreislauf. Die Eigenschaften der Flamme bedürfen bei der Verwendung von Abfallbrennstoffen einer kritischen Überprüfung, da die Flammenlänge und -form sich deutlich mit der Qualität und der Quantität der Abfallbrennstoffe ändern (insbesondere der Gehalt an flüchtigen Bestandteilen). Hierbei ist anzumerken, daß Abfallbrennstoffe mit hohen Gehalten an flüchtigen Bestandteilen mit Hilfe von Petrolkoks und anderen schwerer flüchtigen Kohlen an die Anforderungen angepaßt werden können. Eine sehr genaue Überwachung der Abgasemission und des Prozesses ist unbedingt erforderlich. In bezug auf Veränderungen der Betriebsparameter und der Klinkerqualität bei Einsatz verschiedener Brennstoffe wurde über verschiedene Erfahrungen an Zementöfen berichtet. Während in einigen Fällen die Erfahrung gemacht wurde, daß die Verwendung von Brennstoffen auf Abfallbasis einen erhöhten Wärmeverbrauch, eine geringere Klinkerleistung und höhere Anforderungen an die Aufmerksamkeit des Anlagenpersonals nach sich zog, wird dies von anderen Quellen nicht bestätigt. Dies zeigt klar, daß jeder Ofen für sich optimiert werden muß. In jedem Fall enthalten die Flüssigbrennstoffe gewisse Wassergehalte, so daß die Wärme, die zu deren Verdampfung erforderlich ist, berücksichtigt werden muß.

An einem Vorcalcinierofen des Dalmia-Werkes wurden Versuche mit Abfallbrennstoffen auf Erdölbasis (OWF – Oil based Waste Fuels) als teilweisem Ersatz des herkömmlichen Brennstoffs mit den folgenden Merkmalen durchgeführt:

Heizwert: 12960 kJ/kg
Wassergehalt: 10 %
Flüchtige Stoffe: 64 %
Asche: 14 %
Fester Kohlenstoff: 22 %.

Die Zusatzbrennstoffe wurden manuell mit einer Leistung von 4 t/h mittels Schneckenförderer zugegeben. Dabei wurden auffällige Schwankungen der Kohlenmonoxidgehalte im Vorwärmerabgas beobachtet. Die Regelung der Brennstoffzufuhr zum Calcinator, deren Grundlage ein Temperaturregelkreis am Ausgang der Zyklonstufe IV war, arbeitete zufriedenstellend. Da die Gehalte an Schwefel, Chloriden und alkalischen Stoffen der OWF nicht deutlich höher liegen, wurde kein Bypass benötigt. Auch andere Parameter blieben deutlich innerhalb der Grenzwerte für den Betrieb. Die Qualität des Klinkers hat sich durch die Verwendung der OWF verbessert. Den Einfluß auf die Prozeßparameter zeigt **Tabelle 1** [13].

TABELLE 1: Prozeßparameter mit und ohne Einsatz von von Abfallbrennstoffen auf Ölbasis Zementwerk Dalmia

Bezeichnung		im Versuch	vor dem Versuch
Abgastemperatur Vorcalcinator	(°C)	883	887
Temperatur 4. Stufe	(°C)	875	878
Temperatur 3. Stufe	(°C)	769	765
Temperatur 2. Stufe	(°C)	586	570
Temperatur WT-Gebläse	(°C)	365	353
Druck vor WT-Gebläse	(mm Hg)	−810	−838
Sauerstoff im Rohgas	(%)	5,1	4,4
Kohlenmonoxid im Rohgas	(%)	0,3	0,09
Drehzahl WT-Gebläse	(min⁻¹)	910	920
Rohmehlaufgabe	(t/h)	135	137
Wärmeverbrauch	(kJ/kg Klinker)	3574	3611

Trials have been conducted at Dalmia plant's precalciner kiln using oil based waste fuels (OWF) as a part replacement of the conventional fuel with following characteristics:

Calorific value: 12,960 kJ/kg
Moisture: 10 %
Volatile matter: 64 %
Ash: 14 %
Fixed carbon: 22 %.

OWF was fed manually through the screw conveyor at a rate of 4.0 tonnes/h. It was observed that, with the use of OWF, the CO fluctuations of the tower outlet were more prominent. The precalciner fuel control, based on stage IV outlet temperature loop, was found to work satisfactorily. Since the sulfur, chlorine and alkali content of OWF is not significantly higher, no bypass system was needed. Other parameters have also remained well within operating limits. The quality of clinker produced has improved after using OWF. The influence on process parameters is given in **Table 1** [13].

TABLE 1: Process parameters during blank as well as OWF trial period at Dalmia Cement Plant

Description		During trial	Before trial
PC Exit gas temperature	(degree C)	883	887
4th Stage temperature	(degree C)	875	878
3rd Stage temperature	(degree C)	769	765
2nd Stage temperature	(degree C)	586	570
PH Fan inlet temperature	(degree C)	365	353
PH Fan inlet draught	(mm Hg)	−810	−838
PH Outlet Oxygen	(%)	5.1	4.4
PH Outlet Carbon monoxide	(%)	0.3	0.09
PH Fan speed	(rpm)	910	920
Kiln Feed Rate	(t/h)	135	137
Heat Consumption	(kJ/kg of clinker)	3574	3611

4. Effect on the chemistry of cement

Components of the Portland cement, i.e. C_3S, C_2S, C_3A and C_4AF, undergo changes in their crystal structure, reactivity and performance with the incorporation of foreign ions/minor oxides which are inherent in waste fuel being used. The main oxides which have direct influence on the properties of cement are SO_3, alkalies and fluorides. Alkalies in the raw meal affect both the clinker phase formation and properties of cement and try to combine preferentially with each other to form alkali or potassium calcium sulfate. While the process parameters are very sensitive to alkali and sulfates, the same is not the case with chlorides where high levels can be tolerated. However the same should not be permitted to go in the clinker as it will lead to alinite clinker formation. The alinite type of cement is not yet free from application hazards due to leaching out of chlorides. While much higher sodium oxide/potassium oxide can be incorporated in the C_3A, their influence on the characteristics of clinker is felt even at much lower levels. For example, K_2O in C_3A accelerates hydration of pure C_3A phase. Woermann determined that up to 1.4 % (by mass) of Na_2O and K_2O each can be incorporated in C_3S. In case of higher K_2O contents, free lime and alkali belite are formed apart from C_3S. Fluorides coming from waste fuels have both fluxing and mineralising effect on cement making. There is an optimum level, say about 1 % for CaF_2, at which the surface tension and viscosity are reduced significantly. The other effects are the reduction in average crystal sizes and additional alite formation. With reduction of C_3A which is replaced by $C_{11}A_7 \cdot CaF_2$, resulting in extra lime for alite formation in the early stage. Reports indicate that the use of waste-derived fuels generally leads to the formation of alite crystal size which is lower, and that no element is carried to the clinker and therefore has no adverse effect [14–16].

4. Auswirkungen auf die Zementchemie

Die Bestandteile von Portlandzementklinker, also C_3S, C_2S, C_3A und C_4AF, unterliegen nach der Aufnahme von fremden Ionen/Oxiden, die unvermeidliche Bestandteile des verwendeten Brennstoffs sind, Veränderungen in ihrer Kristallstruktur, der Reaktivität und der Eigenschaften. Die wichtigsten Oxide, die direkt die Eigenschaften des Zements beeinflussen, sind SO_3, Alkalien und Fluoride. Alkalien im Rohmehl beeinflussen sowohl die Klinkerphasenbildung als auch die Eigenschaften des Zements und reagieren vorzugsweise miteinander zu Alkali- oder Kalium-Kalzium-Sulfaten. Während die Betriebsbedingungen des Ofens gegenüber Alkalien und Sulfaten sehr empfindlich reagieren, ist dies gegenüber Chloriden nicht gleichermaßen der Fall, so daß von diesen demzufolge höhere Gehalte toleriert werden können. Eine Einbindung in den Klinker sollte jedoch möglichst verhindert werden, da diese zur Bildung

5. Effect on lining life of rotary kiln

The conventional refractories consist of pure oxides or oxide combinations which, from thermodynamic point of view, are usually not in an energetically stable state of equilibrium. As is obvious from **Table 2**, the carbon content of most of the waste-derived fuels is in the range of 25 to 90 % which is quite high. Under the catalytic influence of the iron, carbon deposition may occur preferentially in the brick pores in the presence of CO and, due to crystallisation pressure, high stress can occur in the matrix of the brick leading to a complete destruction of the brickwork. This phenomenon is more obvious in the transition and calcining zone of the kiln where the brick surfaces are open for reaction whereas in the burning zone the coating may resist the interaction due to relatively low porosity.

Refractories having compounds such as mullite decompose to form corundum, silicon monoxide and a glassy phase in

TABELLE 2: Charakteristische Eigenschaften von Abfallbrennstoffen
TABLE 2: Characteristics of waste derived fuels

Parameter	Altreifen/ Tyre derived fuels	Holzabfälle/ Wood based fuels	Landwirtschaftliche Abfälle/ Agriculture derived fuels	Ölschiefer/ Oil shale	Hausmüll/ Refuse derived fuels	Kommunale Abfälle/ Mixed municipal waste fuels	Medizinische Abfälle/ Mixed medical waste fuels	Sondermüll/ Hazardous waste
Quellen/ Sources	Reifen Automobil-industrie/ Tyre Automobile industry	Forst/ Forest	Zucker-, Wein-, Baumwoll-, Obst- und Samen-verarbeitung/ Sugar, wine, cotton fruits and seeds processing ind.	Natürliches Vorkommen/ Natural occurrence	Städtische Müllkippe/ Municipal waste depot	Städtische und ländliche Kommunen/ Urban and rural society	Krankenhäuser; Pathologische Laboratorien, Kliniken/ Hospitals, pathological labs, clinics,	Stoffe nach RCRA*), TSCA**), CERCLA***)/ Materials listed in RCRA*), TSCA**), CERCLA***)
Feuchtigkeitsgehalt	–	–	–	–	–	10–60	9,0	7,52–26,43
Flüchtige Bestandteile	–	73–88	68–85	–	–	–	–	–
Fester Kohlenstoff	–	11–25	0,1–22,0	–	–	–	–	–
Asche	–	0,4–3,0	1,1–22,0	–	7,4–23,0	1,5–22,5	7,62	11,77–45,22
Elementaranalyse								
Kohlenstoff	88,3	49–59	38–55	4,1–22,6	34–60	18–56	51,1	19,35–38,94
Wasserstoff	7,1	6–7	4,4–6,3	0,5–3,0	4,3–8,9	2,0–7,8	6,23	3,07–6,81
Sauerstoff	–	38–44	32,1–43,5	0,3–1,8	13–35,7	8–35	21,31	3,87–10,97
Stickstoff	–	0,1–0,4	0,2–2,1	0,1–0,7	0,6–1,58	0,1–3,1	0,45	0,01–0,36
Chlor	–	–	–	–	–	0,11–5,0	0,01–0,07	7,31–12,24
Schwefel	2,10	–	–	0,02–0,55	0,6–2,0	0,06–1,17	0,25–0,57	0,10–3,00
Heizwert (btu/lb)	10400–16000	8000–10400	6700–9000	840–6700	6300–13300	3200–11600	9240	4050–8560
Spurenmetalle (mg/kg)								
Ag	–	<0,08	<0,8	–	–	–		
B	–	–	–	–	–	–	6–43,5	
Ba	–	130	41–220	–	49–285	19,8–675		
Cd	–	1,5–16,0	0,36–1,1	–	1,1–17,3	0,05–2,5		
Cr	–	16–25	11–20	–	23–95	3–375		
Cu	–	40–77	14–31	–	12–2740	10–1475		
Hg	–	<0,05	<0,5	–	0,3–1,0	4,38		
Mg	–	–	–	–	–	175–4000		
Mn	–	–	–	–	61–367	3,75–782,5		
Mo	–	3–14	2–16	1–10	–	0,6–72,5		
Ni	–	11–50	4,4–5,8	1–10	10,4–170	3,75–3228		
Pb	–	38–70	21–55	–	88–836	7,75–9,15		
Sb	–	10	10	–	–	–		
Se	–	5	<0,02	1–10	2,0–3,3	0,03–12,5		
Zn	–	130–560	40–190	1–10	164–2494	23–11500		

*) RCRA – Gesetz zur Wiedergewinnung und Bewahrung der natürlichen Ressourcen
*) RCRA – Resource recovery & consvr. act
**) TSCA – Gesetz über giftige Substanzen
**) TSCA – Toxic substances control act
***) CERCLA – Gesetz über Superfund-Gesetzgebung
***) CERCLA – Superfund legislation act

Feuchtigkeitsgehalt	= Moisture	Kohlenstoff	= Carbon	Schwefel	= Sulfur
Flüchtige Bestandteile	= Volatile matter	Wasserstoff	= Hydrogen	Heizwert (btu/lb)	= Heat value (btu/lb)
Fester Kohlenstoff	= Fixed carbon	Sauerstoff	= Oxygen	Spurenmetalle (mg/kg)	= Trace metal (mg/kg)
Asche	= Ash	Stickstoff	= Nitrogen		
Elementaranalyse	= Ultimate Ana.	Chlor	= Chlorine		

von Alinitklinker führt. Die Verwendung von Alinitzement ist wegen der Auslaugung von Chloriden nicht ohne Risiko. Obwohl der Klinker sehr viel mehr Natrium- bzw. Kaliumoxid im C_3A aufnehmen kann, wird deren Einfluß auf die Eigenschaften des Klinkers bereits bei niedrigen Gehalten spürbar. So beschleunigt zum Beispiel K_2O im C_3A die Hydratation der reinen C_3A-Phase. Woermann hat ermittelt, daß bis zu 1,4 % Na_2O und K_2O im C_3S aufgenommen werden können. Im Falle höherer K_2O-Gehalte bilden sich neben C_3S Freikalk und alkalischer Belit. Fluoride, die aus den Abfallbrennstoffen stammen, wirken bei der Klinkerbildung sowohl als Flußmittel als auch als Mineralisator. Es gibt einen optimalen Gehalt bei ungefähr 1 % CaF_2, bei dem die Oberflächenspannung und die Viskosität der Klinkerschmelze deutlich vermindert werden. Andere Auswirkungen sind die Verminderung der durchschnittlichen Kristallgröße und die zusätzliche Alitbildung. Dabei resultiert die Verminderung von C_3A, das durch $C_{11}A_7 \cdot CaF_2$ ersetzt wird, in einem höheren Kalkangebot zur Alitbildung in einem frühen Stadium. Berichte weisen darauf hin, daß die Verwendung von Abfallbrennstoffen allgemein zu kleineren Alitkristallen führt und daß kein Element in den Klinker eingebracht wird und daher auch keinen negativen Einfluß hat [14–16].

5. Einfluß auf die Lebensdauer der Ofenausmauerung

Die herkömmlichen Ausmauerungen bestehen aus reinen Oxiden oder Oxidverbindungen, die aus dem Blickwinkel der Thermodynamik für gewöhnlich nicht im energetisch stabilen Gleichgewicht sind. Wie aus **Tabelle 2** offensichtlich wird, liegt der Kohlenstoffgehalt der meisten Abfallbrennstoffe im Bereich von 25–90 %. Unter dem katalytischen Einfluß von Eisen kann sich Kohlenstoff vorzugsweise in Gegenwart von CO in den Poren der Steine absetzen. Infolge des Kristallisationsdruckes können hohe Spannungen in der Steinmatrix auftreten, die zu einer vollständigen Zerstörung der Ausmauerung führen. Diese Erscheinung tritt vorwiegend in der Übergangs- und Calcinierzone des Ofens auf, da in diesem Bereich die Ausmauerungen nicht durch Ansätze geschützt sind, wohingegen die Ansätze in der Sinterzone aufgrund ihrer relativ geringen Porosität diese Wechselwirkung verhindern.

Ausmauerungen mit Bestandteilen wie z. B. Mullit zersetzen sich in Gegenwart hoher Partialdrücke von Wasserstoff (H_2) und Kohlenmonoxid (CO) zu Korund, Siliziummonoxid und einer glasigen Phase. Feuerfestmaterial, das leichtreduzierbare Oxide wie z. B. Eisen enthält, wird zu dem entsprechenden Metall reduziert. In einer stark mit H_2 und CO angereicherten Ofenatmosphäre sollten deshalb Ausmauerungen mit einem minimalen Eisengehalt verwendet werden.

In Gegenwart genügend großer Mengen von Wasserstoff und Fluor aus den Abfallbrennstoffen besteht die Möglichkeit der Bildung von Flußsäure (HF). HF ist ein sehr gutes Lösungsmittel für das SiO_2 aus dem Feuerfestmaterial, wobei gasförmiges SiF_4 gebildet wird. Daher sollten unter solchen Bedingungen SiO_2-freie und korundreiche Feuerfest-Steine verwendet werden. Wie aus Tabelle 2 hervorgeht, ist der Wassergehalt der meisten Abfallbrennstoffe hoch. Hohe Wasserdampfpartialdrücke tragen aber wahrscheinlich zur Verdampfung des SiO_2 bei. Bei Taupunktunterschreitung greifen die Säuren, also HCl/H_2SO_3, die Mikrostruktur der Ausmauerung an, lösen einzelne Bestandteile heraus und verursachen so den Verschleiß der Steinoberflächen. Hohe Alkalikonzentrationen führen nicht nur zur Ausbildung von Kreisläufen, sondern haben darüber hinaus schädliche Auswirkungen auf die Ausmauerungen. Unter Berücksichtigung des Al_2O_3-SiO_2-Komplexes bilden sich Alkalisilikate und Alkalialuminiumsilikate. In Gegenwart von Al_2O_3-haltigen Materialien wird unter Zerstörung der Struktur β-Al_2O_3 gebildet. Bekanntlich hat die infolge der erhöhten Sauerstoffgehalte der Abfallbrennstoffe zu erwartende oxidierende Atmosphäre im Ofen einen positiven Einfluß auf die Lebensdauer der Ausmauerungen. Allerdings ist eine sehr genaue Überwachung des Sauerstoffgehalts erforderlich, um lokale reduzierende Bedingungen zu ver-

presence of hydrogen and CO, especially at high gaseous pressure. Refractory materials which contain easily reducible oxides such as iron, are reduced to the corresponding metal. For application under highly hydrogen and CO rich atmosphere the refractories with the lowest iron content should be used.

Under the situation where both hydrogen and fluorine are coming from the waste-derived fuels in sufficient amount, then there is a possibility of the formation of hydrofluoric acid. HF is very much prone to dissolve the SiO_2 of the refractory and form SiF_4 which is in gaseous form. Therefore under such environment, refractories free from SiO_2 and rich in corundum should be used. The water content of most of the waste-derived fuels is high as is obvious from Table 2. Water vapour is likely to contribute to volatilisation of the SiO_2 in presence of high partial pressure of the water vapour. When the temperature is below the dew point the acids, i.e. HCl/H_2SO_3, attack the microstructure of the refractory and dissolve out single components causing wearing of the refractory brick surface. Alkalies, apart from their characteristic cyclic phenomena, also have a detrimental effect on the refractory lining. While considering Al_2O_3-SiO_2 system, alkali silicates and alkali alumosilicate are formed. In the presence of the Al_2O_3 containing materials b-alumina will be formed with destruction of the structure. Oxidizing atmosphere is known to have a positive effect on the refractory lining which is expected to prevail inside the kiln due to relatively higher content of the oxygen in most of the waste-derived fuels. However a precise control of oxygen content is required in order to avoid local alternating of the atmosphere from oxidizing to reducing. The alternating atmosphere is detrimental to the chrome based refractories which may cause a phenomenon known as bursting with subsequent destruction of the structure [17].

Some of the plants have reported early failure of high alumina refractories and have gone for basic refractories while using waste-derived fuels. It was also reported that magnesia spinels perform the best in the burning zone. There is no reported data on the kiln shell corrosion. This aspect requires long term studies and should not be ignored.

Literature

[1] Huhta, R.S.: Waste fuel survey report-I. Rock Products 88 (1985) No. 4, pp. 40–43.

[2] Huhta, R.S.: Waste fuel survey report-II. Rock Products 88 (1985) No. 5, pp. 46–49, 67.

[3] Rose, D., and Kupper, D.: Ecological and economic aspects of cement production when using waste derived fuels. Zement-Kalk-Gips 44 (1991) No. 11, pp. 554–559.

[4] Narang, K.C.: Tests relating to the use of solid fuels with low volatile content in precalciner kilns. Zement-Kalk-Gips 38 (1987) No. 8, pp. 460–462.

[5] Tillman, D.A.: The combustion of solid fuels and waste. Handbook, Academic Press Inc., 1991, pp. 81–257.

[6] Siemering, W.H., Parsons, L.S., and Lochbrunner, P.: Experience with burning waste. Rock Products 94 (1991) No. 4, pp. 36–46.

[7] Blumenthal, M.: The rationale for using whole tires. Rock Products (1992) No. 7, pp. 48–50.

[8] Tresouthick, S.W.: Spent pot lining as a supplementary fuel in cement production, proc. of 22 int. cement seminar. Rock Products, 1986, pp. 240–243.

[9] Obrist, A.: Burning sewage sludge in cement kilns. World Cement 18 (1987) No. 2, pp. 57–64.

[10] Maury, H.D., and Pavenstedt, R.G.: Chlorine bypass for increasing the fuel input from refuse in clinker burning. Zement-Kalk-Gips 41 (1988) No. 9, pp. 540–543.

[11] Mantus, E.K.: All fired up-burning hazardous waste in cement kilns. Environmental toxicology Int. pub, 1992.

meiden. Lokale reduzierende Atmosphäre ist schädlich für Ausmauerungen auf Chrombasis und kann eine als Bersten bekannte Erscheinung mit nachfolgender Zerstörung der Steinstruktur hervorrufen [17].

Berichten zufolge haben Ausmauerungen mit hohem Tonerdegehalt in einigen Anlagen bei Verwendung von Abfallbrennstoffen früh versagt und mußten durch basische Materialien ersetzt werden. Weiterhin wurde berichtet, daß Magnesiaspinellsteine in der Sinterzone am besten geeignet sind. Über die Korrosion des Ofenmantels liegen keine veröffentlichten Daten vor. Dieser Aspekt erfordert Langzeitstudien und sollte nicht vernachlässigt werden.

[12] Weisweiler, W., Dallibor, W., and Luck, M.P.: Lead, cadmium and the alkali balances of a cement kiln plant with grate preheater, operating with increased chloride input. Zement-Kalk-Gips (1987) No. 11, pp. 571–573.

[13] Trial study report on use of alternate waste derived fuels, Dalmia cement (B) Ltd. Dalmiapuam, India, 1992.

[14] Hahn, T., Eysel, W., and Woermann, E.: Vth Int. symposium on chemistry of cement, Tokyo, Japan, 1 (1968), pp. 61.

[15] Skalny, J.K., and Klemm, W.A.: Alkalies in clinker-origin, chemistry, effects. Conference on alkali aggregate reaction in concrete, Capetown, South Africa, 1981.

[16] Narang, K.C.: Portland and blended cement. IX Int. congress on chemistry of cement, New Delhi, 1 (1992), pp. 218–220.

[17] Kronert, W., and Krebs, R.: The effect of gas, liquid and solids on the refractory lining of processing furnaces of the chemical and of incinerators, XXXth Int. Colloquium on refractories, Aachen, 1987, pp. 13–17.

Betriebserfahrungen mit IKN-Kühlern im Werk Göllheim der Dyckerhoff AG

Operating experience with IKN coolers in Dyckerhoff AG's Göllheim works

Expérience acquise en exploitation avec les refroidisseurs IKN à l'usine Göllheim de la Dyckerhoff AG

Experiencias adquiridas durante el servicio con enfriadores IKN en la factoría de Göllheim, de Dyckerhoff AG

Von **H. Reckziegel**, Göllheim/Deutschland

Zusammenfassung – Der Durchsatz der beiden Drehofenanlagen im Werk Göllheim der Dyckerhoff AG, Wiesbaden, sollte aufgrund der günstigen Marktentwicklung gesteigert werden. Als erste wurde die Drehofenanlage 2 umgebaut. Dieser Drehofen (4,4 × 71 m) ist mit einem 4stufigen Schwebegas-Wärmetauscher und einem Kombi-Rostkühler (820/1044) ausgestattet. Bei diesem Kühler wurde der erste Rost mit 13 Reihen von der Firma IKN neu gestaltet und pendelnd aufgehängt. 1992 folgte ein kompletter Umbau des Horizontal-Rostkühlers (Größe 1244) der 4stufigen Zyklonvorwärmer-Drehofenanlage 1 (4,0 × 68 m). Der Kühler wurde an 6 Pendelstützen aufgehängt. Die umgebauten Roste erhielten hydraulische Antriebe. Bei den Kühlern wird der Volumenstrom der Kühlluftgebläse geregelt. Die Sollwerte werden in Abhängigkeit von der Sekundärluft-, Abluft- und Kaltklinkertemperatur festgelegt. Die Rostgeschwindigkeit ist abhängig vom Ofendurchsatz. Die verfahrenstechnischen Verbesserungen waren nach dem Umbau an höheren Sekundärlufttemperaturen, niedrigeren Kaltklinkertemperaturen und an einem gleichmäßigeren Ofenlauf zu erkennen. Die Garantiedaten wurden erreicht. Stillstände durch defekte Rostplatten traten bisher nicht ein. Demgegenüber wurden die Ofenauslaufsegmente deutlich höher beansprucht.

Summary – Because of the favourable market trend the outputs of the two rotary kiln plants at Dyckerhoff AG's Göllheim works in Wiesbaden had to be increased. Rotary kiln plant 2 was modified first. This rotary kiln (4.4 × 71 m) has a 4-stage suspension gas preheater and a Kombi grate cooler (820/1044). The first grate in this cooler with 13 rows was redesigned by IKN and suspended so that it could oscillate. This was followed in 1992 by complete conversion of the horizontal grate cooler (model 1244) of the 4-stage cyclone preheater rotary kiln plant 1 (4.0 × 68 m). The cooler was suspended on 6 oscillating supports and the modified grate was given hydraulic drives. The volume flow from the cooling air fans is controlled at the coolers. The setpoint values are set as a function of the secondary air, exhaust air, and cold clinker temperatures. The grate speed depends on the kiln output. The process engineering improvements after the conversion were apparent in higher secondary air temperatures, lower cold clinker temperatures, and more uniform kiln running. The guaranteed data were achieved, and so far there have been no stoppages caused by defective grate plates. However, the kiln outlet segments are exposed to significantly higher stresses.

Résumé – Les capacités des deux lignes de four à l'usine Göllheim de la Dyckerhoff AG, Wiesbaden devaient être accrues en raison des conditions favorables du marché. La première ligne modifiée a été la 2. Ce four (4,4 × 71 m) est équipé d'un préchauffeur à suspension de 4 étages et d'un refroidisseur à grille Kombi (840/1044). Sur ce refroidisseur, la première grille à 13 rangées a été reconfigurée par la firme IKN et suspendue en pendule. En 1992 a été entreprise une reconstruction complète du refroidisseur à grille horizontale (taille 1244) de la ligne de four rotatif 1 (4,0 × 68 m) à 4 étages de cyclones. Le refroidisseur a été suspendu à 6 supports pendulaires. Les grilles reconfigurées ont reçu des entraînements hydrauliques. Dans les refroidisseurs est régulé le flux volumique des soufflantes d'air de refroidissement. Les valeurs de consigne sont fixées en fonction des températures de l'air secondaire, de l'air d'exhaure et du clinker refroidi. La vitesse de la grille dépend du débit du four. Les améliorations apportées à la technique du process pouvaient être jugées, après la reconstruction, sur la base de températures d'air secondaire plus élevées, de températures plus basses du clinker refroidi et d'une marche du four plus régulière. Les valeurs garanties ont été atteintes. Jusqu'à présent, des arrêts pour plaques de grille défectueuses ne sont pas à noter. Les segments de sortie du four, par contre, sont nettement plus sollicités.

Resumen – Debido a la evolución favorable del mercado, surgió la necesidad de aumentar la capacidad de producción de las dos líneas de hornos rotatorios en la factoría de Göllheim, perteneciente a la Dyckerhoff AG, de Wiesbaden. Se convirtió primero la línea 2. Este horno (4,4 × 71 m) va equipado con un intercambiador de suspensión del material en los gases calientes, de 4 etapas, y un enfriador de parrilla combinado (820/1044).

En este enfriador, la primera parrilla, de 13 hileras, fue concebida por la firma IKN, con suspensión pendular. En 1992 se procedió a la reconversión total del enfriador de parrilla horizontal (tamaño 1244) de la instalación de horno rotatorio con precalentador de ciclones, de 4 etapas (4,0 × 68 m). El enfriador fue suspendido mediante 6 apoyos pendulares. Las parrillas modificadas fueron equipadas con dispositivos de accionamiento hidráulico. En los enfriadores se regula el caudal volumétrico de las soplantes de aire de enfriamiento. Se fijan los valores teóricos en función de la temperatura del aire secundario, del aire de escape y del clínker enfriado. La velocidad de la parrilla depende del caudal del horno. La mejoras tecnológicas se notaron después de la transformación, por las temperaturas más elevadas del aire secundario, las temperaturas más bajas del clínker enfriado así como por la marcha más uniforme del horno. Se alcanzaron los datos garantizados. No se han producido, hasta la fecha, paradas debidas a averías en las placas de la parrilla. Pero, por otro lado, los segmentos de la salida del horno han sido sometidos a mayores esfuerzos.

Vor 30 Jahren ging im Werk Göllheim der Dyckerhoff AG die erste Drehofenanlage in Betrieb, 1965 die zweite, gebaut von KHD und Krupp-Polysius.

In den letzten Jahren wurde die Leistung beider Ofenanlagen durch Umbaumaßnahmen deutlich gesteigert.

Durch die Reduzierung der spezifischen Luftmenge der Klinkerkühler wurde die geplante Leistungssteigerung erreicht, ohne die absoluten Gasvolumenströme zu erhöhen. Alle anschließenden Aggregate wie Trockner, EGR und EGR-Gebläse konnten erhalten bleiben. Nur das Abgasgebläse der Ofenanlage 2 mußte vergrößert werden.

Außer dem Ofenabgas wird in Göllheim auch die Kühlerabluft zur Rohmaterialtrocknung verwendet. Daher war es neben der Erhöhung des Rekuperationsgrades des Kühlers erstrebenswert, den Gesamtwirkungsgrad und die Temperatur der Kühlerabluft zu erhöhen und den Kühlluftbedarf zu senken.

Als erste Anlage wurde 1990 die Drehofenanlage 2 umgebaut. Der Drehofen dieser Anlage mit einem Durchmesser von 4,4 Metern und einer Länge von 71 Metern besitzt einen 4stufigen Schwebegas-Wärmetauscher, einen Kombi-Rostkühler der Größe 8 × 20 Fuß für Rost 1 und 10 × 44 Fuß für Rost 2. Der erste schräge Rost mit 20 Reihen wurde von IKN neu gestaltet und pendelnd aufgehängt. Der umgebaute Rost erhielt, wie heute üblich, einen hydraulischen Antrieb.

Blickt man in den Kühler bei Vollastbetrieb, so ist deutlich der rotglühende Bereich des Kühlereinlaufes zu erkennen. Der Klinker wird schnell dunkel und dann gleichmäßig über die gesamte Rostbreite zum Kühlerauslauf transportiert — es sind keine roten Klinkersträhnen zu erkennen.

Durch diese Umbaumaßnahme wurde der Kühlerwirkungsgrad bei gesteigerter Leistung der Ofenanlage von 63 % auf 72 % verbessert. Die spezifische Kühlluftmenge sank von 2,27 Nm³/kg Klinker auf unter 1,9 Nm³/kg Klinker und erreichte inzwischen 1,7 Nm³/kg Klinker.

Bild 1 zeigt an einem Beispiel den Kühlerbetrieb über 24 Stunden. Dargestellt ist der Geschwindigkeitsverlauf der beiden Schubroste, die Sekundärlufttemperatur und die Kaltklinkertemperatur. Man erkennt die konstante Rostgeschwindigkeit, die Sekundärlufttemperatur — gemessen

Dyckerhoff AG's first rotary kiln plant at the Göllheim works went into operation 30 years ago and the second, built by KHD and Krupp-Polysius, in 1965.

The output of both plants has been considerably increased in recent years by conversion work.

By reducing the specific volume of air in the clinker coolers the planned improvement of performance was obtained without increasing the absolute volume flow of gas. It was possible to retain all the adjoining assemblies such as dryers, electrostatic precipitators and electrostatic precipitation fans. Only the exhaust gas fan in kiln plant 2 had to be enlarged.

Apart from the exhaust gas from the kiln the exhaust air from the cooler is also used for drying raw material at Göllheim. So it was desirable not only to increase the cooler's recuperation rate but also to improve overall efficiency, raise the temperature of the exhaust air from the cooler and lower the cooling air requirement.

Rotary kiln plant 2 was the first plant to be converted, in 1990. The kiln, which is 4.4 m in diameter and 71 m long, has a 4-stage suspension gas preheater and a combined grate cooler measuring 8 × 20 ft for grate 1 and 10 × 44 ft for grate 2. The first inclined grate with 20 rows was redesigned by IKN and suspended so that it could oscillate. The converted grate was given a hydraulic drive as is now the practice.

If the cooler is inspected when it is operating at full load a red glow can be seen clearly at the inlet. The clinker soon darkens and is then transported evenly over the whole width of the grate to the outlet. No red streaks of clinker can be seen.

This conversion work increased cooling efficiency from 63 to 72 % and raised the output of the kiln plant. The specific volume of cooling air dropped from 2.27 to less than 1.9 m³(stp)/kg clinker and has since reached 1.7 m³(stp)/kg.

Fig. 1 shows an example of cooler operation over 24 hours. The speed curve for the two reciprocating grates, the secondary air temperature and the cold clinker temperature are plotted. The grate speed can be seen to be constant. The secondary air temperature, measured with thermocouples, is between 1000 and 1150 °C and the clinker temperature between 80 and 100 °C.

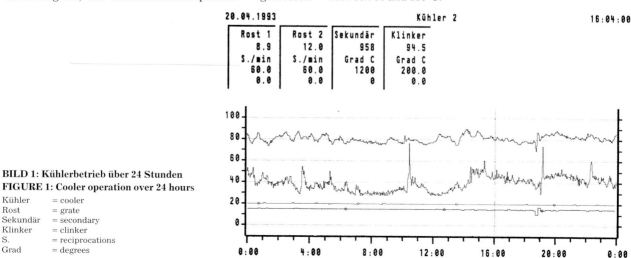

BILD 1: Kühlerbetrieb über 24 Stunden
FIGURE 1: Cooler operation over 24 hours

Kühler = cooler
Rost = grate
Sekundär = secondary
Klinker = clinker
S. = reciprocations
Grad = degrees

Kühler der Drehofenanlage 1	
Sekundärlufttemperatur gemessen mit Thermoelement	1100–1200 °C
Kühlerablufttemperatur	250–300 °C
Kaltklinkertemperatur	80–110 °C
spezifische Kühlluftmenge	1,5–1,7 Nm³/kg Kli
Kühlerwirkungsgrad	ca. 74 %

BILD 2: Betriebsmeßergebnisse nach Umbau von Kühler 1

Cooler for kiln plant 1	
Secondary air temperature measured with thermocouple	1100–1200 °C
Temperature of exhaust air from cooler	250–300 °C
Cold clinker temperature	80–110 °C
Specific volume of cooling air	1.5–1.7 m³(stp)/kg clinker
Cooler efficiency	approx. 74 %

FIGURE 2: Operating results after conversion of cooler 1

Kühler der Drehofenanlage 2	
Quelle: 24h-Messung des VDZ und Betriebsmessungen	
Sekundärlufttemperatur gemessen mit Thermoelement	1000–1150 °C
Kühlerablufttemperatur	ca. 250 °C
Kaltklinkertemperatur	80–100 °C
spezifische Kühlluftmenge	ca. 1,7 Nm³/kg Kli
Kühlerwirkungsgrad	ca. 72 %

BILD 3: Ergebnisse der VDZ-Messung Kühler 2

Cooler for kiln plant 2	
Source: 24 h measurement by the VDZ and plant measurements	
Secondary air temperature measured with thermocouple	1000–1150 °C
Temperature of exhaust air from cooler	approx. 250 °C
Cold clinker temperature	80–100 °C
Specific volume of cooling air	1.7 (stp)/kg clinker
Cooler efficiency	approx. 72 %

FIGURE 3: Results of VDZ measurement, cooler 2

mit Thermoelementen – liegt zwischen 1000 und 1150 Grad Celsius, die Klinkertemperatur zwischen 80 und 100 Grad Celsius.

Geregelt werden bei dem Kühler lediglich die Kühlluftgebläse. Sie sind in ihrer Kapazität so ausgelegt, daß auch bei höheren Klinkerbetten der Volumenstrom konstant gehalten werden kann. Die Geschwindigkeit des Kühlrostes 1 wird in Abhängigkeit der Ofenleistung eingestellt und über lange Zeiten konstant betrieben. Der Rost 2 läuft im gleichen Maße mit. Von der gleichmäßigen Klinkerverteilung am ersten Rost profitiert auch der nicht umgebaute Rost 2, der nun mit wesentlich geringerer Schubzahl und geringerem Verschleiß als früher betrieben werden kann.

Ofenstillstände infolge defekter Rostplatten sind seit der Inbetriebnahme vor 3 Jahren nicht notwendig gewesen.

Schwierigkeiten gab es mit der Haltbarkeit der Auslaufsegmente. Durch die hohe Sekundärlufttemperatur werden die Auslaufsegmente auch bei guter Kühlung sehr heiß, d. h. rotglühend. Damit wird der Werkstoff in erheblicher Weise beansprucht und verliert einen großen Teil seiner Festigkeit. Die Segmente müssen deshalb ausreichend gekühlt werden.

Nach den positiven Erfahrungen des ersten Umbaus wurde 1992 der Kühler der Ofenanlage 1, ein Horizontalrostkühler, über seine gesamte Länge umgebaut. Der Drehofen 1 mit einem Durchmesser von 4 Metern und einer Länge von 68 Metern besitzt ebenfalls einen 4stufigen Schwebegaswärmetauscher. Der Rost der Größe 12 × 44 Fuß wurde an 3 Pendelstützenpaaren aufgehängt. An Stelle der vorher vorhandenen Wasserschurre wurden 6 feststehende Reihen montiert. Der gesamte Rost wird mit Hilfe eines einzigen Hydraulikzylinders angetrieben.

Die in **Bild 2** zusammengestellten Betriebsdaten zeigen, daß auch an diesem Kühler durchweg befriedigende Ergebnisse erreicht wurden. Die Sekundärlufttemperatur liegt nun zwischen 1100 und 1200 Grad Celsius, die Kühlerabluft zwischen 250 und 300 Grad Celsius, die Klinkertemperatur nach Kühler zwischen 80 und 100 Grad Celsius und die spezifische Luftmenge zwischen 1,6 und 1,7 Nm³/kg Klinker. Der Kühlerwirkungsgrad beträgt etwa 74 %.

Im **Bild 3** sind die Betriebsdaten nach dem Umbau des Kühlers 2 dargestellt. Der Kühlerwirkungsgrad betrug bei dem betrachteten 24-Stundenversuch 72 %.

Zusammenfassend kann gesagt werden, daß die Garantiedaten bei beiden Kühlern eingehalten und die Erwartungen erfüllt wurden.

The only items controlled in the cooler are the cooling air fans. Their capacity is such that volume flow can be kept constant even at deeper beds of clinker. The speed of cooler grate 1 is set according to kiln output and kept constant over long periods. Grate 2, which has not been converted, operates at the same rate. It benefits from the uniform clinker distribution on the first grate and can now be run at a far lower reciprocating rate and with far less wear than previously.

It has not been necessary to stop the kiln because of defective grate plates since the conversion came into operation three years ago.

There were difficulties about the durability of the outlet segments. Owing to the high temperature of the secondary air they become very hot, i.e. red hot, even with effective cooling. This puts a considerable strain on the material, which loses a great deal of its strength. The segments must therefore be cooled adequately.

After the positive experience with the first conversion the cooler of kiln plant 1, a horizontal grate cooler, was converted over its whole length in 1992. Kiln 1, which is 4 m in diameter and 68 m long, also has a 4-stage suspension gas preheater. The grate measuring 12 × 44 ft was suspended from six oscillating supports. Six stationary rows were installed instead of the previous water-cooled chute. The whole grate is driven by one hydraulic cylinder.

The operating data summarised in **Fig. 2** shows that perfectly satisfactory results were also obtained with this cooler. The secondary air temperature is between 1100 and 1 200 °C, the exhaust air from the cooler between 250 and 300 °C, the clinker temperature after the cooler between 80 and 110 °C and the specific volume of air between 1.6 and 1.7 m³(stp)/kg clinker. Cooling efficiency is approx. 74 %.

Fig. 3 gives the operating data after conversion of cooler 2. Cooling efficiency was 72 % after the 24 hour test illustrated.

In summarising it can be said that the guaranteed data were achieved by both coolers and expectations fulfilled.

Magnesia-Spinell-Steine – eine praxisbezogene Betrachtung

Magnesia spinel bricks – a practical view

Briques de spinelle-magnésie – aspect pratique

Ladrillos de espinela magnésica – aspectos prácticos

Von **A. Rodrigues** und **W. S. Oliveira,** Contagem/Brasilien

Zusammenfassung – Der Zement-Drehrohrofen hat verschiedene Bereiche mit besonderen Anforderungen an eine geeignete feuerfeste Ausmauerung. Die in letzter Zeit erfolgten Modifikationen der Verfahrens- und Brenntechnologie haben zu Veränderungen im Verschleißprofil und in den Eigenschaften feuerfester Ausmauerungen geführt; das gilt besonders für die Sinter- und Übergangszone. In diesen Bereichen wurden bisher basische Steine verwendet. Die Magnesia-Chrom-Ausmauerungen sind in der Zwischenzeit fast vollständig durch chromfreie Produkte, hauptsächlich Magnesia-Spinell-Steine, ersetzt worden. Verschiedene Verfahrensparameter, wie Brennbarkeit des Rohmehls, Einsatz alternativer Brennstoffe, Anwesenheit alkalischer Verbindungen und Herstellung verschiedener Klinkerarten, beeinflussen die Menge und die charakteristischen Eigenschaften der Schmelzphase, die Bildung von Ansätzen wie auch deren Stabilisierung und Position innerhalb des Ofens. Dementsprechend hat MAGNESITA spezielle Magnesia-Spinell-Steine entwickelt, die mit oder ohne Ansatzbildung verwendbar und gegenüber chemischer Korrosion und Redox-Bedingungen resistent sind. Das ist das Ergebnis verschiedener praktischer Untersuchungen in Verbindung mit der Anwendung von Magnesia-Spinell-Steinen.

Summary – The cement rotary kiln has different regions with special requirements for suitable refractory lining. The recent modifications in the process and burning technologies have caused changes in the wear profile and in the properties of refractory linings, especially in the sintering and transition zones. In these regions, basic bricks have been used. The magnesia chrome refractories were nearly completely replaced in the meantime by chrome-free products, mainly by magnesia spinel bricks. Different process parameters, such as burnability of the raw meals, use of alternative fuels, presence of alkali compounds and production of different types of clinker, affect the amount and characteristic properties of the liquid phase, the formation of coatings, as well as its stabilization and position inside the kiln. Therefore MAGNESITA has developed specific magnesia spinel products which work with or without coating and which are resistant to chemical corrosion and redox conditions. This was the result of investigations into different practical cases of specification and use of magnesia spinel bricks.

Résumé – Le four tubulaire rotatif de cimenterie présente différentes zones particulièrement exigeantes au niveau du garnissage réfractaire. Les modifications de la technologie de process et de cuisson effectuées dernièrement ont entraîné des changements au point de vue profil d'usure et propriétés des garnissages réfractaires; ceci s'applique essentiellement pour la zone de cuisson et de transition; ces zones font appel jusqu'à présent à des briques basiques. Les garnissages au chrome-magnésie sont, entre-temps, presque entièrement remplacés par des produits exempts de chrome, notamment les briques de spinelle-magnésie. Divers paramètres d'exploitation comme la broyabilité, l'utilisation de combustibles alternatifs, la présence de composés alcalins et la fabrication de divers types de clinker influent sur le volume et les caractéristiques de la phase de fusion, la formation de croûtages et leur stabilisation et emplacement dans le four. Pour ce faire, MAGNESITA a développé des briques spéciales de spinelle-magnésie qui sont utilisables avec ou sans formation de croûtage et résistantes à la corrosion chimique et aux conditions rédox. C'est là le résultat de diverses recherches pratiques en liaison avec l'utilisation de briques de spinelle-magnésie.

Resumen – El horno rotatorio de cemento tiene varias zonas especialmente exigentes en cuanto a un revestimiento refractario apropiado. Las modificaciones introducidas últimamente en la tecnología de los procesos y en la cocción hacen que se hayan producido cambios en el perfil de desgaste y en las características de la mampostería refractaria, sobre todo en las zonas de sinterización y de transición. En dichas zonas se han utilizado hasta ahora ladrillos básicos. Entretanto, los revestimientos de cromomagnesita han sido sustituidos casi por completo por productos exentos de cromo, en primer lugar por ladrillos de espinela magnésica. Existen varios parámetros, tales como la aptitud a la cocción del crudo, la utilización de combustibles alternativos, la presencia de combinaciones alcalinas y la fabricación de diferentes tipos de clínker, que influyen en la cantidad y características de la fase fundida, en la formación de adherencias y también en la estabilización y situación de estas últimas dentro del horno. Por esta razón, MAGNESITA ha desarrollado ladrillos especiales de espinela

magnésica que pueden emplearse con o sin adherencias y que son resistentes frente a
la corrosión por agentes químicos y a las condiciones redox. Esto es el resultado de
diferentes estudios prácticos efectuados en relación con la aplicación de ladrillos de
espinela magnésica.

1. Einleitung

Die Prozeß- und Brenntechnik bei der Zemenherstellung hat sich in den letzten Jahren verändert. Dies zog Änderungen im Verschleißprofil und in den Anforderungen an die feuerfeste Ausmauerung nach sich, insbesondere in der Sinter- und Übergangszone [1].

Um der gegenwärtigen Nachfrage gerecht zu werden, und auch infolge der Forderungen des Umweltschutzes im Hinblick auf die Toxizität sechswertiger Chrombestandteile in feuerfesten Steinen hat die Firma Magnesita eine Palette chromfreier Steine entwickelt.

Das Verhalten dieser Steine im Zementofen hängt vom Herstellungsverfahren und von den Rohstoffen ab, insbesondere von der Herkunft der verwendeten Rohmaterialien.

Der vorliegende Artikel beschreibt die Eigenschaften chromfreier Steine und erörtert darüber hinaus auch die dargestellten Anforderungen und die Verwendung in der Praxis anhand von Untersuchungen an zwölf Ofenanlagen.

2. Magnesia-Spinell-Steine

2.1 Rohstoffe

Die Magnesia-Spinell-Steine werden im wesentlichen aus Magnesia (MgO) und Spinell ($MgAl_2O_3$) hergestellt. Sie bestehen aus zwei Arten gesinterter Magnesia. Beide sind totgebrannte Magnesite aus brasilianischer Produktion. Die wichtigsten Eigenschaften der Sinterstoffe zeigt **Tabelle 1**.

Die Art des jeweiligen Sinterwerkstoffes teilt die Steine in zwei Gruppen ein. Hauptbestandteil der ersten Gruppe ist Sintermagnesia I und der zweiten Sintermagnesia II. Daneben gibt es zwei Spinellquellen: elektrisch erschmolzenen Spinell, der zur körnigen Magnesiamischung zugegeben wird, und elektrisch erschmolzenes Aluminiumoxid, das zur Mischung zugegeben wird und beim Brennen mit dem Periklas der Matrix der Steine unter Bildung von Spinell reagiert.

Die Rohstoffzusammensetzung der Steine ist in **Tabelle 2**, deren Eigenschaften in **Tabelle 3** aufgeführt.

2.2 Eigenschaften

Zur Erfüllung der Anforderungen in der Sinter- und Übergangszone sollten die Steine spezifische Eigenschaften wie

1. Introduction

The process and burning technologies have been modified in the last years, causing changes in the wear profile and in the requirements of refractory lining, specially in the sintering and transition zones [1].

In order to attend the actual demand, also due to the environmental protection regards to toxic hexavalent chromium compounds, Magnesita has developed a variety of chrome-free bricks.

The behaviour of these bricks inside the kiln depends on the fabrication procedures and on the raw materials, specially on the genesis of the raw materials used.

In this paper the chrome-free bricks are described, also practical cases of specification and use of these bricks in twelve real plants are discussed.

2. Periclase-Spinel bricks

2.1 Raw materials

The periclase-spinel bricks are mainly composed of magnesia (MgO) and spinel ($MgAl_2O_4$). Our bricks are made using two types of sintered magnesia. Both are dead-burned magnesite, produced in Brazil. The main characteristics of the sinters are shown in **Table 1**.

The type of sinter classifies the bricks in two groups. The first has sintered magnesia I and the second has sintered magnesia II as the major component. There are also two sources of spinel: electrofused spinel which is added to the magnesia granular mix, and electrofused alumina which is added to the mix, and during the firing it reacts with the periclase of the brick's matrix forming spinel.

The composition of the bricks in terms of raw materials is illustrated in **Table 2** and the brick's properties are described in **Table 3**.

2.2 Properties

To attend the requirements in the sintering and transition zones, the bricks should have specific properties such as chemical corrosion resistance (by alkali or liquid phase), abrasion resistance, flexibility, adhesion and stabilization of the coating, thermal shock resistance, stability under redox conditions and a low permeability.

TABELLE 1: Hauptmerkmale gesinterter Magnesia-Steine

Eigenschaften	Sintermagnesia I	II
Chemische Analyse (M.-%)		
SiO_2	1,30	0,19
Al_2O_3	0,33	0,03
CaO	0,51	0,81
Fe_2O_3	2,37	0,37
MnO	0,84	0,11
MgO	94,64	98,48
CaO/SiO_2 (Molverhältnis)	0,42	4,56
Rohdichte (g/cm³)	3,02	3,31
Porosität (%)	16	3
Mittlerer Kristalldurchmesser des Periklas (μm)	65	125
XRD – mineralische Phasen*)	M, M_2S, CMS, MF	M, C_3S, C_2S

*) M = MgO; S = SiO_2; C = CaO; F = Fe_2O_3

TABLE 1: Main characteristics of sintered magnesia bricks

Characteristics	Sintered magnesia I	II
Chemical analysis (wt.%)		
SiO_2	1.30	0.19
Al_2O_3	0.33	0.03
CaO	0.51	0.81
Fe_2O_3	2.37	0.37
MnO	0.84	0.11
MgO	94.64	98.48
CaO/SiO_2 (molar ratio)	0.42	4.56
Bulk Density (g/cm³)	3.02	3.31
Apparent Porosity (%)	16	3
Average Periclase Crystal Diameter (μm)	65	125
XRD – mineralogical phases*)	M, M_2S, CMS, MF	M, C_3S, C_2S

*) M = MgO; S = SiO_2; C = CaO; F = Fe_2O_3

Widerstandsfähigkeit gegen chemische Korrosion (durch Alkali- oder Schmelzphasen), Abriebfestigkeit, Flexibilität, Adhäsion und Stabilisierung des Ansatzes, Unempfindlichkeit gegen Temperaturwechsel, Stabilität unter Redoxbedingungen und geringe Durchlässigkeit aufweisen.

Die Eigenschaften der Steine hängen bei gleichen Produktionsbedingungen von den Rohmaterialien ab, insbesondere von deren Herkunft. Die Maximierung aller geforderten Eigenschaften ist eine sehr schwierige Aufgabe, da die Verbesserung einer Eigenschaft die Verschlechterung einer anderen nach sich ziehen kann. Daher ist es für die Verbesserung eines Steins erforderlich, bestimmte Eigenschaften zu optimieren und die anderen dabei auf akzeptablem Niveau zu halten. Schließlich bedeutet die Festlegung der richtigen Steine für eine bestimmte Ofenzone oder eine bestimmte Anforderung die Verbesserung der Lebensdauer der Ausmauerung für die gesamte Ofenreise. **Tabelle 4** zeigt die wichtigsten Anforderungen an das Feuerfestmaterial im

The properties of the bricks depend on the raw materials, specially on the genesis of the raw materials used, for the same fabrication procedures. To maximize all the properties it is a very difficult task since the improvement of a property may reduce another one. Therefore, it is necessary to optimize specific properties of a brick keeping the others in good levels, in order to improve the brick's performance. Finally, the specification of a correct brick for a specific kiln region or requirement, means the improvement of the refractory lining campaign in the whole kiln. **Table 4** shows the most important demands inside the kiln with the indication of suitable bricks and their characteristics.

Emphasis should be given to the coating adhesion or stability since it influences a great number of wear mechanisms, also because it seems to be a usual deficiency of ordinary magnesia spinel bricks [2].

Concerning the brick, it was found that the adherence and stability of the coating could be influenced by two main

TABELLE 2: Zusammensetzung von Magnesia-Spinell-Steinen

	Gruppe I MAGKOR	MAGKOR-A	MAGKOR-D	Gruppe II MAGKOR-B	MAGKOR-B-RA
Sintermagnesia I	*	*	*		
Sintermagnesia II				*	*
elektrisch erschmolzener Spinell		*	*	*	*
elektrisch erschmolzenes Aluminium	*				

TABLE 2: Composition of periclase-spinel bricks

	First Group MAGKOR	MAGKOR-A	MAGKOR-D	Second Group MAGKOR-B	MAGKOR-B-RA
Sintered Magnesia I	*	*	*		
Sintered Magnesia II				*	*
Electrofused Spinel		*	*	*	*
Electrofused Alumina	*				

TABELLE 3: Eigenschaften von Magnesia-Spinell-Steinen
TABLE 3: Properties of periclase-spinel bricks

Eigenschaften	MAGKOR	MAGKOR-A	MAGKOR-D	MAGKOR-B	MAGKOR-B-RA
Chemische Analyse (M.-%)					
MgO	90,00	81,00	87,00	83,00	90,00
Al_2O_3	3,00	12,50	6,50	12,00	6,00
Fe_2O_3	2,50	2,00	2,20	0,60	0,70
CaO	0,60	0,60	0,60	1,10	1,10
SiO_2	1,50	1,30	1,30	0,70	0,50
Rohdichte (g/cm³)	2,80	2,87	2,85	2,94	2,95
Rohporosität (%)	21	18	20	16	16
Kaltdruckfestigkeit (MPa)	45	50	50	55	60
Temperaturwechselbeständigkeit (Zyklen)	>50	>100	>100	>100	>100
Ausmauerung Teillast (T5)	>1762	>1762	>1762	>1762	>1762
Reversible lineare Wärmeausdehnung bis 1000/1500 °C	1,2/1,9	1,0/1,7	1,0/1,6	1,0/1,7	1,0/1,7
Permeabilität (CD)	51	38	29	24	14

Eigenschaften	= Characteristics
Chemische Analyse (M.-%)	= Chemical Analysis (wt.%)
Rohdichte (g/cm³)	= Bulk Density (g/cm³)
Rohporosität (%)	= Apparent Porosity (%)
Kaltdruckfestigkeit (MPa)	= Cold Crushing Strength (MPa)
Temperaturwechselbeständigkeit (Zyklen)	= Thermal Shock Resistance-Cycles
Ausmauerung Teillast (T5)	= Refractories Unterload (T5)
Reversible lineare Wärmeausdehnung bis 1000/1500 °C	= Reversible Linear Thermal Expansion up to 1000/1500 °C
Permeabilität (CD)	= Permeability (CD)

TABELLE 4: Eignung verschiedener Magnesia-Spinell-Steine

Anforderungen	Produkt	Rohmaterial und Eigenschaften
Ansatzstabilität	MAGKOR MAGKOR-A MAGKOR-D	Sinterwerkstoff aus Magnesia I: porige Struktur, Eisengehalt
Redoxbedingungen	MAGKOR-B MAGKOR-B-RA	Sinterwerkstoff aus Magnesia II: niedriger Eisengehalt
Korrosion durch Alkali und Flüssigphase	MAGKOR-D MAGKOR-B-RA	Elektrisch erschmolzener Spinell: weniger Spinell; geringere Permeabilität
Abriebfestigkeit	MAGKOR-B MAGKOR-B-RA	Sinterwerkstoff aus Magnesia II: reiner und dichter

TABLE 4: Suitability of different periclase-spinel bricks

Requirements	Brand	Raw materials and characteristics
Coating stability	MAGKOR MAGKOR-A MAGKOR-D	Sinter of magenisa I: pore structure and iron content
Redox condition	MAGKOR-B MAGKOR-B-RA	Sinter of magnesia II: low iron content
Alkali and liquid phase corrosion	MAGKOR-D MAGKOR-B-RA	Electrofused spinel; Lower spinel amount; Lower permeability
Abrasion resistance	MAGKOR-B MAGKOR-B-RA	Sinter of magnesia II: purer and denser

Ofen mit Angabe der geeigneten Steine und deren Merkmale. Das Schwergewicht sollte auf der Haftung bzw. der Stabilität des Ansatzes liegen, da dies eine Vielzahl von Verschleißmechanismen beeinflußt und auch, weil dies eine übliche Unzulänglichkeit bei Magnesia-Spinell-Steinen zu sein scheint [2].

Untersuchungen haben ergeben, daß die Adhäsion und Stabilität des Ansatzes von zwei Haupteigenschaften der Steine beeinflußt werden könnte. Die eine ist der Eisengehalt des Steins. Das Eisen kann mit dem Brenngut unter Bildung von 4 CaO · Al$_2$O$_3$ · Fe$_2$O$_3$ reagieren, was einen stabilen Ansatz bewirkt [2].

Die andere Haupteigenschaft ist die Porenstruktur des gesinterten Magnesia. Es hat sich herausgestellt, daß eine Porenstruktur, wie in **Bild 1** dargestellt, eine bessere physikalische Haftung zwischen Klinker und Stein zu fördern scheint, wenn die Flüssigphase des Klinkers in die Poren eindringt. Diese Struktur ergibt sich aus den Eigenschaften des für die Steine verwendeten Rohmaterials, das ja Magnesit ist. Diese Struktur bleibt während des Sinterprozesses erhalten. Das heißt, daß nur Magnesit diese Porenstruktur bieten kann, niemals jedoch synthetisches Magnesia, das aus Meerwasser oder Sole gewonnen wird.

Das gesinterte Magnesia I hat den geeigneten Eisengehalt und die Porenstruktur, die die exzellente Fähigkeit zur Ansatzbildung der Steine MAGKOR, MAGKOR-A und MAGKOR-D fördert.

Eine weitere wichtige Anforderung an die feuerfesten Steine im Ofen ist die Widerstandsfähigkeit gegen Alkalikorrosion. Es stellte sich heraus, daß der elektrisch erschmolzene Spinell widerstandsfähiger ist als der gesinterte Spinell [3]. Jedoch muß der Spinellanteil begrenzt werden, da sogar der elektrisch erschmolzene Spinell weniger widerstandsfähig ist als die Magnesiakomponente.

2.3 Praktische Überlegungen

In diesem Teil werden das Verhalten und die Leistung chromfreier Ausmauerungen in zwölf in Betrieb befindlichen Anlagen und die praktischen Überlegungen in Zusammenhang mit den Anforderungen an Spinell-Steinen diskutiert. Die chemische Zusammensetzung des Klinkers aus den zwölf untersuchten Anlagen zeigt **Tabelle 5**.

Für die Auswahl von Feuerfestmaterialien ist es wichtig, die Eigenschaften und das Verhalten der flüssigen Phase des Klinkers sowie deren Menge und den Beginn der Schmelzbildung zu kennen.

Diese Daten liefern Informationen über die folgenden Faktoren:

— Länge der Übergangs- und Sinterzone,

— Stabilität des Ansatzes,

— Grad der Infiltration der Ausmauerung durch die flüssige Phase.

Daneben haben auch das Temperaturprofil, die Flammenform, die Reaktivität und die Brennbarkeit des Rohmehls ebenso wie die Klinkerzusammensetzung Einfluß auf diese

characteristics. One is the iron content of the brick that can react with cement raw meal forming 4CaO·Al$_2$O$_3$·Fe$_2$O$_3$, which provides a stable coating [2].

The other characteristic is the pore structure of the sintered magnesia. It was found that the pore structure such as illustrated in **Fig. 1** seems to promote a better physical adherence between the clinker and the brick when the liquid phase of the clinker penetrates through those pores. This structure is due to the genesis of the raw material used which is a magnesite. It was preserved during the sintering process. In other words just a magnesite can provide this pore structure, not synthetic magnesia extracted from seawater or brines.

The sintered magnesia I has an adequate iron content and the pore structure, promoting an excellent "coatability" of the bricks MAGKOR, MAGKOR-A and MAGKOR-D.

Another important demand inside the kiln is the alkali corrosion. It was found that the electrofused spinel is more resistant that the sintered spinel [3]. Eventhough the amount of spinel must be limited because even the electrofused type is less resistant than the magnesia component.

2.3 Practical considerations

In this part, the behaviour and performance of chrome free lining in twelve real plants and the practical considerations related to the specification of spinel bricks are discussed. The clinker chemical composition of the twelve plants are listed in **Table 5**.

In the point of view of refractories, it is important to know the characteristics and behaviour of the clinker liquid phase, its amount and initial point of liquidus formation.

These data provide informations about the following factors:

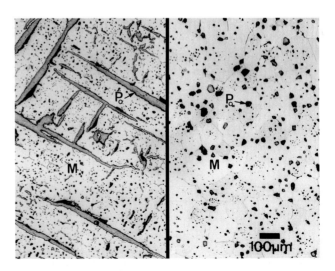

BILD 1: Porenstruktur von Sintermagnesia
FIGURE 1: Pore structure of sintered magnesia

Faktoren. In der Tat bedeutet ein stabiler Ofenbetrieb gleichzeitig eine höhere Leistungsfähigkeit von Ausmauerung und Ofen. Andere Aspekte wie der Gehalt an Alkalien, mechanische Spannungen, Redoxbedingung usw. sind ebenfalls zu berücksichtigen.

Zur besseren Analyse werden entsprechend dem Zusammenhang zwischen Silikatmodul, Kalksättigungsgrad und Brennbarkeit des Klinkers jeweils vier leicht-, normal- und schwerbrennbare Klinker vorgestellt [4] (vgl. **Bilder 2, 3, 4 und 5**). Die Werte in Tabelle 5 sind Mittelwerte. Aus der Aufzeichnung der Tageswerte können Informationen über die Homogenität des Prozesses entnommen werden. Die Bilder 2, 3, 4 und 5 stellen diese Situation für die Anlagen 3, 4, 7 und 10 dar. Die Parameter der Übergangs- und Sinterzonen aller Anlagen und die jeweils empfohlenen Arten von Spinell-Steinen sind wie folgt:

- length of transition and burning zones,
- stability of coating,
- degree of liquid phase infiltration in the lining.

Also, the profile of the temperature, flame shape, reactivity and burnability of raw materials and as well as clinker composition act on these factors.

Actually, for refractory and also for kiln operator, higher operational stability means higher performance of lining and kiln.

Other aspects such as alkalis, mechanical stresses, redox condition, ect. must be taken into account.

For a better analysis, there are presented four clinkers easy to burn, four normal and four difficult to burn, according to the curve which relates the Silica Ratio with Line Saturation Factor and the clinker burnability condition [4] (See **Figs. 2,**

TABELLE 5: Klinkerzusammensetzung der untersuchten Anlagen
TABLE 5: Clinker composition of investigated kilns

Anlage/Plants	1	2	3	4	5	6	7	8	9	10	11	12
SiO_2	21,60	21,88	20,91	21,66	21,73	21,28	21,51	22,96	21,77	19,89	21,30	21,12
CaO	65,30	66,44	66,39	65,69	66,65	64,41	65,02	66,58	66,26	63,86	66,10	67,60
Al_2O_3	5,77	5,36	7,14	5,21	5,26	4,77	5,46	4,85	5,56	6,12	5,20	5,36
Fe_2O_3	3,41	4,41	3,13	4,88	3,45	4,10	2,79	2,50	2,83	3,77	3,20	3,04
MgO	3,08	1,14	1,05	1,52	0,87	2,60	4,53	1,38	1,47	3,28	1,50	0,80
K_2O	0,70	0,51	0,21	0,34	0,54	0,80	0,50	0,40	0,45	0,98	0,90	0,50
Na_2O	–	0,10	0,10	0,19	0,00	0,10	0,20	0,10	0,30	0,36	0,18	0,10
SO_3	0,70	0,00	0,75	0,33	0,79	1,21	0,25	0,48	–	0,96	0,80	0,45
Freier Kalk/free lime	2,26	0,48	2,96	0,36	0,83	1,55	1,55	1,50	0,98	0,36	0,70	1,41
C_3S	46,84	59,92	44,73	58,42	60,28	52,82	53,53	52,93	58,90	57,98	62,57	67,31
C_2S	26,61	17,55	26,23	18,05	16,85	21,19	21,31	25,93	18,00	13,31	13,89	9,80
C_3A	9,52	6,74	13,63	5,55	8,10	5,70	9,75	8,62	9,95	9,84	8,37	9,06
C_4AF	10,37	13,41	9,52	14,84	10,49	12,46	8,48	7,60	8,60	11,46	9,73	9,24
MS	2,35	2,24	2,04	2,15	2,49	2,40	2,61	3,12	2,59	2,01	2,54	2,51
MA	1,69	1,22	2,28	1,07	1,52	1,16	1,96	1,94	1,96	1,62	1,63	1,76
SAT	91,44	93,38	91,67	93,55	94,45	93,35	93,40	90,84	94,23	97,91	96,52	97,51
AW	32,03	32,48	34,65	33,75	28,86	30,60	28,07	26,41	27,81	31,55	27,27	26,34
% FL 1450 °C	29,47	27,75	30,57	28,99	25,74	28,25	28,14	22,54	25,27	32,42	26,18	24,77
% FL 1338 °C	25,28	22,68	21,10	19,63	23,25	22,42	22,30	17,51	19,33	28,22	22,72	20,29
IQ	2,36	2,97	1,93	2,87	3,24	2,91	2,94	3,26	3,18	2,72	3,46	3,68
RAS	1,00	–	2,88	0,76	1,46	1,42	0,42	1,07	–	0,83	0,81	0,82
EA	0,46	0,40	0,24	0,41	0,36	0,63	0,53	0,36	0,60	1,01	0,77	0,43

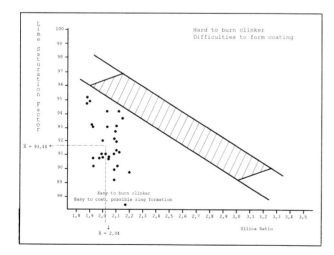

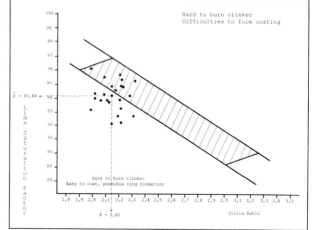

BILD 2: Kalksättigungsgrad in Abhängigkeit des Silikatmoduls, Anlage 3
FIGURE 2: Lime Saturation Factor vs Silica Ratio, Plant 3

Kalksättigungsgrad	= Lime saturation factor
Schwer brennbarer Klinker	= Hard to burn clinker
Geringe Neigung zur Ansatzbildung	= Difficulties to form coating
leicht brennbarer Klinker	= Easy to burn clinker
leichte Ansatzbildung, Ringbildung möglich	= Easy to coat, possible ring formation
Silikatmodul	= Silica Ratio

BILD 3: Kalksättigungsgrad in Abhängigkeit des Silikatmoduls, Anlage 4
FIGURE 3: Lime Saturation Factor vs Silica Ratio, Plant 4

Kalksättigungsgrad	= Lime saturation factor
Schwer brennbarer Klinker	= Hard to burn clinker
Geringe Neigung zur Ansatzbildung	= Difficulties to form coating
leicht brennbarer Klinker	= Easy to burn clinker
leichte Ansatzbildung, Ringbildung möglich	= Easy to coat, possible ring formation
Silikatmodul	= Silica Ratio

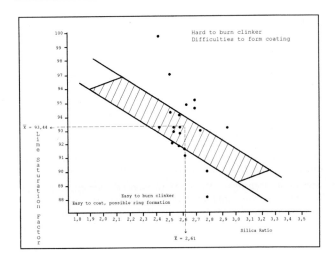
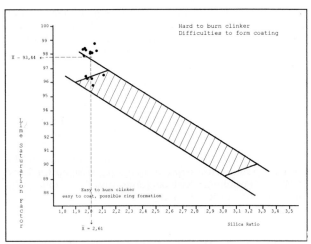

BILD 4: Kalksättigungsgrad in Abhängigkeit des Silikatmoduls, Anlage 7
FIGURE 4: Lime Saturation Factor vs Silica Ratio, Plant 7

Kalksättigungsgrad	= Lime saturation factor
Schwer brennbarer Klinker	= Hard to burn clinker
Geringe Neigung zur Ansatzbildung	= Difficulties to form coating
leicht brennbarer Klinker	= Easy to burn clinker
leichte Ansatzbildung, Ringbildung möglich	= Easy to coat, possible ring formation
Silikatmodul	= Silica Ratio

BILD 5: Kalksättigungsgrad in Abhängigkeit des Silikatmoduls, Anlage 10
FIGURE 5: Lime Saturation Factor vs Silica Ratio, Plant 10

Kalksättigungsgrad	= Lime saturation factor
Schwer brennbarer Klinker	= Hard to burn clinker
Geringe Neigung zur Ansatzbildung	= Difficulties to form coating
leicht brennbarer Klinker	= Easy to burn clinker
leichte Ansatzbildung, Ringbildung möglich	= Easy to coat, possible ring formation
Silikatmodul	= Silica Ratio

a) Obere Übergangszone

Die Anlagen 1, 2, 3, 4 und 10 arbeiten mit Überschuß an Schmelzphase und damit auch an Ansatz. Dies läßt sich aus den Werten für das Silikatmodul (SR), den Ansatzbildungsindex (AW) und den Anteil der flüssigen Phase bei 1450°C ablesen. Jedoch lassen sich anhand des Aluminiumoxidverhältnisses (AR) dieser Öfen deutliche Unterschiede feststellen, die Einfluß auf den Schmelzbeginn des Brennguts und auf die Menge und die Viskosität der Schmelzphase bei 1330°C haben.

Als Beispiel liegen für die Öfen 3 und 4 die AR-Werte bei 2,28 bzw. 1,07, außerdem beträgt bei 1338°C der Anteil der Schmelzphase 21,10% und 19,63% gegenüber 30,57% und 28,99% bei 1450°C. Dieser große Unterschied ist typisch für eine kurze Übergangszone mit wenig Ansatzbildung verglichen mit der Sinterzone. In diesen Anlagen ist eher die Verwendung von MAGKOR-A angezeigt, da dieser Stein für eine gute Haftung des Ansatzes sorgt.

Ebenso wird für Anlage 8, in der die Menge der Schmelzphase in der Übergangszone ebenfalls niedrig ist, auch MAGKOR-A empfohlen.

Im Gegensatz dazu zeigen die Anlagen 1 und 10 eine sehr ausgedehnte Übergangszone mit reichlich Flüssigphase und Ansatzbildung. Ringbildung und Instabilität des Ansatzes kommen hier ziemlich häufig vor. Daher ist die Verwendung eines Steins mit geringer Haftfähigkeit für Ansätze wie etwa MAGKOR-B wünschenswert.

Ein weiterer Punkt, der in den Anlagen 1, 10 und 11 überprüft wurde, ist der hohe Alkaligehalt im Rohmehl, der wiederum nach MAGKOR-B mit seiner ausgezeichneten Widerstandsfähigkeit gegenüber Eindringen und Angriff von Alkalien verlangt [4, 5].

Zusammengefaßt gilt, daß die obere Übergangszone grundsätzlich anfällig für instabilen Ansatz ist. Daher wird die Verwendung von Magnesia-Spinell-Steinen wie etwa MAGKOR-A empfohlen. Im Falle starken Angriffs durch Alkalien und übermäßig starke Ansatz- oder Ringbildung kann ein reinerer Spinell-Stein wie MAGKOR-B bessere Ergebnisse liefern und wird deshalb empfohlen.

b) Sinterzone

Ein Ersatz für Magnesia-Chrom-Steine in der Sinterzone, in der die Bildung und Adhäsion eines stabilen Ansatzes erwünscht ist, war die größte Herausforderung. Dolomit-

3, 4, and 5). The values listed in Table 5 are averages and when the daily values are plotted some informations about the homogeneity of the process can be observed. The Figs. 2, 3, 4, and 5 present this situation for plants 3, 4, 7, and 10. The characteristics of the transition and burning zones in each plant and what kind of spinel brick has been recommended for that condition, are presented as follows:

a) Upper Transition Zone

The plants 1, 2, 3, 4 and 10 work with an abundant amount of liquid phase and consequently, of coating as well. This can be observed by the values of Silica Ratio (SR), coatability index (AW) and percentage of liquid phase at 1450°C (% LP 1450°C). However through the Alumina Ratio (AR) of these kilns, significative differences can be noticed which will act on the initial temperature of liquid phase formation, on its amount at 1338°C and also on its viscosity.

As an example, for kilns 3 and 4, the value of AR is respectively 2.28 and 1.07, and the amount of liquid phase at 1338°C is 21.01% and 19.63% against 30.57% and 28.99% at 1450°C. This great difference is typical of a short transition zone with scarce coating, which is an opposite feature when compared to the burning zone. In these plants, the use of MAGKOR-A has been more indicated, since it provides a high adherence of coating.

In plant 8, where the amount of liquid phase in transition zone is also low, MAGKOR-A is also indicated.

On the other hand, the plants 1 and 10 present a very extensive transition zone with excessive amounts of liquid phase and coating formation. The ring formation and coating instability occur with a certain frequency and the use of a brick which shows a very low adherence of coating, such as MAGKOR-B is desired.

Another point verified in plants 1, 10, and also 11 is the high alkalis content that again demands the MAGKOR-B by its excellent resistance to alkalis infiltration and attack [4, 5].

In short, the upper transition zone is always liable to a coating instability and the utilization of magnesia spinel bricks, such as MAGKOR-A, is recommended. In case of strong alkalis attack and excessive formation of coating or ring, a purer spinel brick as MAGKOR-B may present better results, therefore being the recommended brick.

steine sind nur begrenzt einsetzbar, und die ersten konventionellen Spinell-Steine versagten, sobald die thermische Belastung höher wurde. Daher entwickelte MAGNESITA neue Magnesia-Spinell-Steine mit verbessertem Haftungsvermögen für eine starke, stabile Beschichtung.

Derzeit ist in Öfen, in denen mit Magnesia-Chrom-Silikat gebundene Steine wie etwa MAGNEFOR gute Leistungsfähigkeit und Ansatzverhalten zeigten, die Substitution durch MAGKOR abgeschlossen. Dies ist der Fall in den Anlagen 1, 3, 5, 7 und 9, in denen die Klinkereigenschaften eine stabile Ansatzbildung unterstützen.

Die niedrigen AR-Werte der Anlagen 2, 4 und 6 zeigen sofort die Gegenwart einer Flüssigphase sehr niedriger Viskosität, niedriger Oberflächenspannung und auch niedriger Hitzebeständigkeit an, was einem dichten, schmelzflüssigen Ansatz entspricht. Diese Art von Flüssigphase mit hohem C_4AF-Gehalt, die sehr aggressiv und infiltrierend wirkt, macht die basische Ausmauerung der Brennzone anfällig für Überhitzung. In diesen Fällen ist die Forderung nach einem Spinell-Stein in der Art des MAGKOR-A oder seiner alkalibeständigeren Version MAGKOR-D die beste Alternative [4].

Im Vergleich zu den Magnesia-Chrom-Steinen zeigen MAGKOR-A und MAGKOR-D ähnliche Ansatzhaftung, jedoch mit geringerer Infiltrationsneigung. Dies bedeutet, daß die Spinell-Steine von der flüssigen Phase weniger stark durchdrungen und daher weniger von strukturellen Abplatzungen verschlissen werden [6]. **Bild 6** zeigt den Vergleich.

In den Anlagen 8, 11 und 12 ist der Anteil an Schmelzphase infolge der niedrigen Reaktivität des Rohmehls und der Herstellung von Spezialklinker sehr niedrig. Die Klinker sind sicher brennbar (hoher Brennbarkeitsindex), und zur Vermeidung höherer Freikalkgehalte wird das Risiko der Ofenüberhitzung noch erhöht. Wiederum ist MAGKOR-A eher geeignet aufgrund seiner höheren Hitzebeständigkeit und größerem Widerstand gegen das Eindringen der Schmelzphase.

Als Ergebnis empfiehlt MAGNESITA für die Sinterzone als chromfreie Alternativen MAGKOR für normale Betriebsbedingungen und MAGKOR-A bzw. MAGKOR-D für Bedingungen, bei denen das Risiko von Überhitzung oder starker Infiltration von Schmelze durch Überhitzung und/oder Flüssigphase niedriger Viskosität besteht.

c) Untere Übergangszone

Die untere Übergangszone unterliegt mechanischer Abrasionsbeanspruchung, Ansatzinstabilität, Temperaturwechseln und chemischen Angriffen. Die Temperaturgradienten

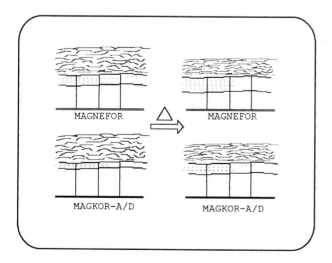

BILD 6: Infiltrationsgrad für MAGNEFOR- und MAGKOR-A/D-Steine
FIGURE 6: Degree of infiltration for MAGNEFOR and MAGKOR-A/D bricks

b) Burning Zone

The substitution of chrome-magnesia bricks in the burning zone, where the formation and adhesion of a stable coating are desired, have been the major challenge. The dolomite brick presents some limitations and the first conventional spinel bricks have failed whenever the thermal load increased. As a result, MAGNESITA developed new magnesia spinel bricks with better capacity to fix a strong, stable coating.

Actually, in kilns where magnesia chrome silicate bonded brick, as MAGNEFOR, has presented an adequate performance and coating, the substitution has been occurred by MAGKOR. This is the case of plants 1, 3, 5, 7 and 9, where the clinker characteristics support a stable coating formation.

At once, in plants 2, 4 and 6 the low values of AR indicate the presence of a liquid phase of very low viscosity, low surface tension and also low refractoriness, corresponding to a dense, fused coating formation. This type of liquid phase, rich in C_4AF which are very aggressive and penetrating, turns the basic lining at burning zone susceptible to thermal overload. In these cases, the specification of a spinel brick of type MAGKOR-A or its version more resistant to alkalis, MAGKOR-D, will be the best alternatives [4].

Comparatively to magnesia chrome bricks, MAGKOR-A or MAGKOR-D show a similar coating attachment but with less degree of infiltration. It means that the spinel bricks are less penetrated by liquid phase and, hence, less worn by structural spalling [6]. **Fig. 6** shows this comparative situation.

For plants 8, 11 and 12, due to low reactivity of raw meal and production of special clinkers, the quantity of the liquid phase is very low. The clinkers are hard to burn (high Burnability Index) and in order to avoid high free lime content, the risk of kiln overheatings is increased. Once more the use of MAGKOR-A is more adequate by its higher refractoriness and less liquid phase penetration.

In resume, for burning zone, MAGNESITA advices, as chrome free alternatives, MAGKOR for normal operational condition; and MAGKOR-A or MAGKOR-D for conditions where the risk of thermal overload or deep liquid phase penetration, by overheating and/or liquid phase of low viscosity, become the operational condition.

c) Lower Transition Zone

The lower transition zone is subject to mechanical abrasion, coating instability, thermal shock and chemical attack. The temperature gradients are high in this region and the chemical attack is caused by alkali compounds infiltration followed by reactions or just solidification. The alternation of the kiln atmosphere (reduction/oxidation-redox cycles) can also occur in this region, which can lead to the destruction of the lining due to the crack formation, increasing the alkali infiltration as well.

The instability of the coating, which leads to thermal shock and structural spalling, is also very critical in this region. The instability might be caused by alterations on the flame shape, burner position and amount and characteristics of liquid phase. These factors change the length of the burning zone very frequently.

It means that the coating formation in the lower transition zone is eventual with cycles of formation and falling down. In each loss of coating the infiltrated portion of brick is spalled in layers (structural spalling).

In opposite to the burning zone, where an adequate coating formation and mainly adhesion is required, in the lower transition zone the attachment of coating should be avoided, meaning that is necessary to line the zone with a brick which is "repellent" to coating formation and stabilization.

For lower transition zone, the best results have been obtained with a spinel brick as MAGKOR-B, which besides the very low adherence of coating, presents a good resistance to alkali attack and to redox condition, provided by its chemical and mineralogical composition and by its low

in diesem Bereich sind groß. Der chemische Angriff wird hervorgerufen durch Infiltration von Alkaliverbindungen, gefolgt von chemischen Reaktionen oder einfach nur Erstarrung. In diesem Bereich kann sich auch die Ofenatmosphäre (Reduktion/Oxidation – Redoxyklen) verändern. Dies kann zur Zerstörung der Ausmauerung durch Rißbildung führen. Dies verstärkt wiederum auch die Infiltration von Alkalien.

Die Instabilität des Ansatzes, die zu Temperaturwechseln und örtlichen Abplatzungen führt, ist in diesem Bereich ebenfalls sehr kritisch. Die Instabilität kann durch Veränderungen der Flammenform, der Brennerposition und der Menge und Eigenschaften der Schmelzphase hervorgerufen werden. Diese Faktoren verändern sehr häufig die Länge der Sinterzone.

Dies bedeutet, daß in der unteren Übergangszone Zyklen von Ansatzbildung und Ansatzfall entstehen. Bei jedem Ansatzfall platzt der infiltrierte Teil der Steine schichtweise ab (strukturelles Abplatzen).

Im Gegensatz zur Sinterzone, wo eine adäquate Ansatzbildung und vor allem Adhäsion erforderlich ist, sollte das Festsetzen von Ansätzen in der unteren Übergangszone vermieden werden. Dies bedeutet, daß es erforderlich ist, diese Zone mit Steinen auszumauern, die sich gegenüber Ansatzbildung und -stabilisierung „abweisend" verhalten.

In der unteren Übergangszone erzielten die Spinell-Steine wie MAGKOR-B, die neben der sehr geringen Haftwirkung des Ansatzes aufgrund ihrer chemischen und mineralogischen Zusammensetzung und der geringen Porosität und Durchlässigkeit auch einen guten Widerstand gegen Alkaliangriffe und Redoxbedingungen bieten, die besten Ergebnisse. Für Öfen mit Satellitenkühler, bei denen der mechanische Abrieb in der unteren Übergangszone stärker ist, empfiehlt MAGNESITA die Verwendung von MAGKOR-B-RA, einer gegenüber MAGKOR-B abrasionsbeständigeren Ausführung [6].

3. Schlußbemerkungen

Grundsätzlich kann die Zusammensetzung des Klinkers die Leistungsfähigkeit der Ausmauerung beeinflussen, da sie die Eigenschaften der gebildeten Schmelzphase und damit die Länge und die Bedingungen aller Ofenbereiche verändert. In Verbindung mit anderen Prozeßbedingungen bestimmt sie die Stabilität des schützenden Ansatzes. Die Spezifikation eines Magnesia-Spinell-Steines für Sinter- und Übergangszone muß die wichtigsten Belastungen und Anforderungen berücksichtigen.

Aus den bisher erzielten Ergebnissen läßt sich ableiten, daß – abgesehen von kleineren Problemen – ein Wechsel von Magnesia-Chrom-Steinen zu Magnesia-Spinell-Steinen mit der MAGKOR-Palette erfolgversprechend ist.

porosity and permeability. For kilns with satellite cooler where the mechanical abrasion in lower transition zone is more severe, MAGNESITA S.A. advices the use of MAGKOR-B-RA that is a more resistant version to abrasion of MAGKOR-B [6].

3. Final comments

Intrinsically, the clinker composition may influence the refractory lining performance by modifying the characteristics of the liquid formed, the length and behaviour of each zone of the kiln. Allied to other process conditions it determines the stability of a protective coating. The specification of a magnesia spinel brick for burning and transition zones must take into account the main stresses and requirements.

From the results achieved up to now, so far it is possible to conclude that, apart from minor problems, the changeover from magnesia chrome to magnesia spinel bricks can be carried out successfully with the MAGKOR line.

Literature

[1] Barthel, H., and Kaltner, E.: The basic refractory lining of cement rotary kilns to conform to changed requirements. Proceedings of 2nd International Conference on Refractories, Tokyo, TAR-J, Vol. 2 (1987), pp. 623–639.

[2] Tokunaga, K., Kozuda, H., Honda, T., and Tanemura, F.: Further improvement in high temperature strength, coating adherence and corrosion resistance of magnesia-spinel bricks for rotary cement kiln. UNITCER Congress, Aachen, Sept. (1991), pp. 431–435.

[3] Gonsalves, G.E., Duarte, A.K., and Brant, P.O.R.C.: Magnesia spinel brick for cement rotary kiln. American Ceramic Society Bulletin 72 (1993), No. 2, pp. 49–54.

[4] Oliveira, W.S.: Reactivity and burnability of raw meals – Their influence on the specification of refractories. World Cement, April (1991), pp. 146–154.

[5] Oliveira, W.S.: Refractories for cement industry – Newly concepts. Magnesita SA (1992).

[6] Gontijo, L.M.: Refratarios isentos de cromo – Uma abordagem pratica (in Portuguese). Magnesita SA (1992).

PYROSTEP® Rostkühler – Weiterentwicklungen in der Rostkühlertechnik

PYROSTEP® grate cooler – further developments in grate cooler technology

Refroidisseurs à grille PYROSTEP® – Nouveaux développements de la technique des refroidisseurs à grille

Enfriador de parrilla PYROSTEP® – Nuevos desarrollos en la técnica de los enfriadores de parrilla

Von **R. Schneider**, Köln/Deutschland

Zusammenfassung – Wesentliche Ziele bei der Optimierung von Anlagen zur Herstellung von Zementklinker sind die Senkung des Brennstoffenergieverbrauchs sowie die Minderung der Emissionen. Ein moderner Schubrostkühler kann hierzu einen entscheidenden Beitrag leisten. Neue Rostplattenformen, geänderte Luftzuführungen und eine gezielte Kühlung des Klinkers tragen dazu bei, den Wärmeaustausch zwischen Klinker und Kühlluft zu verbessern. Durch einen feststehenden Treppenrost, welcher mit Rostplatten, die horizontale Luftaustrittsöffnungen aufweisen, bestückt ist, wird die Klinkerschicht in Sektionen pulsierend belüftet. Die Luftmengen können für jede Sektion individuell auch während des Betriebs eingestellt werden. Da Treppenroste aus verfahrenstechnischen Gründen nicht beliebig lang gebaut werden können, wird der Treppenrost für hohe Durchsätze durch einen Schubrostbereich unterbrochen. Darüber hinaus werden die Kühlluftmengen durch erhöhte Klinkerschichten auf dem Rostsystem sowie durch weitere Änderungen in der Konstruktion vermindert und somit weniger Abluft erzeugt.

Summary – Important objectives when optimizing plants for producing cement clinker are to lower the consumption of fuel energy and to reduce the emissions. A modern reciprocating grate cooler can make a decisive contribution towards this. New grate plate shapes, changed air supply systems, and carefully controlled cooling of the clinker all help to improve the heat exchange between clinker and cooling air. The clinker bed is pulse-aerated in sections through a fixed stepped grate which has grate plates with horizontal air outlet openings. The quantities of air can be adjusted individually for each section, even during operation. For process engineering reasons, stepped grates cannot be built to just any length, so for high throughput capacities the stepped grate is interrupted by a reciprocating grate section. The quantities of cooling air are also reduced through deepened clinker beds on the grate system and by other changes in the design, so less exhaust air is produced.

Résumé – Les objectifs essentiels de l'optimisation d'installations de production de clinker à ciment sont l'abaissement de la consommation d'énergie fossile et aussi la réduction des émissions. Un refroidisseur moderne à grille poussée peut y contribuer de manière décisive. Des formes nouvelles des plaques de la grille, des ventilations modifiées et un refroidissement ciblé du clinker contribuent à améliorer l'échange de chaleur entre clinker et air de refroidissement. A travers une grille fixe à gradins, équipée de plaques à ouvertures de sortie d'air horizontales, la couche de clinker est ventilée en alternance par sections. Pour chaque section, les quantités d'air peuvent être réglées individuellement, même pendant la marche. Pour des raisons de technique de process, les grilles à gradins ne peuvent pas être configurées pour n'importe quelle longueur; la grille à gradins est donc interrompue, pour les grandes capacités de débit, par une zone à grille poussée. En outre, les quantités d'air de refroidissement sont diminuées par des épaisseurs plus grandes de la couche de clinker sur le système de grille et par d'autres modifications de la construction, ce qui réduit la production d'air d'exhaure.

Resumen – Los objetivos esenciales de la optimización de las instalaciones destinadas a la fabricación de clínker de cemento son la reducción del consumo de energía y de las emisiones. Un moderno enfriador de parrilla de vaivén puede contribuir a ello de forma decisiva. Nuevas formas de placas de parrilla, conductos de llegada de aire modificados y un enfriamiento bien enfocado del clínker contribuyen a mejorar el intercambio de calor entre el clínker y el aire de enfriamiento. Mediante una parrilla escalonada, equipada con placas de parrilla que presentan aberturas horizontales de salida de aire, la capa de clínker es aireada por secciones, de forma pulsátil. Los caudales de aire pueden ser regulados individualmente para cada sección, incluso durante el servicio. Puesto que las parrillas escalonadas no pueden ser construidas con cualquier longitud, por razones tecnológicas, para los grandes caudales la parrilla escalonada queda interrumpida por una zona de parrilla de vaivén. Además, las cantidades de aire de enfriamiento se reducen, debido al aumento de las capas de clínker que descansan sobre el sistema de parrillas así como a otras modificaciones constructivas, de modo que se produce menos cantidad de aire de escape.

KHD Humboldt Wedag AG have had extensive know-how on the construction of grate coolers for decades, having supplied a large number of such coolers and gained operating experience with over 370 rotary kiln plants supplied to the cement industry. Their first grate cooler patent was granted as early as 1936 and their first grate cooler was manufactured in the fifties (**Fig. 1**).

KHD grate coolers with a throughput of up to 5 000 t/d work to the full satisfaction of plant operators.

The purpose of continued development has been to lower the heat consumption of the kiln plant, improve cooler availability and diminish dust emission. Extensive trials have been carried out to test grate plates with pockets, individual grate plate aeration and a sequentially aerated, fixed stepped grate with nearly horizontal air outlet openings in the kiln discharge area.

All the positive test results and other operating experience have led to a fundamental revision of the construction of the grate cooler, so that KHD's new, third-generation grate cooler, the PYROSTEP grate cooler, is now available.

The various sections of the PYROSTEP grate cooler are illustrated in the system diagram (**Fig. 2**).

Sections I – IV are aerated directly, i.e. an adjustable volume of air is blown onto each row individually.

The size and design of sections I – IV are determined by the throughput and the amount of air recovered.

Section I consists of a fixed stepped grate where the grate plates (**Fig. 3**) have nearly horizontal air outlet openings. Aeration of the individual rows is subdivided crosswise, and air is supplied with a pulsating action and can be adjusted from outside.

Thus the classic "horse shoe" covering of individual plates is discontinued and red river formation is much reduced by the targeted supply of air.

Section II is fitted with newly developed Omega grate plates (**Fig. 4**). These have pockets in which clinker collects, thereby preventing air from breaking vertically through the layer of clinker. Furthermore, there is specific cooling of the end wall to give Omega grate plates a still longer service life. In the moving rows air is supplied through pipes which slide into each other and have a newly developed seal. In small coolers the chamber-aerated section V adjoins section II.

Section III corresponds in essence to section I but the air is supplied without pulsation. In this way the advantages of the stepped grate are obtained for large throughputs, without the risk of uncontrolled clinker slippage (avalanche effect) and the indifferent operating behaviour which would result.

Section IV is a repetition of section II of the Omega grate plates. It varies in length according to processing requirements and plant size.

BILD 1: Rostkühler der 2. Generation
FIGURE 1: Second generation grate cooler

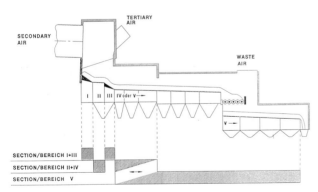

BILD 2: Systemskizze des PYROSTEP-Rostkühlers
FIGURE 2: System sketch of PYROSTEP grate cooler

Sekundärluft = Secondary Air
Tertiärluft = Tertiary Air
Abluft = Waste Air

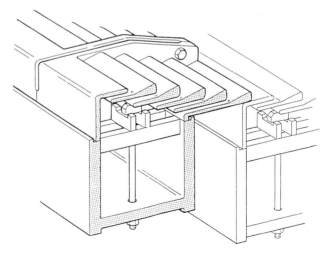

BILD 3: Stationäre Treppen-Rostplatte
FIGURE 3: Stationary stepped grate plate

Bereich I besteht aus einem feststehenden Treppenrost, dessen Rostplatten (**Bild 3**) nahezu horizontale Ausströmöffnungen aufweisen. Die Belüftung der einzelnen Reihen ist in Querrichtung unterteilt, wobei die Luft pulsierend zugeführt wird und sich von außen verstellen läßt.

Somit ist die klassische Abdeckung einzelner Platten als „horse shoe" entfallen, und die Entstehung eines „red river" ist durch gezielte Luftmengenzuführung erheblich reduziert worden.

Bereich II ist mit neu entwickelten Omega-Rostplatten (**Bild 4**) bestückt. Diese Rostplatten besitzen Taschen, in denen sich Klinker sammelt, wodurch ein senkrechtes Durchbrechen der Luft durch die Klinkerschicht vermieden wird. Weiterhin ist auch die Stirnwand der Rostplatte gezielt gekühlt, um eine noch längere Standzeit der Omega-Rostplatten zu erreichen. Die Luft wird in den beweglichen Reihen über Rohrleitungen zugeführt, die sich ineinander schieben und eine neu entwickelte Dichtung aufweisen. Bei kleinen Kühlertypen beginnt im Anschluß an den Bereich II der kammerbelüftete Bereich V.

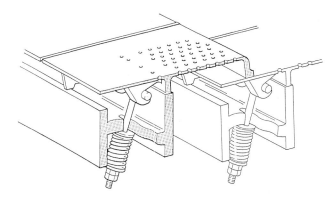

BILD 5: Standard-Rostplatte
FIGURE 5: Standard grate plate

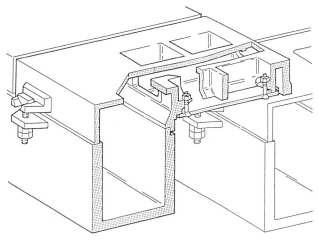

BILD 4: Omega-Rostplatte
FIGURE 4: Omega grate plate

Bereich III entspricht im wesentlichen dem Bereich I, jedoch wird die Luft ohne Pulsation zugeführt. Hierdurch werden für große Durchsätze die Vorteile des Treppenrostes – ohne die Gefahr des unkontrollierten Nachrutschens des Klinkers (Lawineneffekt) und die damit verbundenen indifferenten Betriebsverhältnisse – erreicht.

Bereich IV ist eine Wiederholung des Bereichs II der Omega-Rostplatten. Er variiert in der Länge entsprechend den verfahrenstechnischen Anforderungen und der Anlagengröße.

Bereich V stellt den neuen konventionellen, kammerbelüfteten Bereich dar (**Bild 5**). Die spezifische Luftbeaufschlagung pro Rostplatte liegt am Anfang dieses Bereichs bereits so niedrig, daß Luftdurchbruchsfontänen nahezu ausgeschlossen sind.

Auch ist der Durchfall von Klinker in das Gehäuseunterteil durch die modifizierte Konstruktion (Reduzierung der Spalte zwischen den Rostplatten) vermindert.

Der kammerbelüftete Bereich läßt sich auch vollständig mit den neuen Omega-Rostplatten ausführen. Es entstehen dadurch verfahrenstechnische Vorteile, jedoch sind deutlich höhere Investitions- und Energiekosten sowie eine aufwendigere Wartung zu berücksichtigen. Die Beplattung in diesem Bereich des PYROSTEP-Rostkühlers kann individuell angepaßt werden.

Um die geringen Spalte zwischen den Rostplattenreihen zueinander herstellen zu können, wurde die Durchbiegung des Rostsystems (fester und beweglicher Teil) durch zielgerechte Dimensionierung der tragenden Teile und durch eine Vielzahl von Abstützungen minimiert. Bewährte Konstruktionselemente, wie reale Mittenunterstützung und Laufrollenböcke im belüfteten Bereich, die keine Abdichtung zur Umgebung und damit keine Sperrluft erfordern, tragen hierzu bei.

Section V is the new conventional, chamber-aerated section (**Fig. 5**). At the beginning of the section specific air impingement per grate plate is already so low that „fountains" caused by air channelling are almost impossible. The amount of clinker dropping through into the bottom of the housing is also lessened by the modified construction (reduction of the gaps between grate plates).

The chamber-aerated section may be made entirely with the new Omega grate plates. This has processing advantages but the far higher capital and energy costs and more expensive maintenance must be taken into account. Fitting of plates in this section of the PYROSTEP grate cooler can be adapted to individual requirements.

Bending of the grate system (the stationary and the moving parts) has been minimised by appropriate dimensioning of load-bearing parts and by a number of supports so that narrow gaps can be formed between the rows of grate plates. Well-tried constructional elements such as true central support and carrying roller frames in the aerated section, which need not be shut off from the environment so do not require sealing air, also help to minimise bending.

The standard construction range includes hydraulic or mechanical drives as well as roller and hammer crushers at the end of the cooler or between the grates. Some characteristic data on the PYROSTEP grate cooler is set out in the following table (**Table 1**).

TABLE 1: Characteristic data for PYROSTEP grate cooler

Characteristic data for standard preheater processes	
Throughput	650 – 11 000 t/d
Grate areas	13 – 260 m^2
Grate area loading	40 – 55 t/d.m^2
Grate inclination	3.5 %
Specific air volume	1.6 – 2.1 m^3(stp)/kg clinker
Number of reciprocating frames	1 – 4

Values recorded for specific loading of grate area and specific quantity of air may be above or below the given ranges according to the processing requirements for the whole plant. With a higher clinker exit temperature, for example, there can be higher grate loading, while porous clinker involves a longer dwell time and a larger specific quantity of cooling air.

The PYROSTEP grate cooler (**Fig. 6**) is expected to give thermal recuperation rates of between 70 and 74 %, depending on the kiln plant.

The PYROSTEP grate cooler can be used in any kiln plant, but for plants belonging to other companies the whole plant must first be submitted to a process engineering check. Thermal efficiency can be improved by at least 5 % under the same processing conditions. The advantages can also be obtained in existing grate coolers by retrofitting the stepped grate.

Das Standardbauprogramm enthält hydraulische oder mechanische Antriebe sowie auch Walzen- und Hammerbrecher am Kühlerende oder zwischen den Rosten. In der folgenden Tabelle sind einige Kenndaten des PYROSTEP-Rostkühlers zusammengestellt (**Tabelle 1**).

TABELLE 1: Kenndaten des PYROSTEP-Rostkühlers

Kenndaten für Standard WT-Verfahren	
Durchsatzleistungen	650–11 000 t/d
Rostflächen	13–260 m²
Rostflächenbelastung	40–55 tato/m²
Rostneigung	3,5 %
spezifische Luftmenge	1,6–2,1 Nm³/kg Kli
Anzahl der Schubrahmen	1–4

BILD 7: Beplattung des Rostanfangs eines PYROSTEP-Rostkühlers
FIGURE 7: Plates fitted at beginning of grate in a PYROSTEP cooler

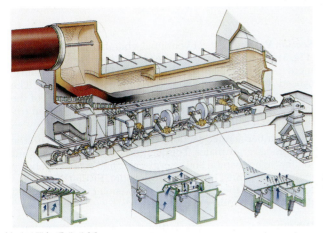

BILD 6: Schnittbild des PYROSTEP-Rostkühlers
FIGURE 6: Section through PYROSTEP grate cooler

Entsprechend den verfahrenstechnischen Anforderungen der Gesamtanlage können die aufgezeigten Werte der spezifischen Rostflächenbelastung und der spezifischen Luftmenge die angegebenen Bereiche unter- bzw. überschreiten. Eine höhere Klinkeraustrittstemperatur erlaubt beispielsweise eine höhere Rostbelastung, poröser Klinker bedingt eine längere Verweilzeit und eine größere spezifische Kühlluftmenge. Für thermische Rekuperationsgrade des PYROSTEP-Rostkühlers (**Bild 6**) werden je nach Ofenanlage Werte zwischen 70 und 74 % erwartet.

Der PYROSTEP-Rostkühler ist in jeder Ofenanlage einsetzbar. Bei Fremdanlagen ist eine vorherige verfahrenstechnische Überprüfung der Gesamtanlage erforderlich. Bei gleichen Prozeßbedingungen sind Verbesserungen des thermischen Wirkungsgrades von mindestens 5 % realisierbar. Ebenso besteht die Möglichkeit, auch bei bestehenden Rostkühlern durch eine Nachrüstung die Vorteile des Treppenrostes zu nutzen. Die letzten zwei Bilder zeigen den PYROSTEP-Rostkühler bei der Werksmontage.

Bild 7 zeigt den Einlaufbereich mit Treppenrost und den anschließenden Omega-Bereich sowie den neuen konventionellen Rostplatten-Bereich.

Bild 8 zeigt einen Rostkühler mit 60 m² Rostfläche für einen Durchsatz von 2 700 t/d.

KHD hat inzwischen mehrere PYROSTEP-Rostkühler in Auftrag genommen, davon befinden sich vier im Bau bzw. in der Fertigung. Über Betriebsergebnisse wird zu gegebener Zeit berichtet.

BILD 8: Werkstatt-Montage eines PYROSTEP-Rostkühlers
FIGURE 8: Workshop assembly of a PYROSTEP grate cooler

The last two figures show a workshop assembly of the PYROSTEP grate cooler. **Fig. 7** shows the inlet section with the stepped grate, the adjoining Omega section and the new conventional grate plate section.

Fig. 8 shows a cooler with a 60 m² grate area for a throughput of 2 700 t/d/

In the meantime KHD have received orders for several PYROSTEP grate coolers, of which four are in the course of construction or manufacture. The operating results will be reported in due course.

Zur Einflußnahme der Mahlfeinheit bei Stein- und Braunkohlen auf die Flammenausbildung in Zementdrehöfen

Influence of the fineness of coal and lignite on the flame formation in rotary cement kilns

Sur l'influence de la finesse de mouture de charbons et lignites sur la formation des flammes dans les fours rotatifs à ciment

El influjo de la finura de molido de hulla y de lignito sobre la formación de la llama en los hornos rotatorios de cemento

Von **G. Seidel,** Weimar/Deutschland

Zusammenfassung – Bisher folgte die Mahlfeinheit von mittel- und hochbituminösen Steinkohlestäuben für den Zementbrennprozeß im allgemeinen der Regel „je weniger flüchtige Bestandteile in der Kohle – desto feiner muß gemahlen werden". Sie führte im Mittel zu Rückstandswerten R90 von 10 – 13%. Die Erweiterung der Brennstoffpalette auf extrem aschereiche Steinkohlen, Magerkohlen und Weichbraunkohlen erfordert eine Präzisierung der Anforderungen. Ausgehend von Verbrennungsversuchen mit einer Vielzahl von Brennstoffen wurde ein Bewertungsverfahren für Brennstoffe mit Hilfe eines mathematischen Modells entwickelt, das auf die besonderen Anforderungen des Klinkerbrennprozesses zugeschnitten ist. Darin werden Einflüsse des Brenners (Luftanteil, Austrittsgeschwindigkeit, Vermischungsintensität), des Brennraums (Luftverhältniszahl, Sekundärlufttemperatur, Brennraumbelastung, Wandverluste) und des Brennstoffs (Zusammensetzung, Heizwert, Korngröße) berücksichtigt. Versuche und Simulationsrechnungen zeigten, daß sich die Ausmahlung der Kohle vor allem auf die Flammentemperaturen, die Flammenform und die Wärmeübertragung auf das Brenngut auswirkt. Die Festlegung der Mahlfeinheit gestattet Rückschlüsse auf den zu erwartenden Ofen-Betriebszustand, insbesondere den spezifischen Brennstoffenergieverbrauch.

Zur Einflußnahme der Mahlfeinheit bei Stein- und Braunkohlen auf die Flammenausbildung in Zementdrehöfen

Summary – Until now the fineness of grinding of medium- and high-volatiles pulverized coals for the cement burning process has usually followed the rule "the lower the volatiles in the coal the finer it must be ground". On average this has led to residues of 10–13% on 90 μm. Extension of the range of fuels to cover extremely ash-rich coals, non-bituminous coals and soft lignites means that the requirements must be more precise. Based on combustion trials with a large number of fuels, a method of evaluating fuels has been developed with the aid of a mathematical model which is tailor-made to the particular requirements of the clinker burning process. It takes account of the effects of the burner (air percentage, outlet velocity, blending intensity), of the combustion space (excess air ratio, secondary air temperature, combustion space loading, wall loss) and of the fuel (composition, calorific value, particle size). Trials and simulation calculations have shown that the main factors affected by the coal grinding are the flame temperature, the flame shape, and heat transfer to the kiln feed. Specifying the fineness means that inferences can be drawn about the anticipated kiln operating state, particularly the specific fuel energy consumption.

Influence of the fineness of coal and lignite on the flame formation in rotary cement kilns

Résumé – Jusqu'à présent, la règle pour la finesse de mouture de charbons moyennement ou hautement bitumineux destinés au process de cuisson du ciment était généralement: „moins il y a de composants volatils dans le charbon, plus il faut broyer fin". Elle avait conduit, en moyenne, à des valeurs de refus R90 de 10–13%. L'élargissement du choix de combustibles, à des charbons extrêmement cendreux, des charbons maigres et des lignites tendres exige de préciser les besoins. A partir d'essais de combustion avec une multitude de combustibles, a été élaboré un procédé d'évaluation à l'aide d'un modèle mathématique taillé sur mesure pour les besoins du process de cuisson du clinker. Y est tenu compte des influences du brûleur (quantité d'air, vitesse de sortie, intensité de mélange), de la zone de combustion (rapport de l'air, température de l'air secondaire, charge de la zone de combustion, pertes par les parois) et du combustible (composition, pouvoir calorifique, granulométrie). Des essais et des calculs de simulation ont montré, que le degré de mouture du charbon influe surtout sur la température et la forme de la flamme et sur le tranfert de chaleur à la matière à cuire. La définition de la finesse de mouture permet de connaître les répercussons sur l'état de marche prévisible du four, surtout côté consommation spécifique d'énergie.

Sur l'influence de la finesse de mouture de charbons et lignites sur la formation des flammes dans les fours rotatifs à ciment

Resumen – Hasta ahora, la finura de molido de hulla mediana y altamente bituminosa, destinada al proceso de cocción del cemento, solía regirse por la regla siguiente: „Cuanto menos componentes volátiles contenga el carbón, más fina tendrá que ser la molienda". De esta forma, se alcanzaban, por término medio, residuos de 10 – 13 % sobre el tamiz de 90 μm. La extensión de la gama de combustibles a hullas extremadamente ricas en cenizas, hulla magra y lignito blando hace necesaria una definición más precisa de los requerimientos. Partiendo de ensayos de combustión realizados con numerosos combustibles, se ha desarrollado un procedimiento de evaluación para combustibles, con ayuda de un modelo matemático adaptado a las exigencias específicas del proceso de cocción de clínker. A este respecto, se tienen en cuenta los influjos del quemador (cantidad de aire, velocidad a la salida, intensidad de mezcla), del recinto de combustión (relación de aire, temperatura del aire secundario, carga del recinto de combustión, pérdidas por la pared) y del combustible (composición, valor calorífico, granulometría). Los ensayos llevados a cabo así como los cálculos de simulación han demostrado que el grado de molido del carbón influye, sobre todo, en las temperaturas y la forma de la llama así como en la transmisión del calor al material a cocer. La fijación de la finura de molido permite sacar conclusiones respecto del futuro estado de marcha del horno y, especialmente, del consumo de energía.

El influjo de la finura de molido de hulla y de lignito sobre la formación de la llama en los hornos rotatorios de cemento

1. Problemstellung

Im Zusammenhang mit der stärkeren Nutzung von Weichbraunkohlen als Brennstoff für den Zementbrennprozeß wurde das Problem der Mahlfeinheit letztmalig Anfang der 80er Jahre in größerem Umfang in der Literatur diskutiert. Seinerzeit wurde die für mittel- und hochbituminöse Steinkohlen gefundene Regel: „je weniger flüchtige Bestandteile – desto feiner muß gemahlen werden" leicht modifiziert auch auf Braunkohlen übertragen.

Es erhebt sich die Frage, ob kleinere Korngrößen bei den Weichbraunkohlen sinnvoll sind und welche Effekte damit erreichbar wären. Da sich die Brennstoffpalette – neben den Weichbraunkohlen – auch in Richtung aschereiche Steinkohlen, Magerkohlen etc. – erweitert hat, konnte eine Präzisierung der o.g. Regel von prinzipiellem Interesse sein.

Um den Zusammenhang zwischen Brennstoffqualität und Flammenausbildung im Drehofen besser zu erkennen, wurde aus Verfeuerungsversuchen mit unterschiedlichen Brennstäuben an kleintechnischen Anlagen ein Bewertungsverfahren entwickelt, das auf die besonderen Anforderungen des Klinkerbrennprozesses zugeschnitten ist. Beurteilt werden die Wirkungen auf die Flammentemperaturen und damit auf die Intensität der Wärmeübertragung von der Flamme auf das Materialbett.

2. Bisherige Erkenntnisse und Erfahrungen zur Festlegung der Kohlemahlfreiheit

Bereits seit den Zeiten, da Steinkohle guter Qualität ausreichend und preiswert als „klassischer" Drehofenbrennstoff zur Verfügung stand, ist es unstrittig, daß die Mahlfeinheit dem jeweiligen Gehalt (besser: „der Ausbeute") an flüchtigen Bestandteilen anzupassen ist. Diese Erkenntnis rührt sowohl aus den Beobachtungen zur Flammenausbildung als auch zur Zündfähigkeit der Kohle. Die Erfahrungen wurden beispielsweise von Ramesohl [1] zusammenfassend formuliert, der dabei auch die Spezifika besonders aschereicher Kohlen berücksichtigte. Faßt man die gegebenen Empfehlungen [2, 3] zusammen, ergeben sich Rückstandswerte von R 200 = 0,5 – 1,5% und R 90 = 10 – 15% je nach Aschegehalt. Für Weichbraunkohlen, in der Zündfähigkeit unproblematisch, wurden z.B. von Schubert [4] die Maximalwerte R 200 = 16% und R 90 = 43% genannt; realisiert wurde der Bereich R 200 = 8 – 15% und R 90 = 35 – 45%.

Zur Charakterisierung der Korngrößenverteilung lassen sich dabei als Entsprechungen angeben:

– Steinkohlen: Der Bereich R 90 = 10–15% entspricht für das RRSB-Netz den Werten d' = 15–30 μm und n = 0,6–0,8. Als mittlere Korngröße (gewogenes Mittel im Bereich 3–250 μm):
$\bar{d}_P = 25-35$ μm.

1. The problem

The last time the problem of the fineness of grinding was widely discussed in the literature was at the beginning of the eighties, in connection with the increasing use of soft lignites as a fuel for the cement burning process. At that time the rule arrived at for medium and high-bitumen pit coal, "the lower the volatiles in the coal the finer it must be ground" was also applied in a slightly modified form to lignites.

The question is whether smaller particle sizes are helpful in soft lignites and what effects can be obtained with them. Extension of the range of fuels to cover extremely ash-rich coals and non-bituminous coals as well as soft lignites means that the requirements must be more precise.

As a means of establishing the connection between fuel quality and flame shape in the kiln, an evaluation method which is tailor-made to the particular requirements of the clinker burning process has been developed, based on combustion trials with different pulverised coals in small industrial plant. The method assesses the effects on flame temperatures and thus on the intensity of heat transmission from the flame to the bed of material.

2. Previous findings and experience in establishing the required fineness of the coal

Even since the times when enough cheap, good-quality pit coal was available as a "classic" rotary kiln fuel, it has never been disputed that the fineness of grinding has to be adapted to the content (or better, the yield) of volatile constituents. This realisation is derived from observations of both the shape of the flame and the flammability of the coal. The experience has been summarised, for example, by Ramesohl [1], who also considered the specific requirements for particularly ash-rich coal. If the recommendations are combined [2, 3] the residue values obtained are R 200 (residue on 200 μm) = 0.5–1.5, R 90 = 10–15% according to ash content. For soft lignites where flammability is no problem Schubert [4], for example, quotes maximum values of R 200 = 16%, R 90 = 43%; the values obtained were in the range R 200 = 8–15%, R 90 = 35–45%.

The particle size distribution can be characterised in terms of the correspondence between the following:

– Pit coal: The R 90 = 10–15% range corresponds, for the RRSB grid, to the values d' = 15–30 μm and n = 0.6–0.8. The mean particle size (weighted mean within the 3–250 μm range) is:
$\bar{d}_P = 25-35$ μm.

— Braunkohlen: Der Bereich R 90 = 35–45 % entspricht für das RRSB-Netz den Werten d' = 90–120 μm, n = 1,0–1,2. Als mittlere Korngröße (gewogenes Mittel im Bereich 3–500 μm):

$\overline{d}_P = 85$–$110\,\mu m$.

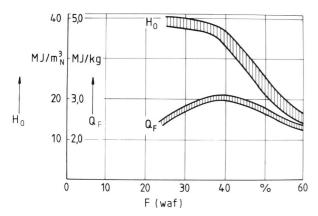

BILD 1: Energetische Kennwerte der flüchtigen Bestandteile bei Kohlen

H_O : Brennwert je m^3_N Flüchtiges
Q_F : Brennwert, umgerechnet auf kg Kohle
F_{waf}: Ausbeute an Flüchtigem, bezogen auf wasser- und aschefreien Zustand

FIGURE 1: Characteristic energy values of volatiles in coal

H_O : calorific value per m^3_{stp} volatiles
Q_F : calorific value, converted to kg coal
F_{waf}: yield of volatiles, relative to water and ashfree state

3. Verbrennung und Flammenausbildung bei Kohlenstaubfeuerungen an Drehöfen

Auf die Qualität des Zementklinkers haben das Temperaturfeld in der Sinterzone und damit die Flammenausbildung entscheidenden Einfluß. Erfahrungen beim Wechsel der Brennstoffart an Zementdrehöfen zeigten, daß insbesondere Alitgehalt und hydraulische Aktivität der Klinkermineralien auf Veränderungen des Temperaturfeldes im Sinterbereich sensibel reagieren [5]. Der Übergang von einer Steinkohlen- bzw. Ölbefeuerung auf Braunkohlenbefeuerung zeigte das in besonderer Weise, so daß die Frage nach einer Intensivierung der Verbrennungsprozesse beim Einsatz von Braunkohlen auch das Problem der Mahlfeinheit aktualisierte.

Werden aus Sicht des Verbrennungsablaufs die Anforderungen an die Korngrößen bei Stein- und Braunkohlenstäuben diskutiert, so wird häufig auf das unterschiedliche Verhalten des Staubkorns während seiner Entgasung und der Frühphase der Verbrennung hingewiesen: Steinkohlepartikel blähen meist auf, Braunkohlepartikel schwinden bzw. zerplatzen. Daraus wird dann der Schluß gezogen, daß bei Braunkohlenstaub wesentlich größere Partikel zulässig wären. Das berücksichtigt jedoch die spezifischen Anforderungen an die Flammenausbildung, insbesondere das Temperaturfeld, in der Sinterzone des Drehofens nicht. Die für die Garbrandphase notwendigen hohen Temperaturen in der Sinterzone, die zu einer kurzen, heißen Flamme führen, sind vorwiegend eine Folge der Verbrennung der zuvor ausgetriebenen flüchtigen Bestandteile und eines Teiles der festen Kohlesubstanz.

Nun ist der Brennwert des Flüchtigen sowohl von der Ausbeute an Flüchtigem als auch ihrem Gehalt an brennbaren Gasen abhängig. **Bild 1** zeigt den hohen Brennwert des Flüchtigen [6] bei gasarmen Steinkohlen und den starken Abfall hin zu den Weichbraunkohlen. Umgerechnet auf kg Kohle zeigt die Wärmefreisetzung ein Maximum für hochbituminöse Steinkohlen. Mit dieser schnellen Wärmefreisetzung aus dem Flüchtigen wird die kurze, mit einem eng begrenzten Temperaturmaximum ausgezeichnete Steinkohlenstaubflamme erreicht — schematisch dargestellt in **Bild 2**. Für die Braunkohlenstaubflamme mindert der hohe

— Lignites: The R 90 = 35–45 % range corresponds, for the RRSB grid, to the values d' = 90–120 μm, n = 1.0–1.2. The mean particle size (weighted mean within the 3–500 μm range) is:

$\overline{d}_P = 85$–$110\,\mu m$.

3. Combustion and shape of flame in pulverised coal firing at rotary kilns

The temperature field in the sintering zone and thus the shape of the flame have a decisive influence on the quality of the cement clinker. Experience in changing the type of fuel in rotary cement kilns has shown that the alite content and the hydraulic activity of the clinker minerals in particular react markedly to changes of temperature field in the sintering region [5]. The transition from pit coal and oil firing to lignite firing has shown this up especially, so the question of intensifying combustion processes for lignites has also updated the problem of the fineness of grinding.

When particle size requirements for pulverised coal and lignites are discussed from the point of view of the progress of combustion, reference is often made to the different behaviour of the particles during degassing and the early phase of combustion; coal particles mostly swell, whereas lignite particles shrink or burst. From this it is concluded that considerably larger particles can be used in the case of lignite. However this reasoning does not take into account the specific requirements as to the flame structure, particularly of the temperature field, in the sintering zone of the kiln. The high temperatures required in the sintering zone for the finishing burning phase (a short, hot flame) are largely the result of combustion of the previously expelled volatile constituents and, naturally, part of the solid coal substance. Now the calorific value of volatiles depends both on the yield of volatiles and their content of combustible gases. **Fig. 1** shows the high calorific value of the volatiles (values taken from [6] inter alia) in non-bituminous coals and the sharp drop to the soft lignites. When converted to kilograms of coal the heat liberated shows a maximum for high-bitumen pit coals. This rapid liberation of heat from the volatiles gives the short pulverised coal flame; it is distinctive in having a strictly limited temperature maximum and is shown diagrammatically in **Fig. 2**. In the case of the pulverised lignite flame the high proportion of inert gas in the volatiles reduces the quantity of heat liberated, so that the maximum temperature in the flame occurs later, aided by a higher degree of solids burnout. Altogether the pulverised lignite flame shows a lower temperature level and transposition of the flame into the kiln (Fig. 2). These disadvantages could be partly offset by the burner design. A slight increase in specific heat consumption with lignite firing is accepted in a number of applications — partly owing to the lower fuel costs. In theory it should be possible to adapt to the temper-

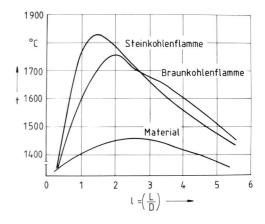

BILD 2: Schematische Temperaturverläufe in der Sinterzone eines Zementdrehofens

FIGURE 2: Diagram showing progress of temperature in the sintering zone of a rotary cement kiln

Steinkohlenflamme = pit coal flame
Braunkohlenflamme = lignite flame
Material = material

TABLE 1: Characteristic fuel values of coals and lignites

		Pit coal	Lignite
Water content (basis for reference)	%	2.0	10.0
Ash content	%	5.0 ... 35.0	3.0 ... 15.0
Volatiles	%	25.0 ... 35.0	45.0 ... 55.0
Calorific value	MJ/kg	22.0 ... 32.0	18.0 ... 23.0
Volatiles (water and ash-free)	%	28.0 ... 38.0	48.0 ... 60.0
Spec. quantity of flue gas	m^3stp/GJ	260.0 ... 290.0	270.0 ... 320.0
Degree of coking (C : 0)	–	4.5 ... 9.0	1.9 ... 4.0

ature profile of the pulverised coal flame by increasing the proportion of fine particles in lignites; the question is whether the effect is great enough to compensate for the extra expenditure on treating the pulverised lignites.

Operating experience with both fuels is naturally influenced by the degree to which the characteristic quality values fluctuate (**Table 1**): the fluctuations lead to overlaps between "bad" coals and "good" lignites. This means that the actual flame temperatures obtainable in medium-size rotary kilns are e. g.:

— for high-grade medium-bitumen pit coals: 1750–1850°C

— for ash-rich pit coals: 1700–1800°C

— for soft lignites: 1650–1750°C.

The basic difference between lignites and coals remains in respect of the position of the sintering zone in the kiln; though their action can be made more similar by the burner design with its influence on mixing and burning intensity.

4. The effect of the fineness of grinding – an attempt at a quantifying evaluation

If the effect of an external parameter, such as the particle composition of the pulverised fuel, on the quality of the kiln feed and the technical/economic characteristics of the burning process is to be reproduced correctly, the complexity of the heat liberation and heat transfer processes in the combustion zone must be mirrored accurately.

Based on combustion trials with a large number of fuels, a method of evaluation of fuels has been developed, which is tailor-made to the particular requirements of the clinker burning process. The mathematical model, which is supported by experimental results from small industrial plants, considers effects from

— the burner (air percentage, outlet velocity, blending intensity),

— the combustion space (excess air ratio, secondary air temperature, combustion space loading, wall loss),

— the fuel (composition, calorific value, particle size)

in their complexity and interaction.

The readings obtained in small industrial plants were converted using the laws of physical similarity to a kiln plant with a sintering zone diameter of 4.2 m and the status data (quantity of fuel etc.) of a 2000 t/d plant.

The effect of the fineness of grinding could then be determined by simulation computations. As a means of formulating this effect on the shape of the flame two criteria are considered:

— The position of the flame: the distance between the burner orifice and the cross-section with the maximum wall temperature was found in the combustion space of each kiln (**Fig. 3**)

— The intensity of heat transfer: for the maximum flame temperature range the effective heat flow density from the flame or the wall of the kiln to the bed of material was calculated from the measured temperatures (**Fig. 4**).

The results are given in Figs. 3 and 4, as functions of the mean particle size of the powder (this is easier to handle in calculations than the characteristic distribution values d'

lichkeit auf eine Ofenanlage mit einem Sinterzonendurchmesser von 4,2 m und den Zustandsdaten (Brennstoffmenge etc.) einer Anlage mit einem Klinkerdurchsatz von 2000 t/d umgerechnet.

Durch Simulationsrechnungen konnte dann der Einfluß der Mahlfeinheit ermittelt werden. Um diesen Einfluß auf die Flammenausbildung formulieren zu können, werden zwei Kriterien betrachtet:

— Die Flammenlage: Im Brennraum des Ofens wurde jeweils der Abstand zwischen Brennermündung und dem Querschnitt mit der maximalen Ofenwandtemperatur festgestellt (**Bild 3**).

— Die Intensität der Wärmeübertragung: Für den Bereich der maximalen Flammentemperatur wurde die effektive Wärmestromdichte von der Flamme bzw. der Ofenwandung an das Materialbett aus den gemessenen Temperaturen berechnet (**Bild 4**).

Die Ergebnisse sind in den Bildern 3 und 4 in Abhängigkeit von der mittleren Korngröße des Staubes (sie ist gegenüber den Verteilungskenngrößen d' und n für Rechnungen leichter handhabbar) dargestellt. Die Flammenlage wird für Braunkohlen auch bei denkbar verminderter Korngröße immer etwas ungünstiger sein als bei Steinkohlen. Für die Wärmeübertragung existieren Überdeckungsbereiche: es ist bei höherer Aufmahlung also möglich, auch für Braunkohlen eine Intensität der Wärmeübertragung in der Sinterzone zu erreichen wie für Steinkohlen mittlerer Qualität.

Damit ist festzustellen, daß ein Absenken der mittleren Korngrößen bei Braunkohlenstäuben von den bisher üblichen 85–110 μm durchaus Effekte zur Anlagenoptimierung bringen kann.

5. Flammenausbildung und spezifischer Wärmeaufwand

Direkt kann dieser Zusammenhang aus den Simulationsrechnungen nicht entnommen werden. Quantitative Einschätzungen sind jedoch möglich. Die o.g. Wärmestromdichte kann als Synonym für die Intensität der Wärmeübertragung im Drehrohr angesehen werden, die wesentlich solche Größen wie Klinkerdurchsatz und spezifischen Wärmeaufwand fixiert.

Die an einer Ofenanlage konkret noch gegebenen Möglichkeiten zur Energieeinsparung sind abhängig vom bereits erreichten technologischen Niveau, also dem erreichten thermischen Wirkungsgrad [7, 8]. Konkret zur Wirkung der Mahlfeinheit kann auf Betriebsuntersuchungen von Ueda u.a. [9] an Trockenöfen verwiesen werden: Etwa 70–100 kJ/kg Klinker werden bei Absenkung der Korngröße von 110 μm auf 70 μm einzusparen sein. Um den betriebswirtschaftlichen Effekt festzustellen, müßte eine Aufrechnung gegen die vermutlich höheren Brennstoffkosten (Energieaufwand für Zerkleinerung) erfolgen.

In anderer Hinsicht ist die Verringerung der Mahlfeinheit noch von Interesse: Nach Endres [10] sinkt die NO_x-Bildung mit abnehmender Feinheit des Kohlenstaubes.

Zusammenfassund läßt sich feststellen: Eine Verringerung der Korngröße bei Braunkohlenstäuben aus dem Bereich von $\bar{d}_P$ = 85–110 μm nach $\bar{d}_P$ = 40–60 μm läßt Vorteile im Brennbetrieb erwarten. Die Effektivität insgesamt hängt dabei von den größeren Aufwendungen für die Aufbereitung einschließlich der sicherheitstechnischen Aspekte ab.

Literaturverzeichnis

[1] Ramesohl, H.: Betriebserfahrungen beim Verbrennen fester Brennstoffe im Zementdrehofen und daraus resultierende Folgerungen. Zement-Kalk-Gips 32 (1979) Nr. 5, S. 227–229.

[2] Braig, H.: Betriebserfahrungen mit der Kohlemahlanlage im Zementwerk Mergelstetten. Zement-Kalk-Gips 35 (1982) Nr. 8, S. 391–397.

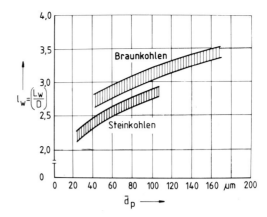

BILD 3: Maximale Wärmestromdichte von der Flamme zum Materialbett, in Abhängigkeit von der Mahlfeinheit der Kohle
FIGURE 3: Maximum heat flow density from flame to bed of material, dependent on fineness to which coal is ground.
Braunkohlen = lignites
Steinkohlen = pit coals

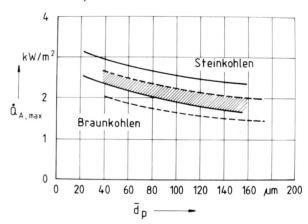

BILD 4: Flammenlage – beurteilt nach Lage des Temperaturmaximums – in Abhängigkeit von der Mahlfeinheit der Kohle
$\bar{d}_P$: Mittlerer Korndurchmesser
L_W : Entfernung Brennermündung – Querschnitt mit maximaler Wandtemperatur
D : Ofendurchmesser

FIGURE 4: Position of flame – assessed according to position of maximum temperature – dependent on fineness to which coal is ground
$\bar{d}_P$: mean particle diameter
L_W : distance from burner orifice to cross-section with maximum wall temperature
D : kiln diameter

and n). The position of the flame will always be somewhat more unfavourable for lignites than for coal even if the particle size is considerably reduced. For heat transfer there are overlapping regions: thus with more grinding the same intensity of heat transfer in the sintering zone can be obtained for lignites as for medium-grade coal.

Hence lowering of mean particle sizes in pulverised lignites from the hitherto customary 85–110 μm may very well have effects in optimising the plant.

5. Flame shape and specific heat consumption

This relationship cannot be derived directly from the simulation computations, but quantitative estimates are possible. The above-mentioned heat flow density may be regarded as synonymous with the intensity of heat transfer in the rotary kiln, which essentially establishes such values as clinker throughput and specific heat consumption.

The concrete energy-saving opportunities which still exist at a plant depend on the technological level already reached, i.e. the thermal efficiency achieved [7, 8]. Concrete evidence of the effect of the fineness of grinding can be obtained from working investigations in dry kilns by Ueda et al [9]: about

[3] Patzke, J.: Sicherheitstechnische Betriebserfahrungen bei der Kohlenmahlung im Zementwerk Lägerdorf. Zement-Kalk-Gips 34 (1981) Nr. 5, S. 238–242.

[4] Schubert, P.: Mehrkanalbrenner für grobkörnige feste Brennstoffe. Zement-Kalk-Gips 35 (1982) Nr. 5, S. 246–249.

[5] Seidel, G.: Kohlequalität – Sinterzonenzustand – Betriebsergebnisse bei Zementdrehöfen. Zement-Kalk-Gips 38 (1985) Nr. 6, S. 289–292.

[6] Gumz, W.: Kurzes Handbuch der Brennstoff- und Feuerungstechnik. Springer-Verlag Berlin/Göttingen/Heidelberg, 3. Aufl., 1962.

[7] Seidel, G.: Zusammenhang zwischen rationeller Energieanwendung und der Führung des Verbrennungsprozesses bei Zementdrehöfen. Silikattechnik 37 (1986) H. 9, S. 297/298.

[8] Scheuer, A. und Ellerbrock, H.-G.: Möglichkeiten der Energieeinsparung bei der Zementherstellung. Zement-Kalk-Gips 45 (1992) Nr. 5, S. 222–230.

[9] Ueda, Y. und Suzuki, Y.: Der Einfluß des Heizwertes der Kohle auf die Zementqualität. Zement-Kalk-Gips 38 (1985) Nr. 2, S. 77–83.

[10] Endres, G.: Aktuelle Brennerentwicklung im Hinblick auf gesteigerte Umweltanforderungen. Vortrag auf der 2. Feuerungstechnischen Fachtagung „Pillard aktuell" am 15. Nov. 1991 in Wiesbaden.

70–100 kJ/kg clinker could be saved by reducing the particle size from 110 to 70 μm. To establish the overall effect on the business this would naturally have to be set off against the presumably higher fuel costs (energy used for grinding).

A reduction in the fineness of grinding is also of interest from another point of view: according to Endres [10] NO_x formation drops with a decrease in the fineness of the pulverised coal.

In summarising it can be said that a reduction in the particle size of pulverised lignites from the $\bar{d}_P = 85-110$ μm range to the $\bar{d}_P = 40-60$ μm range should bring advantages in the combustion process. Total effectiveness would depend on the increased expenditure on preparation including safety aspects.

Innovation bei Schubrostkühlern mit großen Durchsatzleistungen

Innovations with reciprocating grate coolers with large throughput capacities

Innovation dans les refroidisseurs à grille poussée de grande capacité de débit

Innovaciones en el campo de los enfriadores de parrilla de vaivén, de gran capacidad de rendimiento

Von **E. Steffen, G. Koeberer** und **H. Meyer**, Buxtehude/Deutschland

Zusammenfassung – Rostkühler mit Durchsätzen zwischen 5000 und 12000 t/d stellen aufgrund der hohen Massenstromdichte des Klinkers auf den Kühlrosten spezielle Anforderungen an das Belüftungssystem und den Klinkerbrecher. Die Massenstromdichte hat einen deutlich stärkeren Einfluß auf den Wärmeaustausch zwischen Klinker und Luft als bisher angenommen wurde. Entscheidend für die Verbesserung des Wärmeaustauschs bei hoher Massenstromdichte ist die Abstimmung der Kühlluftzufuhr auf die unterschiedliche Verteilung der Klinkerkörnung über dem Rostquerschnitt. Die aus der Praxis gewonnenen Erfahrungen zeigen, daß ein Scale-up von zum Beispiel ca. 3 m (10 ft) breiten Kühlern für 3000 t/d auf knapp 5 m (16 ft) breite Kühler für 10000 t/d nur mit entsprechenden Anpassungen möglich ist. Das führte zur Entwicklung neuartiger Belüftungssysteme mit direkt- und kammerbelüfteten Muldenplatten und eines Hochleistungswalzenbrechers. Die Einsatzmöglichkeiten dieser innovativen Technik werden am Beispiel einer 6000 t/d-Anlage dargestellt. Sie kann auch zur Modernisierung bestehender Rostkühler eingesetzt werden.

Innovation bei Schubrostkühlern mit großen Durchsatzleistungen

Summary – Grate coolers with throughputs between 5000 and 12000 t/d make special demands on the aeration system and the clinker crusher because of the high mass flow density of the clinker on the cooler grates. The mass flow density has assumed a significantly greater influence on the heat exchange between clinker and air than used to be the case. Adaptation of the cooling air supply to the differing distribution of the clinker particle size over the grate cross-section is critical for improving the heat exchange at high mass flow densities. Practical experience shows that it is only possible to scale up from, for example, coolers around 3 m (10 ft) wide for 3000 t/d to coolers close to 5 m (16 ft) wide for 10000 t/d if appropriate adjustments are made. This has led to the development of new types of aeration systems with direct- and chamber-aerated troughed grate plates and heavy duty roller crushers. The possible applications of this innovative technology are explained using the example of a 6000 t/d plant. It can also be used for modernizing existing grate coolers.

Innovations with reciprocating grate coolers with large throughput capacities

Résumé – Les refroidisseurs à grille de débits entre 5000 et 12000 t/j posent, en raison de la haute densité du courant massique du clinker sur les grilles de refroidissement, des exigences spéciales côté système de ventilation et concasseur de clinker. La densité du courant massique exerce une influence nettement plus forte sur l'échange de chaleur entre clinker et air, que celle jusqu'à présent présumée. Décisif pour l'amélioration de l'échange de chaleur avec une haute densité de courant massique est d'accorder l'arrivage de l'air de refroidissement dans la distribution inhomogène des grains de clinker sur la section transversale de la grille. L'expérience acquise dans la pratique montre, qu'une transposition de refroidisseurs de, p. ex. 3 m (10 ft) pour 3000 t/j à des refroidisseurs à peine larges de 5 m (16 ft) pour 10000 t/j n'est possible qu'avec des adaptations conséquentes. Cela a conduit au développement de nouveaux systèmes de ventilation avec des plaques à auges ventilées directement ou par secteurs et d'un concasseur à cylindres de haute performance. Les possibilités de mise en oeuvre de cette technique innovative sont présentées au moyen de l'exemple d'une installation de 6000 t/j. Elle peut aussi être mise à profit pour la modernisation de refroidisseurs à grille existants.

Innovation dans les refroidisseurs à grille poussée de grande capacité de débit

Resumen – Los enfriadores de parrilla con capacidades de 5000 – 12000 t/d exigen del sistema de aireación y de la trituradora de clínker que cumplan determinadas condiciones, debido a la elevada densidad del caudal de masa del clínker sobre las parrillas de enfriamiento. La densidad del caudal de masa tiene un influjo notablemente mayor de lo que se pensaba hasta ahora sobre el intercambio de calor entre el clínker y el aire. Es decisivo para la mejora del intercambio de calor, cuando existe una elevada densidad del caudal de masa, que la llegada de aire de enfriamiento se adapte a la distribución no homogénea de los granos de clínker por la sección transversal de la parrilla. Las experiencias adquiridas en la práctica muestran que un scale-up, por ejemplo de enfriadores de unos 3 m (10 ft) de ancho, para 3000 t/d, a enfriadores de casi 5m (16 ft) de ancho, para 10000 t/d, sólo es posible si se llevan a cabo las adaptaciones necesarias. Esto ha conducido al desarrollo de nuevos sistemas de

Innovaciones en el campo de los enfriadores de parrilla de vaivén, de gran capacidad de rendimiento

aireación, con placas de artesa, de aireación directa o por cámaras, y de una trituradora de cilindros de alto rendimiento. Se explican las posibilidades de aplicación de esta nueva técnica, citando como ejemplo una instalación para 6 000 t/d. Dicha técnica se puede emplear también para la modernización de enfriadores de parrilla ya existentes.

Bei der Umsetzung eines verfahrenstechnischen Prozesses von einer Versuchsanlage auf eine Pilotanlage oder Produktionsanlage ändern sich eine Reihe von Parametern so stark, daß sie den Prozeßablauf negativ beeinflussen. Beim Scale-up der Rostkühlerbreite in Abhängigkeit vom Klinkerdurchsatz treten ähnliche Probleme auf, wobei die Parameter in beeinflußbare und nicht beeinflußbare Größen eingeteilt werden können. Als wesentliche Parameter, die auf den Abkühlungsverlauf des Klinkers im Schubrostkühler einwirken, sind die Klinkerkörnung, die Ofenabwurfbedingungen, die Luftverteilung, die Rostplattenausführung und die Zwischenzerkleinerung zu nennen. Welche Auswirkungen haben nun diese Parameter auf die Kühlung?

Die ofenfallende Klinkerkörnung ist bei den bekannten Brennverfahren nur in geringem Maße zu beeinflussen und tendiert bei großen Durchsatzleistungen mehr zu einem größeren Feinanteil, der höhere Drücke bei der Auslegung der Ventilatoren erforderlich macht.

Beim Austrag aus dem Drehrohrofen stellt sich die bekannte Separation des Klinkers ein. **Bild 1** zeigt die durch unterschiedliche Punktierung markierte Körnungsverteilung im Kühler. Die dunklen Balken unterhalb des Ofens stellen die Größe der Massenstromdichte (Masse pro Flächen- und Zeiteinheit) des Klinkers dar. Im mittleren Bereich ist die Massenstromdichte am größten. Bei hohen Durchsatzleistungen erhöht sich die Massenstromdichte in der Mitte um ein Mehrfaches. Im Aufprallbereich im Kühler wird die Klinkerschüttung dadurch stärker zusammengepreßt und einschließlich des sich darüber bildenden Klinkerberges ergeben sich hohe Luftwiderstände, d.h., daß die Luftgeschwindigkeit in diesem Bereich gegen Null geht. Die mögliche Folge sind Plattenschäden durch Überhitzung.

Bild 2 zeigt die Luftverteilung über den Querschnitt eines Kühlers mit 8 und 14 ft Breite. In einigen Veröffentlichun-

When an industrial process is transferred from an experimental plant to a pilot or production plant a number of parameters change so much that they have a negative effect on the process. Similar problems arise when the width of a grate cooler is scaled up according to the clinker throughput, and here the parameters can be divided into those which it is possible or impossible to influence. The essential parameters affecting the clinker cooling process in a reciprocating grate cooler are the coarseness of the clinker, the kiln discharge conditions, air distribution, the shape of the grate plates and intermediate size reduction. What effects do these parameters have on cooling?

In the familiar combustion processes the coarseness of the clinker dropping from the kiln can only be slightly influenced. With a high throughput the tendency is more to a greater proportion of fines, necessitating higher pressures in fan design.

When the clinker is discharged from the rotary kiln the well-known separation process starts. **Fig. 1** shows the particle size distribution in the cooler, marked by different dotting. The dark bars below the kiln represent the level of mass flow density (mass per unit area and time) of the clinker. Density is greatest in the central section. With a high throughput the mass flow density in the centre becomes several times greater. The granular mass of clinker in the area of impingement in the cooler is thereby more strongly compressed and strong air resistance is created, including the heap of clinker which forms above it, i.e. the air speed in this area approaches zero. This may result in damage to the plates through overheating.

Fig. 2 shows the air distribution over the cross-section of a cooler 8 and 14 ft wide. In some publications the same air speed is indicated at the coarse and the fine side of the flow, with a reference to the use of directly aerated grate plate systems. This indication is only correct if the plates have infi-

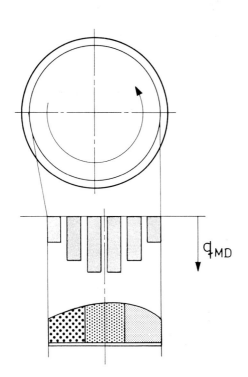

BILD 1: Massenstromdichte im Ofenabwurfbereich
FIGURE 1: Mass flow density in kiln discharge area

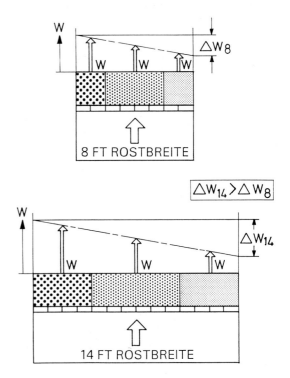

BILD 2: Luftverteilung über den Kühlerquerschnitt
FIGURE 2: Air distribution over cross-section of cooler
Rostbreite = Grate width

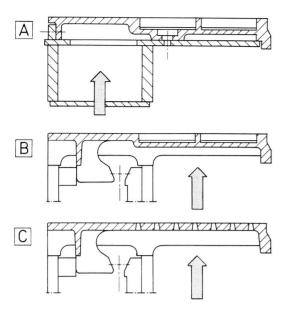

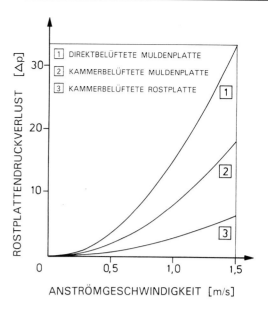

BILD 3: Rostplattensysteme
FIGURE 3: Grate plate systems
a) Direktbelüftete Muldenplatte = Directly aerated troughed plate
b) Kammerbelüftete Muldenplatte = Chamber-aerated troughed plate
c) Kammerbelüftete Rostplatte = Chamber-aerated grate plate

BILD 4: Eigendruckverlustkurven der Rostplattensysteme
FIGURE 4: Inherent pressure loss curves of grate plate systems
1) Direktbelüftete Muldenplatte = Directly aerated troughed plate
2) Kammerbelüftete Muldenplatte = Chamber-aerated troughed plate
3) Kammerbelüftete Rostplatte = Chamber-aerated grate plate
Rostplattendruckverlust = Grate plate pressure loss
Anströmgeschwindigkeit = aeration speed

gen wird unter dem Hinweis auf den Einsatz von direkt belüfteten Rostplattensystemen die Luftgeschwindigkeit auf der Grob- und Feinstromseite in gleicher Größe dargestellt. Diese Darstellung ist nur dann richtig, wenn der Widerstand der Platten unendlich groß ist. Die Reduzierung der spez. Kühlluftmenge mit gleichzeitiger Erhöhung des Druckverlustes durch die Rostplatten sollte jedoch den heutigen elektrischen Energiebedarf in kWh/t nicht wesentlich erhöhen, sofern dadurch nicht eine erhebliche Wärmerückgewinnung aus dem Klinker erzielt werden kann. Auf jeden Fall verbleibt eine Luftgeschwindigkeitsdifferenz zwischen Grob- und Feinstromseite. Bei Kühlern ab 12 ft Breite kann diese Differenz nicht mehr vernachlässigt werden. Die Vergrößerung der Geschwindigkeitsdifferenz von z. B. delta w8 auf delta w14 kann nur durch zusätzliche Maßnahmen verhindert werden. Eine solche Maßnahme kann z. B. die Aufteilung in drei Belüftungsbereiche quer zum Kühler sein. Hieraus ergeben sich folgende Möglichkeiten:

1. Zusätzliche Beeinflussung durch Drosselorgane in der Luftzuführung
2. Jeder Bereich wird mit einem eigenen Ventilator ausgerüstet
3. Einsatz von Rostplatten mit unterschiedlichen Luftwiderständen

Wie bereits ausgeführt, kann die Luftverteilung durch die Plattenausführung gesteuert werden. **Bild 3** zeigt drei Plattenausführungen mit unterschiedlichen Luftwiderständen. Ausführung A zeigt die direkt belüftete Muldenplatte mit hoher innerer Kühlung, so daß geringere Anströmgeschwindigkeiten mit hohem Wärmerückgewinn gewählt werden können. Ausführung C zeigt die bekannte Lochplatte, die nur noch in der Nachkühlzone eingesetzt werden sollte. Ausführung B ist eine kammerbelüftete Rostplatte, eine offene Ausführung der Muldenplatte A. Diese kammerbelüftete Muldenplatte ist in einem Kühler in der Nachkühlzone seit 9 Monaten in Betrieb. Bei gleicher spez. Kühlluftmenge konnte die Klinkeraustrittstemperatur um über 20 °C herabgesetzt werden. Seit einigen Wochen ist diese offene Muldenplatte auch in der Rekuperationszone in Betrieb. Die kammerbelüftete Muldenplatte kann ohne Umbauten des Kühlerrostes gegen die normale Lochplatte ausgetauscht werden. **Bild 4** zeigt den Druckverlust der drei Plattenausführungen in Abhängigkeit der Anströmgeschwindigkeit der Luft. Anhand dieses Diagrammes läßt sich leicht erken-

nite resistance. However, the present power requirement in kWh/t ought not to be increased significantly by reducing of the specific quantity of cooling air with a simultaneous increase in pressure loss by the grate plates unless this achieves considerable heat recovery from the clinker. In any case there is still a difference in air speed between the coarse and the fine side of the flow, and in coolers 12 ft wide and over this difference cannot be ignored. An increase in the difference in speed from e.g. delta w8 to delta w14 can only be prevented by additional measures. One such measure may be e.g. division into three aeration areas across the cooler, opening up the following possibilities:

1. producing an additional effect with restrictors in the air supply
2. equipping each area with its own fan
3. using grate plates with different air resistances

As already mentioned, air distribution can be controlled through plate design. **Fig. 3** shows three plate designs with different air resistance. Version A shows the directly aerated troughed plate with strong internal cooling, so that lower air flow speeds with high heat recovery can be selected. Version C shows the well-known plate containing holes, which should only be used in the secondary cooling zone. Version B is a chamber-aerated grate plate, an open version of

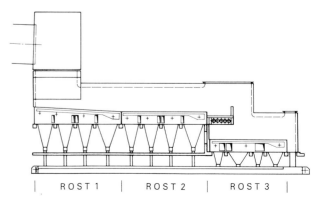

BILD 5: Combi-Stufenkühler mit integriertem Walzenbrecher
FIGURE 5: Combined-stage cooler with integral roller crusher
Rost = Grate

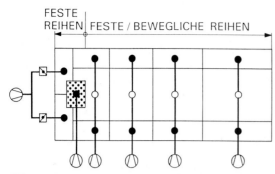

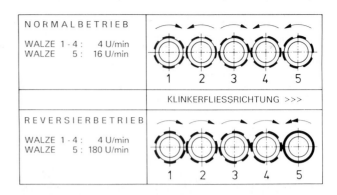

BILD 6: Belüftungsfelder eines Rostes
FIGURE 6: Aeration fields of a grate

Feste Reihen	= Stationary rows
Feste/bewegliche Reihen	= Stationary/moving rows
Individuelle Dosierung der Luft	= Individual air metering
Belüftung des zentralen Ofenabwurfes	= Aeration of central kiln discharge

BILD 8: Funktionsprinzip PETERS-Hochleistungswalzenbrecher
FIGURE 8: Operating principle of PETERS heavy-duty roller crusher

Normalbetrieb	= Normal operation
Walze	= Roller
U/min	= r.p.m.
Klinkerfließrichtung	= Clinker flow direction
Reversierbetrieb	= Reversing operation

nen, daß die normale Lochplatte für Kühler mit großer Breite für eine gute Luftverteilung nicht geeignet ist. Die kammerbelüftete Muldenplatte ist auf Grund des Eigendruckverlustes für Umbauten besonders geeignet.

Die Erfahrungen aus in Betrieb befindlichen Kühlern, die durchgeführten Versuche und die hier aufgezeigten Einflußgrößen sind die Grundlage für die Auslegung moderner Schubrostkühler insbesondere mit hohen Durchsatzleistungen. Die Einsatzmöglichkeiten der verschiedenen Rostplattensysteme und deren sinnvolle Aufteilung in Rostflächen-Belüftungsfelder werden am Beispiel einer Anlage für 6000 t/d dargestellt. **Bild 5** zeigt einen Combi-Stufenkühler mit drei Rosten und einem Walzenbrecher. Rost 1 und 2, das ist der Bereich der Rekuperation und der heißen Nachkühlzone, sind mit direkt belüfteten Muldenplatten ausgerüstet, während Rost 3 mit kammerbelüfteten Muldenplatten belegt ist. Die Anordnung des Walzenbrechers nach dem zweiten Rost hat den entscheidenden Vorteil, daß der gesamte Klinker auf eine konstante Endtemperatur gekühlt werden kann.

In **Bild 6** sind nun die einzelnen Belüftungsfelder speziell für Rost 1 dargestellt. Die Ausführung betrifft einen 14 ft breiten Kühler, der in drei Sektionen in Querrichtung unterteilt ist. Die Breite der mittleren Belüftungsfelder entspricht einem 8 ft breiten Kühler. Die Kühlluftversorgung erfolgt über je einen Ventilator pro Kühlabschnitt. Die Luftmengendosierung wird durch Drosselorgane in den Luftzuführungskanälen eingestellt. Fünf Reihen am Anfang des

troughed plate A. This chamber-aerated troughed plate has been in operation for 9 months in the secondary cooling zone of a cooler. It has been possible to lower the clinker exit temperature by over 20 °C with the same specific quantity of cooling air. This open troughed plate has also been operating in the recovery zone for some weeks. The chamber-aerated troughed plate may be exchanged for the normal plate containing holes without converting the cooler grate. **Fig. 4** shows the pressure loss of the three versions of the plate versus the aeration speed. It can easily be seen from this graph that the normal plate containing holes is not suitable for wide coolers requiring good air distribution. The chamber-aerated troughed plate is particularly appropriate for conversions owing to the inherent pressure loss.

Experience with operating coolers, the trials carried out, and the influencing factors noted here provide a basis for the design of modern reciprocating grate coolers, especially those with a high throughput. The possible applications of the various grate plate systems and their appropriate division into grate surface aerating fields are illustrated using the example of a 6000 t/d plant. **Fig. 5** shows a combined-stage cooler with three grates and a roller crusher. Grates 1 and 2, which are the recovery area and hot secondary cooling zone, are fitted with directly aerated troughed plates, while grate 3 has chamber-aerated troughed plates. The arrangement of the crusher after the second grate has the decisive advantage that all the clinker can be cooled to a constant end temperature.

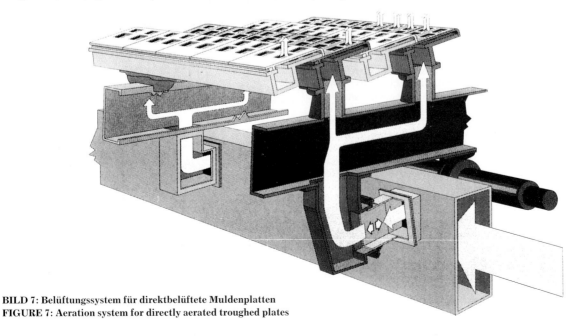

BILD 7: Belüftungssystem für direktbelüftete Muldenplatten
FIGURE 7: Aeration system for directly aerated troughed plates

Rostes sind feste Reihen, wobei die ersten drei Reihen mit einer Luftpulsation beaufschlagt werden. Das punktiert dargestellte Belüftungsfeld ist der Bereich des zentralen Ofenabwurfes mit der höchsten Massenstromdichte. Dieses Belüftungsfeld besitzt einen eigenen Ventilator, um zu gewährleisten, daß ausreichend Luft zugeführt wird. **Bild 7** zeigt die Ausführung des Belüftungssystemes. Die Luft zweigt vom Hauptluftkanal über feste Verbindungen für feste Reihen und über Schiebekompensatoren für bewegliche Reihen zu den Schwingrahmenholmen ab und wird dann durch die Rostträger den einzelnen Platten zugeführt.

Bei großen Anlagen treten häufig kurzzeitige Durchsatzschwankungen auf, wenn z. B. Ofenansatz fällt. Der erhöhte Anfall an großen Brocken kann mit herkömmlichen Hammerbrechern nicht mehr einwandfrei bewältigt werden, so daß Walzenbrecher mit Speichervolumen eingesetzt werden müssen. Der PETERS-Hochleistungswalzenbrecher besteht aus mehreren modular aufgebauten Brechwalzen, die sich in Drehrichtung und Drehzahl den Betriebsbedingungen anpassen. **Bild 8** oben zeigt den normalen Betriebszustand. Bei starkem Grobanfall, z. B. Ofenansatz, wird automatisch über eine Schichthöhenregelung die Drehrichtung der Walzen gemäß der unteren Darstellung geändert. Walze 5 wird dann zur Schnelläuferwalze mit Drehzahlen bis 180 U/min. Innerhalb von Sekunden werden große Brocken zerkleinert. Nach erfolgter Zerkleinerung stellt sich der Reversierbetrieb automatisch wieder auf Normalbetrieb um.

Bei Schubrostkühlern mit großen Durchsatzleistungen gewährleisten die Einteilung der Roste in Belüftungsfelder, die richtige Auswahl des Rostplattensystems und die sinnvolle Anordnung des Hochleistungs-Walzenbrechers eine hohe Wärmerückgewinnung mit niedriger Klinkerendtemperatur bei minimaler spez. Kühlluftmenge.

The individual aerating fields specifically for grate 1 are illustrated in **Fig. 6**. The construction is for a cooler 14 ft wide, divided crosswise into three sections. The width of the central aerating fields corresponds to an 8 ft wide cooler. The cooling air is supplied through one fan per cooling section. Air metering is adjusted by restrictors in the air supply ducts. Five rows at the beginning of the grate are stationary rows, and air is pulsated at the first three. The dotted aeration field in the figure is the central kiln discharge area with the highest mass flow density. It has its own fan, to ensure that enough air is supplied. **Fig. 7** shows the construction of the aeration system. The air branches off from the main air duct, through stationary connections for stationary rows and through sliding compensators for mobile rows, to the bars of the movable frames and is then fed through the grate supports to the individual plates.

Short-term fluctuations in throughput occur frequently in large plant, e.g. if there is a fall of kiln coating. The increased amount of large lumps cannot be dealt with satisfactorily by conventional hammer crushers, so roller crushers with a hold-up capacity have to be used. The PETERS heavy-duty roller crusher consists of several crushing rolls of modular construction, which adapt to the operating conditions in direction of rotation and speed. The normal operative state is shown at the top of **Fig. 8**. On the appearance of a large amount of coarse material such as kiln coating, the direction of rotation of the rollers is changed automatically by a layer height control as shown at the bottom of the figure. Roller 5 then becomes the high-speed roller with speeds of up to 180 rpm. Large lumps are pulverised in seconds. When this is complete the system is automatically returned to normal operation.

In reciprocating grate coolers with a large throughput, division of the grates into aerating fields, correct choice of grate plate system and appropriate arrangement of the heavy-duty roller crusher ensure good heat recovery with a low final clinker temperature and a minimum specific quantity of cooling air.

Entwicklungsstand des PYRO-JET®-Brenners
Current state of development of the PYRO-JET® burner

Etat actuel du développement du brûleur PYRO-JET®

Estado de desarrollo del quemador PYRO-JET®

Von **E. Steinbiß, C. Bauer** und **W. Breidenstein,** Köln/Deutschland

Entwicklungsstand des PYRO-JET®-Brenners

Zusammenfassung – 10jährige Betriebserfahrungen mit PYRO-JET®-Brennern an Drehöfen mit unterschiedlichen Durchsätzen haben zu einer gesicherten Auslegung und verläßlichen Konstruktion dieses Brenners geführt. Die Axialluft tritt beim PYRO-JET®-Brenner über mehrere am Umfang verteilte Einzelstrahlen mit hoher Geschwindigkeit aus. Dadurch wird der Primärluftanteil gegenüber herkömmlichen Brennertypen deutlich verringert und der Brennstoffenergieverbrauch um bis zu 160 kJ/kg Kl gesenkt. Durch die sogenannten Jetstrahlen entsteht eine straffe Flamme mit gleichmäßiger Wärmeentwicklung. Die frühe Zündung des Brennstoffs am Austritt des PYRO-JET®-Brenners bewirkt eine deutliche Senkung der Stickoxid-Emissionen, wie Messungen gezeigt haben. Aufgrund seiner hervorragenden Verbrennungseigenschaften ist der Brenner für schwierige feste Brennstoffe mit geringen Anteilen flüchtiger Bestandteile, wie z. B. Anthrazit oder Petrolkoks, sowie auch zur Mischfeuerung mit Öl und Gas geeignet. Die Weiterentwicklung konzentriert sich auf Verbesserungen der Konstruktion sowie der Standzeiten.

Current state of development of the PYRO-JET® burner

Summary – Ten years' operating experience with PYRO-JET® burners in rotary kilns of varying capacities has led to an established design and dependable construction for this burner. In the PYRO-JET® burner the axial air emerges at high velocity through several individual jets distributed around the circumference. This means that the percentage of primary air can be significantly reduced compared with traditional types of burners, and the fuel energy consumption is lowered by up to 160 kJ/kg clinker. The so-called jet-streams produce a fierce flame with uniform heat development. Early ignition of the fuel at the outlet of the PYRO-JET® burner produces a significant lowering of nitrogen oxide emissions, as has been demonstrated by measurements. Due to its excellent burning characteristics the burner is suitable for difficult solid fuels with lower proportions of volatiles, such as anthracite or petroleum coke, as well as for mixed firing with oil and gas. Further development work is being concentrated on improving the construction and service life.

Etat actuel du développement du brûleur PYRO-JET®

Résumé – 10 années d'expérience industrielle avec le brûleur PYRO-JET® dans des fours rotatifs de capacités différentes ont conduit à une conception confirmée et une construction fiable de ce brûleur. Dans le brûleur PYRO-JET®, l'air axial sort à haute vitesse par plusieurs orifices distincts disposés à la périphérie. Cela permet de réduire nettement, par rapport aux types de brûleurs conventionnels, la part d'air primaire et d'abaisser la consommation d'énergie de combustible, d'environ 160 kJ/kg KL. A l'aide des courants jet ainsi nommés se forme une flamme tendue avec un dégagement de chaleur régulier. L'ignition précoce du combustible, à la sortie du brûleur PYRO-JET®, a comme conséquence un abaissement remarquable des émissions d'oxydes d'azote, ce qui est vérifié par des mesures. En raison de ses propriétés de combustion excellentes, ce brûleur convient pour les combustibles solides difficiles à faible teneur en composants volatils, tels que anthracite ou coke de pétrole, ou encore au chauffage mixte avec mazout ou gaz. Le développement ultérieur est axé sur l'amélioration de la construction et de la durée de service.

Estado de desarrollo del quemador PYRO-JET®

Resumen – 10 años de experiencias adquiridas con quemadores PYRO-JET® en hornos rotatorios de diferentes capacidades de producción han permitido llegar a un dimensionamiento seguro y a una construcción fiable de este quemador. El aire axial sale a gran velocidad del quemador PYRO-JET®, a través de diferentes orificios repartidos por la periferia del mismo. De esta forma, se reduce notablemente la proporción de aire secundario en comparación con los tipos de quemadores convencionales, disminuyendo al mismo tiempo el consumo de energía, que puede llegar a ser hasta 160 kJ/kg más bajo. Debido a los llamados chorros „jet", se produce una llama rígida, con desarrollo uniforme del calor. La ignición precoz del combustible a la salida del quemador PYRO-JET® conduce a una reducción notable de las emisiones de óxido de nitrógeno, según se desprende de las mediciones efectuadas. Gracias a sus excelentes propiedades de combustión, el quemador es apto para combustibles sólidos difíciles, con reducida proporción de componentes volátiles, tales como la antracita o el coque de petróleo, así como para mezclas de petróleo y gas. El futuro desarrollo se centra en las mejoras constructivas y de la duración de vida.

1. Einleitung

KHD Humboldt Wedag hat 1980 einen völlig neuartigen Kohlenstaubbrenner für Zementdrehöfen entwickelt [1]. Diese patentierte Entwicklung basierte auf dem KHD-Hochdruck-Gasbrenner, der sich bereits mehrfach im Drehofenbetrieb bewährt hatte [2]. Seit Einsatz des ersten PYRO-JET-Brenners für Kohlenstaub sind mehr als 12 Jahre vergangen, und über 100 Drehöfen sind damit ausgerüstet worden. Aus diesem Grunde soll hier über Erfahrungen mit diesem Brenner zusammenfassend berichtet werden, insbesondere auch, weil er heute als Brenner zur Senkung der Stickoxid-Emission verwendet wird.

Der Anlaß für die Entwicklung des PYRO-JET-Brenners war der Wunsch, den Wärmeverbrauch der Ofenanlage zu senken, indem man den kalten Primärluftanteil reduziert zugunsten heißer Sekundärluft. Der Gewinn lag zu Beginn der Anwendung immerhin schon bei 60–80 kJ/kg (bezogen auf Klinker) und ließ sich bis heute im günstigen Fall noch einmal um den gleichen Betrag verbessern.

Dies sind beträchtliche Einsparungen an wertvoller Brennstoffenergie, die außerdem zur Entlastung der Umwelt beitragen. Darüber hinaus trägt der PYRO-JET-Brenner aufgrund seiner besonderen Konstruktion zu einer weiteren Entlastung der Umwelt bei, da er wesentlich weniger Stickoxid in der Flamme entstehen läßt als vergleichbare 3-Kanal-Brenner [3].

Wegen dieser Vorzüge und der Möglichkeit, ihn für alle verfügbaren Brennstoffe einsetzen zu können, wurde der PYRO-JET-Brenner auch in Ofenanlagen anderer Lieferanten eingesetzt. Damit hat sich dieser Brenner einen bemerkenswerten Anteil im Markt gesichert.

Im Laufe der letzten Jahre erfuhr der Brenner eine Weiterentwicklung, um ihn an die veränderten Anforderungen des Brennprozesses anzupassen.

Der PYRO-JET-Brenner wird heute üblicherweise als Kombinationsbrenner für die gleichzeitige oder alternative Verbrennung von festen, flüssigen und gasförmigen Brennstoffen gebaut, wobei häufig noch zusätzliche Kanäle für die Zuführung verschiedenster Substitutionsbrennstoffe vorgesehen sind.

Tabelle 1 zeigt eine Auswahl in Betrieb befindlicher PYRO-JET-Brenner, aufgelistet nach ihrer Kapazität.

2. Konstruktion und Funktionsweise

Konstruktion und Funktionsweise des Brenners wurden bereits früher ausführlich beschrieben [1] und haben sich im Prinzip nicht verändert (**Bild 1**).

Der wesentliche Unterschied zu den herkömmlichen Brennern besteht in der patentierten Ausführung des Axialluftdüsensystems. Die Axialluft tritt beim PYRO-JET-Brenner

1. Introduction

KHD Humboldt Wedag developed a completely new type of pulverized coal burner for rotary cement kilns in 1980 [1]. This patented development was based on the KHD high-pressure gas burner, which had already been used successfully in several cases for rotary kiln operation [2]. Over 12 years have passed since the first PYRO-JET burner was used for pulverized coal, and over 100 rotary kilns have been fitted with it. A summary of the experience with this burner will therefore be reported here, particularly as the burner is used nowadays for lowering nitrogen oxide emissions.

The reason for developing the PYRO-JET burner was the wish to diminish the heat consumption of the kiln plant by reducing the proportion of cold primary air in favour of hot secondary air. There was a gain of 60 to 80 kJ/kg (relative to clinker) when the burner was first used, and this gain has now doubled in favourable cases.

These are considerable savings in valuable fuel energy, which also help to relieve the pressure on the environment. The special design of the PYRO-JET burner further relieves pressure on the environment since its flame produces far less nitrogen oxide than comparable 3-channel burners [3].

Because of these advantages and the fact that it can be used for all available fuels the PYRO-JET burner was also installed in other suppliers' kiln plants. In this way it has taken over a remarkably large share of the market.

The burner has been further developed in recent years to adapt it to the changed requirements of the burning process.

The PYRO-JET burner is now normally constructed as a combination burner for simultaneous or alternative combustion of solid, liquid and gaseous fuels, and additional channels are often provided for supplying a wide variety of substitute fuels.

Table 1 shows a selection of PYRO-JET burners which are in operation, listed according to their capacity.

2. Design and mode of operation

The design and mode of operation of the burner have already been described in detail [1] and have not changed in principle (**Fig. 1**).

The essential difference from conventional burners lies in the patented construction of the axial air jet system. In the PYRO-JET burner the axial air passes out of the outermost burner channel through several individual round nozzles distributed around the circumference. Owing to their high velocity the resultant individual jet streams draw in the slow-flowing but hot secondary air from the surroundings and lead to particularly fast ignition of the fuel.

Early ignition is also ensured by internal recirculation of flame gases, as a result of the swirl air emerging from the inner part of the burner.

TABELLE 1: Auswahl in Betrieb befindlicher PYRO-JET-Brenner

Anlage	Jahr	Brennstoff[1])	Ofen Ø × L m	Klinkerprod. t/d	Brennerkapazität GJ/h	t/h [2])
1	1981	K, Ö	3,6 × 49	700	126	5,0
2	1982	K, Ö	3,8 × 50	1500	167	10,0
3	1981	K	3,8 × 60	1200	183	7,3
4	1981	K, Ö	4,8 × 72	3000	201	8,0
5	1992	K, Ö, G	4,0 × 58	1100	230	8,6
6	1991	K, Ö, P	4,1 × 65	1700	250	9,9
7	1983	K, Ö	4,5 × 68	2000	266	12,0
8	1981	K	4,6 × 70	1600	298	14,2
9	1984	K, Ö	5,0 × 78	4600	312	12,4
10	1981	K, Ö	4,8 × 72	2200	333	13,3
11	1990	K, Ö	5,0 × 55	4800	390	15,5
12	1985	K, BK, Ö	5,3 × 80	2800	444	20,0
13	1990	K, Ö	5,75 × 92	3400	510	20,3
14	1992	K, Ö	5,6 × 87	7600	530	20,1

[1]) BK Braunkohle   [2]) für Kohle
    G   Gas
    K   Steinkohle
    Ö   Öl
    P   Petrolkoks

TABLE 1: Selection of PYRO-JET burners now in operation

Plant	Year	Fuel[1]	Kiln Ø × L m	Clinker produced t/d	Burner capacity GJ/h	t/h [2]
1	1981	K,Ö	3.6 × 49	700	126	5.0
2	1982	K,Ö	3.8 × 50	1500	167	10.0
3	1981	K	3.8 × 60	1200	183	7.3
4	1981	K,Ö	4.8 × 72	3000	201	8.0
5	1992	K,Ö,G	4.0 × 58	1100	230	8.6
6	1991	K,Ö,P	4.1 × 65	1700	250	9.9
7	1983	K,Ö	4.5 × 68	2000	266	12.0
8	1981	K	4.6 × 70	1600	298	14.2
9	1984	K,Ö	5.0 × 78	4600	312	12.4
10	1981	K,Ö	4.8 × 72	2200	333	13.3
11	1990	K,Ö	5.0 × 55	4800	390	15.5
12	1985	K,BK,Ö	5.3 × 80	2800	444	20.0
13	1990	K,Ö	5.75 × 92	3400	510	20.3
14	1992	K,Ö	5.6 × 87	7600	530	20.1

[1] BK lignite
G gas
K coal
Ö oil
P petroleum coke

[2] for coal

über mehrere über den Umfang verteilte, runde Einzeldüsen aus dem äußersten Brennerkanal aus. Die dadurch entstehenden einzelnen Freistrahlen (Jets) saugen infolge ihrer hohen Geschwindigkeit die langsam strömende, aber heiße Sekundärluft aus der Umgebung an und führen zu einem besonders schnellen Zünden des Brennstoffs.

Eine frühzeitige Zündung wird darüber hinaus durch die innere Rezirkulation von Flammengasen sichergestellt, die durch die im inneren Bereich des Brenners austretende Drallluft entsteht.

Diese Maßnahmen zur Annäherung der Zündfront an den Brenner verbessern den Ausbrand und verhindern zum Teil die unerwünschte Bildung von Stickoxid. In Verbindung mit den Jetstrahlen entsteht eine straffe, heiße und leuchtende Flamme mit gleichmäßiger Wärmedarbietung.

Der PYRO-JET-Brenner kann mit den üblichen Luftdrükken von 150 mbar betrieben werden. Seine Funktion läßt sich damit bereits mit nur 6–8 % Primärluft einschließlich Förderluft voll erreichen. Die Austrittsgeschwindigkeiten der Drall- und Jetluft liegen dann bei ca. 150 m/s.

Der besondere Vorteil dieses Brenners, d.h. ein noch niedrigerer Primärluftanteil, läßt sich besonders dann erzielen, wenn der Druck der Primärluft erhöht wird. Dadurch steigt der Impuls der Primärluft, welcher für eine gute Vermischung und Flammenformung entscheidend ist. Mit einem Druck von 800 mbar läßt sich die Geschwindigkeit der Jetluft auf etwa 350 m/s erhöhen, wobei der Primärluftanteil bis auf 4 % (bzw. 6 % einschließlich der Kohle-Förderluft) sinken kann (**Bild 2**). Dieser erhöhte Druck wird also lediglich für den Jetluftanteil benötigt, der in der Größenordnung von nur 1,6 % liegt. Das hierfür benötigte Kapselgebläse ist wegen der geringen Luftmenge so klein, daß die Kosten des erhöhten elektronischen Energieaufwandes durch den Gewinn an eingespartem Wärmeaufwand mehr als kompensiert werden (**Tabelle 2**).

Die Jetluftdüsen sitzen in einer Düsenscheibe, die zwölf oder mehr Bohrungen aufweist. Austrittswinkel und Düsenquerschnitt lassen sich durch Austausch der Düsen den Gegebenheiten des Brennprozesses anpassen.

Der Brenner besitzt normalerweise keine beweglichen Teile im feuerseitigen Bereich. Der Kohlenstaub tritt über eine Ringspaltdüse mit geringer Divergenz in den Brennraum. Die Austrittsgeschwindigkeit läßt sich in Abhängigkeit von den Kohleeigenschaften einstellen und sollte aus Verschleißgründen möglichst niedrig liegen, d.h. im Bereich von 20 bis 35 m/s.

Die wesentlichen Brennerparameter zur Beeinflussung der Flamme, wie Luftmengen und -geschwindigkeiten, lassen sich den individuellen Gegebenheiten und Anforderungen während des Betriebes und im Stillstand anpassen.

These measures to bring the ignition front closer to the burner improve burn-out and to some extent prevent the undesirable formation of nitrogen oxide. In conjunction with the jet streams a fierce, hot, luminous flame is formed with uniform heat development.

The PYRO-JET burner can be operated at the usual air pressure of 150 mbar. Thus it is fully functional with only 6 to 8 % primary air including transport air. The exit velocities of the swirl air and jet stream are then approx. 150 m/s.

The special advantage of this burner, i.e. an even lower proportion of primary air, is best achieved by raising the pressure of the primary air. This increases its momentum, a decisive factor in good mixing and flame formation. At a pressure of 800 mbar the velocity of the jet stream can be increased to about 350 m/s and the proportion of primary air can be reduced to 4 % (or 6 % including the coal transport air) (**Fig. 2**). The increased pressure is only required for the jet stream component, which is only of the order of 1.6 %. Owing to the small amount of air the positive displacement blower

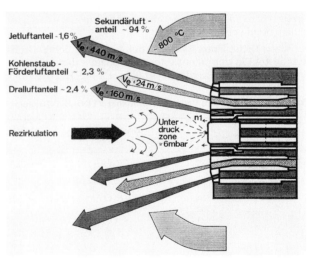

BILD 1: Funktionsweise des PYRO-JET-Brenners
FIGURE 1: Mode of operation of the PYRO-JET burner

Sekundärluftanteil	= proportion of secondary air
Jetluftanteil	= proportion of jet air
Kohlenstaub-Förderluftanteil	= proportion of transport air for pulverized coal
Drallluftanteil	= proportion of swirl air
Rezirkulation	= recirculation
Unterdruckzone	= low pressure zone
Zentralluftanteil	= proportion of central air

TABLE 2: Total heat consumption and power requirement dependent on the quantity of primary air

Parameter	Unit	Conventional burner 13% primary air	PYRO-JET burner 8% primary air
Production	t/d	2000	2000
Specific heat requirement	kJ/kg	3350	3280
Specific quantity of exhaust gas after preheater	m³(stp)/kg	1.53	1.51
Pressure drop, cyclone preheater	mbar	50	48
Calorific value of coal	kJ/kg	25120	25120
Minimum air requirement	m³(stp)/kg	6.56	6.56
Quantity of coal	t/h	11.11	10.89
Saving of coal	t/a		1760
Saving in cost of coal	DM/a	(110 DM per t coal)	193600
Quantity of primary air	m³(stp)/h	10610	6400
Primary air pressure	mbar	130	160/800
Fan capacity			
– primary air	kW	48	73
– exhaust gas	kW	575	548
– total	kW	**623**	**621**

required is so small that the additional power cost is more than compensated for by the saving in heating (**Table 2**).

The nozzles are seated in a nozzle disc containing twelve or more holes. The exit angle and the nozzle cross-section can be adapted to the requirements of the burning process by exchanging the nozzles.

The burner does not normally have any movable parts near the hot face. The pulverized coal enters the combustion chamber through an annular nozzle with little divergence. The exit velocity can be adjusted to the properties of the coal and should be kept as low as possible, i.e. within the 20 to 35 m/s range, for reasons of wear.

The essential burner parameters for controlling the flame, such as air quantities and velocities, may be adapted to individual conditions and requirements during operation or when stopped.

3. Design development

A modern version of the PYRO-JET burner is shown in the photograph in **Fig. 3** and a section through the nozzle of a typical PYRO-JET burner is given in **Fig. 4**. The attachment of the jet nozzle ring is regarded as a critically important new development. The six screws are now only used to seal off this ring securely from the inner and outer pipes. The two concentric pipes of the annular jet stream channel are braced by the screw fitting behind the ring, which is thus in a heat-protected zone. This design also facilitates the changing of the jet nozzle ring.

Another change in design, to reduce wear on the coal nozzle, has been developed and is being patented. It is a ceramic nozzle insert (Fig. 4). This closed, tapering insert has been in operation in a burner for a year for firing normally abrasive coal but as yet shows no visible wear.

The second point of possible wear, i.e. when the coal enters the burner channel, has already been protected with ceramic plates for some years. The plates are stuck to the piece of pipe requiring protection, thus giving it a service life of several years.

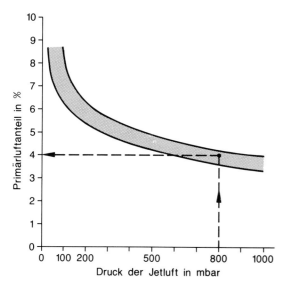

FIGURE 2: Effect of pressure of jet air on the required quantity of primary air

inneren und äußeren Rohr. Die Verspannung der beiden konzentrischen Rohre des Jetluftringkanals übernimmt die dahinterliegende Verschraubung, sie liegt somit im temperaturgeschützten Bereich. Diese Konstruktion erleichtert überdies einen Wechsel des Jetdüsenrings erheblich.

Eine weitere konstruktive Änderung wurde zur Senkung des Verschleißes der Kohledüse entwickelt und zum Patent angemeldet. Es handelt sich um einen keramischen Düseneinsatz (Bild 4). Dieser geschlossene konische Einsatz ist in einem Brenner seit einem Jahr in Betrieb und weist bei Verfeuerung normal schleißender Steinkohle bisher keinen sichtbaren Verschleiß auf.

Die zweite Stelle möglichen Verschleißes, d. h. der Eintritt der Kohle in den Brennerkanal, wird schon seit einigen Jahren mit einer keramischen Beplattung geschützt. Die Plättchen werden auf das zu schützende Rohrstück geklebt, wodurch Standzeiten von mehreren Jahren erreicht wurden.

Eine dritte Ursache für vorzeitigen Verschleiß eines Brenners liegt häufig in seiner ff-Ummantelung. Dieses Problem tritt bevorzugt dann auf, wenn der Verarbeitung der feuerfesten Masse nicht die erforderliche Aufmerksamkeit geschenkt wird. Nur die einwandfreie Befolgung der Verarbeitungsrichtlinien insbesondere der Austrocknungs- und Aufheizzeiten führt zu brauchbaren Standzeiten. Als sinnvoll für schnellen Wechsel hat sich der vorbereitete Reservebrenner erwiesen.

Für relativ kurze Brenner verwendet man auch vorbereitete, d. h. mit ff-Masse überzogene Hüllrohre für den PYRO-JET-Brenner, so daß man anstatt eines Brennerwechsels nur noch einen Wechsel seines äußeren Mantelrohres vornehmen muß. Für den Betrieb ergibt sich damit eine preiswerte Bevorratung nur eines oder mehrerer feuerfester Mantelrohre. Die Zeit für das Auswechseln läßt sich durch entsprechende konstruktive Ausbildung erheblich verkürzen. Bild 4 zeigt ein solches Mantelrohr, das auf den Brenner aufgeschoben und angeflanscht wird.

4. Flammenform und Verbrennungseigenschaften

Der PYRO-JET-Brenner liefert eine ausreichend heiße Flamme für alle Klinkerarten. Die Flammenform unterscheidet sich jedoch bei normalem Luftüberschuß von der eines konventionellen 3-Kanal-Brenners ganz wesentlich.

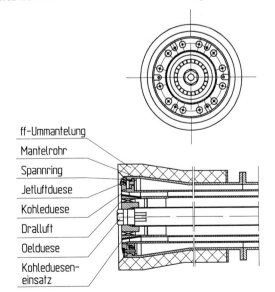

BILD 4: PRO-JET-Brenner für Kohle und Öl
FIGURE 4: PYRO-JET burner for coal and oil

ff-Ummantelung	=	refractory jacket
Mantelrohr	=	tubular jacket
Spannring	=	clamping ring
Jetluftdüse	=	jet air nozzle
Kohledüse	=	coal nozzle
Drallluft	=	swirl air
Öldüse	=	oil nozzle
Kohledüseneinsatz	=	coal nozzle insert

BILD 3: PYRO-JET-Brenner für Kohle, Öl und Gas
FIGURE 3: PYRO-JET burner for coal, oil and gas

A third cause of premature wear on a burner is often its refractory jacket. The problem arises particularly when insufficient attention is paid to the processing of the cast refractory material. A satisfactory service life can only be obtained by strictly following the processing instructions, especially with regard to drying-out and heating-up times. It has been found helpful to have a reserve burner in readiness so that a quick change can be made.

Prepared covering pipes, i.e. pipes coated with refractory material, are also used for relatively short PYRO-JET burners, so that only the outer tubular jacket has to be changed instead of the burner. This enables the operator to economise by stocking only one or more refractory jackets. The time taken to change the parts can be shortened considerably if they are appropriately designed. Fig. 4 shows such a tubular jacket, which is slid onto the burner and flange mounted.

4. Flame shape and burning characteristics

The PYRO-JET burner provides a hot enough flame for all types of clinker. However, when there is a normal excess of air the shape of the flame is very different from that of a conventional 3-channel burner.

The flame of the PYRO-JET burner has both an internal recirculation zone and a large external one, resulting in substantially uniform heat distribution. Thus there are no sharp temperature peaks, as shown in **Fig. 5**. This is confirmed by tests carried out by operators at various rotary cement kilns [4].

This flame shape is particularly desirable since it allows stable coat formation and thus protects the refractory lining in the sintering zone owing to its uniform heat development. It should be mentioned in this connection that the defined path of the jet streams is maintained throughout the operating period and the flame always keeps its set shape, to the benefit of the coating and refractory lining in the sintering zone.

In addition, the thorough mixing in the flame ensures extremely good burn-out, as evidenced by the CO values measured at the kiln inlet (less than 0.1% with 1.5% O_2 in normal operation).

The flame can be set manually or by a remote-controlled air volume adjustment device to meet the special requirements for combustion of different grades of clinker. It also clearly fulfils the requirements for producing clinker for very high grades of cement. This can also be done with the usual free lime content of 1% or even higher, that is to say, the clinker need not be over-burnt, so fuel can be saved.

The turndown ratio of the burner is 10 : 1 with pure coal firing, so the PYRO-JET burner can take the kiln from the pre-

Die Flamme des PYRO-JET-Brenners besitzt neben der inneren Rezirkulationszone auch eine große äußere Rezirkulationszone, wodurch sich eine weitgehend gleichmäßige Temperaturverteilung ergibt. Somit treten keine scharfen Temperaturspitzen auf, wie **Bild 5** zeigt. Dies wird durch Untersuchungen von Betreibern an verschiedenen Zementdrehöfen bestätigt [4].

Diese Flammenform ist besonders erwünscht, da sie wegen ihrer gleichmäßigen Wärmedarbietung eine stabile Ansatzbildung ermöglicht und damit die ff-Auskleidung in der Sinterzone schont. In diesem Zusammenhang ist zu erwähnen, daß die definierte Führung der Jetstrahlen über die gesamte Betriebszeit erhalten bleibt und die Flamme ihre eingestellte Form immer beibehält, was insbesondere dem Ansatz und der ff-Auskleidung in der Sinterzone zugute kommt.

Darüber hinaus sorgt die gute Vermischung in der Flamme für einen äußerst guten Ausbrand, der sich anhand der gemessenen CO-Werte am Ofeneinlauf (kleiner 0,1 % bei 1,5 % O_2 im Normalbetrieb) nachweisen läßt.

Den besonderen Erfordernissen beim Brennen verschiedener Klinkerqualitäten entsprechend läßt sich die Flamme von Hand oder durch eine ferngesteuerte Luftmengenregelung einstellen. Auch der Wunsch, Klinker für Höchstwert-Zemente zu erzeugen, wird nachweisbar erfüllt. Dies gelingt auch mit den üblichen Freikalkgehalten von 1 % oder auch darüber, d.h. der Klinker muß nicht überbrannt werden, so daß sich dadurch Brennstoff einsparen läßt.

Der Regelbereich des Brenners beträgt 10 : 1 bei reiner Kohlefeuerung. Damit läßt sich der Ofen mit dem PYRO-JET-Brenner schnell und energiesparend aus dem vorgewärmten Zustand in den heißen Normalbetrieb fahren.

Zur Verfeuerung extrem unterschiedlicher Kohlesorten läßt sich die Kohledüse ohne Schwierigkeiten anpassen.

Wegen seiner hervorragenden Verbrennungseigenschaften eignet sich der PYRO-JET-Brenner besonders für „schwierige", feste Brennstoffe mit geringen flüchtigen Bestandteilen, wie Anthrazit und Petrolkoks. Damit ergeben sich erhebliche wirtschaftliche Vorteile bezüglich der Brennstoffkosten. Es befinden sich bereits mehrere Brenner seit Jahren mit 100 % Anthrazit sowie auch mit 100 % Petrolkoks im Dauerbetrieb im Einsatz.

5. Stickoxid-Emission

Der PYRO-JET-Brenner läßt sich auch zur Senkung der Stickoxid-Emission einsetzen, da er im Vergleich zu herkömmlichen Brennern die Stickoxidbildung deutlich dämpft. Ein Grund für diese positive Eigenschaft liegt darin, daß der Brenner mit vielen einzelnen Jetstrahlen hoher Geschwindigkeit arbeitet, die für eine rasche Zündung kurz vor der Brennermündung sorgen und keine Temperaturspitze der Flamme entstehen lassen. Die Vermeidung einer Temperaturspitze behindert die thermische Stickoxid-Bildung, und die definierten Strömungsbedingungen schaffen reduzierende Flammenbereiche, wo bereits gebildetes Stickoxid teilweise wieder reduziert wird.

Modellrechnungen an der Universität Bochum haben dies bestätigt. **Bild 6** zeigt das Ergebnis dieser Rechnungen als

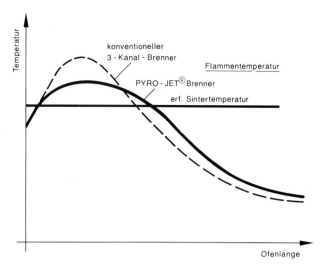

BILD 5: Qualitativer Verlauf der Flammentemperatur beim PYRO-JET-Brenner
FIGURE 5: Qualitative development of flame temperature with the PYRO-JET burner

Temperatur	= temperature
konventioneller 3-Kanal-Brenner	= conventional 3-channel burner
Flammentemperatur	= flame temperature
Brenner	= burner
erf. Sintertemperatur	= required sintering temperature
Ofenlänge	= distance along kiln

heated state to normal hot operation quickly and with low energy consumption.

The coal nozzle can be adapted without difficulty to fire widely differing types of coal.

Due to its excellent burning characteristics the PYRO-JET burner is particularly suitable for "difficult" solid fuels with lower proportions of volatiles, such as anthracite or petroleum coke. It therefore has considerable economic advantages in respect of fuel costs. Several burners have already been in continuous operation for years with 100% anthracite and 100% petroleum coke.

5. Nitrogen oxide emissions

The PYRO-JET burner can be used to reduce nitrogen oxide emissions, since it suppresses nitrogen oxide formation to a significantly greater extent than conventional burners. One reason for this positive characteristic is that the burner operates with many individual high-velocity jet streams, which cause early ignition close to its outlet and do not allow temperature peaks to form in the flame. Avoidance of a temperature peak prevents thermal nitrogen oxide formation, and the defined flow conditions produce reducing zones in the flame, where the nitrogen oxide already formed is partly reduced.

This has been confirmed by model analysis at the University of Bochum. **Fig. 6** shows the result of these computations in the form of the distribution of temperature, oxygen and

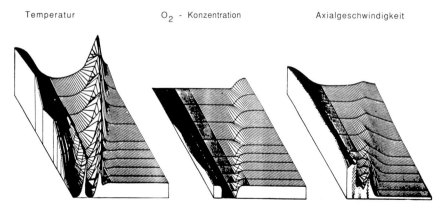

BILD 6: Verteilung von Temperatur, Sauerstoffkonzentration und Axialgeschwindigkeit im Strömungsgebiet eines PYRO-JET-Brenners
FIGURE 6: Distribution of temperature, oxygen concentration and axial velocity in the flow area of a PYRO-JET burner

Temperatur	= temperature
O_2-Konzentration	= O_2 concentration
Axialgeschwindigkeit	= axial velocity

TABELLE 3: Betriebsdaten und NO$_x$-Werte von Zementöfen vor und nach Einbau eines PYRO-JET-Brenners

Werk		A vor	A nach	B vor	B nach	C vor	C nach	D vor	D nach	E vor	E nach
Klinkerproduktion	t/d	2450	2450	1650	1700	1100	1250	1700	1700	2800	2800
Calcinator-Brennstoff	%	50	45	–	–	–	–	20	20	–	–
Ofenbrennstoff	t/h	–	–	8600	8900	6500	7300	7300	7300	16500	16500
Heizwert der Kohle	kJ	29300	29300	26000	26000	25500	25500	25500	25500	23000	23000
Primärluftanteil einschl. Förderluft	%	19	8	13	9	12	6	12	9	17	10
Petrolkoksanteil	%	20	50	–	–	–	–	–	–	–	–
Ofeneinlauftemperatur	°C	1150	1080	1200	1200	–	–	–	–	1200	1200
Sinterzonenlänge	m	21	18	–	–	24	21	–	–	–	–
NO$_x$-Gehalt	ppm	970	650	X	X – 30 %	650	460	900	730	X	X – 15 %
Litergew.	g/l	1350	1200	1350	1250	1500	1350	1450	1400	1230	1230
Freikalkgeh.	%	0,5–1,0	1,0–1,5	1,0–1,5	1,0–1,5	0,5	1,0	0,4	0,4–1,0	0,8–1,5	0,8–1,5

Verteilung von Temperatur, Sauerstoff und Axialgeschwindigkeit im Strömungsgebiet eines PYRO-JET-Brenners. Die Mündung des Brenners liegt bei dieser Darstellung jeweils im Achsenkreuz, d. h. links unten.

Auch Untersuchungen der Internationalen Flammenforschungsgemeinschaft (IFRF), Ijmuiden, haben in einem Vergleichstest verschiedener Versuchsbrenner für 2,5 MW gezeigt, daß der PYRO-JET-Brenner bei unterschiedlichen Kohlesorten und auch für Petrolkoks hervorragende Zünd- und Ausbrandeigenschaften aufwies und hohe Flammentemperaturen bei gleichzeitig niedrigsten NO$_x$-Werten erreichte [5].

An einer Vielzahl von Betriebsöfen durchgeführte Messungen liefern ebenfalls eindeutige Absenkungen der Stickoxid-Werte nach Einsatz des PYRO-JET-Brenners anstelle anderer Brennertypen. **Tabelle 3** zeigt die Absenkung der Stickoxid-Emissionen verschiedener Zementöfen nach Einbau eines PYRO-JET-Brenners. Zum Teil sind die absoluten Werte für den NO$_x$-Gehalt angegeben. Dabei wurde auf die Angabe des korrespondierenden Sauerstoffgehalts bewußt verzichtet, weil für den Vergleich der jeweils gleiche axial velocity within the flow area of a PYRO-JET burner. In each case the outlet of the burner lies the coordinate axes, i.e. at the bottom left-hand corner.

Investigations by the International Flame Research Association (the IFRF) of Ijmuiden, including a comparative test of different burners for 2.5 MW, revealed that the PYRO-JET burner had excellent ignition and burnout properties with different types of coal and also petroleum coke, and reached high flame temperatures with extremely low NO$_x$ levels [5].

Readings taken at many industrial kilns also indicate a definite drop in nitrogen oxide levels when the PYRO-JET burner was used instead of other types. **Table 3** shows the drop in nitrogen oxide emissions from various cement kilns following the installation of a PYRO-JET burner. The absolute values for the NO$_x$ content are given in some cases. The corresponding oxygen content is deliberately not given, because an identical oxygen content is used for the comparison. A clear drop in NO$_x$ emissions of up to 30% or more was obtained in all cases.

However it should be noted that the mode of operation of the kiln has a decisive effect on the absolute level of NO$_x$ emis-

TABLE 3: Operating data and NO$_x$ levels at cement kilns before and after installation of a PYRO-JET burner

Works		A before	A after	B before	B after	C before	C after	D before	D after	E before	E after
Clinker production	t/d	2450	2450	1650	1700	1100	1250	1700	1700	2800	2800
Calciner fuel	%	50	45	–	–	–	–	20	20	–	–
Kiln fuel	t/h	–	–	8600	8900	6500	7300	7300	7300	16500	16500
Calorific value of coal	kJ	29300	29300	26000	26000	25500	25500	25500	25500	23000	23000
Proportion of primary air incl. transport air	%	19	8	13	9	12	6	12	9	17	10
Proportion of petroleum coke	%	20	50	–	–	–	–	–	–	–	–
Kiln inlet temperature	°C	1150	1080	1200	1200	–	–	–	–	1200	1200
Length of sintering zone	m	21	18	–	–	24	21	–	–	–	–
NO$_x$ content	ppm	970	650	X	X – 30 %	650	460	900	730	X	X – 15 %
Litre weight	g/l	1350	1200	1350	1250	1500	1350	1450	1400	1230	1230
Free lime content	%	0.5–1.0	1.0–1.5	1.0–1.5	1.0–1.5	0.5	1.0	0.4	0.4–1.0	0.8–1.5	0.8–1.5

Sauerstoffgehalt herangezogen wurde. In allen Fällen konnte eine deutliche Senkung der NO_x-Emission bis zu 30% oder darüber erzielt werden.

Hierzu ist jedoch anzumerken, daß die Betriebsweise der Öfen einen erheblichen Einfluß auf die absolute Höhe der NO_x-Emission ausübt. Die entscheidenden Parameter, wie Luftüberschuß, Sekundärlufttemperatur und Kohlequalität müssen in jedem Fall bei der Beurteilung solcher Ergebnisse mit herangezogen werden.

6. Zusammenfassung

10jährige Betriebserfahrung mit PYRO-JET-Brennern an Drehöfen bis zu größten Leistungen haben zu einer gesicherten Auslegung und verläßlichen Konstruktion dieses Brenners geführt. Der PYRO-JET-Brenner nutzt das Prinzip einer Vielzahl von Einzelstrahlen mit hoher Geschwindigkeit und kommt deshalb mit wesentlich weniger Primärluft aus als herkömmliche Brennertypen. Dadurch lassen sich bis zu 160 kJ/kg Klinker an Brennstoffwärme einsparen.

In Verbindung mit den Jetstrahlen entsteht eine straffe Flamme mit gleichmäßiger Wärmedarbietung.

Besondere Maßnahmen zur Annäherung der Zündfront an den Brenner führten dazu, daß sich der PYRO-JET-Brenner zur deutlichen Senkung der Stickoxid-Emission nutzen läßt.

Aufgrund seiner hervorragenden Verbrennungseigenschaften eignet sich der Brenner besonders für schwierige feste Brennstoffe mit geringen Anteilen flüchtiger Bestandteile, wie z.B. Anthrazit oder Petrolkoks sowie auch zur Mischfeuerung mit Öl und Gas.

sions. The decisive parameters such as excess air, secondary air temperature and coal quality have to be considered in each case when judging such results.

6. Summary

Ten years' experience in working with PYRO-JET burners in rotary kilns of up to the highest capacity has led to the development of an established and dependable design. The PYRO-JET burner uses the principle of a number of individual high-velocity jet streams and can therefore operate with considerably less primary air than conventional types of burner. A saving of up to 160 kJ/kg of clinker can thus be made in fuel heat.

The jet streams form a fierce flame with uniform heat development.

As a result of special measures to bring the ignition front close to the burner, the PYRO-JET can be used to obtain a significant reduction in nitrogen oxide emissions.

Due to its excellent burning characteristics the burner is especially suitable for difficult solid fuels with lower proportions of volatiles, such as anthracite or petroleum coke, as well as for mixed firing with oil and gas.

Literatur

[1] Steinbiß, E.: Mehrstrahlbrenner für die besonderen Anforderungen bei Kohlenstaubfeuerungen. Zement-Kalk-Gips 5 (1982), S. 250–252.

[2] Herchenbach, H.: Hochdruck-Gasbrenner für Drehöfen. Zement-Kalk-Gips 10 (1973), S. 491–496.

[3] Bauer, C.: PYRO-JET burners to reduce NO_x emissions-current developments and practical experience. World Cement 4 (1990), S. 118–124.

[4] Schraemli, W.: Experiences with measures to reduce SO_2 and NO_x emissions from cement kilns. PCA Seminar, Dallas, USA, Sept. 1990.

[5] Forschungsbericht der IFRF, Ijmuiden (NL), über CEMFLAME 1, April 1992.

Betriebsergebnisse von Pendelrostkühlern mit horizontaler Anströmung des Klinkers

Operational experience with pendulum clinker coolers with horizontal air jets

Résultats obtenus en exploitation avec des refroidisseurs à grille pendulaire avec soufflage horizontal du clinker

Resultados obtenidos durante el servicio con enfriadores de parrilla pendulares con chorros de aire horizontales

Von **K. von Wedel**, Neustadt/Deutschland

Zusammenfassung – In Pendelkühlern für Zementklinker werden Rostplatten eingesetzt, die die Kühlluft als horizontal gerichtete Luftstrahlen mit hoher Geschwindigkeit in das Klinkerbett einleiten. Der erste Pendelkühler war die Modifikation eines Standard-Rostkühlers. Er wurde im März 1989 in Betrieb genommen. Dieser Kühler demonstrierte, daß sowohl horizontale Luftstrahlen als auch die Pendelaufhängung des beweglichen Teils der Konstruktion entscheidende Schritte zur Verbesserung der Kühlung und zur Minimierung des Verschleißes darstellen. Zur Zeit sind mehr als 12 Pendelkühler in Betrieb. Um das Potential dieses neuen Kühlertyps auszuschöpfen, wurden die Parameter Luftanströmung, Betthöhe und spezifische Klinkerleistung verändert.

Summary – Pendulum coolers for cement clinker use grate plates which introduce the cooling air into the clinker bed at high velocity as horizontal air jets. The first pendulum cooler was a modification of a standard type grate cooler. It was put into operation in March 1989 and demonstrated that the horizontal air jets and the pendulum suspension of the mobile frame are both decisive steps towards increasing the cooling efficiency and minimizing the wear. Today more than 12 pendulum coolers are in operation. The air flows, bed depths and specific clinker loadings were varied to make the most of the potential of this new type of cooler.

Résumé – Dans les refroidisseurs de clinker à ciment pendulaires sont utilisées des plaques de grille conduisant, à haute vitesse, l'air de refroidissement sous forme de jets orientés horizontalement dans le lit de clinker. Le premier refroidisseur pendulaire était une modification d'un refroidisseur de clinker conventionnel. Il a été mis en service en Mars 1989. Ce refroidisseur a fourni la démonstration, que des jets d'air horizontaux aussi bien qu'une suspension pendulaire de la partie mobile de la construction constituaient des pas décisifs pour l'amélioration du refroidissement et pour réduire l'usure au minimum. Plus de 12 refroidisseurs pendulaires sont actuellement en service. A fin d'exploiter au maximum le potentiel de ce nouveau type de refroidisseur, les paramètres réglés sont le soufflage d'air, la hauteur du lit et le débit spécifique de clinker.

Resumen – En los enfriadores pendulares para clínker de cemento se emplean placas de parrilla que introducen el aire de enfriamiento con gran velocidad en el lecho de clínker, en forma de chorros de aire horizontales. El primer enfriador pendular fue la versión modificada de un enfriador de parrilla standard. Se puso en servicio en marzo de 1989. Este enfriador demostró que tanto los chorros de aire horizontales como la suspensión pendular de la parte móvil de la estructura constituyen un paso decisivo para mejorar el enfriamiento y reducir el desgaste a un mínimo. Actualmente, se encuentran en servicio más de 12 enfriadores pendulares. Con el fin de aprovechar al máximo el potencial de este nuevo tipo de enfriador, se variaron los siguientes parámetros: el soplado de aire, la altura del lecho y el caudal específico de clínker.

1. Rostverschleiß und Rostwiderstand

Pendelroste mit horizontaler Anströmung des Klinkers sind Widerstandsroste [1]. Ein günstiger Widerstand gegen den Durchtritt der Kühlluft wird durch 2,5 % offene Rostfläche erreicht [2]. Diese günstige offene Rostfläche darf sich durch Verschleiß nicht vergrößern. An Standardrosten wurden Verschleißraten beobachtet, die einem jährlichen Zuwachs an offener Rostfläche von 4 % entsprechen. Eine solche Verschleißrate ist mit dem Widerstandskonzept nicht vereinbar.

Die Betriebsergebnisse lassen erkennen, daß IKN-Roste die wesentlichen Ursachen von Roststillständen ausschließen.

1. Grate wear and grate resistance

Pendulum coolers with horizontal aeration of the clinker use high resistance grates [1]. A favourable resistance against the passing of cooling air is achieved with 2.5 % of open grate area [2]. This favourable open grate area must not increase because of wear. On standard grates wear rates have been observed equal to an annual increment of 4 % of open grate area. Such a wear rate is contradictory to the resistance concept.

The operating results show that IKN Coanda Nozzles eliminate the main reasons for cooler down times. The chart in

Ursachen von Verschleiß an Standard-Rostkühlern	IKN Lösungen
verbrannte Platten	gleichmäßige Luftverteilung
Spannungsrisse in Rostplatten	bereits unterteilte Rostelemente
Sandstrahleffekts	horizontale Belüftung
Klinkertransport	Bimetallische Rostelemente
Absinken des Schwingrahmens	verschleißfreie Pendelaufhängung

BILD 1: Tabelle Verschleißursachen und IKN-Lösungen

Die Tabelle in **Bild 1** zeigt die wesentlichen Verschleißursachen an Standardrosten und die IKN-spezifischen Lösungen:

1. Es gibt bei IKN-Rosten keine verbrannten Rostplatten, weil der hohe Rostwiderstand der Coanda-Düsen die kühlende Luft gleichmäßig verteilt.
2. Die standardmäßig einen Quadratfuß großen Rostplatten zeigen bei Erwärmung Spannungen und diagonale Risse. Der IKN-Rost ist aus schmalen, mehrteiligen Coanda-Düsen aufgebaut, die sich auch bei Erwärmung nicht verwerfen.
3. Es gibt keine Sandstrahleffekte rings um die Rostplatte, weil die horizontale Anströmung nach dem Coanda-Effekt den fluidierbaren Feinanteil an die Bettoberfläche treibt [2].
4. Die erreichten Standzeiten von bis zu 8 Jahren und mehreren Millionen Tonnen Klinker zeigen, daß der verbleibende Verschleiß durch den Vorschub des groben Klinkers vernachlässigbar ist. An einem Rost mit hohem Rostbett und niedrig legiertem Rostbelag führte eine Vertiefung nach jeder schiebenden Rostreihe nach 2 Jahren zu einem Wechsel. Hier haben wir Rostelemente (**Bild 2**) eingesetzt, deren Kästen mit harten, auswechselbaren Lamellen bestückt sind.
5. Bei Einsatz von Rostkästen mit geschlossenen Böden kommt es zu Verschleiß an der Unterseite der Rostelemente. Die verschleißfreie Aufhängung des Schwingrahmens nach Art eines Pendels **Bild 3** schaltet diese Verschleißursache aus. Wir haben es abgelehnt, bewegliche Rostreihen ohne Pendelaufhängung zu liefern.

Die Pendelaufhängung ist darüber hinaus selbstzentrierend, so daß die festen Seitenborde des Rostes nicht berührt werden. So konnte die Standzeit der Seitenborde der des Rostbelages angepaßt werden.

Der erste komplette IKN-Rost ist seit Mai 1991 bei Adelaide Brighton Cement Co. in Südaustralien in Betrieb. Die Leistung beträgt 4000 t/d. Nach anfänglichen Schäden und Nachbesserungen arbeitet dieser Rost äußerst verschleiß-

Reasons for Wear of Standard Grate Coolers	IKN solutions
burnt plates	cooling by air distribution
cracked plates	grate elements 'cracked' already
sandblasting effects	horizontal aeration
clinker transport	bimetallic elements
sinking of mobile frame	pendulum suspension

FIGURE 1: Solution of the Wear Problems

Fig. 1 shows the most important reasons for wear of standard grates and IKN's special solutions:

1. IKN coolers don't have burnt grate plates because the high resistance of the coanda nozzles distributes the cooling air evenly.
2. The standard grate plate with 1 square foot grate area develops internal thermal expansion while heating up which often ends up in diagonal cracks. The IKN grate is made of small coanda nozzles, consisting of multiple parts, which keep without tension when getting hot.
3. There are no sand blasting effects around the coanda nozzle because the horizontal aeration, based on the Coanda effect, sweeps the fluidizable fines to the bed surface.
4. The achieved lifetime of up to 8 years and several million tons of clinker demonstrate that the remaining wear caused by the transport of the coarse clinker is neglectable. An existing grate with a deep clinker bed and a base alloy grate surface showed cavities in each mobile grate row which caused a replacement of grate plates after two years. In this case we used grate elements (**Fig. 2**) consisting of a box with replaceable hard blades.
5. While using boxes as grate elements, their closed bottom was exposed to wear from down side. The wearless suspension of the mobile frame as a pendulum (**Fig. 3**) eliminated this effect. We refused to supply mobile rows without pendulum suspension.

Furthermore the pendulum suspension is self-centering, so that the fixed side castings of the grate are not in touch with the mobile rows. Thus the lifetime of the side castings can be extended same as the lifetime of the grate plates. The first complete IKN Pendulum Cooler is in operation since early 1991 at Adelaide Brighton Cement Co. in South Australia. The capacity is 4000 t/d. After some initial problems, damages and modifications in the early stages, the cooler is running now with extreme little wear [3]. The company is now working to have scheduled kiln stops every two years only.

As we have solved the wear problems we guarantee the pro-

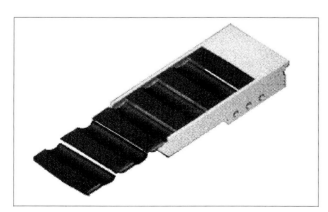

BILD 2: Bimetallische Rostelemente
FIGURE 2: Bi-metallic Grate Element

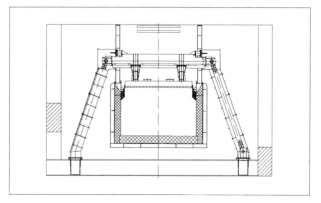

BILD 3: Typischer Querschnitt der Pendelaufhängung
FIGURE 3: Typical Cross Section of Pendulum Suspension

arm [3]. Das Werk arbeitet jetzt darauf hin, Stillstände der Ofenanlage nur noch alle 2 Jahre einzuplanen.

Nach Lösung des Verschleißproblems garantieren wir die verfahrenstechnischen Daten für das Ende der 2jährigen Gewährleistungsfrist des Rostes. Wir möchten damit der Tendenz der Betreiber entgegenwirken, die besseren Leistungsdaten des IKN-Rostes durch Zuschläge zu verwischen, zum Beispiel für künftigen Verschleiß oder zur allgemeinen Sicherheit.

2. Betriebsergebnisse der 3000-t/d-Rostanlage bei Goliath Portland Cement, Tasmanien, Australien

Zum Zeitpunkt dieses Berichts lieferte der 3000-t/d-Kühler der Goliath Portland Cement Co. die neuesten Betriebsergebnisse. In der neutralen Zone zwischen Abluft und Sekundärluft ist der Kühler mit einer sogenannten Stauwand ausgestattet (**Bild 4**). Sie ist luftgekühlt, hydraulisch schwenkbar und reicht bis auf das Rostbett. Der Wandfuß wird mit einer Farbkamera beobachtet.

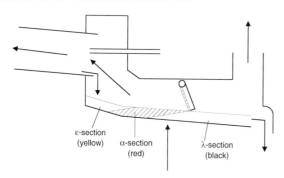

BILD 4: Schema der Stauwand
FIGURE 4: Arrangement of Partition Wall
ε-Zone (gelb) = ε-section (yellow)
α-Zone (rot)   = α-section (red)
λ-Zone (schwarz) = λ-section (black)

In die Kühlerbilanz **Bild 5** sind die Mittelwerte des Abnahmeversuchs über 72 Stunden eingegangen. Als gesichert sind zunächst die gemessenen Temperaturen anzusehen. Die Temperatur der Tertiärluft wird in der Rohrleitung nach einer rückläufig über dem Kühler angeordneten Staubabsetzkammer von KHD gemessen. Die Temperatur der Abluft wird durch die heißen Stäube dieser Absetzkammer erhöht. Andernfalls wäre die Rekuperation noch höher. Die Temperatur des Kaltklinkers wird von einem Infrarotgerät auf dem Plattenband von AUMUND unmittelbar nach der Aufgabe angezeigt. Die Kühlluftmenge wurde von uns anhand der aus Deutschland gelieferten und geeichten Einlaufmeßdüsen der Ventilatoren zu 1.44 Nm³/kg Klinker bestimmt.

Die Bilanzwerte (Bild 5) haben wir unter Annahme von heißerem und kälterem ofenfallenden Klinker variiert. Auch haben wir weniger Kohleverbrauch und weniger Verbrennungsluft angenommen. Keine der Variationen führte zu mehr als 1.44 Nm³/kg an Kühlluft und zu weniger als 75 % Rekuperation. Der geringe Kühlluftverbrauch als die entscheidende Kennzahl für den Wirkungsgrad bestätigt Angaben, die auch schon der Dyckerhoff AG für die Werke Amöneburg und Göllheim gemacht wurden. Die Rekuperation dürfte bisher nicht erreicht worden sein.

Der Beitrag der Stauwand Bild 4 zu der Rekuperation soll noch theoretisch angesprochen werden. Erwartet haben wir zunächst, daß sich die Rekuperation gewissermaßen auf Knopfdruck durch Ein- und Ausschwenken verändern ließe. Dem war nicht so. Zu Versuchen über mehrere Stunden mit und ohne Stauwand ist es bis zu diesem Bericht nicht gekommen.

Mit der Stauwand läßt sich die Betthöhe im Bereich kräftiger Belüftung genau messen. Im Vergleich zu deren Meßwert von 600 mm Betthöhe überraschte die hohe Schubzahl. Bei der durch Versuch an gebrochenem Klinker bestätigten

cess data and the heat recuperation of the cooler to be verified at the end of the 2 years warrantee period.

2. Operating results with a 3000 t/d Pendulum Cooler at Goliath Portland Cement Co., Tasmania, Australia

When this paper was written the 3000 t/d Pendulum Cooler at Goliath Cement Co. delivered the latest operation data. The cooler is equipped with a so called partition wall, hanging in the neutral zone where exhaust air seperates from secondary air (**Fig. 4**). The partition wall is air cooled and can be hydraulically driven into its working position close above the clinker bed. The gap between the wall and the clinker bed is observed by a colour video camera.

The cooler balance (**Fig. 5**) shows average data from the 72 hours performance test. The temperatures are reliable so far. The temperature of the tertiary air is measured in an air duct located after the dust settling chamber of KHD which is installed reversed to the cooler. The vent air temperature is increased by the hot clinker dust out of this settling

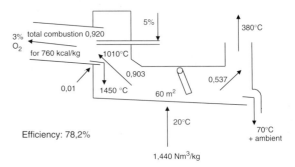

BILD 5: Abnahmebilanz bei Goliath Cement
FIGURE 5: Measured Cooler Data at Goliath Cement
Verbrennungsluft = combustion air
für 760 kcal/kg = for 760 kcal/kg
Wirkungsgrad   = efficiency
Umgebung   = ambient

chamber, otherwise the recuperation would be much higher. The temperature of the cold clinker is indicated by a pyrometer above the AUMUND conveyor close to the clinker discharge. The cooling air volume is measured by calibrated inlet flow nozzles from Germany which are attached to the fans. The total cooling air volume is 1.44 Nm³/kg clinker.

We have calculated the cooler balance for various data sets, such as hotter and colder clinker inlet temperature, more and less fuel consumption and more and less combustion air. None of the variations showed more than 1.44 Nm³/kg of cooling air and less than 75 % recuperation. The small cooling air consumption is indication for an excellent heat recuperation and is verified by results from Dyckerhoff AG in the Amöneburg and Göllheim plants. The heat recuperation at Goliath is the best achieved so far.

The contribution of the partition wall (figure 4) will be shortly discussed here: The recuperation could not be varied by simply pressing the button to drive the wall into a different working position. Trials over several hours with the wall lifted and the wall down were not carried out until the writing of this report. With the partition wall the bed depth can be measured easily and exactly. The 600 mm bed depth did not fit to the actual grate speed. With the verified density of the bulk clinker after the crusher at 1.5 t/m³, the grate width of 2.7 m, 100 mm stroke length and the bed depth of 600 mm the theoretically grate speed is calculated to 8.65 strokes/min. Actually 13.5 spm are necessary. 10 % slippage was found in other installations. We explain this grate speed with a significant lower density of the bulk clinker caused by the expansion of the clinker bed. The partial fluidisation of the fines expands the bed. A clinker bed with coarse particles contains 50 % pore volume in which additional mass, here fine clinker particles, can be stored without increasing the total volume. Such a clinker bed could be expanded to a

$$\text{Wärmeleitung (Heat conduction):} \quad \dot{Q} = \lambda \cdot A \cdot d\vartheta/dx$$

$$\text{Konvektion (Convection and heat transfer):} \quad \dot{Q} = \alpha \cdot A \cdot \Delta\vartheta_m$$

$$\text{Wärmestrahlung (Radiative heat transfer):} \quad \dot{Q} = \varepsilon \cdot A \cdot \sigma \cdot (T_1^4 - T_2^4)$$

BILD 6: Formeln der Wärmeübertragung
FIGURE 6: Formulas of heat transfer

BILD 7: IKN-Firmen-Logo
FIGURE 7: IKN company logo

Schüttdichte von 1,5 t/m³, der Rostbreite von 2,7 m, 100 mm Hub und der Betthöhe von 600 mm ergab sich die theoretische Schubzahl von 8,65 Hüben/min. Es werden jedoch 13,5 Hübe/min benötigt. Aufgrund von Beobachtungen an anderer Stelle nehmen wir 10 % Schlupf an.

Wir erklären uns die Schubzahl mit wesentlich geringerer Schüttdichte aufgrund der Expansion des Bettes durch die partielle Fluidierung des Feinanteils. Ein Bett aus Grobkorn kann bis zu 50 % Zwickelvolumen aufweisen, in das zusätzliche Masse als Feinkorn unterbracht werden kann, ohne daß sich der Raumbedarf vergrößert. Dieses Klinkerbett ließe sich durch Ausblasen des Feinanteils auf eine geringere Dichte expandieren. Nach 10 % Schlupf ergibt die Schubzahl im Fall Goliath, daß das Bett infolge der Belüftung auf die Dichte 1,069 expandiert ist und der Feinanteil zur Oberfläche ausgetragen wurde.

3. Arten der Wärmeübertragung bei der Klinkerkühlung

Die Stauwand ist das Ergebnis theoretischer Überlegungen zur optimalen Betthöhe. Wärmetechnisch wird ein zu hohes Bett dadurch angezeigt, daß die Luft sich nicht weiter aufheizt, sondern an der durch Strahlung erkalteten Bettoberfläche rückgekühlt wird. Manchmal kann man die Abbruchkante eines Rostbettes beobachten, das im Innern noch glüht während die Oberfläche schwarz ist. Noch deutlicher wird die voreilende Kühlung der Bettoberfläche an Luftdurchbrüchen bei zu hoch eingestelltem Klinkerbett: Hier wird hellgelber Klinker an die schon dunkelrote Bettoberfläche gefördert. Es hat auch Schäden an der Ausmauerung und am Brenner durch zu schwach belüftete Betten am Einlauf des Rostes gegeben. Auch wir waren zu Beginn unserer Entwicklung der optimistischen Ansicht, man müsse nur die Luftbeaufschlagung weit genug zurücknehmen, um die Temperatur der Sekundärluft an die des Klinkers heranzuführen. Die voreilende Kühlung der Bettoberfläche durch Strahlung hat diese Erwartung zunächst zunichte gemacht. Der Erfolg der Stauwand bestätigt diese Überlegungen.

Bei der Klinkerkühlung wird Wärme durch Strahlung, Konvektion und Leitung übertragen, wobei Konvektion als Sonderfall der Leitung für ein strömendes und ein festes Medium gilt. **Bild 6** zeigt die entsprechenden Formeln mit den griechischen Buchstaben für die Stoffwerte [4]. Firmenintern sprechen wir von der Epsilon-, Alpha- und Lambdazone und meinen damit die Strahlungs-, die Konvektions- und die Nachkühlzone des Kühlers. Die entsprechenden Klinkerfarben gold/rot/schwarz haben wir in unser Firmenlogo aufgenommen (**Bild 7**). In der Epsilonzone strahlt der Klinker, ohne die Luft aufzuheizen, da diese keine Strahlung absorbiert. In der Alphazone findet die Aufheizung der Sekundärluft am Klinkerkorn durch Konvektion statt. In der Lambdazone kühlt der Klinker durch Wärmeleitung. Mit unserer Kühlerentwicklung sind wir an der Epsilonzone angelangt, ohne diese zu beherrschen. Die Rolle der Strahlung an der Schnittstelle zwischen Drehofen, Brenner und Kühler ist bisher wenig definiert.

lower density by sweeping the fines to the bed surface. With 10 % slippage the grate speed in Goliath indicates that the clinker bed has expanded to a density of 1.069 t/m³ caused by aeration and that the fines were swept to the surface.

3. Process of heat transfer of the clinker cooling

The partition wall is the result of theoretical reflections about the optimum bed depth. In terms of heat transfer a clinker bed which is too high, would be indicated by air which stops heating up during its way through the bed and is then cooled down while passing through the bed surface. The surface is always colder than the clinker layer below because of radiation losses. Sometimes this can be observed at a shear zone of a clinker bed where the pit is still red hot while the surface is black. Better still the faster cooling of the surface is indicated by air channelling caused by a too high clinker bed: light yellow clinker pieces from the layer below are spattering out of the air channel to the dark red surface. Damages have occurred to the refractory and at the burner caused by too less aeration of the clinker inlet area. At the initial stage of our development we had the optimistic opinion that to increase the secondary air temperature closer to the clinker temperature we just need to decrease the aeration. However due to cooling of the bed surface by radiation this proved contrary. The success of the partition wall verifies our reflections.

During the clinker cooling heat is transferred through radiation, convection and conduction, whereas convection is a special case of conduction of a flow and a fixed medium. **Fig. 6** shows the corresponding formulas with the greek letters of the constants [4]. Internally we are talking about the Epsilon-, Alpha and Lambda-Section which means radiation section, the convection section and the retention section. We composed the corresponding colours of the clinker, gold/red/black, in our company logo (**Fig. 7**). In the epsilon zone the radiation of the clinker does not heat up the air because it does not absorb radiation. In the Alpha zone the heat transfer to the secondary air is caused by convection at the clinker particle. In the Lambda zone the clinker is cooled by conduction. With our cooler development we have reached the epsilon zone without being able to control it properly yet. The rule of radiation at the intersection of kiln, burner and clinker cooler is up to now little defined.

Literature

[1] Wagner, R., und von Wedel, K.: Sind Kühlroste Klinkerkühler oder Wärmerekuperatoren? Zement-Kalk-Gips 5 (1984), S. 244.

[2] von Wedel, K.: Pendelrostkühler mit horizontaler Anströmung des Klinkers. Zement-Kalk-Gips 4 (1992), S. 171–176.

[3] Manias, C. G.: A New Generation IKN Clinker Cooler – A Reflection on the First Year of Operation. Portland Cement Association of Canada, Technical Conference, Toronto, September 1992.

[4] VDI Wärmeatlas, 1984.

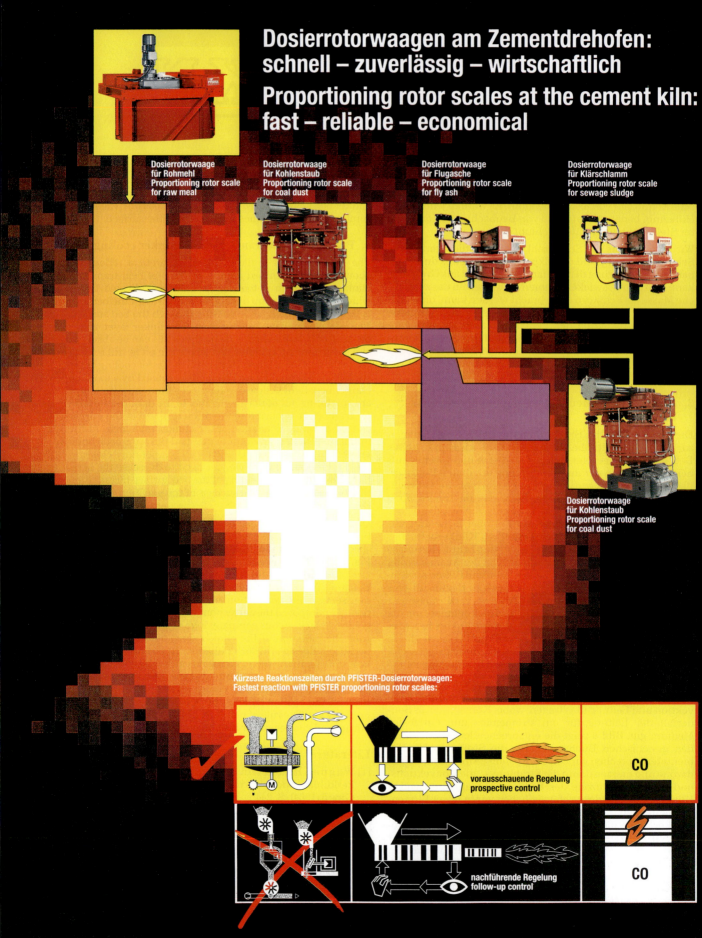

ACT-PASEC® Precalcining System

AUSTRIAN CEMENT TECHNOLOGY GmbH

YOUR PARTNER FOR MODERN CEMENT PLANTS

"Conversion of an existing 4-stage conventional preheater to a 4-stage ACT - PASEC® precalcining system"

PASEC® - THE BASIS FOR HIGH QUALITY CLINKER PRODUCTION

The outstanding advantages of the ACT-PASEC® System are as follows:

- 690 kcal/kg clinker (5-stage type) - the lowest heat consumption world-wide
- Optimum heat utilization - exhaust gas temperature lower than 260° C (5-stage type)
- Low electric power consumption
- Cyclones with low pressure loss
- Low grade fuels applicable for calciner
- Drying of raw materials with preheater exhaust and cooler excess air up to 10% moisture content
- Optimum burning conditions for lowest emission rates
- High degree of automation for reduced personnel costs
- High degree of operating reliability - proven plant availability >95%

These advantages especially apply for the modernization of existing plants!

EXPERIENCE IS A GREAT ASSET!

Assistance of ACT experts is available to our clients for every phase of the project:

- Raw Material Investigations
- Economical Assessment
- Financing Assistance
- Thoroughly Engineered Plant Concepts
- Expert Project Management
- Start-up and Operational Assistance
- Comprehensive Training Service

AUSTRIAN CEMENT TECHNOLOGY GmbH
Reuchlinstr. 6   A-4020 Linz/Austria
P. O. Box 56, A-4024 Linz

Phone: **43/732/60 42 20
Fax : ** 43/732/66 98 88

PATO AG
Aegeristrasse 24, P. O. Box 415, 6300 Zug/Switzerland
Tel. (42) 223 335, Fax (42) 217 620

Ihr Partner in:
- Weltweitem „after sales service" (sämtliches Material) für Zementwerke, Elektrizitätswerke und andere Industrieanlagen.

  Wir arbeiten nach drei Grundsätzen:
  - Die Sicherheit ausgeglichener hoher Qualität.
  - Den Vorteil kurzfristiger Auslieferung; JUST IN TIME
  - Den Nutzen kompetenter Unterstützung.

Your partner in:
- Worldwide „after sales service" (the whole range of material) for cement works, industrial plants and electric power stations etc.

  We work according to three principles:
  - The security of a well-balanced high quality.
  - The advantage of short-term delivery, JUST IN TIME.
  - The benefit of qualified support.

Ser su Cocio para:
- Servicio post ventas a nivel mundial para plantas de cemento, instalaciones industriales, plantas generadores de energía eléctrica, etc.

  Trabajamos de acuerdo a los siguientes tres principios:
  - La seguridad de tener una alta y constante calidad.
  - La ventaja de entregas a corto plaza.
  - El beneficio del soporte calificado y constante.

Industrie- und Hochbau
Feuerfest- und Schornsteinbau
Schlüsselfertigbau
Siloanlagenbau

Niederlassungen
Berlin
Braunschweig
Köln
Magdeburg
Mannheim

Geschäftsbereich
Feuerfest- und Schornsteinbau
Gocher Straße 15
50733 Köln
Tel. 02 21. 77 56-0
Fax 02 21. 7 75 61 01

Feuerfest- und Schornsteinbau

Lucks+Co zählt seit seiner Gründung zu den führenden deutschen Spezialbau-Unternehmen im Feuerfest- und Schornsteinbau.

Wir konstruieren und erstellen feuerfeste Auskleidungen für die Kalk-, Gips- und Zementindustrie:

- Mehrschachtkammeröfen
- Ringschachtöfen
- Gleichstromschachtöfen
- Schachtöfen konventioneller Bauart
- Drehrohröfen
- Wärmetauscher
- Klinkerkühler
- Brennkammern jeglicher Art

If the grate cooler has you in a sweat, let us recover your cool...

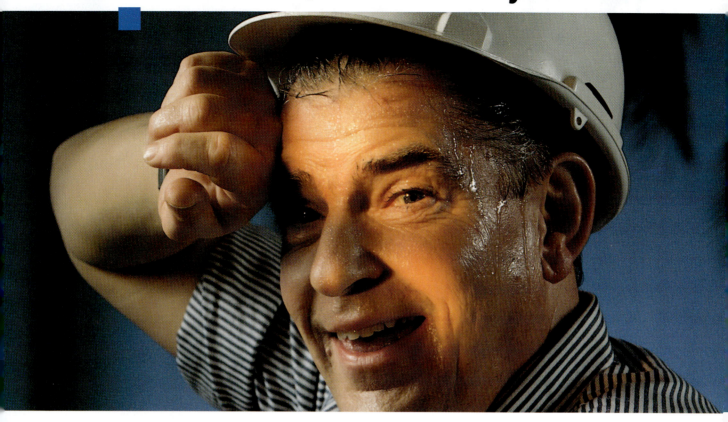

...with the FLS Coolax CFG retrofit

In addition to the operating advantages, there are compelling economic considerations like heat savings of 30-83 kcal/kg clinker, a yearly capacity increase typically in the neighbourhood of 5% and a payback time of less than 1.5 years.

With the CFG system - movable grates from the very start of the cooler - and reduced maintenance requirements, our CFG system offers some evident operational benefits, including:

- Reduced dust circulation
- Reduced wear on grate plates
- Reduced air consumption
- Reduced tendency to snowman and red river formation
- More stable kiln and cooler operation

Do not compromise by modifying less than the complete first grate if you want to solve your cooler problems once and for all.

In our CFG system there is no formation of ugly snowmen and no need for airblasters which will have a negative effect on the operation.

Installed in less than 3 weeks!

You can have our Coolax CFG retrofit installed and ready to go during one of your regularly scheduled plant shutdowns.

So if you, like us, believe it makes sense to consider upgrading now, ask for further details from the:

FLS Cooler Systems Department

From upgrades to complete plants.
FLS know-how: Nearly a third of the world's cement production comes from plants and machinery provided by F.L.Smidth.

F.L.Smidth & Co. A/S. Vigerslev Allé 77. DK-2500 Valby-Copenhagen, Denmark. Tel. + 45 3618 1000. Fax + 45 3117 4724

Wülfrath SINDOFORM.
Rapid Forming Stable Coating, High Durability Lining.

Tailor-made lining designs are required for each zone, in every rotary kiln. The vast range of refractory products available from the Wülfrath Group enables optimal selection for all requirements. For the actual sintering zone we supply our highly efficient dolomite refractory Wülfrath SINDOFORM to meet the special demands in this region.

The main advantages are:

- Rapid forming stable coating, due to chemical compatibility with the cement clinker.
- High thermo-elasticity from the addition of Zirconium Dioxide.
- A chrome free refractory assuring no toxic waste problems.

All in all:

With Wülfrath SINDOFORM, low cost long campaigns are possible. High kiln availability will give you the competitive edge.

And do not forget, the exclusive Wülfrath lining design and on site installation service is available to you.

Interested? We will be happy to supply detailed information to suit your requirements.

Wülfrath Refractories/Ceramics Group

In Zukunft Wülfrather

Dolomitwerke GmbH, Wilhelmstrasse 77, D-42489 Wülfrath, Telephone: (20 58) 17-0, Fax (20 58) 17 22 10, Telex: 8 592 082 dolo d
Wulfrath Refractories (U. K.) Ltd., 24 Lambourne Crescent, Cardiff Business Park, Llanishen · Cardiff CF4 5 GG, Telephone: (02 22) 76 57 57, Fax: (02 22) 76 13 42, Telex: 4 97 799 w
Wulfrath Refractories Inc., Tarentum/PA 15084 USA, Tel.: 4 12-2 24-88 00, Fax: 4 12-2 24-33 53

Single Source for Materials Handling in Cement Plants

Claudius Peters has been designing and building tailor-made plants for the cement industry for more than 80 years. Our clinker coolers, silos and pneumatic conveying systems have become synonymous with progress, reliability and quality.

Claudius Peters Cement represents the combined products and technologies within BMH, of Siwertell, PHB Someral, Pacpal and PETERS systems, which can be integrated to meet specific requirements for new cement plants and modifications of existing lines.

Our system capabilities in the Cement sector include:

- raw materials mixing beds and transport systems
- homogenising of raw meal
- feeding of pre-heaters
- coal mixing beds and preparation
- clinker cooling
- cement storage and transport
- packers and palletizers
- ship loaders and unloaders

BMH Division

Babcock Materials Handling
Division GmbH
Bahnhofstraße 24
D-21614 Buxtehude
Phone: 04161/706-10
Fax: 04161/70 62 20

Brand names of BMH in the Cement sector.

We have the experience to engineer your special technology process

Our wide experience covers the whole range of

- white cement technology employing own patented processes,
- portland cement plants supply and optimizing, applying precalcining technology,
- conversion of wet process cement plants
- grinding technology for cement and raw materials.

BKMI, who are now a fully integrated division of the Babcock-BSH AG, employs a competent team of engineers and scientists ensuring that all design criteria will be solved to the full satisfaction of the customer. Intensive research work is applied to ensure a further development of all applied processes utilising feed back of experience from executed plants.

White cement plant with PRE-AXIAL preheater

Division BKMI, Parkstraße 10, P.O. Box 4/6, D-47829 Krefeld 11, Phone (0 21 51) 4 48-0, Telefax (0 21 51) 44 82 44

Top Performance
Chrome Ore-Free

environmentally safe refractory bricks for
the burning and transition zones of cement rotary kilns.

A new generation of refractory bricks
for use in sintering and transition zones of cement rotary kilns.
These bricks are chrome ore-free, and feature excellent
thermal shock resistance despite high MgO content.

- chrome ore-free – no formation of highly toxic, watersoluble alkali chromates
- chrome ore-free – resistance to thermo-chemical alkali attacks
- optimised structural flexibility – the result is an excellent thermal shock resistance
- thermal conductivity – comparable to magnesia chromite products

Almag 85

Magpure 95

Magpure 93

Thermo-chemical load	Lower transition zone	Central sintering zone	Upper transition zone
Normal	Magpure 93	Magpure 93	Magpure 93
High	Almag 85	Magpure 93	Almag 85
Extreme	Magpure 95	Magpure 93	Magpure 95

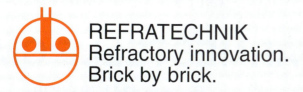

REFRATECHNIK
Refractory innovation.
Brick by brick.

REFRATECHNIK GmbH
Rudolf-Winkel-Straße 1
D-37079 Göttingen
Germany
Phone: (0) 551-69410
Telex: 96863 + 96811
Telefax: (0) 551-6941-104

BECKENBACH LIME SHAFT KILNS (PATENTS, PEND.)

**FOR STEEL-, SODA-, SUGAR-, ENVIRONMENTAL- AND BUILDING INDUSTRIES. ALREADY 607 BECKENBACH KILNS INCL. 305 ANNULAR SHAFT KILNS SOLD. NOBODY HAS EVER BUILT MORE.
IN 1993 ANNULAR SHAFT KILNS ARE STARTED UP WITH TWO UNITS EACH 500 TPD ONE UNIT 300 TPD, THREE UNITS EACH 200 TPD AND TWO UNITS EACH 150 TPD. BECKENBACH KILNS PRODUCE FIRST QUALITY LIME AND DOLOMITE. FUELS ACCORDING TO SPECIAL REQUIREMENTS. HIGHEST LONGTERM EFFICIENCY. LONGEST REFRACTORY LIFE. N E W: ANNULAR SHAFT KILNS
FOR SMALL SIZE AND/OR DENSELY BURNT LIME.**

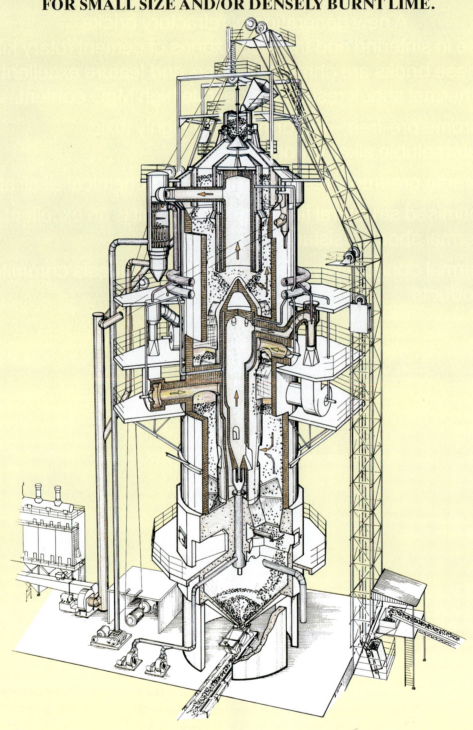

RINGSCHACHTÖFEN/ANNULAR SHAFT KILNS 70 – 900 TPD

BECKENBACH WÄRMESTELLE GMBH

Telefax 0211 4910077   Telefon 0211 4931019
P. O. Box/Postfach 32 12 25   D-40427 Düsseldorf Germany
Klever Straße 66   D-40477 Düsseldorf Germany

Horizontal Aeration by Coanda Effect!

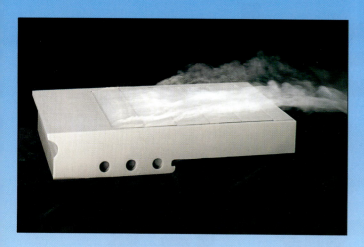

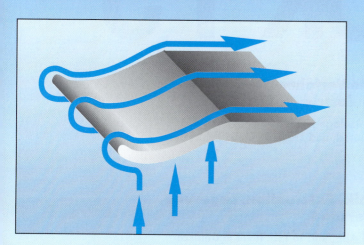

You want less:
- ❏ heat consumption
- ❏ air consumption
- ❏ electrical consumption
- ❏ dust recirculation back to the kiln
- ❏ maintenance costs
- ❏ clinker temperature

You want more
- ❏ heat recuperation
- ❏ production
- ❏ life of your grate plates

KIDS is simple, efficient and effective.
There is hardly any maintenance up to 9 years of operation and 7 Million tons of clinker. The best part is, KIDS takes only few days to install!

Can you think of any better return on your investment?

Let us analyze your cooler operating data and show you how you will benefit.

Join the 120 + IKN customers
in 34 countries who are enjoying the benefits.
For our clients it is no longer a dream.

Offices in:

US, Allentown Pa.
Telephone (1) 610 266-5441
Telefax   1) 610 266-9773

France, Paris
Telephone (33) 1 4963-2592
Telefax   (33) 1 4860-7261

IKN GmbH
P.O.Box 1121 · D-31519 Neustadt
Telephone (49) 50 32-895-0
Telefax   (49) 50 32-895-95
Telex   924 527

 BAUVERLAG

Dipl.-Ing. Walter H. Duda

cement data-book 1-3

Band 1: Internationale Verfahrenstechniken der Zementindustrie

3., neubearb. und erweiterte Auflage 1985. 656 Seiten mit 413 Abb. und 150 Tab. Texte in Deutsch und Englisch. Geb. DM 290,– öS 2.262,– / sFr 290,–
ISBN 3-7625-2137-9

Band 2: Elektrotechnik, Automation, Lagerung, Transport, Versand

1984. XII, 456 Seiten mit 286 Abb. und 16 Tab. Texte in Deutsch und Englisch. Geb. DM 240,– / öS 1.872,– / sFr 240,–
ISBN 3-7625-2042-9

Band 3: Rohmaterial für die Zementherstellung

1988. V, 188 Seiten mit 224 Abb. und Tab. Texte in Deutsch und Englisch. Geb. DM 148,– / öS 1.155,– / sFr 148,–
ISBN 3-7625-2286-3

Das dreibändige „Cement-Data-Book" ist eines der erfolgreichsten und umfassendsten Werke zum Thema Zementherstellung und Verarbeitung. In Band 1 werden die in der Zementindustrie international bekannten Verfahren und maschinellen Einrichtungen in Kurzdarstellungen beschrieben. In Band 2 werden solche Arbeitsbereiche der Zementindustrie behandelt, in denen in den letzten Jahren Neuerungen in der Anlagentechnik zu verzeichnen waren und die der Kostendämpfung in der Produktion und der Qualitätssicherung dienen. Band 3 befaßt sich mit den Zementrohmaterialien, deren Exploration, Gewinnung und Beförderung zum Zementwerk.

BAUVERLAG GMBH · D-65173 Wiesbaden

Great success!

Asia Cement successfully started operation of their PYRORAPID® cement kiln with a capacity in excess of 5000 tons per day. Even with the difficult raw materials, heat consumption of 3050 kJ/kg and electrical energy demand of 8.4 kWh/t are below the guaranteed values.

Short rotary kiln 5 m x 55 m ■ Capacity, guaranteed 4,800 t/d ■ Availability 350 d/a ■ Production 1.78 mill t/a

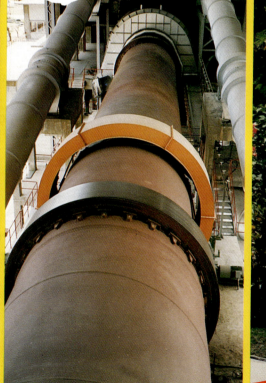

These achievements
are the result of a successful combination of
- PYRORAPID® short rotary kiln
- 4-stage cyclone preheater of the latest design
- PYROCLON-LowNOx®-calciner
- PYRO-JET® burner
- Grate cooler with maximum recuperation

HUMBOLDT WEDAG: the professional solution.

KHD Humboldt Wedag AG · D-51057 Cologne, Phone +221-822-6323, Fax +221-822-6627
HUMBOLDT WEDAG ZAB GmbH · Brauereistr. 13, D-06847 Dessau, Phone +340/7310-0, Fax +340/7310-676
Humboldt Wedag, Inc. · 3883 Steve Reynolds Blvd., Norcross, GA. 30093, USA
Phone +404 564 7300, Fax +404 564 7333
International sales offices: Beijing · Hongkong · Madrid · Melbourne · Mexico · Moscow · São Paulo · Shanghai

unitherm

NEU / **NEW**

M.A.S.
Mono Airduct System

Drehofenbrenner / *Rotary kiln burners*

Die logische Konsequenz zu herkömmlichen Brennersystemen mit zwei Primärluftkanälen
The logical conclusion from conventional burner systems with two primary air ducts

Rasche Amortisation durch Weiterverwendung der vorhandenen Brennerperipherie
Rapid repayment of investment through further use of burner periphery

- **Geringerer Primärluftbedarf**
  Less primary air
- **Optimale Flammenformbarkeit**
  Better flame adjustment
- **Geringere Schadstoffemissionen**
  Less pollution
- **Einfachere Bedienung**
  Easier handling
- **Längere Düsenstandzeiten**
  Longer nozzle life
- **Bis zu 30% Gewichtseinsparung**
  Up to 30% weight reduction

Ausströmsystem eines M.A.S. Dreistoffbrenners für Kohle / Gas / Öl
Nozzle system of an M.A.S. three-fuel burner for coal / gas / oil

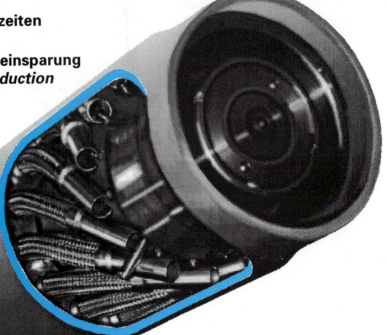

UNITHERM - A-1110 Wien, Nemelkagasse 9 - Tel. (+43)/1/74 041-0 - Fax (+43)/1/74 041/285 - Telex 131204 unith a - Österreich/Austria

stelma

Kohlenstaub
Silo-, Dosier- und Verladeanlagen

*Neubau
Sicherheitsüberprüfungen
Modernisierung*

STELMA-Kohlenstaubanlagen sind unser praxisbewährter Beitrag, Energiequellen wirtschaftlich bei höchstem Sicherheitsstandard zu nutzen.

Die Staubgutlagerung erfolgt in druckstoßfesten Silos bis 1500 m³ Inhalt.

Das STELMA-Grados-Dosiersystem – gravimetrisch geregelt – gibt es für Leistungen von 150 kg/h bis 15 t/h bei einer Dosiergenauigkeit von ± 2% der jeweils gefahrenen Leistung.

6 Punkte, die Prüfsteine Ihrer Entscheidung werden sollten:
- Extrem geringer Verschleiß
- Hohe Kurzzeitgenauigkeit
- Völlig impulsfreier Lauf
- Rückdruck-Unabhängigkeit
- Unempfindlichkeit gegen Materialfließverhalten
- Mehrere einzeln regelbare Brennstellen mit nur einem Dosiersystem

STELMA Industrieanlagen GmbH & Co. KG · D-59 269 Beckum
Daimlerring 22 · Telefon 02521/9341-0 · Fax 02521/9341-30

Explosion Venting without flame-propagation

ROHR
Int. pat. pend.

Consequent further development of 100.000 times proved Triple Section Bursting Disc Technology

Safety Experts

Rembe® GmbH · D-59918 Brilon
Germany · Fax (02961) 50714
Phone (02961) 7405-10
Teletex 296134 = REMBE
Telex 17296134

e121

Unshaken reliability automatically...

Air-Cushioned Explosion Doors

for coal- and lignite silos and for cyclones in coal grinding plant.

- extremely high venting efficiency resulting in smaller venting areas
- low torque impact on support due to smoothly dampened stopping of the door's lid (air-cushion and inertia, patented system)
- automatically closing after venting
- built-in optional vacuum breakers
- stainless steel and heated versions for cold climates available

Pressure-relief as a safety measure for dust explosions

Also available: SIT Air-Cushioned Explosion Doors for use on (electrostatic) filters, dryers etc., as well as for the protection of combustion boilers. Use our know-how in silo engineering!

SILO-THORWESTEN GmbH
D-59 269 Beckum · Daimlerring 39
Telefon 02521/9333-0 · Fax 02521/9333-33

Does a payback time of less than 1.5 years sound like magic?

No, it sounds like an FLS Grate Cooler retrofit!

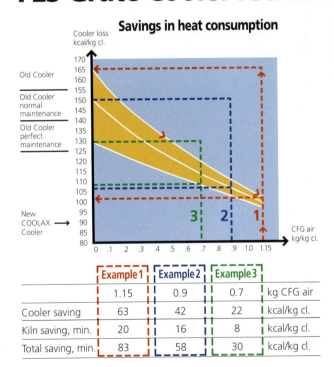

Upgrading with the CFG Coolax system gives you many stunning advantages.

By retrofitting the complete first grate you will have maximum benefits since this grate in most cases will cover the complete heat recuperation zone. If we compare the investment to the savings in specific heat and power, maintenance and a higher production availability we find in most cases a payback time of 0.6 to 1.0 year for units between 2000 and 5000 tpd, and 1.5 years for a 1000 tpd unit.

If you are interested in further details, contact the:

FLS Cooler Systems Department

From upgrades to complete plants.
FLS know-how: Nearly a third of the world's cement production comes from plants and machinery provided by F.L.Smidth.

F.L.Smidth & Co. A/S. Vigerslev Allé 77. DK-2500 Valby-Copenhagen, Denmark. Tel. + 45 3618 1000. Fax + 45 3117 4724

Fachbereich 4

Maßnahmen zum Schutz der Umwelt

(Emissionen und Immissionen von Staub, Gas, Lärm und Erschütterungen, verfahrenstechnische Minderungsmaßnahmen, Auswahl von Einsatzstoffen und Emissionsprognose, Gasreinigung, Lärmschutz, Rekultivierung, Renaturierung)

Subject 4

Measures to protect the environment

(Emission and immission of dust, gas, noise and vibrations, process-technological measures to reduce annoyances from industrial processes, selection of constituents and emission prognosis, gas cleaning, noise protection, recultivation, renaturalization)

Séance Technique 4

Mesures de protection de l'environnement

(Emissions et immissions de poussière, gaz, bruit et vibrations, réduction par des mesures de la technologie des procédés, sélection des matières et pronostic d'émission, épuration des gaz, protection contre les bruits, replantation, restauration de la végétation)

Tema de ramo 4

Medidas de protección del medio ambiente

(Emisiones e inmisiones de polvo, gas, ruido y trepidaciones, medidas de reducción usando la tecnología del proceso, elección de material y prognóstico de emisión, instalaciones de tratamiento de gases, control de niveles ruidosos, replantación, restauración de paisajes)

Maßnahmen zum Schutz der Umwelt*)
Measures to protect the environment*)
Mesures de protection de l'environement

Medidas de protección del medio ambiente

Von **J. Kirsch,** Wiesbaden/Deutschland

Generalbericht 4 · Zusammenfassung – In der Bundesrepublik Deutschland und in der Europäischen Gemeinschaft wurden die immissionsschutzrechtlichen Regelungen weiterentwickelt. Nach wie vor besteht jedoch die Notwendigkeit einer Harmonisierung in der Durchführung und im Vollzug. Die der Zementindustrie gestellte Aufgabe, Staubemissionen weitgehend zu vermeiden, kann in Deutschland und auch in anderen Ländern als technisch gelöst angesehen werden. Entsprechend gering ist die Immission von Staub und staubgebundenen Spurenstoffen in der Nachbarschaft der Zementwerke. Die im Zuge der Dynamisierung weiter herabgesetzten Grenzwerte für die beim Brennprozeß zwangsläufig entstehenden Stickstoffoxide und ihre Einhaltung stellen für die deutsche Zementindustrie ein aktuelles Problem dar. Derzeit werden verschiedene primäre und sekundäre Maßnahmen im industriellen Maßstab in Langzeitversuchen erprobt, um die Erfolgsaussichten für eine Anwendung im Dauerbetrieb auf einer gesicherten Datengrundlage beurteilen zu können. Einen anderen Schwerpunkt bildet unter anderem die Entwicklung von betriebstechnischen Maßnahmen zur Vermeidung von Emissionen durch SO_2 aus den Rohstoffen. Darüber hinaus gewinnt der Lärmschutz immer mehr an Bedeutung. Das erfordert die Weiterentwicklung bewährter wie auch neuartiger Minderungsmaßnahmen. Grundlage der Zementherstellung ist die Rohstoffgewinnung. Vielfältige Möglichkeiten zur späteren Rekultivierung bzw. Renaturierung sind Grundlage für die abbaurechtliche Genehmigung. Die Zementrezeptur erlaubt es außerdem, geeignete Reststoffe aus anderen Industrien als sekundäre Roh- bzw. Brennstoffe ohne Nachteile für Umwelt und Produktqualität einzusetzen und damit einen aktiven Beitrag zur Entlastung der Umwelt zu leisten. Die gesamten Kosten für Umweltschutzmaßnahmen sind in der Zementindustrie der Bundesrepublik mit mehr als 10% des Umsatzes außergewöhnlich hoch. Nicht allein die technische Machbarkeit, sondern die Abwägung der notwendigen Vorsorge einerseits und die daraus entstehenden wirtschaftlichen Belastung andererseits sollten den Maßstab für die Umweltpolitik der Zukunft bilden.

General report 4 · Summary – There have been further developments in legal immission protection regulation in the Federal Republic of Germany and in the European Community but now, as in the past, their execution and performance must be harmonized. The task given to the cement industry of making extensive reductions in dust emission can now be regarded as technically solved in Germany and in other countries. The immission of dust and of trace materials associated with the dust in the neighbourhood of cement works is correspondingly low. Production of nitrogen oxides in the burning process is unavoidable, and their limits were reduced further when the regulations were made more effective. Adherence to these limits is a current problem for the German cement industry. Various primary and secondary measures are at present being tested in long-term trials on an industrial scale so that the prospects of their successful application in continuous operation can be assessed on the basis of confirmed data. One more problem, among others, is the development of process engineering measures to avoid emissions of SO_2 from the raw materials. Noise protection is also gaining increasing importance. This requires new types of abatement measures as well as the further development of proven ones. Cement production is based on the excavation of raw materials. Legal approval for excavation is based on diverse options for later recultivation or return to nature. The cement mix formulation also makes it possible to use suitable residual materials from other industries as secondary raw materials or fuels without any disadvantages for the environment or the product quality, and therefore to make an active contribution towards relieving the load on the environment. At more than 10% of turnover the total costs for environmental protection measures in the cement industry in Germany are extremely high. Environmental policies of the future should not take technical feasibility as their sole yardstick; they should also weigh up the necessary precautions against the resulting financial burden.

Maßnahmen zum Schutz der Umwelt

Measures to protect the environment

---

*) Überarbeitete Fassung eines Vortrages zum VDZ Kongreß '93, Düsseldorf (27. 9. – 1. 10. 1993)
Revised text of a lecture to the VDZ Congress '93, Düsseldorf (27.9. – 1.10.1993)

Fachbereich 4 · Subject 4 · Séance Technique 4 · Tema de ramo 4

Mesures de protection de l'environement

Rapport général 4 · Résumé – En République fédérale d'Allemagne et dans la Communauté Européenne les réglementations en matière de protection contre les émissions ont été améliorées. Cependant, comme par le passé la nécessité d'une harmonisation au niveau de l'exécution et de la réalisation s'impose. La tâche qui revient à l'industrie du ciment consistant à réduire substantiellement les émissions de poussière, peut être considérée comme techniquement résolue en Allemagne mais aussi dans les autres pays. Les émisions de poussière et de matières en trace combinées à la poussière sont donc faibles à proximité des cimenteries. Les valeurs-limites d'oxyde d'azote se formant nécessairement lors de la cuisson – encore abaissées dans le cadre de la dynamisation – et leur respect, posent actuellement un problème pour l'industrie cimentière allemande. A l'heure actuelle, diverses mesures primaires et secondaires à l'échelle industrielle sous forme de test de longue durée sont à l'étude, pour pouvoir apprécier les chances de succès pour une utilisation continue reposant sur une base de données garanties. Une autre priorité est accordée, entre autres, au développement de mesures d'ordre technique visant à supprimer les émissions de SO_2 provenant des matières premières. De plus, l'importance de la lutte contre le bruit prend de plus en plus d'importance. Cela suppose le développement de nouvelles mesures d'insonorisation comme de celles qui ont fait leur preuve jusqu'ici. La fabrication du ciment repose sur la production de matières premières. Les multiples possibilités de reconditionnement et de récupération ultérieure sont la base d'une autorisation réglementant l'exploitation. La recette du ciment autorise, en outre, l'utilisation de produits résiduels appropriés issus d'autres industries comme matières premières ou combustibles secondaires sans préjudice pour l'environnement et la qualité du produit, et apporte ainsi sa contribution active à la sauvegarde de l'environnement. L'ensemble des frais de lutte pour la protection de l'environnement dans l'industrie cimentière de la République fédérale d'Allemagne avec plus de 10% du chiffre d'affaires, est extrêmement élevé. La faisabilité technique mais aussi l'appréciation des mesures préventives nécessaires, d'un côté, et de la charge économique en résultant, de l'autre, devraient constituer la mesure de l'enjeu de la future politique de lutte pour la protection de l'environnement.

Medidas de protección del medio ambiente

Informe general 4 · Resumen – En la República Federal de Alemania y en la Comunidad Europea se han seguido desarrollando las prescripciones referentes a la protección contra las emisiones. Sin embargo, es todavia necesaria una armonización en cuanto a su ejecución y realización. La tarea que se le plantea a la industria del cemento, o sea evitar en lo posible las emisiones de polvo, puede considerarse en Alemania, lo mismo que en otros países, como técnicamente solucionada. Es muy reducida la emisión de polvo y de elementos traza, contenidos en el mismo, a proximidad de las plantas cementeras. Los valores limite, que han sido reducidos aún más para los óxidos de nitrógeno, producidos inevitablemente durante el proceso de cocción, y su observación constituyen actualmente un problema para la industria del cemento. Se están estudiando, a escala industrial, diferentes medidas primarias y secundarias, para lo cual se llevan a cabo ensayos a largo plazo, con el fin de poder evaluar, con datos suficientemente fiables, las perspectivas de éxito de estas medidas cuando se utilicen en el servicio continuo. Otro punto importante lo constituye el desarrollo de medidas técnicas destinadas a evitar las emisiones de SO_2 de las materias primas. Además, la protección contra los ruidos adquiere cada vez mayor importancia, lo cual requiere el desarrollo de medidas de reducción nuevas y también el perfeccionamiento de las que ya han probado su eficacia. La base de la fabricación del cemento es la obtención de materias primas. Existen múltiples posibilidades de restauración y recuperación del paisaje, que son requisitos imprescindibles para obtener la autorización para la explotación del yacimiento. La fórmula de mezcla del cemento permite, además, emplear materias residuales apropiadas, provenientes de otras industrias, como materia prima o combustibles secundarios, sin perjuicio para el medio ambiente o para la calidad del producto, contribuyendo asi activamente a la conservación del medio ambiente. El coste total de las medidas destinadas a la protección ambiental son extremadamente altas en la República Federal de Alemania, alcanzando más del 10% de la cifra de ventas. No solamente la factibilidad técnica, sino también la evaluación de las necesarias medidas de prevención, por una parte, y las cargas económicas de ahí resultantes, por otra parte, deben servir de criterios para fijar la futura política de protección del medio ambiente.

1. Einleitung

Die Herstellung von Zement ist zwangsläufig mit Eingriffen in die Umwelt verbunden. Die Zementindustrie bemüht sich nach Kräften, die Auswirkungen auf die Umwelt so gering wie möglich zu halten. Dabei haben sich die Schwerpunkte im Laufe der letzten Jahrzehnte verschoben. Nachdem die Rückhaltung der Produktionsstäube inzwischen als technisch gelöst und praktisch weitgehend realisiert betrachtet werden darf, galt die Aufmerksamkeit den als Staubinhaltsstoffe emittierten Schwermetallen und in jüngster Zeit den gasförmigen Emissionen. Hier sind es insbesondere die Stickstoffoxide, deren Reduzierung eine erhebliche Herausforderung darstellt [1, 2].

1. Introduction

The manufacture of cement necessarily involves interference with the environment, though the cement industry is doing its best to minimize adverse effects on it. The focus has shifted several times in recent decades. Once the technical problem of collecting the dust created during manufacture could be regarded as solved and substantially implemented, attention then switched to the constituent heavy metals emitted with the dust, and more recently to gaseous emissions. Nitrogen oxides are the main concern, and their reduction is a real challenge [1, 2].

This trend corresponds to the general development in environmental protection: starting with local questions

Dieser Trend entspricht der generellen Entwicklung im Umweltschutz, die, von der lokalen Betrachtung ausgehend, mittlerweile auf eine überregionale und, wenn an den Schutz der Atmosphäre gedacht wird, sogar globale Sicht übergreift. Eine entscheidende Voraussetzung für die Ausdehnung des Umweltschutzes auf immer weitere Stoffe war die Entwicklung der Meß- und Analysentechnik [3]. Dieses Sachgebiet wird ausführlich in [4] behandelt. Auch die integrierte Betrachtung bei der Umweltvorsorge entspricht einer eigenen Thematik. In [5–7] wird der Umweltinanspruchnahme die Entlastung und der Umweltnutzen von Zement und Beton gegenübergestellt.

Die folgenden Ausführungen befassen sich mit den Themenschwerpunkten Immissionsschutz, Rohstoffgewinnung und Wiederherstellung der Landschaft sowie Umweltentlastung durch Einsatz von Sekundärstoffen. Auf die Umweltbereiche Wasser und Abfall wird nicht eingegangen, da sie im Normalfall bei der Zementherstellung keine produktionsspezifische Bedeutung haben.

2. Entwicklung der Umweltgesetzgebung in der Bundesrepublik Deutschland und in der Europäischen Gemeinschaft

Die Errichtung und der Betrieb von Anlagen zur Herstellung von Zement unterliegen in Deutschland dem Bundes-Immissionsschutz-Gesetz mit den entsprechenden Verordnungen und Verwaltungsvorschriften. Die Änderung des BImSchG im April 1986 brachte mit § 5 eine Verstärkung der Wiederverwertungspflicht von Reststoffen und die Regelung eines generellen Wärmenutzungsgebotes [8, 9].

Der Klinkerherstellungsprozeß kann in der Regel so gesteuert werden, daß keine Reststoffe anfallen. Voraussetzung hierfür ist, daß weder durch Einsatzstoffe noch durch überzogene Sicherheitsüberlegungen bei der Anwendung des Zementes Reststoffprobleme geschaffen werden. Das gilt insbesondere für die im Zusammenhang mit der Frage der Alkalizuschlagreaktion diskutierten Forderung nach einem alkaliarmen Zement.

Die Konkretisierung des Wärmenutzungsgebotes in Form einer Wärmenutzungsverordnung liegt noch nicht in verbindlicher Form vor [10]. Das Gebot kann jedoch bei der Zementherstellung zu keiner weiteren Verbesserung führen, da bereits jetzt eine Ausnutzung der thermischen Energie in der Größenordnung von 70 % erreicht wird. Dagegen haben selbst moderne Kraftwerke heute noch einen Wirkungsgrad, der in der Größenordnung von nur 40 % liegt.

Sehr bedeutsam war ferner der § 7 der Neufassung des BImSchG von 1986, der die Grundlage für die sog. Altanlagensanierung darstellt. Durch die neue Anleitung zur Reinhaltung der Luft (TA Luft), die Ende Februar 1986 in Kraft trat, wurden die Voraussetzungen für die Anpassung aller bestehenden Anlagen, der sog. Altanlagen, an den Stand der Technik geschaffen [11–13]. Das war die Grundlage für eine deutliche Emissionsminderung, die Mitte dieses Jahrzehnts in Westdeutschland und Ende der 90er Jahre auch in Ostdeutschland abgeschlossen sein wird [14]. Allerdings hat diese Anpassung bestehender Anlagen an den Stand der Technik erhebliche finanzielle Mittel beansprucht und damit die internationalen Wettbewerbsvoraussetzungen belastet.

Mit Wirkung vom 1. 9. 1990 hat das Bundes-Immissionsschutzgesetz erneut eine Reihe von Änderungen erfahren. Zu nennen ist hier insbesondere die mit § 52a neu eingeführte Mitteilungspflicht zur Betriebsorganisation, die eine organisatorische Maßnahme zum Schutz der Umwelt darstellt.

Im Rahmen der Aktionsprogramme der Europäischen Gemeinschaft wurde eine Reihe von Verordnungen und Richtlinien erlassen, welche die gleiche Zielsetzung wie das deutsche Bundes-Immissionsschutzgesetz verfolgen [15].

there is a move to supra-regional ones and even to global ones in terms of protecting the atmosphere. An important pre-condition for the expansion of environmental protection to more and more materials was the development of measuring and analytical technology [3]. This field is treated in detail in [4]. Integrated consideration of environmental precautions is a subject in itself. In [5–7] the relief measures and the environmental benefit of cement and concrete are contrasted with the utilization of the environment.

This article deals with the important topics of immission protection, quarrying of raw materials, landscape reclamation and relief of the environment through the use of secondary materials. The subjects of water and waste will not be discussed, as these do not normally have a specific role in cement production.

2. Development of environmental legislation in the Federal Republic of Germany and the European Community

The installation and operation of cement manufacturing plant in Germany are subject to the BImSchG (Federal Immission Protection Act) with the corresponding rules and administrative regulations. With the amendment of the Act in 1986 clause 5 reinforced the duty to re-use residual materials and introduced a general heat recovery requirement [8, 9].

The clinker manufacturing process can generally be controlled so that there are no residual materials. This presupposes that no residual material problems are created either by the materials used or by exaggerated safety considerations in the use of the cement. This applies particularly to the requirement for low-alkali cement, which is discussed in connection with the question of the alkali-aggregate reaction.

The heat recovery requirement has not yet come into force as a binding heat recovery order [10]. However it cannot lead to any further improvement in cement production, since the process already utilizes about 70 % of the heat energy. By comparison even modern power stations are still only 40 % efficient.

Clause 7 of the amended BImSchG of 1986 was also very important, laying the foundation for the renovation of old plant. The new Clean Air Regulations (TA Luft) which came into force in February 1986 laid down the conditions for bringing all existing plant, so-called old plant, up to the state of the art [11–13]. This provided the basis for a marked reduction in emissions, which is to be completed by the middle of this decade in West Germany and by the end of the nineties in East Germany [14]. The modernization of existing plant has required a considerable financial input though, thus putting a strain on international competitiveness.

Another series of amendments to the BImSchG took effect from 1.9.1990. Clause 52a should be mentioned particularly. It gave works management a new duty to provide information and was an organizational measure to protect the environment.

A number of orders and directives following the same aims as the BImSchG have been issued under the action programmes of the European Community [15]. The most important ones for the cement industry are:

– the Industrial Licensing Directive of 28.6.1984 [16],

– the Directive of 27.6.1985 on Testing for Compatibility with the Environment [17–19],

– the Environmental Information Directive of 7.6.1990 [20],

– the Ecological Audit Regulation of 26.9.1993 [21],

– the outline proposal on "Integrated Pollution, Prevention and Control" (IPPC) (from 1.7.1995) [22].

The EC sees limitation of carbon dioxide emission through a CO_2/energy tax as an important new feature of their clean air policy. Together with other trace gases carbon dioxide is held responsible for the feared future climatic heating known as the "greenhouse effect". So as early as 1990 the EC Commission set the target of stabilizing CO_2 emissions at the

Die wichtigsten für die Zementindustrie sind

— die Industriezulassungsrichtlinie vom 28. 6. 1984 [16],

— die Richtlinie über die Umweltverträglichkeitsprüfung vom 27. 6. 1985 [17–19],

— die Umweltinformationsrichtlinie vom 7. 6. 1990 [20],

— die Öko-Audit-Verordnung vom 26. 9. 1993 [21],

— die Rahmenrichtlinie „Integrated Pollution, Prevention and Control" (IPPC) (ab 1. 7. 1995) [22].

Einen neuen Schwerpunkt ihrer Luftreinhaltungspolitik sieht die EG in der Begrenzung der Kohlendioxid-Emission durch eine CO_2/Energie-Steuer. Zusammen mit anderen Spurengasen wird Kohlendioxid bekanntlich unter dem Stichwort „Treibhauseffekt" für eine befürchtete zukünftige Klimaerwärmung verantwortlich gemacht. Daher hat sich die EG-Kommission bereits 1990 das Ziel gesteckt, die EG-weiten CO_2-Emissionen bis zum Jahr 2000 auf dem Stand von 1990 zu stabilisieren. Hierzu soll eine kombinierte CO_2-/Energie-Steuer dienen. Im Juni 1992 wurde von der Kommission ein Richtlinien-Vorschlag vorgelegt [23–26]. Danach sollen dieser Steuer sowohl der Energieverbrauch als auch die CO_2-Emissionen je zur Hälfte zu Grunde gelegt werden. Die Steuer soll mit 30% der endgültig geplanten Höhe beginnen und dann über 7 Jahre um jeweils 10 Prozentpunkte angehoben werden. Nach dieser Berechnungsbasis würde eine Tonne Zement aufgrund des für die Herstellung benötigten Brennstoffbedarfs in der Endstufe mit ca. 11,— DM belastet. Das macht, bezogen auf den Umsatzerlös, ca. 10% aus. Hinzu käme noch die Vorbelastung aus Strom und sonstigen Bezügen. Diese immense zusätzliche finanzielle Belastung würde z.B. für die deutsche Zementindustrie jährlich ca. 400 Mio. DM betragen.

Die Folge dieser erheblichen Verteuerung der Zementherstellung in der Europäischen Gemeinschaft wäre eine weitere Wettbewerbsverzerrung gegenüber Einlieferungen in den EG-Wirtschaftsraum. Einer Klimabeeinträchtigung durch Spurengase wie CO_2 kann nur durch eine globale Anstrengung begegnet werden. Lokal begrenzte Bemühungen würden nur Produktionsverlagerungen an andere Standorte zur Folge haben und hätten wegen der dort zu erwartenden geringeren Energieeffizienz sogar einen gegenteiligen Effekt.

Werden bei der Zementproduktion auch sog. „gefährliche Abfallstoffe" als Sekundärbrennstoff eingesetzt, so wird in Zukunft auch eine entsprechende EG-Richtlinie hierzu berücksichtigt werden müssen [27]. Ihr Anwendungsbereich soll auf Sekundärbrennstoffe mit kritischen Inhaltsstoffen begrenzt werden. Dagegen unterliegen der entsprechenden deutschen 17. Durchführungsverordnung zum BImSchG, der sog. Abfallverbrennungsanlagenverordnung vom 30. 11. 1990, mit Ausnahme der enumerativ benannten fossilen Brennstoffe, alle anderen brennbaren Stoffe [28].

Um die bei der Entsorgung von Sonderabfällen bestehenden Engpässe zu entschärfen, hat der Gesetzgeber in Deutschland im Mai 1990 grundsätzlich die Mitverbrennung von Abfällen in immissionsschutzrechtlich genehmigungspflichtigen Anlagen zugelassen. Das Abfallbeseitigungsgesetz nimmt auf diese neue Regelung ausdrücklich Bezug. Damit entfällt die Notwendigkeit einer abfallrechtlichen Ausnahmegenehmigung.

Andererseits wird jedoch durch die Verordnung über Verbrennungsanlagen für Abfälle und ähnliche brennbare Stoffe (17. BImSchV) und deren Anwendung auf alle Sekundärbrennstoffe diese Möglichkeit der thermischen Entsorgung erheblich behindert. Eine unabdingbare Voraussetzung für den Einsatz von Ersatzbrennstoffen in der Zementindustrie ist die Berücksichtigung strenger Auswahlkriterien, die eine Erhöhung der Umweltbelastung bzw. Nachteile für die Qualität ausschließen.

In den USA sind am 21. 2. 1991 spezielle Richtlinien für die Verbrennung gefährlicher Abfallstoffe in Kraft getreten, die auch für die Zementindustrie Gültigkeit haben [29, 30].

1990 level throughout the EC by the year 2000. This is to be done by imposing a combined CO_2/energy tax. The Commission put forward a proposed directive in June 1992 [23–26], to the effect that the tax should be based half on energy consumption and half on CO_2 emissions. The tax is to start at 30% of its final projected level, then be raised by 10 percentage points each time over 7 years. On this basis a tonne of cement would carry a tax of about DM 11.— at the final stage due to the fuel required for production. This would raise the selling price by about 10%. Prepaid tax would also be included in the cost of power and other purchases. For the German cement industry, for example, this enormous extra financial burden would amount to about 400 million DM p.a.

The result of this considerable increase in the cost of cement production in the European Community would be further distortion of competition with imports into the EC economic area. Damage to the climate by trace gases such as CO_2 can only be dealt with by global action. Local efforts would only result in production being moved to other locations and would even have a detrimental effect owing to the lower energy efficiency which would be expected there.

If so-called "hazardous waste materials" are used as secondary fuels in cement production, an EC directive on the subject will have to be complied with in future [27]. Its application is to be limited to secondary fuels with critical constituents. On the other hand the corresponding German legislation, the 17th Order implementing the BImSchG, known as the waste incineration plant order of 30.11.1990, covers all combustible materials except for the fossil fuels enumerated [28].

To open up existing bottle-necks in the disposal of hazardous waste, German legislation enacted in May 1990 allows waste to be incinerated in plant licensed under the immission protection laws. The Waste Disposal Act expressly refers to this new arrangement, and hence there is no need to obtain exemption under the present legislation.

On the other hand the Order on Incineration Plant for Waste and other Combustible Materials (17th Order under the BImSchG) and its application to all secondary fuels considerably limits this option for disposal by burning. An essential prerequisite for the use of alternative fuels in the cement industry is the application of strict selection criteria, avoiding any increase in environmental pollution or lowering of quality.

In the USA special directives on the incineration of hazardous waste materials came into force on 21.2.1991. These also apply to the cement industry [29,30].

3. Clean air

3.1 Emissions

In view of the existing differences in the implementation and particularly the enforcement of Community law, the urgent aim should be to harmonize the implementation and observation of the set standard within the Community. As a step in this direction the EC Council has laid down requirements for reducing emissions from industrial plant, and thus also for cement-making plant, in the EC in its Directive 84/360 of 28 June 1984 [16]. Article 7 of this Directive calls for exchange of information on **B**est **A**vailable **T**echnology — not entailing excessive cost (BAT). As a result of a pilot project started in 1989 there is now a draft study for the cement industry, in which the BAT limits set out in **Table 1** are proposed for dust, SO_2 and NO_x [31]. However, the BAT values cannot yet ensure any harmonization of immission protection in the EC. They are not mandatory limits, merely recommendations for Member States and the EC Commission itself. In future these limits are to be observed by new plants in normal continuous operation.

A uniform EC environmental protection level requires not only uniform limits but also harmonization of measuring and monitoring regulations which still differ considerably at present. Periods of assessment, times over which the average is taken, and frequency of compliance have an impor-

3. Luftreinhaltung

3.1 Emissionen

Angesichts der bestehenden Unterschiede in der Umsetzung von Gemeinschaftsrecht sowie insbesondere im Vollzug sollte das vordringliche Ziel in einer Harmonisierung bei der Umsetzung und Durchführung des gesetzten Standards innerhalb der Gemeinschaft gesehen werden. Um diesem Ziel näher zu kommen, hat der Rat der EG in seiner Richtlinie 84/360 vom 28. Juni 1984 die Anforderungen zur Minderung der Emissionen aus Industrieanlagen und damit auch aus Anlagen zur Herstellung von Zement in der EG festgelegt [16]. In Artikel 7 dieser Richtlinie wird ein Informationsaustausch über den bestmöglichen technologischen Stand unter Beachtung angemessener Kosten gefordert (**B**est **A**vailable **T**echnology — not entailing excessive cost) (BAT). Als Ergebnis eines 1989 gestarteten Pilotprojektes liegt inzwischen für die Zementindustrie der Entwurf einer Studie vor, in der die in **Tabelle 1** zusammengestellten BAT-Grenzwerte für Staub sowie SO_2 und NO_x vorgeschlagen werden [31]. Die BAT-Werte können allerdings noch keine Harmonisierung des Immissionsschutzes in der EG sicherstellen. Sie sind keine rechtskräftigen Grenzwerte, sondern lediglich Empfehlungen für die Mitgliedsstaaten und die EG-Kommission selbst. Die Werte sollen zukünftig von Neuanlagen im kontinuierlichen Normalbetrieb eingehalten werden.

TABELLE 1: Emissionsgrenzwerte nach BAT

Emission	Quelle	Zahlenwert mg/m$^3_{Nf}$
Staub	Ofen/Mahltrocknung	100
	Klinkerkühler	100
	Zementmahlung	80
	Andere Quellen	50
SO_2		400– 700
NO_x		1300–1800

Für ein EG-einheitliches Umweltschutzniveau sind darüber hinaus nicht nur einheitliche Grenzwerte erforderlich, sondern auch eine Harmonisierung der Meß- und Überwachungsvorschriften, die z. Z. noch erhebliche Unterschiede aufweisen. Beurteilungszeiträume, Mittelungszeiten und Einhaltungshäufigkeiten sind insbesondere bei den Emissionswerten, die prozeßbedingt stärkeren Schwankungen unterworfen sind, von nicht zu unterschätzender Bedeutung.

Ein Vergleich der Grenzwerte für Staub zeigt, daß die BAT-Werte mit bis zu 100 mg/m^3*) z. T. doppelt so hoch sind wie die TA Luft-Grenzwerte. Das hat einen Einspruch der deutschen Behörden ausgelöst.

Im Vergleich zu dem BAT-Vorschlag für die SO_2-Begrenzung von 400–750 mg/m^3 (feucht) liegt der Grenzwert in Deutschland mit 400 mg/m^3 noch unter dieser Spanne, da er sich auf den Normzustand nach Abzug des Feuchtegehaltes an Wasserdampf bezieht.

Wie **Tabelle 2** zu entnehmen ist, entsprechen die BAT-Grenzwerte für NO_x, abgesehen von der Einbeziehung des Wasserdampfanteils in das Gasbezugsvolumen, den z. Z. in Deutschland noch bis 28. 2. 1994 gültigen Werten. Die Gültigkeit dieser Grenzwerte ist in der Technischen Anleitung zur Reinhaltung der Luft von 1986 jedoch mit dem nachstehenden, als Dynamisierungsklausel bezeichneten Zusatz eingeschränkt:

„Die Möglichkeiten, die Emissionen durch feuerungstechnische oder andere dem Stand der Technik entsprechende Maßnahmen weiter zu vermindern, sind auszuschöpfen" [11].

Zur Konkretisierung dieser Dynamisierungsklausel hat der Länderausschuß für Immissionsschutz (LAI) im Mai 1991

---

*) Sofern keine anderslautenden Angaben gemacht werden, beziehen sich alle Angaben des Abgasvolumens auf Normzustand (0°C; 1013 mbar) nach Abzug des Feuchtegehaltes an Wasserdampf.

TABLE 1: Emission limits using Best Available Technology

Emission	Source	Numerical value mg/m$^3_{stp,moist}$
Dust	kiln/drying and grinding	100
	clinker cooler	100
	cement grinding	80
	other sources	50
SO_2		400– 700
NO_x		1300–1800

tance which should not be under-estimated, particularly in the case of emissions which are subject to quite large fluctuations due to the nature of the process.

Comparison of the limits for dust shows that the BAT levels of up to 100 mg/m^3*) are sometimes twice as high as those in the TA Luft. This has caused protest by the German authorities.

Compared to the BAT proposal for an SO_2 limit of 400–750 mg/m^3 (moist) the German limit of 400 mg/m^3 is still below that range, since it relates to the standard state after deduction of the water vapour moisture content.

As will be seen from **Table 2**, the BAT limits for NO_x correspond to the levels still valid in Germany to 28.2.1994, apart from the inclusion of the steam component in the reference volume of gas. However, the TA Luft of 1986 qualifies the validity of these limits in the following words, known as the adjustment clause:

"Possible ways of further reducing emissions by firing methods or other state of the art measures are to be exploited to the full" [11].

To put this adjustment clause in concrete terms the State Committee on Immission Protection (LAI) made a recommendation to the governments of the German States in May 1991. This was that the new limits to be observed should be 0.50 g NO_x/m^3 for new plant and 0.80 g NO_x/m^3 for old, in each case in the form of NO_2 with a reference oxygen value of 10%. The provincial governments instructed the competent authorities to implement this recommendation from 1.3.1994 [32].

TABLE 2: Emission levels for nitrogen oxides (as specified in TA Luft 1986)

Emission source	Numerical value g NO_2/m$^3_{stpd}$
cement kilns with	
— grate preheater	1.5
— cyclone preheater and exhaust gas heat utilization	1.3
— cyclone preheater without waste gas heat utilization	1.8
After final drafting of adjustment clause (with a reference oxygen content of 10%)	
— new plants	0.50
— old plants	0.80

However most kiln plants in the German cement industry will not be able to meet these drastically lowered limits in conjunction with the strict evaluation rules for continuous measurement by the set date. The waste incineration plant order (17th BImSchV) also lowers the limits for some exhaust gas constituents at cement kilns using secondary fuels. In these cases the limit to be met can be calculated using the proportional calculation described in Clause 5 of this Order [33]. **Fig. 1** takes dust as an example and shows how the limit to be applied can be determined from the weighted proportions of the limits from the TA Luft and the

---

*) Unless otherwise stated, any mention of the volume of exhaust gas refers to the standard state (0°C; 1013 mbar) after deduction of the water vapour moisture content.

TABELLE 2: Emissionswerte für Stickstoffoxide (nach TA Luft 1986)

Emissionsquelle	Zahlenwert g $NO_2/m^3_{N.tr.}$
Zementöfen mit	
— Rostvorwärmer	1,5
— Zyklonvorwärmer und Abgaswärmenutzung	1,3
— Zyklonvorwärmer ohne Abgaswärmenutzung	1,8
Nach Konkretisierung der Dynamisierungsklausel (bei einem Bezugssauerstoffgehalt von 10%)	
— Neuanlagen	0,50
— Altanlagen	0,80

eine Empfehlung an die Landesregierungen der deutschen Bundesländer gegeben. Diese Konkretisierung besagt, daß als neue Grenzwerte für Neuanlagen 0,50 g NO_x/m^3 und für Altanlagen 0,80 g NO_x/m^3, jeweils als NO_2 und bei einem Sauerstoffbezugswert von 10% angegeben, eingehalten werden sollen. Dieser Empfehlung folgend haben die Landesregierungen durch Erlaß die zuständigen Behörden zur Umsetzung ab 1. 3. 94 angewiesen [32].

Bei den meisten Ofenanlagen der deutschen Zementindustrie werden diese drastisch verschärften Grenzwerte in Verbindung mit den strengen Auswertevorschriften für kontinuierliche Messungen bis zum genannten Termin jedoch nicht eingehalten werden können. Bei Zementöfen, die Sekundärbrennstoffe einsetzen, ergeben sich zusätzlich für einige Abgasbestandteile noch Grenzwertherabsetzungen durch die Abfallverbrennungsanlagenverordnung (17. BImSchV). Der zu beachtende Grenzwert errechnet sich in diesen Fällen durch Anwendung der in §5 dieser Verordnung beschriebenen Anteilsrechnung [33]. **Bild 1** zeigt am Beispiel Staub die Ermittlung des anzuwendenden Grenzwertes aus den gewogenen Anteilen der Grenzwerte aus TA Luft und 17. BImSchV. Für einen Sekundärbrennstoffanteil von 25% am Gesamt-Brennstoffenergiebedarf erhält man bei diesem Beispiel als Mischrechenwert aus 50 mg/m³

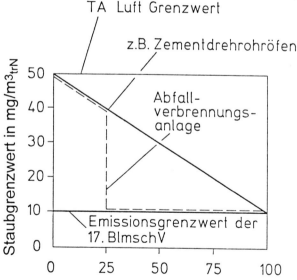

BILD 1: Emissionsgrenzwert für Staub gemäß 17. BImSchV [34]
FIGURE 1: Dust emission limit under 17th Federal Immission Protection Order [34]

Staubgrenzwert in mg/m³$_{trN}$	= dust limit in mg/m³stpd
TA Luft Grenzwert	= TA Luft limit
z.B. Zementdrehrohröfen	= e.g. rotary cement kilns
Abfallverbrennungsanlage	= waste incineration plant
Emissionsgrenzwert der 17. BImSchV	= emission limit under 17th BImSchV
Anteil an anderen brennbaren Stoffen am Gesamtbrennstoffanteil in %	= Percentage of other combustible substances in total fuel in %

17th BImSchV. A secondary fuel fraction of 25% of the total fuel energy requirement gives a new limit of 40 mg/m³ in this example, obtained by the rule of alligation from 50 mg/m³ under the TA Luft and 10 mg/m³ under the 17th BImSchV [34].

A corresponding calculation, which is also required by the authorities for the dioxin limit in Germany, cannot be carried out though as the measuring technology for such low concentrations has not yet been validated. This is taken into account by the draft EC directive, in which the value of 0.1 ng/m³ for dioxin is only given as a recommendation until 1996. The actual limit will not be stipulated until the measurement instructions have been harmonized within the European Community and a further preliminary period of 6 months has elapsed [27].

At the request of the German cement industry the State Committee on Immission Protection has recommended that there should be no limit on emissions of total carbon or carbon monoxide, as an exceptional arrangement which is possible under clause 19 of the 17th BImSchV. As NO_x emissions from cement kilns are not significantly affected by the type of fuel, the possibility of an exceptional arrangement for NO_x is also conceded [35].

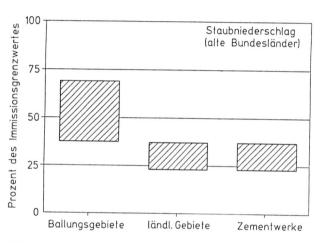

BILD 2: Vergleich des Staubniederschlags in verschiedenen Gebieten [36]
FIGURE 2: Comparison of dust deposition in different areas [36]

Prozent des Immissionsgrenzwertes	= percentage of immission limit
Staubniederschlag (alte Bundesländer)	= dust deposition (former West Germany)
Ballungsgebiete	= conurbations
ländl. Gebiete	= rural areas
Zementwerke	= cement works

3.2 Immissions

With the enormous reduction in total dust emission by the German cement industry from 35 kg/t cement in 1950 to less than 0.5 kg/t, pollution by immissions in the vicinity of cement works has been reduced to 0.18 g/m² × d dust deposit; as shown in **Fig. 2** this corresponds to the pollution generally found in rural areas [36]. Owing to the high degree of dust removal directional dust emission sources only cause additional pollution which is virtually indistinguishable from the existing background pollution. The effects of these emission sources cannot therefore be measured. Consequently any further reductions in the limit as a preventive immission protection measure would not be reasonable.

On the other hand particular attention should still be paid to combatting diffuse dust sources, since these may be significant in the deposition of dust in the immediate vicinity of the works owing to the low level of the source and the usually coarser particle size distribution.

There is also far less heavy metal pollution in the catchment area of cement works, corresponding to the sharply declining dust emission. Measurements taken by the State Immission Protection Institute (LIS) for the state of North Rhine-Westphalia can be quoted as an example; the maximum level for thallium and its compounds in the vicinity of

gemäß TA Luft und 10 mg/m³ gemäß 17. BImSchV einen neuen Grenzwert von 40 mg/m³ [34].

Eine entsprechende Mischungsrechnung, die von den Behörden auch für den Dioxin-Grenzwert in der BRD gefordert wird, ist allerdings nicht durchführbar, da die Meßtechnik für derartig geringe Konzentrationen noch nicht validiert ist. Dem trägt der Entwurf der EG-Richtlinie Rechnung, in dem der Wert von 0,1 ng/m³ für Dioxin bis 1996 lediglich als Richtwert angegeben wird. Die Fixierung als Grenzwert soll erst erfolgen, wenn in der Europäischen Gemeinschaft harmonisierte Meßvorschriften vorliegen und eine weitere Vorlaufzeit von 6 Monaten abgelaufen ist [27].

Auf Antrag der deutschen Zementindustrie hat der Länderausschuß für Immissionsschutz im Rahmen der nach §19 der 17. BImSchV möglichen Ausnahmeregelung einen Verzicht auf eine Emissionsbegrenzung für Gesamtkohlenstoff und Kohlenmonoxid empfohlen. Da die NO_x-Emissionen von Zementöfen nicht signifikant von der Brennstoffart beeinflußt werden, wird auch die Möglichkeit einer Ausnahmeregelung für NO_x eingeräumt [35].

3.2 Immissionen

Entsprechend der enormen Reduzierung der Gesamt-Staubemission der deutschen Zementindustrie von 35 kg/t Zement im Jahre 1950 auf unter 0,5 kg/t hat sich auch die Immissionsbelastung in der Nachbarschaft von Zementwerken auf 0,18 g/m² × d Staubniederschlag reduziert und entspricht damit, wie in **Bild 2** dargestellt, der Belastung, wie sie in ländlichen Gebieten allgemein festzustellen ist [36]. Gerichtete Staubemissionsquellen führen aufgrund des hohen Entstaubungsgrades nur noch zu einer Zusatzbelastung, die von der vorhandenen Hintergrundbelastung praktisch nicht mehr zu unterscheiden ist. Diese Emissionsquellen wirken sich daher nicht meßbar aus. Demzufolge sind weitere Grenzwertherabsetzungen im Sinne eines vorbeugenden Immissionsschutzes nicht zumutbar.

Dagegen sollte der Bekämpfung von diffusen Staubquellen auch weiterhin besondere Aufmerksamkeit gewidmet werden, da diese aufgrund der niedrigen Quellhöhe und der meist gröberen Korngrößenverteilung für den Staubniederschlag in unmittelbarer Werksumgebung von Bedeutung sein können.

Auch die Schwermetallbelastung im Einzugsbereich von Zementwerken hat sich entsprechend der stark rückläufigen Staubemission erheblich reduziert. Als Beispiel seien die Messungen der Landesanstalt für Immissionsschutz (LIS) des Landes Nordrhein-Westfalen zitiert, die für Thallium und seine Verbindungen im Bereich von Zementwerken bei maximal 2 % der Grenzkonzentrationen der TA-Luft lagen [37].

Die Zementindustrie der ehemaligen DDR war durch eine Ballung der Produktion an den 4 Standorten Bernburg, Deuna, Karsdorf und Rüdersdorf gekennzeichnet. In diesen 4 Großbetrieben wurden jährlich jeweils 2,5–3 Mio. t Zement produziert. Inzwischen wurde für diese Standorte eine Konzentration der Klinkerproduktion auf wenige Ofenlinien eingeleitet. **Bild 3** zeigt den Verlauf der Emission am Beispiel des Zementwerkes Rüdersdorf in den letzten 20 Jahren [38]. Der Erfolg der Ende der 70er Jahre eingeleiteten Bemühungen zur Emissionsminderung ist deutlich erkennbar. Mitte der 80er Jahre trat jedoch eine Stagnation des erreichten Standes ein, der insbesondere auf die Engpaßsituation bei den Ersatzinvestitionen zurückzuführen war. Mit der Wiedervereinigung Deutschlands konnte ab 1990 durch Stillegung veralteter Anlagen und Realisierung von Neuinvestitionen mit modernen Entstaubungsanlagen eine deutliche Reduzierung der Staubemissionen und entsprechend auch der Immissionen erreicht werden.

Bodenuntersuchungen, die vom Forschungsinstitut der Zementindustrie in der Umgebung eines anderen Zementwerkes in den neuen Bundesländern durchgeführt wurden, haben trotz der in früheren Jahren extrem hohen Staubniederschlagsbelastung keine überhöhten Schwermetallge-

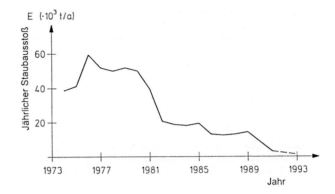

BILD 3: Entwicklung der Staub-Emission im Zementwerk Rüdersdorf
FIGURE 3: Development of dust emission at Rüdersdorf cement works

Jährlicher Staubausstoß = annual dust discharge
Jahr = year

cement works was 2 % of the concentration limits set by the TA Luft [37].

The cement industry in the former East Germany was characterized by clustering of production at the four locations of Bernburg, Deuna, Karsdorf and Rüdersdorf. 2.5–3 million t cement p.a. were produced at each of these large-scale works. Concentration of clinker production in only a few kiln lines has now been introduced at these locations. Fig. 3 shows the emission curve for the Rüdersdorf cement works over the last 20 years [38]. The success of the efforts to decrease emissions at the end of the seventies is clearly visible. However stagnation set in the mid-eighties, due mainly to the critical situation with regard to replacement expenditure. With the reunification of Germany a clear reduction in dust emission and a corresponding reduction in immissions was obtained from 1990 by shutting down old plant and bringing in new investment with modern de-dusting equipment.

Soil tests conducted by the Research Institute of the Cement Industry in the vicinity of another cement works in the former East Germany have not revealed any excessive heavy metal content in spite of the very severe dust pollution in earlier years. The soil values found for the heavy metals studied, cadmium, lead and mercury, were well below the recommended levels [34].

3.3 Measures to reduce emissions

3.3.1 Total dust

Of the 10–15 m³ exhaust gas discharged in producing 1 tonne of cement by far the largest part is exhaust air which is discharged from mills, clinker coolers, dryers and conveying equipment and is de-dusted in fabric filters. Exhaust gases from kilns on the other hand are largely de-dusted by electrostatic precipitators because of the prevailing temperatures. The areas of use of the two types of collector overlap in the de-dusting of exhaust air from clinker coolers and exhaust gases from dryers [39].

Cleaning of fabric filters by mechanical rapping or scavenging air has now largely been superseded by the use of compressed air. Fig. 4 shows the design of such a de-dusting system in section. The use of compressed air increases availability and at the same time reduces the required maintenance. Compressed air cleaning also enables very dense filter materials based on needle felt to be used. These advantages have even made it possible to convert existing filters to modern cleaning technology. A cleaning control system based on differential pressure has been developed as an energy-saving measure.

With the filter designs available on the market and the experience which has been gained of correct dimensioning of the filter area, it is possible nowadays to comply with the TA Luft limit of 50 mg/m³ even with the strict evaluation criteria of the TA Luft 1986 [40–42].

halte ergeben. Die für die untersuchten Schwermetalle Cadmium, Blei und Quecksilber gefundenen Bodenwerte lagen deutlich unter den jeweiligen Richtwerten [34].

3.3 Maßnahmen zur Minderung der Emissionen

3.3.1 Gesamtstaub

Bei dem Abgasaufkommen von 10–15 m³, das mit der Produktion von 1 t Zement verbunden ist, handelt es sich zum weitaus überwiegenden Teil um Abluft, die an Mühlen, Klinkerkühlern, Trocknern und insbesondere Fördereinrichtungen anfällt und in Faserstoffiltern entstaubt wird. Dagegen ist die Entstaubung der Ofenabgase mit Rücksicht auf die zu beherrschenden Temperaturen weitgehend den Elektrofiltern vorbehalten. Bei der Entstaubung von Klinkerkühlerabluft und Trocknerabgasen überschneiden sich die Einsatzgebiete beider Abscheider-Bauarten [39].

Bei den Faserstoffiltern wurde die Abreinigung durch mechanische Klopfung bzw. Spülluft inzwischen weitgehend durch den Einsatz von Druckluft abgelöst. **Bild 4** zeigt den Aufbau einer derartigen Entstaubungsanlage im Schnitt. Durch den Drucklufteinsatz wurde die Verfügbarkeit erhöht und der Wartungsaufwand gleichzeitig verringert. Darüber hinaus macht die Druckluftabreinigung den Einsatz sehr dichter Filtermedien auf Nadelfilzbasis möglich. Aufgrund dieser Vorteile wurden auch bestehende Filter auf die moderne Abreinigungstechnik umgerüstet. Als energiesparende Maßnahme wurde die differenzdruckabhängige Abreinigungssteuerung entwickelt.

Die auf dem Markt verfügbaren Filterkonstruktionen, und die inzwischen vorliegenden Erfahrungen über die richtige Filterflächendimensionierung machen es heute möglich, den TA Luft-Grenzwert von 50 mg/m³ auch mit den strengen Auswertekriterien der TA Luft 1986 gesichert einzuhalten [40–42].

Auch bei Elektrofiltern sind in den letzten Jahren Weiterentwicklungen zu verzeichnen [43–46]. Hierzu gehört die Vergrößerung des Gassenabstandes zwischen den Niederschlagsplatten von 300 auf 400 mm in Verbindung mit einer angepaßten Geometrie zwischen Sprüh- und Niederschlagselektrode. Fortschritte sind aber auch dem Einsatz der Mikro-Elektronik zur Aussteuerung der Hochspannung zu verdanken. Im Gegensatz zu den früheren Analog-Steuerungssystemen paßt sich die computergesteuerte Spannungsumsetzanlage optimal den veränderlichen Betriebsbedingungen an. Wie in **Bild 5** dargestellt, sind derartige Steuerungen in der Lage, Schwankungen in der Durchschlagsgrenze zu erfassen und darauf mit variablen Abtastzeiten und -stufen zu reagieren. Ein weiterer Vorteil der Computersteuerung ist die weitgehende Vermeidung von Staubstößen am Kamin durch die automatische Anpassung der Spannung an den Klopftakt (**Bild 6**).

Auch bei der notwendigen Anpassung an die aktuellen Betriebsbedingungen des Ofensystems können die modernen Mikroelektronik-Regelsysteme Verbesserungen bringen.

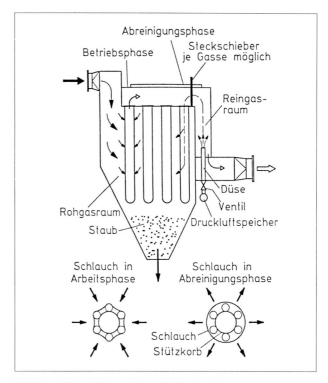

BILD 4: Schlauchfilter mit Druckluft-Abreinigung
FIGURE 4: Bag filter with compressed air cleaning

Abreinigungsphase	= cleaning phase
Steckschieber je Gasse möglich	= slide can be inserted in each duct
Betriebsphase	= operating phase
Reingasraum	= clean gas chamber
Düse	= nozzle
Ventil	= valve
Rohgasraum	= raw gas chamber
Staub	= dust
Druckluftspeicher	= compressed air receiver
Schlauch in Arbeitsphase	= bag in working phase
Schlauch in Abreinigungsphase	= bag in cleaning phase
Schlauch	= bag
Stützkorb	= supporting cage

There have also been developments in electrostatic precipitators in recent years [43–46]. One of these is the enlargement of the spacing between the collector plates from 300 to 400 mm, in conjunction with adapted geometry between the discharge and collector electrodes. Progress has also been achieved through the use of micro-electronics for controlling the high voltage system. Unlike the earlier analog control systems, the computer-controlled voltage converter can adapt optimally to changing operating conditions. As indicated in **Fig. 5**, such controls can detect fluctuations in the disruptive discharge limit and react to them using variable scanning times and stages. Another advantage of computer control is that it largely avoids dust surges at the stack by automatically adapting the voltage to the rapping cycle (**Fig. 6**).

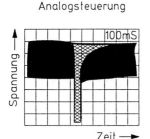

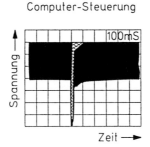

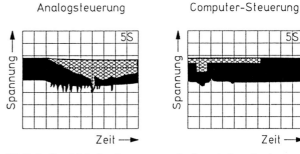

BILD 5: Spannungsverlauf bei Reaktion auf einen Durchschlag bei Analog- bzw. Computer-Steuerung
FIGURE 5: Voltage behaviour in reacting to a discharge with analog and computer control

Spannung	= voltage
Analogsteuerung	= analog control
Computer-Steuerung	= computer control
Zeit	= time

BILD 6: Vergleich des Spannungsverlaufs zwischen Analog- und Computer-Steuerung beim Abklopfvorgang
FIGURE 6: Comparison of voltage behaviour with analog and computer control during the rapping procedure

Spannung	= voltage
Analogsteuerung	= analog control
Computer-Steuerung	= computer control
Zeit	= time

3.3.2 Dust constituents

Most of the trace elements which pass into the burning process with the main and secondary constituents of the materials used remain in the clinker. The extent to which they are retained depends on the processing conditions and the volatility of the particular elements. Even under unfavourable conditions the exhaust gas from the kiln comes within the limits set by the TA Luft [47–51].

3.3.3 Gaseous emissions

As a means of reducing NO_x emissions the State Committee for Immission Protection (LAI) has defined the following technical measures in connection with the drafting of the expediency clause, as the state of the art which could be adopted in the cement industry:

— smoothing the kiln operation,
— low NO_x firing technology,
— low NO_x secondary firing system with tertiary air supply,
— selective, non-catalytic reduction (SNCR).

The first point – smoothing the kiln operation – is undoubtedly an appropriate and feasible method of reducing NO_x emissions which was quite commonly carried out in the past for quality reasons alone.

Low NO_x firing technology refers to the use of special burners which have been developed with the aim of reducing NO_x formation. They differ from the older type of burner in having a novel design for the swirl, axial and conveying air supplies and a modified flame shape. The most important aim in optimizing the flame for NO_x reduction is to obtain the same concentration profiles for the various components of the exhaust gas [52–56].

An example of such a design is the Pyro-Jet burner [57, 58], and another new development aimed at reducing NO_x is the Rotaflam burner [60–63]. Working outwards in **Fig. 7** there are the central flame stabilizer and the ducts for the pulverized coal/air mixture, swirl air and axial air. The basis for the NO_x reduction strategy is the central pulverized coal feed surrounded by the swirl air. This arrangement counteracts the combustion of pulverized coal particles under substoichiometric conditions in the hot secondary air. An NO_x emission reduction rate of about 10–30 % of that for conventional burners is quoted for this so-called low NO_x burner. However, the introduction of these new developments has been delayed, by the fact that there were problems with coating and wear in some cases as well as difficulties with burning. In addition, the emission reduction rates obtained at some plants could not always be obtained at others.

There is not yet any long-term experience of the secondary measures recommended by the State Committee for Immission Protection for reducing NO_x emissions, so they have not yet been adopted [64–70]. However, large-scale projects sponsored by the Federal Ministry of the Environment are at present being carried out at three German kiln plants [71]. Only when reliable reports are available on a series of still unknown long-term effects in these projects can such extensive retrofitting be undertaken at other cement works. The first results of experiments in reducing NO_x by pre-calcining in stages are coming in from abroad, but these need to be followed up further [72].

In the SNCR process NO reduction can be obtained without a catalyst by adding NH_3 in the 800–1000°C temperature range, but NO reduction by NH_3 in the clean gas at 80–100°C requires an activated coke catalyst [73]. This very expensive process has only been considered once so far in connection with the use of dried sewage sludge as an alternative fuel [74].

The NO_x reduction rate in the activated coke catalyst process is similar to that in the non-catalytic SNCR method,

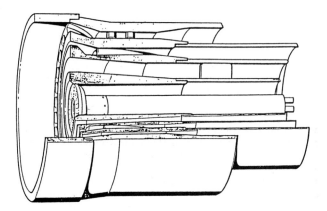

FIGURE 7: Rotaflam kiln burner [60–63]

Über die vom Länderausschuß für Immissionsschutz zur Reduzierung der NO_x-Emission empfohlenen sekundären Minderungsmaßnahmen liegen noch keine Langzeiterfahrungen vor. Sie entsprechen daher noch nicht dem Stand der Technik [64–70]. Zur Zeit werden jedoch an drei deutschen Ofenanlagen vom Bundesministerium für Umwelt geförderte Großprojekte durchgeführt [71]. Erst wenn gesicherte Aussagen über eine Reihe noch unbekannter Langzeitauswirkungen bei diesem Vorhaben vorliegen, können in weiteren Zementwerken derartig umfangreiche Nachrüstungen in Angriff genommen werden. Auch aus dem Ausland sind erste Untersuchungsergebnisse über die NO_x-Minderung durch gestufte Vorcalcination bekannt geworden, die jedoch noch einer weiteren Verfolgung bedürfen [72].

Während beim SNR-Verfahren durch NH_3-Zugabe im Temperaturbereich von 800–1000°C eine NO-Reduzierung ohne Katalysator erreicht werden kann, setzt die Anwendung der NO-Minderung durch NH_3 im Reingas bei 80–100°C einen Aktivkoks-Katalysator voraus [73]. Bisher wurde dieses sehr aufwendige Verfahren allerdings nur in einem Fall in Verbindung mit dem Einsatz von Trockenklärschlamm als Alternativbrennstoff in Erwägung gezogen [74].

Die NO_x-Minderungsrate des Aktivkoks-Katalysatorverfahrens liegt in der gleichen Größenordnung wie beim nichtkatalytischen SNR-Verfahren, wobei allerdings die spezifischen Kosten für das Aktivkoksverfahren etwa das Zehnfache betragen [75].

Die beim Zementbrennprozeß eingesetzten Roh- und Brennstoffe enthalten schwefelhaltige Verbindungen, durch deren Dissoziation und Verbrennung Schwefeldioxid (SO_2) entsteht. Dieses SO_2 bildet mit den durch das Rohmaterial eingebrachten Alkalien in der Ofengasphase direkt, oder über den Umweg einer Bindung als Calciumsulfat, Alkalisulfate [76]. Die systemimmanente Schwefeleinbindung in das Brenngut führt in der Regel zu einer deutlichen Unterschreitung des in der TA-Luft (1986) festgelegten Emissionsgrenzwertes von 400 mg SO_2/m^3. Durch Optimierung der Mehlverteilung im Vorwärmer und Erhöhung des Luftüberschusses kann die Schwefeleinbindung noch unterstützt werden [77, 78]. Enthält das Rohmaterial jedoch Schwefel in Form von Sulfiden, so wird dieser Schwefel nicht im Klinker eingebunden, und es können erhöhte SO_2-Konzentrationen im Abgas auftreten. Das ist darauf zurückzuführen, daß aus den Sulfiden bei vergleichsweise niedrigen Temperaturen beim Eintritt in den Vorwärmer SO_2 entsteht, das in diesem Bereich nur unvollkommen eingebunden wird. In derartigen Fällen kann durch Zugabe von trokkenem oder in Wasser dispergiertem Kalkhydrat $Ca(OH)_2$ eine Bindung des Schwefels erreicht werden [79–82].

Nach Laboruntersuchungen liegt das Reaktionsmaximum im Bereich des Wassertaupunktes. Einfacher ist jedoch die betriebliche Handhabung im Temperaturbereich von 400–600°C, entsprechend dem rechten Kurvenast in **Bild 8**. Bei diesen Temperaturen, die in den oberen Stufen von Zyklonvorwärmern anzutreffen sind, wird auch das SO_2 aus den Sulfiden freigesetzt. Die in Abhängigkeit vom eingesetzten Ca/S-Molverhältnis erreichbare Schwefeleinbindung nimmt allerdings mit zunehmender Einsatzmenge ab. Hierdurch wird die Anwendbarkeit dieses Verfahrens auf nicht allzu hohe SO_2-Konzentrationen begrenzt [83].

Im Zementofenabgas können auch organische Stoffe enthalten sein. Die bisher vorliegenden Untersuchungen lassen vermuten, daß bei der Emission von Gesamt-Kohlenstoff den organischen Bestandteilen im Rohmaterial besondere Bedeutung zukommt. Die Grenzwerte der TA Luft von 1986 werden jedoch mit Sicherheit eingehalten. Die gemessenen Emissionen an polyzyklischen aromatischen Kohlenwasserstoffen (PAH) liegen im Bereich von 0,1–1 % der entsprechenden Grenzwerte. Im Zusammenhang mit dem Einsatz von Sekundärbrennstoffen wurden an einer Reihe von Zementofenanlagen auch Messungen der Dioxine und Furane durchgeführt. Diese haben ergeben, daß der Grenzwert der 17. BImSchV von 0,1 ng TE/m^3 eingehalten wird [84]. Die bei unterschiedlichen Betriebsbedingungen gefundenen Meßergebnisse ließen keine Abhängigkeit von den eingesetzten Brenn- bzw. Rohstoffen erkennen [34].

though the specific costs for the activated coke process are about ten times as high [75].

The raw materials and fuels used in the cement burning process contain sulphur compounds which produce sulphur dioxide (SO_2) when dissociated and burnt. The SO_2 forms alkali sulphates with the alkalis introduced with the raw material, either directly in the kiln gas phase or indirectly after being bonded as calcium sulphate [76]. The bonding of sulphur with the kiln feed, which is inherent in the system, generally leads to levels well below the emission limit of 400 mg SO_2/m^3 laid down in the TA Luft (1986). Sulphur bonding can be further assisted by optimizing meal distribution in the preheater and increasing the excess of air [77, 78]. However, if the raw material contains sulphur in sulphide form this sulphur will not become combined in the clinker and there may be higher SO_2 concentrations in the exhaust gas. This is due to the fact that SO_2 is formed from the sulphides at comparatively low temperatures on entering the preheater and is only incompletely bonded in this zone. The sulphur can be bonded in such cases by adding calcium hydrate $Ca(OH)_2$ either dry or dispersed in water [79–82].

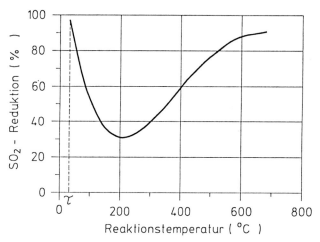

BILD 8: SO_2-Reduktion durch $Ca(OH)_2$ in Funktion der Temperatur
FIGURE 8: SO_2 reduction by $Ca(OH)_2$ as a function of temperature

SO_2-Reduktion (%) = SO_2 reduction (%)
Reaktionstemperatur (°C) = reaction temperature (°C)

According to laboratory tests the reaction is at its maximum near the dewpoint of water. However handling is simpler within the 400–600°C temperature range, corresponding to the right-hand branch of the curve in **Fig. 8**. At these temperatures, which occur in the upper stages of cyclone preheaters, the SO_2 is also liberated from the sulphides. The sulphur bonding which can be achieved is dependent on the Ca/S molar ratio used, though it decreases where larger quantities are introduced. This restricts the applicability of the process to SO_2 concentrations which are not excessively high [83].

The exhaust gas from cement kilns may also contain organic substances. Investigations so far carried out suggest that the organic constituents of the raw material are particularly significant in the emission of total carbon. However the limits set in the TA Luft of 1986 are definitely observed. Measured emissions of polycyclic aromatic hydrocarbons (PAH) come within 0.1–1 % of the corresponding limits. Measurements of dioxins and furans have been taken at a number of cement kiln plants in connection with the use of secondary fuels. They show that the limit of 0.1 ng toxic equivalent/m^3 in the 17th BImSchV is met [84]. Readings taken under differing operating conditions do not show any dependence on the fuels or raw materials used [34].

4. Blasting vibrations

Owing to their considerable hardness the limestone or lime marl strata required as a raw material sometimes have to be broken up by blasting. DIN 4150 Part 2 "Effects of vibrations on people in buildings" and Part 3 "Effects of vibrations on

4. Sprengerschütterungen

Die als Rohmaterial benötigten Kalkstein- bzw. Kalkmergelschichten müssen wegen ihrer erheblichen Festigkeit z.T. durch Sprengung gelöst werden. Zur Beurteilung von Erschütterungs-Immissionen in der Nachbarschaft werden die DIN 4150, Teil 2 „Erschütterungseinwirkungen auf Menschen in Gebäuden" und Teil 3 „Erschütterungseinwirkungen auf bauliche Anlagen" herangezogen. Sofern ein geringer Abstand zwischen Standort der Sprengung und Bebauung dies erforderlich macht, kann durch Einflußnahme auf die Lademenge je Sprengbohrloch und die Abstufung der Zündzeitstufen die Erschütterung gemindert werden [85].

Dank der gerätetechnischen Entwicklung konnte in Lagerstätten mit mäßiger Gesteinsfestigkeit das Abbauverfahren von der Gewinnungssprengung auf das Lösen mit entsprechend stark ausgelegten Hydraulik-Baggern umgestellt werden [86].

5. Lärmschutz

Die wichtigste Maßnahme für einen vorbeugenden Lärmschutz ist die Einhaltung ausreichender Abstände zwischen Zementwerk und Wohnbebauung bei der Bauleitplanung. Da dieser „natürliche" Lärmschutz in zahlreichen Fällen jedoch nicht in ausreichendem Maße eingehalten werden konnte, sind z.T. erhebliche Schallschutz-Maßnahmen erforderlich, um die Immissionswerte für Geräusche nach TA-Lärm einzuhalten [87].

Mögliche Lärmminderungsmaßnahmen beziehen sich nach [45, 88] auf

— den Einbau von Schalldämpfern,

— die lärmdämmende Einhausung hochliegender Antriebe,

— das Schließen von Gebäudeöffnungen,

— die Verwendung lärmdämmender Beläge an Fördergut-Übergabestellen,

— das Ausrüsten von Gebläsen und Kompressoren mit Ansaug- und Austrittsschalldämpfern sowie einer schalldämmenden Umhausung,

— die schwingungsisolierte Maschinenaufstellung zur Vermeidung von Körperschallübertragung.

Ausgehend von Werkslärmkarten, die den Ist-Zustand beschreiben, können die Auswirkungen geplanter Minderungsmaßnahmen auf die Lärmimmissionssituation in der Umgebung eines Zementwerkes berechnet werden [88]. Zur Minderung des Mündungsgeräusches beim Austritt von Gasströmen aus Rohrleitungen bzw. Kaminen werden Schalldämpfer eingesetzt. Dabei sind nach dem Wirkungsprinzip Absorptionsschalldämpfer und Resonatorschalldämpfer zu unterscheiden [89]. Die Wirkung eines neuentwickelten Kulissenschalldämpfers aus Membran-Absorbern, der sich durch weitgehende Unempfindlichkeit gegenüber Staubablagerungen auszeichnet, zeigt das **Bild 9** [90, 91].

6. Rohstoffgewinnung und Wiederherstellung der Landschaft

Die Sicherung des Rohstoffabbaues ist für die Zementindustrie im Wettbewerb mit anderen Nutzungsmöglichkeiten der Erdoberfläche ein lebenswichtiges Anliegen. Während in der BRD aber z.B. Landschaft, Natur und Trinkwasser durch Gesetze geschützt werden, fehlt ein spezielles Rohstoffsicherungsgesetz. Das hat eine ständige Verringerung der verbleibenden Lagerstättenvorräte durch Überplanung zugunsten anderer Nutzungsarten zur Folge [92–94].

Wesentlicher Bestandteil zur Genehmigung des Abbaus einer Lagerstätte ist der Rekultivierungsplan, in dem die spätere Folgenutzung festgeschrieben wird. Wie vielfältig die Nutzungsmöglichkeiten sind, geht aus einer vom VDZ 1984 herausgegebenen und 1990 überarbeiteten Broschüre „Alte Steinbrüche — Neues Leben" hervor. Die Broschüre enthält Bildbeispiele für die Nutzung alter Steinbrüche durch die Land- und Forstwirtschaft sowie Beispiele von

structural installations" are used for assessing vibration immissions in the vicinity. Where the closeness of the blasting location to a built-up area makes it necessary, the vibration can be lessened by adjusting the quantity of charge per blasting hole and graduating the ignition time intervals [85].

With the development of equipment technology it has been possible to switch over the quarrying process from blasting to loosening with heavy-duty hydraulic excavators in deposits with moderately hard rock [86].

5. Noise protection

The most important measure for preventive noise protection is to leave adequate distances between cement works and dwellings in the general development plan. In many cases however this "natural" protection could not be fully maintained, so extensive measures are sometimes required in order to reach the noise immission levels in the Noise Directive (TA Lärm) [87].

According to [45 and 88] possible noise abatement measures include:

— fitting sound absorbers,

— sound-insulating encasing of high-level drives,

— closing of apertures in buildings,

— use of sound-insulating lagging at transfer points for conveyed material,

— equipping fans and compressors with suction and outlet sound absorbers and a sound-insulating housing,

— vibration-isolated machine mounting to avoid structure-borne noise transmission.

The effects of planned abatement measures on the noise immission situation around a cement works can be calculated starting from works noise maps which describe the current situation [88]. Sound absorbers are fitted to reduce the noise created by gas emerging from pipes or chimneys. A distinction has to be made between absorption and resonator sound absorbers, which operate on different principles [89]. The action of a newly developed link sound absorber, which is made up of membrane absorbers and is largely insensitive to dust deposits, is shown in **Fig. 9** [90, 91].

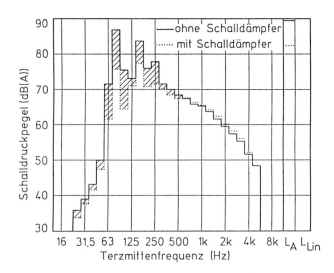

BILD 9: Lärmminderung durch einen neuartigen Kulissenschalldämpfer aus Membran-Absorbern [90, 91]

FIGURE 9: Noise abatement by a novel splitter silencer made up of membrane absorbers [90, 91]

Schalldruckpegel (dB(A))	= sound pressure level (dB(A))
ohne Schalldämpfer	= without sound absorber
mit Schalldämpfer	= with sound absorber
Terzmittenfrequenz (Hz)	= Third-band mean frequency (Hz)

Deponien aber auch von Freizeitanlagen und Naherholungsgebieten [95, 96].

7. Umweltentlastung durch Einsatz von Sekundärstoffen in der Zementindustrie

Die Rezeptur der Zementrohmischung erlaubt es, in einigen Fällen auch Reststoffe aus anderen Industrien als Rohmischungsbestandteil beim Zementbrennprozeß oder als Zumahlstoff zuzusetzen. Voraussetzung dafür ist, daß diese Reststoffe keine Nebenbestandteile enthalten, die eine Erhöhung der Umweltbelastung oder Nachteile für die Qualität zur Folge haben könnten [97–98].

Ein wichtiges Beispiel für die stoffliche Verwertung ist der Einsatz von Rauchgas-Entschwefelungsgips als Sulfatträger bei der Zementmahlung zur Regulierung des Erstarrungsprozesses.

Die bekannten, besonders günstigen Brennraumbedingungen des Drehrohrofens mit einer ausreichend langen Verweildauer der Verbrennungsgase im Temperaturbereich über 1200 °C in Verbindung mit einer oxidierenden Ofenatmosphäre bieten vielfältige Möglichkeiten, die konventionellen Brennstoffe z. T. durch brennbare Reststoffe zu ersetzen. Dabei können auch toxisch relevante Verbindungen umweltverträglich abgebaut werden. Die Sekundärbrennstoffe werden damit sowohl einer thermischen Verwertung als auch gleichzeitig einer Entsorgung zugeführt [99–114].

8. Schlußbetrachtung

Die Zementindustrie verfügt heute über Maßnahmen, mit deren Hilfe Umweltbeeinträchtigungen vermieden bzw. auf ein umweltverträgliches Maß gemindert werden können. Auch in der Zementindustrie der neuen Bundesländer konnten bei der Einführung dieser Maßnahmen bereits deutliche Fortschritte erzielt werden. Die dadurch erreichte Umweltentlastung ist beachtlich.

Die Maßnahmen zum Schutz der Umwelt haben allerdings auch ihren Preis. Etwa 15 % der Investitionskosten für Anlagen zur Herstellung von Zement entfallen in der Bundesrepublik Deutschland inzwischen auf Umweltschutzeinrichtungen. Der daraus resultierende Kapitaldienst macht zusammen mit den Kosten für den Betrieb dieser Umweltschutzeinrichtungen mehr als 10 % vom Umsatz aus. Weitere Maßnahmen, die z. Z. noch in der Entwicklung bzw. Langzeiterprobung sind, wie das SNR-Verfahren zur NO_x-Minderung, werden diese Kostenanteile noch erhöhen.

Von besonderer Bedeutung für die energieintensive Zementindustrie ist auch der Preis der elektrischen Energie. **Bild 10** läßt die erhebliche Differenzierung in einer Auswahl wichtiger Industrieländer erkennen. Dabei ist der Strompreis in Deutschland seinerseits bereits stark durch Umweltschutzkosten belastet.

Unterschiedliche Belastungen durch kostenrelevante Umweltauflagen bzw. -abgaben und -steuern, wie das im Zusammenhang mit dem Klimaschutz diskutiert wird, dürfen in einem zusammenhängenden Wirtschaftsraum nicht zu Wettbewerbsverzerrungen führen. Die zwangsläufige Folge wäre eine Produktionsverlagerung an Standorte mit geringerer Kostenbelastung für die Umweltvorsorge, aber mit wahrscheinlich höherer Belastung der Umwelt infolge einer weniger modernen Anlagentechnik und geringerer Effizienz des Energieeinsatzes.

Wie differenziert die einwohnerbezogenen Umweltausgaben in der OECD inzwischen sind, zeigt **Bild 11**. Danach werden in den aufgeführten OECD-Ländern einwohnerbezogen im Durchschnitt jährlich etwa 180 US$ für den Umweltschutz aufgewendet. Vordringlich erscheint es nun, auch in anderen Ländern, z. B. in den ehemaligen Ostblockstaaten, eine Angleichung an dieses Niveau zu erreichen. Auf diese Weise könnte ein deutlich höherer, aufwandbezogener Nutzen und damit auch ein größerer Beitrag zur Umweltentlastung geleistet werden als das durch weitere kostenträchtige Maßnahmen in der Europäischen Gemein-

6. Quarrying the raw materials and restoring the landscape

Safeguarding raw material extraction is a vital concern for the cement industry in competition with other possible uses of the earth's surface. But whereas there are laws to protect e.g. the landscape, nature and drinking water in the Federal Republic, there is no particular law to safeguard raw materials. As a result the remaining deposits are constantly being reduced through over-planning in favour of other uses [92–94].

The recultivation plan laying down the subsequent use of the area plays an important part in obtaining approval for quarrying a deposit. The variety of possible uses can be seen from a brochure published by the VDZ in 1984 and revised in 1990 entitled "Old quarries – new life". The brochure contains illustrations giving examples of the use of old quarries for agriculture and forestry and examples of landfill sites, and also of leisure amenities and recreation areas [95, 96].

7. Relieving environmental pollution by using secondary materials in the cement industry

The formulation of the raw cement mix allows residual materials from other industries to be added in some cases, either as a component of the raw mix in the cement-burning process or as an interground additive. The prerequisite for this is that the residual materials should not contain any secondary constituents which might cause an increase in environmental pollution or be detrimental to quality [97–98].

An important example of the re-use of materials is the use of flue gas desulphogypsum as a sulphate agent in cement grinding to regulate the setting process.

The well-known and particularly favourable combustion conditions in a rotary kiln, with a sufficiently long residence time for the combustion gases in the temperature range above 1200°C combined with an oxidizing kiln atmosphere, offer various opportunities for replacing some conventional fuels with combustible residual materials. Toxic compounds may also be decomposed in an environmentally friendly manner. This not only uses the secondary fuels for heating but also disposes of them [99–114].

8. Conclusions

The cement industry now has measures whereby damage to the environment can be avoided or reduced to an environmentally tolerable level. The cement industry in the former East Germany has also made clear progress in introducing these measures. The relief to the environment is considerable.

However, there is a price to pay for environmental protection measures. About 15 % of the capital cost for German cement-producing plant is now allotted to environmental protection equipment. The resultant capital charges together with the cost of operating the equipment account for more than 10 % of the turnover. Other measures which are at present still being developed or undergoing long-term testing, such as the SNCR process for reducing NO_x, will further increase this percentage.

A particularly important factor for the energy-intensive cement industry is the price of electricity. **Fig. 10** shows the considerable differences between a number of important industrial countries. The price of power in Germany is already greatly inflated by environmental protection costs.

Different burdens imposed by cost-relevant environmental requirements or environmental taxes and duties, such as are discussed in connection with the protection of the climate, must not be allowed to distort competition in a unified economic area. This would necessarily lead to the transfer of production to locations where costs are less inflated by environmental provisions but where there is probably more

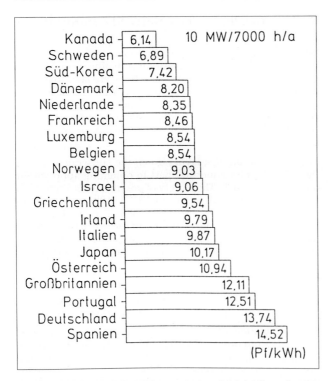

BILD 10: Internationaler Industriestrompreisvergleich nach UNI-PEDE, Paris (Preise einschl. Steuern und Abgaben ohne MwSt, Preisstand 1. 1. 1991)
FIGURE 10: Comparison of international industrial power prices by UNIPEDE, Paris (prices include taxes and duties except VAT, as at 1.1.1991 level)

Kanada	= Canada	Griechenland	= Greece
Schweden	= Sweden	Irland	= Ireland
Süd-Korea	= South Korea	Italien	= Italy
Dänemark	= Denmark	Japan	= Japan
Niederlande	= Netherlands	Österreich	= Austria
Frankreich	= France	Großbritannien	= Great Britain
Luxemburg	= Luxembourg	Portugal	= Portugal
Belgien	= Belgium	Deutschland	= Germany
Norwegen	= Norway	Spanien	= Spain
Israel	= Israel	(Pf/kWh)	= (German Pfennig/kWh)

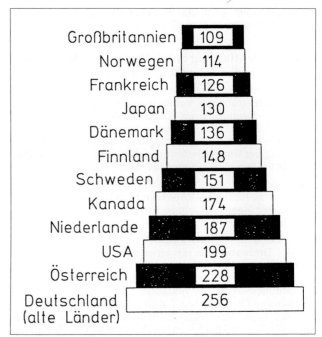

BILD 11: Einwohnerbezogene Umweltausgaben in der OECD (Öffentliche und private Ausgaben nach Preisen von 1980)
FIGURE 11: Expenditure on the environment in the OECD, related to inhabitants (public and private expenditure at 1980 prices)

Großbritannien	= Great Britain
Norwegen	= Norway
Frankreich	= France
Japan	= Japan
Dänemark	= Denmark
Finnland	= Finland
Schweden	= Sweden
Kanada	= Canada
Niederlande	= Netherlands
USA	= USA
Österreich	= Austria
Deutschland (alte Länder)	= Germany (former West Germany)

schaft möglich ist. Dort stellen die Aufwendungen für den Umweltschutz, zumindest in einigen Wirtschaftszweigen, zu denen auch die Zementindustrie gehört, bereits einen erheblichen Anteil der Gesamtkosten dar. Für die Zukunft darf nicht allein das technisch Machbare als Maßstab gelten. Vielmehr ist Umweltschutz-Vorsorge einerseits, gegen wirtschaftliche Belastung andererseits, abzuwägen.

Literaturverzeichnis

[1] Kroboth, K., und Xeller, H.: Entwicklungen beim Umweltschutz in der Zementindustrie. Zement-Kalk-Gips 39 (1986) Nr. 1, S. 1–14.

[2] Locher, F.W.: Entwicklung des Umweltschutzes in der Zementindustrie. Zement-Kalk-Gips 42 (1989) Nr. 3, S. 120–127.

[3] Kontinuierliche Gasanalyse in Zementwerken, Merkblatt Vt 9 des Vereins Deutscher Zementwerke Düsseldorf, Juni 1990.

[4] Xeller, H.: Messen und Verarbeiten von umweltrelevanten Daten. Zement-Kalk-Gips 47 (1994) Nr. 1, S. 13–20.

[5] Kuhlmann, K.: Ökobilanz für Baustoffe am Beispiel von Zement und Beton. Zement-Kalk-Gips 47 (1994) Nr. 1, S. 25–28.

[6] Wischers, G.: Beton und Umwelt — Ökobilanz für Beton. Betonwerk + Fertigteil-Technik 58 (1992) Nr. 4, S. 50–61.

[7] Wischers, G., und Kuhlmann, K.: Ökobilanz von Zement und Beton. Zement-Kalk-Gips 44 (1991) Nr. 11, S. 545–553.

pollution as a result of less modern plant technology and less efficient use of energy.

The difference between environmental expenditure related to inhabitants in the OECD is shown in **Fig. 11**. The average annual expenditure on environmental protection per inhabitant of the OECD countries listed is approximately 180 US$. There would seem to be an urgent need to bring other countries, such as those in the former East block, to this level. This could obtain far greater benefit, relative to expenditure, and thus a greater contribution could be made towards relieving the environment than is possible through further costly measures in the European Community. In the Community expenditure on environmental protection, at least in some branches of the economy including the cement industry, already accounts for a considerable proportion of total costs. Technical feasibility must not form the sole yardstick in the future; it is more important to balance provision for environmental protection against the economic burden.

[8] Bundes-Immissions-Schutzgesetz (BImSchG) in der Fassung vom 14. 5. 1990 (BGBl. S. 880), zuletzt geändert am 10. 12. 1990 (BGBl. IS 2634).

[9] Büge, D.: Die 3. Novelle zum Bundes-Immissions-Schutzgesetz und ihre Bedeutung für die Betreiber genehmigungspflichtiger Anlagen. Wirtschaftsrecht, H. 48 vom 30. 11. 1990, S. 2408–2412.

[10] Entwurf Wärmenutzungsverordnung. Energiespektrum Mai 1992, S. 39–50.

[11] 1. Allgemeine Verwaltungsvorschrift zum Bundes-Immissionsschutzgesetz (Technische Anleitung zur Reinhaltung der Luft – TA Luft – vom 27. 2. 1986 (GMBL. S. 95, 202).

[12] Hinz, W.: Probleme der Altanlagensanierung in der Zementindustrie. Zement-Kalk-Gips 42 (1989) Nr. 3, S. 136–141.

[13] Keinhorst, H.: Anforderungen an den Umweltschutz in der Zementindustrie. Zement-Kalk-Gips 41 (1988) Nr. 1, S. 30–31.

[14] Davids, P.: Den Stand der Technik konkretisieren. Umwelt 23 (1993) Nr. 4, S. 202–206.

[15] Feldhaus, G.: Luftreinhaltungsgesetzgebung in Europa, Vortragsband der VGB-Konferenz Kraftwerke und Umwelt 1993. VGB-Kraftwerkstechnik GmbH, Verlag technisch-wissenschaftlicher Schriften, S. 27–31.

[16] EG-Richtlinie 84/360 v. 28. 6. 1984, Richtlinie zur Bekämpfung der Luftverunreinigung durch die Industrieanlagen, sogenannte Industriezulassungsrichtlinie. Amtsblatt der Europäischen Gemeinschaft, 16. 7. 1984, Nr. L 188/20.

[17] Richtlinie 85/337 vom 27. 6. 1985, Richtlinie über die Umweltverträglichkeitsprüfung bei bestimmten öffentlichen und privaten Projekten.

[18] Gesetz zur Umsetzung der Richtlinie des Rates vom 27. 6. 1985 über die Umweltverträglichkeitsprüfung bei bestimmten öffentlichen und privaten Projekten (85/337/EWG) vom 12. 2. 1990 (BGBl. I. S. 205), zuletzt geändert am 20. 6. 1990 (BGBl. I. 1080).

[19] Schwab, J.: Die Umweltverträglichkeitsprüfung in der Genehmigungspraxis. Abfall-Wirtschaftsjournal 4 (1992) Nr. 6, S. 501–504.

[20] EG-Richtlinie 90/313 (EWG) vom 7. 6. 1990, Richtlinie über den freien Zugang zu Umweltinformationen, sog. Umweltinformationsrichtlinie.

[21] Verordnung (EWG) Nr. 1836/93 des Rates v. 29. 6. 1993 über die freiwillige Beteiligung gewerblicher Unternehmen an einem Gemeinschaftssystem für das Umweltmanagement und die Umweltbetriebsprüfung, Amtsblatt der Europäischen Gemeinschaft Nr. L 168/1–14 sog. Öko-Audit-Verordnung.

[22] Vorschlag für eine Richtlinie des Rates über die integrierte Vermeidung und Verminderung der Umweltverschmutzung (Integrated Pollution, Prevention and Control) (IPPC) KOM (93) 423 endg. 14. 9. 1993.

[23] Vorschlag für eine Richtlinie des Rates zur Einführung einer Steuer auf Kohlendioxidemissionen und Energie, 92/C 196/01, vorgelegt am 2. 6. 1992, Amtsblatt der Europäischen Gemeinschaft Nr. C 196/1–8.

[24] Natürlicher Treibhauseffekt lebensnotwendig, VDZ-Mitteilungen 79, Mai 1989, S. 1–2.

[25] Comments of the European Cement Industry on Taxation of CO_2 Emissions. Ciments, Bétons, Plâtres, Chaux, Nr. 792-5/91, S. 294–295.

[26] Beschluß der Wirtschaftsministerkonferenz zu den Vorschlägen der EG-Kommission zur Einführung einer CO_2-/Energie-Steuer. Energiespektrum 11 (1992) November, S. 6–7.

[27] Entwurf für EG-Richtlinie über Verbrennung gefährlicher Abfälle (COM (92)) (9 final) Sitzung des Rates vom Juni 1993.

[28] 17. Verordnung zur Durchführung des Bundes-Immissions-Schutzgesetzes (Verordnung über Verbrennungsanlagen für Abfälle und ähnliche brennbare Stoffe) vom 23. 11. 1990 (BGBl., I. S. 2545, berichtigt S. 2832/BGBl. III 2129-8-1-17) 5.

[29] Mikols, E. H.: Staatliche Umweltschutzregelungen in den USA für die Verbrennung von Sondermüll in Portlandzementöfen. Vortrag zum VDZ-Kongreß '93, Fachbereich 4.

[30] von Seebach, H.M., und Tseng, H.: Herstellung von Zementen unter Verwendung von Sondermüll und von fossilen Brennstoffen und ihre Eignung bei Anwendung im Trinkwasserbereich. Vortrag zum VDZ-Kongreß '93, Fachbereich 1.

[31] Kroboth, K., und Kuhlmann, K.: Stand der Technik der Emissionsminderung in Europa. Zement-Kalk-Gips 43 (1990) Nr. 3, S. 122–131.

[32] Erlaß des Ministeriums für Umwelt, Raumordnung und Landwirtschaft des Landes Nordrhein-Westfalen (MURL) vom 10. 7. 1991 über die Dynamisierung des NO_x-Grenzwertes.

[33] Neisecke, P., und Kamm, K.: Thermische Altreifenverwertung – Mischgrenzwertberechnung nach der 17. BImSchV am Beispiel der Zementindustrie, WLB Envitec-Report 1992, S. 52–56.

[34] Verein Deutscher Zementwerke e. V., Forschungsinstitut der Zementindustrie: Tätigkeitsbericht 1990–1993. Beton-Verlag GmbH, Düsseldorf.

[35] Länderausschuß für Immissionsschutz (LAI): Zweifelsfragen bei der Auslegung und Anwendung der 17. BImSchV (Abfallverbrennungsanlagen-Verordnung), herausgegeben vom Ministerium für Umwelt, Raumordnung und Landwirtschaft des Landes Nordrhein-Westfalen, Düsseldorf 1992.

[36] Kuhlmann, K., und Kirchartz, B.: Beitrag der Emissionen von Zementwerken zur regionalen Immission. Vortrag anläßlich der Technisch-Wissenschaftlichen Zementtagung 1991 in München.

[37] Winkler, H.-D.: Thallium-Emissionen bei der Zementherstellung. LIS-Berichte Nr. 64, 1986, Landesanstalt für Immissionsschutz des Landes Nordrhein-Westfalen.

[38] Scur, P.: Entwicklung der Immissionen in der Umgebung der Zementwerke in den neuen Bundesländern am Beispiel Rüdersdorf. Vortrag zum VDZ-Kongreß '93, Fachbereich 4.

[39] Binn, F.J.: Modernisierung bestehender Filteranlagen in Zementwerken. Zement-Kalk-Gips 41 (1988) Nr. 4, S. 183–187.

[40] Funke, G.: Beurteilung von Faserstoff-Filtern. Zement-Kalk-Gips 40 (1987) Nr. 5, S. 273–274.

[41] Dietrich, H.: Wirtschaftliche Filtermedien für Staubabscheider in der Zementindustrie. Zement-Kalk-Gips 40 (1987) Nr. 11, S. 543–548.

[42] Adlhoch, H.-J.: Fortschritte bei der Konstruktion und Bewertung von Filteranlagen. Zement-Kalk-Gips 46 (1993) Nr. 5, S. 256–260.

[43] Werner, G.: Electrostatic Precipitators in Cement Plants. International Cement Review 1991, August, S. 61–65.

[44] Xeller, H.: Modernisierung bestehender Verdampfungskühler in der Zementindustrie. Zement-Kalk-Gips 41 (1988) Nr. 4, S. 176–182.

[45] Verein Deutscher Zementwerke e. V., Forschungsinstitut der Zementindustrie: Tätigkeitsbericht 1984–1987, Beton-Verlag GmbH, Düsseldorf.

[46] Braig, H.: Maßnahmen zum optimierten Betrieb von Elektrofiltern. Vortrag zum VDZ-Kongreß '93, Fachbereich 4.

[47] Sprung, S.: Spurenelemente. Zement-Kalk-Gips 41 (1988) Nr. 5, S. 251–256.

[48] Kirchner, G.: Das Verhalten des Thalliums beim Brennen von Zementklinker. Schriftenreihe der Zementindustrie, Nr. 47/1986, Beton-Verlag GmbH, Düsseldorf.

[49] Kirchner, G.: Thallium-Kreisläufe und Thallium-Emissionen beim Brennen von Zementklinker. Zement-Kalk-Gips 40 (1987) Nr. 3, S. 134–144.

[50] Kirchner, G.: Reaktionen des Cadmiums beim Klinkerbrennprozeß. Zement-Kalk-Gips 38 (1985) Nr. 9, S. 535–539.

[51] Kirchartz, B.: Abscheidung von Spurenelementen beim Klinkerbrennprozeß. Vortrag zum VDZ-Kongreß '93, Fachbereich 4.

[52] Scheuer, A., und Gardeik, H.O.: Bildung und Abbau von NO in Zementofenanlagen. Zement-Kalk-Gips 38 (1985) Nr. 2, S. 57–66.

[53] Scheuer, A., Schmidt, K.D., Gardeik, H.O., und Rosemann, H.: Einflüsse auf die Entstehung von gasförmigen Schadstoffen bei Drehofenanlagen der Zementindustrie und Primärmaßnahmen zu ihrer Minderung. Sonderdruck aus „IF – Die Industriefeuerung", H. 38, S. 65–78, Vulkan-Verlag, Essen.

[54] Scheuer, A.: Minderung der NO_x-Emission beim Brennen von Zementklinker. Zement-Kalk-Gips 41 (1988) Nr. 1, S. 37–42.

[55] Kirsch, J., und Scheuer, A.: Tätigkeit der VDZ-Kommission „NO_x-Minderung". Zement-Kalk-Gips 41 (1988) Nr. 1, S. 32–36.

[56] Xeller, H.: NO_x-Minderung durch Einsatz eines Stufenbrenners mit Rauchgasrückführung vom Vorwärmer. Zement-Kalk-Gips 40 (1987) Nr. 2, S. 57–63.

[57] Bauer, C.: Pyro-Jet burners to reduce NO_x emissions – current developments and practical experience. World Cement, April 1990.

[58] Steinbiß, E., Bauer, C., und Breidenstein, W.: Entwicklungsstand des Pyro-Jet-Brenners. Vortrag zum VDZ-Kongreß '93, Fachbereich 3.

[59] Rosemann, H., und Künne, P.: Betriebserfahrungen mit einem neuartigen Drehofen-Brenner. Zement-Kalk-Gips 43 (1990) Nr. 9, S. 421–424.

[60] Breitenbaumer, C.: Betriebsergebnisse und NO_x-Emissionen mit dem ROTAFLAM-Brenner im Zementwerk Retznei. Zement-Kalk-Gips 45 (1992) Nr. 5, S. 239–242.

[61] Endres, G., und Gauthier, J.-C.: Réduction des émissions et de la consommation d'énergie des fours rotatifs au moyen d'une technologie de combustion de pointe. Ciments, Bétons, Plâtres, Chaux, No. 800, 1/93.

[62] Endres, G.: Utilization of advanced combustion technology in kilns. World Cement, März 1993.

[63] Endres, G., Gauthier, J.-C., und Tejuca, J.A.: Reducción de emisiones y consumo de energía en hornos rotativos, por medio de una tecnología punta de combustión. Cemento-Hormigón No. 718, April 1993, S. 413–426.

[64] Scheuer, A.: Nichtkatalytische Reduktion des NO mit NH_3 beim Zementbrennen. Zement-Kalk-Gips 43 (1990) Nr. 1, S. 1–12.

[65] Rother, W., und Kupper, D.: Brennstoffstufung – ein wirksames Mittel zur NO_x-Emissionsminderung. Zement-Kalk-Gips 42 (1989) Nr. 9, S. 444–447.

[66] Gardeik, H.O.: Optimierung von Drehöfen der Zementindustrie im Hinblick auf Produktqualität, Energieeinsatz und Schadstoffemission. Zement-Kalk-Gips 44 (1991) Nr. 3, S. 105–109.

[67] Kupper, D., und Brentrup, L.: SNCR technology for NO_x reduction in the cement industry. World Cement, März 1992, S. 4–8.

[68] Kupper, D., und Adler, K.: Multi-stage combustion minimises NO_x emissions. International Cement Review, Juni 1993, S. 61–69.

[69] Høidalen, Ø.: Modernisierung und Produktionssteigerung der Drehofenanlage 6 im Werk Dalen der Norcem A/S. Sonderdruck aus Zement-Kalk-Gips 43 (1990) Nr. 3, S. 132–138.

[70] Jorget, S.: The new CF/FCB Low NO_x-precalciner. Zement-Kalk-Gips 46 (1993) Nr. 4, S. 193–196.

[71] Ruhland, W., und Hoppe, H.: Die betriebliche Umsetzung von NO_x-Minderungsmaßnahmen (Bericht des VDZ-Arbeitskreises „NO_x-Minderung"). Vortrag zum VDZ-Kongreß '93, Fachbereich 4.

[72] Syverud, T.S., Thomassen, A., und Høidalen, Ø., : NO_x-Reduzierung im norwegischen Zementwerk Brevik – Versuche mit gestufter Brennstoffzufuhr zum Calcinator. Zement-Kalk-Gips 47 (1994) Nr. 1, S. 40–42.

[73] Rose, D.: Technik der Emissionsminderung für Drehofenabgase. Vortrag zum VDZ-Kongreß '93, Fachbereich 4.

[74] de Quervain, B.: Weitergehende Rauchgasreinigung für den Einsatz von Trockenklärschlamm als Alternativ-Brennstoff bei der Portland-Cement-Werk Würenlingen-Siggenthal AG. Vortrag zum VDZ-Kongreß '93, Fachbereich 4.

[75] Sadowsky, U., und Söllenböhmer, F.: Relevanz organischer Emissionen in der Zementindustrie. Vortrag zum VDZ-Kongreß '93, Fachbereich 4.

[76] Sprung, S.: Technologische Probleme beim Brennen des Zementklinkers, Ursache und Lösung. Schriftenreihe der Zementindustrie, Heft 43 (1982), Beton-Verlag GmbH, Düsseldorf.

[77] Schmidt, K.D., Gardeik, H.O., und Ruhland, W.: Verfahrenstechnische Einflüsse auf die SO_2-Emission aus Drehofenanlagen. Zement-Kalk-Gips 39 (1986) Nr. 2, S. 93–101.

[78] Rodenhäuser, F., und Herchenbach, H.: Verringerung der gasförmigen Schadstoff-Emissionen aus Zementanlagen. Zement-Kalk-Gips 39 (1986) Nr. 10, S. 575–577.

[79] Nielsen, P.B.: Die SO_2- und NO_x-Emissionen bei modernen Zementdrehofenanlagen mit Blick auf zukünftige Verordnungen. Zement-Kalk-Gips 44 (1991) Nr. 9, S. 449–456.

[80] Kühle, K.: Verfahren zur trockenen Schadgasabscheidung aus Rauchgas. Zement-Kalk-Gips 11 (1987) Nr. 11, S. 537–542.

[81] Bonn, W., und Hasler, R.: Verfahren und Erfahrung einer rohstoffbedingten SO_2-Emission im Werk Untervaz der Bündner Cementwerke. Zement-Kalk-Gips 43 (1990) Nr. 3, S. 139–143.

[82] Kupper, D., Rother, W., und Unland, G.: Trends in desulphurisation and denitration techniques in the cement industry. World Cement, März 1991. S. 94–103.

[83] Boes, K.H.: Maßnahmen zur Minderung der SO_2-Emission beim Klinkerbrennen im Werk Höver der Nordcement AG. Vortrag zum VDZ-Kongreß '93, Fachbereich 4.

[84] Bolwerk, R.: The reduction of dioxin in connection with the use of wastederived fuels in the cement industry. Tagungsband 9. Internationaler Zement-Chemie-Kongreß in Delhi vom 23. 11.–28. 11. 1992, S. 217–223.

[85] Küllertz, P., und Schneider, M.: Der Einfluß von Gewinnungssprengungen auf die Erschütterungsimmission. Vortrag zum VDZ-Kongreß '93, Fachbereich 4.

[86] Boßmann, W.: Abbau mit dem Hydraulik-Bagger als Alternative zur Gewinnungssprengung. Vortrag zum VDZ-Kongreß '93, Fachbereich 4.

[87] Allgemeine Verwaltungsvorschrift über genehmigungsbedürftige Anlagen nach § 16 der Gewerbeordnung. Technische Anleitung zum Schutz gegen Lärm

(TA Lärm) vom 16. 7. 1968 (Beilage zum BAnz. Nr. 137 v. 26. 7. 1968).

[88] Schneider, M., und Küllertz, P.: Gezielte Lärmminderungsprogramme für Zementwerke. Vortrag zum VDZ-Kongreß '93, Fachbereich 4.

[89] Funke, G.: Auslegung und Betrieb von Abgasschalldämpfern in Zementwerken. Zement-Kalk-Gips 41 (1988) Nr. 1, S. 42–47.

[90] Fuchs, H. V.: Schalldämpfer für stark verschmutzende Abgasanlagen. Zement-Kalk-Gips 46 (1993) Nr. 5, S. 261–267.

[91] Fuchs, H. V., und Eckoldt, D.: Entwicklung von Schalldämpfern für die Zementindustrie. Vortrag zum VDZ-Kongreß '93, Fachbereich 4.

[92] Bundesverband der Deutschen Zementindustrie e. V.: Leitfaden zur Rohstoffsicherung für die Zementindustrie. Hausdruckerei BDZ, Köln, 1981.

[93] Becker-Platen, J. D., und Pauly, E.: Rohstoffsicherung und Kategorisierung oberflächennaher Rohstoffe in den Ländern der BRD, Geologisches Jahrbuch A 75, Hannover 1984.

[94] Mattig, U.: Der Einsatz von Naturraumpotential-Karten als Beitrag zur raumplanerischen Sicherung oberflächennaher Rohstoffe Sand und Kies in der BRD und in Norwegen. Dissertation Erlangen (1991).

[95] Verein Deutscher Zementwerke e. V.: Alte Steinbrüche – Neues Leben, 2. überarbeitete Auflage 1990, Beton-Verlag GmbH, Düsseldorf.

[96] Stein, V.: Anleitung zur Rekultivierung von Steinbrüchen und Gruben der Steine- und Erden-Industrie. Deutscher Instituts-Verlag (1985).

[97] Sprung, S.: Umweltentlastung durch Verwertung von Sekundärrohstoffen. Zement-Kalk-Gips 45 (1992) Nr. 5, S. 213–221.

[98] Schmidt, M.: Zement mit Zumahlstoffen – Leistungsfähigkeit und Umweltentlastung, Teil 2. Zement-Kalk-Gips 45 (1992) Nr. 6, S. 296–301.

[99] Kirsch, J.: Umweltentlastung durch Verwertung von Sekundärbrennstoffen. Zement-Kalk-Gips 44 (1991) Nr. 12, S. 605–610.

[100] Bolwerk, R.: Behördliche Umweltschutzanforderungen beim Einsatz alternativer Brenn- und industrieller Reststoffe im Zementdrehrohrofen. Vortrag zum VDZ-Kongreß '93, Fachbereich 4.

[101] Lawton, J. M.: The European cement industry's approach to the use of secondary raw materials and fuels. Ciments, Bétons, Plâtres, Chaux No. 799, 6/92, S. 357–359.

[102] Rose, D., und Kupper, D.: Ökologische und ökonomische Aspekte der Zementherstellung bei Einsatz von Abfallbrennstoffen. Zement-Kalk-Gips 44 (1991) Nr. 11, S. 554–559.

[103] Campbell, R. L.: What you should know about sourcing waste fuels. Rock Products, April 1993, S. 28–32.

[104] Chahine, G.: Incidence de l'incinération sur les émissions conventionelles du four à ciment. Ciments, Bétons, Plâtres, Chaux No. 791 4/91, S. 215–220.

[105] Dégre, J.-P.: Incineration of waste in a cement kiln. International Cement Review, Dezember 1991, S. 25–32.

[106] Leveque, E.: De l'énergie avec des déchets: le combsu de scori. Ciments, Bétons, Plâtres, Chaux No. 799, 6/92, S. 388–389.

[107] Krogbeumker, G.: Sicherheitseinrichtungen bei zusätzlicher Verbrennung PCB-haltiger Altöle im Zementdrehofen. Zement-Kalk-Gips 41 (1988) Nr. 4, S. 188–192.

[108] Dambrine, E.: Incinération des déchets et destruction des composés organiques dans le four à ciment. Ciments, Bétons, Plâtres, Chaux No. 800, 1/93, S. 37–42.

[109] Gerger, W., und Liebl, P.: Thermische Verwertung von Sekundärbrennstoffen bei den Gmundner Zementwerken. Zement-Kalk-Gips 44 (1991) Nr. 9, S. 457–462.

[110] Neumann, E.: Energy alternatives. International Cement Review, Mai 1992, S. 61–67.

[111] Dawson, B.: Emerging technologies for utilizing waste in cement production. World Cement, Dezember 1992, S. 22–24.

[112] Blumenthal, M.: The use of scrap tyres in the US cement industry. World Cement, Dezember 1992, S. 14–20.

[113] Siemering, W., Parsons, L. J., und Lochbrunner jr., P.: Experiences with burning waste. Rock Products, April 1991, S. 36–47.

[114] Kelly, K. E., und Beahler, C. C.: Burning hazardous waste in cement kilns. Vortragsmanuskript anl. der Kilnburn '92 Conference, Brisbane, am 10. 9. 1992.

Messen und Verarbeiten von umweltrelevanten Daten*)
Measuring and processing environmentally relevant data*)

Mesure et traitement des données concernant l'environnement

Medición y tratamiento de datos relevantes para el medio ambiente

Von **H. Xeller,** Heidelberg/Deutschland

Fachbericht 4.1 · Zusammenfassung – Das Messen und Auswerten umweltrelevanter Daten spielt eine zentrale Rolle in der Umweltdiskussion. Ohne eine leistungsfähige Umweltanalytik ist wirkungsvolle Umweltvorsorge nicht denkbar. Durch moderne Analysenmethoden können immer kleinere Schadstoffmengen aufgespürt und kontrolliert werden. Mit zunehmender Empfindlichkeit steigt aber auch die Fehleranfälligkeit, insbesondere bei der Probenahme und Aufbereitung. Höhere Anforderungen an die Qualitätsüberwachung bei der Ermittlung umweltrelevanter Daten haben jedoch erheblich steigende Kosten zur Folge. Für die Beurteilung der Umweltsituation bei der Zementherstellung bildet die Reinhaltung der Luft den Schwerpunkt. Daneben können aber auch Lärm und Erschütterungen von Bedeutung sein. Zur Überwachung der Emissionen und Immissionen steht ein umfangreiches Instrumentarium leistungsfähiger Meßverfahren für kontinuierliche Messungen und Einzelmessungen zur Verfügung. Darüber hinaus gewinnen die Meßdatenaufzeichnung, -verarbeitung und -auswertung, angefangen beim Emissionsrechner für die Einzelanlage bis hin zur zentralen landesweiten Fernüberwachung, sowie der Aufbau bereichsübergreifender Informationssysteme zunehmend an Bedeutung. Die Meß- und Verarbeitungsmethoden für umweltrelevante Daten sind international noch nicht vereinheitlicht. Die Art und Weise, wie die Daten erfaßt und ausgewertet werden, kann sich daher auf das Ergebnis auswirken. Hinzu kommt, daß die Forschung, die sich mit der Wirkung von Emissionen und Immissionen befaßt, nicht mehr mit der Fortentwicklung der Meßtechnik Schritt halten kann. Das führt vielfach zu Verunsicherungen und damit zu einer teilweise chaotischen Informationspolitik. Zur Verbesserung der Transparenz und Vergleichbarkeit in der Umweltdiskussion sind klare Vorgaben für das Messen und Auswerten und die Bewertung umweltrelevanter Daten auf internationaler Ebene dringend erforderlich.

Special report 4.1 · Summary – Measurement and evaluation of environmentally relevant data play a central role in the environmental debate. Effective environmental precautions are not possible without an efficient method of environmental analysis. Ever smaller quantities of harmful materials can be tracked down and checked with modern analytical methods. However, with increasing sensitivity there is also an increase in the susceptibility to errors, especially when taking and processing the samples. Higher requirements for quality monitoring when measuring environmentally relevant data have resulted in greatly increased costs. Reducing air-borne pollution forms the focus for assessing the environmental situation in cement production, but noise and vibration can also be important. There is an extensive range of efficient test methods available for continuous measurements and spot measurements for monitoring emissions and immissions. Increasing importance is also being placed on the display, processing and evalution of the measured data, ranging from the emission computer for individual plants right up to centralized regional remote monitoring, and on setting up information systems covering all spheres. The methods of measuring and processing environmentally relevant data are not yet standardized internationally. The way in which the data is collected and evaluated can affect the result. Added to this is the fact that research which is concerned with the effect of emissions and immissions can no longer keep pace with the onward development of measurement technology. This often leads to uncertainties and in some cases to chaotic information policy. Clear objectives for the measurement, analysis and evaluation of environmentally relevant data are absolutely essential on an international level to improve transparency and comparability in the environmental debate.

Rapport spécial 4.1 · Résumé – La mesure et l'analyse des données sur l'environnement jouent un rôle capital dans les discussions. Sans une analyse performante des données, il est impensable de mettre en place une politique de protection de l'environnement efficace. Les méthodes d'analyse moderne permettent de déceler et de contrôler des quantités de matières toxiques toujours plus fines. Cependant au fur et à mesure qu'augmente la sensibilité, le risque d'erreurs augmente aussi, notamment pour la pri-

---

*) Überarbeitete Fassung eines Vortrages zum VDZ Kongreß '93, Düsseldorf (27.9.–1.10.1993)
Revised text of a lecture to the VDZ Congress '93, Düsseldorf (27.9.–1.10.1993)

se d'échantillons et le traitement. Les exigences accrues en matière de surveillance de la qualité notamment pour le calcul de donnée concernant l'environnement, entraînent des frais croissants considérables. Pour apprécier la situation de l'environnement durant la fabrication du ciment, le maintien de la pureté de l'air constitue une priorité, à côté de cela on peut ranger par ordre d'importance également le bruit et les vibrations. Pour surveiller les émissions on dispose de nombreux outils de méthodes de mesure performantes pour mesures continues et mesures ponctuelles. De plus, l'enregistrement, le traitement et l'analyse des données mesurées, à commencer par le calculateur d'émissions pour une installation isolée jusqu'à la télésurveillance centrale étendue à tout un pays, ainsi que la mise en place de systèmes d'informations inter-secteurs, prennent une importance accrue. Les méthodes de traitement et de mesure des données concernant l'environnement ne sont pas encore uniformisées à l'échelon international. Le mode de saisie et d'analyse des données peut donc influer sur le résultat. A cela s'ajoute le fait que la recherche, qui se penche sur les effets des émissions, ne parvient plus à avancer au même rythme que les progrès réalisés par la technique des mesures. Cela entraîne fréquemment des données peu sûres et par là même une politique d'information chaotique par moments. Pour améliorer la transparence et les moyens de comparaison lors des discussions portant sur l'environnement, il est impératif de pouvoir disposer de données claires pour effectuer les mesures et analyses, et formuler une appréciation au plan international sur les données touchant l'environnement.

Medición y tratamiento de datos relevantes para el medio ambiente

Informe de ramo 4.1 · Resumen – La medición y evaluación de datos relevantes para el medio ambiente juega un papel muy importante en las discusiones sobre el medio ambiente. Sin un análisis apropiado del medio ambiente, es imposible pensar en una prevención eficaz. Los modernos métodos de análisis permiten detectar y controlar cantidades cada vez menores de sustancias nocivas. Pero a medida que aumenta la sensibilidad, crecen también las posibilidades de incurrir en errores, sobre todo en cuanto a la toma y preparación de muestras. El aumento de las exigencias en materia de control de la calidad, en relación con la obtención de datos relevantes para el medio ambiente, hace que los costes crezcan de forma considerable. Para evaluar la situación del medio ambiente en la fabricación del cemento, la prevención de la contaminación atmosférica es un aspecto muy importante, pero aparte de ello pueden adquirir cierta importancia también los ruidos y las vibraciones. Para el control de las emisiones se dispone de un amplio abanico de métodos, tanto para la medición continua como para mediciones individuales. Además, adquieren creciente importancia el registro, tratamiento y evaluación de datos de medición, desde el ordenador para determinar las emisiones en instalaciones individuales hasta el control remoto, centralizado, que abarca todo el país, así como la creación de sistemas de información que abarcan diferentes sectores. Los métodos de medición y tratamiento de datos relevantes para el medio ambiente todavía no están armonizados a nivel internacional. La forma en que estos datos son recogidos y evaluados puede influir, por lo tanto, en el resultado obtenido. Hay que añadir a ello el hecho de que la investigación que se ocupa de los efectos de las emisiones ya no puede seguir el ritmo de desarrollo que toma la técnica de medición. Esto conduce, en muchos casos, a una actitud de inseguridad y, debido a ello, a una política de información en parte caótica. Para mejorar la transparencia y poder hacer comparaciones, en cuanto a las discusiones sobre el medio ambiente, se requieren urgentemente unas normas claras para la medición y evaluación así como para la apreciación, a nivel internacional, de los datos relevantes para el medio ambiente.

1. Vorbemerkungen

Churchill soll einmal gesagt haben: „Traue keiner Statistik, die Du nicht selbst gefälscht hast". Zu Lebzeiten von Churchill waren umweltrelevante Daten noch kein Thema, sonst hätte er seine Aussage vielleicht wie folgt erweitert: „Traue keiner auf der Grundlage umweltrelevanter Daten aufgestellten Aussage, wenn Du die Randbedingungen für das Messen und Verarbeiten nicht selbst vorgegeben hast". Diese Einschränkung der „absoluten Wahrheit" gilt insbesondere unter dem Aspekt der internationalen Vergleichbarkeit.

Selbst auf der EU-Ebene scheiterte bisher eine angestrebte Harmonisierung, obwohl in der Brüsseler Umweltpolitik umweltrelevante Daten zunehmend an Bedeutung gewinnen, wie die EG-Verordnungen zum Öko-Audit [1] und „Freier Zugang zu Daten der Umwelt" [2] zeigen.

Ohne daß die Wirkungsforschung damit schritthalten kann, wird die Meßtechnik von Jahr zu Jahr verfeinert und ermöglicht inzwischen die Erfassung von Konzentrationen bis in den Spurenbereich. Angaben in ng/kg (Nanogramm pro kg = 10^{-9} g/kg), pg/kg (Picogramm pro kg = 10^{-12} g/kg), fg/kg (Femtogramm pro kg = 10^{-15} g/kg) sind keine Seltenheit mehr [3].

Damit werden der Wissenschaft zwar wertvolle Hilfsmittel an die Hand gegeben, um umweltrelevante Einflüsse zuneh-

1. Preliminary comments

Churchill is said once to have remarked: "Don't believe any statistics which you haven't doctored yourself". During Churchill's lifetime environmentally relevant data were still not a topic for discussion, otherwise he might perhaps have expanded his remark to: "Don't believe any statements made on the basis of environmentally relevant data, unless you yourself have specified the constraints on the measurements and processing." This qualification of the "absolute truth" is particularly valid from the aspect of international comparability.

Even at the EU level an attempt at harmonization has failed so far although environmentally relevant data is becoming increasing important in Brussels' environmental policy, as is shown by the EC regulations for the Eco Audit [1] and "Free access to environmental data" [2].

Research into effects has not been able to keep pace with measurement technology which is becoming more refined every year and which now enables concentrations to be measured down to the trace range. Data in ng/kg (nanograms per kg = 10^{-9} g/kg), pg/kg (picograms per kg = 10^{-12} g/kg), and fg/kg (femtograms per kg = 10^{-15} g/kg) are no longer rare [3].

This does indeed provide science with a valuable tool for ever better assessment of environmentally relevant effects

mend besser beurteilen und sichere Vorsorgemaßnahmen im Umweltschutz treffen zu können, der Normalbürger wird jedoch häufig, oft sogar vorsätzlich, verunsichert, insbesondere wenn Meßwerte ohne Bezug zu Grenz- oder Vergleichswerten [4-9] in den Raum gestellt werden und eine alte Erkenntnis von Paracelsus nicht beachtet wird, der da sagte: „Kein Ding an sich ist Gift. Die Dosis macht's, ob ein Ding Gift ist oder nicht".

Politiker, Wissenschaftler und Journalisten sind hier in der Verantwortung [10]. Politikverdrossenheit, fehlende Glaubwürdigkeit und mangelnde Akzeptanz rühren vielfach aus dem leichtfertigen Umgang mit Daten, Objektivität ist im Bereich Umweltschutz praktisch nicht erreichbar. Die Subjektivität beginnt bereits beim Begriff „umweltrelevant" bzw. bei der Auswahl der Daten, die gemessen und verarbeitet werden. Die sogenannte Relevanz ändert sich nicht nur fortlaufend durch neue wissenschaftliche Erkenntnisse, sondern in noch stärkerem Maße von Land zu Land, ja von Genehmigungsbehörde zu Genehmigungsbehörde, je nach politischem Umfeld. Insbesondere Grenzwerte sind zunehmend politische Werte. Das gilt entsprechend auch für den im Umweltschutz besonders bedeutsamen Begriff „Stand der Technik". Beispielsweise hat der Rat der Europäischen Gemeinschaften durch eine Expertengruppe den Stand der Technik bei der Luftreinhaltung unter besonderer Beachtung von Zementanlagen erarbeiten lassen, um ihn in die Direktive No 84/360 aufzunehmen [11]. Während die Französische Gesetzgebung die daraus resultierenden Vorgaben für die Grenzwerte in ihrem jüngsten Erlaß von 1993 [12] umgesetzt hat, versuchen in Deutschland einzelne Genehmigungsbehörden, darüber hinausgehende neue schärfere Grenzwerte durchzusetzen [13]. Als Begründung dienen Einzelmessungen an Einzelanlagen [14]. Unterschiedliche Meß- und Auswertebedingungen [15] bei Einzelmessungen und kontinuierlichen Dauermessungen sowie eine gesamtökologische Betrachtungsweise [16, 17] werden nicht berücksichtigt. Wirtschaftliche Gesichtspunkte bleiben unbeachtet.

2. Zielsetzung

Ein verantwortungsvoller Umgang mit umweltrelevanten Daten ist es, für zu planende Maßnahmen im Bereich Umweltschutz und für eine langfristig tragfähige Entwicklung notwendig.

Ziel beim Messen und Verarbeiten von umweltrelevanten Daten ist es, die Umweltsituation, den Status quo objektiv festzustellen und die Umweltbeeinflussungen durch Emissionsmessungen zu ermitteln sowie die Umweltauswirkungen durch Immissionsmessungen zu untersuchen.

Emissionsbezogene Daten sind wichtig zur Überwachung und Beurteilung, d.h. zur Qualitätskontrolle von Anlagen, Stoffen (Einsatz-, Reststoffe) und Produkten (Zwischen-, Endprodukte); immissionsbezogene Daten zur Bestimmung der Auswirkungen auf Mensch und Biosphäre, Luft, Wasser und Boden.

Bei allen Herstellprozessen, nicht zuletzt auch beim Zementherstellungsprozeß, ist heute die Ausrichtung auf eine umweltverträgliche Prozeßführung selbstverständlich. Dabei werden eine Vielzahl von Meßwerten und eine zielgerichtete Auswertung erforderlich, um den Prozeß im Sinne eines integrierten Umweltschutzes mit minimalem Energie- und Rohstoffeinsatz zu optimieren und die Abscheideanlagen umweltgerecht zu regeln und zu steuern. Die anlagenbezogene Emissionsüberwachung wird ergänzt durch standortbezogene oder globale Immissionsbetrachtungen. Emissions- und Immissionsuntersuchungen zusammen ermöglichen: die Simulation komplexer Systeme sowie Abläufe [18, 19] und Ausbreitungs- und Klimamodelle. Zutreffende Auswirkungsmodelle setzen voraus, daß sich die Verarbeitung emissions-, immissions- und stoffbezogener Daten nicht in einer isolierten Betrachtungsweise erschöpft, sondern eine anlagen- und medienübergreifende Ganzheitsbetrachtung im Sinne einer Öko-Bilanz zuläßt [20-22]. Die Aufgabe der Forschung und Entwicklung ist es, hier Hilfe anzubieten und nicht den Stand der Technik einseitig nur in der Analytik voranzutreiben.

and means that reliable precautionary measures can be taken for environmental protection, but the normal citizen is frequently, often even deliberately, baffled. This is especially true if measurements are given without reference to limiting or comparative values [4-9] and if an old saying of Paracelsus is disregarded. He said "Nothing is in itself a poison. The dose determines whether or not something is poisonous".

Politicians, scientists and journalists have a responsibility here [10]. Tiredness with politics, failure of credibility, and lack of acceptance frequently stem from superficial handling of the data, and objectivity is virtually unattainable in the field of environmental protection. Subjectivity begins right with the term "environmentally relevant" and in the selection of the data to be measured and processed. The so-called relevance changes continuously not only with new scientific findings, but to an even greater extent from country to country, even from licensing authority to licensing authority and according to political milieu. Limiting values in particular are increasingly becoming political values. This also applies to the term "state of the art" which is particularly important in environmental protection. For example, the council of the European Community arranged for a group of experts to study the state of the art for air pollution control with special attention to cement plants so that it could be included in Directive No. 84/360 [11]. While the French legislature have put the resulting guidelines for the limiting values into action in their latest decree of 1993 [12], individual licensing authorities in Germany are trying to go beyond this and enforce new tighter limits [13]. This is based on single measurements on individual plants [14]. No consideration is given to differing measurement and evaluation conditions [15] for single measurements and continuous, long-term, measurements, nor to a total-ecology approach [16, 17], and economic considerations continue to be disregarded.

2. Objectives

Responsible dealings with environmentally relevant data are essential for planning measures in the environmental protection sector and for long-term productive development work.

The target when measuring and processing environmentally relevant data is to establish the environmental situation, the status quo, objectively, to determine the influences on the environment by emission measurements, and to investigate the environmental effects by immission measurements.

Emission-related data are important for monitoring and evaluating, i.e. for quality controls on plants, substances (feed materials, residues) and products (intermediate and end products); immission-related data are important for determining the effects on human beings and on the biosphere, air, water and soil.

As a matter of course, all manufacturing processes, not least the cement manufacturing process, now aim for environmentally compatible process management. This requires a large number of measurements and carefully directed evaluation in order to optimize the process in the sense of integrated environmental protection with minimum use of energy and raw materials and to control the collecting equipment in a manner appropriate to the environment. Plant-related emission monitoring is complemented by local or global immission considerations. Emission and immission investigations together make it possible to simulate complex systems and processes [18, 19] and produce dispersion and climatic models. Appropriate models of effects presuppose that the processing of emission-, immission- and material-related data is not wasted in an isolated approach, but permits a holistic examination in the sense of an ecology balance [20-22] embracing the plants and media. The task of research and development is to provide help here, and not just to advance the state of the art on the analytical side alone.

3. Grundsätzliche Anforderungen

Das Messen und Verarbeiten umweltrelevanter Daten ist zumindest in den Industriestaaten weitgehend gesetzlich geregelt. Die Anforderungen und Vorgaben sowie insbesondere die Umsetzung der Vorschriften sind jedoch von Land zu Land unterschiedlich [23] und weitgehend vom Wohlstandsgefälle abhängig. Die Abkehr von der reinen Gefahrenabwehr hin zum Vorsorgeprinzip verstärkt weltweit den Trend nach einer immer intensiveren und lückenloseren Überwachung und immer umfangreicheren Anforderungen an das Messen und Auswerten. Das hat bei der Analytik zu einer stürmischen Entwicklung geführt. Der analytische Aufwand muß jedoch in einem gesunden Verhältnis zum Nutzen stehen. Deshalb muß obenan im Anforderungskatalog die Auswahl der geeigneten Meßkomponente eingeordnet werden. Dabei sind Relevanz und Aufwand maßgebend. Bei der Auswahl der geeigneten Meßgeräte sind Genauigkeit und Reproduzierbarkeit von Bedeutung.

Der Meßwert allein ist nicht ausreichend aussagekräftig. Er bedarf der Beschreibung der Randbedingungen. Probenahme und Probenvorbereitung müssen so eindeutig wie möglich festgelegt werden. Genormte Analysen- und einheitliche Auswerteverfahren [24, 25] sind Voraussetzung für vergleichbare Ergebnisse. Insbesondere für kontinuierliche Messungen sollten nur eignungsgeprüfte Geräte [26, 27] verwendet werden, damit die notwendigen Anforderungen an die Wartungsfreundlichkeit, die Ansprechzeit, Zuverlässigkeit und Kalibrierbarkeit erfüllt werden. International klaffen die Anforderungen bei der Erfassung und Auswertung umweltrelevanter Daten enorm auseinander. Die staatliche Regelungsdichte ist in der Bundesrepublik besonders groß. Beispielsweise gibt es für die Bestimmung des Reingas-Gesamtstaubes in der Bundesrepublik genaue behördliche Vorgaben von der Meßstellenplanung über die Verwendung amtlich zugelassener Meßgeräte, die Meßausführung durch amtlich zugelassene Institute bis hin zum Meßbericht. Zusätzlich zu Einzelmessungen werden bei allen relevanten Quellen kontinuierliche Messungen mit kalibrierten Meßgeräten und eine Auswertung mit Emissionsrechner verlangt. In anderen Ländern [28], beispielsweise auch in einigen Staaten der USA, werden nur Einzelmessungen vorgeschrieben, und selbst nach Ofenanlagen erfolgt die Überwachung der Staubemissionen teilweise noch durch nicht kalibrierte Rauchdichte-Meßgeräte nach der „black smoke"-Methode, welche nur die Lichtschwächung der emittierten Partikel erfaßt, oder gar nur stichprobenartig durch sogenannte Smoke-Reader. Das sind speziell ausgebildete Fachleute, welche die Abgasfahne nach ihrem optischen Aussehen beurteilen.

Auch der Umfang der Emissionsüberwachungswerte ist international nicht einheitlich. So fehlen bisher auch in den meisten Ländern, im Gegensatz zu Deutschland, staatliche Vorgaben für die kontinuierliche Überwachung von gasförmigen Emissionen. Die Beispiele machen deutlich, wie unterschiedlich die Anforderungen an das Messen und das Aufarbeiten umweltrelevanter Daten im internationalen Vergleich sind, und daß ein reiner Grenzwertvergleich für eine Beurteilung des Umweltschutzes nicht ausreicht. Parallel zu den Anforderungen an den Umfang und die Qualität der umweltrelevanten Daten steigen die Kosten für die Gesamtheit der Umweltschutzmaßnahmen.

4. Emissionsmessungen

Schwerpunktmäßig umfassen die Messungen bei der Zementherstellung den Bereich Luftreinhaltung. Daneben sind aber auch die Umweltauswirkungen wie Lärm und Erschütterungen sowie Beeinflussungen von Wasser und Boden zu bestimmen und zu beurteilen.

4.1 Luftreinhaltung

4.1.1 Kontinuierliche Messungen

Aufwands- und kostenmäßig spielen kontinuierliche Messungen [29] die größte Rolle. Hier gibt es auch die deutlichsten Unterschiede im internationalen Vergleich.

3. Basic requirements

The measurement and processing of environmentally relevant data is, at least in the industrialized countries, largely governed by law. The requirements and guidelines, and in particular, the translation of the directives into action, do however differ from country to country [23] and are substantially dependent on the degree of affluence. The change from simple defence against danger to the principle of precautionary measures is strengthening the trend throughout the world towards ever more intensive and uninterrupted monitoring and ever more extensive demands on the measurement and evaluation. This has led to a brisk development in analysis. However, the expenditure on analysis must stand in a healthy relationship with the benefits. The selection of suitable measured components must therefore be placed at the top of the list of requirements. Relevance and expenditure are decisive factors. Accuracy and reproducibility are important when selecting suitable measuring equipment.

The measured value by itself does not have sufficient informative value; it is necessary to describe the constraints. Sampling and sample preparation must be defined as clearly as possible. Standardized methods of analysis and consistent methods of evaluation [24, 25] are requirements for comparable results. Only equipment which has been tested for suitability [26, 27] should be used, especially for continuous measurements, to ensure that the necessary requirements for ease of maintenance, response time, reliability and calibratability are fulfilled. Internationally, there is an enormous divergence in the requirements for the measurement and evaluation of environmentally relevant data. There is a particularly large quantity of governmental regulations in Germany. For the measurement of the total clean gas dust content, for example, there are exact official guidelines in Germany from the design of the measuring point, through the use of officially approved measuring equipment and performance of the measurements by officially approved institutes, right up to the test report. Continuous measurements with calibrated measuring equipment and evaluation with emission computers are required at all relevant sources, in addition to single measurements. In other countries [28] for example, including some states in the USA, only single measurements are specified. Even after kiln plants the dust emission is sometimes only monitored by uncalibrated smoke-density meters using the "black smoke" method, which only measure the light attenuation caused by the emitted particles, or even only on a spot sample basis by so-called smoke readers. These are specially trained experts who assess the exhaust gas plume by its visual appearance.

The extent of the emission monitoring values is also not internationally consistent. Unlike Germany, most countries so far have no government guidelines for continuous monitoring of gaseous emissions. These examples illustrate the extent of the differences in the requirements for measuring and processing environmentally relevant data in the international scene, and show that a pure limit comparison is not sufficient for assessing environmental protection. The costs of all the environmental protection measures increase in parallel with the demands on the extent and quality of the environmentally relevant data.

4. Emission measurements

Measurements during cement manufacture are concentrated on the area of air pollution control. However, environmental effects such as noise and vibration, as well as the effects on water and soil, also have to be measured and assessed.

4.1 Air pollution control

4.1.1 Continuous measurements

Continuous measurements [29] play the greatest part as far as effort and costs are concerned. Here again there are significant differences internationally.

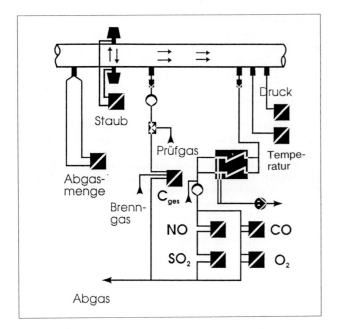

BILD 1: Beispiel für kontinuierliche Messungen im Abgas eines Zementdrehofens
FIGURE 1: Example of continuous measurements on the exhaust gas from a rotary cement kiln

Staub	= dust	Brenngas	= fuel gas
Abgasmenge	= exhaust gas volume	Druck	= pressure
Abgas	= exhaust gas	Temperatur	= temperature
Prüfgas	= test gas	C_{ges}	= C_{total}

Continuous monitoring of mass concentrations and mass flows of certain harmful materials is always specified in Germany if they are regarded as particularly environmentally relevant and if the emission mass flows exceed predetermined values [15] (**Fig. 1**).

Continuous measurement of the dust and nitrogen oxide components as stipulated in [30] is almost always specified in Germany. Measurement of the sulphur dioxide emissions is often also required.

The emissions in Germany have to be expressed relative to dry exhaust gas in the standard state. An oxygen reference value and a mass flow measurement are sometimes also required. Various other auxiliary variables therefore also have to be measured continuously in addition to the harmful substances [31].

Continuous measurements [32–35] take place using either the extractive or the in-situ measurement strategies (**Fig. 2**).

In the extractive method of measurement a partial gas stream is taken from the process gas and then analyzed, while in the in-situ method the process gas is measured without treatment. This method requires very little maintenance work because no preparation of the test gas is required. The advantage of the extractive method is that it also works under critical conditions, such as high dust loadings and high temperatures, and can therefore be used for process control, while the in-situ method can essentially only be applied to clean gas measurements. The test methods used for continuous extractive measurements are very varied [36–41]. The photometric methods, with which carbon monoxide, nitrogen oxide and sulphur dioxide can be measured, are of particular importance in the non-dispersive infra-red and ultra-violet ranges. The appropriate measuring equipment and the required gas processing systems, such as gas cooler and filter, can now be obtained in compact, easily maintained, designs.

Optical in-situ measuring equipment consists of a transceiver unit and a reflector unit which are located outside the chimney on opposite sides (**Fig. 3**).

The in-situ equipment shown [42–45] for the simultaneous detection of dust, NO and SO_2 works on the absorption method and has, in addition to the detectors for the reference and dust ducts, additional detectors for SO_2 and NO.

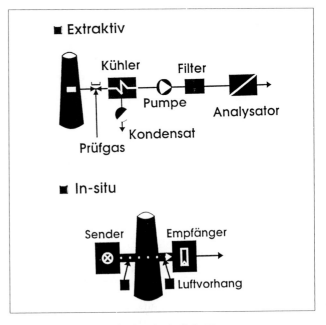

BILD 2: Meßstrategien für kontinuierliche Messungen
FIGURE 2: Measurement strategy for continuous measurements

Extraktiv	= extractive		
Kühler	= cooler	Kondensat	= condensate
Filter	= filter	In-situ	= in-situ
Pumpe	= pump	Sender	= transmitter
Analysator	= analyzer	Empfänger	= receiver
Prüfgas	= test gas	Luftvorhang	= air curtain

In Deutschland ist die kontinuierliche Überwachung der Massenkonzentration und des Massenstromes bestimmter Schadstoffe immer dann vorgeschrieben, wenn sie als besonders umweltrelevant angesehen werden und die Emissionsmassenströme vorgegebene Werte [15] überschreiten (**Bild 1**).

Entsprechend der Festlegung [30] wird an Zementofenanlagen die kontinuierliche Messung der Komponenten Staub und Stickstoffoxide in Deutschland fast immer vorgegeben. Zusätzlich wird häufig auch noch die Messung der Schwefeloxidemissionen verlangt.

Die Emissionen sind in Deutschland auf trockenes Abgas im Normzustand zu beziehen. Bisweilen wird auch noch ein Sauerstoffbezug und eine Massenstrombestimmung gefordert. Entsprechend müssen zusätzlich zu den Schadstoffen vielfach noch verschiedene Hilfsgrößen ebenfalls kontinuierlich gemessen werden [31].

Kontinuierliche Messungen [32–35] erfolgen entweder nach der extraktiven oder nach der In-Situ-Meßstrategie (**Bild 2**).

Bei der extraktiven Meßmethode wird ein Teilgasstrom des Prozeßgases entnommen und dann analysiert, während das Prozeßgas bei der In-Situ-Methode unbehandelt gemessen wird. Wegen der nicht benötigten Meßgasaufbereitung erfordert diese Methode nur wenig Wartungsaufwand. Vorteilhaft beim extraktiven Verfahren ist, daß es auch unter kritischen Bedingungen, wie beispielsweise bei hohen Staubbeladungen und hohen Temperaturen, arbeiten und dadurch eine Nutzung zur Prozeßführung ermöglicht, während die In-Situ-Methode im wesentlichen nur bei Messungen im Reingas angewandt werden kann. Die eingesetzten Meßverfahren für kontinuierlich extraktive Messungen sind sehr unterschiedlich [36–41]. Von besonderer Bedeutung sind insbesondere die photometrischen Verfahren im nicht dispersiven infraroten und ultravioletten Bereich, mit denen Kohlenmonoxid, Stickoxid und Schwefeldioxid gemessen werden können. Entsprechende Meßgeräte und die notwendige Probegasaufbereitung, wie Gaskühler und Filter, sind heute in kompakter und wartungsfreundlicher Ausführung erhältlich.

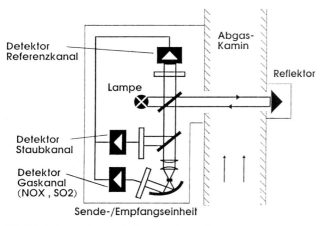

BILD 3: Schema eines optischen In-Situ-Meßgerätes zum gleichzeitigen Nachweis von Staub, NO und SO$_2$
FIGURE 3: Diagram of optical in-situ measuring equipment for simultaneous detection of dust, NO and SO$_2$

Detektor Referenzkanal	= detector reference duct
Detektor Staubkanal	= detector dust duct
Detektor Gaskanal (NO$_x$, SO$_2$)	= detector gas duct (NO$_x$, SO$_2$)
Sende-/Empfangseinheit	= transceiver unit
Lampe	= lamp
Abgas-Kamin	= exhaust gas chimney
Reflektor	= reflector

Optische In-Situ-Meßgeräte bestehen aus einer Sende- und Empfangs- sowie einer Reflektoreinheit, die jeweils außerhalb des Kamines gegenüberliegend angeordnet sind (**Bild 3**).

Das dargestellte In-Situ-Gerät [42–45] für den simultanen Nachweis von Staub, NO und SO$_2$ arbeitet nach dem Absorptionsverfahren und besitzt neben dem Detektor für den Referenz- und Staubkanal noch zusätzliche Detektoren für SO$_2$ und NO.

Für die kontinuierliche Emissionsmessung sind in Deutschland nur solche Meßeinrichtungen zugelassen, welche einer Eignungsprüfung unterzogen wurden. Eine branchenbezogene Eignungsbekanntgabe [26] im Zementbereich liegt bisher nur für Staubmeßgeräte und ein SO$_2$-Gerät nach dem In-Situ-Verfahren vor. Soweit eine Überwachung anderer Komponenten erfolgt, werden die verwendeten Geräte aufgrund von freiwilligen Vereinbarungen betrieben.

In Eignungsprüfungen werden die Meßeinrichtungen umfassenden Labor- und Praxistests unterzogen [46]. Durch dieses Vorgehen soll sichergestellt werden, daß die Meßgeräte für den Einsatzzweck wirklich geeignet sind und an verschiedenen Anlagen mit vergleichbaren Verfahren gemessen wird [47].

Das Netzwerk zur Qualitätssicherung in der Bundesrepublik ist sehr umfangreich (**Bild 4**).

In die Qualitätssicherung als Eckpfeiler eingebunden sind: der Betreiber, der Meßgerätehersteller, eine sachverständige Stelle [48, 49] und vor allem die Aufsichtsbehörde. Betreiber und Meßgerätehersteller sind insbesondere durch die Wartung verknüpft.

Die sachverständige Stelle erhält vom Betreiber Prüfaufträge und erstellt Prüfberichte für die Aufsichtsbehörde, wird von dieser aber selbst auch wieder geprüft. Meßgerätehersteller und sachverständige Stelle arbeiten bei der Eignungsprüfung zusammen. Die Aufsichtsbehörde erteilt dem Betreiber Auflagen und führt Betriebsinspektionen durch. Sie verlangt vom Betreiber Meßberichte.

Bei der Einrichtung der Meßplätze, der Auswahl und dem Einbau der Geräte arbeiten sachverständige Stelle, Gerätehersteller und Betreiber eng zusammen. Nach der Inbetriebnahme müssen die Meßeinrichtung kalibriert [50] und die Auswerteanlage von der sachverständigen Stelle geprüft werden.

Bei den Funktionsprüfungen werden die wesentlichen Geräte-Parameter kontrolliert. Funktionsprüfungen sind in Deutschland jährlich durchzuführen. Kalibrierungen, das

In Germany the only items of test equipment permitted for continuous emission measurement are those which have been submitted to a suitability test. So far, the only available industry-related suitability announcement [26] in the cement sector is for dust measuring equipment and an SO$_2$ device using the in-situ method. Where other components are monitored the equipment used is operated on the basis of voluntary agreements.

During suitability testing the measuring equipment is submitted to comprehensive laboratory and practical tests [46]. This procedure is intended to ensure that the measuring equipment is truly suitable for the intended application and that measurements are carried out in different plants with comparable methods [47].

The quality assurance network in Germany is very extensive (**Fig. 4**).

The cornerstones of the quality assurance system are: the operator, the equipment manufacturer, a specialist agency [48, 49] and, above all, the regulatory authority. Maintenance forms a special link between the operator and the equipment manufacturer.

The specialist agency is awarded test contracts by the operator and draws up test reports for the regulatory authority, by whom it is itself also checked. The measuring equipment manufacturer and specialist agency cooperate in the suitability testing. The regulatory authority gives directions to the operator and carries out plant inspections. It requires the operator to provide test reports.

The specialist agency, equipment manufacturer and operator work closely together when establishing the test positions, and selecting and installing the equipment. After the commissioning, measuring equipment must be calibrated [50] and the evaluating system must be checked by the specialist agency.

The essential equipment parameters are checked during the function tests. Function tests have to be carried out annually in Germany. Calibrations, i.e. comparison measurements

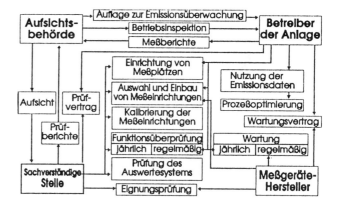

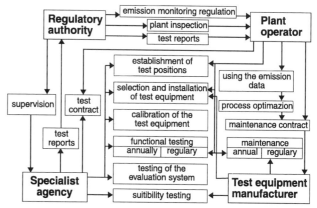

BILD 4: Netzwerk der Qualitätssicherung für das Messen und Auswerten umweltrelevanter Daten
FIGURE 4: Network of the quality assurance scheme for measuring and evaluating environmentally relevant data

heißt die Durchführung von Vergleichsmessungen mit einem unabhängigen Analyseverfahren, sind alle 5 Jahre zu wiederholen.

Die Kalibrierung der registrierenden Staubgehaltsmeßgeräte erfolgt mit Hilfe von gravimetrischen Einzelmessungen [51].

In einer Kalibrierkurve (**Bild 5**) wird der Zusammenhang zwischen der Konzentration des Schadstoffes im Abgas und der Analysefunktion der Betriebsmeßeinrichtung dargestellt. Neben der Regressionsgeraden wird der Toleranzbereich der Einzelwerte und der Vertrauensbereich berechnet. In Einzelfällen ergibt sich eine gute Übereinstimmung mit engen Vertrauens- und Toleranzbereichen. Insbesondere bei Gesamtstaub werden jedoch wegen der prozeßbedingten Schwankungen in der Feinheit der Stäube überwiegend breite Vertrauens- und Toleranzbereiche gefunden. Die Werte der Kalibrierkurve sind für die Parametrierung des Emissionsrechners wichtig. Dort werden sie beim maßgebenden Ausdruck für die Grenzwertbeurteilung berücksichtigt. Es kann der Toleranzbereich für die Halbstundenmittelwerte, d. h. die Einzelwertbetrachtung, bzw. der Vertrauensbereich für die Summenwerte, z. B. die Tagesmittelwerte, ausgeschöpft werden.

4.1.2 Einzelmessungen

Soweit keine kontinuierlichen Messungen vorgeschrieben sind, sehen die behördlichen Anforderungen in Deutschland zur Überwachung von Emissionen Einzelmessungen derjenigen Komponenten vor, die für die entsprechende Anlage als relevant angesehen werden und für die es gesetzliche Grenzwerte gibt.

Die Einzelmessungen sind in der Regel alle 3 Jahre zu wiederholen. Grenzwerte im Bereich Luftreinhaltung gibt es in der Bundesrepublik für 35 anorganische Stoffe, für etwa 150 organische Stoffe und für etwa 20 canzerogene Stoffe. Prozeßbedingt sind nur wenige dieser Stoffe für Zementwerke relevant.

Unter einer Einzelmessung ist eine stichprobenartige Messung zu verstehen, bei welcher nur über einen begrenzten Zeitraum Messungen durchgeführt werden. Dafür können sowohl diskontinuierliche Verfahren als auch kontinuierliche Verfahren eingesetzt werden.

Beispielsweise wird zur Bestimmung des Gehaltes an Gesamtkohlenwasserstoffen im Abgas von Zementbrennöfen das kontinuierlich arbeitende Flammenionisationsverfahren angewendet, da Einzelmessungen nach der Silikagelmethode zu Fehlmessungen führen.

Für die eigentlichen Messungen sind als Probenahmetechniken die Abscheidung auf Filtermaterial (Staub), die Absorption in Lösung (SO_2), die Adsorption an Adsorbentien (organische Einzelstoffe), die Entnahme mit Gassammelgefäß (Stickstoffoxide) und kombinierte Meßverfahren (Dioxin) von Bedeutung [53, 54].

Die häufigste Anwendung für die Abscheidung auf Filtermaterialien ist bei der Messung von Staub und Staubinhaltsstoffen [55, 56]. Grundlegende Voraussetzung ist die isokinetische Probenahme. Der Staub wird auf einem in der Absaugsonde integrierten Filter abgeschieden. Die Bestimmung der gesammelten Staubmasse erfolgt gravimetrisch (**Bild 6**).

Je nach verwendetem Gerät schwanken bei den zum Einsatz in der Zementindustrie geeigneten Verfahren die Nachweisgrenze und die Standardabweichung. Zur Bestimmung der Staubinhaltsstoffe und der Schwermetalle werden die Stäube durch Säure aufgeschlossen. Die Analyse der Metalle erfolgt dann vorwiegend mit der Atom-Absorptions-Spektroskopie [57–59].

Die Nachweisgrenze und die Standardabweichung sind für die einzelnen Spurenelemente unterschiedlich und abhängig von der Probenmenge [60].

Bei im Abgaskamin durchgeführten Ringversuchen, an denen auch das Forschungsinstitut der Zementindustrie teilgenommen hat, ergaben sich bei der Gesamtstaubermitt-

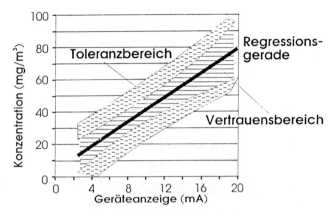

BILD 5: Beispiel einer Kalibrierkurve für Gesamtstaub
FIGURE 5: Example of a calibration curve for total dust

Konzentration	= concentration
Toleranzbereich	= tolerance range
Regressionsgerade	= regression line
Vertrauensbereich	= confidence region
Geräteanzeige	= equipment indication

carried out with an independent method of analysis, have to be repeated every 5 years.

Recording dust measuring equipment is calibrated with the aid of single gravimetric measurements [51].

The relationship between the concentration of harmful substance in the exhaust gas and the analysis function of the plant test equipment is shown in a calibration curve (**Fig. 5**). The tolerance range of the individual values and the confidence range are calculated in addition to the regression line. In individual cases there is good correspondence with narrow confidence and tolerance ranges. However, in most cases wide confidence and tolerance ranges are found, especially when measuring total dust, because of the fluctuations in fineness of the dust due to the nature of the process. The values of the calibration curves are important for parameterization of the emission computer where they are taken into account in the decisive print-out for the limit value assessment. The tolerance range for the half-hour average values, i.e. single value examination, and the confidence range for the total values, e. g. the daily average values, can be fully utilized.

4.1.2 Single measurements

Where continuous measurements are not prescribed, the official requirements in Germany schedule single measurements for monitoring emissions of those components which are regarded as relevant for the corresponding plants and for which there are legal limits.

As a rule the single measurements have to be repeated every 3 years. In the field of air pollution control there are limiting values in Germany for 35 inorganic substances, for about 150 organic substances and for about 20 carcinogens. Because of the nature of the process only a few of these substances are relevant to cement works.

A single measurement is taken to mean a measurement of the spot sample type in which measurements are only carried out over a limited period. Discontinuous and continuous methods can both be used for this purpose.

For example, the continuous flame ionization method is used for measuring the total hydrocarbon content of the exhaust gas from cement-burning kilns because single measurements based on the silica gel method give false readings.

Collection on filter material (dust), absorption in solution (SO_2), adsorption on adsorbents (individual organic substances), removal with gas collecting vessels (nitrogen oxides), and combined measuring methods (dioxins) are important sampling techniques for the actual measurements [53, 54].

The most frequent application for collection on filter materials is during the measurement of dust and of substances

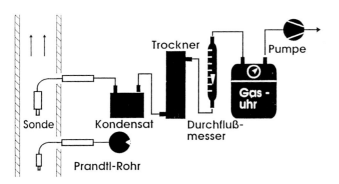
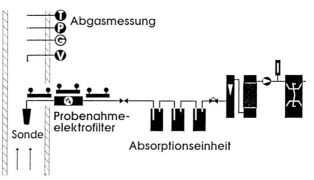

BILD 6: Schema für Abscheidung auf Filtermaterial, Staubprobenahme
FIGURE 6: Scheme for collection on filter material; dust sampling

Sonde	= probe
Trockner	= dryer
Pumpe	= pump
Kondensat	= condensate
Durchflußmesser	= flow meter
Gasuhr	= gas totalizer
Prandtl-Rohr	= Prandtl tube

BILD 7: Schema für eine Probenahme nach dem Absorptionsverfahren
FIGURE 7: Scheme for sampling using the absorption method

Sonde	= probe
Abgasmessung	= exhaust gas measurement
Probenahmeelektrofilter	= sampling-electrostatic precipitator
Absorptionseinheit	= absorption unit

lung Einzelwerte, die bis zu 10 mg/m³ unterschiedlich waren [61]. Eine noch weitaus größere Streuung zeigte sich bei der Bestimmung der Spurenelementkonzentrationen. Hier traten Abweichungen bei den Einzelwerten bis 300 % auf [62].

Bei den Absorptionsverfahren [63–65] wird die Löslichkeit eines Stoffes in einem Absorptionsmedium zur Probenahme ausgenutzt. Die Probenahme erfolgt über Quarzglassonden mit nachgeschalteten Staubabscheidern und Waschflaschen. Die Apparatur ist von der Entnahme bis zu den Waschflaschen zu beheizen. Mit diesem Verfahren werden beispielsweise SO$_2$ und flüchtige anorganische Spurenelemente im Abgas von Zementöfen bestimmt (**Bild 7**).

Das Ergebnis der Analyse ist sehr stark davon abhängig, wie gut die Probenahmeeinrichtung arbeitet. Auf keinen Fall dürfen Feinststäube in die Waschflaschen gelangen, da ansonsten von einer Fehlmessung durch chlor- und fluorhaltige Stäube auszugehen ist.

Für flüchtige organische Verbindungen erfolgt im Spurenbereich die Probenahme durch Adsorption. Als Adsorptionsmittel wird im allgemeinen Aktivkohle verwendet. Der Aufbau der Probenahmeeinrichtung entspricht sonst im Prinzip der Apparatur, wie sie beim Absorptionsverfahren verwendet wird.

Die Analyse selbst wird mit der Gaschromatographie und einem Massenspektrometer durchgeführt [66–70].

Gassammelgefäße kommen zum Einsatz, wenn es keine geeigneten Absorptions- und Adsorptionsbasen gibt, wie etwa bei Stickoxiden, oder wenn eine Kurzzeitprobenahme ausreichend ist (**Bild 8**).

Die Probenahme von Dioxin, die im Schema stark vereinfacht dargestellt ist, wird nach dem kombinierten Verfahren unter Verwendung der Kondensationsmethode durchgeführt und ist sehr aufwendig. Die Analyse selbst erfolgt nach der Probenaufbereitung, die eine Reihe von Extraktionen und Vorreinigungsschritte umfaßt, mit einem Gaschromatographen, der mit einem hochauflösenden Massenspektrometer als Detektor kombiniert ist [71–74] (**Bild 9**).

Neuerdings wird in Pilotuntersuchungen auch die Lasertechnik dazu verwendet, vom Boden oder von Flugzeugen aus, Abgasfahnen auf Inhaltsstoffe zu überprüfen [75].

4.2 Lärm

Für die Zementwerke ist es wichtig, neben den Emissionsquellen, welche die Luftreinhaltung beeinflussen, auch die lärmverursachenden Quellen zu kontrollieren [76–79]. Dazu werden Schallmessungen durchgeführt. Die Bestimmung der umweltrelevanten Lärmdaten erfolgt auf der Emissions- und Immissionsseite im allgemeinen mit gleichen Geräten. Gemessen wird dabei mit unterschiedlichen

contained in the dust [55, 56]. A fundamental requirement is isokinetic sampling. The dust is collected on an integral filter in the suction probe. The mass of dust collected is determined gravimetrically (**Fig. 6**).

The limits of detection and the standard deviation for the methods suitable for use in the cement industry vary depending on the equipment used. In order to measure the substances contained in the dust and the heavy metals the dusts are digested with acids. The metals are then normally analyzed by atom absorption spectroscopy [57–59].

The limits of detection and the standard deviation are different for the individual trace elements and are dependent on the sample size [60].

In inter-laboratory tests carried out in the exhaust gas chimney, in which the Research Institute of the Cement Industry took part, single values which differed by up to 10 mg/m³ [61] were obtained when measuring the total dust. An even greater spread was found when measuring the trace element concentrations. In this case there were deviations in the single values of up to 300 % [62].

In the absorption method [63–65] the sampling procedure makes use of the solubility of a substance in an absorption medium. The sampling takes place through quartz glass probes connected to dust collectors and wash bottles. The apparatus from the sampling point to the wash bottles has to be heated before sampling. This method is used, for example, for measuring SO$_2$ and volatile inorganic trace elements in the exhaust gas from cement kilns (**Fig. 7**).

The result of the analysis is heavily dependent on how well the sampling equipment works. It is essential that very fine

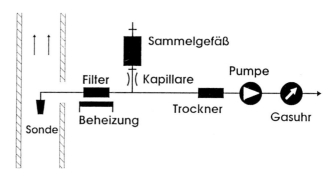

BILD 8: Schema für eine Probenahme mit einem Gassammelgefäß
FIGURE 8: Scheme for sampling with a gas collecting vessel

Sonde	= probe
Filter	= filter
Beheizung	= heating
Sammelgefäß	= collecting vessel
Kapillare	= capillary
Pumpe	= pump
Trockner	= dryer
Gasuhr	= gas totalizer

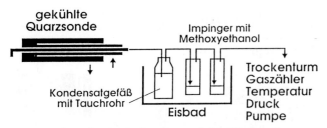

BILD 9: Schema für eine Probenahme nach dem Kondensationsverfahren
FIGURE 9: Scheme for sampling using the condensation method

gekühlte Quarzsonde	= cooled quartz probe
Impinger mit Methoxyethanol	= impinger with methoxyethanol
Kondensatgefäß mit Tauchrohr	= condensate vessel with dip tube
Eisbad	= ice bath
Trockenturm, Gaszähler, Temperatur, Druck, Pumpe	= drying tower, gas counter, temperature, pressure, pump

Meßbereichen. Dazu werden Schallmessungen durchgeführt. Für orientierende Messungen ist ein Schallpegelmesser mit Stativ, Windschirm und Kalibriereinheit erforderlich (**Bild 10**).

Neben dem bewerteten Schallpegel muß der energieäquivalente Dauerschallpegel und der Taktmaximalpegel für Taktdauern bis fünf Sekunden ablesbar sein. Als Zusatzeinrichtungen sind wichtig, ein Frequenzanalysator sowie Speicher- und Registriergeräte, die den Schallpegelverlauf über die Zeit dokumentieren können. Zur Darstellung der Lärm-Emissionssituation haben eine Reihe von Zementwerken mit Hilfe des Schallpegelmessers sogenannte Lärmkarten aufgestellt, in welchen das Werksgebiet in Bereiche gleicher Lärmpegel aufgeteilt ist. Diese Karten sagen jedoch noch nichts über die Einzelmaschinen aus und über die Lästigkeit des von ihnen ausgehenden Lärms. Daher sind mit Hilfe des Frequenzanalysators quellenbezogene Frequenzanalysen wichtig.

4.3 Erschütterungen

Neben Luftschwingungen können von Zementwerksanlagen auch Bodenschwingungen ausgehen, die dann als Erschütterungen lästig werden. Solche Erschütterungen können insbesondere von Sprengungen hervorgerufen werden [80–82]. Sie können aber beispielsweise auch von Maschinen, ja sogar von Silos, verursacht werden (**Bild 11**).

Moderne Meßeinrichtungen sind meist zur gleichzeitigen Verarbeitung von 6 Kanälen ausgerüstet und besitzen mehrere Schwingungsaufnehmer, einen Dreikomponenten-Schwingungsaufnehmer für die x, y und z-Richtung, einen Zweikomponenten-Aufnehmer für die x, y-Richtung und einen Einkomponenten-Aufnehmer für die z-Richtung.

Die Schwingungsaufnehmer arbeiten nach dem elektrodynamischen Prinzip und werden über Kabel mit dem Steuer- und Auswerterechner verbunden.

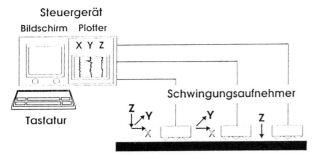

BILD 11: Schema für eine Erschütterungsmeßeinrichtung
FIGURE 11: Scheme for vibration measuring equipment

Steuergerät	= control equipment
Bildschirm	= screen
Plotter	= plotter
Tastatur	= keyboard
Schwingungsaufnehmer	= vibration sensor

dust is not allowed to reach the wash bottles as otherwise a false reading can be expected due to the chlorine and fluorine contained in the dust.

Sampling in the trace range for volatile organic compounds takes place by adsorption. Activated carbon is generally used as the adsorbent. In other respects the design of the sampling equipment is the same in principle as the apparatus used in the absorption process.

The analysis itself is carried out by gas chromatography and a mass spectrometer [66–70].

Gas collecting vessels are used if there are no suitable absorption and adsorption bases, such as is the case with nitrogen oxides, or if short-term sampling is sufficient (**Fig. 8**).

The sampling system for dioxin, which is shown in greatly simplified form in the diagram, is carried out by the combined process using the condensation method, and is very laborious. After sample preparation, which includes a series of extractions and preliminary purification stages, the analysis itself takes place with a gas chromatograph which is combined with a high-resolution mass spectrometer as the detector [71–74] (**Fig. 9**).

Laser technology has also been used recently in pilot investigations for testing the substances contained in exhaust gas plumes from the ground or from aircraft [75].

BILD 10: Beispiel für eine Schallpegelmessung
FIGURE 10: Example of a system for measuring sound level

4.2 Noise

It is important for a cement works to check not only sources of emission which affect air pollution control, but also the sources of noise [76–79], so sound measurements are carried out. The environmentally relevant noise data are generally measured with the same equipment on the emission and immission sides, but the measurements are made in different test ranges. A sound level meter with stand, windshield and calibration unit is necessary for preliminary investigative measurements (**Fig. 10**).

In addition to the evaluated sound level it must also be possible to read off the energy-equivalent continuous sound level and the cyclic maximum level for cycle times of up to five seconds. Important auxiliary equipment includes a frequency analyzer as well as storage and recording equipment which can document the sound level behaviour with time. A number of cement works have used sound level meters to draw up so-called noise maps – in which the works area is divided up into sectors of equal noise level – in order to represent the noise emission situation. However, these maps do

Der Rechner ist mit einer Tastatur und einem Bildschirm ausgerüstet. Die notwendigen Daten und die gewünschte Auswertung können vor der Messung eingegeben werden. Die Messung selbst läuft dann vollautomatisch, etwa bei Erreichen eines bestimmten Schwellenwertes, ab. Die aufgezeichneten Kurven und Maxima mit deren Frequenzen können nach der Messung mit einem eingebauten Plotter ausgegeben werden.

5. Immissionsmessungen

Der Übergang von der Emissions- zur Immissionsmessung ist fließend, was nicht nur auf die Erschütterungs- und Lärmmessungen zutrifft. Vielfach werden für Emissions- und Immissionsmessungen die gleichen Analyse- und Meßverfahren mit entsprechend angepaßten Meßbereichen angewendet. Gemessen werden vor allem partikelförmige und gasförmige Stoffe [83–85]. Zusätzlich werden Bioindikatoren zur Beurteilung herangezogen [86, 87].

Für die Immissionen in der Nachbarschaft von Zementwerken sind neben den Emissionen des Werkes auch Emissionen anderer Quellen, die Entfernung und die Lage der Emissionsquellen sowie vor allem auch die Witterungseinflüsse maßgebend. Das gilt insbesondere für die Komponenten aus dem Bereich der Luftreinhaltung. Die Übertragungsverhältnisse sind sehr komplex (**Bild 12**). Daher erweisen sich Immissionsmessungen als sehr kosten- und zeitaufwendig.

Darüber hinaus ist es schwierig, Rückschlüsse zu ziehen, von welchen Emittenten die Immissionen maßgebend beeinflußt werden. Deshalb wird vielfach versucht, über Ausbreitungsrechnungen den Immissionsanteil zu bestimmen. Es gibt inzwischen sehr viele und ausgefeilte Verfahren, die von der Rechnung her einen hohen Zuverlässigkeitsgrad aufweisen, aber voraussetzen, daß die örtlichen Witterungsbedingungen sehr genau bekannt sind [88–91]. Vielfach herrschen in Bodennähe andere Windbedingungen als in Höhe des Kaminaustritts. Im Hinblick auf mögliche Nachbarschaftsbeschwerden und Haftungsansprüche ist jedes Zementwerk gut beraten, eine eigene Windmeßstation zu betreiben. Eine Anzahl von Zementwerken führt in Eigeninitiative darüber hinaus ständig Staubniederschlagsmessungen in der Nachbarschaft durch. Hierfür werden üblicherweise einfache Auffangbehälter nach Bergerhoff verwendet [92, 93].

Das südafrikanische Werk Hercules setzt sogenannte Zwei-Behälter-Meßgeräte ein [94]. Bei diesen wird in Abhängigkeit von der Windrichtung entweder nur der eine Behälter oder der andere geöffnet. Weiterhin sind auch Meßgeräte für Spezialuntersuchungen im Einsatz, die nur die sogenannte Naßdeposition erfassen.

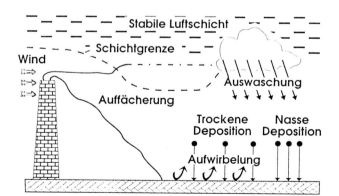

BILD 12: Beeinflussungsfaktoren für die Ausbreitung von Emissionen im Bereich Luftreinhaltung

FIGURE 12: Factors affecting the dispersion of emissions in the air pollution control sector

Stabile Luftschicht	= stable air layer
Schichtgrenze	= boundary layer
Wind	= wind
Auffächerung	= dispersion
Auswaschung	= washing out
Trockene Deposition	= dry deposition
Nasse Deposition	= wet deposition
Aufwirbelung	= swirling

not provide any information about the individual machines and about the nuisance of the noise which they produce. Source-related frequency analyses made with the aid of the frequency analyzer are therefore important.

4.3 Vibrations

Cement works plants can be the source of ground oscillations, which then cause nuisance in the form of vibrations, as well as of air-borne oscillations. Blasting, in particular, can cause these vibrations [80–82], but they can also be caused by, for example, machines or even by silos (**Fig. 11**).

Modern measuring equipment is usually designed for simultaneous operation on 6 channels and has several oscillation sensors, a three-component oscillation sensor for the x, y and z directions, a two-component oscillation sensor for the x and y directions, and a single-component oscillation sensor for the z direction.

The oscillation sensors work on the electrodynamic principle and are linked by cables to the control and evaluation computer.

The computer has a keyboard and a display screen. The necessary data and the desired evaluation can be input before the measurement. The measurement itself then runs fully automatically, possibly on reaching a certain threshold value. The recorded curves and maxima with their frequencies can be output after the measurements with a built-in plotter.

5. Immission measurements

There is a smooth transition from emission to immission measurement, and this does not apply just to vibration and noise measurements. The same methods of analysis and measurement, with correspondingly adapted test ranges, are often used for the emission and immission measurements. Most of the measurements are on particulate or gaseous substances [83–85], but bio-indicators are also used in the assessment [86, 87].

For immissions in the neighbourhood of cement works the important factors are not only the emissions from the works, but also emissions from other sources, the distance and position of the emission sources and, above all, the influence of the weather. This is particularly true of the components from the air pollution control sector. The transfer conditions are very complex (**Fig. 12**), which is why immission measurements prove to be very cost- and time-intensive.

It is also difficult to draw conclusions about which emission sources have the main influence on the immissions. The attempt is therefore often made to determine the amount of immission from dispersion calculations. There are now a great many elaborate methods which are highly reliable from the calculation side, but presuppose that the local weather conditions are very accurately known [88–91]. The wind conditions at ground level are often different from those at the level of the chimney outlet. In view of the possible complaints from the neighbourhood and claims for liability, every cement works is well advised to operate its own wind measuring station. A number of cement works also carry out permanent dust deposition measurements in the neighbourhood on their own initiative. Simple Bergerhoff collectors are normally used for this purpose [92, 93].

The South African Hercules works uses so-called twin-container test equipment [94] in which only one or other of the containers is open depending on the wind direction. There is also test equipment in use for special investigations which only measures the so-called wet deposition.

In addition to this, mobile test equipment which can be installed in vehicles or containers is also used in the neighbourhood of cement works for special investigations of restricted duration [95–97] (**Fig. 13**). They can be used for measuring air-borne dust and gaseous immissions.

Bio-indicators such as ryegrass and lichens are generally only used for particular investigations and are of only limited informative value.

Für spezielle zeitlich beschränkte Untersuchungen werden darüber hinaus in der Nachbarschaft von Zementwerken mobile Meßgeräte, die in Fahrzeugen oder Containern installiert sein können, zu Untersuchungen herangezogen [95-97] (**Bild 13**). Damit können Schwebstaub und gasförmige Immissionen gemessen werden.

Bioindikatoren wie Weidelgras und Flechten werden im allgemeinen nur für besondere Untersuchungen eingesetzt und haben nur eine beschränkte Aussagekraft.

Mit Weidelgras konnte beispielsweise in der Nachbarschaft eines Zementwerkes nachgewiesen werden, daß ein Zusammenhang zwischen Staubniederschlag und Thalliumgehalt in den Pflanzen nicht besteht [98]. Flechtenwachstum läßt Rückschlüsse auf die Belastung durch saure Luftschadstoffe zu. Im Umkreis von Zementwerken sind vor allem Flechten zu finden, die auf alkalischem Milieu gedeihen [99]. Zur Bestimmung von Schwebstaub und der gasförmigen Komponenten wurde ein über ganz Deutschland verteiltes Meßnetz aufgebaut [100-102]. Die ermittelten Werte werden in Fachzeitschriften und auch in Tageszeitungen wöchentlich veröffentlicht. Ständige Meßstationen in der Nähe von Zementwerken können eine wertvolle vertrauensbildende Maßnahme sein.

6. Untersuchungen an Böden, Pflanzen, Produkten, Einsatz- und Reststoffen

Neben der Emissions- und Immissionsanalyse sind heute zunehmend auch Stoffanalysen wichtig. Hier geht es um organische und anorganische Bestandteile. Solche Untersuchungen werden sowohl von den Herstellern, den Lieferanten, den Kunden als auch von den Behörden durchgeführt. Gestützt auf das Bodenschutzgesetz ist in der Bundesrepublik vor allem Baden-Württemberg [103] sehr aktiv im Bereich der Bodenuntersuchungen [104]. Dabei werden beispielsweise bei Thallium verschiedene Aufschlußverfahren angewendet, um nicht nur die Gesamtkonzentration bestimmen zu können, sondern auch insbesondere den mobilen Anteil, der für den Pflanzentransfer zur Verfügung steht [105]. Zur Überwachung von Anbauempfehlungen werden in Deutschland auch regelmäßig die Nutzpflanzen bestimmter Gebiete untersucht. Die Analyse auf umweltrelevante Bestandteile bei Sekundärstoffen ist ein wichtiger Bestandteil des Qualitätsüberwachungssystems [106, 107], in das Lieferanten durch Qualitätszertifikate, Verarbeiter durch Eingangsuntersuchungen und Behörden durch Kontrolluntersuchungen eingebunden sind.

7. Meßdatenaufzeichnung, -verarbeitung, -auswertung

Die umweltrelevanten Daten erhalten erst Aussagekraft, wenn die Randbedingungen dazu bekannt sind und der Gesamtzusammenhang hergestellt wird. Dazu ist eine zielgerichtete Meßwertaufzeichnung, -verarbeitung und -auswertung notwendig. Wichtige Vorgaben, welche an ein Protokoll für z.B. Einzeluntersuchungen gestellt werden, sind [108]: Zweck, Anlagenbeschreibung, Probenahmestelle, Meßverfahren, Betriebszustand und Ergebnisse. Sehr genau ist auch die Aufzeichnung von kontinuierlichen Messungen geregelt. Hier reicht eine Dokumentation mit Schreibern nicht mehr aus, sondern fast immer ist der Anschluß an einen Emissionsrechner vorgeschrieben [109]. Emissionsrechner bestehen aus Datenerfassungs-, Auswerte-, Eingabe- und Ausgabeeinheiten.

Mit Hilfe von Emissionsrechnern ist eine lückenlose Erfassung und Dokumentation über die gesamte Betriebszeit einer Anlage möglich (**Bild 14**).

Neben den eigentlichen Überwachungswerten, wie SO_2, NO und Staub, werden auch die Bezugswerte, wie O_2, Temperatur und Volumenstrom, berücksichtigt.

Gleichzeitig erfolgt die Überwachung auf Grenzwertüberschreitungen, und es werden besondere Ereignisse, wie z.B. Störungen der Aufzeichnungsgeräte oder CO-Abschaltungen der Anlage, festgehalten. Die Emissionsrechner können bei besonderen Ereignissen Alarme auslösen und diese

BILD 13: Schwebstaub- und Gasimmissionsmessungen mit einem Meßfahrzeug
FIGURE 13: Measurements of air-borne dust and gas immissions with a test vehicle

With ryegrass, for example, it has been possible to prove that there is no connection between dust deposition in the neighbourhood of a cement works and the thallium content in the plants [98]. Inferences about pollution by harmful acid airborne materials can be drawn from lichen growth. Most of the lichens found in the areas surrounding cement works are those which thrive in an alkaline milieu [99]. A test network distributed over the whole of Germany has been set up for measuring air-borne dust and the gaseous components [100-102]. The measured values are published in technical journals and also weekly in daily papers. Permanent test stations close to cement works can be a valuable means of building up trust.

6. Investigations on soils, plants, products, feed materials and residues

Nowadays, increasing importance is given not only to emission and immission analyses but also to analysis for substances. This involves organic and inorganic constituents. Such investigations are carried out by the manufacturers, the suppliers and the clients as well as by the authorities. Based on the law for soil protection there is great activity in Germany, especially in Baden-Württemberg [103], in the field of soil investigations [104]. These use various digestion methods for thallium, for example, so that not only can the total concentration be measured, but also, and in particular, the mobile fraction which is available for plant transfer [105]. The economically useful plants in certain areas are also investigated regularly in Germany to monitor land cultivation recommendations. Analysis for environmentally relevant constituents in secondary substances is an important part of the quality monitoring system [106, 107], in which suppliers are involved through quality certificates, the users through incoming goods tests, and the authorities through check tests.

7. Recording, processing and evaluation of measured data

Environmental data only have informative value if the associated constraints are known and all the circumstances are established. This requires purposeful recording, processing and evaluation of the measured data. For example, the important guidelines set out in a report for single investiga-

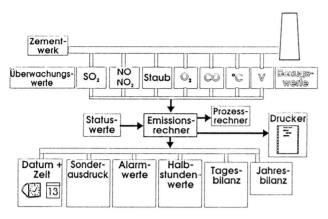

BILD 14: Beispiel für Eingangs- und Ausgangswerte bei einem Emissionsrechner
FIGURE 14: Example of input and output values for an emission computer

Zementwerk	= cement works
Überwachungswerte	= monitored values
Staub	= dust
Bezugswerte	= reference values
Statuswerte	= status values
Emissionsrechner	= emission computer
Prozeßrechner	= process computer
Drucker	= printer
Datum + Zeit	= date + time
Sonderausdruck	= special printout
Alarmwerte	= alarm values
Halbstundenwerte	= half-hour values
Tagesbilanz	= daily balance
Jahresbilanz	= annual balance

Ereignisse auch sofort dokumentieren. Damit ist eine schnelle Ursachenverfolgung möglich. Täglich können weiterhin Protokolle für die Halbstundenmittelwerte für die Tages- und Jahresklassierungen abgerufen werden. Die Emissionsausdrucke stellen daher wertvolle Hilfsmittel für die Verantwortlichen dar. Am Jahresende kann mit Hilfe des Jahresausdruckes ein Jahresbericht erstellt werden.

Den Beurteilungskriterien liegen in Deutschland die Halbstundenmittelwerte zugrunde. Nach Statuskontrollen und Plausibilitätsprüfung sowie Umrechnung auf Bezugsgrößen sind Tagesmittelwerte zu bilden und diese sowie die 1/2-Stundenmittelwerte zu klassieren.

Die behördlichen Vorgaben sind nur erfüllt, wenn sämtliche Tagesmittelwerte unter dem einfachen Grenzwert, 97 % der Halbstundenmittelwerte unter dem 6/5tel-Grenzwert sowie alle Halbstundenmittelwerte unter dem doppelten Grenzwert liegen.

Die Bedeutung dieser Bewertung wird am Beispiel einer NO_x-Auswertung deutlich (**Bild 15**).

Obwohl die Halbstundenwerte eines Jahres im Mittel bei nur 65 % des zulässigen Grenzwertes liegen, erfüllt die Anlage nicht die behördlichen Vorgaben. Wegen der prozeßbedingten Schwankungen ergibt sich eine breite Klassenverteilung, und es treten beim Tagesmittelwert und dem doppelten Grenzwert Überschreitungen auf.

Über Verbindungen des Emissionsrechners mit Prozeßrechnern können die Emissionsdaten mit den Prozeßdaten korreliert werden. Auch Verknüpfungen mit den Winddaten sind denkbar, so daß beispielsweise Emissionsmassenströme auch nach Windsektoren klassiert werden können [110].

Eine Dokumentation von relevanten Umweltdaten mit dem Werksrechner, erlaubt dem Betreiber den Nachweis des bestimmungsgemäßen Betriebes und stellt damit ein Hilfsmittel zur Abwehr von Umwelthaftungsansprüchen dar [111–113].

Dagegen kann die staatliche Fernüberwachung als besonders gravierender Auswuchs einer Emissionskontrolle angesehen werden. Hiermit werden Orwell'sche Visionen umgesetzt und erste Anfänge für eine „gläserne Fabrik" mit „Big Brother Überwachung" geschaffen (**Bild 16**).

tions are [108]: objective, plant description, sampling position, method of measurement, operating conditions and results. The recording of continuous measurements is also very precisely controlled. Documentation with chart recorders is no longer sufficient in this case, and connection to an emission computer is almost always stipulated [109]. Emission computers consist of data acquisition, evaluation, input and output units.

Uninterrupted measurement and documentation over the entire operating period of a plant can be achieved with the aid of emission computers (**Fig. 14**).

Reference values such as O_2, temperature and volume flow are taken into account in addition to the actual monitored values, such as SO_2, NO and dust.

Monitoring for exceeded limits takes place at the same time, and any special events, e.g. faults in the recording equipment or system shutdowns for high CO, are noted. For special occurrences the emission computers can initiate alarms and also immediately document these events, making it possible to track down the cause quickly. Logs of the half-hour mean values can also be retrieved for the daily and annual classification. The emission print-outs therefore represent valuable aids for the responsible persons. An annual report can be drawn up at the end of the year with the aid of the annual print-out.

In Germany the assessment criteria are based on the half-hour mean values. Daily mean values have to be constructed after status checks and plausibility testing and conversion using reference variables, and these and the 1/2-hour values have to be classified.

The official guidelines are only fulfilled if all the daily average values lie under the plain limit, 97 % of the half-hour mean values lie under 6/5 of the limit, and all the half-hour mean values lie under twice the limit.

The significance of this evaluation is illustrated using the example of an NO_x evaluation (**Fig. 15**).

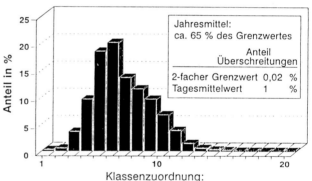

BILD 15: Jahresauswertung am Beispiel Stickstoffoxid (NO als NO_2 gerechnet)
FIGURE 15: Annual evaluation, using the example of nitrogen oxide (NO calculated as NO_2)

Anteil in %	= proportion in %
Jahresmittel: ca. 65 % des Grenzwertes	= yearly average: approx. 65 % of the limit
Anteil Überschreitungen	= proportion over limit
2facher Grenzwert 0,02 %	= 2-times limit value 0.02 %
Tagesmittelwert 1 %	= daily mean value 1 %
Klassenzuordnung	= classification
Klasse 10 ≙ Grenzwert NO_x als NO_2 in mg/m³ (i.N.tr.)	= class 10 ≙ limit for NO_x as NO_2 in mg/m³ (stp, dry)

Although on average the half-hour values for a year were only 65 % of the permitted limit, the plant did not meet the official guidelines. Because of the fluctuations caused by the process there is a wide distribution of the classes, and the daily mean value and the doubled limit are both exceeded.

The emission data can be correlated with the process data through connections between the emission computer and the process computers. The results can also be linked with wind data so that, for example, emission mass flows can also be classified according to wind sector [110].

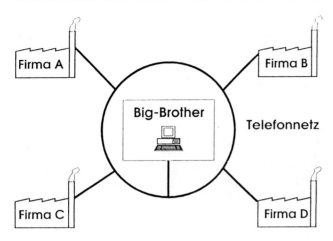

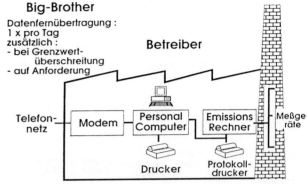

BILD 16: Gesamtkonzept einer Big-Brother-Überwachung
FIGURE 16: Total concept of a Big Brother monitoring system

Firma A	=	Firm A
Firma B	=	Firm B
Big-Brother	=	Big Brother
Telefonnetz	=	telephone network
Firma C	=	Firm C
Firma D	=	Firm D

Bei dem staatlichen Fernüberwachungssystem wird der Werksemissionsrechner über ein Modem an das Telefonnetz angeschlossen. Die Datenfernübertragung erfolgt einmal täglich und zusätzlich auf Anforderung oder automatisch bei Grenzwertüberschreitungen (**Bild 17**).

Damit ist eine quasi On-line-Auswertung der verschiedenen Emissionsquellen eines Gebietes bei einer Landesbehörde möglich. Ein derartiges Netz mit der Bezeichnung EFÜ (Emissions-Fern-Überwachung) [114, 115] wurde beispielsweise in Niedersachsen aufgebaut und ein Zementwerk an dieses Netz angeschlossen.

Das Land Niedersachsen strebt eine flächendeckende Einführung dieses Systems an und begründet die Anschlußpflicht der Industriebetriebe mit dem Verweis auf §§ 26 bis 31 BImSchG [116].

Das an die EFÜ angeschlossene Unternehmen muß gleichzeitig mit den eigentlichen Emissionsdaten auch zur Beurteilung der Emissionssituation erforderliche Werte wie Temperaturen, Sauerstoffgehalt, Feuchte, Abgasvolumen und Betriebszeiten und damit wettbewerbsrelevante Betriebsdaten offenlegen.

Die Anschlußpflicht von genehmigungsbedürftigen Anlagen ist umstritten. Eine derartige Überwachung sowie die als Zielsetzung u. a. genannte „Beweissicherung" ist rechtsstaatlich äußerst bedenklich und läßt sich nicht mit der sonst in der Bundesrepublik gepflegten Einstellung zum Datenschutz und zum sogenannten „großen Lauschangriff" vereinbaren. Sie steht nach Auffassung zahlreicher namhafter Juristen nicht im Einklang mit deutschem Recht.

Neben dem staatlichen Fernüberwachungssystem wird derzeitig auf Länder-, Bundes- und EU-Ebene ein umfassendes Umweltinformationssystem diskutiert. Es soll eine bereichsübergreifende Datenbank aufgebaut werden, in welcher nicht nur Luftemissionen und -immissionen erfaßt, sondern auch Pflanzen-, Boden- und Gewässeruntersuchungen einbezogen werden [117].

8. Ausblick

Die Bedeutung der Messung und Verarbeitung umweltrelevanter Daten wächst. Zunehmend zeigt sich, daß Umweltprobleme nicht mehr nur isoliert auf Einzelmedien oder standortbezogen angegangen werden dürfen, sondern eine Ganzheitsbetrachtung und ein globaler Ansatz notwendig sind. Die Informationen müssen aber vergleichbar sein. Die Art und Weise, wie Daten erfaßt und ausgewertet werden, beeinflußt das Ergebnis. Zur Verbesserung der Transparenz und Vergleichbarkeit in der Umweltdiskussion sind klare Vorgaben für das Messen und Auswerten sowie die toxische

BILD 17: Schema einer Fernübertragung von Emissionswerten
FIGURE 17: Scheme for remote transmission of emission values

Big-Brother	=	Big Brother
Datenfernübertragung:	=	remote data transmission
1 × pro Tag	=	1 × per day
zusätzlich:	=	and also:
– bei Grenzwertüberschreitung	=	– on exceeding limit
– auf Anforderung	=	– on request
Betreiber	=	operator
Telefonnetz	=	telephone network
Modem	=	modem
Personal-Computer	=	personal computer
Emissions-Rechner	=	emission computer
Meßgeräte	=	measuring equipment
Drucker	=	printer
Protokolldrucker	=	report printer

Documentation of relevant environmental data with the works computer provides the operator with evidence of correct operation, and represents an aid for defending against claims of environmental liability [111–113].

On the other hand the governmental remote monitoring system can be regarded as a particularly serious abuse of emission checking. This is turning Orwellian visions into reality and is creating the beginnings of a "glass factory" with "Big Brother monitoring" (**Fig. 16**).

In the governmental remote monitoring system the works emission computer is connected by modem to the telephone network. Remote data transmission takes place once a day and additionally on request, or automatically if a limit is exceeded (**Fig. 17**).

This enables a state authority to have quasi on-line evaluation of the different emission sources in an area. A network of this type, designated EFÜ (remote emission monitoring) [114, 115] has, for example, been set up in Lower Saxony and a cement works has been connected to this network.

The state of Lower Saxony is trying to introduce this system across the board and is basing the obligation of the industrial operations to participate on a reference to §§ 26 to 31 BImSchG [Federal Immission Protection Act] [116].

The companies connected to the EFÜ must also, simultaneously with the actual emission data, publish values necessary for assessing the emission situation, such as temperatures, oxygen content, moistures, exhaust gas volumes and operating times, and hence operating data relevant to competition.

The obligation to participate on the part of plants which require authorization is disputed. This type of monitoring and the "preservation of evidence" mentioned among other things as the objective is extremely questionable from the constitutional point of view and is incompatible with the attitude towards data protection and to the so-called "great surveillance offensive" otherwise maintained in Germany. In the opinion of a great many noted lawyers it is not in accord with German law.

In addition to the governmental remote monitoring system a comprehensive environmental information system is at present under discussion at the state, federal, and EU levels. A data base which includes not only air emissions and immissions but also investigations of plants, soil and water is to be set up covering all areas [117].

Bewertung umweltrelevanter Daten auf internationaler Ebene dringend erforderlich. Der verantwortungsvolle Umgang mit umweltrelevanten Daten ist eine große Herausforderung für die Industrie, die Behörden und die Politik.

8. Outlook

The importance of measuring and processing environmentally relevant data is growing. It is becoming increasingly apparent that environmental problems must no longer be tackled in isolation in individual media or on a local basis; there is need for a holistic examination and a global approach. However, the information must be comparable. The manner in which data is collected and evaluated affects the result. There is a pressing need for clear guidelines at the international level for measuring and evaluation and for the toxic evaluation of environmentally relevant data in order to improve the transparency and comparability of the environmental debate. Responsible dealings with environmentally relevant data is a great challenge for industry, the authorities, and the politicians.

Literaturverzeichnis

[1] Verordnung (EWG) Nr. 1836/93 des Rates vom 29. Juni 1993 über die freiwillige Beteiligung gewerblicher Unternehmen an einem Gemeinschaftssystem für das Umweltmanagement und die Umweltbetriebsprüfung. Amtsblatt der Europäischen Gemeinschaften Nr. L 168/1, 10. 7. 1993.

[2] Richtlinie des Rates vom 7. Juni 1990 über den freien Zugang zu Informationen über die Umwelt (UIRL, 90/313/EWG). Amtsblatt der Europäischen Gemeinschaften Nr. L 158/56

[3] Ballschmiter, K.-H.: Analytische Grundlagen zur Ermittlung von Grenzwerten. Umwelt-Magazin, Juli 1990, S. 7–10.

[4] Montanus, K.: Große Angst vor kleinen Mengen. Gesellschaft für Strahlen und Umweltforschung München.

[5] Reiter, J.: Die Grenzen der Grenzwerte. Umwelt-Magazin, Juli 1990, S. 33–36.

[6] Grein, H.: Toxikologische Voraussetzungen für die Festlegung von Grenzwerten. Umwelt-Magazin, Juli 1990, S. 24–27.

[7] Müller, P.: Ökologische Relevanz kleiner Stoffmengen. Umwelt-Magazin, Juli 1990, S. 15–20.

[8] Peine, H. G.: Zur Vielfalt bestehender Grenzsysteme. Umwelt-Magazin, Juli 1990, S. 5–6.

[9] Ditbern, D.: Grenzwertdiskussion heute. Umwelt-Magazin, Juli 1990, S. 4.

[10] Repenning, K.: Risikodiskussion – politische Aufgabe. Umwelt-Magazin, November 1992, S. 82–84.

[11] Interne VDZ-Mitteilungen Nr. 81.

[12] Prévention de la Pollution dans les Cimenteries (Arrêté du 3 mai 1993). Publié au Journal Officiel de la République Française le 15 juin 1993.

[13] Davids, P.: Den Stand der Technik konkretisieren. Wann soll der Staat tätig werden? Machbarkeit kontra Verhältnismäßigkeit. Umwelt Bd. 23 (1993) Nr. 4, S. 202–206.

[14] Verwaltungsgericht Karlsruhe, Urteil im Namen des Volkes in der Verwaltungsrechtssache, AZ.: 4 K 26/91.

[15] Erste Allgemeine Verwaltungsvorschrift zum Bundes-Immissionsschutzgesetz (Technische Anleitung zur Reinhaltung der Luft) vom 27. 2. 1986 (GMBL, S. 95).

[16] Kuhlmann, K.: Produktökobilanz – Verfahren für eine ganzheitliche Beurteilung, beispielsweise von Bauprodukten aus Zement und Beton. Zement-Kalk-Gips 47 (1994) Nr. 1, S. 25–30.

[17] Interne VDZ-Mitteilungen Nr. 90.

[18] Thudium, J., Törnerik, H., und Förderer, L.: Modellierung der Ausbreitung von Luftschadstoffen in einem Alpental. WLB, Wasser, Luft, Boden 3/1994, S. 47–50.

[19] Dreizeitl, M.E., und Stöhr, D.: Emissionen – Meteorologie – Immissionen, Transitstudie, Institut für Meterologie, Universität Insbruck, 1993.

[20] Kuhlmann, K., und Kirchartz, B.: Beitrag der Emissionen von Zementwerken zur regionalen Immission. (Veröffentlichung in Vorbereitung).

[21] Meyer, J.: Verantwortbares Wachstum. Umwelt 3/92, S. 139–146.

[22] Enquete-Kommission des Deutschen Bundestages „Schutz des Menschen und der Umwelt – Bewertungskriterien und Perspektiven für umweltverträgliche Stoffkreisläufe in der Industriegesellschaft". Drucksache 12/1951 vom 16. Januar 1992.

[23] Kroboth, K., Kuhlmann, K., und Xeller, H.: Stand der Technik der Emissionsminderung in Europa. Zement-Kalk-Gips 43 (1990) Nr. 3, S. 121–131.

[24] Schwarz, O., und Grefen, K.: Die Richtlinienarbeit der VDI-Kommission. Reinhaltung der Luft, Band 47 (1987) Nr. 3/4, S. 49–57.

[25] Schnabel, W.: Zuverlässig, Probenahme in der Ablufttechnik. Entsorga-Magazin 3 (1989), S. 31–34.

[26] Bundeseinheitliche Praxis bei der Überwachung der Emissionen; Richtlinien über die Eignungsprüfung, den Einbau, die Kalibrierung und die Wartung von Meßeinrichtungen für kontinuierliche Emissionsmessungen, Rundschreiben des BMU vom 1. 3. 1990; IG I 2-556 134/4 im gemeinsamen Ministerialblatt (GMBl), 1990, S. 226.

[27] Görgen, R.: Rechtliche Rahmenbedingungen der Meßtechnik. Umwelt (1992) Nr. 5, S. 300–303.

[28] N. N.: Gemessen wird viel, richtig ausgewertet wenig. VDI-Nachrichten, Nr. 4, 25. Januar 1991.

[29] Luftreinhaltung, Leitfaden zur kontinuierlichen Emissionsüberwachung. Hrsg. Umweltbundesamt. UBA-Berichte, Band 1/86, Berlin: Erich Schmidt Verlag 1986.

[30] Anlage zum Rundschreiben Nr. 7/1986, Forschungsinstitut der Zementindustrie Düsseldorf. Hinweise für den Einsatz von Meßeinrichtungen zur kontinuierlichen Überwachung von gasförmigen Emissionen in Zementwerken.

[31] Fabinski, W.: Rauchgase kontinuierlich messen. Umwelt, Bd. 22 (1992) Nr. 4, S. 205–208.

[32] Scheuer, A.: Kontinuierliche Gasanalyse bei Zementöfen. Zement-Kalk-Gips 38 (1985) Nr. 5, S. 229–232.

[33] Nyffenegger, H.: Die Gasanalyse in der Zementproduktion. Vortrag 20. März 1985 in Boden/Schweiz.

[34] Gromadzki, D.: Methoden zur Rauchgasanalyse. TIZ International, Vol. 113, No. 7, 1989, S. 567–571.

[35] Brandt, A., und Fabinski, W.: Schadgase kontinuierlich messen. BWK/TÜ/Umwelt-Special, März 1993, L 61–68.

[36] Alshagen, J., Lietenow, D., und Stahl, H.: Automatisierte Meßsysteme zur Erfassung von Luftschadstoffen. Umwelt 1988, M 4–M 12.

[37] Geneken, K.-J.: Emissionsmessungen – Aufgaben, Struktur, Durchführungen. Umwelt (1992) Nr. 6, S. 354–358.

[38] Kontinuierliche Gasanalyse in Zementwerken. Verein Deutscher Zementwerke, Ausschuß Verfahrenstechnik, Merkblatt VT 9, Juni 1990.

[39] N.N.: Analysensysteme zur kontinuierlichen Luftüberwachung, Umwelt und Technik 4/1989, S. 52–56.

[40] Lützke, Kl., und Burk, H.-D.: Meß- und Überwachungstechnik bei der Emissionsminderung von Stickoxiden. Umwelt, März 1988, M 14–M 19.

[41] Interne VDZ-Mitteilungen Nr. 88.

[42] Verein Deutscher Zementwerke e.V., Forschungsinstitut der Zementindustrie, Tätigkeitsbericht 1990–93. Beton-Verlag, Umweltschutz bei der Zementherstellung, S. 78–80.

[43] Interne VDZ-Mitteilungen Nr. 78.

[44] Interne VDZ-Mitteilungen Nr. 76.

[45] Biniuris, S.: Entwicklung der kontinuierlichen Staubemissionsmessung. VGB Kraftwerkstechnik 71 (1991) Nr. 5, S. 490–495.

[46] Stahl, H.: Maßnahmen zur Qualitätssicherung bei der kontinuierlichen Emissionsüberwachung. WLB Wasser, Luft und Beton 3/1992, S. 58–63.

[47] Bröker, G.: Qualitätssicherung bei Emissionsmessungen. VDI-Berichte Band 838, VDI-Verlag, Düsseldorf 1990, S. 633–642.

[48] Lietenow, D., und Stahl, H.: Stellen nach § 26 BImSchG für Emissions- und Immissionsmessungen von Luftverunreinigungen. WLB-Handbuch, Umwelttechnik 1990/91, S. 104 ff.

[49] Richtlinie zur Bekanntgabe von Stellen zur Ermittlung von Emissionen und Immissionen nach §§ 26 BImSchG. Ministerialblatt für das Land Nordrhein-Westfalen 1986, S. 525–528.

[50] VDI Richtlinie 3950, Entwurf, Kalibrierung automatischer Meßeinrichtungen.

[51] VDI Richtlinie 2066, Blatt 1, 2E, 3E, 7E, Messung von Partikeln, Staubmessen in strömenden Gasen; manuelles Verfahren.

[52] VDI Richtlinie 2448, Blatt 1, Planung von stichprobenartigen Emissionsmessungen an geführten Quellen.

[53] Brandl, A.: Emissionsüberwachung bei der Glasherstellung. Unveröffentlichter Vortrag anläßlich einer Tagung der Hüttentechnischen Vereinigung der Deutschen Glasindustrie 1991.

[54] Emissionsmessungen, Verein Deutscher Zementwerke e.V., Forschungsinstitut der Zementindustrie. Tätigkeitsbericht 1990–1993, S. 78–80.

[55] Kuhlmann, K., Kirchartz, B., Rechenberg, W., und Bachmann, G.: Probenahme von Spurenbestandteilen im Reingas von Zementdrehrohröfen. Zement-Kalk-Gips 44 (1991) Nr. 5, S. 209–216.

[56] Hahn, U.: Bestimmung der Staubemission mit Hilfe der gravimetrischen Staubmeßtechnik. Meßtechnik für Stäube und Schadgase, Hüttentechnische Vereinigung der Deutschen Glasindustrie. Fortbildungskurs 1986, S. 11–24.

[57] Interne VDZ-Mitteilungen Nr. 78.

[58] Methodensammlung für Spurenelemente. VDZ-Arbeitskreis Analytische Chemie, Aufschluß und Bestimmungsverfahren für Schwermetalle, 1993.

[59] Schwedt, G.: Computergesteuerte Spurenanalyse von Schwermetallen. Umwelt-Magazin 1991, S. 100–102.

[60] Interne VDZ-Mitteilungen Nr. 78.

[61] Interne VDZ-Mitteilungen Nr. 89.

[62] Interne VDZ-Mitteilungen Nr. 89.

[63] Interne VDZ-Mitteilungen Nr. 89.

[64] VDI-Richtlinie 2462, Blatt 1, 2, 3, 7 und 8.

[65] Interne VDZ-Mitteilungen Nr. 83.

[66] Gerger, W.: Erfahrungen mit spezialisierten Kontrolltechniken beim Einsatz von Sekundärstoffen. Vortrag zum VDZ-Kongreß '93, Fachbereich 1 (Veröffentlichung vorgesehen).

[67] VDI-Richtlinie 2546, Blatt 1, Blatt 2, Blatt 8.

[68] VDI-Richtlinie 3481, Blatt 2, Bestimmung des durch Adsorption an Kieselgel erfaßbaren organisch gebundenen Kohlenstoffs in Abgasen.

[69] Interne VDZ-Mitteilungen Nr. 86.

[70] Interne VDZ-Mitteilungen Nr. 91.

[71] Emissionsminderung, Dioxine und Furane. Verein Deutscher Zementwerke e.V., Forschungsinstitut der Zementindustrie, Tätigkeitsbericht 1990–93, S. 82–84.

[72] Interne VDZ-Mitteilungen Nr. 85.

[73] Düwel, U.: Einschneidende Konsequenzen. Entsorga-Magazin, Mai 1992, S. I–XVII.

[74] Messen von polychlorierten Dibenzo-p-Dioxinen und Dibenzofuranen an industriellen Anlagen. VDI 3499, Blatt 2, Blatt 3 (1990).

[75] Schäfer, K., Wehner, D., und Haus, R.: Remote Sensing of Smoke Stack Emissions Using a Mobile Environmental Laboratory. SPIE Proceedings on the "Environmental Sensing" conference in Berlin, June 1992.

[76] Lärm und Erschütterungen, Verein Deutscher Zementwerke e.V., Forschungsinstitut der Zementindustrie. Tätigkeitsbericht 1990–93, S. 88–90.

[77] Interne VDZ-Mitteilungen Nr. 88.

[78] Interne VDZ-Mitteilungen Nr. 87.

[79] Interne VDZ-Mitteilungen Nr. 80.

[80] Interne VDZ-Mitteilungen Nr. 90.

[81] Interne VDZ-Mitteilungen Nr. 91.

[82] DIN 4150, Teil 2. Erschütterungen im Bauwesen.

[83] Staub- und gasförmige Immissionen. Verein Deutscher Zementwerke e.V., Forschungsinstitut der Zementindustrie, Tätigkeitsbericht 1990–1993, S. 85–88.

[84] Bestimmung des Staubniederschlags. VDI-Richtlinie 2119, Blatt 2.

[85] Staubinhaltsstoffe im Staubniederschlag. VDI-Richtlinie 2267, Blatt 2.

[86] Flechten geben Auskunft über Luftqualität. Thema des Monats, April 1993, TÜV-Südwest, Filderstadt.

[87] Pflanzen eignen sich als Indikator für Schwermetallbelastungen. VDI-Nachrichten Nr. 45, 12. November 1993, S. 36.

[88] Interne VDZ-Mitteilungen Nr. 90.

[89] Interne VDZ-Mitteilungen Nr. 86.

[90] Interne VDZ-Mitteilungen Nr. 86.

[91] Lung, T.: Zur Erstellung standortbezogener Ausbreitungsklassenstatistiken. WLB Wasser, Luft und Boden 11–12/1992, S. 55–58.

[92] Joel, H.: Staubniederschlagsmessungen in der Umgebung von Zementwerken. Zement-Kalk-Gips 18 (1965) Nr. 3, S. 114–121.

[93] VDI 2119, Blatt 1, Blatt 2, Messung partikelförmiger Niederschläge.

[94] Riekert, J.A., van Doornum, A.D., und Gaylard, J.M.: Environmental improvement programme at a cement plant in the Pretoria urban area. Vortrag zum VDZ-Kongreß '93, Fachbereich 4.

[95] Depositionsnetz Zementwerke 1992/1993. Landesanstalt für Umweltschutz Baden-Württemberg. UMEG-Bericht Nr. 32–6/1993.

[96] Jahresbericht 1992 – Luftschadstoffmessungen. Umweltministerium Baden-Württemberg, Bericht Nr. 31–07/1993.

[97] Mindt, J.: Mobiles und modulares Labor. Elektronik Journal 10/92, S. 47–50.

[98] Schweiger, D.: Untersuchungen zur Ermittlung der Tl- und Cd-Belastung in der Umgebung eines Zementwerkes mittels Bioindikatoren. Staatliche Landwirt-

schaftliche Untersuchungs- und Forschungsanstalt Augustenberg. Unveröffentlichter Bericht, Dezember 1985.

[99] Bartholmeß, E.: Bioindikation der Immissionsbelastung mit Hilfe von Flechtenkartierungen. Juli 1991, unveröffentlichter Bericht TÜV Südwest.

[100] Luftqualität in der Bundesrepublik, Daten zur Umwelt 1990/91. Umweltbundesamt Berlin, Erich Schmidt Verlag, S. 182–245.

[101] Luft, Meßnetz, Immission, Umweltdaten 91/92. Landesanstalt für Umweltschutz Baden-Württemberg. Umweltministerium Baden-Württemberg 1992, B/12-B/28.

[102] Index und Karte geben Durchblick bei „dicker Luft". Umweltindex der VDI-Nachrichten erhält Unterstützung durch Immissionskarte. Umwelt Bd. 23 (1993) Nr. 5, S. 246.

[103] Staatsministerium Baden-Württemberg (1991): Gesetz zum Schutz des Bodens (Bodenschutzgesetz BodSchG) – Gesetzblatt für Baden-Württemberg 16: 434–440, Stuttgart.

[104] Umweltministerium Baden-Württemberg (1993): Zweite Verwaltungsvorschrift des Umweltministeriums zum Bodenschutzgesetz über die Probenahme und -aufbereitung (VwV Bodenproben) vom 24. August 1993 – Az.: 44-8810.30-1/46.

[105] Umweltministerium Baden-Württemberg (1993): Dritte Verwaltungsvorschrift des Umweltministeriums zum Bodenschutzgesetz über die Ermittlung und Einstufung von Gehalten anorganischer Schadstoffe in Baden (VwV Anorganische Schadstoffe).

[106] Liebl, P., und Gerger, W.: Nutzen und Grenzen beim Einsatz von Sekundärstoffen. Zement-Kalk-Gips 46 (1993) Nr. 10, S. 632–638.

[107] Gerger, W.: Erfahrungen mit spezialisierten Kontrolltechniken beim Einsatz von Sekundärstoffen. Vortrag VDZ-Kongreß 1993, Düsseldorf.

[108] Interne VDZ-Mitteilungen Nr. 88.

[109] Interne VDZ-Mitteilungen Nr. 78.

[110] Mattern, E., Eberbach, W., und Schulz, V.: Werksweites Netzwerk für die Erfassung, Verarbeitung und behördengerechte Protokollierung von Umweltwerten. VGB Kraftwerkstechnik 72 (1992) Heft 2, S. 116–124.

[111] Interne VDZ-Mitteilungen Nr. 90.

[112] Interne VDZ-Mitteilungen Nr. 88.

[113] Gesetz über die Umwelthaftung. Bundesgesetzblatt, Jahrgang 1990, Teil 1, 10. Dezember 1990.

[114] Emissionsfernüberwachung in Niedersachsen. Sonderdruck des Niedersächsischen Umweltministeriums, 1992.

[115] Liebe, H.-G., und Thiele, W.: Automatische Erfassung und Übertragung von Emissionsmeßwerten. WLB Wasser, Luft und Boden 7–8, 1989, S. 40–51.

[116] Bundes-Immissionsschutzgesetz – BImSchG, zuletzt geändert durch Gesetz vom 22. April 1993, BGBl I, S. 466.

[117] UIS Das Umweltinformationssystem des Landes Baden-Württemberg. Umweltministerium Baden-Württemberg 1992.

Produktökobilanz –
Verfahren für eine ganzheitliche Beurteilung, beispielsweise von Bauprodukten aus Zement und Beton*)

Product ecobalance –
a method for complete assessment of, for example, building products made of cement and concrete*)

Bilan écologique du produit –
Procédé pour une évaluation globale, par l'exemple de produits de construction à base de ciment et de béton

Balance ecológico del producto –
Procedimiento para una evaluación i ntegral, por ejemplo de productos de cemento y de hormigón

Von **K. Kuhlmann**, Forschungsinstitut der Zementindustrie, Düsseldorf/Deutschland

Fachbericht 4.2 · Zusammenfassung – Seit einigen Jahren arbeiten staatliche Organisationen, Industrie, Verbraucher- und Umweltverbände an Verfahren, mit denen analog zu betriebswirtschaftlichen Bilanzen die bei der Herstellung von Produkten entstehenden Umweltbelastungen erfaßt und mit dem Ziel bewertet werden, Produkte ökologisch vergleichbar zu machen. Angewendet wurden bislang Bilanzmodelle und Begriffe mit unterschiedlichstem Sachinhalt, wie „Ökobilanz", „Produktlinienanalyse" (PLA) oder im amerikanischen Sprachraum „Product-Life-Cycle-Assessment" (LCA). In der PLA sollten alle umweltbelastenden Faktoren während des Lebenswegs eines Produkts zusammengefaßt und beurteilt werden. Die Bewertung soll allerdings über den Umweltbereich hinausgehen und insbesondere ökonomische, soziale und gesellschaftliche Kriterien mit einbeziehen. Ein konsensfähiges Konzept liegt bislang hierfür jedoch noch nicht vor. Sowohl in der LCA als auch in der Ökobilanz wird innerhalb eines exakt definierten Bilanzrahmens der Lebensweg des Produkts von der Rohmaterialgewinnung bis zur Entsorgung dargestellt. Auf der Einnahmenseite steht der Verbrauch an Energie, Rohmaterial, Wasser und Luft ebenso wie die Inanspruchnahme von Grund und Boden. Die Ausgabenseite enthält die luftgetragenen Emissionen, die Lärmemissionen, den Anfall an Abfall, Abwasser und Abwärme sowie an verwertbaren Rohstoffen. Die Bewertung soll auf die Umweltinanspruchnahme beschränkt bleiben; sie wird im Vergleich zu derjenigen bei der komplexen PLA allerdings nur teilweise einfacher, da es derzeit noch keine allgemein akzeptierten „ökologischen" Verrechnungseinheiten gibt. In der Ökobilanz für Beton werden die Normierung unterschiedlicher Umwelteinheiten in Anlehnung an das Vorgehen des schweizerischen Bundesamtes für Umwelt, Wald und Landschaft (BUWAL) durchgeführt und dimensionslose Verhältniszahlen (Grenzwerte) auf der Basis von tatsächlichen und maximal möglichen Belastungen (Grenzwerte) zusammengefaßt. Die Abwägung zwischen ökologischer Belastung durch das Produkt und dessen Nutzen für die Umwelt erfolgt in Form einer Nutzwertanalyse. Diese vor allem aus Planungsprozessen bekannte Entscheidungstechnik erlaubt es, mehrere Gesichtspunkte gleichzeitig in die Entscheidungsfindung mit einzubeziehen. Dabei können unterschiedliche quantitative, aber auch qualitative Daten in Form gewichteter dimensionsloser Nutzwerte berücksichtigt werden.

Special report 4.2 · Summary – For some years government organizations, industry, and consumer and environmental organizations have been working on methods with which, like industrial management balances, the environmental burden resulting from the manufacture of products are measured and assessed with the object of making products ecologically comparable. Balance models and terms with widely differing factual content have been used, such as "ecobalance", "product line analysis" (PLA) or in the American language sphere "Product-Life-Cycle-Assessment" (LCA). The PLA is supposed to combine and assess all the factors affecting the environment during the life of a product. The evaluation is intended to extent beyond the environmental sphere and also incorporate economic, social and communal criteria. However, no scheme has as yet been agreed for this. Both in the LCA and in the ecobalance the life of the product is described from raw material extraction to final disposal within an exactly defined balance framework. On the input side are the consumption of energy, raw materials, water and air, as well as the demands on land. The output side contains the airborne emissions, the noise immission, the waste material, waste water and waste heat, as well as useful raw materials produced. The evaluation is to be confined to environmental demands; however, it is only in some respects that it is simpler than that of the complex

---

*) Überarbeitete Fassung eines Vortrages zum VDZ Kongreß '93, Düsseldorf (27. 9.–1. 10. 1993)
Revised text of a lecture to the VDZ Congress '93, Düsseldorf (27. 9.–1. 10. 1993)

PLA, as at present there is still no generally accepted "ecological" calculation unit. In the ecobalance for concrete various environmental units are standardized on the basis of the procedure used by the Federal Swiss Office for Environment, Forests and Land, and dimensionless proportional numbers are compiled on the basis of actual and maximum possible loadings (limit values). The ecological loading caused by the product is weighed against its benefit to the environment in the form of a utility analysis. This decision technique familiar mainly from planning processes makes it possible to include several points of view at once arriving at a decision. Differing quantitative, and also qualitative, data can be taken into account in the form of weighted dimensionless utility values.

Rapport spécial 4.2 · Résumé – Depuis quelques années, organismes d'Etat, industrie, confédérations de consommateurs et associations de protection de l'environnement travaillent conjointement sur des procédés, permettant, de manière analogue aux bilans d'économie industrielle, d'enregistrer les charges de l'environnement se produisant lors de la fabrication de produits, et de les évaluer en rendant comparables les produits sur le plan écologique. Jusqu'à ce jour, il a été mis en oeuvre des modèles de bilan et des concepts d'une teneur des plus variées comme „bilan écologique", „analyse de ligne de produit" (PLA) ou ce qu'appellent les américains „Product-Life-Cycle-Assessment" (LCA). Le PLA se propose de regrouper et d'apprécier tous les facteurs de pollution de l'environnement durant toute la vie d'un produit. L'appréciation, quant à elle, doit cependant dépasser le cadre de l'environnement et intégrer notamment des critères économiques, sociaux et de societé. Un concept consensuel n'est cependant pas encore disponible jusqu'ici. Tant dans le LCA que dans le bilan écologique on trouve, à l'intérieur d'un cadre de bilan extrêmement précis, le „curriculum vitae" du produit de l'extraction de la matière première jusqu'au traitement des déchets. Côté recette on trouve la consommation d'énergie, la matière première, l'eau et l'air ainsi que la prise en compte du terrain et du sol. Côté dépenses on range les émissions polluant l'air, les nuisances sonores, la formation de déchets, d'eaux résiduaires, de chaleur perdue ainsi que de matières premières récupérables. L'appréciation reste limitée à la prise en compte des questions d'environnement. Elle est en comparaison avec celle de PLA qui est complexe, plus simple du fait qu'il n'existe actuellement encore aucune unité de décompte „écologique" reconnue par tous. Dans le bilan écologique du béton il est procédé à une normalisation des unités les plus variées en matière d'environnement s'appuyant sur la procédure de l'Office fédéral Suisse pour l'environnement, les forêts et le paysage (BUWAL), et des coefficients sans dimension basés sur des charges (valeurs-limites) effectives, maximales possibles y sont regroupées. L'appréciation entre la charge écologique due au produit et son exploitation pour l'environnement est réalisée sous forme d'analyse de rentabilité des coûts. Cette technique de prise de décision connue surtout dans les processus de planification permet d'intégrer simultanément plusieurs points de vue dans la recherche de décision. Pour cela on peut prendre en compte diverses données quantitatives mais aussi qualitatives sous forme de valeurs utiles pondérées sans dimension.

Bilan écologique du produit –
Procédé pour une évaluation globale, par l'exemple de produits de construction à base de ciment et de béton

Informe de ramo 4.2 · Resumen – Desde hace varios años, algunos organismos estatales, la industria, los consumidores y las organizaciones de protección del medio ambiente están elaborando métodos que permitan – analogamente al balance económico – registrar y evaluar las repercusiones de los productos sobre el medio ambiente, con el fin de conseguir que tales productos sean comparables desde el punto de vista ecológico. Hasta ahora se han aplicado modelos y términos de muy diferente significado, tales como „balance ecológico", „análisis de lineas de productos" (ALP) o, en el área norteamericano, „Product-Life-Cycle-Assessment" (LCA). Mediante el ALP se pretendia reunir y evaluar todos los factores perjudiciales para el medio ambiente durante la vida útil del producto. Pero la evaluación tenia que pasar más allá del ámbito medioambiental, incluyendo, especialmente, criterios económicos y sociales. Para ello no se dispone, hasta la fecha, de ningún concepto susceptible de ser generalmente aceptado. Tanto en el LCA como el balance ecológico se representa la vida del producto, dentro del marco de un balance exactamente definido, desde la obtención de las materias primas hasta la eliminación de los desechos. En el lado „entradas" figura el consumo de energía, materias primas, agua y aire, lo mismo que la utilización del suelo. En el lado „salidas" se apuntan las emisiones que afectan al aire, las emisiones sonoras, las emisiones en forma de desechos, aguas residuales y calor residual así como las materias primas aprovechables. La evaluación debe quedar limitada al medio ambiente, pero sólo será en parte más sencilla, en comparación con el ALP, que es bastante complejo, ya que de momento todavía no existen unidades de cuenta „ecológicas", generalmente aceptadas. En el balance ecológico del hormigón, se lleva a cabo la normalización de diferentes unidades del medio ambiente, siguiendo el ejemplo de la Oficina Suiza del Medio Ambiente, Bosques y Paisaje, mediante la combinación de cifras proporcionales, adimensionales, sobre la base de las cargas efectivas y de las cargas máximas posibles (valores límite). La ponderación de la carga ecológica, debida al producto, y su utilidad para el medio ambiente se efectua en forma de análisis del valor útil. Esta técnica, conocida sobre todo de los procesos de planificación, permite incorporar varios aspectos a la vez, para llegar a una decisión. A este respecto, se pueden tener en cuenta diferentes datos cuantitativos y también cualitativos, en forma de valores utiles ponderados, adimensionales.

Balance ecológico del producto –
Procedimiento para una evaluación integral, por ejemplo de productos de cemento y de hormigón

1. Einleitung

Jährlich werden in der Bundesrepublik Deutschland mehr als 600 Mio. Tonnen mineralische Baustoffe produziert. Das Bauvolumen in Deutschland betrug 1992 488 Mrd. DM. Dabei entfielen 412 Mrd. DM auf die alten Bundesländer. Sie verteilten sich zu 43 % auf das Bauhauptgewerbe, zu 29 % auf das Ausbaugewerbe und zu 28 % auf andere Produzentengruppen. Im Bauhauptgewerbe ist im wesentlichen die Rohbautätigkeit (86 %) enthalten, in der der Beton- und Mauerwerksbau dominieren. Wertmäßig entfallen rund zwei Drittel des Rohbaus auf die Betonbauart (**Tabelle 1**). Insgesamt stellt die Bautätigkeit einen erheblichen Anteil von knapp einem Fünftel am Bruttosozialprodukt der Bundesrepublik Deutschland dar [1].

Unbestritten trägt das Bauen dazu bei, wesentliche Bedürfnisse unserer Gesellschaft zu befriedigen, wie etwa das Wohnen, das Erzeugen von Nahrungsmitteln und Gebrauchsgütern sowie die Mobilität. Aber auch Ver- und Entsorgung sind in einer modernen Industriegesellschaft ohne Bauen mit Beton nicht möglich. Eine entsprechende Anerkennung findet das Bauen, insbesondere der Wohnungsbau, auch in der Bevölkerung, wie Ergebnisse einer 1992 durchgeführten Meinungsumfrage zeigen (**Tabelle 2**). Das Bauen rangiert danach in der Prioritätenliste auf den vorderen Plätzen [2]. Aus der Zusammenstellung geht hervor, daß Umweltschutz und Wohnungsbau als etwa gleichrangige politische Aufgaben und Ziele in der Öffentlichkeit eingestuft werden. Daraus ergibt sich zwangsläufig ein Spannungsfeld, denn das Bauen und die Herstellung von Baustoffen sind unweigerlich mit Eingriffen in Natur und Umwelt verbunden. Die Frage, inwieweit diese Eingriffe jedoch gerechtfertigt sind, bildet oftmals Gegenstand von Auseinandersetzungen mit der Bevölkerung, die Bauen für unabdingbar hält.

Was bislang fehlt, sind allgemein akzeptierte Regeln, um komplexe Prozesse oder Produkte unter ökologischen Gesichtspunkten ganzheitlich zu bewerten. In diesem Zusammenhang werden Ökobilanzen zukünftig besondere Bedeutung erlangen. Allerdings darf sich eine solche Bilanz nicht nur auf eine Zusammenstellung der Umweltbelastungen beschränken, sondern es muß auch deren Bewertung erfolgen [3, 4]. Lediglich eine Liste mit quantitativ ermittelten Umweltbelastungen gilt nicht als Ökobilanz. Zudem muß Einigkeit darüber bestehen, daß Ökobilanzen keine Instrumente sind, mit denen man Materialien gegeneinander ausspielen kann. Plakative Aussagen wie „Baustoff A ist besser als Baustoff B" sind auf Basis seriöser Ökobilanzen nicht zu erwarten.

1. Introduction

More than 600 million tonnes of mineral building materials are produced every year in Germany. The volume of construction in Germany in 1992 amounted to 488 billion DM, of which 412 billion DM were accounted for by the former West Germany. This broke down into 43 % for the building industry proper, 29 % for the finishing trade, and 28 % for other producer groups. Most of the carcass work (86 %), in which concrete and masonry construction are predominant, is covered by the building industry proper. Concrete construction accounts for about two-thirds of the carcass work in terms of value (**Table 1**). In total, construction work represents the considerable proportion of just under one fifth of Germany's gross national product [1].

It is undisputed that the construction industry contributes to satisfying some of the essential needs of our society such as, for example, accommodation and the production of food and consumer durables, as well as mobility. Nor are supply and waste disposal possible in a modern industrial society without concrete construction work. Construction, especially of housing, is also duly recognized by the population, as is shown by the results of an opinion poll carried out in 1992 (**Table 2**). This shows that building occupies one of the top places in the list of priorities [2]. It can be seen from the table that public opinion classifies environmental protection and the building of housing as political duties and objectives of roughly equal rank. This inevitably leads to a conflict of interests, as building and the manufacture of building materials are unavoidably associated with interference in nature and the environment. However, the question as to how far this interference is justified often causes arguments among the population which considers construction to be indispensable.

What has been lacking so far are generally acceptable rules for evaluating complex processes or products as a whole from the ecological point of view. Ecobalances are going to become particularly important in this connection in the future. However, such a balance must not be confined just to compiling the adverse environmental effects; it must also evaluate them [3, 4]. A mere list with quantitatively measured environmental damage does not constitute an ecobalance. There must also be general agreement that ecobalances are not instruments for playing off materials against each other. Serious ecobalances should not be expected to produce simplistic statements such as "Construction material A is better than construction material B".

TABELLE 1: Aufteilung des Bauvolumens auf die wesentlichen Baugewerbe (1992)

Bezeichnung	Bauvolumen in Mrd. DM	%
Bauvolumen 1992 (alte Bundesländer)	412	100
Bauhauptgewerbe	177	43
Ausbaugewerbe	121	29
Andere Produzentengruppen	114	28

TABLE 1: Breakdown of the volume of construction work between the main buildings trades (1992)

Designation	Construction volume in billion DM	%
Construction volume 1992 (former West Germany)	412	100
Building industry proper	177	43
Finishing trade	121	29
Other producer groups	114	28

TABELLE 2: Ergebnis einer Umfrage nach den wesentlichen politischen Aufgaben und Zielen in der Bundesrepublik (1992)

Bezeichnung	Angaben in %
Umweltschutz	72
Wohnungsbau	68
Renten	70
Asylmißbrauch	64
Arbeitsplätze	68
Wirtschaftsstabilität	65

TABLE 2: Results of an opinion poll about essential political tasks and objectives in Germany (1992)

Designation	Data in %
Environmental protection	72
Housing	68
Pensions	70
Misuse of asylum	64
Jobs	68
Economic stability	65

2. Verfahren der Bilanzierung

Die Ökobilanz (international: Life Cycle Assessment, LCA) ist keine neue Methode. Schon in den 70er Jahren wurden umweltbezogene Systemanalysen von Verpackungsmaterialien und chemischen Prozessen durchgeführt [5–9]; Ökobilanz als Begriff tauchte erstmals in den 80er Jahren auf [10–13]. Fraglos handelt es sich dabei um eine fast geniale Wortschöpfung, die sprachlich geschickt Ökologie und Ökonomie miteinander verbindet. Zudem entsteht der Eindruck, daß bislang mehr qualitativ behandelte Aspekte quantifizierbar und bilanzierbar geworden sind. Es erstaunt also nicht, daß Ökobilanzen als umfassendes ökologisches Beurteilungsmittel bei allen gesellschaftlichen Gruppen sofort hoch im Kurs standen und auch noch stehen, ohne daß anfangs einheitliche Methoden für ihre Erstellung vorlagen. Als Folge davon ist ein Teil der bislang erstellten Ökobilanzen hinsichtlich ihrer Aussage einseitig geprägt und für eine akzeptable Beurteilung ungeeignet.

Inzwischen besteht jedoch weitgehend Einigkeit, wie die ökologische Bewertung von Produkten international erfolgen soll. Das Standardmodell [14] einer solchen Ökobilanz umfaßt vier Schritte: die Zielsetzung (Scoping), die Sachbilanz (Inventory), die Wirkungsanalyse (Impact assessment) und die Bewertung (Evaluation) (**Bild 1**).

2.1 Zielsetzung

Im Rahmen der Zielsetzung der Ökobilanz muß zunächst eine Aufgabe formuliert werden. Beispielsweise muß vorher klar sein, ob ein Vergleich zweier oder mehrerer Produkte für einen bestimmten, genau definierten Anwendungsfall oder eine Optimierung eines Produktes unter ökologischen Gesichtspunkten durchgeführt werden soll. Von besonderer Bedeutung ist dabei die Definition einer sogenannten „Funktionalen Einheit", die verhindert, daß Baustoff A mit Baustoff B verglichen wird, ohne auf gleiche Leistungsmerkmale zu achten. Gegenüberzustellen sind stattdessen z. B. zwei Bauteile oder Bauwerke gleicher Funktion und Leistungsfähigkeit aus unterschiedlichen Baustoffen für einen ganz konkreten Einsatz. Schließlich ist im Rahmen der Zielsetzung auch die Genauigkeit vorzugeben, mit der die Bilanz erhoben werden soll. Meßwerte führen zu gänzlich anderen Ergebnissen als mittlere Emissionsfaktoren, so daß in einer Bilanz nicht beides Eingang finden kann.

2.2 Sachbilanz

In der Sachbilanz werden der Bedarf an Rohstoffen, an Energie, die Emissionen sowie die Abfall- und Abwasserströme quantitativ beschrieben. Dabei ist es erforderlich, den Lebensweg des Produktes exakt zu kennen, um sachgerechte Bilanzgrenzen festlegen zu können (**Bild 2**). Energie, Rohmaterial, Luft, Wasser und Landschaft können sowohl bei der Gewinnung und Aufbereitung der Produktausgangsstoffe, bei seiner Produktion, seiner Distribution und seinem Gebrauch als auch bei seiner Entsorgung benötigt werden. Zu definieren ist, welche Stoff- oder Energieströme bzw. Lebenslaufphasen von untergeordneter Bedeutung und nicht zu betrachten sind.

Auf der Ausgabenseite der Sachbilanz stehen die Emissionen aller relevanten Schadstoffe, angefangen von Staub und Staubinhaltsstoffen über gasförmige anorganische und organische Abgaskomponenten bis hin zu hochtoxischen Stoffen wie Dioxinen und Furanen. An dieser Stelle tritt in

- Zielsetzung der Ökobilanz
  (= Scoping)
- Sachbilanz
  (= Life Cycle Inventory)
- Wirkungsanalyse
  (= Environmental Impact Assessment)
- Bewertung
  (= Life Cycle Evaluation)

Bild 1: Teilschritte einer Ökobilanz

2. The procedure for drawing up the balance

The ecobalance (internationally: Life Cycle Assessment, LCA) is not a new method. Environmentally related system analyses were being carried out on packaging materials and chemical processes as early as the 70s [5–9]; the term "ecobalance" first emerged in the 80s [10–13]. Without doubt this is an almost inspired piece of word coinage which skilfully links ecology with economy. It also gives the impression that it has become possible to quantify and balance aspects which could previously only be treated qualitatively. It is not surprising that ecobalances immediately became, and still are, highly regarded in all social groups as a tool for comprehensive ecological assessment, without at first there being any consistent methods for drawing them up. As a result, some of the ecobalances which have been drawn up so far are markedly one-sided in their message and are unsuitable for acceptable assessment.

However there is now substantial agreement as to how the ecological evaluation of products is to be carried out internationally. The standard model [14] of such an ecobalance encompasses four steps: setting the objective (Scoping), the material balance (Inventory), the analysis of effects (Impact Assessment) and the Evaluation (**Fig. 1**).

2.1 Setting the objective (Scoping)

When setting the objective for the ecobalance it is first necessary to formulate the problem. For example, it must be clear in advance whether a comparison of two or more products is to be carried out for a certain, precisely defined, application or for optimization of a product from the ecological point of view. Of particular importance is the definition of a "functional unit"; this prevents construction material A from being compared with constructional material B without taking the same performance features into account. The comparison should instead be made between, for example, two components or two structures of the same function and capabilities made of different construction materials for an entirely specific use. As a part of the objective it is also necessary to specify the accuracy with which the balance is to be made. Measured values lead to very different results from average emission factors, so they cannot both be included in one balance.

2.2 The material balance (Life Cycle Inventory)

The material balance provides a quantitative description of the consumption of raw materials and energy, the emissions, and the waste and waste water flows. It is necessary to have precise knowledge of the life cycle of the product in order to be able to specify proper boundaries for the balance (**Fig. 2**). Energy, raw material, air, water, and land can be required during the extraction and preparation of the basic materials for the product, in its production, its distribution and its use, as well as in its final disposal. It is necessary to define which material and energy flows, or life cycle phases, are of secondary importance and are not to be taken into account.

On the output side of the material balance are the emissions of all relevant pollutants, starting from the dust and the substances contained in it, through gaseous inorganic and organic exhaust gas components, right up to highly toxic substances like dioxins and furans. In most cases it is at this point that the first difficulty appears, as it is only in exceptional cases that these data are available. Noise, wastes,

- Setting the objectives for the ecobalance
  (= Scoping)
- Material balance
  (= Life Cycle Inventory)
- Analysis of effects
  (= Environmental Impact Assessment)
- Evaluation
  (= Life Cycle Evaluation)

Figure 1: Stages in an ecobalance

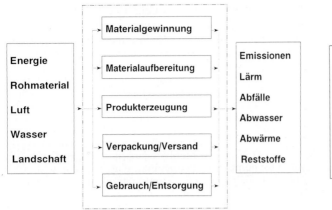

BILD 2: Bilanzrahmen für die Sachbilanz einer Ökobilanz

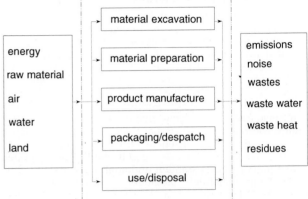

FIGURE 2: Framework of the material balance within an ecobalance

den meisten Fällen bereits die erste Schwierigkeit auf, da diese Daten nur in Ausnahmefällen vorliegen. Ferner sind zu erfassen: Lärm, Abfälle, Abwasser, Abwärme und Reststoffe. Diese Reststoffe können im eigenen oder in anderen Produktionsprozessen wiederverwertbar sein. Ihre Behandlung innerhalb einer Ökobilanz bedarf daher einer Konvention [14]. Handelt es sich um ein Recycling, also eine produktionsinterne Wiederverwertung, kann Rohmaterial eingespart werden, und die damit verbundene Einsparungsmöglichkeit auf der Einnahmen- und Ausgabenseite muß gutgeschrieben werden. Bei der Verwertung von sekundären Stoffen aus anderen Prozessen müssen die Umweltbelastungen des Prozesses, bei dem sie entstehen, in geeigneter Form auf den Reststoff verteilt werden. Die Aufteilung kann z.B. proportional zum Erlös erfolgen, der aus Haupt- und Abfallprodukten erzielt wird.

Zielsetzung und Sachbilanz sind vergleichsweise einfache Teilschritte der Ökobilanz, da man sich dabei auf weitgehend abgesicherte Grundlagen der Bilanzierung abstützen kann; das gilt nicht mehr für die Wirkungsanalyse und die Bewertung. Um jedoch ein handhabbares Ökobilanzergebnis zu erzielen, sind zusammenfassende Rechenoperationen unumgänglich. Im Idealfall ist die Basis dafür eine ökologische Verrechnungseinheit, wie die Währung in betriebswirtschaftlichen Bilanzen. Allerdings muß die Herleitung nachvollziehbar und die Bewertung transparent sein, um eine breite Akzeptanz beim Adressaten der Ökobilanz zu erreichen.

2.3 Wirkungsanalyse

In der Wirkungsanalyse der Ökobilanz muß die Relevanz der Umweltbelastung, die mit dem Produkt oder dem Produktionsprozeß verbunden ist, im einzelnen eingeschätzt und beurteilt werden. Ziel ist es, die Hauptbelastungsfaktoren herauszuarbeiten. Eine vergleichbare Fragestellung ergibt sich auch im Rahmen von Umweltverträglichkeitsprüfungen, die gemäß EG-Richtlinie vom 27. Juli 1985 bei Vorhaben durchzuführen sind, die erhebliche Umweltauswirkungen haben können. Dazu zählen unter bestimmten Voraussetzungen auch der Bau bzw. die Änderung des Betriebs von Anlagen zur Herstellung von Zementklinker. Insbesondere bei der Festlegung des Untersuchungsrahmens der Umweltverträglichkeitsprüfung, d.h. beispielsweise bei der Auswahl relevanter Emissionskomponenten, kommt dieser Analyse besondere Bedeutung zu.

In der Praxis hat sich eine Vorgehensweise zur Beurteilung der Erheblichkeit von anlagenbedingten Umweltauswirkungen bewährt. Danach wird eine zusätzliche Umweltbelastung nicht als relevante Risikoerhöhung und damit auch nicht als ursächlicher Beitrag zu schädlichen Umwelteinwirkungen angesehen, wenn sie 1% eines schadstoffspezifischen Schwellenwertes unterschreitet. Die Umweltrelevanz ist wie folgt definiert:

$$R_E = \frac{I1Z}{Sw} \cdot 100 \text{ in }\% \qquad (1)$$

waste water, waste heat and residues also have to be covered. These residues may be re-usable in the same or other production processes. Their treatment within an ecobalance therefore requires the use of a convention [14]. If recycling is involved, i.e. re-use within the production system, then raw material can be saved, and the associated possible saving must be credited on the input and output sides. When utilizing secondary materials from other processes the environmental pollution caused by the processes in which they are produced must be allocated in appropriate form to the residues. The allocation can, for example, be made proportional to the profit which is obtained from the main and waste products.

Setting the objective and carrying out the material balance are comparatively simple stages in the ecobalance, as they are based on substantially factual foundations; this no longer applies to the analysis of effects and the evaluation. All-embracing computation operations are unavoidable if a manageable ecobalance is to be achieved. In the ideal case the basis for this is an ecological accounting unit, like the currency in an industrial management balance. However, its derivation must be reconstructible, and the evaluation must be transparent, to achieve broad acceptance with those for whom the ecobalance is intended.

2.3 Analysis of effects (Environmental Impact Assessment)

In the ecobalance's analysis of effects the relevance of the environmental impact which is associated with the product or production process must be estimated and assessed in detail. The aim is to bring out the principal pollution factors. A comparable problem also arises in the context of environmental compatibility tests which have to be carried out in accordance with EC guidelines of 27th July 1985 for projects which could have considerable environmental effects. This also includes, with certain preconditions, the construction or the change of operation of plants for manufacturing cement clinker. This analysis is of particular importance, especially when the investigative framework of the environmental compatibility testing is being specified, i.e. for example when selecting relevant emission components.

A procedure for assessing the relevance of environmental effects caused by industrial plants has proved successful in practice. According to this, additional environmental pollution is not regarded as a relevant increase in risk, and therefore also not as a causal contribution to harmful environmental effects, if it is less than 1% of a threshold value specific to the pollutant. The environmental relevance is defined as follows:

$$R_E = \frac{I1Z}{Sw} \cdot 100 \text{ in }\% \qquad (1)$$

In Gleichung (1) bedeuten:

R_E: Umweltrelevanz
Sw: Schwellenwert
I1Z: Zusatzbelastung

Für $R_E \leq 1\%$ ist die Zusatzbelastung vernachlässigbar.

Dieses Schwellenwertkonzept zur Umweltrelevanz läßt sich auf alle Emissionen in die Umweltmedien Luft, Boden und Wasser übertragen. Anerkannte Schwellenwerte für die wichtigsten Stoffe sind im Schrifttum enthalten. Die Zusatzbelastung im Quotienten zur Ermittlung der Umweltrelevanz kann rechnerisch ermittelt werden. Überträgt man dieses Schwellenwertkonzept auf die Wirkungsanalyse der Ökobilanz, kann damit auch die Relevanz der Umweltbelastung eines Produktes eingeschätzt werden.

Hilfreich für die weitere Bewertung der Ergebnisse der Sachbilanz ist die Aggregation der relevanten Umweltbelastungen in normierter Form. Damit hat sich sehr eingehend das schweizerische Bundesamt für Umwelt, Wald und Landschaft (BUWAL) in Bern befaßt. Die Arbeiten haben zu einer pragmatischen Konvention geführt, die einer einheitlichen Ökoeinheit sehr nahe kommt. Entwickelt wurden zwei Modelle: das „Kritische Volumen" und die „Ökologische Knappheit". Bei der Berechnung ökologischer Knappheiten werden die Öko-Punkte ermittelt, wobei dem Verhältnis von tatsächlicher Belastung und der maximal tolerierbaren Belastung besondere Bedeutung zukommt [15]. Eine breitere Anwendung, nicht nur in der Schweiz, sondern auch in den Niederlanden und den skandinavischen Ländern, findet die Berechnung von kritischen Volumen [16]. Danach wird für jeden Schadstoff der Quotient aus seiner in ein Umweltmedium abgegebenen Menge und einem Grenzwert (z.B. Emissionsgrenzwert, MIK- oder MAK-Wert) gebildet. Aufsummiert über alle Schadstoffe ergibt sich das sogenannte „Kritische Volumen", das nötig ist, um das Umweltmedium, beispielsweise Luft oder Wasser, bis zu einer unbedenklichen Konzentration zu verdünnen. Je größer dieses Volumen ist, um so höher muß die Umweltbelastung eingeschätzt werden.

Mathematisch ist das „Kritische Volumen" wie folgt definiert:

$$V_K = \sum_{i=1}^{N} \frac{m_i}{G_i} \quad (2)$$

Darin bedeuten:

V_K: kritisches Volumen
m_i: Menge des Schadstoffs i
G_i: Grenzwert des Schadstoffs i

2.4 Bewertung

Abschließender Baustein der Ökobilanz ist die zusammenfassende Bewertung aller wesentlichen Merkmale des betrachteten Produkts. Die ökologische Belastung, die sich aus der Sachbilanz ergibt, stellt dabei lediglich einen Teil der Gesamtbilanz dar; andere Merkmale müssen ebenso mit berücksichtigt werden. Bei einem Betonbauteil können das beispielsweise die Dauerhaftigkeit, die Feuerwiderstandsfähigkeit oder Tragfähigkeit, bei einem Klärwerk aber auch die Abwasserreinigungskapazität sein. Wesentlich ist, daß diese Merkmale in der Zieldefinition zu Beginn der Ökobilanz festgelegt werden. Sie bestimmen entscheidend die funktionale Einheit, für die die Ökobilanz erstellt wird.

Als Bewertungsmaßstab für eine Entscheidung bei mehreren gleichrangigen Merkmalen, die nicht gegeneinander abwägbar sind, also nicht die gleiche Verrechnungseinheit haben, wurde in der Betriebswirtschaftslehre die Theorie der Nutzwertanalyse entwickelt, die auch im Rahmen von Ökobilanzen anwendbar ist [17–20]. Dabei werden die Merkmale anhand einer Werteskala eingeordnet. Das Produkt mit der höchsten Umweltbelastung erhält den Zielerfüllungsgrad 0, für die niedrigste Belastung wird der Wert 1 vergeben. Dazwischen wird proportional abgestuft. Ferner werden die Merkmale nach ihrer Bedeutung gewichtet,

In Equation (1)

R_E: environmental relevance
S_W: threshold value
I1Z: additional pollution

The additional pollution can be neglected for $R_E \leq 1\%$.

The threshold concept for environmental relevance can be applied to all emissions in the environmental media of air, soil and water. Recognized threshold values for the most important substances are contained in the literature. The additional pollution in the quotient for determining the environmental relevance can be determined by calculation. If this threshold concept is applied to the analysis of effects in the ecobalance then it is also possible to estimate the relevance of the environmental pollution of a product.

The aggregate of the relevant environmental pollutions in normalized form is helpful for further evaluation of the results of the material balance. The Swiss Federal Office for the Environment, Forests and Land in Bern have studied this very thoroughly. The work has led to a pragmatic convention which comes very close to being a consistent eco-unit. Two models were developed: the "Critical volume" and the "Ecological scarcity". Eco points are determined when calculating ecological scarcity, particular importance being given to the ratio of actual pollution to maximum tolerable pollution [15]. The calculation of critical volumes is being applied more widely, not only in Switzerland but also in the Netherlands and the Scandinavian countries [16]. A quotient is formed for each pollutant from the quantity discharged into an environmental medium and a limit value (e.g. emission limit, maximum immission concentration or maximum workplace concentration). The sum for all the pollutants produces the "critical volume" which is required to dilute the environmental medium, e.g. air or water, down to a harmless concentration. The greater this volume the higher is the estimate of the environmental pollution.

The "critical volume" is defined mathematically as follows

$$V_K = \sum_{i=1}^{N} \frac{m_i}{G_i} \quad (2)$$

in which

V_K: critical volume
m_i: quantity of pollutant i
G_i: limit value for pollutant i

2.4 Evaluation

The final building block in the ecobalance is the comprehensive evaluation of all essential features of the product under consideration. The ecological pollution given by the material balance represents only one part of the overall balance; other features must also be taken into account. For a concrete component this could, for example, be the durability, the fire resistance, or the load-bearing capacity, and for a sewage works the waste water purifying capacity. It is important that these features are specified in the target definition at the start of the ecobalance. They are decisive in determining the functional unit for which the ecobalance is being drawn up.

The theory of utility analysis, which can also be applied in ecobalances, was developed in business management science as an evaluation criterion for reaching a decision where there are several equal-ranking features which cannot be weighed against one another, i.e. which are not in the same accounting units [17–20]. The features are classified with the aid of a value scale. The product with the highest environmental impact is given the target-fulfilment coefficient of 0, and the value 1 is awarded to the lowest impact. Proportional values are allocated between them. The features are also weighted in accordance with their importance, usually with factors between 0 and 100. For a component where fire resistance is absolutely essential, this feature is given the factor 100. The other features are given lower graduated values.

üblicherweise mit Faktoren zwischen 0 und 100. Bei einem Bauteil, dessen Feuerwiderstandsfähigkeit unabdingbar notwendig ist, wird dieses Merkmal mit dem Faktor 100 bewertet. Die anderen Merkmale werden abgestuft geringer bewertet.

Auf diese Art und Weise erhält man eine Matrix aus Zielerfüllungsgraden einzelner Merkmale, die entsprechend ihrer Bedeutung für das Produkt gewichtet sind. Die Summe der einzelnen, gewichteten Zielerfüllungsgrade, also der Wert der Matrix, ist der Nutzwert. Vorteil dieser Vorgehensweise ist es, eine subjektiv geprägte Entscheidung nachvollziehbar zu strukturieren.

Der Nutzwert eines Produktes ist definiert nach:

$$N = \sum_{i=1}^{M} n_i \qquad (3)$$

Der Teilnutzen einzelner Merkmale nach:

$$n_i = \sum_{i=1}^{Q} Z_i \cdot g_i \qquad (4)$$

Darin bedeuten:
N: Nutzwert des Produktes
n_i: Teilnutzen einzelner Merkmale
Z_i: Zielerfüllungsgrad des Merkmals i
g_i: Gewichtungsfaktoren für das Merkmal i

3. Schlußbemerkungen

Das Handwerkszeug für Produktökobilanzen liegt vor. Es besteht aus den vier Teilschritten Zieldefinition, Sachbilanz, Wirkungsanalyse und Bewertung. Es fehlen jedoch bislang für die meisten Produktausgangsstoffe überprüfbare Angaben über Emissionen, Energieverbräuche und andere Grundlagen der Sachbilanz. Einigkeit muß darüber herrschen, daß Produktökobilanzen immer für einen speziellen Anwendungsfall, eine „Funktionale Einheit", erstellt werden. Aussagen gelten ausschließlich für diesen Fall, eine Verallgemeinerung ist nicht zulässig. Deshalb sind von sachgerechten Produktökobilanzen keine plakativen Aussagen zu erwarten, wie „Baustoff A ist besser als Baustoff B".

Bei der Bewertung der Umweltinanspruchnahme müssen ökologische Aspekte im Vergleich zu anderen Merkmalen des Produkts beurteilt werden. Ein geeignetes Hilfsmittel stellt die Nutzwertanalyse dar. Auf dieser Basis kann eine umfassende Beurteilung eines Produkts unter Berücksichtigung der Ökologie erfolgen, ohne sie jedoch zum alleinigen Maßstab zu erheben.

This produces a matrix consisting of target-fulfilment coefficients for single features which are weighted according to their importance for the product. The total of the individual, weighted, target-fulfilment coefficients, i.e. the value of the matrix, gives the utility. The advantage of this procedure is that it gives a retraceable structure to a subjective type of decision.

The utility of a product is defined by:

$$N = \sum_{i=1}^{M} n_i \qquad (3)$$

and the constituent benefits of individual features by:

$$n_i = \sum_{i=1}^{Q} Z_i \cdot g_i \qquad (4)$$

in which
N: utility of the product
n_i: constituent benefit of individual features
Z_i: target-fulfilment coefficient of feature i
g_i: weighting factor for feature i

3. Final comments

The tool for product ecobalances is available. It consists of the four stages of definition of the objective, material balance, analysis of effects, and evaluation. However, so far for the majority of basic product materials there is a lack of checkable data on emissions, energy consumptions and other fundamental requirements for the material balance. There must be agreement that product ecobalances are always set up for one specific application, one "functional unit". Statements apply to this case only, and generalization is not permissible. Simplistic statements such as "Construction material A is better than construction material B" are not to be expected from objective ecobalances.

In the evaluation of the environmental demands the ecological aspects must be assessed in comparison with other features of the product. Utility analysis is a suitable tool for this purpose. On this basis it is possible to carry out a comprehensive assessment of a product in which the ecology is taken into account, but without it being adopted as the sole criterion.

Literaturverzeichnis

[1] Hauptverband der Deutschen Bauindustrie: Jahresbericht 1992, Wiesbaden 1993.

[2] Umweltbundesamt: Umweltdaten – kurzgefaßt, Ausgabe 1993, Berlin 1993.

[3] Wischers, G., und Kuhlmann, K.: Ökobilanz von Zement und Beton – Abwägende Gegenüberstellung von ökologisch entlastenden und belastenden Einwirkungen auf die Umwelt. Zement-Kalk-Gips 44 (1991) Nr. 11, S. 545–553.

[4] Wischers, G.: Beton und Umwelt – Ökobilanz für Beton. Betonwerk + Fertigteil-Technik 57 (1991) Nr. 11, S. 33–40.

[5] Oberbacher, B., et al.: Abbaubare Kunststoffe und Müllprobleme. Beiträge zur Umweltgestaltung, E. Schmidt, Berlin, Heft A 23 (1974).

[6] Hunt, R. G., et al.: Resource and environmental profile analysis of nine beverage container alternatives. Report of Midwest Res. Inst. to U.S. Environmental Protection Agency (EPA) Washington, D.C. 1974.

[7] Boustead, I., und Hancock, G.F.: Handbook of Industrial Energy Analysis. Ellis Horwood Ltd. Chichester, England 1979.

[8] Kindler, H., und Nikles, A.: Energiebedarf bei der Herstellung und Verarbeitung von Kunststoffen. Chem.-Ing.-Tech. 51 (1979) S. 1–3.

[9] Kindler, H., und Nikles, A.: Energieaufwand zur Herstellung von Werkstoffen – Berechnungsgrundsätze und Energieäquivalenzwerte von Kunststoffen. Kunststoffe 70 (1980) S. 802–807.

[10] Franke, M.: Umweltauswirkungen durch Getränkeverpackungen. Systematik zur Ermittlung der Umweltauswirkungen von komplexen Prozessen am Beispiel von Einweg- und Mehrweg-Getränkebehältern. E.F.-Verlag für Energie- und Umwelttechnik GmbH, Berlin 1984.

[11] Bundesamt für Umweltschutz, Bern (Ed.): Ökobilanzen von Packstoffen. Schriftenreihe Umweltschutz, Nr. 24 (1984).

[12] Lundholm, M.P., und Sundström, G.: Ressourcen- und Umweltbeeinflussung – Tetrabrik Aseptic Kartonpackungen sowie Pfandflaschen und Einwegflaschen aus Glas, Malmö 1985.

[13] Lundholm, M.P., und Sundström, G.: Ressourcen- und Umweltbeeinflussung durch zwei Verpackungssysteme für Milch, Tetra Brik und Pfandflasche, Malmö 1986.

[14] Umweltbundesamt: Ökobilanzen für Produkte – Bedeutung/Sachstand/Perspektiven, Berlin 1992.

[15] Ahbe, S., Braunschweig, A., und Müller-Wenk, R.: Methodik für Ökobilanzen auf der Basis ökologischer Optimierung. Schriftenreihe Umwelt 133 des Bundesamtes für Umwelt, Wald und Landschaft, Bern 1990.

[16] Haberstatter, K.: Ökobilanzen für Packstoffe. Schriftenreihe Umwelt Nr. 132 des Bundesamtes für Umwelt, Wald und Landschaft, Bern 1991.

[17] Beckmann, A.: Nutzwertanalyse, Bewertungstheorie und Planung, Haupt, Bern und Stuttgart 1978.

[18] Lillich, L.: Nutzwertverfahren, Physica-Verlag Heidelberg 1992.

[19] Deutscher Verband für Wasserwirtschaft und Kulturbau (DVWK): Nutzwertanalytische Ansätze zur Planungsunterstützung und Projektbewertung, DVWK-Mitteilungen 19, Bonn 1989.

[20] Deutscher Verband für Wasserwirtschaft und Kulturbau (DVWK): Pilotstudie zur Anwendung nutzwertanalytischer Verfahren, DVWK-Mitteilungen 22, Bonn 1991.

Emissionsminderung von Dioxinen und Furanen
Reduction in the emission of dioxins and furans

Réduction des émissions de dioxines et de furanes

Reducción de emisiones de dioxinas y de furanos

Von **J. Blumbach** und **L.-P. Nethe,** Harburg/Deutschland

Zusammenfassung – In Müllverbrennungsanlagen werden heute hochwirksame Techniken zur Emissionsminderung, insbesondere für Dioxine und Furane sowie für flüchtige Schwermetalle, eingesetzt. Das Flugstromverfahren ist eines von den zwei großtechnisch zum Einsatz kommenden Verfahren. Dabei werden die ökotoxischen Schadstoffe adsorptiv gebunden und an einem Gewebefilter abgeschieden. Am Beispiel einer Sondermüllverbrennungsanlage kann gezeigt werden, daß die in der 17. BImSchV vorgeschriebenen Grenzwerte deutlich unterschritten werden. Das Verfahren läßt sich auch bei anderen Industriefeuerungen anwenden. Als Beispiel dafür dient eine Leichtmetallumschmelzanlage. Die Erkenntnisse aus dem Einsatz des Flugstromverfahrens zeigen, daß im Wärmetauscher von Zementdrehöfen ein vergleichbarer Adsorptionsprozeß systemimmanent abläuft. Das bietet eine Erklärung für die Tatsache, daß im Abgas von Ofensystemen der Zementindustrie extrem niedrige Dioxin- und Furankonzentrationen festgestellt werden.

Emissionsminderung von Dioxinen und Furanen

Summary – Highly efficient techniques are employed nowadays in waste incineration plants to reduce emissions, particulary of dioxins and furans and also of volatile heavy metals. The suspension flow process is one of two full-scale procedures being used in which the ecotoxic pollutants are fixed by adsorption and then removed in a fabric filter. Using a hazardous waste incineration plant as an example it can be shown that the emissions fall substantially below the limit values laid down in the 17th Federal Immissions Control Law. The process may also be applied to other industrial combustion systems, an example of which is a remelt plant for light metals. The knowledge gained from the use of the suspension flow process shows that a comparable adsorption process which is inherent in the system takes place in the preheaters of cement rotary kilns. This explains the fact that extremely low concentrations of dioxins and furans are found in the exhaust gases from the kiln systems in the cement industry.

Reduction in the emission of dioxins and furans

Résumé – Les stations d'incinération de déchets emploient, de nos jours, des techniques sophistiquées pour réduire le taux des émissions, notamment les dioxines et furanes mais aussi les métaux lourds volatils. La technique du flux volant est un des deux procédés mis en oeuvre à l'échelle industrielle. A cette occasion, les déchets écologiquement toxiques sont combinés par adsorption et séparés sur un filtre à toile. L'exemple d'une station d'incinération de déchets spéciaux montre que les valeurs-limites prescrites par la 17e BImSchV sont nettement inférieures aux valeurs de consigne. Et ce procédé est utilisable aussi pour d'autres systèmes de chauffe industriels. Comme exemple citons ici une installation de fusion d'alliage léger. Les conclusions établies à partir du procédé de flux volant montrent qu'un procédé d'adsorption comparable, se déroule de manière immanente au système dans l'échangeur de température des fours rotatifs de ciment. Ceci explique le fait que les effluents gazeux des systèmes de four de l'industrie cimentière accusent des concentrations de dioxine et de furane extrêmement faibles.

Réduction des émissions de dioxines et de furanes

Resumen – En las instalaciones de incineración de basuras se emplean, hoy en día, técnicas eficaces que permiten reducir las emisiones, especialmente las de dioxinas y de furanos, pero también los metales pesados volátiles. El procedimiento por suspensión es uno de los dos procedimientos empleados a escala industrial. Permite combinar por adsorción las sustancias ecológicamente tóxicas, las cuales son separadas por medio de un filtro de tejido. Citando el ejemplo de una instalación de incineración de basuras especiales, se puede demostrar que los valores obtenidos son notablemente inferiores a los prescritos en la Directiva 17. BImSchV. Este procedimiento puede aplicarse también a otros hogares industriales. Sirve de ejemplo una instalación de fusión de metal ligero. Los conocimientos adquiridos con el procedimiento por suspensión muestran que en el intercambiador de calor de los hornos rotatorios de cemento se desarrolla un proceso de adsorción similar, inmanente del sistema. Esto explica el hecho de que en los gases de escape de los sistemas de hornos empleados en la industria del cemento se hayan registrado concentraciones de dioxina y de furano extremadamente bajas.

Reducción de emisiones de dioxinas y de furanos

1. Introduction

There has been a surprising change in public opinion recently. A few years ago, demonstrations were still being staged against waste incineration plants, whereas nowadays even dedicated conservationists and opponents of waste incineration plants are debating about whether 50 or 60 additional waste incineration plants should be built in Germany. There are certainly many reasons for this change in attitude, but there is one aspect which stands out above the rest in the current debate: Waste incineration plants which were formerly condemned for "fouling up the nation" have now been equipped or retrofitted with highly efficient technology for minimizing emissions. This also has a tremendous impact on the removal of critical compounds, such as dioxins and furans, for which the concentrations in the clean gas flow may not exceed 0.1 ng TE/Nm^3 (stp, dry) [TE = toxicity equivalents] [1]. Similar limits are applied to other organic compounds and to the removal of volatile heavy metals, such as mercury. Waste incineration plant operators have also realized the necessity of educating the public about such new technology, and of documenting the effectiveness of the individual measures taken.

Two processes are primarily involved in minimizing these emissions:

– the fixed bed filter, and
– the suspension flow process.

These have now been widely used in industry and have proven their suitability. The suspension flow process will be described in detail below. It is mainly used after waste incineration plants and is technically proven. Experience with this technology has also explained the fact that no relevant quantities of dioxins and furans have been observed in the exhaust gases from rotary cement kilns, even where secondary fuels are used.

2. Suspension flow process

The suspension flow process consists essentially of two components:

– an adsorbent for ecotoxic pollutants, and
– a reaction unit in the exhaust gas stream which permits adequate mixing of the exhaust gas and the adsorbent and subsequent separation on a fabric filter.

The composition of the adsorbent enables other pollutants to be removed in addition to neutralizing the acid exhaust gas components which is often the main requirement. These pollutants consist predominantly of:

– volatile heavy metals; e.g. mercury, cadmium, thallium, selenium and arsenic,
– chlorinated dibenzodioxins and dibenzofurans (abbreviated to dioxins and furans),
– low-volatility chlorinated hydrocarbons, e.g. hexachlorobenzene and hexachlorocyclohexane,
– polychlorinated biphenyls (abbreviated to PCBs),
– polycyclic aromatic hydrocarbons (abbreviated to PAHs).

Calcium hydroxide has proved to be a particularly suitable base material. The adsorptive capacity is generally provided by different forms of activated carbon. Calcium hydroxide has been employed for the removal of acid pollutants from flue gases downstream of waste incineration plant processes for many decades. However, it does not remove other ecotoxic pollutants, such as heavy metals and organic components. On the other hand, these pollutants can be reduced at reasonable cost and high separation efficiencies, even when present in extremely low concentration, by using adsorptive processes on activated carbon. Catalytic oxidation and chemical combination with the basic constituents in the lime take place as secondary reactions in addition to the pure adsorption [3]. Previously, the standard technology was to connect the various process stages for removing different ecotoxic pollutants in series. However, this was not as successful as anticipated; nor was such a concept economi-

Nebenreaktionen die katalytische Oxidation und die chemische Bindung an den basischen Bestandteilen des Kalkes ab [3]. Bisher bestand die übliche Technik darin, verschiedene Verfahrensstufen für die Abscheidung unterschiedlicher ökotoxischer Schadstoffe hintereinander zu schalten. Dies brachte aber nicht den gewünschten Erfolg. Zudem war eine solche Konzeption auch aus ökonomischer Sicht nicht sinnvoll.

Deshalb wurde ein Weg gewählt, die Wirkung der Einzelkomponenten zu kombinieren und Weißkalkhydrat mit aktivierten Kohlen zu mischen. Dadurch entstand ein System, das einerseits als Breitbandadsorbens wirkt und andererseits durch einfache Modifizierungen zur Abscheidung spezieller Schadstoffe genutzt werden kann. Entscheidend war hier, ein homogenes Gemisch aus den Ausgangsstoffen zu erhalten, das sich auch im Einsatz nicht entmischt. Dazu wurde ein neuartiges Mischverfahren entwickelt. Eine typische Zusammensetzung des Adsorbens für die Trockensorption in Müllverbrennungsanlagen ist etwa 95% Weißkalkhydrat und 5% Herdofenkoks.

Das Flugstromverfahren bietet erhebliche Vorteile. Sie bestehen darin, daß bewährte Anlageteile, angefangen vom Bevorratungssilo über die Fördereinrichtung bis zu den Mischstrecken sowie Kontrolleinrichtungen und Gewebefilter, zur Verfügung stehen. Das Gewebefilter ist eine massive Sperre gegenüber den Emissionen von Feststoffen. Ohne Probleme können Reststaubgehalte < 10 mg/m^3 erreicht werden. Werte von 1 mg/m^3 sind durchaus realistisch. Dieser geringe Reststaubgehalt ist von erheblicher Bedeutung, da damit die Emission von mit toxischen Schadstoffen beladener Kohle unterbunden wird.

Ein weiterer Vorteil ist darin zu sehen, daß sich das Flugstromverfahren problemlos in bestehende Anlagen integrieren läßt. Damit ist ein sehr breites Anwendungs- und Wirkungsspektrum für dieses System gegeben.

3. Sondermüllverbrennungsanlage (SVA) Schöneiche

Als erstes Beispiel für das Flugstromverfahren wurde die SVA Schöneiche gewählt. Die Anlage ist auf eine Müllmenge von 2,3 t/h ausgelegt, was einer Abgasmenge von 31 000 m^3/h entspricht. Das **Bild 1** zeigt ein stark vereinfachtes Verfahrensschema dieser Anlage. Die Abgase werden in einem Kühler mit Wasser auf 140 °C abgekühlt und gelangen danach in einen Reaktor, in dem das Adsorptionsmittel, bestehend aus Weißfeinkalk und aktivierter Kohle, dem Abgasstrom zugemischt und in ihm verwirbelt wird. Das mit Adsorptionsmittel beladene Abgas wird in einem Gewebefilter gereinigt. Auf dem Weg bis zum Filter und in dem Filterkuchen geschieht die Adsorption der ökotoxischen Schadstoffe. Ein Teil des abgeschiedenen Staubes wird rezirkuliert, ein anderer Teil geht zur Deponie. Die Anlage wird seit Dezember 1989 im Dauerbetrieb gefahren. Die Versuche wurden vom TÜV Berlin meßtechnisch begleitet [4].

cally feasible. A method was therefore chosen which combined the effects of the individual components using a mix of calcium hydroxide and activated carbon. This produced a system which not only acted as a wide-band adsorbent but also, by a simple modification, could be used for the removal of specific pollutants. It was critical here that a homogenous mix which did not segregate during use should be produced from the original materials. A new type of mixing process was developed for this purpose. A typical composition for the adsorbent for dry sorption in waste incineration plants is approx. 95% calcium hydroxide and 5% open-hearth coke.

The suspension flow process offers substantial advantages. These include the fact that tried and tested plant components, starting from the storage silo and the conveying equipment and extending to the mixing sections, as well as the control equipment and fabric filter, are all available. The fabric filter provides an excellent barrier against the emission of solids. Residual dust levels of < 10 mg/m^3 are easily achieved, and values of 1 mg/m^3 are entirely realistic. This low residual dust level is extremely important as this prevents the emission of carbon laden with toxic pollutants.

The fact that the suspension flow process is easily integrated into existing plants is a further advantage. This system therefore offers a very wide range of applications and effects.

3. Schöneiche hazardous waste incineration plant

The Schöneiche hazardous waste incineration plant was chosen as the first example of a suspension flow process. The plant is designed to process 2.3 t/h of waste, corresponding to an exhaust gas volume of 31 000 m^3/h. **Fig. 1** shows a greatly simplified process flow diagram of this plant. The exhaust gas is cooled with water to 140 °C in a cooler and then led into a reactor where the adsorbent, consisting of fine calcium hydroxide and activated carbon, is added to the exhaust gas stream and entrained in it. The exhaust gas containing the adsorbent is then cleaned in a fabric filter. The adsorption of the ecotoxic pollutants takes place on the way to the filter and in the filter cake. Part of the separated dust is recirculated and the remainder is landfilled. The plant has been in continuous operation since December 1989. The TÜV Berlin [Berlin Technical Inspectorate] assisted with the measurements during the trials [4].

The following diagrams show the success of the adsorption process. For the sake of clarity only two pollutant components are shown by way of example in each case. The clean gas values for mercury are well below the limit of 50 μg/Nm^3 (**Fig. 2**). Comparable success in reducing levels were also obtained with dioxins and furans (**Fig. 3**), and it can be seen from the example that in some cases there were substantial concentrations of dioxins in the raw gas. The engineering company, Technischer Umweltschutz GmbH, was appointed as the independent expert to monitor and check

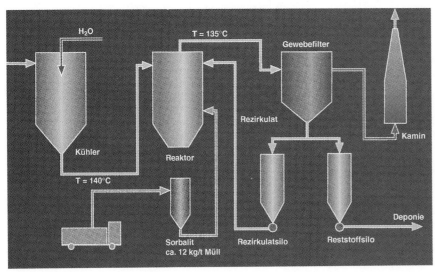

Bild 1: Konditionierte Trockensorption – SVA Schöneiche
FIGURE 1: Conditioned dry sorption process – Schöneiche hazardous waste incineration plant

Kühler	= cooler
Reaktor	= reactor
Sorbalit	= Sorbalit
ca. 12 kg/t Müll	= approx. 12 kg/t waste
Gewebefilter	= fabric filter
Rezirkulat	= recycled material
Rezirkulatsilo	= silo for recycled material
Kamin	= stack
Reststoffsilo	= silo for residues
Deponie	= landfill

Rohgas [µg/Nm³]	Reingas [µg/Nm³]
180,0	2,5
250,9	29,0

Bild 2: SVA Schöneiche – Quecksilberemissionen

Rohgas [ng TE/Nm³]	Reingas [ng TE/Nm³]
0,34	n.n.
1,74	0,0022

Bild 3: SVA Schöneiche – Dioxinemissionen

Die folgenden Bilder zeigen den Adsorptionserfolg, wobei der besseren Übersicht wegen nur jeweils 2 Schadstoffkomponenten beispielhaft ausgewählt wurden. Aus den Reingaswerten für Quecksilber ergibt sich, daß der Grenzwert von 50 µg/Nm³ deutlich unterschritten wird (**Bild 2**). Bei den Dioxinen und Furanen ergeben sich vergleichbare Minderungserfolge (**Bild 3**), wobei aus dem Beispiel ersichtlich ist, daß zum Teil erhebliche Dioxinkonzentrationen im Rohgas vorliegen. Die Ingenieurgemeinschaft Technischer Umweltschutz GmbH als neutraler Sachverständiger hat die Messungen des TÜV Berlin begleitet und kontrolliert und kommt zu folgendem Ergebnis:

„Auch für diese Emissionen sind nach der Modifikation der Rauchgasreinigung Meßwerte ermittelt worden, die erkennen lassen, daß der vorgeschlagene Grenzwert von 0,1 ng TE/Nm³ unter den Versuchsbedingungen erreicht werden kann. Damit ist die Sondermüllverbrennungsanlage Schöneiche die erste Anlage in Deutschland, die diesen Standard demonstriert hat" [5].

Die beschriebene Anlage ist wohl inzwischen die am besten untersuchte Anlage der Welt, es liegen mehr als hundert Dioxinmessungen vor. In keiner Messung wurde der Grenzwert der 17. BImSchV überschritten. Hervorzuheben ist besonders, daß diese Reingaswerte ohne Anlagenänderung erreicht werden konnten.

Für PCB existieren bisher keine Grenzwerte. Aus dem folgenden **Bild 4** wird aber deutlich, daß mit dem Flugstromverfahren auch hier eine erhebliche Adsorption erreicht wird. Die PCB-Konzentrationen im Reingas liegen in der Größenordnung der Nachweisgrenze des Meßverfahrens.

PCB-Art	Rohgasgehalt [ng/Nm³]	Reingasgehalt [ng/Nm³]
Monochlorbiphenyle	<1	<1
Dichlorbiphenyle	<1	<1
Trichlorbiphenyle	10	<2
Tetrachlorbiphenyle	18	<3
Pentachlorbiphenyle	30	<3
Hexachlorbiphenyle	38	<2
Heptachlorbiphenyle	34	<2
Octachlorbiphenyle	30	<1
Nonachlorbiphenyle	24	<1
Decachlorbiphenyle	14	<1
Summe PCB's	**130**	

Bild 4: SVA Schöneiche – PCB-Emissionen

Raw gas [µg/Nm³]	Clean gas [µg/Nm³]
180.0	2.5
250.9	29.0

FIGURE 2: Schöneiche hazardous waste incineration plant – mercury emissions

Raw gas [TE/Nm³]	Clean gas [TE/Nm³]
0.34	not detectable
1.74	0.022

TE = toxicity equivalents

FIGURE 3: Schöneiche hazardous waste incineration plant – dioxin emissions

measurements by the TÜV Berlin. They reached the following conclusions:

"Measurements have been obtained for these emissions after the modification to the flue gas cleaning process which demonstrate that the proposed limit of 0.1 ng TE/Nm³ can be achieved under the trial conditions. The Schöneiche hazardous waste incineration plant is therefore the first plant in Germany to have demonstrated this standard" [5].

The plant in question must now be one of the most tested plants in the world. There have been more than a hundred individual dioxin measurements. The limit specified in the 17th BImSchG was not exceeded in any of these measurements. It should be particularly emphasized that these clean gas values were achieved without changing the plant.

There are no limits as yet for PCBs, but **Fig. 4** clearly shows that here again excellent adsorption can be achieved with the suspension flow process. The concentrations of PCBs in the clean gas are of the same order of magnitude as the limit of detection of the method of measurement.

4. Pilot filter in a light metal remelt plant

The exhaust gases from an induction furnace, a holding furnace and a foundry furnace were combined and conveyed to this system. The temperature of the exhaust gas is reduced to 150°C through the addition of false air. An adsorbent, which is a modification of the one used in Example 1, is added to the flue gas stream. The used Sorbalit is removed in a pilot filter system (fabric filter with membrane filter bags). The final design of the dedusting plant is to be decided in the light of the experience gained from the pilot trial. This will

PCB type	Raw gas [ng/Nm³]	Clean gas [ng/Nm³]
Monochlorobiphenyls	<1	<1
Dichlorobiphenyls	<1	<1
Trichlorobiphenyls	10	<2
Tetrachlorobiphenyls	18	<3
Pentachlorobiphenyls	30	<3
Hexachlorobiphenyls	38	<2
Heptachlorobiphenyls	34	<2
Octachlorobiphenyls	30	<1
Nonachlorobiphenyls	24	<1
Decachlorobiphenyls	14	<1
Total PCBs	**130**	

FIGURE 4: Schöneiche hazardous waste incineration plant – PCB emissions

4. Pilotfilter an einer Leichtmetall-Umschmelzanlage

Die Abgase eines Induktionsofens, eines Warmhalteofens und eines Gießofens wurden bei dieser Anlage zusammengeführt. Über Falschluftzufuhr wird die Temperatur der Abgase auf 150°C gesenkt. Dem Rauchgasstrom wird ein gegenüber Beispiel 1 modifiziertes Adsorbens eingesetzt. Die Abscheidung des verbrauchten Sorbalit erfolgt an einer Pilot-Filteranlage (Gewebefilter mit Membran-Filterschläuchen). Mit Hilfe der aus dem Pilotversuch gewonnenen Erkenntnisse soll die endgültige Ausführung der Entstaubungsanlage festgelegt werden. Dabei wird die Rezirkulation des abgereinigten Filterstaubes einbezogen, um den Ausnutzungsgrad zu erhöhen und die Deponiekosten in Grenzen zu halten. Ferner soll aufgrund der Ergebnisse die Adsorbensmenge weiter reduziert werden [6].

In folgender Tabelle (**Bild 5**) werden die Emissionsdaten dargestellt. Dabei handelt es sich um Maximalwerte. Die mittleren Emissionswerte lagen zum Teil erheblich unter den oben genannten Ergebnissen. Deutlich erkennbar ist der ausgezeichnete Abscheidegrad von > 99 % bei Dioxinen und Furanen, wobei auch hier mit nur einem einzigen Adsorptionsmittel geeigneter Zusammensetzung die vorgeschriebenen Grenzwerte sicher eingehalten werden können.

5. Schlußfolgerungen

Die bisher durchgeführten Untersuchungen und die Erprobung des Verfahrens in der Praxis haben gezeigt, daß das Flugstromverfahren dazu geeignet ist, dampfförmige organische Verbindungen hoher Umweltrelevanz, wie z.B. Dioxine, Furane, PCB, PAK, aber auch flüchtige Schwermetallverbindungen, wirksam abzuscheiden. Bewährt hat sich das Abscheideverfahren insbesondere bei Müll- und Sondermüllverbrennungsanlagen sowie sonstigen Feuerungen, die der Gewinnung von Prozeßwärme dienen.

Im Gegensatz hierzu handelt es sich bei den Verfahren zum Brennen von Zementklinker um Stoffumwandlungsprozesse. Kennzeichnend hierfür sind ein direkter Kontakt zwischen Flamme und Brenngut bei Sauerstoff-Überschuß und die hierdurch hervorgerufenen chemischen Reaktionen, hohe Brennraumtemperaturen von bis zu 2000°C und eine Verweilzeit der Gase von mehreren Sekunden im Hochtemperaturbereich von über 1000°C. Unter diesen Bedingungen werden die polychlorierten Kohlenwasserstoffe zerstört. Das hat dazu geführt, daß z.B. PCB-haltige Altöle als teilweiser Ersatz von natürlichen Primärbrennstoffen ohne Gefahr für die Umwelt vollständig im Drehrohrofen verbrannt werden können.

Demgegenüber können im Vorwärmer Bedingungen herrschen, die eine denovo-Synthese von Dioxinen und Furanen trotz O_2-Überschuß bei der Gasanalyse denkbar erscheinen lassen. Maßgebend hierfür sind Gas- und Brenngutemperaturen von unter 1000°C und ein je nach Rohstoffvorkommen unterschiedlich hoher Gehalt an natürlicher organischer Substanz in den Kalksteinen, Kalkmergeln und Tonen, der jedoch auch Werte von 0,5 Gew.-% TOC nicht übersteigt.

Aus den Erfahrungen bei der Einführung des Flugstromverfahrens für MVA's sind allerdings Faktoren sichtbar geworden, die auch auf das Klinkerbrennen übertragbar sind und eine mögliche Erklärung dafür liefern, warum in Zementwerken, unabhängig von der Art des verwendeten Brennstoffes, Dioxin- und Furankonzentrationen weit unter 0,1 ng/Nm³ festgestellt werden und somit keine Emissionsprobleme darstellen. Bei MVA's stellte sich heraus, daß die Konzentration der Dioxine und Furane im Reingas anstieg, wenn die Anlagen in bezug auf die Feuerung optimiert und hierdurch die Gehalte an unverbranntem Kohlenstoff im Flugstaub vermindert wurden. Die unverbrannten Kohlenstoffteilchen wirkten ähnlich adsorptiv wie Aktivkohle im Flugstromverfahren. Nachdem der Feuerungswirkungsgrad verbessert worden war, fehlten diese Adsorbentien oder ihr Anteil war deutlich vermindert, und die Dioxinemissionen stiegen an. Da aber der Grundpegel an Dioxinen und Furanen im MVA's um Größenordnungen höher als bei

Gaskomponente	Rohgas	Reingas	zulässiger Wert	Maßeinheit	Abscheidegrad
Staub	167,0	<0,8	20,0	mg/m³	>99,52%
anorg. gasf. Cl als HCl	133,0	0,3	30,0	mg/m³	99,77%
anorg. gasf. F als HF	0,07	0,02	0,6	mg/m³	71,43%
Summe SO als SO_2	62,0	2,1	30,0	mg/m³	96,61%
PCDD/PCDF (TE nach NATO)	3,2	0,015	0,1	ng/m³	99,53%

Bild 5: Pilotfilter einer Leichtmetall-Umschmelzung

Gas components	Raw gas	Clean gas	Permissable value	Units	Separating efficiency
Dust	167.0	<0.8	20.0	mg/m³	>99.52%
Inorganic gaseous Cl as HCl	133.0	0.3	30.0	mg/m³	99.77%
Inorganic gaseous F as HF	0.07	0.02	0.6	mg/m³	71.43%
Total SO as SO_2	62.0	2.1	30.0	mg/m³	96.61%
PCDD/PCDF (TE as per NATO)	3.2	0.015	0.1	ng/m³	99.53%

FIGURE 5: Pilot filter for a light metal remelt plant

incorporate the recirculation of the treated filter dust to increase the level of utilization and limit waste disposal costs. Further reductions in the quantity of adsorbent are anticipated in the light of these results [6].

The following table (**Fig. 5**) shows the emission data. These are maximum values. In some cases the average emission values were substantially less than the results mentioned above. The excellent removal rate of > 99 % for dioxins and furans is particularly notable, and here again the prescribed limits can be met reliably with just a single adsorbent of suitable composition.

5. Conclusions

The investigations conducted so far and the tests of the process under practical conditions have shown that the suspension flow process is suitable for effective removal of vaporous organic compounds which are of major environmental impact, such as dioxins, furans, PCBs, PAHs, as well as volatile heavy metal compounds. The removal process has proved particularly effective in waste and hazardous waste incineration plants and other combustion systems designed for recovering process heat.

In contrast to this, the process for burning cement clinker involves materials conversion processes. It is characterized by direct contact between the flame and kiln feed in the presence of excess oxygen, and by the resulting chemical reactions, high burning zone temperatures of up to 2000°C and a residence time for the gases of several seconds in the high temperature range of over 1000°C. The polychlorinated hydrocarbons are destroyed under these conditions. That means, for example, that used oil containing PCBs can be completely burnt in a rotary tube kiln as a partial replacement for natural primary fuels without danger to the environment.

On the other hand, there may be conditions in the preheater under which "denovo" synthesis of dioxins and furans could theoretically occur in spite of excess O_2 in the gas analysis. The decisive factors here are gas and kiln feed temperatures of less than 1000°C, and levels of naturally occurring organic substances in the limestones, lime marls and clays which differ according to the raw material deposit but do not exceed 0.5 wt.% TOC.

The experience gained from the introduction of the suspension flow process for waste incineration plants has revealed

Zementwerken liegt, war grundsätzlich der Einsatz von zusätzlichen Emissionsminderungstechniken erforderlich.

Demgegenüber ist im Wärmetauscherturm eines Drehrohrofens eher von einem Gleichgewicht zwischen Neubildung und Adsorption auszugehen, mit deutlicher Verschiebung zur Adsorption hin und auf einem wesentlich niedrigeren Konzentrationsniveau. Das führt dazu, daß im Wärmetauscherturm das Flugstromverfahren in der Reaktion zwischen Abgasen und dem Zementrohmehl systemimmanent praktiziert wird. Die Folgen sind extrem niedrige Dioxin- und Furankonzentrationen im Reingas der Öfen der Zementindustrie.

Literaturverzeichnis

[1] 17. Verordnung zur Durchführung des Bundesimmissionschutzgesetzes vom 23. 11. 1990 BGBL Teil 1 vom 30. 11. 1990, S. 2545 ff.

[2] L.-P. Nethe: Sorbalit — Modifiziertes Calciumhydroxid zur Rauchgasreinigung hinter Verbrennungsanlagen, „Müllverbrennung und Umwelt" Band 4, EF-Verlag, Berlin, 1990, S. 299–314.

[3] W. Esser-Schmittmann, u. a.: Einsatzmöglichkeiten und Sicherheitskriterien für Herdofenkoks in der Rauchgasreinigung, VDI Bildungswerk BW 797.

[4] TÜV Berlin, Technischer Bericht D 90/064 vom 28. 2. 1991 Emissionsmessungen im Roh- und Reingas während des Probebetriebes II Sondermüllverbrennungsanlage Schöneiche.

[5] SVA Schöneiche — Verfahrenstechnisches und Emissionstechnisches Gutachten der ITU — Ingenieurgemeinschaft Technischer Umweltschutz GmbH, Dr. Jager, Dr. Obermeier, Berlin, März 1990.

[6] A. Brünger: „Reduzierung von Schadstoffemissionen mit filternden Abscheidern", 28. Metallurgisches Seminar 24. – 26. 3. 1993.

a number of factors which can also be applied to clinker burning. These may explain why dioxin and furan concentrations of well below 0.1 ng/Nm^3 are found in cement plants, irrespective of the type of fuel used, and therefore do not present any emission problems. It has been established that dioxin and furan concentrations in the clean gas rose in waste incineration plants when the plant firing systems were optimized with a resulting reduction in the content of the unburnt carbon in the flue dust. The unburnt carbon particles had a similar adsorptive action to that of the activated carbon in the suspension flow process. After the firing efficiency had been improved, these adsorbents were either absent or their proportion was substantially reduced, and dioxin emissions rose. The base level for dioxins and furans in waste incineration plants was orders of magnitudes higher than in cement plants so there was a fundamental need for additional techniques for minimizing emissions.

In the preheater tower of a rotary kiln, on the other hand, there is more likely to be an equilibrium between new formation and adsorption, with a significant shift towards adsorption and a substantially lower level of concentration. The suspension flow process therefore occurs as an inherent part of the system in the reaction between exhaust gases and the cement raw meal in the preheater tower. This results in extremely low dioxin and furan concentrations in the clean gas from kilns in the cement industry.

Maßnahmen zur Minderung der SO₂-Emission beim Klinkerbrennen im Werk Höver der Nordcement AG

Measures to reduce the SO₂ emission during clinker burning at Nordcement AG's Höver works

Contremesures pour réduire l'émission SO₂ dans la cuisson du clinker à l'usine Höver de la Nordcement AG

Medidas para reducir las emisiones de SO₂ durante la cocción del clínker en la fábrica de Höver, de Nordcement AG

Von **K.-H. Boes**, Hannover/Deutschland

Zusammenfassung – Bei Emissionsmessungen waren bereits in den 80er Jahren erhöhte, rohstoffbedingte SO_2-Werte im Reingas der Drehofenanlage festgestellt worden. Ursache dafür waren Schwefelverbindungen im Rohmaterial (Pyrit und Markasit, Gehalte zwischen 0,05 und 0,45Gew.-%), die im Wärmetauscher freigesetzt und nicht im Brenngut eingebunden werden. Nach Einführung einer Dauermessung von SO_2 stellte sich heraus, daß die gemessenen Werte die Emissionsbegrenzungen der TA Luft zeitweilig überschritten. In Zusammenarbeit mit der Technischen Stelle Holderbank wurde ein in einem Zementwerk in der Schweiz erprobtes Minderungsverfahren auf der Basis von Kalkhydratzugabe an die speziellen Bedingungen im Werk Höver der Nordcement AG angepaßt. Mit weitgehend vorhandener Anlagentechnik wurde bei vertretbarem finanziellem Aufwand Kalkhydrat mit den Rohmehlbecherwerken in die erste Zyklonstufe des Wärmetauschers aufgegeben und so das entstehende SO_2 gebunden. Nach anfänglichen Problemen bei der Inbetriebnahme der Anlage im Bereich der Dosierung wird nach einer inzwischen installierten SO_2-konzentrationsabhängigen Kalkhydratdosierung der SO_2-Grenzwert der TA Luft sicher eingehalten.

Summary – As early as the eighties, emission measurements had shown increased SO_2 values in the clean gas from the rotary kiln plant caused by the raw materials. This was due to sulphur compounds in the raw material (pyrites and marcasite, levels between 0.05 and 0.45 wt.%) which are liberated in the preheater and not retained in the kiln feed. After introduction of continuous SO_2 measurement it was found that the measured values occasionally exceed the emission limits set by the German Clean Air Regulations. Working in conjunction with the technical agency Holderbank, a method for reducing the emissions, tested in a cement works in Switzerland and based on the addition of slaked lime, was adapted to the specific conditions in Nordcement AG's Höver works. Extensive use was made of existing plant technology to feed slaked lime, at a reasonable cost, into the first cyclone stage of the preheater using the raw meal bucket elevator; this then fixed the SO_2 produced. After initial problems with the metering system during plant commissioning and after installing a slaked lime metering system which is governed by the SO_2 concentration, the SO_2 limit set by the Clean Air Regulations can now be met reliably.

Résumé – Déjà dans les années 80, des mesures d'émission ont permis de constater des valeurs SO_2 plus fortes, dues à la matière première, dans les gaz épurés de la ligne de four rotatif. La cause en étaient des composants soufrés de la matière première (pyrite et marcassite, teneurs entre 0,05 et 0,45% pds.), libérés dans le préchauffeur et non liés dans la matière à cuire. Après mise en service d'une mesure en continu du SO_2, s'est révélé que les valeurs mesurées dépassaient parfois les limites d'émission de la TA Luft. En collaboration avec l'Instance Technique Holderbank a été alors adapté, aux conditions spécifiques de l'usine Höver de la Nordcement AG, un procédé de réduction basé sur addition d'hydrate de chaux expérimenté avec succès dans une cimenterie en Suisse. La configuration technique de l'installation étant largement disponible, peu d'efforts financiers ont été nécessaires pour introduire, au moyen des norias à farine crue, l'hydrate de chaux dans le premier étage de cyclones et de lier ainsi le SO_2 naissant. Après quelques problèmes au début, lors de la mise en service de l'installation, dans le domaine du dosage, la valeur limite SO_2 de la TA Luft est respectée de manière sûre, grâce à un dosage de l'hydrate de chaux en fonction de la concentration SO_2, installé récemment.

Resumen – Al medirse las emisiones en los años 80, ya se registraron valores demasiado altos de SO_2 en los gases depurados de la planta de horno rotatorio, causados por las materias primas. Se trataba, concretamente, de combinaciones de azufre contenidas en la materia prima (pirita y marcasita, contenidos entre 0,05 y 0,45 % en peso), las cuales son liberadas en el intercambiador de calor, sin ser incluidas en el material a cocer. Después de proceder a una medición permanente de SO_2, resultó que los valores obtenidos eran a veces superiores a los límites fijados en la TA Luft (Instrucción

técnica para el mantenimiento de la limpieza del aire). En colaboración con la Technische Stelle Holderbank se adaptó un procedimiento de reducción de emisiones, basado en la adición de hidrato de cal y que había probado su eficacia en una fábrica de cemento en Suiza, a las condiciones especiales reinantes en la fábrica de Höver, perteneciente a Nordcement AG. Aprovechando una técnica mayormente disponible, y con unos gastos económicos razonables, se introducía hidrato de cal en la primera etapa de ciclones del intercambiador de calor, mediante los elevadores de cangilones para harina cruda, combinando de esta forma el SO_2 que se formaba. Tras los problemas iniciales surgidos durante la puesta en servicio de la instalación, en la sección de dosificación, y después del montaje de una instalación de dosificación de hidrato de cal, que trabaja en función de la concentración de SO_2, se respeta ahora de forma segura el valor límite de SO_2, fijado por la TA Luft.

1. Werk Höver

Im Werk Höver bei Hannover betreibt die NORDCEMENT AG einen 3000t/d Wärmetauscherofen als zweisträngigen 4-Stufen-Zyklonvorwärmer mit Satellitenkühler der Fa. KHD (Bj. 1972). Als Sekundärbrennstoff werden im Ofeneinlauf Altreifen verbrannt (max. 10% der Gesamtwärmemenge). Die Rohmahlung erfolgt in einer Walzenschüsselmühle der Fa. Loesche unter Nutzung der Ofenabgase zur Rohmaterialtrocknung.

2. SO_2-Emission

Im Werk Höver wurden im Jahre 1984 bei Messungen des Forschungsinstituts der Zementindustrie im Reingas der Drehofenanlage erhöhte SO_2-Konzentrationen nachgewiesen. Zunächst fand sich keine Erklärung dafür, da die Feuerung mit schwefelarmen Brennstoffen betrieben wurde und eine Schwefelbilanz keine Hinweise für das Auftreten hoher SO_2-Konzentrationen lieferte.

Untersuchungen der Holderbank hatten gezeigt, daß SO_2-Emissionen an Wärmetauscheröfen häufig auf Sulfidverbindungen im Rohmaterial zurückzuführen sind. Es handelt sich dabei im wesentlichen um Pyrit- und Markasitanteile (FeS_2), die in den oberen Zyklonstufen des Wärmetauschers zu erhöhten SO_2-Konzentrationen im Abgas führen. Ergebnisse von Holderbank-Gutachten belegen, daß die Emissionsbegrenzung bereits durch geringe Pyrit- und Markasitanteile im Rohmaterial überschritten wird, auch wenn davon bis zu 70% in der dem Wärmetauscher nachgeschalteten Mahltrocknung wieder abgeschieden werden.

Analysen des Höver'schen Rohmehls bestätigen den Verdacht, daß diese Stoffe für die erhöhten Emissionen verantwortlich sind. Die Pyrit- und Markasitkonzentrationen betrugen hier zwischen 0,05 M.-% und 0,45 M-%, so daß bei Messungen der SO_2-Konzentrationen im Abgas Werte gefunden wurden, die im Verbundbetrieb bis zu 900 und bei Stillstand der Rohmühle bis zu 1300 mg/m³ erreichten. Daraus ergab sich die Notwendigkeit zur Entschwefelung der Rauchgase.

3. Grundlagen

Grundlage für das Verfahren stellt die Reaktion von $Ca(OH)_2$ mit SO_2 zu Calciumsulfit dar, das in Gegenwart von Sauerstoff zu Calciumsulfat oxidiert werden kann.

$Ca(OH)_2 \rightarrow CaO + H_2O$ (1)
$CaO + SO_2 \rightarrow CaSO_3$ (2)

Diese Reaktion führt jedoch nur dann zu einer optimalen SO_2-Bindung im Wärmetauscher, wenn Kalkhydrat im Überschuß aufgegeben wird.

Der Überschuß beträgt dabei 400 bis 700%. Der Ablauf der Reaktion ist außerdem sehr stark von der herrrschenden Temperatur und der Feinheit sowie Menge des Kalkhydrats abhängig. Zusätzlich hat die Feuchte einen Einfluß auf den Reaktionsablauf. Diese ist aber im System normalerweise nicht zu beeinflussen. Deshalb kann die Aufgabestelle für das Entschwefelungsmittel ($Ca(OH)_2$) im Wärmetauscher nur über die dort herrschende Temperatur ausgewählt werden. Hierfür kommt praktisch der Temperaturbereich von 200 bis 400 °C in Frage. Für eine optimale Bindung des SO_2 ist entscheidend, daß die Entschwefelungsreaktion möglichst dort abläuft, wo das SO_2 entsteht. Bei den Untersu-

1. The Höver works

At its Höver works near Hannover, NORDCEMENT AG operates a 3000 t/d preheater kiln in the form of a twin-line four-stage cyclone preheater with planetary cooler (built by KHD in 1972). Old tyres are burnt in the kiln inlet as secondary fuel (max. 10% of the total quantity of heat). Raw grinding is carried out in a Loesche roller grinding mill, using the kiln flue gases for raw material drying.

2. SO_2 emissions

During measurements by the Cement Industry Research Institute at the Höver works in 1984, increased SO_2 concentrations were detected in the clean gas from the rotary kiln plant. At first no explanation could be found for this, as the firing system was operated with low-sulphur fuel and a sulphur balance gave no indication of why high SO_2 concentrations should occur.

Investigations by Holderbank had shown that SO_2 emissions from preheater kilns could often be traced to sulphide compounds in the raw materials. These compounds are mainly pyrites and marcasite (FeS_2), which in the upper cyclone stages of the preheater result in increased SO_2 concentrations in the flue gas. Results of audits by Holderbank confirm that even small proportions of pyrites and marcasite cause the emission limits to be exceeded, even though up to 70% of the content of these substances are removed again in the drying and grinding stage downstream of the preheater.

Analyses of the raw meal from the Höver works confirmed the suspicion that these substances are responsible for the increased emission levels.

In this case the pyrites and marcasite concentrations ranged from 0.05 to 0.45 wt.%, with the result that measurements of the SO_2 concentration in the flue gas showed levels of up to 900 mg/m³ (NTP) in interconnected operation and up to 1300 mg/m³ (NTP) with the raw-grinding mill off-line. Desulphurisation of the flue gases was therefore necessary.

3. Basic principles

The desulphurisation technology is based on the reaction of $Ca(OH)_2$ with SO_2 to form calcium sulphite, which can be oxidised to calcium sulphate in the presence of oxygen.

$Ca(OH)_2 \rightarrow CaO + H_2O$ (1)
$CaO + SO_2 \rightarrow CaSO_3$ (2)

However, this reaction only leads to optimum SO_2 fixation in the preheater if slaked lime is fed in excess.

The excess amounts to approx. 400–700%. The progress of the reaction also depends very greatly on the prevailing temperature and on the fineness and quantity of the slaked lime. In addition, the progress of the reaction is affected by the moisture content. This, however, cannot normally be influenced in the system. Therefore the feed location for the desulphurisation agent ($Ca(OH)_2$) in the preheater can only be chosen on the basis of the temperature prevailing there. In practice, the temperature range 200–400 °C is suitable for this. For optimum SO_2 fixation it is vital, if at all possible, that the desulphurisation reaction takes place at the point where the SO_2 forms. The tests in the cyclone preheater plant at Höver revealed that the only suitable place for feed-

chungen an der Zyklonvorwärmeranlage in Höver hat sich gezeigt, daß für die Aufgabe des Kalkhydrates nur die oberste Zyklonstufe mit einer Gastemperatur von etwa 360 °C in Frage kam. Weitere Versuche zeigten, daß die Aufgabe in Zyklon 1 am wirksamsten war. Aufgrund dieser Erfahrungen erscheint es daher notwendig, bei Einführung des Verfahrens an einem anderen Ofen die optimale Aufgabestelle für das Kalkhydrat anhand von Versuchen auszuwählen.

4. Verfahren und Betriebsergebnisse

Grundlage des Entschwefelungsverfahrens waren erste Erfahrungen der Holderbank in einem Schweizer Zementwerk. Dort wurde feines Kalkhydrat mit einem Airlift in die 1. Zyklonstufe des Wärmetauschers aufgegeben, um so das entstehende SO_2 zu binden. Dieses Verfahren erwies sich als zuverlässig und wurde auch in Höver getestet. Die physikalischen und chemischen Daten des verwendeten Kalkhydrates sind der **Tabelle 1** zu entnehmen. Für die Lagerung des Kalkhydrates wurde ein Kohlenstaubsilo (900 m^3) verwendet, das zur damaligen Zeit nicht benötigt wurde. Die Kapazität reichte bei voller Leistung für etwa 10 Tage. Im Unterschied zum Schweizer Verfahren wird das Kalkhydrat in Höver über die Rohmehlbecherwerke in den Zyklon 1 gefördert. Die Dosierung erfolgt über drehzahlgesteuerte Zellenradschleusen mit drei unabhängigen Blasleitungen, die bei Bedarf auch gleichzeitig genutzt werden können. So läßt sich die erforderliche Kalkhydratmenge genau und sparsam dosieren.

TABELLE 1: Chemische und physikalische Eigenschaften des Kalkhydrats

Weißkalkhydrat „Blütenweiß" Fels-Werke GmbH
Chemische und physikalische Kenndaten

CaO	71,99 %
MgO	0,65 %
SiO_2	2,00 %
Fe_2O_3	0,45 %
Al_2O_3	0,47 %
SO_3	0,27 %
CO_2	1,23 %
geb. H_2O	21,60 %
Feuchte	0,90 %
BET	ca. 10 m^2/g
$Ca(OH)_2$	90,0 %
Feinheit:	
R > 0,063 mm	4,0 %

Die Aufgabe des Kalkhydrates in die Rohmehlbecherwerke hat sich als technisch einfach und äußerst wirksam erwiesen, wie aus **Bild 1** ersichtlich wird. Die Darstellung zeigt die Abhängigkeit der SO_2-Emission von der Kalkhydrataufgabe. Es wurde bei einer SO_2-Konzentration von etwa 250 ppm nach 3 Stunden zunächst über zwei Stränge die gleiche Kalkhydratmenge von etwa 200 kg/h aufgegeben. Nach 5 Stunden wurde ein Strang abgeschaltet. Deutlich ist die SO_2-Minderung nach etwa 3 Stunden auf etwa 150 ppm zu erkennen. Nach Abschalten des einen Stranges stieg sie wieder auf 200 ppm, was die Wirksamkeit des Verfahrens belegt.

Aus den Betriebsversuchen wurde eine Kurve abgeleitet, die die Abhängigkeit der SO_2-Minderung von der aufgegebenen Kalkhydratmenge zeigt (**Bild 2**). Deutlich ist zu erkennen, daß die Kurve von einem anfangs linearen Bereich mit zunehmender Kalkhydratmenge eine immer geringer werdende SO_2-Abscheidung aufweist und sich stark abflacht. Daraus geht hervor, daß das Verfahren – gemessen am Aufwand – nicht zu beliebig geringen SO_2-Konzentrationen im Abgas führt. Bei einer durchschnittlichen Konzentration von etwa 700 mg/m^3 ergibt sich außerdem ein Kalkhydratbedarf von etwa 900 kg/h. Erreicht der SO_2-Wert jedoch 1000 mg/m^3, so sind bei gleicher Endkonzentration bereits ca. 1200 kg/h aufzugeben. Damit ist die Grenze des Verfahrens nahezu erreicht.

ing the slaked lime was the top cyclone stage, where the gas temperature is about 360 °C. Further experiments showed that feeding into cyclone 1 gave the best results. Based on this experience, it appears that when this technology is introduced in a different kiln, it will be necessary to determine the optimum feed location for the slaked lime by means of experiments.

4. Method and operating results

The desulphurisation method was based on initial experience gained by Holderbank in a Swiss cement works. There, fine slaked lime was fed by means of an airlift into the first cyclone stage of the preheater to fix the SO_2 generated. This method proved reliable and was also tested at Höver. The physical and chemical characteristics of the slaked lime used are shown in **Table 1.** The slaked lime was stored in a pulverised coal storage bin (900 m^3) which was not needed at the time. Its capacity was sufficient for about 10 days at full production. In contrast to the process used in Switzerland, at Höver the slaked lime is fed into cyclone 1 by means of the bucket elevators for raw meal. It is proportioned by means of speed-controlled rotary valves with three independent pneumatic injection pipes which can also be used simultaneously when required. Thus the necessary quantity of slaked lime can be proportioned precisely and economically.

TABLE 1: Chemical and physical properties of the slaked lime

"Blütenweiß" slaked white lime from Fels-Werke GmbH
Chemical and physical data

CaO	71.99 %
MgO	0.65 %
SiO_2	2.00 %
Fe_2O_3	0.45 %
Al_2O_3	0.47 %
SO_3	0.27 %
CO_2	1.23 %
Combined H_2O	21.60 %
Moisture	0.90 %
BET	approx. 10 m^2/g
$Ca(OH)_2$	90.0 %
Fineness:	
R > 0.063 mm	4.0 %

Feeding the slaked lime into the raw meal bucket elevators has proved to be technically simple and extremely effective, as is evident from **Figure 1.** This chart shows the relationship between SO_2 emissions and slaked lime feeding. After about 3 hours at an SO_2 concentration of about 250 ppm, the slaked lime was initially fed at the same rate, about 200 kg/h, to two lines. After five hours, one line was shut off. The reduction in SO_2 to about 150 ppm after about three hours can be clearly seen. After shutting off one of the lines, it rose again to 200 ppm, confirming the effectiveness of the process.

A curve was plotted on the basis of the operating trials which shows the relationship between the SO_2 reduction and the slaked lime feed rate (**Figure 2**). It is clear that as the slaked lime feed rate increases, the curve, after an initial linear section, exhibits a declining rate of SO_2 removal and becomes much flatter. This shows that, measured against the cost, this technology cannot be used to achieve any desired low flue gas SO_2 concentration. In addition, for an average concentration of approx. 700 mg/m^3 about 900 kg/h of slaked lime are required. However, if the SO_2 level reaches 1000 mg/m^3, a feed rate of about 1200 kg/h is needed to achieve the same final concentration. At this point the technology has practically reached its limits.

After commissioning the slaked lime feeding system at Höver in November 1991, problems were initially encoun-

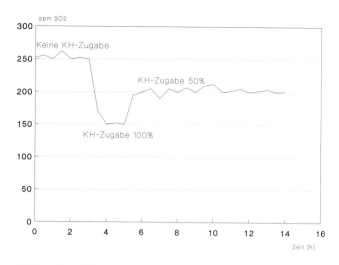

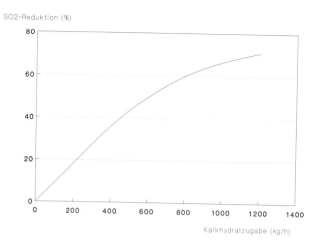

BILD 1: Erreichbare SO$_2$-Emissionsminderung durch Kalkhydratzugabe
FIGURE 1: Reduction in SO$_2$ emissions which can be achieved by adding slaked lime

Keine KH-Zugabe = No slaked lime feed
KH-Zugabe = Slaked lime feed
Zeit = Time

BILD 2: Zusammenhang zwischen benötigter Kalkhydratzugabe und SO$_2$-Reduktionsgrad
FIGURE 2: Relationship between required slaked lime addition and level of SO$_2$ reduction

SO$_2$-Reduktion = SO$_2$ reduction
Kalkhydratzugabe = Slaked lime feed rate

Nach Inbetriebnahme der Kalkhydrataufgabe in Höver im November 1991 traten zunächst Probleme bei der Dosierung des Kalkhydrates als Folge von Leitungsverstopfern auf. Diese konnten jedoch nach Vergrößerung des Leistungsquerschnitts behoben werden. Danach lief die Anlage weitgehend störungsfrei. Eine Verbesserung bei den Emissionswerten konnte zusätzlich noch durch Einbau einer Regelung der Kalkhydrataufgabe abhängig vom Emissionswert und dem Betriebszustand (Verbund oder Direkt) erreicht werden. Somit konnte der Leitstandfahrer von einer häufigen manuellen Nachregelung der Kalkhydratzugabe befreit werden.

5. Verwendung anderer Entschwefelungsmittel

Um die hohen Kosten für das feine Kalkhydrat einzusparen, wurden andere Entschwefelungsmittel auf ihre Verwendbarkeit untersucht. Dies waren Kalkhydratgrieße, Kreidemehl und teilentsäuertes Ofenmehl. Die Ergebnisse dieser Versuche waren aber überwiegend unbefriedigend. So konnte in keinem Fall auch nur annähernd die Wirksamkeit des z. Zt. verwendeten Kalkhydrates erreicht werden. Kreidemehl wies die geringste Wirksamkeit auf. Die Reaktivität der Kalkhydratgrieße war ebenfalls aufgrund der geringen Oberfläche kaum nachweisbar. Beim teilentsäuertene Rohmel, das aus der untersten Zyklonstufe eines Wärmetauschers abgesaugt worden war, ließ sich eine geringe entschwefelnde Wirkung feststellen. Die technischen Probleme beim Handling waren aber so groß, daß von einem weiteren Einsatz abgesehen wurde. Zudem waren für eine ausreichende Aufgabemenge die Blasleitungen zu klein dimensioniert, so daß ein Einsatz nicht möglich ist.

Versuche mit ungelöschtem Kalk (CaO) unterblieben, weil aufgrund seiner geringeren Reaktivität im Vergleich zu Kalkhydrat die Kostendifferenz den Mehreinsatz nicht rechtfertigt. Eine Optimierung, z.B. auch zur Absorption größerer SO$_2$-Mengen, wäre wohl nur über die Wahl feinerer Kalkhydrates möglich. So könnte ein Material mit einer BET-Oberfläche von 18 oder 36 m^2/g durchaus auch zu einer Bindung größerer SO$_2$-Mengen führen.

6. Zusammenfassung

Die trockene Rauchgasentschwefelung mit feinem Kalkhydrat in den oberen Zyklonstufen des Wärmetauschers stellt ein preiswertes, wirksames Verfahren dar, um die Abgase von Drehrohrofenanlagen zu entschwefeln. Allerdings hat dieses Verfahren Grenzen. So können damit nur Konzentrationen gemindert werden, die etwa 1 000 mg SO$_2$/Nm^3 (Verbundbetrieb) nicht übersteigen. Gleichzeitig sollte die

tered with slaked lime proportioning as a result of pipe blockages. However, they were cured by enlarging the pipe cross-section. After this, operation of the system was largely trouble-free. An improvement in emission levels was additionally achieved by installing a slaked lime feed controller operating in relation to the emission level and the operating mode (interconnected or direct). This freed the control room operator from having to make frequent manual adjustments to the slaked lime feed.

5. Use of other desulphurisation agents

To save on the high cost of the fine slaked lime, other desulphurisation agents were tested for their suitability. These were coarse slaked lime, powdered chalk and partly calcined kiln meal. However, the results of these experiments were mostly unsatisfactory. In no case was the effectiveness of the slaked lime currently in use anywhere near achieved. Powdered chalk had the lowest effectiveness. The reactivity of the coarse slaked lime was also barely detectable due to the small surface area. In the case of the partly calcined raw meal, which had been pneumatically withdrawn from the lowest cyclone stage of a preheater, a slight desulphurising effect could be discerned. However, the technical problems in handling were so great that its further use was abandoned. Furthermore, the pneumatic injection tubes were too small to deliver an adequate feed quantity; thus they cannot be used.

Tests with unslaked lime (CaO) were not carried out, since on account of its lower reactivity compared with slaked lime, the difference in cost does not justify the increased quantities required. Optimisation, e.g. also for absorption of larger quantities of SO$_2$, is only likely to be achieved by selecting a finer slaked lime. Thus a material with a BET specific surface of 18 or 36 m^2/g could certainly lead to fixation of larger quantities of SO$_2$.

6. Summary

Dry flue gas desulphurisation with fine slaked lime in the upper cyclone stages of the preheater is an inexpensive and effective method of desulphurising the flue gases of rotary kilns. However, this technology has its limits. For instance, only concentrations not exceeding about 1 000 mg SO$_2$/m^3 (NTP) in interconnected operation can be reduced. At the same time, the emission limit value should be less than 400 mg SO$_2$/m^3 (NTP). If finer slaked lime were used, it would probably be possible to fix larger quantities of SO$_2$. The SO$_2$ emission limits can be met if the outlets are also appropriately sized and the slaked lime can be precisely proportioned.

Emissionsbegrenzung bei Werten unter 400 mg SO_2/Nm^3 liegen. Beim Einsatz feineren Kalkhydrats könnten vermutlich auch größere SO_2-Mengen gebunden werden. Bei ebenfalls entsprechender Dimensionierung der Austräge und einer genauen Dosierungsmöglichkeit des Kalkhydrates läßt sich die Emissionsbegrenzung für SO_2 einhalten.

Literaturverzeichnis

[1] Nielsen, P. B.: Die So_2- und No_x-Emissionen bei modernen Zementdrehofenanlagen mit Blick auf zukünftige Verordnungen. Zement-Kalk-Gips (1991) 44, Nr. 9, S. 449–456.

[2] Bonn, W., und Hasler, R.: Verfahren und Erfahrung einer rohstoffbedingten SO_2-Emission im Werk Untervaz der Bündner Cementwerke. Zement-Kalk-Gips (1990), 44, Nr. 3, S. 139–143.

[3] Hünlich, Th., Jeschar, R., und Scholz, R.: Sorptionskinetik von SO_2 aus Verbrennungsabgasen bei niedrigen Temperaturen. Zement-Kalk-Gips (1991), 44, Nr. 5, S. 228–237.

[4] Hennecke, H. P., König, W., Roeder, A., Schmitz, F., und Stumpf, T.: Trockengelöschtes Kalkhydrat mit großer Oberfläche – Ein wirksames Reagenz zur Abgasreinigung. Zement-Kalk-Gips (1986), 39, Nr. 5, S. 251–259.

[5] Sprung, S.: Das Verhalten des Schwefels beim Brennen von Zementklinker. Schriftenreihe der Zementindustrie, Heft 31/1964.

[6] Sprung, S.: Technologische Probleme beim Brennen des Zementklinkers, Ursache und Lösung. Schriftenreihe der Zementindustrie, Nr. 43/1982.

[7] Schütte, R.: Möglichkeiten der Entstehung und Minderung von SO_2-Emissionen in Zementwerken. Zement-Kalk-Gips (1989), 42, Nr. 3, S. 128–133.

[8] Kühle, K.: Verfahren zur trockenen Schadgasabscheidung aus Rauchgas. Zement-Kalk-Gips (1987). Nr. 11, S. 537–542.

[9] Schmidt, K. D., Gardeik, H. O., und Ruhland, W.: Verfahrenstechnische Einflüsse auf die SO_2-Emission an Drehofenanlagen. Zement-Kalk-Gips (1986), 39, Nr. 2, S. 93–101.

Behördliche Umweltschutzanforderungen beim Einsatz alternativer Brenn- und industrieller Reststoffe im Zementdrehrohrofen

Official environmental protection requirements when using alternative fuels and industrial residues in cement rotary tube kilns

Exigences de protection de l'environnement posées par le Ministère Public pour l'utilisation de combustibles alternatifs et de résidus industriels dans le four rotatif à ciment

Prescripciones oficiales en materia de protección del medio ambiente, en relación con la utilización de combustibles alternativos y de residuos industriales en los hornos rotatorios de cemento

Von **R. Bolwerk,** Münster/Deutschland

Behördliche Umweltschutzanforderungen beim Einsatz alternativer Brenn- und industrieller Reststoffe im Zementdrehrohrofen

Zusammenfassung – Nach derzeitigem Erkenntnisstand kann man grundsätzlich davon ausgehen, daß der Einsatz von Reststoffen im Zementdrehrohrofen möglich ist. Die Voraussetzungen sind jedoch bei jeder Anlage unterschiedlich und müssen in der Regel im Rahmen eines Genehmigungsverfahrens nach dem Bundes-Immissionsschutzgesetz (BImSchG) festgelegt werden. Die im Rahmen des Genehmigungsverfahrens durchzuführenden Prüfungen, Betrachtungen und Abschätzungen wie z.B. eine Immissionsprognose, toxikologische Bewertung, Sicherheitsprüfung, Stellungnahme anderer Behörden sowie öffentliche Erörterungen, müssen zu dem Ergebnis kommen, daß der Einsatz von Reststoffen unbedenklich ist. Zur Umsetzung unbestimmter Rechtsbegriffe des BImSchG ist die Technische Anleitung zur Reinhaltung der Luft – TA Luft 86 – heranzuziehen. Dabei sind die Dynamisierungsklauseln sowie der fortgeschrittene Stand der Technik zu berücksichtigen. Unterliegt die Anlage aufgrund der Änderung des Einsatzstoffes der Störfall-Verordnung (12. BImSchG), so sind Art und Ausmaß möglicher Gefahren und Vorkehrungen zu beschreiben, um Störfälle zu verhindern. Künftig werden beim Einsatz von Reststoffen im Zementdrehrohrofen die Vorschriften der 17. BImSchV hinsichtlich der Überwachung und Begrenzung der Emissionen anzuwenden sein. Hierbei kann jedoch aufgrund der besonderen Verfahrenstechnik der §19 der VO Ausnahmen zulassen. Wird eine Genehmigung in einem Verfahren unter Einbeziehung der Öffentlichkeit durchgeführt, so kommt auch eine Umweltverträglichkeitsprüfung in Betracht. Sie ist jedoch ein unselbständiger Teil des immissionsschutzrechtlichen Genehmigungsverfahrens.

Official environmental protection requirements when using alternative fuels and industrial residues in cement rotary tube kilns

Summary – Basically, it can be assumed from current findings that it is possible to use residual materials in cement rotary tube kilns. However, the requirements differ for each plant and, as a rule, have to be established within the framework of an authorization process in accordance with the Federal Emission Protection Act (BImSchG). It is necessary for the tests, inspections, and estimates which have to be carried out in the context of the licensing procedure, such as an emission prediction, toxicological evaluation, safety testing, views of other authorities, and public discussions, to come to the conclusion that the use of residual materials is absolutely safe. The German Clean Air Regulations – TA Luft 86 – is used for translating the indefinite legal terms in the BImSchG. The adjustment clauses and the advanced state of technology have to be taken into account. If the plant becomes subject to the Industrial Incident Regulations (12th BImSchG) due to the change in the material used, then it is necessary to describe the type and extent of possible dangers and precautions to prevent industrial incidents. In future, the regulations in the 17th BImSchV will apply to the use of residues in cement rotary tube kilns as far as monitoring and limiting of emissions are concerned. However, Section 19 of the regulations can permit exceptions in this case due to the particular process technology. If an authorization is carried out in a process involving the public, then an environmental compatibility test is also a possibility. It is, however, a dependent part of the licensing process governed by the immission protection laws.

Résumé – *D'après l'état actuel des connaissances, il peut en principe être admis que l'utilisation de résidus est possible dans le four rotatif à ciment. Les conditions sont, par contre, différentes pour chaque installation et doivent être fixées au cours d'une procédure d'autorisation selon la Loi fédérale de protection contre les immissions (BImSchG). Les vérifications, considérations et estimations à réaliser dans le cadre de la procédure d'autorisation, comme p. ex. un pronostic d'immission, une évaluation toxicologique, des essais de sécurité, l'avis d'autres instances officielles et des discussions publiques doivent conduire au résultat, que l'utilisation de résidus est sans risque. Afin de transposer des notions législatives non définies dans la BImSchG, il convient de consulter le Guide Technique de conservation de la pureté de l'air – TA Luft 86. Y sont à considérer les clauses à effet dynamique et l'état avancé de la technique. Si l'installation, à cause du changement de la substance utilisée, est concernée par le décret régissant les incidents (12. BImSchG), il faut décrire la nature et l'étendue des risques et précautions, afin d'éviter les incidents. A l'avenir seront à observer, lors de l'utilisation de résidus dans le four rotatif à ciment, les préscriptions de la 17. BImSchV au sujet de la surveillance et la limitation des émissions. Néanmoins et en raison de la technique de process particulière, le § 19 de la VO peut autoriser des exceptions. Si une autorisation est examinée dans une procédure incluant le public, il faut aussi tenir compte d'une vérification de compatibilité avec l'environnement. Celle-ci n'est, par contre, pas une partie indépendante de la procédure d'autorisation relevant de la législation pour la protection contre les immissions.*

Exigences de protection de l'environnement posées par le Ministère Public pour l'utilisation de combustibles alternatifs et de résidus industriels dans le four rotatif à ciment

Resumen – *Según el estado actual de los conocimientos, se puede partir, en principio, de la base de que es posible la utilización de sustancias residuales en los hornos rotatorios de cemento. Sin embargo, las condiciones para ello son diferentes en cada caso y deben ser fijadas, por regla general, dentro del marco de un procedimiento de autorización según la Ley federal de protección contra las emisiones (BImSchG). Los ensayos, observaciones y evaluaciones a efectuar durante el procedimiento de autorización, tales como el pronóstico de las emisiones, la evaluación toxicológica, el control de seguridad, la obtención del parecer de otras Autoridades así como la discusión pública del problema, tienen que llegar a la conclusión de que el empleo de las sustancias residuales no encierra ningún riesgo. En cuanto a la aplicación de términos jurídicos imprecisos, contenidos en la BImSchG, hay que consultar la Instrucción técnica para el mantenimiento de la limpieza del aire – TA Luft 86. A este respecto, hay que tener en cuenta las cláusulas de ajuste así como el estado actual de la técnica. En el caso de que la instalación esté sujeta a la Directiva sobre incidentes (12. BImSchG), se decribirá la naturaleza y el alcance de posibles riesgos así como las medidas a tomar para prevenir tales incidentes. En el futuro habrá que aplicar, para la utilización de residuos en los hornos rotatorios de cemento, las prescripciones de la 17. BImSchV, en cuanto se refiere al control y limitación de las emisiones. Sin embargo, debido al carácter especial de esta tecnología, el 19 de la VO puede hacer excepciones. Si se lleva a cabo una autorización, en un procedimiento público, hay que tener en cuenta también un examen de inocuidad para el medio ambiente. Pero este examen forma parte del procedimiento de autorización según la Ley de protección contra las emisiones.*

Prescripciones oficiales en materia de protección del medio ambiente, en relación con la utilización de combustibles alternativos y de residuos industriales en los hornos rotatorios de cemento

Erfahrungen der Vergangenheit haben gezeigt, daß die Zementindustrie zur Verwertung von Sekundärbrenn- und -rohstoffen einen wichtigen Beitrag leisten kann. Die günstigen Bedingungen im Drehrohrofen, eine optimierte Verfahrens- und Sicherheitstechnik und verbesserte Abgasreinigungsmaßnahmen sind dabei wichtige Faktoren. Die Voraussetzungen sind jedoch bei jeder Anlage unterschiedlich und müssen im Rahmen eines Genehmigungsverfahrens nach dem Bundes-Immissionsschutzgesetz (BImSchG) geprüft und festgelegt werden.

Dabei wird eine Umweltverträglichkeitsprüfung (UVP) erforderlich, wenn das Vorhaben öffentlich bekannt gemacht werden muß und nachteilige Auswirkungen auf Menschen, Tiere und Pflanzen, Boden, Wasser, Luft, Klima und Landschaft einschließlich der jeweiligen Wechselwirkungen sowie Kultur und Sachgüter haben kann. Dem Genehmigungsantrag ist in diesen Fällen zusätzlich eine Beschreibung der Umwelt und ihrer Bestandteile sowie die zu erwartenden Auswirkungen des Vorhabens auf die v. g. Schutzgüter beizufügen.

Prüfgegenstand ist die Errichtung und der Betrieb der Anlage bzw. der Maßnahme. Beide Begriffe – Errichtung und Betrieb – sind in einem umfassenden Sinne zu verstehen. Errichtung ist nicht allein das Stadium des Aufbaus, sondern auch die Einrichtung der Anlage, so daß die gesamte technisch-konstruktive Beschaffenheit der Anlage, einschließlich ihrer Funktionsweise, zu prüfen ist.

Betrieb ist nicht nur die Produktion im engeren Sinne, sondern die gesamte Betriebsweise unter Einschluß der Wartung und Unterhaltung. Somit sind im Genehmigungsantrag die Anforderungen an den Betrieb der Zementofenan-

Past experience has shown that the cement industry can play an important part in the utilization of secondary fuels and raw materials. Key factors include favourable conditions inside rotary tube kilns, optimized process and safety technology, and improved exhaust gas cleaning systems. The requirements differ for each plant and these must be examined and defined as part of the licensing procedure in accordance with the Federal Immission Protection Act (BImSchG).

An environmental compatibility test is compulsory if the project has to be made public and could have disadvantageous effects on human beings, animals or plant life, soil, water, air, the climate or the landscape – including any interactive effects – or on cultivation and property. In these cases the application for a licence must be accompanied by a description of the local environment and its features including the anticipated effects of the project on the above-mentioned factors which require protection.

The object under examination in this case is the installation and operation of the plant or process. Both terms – installation and operation – apply in a comprehensive sense. Installation is not just the construction phase of the plant, but also the equipment of the plant itself, so that the entire technical and constructional nature of the plant including its mode of operation has to be examined.

Operation is not just production in its narrowest sense, it covers the entire mode of operation including maintenance and repair. In other words, the application for a licence must give a comprehensive specification of the operating requirements for the cement kiln plant for ensuring safe combustion of the residues, together with a description of the neces-

lage zur Gewährleistung einer sicheren Verbrennung der Reststoffe umfassend darzulegen und die erforderlichen betrieblichen Maßnahmen unter Beachtung folgender Kriterien zu beschreiben:

- Menge des zu ersetzenden Primärbrenn- bzw. Rohstoffes
- Heizwert und Zugabemenge (kg/h) des Ersatzstoffes
- Gesamtchlorgehalte/Chlorfracht pro Zeiteinheit
- Schadstoffgehalte (PCB, PCDM, Schwermetalle usw.)
- Angaben zur Identität des Einsatzstoffes
- chemische und physikalisch-chemische sowie toxische und ökotoxische Eigenschaften des Stoffes
- Verbrennungsbedingungen und Zerstörungseffizienz
- Verfahrenskreisläufe, die zu Anreicherungen führen
- Möglichkeiten der Stoffausschleusung, Kreislaufentlastungen
- Betriebsvorgänge mit Abschaltungen (CO-Abschaltungen)
- Wirkung und Art der Abgasreinigungsverfahren

Einzugehen ist insbesondere auf mögliche Störungen des Brennbetriebes, die eine Unterbrechung bzw. Änderung des Materialflusses durch die Zementofenanlage bewirken können, sowie die Auswirkungen auf die Zerstörungseffizienz im Drehrohrofen. Sind Verfahrenskreisläufe notwendig, so müssen Maßnahmen zur Vermeidung erhöhter Emissionen beispielsweise durch gezielte Stoffausschleusung plausibel dargelegt werden. Betriebsvorgänge, die mit Abschaltungen (CO-Abschaltung) verbunden sind, müssen im Hinblick auf geringe Emissionen ausgelegt sein und durch Aufzeichnung geeigneter Prozeßgrößen besonders überwacht werden. Für den Ausfall von Einrichtungen zur Emissionsminderung müssen Maßnahmen vorgesehen sein, um die Emissionen unverzüglich soweit wie möglich zu vermindern.

1. Einsatzbedingungen

Wesentliche Voraussetzungen für niedrige Emissionen sind ein gleichbleibender Ofenbetrieb und konstante Bedingungen beim Einsatz von Sekundärroh- und brennstoffen. Hieraus ergibt sich:

- Der Brennprozeß muß kontinuierlich mit moderner Prozeßleittechnik überwacht werden.
- Abfallstoffe bedürfen einer ständigen Eingangskontrolle und aufwendiger Vorhomogenisierung. Flüssige Stoffe werden über Tropfleitungen zur Qualitätskontrolle kontinuierlich entnommen.
- Die wesentlichen Analysenparameter der Abfallstoffe (Heizwert, chemische Zusammensetzung etc.) müssen quasi kontinuierlich in die Prozeßsteuerung eingehen.
- Die Aufgabelanze muß so gestaltet sein, daß der Abfallbrennstoff zentral eingeblasen wird und in der Flammenfront des Hauptbrennstoffes zur Zündung kommt.
- Die Regelaggregate müssen eine vom Hauptbrennstoff unabhängige Zugabe von Abfallbrennstoffen ermöglichen.
- Die Zufuhr von Abfallbrennstoffen darf nur bei störungsfreiem Dauerbetrieb im Bereich der Nennlast möglich sein.

Eingriffe in den Betrieb einer Ofenanlage aufgrund unterschiedlicher Störungen haben nicht nur Einfluß auf die Temperaturen im Ofen, sondern auch auf andere Betriebsparameter. Über kontinuierliche Messungen müssen die hierdurch bedingten Abweichungen vom Normalbetrieb registriert und dokumentiert werden. Dabei ist die Aufzeichnung folgender Daten erforderlich:

Sauerstoff (O_2), Kohlenmonoxid (CO), Kohlendioxid (CO_2)
Summe organisch gebundener Kohlenstoffe (Gesamt C)
Temperatur des Klinkers am Ofenausgang
Temperatur des Ofengases am Ofeneingang
Unterdruck vor dem Abgasgebläse
Schwefeldioxid (SO_2), Staub, Stickoxide (NO_x)

sary operational measures with regard to the following criteria:

- Quantity of primary fuel and/or raw material to be replaced
- Calorific value and added quantity (kg/h) of substitute material
- Total chlorine content/loading per unit time
- Pollutant content (PCB, PCDM, heavy metals, etc.)
- Information on the identity of the material used
- Chemical, physico-chemical, toxic and ecotoxic properties of the material
- Combustion conditions and destruction efficiency
- Recirculation systems leading to concentration
- Possible ways of purging material and relieving the recirculation systems
- Operating processes with cut-offs (CO cut-off)
- Effect and type of exhaust gas cleaning processes

Potential faults in the combustion process which could interrupt or alter the flow of material through the cement rotary tube kiln, as well as their effects on the destruction efficiency in the rotary tube kiln, are to be documented in detail. Where recirculation processes are necessary, plausible descriptions must be given of measures for preventing increased emissions, e.g. by way of strategic material purging. Operating processes associated with cut-off systems (CO cut-off) must be designed specifically for low emissions and be monitored particularly carefully by recording suitable process variables. Special measures must be provided in case of failure of the emission reduction equipment to minimize emissions as rapidly as possible.

1. Operating conditions

The main requirements for low emissions are uniform kiln operation and constant operating conditions when using secondary raw materials and fuels. From this it follows that:

- The burning process must be monitored continuously using modern process control technology.
- Waste materials require fixed inspections on arrival and comprehensive preliminary homogenization. Liquid media are sampled continuously through trickle tubes for quality control.
- The main parameters for analysis of the waste material (calorific value, chemical composition, etc.) must be input into the process control system on a semi-continuous basis.
- The feed lance must be designed so that the waste fuel is injected centrally and is ignited at the flame front of the main fuel.
- The control units must allow the waste fuel to be supplied independently of the main fuel.
- Waste fuel may only be supplied during normal continuous operation within the rated output range.

Intervention in the operation of a kiln plant as a result of a variety of faults affects the temperatures in the kiln as well as a number of other operating parameters. The resulting deviations from normal operation must be recorded and documented by continuous measurement. The following data should be recorded:

Oxygen (O_2), carbon monoxide (CO), carbon dioxide (CO_2)
Total organically combined carbon (Total C)
Temperature of clinker at kiln outlet
Temperature of kiln gas at kiln inlet
Negative pressure before the exhaust gas fan
Sulphur dioxide (SO_2), dust, nitrogen oxides (NO_x)

2. Monitoring safe combustion

One of the main preconditions for granting a licence under the Federal Immission Protection Act (BImSchG) to use substitute materials is the provision of a safety system which

2. Überwachung einer sicheren Verbrennung

Wesentliche Voraussetzung für die Erteilung einer Genehmigung nach dem Bundes-Immissionsschutzgesetz zum Einsatz von Ersatzstoffen ist die Installation eines als Sicherheitskette arbeitendes Sicherheitssystems, welches eine lückenlose Überwachung gewährleistet. Es soll Störungen des Normalbetriebes schnell erkennen und durch geeignete Ansprechsysteme eine unkontrollierte Verbrennung von Reststoffen verhindern.

Mittels einer rechnergesteuerten Logik müssen mindestens die nachfolgend genannten Parameter der „10-Punkte-Sicherheitskette" so miteinander verknüpft werden, daß in Abhängigkeit von der Größe der Abweichung vom Sollwert bzw. der Ausfalldauer eines Aggregates die Auswirkung auf die Emission ermittelt wird und bei festgelegten Grenzen eine Abschaltung erfolgt:

1. Unterschreitung einer Gastemperatur von 900 °C am Ofeneinlauf.
2. Unterschreitung einer Brennguttemperatur am Ofenauslauf von 1250 °C.
3. Überschreitung eines im Versuch festzulegenden CO-Gehaltes (Vol. 1 %).
4. Unzulässige Regelabweichungen des Soll-/Ist-Wertevergleiches an der Primärbrennstoffaufgabe.
5. Unterschreitung einer Rohmehlaufgabemenge von 75 % der max. möglichen Menge.
6. Unterschreitung des bei Nennleistung erforderlichen Unterdrucks vor dem Abgasgebläse.
7. Ausfall Elektro-Filter.
8. Ausfall Abgasventilator.
9. Ausfall Brenner.
10. Überschreitung des zulässigen Staubgrenzwertes.

Zur genauen Festlegung der Schwellwerte in der 10-Punkte-Sicherheitskette ist vorab die Kalibrierung der Temperaturfühler im Ofenein- und -auslauf unter Variation der Nennlast notwendig, dabei ist auch die Repräsentanz der Meßpunkte zu beurteilen. Aus diesen Ergebnissen kann dann der Lastbereich, in dem die Verbrennung von Afballbrennstoffen zu vertreten ist, festgelegt werden. Dabei ergeben sich auch die notwendigen Schwellwerte der Überwachungstemperaturen.

3. Untersuchung der Emissionen

Im Hochtemperaturbereich des Drehrohrofens wird aufgrund vorliegender Erfahrungen gewährleistet, daß insbesondere chlorierte Stoffe dort sicher zerstört und unschädlich gemacht werden können. Im Vergleich hierzu sind die Brennbedingungen an der Sekundärseite einer Anlage für eine Verbrennung halogenierter Kohlenwasserstoffe ungünstiger, und es bedarf sorgfältiger Untersuchungen, wenn dort Stoffe eingesetzt werden sollen.

In jedem Fall ist vor Aufnahme des Betriebes mit Reststoffen die Zuverlässigkeit der Anlage mittels Simulation von „Störungen" zu testen und die Ermittlung der Emissionen im „Nullzustand" vorzunehmen, die bei herkömmlicher Betriebsweise bisher nicht ermittelt wurden. Zusätzlich mit diesen Ermittlungen kommen dann in Abhängigkeit von der Zusammensetzung des neu einzubringenden Reststoffes für die Reingasmessung ingesamt folgende Abgasinhaltsstoffe und Betriebsparameter für die Emissionsüberwachung in Betracht:

Staub; HCl; Gesamt-C; CO; SO_2; NO_x; Schwermetalle (Cd, Hg, Tl, As, Co, Ni, Se, Te, Sb, Pb, Cr, Cu, Mn, Ni, V, Sn, Be);
Benzol, Toluol, Xylol (BTX); Chlorbenzole; Chlorphenole; PAH; PCDD/F; PCB; TCDM;
Abgaszustand (O_2, CO_2, Temperatur, Feuchte, Volumenstrom)

operates as a safety chain and ensures uninterrupted monitoring. This should ensure rapid detection of any disruption to normal operation and use appropriate response systems to prevent uncontrolled combustion of residues.

As a minimum, the parameters of the "10-point safety chain" listed below must be linked to one another by a computer-controlled logic system so that their effect on emissions can be ascertained and the operation can be shut down at predetermined limits as a function of the degree of deviation from the setpoint value or the plant stoppage time.

1. Gas temperature less than 900 °C at kiln inlet.
2. Temperature of material at kiln outlet less than 1250 °C.
3. CO level above a value to be established by trial (Vol. %).
4. Inadmissible control deviations in the setpoint/actual-value comparison for the primary fuel feed.
5. Raw meal feed of less than 75% of the max. possible quantity.
6. Negative pressure before the exhaust gas fan below the value required at rated output.
7. Failure of electrostatic precipitator.
8. Failure of exhaust gas fan.
9. Failure of burner.
10. Dust level above permissible limit.

The temperature sensors in the kiln inlet and outlet have to be calibrated at different outputs before the threshold values for the 10-point safety chain can be specified accurately. It is also necessary to assess the representativeness of the measuring points. The output range over which waste fuel combustion can reasonably be carried out can be established from these results. They also give the required threshold values for the monitored temperatures.

3. Emission testing

Previous experience shows that chlorinated materials in particular can be safely broken down and rendered harmless in the high temperature section of a rotary kiln. On the other hand, the combustion conditions on the secondary side of a combustion plant are less favourable for burning halogenated hydrocarbons and careful investigations are required if these materials are to be used there.

The reliability of the plant must be tested in every case by "fault" simulation before starting operation with residues. Those emissions which were not measured previously during conventional operation must be measured under "zero input" conditions. In addition to these measurements, and depending on the composition of the residue which is about to be introduced, all the following exhaust gas components and operating parameters for the emission monitoring system also have to be taken into account for the clean gas measurement:

dust; HCl; total C; CO; SO_2; NO_x; heavy metals (Cd, Hg, Tl, As, Co, Ni, Se, Te, Sb, Pb, Cr, Cu, Mn, Ni, V, Sn, Be);
benzene, toluene, xylene (BTX); chlorobenzenes; chlorophenols; PAH; PCDD/F; PCB; TCDM;
Exhaust gas conditions (O_2, CO_2, temperature, moisture, volumetric flow)

4. Application of German Clean Air Regulations (TA-Luft)

When the requirements for a licence to operate are being checked the German Clean Air Regulations are mainly applied for assessing whether harmful environmental effects or other hazards, major disadvantages, or substantial annoyance to the general public or the neighbourhood have been eliminated.

Where immission values are specified for emitted pollutants to provide protection against health hazards, the suitability of a project for approval basically depends on whether the prescribed immission values (IW 1, IW 2) are complied with

4. Anwendung der Technischen Anleitung zur Reinhaltung der Luft (TA-Luft)

Bei der Prüfung der genehmigungsrechtlichen Voraussetzungen ist die TA-Luft maßgeblich für die Beurteilung der Frage, ob schädliche Umwelteinwirkungen oder sonstige Gefahren, erhebliche Nachteile oder erhebliche Belästigungen für die Allgemeinheit oder die Nachbarschaft ausgeschlossen sind.

Soweit für die emittierten Schadstoffe Immissionswerte zum Schutz vor Gesundheitsgefahren festgelegt sind, hängt die Genehmigungsfähigkeit des Vorhabens grundsätzlich davon ab, ob die dort festgelegten Immissionswerte (IW 1, IW 2) auf allen Beurteilungsflächen des entsprechend der Schornsteinhöhe festzulegenden Beurteilungsgebietes eingehalten sind.

Bei der Bewertung von Schadstoffen, für die keine Immissionswerte festgelegt sind, sind die von der Wissenschaft anerkannten Beurteilungsmaßstäbe mit den zu erwartenden Immissionsbelastungen in Beziehung zu setzen. In Abhängigkeit von den möglichen Folgen der Umwelteinwirkungen ist zu entscheiden, ob die Immissionsbelastung hingenommen werden kann und welche Schutzmaßnahmen ggf. ergriffen werden müssen.

5. Krebserzeugende Stoffe

Beim Einsatz von Reststoffen ist den krebserzeugenden Stoffen im Genehmigungsverfahren und bei der Überwachung besondere Aufmerksamkeit zu widmen. Für krebserzeugende Stoffe gilt der stets anzuwendende Grundsatz als Genehmigungsvoraussetzung, daß die Emissionen unter Beachtung des Grundsatzes der Verhältnismäßigkeit soweit wie möglich zu begrenzen sind.

Mit dem Minimierungsgebot in Nr. 2.3 TA-Luft (Emissionsbegrenzung soweit wie möglich) wurde bewußt nicht an die sonst übliche generelle Anforderung zur Emissionsbegrenzung nach dem Stand der Technik angeknüpft. Dies ergibt sich aus dem besonderen Wirkungscharakter der Stoffe und der Doppelfunktion der emissionsbegrenzenden Vorschriften. Die Anforderung „soweit wie möglich" hebt hervor, daß bei krebserzeugenden Stoffen die Möglichkeiten zur Emissionsminderung und Vermeidung voll ausgeschöpft werden müssen. Maßstab für die Pflicht zur Emissionsbegrenzung ist das in den Grenzen der praktischen Vernunft technisch Mögliche.

Aufgrund des Risikos der Gesundheitsgefährdung durch krebserzeugende Stoffe sind Emissionen (wenn möglich) zu vermeiden bzw. die Emissionen durch prozeßtechnische Maßnahmen und Abgasreinigungsmaßnahmen zu minimieren. Weiterhin ergibt sich aus dem Minimierungsgebot die besondere Verpflichtung, den Entwicklungs- und Anwendungsstand von Emissionsminderungstechniken für krebserzeugende Stoffe zu verbessern. Wichtig ist neben der Einhaltung niedriger Massenkonzentrationen dabei auch die Minimierung der Massenströme.

6. Anwendung der Störfall-Verordnung – 12. BImSchV –

Werden Reststoffe in Anlagen der Zementindustrie eingesetzt, so ist natürlich zu fragen, ob die Genehmigungsvoraussetzungen im Sinne der Störfall-Verordnung geprüft werden müssen.

Anlage im Sinne der Störfall-Verordnung ist die genehmigungsbedürftige Anlage im Sinne des § 4 BImSchG in Verbindung mit der Verordnung über genehmigungsbedürftige Anlagen. Für die Anwendung der Störfall-Verordnung genügt es, wenn Stoffe nach den Anhängen II, III oder IV zur 12. Verordnung in bestimmungsgemäßen Betrieb vorhanden sein oder bei einer Störung des bestimmungsgemäßen Betriebes entstehen können.

Die Verordnung gilt nicht, soweit Stoffe nur in so geringen Mengen vorhanden sein oder entstehen können, daß der Eintritt eines Störfalls offensichtlich ausgeschlossen ist.

at all measuring surfaces in the assessment area, which is specified according to the stack height.

When assessing pollutants for which no prescribed immission values are set the anticipated immission levels must be related to scientifically recognized assessment standards. The decision as to whether these immission levels are acceptable and what protective measures, if any, must be taken will depend on the possible consequences of the environmental effects.

5. Carcinogens

When residual materials are used particular attention must be paid to carcinogens in the licensing procedure and during monitoring. The basic principle which is always applied to carcinogens as a requirement for issuing a licence states that emissions are to be restricted as far as possible in accordance with the principle of proportionality.

The directive for minimization in No. 2.3 of the German Clean Air Regulations (Restricting emissions as far as possible) was intentionally not linked to the otherwise general requirement for state-of-the-art emission limitation. This is due to the specific nature of the effect of the materials and the dual function of the regulations which restrict emissions. The requirement "as far as possible" emphasizes that for carcinogens the alternatives available for reducing and preventing emissions must be utilized to the full. The general rule concerning the obligation to restrict emissions is to do all that is technically possible within the limits of practical common sense.

Given the health risk posed by carcinogens, emissions are to be avoided (if possible) or minimized by process engineering means and flue gas cleaning systems. Also arising from the directive on minimization is the specific undertaking to improve the level of development and application of emission reduction technology for carcinogens. In addition to maintaining low mass concentrations it is also important to minimize the mass flows.

6. Application of the Regulation on Industrial Incidents (12th BImSchV)

If residual materials are used in plants in the cement industry the question naturally arises as to whether the licensing requirements are subject to compliance with the Regulation on Industrial Incidents.

A plant – for the purposes of the Regulation on Industrial Incidents – is one which requires a licence to operate under the terms of §4 BImSchG (Federal Immission Protection Act) in conjunction with the Regulation relating to plants requiring a licence to operate. The Regulation on Industrial Incidents applies where materials, as specified in Appendix II, III or IV of the 12th Regulation, are present during designated operation, or where they can be produced in the event of a fault in the designated operation.

This regulation shall not apply where materials are only present or can only be produced in such small quantities that the occurrence of an incident is clearly ruled out.

7. Application of the Regulation relating to incineration plants for waste materials and similar combustible substances (17th BImSchV)

The 17th Regulation contained in the Federal Immission Protection Act (17th BImSchG) also applies for assessing the licensing requirements when using residual materials. This contains comprehensive requirements for state-of-the-art technology and supplements the Directive on Minimization in conjunction with No. 3.1.7 of the German Clean Air Regulations (e.g. polyhalogenated dibenzodioxins, polyhalogenated dibenzofurans or polyhalogenated biphenyls) and No. 2.3 of the German Clean Air Regulations (carcinogens), providing that state-of-the-art technology does not change substantially.

7. Anwendung der Verordnung über Verbrennungsanlagen für Abfälle und ähnliche brennbare Stoffe – 17. BImSchV –

Zur Beurteilung der genehmigungsrechtlichen Voraussetzungen beim Einsatz von Reststoffen ist auch die 17. Verordnung zum Bundes-Immissionsschutzgesetz (17. BImSch) heranzuziehen. Sie enthält umfassende Anforderungen zum Stand der Technik und füllt beispielsweise das Minimierungsgebot im Hinblick auf Nr. 3.1.7 TA-Luft (z. B. polyhalogenierte Dibenzodioxine, polyhalogenierte Dibenzofurane oder polyhalogenierte Biphenyle) und Nr. 2.3 TA-Luft (krebserzeugende Stoffe) aus, solange sich der Stand der Technik nicht wesentlich ändert.

Bei der Verwertung von Sekundärbrennstoffen sind die Emissionsgrenzwerte der 17. BImSchV auf den Teil des Abgasstromes anzuwenden, der bei der Verbrennung des höchstzulässigen Anteils der Reststoffe und des für die Verbrennung von Abfällen zusätzlich benötigten Brennstoffes entsteht. Für den übrigen Teil des Abgasstromes gelten die hierfür verbindlichen Emissionsbegrenzungen der Genehmigungen oder der bestehenden Vorschriften (z. B. TA Luft). Die Gesamtbegrenzung ist daher im Einzelfall von der Genehmigungsbehörde zu ermitteln und im Genehmigungsbescheid unter Berücksichtigung der Ausnahmevorschriften festzulegen.

When secondary fuels are used the emission limits contained in the 17th BImSchV shall apply to the part of the exhaust gas flow generated during the combustion of the maximum admissible proportion of residual materials and of the additional fuel required for combustion of the wastes. The binding emission limits contained in the licence to operate or other existing regulations (e. g. German Clean Air Regulations) shall apply to the remainder of the exhaust gas flow. The overall restriction therefore has to be determined by the licensing authorities for the individual case and be specified in the licence to operate, taking any exemptive provisions into account.

Abbau mit dem Hydraulikbagger als Alternative zur Gewinnungssprengung

Quarrying with a hydraulic excavator as an alternative to primary blasting

Extraction avec pelle hydraulique comme alternative au tir d'abattage

Extracción por medio de la excavadora hidráulica, como alternativa de la explotación por voladura

Von **W. Boßmann**, Beckum/Deutschland

Zusammenfassung – Der Beckumer Kalkmergel, mit einer durchschnittlichen Mächtigkeit von 15 m, wird bis in die unmittelbare Nähe der Wohnbebauung abgebaut. Die gebotenen Sicherheitsabstände aufgrund von Umweltanforderungen führten bei der Durchführung der Sprengarbeiten zu hohen Abbauverlusten in Form von nicht abgebautem steinführenden Areal. Erste Abbauversuche mit einem Hydraulikbagger der unteren Gewichtsklasse (ca. 50 t) zu Beginn der 70er Jahre zeigten eine Alternative zur Gewinnungssprengung auf. In Zusammenarbeit mit den Herstellern wurden besonders lärmgedämpfte Bagger bis in die Gewichtsklasse ≥ 150 t entwickelt. Der Kostenvergleich zeigt, daß sich der bankig abgelagerte Beckumer Kalkmergel wirtschaftlich und umweltfreundlich mittels Hydraulikbagger gewinnen läßt.

Summary – The Beckum lime marl with an average thickness of 15 m is being quarried right up to the immediate vicinity of residential buildings. The safety distances required by environmental demands when blasting led to high quarrying losses in the form of unquarried areas containing rock. Initial quarrying trials with a hydraulic excavator of the lower weight class (approx. 50 t) at the start of the 70s pointed to an alternative to primary blasting. Special low-noise excavators were developed up to the weight class ≥ 150 t in co-operation with the manufacturers. Cost comparisons show that the Beckum lime marl deposited in seams can be quarried economically by hydraulic excavators without causing environmental problems.

Résumé – La marne calcaire de Beckum, d'une épaisseur de 15 m, est extraite jusqu'au voisinage immédiat des constructions d'habitation. Les distances de sécurité imposées par le respect de l'environnement avaient conduit, lors de l'exécution des tirs, à des pertes d'extraction importantes sous forme de zones de gisement non extractables. Les premiers essais avec une pelle hydraulique de classe de masse faible (env. 50 t), au début des années 70, ont révélé une alternative possible au tir d'abattage. En collaboration avec les constructeurs ont été développées des pelles particulièrement isolées côté bruit, allant jusqu'à la classe de ≥ 150 t de masse. La comparaison des coûts a montré, qu'il est possible d'extraire avec une pelle hydraulique, de façon rentable et en respectant l'environnement, la marne calcaire de Beckum disposée en bancs.

Resumen – La extracción de la marga calcárea de Beckum, que se presenta con un espesor medio de 15 m, se efectúa muy cerca de las zonas residenciales. Las distancias de seguridad a guardar, de acuerdo con las exigencias medioambientales, han traído consigo importantes pérdidas durante los trabajos de voladura, puesto que algunas áreas explotables no han podido ser aprovechadas. Los primeros intentos de extracción por medio de una excavadora hidráulica ligera (unas 50 t), a principios de los años 70, han mostrado que existe una alternativa de la explotación por voladura. En colaboración con los fabricantes, se han desarrollado excavadoras especialmente silenciosas, con un peso de hasta ≥ 150 t. De una comparación de costes resulta que es posibles explotar las bancadas de marga calcárea de Beckum, de forma económica y sin afectar al medio ambiente, mediante una excavadora hidráulica.

Der Beckumer Kalkmergel stammt aus der oberen Kreide und ist ca. 70 Mio Jahre alt. Er ist in horizontalen Schichten abgelagert und hat eine durchschnittliche Mächtigkeit von 20 m. Auf dem **Bild 1** sind die einzelnen Steinbänke, die im Kalkgehalt unterschiedlich sind, zu erkennen. Im oberen Bereich der Lagerstätte sind die Kalkmergelbänke eher geringmächtig und kurzbrüchig, im unteren Bereich sind die Bänke mächtiger und auch härter.

Bis in die Mitte der siebziger Jahre wurde das Rohmaterial durch Großbohrlochsprengung gewonnen und mittels

The Beckum lime marl comes from the Upper Chalk and is approximately 70 million years old. It is deposited in horizontal beds and has an average thickness of 20 m. The individual rock strata of varying lime content can be seen in **Fig. 1**. The lime marl strata in the upper part of the deposit tend to be shallower and brittle, while in the lower part the strata are thicker and also harder.

Until the mid 70s the raw material was quarried by large-diameter hole blasting and loaded onto trucks with rope shovels. However, this method of quarrying restricted the

Hochlöffel-Seilbagger auf Lastkraftwagen verladen. Diese Abbaumethode begrenzte jedoch die Nutzung der zur Verfügung stehenden Abbauflächen. Randgebiete, die direkt an die Wohnbebauung angrenzen, konnten wegen der Gefahr von Sprengschäden, Gebäudeerschütterungen und auch wegen der Lärmbelästigung nicht abgebaut werden. Selbst bei äußerster Optimierung von Bohrlochdurchmesser, Seitenabstand, Vorgabe, Sprengstoffmenge je Bohrloch, Zündfolge und Wandhöhe traten Sprengerschütterungen auf, die in der Nachbarschaft zu Beschwerden führten.

Im Genehmigungsbescheid war eine maximale Schwinggeschwindigkeit von 4 mm/s festgelegt worden. Um den ständigen Beschwerden der Anlieger entgegentreten zu können, wurden von 1968 bis 1975 ca. 2000 Schwinggeschwindigkeitsmessungen durchgeführt. Mit den Messungen, den Auswertungen und der Dokumentation war ein Mitarbeiter voll beschäftigt. Um zukünftig Abbauverluste und Nachbarschaftsbelästigungen zu vermeiden, wurde zu Beginn der 70er Jahre über eine neue Abbaumethode nachgedacht.

Bei Erdbaustellen waren zu dieser Zeit schon leichtere Hydraulikbagger im Einsatz. In einigen Steinbrüchen wurden Reißraupen eingesetzt, die große Abbauflächen benötigten. Die Reißraupen verursachten großen Lärm. Der Abbau in unmittelbarer Nachbarschaft der Wohnbebauung ließ dieses Verfahren nicht zu.

Nach intensiven Untersuchungen und Einsatzbesichtigungen anderer Abbaumethoden wurde 1972 der erste Hydraulikbagger mit einem Dienstgewicht von 56 t, einer 2 m³-Klappschaufel und 210 PS Antriebsleistung eingesetzt. Da die Abbauwand wegen der Minimierung der Sprengerschütterungen in zwei Sohlen unterteilt war, wurde der Hydraulikbagger zunächst auf der oberen Sohle eingesetzt. Die Wandhöhe betrug max. 5 m. Schon bald beherrschten die Baggerführer die richtige Lösetechnik. Mit der Schaufel werden die einzelnen Schichtenpakete von oben nach unten abgegraben. Die obere Sohle mit den kurzbrüchigen und weniger harten Bänken bereitete dem Bagger keine Schwierigkeit.

Mit der 2 m³-Schaufel (**Bild 2**) füllte der Hydraulikbagger einen 25 t fassenden LKW mit 7 bis 8 Förderspielen in etwa 8 Minuten. Insgesamt wurde eine Löse- und Ladeleistung von 185 t/h erreicht. Mit dem Hydraulikbagger ist es möglich, einzelne Bänke abzugraben. So wurde schon in den ersten Anfängen des Hydraulikbaggerbetriebes eine im Kalkgehalt sehr niedrige, 1 m mächtige Schicht separat abgebaut und als Abraum verworfen. Das Selektieren von einzelnen Kalkmergelschichten ist mittels Hydraulikbagger erst möglich geworden und wird seit der Zeit durchgeführt. Ein weiterer Vorteil des Hydraulikbaggers ist seine Wendigkeit. Da er im Gegensatz zu einem Seilbagger nicht mit einem Plattenlaufwerk, sondern mit einem Traktorenlaufwerk ausgerüstet ist, kann er schnell an eine andere Abbaustelle umgesetzt werden. Die ersten Hydraulikbagger waren damals sicherlich nicht für den harten Steinbrucheinsatz zum Lösen und Laden des Materials konzipiert.

Die Schwachpunkte waren:

Reißen der Hydraulikleitungsschweißnähte
Platzen der Hydraulikschläuche
Vorzeitiger Verschleiß der Hydraulikpumpe
Frühzeitiger Verschleiß des Dieselmotors
Starker Verschleiß an den Schaufelzahnhaltern
Zu hohe Hydrauliköltemperaturen

TABELLE 1: Kostenvergleich Hydraulikbagger/Seilbagger

	Hydraulikbagger 2 m³-Schaufel	Seilbagger 2,5 m³-Schaufel
Ladeleistung t/h	186	177
Treibstoffverbrauch l/h	21,3	21,6
Kosten ohne Abschreib. DM/t	0,45	0,53

Beim Seilbaggerbetrieb erhöhen sich die Gewinnungskosten um ca. 0,30 DM/t für Bohren und Sprengen.

BILD 1: Horizontal abgelagerter Beckumer Kalkmergel
FIGURE 1: Horizontally stratified Beckum lime marl

BILD 2: Abgraben der Kalkmergelbänke von oben nach unten
FIGURE 2: Digging the lime marl strata from top to bottom

use of the available quarrying areas. The boundary areas directly adjacent to residential building could not be quarried because of the danger of blasting damage and building vibration, as well as the noise nuisance. Blasting vibration, which led to complaints in the neighbourhood, occurred even with the best possible drill hole diameter, lateral spacing, length of shot, quantity of explosive per drill hole, ignition sequence and bench height.

A maximum oscillation speed of 4 mm/s had been stipulated in the licensing approval. Approximately 2000 oscillation speed measurements were carried out from 1968 to 1975 in order to counter the continuous complaints from the neighbours. One member of staff was fully occupied with the measurements, assessments and documentation. At the start of the 70s a new method of quarrying was considered to avoid future quarry losses and annoyance to the neighbourhood.

Fairly light-weight hydraulic excavators were already in use by then on earthmoving sites. Crawler-mounted rippers, which needed large working areas, were used in some quarries. However, they made a great deal of noise, so this method was ruled out for quarrying in the immediate vicinity of residential buildings.

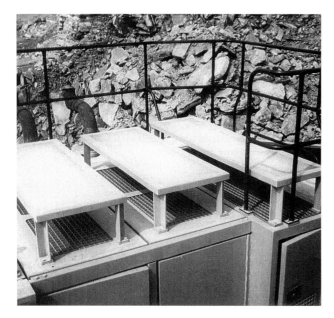

BILD 3: Schalldämmung des Motorraumes
FIGURE 3: Noise suppression in the engine compartment

BILD 4: Hydraulikbagger mit 140 t Gewicht
FIGURE 4: 140 t hydraulic excavator

Beim Einsatz des Gerätes auf der unteren Sohle erwies es sich als zu schwach, die Standfestigkeit war aufgrund des niedrigen Baggergewichtes zu gering. Der Bagger rutschte beim Lösen des Materials von der Wand weg.

Die dreijährigen Betriebserfahrungen mit Hydraulikbagger führten dazu, im Jahre 1975 einen zweiten, jedoch schwereren Hydraulikbagger einzusetzen.

Gewicht:	75 t
Schaufelinhalt:	2,7 m³
Löse- und Ladeleistung:	275 t/h
Antriebsleistung:	320 PS

Vom Hersteller wurde verlangt, daß der Bagger besonders schallgedämpft sein müsse. Durch Kapselung des Motorenraumes, des Ansaugkanals und der Auspuffanlage mit hochwertigen Dämmaterialien konnte der Schallpegel in 7,5 m Abstand vom Gerät von 89 auf 75 dB (A) und in der Kabine von 85 auf 73 dB (A) gesenkt werden (**Bild 3**). Um den Wärmehaushalt des Gerätes durch diese Sondermaßnahmen nicht zu beeinträchtigen, wurde der Kühlkreislauf für das Hydrauliköl vergrößert. Neben dem erschütterungsfreien Abbau ist die zusätzliche Lärmdämmung ein weiterer Schritt hin zu einer verbesserten Umweltverträglichkeit.

The first hydraulic excavator, with an operating weight of 56 t, a 2 m³ clam bucket and 210 hp drive rating, was brought into use in 1972 after intensive investigations and site visits to other methods of quarrying. As the working face was divided into two levels to minimize blast vibration the hydraulic excavator was at first used on the upper level. The maximum face height was 5 m. The excavator drivers very soon mastered the correct digging technique. The individual sedimentary complexes were dug from top to bottom with the bucket. The upper level with the brittle and less hard strata caused the excavator no problems.

With the 2 m³ bucket (**Fig. 2**) the hydraulic excavator filled a 25 t truck in about 8 minutes with 7 or 8 loading cycles. An overall digging and loading rate of 185 t/h was achieved. With the hydraulic excavator it is possible to dig individual strata. At the very start of operation with the hydraulic excavator a 1 m thick bed with a very low lime content was dug separately and discarded as overburden. For the first time it became possible with the hydraulic excavator to select individual lime marl strata, and this has been carried on ever since. Another advantage of the hydraulic excavator is its manoeuvrability. Unlike a rope shovel it is mounted on a tractor unit rather than a crawler unit so it can be shifted rapidly to another digging position. At that time the first hydraulic excavators were certainly not designed for tough quarry usage for digging and loading the material.

The weak points were:

Splitting of the weld seams in the hydraulic lines
Bursting of the hydraulic hoses
Premature wear of the hydraulic pump
Early wear of the diesel motor
Severe wear of the bucket tooth sockets
Excessive hydraulic oil temperatures

TABLE 1: Cost comparison for hydraulic excavators and rope shovels

	Hydraulic excavator 2 m³ bucket	Rope shovel 2,5 m³ bucket
Loading capacity t/h	186	177
Fuel consumption l/h	21.3	21.6
Costs without depreciation DM/t	0.45	0.53

For operation with a rope shovel the quarrying costs are increased by approximately 0.30 DM/t for drilling and blasting.

When the machine was used on the lower level it proved to be too weak, and it was not stable enough because of the low weight of the excavator. The excavator slid away from the face when digging material.

The three years of operation experience with a hydraulic excavator led in 1975 to the use of a second, but heavier, hydraulic excavator.

Weight:	75 t
Bucket capacity:	2.7 m³
Digging and loading rate:	275 t/h
Drive rating:	320 hp

The manufacturer was requested to make the excavator with a particularly effective noise suppression system. Enclosure of the engine compartment, intake duct and exhaust systems with high-grade insulating materials reduced the sound level at a distance of 7.5 m from the machine from 89 to 75 dB (A), and in the cab from 85 to 73 dB (A) (**Fig. 3**). The cooling circuit for the hydraulic oil was enlarged to prevent these special measures from having a detrimental affect on the machine's heat balance. In addition to the vibration-free excavation, the extra noise suppression is another step towards environmental compatibility.

Auf Dauer erwies sich auch das 75-t-Gerät als zu schwach. In den Jahren 1988 und 1992 wurde je ein 140-t-Gerät mit einer 6,5 m^3-Schaufel und einer Antriebsleistung von 782 PS angeschafft (**Bild 4**). Die Löse- und Ladeleistung beträgt ca. 400 t/h. Als Hydrauliköl ist Rapsöl „Plantohyd 40" im Einsatz. Die Betriebskosten dieser Geräte betrugen im Jahre 1992 0,53 DM/t.

Zusammenfassend kann nach 20jährigem Hydraulikbaggerbetrieb gesagt werden: Der Löse- und Ladebetrieb mit den Hydraulikbaggern ist da, wo er aufgrund der Gesteinsart möglich ist, kostengünstiger als der Bohr-, Spreng- und Ladebetrieb. Darüberhinaus ist er deutlich umweltfreundlicher.

In the long term the 75 t machine also proved to be too weak. 140 t machines with 6.5 m^3 buckets and 782 hp drive ratings were purchased in 1988 and 1992 (**Fig. 4**) with digging and loading capacities of approximately 400 t/h. „Plantohyd 40" rape seed oil is used as the hydraulic oil. In 1992 the operating costs of these machines amounted to 0.53 DM/t.

To summarize after 20 years of hydraulic excavator operation it can be said that: Wherever the type of rock makes it possible, it is less expensive to dig and load with hydraulic excavators than to drill, blast and load. It is also significantly more environmentally friendly.

Maßnahmen zum optimierten Betrieb von Elektrofiltern

Measures for optimized operation of electrostatic precipitators

Règles pour l'exploitation optimale des électrofiltres

Medidas para optimizar el servicio con filtros eléctricos

Von **H. Braig**, Mergelstetten, und **B. Kirchartz**, Düsseldorf/Deutschland

Zusammenfassung – Zur Entstaubung der Abgase von Drehrohröfen mit Zyklon- oder Rostvorwärmer haben sich Elektrofilter bewährt. Für den stationären Dauerbetrieb werden die Anforderungen an die Entstaubungsleistung bzw. den Reingasstaubgehalt erfüllt. Die strengen Auswertevorschriften der TA Luft erfordern jedoch auch Maßnahmen zum optimierten Betrieb von Elektrofiltern bei instationären Betriebszuständen, wie z. B. im Umschaltbetrieb. Besondere Bedeutung kommt dabei einer angepaßten Regelung des Verdampfungskühlers sowie der Spannungsversorgung zu. Aus sicherheitstechnischen Gründen müssen Elektrofilter bei zu hohem CO-Gehalt im Abgas abgeschaltet werden. Die Praxis zeigt, daß die Zahl dieser Abschaltungen reduziert werden kann, wenn bei Auftreten einer erhöhten CO-Bildung abgestufte Eingriffe in die Brennstoffdosierung vorgenommen werden. Voraussetzung hierfür ist im wesentlichen eine geeignete Auslegung, Regelung und Überwachung der Brennstoffdosierung sowie eine schnelle CO-Meßtechnik.

Summary – Electrostatic precipitators have proved successful for dedusting the exhaust gases from rotary tube kilns with cyclone or grate preheaters. The requirements for dedusting performance and clean gas dust content are fulfilled for steady-state continuous operation. However, the strict evaluation instructions of the German Clean Air Regulations also require measures for optimized operation of electrostatic precipitators in unsteady-state operating conditions, such as changeover operation. Particular importance is given in this case to suitable control of the evaporative cooler and of the voltage supply. For safety reasons electrostatic precipitators have to be switched off if the CO content in the exhaust gas is too high. Practical experience shows that the number of these disconnections can be reduced if stepped interventions are made in the fuel metering system on the appearance of increased CO. The prerequisite for this is essentially a suitably designed, controlled and monitored fuel metering system and a rapid method of measuring the CO.

Résumé – Les électrofiltres ont fait leur preuve dans le dépoussiérage des gaz d'exhaure de fours rotatifs à préchauffeur à grille ou à cyclones. Les exigences concernant la capacité de dépoussiérage et la teneur en poussière des gaz épurés sont satisfaites en service continu régulier. Les préscriptions d'exploitation strictes de la TA Luft exigent néanmoins des mesures prises pour assurer le fonctionnement optimal des électrofiltres dans des conditions d'exploitation instables, comme p. ex. la marche alternée. D'une importance particulière y est une régulation adaptée de la tour de conditionnement et de l'alimentation haute tension. Pour des raisons de sécurité, les électrofiltres doivent être débranchés lors d'une teneur CO trop forte des gaz de sortie. La pratique montre, que le nombre de ces débranchements peut être réduit quand, lors de l'apparition d'une formation CO accrue, sont effectuées des interventions graduées dans le dosage du combustible. Pour cela, les conditions préalables sont essentiellement une conception, une régulation et une surveillance adaptées du dosage du combustible et une technique rapide de mesure CO.

Resumen – Para el desempolvado de los gases de escape de los hornos rotatorios con precalentadores de ciclones o de parrilla, han probado su eficacia los electrofiltros. En cuanto al servicio continuo, estacionario, se cumplen las exigencias respecto de la capacidad de desempolvado y del contenido de polvo en los gases depurados. Sin embargo, los requisitos bastante rigorosos de la TA Luft (Instrucción técnica para el mantenimiento de la limpieza del aire) exigen, además, unas medidas destinadas a la optimización del servicio bajo condiciones no estacionarias, por ejemplo durante los cambios de servicio. A este respecto, tiene especial importancia la regulación adecuada del refrigerador por evaporación y el suministro de tensión. Por razones de seguridad, los electrofiltros tienen que ser desconectados, cuando el contenido de CO en los gases de escape llega a ser demasiado alto. Se ha podido comprobar en la práctica que es posible reducir el número de desconexiones, mediante intervenciones escalonadas en la dosificación del combustible, siempre y cuando se presente un exceso de CO. Para ello se requiere, esencialmente, un dimensionamiento apropiado, una regulación y control de la dosificación de combustibles y un método de medición rápida del CO.

Durch die Herabsetzung des Emissionswertes für Gesamtstaub auf 50 mg/m$^3_{N.tr.}$ sind die Anforderungen an die Abscheideleistung von Elektrofiltern an Drehrohröfen der Zementindustrie weiter gestiegen. Zu beachten ist dabei, daß die Einhaltung dieses Wertes unter Berücksichtigung der strengen Auswertevorschrift der TA-Luft erfahrungsgemäß ein Vorhaltemaß im stationären Betrieb von mindestens 20 mg/m^3 erfordert. Zudem müssen die Staubemissionen in instationären Betriebszuständen minimiert werden.

Die Abscheideleistung eines Elektrofilters kann bei vorgegebener Auslegung maßgeblich von wechselnden Eigenschaften der abzuscheidenden Stäube sowie den Abgasrandbedingungen abhängen. In erster Linie ist dabei der spezifische elektrische Widerstand der sich auf der Niederschlagselektrode bildenden Staubschicht, die Temperatur, der Volumenstrom sowie der Wassertaupunkt des Abgases von Bedeutung.

Um einen optimalen Elektrofilterbetrieb auch in verschiedenen Betriebszuständen sicherzustellen, müssen geeignete Maßnahmen ergriffen werden, wie z. B. durch

— konstruktive Ausführung

— regelungstechnische Optimierung der Hochspannungsversorgung mittels Microprozessoren

— Verwendung von Impulsgeneratoren

— geregelte Abgaskonditionierung.

Während die konstruktive Ausführung und die elektrische Spannungsregelung im Normalbetrieb eine hohe Abscheideleistung sicherstellen, ist dies bei instationären Betriebszuständen nicht immer befriedigend möglich. Probleme ergeben sich vorwiegend beim An- und Abfahren, beim Wechsel zwischen verschiedenen Betriebszuständen sowie bei Betriebsstörungen.

Aufgrund der Häufigkeit ihres Auftretens ist dem Umschalten zwischen dem Betrieb mit und ohne Abgasverwertung besondere Aufmerksamkeit zu widmen. Hierbei muß die mit der Trocknung des Rohmaterials verbundene Aufnahme von Wasserdampf bei Betrieb ohne Abgasverwertung über eine entsprechende Dosierung von Wasser in einem Verdampfungskühler ausgeglichen werden. In einigen Fällen gestaltet sich diese Maßnahme als schwierig, da sie nicht ausreichend schnell auf die geänderten Betriebsverhältnisse abgestimmt werden kann.

Die mit dieser Abgaskonditionierung verbundene regelungstechnische Aufgabe geht aus dem folgenden **Bild 1** hervor. Maßgebende Regelgröße ist die Gastemperatur nach Verdampfungskühler, deren Abweichung vom vorgegebenen Sollwert über die Stellgröße eine Änderung der Ventilstellung und somit der eingedüsten Wassermenge erzwingt. Um die dynamische Stabilität des Regelkreises aufrechtzuerhalten, arbeitet dieser Regelkreis im stationären Fall vergleichsweise träge. Kurzfristige Änderungen des Lastzustandes, wie sie bei Umschaltzuständen auftreten, erfordern jedoch ein schnelleres Regelungsverhalten. Als vorteilhaft hat es sich daher erwiesen, bei kurzfristigen Änderungen des Lastzustands die vorhandene Regelung für die Abgaskonditionierung zu umgehen und während der Umschaltphase durch eine Zusatzsteuerung zu ersetzen.

Wie aus **Bild 2** hervorgeht, wird hierzu der Ausgang des Temperaturreglers stillgelegt und die Rücklaufventile der Bedüsungseinrichtung mit Hilfe einer Zusatzsteuerung in eine definierte Stellung gebracht. Für jede mögliche Umschaltphase kann für die Zusatzsteuerung ein vorab ermittelter optimaler Wert für die einzudüsende Wassermenge vorgegeben werden (Expertensystem). Somit kann das ungünstige Verhalten des Regelkreises bei kurzfristigen Änderungen des Lastzustandes umgangen werden. Nach Beendigung der Umschaltphase übernimmt der Regler wieder die Verstellung der Regelventile. Steht zusätzlich eine zuverlässige und schnelle Volumenstrommessung zur Verfügung, kann die Einstellung der Konditionierung wieder vom Regler übernommen werden, wenn das Meßergebnis

With the reduction in permitted total dust emission to 50 mg/m$^3_{stp,dry}$ still higher demands are made on the precipitating performance of electrostatic precipitators on rotary tube kilns in the cement industry. To reach this level an allowance of at least 20 mg/m^3 is required in steady-state operation in view of the strict evaluation instructions of the German Clean Air Regulations. Dust emissions also have to be minimised in unsteady-state operating conditions.

The precipitating performance of a given design of electrostatic precipitator may depend essentially on changing properties of the dusts and marginal exhaust gas conditions. The specific electrical resistance of the dust layer forming on the collecting electrode, the temperature, the volume flow and the water dew point of the exhaust gas are of primary importance.

To ensure optimum electrostatic precipitator operation in different operative states suitable measures must be taken such as

— appropriate design

— optimum automatic control of the high voltage supply by microprocessors

— use of pulse generators

— controlled exhaust gas conditioning.

Although appropriate design and voltage control ensure good precipitation during normal operation, performance is not always satisfactory in unsteady-state operating conditions. Problems arise chiefly during start-up and shut-down, changes between different operating states and breakdowns in production.

Change-over between operation with and without exhaust gas utilisation merits special attention owing to the frequency of its occurrence. The take-up of water vapour associated with the drying of the raw material has to be compensated for during operation without exhaust gas utilisation by corresponding metered addition of water in an evaporative cooler. In some cases this measure proves difficult as it cannot be adapted fast enough to the changed operating conditions.

The control task involved in this waste gas conditioning is indicated in **Fig. 1**. The decisive controlled variable is the gas temperature after the evaporative cooler; its deviation from the predetermined desired value caused by the actuating variable brings about a change in the valve position and thus of the quantity of water injected. To maintain the dynamic stability of the control loop, it has comparatively sluggish operating characteristics in the steady state. However temporary changes in load condition such as occur in changeover states require a quicker control action. It has therefore been found advantageous to bypass the existing control for exhaust gas conditioning in the event of temporary changes in load condition and to replace it with an auxiliary control during the changeover phase.

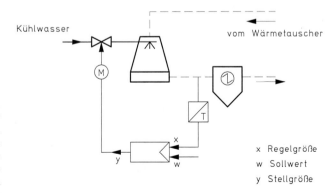

BILD 1: Verdampfungskühler- Regelkreis (Normalbetrieb)
FIGURE 1 : Evaporative cooler control circuit (normal operation)

Kühlwasser	= cooling water
vonWärmetauscher	= from heat exchanger
Regelgröße	= controlled variable
Sollwert	= desired value
Stellgröße	= actuating variable

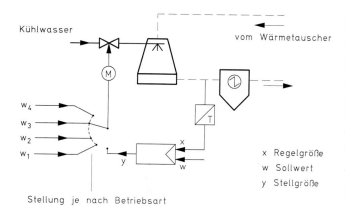

BILD 2: Zusatzsteuerung (Umschaltbetrieb)
FIGURE 2 : Auxiliary control (changeover)

Kühlwasser	= cooling water
von Wärmetauscher	= from heat exchanger
Regelgröße	= controlled variable
Sollwert	= desired value
Stellgröße	= actuating variable
Stellung je nach Betriebsart	= position according to type of operation

für den Volumenstrom als Störgröße im Regler verarbeitet wird. Erste Anwendungen werden derzeit im Dauerbetrieb erprobt.

Eine erhebliche und deutlich sichtbare Staubemission tritt bei einer sicherheitsbedingten Abschaltung des Elektrofilters hinter einem Drehrohrofen infolge erhöhter CO-Bildung auf. Einige wesentliche Ursachen für die Entstehung von CO sind:

— Störungen des Gasdurchsatzes
— Einstürzen von Ansatzringen
— Schäden an der Brennerdüse
— Mängel bei der Brennstoffdosierung
— Handhabungsfehler.

Zur Vermeidung der sicherheitsbedingten Abschaltung von Elektrofiltern haben die Betreiber von Drehrohröfen der Zementindustrie in der Vergangenheit zusätzliche Anstrengungen unternommen. Die ungleichmäßige Zufuhr von Brennstoffen durch Handhabungsfehler oder Störungen an den Dosiersystemen als Ursache für erhöhte CO-Konzentrationen kann beispielsweise über eine entsprechende Sicherheitskette beim Dosiervorgang weitgehend vermieden werden.

Bewährt haben sich ferner gezielte Eingriffe in den Prozeß, um das Auftreten von CO-Spitzen zu unterbinden. Voraussetzung ist eine redundante Auslegung der CO-Überwachung, wie sie beispielhaft aus dem folgenden **Bild 3** hervorgeht. In nachfolgender **Tabelle 1** sind die abgestuften Maßnahmen aufgeführt, die in Abhängigkeit vom Meßort sowie vom zugehörigen CO-Grenzwert eingeleitet werden, um CO-Spitzen zu vermeiden. Im wesentlichen handelt es sich hierbei um präventive Eingriffe in die Brennstoffdosierung, ohne daß der Prozeß zunächst vollständig abgeschaltet werden muß.

As shown in **Fig. 2**, the output from the temperature controller is shut down for this purpose and the return valves of the injector moved to a defined position by an auxiliary control. A previously determined optimum value for the quantity of water to be injected during any possible changeover phase may be stipulated for the auxiliary control (expert system). In this way the unfavourable action of the control loop during temporary changes of load condition can be avoided. When the changeover phase is completed the main control takes over adjustment of the control valves again. If quick, reliable volume flow measurement is also available the main control can take over adjustment of the conditioning again if the volume flow reading is processed in the controller as a disturbance variable. Initial applications are at present being tried in continuous operation.

There is considerable, clearly visible dust emission when an electrostatic precipitator after a rotary tube kiln is switched off for safety reasons owing to increased CO formation. Some of the main causes of the appearance of the CO are:

— disturbances in gas throughput
— collapse of clinker rings
— damage to the firing nozzle
— faults in the fuel metering system
— incorrect operation.

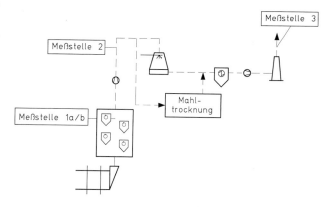

BILD 3: Meßorte für die CO-Überwachung
FIGURE 3 : Measuring points for CO monitoring

| Meßstelle | = measuring point |
| Mahltrocknung | = drying and grinding |

In the past, operators of rotary tube kilns in the cement industry have made additional efforts to avoid switching off electrostatic precipitators for safety reasons. Fluctuations in fuel supply due to incorrect operation or malfunctioning metering systems, causing raised CO concentrations, may largely be avoided e. g. by introducing an appropriate safety chain in the metering process. Specific interventions in the process to avoid the appearance of CO peaks have also been successful. This requires redundant design of the CO monitoring system as shown e. g. in **Fig. 3** below. The stepped measures which are taken to avoid CO peaks, dependent on the measuring point and the associated CO limit, are

TABELLE 1: Maßnahmen bei CO-Überschreitung

Meßstelle	MS[1]	Maßnahme
1a, b	1	Warnung, Zweitfeuerung stop
	2	Hauptfeuerung stop
	3	EGR-Abschaltung
2	1	Warnung, Zweitfeuerung stop
	2	Hauptfeuerung stop
	3	EGR-Abschaltung
3	—	Meßwertkontrolle

[1]) Maßnahmenstufe

TABLE 1: Measures when CO level exceeded

Measuring point	Action step	Action taken
1a, b	1	Warning, stop secondary firing system
	2	Stop main firing system
	3	Switch off electrostatic precipitation
2	1	Warning, stop secondary firing system
	2	Stop main firing system
	3	Switch off electrostatic precipitation
3	—	Measured value check

Beispielhaft seien die CO-Schwellenwerte für eine 4-stufige Zyklonvorwärmeranlage genannt, bei der 90 % des Wärmebedarfs über den Hauptbrenner und 10 % über die Vorcalcinierung bereitgestellt werden.

— Meßstelle 1 a,b 1 : 0,5 Vol-% CO
　　　　　　　　　　Abschaltung Vorcalcinierung
　　　　　　　　　　2 : 2,5 Vol.-% CO;
　　　　　　　　　　Abschaltung Hauptbrenner
　　　　　　　　　　3 : 3,5 Vol.-% CO
　　　　　　　　　　Abschaltung Filter
— Meßstelle 2　　　1 : 0,5 Vol-% CO
　　　　　　　　　　Abschaltung Vorcalcinierung
　　　　　　　　　　2 : 2,5 Vol.-% CO;
　　　　　　　　　　Abschaltung Hauptbrenner
　　　　　　　　　　3 : 2,6 Vol.-% CO
　　　　　　　　　　Abschaltung Filter
— Meßstelle 3　　　keine Maßnahmen
　　　　　　　　　　Meßwertkontrolle

Diese Werte gelten nur für den zitierten Einzelfall. An anderen Anlagen können abhängig von der Verfahrensführung, dem individuellen Sicherheitsbedürfnis sowie der Verzögerung der Meßwertanzeige andere Werte notwendig sein.

Bei Berücksichtigung der in jedem Einzelfall vorliegenden meß- und anlagentechnischen Randbedingungen lassen sich erfahrungsgemäß Dauer und Häufigkeit der sicherheitsbedingten Abschaltung von Elektrofiltern über abgestufte Eingriffe in die Prozeßführung drastisch reduzieren.

listed in **Table 1**. They are essentially preventive interventions in the fuel metering system and do not initially involve stopping the process completely.

The example given in the table concerns the CO threshold values for a four-stage cyclone preheater plant where 90% of the heat requirement is provided by the main burner and 10% by pre-calcination.

— Measuring point 1 a,b 1: 0.5 vol.% CO
　　　　　　　　　　　　　Switch off pre-calcination
　　　　　　　　　　　　　2: 2.5 vol.% CO
　　　　　　　　　　　　　Switch off main burner
　　　　　　　　　　　　　3: 3.5 vol.% CO
　　　　　　　　　　　　　Switch off precipitator
— Measuring point 2　　　1: 0.5 vol.% CO
　　　　　　　　　　　　　Switch off pre-calcination
　　　　　　　　　　　　　2: 2.5 vol.% CO
　　　　　　　　　　　　　Switch off main burner
　　　　　　　　　　　　　3: 2.6 vol.% CO
　　　　　　　　　　　　　Switch off precipitator
— Measuring point 3　　　No action
　　　　　　　　　　　　　Measured value check

These values only apply to the case mentioned above. Different values may be needed at different plant, depending on the process, the individual safety requirement and the delay in displaying readings.

Experience shows that the duration and frequency of electrostatic precipitator disconnections for safety purposes can be drastically reduced by stepped interventions in the process, taking into account the marginal metering and plant engineering conditions in each individual case.

Das Rekultivierungsprogramm der Rüdersdorfer Zement GmbH

The recultivation programme at Rüdersdorfer Zement GmbH

Le programme de recultivation de la Rüdersdorfer Zement GmbH

El programa de restauración de paisajes de la Rüdersdorfer Zement GmbH

Von **S. Brandenfels,** Münster, **A. Koszinski** und **V. Rahn,** Rüdersdorf/Deutschland

Zusammenfassung – Bei der Rekultivierung von Tagebauen, die zur Kalksteingewinnung genutzt werden, kommt der Schaffung ökologisch hochwertiger, vielfältiger Areale unter Berücksichtigung der natürlichen Standortmerkmale besondere Bedeutung zu. Bedingt durch die Besonderheiten des Rüdersdorfer Kalksteinvorkommens bietet sich bereits heute die Möglichkeit, alle Randbereiche, vornehmlich flache Böschungen und sämtliche Außenhalden des Tagebaus zu renaturieren, da der weitere Abbau auf den tieferen Sohlen erfolgen wird und keine Inanspruchnahme zusätzlicher Landschaftsteile erforderlich ist. Geschaffen werden ökologisch ausgeglichene Landschaften, die auch zur Naherholung mit naturnahen Erlebnisbereichen genutzt werden sollen. Ökologische Besonderheiten sollen erhalten bleiben. Dabei werden gezielt angelegte Baumgruppen, Hecken und Alleen, Hangbepflanzungen sowie Wassergräben die Besiedlung der neu entstehenden Biotope begünstigen. Darin eingebunden ist der „Museumspark Rüdersdorf", in dem die Historie von Bergbau und Baustoffindustrie präsentiert wird. Ein Aussichtsturm sowie Ausblicke von markanten Punkten erhöhen die Attraktivität der Bergbaufolgelandschaft ebenso wie an die Morphologie (Tagebaurand, Haldenkonturen) angepaßte Wegeführungen mit Rad-, Wanderwegen und Lehrpfaden.

Summary – In the recultivation of quarries used for excavating limestone great importance is given to the creation of varied, ecologically sound, areas taking the natural features of the location into account. Because of the peculiarities of the Rüdersdorf limestone deposit all the edge regions, mainly shallow slopes and all the external banks of the quarry, can already be returned to nature. This is because future excavation will take place at lower levels and no additional areas of the landscape are needed. Ecologically balanced landscapes are being created which will also be used for providing readily accessible recreational facilities which are close to nature. Special ecological features are to be retained. Carefully placed groups of trees, hedges and avenues, planting of the slopes, and ditches will encourage colonization of the newly produced biotopes. Incorporated in this is the "Rüdersdorf Open-air Museum" which presents the history of mining and the building materials industry. The attractiveness of the landscape, which follows the lines of the quarry, is enhanced by an observation tower and views from prominent points, as well as by routes with bicycle paths, footpaths and nature trails adapted to the edge of the quarry and the contours of the slopes.

Résumé – La création de sites variés, d'une grande valeur écologique tout en respectant le caractère naturel du lieu, est d'une importance particulière lors de la recultivation de carrières d'extraction de calcaire. Suite aux particularités du gisement de calcaire de Rüdersdorf, s'offre déjà aujourd'hui la possibilité de réhabiliter toutes les zones périphériques, essentiellement des talus à pente douce et la totalité des coteaux extérieurs de la carrière, l'extraction ultérieure devant s'effectuer à des niveaux plus bas et ne nécessitant pas l'empiètement sur des parties supplémentaires du paysage. Des paysages équilibrés écologiquement ont été créés, utilisables aussi comme parcs de loisirs de proximité, avec des domaines d'activités proches de la nature. Les particularités écologiques devaient être conservées. Dans ce but, des groupes d'arbres, bosquets et sentiers, coteaux plantés et cours d'eau favoriseront la colonisation du biotope nouvellement formé. Y est inclus le „Parc-musée Rüdersdorf", présentant l'histoire de l'extraction et de l'industrie des matériaux de construction. Une tour d'observation et des points de vue des lieux privilégiés accroissent encore l'attrait du paysage qui remplace la carrière, comme le font aussi les dessins des sentiers avec des voies cyclistes, de randonnée et des parcours éducatifs respectant la morphologie (lisière de la carrière, contours des coteaux).

Resumen – En la restauración del paisaje de las áreas de explotación a cielo abierto de piedra caliza, adquiere especial importancia la creación de zonas variadas, de gran valor ecológico, teniendo en cuenta las características naturales del emplazamiento. Debido a las propiedades del yacimiento calizo de Rüdersdorf, existe desde ahora la posibilidad de rehabilitar todas las zonas marginales del mismo, sobre todo los taludes de poca pendiente y la totalidad de las pilas exteriores de la explotación a cielo abier-

to, ya que la explotación posterior se efectuará en las plataformas inferiores, sin que sea necesario utilizar otras áreas adicionales. Lo que se creará son paisajes ecológicamente equilibrados, que han de ser aprovechados para el recreo a poco distancia, en un entorno que permita el contacto con la naturaleza, conservando las particularidades ecológicas. A este respecto, se crearán sistemáticamente grupos de árboles, setos y avenidas, pendientes pobladas así como fosos de agua, favoreciendo de esta manera la colonización de los nuevos biótopos. Y en medio de todo ello estará ubicado el nuevo „Parque-museo de Rüdersdorf", en el que se presentará la historia de la minería y de la industria de materiales para la construcción. Una torre de observación así como las posibilidades de observación desde puntos destacados contribuirán a aumentar el atractivo del paisaje restaurado, lo mismo que los caminos adaptados a la morfología del entorno (zonas marginales del área de explotación a cielo abierto, contornos de las pilas), con recorridos para bicicletas y pedestrismo y senderos educativos.

1. Planungsraum

Rüdersdorf liegt 30 km östlich von Berlin. Der betrachtete Raum dient seit dem 17. Jh. der Gewinnung von Baustoffen. Zunächst wurde der Kalkstein in Blöcken gebrochen und verarbeitet, später folgte die Produktion von Branntkalken. Im Rahmen der heutigen großindustriellen Produktion werden verschiedenste Produkte auf der Basis von Kalk hergestellt. Mit der industriellen Entwicklung beschleunigte sich auch die Entwicklung der Abbaumethode und -geschwindigkeit. In immer kürzeren Intervallen wurden verschiedene Techniken eingesetzt und wieder verworfen. Diese technisch-industrielle Evolution mit ihren im Gelände manifestierten topographischen, architektonischen und landwirtschaftlichen Relikten prägt die heutige Situation.

2. Rekultivierung

Basis der Rekultivierungsplanung ist eine gründliche Bestandserhebung und Analyse der vorhandenen Arten und Lebensgemeinschaften der Pflanzen- und Tierwelt. Zu ihrer Erfassung gehört auch die Beschreibung der unterschiedlichen Standorte, wie z. B. warme, trockene Kalkhänge, feuchte Senken (in denen sich z. T. kleine Kalkmoore gebildet haben) und die sog. Ruderalstandorte. Weiterhin erstreckt sich die Bestandserhebung auf die durch die Industrie geschaffenen Bauwerke und technischen Einrichtungen, die man kurz als das „kulturelle Erbe" bezeichnen kann.

In der Wahrnehmung des Spannungsfeldes zwischen den durch die natürliche Sukzession geprägten Abbaubereichen und den durch die menschliche Tätigkeit beeinflußten Entwicklungen liegt der eigentliche Reiz des Rüdersdorfer Tagebaus. Damit ein ausgewogenes Verhältnis zwischen natürlicher Entwicklung und kulturellem Erbe in der Zukunft gewahrt bleibt bzw. hergestellt wird, werden zunächst generelle Ziele formuliert.

3. Rekultivierungsziele

Folgende Rekultivierungsziele können angeführt werden:

- Die bereits vorhandenen wertvollen Biotop-Strukturen müssen vor nachteiligen Beeinträchtigungen geschützt werden,
- bedeutsame Zeugnisse der Industriearchitektur müssen so rekonstruiert und erhalten werden, daß sie zukünftigen Besuchern vermitteln, wie der Kalkstein in Rüdersdorf abgebaut und verarbeitet worden ist,
- weite Bereiche im Umfeld des Tagebaus sollen für die Naherholung der Rüdersdorfer Bevölkerung hergerichtet werden,
- bei der Rekultivierung müssen die auf die Zukunft ausgerichteten betrieblichen Belange berücksichtigt werden.

4. Naturfaktoren

Die Naturfaktoren werden üblicherweise in die abiotischen und biotischen unterteilt. Zu den abiotischen Faktoren gehören die Gesteinsformationen und die sich daraus entwickelnden Böden, das Grundwasser, die klimatischen

1. The site

Rüdersdorf is located 30 km east of Berlin. The site has been used since the 17th century for the extraction of building materials. At first limestone was broken into boulders and worked, then later quick lime was produced. The present industrial production programme includes a very wide range of limestone-based products.

With the advance of industrialization, methods of extraction and speed of extraction also developed rapidly. Techniques were tried and rejected in favour of others in increasingly rapid succession. The present situation is shaped by this process of industrial evolution, which has left its mark on the topographical, architectural and agricultural features of the landscape.

2. Recultivation

The recultivation plan is based on a thorough stocktaking and analysis of the existing types of fauna and flora and their symbioses. This list also included descriptions of the various locations, for example warm, dry limestone slopes, moist hollows (in which small lime bogs have formed), and ruderal locations. The survey also covered industrial structures and equipment which can be summarized as "industrial heritage".

The real fascination of the Rüdersdorf open cast pit lies in observing the interaction between workings marked by natural succession, and developments influenced by human activity. Initial guidelines are being formulated to ensure a balanced relationship is maintained or established between natural development and industrial heritage.

3. Recultivation goals

The following recultivation goals have been identified:

- to protect the existing valuable biotope structures against degradation,
- to reconstruct and preserve significant examples of industrial architecture in such a way as to convey to future visitors how limestone was extracted and processed at Rüdersdorf,
- to reinstate large areas surrounding the pit as a local recreational amenity for the population of Rüdersdorf,
- to take account of the future operational needs of the plant.

4. Natural factors

Natural factors are normally divided into the biotic and the abiotic. Abiotic factors include rock formations and the earth derived from them, groundwater, climatic conditions, and physical features. Biotic factors include fauna and flora and the physical presence of human beings.

A salt block below the substratum has caused the lower Triassic formation to curve upwards in the Tertiary geological formation. As a result the shell limestone of this formation is near the surface, an exceptional situation in the surrounding moraine landscape. Rüdersdorf owes its fame to this geological peculiarity. All ecological factors are determined by the presence of calcium carbonate.

Gegebenheiten und die Oberflächenformen. Zu den biotischen Faktoren zählen dagegen die Pflanzen- und Tierwelt und der Mensch in seiner physischen Existenz.

Durch einen Salzblock im Untergrund ist in der geologischen Formation des Tertiärs die tieferliegende Triasformation aufgewölbt worden. Somit steht der Muschelkalk dieser Formation oberflächennah an, eine singuläre Situation in der umliegenden Moränenlandschaft. Dieser geologischen Besonderheit verdankt Rüdersdorf seine Berühmtheit. Alle ökologischen Faktoren sind durch die Präsenz des Kalziumcarbonats bestimmt.

Als durch die moderne Technik die Sümpfung so weit fortgeschritten war, daß der Abbau auch im Grundwasserbereich stattfinden konnte, sind Abbauwände von beträchtlicher Höhe entstanden. In der Formation des Unteren Muschelkalks (Schaumkalk) wurden zur Kalkgewinnung Stollen vorgetrieben. Die dazwischen verbliebenen Säulen wurden anschließend gesprengt, so daß das Hangende in dramatischer Weise hinunterstürzte (sog. Pfeiler-Bruch-Sturz-Verfahren). Einige Stollen sind noch vorhanden. Diese sind heute ein außerordentlich wertvolles Winter-Biotop für Fledermäuse unterschiedlicher Gefährdungskategorien (bedeutsames Fledermaus-Vorkommen in Mitteleuropa). Im Rahmen der Rekultivierung werden jetzt schon Vorkehrungen für Ersatzbiotope getroffen, damit diese empfindlichen Kleinsäuger sich rechtzeitig umorientieren können. Ähnlich wird mit der Schaffung von Ersatzbiotopen (Second-Hand-Biotope) für die Pflanzenwelt verfahren. Auch hier ist festzustellen, daß durch das Aufdecken der Trias-Formation, die Schaffung von oft schroffen Geländeformen, die Veränderung des Wasserhaushaltes und weitere Eingriffe, extreme Standortbedingungen entstanden sind, die bedrohten Pflanzenarten (sog. stenöken Arten) einen Lebensraum bieten.

War in der Vergangenheit die Nivellierung dieser extremen Standortverhältnisse die Regel, so verfährt man heutzutage gerade umgekehrt. Die hohe Diversität der Standorte wird erhalten. Das bedeutet in der Praxis, daß z. B. nur auf einigen wenigen Standorten nährstoffhaltiger Oberboden aufgetragen wird, z. B. nur dort, wo Bäume und Gehölze gepflanzt werden sollen. Bei dieser Vorgehensweise geht es

With the help of modern technology, draining advanced to the point that extraction could proceed below ground water level, creating impressively tall working faces. In the lower shell limestone formation (foam limestone), tunnels were driven to extract limestone. The columns left in between were then blown up, so that the hanging roof collapsed dramatically (so-called "pillar mining". Some tunnels still exist. They are now a valuable winter biotope for various endangered species of bat (major bat population in Central Europe). Measures are being taken as part of the recultivation program to establish replacement biotopes so that these sensitive small mammals can readapt in good time. Work is also in hand to establish replacement biotopes for the flora. Here too it is evident that uncovering the Triassic formation, creating often precipitous terrain, changing the water economy and other activities have created extreme local conditions affording habitats for threatened species of plant (species with limited range of habitat).

Whereas in the past these extreme local conditions were made to conform to a standard pattern, today by contrast the approach is precisely the opposite, with the emphasis on maintaining the great diversity of habitats. This means in practice that for example nutrient-rich top-soil is applied in a few locations only, e.g. only where trees and undergrowth are to be planted. The key consideration in this programme is to apply the recultivation resources available in the most efficient and environmentally appropriate manner.

5. Location analysis

The three-dimensional system of coordinates (ecological coordinates – **Fig. 1**) illustrates how a location can be defined in terms of the following variables:
– lime content
– humidity, and
– slope.

In the case of a steep south-facing limestone scree for example the environment can be defined as having a high slope factor, low moisture content and high lime content. The necessary recultivation measures can be defined by applying this analysis to the various sectors to be recultivated.

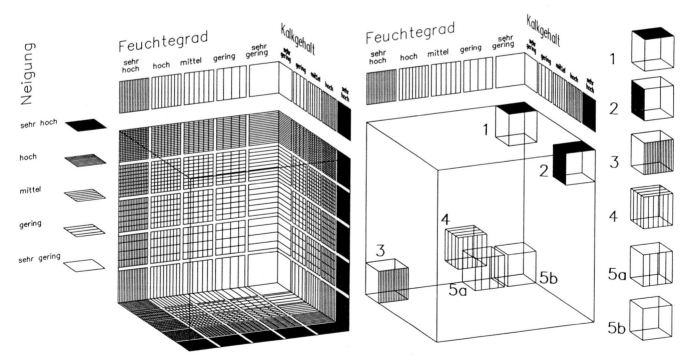

BILD 1/FIGURE 1: Ökokoordinaten / Ecological coordinates

Feuchtegrad	= Humidity	Kalkgehalt	= Lime content	Neigung	= Slope
sehr hoch	= very high	sehr hoch	= very high	sehr gering	= very gentle
hoch	= high	hoch	= high	gering	= gentle
mittel	= medium	mittel	= medium	mittel	= moderate
gering	= low	gering	= low	hoch	= steep
sehr gering	= very low	sehr gering	= very low	sehr hoch	= very steep

vor allem darum, die Mittel, die für die Rekultivierung vorgesehen sind, effizient und situationsgerecht einzusetzen.

5. Standortanalyse

Anhand eines dreidimensionalen Koordinatensystems (Ökokoordinaten) (**Bild 1**) kann veranschaulicht werden, daß ein Standort beispielsweise durch

— den unterschiedlichen Kalkgehalt,

— den unterschiedlichen Feuchtegrad und

— die unterschiedliche Neigung definiert werden kann.

Nimmt man beispielsweise einen nach Süden stark geneigten Geröllhang aus Kalkstein an, so kann der Standort durch einen hohen Neigungsfaktor, einen geringen Feuchtegrad und einen hohen Kalkgehalt definiert werden. Unterzieht man die einzelnen Teilbereiche, die zu rekultivieren sind, einer derartigen Standortanalyse, so lassen sich daraus die geeigneten Rekultivierungsmaßnahmen ableiten.

Würde der oben beschriebene Standort z. B. sich selbst überlassen bleiben (natürliche Sukzession), könnten sich Eidechsen ansiedeln oder wärmeliebende Wildrosen entfalten. Im Fall von nährstoffreicheren Standorten, deren Bodensubstrat im Pleistozän entstand, ließe sich demgegenüber die Vegetationsentwicklung mittels Pflanzungen beschleunigen. Für den Raum Rüdersdorf kommen Großbäume wie Bergahorn (Acer pseudoplatanus), Stieleiche (Quercus robur) und Esche (Fraxinus excelsior) sowie Gehölze wie Hasel (Corylus avellana), Weißdorn (Crataegus monogyna), Liguster (Ligustrum vulgare) oder die bereits zitierten Wildrosen (Rosa spec.) und viele andere Arten in Frage. In Bereichen, die überwiegend der Naherholung dienen, können auch Wildapfel (Malus sylvestris), Flieder (Syringa vulgaris) oder Schmetterlingsstrauch (Buddlia alternifolia) gepflanzt werden.

Das gewählte Pflanzschema sollte so aufgebaut sein, daß sich eine Verzahnung von höheren Bäumen (im zentralen Pflanzbereich) mit solchen von geringerer Wuchshöhe (in äußeren Gehölzstreifen) ergibt (**Bild 2**). Dadurch werden wertvolle Säume geschaffen, die als sog. Übergangsbiotope (Ökotone) für die Vernetzung von ökologischen Systemen eine wichtige Rolle spielen.

Eine weitere Form der Vegetationsentwicklung, die für nicht erschlossene Bereiche sinnvoll ist, kann durch Ausbringung von abgeschlagenen Ästen (Totholz) initiiert werden. Diese Zweige dienen als Ansitz für die Vogelwelt, dadurch erfolgt ein Sameneintrag von Gehölzen. Die so entstandenen natürlichen Gehölze sind außerordentlich ästhetisch und geeignet, Leitlinien für die Landschaft darzustellen.

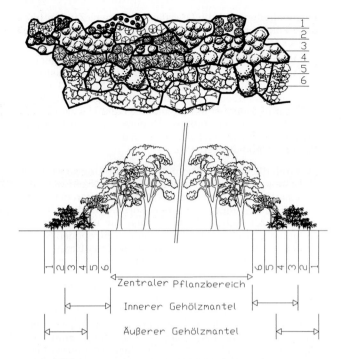

BILD 2/FIGURE 2: Pflanzschema / Scheme of plantation

Zentraler Pflanzbereich = Central plant area
Innerer Gehölzmantel = Inner wooded covering
Äußerer Gehölzmantel = Outer wooded covering

If the location described above were for example left undisturbed (natural succession), lizards or warmth-loving wild roses could establish themselves. In the case of nutrient-rich locations with a substratum laid down in the Great Ice Age, on the other hand, the development of vegetation could be accelerated by plantation. In the case of Rüdersheim, large trees are appropriate, such as mountain maple (Acer pseudoplanatanus), oak (Quercus robur) and ash (Fraxinus excelsior) and undergrowth such as hazel (Corylus avellana), hawthorn (Crataegus monogyna), privet (Ligustrum vulgare) or the wild rose (Rosa spec.) already mentioned, as well as many other species. In areas dedicated principally to local recreational use, crab apple (Malus sylvestris), lilac (Syringa vulgaris) or buddleia (Buddlia alternifolia) can be planted.

The plants should be selected to create a structure whereby taller trees (in the central planting area) combine with those of lesser stature (in the fringe undergrowth) (**Fig. 2**). This creates valuable fringe areas which serve as transitional biotopes (ecotones) for interlinking ecological systems.

A further form of vegetation appropriate for undeveloped areas can be fostered by spreading tree cuttings (dead wood). These twigs serve as a habitat for birds which in turn bring in seeds for the undergrowth. This leads to natural woody growth of outstanding aesthetic quality, well suited for profiling the landscape.

Fachbereich 4 · Subject 4 · Séance Technique 4 · Tema de ramo 4

Entwicklung von Schalldämpfern für die Zementindustrie

Development of sound absorbers for the cement industry

Développement de silencieux pour l'industrie cimentière

Desarrollo de silenciadores para la industria del cemento

Von **D. Eckoldt** und **H. V. Fuchs,** Stuttgart/Deutschland

Zusammenfassung – *Die Ausblasöffnungen hinter den Entstaubungsanlagen gehören zu den dominanten Lärmquellen. Als besonders lästig werden, auch noch in großer Entfernung, die tieffrequenten Anteile des Drehklangs der Ventilatoren, Verdichter und Pumpen wahrgenommen. Am Modellfall der Vakuumanlagen einer Papierfabrik werden neuartige Schalldämpfer beschrieben, die mit völlig glatten metallischen Oberflächen unempfindlich sind gegenüber Ablagerungen aus dem Fördermedium. Die Resonanz-Absorber können, wenn sie z.B. ganz aus hochwertigem Edelstahl gefertigt werden, mechanisch und chemisch außerordentlich resistent sein. Wenn sie gut zugänglich in den Abgasstrang eingebaut werden, lassen sich die sogenannten Membran-Absorber auch leicht, z.B. durch Waschen, Bürsten oder Dampfstrahlen, in regelmäßigen Intervallen reinigen. Dadurch kann verhindert werden, daß „Anbackungen" auf den Schalldämpfer-Kulissen ihre akustische Wirksamkeit schmälern und die Energie verzehrenden Druckverluste durch allmähliches „Zuwachsen" der Kulissen-Spalte unnötig ansteigen. Ein neues Auslegungs-Programm für Schalldämpfer gestattet es, neben der Einfügungsdämpfung auch das Eigengeräusch sowie die Energiekosten abzuschätzen.*

Entwicklung von Schalldämpfern für die Zementindustrie

Summary – *The outlets from dedusting plants are among the dominant sources of noise. The low-frequency range of the rotational noise from fans, compressors and pumps is perceived as particularly annoying, even at great distances. Using the case of the vacuum pumps in a paper factory a description is given of new types of sound absorbers which have completely smooth metal surfaces and are not sensitive to deposits from the conveying medium. The resonance absorbers can be extremely mechanically and chemically resistant if, for example, they are made entirely of high-grade special steel. The so-called membrane absorbers are also easy to clean at regular intervals by washing, brushing or steam jets if they are installed in the exhaust gas ducts with suitable access. This prevents any build-up on the splitter silencers from encroaching on their acoustic effectiveness and the energy-consuming pressure drop from rising unnecessarily due to gradual blockage of the gaps between the panels. A new design program for sound absorbers makes it possible to estimate not only the insertion attenuation but also the residual noise and the power costs.*

Development of sound absorbers for the cement industry

Résumé – *Les ouvertures d'exhaure derrière les installations de dépoussiérage comptent parmi les sources de bruit dominantes. Comme particulièrement dérangeant sont perçues, même encore à grande distance, les composantes basse fréquence du son de rotation des ventilateurs, compresseurs et pompes. A l'exemple des installations à vide d'une usine à papier sont décrits des silencieux nouveaux, complètement immunisés contre les concrétions à partir du médium de transport, grâce à leurs surfaces métalliques absolument lisses. Les absorbeurs de résonance peuvent, quand ils sont p. ex. construits totalement en acier inoxydable de haute qualité, être mécaniquement et chimiquement extrêmement résistants. S'ils sont montés bien accessibles dans le trajet des gaz d'exhaure, les absorbeurs à membrane ainsi nommés peuvent aussi être facilement nettoyés à intervalles réguliers, p. ex. par lavage, brossage ou jet de vapeur. Ainsi peut être évité, que des concrétions sur les chicanes des silencieux amoindrissent leur efficacité acoustique et que des pertes de charge énergivores augmentent inutilement par un colmatage insidueux. Un nouveau programme de dimensionnement de silencieux permet d'estimer à la fois l'atténuation apportée, le bruit propre et les dépenses d'énergie.*

Développement de silencieux pour l'industrie cimentière

Resumen – *Las aberturas de expulsión detrás de las instalaciones de desempolvado forman parte de las fuentes de ruidos predominantes. Se perciben, y son particularmente molestos, incluso a gran distancia, los ruidos de baja frecuencia causados por el giro de los ventiladores, compresores y bombas. Citando el ejemplo de las bombas al vacío de una fábrica de papel, se describen nuevos tipos de silenciadores, que tienen unas superficies metálicas completamente lisas y que son insensibles a las adherencias causadas por el material transportado. Los dispositivos de absorción de resonancias, fabricados p. ej. en su totalidad de acero fino de alta calidad, pueden ser extremadamente resistentes a las acciones mecánicas y químicas. Siempre y cuando se mon-*

Desarrollo de silenciadores para la industria del cemento

ten en el tramo de gases de escape, bien accesibles, resulta fácil limpiar periódicamente los llamados absorbedores de membrana, por ejemplo lavándolos, pasándoles un cepillo o mediante chorros de vapor. Con ello se puede impedir que las „adherencias" formadas en las colisas de los silenciadores disminuyan la eficacia acústica de los mismos y que aumenten innecesariamente las pérdidas de presión, responsables de un mayor consumo de energía, por „cerrarse" poco a poco los intersticios entre las mencionadas colisas. Un nuevo programa de dimensionamiento de silenciadores permite evaluar no sólo la amortiguación de inserción, sino también los ruidos propios y los gastos de energía.

1. Einleitung

Die Ausblasöffnung hinter Entstaubungsanlagen gehören zu den herausragenden Lärmquellen in Zement- und Beton-Werken. Als besonders lästig werden, auch noch in großer Entfernung, die tieffrequenten Anteile des Drehklangs der Ventilatoren wahrgenommen. Angesichts der zunehmenden Anforderungen im Bereich des Lärmschutzes stellt sich für die Zementindustrie in zunehmendem Maße die Frage nach kostengünstigen und effektiven Maßnahmen der Lärmemissionsminderung. Das Forschungsinstitut der Zementindustrie stellte in seiner Arbeit für die Zementwerke fest [1], daß insbesondere der Einsatz konventioneller Schalldämpfertechnik zur Minimierung der von Ausblasöffnungen und Kaminen abgestrahlten Geräusche nur begrenzt möglich ist. Standzeit und Wartungsintervalle dieser Schalldämpfer sind aufgrund der zementspezifischen Zusammensetzung der Stäube in den Abgasen in vielen Fällen unbefriedigend. Aus diesem Grunde sollen hier Erfahrungen des IBP mit einer neuartigen Membran-Absorber-Technologie beim großtechnischen Einsatz an Vakuum-Pumpen von Papier-Maschinen vorgestellt werden. Da die harten Anforderungen und aggressiven Fluiden überall ähnlich sind, lassen sich die dargestellten Schalldämpfer-Protoypen und die akustiscchen Ergebnisse auch auf Anlagen der Zementindustrie übertragen.

2. Schalldämpfer für staubbeladene Fluide

Es hat in der Vergangenheit immer wieder Versuche gegeben, konventionell mit porösen oder faserigen Dämpfungs-Materialien ausgestattete Schalldämper gegenüber Staubablagerungen resistent zu machen. Sie zielten darauf ab, die Dämpfungs-Materialien entweder durch alle möglichen Abdeckungen zu schützen, die Ablagerungen in Wartungs-Intervallen wieder abzutragen oder – als ultima ratio – die Dämpfungs-Einlagen regelmäßig zu erneuern. Letzteres setzt bereits voraus, daß z.B. Schalldämpfer-Kulissen als Ganzes aus den in die Anlagen integrierten Schalldämpfer-Gehäusen ohne viele Umstände und, vor allem, ohne lange Stillstandszeiten herausgenommen werden können. An diese bisher nicht sehr erfolgreichen Versuche, ein altes Problem zu lösen, soll hier angeknüpft werden. Zum Einsatz kommt eine neuartige Schalldämpfer-Technologie, bei welcher

– alle Oberflächen absolut eben und glatt ausgeführt sind,
– die innerhalb der Kulissen angeordneten absorbierenden Bauelemente durch die außen angeordneten, schwingfähigen Membranen rundum hermetisch gegenüber dem jeweiligen Fördermedium abgeschlossen werden,
– die äußeren, akustisch aktiven Membranen z.B. durch Bürsten, Dampfstrahlen oder andere mechanische Verfahren nach Bedarf gereinigt werden können.

Dabei erscheint eine Ausführung der Schalldämpfer aus nur einem Werkstoff, im Hinblick auf ihre Rückführbarkeit, wünschenswert und, unter Berücksichtigung ihrer jeweiligen Einsatzbedingungen, auch realisierbar.

Als Resonanz-Dämpfer läßt sich der Membran-Absorber sehr gut auf mittlere und vor allem auf tiefe Frequenzen (auch unter 100 Hz) abstimmen. Er kommt dabei mit einem minimalen Bauvolumen aus. Auf seinen ganz glatten und geschlossenen Oberflächen können sich Feststoffe kaum ablagern. Da der Membran-Absorber vollständig aus hoch-

1. Introduction

The outlets from dedusting plants are among the dominant sources of noise in cement and concrete works. The low-frequency range of the rotational noise from fans is perceived as particularly annoying, even at great distances. In view of the increasing demand for noise control the cement industry is increasingly looking for favourably priced and effective means of reducing noise emission. In its research on cement works the Cement Industry Research Institute has found [1] that conventional sound absorption technology can only be used to a limited extent for minimising noise radiated from discharge outlets and chimneys. The durability and servicing frequency of conventional sound absorbers is often unsatisfactory owing to the specific composition of the dusts in the exhaust gases from cement works. For this reason we shall now describe the experience of the Building Physics Institute with a novel "membrane absorber" technology in a large-scale industrial application to vacuum pumps for paper machines. Since the strict requirements for process air plant with severely polluting and aggressive fluids are the same everywhere, the sound absorber prototypes shown and the acoustic results can also be applied to plants in the cement industry.

2. Sound absorbers for dust-laden fluids

There have been frequent attempts in the past to make conventional sound absorbers fitted with porous or fibrous materials resistant to dust deposits. Their aim was either to protect the sound-absorbing materials with all possible types of coverings, to remove the deposits during the maintenance period or, as a last resort, to renew the damping inserts regularly. This last method presupposes e.g. that silencer splitters can be removed as an entity from the sound absorbing housings integral with the installation, without any trouble and above all without causing long stoppages. We shall take these not very successful attempts to solve an old problem as our starting point. A novel sound absorption technology is being applied, in which

– all surfaces are absolutely level and smooth
– the absorbing components inside the splitters are hermetically sealed from the conveying medium by the external vibrating membranes
– the external, acoustically active membranes can be cleaned as required, e.g. by brushing, steam jet or other mechanical processes.

It seems desirable for the sound absorber to be made of one material only for recycling purposes, and this is possible in view of its conditions of use. As a resonance absorber the membrane absorber can be tuned very well to medium frequencies and especially to low ones (even below 100 Hz). It only requires a minimal unit volume. It is almost impossible for solids to be deposited on its smooth, closed surfaces and, since the whole membrane absorber may be made of high-alloy special steels, good durability can be obtained. For example it has been tested successfully in a flue-gas dust collector at a combined heating and power station for two and a half years [2].

3. Testing on paper machines

Vacuum pumps on paper machines are the dominant source of the noise emitted from paper factories. They radiate a low-frequency humming ("rumbling") through the chimney out-

legierten Edelstählen aufgebaut werden kann, sind hohe Standzeiten erreichbar. Er wurde z. B. über 2½ Jahre in einer Rauchgasreinigungsanlage eines Heizkraftwerks erfolgreich getestet [2].

3. Erprobung an Papier-Maschinen

Bei der Lärmemission von Papierfabriken sind die Vakuumpumpen der Papiermaschinen dominierende Schallquelen. Sie strahlen ein tieffrequentes Brummen („Wummern") im Frequenzbereich um 100 Hz über die Kaminöffnungen ab. Da dies auf dem Ausbreitungsweg kaum gedämpft wird, trägt es häufig auch zum A-bewerteten Gesamtpegel bei, wobei dieser Beitrag in großer Entfernung größer wird. Bei der schalltechnischen Sanierung einer Papierfabrik bestand die Aufgabe darin, den A-bewerteten Immissionspegel um mindestns 5 dB zu senken. Dies konnte nur erreicht werden, indem die Pegel zweier Töne bei 80 Hz und 160 Hz vermindert wurden. Für die Lösung dieses Lärmproblems wurden spezielle Schalldämpfer aus Membran-Absorbern entwickelt und eingesetzt [3].

Zur nachhaltigen Lärmminderung wurden bei einer Papiermaschine Kulissen-Schalldämpfer in den rechteckigen Abluftkamin eingebaut und bei einer zweiten Rohr-Schalldämpfer auf die runden Ausblasöffnungen aufgesetzt. Die Problemlösung erfolgte in folgenden Schritten:

1. Aufnahme des Ist-Zustandes durch Schallpegel-Messungen mit Teilabschaltung der Papierfabrik zum einen an ausgewählten Immissionspunkten und zum anderen auf einer Hüllfläche an den Kaminöffnungen,
2. Auslegung der neuartigen Resonanz-Schalldämpfer auf das Immissions-Spektrum und Bau von Prototypen aus Aluminium,
3. Test der Prototypen in einem speziellen Kleinprüfstand (Einzelelemente) sowie im Schalldämpfer-Prüfstand nach DIN 45 646/ISO 7235 (Kulissen) und im Rohr-Schalldämpfer-Prüfstand in Anlehnung an die neue Norm,
4. Test der Aluminium-Prototypen an den Kaminöffnungen bei normalem Betrieb der Papiermaschinen,
5. Fertigung der Schalldämpfer, vollständig aus Edelstahl Nr. 1.4571, durch den Schalldämpfer-Hersteller und Einbau mit Hilfe einer Montagefirma auf dem Dach der Papierfabrik.

4. Ergebnisse mit Kulissen-Schalldämpfern

Das Kammervolumen der Membran-Absorber beträgt 4 Liter, die Kammerwände bestehen aus 1,5 mm dickem Edelstahl. Schlitz- und Abdeckmembranen sind 0,3 mm dick. In die 0,6 m × 5,8 m große Öffnung des Abgaskamins der Papiermaschine werden insgesamt acht beidseitig absorbierende Kulissen von oben eingehängt (**Bild 1**). Jede Kulisse ist 0,3 m breit, 0,5 m hoch und 3 m lang. Damit ergibt sich ein Kulissenabstand von 0,43 m. Die Kulissen sind aus je drei 1 m langen Segmenten gefertigt. Zur Verringerung des Druckverlustes sind die Halterungen mit halbkreisförmigen Anströmprofilen versehen. Der Einbau der Schalldämpfer reduziert den A-bewerteten Hüllflächen-Pegel um 7 dB (**Bild 2**). Die Auswertung einzelner Meßpunkte auf der Hüllfläche zeigt aber, daß bei den immissionsrelevanten flachen Winkeln sogar Dämpfungswerte bis 15 dB erreicht werden. Die Wirkung der Schalldämpfer an den Immissionsorten ist demnach sehr viel größer als nach dem Mittelwert auf der Hüllfläche und der Messung im Prüfstand zu erwarten ist.

5. Alterungsverhalten

Am 15. 11. 1990 wurden die Kulissenschalldämpfer aus Membran-Absorbern nach fast 1 Jahr Einsatz im Abgaskamin akustisch und mechanisch überprüft. Ein Vergleich des mittleren Hüllflächenschalldruckpegels mit dem Ergebnis vom 6. 2. 1990 zeigte keine Verschlechterung der Dämpfung.

Um das akustische Alterungsverhalten auch unter den reproduzierbaren Meßbedingungen im Schalldämpferprüfstand bestimmen zu können, wurde ein Kulissensegment gegen ein Reserveelement bei laufender Papiermaschine

lets, in the range around 100 Hz. As the humming is hardly damped during its propagation it often contributes to the A-weighted total level, and its contribution increases with the distance covered. When sound control at a paper factory was being improved an attempt was made to reduce the A-weighted immission level by at least 5 dB. This could only be achieved by lowering the level of two sounds at 80 Hz and 160 Hz. Special membrane-type sound absorbers were developed and used to solve the noise problem [3].

For lasting reduction of noise from a paper machine splitter silencers were fitted in the rectangular exhaust air stack, and in the case of a second tubular sound absorber they were placed on the round discharge outlets. The problem was solved in the following steps:

1. recording the current situation by taking sound level readings with partial shutdown of the paper factory (a) at selected immission points and (b) on an enveloping surface at the stack outlets;
2. designing the novel resonance sound absorbers on the basis of the immission spectrum and constructing aluminium prototypes;
3. testing the prototypes in a special miniature test stand (single elements) as well as in the sound absorber test stand to DIN 45 646/ISO 7235 (splitters) and the tubular sound absorber test stand on the lines of the new standard;
4. testing the aluminium prototypes at the stack outlets with the paper machines running normally, and
5. having the sound absorbers made by an appropriate manufacturer, entirely of special steel no. 1.4571, and fitted on the roof of the paper factory by an installation company.

4. Results with splitter silencers

The chamber volume of the membrane absorbers is 4 litres and the chamber walls are made of 1.5 mm thick special steel. The slotted and covering membranes are 0.3 mm thick. A total of eight splitters, absorbing on both sides, are inserted from above in the outlet of the exhaust gas stack of the paper machine, measuring 0.6 m × 5.8 m (**Fig. 1**). Each panel is 0.3 m wide, 0.5 m high and 3 m long, giving a gap of 0.43 m between them. The panels are each made up of three

BILD 1: Kulissen-Schalldämpfer aus Membran-Absorbern in der Kaminöffnung

FIGURE 1: Splitter silencer made up of membrane absorbers in the stack outlet

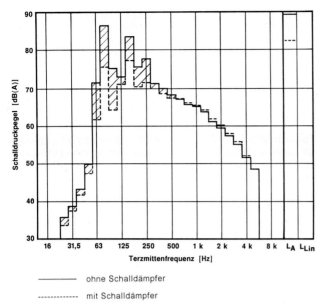

BILD 2: Minderung des A-bewerteten Emissionsspektrums der Vakuumanlage einer Papiermaschine durch Einbau der Kulissen-Schalldämpfer aus Membran-Absorbern
FIGURE 2: Reduction in the A-weighted emission spectrum of the vacuum system of a papermachine by installation of splitter silencer made up of membrane absorbers

Schalldruckpegel	= sound-pressure level
Terzmittenfrequenz	= third-band mean frequency
ohne Schalldämpfer	= without sound absorbers
mit Schalldämpfer	= with sound absorbers

ausgetauscht. **Bild 3** zeigt das ausgewählte Segment direkt nach dem Ausbau. Es hat eine trockene Oberfläche, die von einer dünnen grauen Schicht bedeckt ist. Die Schicht läßt sich leicht entfernen, und darunter kommt wieder die sauberglänzende Metalloberfläche zum Vorschein. Papierreste, wie an der Kaminwand in **Bild 4** zu erkennen, haben sich auf der Deckmembran nicht abgesetzt. Da von allen Kulissensegmenten vor Ihrem Einbau in den Abluftkamin die Einfügungsdämpfung im Schalldämpfer-Prüfstand bestimmt wurde, ist das akustitsche Alterungsverhalten jedes Ele-

BILD 4: Blick vom Dach in den Abluftkamin der Vakuumpumpen einer Papiermaschine vor dem Einbau der Schalldämpfer
FIGURE 4: View from the roof into the exhaust air stack of the vauum pumps of a paper machine before installation of the sound absorbers

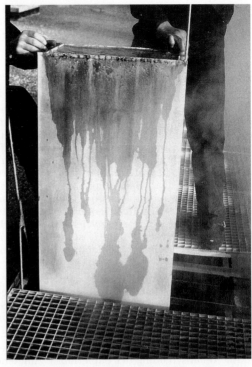

BILD 3: Austausch eines Kulissensegments nach einem Jahr Betriebszeit zur Überprüfung im Schalldämpfer-Prüfstand
FIGURE 3: Exchanging a splitter section after one year's operation for checking on the sound absorber test stand

sections 1 m long. The supports have semi-circular inlet profiles to reduce pressure loss. Installation of the sound absorbers lowers the A-weighted enveloping surface level by 7 dB (**Fig. 2**). However, evaluation of individual measuring points on the enveloping surface shows that absorption values of up to 15 dB are reached at the flat angles relevant to immission. The effect of the sound absorbers at the immission points is accordingly far greater than would be expected from the mean value on the enveloping surface and measurement on the test stand.

5. Deterioration

On 15. 11. 1990 acoustic and mechanical tests were carried out on the splitter silencers made up of membrane absorbers after they had been installed in the exhaust gas stack for nearly a year. When the mean sound-pressure level of the enveloping surface was compared with the result obtained on 6. 2. 1990 there was found to be no loss of absorption.

To determine acoustic deterioration under reproducible measuring conditions on the sound absorber test stand, a splitter section was exchanged for a reserve element with the paper machine running. **Fig. 3** shows the selected section immediately after its removal. It has a dry surface covered with a thin grey film. The film is easily removed and the clean shiny metal surface reappears from below it. No paper residues, such as can be seen on the stack wall in **Fig. 4**, have been deposited on the covering membrane. Since the insertion attenuation of all the splitter sections is recorded in the test stand before they are installed in the exhaust air chimney, the acoustic deterioration of each individual element can be determined accurately. **Fig. 5** compares the insertion attenuation of the exchanged section before installation and after removal. Nor does measurement on the test stand reveal any loss of absorption. Even after 4 years under extremely severe conditions on the paper machine all the splitter silencers in Fig. 1 are still fully operative and there have been no complaints from the operator.

6. Conclusions

Membrane-type sound absorbers fitted in the vacuum systems of paper machines have fulfilled expectations as far as sound control is concerned. Emission readings at the outlets

ments genau zu bestimmen. In **Bild 5** wird die Einfügungsdämpfung des ausgetauschten Segments vor dem Einbau und nach dem Ausbau verglichen. Auch die Prüfstandsmessung zeigt keine Verschlechterung der Dämpfung. Auch nach 4 Jahren unter den extrem harten Einsatzbedingungen an der Papiermaschine sind alle Kulissen-Schalldämpfer in Bild 1 noch voll funktionsfähig und ohne irgendwelche Beanstandungen durch den Betreiber geblieben.

6. Schlußfolgerungen

Die an den Vakuumanlagen der Papiermaschinen eingebauten Schalldämpfer aus Membran-Absorbern haben die in sie gesetzte Erwartung schalltechnisch erfüllt. Die Ergebnisse der Emissionsmessungen an den Öffnungen der Abgaskamine wurden durch die Befragung von einigen Mitarbeitern der Papierfabrik untermauert, die in der Nähe des Betriebes wohnen. Die Mitarbeiter der Papierfabrik in den Arbeitszimmern nahe den Öffnungen der Abgaskamine erklärten übereinstimmend, daß auch die Schalldruckpegel in diesen Räumen deutlich abgenommen haben. Damit konnte ein für alle Beteiligten sehr entscheidendes Vorhaben zur Umsetzung einer neuartigen Schalldämpfer-Technologie aus der Prototyp-Phase im Labor in die industrielle Anwendung im großtechnischen Maßstab entsprechend dem Projektplan in enger Kooperation zwischen dem Forschungs-Institut, dem Betreiber und dem Schalldämpfer-Hersteller erfolgreich abgewickelt werden. Regelmäßige Überprüfungen des akustischen und mechanischen Langzeitverhaltens der Membran-Schalldämpfer werden zeigen, ob das Problem tiefer Frequenzen bei Vakuumpumpen nach dem dargestellten technischen Konzept dauerhaft gelöst werden kann.

Bereits heute läßt sich aber konstatieren, daß die in [4] ausführlicher beschriebenen Membran-Schalldämpfer alle wesentlichen Eigenschaften für den sicheren und dauerhaften Einsatz auch in Abgasanlagen der Zementindustrie aufweisen. Im Rahmen eines Verbund-Projektes, an dem sich Betreiber von Zementwerken beteiligen, sollte die Eignung dieser Schalldämpfer auch an Abgasanlagen der Zementindustrie geprüft werden.

Literatur

[1] Funke, G.: Auslegung und Betrieb von Abgasschalldämpfern in Zementwerken. Zement-Kalk-Gips ZKG International 41 (1988), H. 1, S. 42–47.

[2] Fuchs, H.V., Ackermann, U., und Rambausek, N.: Nichtporöser Schalldämpfer für den Einsatz in Rauchgasreinigungsanlagen. VGB Kraftwerkstechnik 69 (1989), H. 11, S. 1102–1110.

[3] Fuchs, H.V., Ackermann, U., und Rambausek, N.: Membran-Absorber für den technsichen Schallschutz. Fortschritte der Akustik DAGA '87, DPG-GmbH, Bad Honnef 1987, S. 741–744.

[4] Fuchs, H.V., Ackermann, U., und Neemann, W.: Neuartige Membran-Schalldämpfer an Vakuumanlagen von Papiermaschinen. Das Papier 46 (1992), H. 5, S. 219–231.

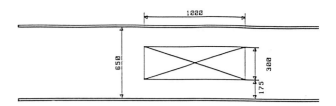

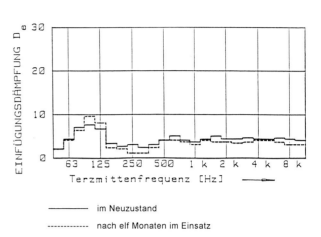

BILD 5: Vergleich der Einfügungsdämpfung eines Kulissensegments im Neuzustand und nach elf Monaten im Einsatz
FIGURE 5: Comparison of the insertion attenuation of one splitter section when new and after 11 months' use

Einfügungsdämpfung	= insertion attenuation
Terzmittenfrequenz	= third-band mean frequency
im Neuzustand	= when new
nach elf Monaten im Einsatz	= after 11 months' use

of exhaust gas stacks have been confirmed by questioning some company employees who lived near the factory. Employees who worked in rooms close to the stack outlets were unanimous in declaring that there had also been a clear drop in sound-pressure levels in those rooms. Hence a very important plan for all parties involved, to convert a novel sound absorber technology from the prototype phase in the laboratory to a large-scale industrial application was carried out successfully in close cooperation between the Research Institute, the operator and the sound absorber manufacturer. Regular checking of the long-term acoustic and mechanical action of the membrane-type sound absorbers will show whether the problem of low frequencies in vacuum pumps can be permanently solved by the technology described.

Even today the membrane-type sound absorbers described more fully in [4] can be said to have all the essential properties for safe, permanent use in exhaust gas systems in the cement industry. The suitability of these sound absorbers for such systems is to be tested in a combined project in which cement works operators will take part.

Ersatz eines 3feldrigen Elektrofilters nach 20jährigem Betrieb

Replacement of a 3-compartment electrostatic precipitator after 20 years' operation

Remplacement d'un électrofiltre à trois champs après 20 ans de service

Sustitución de un electrofiltro de 3 compartimientos, tras 20 años de servicio

Von **H. Fleck,** Harburg/Deutschland

Zusammenfassung – Die Abscheideleistung des Filters, das mit dem 3000 t/d-Ofen 1974 in Betrieb gegangen war, hatte sich aufgrund von Undichtigkeiten im Gehäuse, die durch Korrosion entstanden waren, so verschlechtert, daß die speziellen Auswertebedingungen der TA Luft (Nr. 2.1.5) nicht mehr sicher erfüllt werden konnten. Ein Austausch ganzer Felder, wie bereits vor einigen Jahren durchgeführt, erschien nicht sinnvoll, da die durch Undichtigkeiten hervorgerufene Korrosion die neu bestückten Teile weiter gefährdet hätte. Das Werk entschied sich daher zum Komplettaustausch des Filters, bei dem der neueste Stand der Erkenntnisse berücksichtigt werden konnte. Bedingung war zudem, daß die Grundfläche und die Abstützpunkte des Filters sich nicht ändern durften. Außerdem stand für den Austausch und Anschluß der Leitungen sowie für die Inbetriebnahme nur ein Zeitrahmen von wenigen Wochen zur Verfügung. Realisiert wurden diese bau- bzw. produktionstechnischen Bedingungen dadurch, daß das alte Filter komplett ausgebaut, weggehoben und abtransportiert wurde. Das vormontierte neue Filter mit einem Gewicht von 620 t konnte unmittelbar im Anschluß daran einschließlich Isolierung vom Montageort mit 2 Kränen auf den Platz des alten Filters gehoben werden. Das Abheben des alten Filters und Einheben des neuen Filters erfolgte planmäßig innerhalb von 2 Tagen. Die Abnahmemessungen und der anschließende Betrieb zeigen, daß die Anforderungen der TA Luft nun voll erfüllt werden.

Ersatz eines 3feldrigen Elektrofilters nach 20jährigem Betrieb

Summary – The collecting efficiency of the filter which had gone into operation in 1974 with the 3000 t/d kiln had deteriorated to such an extent due to leaks in the casing caused by corrosion that the special evaluation conditions of the German Clean Air Regulations (No. 2.1.5) could no longer be fulfilled with certainty. Replacement of complete compartments, as had already been carried out a few years ago, did not appear practical as the corrosion resulting from the leaks would have put the newly fitted parts at risk again. The works therefore decided on a complete replacement of the precipitator, which would also allow the use of the latest technology. It was a condition that the base area and support points of the filter could not be changed. There was also a period of only a few weeks available for the replacement, for making the connections, and for commissioning. These structural and production conditions were met by completely dismantling the old precipitator, lifting it out, and transporting it away. Immediately after this the new preassembled precipitator with a weight of 620 t was then lifted from where it had been assembled, complete with insulation, into the position of the old precipitator using 2 cranes. The lifting out of the old precipitator and the lifting in of the new one took place according to plan within two days. The acceptance measurements and the subsequent operation show that the requirements of the German Clean Air Regulations are now completely fulfilled.

Replacement of a 3-compartment electrostatic precipitator after 20 years' operation

Résumé – Le rendement de séparation du filtre, qui a été mis en service avec le four de 3000 t/j en 1974, s'était tellement dégradé suite à des fuites dans l'enceinte causées par corrosion, que les conditions spécifiques d'exploitation de la TA Luft (N° 2.1.5) ne pouvaient plus être remplies avec certitude. Un remplacement de champs entiers, comme déjà fait il y a quelques années, ne paraissait pas justifié, parce que la corrosion due aux fuites aurait, par la suite, mis en danger les parties nouvellement équipées. L'usine a donc décidé le remplacement complet du filtre, en tenant compte de l'état le plus récent des connaissances. En outre, le préalable était, que la surface au sol et les points d'appui du filtre ne devaient pas être changés. De plus, un intervalle de peu de semaines était disponible pour le remplacement, le branchement des conduits et la mise en service. Ces conditions posées aux techniques de construction et de production ont été remplies grâce au démontage, au soulèvement et à l'enlèvement du vieux filtre entier. Immédiatement après, le nouveau filtre prémonté, d'un poids de 620 t, y compris l'isolation, a pu être transplanté, à l'aide de deux grues, du lieu de montage et posé à l'emplacement de l'ancien filtre. L'enlèvement du filtre ancien et la mise en place du nouveau filtre a été effectué, comme prévu, en deux jours. Les mesures de réception et le fonctionnement après montrent, que les exigences de la TA Luft sont maintenant pleinement satisfaites.

Remplacement d'un électrofiltre à trois champs après 20 ans de service

Sustitución de un electrofiltro de 3 compartimientos, tras 20 años de servicio

Resumen – *La capacidad de separación del filtro puesto en servicio en 1974, junto con el horno para 3.000 t/d, había empeorado a causa de fugas producidas en el cuerpo del mismo por la corrosión, de modo que ya no era posible cumplir, de forma segura, las condiciones particulares de evaluación fijadas en la TA Luft - Instrucción técnica para el mantenimiento de la limpieza del aire (No. 2.1.5). Una sustitución de compartimientos completos, como ya se había hecho hace años, no tenía mucho sentido, ya que la corrosión causada por las fugas hubiera puesto en peligro también las piezas nuevas. Por esta razón, la empresa decidió proceder al cambio completo del filtro, teniendo en cuenta el estado actual de la técnica. Además, se puso como condición que la superficie ocupada por el filtro así como los puntos de apoyo del mismo no debían cambiar. Para sustituir el filtro y realizar las necesarias conexiones así como para la puesta en servicio del nuevo filtro, se disponía sólo de unas pocas semanas. Los requerimientos de tipo técnico y constructivo se cumplieron de la forma siguiente: El filtro antiguo fue desmontado por completo, levantado y sacado de su sitio. Luego, el nuevo filtro, premontado y con un peso de 620 t, incluyendo el aislamiento, pudo ser colocado directamente en el mismo lugar, con ayuda de 2 grúas. Para levantar el filtro antiguo y poner en su sitio el filtro nuevo fueron necesarios dos días, tal como estaba previsto. Las mediciones de recepción y el subsiguiente servicio dieron como resultado que los requerimientos de la TA Luft se cumplían satisfactoriamente.*

1974 wurde im Märker Zementwerk, Harburg, eine Wärmetauscherofenanlage mit 3000 t/d in Betrieb genommen. Zur Abscheidung des Rohmehls im Verbund- und Direktbetrieb wurde ein 3feldriges Elektrofilter eingebaut. Die damit erzielten Abscheidegrade lagen im 3-Feld-Betrieb deutlich unter den genehmigten Werten von 75 mg/Nm^3 trocken. Selbst mit 2 Feldern konnten die neueren Auflagen von 50 mg/Nm^3 trocken im Normalbetrieb sicher eingehalten werden. Die großzügige Auslegung des Filters bewährte sich in den letzten Jahren, obwohl unregelmäßig auftretende Verschlechterungen der Normalwerte zu beobachten waren. Ursache hierfür waren durch Korrosion verursachte Undichtigkeiten des Gehäuses. Der Austausch ganzer Felder mit Reparatur des Gehäuses, wie bereits vor einigen Jahren durchgeführt, erschien nicht sinnvoll, da undichte Stellen an den anderen Feldern zu erneuter, schneller Korrosion der reparierten Bereiche führten. Die Entscheidung zur Generalreparatur mit Austausch des Filters erfolgte unter Berücksichtigung der neuesten Erkenntnisse hinsichtlich Filtersteuerung und Gasführung. Voraussetzung war, daß sich die Grundfläche und die Abstützpunkte des Filters nicht veränderten. Nach der ersten Planung sollte der alte Filter vor Ort verschrottet und danach der neue Filter montiert werden. Dies hätte einen Stillstand der Ofenanlage von mindestens 6 Monaten bedeutet. Angeregt durch einen Film eines Anbieters wurde daher der Plan verfolgt, den alten Filter abzuheben und den neuen Filter in wenigen Tagen einzuheben. Schwierigkeiten ergaben sich jedoch aus der Tatsache, daß im Gegensatz zu den bisherigen Erfahrungen nicht nur enorme Lasten, sondern auch extreme Abmessungen bewältigt werden mußten. **Bild 1** zeigt den Größenvergleich der beiden Filter alt und neu.

Gelöst wurde die Aufgabe in Zusammenarbeit mit zwei deutschen und einer niederländischen Kran- sowie einer belgischen Transportfirma. Die Arbeiten wurden in die folgenden Teilabschnitte aufgeteilt:

1. Vorbereitung der Montage- und Abstellflächen
   Dazu mußte eine kleinere Rohmehlmahlanlage komplett und etwa 50 m Rollenbahn der Reifenaufgabe demontiert werden.
2. Montage eines neuen Filters
   Auf der Transportkonsole wurde der Filter ab 1. September montiert; Montageende war der 10. Dezember.
3. Vorbereitung der Demontage des alten Filters
   Am 30. November wurde die Ofenanlage abgestellt. Die Aussteifungen wurden montiert. Parallel dazu wurden die Rohrleitungen demontiert und die Filtertrichter abgeschnitten und auf die Bühne abgelassen. Die Transportkonsole wurde montiert. Die Fertigstellung war zum 15. Dezember vorgesehen.
4. Austausch der Filter
   Für den Austausch der Filter waren 3 Kräne und 2 selbstfahrende Kamag Modultransporter im Einsatz, deren

A 3000 t/d preheater kiln plant was brought into operation at the Märker cement works in Harburg in 1974. A 3-compartment electrostatic precipitator was installed for removing the raw meal in interconnected and direct operation. In 3-compartment operation it achieved collecting efficiencies which gave values significantly below the approved value of 75 mg/m^3 (stp, dry). It also met the more recent conditions of 50 mg/m^3 (stp, dry) reliably, even with 2 compartments. The generous design of the precipitator proved its worth in later years although the normal values sometimes deteriorated at irregular intervals. This was due to leaks in the casing caused by corrosion. Replacement of entire compartments with repairs to the casing, as had already been carried out a few years previously, did not appear practical as leaks in the other compartments would lead to renewed, rapid, corrosion in the repaired section. The decision to carry out general repairs and replace the precipitator was taken with due regard to the latest findings on control of the precipitator and gas flow. One precondition was that the base area and support points of the precipitator should not be changed. The first plan was to scrap the old precipitator on site and then erect the new one. This would have meant stopping the kiln plant for at least 6 months. Inspired by a film from one of the suppliers the plan was therefore followed to lift out the old precipitator and lift in the new one within a few days. However, difficulties arose from the fact that in contrast to previous experience it was necessary to deal with not only enormous loads but also extreme dimensions. **Fig. 1** shows a size comparison of the old and new precipitators.

The problem was solved in cooperation with one Dutch and two German crane firms and a Belgium transport firm. The work was divided into the following sections:

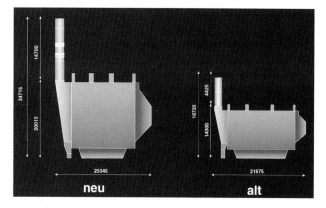

BILD 1: Größenvergleich alter und neuer Filter
FIGURE 1: Size comparison of the old and new precipitators

Elektrofilter:	=	electrostatic precipitators:
Größenvergleich		size comparison
neu	=	new
alt	=	old

Kräne	Tragkraft t	Hauptausleger m	Gegenausleger m	Ausladung m	Ballast t	Einsatzgewicht t	Last t
A	850 / 1100	77	43	41	420	950	300
B	500 / 800	72	30	22	175	520	160
C	500 / 800	72	30	22	175	520	160
D	Transportfahrzeuge					150	660

TABELLE 1: Daten der Kräne und Transportfahrzeuge
TABLE 1: Data for the cranes and transporters

Eletrofilter: Kräne und Transportfahrzeuge	=	Electrostatic precipitators: cranes and transporters
Kräne	=	Cranes
Tragkraft	=	Lifting capacity
Hauptausleger	=	Main jib
Gegenausleger	=	Counterweight jib
Ausladung	=	Reach
Ballast	=	Ballast
Einsatzgewicht	=	Service weight
Last	=	Load
Transportfahrzeuge	=	Transporters

BILD 2: Abheben des alten Filters
FIGURE 2: Lifting of the old precipitator

1. Preparation of the assembly and standing surfaces
   This required the complete removal of a fairly small raw meal grinding plant and about 50 m of the roller conveyor for the tyre feed.
2. Assembly of a new precipitator
   The assembly of the precipitator on the transport support started on 1st September and was completed on 10th December.
3. Preparation for the removal of the old precipitator
   The kiln plant was shut down on 30th November. The stiffening members were installed, and at the same time the ducts were removed and the precipitator hopper cut off and lowered onto the working platform. The transport support was assembled. Completion was planned for 15th December.
4. Exchange of the precipitators
   3 cranes and 2 selfpropelled Kamag module transporters were used for exchanging the precipitators; their data are given in **Table 1**. 72 heavy transporters were used for delivering and for removing the lifting and transporting equipment. Large numbers of telescopic mobile cranes with lifting capacities from 70 to 200 t were required for about 150 working hours for rigging and stripping the equipment. The operating personnel consisted of 2 coordinators and 9 crane drivers, as well as 24 fitters for the assembly and dismantling work.

The old precipitator was lifted out on 14th December (**Fig. 2**) and taken on the transport trolley to the storage position. The transport trolley consisted of two units with a total of 34 axles which could be rotated through ± 90°. Each axle could carry a load of 17 t. The entire vehicle was remote-controlled and could be lifted by 700 mm so that each axle could compensate individually for unevennesses of ± 350 mm.

Fig. 3 shows the two precipitators next to one another. On 15th December the new precipitator was driven to the lifting position; the inlet connection piece weighing 65 t had to be

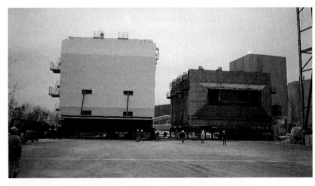

BILD 3: Alter und neuer Filter nebeneinander
FIGURE 3: Old and new precipitators next to each other

Daten der **Tabelle 1** zu entnehmen sind. Für den An- und Abtransport der Hebe- und Transportmittel wurden je 72 Schwertransporter eingesetzt. Für das Auf- und Abrüsten der Geräte wurden zahlreiche Teleskop-Autokräne mit einer Tragfähigkeit von 70–200 t für ca. 150 Arbeitsstunden benötigt. Als Bedienpersonal waren 2 Koordinatoren, 9 Kranfahrer und für den Auf- und Abbau 24 Monteure im Einsatz.

Am 14. Dezember wurde der alte Filter abgehoben (**Bild 2**) und auf dem Transportwagen zum Lagerplatz gefahren. Der Transportwagen bestand aus zwei Einheiten mit insgesamt 34 Achsen, die um +/− 90° gedreht werden können. Jede Achse konnte 17 t Last tragen. Das gesamte Fahrzeug war ferngesteuert und konnte um 700 mm abgehoben werden, wobei jede Achse einzeln Unebenheiten von +/− 350 mm ausglich.

Bild 3 zeigt die beiden Filter nebeneinander. Am 15. Dezember wurde der neue Filter zum Verhebeplatz gefahren, wobei der Eintrittsstutzen mit 65 t noch abgenommen werden mußte, um den Filter überhaupt heben zu können (**Bild 4**). Da sich die Hebekonzeption geändert hatte, mußte der Filter vor dem Hub noch um 180° gedreht werden. **Bild 5** zeigt das Abheben des neuen Filters vom Transportfahrzeug.

Am 16. Dezember wurde der neue Filter bei zunehmend schlechter werdenden Wetterbedingungen mit Temperaturen knapp über 0°C und immer stärker aufkommendem

BILD 4: Trennen des Eintrittskonus vom vormontierten Filter
FIGURE 4: Removal of the inlet cone from the preassembled precipitator

TABELLE 2: Kenndaten des neuen und alten Elektrofilters
TABLE 2: Characteristic data for the old and new electrostatic precipitators

	Alt	Neu
Abmessungen in mm		
Länge	21675	23345
Breite	16800	16800
Höhe	14300	20015
Gassenzahl	56	40
Gassenabstand	300	400
Feldhöhe in mm	7850	13950
Feldlänge in mm	4320	4320
Abscheidefläche in m²	11249	14463
inst. Leistung in KW	179	357
Gasmenge in m³/s	147	147
Garantie für		
2 Felder in mg/Nm³	< 50	< 30

Elektrofilter: Kenndaten = Electrostatic precipitators: characteristic data
Alt = Old
Neu = New
Abmessungen in mm = Dimensions in mm
Länge = Length
Breite = Width
Höhe = Height
Gassenzahl = Number of gas passages
Gassenabstand = Collector spacing
Feldhöhe in mm = Compartment height in mm
Feldlänge in mm = Compartment length in mm
Abscheidefläche in m² = Collecting area in m²
Inst. Leistung in kW = Installed rating in kW
Gasmenge in m³/s = Gas flow in m³/s
Garantie für = Guarantee for
2 Felder in mg/Nm³ = 2 compartments in mg/m³(stp)

Nebel eingehoben, am 17. Dezember folgte der Eintrittsstutzen. Das Umsetzen und die dazu notwendigen Hilfskonstruktionen haben ca. 1,2 Mio. DM gekostet. Verteilerkonen und die Verbindungsrohrleitungen zur neuen Mahlanlage waren am 29. März 1993 fertiggestellt. Am 30.03.1993 begannen die Probeläufe der neuen Mahlanlage mit Feuerung. Am 15. April ging die Ofenanlage in Betrieb. **Bild 6** zeigt den neuen Filter mit den Rohrleitungen. Die technischen und verfahrenstechnischen Daten von altem und neuem Filter sind aus der **Tabelle 2** zu nehmen.

Die Inbetriebnahme des Filters war bis auf kleine, schnell zu behebende mechanische Störungen an den Austragsorganen unproblematisch. Sowohl im Direktbetrieb als auch im Verbundbetrieb werden die garantierten Werte im 3-Feld-Betrieb sicher eingehalten. Beim Umschalten vom Verbundbetrieb zum Direktbetrieb steigt der Reingasstaubgehalt kurzzeitig über den garantierten Wert, da die Gastemperatur von 90 – 95°C auf 180 – 185°C zunimmt. Die Rückkehr zu normalen Betriebswerten dauert eine halbe Stunde. Die Umschaltung vom Direktbetrieb zum Verbundbetrieb hat eine Einschwingzeit von etwa 25 Minuten. Durch Umstellung des Sprühsystems in den Kühltürmen soll die Temperatur im Direktbetrieb auf < 120°C herabgesetzt werden. Beim Umschalten wird sich die Gastemperatur dann nur noch um ca. 60°C ändern, was zu einer sehr schnellen Umschaltphase führt. Die momentan auftretenden Spitzen im Reingasstaubgehalt sollen damit vermieden werden.

BILD 5: Beginn des Hubes
FIGURE 5: Start of the lift

Bei Betrieb mit 2 Feldern stellen sich, je nachdem, welches Feld ausgeschaltet wurde, verschiedene Reingasstaubgehalte ein, die jedoch immer noch unter dem garantierten Wert liegen. Nach einer relativ kurzen Betriebszeit von nur 7 Monaten ist bereits festzustellen, daß der neue Filter den Erwartungen entspricht und Unregelmäßigkeiten im Ofenbetrieb nicht zu einer Überschreitung des Halbstundenmittelwertes führen.

removed again for it to be possible to lift the precipitator at all (**Fig. 4**). As the lifting plan had changed the precipitator had to be rotated through 180° before it was lifted. **Fig. 5** shows the new precipitator being lifted from the transport vehicle.

On 16th December the new precipitator was lifted into position under ever worsening weather conditions with temperatures just above 0°C and increasing fog; this was followed on 17th December by the inlet connection piece. The transfer and the necessary auxiliary construction work have cost about 1.2million DM. The distributor cones and the connecting ducts to the new grinding plant were completed on 29th March 1993. The test runs of the new grinding plant and combustion system started on 30.03.1993. The kiln plant went into operation on 15th April. **Fig. 6** shows the new precipitator and the ducts. The engineering and process data for the old and new precipitators are given in **Table 2**.

BILD 6: Gesamtansicht des neuen Elektrofilters
FIGURE 6: Overall view of the new electrostatic precipitator

The commissioning of the precipitator went smoothly with the exception of small mechanical faults in the extraction equipment which were quickly remedied. The guaranteed values are met reliably during 3-compartment operation both in direct and interconnected operation. When changing over from interconnected operation to direct operation the clean gas dust content rises above the guaranteed value for a short time because the gas temperature increases from 90 – 95°C to 180 – 185°C. It takes about half an hour to return to normal operating values. The changeover from direct to interconnected operation has a response time of about 25 minutes. The temperature during direct operation is to be reduced to less than 120°C by modifying the spray system in the cooling towers. The gas temperature will then only change by about 60°C during changeover, which will lead to a very rapid transition phase. This should avoid the brief peaks which appear in the clean gas dust content.

Different clean gas dust contents are obtained when operating with 2 compartments depending on which compartment had been switched off, but they are still all below the guarantee value. After a relatively short operating time of only 7 months it has already been established that the new precipitator fulfils the expectations and that irregularities in kiln operation do not cause the limits for the halfhour mean values to be exceeded.

Staubfreie Klinkerentladung von Schiffen – eine neue Problemlösung

Dust-free clinker unloading from ships – a new solution to the problem

Déchargement sans poussière de bâteaux transportant du clinker – une nouvelle solution du problème

Descarga del clínker de los barcos, sin producir polvo – una nueva solución del problema

Von **W. Heine**, Rheinberg/Deutschland

Zusammenfassung – Für die Entladung von Klinkerschiffen werden derzeit häufig Hafenkrane mit Seilgreifern eingesetzt. Extreme Staubbelästigung für das Bedienungspersonal und die Umwelt kennzeichnen daher die Schiffsentladeanlagen für Zementklinker. Hydraulisch betriebene Krane mit Gestängeausleger und Hydraulikgreifer ermöglichen dagegen eine starre Greiferführung mit hoher Umschlagleistung. Die sinnvolle Anordnung von Trichter und Abzugsförderern führt zu kurzen Greiferwegen. Der Abwurftrichter wird mit Kompaktfiltern und einer speziellen Rostabdeckung wirksam entstaubt. Eine technische Besonderheit stellt die Vorrichtung für die Entstaubung des Greifers dar. Durch die besondere Bauart des Krans mit Gestängeauslegern ist es möglich, Entstaubungsrohre bis zum Greifer zu führen, um über eine Ringdüse die unmittelbare Umgebung des Greifers zu entstauben. Im Rahmen des Entwicklungsprojekts wurden auch alternative Entstaubungskonzepte wie z.B. die Wasserbedüsung von nicht feuchtigkeitsempfindlichen Staubquellen untersucht.

Summary – At present harbour cranes with cable-operated grabs are often used for unloading clinker ships. Ship unloading systems for cement clinker therefore create a great deal of dust nuisance both for the operators and the environment. Hydraulically driven cranes with rod-linkage jibs and hydraulic grabs facilitate positive grab control with high handling capacities. Suitable positioning of the hopper and extraction conveyors leads to shorter grab travel. The discharge hopper is dedusted efficiently with compact filters and a special grid covering. One special technical feature is the device for dedusting the grab. Due to the particular design of the crane with rod-linkage jib it is possible to run dedusting pipes out to the grab so that the immediate vicinity of the grab can be dedusted by an annular nozzle. Alternative dedusting schemes, such as spraying water on dust sources which are not moisture sensitive, were also investigated as part of the development project.

Résumé – Pour le déchargement de vraquiers à clinker sont souvent utilisées, actuellement, des grues de port à benne prenante sur câbles. Une gêne extrême de poussière, pour le personnel de manutention et pour l'environnement, caractérise ainsi les installations de déchargements de bâteaux à clinker. Des grues à entraînement hydraulique, avec bras-poutre et benne hydraulique permettent, par contre, un guidage rigide de la benne avec une haute performance de manutention. La disposition judicieuse de la trémie et des transporteurs d'évacuation conduit à des trajets courts de la benne. La trémie de déversement est dépoussiérée efficacement à l'aide de filtres compacts et d'un couvercle spécial à grille. Le dispositif de dépoussiérage de la benne constitue une particularité technique. La construction particulière de la grue à bras-poutre permet de conduire des tubes de dépoussiérage jusqu'à la benne, afin de dépoussiérer l'alentour immédiat de la benne au moyen d'une buse annulaire. Dans le cadre du projet de développement, ont aussi été étudiés des concepts alteratifs du dépoussiérage, comme p. ex. l'aspersion d'eau de sources de poussière non sensibles à l'humidité.

Resumen – Para la descarga de los barcos utilizados para el transporte de clínker, se emplean actualmente más a menudo las grúas de los puertos, equipadas con cuchara y cables. La formación de polvo, que constituye una molestia grandísima para el personal de servicio y también para el medio ambiente, es, por lo tanto, característica de la descarga de los barcos que transportan el clínker de cemento. Las grúas de accionamiento hidráulico, con pluma de varillaje y cuchara hidráulica, permiten, por el contrario, un movimiento rígido de la cuchara, con una gran capacidad de transbordo. La disposición adecuada de la tolva y de los extractores hace que los recorridos de la cuchara sean muy cortos. La tolva de evacuación queda eficazmente desempolvada mediante filtros compactos y un recubrimiento especial de parrilla. Una particularidad técnica la constituye el dispositivo de desempolvado de la cuchara. Debido a la construcción de la grúa con plumas de varillaje, es posible prever unos tubos de desempolvado, que llegan hasta la cuchara, y desempolvar, de esta forma, el entorno de la misma, por medio de una tobera anular. Dentro del marco de este proyecto de desarrollo, se han estudiado también otros conceptos alternativos de desempolvado, por ejemplo la inyección de agua en las fuentes de polvo insensibles al agua.

1. Einleitung

Weltweit wird loser Zement mit Schiffen transportiert, wofür bereits staubarme Belade- und Entladevorrichtungen bekannt sind. Für die Entladung haben sich dort Vertikalschnecken oder aber Sauganlagen bewährt. Ähnlich kontinuierlich arbeitende Entladevorrichtungen für Zementklinker gibt es jedoch nicht.

Für die Entladung von Klinkerschiffen werden typische Hafenkrane mit Seilgreifern eingesetzt. Wegen der unterschiedlichen Kornfraktion des Klinkers können keine Sauganlagen zum Einsatz kommen. Der hohe Verschleiß verhindert auch den Einsatz von kontinuierlich arbeitenden Entladevorrichtungen, wie sie für den Erz- und Kohleumschlag üblich sind. Extreme Staubbelästigung für das Bedienungspersonal und die Umwelt kennzeichnen diese Art der Schiffsentladeanlagen für Zementklinker.

Seit Jahren liefert AUMUND staubarme Beladeanlagen für Eisenbahnwaggons, Schiffe und LKW's. Diese Anlagen sind mit einem besonderen Verschleißschutz ausgerüstet, um einen wartungsarmen Betrieb zu ermöglichen. Großzügig dimensionierte Enstaubungsanlagen ermöglichen den staubarmen Beladevorgang, entsprechend den Anforderungen des Umweltschutzes.

2. Prinzip

Für den Entladevorgang von Schiffen werden Greiferkrane (Schiffsentlader) mit einem zugehörigen Abwurftrichter eingesetzt. Dabei muß der Trichter der Greifergröße entsprechen und der Aufgabenstellung für den Weitertransport angepaßt sein.

2.1 Schiffsentlader

Als Schiffsentlader kommt ein hydraulisch betriebener Balancekran zum Einsatz, wie er heute vorzugsweise für den Schrottumschlag eingesetzt wird. Der Balancekran hat seinen Namen von einer Kontergewichtsverlagerung; das dabei eingesetzte Kontergewicht verlagert sich automatisch mit jeder Positionsänderung des Auslegers. Das ist eine wichtige Voraussetzung für die hohe Standfestigkeit des Kranes (**Bild 1**).

Die Krane werden fahrbar oder aber auch speziell für den Klinkerentlader stationär ausgeführt. Die exakte und zielgenaue Steuerung des Greifers ermöglicht eine Steigerung der Umschlagleistung von 20 bis 30 % bei gleicher Greifergröße. **Bild 2** zeigt einen Balancekran in stationärer Ausführung für

1. Introduction

Bulk cement is shipped worldwide, and relatively dust-free loading and unloading equipment is already familiar. Vertical screws, or else suction systems, have been found to work well in unloading, but there is no similar, continuously operating equipment for unloading cement clinker.

Typical harbour cranes with cable-operated grabs are employed to unload clinker ships. Suction systems cannot be used because of the different particle size fractions in the clinker. The heavy wear also prevents the use of continuously operating unloading equipment like that used for handling ore and coal. Ship unloading systems for cement clinker typically create a great deal of dust nuisance both for the operators and the environment.

AUMUND has been supplying low-dust loading systems for rail waggons, ships and lorries for some years. These have special wear protection so that they can operate with little maintenance. Generously dimensioned dust removal systems make it possible to load with little dust in compliance with environmental protection requirements.

2. The basic principle

Grab cranes with associated discharge hoppers are used for unloading ships. The hopper must be of an appropriate size for the grab and must be adapted to suit the subsequent transport system.

2.1 Ship unloaders

The ship unloader is a hydraulically operated balancing crane such as is nowadays used chiefly for handling scrap. It takes its name from the counterweight displacement; the counterweight used is automatically displaced each time the jib changes its position. This is an important prerequisite for good crane stability (**Fig. 1**).

The cranes may be of the travelling type or the stationary type specially for clinker unloaders. Accurate, well-targeted control of the grab enables the unloading performance to be increased by 20 to 30 % with a grab of the same size. **Fig. 2** shows the stationary type of balancing crane for unloading ships of 1000 t load capacity. Larger installations are available for ships with 3 000/10 000/20 000 t load capacities. It is for the larger ships that travelling balancing cranes are used, so that the most favourable working position can be adopted without moving the ship.

BILD 1:
Fahrbarer Balancekran für den Schüttgutumschlag
FIGURE 1:
Mobile balancing crane for handling bulk materials

die Entladung von Schiffen mit einer Ladekapazität von 1000 t. Größere Anlagen sind bestimmt für Schiffe mit einer Ladekapazität von 3000/10000/20000 t. Gerade für die größeren Schiffe werden fahrbare Balancekrane eingesetzt, damit die jeweils günstigste Arbeitsposition ohne Bewegung des Schiffes angesteuert werden kann.

2.2. Abwurftrichter

Je nach Aufgabenstellung gibt es verschiedene Ausführungsformen als

- stationärer Trichter mit Unterflurabzug,
- stationärer Trichter mit Überflurabzug,
- fahrbarer Trichter mit Kailängsband,
- stationärer oder fahrbarer Trichter zur Direktbeschickung von LKW's.

Dabei wird man der Bauart den Vorzug geben, welche die kürzesten Greiferwege vom Schiff zum Trichter ermöglicht. Zur Vermeidung der Staubentwicklung beim Öffnen des Greifers oberhalb des Abwurftrichters werden heute aufwendige Entstaubungsanlagen eingesetzt, die mit einem Luftmengendurchsatz von ca. 100000 m³/h ausgeführt werden.

Das **Bild 3** zeigt das neue Konzept eines Abwurftrichters, der durch den Greifer des Balancekranes beschickt wird. Der Abwurftrichter wird mit vier Kompaktfiltern ausgerüstet, die platzsparend an zwei Stirnseiten angeordnet sind. Der Luftmengendurchsatz für diese Filtereinheiten beträgt 18000 m³/h. Die Einwurffläche ist mit einem Rost abgedeckt, so daß Fremdkörper oberhalb des Rostes gehalten werden. Unterhalb des Rostes ist ein spezielles Klappensystem vorgesehen, welches sich nur bei der Klinkeraufgabe öffnet und nach dem Durchfall des Klinkers automatisch wieder schließt.

Um den Windeinfluß auszuschalten, erhält der Trichter auf allen vier Seiten oberhalb des Rostes eine 3 m hohe Windschutzwand. Die kompakte Bauart dieses Abwurftrichters ermöglicht auch die fahrbare Ausführung in Verbindung mit dem fahrbaren Balancekran.

Unterhalb des Trichters befinden sich Abzugsvorrichtungen, die nach dem Schwerkraftprinzip ohne besondere Staubentwicklung arbeiten. Hierfür eignen sich Plattenbänder oder Gurtbandförderer. Es besteht auch die Möglichkeit, den Trichter so hoch anzuordnen, daß eine direkte Beschickung von LKW's möglich ist. Für den Beladevorgang auf LKW's stehen Teleskopschurren mit Entstaubungsanlage aus dem Standardprogramm zur Verfügung. Für den Weitertransport des Klinkers von dem Entladetrichter bis zu den Klinkersilos werden Gurtbänder, Becherwerke und Trogkettenförderer eingesetzt.

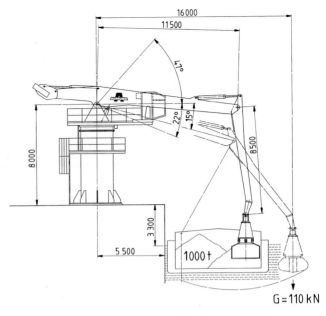

BILD 2: Stationärer Balancekran für die Schiffsentladung
FIGURE 2: Stationary balancing crane for ship unloading

2.2 Discharge hoppers

There are various forms of hopper, according to the application:

- stationary hoppers with underground discharge,
- stationary hoppers with above-ground discharge,
- travelling hoppers with a quayside conveyor,
- stationary or travelling hoppers for direct loading of lorries.

Preference will be given to the type giving the shortest grab travel from ship to hopper. Expensive dedusting systems with air throughputs of approx. 100000 m³/h are used nowadays to avoid creating dust when the grab is opened above the discharge hopper.

Fig. 3 shows the new design for a discharge hopper which is filled by the grab of the balancing crane. The hopper has four compact filters in a space-saving arrangement at two ends. The air throughput for these filter units is 18000 m³/h. The receiving area is covered with a grid to retain foreign bodies. Below the grid there is a special flap system, which opens only when clinker is fed in and recloses automatically when the clinker has dropped through.

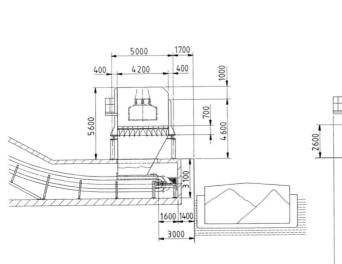
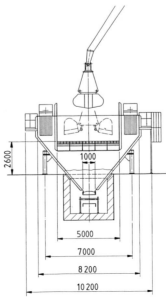

BILD 3:
Stationärer Abwurftrichter mit Kompaktentstaubung

FIGURE 3:
Stationary discharge hopper with compact dedusting system

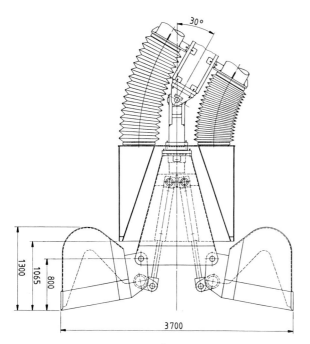

BILD 4: Hydraulische Greifer (4 m³ Inhalt) mit Greiferentstaubung
FIGURE 4: Hydraulic grab (4 m³ capacity) with grab dedusting system

3. Greiferentstaubung

Im Rahmen des Entwicklungsprojektes wurde auch die Möglichkeit einer Greiferentstaubung untersucht. Das **Bild 4** zeigt einen Hydraulikgreifer mit einem Inhalt von 4 m³, welcher für eine Entladeleistung von 300 t/h Zementklinker ausreichend dimensioniert ist. Die kardanische Aufhängung des Greifers ermöglicht eine allseitige Schiefstellung um 30°. Unterhalb des Kardangelenkes ist eine Saugglocke angeordnet, die über eine Ringdüse den Staub im Greiferbereich aufnimmt. Die Saugglocke ist über zwei flexible Rohrleitungen mit den festen Rohrleitungen des starren Auslegersystems des Balancekranes verbunden. Saugglocke und Rohrleitungsdurchmesser sind für einen Luftmengendurchsatz von 15 000 m³/h ausgelegt.

Unterhalb der Saugglocke befindet sich der Drehantrieb, mit welchem der Greifer selbst in die gewünschte Arbeitsposition gebracht werden kann. Zur Vermeidung der Staubentwicklung werden außerdem alternative Systeme untersucht. So wäre es denkbar, die Staubentwicklung durch eine Wasserbedüsung im Greiferbereich zu reduzieren. Diese Methode könnte auch im Bereich des Trichters von Interesse sein. Dabei ist jedoch besonders Wert auf die zuverlässige Arbeitsweise der Wasserdüsen zu legen.

4. Ausblick und Wertung

Durch den Einsatz eines hydraulisch betriebenen Balancekranes können Klinkerentladeanlagen in ihrem Einsatz optimiert werden. Damit sind Entladeleistungen von 300–600 t/h für Schiffsgrößen von 1 000–20 000 t möglich. Die Ausladung des Greifers beträgt dabei maximal 30–35 m.

Für die Module Hydraulikkran, Trichterausführung und staubarme Transportwege wurden inzwischen die technischen Vorklärungen abgeschlossen. Das trifft auch für die vorgestellte Greiferentstaubung zu. Sie soll an einem Prototyp erprobt werden. Dabei wird sowohl die Luftentstaubung als auch die Möglichkeit der Wasserbedüsung untersucht. Der Einsatz des Balancekranes in Verbindung mit dem Abwurftrichter in stationärer oder fahrbarer Ausführung ermöglicht schon jetzt eine erhebliche Leistungssteigerung bei gleichzeitiger Verminderung der Staubabgabe an die Umwelt.

The hopper has a 3 m high protective wall above the grid on all four sides in order to exclude the effects of wind. In this compact form the hopper may be a travelling model for use in conjunction with the travelling balancing crane.

Discharge equipment is provided below the hopper and works on the gravity principle without any particular dust formation. Apron or belt conveyors are suitable for the purpose. Another possibility is to arrange the hopper at a level where it can load lorries directly. Telescopic chutes with a dedusting system are available from the standard range for loading lorries. Belts, bucket elevators and troughed drag chain conveyors are used to carry the clinker from the discharge hopper to the silos.

3. Dedusting the grab

The possibility of dedusting the grab has been studied in a development project. **Fig. 4** shows a hydraulic grab with a capacity of 4 m³, which is large enough to unload 300 t cement clinker per hour. The grab is suspended by a universal joint, enabling it to be inclined by 30° in all directions. Below the universal joint there is a suction bell which picks up the dust in the vicinity of the grab through an annular nozzle. The suction bell is connected by two flexible pipes to the fixed pipes of the rigid jib system of the balancing crane. The suction bell and pipe diameter are dimensioned to give an air throughput of 15 000 m³/h.

The rotating drive by which the grab itself can be brought to the desired working position is located below the suction bell.

Alternative systems for avoiding dust formation are also being studied. Thus it would be possible to reduce dust formation by spraying water around the grab. This method might also be applied around the hopper, but reliable operation of the water nozzles would be particularly important in that case.

4. Prospects and assessment

The use of clinker unloading systems can be optimised by employing a hydraulically operated balancing crane. Unloading rates of 300 to 600 t/h are then possible for 1 000 to 20 000 t ships. The maximum working radius of the grab is 30 to 35 m.

Technical arrangements for the hydraulic crane, hopper construction and low-dust transport path modules have now been completed. This also applies to the above-mentioned dedusting of the grab, which is to be tested on a prototype. Both dedusting of air and the possibility of spraying water are being considered. The use of the balancing crane in conjunction with the discharge hopper of either the stationary or the travelling type is already increasing output considerably, while at the same time reducing dust pollution of the environment.

NO$_x$-Reduzierung im norwegischen Zementwerk Brevik —
Versuche mit gestufter Brennstoffzufuhr zum Calcinator *)

Reducing NO$_x$ at the Brevik cement works in Norway —
Trials with stepped fuel supply to the calciner *)

Réduction NO$_x$ à la cimenterie norvégienne Brevik –
Essais avec alimentation étagée en combustible au calcinateur

Reducción de NO$_x$ en la fábrica de cemento noruega de Brevik –
Ensayos efectuados con la alimentación escalonada de combustible al calcinador

Von **Ø. Hoidalen, A. Thomassen** und **T. Syverud,** Brevik/Norwegen

Zusammenfassung — Im Zementwerk Brevik der Norcem AS ist seit 1987 ein Ofen mit Low NO$_x$-Calcinator in Betrieb. Im Calcinator wird die Verbrennung mehrstufig geführt. Der Brennstoff wird dabei zu etwa 50% auf einen Low NO$_x$-Brenner und auf einen Calcinatorbrenner aufgeteilt. Als Calcinatorbrennstoff dient Kohlenstaub. Im Low NO$_x$-Bereich des Calcinators wurden versuchsweise sowohl Kohlenstaub über den Brenner als auch geschredderte Altreifen über eine Schurre eingesetzt (bis max. 15% des Gesamtwärmeverbrauchs der Ofenanlage). Gemessen wurden die Konzentrationen an NO$_x$, SO$_2$ und CO im Reingas mittels kontinuierlich arbeitender Meßgeräte. Die ersten Untersuchungen zeigten, daß bei Einsatz von Kohle als Reduktionsbrennstoff die NO$_x$-Emission zwar abnahm, die Konzentrationen des CO und SO$_2$ jedoch deutlich anstiegen. Demgegenüber war der Anstieg der CO- und SO$_2$-Konzentration bei Zugabe von geschredderten Reifen geringer. Weitere Versuche sollen Aufschluß über die Ursachen geben, insbesondere auch über den Einfluß der Ofenfahrweise im Dauerbetrieb. Bislang liegt das NO$_x$-Konzentrationsniveau ohne gestufte Verbrennung im Reingas zwischen 0,8 und 1,1 g NO$_2$/m$^3_{trN}$ bezogen auf 10 Vol.-% O$_2$.

NO$_x$-Reduzierung im norwegischen Zementwerk Brevik —
Vesuche mit gestufter Brennstoffzufuhr zum Calcinator

Summary — A kiln with a Low-NO$_x$ calciner for precalcination has been in operation at Norcem AS's Brevik works since 1987. The combustion is carried out in stages in the calciner. About 50% of the fuel is divided between a Low-NO$_x$ burner and a calciner burner. The calciner is fueled with pulverized coal. Only for testing the Low-NO$_x$ part of the calciner uses both pulverized coal through the burner and shredded used tyres through a chute (up to a maximum of 15% of the kiln plant's total energy consumption). The concentrations of NO$_x$, SO$_2$ and CO in the clean gas are measured by continuous emission measuring equipment. Initial investigations showed that when using coal as the reducing fuel the NO$_x$ emission did drop, but there was a significant increase in the concentration of CO and SO$_2$. On the other hand the increases in CO and SO$_2$ concentrations were less with the addition of shredded tyres. Further investigations should provide explanations about the causes, and in particular should also explain the effect on the kiln operating characteristics during continuous operation. So far the NO$_x$ concentration level in the clean gas without stepped combustion has been between about 0.8 and 1.1 g NO$_2$/m^3 (dry, standard state) at 10 vol.% O$_2$.

Reducing NO$_x$ at the Brevik cement works in Norway —
trials with stepped fuel supply to the calciner

Résumé — A l'usine Brevik de la Norcem AS est en service, depuis 1987, un four précalcination avec calcinateur Low-NO$_x$. Dans le calcinateur, la combustion s'effectue par étages. La part du combustible, d'environ 50%, y est partagée entre un brûleur Low-NO$_x$ et un brûleur de calcinateur. Le combustible pour précalcination est du charbon pulvérisé. Dans la zone Low-NO$_x$ du calcinateur ont été utilisés, à titre d'essai, aussi bien du charbon pulvérisé au brûleur, que des vieux pneus déchiquetés introduits par goulotte (jusqu'à 15% de la consommation d'énergie totale de la ligne de four). Les concentrations de NO$_x$, SO$_2$ et CO dans les gaz épurés ont été mesurées au moyen d'appareils de mesure d'émission travaillant en continu. Les premières investigations ont montré, que l'utilisation de charbon comme combustible de réduction abaissait bien l'émission NO$_x$, mais les concentrations CO et SO$_2$ augmentaient nettement. Avec addition de copeaux de pneus, par contre, l'augmentation des concentrations CO et SO$_2$ était moindre. D'autres essais doivent en préciser les causes, surtout aussi expliquer l'influence de la manière de conduire le four en marche continue. Jusqu'à présent, le niveau de concentration NO$_x$ sans combustion étagée se situe, dans les gaz épurés, entre à peu près 0,8 et 1,1 g NO$_2$/m^3 (n., s.) pour 10% vol. O$_2$.

Réduction NO$_x$ à la cimenterie norvégienne Brevik —
Essais avec alimentation étagée en combustible au calcinateur

---

*) Überarbeitete Fassung eines Vortrages zum VDZ Kongreß '93, Düsseldorf (27. 9.–1. 10. 1993)
Revised text of a lecture to the VDZ Congress '93, Düsseldorf (27.9.–1.10.1993)

Reducción de NO$_x$ en la fábrica de cemento noruega de Brevik —
Ensayos efectuados con la alimentación escalonada de combustible al calcinador

Resumen — *En la fábrica de cemento noruega de Brevik, de Norcem AS, está funcionando desde 1987 un horno con Low-NO$_x$-Calcinator. En este calcinador, se lleva a cabo la combustión en varias etapas. En este proceso, el combustible se reparte en un 50% aprox. entre un quemador-Low-NO$_x$ y un quemador de calcinador. Como combustible para el calcinador se emplea carbón pulverizado. En la zona Low-NO$_x$ del calcinador se introdujeron, a título de ensayo, tanto carbón pulverizado, a través del quemador, como neumáticos viejos previamente triturados, a través de una resbaladera (hasta un 15%, como máx., del consumo total de calor de la instalación de horno). Se midieron las concentraciones de NO$_x$, SO$_2$ y CO en el gas depurado, mediante unos aparatos de funcionamiento continuo. Los primeros estudios demostraron que con el empleo de carbón como combustible reductor, la emisión de NO$_x$ disminuía, pero que las concentraciones de CO y SO$_2$ aumentaban notablemente. Sin embargo, el aumento de las concentraciones de CO y SO$_2$ eran más bajas al añadirse neumáticos triturados. Se llevaron a cabo luego otros ensayos, con el fin de averiguar las causas y, en especial, también el influjo de la marcha del horno en servicio continuo. Hasta ahora, el nivel de concentración de NO$_x$, sin combustión escalonada, se sitúa en el gas depurado entre 0,8 y 1,1 g de NO$_2$/m$^3_{trN}$, referido al 10% de O$_2$ en volumen.*

1. Einleitung

Im Jahre 1987 wurde der mit Zyklonvorwärmer ausgerüstete Drehofen 6 im Zementwerk Brevik der NORCEM AS auf das Vorcalcinierverfahren umgestellt. Im Ergebnis von umwelttechnischen Überlegungen war entschieden worden, den zur damaligen Zeit gerade erst neu entwickelten Pyroclon RP Low-NO$_x$-Calcinator zu installieren. Dieses Calcinatorsystem ermöglicht bekanntlich die Anwendung einer gestuften Verbrennung im Calcinator mit Hilfe des sogenannten Low-NO$_x$-Brenners (**Bild 1**). Im folgenden werden einige Ergebnisse aus verschiedenen Untersuchungen an diesem Calcinatorsystem mitgeteilt.

2. NO$_x$-Bildung und Reduktionsmechanismen

2.1 Allgemeines

Es ist bekannt, daß sich NO$_x$ im Zementofen als Kombination von thermischem und Brennstoff-NO$_x$ bildet. Da das thermische NO$_x$ im Hinblick auf die NO$_x$-Bildung von besonderer Bedeutung ist, besitzt ein mit Vorcalcinator ausgerüsteter Drehofen im Vergleich zu einem Drehofen mit Zyklonvorwärmer niedrigere NO$_x$-Emissionen, da wesentlich weniger Brennstoff für den Hochtemperaturprozeß eingesetzt werden muß.

Die NO$_x$-Emissionen von Drehofen 6 im Zementwerk Brevik schwanken normalerweise zwischen 800 und 1100 mg/m^3_N, bezogen auf einen O$_2$-Gehalt von 10%. Die NO$_x$-Emission liegt damit höher als bei einigen nach dem konventionellen Trockenverfahren arbeitenden Öfen, was schließlich die Vermutung nahe legt, daß auch andere Parameter, wie z.B. das Rohmehl-Brennverhalten, das Klinkerkühlersystem usw. eine Bedeutung auf die NO$_x$-Emission haben müssen. Weitergehende Untersuchungen werden deshalb noch erforderlich sein, um die verschiedenen Bildungsbedingungen von NO$_x$ in Zementöfen zu klären.

2.2 NO$_x$-Reduktionsmechanismen

Eine Reihe von Reaktionsabläufen ist bekannt, bei welchen das NO$_x$ zu molekularem Stickstoff (N$_2$) reduziert wird. Die wichtigsten Bildungsmechanismen in einem mit Calcinator ausgerüsteten Zementofen sind die NO/CO-Reaktionen, dargestellt durch die chemische Gleichung:

$$NO + CO \rightarrow \tfrac{1}{2} N_2 + CO_2 \tag{1}$$

und die sog. Nachverbrennungs-Reaktionen, dargestellt durch die chemischen Gleichungen

$$CH_i + NO \rightarrow XN + ... \tag{2}$$

$$XN + NO \rightarrow N_2 + ... \tag{3}$$

In den letzten Gleichungen bedeuten:

CH: Kohlenwasserstoff

XN: Cyanide/Amine

Die Effektivität der NO/CO-Reaktionen wird neben anderen insbesondere durch solche Parameter wie die Temperatur,

1. Introduction

In 1987 the suspension preheater kiln no. 6 at the BREVIK cement works of NORCEM AS was converted to the precalcining process. For environmental reasons it was decided to install the newly developed Pyroclon RP Low-NO$_x$ calciner. This system enables stepped combustion to be carried out in the precalciner with the Low-NO$_x$ burner (**Fig. 1**). This report summarizes some of the findings from different tests carried out on this precalciner unit.

2. NO$_x$ formation and reduction mechanisms

2.1 General

It is known that NO$_x$ is formed in cement kilns as a combination of thermal and fuel NO$_x$. Thermal NO$_x$ is of particular importance in the formation of NO$_x$, so a precalciner kiln has lower overall NO$_x$ emissions than a preheater kiln because less fuel is used for the high temperature process.

The NO$_x$ emission level from kiln no. 6 in Brevik normally varies between 800 and 1100 mg/m$^3_{stp}$ at 10% O$_2$. This is

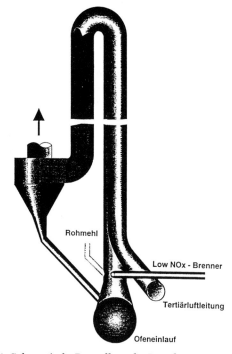

BILD 1: Schematische Darstellung des Pyroclon RP Low-NO$_x$-Calcinators
FIGURE 1: Diagrammatic representation of the Pyroclon RP Low-NO$_x$ calciner

Rohmehl = raw meal
Low-NO$_x$-Brenner = Low-NO$_x$ burner
Tertiärluftleitung = tertiary air duct
Ofeneinlauf = kiln inlet

die CO-Konzentration und die Verweilzeit beeinflußt. Bei den Nachverbrennungs-Reaktionen wird das NO durch Zwischenprodukte des Verbrennungsprozesses reduziert.

3. Testergebnisse und Bewertung

3.1 NO$_x$-Minderung durch die Verbrennung von Kohle mit dem Low-NO$_x$-Brenner

Bei der Bestimmung der Wirksamkeit der NO$_x$-Minderung in Öfen mit Vorcalcinierung besteht das Hauptproblem darin, herauszufinden, in welchem Grade zusätzlich NO$_x$ im Calcinator gebildet wird.

Um die Effektivität der NO$_x$-Minderung im Zementwerk Brevik messen zu können, wurde der Umwandlungsgrad $\emptyset$ eingeführt, der wie folgt definiert ist:

$$\emptyset = \left(\frac{\dot{m}_{NOxOE} - \dot{m}_{NOxOS}}{\dot{m}_{NOxOE}}\right) \cdot 100\% \quad (4)$$

Darin bedeuten:

OE Ofeneinlauf

OS Ofenschornstein

Ein negativer Wert von $\emptyset$ bedeutet, daß NO$_x$ im Calcinator gebildet wurde. Ein positiver Wert von $\emptyset$ bedeutet, daß NO$_x$ im Calcinator reduziert wurde.

Die Gleichung (4) drückt die Veränderung des NO$_x$-Masseflusses vom Drehofen bis hin zum Ofenschornstein in Prozent der im Ofen erzeugten NO$_x$-Menge aus. Der Gasmengenfluß wird dabei berechnet auf der Grundlage der Brennstoff- und Gaszusammensetzung, der Stöchiometrie der Verbrennung sowie der chemischen Reaktionen im Drehofen, da direkte Gasmengenmessungen nicht möglich sind.

Ein Zahlenwert von $\emptyset = -50\%$ bedeutet, daß 50% zusätzliches NO$_x$ im Calcinator gebildet wurde. Wie aus **Bild 2** hervorgeht, ist der NO$_x$-Umwandlungsgrad erwartungsgemäß von der NO$_x$-Konzentration am Ofeneinlauf abhängig. Schlußfolgernd kann deshalb auch angenommen werden, daß die NO$_x$-Erzeugung im Calcinator ziemlich konstant verläuft. Aus dem Diagramm geht weiterhin hervor, daß bis zu 50% des Gesamt-NO$_x$, dargestellt durch einen Umwandlungsgrad von $\emptyset = -100\%$, im Calcinator erzeugt werden. Die durchschnittliche NO$_x$-Emission von Ofen 6 korreliert demzufolge bei ca. 1000 mg/m^3_N mit einem $\emptyset = -50\%$, was bedeutet, daß im Durchschnitt 1/3 des NO$_x$-Gehaltes im Calcinator erzeugt wird.

3.2 NO$_x$-Minderung als Funktion der CO-Konzentration im Calcinator

Um den Zusammenhang zwischen dem NO$_x$-Umwandlungsgrad und der CO-Konzentration im Calcinator zu ermitteln, wurde der Low-NO$_x$-Brenner mit einer Brennstoffmenge von 1,0 bis 2,5 t/h, entsprechend 10 bis 30% des in den Calcinator eingebrachten Brennstoffs betrieben. Das führte zu CO-Konzentrationen im Calcinator zwischen 0,5

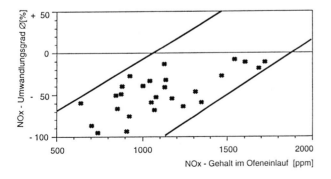

BILD 2: NO$_x$-Umwandlungsgrad $\emptyset$ in Abhängigkeit vom NO$_x$-Gehalt im Ofeneinlauf

FIGURE 2: Degree of conversion $\emptyset$ for NO$_x$ as a function of the NO$_x$ concentration at the kiln inlet

NO$_x$-Umwandlungsgrad $\emptyset$ [%] = NO$_x$ degree of conversion [%]
NO$_x$-Gehalt im Ofeneinlauf [ppm] = NO$_x$ concentration at kiln inlet [ppm]

higher than for some preheater kilns operating with the conventional dry process and suggests that other parameters, such as raw meal burnability, cooler system, etc., must also be of importance for the overall NO$_x$ formation. Further studies are therefore required in order to explain the different conditions under which NO$_x$ is formed in cement kilns.

2.2 NO$_x$ reduction mechanism

A series of reaction paths are known in which the NO$_x$ is reduced to molecular nitrogen (N$_2$). The most important mechanisms likely to occur in a cement kiln with calciner are the NO/CO reactions, represented by the chemical equation:

$$NO + CO \rightarrow \tfrac{1}{2} N_2 + CO_2 \quad (1)$$

and the so-called reburning reactions, represented by the chemical reactions

$$CH_i + NO \rightarrow XN + ... \quad (2)$$

$$XN + NO \rightarrow N_2 + ... \quad (3)$$

In these last equations

CH = hydrocarbon

XN = cyanides/amines

The effectiveness of the NO/CO reactions is to some extent influenced by, among other things, parameters such as temperature, CO concentration and retention time. Reburning reactions are based on the principle that NO is reduced by intermediate products from the combustion process.

3. Test results and evaluation

3.1 NO$_x$ reduction by burning coal with the Low-NO$_x$ burner

A major problem in connection with the determination of the effectiveness of NO$_x$ reduction in precalciner kilns is assessing the extent to which additional NO$_x$ is formed in the precalciner.

In order to measure the effectiveness of NO$_x$ reduction techniques at the Brevik cement works the degree of conversion $\emptyset$ was introduced, which is defined as follows:

$$\emptyset = \left(\frac{\dot{m}_{NOxOE} - \dot{m}_{NOxOS}}{\dot{m}_{NOxOE}}\right) \cdot 100\% \quad (4)$$

in which

OE is the kiln inlet

OS is the kiln stack

A negative value of $\emptyset$ signifies that NO$_x$ is formed in the calciner, and a positive value of $\emptyset$ signifies that NO$_x$ is reduced in the calciner.

Equation (4) expresses the change in mass flow of NO$_x$ from the rotary kiln to the kiln stack as a percentage of the quantity of NO$_x$ generated in the rotary kiln. The gas flow is calculated from the composition of the fuel and gas as well as from the stoichiometry of the combustion and of the chemical reactions in the rotary kiln, since direct measurements of the gas volumes are not possible.

A numerical value of $\emptyset = -50\%$ means that 50% additional NO$_x$ is formed in the precalciner. As can be seen from **Fig. 2**, the degree of conversion of NO$_x$ is, as expected, dependent on the NO$_x$ concentration at the kiln inlet. As a consequence it can be assumed that the NO$_x$ generation in the precalciner is fairly constant. The diagram also shows that up to 50% of the total NO$_x$, represented by a value $\emptyset = -100\%$, is generated in the precalciner. The average NO$_x$ emission from kiln no. 6 is about 1000 mg/m$^3_{stp}$, correlating with $\emptyset = -50\%$, which means that on average 1/3 of the NO$_x$ is formed in the precalciner.

bis 1,8 %, gemessen 5 m oberhalb des Low-NO$_x$-Brenners im Steigrohrkanal der Ofenanlage.

Bild 3 zeigt deutlich, daß eine NO$_x$-Minderung deshalb erreicht wird, weil der Low-NO$_x$-Brenner wie bereits dargestellt bei positiven ∅-Werten arbeitet. Im Bild wurde auch ein Meßwert aufgenommen, bei dem der Low-NO$_x$-Brenner nicht in Betrieb war. Nach Bild 3 hat es den Anschein, als würde kein fester Zusammenhang zwischen dem NO$_x$-Umwandlungsgrad ∅ und dem CO-Niveau im Calcinator bestehen. Die Meßpunkte scheinen auch unabhängig vom NO$_x$-Niveau am Ofeneinlauf zu sein. Wie dem Diagramm auf Bild 2 entnommen werden kann, ist eine Korrelation zwischen ∅-Wert und NO$_x$-Konzentration am Ofeneinlauf durchaus denkbar, wenn der Low-NO$_x$-Brenner nicht in Betrieb ist. Dieser Befund ist in erster Linie darauf zurückzuführen, daß die NO/CO-Reaktion nicht die wichtigste Rolle für die NO$_x$-Minderung im Vorcalcinierofen spielt und andere Reduktionsmechanismen, z.B. die Nachverbrennungs-Reaktion, stattzufinden scheinen.

Der Einsatz eines Low-NO$_x$-Brenners erhöht den NO$_x$-Umwandlungsgrad im Mittel auf ca. +20 %, wie aus Bild 3 entnommen werden kann. Bei diesem Level wird die anfängliche NO$_x$-Emission von 1000 auf 550 mg/m^3_N, oder um annähernd 45 % gesenkt.

3.3 Randeffekte des Low-NO$_x$-Brenners

Während der Versuche mit dem Low-NO$_x$-Brenner wurden kontinuierlich CO- und SO$_2$-Messungen durchgeführt. Der Betrieb mit dem Low-NO$_x$-Brenner ergab zum Teil bedeutend höhere CO- und SO$_2$-Emissionen. Weiterführende Tests mit dem Low-NO$_x$-Brenner verfolgen das Ziel, die optimalen Betriebsparameter zu ermitteln, bei denen eine wirkungsvolle NO$_x$-Minderung ohne Anstieg der CO- und SO$_2$-Emissionen gewährleistet ist. Als eine Konsequenz dieser hier dargestellten Erkenntnisse ist der Low-NO$_x$-Brenner deshalb auch nicht im ständigen Betriebseinsatz gewesen.

4. NO$_x$-Minderung mit Autoreifen

Es ist eine praktische Erfahrung, daß durch die Reifenverbrennung die NO$_x$-Emissionen gesenkt werden. Gestützt auf die Erfahrungen mit dem Low-NO$_x$-Brenner und auf theoretische Kenntnisse, wurden auch Experimente mit zerkleinerten Autoreifen am Ofen 6 des Zementwerkes Brevik durchgeführt. Der Ofen wurde bei diesen Experimenten mit einem Luftüberschuß im Ofeneinlauf zwischen 4 und 7 % betrieben, während dem Calcinator die Autoreifen in einer Menge von bis zu 3 t/h aufgegeben wurden. Das untere Diagramm in **Bild 4** zeigt, wie der Kohleverbrauch des Calcinators durch die Reifenaufgabe reduziert werden konnte, während das obere Diagramm die Entwicklung der NO$_x$-Emission enthält, die bis zu 50 % gesenkt werden konnte. Zur gleichen Zeit konnte im Calcinator kein CO an der Meß-

3.2 NO$_x$ reduction as a function of CO concentration in the precalciner

In order to determine the relationship between the NO$_x$ degree of conversion and the CO concentration in the precalciner, the Low-NO$_x$ burner was operated at fuel feed rates between 1.0 and 2.5 t/h, corresponding to 10 to 30 % of the fuel used in the precalciner. This resulted in CO concentrations in the precalciner between 0.5 and 1.8 % measured 5 m above the Low-NO$_x$ burner in the kiln riser duct.

Fig. 3 shows clearly that a NO$_x$ reduction is obtained because, as already shown, the Low-NO$_x$ burner operates with positive ∅ values. The diagram includes one measurement taken when the Low-NO$_x$ burner was not in operation. There appears, however, to be no fixed relationship between the NO$_x$ degree of conversion ∅ and the CO level in the precalciner. The measurements also seem to be independent of the NO$_x$ level at the kiln inlet. As can be seen from Fig. 2, the correlation between ∅ and NO$_x$ at the kiln inlet is, however, reasonable when the Low-NO$_x$ burner is not in operation. This is primarily attributed to the fact that the NO/CO reaction is not necessarily the most important for NO$_x$ reduction in precalciner kilns, and other reaction mechanisms, e.g. the reburning reactions, appear to take place.

The Low-NO$_x$ burner increases the NO$_x$ degree of conversion to an average of about +20 %, as shown in Fig. 3. This level reduced the initial NO$_x$ emission from about 1000 mg/m$^3_{stp}$ to about 550 mg/m$^3_{stp}$ or by approximately 45 %.

3.3 Side effects of the Low-NO$_x$ burner

During the tests with the Low-NO$_x$ burner, continuous measurements of CO and SO$_2$ emissions were carried out and at times the operation with the Low-NO$_x$ burner resulted in significantly higher CO and SO$_2$ emissions. Further tests with the Low-NO$_x$ burner are being aimed at finding optimum operating parameters, at which efficient NO$_x$ reduction is ensured without an increase in CO and SO$_2$ emissions. As a consequence of the above findings the Low-NO$_x$ burner has not been in continuous operation at the Brevik plant.

4. NO$_x$ reduction with car tyres

Practical experience has shown that NO$_x$ emissions are reduced by burning car tyres. Experiments with shredded car tyres were also carried out at kiln no. 6 at the Brevik cement works based on the experience with the Low-NO$_x$ burner and on theoretical findings. During these experiments the kiln was operated with 4 to 7 % excess air at the kiln inlet, and the car tyres were fed to the precalciner unit at rates of up to 3 t/h. The lower part of **Fig. 4** shows how the coal consumption in the calciner can be reduced by the tyre feed, and the upper part of the figure shows the NO$_x$ emissions, which were reduced by up to 50 %. At the same time,

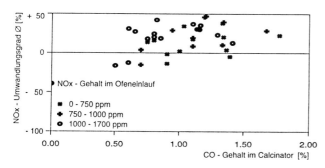

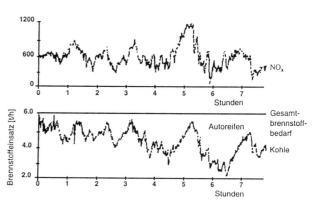

BILD 3: NO$_x$-Minderung in Abhängigkeit vom CO-Gehalt im Calcinator bei Einsatz des Low-NO$_x$-Brenners
FIGURE 3: NO$_x$ reduction as a function of the CO concentration in the calciner when using the Low-NO$_x$ burner

NO$_x$-Umwandlungsgrad ∅ [%]	= NO$_x$ degree of conversion [∅%]
NO$_x$-Gehalt im Ofeneinlauf	= NO$_x$ concentration at kiln inlet
CO-Gehalt im Calcinator [%]	= CO concentration in calciner [%]

BILD 4: NO$_x$-Emissionen bei der Reifenverbrennung
FIGURE 4: NO$_x$ emissions when burning car tyres

Brennstoffeinsatz [t/h]	= fuel feed [t/h]
NO$_2$-Gehalt mg/m^3_N	= NO$_2$ concentration in mg/m^3 stp
Stunden	= hours
Autoreifen	= car tyres
Gesamtbrennstoffbedarf	= total fuel consumption
Kohle	= coal

stelle registriert werden, die schon bei den Versuchen mit dem Low-NO_x-Brenner benutzt worden war. Die wirkungsvolle NO_x-Minderung über die Verbrennung von Autoreifen ist möglicherweise auf die Nachverbrennungs-Reaktion zurückzuführen, unterstützt durch den höheren Gehalt der Autoreifen an flüchtigen Bestandteilen. In Abhängigkeit von den Abmessungen des geschredderten Reifenmaterials können dabei auch lokale Reaktionen mit dem CO stattfinden. Die NO_x-Minderung kann schließlich auch dadurch begünstigt werden, daß Autoreifen im Vergleich zur Kohle die Eigenschaft haben, weniger Brennstoff-NO_x zu bilden. Die zur NO_x-Minderung verwendete Kohle besaß ca. 30% flüchtige Bestandteile mit einem Stickstoffgehalt um 1,9%. Im Vergleich dazu haben Autoreifen einen Anteil von 65% an Flüchtigem und einen Stickstoffgehalt von ca. 0,3%.

Der niedrige Gehalt an Stickstoff in den Autoreifen im Vergleich zur Kohle kann allerdings nicht die NO_x-Minderung von bis zu 50% erklären, die einer NO_x-Emission um 500 mg/m^3_N bei einem O_2-Gehalt von 10% entspricht. Während der Versuche mit Autoreifen wurde kein Anstieg der CO- oder SO_2-Emissionen nachgewiesen. Langzeitversuche haben allerdings gezeigt, daß betriebstechnische Schwierigkeiten, wie z. B. Schwefelkreisläufe im Calcinator, schon bei einer Reifenaufgabe oberhalb 1 t/h auftreten können.

5. Schlußbemerkungen

Experimente mit gestufter Verbrennung von Kohle im Calcinator sowie geschredderten Autoreifen im Bereich der Ofeneinlaufzone haben zu einer beachtlichen Reduzierung der NO_x-Emissionen geführt. Die Mechanismen, die hinter den erreichten Resultaten stehen, werden in den NO/CO- und Nachverbrennungs-Reaktionen vermutet. Gegenwärtig wird den letzteren Reaktionen besondere Aufmerksamkeit gewidmet, indem hochflüchtige Sekundärbrennstoffe zur Anwendung gebracht werden. Der Einsatz solcher Brennstoffe stellt bei heutigem Kenntnisstand einen Weg dar, die NO_x-Emissionen beim Klinkerbrennen wirkungsvoll zu senken.

no CO was detected in the precalciner at the measuring point used during the tests with the Low-NO_x burner. The efficiency of NO_x reduction with car tyres can possibly be attributed to reburning reactions, enhanced by the higher content of volatile components in the car tyres. Local reduction reactions with CO may, however, also take place depending on the dimensions of the shredded tyre material. The NO_x reduction may also be assisted by the reduced tendency of car tyres to form fuel NO_x when compared with coal. The coal used for NO_x reduction has about 30% volatile matter with a nitrogen content of about 1.9%, compared with about 65% volatiles and approximately 0.3% nitrogen in the car tyres.

The lower amount of nitrogen in the car tyres compared to the coal cannot, however, explain the NO_x reduction of up to 50%, corresponding to a NO_x emission of about 500 mg/m^3_{stp} at 10% O_2. During the tests with car tyres, no increase of CO or SO_2 emission was detected. Long-term tests with car tyres have, however, revealed that operational difficulties, e.g. sulphur recirculation in the precalciner, occur at car tyre feed rates above 1 t/h.

5. Conclusion

Experiments with stepped combustion of coal in the precalciner and with shredded car tyres fed into the kiln inlet zone have led to major reductions in NO_x emissions. The mechanisms behind these results are assumed to be NO/CO reactions and reburning reactions. At present, particular attention is being paid to the latter reactions by using high-volatile secondary fuels. Current findings indicate that the use of such fuels represents an effective way of lowering NO_x emissions during clinker burning.

Norwegische Umweltanforderungen an den Einsatz von flüssigen Ersatzbrennstoffen in Drehrohröfen

Norwegian environmental requirements when using liquid substitute fuels in rotary tube kilns

Exigences norvégiennes côté environnement pour l'utilisation de combustibles alternatifs liquides dans les fours rotatifs

Exigencias medioambientales noruegas en cuanto al empleo de combustibles secundarios líquidos en los hornos rotatorios

Von Ø. Høidalen, Brevik/Norwegen

Zusammenfassung – NORCEM verbrennt seit 1980 organische Flüssigabfälle. Wiederholte Versuche haben gezeigt, daß sowohl die Schadstoffemissionen als auch die Produktqualität (bautechnische Eigenschaften und Umweltverträglichkeit) nicht negativ beeinflußt werden. Die gesetzlichen Randbedingungen, die dafür gelten, entsprechen internationalen Anforderungen. Das Werk Brevik der NORCEM AS ist der alleinige staatlich autorisierte Entsorger für bestimmte flüssige organische Reststoffe. Damit leistet die norwegische Zementindustrie einen wesentlichen Beitrag in einem ökologischen Entsorgungskonzept. Der Einsatz dieser sekundären Brennstoffe findet in der Gemeinde Brevik breite Akzeptanz.

Summary – Since 1980 NORCEM has been burning organic liquid wastes. Repeated trials have shown that there is no detrimental effect on the emission of harmful substances or on the product quality (structural properties and environmental compatibility). The legal conditions which apply in this case comply with international requirements. NORCEM AS's Brevik works is the sole government authorized disposer of certain liquid organic residues, so the Norwegian cement industry is making an important contribution in an ecological waste disposal scheme. The use of these secondary fuels is widely accepted in the Brevik community.

Résumé – Depuis 1980, NORCEM brûle des déchets organiques liquides. Des essais répétés ont montré, que les émissions de substances nocives, aussi bien que la qualité du produit (propriétés d'emploi dans la construction et compatibilité avec l'environnement) n'étaient pas influencées négativement. Le cadre législatif applicable correspond aux exigences internationales. L'usine Brevik de la NORCEM est le seul destructeur autorisé par l'Etat pour certains déchets organiques liquides. Ainsi, l'industrie cimentière norvégienne contribue efficacement au projet d'élimination écologique. L'emploi de ces combustibles secondaires a été largement accepté dans la commune de Brevik.

Resumen – NORCEM viene quemando, desde 1980, residuos orgánicos líquidos. Repetidos ensayos han demostrado que ni las emisiones de sustancias nocivas ni la calidad del producto (características técnicas e inocuidad para el medio ambiente) se ven afectados por ello de forma negativa. Los requisitos legales aplicables al respecto cumplen las normas internacionales. La factoría Brevik de NORCEM AS es la única empresa autorizada oficialmente para eliminar determinadas sustancias orgánicas líquidas. Con ello, la industria del cemento noruega contribuye, de forma eficaz, a llevar a efecto un proyecto de eliminación ecológica. El empleo de los mencionados combustibles secundarios ha encontrado una amplia aceptación en el municipio de Brevik.

1. Gegenwärtige Situation

Nach den Informationen der Europäischen Union wird der Gesamtanteil an gefährlichen Abfallstoffen innerhalb der Mitgliedsländer auf 2 bis 20% des insgesamt erzeugten Abfalls geschätzt. Die große Schwankungsbreite von 2 bis 20% erklärt sich dadurch, daß für gefährliche Abfallstoffe unterschiedliche Definitionen bestehen und eine harmonisierte Nomenklatur noch fehlt. Es wird eingeschätzt, daß in Norwegen insgesamt 200 000 t an gefährlichen Abfallstoffen

1. Present situation

Following information from the European Community, the total amount of hazardous wastes within the Member States is estimated to be 2–20% of the total waste generated. The large variations in the amount of "Hazardous Waste" may be a result of different definitions and lack of harmonized nomenclature for this type of waste. It is estimated that in total 200 000 tons of hazardous wastes are generated in Norway every year. This represents approximately 10% of the

jährlich anfallen. Dieser Zahlenwert stellt annähernd 10% des Gesamtabfalls dar, der jährlich in den Gemeinden anfällt.

Außerhalb dieser auf jährlich 200 000 t geschätzten Abfallmenge verfügt die norwegische Industrie über eine eigene Aufbereitungskapazität von ca. 80 bis 90 000 t, so daß jährlich eine Kapazität von ca. 110 bis 120 000 t zur Aufbereitung und/ oder Beseitigung übrig bleibt. Die gegenwärtig als gefährlich eingestuften Abfallstoffe sind in insgesamt 17 verschiedene Arten eingestuft worden (**Tabelle 1**). Mischungen aus verschiedenen Kategorien von organischen Abfällen, wie ölige Emulsionen, Farben, Lösungsmitteln, usw., ausgenommen Abfallöle, sollen nachfolgend als gefährliche Flüssigabfälle bezeichnet werden. In den letzten Jahren betrug die Menge an gefährlichen Abfällen, die an Sammelstellen abgeliefert wurden, nur ca. 50 bis 60 000 t/a, was nur 50% der geschätzten Gesamtmenge entspricht. Eine beträchtliche Menge an gefährlichen Abfällen wurde demzufolge nicht gesammelt und entsprechend behandelt. Deshalb hat die norwegische Regierung erklärt, daß es dringend notwendig sei, ein System zur Ablieferung, Sammlung und Behandlung von gefährlichen Abfällen zu schaffen, um zu verhindern, daß diese Abfälle irgendwo untertauchen. Schon im Jahre 1988 schlug deshalb die Regierung vor, eine zentrale Verbrennungsanlage für gefährliche Abfälle zu errichten. Aus ökonomischen, technischen als auch Umweltzwängen wurde jedoch 1991 entschieden, daß eine derartige Anlage nicht gebaut wird. Den Hintergrund für diese Entscheidung bildeten hauptsächlich neue Umweltauflagen, die zu wesentlich höheren Kosten als ursprünglich angenommen führten. Allerdings wurden auch Unsicherheiten über die in der neuen Anlage zu behandelnde Abfallmenge zur Debatte gestellt. Schließlich stellten sich auch Umweltgruppen der Errichtung von Verbrennungsanlagen entgegen und klagten die Regierung an, die Priorität nicht auf die Minimierung der Abfallentstehung zu setzen. Gegenwärtig ist die Regierung damit befaßt, verschiedene Alternativen zur Behandlung von gefährlichen Abfallstoffen zu sondieren.

Die norwegische Zementindustrie hält die Verbrennung von gefährlichen Flüssigabfällen in Zementdrehöfen für eine realistische Alternative zu anderen Methoden der Abfallbeseitigung.

total waste which annually is delivered to municipal waste facilities.

Out of the 200 000 tons which are generated every year, the Norwegian industry's own treatment capacity is around 80–90 000 tons, which results in the requirement of treatment and/or disposal facility for around 110–120 000 tons per year. Wastes presently classified as hazardous wastes in Norway are grouped into total 17 types as shown in **Table 1**. Mixtures of different categories of organic wastes such as oil emulsions, paints, solvents etc. and excluding waste oil, shall hereafter be called liquid hazardous wastes.

In recent years the amount of hazardous wastes delivered to approved facilities has only been around 50–60 000 tons per year, which is only 50% of total estimates. A considerable amount of hazardous waste is therefore not collected and treated adequately and the Norwegian government states that "there is an urgent need for establishment of a system for delivery, collection and treatment of hazardous wastes to prevent this type of waste from going astray". The Norwegian government therefore in 1988 proposed to establish a central incineration plant for treatment of hazardous wastes. However, due to economical, technical and environmental constraints, it was in 1991 decided not to build such a plant. The background for this decision were mainly new environmental standards which led to considerable higher destruction costs than previously estimated. However, also uncertainties with respect to the total amount of wastes to be treated in the new plant were questioned. Last but not least, environmental pressure groups totally opposed the establishment of the proposed facilities, mainly because they accused the government of failure to give priority to the minimization of the generation of wastes. The government is at present considering various alternatives for treatment of hazardous waste. The Norwegian cement industry considers that the burning of liquid hazardous wastes in cement kilns is a viable alternative to other disposal methods as shown in this article.

2. Burning of liquid waste

NORCEM started with the burning of liquid hazardous wastes at its Slemmestad plant near Oslo in 1980. The background for NORCEM's involvement in the development of

TABELLE 1: Arten gefährlicher Abfälle in Norwegen

	Art des gefährlichen Abfallstoffes	t/a
1	Abfallöl	34 500
2	Ölrückstände	8 000
3	Ölemulsionen	5 700
4	Organische Lösungen	7 000
5	Farben, Klebstoffe, Lacke, etc.	5 700
6	Destillationsrückstände	800
7	Teer	2 600
8	Abfall mit Hg und Cd	300
9	Wasserlösliche Schwermetallverbindungen	20 000
10	Cyanidhaltige Abfälle	20
11	Pestizide	350
12	PCB-haltige Abfälle	250
13	Isocyanate	70
14	Verschiedene organische Abfälle	11 000
15	Stark saure Lösungen	7 000
16	Stark basische Lösungen	3 400
17	Verschiedene anorganische Abfälle	11 000
	Gesamtabfälle pro Jahr	118 000
	Organische Abfälle	81 000
	Anorganische Abfälle	37 000

TABLE 1: Types of hazardous wastes in Norway

	Type of hazardous waste	t/a
1	Waste oil	34 500
2	Oil containing residues	8 000
3	Oil emulsions	5 700
4	Organic solvents	7 000
5	Paints, adhesives, varnish, etc.	5 700
6	Destillation residues	800
7	Tar	2 600
8	Waste with Hg and Cd	300
9	Water soluble heavy metal compounds	20 000
10	Wastes containing cyanides	20
11	Pesticides	350
12	Wastes containing PCB	250
13	Isocyanates	70
14	Sundry organic	11 000
15	Strong acidic solutions	7 000
16	Strong basic solutions	3 400
17	Sundry inorganic	11 000
	Total per year	118 000
	Organic	81 000
	Inorganic	37 000

2. Die Verbrennung von gefährlichen Flüssigabfällen

Der norwegische Zementkonzern NORCEM begann 1980 mit der Verbrennung von gefährlichen Flüssigabfällen im Zementwerk Slemmestad, in der Nähe von Oslo. Für den Einstieg von NORCEM in die Verwertung von Abfällen als Brennstoff für den Zementdrehofen gab es hauptsächlich die folgenden Gründe:

— Die Regierung (das Umweltministerium) war auf der Suche nach Optionen für die Abfallverwertung nach soliden Umwelttechnologien.

— Die Zementindustrie und insbesondere solche Werke, die noch nach dem Naßverfahren arbeiten, sahen sich nach der sogenannten Ölkrise in den 70er Jahren gerade nach Möglichkeiten zur Senkung der Brennstoffkosten um.

Als ein Ergebnis dieser Situation begann NORCEM gemeinsam mit der Regierung im Jahre 1980 im Zementwerk Slemmestad Brennversuche mit gefährlichen Flüssigabfällen. Auf der Grundlage von umfangreichen Messungen und Testversuchen erhielt das Zementwerk die Erlaubnis zur Verbrennung von Flüssigabfällen in Höhe von zunächst 5000 t/a. 1981 wurde mit der kontinuierlichen Abfallverbrennung begonnen. Nach der Entscheidung im Jahre 1985, die nach dem Naßverfahren am Standort Slemmestad arbeitende Ofenanlage stillzulegen, wurden schließlich die Brennereinrichtungen umgesetzt und im Zementwerk Brevik in Südnorwegen installiert.

Wegen der Besorgnis über mögliche Veränderungen der Emissionshöhe, aber auch wegen betrieblicher Schwierigkeiten, die aus der Anpassung der Brennereinrichtungen an die nach dem Trockenverfahren arbeitende Ofenlinie entstanden, besonders aber wegen auftretender Chloridbelastungen aus dem Brennstoff, war es notwendig, auf der Durchführung von neuen Testversuchen zu bestehen. Ergänzend sei in diesem Zusammenhang erwähnt, daß die Zementfabrik direkt an der dörflichen Gemeinde Brevik liegt und daß auch aus einer öffentlichen Sorge um mögliche Emissionen und Vorkommnisse die Forderung nach neuen und umfangreichen Messungen ausgelöst wurde. Als eine Konsequenz dieser Entwicklung führte deshalb NORCEM zusammen mit dem Zentrum für Industrieforschung (SI) neue Versuche zur Verbrennung von gefährlichen Flüssigabfällen in den Trockenöfen durch. Die Genehmigung zur Verbrennung dieser Flüssigabfälle in der norwegischen Zementindustrie basiert auf drei sehr umfangreichen Meßprogrammen, die in den Jahren 1980, 1983 und 1987 sowohl an Drehöfen nach dem Naß- als auch nach dem Trockenverfahren durchgeführt wurden. Auf der Grundlage der Meßergebnisse wurde dem Zementwerk Brevik die Erlaubnis zur Verbrennung von jährlich 5000 t gefährlicher Flüssigabfälle durch die State Pollution Control Authority (SFT) erteilt.

Tabelle 1 enthält eine Zusammenstellung der gefährlichen Abfälle in Norwegen. NORCEM besitzt die Erlaubnis, die Abfallarten 2 bis 7 sowie 11 und 12 zu verbrennen. Bisher hat NORCEM noch nie Pestizide verbrennen müssen. Selbst PCB-haltige Abfälle wurden bisher nur in sehr geringen Mengen verbrannt. Pestizide werden bereits durch die Gesellschaft zurückgewiesen, welche die gefährlichen Flüssigabfälle an NORCEM liefert, und PCB-haltige Öle erfordern keine sehr umfangreiche Behandlung. Annähernd 40% der organischen Abfälle fallen in die Gruppe 1 der Abfallöle. NORCEM besitzt neben seiner Genehmigung, jährlich 5000 t gefährliche Flüssigabfälle zu verbrennen, auch die Erlaubnis zur Verbrennung von 30000 t Altölen. Abfallöle für die Verbrennung in Zementöfen werden in Norwegen über den PCB-Gehalt mit max. 100 ppm, Chlor mit max. 1500 ppm und Blei mit max. 1500 ppm begrenzt.

Die volle Ausschöpfung der Genehmigungen für das Zementwerk Brevik würde bedeuten, daß mit der jährlichen Verbrennung von 30000 t Abfallölen und 5000 t gefährlichen Flüssigabfällen 30 bis 35% des Gesamtwärmebedarfs gedeckt werden könnten. Gleichzeitig würde das Zementwerk Brevik auch mehr als 40% solcher gefährlicher organi-

waste used as fuels in cement kilns was mainly due to the following reasons:

— The government (Ministry of Environment) sought options for recovery facilities according to "sound environmental practices".

— The cement industry and in particular plants with wet process kilns were looking at opportunities to reduce fossil fuel costs following the so-called oil crisis in the nineteenseventies.

As a result of this situation NORCEM and the government jointly started test burning of liquid hazardous wastes at Slemmestad in 1980. Based on extensive measurements and test trials, NORCEM Slemmestad received permission to burn up to 5000 tons of liquid hazardous wastes per year and continuous operation started in 1981. Following the decision to close down the wet process kiln at Slemmestad in 1985, it was decided to relocate and establish the burning facilities to the Brevik-Plant in the south of Norway.

Due to concern about possible changes in emission levels and operational difficulties when shifting from a wet process to a dry process kiln — in particular due to chlorine in the fuel it was decided to carry out new tests. It must in addition be noted that the Brevik Plant is situated closely to the village of Brevik and that public concern about possible emissions and accidents with the proposed facility further triggered the need for new and extensive measurements. As a consequence NORCEM together with SI (Center for Industrial Research) carried out new tests with the burning of liquid hazardous wastes in dry process kilns. The present permission for burning of liquid hazardous waste in the Norwegian cement industry is therefore based on three extensive programmes of measurements carried out in 1980, 1983 and 1987 both on wet and dry process kilns. Based on the test results, the Brevik Plant was given permission to burn up to 5000 tonnes liquid hazardous wastes per year as defined by the Norwegian State Pollution Control Authority (SFT).

With reference to Table 1 showing the nomenclature of hazardous wastes in Norway, NORCEM has permission to burn types no 2 through 7 and types 11 and 12. So far NORCEM has not received pesticides and only minor quantities of wastes containing PCB's have been burnt. Pesticides are rejected by the company which delivers liquid haz. wastes to NORCEM and oils containing PCB have only to a little extent required treatment. Approximately 40% of the organic wastes falls within group no. 1 — waste oils . NORCEM, Brevik, has in addition to the permission to burn 5000 tonnes of liquid haz. wastes, also permission to burn up to 30000 tonnes of waste oils per year. Waste oils for use in the cement industry are in Norway defined according to the levels of PCB (max. 100 ppm), Chlorine (max. 1500 ppm) and Lead (max.1500 ppm).

Full utilization of the permits for the Brevik Plant, i.e. 30000 tonnes waste oil and 5000 tonnes of liquid hazardous waste would represent between 30 and 35% of total heat requirements for the Brevik Plant. At the same time the Brevik Plant would dispose of more than 40% of the organic hazardous wastes which need treatment in Norway. Taking into account similar permissions for the Kjøpsvik Plant in the north of Norway it is clear that the Norwegian cement industry to a considerable degree can contribute with treatment facilities for organic hazardous wastes and it is assumed that similar opportunities exist in other countries.

3. Operation of the waste treatment plant at Brevik

The test results led to the following conclusions:

— The measurements revealed very good destruction efficiency for all types of organic wastes including PCB-containing wastes. The State Pollution Control Authority (SFT) therefore in 1986 stated that "Based on today's knowledge the cement kiln appears to be the best alternative for burning of liquid hazardous waste".

scher Abfälle verwerten, die in Norwegen einer Behandlung bedürfen.

Wird die Erteilung ähnlicher Genehmigungen auch für das Zementwerk Kjøpsvik im Norden Norwegens unterstellt, dann wird deutlich, daß die norwegische Zementindustrie mit ihren eigenen Möglichkeiten im beträchtlichen Umfang dazu beitragen kann, gefährliche organische Abfälle zu verwerten. Es kann sicherlich davon ausgegangen werden, daß ähnliche Voraussetzungen auch in anderen Ländern gegeben sind.

3. Durchführung der Abfallbehandlung in Brevik

Die während der durchgeführten Versuchsreihen erzielten Ergebnisse führten zu den folgenden Schlußfolgerungen:

Die Zerstörungswirkung aller organischer Abfallarten einschließlich PCB-haltiger Abfälle ist sehr gut. Die State Pollution Control Authority konstatierte deshalb schon im Jahre 1986, daß „auf der Basis des heutigen Erkenntnisstandes der Zementdrehofen als die beste Alternative zur Verbrennung von gefährlichen Flüssigabfällen anzusehen ist". Die Höhe der Emission wird mehr durch die Betriebsbedingungen als durch die verwendete Brennstoffart beeinflußt.

Als eine Konsequenz dieser Erkenntnisse wurde deshalb auch die Zustimmung zur Abfallverbrennung an die Ofenparameter gekoppelt, weshalb durch NORCEM in der Folge bestimmte Betriebsparameter, speziell zu den Erfordernissen von Temperatur und Sauerstoffgehalt definiert wurden. Allerdings wurde auch erkannt, daß die wichtigste Vorbedingung der Abfallverbrennung ein normaler und stabiler Ofenbetrieb darstellt. Die gleichen Vorschriften werden heute auch bei der Verbrennung von Abfallölen angewendet. Darüber hinaus sind in der Genehmigung die Betriebsweisen und insbesondere Anlieferung, Abnahme, Brennbedingungen, Abwassereinleitung usw. im einzelnen beschrieben.

NORCEM kontrolliert selbst die Qualität der angelieferten gefährlichen Flüssigabfälle und befindet sich in einer periodischen Berichterstattung gegenüber der State Pollution Control Authority. **Tabelle 2** enthält eine Zusammenstellung der durchschnittlichen Konzentrationswerte, die kontrolliert werden. Wie der Tabelle entnommen werden kann, lag der Chloridgehalt um ca. 1,0 %. Er wurde entsprechend den gemachten Betriebserfahrungen und dem aktuellen Chloridgehalt im Zement schrittweise erhöht. Der Bleigehalt war ziemlich konstant, während der Zinkgehalt wegen eines höheren Anteils von Farben im verfeuerten Flüssigabfall angestiegen war. Wegen seines Wassergehaltes von 26 bis 28 % ist es von besonderer Bedeutung, den zu lagernden Flüssigabfall als eine „homogene Emulsion" so lange, wie es erforderlich ist, im Tanksystem zu belassen. Die Abfall-Aufbereitungsanlage ist mit einer Vorrichtung für die Zerkleinerung von Faserstoffen wie Textilien oder Plastikmaterial ausgerüstet, um Pumpen und Stellventile zu schützen. Die Installationen einschließlich der Brennerdüse sind so kon-

— The emission level seemed to be more influenced by operational conditions than by the type of fuel used. As a consequence the permission to burn was linked with parameters of the kiln. NORCEM has therefore defined certain operational parameters, in particular temperature and oxygen requirements when burning liquid hazardous wastes. However, the most important precondition is that the kiln is in normal, stable condition. The same rule applies to the burning of waste oils.

Furthermore, the permission describes in detail operational procedures and in particular the delivery, reception, burning condition, discharges of waste water etc.

NORCEM is carrying out its own quality control on the received liquid hazardous wastes and reports periodically to the State Pollution Control Authority. **Table 2** shows the average values of the parameters which are monitored. As can be seen from the table the chlorine content has been around 1% and it has been gradually increased following operational experience and level of chlorine in the cement. The lead content has been fairly stable whereas the zinc content has been increased due to higher quantities of paints in the liquid hazardous waste. Due to the water content of 26-28%, it is of particular importance to maintain the stored liquid as an homogeneous "emulsion" as long as it remains in the tank system. The plant is equipped with a cutting unit for size reduction of long fibers such as textiles, plastics etc. in order to protect pumps and valves. It is, however, noted that the installation, including the burning nozzle, is constructed to handle particles up to 4–5 mm in diameter. The system is as a consequence constructed with particular emphasis on avoiding blockages in the tanks, valves, pumps and pipelines. It is noted that the receiving and mixing station is located approximately 500 m away from the main burner of the kiln.

4. Emission levels

The Brevik Plant has up to now been burning around 5 000 tonnes liquid hazardous wastes per year. The permit allows a burning capacity up to 3 t/hour and the plant is therefore only utilized around 2 000 hours per year. Following a troublefree operation over the last years NORCEM has applied for permission to burn up to 15 000 tonnes liquid hazardous wastes per year, i.e. an increase of 10 000 tonnes.

In connection with the establishment of the European Economic Area (EEA – EC/EFTA-countries), Norway will have to adhere to the EC-directives. In this respect the proposed "European Council Directive on the Incineration of Hazardous Waste" is of particular interest. The proposal by the Commission sets new and strict emission limit values for many pollutants which so far have not been measured by the Norwegian cement industry, in particular some of the heavy metals. The dust emission level is thereby reduced from 50 to 10 mg/Nm^3. The proposal allows use in plants not solely destined for the purpose of incineration of hazardous

TABELLE 2: Durchschnittswerte von untersuchten Parametern gefährlicher Flüssigabfälle im Zeitraum von 1989 bis 1992

Bezeichnung	Maßeinheit	1989	1990	1991	1992
Heizwert	kJ/kg	25941	24862	23425	23222
Cl	(%)	0,97	1,09	1,25	1,42
F	(%)	0,27	0,16	0,12	0,03
Pb	(ppm)	841	945	856	931
Zn	(ppm)	2975	3161	5017	4346
Cd	(ppm)	3	4	4	4
Hg	(ppm)	8	5	14	3
Wasser	(%)	26,2	27,7	28,0	23,2
Gesamte gefährliche Abfälle	t	4194	3915	5074	5371
Abfallöl	t	0	5627	4811	10213

TABLE 2: Average values of the analyzed parameters of liquid hazardous wastes in the period from 1989 to 1992

Designation	Unit	1989	1990	1991	1992
Energy content	kJ/kg	25941	24862	23425	23222
Cl	(%)	0.97	1.09	1.25	1.42
F	(%)	0.27	0.16	0.12	0.03
Pb	(ppm)	841	945	856	931
Zn	(ppm)	2975	3161	5017	4346
Cd	(ppm)	3	4	4	4
Hg	(ppm)	8	5	14	3
Water	(%)	26.2	27.7	28.0	23.2
Total hazardous wastes	t	4194	3915	5074	5371
Waste oil	t	0	5627	4811	10213

struiert, daß Feststoffpartikel bis zu 4 bis 5 mm Durchmesser sicher gehandhabt werden können. Sowohl die Empfangs- als auch die Mischstation sind ungefähr 500 m vom Hauptbrenner des Ofens entfernt stationiert.

4. Emissionswerte

Im Zementwerk Brevik werden bis heute gefährliche Flüssigabfälle von jährlich ca. 5000 t verbrannt. Die zur Abfallverbrennung erteilte Genehmigung erlaubt die Verbrennung einer Abfallmenge bis zu 3 t/h. Das Zementwerk ist demzufolge nur über ca. 2000 h/a mit brennbaren Abfällen versorgt. Ausgehend von der Tatsache eines störungsfreien Betriebes über die letzten Jahre hat NORCEM die Genehmigung zur Verbrennung von 15000 t, d.h. eine jährliche Erhöhung um 10000 t, beantragt.

Im Zusammenhang mit der Einrichtung der „European Economic Area" (EEA-EC/EFTA-Länder) wird Norwegen an den EG-Verordnungen festhalten. In dieser Hinsicht ist die vorgeschlagene „Europäische Ratsdirektive für die Verbrennung von gefährlichen Abfällen" von besonderem Interesse. Der Vorschlag der Kommission setzt neue und strikte Emissionsgrenzwerte für viele Schadstoffe fest, wie sie von der norwegischen Zementindustrie bisher noch nicht gemessen worden sind. Insbesondere betrifft dies einige Schwermetalle. Der spezifische Grenzwert für Staubemissionen wird dabei von 50 auf 10 mg/m^{3}N herabgesetzt. Der Vorschlag sieht für die Zementindustrie vor, daß die Emissions-Grenzwerte im Verhältnis zur verwerteten Abfallmenge bestimmt werden sollen. Befreiungen von bestimmten Verfügungen sind für die Zementindustrie von wesentlicher Bedeutung. In diesem Zusammenhang sei besonders erwähnt, daß das geforderte CO-Emissionsniveau von 50 mg/m^{3}N bei den Zementdrehöfen nicht erreicht werden kann. Auch andere Emissionswerte, wie beispielsweise der rohmaterialabhängige TOC-Wert, werden nur schwer zu erreichen sein. Obwohl der Vorschlag von verschiedenen Organisationen, eingeschlossen des Cembureau, angegriffen wird und darin enthaltene Bestimmungen noch verändert werden können, hat NORCEM die Durchführung neuer Versuche beschlossen, um zu prüfen, ob die neuen Vorgaben mit der umgestellten Ofenanlage 6 des Zementwerkes Brevik erfüllt werden können. **Bild 1** zeigt ein vereinfachtes Verfahrensfließbild von Ofenanlage 6, die mit einem Calcinator und einem zweisträngigen Vorwärmer ausgerüstet, eine Klinkerkapazität von ca. 3300 t/d hat. Das Abgas des einen Vorwärmerstranges wird für die Rohmaterialtrocknung verwendet, während das Abgas des anderen Vorwärmerstranges in einem Verdampfungskühler heruntergekühlt wird, bevor es in die Entstaubungsanlage gelangt. Nahezu 50% des Abgases gelangt somit im Direktbetrieb, 50% über die Rohmahlanlage, zu den Entstaubungseinrichtungen. Die Emissionshöhe beider Abgasströme wird an den in Bild 1 dargestellten Punkten A und B gemessen. Einige der wichtigsten Ergebnisse sind in **Tabelle 3** zusammengefaßt. Die Staubemission zeigt bei der Verbrennung von gefährlichen Abfällen einen etwas höheren Wert, als wenn der Ofen nur mit Kohle befeuert wird. Dies ist wesentlich auf übliche Meßunsicherheiten bei den Staubemissionsmessungen zurückzuführen. Die Emissionswerte der Schwermetalle, insbesondere Thallium und Quecksilber, werden vom verwendeten Brennstoff nicht beeinflußt. Die Emissionswerte aller Spurenmetalle liegen unter dem Vorschlag der neuen EG-Verordnung. Für SO$_2$ wurde ein Emissionswert zwischen 100 bis 140 mg/m^{3}N gemessen. Dieser Wertebereich liegt weit unter den gegenwärtig in Europa für Zementwerke geltenden gesetzlichen Regelungen. Die SO$_2$-Emission wird nicht durch den verwendeten Brennstoff beeinflußt, was im übrigen auch für die NO$_x$-Emission zutrifft. Die Ergebnisse der CO-Messung lieferten relativ niedrige Werte. Sie waren ebenfalls nicht durch die Art des verwendeten Brennstoffes beeinflußt, ebenso die Emissionen von HCL und HF, die unter den Konzentrationen der EG-Verordnung von 5 bzw. 1 mg/m^{3}N lagen. Die gemessenen Emissionen von TOC lieferten Werte unterhalb der Vorgaben der neuen EG-Verordnung.

wastes, e g the cement industry, where the emission limit values shall be in proportion to the hazardous waste used. Exemptions from certain articles in the proposal are essential for the cement industry. It is particularly noted that the emission level of 50 mg/Nm^3 CO cannot be met with a cement kiln. Also other emission levels such as raw material derived TOC may be difficult to achieve. Although the proposal is being challenged by various organizations, including Cembureau, and certain of the provisions may still be altered, NORCEM decided to carry out new tests in order to verify if compliance with the new standards could be met with the new converted kiln no VI at Brevik.

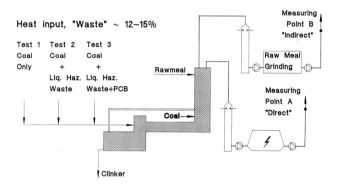

BILD 1: Vereinfachtes Verfahrensfließbild der umgebauten Drehofenanlage im Zementwerk Brevik

FIGURE 1: Simplified flow-sheet of the converted kiln No. 6, Brevik

Wärmeeinnahme durch Abfall ~ 12–15%	= Heat input by waste ~ 12–15%
Meßpunkt B „indirekter Betrieb"	= Measuring point B "indirect"
Meßpunkt A „direkter Betrieb"	= Measuring point A "direct"
Kohle	= coal
Kohle allein	= coal only
gefährlicher Flüssigabfall	= Liq. Haz. waste
Rohmehl	= raw meal
Rohmahlung	= raw meal grinding
Vorwärmer	= preheater
Klinker	= clinker

Figure 1 shows a simplified flow-sheet of the converted kiln no 6 which is a precalciner kiln with twin type preheater and with a capacity of around 3300 tons/day. The exit gas from one of the preheaters is utilized for raw material drying whereas the temperature of the exit gas from the other preheater is reduced in a cooling tower before dedusting. As a consequence approximately 50% of the exit gases is operated in the "direct" system and 50% in the "indirect" system. The emission levels from both gas streams marked A and B in the figure were measured and some of the main findings are summarized in **Table 3**.

The dust emission shows a slightly higher value when burning hazardous wastes as compared with "coal only" tests. It is assumed that this is mainly a result of normal fluctuations in the dust emission measurements. The emission levels of heavy metals, in particular Thallium and Mercury, are not influenced by the fuels utilized. All the metal emissions are below the values being required in the new EC-directive proposal. The emission of SO$_2$ was measured to be 100–140 mg/Nm^3 which is far below present cement works legislations in Europe. The SO$_2$ emission is not influenced by the fuel utilized and the same applies to the emission of NO$_x$. The results of the CO-measurements revealed fairly low values and were not influenced by the fuel utilized. Both the emissions of HCl and HF did not seem to be affected by the type of fuel and the results in general indicated values below the new EC-Directive of 5 mg/Nm^3, resp. 1 mg/Nm$_3$. The measured emissions of TOC showed values below the standards of the new EC-proposal. In total 15 measurements of emissions of dioxin/furans where carried out and the main conclusion are that:

— The emission is not influenced by the fuels utilized.

— The avarage emission level for dioxins and furans is ng/Nm^3 (10% O$_2$) as measured according to German and Norwegian procedures (Total 11 measurements).

TABELLE 3: Testergebnisse an Drehofen 6 im Zementwerk Brevik (1992)
TABLE 3: Test results, kiln No. 6, Brevik

Staub- und Gasemission		Meßstelle A (Direktbetrieb)			Meßstelle B (Verbundbetrieb)			Direkt- und Verbundbetrieb		
		Test 1	Test 2	Test 3	Test 1	Test 2	Test 3	Test 1	Test 2	Test 3
Staub:										
Anorg. Verb.	mg/m³ (N, d)	37,1	51,4	46,6	50,7	47,9	57,2	44,6	49,5	52,4
Cd, Tl		0,025	0,021	0,021	0,001	0,005	0,002	0,013	0,013	0,011
Hg		0,007	0,004	0,007	0,001	0,001	0,0007	0,004	0,002	0,004
Sb, As, Pb, Cr, CO, Cu Mn, Ni, V, Sn		0,024	0,013	0,032	0,017	0,021	0,025	0,020	0,017	0,028
Anorg. Gase:										
SO_2	mg/m³ (N, d)	238	201	181	48	24	24	133	103	94
NO_x		1145	1070	772		1220		1145	1153	772
CO		–	–	–	297	292	152	297	292	152
HCl		7,1	2,8	5,9	7,1	2,6	6,6	7,1	2,7	6,3
HF		0,2	0,18	0,19	0,47	0,17	0,17	0,35	0,17	0,18
Org. Gase:										
Dioxin/Furan	ng/m³ (N, d)	–	–	–	–	–	–	<0,1	<0,1	<0,1
TOC	mg/m³ (N, d)	<3	<3	<3	<1,5	<1,5	<1,5	<2,2	<2,2	<2,2

Staub- und Gasemission	= Dust and gas emission
Meßstelle A (Direktbetrieb)	= Measuring point A (Direct operation)
Meßstelle B (Verbundbetrieb)	= Measuring point B (Operation)
Direkt- und Verbundbetrieb	= Total Direct + Indirect
Staub	= Dust
Anorg. Verb.	= Inorg. comp.
Anorg. Gase	= Inorg. Gases
Org. Gase	= Organic Gases

Aus insgesamt 15 Dioxin/Furan-Messungen konnte die Schlußfolgerung gezogen werden, daß auch diese Emissionen brennstoffunabhängig sind. Nach deutschen bzw. norwegischen Meßverfahren beträgt die Emissionshöhe kleiner 0,1 mg/Nm³. Einige Messungen lieferten allerdings auch Emissionswerte über 0,1 mg/Nm³, was an der Ungenauigkeit der Messungen gelegen haben kann, da die Messungen der Dioxin/Furan-Emissionen auch an veralteten Zementöfen durchgeführt wurden.

5. Ausblick

Das oberste Ziel der Zementindustrie besteht darin, ein Produkt hoher Qualität zu erzeugen, das in jeder Hinsicht den Herausforderungen anderer konkurrierender Baumaterialien gewachsen ist. Dabei sollte nicht vergessen werden, daß die Verwertung von gefährlichen Abfallstoffen sowie anderer Sekundärbrenn- und Rohstoffe durchaus einen Einfluß auf die Qualität und die Umweltverträglichkeit von Zement und Beton haben kann. Den Erfahrungen von NORCEM zufolge bleibt die Zementqualität unbeeinflußt, wenn – wie in diesem Beitrag beschrieben – Abfallstoffe als Brennstoffe verwendet werden. Demnach werden die Forschungsarbeiten auf diesem Gebiet weitergehen.

Im Zusammenhang mit der einhergehenden Diskussion um die Klassifizierung und Behandlung der unterschiedlichen Abfallarten wird es eine Schlüsselfrage sein, in welchem Umfang die Zementindustrie mit entsprechenden Lösungen zur Abfallverwertung beitragen soll. Die neuen Verordnungen räumen der Abfallminimierung durch Wiederverwertung die Priorität ein. Der bevorzugte Einsatz als Brennstoff wird als eine Verwertungsmöglichkeit bei der zukünftigen Abfallbehandlung eine wichtige Rolle spielen. Die Vorteile, die mit der Verbrennung verschiedener Sekundärbrennstoffe in der Zementindustrie verbunden sind, wie etwa die Schonung nicht erneuerbarer Energie-Ressourcen, die Kosteneffektivität etc., sind wohlbekannt. Dazu kommt, daß die Zementindustrie schärfer werdende Umweltauflagen ebenso wie andere amtliche Anforderungen erfüllen muß. Die Regierung, die Umweltschutzgruppen und die allgemeine Öffentlichkeit widmen dem Umgang und der Behandlung von Abfällen mehr und mehr Aufmerksamkeit. In dieser Hinsicht kann die Zementindustrie mit Lösungen beitragen, indem sie sowohl die Qualitätssicherung des Zementes als auch die Emissionen und andere Umweltaspekte in Betracht zieht.

Some measurements gave emissions higher than 0.1 mg/Nm³. This may be a result of inaccuracies in the measuring procedures when determining dioxin/furan-emissions from cement kilns at low levels.

5. Outlook

The cement industry's prime objective is to produce a high quality product which in every respect can meet the challenges of other competing building materials. It should not be forgotten that the utilization of hazardous wastes as well as other secondary fuels and raw materials may have an impact on the quality and environmental performance of cement and concrete. Following NORCEM's experience, the cement quality remains the same when utilizing fuels as described although research work continues to be carried out.

In connection with ongoing discussions related to classification and treatment of different types of wastes, a key question will be to what extent the cement industry shall contribute with solutions for waste recovery. The new waste Directives give priority to waste minimization followed by reuse and recovery. The use principally as a fuel is regarded as a means of recovery and will play an important role in future waste treatment. The benefits from burning various secondary fuels in the cement industry, such as the reduction in the use of non-renewable energy resources, cost effectiveness etc. are well known. In addition the cement industry must comply with increasingly tougher environmental standards as well as other official requirements. The government, environmental pressure groups and the public in general are paying more and more attention to the issue of waste handling and treatment. The cement indusry in this respect can contribute with solutions both taking the cement quality and emissions and other environmental impacts as well into account.

Eigenschaften der Ofenstäube aus verschiedenen mit hoch aschehaltigem Kohlenstaub befeuerten Öfen

Kiln dust characteristics from high ash coal fired kilns of different processes

Propriétés des poussières de four de différents fours chauffés au poussier de charbon riche en cendres

Características de los polvos procedentes de hornos con calefacción por combustión de carbón pulverizado, de elevado contenido de cenizas

Von **S. A. Khadilkar, N. A. Krishnan, D. Ghosh, C. H. Page** und **A. K. Chatterjee**, Thane/Indien

Zusammenfassung – Es wurden die Eigenschaften von Ofenstäuben an Anlagen untersucht, die nach dem Naß-, Halbnaß- (Vorcalcinieranlage mit 2stufigem Vorwärmer), Halbtrocken- (Lepol-Ofen) und dem Trockenverfahren (Vorcalcinieranlage mit 4stufigem Vorwärmer) arbeiten und mit aschereicher (> 30%) Kohle befeuert werden. Die Haupt- und Nebenbestandteile sowie die Spurenelemente im Ofenstaub und ihre Verteilung in den verschiedenen Korngrößenfraktionen wurden bestimmt. Die aus dem Rohmaterial und der Kohleasche stammenden Gehalte an Nebenbestandteilen und Spurenelementen wurden untersucht. Dabei wurde der Alkalien- und Sulphatanreicherung der Ofenstäube und den Schwermetallgehalten besondere Aufmerksamkeit gewidmet. Die zeitabhängige Auslaugung der Ofenstäube wurde in die Untersuchungen einbezogen. Die Rückführung dieser Stäube in den Kreislauf und die Auswirkungen der Staubrückführung sind eingehend untersucht worden. Auf der Grundlage der erarbeiteten Daten werden die Auswirkungen auf die Umwelt beschrieben.

Summary – The kiln dust charcteristics from wet, semi-wet (two-stage preheater precalciner), semi-dry (Lepol kiln), dry (four-stage preheater precalciner) process plants using high-ash (> 30%) coals have been studied. The major, minor and trace elements in the kiln dusts have been determined along with their distribution in different fractions of particele size. The sources of minor and trace elements from feed, fuel ash have been traced. The alkali and sulphate enrichment of kiln dusts have been particularly studied along with the concentration of heavy metals. The time-dependent leachability of kiln dusts has been examined. The process recycling of these dusts and the resultant cyclic effects have been studied in some details. Based on the data, the environmental implications have been discussed.

Résumé – Les études des propriétés des poussières de four ont été conduites dans des installations travaillant selon les procédés humide, semi-humide (ligne à précalcination et préchauffeur à deux étages), semi-sec (four Lepol) et sec (ligne à précalcination et préchauffeur à 4 étages) chauffées au charbon à haute teneur en cendres (> 30%). Les composants primaires et secondaires ainsi que les éléments traces dans la poussière de four et leur distribution dans les différentes fractions granulométriques ont été déterminés. Les teneurs en composants secondaires et éléments traces provenant de la matière première et de la cendre de charbon ont été étudiées. Une attention particulière a été portée à l'enrichissement des poussières de four en alcalis et sulfates et aux teneurs en métaux lourds. La lixiviation en fonction du temps, des poussières de four, avait été incluse dans les études. La recirculation de ces poussières dans le circuit et ses conséquences ont été étudiées de façon approfondie. Sur la base des données recueillies sont décrites les influences exercées sur l'environnement.

Resumen – Se han estudiado las características de los polvos procedentes de hornos de instalaciones que trabajan por vía húmeda, semihúmeda (instalación de percalcinación con precalentador de dos etapas), semiseca (horno Lepol) y seca (instalación de precalcinación con precalentador de cuatro etapas) y que emplean como combustible carbón rico en cenizas (> 30%). Han sido determinados los componentes principales y adicionales así como los elementos-traza, contenidos en el polvo de los hornos, así como su distribución en las distintas fracciones granulométricas. Además, se han estudiado los contenidos de componentes adicionales y elementos-traza, provenientes de las materias primas y de las cenizas del carbón. A este respecto, se ha dedicado mucha atención al enriquecimiento de los polvos de los hornos en álcalis y sulfatos así como a los contenidos de metales pesado. Se ha incorporada a las investigaciones la lixiviación de los polvos de los hornos en función del tiempo, y se ha estudiado detenidamente el retorno de esos polvos al circuito así como las repercusiones de dicho retorno. Sobre la base de los datos elaborados, se describen los correspondientes efectos sobre el medio ambiente.

1. Einführung

Die bei der Zementklinkerherstellung eingesetzten natürlichen Roh- und Brennstoffe enthalten Spurenelemente in unterschiedlichen Konzentrationen. Je nach den physikalisch-chemischen Merkmalen und dem Brennverfahren geraten diese Elemente in den Klinker oder in den Kreislauf des Ofensystems und reichern sich auf diese Weise im Ofenstaub relativ stark an. Über das Verhalten von Ba, Sr, Cr, Mn, Co, Ni, Cu, Zn, Cd, Sb, As und Pb im Ofenkreislauf beim Trocken-, Halbtrocken-, Halbnaß- und Naßverfahren der Zementherstellung wurde bereits berichtet [1]. Die vorliegende Arbeit beschreibt die Ergebnisse weiterer Untersuchungen der zeitabhängigen Auslaugbarkeit dieser Spurenelemente aus den Ofenstäuben und der Verteilung der Spurenelemente in den einzelnen Größenfraktionen der bei den verschiedenen Brennverfahren entstehenden Ofenstäube. Solche Daten könnten zu einem besseren Verständnis der Auswirkungen der aus dem Ofenstaub freigesetzten Spurenelemente auf die Umwelt führen.

2. Methode

Die Untersuchungen wurden in folgenden Zementwerken durchgeführt:

Werk P-1: Nach dem Trockenverfahren arbeitendes Werk mit Zyklonvorwärmer und Vorcalcinierung.

Werk P-2: Nach dem Halbtrockenverfahren arbeitendes Werk (Lepol-Rostvorwärmeranlage) ohne Ofenstaubrückführung.

Werk P-3: Nach dem Halbnaßverfahren arbeitendes Werk mit Zyklonvorwärmer und Vorcalcinierung.

Werk P-4: Nach dem Naßverfahren arbeitendes Werk ohne Ofenstaubrückführung.

Die Gehalte der Spurenbestandteile in den Roh- und Brennstoffen (Asche) wurden in den aufgegebenen (Rohmehl und Kohleasche) (**Tabelle 1**) und ausgetragenen Stoffen

1. Introduction

The natural raw materials and fuels used in cement clinker production contain various concentrations of trace elements. Depending on the physico-chemical characteristics and the type of clinkering process, these elements get incorporated in clinker or recycle in the klin systems and thereby getting relatively concentrated in the kiln dusts. The observations on the process recycling behaviour of Ba, Sr, Cr, Mn, Co, Ni, Cu, Zn, Cd, Sb, As and Pb in dry process, semi-dry process, semi-wet process and wet process of cement manufacture have already been reported [1]. The present paper discusses the results of further investigative studies carried out on time-dependent leachability of the trace elements from kiln dusts and on distribution of these trace elements in the different size fractions of kiln dusts generated from the different processes of cement manufacture. These aspects could be of importance to understand the environmental implications of these trace element emissions through kiln dusts.

2. Basic approach

The investigations were carried out on the following cement plants:

Plant P-1: Dry process plant with suspension preheater and precalciner.

Plant P-2: Semi-dry process plant with Lepol system (grate calciner system) without kiln dust recirculation.

Plant P-3: Semi-wet process plant with suspension preheater and precalciner.

Plant P-4: Wet process plant without kiln dust recirculation.

The sources of trace elements from raw materials and fuel (ash) have been traced in the input (raw meal and coal ash) (**Table 1**) and output material (kiln dust and clinker). The total concentrations of main components in input and output are given in **Table 2**. The enrichment of the elements in

TABELLE 1: Konzentrationen der Elemente in ppm (glühverlustfrei)

Werk	Material	Ba	Sr	Cr	Mn	Co	Ni	Cu	Zn	Cd	Sb	As	Pb
							(ppm)						
P-1	Kalkstein	1295	315	16	350	20	22	7	27	12	50	–	<10
	Laterit	<100	<10	540	234	33	72	8	73	9	40	–	15
	Eisenerz	<100	<10	9	760	25	41	12	25	8	45	–	<10
	Kohle	140	10	58	95	14	33	28	26	<10	30	4,0	<10
P-2	Kalkstein-1	<100	94	20	400	14	20	4	20	10	96	–	<10
	Kalkstein-2	<100	83	22	309	12	14	6	22	9	104	–	<10
	Kohle	<100	<10	47	170	11	28	26	45	<10	30	4,2	30
P-3	Kalkstein	695	89	21	377	11	20	9	39	11	35	–	<10
	Bauxit	<100	<10	3640	245	30	1020	149	131	11	76	–	<10
	Eisenerz	<100	<10	115	295	23	48	28	57	11	76	–	<10
	Kohle	112	<10	69	230	11	39	23	46	<10	30	1,0	<10
P-4	Kalkstein	<100	385	13	16	12	20	8	29	10	84	–	<10
	Laterit	<100	<10	345	121	75	120	131	85	7	78	–	49
	Kohle	<100	<10	37	90	11	22	24	26	<10	20	4,4	<10

TABLE 1: Concentration of Elements in ppm (Loss Free Basis)

Plant	Material	Ba	Sr	Cr	Mn	Co	Ni	Cu	Zn	Cd	Sb	As	Pb
							(ppm)						
P-1	Limestone	1295	315	16	350	20	22	7	27	12	50	–	<10
	Laterite	<100	<10	540	234	33	72	8	73	9	40	–	15
	Iron Ore	<100	<10	9	760	25	41	12	25	8	45	–	<10
	Coal	140	10	58	95	14	33	28	26	<10	30	4.0	<10
P-2	Limestone-1	<100	94	20	400	14	20	4	20	10	96	–	<10
	Limestone-2	<100	83	22	309	12	14	6	22	9	104	–	<10
	Coal	<100	<10	47	170	11	28	26	45	<10	30	4.2	30
P-3	Limestone	695	89	21	377	11	20	9	39	11	35	–	<10
	Bauxite	<100	<10	3640	245	30	1020	149	131	11	76	–	<10
	Iron Ore	<100	<10	115	295	23	48	28	57	11	76	–	<10
	Coal	112	<10	69	230	11	39	23	46	<10	30	1.0	<10
P-4	Limestone	<100	385	13	16	12	20	8	29	10	84	–	<10
	Laterite	<100	<10	345	121	75	120	131	85	7	78	–	49
	Coal	<100	<10	37	90	11	22	24	26	<10	20	4.4	<10

(Ofenstaub und Klinker) bestimmt. Die Gehalte der Hauptbestandteile im Eingangs- und Ausgangsmaterial sind in **Tabelle 2** wiedergegeben. Die Anreicherung der Elemente bei den verschiedenen Prozessen ist aus **Bild 1** ersichtlich. Die Flüchtigkeit der Elemente Ni, Cu, Zn, Cd, Sb, As und Pb bei den verschiedenen Herstellungsprozessen wird graphisch in **Bild 2** dargestellt.

Die Verteilung dieser Spurenbestandteile in den Größenfraktionen der Ofenstäube der oben genannten Werke wurde durch die Bestimmung der Spurenelementgehalte in den Korngrößenfraktionen von über und unter 45 μm bestimmt. Die Konzentrationen der Spurenelemente im Ofenstaub und ihre Fraktionen sind graphisch in **Bild 3** dargestellt.

Die Gehalte der Hauptbestandteile und die Konzentrationen von Na_2O, K_2O, SO_3 und Chloriden im Rohmehl und Ofenstaub wurden ebenfalls analysiert (Tabelle 2), der Flüchtigkeitsgrad von Na_2O, K_2O, SO_3 und Cl ist aus **Bild 4**

different processes is shown in **Fig. 1**. The volatile behaviour of the elements Ni, Cu, Zn, Cd, Sb, As and Pb in different processes is graphically presented in **Fig. 2**.

The distribution of trace elements in size fractions of kiln dusts of the above plants were studied through determination of trace element concentrations in different size fractions above and below 45 microns. The concentractions of elements in kiln dust and their fractions are graphically presented in **Fig. 3**.

The major oxides concentrations and concentrations of Na_2O, K_2O, SO_3 and chlorides in raw meal and kiln dust were also analysed (Table 2), the percent volatility levels of Na_2O, K_2O, SO_3 and Cl are shown in **Fig. 4**. The time-dependent leachability of the trace elements in the kiln dusts was attempted through extraction of a known quantity of kiln dust with distilled water for 15, 30 minutes and 72 hours. The filtrate was then analysed for trace concentrations. The percent leachability of the trace elements is given in **Table 3**.

TABELLE 2: Die Hauptbestandteile, Alkalien, Chloride und Sulphate im Rohmehl und Ofenstaub

Oxide	Werk-1		Werk-2		Werk-3		Werk-4	
	Rohmehl	Ofenstaub	Rohmehl	Ofenstaub	Rohmehl	Ofenstaub	Rohmehl	Ofenstaub
	(M.-%)							
SiO_2	12,0	12,9	12,3	23,7	12,7	9,9	12,1	19,2
Al_2O_3	1,8	3,0	4,0	12,0	2,8	3,4	2,7	6,4
Fe_2O_3	2,6	2,4	1,7	3,5	2,3	2,8	2,1	2,7
CaO	45,0	43,8	43,8	31,4	44,4	45,3	43,6	38,6
MgO	0,9	1,2	1,8	1,6	0,8	0,9	2,4	2,6
LOI	36,6	34,9	35,8	21,8	35,9	36,3	36,7	26,1
Na_2O	0,04	0,07	0,16	0,59	0,09	0,19	0,04	0,08
K_2O	0,24	0,39	0,74	3,41	0,23	0,48	0,35	1,88
Gesamt SO_3	0,6	1,1	0,3	1,7	0,1	0,5	0,10	2,1
Lösliches Cl	0,01	0,05	0,01	0,07	0,01	0,39	0,02	0,14

TABLE 2: Major Oxides, Alkalis, Chlorides and Sulphates in Raw Meal and Kiln Dust

Oxides	Plant-1		Plant-2		Rwa-3		Plant-4	
	Raw Meal	Kiln Dust	Raw Meal	Kiln Dust	Rwa Meal	Kiln Dust	Raw Meal	Kiln Dust
	(wt.-%)							
SiO_2	12.0	12.9	12.3	23.7	12.7	9.9	12.1	19.2
Al_2O_3	1.8	3.0	4.0	12.0	2.8	3.4	2.7	6.4
Fe_2O_3	2.6	2.4	1.7	3.5	2.3	2.8	2.1	2.7
CaO	45.0	43.8	43.8	31.4	44.4	45.3	43.6	38.6
MgO	0.9	1.2	1.8	1.6	0.8	0.9	2.4	2.6
LOI	36.6	34.9	35.8	21.8	35.9	36.3	36.7	26.1
Na_2O	0.04	0.07	0.16	0.59	0.09	0.19	0.04	0.08
K_2O	0.24	0.39	0.74	3.41	0.23	0.48	0.35	1.88
Total SO_3	0.6	1.1	0.3	1.7	0.1	0.5	0.10	2.1
Soluble Cl	0.01	0.05	0.01	0.07	0.01	0.39	0.02	0.14

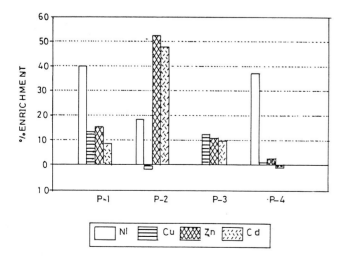

BILD 1: Prozentuale Anreicherung von Ni, Cu, Cd, Sb, As und Pb in den Austragsstoffen verschiedener Herstellungsverfahren

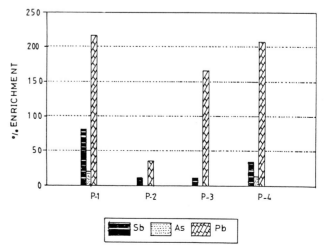

FIGURE 1: % Enrichment of Ni, Cu, Cd, Sb, As, Pb in outputs of different processes

Anreicherung = Enrichment

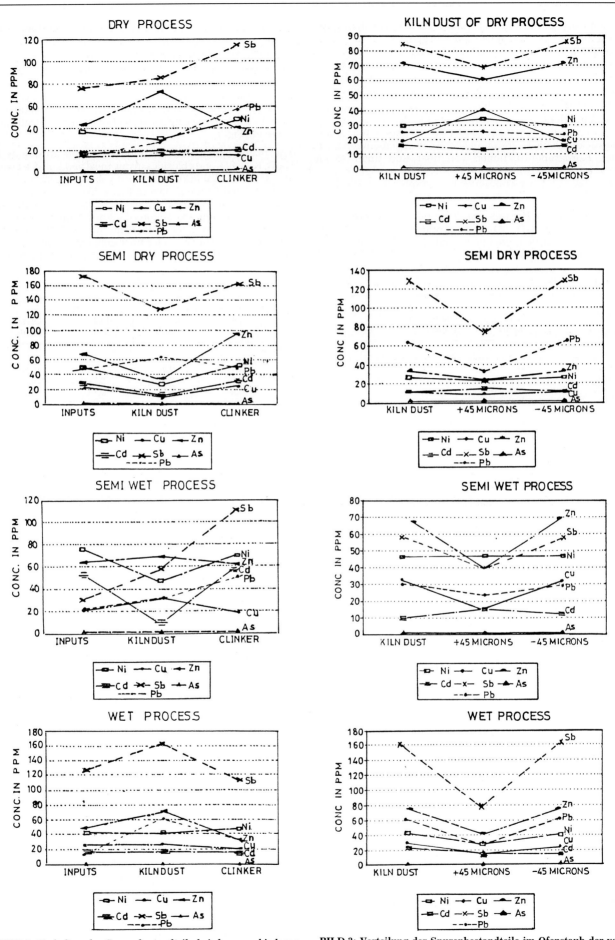

BILD 2: Verhalten der Spurenbestandteile bei den verschiedenen Herstellungsverfahren
FIGURE 2: Behaviour of trace elements in different processes

Trockenverfahren	= Dry process	Ofenstaub	= Kiln Dust
Konz.	= Conc.	Halbnaßverfahren	= Semi wet process
Eingangsstoffe	= Inputs	Naßverfahren	= Wet process

BILD 3: Verteilung der Spurenbestandteile im Ofenstaub der verschiedenen Verfahren
FIGURE 3: Distribution of trace elements in kiln dust of different processes

| Trockenverfahren | = Dry process | Halbnaßverfahren | = Semi wet process |
| Halbtrockenverfahren | = Semi dry process | Naßverfahren | = Wet process |

ersichtlich. Die zeitabhängige Auslaugbarkeit der Spurenbestandteile im Ofenstaub wurde durch Extraktion einer bestimmten Menge Ofenstaub in destilliertem Wasser nach 15 und 30 Minuten sowie 72 Stunden bestimmt. Die Spurenelementkonzentrationen wurden im Filtrat ermittelt. Die prozentuale Auslaugbarkeit ist in **Tabelle 3** angegeben.

Die Analyse der Spurenelemente wurde mit Zweistrahl-Atomabsorptions-Spektrometern (Instrumentation Laboratory, USA) durchgeführt. Die Analyseverfahren waren vergleichbar mit dem in [1] beschriebenen Vorgehen.

3. Diskussion

3.1 Herkunft der Spurenbestandteile

Der in den verschiedenen Werken eingesetzte Kalkstein stammt aus verschiedenen Steinbrüchen, wobei zwischen zwei Arten mit höherem Ba-Gehalt (> 695 ppm) und mit niedrigerem Ba-Gehalt (< 100 ppm) zu unterscheiden ist. Obgleich Barium und Strontium zur gleichen Gruppe des periodischen Systems gehören, stehen ihre Konzentrationen in keiner Beziehung zueinander. Der Mangangehalt im

The analysis of trace elements was carried out on atomic absorption spectrophotometer, double beam, double channel of The Instrumentation Laboratory, USA. The analytical procedures followed were similar to those discussed in [1].

3. Discussions

3.1 Source of trace elements

Limestone of different sources for the plants under study are distinguished between two varieties, one with higher Ba contents (> 695 ppm), and the other with lower Ba contents (< 100 ppm). Although barium and strontium are of the same group of periodic table, their occurrence is not related. The

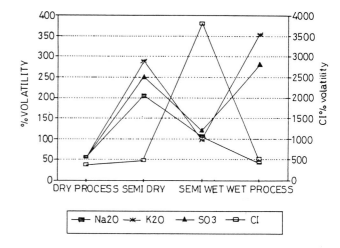

BILD 4: Flüchtigkeit von Na$_2$O, K$_2$O, SO$_3$ und Cl bei den einzelnen Verfahren
FIGURE 4: % volatility levels of Na$_2$O, K$_2$O, SO$_3$ and Cl in different processes

Flüchtigkeit	= Volatility
Trockenverfahren	= Dry process
Halbtrockenverfahren	= Semi dry process
Halbnaßverfahren	= Semi wet process
Naßverfahren	= Wet process

TABELLE 3: Auslaugbarkeit von Spurenbestandteilen

Spuren-element	Werk-1 Trockenverfahren Zeit			Werk-2 Halbtrockenverfahren Zeit			Werk-3 Halbnaßverfahren Zeit			Werk-4 Naßverfahren Zeit		
	15 min	30 min	72 h	15 min	30 min	72 h	15 min	30 min	72 h	15 min	30 min	72 h
	%											
Ba	1,1	1,4	2,1	N.L.	N.L.	N.L.	3,0	4,6	4,6	N.L.	N.L.	N.L.
Sr	6,0	8,3	9,4	10,5	17,6	17,8	11,0	13,7	14,0	6,0	7,4	7,4
Cr	N.L.	N.L.	N.L.	N.L.	N.L.	N.L.	N.L.	N.L.	N.L.	N.L.	N.L.	N.L.
Mn	0,7	0,7	N.L.	N.L.	N.L.	N.L.	0,5	0,5	N.L.	N.L.	N.L.	N.L.
Co	N.L.	N.L.	N.L.	N.L.	N.L.	N.L.	N.L.	N.L.	N.L.	N.L.	N.L.	N.L.
Ni	N.L.	N.L.	N.L.	N.L.	N.L.	N.L.	N.L.	N.L.	N.L.	N.L.	N.L.	N.L.
Cu	N.L.	N.L.	N.L.	N.L.	N.L.	N.L.	N.L.	N.L.	N.L.	N.L.	N.L.	N.L.
Zn	2,0	2,5	2,5	4,0	4,8	4,8	1,8	2,2	2,2	2,8	3,8	3,8
Cd	N.L.	N.L.	N.L.	N.L.	N.L.	N.L.	N.L.	N.L.	N.L.	N.L.	N.L.	N.L.
Sb	20,0	26,0	N.L.	15,0	18,3	N.L.	10,2	13,8	N.L.	10,5	13,8	N.L.
As	N.L.	N.L.	N.L.	N.L.	N.L.	N.L.	N.L.	N.L.	N.L.	N.L.	N.L.	N.L.
Pb	N.L.	N.L.	N.L.	N.L.	N.L.	N.L.	N.L.	N.L.	N.L.	N.L.	N.L.	N.L.

N.L. = vernachlässigbare Auslaugbarkeit

TABLE 3: Leachability of Trace Elements

Trace Element	Plant-1 Dry Process Time			Plant-2 Semi-dry Process Time			Plant-3 Semi-wet Process Time			Plant-4 Wet Process Time		
	15 min	30 min	72 h	15 min	30 min	72 h	15 min	30 min	72 h	15 min	30 min	72 h
	%											
Ba	1.1	1.4	2.1	N.L.	N.L.	N.L.	3.0	4.6	4.6	N.L.	N.L.	N.L.
Sr	6.0	8.3	9.4	10.5	17.6	17.8	11.0	13.7	14.0	6.0	7.4	7.4
Cr	N.L.	N.L.	N.L.	N.L.	N.L.	N.L.	N.L.	N.L.	N.L.	N.L.	N.L.	N.L.
Mn	0.7	0.7	N.L.	N.L.	N.L.	N.L.	0.5	0.5	N.L.	N.L.	N.L.	N.L.
Co	N.L.	N.L.	N.L.	N.L.	N.L.	N.L.	N.L.	N.L.	N.L.	N.L.	N.L.	N.L.
Ni	N.L.	N.L.	N.L.	N.L.	N.L.	N.L.	N.L.	N.L.	N.L.	N.L.	N.L.	N.L.
Cu	N.L.	N.L.	N.L.	N.L.	N.L.	N.L.	N.L.	N.L.	N.L.	N.L.	N.L.	N.L.
Zn	2.0	2.5	2.5	4.0	4.8	4.8	1.8	2.2	2.2	2.8	3.8	3.8
Cd	N.L.	N.L.	N.L.	N.L.	N.L.	N.L.	N.L.	N.L.	N.L.	N.L.	N.L.	N.L.
Sb	20.0	26.0	N.L.	15.0	18.3	N.L.	10.2	13.8	N.L.	10.5	13.8	N.L.
As	N.L.	N.L.	N.L.	N.L.	N.L.	N.L.	N.L.	N.L.	N.L.	N.L.	N.L.	N.L.
Pb	N.L.	N.L.	N.L.	N.L.	N.L.	N.L.	N.L.	N.L.	N.L.	N.L.	N.L.	N.L.

N.L. = Negligible Leachability

manganese content in the limestones is 300–400 ppm with the exception in limestone from plant P-4. The ohter elements are in the range of 13–76 ppm (Table 1). In the corrective materials, viz. laterite, bauxite, iron ore, the presence of chromium, manganese, nickel levels are relatively higher than in the limestone. The coals with high ash content of different sources have significant arsenic contents of 4 to 4.4 ppm. The presence of arsenic was not noticed in the raw materials, at least within the detectable limits of the present analysis (0.1 ppm). Some of the coals have higher manganese levels (around 200 ppm) while others are lower in manganese ($\approx$ 90 ppm).

3.2 Enrichment tendency of trace elements

Trace elements mass balance investigation studies carried out for plants with different processes have already been discussed in an earlier paper by the authors [1]. The present paper discusses further the behaviour of trace elements in the light of further investigations.

The elements Ba, Sr, Cr, Mn and Co have been observed to show negligible enrichment in outputs, while the elements Ni, Cu, Zn, Cd, As, Sb and Pb show measurable enrichments in outputs (Fig. 1). The % enrichment is calculated by the formula

$$\frac{\text{Outputs} - \text{Inputs}}{\text{Inputs}} \times 100$$

where:

Inputs: Conc. in raw meal + conc. in coal consumed.
Outputs: Conc. in klin dust generated + conc. in clinker.

The percent enrichment in outputs in excess of 10% was considered appreciable and worth consideration. Accordingly, the enrichment of the trace elements can be rated in the following order, irrespective of the effect of process.

Highest   Lowest
             Pb, Sb, Ni, Cu, Zn, Cd, As

3.3 Trace elements enrichment in kiln dust

The enrichment of trace elements in the kiln dust could be due to the volatility of trace elements and the concentration of trace elements in very fine particles. Both the above factors would be influenced by the process stream [2–4, 6]. In a cyclone preheater system, the gases loaded with dust come into contact with the relatively colder feed material, and the volatile compounds in gases would condensate and return back to the kiln, such a phenomenon has been known to exist in many preheater systems. The concentrations of trace elements will increase with the number of such cycles.

In a Lepol grate kiln, the bed above the Lepol grate acts as a filter bed for the gases passing through it and condensation of volatiles as found in preheater system is expected, but the finer particles mostly escape. In the present stydy, the Lepol grate plant (P-2) does not feed back its kiln dust, thereby the internal cycles are expected to be limited.

The wet process kiln under study (P-4) was not recirculating its kiln dust, only a certain amount of recycling is expected from the chain zone area.

An attempt to correlate the process aspects with the enrichment of trace elements (Fig. 2) in the kiln dust and their distribution in size fractions of the dust leads to following conclusions:

– The dry process plant P-1 and the semi-wet process plant P-3 show an enrichment of the same trace elements in the kiln dust, viz. Cu, Zn, Sb and Pb with Zn enriching by about 30 ppm in P-1 and Sb enriching to around 37 ppm in P-3. As levels in kiln dusts in both plants are similar to inputs, while Ni concentrations are marginally less than inputs. Cd concentration in P-1 is similar to inputs while in P-3 it is less than inputs.
– In the semi-dry process plant P-2 only Pb shows an enrichment in kiln dust to the tune of 22 ppm, while Ni, Cu, Zn, Cd, Sb concentration in dust is lesser than inputs, As levels being similar to inputs.

etwa 37 ppm im Werk P-3 anreicherte. Die As-Gehalte in den Ofenstäuben beider Werke sind ähnlich denen im Eingangsmaterial, während die Ni-Gehalte marginal geringer als im Eingangsmaterial ausfallen. Die Cd-Gehalte im Werk P-1 sind ähnlich denen im Eingangsmaterial, im Werk P-3 hingegen liegt die Cd-Konzentration unter den Werten des Eingangsmaterials.

– In dem nach dem Halbtrockenverfahren arbeitenden Werk P-2 wies nur Pb eine Anreicherung auf 22 ppm im Ofenstaub auf, während die Elemente Ni, Cu, Zn, Cd und Sb Gehalte im Ofenstaub aufwiesen, die unter denen der Eingangsstoffe lagen. Die As-Werte waren ähnlich wie im Eingangsmaterial.

– Im Werk P-4, das nach dem Naßverfahren arbeitet, reicherten sich Zn, Sb und Pb im Ofenstaub an, wobei für Zn ein Gehalt von 20 ppm, für Sb von 36 ppm und für Pb von etwa 46 ppm nachgewiesen wurde. Die Elemente Ni, Cu, Cd und As erreichten dieselben Werte.

Bei den in den verschiedenen Prozessen entstehenden Ofenstäuben handelt es sich vorwiegend um sehr feine Stäube mit einem Anteil von 90 bis 95 % kleiner 45 μm, und deshalb wird die bevorzugte Verteilung der Spurenelemente in den feineren Fraktionen (< 45 μm) der Ofenstäube der verschiedenen Verfahren in Bild 3 dargestellt.

– Im Werk P-1, das nach dem Trockenverfahren arbeitet, sind die feineren Fraktionen stärker mit den Elementen Zn, Cd und Sb angereichert, während die Anreicherung von Ni, Cu, As und Pb etwa dem Wert des Ofenstaubs insgesamt entspricht.

– Beim Halbtrockenverfahren (Werk P-2) und beim Halbnaßverfahren (Werk P-3) traten Zn, Cd und Pb verstärkt in der feineren Fraktion auf, während es beim Naßverfahren (Werk P-4) vorwiegend Cu und Sb waren, die sich in den Feinanteilen anreicherten.

– In the wet process plant P-4 the elements Zn, Sb and Pb show an enrichment in kiln dust with Zn concentrating by 20 ppm, Sb by 36 ppm and Pb by around 46 ppm. The elements Ni, Cu, Cd, As being at the same levels.

The kiln dusts from the different processes are primarily of very fine nature with 90–95 % being of size less than 45 microns and so the preferential distribution of trace elements in the finer fraction (< 45 microns) of the kiln dusts of different processes has been discussed (Fig. 3).

– In dry process plant (P-1), the finer fraction is richer in the elements Zn, Cd and Sb with its Ni, Cu, As, Pb levels similar to overall kiln dust.

– In semi-dry process (P-2) and in semi-wet process (P-3), the finer fraction is richer in Zn, Cd and Pb while in wet process (P-4), Cu and Sb are observed to concentrate in the fines.

3.4 Effect of volatility on trace element enrichments

The percent volatility of Na_2O, K_2O, SO_3 and Cl for the different processes (Fig. 4) indicates the following:

– In semi-dry process (P-2) and wet process (P-4) plants (both plants without dust recirculation), the % volatility of K_2O, SO_3 and Cl is high, they differ only in their Na_2O volatility.

– In the dry process plant with preheater-precalciner (P-1) the chloride volatility levels are high and the volatility levels of Na_2O, K_2O and SO_3 are low (50–60 %).

– In the semi-wet process plant with preheater-precalciner, the chloride volatility levels are the highest with Na_2O, K_2O and SO_3 volatility levels of 100–110 %.

The effect of percent volatility levels of Na_2O, K_2O, SO_3 and Cl on the tendency of enrichment of trace elements in kiln dust/clinker is shown in **Table 4**.

TABELLE 4: Auswirkung der Flüchtigkeit von Na_2O, K_2O, SO_3 und Cl auf die Spurenelementanreicherung im Ofenstaub und Klinker

ohne Ofenstaubrückführung		mit Ofenstaubrückführung	
Halbtrockenverfahren	Naßverfahren	Trockenverfahren	Halbnaßverfahren
Hohes Cl, Na_2O, K_2O, SO_3	Hohes Cl, K_2O, SO_3 mit niedrigem Na_2O	Hohes Cl mit Na_2O, K_2O, SO_3 (50–60 %)	Sehr hohes Cl. mit Na_2O, K_2O, SO_3 (100–110 %)
angereicherte Elemente (ppm)		angereicherte Elemente (ppm)		angereicherte Elemente (ppm)		angereicherte Elemente (ppm)	
im Ofenstaub	im Klinker	im Ofenstaub	im Klinker	im Ofenstaub	im Klinker	im Ofenstaub	im Klinker
Pb (17)		Pb (46)	Pb (19)	Zn (30)	Sb	Sb (40)	Pb (80)
	Zn (30)	Sb (36)	Ni (7)	Cu	Pb	Cu	Sb (30)
	Cd (91)			Sb (5–15)		Zn (5–10)	Cd (10)
	Ni	Zn (20)	Cd (1–2)	Pb		Pb	
	Cu (3–5)						
	Pb						

Die Zahlen in Klammern geben die Anreicherungskonzentration der Elemente (ppm) im Vergleich zum Eingangsmaterial wieder

TABLE 4: Effect of % volatility levels of Na_2O, K_2O, SO_3 and Cl on enrichment of trace elements in kiln dusts/clinker

Kiln-dust not Recirculated		Kiln-dust Recirulated	
Semi-dry process	Wet process	Dry process	Semi-wet process
High Cl, Na_2O, K_2O, SO_3	High Cl, K_2O, SO_3 with low Na_2O	High Cl with Na_2O, K_2O, SO_3 (50–60 %)	V. High Cl. with Na_2O, K_2O, SO_3 (100–110 %)
Enriched elements (ppm)		Enriched elements (ppm)		Enriched elements (ppm)		Enriched elements (ppm)	
in kiln-dust	in clinker	in klin-dust	in clinker	in kiln-dust	in clinker	in kiln-dust	in clinker
Pb (17)		Pb (46)	Pb (19)	Zn (30)	Sb	Sb (40)	Pb (80)
	Zn (30)	Sb (36)	Ni (7)	Cu	Pb	Cu	Sb (30)
	Cd (91)			Sb (5–15)		Zn (5–10)	Cd (10)
	Ni	Zn (20)	Cd (1–2)	Pb		Pb	
	Cu (3–5)						
	Pb						

Number in bracket indicate enriched concentration of the element (ppm) as compared to total inputs.

3.4 Auswirkung der Flüchtigkeit auf die Anreicherung mit Spurenbestandteilen

Hinsichtlich der Flüchtigkeit von Na_2O, K_2O, SO_3 und Cl bei den verschiedenen Zementherstellungsverfahren (Bild 4) ergab sich folgendes Bild:

— Sowohl beim Halbtrocken- (P-2) als auch beim Naßverfahren (P-4) (beide Werke arbeiten ohne Staubrückführung) ist der flüchtige Anteil von K_2O, SO_3 und Cl hoch, nur hinsichtlich der Flüchtigkeit von Na_2O unterscheiden sich die beiden Verfahren.

— Im Werk P-1, das nach dem Trockenverfahren mit Vorcalcinierung arbeitet, ist die Flüchtigkeit von Chlorid hoch und die Flüchtigkeit von Na_2O, K_2O und SO_3 gering (50–60%).

— Beim Halbnaßverfahren mit Vorcalcinierung ist die Flüchtigkeit von Chlorid am höchsten; Na_2O, K_2O und SO_3 erreichen Werte von 100–110%.

Die Auswirkung der Flüchtigkeit von Na_2O, K_2O, SO_3 und Cl auf die Anreicherung der Spurenelemente im Ofenstaub und Klinker wird in **Tabelle 4** wiedergegeben. Die Daten zeigen, daß

— bei hoher Flüchtigkeit von Na_2O, K_2O, SO_3 und Cl sich nur Pb im Staub anreichert, während im Klinker eine Anreicherung mit Zn, Cd, Ni, Cu und Pb zu beobachten ist,

— bei niedriger Na_2O-Flüchtigkeit die amphoteren Elemente Zn, Cd und Sb eine Tendenz zur Anreicherung im Ofenstaub zeigen, während bei höherer Flüchtigkeit von Na_2O und Chlorid eine stärkere Anreicherung der Elemente Zn, Cd und Sb im Klinker festzustellen ist,

— im allgemeinen höhere Chloridgehalte zu einer Anreicherung der Elemente Pb, Sb, Zn und Cu im Ofenstaub führen.

Die Tendenz der Spurenelemente, sich im Ofenstaub bzw. Klinker anzureichern, wird demnach von der relativen Flüchtigkeit von Na_2O, K_2O, SO_3 und Cl beeinflußt, wobei jedoch der Grad der Anreicherung von der Art des Brennverfahrens abhängig zu sein scheint und ferner davon, ob eine Staubrückführung vorhanden ist oder nicht. Über einen ähnlichen Einfluß der Chloride auf das Verhalten von Pb und Cd wurde auch von W. Weisweiler und W. Dallibor berichtet [4].

3.5 Auslaugung der Spurenbestandteile

Die Ergebnisse der Auslaugversuche von Spurenelementen (Tabelle 3) zeigen folgendes:

— Die Elemente Cr, Co, Ni, Cu, Cd, As und Pb weisen eine sehr geringe Auslaugung aus den Ofenstäuben der verschiedenen Verfahren auf. Die Nachweisgrenze lag bei 3–5 ppm für alle Elemente außer für Arsen mit 0,1 ppm.

— Ofenstaub mit höherem Ba-Gehalt wies eine Auslaugung von 1,4% (P-1) und 4,6% (P-3) nach 30 Minuten bzw. 2,1% und 4,6% nach 72 Stunden auf.

— Sr, Zn und Sb zeigten eine verhältnismäßig starke Auslaugung aus den Stäuben aller Herstellungsverfahren.

Diese Daten weisen darauf hin, daß ein länger andauernder Kontakt mit Wasser zu keiner stärkeren Auslaugung von Sr und Zn führt, Antimon weist nach 72 Stunden eine nur ganz geringe Auslaugung auf. S. Sprung [3] gibt an, daß bei stark alkalischem pH die Thallium-Auslaugung aufgrund chemischer und/oder adsorptiver Bindung an die bei wäßrigen Aufschlämmungen von Ofenstaub und Zement gebildeten Hydratphasen deutlich geringer ist. Das Verhalten von Antimon läßt sich auf ähnliche Weise erklären. Die Untersuchungen zur Auslaugung zeigen, daß die Gehalte der Spurenelemente in Ofenstäuben durch wäßrige Auslaugung nicht vermindert werden können. Bei der Entsorgung von Ofenstaub ist demnach auch nur mit einer geringfügigen Auslaugung zu rechnen. Um Auswirkungen auf die Umwelt und das Ökosystem zu verhindern, könnte eine Einbindung von Ofenstäuben mit Zement erwogen werden; bei der

The data indicate that

— at high volatility levels of Na_2O, K_2O, SO_3 and Cl, only Pb is enriched in dust, while Zn, Cd, Ni, Cu and Pb are enriched in clinker,

— at lower Na_2O volatility levels, the amphoteric elements, viz. Zn, Cd and Sb, show a tendency to enrich in kiln dust, while at higher Na_2O and higher chloride volatility levels, the enrichment of the elements Zn, Cd, Sb in clinker is higher;

— in general, higher chloride levels tend to enrich the elements Pb, Sb, Zn, Cu in the kiln dusts.

The tendency of trace elements to enrich in kiln dust/clinker is thus influenced by the relative volatility levels of Na_2O, K_2O, SO_3 and Cl, however, the extent of enrichment seems to be influenced by the type of process and also by the fact whether the kiln dust is recirculated in the kiln system. A similar influence of chlorides on the behaviour of Pb and Cd has also been reported by W. Weisweiler and W. Dallibor [4].

3.5 Leachability of trace elements

The leachability data of trace elements (Table 3) indicate:

— The elements Cr, Co, Ni, Cu, Cd, As and Pb show negligible leachability from kiln dusts of different processes. The detectability limit amounted to about 3–5 ppm for all elements except for arsenic with 0.1 ppm.

— Kiln dust with higher Ba content shows a leachability of 1.4% (P-1) and 4.6% (P-3) after 30 min. and 2.1% and 4.6% respectively after 72 hours.

— Sr, Zn and Sb exhibit considerable leachability in kiln dust of all processes.

The data indicate that a longer duration of aqueous contact does not increase the leachability of Sr and Zn, antimony shows negligible leachability after 72 hours. S. Sprung [3] had observed that at highly alkaline pH, the aqueous leachability of thallium is very less due to chemical and/or adsorptive bonding to the hydrated phases formed in aqueous suspensions of kiln dust and cement. The behaviour of antimony could be explained on similar lines. The leachability studies reveal that the levels of trace elements in kiln dusts cannot be reduced by aqueous leaching. In case of disposal of kiln dusts, a small amount of leaching is expected. In order to avoid the impact on environment and ecosystem, encapsulation of kiln dusts in cementitious systems can be looked into, the hardening of cement is able to fix heavy metals chemically, adsorptively and by formation of dense texture.

4. Conclusions

The elements Ba, Sr, Cr, Mn and Co do not show an enrichment on the clinkerisation process and their behaviour could be termed similar to calcium.

The elements Ni, Cu, Zn, Sb, As and Pb show a tendency of enrichment in kiln dust/clinker. Their behaviour is influenced by

— the type of process,

— whether the kiln dust is recirculated or not, and

— the percent volatility of Na_2O, K_2O, SO_3 and Cl in the kiln systems.

In the semi-dry process and wet process plant where kiln dust is not recirculated the behaviour of trace elements is similar while the semi-wet and dry process plants (both with preheater-precalciner) exhibit similarities in the enrichment of trace elements. However, the relative extent of enrichment of the elements in kiln dusts and clinker is substantially influenced by the percent volatility levels of Na_2O, K_2O, SO_3 and Cl in the kiln system. Higher levels of chlorides substantially increase the enrichment of trace elements, viz. Cu, Zn, Sb and Pb, in kiln dust and clinker, the extent of enrichment is however influenced by the relative volatility levels of Na_2O, K_2O and SO_3. At higher volatility levels of Na_2O, the incorporation of elements like Zn, Cd, Sb, is more in clinker. Arsenic primarily shows a tendency of incorpora-

Zementerhärtung werden Schwermetalle chemisch, adsorptiv und durch die Bildung eines dichten Gefüges gebunden.

4. Schlußfolgerungen

Die Elemente Ba, Sr, Cr, Mn und Co zeigen keine Anreicherung während der Klinkerbildung; sie verhalten sich ähnlich wie Calcium.

Die Elemente Ni, Cu, Zn, Sb, As und Pb zeigen eine Tendenz zur Anreicherung im Ofenstaub bzw. Klinker. Ihr Verhalten wird von

— der Art des Herstellungsverfahrens,
— dem Vorhandensein bzw. Fehlen einer Staubrückführung und der Flüchtigkeit von Na_2O, K_2O, SO_3 und Cl im Ofensystem beeinflußt.

In den nach dem Halbtrocken- und dem Naßverfahren arbeitenden Werken ohne Ofenstaubrückführung verhalten sich die Spurenbestandteile ähnlich, während in den Werken mit Halbnaß- und Trockenverfahren (in beiden Fällen mit Vorcalcinierung) die Anreicherung der Spurenelemente Ähnlichkeiten aufweist. Das relative Ausmaß der Anreicherung der Spurenelemente im Ofenstaub und Klinker wird jedoch stark von der Flüchtigkeit von Na_2O, K_2O, SO_3 und Cl im Ofensystem beeinflußt. Höhere Cloridgehalte führen zu einer deutlich stärkeren Anreicherung der Spurenelemente Cu, Zn, Sb und Pb im Ofenstaub und im Klinker, wobei der Grad der Anreicherung jedoch von der relativen Flüchtigkeit von Na_2O, K_2O und SO_3 abhängig ist. Bei höherer Flüchtigkeit von Na_2O reichert sich der Klinker stärker mit den Elementen Zn, Cd und Sb an. Arsen wird in erster Linie im Klinker eingebunden, und das ist von der Art des Brennverfahrens oder der Flüchtigkeit anderer Bestandteile unabhängig.

Die Untersuchung der Auslaugbarkeit in wäßrigen Suspensionen von Ofenstäuben zeigt, daß Barium bei einer Konzentration im Ofenstaub von mehr als 1000 ppm verstärkt und Sr, Zn und Sb aus den Ofenstäuben aller Herstellungsverfahren in beachtlichem Umfang ausgelaugt werden.

tion in clinker and is not affected by the type of process or volatility levels of other volatiles.

Aqueous leachability studies on kiln dust reveal that barium exhibits leachability in kiln dusts having a concentration higher than 1000 ppm, while Sr, Zn and Sb show considerable leachability in kiln dusts of all processes.

Literature

[1] Krishnan, N. A., Gore, V. K., Khadilkar, S. A., Hargave, R. V., Page, C, H., and Chatterjee, A. K.: Trace element balances in kiln system of a few Indian cement plants: 9th ICCC, New Delhi, India. 1992, Vol. II, pp. 67–73.

[2] Sprung, S., Kirchner, G., and Rechenberg, W.: Reactions of poorly volatile trace elements in cement clinker burning. Zement-Kalk-Gips 10 (1984), pp. 513–518.

[3] Sprung, S.: Trace elements – Concentration build up and measures of reduction. Zement-Kalk-Gips 7 (1988), pp. 172–176.

[4] Weisweiler, W., Dallibor, W., and Lück, M. P.: Lead, cadmium and thallium balances in a cement kiln with grate preheater operating with increased chloride input. Zement-Kalk-Gips 1 (1988), pp. 27–28.

[5] Weisweiler, W., and Kreçmar, W.: Arsenic and antimony balances of cement kiln plant with grate preheater. Zement-Kalk-Gips 3 (1989), pp. 133–135.

[6] Kirchner, G.: Behaviour of heavy metals during clinker burning. Zement-Kalk-Gips 10 (1986), pp. 555–557.

Abscheidung von Spurenelementen beim Klinkerbrennprozeß

Removal of trace elements in the clinker burning process

Séparation d'éléments traces dans le processus de cuisson du clinker

Separación de elementos-traza durante el proceso de cocción del clínker

Von **B. Kirchartz**, Düsseldorf/Deutschland

Zusammenfassung – Die überwiegende Zahl der Spurenelemente verhält sich im Klinkerbrennprozeß ähnlich wie die Hauptkomponenten des Brennguts. Sie werden daher in den einzelnen Anlagenteilen des Zementbrennprozesses wie die Hauptkomponenten auch abgeschieden. Demgegenüber reagieren die Elemente Blei, Cadmium und Thallium bevorzugt zu leichterflüchtigen Halogeniden und Sulfaten, die jedoch unter üblichen Reingasbedingungen wie alle nichtflüchtigen Elemente praktisch vollständig in kondensierter Form vorliegen. Die Abscheidung der im Ofengasstaub vorliegenden Elementmenge und somit die Spurenelementemission hängt daher im wesentlichen vom Gesamtentstaubungsgrad der Abgasreinigungseinrichtung ab. Eine Ausnahme hiervon stellt das Element Quecksilber dar, dessen Abscheidung aus der Gasphase vorwiegend im Bereich der Mahltrocknung und Abgasreinigung stattfindet. Sie verläuft um so vollständiger, je niedriger die Temperatur im Bereich der Abgasreinigung und je höher das dort vorliegende Oberflächenangebot an Ofengasstaub ist. Der nicht mit dem Ofengasstaub abgeschiedene Hg-Anteil wird in aller Regel dampfförmig emittiert. Aufgrund der geringen Gehalte an Quecksilber in den natürlichen Roh- und Brennstoffen ist jedoch die Hg-Emission von Zementdrehrohröfen im Vergleich zum Emissionsgrenzwert sehr gering. Die hohe Flüchtigkeit des Quecksilbers und seiner Verbindungen erfordert jedoch bei Einsatz von Sekundärstoffen eine strikte Begrenzung der damit verbundenen Hg-Einträge in den Brennprozeß.

Summary – In the clinker burning process the majority of trace elements behave like the main components of the kiln feed. They are therefore also removed in the individual sections of the plant in the cement burning process in the same way as the main components. The elements lead, cadmium and thallium normally react to form more volatile halides and sulphates, but under the usual clean gas conditions these are present virtually entirely in condensed form like all the non-volatile elements. Removal of the elements present in the kiln gas dust, and hence the trace element emission, is therefore heavily dependent on the overall dedusting efficiency of the exhaust gas cleaning equipment. An exception to this is the element mercury which is generally removed in the gas phase in the drying and grinding system and the exhaust gas cleaning system. This is more complete the lower the temperature in the exhaust gas cleaning system and the higher the available surface area of the kiln dust there. The fraction of Hg which is not removed with the kiln gas dust is normally emitted in vapour form, but because of the low levels of mercury in the natural raw materials and fuels the Hg emission from rotary cement kilns is very low when compared with the emission limit. However, when secondary materials are used the high volatility of mercury and its compounds means that there must be strict limits on the associated intake of Hg into the burning process.

Résumé – Le plus grand nombre d'éléments traces se comporte, dans le processus de cuisson du clinker, à peu près comme les composants principaux de la matière à cuire. Ils sont donc séparés, dans les différentes parties d'installation du process de cuisson du clinker, de même façon que les composants principaux. Les éléments plomb, cadmium et thallium, par contre, se transforment préférentiellement en halogénures et sulfates plus fortement volatils, qui néanmoins se présentent pratiquement complètement sous forme condensée, comme tous les éléments non volatils, dans les conditions courantes des gaz épurés. La séparation de la quantité d'éléments présents dans la poussière des gaz du four dépend alors essentiellement du degré total de dépoussiérage de l'équipement d'épuration des gaz d'exhaure. Une exception y est l'élément mercure, dont la séparation de la phase gazeuse s'effectue préférentiellement dans les secteurs broyage-séchage et épuration des gaz d'exhaure. Elle se fait d'autant plus complètement, que la température est basse dans le secteur de l'épuration des gaz d'exhaure et qu'y est plus grand le potentiel de surface de poussière des gaz du four. La part Hg non séparée avec la poussière des gaz du four est de toute façon émise sous forme de vapeur. En raison des faibles teneurs en mercure des matières premières et combustibles naturels, l'émission Hg de fours rotatifs à ciment est néanmoins très faible, comparée à la valeur limite d'émission. La grande volatilité du mercure et de ses composés nécessite néanmoins, lors de l'utilisation de substances secondaires, une limitation stricte des entrées Hg, qui peut en résulter, dans le process de cuisson.

Resumen – La mayoría de los elementos-traza se comporta durante el proceso de cocción del clínker de forma similar a los componentes principales del material sometido a cocción. Por esta razón, son eliminados en las diferentes secciones del proceso de cocción, lo mismo que los componentes principales. Por otro lado, los elementos formados por el plomo, el cadmio y el talio se transforman preferentemente en halogenuros y sulfatos más volátiles, pero bajo las condiciones habituales de gases depurados, están presentes casi de forma completa en estado condensado, como es el caso de todos los elementos no volátiles. La separación de las cantidades de elementos presentes en el polvo de los gases del horno, y por lo tanto las emisiones de elementos-traza, depende esencialmente del grado de desempolvado total de la instalación depuradora de los gases de escape. Una excepción la constituye el mercurio, cuya separación de la fase gaseosa tiene lugar, sobre todo, durante la operación de molienda-secado y la depuración de los gases de escape. Será tanto más completa, cuanto más baja sea la temperatura en la zona de depuración de los gases de escape y cuanto más alto sea la superficie de polvo de los gases del horno. La proporción de Hg no eliminado con el polvo de los gases del horno se emite, por regla general, en forma de vapor. Sin embargo, debido al reducido contenido de mercurio en las materias primas y los combustibles naturales, las emisiones de Hg de los hornos rotatorios de cemento son muy bajas, en comparación con el valor límite. Pero en caso de emplearse materias primas secundarias, la alta volatilidad del mercurio y de sus combinaciones requiere una limitación estricta de la introducción de Hg en el proceso de cocción.

Separación de elementos-traza durante el proceso de cocción del clínker

Anlagen zum Herstellen von Zementen und Zementklinker gehören zu den genehmigungsbedürftigen Anlagen, deren Emissionsbegrenzungen bislang in der Technischen Anleitung zur Reinhaltung der Luft festgelegt waren. Zukünftig wird für Zementwerke, in deren Drehrohröfen sekundäre Brennstoffe eingesetzt werden, die „Verordnung über Verbrennungsanlagen für Abfälle und ähnliche brennbare Stoffe" (17. BImSchV) mit weitaus schärferen Emissionsbegrenzungen maßgebend sein.

Die Emissionsbegrenzungen der TA-Luft sowie der 17. BImSchV für umweltrelevante Spurenelemente gehen aus der nachfolgenden **Tabelle 1** hervor. Danach werden die Elemente in Klassen eingeteilt, für die meist Summengrenzwerte festgelegt worden sind. Lediglich beim Element Quecksilber existiert mit Einführung der 17. BImSchV ein gesonderter Grenzwert. Besondere Bedeutung kommt diesem Element zu, da aufgrund der zu erwartenden hohen Flüchtigkeit des Quecksilbers und seiner Verbindungen nicht ausgeschlossen werden kann, daß ein wesentlicher Teil der in den Einsatzstoffen enthaltenen Quecksilbermenge ganz oder teilweise emittiert wird.

Mit der Abscheidung und Emission umweltrelevanter Spurenelemente an Drehrohröfen der Zementindustrie hat sich das Forschungsinstitut der Zementindustrie in den vergangenen Jahren eingehend beschäftigt. Danach wird die Emission im wesentlichen vom Spurenelementeintrag mit den Roh- und Brennstoffen, der Reaktion der Elemente während des Brennvorgangs sowie den anschließenden Abscheidevorgängen im Ofensystem sowie den nachgeschalteten Anlagenteilen bestimmt. Bei den Abscheidevorgängen kann grundsätzlich zwischen der Kondensation bzw. Adsorption an Staub aus der Gasphase sowie der Abscheidung der an Staub gebundenen Spurenelemente unterschieden werden.

Die Abscheidung aus der Gasphase erfolgt je nach Flüchtigkeit der Verbindungen bei unterschiedlichen Temperaturen

Until now, the emission limits for cement and cement clinker production plants requiring an operating licence have been laid down in the German Clean Air Regulations ("TA-Luft"). In the future, cement works with rotary kilns in which secondary fuels are used will be required to comply with the "Ordinance on Incinerators for Wastes and Similar Combustible Substances" (17th Federal Immission Protection Ordinance: "17. BImSchV") with far more stringent emission limits.

The TA Luft and 17. BImSchV emission limits for environmentally harmful trace elements can be seen in **Table 1** below. The elements are divided into classes for which cumulative limits have been imposed in most cases. Only mercury has been given a separate limit since the introduction of 17. BImSchV. This element is particularly important since, in view of the highly volatile nature of mercury and its compounds, one cannot rule out the possibility that a substantial proportion of the mercury contained in the starting products will be emitted wholly or in part.

In recent years, the Cement Industry Research Institute has examined in detail the separation and emission of environmentally harmful trace elements from rotary kilns used in the cement industry. According to the Institute's findings, the emissions are determined essentially by the trace element intake with the raw materials and fuels, the reaction of the elements during the burning process, and the subsequent separation processes in the kiln system and the downstream parts of the plant. With regard to the separation processes, a fundamental distinction can be made between condensation or adsorption on dust from the gas phase and separation of the trace elements bound to the dust.

Depending on the volatility of the compounds, separation from the gas phase takes place at different temperatures and hence also in different zones of the kiln system. This can be used to assess the different behaviour of trace elements, enabling them to be assigned to different classes. **Table 2**

TABELLE 1: Emissionsgrenzwerte für Spurenelemente nach TA Luft sowie der 17. BImSchV

Stoffklasse	Elemente	TA-Luft 1986	17. BImSchV 1990
		Angaben in mg/m^3	
I	Cd, Tl	0,2	0,05
	Hg		0,05
II	Se, Te As, Co, Ni	1	–
III	Sb, Pb, Cr, Cu, Mn, V, Sn CN, F, Pt, Pd, Rh	5	0,5 –

TABLE 1: Emission limits for trace elements according to TA Luft and 17. BImSchV

Substance class	Elements	TA Luft 1986	17. BImSchV 1990
		Data in mg/m^3	
I	Cd, Tl	0.2	0.05
	Hg		0.05
II	Se, Te As, Co, Ni	1	–
III	Sb, Pb, Cr, Cu, Mn, V, Sn CN, F, Pt, Pd, Rh	5	0.5 –

TABELLE 2: Einordnung von Spurenelementen nach ihrer Flüchtigkeit

Stoffklasse	Elemente	Kondensationspunkt in °C
nichtflüchtig	As, Co, Ni, Sb, Cr, Cu, Mn, V, Sn	–
schwerflüchtig	Cd, Pb	700–900
leichtflüchtig	Tl	450–550
hochflüchtig	Hg	< 250

TABLE 2: Classification of trace elements according to volatility

Substance class	Elements	Condensation point in °C
Non-volatile	As, Co, Ni, Sb, Cr, Cu, Mn, V, Sn	–
Low volatility	Cd, Pb	700–900
Readily volatile	Tl	450–550
Highly volatile	Hg	< 250

und somit auch in unterschiedlichen Bereichen des Ofensystems. Dies kann zur Beurteilung des unterschiedlichen Verhaltens von Spurenelementen herangezogen werden und erlaubt eine Einstufung in unterschiedliche Klassen. Die folgende **Tabelle 2** enthält eine Einordnung, wie sie aufgrund der mittlerweile weitgehend abgeschlossenen Forschungsarbeiten vorgenommen werden kann.

Daraus geht hervor, daß die Mehrzahl der Spurenelemente sich in ähnlicher Weise wie die Hauptelemente verhält. Kennzeichnend für diese Elemente ist unter anderem, daß sie beim Brennvorgang praktisch nicht verdampfen und daher mit dem Klinker nahezu vollständig das Ofensystem verlassen. Bei den Elementen Blei und Cadmium kann es hingegen je nach Reaktionsbedingungen zu Verdampfungs- und Kondensationsvorgängen kommen, die zur Ausbildung von Kreisläufen zwischen Drehofen und Vorwärmer führen. Da die Abscheidung aus der Gasphase bereits bei Temperaturen zwischen 700 °C und 900 °C erfolgt, ist dies bei Drehrohröfen mit Zyklonvorwärmer jedoch aufgrund des hohen Rückhaltevermögens des Vorwärmers für die Emissionen meist ohne wesentliche Bedeutung.

Beim Element Thallium liegt der Kondensationspunkt mit Temperaturen zwischen 450 °C und 550 °C deutlich niedriger. Infolgedessen wandert der Kondensationspunkt in den oberen Bereich des Zyklonvorwärmers. Dementsprechend nimmt der Thalliumabscheidegrad des Zyklonvorwärmers im Vergleich zu den Elementen Blei und Cadmium geringfügig ab. An Drehrohröfen mit Rostvorwärmer kondensieren flüchtige Blei-, Cadmium- und Thalliumverbindungen überwiegend an den Zwischengas- und Elektrofilterstäuben. In den nachgeschalteten Anlagenteilen und Staubabscheidern werden die partikelgebundenen Spurenelemente dann entsprechend dem Gesamtentstaubungsgrad der Staubabscheider abgeschieden.

Umfangreiche Untersuchungen zum Verhalten des Quecksilbers haben demgegenüber gezeigt, daß dessen Verbindungen den Vorwärmer zunächst praktisch vollständig in der Gasphase verlassen und sich dementsprechend im Vorwärmer auch nicht anreichern. Das folgende **Bild 1** enthält hierzu die Abnahme der Hg-Konzentration im Brenngut einer Drehofenanlage mit Zyklonvorwärmer in Abhängigkeit von der Temperatur.

Eine merkliche Abnahme der Konzentration der zunächst dampfförmigen Hg-Verbindungen im Abgas kann unterhalb von 200 °C beobachtet werden. Ursache hierfür ist eine Adsorption von Hg-Verbindungen aus der Gasphase an den Ofengasstäuben, die bevorzugt bei niedrigen Gastemperaturen und hohem Oberflächenangebot abläuft. Aus dem folgenden **Bild 2** geht die an Drehrohröfen mit Zyklon- und Rostvorwärmer auf die Ofeneinnahmen bezogene Quecksilberabscheidung in Abhängigkeit von der Abgastemperatur hervor. Der verbleibende Anteil an Hg-Einnahmen wird mit dem Abgasstrom überwiegend dampfförmig emittiert. Im Unterschied zu allen anderen Spurenelementen, die unter üblichen Reingasbedingungen weitgehend in kondensierter Form vorliegen und aufgrund der effizienten Staubabscheidung praktisch vollständig abgeschieden werden, kommt beim Element Quecksilber daher dem Eintrag in den Brennprozeß besondere Bedeutung zu.

below contains a classification that can be made on the basis of the research work which has now been largely completed.

It can be seen from this table that most of the trace elements behave in a similar way to the main elements. One of the characteristics of these elements is that they hardly evaporate at all during the burning process and therefore leave the kiln system almost entirely with the clinker. In the case of lead and cadmium, however, evaporation and condensation processes may occur, depending on the reaction conditions, leading to the formation of recirculating systems between rotary kiln and preheater. As separation from the gas phase takes place at temperatures as high as 700 °C to 900 °C, this is not usually very important for the emissions with rotary kilns with a cyclone preheater because of the high retaining capacity of the preheater.

The condensation point for thallium is much lower, at temperatures of between 450 °C and 550 °C. Consequently, the condensation point moves to the upper zone of the cyclone preheater. The thallium separating efficiency of the cyclone preheater therefore falls slightly compared with lead and cadmium. In rotary kilns with a grate preheater, volatile lead, cadmium and thallium compounds condense predominantly on the intermediate gas and electrostatic precipitator dusts. In the downstream parts of the plant and dust separators, the particle-bound trace elements are then removed to an extent which depends on the overall dedusting efficiency of the dust separators.

Extensive investigations into the behaviour of mercury, on the other hand, have shown that its compounds initially leave the preheater almost entirely in the gas phase and do not, therefore, accumulate in the preheater. **Fig. 1** below

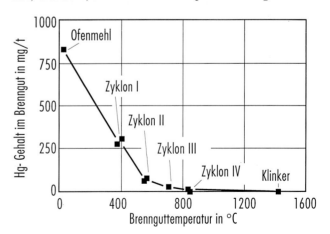

BILD 1: Hg-Gehalte im Brenngut in Abhängigkeit von der Brennguttemperatur
FIGURE 1: Hg levels in the kiln feed as a function of the kiln feed temperature

Hg-Gehalt im Brenngut in mg/t = Hg level in the kiln feed in mg/t
Brennguttemperatur in °C = Kiln feed temperature in °C
Ofenmehl = Kiln meal
Zyklon I = Cyclone I
Zyklon II = Cyclone II
Zyklon III = Cyclone III
Zyklon IV = Cyclone IV
Klinker = Clinker

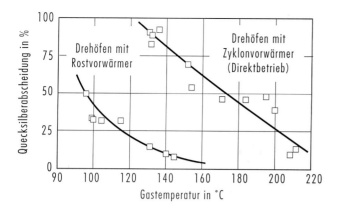

BILD 2: Quecksilberabscheidung in Abhängigkeit von der Gastemperatur
FIGURE 2: Mercury separation as a function of the gas temperature

Quecksilberabscheidung in %	= Mercury separation in %
Gastemperatur in °C	= Gas temperature in °C
Drehöfen mit Rostvorwärmer	= Rotary kilns with grate preheater
Drehöfen mit Zyklonvorwärmer (Direktbetrieb)	= Rotary kilns with cyclone preheater (direct operation)

Nach dem derzeitigen Stand der Untersuchungen des Forschungsinstituts der Zementindustrie liegen die Hg-Einträge jedoch derart niedrig, daß auch unter der ungünstigsten Annahme einer vollständigen Hg-Emission die Emissionsbegrenzungen der TA Luft sowie der 17. BImSchV auch in Summe mit den Elementen Thallium und Cadmium eingehalten werden. Dies trifft unter anderem auch auf den Einsatz sekundärer Brennstoffe zu, wenn deren Gehalt an Thallium, Cadmium und insbesondere Quecksilber innerhalb der Schwankungsbreite der natürlichen Brennstoffe liegt.

shows the fall in the Hg concentration in the kiln feed of a rotary kiln with a cyclone preheater as a function of the temperature.

A marked fall in the concentration of the Hg compounds initially in the vapour state in the exhaust gas may be observed at temperatures below 200°C. The reason for this is the adsorption of Hg compounds from the gas phase on the kiln gas dusts which takes place preferentially at low gas temperatures and when the available surface area of dust is high. **Fig. 2** below shows the mercury separation based on kiln intakes as a function of the exhaust gas temperature, in rotary kilns with cyclone and grate preheaters. The remaining proportion of the Hg intake is emitted predominantly in vapour form with the stream of exhaust gas. Unlike all the other trace elements, which are present largely in the condensed form under conventional clean gas conditions and are almost completely removed because of the efficient dust separation, the mercury intake into the burning process is particularly important.

According to the investigations carried out so far at the Cement Industry Research Institute, the Hg intakes are so low, however, that even under the least favourable assumption of total Hg emission, the emission limits of TA Luft and 17. BImSchV are complied with, even when combined with the figures for thallium and cadmium. This also applies to the use of secondary fuels if their content of thallium, cadmium and particularly mercury lies within the range of variation of natural fuels.

Der Einfluß von Gewinnungssprengungen auf die Erschütterungsimmission

The influence of primary blasting on vibration immission

L'influence des tirs d'abattage sur les immissions d'ébranlement

El influjo de la explotación por voladuras sobre las inmisiones debidas a las vibraciones del suelo

Von **P. Küllertz** und **M. Schneider**, Düsseldorf/Deutschland

Zusammenfassung – Aufgrund der hohen Siedlungsdichte in Deutschland werden an Gewinnungssprengungen und die dabei auftretenden Erschütterungsimmissionen in zunehmendem Maße höhere Anforderungen gestellt. Neben der Erschütterungsmessung kommt der Immissionsprognose für Schwinggeschwindigkeiten in der näheren Umgebung von Steinbrüchen besondere Bedeutung zu. Damit können sprengtechnische Randbedingungen optimiert werden. Bei der Ausbreitung von Sprengerschütterungen im Gelände nimmt die Schwingungsenergie mit zunehmendem Abstand vom Sprengort ab. Die Prognose der Schwinggeschwindigkeit am Immissionsort beruht auf empirischen Zusammenhängen, in die neben der Entfernung zur Sprengstelle die Lademenge pro Zündzeitstufe sowie die sprengtechnischen Bedingungen einschließlich der geologischen Lagerstättenstruktur eingehen. Während eine Erschütterungsprognose Aussagen über die zu erwartende maximale Schwinggeschwindigkeit am Fundament von Gebäuden liefert, ist für die Vorhersage der Immission im Gebäude zusätzlich die Kenntnis der Übertragung einer Fundamentanregung auf eine Geschoßdecke erforderlich. Der Übertragungsfaktor wird in der Regel bei orientierenden Erschütterungsmessungen ermittelt.

Summary – Primary blasting and the resulting vibration immission are being subjected to ever stricter requirements because of the high housing density in Germany. Great importance is attached not only to vibration measurements but also to immission predictions for oscillation velocities in the immediate vicinity of quarries. These make it possible to optimize the conditions associated with blasting. As the blasting vibrations propagate through the ground the oscillation energy decreases with increasing distance from the blasting location. Prediction of the oscillation velocity at the immission location is based on empirical relationships which include not only the distance from the blasting site but also the charge per ignition delay interval and the blasting conditions including the geological structure of the deposit. A vibration prediction provides information about the maximum oscillation velocity to be expected in building foundations, but to predict the immission in the building it is also necessary to understand how an excitation in the foundation is transmitted to a floor. The transmission factor is normally determined during preliminary investigative vibration measurements.

Résumé – En raison de la forte densité des habitations, en Allemagne, les tirs d'abattage et les immissions d'ébranlement qui en résultent, sont de plus en plus dans le collimateur. Outre la mesure de l'ébranlement, le pronostic d'immission des vitesses ondulatoires dans le voisinage immédiat des carrières revêt une importance particulière. Ainsi, les conditions d'accompagnement du point de vue technique des tirs peuvent être optimisées. Lors de la propagation des ébranlements de tirs dans le terrain, l'énergie ondulatoire décroît au fur et à mesure de l'éloignement du lieu de tir. Le pronostic de la vitesse ondulatoire au lieu d'immission relève de circonstances empiriques, incluant à la fois la distance par rapport au lieu de tir, l'importance de la charge par séquence de déclenchement et les conditions techniques du tir, y compris la structure du gisement. Alors qu'un pronostic d'ébranlement fournit des données quant aux vitesses ondulatoires maximales prévisibles aux fondations des bâtiments, il est, de plus, nécessaire de connaître la transmission d'une sollicitation des fondations à un niveau d'étage, pour prévoir l'immission dans le bâtiment. Le facteur de transmission est déterminé, en règle générale, au moyen de mesures d'ébranlement orientées.

Resumen – A causa de la gran densidad demográfica registrada en Alemania, las exigencias respecto de la explotación por voladuras y de las inmisiones debidas a las vibraciones del suelo son cada vez más rigorosas. Adquiere especial importancia, aparte de la medición de las sacudidas, el pronóstico de las inmisiones relacionadas con la velocidad de las vibraciones a proximidad de las canteras. De esta forma es posible optimizar las condiciones técnicas de la operación de voladura. Durante la propagación de las vibraciones del terreno, debidas a las voladuras, la energía de las vibraciones disminuye con la distancia del lugar en que se efectúen las voladuras. El pronóstico de la velocidad de las vibraciones en el lugar de las inmisiones se basa en condiciones empíricas, a las que hay que sumar no solamente las distancias del lugar de las

voladuras, sino también la carga por cada secuencia de ignición y las condiciones técnicas de las voladuras, incluyendo la estructura geológica del yacimiento. Mientras que un pronóstico de las vibraciones nos da detalles acerca de la velocidad máxima de las vibraciones, con las que se puede contar en los fundamentos de los edificios, es necesario, además, para la predicción de las inmisiones producidas dentro del edificio, conocer la transmisión de las excitaciones del fundamento al techo de una planta. El factor de transmisión suele determinarse por medio de mediciones orientativas de las vibraciones.

Aufgrund der hohen Besiedlungsdichte in Deutschland werden an Gewinnungssprengungen hinsichtlich der Erschütterungsimmissionen in zunehmendem Maße höhere Anforderungen gestellt. Auch bei Gewinnungssprengungen, so wie sie beispielsweise in Steinbrüchen und Tagebauen der Zementindustrie durchgeführt werden, werden die Erschütterungsimmissionen in der Umgebung des Abbaugebietes dadurch begrenzt, daß maximal zulässige Schwinggeschwindigkeiten nicht überschritten werden dürfen. Hierzu sind in der Normenreihe DIN 4150 „Erschütterungen im Bauwesen" Anhaltswerte hinsichtlich des Schutzes von Menschen in Gebäuden und von Bauwerken aufgestellt. Dabei sind im Teil 2 dieser Norm Anhaltswerte für die maximal zulässige Erschütterungseinwirkung auf Menschen zusammengefaßt. Dem unterschiedlichen Wahrnehmungsempfinden des Menschen für Erschütterungen verschiedener Frequenzbereiche wird dabei durch das Verfahren der sogenannten KB-Bewertung Rechnung getragen. Im Teil 3 der Norm sind maximale Schwinggeschwindigkeiten dargestellt, die den Schutz von Bauwerken gewährleisten. Dabei wird die unterschiedliche Empfindlichkeit verschiedener Gebäudearten auf Erschütterungseinwirkungen berücksichtigt und auch zwischen verschiedenen Frequenzbereichen differenziert, die jeweils unterschiedliche Auswirkungen auf die zu betrachtenden Gebäude haben können. Aus der Zusammenfassung in **Tabelle 1** geht hervor, daß die maximale Schwinggeschwindigkeit beispielsweise an Fundamenten von Wohngebäuden je nach Frequenzbereich einen Wert von 5–20 mm/s nicht überschreiten darf.

Die Höhe der Erschütterungsimmission wird im wesentlichen bestimmt durch den Abstand des Immissionsortes zur Sprengstelle sowie durch die Stärke der Sprengung selbst. Hierbei ist in erster Linie die Lademenge pro Zündzeitstufe von Bedeutung. Darüber hinaus hängen die Erschütterungsimmissionen in der Umgebung eines Steinbruchs sehr

Primary blasting and the resulting vibration nuisance are being subjected to ever stricter requirements because of the high housing density in Germany. In the case of primary blasting of the kind carried out, for instance, in the cement industry's quarries and open-cast mines, the vibration nuisance in the neighbourhood of the workings is also restricted by prescribing maximum acceptable oscillation velocities which must not be exceeded. Guide values for the protection of people in buildings and of structures are specified in the series of standards DIN 4150 "Vibrations in building". Part 2 of this standard lists guide values for the maximum acceptable effect of vibration on people. The differing sensitivities of people to vibrations of various frequency ranges is taken into account by means of the so-called "KB" evaluation method. Part 3 of the standard gives maximum oscillation velocities which ensure the protection of structures. They take into account the differing sensitivity of different types of buildings to the effects of vibration and also differentiate between various frequency ranges, each of which may have different effects on the buildings under consideration. The summary in **Table 1** shows, for example, that the maximum oscillation velocity at the foundations of residential buildings must not exceed a figure of 5 – 20 mm/s depending on the frequency range.

The magnitude of the vibration nuisance is essentially determined by the distance between the place where the nuisance is felt and the blasting site, and by the intensity of the blast. The most important factor is the charge per ignition delay interval. In addition, the vibration nuisance in the neighbourhood of a quarry depends very greatly on the local geological conditions. The transmission of the vibration may therefore vary considerably depending on the geology and the direction of propagation.

As a general rule, the levels of vibration nuisance are ascertained by measurement; this is done by on-site determination of the oscillation velocity, especially in relation to build-

TABELLE 1: Anhaltswerte für die zulässige Schwinggeschwindigkeit (DIN 4150, Teil 3, Tabelle 1)

Gebäudeart	Schwinggeschwindigkeit v_i am Fundament in mm/s			Deckenebene oberstes Vollgeschoß
	Frequenz <10 Hz	Frequenz 10–50 Hz	Frequenz 50–100 Hz	alle Frequenzen
1. Gewerblich genutzte Bauten und Industriebauten	20	20–40	40–50	40
2. Wohngebäude, Gebäude mit Außenputz	5	5–15	15–20	15 (20)*
3. Besonders empfindliche Bauten	3	3–8	8–10	8

* Anhaltswert für vertikale Richtung

TABLE 1: Guide values for acceptable oscillation velocities (DIN 4150, Part 3, Table 1)

Type of building	Oscillation velocity v_i at foundations in mm/s			Floor level of top storey
	Frequency <10 Hz	Frequency 10–50 Hz	Frequency 50–100 Hz	All frequencies
1. Commercial and industrial buildings	20	20–40	40–50	40
2. Residential buildings, externally rendered buildings	5	5–15	15–20	15 (20)*
3. Particularly sensitive structures	3	3–8	8–10	8

* Guide value for vertical direction

stark von den geologischen Standorteigenschaften ab. Die Weiterleitung der Erschütterungsemissionen kann daher je nach Geologie und Ausbereitungsrichtung stark variieren.

Im allgemeinen werden die Erschütterungsimmissionen meßtechnisch ermittelt, indem vor Ort im Rahmen von Gewinnungssprengungen mit Schwingungsmessern die Schwinggeschwindigkeit, insbesondere an Gebäuden in der Umgebung eines Abbaugebietes, bestimmt wird. Bei einer geplanten Abbauerweiterung ist hingegen häufig eine rechnerische Prognose der zu erwartenden Erschütterungsimmissionen erforderlich, um die sprengtechnischen Randbedingungen hinsichtlich des Abbaus einerseits und der zukünftig zu erwartenden Erschütterungsimmisionen andererseits optimal wählen zu können.

Bild 1 zeigt die sogenannte Abstand-Lademengen-Beziehung, durch die die zu erwartende maximale Schwinggeschwindigkeit am Immissionsort als Funktion der Lademenge und des Abstands zur Sprengstelle dargestellt werden kann. Dabei bezeichnet v_{max} die zu erwartende maximale Schwinggeschwindigkeit in mm/s, L bezeichnet die Lademenge pro Zündzeitstufe in kg und R gibt den Abstand zwischen Sprengort und Meßort in Meter an. Die Konstante k bzw. die Exponenten b, n sind empirisch bestimmte Größen.

$$v_{max} = k \cdot L^b \cdot R^{-n}$$

v_{max}	=	maximale Schwinggeschwindigkeit [mm/s]
L	=	Lademenge pro Zündzeitstufe [kg]
R	=	Abstand zur Sprengstelle [m]
k, b, n	=	geologieabhängige Konstanten

BILD 1: Abstand-Lademengen-Beziehung

Dieser theoretische Zusammenhang zwischen Erschütterungsemission und -immission sowie den empirisch zu bestimmenden Konstanten ergibt sich auf Grundlage der statistischen Auswertung einer Vielzahl von Erschütterungsmessungen bei Gewinnungssprengungen, die im Rahmen eines Forschungsvorhabens im Auftrag des Umweltbundesamtes zur Erstellung eines Erschütterungskatasters für den norddeutschen Raum durchgeführt wurden. Von Lüdeling [1] sind seinerzeit Daten aus ca. 3000 Einzelmessungen bei ca. 150 Sprengungen zusammengetragen worden, um daraus ein Prognosemodell für die Erschütterungsimmissionen zu erarbeiten.

In der Praxis liefern derartige empirische Beziehungen naturgemäß Prognosewerte, die einen entsprechenden Sicherheitsfaktor beinhalten müssen, um unterschiedlichen Ausbreitungsbedingungen Rechnung tragen zu können. Dadurch wird beispielsweise berücksichtigt, daß die vereinfachende Annahme einer kreisförmigen Ausbreitung der Erschütterungswellen in der Praxis nicht gegeben ist, vielmehr führt der geologische Aufbau in der Umgebung des Abbaugebietes im allgemeinen zu unsymmetrischen Abstrahlcharakteristiken, die aber dennoch in einer Prognose mit hinreichender Sicherheit erfaßt werden sollen.

Bild 2 zeigt als sogenanntes Abstands-Lademengen-Diagramm denjenigen Abstand, der bei vorgegebener Lademenge und maximal zulässiger Schwinggeschwindigkeit in der Umgebung von Steinbrüchen der Zementindustrie nicht unterschritten werden darf. Der Zusammenhang gilt mit einer statistischen Sicherheit von 95 %, die empirischen Konstanten k = 447, b= 0,67, n= 1,32 beruhen auf Messungen, bei denen die Ausbreitung von Erschütterungen in Kalkstein bzw. Kalkmergel untersucht wurden.

In vielen Fällen ist eine rein theoretische Betrachtung der zu erwartenden Erschütterungsimmissionen beispielsweise aufgrund komplizierter Abbaufortschritte oder aber wegen der zu geringen Entfernung zur nächsten Bebauung nicht ausreichend, um optimale sprengtechnische Randbedingungen festzulegen zu können.

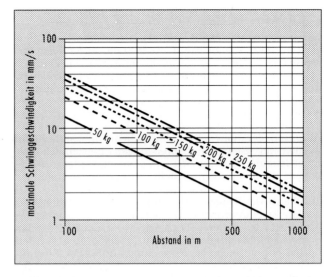

BILD 2: Zusammenhang zwischen maximaler Schwinggeschwindigkeit und Abstand Sprengstelle-Immissionspunkt für unterschiedliche Lademengen pro Zündzeitstufe

FIGURE 2: Relationship between the maximum oscillation velocity and the distance between the blasting site and the point where the nuisance is felt for different charges per ignition delay interval

X-axis: Distance in m
Y-axis: Maximum oscillation velocity in mm/s

ings in the neighbourhood of the workings, with vibration meters while carrying out primary blasting. In contrast, in the case of plans to extend the workings it is often necessary to make a mathematical prediction of the probable vibration nuisance in order to optimise the blasting parameters with regard to extraction on the one hand and the probable future vibration nuisance on the other.

Fig. 1 depicts the so-called distance-charge relationship, by means of which the probable maximum oscillation velocity at the place where the nuisance is felt can be represented as a function of the charge and the distance from the blasting site. Here, v_{max} denotes the probable maximum oscillation velocity in mm/s, L denotes the charge per ignition delay interval in kg and R indicates the distance between the blasting site and the measurement site in metres. The constant k and the exponents b and n are empirically determined variables.

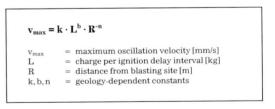

v_{max}	=	maximum oscillation velocity [mm/s]
L	=	charge per ignition delay interval [kg]
R	=	distance from blasting site [m]
k, b, n	=	geology-dependent constants

FIGURE 1: Distance-charge relationship

This theoretical relationship between the vibration emissions, the vibration nuisance and the empirically determined constants is based on a statistical analysis of a large number of vibration measurements carried out during primary blasting in the course of a research project conducted on behalf of the Federal Environmental Agency in order to compile a vibration register for northern Germany. At the time, Lüdeling [1] collected data from some 3000 individual measurements made during 150 or so blasting operations in order to use them to develop a model for predicting vibration nuisance.

In practice, empirical relationships of this kind naturally supply predicted values which must necessarily include an appropriate safety factor in order to allow for different con-

Zuverlässigere Aussagen hinsichtlich der zu erwartenden Erschütterungsimmission bei Gewinnungssprengungen lassen sich dann gewinnen, wenn durch Testsprengungen mit entsprechend geringer Lademenge die empirischen Parameter im Prognosemodell für den konkreten Einzelfall bestimmt werden. Durch den Einsatz von bis zu acht Beschleunigungssensoren während einer Messung kann vom Forschungsinstitut beispielsweise gleichzeitig am Fundament sowie in einzelnen Geschoßebenen eines Gebäudes die jeweilige Schwinggeschwindigkeit normgerecht bestimmt werden. Durch eine derartige Anordnung wird der Einfluß der geologischen Parameter ermittelt. Gleichzeitig kann die Stärke der Übertragung der Erschütterung vom Gebäudefundament auf die entsprechende Geschoßdecke eines Gebäudes bestimmt werden. Die Erfahrung zeigt, daß je nach Bauart des betrachteten Gebäudes die Schwinggeschwindigkeit in Deckenmitte eines Obergeschosses bis zum 5fachen des am Gebäudefundament gemessenen Wertes betragen kann.

Zusammenfassed kann festgestellt werden, daß Gewinnungssprengungen zunehmend im Hinblick auf die Erschütterungsimmisionen optimiert werden müssen. Während in vielen Fällen bei Abbauerweiterungen eine theoretische Prognose möglich ist, kann oft auf Messungen vor Ort nicht mehr verzichtet werden. Im Rahmen von Testsprengungen werden dabei vom Forschungsinstitut beispielsweise während einer einzelnen Messung neben der Schwinggeschwindikeit an Gebäudefundamenten auch die Erschütterungen in einzelnen Geschoßebenen ermittelt. Mit geringem Meßaufwand können so die sprengtechnischen Randbedingungen gleichzeitig im Hinblick auf die Erschütterungsauswirkungen auf Menschen im Gebäude und Bauwerke optimiert werden.

Literaturverzeichnis

[1] Lüdeling, R.: Erschütterungsemissionen von Sprengungen in Norddeutschland, im Auftrag des Umweltbundesamtes, Berichtsnummer UBA-FB 82-105 02 802.

ditions of propagation. This takes account, for example, of the fact that the simplifying assumption that the vibration waves are propagated in a circular pattern is not borne out in practice; instead, the geological formation in the neighbourhood of the workings generally produces non-symmetrical radiation characteristics, which nevertheless have to be taken into account in a prediction with a sufficient degree of certainty.

Fig. 2 shows, in the form of a distance-charge diagram, the minimum distance permitted with a given charge and the maximum acceptable oscillation velocity in the neighbourhood of the cement industry's quarries. The relationship is valid with a statistical certainty of 95%; the empirical constants $k = 447$, $b = 0.67$ and $n = 1.32$ are based on measurements made during a study of the propagation of vibrations in limestone and calcareous marl.

In many cases, a purely theoretical assessment of the expected levels of vibration nuisance is inadequate for establishing optimum blasting parameters – for example, due to complicated progress of the workings or because the nearest housing area is too close.

More reliable data on the probable vibration nuisance arising from primary blasting can be obtained if the empirical parameters in the prediction model can be determined for an individual real-life case by means of test blasting with an appropriately small charge. By using up to eight acceleration sensors in the course of a measurement, the Research Institute can, for instance, simultaneously measure the individual oscillation velocities in the foundations and in individual storeys of a building in accordance with the relevant standards. Such an arrangement enables the influence of the geological parameters to be ascertained. At the same time, the intensity of the transmission of the vibration from the building's foundations to the respective intermediate floor of a building can be determined. Experience shows that, depending on the design of the building in question, the oscillation velocity in the middle of the floor of an upper storey may be as much as five times the value measured in the foundations.

Summarising, it is true to say that there is an increasing requirement for primary blasting operations to be optimised with regard to the vibration nuisance. Whilst theoretical predictions are often feasible when workings are to be extended, on-site measurements cannot always be dispensed with. In such cases, the Research Institute uses a single measurement during test blasting to measure, e.g. not only the oscillation velocity in the building's foundations but also the vibrations in individual storeys. In this way the blasting parameters can be optimised simultaneously with regard to both the effects of vibration on people in the building and on structures while minimising the cost of testing.

Leitlinien der europäischen Zementindustrie zum Schutz der Umwelt

The European cement industry's approach to environmental protection

L'approche de l'industrie cimentière européenne pour la protection de l'environnement

La industria del cemento y la protección del medio ambiente

Von **J. Lawton,** Brüssel/Belgien

Zusammenfassung – Die Zementindustrie in den Mitgliedsländern des Cembureau hat eine Reihe von Stellungnahmen zu allen Umweltfragen abgegeben, die sich auf die Herstellung und den Vertrieb von Zement beziehen und eine entscheidende Rolle für das umweltgerechte Verhalten der Zementindustrie spielen. Diese Stellungnahmen konzentrieren sich auf 10 Bereiche (gesellschaftliche Beziehungen, Genehmigung von Zementwerken und Steinbrüchen, Umweltvorsorge bei Zementwerken und Steinbrüchen, Sekundärrohstoffe, Sekundärbrennstoffe, Kohlendioxid-Abgabe, Staubemissionen, andere Emissionen in die Atmosphäre, Entsorgung von Prozeßrückständen und Transport) und bilden zusammen die „Leitlinien der europäischen Zementindustrie zur Umweltvorsorge". Die Ziele dieser Leitlinien und die ihnen zugrunde liegende Philosophie werden im Beitrag erläutert und die Stellungnahmen dargelegt. Es wird betont, daß die Stellungnahmen bewußt keine Aufzählung vergangener Errungenschaften sind, sondern die zukünftigen Verpflichtungen der Zementindustrie widerspiegeln und ihre Entschlossenheit, vom gegenwärtigen Stand aus weitere Fortschritte zu erzielen.

Summary – The cement industry in the Member countries of Cembureau has taken the initiative to prepare a series of position statements on all environmental issues which relate to the manufacture and distribution of cement and are central to the industry's environmental performance. These statements focus on 10 areas (community relations, permitting of cement plants and quarries, environmental management of plants and quarries, secondary raw materials, secondary fuels, release of carbon dioxide, dust emissions, other atmospheric emissions, management of process residues and transportation) and together form "the European Cement Industry's Approach to the Environment". The objectives of the Approach and the philosophy behind it are outlined and the statements are presented. It is emphasized that the statements are deliberately not a catalogue of past achievements but describe the industry's commitments for the future and its intention of moving forward from its present position.

Résumé – L'industrie cimentière des pays membres de la Communauté Européenne du Cembureau a présenté tout un catalogue de prises de position sur les questions touchant l'environnement, qui ont trait à la fabrication et à la commercialisation du ciment et jouent un rôle décisif sur le respect de l'environnement par cette même industrie. Ces prises de position se concentrent sur 10 secteurs (rapports sociaux, agrément des cimenteries et carrières, respect de l'environnement par les cimenteries et carrières, matières premières secondaires, combustibles secondaires, émissions de dioxyde de carbone, émissions de poussière, autres émissions dégagées dans l'atmosphère, retraitement de résidue de procédé industriels et transport), et regoupent ainsi „L'approche de l'Industrie cimentière européenne pour la protection de l'environnement". Les objectifs de ces directives et la philosophie sur laquelle elles reposent, sont expliqués dans le présent exposé de même que les prises de position. L'accent porte sur le fait que ces prises de position ne veulent en aucun cas être un liste des acquis du passé mais reflètent les engagements futurs de l'industrie cimentière et sa détermination, à partir de la situation actuelle, d'avancer dans le sens du progrès.

Resumen – La industria cementera de los países miembros de Cembureau ha emitido una serie de juicios sobre el conjunto de los problemas ambientales, relacionados con la fabricación y venta de cemento, que desempeñan un papel importante para el comportamiento de esta industria respecto del medio ambiente. Los mencionados juicios se centran en 10 sectores (relaciones sociales, autorizaciones para fábricas de cemento y canteras, medidas de prevención en fábricas de cemento y canteras, materias primas secundarias, combustibles secundarios, emisión de dióxido de carbono, emisión de polvo, otras emisiones que afectan a la atmósfera, eliminación de los residuos de los procesos industriales, transporte). Estos juicios constituyen, en su conjunto, las „Directrices de la industria europea del cemento en materia de protección del medio am-

biente". En el presente artículo se explican los objetivos de estas directrices y la filosofía en que están basadas las mismas, así como los referidos juicios. Se destaca que los juicios emitidos no pueden ser considerados como un catálogo de lo conseguido hasta ahora, sino que reflejan las obligaciones de la industria del cemento y su determinación de hacer progresos también en el futuro, basándose en los logros actuales.

1. Einleitung

Im Jahr 1990 hat sich die europäische Zementindustrie für eine notwendige Ausarbeitung von strukturierten und logisch zusammenhängenden Leitlinien zu Umweltfragen entschieden. Sie beauftragte ihre europäische Organisation, das Cembureau[1]), die Formulierung von umfassenden „Leitlinien zum Schutz der Umwelt" zu koordinieren. Ziel dieser Leitlinien ist es, die verantwortungsvolle Haltung der Zementindustrie gegenüber der Umwelt zu bekräftigen, den nationalen Verbänden der Zementindustrie und den einzelnen Zementherstellern eine Grundlage für die Ausarbeitung eigener Initiativen an die Hand zu geben, der Industrie die Möglichkeit zu einheitlichen Stellungnahmen in Umweltfragen zu eröffnen (z.B. in Diskussionen mit den Europäischen Institutionen und den Behörden der EFTA-Länder) und die Kommunikationsmittel bereitzustellen, die zur Imageverbesserung verwendet werden können. Die Initiative bezieht alle relevanten Aspekte der Herstellung und des Versands von Zement ein, bewertet aber bewußt nicht die Auswirkungen auf die Umwelt durch die Anwendung von Zement und Beton. Dieser Teil soll jedoch in einem zweiten Dokument behandelt werden, das gegenwärtig vorbereitet und dessen Fertigstellung für 1994 erwartet wird.

Die Verantwortung für die Ausarbeitung der Leitlinien wurde von der aus 17 Vertretern der Mitgliedsländer des Cembureau zusammengesetzten Arbeitsgruppe „Umwelt" übernommen. Um Objektivität und Unparteilichkeit zu gewährleisten und schnell voranzukommen, wurde zur Unterstützung der Arbeitsgruppe die Environmental Resources Ltd. als externes Beratungsunternehmen herangezogen. Die zahlreichen Ergebnisse und Überlegungen wurden in 10 Artikeln zusammengefaßt, die sich auf die betroffenen Schlüsselbereiche konzentrieren und zusammen mit einer Einleitung die „Leitlinien der Europäischen Zementindustrie zum Schutz der Umwelt" bilden. Sie wurden im Jahre 1992 von den Entscheidungsgremien des Cembureau sowohl für den internen als auch für den externen Gebrauch verabschiedet.

2. Grundsätze

Die „Leitlinien der Europäischen Zementindustrie zum Schutz der Umwelt" sollen richtungsweisend, offen für Belange von außen und „ausgewogen" sein. Die Industrie ist von der Notwendigkeit überzeugt, ein vorausschauendes Konzept aufzubauen, mit dem versucht werden soll, die Zukunft vorausschauend zu planen. Dies war einer der Hauptgründe für die Entscheidung, die Leitlinien auszuarbeiten. Demnach stellt die Leitlinie ganz bewußt keine Aufzählung der bisher erzielten Fortschritte dar, sondern verdeutlicht das Verantwortungsbewußtsein für die Zukunft und den festen Willen der Industrie, auf der Basis des bisher Erreichten weitere Fortschritte zu erzielen. Als Zeichen ihres Engagements nennt die Industrie in den Stellungnahmen zum Beispiel die Offenlegung ihrer Umweltpolitik, periodische Überprüfungen der zum Schutz der Umwelt erbrachten Leistungen und die Bereitschaft, Meinungen und Vorschläge außenstehender, von den Behörden anerkannter Organisationen mit in Betracht zu ziehen. Auf lokaler oder kommunaler Ebene wird sie sich mit den Vertretern der Behörden und der genannten Organisationen treffen, um umweltrelevante Probleme zu diskutieren. Um ihre Bereitschaft zum Aufbau partnerschaftlicher Beziehungen

1. Introduction

In 1990 the European cement industry decided that it was necessary to develop a structured and coherent approach to environmental matters and requested its representative Association, Cembureau[1]), to co-ordinate the formulation of an overall "Approach to the Environment". The objectives of the Approach are to confirm the responsible attitude which the cement industry takes towards the environment, provide a reference which National Associations and individual Companies may use when considering their own approaches, allow the industry to speak with a common voice on environmental matters (e.g. in discussions with the European Institutions and the EFTA country Authorities), and provide communication messages which may be used to further improve the image of the industry. The Approach addresses all relevant aspects of the manufacture and distribution of cement but deliberately does not consider the effects of the use of cement and of concrete on the environment. It is, however, intended that the latter should be dealt with in a second documentation which is at present being prepared and is expected to be completed during 1994.

Responsibility for the formulation of the Approach was taken by Cembureau's "Environmental Matters Task Force" comprising representatives from 17 of Cembureau's Member countries. It was decided that, in order to ensure objectivity and impartiality and to achieve the rapid progress required, Environmental Resources Ltd. had been appointed as external consultant to assist the Task Force. The numerous findings and the considerations were grouped together in 10 Approach Statements which focus on the key areas involved and, together with the introductory section, form "The European Cement Industry's Approach to the Environment". The approach was approved for both internal and external use by the Executive Bodies of Cembureau in 1992.

2. Principles

"The European Cement Industries Approach to the Environment" is intended to be pro-active, sensitive to external concerns and "balanced". The industry believes that it is necessary to adopt a pro-active approach which seeks to anticipate and plan for the future and this was one of the principal reasons which led to the decision to prepare the Approach Statements. The Statements are consequently and deliberately not a schedule of the considerable progress which has already been made but describe the industry's commitments for the future and its intention of going forward from its present position. In indicating a more pro-active approach, the Statement note, for example, that the industry will make public its environmental policies, carry out periodic reviews of its environmental performance and consider options and proposals put forward by external organizations recognized by the authorities. At the local or community level, it will meet with representatives of the authorities and of these organizations to discuss environmental issues of concern and, in order to demonstrate its commitment to the establishment of good relationships, will operate a policy of open information and will make appropriate information available.

The cement industry recognizes society's aspiration for high and increasing standards of environmental protection and accepts the need for responsiveness and sensitivity to environmental concerns. This is reflected at numerous points in the Approach.

---

[1]) Cembureau repräsentiert als Organisation die Zementindustrie in den Mitgliedsländern der EU und der EFTA sowie der Türkei. Die Verbände der Zementindustrie in den Tschechischen und Slowakischen Republiken, Ungarns und Polens traten am 01.07.1992 als assoziierte Mitglieder bei.

[1]) Cembureau is the representative organization of the cement industry in the EEC and EFTA countries and Turkey. The cement Associations of Czech and Slovakia, Hungary and Poland joined as associate Members on 01.07.1992.

zu demonstrieren, wird sie eine Politik der Offenheit pflegen und angemessene Informationen zur Verfügung stellen.

Die Zementindustrie ist sich der gesellschaftlichen Erwartungen an einen hohen und stets weiter zu verbessernden Stand des Umweltschutzes bewußt und akzeptiert die Notwendigkeit der Verantwortung und Sensibilität in Umweltfragen. Dies spiegelt sich an vielen Stellen der Leitlinien wider.

Die Zementindustrie ist fest davon überzeugt, daß die Vorteile, die Zement der modernen Gesellschaft bringt, in der Gesamtumweltbilanz die Nachteile bei weitem überwiegen. Weiter wird in Betracht gezogen, daß Umweltfragen ganzheitlich angegangen werden sollten und die Notwendigkeit besteht, eine „ausgewogene" umfassende Leitlinie zu entwickeln. In diesem Zusammenhang ist zu beachten, daß die Behörden zunehmend die Notwendigkeit anerkennen, Kosten und Nutzen einer Tätigkeit oder deren Unterlassung in Betracht zu ziehen [2, 3]. Insbesondere wird in den Leitlinien dargelegt, daß

- die Industrie der Überzeugung ist, daß die gegenwärtig in den europäischen Ländern herrschenden Standards für Staubemissionen aus Punktquellen die Umweltbelange zufriedenstellend berücksichtigen und jede weitere deutliche Verringerung auf der Basis von wissenschaftlich nachgewiesenem Nutzen für die Umwelt und unter Berücksichtigung der zunehmenden Anforderungen an die Werke sowie des damit verbundenen erhöhten Verbrauchs an elektrischer Energie vorzunehmen sind.

- es von seiten der europäischen Gesellschaft unverantwortlich wäre, sich darauf zu verlassen, daß ausreichend Zement für die zukünftige Entwicklung verfügbar ist, wenn gleichzeitig dessen Herstellung in Europa durch die Verweigerung von Betriebsgenehmigungen für Steinbrüche und Zementwerke verhindert wird. Die Industrie versucht daher, in Zusammenarbeit mit den Behörden eine Politik zu verfolgen, durch die Rohstoffreserven ermittelt und langfristig gesichert werden.

- die Zementindustrie als „idealer Kohleverwerter" gilt und der Meinung ist, daß es trotz der zunehmenden Freisetzung von Kohlendioxid insgesamt gesehen sinnvoll ist, Kohle zu verbrennen, da Kohlereserven in weit größerem Umfang zur Verfügung stehen als Erdöl oder Erdgas.

Der Wortlaut der „Leitlinien der Europäischen Zementindustrie zum Schutz der Umwelt" berücksichtigt die oben genannten und eine Reihe weiterer Gesichtspunkte; er ist im Abschnitt 4 des Beitrags wiedergegeben.

3. Anwendung der Leitlinien

Die „Industrieleitlinie" wurde 1992 von den Entscheidungsgremien des Cembureau verabschiedet, anschließend – im Einklang mit den in Abschnitt 1 genannten Zielvorstellungen – verteilt und in weiten Kreisen sowohl innerhalb als auch außerhalb der Industrie angewandt. Die Leitlinien wurden zunächst in den beiden offiziellen Sprachen von Cembureau (Englisch und Französisch) gedruckt, danach jedoch von Mitgliedern des Cembureau auch in die Sprachen Deutsch, Griechisch, Italienisch, Portugiesisch, Spanisch und (teilweise) Dänisch übersetzt, um die Anwendung auf nationaler Ebene zu erleichtern. Die englische, französische, deutsche und spanische Fassung wurde zur Erleichterung der Identifizierung in gleichartigem Layout gedruckt, um ein einheitliches Erscheinungsbild zu gewährleisten. Die Stellungnahmen zum Kohlendioxidausstoß und zu Sekundärbrennstoffen dienten als Grundlage für zwei Positionspapiere des Cembureau [4, 5]. Diese wurden bei aktuellen Erörterungen mit europäischen Institutionen und nationalen Behörden im Zusammenhang mit der vorgeschlagenen EG-Energie/CO_2-Steuer sowie dem Vorschlag für eine EG-Ratsdirektive zur Verbrennung von Sondermüll verwendet.

Auskünfte über die Art der Verwendung der Leitlinien auf nationaler Ebene und die erzielten Fortschritte sind über die Arbeitsgruppe „Umwelt" des Cembureau zu erhalten. Für

The cement industry strongly believes that the benefits which cement brings to modern society substantially outweigh any disadvantages in the overall environmental balance. It is further considered that environmental issues should be addressed in their entirety and that it is necessary to develop a "balanced" coherent approach. In this context, the growing recognition by the authorities of the need to take into account the costs and benefits of action or lack of action (see e. g. [2] [3]) is noted. Specifically in the Approach, it is indicated that:

- the industry believes that the standards currently prevailing for dust emissions from point sources in European countries satisfactorily address environmental concerns and that any further significant reductions need to be adopted on the basis of scientifically proven environmental benefits and taking into account the resultant increases in plant requirements and electricity consumption.

- it would be irresponsible of society in Europe to rely upon there being supplies of cement available for its future development, whilst precluding its manufacture in Europe through the refusal of quarrying and plant permits. The industry therefore seeks to pursue, in partnership with the authorities, a policy whereby reserves of raw materials are identified and secured for the long term.

- the cement industry is considered to be an "ideal coal user" and believes that, despite the increased release of carbon dioxide, it makes overall sense for it to burn coal, a resource which is in far greater abundance than natural gas or oil.

The text of "The European Cement Industry's Approach to the Environment" considers the above and many other points and is included in section 4 of this paper.

3. Use of the Approach

The "Industry's Approach" was approved by the Executive Bodies of Cembureau in 1992 and has subsequently been distributed and used widely both inside and outside the industry in line with the objectives described in section 1 above. The Approach was initially printed in the two official languages of Cembureau (English and French) but has subsequently been translated by Cembureau's Members into German, Greek, Italian, Portuguese, Spanish and (in part) Danish to facilitate use nationally. The English, French, German and Spanish versions have been printed using common art-work in order to aid identification and provide consistency of appearance. The Statements on the release of carbon dioxide and on secondary fuels have been used as the basis for the preparation of two Cembureau Position Papers [4] [5]. These have been used in the active representation actions taken towards the European institutions and the national authorities in relation to the proposed European Community energy/carbon tax and the Proposal for a European Council Directive on the Incineration of Hazardous Waste.

Information on the way in which the Approach is being used nationally and on the progress made is obtained through Cembureau's Environmental Matters Task Force. New Members of Cembureau are required, as a condition of Membership, to endorse the Approach on behalf of the industry which they represent and this condition was accepted by the cement Associations of Czech and Slovakia, Hungary and Poland when they joined Cembureau on 1.7.92. Finally, it should be noted that it has been proposed that the Approach should be reviewed in 1995 (i.e. in the light of 3 years experience) and that any appropriate changes should be made at that time.

4. The European Cement Industry's Approach to the Enviroment

Introduction

Cement is an essential component of the continued development and renewal of buildings and infrastructure of all kinds, and is thus meeting a basic need of modern society.

neue Mitglieder des Cembureau wird als Aufnahmebedingung die Billigung der Leitlinien im Namen des durch sie vertretenen Industriezweiges gefordert. Diese Bedingung wurde von den Verbänden der tschechischen, slowakischen, ungarischen und polnischen Zementindustrie akzeptiert, als sie am 1.7.1992 dem Cembureau beitraten. Schließlich sollte noch der Vorschlag erwähnt werden, die Leitlinien im Jahr 1995 zu überarbeiten (d. h. nach 3jähriger Erfahrung) und entsprechende Änderungen vorzunehmen.

4. Leitlinien der Europäischen Zementindustrie zum Schutz der Umwelt
Einführung

Zement kommt bei der Errichtung und der Instandhaltung von Bauwerken aller Art eine besondere Bedeutung zu. Erst durch seine Anwendung können die Grundbedürfnisse einer modernen Gesellschaft befriedigt werden. Darüber hinaus tragen die Herstellung und die Anwendung von Zement zur Lösung einer Reihe von Umweltproblemen bei. Nach Auffassung der europäischen Zementindustrie ist deshalb die Gesamt-Umweltbilanz des Zementes positiv.

Ungeachtet dessen ist festzustellen, daß die Erwartungen der Gesellschaft hinsichtlich einer Umweltverträglichkeit der Industrie immer größer werden. Von den Unternehmen wird deshalb erwartet, daß sie bei der Planung und Durchführung ihrer Aktivitäten diese gesellschaftlichen Forderungen zu berücksichtigen haben. Die Zementindustrie ist sich durchaus bewußt, daß auch ihre Produktion die Umwelt beeinflußt. Sie ist deshalb nachhaltig bemüht, schädliche Einflüsse zu vermeiden oder wenigstens zu vermindern.

Die ständige Modernisierung und Nachrüstung mit neuen und umweltfreundlichen Techniken hat in der Zementindustrie eine lange Tradition. Damit haben die Unternehmen bereits, speziell im Bereich der Luftreinhaltung, große Fortschritte erzielt. Angesichts des Zusammenhanges zwischen Energieeinsatz und Umweltbelastung reduzierte die Zementindustrie vor allem ihren Verbrauch an fossilen Brennstoffen sowie den Einsatz elektrischer Energie in den vergangenen Jahrzehnten erheblich. Im Rahmen ihrer technischen und wirtschaftlichen Möglichkeiten wird sie das vorhandene Minderungspotential auch weiterhin ausschöpfen. Diesem Ziel dienen eigene Forschungsprogramme zur optimalen Energieausnutzung und Umweltvorsorge bei der Zementherstellung.

Die europäische Zementindustrie ist weiterhin der Meinung, daß sie eine aktive Rolle bei der Bewältigung von Umweltproblemen übernehmen muß, um zukünftige Entwicklungen frühzeitig zu erkennen und zu steuern. Aus diesem Grund hat sie eine Reihe von Stellungnahmen zu allen umweltrelevanten Themen erarbeitet, die sich auf die Herstellung und den Versand von Zement beziehen und die für den Umweltschutz dieses Industriezweiges Bedeutung haben.

Die nachfolgenden Feststellungen sind selbstverständlich keine Aufzählung von erreichten Erfolgen. Sie verdeutlichen vielmehr den festen Willen der Zementindustrie, auf der Basis des bereits Erreichten weitere Fortschritte zu erzielen. Sie ergreift die Initiative, indem sie ihre Umweltpolitik offenlegt und in Zukunft regelmäßig über Stand und Entwicklung des Umweltschutzes berichtet.

Angesichts des technischen Fortschritts, des Wandels der Gesetzgebung und auch der eigenen, sich ständig verändernden Rolle, wird sich die Zementindustrie regelmäßig kritisch überprüfen. Um selbst über Neuerungen und Änderungen informiert zu sein, unterhält sie enge Kontakte mit den zuständigen Behörden und politischen Institutionen. Darüber hinaus setzt sie sich mit der Kritik und Meinung anerkannter Organisationen auseinander und wird deren Vorschläge, sofern sie angemessen und praktikabel erscheinen, nach Kräften in die Tat umsetzen.

Beziehungen auf kommunaler Ebene

Ziel der Zementindustrie ist es, die durch sie verursachte Inanspruchnahme der Umwelt mit den sozialen und wirt-

Further, a number of society's environmental problems are solved by the manufacture and use of cement. For these reasons, the European cement industry believes that the benefits of cement substantially outweigh any disbenefits in the overall environmental balance.

However it is recognized that society's aspirations regarding the environmental performance of industry are increasing and that it is a responsibility of industry to take these aspirations into account in the planning and operation of its activities. The cement industry acknowledges the environmental implications of it activities and seeks to avoid, reduce or mitigate adverse effects. Though long established, it has undertaken the step-by-step modernisation of its facilities and operations and has achieved major improvements in its environmental performance, in particular with regard to atmospheric emissions. The industry recognises the relationship between environmental impacts and energy use and has substantially reduced its consumption of thermal energy over recent decades. It has carefully controlled its usage of electricity and is striving to make more reductions in energy usage given the technical and economic constraints within which it operates. It also undertakes research to assist in making further improvements to the efficiency of its operations and to its environmental performance.

Furthermore the European cement industry believes that a pro-active approach to environment is necessary which seeks to anticipate and plan for the future. In order to assist the achievement of this, the industry has taken the initiative of preparing a series of position statements on all environmental issues which relate to the manufacture and distribution of cement and are central to the industry's environmental performance.

These statements, which are presented on the following pages, are deliberately not a catalogue of past achievements but indicate the industry's intention in going forward from its present position. In adopting a more pro-active approach, the industry will, for example, make public its environmental policies and carry out periodic reviews of its environmental performance.

The industry will also periodically review its approach to the environment in the light of changes in attitudes, technology and legislation. To keep itself informed of such changes, the industry maintains close contacts with relevant authorities and assists in the development of appropriate environmental policies. In reviewing its approach, the industry will consider the options and proposals put forward by organizations recognized by the authorities, and whenever appropriate and practical will seek to adopt such proposals.

Community Relations

The cement industry strives to achieve the right balance between environmental performance and the social and economic benefits that cement brings to society. In order that the industry can better aim to meet local aspirations, it will meet with local representatives of organizations recognized by the authorities to discuss environmental issues of concern.

In order to demonstrate its commitment to establishing good relationships in the community, the cement industry will have a policy of open information and will make appropriate information available to the authorities and organizations recognized by them.

Management, acting at the local level in accordance with company procedures and guidelines, will deal with environmental problems that arise in the community and ensure that these are given proper and due attention.

Permitting of Cement Plants and their Quarries

Cement is an essential construction material and is manufactured by an industry with a proven record in the environmental management of its quarrying and processing activities. It would be irresponsible of society in Europe to rely upon there being supplies of cement available for its future development, whilst precluding its manufacture in

schaftlichen Vorteilen, die der Einsatz von Zement für die Gesellschaft mit sich bringt, im Gleichgewicht zu halten. Damit sie den Erwartungen, die in ihrem direkten Umfeld an sie herangetragen werden, besser entsprechen kann, hält sie engen Kontakt mit den lokalen Vertretern der Organisationen, die von den Behörden als kompetente Diskussionspartner für Umweltfragen autorisiert wurden.

Die Zementindustrie legt großen Wert auf gute Beziehungen zu ihren Nachbarn, Anwohnern und dem kommunalen Umfeld. Um ihre Bereitschaft zum Engagement zu demonstrieren, pflegt sie eine Politik der Offenheit. Sowohl den Behörden als auch den von den Behörden anerkannten Institutionen erteilt sie bereitwillig Auskunft.

Umweltprobleme auf lokaler Ebene werden von den Verantwortlichen der Werke in Übereinstimmung mit den generellen Leitlinien des Unternehmens behandelt. Es ist selbstverständlich, daß ihre Lösung mit der gebotenen Aufmerksamkeit angestrebt wird.

Genehmigungsverfahren für Zementwerke und deren Steinbrüche

Zement ist ein wichtiger Baustoff und wird von Unternehmen produziert, die nachweislich erhebliche Erfolge zum Schutz der Umwelt bei der Gewinnung des Rohmaterials und bei der Verfahrenstechnik im Zementwerk erzielt haben. Es wäre deshalb unrealistisch und unverantwortlich, wenn die Gesellschaft in Europa wie selbstverständlich eine ausreichende Verfügbarkeit von Zement zu ihrer Weiterentwicklung erwarten, andererseits den Unternehmen in Europa die Genehmigung zum Abbau der dafür benötigten Rohstoffe und zu deren Verarbeitung in entsprechenden Anlagen verweigern würde.

Die Industrie strebt deshalb – in Übereinkunft mit den zuständigen Behörden – eine Entwicklung an, die allgemeine Akzeptanz findet, zukünftige Lagerstätten ausweist und deren langfristige Nutzung ermöglicht. Sie erkennt den Einfluß ihrer Aktivitäten auf die Umwelt an und trägt diesem Umstand sowohl beim Betrieb ihrer Anlagen als auch bei der Renaturierung nicht mehr genutzter Lagerstätten Rechnung.

Innerhalb der jeweiligen nationalen gesetzlichen Rahmenbedingungen unternimmt die Zementindustrie deshalb folgende Schritte:

- Sie erarbeitet langfristige Pläne über ihren Bedarf und die Gewinnung von Rohstoffen.
- Sie bezieht bereits bei der Standortsuche für neue Zementwerke und der Erschließung neuer Rohmaterialvorkommen neben technischen und ökonomischen Gesichtspunkten auch Umweltschutzaspekte ein.
- Bei der Errichtung, Erweiterung und wesentlichen Änderungen von Werken und Lagerstätten werden entsprechende Umweltstudien durchgeführt.
- Für nicht mehr genutzte Steinbrüche werden in Zusammenarbeit mit den entsprechenden Behörden Renaturierungspläne erstellt.

Es ist das Ziel der Zementindustrie, daß sowohl die Aufsichtsbehörden als auch die breite Öffentlichkeit die Bedeutung der Branche für die Volkswirtschaft erkennen. Außerdem wird verdeutlicht, welche Anstrengungen die Werke für den Schutz der Umwelt unternehmen. Damit soll, speziell wenn Auflagen für den Abbaubetrieb oder für den Produktionsprozeß neu festgelegt werden, eine sachliche Grundlage für die Diskussion geschaffen werden.

Umweltschutz in Zementwerken und deren Steinbrüchen

Die Industrie akzeptiert und unterstützt das Bedürfnis einer demokratischen Gesellschaft, sich selbst und ihre Umwelt vor nachteiligen Einflüssen industrieller Tätigkeit durch gesetzliche Vorschriften zu schützen. Als ein verantwortlicher Teil dieser Gesellschaft bekennt sich die Zementindu-

Europe through the refusal of quarrying and plant permits. The industry therefore seeks to pursue, in partnership with the relevant authorities, a policy of sustainable development, within an agreed planning framework, whereby future reserves of raw materials are identified and secured for the long term.

The cement industry acknowledges the impact of its activities on the environment and has responded through the improved management of operations and the rehabilitation of disused quarries.

Operating within national legislative frameworks, the industry therefore undertakes:

- to present, at the appropriate level, long term prospects for the development of resources;
- to employ environmental, as well as technical and economic, criteria in the selection of sites for new plants and quarries;
- to carry out environmental studies concerning new plant and quarries, and major extensions to, or modifications of, existing plants and quarries;
- to negotiate with the appropriate authority a rehabilitation plan for quarries following the cessation of activities.

The industry will ensure that the regulatory authorities and the general public recognise the importance of the cement industry to the national economy and are aware of the efforts made by the industry in protecting the environment. The industry will advocate for these factors to be taken into consideration when drafting legislation on the use of mineral resources and in setting the procedural arrangements for granting quarrying and manufacturing permits.

Environmental Management of Cement Plants and their Quarries

The industry understands and supports the desire of a democratic society to protect itself and the environment from the potentially adverse effects of industrial activities through the development and implementation of legislation. As a responsible sector of society, the cement industry is committed to operating within the terms of all such legislation.

The cement industry also recognises that there is often a public perception that it, along with industry in general, has associated environmental impacts on the local community.

The industry strives to initially address local environmental issues to local authorities and organizations recognized by them, and knows from experience that strong environmental management can help reduce potential impacts to acceptable levels.

Training of its employees in environmental matters will be actively pursued in order to ensure that they are competent to achieve the high standards required.

The nature and extent of local impacts often depend on the distance to, and the sensitivity of, the receptors; the industry, in order to respond to local circumstances will adopt appropriate environmental management procedures. These procedures will generally include control of noise, vibration, dust emissions, water discharges, traffic impacts and impacts to visual amenity.

Secondary Raw Materials

The cement industry uses large quantities of secondary raw materials (e.g. granulated blast furnace slag and power station fly ash), much of which would otherwise become waste and would require disposal. Use of secondary raw materials in this way is a benefit to society, releasing disposal capacity and eliminating the need for further treatment.

In addition, the utilisation of these materials by the cement industry provides an opportunity to reduce the requirements for the quarrying of primary raw materials, and together with the use of natural pozzolanas and fillers, to reduce energy consumption and overall emissions of dust, NO_x, SO_2 and CO_2.

strie selbstverständlich dazu, die entsprechenden Gesetze zu achten und einzuhalten.

Die Zementindustrie ist sich bewußt, daß sie selbst, wie die Industrie generell, von der Öffentlichkeit für Umwelteinflüsse auf ihre unmittelbare Nachbarschaft verantwortlich gemacht wird.

Aus diesem Grund streben die Unternehmen an, von sich aus Umweltschutzthemen mit lokalen Behörden sowie mit von den Behörden anerkannten Organisationen zur Sprache zu bringen. Sie weiß aus Erfahrung, daß mögliche Umweltbelastungen durch rechtzeitiges und effektives Handeln auf ein erträgliches Maß reduziert werden können.

Durch die Aus- und Weiterbildung der Mitarbeiter soll erreicht werden, daß diese den hohen Anforderungen auf dem Gebiet des Umweltschutzes fachlich entsprechen.

Art und Umfang örtlicher Belastungen stehen oft in direkter Beziehung zu der Entfernung einer Anlage von den Betroffenen sowie deren Sensibilität. Um den daraus folgenden unterschiedlichen Anforderungen zu begegnen, werden die jeweils geeigneten Maßnahmen ergriffen. Hierzu gehören im allgemeinen: Verminderung der Emission von Lärm, Erschütterungen und Staub, des Verkehrs, die Verbesserung der Abwasserreinigung sowie des optischen Erscheinungsbildes.

Sekundärrohstoffe

Die Zementindustrie setzt erhebliche Mengen an Sekundärrohstoffen ein, wie beispielsweise granulierte Hochofenschlacke und Flugasche aus Kraftwerken. Diese Materialien sind Reststoffe anderer Industrien und müßten ansonsten fast ausnahmslos auf Deponien verbracht werden. Die sinnvolle Nutzung von Sekundärrohstoffen ist also von Vorteil für die Gesellschaft und unsere Umwelt. Sie entlastet die Deponien und macht eine weitere Behandlung dieser Stoffe überflüssig.

Zusätzlich hat die Verwertung von Sekundärrohstoffen bei der Zementherstellung den Vorteil, daß weniger Primärrohstoffe abgebaut werden müssen. Außerdem trägt ihr Einsatz, zusammen mit der Verwendung von Puzzolanerde und Füllern dazu bei, den Energieverbrauch sowie den Gesamtausstoß an Staub, Stickoxiden, Schwefeldioxid und Kohlendioxid zu vermindern.

Die Zementindustrie ist deshalb der Meinung, daß durch entsprechende gesetzgeberische Maßnahmen ein rechtliches Umfeld geschaffen werden sollte, damit der Einsatz von Sekundärrohstoffen nicht durch unnötige restriktive Regelungen behindert wird.

Angesichts der Notwendigkeit, die kontinuierliche Produktion von Zement sicherzustellen, sucht die Zementindustrie im Rahmen einer solchen Gesetzgebung weiterhin nach Wegen, die den Einsatz solcher Stoffe ermöglichen. Bedingung dafür ist, daß die erforderlichen Kriterien hinsichtlich Qualität der Produkte und Leistungsfähigkeit der Anlagen erfüllt sind.

Sekundärbrennstoffe

Die europäische Zementindustrie hat erkannt, daß eine Reihe von Stoffen, die gemeinhin als Reststoffe bezeichnet werden, als Zusatzbrennstoffe in Zementöfen eingesetzt werden können. Durch ihren Einsatz leistet die Zementindustrie einen Beitrag zur Verringerung des Abfallvolumens, das ansonsten deponiert oder anderweitig entsorgt werden müßte. Die Verwendung von Sekundärbrennstoffen in Drehrohröfen der Zementindustrie ist also ein Beitrag zu aktivem Umweltschutz.

Die technischen Besonderheiten und die Umweltrelevanz der Verwendung von Sekundärbrennstoffen sind bekannt und werden ausreichend berücksichtigt, so daß dieses Verfahren eine sichere, endgültige und wirtschaftliche Methode der Abfallbeseitigung darstellt.

Das oberste Ziel der Zementindustrie ist jedoch, ihre Abnehmer mit kostengünstigen Produkten hoher Qualität zu beliefern. Der Einsatz von Sekundärbrennstoffen darf mit diesem Grundsatz niemals in Konflikt geraten.

The industry considers that an appropriate legal climate is required in order to ensure that the use of secondary raw materials in the manufacture of cement is not hampered by unnecessarily restrictive regulations.

Subject to the provision of such a legal climate and the need to ensure the continued production of cement which meets the necessary quality and performance criteria, the cement industry will continue to seek opportunities to extend its usage of these materials.

Secondary Fuels

The European cement industry recognises that many materials which may be classified as waste can be used as secondary fuels in cement kilns. By using the materials as kiln fuels, the industry is fulfilling a societal need, reducing the quantities which have to be used/disposed of in other ways, and making an effective contribution to the protection of the environment.

The technical and environmental issues associated with the combustion of secondary fuels in cement kilns are well understood and combustion of suitable fuels can represent a safe, final and cost effective method of utilisation.

A prime objective of the cement industry is, however, to supply its customers with good quality cement manufactured at the lowest possible cost. The use of secondary fuels must never be allowed to interfere with this objective.

The industry will only consider the use of a particular secondary fuel if:

– it has been demonstrated that previously agreed technical criteria can be satisfied e.g. with regard to possible effects on emissions; plant operations; product quality; and the health and safety of employees and residents in surrounding areas.

– a comprehensive assessment indicates that use of the fuels is beneficial taking into account economic factors (e.g. effects on fuel costs, possible loss of output ect.), legal constraints, social implications, and possible market impacts.

An acceptable legal climate is essential if the industry is to be encouraged to use secondary fuels. In particular, the industry considers that for normal products of combustion and for dust, the emission limits applied to cement plants using secondary fuels should be the same as those for all other cement plants. For emissions associated with the combustion of secondary fuels, emission limits should be similar to those applied in other industrial sectors but taking into account the characteristics of the cement making process.

Release of Carbon Dioxide

Since the 1960's, the European cement industry has achieved significant reduction in the release of CO_2 through a reduction in fuel usage arising from the adoption of energy efficient technology and the increased use of secondary raw materials.

The cement industry will continue to make future improvements in energy efficiency within the cement making process. However, in the future, the potential for further improvement is limited. The industry will continue to seek opportunities to increase the use of natural pozzolanas, fillers and secondary raw materials to partially replace cement clinker, with pro-rata reduction in the release of CO_2, provided that cement and concrete quality are maintained.

A wide range of primary fuels including natural gas, petroleum coke and oil can be used by the cement industry. However, the principal primary fuel used to fire cement kilns is coal. This can be justified on economic grounds but there are also environmental benefits from using coal despite the fact that it generates more CO_2. The chemical composition of the raw materials utilised by much of the industry is such that the cement process can use coals with relatively high contents of sulphur and ash, nearly all of which are absorbed upon combustion into the cement

Die Industrie wird die Einsatzmöglichkeit von Sekundärbrennstoffen nur dann in Betracht ziehen, wenn:

- sichergestellt ist, daß bewährte technische Standards, beispielsweise in bezug auf Betriebssicherheit, Emissionen und Produktqualität, eingehalten werden. Außerdem dürfen die Gesundheit und die Sicherheit der Beschäftigten und natürlich die der Anwohner keinen Schaden nehmen;
- eine umfassende Analyse ergibt, daß der Einsatz des Brennstoffes von Nutzen ist. Dabei müssen wirtschaftliche Faktoren, wie beispielsweise Brennstoffkosten und eine mögliche Verminderung der Produktionsleistung, ebenso berücksichtigt werden wie die gesetzlichen Rahmenbedingungen, gesellschaftliche Zusammenhänge und gegebenenfalls auch Markteinflüsse.

Nur akzeptable gesetzliche Rahmenbedingungen können die Zementindustrie ermutigen, Sekundärbrennstoffe einzusetzen. Insbesondere dürfen die Grenzwerte für solche Werke, die Sekundärbrennstoffe verwenden, nicht anders festgesetzt werden als für die übrigen Zementwerke. Bei Emissionen, die speziell durch die Verbrennung von Sekundärbrennstoffen entstehen, müssen vergleichbare Grenzwerte gelten, wie sie in anderen industriellen Bereichen festgelegt werden. Zu berücksichtigen sind dabei jedoch die spezifischen Eigenheiten des Zementherstellungsprozesses.

Kohlendioxid

Seit den 60er Jahren konnte die europäische Zementindustrie den Ausstoß von Kohlendioxid erheblich verringern. Beigetragen haben dazu die Anwendung energiesparender Techniken und der Einsatz von Sekundärrohstoffen.

Die Zementindustrie wird ihre Anstrengungen fortsetzen, den Energiewirkungsgrad ihrer Anlagen weiter zu optimieren. Das Potential zusätzlicher Verbesserungen ist jedoch begrenzt. Die Industrie wird auch zukünftig nach Möglichkeiten suchen, Zementklinker durch natürliche Puzzolane, Füller und Sekundärrohstoffe zu ersetzen; der Ausstoß von Kohlendioxid wird sich dadurch anteilmäßig verringern. Grundvoraussetzung ist jedoch, daß die hohe Qualität des Zementes – und die des daraus hergestellten Betons – unverändert bleibt.

In der Zementindustrie kann eine breite Palette von Brennstoffen eingesetzt werden. Dazu zählen beispielsweise Erdgas, Petrolkoks und Öl. Der wichtigste Brennstoff ist jedoch Kohle. Dies hat wirtschaftliche Gründe, aber trotz des höheren CO_2-Anteils auch ökologische Vorteile. Die chemische Zusammensetzung der bei der Zementherstellung verwendeten Rohstoffe ist derart, daß Kohle mit einem relativ hohen Anteil an Schwefel und Asche eingesetzt werden kann. Alle diese Komponenten sind zum Brennen des Klinkers erforderlich. Deshalb kann die Zementindustrie für Feuerungsanlagen auch als minderwertig geltende Kohle nutzen und hält die Verbrennung von Kohle für sinnvoll, da deren Reserven bei weitem größer sind als die von Erdgas und Erdöl.

Die Frage, inwieweit Sekundärbrennstoffe verstärkt in Zementöfen technisch und umweltgerecht eingesetzt werden können, wird weiter geprüft. Der Einsatz solcher Sekundärbrennstoffe ermöglicht die Einsparung fossiler Brennstoffe und global eine Verminderung des Ausstoßes an CO_2. Dies gilt besonders deshalb, da an den speziell für den Einsatz von Sekundärbrennstoffen gebauten Anlagen teilweise Erdgas, Öl oder Kohle zugefeuert werden muß. Die Zementindustrie ist deshalb bereit, solche Brennstoffe zu verwenden, vorausgesetzt die Produktqualität wird nicht beeinträchtigt und die technischen, wirtschaftlichen, gesellschaftlichen und gesetzlichen Anforderungen sind erfüllt.

Bisher ist kein alternativer Prozeß zur Zementherstellung bekannt, der ohne den Einsatz von fossilen Brennstoffen als Energieträger auskommt. Solche alternativen Verfahren wurden in der Vergangenheit geprüft, ohne daß sie bislang Erfolge brachten. Dennoch werden sie auch in Zukunft Gegenstand weiterer Untersuchungen sein.

clinker. Therefore the cement industry burns low grade coals efficiently. Consequently, the industry believes that it makes overall sense for it to burn coal, a resource which is in far greater abundance than natural gas and oil.

The technical and environmental feasibility of increasing the use of secondary fuels in cement kilns will continue to be investigated. By using secondary fuels in this way, primary fossil fuels can be replaced and the release of CO_2 may be reduced overall, particularly as incinerators built specifically to dispose of these materials are partially coal, oil or gas fired. The industry will therefore consider and accept the use of these fuels provided quality, technical, economic, social and legal conditions are met.

There is no known viable alternative process for the manufacture of cement which does not and have, use carbon or hydrocarbon fuels. Alternative methods of making cement have, however, been investigated without success in the past and will continue to be examined in the future.

The industry considers that any regulatory, economic or fiscal measures to limit the release of CO_2 should take into account the particular situation of the cement industry and its existing levels.

Dust Emissions

Point Sources

The European cement industry fully accepts and shares the widespread concern regarding the effect of particulate emissions on the environment. By controlling dust from their operations to meet the increasingly lower regulatory emission limits, cement companies have demonstrated their high level of commitment to resolving the problem.

The industry believes that the standard currently prevailing in European countries satisfactorily address environmental concerns, and that any further significant reductions need to be adopted on the basis of scientifically proven environmental benefits and taking into account the resultant increases in plant requirements and electricity consumption.

In order to demonstrate to the authorities its commitment to minimising dust emissions to the surrounding environment, the industry will keep records of their regular monitoring of stack dust emissions and of any outages of its dust abatement equipment and will continually seek to reduce the number and effect of such outages.

Fugitive Dust

In addition to improving the control of dust emissions from point sources, the industry has addressed the need to control fugitive dust emissions. The industry will continue to improve the control of fugitive dust emissions through the employment of good management practices, by careful design, containment of inherently dusty activities, continual attention to the cleanliness of the site, the selective application of water and the use of screening techniques.

Other Atmospheric Emissions

The European cement industry is striving to play its part in meeting society's goal for lower emissions of sulphur dioxide (SO_2) and oxides of nitrogen (NO_x).

In general, the cement industry does not emit significant quantities of SO_2. However, due to the sulphur content of raw material used at a few specific plants, elevated levels of SO_2 may result. The industry has introduced appropriate technology at a number of these plants and will continue to respond on a site-by-site basis.

Significant reductions in NO_x emissions have already been made. The industry continues to fund research and is committed to seeking ways of further reducing NO_x emissions. In order to meet NO_x limits, technical solutions such as low NO_x burners are introduced. Further optimisation of burning techniques and other technical solutions such as ammonia injection are the subject of further development.

The industry recognises concerns about other possible emissions such as heavy metals and organics. Where neces-

Alle Maßnahmen, seien sie wirtschaftlicher oder fiskalischer Art, die den Ausstoß an CO_2 durch Vorschriften vermindern sollen, müssen die spezielle Situation der Zementindustrie und den bereits erreichten hohen Wirkungsgrad des Zementherstellungsprozesses berücksichtigen.

Staubemissionen

Punktquellen

Die europäische Zementindustrie akzeptiert und teilt die verbreitete Beachtung der Auswirkungen von Staubemissionen auf die Umwelt. Durch wirksame Maßnahmen ist es ihr gelungen, die ständig verschärften Grenzwerte für solche Emissionen einzuhalten. Sie hat damit ihre Bereitschaft und ihr großes Engagement bewiesen, die anstehenden Probleme zu lösen.

Die Industrie steht nun auf dem Standpunkt, daß der in Europa erreichte Standard die Interessen des Umweltschutzes zufriedenstellend berücksichtigt. Jede weitere Verringerung der Staubemissionen ist nur dann sinnvoll, wenn ihr Nutzen für die Umwelt mit wissenschaftlich anerkannten Methoden nachgewiesen wird. Fragen der aufwendigeren Verfahrenstechnik sowie der gegebenenfalls höhere Einsatz elektrischer Energie müssen dabei berücksichtigt werden.

Um den Behörden ihre Bereitschaft zur weiteren Verminderung der Emissionen unter Beweis zu stellen, wird die Zementindustrie die Staubemissionsmessungen an den Schornsteinen sowie die eventuellen Ausfälle von Abgasreinigungsanlagen dokumentieren und archivieren. Sie ist selbstverständlich bemüht, die Anzahl und die Auswirkungen solcher Ausfälle weiter zu reduzieren.

Diffuse Staubquellen

Neben der Verringerung von Staubemissionen aus gerichteten Quellen beschäftigt sich die Zementindustrie auch damit, die Emissionen von Flugstaub zu minimieren. Die Industrie wird dazu die Produktionstechniken verbessern und geschlossene Systeme in den Bereichen einführen, in denen die Gefahr von Staubemissionen latent ist. Außerdem wird sie weiterhin vermehrt auf Sauberkeit der Werksgelände achten, durch gezielten Einsatz von Wasser Flugstaub binden sowie entsprechende Inspektionen durchführen.

Andere atmosphärische Emissionen

Die europäische Zementindustrie hat sich zum Ziel gesetzt, ihren Teil zur Erfüllung der gesamtgesellschaftlichen Aufgabe beizutragen, die Emissionen von Schwefeldioxid (SO_2) und Stickoxiden (NO_x) zu reduzieren.

Grundsätzlich gehört die Zementindustrie nicht zu den Emittenten von größeren Mengen an Schwefeldioxid. Trotzdem kann es durch den Schwefelgehalt der in einigen Anlagen verwendeten Rohstoffe zu einem erhöhten Ausstoß an SO_2 kommen. In einer Reihe der betroffenen Anlagen wurden bereits entsprechende Technologien zur Reduzierung dieser Emissionen installiert. Die Industrie wird bei betroffenen Anlagen diese technische Verbesserung konsequent fortführen.

Bei den NO_x-Emissionen wurden bereits wesentliche Verminderungen erreicht. Die Industrie wird ihre Anstrengungen auf diesem Gebiet weiter fortsetzen und auch in Zukunft den Ausstoß an Stickoxiden verringern. Um die NO_x-Grenzwerte einzuhalten, werden als verfahrenstechnische Maßnahmen unter anderem NO_x-arme Brenner installiert. Die darüber hinausgehende Optimierung der Brenntechniken sowie andere technische Lösungen, hierzu gehört beispielsweise die Eindüsung von Ammoniak, sind Gegenstand der weiteren Entwicklung.

Die Industrie kennt die Besorgnis um die Emission anderer möglicher Schadstoffe wie zum Beispiel Schwermetalle und organischer Substanzen. Wo es erforderlich ist, wird der Produktionsprozeß sorgfältig überwacht und gesteuert, um sicherzustellen, daß die Umwelt nicht belastet wird. In vielen Fällen genügt zur Überprüfung eine einmalige Messung. Falls erforderlich, werden wiederholt Messungen durchge-

sary, it is closely monitoring and controlling activities to check that these do not give rise to significant environmental impacts. In many cases, the position is easily established on the basis of initial measurements. More frequent measurements are carried out and appropriate actions taken if results require it.

Management of Process Residues

The cement industry normally produces only small quantities of process residues and liquid and solid wastes. Nevertheless, by the use of actively managed waste reduction programmes, it intends to continue to reduce further the quantities of waste that are produced.

- The use of refractory bricks containing hazardous materials such as chromium will continue to be reduced.
- The amount of collected dust which is not recycled will be minimised. Where dust cannot be recycled, alternative uses will continue to be developed.

Where process wastes are produced, their disposal will be managed in an environmentally responsible way. Where on-site disposal of process or general factory waste takes place, disposal will be in accordance with the relevant regulations relating to, amongst other things, the avoidance of contamination of the air, soil, surface or groundwater.

When disposing of process wastes off-site, cement plants will always use permitted or registered transporters and waste contractors. From time to time, they will ensure that these operators are disposing of the wastes in a satisfactory manner according to the relevant regulations and good operating practices.

Transportation

The cement industry requires the bulk transportation of substantial quantities of raw materials, fuel and cement products. The industry is aware of the public's concerns regarding the transport of such materials and, in order to accommodate these concerns, adopts the following practices:

- Manufacturing plants are normally sited near to reserves of main raw materials in order to minimise the need for transportation.
- The majority of products leave manufacturing plants by road. This is determined by the limited availability of alternative facilities either at the plants and/or at the customers' premises. Rail and water are, however, used where it is economically feasible and there are suitable facilities.
- Cement is transported either in bags or in fully enclosed containers that cannot leak. When cement is sold in bags, these are cleaned to minimise the risk of dust.
- Vehicles owned or contracted by the industry are maintained to the highest standards required in the country of operation and are kept in a clean condition.
- When appropriate, discussions are held with the local authorities to address problems associated with traffic.

Literature

[1] Cembureau, the European Cement Association: The European Cement Industry's Approach to the Environment. 1992.
[2] Commission of European Communities: Proposal for a Council Directive on Integrated Pollution Prevention and Control. 5th draft, November, 1992.
[3] European Council: Resolution on Industrial Competitiveness and Environmental Protection. 24.11.1992.
[4] Cembureau: Position Paper on Fiscal Instruments and Environmental Policy: Energy/CO_2 Tax. September, 1991.
[5] Cembureau: Position Paper on the Proposal for a Council Directive on the Incineration of Hazardous Waste, June 1992.

führt und gegebenenfalls notwendige Maßnahmen eingeleitet.

Umgang mit Reststoffen

Grundsätzlich entstehen in der Zementindustrie nur geringe Mengen an Reststoffen, flüssigem oder festem Abfall. Trotzdem soll durch geeignete Techniken und Vorsorgemaßnahmen der Anfall an Reststoffen weiter minimiert werden.

- Der Einsatz von Feuerfest-Materialien, die kritische Stoffe wie beispielsweise Chrom enthalten, wird weiterhin reduziert werden.
- Die Menge an abgeschiedenem Staub, der nicht wieder in den Produktionsprozeß zurückgeführt werden kann, wird verringert werden. Dort, wo der Staub nicht recycelt werden kann, werden alternative Verwendungsmöglichkeiten weiterentwickelt.

Wo Produktionsabfälle entstehen, geschieht ihre Entsorgung auf umweltverträgliche Weise. Wenn Abfälle oder Reststoffe auf dem Werksgelände gelagert oder weiterbehandelt werden, geschieht dies unter Beachtung der entsprechenden Auflagen. Es wird sichergestellt, daß unter anderem weder Luft, Boden, Grund- oder Oberflächenwasser belastet werden.

Bei der Entsorgung von Produktionsabfällen außerhalb des Werksgeländes bedient sich die Zementindustrie nur zugelassener Transportunternehmen und registrierter Entsorgungsfirmen. Sie wird sich von Zeit zu Zeit davon überzeugen, daß diese Unternehmen die Abfälle den gesetzlichen Anforderungen entsprechend und unter Beachtung der Vorschriften deponieren.

Transport

Die Zementindustrie benötigt für die Beförderung ihrer Rohstoffe, Brennstoffe und Produkte erhebliche Mengen an Transportkapazität. Sie nimmt die Sorgen der Öffentlichkeit, die angesichts des Transportes dieser Stoffe bestehen, ernst und stellt dazu fest:

- Die Produktionsanlagen sind in der Regel nahe an den Rohmaterialvorkommen gelegen, um lange Transportwege zu vermeiden.
- Der überwiegende Teil des Versandes wird über die Straße abgewickelt, da sowohl die Werke als auch die Abnehmer kaum über alternative Transporteinrichtungen verfügen. Schienen- und Wasserwege werden jedoch immer dann benutzt, wenn es wirtschaftlich vertretbar ist und entsprechende Einrichtungen vorhanden sind.
- Zement wird entweder als Sackware verpackt oder in hermetisch dichten Silobehältern transportiert. Die Säcke werden vor dem Versand äußerlich entstaubt, um Belästigungen zu verhindern.
- Fahrzeuge, die den Unternehmen selbst oder den von den Unternehmen unter Vertrag genommenen Transportfirmen gehören, werden entsprechend den strengsten Vorschriften des jeweiligen Landes gewartet und überprüft. Sie werden darüber hinaus in sauberem Zustand gehalten.
- Falls erforderlich, werden Verkehrsprobleme im Einflußbereich des Werkes mit den lokalen Behörden erörtert.

Staatliche Umweltschutzregelungen in den USA für die Verbrennung von Sondermüll in Portlandzementöfen

Federal environmental regulations on portland cement kilns that burn hazardous wastes in the United States

Réglementations d'Etat en matière de protection de l'environnement appliquées aux USA pour l'incinération de déchets spéciaux dans les fours à ciment de Portland

Normas estatales en materia de protección ambiental, establecidas en EE.UU. para la incineración de basuras especiales en los hornos de cemento Portland

Von **E. Mikols,** Allentown/USA

Zusammenfassung – In den Vereinigten Staaten von Amerika wird seit mehr als zehn Jahren Sondermüll als Ergänzungsbrennstoff für Kessel und Industrieöfen verwendet. Bei geeigneten Bedingungen stellen diese Verbrennungsanlagen eine für die Umwelt unbedenkliche Methode zur thermischen Beseitigung großer Mengen gefährlicher organischer Abfallstoffe dar. Dank der inhärenten Sicherheitsfaktoren, wie ihre große thermische Trägheit, galten für die entsprechenden Anlagen viele Jahre lang minimale Regelungen. Als Reaktion auf den Druck von seiten der Umweltschützer und Firmen, die private Verbrennungsanlagen ohne Wärmerückgewinnung aus dem Müll betreiben, hat die US-Umweltbehörde EPA Regelungen für solche Kessel und Industrieöfen erlassen, die sogenannten BIFs. Diese Regelungen traten am 21. Februar 1991 in Kraft. Die wichtigsten in ihnen enthaltenen Auflagen und Grenzwerte betreffen in erster Linie die Zementindustrie. Die Auflagen umfassen die Forderung nach Emissionsüberwachung, Emissionsgrenzwerte für Schwermetalle, Kohlenmonoxid und Kohlenwasserstoffe und machen es zur Pflicht, betriebliche Unterlagen zu führen; außerdem enthalten diese Regelungen bestimmte Vorschriften bezüglich der Auslegung der Sondermüll-Zuführeinrichtungen. Auch müssen besondere Verfahren zur Erlangung einer Genehmigung eingehalten werden.

Staatliche Umweltschutzregelungen in den USA für die Verbrennung von Sondermüll in Portlandzementöfen

Summary – Hazardous wastes have been used as a supplementary fuel in boilers and industrial furnaces for well over a decade in the United States of America. Under proper conditions, these combustion units provide an environmentally sound way to thermally destroy a significant amount of the hazardous organic waste. For several years, this practice was subject to minimal regulations, due to the inherent safeguards associated with these facilities, such as a large thermal inertia. In response to pressure from environmental activists and companies which operate private incinerators which do not reclaim the heat from the waste, the federal Environmental Protection Agency (EPA) passed regulations for these boilers and industrial furnaces (BIF's). This rule, known as "BIF", went into effect on February 21, 1991. The major restrictions contained in the BIF Rule with their new limits affect specifically the cement industry. These restrictions include emissions monitoring requirements, emission limits for heavy metals, CO and hydrocarbons, operational record-keeping, and waste feed design. Furthermore, special procedures are required to obtain permits.

Federal environmental regulations on portland cement kilns that burn hazardous wastes in the United States

Résumé – Aux Etats Unis, depuis plus de 10 ans, les déchets spéciaux sont utilisés comme combustibles d'appoint dans les chaudières et fours industriels. Quand les conditions sont conformes, ces installations d'incinération représentent pour l'environnement une solution satisfaisante pour l'élimination thermique de grandes quantités de déchets organiques dangereux. En raison des facteurs de sécurité inhérents notamment leur grande inertie thermique, ces installations ont durant de nombreuses années bénéficié de régulations minimales. Face à la pression exercée par les écologistes et du fait des entreprises qui exploitent des installations d'incinération privées à partir de déchets sans système de récupération de chaleur, le Ministère de L'Environnement des Etats Unis, EPA, a publié des réglementations, les BIFs, applicables pour chaudières et fours industriels. Ces règles sont entrées en vigueur le 21 février 1991. Les obligations et valeurs-limites les plus importantes qu'elles comportent concernent, en premier chef, l'industrie du ciment. Ces obligations concernent l'exigence de surveillance des émissions, mais aussi valeurs-limites des émissions pour métaux lourds, monooxyde de carbone et hydrocarbures, et imposent la gestion régulière de documents d'exploitation justificatifs; de plus, ces réglementations comportent des directives sur le dimensionnement des installations d'alimentation de déchets spéciaux. Là aussi, le respect de procédés spéciaux visant à l'obtention d'une autorisation s'impose.

Réglementations d'Etat en matière de protection de l'environnement appliquées aux USA pour l'incinération de déchets spéciaux dans les fours à ciment de Portland

Resumen – En los Estados Unidos de América se viene empleando, desde hace más de diez años, la basura especial para ser quemada como combustible complementario en las calderas y los hornos industriales. En condiciones normales, estas instalaciones de incineración constituyen un método, inofensivo para el medio ambiente, de eliminar grandes cantidades de residuos orgánicos peligrosos. Gracias a los factores de seguridad inherentes a estas instalaciones, tales como su gran inercia térmica, durante muchos años sólo hubo una reglamentación mínima para las mismas. Reaccionando a las presiones de los protectores del medio ambiente y de las firmas que explotan plantas privadas de incineración de basuras, sin recuperación de calor, el ministerio estadounidense del Medio Ambiente EPA emitió reglas para calderas y hornos industriales, las llamadas BIF's, las cuales entraron en vigor el 21 de febrero de 1991. Los principales requerimientos y valores límites contenidos en ellas afectan en primer lugar a la industria del cemento. Incluyen la exigencia de un control de las emisiones y la observación de valores límite para emisiones de metales pesados, monóxido de carbono e hidrocarburos, y obligan a las empresas a llevar libros de control. Además, estas normas contienen ciertas prescripciones respecto del dimensionamiento de las instalaciones de alimentación de basuras especiales. Asimismo, hay que seguir determinados trámites para la obtención de las autorizaciones correspondientes.

Normas estatales en materia de protección ambiental, establecidas en EE.UU. para la incineración de basuras especiales en los hornos de cemento Portland

Am 21. Februar 1991 erließ die US-Umweltbehörde (USEPA) ihre endgültigen Regelungen für die Verbrennung von Sondermüll in Kesseln und Industrieöfen (Boilers and Industrial Furnaces = BIF). Deshalb wird von den „BIF-Regelungen" gesprochen. Portlandzementöfen sind Industrieöfen und fallen darum unter die BIF-Regelungen, wenn in ihnen Sondermüll als Brennstoff eingesetzt wird. **Bild 1** zeigt die 26 BIF-Öfen, die zur Zeit den BIF-Regelungen genügen.

Mit den BIF-Regelungen wurden mehrere neue Normen für Emissionsgrenzwerte bei Sondermüll verfeuernden BIF-Anlagen erlassen. Da die BIF-Regelungen eine große Vielzahl verschiedener Verbrennungsanlagen abdecken mußten, führte die USEPA drei Methoden zur Zertifizierung der Einhaltung der Vorschriften ein: Stufe I, Stufe II und Stufe III. Die Zementofenbetreiber entschieden sich für eine Reglementierung nach Stufe II und Stufe III, die sich dadurch unterscheiden, wie die Umweltbelastung durch eine BIF-Anlage analysiert wird. Die Emissionsgrenzwerte für Metalle, HCl, Cl_2 und die kanzerogenen Metalle (As, Cr(VI), Be und Cd) beruhen sämtlich auf der jeweiligen Gefährdung. Da diese Normen von den besonderen Kenndaten des Ofens und seiner Betriebsweise abhängen, sind diese Emissionsgrenzwerte nicht spezifisch formuliert. Vielmehr werden sie für jede Anlage nach einer geeigneten Analyse der Randbedingungen festgelegt.

Nach den BIF-Regelungen sind folgende Normen einzuhalten (**Bilder 2-4**):

BILD 2: Voraussetzungen für die weitere Genehmigung der Verbrennung von Abfallstoffen in Zementöfen

- Erlangung eines Interimsstatus
  Finanzielle Absicherung
  Keine Dioxinabfälle
  Notfallbereitschaft

- Vorläufige Bescheinigung der Einhaltung der Bestimmungen
  21. August 1991

- Bescheinigung der Einhaltung der Bestimmungen
  21. August 1992

BILD 3: Numerische Emissionsgrenzwerte nach den BIF-Regelungen

- Kohlenmonoxid (Kohlenwasserstoff-Alternative)
  100 ppmv CO — auf 7 % Sauerstoff korrigiert
  oder
  20 ppmv KW — auf 7 % Sauerstoff korrigiert
  — als Propan ausgedrückt
  — unter Ausschluß von Methan

- Zerstörungs- und Beseitigungswirkungsgrad (DRE)
  99,99 % (99,9999 % bei Dioxinabfällen)

- Feststoffausstoß
  0,08 grain/dscf — auf 7 % Sauerstoff korrigiert
  (183 mg/Nm³)

On February 21, 1991, the United States Environmental Protection Agency (USEPA) issued its final rules for the burning of hazardous wastes in Boilers and Industrial Furnaces. Hence, the rules are called "the BIF Rule". Portland cement kilns are industrial furnaces, so they are covered by the BIF Rule if they burn hazardous waste derived fuels. **Fig. 1** shows the 26 BIF kilns which currently comply with the BIF Rule.

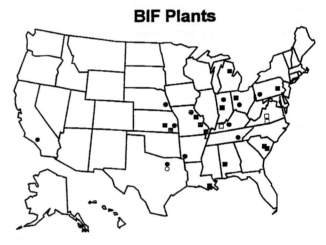

BILD 1: Sondermüll verbrennende Zementöfen
FIGURE 1: Cement kilns burning hazardous waste

The BIF Rule imposed several new standards for the permissible emissions from BIF units that burn hazardous wastes. Because the BIF Rule had to cover so many kinds of combustion units, USEPA created three ways to certify compliance: Tier I, Tier II and Tier III. Cement kiln operators opted to be regulated under Tier II and III, which differ in the way the impact from a BIF unit is analyzed. The emissions for metals, HCl, Cl_2 and the carcinogenic metals (As, Cr(VI), Be and Cd) are all riskbased standards. Because these standards depend on the particular characteristics of the kiln and its setting, those emission rates are not specific limits. Rather, they are determined for each facility after the appropriate analysis which is performed for that facility. The BIF Rule requires compliance with the following standards (**Fig. 2-4**):

BILD 4: Auf den Gefährdungen beruhende Emissionsgrenzwerte nach den BIF-Regelungen

- Metalle
  Kanzerogen (As, Be, Cr, Cd)
    - Gefährdungssumme für alle 4 Metalle = 10^{-5}
  Nichtkanzerogen (Sb, Ba, Pb, Hg, Ag, Tl)
    - Gefährdungsbewertung aufgrund der oralen Dosis

- HCl und Cl_2
  Nach der Gefährdung festgesetzte Grenzwerte;
  Angaben in den Regelungen

- Dioxine und Furane
  Ausgedrückt als toxisches Äquivalent von 2,3,7,8-TCDD
  MEI-Risiko = 10^{-5}

1. Kohlenmonoxid oder Kohlenwasserstoffe

Zur Sicherstellung der vollständigen Verbrennung wurden höchstens 100 ppmv CO erlaubt. Da viele Öfen diesen CO-Grenzwert nicht einhalten können, wurde als Alternative ein Grenzwert von 20 ppmv Kohlenwasserstoffe (KW) — unter Ausschluß von Methan und als Propan ausgedrückt — erlaubt. Beide Grenzwerte gelten für auf 7 % Sauerstoff korrigiertes Trockengas. Für die Messung von CO oder KW und O_2 sind kontinuierlich registrierende Meßgeräte erforderlich.

2. Zerstörungs- und Beseitigungswirkungsgrad (DRE = Destruction and Removal Efficiency)

Ein DRE von 99,99 % muß für einen oder mehrere schwer zu zerstörende organische Bestandteile des Mülls nachgewiesen werden. Bei dioxinhaltigen Abfällen ist ein DRE von 99,9999 % erforderlich.

3. Feststoffausstoß

Die Emissionsgrenze liegt nach einer Korrektur auf 7 % Sauerstoff bei 0,08 grain je trockenen Standard-Kubikfuß (183 mg je trockenen Kubikmeter). Dieser Grenzwert ist die für Müllverbrennungsanlagen geltende Norm. Für Öfen gelten häufig strengere Normen der NSPS oder des jeweiligen Bundesstaats.

4. Metalle (Sb, As, Be, Ba, Cr, Cd, Pb, Hg, Ag, Tl)

Die Emissionsraten werden je nach Gefährdung angegeben. Das Gesamtrisiko beträgt bei kanzerogenen Metallen (As, Be, Cd und Cr(VI)) 10^{-5}. Die höchstzulässigen Konzentrationen nichtkanzerogener Metalle (Sb, Ba, Pb, Hg, Ag, Tl) beruhen auf der oralen Bezugsdosis und sind in den Regelungen aufgeführt.

5. HCl und Cl_2

Die Grenzwerte sind nach der jeweiligen Gefährdung in den Regelungen aufgeführt.

6. Dioxine und Furane

Die Emissionen werden als toxisches Äquivalent (TEQ) von 2,3,7,8-TCDD aufgeführt. Die Gefährdung der am stärksten exponierten Personen (MEI = Most Exposed Individuals) darf nach dem toxischen Äquivalent nicht mehr als 10^{-5} betragen. Bei Öfen muß die Gefährdung in der Regel berechnet werden, da das Staubmeßgerät bei 240–400 °C betrieben wird.

Die obigen Grenzwerte können punktuell oder als jeweilige Stundendurchschnitte ausgedrückt werden. Die BIF-Zementöfen haben kaum Probleme mit der Einhaltung der einzelnen Grenzwerte.

Zu den Zukunftsproblemen der BIF-Zementöfen gehören:

- der Zementofenstaub, der regelmäßig untersucht werden muß, um nachzuweisen, daß er sich statistisch nicht vom Ofenstaub ohne Sondermüllverbrennung unterscheidet;

- die Erhaltung von „Gleichgewichtsbedingungen" entsprechend der Definition der USEPA im Hinblick auf die Einhaltungsprüfung;

FIGURE 2: Requirements to allow cement kilns to continue burning wastes

- Obtain interim status
  Financial assurance
  No dioxin wastes
  emergency preparedness

- Certification of precompliance
  21. August 1991

- Certification of compliance
  21. August 1992

FIGURE 3: Numerical emission limits under BIF

- Carbon monoxide (Hydrocarbon alternative)
  100 ppmv CO   — corrected to 7 % oxygen
  or
  20 ppmv HC   — corrected to 7 % oxygen
                 — expressed as propane
                 — excluding methane

- Destruction and removal efficiency (DRE)
  99,99 % (99,9999 % for dioxin wastes)

- Particulate matter
  0,08 grain/dscf   — corrected to 7 % oxygen
  (183 mg/Nm^3)

FIGURE 4: Risk-based emission limits under BIF

- metals
  Carcinogenic (As, Be, Cr, Cd)
    — Summed risk for all 4 metals = 10^{-5}
  Non-carcinogenic (Sb, Ba, Pb, Hg, Ag, Tl)
    — Risks based on oral dose

- HCl and Cl_2
  Limits based on risk; specified in rule

- Dioxins and furans
  Expressed as toxic equivalent for 2,3,7,8-TCDD
  Risk to MEI = 10^{-5}

1. Carbon Monoxide or Hydrocarbons

A maximum of 100 ppmv CO was allowed to insure complete combustion. Since many kilns can not meet the CO limit, an alternative limit of 20 ppmv hydrocarbons (HC), excluding methane and expressed as propane, was allowed. Both limits are for dry gas corrected to 7 % oxygen. A continuous monitor is required to measure CO or HC and O_2.

2. Destruction and Removal Efficiency (DRE)

A 99.99 % DRE has to be demonstrated for one or more organic components in the waste which are difficult to destroy. Wastes containing dioxin require a 99.9999 % DRE.

3. Particulate Matter

Emission limit equals to 0.08 grain per dry standard cubic feet (183 mg per dry cubic meter), corrected to 7 % oxygen. This limit is the incinerator standard; more stringent NSPS or state standards often apply to kilns.

4. Metals (Sb, As, Be, Ba, Cr, Cd, Pb, Hg, Ag, Tl)

Emission rates are specified, based on risk. The total risk for carcinogenic metals (As, Be, Cd and Cr(VI)) is 10^{-5}. The allowable concentrations for noncarcinogenic metals (Sb, Ba, Pb, Hg, Ag, Tl) are based on Oral ReferenceDose and are listed in the Rule.

5. HCl and Cl_2

The limits are specified in the Rule, based on risk.

6. Dioxins and Furans

Emissions are expressed as the toxic equivalency to 2,3,7,8-TCDD (TEQ). Risk to the most exposed individual (MEI) cannot exceed 10^{-5}, based on the toxic equivalency. Kilns generally must calculate the risk, since their particulate matter control device is operated at 450–750 °F (240–400 °C).

— die Messung von Kohlenwasserstoffen, bei der noch Meinungsverschiedenheiten über beheizte oder unbeheizte Sonden bestehen.

Eine weitere Entwicklung aus jüngerer Zeit ist der neue „Entwurf einer Verbrennungsstrategie", der von der Clinton-Administration am 18. Mai 1993 vorgelegt wurde. In diesem Entwurf wird eine neue Feststoffnorm vorgeschlagen (0,015 gr/dscf = 34,3 mg/Nm^3), was der Hälfte der Bundesnorm für neue Zementöfen entspricht. Darüber hinaus beschränkt ein neuer Dioxin-Grenzwert die Summe aller Dioxine (nicht das toxische Äquivalent) auf 30 ng/Nm^3. Auch die Emissionsgrenzwerte werden auf der Summe der Gefährdungen über sämtliche Expositionswege wie die Nahrungskette und die Haut festgelegt. Gegenwärtig werden nur die inhalative oder orale Exposition berücksichtigt.

Vor den BIF-Regelungen stellten die Portlandzementwerke in den Vereinigten Staaten eine brauchbare Alternative für die Verwertung von energiereichem Sondermüll dar. Die Abfälle konnten problemlos und sicher ohne Umweltbelastung anstelle herkömmlicher Brennstoffe eingesetzt werden. Mit den BIF-Regelungen wird der reglementierte Bereich erweitert, und der Betreiber ist nun verpflichtet, den betrieblichen Gesamtrahmen zu beschreiben, in dem er tätig werden und für die Einhaltung der auf den Gefährdungen beruhenden Emissionsgrenzwerte sorgen wird. Angesichts der mittlerweile langen Liste wichtiger Betriebsparameter ist eine solche Beschreibung des betrieblichen Gesamtrahmens überaus schwierig geworden. Noch schwerer fällt dem Betreiber der laufende Nachweis, daß der Ofen unter zulässigen Betriebsbedingungen gefahren wird. Angesichts dieser praktischen Schwierigkeiten und der Kosten des Nachweises der Einhaltung der Bestimmungen ist die künftige Wirtschaftlichkeit der Verbrennung von Sondermüll in Zementöfen fraglich geworden.

The above limits may be expressed as an instantaneous or on an hourly rolling averge basis. Very few problems meeting each of these limits have been reported by BIF cement kilns.

Issues facing BIF cement kilns in the future include:

- Cement kiln dust which has to be regulary tested to show that is not statistically different from kiln dust before it burned hazardous waste,
- maintaining "equilibrium conditions" as defined by USEPA for the compliance test,
- measuring hydrocarbons, where disagreement still exists regarding heated or non-heated probes.

Another recent development is the new "Draft Combustion Strategy", issued by the Clinton Administration on May 18, 1993. In this draft a new particulate matter standard is proposed (0.015 gr/dscf (34.3 mg/Nm3)), which is one-half the federal standard for new cement kilns. In addition a new Dioxin standard limits the sum of all congeners (not the TEQ) to 30 ng/Nm^3. Also the emission limits will be based on the summed risks from all pathways of exposure, such as food chain and dermal methods. Now only an inhalation or oral exposure is used.

Prior to the BIF Rule, Portland cement plants in the United States posed a viable alternative to the destruction of energy rich hazardous wastes. The wastes could be easily and safely substituted for conventional fuels with no adverse effects to the environment. The BIF Rule expands the regulatory focus and requires the operator to describe the operational envelope within which he will operate and maintain compliance with risk based emissions limits. Because the list of important operating variables is lengthy, the task of defining the operating envelope has become exceedingly difficult. It is even more difficult for the operator to continually prove that the kiln is operating within the permissible conditions. Because of these practical difficulties and the costs associated with demonstrating compliance, the future economic advantage of burning hazardous wastes in cement kilns has become questionable.

Optimale Ofenführung zur Einhaltung von Umweltschutzvorschriften

Optimal kiln control under environmental regulations

Conduite de four optimale pour respect des réglementations sur la protection de l'environnement

Conducción óptima del horno para cumplir las normas de protección ambiental

Von **J.-J. Østergaard,** Valby/Dänemark

Zusammenfassung – Es wird eine Überwachungsstrategie für den optimalen Ofenbetrieb zur Einhaltung von Umweltschutzvorschriften vorgestellt. Bis jetzt waren maximale Produktion, minimaler Brennstoff- und Energieeinsatz und gute Produktqualität die Hauptziele einer optimalen Ofenführung. Diese Ziele sind für eine wirtschaftliche Zementproduktion natürlich nach wie vor sehr wichtig. Daneben gewinnt aber eine Ofenführungsstrategie für den optimalen Ofenbetrieb zur Einhaltung von Umweltschutzvorschriften immer mehr an Bedeutung. Die Einführung strenger Vorschriften bezüglich der Emission umweltrelevanter Bestandteile aus industriellen Prozessen hat eine Reihe neuer Ziele entstehen lassen, die oft zu den traditionellen Anforderungen an einen optimalen Ofenbetrieb in Widerspruch geraten. Die Ofenführungsstrategie basiert auf einem Überwachungskonzept, welches den optimalen Ofenbetrieb gewährleistet. Dieser ist als die größtmögliche fortlaufende Produktion hochwertigen Klinkers bei NO_x- und CO-Emissionen unterhalb der oberen Grenzwerte definiert. Diese Überwachungsstrategie wurde mit Hilfe des Hochleistungs-Steuerungssystems FLS Fuzzy II realisiert.

Optimale Ofenführung zur Einhaltung von Umweltschutzvorschriften

Summary – A control strategy for optimal kiln operation under environmental regulations is presented. Up to now, maximum production, minimum fuel and energy consumption and good product quality have been the main objectives for optimal kiln control. These goals, of course, are still of decisive importance for profitable cement production. Besides this a control strategy for optimal kiln operation under environmental regulations gains an increasing importance. The introduction of tough regulations with reference to the emission of environmentally relevant components from industrial processes has added a number of new objectives, which are often in conflict with the traditional requirements for optimal kiln operation. The control strategy is based on a control scheme which produces optimal kiln operation defined as the largest possible continuous production of good quality clinker, and with NO_x – and CO-emission below the high limits. The control strategy has been implemented in the FLS Fuzzy II high level control system.

Optimal kiln control under environmental regulations

Résumé – Une stratégie de surveillance pour conduite de four optimale respectant les réglementations antipollution a été présentée. Jusqu'ici, la production maximale, l'utilisation de combustible et d'énergie minimale et la bonne qualité du produit étaient les objectifs primordiaux d'une conduite de four optimale. Ces objectifs restent bien entendu comme par le passé d'une importance vitale pour une production de ciment rentable. A côté de cela cependant, une stratégie de gestion de four pour conduite optimale respectant les normes de protection de l'environnement s'avère être de plus en plus importante. L'introduction de directives sévères en matière d'émissions de composants polluants provenant de procédés industriels a créé un grand nombre de nouveaux objectifs, qui sont souvent en contradiction avec les exigences traditionnelles imposées à une conduite de four optimale. La stratégie de gestion de four repose sur un concept de surveillance garantissant la meilleure conduite de four possible. Celle-ci est définie comme étant, en continu, la production de clinker de qualité supérieure la plus élevée possible avec émissions de NO_x et de CO inférieures aux valeurs limites supérieures. Cette stratégie de surveillance a été réalisée avec le concours du système de conduite haute puissance FLS Fuzzy II.

Conduite de four optimale pour respect des réglementations sur la protection de l'environnement

Resumen – Se presenta una estrategia de control, destinada a conseguir un servicio óptimo del horno que permita cumplir las normas de protección ambiental. Hasta ahora, los objetivos principales de la conducción óptima del horno han sido: una producción máxima, un consumo mínimo de combustible y de energía y una buena calidad del producto. Lo cierto es que estos objetivos siguen siendo muy importantes, para que la fabricación del cemento sea rentable. Pero a parte de ello, adquiere cada vez mayor importancia la estrategia de conducción óptima del horno, que permita cumplir las prescripciones en materia de protección del medio ambiente. El establecimiento de reglas rigorosas referentes a la emisión de sustancias relevantes para el medio ambiente, y procedentes de los procesos industriales, hace que hayan surgido una serie de objetivos

Conducción óptima del horno para cumplir las normas de protección ambiental

nuevos, los cuales están a menudo reñidos con los requerimientos tradicionales de un servicio óptimo del horno. La estrategia de conducción del horno se basa en un método de control que garantiza el servicio óptimo del horno. Este queda definido como producción máximo y continua de clínker de alta calidad, con emisiones de NO_x y CO inferiores a los valores límite superiores. La citada estrategia de control se realizó con ayuda del sistema de mando de alto rendimiento FLS Fuzzy II.

1. Emissionskontrolle

Die Emission von Zementöfen hängt in starkem Maße von ihrer Kontrolle ab. Dabei stellen die Umweltvorschriften zunehmend höhere Anforderungen an den Leitstandsfahrer. Er muß die Emissionen unterhalb der festgelegten Grenzwerte halten und gleichzeitig soll er einen maximalen Durchsatz hochwertigen Klinkers bei möglichst niedrigem Brennstoff- und Energieeinsatz erzielen. High Level Steuerungsstrategien von Expertensystemen, wie dem Fuzzi II System, haben sich für die Gewährleistung eines optimalen Ofenbetriebes unter Einhaltung der Umweltschutzvorschriften als ein effizientes Werkzeug für den Leitstandsfahrer bewährt. Das Fuzzi II System selbst kann zwar nicht das Emissionsverhalten der Ofenanlage verändern, aber es sorgt dafür, daß die Emissionen so niedrig wie möglich gehalten werden. Mit der Installation eines Fuzzi II Systems ist eine Basis zur Berücksichtigung möglicher Prozeßmodifikationen gegeben, etwa durch Einsatz eines Low NO_x-Brenners, der Eindüsung von Ammoniakwasser, usw.

Dieser Beitrag konzentriert sich auf die Fuzzi II Strategie, die zur Kontrolle von CO und dem in der Brennzone aus der Reaktion zwischen Sauerstoff und dem atmosphärischen Stickstoff erzeugten thermischen NO_x getestet worden ist.

2. Die Fuzzi II Kontrollstrategie

Generell kann die NO_x-Bildung durch den Sauerstoffgehalt und durch die Herabsetzung der Brennzonentemperatur beeinflußt werden. Wie in **Bild 1** dargestellt, wird die Fuzzi II Strategie zur Überwachung von NO_x und CO in zwei Gruppen unterteilt: in die Brennzonen- und Verbrennungssteuerung. Das Überwachungskonzept für NO_x setzt sich wiederum zusammen aus einer Langzeitstrategie (LTS), die Bestandteil der Brennzonensteuerung ist, und aus einer Kurzzeitstrategie (STS), die zusammen mit der Überwachung des Sauerstoffgehaltes in der Gruppe der Verbrennungssteuerung verankert ist. Die Prioritäten zeigen den Grad der Wichtigkeit an. Mit der Priorität 1 ist demzufolge das wichtigste Steuerziel bezeichnet. Das Fuzzi II Prioritäts-Management-System (PMS) stellt sicher, daß in aktuellen Betriebssituationen die Steuerungsziele in der Rangordnung ihrer Bedeutung erfüllt werden.

3. Die NO_x-Kurzzeitstrategie

Die NO_x-Kurzzeitstrategie basiert auf der Tatsache, daß um so mehr NO_x erzeugt wird, je höher der Sauerstoffgehalt ist. **Bild 2** zeigt den grundlegenden Programmablauf zur NO_x-Kurzzeitstrategie. Wenn der NO_x-Gehalt zu hoch ist, aktiviert das Prioritäts-Management-System automatisch die Kurzzeitstrategie und deaktiviert die Sauerstoffüberwachung. Der Zweck dieser Steuerungsvorgänge ist, den Sauerstoffgehalt soweit zu senken, daß es zu einer beginnenden CO-Spitze kommt, die unverzüglich die NO_x-Konzentration reduziert. Das Prioritäts-Management-System stellt sicher, daß der CO-Wert nicht zu hoch ansteigt. Wenn der CO-Wert seinen obersten Grenzwert überschritten hat, erhöht das Ofengebläse seine Geschwindigkeit. Diese Balance zwischen dem CO-und NO_x-Wert ist kritisch und durch den Leitstandsfahrer sehr schwer aufrecht zu erhalten.

Wenn die NO_x-Kurzzeitstrategie erfolgreich ihre Funktion erfüllt hat, d.h. die NO_x-Konzentration durch die Zurücknahme des Sauerstoffgehaltes ohne Veränderung der Brennzonentemperatur reduziert werden konnte, dann wird der Sollwert für die O_2-Kontrolle entweder automa-

1. Emission Control

The emission from a cement kiln depends very much on how it is controlled. Environmental regulations put increasingly tough demands on the kiln operator. He must keep emissions below the specified limits and simultaneously, he shall produce maximum output of good quality clinker by using as little fuel and energy as possible. High level control strategies of the expert system type, like the Fuzzy II System, have proven as efficient tools for the operator to achieve optimal kiln operation under environmental regulations. Fuzzy II does not enhere the kiln system itself with low emission characteristics, but it ensures that the emission is as low as possible. With Fuzzy II installed, the basis is thus established for consideration of possible process design modifications, like low NO_x burners, injection of NH_3 ect., if such are necessary. This presentation will focus on the Fuzzy II strategy, which has been tested in practice, for control of CO and thermal NO_x, i.e. the NO generated in the burning zone from the reaction between oxygen and atmospheric nitrogen.

2. Fuzzy II Control Strategy

From a control point of view, the formation of NO_x can be influenced through the oxygen level and by lowering the burning zone temperature. As shown in **Fig. 1**, the Fuzzy II control strategy, which is relevant for NO_x and CO control, is divided into two Control Groups, i.e. Burning Zone Control and Combustion Control.

The control scheme for NO_x is composed of a Long Term Strategy (LTS), which is part of the Burning Zone Control group, and a Short Term Strategy (STS) in the Combustion Control group together with control of the oxygen level. The priorities indicate the order of importance, with priority 1 being the most important control objective. The Fuzzy II Priority Management System (PMS) ensures that in actual operations the control objectives are fulfilled in order of importance.

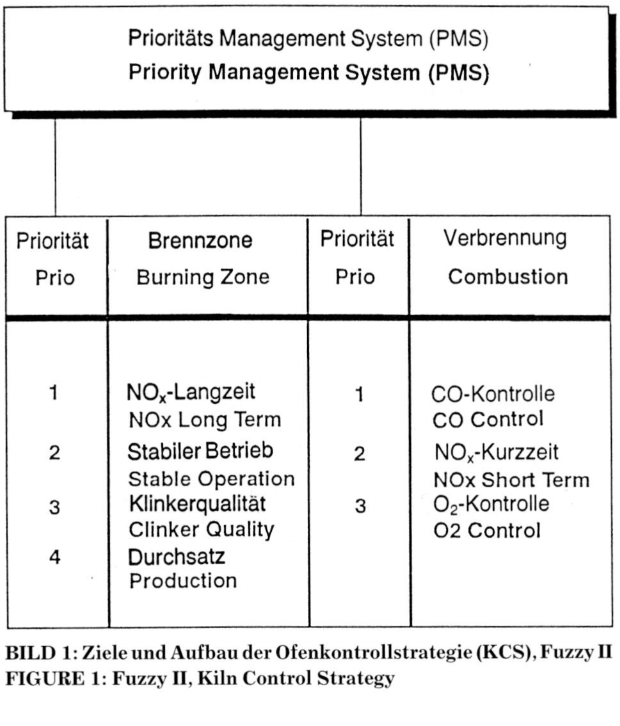

Prioritäts Management System (PMS) Priority Management System (PMS)			
Priorität Prio	Brennzone Burning Zone	Priorität Prio	Verbrennung Combustion
1	NO_x-Langzeit NOx Long Term	1	CO-Kontrolle CO Control
2	Stabiler Betrieb Stable Operation	2	NO_x-Kurzzeit NOx Short Term
3	Klinkerqualität Clinker Quality	3	O_2-Kontrolle O2 Control
4	Durchsatz Production		

BILD 1: Ziele und Aufbau der Ofenkontrollstrategie (KCS), Fuzzi II
FIGURE 1: Fuzzy II, Kiln Control Strategy

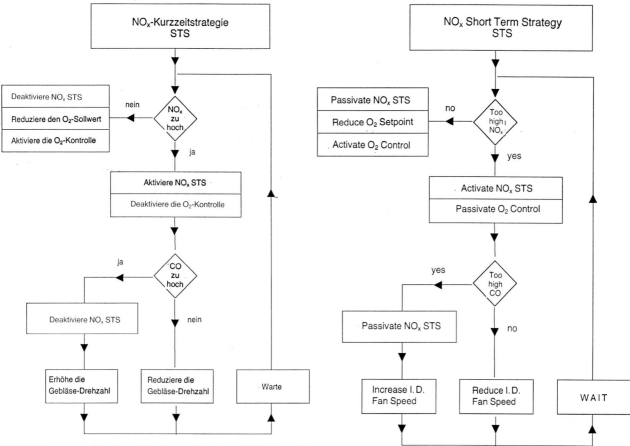

BILD 2: Programmablauf zur NO$_x$-Kurzzeitstrategie (STS), Fuzzy II
FIGURE 2: Fuzzy II, NO$_x$ Short Term Strategy (STS)

tisch oder als ein Hinweis für den Leitstandsfahrer geringfügig reduziert. Wenn der O$_2$-Sollwert nicht herabgesetzt wird, dann ist es sehr wahrscheinlich, daß der NO$_x$-Gehalt wieder ansteigt, wenn die O$_2$-Steuerung den Sauerstoffgehalt wieder auf den alten Sollwert zurückbringt. Wenn der Sauerstoffgehalt die Ursache dafür war, daß die NO$_x$-Konzentration zu hoch ist, dann reduziert die NO$_x$-Kurzzeitstrategie von Fuzzi II den O$_2$-Sollwert in solch kleinen Schritten, bis der Grenzwert für CO erreicht worden ist.

4. Die NO$_x$-Langzeitstrategie

Die NO$_x$-Langzeitstrategie basiert auf der Tatsache, daß die Brennzonentemperatur um so höher ist, je mehr NO$_x$ erzeugt wird. **Bild 3** zeigt die Anwendung der Langzeitstrategie nach einem Abfrageprogramm. Die Langzeitstrategie wird nicht wirksam, solange die Kurzzeitstrategie die Möglichkeit hat, den NO$_x$-Gehalt durch Herstellen des Gleichgewichts zwischen CO und NO$_x$ zu reduzieren. Das entspricht einer Vorzugsvariante, da das Prioritäts-Management-System die normale Ofenoptimierung deaktiviert, wenn die NO$_x$-Langzeitstrategie wirksam ist.

Zunächst wird die Brennzonentemperatur reduziert, während der Ofendurchsatz sich erhöht. Die Zurücknahme des Brennstoffs als ein erster Versuch würde das Ergebnis zunichte machen, das schon durch die Anwendung der NO$_x$-Kurzzeitstrategie erreicht worden ist. Im nächsten Schritt, wenn der maximale Durchsatz erreicht worden ist und der NO$_x$-Wert immer noch zu hoch liegt, muß die Brennstoffzufuhr reduziert werden.

Die Vorgehensweise ist geeignet, die Balance zwischen dem CO- und NO$_x$-Gehalt zu stören, wobei die NO$_x$-Kurzzeitstrategie erneut aktiviert wird, bis ein neues Gleichgewicht erreicht worden ist. Auf diese Weise kommt es zu einer on/off-Kooperation zwischen den beiden NO$_x$-Strategien, bis der NO$_x$-Wert unterhalb des oberen Grenzwertes gesunken ist.

Warum können die Vorschriften zum NO$_x$-Wert zu einer Durchsatzminderung führen, wenn der Ofen nicht ord-

3. NO$_x$ – Short Term Strategy

The NO$_x$ Short Term Strategy (STS) is based on the fact that more NO$_x$ is generated the higher the oxygen level. **Fig. 2** shows the basic diagram for the NO$_x$ STS. When NO$_x$ is too high, the Priority Management System automatically activates the STS and passivates the oxygen control. The purpose of the control actions is to lower the oxygen so that CO starts to spike, which immediately reduces the NO$_x$. The Priority Management System ensures that CO will not grow too high as the I.D. Fan speed is increased again, if CO exceeds its high limit. This balance between CO and NO$_x$ is critical and very difficult to maintain for an operator. If the NO$_x$ Short Term Strategy exits successfully, i.e. NO$_x$ has been reduced by lowering the oxygen level and without reducing the burning zone temperature, then the setpoint for the O$_2$ control objective is reduced slightly, either automatically or as an advise to the operator.

If the oxygen setpoint is not reduced, it is likely that NO$_x$ will increase again when the O$_2$ control brings the oxygen back to the old setpoint. If the oxygen level is the reason for the NO$_x$ being too high, the Fuzzy II NO$_x$ Short Term Strategy will reduce the O$_2$ setpoint in small steps until the limit for CO has been reached.

4. NO$_x$ – Long Term Strategy

The NO$_x$ Long Term Strategy is based on the fact that more NO$_x$ is generated the higher the burning zone temperature. **Fig. 3** shows the basic diagram for the NO$_x$-LTS. The Long Term Strategy is not activated until the Short Term Strategy has had the opportunity to reduce NO$_x$ by establishing the balance between CO and NO$_x$. This is preferable as the Priority Management System passivates the normal kiln optimization when NO$_x$-LTS is activated. First, the burning zone temperature is reduced by increase of the production. Cutting fuel as the first attempt would spoil what the NO$_x$ Short Term Strategy had already attained. Next, when maximum production has been reached, and if NO$_x$ is still too high,

nungsgemäß gesteuert wird? Vom Standpunkt der NO_x-Langzeitstrategie wird eigentlich das Gegenteil angestrebt, da Produktionssteigerungen ein Bestandteil der Strategie sind. Tatsache ist allerdings, daß Leitstandsfahrer im allgemeinen darauf bedacht sind, den Ofen heiß zu betreiben, da er unter diesen Bedingungen stabiler läuft bzw. Instabilitäten besser abfängt. Wenn die Temperatur der Brennzone aus NO_x-Gründen reduziert wird, wird der Ofen gegenüber Störungen anfälliger, und seine Steuerung wird wesentlich schwieriger. Gerade auch die kleineren Instabilitäten können den Ofen in einen Zustand versetzen, wo er außer Kontrolle gerät und es notwendig wird, den Durchsatz zurückzunehmen, um ihn wieder in einen normalen Betriebszustand zu versetzen.

Ähnlich der CO-NO_x-Balance ist es notwendig, den Ofen nahe seiner Stabilitätsgrenze zu betreiben. Wenn er nicht sorgfältig gesteuert wird, kann sich daraus ein Durchsatzabfall ergeben. Wenn der Ofen auf seinen Normaldurchsatz zurückgebracht wird, ist das gewöhnlich mit einer vorübergehenden Aufheizung und damit wiederum mit zu hohen NO_x-Emissionen verbunden. Nachdem der NO_x-Gehalt durch die Zurücknahme der Brennzonentemperatur reduziert wurde, ist es wichtig, den Ofen auf eine neue Instabilität vorzubereiten. Das ist damit getan, wenn der Durchsatz um einen bestimmten Prozentsatz der Steigerungen zurückgenommen wird, die während der letzten NO_x-Gegensteuerung erreicht worden sind. Die Brennstoffzufuhr muß dabei ebenfalls um einen bestimmten Prozentsatz erhöht werden.

5. Beispiel aus der Solnhofer Portland Zementwerke GmbH

Bild 4 zeigt beispielhaft an einem Meßschrieb aus der Solnhofer Portland Zementwerke GmbH vom Juli 1992 die Arbeitsweise der NO_x-Überwachung durch Fuzzy II. Die

then fuel must be reduced. This is likely to disturb the balance between CO and NO_x by which the NO_x Short Term Strategy is activated again until a new balance has been obtained. Like this, an ON/OFF cooperation between the two NO_x strategies may be established until NO_x is below the high limit.

Why is it that regulations on NO_x may result in lower production if the kiln is not properly controlled? From the NO_x Long Term Strategy the opposite seems to be the case as production increases are parts of the strategy. The fact, however, is that operators tend to operate the kiln on the hot side where it runs more stable and where the kiln itself can cope with minor upsets. When the burning zone temperature is reduced, for NO_x reasons, then the kiln becomes more sensitive to disturbances, and it is much more difficult to control. Even minor upsets may take the kiln over the edge where it is out of control, and where it is necessary to cut production to bring the kiln back to normal operation. Like for the CO-NO_x balance, it is very demanding to operate the kiln close to the cutting edge, and if not controlled carefully, it will result in an overall decrease in production. And furthermore, bringing the kiln back from low production normally involves a temporary state of overburning and by this, in too high NO_x emission. Having reduced NO_x by lowering the burning zone temperature, it is important to prepare the kiln for a possible upset. This is done by taking the production back a certain percentage of the increases that were accumulated during the last campaign against NO_x. The fuel is also increased a certain percentage of the accumulated reduction.

5. Example

Fig. 4 shows an example registered at Solnhofer Portland Zementwerke GmbH in July 1992. The curves cover a period of 8 hours and they show NO_x, oxygen and CO at the stack

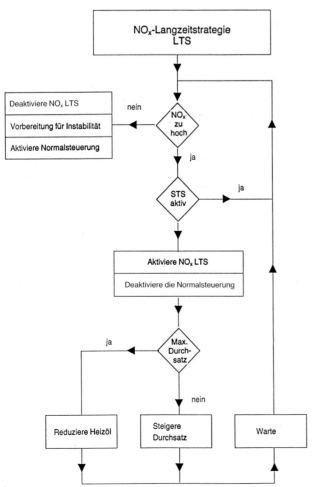

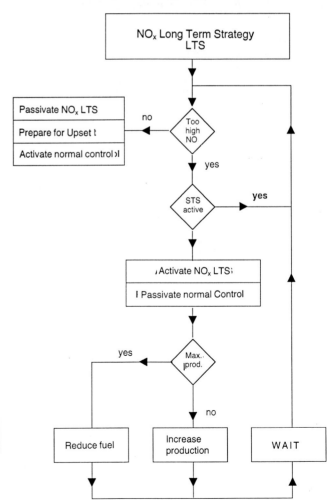

BILD 3: Programmablauf zur NO_x-Langzeitstrategie (LTS), Fuzzy II
FIGURE 3: Fuzzy II, NO_x Long Term Strategy (LTS)

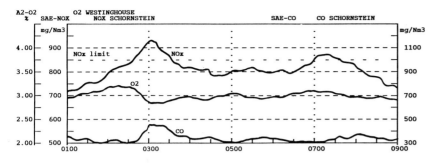

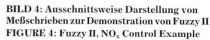

BILD 4: Ausschnittsweise Darstellung von Meßschrieben zur Demonstration von Fuzzy II
FIGURE 4: Fuzzy II, NO_x Control Example

O_2-Gehalt	= O_2 content
NO_x-Konzentration	= NO_x concentration
CO-Konzentration	= CO concentration
NO_x und CO gemessen	= NO_x and CO measured
am Schornstein	= at the stack
Zeit h	= Time h
Heizölmenge l/h	= Fuel oil quantity l/h
Rohmehlaufgabe t/h	= Raw meal feed t/h
Gebläsedrehzahl min^{-1}	= Fan speed min^{-1}

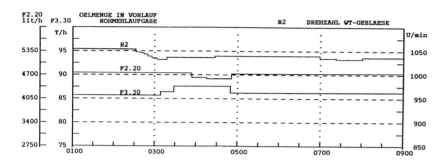

Meßschriebe über eine Dauer von 8 Stunden zeigen den NO_x-, O_2- und CO-Verlauf, gemessen am Schornstein im Vergleich zu Parametern wie der Rohmehlaufgabe auf den Vorwärmer, der Heizölzufuhr zum Ofen und der Drehzahl des Wärmetauschergebläses. In Phase 1 wird der Sauerstoffgehalt durch die NO_x-Kurzzeitstrategie soweit gesenkt, bis der CO-Wert zu hoch ansteigt. In der Phase 2 beginnt die NO_x-Langzeitstrategie die Rohmehlaufgabe zu erhöhen, bis die maximale Aufgabe von 88 t/h erreicht worden ist. In Phase 3 wird das Heizöl in on/off-Kooperation mit der Kurzzeitstrategie reduziert, bis die NO_x-Konzentration unter 800 mg/Nm³ zurückgegangen ist, während mit der Langzeitstrategie der Ofen auf einen möglichen labilen Zustand durch erhöhte Heizölzugabe und Zurücknahme der Rohmehlaufgabe vorbereitet wird. Gegen 7 Uhr morgens tritt nach Bild 4 die NO_x-Kurzzeitstrategie für eine kurze Zeitperiode wieder in Aktion.

together with the control parameters, i.e. kiln feed, oil, and I.D. fan speed. In phase 1, the NO_x Short Term Strategy lowers the oxygen until CO becomes too high. In phase 2, the NO_x Long Term Strategy starts to increase the kiln feed until maximum feed of 88 t/h has been reached. In phase 3, oil is reduced in ON/OFF cooperation with STS until NO_x is below 800 mg/Nm³, where LTS prepares the kiln for a possible upset by increasing oil and reducing kiln feed. Around 7:00 a.m., the NO_x Short Term Strategy is active again for a short period.

Weitergehende Rauchgasreinigung für den Einsatz von Trockenklärschlamm als Alternativ-Brennstoff bei der „Holderbank" Cement und Beton, Siggenthal*)

Extensive flue gas cleaning when using dry sewage sludge as an alternative fuel at the "Holderbank" Cement und Beton, Siggenthal*)

Epuration des gaz de fumée encore améliorée pour l'utilisation de boue de décantation séchée comme combustible alternatif chez „Holderbank" Cement und Beton, Siggenthal

Depuración mejorada de los gases de humo, en relación con el empleo de los lodos secos como combustible alternativo en „Holderbank" Cement und Beton, Siggenthal

Von **B. de Quervain**, Siggenthal/Schweiz

Zusammenfassung – Das in der Zementindustrie weltweit erstmalig zum Einsatz kommende Aktiv-Koks-Festbett-Filter „POLVITEC" von Krupp Polysius, erlaubt die umweltfreundliche Nutzung der Verbrennungsenergie aus Trockenklärschlamm. Dabei werden flüchtige Schwermetalle wie z.B. Quecksilber zurückgehalten. Der hohe Ascheanteil im TKS dient als Ersatz von Rohmaterial für die Zementherstellung. Die Inbetriebnahme der Anlagen ist für das erste Quartal 1994 vorgesehen, nachdem die Baubewilligung im Mai 1992 erteilt wurde und mit dem Bau im August 1992 begonnen werden konnte. Aufgrund der Planungsvorgaben wird erwartet, daß die SO_2-Emission um das 7fache, die NO_x-Emission um das 4fache vermindert werden können. Die zur Verbrennung gelangenden 25000 t Trockenklärschlamm werden jährlich 8000 t Kohle und 12000 t Rohmaterial substituieren sowie den globalen Ausstoß an CO_2 um ca. 25000 t verringern. Die Investitions- und Betriebskosten sollen aus Entsorgungsgebühren finanziert werden.

Weitergehende Rauchgasreinigung für den Einsatz von Trockenklärschlamm als Alternativ-Brennstoff bei der „Holderbank" Cement und Beton, Siggenthal

Summary – The "POLVITEC" activated coke fixed-bed filter from Krupp Polysius used for the first time in the world in the cement industry allows the combustion energy from dry sewage sludge to be utilized without harm to the environment. It retains volatile heavy metals like mercury. The high ash content in the dry sewage sludge acts as raw material replacement in the production of cement. The plant is due to be commissioned in the first quarter of 1994 after the building permit had been granted in May 1992 and building started in August 1992. From the design data it is expected that the SO_2 emission can be reduced by a factor of 7 and the NO_x emission by a factor of 4. The 25000 t dry sewage sludge used for combustion will replace 8000 t coal and 12000 t raw material annually and reduce the global emission of CO_2 by about 25000 t. The capital and operating costs are to be financed from disposal charges.

Extensive flue gas cleaning when using dry sewage sludge as an alternative fuel at the "Holderbank" Cement und Beton, Siggenthal

Résumé – Le filtre POLVITEC à lit solide de coke actif, utilisé pour la première fois au monde dans l'industrie cimentière, permet l'emploi conforme à l'environnement de l'énergie thermique de la boue de décantation. Dans ce procédé, des métaux lourds comme p. ex. mercure, sont captés. La forte teneur en cendres de la boue remplace des matières premières pour la production du ciment. La mise en service des installations est prévue pour le premier trimestre 1994; le permis de construire a été délivré en Mai 1992 et la construction a pu être lancée en Août 1992. Compte tenu des bases du projet, il est prévu de réduire de 7 fois l'émission SO_2 et de 4 fois l'émission NO_x. Les 25000 t de boue de décantation à brûler substitueront chaque année 8000 t de charbon et 12000 t de matière première et réduiront l'émission de CO_2 d'environ 25000 t. Les dépenses d'investissement et d'exploitation doivent être financées par des contributions d'élimination.

Epuration des gaz de fumée encore améliorée pour l'utilisation de boue de décantation séchée comme combustible alternatif chez „Holderbank" Cement und Beton, Siggenthal

---

*) Früher/formerly: Portland-Cement-Werk Würenlingen-Siggenthal AG

Resumen – *El filtro de lecho sólido, de coque activo, „POLVITEC", fabricado por Krupp Polysius y empleado por primera vez en la industria mundial del cemento, permite utilizar la energía procedente de la incineración de lodos activos secos, sin afectar al medio ambiente. En este proceso, los metales pesados volátiles, tales como el mercurio, son retenidos. El elevado contenido de cenizas en los lodos activos secos sirve para sustituir a las materias primas para la fabricación del cemento. La puesta en servicio de la instalación está prevista para el primer trimestre de 1994, una vez que la autorización para la construcción fue concedida en mayo de 1992 y que se iniciaron los trabajos en agosto de 1992. De acuerdo con los valores prefijados, se cuenta con que las emisiones de SO_2 resulten 7 veces más bajas y las de NO_x, 4 veces más bajas. Las 25000 t de lodos activos secos que hay que quemar permitirán sustituir 8 000 t de carbón y 12000 t de materias primas al año, reduciendo la emisión global de CO_2 en 25 000 t aprox. Los gastos de inversión y de explotación serán financiados con tasas cobradas por la eliminación de residuos.*

Depuración mejorada de los gases de humo, en relación con el empleo de los lodos secos como combustible alternativo en „Holderbank" Cement und Beton, Siggenthal

1. Einleitung

Am Beispiel der in **Bild 1** dargestellten Verteilungskurve der NO_x-Emission der vom Verein Deutscher Zementwerke (VDZ) untersuchten Werke wird deutlich, welcher Handlungsbedarf zur Stickoxidminderung für „Holderbank" Cement und Beton, Siggenthal, früher Portland-Cement-Werk Würenlingen-Siggenthal AG (PCW), besteht. Die SO_2-Emissionssituation, verursacht durch pyritischen Schwefel aus dem Rohmaterial, verlangt ebenso drastische Minderungsmaßnahmen. Herkömmliche Techniken zur Emissionsminderung hätten Investitionen in der Größenordnung von 7 Mio Sfr. erfordert.

Bei der zu fällenden Entscheidung für eine neue Abgasreinigung muß der Standort des Zementwerkes in der Gemeinde Würenlingen miteinbezogen werden. So sind auf dem Gemeindegebiet eine grundwasserverschmutzende Altlast, ein Zwischenlager für radioaktive Abfälle, ein Reaktorforschungsinstitut sowie im Umkreis von nur wenigen Kilometern auch noch zwei Kernkraftwerke zu finden. Insgesamt kam daher bei Einsatz von Trockenklärschlamm als Brennstoff nur eine Lösung in Betracht, bei der über die Entsorgungsgebühren die erforderliche Rauchgasreinigungsanlage mitfinanziert werden sollte.

Mit neuer Abgasreinigungstechnologie ist eine Änderung der Emissionssituation, wie im **Bild 2** dargestellt, zu erwarten. Die Emissionen an Schwefeldioxid werden um das 7fache, jene der Stickoxide um das 4fache reduziert. Minderungen ergeben sich darüber hinaus beim Ammoniak sowie beim Staub.

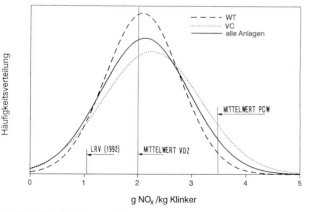

BILD 1: Häufigkeitsverteilung der NO_x-Emissionen
FIGURE 1: Frequency distribution of NO_x emissions

Häufigkeitsverteilung	= Frequency distribution
Klinker	= Clinker
LRV	= Clean Air Order
Mittelwert VDZ	= mean value – VDZ study
Mittelwert PCW	= mean value – PCW
WT	= preheater
VC	= precalciner
alle Anlagen	= all plants

1. Introduction

From the example of the distribution curve of the NO_x emissions from the works investigated by the German Cement Works' Association (VDZ) shown in **Fig. 1**, it is clear that "Holderbank" Cement und Beton, Siggenthal, formerly Portland-Cement-Werk Würenlingen-Siggenthal AG (PCW), needs to take action to reduce its nitrogen oxide emissions. The SO_2 emission situation, caused by pyritic sulphur from the raw material, demands equally drastic control measures. Conventional emission control technologies would have necessitated capital expenditure of the order of SFr. 7 million.

The location of the cement works in the municipality of Würenlingen must be taken into account in the decision to be taken on a new flue gas cleaning system. Within the municipality's boundaries there are a contaminated site

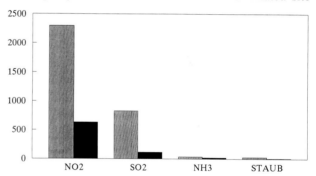

BILD 2: Änderung der Emissionssituation bei Einsatz eines Aktivkoksfilters
FIGURE 2: Changes in emission levels with the use of an activated coke fixed-bed filter

Staub	= dust
heutige/zukünftige Emissionen (t/Jahr)	= Present/future emission levels (t/yr)

which causes groundwater pollution, an intermediate store for radioactive waste, a nuclear research institute and, within a radius of only a few kilometres, two nuclear power stations. Overall, therefore, the only feasible solution using dry sewage sludge as fuel was one in which the disposal fees would help to fund the necessary flue gas cleaning unit.

The new flue gas cleaning technology is expected to produce a change in the emission situation as shown in **Fig. 2**. Sulphur dioxide emissions will be reduced by a factor of 7 and nitrogen oxide emissions by a factor of 4. Emissions of ammonia and dust will also be cut.

2. Implementation of the project

Pilot tests lasting about six months from the beginning of 1990 laid the foundations for a flue gas cleaning system based on activated coke. During these tests it turned out that

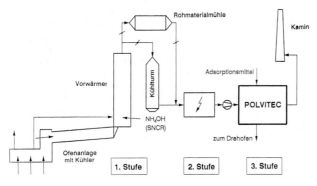

BILD 3: 3stufige Abgasreinigung bei der Zementherstellung
FIGURE 3: Three-stage flue gas cleaning system in cement production

Ofenanlage mit Kühler	=	Kiln system with cooler
Vorwärmer	=	Preheater
Kühlturm	=	Cooling tower
Rohmaterialmühle	=	Raw mill
Adsorptionsmittel	=	Adsorbent
zum Drehofen	=	to rotary kiln
Kamin	=	Stack

(underneath diagram:)
1st stage: NO_x reduction by the SNCR method
2nd stage: Dust collection and mercury retention in ESP
3rd stage: Adsorption of pollutants (SO_2, heavy metals, organics, etc.) by POLVITEC

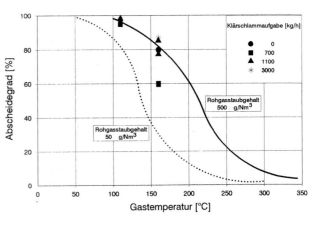

BILD 5: Quecksilberadsorption an Ofengasstaub
FIGURE 5: Adsorption of mercury on kiln flue gas dust in terms of mercury output

Abscheidegrad	=	Collection efficiency
Rohgasstaubgehalt	=	Raw gas dust content
Klärschlammaufgabe	=	Sewage sludge feed rate
Gastemperatur	=	Gas temperature

BILD 4: Abscheideleistung des POLVITEC-Verfahrens

Schwefeldioxidreduktion auf:	< 130 mg/m$^3_{n,tr}$
Ammoniakreduktion auf:	< 10 mg/m$^3_{n,tr}$
Quecksilber:	
Messung Pilotanlage (1 mg Hg/m^3 Zugabe)	99 % ± 1 % Abscheidung
Messung Labor (2 mg Hg/m^3 Zugabe)	99,9 % ± 1 % Abscheidung
Dioxine und Furane:	
Literatur:	Sichere Einhaltung bzw. Unterschreitung des BImSchG-Grenzwertes 0,1 ng/m^3 Toxizitätsäquivalent nach BGA
Staubreduktion:	
Literatur:	< 10 mg/m$^3_{n,tr}$
Messungen Pilotanlage:	< 1 mg/m$^3_{n,tr}$

FIGURE 4: Collection efficiencies of the POLVITEC process

Sulphur dioxide reduced to:	< 130 mg/m^3 (NTP, dry)
Ammonia reduced to:	< 10 mg/m^3 (NTP, dry)
Mercury:	
Pilot plant measurement (1 mg Hg/m^3 addition)	99 % ± 1 % removal
Laboratory measurement (2 mg Hg/m^3 addition)	99.9 % ± 1 % removal
Dioxins and furans:	
Literature:	Values safely comply with or lie below the limit of 0.1 ng/m^3 toxicity equivalent according to Federal Health Office stipulated by Federal Control of Pollution Act
Dust reduction:	
Literature:	< 10 mg/m^3 (NTP, dry)
Pilot plant measurements:	< 1 mg/m^3 (NTP, dry)

2. Realisierung des Projektes

Mit Pilotversuchen, die sich Anfang 1990 über eine Zeitdauer von ca. 6 Monaten erstreckten, wurden die Grundlagen für eine Abgasreinigung mittels Aktivkoks erarbeitet. Dabei zeigte sich, daß der Koksreaktor neben dem Schwefeldioxid und den Ammoniumverbindungen, die zum einen Teil aus dem Rohmaterial, zum anderen aus der SNCR-Entstickungsstufe im unteren Teil des Wärmetauschers stammen, auch Quecksilber adsorbiert. **Bild 3** zeigt schematisch die 3stufige Rauchgasreinigung, wobei der zweiten Stufe, dem konventionellen Elektrofilter, die Aufgabe der Quecksilbersenke zukommt. **Bild 4** zeigt zusammenfassend die angestrebte Abscheideleistung des Koksreaktors. Nach Erreichen der Grenzbeladung im Koks, die durch die Adsorption von SO_2 bestimmt wird, wird dieser ausgeschleust und im Ofen verbrannt. Das abgeschiedene SO_2 aus dem Rauchgas reagiert mit dem im Ofen vorhandenen Calciumoxid zu Gips. Die übrigen Schadstoffe werden (mit Ausnahme des Quecksilbers) thermisch zersetzt. Quecksilber gelangt zurück in den Koksreaktor und baut einen Kreislauf auf. Untersuchungen bei „Holderbank" Cement und Beton, Siggenthal, haben die Angaben in der Literatur bestätigt, wonach das Quecksilber am E-Filterstaub kondensiert. **Bild 5** zeigt die Hg-Adsorption in Abhängigkeit von der Rauchgastemperatur. Wird nun ein Teil des E-Filter-

the coke reactor adsorbs not only sulphur dioxide and the ammonium compounds which originate on the one hand from the raw material and on the other from the SNCR nitrogen oxide removal stage in the lower part of the preheater, but also mercury. **Fig. 3** shows a schematic diagram of the three-stage flue gas cleaning system, in which the second stage, the conventional electrostatic precipitator, functions as the "mercury sink". **Fig. 4** summarises the target collection efficiency of the coke reactor. When the coke has reached its maximum contamination level, which is determined by the adsorption of SO_2, it is discharged and burnt in the kiln. The SO_2 removed from the flue gas reacts with the calcium oxide present in the kiln to form gypsum. The remaining pollutants are thermally decomposed, except for the mercury, which returns to the coke reactor and forms a recirculating system. Investigations at "Holderbank" Cement und Beton, Siggenthal, have confirmed the data in the literature, according to which the mercury condenses on the dust in the precipitator. **Fig. 5** illustrates the adsorption of mercury as a function of the flue gas temperature. An effective "sink" is created if part of the dust collected in the precipitator is then discharged and conveyed to the cement mill. If the mercury contamination rises above a predetermined level then thermal treatment of the dust in an indirect kiln is an option.

BILD 6: Projektkosten Rauchgasreinigung/Klärschlammverbrennung

Projektkosten	M SFr
Vertrag Konsortium: Polysius – Meto Bau	29,4
PCW Eigenleistungen	6,8
Total Projektkosten	36,2
Betriebsmittelkosten	**SFr/t Klinker**
Koks (Heizwert korrigiert)	1,75
Ammoniak	1,23
Strom	1,15
Total Betriebsmittelkosten	4,13

staubes ausgeschleust und der Zementmühle zugegeben, entsteht eine wirkungsvolle Senke. Sollten die Frachten an Quecksilber über ein vorgegebenes Maß ansteigen, so besteht die Möglichkeit, den Staub in einem Indirektofen thermisch zu behandeln.

3. Kosten-Aspekte

Bild 6 zeigt die Kosten für Investitionen sowie die Betriebskosten, wie sie dem Projekt zugrunde liegen. Da dies ein Mehrfaches an Investitionskosten bedeutet, als für die Minimallösung aufgewendet werden müßte, willigten die beteiligten Instanzen ein, daß hier Sonderbrennstoffe, wie z. B. Trocken-Klärschlamm (TSK) eingesetzt werden dürfen. Ihre Entsorgungskosten helfen die Investitionen tragen. Im **Bild 7** sind die Kosten für die heute üblichen Verfahren der Klärschlammentsorgung dargestellt. Bei „Holderbank“ Cement und Beton, Siggenthal, werden ca. Sfr. 400/t veranschlagt, wobei der Abgeber die Möglichkeit hat, durch eine anteilsmäßige Mitfinanzierung der Rauchgasreinigungsanlage die Entsorgungskosten auf etwa die Hälfte zu reduzieren. Die Stadt Zürich hat soeben Sfr. 9,5 Mio. bewilligt und wird ab 1994 jährlich 7 000 Tonnen TKS zur thermischen und stofflichen Verwertung liefern. Die Grundsätze dieser Kostenbeteiligung lassen sich wie folgt darstellen:

- PCW finanziert die Rauchgasreinigung, die von der LRV gefordert wird.
- TKS-Lieferanten übernehmen die Kosten, die durch den Einsatz von TKS entstehen (Bahn/Straßen-Entladung, Silierung, Transport).
- TKS-Lieferanten finanzieren die zusätzliche Reinigungsstufe (POLVITEC).

BILD 7: Geschätzte Kosten der Klärschlammentsorgung in Deutschland (Basis 1991)

Entsorgungsverfahren	Kosten (DM/t TS)
Verwerten von Naßschlamm ohne zusätzliche Lagerung	150– 400
Verwerten von Naßschlamm mit zusätzlicher Lagerung	200– 600
Verwerten von entwässertem Schlamm mit zusätzlicher Lagerung	350–1000
Deponieren von entwässertem Schlamm (35 % TS)	500–1000
Verbrennen von getrocknetem Schlamm + Deponierung der Asche	900–2000

TABELLE 1: Hauptbestandteile von Zementklinker im Vergleich mit Klärschlammasche

	Zementklinker (%)	Klärschlammasche (% bez. Asche)
SiO_2	21–24	30–49
Al_2O_3	4,5– 6	8–15
Fe_2O_3	3– 4	5–23
CaO	64–66	9–22
MgO	1,5	1– 2

FIGURE 6: Cost of flue gas cleaning/sewage sludge burning

Project costs	M SFr
Contract consortium: Polysius – Meto Bau	29.4
Work by PCW in-house	6.8
Total project costs	36.2
Material and utility costs	**SFr/t clinker**
Coke (calorific value adjusted)	1.75
Ammonia	1.23
Electricity	1.15
Total material and utility costs	4.13

3. Cost aspects

Fig. 6 illustrates the capital expenditure costs and the running costs on which the project is based. Since this means capital expenditure of several times what would be required for a minimum solution, the relevant authorities agreed that special fuels such as dry sewage sludge (DSS) could be used here. The disposal fees received help to meet the capital expenditure costs. The costs of the current conventional methods of sewage sludge disposal are shown in **Fig. 7**. At “Holderbank” Cement und Beton, Siggenthal, the cost is put at approx. SFr. 400/t; the provider has the option of reducing the disposal charges by about 50 % through a pro rata contribution to the cost of the flue gas cleaning system. The city of Zurich has just approved expenditure of SFr. 9.5 million and from 1994 will deliver 7 000 tonnes/year of DSS for thermal processing and resource recovery. The principles of this cost-sharing are as follows:

- PCW finances the flue gas cleaning system required by clean air legislation.
- The providers of DSS bear the costs arising from the use of DSS (rail/road unloading, storage, transport).
- The providers of DSS finance the additional cleaning stage (POLVITEC).

4. Resource recovery

Besides the recirculation of sulphur, there are other important effects. **Table 1** shows the composition of the DSS ash compared with cement clinker. It is combined in the clinker. It is therefore appropriate to speak here of “resource recovery”. The fact that, with CO_2-neutral combustion of the

FIGURE 7: Estimated costs for sewage sludge disposal in Germany (1991)

Disposal method	Cost (DM/t DS)
Utilisation of wet sludge without additional storage	150– 400
Utilisation of wet sludge with additional storage	200– 600
Utilisation of dewatered sludge with additional storage	350–1000
Landfilling of dewatered sludge (35% DS)	500–1000
Incineration of dried sludge + landfilling of the ash	900–2000

TABLE 1: Main constituents of cement clinker compared with sewage sludge ash

	cement clinker (%)	Sewage sludge ash (% content in ash)
SiO_2	21–24	30–49
Al_2O_3	4.5– 6	8–15
Fe_2O_3	3– 4	5–23
CaO	64–66	9–22
MgO	1.5	1– 2

4. Stoffliche Verwertung

Neben dem Kreislauf des Schwefels gibt es andere Effekte, die von Bedeutung sind. **Tabelle 1** zeigt die Zusammensetzung der TKS-Asche im Vergleich zu Zementklinker. Sie wird im Klinker gebunden. Daher kann hier von einer stofflichen Wiederverwertung gesprochen werden. Daß bei der CO_2-neutralen Verbrennung der organischen Bestandteile des TKS auch noch jährlich 25 000 Tonnen weniger von diesem Treibhausgas emittiert werden, ist für das Unternehmen von ganz besonderer Bedeutung.

Mit dieser Rauchgasreinigung wird der Einsatz einer ganzen Reihe von weiteren Reststoffen ermöglicht. So könnten PAK- und Mineralöl-verunreinigte Böden unter der Voraussetzung der mineralogischen Eignung als Rohstoffsubstitute eingesetzt werden. Ihre dosierte Aufgabe zur Rohmühle verursacht keine Emissionen mehr an Kohlenwasserstoffen, die über den Kamin in die Atmosphäre gelangen. Auch der Einsatz von kontaminiertem Wasser im Kühlturm kann nun einen ökologisch sinnvollen Beitrag im Bereich der Entsorgung liefern. Der Aktivkoksfilter sorgt dafür, daß problematischen Schadstoffen der Weg ins Freie versperrt bleibt und sie zur thermischen Zersetzung in den Hochtemperaturbereich des Ofens gelangen.

Die Ziele des „Aktionsprogrammes Energie 2000" lassen sich wie folgt zusammenfassen:

- Stabilisierung des Gesamtverbrauchs von fossiler Energie und der CO_2-Emission zwischen 1990 und 2000,
- zunehmende Dämpfung der Verbrauchszunahme von Elektrizität und Stabilisierung der Nachfrage ab 2000,
- Beiträge der erneuerbaren Energien im Jahre 2000: 0,5 % zur Stromerzeugung und 3 % des Verbrauchs fossiler Energien als Wärme,
- Ausbau der Wasserkraft um 5 % und der Leistung der KKW um 10 %.

Eine Argumentation gegen das Projekt wird vor diesem Hintergrund wohl schwierig werden. So haben auch alle unsere Informationsrunden mit der Bevölkerung, den Behörden sowie auch den Umweltschutzorganisationen ein positives Bild hinterlassen.

organic constituents of the DSS, the emissions of this greenhouse gas are also reduced by 25 000 tonnes per annum is extremely important for the firm.

This flue gas cleaning system also enables a whole range of other waste materials to be used. For instance, PAH- and petroleum-contaminated soils can be used as raw material substitutes, provided they are mineralogically suitable. Their controlled feeding to the raw grinding mill does not cause any increased emission of hydrocarbons discharged to atmosphere via the stack. Even the use of contaminated water in the cooling tower can now make an ecologically worthwhile contribution in the field of waste management. The activated coke filter ensures that troublesome pollutants are prevented from escaping into the atmosphere and that they enter the high-temperature zone of the kiln for thermal decomposition.

The goals of the "Energy 2000 programme" can be summarised as follows:

- stabilisation of total fossil energy consumption and CO_2 emissions between 1990 and 2000,
- an increasing slowdown in the growth of electricity consumption and stabilisation of demand from the year 2000,
- contributions by renewable energy sources in the year 2000: 0.5 % to power generation and 3 % of the consumption of fossil fuels as heat,
- expansion of hydroelectric power by 5 % and the output of nuclear power stations by 10 %.

Against this background, it will be quite hard to argue against the project, and all our briefing meetings with the public and the authorities as well as with the environmental organisations have left a positive impression.

Programm zur Umweltverbesserung bei einem Zementwerk im Stadtgebiet von Pretoria

Environmental improvement programme at a cement plant in the Pretoria urban area

Programme visant à l'amélioration des normes antipollution dans une cimenterie dans la banlieue de Prétoria

Programa medioambiental de una fábrica de cemento situada en el término municipal de Pretoria

Von **J. A. Riekert, A. D. Doornum** und **J. M. Gaylard,** Pretoria/Südafrika

Zusammenfassung – Das Hercules-Werk von PPC in Pretoria lag außerhalb der Stadtgrenze, als 1892 die Produktion aufgenommen wurde. Heute liegt das Werk, das eine Kapazität von 640000 Tonnen pro Jahr hat, mitten im Stadtgebiet, und der Umweltschutz ist deshalb zu einem Schlüsselfaktor geworden. Im Verlauf der letzten vier Jahre wurde ein umfangreiches Programm zur Sanierung des Werksumfeldes und zur Quantifizierung der vom Werk und anderen Quellen verursachten Staubimmission durchgeführt. Ein Netz von zweiseitig ausgerichteten Probenahmegeräten, die von der Universität von Witwatersrand entwickelt wurden, wurde errichtet und erstreckte sich stellenweise über eine Entfernung von bis zu 2 km vom Werk. Im Werk wurde das gesamte System des innerbetrieblichen Stoffumschlags und der Prozesse eingehend überprüft. Auf dieser Grundlage entstand ein Umweltsanierungsplan, der sich auf eine Verbesserung der betrieblichen Verfahrensweisen für einen stabilen Betrieb, zusätzliche Entstaubungsanlagen und eine Ausweitung der gepflasterten und der als Grünanlagen gestalteten Bereiche erstreckte. Die Umsetzung dieses Programms hat zur stetigen Abnahme der Staubimmission auf Werte nahe den natürlichen Staubwerten geführt. Das 36 ha große Werksgelände mit ausgedehnten Gärten und Wiesen ist so zu einer Grünfläche und zu einer umweltfreundlichen Anlage der Stadt Pretoria geworden.

Programm zur Umweltverbesserung bei einem Zementwerk im Stadtgebiet von Pretoria

Summary – The Hercules plant of PPC in Pretoria was outside the city limits when production started in 1892. Today the plant, with a capacity of 640000 tons per year, is surrounded by the city and plant environmental control is therefore a key factor. During the past four years an intensive programme has been followed to upgrade the plant environment and to quantify dust fallout from the plant and other sources. A network of bi-directional samplers, designed by the University of the Witwatersrand, was set up, encircling the factory at distances up to 2 km. In the plant, a close examination of the entire material handling and processing system was made. From this an environmental upgrade plan was formulated covering refinement of the operating procedures to promote stable operation, additional dust collection equipment and extension of the grassed, paved, and planted areas. The implementation of this programme has resulted in a steady decrease in dust fallout to levels which are close to background levels. The 36 hectare factory site with extensive gardens and grassed areas has effectively become a green area and an environmental asset to the city of Pretoria.

Environmental improvement programme at a cement plant in the Pretoria urban area

Résumé – L'usine Hercules de PPC à Prétoria était située hors des limites de la ville quand la production fut mise en service en 1892. Aujourd'hui l'usine qui a une capacité annuelle de 640000 t, est au centre de la ville, et la protection de l'environnement est de ce fait devenue d'une importance capitale. Au cours des quatre dernières années, un vaste programme d'assainissement de l'environnement de l'usine et de quantification des émissions de poussière provenent de l'usine mais aussi d'autres sources, a été lancé. Un réseau d'appareils pour prélèvements d'échantillons établis bilatéralement et mis au point par l'Université de Witwatersrand, a été mis en place, par endroits, sur une distance atteignant 2 km à partir de l'usine. Et dans l'usine, l'ensemble du système des transferts de matières internes et de process a fait l'objet d'une étude approfondie. C'est sur ces bases qu'a été établi un plan d'assainissement, comprenant à la fois une amélioration des procédures d'exploitation en vue d'un fonctionnement stable, des installations de dépoussiérage complémentaires et une extension des zones dallées et espaces verts. La concrétisation de ce programme a entraîné une diminution constante des émissions de poussière, ramenées à des valeurs avoisinant les valeurs naturelles. Le terrain de l'usine couvrant une superficie de 36 ha constitué de vastes jardins et de prés est devenu un espace vert et une installation respectueuse de l'environnement à l'intérieur de la ville de Prétoria.

Programme visant à l'amélioration des normes antipollution dans une cimenterie dans la banlieue de Prétoria

Resumen – La planta Hercules de PPC en Pretoria se encontraba fuera del casco urbano, cuando se inició la producción en 1892. Actualmente, la fábrica, que tiene una capacidad de producción de 640.000 t anuales, queda en el centro de la ciudad, por lo que la protección del medio ambiente se ha convertido en un factor clave. En los últimos 4 años, se ha llevado a cabo un amplio programa de saneamiento del entorno de la fábrica y de determinación cuantitativa de las emisiones de polvo, caudadas por la mencionada fábrica y por otras fuentes de contaminación. Se ha instalado una red de tomamuestras bidireccionales, desarrollados por la Universidad de Witwatersrand, que se extiende en algunos puntos hasta una distancia de 2 km de la fábrica. Dentro de la fábrica, se ha sometido a revisión todo el sistema de procesos y de manutención interna de materiales. En base a estos trabajos, se ha establecido un plan de saneamiento del medio ambiente que abarca la mejora de los procesos operativos necesarios para un servicio estable, instalaciones de desempolvado adicionales y una ampliación de las áreas pavimentadas y de las zonas verdes. La realización de este programa ha conducido a una constante disminución de las emisiones de polvo, alcanzándose valores próximos a los valores naturales. La fábrica, que ocupa una superficie de 36 ha dentro de la ciudad de Pretoria, se ha convertido, con sus extensos jardines y prados, no solamente en una zona verde, sino también en una instalación inocua para el medio ambiente.

Programa medioambiental de una fábrica de cemento situada en el término municipal de Pretoria

1. Einleitung

Als die Produktion vor 100 Jahren aufgenommen wurde, lag das Hercules-Werk von Pretoria Portland Cement Company Ltd. (PPC) außerhalb der Stadtgrenzen von Pretoria. Heute ist das Werk von den nordwestlichen Vorstädten Pretorias umgeben. In der Nähe liegen aber auch noch landwirtschaftlich genutzte Flächen. Zwei kohlengefeuerte Öfen von F. L. Smidth, die nach dem Trockenverfahren arbeiten, sind in Betrieb:

– Ofen 4 mit 1stufigem Vorwärmer mit 720 t/d (Baujahr 1966)

– Ofen 5 mit 4stufigem Vorwärmer mit 1250 t/d (Baujahr 1974).

Eine einzige Rohmühle liefert das Rohmehl für beide Öfen. Sie kann die Abgase beider Öfen nutzen. Die Abgase der zwei Öfen werden in einem gemeinsamen Kamin abgeleitet.

Vor rund vier Jahren wurde im Zusammenhang mit zunehmendem Umweltbewußtsein ein umfangreiches Programm zur Sanierung des Werksumfeldes durchgeführt.

2. Staubsammeleinrichtung

Um Veränderungen zu messen und den vom Hercules-Werk und anderen Quellen in der Nähe verursachten Staub zu quantifizieren, wurde ein Netz von Probenahmegeräten in einer Entfernung von bis zu 2 km errichtet. **Bild 1** zeigt die Standorte. Die Standorte der Probenahmegeräte wurden nach Zugänglichkeit und Sicherheit entsprechend den folgenden Gesichtspunkten gewählt:

Haus: An einem privaten Wohnhaus im Vorort „Pretoria Gardens".

Schule: Auf dem Gelände der Hillsview High School.

Umspannwerk und Klärwerk: Städtisches Gelände der Stadt Pretoria.

Werksgelände: Auf dem Hercules-Werksgelände.

Capital Park: Ein Wohnheim der Eisenbahnverwaltung im Vorort „Capital Park".

Signalhof: Auf Eisenbahngelände.

Mutual Park: In einem Gewerbepark.

Doppelprobenahmegeräte wurden von der Universität Witwatersrand in Johannesburg für normale Messungen von Staubniederschlag entwickelt und gebaut [1, 2]. Die Probenahmegeräte haben zwei Gefäße zur Aufnahme des Staubniederschlags und einen Abdeckmechanismus, der durch eine Wetterfahne aktiviert wird. In Abhängigkeit von der Windrichtung ist jeweils nur ein Gefäß geöffnet. Den Aufbau des Probenahmegerätes zeigt **Bild 2**. Probenahmegeräte dieser Art wurden zur Überwachung der Staubentwicklung bei der Sanierung von Goldschlammdeponien der Crown-Minen nahe dem Geschäftsviertel von Johannesburg verwendet.

1. Introduction

When production started 100 years ago the Hercules plant of PPC was situated outside the city limits of Pretoria. Today the plant is surrounded by the north western suburbs of the city but still has a small area of agricultural land nearby. Two dry, coal-fired kilns, supplied by F. L. Smidth, are in operation:

Kiln 4   1-stage-preheater   720 tons/day (1966)
Kiln 5   4-stage preheater   1250 tons/day (1974).

A single raw mill provides meal for both kilns and can operate using exhaust gas from either kiln. A common kiln stack serves both kilns.

Some four years ago, in line with increased environmental awareness, an intensive programme to upgrade the plant environment was started.

2. Dust sampling system

To measure progress and to quantify dust arising from the Hercules plant and from other sources in the area, a network of samplers was installed at distances up to 2 km and located as shown in **Fig. 1**. The sampler sites were selected taking account of accessibility and security, as follows:

House site: At a private home in the suburb of Pretoria Gardens.

School site: On the premises of Hillsview High School.

Sub-station and sewage work sites: On property owned by the Pretoria Municipality.

Factory site: At the Hercules cement plant.

Capital Park site: At a hostel owned by the Railways Administration in the suburb of Capital Park.

Signal Yard site: On railway property.

Mutual park site: In an industrial business park.

The bi-directional samplers were designed and built by Witwatersrand University in Johannesburg, in line with standard practice for particle fallout sampling [1–2]. The samplers have two buckets to collect fallout dust and have a cover mechanism activated by a wind vane. Only one bucket, depending on wind direction, is open at a time. The sampler arrangement is illustrated in **Fig. 2**. Samplers of this type were used to monitor dust arising from the reclamation of the Crown Mines gold tailings dumps near the Johannesburg central business district.

The time during which each bucket is open is recorded, enabling the average rate of dust fallout arising from the direction of the plant and from all other directions to be determined. Changes in wind direction resulting in the lid mechanism moving are also recorded.

The buckets initially contain 2 litres of distilled water to prevent the loss of collected dust. At monthly intervals they are replaced with fresh buckets.

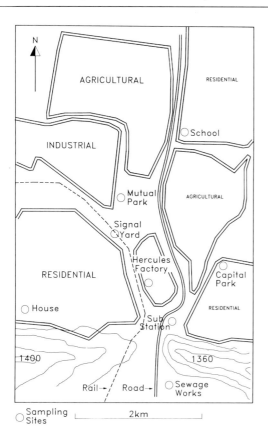

BILD 1: Flächenplan mit Lage des Werks und der Meßstellen sowie Nutzung durch Wohnen, Industrie und Landwirtschaft
FIGURE 1: Area plan indicating plant and sampler location and division into residential, industrial, and agricultural zones

Agricultural	= Landwirtschaft
Residential	= Wohnen
Industrial	= Industrie
School	= Schule
Signal yard	= Signalhof
Sub-station	= Umspannwerk
Rail	= Bahn
Road	= Straße
Sewage Works	= Klärwerk
Sampling sites	= Meßstelle

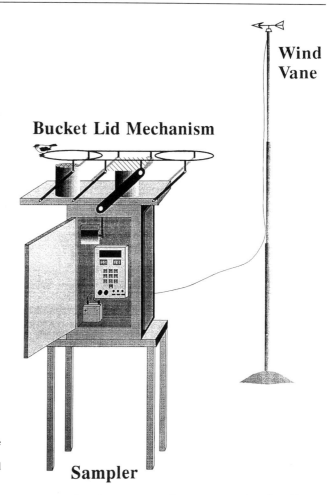

BILD 2: Probenahmegerät zur Erfassung von Staubniederschlag
FIGURE 2: Bi-directional sampler for collecting fallout dust

Wind Vane	= Wetterfahne
Bucket Lid Mechanism	= Mechanismus der Gefäßdeckel
Sampler	= Probenahmegerät

Die zeitliche Dauer, in der die beiden Gefäße geöffnet sind, wird aufgezeichnet. Dadurch kann der durchschnittliche Staubniederschlag aus Richtung des Werks und aus anderen Richtungen ermittelt werden. Jede Änderung der Windrichtung, die den Deckelmechanismus aktiviert, wird ebenfalls aufgezeichnet.

Anfänglich enthalten die Gefäße 2 l destilliertes Wasser, um Staubverluste zu verhindern. Die Gefäße werden im monatlichen Abstand gewechselt.

3. Probenbehandlung

Die Flüssigkeit in den Gefäßen wird zu 2 l aufgefüllt. Eine Teilprobe von 200 ml wird analysiert. Der Rückstand in den Gefäßen wird unter leichtem Vakuum abgesaugt, gewogen und mit Röntgenstrahlen analysiert. Aus der Masse und der Zusammensetzung des Staubniederschlags wird die Staubmenge aus Richtung des Zementwerks und aus anderen Quellen berechnet.

Der Staub hydratisiert in den Gefäßen. Die Zuordnung des Staubs kann dadurch erschwert werden. In den meisten Fällen ist der Gehalt an Calcium, Kieselsäure, Tonerde und Sulfat jedoch ausreichend, um Staub ggf. dem Zementwerk zuzuordnen.

4. Verbesserungen im Werk

Eine weitgehende Untersuchung der gesamten innerbetrieblichen Stoffströme und des Prozesses bildete die Grundlage eines Verbesserungsprogramms. Besonderer

3. Sample processing

The volume in the buckets is made up to 2 litres and a 200 ml sample of the clear solution is taken for analysis. The remaining bucket contents are filtered under low suction, and the residue weighed and analysed by X-ray techniques. From the mass and composition of the fallout dust thus obtained, the amount of dust arising from the direction of the cement plant and from other sources is calculated.

Hydration of dust occurs in the buckets and can present a problem in identifying dust sources. In most cases, however, the content of calcium, silica, alumina, and sulphate is sufficient to determine whether the dust source is the cement plant.

4. Plant improvement

A close examination of the entire material handling and process system formed the basis of the plant improvement programme. Emphasis was placed on equipment, operating procedures, and the general plant environment, as illustrated by the following:

– Modification of transfer points.

– Installation of additional dust collector units, an increase from 26 to 51.

– Upgrade drive on conditioning tower dust screw to eliminate trips.

– Upgrade of kiln 4 coal feed to achieve greater stability and reduce precipitator trips caused by O_2 and CO limits.

Wert wurde auf die Ausrüstung, die Vorgehensweisen beim Betrieb und die allgemeine Umweltsituation des Werks gelegt, wie im folgenden dargestellt wird:

- Modifizierung der Übergabestellen,
- Installation zusätzlicher Filter, die von 26 auf 51 erhöht wurden,
- Verbesserung des Schneckenantriebs im Verdampfungskühler zur Minderung von Undichtigkeiten,
- Modernisierung der Kohledosierung am Ofen 4 zur Verminderung von Leistungseinbrüchen des Staubabscheiders infolge von CO-Abschaltungen,
- Verhindern von Ansätzen in den Vorwärmerzyklonen durch Installation zusätzlicher Druckstoßgeräte und Veränderung der Falleitungen,
- Reduzierung der Anfahrzeit der Rohmühle zur Vermeidung kurzzeitiger Störungen im Betrieb des Staubabscheiders.

Aus der Studie ging ebenfalls hervor, daß Staub vom Werksgelände durch den Wind und Fahrzeuge aufgewirbelt wurde. Dem wurde durch ausgedehnte künstlich bewässerte Garten- und Rasenflächen begegnet, in die Bäume und Büsche gepflanzt wurden. Die geteerten Flächen wurden vergrößert; sie werden jetzt mit Kehrmaschinen gereinigt.

Insgesamt 90 % des 36 ha großen Werksgeländes sind jetzt umgestaltet. Davon entfallen auf geteerte Flächen 57 000 m² und 40 000 m² auf Rasen.

5. Ergebnisse

Die Ergebnisse der Niederschlagsmessungen sind in **Tabelle 1** wiedergegeben. Während des ersten Zeitraumes wurden an den Entnahmestellen Umspannwerk und Klärwerk abnorm hohe Staubniederschläge gemessen. Eine detaillierte Untersuchung zeigte einen häufigen Wechsel der Windrichtung an beiden Stellen infolge einer nahe gelegenen Straße zwischen zwei Hügeln. Diese beiden Entnahmegeräte wurden umgesetzt. Aus den Zahlen der Tabelle ergibt sich ein Hintergrund- oder minimaler Staubniederschlag von 0,2–0,4 g/m² pro Tag.

In **Bild 3** ist das mittlere Niederschlagsverhältnis (Hercules/andere Richtungen) für die Jahre 1991 und 1992 in Halbjahreszeiträumen dargestellt. Aus dem Bild geht ein allgemeiner Abwärtstrend hervor. Die Ergebnisse an den Meßstellen Capital Park, Mutual Park und Schule (1992) wurden um einen Bodenanteil in der Probe korrigiert. Ein hoher Anteil an Silizium (> 40 %) und Tonerde (> 10 %) in Verbindung mit wenig Calcium (< 10 %) im gesammelten Staub zeigen Erdreich an. Dagegen würden Klinker, Zement oder Rohmehl durch eine entgegengesetzte Zusammensetzung angezeigt.

- Elimination of preheater cyclone chokes by installing additional air blasters and modifying cyclone discharge pipes.
- Reduction in raw mill start-up time to eliminate a short-term disturbance in precipitator operation.

The study also showed that dust settling in the plant area was being re-entrained by wind and vehicle movement. This was countered by establishing extensive irrigated garden and grassed areas, and by planting trees and shrubs. Paved areas were extended and are cleaned by mechanical sweepers.

A total of 90 % of the factory site of 36 ha is now covered including paving (57 000 m²) and grass (40 000 m²).

5. Results

Results from the fallout sampling programme are set out in **Table 1**. During the first period, abnormally high dust fallout was measured by the samplers at the Sub-station and Sewage works sites. Detailed examination showed a high frequency of wind changes at both sites brought on by their proximity to a road passing between two hills. These two samplers were then re-positioned at other sites. From figures in the table, a background of minimum dust fallout level of 0.2–0.4 g/m² per day is indicated.

In **Fig. 3**, the average fallout ratio (Hercules/other directions) is illustrated for the years 1991 and 1992, in six month periods, and shows a general downward trend. The results for 1992 for the Capital Park, Mutual Park, and School sites have been corrected for the presence of soil in the collected dust by means of a chemical mass balance. High silica (>40 %) and alumina (>10 %), together with low calcium (<10 %) in the collected dust indicate soil, whilst the reverse would indicate the presence of clinker, cement, or raw meal. Industrial activities close to the Hercules cement plant include a gypsum board manufacturer and a steel scrap merchant. Dust from these sources would clearly affect the composition of the collected dust. There is also some agricultural activity in the area, and through analytical work, it has been possible to prove that some of the fallout initially attributed to the cement plant originated from this source.

6. Current status

Fig. 4 illustrates a section of the plant showing extensive grassing and garden development which makes an attractive work environment. This has contributed to the development by plant personnel of a great sense of pride in their operation, resulting in motivation, good maintenance, and a high standard of housekeeping. At the Hercules plant, the mean annual rainfall is 679 mm. During the rainy summer

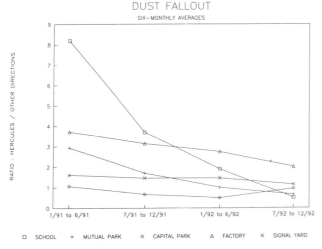

BILD 3: Staubniederschlagsverhältnis: Halbjahresdurchschnitt
FIGURE 3: Dust fallout ratio: six-monthly averages

Ratio: Hercules/other directions = Verhältnis: Hercules/andere Richtungen
School = Schule
Factory = Werk
Signal yard = Signalhof

BILD 4: Teilansicht des Werkes, die die Erweiterung von Rasen- und Gartenflächen veranschaulicht
FIGURE 4: A section of the plant illustrating extensive grassing and garden development

TABELLE 1: Messungen des Staubniederschlags (g/m² pro Tag)

| | Mai 1989 – Dezember 1991 | | | | Januar 1992 – Juni 1992 | | | |
| | Von Hercules | | Aus anderen Richtungen | | Von Hercules | | Aus anderen Richtungen | |
Meßort	Mittelwert	Maximum	Mittelwert	Maximum	Mittelwert	Maximum	Mittelwert	Maximum
Haus	0,2	0,3	0,3	0,4				
Schule	1,2	7,8	0,2	1,3	0,9	3,6	0,2	0,3
Umspannwerk	16	>50	0,3	0,3				
Klärwerk	19	>50	0,2	0,8				
Hercules					0,8	2,5	0,2	0,6
Capital Park					0,2	0,5	0,2	0,8
Signalhof					0,5	1,4	0,4	0,6
Mutual Park					1,0	5,2	0,4	0,8

TABLE 1: Measured dust fallout (grams per square meter per day)

| | May 1989 – December 1991 | | | | January 1992 – June 1992 | | | |
| | From Hercules | | From other directions | | From Hercules | | From other directions | |
Site	Average	Maximum	Average	Maximum	Average	Maximum	Average	Maximum
House	0.2	0.3	0.3	0.4				
School	1.2	7.8	0.2	1.3	0.9	3.6	0.2	0.3
Sub-station	16	>50	0.3	0.3				
Sewage Works	19	>50	0.2	0.8				
Hercules					0.8	2.5	0.2	0.6
Capital Park					0.2	0.5	0.2	0.8
Signal Yard					0.5	1.4	0.4	0.6
Mutual Park					1.0	5.2	0.4	0.8

In der Nähe des Hercules-Werks liegen ein Gipsplattenhersteller und ein Schrotthändler. Staub aus diesen Quellen würde die Zusammensetzung des gesammelten Staubes deutlich beeinflussen. In der Umgebung wird auch Landwirtschaft betrieben. Durch analytische Untersuchungen konnte gezeigt werden, daß ein Teil des Niederschlages, der ursprünglich dem Zementwerk zugeschrieben worden war, aus dieser Quelle stammt.

6. Gegenwärtige Situation

Bild 4 zeigt einen Teil des Werkes mit ausgedehnten Rasen- und Gartenflächen, die eine attraktive Arbeitsumgebung gestalten. Dies hat dazu beigetragen, daß das Werkspersonal gute Arbeitsmoral entwickelt hat, was in hoher Motivation, pfleglichem Umgang mit den Arbeitsmitteln und großer Sauberkeit resultiert. Um das Hercules-Werk fallen jährlich 679 mm Regenniederschlag. In den regnerischen Sommermonaten liegen die Temperaturen zwischen 17,4 und 28,5 °C mit einem Maximum von 35 °C. Die Winter sind trocken mit erfaßten Niedrigsttemperaturen von etwa 4,6 °C.

Die Überwachung wird langfristig fortgesetzt, und zur weiteren Verbesserung der Entnahme werden weitere Geräte installiert. Verfahren zur Messung des Trockenniederschlags in Echtzeit werden angewandt. Dabei wird die Staubmasse auf einem laufenden Filterband mit β-Strahlen bestimmt. Die Einrichtungen zur Aufzeichnung der Windrichtung werden ebenfalls auf Echtzeit umgestellt. Dadurch wird es möglich, den Staubniederschlag zeitgerecht mit den Wind- und Wettermustern in Beziehung zu setzen.

Insgesamt gesehen läßt sich sagen, daß das Hercules-Werk zu einer Grünfläche und einem Aktivposten für die Umwelt in der Stadt Pretoria geworden ist.

months temperatures range from 17.4 to 28.5 with maximums of 35 °C. Winters are dry with minimum temperatures of about 4.6 °C being recorded.

The project will continue on a long-term basis and additional equipment will be introduced to further refine the sampling procedure. Dry deposition techniques on a real time basis will be applied, using mass determination by means of beta radiation across a moving filter strip. A real-time wind logging system will also be added, enabling dust deposition at a given time to be related to wind and weather patterns.

Overall, it can be said that the Hercules factory has become a green area and an environmental asset to the city of Pretoria.

References:

[1] Lodge, J. P. (Ed): Methods of air sampling and analysis. Lewis Publishers, Chelsea MI USA (1989), Method 502, p. 440–445.

[2] ASTM D 1739-82: Standard method for collecting and analysis of dustfall (settleable particulates). Annual book of ASTM Standards Vol. 11.03, American Society for Testing and Materials, Philadelphia PA USA (1982).

[3] Annegarn, H. J., Surridge, A. D., Hlapolosa, H. S. P., Swanepoel, D. J. de V., and Horne, A. R.: A review of 10 years of environmental monitoring at Crown Mines. Ist. IUAPPA Regional Conference on Air Pollution, Pretoria (1990), paper No. 69.

Technik der Emissionsminderung für Drehofenabgase
Technology of emission reduction for rotary kiln exhaust gases

Technique de réduction des émissions pour gaz d'exhaure de fours rotatifs

Técnica de reducción de las emisiones de los gases de escape de los hornos rotativos

Von **D. Rose**, Neubeckum/Deutschland

Zusammenfassung – Für den Einsatz zur Emissionsminderung im Klinkerbrennprozeß sind von Polysius in den letzten Jahren neben verschiedenen anderen Verfahren auch die Eindüsung von Additiven zur Stickstoffoxidreduktion sowie die multifunktionale Rauchgasreinigung auf Aktivkoksbasis entwickelt worden. Untersuchungen zeigten, daß die gewählten Maßnahmen zu hohen Abscheidegraden bei Staub, Schwermetallen und gasförmigen Abgasbestandteilen führen können.

Summary – In addition to various other methods for reducing emissions in the clinker burning process Polysius have also in the last few years developed the injection of additives for nitrogen oxide reduction and the multi-functional flue gas cleaning system based on activated coke. Investigations have shown that the selected measures can lead to high levels of removal of dust, heavy metals and gaseous exhaust gas constituents.

Résumé – Afin de contribuer à la réduction des émissions dans le process de cuisson du clinker, Polysius a développé au cours des dernières années, à côté d'autres procédés divers, aussi l'injection d'additifs pour la réduction des oxydes d'azote ainsi que l'épuration des gaz de fumée multifonctionnelle à base de coke actif. Les recherches ont montré, que les procédés choisis peuvent conduire à des hauts degrés d'élimination de poussière, de métaux lourds et de composants gazeux des gaz d'exhaure.

Resumen – Para la reducción de las emisiones en el proceso de cocción de clínker, Polysius ha desarrollado en los últimos años, además de otros procedimientos, también la inyección de aditivos, destinados a reducir los óxidos de nitrógeno, así como la depuración multifuncional de los gases de humo, sobre la base de carbón activo. De los estudios realizados se desprende que las medidas tomadas pueden contribuir a elevados grados de eliminación de polvo, metales pesados y componentes gaseosos de los gases de escape.

Fragen der Emissionsminderung gewinnen auch in der Zementindustrie durch die ständig steigenden Anforderungen an die Reinhaltung der Luft zunehmend an Bedeutung. In den letzten Jahren sind neben Primärmaßnahmen auch verschiedene sekundäre Verfahren zur Vermeidung von Emissionen für den Einsatz im Zementwerk entwickelt worden.

Bild 1 zeigt schematisch eine Drehofenanlage mit Zyklonvorwärmer, in der die Stickstoffoxide durch das sogenannte SNCR-Verfahren gemindert werden. Die Reduzierung weiterer Schadstoffe erfolgt durch das POLVITEC® Schüttschichtfilter, in dem als Filtermedium Herdofenkoks aus rheinischer Braunkohle eingesetzt wird. Wie dem Schema zu entnehmen ist, erfolgt beim SNCR-Verfahren die Eindüsung des Ammoniakwassers in die unterste Gassteigleitung des Vorwärmers in einem Temperaturbereich zwischen 900°C bis 1100°C, während das POLVITEC®-Filter im staubarmen Abgas nach Elektrofilter angeodnet wird. Die Kombination beider Verfahren ermöglicht hierbei einen optimalen Betrieb der gesamten Rauchgasreinigungsanlage.

Bild 2 zeigt die Ergebnisse, die mit der Eindüsung von Ammoniakwasser in den Vorwärmer einer Produktionsanlage mit einer Kapazität von 2000 Tagestonnen Klinker erzielt wurden. Als Additiv für die Optimierung des Verfahrens wurde Ammoniakwasser ausgewählt, da dies im Vergleich zu druckverflüssigtem Ammoniak genehmigungsrechtlich erheblich einfacher zu behandeln ist und im

Emission reduction is becoming an increasingly important issue in the cement industry due to the increasingly stringent clean air requirements. In addition to primary measures, various secondary processes for preventing emissions have also been developed in recent years for use in cement works.

Fig. 1 is a diagram of a rotary kiln system with a cyclone preheater, in which the nitrogen oxides are reduced by the SNCR process. Other pollutants are reduced by the POLVITEC® granular bed filter, in which open-hearth coke from Rhenish lignite is used as the filter medium. In the SNCR process, as shown in the diagram, the ammonia water is injected into the lowest gas riser pipe of the preheater in a temperature range between 900°C and 1100°C, whilst the POLVITEC® filter is located in the low-dust exhaust gas after the electrostatic precipitator. The combination of both processes enables the entire flue gas purification plant to operate in an optimum manner.

Fig. 2 shows the results that were obtained by injecting ammonia water into the preheater of a production plant with a capacity of 2000 tonnes per day of clinker. Ammonia water was chosen as the additive for optimising the process because it is much easier to handle correctly in conformity with the licensing laws compared with ammonia liquefied under pressure; moreover, in contrast to a release of urea, it does not lead to any secondary emissions of carbon monoxide and dinitrogen monoxide and is much more effi-

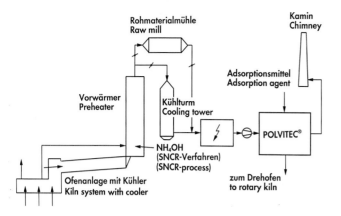

BILD 1: Rauchgasreinigungsverfahren im Zementherstellungsprozeß
FIGURE 1: Flue gas purification process in cement manufacture

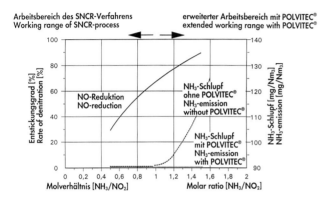

BILD 2: NO-Reduktion und NH_3-Schlupf bei Ammoniakwassereindüsung
FIGURE 2: NO reduction and NH_3 emission with ammonia water injection

BILD 3: Funktionsprinzip der POLVITEC®-Rauchgasreinigungsanlage
FIGURE 3: Operating principle of the POLVITEC® flue gas purification plant

① Rohgas / Dirty gas
② Reingas / Clean gas
③ Systemventilator / System fan
④ Frischkokssilo / Fresh coke silo
⑤ Frischkoksverteilung / Fresh coke distribution
⑥ Koksabzug / Coke discharge
⑦ Koksfilterbetten / Coke filter bed

Gegensatz zu einer Harnstoffauflösung bei deutlich besseren Wirkungsgraden keine Sekundäremissionen an Kohlenmonoxid und Distickstoffoxid verursacht. In Bild 2 ist auf der Abzisse das Molverhältnis von NH_3/NO_2 aufgetragen, mit dem das Verhältnis der beiden Mengenströme dargestellt wird. Die linke Ordinate zeigt den erzielten NO-Reduktionsgrad, während auf der rechten Ordinate der hiermit korrespondierende Ammoniakschlupf, d. h. die nicht umgesetzte und somit emittierte Ammoniakmenge, dargestellt ist. Die im Bild 2 eingezeichnete, durchzogene Kurve zeigt die NO-Reduktion in Abhängigkeit vom eingestellten Molverhältnis. Es wird deutlich, daß mit diesem Verfahren sehr gute Entstickungsgrade erzielt werden können, die bei Molverhältnissen oberhalb von 1.3 bei über 80 % liegen können. Aus dem Bild geht jedoch auch hervor, daß bei Molverhältnissen über 1.0 die eingedüste NH_3-Menge nicht mehr vollständig reagieren wird, sondern zu zusätzlichen Ammoniakemissionen führt, die aufgrund der Toxizität des Ammoniaks unerwünscht sind. So wird der Arbeitsbereich des SNCR-Verfahrens und somit auch dessen Wirkungsgrad durch den Ammoniakschlupf auf Molverhältnisse von maximal ca. 1.0 und hiermit korrespondierende Reduktionswerte von ca. 60 % begrenzt.

Diese Begrenzung wird jedoch aufgehoben, wenn, wie in Bild 1 gezeigt, eine weitergehende Rauchgasreinigung auf Basis eines Aktivkokses nachgeschaltet wird. Das Aktivkoksfilter ist in der Lage, die Ammoniakemissionen aus dem Rauchgasstrom zu entfernen. Somit können auch NH_3/NO_2-Molverhältnisse größer 1.0 vorgewählt werden. Bild 2 zeigt anhand der gepunkteten Linie den Verlauf des Schlupfes hinter einem POLVITEC®-Filter. Tatsächlich wird nicht nur die aus dem SNCR-Verfahren stammende Ammoniakemission, sondern auch die in diesem Fall relativ hohe Ba-

cient. In Fig. 2, the molar ratio of NH_3/NO_2 is plotted on the abscissa, representing the ratio of the two mass flows. The left ordinate shows the degree of NO reduction achieved, whilst the right ordinate shows the corresponding ammonia emission, i.e. the amount of ammonia which did not react and was thus emitted. The continuous line drawn in Fig. 2 shows NO reduction as a function of the molar ratio. It is clear that very good levels of nitrogen removal can be achieved with this process; these may exceed 80 % with molar ratios above 1.3. It is also evident from the figure, however, that at molar ratios above 1.0, the amount of NH_3 injected will no longer react completely and leads to additional ammonia emissions which are undesirable because of the toxicity of ammonia. Consequently, the working range of the SNCR process and hence its efficiency is limited by the ammonia emissions to molar ratios not exceeding about 1.0 and to corresponding reduction values of about 60 %.

This restriction is lifted, however, if a thorough flue gas purification system based on activated coke is installed downstream, as shown in Fig. 1. The activated coke filter is capable of removing the ammonia emissions from the flue gas stream. NH_3/NO_2 molar ratios in excess of 1.0 can thus be chosen. The dotted line in Fig. 2 shows the course of the emission downstream of a POLVITEC® filter. In fact, not only the ammonia emissions originating from the SNCR process but also the basic emissions from the clinker burning process, which are relatively high in this case and amount to 90 mg/Nm³ in this example, are almost completely removed.

The operating principle of such a POLVITEC® system is shown in **Fig. 3**. Essentially, it comprises granular beds of activated coke which are combined on a modular basis to

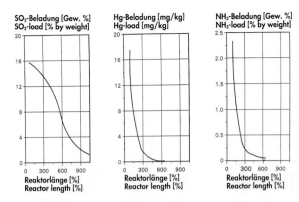

NO$_x$-Reduktion NO$_x$-reduction	bis > 80 % up to > 80 %
SO$_2$-Adsorption SO$_2$-adsorption	> 80 %
Hg-Abscheidung Hg-extraction	> 98 %
NH$_3$-Abscheidung NH$_3$-extraction	> 98 %

BILD 4: Schadstoffkonzentrationsprofile auf Herdofenkoks in der POLVITEC®-Rauchgasreinigungsanlage
FIGURE 4: Pollutant concentration profile on open-hearth coke in the POLVITEC® flue gas purification plant

BILD 5: Abscheideleistungen der Rauchgasreinigung POLVITEC® mit SNCR-Verfahren
FIGURE 5: Separating efficiencies of the POLVITEC® flue gas purification system with the SNCR process

sisemission aus dem Klinkerbrennprozeß, die in diesem Beispiel bei 90 mg/Nm³ liegt, nahezu vollständig abgeschieden.

Das Funktionsprinzip einer solchen POLVITEC®-Anlage zeigt **Bild 3**. Es handelt sich hierbei im wesentlichen um Aktivkoksschüttschichten, die modulartig zu einer Filteranlage kombiniert werden. Da der Aktivkoks durch die abgeschiedenen Schadstoffe mit der Zeit deaktiviert wird, muß er quasi kontinuierlich ausgetauscht werden. Dies geschieht, indem Koks unten aus den Schüttschichten abgezogen und durch frisches Material aus den Vorratssilos ersetzt wird. Der Koks durchwandert die Anlage insgesamt zweimal, wodurch eine optimale Ausnutzung des Adsorptionsmittels und somit der kostengünstige Betrieb der Anlage erreicht wird. Das frische Material wird zunächst durch die stärkere Schüttschicht transportiert, um dann in die vordere dünne Schicht aufgegeben zu werden. Hieraus abgezogener, deaktiver Aktivkoks wird in einem Altkokssilo zwischengelagert und anschließend im Drehofenhauptbrenner als Heizwertspender verbrannt. Das aus dem Elektrofilter austretende Gas wird vom POLVITEC®-Systemventilator angesaugt und auf die Filtermodule verteilt. Es wird dann durch die dünne und dicke Koksschicht gedrückt und über den Gassammelkanal abgeführt. Gas und Material werden somit also im Kreuzgegenstrom geführt. Das weitgehend von Schadstoffen befreite Gas wird anschließend über den Kamin in die Atmosphäre geleitet. Die Betriebstemperaturen des Schüttschichtfilters können im Bereich zwischen 80 °C und 140 °C variiert werden.

Daß mit diesem Verfahren zur Rauchgasreinigung tatsächlich hervorragende Abscheideleistungen erzielt werden können, belegen die in **Bild 4** dargestellten Beladungsprofile des Aktivkokses, die mit einer Pilotanlage während des Einsatzes im Zementwerksbetrieb gemessen wurden. In dem Bild sind über die Länge der Adsorptionsschicht die Konzentrationen an Schwefeldioxid, Quecksilber und Ammoniumverbindungen, berechnet als Ammoniak, aufgetragen. Wie aus dem SO$_2$-Beladungsprofil hervorgeht, ist die bei der Erläuterung des POLVITEC®-Filters angesprochene Rückführung von Material aus dem hinteren Anlageteil in vordere Schichten überaus sinnvoll, da der Koks in den hinteren Schichten nicht seine maximale Beladungskapazität erreicht. Bei hohen SO$_2$-Eingangskonzentrationen ist die Anlagerung des Schwefeldioxids der geschwindigkeitsbestimmende Schritt für den Austausch des beladenen Materials. Hierbei können die Größen SO$_2$-Beladung bzw. SO$_2$-Austrittskonzentration so eingestellt werden, daß der gewünschte Betriebspunkt erreicht ist. Ist nur wenig Schwefeldioxid im Rauchgas enthalten, kann auch die Beladung mit anderen Schadstoffen oder ein durch Staubeintrag verursachter, steigender Druckverlust in den Schüttschichten zum Materialabzug führen.

In den nachfolgenden beiden Beladungsprofilen zeigt sich deutlich, daß die Konzentrationen an Quecksilber und

form a filter unit. As the activated coke becomes deactivated with time due to the separated pollutants, it has to be exchanged almost continuously. This is carried out by removing coke from the granular beds from below and replacing it by fresh material from the storage bins. The coke passes through the system twice altogether, thus ensuring optimum use of the adsorbent and hence economic operation of the plant. The fresh material is transported first through the thicker granular bed, and then fed into the upper thin bed. Deactivated coke drawn off from this bed is stored temporarily in a spent coke silo and then incinerated in the main burner of the rotary kiln as a source of calorific heat. The gas leaving the electrostatic filter is drawn in by the POLVITEC® system fan and distributed over the filter module. It is then forced through the thin and the thick bed of coke and discharged via the gas collecting duct. Gas and material are thus conveyed in cross-current. The gas, largely free of pollutants, is then discharged into the atmosphere through the chimney. The operating temperatures of the granular bed filter can range from 80 °C to 140 °C.

The fact that outstanding separation efficiencies can actually be achieved with this flue gas purification process is confirmed by the loading profiles of activated coke shown in **Fig. 4**, which were measured with a pilot plant whilst the process was in use in the cement works. The sulphur dioxide, mercury and ammonium compounds, calculated as ammonia, are plotted in this figure along the adsorption bed. As can be seen from the SO$_2$ loading profile, the return of material from the rear of the plant to the front beds, mentioned in the description of the POLVITEC® filter, is an extremely practical measure because the coke in the rear beds does not reach its maximum loading capacity. At high SO$_2$ input concentrations, the deposition of sulphur dioxide is the rate-determining step for the exchange of the loaded material. The SO$_2$ loading and SO$_2$ discharge concentration parameters can be adjusted such that the desired operating point is reached. If the flue gas contains only a little sulphur dioxide, loading with other pollutants or a rising pressure drop in the granular beds caused by incoming dust may also lead to material discharge.

The next two loading profiles show clearly that the mercury and ammonia concentrations fell to values approaching zero after only one third of the coke bed length; this suggests a great affinity of the adsorbent to both these pollutants. This assumption is borne out by the determination of the gaseous concentration profiles in the granular bed and the emission values after the activated coke filter. In all cases in which emission values were determined, efficiencies of more than 98 % for NH$_3$ and mercury were determined. It should also be borne in mind that, in addition to mercury, other heavy metals are also separated with similar efficiency on open-hearth coke, this being of particular importance when waste fuels are used. Moreover, it was proved in various measurements that the dust concentrations after the activated coke granu-

Ammoniak schon nach einem Drittel der Koksschichtlänge auf Werte nahe Null zurückgegangen sind, was auf die hohe Affinität des Adsorbens bezüglich dieser beiden Schadstoffe hindeutet. Belegt wird diese Annahme durch die Bestimmung der gasförmigen Konzentrationsprofile in der Schüttschicht sowie der Emissionswerte nach Aktivkoksfilter. In allen Fällen, in denen Emissionswerte bestimmt wurden, konnten Abscheidegrade für NH_3 und Quecksilber von über 98 % ermittelt werden. Es ist weiterhin zu berücksichtigen, daß neben Quecksilber auch andere Schwermetalle mit ähnlicher Effektivität auf dem Herdofenkoks abgeschieden werden, was insbesondere bei Einsatz von Abfallbrennstoffen von Bedeutung ist. Weiterhin konnte in verschiedenen Messungen belegt werden, daß die Staubkonzentrationen nach Aktivkoksschüttschicht in allen Fällen bei Werten unter 10 mg/Nm^3 lagen, also auch in dieser Beziehung bei den durchgeführten Versuchen eine deutliche Verbesserung der Emissionssituation auftrat.

Die hier geschilderten Ergebnisse der Abgasreinigung mit Aktivkoks bestätigen die Multifunktionalität des POLVITEC®-Verfahrens. Neben den hier angeführten Anlagerungsmechanismen ist mit Braunkohleherdofenkoks auch eine katalytische Umsetzung der Stickstoffoxide bei Temperaturen um 100 °C möglich. **Bild 5** resumiert abschließend die wesentlichen Abscheideleistungen, die bei Kombination des SNCR-Verfahrens mit einem POLVITEC®-Filter erzielt werden. Hieraus geht hervor, welche deutliche Entlastung der Emissionssituation sich bei Einsatz der beschriebenen Variante zur Abgasreinigung ergibt, wobei auch die Symbiose zwischen Ökologie und Ökonomie durch die Übernahme der Entsorgung von umweltrelevanten Sekundärroh- und Brennstoffen gegeben ist.

lar bed were below 10 mg/Nm^3 in all cases, so a marked improvement in this emission situation also occurred during the tests.

The results of exhaust gas purification with activated coke described here therefore confirm the multifunctional nature of the POLVITEC® process. Apart from the deposition mechanisms mentioned here, catalytic conversion of nitrogen oxides at temperatures of about 100 °C is also possible with open-hearth lignite coke. **Fig. 5** summarises the main separation efficiencies that can be achieved with a combination of the SNCR process and a POLVITEC® filter. This shows the marked improvement in the emission situation when using the system of exhaust gas purification described. There is also a symbiotic relationship between ecology and economy since this system takes care of the disposal of environmentally harmful secondary raw materials and fuels.

Die betriebliche Umsetzung von NO_x-Minderungsmaßnahmen
(Bericht des VDZ-Arbeitskreises „NO_x-Minderung")

Practical application of measures for reducing NO_x
(Report from the VDZ Working Group "NO_x reduction")

La réalisation industrielle d'actions de réduction de NO_x
(Contribution du groupe de travail VDZ „Réduction de NO_x")

Aplicación práctica de medidas para la reducción de NO_x
(Informe del grupo de trabajo VDZ „Reducción NO_x")

Von **W. Ruhland**, Wiesbaden und **H. Hoppe**, Düsseldorf/Deutschland

Zusammenfassung – Die Spannweite der mittleren NO_x-Emission von Drehöfen der Zementindustrie reicht derzeit noch etwa von 0,7 bis 3,0 g/m³. Ursache dafür ist der exponentielle Anstieg der thermischen NO-Bildung bei steigenden Brenntemperaturen, die auf die Zusammensetzung der natürlichen Rohstoffe sowie die Qualitätsanforderungen an das Produkt abgestimmt sein müssen. Dem „Stand der Technik" entsprechende Maßnahmen, mit denen der vom Gesetzgeber drastisch gesenkte Emissionsgrenzwert von 0,8 g/m³ bei bestehenden Anlagen und 0,5 g/m³ bei Neuanlagen sicher eingehalten werden kann, gibt es derzeitig noch nicht. Primär- und Sekundärmaßnahmen, die aus Pilotuntersuchungen bekannt sind, werden gegenwärtig an ersten Anlagen versuchsweise mit erheblichem Aufwand umgesetzt. Ein Teil der Umbauten wird vom BMU gefördert. Erprobt werden dabei verfahrenstechnische Maßnahmen, durch die eine NO-Bildung behindert wird, Minderungsmaßnahmen nach dem SNCR-Verfahren sowie die mehrstufige Vorcalcination, die jedoch nur bei Öfen mit Tertiärluftleitung möglich ist. Bei der Eindüsung von NH_3-Wasser zur NO-Minderung kommt einem geringen NH_3-Schlupf und dessen Messung besondere Bedeutung zu. Darüber hinaus müssen erhebliche Anstrengungen unternommen werden, um NO_x-Spitzen zu vermeiden, damit selbst bei niedrigen mittleren NO_x-Emissionen die Auswertungskriterien der TA Luft sicher eingehalten werden können.

Summary – At present the range of average NO_x emissions from rotary kilns in the cement industry extends from about 0.7 to 3.0 g/m³. The reason for this is the exponential rise in the thermal formation of NO with increasing combustion temperature, which has to be matched to the composition of the natural raw materials and the quality requirements for the product. At present there are still no "state of the art" measures which can comply reliably with the emission limits (0.8 g/m³ for existing plants and 0.5 g/m³ for new plants) which have been drastically lowered by the legislature. Primary and secondary measures which are known from pilot investigations are at present being put into practice for the first time industrially at great expense on a trial basis. Some of the conversion work is being supported by the Federal Ministry for the Environment. Trials are being carried out with process measures which hinder the formation of NO and with measures to reduce emission based on selective non-catalytic reduction, as well as with multi-stage precalcination which, however, is only possible in kilns with tertiary air ducts. When injecting NH_3-water to lower the NO great importance is attached to a low level of NH_3 bypass and to its measurement. Great efforts also have to be made to avoid NO_x peaks so that even where the average levels of NO_x emissions are low the evaluation criteria of the German Clean Air Regulations can be met reliably.

Résumé – La plage des émissions moyennes NO_x de fours rotatifs de l'industrie cimentière s'étend actuellement encore d'environ 0,7 à 3,0 g/m³. La raison en est l'augmentation exponentielle de la formation thermique du NO avec des températures de cuisson croissantes, qui doivent être ajustées en fonction de la composition des matières premières naturelles ainsi qu'aux critères de qualité du produit. Des procédés correspondant à „l'état de la technique", permettant d'assurer la limite d'émission abaissée outre mesure par les législateurs, de 0,8 g/m³ dans les installations existantes et de 0,5 g/m³ dans les installations nouvelles, n'existent pas encore. Des actions primaires et secondaires, révélées par des essais pilote, sont actuellement entreprises à titre d'essai avec des fortes dépenses, dans quelques premières installations. Une partie des modifications est subventionnée par le BMU (Ministère Fédéral de l'Environnement). Dans ce contexte, sont expérimentés de procédés inhibant la formation de NO, des procédés de réduction basés sur la méthode SNCR ainsi que la précalcination à plusieurs étages, possible pourtant que sur des fours à conduit d'air tertiaire. Dans

Fachbereich 4 · Subject 4 · Séance Technique 4 · Tema de ramo 4

l'injection d'eau ammoniaquée pour la réduction de NO, une faible dérive NH_3 et sa mesure ont une importance particulière. De plus des efforts notables doivent être consentis pour éviter des pointes NO_x, afin que les critères d'application de la TA Luft puissent être respectés avec certitude même lors de faibles émissions moyennes de NO_x.

Resumen – *El promedio de emisiones de NO_x de los hornos rotatorios empleados en la industria del cemento oscila actualmente aún entre 0,7 y 3,0 g/m³. La causa de ello es el aumento exponencial de la formación térmica de NO a medida que aumentan las temperaturas de cocción, las cuales tienen que ser adaptadas a la composición de las materias primas naturales y a la calidad requerida del producto. Actualmente, no existen todavía, según el estado actual de la técnica, medidas que permitan alcanzar, de forma segura, el valor de emisión que ha sido bajado drásticamente por el legislador a 0,8 g/m³ para instalaciones existentes y 0,5 g/m³ para instalaciones nuevas. Se están aplicando, a título de prueba, algunas medidas primarias y secundarias, conocidas de ensayos piloto, a las primeras instalaciones, pero con gastos bastante elevados. Una parte de las reformas es financiada por el Ministerio Federal del Medio Ambiente. A este respecto, se someten a ensayo las medidas tecnológicas capaces de frenar la formación de NO_x, las medidas de reducción mediante el procedimiento SNCR así como la precalcinación en varias etapas, la cual es posible solamente con hornos equipados con tubería de aire terciario. Al inyectar agua de NH_3 para la reducción de NO, adquiere especial importancia un bajo nivel de NH_3 y su medición. Además, hay que hacer grandes esfuerzos para evitar puntas de NO_x, con el fin de poder respetar sin problemas los criterios de evaluación de la TA Luft (Instrucción técnica para el mantenimiento de la limpieza del aire), aun cuando el promedio de emisiones de NO_x es bajo.*

Aplicación práctica de medidas para la reducción de NO_x
(Informe del grupo de trabajo VDZ „Reducción NO_x")

Die dynamisierten TA Luft-Grenzwerte für NO_x von 0,80 g NO_2/m^3 (für Altanlagen) bzw. 0,50 g NO_2/m^3 (für Neuanlagen) [1, 2] sind von den meisten Drehofenanlagen der Zementindustrie ohne umfangreiche zusätzliche Minderungsmaßnahmen für Stickoxide nicht einzuhalten. Um die vorliegenden Erkenntnisse aus Pilotuntersuchungen [3, 4], die noch nicht dem Stand der Technik entsprechen, in Einzelmaß-

The revised limits for NO_x imposed by the German Clean Air Regulations, namely 0.80 g NO_2/m^3 (for existing plants) and 0.50 g NO_2/m^3 (for new plants) [1, 2] cannot be met by most rotary kilns in the cement industry without extensive additional measures for reducing nitrogen oxides. Demonstration projects in which NO_x reduction measures are being put into practice on a trial basis are being carried out on three kiln systems in the cement industry in order to apply the knowledge available from pilot investigations [3, 4], which are not yet state of the art, by taking specific measures in different plants. The „NO_x reduction" Working Group of the VDZ „Environment" committee is responsible for monitoring the projects from a technical angle. The experience gained in connection with the reduction measures and the associated measuring techniques will be collated by the Working Group and made available to all the other member companies. The projects are being funded by the Federal Minister for the Environment. The necessary alterations to the rotary kilns have since been carried out and the kiln systems are in operation.

In the first kiln system (**Figs. 1a** and **1b**) with a four-stage, two-string cyclone preheater, NO_x reduction is carried out by the SNCR process (Selective, Non-Catalytic Reduction). To this end, ammonia water is sprayed into the hot gas duct at a temperature of about 900 °C. Not until after this is the raw meal fed in from cyclone stage 3, as a result of which the temperature falls spontaneously to about 820 °C. The ammonia water is atomised and distributed with compressed air by dual-fluid atomizers with a control range of 10:1. The atomizers are built into the corners of the hot gas duct flush with the brickwork. Scavenging air is blown in around the atomizer connections.

The second kiln system (**Figs. 2a** and **b**), in which the SNCR process is applied, also has a two-string, four-stage cyclone preheater, but with tertiary air duct and a tubular calciner. In this project, the introduction of tertiary air and calcination fuel was staggered such that the ammonia water is first

Zyklon 4	= cyclone 4
Mehl von Zyklon 3	= meal from cyclone 3
Brennstoff	= fuel
Staub-Absetz-kammer	= dust settling chamber
Tertiärluft	= tertiary air
Drehofen	= rotary kiln

BILD 1a: NO_x-Minderung nach dem SNCR-Verfahren an der Ofenanlage 1
FIGURE 1a: NO_x reduction by the SNCR process in kiln system 1

4stufiger Zyklonvorwärmer mit Rohrkalzinator

Klinkerproduktion: 1 200 t/d

Selective Non Catalytic Reduction (SNCR)

Ammoniakwasser mit 25 % NH_3

BILD 1b: Technische Angaben zur Ofenanlage 1

4-stage cyclone preheater with tubular calciner

Clinker production: 1 200 t/day

Selective Non-Catalytic Reduction (SNCR)

Ammonia water with 25 % NH_3

FIGURE 1b: Technical information relating to kiln system 1

nahmen an unterschiedlichen Anlagen anzuwenden, werden an drei Ofenanlagen der Zementindustrie Demonstrationsprojekte durchgeführt, bei denen die Umsetzung von NO_x-Minderungsmaßnahmen in die betriebliche Praxis erprobt wird. Die fachliche Begleitung der Projekte nimmt der Arbeitskeis „NO_x-Minderung" des VDZ-Ausschusses „Umwelt" wahr. Im Arbeitskreis werden die Erfahrungen über die Minderungsmaßnahmen und die zugehörige Meßtechnik zusammmengefaßt, um sie allen Mitgliedsunternehmen zugänglich zu machen. Die Vorhaben werden mit Mitteln des Bundesministers für Umwelt gefördert. Die erforderlichen Umbauten an den Drehöfen sind mittlerweile durchgeführt worden und die Anlagen in Betrieb gegangen.

Bei der ersten Ofenanlage (**Bilder 1a** und **1b**) mit vierstufigem, zweisträngigem Zyklonvorwärmer wird eine NO_x-Minderung nach dem SNCR-Verfahren (selektive, nichtkatalytische Reduktion) durchgeführt. Dazu wird Ammoniakwasser bei einer Temperatur von ca. 900°C in den Heißgaskanal eingedüst. Erst danach wird das Rohmehl von der Zyklonstufe 3 aufgegeben, wodurch die Temperatur spontan auf rund 820°C absinkt. Die Zerstäubung und Verteilung des Ammoniakwassers erfolgt mit Druckluft durch Zweistoffdüsen mit einem Regelbereich von 10 : 1. Die Düsen sind in den Ecken des Heißgaskanals bündig mit dem Mauerwerk eingebaut. Um die Düsenstöcke wird Spülluft eingeblasen.

Die zweite Anlage (**Bilder 2a** und **2b**), an der mit dem SNCR-Verfahren gearbeitet wird, hat ebenfalls einen zweisträngigen, vierstufigen Zyklonvorwärmer, jedoch mit Tertiärluftleitung und Rohrcalcinator. Bei diesem Projekt wurde die Zuführung der Tertiärluft und des Calcinationsbrennstoffs so verlegt, daß das Ammoniakwasser zuerst in das noch nicht verdünnte Drehofenabgas eingedüst wird und somit eine kurze Reaktionsstrecke für den NO-Abbau verbleibt, bevor dann die Tertiärluft und der Brennstoff hinzutreten. Ferner wird an dieser Stelle ein Teilstrom des Rohmehls von der Zyklonstufe 3 dem Ofenabgas zugegeben. Die Menge wird so bemessen, daß sich eine Temperatur zwischen 900 und 1 000°C einstellt. Die Eindüsung des Ammoniakwassers erfolgt über sechs Flachstrahl-Druckzerstäuberdüsen, die in den Ecken bzw. an den Längsseiten des Heißgaskanals angeordnet sind. Die Düsenstöcke sind auch hier bündig mit dem Mauerwerk eingesetzt und werden mit Spülluft umspült. Eine verminderte Bildung von Brennstoff-NO in der Zweitfeuerung soll durch die Verwendung von Braunkohlenstaub erreicht werden, der einen geringeren Stickstoffgehalt als der bisher verwendete Steinkohlenstaub aufweist.

Bei der dritten Anlage (**Bilder 3a** und **3b**) handelt es sich um einen Kurzdrehofen mit fünfstufigem Zyklonvorwärmer, Low-NO_x-Calcinator und Tertiärluftleitung. Durch eine Brennstoffzugabe im Calcinator wird das Ofenabgas in eine reduzierende Atmosphäre überführt und damit ein NO-Abbau erreicht.

Damit aber die durch die Kohlenstaubverbrennung freigesetzte Wärmeenergie nicht zu überhöhten Gastemperaturen führt, muß auch bei diesem Verfahren ein entsprechender Rohmehlteilstrom aus der zweituntersten Zyklonstufe an der Brennstoffzugabestelle zugeführt werden. Ein Teil des für die Vorcalcination erforderlichen Kohlenstaubs wird gemeinsam mit dem restlichen Rohmehl in die Tertiärluft aufgegeben.

injected into the still undiluted rotary kiln waste gas and there is a short reaction section for NO reduction before the tertiary air and the fuel are added. At this point, a partial stream of the raw meal from cyclone stage 3 also is added to the kiln exhaust gas. The quantity is determined such that the resulting temperature is between 900 and 1000°C. The ammonia water is injected by means of six fan-jet pressure atomizers arranged in the corners and on the long sides of the hot gas duct. Here again, the atomizers are inserted with the brickwork and have scavenging air blown round them. The purpose of using pulverised lignite is to reduce the formation of fuel-NO in the secondary firing system, since it has a lower nitrogen content than the pulverised coal used in the past.

The third kiln system (**Figs. 3a** and **b**) is a short rotary kiln with a five-stage cyclone preheater, low-NO_x calciner and tertiary air duct. NO reduction is achieved by adding fuel to the calciner and converting the kiln exhaust gas to a reducing atmosphere.

However, in order to ensure that the heat energy released by burning pulverised coal does not lead to excessively high

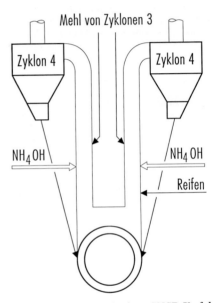

BILD 2a: NO_x-Minderung nach dem SNCR-Verfahren an der Ofenanlage 2
FIGURE 2a: NO_x reduction by the SNCR process in kiln system 2

Zyklon 4 = cyclone 4
Mehl von Zyklonen 3 = meal from cyclones 3
Reifen = tyres

gas temperatures, this process also requires that a corresponding partial stream of raw meal from the second lowest cyclone stage be fed into the fuel delivery point. Part of the pulverised coal required for precalcination is fed together with the remaining raw meal into the tertiary air.

The streams of oxidising tertiary air exhaust gas and reducing rotary kiln exhaust gas must then be mixed intensively so as to oxidise the CO in the flue gas stream. This takes place in a swirl chamber which is installed in the calciner instead of a simple 180° elbow. Raw meal separating out in this chamber can be recycled to either the rising or falling pipe line of the calciner, depending on the proportion of combustible materials (coal) still present in the raw meal.

4stufiger Zyklonvorwärmer ohne Vorkalzinator

Klinkerproduktion: 2 000 t/d

Selective Non Catalytic Reduction (SNCR)

Ammoniakwasser mit 25 % NH_3

BILD 2b: Technische Angaben zur Ofenanlage 2

4-stage cyclone preheater without precalciner

Clinker production: 2 000 t/day

Selective Non-Catalytic Reduction (SNCR)

Ammonia water with 25 % NH_3

FIGURE 2b: Technical information relating to kiln system 2

Kurzdrehofen	Short rotary kiln
5stufiger Zyklonvorwärmer mit Rohrkalzinator und Wirbelkammer	5-stage cyclone preheater with tubular calciner swirl chamber
Klinkerproduktion: 2 000 t/d	Clinker production: 2 000 t/day
NO-Abbau in reduzierender Rauchgasatmosphäre durch Brennstoffzugabe	NO reduction in reducing flue gas atmosphere due to addition of fuel

BILD 3b: Technische Angaben zur Ofenanlage 3

FIGURE 3b: Technical information relating to kiln system 3

Anschließend müsen die Strähnen des oxidierend wirkenden Tertiärluftabgases und des reduzierend wirkenden Drehofenabgases intensiv gemischt werden, um das CO im Rauchgasstrom zu oxidieren. Dies geschieht in einem Wirbeltopf, der anstelle eines einfachen 180°-Krümmers im Calcinator eingebaut ist. In diesem Topf ausfallendes Rohmehl kann alternativ in die aufsteigende oder abfallende Rohrleitung des Calcinator zurückgeführt werden, je nach dem Anteil an Brennbarem (Kohle), der noch im Rohmehl vorhanden ist.

Im Zusammenhang mit der NO_x-Minderung wird vom Arbeitskreis auch die zugehörige Gasmeßtechnik bewertet. Im Rahmen von umfangreichen Versuchen und Kalibriermessungen wurden von der Zementindustrie auf freiwilliger Basis und auf eigene Kosten NO_x-Meßgeräte erprobt und auf die speziellen Anforderungen für eine Messung im Drehofenabgas abgestimmt. Nach Abschluß dieser Arbeiten wird die Emission von Stickoxiden in fast allen Werken kontinuierlich überwacht, obwohl die branchenbezogene Eignungsbekanntgabe für diese Geräte nach wie vor fehlt.

Bei Einsatz des SNCR-Verfahrens kann es zur Emission von nicht umgesetztem Ammoniak kommen („NH_3-Schlupf"). Eine kontinuierliche Überwachung des Schlupfs mit NH_3-Meßgeräten bereitet aber Schwierigkeiten, da zur Zeit kein eignungsgeprüftes Gerät für die Messung von NH_3 an Zementofenanlagen auf dem Markt ist. Die Erfahrungen mit den NO_x-Meßgeräten zeigen, daß sich Erkenntnisse von anderen Feuerungsanlagen nicht direkt auf die Verhältnisse an Zementofenanlagen übertragen lassen. Insbesondere die Kalibrierfähigkeit dieser Meßgeräte müßte erst noch unter den besonderen Bedingungen des Klinkerbrennprozesses nachgewiesen werden. Deshalb sollten die Genehmigungsbehörden bis zum Abschluß der Demonstrationsversuche von der Forderung nach einer kontinuierlichen Schlupfmessung absehen.

Literaturverzeichnis:

[1] Erste Allgemeine Verwaltungsvorschrift zum Bundes-Immissionsschutzgesetz (Technische Anleitung zur Reinhaltung der Luft – TA Luft), 27.2.1986 (GMBl. S. 95).

[2] Gem. RdErl. d. Ministeriums für Umwelt, Raumordung und Landwirtschaft u. d. Ministeriums für Wirtschaft, Mittelstand und Technologie v. 6.2.1992: Durchführung der Technischen Anleitung zur Reinhaltung der Luft. Ministerialblatt für das Land Nordrhein-Westfalen, Nr. 18, 17.3.1992, S. 452.

[3] S c h e u e r, A.: Minderung der NO_x-Emission beim Brennen von Zementklinker. Zement-Kalk-Gips 41 (1988), H. 1, S. 37–42.

[4] S c h e u e r, A.: Nichtkatalytische Reduktion des NO mit NH_3 beim Zementbrennen. Zement-Kalk-Gips 43 (1990), H. 1, S. 1–12.

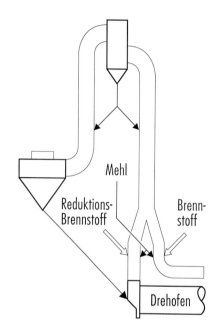

BILD 3a: Low-NO_x-Calcinator an der Ofenanlage 3
FIGURE 3a: Low-NO_x calciner in kiln system 3

Reduktionsbrennstoff	= reducing fuel
Mehl	= meal
Brennstoff	= fuel
Drehofen	= rotary kiln

The Working Group also evaluated the associated method of gas measurement in connection with NO_x reduction. As part of extensive tests and calibration measurements, NO_x measuring instruments were tested voluntarily by the cement industry at its own expense, and matched to the special requirements for measuring NO_x in rotary kiln exhaust gas. When this work is completed, the emission of nitrogen oxides will be monitored continuously in nearly all the works, although the industry itself has not yet made any announcement as to the suitability of these instruments.

Unreacted ammonia may be emitted („NH_3 escape") when the SNCR process is used. Continuous monitoring of the emission with NH_3 measuring instruments is difficult, however, because there is no instrument on the market at present that has been tested for its suitability for measuring NH_3 in cement kilns. Experience with NO_x measuring instruments shows that knowledge about other combustion systems cannot be transferred directly to conditions in cement kiln plants. In particular, the ability of these instruments to be calibrated would have to be proved under the special conditions of the clinker burning process. For this reason, the licensing authorities should not demand continuous measurement of escape emissions until the demonstration trials have been completed.

Die Immissionsverteilung staubförmiger Emissionen aus Zementofenanlagen

The immission distribution of dust emissions from cement kiln plants

La dispersion en immission des émissions sous forme de poussière des installations de fours à ciment

La distribución de las inmisiones en forma de polvo, procedente de plantas de hornos de cemento

Von **M. Schneider,** Düsseldorf/Deutschland

Zusammenfassung – Eine Zuordnung gemessener Staubniederschläge in der Umgebung von Zementwerken zu Quellen des Zementproduktionsprozesses erfolgte bisher nicht. Die Herstellung des Quellenbezuges erfordert die Kenntnis der Korngrößenverteilung sowie der morphologischen und optischen Eigenschaften sowohl der emittierten als auch der niedergeschlagenen Stäube. Im Rahmen von Staubimmissionsmessungen in der Umgebung von Zementwerken werden die stoffspezifischen Parameter der Stäube bestimmt. Die Probenahme erfolgt mittels Haftfolien, die Auswertung wird für zementspezifische Stäube weiterentwickelt. Die wichtigsten Ausbreitungsparameter werden untersucht, um Ausbreitungsmodelle im Hinblick auf ihre Eignung zur Staubniederschlagssimulation zu überprüfen und gegebenenfalls zu überarbeiten.

Summary – So far the dust precipitation measured near cement works has never been assigned to the sources in the cement production process. Making this connection requires knowledge of the particle size distribution as well as the morphological and optical properties of the dusts emitted and precipitated. The material-specific parameters of the dusts are being determined as part of dust emission measurements near cement works. The samples are collected with adhesive foil, and further development work is being carried out on the evaluation of cement-specific dusts. The most important propagation parameters are being investigated in order to check propagation models for their suitability for simulating dust precipitation, and if necessary to revise them.

Résumé – Jusqu'à présent, l'implication des précipitations de poussières mesurées dans le voisinage de cimenteries, ayant comme source le process de production du ciment, n'a pas été établie. La preuve de l'origine exige la connaissance de la distribution granulométrique ainsi que des propriétés morphologiques et optiques des poussières émises et précipitées. Dans le cadre de mesures d'immission de poussières dans le voisinage de cimenteries, sont définis les paramètres spécifiques de la nature des poussières. La prise d'échantillon est effectuée au moyen de films adhésifs, l'exploitation pour les poussières concernant spécifiquement le ciment est en cours de développement. Les paramètres de dispersion les plus importants sont étudiés, afin de vérifier et, le cas échéant, redéfinir des modéles de dispersion en vue de leur aptitude à la simulation de précipitations de poussières.

Resumen – Las precipitaciones de polvo medidas a proximidad de las plantas cementeras no han sido asignadas, hasta la fecha, a los diferentes fuentes del proceso de producción de cemento. Dicha asignación requiere el conocimiento de la distribución granulométrica así como de las características morfológicas y ópticas, tanto de los polvos emitidos como de los precipitados. Al medirse las inmisiones de polvo a proximidad de las plantas cementeras, se determinan los parámetros específicos de los polvos. La toma de muestras se efectúa mediante láminas adhesivas, el método de evaluación se sigue perfeccionando para polvos específicos de cemento. Se estudian los principales parámetros de propagación, con el fin de comprobar y, en caso necesario, revisar los modelos de propagación respecto de su aptitud para la simulación de las precipitaciones de polvo.

Die Staubniederschlagsbelastung, die 1950 in Gebieten mit verstärkter Zementproduktion im Mittel noch bei 0,75 g/m^2 · d lag, nahm auf Werte ab, die heute in der Größenordnung von 0,1 g/m^2 · d und darunter liegen. Der in **Bild 1** dargestellte Verlauf zeigt aber auch, daß die Staubimmission im Gegensatz zu den 50er und 60er Jahren nicht oder nur noch geringfügig abnimmt und sich damit einer Hintergrundbelastung annähert.

Eine weitere Reduzierung der Emissionen aus Anlagen der Zementindustrie ist mit hohen Kosten verbunden, so daß

Pollution by dust precipitation, which in 1950 still amounted to an average of 0.75 g/m^2 · d in areas with high cement production levels, has fallen to present-day values of around 0.1 g/m^2 · d and below. The curve shown in **Fig. 1** also shows, however, that the dust immission is falling only slightly, if at all, in contrast to the 50s and 60s and is thus akin to background pollution.

A further reduction in the emissions from plants in the cement industry would entail substantial costs, so the

eine Verringerung der Immissionskonzentrationen nur erfolgen kann, wenn gezielte Maßnahmen an Einzelquellen durchgeführt werden. Fehlinvestitionen durch oft geforderte pauschale Maßnahmen werden dadurch vermieden.

Eine gezielte Beurteilung von Einzelquellen kann durch Berechnung oder auch durch partikelspezifische Immissionsmessungen erfolgen. Die mathematische Beschreibung desjenigen Anteils an der Immission, der durch eine einzelne Emissionsquelle, wie beispielsweise einer Drehofenanlage, verursacht wird, erfolgt im allgemeinen mit Hilfe Gauß'scher Ausbreitungsmodelle. Allerdings ist es in der Praxis oft schwierig, die für diese Modelle benötigten Ausbreitungsparameter zu erhalten. So ist es z. B. erforderlich, exakte Angaben über die meteorologische Situation zum Ausbreitungszeitpunkt zu haben. Hierzu zählen insbesondere Windgeschwindigkeit und Windrichtung. Die Ausbreitungsmodelle benötigen ferner als Eingangsdaten die Depositions- und Sinkgeschwindigkeit, insbesondere ist hierzu die Kenntnis der Korngrößenverteilung und der zugehörigen mittleren Partikeldichte der emittierten Stäube erforderlich.

Derart rechnerisch ermittelte Konzentrationen stellen langjährige Mittelwerte dar, die sich aufgrund der mittleren meteorologischen Situation am Standort errechnen lassen. Zur Modellierung der Immissionssituation für kürzere Zeiträume sind diese gemittelten Häufigkeitsverteilungen der meteorologischen Ausbreitungsbedingungen ungeeignet. In diesem Fall werden die Windverhältnisse und der Turbulenzzustand der Atmosphäre vom Forschungsinstitut der Zementindustrie direkt in unmittelbarer Umgebung des Werkes ermittelt.

Eine Berücksichtigung von Auswascheffekten durch Niederschlag kann im Ausbreitungsmodell ebenfalls berücksichtigt werden, wenn die partikelspezifischen Auswaschkoeffizienten bekannt sind. Erfahrungsgemäß beträgt dieser Anteil der nassen Deposition etwa das Doppelte des ansonsten ermittelten Anteils des Staubniederschlags.

Die Sinkgeschwindigkeit, die maßgeblich für die Verweilzeit des Staubes in der Luft und damit für die Entfernung, über die er transportiert wird, verantwortlich ist, hängt von der Form, Größe, Dichte und Hygroskopizität eines Partikels ab. Diese Eigenschaften wiederum sind sehr stark anlagenspezifisch. Die Wahl der eingesetzten Rohstoffe und der Fertigungs- und Feuerungsprozeß sowie nachgeschaltete Abgasreinigungsanlagen bestimmen die Art und die Inhaltsstoffe der emittierten Stäube sowie deren Korngrößen- und Massenverteilungen. Während Korngrößenverteilungen und morphologische Eigenschaften von Reingasstäuben für z. B. Kohlekraftwerke, Müllverbrennungsanlagen oder Bleiglashütten dokumentiert sind, liegen keine entsprechend umfangreichen Angaben für unterschiedliche Einzelquellen aus dem Bereich der Zementindustrie vor.

Zur Messung partikelförmiger Niederschläge werden unterschiedliche Geräte eingesetzt; im Bereich der Zementindustrie kommen hierzu erstmalig Haftfoliengeräte zum Einsatz. Während Sammelgefäße, die z.B. nach dem Bergerhoff-Verfahren arbeiten, insbesondere zur Bestimmung der Massendepositionsrate verwendet werden, erlauben Haftfoliengeräte die partikelspezifische Auswertung des Staubniederschlags. Die ca. 7 cm x 7 cm großen Folien werden entsprechend den Vorgaben der Richtlinie VDI 2119, Blatt 4 Entwurf, im Freien aufgestellt. Der Schutztubus schirmt die Fläche weitestgehend vor Wind und Regeneinflüssen ab. Die Ablagerung von Grobfraktion des Staubes sowie Blütenpollen oder anderen organischen Partikeln wird dadurch wirkungsvoll unterdrückt. Aufgrund der im Vergleich zu Sammelgeräten kurzen Expositionsdauer liegen die Partikel auf der Haftfolie überwiegend isoliert vor. Eine routinemäßige Bestimmung der Anzahl- und Massendepositionsrate durch lichtmikroskopische Untersuchungen, insbesondere mit Methoden der modernen Bildanalyse, ist dadurch möglich. Partikelgrößenanalysen sowie die Untersuchungen der morphologischen und optischen Eigenschaften des niedergeschlagenen Staubes erlauben insbesondere die

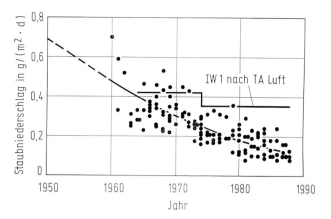

BILD 1: Entwicklung der Staubimmission in der Umgebung von Zementwerken zwischen 1950 und 1989 im Vergleich zum Immissionswert IW1 nach TA Luft
FIGURE 1: Development of the dust immission in the vicinity of cement works between 1950 and 1989 in comparison with the IW1 immission value specified in the German air pollution regulations

Staubniederschlag in g/(m²·d) = Dust precipitation in g/(m²·d)
Jahr = Year
IW1 nach TA Luft = IW1 immission value specified in German air pollution regulations

immission concentrations can be reduced only if specific measures are taken at particular sources. This avoids bad investments by blanket measures of the kind often called for.

A specific assessment of individual sources can be made by calculation or by particle-specific immission measurements. A mathematical description of that proportion of the immission that is caused by a particular source of emission, for example, a rotary kiln plant, is generally obtained with the aid of Gaussian dispersion models. In practice, however, it is often difficult to obtain the dispersion parameters required for these models. For example, exact details of the meteorological conditions at the time of dispersion are required. These include, in particular, the wind speed and direction. Dispersion models also require input data on the rate of deposition and settling velocity, for which a knowledge of the particle size distribution and the associated mean particle density of the dusts emitted is required, in particular.

Concentrations determined mathematically in this way are averages obtained over several years which can be calculated on the basis of the average meteorological conditions at the site. These averaged frequency distributions of the meteorological dispersion conditions are unsuitable for modelling the immission situation for relatively short periods. In this case, the wind conditions and turbulence of the atmosphere have to be determined directly by the Research Institute of the Cement Industry in the immediate vicinity of the works.

The dispersion model can also take account of wash-out effects by precipitation if the particle-specific wash-out coefficients are known. Experience has shown that this proportion of wet deposition is about double the proportion of dust precipitation otherwise determined.

The settling velocity, which is crucial for the residence time of the dust in the air and thus responsible for the distance over which it is transported, depends on the form, size, density and hygroscopic nature of a particle. These properties are, in turn, extremely plant-specific. The choice of the raw materials used and the production and heating process and downstream waste gas cleaning facilities determine the nature and constituents of the dusts emitted and their particle size and mass distributions. Whilst the particle size distributions and morphological properties of clean gas dusts for e.g. coal-fired power stations, waste incineration plants or lead glassworks have been documented, there are no correspondingly comprehensive data for different individual sources from the cement industry.

Einteilung in Stäube biogenen oder anthropogenen Ursprungs.

Darüber hinaus liefern rasterelektronenmikroskopische (REM) Untersuchungen detaillierte Informationen über die Partikeloberflächenstruktur. Phasen- und Spurenelementuntersuchungen an Reingasstäuben von Kohlekraftwerken und erstmalig auch Müllverbrennungsanlagen sind dokumentiert. Mittels des energiedispersiven Röntgennachweises konnte dabei die Schwermetallbelegung der emittierten Stäube mikroanalytisch untersucht werden.

Rasterelektronenmikroskopische Untersuchungen an Reingasstäuben oder Staubniederschlagsproben wurden im Bereich der Zementindustrie bisher nur vereinzelt durchgeführt, um einen quellenspezifischen Zusammenhang zwischen Emission und Immission zu untersuchen. Es hat sich aber gezeigt, das durch REM-Aufnahmen emittierter Stäube, die Aufnahme des Röntgenbeugungsspektrums und des EDX-Spektrums zementproduktionsspezifische Merkmale wiedergefunden werden können, die die Zuordnung zu einzelnen Quellen im Werk ermöglichen.

Das Immissionsverhalten von Stäuben aus dem Zementproduktionsprozeß kann zukünftig stärker quellenbezogen untersucht werden. Der Beitrag einzelner Emissionsquellen eines Zementwerkes zur Immissionssituation wird dadurch darstellbar. Fehlinvestitionen durch oft geforderte pauschale Minderungsmaßnahmen werden dadurch vermeidbar. Darüber hinaus kann ein Betreiber angesichts der verschärften Umwelthaftung zukünftig leichter den Einfluß der Emissionen auf die Immissionssituation in der Umgebung des Werkes darstellen.

Various instruments are used to measure particulate deposits; in the cement industry, adhesive foil instruments are being used for this purpose for the first time. Whilst collecting devices e.g. those operating by the Bergerhoff method are used especially for determining the mass deposition rate, adhesive foil instruments permit a particle-specific evaluation of the dust deposit. The foil squares approx. 7 cm × 7 cm in size are placed in the open air in accordance with the instructions of the VDI Code of Practice 2119, sheet 4, draft. The protective tube screens the surface to a large extent from the effects of wind and rain. The deposition of coarse dust fractions, flower pollen or other organic particles is thereby effectively prevented. In view of the short exposure time compared with collecting devices, most of the particles on the adhesive foil are separate. A routine determination of the number and mass deposition rate can therefore be carried out by light microscopy examinations, particularly with modern methods of image analysis. By analysing the particle size as well as the morphological and optical properties of the precipitated dusts, it is possible to classify them as being of biogenic or anthropomorphic origin.

Investigations by scanning electron microscopy (SEM) also provide detailed information about the surface structure of the particles. Phase and trace element analyses of clean gas dusts from coal-fired power stations and, for the first time, waste incineration plants, have been documented. The heavy metal content of the dusts emitted has been examined microanalytically using energy-dispersive X-ray methods.

Until now, analyses of clean gas dusts or dust deposit samples by scanning electron microscopy have been carried out only sporadically in the cement industry to examine the source-specific correlation between emission and immission. It has become apparent, however, that cement production-specific characteristics have been identified by SEM photographs of emitted dusts, photographs of the X-ray diffraction spectrum and the EDX spectrum, as a result of which dusts can be assigned to individual sources in the works.

The immission behaviour of dusts from the cement production process can be examined on a more source-related basis in the future. As a result, it will be possible to show the contribution made by individual emission sources of a cement works to the immission situation, thereby avoiding bad investments by blanket reduction measures. Moreover, faced with stricter environmental liability, an operator will be able to show the effect of emissions on the immission situation in the vicinity of the works more easily in the future.

Fachbereich 4 · Subject 4 · Séance Technique 4 · Tema de ramo 4

Gezielte Lärmminderungsprogramme für Zementwerke
Specific noise reduction programme for cement works

Programmes ciblés pour la réduction du bruit des cimenteries

Programas bien enfocados para la reducción de los ruidos producidos en las fábricas de cemento

Von **M. Schneider** und **P. Küllertz,** Düsseldorf/Deutschland

Zusammenfassung – Bei Geräuschprognosen für Zementwerke wird die Geräuschimmission in der Nachbarschaft aus der Geräuschemission der Anlagen berechnet. Grundlage sind umfangreiche Lärmmessungen im Nahbereich der Hauptlärmquellen sowie Kenndaten (Schalldruckpegel, Frequenzspektren) aus einer 30jährigen Meßpraxis. Bei der Durchführung effizienter Lärmminderungsmaßnahmen haben sich in letzter Zeit Werkslärmkarten bewährt, die zunächst nur den Ist-Zustand der Lärmsituation beschreiben. Anschließend werden auf der Basis dieser meßtechnisch ermittelten Lärmverteilung die Auswirkungen geplanter Minderungsmaßnahmen auf die Lärmimmissionssituation in der Umgebung des Werks errechnet. Der Schallausbreitungsrechnung liegt ein Modell nach VDI-Richtlinie 2714 zugrunde, das die verschiedenen Einflüsse der Schallausbreitung berücksichtigt. Sämtliche Lärmemissionsparameter, die verwendet werden, sind variierbar und erlauben die Ausarbeitung gezielter, kostengünstiger Pläne zur Reduzierung der Lärmimmission. Abschirmung durch Gebäude werden ebenso berücksichtigt wie sekundäre Schallminderungsmaßnahmen. Gestalt und Position von Schallschirmen, z.B. Erdwällen, können so den vorgegebenen Emissionsbedingungen angepaßt werden.

Gezielte Lärmminderungsprogramme für Zementwerke

Summary – In sound predictions for cement works the noise immission in the neighbourhood is calculated from the noise emission from the plants. The basis is extensive noise measurements close to the main sources of noise and characteristic data (sound pressure levels, frequency spectra) from 30 years of practical measurements. Works noise maps, which initially only describe the current noise situation, have recently proved their value when carrying out efficient noise reduction measures. The effects of planned reduction measures on the noise immission situation near the works are then calculated on the basis of this measured noise distribution. The noise propagation calculation is based on a model given in VDI guidelines 2714 which takes account of the various factors affecting the sound propagation. All the noise emission parameters which are used are variable and make it possible to work out specific plans for reducing noise immission at reasonable cost. Screening by buildings is taken into account, as are secondary sound reduction measures. In this way the shape and position of sound barriers, e.g. earth embankments, can be adapted to suit the given emission conditions.

Specific noise reduction programme for cement works

Résumé – Lors de pronostics de bruit de cimenteries, est calculée l'immission du bruit dans le voisinage, à partir de l'émission de bruit des installations. Les bases en sont des mesures de bruit exhaustives à proximité immédiate des sources principales, ainsi que les valeurs caractéristiques (niveau d'intensité du bruit, spectres de fréquences) connues suite à une expérience de 30 années de mesures. Dans l'appliquation de mesures efficaces de réduction du bruit, se sont révélées excellentes, ces derniers temps, des cartes du bruit de l'usine ne décrivant, en premier lieu, que l'état actuel de la situation du bruit. Ensuite sont calculées, à partir de cette distribution du bruit déterminée par des mesures, les conséquences d'actions de réduction projetées, sur la situation d'immission de bruit dans le voisinage de l'usine. Le calcul de la propagation de bruit est basé sur un modèle selon la directive VDI 2714, qui tient compte des différentes influences de la propagation du bruit. L'ensemble des paramètres d'émission de bruit pris en compte peut être modulé et permet l'élaboration de projets ciblés, économiques, pour la réduction de l'émission de bruit. Des écrans offerts par des constructions sont considérés au même titre que des mesures secondaires de réduction du bruit. La configuration et la position d'écrans anti-bruit, comme p. ex. des talus de terre, peuvent ainsi être adaptées aux conditions d'émission existantes.

Programmes ciblés pour la réduction du bruit des cimenteries

Resumen – En los pronósticos de los ruidos producidos en las fábricas de cemento se calcula la inmisión sonora a proximidad de la fábrica, basándose en los ruidos producidos dentro de la misma, los cuales se miden cerca de las principales fuentes de ruidos, así como en los datos característicos (nivel de presión sonora, espectros de frecuencias) obtenidos en 30 años de experiencias con esta clase de mediciones. Durante la aplicación de medidas eficaces de reducción de ruidos, han probado su eficacia últimamente las tarjetas sonoras de las fábricas, las cuales describen de momento sólo el estado actual de la situación de ruidos. Sobre la base de esta distribución de ruidos, determinada mediante las técnicas de medición, se calculan a continuación las re-

Programas bien enfocados para la reducción de los ruidos producidos en las fábricas de cemento

percusiones de las medidas de reducción previstas sobre las inmisiones sonoras a proximidad de la fábrica. El cálculo de propagación de los ruidos se basa en un modelo de la Directiva VDI 2714, que tiene en cuenta los diferentes influjos de la propagación de ruidos. Todos los parámetros de emisión de ruidos que se utilizan son variables y permiten la elaboración de proyectos, bien enfocados y económicos, de reducción de los ruidos. Se tienen en cuenta tanto los edificios que puedan servir de pantalla como también las medidas secundarias de reducción de los ruidos. De esta manera, la forma y posición de las pantallas sonoras, por ejemplo terraplenes, pueden ser adaptadas a las condiciones existentes.

Die Schallimmissionssituation in der Umgebung eines Zementwerkes ist auf eine Vielzahl von Lärmemissionsquellen zurückzuführen. Dabei tragen die verschiedenen Schallquellen auf dem Werksgelände jeweils unterschiedlich zur Schallimmission in der Umgebung des Werkes bei.

Lärmminderungsmaßnahmen in Zementwerken erfordern daher eine detaillierte Untersuchung der Immissions- und Emissionssituation. Nur so kann eine gezielte Verringerung der Schallabstrahlung erreicht werden. Pauschale Maßnahmen hingegen sind im allgemeinden teuer und führen oft zu unbefriedigenden Ergebnissen.

Im Rahmen von Lärmminderungsprogrammen haben sich bei der Arbeit des Forschungsinstituts der Zementindustrie daher in letzter Zeit sogenannte Werkslärmkarten bewährt, die zunächst nur den Istzustand der Schallsituation auf dem Werksgelände beschreiben. Hierzu werden an sogenannten Rasterpunkten auf dem Werksgelände im Abstand von ca. 25 m jeweils Schallpegelmessungen durchgeführt, die die Lärmzentren und damit auch die bestimmenden Lärmquellen aufzeigen.

The noise immission situation in the neighbourhood of a cement works is due to a multitude of noise emission sources, the various sources of sound on the works site each contributing to a different extent to the noise immission near the works.

Noise reduction measures in cement works therefore necessitate a detailed study of the immission and emission situations. Only in this way can a specific reduction of the sound radiation be achieved. Across-the-board measures, on the other hand, are generally expensive and often produce unsatisfactory results.

In the context of noise reduction programmes, works noise maps, which initially only describe the current noise situation on the works site, have therefore recently proved their value in the work of the Research Institute of the Cement Industry. For this purpose, sound level measurements which indicate the centres of noise and thus the dominant noise sources are taken at grid points spaced approx. 25 m apart on the works site.

Fig. 1 shows the noise situation in the form of a works noise map of this kind, such as results from the grid measurements on the works site. The lines of equal sound pressure, also termed "isobars", show the areas of the works site in which dominant noise emission sources are present. In the case in question, relevant noise emissions stem from the area of the rotary kiln, where sound pressure levels of over 80 dB(A) were measured.

The plotting of a works noise map should be followed by sound measurements at the relevant individual sources during a tour of the site in the area of the centres of noise. These emission measurements are then used as the basis for a mathematical representation of the immission situation in association with a noise propagation calculation.

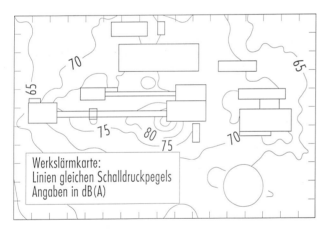

BILD 1: Werkslärmkarte
FIGURE 1: Works noise map
(Lines of equal sound pressure level, figures in dB(A))

Bild 1 zeigt die Lärmsituation in Form einer derartigen Werkslärmkarte, so wie sie sich aufgrund der Rastermessungen auf dem Werksgelände ergibt. Die Linien gleichen Schalldrucks, auch Isobaren genannt, zeigen, in welchen Bereichen auf dem Werksgelände dominante Schallemissionsquellen vorliegen. Im vorliegenden Fall gehen relevante Schallemissionen vom Bereich der Drehofenanlage aus, wo Schalldruckpegel von über 80 dB(A) ermittelt wurden.

Im Anschluß an diese Aufnahme einer Werkslärmkarte sind im Rahmen einer Begehung im Bereich der Lärmzentren Schallmessungen an den relevanten Einzelquellen durchzuführen. Diese Emissionsmessungen dienen anschließend als Grundlage für eine mathematische Darstellung der Immissionssituation im Rahmen einer Ausbreitungsrechnung.

Bild 2 zeigt an einem weiteren Beispiel die auf rechnerischem Wege bestimmten Anteile der Lärmquellen des betrachteten Werkes an der Schallimmission. Im vorliegenden Fall wurde die Schallimmission für einen Beurteilungs-

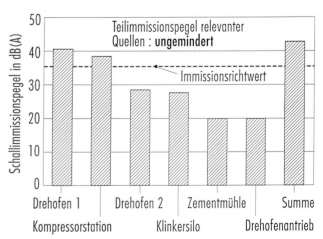

BILD 2: Rechnerisch ermittelte Teilimmissionspegel relevanter Schallquellen (ohne Minderungsmaßnahmen)
FIGURE 2: Component immission levels from relevant sources: before noise reduction

Schallimmissionspegel in dB(A) = Noise immission level in dB(A)
Immissionsrichtwert = Standard value for noise immission
Drehofen 1 = Rotary kiln no. 1
Drehofen 2 = Rotary kiln no. 2
Kompressorstation = Compressor station
Klinkersilo = Clinker silo
Zementmühle = Cement mill
Drehofenantrieb = Rotary kiln drive system
Summe = Overall level

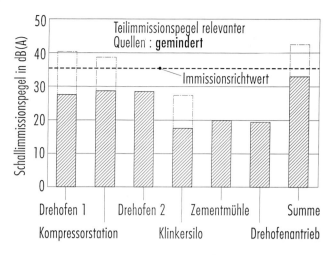

BILD 3: Rechnerisch ermittelte Teilimmissionspegel relevanter Schallquellen (mit Minderungsmaßnahmen)
FIGURE 3: Component immission levels from relevant sources: after noise reduction

Schallimmissionspegel in dB(A)	= Noise immission level in dB(A)
Immissionsrichtwert	= Standard value for noise immission
Drehofen 1	= Rotary kiln no. 1
Drehofen 2	= Rotary kiln no. 2
Kompressorstation	= Compressor station
Klinkersilo	= Clinker silo
Zementmühle	= Cement mill
Drehofenantrieb	= Rotary kiln drive system
Summe	= Overall level

punkt in der Umgebung des Werkes ermittelt. Aus dem Bild ist ersichtlich, daß der größte Teilimmissionspegel von etwa 40 dB(A) vom Hauptkamin der Drehofenanlage 1 verursacht wurde.

Die nächst-dominanten Schallquellen in diesem Beispiel stellen die Außenwände der Kompressorenstation dar. Insbesondere über die Ansaugöffnungen kann hier ein wesentlicher Anteil des Rauminnenpegels in die Umgebung abgestrahlt werden.

Darüber hinaus existieren eine Reihe weiterer Lärmquellen, die relevante Anteile zum Summenpegel in der Umgebung beitragen. Der Summenpegel, als Balken im rechten Teil des Bildes 2 dargestellt, beträgt im vorliegenden Fall fast 42 dB(A). Er liegt damit deutlich über dem hier einzuhaltenden gebietsbezogenen Immissionsrichtwert von 35 dB(A).

Bild 3 zeigt die gleiche Situation. Hierbei werden zunächst rechnerisch gezielte Maßnahmen zur Verringerung der Schallabstrahlung abgeschätzt. Durch den Einbau von Schalldämpfern im Abgaskamin der Drehofenanlage 1 sowie in den Abluftkaminen der Klinkersiloentstaubung wird die Schallabstrahlung verringert. Durch weitere konstruktive Maßnahmen an den Außenwänden der Kompressorstation kann der Schallimmissionspegel am betrachteten Beurteilungspunkt auf unter 35 dB(A) reduziert werden. Die Rechnungen zeigen ferner, daß der Einbau eines Schalldämpfers im Abgaskamin des Drehofenanlage 2 und im Abluftkamin der Zementmühle hier nicht erforderlich ist, um die Einhaltung des gebietsbezogenen Immissionsrichtwertes zu gewährleisten.

Fig. 2 shows, with the aid of a further example, the mathematically determined contributions by the noise sources in the works in question to the level of noise immission. In the case under study, the level of noise immission was determined for an assessment point in the neighbourhood of the works. It is clear from the chart that the largest component of the noise immission level of approx. 40 dB(A) was caused by the main stack of rotary kiln no. 1.

The next most dominant sources of sound in this example are the outside walls of the compressor station. Here, a substantial proportion of the internal noise level may be radiated into the environment, especially via the air intake ports.

In addition, there are a range of other noise sources which contribute relevant proportions to the overall level in the neighbourhood. The overall level, shown as a bar at the right of Fig. 2, is in this case almost 42 dB(A). It is thus well above the standard value for noise immission for the area of 35 dB(A) which has to be met here.

Fig. 3 shows the same situation. Here, specific measures to reduce the sound radiation are assessed, initially on a mathematical basis. The sound radiation is reduced by fitting silencers to the flue gas stack of rotary kiln no. 1 and to the exhaust air stacks of the clinker silo dedusting system. The noise immission level at the assessment point in question can be reduced to less than 35 dB(A) by further structural work on the outside walls of the compressor station. The calculations also show that in this case it is not necessary to fit a silencer to the flue gas stack of rotary kiln no. 2 or to the

BILD 4: Schallpegelverteilung auf dem Werksgelände (Isoliniendarstellung) vor Durchführung gezielter Lärmminderungsmaßnahmen
FIGURE 4: Sound level distribution on the works site (showing isolines) before carrying out specific noise reduction measures

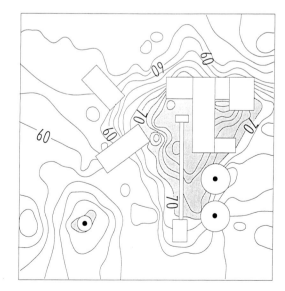

BILD 5: Schallpegelverteilung auf dem Werksgelände (Isoliniendarstellung) nach Durchführung gezielter Lärmminderungsmaßnahmen
FIGURE 5: Sound level distribution on the works site (showing isolines) after carrying out specific noise reduction measures

Die beiden **Bilder 4** und **5** zeigen in Form einer Werkslärmkarte die gemessene Schalldruckpegelverteilung auf dem Werksgelände vor und nach Durchführung gezielter Lärmminderungsmaßnahmen.

Bei Vergleich der 70-dB(A)-Isobaren in beiden Werkslärmkarten ist eine deutliche Verbesserung der Lärmsituation durch die Minderungsmaßnahmen zu erkennen. Die Reduzierung der von diesen 70-dB(A)-Isobaren eingeschlossenen, schraffiert dargestellten Fläche ist dabei ein Maß für die Verringerung der Schallabstrahlung. Im vorliegenden Fall wurden in einem ersten Schritt Minderungsmaßnahmen im Bereich der Zementmahlanlage und des Klinkerkühlers durchgeführt. Dadurch ergibt sich eine Halbierung der insgesamt von den Werksanlagen abgestrahlten Schalleistung, entsprechend einer Verringerung um 3 dB.

Zusammenfassend ist festzustellen, daß mit Hilfe von Werkslärmkarten eine gezielte und damit kostengünstige Durchführung von Lärmminderungsmaßnahmen möglich ist. Durch eine einmalige Aufnahme der Schallsituation auf dem Werksgelände und der anschließenden rechnerischen Umsetzung im Rahmen eines Schallausbreitungsmodells können die Minderungseffekte simuliert und Minderungsmaßnahmen optimal eingesetzt werden.

Neben einer gezielten Verbesserung der bestehenden Lärmsituation können auf diese Weise auch zukünftig Veränderungen der Schallimmission untersucht werden, wenn beispielsweise im Rahmen einer Änderungsgenehmigung Anlagen bestehender Betriebsteile modernisiert oder ergänzt werden sollen.

exhaust air stack of the cement mill in order to ensure compliance with the standard value for noise immission in the area.

Figs. 4 and **5** show the measured sound pressure level distribution on the works site before and after the implementation of specific noise reduction measures in the form of a works noise map.

It is evident from a comparison of the 70-dB(A) isobars on the two works noise maps that a substantial improvement in the noise situation has been achieved by the reduction measures. The reduction in the shaded area enclosed by these 70-dB(A) isobars is a measure of the reduction of the sound radiation. In the case in question, the first step was to carry out reduction measures in the area of the cement grinding plant and the clinker cooler. These measures have halved the total sound power radiated from the plant and machinery, which corresponds to a reduction of 3 dB.

Summarising, it can be stated that works noise maps enable noise reduction measures to be implemented in a specific and thus cost-effective manner. By making a one-off record of the noise situation on the works site and then mathematically converting it with the aid of a sound propagation model, the reduction effects can be simulated and reduction measures can be applied in an optimum manner.

Besides a specific improvement in the existing noise situation, this method also enables future changes in noise immission levels to be investigated if, for example, the plant and machinery of existing parts of the works is to be modernised or added to under the terms of planning permission for alterations.

Entwicklung der Immissionen in der Umgebung der Zementwerke in den neuen Bundesländern am Beispiel Rüdersdorf

Development of the immissions in the vicinity of the cement works in the new Federal Länder using Rüdersdorf as an example

Evolution des immissions dans le voisinage des cimenteries dans les nouveaux Länder Fédéraux, à l'exemple de Rüdersdorf

Evolución de las inmisiones a proximidad de las fábricas de cemento en los 5 nuevos Estados federados, citando el ejemplo de Rüdersdorf

Von **P. Scur**, Rüdersdorf/Deutschland

Zusammenfassung – Die Zementindustrie der ehemaligen DDR war gekennzeichnet durch eine hohe Konzentration der Produktion im wesentlichen an vier Standorten. In den Werken Bernburg, Deuna, Karsdorf und Rüdersdorf betrugen die Jahreskapazitäten jeweils 2,5 bis 3 Mio. t. Infolge unzureichender Entstaubungsanlagen war die Umgebung aller Werke stark durch hohe Staubbelastungen geprägt. Am Standort Rüdersdorf befanden sich nach Inbetriebnahme des zuletzt gebauten Werkes 4 im Jahre 1966 drei Betriebsteile mit insgesamt zwölf Ofenanlagen. Deren Leistung lag jeweils zwischen 300 und 1000 t/d. Meßwerte aus der unmittelbaren Betriebsnachbarschaft aus den 70er Jahren lagen bis zum Faktor 20 über den jetzt gültigen Immissionsrichtwerten. Mittlerweile wurden in allen Zementwerken der neuen Bundesländer Konzepte zur Konzentration der Produktion auf wenige Ofenlinien erarbeitet und teilweise bereits realisiert. Das Rüdersdorfer Konzept sieht die Sanierung von zwei Lepolöfen und den Bau eines neuen Ofens mit einer Kapazität von 5000 t/d vor. Damit sollen insbesondere die staubförmigen Emissionen bis 1995 gegenüber 1989 um 95% gesenkt werden.

Entwicklung der Immissionen in der Umgebung der Zementwerke in den neuen Bundesländern am Beispiel Rüdersdorf

Summary – The cement industry in the former GDR was characterized by a high concentration of the production in essentially four locations. The yearly capacities of the Bernburg, Deuna, Karsdorf and Rüdersdorf works were each between 2.5 and 3 million t. As a result of inadequate dedusting systems the surroundings of each of the works was characterized by high dust levels. After the commissioning of the most recently built works 4 in 1966 there were three operating sections at the Rüdersdorf site with a total of twelve kiln plants. Their outputs lay between 300 and 1000 t/d. Values measured in the immediate vicinity of the plant in the 70s were 20 times higher than the current immission guide values. Schemes for concentrating the production in a few kiln lines have now been worked out for all the cement works in the new Federal Provinces, and some have already been put into practice. The Rüdersdorf scheme envisages the renovation of two Lepol kilns and the construction of a new kiln with a capacity of 5000 t/d. The intention is that by 1995 the dust emissions in particular will be lowered by 95% when compared with the 1989 levels.

Development of the immissions in the vicinity of the cement works in the new Federal Länder using Rüdersdorf as an example

Résumé – L'industrie cimentière de l'ancienne RDA était caractérisée par une forte concentration des lieux de production, essentiellement sur quatre sites. Aux usines Bernburg, Deuna, Karsdorf et Rüdersdorf, les capacités annuelles étaient respectivement de 2,5 à 3 Mt. A cause des installations de dépoussiérage déficientes, le voisinage de toutes ces usines était soumis à des fortes précipitations de poussières. Sur le site de Rüdersdorf se trouvaient, après la mise en service de la dernière usine N° 4, construite en 1966, trois secteurs de production avec, au total douze lignes de four. Leur débit était respectivement de 300 à 1000 t/j. Les mesures effectuées dans le voisinage immédiat, au cours des années 70, étaient jusqu'à 20 fois supérieures aux valeurs d'immission tolérées actuellement. Entretemps et dans toutes les cimenteries des nouveaux Länder Fédéraux, ont été élaborés et déjà partiellement réalisés des projets de concentration de la production sur un minimum de lignes de four. Le projet de Rüdersdorf prévoit l'assainissement de deux fours Lepol et la construction d'un nouveau four, d'une capacité de 5000 t/j. Conjointement, les émissions surtout sous forme de poussières doivent être réduites, en 1995, de 95% par rapport au niveau de 1989.

Evolution des immissions dans le voisinage des cimenteries dans les nouveaux Länder Fédéraux, à l'exemple de Rüdersdorf

Resumen – *La industria del cemento de la antigua RDA estaba caracterizada, esencialmente, por una elevada concentración de la producción en cuatro lugares. En cada una de las factorías de Bernburg, Deuna, Karsdorf y Rüdersdorf, las capacidades de producción anuales ascendían a 2,5 - 3 millones de toneladas. Puesto que las instalaciones de desempolvado eran insuficientes, los alrededores de todas las fábricas estaban caracterizadas por una elevada precipitación de polvo. En Rüdersdorf había, después de la última planta 4, construida en el año 1966, tres secciones con un total de doce instalaciones de hornos. La capacidad de cada una de ellas ascendía a 300 – 1.000 t/d. Los valores de medición de las inmisiones obtenidos en los años 70 a proximidad de las plantas, eran hasta 20 veces superiores a los índices vigentes en la actualidad. Entretanto, se han elaborado, y parcialmente realizado, en todas las fábricas de cemento de los nuevos Estados federados, conceptos que prevén la concentración de la producción en un reducido número de líneas de hornos. El concepto de Rüdersdorf prevé el saneamiento de dos hornos Lepol y el montaje de un horno nuevo, con una capacidad para 5.000 t/d. Con ello, se pretende reducir, sobre todo, las emisiones en forma de polvo hasta 1995 en un 95 %, en comparación con el año 1989.*

Evolución de las inmisiones a proximidad de las fábricas de cemento en los 5 nuevos Estados federados, citando el ejemplo de Rüdersdorf

Die Zementindustrie der ehemaligen DDR war gekennzeichnet durch eine hohe Konzentration der Produktion in 4 Großbetrieben. An den Standorten in Bernburg, Deuna, Karsdorf und Rüdersdorf wurden jährlich jweils 2,5–3 Millionen t Zement produziert. Die Produktion erfolgte generell in vielen kleineren Anlagen mit entsprechend hohem Wartungsaufwand. Infolge der unzureichenden Entstaubungsanlagen ist die Umgebung aller Werke stark durch die hohen Staubbelastungen der letzten 20 bis 30 Jahre geprägt, so daß die Entwicklung in Rüdersdorf stellvertretend für alle Betriebe dargestellt werden kann.

In Rüdersdorf befanden sich nach Inbetriebnahme des Werkes 4 im Jahre 1966 drei Zementwerke mit insgesamt 12 Ofenanlagen. Folgende Situation bei der Abgasreinigung der Ofenabgase in den 70er Jahren soll die Umweltprobleme verdeutlichen:

Zementwerk 2: keine Abgasreinigung.

Zementwerk 3: Zyklonanlagen, die selten betrieben wurden.

Zementwerk 4: Vertikal-Elektroabscheider mit zeitlichen Verfügbarkeiten unter 50 %.

Um die Entwicklung der Immission analysieren zu können, muß man sich zuerst mit der Ursache, der Emission, auseinandersetzen. Das **Bild 1** zeigt die Staubemissionen der Rüdersdorfer Zementwerke in den letzten 20 Jahren. Aus der Grafik lassen sich 4 Etappen ableiten:

1. Mitte der siebziger Jahre – maximale Emissionen durch Produktion ohne Rücksicht auf die Umwelt.

2. Ende der siebziger Jahre – Wirksamwerden der 5. Verordnung zum Landeskulturgesetz mit der 1. Durchführungsbestimmung aus dem Jahre 1973. In den Betrieben wurden Umweltschutzabteilungen gebildet, die behördliche Kontrolle wurde verstärkt, und Umweltschutzmaßnahmen bekamen eine größere Bedeutung. In Rüdersdorf wurde die Zyklonentstaubung an den Öfen im Zementwerk 3 verbessert, an den Öfen des Zementwerkes 2 wurden Heißgasgewebefilter errichtet, im Zementwerk 4 wurden an den Elektrofiltern Verfügbarkeiten über 70 % erreicht. Von den Behörden wurden Grenzwertbescheide für Hauptemissionsquellen erstellt.

3. Mitte der 80er Jahre – Stagnation des erreichten Standes, da falsche Investitionsstrategien verfolgt wurden und die Kapazitäten des Maschinenbaus die notwendigen Ersatzinvestitionen nicht erlaubten.

4. Ab 1989 – drastische Emissionssenkung nach der Wiedervereinigung Deutschlands infolge Anlagestillegung und Realisierung von Neuinvestitionen mit hohen Umweltschutzansprüchen.

An 3 weiteren Bildern soll der Einfluß der Emissionen der 3 Betriebsteile auf den Staubniederschlag in unmittelbarer Betriebsnähe dargestellt werden. An den beiden Altwerken 2 und 3 ist der unmittelbare Zusammenhang sehr gut zu erkennen. Die Absenkung der Emission führt zu einer deutlich sichtbaren Abnahme der Immission. Es muß allerdings hervorgehoben werden, daß der Ausgangswert von über 6 g/m² · d im Zemenmtwerk 2 eine über 20fache Überschrei-

The cement industry in the former East Germany was characterised by a high concentration of production in four large plants. The annual cement capacities of the Bernburg, Deuna, Karsdorf and Rüdersdorf works were each between 2.5 and 3 million tonnes. Production was generally carried out in a large number of fairly small plants with correspondingly high maintenance costs. As a result of inadequate dust separation systems, the surroundings of all the works have had high dust levels for the past 20 to 30 years, so developments in Rüdersdorf can be regarded as representative of all the plants.

After works 4 came on stream in 1966 in Rüdersdorf, there were three cement works with a total of 12 kilns. The following situation regarding the purification of kiln waste gases in the 70s will illustrate the environmental problems:

Cement works 2: no waste gas purification.

Cement works 3: cyclones, which were seldom in operation.

Cement works 4: vertical electrostatic precipitator available for less than 50 % of the time.

The cause of the emissions must be understood before the development of the immissions can be analysed. **Fig. 1** shows the dust emissions from the Rüdersdorf cement works over the past 20 years. Four stages can be derived from the graph:

1. Mid-seventies: maximum emissions due to production without regard for the environment.

2. End of the seventies: The 5th Ordinance relating to the German Soil Improvement Act takes effect with the 1st implementation regulation stemming from 1973. Environmental protection departments were set up in the plants, official controls were stepped up and environmental protection measures took on a greater significance. In Rüdersdorf, the cyclone dust separator system for the kilns in cement works 3 was improved, hot gas fabric filters were installed in the kilns of cement works 2, and availabilities of more than 70 % were achieved for the electrostatic filters in cement works 4. The authorities laid down limits for the main sources of emissions.

BILD 1: Entwicklung der Staubemissionen des Zementwerks Rüdersdorf

FIGURE 1: Development of dust emissions from the Rüdersdorf cement works

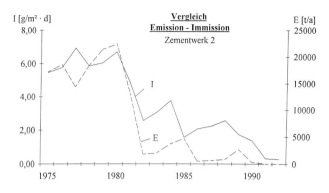

BILD 2: Entwicklung der Emissionen und Immissionen des Zementwerks 2 im Vergleich
FIGURE 2: Comparative development of emissions and immissions for cement works 2

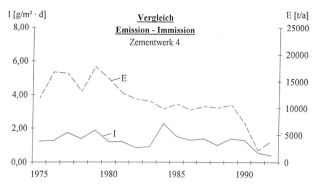

BILD 4: Entwicklung der Emissionen und Immissionen des Zementwerks 4 im Vergleich
FIGURE 4: Comparative development of emissions and immissions for cement works 4

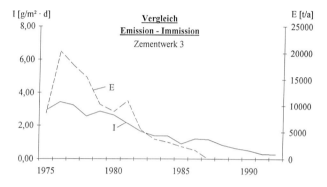

BILD 3: Entwicklung der Emissionen und Immissionen des Zementwerks 3 im Vergleich
FIGURE 3: Comparative development of emissions and immissions for cement works 3

3. Mid-eighties: Stagnation at the level reached by that time, because the wrong investment strategies were followed and the capacities of the mechanical engineering industry did not permit the necessary replacement investment.
4. 1989 onwards: drastic reduction in emissions after the reunification of Germany when plants were shut down and new investment was made with stringent environmental requirements.

The 3 other diagrams show the effect of the emissions from the 3 sectors of the plant on the deposition of dust in the immediate vicinity. The direct correlation can be seen clearly in the case of old works 2 and 3. The fall in emissions leads to an unmistakable reduction in immissions, though it should be emphasised that the starting value of more than 6 g/m² · d in cement works 2 meant that emissions were 20 times higher than the limit for dust deposits. In cement works 4, where the chimneys are much higher, the correlation is not so obvious, but the same trend is discernable.

In the meantime, schemes for concentrating production on just a few kiln lines with appropriate environmental protection facilities have been developed in all the cement works of the new Federal Provinces and some have already been put into practice. The Rüdersdorf cement works has made responsibility for the environment one of four company objectives.

The scheme envisages the renovation of 2 Lepol kilns and the construction of a new 5000 t/day kiln in cement works 4. The other kilns will be shut down. The existing cement mill in works 4 will be thoroughly modernised, as will all the filter units to be kept in operation. As a result of these measures, the emissions of about 900 kg/h in 1989 will be reduced by 95% to 45 kg/h in 1995. The first positive results can be seen from the fact that this is the first year when there has been no month so far when the dust immissions have exceeded the limit of 0.35 g/m² · day.

tung des Grenzwertes für Staubniederschlag bedeutete. Im Zementwerk 4, in dem die Schornsteine wesentlich höher sind, ist der Zusammenhang nicht so deutlich. Ein gleicher Trend ist aber erkennbar.

Mittlerweile wurden in allen Zementwerken der neuen Bundesländer Konzepte zur Konzentration der Produktion auf wenige Ofenlinien mit entsprechenden Umweltschutzkonzepten erarbeitet und teilweise bereits realisiert. Die Rüdersdorfer Zement GmbH haben die Verantwortung für die Umwelt als eines von vier Unternehmenszielen festgeschrieben.

Das Konzept sieht die Sanierung von 2 Lepolöfen und den Bau eines 5000-t/d-Ofens im Zementwerk 4 vor. Die restlichen Ofenanlagen werden stillgelegt. Die vorhandene Zementmahlanlage im Werk 4 wird grundlegend modernisiert. Dies betrifft auch alle weiterzubetreibenden Filteranlagen. Durch die Maßnahmen werden die Emissionen von ca. 900 kg/h im Jahre 1989 auf 45 kg/h im Jahre 1995, also um 95%, gesenkt. Daß sich die ersten Erfolge einstellen ist daran zu sehen, daß erstmalig in diesem Jahr bisher kein Monatswert des Staubimmissionsmeßnetzes über dem Grenzwert von 0,35 g/m² · d liegt.

NO_x-Minderung – ein systematisches Konzept
NO_x reduction – a systematic approach

Diminution de NO_x – un concept systématique

La reducción de NO_x – un concepto sistemático

Von **K. Thomsen,** Valby/Dänemark

Zusammenfassung – *In einer Welt mit sich ständig verändernden Emissionsauflagen ist es unerläßlich, nicht nur die heute gültigen Anforderungen zu erfüllen, sondern auch für die morgen zu erwartenden gerüstet zu sein. Das FLS-Konzept umfaßt eine Werksstudie, die zur Auswahl einer einzelnen Lösung oder eines Lösungspakets zur NO_x-Minderung führt. Neue Anlagen werden als NO_x-arme Systeme ausgelegt oder für die Umstellung auf NO_x-armen Betrieb vorbereitet. Die verschiedenen Optionen, wie Veränderungen an der Konstruktion der Calcinatoranlage, NO_x-Fuzzy-Steuerung, NH_3-Eindüsung, Brenneraustausch usw., die vom Ofentyp abhängen, müssen geprüft und unter Berücksichtigung der vorhandenen Geräte, des Ofenbetriebs, der Rohstoffsituation, des verwendeten Brennstoffs und schließlich der Wirtschaftlichkeits- und Qualitätsanforderungen an den Einzelfall angepaßt werden.*

Summary – *In a world with ever changing emission regulations, it is of vital importance not only to meet today's requirements, but also to be prepared for those expected of tomorrow. The FLS approach comprises a plant study resulting in selecting one or a combined solution from a NO_x-reduction package. New installations are designed as low-NO_x systems or prepared for conversion to low-NO_x systems. The different options, e.g. change in calciner design, NO_x-fuzzy control, NH_3 injection, replacement of burner, etc. related to the type of kiln system, have to be examined and adjusted to each individual case taking into consideration the present equipment, the kiln operation, the raw materials situation, the fuel used, as well as the financial and quality requirements.*

Résumé – *Dans un monde où les obligations en matière d'émissions ne cessent de changer, il est indispensable non seulement de satisfaire aux exigences d'aujourd'hui, mais il faut être armé pour répondre aux normes de demain. Le concept FLS comporte une étude d'usine visant à sélectionner une solution isolée ou un ensemble de solutions pour réduire les émissions de NO_x. De nouvelles installations seront conçues comme systèmes à faible taux de NO_x ou pour la conversion en un service à faible taux de NO_x. Les différentes options, notamment modifications de la construction du calcinateur, commande Fuzzy pour NO_x, gicleur à NH_3, changement du brûleur etc., facteurs qui dépendent du four, doivent être contrôlés et adaptés au cas par cas en fonction des appareils existants, de la conduite du four, des matières premières, du combustible utilisé et, enfin, des impératifs de rentabilité et de qualité.*

Resumen – *En un mundo, en el que las prescripciones en materia de emisiones cambian constantemente, resulta imprescindible cumplir no solamente los requerimientos actualmente vigentes, sino estar preparado también para afrontar las demandas futuras. El concepto FLS comprende un estudio de fábrica, destinado a elegir una solución única o un paquete de soluciones para la reducción de las emisiones de NO_x. Las instalaciones nuevas son dimensionadas como sistemas de bajo nivel de NO_x o preparadas para su transformación en sistemas de bajo nivel de NO_x. Las distintas opciones, tales como cambios constructivos del calcinador, mando NO_x-fuzzy, inyección de NH_3, sustitución del quemador, etc., que dependen del tipo de horno, tienen que ser comprobadas y adaptadas a cada caso, teniendo en cuenta los aparatos existentes, el servicio del horno, la situación de las materias primas, el combustible utilizado y, finalmente, las exigencias de rentabilidad y de calidad.*

1. Einleitung

Zur Unterstützung der Zementhersteller in ihrer Auseinandersetzung mit den mehr und mehr sich verschärfenden Vorschriften zur NO_x-Begrenzung, bietet FL Smidth eine Reihe von Verfahren zur Lösung dieses Problems an. Um die sowohl beste als auch wirtschaftlichste Lösung unter Berücksichtigung der gegenwärtigen als auch der zu erwartenden Erfordernisse zur NO_x-Reduzierung in der Zukunft auswählen zu können, hat es sich aus Erfahrung als notwen-

1. Introduction

When contacted by cement producers asking for help to battle the more and more strict NO_x emission limits, FLS offers a number of methods to solve the problem. To select the best and most economic solution, taking the present and also possible future necessary NO_x reduction into account, experience tells that a plant study, examining the kiln operation, raw materials, fuel, and present equipment, is necessary in each individual case.

dig erwiesen, in jedem Einzelfall eine Anlagenstudie zur Untersuchung des Ofenbetriebes, der Roh- und Brennstoffe sowie der eingesetzten Ausrüstungen durchzuführen.

2. Verfahren zur NO_x-Minderung.

Tabelle 1 enthält eine Zusammenstellung verschiedener Verfahren zur NO_x-Minderung.

— Veränderung der Brennbarkeit der Rohmaterialmischung

Die Herabsetzung der Brennbarkeit einer Rohmaterialmischung führt zu einer niedrigeren Brennzonentemperatur und damit zwangsläufig auch zu einer verminderten NO_x-Bildung. Die zu erwartende NO_x-Minderung beträgt in einigen speziellen Fällen bis zu 30%, im Normalfall jedoch 5 bis 10% des nach einem Drehofen auftretenden NO_x-Gehaltes.

TABELLE 1: Verfahren zur NO_x-Minderung

Veränderung der Brennbarkeit der Rohmaterialmischung
Einsatz des automatischen Ofenüberwachungssystems Fuzzi II
Niedrige Primärluftmenge durch Einsatz des Centrax-Ofenbrenners
Änderung der Brennstoffart im Calcinator
Hochtemperaturverbrennung im Calcinator
Stufenverbrennung im Calcinator
Eindüsung von Ammoniakwasser

— Fuzzi II

Das automatische Ofenüberwachungssystem Fuzzi II kontrolliert die NO_x-Bildung im Drehofen, indem es die Prozeßluft auf einem Minimum hält und eine unakzeptabel hohe CO-Emission über den Vergleich der im Ofeneinlaufbereich gemessenen CO-, NO- und O_2-Konzentrationen vermeidet. Fuzzi II verhindert ein Überbrennen des Zementklinkers und hält die Brennzonentemperatur in einer Höhe, daß sie zur Gewährleistung der gewünschten Klinkerqualität gerade ausreicht. Die daraus resultierende niedrigere Prozeßluft und Brennzonentemperatur führen zu einem NO_x-Wert, der bis zu 30% niedriger liegt.

— Drehofenbrenner mit niedriger Primärluftmenge

Die niedrigere NO_x-Bildung bei Anwendung eines derartigen Ofenbrenners ist einerseits auf gleichmäßigere Strömungsverhältnisse ohne große Temperaturspitzen und andererseits auf die besondere Ausbildung der Ofenflamme zurückzuführen, die lokal eine reduzierende Atmosphäre erzeugt. Auch die niedrigere spezifische Feuerungsleistung trägt zur NO_x-Reduzierung bei. Die NO_x-Minderung im Drehofen liegt zwischen 5 und 15%, verglichen mit einem konventionellen Ofenbrenner.

— Veränderung der Brennstoffart

In Ofensystemen mit Vorcalcinierung kann eine beträchtliche NO_x-Senkung auf einfache Weise erreicht werden, wenn der Brennstoff verändert wird. Die NO_x-Konzentration nach dem Calcinator steht in unmittelbarer Beziehung zum eingesetzten Brennstofftyp. Die NO_x-Bildung verringert sich in der Reihenfolge der Brennstoffe: Petrolkoks, Kohle, Heizöl und Gas. Der NO_x-Wert nach einem Calcinator kann bis über 60% gesenkt werden, wenn in Abhängigkeit vom Calcinatorofen von Petrolkoks auf Heizöl oder Gas umgestellt wird.

— Hochtemperaturverbrennung im Calcinator

Im Grunde vollziehen sich im Calcinator zwei gegenläufige Reaktionen. Bei der einen Reaktion wird NO_x gebildet, bei der anderen NO_x abgebaut. Da die letztere Reaktion mit der Temperatur wesentlich schneller ansteigt als die Bildungsreaktion von NO_x, verursacht eine höhere Verbrennungstemperatur im Calcinator zwangsläufig eine Herabsetzung des NO_x-Wertes. Wenn die Temperatur im Calcinator nur um ca. 100°C erhöht wird, ist es möglich, den NO_x-Wert im Bereich von 10 bis 15% zu senken.

2. NO_x reduction methods

Table 1 shows the different NO_x reduction methods.

— Change of raw mix burnability

Lowering the raw mix burnability will cause a lower necessary burning zone temperature and thus lower the NO_x formation. The expected NO_x reduction is in some special cases up to 30% but normally 5–10% of the NO_x from a rotary kiln.

— Fuzzy II

The automatic kiln control, Fuzzy II, controls the NO_x formation in the rotary kiln by keeping the excess air down at a minimum, avoiding unacceptable CO emission by comparison of CO, NO, and O_2 measured in the kiln inlet. It prevents overburning of the clinker by keeping the burning zone temperature just adequate to ensure good clinker quality. The

TABLE 1: NO_x reduction methods

Change of raw mix burnability
Fuzzy II
Centrax burner — low primary air burner
Change of fuel type in the calciner
High temperature combustion in the calciner
Stage combustion in the calciner
Ammonia (NH_3) injection

resulting lower average excess air and burning zone temperature cause the NO_x from the rotary kiln to be reduced up to 30%.

— Low primary air burner

The lower NO_x formation with this burner type is partly due to a more uniform flow without high temperature peaks and partly due to the flow pattern in the flame creating reducing atmosphere. Also the lower specific firing rate helps to lower the NO_x. The NO_x reduction from the rotary kiln compared with a conventional burner is 5–15%.

— Change of fuel type

In precalciner kiln systems a considerable NO_x reduction is obtainable simply by changing fuel. The NO_x from the calciner is related to the fuel type in the following way with decreasing NO_x formation: Pet coke, coal oil, and gas. The NO_x from the calciner may be reduced to about 60% by changing from petcoke to oil or gas, dependent on the precalciner kiln in question.

— High temperature combustion in calciner

Basically two different gross reactions concerning NO_x are going on in the calciner. One forming NO_x and another removing NO_x. As the rate of the latter reaction is increasing more rapidly with the temperature than the first, a higher combustion temperature in the calciner causes the NO_x to decrease. If the temperature is increased by approximately 100°C, it is possible to lower the NO_x from the calciner by 10–15%.

— Stage combustion in the calciner

The FLS low NO_x calciner is divided into two sections. All the fuel but only a part of the tertiary air is supplied to the lower calciner sections, thus creating a reducing zone. The remaining tertiary air is supplied to the upper section, creating an oxidizing zone. When buying a new kiln system with a low NO_x calciner or converting the existing calciner, the NO_x from the calciner will drop by 30–50%.

— Ammonia (NH_3) injection

NH_3 injection is probably the most effective way to reduce the NO_x emission. The NH_3 is injected in a suitable place in the kiln system where the temperature is between 900°C and

1000°C. The degree of NH_3 utilization is about 70% in a calciner kiln system and the NO_x emission is reduced about 75%. In SP kilns the NH_3 utilization and the possible obtainable reduction will be somewhat smaller since the optimum temperature is hard to establish.

3. Choice of NO_x reduction method

Since, as outlined in the following, it is not possible to choose favourably between the different methods and the methods do not have the same NO_x reducing ability in the different types of kiln systems, the kilns are placed in one of the four categories as **Table 2** shows.

The applicability of the different NO_x reducing methods and estimated percentage of the NO_x reduction in the different types of kiln systems are shown in **Table 3**.

"AC" means that the method is applicable if some minor changes are made and "NA" means that the method is not applicable in the type of kiln system in question. All the estimations are based on full scale tests or on measurements made at kilns where the different methods are being used.

4. Examples of results

The Centrax burner was installed on a 4000 tpd precalciner kiln and a 1200 tpd suspension preheater kiln. The main figures are prescribed in **Table 4**. As can be seen, the NO_x emission reduction is 15% for the SP kiln and 8% for the calciner kiln without any significant change in the clinker quality. In addition, there is a small fuel saving in both kilns.

The ammonia injection system was installed in a 5000 tpd SLC-S kiln. **Fig. 1** shows the effect on the NO_x measured in the preheater exit when stopping and starting the ammonia injection into the calciner. The NO_x is being reduced from

TABLE 2: Categories of kilns

Long dry and wet kilns

Suspension preheater (SP) kilns

SP kilns with riser duct firing and calciner kilns without tertiary air duct

Precalciner kilns with tertiary air duct

TABLE 3: Guidelines for NO_x reduction

Kiln type	Change of raw mix burnability	Fuzzy II	Centrax burner	Change of fuel type in calciner	High temperature combustion in calciner	Stage combustion	Ammonia injection
1	10	20	15	NA	NA	NA	NA
2	10	20	15	NA	NA	NA	AC
3	5–10	15	10	5–15	5	AC	25
4	5	10	5–10	5–60	15	30–50	25–75

NA: not applicable
AC: applicable if some minor changes are made

TABELLE 4: NO$_x$-Minderung durch Anwendung des Centrax-Ofenbrenners

Parameter	Maß-einheit	konv. Vorwärmerofen		Vorcalcinierofen	
		vorher	nachher	vorher	nachher
Durchsatz	t/d	1200	1275	4000	4000
Primärluftanteil	%	11,3	3,7	10,5	3,6
Gesamtbrennstoffverbrauch	kcal/kg	813	790	1030	1020
NO$_x$-Gehalt im Schornstein bei 10% O$_2$	ppm	440	375	545	500
Freikalkgehalt im Klinker	%	1,3	1,3	1,53	1,65

In dieser Tabelle bedeutet „AC", daß das Verfahren mit einigen wenigen Veränderungen anwendbar ist, während die Abkürzung „NA" die Nichtanwendbarkeit des Verfahrens bei dem betreffenden Ofensystem angibt. Die in der Tabelle enthaltenen Schätzwerte basieren auf großtechnischen Versuchen oder Messungen, die an Öfen unter Anwendung der unterschiedlichen Verfahren zur NO$_x$-Minderung durchgeführt wurden.

4. Beispielhafte Ergebnisse

Tabelle 4 enthält die wichtigsten Ergebnisse, die mit dem Einsatz des Centrax-Brenners an einer Ofenanlage mit Vorcalcinierung und einem Klinkerdurchsatz von 4000 t/d sowie an einer konventionellen Ofenanlage nach dem Trockenverfahren mit einem Klinkerdurchsatz von 1200 t/d erreicht wurden. Wie aus der Tabelle hervorgeht, konnte die NO$_x$-Emission an der konventionellen Ofenanlage um 15% und an der Ofenanlage mit Vorcalcinierung um 8% ohne jegliche signifikante Veränderungen in der Klinkerqualität gesenkt werden. Weiterhin konnten bei beiden Ofensystemen auch geringe Einsparungen beim spezifischen Wärmeverbrauch erreicht werden.

Bild 1 zeigt die Auswirkungen auf die NO$_x$-Emission am Vorwärmeraustritt einer Ofenanlage SLC-S mit einem Klinkerdurchsatz von 5000 t/d, in deren Calcinator Ammoniakwasser eingedüst wurde. Die Meßschriebe in Bild 1 zeigen die Entwicklung der NO$_x$-Konzentration in Abhängigkeit von der Eindüsung des Ammoniakwassers, verdeutlicht durch einen Stop und einen Start der Eindüsung. Der NO$_x$-Gehalt konnte bei der Eindüsung von Ammoniakwasser in einer Menge von 70 l/h von ca. 790 auf 475 ppm gesenkt werden, was einem stöchiometrischen Verhältnis von NH$_3$/NO = 0,56 und einer Ammoniakausbeute um 0,70 entspricht.

TABLE 4: NO$_x$ reduction by the Centrax burner

Parameter	Unit	SP-kiln		SLC-S-kiln	
		Before	After	Before	After
Production	tpd	1200	1275	4000	4000
Primary air ratio	%	11.3	3.7	10.5	3.6
Total fuel consumption	kcal/kg	813	790	1030	1020
NO$_x$ content in stack at 10% O$_2$	ppm	440	375	545	500
Free content in clinker	%	1.3	1.3	1.53	1.65

some 790 ppm to 475 ppm injecting 70 l/h ammonia corresponding to a stoichiometric NH$_3$/NO ratio of 0.56 and an ammonia utilization of about 0.7.

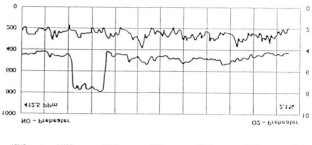
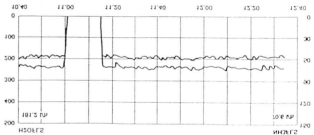

BILD 1: Der Einfluß der Eindüsung von Ammoniakwasser im Calcinator auf die NO$_x$-Konzentration

FIGURE 1: Effect on the NO$_x$ reduction by ammonia injection in the calciner

TECHNIK FÜR REINE LUFT

Alle reden von reiner Luft – wir arbeiten dafür

Entstaubung für Flugasche-Dosierung in einem Zementwerk durch SCHEUCH-Impuls-Schlauchfilter mit einer Absaugleistung von 2500 m³/h.

- Mit Absaug-, Entstaubungs- und Förderanlagen aus eigener Fertigung.
- In der Metall- und Steine-Erden-Industrie, in der Holzverarbeitung und in der Plattenerzeugung.
- Mit jahrzehntelanger Erfahrung und innovativer Technologie.
- Wir bieten Komplettlösungen von der Beratung und Projektierung bis zur Montage und Inbetriebnahme.
- Optimale Auslegung unserer Anlagen auf betriebsspezifische Bedürfnisse garantiert Energieeinsparung, Wirtschaftlichkeit, Betriebssicherheit und Wartungsfreundlichkeit.
- Durch eigene Forschung und innovative Entwicklung ist die sichere Einhaltung gesetzlicher Auflagen gewährleistet.

Aus Verantwortung gegenüber Mensch und Umwelt.

Alois Scheuch Gesellschaft m.b.H.
A-4910 Ried im Innkreis, OÖ., Am Burgfried 14
Tel. 07752/905-0, Fax-37, Telex 027-606

SCHEUCH Entstaubungstechnik GmbH
W-3063 Obernkirchen, Bornemannstr. 1, Postf. 1248
Telefon 05724/40 77, Telefax 05724/51 4 51

Die Zukunft im Blick. Elektrofilter.

Minimierung von Anlagengrößen bei gleichzeitiger Erhöhung des Wirkungsgrades, größer als 99,99%.
WALTHER-Elektrofilter mit Mikropulstechnik oder Gaskonditionierung.

WALTHER & CIE. AG, Postfach 85 05 61, D-51030 Köln,
Tel. (02 21) 67 85-0, Telex 8 873 309, Fax (02 21) 67 85-6 95

Staubfreie Luft für Ihr Verfahren	Dust - free air for your process
◆ Klinkerkühler	◆ Clinker cooler
◆ Siloanlagen	◆ Silo works
◆ Rohmehlmahl-anlage	◆ Raw meal mill
◆ Kalksteinbrecher	
◆ Zementmühle	◆ Crusher
◆ Zementverladung und Verpackung	◆ Cementmill
	◆ Cement loading and packing

Heißluftkühler für Klinkerkühlerabluft
120 000 m3n trocken/h, max. 450 °C auf 110 °C
Cooler for clinker cooler waste air
120 000 m3n dry/h, max. 450 °C down to 110 °C

KG FAVORIT Filterbau GmbH & Co.
Postfach 13 83
D-23503 Lübeck
Telefon (04 51) 4 70 03-0
Telefax (04 51) 4 11 09

Niederlassung West
KG FAVORIT Filterbau GmbH & Co.
Keplerstraße 69
D-45147 Essen
Telefon (02 01) 73 00 24
Telefax (02 01) 73 00 27

Niederlassung Polen
KG FAVORIT Filterbau GmbH & Co.
ul. Wolnosci 31/2
PL-57-100 Strzelin
Telefon/Telefax
(00 48 725) 2 03 99

Vertretung Schweiz
NAVECA AG
R. Schreier
Postfach 119
CH-1784 Courtepin
Telefon (00 41 37) 34 14 44
Telefax (00 41 37) 34 22 66

Technologie nach Maß ◆ Technology made to measure

GOOD - FLEXIBLE - DEPENDABLE - INEXPENSIVE - are not just slogans for us!

Our delivery programme:
WOKU - filter bags
WOKU - filter pockets
WOKU - filter cloths
WOKU - multifilter pockets
WOKU - filter sacks
WOKU - filter press cloths
WOKU - disc filter segments
WOKU - air-conditioning products
WOKU - supporting baskets, supporting frames
WOKU-LIGHT - leak indicator system
WOKU-CLEAN - tubular filtering bag cleaning machine
WOKUTEX - air slide fabrics
Accessories like: covers, supports, tightening scraps and much more;
Emission measurements
Advice in problem cases
Cleaning and repair of used filter materials
Filter alterations
Assembly - disassembly
Maintenance and overhaul of your filtering installation

WOKU-Filtermedien Ost
Schulstraße 4
D-06179 Teutschenthal, Germany
Tel.: (03 46 01) 2 28 30;
Fax: (03 46 01) 2 28 30

Wolfgang Kupke Filtermedien GmbH & Co. KG
Kaiser-Wilhelm-Straße 90
D-59269 Beckum-Neubeckum, Germany
Tel.: (0 25 25) 20 57-59; Fax: (0 25 25) 44 38

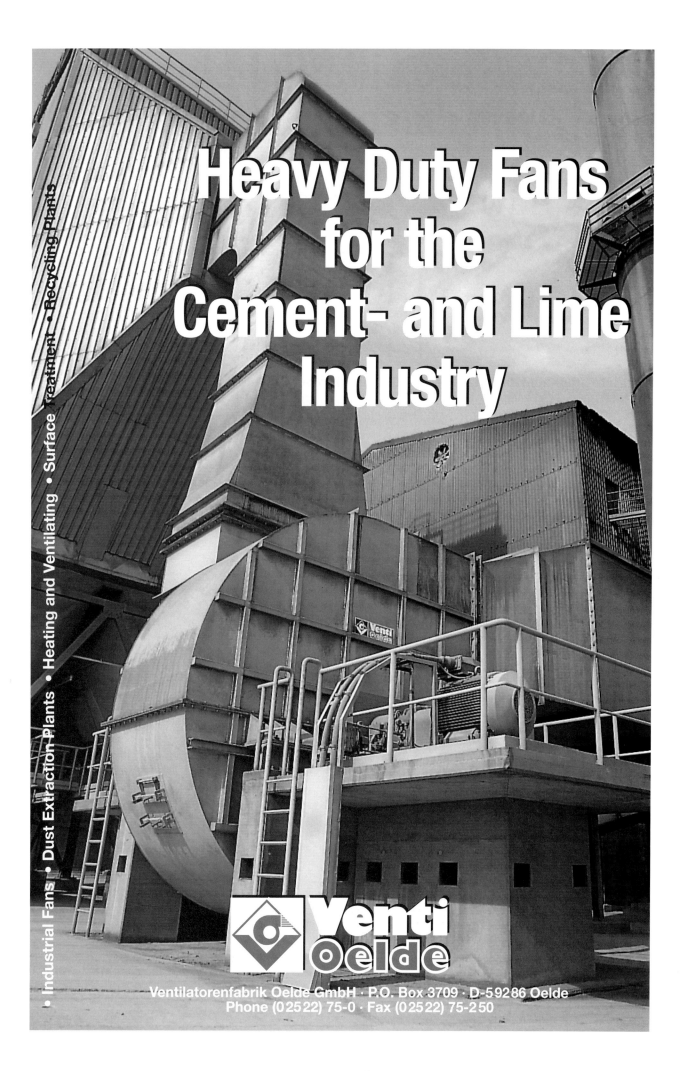

Bessere Luftreinhaltung mit diesen Etiketten

Bei Filterschläuchen mit Filtermedien aus DuPont Fasern können Sie sich auf Qualität verlassen

Dieses Etikett gibt Ihnen die Sicherheit, dass bei der Herstellung der Hochtemperatur-Filtermedien nur 100% DuPont Qualitätsprodukte verwendet werden.

Nähere Informationen erhalten Sie auf Anfrage von DuPont oder von einem autorisierten Filtermedien-Hersteller.

Du Pont de Nemours (Deutschland) GmbH
P.O. Box 1365   D-61343 Bad Homburg v.d.H.
Tel. ++49/6172/87 19 56   Fax ++49/6172/87 19 77

Du Pont de Nemours International S.A.
P.O. Box 50 CH-1218 Le Grand-Saconnex
Tel. ++41/22/717 58 70   Fax ++41/22/717 61 31

Engineering Fiber Systems

International Process Engineering in the Cement Industry

BAUVERLAG

The "Cement-Data-Book", in three volumes, is one of the most successful and extensive works on cement processing and manufacture. Volume 1 presents versatile condensed descriptions of internationally known production processes and machinery arrangements.

Contents:
Raw materials · Calculation of raw mix composition · Coarse size reduction of raw materials · Drying of raw materials · Cement production and grinding · Grinding work index according to Bond · Grinding ball data · Mill drives · The mill shell – optimum dimensions · Finish grinding · Specific mill volume and power demand · Grinding in closed circuit · Roller mills · Grinding methods in the state of development · Air separators · Wet grinding in closed circuit · Prehomogenization · Pneumatic homogenization of raw mix · Fuels in the cement industry · The rotary kiln · The raw mix suspension preheater · Clinker cooling · Kiln lining · Dust collection

By Dipl.-Ing. Walter H. Duda. 3rd revised and enlarged edition 1985. 656 pp. with 413 illustrations and 150 tables. Text in English and German. Size 21 x 29.7 cm. Hardcover DM 290,– (plus postage)
ISBN 3-7625-2137-9

BAUVERLAG GMBH · D-65173 Wiesbaden
Tel. 0 61 23 / 700-0 · Fax 0 61 23 / 700-122

Luftreinhaltung in Zementwerken auf dem Stand der Technik

Aufgaben:
Anlage 1:
Entstaubung der Abgase
einer Trockentrommel für Hüttensand
Anlage 2:
Entstaubung für
Hüttensandtransport und Siloentlüftung

Auslegungsdaten:
Anlage 1: V_A: 62 700 m³/h
T_A: 100 °C, max. 150 °C
m_S: 50 g/m³

Anlage 2: V_A: 9 500 m³/h
T_A: 40 °C
m_S: 5 g/m³

Anlagedaten:
Anlage 1: Filterfläche: 592 m²
Ventilator: 90 kW

Anlage 2: Filterfläche 120 m²
Ventilator: 15 kW

Gewährleistungen:
Standzeit des
Filtertaschenmaterials: 2 Jahre
Reingasstaubgehalt: 20 mg/m³

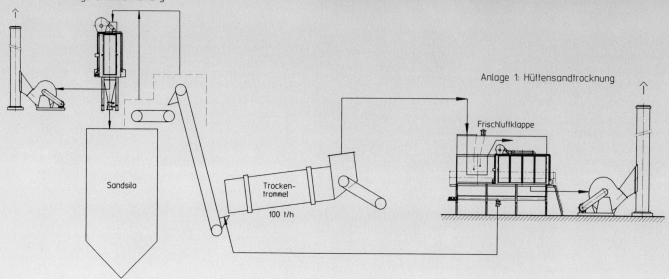

Ideen und Erfahrung...

BMD-GARANT Entstaubungstechnik GmbH
Postfach 27 · 77944 Friesenheim
Tel.: 07821/966-0 · Fax: 07821/966-245

Start into a New Quality Dimension

The new Loesche Roller Mill for cement clinker and interground additives is now on the market. The new Loesche Roller Mill System LM 2+2C/S grinds cement clinker and interground additives to consistent quality, efficiently and at high availability.

The grinding elements of the roller mill will achieve a wearlife of more than 10.000 hours at capacities of up to 140 t/h and finenesses of 3.300 to 4.500 cm^2/g according to Blaine. The specific power consumption is significantly below that of conventional systems. The new Loesche Roller Mill System LM 2+2C/S is of a simple design.

Our references will convince you.

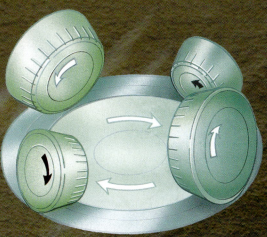

Loesche Roller Mill System, LM 2+2C/S (2 grinding rollers, 2 grinding bed preparation rollers)

Loesche GmbH
P.O. Box 11 07 36
D-40507 Düsseldorf
Telephone: (+49-211) 5353-0
Fax: (+49-211) 5353-499

Fachbereich 5

Zerkleinerungstechnik und Energiewirtschaft

(Brechen, Mahlen, Trocknen, Klassieren, Kühlen, Energiebedarf, Energieausnutzung, Energiemanagement)

Subject 5

Comminution technology and energy management

(Crushing, grinding, drying, classifying, cooling, energy consumption, energy utilization and management)

Séance Technique 5

Technique de broyage et gestion d'énergie

(Broyage, mouture, séchage, classification, refroidissement, consommation d'énergie, utilisation et gestion d'énergie)

Tema de ramo 5

Técnica de trituración y economía energética

(Trituración, molienda, secado, tamizado, enfriamiento, consumo energético, utilización de energía, management de energía)

Zerkleinerungstechnik und Energiewirtschaft*)

Comminution technology and energy management*)

Technique de broyage et gestion d'énergie

Técnica de trituración y economía energética

Von **H.-G. Ellerbrock**, Düsseldorf, und **H. Mathiak**, Dortmund/Deutschland

Generalbericht 5 · Zusammenfassung – Zu den Zerkleinerungsanlagen eines Zementwerks gehören die Anlagen zur Fertiggut- und Halbproduktherstellung. Die grundlegenden Produktionsziele, wie Mengen- und Sortenbedarf, Produktqualität und erzielbare Erlöse, werden vom Markt bestimmt. Im Zementwerk müssen die Anforderungen des Marktes mit minimalen Kosten in produktionstechnische Maßnahmen umgesetzt werden. Dabei spielen heute die Datenerfassung und -verarbeitung, die Bewertung der Meßwerte und die Prozeßoptimierung eine wichtige Rolle. In den letzten Jahren stand die Entwicklung der Gutbett-Walzenmühlen und neuer Hochleistungssichter, insbesondere deren Konstruktion und Anwendung bei der Mahlung von Zementrohmehl und Zement, im Vordergrund des Interesses. Bei der Fertiggutproduktion werden in größerem Umfang die zweistufige Zerkleinerung, d. h. Vorzerkleinerung in Wälz- oder Gutbett-Walzenmühlen und Feinmahlung in Kugelmühlen, angewendet. In vielen Werken ist bereits die Halbproduktmahlung mit anschließender Mischung der Halbprodukte zu Zementen und zementhaltigen Baustoffen eingeführt. Durch den Einsatz der neuen Zerkleinerungsmaschinen und -verfahren konnte der elektrische Energieverbrauch deutlich gesenkt werden. Ein Teil der eingesparten Energie wird jedoch durch die sich laufend verschärfenden Umweltschutzauflagen wieder aufgezehrt.

Zerkleinerungstechnik und Energiewirtschaft

General report 5 · Summary – The comminution systems in a cement works include the systems for producing finished products and intermediate products. The fundamental production targets, such as quantities and types required, product quality, and achievable profit, are determined by the market. In a cement works the market requirements have to be translated into production engineering measures at minimum costs. Data acquisition and processing, evaluation of measurements and process optimization now play an important part. In recent years there has been a great deal of interest in the development of high-pressure grinding rolls and new high-efficiency classifiers, especially in their design and application for grinding cement raw meal and cement. Two-stage comminution, i. e. primary comminution in roller grinding mills or high-pressure grinding rolls and finish grinding in ball mills, is used quite extensively in the production of finished products. Many works are now introducing intermediate product grinding followed by mixing of the intermediate products to make cements and building materials containing cement. The new comminution machinery and methods have resulted in significant reductions in the use of electrical energy. However, part of the energy saved is consumed by the ever stricter environmental regulations.

Comminution technology and energy management

Rapport général 5 · Résumé – Les installations de produit fini et de semi-produit font partie des ateliers de broyage d'une cimenterie. Les objectifs fondamentaux visés en matière de production notamment besoins en quantité et catégorie, qualité du produit et ventes de produit réalisables, sont déterminés par le marché. Dans une cimenterie il faut que les exigences du marché se concrétisent en terme de production, et ce avec un minimum de coûts. Pour cela la saisie et le traitement des données, l'appréciation des valeurs mesurées et l'optimisation du process, jouent un rôle capital. Durant les dernières années, le développement des broyeurs à galets à lit de matière et de nouveaux séparateurs à haut rendement a été au centre de l'intérêt, et ce notamment au point de vue conception et utilisation pour broyage de farine crue et de ciment. Pour la production de produit fini on fait de plus en plus appel au broyage en deux étapes, c'est-à-dire prébroyage dans des broyeurs à cylindres ou à galets à lit de matière puis broyage fin dans des broyeurs à boulets. L'utilisation de nouvelles machines et méthodes de broyage permet de réaliser des économies d'énergie électrique substantielles. Une partie de cette énergie économisée est cependant réutilisée en raison des réglementations toujours plus sévères en matière de protection de l'environnement.

Technique de broyage et gestion d'énergie

Informe general 5 · Resumen – Forman parte de las instalaciones de una fábrica de cemento los equipos destinados a la producción de material acabado y semiacabado. Los objetivos fundamentales de la producción, tales como cantidades y tipos, calidad del producto y beneficios realizables, los determina el mercado. En una fábrica de cemento, es necesario que los requerimientos del mercado se traduzcan en términos de producción, con un mínimo de costes. A este respecto, la recogida y tratamiento de da-

Técnica de trituración y economía energética

---

*) Überarbeitete Fassung eines Vortrages zum VDZ-Kongreß '93, Düsseldorf (27.9.–1.10.1993)
   Revised text of a lecture to the VDZ Congress '93, Düsseldorf (27.9.–1.10.1993)

tos, la evaluación de los valores de medición y la optimización del proceso juegan, hoy en día, un papel muy importante. En los últimos años, ha sido de gran interés el desarrollo del molino de cilindros y lecho de material así como de nuevos separadores de elavado rendimiento, y especialmente la construcción y aplicación de los mismos para la molienda de crudo de cemento y de cemento. Para de producción de material acabado se emplea, en gran medida, la trituración en dos etapas, es decir la trituración previa por medio de molinos de pista de rodadura o de molinos de cilindros y lecho de material, y la trituración fina mediante molinos de bolas. En muchas fábricas se ha introducido ya la molienda de productos semiacabados, con subsiguiente mezcla de los productos semiacabados, para obtener cementos y materiales para la construcción que contengan cemento. Gracias al empleo de las nuevas máquinas trituradoras, se ha podido reducir notablemente el consumo de energía eléctrica. Sin embargo, una parte de la energía ahorrada se vuelve a consumir, debido a las exigencias cada vez mayores en materia de protección del medio ambiente.

1. Einleitung

Zu den wesentlichen Anforderungen des Marktes an die Herstellung von Zement gehören Angaben darüber, welche Zementsorten und -mengen in bestimmten Zeiträumen bereitgestellt werden müssen, welche Anforderungen und Ansprüche an die Qualität gestellt werden und welche Preise am Markt erzielt werden können. Daraus ergeben sich die Produktionsziele und die Aufgaben für die Produktionsanlagen im allgemeinen und für die Zerkleinerungsanlagen im besonderen.

1.1 Produktionsziele

Die vom Verbraucher gewünschten Zementmengen lassen sich nur anbieten, wenn die Produktionsanlagen auch die geforderten Produktions-Leistungen erbringen. Hierbei handelt es sich nicht nur um durchschnittliche Produktmengen, sondern es müssen kurzfristig auch deutlich größere „Spitzenmengen" lieferbar sein. Die geforderten Produktmengen können ferner nur bereitgestellt werden, wenn die Produktionsanlagen auch verfügbar sind. Das bedeutet, daß die Verfügbarkeit ein besonders wichtiges Beurteilungskriterium für eine Produktionsanlage ist. Da in vielen Fällen eine Zerkleinerungsanlage nicht nur zur Herstellung einer sondern meist mehrerer Zementsorten genutzt wird, ist die Fähigkeit zur schnellen Anpassung der Einstellgrößen an die unterschiedlichen Produktionsbedingungen für die einzelnen Sorten von großer Bedeutung. Schließlich erwarten die Verbraucher, daß die Zementeigenschaften auch den Qualitätsanforderungen, insbesondere an Homogenität und Gleichmäßigkeit, gerecht werden.

Außer der Verfügbarkeit, Durchsatzleistung und Anpassungsfähigkeit einer Produktionsanlage sowie dem gleichbleibend hohen Niveau der Produktqualität gehört die Minimierung der Herstellkosten zu den wichtigsten Produktionszielen.

1.2 Herstellkosten

Den größten Anteil an den gesamten Herstellkosten machen die Kosten der Einsatzstoffe aus. Daher hat bei Zementen mit mehreren Hauptbestandteilen die Wahl der Rezeptur einen großen Einfluß auf die Minderung der Herstellkosten, während es bei der Herstellung von Portlandzement unter Verwendung von Klinker aus dem eigenen Unternehmen nur wenig Spielraum gibt.

Ein großes Gewicht bei der Bewertung der Herstellkosten haben die sog. Fixkosten. Hierzu zählen die Kosten für das anlagen- und das durch Ersatzteile gebundene Kapital sowie die Kosten für die Bereitstellung der Energie.

Zu dem beeinflußbaren Kostenblock gehören der Aufwand für Instandhaltung und Reparatur sowie die Energiekosten, deren Anteil an den gesamten Herstellkosten für Zement zwischen 30 und 40 % liegt und sowohl in den einzelnen Werken als auch regional sehr unterschiedlich ist. Angesichts dieses sehr großen Anteils der Energiekosten hat die Zementindustrie bereits in den letzten 40 Jahren große Anstrengungen unternommen, die Energiekosten und den Energieverbrauch zur Herstellung von Zement zu senken.

1. Introduction

One of the market's most important requirements in the manufacture of cement is information about the types and quantities of cement which have to be provided within given periods of time, about the requirements and demands made on the quality, and about the price that can be obtained in the market. From these are derived the production targets and tasks for the production plants in general and for the comminution plants in particular.

1.1 Production targets

The quantities of cement required by the consumer can only be offered if the production plants have the necessary production capacities. This does not just apply to average production quantities; it must also be possible to supply significantly greater "peak quantities" in the short term. What is more, the required production quantities can only be provided if the production plants are available for operation. This means that availability is a particularly important assessment criteria for a production plant. In many cases a comminution plant is used for the production of not one but several types of cement, so the ability to adjust the settings rapidly to the different production conditions for the individual types is also very important. Finally, the consumer also expects the cement properties to meet the quality requirements, especially with respect to homogeneity and uniformity.

One of the most important production objectives, along with availability, throughput capacity and adaptability of a production plant, and a consistently high level of product quality, is to minimize production costs.

1.2 Manufacturing costs

The costs of the feed materials make up the greatest proportion of the total manufacturing costs. In cements with several main constituents the selection of mix formulation therefore has a great influence on minimizing the manufacturing costs, but when producing Portland cement using clinker produced by the same company there is little room for manoeuvre.

The so-called fixed costs are very important when assessing the manufacturing costs. These include the costs of the plant and the capital tied up in the spare parts as well as the cost of providing the energy.

The part of the costs which can be influenced includes expenditure on maintenance and repair and the energy costs, which for cement amount to between 30 and 40 % of the total manufacturing costs and differ very widely both regionally and between individual works. In view of the very large proportion taken by the energy costs the cement industry has been making great efforts in the last 40 years to lower the energy costs and amount of energy consumed when manufacturing cement.

1.3 Energy costs, energy consumption

The total energy costs for cement manufacture are divided approximately equally between the costs for fuel energy and

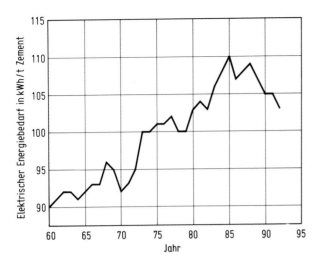

BILD 1: Massebezogener elektrischer Energieverbrauch der deutschen Zementindustrie (Alte Bundesländer) von 1960 bis 1992
FIGURE 1: Specific electrical energy consumption in the German cement industry (former West Germany) from 1960 to 1992

Elektrischer Energiebedarf in kWh/t Zement	= Electrical energy consumption in kWh/t cement
Jahr	= year

1.3 Energiekosten, Energieverbrauch

Die gesamten Energiekosten der Zementherstellung teilen sich zu etwa gleichen Teilen auf die Kosten für Brennstoffenergie und für elektrische Energie auf. Während durch verbesserte Brenntechnik und rationelle Abwärmenutzung der Brennstoffenergieverbrauch je kg Zement in Deutschland inzwischen etwa halbiert [1] und durch den Einsatz preisgünstiger Sekundärbrennstoffe die spezifischen Brennstoffenergiekosten bei gleichem Brennstoffenergieverbrauch vermindert werden konnten, stieg der spezifische elektrische Energiebedarf z.B. in den westdeutschen Zementwerken von etwa 80 kWh/t Zement im Jahr 1960 auf rd. 110 kWh/t Zement im Jahr 1985 an (**Bild 1**). Dieser Anstieg ist auf einen erhöhten Aufwand für den Umweltschutz, Maßnahmen zur Absenkung des Brennstoffenergieverbrauchs, auf die Inbetriebnahme moderner Mahlanlagen für die anstelle von Heizöl eingesetzte Kohle, die Automatisierung des Betriebs von Zementwerksanlagen und schließlich die höhere Mahlfeinheit einzelner Produkte, die insbesondere bei Zementen mit mehreren Hauptbestandteilen zur Bewahrung des Leistungsniveaus erforderlich ist, zurückzuführen. Die in vielen Ländern stark gestiegenen Preise für elektrische Energie verbunden mit einem Anstieg des Verbrauchs an elektrischer Energie einerseits und die Möglichkeiten des Einsatzes preisgünstiger Sekundärbrennstoffe andererseits führten sogar dazu, daß in einzelnen Werken in Europa aber auch in anderen Ländern die Kosten für Elektroenergie die Brennstoffkosten überstiegen. Unter der Voraussetzung, daß für die Zementherstellung ein Brennstoffenergieaufwand von etwa 3,0 GJ/t und ein elektrischer Energieaufwand von rd. 0,4 GJ/t erforderlich sind, bedeutet das, daß Elektroenergie heute mehr als 7 bis 8 mal so teuer wie Brennstoffenergie sein kann. Daher konzentrierten sich die Anstrengungen der Zementindustrie in den letzten 10 Jahren vor allem auf die Senkung der Kosten für elektrische Energie und damit auch auf Maß-

those for electrical energy. Through improved combustion technology and efficient use of waste heat the consumption of fuel energy per kg clinker has now been roughly halved in Germany [1] and the specific fuel energy costs have been reduced for the same fuel energy consumption through the use of cheap secondary fuels. However, the specific electrical energy consumption in, for example, West German cement works rose from about 80 kWh/t cement in 1960 to approximately 110 kWh/t cement in 1985 (**Fig. 1**). This increase can be attributed to increased expenditure on environmental protection, measures for lowering the fuel energy consumption, commissioning of modern grinding plants for coal instead of fuel oil, automation of the operation of cement works plants, and finally to the greater fineness of individual products which is necessary to maintain the performance level, especially for cements with several main constituents. The price of electrical energy, which has risen sharply in many countries, linked with the increase in consumption of electrical energy on the one hand and opportunities for using cheap secondary fuels on the other have even led to the situation where in individual works in Europe, and also in other countries, the costs of electrical energy exceed the fuel costs. On the assumption that cement manufacture requires a fuel energy consumption of about 3.0 GJ/t and an electrical energy consumption of about 0.4 GJ/t this means that electrical energy can now be more than 7 to 8 times as expensive as fuel energy. The cement industry's efforts over the last 10 years have therefore been concentrated primarily on lowering the costs for electrical energy, and therefore also on measures to save electrical energy. By way of example **Table 1** shows how the electrical energy is divided between the process stages in cement manufacture [1]. This shows that the grinding plants for raw material and cement with a total of about 62% are the largest consumers of electrical energy in a cement works, followed by the kiln plant with about 22%. Efforts to save electrical energy have therefore been concentrated on these areas.

2. Saving electrical energy

A series of innovative devices, most of which have been developed in the last 10 years, contribute to the saving of electrical energy [2]. These include all the measures and devices for energy management [3, 4], plants sections [2, 5] and dust separators with low flow pressure drops as well as low energy blending plants, motors and power transmission devices with high levels of efficiency and, above all, the development and operation of grinding plants with high levels of energy utilization.

2.1 Energy management

The important ongoing tasks for efficient operation of a cement works include not only quality management, personnel management, and maintenance and production management but also energy management. A precondition for successful energy management is continuous availability of all the necessary information. This is not difficult to achieve nowadays with the aid of computer-assisted systems for acquisition and processing of the measured data. Application of these systems makes it necessary to set up a continuous energy management organization with suitable working methods and motivated, energy-aware, employees [3, 4]. The work of this organization is aimed at certain energy productivity targets at different cost centre levels. Important areas of action in energy management include

TABELLE 1: Aufteilung des elektrischen Energiebedarfs auf die Verfahrensschritte der Zementherstellung

Verfahrensschritte	Elektr. Energiebedarf in %
Tagebau- und Mischbettbetrieb	5
Rohstoffmahlung	24
Rohmehlhomogenisierung	6
Brennen und Kühlen des Klinkers	22
Zementmahlung	38
Fördern, Verpacken, Verladen	5

TABLE 1: Allocation of the electrical energy consumption between the process stages in cement manufacture

Process stage	Electrical energy consumption in %
quarrying and blending bed operation	5
raw material grinding	24
raw meal blending	6
burning and cooling the clinker	22
cement grinding	38
conveying, packing, loading	5

nahmen zur Einsparung an elektrischer Energie. **Tabelle 1** zeigt beispielhaft die Aufteilung der elektrischen Energie auf die Verfahrensschritte der Zementherstellung [1]. Danach sind die Mahlanlagen für Rohmaterial und Zement mit insgesamt rd. 62 % die größten Verbraucher elektrischer Energie in einem Zementwerk, gefolgt von den Ofenanlagen mit rd. 22 %. Daher waren die Bemühungen zur Einsparung elektrischer Energie insbesondere auf diese Bereiche gerichtet.

2. Einsparung von elektrischer Energie

Zur Einsparung elektrischer Energie tragen eine Reihe innovativer Einrichtungen bei [2], die vor allem in den letzten 10 Jahren entwickelt wurden. Hierzu gehören alle Maßnahmen und Einrichtungen des Energiemanagements [3, 4], druckverlustarme strömungstechnische Anlagenteile [2, 5] und Staubabscheider sowie energieextensive Homogenisieranlagen, Motoren und Energieübertragungseinrichtungen mit hohem Wirkungsgrad und schließlich vor allem Entwicklung und Betrieb von Mahlanlagen mit hoher Energieausnutzung.

2.1 Energie-Management

Für den effizienten Betrieb eines Zementwerks gehört neben einem Qualitäts-Management, Personal-Management, Unterhalts- und Produktions-Management auch das Energie-Management zu den wichtigen ständigen Aufgaben. Voraussetzung für ein erfolgreiches Energie-Management ist die kontinuierliche Bereitstellung aller dafür erforderlichen Informationen. Dies ist heute ohne Schwierigkeiten mit Hilfe computergestützter Systeme zur Meßdatenerfassung und -verarbeitung möglich. Die Anwendung der Systeme erfordert es, eine dauerhaft funktionierende Energie-Management-Organisation mit geeigneten Arbeitsmethoden und motivierten, energiebewußten Mitarbeitern aufzubauen [3, 4]. Die Arbeit dieser Organisation ist nach bestimmten Energie-Produktivitätszielen auf verschiedene Kostenstellenebenen ausgerichtet. Wesentliche Handlungsbereiche im Energie-Management umfassen die

– Rationelle Energienutzung und die
– Nutzung von Sekundärstoffen.

Zur rationellen Energienutzung gehören die

– Optimierung der Betriebsanlagen,
– Einsatz neuer Technologien,
– Optimierung der Stromverträge.

Diese letztgenannten Maßnahmenbereiche sind meist sehr wirkungsvoll und führen schnell zu meßbaren Erfolgen. Demgegenüber sind bei der Nutzung von Sekundärstoffen vor allem auch die Anforderungen an die Produktqualität bei der Herstellung zu berücksichtigen [6].

Die Optimierung der Betriebsanlagen beginnt bei der Einrichtung oder Modernisierung des Energieverteilungssystems. Allein durch Einsatz moderner Schaltanlagen, Motoren mit hohem Wirkungsgrad und Überwachungseinrichtungen für Motoren läßt sich eine deutliche Energieeinsparung erreichen [2, 7]. Aus den kontinuierlichen Meßdaten-Aufzeichnungen können außer Strategien für Instandhaltungsmaßnahmen auch die Planung der Produktion mit möglichst geringem Energieaufwand abgeleitet werden. Die ständige Registrierung des Elektroenergie-Verbrauchs kann z. B. dazu beitragen, durch Vermeiden unnötigen Leerlaufs von Maschinen oder durch zweckmäßigen Lastwechsel oder geeignete Betriebsumstellung die Spitzen im Energieverbrauch zu vermindern. Beispiele aus der Praxis haben gezeigt, daß es meist insgesamt kostengünstiger ist, den Produktionsplan zu ändern und auf diese Weise die Energiekosten zu senken [8].

2.2 Energieübertragung

Insbesondere bei großen Antriebsleistungen ist neben dem Vergleich üblicher Kenngrößen vor allem der Wirkungsgrad der Energieübertragung mit in die Bewertung einzube-

– efficient use of energy, and
– use of secondary materials.

Efficient use of energy covers

– optimization of operational plants,
– use of new technology,
– optimization of contractual arrangements for the use of electricity.

These last-named areas are usually very effective and quickly lead to measurable success. On the other hand particular attention has to be paid to the requirements of product quality during manufacture when using secondary materials [6].

Optimization of plant systems starts with the installation or modernization of the power distribution system. Only by the use of modern switch-gear, high-efficiency motors and motor monitoring devices is it possible to achieve any significant energy saving [2, 7]. From the continuous display of measured data it is also possible to work out strategies for maintenance measures and production planning involving the least possible consumption of energy. Continuous recording of electrical power consumption can, for example, contribute to reducing peaks in the power consumption by avoiding unnecessary noload operation of machinery or by appropriate load changes or by suitable plant changeover. Practical examples have shown that it is usually more cost-effective overall to change the production plan and reduce energy costs in this way [8].

2.2 Power transmission

It is particularly important that the efficiency of the power transmission system should be included in the evaluation alongside the comparison of the usual parameters, especially with large drive ratings. Experience in the USA has shown that gearless drives can be counted among the systems with high transmission efficiencies. An energy saving of between 2 and 5 % can be achieved when using a gearless drive as compared with the use of conventional gear unit drives [9, 10]. Maintenance costs are also said to be relatively low, with a high operational reliability. However, the capital costs are significantly higher than for power transmission systems with conventional gear units. Gearless drives also make it possible to optimize plant operation by changing the rotational speed. With large fan drives, for example, it is possible to achieve a very low power consumption by controlling the speed [11–13]. Although there are no mechanical transmission losses gearless drives are generally only considered an advantage at fairly high ratings. Investigations on industrial mills with gearless drives with speed control have, for example, shown that for the same fineness of material discharged from the mill it was possible to influence the mass throughput as well as the electrical power consumption by changing the mill speed within certain limits while the specific power consumption for grinding remained constant regardless of the mill speed [14].

3. Comminution technology

On average, a specific power consumption of about 62 kWh/t is needed for grinding raw materials and cement-making materials. Most fine grinding is still carried out in plants with ball mills; these are operated either as open-circuit mills or in closed circuit with an air classifier. The majority of the energy used for driving the mill is lost as heat [15, 16]. All process engineering measures for reducing the consumption of electrical energy are therefore aimed at limiting the losses as far as possible.

The energy utilization in a ball mill depends chiefly on the grindability and moisture content of the mill feed components as well as on the fineness of the product. The energy utilization is also heavily influenced by the operating conditions in the grinding plant [17]. It is therefore particularly important to make an appropriate choice of the grinding process and of the optimum settings for the mill and classifier.

ziehen. Dabei können aufgrund von Erfahrungen aus den USA getriebelose Antriebe zu den Systemen mit einem hohen Übertragungs-Wirkungsgrad gezählt werden. Bezogen auf den Einsatz konventioneller Getriebe-Antriebe ließ sich bei Verwendung eines getriebelosen Antriebs eine Energieeinsparung zwischen 2 und 5% erzielen [9, 10]. Außerdem sollen die Unterhaltskosten relativ niedrig und die Betriebssicherheit hoch sein. Demgegenüber sind jedoch die Investitionskosten deutlich höher als bei den Energieübertragungssystemen mit konventionellen Getrieben. Getriebelose Antriebe bieten ferner die Möglichkeit, durch Verändern der Drehzahl den Anlagenbetrieb zu optimieren. Bei großen Ventilatorantrieben ließ sich z. B. durch Drehzahlregelung ein sehr geringer Energieverbrauch erzielen [11–13]. Obwohl keine mechanischen Übertragungsverluste auftreten, sollen getriebelose Antriebe generell aber nur bei größeren Leistungen vorteilhaft sein. Untersuchungen an Betriebsmühlen mit drehzahlregelbarem, getriebelosen Antrieb haben z. B. gezeigt, daß bei gleicher Mahlfeinheit des Mühlenaustragsguts durch Verändern der Mühlendrehzahl in bestimmten Grenzen außer der elektrischen Leistungsaufnahme auch der Durchsatzmassenstrom zu beeinflussen war, während der massebezogene Energieaufwand zum Mahlen unabhängig von der Mühlendrehzahl konstant blieb [14].

3. Zerkleinerungstechnik

Zum Mahlen von Rohmaterial und Zement-Einsatzstoffen ist im Mittel ein massebezogener Energieaufwand von rd. 62 kWh/t erforderlich. Zum Feinmahlen werden heute noch überwiegend Anlagen mit Kugelmühlen eingesetzt, die als Durchlaufmühlen oder im Umlauf mit einem Windsichter betrieben werden. Der größte Teil der zum Betrieb der Mühlen aufgewendeten Energie geht als Wärme verloren [15, 16]. Daher zielen alle verfahrenstechnischen Maßnahmen zur Verminderung des Elektroenergieverbrauchs darauf ab, die Verluste so weit wie möglich zu begrenzen.

Die Energieausnutzung in einer Kugelmühle hängt vor allem von der Mahlbarkeit und der Feuchte der Mahlgutkomponenten sowie von der Mahlfeinheit des erzeugten Produkts ab. Außerdem wird die Energieausnutzung in starkem Maß von den Betriebsbedingungen der Mahlanlage beeinflußt [17]. Daher kommt es vor allem auf eine zweckmäßige Wahl des Mahlverfahrens und der optimalen Einstellgrößen von Mühle und Sichter an.

3.1 Kugelmühlen-Betrieb

In Anlagen mit Kugelmühlen wird der Energieaufwand zum Mahlen in starkem Maß vom Mahlkörperfüllungsgrad, der Ausbildung der Panzerung sowie von der dem gewünschten Zerkleinerungsfortschritt entsprechenden Mahlkörperklassierung in der Feinmahlkammer beeinflußt.

Bild 2 zeigt als Ergebnis von Betriebsuntersuchungen an Zementmühlen unterschiedlicher Größe beim Mahlen von Portlandzement Z 35 F den auf den Mahlgutmassestrom durch die Mühle (Umlaufmassestrom) bezogenen Arbeitsbedarf in Abhängigkeit vom mittleren Mahlkörperfüllungsgrad [17, 18]. Die Mühlen wurden mit relativen Drehzahlen von 55 bis 75 % der kritischen Drehzahl betrieben. Die Feinheit der Mühlenaustragsgüter wurde zur einheitlichen Darstellung der Meßwerte des Arbeitsbedarfs proportional auf eine Mahlfeinheit von 1800 cm²/g normiert. Aus dem Bild geht hervor, daß die auf den Umlaufmassestrom bezogene Energiebedarf einer Kugelmühle zum Mahlen von PZ 35 F bei einem mittleren Mahlkörperfüllungsgrad von etwa 26 % ein Minimum aufweist. Die Lage der Meßpunkte auf einer einheitlichen Kurve und die geringe Streuung lassen den Schluß zu, daß z.B. weder die Mühlendrehzahl noch die Mahlkörpergrößenverteilung in den untersuchten Grenzen einen wesentlichen Einfluß auf den massebezogenen Energieaufwand haben. Mit zunehmender Mühlendrehzahl stieg aber der Fertiggutmassestrom etwa proportional an und war um so höher, je größer der Mahlkörperfüllungsgrad war (**Bild 3**).

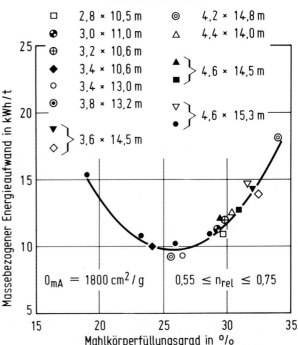

BILD 2: Auf den Umlaufmassenstrom bezogener Energiebedarf von Kugelmühlen in Abhängigkeit vom Mahlkörperfüllungsgrad aufgrund von Untersuchungen an Zementmühlen

n_{rel} = relative Mühlendrehzahl
O_{mA} = massebezogene Oberfläche des Mühlenaustragsguts

FIGURE 2: Specific power consumption, relative to the recirculating mass flow, of ball mills as a function of the grinding media filling ratio based on investigations into cement mills

n_{rel} = relative mill rotational speed
O_{mA} = specific surface area of the mill discharge material

Massebezogener Energieaufwand in kWh/t	= Specific power consumption in kWh/t
Mahlkörperfüllungsgrad in %	= Grinding media ratio in %
Mühlenabmessungen	= mill dimensions

3.1 Ball mill operation

In plants with ball mills the energy consumption for grinding is heavily influenced by the grinding media filling ratio, by the design of the protective lining, and by the grinding media classification in the fine grinding chamber which is matched to the desired progress of comminution.

Fig. 2 is based on industrial investigations in cement mills of different sizes when grinding Z 35 F Portland cement, and shows the specific power consumption relative to the mass flow of material through the mill (recirculating mass flow) as a function of the average grinding media filling ratio [17, 18]. The mills were operated with relative rotational speeds of 55 to 75 % of the critical speed. The fineness of the mill discharge material was normalized proportionally to a fineness of 1800 cm²/g for the sake of consistent representation of the measurements of the power consumption. It can be seen from the diagram that the specific power consumption, relative to the recirculating mass flow, of a ball mill for grinding PZ 35 F Portland cement has a minimum value at an average grinding media filling ratio of about 26%. The positions of the points on a consistent curve and the small amount of scatter lead to the conclusion that, for example, neither the mill speed nor the grinding media size distribution have an important influence on the specific power consumption within the limits investigated. However, with increasing mill speed there was an approximately proportional increase in the finished product mass flow, and this was higher the greater the grinding media filling ratio (**Fig. 3**).

In the fine grinding chambers of the ball mills investigated the grinding media diameters lay between 20 and 60 mm and the bulk densities of the grinding media charges fluctuated between 4.3 and 4.6 t/m³. Provided the grinding media charge gradings remained within these limits their influence

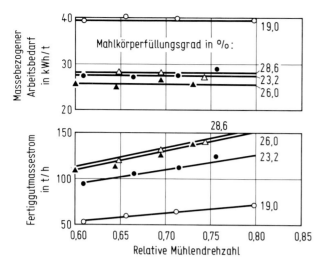

BILD 3: Fertiggutmassenstrom und massebezogener Energiebedarf von Kugelmühlen in Abhängigkeit von der relativen Mühlendrehzahl und vom Mahlkörperfüllungsgrad
massebezogene Oberfläche des Mühlenaustragsguts
$O_{mA} = 1800\ cm^2/g$
massebezogene Oberfläche des Fertigguts
$O_m = 3180\ cm^2/g$
ungesichteter Anteil $\tau = 20\%$

FIGURE 3: Finished product mass flow and specific power consumption of ball mills as a function of the mill relative rotational speed and of the grinding media filling ratio
specific surface area of the mill discharge material
$O_{mA} = 1800\ cm^2/g$
specific surface area of the finished product
$O_m = 3180\ cm^2/g$
unclassified fraction $\tau = 20\%$

Massebezogener Arbeitsbedarf in kWh/t	=	Specific power consumption in kWh/t
Fertiggutmassenstrom in t/h	=	Finished product mass flow in t/h
Relative Mühlendrehzahl	=	Mill relative rotational speed
Mahlkörperfüllungsgrad in %	=	grinding media filling ratio in %

Die Mahlkörperdurchmesser in den Feinmahlkammern von untersuchten Kugelmühlen lagen zwischen 20 und 60 mm. Die Schüttdichten der Mahlkörperfüllungen schwankten zwischen 4,3 und 4,6 t/m³. Solange die Mahlkörpergattierungen innerhalb dieser Grenzen blieben, konnte deren Einfluß auf den Mühlenbetrieb vernachlässigt werden [11]. Demgegenüber kann bei einer ungenügenden Mahlkörperklassierung längs der Mahlbahn der Feinmahlkammer der Zerkleinerungsfortschritt deutlich abnehmen und demzufolge der massebezogene Energieverbrauch zum Mahlen um bis zu 20 % ansteigen.

Auch die Temperatur im Mahlraum spielt für den Energieaufwand beim Mahlen eine große Rolle [15–17]. Bei höheren Mahlguttemperaturen besteht die Gefahr, daß der Zerkleinerungsfortschritt in der Mühle durch Agglomeratbildung, Anbackungen und Verpelzungen stark vermindert wird (**Bild 4**). Dadurch kann die Durchsatzleistung stark abnehmen und der massebezogene Energieaufwand entsprechend ansteigen. Demnach ist es häufig zweckmäßig, den Zement z. B. während des Mahlens oder im Mahlkreislauf zu kühlen. Auch durch den Zusatz oberflächenaktiver Stoffe, der sog. Mahlhilfsmittel, läßt sich die Agglomerationsneigung des Mahlguts in der Mühle erheblich vermindern und der massebezogene Energieaufwand um so mehr herabsetzen, je feiner gemahlen wird. Daher ist der Einsatz von Mahlhilfsmitteln meist erst beim Mahlen von Zementen höherer Mahlfeinheit wirtschaftlich günstig.

Von besonderer Bedeutung sowohl für die Energieausnutzung beim Mahlen als auch für den Verschleiß der Mühleneinbauten ist die geeignete Ausfüllung der Zwischenräume zwischen den Mahlkörpern mit Mahlgut, d. h. der Mahlgutfüllungsgrad in den Mahlkammern, und die den Mahlguttransport unterstützende Mühlenbelüftung. Das Problem tritt vor allem in der Grobmahlkammer auf, die üblicherweise mit Mahlkugeln von 60 bis 90 mm Durchmesser gefüllt

on the mill operation could be neglected [11]. On the other hand, the progress of comminution decreases significantly where there is inadequate grinding media classification along the mill in the fine grinding chamber, and as a result the specific power consumption for grinding increases by up to 20 %.

The temperature in the grinding chamber also plays an important part in the power consumption during grinding [15–17]. At fairly high grinding temperatures there is the danger that the progress of grinding in the mill will be heavily impeded by agglomeration, buildup, and coating of the grinding media (**Fig. 4**). This can cause a sharp drop in the throughput capacity with a corresponding rise in the specific power consumption. It is therefore often advisable to cool the cement during grinding or in the grinding circuit. By the addition of surface-active substances, so-called grinding aids, it is also possible to make a great reduction in the tendency of the material to agglomerate in the mill. This reduces the specific power consumption to a greater extent the finer the grinding, so the use of grinding aids is usually only economically advantageous when grinding cements with fairly high levels of fineness.

The correct filling of the interstitial spaces between the grinding media with the material being ground, i.e. the material filling ratio in the grinding chambers, and the mill ventilation which assists the material transport are of particular importance for the energy utilization as well as for the wear on the mill internals. Problems occur mainly in the coarse grinding chamber which is normally filled with grinding balls with diameters from 60 to 90 mm, and in which the level

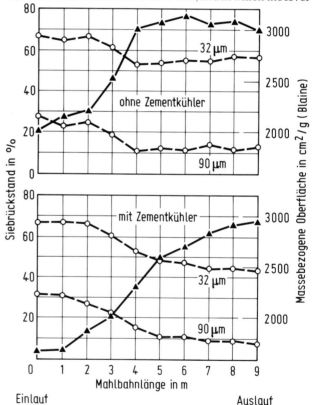

BILD 4: Mahlfeinheit von Mahlgutproben längs der Mahlbahn in der Feinmahlkammer einer 2-Kammer-Kugelmühle 4,4 × 15,5 m beim Mahlen von Portlandzement Z 45 ohne (oberes Diagramm) und mit Kühlung (unteres Diagramm) des Mühlenaustragsguts

FIGURE 4: Fineness of samples of material taken along the length of the mill in the grinding fine chamber of a 4.4 × 14.5 m two-chamber mill when grinding PZ Portland cement without (upper diagram) and with (lower diagram) cooling of the mill discharge material

Siebrückstand in %	=	Sieve residue in %
Mahlbahnlänge in m	=	Distance along mill in m
Einlauf	=	Inlet
Auslauf	=	Outlet
ohne Zementkühler	=	without cement cooler
mit Zementkühler	=	with cement cooler
Massebezogene Oberfläche in cm²/g (Blaine)	=	Specific surface area in cm²/g (Blaine)

BILD 5: Neue Trennwand „AIRFEEL" (Werkfoto der FM Magotteaux S. A., Vaux-Sous-Chèvremont, Belgien) [20]
FIGURE 5: New "AIRFEEL" diaphragm (works photograph from FM Magotteaux S.A., Vaux-Sous-Chèvremont, Belgium) [20]

ist und in der sich meist ein zu niedriger Mahlgutspiegel einstellt, wenn keine geeigneten Gegenmaßnahmen ergriffen werden. Um den Mahlgutfüllstand in der Grobmahlkammer günstig zu beeinflussen, wurden einstellbare Trennwände entwickelt [19, 20]. Eine neue Trennwand zur Regelung des Mahlgutfüllungsgrads kommt ohne bewegliche Teile aus und nutzt die Wirkung des Luftstroms durch die Mühle (**Bild 5**). Durch eine geeignete Mahlgutfüllung in der Grobmahlkammer kann die Durchsatzleistung einer Mühle um 5 bis 8 % gesteigert werden. Bei optimaler Einstellung liegt das Masseverhältnis von Mahlgut zu Mahlkörpern in der Grobmahlkammer bei etwa 0,17 bis 0,19. In der Feinmahlkammer beträgt der optimale Wert etwa 0,08 bis 0,10. Hier kann er leicht durch Verändern der Mahlkörpergattierung eingestellt werden.

Beim Zement-Herstellungsprozeß sind außer wirtschaftlichen Gesichtspunkten und der Umweltverträglichkeit der Herstellung und des Produkts die Sicherung einer marktorientierten Produktqualität unter den gegebenen verfahrenstechnischen Bedingungen von besonderer Bedeutung. Hierzu dienen alle Maßnahmen, die zur Qualitätssteuerung ergriffen werden. Dazu gehört zunächst die Dokumentation des Herstellungsprozesses und der Leistungsfähigkeit des Produkts, wozu auch die Gleichmäßigkeit der Zementeigenschaften zählt [21].

Die Zementeigenschaften werden von den Eigenschaften der Einsatzstoffe sowie in besonderem Maß von den Herstellbedingungen bestimmt. Daher ist für eine dauerhafte Sicherung der Gleichmäßigkeit eine Kopplung von Qualitätsprüfung und Herstellungsprozeß erforderlich. Mit Hilfe der bekannten Zusammenhänge zwischen den wichtigsten Zementeigenschaften und den bestimmenden Prozeßkenngrößen sind rechtzeitige Eingriffe in den Prozeß und damit eine Regelung des Prozesses möglich. Da die Prozeß- und Qualitätskenngrößen meist nur über komplexe Zusammenhänge miteinander verknüpft sind und die Qualitätsgrößen häufig nur mit großer Zeitverzögerung und als zeitlich gemittelte Werte vorliegen, bietet z. B. die SPC(Statistical-Process-Control)-Technik ein geeignetes Mittel für die statistische Aus- und Bewertung [22], um auf eine Änderung der Qualitätsmerkmale durch eine Steuerung des Produktionsprozesses zu reagieren.

Für die Steuerung und Regelung von Zementmahlanlagen wurden verschiedene Systeme entwickelt [5, 23–26]. Durch deren Einsatz ließen sich die Mahlgutmassenströme stabilisieren und die Gleichmäßigkeit der Zementeigenschaften, insbesondere der Feinheit, deutlich verbessern. Für den Mahlprozeß sind außerdem die gewählten Anteile der Mahlgutkomponenten präzise einzuhalten. Daher muß die Mahlgutdosierung in die Regelung des Mahlprozesses einbezogen werden [27]. Die Regelung von Mahlanlagen mit dem Ziel, einen gleichmäßigen und stabilen Anlagenbetrieb zu gewährleisten, führt meist auch zu einer deutlichen Erhö-

of material is usually too low unless suitable counter-measures are taken. Adjustable diaphragms were developed to assist the material filling levels in coarse grinding chambers [19, 20]. One new diaphragm for controlling the material filling level does without any moving parts and makes use of the action of the air flow through the mill (**Fig. 5**). The throughput capacity of a mill can be increased by 5 to 8 % by a suitable material filling ratio in the coarse grinding chamber. At the optimum adjustment the mass ratio of material to grinding media in the coarse grinding chamber is about 0.17 to 0.19, and in the fine grinding chamber the optimum value is around 0.08 to 0.10. In this case it is easily adjusted by changing the grinding media charge grading.

Apart from the economic points of view and the environmental compatibility of the production process and of the product, it is also of particular importance to ensure that the cement manufacturing process produces a market-orientated product quality under the given process engineering conditions. This involves all the measures which are taken to control the quality, including documentation of the manufacturing process and of the performance of the product, which also covers the uniformity of the cement properties [21].

The cement properties are determined by the properties of the input materials used and, in particular, by the manufacturing conditions. To secure long-term consistency it is therefore necessary to link quality testing to the manufacturing process. The known relationships between the most important cement properties and the measured process parameters make it possible to make timely interventions in the process and therefore to control it. The process and quality parameters are normally linked to one another through complex relationships and the quality parameters are frequently only available with large time delays and as intermittently measured values, so the SPC (statistical process control) technique, for example, offers a suitable means of statistical analysis and evaluation [22] for reacting to a change in the quality characteristics by controlling the production process.

Various systems have been developed for open- and closed-loop control of cement grinding plants [5, 23–26]. Through their use it is possible to stabilize the material mass flows and to make significant improvements in the uniformity of the cement properties, especially the fineness. The selected proportions of the mill feed components in the grinding process also have to be maintained accurately. The mill feed metering system must therefore be incorporated into the control system for the grinding process [27]. Control systems for grinding plants aimed at ensuring uniform and stable plant operation usually also lead to significant increases in throughput capacity and to reductions in specific energy consumption.

3.2 Air classifier in the grinding circuit

The material discharged from the mill is separated in the classifier into tailings and fines, and the tailings are fed back to the mill with the fresh material. This relieves the mill of cement fractions which are already finely ground. The related energy-saving effect is greater the more completely the finish-ground fines are removed. The power consumption for grinding is therefore also heavily dependent on the design and mode of operation of the air classifier. This can be characterized by the Tromp separation curve which shows the dependence of the selectivity T on the particle size x. The selectivity T(x) is the percentage of classifier feed material with a certain particle size which passes into the tailings. The quality of separation can be read from the separation curve [28].

During the classifying process every attempt is made to ensure that fine particles do not pass into the tailings and hence back into the mill. The shape of a real separation curve is shown diagrammatically in **Fig. 6** as an example. The distance of the separation curve from the abscissa in the fine range shows that in this real classification a proportion of the fine particles did pass into the tailings. It represents the so-

hung der Durchsatzleistung und zu einer Verminderung des massebezogenen Energieaufwands.

3.2 Windsichter im Mahlkreislauf

Das Mühlenaustragsgut wird im Sichter in Grob- und Feingut getrennt und das Grobgut der Mühle mit dem Frischgut erneut aufgegeben. Dadurch wird die Mühle von bereits feingemahlenen Zementanteilen entlastet. Der damit verbundene Energiespareffekt ist um so größer, je vollständiger das schon fertiggemahlene Feingut abgetrennt wird. Der Energieaufwand zum Mahlen hängt daher auch maßgebend von der Bauform und der Arbeitsweise des Windsichters ab. Diese läßt sich durch die Trennkurve nach Tromp kennzeichnen. Darunter versteht man die Abhängigkeit des Trenngrads T von der Korngröße x. Der Trenngrad T(x) ist der Anteil des Sichteraufgabeguts mit einer bestimmten Korngröße in %, der in das Grobgut gelangt. Aus der Trennkurve läßt sich die Qualität der Trennung ablesen [28].

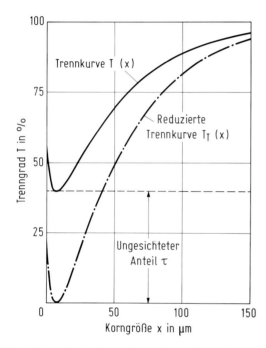

BILD 6: Schematische Darstellung der realen und reduzierten Trennkurve nach Tromp eines Sichters
FIGURE 6: Diagrammatic representation of the real and reduced Tromp separation curves of a classifier

Trenngrad T in % = Selectivity T in %
Korngröße x in µm = Particle size x in µm
Trennkurve T(x) = separation curve T(x)
Reduzierte Trennkurve $T_T(x)$ = reduced separation curve $T_T(x)$
Ungesichteter Anteil τ = unclassified fraction τ

Beim Sichtprozeß wird angestrebt, daß feine Partikel möglichst nicht ins Grobgut und damit zurück in die Mühle geführt werden. Im **Bild 6** ist als Beispiel der Verlauf einer realen Trennkurve schematisch dargestellt. Der Abstand der Trennkurve im Feinbereich von der Abszisse zeigt, daß bei dieser realen Sichtung auch ein Teil der feinen Partikel ins Grobgut gelangte. Er stellt den sog. ungesichteten Anteil τ dar. Höhere Werte von τ bedeuten daher eine schlechtere Aussichtung des Mahlguts. Durch geeignete Wahl der Sichterbauform und/oder der Sichtereinstellung kann der ungesichtete Anteil vermindert, die Durchsatzleistung einer Mahlanlage erhöht und der auf den Fertiggutmassestrom bezogene Energieaufwand dementsprechend gesenkt werden.

Unter dem Trennschärfemaß ϰ versteht man das Verhältnis von zwei Korngrößen mit bestimmtem Trenngrad, z. B. ist:

$$\varkappa = x_{T(x) = 25\%} / x_{T(x) = 75\%}.$$

Das Verhältnis läßt sich nur an der sog. reduzierten Trennkurve ermitteln, bei der der ungesichtete Anteil τ = 0 ist. **Bild 7** zeigt schematisch zwei reduzierte Trennkurven

called unclassified fraction τ. Higher values of τ therefore mean worse classification of the ground material. By suitable selection of the classifier design and/or the classifier settings it is possible to reduce the unclassified fraction, raise the throughput capacity of a grinding plant, and make a corresponding reduction in the specific power consumption in relation to the finished product mass flow.

The size of the separation effect ϰ is taken as the ratio of two particle sizes with given selectivities, for example

$$\varkappa = x_{T(x) = 25\%} / x_{T(x) = 75\%}.$$

The ratio can only be determined on a so-called reduced separation curve in which the unclassified fraction is τ = 0. **Fig. 7** shows two reduced separation curves in diagrammatic form with different separation effects. The greater the separation of a classifier at a given cut size the greater is the size of the separation effect ϰ and the narrower is the particle size distribution of the classifier fines. The smaller the separation the smaller is the proportion of coarser ground material which passes into the tailings and the larger is the classifier fines mass flow and hence the finished product output. The following conclusion can be drawn from this: If permitted by the demands on the cement properties or if, for example, required to ensure the workability in the concrete, the selectivity in the range of coarser particle sizes, and therefore the separation effect, should be reduced in favour of a wider particle size distribution and a greater throughput capacity, and therefore a lower specific power expenditure relative to the finished product mass flow.

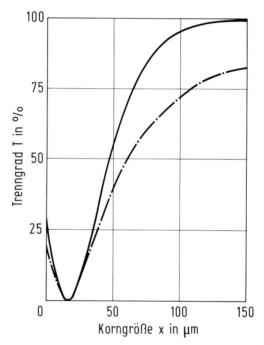

BILD 7: Schematische Darstellung von zwei reduzierten Trennkurven für unterschiedliche Trennschärfe
FIGURE 7: Diagrammatic representation of two reduced separation curves for different separation effects

Trenngrad T in % = Selectivity T in %
Korngröße x in µm = Particle size x in µm

For the same clinker properties, especially for clinker of high reactivity, and for the same specific surface area, cements with narrower particle size distributions do in fact have higher standard compressive strengths than cements with wide particle size distributions [18, 29]. Cements with narrower particle size distributions could therefore achieve the required standard strength at a lower specific surface area. As cements with lower fineness can be ground with a lower expenditure of energy it would be expected that grinding energy could be saved by controlling the particle size distribution in the classifier. However, it has been found in practice that cements with narrower particle size distributions produced by classifying are always associated with a

mit unterschiedlicher Trennschärfe. Je größer bei einer bestimmten Trenngrenze die Trennschärfe eines Sichters ist, desto größer ist das Trennschärfemaß $\varkappa$ und um so enger ist die Korngrößenverteilung des Sichterfeinguts. Je geringer die Trennschärfe ist, desto kleiner ist der Anteil gröberen Mahlguts, das ins Grobgut gelangt und um so größer ist der Sichterfeingutmassestrom und damit die Fertiggutleistung. Daraus ist folgender Schluß zu ziehen: Wenn es die Anforderungen an die Zementeigenschaften zulassen bzw., wie z. B. zur Sicherung der Verarbeitbarkeit im Beton, verlangen, sollte der Trenngrad im Bereich gröberer Korngrößen und damit die Trennschärfe des Sichters zugunsten einer breiteren Kornverteilung und größeren Durchsatzleistung und damit eines geringeren, auf den Fertiggutmassestrom bezogenen Energieaufwands vermindert werden.

Bei gleichen Klinkereigenschaften, insbesondere bei Klinker mit hoher Reaktivität, und bei gleicher massebezogener Oberfläche weisen Zemente mit einer engeren Korngrößenverteilung zwar eine höhere Normdruckfestigkeit auf als Zemente mit breiter Korngrößenverteilung [18, 29]. Demnach könnten Zemente mit engerer Korngrößenverteilung die angestrebte Normfestigkeit bei geringerer massebezogener Oberfläche erreichen. Da Zemente mit geringerer Mahlfeinheit mit einem niedrigeren Energieaufwand ermahlen werden können, wäre zu erwarten, daß sich durch Beeinflussen der Korngrößenverteilung im Sichter Mahlenergie einsparen läßt. In der Praxis hat sich aber gezeigt, daß durch Sichten erzeugte engere Korngrößenverteilungen des Zements stets mit einer Einbuße an Fertiggutleistung und dementsprechend mit einem Anstieg des massebezogenen Energiebedarfs verbunden sind. Außerdem sind in diesem Zusammenhang die Anforderungen an Wasserbedarf und Betonfestigkeit bei gleichem Konsistenzmaß zu beachten. Darüber hinaus ist der massebezogene Energiebedarf der modernen trennschärferen Sichter auch häufig höher als der der früher üblichen Sichterbauten. Eine Einsparung an Mahlenergie erscheint daher nur dann möglich, wenn die Mühle bereits ein Mühlenaustragsgut mit einer engeren Korngrößenverteilung liefert [18].

loss of finished product output and hence with an increase in specific power consumption. In this context it is also necessary to pay attention to the requirements for water demand and concrete strength at equal consistency. In addition to this the specific power consumption of the modern, more selective, classifier is also frequently higher than that of the classifiers used previously. A saving in grinding energy therefore only appears possible when the mill already supplies a mill discharge material with a fairly narrow particle size distribution [18].

The efforts to grind cements with narrower particle size distributions are limited because of the demands on workability characteristics [30, 31]. During attempts to save grinding energy through a narrower cement particle size distribution it is always also necessary to take account of the influence of the process measures on the cement properties, especially the workability and strength development in the concrete.

There have been extensive reports on the influence of different classifier designs and different modes of classifier operation on the specific power consumption of a grinding plant and on the cement properties [18, 28]. The present state of knowledge is summarized in Special Report 5.2 [32]. This shows that, among other things, the separation can be significantly improved by replacing the countervane system by a caged rotor or bladed rotor, by separating the classifier fines in external cyclones, and by operating the classifier with fresh air. As an example taken from practice, **Fig. 8** shows the measured separation curves of two classifiers of different design [33] – an older recirculating air classifier with countervane system and a modern cyclone recirculating air classifier of comparable size with rotor internals – when grinding the same type of PZ 45 F Portland cement with the same ball mill and the same clinker. This shows that the fines fraction in the material discharged from the mill is a great deal more effectively separated out in the caged-rotor classifier than in the conventional recirculating air classifier; this can be seen by the low unclassified fraction of only about 1% during separation in the caged-rotor classifier as opposed to about 52% in the older recirculating air classifier. This effect raised the throughput capacity of the grinding plant. On the other hand the somewhat greater separation effect of the caged-rotor classifier tended to reduce the throughput. These opposing effects resulted in an overall increase of about 15% in output capacity for the grinding plant when producing PZ45F Portland cement with a fineness of about 3500cm^2/g Blaine. However, the specific power consumption of the caged-rotor classifier by itself was significantly greater than that of the conventional recirculating air classifier, so in this case the replacement of the recirculating air classifier by a modern caged-rotor classifier accompanied by adjustment of the ball mill operation to suit the classifier's mode of operation only reduced the specific power consumption of the entire plant by just under 5%. Greater power saving can be obtained when manufacturing cements of higher fineness.

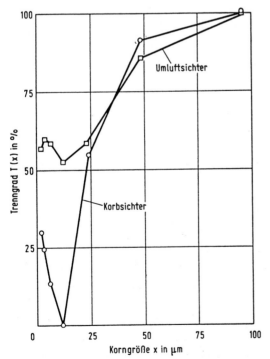

BILD 8: Reale Trennkurven eines konventionellen Umluftsichters und eines Korbsichters vergleichbarer Größe beim Mahlen von Portlandzement Z 35 F
FIGURE 8: Real separation curves for a conventional recirculating air classifier and a cage-rotor classifier of comparable size when grinding Z 35 F Portland cement

Trenngrad T(x) in %	=	Selectivity T(x) in %
Korngröße x in μm	=	Particle size x in μm
Umluftsichter	=	recirculating air classifier
Korbsichter	=	cage-rotor classifier

4. Alternative methods of comminution

The energy utilization in a ball mill is very low [17]. Significantly higher energy utilization is achieved during comminution by pressure stressing in a bed of material and by impact stressing than by comminution by percussion and frictional stressing of the material in a ball mill [34, 35]. Appreciable energy savings can therefore be achieved by plant engineering, mainly by the use of alternative methods of comminution. At present the competition is between impact crushers, high-pressure grinding rolls, and roller grinding mills – either without integral classifier and with mechanical material discharge or with pneumatic material discharge by a system fan and with a classifier integral with the mill housing. Impact crushers are used mainly for primary size reduction while high-pressure grinding rolls are used in a variety of plant configurations for primary, semi-finish and finish grinding, and roller grinding mills for primary and finish grinding.

Dem Bestreben, Zemente mit engerer Korngrößenverteilung zu ermahlen, sind demnach aufgrund der Anforderungen an die Verarbeitungseigenschaften Grenzen gesetzt [30, 31]. Daher ist bei den Bemühungen, durch eine engere Korngrößenverteilung des Zements Mahlenergie einzusparen, stets auch der Einfluß der verfahrenstechnischen Maßnahmen auf die Zementeigenschaften, vor allem auf die Verarbeitbarkeit und die Festigkeitsentwicklung im Beton, zu beachten.

Über den Einfluß verschiedener Sichterbauformen sowie unterschiedlicher Sichterbetriebsweise auf den massebezogenen Energiebedarf einer Mahlanlage und die Zementeigenschaften wurde ausführlich berichtet [18, 28]. Der Stand des Wissens ist im Fachbericht 5.2 zusammengefaßt [32]. Daraus geht u.a. hervor, daß durch Ersetzen des Gegenflügelsystems durch einen Stabkorb oder Flügelkorb, Abscheiden des Sichterfeinguts in außenliegenden Zyklonen und durch Betrieb des Sichters mit Frischluft die Trennung deutlich verbessert werden konnte. **Bild 8** zeigt als Beispiel aus der Praxis die gemessenen Trennkurven von zwei Sichtern unterschiedlicher Bauart [33], und zwar eines älteren Umluftsichters mit Gegenflügelsystem und eines modernen Zyklonumluftsichters mit Korbeinbauten vergleichbarer Größe beim Mahlen der gleichen Zementsorte PZ 45 F mit der gleichen Kugelmühle und dem gleichen Klinker. Danach wurde im Korbsichter der Feingutanteil aus dem Mühlenaustraggut erheblich wirkungsvoller abgetrennt als in dem konventionellen Umluftsichter, kenntlich an dem geringeren ungesichteten Anteil von nur etwa 1 % bei der Trennung im Korbsichter gegenüber etwa 52 % im älteren Umluftsichter. Dadurch wurde die Durchsatzleistung der Mahlanlage erhöht. Andererseits wirkte sich die etwas größere Trennschärfe des Korbsichters durchsatzleistungsmindernd aus. Diese gegenläufigen Einflüsse hatten bei der Herstellung von PZ 45 F mit einer Mahlfeinheit von rd. 3500 cm^2/g nach Blaine insgesamt eine um etwa 15 % größere Durchsatzleistung der Mahlanlage zur Folge. Da der massebezogene Energiebedarf des Korbsichters allein jedoch deutlich größer als der des konventionellen Umluftsichters war, ließ sich in diesem Fall mit dem Ersatz des Umluftsichters durch einen modernen Korbsichter und eine gleichzeitige Abstimmung des Betriebs der Kugelmühle auf die Sichterarbeitsweise der massebezogene Energiebedarf der Gesamtanlage nur um knapp 5 % vermindern. Bei der Herstellung von Zementen höherer Mahlfeinheit kann sich eine größere Energieeinsparung ergeben.

4. Alternative Zerkleinerungsverfahren

Die Energieausnutzung in einer Kugelmühle ist sehr gering [17]. Gegenüber der Zerkleinerung durch Schlag- und Reibbeanspruchung des Mahlguts in Kugelmühlen wird bei der Zerkleinerung durch Druckbeanspruchung im Gutbett und durch Prallbeanspruchung eine deutlich höhere Energieausnutzung erzielt [34, 35]. Daher können bemerkenswerte Energieeinsparungen anlagentechnisch vor allem durch den Einsatz alternativer Zerkleinerungsverfahren erreicht werden. Als solche konkurrieren heute der Prallbrecher, die Gutbett-Walzenmühle sowie die Wälzmühle, und zwar einerseits die Bauart ohne integrierten Sichter und mit mechanischem Mahlgutaustrag und andererseits die Bauart mit pneumatischem Mahlgutaustrag durch einen Systemventilator und mit einem in das Mühlengehäuse integrierten Sichter. Prallbrecher werden vor allem zur Vorzerkleinerung, Gutbett-Walzenmühlen in verschiedenen Anlagen-Konzepten sowohl zur Vor- als auch zur Teilfertig- und Fertigmahlung und Wälzmühlen zur Vor- und Fertigmahlung eingesetzt.

4.1 Vorzerkleinerung mit Prall- und Hammerbrecher

Wegen der sehr geringen Energieausnutzung in der Grobmahlkammer einer Kugelmühle [36] und des großen Investitionsaufwandes für neue Mahlanlagen ist es häufig wirtschaftlich günstiger, bestehende Mahlanlagen mit Kugelmühlen zu optimieren und Einrichtungen zur Vorzerkleinerung des Mahlguts mit höherer Energieausnutzung

4.1 Primary size reduction with impact and hammer mills

Because of the very low energy utilization in the coarse grinding chamber of a ball mill [36] and the high capital cost of new grinding plants it is often more cost-effective to optimize existing grinding plants with ball mills and to position equipment with higher levels of energy utilization upstream of the ball mill for primary size reduction of the mill feed. In this case the fine grinding takes place in the ball mill, so the cement properties remain unaffected by the plant optimization.

The energy utilization in a grinding plant can be significantly improved by primary comminution of the coarse mill feed in a hammer crusher or impact crusher and fine grinding in a ball mill [37 to 40]. The upstream impact crusher can also be operated in circuit with a screening plant. The coarse fraction (e.g > 4mm) in the crushed material is returned to the crusher and the material passing the screen is fine-ground in the following closedcircuit grinding system. If the grinding media charge grading in the ball mill is also adjusted to suit the precomminuted mill feed it is possible to achieve an increase in throughput of 15 to 28 % at a cement specific surface area of about 3300cm^2/g, and a reduction in the overall specific power consumption of 5 to 10 % relative to the finished product mass flow. The relatively low energy saving is mainly due to the fact that hardly any fines are produced in the impact crusher itself.

4.2 Primary grinding in highpressure grinding rolls and roller grinding mills

Significantly more fines are produced during interparticulate comminution; this is carried out on an industrial scale by highpressure grinding rolls and roller grinding mills.

The use of highpressure grinding rolls for grinding cement raw materials, clinker and various additives has been developed in a variety of plant configurations in recent years, mainly in Germany. By high pressure stressing in a gap between two counterrotating grinding rollers (**Fig. 9**) with circumferential velocities of about 0.9 to 1.8 m/s the mill feed is pressed into compacted cake consisting of over 70 % solids by volume and containing up 40 % fines smaller than 90 mm. The fines have to be recovered by disagglomerating the compacted cake. A total of well over 300 sets of highpressure grinding rolls of German manufacture are already in operation, more than half of which are used for the manufacture of cement. An extensive report of the advantages and disadvantages associated with the use of highpressure grinding rolls has already been given in Special Report 5.1 [41]. By using highpressure grinding rolls for primary comminution of the mill feed the specific energy expenditure can be reduced by up to 30 % (depending on the proper-

BILD 9: Mahlwalze mit Lagerung für den Einsatz in einer Gutbett-Walzenmühle (Werkfoto der KHD Humboldt Wedag AG, Köln)
FIGURE 9: Grinding roller with support system for use in high-pressure grinding rolls (works photograph from KHD Humbolt Wedag AG, Cologne)

vorzuschalten. Da die Feinmahlung hierbei in der Kugelmühle erfolgt, bleiben die Zementeigenschaften von der Anlagenoptimierung unbeeinflußt.

Durch Vorzerkleinerung des groben Mahlguts in einem Hammerbrecher oder Prallbrecher und Feinmahlen in einer Kugelmühle läßt sich die Energieausnutzung einer Mahlanlage deutlich verbessern [37 bis 40]. Dabei kann der vorgeschaltete Prallbrecher auch im Kreislauf mit einer Siebanlage betrieben werden. Der Grobanteil (z.B. > 4 mm) des gebrochenen Guts wird wieder dem Brecher zugeführt, der Siebdurchgang in der nachgeschalteten Umlaufmahlanlage feingemahlen. Wenn gleichzeitig die Mahlkörpergattierung der Kugelmühle auf das vorzerkleinerte Mahlgut abgestimmt wird, kann bei einer massebezogenen Oberfläche des Zements von rd. 3300 cm^2/g eine Durchsatzsteigerung von 15 bis 28 % und eine Verminderung des gesamten auf den Fertiggutmassestrom bezogenen Energieverbrauchs von 5 bis 10 % erreicht werden. Die verhältnismäßig geringe Energieeinsparung ist im wesentlichen darauf zurückzuführen, daß im Prallbrecher selbst wenig Feingut erzeugt wird.

4.2 Vormahlung in Gutbett-Walzenmühlen und Wälzmühlen

Deutlich mehr Feingut wird bei der Druckzerkleinerung im Gutbett erzeugt, die großtechnisch in Gutbett-Walzenmühlen und Wälzmühlen verwirklicht wird.

Vor allem in Deutschland wurde in den letzten Jahren der Einsatz von Gutbett-Walzenmühlen zum Mahlen von Zementrohmaterial, Klinker und verschiedenen Zumahlstoffen in unterschiedlichen Anlagenkonzepten entwickelt. Durch Hochdruckbeanspruchung in einem Spalt zwischen zwei sich gegensinnig drehenden Mahlwalzen (**Bild 9**) mit Umfangsgeschwindigkeiten von etwa 0,9 bis 1,8 m/s wird das Mahlgut zu sogenannten Schülpen mit Feststoff-Volumenanteilen von über 70 % gepreßt, die bis zu 40 % Feingutanteil < 90 μm enthalten. Das Feingut muß durch Deglomeration der Schülpen gewonnen werden. Insgesamt sind bereits weit über 300 Gutbett-Walzenmühlen deutscher Hersteller in Betrieb, von denen mehr als die Hälfte für die Herstellung von Zement eingesetzt werden. Über die mit der Nutzung von Gutbett-Walzenmühlen verbundenen Vor- und Nachteile wurde bereits ausführlich im Fachbericht 5.1 berichtet [41]. In Abhängigkeit von den Mahlguteigenschaften und dem aufgebrachten Druck kann durch Einsatz von Gutbett-Walzenmühlen zur Vorzerkleinerung des Mahlguts der massebezogene Energieaufwand gegenüber dem Mahlen in Anlagen mit Kugelmühlen um bis zu 30 % vermindert werden.

Wälzmühlen werden wegen ihres hohen Entwicklungsstandes bereits seit langer Zeit in großem Umfang vor allem zur Rohmaterialmahlung eingesetzt. Ihre Zerkleinerungsleistung hängt von den Einstellgrößen und Betriebsbedingungen, wie Umfangsgeschwindigkeit des Mahltellers, Gestaltung der Mahlrollen und des Mahltellers, Mahlbettdicke und Anpreßdruck der Mahlrollen, sowie von den Mahlguteigenschaften, insbesondere der Mahlbarkeit und der Korngrößenverteilung des Mahlguts, ab.

Vor allem in Japan wurden Vormahl-Verfahren entwickelt, bei denen eine Wälzmühle mit mechanischem Mahlgutaustrag als Vormühle einer bestehenden Kugelmühle vorgeschaltet wird (**Bild 10**). Betriebserfahrungen haben inzwischen gezeigt, daß durch geeignete Maßnahmen auch bei maximal eingestelltem Durchsatz ein stabiler und kontinuierlicher Betrieb einer im Durchlauf betriebenen vorgeschalteten Wälzmühle zu erreichen ist. Dabei wurde eine Steigerung der Durchsatzleistung um bis zu 50 % und eine Energieeinsparung von bis zu 17 % gemessen [40, 42].

Ein Betrieb der vorgeschalteten Wälzmühle mit Mahlgutrückführung (**Bild 10**, gestrichelter Verfahrensschritt) brachte deutliche Verbesserungen in der Betriebsstabilität. Durch Rückführung von 30 bis 40 % des die Wälzmühle verlassenden Mahlguts ohne Absiebung zum Mühleneingang erhöht sich die Schüttdichte des auf dem Mahlteller beanspruchten Mahlbetts, wodurch sowohl die Schwankungen in der Leistungsaufnahme der Mühle als

ties of the mill feed and the pressure applied) when compared with grinding in plants with ball mills.

Because of their high level of development roller grinding mills have already been used very widely for a long time, mainly for grinding raw materials. The comminution capacity depends on the settings and operating conditions such as circumferential velocity of the grinding table, configuration of the grinding rollers and of the grinding table, grinding bed thickness and pressure applied by the grinding rollers, as well as on the mill feed properties, especially grindability and particle size distribution.

Primary grinding processes have been developed, predominantly in Japan, in which a roller grinding mill with mechanical material discharge is positioned as a primary mill upstream of an existing ball mill (**Fig. 10**). Operating experience has now shown that suitable measures can achieve stable and continuous operation of an upstream roller grinding mill operated in open circuit, even at maximum throughput. An increase in the throughput capacity of up to 50 % has been measured with an energy saving of up to 17 % [40, 42].

Operation of the upstream roller grinding mill with material return (process step shown by broken line in **Fig. 10**) brought significant improvements in the operating stability. By returning 30 to 40 % of the material leaving the roller grinding mill to the mill inlet without screening raises the bulk density of the grinding bed which is stressed on the grinding table; this moderates the fluctuations in the power consumption of the mill as well as the oscillations resulting from changes in the grindability of the mill feed and its particle size distribution. Under these conditions the pressure forces of the grinding rollers can be increased and the mill operated stably at maximum power input [43]. This allowed the efficiency of the primary grinding system to be improved significantly, so that increases in throughput of up to 60 % and energy savings of up to 20 % became possible when grinding clinker, provided that the grinding media charge in the ball mill was also adapted to suit the finer feed material.

Increases in throughput of 80 to 120 % and reductions in specific power consumption by 25 to 30 % have been achieved when grinding raw material by primary grinding in a roller grinding mill with material return. In this case most of the raw material drying takes place in the roller grinding mill. The existing ball mill can also be brought into use for drying very moist raw materials. However, in some cases it can be more cost-effective, especially in new plants, to produce the raw meal just by drying and grinding in a suitably dimensioned roller grinding mill with integral classifier and pneumatic removal of the ground material.

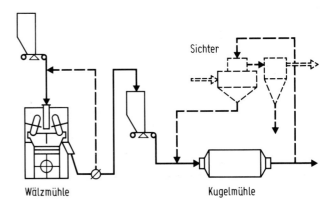

BILD 10: Umlaufmahlanlage mit vorgeschalteter Wälzmühle mit mechanischem Austrag und Mahlgutrückführung; schematische Darstellung
FIGURE 10: Closed-circuit grinding plant with upstream roller grinding mill with mechanical discharge and material return system; diagrammatic representation

Wälzmühle = roller grinding mill
Sichter = classifier
Kugelmühle = ball mill

auch die Schwingungen infolge der Veränderungen in der Mahlbarkeit des Mahlguts und seiner Korngrößenverteilung abgebaut werden. Unter diesen Bedingungen können die Anpreßkräfte der Mahlrollen erhöht und die Mühle bei maximaler Leistungsaufnahme stabil betrieben werden [43]. Damit ließ sich die Effizienz der Vormahlung deutlich verbessern, so daß bei der Klinkermahlung Durchsatzsteigerungen bis zu 60% und Energieeinsparungen bis 20% möglich wurden, wenn auch die Mahlkörperfüllung in der Kugelmühle dem feineren Aufgabegut angepaßt wurde.

Bei der Rohmaterialmahlung wurden durch Vormahlung in der Wälzmühle mit Mahlgutrückführung Durchsatzsteigerungen von 80 bis 120% und eine Verminderung des massebezogenen Energieaufwands um 25 bis 30% erreicht. In diesem Fall erfolgte die Trocknung des Rohmaterials vor allem in der Wälzmühle. Für die Trocknung von sehr feuchtem Rohmaterial kann die vorhandene Kugelmühle zusätzlich mit einbezogen werden. Jedoch kann es in vielen Fällen, insbesondere bei Neuanlagen, wirtschaftlich günstiger sein, das Rohmehl allein durch Mahltrocknung in einer entsprechend dimensionierten Wälzmühle mit integriertem Sichter und pneumatischer Abförderung des gemahlenen Guts zu erzeugen.

Wälzmühlen haben sich seit vielen Jahren auch in den größten Baugrößen als mechanisch zuverlässig bewährt. Das gilt nicht nur für den Einsatz zur Rohmaterialmahlung sondern auch für den Einsatz als Vormühle ohne und mit Mahlgutrückführung zur Klinkermahlung. Außer der Zuverlässigkeit sind die kompakte Bauweise, die einfache und wirtschaftliche Betriebsführung sowie die Möglichkeit der Mahltrocknung wesentliche Vorteile dieses Mühlentyps. Der Verschleißschutz von Mahlwalzen und Mahlteller wird heute aus hochlegiertem Chromstahlguß hergestellt und kann durch eine besondere Wärmebehandlung hohe Standzeiten von mehr als 20000 h erreichen [40, 44, 45]. Die Modifikation einer bestehenden Umlaufmahlanlage durch Vorschalten einer Wälzmühle zur Vorzerkleinerung erfordert geringere Investitionskosten als im Normalfall für eine komplette neue Wälzmühlenanlage notwendig sind [40]. Sie werden sich aufgrund der hohen Durchsatzsteigerungen und Energieeinsparungen meist in verhältnismäßig kurzen Zeiten von wenigen Jahren amortisieren. Das ist vor allem dann der Fall, wenn die Mahlanlage nicht mehr in der Hoch- sondern ausschließlich in der Niedertarifzeit betrieben werden kann. Größere Investitionen rechnen sich in der Regel nur, wenn außerdem ältere Kleinanlagen stillgelegt werden können. Außerdem sollen durch die Installation einer Wälzmühle als Vormühle meist die gesamten Unterhaltskosten geringer sein [43, 44]. Durch den deutlich verbesserten Verschleißschutz in der Wälzmühle wurde erreicht, daß das Zerkleinerungsergebnis weitgehend unabhängig von der Betriebszeit der Mühle ist. Wenn noch ein Teil der Fertigmahlung des Zements in einer Kugelmühle erfolgt, werden außerdem die Zementeigenschaften bei Anwendung der Wälzmühle als Vormühle nicht beeinflußt.

4.3 Fertigmahlung in Gutbett-Walzenmühlen und Wälzmühlen

Je größer der durch Druckbeanspruchung erzeugte Feingutanteil ist, um so größer ist die Energieausnutzung beim Mahlen und damit die Energieeinsparung gegenüber der Mahlung in Kugelmühlen. Daher wurde in den letzten Jahren angestrebt, die gesamte Feinmahlung ausschließlich in einer Gutbett-Walzenmühle oder einer Wälzmühle durchzuführen und auf den Einsatz einer Kugelmühle ganz zu verzichten.

Untersuchungen haben gezeigt, daß bei der Fein- und Feinstmahlung spröder Stoffe in Gutbett-Walzenmühlen eine Energieeinsparung gegenüber Anlagen mit Kugelmühlen um bis zu mehr als 50% möglich ist [41]. In der Zementindustrie wurde die Fertigmahlung bei der Rohmehl- und Hüttensandmahlung inzwischen erfolgreich eingesetzt. Bei Rohmaterial mit relativ geringer Feuchte ist die Rohmehlerzeugung in Gutbett-Walzenmühlen meist noch effizienter als in Wälzmühlen. Während sich die Fertigmahlung in Gut-

Roller grinding mills have for a long time proved themselves to be mechanically reliable, even in the largest models. This is true not only when used for raw material grinding but also when used as primary mills for clinker grinding, with or without material return. Apart from reliability the compact design, the simple and economical operation, and the option of drying and grinding are important advantages of this type of mill. The wear protection on the grinding rollers and grinding table is now made of high-alloy chromium steel casting and, with special heat treatment, can achieve high service lives of over 20000 h [40, 44, 45]. The modification of an existing closed-circuit plant by positioning a roller grinding mill upstream for primary comminution requires less capital expenditure than is normally necessary for a complete, new, roller grinding mill plant [40]. Because of the high increases in throughput and energy savings it usually pays for itself in the relatively short time of a few years. This is especially the case if the grinding plant no longer has to run in the high tariff period and can be operated entirely in the low tariff period. As a rule, higher capital costs are only justified if it is also possible to shut down small obsolete plants. Installation of a roller grinding mill as a primary mill should also normally reduce the total maintenance costs [43, 44]. The significantly improved wear protection in the roller grindng mill means that the comminution results are largely independent of how long the mill has been in operation. Provided some of the finish grinding of the cement still takes place in a ball mill the cement properties are not affected by the use of a roller grinding mill as a primary mill.

4.3 Finish grinding in high-pressure grinding rolls and roller grinding mills

The larger the proportion of fines produced during the pressure stressing the greater is the utilization of energy during grinding, and therefore the saving in energy when compared with grinding in ball mills. Efforts have therefore been made in recent years to carry out all the fine grinding solely in high-pressure grinding rolls or a roller grinding mill and to dispense entirely with the use of a ball mill.

Investigations have shown that during fine and very fine grinding of brittle materials in high-pressure grinding rolls it is possible to achieve energy savings of more than 50% when compared with plants with ball mills [41]. Finish grinding has now been introduced successfully into the cement industry for grinding raw meal and granulated blastfurnace slag. For raw materials with relatively low moisture levels it is usually more efficient to produce raw meal in high-pressure grinding rolls than in roller grinding mills. Finish grinding in high-pressure grinding rolls does not have any adverse effect on the service properties of raw meal or granulated blastfurnace slag. However, the properties of cements which have been finish ground in high-pressure grinding rolls differ significantly in many cases from those ground in ball mills. Depending on the reactivity of the clinker used the water demand to achieve standard stiffness is sometimes significantly higher, and the setting times are drastically shortened. According to current findings this can be attributed to a great extent to excessively narrow particle size distributions of the cement and to inadequate matching of the type and quantity of the sulphate agent to the reactivity of the clinker. Various constructional and process engineering remedies are being tested at the moment.

Initial investigations [46 to 48] into finish grinding of cement in roller grinding mills confirmed the expectation that the specific power consumption of the roller grinding mill system was up to 30% lower than that of a grinding plant with ball mill and classifier. On the other hand, the investigations with cements showed that grinding in a roller grinding mill which was adjusted for raw meal grinding had a very detrimental effect on the technological properties of the cement and concrete. The setting times were shorter and the water demand to achieve standard stiffness was significantly larger than with cements from ball mills. As with cements from high-pressure grinding rolls, the cause was recognized as being due mainly to a much narrower particle size distribution of the cements from roller grinding mills.

bett-Walzenmühlen nicht nachteilig auf die Gebrauchseigenschaften von Rohmehl oder Hüttensand auswirkt, unterscheiden sich demgegenüber die Eigenschaften der in Gutbett-Walzenmühlen fertiggemahlenen Zemente in vielen Fällen noch deutlich von denen aus Kugelmühlen. Je nach der Reaktivität des verwendeten Klinkers sind die Wasseranspruch zur Erzielung der Normsteife z. T. deutlich höher und die Erstarrungszeiten drastisch verkürzt. Dies kann nach bisher vorliegenden Erkenntnissen zum großen Teil auf eine zu enge Korngrößenverteilung des Zements sowie auf eine unzureichende Abstimmung von Art und Menge des Sulfatträgers auf die Reaktivität des Klinkers zurückgeführt werden. Derzeit werden verschiedene konstruktive und verfahrenstechnische Abhilfemaßnahmen erprobt.

Erste Untersuchungen [46 bis 48] beim Fertigmahlen von Zement in Wälzmühlen bestätigten die Erwartung, daß der massebezogene Energieverbrauch des Wälzmühlensystems um bis zu 30 % geringer als der einer Mahlanlage mit Kugelmühle und Sichter war. Andererseits ergaben die Untersuchungen an den Zementen, daß sich die Mahlung auf einer Wälzmühle, die für Rohmehlmahlung eingestellt war, sehr ungünstig auf die zement- und betontechnischen Eigenschaften auswirkte. Die Erstarrungszeiten waren kürzer, und der Wasseranspruch zur Erzielung der Normsteife war deutlich größer als bei Zementen aus Kugelmühlen. Als Ursache wurde wie bei Gutbett-Walzenmühlenzementen vor allem eine erheblich engere Korngrößenverteilung der Wälzmühlen-Zemente erkannt.

In den vergangenen Jahren wurden jedoch von den verschiedenen Mühlenherstellern eine Reihe unterschiedlicher konstruktiver und verfahrenstechnischer Änderungen am Mahl- und Sichtsystem eingeführt, wie z. B. Erhöhen des Staurings und gleichzeitiges Optimieren der Form der Mahlwerkzeuge und Erhöhen des Anpreßdrucks, den Mahlwalzen vorgeschaltete Glättwalzen (**Bild 11**), Verändern des Düsenrings und der Gasströmungsgeschwindigkeit und Anordnen von Bypass-Rohren, um ungesichtetes Mahlgut am Sichter vorbei ins Sichterfeingut zu führen. Mit Hilfe der Glättwalzen soll ein höher verdichtetes Mahlgutbett erreicht werden, das für eine Beanspruchung durch die Mahlwalzen mit höherem Anpreßdruck besser vorbereitet ist. Diese Maßnahmen haben dazu beigetragen, daß beim Mahlen auf Wälzmühlen eine breitere Korngrößenverteilung der Zemente eingestellt werden konnte und die Zementeigenschaften den baupraktischen Anforderungen besser entsprachen und denen der Zemente aus einer Kugelmühlenanlage mindestens gleichwertig waren [44, 48–50]. Durch eine zweckmäßige Gestaltung und einen verbesserten Verschleißschutz haben die Mahlwerkzeuge eine hohe Wirksamkeit und eine lange Lebensdauer erreicht. Daher bleibt darüber hinaus die Produktqualität vom Verschleiß der Mahlwerkzeuge weitgehend unbeeinflußt und während der gesamten Lebensdauer gleich [44]. Mit modernen Wälzmühlen-Anlagen sollen heute beim Mahlen von Klinker bis zu 40 % und beim Mahlen von Hüttensand bis zu 50 % des Energieverbrauchs einer Kugelmühle vergleichbarer Durchsatzleistung eingespart werden können. Mit den unterschiedlichen konstruktiven und verfahrenstechnischen Maßnahmen wurde ein stabiles Mahlbett aufgebaut und damit ein stabiler Mühlenbetrieb erreicht [51–53].

Etwa die Hälfte des gesamten massebezogenen elektrischen Energieverbrauchs einer Wälzmühlenanlage wird von dem Prozeßventilator für die pneumatische Förderung des Mahlguts verbraucht. Daher kann durch Vermindern des Druckverlustes und der Gasgeschwindigkeit im Mahlraum der Energieverbrauch des Mahlsystems erheblich reduziert werden. Eine deutliche Verminderung des Druckverlustes läßt sich bereits durch konstruktive Änderungen des Düsenrings erreichen. Erweiterte Strömungsquerschnitte führen zu niedrigeren Gasgeschwindigkeiten. Dadurch wird allerdings ein zusätzlicher äußerer Mahlgutkreislauf (**Bild 12**) mit mechanischen Förderern benötigt, damit die gröberen Mahlgutpartikel für eine weitere Zerkleinerung wieder auf die Mahlbahn zurückgeführt werden können [44, 54]. Obwohl der Energieverbrauch der Mühle allein dadurch

However, in the past few years the different mill manufacturers have introduced a series of different structural and process engineering changes to the grinding and classifying system, such as raising the dam ring and at the same time optimizing the shape of the grinding elements and raising the applied pressure, placing smoothing rollers in front of the grinding rollers (**Fig. 11**), changing the louvre air ring and the gas flow velocity, and positioning bypass pipes in order to carry unclassified ground material around the classifier and into the classifier fines. The smoothing rollers are intended to produce a more highly compacted bed of material which is better prepared for stressing by the grinding rollers with higher applied pressures. These measures mean that it has become possible to produce a wider cement particle size distribution when grinding in roller grinding mills, and that the cement properties correspond better to the requirements of building practice and are at least equivalent to those of cements from a ball mill plant [44, 48–50]. Appropriate configuration and improved wear protection have made the grinding elements highly effective, and long service lives are achieved. The product quality therefore remains largely unaffected by the wear to the grinding elements and is the same throughout their entire service life [44]. Modern roller grinding mill plants should now save up to 40 % of the energy consumption of a ball mill of comparable output when grinding clinker and up to 50 % when

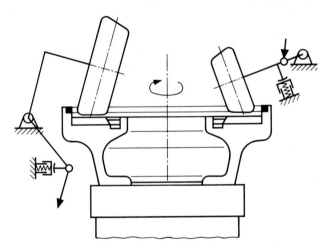

BILD 11: Schematische Darstellung von Glättwalze und Mahlwalze einer Wälzmühle für Zement [50]
FIGURE 11: Diagrammatic representation of smoothing roller and grinding roller in a roller grinding mill for cement [50]

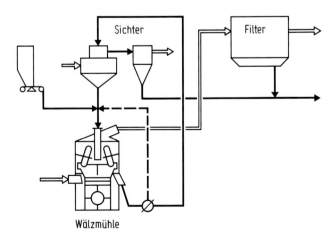

BILD 12: Mahlanlage mit Wälzmühle und zusätzlichem äußeren Mahlgutkreislauf und äußerem Sichter
FIGURE 12: Grinding plant with roller grinding mill and additional external grinding circuit and external classifier

Sichter = classifier
Wälzmühle = roller grinding mill
Filter = filter

leicht ansteigen kann und der zusätzliche äußere Mahlgutkreislauf höhere Betriebskosten erfordert, nimmt der gesamte Energieverbrauch des Wälzmühlensystems mit zunehmendem äußeren Materialkreislauf ab. Insgesamt werden durch den äußeren Mahlgutumlauf die Druckverluste gesenkt, die Stabilität des Mühlenbetriebs verbessert und Energieeinsparungen erzielt.

4.4 Hybrid- und Teilfertigmahlung in Gutbett-Walzenmühlen und Wälzmühlen

Soll aus Gründen der Produktqualität auf die Kugelmühle zur Feinmahlung nicht verzichtet werden, jedoch der Durchsatz einer vorhandenen Mahlanlage mit Kugelmühle stärker gesteigert werden als es mit dem Anlagenkonzept der Vormahlung möglich ist, können Gutbett-Walzenmühlen und Wälzmühlen hybrid betrieben werden [41]. Dabei wird ein Teil des Sichtergrobguts des Kugelmühlen-Sichter-Kreislaufs zurückgeführt und gemeinsam mit dem Frischgut der Vormühle aufgegeben. Die Vormühle kann außerdem auch mit eigener Mahlgutrückführung betrieben werden.

Die erzielten Leistungen und Energieeinsparungen bei Hybrid- und Teilfertigmahlung mit Gutbett-Walzenmühlen wurden ausführlich im Fachbericht 5.1 erläutert [41]. Beim Mahlen von Zement sind auf diese Weise Energieeinsparungen von bis zu 50 % möglich [2, 55].

Bei Wälzmühlen in Hybridschaltung mit einer Kugelmühlen-Umlaufmahlanlage war der gesamte Energieverbrauch höher als bei dem Anlagenkonzept der Vormahlung [43, 44]. Außerdem ist die Steuerung der Mahlanlage in Hybridschaltung gegenüber der bei Schaltung der Wälzmühle als Vormühle deutlich erschwert. Daher kam dieses Konzept bisher nicht zum Einsatz. Auch das System der Teilfertigmahlung in einer Wälzmühle mit anschließender Fertigmahlung in einer Kugelmühlenanlage ist wirtschaftlich ungünstig.

4.5 Folgerungen

Aus den Darstellungen geht ingesamt hervor, daß vor allem durch den zunehmenden Einsatz energiesparender Mahlsysteme und Hochleistungssichter der massebezogene elektrische Energieverbrauch der westdeutschen Zementwerke seit 1985 deutlich zurückgegangen ist (s. Bild 1). Dabei wurde allerdings ein Teil der durch neue Maschinen und Verfahren zur Zerkleinerung eingesparten Energie durch verschärfte Umweltschutzauflagen für den gesamten Herstellungsprozeß wieder aufgezehrt.

In steigendem Umfang wurden Wälz- und Gutbett-Walzenmühlen unter vergleichbaren Bedingungen eingesetzt. In Deutschland wurden vor allem die Mahlkonzepte mit Gutbett-Walzenmühlen weiterentwickelt, während insbesondere in Japan die Systeme mit Wälzmühlen weitere Verbreitung fanden. Aus einzelnen Untersuchungen geht hervor, daß die Mahlung in Gutbett-Walzenmühlen mit einer höheren Energieausnutzung und daher mit einer größeren Energieeinsparung gegenüber der Mahlung in Wälz- und Kugelmühlen verbunden ist [56]. Bei hartem, sprödem Mahlgut hat der Einsatz von Gutbett-Walzenmühlen Vorteile, während Wälzmühlen vorteilhafter für das Mahlen weicherer und vor allem sehr feuchter Stoffe eingesetzt werden können. Inzwischen ist es bei beiden Mühlentypen möglich, durch konstruktive und verfahrenstechnische Maßnahmen in begrenztem Umfang die Korngrößenverteilung des Produkts so zu beeinflussen, daß die angestrebten Produkteigenschaften in vielen Fällen sicher eingestellt werden können. Bei zweckmäßiger Dimensionierung, sachgerechter Betriebsweise und Anwendung modernsten Verschleißschutzes läßt sich heute mit beiden Mühlentypen eine zufriedenstellende Standzeit der Mahlwerkzeuge und damit eine ausreichende Betriebssicherheit erreichen. Demgegenüber sind konventionelle Mahlanlagen mit Kugelmühle und Sichter vor allem unempfindlicher gegenüber Fremdkörpern und haben deshalb meist eine höhere Betriebssicherheit.

grinding granulated blastfurnace slag. The various structural and process engineering measures form a stable grinding bed, which also achieves stable mill operation [51-53].

About half of the total specific electrical energy consumption of a roller grinding mill plant is taken by the process fan used for pneumatic transport of the ground material. The power consumption of the grinding system can therefore be greatly reduced by lowering the pressure drop and the gas velocity in the grinding chamber. A significant reduction in the pressure drop can be achieved just by structural alterations to the louvre air ring. Enlarged flow cross-sections lead to lower gas velocities. This means, however, that an additional external material circuit with mechanical transport is needed (Fig. 12) so that the coarser material particles can be returned to the grinding track for further comminution [44, 54]. This may raise the power consumption of the mill itself slightly and the additional external material circuit involves higher operating costs, but the overall power consumption of the roller grinding mill alone decreases with increasing external material recycling. In all, the pressure losses are lowered by the external material recycling, the stability of the mill operation is improved and energy savings are achieved.

4.4 Hybrid and semi-finish grinding in high-pressure grinding rolls and roller grinding mills

If, for reasons of product quality, the ball mill is not to be discarded for fine grinding but the throughput of an existing grinding plant with ball mill is to be increased by more than can be achieved by a plant configuration with primary grinding then high-pressure grinding rolls and roller grinding mills can be operated in hybrid configurations [41]. Part of the classifier tailings from the circuit consisting of ball mill and classifier is returned and fed to the primary mill together with the fresh material. The primary mill can also be operated with a suitable system for returning ground material.

The performance and energy saving which can be achieved with hybrid and semi-finish grinding with high-pressure grinding rolls have been examined in detail in Special Report 5.1 [41]. Energy savings of up to 50 % are possible when grinding cement by this method [2, 55].

With roller grinding mills in a hybrid configuration with a closed-circuit ball mill grinding plant the total power consumption is higher than in the plant arrangement with primary grinding [43, 44]. Controlling the grinding plant with a hybrid configuration is also significantly harder than in the configuration with a roller grinding mill as a primary mill, so this scheme has not yet come into use. The system for semi-finish grinding in a roller grinding mill followed by finish grinding in a ball mill plant is also economically unfavourable.

4.5 Consequences

From the descriptions it can be seen that, taken as a whole, there has been a significant reduction in the specific electrical energy consumption of West German cement works since 1985 (see Fig. 1), chiefly due to the increasing use of energy-saving grinding systems and high-efficiency classifiers. However, part of the energy saved by the new machinery and processes used for comminution has been consumed by intensified environmental protection requirements for the entire production process.

Roller grinding mills and high-pressure grinding rolls are being increasingly used under comparable conditions. In Germany the continued development has been mainly in the grinding schemes with high-pressure grinding rolls, but in Japan in particular the systems with roller grinding mills have become more widespread. From individual investigations it has emerged that grinding in high-pressure grinding rolls is associated with a higher energy utilization and therefore with greater energy saving than grinding in roller grinding mills and ball mills [56]. The use of high-pressure grinding rolls has advantages with hard, brittle, mill feed, while

4.6 Walzen-Rohrmühle (Horizontal-Rollenmühle)

Im Jahr 1993 wurde erstmals eine völlig neu entwickelte Mühle vorgestellt, die die hohe Zuverlässigkeit einer Mahlanlage mit Kugelmühle und die guten Produkteigenschaften mit der hohen Energieausnutzung und der Flexibilität eines Mahlsystems mit Gutbett-Walzenmühle verbinden soll. Die Mühle besteht aus einem waagerecht gelagerten Mühlenrohr mit einem Durchmesser/Länge-Verhältnis von etwa 1. Im unteren Bereich des Mühlenrohrs wird eine zylindrische Mahlwalze an die Innenwand des sich drehenden Mühlenrohrs gepreßt (**Bild 13**). Das Mahlgut wird zwischen Mühlenrohr und Mahlwalze vorwiegend durch Druck zerkleinert [57].

Aufgrund erster Untersuchungen soll der Anpreßdruck der Mahlwalze deutlich geringer als bei Gutbett-Walzenmühlen sein, der Druck in der Mahlzone demgegenüber erheblich höher. Die Mühle arbeitet im geschlossenen Kreislauf mit einem Sichter. Beim Mahlen von Portlandzement Z 35 wurde bei einem sehr stabilen und ruhigen Mühlenbetrieb eine Energieeinsparung von etwa 40 % gegenüber dem Mahlen in einer optimierten Kugelmühle erreicht. Es wird erwartet, daß die Mühle auch zum Mahlen von Stoffen mit verschiedenen Eigenschaften und in unterschiedlichen Anlagen-Konzepten vorteilhaft eingesetzt werden kann.

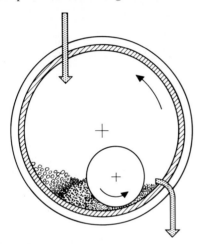

BILD 13: Schematische Darstellung des Aufbaus der Walzen-Rohrmühle „HOROMILL" (Horizontal Roller Mill) [57]
FIGURE 13: Diagrammatic representation of the structure of the "HOROMILL" (horizontal roller mill) roller tube mill [57]

5. Mahlbarkeit der Zementbestandteile

Maßgebend für den Energieaufwand zum Mahlen sind außer dem Mahlverfahren und den Betriebsbedingungen von Mühle und Sichter vor allem auch die Mahlbarkeit der Zementbestandteile und die geforderte Mahlfeinheit. Der Energieaufwand wird um so geringer, je leichter mahlbar das Mahlgut ist. Dabei versteht man unter der Mahlbarkeit eines Stoffs den Energiebedarf einer Labormühle, der für das Feinmahlen einer Probe von einer bestimmten Ausgangskörnung auf eine bestimmte Endfeinheit erforderlich ist [58]. Als Maß der Mahlfeinheit wurde in vielen Fällen die massebezogene Oberfläche benutzt. Da sich die Korngrößenverteilungen verschiedener gemahlener Stoffe bei gleicher massebezogener Oberfläche aber erheblich unterscheiden können, ist es sowohl für den Vergleich der Mahlbarkeitswerte verschiedener Stoffe und Stoffkombinationen untereinander als auch für den Zusammenhang zwischen den Mahlbarkeitswerten und dem Energieaufwand in Betriebsmühlen erforderlich, die Korngrößenverteilungen der Komponenten zu berücksichtigen. Daher wurde in einem modifizierten Mahlbarkeitsprüfverfahren der massebezogene Energieaufwand des Prüfgeräts auf die Parameter der Korngrößenverteilung des erzeugten Mehls bezogen [59–61]. Mit diesem Verfahren können auch kleinere Mahlbarkeitsunterschiede und die Mahlbarkeit harter und weicher Stoffe reproduzierbar ermittelt werden. Das Prüfer-

roller grinding mills can be used to greater advantage for grinding soft and, above all, very moist materials. With both types of mill it has now become possible, through constructional and process engineering means, to influence the particle size distribution of the product to a limited extent so that in many cases the desired product properties can be obtained. With appropriate dimensioning, correct mode of operation, and application of the latest wear protection it is now possible with both types of mill to achieve satisfactory service lives for the grinding elements and therefore also sufficient operational reliability. On the other hand, it is a particular feature of grinding plants with ball mills and classifiers that they are less sensitive to foreign bodies and therefore usually have greater operational reliability.

4.6 Roller tube mill (horizontal roller mill)

A completely newly developed mill was introduced for the first time in 1993; it is intended to combine the high reliability of a ball mill grinding plant with the good production characteristics, high energy utilization, and flexibility of a grinding system with high-pressure grinding rolls. The mill consists of a horizontally supported mill tube with a diameter/length ratio of approximately 1. In the lower part of the mill tube a cylindrical grinding roller is pressed against the inner wall of the rotating mill tube (**Fig. 13**). The mill feed is comminuted between the mill tube and the grinding roll, predominantly by pressure [57].

Initial investigations show that the applied pressure of the grinding roll should be significantly lower than in high-pressure grinding rolls, but that the pressure in the grinding zone is considerably higher. The mill operates in closed circuit with a classifier. When grinding Z 35 Portland cement the mill operation was very stable and smooth and an energy saving was achieved of approximately 40 % when compared with grinding in an optimized ball mill. It is expected that it will also be possible to use the mill advantageously for grinding materials with a variety of properties and in different plant configurations.

5. Grindability of the cement constituents

Apart from the grinding process and the operating conditions of the mill and classifier the power consumption for grinding is also greatly affected by the grindability of the cement constituents and the required fineness of grinding. The power consumption is lower the more easily the mill feed can be ground. The grindability of a material is taken to mean the power consumption of a laboratory mill needed for fine grinding a sample of a given initial particle size composition to a given end fineness [58]. In many cases the specific surface area is used as the measure of fineness. However, the particle size distributions of different ground materials of the same specific surface area can vary considerably, so both for comparing grindability values of different materials and material combinations with one another and for finding the relationship between grindability values and power consumption in industrial mills it is necessary to take the particle size distributions of the components into account. In a modified grindability test the specific power consumption of the test equipment was therefore related to the parameters of the particle size distribution of the meal produced [59–61]. With this test method it is also possible to determine reproducibly the grindabilities of hard and soft materials and quite small differences in grindability. The test results are affected by the type of mechanical stressing in the test equipment. The measurements show that the relationship between the specific power consumption of the test equipment and of industrial grinding plants with ball mills measured with different materials is practically linear and has a high coefficient of determination.

6. Power consumption for grinding cements with several main constituents

The hydraulic or pozzolanic activity of interground additives is generally significantly lower than that of clinker or, in the case of inert constituents, is almost entirely absent, so

gebnis wird von der Art der mechanischen Beanspruchung im Prüfgerät beeinflußt. Die Messungen zeigen, daß der an unterschiedlichem Mahlgut ermittelte Zusammenhang zwischen dem massebezogenen Energiebedarf des Prüfgeräts und von Betriebsmahlanlagen mit Kugelmühlen praktisch linear ist und ein hohes Bestimmtheitsmaß aufweist.

6. Energieaufwand zum Mahlen von Zementen mit mehreren Hauptbestandteilen

Da die hydraulische bzw. puzzolanische Aktivität von Zumahlstoffen im allgemeinen deutlich geringer als die des Klinkers ist oder im Fall von inerten Bestandteilen praktisch ganz fehlt, müssen Zemente mit Zumahlstoffen generell feiner gemahlen werden als ein Portlandzement derselben Festigkeitsklasse und Leistungsfähigkeit [61, 62]. Außerdem kann die Feuchte der Zumahlstoffe den Mahlfortschritt in der Mühle beeinträchtigen und beim gemeinsamen Feinmahlen mit Klinker zu einer Vorhydratation des Klinkers führen. Gerade in solchen Fällen ist dann eine höhere Mahlfeinheit des Zements zum Erreichen der gewünschten Festigkeit erforderlich, die einen zusätzlichen Energieaufwand beim Mahlen von Zement mit feuchten Zumahlstoffen zur Folge hat. Diese Maßnahme läßt sich allerdings nur in eng begrenztem Umfang anwenden. Durch Vortrocknen der Zumahlstoffe kann demgegenüber die Durchsatzleistung der Mühle deutlich gesteigert, der massebezogene Energieaufwand zum Mahlen gesenkt und eine Beeinträchtigung der Produkteigenschaften vermieden werden.

Bei der Herstellung von Zementen mit mehreren Hauptbestandteilen durch gemeinsames Feinmahlen wird die entstehende Korngrößenverteilung des Zements in starkem Maß von der Mahlbarkeit und dem Mengenanteil des jeweiligen Zumahlstoffs bestimmt. Versuche haben gezeigt, daß die Korngrößenverteilung des jeweils schwerer mahlbaren Stoffs enger, die des leichter mahlbaren Stoffs breiter ausfällt als bei einer getrennten Mahlung [62]. Daher sind bei einer zielsicheren Herstellung von Zement mit bestimmten Eigenschaften sowohl durch gemeinsames Feinmahlen als auch durch getrenntes Mahlen der Komponenten und anschließendes Mischen der mehlfeinen Halbprodukte die beim Mahlen erzeugten Korngrößenverteilungen der Bestandteile zu berücksichtigen. Vor allem bei größeren Unterschieden in der Mahlbarkeit und/oder der Aufgabekorngröße, wie z. B. bei Klinker und Kalkstein oder bei Klinker und Hüttensand, kann es aus Gründen der Energieeinsparung gegenüber dem Verfahren der gemeinsamen Fertigmahlung vorteilhaft sein, die Hauptbestandteile getrennt auf eine bestimmte Mahlfeinheit oder Korngrößenverteilung vor- bzw. feinzumahlen und den Zement anschließend durch gemeinsames Feinmahlen oder auch durch Mischen der mehlfeinen Bestandteile herzustellen. Wesentliche Voraussetzung für das Vorgehen sind die Konformität der Zementeigenschaften mit den Festlegungen der jeweils gültigen Normen und deren Gleichmäßigkeit. Ein großer Vorteil des getrennten Feinmahlens und anschließenden Mischens wird auch darin gesehen, daß unter Nutzung optimierter Zusammensetzungen und Mahlfeinheiten von getrennt gemahlenen Halbprodukten die Zementeigenschaften entsprechend den Anforderungen des Marktes zielsicher, gleichmäßig und kostengünstig hergestellt werden können.

Der Einsatz von Zumahlstoffen bei der Zementmahlung anstelle von Klinker kann darüber hinaus generell zu einer Einsparung an Brennstoffenergie und elektrischer Energie bei der Klinkererzeugung führen. Das effektiv erreichbare Einsparpotential läßt sich jedoch nur aus einer im Einzelfall zu ermittelnden Energiebilanz des Gesamtprozesses bestimmen. Dabei sind außer dem Energieaufwand zur Herstellung des Klinkers auch die Energiemengen, die zur Herstellung der Vorprodukte erforderlich sind, zu berücksichtigen. Zusätzlich vermindern die Anforderungen an eine marktgerechte hohe und vergleichbare Leistungsfähigkeit des Zements im Beton die rechnerisch mögliche Senkung des Energieaufwands, die sich aus dem Ersatz von Klinker durch Zumahlstoffe ergibt.

cements with interground additives generally have to be more finely ground than a Portland cement of the same strength class and performance [61, 62]. In addition to this the moisture in the interground additives can impair the progress of grinding in the mill and, during fine intergrinding with clinker, can lead to prehydration of the clinker. In such cases a higher fineness of the cement is then needed to achieve the required strength, which results in additional energy expenditure when grinding cements with moist interground additives. However, this approach can only be used to a limited extent, and in contrast the throughput capacity of the mill can be increased significantly, the specific energy consumption for grinding reduced, and any impairment of the product properties avoided by predrying the interground additives.

When manufacturing cements with several main constituents by fine intergrinding the resulting particle size distribution of the cement is determined to a great extent by the grindabilities and proportions of the particular interground additives. Trials have shown that in any particular case the particle size distribution of the material which is harder to grind is narrower, and that of the more easily ground material is wider, than with separate grinding [62]. The particle size distributions of the constituents produced during grinding therefore have to be taken into account during the precisely controlled manufacture of a cement with specific properties, whether by fine intergrinding or by separate grinding of the components followed by mixing of the mealfine intermediate products. Especially where there are large differences in grindability and/or in feed particle size, such as with clinker and limestone or with clinker and granulated blastfurnace slag, energy can be saved when compared with the intergrinding method by pre-grinding or fine grinding the main constituents separately to a given fineness or particle size distribution. The cement is then manufactured by fine intergrinding or by mixing the mealfine constituents. It is essential for this procedure are that the cement characteristics are consistent and that they conform to the requirements of the relevant standards. One great advantage of separate fine grinding followed by mixing is regarded as the fact that by utilizing optimized compositions and finenesses of separately ground intermediate products the cement characteristics can be produced accurately, uniformly and at reasonable cost to meet the demands of the market.

The use of interground additives in cement grinding instead of clinker also generally leads to a saving in fuel energy and electrical energy in clinker production. The potential saving which is effectively achievable can, however, only be determined from an energy balance of the entire process which has to be carried out in the individual instance. In addition to the energy expenditure for manufacturing the clinker it is also necessary to take into account the amount of energy which is required to manufacture the intermediate products. The theoretical reduction in energy expenditure which could be achieved from the replacement of clinker by interground additives is also reduced by the requirement for the cement in the concrete to have a high and comparable performance which meets market requirements.

Literaturverzeichnis

[1] Scheuer, A., und Ellerbrock, H.-G.: Möglichkeiten der Energieeinsparung bei der Zementherstellung. Zement-Kalk-Gips 45 (1992) 5, S. 222–230.

[2] Fujimoto, S.: Reducing specific power usage in cement plants. World Cement (1993) 7, S. 25–35.

[3] Blanck, M.: Neue Wege im Energiemanagement zur Steigerung der Energieproduktivität im Betrieb. Zement-Kalk-Gips 44 (1991) 11, S. 565–570.

[4] Blanck, M.: Energiedatenanalyse – eine Voraussetzung für die rationelle Elektrizitätsverwendung. Zement-Kalk-Gips 42 (1989) 6, S. 288–290.

[5] Ritzmann, H.: Neuentwicklungen von Krupp Polysius für die Zementindustrie. Zement-Kalk-Gips 46 (1993) 4, S. 173–180.

[6] Albeck, J., und Kirchner, G.: Einfluß der Verfahrenstechnik auf die Herstellung marktorientierter Zemente. Zement-Kalk-Gips 46 (1993) 10, S. 615–626.

[7] Pendo, M. C.: A plan for upgrading electrical power distribution systems at South Dakota Cement. 33rd IEEE Cement Industry Technical Conference, May 1991, S. 385–405.

[8] Barreiro, C., Ferreira, B., Abreu, C., und Blanck, M.: Energy Management for rational electricity use. 32nd IEEE Cement Industry Technical Conference, May 1990, S. 207–236.

[9] Hamdani, R., und Zarif, Z. K.: 20 Years Operating Experience with gearless drives. 32nd IEEE Cement Industry Technical Conference, May 1990, S. 57–80.

[10] Thomas, P. F.: Development of mill drives for the Cement Industry. 33rd IEEE Cement Industry Technical Conference, Mai 1991, S. 171–189.

[11] Ranze, W.: Drehzahlvariable Antriebe in der Zementindustrie. Zement-Kalk-Gips 37 (1984) 11, S. 593–598.

[12] Errath, R. A.: Integrated Variable Speed Drives – The Energy Savers. 33rd IEEE Cement Industry Technical Conference, May 1991, S. 135–149.

[13] Godichon, A.: Variable Speed Drives on large centrifugal Fans. 34th IEEE Cement Industry Technical Conference, May 1992, S. 96–107.

[14] Kuhlmann, K., und Ellerbrock, H.-G.: Energy input for the grinding of cement in grinding plants with ball mills. 1. World Congress Particle Technology, Nürnberg, (1986), S. 265–281.

[15] Ellerbrock, H.-G.: Einflüsse auf die Temperatur beim Mahlen von Zement. Zement-Kalk-Gips 35 (1982) 2, S. 49–57.

[16] Ellerbrock, H.-G., und Deckers, R.: Mühlentemperatur und Zementeigenschaften. Zement-Kalk-Gips 41 (1988) 1, S. 1–12.

[17] Ellerbrock, H.-G., und Schiller, B.: Energieaufwand zum Mahlen von Zement. Zement-Kalk-Gips 41 (1988) 2, S. 57–63.

[18] Kuhlmann, K.: Verbesserung der Energieausnutzung beim Mahlen von Zement. Schriftenreihe der Zementindustrie, Heft 44, Verein Deutscher Zementwerke e. V., Düsseldorf 1985.

[19] Gudat, G., und Albers, J.: Übertragungstrennwände der neuen Generation in Rohrmühlen für die Zementmahlung. Zement-Kalk-Gips 45 (1992) 1, S. 26–31.

[20] Thomart, F.: „Airfeel" – Eine neue Trennwand zur Regelung der Mahlgutfüllung in Kugelmühlen. Kurzbeitrag zum VDZ-Kongreß '93, Düsseldorf.

[21] Sprung, S.: Maßnahmen und Möglichkeiten zur Qualitätssteuerung im Zementwerk. Zement-Kalk-Gips 43 (1990) 7, S. 340–346.

[22] Geibig, K.-F., Koppermann, C., und Scheiding, W.: „Statistical Process Control" – Techniken in der Chemischen Industrie. Automatisierungstechnische Praxis atp 31 (1989) 6, S. 253–258.

[23] Gilbert, S., Coromina, J., Gomez, J., Recio, G., und Pedersen, M.: Fuzzy Logic optimizes cement grinding at Sanson Cement Plant, Spain. Zement-Kalk-Gips 42 (1989) 6, S. 286–287.

[24] Cigánek, A., und Kreysa, K.: Zweiparametrische Steuerung eines Zementmahlprozesses. Zement-Kalk-Gips 44 (1991) 7, S. 364–370.

[25] Hepper, R.: Wissensbasierte Prozeßführung von Mahlanlagen. Zement-Kalk-Gips 47 (1994) 4, S. 212–218.

[26] Barton, F. J.: Einsatz eines Prozeßleitsystems für die neue Zementmahlanlage im Werk Neubeckum der Dyckerhoff AG. Zement-Kalk-Gips 45 (1992) 12, S. 642–647.

[27] Allenberg, B.: Energiesparpotentiale durch optimierte Regelung der Schüttgutflüsse an Kugelmühlen und Gutbett-Walzenmühlen. Kurzbeitrag zum VDZ-Kongreß '93, Düsseldorf.

[28] Kuhlmann, K.: Bedeutung des Sichtens beim Mahlen von Zement – Ergebnisse einer Bilanzierung des Mahlkreislaufs. Zement-Kalk-Gips 37 (1984) 9, S. 474–480.

[29] Kuhlmann, K., Ellerbrock, H.-G., und Sprung, S.: Korngrößenverteilung und Eigenschaften von Zement. Teil 1: Festigkeit von Portlandzement. Zement-Kalk-Gips 38 (1985) 4, S. 169–178.

[30] Sprung, S., Kuhlmann, K., und Ellerbrock, H.-G.: Korngrößenverteilung und Eigenschaften von Zement. Teil 2: Wasseranspruch von Portlandzement. Zement-Kalk-Gips 38 (1985) 9, S. 528–534.

[31] Dauphinee, E., and Breitschmid, K.: Problems with High Efficiency Separators. 33rd IEEE Cement Industry Technical Conference, May 1991, S. 339–354.

[32] Onuma, E., und Ito, M.: Sichter in Mahlkreisläufen. Fachbericht zum VDZ-Kongreß '93, Düsseldorf.

[33] Tätigkeitsbericht 1987 – 1990. Verein Deutscher Zementwerke e. V., Düsseldorf, 1990.

[34] Stairmand, C. J.: The energy efficiency of milling processes. A review of some fundamental investigations and their application to mill design. Dechema Monographien Bd. 79 (1976) Teil A/1, S. 1–17.

[35] Schönert, K.: Energetische Aspekte des Zerkleinerns spröder Stoffe. Zement-Kalk-Gips 32 (1979) 1, S. 1–9.

[36] Ellerbrock, H.-G.: Energieausnutzung beim Mahlen von Zement. Zement-Kalk-Gips 35 (1982) 2, S. 75–82.

[37] Müller, G., und Weigel, T.: Einsatz von Vertikal-Prallbrechern als Alternative zu herkömmlichen Zerkleinerungsverfahren. Aufbereitungstechnik (1988) 1, S. 26–31.

[38] Binn, F. J., und Beese, W.: Einsatz eines Prallbrechers zur Vorzerkleinerung von Zementausgangsstoffen. Zement-Kalk-Gips 42 (1989) 4, S. 170–174.

[39] Stoiber, W., Trenkwalder, J., Pernkopf, H., und Beigl, J.: Modernisierung der Zementmahlanlage im Zementwerk Kirchbichl durch Einsatz eines MFL-Vorbrechers. Zement-Kalk-Gips 47 (1994) 1, S. 43–45.

[40] Sekine, K., Sutoh, K., Ichikawa, M., Hashimoto, I., Sawamura, S., und Ueda, H.: Durchsatzsteigerung von Mahlanlagen durch Anwendung des CKP-Systems. Zement-Kalk-Gips 44 (1991) 7, S. 337–341.

[41] Ellerbrock, H.-G.: Gutbett-Walzenmühlen. Zement-Kalk-Gips 47 (1994) 2, S. 75–82.

[42] Sutoh, K., Murata, M. J., Hashimoto, S., Hashimoto, I., Sawamura, S., und Ueda, H.: Gegenwärtiger Stand der Vormahlung von Klinker und Zementrohmaterialien nach dem CKP-System. Zement-Kalk-Gips 45 (1992) 1, S. 21–25.

[43] Sutoh, K., Murata, M., Hashimoto, S., Hashimoto, I., Sawamura, S., and Ueda, H.: Operation control of the CKP-R pregrinding system. World Cement (1992) 7, S. 14–20.

[44] Sawamura, S., and Uchiyama, S.: CKP Pregrinding System and CK Fine Grinding System. Taiwan Seminar 1992.

[45] Wahl, W.: Verbesserung des Verschleißschutzes der Mahlwalzen von Walzenschüsselmühlen. Zement-Kalk-Gips 47 (1994) 4, S. 206–210.

[46] Schauer, S.: Zementmahlung mit MPS-Walzenschüsselmühlen. Zement-Kalk-Gips 30 (1977) 11, S. 576–578.

[47] Wessel, H. F.: Qualitätseigenschaften von auf einer Walzenschüsselmühle gemahlenen Zementen. Zement-Kalk-Gips 35 (1982) 8, S. 425–431.

[48] Leyser, W., und Sillem, H.: Planung, Bau und Betrieb einer Zementmahlanlage mit einer MPS-Walzenschüsselmühle. Zement-Kalk-Gips 37 (1984) 2, S. 83–86.

[49] F. L. Smith: OK Vertical Mills for clinker and slag. Zement-Kalk-Gips 47 (1994) 4, S. 221.

[50] Brundiek, H.: Die Loesche-Mühle für die Zerkleinerung von Zementklinker und Zumahlstoffen in der Praxis. Zement-Kalk-Gips 47 (1994) 4, S. 179–186.

[51] Ahlkvist, B., und Lohnherr, L.: Durchsatzsteigerungen der Rollenmühle im Zementwerk Slite. Zement-Kalk-Gips 45 (1992) 9, S. 463–466.

[52] Ahlkvist, B.: Increasing the roller mill capacity in line 8 of Cementa AB's Slite plant. World Cement (1992) 10, S. 5–9.

[53] Lohnherr, L.: Steigerung der Mahleffizienz durch verbesserte Kinematik in der Rollenmühle. Kurzbeitrag zum VDZ-Kongreß '93, Düsseldorf.

[54] Keefe, B. P.: Die Vorteile des äußeren Materialkreislaufs beim Betrieb von Wälzmühlen. Zement-Kalk-Gips 45 (1992) 7, S. 355–359.

[55] Strasser, S.: Teilfertigmahlung von Zement-Klinker. KHD-Symposium „Moderne Rollenpressen-Technik", Köln, 1990.

[56] Atzl, R.: Umbau einer Rohmehl-Mahlanlage zu einer Kombimahlanlage mit Gutbett-Walzenmühle, – Vergleich des Energieaufwands mit dem einer Walzenschüsselmühle. Kurzbeitrag zum VDZ-Kongreß '93, Düsseldorf.

[57] Buzzi, S.: The HOROMIL TM – A New Grinding System. Kurzbeitrag zum VDZ-Kongreß '93, Düsseldorf.

[58] Ellerbrock, H.-G.: Mahlbarkeit von Klinker und Hüttensand. Zement-Kalk-Gips 30 (1977) 11, S. 572–575.

[59] Schiller, B., und Ellerbrock, H.-G.: Mahlbarkeit von Zementbestandteilen und Energiebedarf von Zementmühlen, Teil 1. Zement-Kalk-Gips 42 (1989) 11, S. 553–557.

[60] Schiller, B., und Ellerbrock, H.-G.: Mahlbarkeit von Zementbestandteilen und Energiebedarf von Zementmühlen, Teil 2. Zement-Kalk-Gips 47 (1994) 4, S. 200–203.

[61] Schiller, B.: Mahlbarkeit der Hauptbestandteile des Zements und ihr Einfluß auf den Energieaufwand beim Mahlen und die Zementeigenschaften. Schriftenreihe der Zementindustrie, Heft 54, Verein Deutscher Zementwerke e. V., Düsseldorf 1992.

[62] Schiller, B., und Ellerbrock, H.-G.: Mahlung und Eigenschaften von Zementen mit mehreren Hauptbestandteilen. Zement-Kalk-Gips 45 (1992) 7, S. 325–334.

Gutbett-Walzenmühlen*)
High-pressure grinding rolls*)
Broyeurs à cylindres à lit de matière

Molinos de cilindros y lecho de material

Von **H.-G. Ellerbrock,** Düsseldorf/Deutschland

Gutbett-Walzenmühlen

Fachbericht 5.1 · Zusammenfassung – Durch Einführung der Gutbett-Walzenmühlen konnte der elektrische Energiebedarf der Zementmahlung in den letzten Jahren deutlich vermindert werden. Gutbett-Walzenmühlen werden in unterschiedlicher Anordnung in bestehende und neue Mahlanlagen mit Kugelmühlen integriert. Bei Anordnung einer Gutbett-Walzenmühle zur Vorzerkleinerung des Mahlguts einer Umlaufmahlanlage mit Kugelmühle kann z.B. der Durchsatz um bis zu 20% gesteigert und eine Energieeinsparung von 7 bis 15% erreicht werden. Je größer der mit der Gutbett-Walzenmühle erzeugte Feingutanteil ist, um so größer ist die Energieeinsparung. Daher wird angestrebt, möglichst auch die gesamte Feinmahlung ausschließlich in Gutbett-Walzenmühlen durchzuführen. Ohne eine nachgeschaltete Kugelmühle können Gutbett-Walzenmühlen für die Fertigmahlung von Branntkalk, Kalkstein und Zementrohmaterial verwendet werden. Dabei ist eine Energieeinsparung gegenüber Anlagen mit Kugelmühlen um mehr als 50% möglich. Da sich die Eigenschaften von in Gutbett-Walzenmühlen feingemahlenen Zementen deutlich von denen der Zemente aus Kugelmühlen unterscheiden, gibt es noch keine Fertigmahlanlage für Portlandzement. Die Wirtschaftlichkeit von Gutbett-Walzenmühlen wird jedoch derzeitig durch die immer noch zu geringe Betriebssicherheit und den großen Aufwand für Instandhaltung und Reparatur eingeschränkt. Die meisten und schwersten Schäden sind an den hochbeanspruchten Walzenoberflächen aufgetreten. Für einen störungsfreien Betrieb sollten der mittlere Mahldruck nicht zu hoch gewählt und Druckspitzen durch entsprechende Betriebsbedingungen vermieden werden.

High-pressure grinding rolls

Special report 5.1 · Summary – The consumption of electrical energy in cement grinding has been significantly reduced in recent years by the introduction of high-pressure grinding rolls. They are integrated in various configurations into existing and new grinding plants with ball mills. In an arrangement with high-pressure grinding rolls for preliminary comminution of the material in a closed-circuit grinding plant with a ball mill it is possible to increase the throughput by up to 20% and achieve an energy saving of 7 to 15%. The larger the proportion of fines produced with the high-pressure grinding rolls the greater is the energy saving. Attempts are therefore being made wherever possible to carry out the entire fine grinding exclusively in the high-pressure grinding rolls. High-pressure grinding rolls can be used for finish grinding of quick lime, limestone and cement raw material without a downstream ball mill. It is possible to achieve energy savings of more than 50% when compared with plants with ball mills. The properties of cements which are fine-ground in high-pressure grinding rolls differ significantly from those of cements from ball mills, so there are still no finish grinding plants for Portland cement. At present the cost-effectiveness of high-pressure grinding rolls is heavily restricted by the operational reliability, which is still too low, and the great cost of maintenance and repairs. The majority, and the most serious, of the damage occurs at the highly stressed surfaces of the rollers. For trouble-free operation the average grinding pressure selected should not be too high, and pressure peaks should be avoided by appropriate operating conditions.

Broyeurs à cylindres à lit de matière

Rapport spécial 5.1 · Résumé – Au cours des dernières années, l'introduction de broyeurs à cylindres à lit de matière a permis de réduire sensiblement les besoins énergétiques électriques. Les broyeurs à cylindres à lit de matière sont intégrés avec des broyeurs à boulets, selon des dispositions très variées, dans des ateliers de broyage existants et nouveaux. La mise en place d'un broyeur à cylindres à lit de matière pour prébroyage de matière d'un atelier de broyage en circuit fermé avec broyeur à boulets, permet p. e., d'accroître le débit d'un pourcentage allant jusqu'à 20% et de réaliser des économies d'énergie de 7 à 15%. Plus la proportion de fines produites dans le broyeur à cylindres à lit de matière est grande, plus l'économie d'énergie est grande également. Le but est d'effectuer autant que possible l'ensemble du broyage fin exclusivement dans des broyeurs à cylindres à lit de matière. Sans broyeur à boulets monté en aval, les broyeurs à cylindres à lit de matière peuvent être utilisés pour le broyage fin de chaux vive, de calcaire et de matière première de ciment. Par rapport à des installations avec broyeurs à boulets, une économie d'énergie de plus de 50% est alors possible. Etant donné que les propriétés des ciments finement broyés dans les broyeurs à cylindres à lit de matière sont sensiblement différentes de celles des ciments de broyeurs à boulets, il n'existe pas encore d'atelier de broyage fin pour ciment de Portland.

---

*) Überarbeitete Fassung eines Vortrages zum VDZ Kongreß '93, Düsseldorf (27.9.–1.10.1993)
Revised text of a lecture to the VDZ Congress '93, Düsseldorf (27.9.–1.10.1993)

Mais la rentabilité des broyeurs à cylindres à lit de matière est, à l'heure actuelle, considérablement limitée par la sécurité de fonctionnement encore trop faible et les dépenses élevées pour la maintenance et les réparations. La plupart des dommages et les plus graves d'entre eux se sont produits à la surface des cylindres fortement sollicités. Pour un fonctionnement sans problème, il faut que la pression de broyage moyenne sélectionnée ne soit pas trop élevée et aussi éviter des pointes de pression par la mise en place de conditions de fonctionnement appropriées.

Molinos de cilindros y lecho de material

Informe de ramo 5.1 · Resumen – Gracias a la introducción de los molinos de cilindros y lecho de material se ha podido reducir notablemente, en los últimos años, el consumo de energía eléctrica en la molienda del cemento. Los molinos de cilindros y lecho de material son integrados en las instalaciones de molienda ya existentes o nuevas, que trabajan con molinos de bolas, variando su disposición dentro de la instalación. Al utilizarse el molino de cilindros y lecho de material para la molienda previa del material de una instalación de molienda a circuito abierto, con molino de bolas, se puede aumentar, por ejemplo, al rendimiento hasta un 20%, lográndose un ahorro de energía del 7 al 15%. Cuanto mayor sea la proporción de finos alcanzada con el molino de cilindros y lecho de material, tanto mayor será el ahorro de energía. Por esta razón, se pretende afectuar también, en lo posible, toda la molienda fina exclusivamente por medio de molinos de cilindros y lecho de material. Estos últimos pueden emplearse, sin molino de bolas trasconectado, para la molienda de acabado de cal viva, piedra caliza y materias primas para la fabricación de cemento. Con ello se puede conseguir un ahorro de energía de más del 50%, en comparación con las instalaciones que trabajan con molinos de bolas. Puesto que las propiedades de los cementos molidos finamente en los molinos de cilindros y lecho de material se diferencian claramente de los obtenidos con molinos de bolas, no existe todavía ninguna planta de molienda de acabado para cemento Portland. Actualmente, la rentabilidad de los molinos de cilindros y lecho de material queda, sin embargo, muy limitada por la deficiente seguridad de marcha y los elevados gastos de mantenimiento y reparación. La mayor parte de las averías, y las más graves, se han producido en las superficies de los cilindros, expuestos a elevados esfuerzos. Con el fin de garantizar una marcha exenta de perturbaciones, la presión de molienda media no debe ser demasiado alta. Además, hay que evitar puntas de presión, creando las condiciones de servicio necesarias.

1. Einleitung

Der Wirkungsgrad der Zerkleinerung in Kugelmühlen ist vergleichsweise sehr gering [1, 2]. Bei ansteigenden Kosten für elektrische Energie war daher schon immer der Anreiz sehr groß, die Energieausnutzung beim Mahlen zu steigern. Gegenüber der Reib- und Schlagbeanspruchung des Mahlguts in Kugelmühlen ist die Hochdruckbeanspruchung in einem Gutbett mit einer wesentlich größeren Energieausnutzung verbunden [3, 4]. Die Druckbeanspruchung eines Mahlgutbetts läßt sich technisch bei Pressungen von mehr als 50 MPa bis zu 400 MPa und mit hohen Beanspruchungsgeschwindigkeiten in Gutbett-Walzenmühlen verwirklichen [4, 5].

Das Mahlgut wird in einem Spalt zwischen zwei sich gegensinnig drehenden Mahlwalzen (**Bild 1**) mit Umfangsgeschwindigkeiten von etwa 1,0 bis 1,8 m/s zu sogenannten Schülpen mit Feststoff-Volumenanteilen von über 70% gepreßt, die in Abhängigkeit vom Zerkleinerungsverhalten

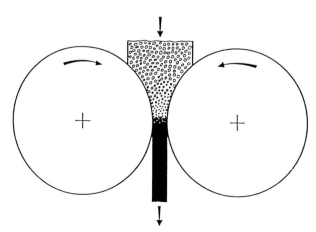

BILD 1: Prinzipielle Darstellung der Hochdruckzerkleinerung in einer Gutbett-Walzenmühle
FIGURE 1: Diagram showing the principle of high pressure comminution in high-pressure grinding rolls

1. Introduction

The efficiency of comminution in ball mills is very low when compared with other methods [1, 2]. With the rising cost of electrical power there has therefore always been a great incentive to increase the utilization of energy during grinding. High compression stressing in a bed of material is associated with substantially greater energy utilization than the frictional and impact stressing of the material in ball mills [3, 4]. Compressive stressing of a bed of material can be carried out industrially in high-pressure grinding rolls at pressures of more than 50 MPa up to 400 MPa and with high-stressing rates [4, 5].

The material to be ground is compressed in a gap between two counter-rotating grinding rolls (**Fig. 1**) with circumferential speeds of about 1.0 to 1.8 m/s to form compacted cake with a volumetric proportion of solids of over 70%. Depending on the comminution behaviour of the material and the pressures applied this can contain up to about 40% fines ($<90\,\mu$m) [6]. The fines have to be obtained by disagglomeration of the compacted cake. The compacted cake also contains coarser particles with large numbers of incipient cracks and weak points which greatly reduce the energy expenditure during further comminution. **Fig. 2** shows the example of a roller with its support system.

High-pressure grinding rolls are now integrated in various arrangements into existing and new grinding plants with ball mills where the throughput of the plant is to be raised by a significant extent and the energy expenditure reduced. High-pressure grinding rolls without a downstream ball mill are already used for the fine grinding of quick lime, limestone, cement raw material and coal. **Fig. 3** shows that nowadays more than half of the total of over 300 sets of high-pressure grinding rolls from German manufacturers are used for grinding Portland cement and cement with interground additives. A large proportion of approximately 20% is accounted for by raw meal production, for which special plant systems have been developed to suit the properties of the raw materials. Just under 10% of the plants are used for ore and coal grinding and somewhat more than 10% for grinding limestone, quick lime and blast furnace slag.

des Mahlguts und der aufgewendeten Pressungen bis zu etwa 40 % Feingutanteil (< 90 μm) enthalten [6]. Das Feingut muß durch Deglomeration der Schülpen gewonnen werden. Daneben enthalten die Schülpen gröbere Körner mit zahlreichen Anrissen und Schwachstellen, die den Energieaufwand bei der weiteren Zerkleinerung erheblich vermindern. **Bild 2** zeigt das Beispiel einer Walze mit Lagerung.

BILD 2: Walze mit Lagerung
(Werkfoto der KHD Humboldt Wedag AG, Köln)
FIGURE 2: Roller with support system
(works photograph from KHD Humboldt Wedag AG, Cologne)

Gutbett-Walzenmühlen werden heute in verschiedener Anordnung in bestehende und neue Mahlanlagen mit Kugelmühlen integriert, wenn der Durchsatz der Anlage deutlich erhöht und der Energieaufwand vermindert werden soll. Ohne eine nachgeschaltete Kugelmühle werden Gutbett-Walzenmühlen bereits für die Feinmahlung von Branntkalk, Kalkstein, Zementrohmaterial und Kohle genutzt. **Bild 3** zeigt, daß heute mehr als die Hälfte der insgesamt über 300 Gutbett-Walzenmühlen deutscher Hersteller für die Mahlung von Portlandzement und Zement mit Zumahlstoffen eingesetzt werden. Mit ca. 20 % entfällt ein großer Anteil auf die Rohmehlerzeugung, für die je nach den Eigenschaften des Rohmaterials besondere Anlagen-Konzepte entwickelt wurden. Knapp 10 % der Anlagen werden für die Erz- und Kohlemahlung und etwas mehr als 10 % für die Kalkstein-, Branntkalk- und Hüttensandmahlung verwendet.

2. Anlagen-Konzepte und Produkteigenschaften

2.1 Vorzerkleinerung

Bild 4 zeigt die Anordnung einer Gutbett-Walzenmühle, die zur Vorzerkleinerung des Mahlguts einer Umlaufmahlanlage mit Kugelmühle vorgeschaltet wurde. Diese Anordnung wird bei der Rohmaterial- und bei der Zementmahlung verwendet. Die in der Gutbett-Walzenmühle erzeugten Schülpen werden z.T. der nachgeschalteten Kugelmühle aufgegeben. Ein Teil der Schülpen kann zur Verbesserung des Guteinzugsvermögens und zur weiteren Verminderung des Gesamtenergiebedarfs zur Gutbett-Walzenmühle rezirkuliert werden. So können auch drehzahlregelbare Antriebe und Zwischenbunker für Schülpen entfallen. Obwohl die Grobzerkleinerung des Mahlguts zum größten Teil in der Gutbett-Walzenmühle erfolgt, kann auf die Grobmahlkammer der Kugelmühle im allgemeinen nicht verzichtet, jedoch der Mahlkörperfüllungsgrad deutlich vermindert werden [6, 7]. Das im nachgeschalteten Sichter abgetrennte Grobgut wird ausschließlich der Kugelmühle zugeführt. Die Endfeinheit des Fertigguts und damit die Produktqualität wird vom Betrieb der Kugelmühle bestimmt. Daher unterscheiden sich die Eigenschaften der so erzeugten Zemente nicht von denen der Zemente aus üblichen

2. Plant design and product properties

2.1 Preliminary comminution

Fig. 4 shows the layout for high-pressure grinding rolls which are connected upstream of a closed-circuit grinding plant with ball mill for preliminary comminution of the mill feed. This configuration is used for grinding raw material and cement. Some of the compacted cake produced in the high-pressure grinding rolls is fed to the downstream ball mill. Part of the compacted cake can be recirculated to the high-pressure grinding rolls to improve the intake conditions for the mill feed and for further reduction in the total energy consumption. This makes it possible to dispense with variable speed drives and an intermediate hopper for compacted cake. It is not generally possible to dispense with the coarse grinding chamber in the ball mill although most of the coarse comminution of the mill feed takes place in the high-pressure grinding rolls, but the grinding media filling ratio can be significantly reduced [6, 7]. The coarse material separated in the following classifier is all fed to the ball mill. The ultimate fineness of the finished product, and hence the product quality, is determined by the operation of the ball mill. The properties of the cements produced in this way do not therefore differ from those of cements from normal cement grinding plants [7]. In this case, the proportion of fines generated in the high-pressure grinding rolls is small, so with this configuration of the grinding plant, it is only possible to achieve an increase in throughput of about 20 % and an energy saving of only about 7–15 % [7, 8]. This plant system is now only used in special cases.

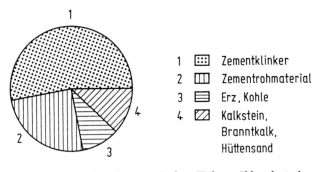

BILD 3: Einsatz der über 300 Gutbett-Walzenmühlen deutscher Hersteller für verschiedene Mahlgüter
(Stand 4/1993)
FIGURE 3: Use of the over 300 sets of high-pressure grinding rolls from German manufacturers for grinding different materials
(as at 4/1993)

Zementklinker	= cement clinker
Zementrohmaterial	= cement raw material
Erz, Kohle	= ore, coal
Kalkstein, Branntkalk, Hüttensand	= limestone, quick lime blast furnace slag

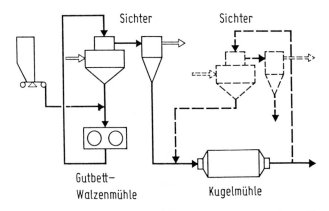

BILD 4: Konzept einer Mahlanlage mit Vorzerkleinerung des Mahlguts in einer Gutbett-Walzenmühle
FIGURE 4: Configuration of a grinding plant with preliminary comminution of the feed material in high-pressure grinding rolls

Gutbett-Walzenmühle	= high-pressure grinding rolls
Sichter	= classifier
Kugelmühle	= ball mill

2.2 Finish grinding

The greater the proportion of fines in the compacted cake produced by the high-pressure grinding rolls, the greater is the energy utilization, and hence the energy saving, when compared with grinding in ball mills [4]. Attempts are therefore made to carry out all the fine grinding in the high-pressure grinding rolls and dispense entirely with the use of a ball mill [9, 10]. **Fig. 5** shows the arrangement of high-pressure grinding rolls used as a fine or finish mill. The compacted cake is broken up in a downstream disagglomerator. The fines are separated in a classifier. It is possible, for example, to use a hammer mill or a vertical impact mill as the disagglomerator. This can also be integrated in the classifier housing. The tailings from the classifier are returned and fed to the high-pressure grinding rolls together with the fresh material. Investigations have shown that, for fine and very fine grinding of brittle materials, this plant system can achieve an energy saving of up to more than 50% when compared with plants with ball mills [9–12]. Finish grinding is now used successfully in the cement industry for grinding raw meal [13–15] and blast furnace slag [16], and in the lime industry for grinding quick lime and limestone [17–19]. In the majority of cases, raw meal production with high-pressure grinding rolls is more efficient than grinding in roller grinding mills. However, the roller grinding mill has advantages with soft and very moist raw material [20]. The particle size distribution of the products from high-pressure grinding rolls is narrower than that of meals from plants with ball mills [9, 10, 21–24]. This has advantages, for example, for the burnability of the raw meal [25]. For the finish grinding of granulated blast furnace slag it has been shown that the particle size distribution of the blast furnace slag meal produced can be governed within certain limits by recycling different quantities of compacted cake [26, 27].

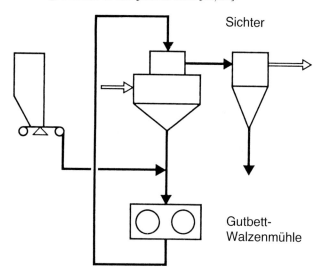

BILD 5: Konzept einer Mahlanlage zur Fertigmahlung des Mahlguts in einer Gutbett-Walzenmühle
FIGURE 5: Configuration of a grinding plant for finish grinding the feed material in high-pressure grinding rolls

Sichter = classifier
Gutbett-Walzenmühle = high-pressure grinding rolls

The service properties of Portland cement which had been finish ground with high-pressure grinding rolls differed from those of the cements from ball mills to an extent which depended on the reactivity of the clinker [9, 10, 21–24, 28]. For the same specific surface area the standard compressive strength was somewhat higher but, when compared with the values for ball mill cements, the water demand to achieve standard stiffness increased by up to 25% depending on the reactivity of the clinker used, and the setting time was shortened drastically to a few minutes. In some cases this affects the workability characteristics so badly that so far the practical use of cements which have been fine ground entirely in high-pressure grinding rolls has been severely limited. At

2.3 Hybridmahlung

Kann einerseits aus Gründen der Zementqualität auf die Kugelmühle nicht verzichtet werden, soll aber andererseits der Durchsatz einer vorhandenen Mahlanlage mit Kugelmühle stärker gesteigert werden als es mit der Vormahlung in der Gutbett-Walzenmühle möglich ist, kann die Gutbett-Walzenmühle hybrid betrieben werden [30, 31]. Dieses Anlagenkonzept der Hybridmahlung, eine kombinierte Vor- und Fertigmahlung, ist schematisch im **Bild 6** dargestellt. Dabei wird ein Teil des Sichtergrobguts des Kugelmühlen-Sichter-Kreislaufs zurückgeführt und gemeinsam mit dem Frischgut der Gutbett-Walzenmühle aufgegeben. Dadurch wird ein größerer Teil der Fertigmahlung von dieser übernommen [32]. Da aber ein großer Teil der Feinmahlung noch in der Kugelmühle erfolgt, werden die Zementeigenschaften nicht beeinträchtigt.

Eine Betriebsweise der Gutbett-Walzenmühle mit größeren rezirkulierenden Sichtergrobgutmengen kann erforderlich werden, wenn in der Hybridmahlanlage höhere Zementfeinheiten und/oder höhere Durchsätze erreicht werden sollen [33]. Durchsatzerhöhungen von 30–70 % sind auf diese Weise möglich. Durch die Rückführung des Sichtergrobguts zur Gutbett-Walzenmühle verändern sich jedoch die Korngrößenverteilung und die Fließeigenschaften des Mühlenaufgabeguts, was sich mehr oder weniger nachteilig auf das Einzugsverhalten auswirkt [34, 35]. Überschreitet der zurückgeführte Massestrom relativ feinen Sichtergrobguts bestimmte Grenzwerte, können sich instabile Betriebszustände einstellen, die durch starke Vibrationen gekennzeichnet sind. Daher muß bei diesem Konzept die zur Gutbett-Walzenmühle rückgeführte Sichtergrobgutmenge begrenzt werden. Der Grenzwert hängt von der Korngrößenverteilung des Frischguts und von der gewünschten Endfeinheit des Produkts ab [36]. Im **Bild 7** ist dargestellt, daß die rückführbare Sichtergrobgutmenge um so geringer ist, je höher die Feinheit des Produkts ist. Zum Beispiel kann bei einem Portlandzement Z 35 F das Verhältnis von Sichtergrobgut/Frischgut etwa 1,0, bei einem PZ 55 jedoch nur noch etwa 0,25 betragen. Bei Rückführung von Schülpen kann das Verhältnis von Sichtergrobgut-/Frischgutmassestrom erhöht werden [26, 36, 37].

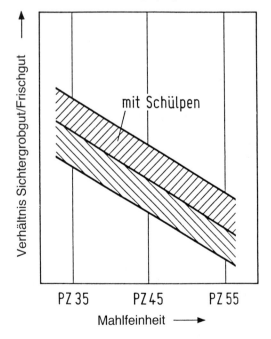

BILD 7: Einfluß von Mahlfeinheit des Produkts und Schülpenrückführung auf das Verhältnis von Sichtergrobgut- zu Frischgutmassestrom [26, 36]
FIGURE 7: Influence of the product fineness and recycling of compacted cake on the ratio of the mass flows of classifier tailings to fresh material [26, 36]

Verhältnis Sichtergrobgut/Frischgut	= ratio of classifier tailings to fresh material
mit Schülpen	= with compacted cake
Mahlfeinheit	= fineness

present there are therefore still no pure finish grinding systems for Portland cement in high-pressure grinding rolls in Germany. An industrial grinding plant using the concept of finish grinding has been in operation in France for about 3 years [11, 12]. However, this is evidently used largely for the production of cements with interground additives. Another plant is to start operation in Belgium in 1994 [29].

According to the findings available at present, the increased water demand of cements from high-pressure grinding rolls can, to a great extent, be attributed to the narrower particle size distribution. A further possible cause is that, as a result of the usually lower temperatures of the material being ground and the shorter residence times of the material in the high-pressure grinding rolls, the type and quantity of the sulphate agent is inadequately matched to the reactivity of the clinker [10, 21–24].

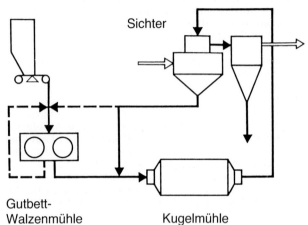

BILD 6: Konzept einer Mahlanlage für eine kombinierte Vor- und Fertigmahlung in einer Gutbett-Walzenmühle (Hybridmahlung)
FIGURE 6: Configuration of a grinding plant for combined pre-finish grinding in high-pressure grinding rolls (hybrid grinding)

Sichter	= classifier
Gutbett-Walzenmühle	= high-pressure grinding rolls
Kugelmühle	= ball mill

2.3 Hybrid grinding

If, for reasons of cement quality, it is not possible to dispense with the ball mill, but the throughput of an existing grinding plant with ball mill is to be increased to a greater extent than is possible with pregrinding in high-pressure grinding rolls, then the high-pressure grinding rolls can be operated in a hybrid system [30, 31]. This hybrid grinding plant scheme – combined pregrinding and finish grinding – is shown diagrammatically in **Fig. 6**. Part of the classifier tailings from the ball mill classifier circuit are returned and fed, together with the fresh material, to the high-pressure grinding rolls which then undertake a greater proportion of the finish grinding [32]. As, however, a large part of the fine grinding still takes place in the ball mill, the cement properties are not impaired.

A mode of operation of the high-pressure grinding rolls with quite large quantities of recirculating classifier tailings may be necessary where greater cement finenesses and/or higher throughputs are to be achieved in the hybrid grinding plant [33]. This can achieve increases in throughput of 30–70 %. However, the return of the classifier tailings to the high-pressure grinding rolls changes the particle size distribution and flow properties of the material being fed to the grinding rolls, which has a certain disadvantageous effect on the in-take characteristics [34, 35]. If the returned mass flow of relatively fine classifier tailings exceeds certain limits then unstable operating conditions can occur, which are characterized by severe vibrations. The quantity of classifier tailings returned to the high-pressure grinding rolls must therefore be restricted in this system. The limit depends on the particle size distribution of the fresh material and on the required ultimate fineness of the product [36]. **Fig. 7** shows that the quantity of classifier tailings which can be returned, is lower the higher the fineness of the product. For example,

2.4 Teilfertig-Mahlung

Noch größere Werte der Durchsatzsteigerung und Energieeinsparung als mit der Hybrid-Mahlung lassen sich mit dem Konzept der sog. Teilfertig-Mahlung oder Kombi-Mahlung erreichen [38]. Jedoch kann bei Anwendung dieses Konzeptes der Investitionsaufwand auch größer sein. Die Kombi-Mahlung wird ebenfalls dann gewählt, wenn aufgrund der geforderten Zementqualität die Fertigmahlung von Zement in der Gutbett-Walzenmühle nicht möglich ist und ein Teil der Feinmahlung in der Rohrmühle erfolgen muß.

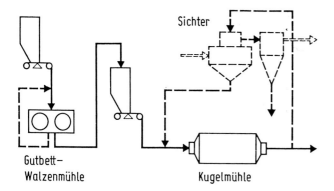

BILD 8: Konzept einer Mahlanlage für eine Teilfertigmahlung in einer Gutbett-Walzenmühle (Kombimahlung)
FIGURE 8: Configuration of a grinding plant for semi-finish grinding in high-pressure grinding rolls (combined grinding)

Sichter = classifier
Gutbett-Walzenmühle = high-pressure grinding rolls
Kugelmühle = ball mill

Bild 8 zeigt das Anlagen-Konzept der Teilfertig-Mahlung. Das Frischgut wird der Gutbett-Walzenmühle aufgegeben. Die Schülpen werden dem Deglomerator zugeführt, der sich auch im Sichtergehäuse oberhalb der Sichtzone befinden kann [39, 40]. Das im Sichter aus den deglomerierten Schülpen abgetrennte Sichtergrobgut wird vollständig zur Gutbett-Walzenmühle zurückgeführt. Das in den Zyklonen des Sichters abgeschiedene Sichterfeingut enthält etwa 50 bis 80 % Fertiggut. Dieses sog. „teilfertige" Produkt aus dem Primär-Mahlkreislauf wird einem zweiten Mahlvorgang in einer Kugelmühle zugeführt, in der die gewünschte Produktfeinheit bzw. Qualität erzeugt wird. Wird das Mahlgut im Gutbett-Walzenmühlen-Kreislauf z.B. auf eine Feinheit von 1800 cm^2/g (Blaine) vorgemahlen und in der nachgeschalteten Kugelmühle auf eine Fertiggutfeinheit von etwa 3500 cm^2/g (Blaine) nachgemahlen, kann die Kugelmühle vorteilhaft als Durchlaufmühle betrieben werden [10]. Bei höheren Produktfeinheiten kann es zweckmäßig sein, das teilfertige Produkt auf eine höhere Feinheit vorzumahlen und die Produktfeinheit in einer nachgeschalteten Umlaufmahlanlage mit Kugelmühle und Sichter zu erzeugen.

Die Kugelmühle erhält bei dieser Anordnung ein definiertes feines Aufgabegut, so daß sie nur noch eine Mahlkammer benötigt. Außerdem können die Abmessungen der Kugelmühle kleiner gewählt und die Mahlkörpergattierung optimal auf das feine Mahlgut abgestimmt werden. Ein weiterer Vorteil ist, daß bei der Produktion von mehreren Zementsorten auf einer Mahlanlage die Feinheit der vorgemahlenen Produkte frei bestimmbar ist. Damit sind die Produktqualität und die Höhe der Energieeinsparung jeweils optimierbar. Vorhandene Mahlkapazitäten können mit dem Anlagen-Konzept der Teilfertig-Mahlung verdreifacht werden [38]. Da das aus den deglomerierten Schülpen abgetrennte Sichtergrobgut wesentlich gröber ist, braucht bei der Teilfertig-Mahlung das zur Gutbett-Walzenmühle zurückgeführte Sichtergrobgut im Gegensatz zur Hybrid-Mahlung nicht begrenzt zu werden. Die Systeme der Vor- und Nachmahlung sind getrennt. Deshalb sind die Anlagen regelungstechnisch einfacher zu betreiben.

with a Z 35 F Portland cement the ratio of classifier tailings to fresh material can be about 1.0 but with a PZ 55 Portland cement, it can only be about 0.25. The ratio of the mass flow of classifier tailings to fresh material can be increased if compacted cake is recycled [26, 36, 37].

2.4 Semi-finish grinding

Even greater values of throughput increase and energy saving than with hybrid grinding can be achieved with the so-called semi-finish grinding or combined grinding system [38]. However, the capital expenditure can also be larger when using this scheme. Combined grinding is also chosen when finish grinding of cement in the high-pressure grinding rolls is not possible because of the required cement quality and part of the finish grinding has to take place in the tube mill.

Fig. 8 shows the plant layout for semi-finish grinding. The fresh material is fed to the high-pressure grinding rolls. The cake is transferred to the disagglomerator, which can also be located in the classifier housing above the classifying zone [39, 40]. All the classifier tailings separated in the classifier from the disagglomerated compacted cake are returned to the high-pressure grinding rolls. The classifier fines separated in the cyclones of the classifier contain about 50–80 % finished product. This so-called "semi-finished" product from the primary grinding circuit is fed into a ball mill in a second grinding process in which the required product fineness and quality is generated. If the material ground in the high-pressure grinding roll circuit is, for example, preground to a fineness of 1800 cm^2/g (Blaine) and ground further in the downstream ball mill to a finished product fineness of about 3500 cm^2/g (Blaine), then the ball mill can be operated to advantage as an open circuit mill [10]. For higher product finesses it can be advisable to pregrind the semi-finished product to a higher fineness and then to generate the product fineness in a downstream closed circuit grinding system with ball mill and classifier.

In this configuration the ball mill receives a feed material of defined fineness, so only one grinding chamber is required. The ball mill can also have smaller dimensions and the grinding media charge grading can be optimally matched to the fine material being ground. A further advantage is that when producing several types of cement in one grinding plant, the fineness of the preground product is freely definable. This means that the product quality and the level of energy saving can be optimised in each instance. Existing grinding capacities can be tripled with the semi-finish grinding plant concept [38]. Unlike the situation in hybrid grinding there is no need to restrict the classifier tailings returned to the high-pressure grinding rolls in semi-finish grinding as the classifier tailings separated from the disagglomerated compacted cake are substantially coarser. The pregrinding and secondary grinding systems are separate, so the plants are simpler to operate from the aspect of control technology.

3. Operational reliability and operating conditions

However, apart from the considerable energy saving by the use of high-pressure grinding rolls, not only the capital expenditure, but also the operational reliability and availability of the plant as well as the expenditure on maintenance and repair, are at least of equal importance for the overall economic calculations. In past years the operational reliability of high-pressure grinding rolls was in some cases still very unsatisfactory [41–44]. The expenditure on maintenance and repair of wear damage was so high in some cases that the overall cost-effectiveness of the high-pressure grinding rolls was greatly curtailed. A VDZ working group was even set up to investigate the causes of wear and to work out design and process technology measures to reduce it. The working group has investigated the design data, the operational behaviour, and the operating faults and damage of over 25 sets of high-pressure grinding rolls. In the first place, it emerged from the investigations that damage to the

3. Betriebssicherheit und Betriebsbedingungen

Außer der erheblichen Energieeinsparung durch den Einsatz von Gutbett-Walzenmühlen haben jedoch für gesamtwirtschaftliche Rechnungen neben dem Aufwand an Kapital die Betriebssicherheit und Verfügbarkeit der Anlagen sowie der Aufwand für Instandhaltung und Reparatur eine mindestens ebenso große Bedeutung. In den vergangenen Jahren war die Betriebssicherheit der Gutbett-Walzenmühlen z. T. noch sehr unbefriedigend [41–44]. Der Aufwand für die Instandhaltung und Reparaturen von Verschleißschäden war in einigen Fällen so groß, daß die Gesamtwirtschaftlichkeit der Gutbett-Walzenmühlen erheblich eingeschränkt war. Daher wurde auch ein VDZ-Arbeitskreis eingesetzt, der den Ursachen für den Verschleiß nachging und zur Erarbeitung konstruktiver und verfahrenstechnischer Maßnahmen zu dessen Minderung beitragen sollte. Dazu hat der Arbeitskreis die Auslegungsdaten, das Betriebsverhalten sowie die Betriebsstörungen und Schäden von über 25 Gutbett-Walzenmühlen untersucht. Aus den Untersuchungen geht zunächst hervor, daß Antriebsschäden nur vereinzelt auftraten. Sie waren vor allem auf konstruktive Mängel zurückzuführen.

3.1 Walzen-Lagerung

Einige Schäden traten an den Walzenlagern auf. Grundlage für die Berechnung der Lebensdauer eines Lagers ist die maximal projektierte Preßkraft der Maschine, in der Regel mit einem 10%igen Sicherheitszuschlag. Die mittlere Grundbelastung wird von einer dynamischen Belastung aus Druck-Schwellkräften überlagert, die durch die Bruchereignisse im Mahlspalt hervorgerufen werden [45, 46]. Zum Beispiel kann bei Klinker der mittlere Druck mehr als 100 MPa betragen und der maximale Druck auf die Walze den vierfachen Wert des mittleren Drucks, also mehr als 400 MPa erreichen.

Für einen störungsfreien Betrieb sollten hohe Druckspitzen vermieden werden. Dafür sind die Betriebsbedingungen der Gutbett-Walzenmühle von ausschlaggebender Bedeutung [47]. Eine gleichmäßige Beschickung und Materialverteilung über die Spaltlänge, gleichbleibende Korngrößenverteilung des Aufgabeguts und Vermeidung des Einziehens von Fremdkörpern sind Voraussetzung für eine gleichmäßige stoßarme Lagerbelastung. Darüber hinaus kann ein Betrieb der Gutbett-Walzenmühle bei etwa 20 % unterhalb der maximalen Preßkraft die Lagerlebensdauer praktisch verdoppeln. Das hat zur Folge, daß Lagerschäden infolge Überlastung nicht mehr auftreten.

Im praktischen Betrieb wurden Lagerschäden meist durch Eindringen von Schmutz verursacht. **Bild 9** zeigt beispielhaft einen solchen Lagerschaden. Die heute verwendete Walzenlagerung wird mit Fettschmierung betrieben. Das Frischfett wird von der Maschinenaußenseite quer durch das Lager gedrückt und tritt an der Maschineninnenseite durch das Dichtungslabyrinth aus. Daher hat sich als geeignete Maßnahme zum Schutz vor Lagerschäden ein Schmierbetrieb mit Fettüberschuß bewährt. Frischer Fettaustritt im Abdichtungsbereich ist ein sicheres Zeichen für die gute Funktion von Schmierung und Abdichtung. Die Abdichtungen der Walzenlager dürfen nicht durch unzulässige Betriebszustände mechanisch beschädigt werden.

3.2 Walzen-Panzerung

Die meisten und vor allem die schwersten Schäden sind an den Walzenoberflächen aufgetreten. **Bild 10** zeigt als Beispiel die Ausbruchstelle einer Panzerung. Die Walzenoberflächen sind die am höchsten beanspruchten Bauteile einer Gutbett-Walzenmühle. Sie werden daher gegen Verschleiß geschützt. Die konstruktive Weiterentwicklung hatte vor allem eine höhere Betriebssicherheit der Walzenoberfläche zum Ziel [43, 44]. Dabei wurden inzwischen beachtliche Verbesserungen erzielt [48, 49]. Für den Aufbau der Panzerung gibt es heute mehrere konkurrierende Möglichkeiten:

— Die Verbundwerkstoff-Lösung, bei der auf den Walzenkörper eine oder mehrere Zwischenschichten als Pufferlagen und darauf eine verschleißfeste Hartschicht aufge-

drives only occurred in isolated cases. They were primarily attributable to design deficiencies.

3.1 Roller support system

Some cases of damage occurred in the roller bearings. The basis for calculating the service life of a bearing is the maximum planned pressure force from the machine, as a rule with a 10 % safety margin. Superimposed on the mean base loading is a dynamic loading caused by pulsating compressive forces which are caused by the fracture events in the grinding gap [45, 46]. With clinker, for example, the mean pressure can be more than 100 MPa and the maximum pressure on the roller can reach four times the value of the mean pressure, i.e. more than 400 MPa.

High pressure peaks should be avoided for trouble-free operation, for which the operating conditions of the high-pressure grinding rolls are of prime importance [47]. Uniform feeding and material distribution over the length of the gap, consistent particle size distribution of the feed material, and avoidance of foreign bodies, are essential for uniform low-impact bearing loading. Moreover, operation of high-pressure grinding rolls at about 20 % below the maximum pressure force can practically double the service life of the bearing because bearing damage resulting from overloading no longer occurs.

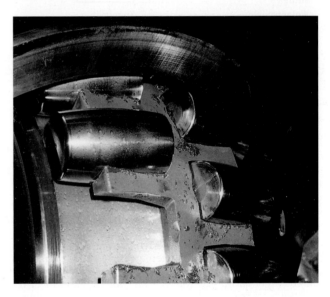

BILD 9: Beispiel eines geschädigten Lagers einer Gutbett-Walzenmühle
FIGURE 9: Example of a damaged bearing in high-pressure grinding rolls

In practical operation bearing damage is usually caused by the penetration of dirt. **Fig. 9** shows an example of such bearing damage. The roller bearing system used nowadays is operated with grease lubrication. The fresh grease is forced transversely through the bearing from the outside of the machine and emerges at the inner side of the machine through the bearing labyrinth. Lubricating with excess grease has therefore proved to be a suitable means of preventing bearing damage. Emergence of fresh grease at the sealing area is a sure sign of efficient functioning of the lubrication system and seal. The seals of the roller bearing must not be damaged mechanically through unacceptable operating conditions.

3.2 Armour protection of the rollers

The majority of, and the most severe, damage occurs at the roller surfaces. **Fig. 10** shows an example of a point where material has broken out of the armouring. The roller surfaces are the most highly stressed components in high-pressure grinding rolls. They are therefore protected against wear. The main aim of continued design development has been to achieve higher operational reliability of the roller

BILD 10: Ausbruchstelle auf der Mantelfläche einer Mahlwalze
FIGURE 10: Surface of a grinding roller where material has broken out

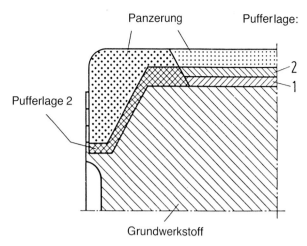

BILD 12: Beispiel für den Aufbau der aufgeschweißten Panzerung einer Mahlwalze; schematische Darstellung nach [50]
FIGURE 12: Example of the construction of the welded armouring on a grinding roller; diagrammatic representation according to [50]

Panzerung = armouring
Pufferlage = buffer layer
Grundwerkstoff = basic material

schweißt wurde, ist inzwischen bei Neuanlagen und Ersatz-Investitionen durch eine zweiteilige Konstruktion aus Walzengrundkörper und aufgeschrumpfter Bandage mit aufgeschweißter Hartschicht als Panzerung abgelöst worden [49–51]. Diese im **Bild 11** schematisch dargestellte Konstruktion gilt heute als bewährter Aufbau der Walzen für den üblichen Einsatzfall, z.B. die Klinkermahlung. Für die Auftragsschweißung sind die am besten geeigneten Werkstoffe auszuwählen und das Fertigungsverfahren im Hinblick auf Wärmeführung und Einstellung der Schweißparameter zu optimieren. Die aufgeschweißte Panzerung (**Bild 12**) ist im Vergleich mit anderen Konstruktionen des Verschleißschutzes meist kostengünstiger.

— Vor allem für die Mahlung stark schleißender Stoffe werden derzeitig Alternativen mit Verschleißteilen aus Hartgußwerkstoffen [49] erprobt, die entweder als Bandage oder Segmente ausgebildet werden, wie beispielhaft im **Bild 13** dargestellt. Die Bandagen sind formschlüssig mit der Walzenwelle verbunden, die Segmente können geschraubt oder — wie in dem dargestellten Fall — geklemmt sein. Der wesentliche Vorteil der Segmente ist, daß der Ersatz des ganzen Verschleißschutzes oder die Reparatur eines Oberflächenschadens mit deutlich weniger Aufwand direkt an der Mühle erfolgen kann. Die neuen Entwicklungen lassen eine deutlich höhere Betriebssicherheit erwarten.

— Der autogene Verschleißschutz durch Materialeinlagerungen zwischen Rasterelementen aus Hartmetall in der Walzenoberfläche [48, 52], die sog. Rasterpanzerung, ist in den **Bildern 14** und **15** dargestellt. Wie erste Betriebserfahrungen zeigen, ist diese Panzerung für alle zu mahlen-

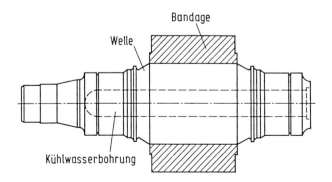

BILD 11: Zweiteilige Mahlwalze, bestehend aus einer Welle mit aufgeschrumpfter Bandage; schematische Darstellung nach [50]
FIGURE 11: Two-piece grinding roller, consisting of a shaft with shrunk-on tyre; diagrammatic representation according to [50]

Bandage = tyre
Welle = shaft
Kühlwasserbohrung = cooling water boring

BILD 13: Mahlwalzen mit Verschleißteilen aus Hartgußwerkstoff in Form von Segmenten [49]
(Werkfoto der Krupp-Polysius AG, Beckum)
FIGURE 13: Grinding rollers with wearing parts made of chilled cast material in the form of segments [49]
(works photograph from Krupp-Polysius AG, Beckum)

surface [43, 44], and very considerable improvements have now been achieved [48, 49]. There are several rival methods for building up the armouring:

— The composite material solution, in which one or more intermediate layers are welded onto the roller body as buffer layers and a wear-resistant hard layer is welded on top, has now been replaced in new plants and replacement projects by a two-part design of basic roller body and shrunk-on tyre with welded hard layer as armouring [49–51]. This design, shown diagrammatically in **Fig. 11**, now rates as a proven method of construction for the rollers for the usual application, e.g. clinker grinding. The most suitable materials should be selected for the build-up welding, and the fabrication process should be optimized with respect to heat control and setting of the weld parameters. The welded armouring (**Fig. 12**) is usually more economical than other wear protection designs.

— Alternative methods with wearing parts made of chilled cast materials [49] which are formed either as tyres or seg-

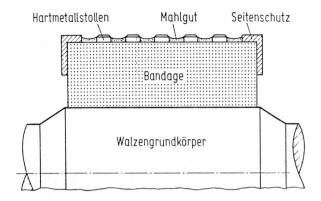

BILD 14: Rasterpanzerung einer Mahlwalze, schematische Darstellung nach [52]
FIGURE 14: Studded armouring of a grinding roller, diagrammatic representation according to [52]

Hartmetallstollen	= hard metal alloy studs
Mahlgut	= material being ground
Seitenschutz	= side protection
Bandage	= tyre
Walzengrundkörper	= basic roller body

den Stoffe geeignet, auch für nahezu trockenen und sehr feuchten Hüttensand mit bis zu 10 % Wassergehalt. Das Einzugsverhalten der Walzen und damit das Zerkleinerungsverhalten entspricht etwa dem bei auftragsgeschweißten Walzen. Schwierigkeiten bei der Befestigung der Raster auf der Walzenoberfläche sollen inzwischen behoben worden sein. Daher werden hohe Standzeiten erwartet. Die Rasterpanzerung kann als Bandage oder als Segmente ausgeführt werden.

Die Beanspruchung der Walzenlagerung und der Walzenoberfläche wird von der zur Zerkleinerung erforderlichen Mahlkraft bestimmt. Außer den Normalkräften wirken an der Mahlwalze auch Umfangskräfte, die als Schub- oder Reibungskräfte auftreten. Je nach dem Reibungsbeiwert rufen sie eine Relativbewegung und damit Abrasivverschleiß hervor. Bei ungleicher Umfangsgeschwindigkeit der beiden Mahlwalzen können Umfangskräfte jedoch sogar Ausbrüche der Walzenpanzerung hervorrufen. Daher ist die Verwirklichung gleicher Walzenumfangsgeschwindigkeiten zwingend erforderlich [53]. Bei Gleichlauf beider Walzen machen die Umfangskräfte nur etwa 3–4 % der auftretenden Normalkräfte aus. In diesem Fall sind sie für die Belastung der Mahlwalzen von untergeordneter Bedeutung. Die Beanspruchung der Walzenoberfläche kann dann vor allem anhand der maximalen Normalkräfte beurteilt werden. Diese hängen außer von den Walzenabmessungen vom Mahlgut ab, da dieses bei der gleichen aufgebrachten Mahlkraft und bei fester Drehzahl und gegebener Walzenprofilierung die Einzugsverhältnisse bestimmt. Ein größerer Einzugswinkel kann zum Abbau der Druckspitzen führen. Bessere Einzugsbedingungen führen aber in der Regel zu einer dickeren Schülpe, wodurch auch die Gesamtmahlkraft entsprechend der Federkennlinie der Anpreßhydraulik ansteigt. Zur Beurteilung der Beanspruchung der Walzenoberfläche ist deshalb die Einstellung des hydropneumatischen Federsystems von großer Bedeutung. Weitere wesentliche Betriebsbedingungen sind ferner die Korngrößenverteilung des Aufgabeguts und die Mahlgutverteilung über die Länge des Mahlspaltes [47].

Die maximale Aufgabegutkorngröße sollte etwa die 2fache Schülpenstärke nicht überschreiten, um eine Einzelkornzerkleinerung weitgehend zu vermeiden. Sowohl ein zu großer Anteil groben Mahlguts als auch zu feinen Mahlguts wirken sich ungünstig auf die Leistungsaufnahme der Mühle und ihr Vibrationsverhalten aus. Daher ist eine möglichst gleichmäßige Mahlgutmischung anzustreben.

Ferner ist für eine möglichst gleichmäßige Mahlgutzufuhr und ebenso gleichmäßige Verteilung des Mahlguts auf die beiden Walzen und über die Länge des Walzenspalts zu sorgen. Die Verteilung des Mahlguts auf die beiden Walzen kann z. B. durch verstellbare Schachtwände beeinflußt werden.

ments are currently being tested, chiefly for grinding highly abrasive materials; an example is shown in **Fig. 13**. The tyres are joined positively to the roller shaft but the segments can be bolted or – as in the case shown – clamped. The important advantage of segments is that replacement of the complete wear protection, or repair of surface damage, can be carried out in situ with significantly less expenditure. The new developments are expected to provide significantly higher operational reliability.

– The autogenous wear protection system by inclusions of material between a pattern of studs made of hard metal alloy on the roller surface [48, 52], so-called studded armouring, is shown in **Figs. 14** and **15**. Initial operating experience indicates that this armouring system is suitable for all materials to be ground, even for almost dry and very wet blast furnace slag with a water content up to 10 %. The intake characteristics of the rollers, and therefore the comminution behaviour, corresponds approximately to that of rollers with build-up welding. Difficulties with attaching the studs to the roller surface are now said to have been eliminated, so long service lives are expected. The studded armouring system can be applied in the form of tyres or segments.

BILD 15: Rasterpanzerung einer Mahlwalze
(Werkfoto der KHD Humboldt Wedag AG, Köln)
FIGURE 15: Studded armouring of a grinding roller
(works photograph from KHD Humboldt Wedag AG, Cologne)

The stressing of the roller support system and of the roller surfaces is determined by the grinding force needed for the comminution. The grinding rolls are subjected not only to the normal forces, but also to circumferential forces which occur as shear or frictional forces. They cause a relative movement, and hence abrasive wear, which depends on the coefficient of friction. With unequal circumferential velocities of the two grinding rolls the circumferential forces can even cause material to break out of the roller armouring. It is therefore absolutely essential to achieve equal roller circumferential velocities [53]. If the two rollers are running uniformly the circumferential forces only make up about 3–4 % of the normal forces which occur. In this case they are of secondary importance for the loadings on the grinding rollers. The stressing of the roller surface can therefore be assessed primarily on the basis of the maximum normal forces. These depend not only on the roller dimensions but also on the material being ground, as for the same applied grinding force and with fixed speed and given roller profiling, these determine the intake conditions. A greater angle of nip can reduce the pressure peaks. However, better intake conditions lead, as a rule, to a thicker compacted cake; this

Bei inhomogener Materialverteilung über der Spaltlänge kann das Durchschießen der feineren Mahlgutanteile durch den Spalt ohne eine Zerkleinerung auch zu einer örtlichen Überpressung der anderen Mahlgutteile, der sog. Strangbildung führen, die eine unterschiedliche Verschleißschädigung der Walzenoberfläche längs der Walzen zur Folge hat.

Eine sowohl nach der Korngröße als auch nach Komponenten inhomogene Materialverteilung im Schacht kann eine Schiefstellung der Loswalze hervorrufen. Entsprechend der eingestellten Federkennlinie der Anpreßhydraulik führt das zu einem einseitig größeren Reaktionsdruck aus dem Gutbett. Dadurch werden nicht nur die Druckkräfte ungleichmäßig über der Walzenlänge verteilt, sondern auch die Axialkräfte einseitig deutlich erhöht. Bei Maschinen mit Pendelrollenlagern wird außerdem die Lagerdichtung der Loswalze erheblich beansprucht. Deshalb wird die Schiefstellung der Loswalze ständig gemessen und in engen Grenzen toleriert. Sie kann durch Umverteilung des Mahlgutstromes über der Walzenbreite wirksam beeinflußt werden.

Ein Teil der Schäden an Walzenoberflächen ist nachweislich auf den Einzug von Fremdkörpern zurückzuführen. Ein wirksamer Schutz vor metallischen Fremdkörpern kann in einer Kombination von Überbandmagnetabscheidern und nachgeschalteter Suchspule bestehen. Zur Ausschleusung von Fremdkörpern bietet sich auch der Einbau einer schnell schaltenden Materialumstellklappe nach dem Bandabwurf an, die für wenige Sekunden das Aufgabegut mit dem Fremdkörper sicher an der Gutbett-Walzenmühle vorbeileitet [54].

4. Ausblick

Die mit hoher Energieausnutzung verbundene Hochdruckzerkleinerung in Gutbett-Walzenmühlen wird erst seit 1985 in der Zementindustrie genutzt. Vor allem in den ersten Jahren haben häufige mechanische Schäden die Betriebssicherheit und Verfügbarkeit von Gutbett-Walzenmühlen erheblich eingeschränkt. Durch eine Kombination mehrerer Maßnahmen, Verbesserung des Verschleißschutzes, gleichmäßigere Betriebsbedingungen und eine Verminderung des mittleren Anpreßdrucks konnte die Verfügbarkeit inzwischen deutlich gesteigert werden. Während diese Mühlen zunächst zur Steigerung des Durchsatzes und Verminderung des Energieaufwands bestehender Mahlanlagen eingesetzt wurden, haben sie inzwischen auch bei neuen Mahlanlagen für Zement, Zementrohmaterial und Hüttensand einen festen Anteil erreicht. Durch eine weitere Steigerung der Betriebssicherheit und durch eine erfolgreiche Weiterentwicklung der Fertigmahlung von Zement mit dem Ziel, marktorientierte Zementeigenschaften zu erzeugen, könnte dieser Anteil in Zukunft noch erhöht werden.

Literaturverzeichnis

[1] Stairmand, C.J.: The energy efficiency of milling processes. 4. Europäisches Symposium Zerkleinern, DECHEMA – Monographien Nr. 79, 1975, S. 1–17.

[2] Schönert, K.: Limits of Energy Saving in Milling, 1. World Congress Particle Technology, Part II Comminution. 6. Europäisches Symposium Zerkleinern, Preprints, Nürnberg, 1986, S. 1–21.

[3] Schönert, K.: Energetische Aspekte des Zerkleinerns spröder Stoffe. Zement-Kalk-Gips 32 (1979) Nr. 1, S. 1–9.

[4] Schönert, K.: Verfahren zur Fein- und Feinstzerkleinerung von Materialien spröden Stoffverhaltens. DP 2708053, 19. 7. 79 (und zahlreiche zugehörige Auslandsanmeldungen bzw. -patente).

[5] Beisner, K., Gemmer, L., Zisselmar, R., und Kellerwessel, H.: Verfahren und Anlagen zur kontinuierlichen Druckzerkleinerung spröden Mahlgutes. EP 0 084 383, 12. 4. 1983.

[6] Conroy, G.H., und Wüstner, H.: Industrial experience with high pressure comminution of cement clinker. World Cement (1986) Nr. 7, S. 297–307.

also increases the total grinding force governed by the spring characteristic of the hydraulic system which provides the pressure force. The setting of the hydropneumatic spring system is therefore of great importance for assessing the stress on the roller surfaces. Other important operating conditions are the particle size distribution of the feed material and the distribution of the mill feed over the length of the grinding gap [47].

In order to avoid single particle comminution as far as possible, the maximum feed particle size should not exceed about twice the thickness of the compacted cake. Too large a proportion of coarse feed material, and also excessively fine feed material, have unfavourable effects on the power consumption of the grinding rolls and on their vibration behaviour. The aim should therefore be to achieve as uniform a mix of feed material as possible.

Care should also be taken to ensure as uniform a feed of material as possible as well as uniform distribution of the feed to the two rollers and over the length of the roller gap. The distribution of the feed material to the two rollers can be governed, for example, by adjustable shaft walls.

Inhomogeneous material distribution over the length of the gap can lead to the finer fraction of the feed material flushing through the gap without comminution and to local overcompression of the other fractions of the feed material, so-called strand formation, which leads to differing wear damage of the roller surface along the roller.

A distribution of material in the feed shaft which is inhomogeneous either in particle size or in components can cause skewing of the free roller. This leads to a greater reaction force from the bed of material at one end which is governed by the set spring characteristic of the hydraulic pressure system. This means that not only are the pressure forces distributed unevenly over the length of the roller, but that the axial forces are also significantly increased at one end. The bearing sealing system of the free roller is also submitted to great stresses in machines with self-aligning roller bearings. The skewing of the free roller is therefore measured continuously and only tolerated within narrow limits. It can be governed effectively by redistributing the flow of feed material over the width of the roller.

Some of the damage to the roller surfaces is clearly due to the intake of foreign bodies. A combination of suspended magnetic belt separator followed by a detector coil can constitute an effective protection from metallic foreign bodies. The installation of a rapid action material diversion flap after the belt discharge point, which, for a few seconds, safely diverts the feed material containing the foreign body past the high-pressure rolls, is a suitable way of removing foreign bodies from the system [54].

4. Outlook

High compression comminution in high-pressure grinding rolls with its associated high level of energy utilization has only been used in the cement industry since 1985. Frequent mechanical damage greatly curtailed the operational reliability and availability of high-pressure grinding rolls, especially in the first few years. The availability has now been significantly increased by a combination of several measures, improvement of the wear protection, more uniform operating conditions and a reduction of the mean applied pressure force. Although these high-pressure grinding rolls were used at first to increase the throughput and reduce the power consumption of existing grinding systems, they now also account for a steady proportion of new grinding plants for cement, cement raw material and blast furnace slag. This proportion could be increased still further in the future through a further increase in operational reliability and successful onward development of finish grinding of cement aimed at producing cement properties adapted to market needs.

[7] Schneider, G.: Die Gutbett-Walzenmühle im Zementwerk Leimen. Zement-Kalk-Gips 40 (1987) Nr. 2, S. 86–89.

[8] Schneider, G., Gudat, G., und Schneider, V.: Betriebserfahrungen mit Gutbett-Walzenmühlen bei der Zementmahlung. Zement-Kalk-Gips 42 (1989) Nr. 4, S. 175–178.

[9] Rosemann, H.: Energieverbrauch und Betriebsverhalten der Gutbett-Walzenmühle bei der Feinmahlung von Zement. KHD-Symposium „Fortschritte der Rollenpressen-Technik", Köln, 1989.

[10] Rosemann, H., Hochdahl, O., Ellerbrock, H.-G., und Richartz, W.: Untersuchungen zum Einsatz einer Gutbett-Walzenmühle zur Feinmahlung von Zement. Zement-Kalk-Gips 42 (1989) Nr. 4, S. 165–169.

[11] Paliard, M., und Cochet, F.: CLE's and Ciments Lafarge's latest developments in roller press. Zement-Kalk-Gips 43 (1990) Nr. 2, S. 71–76.

[12] Cochet, F., et Paliard, M.: Broyage intégral en presse à rouleaux. Ciments, Bétons, Plâtres, Chaux, No. 779, 1989, No. 4.

[13] Beese, W.: Fertigmahlung von Rohmaterial mit Rollenpressen. KHD-Symposium „Moderne Rollenpressen-Technik", Köln, 1990.

[14] Beese, W., und Höse, H.: Rohmehlerzeugung mit der Rollenpresse – ein Bericht über zweijährige Betriebserfahrungen. Zement-Kalk-Gips 45 (1992) Nr. 9, S. 456–462.

[15] Thurat, B., und Wolter, A.: Gutbett-Walzenmühlen zur Feinmahlung von Zementrohmehl, Kohle und Kalk. Zement-Kalk-Gips 42 (1989) Nr. 4, S. 179–183.

[16] Patzelt, N.: Moderne Systeme der Hüttensand-Mahlung. Zement-Kalk-Gips 45 (1992) Nr. 7, S. 342–347.

[17] Plank, F.W.: Die Rollenpresse zur Feinmahlung von Kalk im Sichterkreislauf. KHD-Symposium „Fortschritte der Rollenpressen-Technik", Köln, 1989.

[18] Plank, F.W., Bauerochse, M., und Oberheuser, G.: Erkenntnisse mit der Gutbett-Druckzerkleinerung von Branntkalk und Kalkstein – Praxis-Erprobung einer Rollenpresse. Zement-Kalk-Gips 40 (1987) Nr. 1, S. 1–6.

[19] Paulsen, H.: Erste Betriebsergebnisse beim Einsatz einer Rollenpresse zur Kalkmahlung. Zement-Kalk-Gips 40 (1987) Nr. 12, S. 598–601.

[20] Strasser, S., und Wolter, A.: Zukunftspotentiale der Mahltechnik mit der Rollenpresse. Zement-Kalk-Gips 44 (1991) Nr. 7, S. 345–350.

[21] Wüstner, H., Dreizler, I., und Oberheuser, G.: Einsatz von Rollenpressen in Mahlanlagen für Kohle, Zementrohstoffe und Zement. Zement-Kalk-Gips 40 (1987) Nr. 7, S. 345–353.

[22] Wolter, A., und Dreizler, I.: Einflüsse des Einsatzes der Rollenpresse auf die Zementeigenschaften. KHD-Symposium „Erfahrungen beim Rollenpressen-Einsatz", Köln, 1987.

[23] Odler, I., und Chen, Y.: Einfluß des Mahlens in einer Gutbett-Walzenmühle auf die Eigenschaften des Portlandzements. Zement-Kalk-Gips 43 (1990) Nr. 4, S. 188–191.

[24] Kupper, D., und Knobloch, O.: Fertigmahlung von Zement mit der Gutbett-Walzenmühle POLYCOM. Teil 1: Untersuchungen an Mischungen von Klinkermehlen und Sulfatträgern. Zement-Kalk-Gips 44 (1991) Nr. 1, S. 21–27.

[25] Strasser, S.: Die Fertigmahlung von Rohmaterial unter Druck. Vortrag zum VDZ-Kongreß '93, Fachbereich 5.

[26] Strasser, S.: Verfahren und Einrichtung zur Zerkleinerung bzw. Mahlung spröden Mahlguts. Deutsches Patentamt DE 3518543 C2, 23. 5. 1985.

[27] Patzelt, N.: Beispiele erfolgreicher Integration der Hüttensand-Aufbereitung. Vortrag zum VDZ-Kongreß '93, Fachbereich 5.

[28] Albeck, J., und Kirchner, G.: Einfluß der Verfahrenstechnik auf die Herstellung marktorientierter Zemente. Zement-Kalk-Gips 46 (1993) Nr. 10, S. 615–626.

[29] Maréchal, F., und Melis, G.: New finish grinding facilities at the CBR Group Lixhe and Maastricht plants. World Cement (1993) Nr. 1, S. 25–26.

[30] Kellerwessel, H.: Betriebsergebnisse von Hochdruck-Rollenpressen. Aufbereitungs-Technik 27 (1986) Nr. 10, S. 555–559.

[31] von Seebach, H.M., und Patzelt, N.: Betrieb von Mahlanlagen mit Gutbett-Walzenmühlen für Rohmaterial und Zement. Zement-Kalk-Gips 40 (1987) Nr. 7, S. 337–344.

[32] Oberheuser, G., und Strasser, S.: Mehrfachpressen im Umlaufverfahren. KHD-Symposium „Erfahrungen beim Rollenpressen-Einsatz", Köln, 1987.

[33] Strasser, S.: Generalbericht zum Stand der Rollenpressen-Technik. KHD-Symposium „Fortschritte der Rollenpressen-Technik", Köln, 1989.

[34] Schmitz, T., und Kupper, D.: Einfluß rheologischer Parameter auf den Prozeß der Gutbettbeanspruchung in der Walzenmühle, Teil 1. Zement-Kalk-Gips 45 (1992) Nr. 2, S. 79–85.

[35] Schmitz, T.: Vermeidung von Vibrationen beim Betrieb der Gutbett-Walzenmühle. Vortrag zum VDZ-Kongreß '93, Fachbereich 5.

[36] Patzelt, N.: Hybrid-, Kombi- und Fertigmahlung mit der Gutbett-Walzenmühle POLYCOM. Vortrag zum 24. Polysius-Zementtag, Hannover, 1990.

[37] Conroy, G.: Erfahrungen mit COPLAY's Rollenpressen-Mahlanlage in Nazareth, Pennsylvania. KHD-Symposium „Fortschritte der Rollenpressen-Technik", Köln, 1989.

[38] Strasser, S.: Teilfertigmahlung von Zement-Klinker. KHD-Symposium „Moderne Rollenpressen-Technik", Köln, 1990.

[39] Süßegger, A.: Auswirkung der Hochdruckzerkleinerung auf das Design und den Betrieb von Sichtern. KHD-Symposium „Fortschritte der Rollenpressen-Technik", Köln, 1989.

[40] Süßegger, A.: Sepmaster Report. KHD-Symposium „Moderne Rollenpressen-Technik", Köln, 1990.

[41] Ünal, T.: Erfahrungen mit der Rollenpresse aus maschinentechnischer Sicht und ihr entscheidender Einfluß auf die Verbesserung des Zerkleinerungs-Wirkungsgrads. KHD-Symposium „Fortschritte der Rollenpressen-Technik", Köln, 1989.

[42] Partz, K.-D.: Verschleißschutz an Oberflächen von Walzen für Rollenpressen – Stand der Legierungsentwicklung und der Schweißtechnik, Januar 1989. KHD-Symposium „Fortschritte der Rollenpressen-Technik", Köln, 1989.

[43] Patzelt, N., und Tiggesbäumker, P.: Konstruktive Lösungen für betriebssichere Mahlwalzen von Gutbett-Walzenmühlen. Zement-Kalk-Gips 44 (1991) Nr. 2, S. 88–92.

[44] Brachthäuser, M., und Wollner, M.: Erhöhung der Verfügbarkeit von Gutbett-Walzenmühlen durch konstruktive Maßnahmen. Vortrag zur Fachtagung „Zement-Verfahrenstechnik" des Vereins Deutscher Zementwerke e.V., Düsseldorf, 1991.

[45] Gehlken, C., Fischer-Hellwig, F., und Schneider, R.: Weiterentwicklung an Rollenpressen und Betriebsmessungen. KHD-Symposium „Moderne Rollenpressen-Technik", Köln, 1990.

[46] Gehlken, C., und Zenner, H.: Mechanische Beanspruchungen der Gutbett-Walzenmühlen. Vortrag zur Fachtagung „Zement-Verfahrenstechnik" des Vereins Deutscher Zementwerke e. V., Düsseldorf, 1991.

[47] Kellerwessel, H.: Praxis der Hochdruck-Gutbettzerkleinerung. Zement-Kalk-Gips 43 (1990) Nr. 2, S. 57–64.

[48] Strasser, S.: Aktueller Stand der Rollenpressen-Technik. KHD-Symposium „Moderne Rollenpressen-Technik", Köln, 1992.

[49] Patzelt, N.: Verschleißschutzalternativen für die Beanspruchungsflächen von Gutbett-Walzenmühlen. Zement-Kalk-Gips 47 (1994) Nr. 2, S. 83–86.

[50] Sommer, E.: Konstruktiver Stand der KHD-Rollenpresse. KHD-Symposium „Fortschritte der Rollenpressen-Technik", Köln, 1989.

[51] Stojan, J.: Generalbericht zum Stand der Rollenpressen-Technik. KHD-Symposium „Moderne Rollenpressen-Technik", Köln, 1990.

[52] Kellerwessel, H.: Hochdruck-Gutbett-Zerkleinerung von mineralischen Rohstoffen. Aufbereitungs-Technik 34 (1993) Nr. 5, S. 243–249.

[53] Herchenbach, H.: Einfluß des Antriebes auf die Wirtschaftlichkeit einer Hochdruck-Rollenpresse. Vortrag zum VDZ-Kongreß '93, Fachbereich 5.

[54] Readymix Hüttenzement GmbH: Mahlanlage sowie Verfahren zum Mahlen von stückigem Gut. Deutsches Patent DE 4016262 C1, 19. 5. 1990.

Sichter in Mahlkreisläufen*)
Separators in grinding circuits*)
Séparateurs dans les circuits de broyage

Separadores en los circuios de molienda

Von **E. Onuma** und **M. Ito,** Chiba, Japan

Fachbericht 5.2 · Zusammenfassung – Die konventionellen Sichter in Mahlanlagen für Rohmehl, Kohle und Zement hatten oft nur eine mäßige Trennwirkung. Die in den letzten Jahren entwickelten sog. Hochleistungs-Sichter zeichnen sich durch eine deutlich verbesserte Trennwirkung und Trennschärfe aus. Die Arbeitsweise eines Sichters kann vorteilhaft durch den Verlauf der Trennkurve nach Tromp gekennzeichnet werden, die den Trenngrad in Abhängigkeit von der Korngröße wiedergibt. Die modernen Sichter werden in Mahlanlagen mit Kugel-, Wälz- und Gutbett-Walzenmühlen eingesetzt. Ihre deutlich bessere Trennwirkung führt vor allem bei höheren Produktfeinheiten zu größeren Durchsätzen und zu einem geringeren massebezogenen Energieverbrauch. Jedoch werden auch die Produkteigenschaften durch die Sichterarbeitsweise beeinflußt. Wegen der höheren Trennschärfe ist die Korngrößenverteilung des Produkts enger und daher vor allem die 3- bis 28-d-Festigkeit von Zementen bei gleicher massebezogener Oberfläche höher. Demgegenüber kann hierdurch die Verarbeitbarkeit des Zements beeinträchtigt werden.

Sichter in Mahlkreisläufen

Special report 5.2 · Summary – Conventional separators in grinding plants for raw meal, coal and cement have often shown rather poor separation performances. The recently developed high-efficiency separators are especially noted for their markedly improved separation performance and sharpness of separation. The mode of operation of a separator may be characterized advantageously by the shape of the Tromp separation curve, which shows the selectivity as a function of the particle grain size. Modern separators are used in grinding plants with ball mills, roller mills and high-pressure grinding rolls. Their substantially improved separation performance, especially with higher product finenesses, leads to a higher output and a lower specific energy consumption. However, the properties of the product are also affected by the separator's mode of operation. Due to the greater sharpness of separation the grain size distribution is narrower and the strengths, especially the 3 to 28 day strengths, of cements with the same specific surface area are therefore higher. On the other hand, the workability of the cement may be impaired.

Separators in grinding circuits

Rapport spécial 5.2 · Résumé – Les séparateurs classiques utilisés dans les ateliers de broyage pour farine crue, charbon et ciment n'avaient souvent qu'un effet de séparation moyen. Les séparateurs à haut rendement mis au point ces dernières années se caractérisent par une nette amélioration de l'effet de séparation mais aussi de son degré de séparation. Le mode de fonctionnement d'un séparateur se distingue par le profil de sa courbe de séparation selon Tromp, qui reflète le degré de séparation en fonction de la granulométrie. Les séparateurs modernes sont utilisés dans les ateliers de broyage comportant des broyeurs à boulets, des broyeurs à cylindres et des broyeurs à galets à lit de matière. Leur effet de séparation nettement améliorée entraîne surtout pour des finesses de produit élevées un plus grand rendement et une consommation d'énergie par rapport à une masse plus réduite. Le mode de fonctionnement du séparateur influe néanmoins sur les propriétés du produit. Etant donné le plus grand degré de séparation, la répartition granulométrique du produit est plus étroite, de ce fait, la résistance de 3 à 28-d des ciments est accrue en présence d'une surface identique par rapport à la masse. En revanche, la possibilité de mise en oeuvre du ciment peut, de ce fait, être altérée.

Séparateurs dans les circuits de broyage

Informe de ramo 5.2 · Resumen – Los separadores convencionales, empleados en las instalaciones de molienda de crudo, carbón y cemento, muchas veces tienen un reducido efecto de separación. Los llamados separadores de alta eficacia, desarrollados en los últimos años, se distinguen por un efecto y nitidez de separación notablemente mejorados. El funcionamiento de un separador se puede representar, con ventaja, mediante el trazado de la curva de separación de Tromp, la cual nos da el grado de separación en función de la granulometría. Los separadores modernos se utilizan en las instalaciones de molienda que trabajan con molinos de bolas, molinos de pista de roda-

Separadores en los circuitos de molienda

---

*) Überarbeitete Fassung eines Vortrages zum VDZ-Kongreß '93, Düsseldorf (27.9.–1.10.1993)
Revised draft of a lecture of the VDZ Congress '93, Düsseldorf (27.9.–1.10.1993)

dura y molinos de cilindros y lecho de material. Su efecto de separación, notablemente mejor, conduce a mayores rendimientos y a un menor consumo de energía referido a la masa, sobre todo cuando se trata de productos de mayor finura. Sin embargo, en las propiedades del producto influye también la forma de trabajar de los separadores. Debido a la mayor nitidez de separación, la distribución granulométrica del producto resulta más estrecha, y por ello, sobre todo, la resistencia de los cementos a los 3–28 días resulta más alta, siempre y cuando las superficies referidas a la masa sean iguales. Pero, por otro lado, puede verse afectada la trabajabilidad del cemento.

1. Einleitung

Unter den zahlreichen Bemühungen zur Energieeinsparung bei der Zementherstellung war die Entwicklung des sogenannten Hochleistungssichters eine der herausragendsten und zugleich erfolgreichsten. In diesem Beitrag wird ein kurzer Überblick über diese relativ neue Technik mit ihrer aktuellen Bedeutung und den zukünftigen Möglichkeiten gegeben.

2. Entwicklung von Hochleistungssichtern

Die bei der Zement- und Rohmehlproduktion bislang eingesetzten Sichter besaßen im allgemeinen keine besonders günstigen Trenneigenschaften, vor allem da

— das Aufgabegut auf den Sichter nur unvollständig dispergiert wurde, was zu einem Bypass der Sichtzone und zum Eintritt ungesichteten Materials in das Sichtergrobgut führte,

— die am Sichtvorgang beteiligten Kräfte sowie die Bewegungsbahnen der Partikel im Sichter undefiniert waren, wodurch sich die Trennkorngröße in Abhängigkeit vom Ort in einem weiten Bereich veränderte,

— die Feingutabscheidung in den im Sichtergehäuse untergebrachten Abscheidevorrichtungen unvollständig war, wodurch ein großer Feingutkreislauf innerhalb des Sichters verursacht wurden.

Während mit der Einführung des Zyklonumlaufsichters [1] die Vorgänge der Sichtung und Feingutabscheidung apparativ getrennt wurden, blieben die beiden erstgenannten Probleme im wesentlichen ungelöst. Um auch diese Probleme einer technischen Lösung zuzuführen, war ein neuer Trennmechanismus erforderlich. Dazu wurde das Konzept eines Hochleistungssichters geboren. Alle neuen Sichterkonstruktionen basieren heute auf folgenden Grundgedanken:

— Der Sichtmechanismus sollte definiert sein. Die am Sichtvorgang beteiligten Kräfte sollten klar definiert und vollständig berechenbar sein.

— Die Sichtzone sollte genügend groß dimensioniert sein, um eine gute Trennwirkung zu gewährleisten, und in einer kompakten Anlage untergebracht sein.

— Das Sichtprinzip sollte bei der Maßstabsvergrößerung nicht verletzt werden.

— Die Partikeldispergierung sollte ausreichen, um die hohe Leistungsfähigkeit des Sichters auch unter den Bedingungen hoher Feststoff/Luft-Beladungen zu gewährleisten.

Nach grundlegenden Arbeiten von Quittkat [2], Hukki [3], Knobloch und anderen [4] tauchte der neue Sichter in den frühen 80er Jahren plötzlich auf dem Markt auf, zuerst der O-SEPA-Sichter [5], später gefolgt von einer Reihe weiterer Sichterkonstruktionen.

3. Philosophie, Trennvorgang und Konstruktionen

3.1 Sichter für Zement-Rohrmühlen

Repräsentativ für den neuen Sichtertyp wird zunächst eine kurze Erläuterung des O-SEPA-Sichters (**Bild 1**) gegeben. Im Anschluß daran werden einige andere Sichterbauformen erklärt. Wie aus Bild 1 hervorgeht, wird das zu sichtende Mahlgut über zwei Zuführungen (15) dem Sichter auf-

1. Introduction

Among so many efforts in energy saving in cement manufacturing technology, the development of so-called high-efficiency separators was one of the most prominent and successful ones. A brief survey on this rather new technical field is made in this paper, along with its recent topics and future possibilities briefly introduced.

2. Development of high-efficiency separators

Conventional separators used in cement or raw-meal production [1] did not show very good separation characteristics generally, chiefly because of

— Incomplete dispersion of the feed, causing a by-passing of the classification zone and going into rejects.

— Indefiniteness of the forces participating in the classification and the trajectory of particles in the separator, making the cut-size of classification widely diversified depending on the location in the separator.

— Incomplete capturing of fines in the in-house equipped cyclones, causing a large amount of fines recirculating in the separator.

This third problem was solved by the introduction of cyclone-air-separators [1], while the first and second problems were not completely solved.

To solve all these problems, a new classification mechanism was definitely needed. The concept of high-efficiency separators has been born to meet these requirements, and almost all of these new kinds of separators are now based on the following basic ideas:

— The classification mechanism should be the definite one. Forces participating in the classification should be clear, well defined, and perfectly determined.

— An effective classification zone with space enough to ensure a good classification performance should be well furnished within a compact equipment.

— The classification mechanism should not be distorted when scaled-up for large volume equipments.

— The dispersion of particles should be good enough to secure the good performance of the separator also under high solid-air concentration conditions.

After the pioneering activities by Quittkat [2], Hukki [3] and Knobloch et al. [4], such type of separator emerged on the market in the early eighties, O-SEPA [5] as the first runner and followed by several successors.

3. Philosophy-Mechanism-Constructions

3.1 Separators for clinker grinding ball mills

As representative for this type of separator, a brief explanation of the O-SEPA separator (**Fig. 1**) is made at first, and then several other variants are illustrated. As shown in this illustration, the particles to be classified are fed through the ducts (15), are highly dispersed by means of the rotating dispersion plate (2) and the buffer plate (9), and are thrown into the classifying air which is sucked into the separator through the tangentially extended ducts (5, 11). After passing through the fixed guide vanes (10), the classifying air forms a precise horizontal vortex by means of vortical-flow adjusting rotating blades (3) and horizontal partition plates (4). Particles are at first roughly classified on entering the classifying zone. The principal classification is made in the

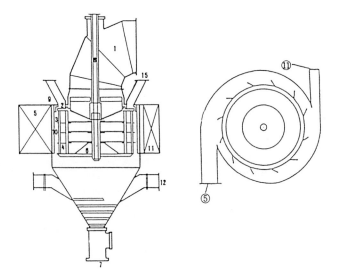

BILD 1: O-SEPA-Sichter
FIGURE 1: O-SEPA-separator

2 Dispergierteller	= dispersion plate
3 Drehzahlregelbarer Stabkorb	= vortical flow adjusting blade
4 Verteilungsscheibe	= partition plate
5 Luftzuführungsleitung	= primary air inlet duct
9 Prallwand	= buffer plate
10 Leitschaufelkranz	= guide vane
11 Luftzuführungsleitung	= secondary air inlet duct
12 Luftzuführungsleitung	= tertiary air inlet duct

gegeben, durch einen rotierenden Dispergierteller (2) und eine Prallwand (9) gut dispergiert und anschließend in die Sichtluft geworfen, die über tangential erweiterte Kanäle (5, 11) in den Sichtraum gesaugt wird. Nachdem die Sichtluft den fest eingestellten Leitapparat (10) durchströmt hat, bildet sie eine genau horizontal gelagerte Wirbelströmung mit Hilfe eines in seiner Drehzahl einstellbaren rotierenden Stabkorbes (3) mit horizontalen Verteilungsscheiben (4). Beim Eintritt in die Sichtzone erfahren die Partikel zunächst eine Grobsichtung. Die eigentliche Sichtung erfolgt in der Wirbelströmung entsprechend dem Gleichgewicht von Zentrifugal- und Schleppkräften durch die nach innen strömende Luft. Nach diesem Sichtvorgang wird das Feingut mit der Sichtluft aus dem Sichter ausgetragen und abgeschieden. Die groben Partikel beschreiben eine drallförmige Abwärtsbewegung auf der Innenseite des Leitschaufelkranzes, werden von der durch den unteren Teil der Luftleitungen im Sichtkonus (5, 11) zugeführten Luft umspült und nachgesichtet (12). Schließlich werden die Grieße über den Sichterkonus aus dem Sichter herausgeführt.

Der neue Sichtertyp mit seinem einmaligen Wirkmechanismus ist durch folgende Merkmale gekennzeichnet:

— Für jedes Partikel besteht die mehrfache Chance, durch ein klar definiertes Kräftegleichgewicht am Trennvorgang teilzunehmen, was zu einer sehr scharfen Sichtung führt.

— Die tangentiale Geschwindigkeitsdifferenz zwischen den Partikeln und dem Stabkorb ist so klein, daß die laufenden Instandhaltungskosten und der Energieverbrauch des Sichters minimiert werden.

— Praktisch der gesamte Innenraum des Sichters wird für die Trennzone genutzt. Das macht die Maschine sehr kompakt.

— Durch die Wirkung des drehzahlregelbaren Stabkorbes und die horizontalen Verteilungsscheiben kann sich auch in einem großräumigen Sichter eine exakt horizontal gelagerte Wirbelströmung ausbilden.

— Die bei der Sichtung angestrebte Trennkorngröße kann leicht durch eine Veränderung der Drehzahl des rotierenden Stabkorbes eingestellt werden. Ebenso kann die Korngrößenverteilung des Fertigproduktes durch eine entsprechende Wahl der Betriebsparameter in einem verhältnismäßig großen Bereich beeinflußt werden.

uniform vortex flow according to the balance of two forces – the centrifugal and the drag force – by inwardly flowing air. After this classification, fine products are carried away and collected. The coarse particles, as they swirl down inside the guide vanes, are rinsed by the air coming inside through the lower parts of ducts (5, 11), and then reclassified by the tertiary air (12). Finally they flow out from the bottom part of the separator.

By these unique mechanisms, the following features of this type of separator are drawn out:

— Repeated chance of classification for every particle by well-defined balance of forces lead to very sharp classification.

— The tangential velocity difference between particles and the rotor blades is so small that the occurring maintenance problems and the power consumption of the separator are minimized.

— Practically all of the inner space of the separator is utilized as classification zone. This makes the apparatus very compact.

— By virtue of the vortical-flow adjusting blades and horizontal partition plates, a precise horizontal vortex can be formed even in large volume apparatus.

— The cut size of classification can be easily adjusted through the adjustment of the rotation speed of the separator. The particle size distribution of the product can be adjusted in fairly wide range through the adjustment of operating variables.

After the O-SEPA separator, several new types of high-efficiency separators based on similar concepts, but having each a feature, have come into the market. The SEPAX separator [6] (**Fig. 2**) is especially suited for air-swept ball mills, since the separator feed is strongly dispersed by means of a special device in the course of its upward pneumatic transportation. This procedure has another advantage. Tramp metals which could stray into the separator are completely separated before entering the separator, which might have caused some maintenance problems.

In SD-classifiers [7] (**Fig. 3**), fine products are sucked out horizontally through the ducts arranged at the side wall of the separator, while inclined deflectors [7] are used instead of vertical guide vanes like in the O-SEPA separator. The former design may reduce the complicated sealing problems at the uppermost part of the separator, while the latter will be helpful in controlling the width of the descending material curtain at the periphery of the classification zone. In addition, this type of separator uses rods instead of blades which adds mechanism of collision to the classification mechanism (O-SEPA) improving the classification efficiency. The SEPOL [8] (**Fig. 4**) and O & K cross stream separators [9] (**Fig. 5**) are characterized by drawing fine products out downwardly from the bottom part of the separator. By adopting this construction a good dispersion of particles by feeding them at the central position of the dispersion plate will be ensured in the case of the SEPOL separator. In O & K separators, the amount of classifying air in the upper part can be adjusted independently from the lower part. With air fans incorporated in cyclone housing, the machinery is enabled to be designed very compactly [10].

3.2 Separators for raw-meal grinding ball mills

Along with the energy saving or capacity increase of existing plants, high-efficiency separators are needed in raw-meal grinding circuits because of their possibilities in the improvement of raw mix burnability by minimizing the coarser fraction of components with poor burnability. Although each type of separator introduced in the former section can be used in raw material grinding circuits, several special types of separators have been developed in this field. These separators are characterized by accepting the feed with a large volume of air swept out from the mill, just as in the case of the SEPAX separator.

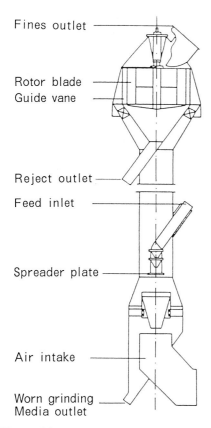

BILD 2: SEPAX-Sichter
FIGURE 2: SEPAX-separator

Feingutabführung	= fines outlet
Stabkorb	= rotor blade
Leitschaufelkranz	= guide vane
Sichtergrobgut	= reject outlet
Sichteraufgabegut	= feed inlet
Dispergierteller	= spreader plate
Luftzufuhr	= air intake
Abführung von verschlissenen Mahlkörpern	= worn grinding media outlet

Nach dem O-SEPA-Sichter sind mehrere Bauarten von Hochleistungssichtern auf den Markt gekommen, die alle auf einem ähnlichen Sichtprinzip beruhen, jedoch durch eigene Merkmale gekennzeichnet sind. So ist der sogenannte SEPAX-Sichter [6] besonders für Rohrmühlen geeignet (**Bild 2**), die nach dem Luftstromprinzip betrieben werden, da das Sichteraufgabegut mit einer besonderen Einrichtung während des nach oben gerichteten pneumatischen Transports intensiv dispergiert wird. Diese technische Lösung hat auch noch einen anderen Vorteil. Metallische Fremdkörper, die in den Sichtraum gelangen und dort Instandhaltungsprobleme verursachen können, werden schon vor ihrem Eintritt in den Sichter vollständig entfernt.

Im sogenannten SD-Sichter [7] wird das Fertigprodukt horizontal durch an der Sichterwand angeordnete Kanäle abgesaugt (**Bild 3**), wobei geneigte Leitbleche anstelle von vertikalen Leiteinrichtungen wie im O-SEPA-Sichter eingesetzt werden. Die frühere Ausführung verminderte bereits die komplizierten Abdichtungsprobleme am oberen Teil des Sichters zwischen dem rotierenden Stabkorb und der Sichtluftabsaugung. In neueren Konstruktionen kann die Größe des sich an der Peripherie der Sichtzone ausbildenden Materialschleiers besser kontrolliert werden. Außerdem werden im rotierenden Teil des SD-Sichters Rundstäbe anstelle von ebenen Flügeln verwendet, wodurch dem Klassierungsmechanismus ein Kollisionsmechanismus überlagert und auf diese Weise die Sichtwirkung verbessert wird. Der SEPOL-Sichter [8] (**Bild 4**) und der O&K-Querstrom-Sichter [9] (**Bild 5**) sind dadurch gekennzeichnet, daß das Sichterfeingut nach unten aus dem Sichter ausgetragen wird. Da bei diesen Sichtern das Aufgabegut zentral auf einen Streuteller oberhalb des Stabkorbes aufgegeben wird, wird für eine gute Dispergierung des Aufgabegutes gesorgt. Im O&K-Sichter kann die Sichtluftmenge im oberen und unteren Teil

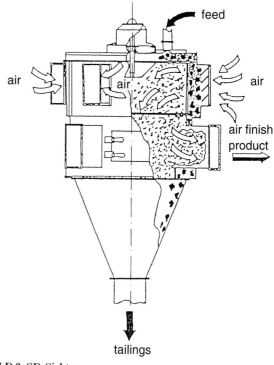

BILD 3: SD-Sichter
FIGURE 3: SD-classifier

Sichteraufgabegut	= feed
Luft	= air
Sichterfeingut	= air finish product
Sichtergrobgut	= tailings

The TSV separator [11] (**Fig. 6**) and SEPMASTER SKS-L separator [12] (**Fig. 7**) are two of such examples, and their typical variants can receive part of the separator feed with the mill-swept air while the remaining part of the separator feed comes through the bucket-elevator.

3.3 Separators in roller mills

The concept of high-efficiency separators can also be successfully realized in roller mills used in raw meal grinding or solid fuel grinding. In roller mills with conventional separators [13], the following problems were almost inevitable:

— Inefficient classification causes the accumulation of fine particles in the mill, which prevents the formation of a good material bed on the table and thus decreases the grinding efficiency.

— Accumulation of fine particles in mill-separator circuits causes an increase of pressure loss in the circuit, thus increasing the necessary power to drive the exhauster fan.

— Incomplete separation of coarser particles often causes the poor burnability of raw meal or a reduction of the combustion rate of solid fuel produced.

All these problems have successfully been solved through the introduction of the high-efficiency separator concept [13] (**Fig. 8**). One more example is the case of re-building ring-ball mills by using TSV or O-SEPA type separators [14] (**Fig. 9**). In addition to an output increase of 70% for the same $R_{0,10}$ residue, the decrease of coarse fraction in the prepared solid fuel for the same output ($R_{0,10}$: 10% → 2%) gave a good promise in the possibility of using low volatile fuels in pre-calciners. The application of roller mills in cement clinker grinding presented another problem in the performance of separators. Classification itself should be excellent, while the particle size distribution should be appropriate from the viewpoint of cement qualtity. The OK mill [15] is one solution of this problem with a sufficiently high amount of fine particles in the product by careful and elaborate choice of roller-table arrangements and operating conditions. At the same time the high-efficiency of classification avoids the straying in of coarser particles to the fine product.

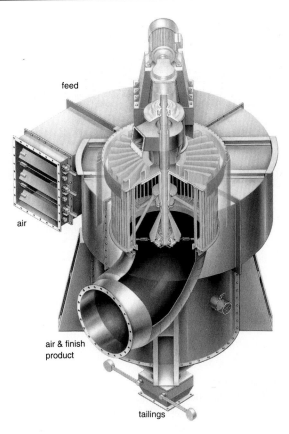

BILD 4: Sepol SV-Sichter
(Werksfoto von Krupp Polysius AG, Beckum)
FIGURE 4: SEPOL-SV separator
(Photograph by Krupp Polysius AG, Beckum)
Sichteraufgabegut = feed
Luft = air
Luft und Sichterfeingut = air & finish product
Sichtergrobgut = tailings

des Sichters unabhängig voneinander eingestellt werden. Mit Ventilatoren, die in die außenliegenden Zyklonabscheider integriert sind, kann der Sichter sehr kompakt ausgeführt werden [10].

3.2 Sichter für die Rohmehlerzeugung in Rohrmühlen

Zur Energieeinsparung oder Kapazitätssteigerung vorhandener Mahlanlagen werden Hochleistungssichter auch in Mahlkreisläufen zur Rohmehlerzeugung benötigt. Die neuen Sichter bieten die Möglichkeit, die Brennbarkeit der Rohmischung zu verbessern, indem sie die gröberen Korngrößenfraktionen der Komponenten mit schlechter Brennbarkeit minimieren. Obwohl jeder der im vorigen Abschnitt beschriebenen Sichter auch bei der Rohmehlerzeugung verwendet werden kann, sind für diesen Einsatz mehrere spezielle Sichtertypen entwickelt worden. Diese Sichter sind dadurch gekennzeichnet, daß sie mit einem großen materialbeladenen Luftstrom aus der Mühle beaufschlagt werden, ähnlich wie im Falle des SEPAX-Sichters.

Der TSV-Sichter [11] (**Bild 6**) und der als SEPMASTER SKS-L bezeichnete Sichter [12] (**Bild 7**) sind zwei solcher Beispiele. Diese Sichter können einen Teil des Sichteraufgabeguts mit dem Luftstrom aus der Mühle erhalten, während der andere Teil mechanisch mit einem Becherwerk zugeführt wird.

3.3 Sichter in Wälzmühlen

Das Konzept des Hochleistungssichters wird auch erfolgreich in Wälzmühlen realisiert, die zum Mahlen von Rohmehl oder festen Brennstoffen eingesetzt werden. In Wälzmühlen mit konventionellen Sichtern [13] gab es immer folgende Probleme:

— Eine wenig wirksame Trennung verursacht eine Anreicherung von Feingut im Mahlraum, wodurch die

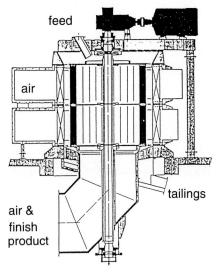

BILD 5: O & K Querstromsichter
FIGURE 5: O & K Cross Stream Separator
Sichteraufgabegut = feed   Luft und Sichterfeingut = air & finish product
Luft = air   Sichtergrobgut = tailings

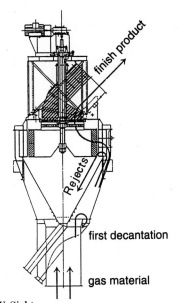

BILD 6: TSV-Sichter
FIGURE 6: TSV-separator
Sichterfeingut = finish product   erste Abscheidung = first decantation
Sichtergrobgut = rejects   Gas und Feststoff = gas material

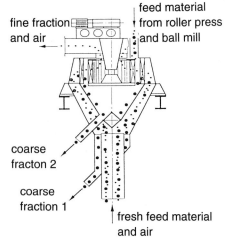

BILD 7: Sepmaster SKS-L-Sichter
FIGURE 7: SEPMASTER SKS-L separator
Sichteraufgabegut = feed material
Luft und Sichterfeingut = fine fraction and air
Grobgutfraktion 2 = coarse fraction 2
Grobgutfraktion 1 = coarse fraction 1
Luft und Frischgut = fresh feed material and air

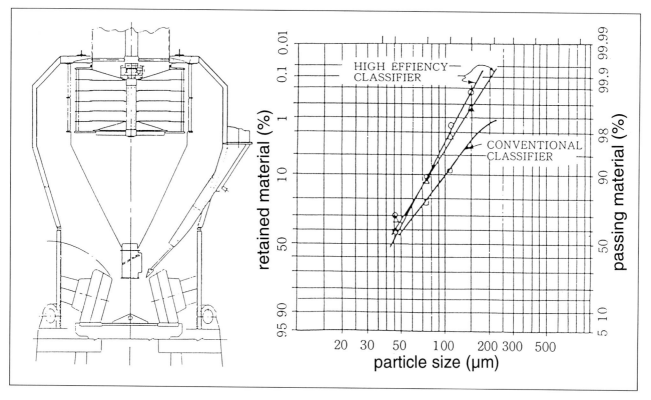

BILD 8: Wälzmühle mit Hochleistungssichter
FIGURE 8: Roller mill with high-efficiency separator

Rückstand	= retained material
Durchgang	= passing material
Korngröße	= particle size
Hochleistungssichter	= high-efficiency classifier
konventioneller Sichter	= conventional classifier

Ausbildung eines stabilen Mahlbetts auf dem Mahlteller verhindert und auf diese Weise die Zerkleinerungseffizienz herabgesetzt wurde.

— Die Feingutanreicherung im Mühlen-Sichter-Kreislauf verursacht einen Anstieg des Druckverlustes und damit eine Erhöhung des erforderlichen elektrischen Leistungsbedarfs des Prozeßventilators.

— Durch eine unvollständige Trennung der gröberen Partikelfraktionen wird die Brennbarkeit des Rohmehls häufig ungünstig beeinflußt oder die Verbrennungsgeschwindigkeit und der Ausbrand des erzeugten Brennstaubs herabgesetzt.

Alle diese aufgezeigten Probleme werden durch die Einführung des Hochleistungssichters [13] (**Bild 8**) erfolgreich gelöst. Ein weiteres Beispiel dafür sind die Ringkugelmühlen, die unter Einsatz von TSV- oder O-SEPA-Sichtern [14] (**Bild 9**) umgebaut werden. Sowohl eine Steigerung der Durchsatzleistung von 70% für den gleichen Siebrückstand R 0,1 mm als auch eine Verminderung der groben Korngrößenfraktionen des erzeugten Brennstaubs von R 0,1 mm = 10% auf 8% für die gleiche Durchsatzleistung gaben einen guten Hinweis auf die Möglichkeit zur Nutzung von niederflüchtigen Brennstoffen für den Vorcalcinatorbetrieb.

Der Einsatz von Wälzmühlen zur Klinkermahlung ließ ein weiteres Problem in der Leistungsfähigkeit des Sichters erkennen. Während einerseits die Sichtung möglichst vollständig sein soll, soll andererseits die Korngrößenverteilung der geforderten Zementqualität entsprechend angepaßt sein. Die O&K-Wälzmühle [15] löst dieses Problem dadurch, daß ein genügend hoher Anteil an feinen Partikelfraktionen im Produkt durch eine sorgfältige Wahl der Geometrie von Mahlteller und Mahlwalzen sowie der Betriebsbedingungen erzeugt wird. Gleichzeitig wird durch die hochwirksame Trennung vermieden, daß gröbere Partikel in das Feingut gelangen.

4. Separation efficiency — its assessment and influencing factors

Amoung various descriptions of the separation efficiency of industrial separators, the Tromp curve or the grade efficiency curve (**Fig. 10**) are the most descriptive and comprehensive. In the abscissa x denotes the particle size and the ordinate $\Phi(x)$ denotes the probability for being separated as fines. In actual cases, $\Phi(x)$ does not reach 100% even in the finest size region, and hence the by-pass model (**Fig. 11**) has been proposed. In this model a considerable amount of separator feed (β part) is assumed to by-pass the classification zone, going directly to the reject, without

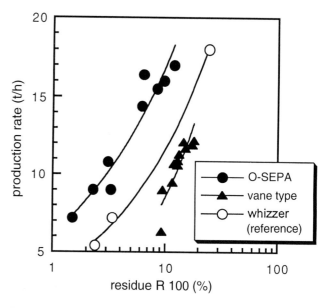

BILD 9: Anwendung von Hochleistungssichtern (O-SEPA-Sichter) in Mahlkreisläufen mit Ring-Kugelmühlen
FIGURE 9: Application of high-efficiency separators (O-SEPA) in ring-ball mills

Durchsatzleistung	= production rate
Rückstand R 0,1 mm	= residue R 100 (%)
O-SEPA	= O-SEPA
statischer Sichter mit Leitflügelsystem	= vane type
dynamischer Sichter mit Horizontalflügelsystem	= whizzer

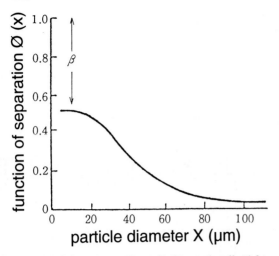

BILD 10: Beispiel einer Tromp-Kurve für konventionelle Sichter
FIGURE 10: Example of Tromp curve (Grade efficiency curve) in conventional separators
Trenngrad = function of separation
Partikeldurchmesser = particle diameter

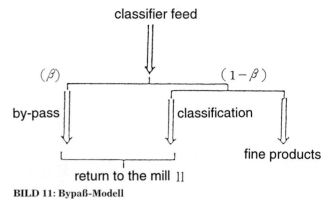

BILD 11: Bypaß-Modell
FIGURE 11: By-pass model
Sichteraufgabegut = classifier feed
Bypaß = by-pass
Sichtung = classification
Sichterfeingut = fine products
Rücklauf zur Mühle (Sichtergrobgut) = return to the mill

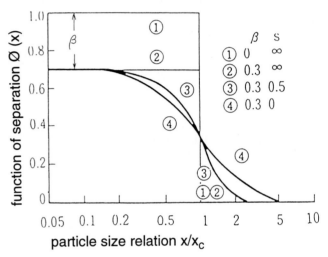

BILD 12: Wirkungsgradkurve; Modell zur Erläuterung der Trennschärfe
FIGURE 12: Grade efficiency curve models
Partikelgrößenverhältnis = particle size relation
Trenngrad $\Phi(x)$ = function of separation

4. Sichteffizienz — Ihre Bewertung und Einflußfaktoren

Unter den verschiedenen Darstellungen über die Sichteffizienz industrieller Sichter ist die Tromp-Kurve oder die Wirkungsgrad-Kurve (**Bild 10**) die anschaulichste und umfassendste. Im Bild 10 ist auf der Abszisse die Partikelgröße x und auf der Ordinate die Wahrscheinlichkeit Φ (x) für die Abscheidung als Feingut aufgetragen. Im praktischen Fall erreicht die Funktion Φ (x) selbst in dem Bereich feinster Korngrößen nicht den Wert von 100%. Daher wurde ein Bypass-Modell (**Bild 11**) vorgeschlagen. Bei diesem Modell wird angenommen, daß ein beträchtlicher Teil des Aufgabegutes (ungesichteter Anteil β) an der Trennzone vorbeigeführt wird und direkt ins Sichtergrobgut gelangt, ohne an der Trennung teilzunehmen. Dieser Modellvorstellung entsprechend kann die Wirkungsgrad-Kurve in zwei Bereiche eingeteilt werden. Der erste Bereich betrifft den Bypass, gekennzeichnet durch den Anteil β, und der zweite Bereich betrifft die Trennung ohne den Bypass, die durch die Trennkorngröße x_c und die Trennschärfe mit dem Index s (**Bild 12**) gekennzeichnet wird. Ergebnisse der Computersimulation [16] (**Bild 13**) zeigten, daß die Größe des Bypasses β bei Zementkreisläufen eine bedeutendere Rolle spielt als die Trennschärfe mit dem Index s, da bei großen Werten von β ein Übermahlen eintreten kann, wobei ein zu großer Anteil sehr feinen Mahlguts erzeugt wird und wodurch zuviel Energie verschwendet wird.

Es ist auch bekannt, daß der Logarithmus von $(1-\beta)$ in industriellen Sichtern [17] linear von der Feststoff/Luft-Konzentration Q_p/Q_a abhängt. Daraus entstand der Gedanke, die Arbeitsweise verschiedener Sichter durch die Lage ihres Betriebspunkts in der Kurve $(1-\beta)$ über Q_p/Q_a miteinander vergleichen zu können [18] (**Bild 14**). In dem Diagramm dieses Bildes liegen die schwarz ausgefüllten Meßpunkte von Hochleistungssichtern oberhalb der nicht ausgefüllten Meßpunkte von konventionellen Sichtern. Das bedeutet, daß für das gleiche Feststoff/Luft-Verhältnis Q_p/Q_a der ungesichtete Anteil β in konventionellen Sichtern größer ausfällt. Daraus geht hervor, daß die Trennung in konventionellen Sichtern mit einer geringeren Leistungsfähigkeit verbunden ist als die Trennung in Hochleistungssichtern.

Es muß darauf hingewiesen werden, daß auch in Hochleistungssichtern eine unzureichende Sichtluftmenge oder eine zu hohe Sichterbelastung ebenfalls zu einer geringeren Sichterleistung führen [19], jedoch bei viel größeren Verhältnissen von Q_p/Q_a als in konventionellen Sichtern. Im krassen Gegensatz zu konventionellen Sichtern hat sich der Einfluß der massebezogenen Oberfläche des Zements auf den ungesichteten Anteil β bei Hochleistungssichtern als vernachlässigbar klein erwiesen [20].

receiving any classification action. According to this assumption the grade efficiency curve can be divided into two parts, the first corresponding to the by-passing (characterized by β) and the second corresponding to the classification without by-passing (characterized by cut size x_c and sharpness index s) (**Fig. 12**). The computer simulation results [16] (**Fig. 13**) show that the ratio of by-passing β plays a more important roll than the sharpness index s in cement grinding circuits since overgrinding occurs at large β, which wastes much energy and produces too much waste superfine particles.

It is also known that the logarithm of $(1-\beta)$ depends linearly on the particle-air concentration (Q_p/Q_a) in industrial separators [17] and so there arises the idea that the separator performance could be compared among each other by the position of the performance point in the $(1-\beta)$ vs. Q_p/Q_a chart [18] (**Fig. 14**). In this chart, the black circles of high-efficiency separators lie more upwards than the white circles of conventional separators which shows that β is larger in conventional separators for the same Q_p/Q_a and that this kind of separation gives a poorer performance than high-efficiency separation.

It must be noted that even in high-efficiency separators an insufficient air flow-rate or a too high circulating load leads to poor separation performance [19], but at quite larger Q_p/Q_a than in conventional separators. The effect of specific surface area of cement on β has been shown to be almost negligible in high-efficiency separators [20] with a striking contrast to conventional separators.

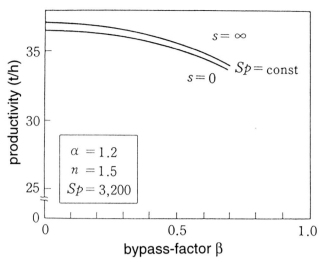

BILD 13: Ergebnisse der Computersimulation für Zement-Mahlkreisläufe mit Kugelmühle und Sichter
FIGURE 13: Simulation results of a cement grinding ball mill
Sp: Blaine (cm²/g)
Bypaß-Faktor = bypass factor
Durchsatzleistung = productivity

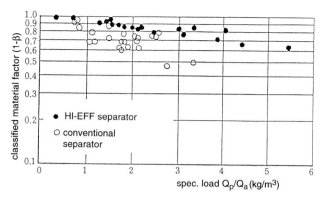

BILD 14: $(1-\beta)$ als Funktion der Feststoff/Luft-Beladung im Sichter
FIGURE 14: $(1-\beta)$ vs. Q_p/Q_a chart of separators
gesichteter Anteil = classified material factor
Feststoff/Luft-Beladung = spec. load
Hochleistungssichter = HI-EFF separator
konventioneller Sichter = conventional separator

5. Einfluß der Sichteffizienz auf den Energieverbrauch

Obwohl allgemein anerkannt wird, daß ein Hochleistungssichter die Leistungsfähigkeit von Mahlkreisläufen günstig beeinflußt und zu einer bedeutenden Energieeinsparung führt, scheint es sowohl über den Mechanismus als auch den Umfang möglicher Verbesserungen noch Fehlvorstellungen zu geben. Die Ursachen für einen Anstieg der Durchsatzleistung infolge einer Verbesserung der Sichteffizienz sind folgende:

— Infolge der verbesserten Trenneigenschaften wird der Feingutanteil, der mit dem Sichtergrobgut in die Mühle zurückgeführt wird, deutlich vermindert. Dadurch wird ein Übermahlen des Mahlguts vermieden und der Energieverbrauch der Mühle gesenkt. Dieser Mechanismus wird durch eine bessere Mühlenbelüftung und durch Vermindern der Mahlraumtemperatur unterstützt. Das kann man oft beim Einsatz moderner Hochleistungssichter beobachten.

— Durch Modifikation der Korngrößenverteilung ergibt sich eine mögliche Verbesserung der Zementqualität, welche die Möglichkeit eröffnet, die massebezogene Oberfläche des Zements zu vermindern.

Die Steigerung der Durchsatzleistung durch eine Verbesserung der Sichterarbeitsweise hängt ab von:

— der Abnahme des ungesichteten Anteils β infolge des verbesserten Sichterbetriebes,

— der Höhe der massebezogenen Oberfläche des Zements,

— der konstruktiven Gestaltung des Mühleninnenraums bzw. den Bedingungen für die Gewährleistung eines genügend hohen Mahlgutdurchflusses durch die Mühle,

— dem Spielraum, der zur Herstellung eines Zements mit engerer Korngrößenverteilung bei geringerer massebezogener Oberfläche unter Berücksichtigung der in der Norm festgelegten Zementeigenschaften verbleibt. Bei konstanter Betonkonsistenz kann der Wasseranspruch ansteigen, was oft zu einer verminderten Betonfestigkeit führen kann.

In bezug auf den ersten Punkt ist es vorstellbar, daß ein Zusammenhang zwischen der Durchsatzleistung und dem Bypass-Faktor β (**Bild 15**) besteht. Die Steigerung der Durchsatzleistung kann bei einem Sichterbetrieb mit hohem Faktor β beträchtlich sein, während sie sehr klein ausfällt, wenn der ungesichtete Anteil β vor Modifikation des Sichters schon nahe Null ist.

5. Influence of separation efficiency on energy consumption

Although it is generally accepted that high-efficiency separators lead to a good performance of grinding circuits and considerable saving of energy, there seems to remain some kinds of confusion as to the mechanism and the extent of improvement in such cases. The cause for a production-rate increase through the improvement of separator-efficiency can be analyzed as follows:

— Owing to the improved separation characteristics, the amount of fines returned to the mill with the rejects decreases considerably which avoids overgrinding and wasting of grinding energy at the mill. This mechanism is facilitated by the intensification of air-sweeping through the mill and the temperature reduction inside the mill which can be often observed when using modern high-efficiency separators.

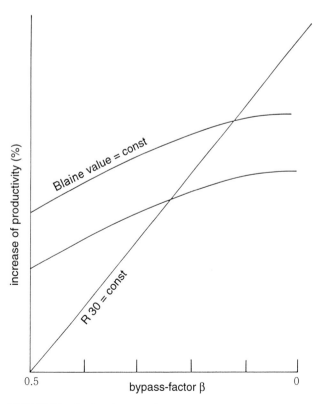

BILD 15: Durchsatzsteigerung als Funktion des Bypaßfaktors β
FIGURE 15: Productivity increase vs bypass factor β
Bypaß-Faktor = bypass factor
Steigerung der Durchsatzleistung = increase of productivity

In bezug auf den zweiten Punkt eröffnet auch eine hohe massebezogene Oberfläche des Zements die Möglichkeit, die Durchsatzleistung zu steigern, da die Gefahr eines stärkeren Übermahlens besteht, besonders wenn keine Mahlhilfsmittel eingesetzt werden.

Der dritte Punkt ist von Bedeutung, wenn die Gestaltung des Mühleninnenraums schon vor Modifikation des Sichters auf einen besonders niedrigen Mahlgutumlaufmassenstrom abgestimmt ist. In einem solchen Fall sollten die Mühleneinbauten wie Trennwände und Stauringe usw. verändert werden, um einen höheren Umlaufmassenstrom zu erreichen, der für eine Verbesserung der Produktivität der Mühle unerläßlich ist. Dies ist besonders dann wichtig, wenn keine Mahlhilfsmittel verwendet werden.

Die nach Einführung von Hochleistungssichtern erreichten guten Betriebsergebnisse, über die viele Autoren [5, 8, 9, 21–29] berichteten, beruhen auf sorgfältigen Untersuchungen. Sie sind zusammengefaßt in **Bild 16** und **Tabelle 1** [28] dargestellt. Tabelle 1 enthält einen Vergleich der Betriebsergebnisse von drei Mahlkreisläufen, von denen jeder mit dem gleichen Mühlentyp, aber mit einem anderen Sichtertyp ausgerüstet ist. **Tabelle 2** [30] zeigt anhand von Betriebsergebnissen, wie durch Austausch eines konventionellen Sichters gegen einen Hochleistungssichter der massebezogene Energieaufwand des Mahlkreislaufs verbessert werden konnte. Alle in der Tabelle dargestellten Ergebnisse zeigen, daß die Durchsatzleistung zwischen 12 und 43 % zunahm und der massebezogene Energieverbrauch der Mahlkreisläufe zwischen 6 und 21 % gesenkt werden konnte.

6. Einfluß der Sichteffizienz auf die Zementeigenschaften

Die Korngrößenverteilung von Zement muß folgende Anforderungen erfüllen:

— Sie soll frei von Überkorngrößen > 90 μm sein,

— sie soll einen möglichst großen Anteil an Korngrößen im Bereich von 3 bis 30 μm haben,

— sie soll nur einen mäßigen Anteil an Korngrößen < 3 μm enthalten.

Durch den Einsatz von Hochleistungssichtern können die ersten beiden Anforderungen leicht erfüllt werden [31] (**Tabelle 3**). Die letztgenannte Anforderung bezieht sich auf das Problem der Verarbeitbarkeit und/oder der Frühfestigkeitseigenschaften. Sie kann auf verschiedene Weise erfüllt werden. Eine Möglichkeit besteht in der Einstellung der Sichter auf einen bestimmten ungesichteten Anteil β, was zu

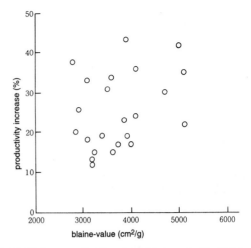

BILD 16: Steigerung der Durchsatzleistung in Hochleistungssichtern

FIGURE 16: Productivity increase in high-efficiency separators

— By virtue of the improvement in cement quality, which is caused by the modification of particle size distribution, the possibility for reducing the specific surface area of cement arises.

Accepting this as a fundamental mechanism, the amount of production-rate increase through the improvement of separators can be considered as dependent from the following factors:

— The amount of decrease of the by-pass factor β through the improvement of separators,

— the present level of the specific surface area of cement,

— whether the mill-inside structures or conditions are appropriate for passing enough amount of particles through the mill or not,

— the room remaining for a sharper particle size distribution at lower specific surface area of cement from the standpoint of standard cement properties. But in the case of constant concrete consistence the water demand may increase, often leading to a diminished concrete strength.

In relation to the first point, a certain kind of relationship could be imagined to exist between the production rate and the by-pass factor β (**Fig. 15**). The production-rate increase would be considerable in high β regions, while the production-rate increase would be very small when β is already rather close to zero before modification of separators.

TABELLE 1: Vergleich der Betriebsdaten von 3 Mahlkreisläufen mit unterschiedlichen Sichtern

Mahlkreislauf Nr.	Verwendeter Sichter		Durchsatz (t/h)	Spez. Energieverbrauch (kWh/t)	Feinheit		
	Typ	Anzahl			Blaine (cm²/g)	Rückstand (%)	
						30 μm	88 μm
11	O-SEPA	1	105	34,8	3240	20	0,2
12	Konventionell	2	85	42,6	3400	18	1,2
13	Konventionell	1	98	38,0	3300	18	0,6

Anmerkung: Mahlhilfen 0,015 (%)

TABLE 1: Comparison of operation data of 3 mill circuits with different types of separators

Circuit No.	Classifiers used		Production rate (t/h)	Specific power consumption (kWh/t)	Fineness		
	Type	Number of equipment			Blaine (cm²/g)	Residue (%)	
						30 μm	88 μm
11	O-SEPA	1	105	34.8	3240	20	0.2
12	Conventional	2	85	42.6	3400	18	1.2
13	Conventional	1	98	38.0	3300	18	0.6

Note: Grinding aids 0.015 (%)

TABELLE 2: Spezifischer Energiebedarf eines O-SEPA-Sichters im Zementwerk Fujiwara

Sichtertyp	Blaine (cm²/g)	Spez. Energieverbrauch (kWh/t)				
		Mühle	Sichter	Becherwerk	Filterventilator	Gesamt
Konventionell*)	3400***)	36,9	4,0	0,6	1,7	43,2
O-SEPA**)	3240***)	31,0	0,7	0,5	2,6	34,8

Anmerkung: Kugelmühle ⌀ 4,3 m × 12,5 m, 3300 kW
 *) Zwei Stück
 **) N-2500, System II, 2000 m³/min.
 ***) Die Zementqualität entspricht derjenigen mit der verbesserten Korngrößenverteilung des O-SEPA-Sichters.

TABLE 2: Power consumption in O-SEPA system in Fujiwara plant

Type of classifier	Blaine (cm²/g)	Spec. energy consumption (kWh/t)				
		Mill	Classifier	Bucket elevator	Bag filter fan	Total
Conventional*)	3400***)	36.9	4.0	0.6	1.7	43.2
O-SEPA**)	3240***)	31.0	0.7	0.5	2.6	34.8

Note: ball mill ⌀ 4.3 m × 12.5 m, 3000 kW
 *) Two sets
 **) N-2500, System II, 2000 m³/min.
 ***) Cement quality is on the same level due to the improved particle size distribution by O-SEPA.

TABELLE 3: Korngrößenverteilung vor und nach der Umstellung auf einen Hochleistungssichter
TABLE 3: Particle size distribution before and after conversion to high-efficiency separators

Plant		PSD*) slope R–R	Blaine (cm²/g)	3–32 μm (%)	−10 μm (%)	−3 μm (%)
A	before	1.01	3562	66.4	41.6	16.8
	after	1.24	3333	74.8	40.4	17.4
B	before	1.15	3510	71.8	39.7	14.9
	after	1.27	3176	70.8	34.2	13.9
C	before	1.07	3538	71.1	39.2	13.5
	after	1.14	3403	69.0	44.8	21.9
D	before	1.03	3879	69.7	45.0	18.1
	after	1.14	3357	70.7	38.2	15.0
E	before	1.01	3900	67.9	43.0	18.6
	after	1.14	4005	69.0	44.8	21.9
F	before	1.03	3796	68.9	42.7	17.0
	after	1.16	3749	71.0	43.0	19.9
G	before	1.16	3588	72.0	42.6	16.9
	after	1.29	3613	78.5	42.4	16.5
H	before	0.95	3276	62.9	35.0	14.0
	after	1.03	3265	64.0	35.8	13.2

*) PSD means: Particle size distribution

Plant = Anlage
PSD*) slope = Steigungsmaß der RRSB-Korngrößenverteilung
*) PSD bedeutet: Korngrößenverteilung
Blaine = Blaine
before = vorher
after = nachher

einer breiteren Korngrößenverteilung führt. Empfehlenswerter dürfte aber die Modifikation des Mahlkreislaufs bzw. eine Veränderung der Mahlraumeinbauten sein.

Ein weiterer Einfluß auf die Zementeigenschaften, der auf die Einführung der Hochleistungssichter zurückzuführen ist, besteht in der Verminderung der Mahlraumtemperatur. In konventionellen Mahlkreisläufen waren die Mahlraumtemperaturen im allgemeinen zu hoch, so daß verschiedene Kühleinrichtungen in den Mahlkreisläufen installiert wurden. Beim Einsatz moderner Hochleistungssichter kann ein großes Kühlluftvolumen sehr wirkungsvoll in den Mahlkreislauf eingeleitet werden, so daß das Problem der Zementkühlung [30, 32] nicht mehr besteht (**Tabelle 4**). Jedoch hat in einigen Fällen eine zu intensive Kühlung eine unvollständige Gipsdehydratation bewirkt. Entsprechend den chemischen und mineralogischen Eigenschaften des Zements führte das zu schnellem Ansteifen des Zements oder zu Konsistenzproblemen, wenn nur Gips und Anhydrit zur Sulfatoptimierung des Zements verwendet wurden [33].

In relation to the second point, a high specific surface area of cement promises the possibility of an increase in the production rate. This is due to the possibility of severer overgrinding in the mill, especially when no grinding aids are used.

The third point is important when the mill-inside conditions before the modification of separators were arranged to fit for exceptionally low level of circulating load. In such a case, mill-inside conditions, i.e. screening plates, retainer rings, etc., should be modified to fit for ordinary level of circulating load, which is essential to the productivity increase of the mill. This is especially important when no grinding aids are used. The successful operational results, caused by the introduction of high-efficiency separators reported by many authors [5, 8, 9, 21–29], are the results of careful examinations. They are summarized in **Fig. 16** and **Table 1** [28]. Table 1 shows the comparison of operating results of three mill circuits, each equipped with different types of separators for the same design of mills. **Table 2** [30] contains the operation data showing how the replacement of a conventional separator by a high-efficiency separator improved the energy consumption of the grinding circuit. All results show that the production rate increased by between 12 and 43 %, and the specific power consumption of the grinding circuits decreased by between 6 and 21 %.

6. Influence of separation efficiency on cement properties

The most desirable particle size distribution of cement has to meet the following requirements:

— Absence of oversize (> 90 μm),

— rich in 3 to 30 μm fraction,

— presence of moderate amount of minus 3 μm fraction.

By using high-efficiency separators the first two requirements can be easily satisfied [31] (**Table 3**). The last item of the above mentioned requirements which is related to the workability problem and/or very early-age hardening properties, can be met by several ways. One would be the adjustment of the β-value of separators, leading to a wider particle size distribution, but more advisable would be the modification of the grinding circuit design, i.e. the modification of mill-inside conditions.

One more influence which arises from the introduction of high-efficiency separators on the properties of cement would be the reduction of mill-inside temperature. In conventional milling circuits mill-inside temperatures were generally too high, and several types of cooling procedures were introduced in milling circuits. In modern type high-efficiency separators a large amount of cooling air can be

TABELLE 4: Senkung der Fertiguttemperatur durch Einsatz eines O-SEPA-Sichters im Zementwerk Ofunato
TABLE 4: Temperature reduction by O-SEPA system Ofunato plant

Type of cement	Type of classifier	Blaine (cm²/g)	Temperature (°C)	
			Mill outlet	Cement
High-early-strength Portland cement	Conventional*)	4100	135	125
	O-SEPA**)	3800	80	55
Ordinary Portland cement	Conventional*)	3100	130	120
	O-SEPA**)	2900	75	50

Note: (1) Water spray 0.1 %
(2) Clinker temperature 80 °C
*) One cyclone air-separator in two mills
**) N-1000, System II, max. 1000 m³/min

Type of cement	= Zementart
Type of classifier	= Sichtertyp
Blaine	= Blaine
Temperature	= Temperatur
Mill outlet	= Mühlenaustragsgut
Cement	= Zement
High-early-strength Portland cement	= Frühhochfester Portlandzement
Ordinary Portland cement	= herkömmlicher Portlandzement
Conventional*)	= konventionell*)
O-SEPA**)	= O-SEPA**)
Note: (1) Water spray 0.1 %	= Anmerkung: (1) eingespritztes Wasser 0,1 M.-%
(2) Clinker temperature 80 °C	(2) Klinkertemperatur 80 °C

*) Ein Zyklon-Umluftsichter für beide Mühlen
**) N-1000, System II, max. 1000 m³/min

In solchen Fällen kann das Temperaturprofil im Mahlkreislauf leicht durch Abgasrückführung beeinflußt werden, besonders dann, wenn nur Gips als Erstarrungsverzögerer zur Verfügung steht [19]. Allgemein können beim Einsatz von Hochleistungssichtern [8, 31] sowohl die Zement- als auch die Betoneigenschaften konstant gehalten oder sogar leicht verbessert werden (**Tabelle 5, Bild 17**). Jedoch hängen die Produkteigenschaften nicht nur von der Sichter-Arbeitsweise ab. Vor allem die Klinkerreaktivität und auch die Optimierung des Calciumsulfatzusatzes müssen dabei sorgfältig betrachtet werden.

7. Aktuelle Probleme und zukünftige Entwicklungen

Seit der Einführung des Hochleistungssichters mit einfacher Luftführung (single-pass) ist die Gestaltung des Mahl-

efficiently introduced into the circuit, and the cooling problem [30, 32] has disappeared (**Table 4**).

In some cases, however, too much cooling caused a very incomplete dehydration of gypsum. This provoked abnormal stiffening of cement or consistency problems according to the chemical and mineralogical character of cement, if only gypsum and anhydrite are employed in an optimized mixture [33]. In such a case, the temperature profile in the grinding circuit can be easily modified through the recirculation of exhaust gas into the circuit, even if only gypsum is available as retarder [19]. As a general result cement properties and also concrete properties can be kept constant or slightly improved by the introduction of high-efficiency separators (**Fig. 17, Table 5**) [8, 31] but this does not only depend on separator operation. Clinker reactivity and calcium sulfate optimization should be thoroughly observed.

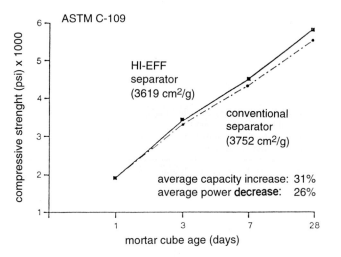

BILD 17: Entwicklung der Normdruckfestigkeit für unterschiedliche Sichtersysteme
FIGURE 17: Mortar compressive strength and type of separators

Alter des Prüfkörpers (in Tagen)	= mortar cube age (days)
Normdruckfestigkeit	= compressive strength
Hochleistungssichter	= HI-EFF separator
konventioneller Sichter	= conventional separator
mittl. Steigerung der Durchsatzleistung	= average capacity increase
durchschnittl. Abnahme des Energieverbrauchs	= average power decrease

TABELLE 5: Betoneigenschaften und Sichtertyp

Zementart	PZ 35F		PZ 45F		PZ 45F HS C₃A-frei	
Sichtertyp	SEPOL	konventionell	SEPOL	konventionell	SEPOL	konventionell
Feinheit cm²/g	2808	2835	3498	3521	3976	4000
Normdruckfestigkeit						
7 Tage N/mm²	33	28	42	38	40	35
28 Tage N/mm²	42	42	54	50	49	49
Frischbeton Ausbreitmaß a_0-a_{60} cm	8	10	7	10	8	8

TABLE 5: Standard concrete properties and type of separators

Cement type	PC 35F		PC 45F		PC 45F HS C₃A-free	
Separator type	SEPOL	conventional sep.	SEPOL	conventional sep.	SEPOL	conventional sep.
Fineness cm²/g	2808	2835	3498	3521	3976	4000
Compressive strength						
7 days N/mm²	33	28	42	38	40	35
28 days N/mm²	42	42	54	50	49	49
Fresh concrete Slump a_0-a_{60} cm	8	10	7	10	8	8

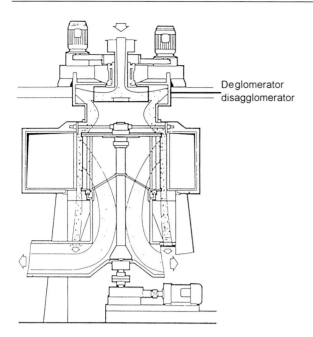

BILD 18: SEPOL-IP-Sichter mit Deglomerator
FIGURE 18: SEPOL-IP separator with disagglomerator

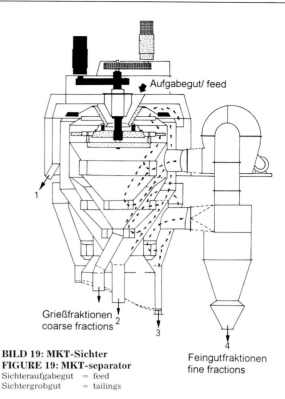

BILD 19: MKT-Sichter
FIGURE 19: MKT-separator
Sichteraufgabegut = feed
Sichtergrobgut = tailings

kreislaufs erheblich einfacher geworden, weil er nur noch aus einer Mühle und einem Sichter mit einem oder meist zwei Staubfiltern [32] besteht. Die kompakte Ausführung des Hochleistungssichters macht es möglich, die Produkte aus drei verschiedenen Mühlen ohne Schwierigkeit nur einem Sichter zuzuführen [30]. Alle diese Entwicklungen haben im hohem Maße dazu beigetragen, daß die Installations- und Betriebskosten von Mahlkreisläufen vermindert werden konnten. Dies wird auch durch die Entwicklung von Sichtern mit integrierten Deglomeratoren unterstrichen, die von mehreren Herstellern in Mahlkreisläufen mit Gutbett-Walzenmühlen eingesetzt werden [35]. Diese Sichter (**Bild 18**) erfüllen mehrere verfahrenstechnische Funktionen und vereinfachen die Mahlkreislaufkonzepte in erheblichem Maß. Eine ganz neue Entwicklung ist die Einführung von neuen Ausführungsformen von Hochleistungssichtern, die besonders für die Sichtung von sehr feinen bzw. ultrafeinen Materialien geeignet sind. Der CLASSIEL-Sichter von Onoda [36] ist solch ein Beispiel, der seine Anwendung bei der Herstellung von sehr feinen Zementen mit einer maximalen Partikelgröße von $10\,\mu m$ oder $5\,\mu m$ in verhältnismäßig großem Maßstab hat. Die Verarbeitung von Flugasche zu einem Produkt mit einer begrenzten Korngrößenverteilung [37] könnte ein weiteres Anwendungsziel des Hochleistungssichters sein. Schließlich sollte in diesem Zusammenhang erwähnt werden, daß einige Sichterkonstruktionen [38] in der Lage sind, gleichzeitig drei oder mehr verschiedene Produktqualitäten zu erzeugen (**Bild 19**). Ein Sichter mit diesen Eigenschaften kann dann besonders nützlich sein, wenn Produkte mit sehr begrenzten bzw. ungewöhnlichen Korngrößenverteilungen benötigt werden.

7. Recent topics and future developments

Since the introduction of so called "single pass" high-efficiency separators, the grinding circuit design has been so much simplified because the circuit needs only one mill and one separator with one or at most two dust collectors [32].

The compactness of high-efficiency separators made it possible to treat the products from three different mills in one separator without much difficulty [30]. All these measures contributed to a high extent to the reduction of installation and operation costs of grinding circuits, and this was again emphasized by the recent introduction of disagglomerator-separators in the roller press circuits by several manufacturers [35]. In these new cases, disagglomerator-separators (**Fig. 18**) play multi-functional roles, which simplifies the grinding circuit design to a great extent. One more topic on high-efficiency separators would be the introduction of their new versions especially suited for the classification of very fine or ultra-fine materials. CLASSIEL by Onoda [36] is such an example, seeking its application for manufacturing microfine cements under a maximum particle size of $10\,\mu m$ or $5\,\mu m$ in fairly large scale. Treatment of fly-ash to deliver a product with considerably limited particle size range [37] would be another target of the application. It should be mentioned finally that some separators [38] have the capability of producing three or more kinds of products at the same time (**Fig. 19**). This would be valuable when products with very limited size range or products with very unusual size distributions are needed.

Literature

[1] Herrmann, C.: Increased Cement grinding efficiency by using high efficiency separators. IEEE Cement Ind. Tech. Conf. 27 (1985).

[2] Quittkat, W.: Der Strahlwindsichter, ein leistungsfähiger, raumsparender Sichter für Feinsttrennung. Zement-Kalk-Gips 26 (1973) No. 7, pp. 326–330.

[3] Hukki, R. T.: Zweistufige Windsichtung im geschlossenen Mahlkreislauf. Zement-Kalk-Gips 30 (1977) No. 5, pp. 199–205.

[4] Knobloch, O., Müller, M., and Eickholt, H.: Entwicklungsstand von Streuteller- und Kanalradsichtern. Zement-Kalk-Gips 31 (1978) No. 8, pp. 413–417.

[5] Furukawa, T., Onuma, E., and Misaka, T.: A new large-scale air classifier O-SEPA – Its principle and operating characteristics. Proc. Int. Symp. on Powder Technology '81 (Kyoto), pp. 750–757.

[6] Cleemann, J. O.: Evaluation of the new high efficient air separators. Zement-Kalk-Gips 39 (1986) No. 6, pp. 295–304.

[7] Klumpar, I. V., Saverse, R. R., Currier, F. N., and Slavsky, S. T.: Air classifier with optimum design and operation. Zement-Kalk-Gips 39 (1986) No. 6, pp. 305–311.

[8] Schmidt, D.: Hochleistungs-Sichter SEPOL – Erfahrungen und Betriebsergebnisse im Zementwerk Hardegsen. Zement-Kalk-Gips 41 (1988) No. 10, pp. 506–510.

[9] Binder, U.: Der O & K-Querstromsichter – Entwicklung und Betriebsergebnisse. Zement-Kalk-Gips 41 (1988) No. 5, pp. 237–242.

[10] Binder, U.: The O & K cross stream separator – development and operating results. World Cement 22 (1991) No. 11, pp. 18–23.

[11] Marchal, G.: The utilisation of high efficiency classifiers in raw grinding. Ciments, Bétons, Plâtres, Chaux (1989) No. 781, pp. 370–376.

[12] Bales, P.: New raw grinding plant with integral roller press at the Hualien cement works in Taiwan. Zement-Kalk-Gips 46 (1993) No. 2, pp. 71–76.

[13] Schonbach, B. H.: High efficiency separators in roller mills. World Cement 19 (1988) No. 11, pp. 436–444.

[14] Ito, M., Miyabe, Y., Takeuchi, M., and Furukawa, T.: Development of "E-SEPA" air classifier for solid fuels. The Cement Manufacturing Technology Symp. (JAPAN) 50 (1993) pp. 66–71.

[15] Shimojima, K., Hamaguchi, M., Obana, H., and Fukuyama, K.: Newly developed roller mill for cement clinker grinding. World Cement 15 (1984) No. 9, pp. 230–232.

[16] Onuma, E., Asai, N., and Jimbo, G.: Analysis of the operating characteristics of steady-state closed-circuit ball mill grinding. 4th European Symp. on Comminution (1975) pp. 559–573.

[17] Onuma, E.: An analysis of the fractional recovery curves of Sturtevant-type air classifiers in high solid concentration conditions. J. Chem. Eng. Japan 6 (1973) No. 6, pp. 527–531.

[18] Onuma, E., and Furukawa, T.: On the criteria for the assessment of performance quality of air classifiers in closed-circuit tube-mill systems. Proc. Int. Symp. on Powder Technology '81 (Kyoto), pp. 412–419.

[19] Dauphinee, E., and Breitschmid, K.: Problems with high efficiency separators. Ciments, Bétons, Plâtres, Chaux (1992) No. 795, pp. 98–102.

[20] Ito, M., and Sota, Y.: Characteristics and application of high efficiency vortex-type classifier "O-SEPA". J. Res. Onoda Cement Company 40 (1988) No. 1, pp. 18–28.

[21] Onuma, E., Furukawa, T., and Fukuyama, K.: The effect of classifier performance on the energy consumption of closed-circuit ball-mill grinding system. J. Res. Onoda Cement Company 34 (1982) No. 2, pp. 56–63.

[22] Knoflicek, M. J.: Betriebserfahrungen mit O-Sepa-Windsichtern in Nordamerika. Zement-Kalk-Gips 39 (1986) No. 6, pp. 335–336.

[23] Clarke, M. B.: Progress report on continuous fracture mill installed at Glens Falls Portland Cement Co. I. C. S. Proceedings 22 (1986), pp. 115–125.

[24] Bernutat, P., and Schroter, H.: Erste Betriebsergebnisse mit einem neuartigen Sichter in Zementmahlanlagen. Zement-Kalk-Gips 43 (1990) No. 4, pp. 192–194.

[25] Bouquelle, J.F.: Modernisation of Ciments d'Obourg's cement grinding plant no. 2. World Cement 20 (1989) No. 4, pp. 115–118.

[26] Henz, F.: Upgrading of a finish mill circuit with a high efficiency single pass separator. I. C. S. Proceedings 21 (1985), pp. 244–262.

[27] Eickholt, H.: Influence of separators for cost-efficient grinding plants. Ciments, Bétons, Plâtres, Chaux (1991) No. 788, pp. 40–47.

[28] Onuma, E.: A new high-efficiency classifier as applied to the cement industry. I. C. S. Proceedings 19 (1983), pp. 40–43.

[29] Toyooka, S., Ohya, S., and Ogawa, K.: Cement mill performance with use of high efficiency separator O-SEPA. The Cement Manufacturing Technology Symp. (JAPAN) 45 (1988), pp. 31–36.

[30] Onuma, E., and Furukawa, T.: O-SEPA – A new high-performance air-classifier. World Cement 15 (1984) No. 1/2, pp. 13–24.

[31] Brugan, J. M.: High efficiency separators – Problems and solutions. Zement-Kalk-Gips 41 (1988) No. 7, pp. 350–355.

[32] Ito, M., Misaka, T., Furukawa, T., Sota, Y., and Onuma, E.: Cooling effect of the O-Sepa air separator in cement grinding. Zement-Kalk-Gips 41 (1988) No. 5, pp. 214–223.

[33] Sumner, M. S., Hepher, N. M., and Moir, G. K.: The influence of a narrow cement particle size distribution on cement paste and concrete water demand. Ciments, Bétons, Plâtres, Chaux (1989) No. 778, pp. 164–168.

[34] Folsberg, J.: A new generation of high efficiency separators for ball mills and roller presses. Zement-Kalk-Gips 44 (1991) No. 1, pp. 37–41.

[35] Disagglomeration and classification with high-efficiency separator. Ciments, Bétons, Plâtres, Chaux (1990) No. 782, p. 59.

[36] Tamashige, T., Ninomiya, H., Fujii, S., and Furukawa, T.: Performance of a newly designed air classifier. 1992 IEEE Cement Ind. Tech. Conf. 34 (1992), pp. 426–439.

[37] Tamashige, T., Kondou, A., Nakamura, S., Takayama, A., and Kouno, I.: Operating results of a newly designed air classifier for fly ash. The Cement Manufacturing Technology Symp. (JAPAN) 49 (1992), pp. 52–57.

[38] Leistungssteigerung durch mehrstufige Feingutrennung mit MKT-Sichtern. Zement-Kalk-Gips 39 (1986) No. 6, pp. 343–344.

Energiesparpotentiale durch optimierte Regelung der Schüttgutflüsse an Kugelmühlen und Gutbett-Walzenmühlen

Potential for saving energy through optimized control of bulk material flows at ball mills and high-pressure grinding rolls

Potentiels d'économie d'énergie dans la régulation optimale des courants de matière sur broyeurs à boulets et broyeurs à rouleaux

Potenciales de ahorro de energía por medio de la regulación optimizada de los flujos de material a granel en los molinos de bolas y los molinos de cilindros y de lecho de material

Von **B. Allenberg,** Darmstadt/Deutschland

Zusammenfassung – *Im Gegensatz zu den zentralisierten Leitsystemen der 80er Jahre sind moderne Leitsysteme modular und dezentral aufgebaut. Dabei werden an einen Leitrechner selbständig arbeitende Teilsysteme angeschlossen, die Anlagenfahrer und Leitsystemzentrale entlasten. Durch die Entwicklung der Mikroelektronik werden Rechenleistungen auch in dezentralen Einheiten verfügbar und ermöglichen dort die Einrichtung neuer selbstoptimierender Regelalgorithmen. Die Schüttgutdosierung stellt im Mahlprozeß ein solches Teilsystem dar, das sich optimal zur Automatisierung eignet. Während die übergeordnete Steuerung und die Bedienung Bestandteile des Zentralsystems bleiben, umfaßt das Teilsystem neben der Steuerung der Dosiereinrichtungen verschiedene Regelschleifen über einen Gruppenregler, der außer den Signalen der Frischgutwaagen die der Rückgutwaage und ggf. anderer Hilfsgrößen verarbeitet. Als solche können z.B. die elektrische Leistungsaufnahme des Becherwerks oder der Schalldruckpegel an der Grobmahlkammer dienen. Kaskadenregler für den Aufgabegutmassenstrom einer Kugelmühle ermöglichen einen stabilen Mahlgutfüllungsgrad als Grundlage für eine Maximierung der Fertiggutleistung und damit eine Einsparung an Energie, bezogen auf die produzierte Fertiggutmenge. Auch die Schüttgutflüsse zur Gutbett-Walzenmühle werden entsprechend geregelt, um kritische Betriebszustände auszuschließen und gleichzeitig den Bauraum für den Vorbunker zu minimieren.*

Summary – *Unlike the centralized control systems of the 80s, modern control systems have modular decentralized structures. Independently operating sub-systems which take the load off the plant operator and the control system centres are connected to a control computer. The development of micro-electronics means that computing capacity is also available in decentralized units where it is possible to draw up new self-optimizing control algorithms. In the grinding process the metered feeding of bulk material represents one such sub-system which is highly suitable for automation. The higher-level control system and the operation remain components of the central system, but the sub-system covers not only the control of the metering equipment but also various control loops through a group controller. This processes the signals from the fresh material weighers as well as those from the return material weigher and, where appropriate, other auxiliary variables. These could be, for example, the electrical power consumption of the bucket elevator, or the noise level in the coarse grinding chambers. Cascade controllers for the mass flow of feed material to a ball mill make it possible to achieve a stable material filling ratio as the basis for maximizing the finished product output, thereby saving energy relative to the quantity of finished product produced. The flows of bulk material to highpressure grinding rolls are also controlled in a corresponding manner to eliminate critical operating conditions and at the same time to minimize the space required for the feed hopper.*

Résumé – *Contrairement aux systèmes de conduite centralisés des années 80, les systèmes de conduite modernes ont une conception modulaire et décentralisée. Dans cette conception, des systèmes partiels travaillant indépendamment et assistant les conducteurs des équipments et la centrale du système de conduite, sont connectés à un ordinateur central. Les développements de la microélectronique permettent de disposer aussi de capacités de traitement dans les unités décentralisées et y autorisent l'implantation de nouveaux algorithmes de régulation auto-optimisants. Dans le process du broyage, le dosage des matières en vrac constitue ainsi un système partiel se prêtant de manière optimale à l'automatisation. Alors que le contrôle supérieur et la conduite restent parts entières du système central, le système partiel regroupe, outre la conduite*

des dispositifs de dosage, différentes boucles de régulation au moyen d'un régulateur de groupage qui traite, en plus des signaux des bascules de matière fraiche, ceux des bascules de recirculation et, le cas échéant, d'autres grandeurs utiles. Comme telles, peuvent servir la puissance électrique absorbée par la noria ou le niveau d'intensité de bruit au compartiment de broyage primaire. Des régulateurs à cascade pour le courant massique de matière d'alimentation d'un broyeur à boulets apportent un taux de remplissage stable de matière à broyer, comme base pour atteindre le maximum de débit de produit fini et, ainsi, une économie d'énergie en fonction de la quantité de matière finie produite. Les courants de matière vers le broyeur à rouleaux sont aussi régulés en conséquence, afin d'éviter des états de fonctionnement critiques et de réduire en même temps l'encombrement de la trémie d'alimentation.

Resumen – *Contrariamente a los sistemas de control centralizados de los años 80, los sistemas de control modernos tienen una estructura modular y descentralizada. A un ordenador de control se conectan otros sistemas parciales, de funcionamiento independiente, que alivian el trabajo del operador y de la central del sistema de control. Gracias al desarrollo de la microelectrónica, las capacidades de tratamiento del ordenador son disponibles también en las unidades descentralizadas, permitiendo el establecimiento de nuevos algoritmos de control autooptimizantes. La dosificación del material a granel constituye uno de estos sistemas parciales en el proceso de molienda, el cual se presta muy bien para la automatización. Mientras que el mando de orden superior así como la operación siguen siendo componentes del sistema central, el sistema parcial abarca, además del mando de los dispositivos de dosificación, diferentes bucles de regulación a través de un regulador de grupo, que trata no solamente las señales de las básculas para materias primas de primer ingreso en el circuito, sino también las de la báscula para material de retorno y, eventualmente, otras magnitudes auxiliares. Estas pueden ser, por ejemplo, la potencia eléctrica absorbida por el elevador de cangilones o el nivel de presión sonora en la cámara de molienda gruesa. Los reguladores en cascada para el flujo de masas de material de alimentación de un molino de bolas permiten obtener un grado de llenado de material estable, como base de una maximización del caudal de producto acabado, y con ello un ahorro de energía, referido a la cantidad de producto acabado obtenido. También los flujos de material a granel hacia el molino de cilindros y de lecho de material son regulados, con el fin de excluir estados de servicio críticos y minimizar al mismo tiempo el espacio requerido para la tolva de alimentación.*

Potenciales de ahorro de energía por medio de la regulación optimizada de los flujos de material a granel en los molinos de bolas y los molinos de cilindros y de lecho de material

1. Die Schüttgutdosierung als autarkes Subsystem

In der Prozeßleittechnik ist ein Trend hin zum dezentralisierten Automatisierungssystem zu verzeichnen. **Bild 1** zeigt den typischen hierarchischen Aufbau [1]. Ziel des Entwurfs sind Ebenen selbständiger Einheiten, entwurfstechnisch auch als Objekte bezeichnet, die ihre Aufgaben möglichst selbstverantwortlich ausführen und deren Schnittstellen zu den anderen Partnern der gleichen Ebene weitgehend reduziert sind.

Bei der Automatisierung von Mühlen im Zementbereich bietet sich die Schüttgutdosierung als ein solches Objekt an. Es ist der Wunsch des Anlagenfahrers, ein vollautomatisch arbeitendes System zu erhalten, bei dem er die Anteile der Rohstoffe, die Feinheit des Endproduktes und Start/Stop-Befehle vorgibt. Eine auf der Basis der MULTICONT-Elektronik realisierte Dosiergruppe fährt dabei automatisch und schnell vom Stillstand in den Optimalzustand des Mühlenkreislaufs (**Bild 2**).

Mit einer Unterbedienebene versehen kann das Subsystem sogar dann betrieben werden, wenn andere Teile der Anlagenautomatisierung nicht oder noch nicht in Betrieb sind. Dies sichert eine höhere Ausnutzung der getätigten Investition insbesondere bei überschrittenen Lieferzeiten anderer Komponenten oder unerwartet langer Inbetriebnahmedauer.

Das Dosier-Subsystem stellt das Rechenleistungsäquivalent mehrerer moderner PCs zur Verfügung und entlastet übergeordnete Automatisierungsebenen durch Übernahme rechenintensiver, zeitkritischer Aufgaben und eine konsequente Verdichtung der Prozeßdaten der Gruppe. Die Kommunikation der Komponenten des Subsystems untereinander kann dabei völlig separat und auf die Regelungsaufgabe hin optimiert ausgelegt werden. Die Anbindung an die übergeordnete Ebene erfolgt über leistungsfähige und standardisierte Kopplungen.

2. Konzepte zur Aufgaberegelung bei Kugelmühlen

Die von der Dosierung zu beeinflussenden Bewertungskriterien für den Mahlprozeß sind die präzise Einhaltung der

1. Metered feeding of bulk material as a self-contained subsystem

In process control technology there is a tendency to decentralise the automation system. **Fig. 1** shows the typical hierarchic structure [1]. It consists of levels of self-contained units, also described as "entities", which carry out their tasks as self-sufficiently as possible and have a substantially reduced number of interfaces with other units at the same level.

In the automation of mills in the cement industry metered feeding of bulk material is one such object. The plant operator wants to have a fully automatic system where he can control the proportions of raw materials, the fineness of the end product and the start/stop commands. A metered feeding group based on MULTICONT electronics moves quickly and automatically from a standstill to the optimum state of the mill recirculating system (**Fig. 2**).

The subsystem has a lower operating level and can work even when other parts of the plant automating system are

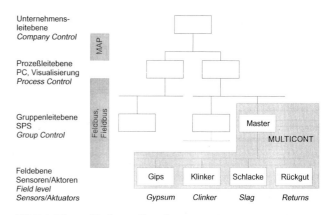

BILD 1: Hierarchischer Aufbau der Unternehmensautomatisierung
FIGURE 1: Hierarchical structure of the company automation system

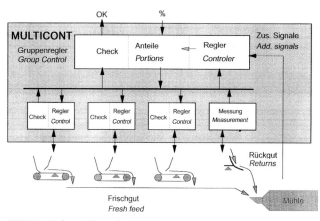

BILD 2: Sicherstellung der Fertiggutzusammensetzung
FIGURE 2: Safeguarding the finished product composition

gewählten Anteile der Komponenten, eine stabile Regelung der Umlaufmenge im Mühlenkreislauf und die Maximierung des Fertiggutstroms bei vorgegebener Feinheit.

Dosierbandwaagen auf der Basis einer ausgereiften Mechanik und der digitalen Elektronik MULTICONT mit BIC (Belt Influence Compensation) garantieren höchste Genauigkeit der einzelnen Dosierkomponente [2]. Die Waage als selbständiges Objekt arbeitet eigenverantwortlich und meldet ungewöhnliche Zustände automatisch. Dem Dosier-Subsystem werden nur noch die Anteile der Schüttgutkomponenten vorgegeben, wobei es die Einhaltung dieser Vorgaben selbst überwacht. Dadurch ist sichergestellt, daß die Zusammensetzung stets den Erwartungen entspricht.

2.1 Regelungskonzepte für den Schüttgutumlauf

Je nach den bauseitigen Gegebenheiten steht eine Teilmenge der in **Bild 3** dargestellten Signale aus dem Prozeß zur Regelung zur Verfügung. Der Regelkreis hat allgemein eine sehr langsame Dynamik, so daß die manuelle Einstellung der Regler ein hohes Maß an Zeit und Erfahrung erfordert.

In der Praxis bewährt sich die Beschränkung der Signale auf die Fördergrößen von Frischgut und Rückgut, die ohnehin zur Verfügung stehen oder mit geringem Aufwand nachgerüstet werden können. Zusatzsignale aus akustischen Aufnehmern an verschiedenen Mahlkammern („elektrisches Ohr") können die Regeldynamik beschleunigen. Oft sind jedoch Beeinflussungen des Aufnehmers durch die Arbeit anderer Mühlen im Umfeld nicht zu vernachlässigen. In diesen Fällen kann das Signal des „elektrischen Ohrs" unzuverlässig sein. Das Signal der Becherwerksleistung sollte nur dann zur Regelung herangezogen werden, wenn ein stabiler und möglichst proportionaler Zusammenhang zwischen dem geförderten Massestrom und der Leistungsaufnahme besteht. Bei Überdimensionierung des Becherwerkantriebs ist dies oft nicht der Fall. Die Erfassung des Fertiggutstroms führt nach derzeitigem Forschungsstand meist nicht zu einer signifikanten Verbesserung der Regelgüte.

Der Regler liefert die Stellgröße für die Frischgutgruppe und hat Kaskadenstruktur, um das unterschiedliche Zeitverhalten der Signale optimal zu einer schnellen Regelung auszunutzen. Die einzelnen, auf der Basis der bekannten PID-Struktur realisierten Regler sind besonders einfach und mit wenigen Einstellern in Betrieb zu nehmen.

Aus der Vergangenheit sind verschiedene Regeldirektiven bekannt. Für Umlaufmahlanlagen haben Untersuchungen den besonders stabilen Betrieb einer Regelung auf konstante, von der Vorgabe der Zementsorte abhängige Rückgutmengen erwiesen. Diese Struktur ist in Bild 3 durchgezogen dargestellt. Regelungstechnisch stellt das Verfahren eine Verallgemeinerung des Prinzips einer konstanten Mühlengesamtaufgabe „Frischgut + Rückgut = konst." dar.

Die Vorteile des vorgestellten Regelungskonzepts konnten in vielen Applikationen belegt werden, so z.B. an der

not, or not yet, in operation. This ensures better utilisation of the capital outlay, especially when other components have not met their delivery dates or when commissioning takes an unexpectedly long time.

The metered feeding subsystem provides computing power equivalent to several modern PCs and relieves the automation levels above it by taking on computationally intensive, real-time tasks and consistent compression of the group's processing data. Communication between components of the subsystem may be designed quite separately, and optimized for the purposes of the control function. Efficient, standardised connections form the link with the higher order level.

2. Designs for controlling ball mill feed

The criteria for assessing the grinding process which are affected by the feed are: precise maintenance of the selected proportions of constituents, stable control of the quantity recycled in the mill recirculating system, and maximization of the flow of finished material with the prescribed fineness.

Weigh-belt feeders based on perfected mechanical systems and MULTICONT digital electronics with BIC (belt influence compensation) guarantee maximum accuracy in the metering of individual constituents [2]. As a self-contained entity the weigher is self-optimising and automatically signals unusual states. The metering subsystem only has the proportions of bulk material constituents specified for it, and it monitors this function itself. This guarantees that the expected composition will always be produced.

2.1 Designs for controlling circulation of bulk material

Depending on the type of construction, some of the signals from the process, shown in **Figure 3**, will be available for control purposes. The control loop generally has a very slow dynamic response, so manual adjustment of the controller takes time and experience.

In practice it pays to limit the signals to the quantities of fresh feed and recycled material, since these are either available anyway or can be obtained at low cost. Additional signals from acoustic sensors to various grinding chambers ("electric ears") may speed up the control dynamics. But other mills operating in the neighbourhood often have effects on the sensor which cannot be ignored, and the signal from the "electric ear" may be unreliable in such cases. The signal for bucket elevator output should only be used for control purposes when there is a stable and as far as possible proportional relationship between the conveyed flow of material and the power consumption. This is often not the case when the elevator drive is too powerful. Research carried out so far indicates that the quality of control is not significantly improved by detecting the flow of finished product.

The controller provides the actuating variable for the fresh feed group and has a cascade structure, so that the different timing of the signals can be used to the optimum for a quick

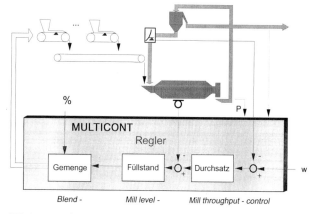

BILD 3: Regelung einer Kugelumlaufmühle
FIGURE 3: Control system for a closed-circuit ball mill

Zementmühle 2 der Firma Ssang-Yong, Korea, Dong-Hae-Plant, an der eine Durchsatzerhöhung von 2,5 % gegenüber einem erfahrenen Mühlenfahrer im Garantielauf nachgewiesen wurde. Die automatisch geregelte Betriebsweise führte zu einer wesentlichen Beruhigung des Mühlenkreises und damit letztendlich zu einer gleichmäßigen Qualität des Endproduktes.

Die Anfahrphasen der Schüttgutflüsse in der Mühle steuert ein Expertensystem (**Bild 4**), in dem die Struktur der Wissensbasis weitgehend vorkonfiguriert ist und das damit eine schnelle und sichere Inbetriebnahme ermöglicht. Der Anfahrvorgang basiert dabei auf einer „intelligenten Zustandsmaschine", wobei die Übergänge der Zustände der Steuerung aus den gemessenen Prozeßzuständen abgeleitet werden.

2.2 On-line-Adaption

Die Wahl des Rückgutsollwerts wird durch die Direktive eines möglichst großen Fertigproduktausstoßes bestimmt. Diese Betriebsweise minimiert die pro Menge Fertigprodukt verbrauchte Energie.

Der optimale Füllstand der Mühle liegt genau dann vor, wenn der Fertigutausstoß im stationären Zustand maximal wird (**Bild 5**). Dosier-Subsysteme auf der Basis MULTICONT beinhalten einen Algorithmus, der im laufenden Betrieb, basierend auf der dabei bestimmten Mühlenkennlinie, den Optimalpunkt durch Veränderung des Sollwerts in Schritten sucht, wobei die Schrittweite an den Fortschritt des Optimierprozesses angeglichen ist. Verschleiß der Mahlkugeln oder Veränderungen der Mahlbarkeit der Rohstoffe werden dabei automatisch mitberücksichtigt. Das adaptive Verfahren entlastet den Mühlenfahrer entscheidend und sichert auch bei weniger geschultem Personal einen optimierten Betrieb.

3. Schüttgutdosierung für Gutbett-Walzenmühlen (Rollenpressen)

In den letzten Jahren kommen vermehrt Rollenpressen allein (Fertigvermahlung) oder in Verbindung mit Kugelmühlen (Vor- oder Hybridvermahlung) zum Einsatz. Das Potential zur Energieeinsparung wurde mehrfach belegt, z. B. [3].

Um ein besseres Einzugsverhalten der Rollenpresse zu erreichen, muß dem Frischgut der Pressenaufgabe ein Anteil Feinmaterial in Form von Schülpen, Grießen oder von beiden gemeinsam beigemischt werden. **Bild 6** zeigt das Beispiel einer Hybridvermahlung.

Rollenpressen benötigen stets einen Schüttgutpuffer im Aufgabebereich, wobei der Wunsch nach möglichst kleinen Behältern aus der Forderung geringer Investitionskosten resultiert. Andererseits kann der Materialeinzug der Presse bei konstanter Drehzahl stark schwanken. Ein Regler für den Füllstand des verwogenen Behälters soll die Durchsatzschwankungen durch Variation der Schülpen oder Grießemenge ausgleichen und so einen gleichmäßigen Puffer auf dem Einzug der Presse gewährleisten. Die Frischgutmenge

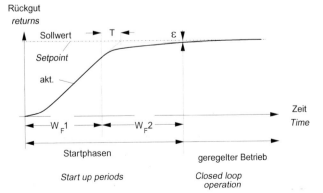

BILD 4: Beispiel eines Anfahrvorgangs
FIGURE 4: Example of a start-up process
akt. = current value

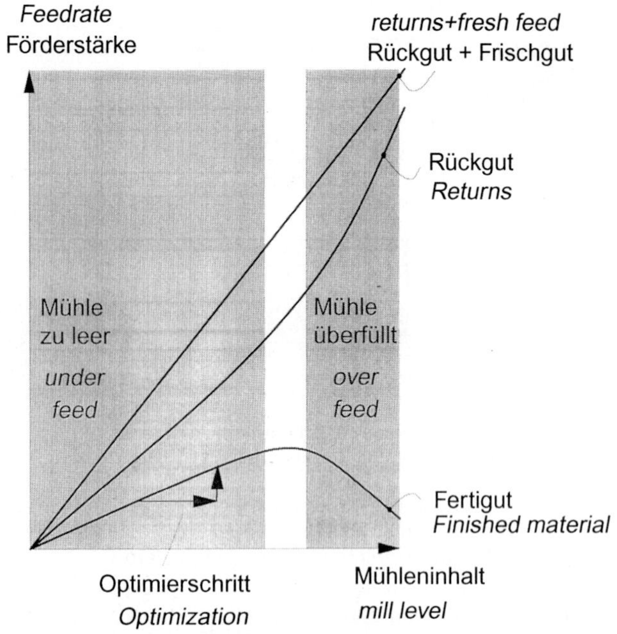

BILD 5: Kennlinie einer Kugelumlaufmühle im stationären Zustand
FIGURE 5: Characteristic curve for a closed-circuit ball mill in steady-state operation

control action. The individual controls based on the well-known PID structure are particularly simple to operate and need very few setting devices.

Various control directives are known from the past. Studies of closed-circuit grinding plants have shown that a control system giving a constant amount of recycled material (the quantity depends on the type of cement specified) is particularly stable in operation. This structure is shown by continuous lines in Figure 3. In terms of control technology the method is a generalisation of the principle of constant total mill feed: "fresh feed + recycled material = const.".

The advantages of the control plan described above have been demonstrated in many applications, e. g. at cement mill 2 of Ssang-Yong's Dong-Hae plant in Korea, where it gave a throughput during the guarantee run which was 2.5 % higher than that achieved by an experienced mill operator. The automatic control of the operation considerably calmed the mill circuit resulting ultimately in an end product of uniform quality.

The start-up periods of bulk material flows in the mill are controlled by an expert system (**Figure 4**), in which the structure of the knowledge base is largely preconfigured, allowing the system to be commissioned quickly and reliably. The start-up operation is based on an "intelligent status machine", with the transitions from one control state to another being derived from the measured states of the process.

2.2 On-line adaptation

The setpoint for the recycled material is selected to achieve the highest possible output of finished product. This mode of operation minimises the energy consumed per quantity of finished product.

The mill is at its optimum filling level precisely when there is maximum output of finished product in steady-state operation (**Figure 5**). Metered feeding subsystems based on MULTICONT contain an algorithm which seeks the optimum point during operation, on the basis of the characteristic curve of the mill determined during the process. It does this by changing the setpoint in steps, in which the width of the step matches the progress of the optimisation process. Wear on the grinding balls or variations in the grindability of the raw materials are automatically taken into account. The adaptive method relieves the mill operator of a considerable burden and ensures optimum operation even with less skilled personnel.

STECKERT AUFBEREITUNGSTECHNIK

DAVID-5-DECK-SIEBMASCHINE

- lineare Schwingung
- hohe Durchsatzleistung
- hohe Trennschärfe bei mehreren Kornfraktionen
- verstopfungsfreie Absiebung
- einfacher Siebbelagwechsel
- geringe Wartung
- Siebbeheizung zur Absiebung von feuchtem Material möglich!

IHR PARTNER SEIT 1965

MARTIN STECKERT
ANLAGEN- UND VERSCHLEISSTECHNIK GMBH

HOHENBRUNNER STR. 59 · D-85521 OTTOBRUNN · TEL. (089) 6 09 80 01-02 · FAX (089) 6 09 80 03

STECKERT AUFBEREITUNGSTECHNIK

MASCHINEN-CHARAKTERISTIK: Das Prinzip sowie der Aufbau der David-5-Deck-Siebmaschine sind vollkommen anders als bei herkömmlichen Siebmaschinen. Die hochfrequenten gerichteten Schwingungen werden durch 2 Stück gegenläufig laufenden Unwuchtmotoren erzeugt. Die Trennung des Aufgabegutes erfolgt durch eine wiederholte Absiebung durch 5 Stück übereinander angeordnete Siebdecken. Die Neigungen dieser Siebdecken nehmen von oben nach unten zu. Dadurch ist die Maschenweite der Siebe in jedem Fall größer als die Größe des gewünschten Endkorns. Aufgrund dieser Tatsache kann mit dem David-5-Deck-Sieb eine wesentlich größere Fertiggutleistung gegenüber herkömmlichen Siebmaschinen erzielt werden. Das Siebgehäuse ist spannungsfrei geglüht.

DAVID-5-DECK-SIEBMASCHINE

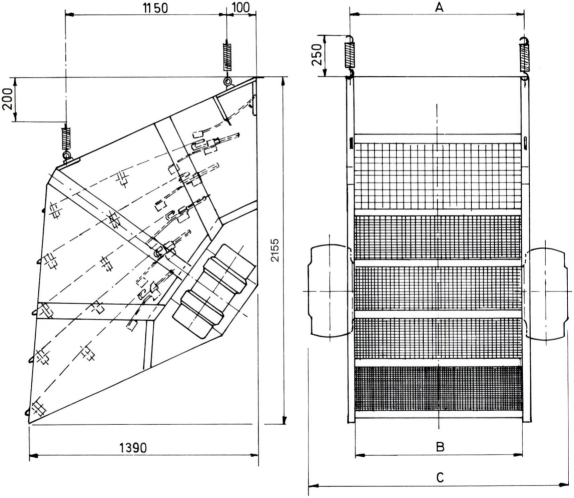

Typ	Leistung to/h 2 mm	Leistung to/h 20 mm	Antriebsleistung KW	Gewicht kp	A	B	C
750	14	80	3,0	890	800	750	1300
1000	20	120	4,0	1080	1050	1000	1600
1250	27	160	4,0	1340	1300	1250	1860
1500	35	200	7,0	1570	1550	1500	2280
1750	40	240	8,0	1710	1800	1750	2550

Techn. Änderungen vorbehalten

DER SPEZIALIST FÜR HORIZONTAL-BRECHER - SYSTEM DAVID

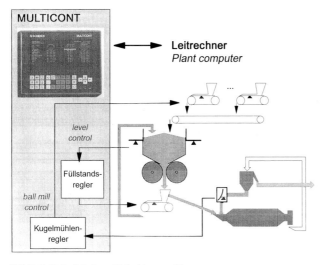

BILD 6: Beispiel einer Hybridvermahlung
FIGURE 6: Example of a hybrid grinding system

3. Metered feeding of bulk material for high-pressure grinding rolls (roller presses)

High-pressure grinding rolls have been used increasingly over recent years, either alone (for finish grinding) or in conjunction with ball mills (for preliminary or hybrid grinding). Their energy-saving potential has been demonstrated in many cases, e. g. [3].

To obtain a better intake behaviour for the high-pressure grinding rolls a proportion of fine material in the form of compacted cake or tailings, or a mixture of the two, must be blended with the fresh feed. **Figure 6** shows an example of a hybrid grinding system.

High-pressure grinding rolls always require a buffer of bulk material in the feed area, and the desire to use the smallest possible hoppers results from the need to keep down investment costs. On the other hand the amount of material drawn into the rolls may fluctuate greatly at constant rotational speed. A device controlling the filling level of the weighed container has to compensate for the fluctuations in throughput by varying the amount of compacted cake or tailings, and thus ensure a uniform buffer at the intake to the rolls. The amount of fresh feed entering the rolls is specified by the ball mill control system described above. The output from the high-pressure grinding rolls circuit follows this signal with a time lag which is negligible compared to the dynamic response of the ball mill.

With the conveyor belts running at a constant speed, the lag time between the metered feeding point and the hopper limits the minimum permissible volume of the feed hopper for the high-pressure grinding rolls. The hopper should hold at least twice the amount which would result from the expected change in throughput of the high-pressure grinding rolls for the duration of the lag. If this condition is not met then consideration should be given to installing variable speed belts which can control the feed with virtually no lag time.

Metered feeding systems based on MULTICONT with an integral group control level contain not only the control system for the flow of bulk material to the ball mill, as described above, but also the system for controlling the flow of compacted cake and grits to the high-pressure grinding rolls. Special start-up circuits for the flows of bulk material increase operating reliability and allow a quick transition to stable operation. The effectiveness of the system has already been demonstrated in over 100 grinding plants.

4. Prospects

The idea of decentralisation is gaining more and more ground in automation technology. The foundation for this is now being laid by general and open communication schemes such as PROFIBUS/MAP. "Intelligent" functions are increasingly moving to the periphery, where adequate computing power must be provided.

Continuous weighing and metered feeding technology meets the requirements by a control concept for the flows of bulk materials to the mills in the form of a package. Even now the modular MULTICONT system offers a basis for later changes in control technology and new installations. The functionality of the system is being extended to cater for quality assurance requirements. Diagnostic functions covering even the details of weighing and convenient methods of parameterisation on PCs will further shorten the time taken up in maintenance [4].

Literaturverzeichnis

[1] Hils, F., und Linder, K.-P.: PROFIBUS – Der Feldbus für die Verfahrenstechnik wird erwachsen, In: atp 34 12, Oldenbourg Verlag 1992, S. 661–667.

[2] Allenberg, B., und Jost, G.: Kurzzeitgenauigkeit und Betriebssicherheit bei der Schüttgutdosierung, In: wägen + dosieren, 22. Jg., 3/92, S. 12–17.

[3] Strasser, S., und Wolter, A.: Zukunftspotentiale der Mahltechnik mit der Rollenpresse, In: ZKG Nr. 7/1991, S. 345–356.

[4] Allenberg, B.: Rationalisierte Instandhaltung und Qualitätssicherung am Beispiel der kontinuierlichen Schüttgutdosierung, Kongreßband zum VDZ Kongreß 27. 9.–1. 10. 93.

Energieeinsparung bei Mahlsystemen
Energy economics in grinding systems

Economie d'énergie dans les systèmes de broyage

Ahorro de energía en sistemas de molienda

Von **R. Anantharaman,** New Delhi/Indien

Zusammenfassung – Die allein in Mahlsystemen verbrauchte elektrische Energie macht fast zwei Drittel der gesamten im Zementwerk verbrauchten elektrischen Energie aus. Der Zerkleinerungseffekt in einer Kugelmühle ist auf die kombinierte Wirkung von Stoß-, Druck- und Scherkräften, die durch den Aufprall der Mahlkörper erzeugt werden, zurückzuführen. Beim Mahlen auf höhere Feinheit und der dadurch bedingten Vergrößerung der Oberfläche insgesamt bedarf es vor allem der Scherkräfte und kleinerer Mahlkugeln sowie eines optimalen Größenverhältnisses zwischen Mahlkugeln und Mahlgut. Nach dem gegenwärtigen Wissensstand gibt es keine einzige Theorie, mit der das Bruchverhalten eines einzelnen Partikels unter einfacher Belastung auch nur einigermaßen genau vorausgesagt werden könnte. Deshalb ist es wichtig zu wissen, nach welchen Prinzipien die Zerkleinerung vorgenommen werden soll, um Methoden für einen wirtschaftlicheren Umgang mit Energie vorschlagen zu können. Die für den Bruch der Partikel erforderliche Energie ist nicht gleichmäßig im Volumen verteilt. Es gibt vielmehr Bereiche mit Energiekonzentrationen aufgrund mikroskopisch kleiner Risse, die zur Spannungserhöhung beitragen. Solche Defekte vermindern die Festigkeit der Bindung innerhalb eines Partikels ganz erheblich. Deshalb liegt die Vermutung nahe, daß die Mahlbarkeit mit zunehmender Partikelgröße besser wird, da die Wahrscheinlichkeit des Vorhandenseins derartiger Schwachstellen steigt. Der für die Mahlung theoretisch errechnete Energiebedarf (bei optimaler Energienutzung ohne Verlust) beträgt demnach 9 bis 10 kWh/t, ein Wert, der deutlich unter den heutigen Verbrauchswerten liegt.

Summary – The electrical power consumed in the grinding section alone accounts for nearly two thirds of the total electric power used in the cement industry. Size reduction inside the ball mill is due to the combined effect of impact, compression and shear forces imparted by the falling of grinding media. As to the grinding progress towards the finer range, shear forces predominate in generating additional surface area requiring smaller grinding media as well as optimum ball to particle size ratio. The present state of knowledge is such that there is no single theory by means of which the behaviour during fracture of even a single particle under the simplest mode of loading can be predicted with moderate accuracy. A thorough understanding of the principle underlying the art of comminution is therefore very essential to suggest methods for improving energy economics. The energy required for fracturing particles is not uniformly distributed over the volume and there are regions of energy concentration produced by microscopic cracks which act as stress raisers. The presence of these imperfections greatly reduces the bond strength of the particles. It is therefore logical to expect that the grindability will become better with an increase in the dimension of particles as the probability of having weak spots is higher. The energy required for grinding on the basis of theoretical calculation (optimum utilisation without wastage of power) is found to be 9 to 10 kWh/t, which is considerably below the present consumption level.

Résumé – Les systèmes de broyage absorbent à eux seuls près des deux tiers de l'énergie électrique consommée par une cimenterie. L'effet de broyage dans un broyeur à boulets est dû à une action combinée de forces de poussée, de pression et de cisaillement, engendrées par l'impact des corps broyants. Un broyage plus fin et l'augmentation de la surface totale en résultant nécessitent surtout des forces de cisaillement, de petits boulets broyants ainsi que des proportions optimales entre les boulets broyant et la matière à broyer. D'après les connaissances actuelles il n'existe pas une seule théorie capable de prévoir tant soit peu exactement le phénomène de broyage d'une seule particule dans des conditions de charge simple. C'est la raison pour laquelle il importe de savoir d'après quels principes s'effectue le broyage pour pouvoir proposer des méthodes d'économie d'énergie satisfaisantes. L'énergie nécessaire au broyage des particules n'est pas répartie uniformément en volume. Il existe au contraire des zones de concentration d'énergie basées sur de petites fissures microscopiques, qui entraînent une augmentation de tension. De pareils défauts diminuent considérablement la résistance de la liaison à l'intérieur de la particule. C'est pourquoi on peut supposer que la broyabilité s'améliore au fur et à mesure de l'augmentation de la taille des particules du fait que la probabilité d'existence de ces points faibles croît. Les besoins énergétiques calculés théoriquement pour le broyage (pour une utilisation d'énergie optimale sans perte) sont, de ce fait, de 9 à 10 kWh/t, valeur qui se situe nettement en-dessous des valeurs de consommation actuelles.

Ahorro de energía en sistemas de molienda

Resumen – *La energía eléctrica consumida tan sólo en los sistemas de molienda supone casi dos terceras partes de toda la energía consumida en una fábrica de cemento. El efecto de desmenuzamiento dentro de un molino de bolas se debe a la acción combinada de fuerzas de choque, de compresión y de cizallamiento, producidas por el impacto de los cuerpos moledores. Una molienda más fina y, con ello, el aumento de la superficie total requieren, sobre todo, fuerzas de cizallamiento y bolas de molienda más pequeñas así como una relación óptima entre el tamaño de las bolas de molienda y el material a moler. De acuerdo con el estado actual de los conocimientos, no existe ninguna teoría que permita predecir, con suficiente aproximación, el comportamiento a la ruptura de una partícula individual normalmente cargada. Por esta razón, es importante saber según qué principios debe efectuarse el desmenuzamiento, con el fin de poder proponer métodos que permitan un empleo más racional de la energía. La energía necesaria para la ruptura de la partícula no está repartida uniformemente dentro de su volumen. Existen, más bien, zonas de concentración de energía, debido a grietas microscópicas, que contribuyen al aumento de las tensiones. Estos defectos reducen considerablemente la cohesión dentro de la partícula. Por eso se supone que a medida que aumenta el tamaño de la partícula, mejora su molturabilidad, puesto que es muy probable que aumente el número de esos puntos débiles. La demanda teórica de energía necesaria para la molienda (con un aprovechamiento óptimo de la energía, sin pérdidas) asciende a 9 – 10 kWh/t, un valor notablemente inferior al nivel de consumo actual.*

Fast zwei Drittel des gesamten elektrischen Energieverbrauchs in der Zementindustrie entfallen auf das Mahlen. Da es an einem umfassenden Verständnis der dabei ablaufenden Vorgänge fehlt, beruhen nur wenige Teilbereiche der herkömmlichen Zerkleinerungstheorien auf theoretischen Betrachtungen. Ohne fundierte theoretische Grundlage lassen sich deshalb nur schwer Leistungsziele festlegen, da die Verbesserungen des Betriebs von Mahlanlagen rein empirisch erfolgen. Obwohl die Brech- und Mahlanlagen einen hohen Entwicklungsstand erreicht haben, blieben entsprechende theoretische Fortschritte aus. Dies könnte an Meinungsverschiedenheiten über die Hypothesen liegen, die von verschiedenen an diesem Thema arbeitenden Autoren vorgetragen wurden.

Die Bond-Theorie betrifft die Beziehung zwischen der Energiezufuhr und der Partikelgröße von aus bestimmten Ausgangsprodukten hergestellten Erzeugnissen. Sie beruht auf der Annahme, daß die Energiezufuhr der Länge der bei einem Zerbrechen von Teilchen entstehenden Risse proportional ist. Die Rißansatzlänge wird der Quadratwurzel aus der Hälfte der neuen Oberfläche gleichgesetzt, und die neue Rißlänge ist proportional

$$\frac{1}{\sqrt{d_1}} - \frac{1}{\sqrt{d_2}} \qquad (1)$$

wobei d_1 und d_2 die Durchmesser in μm für einen Siebdurchgang von 80 % des Produkts (Index 1) bzw. des Aufgabeguts (Index 2) sind. Hieraus ermittelte Bond die für die Zerkleinerung erforderliche Energiezufuhr (in kWh/t) empirisch:

$$W = 10 \left(\frac{W_i}{\sqrt{d_1}} - \frac{W_i}{\sqrt{d_2}} \right) \qquad (2)$$

W_i ist der sogenannte „Arbeitsindex", der von den physikalischen Eigenschaften des zerkleinerten Stoffes abhängt.

Rittinger betrachtete die Zerkleinerung von Würfeln und postulierte, daß die erforderliche Energiezufuhr der neu geschaffenen Oberfläche proportional ist. Obwohl sich diese Beziehung als nützlich erwiesen hat, muß darauf hingewiesen werden, daß sie nur die Grobzerkleinerung zu erklären vermag. Entsprechend der Kick-Theorie wird angenommen, daß die Energiezufuhr der Volumen- oder Gewichtsverminderung der Teilchen proportional ist. Daraus ergibt sich, daß die Energie bei einer Verformung unter Zug oder Druck absorbiert wird.

Keine der oben erwähnten Zerkleinerungstheorien entspricht den Ergebnissen der industriell betriebenen Brech- und Mahlanlagen. In diesem Artikel wird ein umfassender Ansatz auf der Grundlage theoretischer Untersuchungen angeregt, der versucht, die verschiedenen früher vorgeschlagenen Theorien miteinander zu verbinden. Bekanntermaßen wird nur ein sehr kleiner Teil der Zerkleinerungsenergie tatsächlich für das Mahlen verbraucht. Der größte

Nearly two thirds of total electrical power consumption in the cement industry are used for grinding. Due to a lack of complete understanding of the processes involved only few parts of the traditional comminution theories are based on theoretical considerations. Therefore, without a sincere theoretical basis, it is very difficult to set targets of performance because improvements in operation of grinding equipment are mainly based on empiricism. Although crushing and grinding machinery had been brought to a high standard of mechanical development, corresponding advances in theory had not been made. A possible reason for that may be due to the disagreement in the hypothesises proposed by various authors working on this subject.

Bond's theory concerns with a relationship between energy input and particle size of product made from a given feed. It is based on the assumption that energy input is proportional to the tip length of the cracks produced in particle breakage. The crack tip length is taken equivalent to the square root of one half of the new surface area and the new crack length is proportional to

$$\frac{1}{\sqrt{d_1}} - \frac{1}{\sqrt{d_2}} \qquad (1)$$

where d_1 and d_2 are the diameters for 80 % passing, notated in μm, of the product (index 1) and the feed (index 2) respectively.

From that Bond related the necessary energy input for comminution measured in kWh/t empirically.

$$W = 10 \left(\frac{W_i}{\sqrt{d_1}} - \frac{W_i}{\sqrt{d_2}} \right) \qquad (2)$$

W_i is the so called "Work Index", which depends on physical properties of the comminuted material.

Rittinger dealing with the comminution of cubes postulated that the necessary energy input is proportional to the new created surface. Although his relationship has been found useful, it has to be pointed out that it is able to explain coarse comminution only. According to Kick's theory the energy input is supposed to be proportional to the reduction in volume or weight of particles. This implies that the energy is absorbed in deformation under tension or compression.

All above mentioned traditional communition theories do not agree with results taken from industrial scale crushing and grinding plants. In this paper a comprehensive approach based on theoretical investigations is suggested, which tries to reconcile the various theories proposed earlier. It is well known that only a very small amount of the energy used in comminution is actually required for grinding. Most of it is converted into useless heat. The energy required for breakage varies greatly with the rate at which stress is applied to the particle. A light high velocity load is

Teil wird in nutzlose Wärme verwandelt. Die erforderliche Bruchenergie ist stark von der Geschwindigkeit abhängig, mit der das Teilchen beansprucht wird. Eine leichte Belastung mit hoher Geschwindigkeit übt eine stärkere Zerkleinerungswirkung aus als eine vergleichbare Belastung mit verhältnismäßig niedriger Geschwindigkeit.

Der größte Teil der Zerkleinerungsenergie wird für die plastische Verformung des Materials verwendet. Diese Energie ist um ein Vielfaches größer als die benötigte Zerkleinerungsenergie. Das bedeutet, daß der Energieaufwand für die Zerkleinerung des Gesteins im wesentlichen dem zur Verformung des Materials über die kritische Spannung hinaus und zur Bildung von Rißansätzen entspricht. Diese gespeicherte Belastungsenergie wird als Wärmeenergie freigesetzt.

Ein Hauptmerkmal der Gesteinsmassen ist ihre inhomogene Struktur. Die Segregation von Bindemitteln zwischen den Körnern, Mikrorissen und intergranularen Poren bewirken eine Verformung, die zu einer Spannungskonzentration und Bruchausbreitung führt. Da das Feststoffvolumen zwischen diesen Fehlstellen gering ist, bleibt auch die kritische Bruchspannungsenergie niedrig. Der Bruchwiderstand hängt von der Zeit und der Geschwindigkeit ab, mit der die Belastung aufgebracht wird. Der Bruch beginnt an Fehlstellen und verläuft mit einem schnellen Energiefluß in die Risse hinein, die sich aus dem Feld der freigesetzten Spannungen bilden. Nur eine kleine Energiemenge wird dann noch für die Bildung der neuen Oberfläche benötigt. Diese Prozesse können mit Verfahren der computergestützten Finite-Elemente-Methode (FEM) beurteilt werden. Der Riß wird instabil und neigt zu spontaner Beschleunigung. Steht ausreichend Energie zur Verfügung, nähert sich die Rißfortpflanzungsgeschwindigkeit der Schallgeschwindigkeit, und es kommt zu einer Verzweigung.

Finite-Elemente-Methode

Die Finite-Elemente-Methode wurde zur Behandlung von Fällen gewählt, die sich nicht auf dem klassischen Wege lösen lassen. Außerdem kann man sich in diesem Fall von der Annahme eines elastischen Kontinuums lösen und Veränderungen einführen, mit denen man den geometrischen und mechanischen Eigenschaften einer typischen inhomogenen Gesteinsmasse nahekommt.

Wird das Kontinuum in ein System von Elementen mit jeweils gleichförmiger Belastung unterteilt, so wird die Formulierung einer klassischen Grenzwertaufgabe überflüssig. Das Feststoffteilchen im Gleichgewichtszustand existiert als Mikrokontinuum. Wird das Gleichgewicht durch Erhöhung der auf das Teilchen aufgebrachten Belastung gestört, wird das Mikrokontinuum elastisch bis zum Bruch verformt. Finite Verschiebungen herrschen in dem Anpassungsprozeß bald vor und bilden eventuell ein globales Diskontinuum. Jedes Differentialelement unterliegt den drei Spannungsarten σ_x, σ_y und τ_{xy}, die bei dem Zerkleinerungsvorgang auftreten. Die entsprechenden Dehnungen ϵ_x, ϵ_y und v_{xy} sind bis zum Bruch kompatibel und auf das gleichzeitige Auftreten der Spannungen zurückzuführen.

Für in alle Richtungen anisotrope Stoffe läßt sich die Beziehung zwischen Spannungen und Dehnungen wie folgt wiedergeben:

$$\begin{pmatrix} \sigma_x \\ \sigma_y \\ \tau_{xy} \end{pmatrix} = \begin{pmatrix} E_{11} & E_{12} & 0 \\ E_{21} & E_{22} & 0 \\ 0 & 0 & E_{33} \end{pmatrix} \cdot \begin{pmatrix} \epsilon_x \\ \epsilon_y \\ v_{xy} \end{pmatrix} \quad (3)$$

wobei $E_{11} \ldots E_{33}$ von den physikalischen Eigenschaften des Stoffes abhängende Konstanten sind (Elastizitätsmoduln).

Kurz vor dem Bruch übersteigt die Dehnung die oben angegebene lineare Proportionalität, und an stark beanspruchten Stellen kommt es zu ersten Rißbildungen. Die an den Bruchstellen gespeicherte und freigesetzte Energie entspricht folgendem Ausdruck:

$$U = \int \left(\frac{1}{2} \cdot \sigma_x \cdot \epsilon_x + \frac{1}{2} \cdot \sigma_y \cdot \epsilon_y + \frac{1}{2} \tau_{xy} \cdot v_{xy} \right) \cdot d_{vol} \quad (4)$$

more effective in crushing than an equivalent load with a comparable low velocity.

Bulk of energy applied in comminution is used in deforming the material plastically. This energy is many times greater than that required for fracturing. This means that the energy input necessary to break the rock is essentially that which is required to deform the mass beyond the critical strain and to form crack tips. This stored strain energy is released as heat energy.

A main characteristic of rock mass is their inhomogeneous structure. The segregation of bonding material between the grains, micro-cracks and intergranular pores induce deformation which results in stress-concentration and fracture propagation. Since the volume of solid bound inbetween these imperfections is small, the critical strain energy for rupture is small, too. The resistance to fracture is a function of time and of the rate at which stress is applied. Fracturing begins at flaws and with fast rate of energy flow into the cracks from the released stress field. Only a small amount of energy is further required for the formation of new surface. This processes can be assessed by computer aided Finite Element Method (FEM) techniques. The crack becomes unstable with spontaneous acceleration. If sufficient energy is available, the velocity of crack movement approaches that of sound and branching occurs.

Finite Element Method

The finite element method has been chosen to handle the cases that cannot be solved in classical manner. In addition, it becomes possible to depart from the assumption of an elastic continuum by introducing modifications that come close to approximating the geometrical and mechanical properties of typical inhomogeneous rock mass.

By dividing the continuum into a system of elements each having a uniform strain, the formulation of a classical boundary problem becomes unnecessary. The solid particle in the state of equilibrium exists as a micro-continuum. If equilibrium becomes disturbed by increasing the stress applied to the particle the micro-continuum is deformed elastically up to the point of failure. Finite displacements soon dominate in the adjustment process so forming eventually a global discontinuum. Every differential element is subjected to the three kinds of stresses σ_x, σ_y and τ_{xy} occurring in the process of comminution. The corresponding strains ϵ_x, ϵ_y and v_{xy} are mutually compatible up to the point of rupture and are due to the concommittal occurance of the stresses.

For in all directions anisotropic material the relationship between stresses and strains can be written as follows:

$$\begin{pmatrix} \sigma_x \\ \sigma_y \\ \tau_{xy} \end{pmatrix} = \begin{pmatrix} E_{11} & E_{12} & 0 \\ E_{21} & E_{22} & 0 \\ 0 & 0 & E_{33} \end{pmatrix} \cdot \begin{pmatrix} \epsilon_x \\ \epsilon_y \\ v_{xy} \end{pmatrix} \quad (3)$$

Where $E_{11} \ldots E_{33}$ are constants depending on physical properties of the material (Moduli of elasticity).

Just prior to the fracture the strain exceeds the linear proportionality expressed above, resulting in inceptive cracks at points of high stress. The strain energy stored and liberated at the points of rupture is expressed as

$$U = \int \left(\frac{1}{2} \cdot \sigma_x \cdot \epsilon_x + \frac{1}{2} \cdot \sigma_y \cdot \epsilon_y + \frac{1}{2} \tau_{xy} \cdot v_{xy} \right) \cdot d_{vol} \quad (4)$$

Strain energy is reduced by

$$\frac{\pi}{4} \cdot l^2 \cdot \sigma^2 \cdot E \quad (5)$$

due to the presence of microscopic cracks, where l is the length of the crack, σ stress and E the modulus of elasticity. The actual stress can be determined in every particle if displacements associated with the elements are identified and saved in a vector.

Die Dehnungsenergie wird wegen des Vorhandenseins mikroskopischer Risse um

$$\frac{\pi}{4} \cdot l^2 \cdot \sigma^2 \cdot E \qquad (5)$$

vermindert, wobei l die Länge des Risses, σ die Spannung und E der Elastizitätsmodul ist. Die tatsächliche Spannung kann in jedem Teilchen bestimmt werden, wenn die mit den Elementen verbundenen Verschiebungen ermittelt und in einem Vektor abgespeichert werden.

$$(\Delta)_e = \begin{pmatrix} u_1 \\ v_1 \\ u_2 \\ v_2 \\ u_3 \\ v_3 \end{pmatrix} \qquad (6)$$

Zur Spannungsermittlung kann der Ausdruck

$$(\sigma) = (E) \cdot (\epsilon) \qquad (7)$$

herangezogen werden.

Die Dehnungen ergeben sich nach

$$(\epsilon) = (B) \cdot (\Delta)_e$$
$$\therefore (\sigma) = (E) \cdot (B) \cdot (\Delta)_e = (S) \cdot (\Delta)_e \qquad (8)$$

Die Matrix (S) ist die Spannungsmatrix eines linearen Dreiecks.

$$(S) = \frac{1}{2A} \cdot \begin{pmatrix} E_{11} & E_{12} & 0 \\ E_{21} & E_{22} & 0 \\ 0 & 0 & E_{33} \end{pmatrix} \cdot \begin{pmatrix} a_1 & 0 & a_2 & 0 & a_3 & 0 \\ 0 & b_1 & 0 & b_2 & 0 & b_3 \\ b_1 & a_1 & b_2 & a_2 & b_3 & a_3 \end{pmatrix} \qquad (9)$$

$$(S) = \frac{1}{2A} \cdot \begin{pmatrix} E_{11} \cdot a_1 & E_{12} \cdot b_1 & E_{11} \cdot a_1 & E_{12} \cdot b_2 & E_{11} \cdot a_3 & E_{12} \cdot b_3 \\ E_{21} \cdot a_1 & E_{22} \cdot b_1 & E_{21} \cdot a_2 & E_{22} \cdot b_2 & E_{21} \cdot a_3 & E_{33} \cdot b_3 \\ E_{33} \cdot b_1 & E_{33} \cdot a_1 & E_{33} \cdot a_1 & E_{33} \cdot b_2 & E_{33} \cdot a_2 & E_{33} \cdot a_3 \end{pmatrix}$$

Für jedes Element ergeben sich die drei Spannungen σ_x, σ_y und τ_{xy}. Daraus lassen sich die entsprechende Hauptspannung und ihre Richtung ermitteln.

In dieser Veröffentlichung wird die Vermutung geäußert, daß die Zugkraft oder Massenkraft des Teilchens durch starke Trägheitskräfte und eine hohe Beschleunigung der Partikel während des Mahlvorgangs verursacht wird. Um die Zugkraft im Rahmen einer FEM-Analyse verwenden zu können, ist eine Menge äquivalenter Modalkräfte zu entwickeln.

$$(F_{Oberfläche}) = \int (N)^T \cdot \begin{pmatrix} T_x \\ T_y \end{pmatrix} ds = \begin{pmatrix} 0 \\ 0 \\ l_{23} \cdot \frac{T_x}{2} \\ l_{23} \cdot \frac{T_y}{2} \\ l_{23} \cdot \frac{T_x}{2} \\ l_{23} \cdot \frac{T_y}{2} \end{pmatrix} \qquad (10)$$

Der größte Vorteil der Verwendung von FEM-Methoden ist die Möglichkeit der Erstellung von 3D-Grafikmodellen auf dem Computer, um die Ergebnisse der Zerkleinerungssimulation zu testen und zu überprüfen.

Als Beispiel zur Erläuterung des Einsatzes von FEM-Methoden wird im folgenden das Vermahlen von Kalkstein zur Herstellung von Rohmehl für den Klinkerbrennprozeß mit einer Partikelgröße der Einsatzstoffe von 30 mm bei 1 Vol.-% Mikrogefügerissen beschrieben. In der **Tabelle 1** sind einige physikalisch-mechanische Eigenschaften von hartem Kalkstein aufgeführt.

$$(\Delta)_e = \begin{pmatrix} u_1 \\ v_1 \\ u_2 \\ v_2 \\ u_3 \\ v_3 \end{pmatrix} \qquad (6)$$

To determine the stress

$$(\sigma) = (E) \cdot (\epsilon) \qquad (7)$$

can be used.

The strains are obtained by

$$(\epsilon) = (B) \cdot (\Delta)_e \qquad (8)$$
$$\therefore (\sigma) = (E) \cdot (B) \cdot (\Delta)_e = (S) \cdot (\Delta)_e$$

The matrix (S) is the stress matrix for linear triangle.

$$(S) = \frac{1}{2A} \cdot \begin{pmatrix} E_{11} & E_{12} & 0 \\ E_{21} & E_{22} & 0 \\ 0 & 0 & E_{33} \end{pmatrix} \cdot \begin{pmatrix} a_1 & 0 & a_2 & 0 & a_3 & 0 \\ 0 & b_1 & 0 & b_2 & 0 & b_3 \\ b_1 & a_1 & b_2 & a_2 & b_3 & a_3 \end{pmatrix} \qquad (9)$$

$$(S) = \frac{1}{2A} \cdot \begin{pmatrix} E_{11} \cdot a_1 & E_{12} \cdot b_1 & E_{11} \cdot a_1 & E_{12} \cdot b_2 & E_{11} \cdot a_3 & E_{12} \cdot b_3 \\ E_{21} \cdot a_1 & E_{22} \cdot b_1 & E_{21} \cdot a_2 & E_{22} \cdot b_2 & E_{21} \cdot a_3 & E_{33} \cdot b_3 \\ E_{33} \cdot b_1 & E_{33} \cdot a_1 & E_{33} \cdot a_1 & E_{33} \cdot b_2 & E_{33} \cdot a_2 & E_{33} \cdot a_3 \end{pmatrix}$$

For each element will be obtained the three stresses σ_x, σ_y and τ_{xy}. From this the resulting principal stress and its direction can be obtained.

As suggested in this paper traction force or body force in the particle is caused by high inertia forces and by rapid accleration of particle in the grinding process. To be able to use the traction force in FEM analysis a set of equivalent modal forces are to be developed.

$$(F_{Surface}) = \int (N)^T \cdot \begin{pmatrix} T_x \\ T_y \end{pmatrix} ds = \begin{pmatrix} 0 \\ 0 \\ l_{23} \cdot \frac{T_x}{2} \\ l_{23} \cdot \frac{T_y}{2} \\ l_{23} \cdot \frac{T_x}{2} \\ l_{23} \cdot \frac{T_y}{2} \end{pmatrix} \qquad (10)$$

The main advantage in using FEM techniques is the possibility to generate 3D graphic models on the computer for testing and verifying the results of the simulation of comminution.

As a demonstrating example for the use of FEM techniques the grinding of limestone for the production of raw meal for

TABLE 1: Physico-mechanical properties of hard limestone

Velocity of longitudinal waves v_l	5400 m/s
Velocity of transversal waves v_t	3100 m/s
Specific gravity	2.53 g/cm^3
Permeability to air	very low
Modulus of Elasticity E	47500 MPa
Poisson ratio ϑ	0.28
Elastic stress limit σ_p	100 MPa
Rupture stress σ_b	140 MPa

the clinker burning process with a particle size of feed of 30 mm with 1 vol.-% of microstructural cracks is described below. In **Table 1** there are given some physico-mechanical properties of hard limestone.

Idealizing the problem, a cubical mass of limestone 3 cm in size with porous microdefects is considered. The element is discretized into 15 nodes (5 and 3 nodal points at the adjacent sides respectively). Therefore there are $2 \cdot (5-1) \cdot (3-1) = 16$ triangular elements in the grid which will be treated in the FEM analysis.

The co-ordinates of the node which forms one of the input data lists is shown in **Table 2**.

The so called SWAP-VI programme was used for FEM analysis. Physical properties of the limestone and geometrical coordinates are input variables of the programme. Results of the calculation show that, e.g., the stress of element 9 at the node point 7 is enhanced by 3.11 times. That indicates that stress concentrates around microdefects as it does in reality which means that the assumption is correct that the rupture stresses can be simulated by using moderate pre-estimated surface forces.

TABLE 1: Physico-mechanical properties of hard limestone

Velocity of longitudinal waves v_l	5400 m/s
Velocity of transversal waves v_t	3100 m/s
Density	2.53 g/cm^3
Air permeability	very low
Modulus of elasticity E	47500 MPa
Poisson ratio ϑ	0.28
Elastic limit σ_p	100 MPa
Rupture stress σ_b	140 MPa

TABLE 2: Co-ordinates for one node (NI = 5, NJ = 3)

Node number	X co-ordinate	Y co-ordinate
1	0.000	3.000
2	0.000	1.500
3	0.000	0.300
4	0.115	0.277
5	0.212	0.212
6	0.277	0.115
7	0.300	0.000
8	1.500	0.000
9	3.000	0.000
10	3.000	1.500
11	3.000	3.000
12	1.500	3.000
13	2.250	2.250
14	0.750	0.750
15	0.375	1.125

In the grinding process proposed here, the energy is imparted by means of a rotating high speed unit which develops sufficient centrifugal force to traject the particles fed at high velocity against rigid baffle structures studded with sharp spikes. The high impact caused by collision of particle with the barrier develops high amplitude forces which are exerted and removed in a micro second rhythm and which initiate stress waves emanating from the contact region that subsequently propagates throughout the entire dominion of impinging bodies.

Impact forces are calculated by using Hertz' theory of impact of spherical bodies. In this theory the magnitude of force is related to the distance through which the bodies approach one another after the first instant of contact. It is expressed as follows:

$$F = c_1 \cdot \delta^{\frac{3}{2}} \quad (11)$$

where δ is the total deformation of the particle and c_1 is

$$c_1 = \frac{4}{\left(\frac{1}{R_1} + \frac{1}{R_2}\right)^{\frac{1}{3}} \cdot \left(\frac{1}{E_1'} + \frac{1}{E_2'}\right)} \quad (12)$$

R_1 and R_2 are the radii of curvature of the bodies and

$$E_1' = \frac{E_1}{1 - \nu^2} \qquad E_2' = \frac{E_2}{1 - \nu^2} \quad (13)$$

ϑ: Poisson ratio
E_1, E_2: Moduli of elasticity.

$$F = c_1 \cdot \delta^{\frac{3}{2}} \qquad (11)$$

wobei δ die Gesamtverformung des Partikels darstellt und c_1 gleich

$$c_1 = \frac{4}{\left(\frac{1}{R_1} + \frac{1}{R_2}\right)^{\frac{1}{3}} \cdot \left(\frac{1}{E_1'} + \frac{1}{E_2'}\right)} \qquad (12)$$

ist. R_1 und R_2 sind die Krümmungsradien der Körper und

$$E_1' = \frac{E_1}{1 - \nu^2} \qquad E_2' = \frac{E_2}{1 - \nu^2} \qquad (13)$$

ϑ: Poissonsche Zahl
E_1, E_2: Elastizitätsmoduln
Ist die Anfangsgeschwindigkeit v_1, so gilt

$$\delta_{max} = \left(\frac{5 \cdot m_1 \cdot m_2}{4\, c_1 (m_1 + m_2)}\right)^{\frac{2}{5}} \cdot V_1^{\frac{4}{5}} \qquad (14)$$

Für die maximale zwischen den Partikeln wirksam werdende Druckkraft gilt der Ausdruck

$$F_{max} = \left(\frac{5 \cdot m_1 \cdot m_2}{4\, (m_1 + m_2)}\right)^{\frac{3}{5}} \cdot c_1^{\frac{2}{5}} \cdot V_1^{\frac{6}{5}} \qquad (15)$$

wonach die Kontaktzeit folgenden Wert besitzt:

$$t_c = \frac{2.9432 \cdot \delta_{max}}{V_1} \qquad (16)$$

Aus den obigen Gleichungen läßt sich die Mindestfluggeschwindigkeit des einzelnen Partikels berechnen. Durch Anpassung der Leistung des Mühlenantriebsmotors kann dem Partikel genau die Energie zugeführt werden, um es auf die erforderliche Geschwindigkeit zu beschleunigen. Die derzeitigen mechanischen Mahlverfahren sind stochastische Prozesse, bei denen die Teilchen mehrmals beansprucht werden. Wiederholtes (und möglicherweise zu häufiges) Vermahlen des Mahlguts und die geringe Wahrscheinlichkeit, daß die Mahlkörper das einzelne Partikel erfassen, tragen zu einem Anstieg des spezifischen Energieverbrauchs bei. Demgegenüber dürfte das in dieser Veröffentlichung vorgeschlagene Mahlverfahren zu optimaler Energieausnutzung führen. Theoretische Berechnungen für das Mahlen und den Transport von Kalkstein, wie beispielhaft beschrieben, ergaben einen spezifischen Energieverbrauch von 9–10 kWh/t. Das ist deutlich weniger als der gegenwärtige Verbrauch der Industrieanlagen. Zum Nachweis dieser theoretischen Ergebnisse sollen noch weitere experimentelle Untersuchungen durchgeführt werden.

If the initial velocity is v_1 then

$$F_{max} = \left(\frac{5 \cdot m_1 \cdot m_2}{4\, (m_1 + m_2)}\right)^{\frac{3}{5}} \cdot c_1^{\frac{2}{5}} \cdot V_1^{\frac{6}{5}} \qquad (14)$$

Maximum compressive force acting between the particles can be expressed as

$$\delta_{max} = \left(\frac{5 \cdot m_1 \cdot m_2}{4\, c_1 (m_1 + m_2)}\right)^{\frac{2}{5}} \cdot V_1^{\frac{4}{5}} \qquad (15)$$

and time of contact is then

$$t_c = \frac{2.9432 \cdot \delta_{max}}{V_1} \qquad (16)$$

From the above equations minimum velocity of trajection of the single particle can be calculated. By adjusting the mill motor power it is possible to impart the energy to the particle to accelerate it to the velocity needed.

The current mechanical methods of grinding are stochastical processes, for that particles are strained several times. Repeated (and may be over grinding) of the material and the low probability of grinding media to strike the single particle tends to increase specific energy consumption. Whereas the grinding method suggested in this paper should lead to an optimum energy utilization. The specific energy consumption for grinding and transporting limestone as described in our example was found to be 9 to 10 kWh/t on the basis of theoretical calculation. This is considerably below the present level of industrial plants. Further experimental work shall be done to prove this theoretical findings.

Umbau einer Rohmehl-Mahlanlage mit Kugelmühle und Sichter zu einer Kombi-Mahlanlage mit Gutbett-Walzenmühle – Vergleich des Energiebedarfs mit dem einer Walzenschüsselmühle

Conversion of a raw material grinding plant with ball mill and classifier to a combination grinding plant with high-pressure grinding rolls – comparison of the energy consumption with that of a roller grinding mill

Modification d'un atelier de broyage du cru avec broyeur à boulets et séparateur en un atelier de broyage tandem avec broyeur à rouleaux – comparaison de la consommation d'énergie avec celle d'un broyeur à galets

Transformación de una planta de molienda de crudo, con molino de bolas y separador, en planta de molienda combinada, con molino de cilindros y lecho de material – Comparación del consumo de energía con el de un molino de cubeta y rodillos

Von **R. Atzl**, Kufstein/Österreich

Zusammenfassung – Im Zementwerk Eiberg, Kufstein in Tirol, wurde das Rohmehl bisher sowohl auf einer Mahlanlage mit Kugelmühle, Sichter und vorgeschaltetem Doppelwellen-Hammerbrecher als auch mit einer Mahlanlage mit Walzenschüsselmühle ermahlen. Die Kugelmühlen-Mahlanlage hatte einen Durchsatz von 42 t/h bei einem massebezogenen Energieaufwand von 17,5 kWh/t, die Walzenschüsselmühle hatte einen Durchsatz von 26 t/h bei einem Energieaufwand von 16,4 kWh/t. Im Winter 1988/89 wurde die Kugelmühlen-Mahlanlage zu einer Kombi-Mahlanlage umgebaut und dabei der Doppelwellen-Hammerbrecher, der einen großen Anteil der Instandhaltungskosten verursachte, durch eine Gutbett-Walzenmühle mit einer Antriebsleistung von 2 x 140 kW für einen Durchsatz von 100 t/h ersetzt. Dadurch wurde der Durchsatz der Kugelmühlen-Mahlanlage auf 60 t/h gesteigert, der Energieaufwand nahm auf 15,7 kWh/t ab. Durch den Umbau der Mahlanlage wurde demnach eine deutliche Leistungssteigerung und damit eine erhebliche Energieeinsparung gegenüber dem Betrieb der Wälzmühle erzielt. Auch ein Vergleich der Instandhaltungskosten aus den Jahren 1988 – 1993 zeigt die Vorteile der Mahlanlage mit Gutbett-Walzenmühle für das Mahlen von Zementrohmaterial.

Summary – In the Eiberg cement works at Kufstein in Tirol the raw meal used to be ground in a grinding plant with ball mill, classifier and upstream double-shaft hammer crusher, and also in a grinding plant with a roller grinding mill. The ball mill grinding plant had a throughput capacity of 42 t/h with a specific energy consumption of 17.5 kWh/t, and the roller grinding mill had a throughput capacity of 26 t/h with an energy consumption of 16.4 kWh/t. In the winter of 1988/89 the ball mill grinding plant was converted to a combination grinding plant and the double-shaft hammer crusher, which was responsible for a large proportion of the maintenance costs, was replaced by high-pressure grinding rolls with a drive rating of 2 x 140 kW for a throughput mass flow of 100 t/h. This increased the throughput capacity of the ball mill grinding plant greatly to 60 t/h and the energy consumption dropped to 15.7 kWh/t. Conversion of the grinding plant therefore achieved a significant increase in output and a considerable saving in energy when compared with the operation of the roller grinding mill. Comparison of the maintenance costs for the years 1988 – 1993 also shows the advantages of the grinding plant with high-pressure grinding rolls for grinding cement raw material.

Résumé – A la cimenterie Eiberg, à Kufstein au Tyrol, la farine crue avait été produite jusqu'à présent dans un atelier à broyeur à boulets, séparateur et concasseur à marteaux à double rotor en amont et, aussi, dans un atelier à broyeur à galets. L'atelier à broyeur à boulets avait un débit de 42 t/h pour une comsommation spécifique d'énergie de 17,5 kWh/t, celui à broyeur à galets avait un débit de 26 t/h pour une consommation spécifique de 16,4 kWh/t. Dans l'hiver 1988/89, l'atelier à broyeur à boulets a été reconstruit en atelier tandem et le concasseur à marteaux à double rotor, qui avait occasionné une grande partie des dépenses de maintenance, a été remplacé par

un broyeur à rouleaux d'une puissance d'entraînement de 2 x 140 kW pour un débit de 100 t/h. Ainsi la performance de débit de l'atelier à broyeur à boulets a été accrue sensiblement, à 60 t/h et la consommation d'énergie est tombée à 15,7 kWh/t. La reconstruction de l'atelier de broyage a donc apporté une nette augmentation des performances et, conjointement, une économie d'énergie substantielle par rapport à l'exploitation du broyeur à galets. Une comparaison des coûts de maintenance des années 1988–1993 a également montré les avantages de l'atelier à broyeur à rouleaux pour le broyage du cru à ciment.

sommation d'énergie avec celle d'un broyeur à galets

***Resumen** – En la fábrica de cemento Eiberg, en Kufstein/Tirol, el crudo se ha molido hasta ahora tanto en una planta de molienda con molino de bolas, separador y trituradora de martillos de dos ejes, preconectado, como en una planta de molienda con molino de cubeta y rodillos. La instalación de molienda con molino de bolas tenía un rendimiento de paso de 42 t/h, con un consumo de energía referido a la masa de 17,5 kWh/t, mientras que el molino de cubeta y rodillos tenía un rendimiento de paso de 26 t/h, con un consumo de energía referido a la masa de 16,4 kWh/t. En invierno de 1988/89, la instalación de molienda con molino de bolas fue transformada en una planta de molienda combinada, sustituyéndose la trituradora de martillos de dos ejes, la cual causaba gran parte de los gastos de mantenimiento, por un molino de cilindros y lecho de material, que contaba con una potencia de accionamiento de 2 × 140 kW y un rendimiento de paso de 100 t/h. Con ello aumentó la capacidad de producción de la planta de molienda con molino de bolas a 60 t/h, y el consumo de energía bajó a 15,7 kWh/t. Por lo tanto, la transformación de la instalación de molienda ha permitido conseguir un notable aumento del rendimiento y un considerable ahorro de energía en comparación con el servicio con molino de pista de rodadura. También la comparación de los gastos de mantenimiento de los años 1988–1993 demuestra las ventajas de la instalación con molino de cilindros y lecho de material para la molienda del crudo de cemento.*

Transformación de una planta de molienda de crudo, con molino de bolas y separador, en planta de molienda combinada, con molino de cilindros y lecho de material – Comparación del consumo de energía con el de un molino de cubeta y rodillos

Im Zementwerk Eiberg standen für die Mahlung der für den Drehofen täglich erforderlichen 1250 t Rohmaterial bis einschließlich 1988 vier Mahlanlagen zur Verfügung. In **Tabelle 1** sind einige technische Daten dieser Anlagen aufgeführt. Bei den Rohmehl-Mahlanlagen 1 und 2 handelt es sich jeweils um eine kleinere Kugelmühle, die im Umlauf mit einem Sichter betrieben wird. Das Aufgabegut von 16 t/h wird mit einem Einwellen-Hammerbrecher vorzerkleinert. Der spezifische Energiebedarf liegt bei diesen beiden Anlagen über 20 kWh/t. Bei der Rohmühle 3 handelt es sich ebenfalls um eine Kugelmühlen-Umlaufmahlanlage. Zur Vorzerkleinerung wird hier jedoch ein Doppelwellen-Hammerbrecher eingesetzt. Das Verfahrensschema dieser Mahlanlage ist in **Bild 1** dargestellt. Sie hatte einen Durchsatz von 42 t/h bei einem elektrischen Energieaufwand von 17,5 kWh/t. Als vierte Rohmehl-Mahlanlage steht noch eine Wälzmühle vom Typ LM 16 der Loesche GmbH, Düsseldorf, mit einer Durchsatzleistung von 26 t/h und einem spezifischen Energieverbrauch von 16,4 kWh/t zur Verfügung (**Bild 2**). In allen Mühlen wird zur Trocknung des Rohmaterials mit anfänglich 3 – 4 M.-% Feuchte Drehofenabgas eingesetzt.

Während des Winterstillstands 1988/89 wurde die Rohmühle 3 zu einer Hybridmahlanlage umgebaut. Dabei wurde der Doppelwellen-Hammerbrecher, der einen erheblichen

Up to and including 1988 there were four grinding plants available at the Eiberg cement works for grinding the 1250 t raw material required daily for the rotary kiln. **Table 1** lists some of the technical data for these plants. Raw meal grinding plants 1 and 2 each have a fairly small ball mill which is operated in closed circuit with a classifier. The feed of 16 t/h is precomminuted in a single-shaft hammer crusher. The specific power consumption in these two plants is over 20 kWh/t. Raw mill 3 is also a closed-circuit ball mill plant, but in this case a double-shaft hammer crusher is used for preliminary comminution. The process flow sheet for this grinding plant is shown in **Fig. 1**. It has a throughput of 42 t/h and an electric power consumption of 17.5 kWh/t. An LM 16 roller grinding mill from Loesche GmbH, Düsseldorf, with a throughput of 26 t/h and a specific power consumption of 16.4 kWh/t is also available as the fourth raw meal grinding plant (**Fig. 2**). Exhaust gas from the rotary kiln is used in all the mills for drying the raw material with its initial moisture content of 3 – 4 wt.%.

During the 1988/89 winter shutdown raw mill 3 was converted to a hybrid grinding plant. The double-shaft hammer crusher, which was responsible for a considerable proportion of the maintenance costs of the entire plant, was replaced by high-pressure grinding rolls (**Fig. 3**). The high-

TABELLE 1: Technische Daten der Mühlen vor dem Umbau der Rohmühle III

Rohmühle I+II	Rohmühle III	Rohmühle IV
Kugelmühle (Büttner)	Kugelmühle (KHD)	Walzenschüsselmühle (Lösche)
D = 2,3 m	D = 2,8 m	Mahlteller d = 1600 mm
Länge = 4,25 m	Länge = 5,0 m	Mahlwalzen d = 1180 mm
Baujahr 1940	Baujahr 1952	Baujahr 1952
Füllung 20 t, 80–30 mm	Füllung 41 t, 80–30 mm	
Antriebsleistung 230 kW	Antriebsleistung 680 kW	Antriebsleistung 380 kW
Einwellenhammerbrecher	Doppelwellenhammerbrecher	
72 Hämmer je 6 kg	2×48 Hämmer je 10 kg	
Antriebsleistung 55 kW	Antriebsleistung 130 kW	
Sichter (Fabrikat Claes)	Sichter (Fabrikat Claes)	aufgesetzter Kreiselsichter
Antriebsleistung 75 kW	Antriebsleistung 90 kW	Antriebsleistung 7,5 kW
16 t/h	40–42 t/h	26 t/h
> 20 kWh/t	17,5 kWh/t	16,4 kWh/t

TABLE 1: Technical data for the mills before conversion of raw mill III

Raw mills I+II	Raw mill III	Raw mill IV
Ball mill (Büttner)	Ball mill (KHD)	Roller grinding mill (Loesche)
D = 2.3 m	D = 2.8 m	grinding table d = 1600 mm
length = 4.25 m	length = 5.0 m	grinding rollers d = 1180 mm
built 1940	built 1952	built 1952
Grinding media 20 t, 80–30 mm	Grinding media 41 t, 80–30 mm	
drive rating 230 kW	drive rating 680 kW	drive rating 380 kW
single-shaft hammer crusher	double-shaft hammer crusher	
72 hammers 6 kg	2×48 hammers 10 kg	
drive rating 55 kW	drive rating 130 kW	
classifier (Fabrikat Claes)	classifier (Fabrikat Claes)	topmounted rotor classifier
drive rating 75 kW	drive rating 90 kW	drive rating 7.5 kW
16 t/h	40–42 t/h	26 t/h
> 20 kWh/t	17.5 kWh/t	16.4 kWh/t

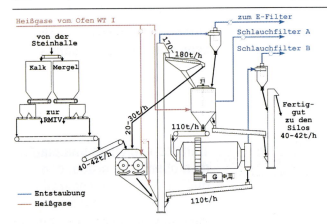
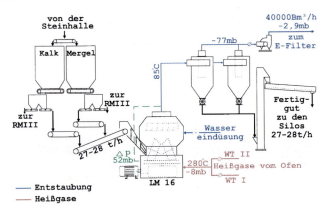

BILD 1: Verfahrensschema der Rohmühle III vor Installierung der Rollenpresse
FIGURE 1: Flow diagram for raw mill III before installation of the high-pressure grinding rolls

Heißgase vom Ofen WT I	=	hot gases from kiln preheater I
von der Steinhalle	=	from the limestone hall
Kalk	=	limestone
Mergel	=	marl
zur RM IV	=	to RM IV
Entstaubung	=	dedusting
Heißgase	=	hot gases
zum E-Filter	=	to electrostatic precipitator
Schlauchfilter	=	bag filter
Fertiggut zu den Silos	=	finished product to the silos

BILD 2: Verfahrensschema der Rohmühle IV
FIGURE 2: Flow diagram for raw mill IV

Heißgase vom Ofen	=	hot gases from kiln
von der Steinhalle	=	from the limestone hall
Kalk	=	limestone
Mergel	=	marl
zur RM III	=	to RM III
Entstaubung	=	dedusting
Heißgase	=	hot gases
40 000 Bm³/h, –2,9 mb zum E-Filter	=	40 000 m³/h (at operating conditions) –2.9 mb to electrostatic precipitator
Wassereindüsung	=	water injection
Fertiggut zu den Silos	=	finished product to the silos
WT II	=	preheater II
WT I	=	preheater I

Anteil der Wartungskosten der Gesamtanlage verursachte, durch eine Gutbett-Walzenmühle ersetzt (**Bild 3**). Die Gutbett-Walzenmühle ist für einen Durchsatz von rd. 100 t/h ausgelegt und besitzt eine installierte Motorleistung von 2 x 140 kW. Wesentliches Ziel des Umbaus war es, eine Erhöhung der Durchsatzleistung von 42 auf rd. 60 t/h zu erreichen und die kleineren Rohmehl-Mahlanlagen 1 und 2 stillzulegen. In Zukunft sollen diese nur noch im Notfall betrieben werden. Außerdem wurde vom Lieferanten eine Verringerung des spezifischen Energiebedarfs von 17,5 auf 15,9 kWh/t in Aussicht gestellt.

In der ersten Ausbaustufe (siehe Bild 3) wurde folgendes Regelkonzept für die Massenströme realisiert. Der Aufgabegutmassenstrom der Kugelmühle wird mittels der drehzahlregelbaren Gleichstromantriebe der Gutbett-Walzenmühle konstant gehalten. Der Füllstand des Aufgabebunkers der Gutbett-Walzenmühle wird mit Druckmeßdosen ermittelt

pressure grinding rolls were designed for a throughput of about 100 t/h and had an installed motor rating of 2 × 140 kW. An important objective of the conversion was to raise the throughput from 42 to about 60 t/h and to close down the small raw meal grinding plants 1 and 2. In future these are only to be operated in emergencies. The suppliers also promised a reduction in specific power consumption from 17.5 to 15.9 kWh/t.

In the first stage of the improvement (see Fig. 3) the following control scheme was used for the mass flows. The feed mass flow to the ball mill is kept constant by the variable speed d.c. drives of the high-pressure grinding rolls. The level in the feed hopper for the high-pressure grinding rolls is measured with pressure cells and controlled by the limestone and marl weighfeeder belts. The compacted cake from the high-pressure grinding rolls is disagglomerated in a small hammer crusher and transported by a bucket elevator to a

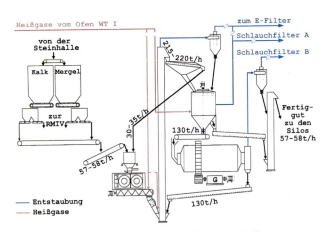
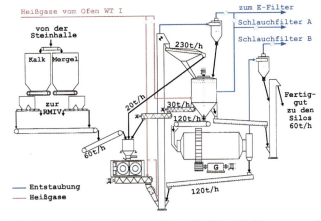

BILD 3: Verfahrensschema der Rohmühle III mit Rollenpresse vom Typ „RPV-100/40"
FIGURE 3: Flow diagram for raw mill III with the RPV-100/40 high-pressure grinding rolls

Heißgase vom Ofen WT I	=	hot gases from kiln preheater I
von der Steinhalle	=	from the limestone hall
Kalk	=	limestone
Mergel	=	marl
zur RM IV	=	to RM IV
Entstaubung	=	dedusting
Heißgase	=	hot gases
zum E-Filter	=	to electrostatic precipitator
Schlauchfilter	=	bag filter
Fertiggut zu den Silos	=	finished product to the silos

BILD 4: Verfahrensschema der Rohmühle III mit Rollenpresse und Grießerückführung
FIGURE 4: Flow diagram for raw mill III with high-pressure grinding rolls and tailings return system

Heißgase vom Ofen WT I	=	hot gases from kiln preheater I
von der Steinhalle	=	from the limestone hall
Kalk	=	limestone
Mergel	=	marl
zur RM IV	=	to RM IV
Entstaubung	=	dedusting
Heißgase	=	hot gases
zum E-Filter	=	to electrostatic precipitator
Schlauchfilter	=	bag filter
Fertiggut zu den Silos	=	finished product to the silos

TABLE 2: Technical data for the mills after conversion to combination grinding

	Mean annual values		optimized over 24 h	
	t/h	kWh/t	t/h	kWh/t
Roller grinding mill RMIV	26	16.4	28	15.2
Combination grinding plant RMIII	60	15.7	61.5	14.3

Moisture of raw material approx. 3–3.5%
Finish product R 200 μm = 3–3.5%,
R 90 μm = 18–10%.

screen with 12 mm mesh. The coarse material which has been screened off is returned directly to the feed hopper for the high-pressure grinding rolls, and the fines are fed to the classifier together with the output from the ball mill. The tailings from the classifier are returned to the ball mill. This configuration gave a throughput of 57–58 t/h and a specific electric power consumption for the entire plant of 16 kWh/t.

In this operating mode the high-pressure grinding rolls can only be operated at 80% of their maximum speed, so the attempt was made to increase the throughput of the entire plant by returning some of the classifier tailings to the high-pressure grinding rolls. Part of the classifier tailings were therefore transferred back to the feed hopper for the high-pressure grinding rolls by a variable speed screw (**Fig. 4**). This process configuration not only reduced the recirculating mass flow in the ball mill/classifier circuit but also increased the mass flow passing through the high-pressure grinding rolls. The flow control system was then modified so that the controlled tailings return to the high-pressure grinding rolls was used to maintain a constant feed of classifier tailings to the ball mill.

The high-pressure grinding rolls are run manually by the operator at the highest possible speed. Returning the tailings to the high-pressure grinding rolls also improves their intake conditions so that a thicker compacted cake is

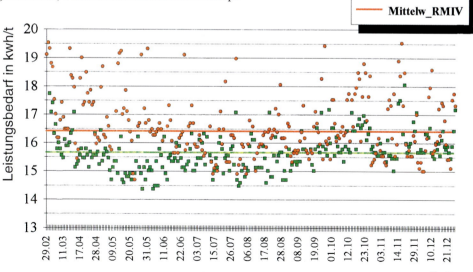

FIGURE 5: Comparison of the specific power consumptions of raw meal grinding plants III (with high-pressure grinding rolls) and IV (roller grinding mill) based on the daily mean values in 1992

Leistungsbedarf in kWh/t	=	Specific power consumption in kWh/t
Zeit	=	Time
RMIII = Kombimahlanlage	=	RMIII = combination grinding plant
RMIV = Mahlanlage mit Walzenschüsselmühle	=	RMIV = grinding plant with roller grinding mill

bilden kann. Durch diese verfahrenstechnische Modifikation konnte die Durchsatzleistung noch einmal um rd. 2 t/h gesteigert werden, so daß heute 60 t/h durchgesetzt werden können. Der spezifische Energiebedarf beträgt nunmehr im statistischen Jahresmittel 15,7 kWh/t.

Um den spezifischen Energieaufwand der Gutbett-Walzenmühlen-Mahlanlage mit dem der Wälzmühle, die beide mit dem gleichen Rohmaterial beschickt werden und dieses auf gleiche Endfeinheit aufmahlen, zu vergleichen, werden am zweckmäßigsten die entsprechenden Tagesmittelwerte beider Systeme herangezogen. Der in **Bild 5** vorgenommene Vergleich zeigt, daß die Gutbett-Walzenmühlen-Mahlanlage sogar noch etwas günstiger als die Wälzmühle liegt. Die Wälzmühle erreichte im Jahr 1992 einen kumulierten Jahresmittelwert von 16,4 kWh/t, während die Gutbett-Walzenmühlen-Mahlanlage mit 15,7 kWh/t um 0,7 kWh/t niedriger lag (**Tabelle 2**). Bei an beiden Anlagen über eine Dauer von 24 h durchgeführten Leistungsversuchen wurden folgende Werte ermittelt. Die Wälzmühle erreichte einen Durchsatz von 27,2 t/h bei einem spezifischen Energieverbrauch von 14,8 kWh/t, die Gutbett-Walzenmühlen-Mahlanlage setzte 61 t/h durch und verbrauchte 14,4 kWh/t.

Sicherlich werden moderne Walzenschüsselmühlen, begründet durch bessere Motor- und Getriebewirkungsgrade und den Einsatz drehzahlregelbarer Mühlenventilatorantriebe, einen geringeren spezifischen Energieverbrauch aufweisen, was den ermittelten Vorteil für die Gutbett-Walzenmühlen-Mahlanlage relativiert. Trotzdem ist der Einsatz einer Gutbett-Walzenmühle eine attraktive Möglichkeit zur Steigerung der Durchsatzleistung einer bestehenden Kugelmühlenaltanlage und der Reduzierung des spezifischen elektrischen Energieaufwands.

Untersuchungen zum Verschleißverhalten und zur Verfügbarkeit der Gutbett-Walzenmühle ergaben, daß erst nach 3 Jahren eine Aufschweißung der Walzen erforderlich war. Dies entspricht einer Standzeit von rd. 18000 Betriebsstunden. Der Verschleiß betrug etwa 150 kg, woraus sich, bezogen auf das Fertiggut, ein spezifischer Verschleiß von rd. 0,1 g/t Rohmehl errechnet. Dies stellt eine wesentliche Zeit- und Kostenersparnis gegenüber dem vorher eingesetzten Doppelwellen-Hammerbrecher dar, bei dem etwa nach 6 Monaten die Hämmer komplett ausgetauscht werden mußten. Inzwischen ist die Gutbett-Walzenmühle rd. 25000 Stunden in Betrieb, ohne daß ein längerer, nicht geplanter Stillstand eingetreten wäre.

formed. This process engineering modification produced a further increase in the throughput capacity of about 2 t/h, so that the output is currently 60 t/h. The mean annual specific power consumption is now 15.7 kWh/t.

In order to compare the specific power consumption of the high-pressure grinding rolls plant with that of the roller grinding mill — both being fed with the same raw material and both grinding it to the same end fineness — the corresponding daily mean values of the two systems were used as being the most appropriate. The comparison made in **Fig. 5** shows that the high-pressure grinding rolls plant is slightly better than the roller grinding mill. In 1992 the roller grinding mill achieved an overall mean annual value of 16.4 kWh/t, while the grinding plant with the high-pressure grinding rolls was 0.7 kWh/t lower with a figure of 15.7 kWh/t (**Table 2**). The values given below were measured in performance trials carried out in the two plants over a period of 24 hours. The roller grinding mill reached a throughput of 27.2 t/h for a specific power consumption of 14.8 kWh/t and the high-pressure grinding rolls plant had a throughput of 61 t/h and consumed 14.4 kWh/t.

Modern roller grinding mills with their better motor and gear unit efficiencies and the use of variable speed mill fan drives will certainly have lower specific power consumptions, which puts the measured advantage of the high-pressure grinding rolls plant into perspective. In spite of this, the use of high-pressure grinding rolls is an attractive option for increasing the throughput of an old existing ball mill plant and for reducing the specific electrical power consumption.

Investigations into the wear behaviour and availability of the high-pressure grinding rolls showed that buildup welding on the rollers was first needed after 3 years. This corresponds to a service life of about 18000 running hours. The wear was about 150 kg, giving a specific wear relative to the finished product of about 0.1 g/t raw meal. This represents a substantial saving in time and costs when compared with the previous double-shaft hammer crusher where all the hammers had to be replaced after about 6 months. The high-pressure grinding rolls have now been in operation for about 25000 hours without the occurrence of any lengthy, unplanned, stoppages.

Die Loesche-Mühle für die Zerkleinerung von Zementklinker und Zumahlstoffen in der Praxis*)

The Loesche mill for comminution of cement clinker and interground additives in practical operation*)

Le broyeur Loesche pour la comminution de clinker à ciment et de matières cobroyées dans la pratique

El molino Loesche para la molienda de clínker de cemento y de adiciones, durante el servicio industrial

Von **H. Brundiek**, Düsseldorf/Deutschland

Zusammenfassung – Die Loesche-Mühle hat sich beim Mahlen von Kohle, Kalkstein und Zementrohmaterial als Mahlsystem mit hoher Effektivität und Verfügbarkeit bei deutlich niedrigerem Energieverbrauch gegenüber Rohrmühlensystemen erwiesen. Bei der Fertigmahlung von Zement in Wälzmühlen wurden bislang nicht alle Qualitätsmerkmale eines Rohrmühlenzements erreicht. Zementmahlversuche mit modifizierten Rohmehl-4-Walzenmühlen haben zu einem neuen Konstruktionskonzept der Loesche-Mühle für Zement geführt. Unter Beibehaltung der bekannten horizontalen Mahlbahn besitzen Loesche-Zementmühlen 4 konische Walzen, von denen jedoch 2 als Mahlwalzen und 2 als sogenannte Präparationswalzen zur Vorverdichtung des Mahlbetts arbeiten. Dadurch können die spezifischen Mahldrücke der Mahlwalzen erheblich gesteigert werden. Im Parallelspalt zwischen den Mahlteilen findet bei internem Grießumlauf eine einstufige Zerkleinerung von Klinker und Zumahlstoffen bei hohen Mahlgeschwindigkeiten statt. Durch diese einfachen konstruktiven Maßnahmen kann die gewünschte Zementqualität bei vibrationsarmem Mühlenlauf erreicht werden. Ein geschlossenes Mühlen-Gebäude ist nicht erforderlich. Standzeiten der Mahlteile von mehr als 10 000 Stunden sorgen bei hoher Mühlenverfügbarkeit für gleichbleibende Produktqualität. Die einfache Anlagenschaltung erlaubt ein einfaches Regelkonzept. In den letzten Jahren wurden in mehreren Werken Loesche-4-Walzenmühlen für die Mahlung von Klinker und Hüttensand optimiert und Durchsätze von bis zu 140 t/h mit Rohrmühlenzementqualität von 3400 Blaine erreicht. Der spezifische Energieverbrauch liegt deutlich unter dem von Rohrmühlen und ist auch günstiger als bei einer Anlage mit Rohrmühle und vorgeschalteter Gutbett-Walzenmühle.

Summary – The Loesche mill has proved to be a grinding system with high levels of efficiency and availability and a significantly lower energy consumption than tube mill systems for grinding coal, limestone and cement raw material. Until now not all the quality features of a tube mill cement have been achieved when finish-grinding cement in roller grinding mills. Cement grinding trials with modified raw meal roller grinding mills with four rollers have led to a new design concept of the Loesche mill for cement. While retaining the familiar horizontal grinding track, the Loesche cement mills have 4 conical rollers of which, however, 2 work as grinding rollers and 2 as so-called preparation rollers for pre-compacting the grinding bed. This has meant that the specific grinding pressures of the grinding rollers can be greatly increased. A single-stage comminution of clinker and interground additives takes place at high grinding speeds in the parallel gap between the grinding elements with internal recirculation of coarse material. These simple design measures mean that the required cement quality can be achieved with low-vibration mill operation. A closed mill building is not necessary. Service lives of the grinding elements of more than 10 000 hours ensure a uniform product quality with high mill availability. The simple plant arrangement allows a simple control concept to be used. In recent years Loesche 4-roller mills for grinding clinker and granulated blast furnace slags in several works have been optimized and throughput capacities of up to 140 t/h have been achieved with tube mill cement quality of 3400 Blaine. The specific energy consumption lies significantly below that of tube mills and is also better than with a plant with a tube mill preceded by high-pressure grinding rolls.

Résumé – Le broyeur Loesche s'est révélé, pour le broyage de charbon, calcaire et matières premières à ciment, comme système de broyage de haute efficacité et disponibilité, avec une consommation d'énergie nettement moindre par rapport à des systèmes avec broyeurs à boulets. Jusqu'à présent, toutes les caractéristiques de qualité d'un ciment de broyeur à boulets n'ont pu être obtenues dans le broyage fini du ciment dans les broyeurs à galets. Des expériences de broyage de ciment, avec des broyeurs de cru à galets, modifiés, ont conduit à un nouveau concept de construction du broyeur Loesche

---

*) Überarbeitete Fassung eines Vortrages zum VDZ-Kongreß '93, Düsseldorf (27.9.–1.10.1993)
Revised text of a lecture to the VDZ Congress '93, Düsseldorf (27.9.–1.10.1993)

pour ciment. Tout en conservant la piste horizontale connue, les broyeurs de ciment Loesche possèdent 4 galets coniques, dont 2 travaillent comme galets de broyage et 2, par contre, comme galets préparateurs pour le précompactage du lit de broyage. Ceci permet d'augmenter nettement les pressions de broyage spécifiques des galets de broyage. Dans l'interstice entre les pièces de broyage s'effectue, avec une recirculation interne des gruaux, une comminution en une seule étape avec des vitesses de broyage élevées. Ces artifices constructives simples permettent d'obtenir la qualité voulue de ciment avec une marche peu vibrante du broyeur. Un bâtiment de broyage fermé n'est pas nécessaire. Des durées de vie des pièces de broyage, de plus de 10 000 heures, garantissent une qualité constante du produit, avec une grande disponibilité du broyeur. La configuration simple de l'atelier permet un concept de régulation simple. Au cours des dernières années, des broyeurs à 4 galets Loesche ont été optimisés dans plusieurs usines pour le broyage de clinker et de laitier granulé. Des performances de débit allant jusqu'à 140 t/h sont atteintes, avec la qualité du ciment de tube broyeur à 3400 Blaine. La consommation spécifique d'énergie se situe nettement en dessous de celle des tubes broyeurs et est aussi plus avantageuse qu'avec un atelier à tube broyeur précédé d'un broyeur à rouleaux.

Resumen – *El molino Loesche ha probado su eficacia en la molienda de carbón, piedra caliza y materias primas para la fabricación del cemento, siendo un sistema de molienda de gran eficacia y alta disponibilidad, que tiene un consumo energético notablemente más bajo que los molinos tubulares. En la molienda de acabado del cemento por medio de molinos de pista de rodadura no se han alcanzado, hasta ahora, todas las características de calidad de un cemento molido en molinos tubulares. Los ensayos de molienda de cemento por medio de molinos de 4 cilindros para crudo, modificados, han conducido a un nuevo concepto de construcción del molino Loesche para cemento. Manteniendo la conocida pista de rodadura horizontal, los molinos Loesche para cemento poseen 4 cilindros cónicos, 2 de los cuales trabajan como rodillos moledores y 2, como rodillos llamados de preparación, que están destinados a la compactación del lecho de molienda. Con ello se pueden aumentar considerablemente las presiones específicas de los rodillos de molienda. En el intersticio paralelo entre los elementos moledores tiene lugar el desmenuzamiento de clínker y de adiciones, a elevadas velocidades de molienda y con recirculación interna de los gruesos. Gracias a estas medidas constructivas sencillas, se puede obtener la calidad de cemento deseada, con una marcha del molino casi exenta de vibraciones. No es necesario un edificio cerrado para los molinos. Una duración de vida de los elementos moledores, de más de 10 000 horas, y una gran disponibilidad de los molinos garantizan una calidad uniforme del producto. La conexión simple de la instalación permite aplicar un concepto de regulación sencillo. En los últimos años se han optimizado, en varias fábricas, molinos Loesche de 4 cilindros, destinados a la molienda de clínker y de escoria siderúrgica, alcanzándose rendimientos de hasta 140 t/h, con calidad de cemento de molino tubular de 3 400 Blaine. El consumo específico de energía se sitúa notablemente por debajo del de los molinos tubulares y resulta también más ventajoso que en las instalaciones con molino tubular y molino de cilindros y lecho de material preconectado.*

El molino Loesche para la molienda de clínker de cemento y de adiciones, durante el servicio industrial

1. Einführung

In der Zementindustrie wird weltweit das Ziel verfolgt, durch eine Verbesserung des spezifischen Energiebedarfs die Herstellkosten bei der Zementfertigmahlung zu reduzieren. Dazu wurden in den letzten Jahren die konventionellen Rohrmühlensysteme weiter optimiert. Weitere Fortschritte brachten Walzenmühlen, die ebenfalls Eingang in diesen Bereich fanden. In diesen Mühlen erfolgt bekanntlich die Zerkleinerung unter Anwendung hoher Drücke im Gutbett. Eingesetzt werden Walzenmühlen bis heute fast ausschließlich zum Vormahlen von Zement, der in einer Rohrmühle fertiggemahlen wird. Hingegen wurden Wälzmühlen bei der Zerkleinerung von Zementklinker und Zumahlstoffen, wie Hochofenschlacken und Puzzolanen, bislang nur in Einzelfällen eingesetzt.

Durch die Loesche GmbH, die weltweit als erster Wälzmühlenhersteller seit 1928 Vertikalmühlen mit dynamischen Sichtern für die Mahlung von Kohle, Kalkstein und Zementrohmaterial eingesetzt hat, konnte wiederholt nachgewiesen werden, daß bei vergleichbarer Produktqualität durch die hohe Effektivität des Mahlsystems der Energieverbrauch gegenüber konventionellen Systemen deutlich niedriger liegt. Die Betriebssicherheit von Wälzmühlen ist hinreichend bekannt; die Wachstumsgesetze und die darauf basierenden Maschinenvergrößerungen sind gesichert, was sowohl für Mühlen mit 2, 3 oder 4 Mahlwalzen gilt.

1. Introduction

The aim of the cement industry throughout the world is to reduce the production costs of cement grinding by improving the specific power consumption. There has been continued optimization of conventional tube mill systems during the last few years, and high-pressure grinding rolls, which were also introduced in this field, have brought further progress. It is well known that the comminution in these rolls takes place in the bed of material with the application of high pressures. Until now high-pressure grinding rolls have been used almost exclusively for preliminary grinding of cement which is then finish ground in a tube mill. Roller grinding mills, on the other hand, have so far only been used in isolated cases for the comminution of cement clinker and of interground additives such as blast furnace slag and pozzolans.

As the first manufacturers of roller grinding mills, Loesche GmbH have installed vertical mills with dynamic classifiers for grinding coal, limestone and cement raw material throughout the world since 1928. They have been able to demonstrate repeatedly that for comparable product quality the power consumption is significantly lower than with conventional systems because of the high efficiency of the grinding system. The operational reliability of roller grinding mills is well known; the scale-up rules and the increased machine sizes based on them are well established, and this applies equally for mills with 2, 3 or 4 grinding rollers.

During recent years the design of the roller grinding mill has been systematically developed for use in grinding cement and blast furnace slag. This continued development work has been carried out in cooperation with UBE Industries in Japan based on modified 4-roller mills used for producing raw meal.

2. The physical problems

An essential requirement in developing the roller grinding mill (used until now almost exclusively for processing cement raw material) into a cement clinker mill was to achieve smooth mill running. In roller grinding mills operating on the airswept principle the very fine grinding to grain sizes between 2 and 50 μm is considerably impaired by imperfect formation of the grinding bed. Due to the strong aeration of the grinding bed, the small grain sizes involved in cement grinding, and the great upward forces, the grinding rollers are not very effective at drawing in large groups of particles. Unlike the situation with the coarser grinding involved with cement raw material and coal, a particle cloud consisting of cement clinker has little tendency to settle unaided on a rotating grinding track. It is well known that aerated dust behaves physically like a liquid. The internal friction and the friction with solid bodies are very low. The particle cloud flows like a liquid, and it flows around the grinding rollers which lie in its path, so only a small mass flow enters the gap between the roller and grinding track. The mass flow must first have formed a solid wedge ahead of the roller before it can be trapped, rolled and comminuted. In the next instant the grinding bed breaks down again because the particle cloud must once more be slowly recompacted into a wedge which will permit the introduction of high pressure forces. Each grinding roller must therefore alternately prepare and comminute its own grinding bed. This leads to a type of slip-stick effect and can generate vibrations.

However, low-vibration mill operation, i.e. carefully controlled and uniform formation of the grinding bed, is a crucial requirement for being able to raise the specific grinding forces above the level used for comminution of cement raw materials, and higher grinding forces are the precondition for achieving the product quality required nowadays for a cement or granulated blast furnace slag.

3. Features of the Loesche roller mill for grinding cement and granulated blast furnace slag

Conversion of the findings from the physical processes into an industrially feasible mechanical design led to a patented solution in which the preparation and comminution of the grinding bed are undertaken by separate elements. A cement mill designed on this basis for the fine grinding of products to high Blaine values is characterized by the following features:

The familiar basic concept of a Loesche mill, the pairing of conical rollers with a flat, horizontal, grinding track, is retained. The rollers are, as before, carried individually on rocker arms. This grinding principle leads in practice to a parallel grinding gap between the grinding elements, which is also a typical feature of all high-pressure grinding rolls. The specific grinding pressure in the comminution gap is significantly higher than when grinding cement raw materials or comparable products and product finenesses. The pressures applied are increased with rising Blaine values.

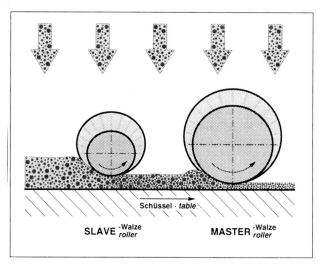

BILD 1: Schematische Darstellung der Mahlbettpräparation
FIGURE 1: Diagrammatic representation of grinding bed preparation

Ein vibrationsarmer Mühlenbetrieb, d. h. eine gezielte und gleichmäßige Mahlbettausbildung, ist aber eine entscheidende Voraussetzung dafür, um die spezifischen Mahlkräfte gegenüber der Zerkleinerung von Zementrohstoffen erhöhen zu können. Höhere Mahlkräfte sind aber auch die Voraussetzung zur Erzielung der Produktqualität, wie sie von einem Zement oder Hüttensand heutzutage gefordert wird.

3. Merkmale der Loesche-Mühle für die Zement- und Hüttensandmahlung

Die Umsetzung der Erkenntnisse aus den physikalischen Vorgängen in eine industriell anwendbare Maschinenkonstruktion führte zu einer patentierten Lösung. Nach dieser Lösung übernehmen getrennte Elemente das Präparieren und das Zerkleinern des Mahlbetts. Eine so konzipierte Zementmühle für die Feinmahlung von Produkten auf hohe Blaine-Werte ist durch folgende Merkmale gekennzeichnet:

Das bekannte Grundkonzept einer Loesche-Mühle, die Paarung von konischen Mahlwalzen mit einer horizontalen, ebenen Mahlbahn, wurde beibehalten. Die Walzen werden wie bisher individuell in Schwinghebeln geführt. Dieses Mahlprinzip führt in der Praxis zu einem parallelen Mahlgutspalt zwischen den Mahlwerkzeugen, der auch ein typisches Merkmal einer jeden Hochdruck-Walzenmühle ist. Der spezifische Mahldruck im Zerkleinerungsspalt liegt deutlich höher als bei der Mahlung von Zementrohstoffen oder vergleichbaren Produkten und Produktfeinheiten. Die angewendeten Drücke werden mit steigenden Blaine-Werten erhöht.

Die Beanspruchung des Mahlgutes erfolgt entsprechend der Definition von Schönert [1] im Gutbett. Das Gutbett wird auf der horizontalen Mahlbahn aufgebaut und kontinuierlich stabil gehalten. Wegen des internen Mahlgutumlaufs im Mühlenraum durchwandern die Mahlgutpartikel den Mahlspalt mehrfach, bis sie als Fertigprodukt die Mühle über den Sichter verlassen. Der für die Zement- und Hüttensandmahlung erforderliche hohe spezifische Mahldruck wird durch gezielte Mahlbettbildung möglich, wozu Walzenpaare dienen, die eine Aktions-Einheit bilden. Jedes Walzenpaar besteht aus einer S-Walze (Slave-roller) und einer M-Walze (Master-roller). Die S-Walze präpariert das Mahlbett. Eine – in Drehrichtung der Schüssel gesehen – hinter der Walze angeordnete M-Walze zerkleinert das Mahlgut. Wegen der paarweisen Arbeitsweise von S- und M-Walze wird eine 4-Walzenmühle LM..2 + 2 C (Cement) bzw. LM..2 + 2 S (Slag) genannt. Bei Durchsätzen beispielsweise über 200 t/h läßt sich unter Verwendung von Baugruppen einer (2 + 2 C)-Mühle auch eine (3 + 3 C)-Mühle realisieren (**Bild 2**). Mechanische Puffer begrenzen den Schwinghebel-Walzenweg nach unten und verhindern so den metallischen Kontakt mit der Mahlbahn. Die Anwendung von Puffern erlaubt den Einsatz hochfester Mahlteile und verhindert Ausbrüche an den harten Beanspruchungsoberflächen.

Der Notwendigkeit, harte und spröde Stoffe möglichst nur unter Druckbeanspruchung zu zerkleinern, wurde konstruktiv dadurch begegnet, daß die M-Walzen schmal und mit großem Durchmesser dimensioniert werden und daß durch die Position der M-Walzen zur Mühlenmitte nur noch geringe Differenzgeschwindigkeiten zwischen dem großen und kleinen Walzendurchmesser entstehen. Die verbliebenen Differenzgeschwindigkeiten an den M-Walzen führen zu Scherbeanspruchungen, die ausreichend sind, um die Kompaktierung von Mahlgut zu vermeiden (**Bild 3**).

Die S-Walzen sind nach den notwendigen Kriterien der optimalen Mahlbettpräparation gestaltet. Sie sind konisch wie die M-Walzen ausgeführt, wobei ihre Breite etwa der Mahlbahnbreite entspricht. Sie sind für eine theoretisch rein rollende Bewegung ausgeführt, d.h. auf der gesamten Walzenbreite sind die Umfangsgeschwindigkeiten aller Rollkreise identisch mit denen der Mahlschüssel. Damit werden keine Scherkräfte auf das Mahlbett übertragen.

Aus **Bild 4** geht hervor, daß auch bei den Zementmühlen die mittlere Mahlgeschwindigkeit mit der Quadratwurzel des

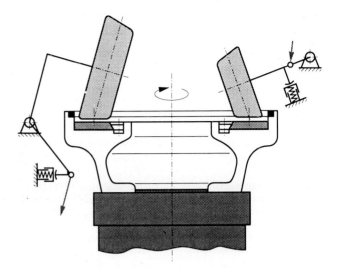

BILD 2: S-Walze und M-Walze auf der Mahlbahn einer Loesche-Mühle
FIGURE 2: S-roller and M-roller on the grinding track of a Loesche mill

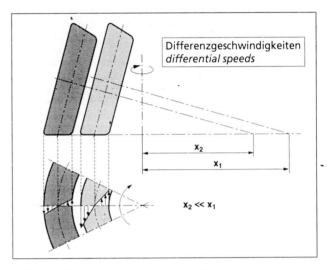

BILD 3: Differenzgeschwindigkeiten der Mahlwalzen als Funktion ihres Abstandes vom Zentrum des Mahltellers
FIGURE 3: Differential speeds of the grinding rollers as a function of their distances from the centre of the grinding table

The stressing of the material being ground takes place in the material bed as defined by Schönert [1]. The material bed is built up on the horizontal grinding track and maintained continuously in a stable condition. Because of the internal recirculation of material in the grinding chamber the particles of material pass through the grinding gap several times before they leave the mill via the classifier as finished product. The high specific grinding pressure needed for grinding cement and blast furnace slag is made possible by carefully controlled formation of the grinding bed; this is carried out by pairs of rollers which form an action unit. Each roller pair consists of an S-roller (slave roller) and an M-roller (master roller). The S-roller prepares the grinding bed. An M-roller positioned behind — in relation to the direction of rotation of the table — the S-roller comminutes the material. Because of the paired mode of operation of the S- and M-rollers a 4-roller mill is called LM..2 + 2 C (Cement) or LM..2 + 2 S (Slag). For throughputs above 200 t/h, for example, it also is possible, using modules from a (2 + 2 C) mill, to construct a (3 + 3 C) mill (**Fig. 2**). Mechanical buffers limit the downward travel of the rocker arm and roller, and so prevent any metallic contact with the grinding track. The use of buffers also makes it possible to employ high-strength grinding elements and prevents pieces from being broken out of the hard working surfaces.

The need for comminuting hard and brittle substances if possible only by pressure stressing is dealt with in the

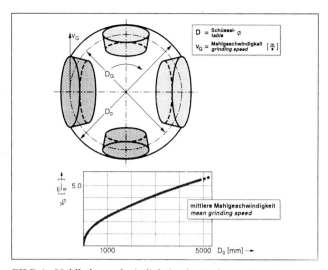

BILD 4: Mahlbahngeschwindigkeit als Funktion des Mahlbahndurchmessers
FIGURE 4: Grinding track speed as a function of grinding track diameter

Mahlbahndurchmessers $D^{0,5}$ ansteigt. Eine Wälzmühle der Baugröße LM 45 C arbeitet beispielsweise mit Mahlbahngeschwindigkeiten, die mehr als 5 m/s betragen. Demnach gilt auch für Loesche-Zementmühlen das Vergrößerungsgesetz, wonach der Durchsatz mit der 2,5-ten Potenz des Mahlbahndurchmessers ansteigt, d. h. es ist $\dot{m} \sim D^{2,5}$.

4. Sichtung

Zur Zementmühle gehört ein Kreiselsichter mit Leistenrotor, der unter den dynamischen Loesche-Sichtern zur zweiten Generation zählt. Der Sichter unterscheidet sich nur marginal von den praxisbewährten Sichtern für andere Mahlgüter. Der Zentraleinlauf für die konzentrische Mahlgutzufuhr durch ein Fallrohr in die Mühle ist vor allem für feuchte Mahlgüter wie z. B. Hüttensand erforderlich. Diese Lösung mit Fallrohrdurchmessern bis 1100 mm ist für Aufgabegüter mit (Misch-)Wassergehalten bis zu 23 % mehrfach praxiserprobt. **Bild 5** zeigt den prinzipiellen Aufbau einer Loesche-Zement- und Hüttensandmühle mit einem integrierten dynamischen Sichter.

Da üblicherweise ein Fertigprodukt mit einer breiten Korngrößenverteilung angestrebt wird, ist bewußt auf die

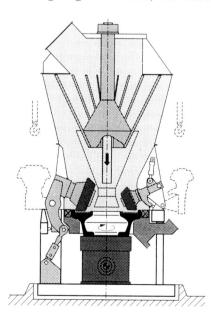

BILD 5: Prinzipieller Aufbau einer Loesche-Zement- und Hüttensandmühle mit integriertem Sichter
FIGURE 5: Basic structure of a Loesche cement and slag mill with integral classifier

design by making the M-rollers narrow and of large diameter and by positioning the M-rollers in relation to the mill centre so that there is only a small differential speed between the large and small diameters of the roller. The residual differential speeds at the M-rollers lead to shear stresses which are sufficient to avoid compacting the material being ground (**Fig. 3**).

The S-rollers are configured to meet the criteria for optimum grinding bed preparation. Like the M-rollers, they are conical, and their width corresponds approximately to the width of the grinding track. They are designed for a theoretically purely rolling movement, i.e. the circumferential speeds of all the rolling circles over the entire width of the roller are identical with those of the grinding table, so no shear forces are transmitted to the grinding bed.

It can be seen from **Fig. 4** that in cement mills the average grinding speed also increases with the square root $D^{0.5}$ of the grinding track diameter. An LM 45 C roller grinding mill, for example, works with grinding track speeds of more than 5 m/s. The scale-up rule by which the throughput increases in proportion to the grinding track diameter raised to the power of 2.5, i.e. $\dot{m} \sim D^{2.5}$, also applies to Loesche cement mills.

4. Classification

The cement mill has a rotor classifier with a bladed rotor which is one of the second generation of dynamic Loesche classifiers. The classifier differs only marginally from the well tried classifiers for other ground materials. The central inlet for concentric material feed into the mill through a drop tube is necessary primarily for moist materials such as granulated blast furnace slag. This solution with drop tube diameters up to 1100 mm has proved successful time and again in practice for feed materials with water contents of up to 23%. **Fig. 5** shows the basic design of a Loesche cement and blast furnace slag mill with integral dynamic classifier.

The object is normally to achieve a finished product with a wide grain size distribution, so the use of a so-called high-effect classifier is intentionally avoided. If, on the other hand, the highest possible slope in the RRSB granulometric diagram in certain grain size ranges is required — e.g. for separate processing of granulated blast furnace slag — because of the product properties to be produced by subsequent mixing with cement clinker, then 2-stage classification is provided with a Loesche high-efficiency classifier. Either the Loesche jalousie classifier (LJKS) [2] or a classifier of the fourth generation which is now available can be used for this purpose. In the latter the static preliminary classifying system can be adjusted during operation.

5. Cement grinding plant

5.1 Structure

Fig. 6 shows an example of the very simple arrangement of a cement grinding plant with an LM.. 2 + 2 C roller grinding mill. Cement clinker, gypsum (and blast furnace slag) are conveyed into the corresponding raw material hoppers. The components are extracted in the correct proportions by metering feeders and transported to the mill by a belt conveyor. A collecting conveyor projects the feed material into the mill through a gate feeder. The mill feed is finish-ground and classified in the mill, which is equipped with a top-mounted dynamic classifier. Oversize material is rejected by the classifier and returned to the centre of the grinding table to be reground. A stream of air passes through the mill and classifier, and takes care of the transport and classification of the material being ground. Tailings which have neither been caught by the grinding rollers for comminution nor reached the classifier with the aid of the air stream are discharged from the mill below the grinding table and transported into a recirculating bucket elevator by a vibrating conveyor. From the bucket elevator this material passes onto a weigh belt with metal separator. The material can be conveyed by diverters either onto the feed belt or into a buf-

Anwendung eines sogenannten Hocheffekt-Sichters verzichtet worden. Wird jedoch – z.B. bei separater Aufbereitung von Hüttensand – wegen der durch die spätere Mischung mit Zementklinker zu erzeugenden Produkteigenschaften in bestimmten Korngrößenbereichen ein möglichst hohes Steigungsmaß im RRSB-Diagramm gefordert, so wird eine 2-Stufen-Sichtung mit einem Loesche-Hocheffekt-Sichter vorgesehen. In Frage kommen dafür entweder der Loesche-Jalousie-Sichter (LJKS) [2] oder ein inzwischen verfügbarer Sichter der vierten Generation. Bei letzterem ist eine Einflußnahme auf die statische Vorsichtung während des Betriebes möglich.

5. Zement-Mahlanlage

5.1 Aufbau

Bild 6 zeigt beispielhaft die sehr einfache Schaltung einer Zementmahlanlage mit einer Wälzmühle der Typenbezeichnung LM..2 + 2 C. Zementklinker, Gips (und Hüttensand) werden in entsprechende Rohmaterialbunker gefördert. Über Dosiereinrichtungen werden die Komponenten proportional abgezogen und über einen Gurtförderer zur Mühle transportiert. Ein Sammelförderer wirft das Aufgabegut über eine Klappenschleuse in die Mühle. In der Mühle, die mit einem aufgesetzten dynamischen Sichter ausgerüstet ist, wird das Mahlgut fertiggemahlen und gesichtet. Überkorn wird vom Sichter abgewiesen und zentral auf die Mahlschüssel zurückgeführt, um nochmals beansprucht zu werden. Mühle und Sichter werden von einem Luftstrom durchzogen, der für den Transport und die Sichtung des Mahlguts sorgt. Grieße, die weder von den Mahlwalzen zum Zerkleinern erfaßt werden, noch mit Hilfe des Luftstroms zum Sichter gelangen, werden unterhalb der Mahlschüssel aus der Mühle ausgeschleust und über einen Schwingförderer in ein Umlaufbecherwerk gefördert. Vom Becherwerk gelangt dieses Material auf ein Wiegeband mit Metallabscheider. Über Förderweichen kann das Material entweder auf das Dosierband oder in einen Pufferbunker mit Schwingrinnenabzug wieder in das Becherwerk gefördert werden. Das Material gelangt so über ein externes Handling System in die Mühle zurück. Metallteile werden aus dem System entfernt.

Das Fertigprodukt wird mit Hilfe des vom Anlagenventilator erzeugten Luftstroms in das Schlauchfilter gefördert und abgeschieden. Die für die Zementmahlung erforderliche Guterwärmung kann durch einen Heißgaserzeuger rea-

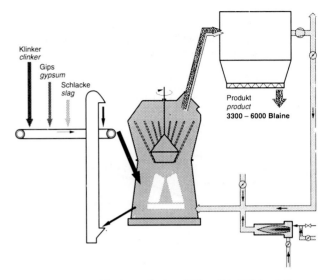

BILD 6: Zementmahlanlage mit einer Wälzmühle LM..2 + 2 C
FIGURE 6: Cement grinding plant with an LM..2 + 2 C roller grinding mill

fer hopper with vibrating conveyor extraction and back into the bucket elevator, so the material is returned to the mill via an external handling system. Any pieces of metal are removed from the system.

The finished product is carried to the bag filter with the aid of the air flow generated by the plant fan, and then separated. The heating of the material which is necessary for cement grinding can be achieved by a hot gas generator. Most of the grinding heat remains in the system as the hot air is returned to the mill through a gas return line. The grinding plant also has metered feeding equipment for water.

5.2 Control loops

The technological control scheme matches the plant configuration and is also simple. The following are controlled:

– the mill differential pressure, by the flow of feed material,
– the volumetric flow after the mill, by the plant fan (rotational speed control),
– the temperature after the mill, with the aid of the returned gas and the heat generator,
– the pressure drop from the atmosphere to the gas inlet into the mill, by a control damper in the air duct to the mill.

6. Properties of cement produced in roller grinding mills

The product quality, including a given grain size distribution, can be set and maintained by varying

– the height of the dam ring on the grinding table (grinding bed thickness),
– the air flow through the mill,
– the product temperature, with a target range of 90 to 100°C,
– the classifier rotor speed, and
– the specific grinding pressure.

The diagram in **Fig. 7** compares typical grain size distribution curves of 3 cements with similar Blaine values. The black TM curve represents tube mill cement, while the light grey LM curve applies to cement from a roller grinding mill with the narrow grain size distribution and pronounced peak normally expected from this mill. The dark grey LM-C curve applies to a cement which was ground in the new Loesche mill. The curve has a slightly different Blaine value but almost matches the continuous TM curve. The match between the curves is achieved by the high specific grinding forces required for grinding cement (and blast furnace slag)

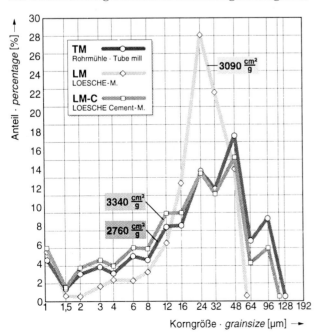

BILD 7: Korngrößenverteilungskurven von Zementen, die in unterschiedlichen Mühlen gemahlen wurden
FIGURE 7: Grain size distribution curves for cements which have been ground in different mills

lisiert werden. Andererseits bleibt die Mahlwärme weitestgehend im System, weil sie über eine Rückgasleitung wieder in die Mühle gelangt. Die Mahlanlage ist auch mit einer Dosiereinrichtung für Wasser ausgestattet.

5.2 Regelkreise

Entsprechend der technologischen Schaltung ist auch das Regelkonzept einfach. Geregelt werden:
— Der Mühlendifferenzdruck über den Aufgabegutstrom,
— der Volumenstrom nach der Mühle über den Anlagenventilator (Drehzahlregelung),
— die Temperatur nach der Mühle mit Hilfe des Rückgases bzw. des Wärmeerzeugers,
— das Druckgefälle von der Atmosphäre zum Gaseintritt in die Mühle über eine Regelklappe im Luftkanal zur Mühle.

6. Eigenschaften des Wälzmühlenzements

Für die Einstellung und Einhaltung der Produktqualität, d.h. auch einer bestimmten Korngrößenverteilung, eignen sich Variationen
— der Stauringhöhe an der Mahlschüssel (Mahlbettdicke),
— des Luftstroms durch die Mühle,
— der Produkttemperatur, die mit 90 bis 100 °C angestrebt wird,
— der Sichterrotor-Drehzahl und
— des spezifischen Mahldrucks.

Das Diagramm auf **Bild 7** zeigt typische Korngrößenverteilungskurven von 3 Zementen mit ähnlichen Blaine-Werten im Vergleich. Dabei gilt die fette TM-Kurve für Rohrmühlenzement, während sich die hellgraue LM-Kurve auf Wälzmühlenzement bezieht, wie man ihn üblicherweise von dieser Mühle mit schmaler Korngrößenverteilung und ausgeprägter Spitze erwartet. Die dunkelgraue LM-C-Kurve gilt für einen Zement, der in der neuen Loesche-Mühle gemahlen wurde. Die Kurve ist bei leicht unterschiedlichem Blaine-Wert fast deckungsgleich mit der durchgezogenen TM-Kurve. Erreicht wird die Deckung der Kurven durch die für die Zement- (und Hüttensand) Mahlung erforderlichen hohen spezifischen Mahldrücke, die bei hohen Mahlgeschwindigkeiten erst durch die paarweise Aktion von S- und M-Walzen voll realisierbar werden.

Elektronenmikroskopische Untersuchungen repräsentativer Proben von in der neuen Loesche-Zementmühle erzeugten Zementen und Rohrmühlenzementen ergaben keine signifikanten Unterschiede.

Neben einer bestimmten Charakteristik der Korngrößenverteilung sind bei einem Zement auch die Anforderungen nach DIN 1164 zu erfüllen. In **Tabelle 1** sind alle charakteristischen Daten eines LM-C-Zementes zusammengestellt. Die Untersuchungen, die auch die mörteltechnischen Prüfungen einschließen, wurden von einem unabhängigen Institut durchgeführt.

7. Betriebserfahrungen

Bei den ersten industriell eingesetzten Zement- und Hüttensandmühlen sind die S- und M-Walzen hinsichtlich ihrer Abmessungen noch identisch ausgeführt. In der Fertigung befindliche Großmühlen für 140 t/h Zement werden bereits mit unterschiedlichen Walzen ausgerüstet.

Dank der Präparation des Mahlbetts vor seiner Hauptbeanspruchung durch die M-Walzen ist der Mühlenbetrieb vibrationsarm. An einer Mühle zur Mahlung von Hüttensand mit einem Durchsatz von 50 t/h und an einer Zementmühle für einen Durchsatz von 130 t/h wurden am Mühlenfuß Schwinggeschwindigkeiten zwischen 1 und 2 mm/s gemessen.

An LM-C-Zementmühlen in Japan und Korea wurden bei den Walzenmänteln Standzeiten – ohne zwischenzeitliche

TABELLE 1: Charakteristische Daten eines LM-C-Zementes
TABLE 1: Characteristic data for an LM-C cement

Zementfeinheit:	Rückstand auf dem Sieb 0,2 DIN 4188: <0,1 % Spezifische Oberfläche nach Blaine: 3440 cm²/g
Cement fineness:	residue on 0.2 DIN 4188 screen: <0.1% specific surface area (Blaine): 3440 cm²/g
Erstarren:	Beginn: 3 h 33 min Ende: 4 h 50 min
Setting times:	initial: 3 h 33 min final: 4 h 50 min
Wasseranspruch:	28,4 %
Water demand:	28.4 %
Biegezugfestigkeit:	2 Tage: 3,7 N/mm² 7 Tage: 6,1 N/mm² 28 Tage: 7,5 N/mm²
Flexural tensile strength:	2 day 3.7 N/mm² 7 day 6.1 N/mm² 28 day 7.5 N/mm²
Druckfestigkeit:	2 Tage: 17,0 N/mm² 7 Tage: 37,6 N/mm² 28 Tage: 52,8 N/mm²
Compressive strength:	2 day 17.0 N/mm² 7 day 37.6 N/mm² 28 day 52.8 N/mm²
Raumbeständigkeit nach Le Chatelier	2,0 mm
Soundness (Le Chatelier):	2.0 mm
SO_3-Gehalt:	2,07 Gew.-%
SO_3 content:	2.07 wt.%
Unlöslicher Rückstand:	0,59 Gew.-%
Insoluble residue:	0.59 wt.%
Glühverlust:	2,98 Gew.-%
Loss on ignition:	2.98 wt.%
Ausbreitmaß:	148 mm (nach alter Norm DIN 1060)
Spread:	148 mm (as specified in the old DIN 1060 standard)

which can only be fully achieved at high grinding speeds by the paired action of the S- and M-rollers.

Electron microscope investigations of representative samples of cements produced in the new Loesche cement mill and tube mill cements did not show any significant differences.

In addition to a certain characteristic of the grain size distribution a cement also has to fulfil the requirements specified in DIN 1164. All the characteristic data of an LM-C cement are summarized in **Table 1**. The investigations, which also include the mortar trials, were carried out by an independent institute.

7. Operating experience

In the first cement and blast furnace slag mill used industrially the S- and M-rollers still had identical dimensions. Large mills for 140 t/h cement currently under manufacture are being equipped with different rollers.

Thanks to the preparation of the grinding bed before its main stressing by the M-rollers there is little vibration in the mill operation. Oscillation speeds of between 1 and 2 mm/s were measured at the foot of a mill for grinding blast furnace slag with a throughput of 50 t/h and of a cement mill with a throughput of 130 t/h.

Service lives of the roller shells — without any intermediate reworking — of more than 8000 h have been achieved in LM-C cement mills in Japan and Korea, and up to 15000 h have

Aufarbeitung – von > 8 000 h erzielt. Bei den Mahlbahnplatten wurden bis zu 15 000 h erreicht. Der Nettoverschleiß an den Mahlteilen beträgt bei der Zementmahlung auf Blaine-Werte von ca. 3 300 cm^2/g zwischen 2 und 4 g/t Fertigprodukt. Kombinationen von neuen Walzenmänteln und teilverschlissenen Mahlplatten haben sich bisher weder negativ auf die Laufruhe noch auf die Produktqualität ausgewirkt. Die Mahlteile von Hüttensandmühlen müssen, speziell bei Produkten mit Blaine-Werten von 4 500 bis 5 000 cm^2/g, in den genannten Zeiträumen zwischendurch partiell aufgepanzert werden. Gegen Strahlverschleiß sind die Mühlen durch Panzerungen geschützt, wie sie bereits seit Jahren beim Mahlen von Kohlen mit Aschegehalten bis zu 45 % eingesetzt werden.

Bei der Zement- und Hüttensandmahlung liegt der notwendige Luftvolumenstrom für den mühleninternen Grießumlauf bzw. den pneumatischen Staubtransport mit < 50 % deutlich unter dem Bedarf der Mahltrocknung von Zementrohstoffen. Dadurch wird der Gesamtenergiebedarf positiv beeinflußt. Für eine komplette Zementmahlanlage nach Bild 6 liegt der spezifische Energiebedarf zwischen 26 und 29 kWh/t bei einer Fertigproduktfeinheit entsprechend einem Blaine-Wert von 3 300 cm^2/g.

8. Schlußbemerkungen

Im Jahre 1935 wurde nachweislich eine Loesche-Wälzmühle LM 11 erstmalig für die Mahlung von Zementklinker an die Companhia Industrias Brasileiras Portella S. A. nach Joao Pessao/Brasilien geliefert. Im Jahre 1993 befanden sich bereits 6 LM-C-Mahlanlagen mit Durchsätzen zwischen 50 und 150 t/h in Betrieb. Die älteste Anlage dieser zuletzt genannten Mahlanlagen läuft seit 1990. Weitere 6 Mühlen LM-S (S = Schlacke) für die Hüttensandmahlung laufen mit Durchsätzen von 25 bis 75 t/h in Ostasien, die älteste Mühle seit 1988. Darüber hinaus befindet sich eine kleine Produktionsanlage mit einer Wälzmühle LM-C für 0,6 t/h zur Herstellung eines Spezial-Aluminiumklinkers seit 4 Jahren in Südafrika in Betrieb. Die mit dieser Anlage erreichte Korngrößenkennlinie sowie der Blaine-Wert von 5 000 cm^2/g entsprechen den Forderungen des Betreibers. Weitere Großmühlen mit Durchsätzen bis zu 150 t/h sind in der Fertigung.

Der Zementindustrie steht heute zur Herstellung marktgerechter Zemente mit einer industriell erprobten Wälzmühle LM..2 + 2 C ein einfaches und energiegünstiges Mahlsystem zur Verfügung.

Literaturverzeichnis

[1] Schönert, K.: Verfahren zur Fein- und Feinstzerkleinerung von Materialien spröden Stoffverhaltens. Auslegeschrift 27 080 53, Offenlegungstag: 7. 9. 1978.

[2] Brundiek, H.: Wälzmühlensichter – Rückblick und gegenwärtiger Entwicklungsstand. Zement-Kalk-Gips 46 (1993) Nr. 8, S. 444–450.

been achieved for the grinding track plates. The net wear on the grinding elements for grinding cement to Blaine values of 3 300 cm^2/g lies between 2 and 4 g/t finished product. Combinations of new grinding roller shells and partially worn grinding plates have so far had no detrimental effects on the smooth running or the product quality. The grinding elements in blast furnace slag mills have to be partially refaced during the periods referred to, especially with products with Blaine values of 4 500 to 5 000 cm^2/g. The mills are protected against abrasive jet wear by armouring of the type which has been used for years when grinding coals with ash levels up to 45 %.

The mills are protected against abrasive jet wear by armouring of the type which has been used for years when grinding coals with ash levels up to 45 %.

When grinding cement and blast furnace slag, the volumetric air flow needed for recycling the tailings within the mill or for pneumatic dust transport is significantly below the requirement for drying and grinding cement raw materials, being less than 50 %. This has a beneficial effect on the overall power consumption. For a complete cement grinding plant of the type shown in Fig. 6 the specific energy consumption lies between 26 and 29 kWh/t for a finished product fineness corresponding to a Blaine value of 3 300 cm^2/g.

8. Final comments

It is known that the first Loesche roller grinding mill for grinding cement clinker was an LM 11 mill supplied in 1935 to Joao Pessao, Brazil, for the Companhia Industrias Brasileiras Portella S.A. In 1993 there were six LM-C grinding plants in operation with throughputs of between 50 and 150 t/h, the oldest of which had been running since 1990. Six more LM-S mills (S = Slag) for grinding blast furnace slag with throughputs from 25 to 75 t/h are operating in East Asia – the oldest mill has been operating since 1988. There is also a small production plant with an LM-C roller grinding mill for 0.6 t/h for producing a special aluminium clinker which has been in operation in South Africa for 4 years. The characteristic grain size curve achieved with this plant and the Blaine value of 5 000 cm^2/g satisfy the operator's requirements. Further large mills with throughputs of up to 150 t/h are being manufactured.

There is now a simple, low-energy, grinding system available to the cement industry for producing cements which comply with the market demands in the form of an industrially proven LM..2 + 2 C roller grinding mill.

BHG-Mühlen – ein neues Mahlverfahren
BHG mill – a new grinding system

Broyeurs BHG – un nouveau procédé de broyage

Molinos BHG – un nuevo procedimiento de molienda

Von **S. Buzzi,** Casale Monferrato/Italien

Zusammenfassung – Die italienische Buzzi-Gruppe hat in Zusammenarbeit mit Five Lille Babcock in Frankreich ein völlig neues Mahlsystem entwickelt, das zu einem Durchbruch in der Zementproduktion führen könnte, weil es nicht nur aus kompakten und robusten Geräten besteht, sondern auch eine ganz erhebliche Reduzierung des Energiebedarfs bei der Fertigmahlung ermöglicht, ohne Qualitätseinbußen beim Endprodukt, z.B. beim Zement, zu verursachen. Die erste BHG-Mühle mit 500 kW und einer Kapazität von 25 t/h Zement wurde vor kurzem im Werk Trino von Cementi Buzzi im Norden Italiens gebaut und untersucht. Obgleich die Mühle in erster Linie für die Zementproduktion konstruiert wurde, gestattet ihr flexibles Konzept nicht nur die Fertigmahlung von Zement, sondern auch das Mahlen von Rohstoffen und Kohle. Das Konzept der BHG-Mühle ermöglicht auch ihren Einsatz zur Vormahlung.

BHG-Mühlen – ein neues Mahlverfahren

Summary – The Buzzi Group, Italy, in cooperation with the Five Lille Babcock Company, France, has developed a complete new grinding system which might lead to a breakthrough in the cement production process. This is mainly due to the fact that, besides compact and sturdy equipment, the energy consumption of final grinding of products can be decreased decisively without quality impairment, e.g. in the case of cement. The first 500 kW BHG mill with the capacity of 25 t/h of cement has been erected and investigated very recently at the Cementi Buzzi Trino Plant in the north of Italy. Although the mill was initially designed for cement production the flexible layout allows the grinding of raw materials and coal as well as finish grinding of cement. The layout means that the BHG mill can also be used as a pre-grinding unit.

BHG mill – a new grinding system

Résumé – Le groupe italien Buzzi a, en coopération avec Five Lille Babcock en France, mis au point un système de broyage de conception entièrement nouvelle, susceptible d'entraîner une véritable percée dans le secteur de la production du ciment, du fait qu'il ne comporte pas seulement des appareils compacts et robustes mais permet aussi une réduction non négligeable des besoins énergétiques nécessaires au broyage de produit fini, sans entraîner de pertes de qualité du produit fini, du ciment en l'occurence. Le premier broyeur BHG de 500 kW d'une capacité de 25 t/h de ciment a été récemment construit et testé à l'usine Trino de Cementi Buzzi au Nord de l'Italie. Bien que ce broyeur soit construit en premier lieu pour la production de ciment, son concept flexible permet non seulement le broyage de produit fini de ciment mais également le broyage de matières premières et de charbon. Le broyeur BHG peut aussi être utilisé pour le prébroyage.

Broyeurs BHG – un nouveau procédé de broyage

Resumen – El grupo italiano Buzzi, en colaboración con Five Lille Babcock, en Francia ha desarrollado un sistema de molienda completamente nuevo, que podría suponer un gran cambio en la producción del cemento, ya que no sólo se compone de aparatos compactos y robustos, sino que permite también una considerable reducción del consumo de energía en el proceso de molienda de acabado, sin mermar la calidad del producto final, por ejemplo del cemento. El primer molino BHG, de 500 kW y una capacidad de producción de 25 t/h de cemento, ha sido montado y estudiado, hace poco, por Cementi Buzzi, en la factoría de Trino, en el Norte de Italia. Aunque el molino se ha construido, en primer lugar, para la fabricación del cemento, su concepto flexible permite no solamente el molido de acabado de cemento, sino también el de materias primas y de carbón. El citado concepto hace posible asimismo su empleo en la molienda previa.

Molinos BHG – un nuevo procedimiento de molienda

Die Veröffentlichung behandelt die neue, technisch hochentwickelte Zementmühle HOROMILL™ mit 500 kW Antriebsleistung und einer Durchsatzleistung von 25–30 t/h Zement, die einen niedrigen Energiebedarf besitzt. Dieses neue Mahlsystem wurde von FCB, Lille, Frankreich, in enger Zusammenarbeit mit der Fratelli Buzzi S.p.A., Casale Monferrato, Italien, entwickelt. Es stellt für die Zementindustrie eine echte Neuheit dar und könnte der Prototyp einer Reihe hocheffizienter Mahlanlagen werden.

Im Rahmen der sorgfältigen Planung von Modernisierungsmaßnahmen an der Mahlanlage des Werks Trino sollte ein einfaches vollständiges Mahlaggregat mit ausgesprochen niedrigem spezifischen Energiebedarf und einer geringen Wahrscheinlichkeit schnellen Veraltens installiert werden.

The paper deals with a new 500 kW (connected load), 25–30 tph (output) High-Tech, Low-Energy cement mill, named HOROMILL™. This new grinding system was developed by FCB, Lille, France, in strict cooperation with Fratelli Buzzi S.p.A., Casale Monferrato, Italy. It is definitely new in the cement industry and could become the prototype of a breed of high efficiency grinding machines.

During the thorough planning of a renovation of the Trino grinding plant, the target was to install a simple integral grinding machine with a definitely low specific energy demand and a minimum risk of getting an equipment which might quickly become obsolete. This may happen, e.g., to ball mills, especially if electrical power costs will continue to rise sharply. A roller press is not designed for integral grind-

Letzteres kann zum Beispiel bei Kugelmühlen eintreten, insbesondere bei weiterhin stark ansteigenden Kosten für elektrische Energie. Eine Gutbett-Walzenmühle ist nicht für die Fertigmahlung ausgelegt und weist darüber hinaus noch schwerwiegende Zuverlässigkeits- und Instandhaltungsprobleme auf. Wälzmühlen scheinen für das Vormahlen ebenso besser geeignet zu sein. Aber der Anteil des gesamten elektrischen Energieverbrauchs, der für den internen pneumatischen Transport des Mahlguts erforderlich ist, ist vergleichsweise hoch.

Darum wurden in Zusammenarbeit mit FCB von Anfang an Forschungsarbeiten mit dem Ziel durchgeführt, ein völlig neues Mahlsystem zu entwickeln. Die neue Mühle sollte folgende Merkmale aufweisen:

— die gleiche Zuverlässigkeit wie eine Kugelmühle und die Eignung zur Herstellung eines Fertigprodukts mit den gleichen Qualitätsmerkmalen wie bei einer solchen Mühle;

— die Energieausnutzung einer Gutbett-Walzenmühle und

— die Flexibilität einer Wälzmühle beim Mahlen verschiedener Arten von Rohstoffen und Kohle.

Für die Planung, die Konstruktion, die Pilotversuche und den Bau waren bis zur Installation der HOROMILL 20 Monate erforderlich. Wegen schwebender Patentverfahren und gewerblicher Schutzrechte kann in dieser Veröffentlichung nur eine allgemeine Beschreibung gegeben werden. Die Mühle besteht aus einem Mantel von 2,2 m Durchmesser und nur 2 m Länge, der von hydrodynamisch-statischen FCB-Gleitlagern getragen wird (**Bild 1**). Der Manteldurchmesser ist etwas geringer als der einer Kugelmühle gleicher Kapazität. Da das Längen-Durchmesser-Verhältnis nur bei etwa 1 liegt, ist der Platzbedarf für die Installation der Mühle verhältnismäßig gering.

BILD 1: Die HOROMILL in Casale Monferrato
FIGURE 1: The HOROMILL at Casale Monferrato

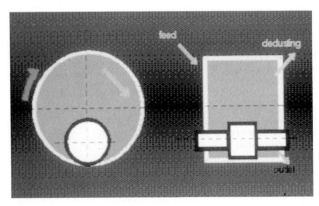

BILD 2: Skizze der HOROMILL
FIGURE 2: Sketch of the HOROMILL

Soweit es den Mahlprozeß betrifft, besteht die HOROMILL prinzipiell aus den drei Teilen Beschickungs-, Mahl- und Austragszone. Wie **Bild 2** zeigt, ist in der Mahlzone ein zylindrischer Satellit angeordnet, der die Energie auf das Mahlgut übertragen soll. Das Mahlgut wird einer kontrollierten Mehrfach-Druckbeanspruchung ausgesetzt (**Bild 3**).

Der Arbeitsdruck beträgt nur etwa ein Fünftel des Drucks in einer Gutbett-Walzenmühle und verursacht daher eine geringere mechanische Beanspruchung als in Walzenmühlen. Jedoch ist der Arbeitsdruck wegen der vergleichsweise kleineren Mahlfläche etwa 4–5mal so hoch. Der Durchgang des Mahlguts durch die Mühle läßt sich leicht steuern, und die Verweilzeit wird entsprechend eingestellt. Das gesamte Mahlsystem arbeitet im geschlossenen Kreislauf. Das Mühlenaustragsgut wird über ein Becherwerk einem Hochleistungssichter zugeführt. Das Sichterfeingut wird als Fertigprodukt in einem Schlauchfilter abgeschieden. Das Sichtergrobgut gelangt wieder in die Mühle zurück. Im Rahmen des Systems kann auch ein sogenannter „interner geschlossener Kreislauf" eingerichtet werden, der es ermöglicht, in be-

ing, and additionally it still has severe reliability and maintenance problems. Roller mills seem to be more suitable for pre-grinding as well. But the portion of total electrical power consumption required for the internal pneumatic transport of materials is comparably high.

Therefore, in cooperation with FCB and from the beginning, research work was started with the target of developing a totally new grinding system. The new mill should have

— the same reliability and produce a finished product with the quality characteristics of a ball mill,

— the energy efficiency of a roller press, and

— the flexibility of a roller mill in the grinding of different kinds of raw material and coal.

It took 20 months of planning, design, pilot tests, and construction to get the HOROMILL installed. Due to pending patents and commercial protection, this paper only gives a description in general terms. The mill consists of a shell of 2.2 m in diameter and only 2 m in length, supported by FCB hydrodynamic-static sliding bearings (**Fig. 1**). The diameter of the shell is somewhat smaller than that of a ball mill of same capacity. Since the L/D-ratio is only about 1 the space requirement for the mill installation is relatively low.

As far as the grinding process is concerned, the HOROMILL in principle consists of the three sections feeding, grinding and discharging zone. As shown in **Fig. 2**, in the grinding zone a cylindrical satellite is arranged which has the task to transmit the energy to the material to be ground. The material undergoes a controlled multiple compression process (**Fig. 3**).

The working pressure is about 5 times lower than that occurring in a roller press and therefore it causes lower mechani-

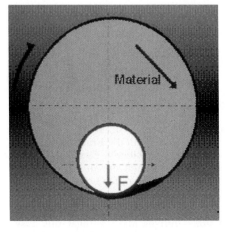

BILD 3: Das Mahlprinzip der HOROMILL
FIGURE 3: Grinding principle realized in the HOROMILL

TABELLE 1: Leistungsfähigkeit und Qualität von Zementen, die in der HOROMILL gemahlen wurden	
Siebrückstand R 32 in %	15–18
Siebrückstand R 40 in %	4
Spez. Oberfläche nach Blaine in cm²/g	3200
Druckfestigkeit nach 1, 3, 28 Tagen in N/mm²	Vergleichbar mit Zementen der Festigkeitsklasse 42.5 R nach ENV 197
Betonverarbeitbarkeit nach DIN 1048, W/C = 0,65	gleich oder besser als Rohrmühlenzement

TABLE 1: Performance and quality of cement ground with the HOROMILL	
Residue on 32 μm sieve (m.-%)	15–18
Residue on 40 μm sieve (m.-%)	4
Specific surface area (Blaine) (cm²/g)	3200
Compressive strength after 1, 3, 28 d (N/mm²)	Complying with cements of strength class 42.5 R according to ENV 197
Concrete workability according to DIN 1048, W/C = 0.65	Equal or better than ball mill finish product

stimmten Grenzen die Abmessungen des Becherwerks, des Sichters und des Filters mit deutlichen Vorteilen zu verringern.

Die rechnergesteuerte HOROMILL läuft bei der Herstellung hochwertigen Portlandzements ruhig und stabil, wie aus den Zahlen der **Tabelle 1** hervorgeht. Die Umlaufzahl liegt zwischen 6 und 7. Es werden Mahlhilfen verwendet. Die ersten Werte des elektrischen Energieverbrauchs für das Mahlen von Portlandzement mit einer spezifischen Oberfläche von 3200 cm²/g nach Blaine (gemessen als Energieverbrauch des Motors) betragen etwa 17–18 kWh/t für die Mühle allein und 7 kWh/t für die Zusatzeinrichtungen, die für die Versuche natürlich überdimensioniert waren, um die größere Flexibilität der halbtechnischen Anlage auszunutzen. Schon während der Testreihen konnte eine größere Feinheit von rund 5000–6000 cm²/g erreicht werden.

Solche Zahlenwerte, die bereits während der allerersten Probeläufe erzielt wurden, belegen eine deutliche Verminderung des elektrischen Energieverbrauchs. Die HOROMILL verbrauchte bei vergleichbarer Qualität des erzeugten Zements nur etwa 60 % der von einer optimierten Kugelmühlen-Anlage benötigten Energie. Damit hat die Mühle ein besseres Ergebnis als die vorhergesagten 70 % gezeigt, die im Vergleich mit der Kugelmühle erwartet wurden. Der Hauptvorteil der Mühle liegt in ihrem ausgesprochen stabilen Betrieb. Es dauert nur einige Minuten, bis sie ihren betrieblichen Gleichgewichtszustand erreicht. Wenn der gewünschte Umlaufmassenstrom durch Anpassung des Sichters eingestellt ist, erreichen alle übrigen Mühlenparameter, z. B. Leistungsaufnahme, Produktfeinheit, Temperatur und Druck konstante Werte, unabhängig davon, ob die Mahlanlage „von Hand" oder ohne Eingreifen des Bedieners automatisch betrieben wird. Diese Leistung wird in hohem Maße von dem TSV-Sichter unterstützt, der sich durch eine hohe Trennschärfe auszeichnet.

Die Schallabstrahlung der HOROMILL ist gering, so daß sogar eine Errichtung im Freien oder in offenen Gebäuden erwogen werden kann. Das noch bestehende Lärmproblem aufgrund des Getriebes wird in absehbarer Zukunft behoben werden. Zum Verschleiß läßt sich noch keine endgültige Aussage machen. Da jedoch zwischen dem Mahlgut und dem Mahlring kein Geschwindigkeitsunterschied besteht und die Walze sich frei bewegen kann, kann keine Reibung auftreten, die eine der Hauptursachen für Verschleißerscheinungen ist.

Die Anlage in Trino konnte in einem Gebäude von 14 × 10 × 26 m untergebracht werden. Sie war so ausgelegt, daß die HOROMILL in verschiedenen Konfigurationen getestet werden konnte, wie:

— Fertigmahlen von Zement,
— Vormahlen von Zement oder Zementklinker,
— Fertigmahlen von Rückgut anderer Mühlen,
— Mahlen von Zementrohstoffen sowie schließlich
— Mahlen von Kohle oder Koks.

Wenn nur eines dieser Mahlverfahren verwirklicht werden soll, kann offensichtlich ein weitaus einfacherer Bauplan erstellt werden, bei dem nur ein Gebäude von 12 × 8 × 12 m

cal stress compared to roller mills. However the working pressure is about 4–5 times higher due to the comparably smaller grinding surface. The passing of the material through the mill can be easily controlled, and the residence time will be adjusted accordingly. The whole grinding system works in a closed circuit. The mill output is fed to a high efficiency separator through an elevator. The separator fines are collected as final product in a bag filter. The separator rejects are recirculated to the mill. The system also allows to install a so-called "internal closed circuit" system which makes it possible to scale down, to a certain extent, the elevator, separator and the filter dimensions as well with significant benefits.

The computer controlled HOROMILL runs quietly and stable when producing Portland cement of high quality, as the figures given in **Table 1** show. The circulation load is in the range of 6–7. Grinding aids are used. First values on power consumption for the grinding of Portland cement with a specific surface of 3200 cm²/g according to Blaine (measured as power consumption of the motor) are in the range of 17–18 kWh/t for the mill alone and 7 kWh/t for the auxiliary equipment which obviously was oversized in the test situation to make use of a greater flexibility of the semi-technical plant. A higher fineness in the range of 5000–6000 cm²/g has been achieved already during the test series.

Such figures, as achieved in the very first test runs, show a significant decrease in the electrical energy consumption. The HOROMILL consumed only about 60 % of the energy required by an optimized ball mill plant at a comparable quality level of the produced cement. So the mill has shown a better result than the forecast 70 % expected with reference to the ball mill. A main advantage of the mill is its definitely stable operation. It only takes a few minutes to reach the operation equilibrium. When the desired circulation load is set by adjusting the separator all other parameters of the mill, e.g. power consumption, fineness of the product, temperature and pressure, become constant even if the grinding plant is operated "by hand" or automatically without operator's intervention. This performance is supported to a great deal by the TSV separator which is characterized by a sharp cut of separation.

The noise generated by the HOROMILL is minimal so that even an open air installation or an installation in open buildings can be envisaged. There still exists a noise problem caused by the gear box which will be solved in the near future. With regard to the wear no final statement can be made. But since there is no speed difference between material and grinding ring and the roller is able to move freely no friction can occur in this case, which is one of the basic reasons for wear abatement.

The layout of the Trino plant could be housed in a 14 × 10 × 26 m building. It was designed for testing the HOROMILL under different configurations like

— integral cement grinding,
— pre-grinding of cement or cement clinker,
— finish-grinding of rejects from other mills,
— grinding of cement raw material, and finally
— grinding of coal or coke.

erforderlich ist. Nach ausreichend langer Untersuchung des Mahlens verschiedener Zementtypen, um mit dem neuen Verfahren vertraut zu werden, sind Versuche zum Mahlen von Rohstoffen und gegebenenfalls auch von Kohle vorgesehen. Nach den Pilot-Untersuchungen wird beim Mahlen des vorliegenden Rohstoffs mit einer Verringerung des Energieverbrauchs um 50 % und der Möglichkeit gerechnet, sogar Materialien mit bis zu 20 Gew.-% Feuchtigkeit zu mahlen. Es ist davon auszugehen, daß die HOROMILL sich für fast jedes Mahlgut eignet, ob es nun grob oder fein, weich oder hart, trocken oder feucht ist. Damit könnte für alle Mahlvorgänge in einem Zementwerk ein einziger Maschinentyp eingesetzt werden.

It is evident that, if only one of this grinding processes shall be realized, a much simpler layout can be designed which requires a building of $12 \times 8 \times 12$ m only. After a sufficiently long time of testing the grinding of different cement types in order to become familiar with the new process, the testing of raw material and eventually coal grinding as well is scheduled. With the given raw material, a 50 % decrease of energy consumption and the ability even to grind material with up to 20 wt.-% of moisture is expected, according to pilot tests. It may be assumed that the HOROMILL is suitable for almost every material, coarse or fine, soft or hard, dry or wet. This would imply that a unique type of machine might be used for all grinding processes of a cement plant.

Sichtersysteme für Fertigmahlung und Halbfertigmahlung auf der Rollenpresse

Separator system for roller press finish and semi-finish grinding

Systèmes de séparateurs pour broyage de produit fini et de produit semi-fini pour presse à rouleaux

Sistemas de separadores para la molienda de acabado y de semiacabado mediante la prensa de rodillos

Von **J. Folsberg** und **O. S. Rasmussen**, Valby/Dänemark

Zusammenfassung – Der Einsatz der Rollenpresse für die Halbfertigmahlung stellt neue Anforderungen an das Sichtersystem zur Desagglomeration von Rollenpressen-Schülpen, an ausreichende Verschleißfestigkeit der Rollenpresse gegenüber dem abrasiven Material, an einen geschlossenen Kreislauf von Rollenpresse und Kugelmühle bei der Halbfertigmahlung und im Falle der Fertigmahlung an die Möglichkeit, bei geringem Energieverbrauch mit hohen Umlaufzahlen zu arbeiten. Um diese neu entstandenen Anforderungen erfüllen zu können, hat FLS ein neues Sichtersystem auf der Basis der Zwei-Stufen-Sichtung entwickelt. Im Vergleich zu einem herkömmlichen Sichtersystem kann eine Zwei-Stufen-Sichteranlage den gesamten spezifischen Energieverbauch und die Kapitalkosten der Anlage verringern und dennoch hohe Verschleißfestigkeit und die Vorteile einer trennscharfen Sichtung sowohl für den Kreislauf der Rollenpresse als auch für den der Kugelmühle gewährleisten. Der Einsatz einer Zwei-Stufen-Hochleistungssichteranlage führt zu einer besseren Gesamtleistung des Mahlsystems. Außerdem kann bei der Halbfertigmahlung die Korngrößenverteilung so angepaßt werden, daß die normale Konsistenz des Zementleims mit dem geforderten niedrigen Wasseranspruch erreicht werden kann. Schließlich sind die Kosten der Anlage für die Halbfertigmahlung insgesamt niedriger, wenn das Zwei-Stufen-Hochleistungssichtersystem eingesetzt wird.

Sichtersysteme für Fertigmahlung und Halbfertigmahlung auf der Rollenpresse

Summary – The application of the roller press for semi-finish grinding has set new demands on the separation system as desagglomeration of roller press flakes, sufficient wear resistance for handling the abrasive material from the roller press, closed circuit operation for both roller press and ball mill for semi-finish grinding, and in the case of roller press finish grinding the ability to work with high circulation factors with low power consumption. To meet these new requirements, FLS has developed a new separator based on the concept of 2-stage separation. A 2-stage separator can reduce the total specific energy consumption and the capital cost for the separator system, compared to a conventional separator system, and provide good wear resistance while maintaining the advantages of high-efficiency separation for both roller press and ball mill circuit. The use of a 2-stage high-efficiency separator results in better overall performance of the grinding system. Furthermore, an adjustment of the particle size distribution can be carried out for the semi-finish system which means that the required low water demand for normal consistency of the cement paste can be obtained. Finally the price for the entire machinery is lower for the semi-finish grinding system when using the 2-stage high-efficiency separator.

Separator system for roller press finish and semi-finish grinding

Résumé – L'utilisation de la presse à rouleaux pour le broyage de produit semi-fini impose de nouvelles exigences au système du séparateur pour la désagglomération des plaquettes de la presse à rouleaux mais aussi à la résistance suffisante à l'usure de la presse à rouleaux vis-à-vis de la matière abrasive, ou au circuit fermé de la presse à rouleaux et du broyeur à boulets pour le broyage de produit semi-fini et, enfin, dans le cas du broyage de produit fini à la possibilité de travailler à des vitesses de rotation élévées avec une faible consommation d'énergie. Afin de pouvoir remplir ces exigences nouvelles, FLS a mis au point un nouveau système de séparateur basé sur le principe de la séparation à deux étages. Comparée au système de séparation classique, l'installation à deux étages permet de réduire la consommation spécifique d'énergie ainsi que les coûts financiers de l'installation et garantit, néanmoins, une résistance à l'usure élevée tout en assurant les avantages d'une sélection fine aussi bien pour le circuit de la presse à rouleaux que pour celui du broyeur à boulets. L'utilisation d'une installation de séparation à haut rendement à deux étages contribue à un meilleur rendement total du système de broyage. De plus, lors du broyage de produit semi-fini la répartition granulométrique peut être adaptée de manière à obtenir une consistance normale de la pâte de ciment avec une demande d'eau faible. Et enfin, les coûts d'une installation de broyage de produit semi-fini sont, globalement, inférieurs quand on fait appel à un système de séparation de haut rendement à deux étages.

Systèmes de séparateurs pour broyage de produit fini et de produit semi-fini pour presse à rouleaux

Resumen — *El empleo de la prensa de rodillos para la molienda de semiacabado supone nuevas exigencias para el sistema de separación, respecto de la desaglomeración de las escamas de la prensa de rodillos, en lo que se refiere a una resistencia suficiente al desgaste de la prensa de rodillos frente al material abrasivo, respecto de un circuito cerrado de prensa de rodillos y molino de bolas, en el caso de la molienda de semiacabado, y de la posibilidad de trabajar con elevadas velocidades de rotación y bajo consumo de energía, en el caso de la molienda de acabado. Para poder cumplir estas nuevas exigencias, FLS ha desarrollado un nuevo sistema de separador, sobre la base de la separación en dos etapas. En comparación con un sistema de separador convencional, una instalación de separación en dos etapas puede reducir el consumo específico de energía total así como el capital de inversión, garantizando, no obstante, la elevada resistencia al desgaste y las ventajas de una separación nítida, tanto para el circuito de la prensa de rodillos como para el del molino de bolas. El empleo de una instalación de separación en dos etapas, de alto rendimiento, conduce a un mejor rendimiento total del sistema de molienda. Además, en el caso de la molienda de semiacabado, se puede adaptar la distribución del tamaño granulométrico de tal forma que se puede lograr la consistencia normal de la pasta de cemento, con una reducida demanda de agua. Finalmente, el costo de la instalación de molienda de semiacabado es, en su conjunto, más bajo, si se emplea el sistema de separación en dos etapas, de alto rendimiento.*

Sistemas de separadores para la molienda de acabado y de semiacabado mediante la prensa de rodillos

1. Einleitung

Im Gegensatz zu dem in einer üblichen Umlaufmahlanlage gemahlenen Gut ist das in einer Rollenpresse beanspruchte Material dadurch gekennzeichnet, daß es aus großen Agglomeraten besteht und eine insgesamt viel gröbere Korngrößenzusammensetzung aufweist. Außerdem sind sowohl die gebrochenen Agglomerate als auch die Einzelpartikel extrem abrasiv, da sie durch die Hochdruckbeanspruchung scharfe Kanten und eine hohe Dichte besitzen. Bei einem spezifischen Energieaufwand im Bereich von 2,5 bis 3,5 kWh/t, bezogen auf einen einmaligen Materialdurchgang zwischen den beiden Rollen, liegt die Umlaufzahl bei der sogenannten Halbfertigmahlung zwischen 3 bis 5, während diese Zahl bei der Fertigmahlung im Bereich von 5 bis 10 oder auch höher liegen kann. Derartig hohe Umlaufzahlen führen normalerweise in konventionellen Sichtern zu einer schlechteren Sichtwirkung, ausgenommen die Materialbeaufschlagung ist gering. Die Anforderungen an einen Sichter in einem Mahlkreislauf mit Gutbett-Walzenmühle sind demzufolge:

— die kombinierte Durchführung von Desagglomeration und Sichtung,

— die verschleißschützende Ausgestaltung des Sichters wegen der in der Rollenpresse realisierten Materialbeanspruchung,

— die Schaffung der technischen Voraussetzungen für eine optimale Grobsichtung,

— die Schaffung der technischen Voraussetzungen für eine optimale Feinsichtung.

2. Beschreibung des Sepax-Sichters und des Mahlanlagenkonzeptes

Die oben aufgeführten Anforderungen werden von dem neuen zweistufigen Sepax-Sichter (**Bild 1**) in vollem Umfang erfüllt. Das neue Sichtersystem besteht aus zwei getrennten Einheiten für die Fein- und Grobsichtung und einem oberhalb des Grobsichters eingebauten Desagglomerator, der mit niedrigen Geschwindigkeiten nach dem Prinzip eines vertikalen Prallbrechers arbeitet. Sowohl bei der Halbfertig- als auch Fertigmahlung wird das aus der Rollenpresse austretende Mahlgut der unteren Sichteraufgabeeinrichtung zugeführt. Dagegen wird das Mahlgut aus der Rohrmühle einer Halbfertigmahlanlage der oberen Aufgabe zugeleitet. Das Hauptmerkmal des zweistufigen Sepax-Sichters ist der verglichen mit konventionellen Sichtern stark verschleißgeschützte Grobsichter, der für hohe Materialbeladungen vorgesehen ist, und ein hochwirksamer oberer Sichter zur Erzeugung von niedrigen Trennkorngrößen. Da die Sichtluft gewöhnlich für beide Sichtstufen bereitgestellt wird, besteht das Endprodukt einer Halbfertigmahlanlage im Normalfall zu 25 bis 50 % aus Partikeln, die direkt aus dem in der Rollenpresse beanspruchten Material ausgesich-

1. Introduction

Contrary to the material ground in ordinary closed ball mill systems the material from the roller press is characterized by large pieces of agglomerates and a much coarser product. Furthermore both the broken agglomerates as well as the individual broken particles are extremely abrasive because of very sharp edges and the high density of the pressed material. As the energy input for one pass through the roller press is in the range of 2.5–3.5 kWh/t, the circulation factor required for semi-finish grinding is 3–5 while for finish grinding the circulation can be 5–10 or even higher. Such high circulation factors in conventional separators normally give a lower separation efficiency unless the material load is low. The demands on a separator operating in closed circuit grinding with roller press are then:

— Combined units for disagglomeration and separation,

— Wear resistant system for roller pressed material,

— New parameters for optimum for the coarse separation,

— Maintained parameters for fine separation.

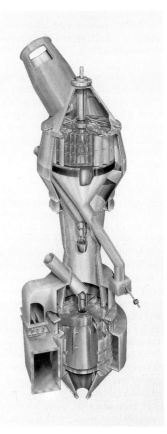

BILD 1:
Zweistufen Sepax-Sichter
FIGURE 1:
Two-stage Sepax separator

tet wurden, und dem Teil von Partikeln, die aus dem in der Rohrmühle gemahlenen Material durch Sichtung gewonnen wurden.

Das Fließschema des zweistufigen Sepax-Sichters ist in **Bild 2** dargestellt. Das Grundprinzip besteht in zwei unabhängig voneinander arbeitenden Sichtern. Beim Fertigmahlsystem setzt sich das Aufgabegut der Rollenpresse aus einem Gemisch von Frischgut, Grießen vom Grobsichter und Grobgut vom Feinsichter zusammen. Beim Halbfertigmahlsystem besteht das Aufgabegut der Rollenpresse aus Frischgut und Grießgut vom Grobsichter, während die Rohrmühle nur mit dem Grobgut des Feinsichters beaufschlagt wird. Dieses System liefert optimale Mahlbedingungen für beide Mahlanlagen-Konzepte, da die Rollenpresse am effektivsten arbeitet, wenn sie mit einem Aufgabegut ohne Feinanteile beaufschlagt wird, während die Einkammer-Rohrmühle ihre größte Leistungsfähigkeit mit einem Aufgabegut erreicht, das frei von agglomerierendem Feingut und hinsichtlich seiner oberen Aufgabekorngröße begrenzt ist.

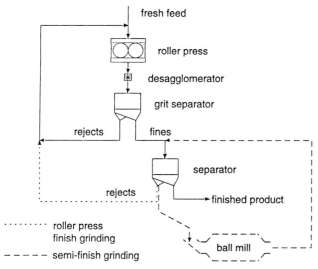

BILD 2: Fließbild eines Mahlanlagenkonzeptes mit zweistufigem Sepax-Sichter
FIGURE 2: Flowsheet of the mill concept with the two-stage Sepax separator

Frischgut	= fresh feed
Rollenpresse	= roller press
Desagglomerator	= desagglomerator
Grobsichter	= grit separator
Sichter	= separator
Kugelmühle	= ball mill
Feingut	= fines
Umlaufgut	= rejects
Fertigmahlung	= finish grinding
Halbfertig-Mahlung	= semi-finish grinding

Da das Fertigprodukt des oberen Sichters aus einem Gemisch von Feingut aus der Rollenpresse und der Rohrmühle besteht, kann der zweistufige Sepax-Sichter zur direkten Einstellung der Korngrößenverteilung des Zementes benutzt werden. Wenn der obere Sichter auf eine höhere Trennkorngröße eingestellt wird und daher mit niedrigerer Rotorgeschwindigkeit läuft, dann gelangt ein höherer Anteil von in der Rollenpresse beanspruchten Mahlgutes direkt in das Feingut. Je weniger Grobgut der Rohrmühle aufgegeben wird, desto feiner ist folglich das die Rohrmühle verlassende Mahlgut. Da unter diesen Bedingungen das Fertigprodukt aus einem Gemisch von relativ groben und relativ feinen Korngrößenfraktionen besteht, wird im Ergebnis eine breitere Korngrößenverteilung erhalten.

3. Die Veränderung der Korngrößenverteilung am Beispiel eines OPC

Zur Herstellung eines gewöhnlichen Portlandzementes (OPC) mit einem Blaine-Wert von 300 m²/kg in der Rohrmühle wurden systematische Untersuchungen über die Entwicklung der Umlaufzahl angestellt. Die Ergebnisse dieser Untersuchungen sind in **Bild 3** dargestellt. Während die

2. Description of the Sepax separator and the mill concept

The new two-stage Sepax separator (**Fig. 1**) is fulfilling the above mentioned requirements. The separator system consists of two separate units for coarse and fine separation respectively and a built-in disagglomerator of the low speed vertical impact type at the top of the grit separator. For both semi-finish and finish grinding the roller pressed material is fed to the lower inlet. For semi-finish grinding the ball mill product is fed to the upper inlet. The main features of the two-stage Sepax separator compared to conventional separator systems are the extensively wearprotected grit separator foreseen for a high material load and a high efficiency top separator module for the low cutsize. As the separation air is common for the two separators, the total fines for semi-finish systems consist normally of at least 25–50% of particles directly separated from the roller pressed material and the remaining part of particles separated from the material ground in the ball mill.

The flowsheet for the two-stage Sepax separator system is shown in **Fig. 2**. The principle is two independently working separators. For the finish grinding system the roller press feed comprises a mix of new feed, coarse particles from the grit separator and rejects from the fine separator. For the semi-finish grinding system the feed to the roller press consists of new feed and coarse material from the grit separator only, while the feed to the ball mill consists of rejects from the fine separator only. This system results in optimum grinding conditions for both grinding systems, as the roller press is performing best with the total feed without any fines, while the one compartment mill is performing best with a feed both limited in top size and free from clustering fines.

As the final product from the top separator consists of a mix of fines from the roller press and of fines from the ball mill, the two-stage Sepax separator can be used for an adjustment of the particle size distribution for the cement. When running the top separator at a higher cutsize (low rotor speed) a higher proportion of the particles ground in the roller press is passed directly to the fines. Consequently, a smaller amount of rejected particles are sent to the ball mill resulting in a correspondingly higher fineness of the material from the ball mill. As the final product now consists of a mix of relatively coarse and relatively fine particle fractions, the result is a flatter particle size distribution.

3. Change of the particle size distribution for OPC

A systematic scan in the circulation factor for the ball mill has been carried out for OPC cement at a Blaine value of 300 m²/kg. The results are given in **Fig. 3**. As the circulation factor for the ball mill is decreased (right to the left) at decreasing rotor speed, the sieve residue on the 32 μm Alpine sieve increase from 12.6% to 15.0% and 17.4% respectively. As can be seen the result for the water demand to normal consistency is a decrease from 30.5% to 29.0% and 28.5% respectively. At even lower ball mill circulation factor the water demand could have been reduced further.

4. Results for blended cement

It has been discussed, whether the cogrinding of easily ground components such as gypsum and limestone with clinker in semi- and finish systems will result in almost straight grinding of the softer components. To verify this problem tests with grinding of blended cement consisting of 73% clinker, 22% of limestone and 5% of natural gypsum (**Fig. 4**) have been carried out at a Blaine value of 340 m²/kg. Chemical analyses have been carried out for determination of the amount of fines produced in the roller press and ball mill and furthermore the circulation factors for the components in the roller press and ball mill system. As shown the amount of fines produced directly in the roller press is 29, 32 and 24% for clinker, limestone and gypsum respectively. The remaining part is fed to the ball mill. The circulation factors for the components in the roller press for both limestone and gypsum are in the same range as the circulation factor

Umlaufzahl für die Rohrmühle mit sinkender Drehzahl des Stabkorbes zurückgeht, steigt der Siebrückstand R 0,032 mm auf dem Alpine-Luftstrahlsieb von 12,6 auf 15,0 % bzw. auf 17,4 % an. Wie man sieht, nimmt der Wasserbedarf zur Erreichung der Normsteife von 30,5 auf 29,0 % bzw. auf 28,5 % ab. Bei noch kleineren Umlaufzahlen könnte der Wasserbedarf noch weiter reduziert werden.

4. Ergebnisse bei der Herstellung von Zementen mit Zumahlstoffen

Es ist wiederholt diskutiert worden, ob die gemeinsame Mahlung von leicht mahlbaren Komponenten wie Gips und Kalkstein zusammen mit Zementklinker nach der Technologie der Halbfertig- bzw. Fertigmahlung das richtige Verfahren für weichere Komponenten ist. Um diese Frage zu beantworten, wurden Mahlversuche an Zement aus mehreren Hauptbestandteilen durchgeführt, der zu 73 % aus Zementklinker, 22 % Kalkstein und 5 % Naturgips bestand (**Bild 4**) und auf einen Blaine-Wert von 340 m^2/kg gemahlen wurde. Zur Bestimmung des Fertigproduktes, das anteilig in der Rollenpresse und in der Rohrmühle produziert wurde, wurden chemische Analysen durchgeführt. Des weiteren wurden auch die Umlaufzahlen für die einzelnen Komponenten sowohl im Kreislauf der Rollenpresse als auch der Rohrmühle untersucht. Wie aus Bild 4 hervorgeht, betrug für die einzelnen Komponenten Zementklinker, Kalkstein und Gips der Feingutanteil, der direkt in der Rollenpresse erzeugt wurde, 29, 32 bzw. 24 %. Der verbleibende Grobgutanteil wurde entsprechend der Rohrmühle aufgegeben. Die Umlaufzahlen für die Komponenten Kalkstein und Gips nach ihrer Beanspruchung in der Rollenpresse lagen in der gleichen Größenordnung wie die Umlaufzahl für Zementklinker. In dem Rohrmühlen-Kreislauf waren die Umlaufzahlen für Kalkstein und Gips etwas niedriger als für Klinker, wie normalerweise aus Mahlsystemen mit Rohrmühle bekannt. Verursacht durch die hohen Umlaufzahlen für alle Komponenten werden schließlich auch homogene Produkte mit Zementeigenschaften erhalten, die denen ähneln, die in üblichen Mahlsystemen mit Rohrmühle und Hochleistungssichter erzeugt werden.

5. Abschätzung des Leistungsvermögens und der Kosten

Für konventionelle Mahlsysteme und solche, die mit einem zweistufigen Sepax-Sichter ausgerüstet sind, wurde ein Leistungs- und Kostenvergleich durchgeführt (**Tabelle 1** und **2**). Wegen der hohen Beladungen des bei relativ großer Trennkorngröße arbeitenden Grobsichters liegt der Ener-

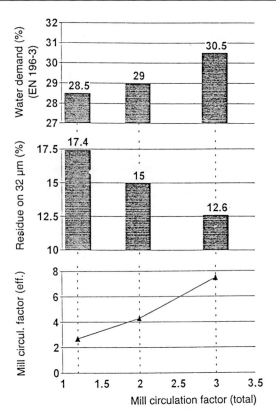

BILD 3: Wasseranspruch und Rückstand R 0,032 mm in Abhängigkeit von der Umlaufzahl (OP-Zement mit einem Blaine-Wert von 300 m^2/kg)
FIGURE 3: Water demand and R32-residue in dependence of the mill circulation factor (OPC cement with a Blaine value of 300 m^2/kg)

Wasseranspruch	= Water demand
Rückstand R 0,032 mm (%)	= Residue on 32 µm (%)
Umlaufzahl	= Circulation factor
Umlaufzahl der Rohrmühle (total)	= Mill circulation factor (total)
Umlaufzahl der Rohrmühle (effektiv)	= Mill circulation factor (effective)

for the clinker. In the ball mill system the circulation factors for limestone and gypsum are somewhat lower than for the clinker as normally seen for ball mill systems. Caused by the high circulation factors for all components also homogeneous products are obtained resulting in cement properties similar to properties obtained in ordinary ball mill systems with a high efficiency separator.

5. Evaluation of performance and costs

A comparison between the conventional grinding systems and systems comprising the two-stage Sepax separator (**Table 1** and **2**) is carried out both in terms of performance and cost. Caused by the allowed higher load for the grit separator working at relative high cutsize, the power consumption of the separator and fan is lower compared to the total power consumption for two individual separators for semi-finish grinding and the same is found for the finish grinding even though only one conventional separator is used. The better grinding performance in the semi-grinding system is caused by the fact that at least 25 % of the product is passed directly from the roller press as fines. Compared to conventional separator systems the use of the two-stage Sepax separator is estimated to result in 7–8 % better overall total specific power consumption. As further shown the use of the two-stage Sepax separator for semi-finish grinding systems results in a cost reduction of 7 % for the machinery.

6. Conclusions

It can be concluded that the use of the two-stage Sepax separator results in better overall performance of the grinding system in terms of specific power consumption. Furthermore, an adjustment of the particle size distribution can be carried out for the semi-finish system which means that the required low water demand for normal consistency of the

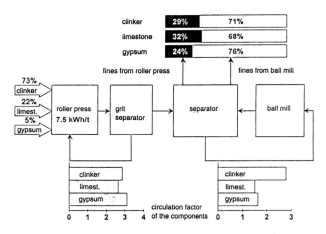

BILD 4: Fertiggutanteil von Rollenpresse und Rohrmühle in einem Halbfertig-Mahlsystem (Zement mit Zumahlstoffen mit einem Blaine-Wert von 340 m^2/kg)
FIGURE 4: Fines proportion of the roller press and the ball mill in a semi-finish grinding system (Blended cement with a Blaine value of 340 m^2/kg)

Fertiggut von der Rollenpresse	= Fines from roller press
Fertiggut von der Rohrmühle	= Fines from the ball mill
Umlaufzahl, bezogen auf die Komponente	= Circulation factor pr. component

TABELLE 1: Vergleich des spezifischen Energiebedarfs
TABLE 1: Comparison of the specific power consumption

Power consumption kWh/t	Semi-finish grinding		Roller press finish grinding	
	2 separators	2-stage separator	Conv. separator	2-stage separator
Ball mill	23.9	21.5	—	—
Roller press	9.8	9.8	25.8	24.3
Desagglomerator	0.8	0.8	1.6	1.5
Grit separator & fan	1.5	—	—	—
Separator & fan	4.2	5.3	9.8	8.5
Total kWh/t	40.2	37.4	37.2	34.3

Energiebedarf	= power consumption	Kugelmühle	= ball mill
Teilfertigmahlung	= semi-finish grinding	Rollenpresse	= roller press
Fertigmahlung auf der Rollenpresse	= roller press finish grinding	Deglomerator	= desagglomerator
2 Sichter	= 2 separators	Grobgutsichter und Ventilator	= grit separator and fan
Zwei-Stufen-Sichter	= 2-stage separator	Sichter und Ventilator	= separator and fan
Konv. Sichter	= conv. separator	gesamt kWh/t	= total kWh/t

gieverbrauch von Sichter und Umluftventilator niedriger, verglichen mit dem Gesamtenergieverbrauch von zwei einzelnen Sichtern im System einer Halbfertigmahlanlage. Zum gleichen Ergebnis gelangt man auch bei Anwendung der Fertigmahlung, obgleich eine derartige Mahlanlagenschaltung nur einen einzigen konventionellen Sichter benötigt. Die bessere Mahlwirkung eines Halbfertig-Mahlsystems ist bedingt durch die Tatsache, daß wenigstens 25 % des Produktes, das die Rollenpresse passiert, als Fertiggut anfällt. Verglichen mit konventionellen Sichtsystemen führt der Einsatz des zweistufigen Sepax-Sichters zu einer Verbesserung des spezifischen Gesamtenergiebedarfs von ca. 7 bis 8%. Außerdem werden durch den Einsatz des Sepax-Sichters bei der Halbfertig-Mahlung die Maschinenkosten um 7 % reduziert.

6. Schlußbemerkungen

Daraus kann geschlossen werden, daß der Einsatz eines zweistufigen Sepax-Sichters insgesamt zu einer Verbesserung der Leistungsfähigkeit des gesamten Mahlsystems, insbesondere hinsichtlich des spezifischen Energieverbrauchs, führt. Weiterhin gestattet der Einsatz dieses Sichtsystems bei der Halbfertig-Mahlung eine Einstellung der Korngrößenverteilung, wodurch der erforderliche niedrige Wasseranspruch zur Erzeugung einer Zementpaste mit Normsteife erreicht werden kann. Bei der Herstellung von Zementen mit Zumahlstoffen werden durch die hohen Umlaufzahlen für die Zementkomponenten im Rollenpressen-Kreislauf normale Zementeigenschaften gewährleistet. Schließlich sind bei der Halbfertig-Mahlung auch die Koşten für die gesamte Maschinenausrüstung niedriger, wenn ein zweistufiger Sepax-Sichter zum Einsatz gelangt.

TABELLE 2: Vergleich der Maschinenkosten
TABLE 2: Comparison of the machinery costs

Comparative prices 9.8 kWh/t in roller press*)	Semi-finish grinding	
	2 separators	2-stage separator
Machines	100 %	93 %

*) Related to a finish product with a Blaine value of 400 m²/kg

Vergleichbare Preise	= comparative prices
Bezogen auf ein Fertigprodukt mit einem Blaine-Wert von 400 m²/kg	= Related to a finish product with a Blaine-value of 400 m²/kg

cement paste can be obtained. For blended cements the high circulation factors for the added components in the roller press ensure normal cement properties. Finally the price for the entire machinery is lower for the semi-finish grinding system when using the two-stage Sepax separator.

Einfluß des Antriebes auf die Wirtschaftlichkeit einer Hochdruck-Rollenpresse

Influence of the drive on the cost-effectiveness of high-pressure grinding rolls

Influence de la motorisation sur la rentabilité d'une presse à rouleaux haute pression

Influjo del mando sobre la rentabilidad de una prensa de rodillos, de alta presión

Von **H. Herchenbach**, Kufstein/Österreich

Zusammenfassung – Verschleißkosten, Verfügbarkeit und Einsparung an elektrischer Energie können bei Einsatz von Gutbett-Walzenmühlen verbessert werden, wenn der Antrieb technisch richtig ausgeführt wird. Dazu muß gefordert werden, daß bei allen Betriebszuständen über einen langen Zeitraum nicht einmal kurzfristige Relativbewegungen zwischen aufgeschweißter Walzenoberfläche und Schülpenoberfläche auftreten können. Diese führen unweigerlich zu Verschleiß bis hin zu Abpellungen, da die Aufschweißung der Walzenoberfläche keine Scherspannungen verkraftet. Die Scherbelastung der Aufschweißung ist um so größer, je fester die Schülpe ist und je größer der Unterschied bei den Umfangsgeschwindigkeiten der Walzen sein kann. Werden diese Zusammenhänge bei der Ausführung der Walzenantriebe berücksichtigt, so können Betriebszeiten von 20000 h erreicht werden, ohne mechanische Teile der Gutbett-Walzenmühle einschließlich der Walzen-Aufschweißung ersetzen zu müssen. Das zeigen Betriebsergebnisse an Anlagen für Rohmaterial, Klinker und Hüttensand, die 4 bzw. 5 Jahre in Betrieb sind. Dabei führen Anpreßdrücke bis zu 150 bar zu optimalen Energieeinsparungen, ohne daß die Verfügbarkeit der Gutbett-Walzenmühle unter die einer Kugelmühle oder einer Walzenschüssel-Mühle vermindert wird.

Einfluß des Antriebes auf die Wirtschaftlichkeit einer Hochdruck-Rollenpresse

Summary – Wear costs, availability, and saving in electrical power can be improved when using high-pressure grinding rolls if the drive has the correct technical design. An essential requirement is that for all operating conditions taken over a long period there must not be any short-term relative movements between the welded surfaces of the rollers and the surface of the compacted cake. These inevitably lead to wear, resulting in peeling, as the build-up welding on the roller surfaces cannot cope with shear stresses. The shear loading on the welding is greater the harder the compacted cake and the greater the possible difference in the circumferential speeds of the rollers. If these relation-ships are borne in mind during the design of the roller drives then operating times of 20000 h can be achieved without having to replace mechanical parts of the high-pressure grinding rolls, including the build-up welding on the rollers. This is demonstrated by operating results in plants for raw material, clinker, and granulated blast furnace slag which have been in operation for four or five years. Surface pressures up to 150 bar lead to optimum energy savings without reducing the availability of the high-pressure grinding rolls to less than that of a ball mill or roller grinding mill.

Influence of the drive on the cost-effectiveness of high-pressure grinding rolls

Résumé – Les coûts d'usure peuvent être réduits et la disponibilité et l'économie d'énergie électrique peuvent être améliorées dans l'exploitation de broyeurs à rouleaux, quand l'entraînement est conçu correctement. Cela exige, que pendant tous les états de marche sur une longue durée d'exploitation, n'apparaissent nullement, même occasionellement, des mouvements relatifs entre la surface stellitée des rouleaux et celle des plaquettes. Ceux-ci conduisent irrémédiablement à l'usure, jusqu'à l'écaillage du stellitage, parce que la surface d'usure rapportée par soudure n'accepte pas les forces de cisaillement. La sollicitation en cisaillement du stellitage est d'autant plus grande, qu'est ferme la plaquette et plus grande la différence des vitesses périphériques de rouleaux. Si ces relations sont prises en compte lors de la conception des entraînements des rouleaux, il est possible d'atteindre des durées de service de 20000 h, sans être obligé de remplacer des pièces mécaniques du broyeur à rouleaux, y compris le stellitage des rouleaux. Ceci est démontré par des résultats d'exploitation d'installations pour matière première, clinker et laitier granulé, en service depuis 4 ou 5 années. Dans ce contexte, des pressions d'appui allant jusqu'à 150 bar ont conduit à des économies d'énergie optimales, sans que la disponibilité du broyeurs à rouleaux tombe en dessous de celle d'un broyeur à boulets ou d'un broyeur à galets.

Influence de la motorisation sur la rentabilité d'une presse à rouleaux haute pression

Resumen – Los gastos originados por el desgaste, la disponibilidad y el ahorro de energía eléctrica pueden mejorarse al utilizar molinos de cilindros y lecho de material, siempre y cuando el dispositivo de accionamiento haya sido diseñado correctamente. A este respecto, hay que exigir que en todos los estados de servicio y durante períodos prolongados no se produzcan siquiera movimientos relativos de corta duración entre la superficie soldada de los rodillos y la superficie del material compactado. Porque

Influjo del mando sobre la rentabilidad de una prensa de rodillos, de alta presión

estos movimientos conducen inevitablemente al desgaste e incluso a fenómenos de desprendimiento, ya que el aporte de soldadura en las superficies de los rodillos no aguanta las tensiones de cizallamiento. El esfuerzo de cizallamiento del aporte de soldadura será tanto mayor, cuanto más duras estén las plaquetas de material y cuanto mayor la diferencia entre las velocidades de rotación de los rodillos. Si se tienen en cuenta estas relaciones a la hora de elegir los sistemas de mando de los rodillos, se podrán alcanzar períodos de servicio de 20.000 h, sin necesidad de sustituir elementos mecánicos del molino de cilindros y lecho de material, incluyendo el aporte de soldadura de los rodillos. Esto se desprende de los resultados obtenidos durante el servicio con instalaciones para materias primas, clínker y escoria siderúrgica, que están en servicio desde hace 4 – 5 años. A este respecto, las presiones de compresión de hasta 150 bar permiten conseguir ahorros de energía, sin que la disponibilidad del molino de cilindros y lecho de material quede inferior a la de un molino de bolas o de un molino de cubeta y rodillos.

Das Prinzip der Gutbettzerkleinerung in einer Hochdruck-Rollenpresse ist das Zerkleinerungsverfahren mit der besten Energieausnutzung. Die Wirtschaftlichkeit einer Maschine oder eines Verfahrens wird jedoch nicht allein durch die Einsparung von Betriebsmitteln bestimmt sondern auch vom Wartungsaufwand und von der Verfügbarkeit. Viele der bis heute produzierenden Hochdruck-Rollenpressen lassen diesbezüglich noch viele Wünsche offen. Es kommt zu erhöhtem Verschleiß bis hin zu Abpellungen an der Walzenoberfläche und den damit verbundenen Ausfallzeiten. Daneben wird von häufigen Lagerschäden, Getriebeschäden, Kupplungsdefekten bis hin zu Schäden in der Basiskonstruktion der Walzenkörper und des Pressengehäuses berichtet. Eine Hauptursache aller dieser Probleme könnte in der technischen Ausführung der Pressenantriebe liegen.

Fast alle Rollenpressen sind mit Drehstrom-Kurzschlußläufer-Motoren ausgerüstet. Diese Maschinen zeichnen sich durch eine stabile Drehzahl über weite Drehmomentbereiche aus.

Interparticle comminution between high-pressure grinding rolls is in principle the most energy-efficient comminution process. Yet the cost-effectiveness of a machine or process does not depend solely on economies with resources; expenditure on maintenance and availability are also important factors. Many of the high-pressure grinding rolls at present in production leave much to be desired in this respect. Increased wear is experienced, including peeling of the roller surfaces with the stoppages which this entails. There are also frequent reports of damage to bearings, damage to gear units, coupling faults and even damage to the basic structure of the roller bodies and roller housing. One main cause of all these problems may be the technical design of the roller drives.

Nearly all high-pressure grinding rolls are fitted with three-phase squirrel cage induction motors. The distinctive feature of these machines is their stable speed over wide torque ranges. The absolute level of the speed depends on individual details in the manufacture of the motor and on the

$Md \leq Md_{max}$		$n_1 <> n_2$
$Md_1 \cong Md_2$	$\rightarrow$ bei $\rightarrow$	$s_1 <> s_2$
$V_{u^1} = V_{u^2}$		$D_1 <> D_2$

BILD 1: Voraussetzungen für den Antrieb von Rollenpressen bei wartungsarmem Betrieb

$Md \leq Md_{max}$		$n_1 <> n_2$
$Md_1 \cong Md_2$	$\rightarrow$ with $\rightarrow$	$s_1 <> s_2$
$V_{u^1} = V_{u^2}$		$D_1 <> D_2$

FIGURE 1: Requirements for the drives of high-pressure grinding rolls in low-maintenance operation

che aus. Die absolute Höhe der Drehzahl ist von individuellen Fertigungsdetails am Motor und vom elektrischen Netz abhängig. Somit ist von Beginn an nicht gewährleistet, daß die Antriebsdrehzahl je Rolle absolut gleich ist (**Bild 1** und **2**). Andererseits ist bekannt, daß Fertigungstoleranzen und ungleicher Verschleiß an den Rollen zu unterschiedlichen Durchmessern führen. Beides führt schließlich zu unterschiedlichen Umfangsgeschwindigkeiten an den Rollen und somit zu Relativbewegungen zwischen Schülpen- und Rollenoberfläche. Ist die Schülpe nicht besonders konsistent, so wird sich lediglich ein erhöhter Verschleiß an der Rollenoberfläche einstellen, weil sich kein schützender „Materialpelz" bilden kann. Ist die Schülpe besonders hart und ist die Oberfläche der Walzen zur Verbesserung des Einzugverhaltens profiliert, so treten Scherkräfte auf, die den Verschleiß erheblich ansteigen lassen bis hin zum Abpellen der Walzenoberfläche. In keinem Fall sollten den druckfesten Walzenoberflächen diese hohen Scherkräfte zugemutet werden. Die eingesetzten superharten Materialien sind dafür nicht geeignet. Darüber hinaus können Drehmoment-Spitzen durch Fremdkörper, Materialüberkorn und ungleiche Materialzuführung zum Einzugsspalt beim Anfahr- und Abstellvorgang entstehen, die vom Kurzschlußläufer-Motor bis zum 5fachen Nennstrom übertragen werden. Dabei sind die Maschinenelemente der Presse, wie Lager, Getriebe, Kupplungen usw., enorm beansprucht [1]. Besonders konsistente Schülpen entstehen beim Pressen von Zementklinker mit Naßschlacke sowie feuchtem Zementrohmaterial und Erzen aller Art.

electric power supply, so there is no assurance from the outset that the drive speed is absolutely the same for each roller (**Figs. 1** and **2**). Moreover, manufacturing tolerances and uneven wear on the rollers are known to lead to different diameters. Both eventually lead to different circumferential speeds of the rollers and thus to relative movement between the surfaces of the compacted cake and the rollers. If the cake is not particularly consistent there will merely be

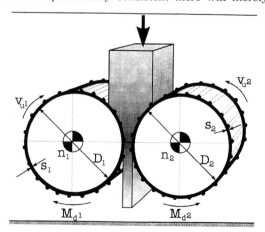

BILD 2: Schematische Darstellung der Zerkleinerung im Gutbett
FIGURE 2: Diagrammatic representation of interparticulate comminution

Drehzahlregelung:	Istwert wird von Hochlaufregler zeitlich so geregelt, daß I-Max nicht überschritten wird
Stromregelung:	Strombegrenzung I-Max von 0–100 % einstellbar
Stromvergleich:	Über I-Ist-Vergleich und n-Zusatzstoff ist $v_{u^1} = v_{u^2}$
Blockschutz:	bei I-Ist > I-Max und n-Ist < n-Min Motor-Abschaltzeit von 0–2000 ms einstellbar
Kriechdrehzahl:	bei Revision möglich

BILD 3: Umsetzung der Voraussetzungen für den wartungsarmen Rollenpressen-Betrieb

Speed control:	actual value is regulated by the acceleration control unit so that I-max is not exceeded
Current control:	current limiting system I-max adjustable from 0–100 %
Current comparison:	Using I-actual value comparison and n-additional setpoint gives $v_{u^1} = v_{u^2}$
Unit protection:	For I-actual value > I-max and n-actual value < n-min motor disconnecting time adjustable from 0–2000 ms
Creep speed:	Possible for inspection

FIGURE 3: Implementation of the requirements for low-maintenance operation of high-pressure grinding rolls

Zwei Wege, die in letzter Zeit eingeschlagen wurden, um diese Probleme in den Griff zu bekommen, sind nicht zielführend. Hierbei handelt es sich um den Einsatz von weicheren Rollenoberflächen und um die Reduktion des Preßdrucks. Im ersten Fall sind aufgrund der kurzen Standzeiten der Rollenpresse Wirtschaftlichkeit und Verfügbarkeit unbefriedigend, im zweiten Fall ist der gesamte Vorteil der Gutbettzerkleinerung im Hochdruckbereich gefährdet. Das Problem kann nur durch drehzahlelastische Antriebe gelöst werden (**Bild 3** und **4**).

Im Zementwerk Eiberg wurden zum Erreichen dieses Zieles Gleichstrom-Nebenschlußmotore an zwei Hochdruck-Rollenpressen von Humboldt-Wedag eingesetzt. Die Schaltung garantiert, daß die Umfangsgeschwindigkeiten bei beiden Rollen gleich bleibt. Gleichzeitig ist die Belastung beider Antriebe in etwa gleich und in der Spitze begrenzt. Daneben bieten die Gleichstrommotore noch die Möglichkeit, bei An- und Abfahrvorgängen der Rollenpresse die variable Drehzahl so einzusetzen, daß die Maschine nicht übermäßig belastet wird [1].

Die Rollenpresse 100/40 für gemeinsames Zerkleinern von Kalkstein und Mergel mit ca. 3,5 M.-% Feuchte ist im 5. Jahr in Betrieb. Die ersten Rollenoberflächen waren 3 Jahre oder ca. 18 000 h im Einsatz. Im Winter 1991/92 wurde eine Reserverolle als Losrolle eingebaut und die Festrolle in der Maschine aufgeschweißt. Bis auf die üblichen Wartungen an Druckzylindern und Herzstücken sind alle anderen Maschinenelemente nach ca. 25 000 Betriebsstunden noch ohne Schaden.

Die Rollenpresse 100/63 mit Randzonen-Schülpenrückführung für getrenntes Pressen von Zementklinker und getrocknetem Hüttensand ist im 4. Jahr in Betrieb. Dabei wurde nach ca. 8000 h (2,5 Jahre) die Losrolle in der Maschine neu aufgeschweißt, während die Festrolle noch nach ca. 10 500 h (> 3 Jahre) ohne Schaden in Betrieb ist. Auch an dieser Presse ist, außer den vorher genannten Wartungen, kein sonstiger Schaden aufgetreten, obschon mit Preßdrücken von 130 bar sowohl bei Klinker als auch bei Schlacke gefahren wird. Die beiden Pressen sind wiederholt durch Fremdkörpereinbringung gestoppt worden, was zu keinerlei Schäden an der Walzenoberfläche oder an sonstigen Maschinenelementen führte.

increased wear on the roller surfaces because no protective "skin of material" can form. If the cake is especially hard and if the roller surfaces are profiled to improve the in-take action, shearing forces will appear and considerably increase the wear, even causing the roller surfaces to peel. The pressure-resistant roller surfaces should definitely not be exposed to these strong shearing forces. The super-hard materials used are not suitable for them. Furthermore, torque peaks may be caused by foreign bodies, oversize grains of material, and uneven feeding of material to the nip during starting or stopping; these are transmitted by the squirrel cage motor at up to 5 times the rated current. There is enormous strain on the mechanical elements of the press such as the bearings, gear units, couplings etc [1]. Particularly consistent compact cake is formed by compressing cement clinker with wet slag, moist raw material for cement and ores of all kinds.

Two ways of tackling these problems which have been adopted recently have not proved successful. These involve the use of softer roller surfaces or reduction in the applied pressure. In the first case cost-effectiveness and availability are unsatisfactory because of the short service life of the high-pressure grinding rolls, and in the second case the whole advantage of grinding the bed of material in the high-pressure range is put at risk. The problem can only be solved by using flexible-speed drives (**Figs. 3** and **4**).

With this end in view, DC shunt motors have been installed on two Humboldt-Wedag high-pressure grinding roll units. The circuit ensures that the circumferential speed remains the same on both rollers. At the same time the load on the two drives is approximately the same and its peak level is limited. The DC motors also provide an opportunity for the variable speed to be used during the starting and stopping of the high-pressure grinding rolls, so that the machine is not overloaded [1].

The 100/40 high-pressure grinding rolls for intergrinding limestone and marl with approx. 3.5 % moisture are in their fifth year of operation. The first roller surfaces were in use for 3 years or approx. 18 000 hours. In winter 1991/92 a reserve roller was fitted as the loose roller and the fixed roller was hard-faced in the machine. Apart from the usual maintenance of pressure cylinders and main components, all the

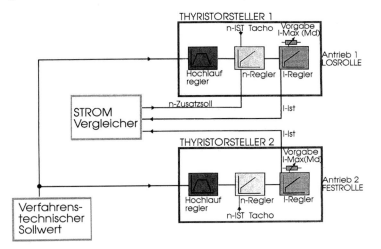

BILD 4: Blockschaltbild der Rollenpressen-Antriebe
FIGURE 4: Block diagram for the high-pressure grinding rolls drives

Thyristorsteller	= thyristor controller
Strom Vergleicher	= current comparator
Verfahrenstechnischer Sollwert	= process setpoint
n-Ist Tacho	= n-actual value tacho
Vorgabe I-Max (Md)	= I-max input
Hochlaufregler	= acceleration control unit
n-Regler	= n controller
I-Regler	= i controller
Antrieb	= drive
Losrolle	= free roller
Festrolle	= fixed roller
n-Zusatzsoll	= n additional setpoint
I-Ist	= I-actual value

Auf der Basis dieser Erfahrung ist davon auszugehen, daß Hochdruck-Rollenpressen bezüglich Wartungsaufwand und Verfügbarkeit nicht hinter Kugelmühlen und Niederdruck-Walzenmühlen zurückstehen müssen. Dies gilt auch dann, wenn mit Anpreßdrücken bis zu 150 bar optimale Energieeinsparungen erzielt werden.

Literaturverzeichnis

[1] Gehlken, Ch., und Zenner, H.: Mechanische Beanspruchungen der Gutbett-Walzenmühlen. Zement-Kalk-Gips 44 (1991) H. 2, S. 84–87.

other mechanical elements are still operating undamaged after about 25 000 hours.

The 100/63 grinding roll unit with edge-zone recycling of the compacted cake is in its fourth year of operation for separate compression of cement clinker and dried granulated blast furnace slag. The loose roller was hard-faced again in the machine after about 8000 hours (2.5 years), while the fixed roll is still working undamaged after about 10 500 hours (over 3 years). There has been no special damage to these high-pressure grinding rolls either, apart from the above-mentioned maintenance, although they run at pressures of 130 bar for both clinker and slag. Both units have been stopped repeatedly through the introduction of foreign bodies, though this did not cause any damage to the roller surfaces or other parts of the machine.

This experience shows that high-pressure grinding rolls are not necessarily inferior to ball mills or low-pressure roller mills in respect of maintenance costs and availability. This also applies when optimum energy saving is obtained at pressures up to 150 bar.

Analyse des Energieeinsatzes in der indischen Zementindustrie – ein Überblick
Energy use analysis in the Indian cement industry – an overview

Analyse de l'utilisation d'énergie dans l'industrie cimentière indienne – un aperçu
Análisis del empleo de energía en la industria del cemento india – un resumen

Von **K. Kumar, A. K. Mullick, J. P. Saxena** und **A. Pahuja,** New Delhi/Indien

Zusammenfassung – In den letzten Jahren sind die Produktionskosten in der Zementindustrie stark angestiegen, weil die Kosten des Energieeinsatzes deutlich höher geworden sind. Wie Untersuchungen gezeigt haben, belaufen sich die Energiekosten auf etwa 45 % der Kosten der Zementproduktion. Um im Bereich der Kosten wettbewerbsfähig zu sein, ist es unerläßlich, Energie in den Zementfabriken wirtschaftlicher einzusetzen. Im Vergleich zu einigen Industrieländern ist der durchschnittliche spezifische Verbrauch an thermischer und elektrischer Energie in der indischen Zementindustrie sehr hoch, wie man aus Untersuchungen in zahlreichen Werken weiß. Neben der Qualität des Energieeinsatzes, der Rohstoffe und des angewandten Verfahrens gibt es viele andere Faktoren, die für einen wirtschaftlichen Betrieb industrieller Anlagen entscheidend sind. Es wurde festgestellt, daß der spezifische Energieverbrauch mit zunehmender Kapazitätsauslastung bzw. bei den einzelnen Werken mit zunehmender Produktion sinkt. Mit zunehmender Ofenkapazität und steigendem technologischen Standard weisen sowohl der spezifische Wärmeverbrauch als auch der spezifische Verbrauch an elektrischer Energie abnehmende Tendenz auf. Die größten Beiträge zur Senkung des spezifischen Energieverbrauchs wurden geleistet durch die Reduzierung der Abgasverluste infolge Optimierung des Ofen- und Kühlerbetriebs, durch Einsatz 5- oder sogar 6-stufiger Vorwärmer mit Zyklonen mit geringem Druckverlust, durch Einbau von Wärmerückgewinnungssystemen, die Verbesserung der Verbrennungstechniken, Senkung der Ofenmantelverluste und durch Einsatz von Wälzmühlen, Gutbettwalzenmühlen und Sichtern mit hoher Energieausnutzung.

Summary – In the cement industry, during the last few years, there has been a steep rise in the cost of production as the cost of energy inputs has increased sharply. Studies have revealed that energy costs constitute about 45% of the cost of cement production. In order to be cost competitive it is essential to improve the energy efficiency of cement plants. In comparison to some countries with advanced development, the average specific thermal and electrical energy consumption of the Indian cement industry is very high, based on the studies carried out in a large number of plants. Apart from the quality of energy inputs and raw materials and the process employed, there are many factors which are decisive for the operating efficiency of industrial plants. The specific energy consumption has been found to decrease with increasing capacity utilization, and for individual plants, with increasing production rates. With the increase in kiln capacity and a higher technology level both the specific heat consumption and the specific power consumption are showing a decreasing trend. The main areas of improvements for minimising the specific energy consumption are the reduction of exhaust gas losses by optimizing the operation of kiln and coolers, the use of 5 or even 6 stage preheater systems with low pressure drop cyclones, incorporation of heat recovery systems, improvement of combustion techniques, reduction of shell losses, and utilization of energy efficient roller mills, roll presses and classifiers.

Résumé – Au cours des dernières années les coûts de production dans l'industrie cimentière ont fortement augmenté du fait de la hausse sensible des coûts énergétiques. Comme le prouvent des études, les coûts énergétiques représentent environ 45% des coûts de la production de ciment. Pour rester compétitif au niveau des coûts, il est indispensable, dans les cimenteries, d'utiliser l'énergie de manière plus rentable. Comparé à d'autres pays industriels, la consommation moyenne spécifique d'énergies thermique et électrique est très élevée dans l'industrie cimentière indienne, comme l'indiquent des études concernant de nombreuses usines. Outre la qualité de l'utilisation de l'énergie, des matières premières et de la technique utilisée, il existe nombre d'autres facteurs qui sont d'une grande importance pour l'exploitation économique des installations industrielles. Il a été constaté que la consommation d'énergie spécifique diminue au fur et à mesure qu'augmente le degré d'utilisation de capacité ou qu' augmente la production pour les différentes usines prises isolément. Quand la capacité du four augmente de même que le niveau technologique, la consommation calorifique spécifique ainsi que la consommation spécifique d'énergie électrique ont tendance à baisser. Parmi les secteurs qui contribuent le plus à baisser la consommation d'énergie spécifique, citons la réduction des pertes des effluents gazeux par optimisation de la conduite du four et du refroidisseur, l'utilisation de préchauffeurs de 5 ou même 6 étages avec cyclones à faible perte de charge, l'installation de systèmes de récupération de chaleur, l'amélioration des techniques de combustion, l'abaissement des pertes de chaleur de la virole du four et, enfin, la mise en oeuvre de broyeurs à cylindres, de broyeurs à cylindres à lit de matière et de séparateurs à rendement énergétique élevé.

Análisis del empleo de energía en la industria del cemento india – un resumen

Resumen – *En los últimos años, los gastos de producción de la industria del cemento han aumentado notablemente, puesto que el coste de la energía ha subido mucho. Se desprende de los estudios realizados que los gastos de energía ascienden aproximadamente al 45 % de los costes de la fabricación del cemento. Para guardar competitividad a nivel de costes, es imprescindible emplear la energía en las fábricas de cemento de forma más razonable. En comparación con algunos países industrializados, el consumo específico medio de energía térmica y eléctrica es muy alta en la industria del cemento india, según se conoce por los estudios llevados a cabo en numerosas fábricas. Aparte de la calidad del empleo de la energía, de las materias primas y del proceso aplicado, existen aún muchos otros factores decisivos para un servicio rentable de instalaciones industriales. Se ha comprobado que el consumo específico de energía disminuye a medida que crece el aprovechamiento de las capacidades o, en las diferentes fábricas, el volumen de producción. A medida que aumenta la capacidad del horno y el standard tecnológico, tanto el consumo específico de calor como el consumo específico de energía eléctrica presentan una tendencia bajista. Lo que más ha contribuido a la reducción del consumo específico de energía ha sido la disminución de las pérdidas de gases de escape, como consecuencia de la optimización del servicio del horno y del enfriador, del empleo de precalentadores de ciclones, de 5 o incluso de 6 etapas, con bajas pérdidas de presión, así como debido al montaje de sistemas de recuperación de calor, a la mejora de las técnicas de combustión, la reducción de las pérdidas por la envolvente del horno y el empleo de molinos de pista de rodadura, molinos de cilindros y lecho de material y separadores con un alto grado de utilización de la energía.*

Der Wirkungsgrad der in der indischen Zementindustrie eingesetzten Energie ist nicht so hoch wie bei den besten Beispielen aus einigen anderen Ländern. Bis zu einem gewissem Maße liegt dies an der vergleichsweise schlechteren Qualität der der indischen Zementindustrie zur Verfügung stehenden Energien (vorwiegend Kohle und Elektrizität). Darüber hinaus gibt es in Indien alte Anlagen mit Naß- oder Halbtrockendrehöfen neben modernen Anlagen nach dem Trockenverfahren mit Zyklonvorwärmern und Vorkalziniereinrichtungen, um den Zementbedarf des Landes zu decken. Bei den Roh- und Brennstoffen muß die indische Zementindustrie mit niedriger Qualität und schwerer mahlbaren Stoffen auskommen, was die Energiekosten verteuert. In Studien konnte jedoch gezeigt werden, daß der Energieverbrauch der indischen Zementindustrie deutlich gesenkt werden kann. Dazu wäre ein umfassendes Vorgehen erforderlich, um die Bemühungen zur Energieeinsparung zu verstärken.

Aus Energieverbrauchsdaten von 48 der 97 Zementwerke aus dem Zeitraum 1991–92 wurde die nachstehende Analyse erarbeitet. Der durchschnittliche thermische Energieverbrauch betrug beim Trockenverfahren 846 kcal/kg Klinker (3540 kJ/kg) und beim Naßverfahren 1408 kcal/kg Klinker (5900 kJ/kg), wobei sich der gewichtete Durchschnitt insgesamt auf 938 kcal/kg Klinker (3930 kJ/kg) belief. Der durchschnittliche elektrische Energieverbrauch lag bei 116,6 kWh/t Zement.

In den teilnehmenden Zementwerken betrugen die Gesamtenergiekosten rund US$ 12,20 je Tonne, wovon die Wärmeenergiekosten US$ 6,30 je Tonne und die Elektrizitätskosten US$ 6 je Tonne ausmachten. Die gewichteten durchschnittlichen Energiekosten entsprachen 38 % der Gesamtproduktionskosten. Die Analyse der verfahrensbedingten Energiekosten ergab, daß die Kosten der thermischen Energie beim Naßverfahren um US$ 2,70 je Tonne höher als beim Trockenverfahren lagen, während die Elektrizitätskosten beim Naßverfahren um US$ 0,21 je Tonne Zement niedriger waren. Damit lagen die spezifischen Gesamtenergiekosten der nach dem Naßverfahren produzierenden Werke um etwa US$ 2,50 je Tonne über denen der nach dem Trockenverfahren arbeitenden Werke.

Neben der Qualität der eingesetzten Energie und Rohstoffe sowie dem angewandten Herstellungsverfahren entscheiden noch viele weitere Faktoren über die Wirtschaftlichkeit des Betriebs einzelner Werke, z. B. die Kapazitätsauslastung, die Größe des Werkes, das Baujahr der Anlagen, der technische Leistungsstand usw.

Die Kapazitätsauslastung und die Produktion der Ausrüstungen wirken sich unmittelbar auf den Energieverbrauch aus. Beide Formen des Energieverbrauchs nehmen mit steigender Kapazitätsauslastung ab. Die Analyse der Daten der nach dem Trockenverfahren arbeitenden Werke mit Produktionskapazitäten von 2000–3850 t/Tag ergab bei einer

The energy performance in the Indian cement industry is not as good as compared to the best results achieved in some other countries. To some extent, this is due to relatively inferior quality of energy inputs (primarily coal and power) available to the Indian cement industry. Further, in the Indian cement industry, old wet and semidry process plants as well as modern dry process plants with suspension preheaters and precalciners are playing their complementary roles in meeting the country's demand for cement. In terms of raw materials and fuels, the Indian cement industry has to cope with low quality and harder to grind materials thus adding to the energy bill. Nevertheless, studies have shown that the energy consumption levels of Indian cement industry can be brought down considerably. For this a comprehensive approach would be necessary to intensify energy conservation efforts.

Based on energy consumption data for 1991–92, of 48 out of the 97 cement plants in the country, the following analysis was obtained. The average thermal energy consumption was 846 kcal/kg clinker (3540 kJ/kg) for dry process and 1408 kcal/kg clinker (5900 kJ/kg) for wet process with an overall weighted average of 938 kcal/kg clinker (3930 kJ/kg). The average electrical energy consumption was 116.6 kWh/t cement.

For the participating cement plants, the overall energy cost was of the order of US$ 12.20 per ton with the component of thermal energy cost of US$ 6.30 per ton and electrical energy cost of US$ 6 per ton. The weighted average cost of energy as a percentage of total cost of production worked out to 38%. The analysis of process-wise energy costs showed that that thermal energy cost in wet process compared to dry process is higher by US$ 2.70 per ton, the electrical energy cost in wet process is lower by US$ .21 per ton of cement. Thus the overall specific energy cost of wet process plants worked out to be higher by about US$ 2.50 per ton in comparison with dry process plants.

Apart from quality of energy inputs and raw materials and the process employed there are many factors which decide the operating efficiency of individual plants such as capacity utilisation, size of the plant, vintage of plants, level of technology etc.

Capacity utilisation and production rates of equipment are found to have a direct bearing on energy consumption. Both the forms of energy consumption are found to decrease with increasing capacity utilisation. The analysis of the data of dry process plants in the capacity range of 2000–3850 tpd showed that for 15% increase in capacity utilisation, the thermal energy consumption is reduced by 15 kcal/kg clinker and power consumption reduces by 5 kWh/t cement. The analysis of production rates vs energy consumption for different sections clearly showed the decreasing trend with

Steigerung der Kapazitätsauslastung um 15 % eine Senkung des thermischen Energieverbrauchs um 15 kcal/kg Klinker und des elektrischen Energieverbrauchs um 5 kWh/t Zement. Die Analyse der Produktionszahlen im Vergleich zum Energieverbrauch in verschiedenen Abschnitten zeigte bei einem Anstieg der Durchsatzleistungen der Anlagen eine abnehmende Tendenz. Daraus ergibt sich die Forderung nach hervorragender Betriebsüberwachung und Instandhaltung der Anlagen und Maschinen, um die höchste Produktivität und auch größte Energieeinsparung zu erzielen.

Mit steigender Ofenleistung nehmen sowohl der spezifische Wärmeverbrauch als auch der spezifische elektrische Energieverbrauch ab. Um den Einfluß der Anlagengröße auf den Energieverbrauch zu untersuchen, wurde außerdem der durchschnittliche Energieverbrauch der Anlagenteile von zwei Werken mit einer Jahreskapazität von einer halben Million Tonnen bzw. von einer Million Tonnen miteinander verglichen. Bei den Großanlagen zeigte sich in allen Anlagenteilen ein niedrigerer Energieverbrauch, sieht man von den Zementmahlanlagen ab, bei denen die Verbrauchszahlen mehr oder weniger vergleichbar waren.

Bis in die siebziger Jahre herrschte das Naßverfahren bei der Klinkerbrenntechnik vor, wobei die Ofenkapazitäten in den letzten fünf Jahrzehnten allmählich auf rund 750 t/Tag stiegen. Der Einsatz des Trockenverfahrens und später der Vorkalziniertechnik brachte in den beiden letzten Jahrzehnten einen schnellen Anstieg der Ofenkapazitäten von 600 auf 3 000 t/Tag und mehr. Dieser schnelle Anstieg der Produktionskapazitäten wurde begleitet von der gleichzeitigen Entwicklung des Standes der Geräte- und Steuerungstechnik sowie Zerkleinerungsverfahren mit hoher Energieausnutzung. Dementsprechend konnte der Energieverbrauch in den neuesten Werken ganz beträchtlich verringert werden. Heute werden einige der modernen Werke aus jüngster Zeit mit einem spezifischen Wärmeverbrauch von rund 750 kcal/kg Klinker (3 140 kJ/kg) und einem spezifischen Verbrauch an elektrischer Energie von etwa 100 kWh/t oder noch weniger betrieben, was mit den besten weltweit erzielten Ergebnissen vergleichbar ist.

Ausgehend von Energiestudien des NCB (National Council for Cement and Building Materials) sind vor allem in folgenden Bereichen Verbesserungen bei der Minimierung der Energieverbrauchswerte zu erzielen:

— Verminderung der Abgasverluste durch Verringerung der Gasaustrittstemperatur und des Volumens. Dies könnte durch Optimierung des Ofen- und Rostkühlerbetriebs und weitestgehende Vermeidung eines Falschlufteintritts in das System ermöglicht werden.

— Einbau von Wärmerückgewinnungssystemen zur Kraft-Wärme-Kopplung. Auf diesem Wege lassen sich bis zu 30 % des gesamten Energiebedarfs decken.

— Verbesserung der Verbrennungstechniken und Verringerung der Primärluft durch Verwendung verbesserter Brenner.

— Senkung der Ofen-Mantelverluste durch bessere Ofenausmauerung. Bei Naßdrehöfen liegen die Wärmeverluste zwischen 7 und 16,3 %.

— Einsatz fünf- oder sogar sechsstufiger Vorwärmersysteme mit Zyklonen mit niedrigem Druckverlust.

— Einsatz von Anlagen mit hoher Energieausnutzung, wie z. B. Wälzmühlen, Gutbett-Walzenmühlen, Hochleistungssichter, Antriebe mit variabler Drehzahl usw.

— Zweckmäßige Steuerung der Belastung des elektrischen Netzes zur optimalen Nutzung der verfügbaren Energie.

— Beseitigung der unzureichenden Auslastung und des Leerlaufs von Anlagen, insbesondere von allgemeinen Betriebseinrichtungen.

— Sachgerechte Steuerung der Aufgabegut-Korngröße und der Produktfeinheit in den Mahlanlagen.

— Verbesserung der Instandhaltungs- und Betriebsführungsmethoden zur Steigerung der Verfügbarkeit der Anlagen.

increase in output rates from equipment. This calls for highest level house-keeping and maintenance of plant and machinery in order to obtain maximum benefits in terms of productivity as well as energy savings.

With the increase in kiln capacity, both the specific heat consumption and specific power consumption show a decreasing trend. To study the effect of equipment size on energy consumption, also the section-wise weighted average energy consumption of half-million tpa (tons per annum) and one million tpa capacity plant were compared. This also showed lower power consumption for large size equipment in all sections except cement grinding where it was more or less comparable.

Wet process technology dominates the cement production in the country up to the seventies, and their kiln capacities gradually increased to around 750 tpd (tons per day) in the last five decades. The advent of dry process kiln technology and later precalciner technology saw a rapid change in kiln capacities from 600 to 3 000 tpd and above in the last two decades. This rapid increase in production capacities was also simultaneously matched with the advanced level of instrumentation, control and production technologies for energy efficient size reduction operations. As a result of this the energy consumption of plants of recent origin was reduced drastically. Today, some of the modern plants of recent origin are operating at a specific heat consumption of around 750 kcal/kg clinker (3 140 kJ/kg) and a specific power consumption level of 100 kWh/t or even below, which is comparable to the best standards of the world.

Based on energy studies by NCB (National Council for Cement and Building Materials), the main areas of improvement for minimising the energy consumption levels are:

— Reduction of exhaust gas losses by reducing the exit gas temperature as well as the volume. This could be achieved by optimising the operation of kiln and grate coolers and minimising the false air entry into the system.

— Incorporation of heat recovery systems for cogeneration of power. Up to 30 % of the total power requirement can be met through cogeneration.

— Improvements in combustion techniques and reduction in primary air through use of improved burners.

— Reduction in shell losses through improved refractory practices. For wet process plants the heat losses vary from 7 to 16.3 %.

— Adoption of five or even six stage preheater systems using low pressure drop cyclones.

— Utilisation of energy efficient equipment like roller mills, roll press, high efficiency classifiers, variable speed drives etc.

— Proper load management for optimum use of available power.

— Eliminating under utilization and idle running of equipment, especially auxilliaries.

— Maintaining proper control of feed and product sizes in grinding mills.

— Improving maintenance and house-keeping techniques to increase plant availability.

Studies carried out by NCB have shown potentials for reduction in power consumption up to 12.5 kWh/t cement and reduction in heat consumption up to 117 kcal/kg clinker (490 kJ/kg) in some plants.

Die vom NCB durchgeführten Studien ergaben in einigen Werken Einsparpotentiale von bis zu 12,5 kWh/t Zement beim elektrischen Energieverbrauch und von bis zu 117 kcal/kg Klinker (490 kJ/kg) beim Brennstoffenergieverbrauch.

Steigerung der Mahleffizienz durch verbesserte Kinematik in der Rollenmühle

Increasing the grinding efficiency through improved kinematics in the roller grinding mill

Augmentation de l'efficacité de broyage par amélioration de la cinématique dans le broyeur à galets

Aumento de la eficiencia de molienda mejorando la cinemática dentro del molino de rodillos

Von **L. Lohnherr**, Neubeckum/Deutschland

Zusammenfassung – Das Wälzmühlenkonzept von Krupp Polysius ist durch das Doppelrollensystem mit seiner guten Anpassungsfähigkeit an die physikalischen Eigenschaften des Mahlguts gekennzeichnet. Für weitere Energieeinsparungen beim Wälzmühlenprozeß ist die Verbesserung der Mahleffizienz, des Sichtsystems und des Transports der sich im Mahlsystem einstellenden Mahlgutmassenströme erforderlich. Durch Untersuchungen wurde festgestellt, daß sich die Stabilität des Mahlrollensystems, d.h. seine Lage und seine Bewegungen im Betrieb, wesentlich auf die Mahleffizienz auswirkt. Die Bewegungsabläufe werden wiederum von den unterschiedlichen Mahlguteigenschaften beeinflußt. Durch rein konstruktive Maßnahmen wurde eine Verbesserung der Stabilität des Mahlrollensystems der Wälzmühle von Krupp Polysius erreicht, die eine wesentliche Steigerung der Mahleffizienz zur Folge hat. Das wird durch höhere Durchsätze und einen niedrigeren massebezogenen Energieaufwand deutlich sichtbar. Außerdem wurde eine verbesserte Laufruhe beobachtet, die die dynamische Beanspruchung der Mühle vermindert.

Summary – Krupp Polysius' roller grinding mill design is characterized by the double roller system with its good adaptability to the physical properties of the material being ground. For further energy savings in the roller grinding mill process it is necessary to improve the grinding efficiency, the classifying system, and the transport in the mass material flows which occur in the grinding system. It was established through investigations that the stability of the grinding roller system, i.e. its position and movements during operation, have an important effect on the grinding efficiency. The movement processes are in turn affected by the differing properties of the material being ground. An improvement of the stability of the grinding roller system in Krupp Polysius' roller grinding mills, which resulted in a substantial increase in grinding efficiency, was achieved by purely structural measures. This is clearly apparent in a higher throughput capacity and lower specific power consumption. Improved smoothness of running, which reduces the dynamic stressing in the mill, was also observed.

Résumé – Le concept du broyeur à galets de Krupp Polysius est caractérisé par le système à galets doubles, avec sa bonne capacité d'adaptation aux propriétés physiques de la matière à broyer. Pour réaliser des économies d'énergie supplémentaires, il est nécessaire d'améliorer l'efficacité du broyage, le système de séparation et le transport des flux de matière à broyer occurrants dans le système de broyage. Des expériences ont permis de constater, que la stabilité du système des galets de broyage, c'est à dire ses positions et mouvements en marche, influe fortement sur l'efficacité du broyage. Les circonstances de mouvement sont, de leur côté, influencées par les variations de propriété de la matière à broyer. Des modifications purement constructives ont permis d'obtenir une amélioration de la stabilité du système des galets broyants du broyeur à piste de Krupp Polysius, apportant un net accroissement de l'efficacité de broyage. Cela est réellement démontré par une meilleure performance de débit et une moindre consommation d'énergie spécifique. De plus, a été constatée une marche plus calme, diminuant la sollicitation dynamique du broyeur.

Resumen – El concepto del molino de rodillos de Krupp Polysius está caracterizado por el sistema de rodillos dobles y su buena adaptabilidad a las propiedades físicas del material a moler. Para ahorrar más energía en el proceso de los molinos de rodillos, hay que mejorar la eficiencia de la molienda, el sistema de separación y el transporte de los flujos de masa del material a moler, formados en el sistema de molienda. Por medio de investigaciones se ha podido comprobar que la estabilidad del sistema de rodillos de molienda, es decir su posición y movimiento durante el servicio, repercute considerablemente en la eficiencia de la molienda. El desarrollo de los movimientos se efectúa bajo el influjo de las diferentes características del material a moler. Tomando medidas de tipo puramente constructivo, se ha conseguido una mejor estabilidad del sistema de rodillos de molienda del molino de rodillos de Krupp Polysius, que ha conducido a un aumento considerable de la eficiencia de la molienda. Esto se ve claramente por el aumento del rendimiento y un consumo de energía más bajo, referido a la masa. Además, se ha podido observar una marcha más tranquila, lo cual reduce la solicitación dinámica del molino.

Introduction

Krupp Polysius' roller grinding mill design is characterised by the double roller system with its good adaptability to the physical properties of the material being ground. In past years the technical efficiency of the design has been constantly improved through extensive studies at both the research centre and industrial mills. Special attention has been paid to the recirculating material systems set up in the grinding plant as a function of grinding efficiency. In one study the roller grinding mill design has been further optimised with the aid of a theoretical description of the grinding process and by practical measurements and carrying out trial grinding.

1. Modelling the grinding process

The pressure profiles obtained were calculated with the aid of a mathematical model in order to clarify the processes which take place in the grinding bed of a roller-type mill. At the same time grinding rollers of an experimental mill at the Polysius research centre were fitted with pressure sensors to check the theoretical results. These are given in **Fig. 1**, which shows the pressure profile of one wide single roller and that of a double roller. Width utilisation can be seen to be considerably more efficient with a double roller than with a wide single one. This is because the rollers adapt better to the material being ground. The maximum grinding pressures for the double roller are approx. 40 MPa, a level which is favourable for grinding. These findings are in accordance with the favourable specific power consumption figures found in the industrial mills.

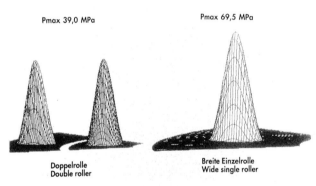

BILD 1: Druckprofile von Einzel- und Doppelrollen
FIGURE 1: Pressure profiles for single and double rollers

2. Material-specific influences on grinding efficiency

In practice the material being ground has been found to have an unfavourable effect on the grinding action in various industrial mills, and the effect is sometimes specific to the material. The stability of the grinding roller system, in respect both of movements about the neutral position and of the absolute neutral position itself, is known to have a great effect on grinding efficiency. **Fig. 2** shows the relative production of surface area in the material being ground plotted against a stability factor. The flow behaviour and internal supporting properties of the materials being ground sometimes vary widely and affect the movements of the roller system. The effect of the material is expressed by the stability factor, which takes into account the flow function of the material being ground and its internal stability derived from the compacted cake test. From this it can be seen that the physical properties of the material being ground influence

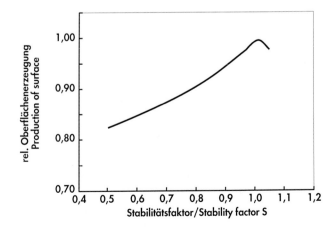

BILD 2: Zerkleinerungsfortschritt in Abhängigkeit vom Stabilitätsfaktor
FIGURE 2: Comminution progress as a function of the stability factor

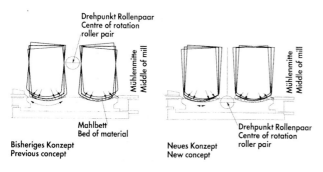

BILD 3: Lage des Rollenpaar-Drehpunkts auf der Mahlbettebene
FIGURE 3: Position on the grinding bed plane of the centre of rotation of the roller pair

3. Konstruktive Lösung

Als eine der möglichen Maßnahmen zur Verbesserung der Kinematik des Rollenpaarsystems erwies sich die Verschiebung des Rollenpaardrehpunktes auf die Mahlbettebene als sehr wirkungsvoll. In **Bild 3** sind das bisherige und das neue Konzept dargestellt. Bei dem bisherigen Konzept traten bei einer Auslenkung des Rollenpaares um den Drehpunkt des Mahlrollensystems im Bereich des Mahlbettes starke relative Bewegungen gegenüber dem Mahlgut auf. Dies führte zur Verringerung der ansonsten gut genutzten Mahlfläche sowie bei kritischen Mahlgütern zur Zerstörung des Mahlbettes und damit zu schlechteren Mahlergebnissen. Um diese Störgrößen auszuschließen, wurde der Drehpunkt des Rollenpaares in die Mahlbettebene gelegt. Aus dem Bild sind die verbesserten Bewegungsabläufe zwischen den Mahlrollen und dem Mahlgut bei der Darstellung des neuen Konzeptes zu erkennen.

4. Verfahrenstechnische Ergebnisse

Mit dem bisherigen und dem neuen Konzept wurde das Kennfeld einer Versuchsmühle aufgenommen. **Bild 4** zeigt den massebezogenen Arbeitsaufwand sowie die Durchsatzleistung in Abhängigkeit von der Mahlkraft. Die verbesserte Mahleffizienz mit dem neuen Konzept wird hierbei durch

the grinding performance of a roller mill system. In the light of this finding it is necessary to stabilise the grinding roller system, but without adversely affecting the good adaptability of the rollers to the material being ground.

3. Design solution

Shifting the pivot point of the pair of rollers to the plane of the grinding bed has proved to be very effective as one of the possible ways of improving the kinematics of the roller pair system. Fig. 3 shows the previous design and the new one. With the previous design there were strong movements relative to the material being ground in the region of the grinding bed when the pair of rollers was swivelled about the pivot point of the roller system. This led to a reduction in the otherwise well utilised grinding surface and, in the case of critical materials, to destruction of the grinding bed and thus to poorer grinding results. In order to avoid these disturbances the pivot point of the pair of rollers has been moved to the plane of the grinding bed. The improved movement between the rollers and the material can be seen from the illustration of the new design in the diagram.

4. Process engineering results

The characteristic diagram of an experimental mill has been plotted with the previous design and the new one. **Fig. 4** shows the mass-related energy expenditure and throughput capacity versus the grinding force. The improved grinding efficiency obtained with the new design is clearly visible from the low mass-related energy expenditure and the higher throughput capacity. The results also confirm the statement that grinding efficiency is reduced when excessive pressures are applied. In addition to the improved process engineering data shown, far smoother running is observed, which has a favourable effect on the dynamic stresses in the machine.

5. Construction of the industrial mill

On the basis of the test results described and the resultant advantages, Krupp Polysius received an order from Portlandzementwerk Dotternhausen Rudolf Rohrbach KG

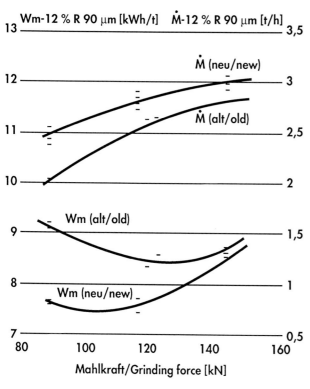

BILD 4: Arbeitsaufwand und Durchsatzleistung einer Versuchsmühle
FIGURE 4: Power consumption and throughput capacity of a test mill

BILD 5: Rollenmühle DOROL zur Mahlung von Ölschiefer
FIGURE 5: DOROL roller mill for grinding oil shale

den niedrigen massebezogenen Arbeitsaufwand und die höhere Durchsatzleistung deutlich sichtbar. Gleichzeitig wird auch die Aussage bestätigt, daß bei überhöhten Mahldrücken die Mahleffizienz vermindert wird. Ergänzend zu den dargestellten besseren verfahrenstechnischen Daten wurde ein deutlich ruhigeres Laufverfahren beobachtet, das sich günstig auf die dynamische Beanpruchung der Maschine auswirkt.

5. Ausführung der Industriemühle

Ausgehend von den beschriebenen Untersuchungsergebnissen und den daraus resultierenden Vorteilen erhielt die Fa. Krupp Polysius im Jahre 1992 von der Portlandzementwerk Dotternhausen Rudolf Rohrbach KG den Auftrag zur Lieferung einer Rollenmühle mit dem neuen Konzept zur Vermahlung von Ölschiefer. Maßgeblich für diese Entscheidung waren technische Gründe – nicht zuletzt die Möglichkeit eines energetisch günstigen Teillastbetriebs. **Bild 5** zeigt die technische Anordnung für die Krafteinleitung und den auf die Mahlbettebene gelegten Drehpunkt des neuen Mühlenkonzeptes. Die Entwicklung stellt einen wichtigen Schritt zur Stabilisierung des Mahlrollensystems und damit zur Steigerung der Mahleffizienz dar.

in 1992 for the supply of a roller grinding mill of the new design for grinding oil shale. The decision was taken mainly on technical grounds – not least because of the possibility of energy-saving operation under partial load. **Fig. 5** shows the power intake arrangement and the pivot point of the new mill design located in the grinding bed plane. The development represents an important step towards stabilising the grinding roller system and thus increasing grinding efficiency.

Verschleißschutzalternativen für die Beanspruchungsflächen von Gutbett-Walzenmühlen*)

Alternative means of protecting the working sufaces of high-pressure grinding rolls against wear*)

Moyens de protection des surfaces des broyeurs à rouleaux, soumises à l'usage

Alternativas de protección de las superficies de los molinos de cilindros y lecho de material, sometidas al desgaste

Von **N. Patzelt**, Neubeckum/Deutschland

Zusammenfassung — Die Verfügbarkeit von Gutbett-Walzenmühlen hängt in hohem Maße von der Standzeit des Verschleißschutzes der Mahlwalzen ab. Für eine hohe Betriebssicherheit sind neben einem zweckmäßigen Anlagenbetrieb, d. h. einem verminderten Mahldruck und einer gleichmäßigen Mahlgutzuführung und -verteilung, verschiedene Lösungen für die Ausbildung des Verschleißschutzes entwickelt worden. Dabei hat die Minenindustrie, in der vorwiegend hochabrasive Mahlgüter verarbeitet werden, entscheidende Impulse zur Weiterentwicklung von Verschleißkonzepten geliefert. Parallel zur Optimierung der Auftragsschweißung wurde ein alternatives Verschleißschutzkonzept auf der Grundlage von Gußwerkstoffen erarbeitet, das nunmehr bei Vollwalzen, Bandagen oder Segmenten eingesetzt wird. Der Einsatz von Vollwalzen und Bandagen ist auf 1,7 m Walzendurchmesser und auf eine maximale Mahlguttemperatur von 110°C begrenzt und besonders auf die Vorzerkleinerung und Hybridmahlung festgelegt. Obwohl profilierte Segmente (PROSEG) höhere Investitionskosten verursachen, führen die kürzeren Montagezeiten und deutlich längeren Standzeiten von weit über 20 000 Betriebsstunden zu einer schnellen Amortisation. Daher können profilierte Segmente die gesamtwirtschaftlich günstigere Lösung sein.

Summary — The availability of high-pressure grinding rolls is heavily dependent on the service life of the wear protection on the grinding rollers. In addition to appropriate plant operation, i. e. reduced grinding pressure and more uniform feed and distribution of the material, various solutions for forming the wear protection have been developed to achieve high operational reliability. The mining industry, which usually processes highly abrasive material, has provided decisive impetus for further development of wear schemes. As well as optimizing wear protection by build-up welding an alternative wear protection scheme was worked out using cast materials which are made as complete rollers, tyres or segments. The use of complete rollers and tyres is limited to a roller diameter of 1.7 m and a maximum temperature of the material being ground of 110°C. They are therefore mainly used for preliminary comminution and hybrid grinding. Although profiled segments (PROSEG) give rise to higher investment costs, their advantages — shorter installation time and significantly longer service life of well over 20 000 operating hours — mean rapid pay-back. Profiled segments can therefore be the better solution when all costs are taken into account.

Résumé — La disponibilité des broyeurs à rouleaux dépend essentiellement de la durée de vie des couches d'usure des rouleaux de broyage. Pour une haute sécurité de fonctionnement ont été définies, outre l'exploitation appropriée de l'atelier, c'est à dire une pression de broyage raisonnable et une alimentation et distribution régulières de la matière à broyer, différentes solutions pour la configuration de la protection anti-usure. Dans ce contexte, l'industrie minière, traitant le plus souvent des matières hautement abrasives, a fourni des indications décisives pour le développement de concepts d'usure. Parallèlement à l'optimisation de la protection anti-usure par apport de soudure, a été mis au point un concept alternatif de protection contre l'usure, reposant sur le principe de matériaux à base de fonte, employés sous forme de rouleau massif, bandage ou segments. L'emploi de rouleaux massifs et de bandage est limité à un diamètre de rouleau de 1,7 m et à une température maximale de la matière à broyer, de 110°C. Ils sont donc utilisés de préférence pour la préfragmentation et le broyage hybride. Bien que les segments profilés (PROSEG) occasionnent des coûts d'investissement plus élevés, leurs avantages tels que temps de montage plus courts et durées de vie nettement prolongées, de loin au-delà de 20 000 heures de marche, signifient un amortissement rapide. Ainsi, les segments profilés peuvent être la solution la plus avantageuse, du point de vue économie en général.

---

*) Überarbeitete Fassung eines Vortrages zum VDZ-Kongreß '93, Düsseldorf (27.9.–1.10.1993)
Revised text of a lecture to the VDZ Congress '93, Düsseldorf (27.9.–1.10.1993)

Alternativas de protección de las superficies de los molinos de cilindros y lecho de material, sometidas al desgaste

Resumen — La disponibilidad de los molinos de cilindros y lecho de material depende en gran medida de la duración del revestimiento protector de los cilindros. Para obtener una elevada seguridad de servicio se han desarrollado, aparte de una adecuada marcha de la instalación es decir, una reducida presión de molienda y una alimentación y distribución más uniforme del material a moler, diferentes soluciones respecto de la formación del revestimiento contra el desgaste. A este respecto, la industria minera, en la que se trata mayormente material altamente abrasivo, ha dado impulsos decisivos para el desarrollo de conceptos relacionados con el desgaste. Paralelamente a la optimización de la protección contra el desgaste, por aporte de soldadura, se ha elaborado un concepto alternativo contra el desgaste, basado en materiales de fundición que se emplean en forma de cilindros macizos, bandajes o segmentos. El empleo de cilindros macizos y bandajes queda limitado a un diámetro de cilindros de 1,7 m y a una temperatura máxima del material a moler de 110°C. Por esta razon, se utilizan éstos principalmente para el desmenuzamiento previo y para la molienda hibrida. Aun que los segmentos perfilados (PROSEG) requieren mayores gastos de inversión, sus ventajas — tiempo de montaje más corto y duración de vida notablemente más larga, o sea de mucho más de 20 000 horas de servicio — residen en una rápida amortización. Por eso, los segmentos perfilados pueden ser, en su conjunto, la solución más ventajosa.

1. Einleitung

Die Verfügbarkeit von Gutbett-Walzenmühlen wird z. Z. noch maßgeblich durch den Verschleiß der Mahlwalzen bestimmt. Während die ersten Gutbett-Walzenmühlen Mitte der 80er Jahre mit Nihard-Bandagen ausgerüstet wurden – diese Anlagen befinden sich heute noch mit Standzeiten von inzwischen 40 000 Stunden in Betrieb – wurden die Walzenoberflächen der nachfolgend gelieferten Anlagen aufgrund der besseren Einzugsbedingungen mit einem geschweißten Verschleißschutz versehen. Eine Auftragsschweißung von großer Härte und hohem Verschleißwiderstand – beansprucht durch hohen oszillierenden Druck – war damals neu. Deshalb konnten die bisweilen auftretenden Probleme nur durch aufwendige Untersuchungen und langwierige Maßnahmen beseitigt werden. Diese Maßnahmen lassen sich in 3 Gruppen einteilen:

– Verbesserung der Schweißqualität,
– alternative Verschleißschutzkonzepte,
– Absenkung des Mahldruckniveaus.

Es ist bekannt, daß die Standzeit eines geschweißten Verschleißschutzes im wesentlichen von der Höhe des Mahldrucks, von der Gleichmäßigkeit der Druckverteilung, der Zusammensetzung des Aufgabematerials sowie der Elimination von Fremdkörpern abhängt.

Eine der wirkungsvollsten Maßnahmen zur Verlängerung der Standzeit ist die Verminderung des Mahldruckes. Während man die Gutbett-Walzenmühlen der ersten Generation mit spezifischen Mahlkräften (F/DxL) von ca. 8 N/mm² betrieb, wurden in der nachfolgenden Generation die Drücke auf ca. 6 N/mm² reduziert. Allein diese Maßnahme hat spürbar zur Senkung des Ausfallrisikos [1] beigetragen. Ein gewisser Durchsatzrückgang, der damit verbunden war, kann bei neuen Anlagen durch eine breitere Walzenausführung kompensiert werden.

Neben der Absenkung des Mahldruckniveaus wurde auch die Qualität der Auftragsschweißung ständig verbessert. Fast rißfreie Unterpulverschweißungen mit hohem Verschleißwiderstand sind heute möglich. Die Standzeitverbesserungen aus der Verminderung des Mahldruckes und die Verbesserung der Auftragsschweißung sind auf **Bild 1** dargestellt. Während bei der ersten Generation Standzeiten von durchschnittlich 5000 bis 7000 h erreicht wurden, liegen diese heute zwischen 12 000 bis 15 000 h.

Parallel zu den o. a. zwei Maßnahmen wurde ein alternatives Verschleißschutzkonzept auf der Grundlage von Gußwerkstoffen entwickelt. Gußwerkstoffe können bei der Herstellung von Vollwalzen, Bandagen oder Segmenten eingesetzt werden (**Bild 2**) und besitzen gegenüber geschweißten Werkstoffen gewöhnlich Festigkeiten, die doppelt so hoch sind.

2. Verschleißschutzkonzepte

Die Ausbildung der Walzenoberfläche beeinflußt die Gutbettzerkleinerung wesentlich. Es gilt als nachgewiesen, daß

1. Introduction

The availability of high-pressure grinding rolls is at present still largely determined by the wear on the grinding rollers. Whereas the first high-pressure grinding rolls in the mid 80s were fitted with Nihard tyres – these plants are still in operation with service lives which have now reached 40 000 hours – the roller surfaces in subsequent installations were provided with welded wear protection because of the improved intake conditions. Build-up welding of great hardness and high wear resistance – subjected to high oscillating pressures – was new at the time, so the problems which occurred from time to time could only be eliminated by expensive investigations and endless improvements. These measures can be divided into 3 groups:

– improving the quality of welding,
– alternative wear protection systems,
– lowering the grinding pressure level.

It is well known that the service life of welded wear protection is essentially dependent on the level of the grinding pressure, the evenness of the pressure distribution, the composition of the feed material, and the elimination of foreign bodies.

One of the most effective ways of extending the service life is to reduce the grinding pressure. High-pressure grinding rolls of the first generation were operated with specific grinding forces (F/DxL) of approximately 8 N/mm², but in the following generation the pressures were reduced to approximately 6 N/mm². This measure alone has made a considerable contribution towards lowering the risk of breakdown [1]. A certain loss in throughput associated with it can be offset in new plants by a wider roller.

Not only was the grinding pressure level lowered but there were also continuous improvements in the quality of the

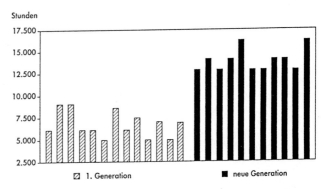

BILD 1: Entwicklung der Standzeiten der Beanspruchungsflächen bei den Gutbett-Walzenmühlen
FIGURE 1: Development of the service life of the working surfaces of high-pressure grinding rolls

Stunden = hours
1. Generation = 1st generation
neue Generation = new generation

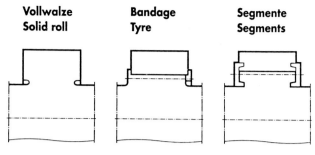

BILD 2: Verschleißschutzkonzepte für die Beanspruchungsflächen von Gutbett-Walzenmühlen
FIGURE 2: Wear protection systems for the working surfaces of high-pressure grinding rolls

Vibrationen an der Gutbett-Walzenmühle durch eine entsprechende Profilierung der Mahlwalzen reduziert werden können [2]. Der Einfluß unterschiedlicher Walzenprofilierungen auf den Durchsatz einer Gutbett-Walzenmühle in Abhängigkeit von der Mahlgutzusammensetzung (Klinker, Grieße, Schülpen) ist auf **Bild 3** dargestellt. Es ist erkennbar, daß bei der Beanspruchung von Zementklinker der Durchsatz durch eine Profilierung (Chev-w, Chev-e) verbessert wird. Bei Frischgut-/Grießgemischen und einer glatten Walzenoberfläche kann es passieren, daß das Mahlgut nicht eingezogen wird. Je enger die Profilierung auf den Walzen (Chev-e) ausgeführt ist, desto besser ist der Materialeinzug gewährleistet und desto höhere Durchsätze können realisiert werden. Außerdem trägt die Profilierung der Beanspruchungsflächen zur Stabilisierung des Mahlbettes bei. Da bei der Hybrid-, Combi- und Fertigmahlung ein erheblicher Anteil an bereits beanspruchtem Material in den Walzenspalt wieder zurückgeführt werden muß, kann auf eine Walzenprofilierung im allgemeinen nicht verzichtet werden.

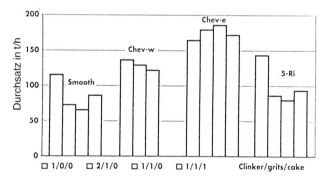

BILD 3: Einzugsverhalten unterschiedlicher Walzenoberflächen
FIGURE 3: Intake characteristics of different roller surfaces

Durchsatz in t/h = throughput in t/h
5-Ri = 5 grooves

Bewährt haben sich auch sogenannte „negative Profile", wie im **Bild 4** dargestellt. Die mit Rillen versehenen Beanspruchungsflächen setzen sich besonders bei grobkörnigen, zähen und teilweise feuchten Mahlgütern zu, verbessern damit die Einzugsverhältnisse und erhöhen die Stabilität des Gutbettes im Mahlspalt. In Hybrid- und Combimahlanlagen übernimmt die Gutbett-Walzenmühle einen erheblichen Anteil der Feinzerkleinerung. Bei Anlagen, die nach der Technologie der sogenannten Fertigmahlung arbeiten, erfolgt die Zerkleinerung ausschließlich in der Gutbett-Walzenmühle. Der Ausfall einer Gutbett-Walzenmühle würde gerade bei diesen Anlagenschaltungen große Produktionsverluste und hohe Betriebskosten verursachen. Der Austausch von Vollwalzen und Bandagen ist im allgemeinen mit einem längeren Betriebsstillstand verbunden. Außerdem ist die gießtechnische Ausführung von Vollwalzen und insbesondere von Bandagen nicht ganz einfach. Für die Weiterentwicklung von Verschleißschutzkonzepten hat die Minenindustrie, in der meist hochabrasives Mahlgut verarbeitet wird, entscheidende Impulse geliefert. Neben hohen

BILD 4: Bandage mit Rillen
FIGURE 4: Tyres with grooves

build-up welding. Virtually crack-free submerged-arc welding with a high wear resistance is possible nowadays. The improvements in service life resulting from the reduction in grinding pressure and the improvement in build-up welding are shown in **Fig. 1**. While average service lives of 5 000 to 7 000 h were achieved with the first generation, they now lie between 12 000 and 15 000 h.

An alternative wear protection system based on cast materials was developed in parallel with the two measures mentioned above. Cast materials can be used in the manufacture of complete rollers, tyres or segments (**Fig. 2**) and are normally twice as strong as welded materials.

2. Wear protection systems

The form of the roller surface has a substantial influence on the interparticulate comminution. It has been shown that vibrations in high-pressure grinding rolls can be reduced by appropriate profiling of the grinding rollers [2]. The effect of different roller profiles on the throughput of high-pressure grinding rolls is shown in **Fig. 3** as a function of the material being ground (clinker, grits, compacted cake). It can be seen that the throughput when comminuting cement clinker is improved by profiling (Chev-w, Chev-e). With mixtures of fresh material and grits and a smooth roller surface it can happen that the feed is not drawn in at all. The closer the profiling is made on the rollers (Chev-e) the better is the material intake and the higher are the throughputs which can be achieved. Profiling the working surfaces also contributes to stabilizing the material bed. In hybrid, combined and finish grinding a considerable proportion of material which has already been comminuted has to be returned to the roller gap, so in general it is not possible to dispense with roller profiling.

So-called "negative profiles" as shown in **Fig. 4** have also proved successful. The grooved working surfaces become clogged, particularly when the materials being ground are coarse, tenacious and partly moist, thereby improving intake conditions and increasing the stability of the material bed in the grinding gap. In hybrid and combined grinding plants a considerable proportion of the fine comminution is carried out by the high-pressure grinding rolls, and in plants operating on the finish grinding system all the comminution takes place in the high-pressure grinding rolls. In these plant configurations any breakdown of the high-pressure grind-

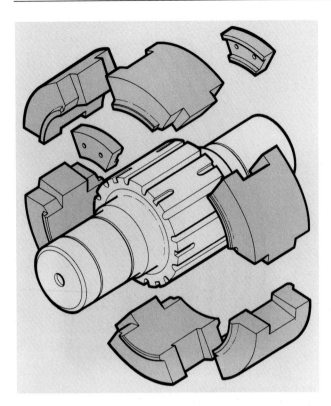

BILD 5: POLYCOM-Segmentwalze
FIGURE 5: POLYCOM® segmented roller

Standzeiten und einer großen Verfügbarkeit werden vor allem kurze Wechselzeiten für den Verschleißschutz gefordert.

Um diesen Forderungen besser nachzukommen, wurden Verschleißschutzsegmente entwickelt, die nicht geschraubt sind, sondern durch Klemmen am Walzenkörper befestigt werden. Damit sind hohe Verschleißdicken von 60 bis 120 mm, d. h. Segmenthöhen bis zu 400 mm, realisierbar. Die im Betrieb auftretenden Kippmomente können nur durch eine derartige Klemmkonstruktion beherrscht werden. Außerdem ist die gießtechnische Herstellung von Segmenten einfacher als die von Vollwalzen und Bandagen. Desweiteren können die Segmente ohne jegliche Vorspannung auf dem Walzenkörper befestigt werden. Die Segmentierung einer Walzenoberfläche ist im **Bild 5** dargestellt. Die auf diesem Bild dargestellten Segmente haben sich seit 3 Jahren in Mühlen mit Walzendurchmessern von 1,2 bis 2,2 m bewährt. Die Austauschzeiten der Segmente liegen bei 1 bis 2 Tagen. In der Minenindustrie sind die Oberflächen der Segmente aufgrund des hohen Verschleißes nicht profiliert. Die laufende Nachbearbeitung der Profilierung würde zu lange Stillzeiten erforderlich machen.

In der Zementindustrie ist jedoch, wie oben beschrieben, für die Mahlung von feinen und trockenen Mahlgütern eine negative oder positive Profilierung der Walzenoberfläche erforderlich. Versuche mit auftragsgeschweißten Profilen auf gegossenen Segmenten, wie sie in **Bild 6** dargestellt sind, waren so erfolgreich, daß der Einsatz dieses als PROSEG bezeichnete Verschleißschutzkonzept auch in der Zementindustrie eingeführt werden konnte. Bis heute wurden mit dem PROSEG-Verschleißschutzkonzept Standzeiten von mehreren 1000 h ohne Probleme erreicht. In der Minenindustrie liegen mit der Aufschweißung von verschlissenen Segmenten umfangreiche Erfahrungen vor, wobei das fertigungstechnisch teuerste Teil der Segmentfuß mit einer Spannvorrichtung darstellt. Die Verschleißraten liegen in der Minenindustrie deutlich höher als in der Zementindustrie, so daß dort ein Segmentwechsel häufiger erforderlich wird. Deshalb ist eine komplette Wiederaufschweißung der Segmente in der Minenindustrie durchaus wirtschaftlich vertretbar. Mit Erfolg wird deshalb auch die Segmentauf-

ing rolls would result in great production losses and high operating costs. Replacing complete rollers and tyres usually involves quite long stoppages. Apart from this, the casting of complete rollers, and especially of tyres, is not entirely straightforward. The mining industry, which usually processes highly abrasive material, has provided decisive impetus for further development of wear protection systems. The main requirements, in addition to a long service life and good availability, are short replacement times for the wear protection.

Wear protection segments which are clamped, rather than bolted, to the roller body were developed to meet these requirements. Large wear thicknesses of 60 to 120 mm can be achieved in this way, i.e. segment heights of up to 400 mm. The tipping moments which occur during operation can only be overcome by a clamped design of this type. Segments are also easier to cast than complete rollers or tyres. What is more, segments can be attached to the roller body without any prestressing. The segmenting of a roller surface is illustrated in **Fig. 5**. The segments shown in this diagram have been used successfully for 3 years in high-pressure grinding rolls with roller diameters from 1.2 to 2.2 m. They take one to two days to replace. In the mining industry the segment surfaces are not profiled because of the heavy wear. Continual refinishing of the profiles would require excessively long stoppages.

Positive or negative profiling of the roller surface is, however, necessary in the cement industry for grinding fine and dry materials, as explained above. Trials with build-up welded profiles on cast segments, as illustrated in **Fig. 6**, were so successful that this wear protection system, known as PROSEG has also been introduced into the cement industry. Service lives of several 1000 h have been achieved to date with the PROSEG system without any problem. The mining industry has extensive experience of build-up welding on worn segments, the segment foot with clamping device being the most expensive part from the manufacturing point of view. The wear rates in the mining industry are significantly higher than in the cement industry so segments have to be replaced more often. Complete renovation of the segments by build-up welding is therefore entirely economically justifiable in the mining industry, so build-up welding on segments has been carried out successfully for some time (**Fig. 7**). Segments have decisive advantages over welded wear protection. It is definitely possible to achieve service lives of far longer than 20 000 h, as opposed to 9 000 to 15 000 h for welded wear protection. At least two spare tyres are needed with welded wear protection (meaning stoppage times of 1 to 2 weeks), but with only two sets of segments the change can be carried out in 1 to 2 days (**Fig. 8**). At least two complete spare rollers with bearings and bearing blocks are needed to achieve equivalent replacement times with welded wear protection. The costs of these spare parts amount to about 35 to 40 % of the machine price so the use of

BILD 6: Profilierte Segmente des Verschleißschutzes PROSEG
FIGURE 6: Profiled segments of the PROSEG® wear protection system

BILD 7: Blick auf ein einzelnes Segment mit Aufschweißung
FIGURE 7: View of a single segment with build-up welding

schweißung seit einiger Zeit durchgeführt (**Bild 7**). Segmente haben gegenüber dem geschweißten Verschleißschutz entscheidende Vorteile. Standzeiten von weit über 20 000 h gegenüber einem geschweißten Verschleißschutz von 9 000 bis 15 000 h sind durchaus erreichbar. Während bei geschweißtem Verschleißschutz mindestens zwei Ersatzbandagen erforderlich sind (das bedeutet Stillstandzeiten von 1 bis 2 Wochen), kann mit nur zwei Segementsätzen der Wechsel in 1 bis 2 Tagen erfolgen (**Bild 8**). Um möglichst gleiche Wechselzeiten beim geschweißten Verschleißschutz zu erreichen, sind mindestens zwei komplette Ersatzrollen mit Lagern und Lagersteinen erforderlich. Da die Kosten für diese Ersatzteile ca. 35 bis 40 % des Maschinenpreises betragen, bietet der Einsatz von Segmenten auch hinsichtlich der Investitionskosten entscheidende Vorteile. Die Nachbearbeitung der Profile kann ohne besondere Vorbehandlung in der Maschine erfolgen.

3. Schlußbemerkung

Der Einsatz von Gußwerkstoffen stellt bei den Gutbett-Walzenmühlen eine interessante Alternative zum auftragsgeschweißten Verschleißschutz von Mahlwalzen dar. Gußwerkstoffe können sowohl bei der Ausführung von Vollwalzen, Bandagen wie auch Segmenten vorteilhaft zur Anwendung gelangen. Aus fertigungstechnischen Gründen ist allerdings der Einsatz von Vollwalzen und Bandagen auf Walzendurchmesser von 1,7 m begrenzt. Bei Vollwalzen darf die maximale Materialtemperatur 110 °C nicht überschreiten. Vollwalzen und Bandagen werden deshalb vor allem bei der Vorzerkleinerung und Hybridmahlung verwendet. Obwohl profilierte Segmente höhere Investitionskosten verursachen, führen die kürzeren Montagezeiten und die deutlich längeren Standzeiten von weit über 20 000 h zu einer schnellen Amortisation. Daher sind profilierte Verschleißsegmente aus Gußwerkstoffen eine betriebssichere und gesamtwirtschaftlich günstige Lösung.

BILD 8: Werkstattmontage von Verschleißschutzsegmenten auf den Walzengrundkörper
FIGURE 8: Workshop assembly of wear protection segments on the basic roller body

segments also offers decisive advantages in respect of the capital costs. The profiles can be refinished in the machine without any special pretreatment.

3. Summary

For high-pressure grinding rolls the use of cast materials offers an attractive alternative to build-up welding wear protection for the grinding rollers. Cast materials can be used advantageously not only as complete rollers and tyres, but also as segments. For manufacturing reasons the use of complete rollers and tyres is, however, limited to roller diameters of 1.7 m, and with complete rollers the maximum material temperature must not exceed 110°C. Complete rollers and tyres are therefore used mainly for preliminary comminution and hybrid grinding. Although profiled segments give rise to higher capital costs, their shorter installation times and significantly longer service lives of well over 20 000 h lead to rapid pay-back. Profiled wear segments made of cast materials therefore represent an operationally reliable and economically advantageous solution.

Literature

[1] Patzelt, N., und Tiggesbäumker, P.: Konstruktive Lösungen für betriebssichere Mahlwalzen von Gutbett-Walzenmühlen. Zement-Kalk-Gips 44 (1991) No. 2, pp. 88–92.

[2] Schmitz, T., und Kupper, D.: Einfluß rheologischer Parameter auf den Prozeß der Gutbettbeanspruchung in der Walzenmühle (Teil 2). Zement-Kalk-Gips 45 (1992) No. 12, pp. 634–639.

Beispiele erfolgreicher Integration der Hüttensandaufbereitung

Examples of successful integration of the granulated blast furnace slag processing system

Exemples d'intégration réussie de la préparation du laitier granulé

Ejemplos de integración bien lograda de la preparación de escoria siderúrgica

Von **N. Patzelt**, Neubeckum/Deutschland

Zusammenfassung – Angesichts steigender Qualitätsanforderungen kommt es bei der Mahlung von Hochofenzement darauf an, mit Hilfe flexibler und energiesparender Mahlsysteme die Produktqualitäten beeinflussen zu können. Als energiesparende Mahlsysteme kommen Gutbett-Walzenmühlen und Wälzmühlen in Betracht. Leider haben nicht beide Mahlsysteme die verlangte Flexibilität zur Einstellung der Korngrößenverteilung und damit der Qualität des Produkts. Die niedrigen Mahlgutverweilzeiten führen zu engen Korngrößenverteilungen im Produkt, die deutlich höhere Werte des Wasseranspruchs zur Folge haben. Beim Betrieb von Gutbett-Walzenmühlen bietet die Schülpenrezirkulation jedoch die Möglichkeit, Hüttensandmehl mit breiterer Korngrößenverteilung zu erzeugen. Auch die Mahlung auf höhere Feinheit ist möglich. Damit kann bei der getrennten Mahlung von Hüttensandmehl und nachfolgender Mischung mit Portlandzement die Qualität des erzeugten Hochofenzements gegenüber der bei gemeinsamer Vermahlung deutlich verbessert werden. Beim Betrieb von Wälzmühlen ist die Korngrößenverteilung des Produkts nicht einstellbar. Die Mahlung auf hohe Feinheiten ist aus betriebstechnischen Gründen nicht möglich.

Summary – In view of increasing quality requirements it is important when grinding blast furnace cement to be able to influence the product quality with the aid of flexible and energy-saving grinding systems. High-pressure grinding rolls and roller grinding mills are possible energy-saving grinding systems. Unfortunately, the two grinding systems do not have the required flexibility for adjusting the particle size distribution and hence the quality of the product. The low residence times of the material being ground lead to narrow particle size distributions in the product, resulting in significantly higher values for the water demand. However, recirculation of the compacted cake when operating high-pressure grinding rolls offers the chance of producing blast furnace slag meal with a wider particle size distribution. It is also possible to grind to a higher level of fineness. By grinding the blast furnace slag meal separately, followed by blending with Portland cement, it is possible to improve the quality of the blast furnace cement significantly over that produced by intergrinding. The particle size distribution of the product cannot be adjusted when operating roller grinding mills, and grinding to high levels of fineness is not possible for process engineering reasons.

Résumé – Compte tenu des besoins croissants de qualité, il est nécessaire, lors de la mouture de ciment de haut fourneau, de pouvoir agir sur les qualités du produit au moyen de systèmes de broyage flexibles et peu gourmands en énergie. Des broyeurs à rouleaux ou des broyeurs à galets s'offrent comme systèmes de broyage économes en énergie. Malheureusement, ni l'un, ni l'autre de ces systèmes de broyage ont la flexibilité demandée pour le réglage de la distribution granulométrique et, ainsi, de la qualité du produit. Les courts séjours de la matière à broyer conduisent à des distributions granulométriques étroites du produit, ayant comme conséquence une demande d'eau nettement plus élevée. Par contre, dans l'exploitation des broyeurs à rouleaux, la recirculation des plaquettes offre la possibilité de fabriquer du laitier finement broyé avec une distribution granulométrique plus large. Il est aussi possible de broyer plus finement. Le broyage séparé du laitier et un mélange ultérieur avec du ciment Portland permet ainsi d'améliorer nettement la qualité du ciment de haut fourneau fabriqué, par rapport à celle obtenue en broyage commun. L'exploitation de broyeurs à galets ne permet pas de régler la distribution granulométrique du produit. Le broyage aux grandes finesses est, pour des raisons techniques du fonctionnement, impossible.

Resumen – Teniendo en cuenta las crecientes exigencias de calidad, es muy importante, en la molienda de escoria siderúrgica, poder influir en la calidad del producto, mediante unos sistemas de molienda flexibles y que ayuden a ahorrar energía. En cuanto al ahorro de energía, son aptos los molinos de cilindros y lecho de material y los molinos de pista de rodadura. Desgraciadamente, los dos sistemas de molienda no poseen la necesaria flexibilidad para poder regular la distribución granulométrica y, con ello, la calidad del producto. Los cortos períodos de permanencia del material a moler conducen a una distribución granulométrica estrecha dentro del producto, con el consi-

guiente aumento de la demanda de agua. Durante el servicio de los molinos de cilindros y lecho de material, la recirculación del material compactado ofrece, sin embargo, la posibilidad de producir harina de escoria siderúrgica, con una distribución granulométrica más ancha. También es posible una molienda más fina. Mediante la molienda separada de harina de escoria siderúrgica y subsiguiente mezcla con cemento portland, se puede mejorar notablemente la calidad del cemento de alto horno, en comparación con la molienda conjunta. Durante el servicio de los molinos de pista de rodadura no se puede regular la distribución granulométrica del producto. Por razones técnicas, no es posible conseguir una molienda finísima.

Bei der separaten Mahlung von Hüttensand und nachfolgender Mischung mit Portlandzement kann die Produktqualität des Hochofenzementes gegenüber der gemeinsamen Mahlung wesentlich verbessert werden. Dies wird u. a. durch eine höhere Feinheit des Hüttensandmehls erreicht. Dabei gelten auch für das Hüttensandmehl Qualitätsanforderungen. Abgesehen von der Zusammensetzung beeinflußt z. B. dessen Korngrößenverteilung u. a. den Wasseranspruch bei Normsteife. Bei neuen Mahlanlagen ist daher neben der energiesparenden Mahlung eine hohe Flexibilität zur Einstellung der Produktqualität sehr wichtig. Energiesparende Mahlsysteme sind Gutbett-Walzenmühlen, die als Fertigmühlen betrieben werden, und Rollenmühlen. Leider erfüllen nicht beide Mahlsysteme die Anforderungen an die Flexibilität. Beide Mahlsysteme haben sehr niedrige Mahlgutverweilzeiten. Eine ausreichend große Menge von Feinstgut wird daher nicht immer erzeugt. Dies führt zu engen Korngrößenverteilungen des Produkts, die hohe Wasseranspruchswerte zur Folge haben.

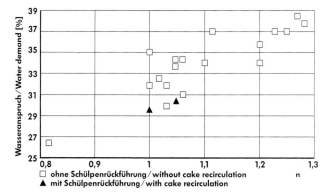

BILD 1: Wasseranspruch als Funktion des Steigungsmaßes n der RRSB-Funktion verschiedener mit einer Gutbett-Walzenmühle fertiggemahlener Hüttensandmehle
FIGURE 1: Water demand as a function of the slope n of the RRSB function of different blast furnace slag meals finish ground in a high pressure grinding roll

Die Gutbett-Walzenmühle bietet Möglichkeiten, durch unterschiedliche Einstellungen im Mahlsystem Hüttensandmehl mit breiter Korngrößenverteilung herzustellen. Die Mahlung auf sehr hohe Feinheiten ist ebenfalls möglich. Um die Auswirkungen der Breite von Korngrößenverteilungen zu demonstrieren, zeigt **Bild 1** den Wasseranspruch bei Normsteife in Abhängigkeit vom Steigungsmaß n der an die Meßwerte der Korngrößenverteilung angepaßten RRSB-Funktion des Hüttensandmehls. Bei einer Schülpenrezirkulation wird die Korngrößenverteilung des Sichtaufgabeguts etwas breiter und damit auch die des Produkts. In **Bild 2** ist dargestellt, wie mit steigendem Anteil der zur POLYCOM zurückgeführten Schülpenmenge am gesamten Aufgabegutmassenstrom der Wasseranspruch bei Normsteife reduziert wird. Die Schülpenrezirkulation hat noch einen weiteren Effekt. Das Mahlgutbett zwischen den Mahlwalzen wird stabilisiert, und die Mahlung auf sehr hohe Feinheiten (größer 5000 cm²/g) ist möglich. Da Hüttensand im Vergleich mit anderen Mahlgütern wesentlich spröder ist, tritt eine dramatische Absenkung der Mahleffizienz bei einer Schülpenrezirkulation nicht ein.

In der Zwischenzeit wurden mehrere Mahlanlagen für Hüttensand in den USA und in England gebaut und sind seit

If granulated blast furnace slag is ground separately and then mixed with Portland cement, the quality of the blast furnace cement produced is considerably better than if the two constituents are interground. This is partly because the slag meal is finer. There are also quality requirements for the slag meal. The water demand at standard stiffness is affected not only by the composition but also by the particle size distribution. Hence it is very important for new grinding plants to have great flexibility in adjusting the quality of the product as well as an energy-saving grinding process. High-pressure grinding rolls which are operated as finishing mills, and roller grinding mills, are both energy-saving grinding systems, but unfortunately neither has the required flexibility. Both grinding systems have very short residence times for the material being ground, and consequently do not always produce enough very fine material. This leads to narrow particle size distributions in the product, resulting in high values for the water demand.

High-pressure grinding rolls provide opportunities for making blast furnace slag meal with a wide particle size distribution through various adjustments to the grinding system. It can also grind to a very high level of fineness. To demonstrate the effects of a wide particle size distribution **Fig. 1** shows the water demand at standard thickness as a function of the slope n of the RRSB function of the blast furnace slag meal adapted to the measured values of the particle size distribution. If the compacted cake is recirculated the particle size distribution of the separator feed, and thus of the product, becomes somewhat wider. **Fig. 2** shows how a rise in the quantity of compacted cake recycled to the POLYCOM as a proportion of the total feed mass flow will bring a reduction in the water demand at standard stiffness.

Recirculation of compacted cake has a further effect. The bed of material between the grinding rollers is stabilised and grinding to very high levels of fineness is possible (> 5000 cm²/g). Since granulated blast furnace slag is considerably

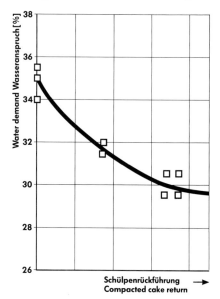

BILD 2: Beeinflussung der Produktqualität: Wasseranspruch von Hüttensandmehlen in Abhängigkeit vom rückgeführten Schülpenanteil
FIGURE 2: Influencing on product quality: Water demand of blast furnace slag meals in dependence of the proportion of the compacted cake recycled to the mill

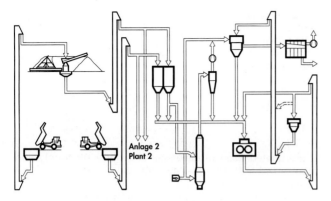

BILD 3: Verfahrenschema einer in Taiwan betriebenen Hüttensand-Mahlanlage
FIGURE 3: Process flow sheet of a grinding plant for blast furnace slag operated in Taiwan

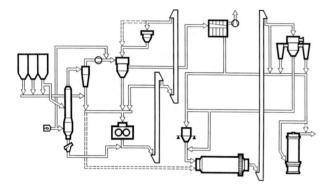

BILD 4: Verfahrenschema der Hüttensand-Mahlanlage im Werk Schwelgern der „Wülfrather Zement GmbH"
FIGURE 4: Process flow sheet of the grinding plant for blast furnace slag in the Schwelgern works of "Wülfrather Zement GmbH"

more brittle than other materials which are ground, there is no dramatic drop in grinding efficiency when compacted cake is recirculated.

In the meantime several grinding plants for granulated blast furnace slag have been built in the USA and England and have been operating successfully for a long period. As an example **Fig. 3** shows the design of a grinding plant which was commissioned in Taiwan in the spring of 1993. It has two parallel lines, each designed for a throughput of 50 t/h blast furnace slag meal with fineness levels of over 4700 cm^2/g. 600 000 t blast furnace slag meal per annum is blended with 600 000 t ordinary Portland cement to produce blast furnace cement.

The grinding plant essentially consists of the granulated blast furnace slag store, the pneumatic conveyor dryer, the POLYCOM 17/12 high-pressure grinding rolls with 2 × 1200 kW drive rating, a SEPOL bladed-rotor classifier, a firing system and a multi-cell silo with 18 cells including mixer. The particle size distribution is adjusted by recirculating the compacted cake. The slope n of the RRSB function is 1.02. It can be varied by ± 0.1. The specific power requirement of the entire plant is 52 kWh/t.

Another grinding plant is in operation at Wülfrather Zement GmbH's works at Duisburg-Schwelgern (**Fig. 4**). Although the plant is used primarily for granulated blast furnace slag, there is a need to switch production to blast furnace cement of the same quality when existing grinding plant 1 is out of action. For quality reasons the new plant has therefore been constructed as a combined plant for grinding blast furnace slag meal and blast furnace cement, with the option of finish grinding the blast furnace slag. In the first circuit the slag is either preground to about 3000 cm^2/g or finish ground to 4000 cm^2/g. The throughput of the combined grinding plant is approx. 55 t/h blast furnace slag meal with a specific power requirement of 56.5 kWh/t. For finish grinding, the throughput is 32.5 t/h with a specific power requirement of only 45 kWh/t.

The saving in electricity obtained with high-pressure grinding rolls in a closed circuit as compared with tube mills in a closed circuit is approx. 40%. However the saving obtained

langer Zeit erfolgreich in Betrieb. **Bild 3** zeigt als Beispiel den Anlagenaufbau einer Mahlanlage, die im Frühjahr 1993 in Taiwan in Betrieb genommen wurde. Diese Mahlanlage besteht aus zwei parallelen Linien für einen Durchsatz von je 50 t/h Hüttensandmehl mit Feinheiten von über 4700 cm^2/g. Pro Jahr werden 600 000 t Hüttensandmehl mit 600 000 t OPC zu Hochofenzement gemischt.

Die Mahlanlage besteht im wesentlichen aus dem Hüttensandlager, dem Steigrohrtrockner, der Gutbett-Walzenmühle POLYCOM 17/12 mit 2 x 1200 kW Antriebsleistung, einem Korbsichter SEPOL, einer Feuerung und einem Mehrzellensilo mit 18 Silozellen incl. Mischer. Die Korngrößenverteilung wird durch eine Schülpenrezirkulation eingestellt. Das Steigungsmaß n der RRSB-Funktion beträgt 1,02. Es läßt sich um ± 0,1 verändern. Der spezifische Arbeitsbedarf der Gesamtanlage liegt bei 52 kWh/t.

Eine weitere Mahlanlage ist bei der „Wülfrather Zement GmbH" in Duisburg-Schwelgern in Betrieb. Obwohl diese Mahlanlage (**Bild 4**) primär für Hüttensand eingesetzt wird, muß bei Stillstand der vorhandenen Mahlanlage 1 eine Umstellung der Produktion auf Hochofenzement bei gleicher Qualität möglich sein. Aus Qualitätsgründen wurde deshalb diese Mahlanlage als Combi-Mahlanlage für Hüttensandmehl und Hochofenzement mit der Option gebaut, Hüttensand auch fertig zu mahlen. Im 1. Kreislauf wird der Hüttensand entweder auf ca. 3000 cm^2/g vor- oder auf 4000 cm^2/g fertiggemahlen. Der Durchsatz der Combi-Mahlanlage liegt bei 55 t/h Hüttensandmehl mit einem spezifischen Arbeitsbedarf von 56,5 kWh/t. Bei der Fertigmahlung beträgt der Durchsatz 32,5 t/h bei einem spezifischen Arbeitsbedarf von nur 45 kWh/t.

Die Einsparung an elektrischer Energie durch Gutbett-Walzenmühlen im geschlossenen Kreislauf gegenüber Rohrmühlen im geschlossenen Kreislauf liegt bei ca. 40 %. Die Einsparung durch geringere Energiekosten wird jedoch zum Teil durch die höheren Verschleißkosten infolge des

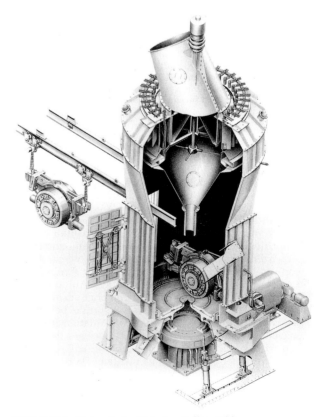

BILD 5: Schnitt durch eine Polysius-Rollenmühle
FIGURE 5: Sectional view of a Polysius roller mill

abrasiven Hüttensandes wieder aufgezehrt. Hier bietet das neue PROSEG-System, ein Verschleißschutz der Mahlwalzen durch Hartguß-Segmente mit aufgeschweißten Raupen, noch ein erhebliches Einsparpotential.

Rollenmühlen (**Bild 5**) sind kompakte Mahlanlagen, in denen die Mahlung und Trocknung gemeinsam erfolgen. Zur Verschleißreduzierung ist eine Absenkung der Gasgeschwindigkeit in der Mühle erforderlich. Diese Maßnahme führt dann zwangsweise zum externen Mahlgut-Kreislauf, bei dem das Mahlgut mechanisch zum Sichter gefördert wird. Bei der Rollenmühle ist die Korngrößenverteilung des Produkts nicht durch Ändern der Mühlen-Einstellgrößen beeinflußbar. Dies kann in einigen Ländern bei geringen Anteilen des Hüttensandmehls im Hochofenzement oder geringeren Anforderungen an die Produktqualität durchaus akzeptiert werden. Die Mahlung auf hohe Feinheiten ist aus Gründen zu starker Vibration nicht möglich. Rollenmühlen benötigen allerdings kein Mühlengebäude, was bei einer POLYCOM nicht fehlen sollte. Dies drückt sich damit auch günstig auf die Investitionskosten einer Rollenmühle aus. Die Verschleißteile sind nachschweißbar, eine Möglichkeit, die Verschleißkosten in Ländern mit niedrigem Lohnniveau zu reduzieren.

through lower energy costs is partly taken up by the higher costs of wear resulting from the abrasive blast furnace slag. Here the new PROSEG system – wear protection for the grinding rollers provided by chilled cast iron segments with treads welded onto them – also has considerable potential for saving.

Roller grinding mills (**Fig. 5**) are compact grinding plants where grinding and drying are combined. The gas velocity in the mill has to be lowered to reduce wear. This measure then of necessity leads to the external material circuit in which the material being ground is conveyed mechanically to the classifier. In roller grinding mills the particle size distribution of the product cannot be governed by varying the mill settings. This may be quite acceptable in some countries where there are only small proportions of slag meal in the blast furnace cement or where lower specifications for the product quality are definitely acceptable. Grinding to high levels of fineness is not possible owing to the excessive vibration. However, roller grinding mills do not need a building – which is essential with a POLYCOM system – and this fact reduces their capital cost. Wearing parts can be re-welded, one option for reducing wear costs in countries with low wages.

Verbesserung einer Rohmehlmahlanlage im ACC-Chanda-Werk

Raw mill upgrading at ACC-Chanda works

Amélioration d'un atelier de broyage cru à l'usine ACC Chanda

Mejoras introducidas en una instalación de molienda de crudo de la factoría ACC-Chanda

Von **K. Ravi Kumar, Y. V. Satyamurthy, M. A. Purohit** und **T. N. Tiwari,** Thane/Indien

Zusammenfassung – Ein Ofenoptimierungsplan im Chanda-Zementwerk der Associated Cement Cos. Ltd. wurde in den Jahren 1989–1990 durchgeführt. Um die erhöhten Anforderungen an das Rohmehl für die Öfen erfüllen zu können, wurde in Verbindung mit der Ofenoptimierung die Modernisierung der in einem geschlossenen Kreislauf arbeitenden Rohmehlmahlanlage geplant. Dazu wurden verschiedene Optionen, wie die Installation einer Gutbett-Walzenmühle, der Umbau der Inneneinbauten der Mühle und der Austausch des vorhandenen dynamischen Sichtersystems durch ein Hochleistungssystem, geprüft. Aufgrund einer eingehenden technisch-wirtschaftlichen Analyse wurde beschlossen, das bestehende Sichtersystem durch ein Hochleistungssystem zu ersetzen. Ausgewählt wurde ein O-Sepa-Sichter. Es war das erste System dieser Art, das in Indien für die Rohmehlmahlung von Fuller-KCP, Indien, geliefert wurde. Nach Aufbau der Anlage und Optimierung des Mahlkreislaufs erhöhte sich der spezifische Durchsatz der Rohmehlmahlanlage von 53 auf 69 t/h bei etwa 15% Energieeinsparung. Diese Leistungsverbesserung wird durch die Optimierung der Mahlanlage erzielt, deren Hochleistungssichter an den Betrieb der bestehenden Mahlanlage angepaßt wurde.

Summary – The kiln optimization plan at the Chanda cement works of The Associated Cement Cos. Ltd., was implemented in the year 1989–90. To meet the enhanced raw meal requirements for the kilns, the upgrading of closed-circuit raw grinding systems was planned along with kiln optimization. Different options such as roller press installation, modifications to mill internals, replacement of existing dynamic separator with a high efficiency separator were considered for this upgrading. After a thorough techno-economic analysis it was decided to replace the existing separator by a high efficiency (H.E.) separator. The system selected was an O-Sepa separator supplied for the first time in India for raw mill applications by Fuller-KCP, India. After the installation and optimisation of the grinding circuit the specific output of the raw grinding system went up from a level of 53 to 69 t/h with energy savings of 15%. The above improvement in performance is achieved through optimising the performance of the grinding system by matching the operation of the H.E. separator with the existing mill system.

Résumé – Le programme d'optimisation du four à la cimenterie de Chanda de Associated Cement Cos. Ltd. a été réalisé durant les années 1989–1990. Pour pouvoir répondre aux exigences plus sévères en matière de farine crue pour les fours, la modernisation de l'atelier de broyage cru travaillant en circuit fermé a été programmée en liaison avec l'optimisation du four. Pour ce faire, diverses options comme l'installation d'un broyeur à cylindres à lit de matière, la modification des équipements internes du broyeur et le changement du système du séparateur dynamique existant par un système à haut rendement, ont été étudiées. A la suite d'une analyse technico-économique approfondie, il a été décidé de remplacer le système du séparateur existant par un système à haut rendement. L'installation choisie était un système O-Sepa. C'était là le premier système de ce genre à être livré en Inde pour le broyage de cru de Fuller-KCP, Inde. Après montage de l'installation et optimisation du circuit de broyage la capacité spécifique de l'atelier de broyage de cru de 53 t/h a été portée à 69 t/h, avec environ 15% d'économie d'énergie. Cette amélioration de la capacité a pu être obtenue grâce à l'optimisation de l'atelier de broyage, et adaptation du séparateur à haut rendement à l'exploitation de l'atelier de broyage existant.

Resumen – El proyecto de optimización de los hornos de la planta cementera de Chanda, perteneciente a la Associated Cement Cos. Ltd. se llevó a efecto en los años 1989–1990. Con el fin de poder cumplir las elevadas exigencias respecto del crudo de alimentación de los hornos, se planeó la modernización de la instalación de molienda de crudo que trabaja en circuito cerrado, junto con la optimización de los hornos. Para ello, se estudiaron diferentes opciones, entre ellas la instalación de un molino de cilindros y lecho de material, la modificación constructiva de los dispositivos internos del molino así como la sustitución del separador dinámico existente por otro sistema de alto rendimiento. Tras un detenido análisis técnico-económico, se acordó sustituir el actual sistema de separación por otro de alto rendimiento, eligiéndose un separador O-Sepa. Fue el primer sistema de este tipo suministrado en la India por Fuller-KCP y destinado a la molienda de crudo. Después del montaje de la instalación y de la optimización del circuito de molienda, el caudal específico de la planta de molienda de crudo pasó de

53 a 69 t/h, con un ahorro de energía del 15 % aprox. Esta mejora del rendimiento se ha conseguido gracias a la optimización de la planta de molienda, cuyo separador de alto rendimiento ha sido adaptado al funcionamiento de la planta de molienda existente.

1. Einführung

Im Chanda-Zementwerk der Associated Cement Cos. Ltd. in Chandrapur, Indien, wurde 1989 ein Optimierungsprogramm durchgeführt, um die Klinkerkapazität der beiden vorhandenen Ofensysteme um 400 Tonnen pro Tag zu erhöhen. Diese Maßnahmen umfaßten Veränderungen der Vorkalzinierung, am Vorwärmer und am Kühler der Ofensysteme. Zur Steigerung der Klinkerproduktion wurde bei den vor- und nachgelagerten Anlagen ein Kapazitätsvergleich vorgenommen. Dabei wurden verschiedene Möglichkeiten ausgearbeitet, um die volle Ausnutzung der Rohmehl-Mahlanlage sicherzustellen. In diesem Beitrag werden die für die Verbesserung der Rohmehl-Mahlanlage gewählte Methode, die dabei aufgetretenen Probleme und die eingeleiteten Gegenmaßnahmen dargestellt.

2. Mahlanlage

Das Flußdiagramm des Rohmehl-Mahlkreislaufs vor der Verbesserung ist im **Bild 1** dargestellt. Technische Einzelheiten über das vor der Verbesserung verfügbare Rohmehl-Mahlsystem enthält die **Tabelle 1**. Die Mühle mit einem Durchmesser von 3,4 m und einer Länge von 7,93 m wird von einem 1200 kW-Motor angetrieben und arbeitet in einem geschlossenen Kreislauf mit einem Heyd-Windsichter von 4,8 m Durchmesser. Der mittlere Durchsatz der Mahlanlage lag bei etwa 53 t/h.

3. Vorgehen bei den Verbesserungsmaßnahmen

Die nach der Ofenoptimierung erforderlich werdende zusätzliche Rohmehlmenge von etwa 600 t/Tag sollte von den beiden vorhandenen Rohmehl-Mahlanlagen bereitgestellt werden. Angesichts der gegenwärtig erreichten Durchsatzleistung und der Verfügbarkeit der Anlagen mußten die Mahlsysteme ganz offensichtlich deutlich verbessert werden, um die der gesteigerten Klinkerproduktion der Öfen entsprechend erforderliche Mahlkapazität zu erreichen. Detaillierte Untersuchungen der Leistungsfähigkeit der Mahlanlagen ergaben, daß die Anlagen die benötigte Zusatzkapazität durchaus erreichen konnten. Als größter Engpaß bei der Verwirklichung dieses Ziels erwiesen sich die vorhandenen dynamischen Sichter. Vor allem der niedrige Wirkungsgrad dieser Sichter bei geringer Umlaufzahl begrenzte die Mahlkapazität.

Zur Ausnutzung der vollen Leistungsfähigkeit der Rohmehl-Mahlanlagen wurden folgende Optionen geprüft:

— Ausstattung der Mühle mit besonders wirksamen Inneneinbauten;

— Austausch des Windsichters gegen einen Hochleistungssichter;

— Einbau einer Gutbett-Walzenmühle zur Vorzerkleinerung (gemeinsam für beide Rohmehl-Mahlanlagen).

Die Option Gutbett-Walzenmühle wurde wegen des sehr hohen Kapitalaufwands für die erforderliche Verbesserung verworfen, da diese Technik außerdem in der indischen Zementindustrie noch nicht verbreitet ist.

Die beiden verbleibenden Optionen waren technisch-wirtschaftlich betrachtet gleichwertig. Es wurde beschlossen, bei einer Mühle den dynamischen Sichter zu ersetzen und bei der anderen die Inneneinbauten gegen eine Hochlei-

1. Introduction

Optimisation of Chanda Cement Works of ACC, located in Chandrapur, India, was carried out in 1989 to enhance the clinkering capacity of two existing kiln systems by 400 tons per day. This exercise involved modifications of the precalciner, preheater and the cooler of the kiln systems. In order to achieve the additional clinker production, capacity balancing was carried out to upstream and downstream systems. Various options for achieving the full potential of the raw grinding systems were worked out. The method adopted for upgrading the raw grinding system, problems encountered and remedial actions taken are presented in this paper.

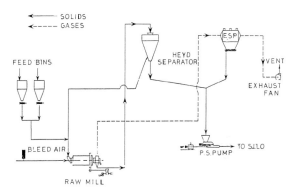

BILD 1: Rohmehl-Mahlkreislauf vor dem Umbau
FIGURE 1: Raw grinding mill circuit before upgradation

Feststoff	= Solids	Heyd-Sichter	= Heyd separator
Gas	= Gases	Elektrofilter	= ESP
Aufgabebunker	= Feed bins	Abgasventilator	= Exhaust fan
Frischluft	= Bleed air	zum Silo	= to silo
Rohmühle	= Raw mill	Schneckenpumpe	= P.S. Pump

2. Grinding system

The flowsheet of the raw grinding circuit before upgradation is shown in **Fig. 1**. The details of the raw grinding system existing before upgradation are furnished in **Table 1**. The mill with the dimensions of 3.4 m in diameter and 7.93 m in length is driven by a 1,200 kW motor and operated in closed circuit with a 4.8 m diameter Heyd air separator. The grinding plant had an average total capacity of approximately 53 t/h.

3. Upgradation approach

The requirement of additional kiln feed of about 600 tpd after kiln optimisation was to be met from the two existing raw mills. Considering the operating output levels and the availability of the mills, it was obvious that the mill systems needed a major improvement to achieve the required grinding capacity which was able to meet the enhanced production levels of the kilns. Detailed investigations on the mills' performance revealed that the mills have a potential to meet the additional capacities required. The major bottleneck for achieving this was found to be the existing dynamic separators. The low efficiency of these separators at low circulating loads was the main factor limiting the mill capacity.

In order to achieve the full potential of the raw grinding systems, following options were considered:

— Installation of high efficiency internals in the mill.

TABELLE 1: Rohmehlanlage vor dem Umbau

Mühlengröße	3,4 m × 7,93 m
Mühlenantriebsleistung	1200 kW
Sichter	Heyd-Sichter (⌀ 4,8 m)
	Antrieb 110 kW 975 Upm
Mühlengebläse	60 000 m³/h @ 500 mm WS

TABLE 1: Raw grinding system details before upgradation

Mill size	3.4 m dia. × 7.93 m long
Mill drive	1,200 kW
Separator	Heyd separator (dia. 4.8 m)
	Drive 110 kW 975 rpm
Mill ventilator fan	60,000 m³/h @ 500 mm wg

- Replacement of the air separator in favour of high efficiency separator.
- Installation of a roller press in pregrinding mode (common for both raw mills).

The roller press option was ruled out due to very high capital cost for the level of upgradation needed and the technology was not well established in the Indian cement industry.

The remaining two options were at par on techno-economic considerations. It was decided to replace the dynamic separator for one mill and change the internals with high efficiency design for the other. This was done to understand the individual merits of each system and assimilate the information. Various high efficiency separators were considered, and, at least, an O-sepa separator was selected for installation owing to its compact design which just suited the existing layout. The O-sepa separator installed was supplied by Fuller-KCP who had already supplied O-sepa separators for several cement grinding mills in India. The engineering and installation of the O-sepa separator and the modified mill circuit was carried out by ACC. Along with the O-sepa installation, the grinding media loading pattern and modifications to diaphragms were carried out as recommended by the supplier.

4. Modified raw grinding circuit

The modified raw grinding circuit after the installation of the O-sepa separator is shown in **Fig. 2**. The details of the installed O-sepa separator and the design parameters are given in **Table 2**. The separator has a case diameter of 3.1 m, a cage diameter of 2.0 m and a cage rotary frequency of 120 to 160 min^{-1}. The fines discharge is precipitated in two cyclones. The operating parameters before and after the mill upgradation are given in **Table 3**. The first grinding compartment was lengthened from 2.5 m to 3.5 m, the second correspondingly shortened from 4.75 to 3.75 m. Accordingly, the grinding media load of the first compartment was increased from 33 to 44 t and that of the second compartment reduced from 60 to 47 t. The air rate in the free mill cross section was raised from 1.0 m/s to 1.5 m/s. In the modified circuit, the mill is vented through the separator. After the modernisation the mill has a capacity of 75 t/h with a product fineness of 15 to 16 % residue at 88 R 0,088 mm.

Detailed investigations were carried out through systematic data collection comprising monitoring of grinding progression inside the mill, mill ventilation rates and separator performance for different operating conditions. The problems encountered during commissioning of the modified mill and

TABLE 2: Details of the newly installed O-sepa separator

Size	N-1500
Diameter, m	3.10
Length, m	5.55
Rotor dia., m	2.0
Rotor speed, rpm	120–160
Drive, kW	110
Dia. of cyclones, m	2.1 × 2
Design performance:	
Mill output, t/h	75
Product fineness	15–16 % on 88 microns

TABLE 3: Mill operating parameters before and after upgradation

	Before	After
Chamber Lengths, m		
Chamber 1	2.5	3.5
Chamber 2	4.75	3.75
Grinding media load, Tonnes		
Chamber 1	33	44
Chamber 2	60	47
Mill ventilation on empty cross section, m/s	1.0	1.5

BILD 2: Rohmehl-Mahlkreislauf nach dem Umbau
FIGURE 2: Raw grinding mill circuit after upgradation

Sekundäre Zusatzluft	= Secondary bleed air	Abgasventilator	= Exhaust fan
Aufgabebunker	= Feed bins	zum Silo	= to silo
Frischluft	= Bleed air	Schneckenpumpe	= P. S. Pump
Hilfsluft	= Auxiliary vent		
Primäre Zusatzluft	= Primary bleed air		
Zyklonabscheider	= Separator cyclone		
O-Sepa-Sichter	= O-Sepa separator		
Feststoff	= Solids		
Gas	= Gases		
Elektrofilter	= ESP		

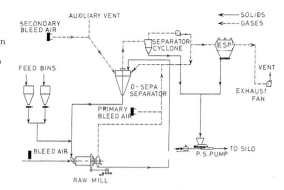

— Erhöhung des Anteils grober Mahlkörper und Ausschluß von Mahlkörpern von 50 mm aus Kammer 1 sowie
— Verkleinerung der Schlitzweite der Trennwand von 16/18 mm auf 12/14 mm

konnte die Mahlleistung beträchtlich verbessert werden, wodurch es zu weniger häufigen Störungen und damit zu einem gleichmäßigeren Mühlenbetrieb mit größerer Durchsatzleistung kam. Die Betriebsparameter des O-Sepa-Sichters, z. B. die Lage der Leitbleche, die Geschwindigkeit des Sichterkorbs usw. – wurden im Hinblick auf eine weitere Steigerung der Leistungsfähigkeit der Mühle und eine Verbesserung der Gleichmäßigkeit des Produkts entsprechend angepaßt.

Mit den oben genannten Veränderungen konnte die Mühlen-Durchsatzleistung gesteigert und während der trockenen Jahreszeit bei 72 t/h stabilisiert werden. Unter Berücksichtigung der jahreszeitlich bedingten Feuchtigkeitsschwankungen des Mahlguts erreichte die Mahlanlage im Jahresdurchschnitt eine Durchsatzleistung von 69 t/h. Ein Vergleich der Mühlenleistung vor und nach der Verbesserung ist in der **Tabelle 5** aufgeführt. Hierbei zeigt sich, daß der massebezogene Energieverbrauch des modifizierten Mahlkreislaufs um 4,3 kWh/t geringer ist, was 15,4 % entspricht.

5. Schlußfolgerung

Die gesteigerte Durchsatzleistung nach der Verbesserung der Mahlanlage entspricht dem Rohmehlbedarf des Ofens. Um jedoch die vorgesehenen Leistungskenndaten des umgebauten Mühlenkreislaufs zu erreichen, werden gegenwärtig noch weitere Anstrengungen unternommen.

remedial actions taken are presented in **Table 4**. With the above changes,
— increased length of chamber 1,
— increased proportion of large grinding media and eliminated 50 mm grinding media from chamber 1, and
— reduced size of the diaphragm slot from 16/18 to 12/14 mm,

the mill performance had been improved substantially resulting in reducing the jamming and hence steady running of mill with higher feed rates. The operating parameters of the O-sepa separator such as dampers' positions, speed of separator were adjusted to improve the mill performance further and have better control over the residues.

With the above changes, the mill output levels have increased and stabilised at 72 t/h during the dry season. Taking into account the seasonal variations in feed moisture, the mill has attained an annual average output of 69 t/h. The comparison of the mill performance before and after the upgradation is given in **Table 5**. It is apparent that the specific power consumption for the modified circuit is lower by 4.3 kWh/t of material, corresponding to 15.4 %.

5. Conclusion

The improved mill output after the upgradation has met the kiln feed requirement. But to achieve the design performance parameters for the modified mill circuit, further efforts are being made.

TABELLE 4: Probleme und Gegenmaßnahmen nach Einbau des O-Sepa-Sichters

Aufgetretene Probleme	Gegenmaßnahmen
Verstopfung der Kammer I und des Mühleneinlaufs	Verlängerung der Kammer I um 1,2 m
Rückgang der Durchsatzleistung auf 40–45 t/h	Verwendung größerer Mahlkörper (90/75 mm) und Entnahme der 50 mm-Kugeln
Geringer Siebrückstand	Schlitzbreite der Trennwand vermindert von 16/18 mm auf 12/14 mm
	Feineinstellung der Rotordrehzahl und Abstimmung der Sichterluftmengen

TABLE 4: Problems encountered and action taken after installation of the O-sepa separator

Problems encountered	Remedial actions
Jamming of I-Chamber and mill inlet	I chamber length increased by 1.2 m
Drop in output to 40–45 t/h	Increase of 90/75 mm size and elimination of 50 mm gr. media
Poor residue control	Diaphragm slot size reduced from 16/18 to 12/14 mm
	Fine tuning of rotor speed and balancing of air flows

TABELLE 5: Vergleich der Mühlenleistung vor und nach Umbau

	vorher	nachher
Mittlere Durchsatzleistung, t/h	53	69
Spezifischer Energieverbrauch, kWh/t	28	23,7
Produktfeinheit:		
R 0,088 mm, %	82	85
R 0,210 mm, %	97,5	98
Sichter:		
Umlaufzahl	1,5	2,3
Sichtergutleistung bezogen auf 88 μm, %	45–50	75–80

TABLE 5: Comparison of mill performance before and after upgradation

	Before	After
Average output, t/h	53	69
Specific power consumption, kWh/t	28	23.7
Product fineness 88 μm	82	85
% passing through 210 μm	97.5	98
Separator performance		
Circulating load, %	1.5	2.3
Sepn. eff. for fines, % @ 88 microns	45–50	75–80

Mahlbarkeit von Zement-Bestandteilen und Energiebedarf von Zementmühlen*)

Grindability of cement components and power consumption of cement mills*)

Broyabilité des composants du ciment et besoin d'énergie des broyeurs à ciment

Molturabilidad de los componentes del cemento y consumo de energía de los molinos de cemento

Von B. **Schiller**, Ulm und **H. G. Ellerbrock**, Düsseldorf/Deutschland

Zusammenfassung – Sowohl die Zementeigenschaften als auch die Energieausnutzung beim Mahlen werden in starkem Maß von der Mahlbarkeit des Mahlguts beeinflußt. Mit einem im Forschungsinstitut der Zementindustrie entwickelten neuen Mahlbarkeitsprüfverfahren lassen sich sowohl kleinere Mahlbarkeitsunterschiede verschiedener Stoffe als auch die Mahlbarkeitskennwerte schwer mahlbarer und leicht mahlbarer Stoffe reproduzierbar ermitteln. Das neue Prüfverfahren wird in Schritten ausgeführt, wobei das erzeugte Feingut < 125 µm durch Frischgut ersetzt und die Mahldauer auf einen Feingutanteil von 50 M.-% eingestellt wird. Gemessen werden die pro Zeit erzeugte Feingutmasse, der darauf bezogene Energiebedarf des Prüfgeräts und die Korngrößenverteilung des gemahlenen Guts. Die für jeden Stoff kennzeichnende Mahlbarkeitskennlinie gibt den Zusammenhang zwischen dem Energieaufwand und dem Lageparameter x' der RRSB-Korngrößenverteilung wieder. Für die Durchführung der Mahlbarkeitsprüfung können das Mahlbarkeitsprüfgerät nach Zeisel, eine Laborkugelmühle oder eine Laborwälzmühle benutzt werden. Der Verlauf der Mahlbarkeitskennlinie hängt dann auch vom verwendeten Prüfgerät ab. Die Korrelation zwischen dem Arbeitsbedarf von Betriebsmühlen und den Laborprüfwerten wurde gegenüber älteren Prüfverfahren deutlich verbessert.

Mahlbarkeit von Zement-Bestandteilen und Energiebedarf von Zementmühlen

Summary – The cement properties and the power consumption during grinding are both heavily influenced by the grindability of the material. With a new grindability test method developed at the cement industry's Research Institute it is possible to make reproducible measurements not only of quite small differences in the grindability of different materials, but also of the characteristic grindability values of materials which are difficult and easy to grind. The new test method is carried out in stages in which the fines < 125 µm which have been produced are replaced by fresh material and the grinding time is set to a proportion of fines of 50 wt.%. The mass of fines produced per unit time, the specific energy consumption of the test equipment, and the particle size distribution of the ground material are measured. The grindability curves characteristic of each material show the relationship between the power consumption and the position parameter x' in the RRSB particle size distribution. The grindability test can be carried out in the Zeisel grindability test equipment, a laboratory ball mill, or a laboratory roller grinding mill. The shape of the characteristic grindability curve then depends on the test equipment used. The correlation between the power consumption of industrial mills and the laboratory test values is significantly better than with older test methods.

Grindability of cement components and power consumption of cement mills

Résumé – Aussi bien les propriétés des ciments, que l'exploitation de l'énergie lors de la mouture, sont grandement influencées par la broyabilité de la matière. Avec un nouveau test de broyabilité mis au point à l'Institut de Recherche de l'Industrie Cimentière, il est possible de déterminer de manière reproductible les faibles différences de broyabilité de diverses matières et, aussi, les grandeurs caractéristiques de matières difficilement ou facilement broyables. La nouvelle méthode d'essai est conduite par étapes, où la matière fine produite < 125 µm est remplacée par de la matière fraiche et où la durée du broyage est réglée en fonction d'une part de 50% masse de matière fine. On mesure la masse de produit fin obtenu en fonction du temps, le besoin d'énergie de l'appareil de test s'y rapportant et la distribution granulométrique du produit broyé. La courbe caractéristique de broyabilité propre à chaque matière donne la relation entre consommation d'énergie et paramètre de position x' de la distribution granulométrique RRSB. Pour effectuer ce test de broyabilité, on peut se servir de l'appareil de mesure de broyabilité selon Zeisel, d'un broyeur à boulets ou d'un broyeur à galets de laboratoire. L'allure de la courbe caractéristique de broyabilité dépend, bien sûr, aussi de l'appareil d'essai utilisé. La corrélation entre l'absorption de travail des broyeurs industriels et des valeurs d'essai de laboratoire a été nettement améliorée par rapport aux méthodes d'essai plus anciennes.

Broyabilité des composants du ciment et besoin d'énergie des broyeurs à ciment

---

*) Überarbeitete Fassung eines Vortrages zum VDZ-Kongreß '93, Düsseldorf (27. 9.–1. 10. 1993).
Revised text of a lecture to the VDZ Congress '93, Düsseldorf, (27.9.–1.10.1993)

Molturabilidad de los componentes del cemento y consumo de energía de los molinos de cemento

Resumen – *Tanto las características del cemento como la utilización de la energía durante el proceso de molienda dependen, en gran medida, de la molturabilidad del matrial. Un nuevo procedimiento de ensayo de la molturabilidad, desarrollado en el Instituto de Investigación de la Industria del Cemento, permite determinar, de forma reproducible, no solamente pequeñas diferencias de molturabilidad entre varios materiales, sino también los valores de molturabilidad de materiales difíciles y fáciles de moler. El nuevo procedimiento se lleva a efecto por etapas, sustituyéndose el producto fino < 125 µm por material de primer ingreso en el circuito y ajustándose la duración de la molienda para una proporción de finos del 50% en masa. Se mide la masa de finos producida en función del tiempo, el correspondiente consumo de energía del aparato de ensayo así como la distribución granulométrica del producto de la molienda. La curva de molturabilidad característica de cada tipo de material nos reproduce la relación entre el consumo de energía y el parámetro de posición x' de la disribución granulométrica RRSB. Para llevar a cabo el ensayo de molturabilidad se puede emplear el aparato de ensayo de la molturabilidad de Zeisel, un molino de bolas, de laboratorio, o un molino de pista de rodadura, también de laboratorio. La forma de la curva de molturabilidad depende entonces también del aparato de ensayo empleado. La correlación entre la exigencia de trabajo de los molinos indutriales y los valores de los ensayos de laboratorio ha mejorado notablemente en comparación con los métodos de ensayo más antiguos.*

1. Introduction

By far the greatest part of the electrical energy needed for cement grinding is not used for comminution but is lost and converted into heat. It is therefore economically essential to select the grinding process with the greatest energy utilization and to adjust the grinding plant so that the energy losses are kept as low as possible.

Energy utilization during grinding is heavily influenced by the grindability of the material being ground. Because of the large number of influencing factors it has not yet been possible to determine the grindability from chemical and physical data, and it is necessary to turn to investigations in laboratory mills for assessing the grinding characteristics of the materials. Grindability is therefore understood to mean the power consumption of a laboratory mill for the comminution of a material from a given initial particle size to a certain end fineness. The relationship between the power consumption of the laboratory mill and the fineness produced is given by the characteristic grindability curve.

A requirement for using the characteristic grindability values for assessing the operation of industrial mills is that their relationship with the power consumption of industrial mills exhibits a very small range of scatter. It is therefore essential that the grindability test method provides very reproducible results.

2. Test method

A new grindability test method has been developed at the Research Institute of the Cement Industry with which it is possible to make reproducible measurements both of quite small differences in grindability of different materials and also of the characteristic grindability values of materials which are difficult and easy to grind. The new test method is carried out in stages. After each grinding stage the fines < 125 µm which have been produced are removed from the grinding chamber and replaced for the next grinding stage by the same amount of fresh material. The grinding time has to be set so that after each grinding stage 50 wt. % of the mill feed can be removed as fines. The mass of fines produced per unit time, the power consumption of the test equipment relative to the mass of fines produced, and the particle size distribution of the ground material are measured [1, 2].

3. Results

The grindability test provides not only the specific power consumption of the test equipment but also a characteristic grindability value M which is defined as the surface area generated per unit time and mass. The meals produced can have very different particle size distributions because of the different grindabilities of the materials [3]; this means that it is only possible to compare the power consumptions when grinding different materials if the pairs of values are con-

haben können [3], ist ein Vergleich des Arbeitsbedarfs beim Mahlen verschiedener Stoffe erst möglich, wenn die Wertepaare auf den gleichen Feinheitsparameter umgerechnet werden. Hierzu wurde ein mathematisches Modell der Zerkleinerung herangezogen. Die im Modell genutzten Zerkleinerungs- und Energieparameter beschreiben die Abhängigkeit der sich einstellenden Korngrößenverteilung und des dazu erforderlichen Energieaufwands von der Mahldauer. Sie berücksichtigen die physikalischen Eigenschaften des Mahlguts und werden vom Beanspruchungsmechanismus, d.h. von Bauart und Betriebsweise des jeweils verwendeten Zerkleinerungsgeräts, beeinflußt. Es hat sich als zweckmäßig erwiesen, für die Darstellung der Ergebnisse den sich einstellenden Lageparameter x' der Korngrößenverteilung des Mahlguts im RRSB-Körnungsnetz anstelle der massebezogenen Oberfläche zu wählen.

Im **Bild 1** ist für verschiedene gemahlene Stoffe mit einem Lageparameter von x' = 16 μm der massebezogene Arbeitsbedarf A_m des Mahlbarkeitsprüfgeräts in Abhängigkeit von der berechneten Mahlbarkeitskennzahl M aufgetragen. Die Kennzahl M nimmt mit abnehmender Härte der untersuchten Stoffe zu.

Die Zusammenstellung in **Tabelle 1** zeigt, daß Hüttensande Mahlbarkeitskennzahlen zwischen 6 und 11 cm²/g · s, Klinker Werte zwischen 16 und 30 cm²/g · s, Traß Werte zwischen 50 und 70 cm²/g · s und Kalkstein Werte zwischen 90 und 120 cm²/g · s annehmen können. Daraus und aus Bild 1 geht hervor, daß die Mahlbarkeitskennzahlen sowohl schwer mahlbarer als auch leicht mahlbarer Stoffe mit dem erforderlichen Arbeitsbedarf des Prüfgeräts durch eine einheitliche Funktion verknüpft sind und daher nach dem gleichen Verfahren ermittelt werden können. Die Wiederholstreuung lag bei den geprüften Stoffen unter 5 %. Es ist demnach möglich, mit diesem Verfahren deutlich kleinere Mahlbarkeitsunterschiede als mit anderen bekannten Prüfverfahren reproduzierbar festzustellen.

TABELLE 1: Mahlbarkeitskennzahlen verschiedener Stoffe für die Zementherstellung

Stoff	M [cm²/g · s]	ΔM [cm²/g · s]
Hüttensand	5 bis 11	± 0,25
Klinker	15 bis 30	± 0,6
Traß	50 bis 70	± 1,5
Kalkstein	90 bis 120	± 2,0

Die Beziehung zwischen dem Energieaufwand A_m und dem Ergebnis der Modellrechnung läßt sich dazu nutzen, um aus dem Versuchsergebnis eine Mahlbarkeitskennlinie für den untersuchten Stoff zu entwickeln, die den Zusammenhang zwischen dem Energieaufwand und einem Feinheitskennwert wiedergibt. Als Ergebnis der Messung und Rechnung sind im **Bild 2** die Mahlbarkeitskennlinien eines Klinkers, d.h. der Arbeitsbedarf des Mahlbarkeitsprüfgeräts in Abhängigkeit vom Lageparameter x', für unterschiedliche Steigungsmaße n des gemahlenen Gutes aufgetragen. Die

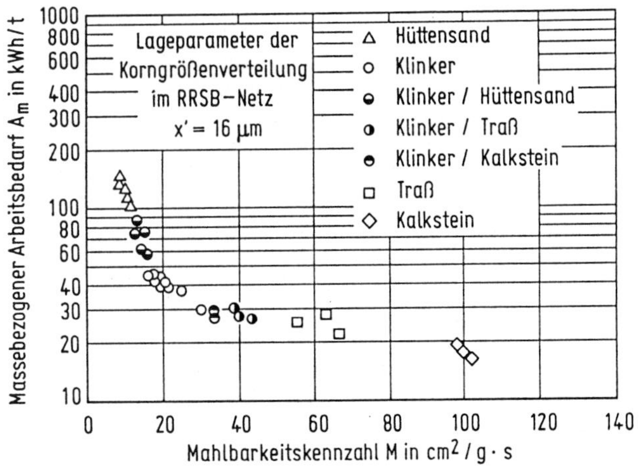

verted to the same fineness parameter. A mathematical comminution model is used for this purpose. The comminution and power parameters used in the model describe the dependence on the grinding time of the particle size distribution produced and the power consumption which this requires. They take account of the physical properties of the material being ground and are affected by the stressing mechanism, i.e. by the design and mode of operation of the particular comminution equipment used. Instead of the specific surface area, it has proved appropriate to choose the position parameter x' obtained in the RRSB granulometric diagram for the particle size distribution of the ground material for the representation of the results.

The specific power consumption A_m of the grindability test equipment is plotted in **Fig. 1** against the calculated characteristic grindability number M for various ground materials with position parameters of x' = 16 μm. The characteristic number M increases with decreasing hardness of the materials under investigation.

The summary in **Table 1** shows that granulated blast furnace slag can have characteristic grindability numbers between 6 and 11 cm²/g · s, clinker between 16 and 30 cm²/g · s, trass between 50 and 70 cm²/g · s and limestone between 90 and 120 cm²/g·s. From this and from Fig. 1 it can be seen that the characteristic grindability numbers of materials which are hard to grind and those which are easy to grind are linked to the required power consumption of the test equipment by a common function, and can therefore be determined by the same method. For the materials tested the spread on repetition was less than 5 %. Significantly smaller differences in grindability can therefore be identified reproducibly with this method than with other familiar test methods.

TABLE 1: Characteristic grindability numbers of various materials for cement manufacture

Material	M [cm²/g · s]	ΔM [cm²/g · s]
Granulated blast furnace slag	5 to 11	± 0.25
Clinker	15 to 30	± 0.6
Trass	50 to 70	± 1.5
Limestone	90 to 120	± 2.0

The relationship between the power consumption A_m and the results of the model calculation can be used to develop from the test result a characteristic grindability curve for the material investigated which describes the relationship between the power consumption and the characteristic fineness value. As a result of the measurement and calculation the characteristic grindability curves for a clinker, i.e. the power consumption of the grindability test equipment as a function of the position parameter x', are plotted in **Fig. 2** for different slopes n of the ground material. The values of the position parameter were determined by calculation for different slopes, in each case at the same specific surface area of

BILD 1: Massebezogener Arbeitsbedarf A_m beim Mahlen verschiedener Stoffe im Mahlbarkeitsprüfgerät in Abhängigkeit von der Mahlbarkeitskennzahl M aufgrund der Prüfung nach dem neuen Verfahren

FIGURE 1: Specific power consumption A_m when grinding different materials in the grindability test equipment as a function of the characteristic grindability number M based on testing by the new method

Massebezogener Arbeitsbedarf A_m in kWh/t	= specific power consumption A_m in kWh/t
Lageparameter der Korngrößenverteilung im RRSB-Netz	= position parameter of the particle size distribution in the RRSB diagram
Hüttensand	= blast furnance slag
Klinker	= clinker
Klinker/Hüttensand	= clinker/slag
Klinker/Traß	= clinker/trass
Klinker/Kalkstein	= clinker/limestone
Traß	= trass
Kalkstein	= limestone
Mahlbarkeitskennzahl M in cm²/g · s	= Characteristic grindability number M in cm²/g · s

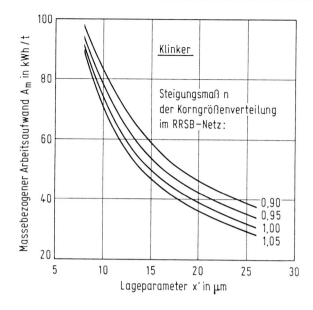

BILD 2: Mahlbarkeitskennlinien eines Klinkers für unterschiedliche Steigungsmaße n der RRSB-Verteilung des erzeugten Klinkermehls

FIGURE 2: Characteristic grindability curves of a clinker for different slopes n of the RRSB distribution of the clinker meal produced

Massebezogener Arbeitsaufwand A_m in kWh/t	= specific power consumption A_m in kWh/t
Klinker	= clinker
Steigungsmaß n der Korngrößenverteilung im RRSB-Netz	= slope n of the particle size distribution in the RRSB diagram
Lageparameter x' in μm	= position parameter x' in μm

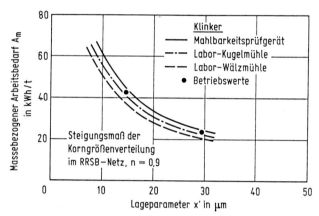

BILD 3: Massebezogener Arbeitsbedarf A_m beim Mahlen eines Klinkermehls im Mahlbarkeitsprüfgerät, in einer Labor-Kugelmühle und in einer Labor-Wälzmühle in Abhängigkeit vom Lageparamter x' der RRSB-Verteilung bei gleichem Steigungsmaß von n = 0,9

FIGURE 3: Specific power consumption A_m when grinding a clinker meal in the grindability test equipment, in a laboratory ball mill and in a laboratory roller mill as a function of the position parameter x' of the RRSB distribution for the same slope of n = 0.9

Massebezogener Arbeitsbedarf A_m in kWh/t	= specific power consumption A_m in kWh/t
Steigungsmaß der Korngrößenverteilung im RRSB-Netz	= slope of the particle size distribution in the RRSB diagram
Klinker	= clinker
Mahlbarkeitsprüfgerät	= grindability test equipment
Labor-Kugelmühle	= laboratory ball mill
Labor-Wälzmühle	= laboratory roller mill
Betriebswerte	= industrial values
Lageparameter x' in μm	= position parameter x' in μm

BILD 4: Massebezogener Arbeitsbedarf beim Mahlen von Portlandzementen unterschiedlicher Mahlfeinheit in verschiedenen Betriebskugelmühlen sowie massebezogener Arbeitsbedarf der entsprechenden Klinkermehle, bezogen auf gleiches Steigungsmaß der RRSB-Verteilung

FIGURE 4: Specific power consumption when grinding Portland cements of different finenesses in various industrial ball mills and the specific power consumption for the corresponding clinker meals relative to the same slope of the RRSB distribution

Symbol	Mühlenabmessungen	Sichter
●	⌀ 4,4 m · 14,0 m	2 Umluftsichter ⌀ 6,0 m
▽	⌀ 3,6 m · 14,5 m	2 Zyklonumluftsichter ⌀ 4,8 m
▼	⌀ 4,6 m · 14,5 m	2 Zyklonumluftsichter ⌀ 3,8 m
■	⌀ 3,4 m · 12,5 m	Korbsichter ⌀ 1,7 m
○	⌀ 3,8 m · 12,9 m	Zyklonumluftsichter ⌀ 4,5 m
□	⌀ 4,4 m · 15,5 m	Umluftsichter ⌀ 8,5 m
△	⌀ 3,4 m · 10,6 m	2 Umluftsichter ⌀ 4,5 m
▲	⌀ 3,8 m · 11,0 m	Korbsichter ⌀ 1,25 m

Symbol	Mill dimensions	Classifier
●	4.4 m dia. × 14.0 m	2 circulating air classifiers, 6.0 m dia.
▽	3.6 m dia. × 14.5 m	2 cyclone circulating air classifiers 4.8 m dia.
▼	4.6 m dia. × 14.5 m	2 cyclone circulating air classifiers, 3.8 m dia.
■	3.4 m dia. × 12.5 m	cage rotor classifier, 1.7 m dia.
○	3.8 m dia. × 12.9 m	cyclone circulating air classifier, 4.5 m dia.
□	4.4 m dia. × 15.5 m	circulating air classifier, 8.5 m dia.
△	3.4 m dia. × 10.6 m	2 circulating air classifiers, 4.5 m dia.
▲	3.8 m dia. × 11.0 m	cage rotor classifier, 1.25 m dia.

Massebezogener Arbeitsbedarf A_{mB} von Betriebskugelmühlen in kWh/t	= specific power consumption A_{mB} of industrial ball mills in kWh/t
Portlandzemente	= Portland cements
Steigungsmaß der Korngrößenverteilung im RRSB-Netz	= slope of the particle size distribution in the RRSB diagram
Massebezogener Arbeitsbedarf A_m des Mahlbarkeitsprüfgeräts in kWh/t	= specific power consumption A_m of the grindability test equipment in kWh/t

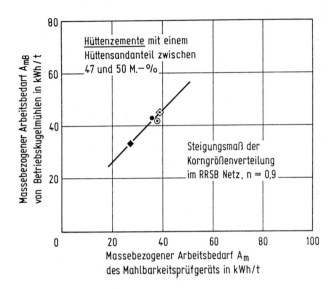

BILD 5: Massebezogener Arbeitsbedarf beim Mahlen von Hüttenzementen unterschiedlicher Mahlfeinheit mit Hüttensandanteilen zwischen 47 und 50 M.-% in verschiedenen Betriebskugelmühlen sowie massebezogener Arbeitsbedarf des Mahlbarkeitsprüfgeräts beim Mahlen der entsprechenden Klinker/Hüttensand-Gemische, bezogen auf gleiches Steigungsmaß der RRSB-Verteilung

FIGURE 5: Specific power consumption when grinding blast furnace cements of different finenesses containing between 47 and 50 wt. % granulated blast furnace slag in various industrial ball mills, and the specific power consumption of the grindability test equipment when grinding the corresponding clinker/slag mixes, relative to the same slope of the RRSB distribution

Symbol	Mühlenabmessungen	Sichter
◇	⌀ 3,2 m · 11,0 m	2 Umluftsichter ⌀ 4,2 m
●	⌀ 4,4 m · 14,0 m	2 Umluftsichter ⌀ 6,0 m
⊙	⌀ 3,8 m · 14,6 m	2 Zyklonumluftsichter ⌀ 5,0 m
◆	⌀ 3,8 m · 11,0 m	Zyklonumluftsichter ⌀ 3,5 m

Symbol	Mill dimensions	Classifier
◇	3.2 m dia. × 11.0 m	2 circulating air classifiers, 4.2 m dia.
●	4.4 m dia. × 14.0 m	2 circulating air classifiers, 6.0 m dia.
⊙	3.8 m dia. × 14.6 m	2 cyclone circulating air classifiers, 5.0 m dia.
◆	3.8 m dia. × 11.0 m	cyclone circulating air classifier, 3.5 m dia.

Massebezogener Arbeitsbedarf A_{mB} von Betriebskugelmühlen in kWh/t	= specific power consumption A_{mB} of industrial ball mills in kWh/t
Hüttenzemente mit einem Hüttensandanteil zwischen 47 und 50 M.-%	= blast furnace slag cements containing between 47 and 50 wt. % slag
Steigungsmaß der Korngrößenverteilung im RRSB-Netz	= slope of the particle size distribution in the RRSB diagram
Massebezogener Arbeitsbedarf A_m des Mahlbarkeitsprüfgeräts in kWh/t	= specific power consumption A_m of the grindability test equipment in kWh/t

Werte des Lageparameters wurden bei jeweils gleicher massebezogener Oberfläche des Mahlguts für verschiedene Steigungsmaße rechnerisch ermittelt. Aus dem Bild geht deutlich hervor, daß für die Ermittlung und Beurteilung der Mahlbarkeit neben dem Lageparameter x' auch das Steigungsmaß n der Korngrößenverteilung des gemahlenen Gutes von Bedeutung ist.

Bild 3 zeigt die Mahlbarkeitskennlinien eines Klinkers, wenn die Mahlbarkeitsprüfung mit unterschiedlichen Prüfgeräten durchgeführt wird. Die durchgezogene Linie gilt für das genannte Mahlbarkeitsprüfgerät, die strichpunktierte Linie für eine Laborkugelmühle und die gestrichelte Linie für eine Labor-Wälzmühle. Aus der Darstellung geht hervor, daß auch die unterschiedliche mechanische Beanspruchung des Mahlguts für die Ermittlung der Mahlbarkeit im Labor von großer Bedeutung ist und bei einem Vergleich von Werten beachtet werden muß.

Die Übertragbarkeit der mit einer satzweise betriebenen Labormühle ermittelten Mahlbarkeitskennzahlen auf Betriebsverhältnisse geht aus den folgenden Bildern hervor. **Bild 4** zeigt den Zusammenhang zwischen den mit dem Prüfgerät ermittelten Mahlbarkeitswerten verschiedener Klinker und dem Arbeitsbedarf optimal eingestellter Betriebsmühlen beim Mahlen von Portlandzementen unterschiedlicher Feinheit. Die massebezogene Oberfläche der Zemente lag im Bereich zwischen 2500 und 5900 cm²/g. Danach ist festzustellen, daß dieser Zusammenhang praktisch linear ist und ein hohes Bestimmtheitsmaß aufweist.

Auch beim Mahlen von Zementen mit Zumahlstoffen ergab sich ein enger Zusammenhang zwischen dem Ergebnis der Mahlbarkeitsprüfung und der Betriebsmahlung. **Bild 5** gilt für die Mahlung von Hochofenzementen mit einem Hüttensandanteil von 47 bis 50 M.-%. Die massebezogene Oberfläche der ermahlenen Zemente lag im Bereich zwischen 3300 und 4050 cm²/g. Lage und Steigung des Zusammenhangs unterscheiden sich jedoch deutlich von dem für Portlandzement gefundenen Zusammenhang. Der im Mahlbarkeitsprüfgerät für Klinker/Hüttensand-Gemische ermittelte Arbeitsbedarf ist im Gegensatz zu den Verhältnissen bei Klinker im gesamten Feinheitsbereich deutlich höher als der von Betriebsmühlen beim Mahlen von Hüttenzementen bzw. Portlandzement. Demnach ist die Energieausnutzung im Mahlbarkeitsprüfgerät beim Mahlen von schwer mahlbaren Stoffen, wie z.B. Hüttensand, insbesondere bei höheren Mahlfeinheiten deutlich geringer als in Betriebsmahlanlagen. **Bild 6** zeigt, daß auch der Vergleich von Betriebs- und Laborwerten beim Mahlen von Portlandkalksteinzementen mit Kalksteinanteilen von 16 bis 20 M.-% ein hohes Bestimmtheitsmaß aufweist. Die massebezogene Oberfläche der untersuchten Betriebszemente lag zwischen 4300 und 4650 cm²/g. Im Gegensatz zum Mahlen von Hüttenzementen ist die Energieausnutzung im Mahlbarkeitsprüfgerät beim Mahlen von Klinker/Kalkstein-Gemischen deutlich besser als in Betriebsmühlen.

the ground material. It is clear from the diagram that not only the position parameter x' but also the slope n of the particle size distribution of the ground material are important for determining and assessing the grindability.

Fig. 3 shows the characteristic grindability curves of a clinker when the grindability testing is carried out with different test equipment. The continuous line applies to the grindability test equipment already mentioned, the dot-dash line applies to a laboratory ball mill, and the broken line to a laboratory roller mill. It can be seen from the diagram that the different mechanical stressing of the ground material is also very important for the measurement of grindability in the laboratory and has to be taken into account when comparing values.

The applicability of the characteristic grindability numbers measured in a batch-operated laboratory mill to industrial conditions can be seen from the following diagrams. **Fig. 4** shows the relationship between grindability values of various clinkers measured with the test equipment and the power consumption of optimally adjusted industrial mills when grinding Portland cements of different finenesses. The cements had specific surface areas lying in the range between 2500 and 5900 cm²/g. It can be seen that this relationship is practically linear and has a high coefficient of determination.

A close relationship between the results of the grindability testing and of the industrial grinding was also found when grinding cements with interground additives. **Fig. 5** applies to the grinding of blast furnace slag cements containing 47 to 50 wt. % granulated blast furnace slag. The ground cements had specific surface areas between 3300 and 4050 cm²/g. However, the position and slope of the relationship differ significantly from the relationship found for Portland cement. The power consumption measured in the grindability test equipment for clinker/slag mixtures is, in contrast to the conditions with clinker in the same fineness range, significantly higher than that of industrial mills when grinding blast furnace slag cement or Portland cement. This shows that the power utilization in the grindability test equipment when grinding materials which are hard to grind, such as granulated blast furnace slag, is significantly lower than in industrial grinding plants, especially at the higher fineness levels. **Fig. 6** shows that the comparison of industrial and labora-

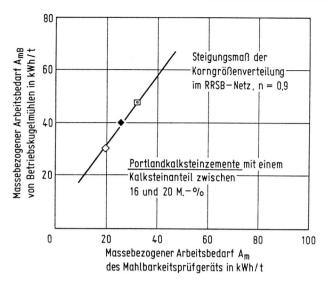

BILD 6: Massebezogener Arbeitsbedarf beim Mahlen von Portlandkalksteinzementen unterschiedlicher Mahlfeinheit mit Kalksteinanteilen zwischen 16 und 20 M.-% in verschiedenen Betriebskugelmühlen sowie massebezogener Arbeitsbedarf des Mahlbarkeitsprüfgeräts beim Mahlen der entsprechenden Klinker/Kalkstein-Gemische, bezogen auf gleiches Steigungsmaß der RRSB-Verteilung

FIGURE 6: Specific power consumption when grinding Portland limestone cements of different finenesses containing between 16 and 20 wt.% limestone in various industrial ball mills, and the specific power consumption of the grindability test equipment when grinding the corresponding clinker/limestone mixes, relative to the same slope of the RRSB distribution

Symbol	Mühlenabmessungen	Sichter
⊡	⌀ 4,0 m · 12,0 m	2 Umluftsichter ⌀ 5,2 m
◇	⌀ 3,8 m · 12,9 m	Zyklonumluftsichter ⌀ 4,5 m
◆	⌀ 3,8 m · 11,0 m	Zyklonumluftsichter ⌀ 3,5 m

Symbol	Mill dimensions	Classifier
⊡	4.0 m dia. × 12.0 m	2 circulating air classifiers, 5.2 m dia.
◇	3.8 m dia. × 12.9 m	cyclone circulating air classifier, 4.5 m dia.
◆	3.8 m dia. × 11.0 m	cyclone circulating air classifier, 3.5 m dia.

Massebezogener Arbeitsbedarf A_{mB} von Betriebskugelmühlen in kWh/t = specific power consumption A_{mB} of industrial ball mills in kWh/t

Steigungsmaß der Korngrößenverteilung im RRSB-Netz = slope of the particle size distribution in the RRSB diagram

Portlandkalksteinzemente mit einem Kalksteinanteil zwischen 16 und 20 M.-% = portland limestone cements containing between 16 and 20 wt.% limestone

Massebezogener Arbeitsbedarf A_{mB} des Mahlbarkeitsprüfgeräts in kWh/t = specific power consumption A_m of the grindability test equipment in kWh/t

tory mills when grinding Portland limestone cements containing 16 to 20 wt.% limestone also has a high coefficient of determination. The specific surface areas of the industrial cements investigated lay between 4300 and 4650 cm²/g. Unlike the situation when grinding blast furnace slag cements, the energy utilization in the grindability test equipment when grinding clinker/limestone mixtures is significantly better than in industrial mills.

Literaturverzeichnis

[1] Schiller, B.: Mahlbarkeit der Hauptbestandteile des Zements und ihr Einfluß auf den Energieaufwand beim Mahlen und die Zementeigenschaften. Schriftenreihe der Zementindustrie 1992, Heft 54, Verein Deutscher Zementwerke e.V., Düsseldorf.

[2] Schiller, B., und Ellerbrock, H.-G.: Mahlbarkeit von Zement-Bestandteilen und Energiebedarf von Zementmühlen. Zement-Kalk-Gips 42 (1989) Nr. 11, S. 553–557.

[3] Schiller, B., und Ellerbrock, H.-G.: Mahlung und Eigenschaften von Zementen mit mehreren Hauptbestandteilen. Zement-Kalk-Gips 45 (1992) Nr. 7, S. 325–334.

Vermeidung von Vibrationen beim Betrieb der Gutbett-Walzenmühle

Avoiding vibrations when operating high-pressure grinding rolls

Comment éviter les vibrations dans la marche du broyeur à rouleaux

Evitando vibraciones durante la marcha del molino de cilindros y lecho de material

Von **T. Schmitz**, Neubeckum/Deutschland

Zusammenfassung – Bei feinem Aufgabegut und gleichzeitig hohen Walzenumfangsgeschwindigkeiten kann es beim Betrieb von Gutbett-Walzenmühlen zu Vibrationen kommen, die durch die rheologischen Mahlguteigenschaften und das Entweichen der Luft aus der Kompressionszone des Mahlguts hervorgerufen werden. Untersuchungen an einer Labor-Gutbett-Walzenmühle ergaben, daß das Auftreten der Vibrationen vor allem von der Korngrößenverteilung des Mühlenaufgabeguts und von der Walzenumfangsgeschwindigkeit abhängt. Bei feinerem Aufgabegut ist die sog. kritische Umfangsgeschwindigkeit geringer. Daher lassen sich bei Walzen mit profilierter Oberfläche höhere kritische Umfangsgeschwindigkeiten einstellen. Für einen vibrationsfreien Betrieb der Gutbett-Walzenmühle sollte der Druck der Schüttgutsäule auf die Walzen möglichst groß sein. Das Aufgabegut sollte vor allem bei erhöhtem Feinanteil der Mühle möglichst homogen zugeführt werden. Die maximale Walzenumfangsgeschwindigkeit ist je nach Aufgabegutfeinheit begrenzt. Bei Fertig- und Kombimahlanlagen für mehrere Produkte unterschiedlicher Feinheit sollten drehzahlvariable Antriebe eingesetzt werden.

Summary – Vibrations can occur when operating high-pressure grinding rolls with fine feed material and high roller circumferential speeds. These vibrations are caused by the rheological properties of the material being ground and by the escape of the air from the compression zone in the material. Investigations with laboratory high-pressure grinding rolls showed that the occurrence of vibrations is dependent mainly on the particle size distribution of the mill feed material and on the roller circumferential speed. The so-called critical circumferential speed is lower with finer feed material. The critical circumferential speeds become higher with rollers with profiled surfaces. For vibration free operation of high-pressure grinding rolls, the pressure of the column of material on the rollers should be as high as possible. The feed material, especially if it has a high proportion of fines, should be as homogeneous as possible when it is fed to the mill. The maximum roller circumferential speed is limited by the fineness of the feed material. Variable speed drives should be used for finish and combination grinding plants for several products with different levels of fineness.

Résumé – Avec une matière fraiche fine et, en même temps, des hautes vitesses périphériques des rouleaux, peuvent apparaître des vibrations sur le broyeur à rouleaux en marche, occasionnées par les paramètres rhéologiques de la matière à broyer et par l'échappement de l'air hors de la zone de compression de la matière à broyer. Des investigations menées sur un broyeur à rouleaux de laboratoire ont révélé, que l'apparition des vibrations dépend surtout de la distribution granulométrique de la matière introduite dans le broyeur et de la vitesse périphérique des rouleaux. Avec une matière fraiche plus fine, la vitesse périphérique critique est plus basse. Ceci explique, que des rouleaux à surface profilée permettent de travailler avec des vitesses périphériques critiques plus élevées. Pour un fonctionnement sans vibration du broyeur à rouleaux, la pression de la colonne de matière en vrac sur les rouleaux devrait être aussi forte que possible. La matière fraiche, surtout avec une haute teneur en fines, devrait être introduite dans le broyeur de manière aussi homogène que possible. La vitesse périphérique maximale des rouleaux est limitée en fonction de la finesse de la matière fraiche. Dans les ateliers de finition ou combinés, pour plusieurs produits de finesses différentes, il est conseillé d'utiliser des entraînements permettant le réglage de la vitesse de rotation.

Resumen – Con materiales de alimentación muy finos y elevadas velocidades circunferenciales de los cilindros, pueden producirse vibraciones durante la marcha de los molinos de cilindros y lecho de material, debido a las características reológicas del material a moler y a que el aire se escapa de la zona de compresión del material a moler. De los estudios realizados con un molino de cilindros y lecho de material, de laboratorio, se desprende que la aparición de vibraciones depende, sobre todo, de la distribución granulométrica del material introducido en el molino así como de la velocidad circunferencial de los cilindros. Cuando el material de alimentación es más fino, la llamada velocidad circunferencial crítica es más reducida. Por esta razón, en

los cilindros de superficie perfilada, se pueden ajustar velocidades circunferenciales críticas más elevadas. Para conseguir una marcha exenta de vibraciones del molino de cilindros y lecho de material, hay que procurar que la presión de la columna de material a granel sea lo más alta posible. Conviene alimentar el molino con material lo más homogéneo posible, sobre todo cuando éste contiene una elevada proporción de finos. La velocidad circunferencial máxima de los cilindros es limitada y depende de la finura del material de alimentación. En las instalaciones de molienda de acabado y combinadas, destinadas a varios productos de diferente finura, se aconseja emplear dispositivos de accionamiento de velocidad variable.

Seit der Einführung der Gutbett-Walzenmühle Mitte der achtziger Jahre sind weltweit mehr als 300 Anlagen in Betrieb genommen worden und haben sich als ein energiesparendes Mahlsystem zur Zerkleinerung spröder Materialien bewährt. In zunehmendem Maß wird Zerkleinerungsarbeit von Kugelmühlen auf Gutbett-Walzenmühlen übertragen. Als Folge davon und auf Grund der vermehrten Rückführung von Sichtergrobgut und/oder Schülpen ändern sich die granulometrische Zusammensetzung des Mühlenaufgabeguts sowie auch die rheologischen Eigenschaften des Fertigguts.

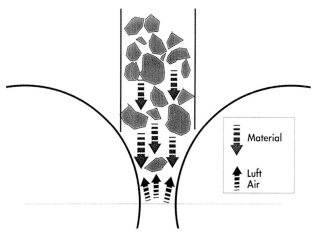

BILD 1: Schematische Darstellung des Materialeinzugs bei Gutbett-Walzenmühlen
FIGURE 1: Diagrammatic representation of the material in-take by high-pressure grinding rolls

Bei hohen Walzenumfangsgeschwindigkeiten und gleichzeitig feinem Aufgabegut kann es zu mehr oder weniger starken Vibrationen der Gutbett-Walzenmühle kommen, die auf ungenügendes Entweichen der Luft aus der Kompressionszone zurückzuführen sind. Die dem Mahlgut entgegenströmende Luft übt eine Kraft aus, die dem Gewicht der Schüttgutsäule entgegensteht und den Anpreßdruck des Mahlguts auf die Walzen vermindert. Dies führt im Bereich direkt oberhalb der Kompressionszone zu Auflockerungen des Schüttgutes. Dieses Phänomen ist auch vom Betrieb der Walzenmühlen in der Brikettiertechnik bekannt. Der Störungsmechanismus ist nicht stationär, sondern tritt periodisch auf. Solange das System ungestört ist, wird das Material durch das Gewicht der Schüttgutsäule auf die Walzenoberfläche gedrückt. Das Schüttgut wird auf die Walzenumfangsgeschwindigkeit beschleunigt und komprimiert, wobei das Lückenvolumen stark vermindert wird. Die verdrängte Luft strömt aus der Kompressionszone heraus und dabei dem Mahlgutfluß entgegen (**Bild 1**). Hierdurch wird auf das Schüttgut eine Kraft übertragen, die die Anpreßkraft vermindert und das Mahlgut auflockert. Aufgrund dieser Behinderungen wird weniger Mahlgut eingezogen, wobei auch der Volumenstrom der entweichenden Luft zurückgeht. Mit dem Rückgang des Luftvolumenstromes vermindern sich auch die Behinderungen, und das Mahlgut wird wieder besser eingezogen. Durch diesen Mechanismus kommt es zu Störungen, wenn entweder bei gegebenem Strömungswiderstand der Luftvolumenstrom einen Grenzwert überschreitet, oder aber bei konstantem Luftvolumenstrom durch Änderung der Korngrößenverteilung der Strömungswiderstand über einem Grenzwert liegt. Das Auftretenh von Vibrationen erscheint somit als ein Problem der

Since the introduction of high-pressure grinding rolls in the mid-eighties more than 300 plants have been commissioned worldwide and have proved to be an efficient energy-saving grinding system for comminution of brittle materials. Comminution work is increasingly being transferred from ball mills to high-pressure grinding rolls. As a result of this and owing to the increased recycling of classifier tailings and/or compacted cake, there is a change in the particle size distribution in the feed material and also in the rheological properties of the finished product.

When there are high peripheral roller speeds accompanied by fine feed material, the high-pressure grinding rolls may vibrate to a varying degree owing to inadequate escape of air from the compression zone. The air in the opposite direction to the material being ground exerts a force which opposes the weight of the column of bulk material and reduces the pressure of the material on the rollers. This leads to loosening of the bulk material in the region directly above the compression zone. The phenomenon is also known from the operation of high-pressure grinding rolls in briquette making. Rather than being a steady-state malfunction mechanism it occurs periodically. Provided that the system is operating smoothly the material is pressed onto the roller surface by the weight of the column of bulk material. The material is accelerated to the peripheral speed of the rollers and compressed, with a great reduction in the void volume. The expelled air flows out of the compression zone and against the flow of material to be ground (**Fig. 1**). This transmits a force to the bulk material which reduces the pressure and loosens the material. As a result of these obstructions

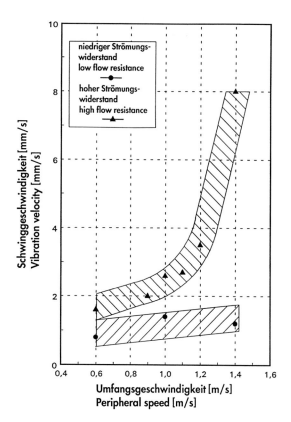

BILD 2: Schwinggeschwindigkeit als Funktion der Umfangsgeschwindigkeit
FIGURE 2: Vibration velocity as a function of peripheral speed

Durchströmung des Schüttgutes. Die Störungen führen zu Schwankungen im Drehmoment an den Wellen und zu Schwankungen der Spaltweite, wodurch die Maschine zu mehr oder weniger starken Vibrationen angeregt werden kann.

Im ungestörten Zustand ist die Menge der ausströmenden Luft in erster Näherung proportional der Walzenumfangsgeschwindigkeit. Somit läßt sich bei Konstanz der Mahlgutparameter die höchste Walzenumfangsgeschwindigkeit bestimmen, bei der noch keine Vibrationen auftreten. Entsprechende Untersuchungen wurden an einer Labor-Gutbett-Walzenmühle im Forschungszentrum Krupp Polysius AG durchgeführt. Die Walzen der Labormühle besitzen einen Durchmesser von 300 mm. Zunächst wurden die Versuche mit glatten Walzenoberflächen durchgeführt. **Bild 2** zeigt exemplarisch die Ergebnisse für ein feines und ein grobes Schüttgut. Als ein Maß für die Stärke der Vibrationen ist die Schwinggeschwindigkeit, die am Rahmen der Labor-Gutbett-Walzenmühle gemessen wurde, in Abhängigkeit der Walzenumfangsgeschwindigkeit aufgetragen. Für das grobe Mahlgut, dem ein niedriger Strömungswiderstand zuzuordnen ist, ergab sich im gesamten Drehzahlbereich ein störungsfreier Betrieb. Beim feinen Mahlgut, das einen vergleichsweise hohen Strömungswiderstand aufweist, traten ab einer bestimmten Umfangsgeschwindigkeit Vibrationen auf. Diese Geschwindigkeit wird im folgenden als kritische Geschwindigkeit bezeichnet.

Als ein Kriterium für die Höhe des Strömungswiderstandes wird der hydraulische Durchmesser herangezogen. Der hydraulische Durchmesser ist ein Maß für den mittleren Durchmesser einer Pore im Haufwerk. Die Haufwerksporen können modellhaft als ein Bündel paralleler Rohre dargestellt werden (**Bild 3**). Der Durchmesser eines Rohres entspricht dabei dem hydraulischen Durchmesser der Haufwerksporen, der eine Funktion der mittleren Partikelgröße und der Packungsdichte der Schüttung ist. Die kritische Umfangsgeschwindigkeit der Labor-Gutbett-Walzenmühle wurde an Schüttungen mit unterschiedlichen hydraulischen Durchmessern ermittelt. Die Ergebnisse dieser Versuche sind in **Bild 4** zusammengefaßt.

Das Bild zeigt die ermittelten kritischen Umfangsgeschwindigkeiten in Abhängigkeit von hydraulischen Durchmessern der Haufwerksporen. Daraus geht hervor, daß die kritische Umfangsgeschwindigkeit bei kleineren hydraulischen Durchmessern stark abnimmt. Eine zweite Versuchsreihe wurde mit profilierten Walzen durchgeführt. Die Ergebnisse zeigen, daß sich bei gleichem hydraulischen Durchmesser mit profilierten Walzen höhere kritische Umfangsgeschwindigkeiten einstellen. Die Laborergebnisse konnten durch Messungen an Industrieanlagen bestätigt werden.

Für den vibrationsfreien Betrieb einer Gutbett-Walzenmühle sollte der aus dem Gewicht der Schüttgutsäule resultierende Druck auf die Walzenoberfläche möglichst groß sein, damit erst bei deutlich erkennbaren Störeinflüssen

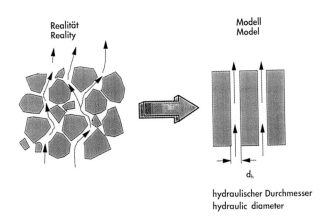

BILD 3: Lückenvolumen und hydraulischer Durchmesser eines Kornhaufwerks
FIGURE 3: Void volume and hydraulic diameter of granular fragmented rock

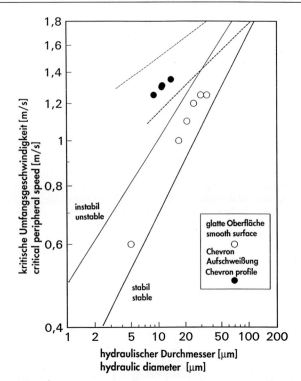

BILD 4: Kritische Umfangsgeschwindigkeit bei unterschiedlichem hydraulischem Durchmesser der Poren eines Kornhaufwerks
FIGURE 4: Critical peripheral speed with different hydraulic diameters of the pores in granular fragmented rock

less material is drawn in and the volume flow of escaping air recedes. With the receding of the airflow the obstructions are also diminished and the material in-take improves again. This mechanism causes trouble either when, for a given resistance to flow, the volume flow of air exceeds a limit, or when at a constant volume flow of air the flow resistance is above a limit owing to a change in particle size distribution. The occurrence of vibrations thus appears to be a problem of the through-flow of bulk material. The malfunctions lead to fluctuations in torque at the shafts and fluctuations in the gap width, whereby the machine can experience varying degrees of vibration.

In the undisturbed state the quantity of escaping air is, to a first approximation, proportional to the peripheral roller speed. So given constant parameters for the material being ground, it is possible to determine the highest peripheral roller speed at which there is still no vibration. Tests along these lines have been carried out on laboratory high-pressure grinding rolls at Krupp Polysius AG's research centre. The laboratory rollers are 300 mm in diameter. The first experiments were with smooth roller surfaces. **Fig. 2** shows the results for a fine and a coarse bulk material by way of example. As a measure of the strength of the vibrations, the vibration velocity measured at the frame of the grinding rolls was plotted against the peripheral roller speed. With coarse material, which has low resistance to flow, operation was found to be trouble-free throughout the whole range of speeds. With fine material, which has a relatively high resistance to flow, vibrations occurred above a certain peripheral speed. This is referred to below as the critical speed.

The hydraulic diameter is taken as a criterion for the level of resistance to flow. It is a measure of the average diameter of a pore in the fragmented rock. The pores in the rock may be modelled as a bundle of parallel tubes (**Fig. 3**). The diameter of a tube corresponds to the hydraulic diameter of the pores in the rock, which is a function of the average particle size and the packing density of the granular mass. The critical peripheral speed of the laboratory grinding rolls was determined for granular masses with different hydraulic diameters. The results of these experiments are shown in **Fig. 4**.

The diagram shows the critical peripheral speeds obtained as a function of the hydraulic diameters of the rock pores. The speed was found to drop sharply with smaller hydraulic

Vibrationen eintreten. Dieser Anforderung ist dementsprechend bei der Gestaltung von Aufgabegutschacht und -bunker Rechnung zu tragen. Das Aufgabegut sollte der Mühle außerdem möglichst homogen zugeführt werden. Insbesondere bei Mahlgut mit erhöhtem Feinanteil sollten Entmischungen vermieden werden. Messungen an Industrianlagen zeigen, daß schon eine geringfügige Erhöhung des Strömungswiderstandes zu instabilen Betriebsverhältnissen führen kann, wenn sich die Walzenumfangsgeschwindigkeit in der Nähe des kritischen Betriebspunktes befindet.

Neben der Gestaltung der Mahlgutzuführung ist darauf zu achten, daß die maximale Walzenumfangsgeschwindigkeit vom Strömungswiderstand der Aufgabegutschüttung begrenzt ist. Im Bereich der Vormahlung liegt die kritische Walzenumfangsgeschwindigkeit in den meisten Fällen deutlich über den üblichen Betriebsgeschwindigkeiten. Nur bei sehr feinem Klinker und starken Entmischungserscheinungen ist eine Überprüfung der maximalen Umfangsgeschwindigkeit notwendig. In Hybridmahlanlagen ist die Sichtergrobgutmenge, die zur Gutbett-Walzenmühle zurückgeführt werden kann, durch das Auftreten von Vibrationen in vielen Fällen begrenzt. Bei Combi- und Fertigmahlanlagen wird demgegenüber die gesamte anfallende Sichtergrobgutmenge zur Gutbett-Walzenmühle zurückgeführt. In Combimahlanlagen mit konstanter Walzenumfangsgeschwindigkeit ist die Feinheit des Produktes aus der Kreislaufmahlung mit Gutbett-Walzenmühlen begrenzt. Für Fertigmahlanlagen und Combimahlanlagen, bei denen der Kreislauf der Gutbett-Walzenmühle mehrere Produkte mit unterschiedlich hohen Feinheiten ermahlen soll, ist es vorteilhaft, drehzahlvariable Antriebe einzusetzen.

diameters. A second series of experiments was carried out with profiled rollers. The results show that, given the same hydraulic diameter, higher critical peripheral speeds are obtained with profiled rollers. The laboratory results are confirmed by readings taken at industrial plants.

For vibration-free operation of high-pressure grinding rolls the pressure on the roller surface, resulting from the weight of the column of bulk material, should be as high as possible so that vibrations only occur when disturbing influences are clearly recognisable. This requirement should be borne in mind when designing the feed shaft and feed hopper. The feed material should be as homogeneous as possible when it is supplied to the mill. Segregation should be avoided, especially if the material has a high proportion of fines. Readings taken at industrial plants show that even a slight rise in resistance to flow may lead to unstable operating conditions if the peripheral roller speed is near the critical point.

Apart from the design of the material feed it should be borne in mind that the maximum peripheral speed is limited by the flow resistance of the granular feed mass. In the preliminary grinding range the critical peripheral roller speed is in most cases well above normal operating speeds. Only when there is very fine clinker and much segregation is it necessary to check the maximum peripheral speed. In hybrid grinding plants the amount of classifier tailings which can be recycled to the high-pressure grinding rolls is in many cases limited by the occurrence of vibrations. In combined and finish grinding plants on the other hand all the tailings obtained are recycled to the rolls. In combined grinding plants with a constant peripheral roller speed there is a limit to the fineness of the product from closed circuit grinding with high-pressure grinding rolls. In finish and combination grinding plants where the high-pressure grinding roll circuit has to grind several products with different levels of fineness it is an advantage to use variable speed drives.

Die Fertigmahlung von Rohmaterial unter Druck
Finish grinding of raw material under pressure

Le broyage fini sous pression de la matière première

La molienda de acabado de la materia prima bajo presión

Von **S. Strasser**, Köln/Deutschland

Die Fertigmahlung von Rohmaterial unter Druck

Zusammenfassung – Während die Gutbettzerkleinerung in den ersten Jahren nach Markteinführung im Verbund mit Kugelmühlen verwendet wurde, wird sie heute zunehmend zur Fertigmahlung eingesetzt. Dabei bildet die Gutbett-Walzenmühle als einziges Mahlaggregat mit dem Sichter und dem Desagglomerator einen geschlossenen Kreislauf. Anhand von Daten aus Betriebsanlagen und gezielten Untersuchungen werden diese Mahlanlagen auf Tauglichkeit hinsichtlich Energieverbrauch, Verschleiß und Produktqualität untersucht. Dabei ergab sich, daß die Feinmahlung von Zementrohmaterial mit einer Gutbett-Walzenmühle besonders vorteilhaft ist. Bezogen auf vergleichbaren Durchsatz und gleiche Fertiggut-Feinheit ist der massebezogene Energieaufwand der Gutbett-Walzenmühle deutlich geringer als der einer Kugelmühlenanlage und der einer Mahlanlage aus Kugel- und Gutbett-Walzenmühle. Trotz des sehr verschleißträchtigen Rohmaterials ist der Walzenverschleiß der Gutbett-Walzenmühle mit nur 0,22 g/t sehr niedrig und der Mühlenbetrieb frei von störenden Vibrationen, da das Aufgabegut zerkleinert und vorgetrocknet wurde.

Finish grinding of raw material under pressure

Summary – In the first years after it was introduced to the market interparticulate comminution was used in conjunction with ball mills, but it is now being used increasingly for finish grinding. In this case, the high-pressure grinding rolls are the sole grinding unit and form a close circuit with the classifier and disagglomerator. The data from industrial plants and specific investigations were used to assess these grinding plants for their suitability with respect to power consumption, wear, and product quality. It was found that high-pressure grinding rolls are particularly advantageous for the fine grinding of cement raw material. Relative to the same throughput capacity and finished product fineness, the specific power consumption of high-pressure grinding rolls is significantly lower than that of a ball mill plant or of a grinding plant consisting of ball mill and high-pressure grinding rolls. In spite of the very abrasive raw material, the roller wear in the high-pressure grinding rolls of only 0.22 g/t is very low and the operation is free from disruptive vibrations as the feed material has been comminuted and predried.

Le broyage fini sous pression de la matière première

Résumé – Alors qu'au cours des premières années après la mise sur le marché, la fragmentation en lit de matière avait été utilisée en tandem avec des broyeurs à boulets, elle est aujourd'hui employée de plus en plus au broyage de finition. Le broyeur à rouleaux forme alors, comme seul appareil de mouture, un circuit fermé avec le séparateur et le désagglomérateur. Ces ateliers de broyage ont été étudiés, au moyen de données d'exploitation et d'investigations ciblées, en vue de leurs performances concernant la consommation d'énergie, l'usure et la qualité du produit. Il s'est révélé, que le broyage de finition des matières premières à ciment dans un broyeur à rouleaux est particulièrement avantageux. Pour les mêmes performances de débit et de finesse du produit fini, la dépense d'énergie rapportée à la masse, du broyeur à rouleaux, est nettement moindre que celle d'un atelier à broyeur à boulets ou celle d'un atelier combiné broyeur à boulets et broyeur à rouleaux. Malgré la matière première très abrasive, l'usure des rouleaux du broyeur à rouleaux est, avec 0,22 g/t, très faible et le fonctionnement est exempt de vibrations gênantes, parce que la matière introduite avait été concassée et préséchée.

La molienda de acabado de la materia prima bajo presión

Resumen – Mientras que la molienda en el lecho de material se utilizaba, en los primeros años de su introducción en el mercado, en combinación con molinos de bolas, se aplica hoy en día cada vez más para la molienda de acabado. El molino de cilindros y lecho de material, como único equipo de molienda, forma circuito cerrado con el separador y el desaglomerador. Con ayuda de datos obtenidos en instalaciones industriales y otras investigaciones específicas, se ha estudiado la aptitud de estas instalaciones de molienda, con respecto a consumo de energía, desgaste y calidad del producto. Resulta que la molienda fina de materias primas destinadas a la fabricación de cemento por medio de un molino de cilindros y lecho de material es especialmente ventajosa. Para el mismo rendimiento y finura del producto acabado, el consumo de energía referido a la masa de un molino de cilindros y lecho de material es notablemente más bajo que el de una instalación de molino de bolas y el de una instalación combinada de molino de bolas y molino de cilindros y lecho de material. A pesar de la gran abrasividad de la materia prima, el desgaste de los cilindros del molino de cilindros y lecho de material es muy bajo, ya que asciende sólo a 0,22 g/t. La marcha del molino es exenta de vibraciones molestas, debido a que el material de alimentación llega previamente desmenuzado y desecado.

The intention is to use the example of a high-pressure grinding rolls unit commissioned in 1992 to show what experience has been gained since then in the grinding of raw material and to indicate the level of the specific power consumption compared to a grinding plant equipped with a ball mill. Tests were also carried out to find whether it is possible to control the wear which the rollers suffer when grinding raw material.

BILD 1: Gutbett-Walzenmühle vom Typ RP20.0-170/130 zur Rohmaterialmahlung
FIGURE 1: RP20.0-170/130 High-pressure grinding rolls for grinding raw materials

Fig. 1 shows the high-pressure grinding rolls or roller press of the grinding plant. The raw material mix to be comminuted consists essentially of limestone in the form of marble with a high proportion of quartz. Approximately 11 wt.-% clay and 1 wt.-% iron ore are added to the limestone. **Table 1** sets out the properties of the main component, limestone, which are important for grinding. The marble has a dense, crystalline structure. It is compact and hard, with a Mohs hardness of about 4.0. The breaking action during coarse size reduction is classified as "slight" to "normal". The raw material feed has a particle size fraction between 0 and 60 mm. The fine grinding action is described as "normal". A wear test carried out suggests that there would be "heavy" wear.

Fig. 2 is a process flow sheet for the grinding plant. The material being ground is dried and pre-comminuted in an impact dryer supplied with gas from the preheater. The gas has a temperature of only 220°C as it has first been used for

TABLE 1: Properties of the limestone components used

Description	hard, compact, dense structure
Mohs hardness	4
Moisture of feed	max. 6 % water
Particle size of feed	0 to 60 mm
Coarse size reduction	slight to normal
Fine grinding	normal
Wear	heavy

TABLE 2: Important technical data for the main equipment in the grinding plant

High-pressure grinding rolls Model RP20.0	⌀ 1.7 × 1.3 m	2000 kW motor
Ball mill with one grinding chamber	4.6 m × 7.5 m	2300 kW motor
Classifier model SKS-L Diameter of caged rotor	SKS-L 4300	600 000 m³/h

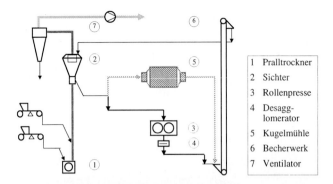

BILD 2: Verfahrenstechnisches Fließbild der Mahlanlage
FIGURE 2: Process flow chart for the grinding plant

1 Pralltrockner	1 Impact dryer
2 Sichter	2 Classifier
3 Rollenpresse	3 High-pressure grinding rolls
4 Desagglomerator	4 Desagglomerator
5 Kugelmühle	5 Ball mill
6 Becherwerk	6 Bucket elevator
7 Ventilator	7 Fan

und eine Breite von 1,3 m. Die maximale Durchsatzleistung der Gutbett-Walzenmühle beträgt 850 t/h. Die Einkammer-Kugelmühle wird von einem Motor mit 2300 kW Anschlußleistung angetrieben. Sie hat einen lichten Durchmesser von 4,6 m und eine Mahlbahnlänge von 7,5 m. Der Querstromsichter vom Typ „SKS-L" hat einen Stabkorbdurchmesser von 4,3 m und wird mit einer Nennluftmenge von 600 000 m³/h betrieben. Der maximale Sichteraufgabegutmassenstrom beträgt 1500 t/h. Davon erreichen rd. 35 % den Sichter durch das Steigrohr und rd. 65 % über das Becherwerk.

In **Tabelle 3** sind die bei den unterschiedlichen Betriebsweisen der Mahlanlage erzielten Betriebsergebnisse einander gegenübergestellt. In allen drei Fällen betrug der Rückstand des Fertigguts auf dem 90-μm-Sieb 10 M.-%. Wurde das Sichtergrobgut allein der Kugelmühle aufgegeben (A), so produzierte die Mahlanlage 175 t/h Fertiggut bei einem spezifischen Gesamtenergieverbrauch von 18,8 kWh/t Rohmehl. Die Leistungsaufnahme der Kugelmühle betrug dann 1700 kW. Bei Betrieb der Gutbett-Walzenmühle in Fertigmahlschaltung (B) wurden 360 t/h Rohmehl hergestellt. Die Leistungsaufnahme der Gutbett-Walzenmühle lag bei 1950 kW. Der spezifische Energieverbrauch betrug bei dieser Betriebsweise nur 12,2 kWh/t Rohmehl. Bei Betrieb der Gutbett-Walzen- und der Kugelmühle im Verbund konnten mit der Mahlanlage 470 t/h Rohmehl mit einem spezifischen Gesamtenergieverbrauch der Mahlanlage von 13,6 kWh/t produziert werden. Die elektrischen Leistungsaufnahmen der beiden Mühlen änderten sich gegenüber den Werten bei Einzelbetrieb nicht.

Die Betriebsergebnisse verschiedener Mühlen- bzw. Mahlsysteme können nur bei gleicher Endproduktfeinheit und gleicher Durchsatzleistung verglichen werden. Da alle Aggregate für eine Fertiggutproduktion von über 500 t/h dimensioniert sind, wird bei geringerer Durchsatzleistung und Einzelbetrieb der Gutbett-Walzenmühle und insbeson-

TABELLE 3: Vergleich der Betriebsergebnisse bei unterschiedlicher Anlagenschaltung

		System A Kugelmühle allein	System B Rollenpresse allein	System C Kugelmühle und Rollenpresse
Produktion	t/h	175	360	470
spezifischer Energieverbrauch	kWh/t	18,8	12,2	13,6
Kugelmühle	kW	1700	0	1700
Rollenpresse	kW	0	1950	1950

generating current. The pre-comminuted material is conveyed pneumatically (by a pneumatic conveyor dryer) to the classifier of the grinding plant by the hot exhaust gas. The high-pressure grinding rolls and ball mill are arranged in parallel downstream of the classifier. The classifier tailings can therefore be fed either to the ball mill, to the grinding rolls or to both simultaneously, so the grinding plant has three possible modes of operation:

— the ball mill working alone without the high-pressure grinding rolls (A)
— the grinding rolls working without the ball mill (B)
— combined operation of grinding rolls and ball mill (C).

In mode C about 50 % of the total electricity requirement is consumed by the ball mill.

The most important technical data on the main units of the grinding plant, the high-pressure grinding rolls, ball mill and classifier, is given in **Table 2**. The high-pressure grinding rolls are each driven by an electric motor with a connected capacity of 1000 kW. The rolls are 1.7 m in diameter and 1.3 m wide. The maximum throughput capacity of the high-pressure grinding rolls unit is 850 t/h. The single-chamber ball mill is driven by a motor with a connected capacity of 2300 kW. Its internal diameter is 4.6 m and the length of the grinding path is 7.5 m. The "SKS-L" cross-flow classifier has a caged rotor diameter of 4.3 m and is operated by a nominal air volume of 600 000 m³/h. The maximum mass flow of classifier feed is 1500 t/h, of which about 35 % reaches the classifier via the pneumatic conveyor and about 65 % via the bucket elevator.

TABLE 3: Comparison of the operating results for different plant configurations

		System A Ball mill alone	System B High-pressure grinding rolls alone	System C Ball mill and high-pressure grinding rolls
Production	t/h	175	360	470
Specific power consumption	kWh/t	18.8	12.2	13.6
Ball mill	kW	1700	0	1700
High-pressure grinding rolls	kW	0	1950	1950

The results obtained with the different modes of operation are compared in **Table 3**. In all three cases the finished material gave a residue on the 90 μm screen of 10 wt.-%. With the classifier tailings fed only to the ball mill (A), the grinding plant produced 175 t/h of finished material with a specific total power consumption of 18.8 kWh/t of raw meal. The power requirement of the ball mill was then 1700 kW. 360 t/h raw meal was produced with the high-pressure grinding rolls operating in the finish grinding configuration (B). The power requirement of the high-pressure grinding rolls was about 1950 kW. The specific power consumption with this mode of operation was only 12.2 kWh/t raw meal. With the grinding rolls and ball mill operating in combination the plant could produce 470 t/h raw meal and had a specific total energy requirement of 13.6 kWh/t. The electrical power requirement of the two mills was no different from when they were operating singly.

The operating results for different mill or grinding systems can only be compared if the fineness of the end product and the throughput capacity are the same. Since all units are dimensioned to produce over 500 t/h of finished material, a low throughput capacity and exclusive operation of the high-pressure grinding rolls or, in particular, exclusive operation of the ball mill will markedly raise the specific power requirement, as the system fan cannot be throttled for lower

TABELLE 4: Rechnerischer Vergleich der 3 verschiedenen Anlagenschaltungen für gleiche Fertiggutleistung

		System A Kugelmühle allein	System B Rollenpresse allein	System C Kugelmühle und Rollenpresse
Produktion	t/h	500	500	500
spezifischer Energieverbrauch	kWh/t	14,4	11,2	13,6
Kugelmühle	kW	4800	0	1700
Rollenpresse	kW	0	2800	2100

dere bei Einzelbetrieb der Kugelmühle der spezifische Energiebedarf deutlich erhöht, da der Systemventilator für geringere Durchsatzleistungen nicht gedrosselt werden kann. Deshalb sind in Tabelle 4 die für eine Rohmehlproduktion von 500 t/h erforderlichen Maschinengrößen sowie der entsprechende spezifische Energiebedarf der drei Mahlsysteme gegenübergestellt.

Sollte die Mahlanlage alleine mit einer Kugelmühle (System A) betrieben werden, so müßte diese einen Innendurchmesser von 5,8 m besitzen und würde voraussichtlich eine elektrische Leistung von rd. 4800 kW aufnehmen. Der spezifische Energiebedarf der Mahlanlage würde dann rd. 14,4 kWh/t betragen. Wenn anstelle der Kugelmühle eine Gutbett-Walzenmühle zum Einsatz käme, so müßten die Mahlwalzen eine Breite von 1,8 m aufweisen. Die elektrische Leistungsaufnahme dieser Mühle läge bei 2800 kW, der spezifische Energiebedarf der Gesamtanlage würde auf 11,2 kW/t Rohmehl sinken.

Die beschriebene Rohmehl-Mahlanlage ist seit mehr als einem Jahr in Betrieb. Wegen der großen, für einen entsprechend dimensionierten Ofen ausreichenden Durchsatzleistung und wegen des äußerst niedrigen spezifischen Energieverbrauchs wird die Mahlanlage meistens ohne die Kugelmühle nur mit der Gutbett-Walzenmühle in Fertigmahlschaltung (System B) betrieben. Auf den Ofenbetrieb hat die Zerkleinerung im Gutbett keinen negativen Einfluß. Die Mahlwalzen dieser Gutbett-Walzenmühle besitzen eine aufgeschweißte Panzerung. Der Walzenverschleiß liegt bei nur 0,22 g/t Rohmehl. Dieser trotz des stark schleißenden Rohmaterials relativ niedrige Walzenverschleiß ist auf das verfahrenstechnische Konzept der Mahlanlage zurückzuführen, bei dem die Gutbett-Walzenmühle ausschließlich bereits vorzerkleinertes und vorgetrocknetes Aufgabegut erhält. Das führt zu einem sehr ruhigen, vibrationsarmen Mühlenbetrieb. Der spezifische Energieverbrauch zur Erzielung einer Rohmehlfeinheit von $R_{0,09}$ = 10 M.-% liegt mit 11,2 bzw. 12,2 kWh/t Rohmehl konkurrenzlos günstig. Aufgrund der inzwischen vorliegenden Betriebserfahrungen ist mit einer heute verfügbaren Maschinengröße eine Rohmehlproduktion von bis zu 500 t/h alleine mit einer Gutbett-Walzenmühle in Fertigmahlschaltung ohne großes Risiko realisierbar.

throughputs. Table 4 therefore sets out the machine sizes required to produce 500 t/h of raw meal and the corresponding specific power requirement of the three grinding systems.

Should the grinding plant be operated only with a ball mill (system A), it would need an internal diameter of 5.8 m and would probably consume about 4800 kW. The specific power requirement of the plant would then be about 14.4 kWh/t. If high-pressure grinding rolls were used instead of the ball mill, the rolls would have to be 1.8 m wide. The electrical power requirement of the mill would be about 2800 kW and the specific power requirement of the whole plant would drop to 11.2 kWh/t of raw meal.

TABLE 4: Calculated comparison of the 3 different plant configurations for the same finished product capacity

		System A Ball mill alone	System B High-pressure grinding rolls alone	System C Ball mill and high-pressure grinding rolls
Production	t/h	500	500	500
Specific power consumption	kWh/t	14.4	11.2	13.6
Ball mill	kW	4800	0	1700
High-pressure grinding rolls	kW	0	2800	2100

The raw meal grinding plant described has been in operation for over a year. Owing to the large throughput capacity demanded by a kiln of corresponding dimensions and owing to the extremely low specific power consumption, the plant is usually operated without the ball mill, solely with the high-pressure grinding rolls in a finish grinding configuration (system B). Interparticle comminution does not have any negative effect on kiln operation. The high-pressure grinding rolls have hard-faced armouring, and the wear is only 0.22 g/t raw meal. The fact that roller wear is relatively low in spite of the highly abrasive raw material is due to the process engineering scheme for the plant, in which the high-pressure grinding rolls are only fed with pre-comminuted and pre-dried material. This leads to very smooth, vibration-free running of the rolls. The specific power consumption required to obtain a raw meal fineness of $R_{0.09}$ = 10 wt.% of 11.2 and 12.2 kWh/t raw meal is unrivalled. The operating experience gained shows that with the size of machines now available a raw meal production rate of up to 500 t/h can be obtained without any great risk, solely with high-pressure grinding rolls in a finished grinding configuration.

Moderne Mahlhilfsmittel-Technologie
Modern grinding additive technology

Technologie moderne avec mise en oeuvre d'additifs de broyage

Tecnología moderna de empleo de coadyuvantes de molienda

Von **M. S. Sumner,** Slough/Großbritannien

Zusammenfassung – Die Wirkung von Mahlhilfsmitteln läßt sich weitgehend dadurch erklären, daß sie die Agglomeration und die Verpelzung an den Mühleneinbauten verhindern und den Füllgrad der Mühle optimieren. Mahlhilfsmittel haben auch einen großen Einfluß auf die Fließeigenschaften des Mahlguts und somit auf die Verweilzeit in der Mühle. Bei richtiger Anwendung können sie im typischen Fall die Mühlenleistung um etwa 10 – 20% erhöhen. Es stehen verschiedene Arten von Mahlhilfsmitteln zur Verfügung, und außer einer erhöhten Mühlenleistung und der damit verbundenen Produktionssteigerung lassen sich mit ihnen auch die Leistungsmerkmale des Zements deutlich verbessern. Es gibt eine Reihe wichtiger Parameter, die berücksichtigt werden müssen, um die optimale Anwendung eines Mahlhilfsmittels zu gewährleisten. Sie beziehen sich sowohl auf die Zement- und Klinkereigenschaften als auch auf die Konstruktions- und Betriebsweise der Mühle.

Summary – The action of grinding aids can largely be explained by the prevention of agglomeration and coating of mill internals and the optimization of mill filling level. Grinding additives also have an important influence on the material flowability characteristics and hence on the mill residence time. With the appropriate application, cement mill efficiency can typically be improved by around 10–20%. Various types of additives are available, and in addition to the improved mill efficiency and the associated increase in production rate it is also possible to significantly enhance the cement performance characteristics. There are a number of important parameters that need to be considered in order to ensure the optimum application of a grinding additive. These concern both the cement and clinker properties as well as the mill design and operating parameters.

Résumé – L'effet des additifs de broyage s'explique largement en ce sens qu'ils préviennent l'agglomération et la formation de croûte sur les équipements intérieurs du broyeur, et optimisent le taux de remplissage du broyeur. Les additifs de broyage ont aussi une grande influence sur les propriétés d'écoulement de la matière à broyer et, par là même, sur le temps de séjour dans le broyeur. S'ils sont utilisés judicieusement, ils peuvent contribuer à un accroissement de la capacité du broyeur de l'ordre de 10 à 20%. Il existe divers types d'additifs de broyage qui, en plus d'un accroissement de la capacité du broyeur et de l'augmentation de la production en résultant, permettent d'améliorer sensiblement les caractéristiques des performances du ciment. Il convient de tenir compte de toute une série de paramètres importants pour garantir l'utilisation optimale d'un additif de broyage. Ceux-ci se réfèrent aussi bien aux propriétés du ciment et du clinker qu'à la conception et au mode de fonctionnement du broyeur.

Resumen – El efecto de los coadyuvantes de molienda se explica, en gran medida, porque éstos impiden la aglomeración y recubrimiento de los dispositivos internos, optimizando el grado de llenado del molino. Los coadyuvantes de molienda tienen también un gran influjo sobre la reología del material a moler y con ello, sobre el tiempo de permanencia del material dentro del molino. Si son correctamente empleados, pueden contribuir a aumentar el rendimiento del molino en un 10 – 20 %. Se dispone de diferentes tipos de coadyuvantes de molienda, los cuales permiten no solamente aumentar el rendimiento del molino, con el consiguiente incremento de la producción, sino mejorar también notablemente las características del cemento. Hay que tener en cuenta una serie de parámetros importantes, con el fin de garantizar un empleo óptimo de los coadyuvantes de molienda. Estos parámetros se refieren tanto a las propiedades del cemento y del clínker como al diseño y modo de funcionamiento del molino.

Die Wirkung von Mahlhilfsmitteln läßt sich weitgehend dadurch erklären, daß die Mahlgut-Agglomeration und die Verpelzung von Mahlkörpern und Mühlenpanzerung verhindert, der Mahlgutfüllungsgrad der Mühle optimiert und evtl. der Zerkleinerungswiderstand des Mahlguts herabgesetzt werden. Nach Rittinger [1] ist die durch die Zerkleinerung erzeugte Oberfläche dem Aufwand an Energie direkt proportional. Somit ist die Energieausnutzung (cm^2/J) unabhängig von der Mahlfeinheit konstant (**Bild 1**). In der Praxis jedoch steigt der Energiebedarf während des Mahlvorgangs

The action of grinding additives can largely be explained by the prevention of agglomeration and mill internals coating, the optimisation of mill filling level and the decrease in the resistance to comminution. According to Rittinger [1] the area of the new surface produced by comminutions is directly proportional to the useful work input. Hence the grinding efficiency notated in cm^2/J is constant for any level of fineness (**Fig. 1**). However in practice the energy input increases by an amount in excess of this as a result of the negative influences of particle agglomeration and surface

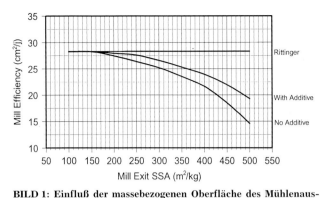

BILD 1: Einfluß der massebezogenen Oberfläche des Mühlenaustragsguts auf die Energieausnutzung in einer Kugelmühle
FIGURE 1: Influence of mill exit fineness on mill efficiency of a ball mill

Energieausnutzung in der Mühle	= Mill Efficiency
mit Mahlhilfsmitteln	= With Additive
ohne Mahlhilfsmittel	= No Additive
Massebezogene Oberfläche des Mühlenaustragsguts	= Mill Exit SSA

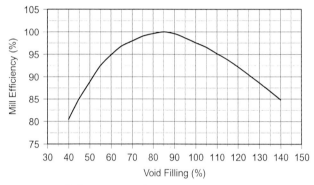

BILD 2: Einfluß des Mahlgutfüllungsgrads auf die Energieausnutzung in einer Kugelmühle
FIGURE 2: Influence of powder filling level on mill efficiency of a ball mill

Energieausnutzung in Kugelmühlen	= Mill Efficiency
Mahlgutfüllungsgrad	= Void Filling

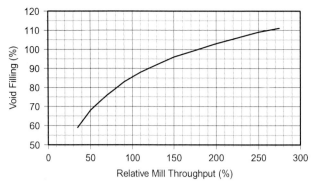

BILD 3: Einfluß des relativen Mühlendurchsatzmassenstroms auf den Mahlgutfüllungsgrad
FIGURE 3: Influence of mill throughput on powder filling level

Mahlgutfüllungsgrad	= Void Filling
Relativer Mühlendurchsatzmassenstrom	= Relative Mill Throughput

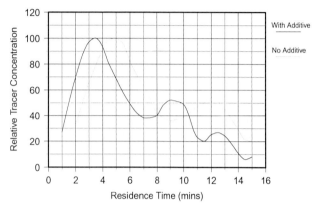

BILD 4: Verweilzeit-Verteilungen des Mahlguts in einer kontinuierlich betriebenen Kugelmühle beim Betrieb ohne und mit Mahlhilfsmittel
FIGURE 4: Residence time distribution of material in a continuously operated ball mill

Relative Konzentration der Markierungssubstanz	= Relative Tracer Concentration
mit Mahlhilfsmitteln	= With Additive
ohne Mahlhilfsmittel	= No Additive
Verweilzeit	= Residence Time

wegen der negativen Auswirkungen der Partikelagglomeration und der Verpelzung stärker an [2, 3]. Deshalb sinkt die Energieausnutzung mit zunehmender Mahlfeinheit. Mahlhilfsmittel enthalten Bestandteile, die die Oberflächenkräfte der Zementpartikel neutralisieren, so daß die Neigung zur Agglomeration und Adhäsion des Mahlguts an den Mahlkörpern und der Panzerung abnimmt und die Mahlwirkung bzw. Energieausnutzung verbessert wird (Bild 1).

Im stationären Zustand besteht ein in Relation zum Volumen der Mahlkörper optimaler Mahlgutfüllungsgrad der Mühle, bei dem die maximale Energieausnutzung gewährleistet ist (d.h. ein optimales Verhältnis von Stahl zu Klinker oder eine optimale Verweilzeit in der Mühle). Dieses Optimum ist nach vorherrschender Meinung dann erreicht, wenn der Leerraum zwischen den Mahlkörpern zu etwa 85 % mit Mahlgut gefüllt ist [4] (**Bild 2**).

In Wirklichkeit jedoch führt ein optimaler Mahlgutmassenstrom durch die Mühle zu einem etwas höheren Mahlgutfüllungsgrad von üblicherweise etwa 110–120 %, weil ein höherer Durchsatzmassenstrom einen höheren Mahlgutspiegel erfordert, um das Material durch die Mahlkörperfüllung bis zum Austragsende hindurchzudrücken [5] (**Bild 3**). Das wird bei kleinen Mahlkörpern besonders deutlich. Mahlhilfsmittel vermindern die Kohäsion des trockenen Pulvers und verbessern die Fließeigenschaften, so daß der Mahlgutfüllungsgrad der Mühle bei konstantem Mahlgutmassenstrom durch die Mühle niedriger ist, wenn Mahlhilfsmittel verwendet werden (d.h. die Mahlgut-Verweilzeit ist kürzer) [2] (**Bild 4**). Das kann zu einem Mahlgutfüllungsgrad führen, der näher am Optimum liegt.

Nach Rehbinders Hypothese [3] werden Mahlhilfsmittel von den Mikrorissen in den Partikeln absorbiert, so daß

coating during grinding [2, 3]. Therefore the grinding efficiency decreases with increasing fineness. Grinding additives contain components which neutralise the surface charges on cement grains and hence the tendency for agglomeration and adhesion to media and linings decreases and thus grinding efficiency is enhanced (Fig. 1).

Under steady-state conditions there exists an optimum powder filling level in the mill relative to the media volume to ensure maximum grinding efficiency (i.e. an optimum steel to clinker ratio or retention time). This is widely believed to occur when approximately 85 % of the media voidage is filled with material [4] (**Fig. 2**).

However, in reality the optimum circulating load results in a void filling somewhat above this, typically around 110–120 %. This results from the fact that a higher mill throughput requires a higher head for material to push itself through the media to reach the discharge end [5] (**Fig. 3**). This is particularly noticeable when small grinding media are present. Grinding additives reduce dry powder cohesion and flowability is improved, thus the filling level at constant mill throughput assumed is lower when additives are present (i.e. the retention time is reduced) [2] (**Fig. 4**). This might lead to a filling level closer to the optimum.

According to Rehbinder's hypothesis [3], grinding additives are absorbed on microcracks, which are present in grains, so eliminating or reducing their valency forces and therefore preventing their rejoining. Modern grinding additives are the result of 50 years of continual research and development acitivity combined with detailed analysis of their influence on the cement production process and on cement quality. These grinding additives combine the "traditional" benefits of increased mill output and improved dry cement flowabil-

deren Valenzkräfte ausgeschaltet oder vermindert werden und sie sich nicht wieder vereinigen können. Moderne Mahlhilfsmittel sind das Ergebnis von 50 Jahren Forschung und Entwicklung verbunden mit einer eingehenden Analyse ihres Einflusses auf den Prozeß der Zementherstellung und die Zementqualität. Diese Mahlhilfsmittel verbinden die „traditionellen" Vorteile höherer Mühlenleistung und besserer Fließeigenschaften des Zements mit zuweilen zusätzlichen Vorteilen einer kürzeren oder längeren Abbindezeit, optimierter Festigkeitsentwicklung und verminderten Wasserbedarfs. Sie werden sorgfältig zusammengesetzt, um die besonderen Anforderungen der Zementwerke zu erfüllen.

Solche Hilfsmittel werden mit dem Ziel eingesetzt, die Herstellungskosten zu senken und gleichzeitig die Zementqualität zu erhöhen. Das erfordert selbstverständlich die Berücksichtigung der Leistungsmerkmale des Zements und die geplante Verwendung verschiedener Zuschlagstoffe im Beton.

Um den optimalen Einsatz von Mahlhilfsmitteln für Zement zu gewährleisten, sind eine Reihe wichtiger Parameter zu beachten. Diese beziehen sich sowohl auf die Zement- und Klinkereigenschaften als auch auf das Mahlanlagen-Konzept und die Betriebsweise der Mühle. Mahlhilfsmittel können ebenso günstig in offenen und geschlossenen Mahlkreislaufsystemen eingesetzt werden einschließlich solcher mit Hochleistungssichtern und Gutbett-Walzenmühlen.

Die Mahlleistung hängt von der Gestaltung und Auslegung des Mühlenkreislaufs und von den Betriebsparametern ab, von denen einige die Reaktion des Mühlensystems auf die Verwendung von Mahlhilfsmitteln stark beeinflussen. Somit ist die erfolgreiche Anwendung von Mahlhilfsmitteln von einer Reihe von Faktoren abhängig, wie die Neigung zu Verpelzung und Partikelagglomeration, Fließeigenschaften des Mahlguts, Mahlgutfüllungsgrad, spezifische Gegebenheiten des Mühlensystems, Zementleistung, Kundenwünsche und besondere wirtschaftliche Faktoren.

Der Hauptvorteil eines erfolgreichen Einsatzes von Mahlhilfsmitteln besteht in der Verminderung des massebezogenen Energiebedarfs aufgrund der höheren Mühlenleistung und der Möglichkeit, die angestrebte Mahlfeinheit zu vermindern, wenn aufgrund der engeren Korngrößenverteilung und/oder verbesserten Hydratationseigenschaften bessere Festigkeitseigenschaften zu erzielen sind. Weitere Einsparungen können durch eine umfangreichere Nutzung von Zumahlstoffen oder Stoffen geringerer Reaktivität (eine wichtige Überlegung hinsichtlich der CO_2-Emissionen) erzielt werden.

Die Produktionssteigerung, die im typischen Fall 10–20 % beträgt, kann auch zu einem höheren Zementabsatz führen oder die Möglichkeit bieten, Nachfragespitzen zu befriedigen oder den Einsatz leistungsfähiger Mühlensysteme zu maximieren, so daß seltener mit älteren, weniger leistungsfähigen Mahlsystemen gearbeitet zu werden braucht.

Die damit einhergehende Senkung der Mühlenlaufzeit für eine bestimmte Jahresproduktion kann weitere Vorteile bieten, wie maximalen Einsatz kostengünstiger Elektrizität und reduzierte Wartungs- und Instandhaltungskosten. Letzteres versteht sich von selbst, da diese Vorteile im wesentlichen eine Funktion der Anlagenlaufzeit und weniger der Anlagendurchsatzleistung sind.

Die sich ergebenden verbesserten Fließeigenschaften können ebenfalls zu deutlichen Vorteilen führen, wie schnelleres Be- und Entladen bei Lagerung und Transport des Zements. Schließlich können Mahlhilfsmittel besser geeignete Leistungsmerkmale des Zements bewirken, so daß dieser die Anforderungen des örtlichen Zementmarkts erfüllt. Alles in allem können Mahlhilfsmittel einen wesentlichen Beitrag zur Senkung der Zementherstellungskosten und zur Verbesserung der Leistungsmerkmale des Zements liefern.

ity, with sometimes additional secondary benefits of reduced or increased setting time, optimised strength development and reduced water demand. They are carefully formulated to meet the specific requirements of the cement manufacturer.

The objective of the use of such additives is to assist in minimising manufacturing cost while maximizing cement quality. This clearly requires consideration of the performance of cement and the likely eventual use of various admixtures in the concrete.

To ensure the optimum application of cement grinding additives there are a number of important parameters that need to be considered. These concern both the cement and clinker properties as well as the mill design and operating characteristics. Additives can work equally well for open and closed-circuit systems, including those with high efficiency separators and roller presses.

The grinding performance depends on the principal mill circuit design and operating parameters and some of these strongly influence the response of a mill system to the utilisation of a grinding additive. Thus the success of the application of an additive will depend on a number of factors such as the propensity for coating and particle agglomeration, material flow characteristics, filling level, mill system specific constraints, cement performance, customer objectives and specific economic factors.

The principal advantages of the successful application of a grinding additive concerns the reduction in specific energy consumption brought about by the enhanced mill efficiency and the possibility to reduce the target fineness where strength properties are improved as a result of a narrower particle size distribution and/or improved hydration characteristics. Other savings can be achieved through the improved utilisation of non-clinker components (an important consideration for CO_2 emissions) or weathered clinker.

The increase in production rate, which is typically of the order of 10–20 %, can also result in increased cement sales or the ability to achieve peak demand or the ability to maximize the use of efficient mill systems and thereby reduce the need to operate older, less efficient grinding systems.

The associated reduction in run time for a given annual capacity can provide further benefits such as maximizing low cost electricity usage and permitting reduced repair and maintenance costs. The latter is evident since these are largely a function of plant run time rather than plant output.

The resultant improved flowability can also provide significant benefits such as an increased rate of loading and unloading operations during cement storage and transport. Finally, additives can provide more appropriate cement performance characteristics to meet the local cement market requirements. Thus, if properly considered, grinding additives can make a significant contribution to the reduction of cement manufacturing costs and to the enhancement of final cement performance.

Literature

[1] Von Rittinger, P.: Ritter — Lehrbuch der Aufbereitungskunde, Berlin, 1867.

[2] Mardulier, F. J.: The mechanism of grinding aids. Amer. Soc. for Testing and Materials. Philadelphia, Pa. Volume 61, 1961.

[3] Massazza, F., and Testolin, M.: Latest developments in the use of admixtures for cement and concrete. Il Cemento 2/1980.

[4] Shoji, K., Austin, L. G., Smaila, F., Brame, K., and Luckie, P. T.: Further studies of ball and powder filling effects in ball milling. Powder Technology 31 (1982), pp. 121–126.

[5] Swaroop, S. H. R., Abouzeid, A-Z. M., and Fuerstenau, D. W.: Flow of particulate solids through tumbling mills. Powder Technology 28 (1981), pp. 253–260.

„AIRFEEL" – eine neue Trennwand zur Regelung der Mahlgutfüllung in Kugelmühlen

The "AIRFEEL" – a new diaphragm for controlling the material level in a ball mill

L'„AIRFEEL" – une nouvelle cloison destinée au contrôle du niveau de matière dans les broyeurs à boulets

„AIRFEEL" – un nuevo tabique destinado al control del nivel de material dentro de los molinos de bolas

Von **F. Thomart**, Louvain-la-Neuve/Belgien

Zusammenfassung – Magotteaux-Slegten hat schon frühzeitig ein Konzept für eine einstellbare Trennwand entwickelt, um in den ersten Mühlenkammern einen für ein besseres Mahlergebnis geeigneten Mahlgutfüllstand zu erzielen. Die bisher eingesetzten Trennwände haben sich nicht automatisch an die Herstellung von sehr verschiedenen Zementen in derselben Mühle anpassen lassen. Die vorgestellte Problemlösung schöpft das Potential des technologischen Fortschritts bei der Regelung der Anlagen aus und führt zu einer höheren Flexibilität. Die neue Trennwand hat den Namen „AIRFEEL". Ihr Einfluß auf den Füllstand der ersten Kammer hängt besonders von der Mühlenluft und dem Mühlendurchsatz ab. Daher führt die in die Regelungsstrategie der Anlage eingebundene Steuerung dieser beiden Parameter auch zu einem korrekten Füllstand der Grobmahlkammer. Die neue Trennwand wurde so konstruiert, daß moderne Gießverfahren für besonders verschleißfesten Stahlguß angewendet werden können.

Summary – Magotteaux-Slegten has already developed a design for a controllable diaphragm for providing a suitable level of material in the first grinding chamber to give better grinding results. The diaphragms used in the past could not be adapted automatically to the manufacture of very different cements in the same mill. The proposed solution to the problem makes use of the potential of the technical progress in controlling the plant and leads to a higher level of flexibility. The new diaphragm is called the "AIRFEEL". Its influence on the level in the first chamber depends in particular on the mill air flow and the mill throughput. The control of these two parameters which is incorporated in the plant's control strategy therefore also leads to a correct level in the coarse grinding chamber. The new diaphragm has been designed so that modern casting methods for particularly wear-resistant cast steel can be used.

Résumé – Très tôt, Magotteaux-Slegten a développé le concept de cloison réglable pour obtenir dans les premiers compartiments des broyeurs le niveau de matière conduisant au meilleur rendement de broyage. Jusqu'à présent, les cloisons réalisées ne s'adaptaient pas automatiquement à la production de ciments très différents dans le même broyeur. Une nouvelle approche du problème a été menée pour accroître la souplesse des réalisations précédentes, exploitant le potentiel des progrès technologiques en matière de régulation des installations. Une nouvelle cloison, l'„AIRFEEL", a été conçue pour que son influence sur le remplissage du compartiment aval soit particulièrement sensible à l'air de ventilation du broyeur et au débit de matière effectif, de telle sorte que le contrôle de ces deux éléments, inclus dans une stratégie de régulation globale de l'installation, aboutisse également au remplissage correct du compartiment préparateur. De plus, le châssis de la cloison a été conçu pour pouvoir appliquer les procédés de fonderie, autorisant l'utilisation de nuances d'acier particulièrement résistantes à l'usure.

Resumen – Magotteaux-Slegden ha desarrollado, hace tiempo ya, un concepto para la construcción de un tabique ajustable, que permita conseguir, en las primeras cámaras del molino, un grado de llenado apropiado, con el fin de mejorar el resultado de la molienda. Los tabiques empleados hasta ahora no han podido ser adaptados automáticamente a la fabricación de cementos muy diferentes en el mismo molino. La solución que el autor nos presenta aprovecha el potencial del progreso tecnológico respecto de la regulación de las instalaciones y conduce a una mayor flexibilidad. El nuevo tabique ha recibido el nombre de „AIRFEEL". Su influjo sobre el nivel de llenado de la primera cámara depende, especialmente, del aire y del caudal del molino. Por esta razón, el dispositivo de mando de estos parámetros, incorporado a la estrategia de regulación de la instalación, conduce también a un nivel de llenado correcto de la cámara de molienda gruesa. El nuevo tabique ha sido concebido de tal manera que se puedan aprovechar los modernos procedimientos de fundición de acero particularmente resistente al desgaste.

1. Introduction

Long experience has shown that it is very important, both for grinding performance and for the wear of the mill internals, that the interstitial spaces between the grinding media in a ball mill should be filled with the material to be ground. This problem is particularly apparent in the coarse grinding chamber which is normally filled with grinding balls with diameters of 60 to 90 mm. The level of material is almost always too low here unless suitable countermeasures are taken.

The performance of an industrial mill can be increased by 5 to 8% by correct material filling in the coarse grinding chamber. The optimum mass ratio of material to grinding media in the coarse grinding chamber is around 17 – 19%. In the fine grinding chamber the optimum value is around 8 – 10%. This ratio can be obtained easily by, for example, changing the charge grading.

2. Adjustable diaphragm

For more than 20 years Magotteaux-Slegten has been developing adjustable diaphragms for regulating the material level in the coarse grinding chamber. The very efficient diaphragms which have been on offer so far have moving parts which require adjustment work to be carried out inside the mill. This normally leads to a compromise in the adjustment to suit the client's grinding programme. By comparison, diaphragms which can be adjusted from outside are much more flexible. Such systems have already been developed but prove to be rather expensive and difficult to maintain.

3. AIRFEEL diaphragm

Magotteaux-Slegten has now developed a new diaphragm known as the "AIRFEEL", which is based on an entirely different control principle without any moving parts. This diaphragm acts as a static control element which responds to the material flow in the mill and, in particular, to the air flow through the mill. It was also designed so that it could be made entirely of cast steel.

The functional principle of the new diaphragm can be described on the basis of the flow of material inside the mill. The flow velocity of the material affects the residence time of the material in the mill, and therefore the level of filling. It is necessary to differentiate between two transport mechanisms:

– the transport effected by the grinding media charge and the diaphragms, and

– the material transport effected by the mill air flowing above the grinding media.

The material being ground is comminuted and fluidized within the grinding media charge, and moves under the

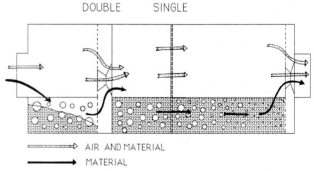

BILD 1: Mahlgutbewegung bei Einsatz einer einfachen und einer doppelten Trennwand
FIGURE 1: Movement of material when using single and double diaphragms

doppelt	=	double
einfach	=	single
Luft und Mahlgut	=	air and material
Mahlgut	=	material

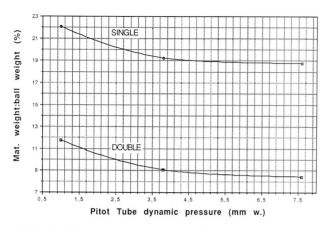

Bild 2: Verhältnis des Gewichts des Mahlguts zum Gewicht der Mahlkörperfüllung einer Kugelmühle als Funktion des Druckverlustes in der Mühle für eine einfache und eine doppelte Trennwand
FIGURE 2: Ratio of the weight of material to the weight of grinding media charge in a ball mill as a function of the pressure drop in the mill for single and double diaphragms

einfach	=	single
doppelt	=	double
Mahlgutgewicht : Mahlkörpergewicht (%)	=	mat. weight : ball weight (%)
dynamischer Druck (Pitot-Rohr) (mm W.S.)	=	Pitot Tube dynamic pressure (mm w.)

tragskonen dargestellt. Hierbei handelt es sich um das Ergebnis von Versuchen an einer Pilotanlage. Eine Erhöhung des Mühlenluftvolumenstroms hat eine Verminderung des Mahlgutspiegels in der Grobmahlkammer zur Folge. Es ist jedoch festzustellen, daß der Mahlgutfüllstand bei einer einfachen Trennwand recht hoch und bei einer doppelten Trennwand zu niedrig ist.

In **Bild 3** ist der prinzipielle Aufbau der „AIRFEEL"-Trennwand dargestellt, **Bild 4** zeigt deren maschinentechnische Ausführung. Wesentliche Merkmale der Trennwand sind:

— Auf der Vorderseite der Trennwand ist der Verschleißbereich mit Schlitzplatten und das Zentrum mit einem Schlitzrost versehen.

— Die Rückseite der Trennwand ist, bis auf das Zentrum, ebenfalls mit Schlitzen versehen. Die Schlitze auf dieser Seite sind größer als auf der Vorderseite, um einen ausreichend großen freien Querschnitt für den Durchgang der Mühlenluft zu erzielen.

— Die radial angeordneten Hubschaufeln sind in der Höhe an die rückseitigen Schlitzplatten angepaßt. Weitere Einbauten für den mechanischen Transport des Mahlguts, wie z.B. ein Kegel oder ähnliches, sind demnach nicht vorhanden.

Wenn die Ventilationsluft gänzlich fehlt, bewegt sich das Mahlgut lediglich im Innern der Mahlkörperfüllung. Die Trennwand wirkt dann praktisch wie eine einfache Schlitzwand, denn sie füllt sich mit Mahlgut und beeinflußt so den Mahlgutspiegel in der Grobmahlkammer. Durchströmt Luft die Trennwand, so entnehmen die Hubschaufeln der Trennwand aus dem Zentrum Mahlgut und geben es dem Luftstrom auf. Das führt zu einer Erleichterung des Mahlguttransports. Durch Einstellen der durch die Mühle gezo-

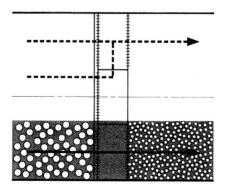

BILD 3: Prinzipskizze der neuen „AIRFEEL"-Trennwand
FIGURE 3: Basic principle of the new "AIRFEEL" diaphragm

action of the pressure difference between the mill inlet and mill discharge ends. The particle size distribution and the type of material, as well as the permeability of the grinding media charge, all influence the velocity of the material. It is the fine, graded, grinding media charges in the fine grinding chamber in particular which normally increase the residence time of the material in the mill.

It is the structural shape of the diaphragm which ultimately determines the movement of material in the mill. With a single diaphragm the retarding effect is transferred to the upstream chamber. This effect does not occur with a double diaphragm with cone and lifters (**Fig. 1**).

Above the ball charge, and in the boundary region of falling balls, the mill air exerts a transport effect on the material being ground which is particularly marked in the first chamber. Depending on the type of the diaphragm the mill air also acts in the region of the diaphragms, as shown by **Fig. 2**. This shows the change in material filling ratio on the ordinate as a function of the mill air velocity on the abscissa, expressed by the mill differential pressure (Pitot tube) in mm of water, for a simple slotted partition and for a double diaphragm with discharge cone. These are the results of trials in a pilot plant. An increase in the mill air volume flow lowers the level of material in the coarse grinding chamber. However, it is found that the level of material is rather high with a single diaphragm, and is too low with a double diaphragm.

The basic principle of the "AIRFEEL" diaphragm is shown in **Fig. 3**, and its mechanical design is shown in **Fig. 4**. Important features of the diaphragm are:

— On the upstream side the diaphragm has slotted plates in the wear region and a slotted grid in the centre.

— The reverse side of the diaphragm also has slots, except for the centre. The slots on this side are larger than on the upstream side in order to provide a sufficiently large free cross-section for the passage of the mill air.

— The levels of the radially positioned lifters are adapted to suit the slotted plates on the reverse side. There are therefore no other internal fittings, such as cones or the like, for mechanical transport of the material being ground.

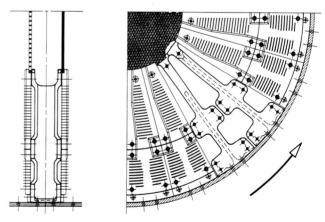

BILD 4: Ausführungszeichnung der „AIRFEEL"-Trennwand
FIGURE 4: Detailed drawing of the "AIRFEEL" diaphragm

If there is absolutely no ventilating air the material only moves within the grinding media charge. The diaphragm then operates in practice as a single slotted partition, as it fills up with material and so affects the level of material in the coarse grinding chamber. If air flows through the diaphragm then the lifters take the material from the centre of the diaphragm and transfer it to the air stream. This assists the transport of material. By adjusting the quantity of air drawn through the mill to suit the operating conditions it is therefore possible to control the material filling level in the diaphragm, and hence also in the coarse grinding chamber.

BILD 5: Verhältnis des Gewichts des Mahlguts zum Gewicht der Mahlkörperfüllung als Funktion des Druckverlustes in der Mühle – Vergleich zwischen einer einfachen und einer doppelten Trennwand und einer „AIRFEEL"-Trennwand

FIGURE 5: Ratio of the weight of material to the weight of grinding media charge as a function of the pressure drop in the mill – comparison of single and double diaphragms with the "AIRFEEL" diaphragm

einfach	=	single
doppelt	=	double
Mahlgutgewicht : Mahlkörpergewicht (%)	=	mat. weight : ball weight (%)
dynamischer Druck (Pitot-Rohr) (mm W.S.)	=	Pitot Tube dynamic pressure (mm w.)

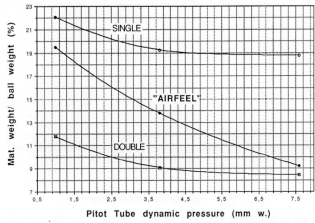

genen Luftmenge in Abhängigkeit von den Betriebsbedingungen kann demnach der Mahlgutfüllstand in der Trennwand und damit auch in der Grobmahlkammer reguliert werden.

In **Bild 5** ist die entsprechende Kennlinie der „AIRFEEL"-Trennwand im Vergleich zu den Kennlinien aus Bild 2 dargestellt. Daraus geht hervor, daß die Wirkung dieser Trennwand in starkem Maße vom Mühlenluftvolumenstrom beeinflußt wird. Das ermöglicht die Einstellung eines optimalen Mahlgutfüllungsgrades in der Mühle.

Die Trennwand eignet sich vor allem für den Einbau in Mühlen und Mahlanlagen, die bereits mit einem guten Steuerungs- und Regelungssystem ausgestattet sind, aber zusätzlich die Implementierung weiterer Regelkreise ermöglicht. Die Mahlanlage muß außerdem auch mit einem Hochleistungssichter ausgerüstet sein, da nur dessen hohe Kühlkapazität die erforderlichen Veränderungen des Mühlenluftvolumenstroms zuläßt.

Fig. 5 compares the corresponding characteristic curve of the "AIRFEEL" diaphragm with the characteristic curves from Fig. 2. From this it can be seen that the effect of this diaphragm is heavily influenced by the volume flow of mill air, making it possible to set an optimum material filling level in the mill.

The diaphragm is particularly suitable for installation in mills and grinding plants which are already provided with good open- and closed-loop control systems, but it also makes it possible to implement other control loops. The grinding plant must also have a high efficiency classifier, as the high cooling capacity is needed to allow the necessary changes to be made to the volume flow of mill air.

Grinding Systems for the nineties...

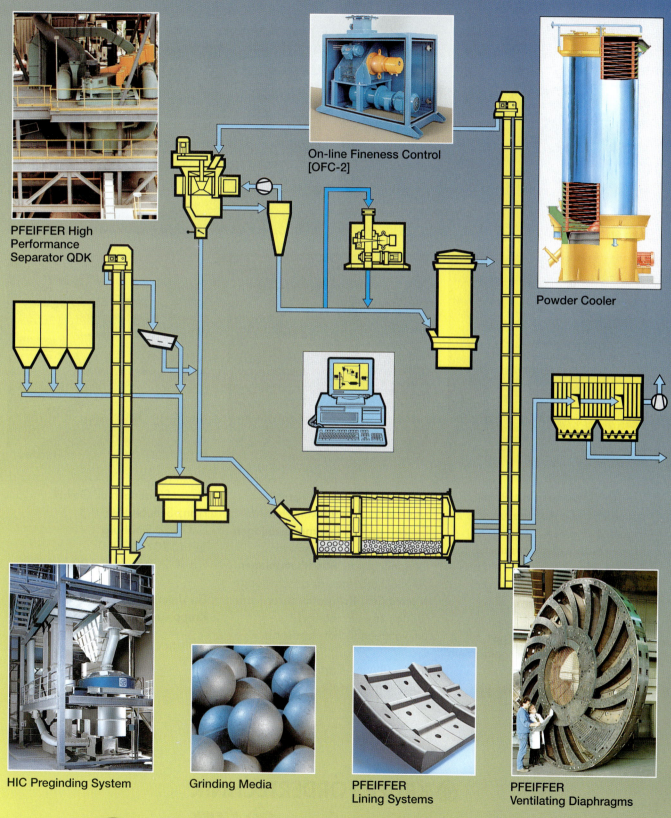

Gewinnen, Aufbereiten, Lagern, Fördern – ein breites Programm.

Von der Gewinnung der Rohmaterialien über die Aufbereitung bis zur Lagerung und zum Transport hat Krupp Fördertechnik zusammen mit PWH Anlagen + Systeme ein breites Programm für viele Stadien des Materialflusses.

Kompakt-Schaufelradbagger mit speziell entwickelten Schneidrädern gewinnen den Kalkstein.
Brechanlagen aller Größen und Leistungsstufen zerkleinern das Material auf die gewünschte Endkorngröße, **Trocknungsanlagen** trocknen und zerkleinern in Doppelfunktion, **Mischbettanlagen** vergleichmäßigen auch schwierige Materialien, **Rollgurtförderer** sorgen für den staubfreien und umweltfreundlichen Transport.

Krupp/PWH – Partner der Zement-, Kalk-, Gipsindustrie.

Krupp Fördertechnik GmbH
Franz-Schubert-Str. 1–3
D-47226 Duisburg
Tel. (0 20 65) 78–0
Fax (0 20 65) 78–28 20

Ein Unternehmen der Gruppe Krupp Anlagenbau

⊗ KRUPP FÖRDERTECHNIK

Krupp Fördertechnik GmbH · Tagebautechnik · D-23554 Lübeck · Telefax (0451) 4501-230
Krupp Fördertechnik GmbH · Aufbereitungstechnik · D-59320 Ennigerloh · Telefax (02524) 2252
PWH Anlagen + Systeme GmbH · Umschlagtechnik · D-66386 St. Inbert-Rohrbach · Telefax (06894) 599468

Ein engagiertes Team, eine halbe Milliarde DM Investitionen und beispielhafte Umweltschutzmaßnahmen – vor den Toren Berlins entsteht derzeit eines der modernsten Zementwerke Europas. Schon heute können wir **Zemente und Beton-Produkte** von hervorragender Qualität und ein perfektes Servicespektrum bieten. Besser läßt sich eine 750jährige Tradition wohl nicht fortsetzen.
Gerne geben wir Ihnen weitere Informationen: Telefon (03 36 38) 54-0.
Rüdersdorfer Zement.
Beste Verbindungen zum Bau.

Weltweit einzigartig: Klinkerofen 5 mit Zirkulierender Wirbelschicht
M 1:4000

Unser Meisterwerk!

RÜDERSDORFER ZEMENT
Ein Unternehmen der Readymix-Gruppe

Sichten ist unser Metier!

SCHMIDT-WINDSICHTER

erfüllen seit Jahrzehnten in vielfältiger Bauart die Anforderungen unserer Kunden

- Korbrotor-Sichter
- Streu-Windsichter
- Zyklon-Umluftsichter

- Luftstrom-Windsichter
- Labor-Windsichter
- Grieß-Abscheider

Mit und ohne stufenlose Feinheitsregelung während des Betriebes, mit hoher Trennschärfe – fehlkornfreien Produkten – robuster Konstruktion.

- Für Feinst-Sichtung
- Mit Trocknung
- Zur Brechsand-Entstaubung

- Für Grob-Sichtung
- Mit Kühlung
- Zur Füller-Gewinnung

Zur Sichtung von Zement, Kalk, Gips, Quarz, Phosphat, Schlacke, Bauxit, Kaolin, Kreide, Kohle, Erz und den anderen Produkten aus dem Sektor der Hartzerkleinerung.

Wir sind Spezialisten für die bei Ihnen anstehenden Sichtprobleme.

Unsere jahrzehntelangen Erfahrungen stehen Ihnen zur Verfügung:

- Bei der Planung von _neuen_ Anlagen,
- Modernisierung _älterer_ Mahl-Sicht-Anlagen,
- Umbau veralteter Sichter nach _heutigen_ Sichtsystemen.

- Nennen Sie uns Ihre Probleme – Wir beraten und helfen

SCHMIDT & CO. MASCHINENFABRIK
BAHNHOFSTRASSE 133 · D-63477 MAINTAL

TELEFON (0 61 81) 49 10 43 · TELEFAX (0 61 81) 49 12 78 · TELEX 4 184 452 schm d

PARTNER DER STEINE- UND ERDEN-INDUSTRIE

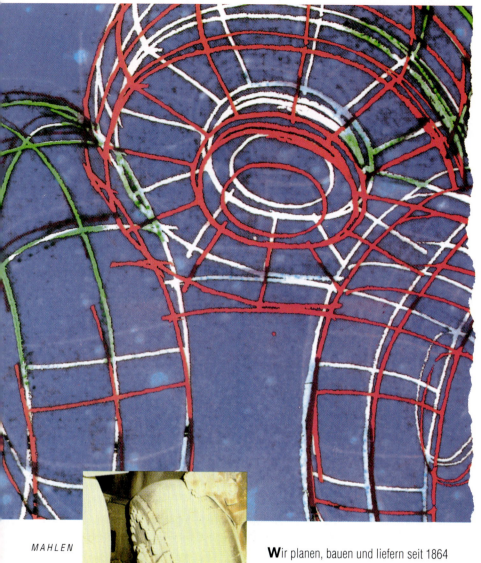

SICHTEN

LÖSCHEN

BRENNEN

MAHLEN

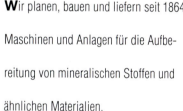

Wir planen, bauen und liefern seit 1864 Maschinen und Anlagen für die Aufbereitung von mineralischen Stoffen und ähnlichen Materialien.

Mit unseren Entwicklungs-, Projekt- und Konstruktionsabteilungen sowie unserem Versuchsfeld und mit eigenen Werkstätten sind wir ein erfahrener, leistungsstarker und zuverlässiger Partner unserer Kunden in der ganzen Welt.

TROCKNEN/KÜHLEN

Zementindustrie

KHD HUMBOLDT WEDAG + GEBR. PFEIFFER: die professionelle Lösung.

GEBR. PFEIFFER AG
Postfach 3080 · D-67618 Kaiserslautern

Rotor-starter for two slip-ring motors

MODERN PLANTS REQUIRE POWERFUL DRIVE UNITS!

Draft-ventilated speed-controllers 1000 + 1500 kW, covers removed

CONSEQUENTLY, YOU DEMAND:

STARTERS

SPEED CONTROLLERS

CONTROL GEAR

RESISTORS

FOR YOUR SLIP-RING INDUCTION MOTORS

WE DELIVER:

STARTERS UP TO 17000 kW

CONTROLLERS
OIL, AIR AND WATERCOOLED
NATURALLY OR SEPARATELY COOLED, DRAFT VENTILATED

CONTROL GEAR

RESISTORS, FOR EXAMPLE
STARTING RESISTORS
SPEED CONTROL RESISTORS,
SLIP RESISTORS
EARTHING RESISTORS
LOAD RESISTORS
CRANE DRIVE RESISTORS

TAPCHANGERS FOR FURNACE-TRANSFORMERS

MAINTENANCE-FREE
FOR MANY YEARS OF OPERATION !

WRITE FOR MORE INFORMATION AND REFERENCE LISTS !

Oil-immersed-water cooled Heavy-duty-Test-Resistor 200 kW/56000 A JET-Project

Group of small oil cooled rotor starters

Load resistor, 6000 kW continuous duty, 11 KV

PAPE & OLBERTZ
FABRIK ELEKTRISCHER SCHALTGERÄTE

Postfach 5149 · D-50354 Hürth · GERMANY
☎ (02233) 65066 · Tx 889338 · Telefax (02233) 65060

All things considered, wouldn't a 30-40% energy advantage stack the odds in your favour?

OK Vertical Mills for clinker and slag

Today, wafer-thin margins are decisive in business just as they are in sports.

An OK Vertical mill from F.L.Smidth offers you a host of competitive advantages. For instance, by reducing your energy requirements for grinding by 30% and 40% for slag, compared with tube mill operations, an OK vertical mill's impact on your bottom line can't be ignored.

But there's more. Like space-saving design, low vibration, less noise and excellent drying, just to highlight a few of the advantages that will put your mill operations in a whole new league.

Winning design and operating advantages:

Easy adjustment of Blaine and particle size distribution for required cement quality.

Grooved roller profile offers high, concentrated grinding pressure on outer path, allowing air to escape in the middle. Inner path evens out and compresses material before transfer to next roller.

Segmented roller wear parts, reversible so full width can be used, can be made of the hardest material without risk of cracking.

OK Vertical mills are just one of many upgrades available from an FLS organisation, dedicated to offering you the latest advances in cement plants - whether it's one machine or the whole works.

For further details, please contact our Vertical Mills Department. We'll get back to you quickly.

FLS Vertical Mills Department

From upgrades to complete plants.
FLS know-how: Nearly a third of the world's cement production comes from plants and machinery provided by F.L.Smidth.

F.L.SMIDTH

F.L.Smidth & Co. A/S. Vigerslev Allé 77. DK-2500 Valby-Copenhagen, Denmark. Tel. + 45 3618 1000. Fax + 45 3646 0277

Trendsetting Technology.
RENK TACKE Industrial Gear Units.

RENK TACKE warrant high equipment availability at low maintenance cost in cement industry plants.

The design and manufacture of gear units for use in the cement industry are a RENK TACKE speciality where the company has gained vast experience. The RENK TACKE know-how is based on over hundred years of involvement in the gear and coupling construction. Due to the extensive use of CAD and CAM methods in design and manufacture maximum flexibility could be achieved. Most advanced technology based on proved and tested design are a guarantee for dependability and economical operation.

RENK TACKE offer full system technology.

1

2

3 Twin planetary gear unit for the central drive of a tube mill

1 REPORA-Gear Units for roller press drives (Gutbett Mills)

2 Bevel Wheel-Planetary Gear Units for vertical roller mills

RT 3.005 e 1f

RENK TACKE GmbH · P.O.Box · D-86013 Augsburg · Telephone 08 21/57 00-0 · Telex 53 781
Telefax 08 21/57 00-4 60

SOMETIMES, GRAFFITI SHOULD BE ENCOURAGED.
Cut your grinding energy by 35% and more.

MAGOTTEAUX specializes in the cement grinding process, and is able to optimize any existing grinding circuit.

	Energy savings:
VERTICAL SHAFT IMPACT CRUSHER	10 to 15 %
MILL INTERNALS	10 to 20 %
STURTEVANT HIGH EFFICIENCY SEPARATOR	10 to 15 %
MILL CONTROL SYSTEM	5 to 10 %

Other MAGOTTEAUX Grinding System benefits are:
Increased output
Improved and stable cement quality
Constant throughput
Reduced transition times
Fineness prediction

MAGOTTEAUX. Always a step ahead.

any information, please contact MAGOTTEAUX INTERNATIONAL, Belgium (phone : 32/41/67.47.04) or your local MAGOTTEAUX sales office.

**Developed to suit tough climatic conditions and perform hard work.
Technical lubricants made by Klüber.**

Again and again trucks are being loaded with stones und ore. Leaving the excavation site, they are heading for the processing plant where the raw materials are crushed in huge mills with a peripheral speed up to 10m/s, driven by gigantic girth gears. In this case, friction is obviously bound to be extreme, and special lubricants are required to cope with it. The cue is "system lubrication". Klüber lubricants cover all fields from pre-start lubrication to running-in and operational lubrication, meeting stringent technical standards and environmental requirements.

Suitable Klüber lubricants are also available for all the other machines and installations used in the base material industry to make it possible to move large things.

Klüber Lubrication München KG, Public Relations, P.O. Box 70 10 47, D-81310 Munich. A member of the Freudenberg Group.

KLÜBER LUBRICATION

KHD Humboldt Wedag AG

All grinding technologies now from a single source. Vertical roller mill complements scope of supply.

KHD Humboldt Wedag AG active worldwide in industrial plant construction and **Gebr. Pfeiffer AG** specialized in grinding systems agreed on exclusive cooperation for grinding installations used by the cement industry. This means that KHD Humboldt Wedag AG complemented their range of tube mills and roller presses by the so far missing vertical roller mill. That technology ensures cost-effective plant operation especially for the grinding of raw materials, coal and clinker.

Gebr. Pfeiffer's vertical roller mill enables universal application for combined grinding and drying. In their position as a full liner, KHD Humboldt Wedag AG offer all popular grinding technologies. Humboldt Wedag and Gebr. Pfeiffer – concentration of expertise and technical competence of two well reputed companies.

HUMBOLDT WEDAG + Gebr. Pfeiffer: the professional solution.

KHD Humboldt Wedag AG · D-51057 Cologne, Phone +221-822-6323, Fax +221-822-6627
HUMBOLDT WEDAG ZAB GmbH · Brauereistr. 13, D-06847 Dessau, Phone +340/7310-0, Fax +340/7310-676
Humboldt Wedag, Inc. · 3883 Steve Reynolds Blvd., Norcross, GA. 30093, USA, Phone +404 564 7300, Fax +404 564 7333
International sales offices: Beijing · Hongkong · Madrid · Melbourne · Mexico · Moscow · São Paulo · Shanghai

DIE GANZE WELT DER AUFBEREITUNGSTECHNIK

Doppelt bricht besser!

Der Barmac-Duopactor - ein Vertikal-Autogenbrecher ohnegleichen: Aus dem Rotor mit bis zu 90 m/s herausgeschleudertes Brechgut prallt auf ein Brechgutbett, kollidiert zuvor jedoch mit einem die Flugbahn in freiem Fall kreuzenden Brechgutstrom! Doppelter Aufprall, mehrfacher Effekt: höherer Wirkungsgrad, verminderter Verschleiß und - einzigartig: effiziente Steuerung der Feinanteile per Freifallstrom-Dosierung. Der Duopactor ist ein äußerst wirtschaftlicher Brecher für alle Mineralstoffe, mit hohem Zerkleinerungsgrad, optimaler Kornform und bis 650 t/h Durchsatz. Wir befriedigen gern Ihre Neugier; bitte Anruf oder Fax!

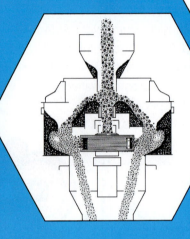

ALLIS MINERAL SYSTEMS GMBH
Theodor-Heuss-Straße 32 • D-61118 Bad Vilbel
Telefon (06101) 52 58 0 • Telefax (06101) 6 41 26

 MITGLIED DER SVEDALA GRUPPE

MOZER-Trommeltrockner

Drum driers

sand	blasting sand	industrial and
gravel	magnesite	domestic studge
limestone	bauxite	
clay	dolomite	**and your**
minerals	iron-powder	**material...**

Zweizügige und einzügige Trockentrommeln. Kombinierte Trocken- und Kühltrommeln. Planung, Fertigung, Montage von kompletten Trocknungsanlagen.

Double Shell and single shell dryers. Combined drying and cooling drums. Engineering, manufacturing, assembly of drying equipments.

Sand
Kies
Kalkstein
Ton
Mineralien
Ferrolegierungen
Strahlmittel
Magnesit
Bauxit
Dolomit
Industrielle und kommunale Schlämme

und Ihr Material...

C. G. MOZER GmbH & Co. KG, Postfach 943
D-73009 Göppingen · Telefon (0 71 61) 67 35-0 · Telefax (0 71 61) 67 35 35

Ein Unternehmen der **ALLGAIER**-Gruppe

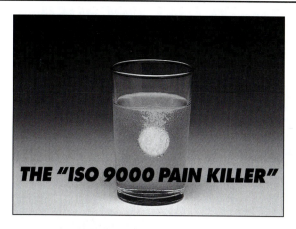

THE "ISO 9000 PAIN KILLER"

THE CILAS PARTICLE SIZE ANALYZER

You think ISO 9000 is a pain? Your certification gives you a headache. Then here is an easy "pain killer"... the CILAS laser diffraction particle size analyzer. When we designed it, we thought of the ISO 9000 certification process. That's why it is good in relieving the pain of your ISO implementation. Why? Because the CILAS particle size analyzer offers exactly the audit control that you require: • Utmost Accuracy, • High Measurement Reproducibility, • Traceability to international industrial and research standards, and above all... • Recalibration by you... To your own standards! So why go on suffering? Just buy the right "pain killer". Buy CILAS and relax.

COMPAGNIE
INDUSTRIELLE
DES LASERS
P.O. Box 27

91460 MARCOUSSIS
FRANCE
PHONE (+33) 1 64 54 48 00
FAX (+33) 1 69 01 37 39

GRAVIT Weighfeeder

HASLER INNOVATION

- 100% digital
- controlled by HF 9 Quadro

Compact and modular
- all desired feeding length available

K-TRON – HASLER
HEAVY DUTY WEIGHFEEDERS

| D K-TRON DEUTSCHLAND GMBH Fax: +49 5481 82978/79 | CH HASLER FRERES SA Fax: +41 38 41 1524 | USA K-TRON NORTH AMERICA, I.N.C. Fax: +1 609 589 8113 | F K-TRON FRANCE Fax: +33 74 57 7884 |
| SINGAPORE K-TRON ASIA PACIFIC PTE LTD Fax: +65 861 5032 | GB K-TRON GREAT BRITAIN LTD Fax: +44 61 6 24 2853 | E K-TRON HASVAN SA Fax: +34 3 308 1460 | BR HASLER FRERES Ind. e Com. Ltda. Fax: +55 11 429 78 32 |

HURRICLON
VOEST-ALPINE KREMS GES.M.B.H.

Weniger = *Mehr*

HURRICLON ® Fliehkraftabscheider setzen neue Maßstäbe.

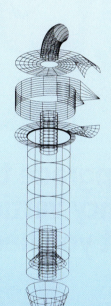

Weniger Druckverlust

Die Halbierung des Druckverlustes gegenüber herkömmlichen Systemen spart wertvolle Energie.

Weniger Reststaub

Höhere Abscheideraten halbieren den Reststaubgehalt und erhöhen die Produktausbeute und die Effizienz von Umweltschutzeinrichtungen.

Weniger Kosten

Das einfache hundertfach erprobte System erhöht die Anlagenlebensdauer. HURRICLON ersetzt Mehrfachzyklone und reduziert so die Investitions- und Wartungskosten bei äußerst wirtschaftlichem Energieeinsatz.

Weniger Energiebedarf

Finanzieren Sie Ihren HURRICLON in wenigen Jahren allein mit der eingesparten Energie.

Können Sie auf diese Vorteile verzichten?
Wir senden Ihnen gerne unseren Spezialkatalog!

IDEEN erobern die WELT

VOEST-ALPINE KREMS
FINALTECHNIK Ges.m.b.H.
DI. Keuschnigg, DW 474
Ing. Wöss, DW 344
Ing. Kaufmann, DW 577
Postfach 42
A-3500 Krems / Austria
Tel. (02732) 885
Telefax (02732) 885 DW 440

You don't replace the whole team just because one player gets tired

Maybe an upgrade to one of our high-efficiency vertical mills is the only replacement you need.

The OK Vertical Mill for cement clinker and slag makes the FLS grinding programme complete.

It means that we can offer you any grinding solution you need, because we have them all.

Just make your choice, and we'll tailor it to meet your exact requirements.

The OK Vertical Mill offers features that can make important contributions to your product quality and profitability:

- It can reduce your energy costs 30-40%
- The compact design requires less space
- Its layout is simple; operation is too

Our OK series of grinding mills are manufactured under patent and licence from Kobe Steel and Onoda Cement Co. It represents a major advance in the finish grinding of cement clinker and slag.

For further details, please contact our Vertical Mills Department. We'll get back to you quickly.

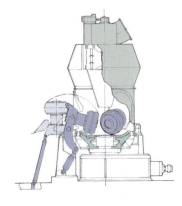

FLS Vertical Mills Department

From upgrades to complete plants.
FLS know-how: Nearly a third of the world's cement production comes from plants and machinery provided by F.L.Smidth.

F.L.Smidth & Co. A/S. Vigerslev Allé 77. DK-2500 Valby-Copenhagen, Denmark. Tel. + 45 3618 1000. Fax + 45 3646 0277

Fachbereich 6

Betriebstechnik

(Wasser-, Druckluft- und Elektrizitätsversorgung, Gewinnen, Homogenisieren und Vergleichmäßigen, Mischen, Fördern, Lagern, Verpacken)

Subject 6

Plant engineering

(Supply of water, compressed air, and electricity, winning, homogenizing and equalizing, blending, materials handling, storage, packaging)

Séance Technique 6

Technique d'exploitation

(Alimentation en eau, en air comprimé et en électricité, extraction, homogénéisation et régulation, mélange, manutention, stockage, emballage)

Tema de ramo 6

Técnica operativa

(Abastecimiento de agua, de aire a presión y de electricidad, explotación, homogeneización, mezcla, transporte, almacenaje, embalaje)

Betriebstechnik*)
Plant engineering*)
Technique d'exploitation

Técnica operativa

Von **F. Guilmin,** München/Deutschland

Betriebstechnik

Generalbericht 6 · Zusammenfassung – Zu den wesentlichsten Aufgaben der Betriebstechnik zählt die Sicherung der Zuverlässigkeit aller derjenigen Einrichtungen, die für den Betrieb der Hauptaggregate und damit für den reibungslosen Ablauf des gesamten Produktionsprozesses erforderlich sind. Diese Anlagen erfordern teilweise einen Anteil von 25 bis 30% der Gesamtinvestitionen. Eine Senkung von Investitionskosten ist nur bei Steigerung der Zuverlässigkeit der Einzelausrüstungen möglich. Betriebskosteneinsparungen lassen sich durch Automatisierung der Anlagen, Einsatz leistungsfähigerer Einzelmaschinen, Verschleißminderung, geplante Wartung, Abbau des Schichtbetriebs und Minderung der Ausfallzeiten durch Arbeitsunfälle erzielen. Schwerpunkte der Entwicklung und Optimierung bilden derzeitig neben einem darauf abgestimmten Layout von Neuanlagen die Verbesserung des Betriebs in Steinbrüchen, Verbesserungen bei der Lagerung und Homogenisierung von Einsatzstoffen und Fertigprodukten, beim Materialtransport zwischen einzelnen Betriebsabteilungen und Anlagenbereichen, bei der Verpackung und beim Versand sowie generell auch bei Hilfsanlagen wie z. B. bei der Behandlung von Kühlwasser oder der Erzeugung und den Einsatz von Druckluft.

Plant engineering

General report 6 · Summary – The main tasks of plant engineering include ensuring the reliability of all the equipment which is necessary for operation of the main units and therefore for the smooth flow of the entire production process. These systems sometimes account for 25 to 30% of the total investment. Investment costs can only reduce by increasing the reliability of the individual items of equipment. Savings in operating costs can be achieved by automating the systems, by the use of more efficient individual machines, by reducing wear, by planned maintenance, by cutting back shift work, and by reducing time lost through industrial accidents. Development and optimization work is currently focused not only on appropriate layouts for new plants, but also on improvement of quarry operation, and on improvements in the storage and blending of primary materials and finished products, in the transport of materials between individual operating departments and plant sections, in packaging and despatch, and also in general in auxiliary systems, such as in the treatment of cooling water or the production and use of compressed air.

Technique d'exploitation

Rapport général 6 · Résumé – Parmi les tâches essentielles de la technique d'exploitation se situe la fiabilité de tous ces équipements, qui sont indispensables pour le fonctionnement des appareils principaux et, ainsi, pour le déroulement sans problème du processus de production dans son ensemble. Ces installations exigent parfois un montant de 25 à 30% des investissements totaux. Un abaissement des coûts d'investissement n'est possible, que par l'amélioration de la fiabilité des équipements distincts. Les économies de coûts d'exploitation peuvent être obtenues par l'automatisation des ateliers, mise en oeuvre de machines spécifiques plus performantes, réduction de l'usure, maintenance planifiée, suppression progressive du travail posté et diminution des temps perdus par accidents de travail. Les points essentiels du progrès et de l'optimisation sont actuellement, avec une conception adéquate des installations nouvelles, l'amélioration de l'exploitation des carrières, un meilleur fonctionnement du stockage et de l'homogénéisation des matières premières et produits finis, du transport des matières entre les différentes zones de l'usine et ateliers, lors de l'ensachage et de l'expédition ainsi que, de manière générale, au niveau des équipements secondaires tels que traitement des eaux de refroidissement ou production et utilisation de l'air comprimé.

Técnica operativa

Informe general 6 · Resumen – Entre las tareas esenciales de la técnica operativa cuenta el aseguramiento de la fiabilidad de todos aquellos elementos que son necesarios para el funcionamiento de los dispositivos principales de la instalación y, con ello, para el buen desarrollo de todo el proceso de producción. Dichos elementos requieren, algunas veces, el 25 a 30% de la totalidad de las inversiones. Una reducción de los gastos de inversión sólo es posible, si se aumenta al mismo tiempo la fiabilidad de los equipos individuales. Un ahorro de gastos de explotación se puede conseguir mediante la automatización de las instalaciones, el empleo de máquinas individuales más efi-

---

*) Überarbeitete Fassung eines Vortrages zum VDZ Kongreß '93, Düsseldorf (27.9.–1.10.1993)
Revised text of a lecture to the VDZ Congress '93, Düsseldorf (27.9.–1.10.1993)

cientes, la reducción del desgaste, la planificación del mantenimiento, la eliminación del trabajo por turnos y la disminución de las paradas debidas a los accidentes de trabajo. Los trabajos de desarrollo y optimización se centran actualmente, aparte del correspondiente diseño de instalaciones nuevas, en el perfeccionamiento de la explotación de la cantera, un mejor almacenamiento y homogeneización de las materias empleadas y de los productos acabados, un transporte mejorado de material entre los distintos departamentos y secciones de la fábrica, en el perfeccionamiento del ensacado y expedición y, de un modo general, de las instalaciones auxiliares, como son el tratamiento de agua de refrigeración y la generación y empleo de aire comprimido.

1. Einleitung

In den vergangenen Jahren war die Zementindustrie stets darauf bedacht, ihre Produktionskosten so optimal wie möglich zu gestalten, und zwar nicht nur über eine Verbesserung der angewandten Verfahren, sondern auch über die Sicherung der Zuverlässigkeit aller zugehörigen Installationen, die den reibungslosen Betrieb zu gewährleisten haben. Solche Anlagen können 25 bis 30 % der Gesamtinvestition einer Produktionslinie ausmachen. Im Verlauf der letzten 10 Jahre trat das Ziel in den Vordergrund, die Investitionskosten für eine Produktionslinie zu senken, ohne dabei die Zuverlässigkeit zu verringern. Weitere Schwerpunktziele in der Zementindustrie sind darüber hinaus der Abbau des Schichtbetriebs in der Produktion, der mit der Automatisierung der Anlagen einhergeht. Hierzu gehört auch der Einsatz immer leistungsfähigerer Geräte und die Beherrschung der entsprechenden Technologien (z. B. zur Verschleißminderung) bei gleichzeitiger Verbesserung der Sicherheit am Arbeitsplatz.

Angesichts dieses weitreichenden Aufgabengebiets und seiner vielfältigen Aspekte muß den Entwicklungen in folgenden Bereichen besondere Aufmerksamkeit geschenkt werden:

— Tagebau,
— Lagerung der Produkte,
— Materialtransport,
— Verpackung und Versand der Endprodukte,
— Hilfsanlagen für Wasser, Luft usw.,
— Wartung und Instandhaltung,
— Anlagenreinigung,
— Layout neuer Zementwerke.

2. Tagebau

Seit dem letzten VDZ-Kongreß im Jahr 1985 sind die Probleme der Umweltbelastung in fast allen Ländern der Welt zu einem der dringlichsten Anliegen geworden. Der Tagebau unterliegt in bezug auf Lärmerzeugung und Erschütterungen durch Sprengungen sowie Staubentwicklung den geltenden Umweltauflagen. Die zuständigen Behörden und Anwohner üben einen beachtlichen Druck aus, damit diese Auflagen eingehalten werden.

Wer einen bereits vorhandenen Steinbruch weiter betreiben will, stößt auf Schwierigkeiten. Einen neuen Steinbruch in Betrieb zu nehmen, erweist sich inzwischen in vielen Ländern als ein langwieriger, manchmal sogar aussichtsloser Prozeß. Deshalb ist die optimale Förderung der vorhandenen Reserven in den bestehenden Steinbrüchen unabdingbar geworden, wenn ein langfristiger Betrieb gesichert werden soll. Die hierbei im allgemeinen angewandte Technik geht zunächst vom Modell des vorhandenen Vorkommens aus, wobei die Daten durch die regelmäßige Analyse von Proben aktualisiert werden, die während des Abbaus entnommen werden. Computergestützte Simulationsverfahren gewährleisten einerseits die fortlaufende Qualität der geförderten Materialien und geben andererseits Aufschluß über die noch vorhandenen Reserven [1].

Neben der klassischen Abbaumethode durch Sprengung setzt sich nach und nach der Abbau durch Schürfen und Fräsen durch, und zwar nicht nur im weichen Gestein, sondern auch in Kalkstein/Mergel-Schichten, in denen der

1. Introduction

In past years the cement industry has always been mindful of optimizing production costs, not only by improving the processes used but also by ensuring the reliability of all the associated installations designed to achieve smooth operation. Such plants can make up 25 to 30 % of the total investment in a production line. During the last 10 years attention has become focused on the aim of lowering the capital costs for a production line without reducing the reliability. Another important objective in the cement industry is to reduce the amount of shift work in production, which goes hand in hand with automation of the plant. This also includes the use of ever more efficient equipment and mastery of the appropriate technologies (e.g. for reducing wear) while at the same time improving industrial safety.

In view of the far reaching range of the task and its manifold aspects, particular attention must be paid to the developments in the following areas:

— quarrying,
— storage of the products,
— materials transport,
— packaging and despatch of the end products,
— auxiliary systems for water, air, etc.,
— maintenance and repair work,
— plant cleaning,
— layout of new cement works.

2. Quarrying

Since the last VDZ Congress in 1985 the problems of environmental pollution have become one of the most pressing concerns in almost all countries. The quarry is subject to relevant environmental regulations in relation to noise and vibrations caused by blasting and the generation of dust. The responsible authorities and the nearby residents exert considerable pressure to ensure that these conditions are met.

Anyone who wants to continue operating an existing quarry runs into difficulties, and bringing a new quarry into operation now proves to be a tedious, sometimes even hopeless, process in many countries. Optimum extraction of the available reserves in the existing quarries has therefore become essential if long-term operation is to be assured. The technique generally used in this situation starts from the model of the existing deposit, in which the data are kept up to date through regular analyses of samples taken during the extraction. Computer-aided simulation methods ensure continued quality of the material extracted and also provide information about the remaining reserves [1].

Extraction by scraping and cutting is gradually gaining acceptance alongside the classical methods of extraction by blasting; this applies not only to soft rock but also to limestone/marl strata in which the pressure resistance can be as high as 20 MPa. In this case blasting explosives are only used for shatter blasting [2, 3].

Another important task is the regular replacement of all existing equipment, such as crushers, loaders, crawlers and transport vehicles, by larger models. As far as transport vehicles are concerned a payload of between 60 and 100 tonnes has proved satisfactory. The manufacturers are anxious to improve the reliability of the mobile equipment. There is

therefore much to recommend equipment with improved facilities for fault diagnosis and reduced downtime for maintenance work. Improvements in efficiency make it possible to lower labour costs; they also favour daytime extraction work, which greatly reduces the disadvantages for the employees and the nuisance to the nearby residents [4].

3. Storage of the products

Storage of the products during the cement manufacturing process essentially fulfils the following functions:

— It smooths out fluctuations in relation to material throughput and makes it possible to avoid 24 hour operation in the auxiliary plants (quarry, despatch, etc.).

— It ensures that the products are homogenized during the respective phases of the process [5].

In the context of environmental protection regulations the cement manufacturers have seen themselves gradually forced to change from open storage to closed stores with dedusting systems. There has been little in the way of development in recent years in the area of plant for prehomogenization of raw materials. At present the market is divided into the two following technologies:

— Linear prehomogenization, in which the blending bed consists of superimposed layers or bands, which is intended to prevent segregation.

— Circular prehomogenization, in which the blending bed is configured to avoid irregularities at the outer regions.

Until now raw meal has been stored in silos in which homogenization is achieved by admitting compressed air and by recirculation. The capital costs and the operating costs (electricity, maintenance, etc.) are both very high for this type of storage. In some plant units built recently the function of homogenization is no longer taken into account, and the storage of raw meal is reduced to a limited buffer (3 to 4 hours' kiln operating time). The reliability of the raw grinding plant must then be just as high as that of the rotary kiln!

Clinker is stored in silos the total capacities of which have been increased in recent years from 25 000 tonnes to 60 000 – 80 000 tonnes. About 80 % of the quantity stored can normally be recovered without using auxiliary equipment [6]. In addition to the usual designs of these silos, e.g. steel or concrete construction, there is now some experience with so-called Dome® silos in which an inflatable plastic dome is lined with gunned concrete.

4. Materials transport

The transport of extensive quantities of products in lump and powder form is an important factor in the cement manufacturing process. Even though the total investment for this sector is limited, selection of the technology requires great care if the reliability of the entire production line is not be thrown off balance. Developments in recent years seem to have been distinguished by two main trends:

— The search for the optimum "layout" of the individual plant sectors, in which the number of conveying systems is restricted to a minimum.

  This concept has two aims, namely to reduce capital costs and improve reliability. It appears that it is primarily in newly erected operating units that these objectives can be achieved, but the search for the optimum solution should also be the rule for every section of the plant in which major modernization work is being carried out.

— Replacement of the pneumatic transport of powdered materials by a mechanical transport system.

  The capital costs of a belt conveyor or a bucket elevator are in fact higher than for an air-lift plant, but the difference in operating costs (electricity, dedusting, heat consumption) offsets the initially higher costs [7].

In view of the increasing automation of the plant and the reduction in the work force it is also necessary to solve the

der gesamten Produktionslinie nicht aus dem Gleichgewicht gebracht werden soll. Zwei Haupttendenzen scheinen die Entwicklung der letzten Jahre geprägt zu haben:

— Die Suche nach dem optimalen „Layout" der einzelnen Werksbereiche, wobei die Zahl der Förderanlagen auf ein Minimum beschränkt wird.

Dieses Konzept verfolgt zwei Ziele, nämlich die Reduzierung der Investitionen und die Verbesserung der Zuverlässigkeit. Diese Zielsetzung scheint in erster Linie für die neu entstehenden Betriebseinheiten realisierbar zu sein. Die Suche nach der optimalen Lösung sollte aber auch für jeden Werksbereich die Regel sein, in dem eine einschneidende Modernisierungsmaßnahme durchgeführt wird.

— Der Ersatz des pneumatischen Transports von pulverförmigen Stoffen durch einen mechanischen Transport.

Die Investitionskosten für ein Gurtförderer- oder ein Becherwerk sind zwar immer noch höher als für eine „Air-lift"-Anlage, aber der Unterschied in den Betriebskosten (Elektrizität, Entstaubung, Wärmeverbrauch) ermöglicht einen rentablen Ausgleich zu den anfänglichen Mehrkosten [7].

Darüber hinaus muß angesichts der zunehmenden Automatisierung der Anlagen und der Reduzierung der Arbeitskräfte das Problem des Überlaufens bei den Gurtband- und Plattenförderern gelöst werden. Positive Ergebnisse werden durch die Optimierung des Verhältnisses Geschwindigkeit/Breite des Bandes erreicht. Einige Lieferanten bieten bereits geschlossene Transportbänder in Form dichter Rohrleitungen an. Diese Anlagen sind vor allem für nicht geradlinig verlaufende Transportwege vorteilhaft.

5. Verpackung und Versand der Endprodukte

Der Bereich Zementversand ist eine der Schnittstellen zwischen Produkthersteller und -anwender und bedarf deshalb der besonderen Aufmerksamkeit. Eigenschaften und Anforderungen an die Qualität der Produkte werden künftig von der europäischen Zementnorm festgelegt, die derzeit als Vornorm (ENV 197, Teil 1) besteht.

Der Absatz von losem Zement setzt sich in Europa zunehmend durch. In einigen Ländern übersteigt der Absatz von losem Zement bereits 80 %. Die angewandten Technologien in diesem Bereich haben sich kaum weiterentwickelt. Dagegen erfreut sich die Selbstbedienung immer größerer Beliebtheit. Die Fahrer der Transportfahrzeuge verfügen über Magnetkarten, die alle erforderlichen Informationen in kodierter Form enthalten. Verwechslungen der Sorten sind damit praktisch ausgeschlossen, und es ist unmöglich, Produkte von mangelhafter Qualität zu liefern.

Obwohl der Versand in Säcken bei einer Reihe von Werken nur einige wenige Prozent des Gesamtabsatzes ausmacht, erfordert gerade dieser Bereich umfangreiche Investitionen. Im allgemeinen werden Rotopack-Maschinen mit einem Durchsatz zwischen 50 und 250 t/h verwendet. Das Verfahren der Gewichtskontrolle der Säcke ist präziser geworden. Die Gewichtskontrolle erfolgt entweder während des Abfüllens oder mittels einer Bandwaage. Neue Wege beschreiten die Hersteller durch den Ersatz von Turbinen mit horizontaler gegen solche mit vertikaler Achse. Der Vorteil solcher Veränderungen besteht darin, daß die relative Geschwindigkeit zwischen Zement und mechanischen Teilen reduziert wird, was einen geringeren Verschleiß zur Folge haben soll [8].

Die Entwicklung der Steckautomaten schreitet kontinuierlich voran. Hier stehen sich zwei Technologien gegenüber. Die erste basiert auf einem mechanischen System der Sackbearbeitung durch Ansaugen, die zweite auf einem System, in dem die Säcke mit einer pneumatischen Einrichtung aufgesteckt werden. Für die Auslieferung an die Zementwerke werden die Säcke auf Paletten oder Rollen abgepackt [9].

Bei den Palettiermaschinen sind keine besonderen technologischen Neuerungen festzustellen. Die Entwicklung hin zum 25-kg-Sack ist bisher langsam verlaufen. Es ist jedoch

problem of overspill at belt and apron conveyors. Good results are being achieved by optimizing the ratio of speed to width of the belt. Some suppliers are already offering closed transport belts in the form of tightly closed tubes. These systems are a particular advantage for transport routes which are not straight.

5. Packaging and despatch of the end products

The cement despatch sector forms an interface between product manufacturer and user, and therefore requires particular attention. Product properties and quality specifications will in the future be specified by the European cement standard, which at the moment exists as a prestandard (ENV 197, Part 1).

Sales of bulk cement are gaining ground in Europe. In some countries the sales of bulk cement already exceed 80 %. There has been very little further development in the technologies used in this area. On the other hand self-service is enjoying ever greater popularity. The driver of the transport vehicle is provided with magnetic cards which contain all the necessary information in coded form. This practically eliminates any confusion over product types and it is impossible to supply products of deficient quality.

Although despatch in bags only makes up a few percent of the total sales in many works this sector requires extensive investment. Rotary packing machines with throughputs between 50 and 250 t/h are generally used. The method of checking the weights of the bags has become more precise. The weight checking is carried out either during the filling process or by means of a belt weigher. The manufacturers are breaking new ground by using impellers with horizontal axes as opposed to vertical ones. The advantage of this change lies in the fact that the relative velocity between the cement and the mechanical parts is reduced, which should result in less wear [8].

Development of automatic applicators continues to make progress. Here there are two contrasting technologies. The first is based on a mechanical system of handling the bag by suction, and the second on a system in which the bags are applied by pneumatic equipment. The bags are packed on pallets or in reels for delivery to the cement works [9].

No particular technological innovations are to be found in the palletizing machines. The development towards 25 kg bags has moved slowly so far, but it is expected that this development will accelerate. The plants are often provided with equipment which draws an elastic film over the pallet. This film does give rise to additional costs, but it improves the stability of the pallet and increases the protection against the ingress of moisture during unfavourable weather conditions.

Until now cement works have been planned so that a limited number of products can be supplied in great quantities. The customers' demands, which have been increasing for several years, for specific products and the opportunities resulting from the new European standard to use greater numbers and quantities of interground additives has led to the situation where more and more cement manufacturers are going over to the installation of mixing plants. Increasing numbers of compartmentalized silos in which the different mixing components and the end products are stored have also been built recently. In some cases the plants are even provided with the option of bulk supply and with mini-pallet packing machines [10].

6. Auxiliary systems for water, air and power

6.1 Water

The general establishment of the dry process in clinker production saw a great decrease in the demand for water in cement works. Nowadays water is used as a coolant for mechanical equipment wherever the use of air is not pos-

zu erwarten, daß sich diese Entwicklung beschleunigen wird. Die Anlagen sind oft mit Geräten ausgestattet, die über die Palette eine elastische Folie ziehen. Diese Folie verursacht zwar Mehrkosten, verbessert jedoch die Stabilität der Palette und erhöht den Schutz vor Feuchtigkeitszutritt bei ungünstigen Witterungsverhältnissen.

Bisher waren Zementwerke so angelegt, daß eine begrenzte Anzahl von Produkten in großen Mengen geliefert werden konnte. Die seit mehreren Jahren steigende Nachfrage der Kunden nach spezifischen Produkten und die auf der Grundlage der neuen europäischen Norm entstandene Möglichkeit, Zumahlstoffe in größerer Zahl und Menge zu verwenden, hat dazu geführt, daß die Zementhersteller mehr und mehr dazu übergehen, Mischanlagen zu installieren. Neuerdings werden auch immer mehr unterteilte Silos aufgestellt, in denen die verschiedenen Mischkomponenten und die Endprodukte gelagert werden. In einigen Fällen werden die Anlagen sogar mit der Möglichkeit zur Loseabgabe sowie durch Mini-Palettpackmaschinen ergänzt [10].

6. Hilfsanlagen für Wasser, Luft und Strom

6.1 Wasser

Mit der Durchsetzung des Trockenverfahrens in der Klinkerproduktion ging der Bedarf an Wasser in den Zementwerken erheblich zurück. Wasser wird heute als Kühlmittel für mechanische Ausrüstungen dort verwendet, wo der Einsatz von Luft nicht möglich oder zu teuer ist. Die meisten Werke sind mit einem doppelten Wasserversorgungsnetz ausgestattet,

— einem Trinkwassernetz, das die Versorgung der Mitarbeiter und einiger sehr anfälliger Geräte sichert, bei denen die Wasserqualität eine wichtige Rolle spielt, und

— einem Brauchwassernetz für den restlichen Bedarf.

Die zunehmenden Umweltschutzauflagen und manchmal hohen Wasserkosten haben eine rentable Investition in ein geschlossenes Brauchwassersystem ermöglicht. Das Wasser wird vor Eintritt in das geschlossene System z.B. mit Anti-Schaum-Mitteln oder Entkalkungsmitteln behandelt. Ein Kühlturm sichert die Abkühlung des Wassers nach Gebrauch. Ein modernes Zementwerk verfügt darüber hinaus über eine Kläranlage für Abwasser und Regenwasser. Diese Anlage muß so bemessen sein, daß sie die anfallende Verschmutzung bewältigen kann, insbesondere bei einem Unfall.

6.2 Druckluft

Der Bedarf an Druckluft in einem Zementwerk kann sehr unterschiedlich sein. Der Bedarf läßt sich in drei Hauptgruppen unterteilen:

— Förderluft (2 bis 3 bar),

— Steuerluft (6 bis 8 bar),

— Reinigungsluft (8 bis 12 bar).

Die Erzeugung von Druckluft ist sehr energieaufwendig. Deshalb sind in einem modernen Werk die Druckluftsysteme getrennt. Die Erzeugung von Druckluft erfolgt dezentral, wodurch der Druckverlust im Netz begrenzt wird. Die Verwendung von leistungsfähigen Schraubenkompressoren hat sich allgemein durchgesetzt. Die Steuerluft muß einer speziellen Behandlung unterzogen werden. Sie wird über ein Kühlsystem gekühlt, getrocknet und zusätzlich entölt. Demgegenüber wird die Förderluft soweit wie möglich durch den Einsatz mechanischer Fördersysteme ersetzt, deren Betriebskosten weniger hoch sind.

6.3 Strom

In den meisten Ländern, in denen ein öffentliches Stromversorgungsnetz mit genügend hoher Kapazität zur Verfügung steht, beziehen die Zementhersteller ihre elektrische Energie von öffentlichen Stromerzeugern. Falls das nicht möglich ist, wird die Versorgung von einer alternativen Einrichtung mit entsprechenden Leistungsmöglichkeiten übersible or is too expensive. The most works are equipped with a dual water supply system,

— a drinking water network, which supplies the employees and some very sensitive equipment in which water quality plays an important part, and

— a service water network for the other requirements.

The increasing environmental protection regulations, and in some cases high water costs, have made it possible for investment in a closed service water system to be profitable. Before entering the closed system the water is treated with, for example, anti-foaming or anti-liming agents. A cooling tower cools the water after use. A modern cement works also has a clarification plant for waste water and rain water. This plant must be dimensioned so that it can deal with the fouling which occurs, especially in the case of an accident.

6.2 Compressed air

The need for compressed air in a cement works can vary considerably. The demand can be divided into three main groups:

— conveying air (2 to 3 bar)

— control air (6 to 8 bar)

— cleaning air (8 to 12 bar).

The production of compressed air is very energy-intensive. The compressed air systems in a cement works are therefore separate, and the production of compressed air is decentralized to limit the pressure drop in the network. The use of efficient screw compressors has become generally established. Control air must be submitted to special treatment. It is cooled in a cooling system, dried and also de-oiled. Conveying air, on the other hand, is as far as possible replaced by the use of mechanical conveying systems with lower operating costs.

6.3 Power

In the majority of countries in which a public power supply system of sufficiently high capacity is available the cement manufacturer obtains his electrical energy from the public power producers. If this is not possible the supply is provided by alternative equipment of appropriate capacity. Electrical energy is being supplied increasingly at high voltage (60 to 110 kV), which reduces the losses in the system.

In some countries the supply contract provides for price adjustments depending on the time of year and of day. It can therefore prove advantageous if some of the equipment or plant (crushers, raw mills, cement mills) are over-dimensioned so that they do not have to be operated at peak tariff times. Another newly developed solution is the purchase of auxiliary generators with capacities of a few megawatts, which in the critical periods take over part of the power supply like emergency generators. As the situation often varies sharply from country to country, and can even be very different from one location to another within a country, it is not possible to give a detailed review of the subject. The rising power costs — in some works they are even higher than the fuel costs — require new schemes to make the best possible use of the supply contracts [11]. Within the works the individual areas are supplied at medium voltage, generally 5 to 6 kV. In modern electrical switch rooms the use of special equipment has further improved the safety of the operating personnel.

7. Wear

Continuous contact between abrasive products, e.g. raw meal, clinker or cement, and mechanical elements leads inevitably to wear. However, for some years the plant manufacturers and university scientists have been carrying out investigations to develop suitable materials for each specific application. In the majority of cases it is now possible to confine the work to preventive maintenance, and the repair costs can be reduced to a minimum [12].

nommen. Die Lieferung von elektrischer Energie erfolgt dabei immer häufiger über Hochspannung (60 bis 110 kV), wodurch die Verluste im Netz verringert werden.

In einigen Ländern sehen die Lieferverträge Preisregelungen in Abhängigkeit von der Jahres- und Tageszeit vor. Es kann sich deshalb als vorteilhaft erweisen, wenn einige Ausrüstungen bzw. Anlagen (Brecher, Rohmühlen, Zementmühlen) größer dimensioniert sind, damit sie zu tariflichen Spitzenzeiten nicht in Betrieb genommen werden müssen. Eine weitere neu entwickelte Lösung ist die Anschaffung von Zusatzgeneratoren mit einer Leistung von einigen Megawatt, die in den kritischen Zeiten einen Teil der Stromerzeugung wie Notstromaggregate übernehmen. Da die Situation von Land zu Land oft stark variiert und selbst innerhalb eines Landes von Standort zu Standort sehr unterschiedlich sein kann, ist es unmöglich, hierzu einen detaillierten Überblick zu geben. Die steigenden Stromkosten, in manchen Werken liegen sie sogar über den Brennstoffkosten, erfordern neue Konzepte, die eine optimale Ausnutzung der Lieferverträge ermöglichen [11]. Innerhalb der Werke erfolgt die Versorgung der einzelnen Bereiche über Mittelspannung mit im allgemeinen 5 bis 6 kV. In modernen Elektroräumen wurde durch besondere Einrichtungen die Sicherheit des Bedienpersonals weiter verbessert.

7. Verschleiß

Der ständige Kontakt zwischen den Produkten mit abrasiven Eigenschaften wie Rohmehl, Klinker oder Zement und Maschinenelementen führt unweigerlich zu Verschleißerscheinungen. Seit einigen Jahren jedoch werden von Anlagenherstellern und Wissenschaftlern an Universitäten Untersuchungen mit dem Ziel durchgeführt, für jeden spezifischen Einsatz das geeignete Material zu entwickeln. Es ist heute in der Mehrzahl der Fälle möglich, sich auf eine vorbeugende Wartung zu beschränken, wodurch die Instandhaltungskosten auf ein Minimum reduziert werden können [12].

Einige der neueren Anlagen wie die Rollenpressen, die sowohl für die Vorzerkleinerung als auch für die Endzerkleinerung des Klinkers verwendet werden, sind noch nicht in allen Fällen ausgereift. Die Betreiber stehen häufig noch vor der Schwierigkeit, die Rollen bei starkem Verschleiß durch Klinker bereits nach wenigen tausend Einsatzstunden neu aufzuschweißen. Bisher wurden verschiedene Methoden und zahlreiche Oberflächenprofile getestet, ohne daß jedoch eine völlig zufriedenstellende Lösung gefunden worden ist. Außerdem werden sehr häufig Risse auf den Beanspruchungsflächen festgestellt, die teilweise sogar zum vollständigen Bruch führten [13]. Durch eine entsprechende Oberflächengestaltung und neue Werkstoffe könnte die künftige Entwicklung positiv beeinflußt werden [14, 15].

8. Wartung

Die Wartung der Anlagen ist zwingend erforderlich, wenn ein regelmäßiger Betrieb unter optimalen wirtschaftlichen Gesichtspunkten gewährleistet werden soll. Deshalb wird der Konzeption und Durchführung einer planmäßigen Wartung immer größere Bedeutung beigemessen [12]. Die Entwicklung neuer Kontroll-, Meß- und Regelungssysteme zwingt die Instandhaltungsabteilungen dazu, sich neue Kenntnisse auf dem Gebiet der Elektronik und Automatisierungstechnik anzueignen. Ein Großteil der modernen Zementwerke ist mittlerweile mit einem elektronischen EDV-Leitsystem ausgerüstet, das genaue Daten über das Verhalten der Anlagen liefert und die Informationen aus den regelmäßigen Inspektionen sinnvoll ergänzt [16]. Für den Betreiber führt das zu einer höheren Zuverlässigkeit der Einzelausrüstungen, was letztendlich zur Qualitätssicherung des Endproduktes beiträgt.

9. Reinigung

Die zunehmend zentrale Steuerung von Maschinen und Anlagen und der damit einhergehende Personalabbau erfor-

Some of the more recent systems, such as high-pressure grinding rolls, which are used both for preliminary comminution and for final comminution of the clinker, are not yet in all cases fully developed. Where there is severe wear from the clinker the operators are still often faced with the problem of hard-facing the rollers again after only a few thousand operating hours. Various methods and large numbers of surface profiles have already been tested, but without finding a fully satisfactory solution. Cracks are also often found in the working surfaces, and in some cases these have led to complete breakage [13]. Appropriate surface configurations and new materials could have a beneficial effect on future development [14, 15].

8. Maintenance

Plant maintenance is absolutely essential from the point of optimum cost-effectiveness if regular operation is to be assured. Increasing importance is therefore being attached to the planning and implementation of planned maintenance [12]. The development of new checking, measuring and control systems is forcing the maintenance departments to make use of new findings in the fields of electronics and automation technology. A great many modern cement works are now equipped with electronic EDP systems which supply accurate data about the behaviour of the plants and provide an appropriate supplement to the information from routine inspections [16]. For the operator this leads to greater reliability of the individual items of equipment, which ultimately contributes towards quality assurance of the end product.

9. Cleaning

The increasingly centralized control of machines and plants and the accompanying reduction in personnel call for logistics which include continuous checking and cleaning of individual parts of the plant. As far as possible, top priority must clearly be given to eliminating the sources of dirt. Production employees must be urged not to accept spillage and overflow of material as an unalterable fact of life. It must be a matter of course for the maintenance personnel to leave the place clean after any maintenance work. In all cases the employees must take part in the analysis of any malfunctions and sources of error and make suggestions as to how they can be remedied.

Compressed air blasting devices, which are sometimes used inappropriately and in excessive numbers, have been developed for cleaning equipment (hoppers, chutes, cyclone preheaters). They are required because of the need for uninterrupted material flow. These needs should be taken into account at the plant's planning stage. There are now also methods available for assessing the flow properties of bulk materials which can be used for continuous monitoring, especially in the area of storage of finished product [17].

A works must be designed so that it is easy to clean. So far little attention has been paid to this aspect of industrial planning, but it must be given more emphasis in the future. The works areas must also be provided with equipment which makes it possible to clean by vacuum. A semi-mobile network of fixed pipes and collecting containers in every part of the works and a diesel-electric vacuum unit appear at the moment to be the most cost-effective solution. The products collected in the cleaning system are reintroduced into the production process at some suitable point. This achieves the dual objective of quality assurance with respect to the end products and limitation of the quantities of waste which have to be disposed of.

10. Layout

A cement works is an industrial plant which must be in tune with its surroundings. For some considerable time now the widely differing parameters such as climatic conditions,

dern eine Logistik, die auch eine laufende Kontrolle und Reinigung einzelner Anlagenteile einschließt. Selbstverständlich muß es das oberste Ziel sein, die Quellen der Verschmutzung so weit wie möglich zu beseitigen. Die Mitarbeiter der Produktion müssen angehalten werden, das Aus- und Überlaufen von Material nicht als unabänderbare Gegebenheit hinzunehmen. Dem Wartungspersonal muß es selbstverständlich sein, nach jeder Wartungsmaßnahme den Einsatzort sauber zurückzulassen. In allen Fällen müssen sich die Mitarbeiter an der Analyse auftretender Fehlfunktionen und Störquellen beteiligen und Vorschläge zu deren Abhilfe vorlegen.

Für die Reinigung der Ausrüstungen (Bunker, Schurren, Zyklonwärmeaustauscher usw.) wurden Druckluftstoßgeräte entwickelt, die teilweise unzweckmäßig und zu zahlreich angewandt werden. Die Notwendigkeit ihres Einsatzes ergibt sich dabei aus der Forderung nach einem ungestörten Materialfluß. Diese Anforderungen sollten bereits im Planungsstadium von Anlagen berücksichtigt werden. Darüber hinaus stehen heute auch Verfahren zur Beurteilung der Fließeigenschaften von Schüttgütern zur Verfügung, die zur laufenden Überwachung, insbesondere im Bereich der Lagerung von Fertiggut, herangezogen werden können [17].

Ein Werk muß so konzipiert werden, daß es leicht zu reinigen ist. Dieser planungstechnische Aspekt hat bislang noch wenig Beachtung gefunden, sollte aber in Zukunft mehr berücksichtigt werden. Außerdem müssen die Werksbereiche mit Ausrüstungen ausgestattet werden, die eine Reinigung durch Absaugen ermöglichen. Ein halb-mobiles Netz, bestehend aus festen Rohren und Sammelbehältern in jedem Werksbereich, sowie ein diesel-elektrisches Absaugaggregat scheinen im Moment die kostengünstigste Lösung zu sein. Die in der Reinigung anfallenden Produkte werden an geeigneter Stelle wieder in den Produktionsprozeß eingeschleust. Damit ist das doppelte Ziel der Qualitätssicherung im Hinblick auf die Endprodukte und Begrenzung der zu entsorgenden Abfallmengen erreicht.

10. Layout

Ein Zementwerk ist eine industrielle Anlage, die sich der Umwelt anpassen muß. Seit geraumer Zeit werden daher in die Planung die unterschiedlichsten Parameter wie Klimabedingungen, Bodenbeschaffenheit und vorhandene Infrastruktur bereits auf dem Reißbrett mit einbezogen. Sie ergänzen die verfahrensspezifischen Anforderungen sowie die Gegebenheiten der Rohstofflagerstätten. Zunehmend gewinnen auch andere Aspekte an Bedeutung, die bisher unberücksichtigt geblieben sind. Hierzu zählen insbesondere Maßnahmen zur Bekämpfung von Emissionen durch Lärm und Erschütterungen.

Bei der Konzeption neuer Produktionseinheiten erfolgt die Anordnung der einzelnen Werksbereiche unter Berücksichtigung des bereits vorgestellten Lastenhefts, wobei der Kostenaufwand bezüglich der Investition sowie des späteren Betriebs so gering wie möglich gehalten werden soll. Die Wegverkürzung zwischen den einzelnen Werksbereichen, die Reduzierung interner Förderwege, die zweckmäßigste Lage der Produktions- und Wartungsgebäude zueinander werden zukünftig noch ausführlicher untersucht werden müssen.

Das Problem der Werksgestaltung ist demgegenüber weitaus komplexer, wenn es um die Modernisierung bereits bestehender Produktionseinheiten geht, da die Möglichkeiten dabei im allgemeinen sehr begrenzt sind. Zudem sind Investitionen aufgrund finanzieller Zwänge oft nur in größeren zeitlichen Abständen möglich. Im Investitionsfalle ist es wichtig, daß vor Beginn der Ausführung ein Gesamtentwurf für den Standort erstellt wird, der als „Grundstruktur" bezeichnet werden könnte. Ausgangsbasis für die Erstellung dieses Entwurfs ist die gemeinsame Erörterung aller technischen und wirtschaftlichen Aspekte durch die zuständigen Abteilungen, was wiederum eine Voraussetzung für die anschließende Festlegung von Qualität und Quantität der herzustellenden Produkte ist. Anschließend folgt die

nature of the soil and existing infrastructure have therefore been included right from the drawing board stage. They supplement the process-specific requirements and the facts relating to the raw material deposits. Other aspects which previously remained unconsidered are also becoming more important. In particular, these include measures to combat the emission of noise and vibration.

When new production units are being designed the individual works areas are laid out to take account of the requirements already mentioned, in which the expenditure in relation to capital investment and subsequent operation should be kept as low as possible. Shortening the distances between the individual works areas, reducing the internal transport distances, and the most appropriate positions for the production and maintenance buildings in relation to one another, will have to be investigated even more thoroughly in the future.

In comparison, the problem of the works configuration is far more complex when it involves the modernization of existing production units, as in this case the options are usually very restricted. In addition, because of financial constraints, investment is often only possible at fairly long time intervals. In the case of capital projects it is important that an overall plan for the site, which could be termed the "basic structure", is drawn up before starting the design. The starting point for drawing up this plan is a joint discussion of all the technical and economic aspects by the responsible departments, which in turn is a precondition for subsequent specification of the quality and quantity of the products to be produced. After completion of the plan it should be possible to identify the development of the works for a period of 5 to 10 years. At the time when it is carried out each capital project must fit in with the objectives of the plan.

11. Conclusion

The plant and equipment installed in a cement works are of great importance for the overall economic results of the cement manufacturing process. It is of fundamental importance for success to have a clear overall concept for the utilization of the site and to stick to the conceptual guide lines drawn up for the site. In addition to harmonious integration of the works into its surroundings, it is also important to comply with relevant environmental protection regulations and to eliminate possible sources of nuisance. Finally, it is also essential to make the best possible use of the new plant and equipment which are available on the market; the general objective should be to increase the capacities and reduce the maintenance work.

Literature

[1] Baumgartner, W.: Computergestützte Rohstoffplanung. Vortrag zum VDZ-Kongreß '93, Fachbereich 6.

[2] Steinberg, H., und Hoffmann, R.: Einsatz eines Schaufelradbaggers im Kalksteintagebau. Zement-Kalk-Gips 47 (1994) No. 2, pp. 104–108.

[3] Boßmann, W.: Abbau mit dem Hydraulikbagger als Alternative zur Gewinnungssprengung. Vortrag zum VDZ-Kongreß '93, Fachbereich 4.

[4] Fleck, H.: Einschichtbetrieb eines Steinbruchs mit einem Durchsatz von 1,8 Mio. t/Jahr. Vortrag zum VDZ-Kongreß '93, Fachbereich 6.

[5] Tschudin, M.: On-line-Kontrolle des Mischbettaufbaus mittels PGNAA in Ramos Arizpe (Mexiko). Vortrag zum VDZ-Kongreß '93, Fachbereich 2.

[6] Heine, W.: Energiesparende Transportsysteme für Zement und Rohmehl – Vergleichende Betrachtung von mechanischen und pneumatischen Transportanlagen. Vortrag zum VDZ-Kongreß '93, Fachbereich 6.

[7] Ehlers, K.: Energieersparnis bei der Ofenbeschickung durch Einsatz von Becherwerken. Vortrag zum VDZ-Kongreß '93, Fachbereich 6.

räumliche Zuordnung der einzelnen Funktionen. Nach Abschluß des Entwurfs sollte für einen Zeitraum von 5 bis 10 Jahren die Entwicklung des Werks erkennbar sein. Jede Investition muß im Moment ihrer Durchführung mit der Vorgabe des Entwurfs übereinstimmen.

11. Schlußfolgerungen

Die in einem Zementwerk installierten Anlagen und Ausrüstungen sind für das wirtschaftliche Gesamtergebnis der Zementherstellung von großer Bedeutung. Grundlegend für den Erfolg sind eine klare Gesamtkonzeption für die Nutzung des Standorts und die Einhaltung der für diesen Standort erstellten Konzeptionslinie. Wichtig ist außerdem die harmonische Einbindung des Werkes in seine Umwelt, die Einhaltung geltender Umweltschutzauflagen sowie die Beseitigung möglicher Belästigungsquellen. Schließlich müssen auch die neuen Anlagen und Ausrüstungen, die auf dem Markt verfügbar sind, optimal genutzt werden, wobei jeweilige Kapazitätserhöhungen und die Reduzierung von Wartungsmaßnahmen als allgemeine Zielstellungen anzusehen sind.

[8] Spiess, J.: Strömungsverhältnisse in Turbinenpackern. Zement-Kalk-Gips 47 (1994) No. 3, pp. 142–145.

[9] Festge, R.: Heutiger Stand der vollautomatischen Verpackungstechnik von Baustoffen aller Art. Vortrag zum VDZ-Kongreß '93, Fachbereich 6.

[10] Thier, B.: Mischtechnik im Werk II der ANNELIESE Zementwerke AG, Ennigerloh. Vortrag zum VDZ-Kongreß '93, Fachbereich 6.

[11] Hasler, R.: Erfahrungen der „Holderbank" mit automatischer Prozeßführung mittels HIGH LEVEL CONTROL. Vortrag zum VDZ-Kongreß '93, Fachbereich 2.

[12] Patzke, J.: Geplante Instandhaltung. Zement-Kalk-Gips 47 (1994) No. 3, pp. 128–132.

[13] Ellerbrock, H.-G.: Gutbett-Walzenmühlen. Zement-Kalk-Gips 47 (1994) No. 2, pp. 75–82.

[14] Wahl, W.: Neue Möglichkeiten zum Verschleißschutz in der Zementindustrie. Vortrag zum VDZ-Kongreß '93, Fachbereich 6.

[15] Patzelt, N.: Verschleißschutzalternativen für die Beanspruchungsflächen von Gutbett-Walzenmühlen. Zement-Kalk-Gips 47 (1994) No. 2, pp. 83–86.

[16] Allenberg, B.: Rationalisierte Instandhaltung von kontinuierlichen Dosiereinrichtungen für Schüttgüter unter Berücksichtigung der Qualitätssicherung. Vortrag zum VDZ-Kongreß '93, Fachbereich 6.

[17] Maltby, L.P., Enstad, G.G., und Stoltenberg-Hansson, E.: „Uniaxial tester" – Ein neues Gerät zur Beurteilung der Fließeigenschaften von Zement. Vortrag zum VDZ-Kongreß '93, Fachbereich 6.

Geplante Instandhaltung*)
Planned maintenance*)
Programmation de la maintenance

Planificación del mantenimiento

Von **J. Patzke** und **K.-A. Krause,** Lägerdorf/Deutschland

Geplante Instandhaltung

Fachbericht 6.1 · Zusammenfassung – Konzept und Aufbau einer geplanten Instandhaltung sowie insbesondere deren Umsetzung im Betrieb erfordern die Auswahl einer geeigneten Software. Von grundsätzlicher Bedeutung ist außerdem, daß Instandhaltung, Materialwirtschaft und Einkauf als integrierte Einheit gesehen und die Schnittstellen zu den übrigen Werksbereichen klar definiert werden. Wesentliche Elemente der geplanten Instandhaltung sind zielgerichtete Inspektionen und systematisierte Wartungsarbeiten. Zielgerichtete Inspektionen führen zur Ermittlung des Istzustands der Maschinen und schaffen damit die wesentliche Voraussetzung für die geplante Instandhaltung als zustandsabhängige und vorbeugende Instandhaltungsstrategie. Systematisierte Wartungsarbeiten dienen der Bewahrung des Sollzustands. Die Erhöhung der Planbarkeit von Instandhaltungsmaßnahmen wird unterstützt durch eine übersichtliche Materialwirtschaft und informative Maschinen-Stücklisten, durch eine gute Werkstattorganisation sowie durch ein klares Auftragsabwicklungsschema mit verschiedenen Klassifizierungskriterien für die Aufträge. Damit können ohne großen Aufwand Großreparaturen geplant und Reparaturprogramme für einen längeren Zeitraum, z.B. als Wochenprogramme, zusammengestellt werden. Die Instandhaltungskostenrechnung als Grundlage der Maschinengeschichte sollte möglichst rasch und detailliert vorliegen, damit sie als weiteres Planungs- und Steuerungselement dienen kann. Mit diesem Konzept wurden bei Alsen-Breitenburg gute Erfahrungen gesammelt. Es führte insgesamt zu einer Erhöhung der Verfügbarkeit der Anlagen sowie zu einer Verbesserung der Kostensituation für die Instandhaltungsmaßnahmen.

Planned maintenance

Special report 6.1 · Summary – Designing and setting up a planned maintenance system, and especially implementing it, require the creation of suitable software. It is also of fundamental importance that maintenance, material management and purchasing are regarded as an integral unit, and that the interfaces with the other works departments are clearly defined. Carefully directed inspections and systematic maintenance work are important elements of the planned maintenance. The inspections determine the current state of the machines and therefore provide the essential requirement for planned maintenance as a condition-related and preventive maintenance strategy. Systematic maintenance work is used to keep the plant in the required state. A clearly laid out material management system and informative machine parts lists, good workshop organization, and a clear order handling scheme with various classification criteria for the orders all help to make the maintenance work easier to plan. This means that large repairs can be planned without great effort, and repair schedules can be compiled for longer periods, e. g. as weekly schedules. Maintenance costing, which forms the basis of machine history, should be available as rapidly and in as great a detail as possible so that it can serve as a further planning and control element. This scheme was a great success at Alsen-Breitenburg. Taken as a whole, it has led to an increase in the availability of the plant and to an improvement in the cost situation for the maintenance work.

Programmation de la maintenance

Rapport spécial 6.1 · Résumé – Le concept et la mise en oeuvre d'un plan de maintenance et notamment sa traduction dans les faits, exigent l'installation d'un logiciel approprié. Il est, en outre, d'une importance fondamentale que la maintenance, la gestion des stocks et le service achat soient considérés comme une unité intégrée, et que les interfaces vers les autres secteurs de l'usine soient clairement définis. Parmi les éléments essentiels de la maintenance envisagée citons les inspections spécifiques et les travaux d'inspection systématique. Les inspections spécifiques mènent au calcul de l'état réel des machines, et créent ainsi les conditions indispensables à l'inspection envisagée en tant que stratégie de maintenance préventive, dépendant de l'état du matérial. Les travaux d'entretien systématique servent à la conservation de l'état des valeurs de consigne. La planification accrue des travaux de maintenance est assistée à la fois par une gestion des stocks de conception claire et des nomenclatures de pièces machine fournies à titre d'information, mais aussi par une bonne organisation de l'atelier et, enfin, par un système de gestion des commandes clairement défini comportant divers critères de classification pour les commandes. Ainsi il est possible sans dépenses élevées de programmer des réparations importantes et de regrouper des programmes de réparation sur une longue période, comme p. ex. un programme hebdomadaire. Il convient de disposer rapidement et en détail du calcul des frais de maintenance comme

---

*) Überarbeitete Fassung eines Vortrages zum VDZ Kongreß '93, Düsseldorf (27. 9.–1. 10. 1993)
 Revised text of a lecture to the VDZ Congress '93, Düsseldorf (27.9.–1.10.1993)

base de l'histoire des machines, afin qu'il puisse servir d'élément de commande et de conduite complémentaires. Un tel concept a été appliqué avec succès chez Alsen-Breitenburg. Il a conduit, globalement, à une augmentation de la disponibilité des installations et à une amélioration de l'état des coûts pour les travaux d'entretien.

Informe de ramo 6.1 · Resumen – La concepción y organización de un mantenimiento planificado y, especialmente, su aplicación al servicio práctico, requiere el desarrollo de un software apropiado. Ademas, es de una importancia fundamental que el mantenimiento, la gestión de stocks y el servico de compras se consideren como unidad integral y que se definan claramente los interfaces hacia los demás sectores de la instalacion. Entre los elementos esenciales de un mantenimiento planificado cuentan las inspecciones específicas y los trabajos de mantenimiento sistemáticos. Las inspecciones específicas conducen a la determinación del estado real de las máquinas, creando así las condiciones necesarias para el mantenimiento planificado, como estrategia preventiva, en función del estado del material. Los trabajos de mantenimiento sistemáticos permiten preservar el estado deseado. La planificación de las medidas de mantenimiento se facilita mediante una buena gestión de stocks y listas detalladas de despiece de las máquinas, una buena organización de los talleres así como mediante un esquema bien claro de gestión de los pedidos, incluyendo diversos criterios de clasificación. De esta forma, se pueden planificar, sin mayores esfuerzos, reparaciones importantes y organizar programas de reparación de larga duración, por ejemplo programas de una semana. El cálculo de gastos de reparación, como base del historial de la máquina, debe estar disponible rápidamente y de forma detallada, con el fin de que pueda servir también de elemento de planificación y guía. Este concepto ha permitido a Alsen-Breitenburg adquirir buenas experiencias. En su conjunto, ha conducido a un aumento de la disponibilidad de las instalaciones y a una mejora de la situación de costes en relación con las medidas de mantenimiento.

Planificación del mantenimiento

1. Einleitung und Zielsetzung

Der wirtschaftliche Betrieb moderner Anlagen erfordert eine hohe Verfügbarkeit. Das trifft insbesondere auf die Steine- und Erden-Industrie zu, deren Produktionslinien im allgemeinen voll ausgelastet sind und der keine Standby-Anlagen zur Verfügung stehen.

Der Herstellprozeß des Zements stellt hohe Anforderungen an die Instandhaltung, weil Rohmaterial und Klinker verschleißintensiv und die Prozeßtemperaturen hoch sind. Zusätzlich haben die Belange der Arbeitssicherheit und des Umweltschutzes verstärkte Bedeutung erlangt.

In früheren Jahren war es gängige Praxis — mit Ausnahme der einmal im Jahr durchgeführten Grundreparatur — die Anlagen nur im Schadensfalle oder bei Ausfall zu reparieren. Das ist jedoch heute kein zufriedenstellendes Konzept mehr. Der hohe Anteil der Instandhaltungskosten an den Betriebskosten und die Notwendigkeit eines möglichst störungsfreien Betriebs zwingen dazu, die Instandhaltung dadurch zu optimieren, daß der Anteil der geplanten Instandhaltung möglichst groß wird. Damit wird die Instandhaltung zu einer Managementaufgabe mit der Zielsetzung, die Kosten zu minimieren und gleichzeitig eine hohe Verfügbarkeit der Anlagen zu gewährleisten.

2. Entwicklung eines Instandhaltungskonzeptes

Dieses Ziel ist nur dadurch zu erreichen, wenn für die Instandhaltung ein klares Konzept erarbeitet wird. Hierfür ist es notwendig, das gesamte Instandhaltungswesen zunächst werksspezifisch von Grund auf zu analysieren, zu überdenken und daraufhin ein verbessertes Konzept zu entwickeln. Bei Alsen-Breitenburg wurde dieser Weg schon frühzeitig beschritten. Ende 1982 wurde die neue Abteilung „Instandhaltungsplanung" (IHP) eingerichtet, deren Aufgabe zunächst darin bestand, die erforderlichen Maßnahmen für die Durchsetzung des Konzeptes gemeinsam mit der Werksleitung, den Instandhaltungsabteilungen, der Materialwirtschaft, der EDV-Abteilung und dem betriebswirtschaftlichen Dienst schrittweise zu koordinieren und durchzusetzen.

Die Strategie bestand darin, die eigenen Mitarbeiter von Anfang an in die Neugestaltung des Instandhaltungskonzeptes voll mit einzubeziehen. Dieses Vorgehen hat sich in der Praxis gut bewährt. Es wurden Teilziele definiert und

1. Introduction and objectives

Modern industrial plant requires a high level of availability to operate efficiently. This applies to the non-metallic minerals industry in particular, where production lines generally operate at full load, and there is no stand-by plant available.

The cement production process is most exacting in terms of maintenance, since the raw material and clinker are highly abrasive, and the process temperatures involved are very high. Health and safety and environmental considerations have also acquired greater significance in recent years.

It used to be normal practice to repair plant only once a year during the general overhaul, and when it broke down or was damaged. This is no longer an acceptable system of work. Since maintenance costs now constitute a large proportion of operating costs, and uninterrupted operation is an overriding priority, maintenance has to be carried out on a planned basis as far as possible. This makes maintenance a management task with the aim of minimizing cost and at the same time ensuring a high level of plant availability.

2. Developing a maintenance system

This aim can only be achieved if there is a clear maintenance strategy. This requires a fundamental analytical review of the individual maintenance system, and the development of an improved system. This course of action was followed at Alsen-Breitenburg at an early stage. A new "maintenance planning" department was set up in late 1982 with the initial aim of gradually coordinating and implementing the necessary action for carrying out the strategy together with works management, the maintenance departments, the materials management function, the EDP department and the industrial administration service.

The strategy was to fully involve our own employees from the start in establishing the new maintenance system. This approach proved very successful in practice. Intermediate objectives were defined and introduced in stages. These were expressed as the following targets for determining the actual condition of the machines:

— Systematization of maintenance work,
 such as lubrication, cleaning filters, drainage precautions.

schrittweise eingeführt. Im einzelnen waren das folgende Zielstellungen zur Ermittlung des Ist-Zustandes der Maschinen.

- Systematisierung von Wartungsarbeiten
  wie Schmierdienst, Filterreinigung, Gewässerschutz,
- Zielgerichtete Inspektionen
  Die Kenntnis des Ist-Zustandes ist die wesentliche Voraussetzung für die geplante Instandhaltung.
- Die geplante Instandhaltung
  versteht sich als zustandsabhängige und vorbeugende Instandhaltungsstrategie. Damit werden ad-hoc-Reparaturen weitestgehend reduziert, und die eigentlichen Instandhaltungsmaßnahmen werden planbar.
- Die effiziente Personaleinsatzplanung
  ermöglicht den wirtschaftlichen Einsatz eigener und fremder Werkstattkapazitäten, was durch die Klassifizierung der IH-Aufträge nach Prioritäten und Überwachung des Arbeitsablaufs unterstützt wird.
- Die termingerechte Ersatzteilbereitstellung
  führt zur Verringerung der Lagerhaltungskosten ohne negative Folgen für die Instandhaltung.
- Die Bereitstellung von Stücklisten
  ermöglicht den schnelleren Zugriff zu Magazinartikeln und Ersatzteilen und verringert Such-, Warte- und Wegezeiten.
- Die hohe Kostentransparenz
  ist wichtig für die Einhaltung des Budgets. Die IH-Kosten sollen schnell und detailliert vorliegen. Sie sind Grundlage für die
- Maschinengeschichte
  als Planungs- und Steuerungsmittel für Instandhaltungsvorhaben. Sie ist auch eine Hilfe bei Ersatzbeschaffungen, Schadensanalysen und bei der Schwachstellenbeseitigung.

3. EDV als Hilfsmittel

Es liegt auf der Hand, daß diese Teilziele nur mit Hilfe der EDV zu bewältigen sind, denn für die geplante Instandhaltung sind eine Vielzahl von Daten für Information und Kommunikation zu verarbeiten, was ohne EDV-Unterstützung einen unverhältnismäßig hohen Personalaufwand erfordern würde. 1983 wurde bei Alsen-Breitenburg mit der Verwirklichung des Konzeptes begonnen. Hierfür waren passende Software-Programme noch nicht auf dem Markt, so daß ein eigenes System geschaffen werden mußte. Diese Situation hat sich inzwischen geändert, da es eine Vielzahl von Software-Anbietern sowohl für PC-Systeme als auch für Großrechner gibt. So sind z.B. in der Zeitschrift „IH-Markt 1993" 36 Anbieter für Instandhaltungssoftware aufgeführt. Die geeignete Auswahl wird wesentlich erleichtert, wenn das jeweilige Unternehmen sich zunächst ein werkspezifisches Instandhaltungskonzept erarbeitet und ein EDV-Pflichtenheft zusammenstellt. Wichtig ist, daß Instandhaltung, Materialwirtschaft und Einkauf als integrierte Einheit gesehen werden – so wie es in **Bild 1** dargestellt ist – und daß die Schnittstellen zu Lohnbuchhaltung, Finanz- und Anlagenbuchhaltung sowie zu Produktion und zur Controlling-Abteilung klar definiert sind, damit es nicht zu Doppel- oder Mehrfacheingaben kommt. Von Anfang an sollte darauf geachtet werden, daß es gerade durch den Einsatz von EDV-Programmen zu einer Minimierung der Papierflut kommt.

Eingesetzt wird eine Zentralrechenanlage – früher IBM 38, jetzt AS/400. 1983 wurde für die Materialwirtschaft das IBM-Softwareprogramm MAS II eingesetzt und dieses um die Module für die Instandhaltung durch Eigenleistung ergänzt. Der Vorteil dieser Vorgehensweise besteht darin, daß die erforderlichen Module den Bedürfnissen direkt angepaßt werden konnten und von Anfang an voll genutzt wurden. Die einzelnen Schritte der geplanten Instandhaltung werden im wesentlichen am Beispiel des Vorgehens bei Alsen-Breitenburg aufgezeigt.

- Targeted inspections
  The key precondition for planned maintenance is to know the actual condition.
- Planned maintenance
  is defined as a context-related, preventive maintenance strategy. This substantially reduces ad hoc repairs, and make it possible to plan maintenance work proper.
- Efficient manpower deployment
  enables in-house and contract workshop capacity to be efficiently used, by means of prioritizing maintenance tasks and monitoring the progress of maintenance work.
- Timely provision of spare parts
  to reduce stockholding costs without reducing the efficiency of maintenance.
- Providing parts lists
  to give more rapid access to supply items and spare parts, and to reduce access time, waiting time and transport time.
- A high level of cost transparency
  is important for keeping to budget. Detailed maintenance costs should be promptly available. These costs are the basis for:
- Machine histories
  as a planning and control aid for maintenance projects. Machine histories also provide a basis for procuring replacements, damage analysis and eliminating weak points.

3. EDP support

It is evident that these objectives can be achieved only with the help of EDP, given the volume of data which has to be processed for information and communication purposes to

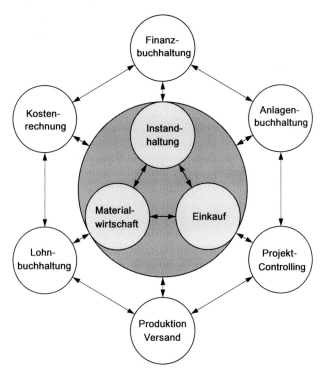

BILD 1: Geplante Instandhaltung: Schnittstellen mit anderen Unternehmensbereichen
FIGURE 1: Planned maintenance: Interfaces with other company departments

Finanzbuchhaltung	= financial accounting
Kostenrechnung	= cost accounting
Anlagenbuchhaltung	= fixed-asset accounting
Instandhaltung	= maintenance
Materialwirtschaft	= materials management
Einkauf	= purchasing
Lohnbuchhaltung	= payroll
Projekt-Controlling	= project control
Produktion Versand	= production despatch

4. Systematische Inspektion und Wartung

Die Planung von Instandhaltungsmaßnahmen ist nur dann möglich, wenn durch gezielte Inspektionen der aktuelle Ist-Zustand der Anlagen bekannt ist. Ein Beispiel für die Organisation von Inspektions- und Wartungsdiensten enthält **Bild 2**.

Nachdem die Maschinencodierung eingegeben ist, werden für jede Maschine die periodischen Arbeitsgänge nach Art und Frequenz festgelegt und am Bildschirm erfaßt. Die wöchentlich ausgedruckten Arbeitskarten werden an die jeweiligen Betriebsabteilungen verteilt, dort bearbeitet und zur Kontrolle an die Instandhaltungsplanung (IHP) rückgemeldet. Werden Schäden festgestellt, so wird ein Arbeitsauftrag geschrieben und an die IHP gegeben. Grundlage für die Festlegung der Arbeitsgänge sind

— Betriebsanweisungen und Herstellervorschriften,

— Erfahrungen der Betriebs- und Handwerkermeister,

— zunehmende eigene Erfahrungen der IHP-Mitarbeiter.

Hierbei ist es wichtig, daß für die einzelnen Arbeitsgänge ein vollständiger unverschlüsselter Text eingegeben werden kann, so daß der Anwender auf einen Blick Art und Umfang der durchzuführenden Arbeit erkennt. Bei Alsen-Breitenburg sind durch dieses System heute etwa 44 000 Jahresstunden für Inspektion, Wartung und Routinearbeiten verplant, davon etwa

— 3 000 Jahresstunden für mechanische Inspektion,
— 8 000 für Schlosser- und Elektrowerkstatt,
— 33 000 für die Produktion.

Es muß ausdrücklich darauf hingewiesen werden, daß die von den Produktionsabteilungen zu erbringenden Leistungen (insbesondere Routinekontrollen und Reinigung) im Rahmen ihrer Aufgaben liegen, d.h. also keinen zusätzlichen Personalaufwand bedeuten und miterledigt werden müssen. Für die systematischen Inspektions- und Kontrollarbeiten sind zwei zuverlässige und sachkundige Handwerker fest abgestellt, welche mit modernen Meßgeräten ausgerüstet sind, z.B.

— Stoßimpulsmeßgerät mit Dokumentations- und Analysensoftware zur Bewertung des Wälzlagerzustandes und der Schmierung,

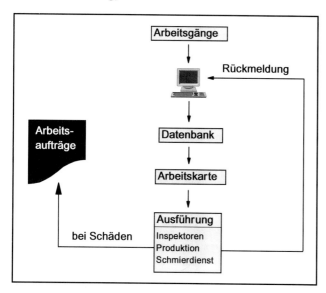

BILD 2: Ablaufschema für gezielte Inspektion und Wartung
FIGURE 2: Flow diagram for targeted inspection and maintenance

Arbeitsaufträge = job orders
Arbeitsgänge = operations
Rückmeldung = feedback
Datenbank = database
Arbeitskarte = job sheet
bei Schäden = if there is damage
Ausführung = implementation
Inspektoren = inspection
Produktion = production
Schmierdienst = lubrication

achieve planned maintenance, which would require inordinate manpower resources without the help of EDP. Implementation of this strategy at Alsen-Breitenburg was started in 1983. Since there were no suitable software programs on the market at that time, we had to develop our own system. This situation has now changed, with a variety of software on offer both for PC systems and also for mainframe computers. There are, for example, 36 suppliers of maintenance software listed in the periodical "IH-Markt 1993". It is much easier for a company to choose suitable software if it first works out a specific maintenance strategy for its own operation, and defines an EDP specification. It is important that maintenance, materials management and purchasing are seen as an integrated whole (as shown in **Fig. 1**), and that there are clearly defined interfaces with payroll, financial accounting and fixed asset accounting and with production and the controllership department, to eliminate duplication of input. It is essential to ensure from the beginning that EDP programs serve to minimize the flood of paperwork.

A central computer facility is used, previously an IBM 38, now AS/400. In 1983, IBM MASII software was used for material management, and the in-house maintenance module was added to it. The advantage of this approach was that the necessary modules could be tailored to our needs, and were fully utilised from the start. The planned maintenance process at Alsen-Breitenburg illustrates in principle the individual steps in planned maintenance.

4. Systematic inspection and maintenance

Maintenance activity can only be planned if the current status of plant is known from targeted inspections. An example of the organisation of inspection and maintenance services is given in **Fig. 2**.

After the machine coding has been entered, the periodic activities are defined for each machine by type and frequency, and input on screen. The work tickets printed out weekly are distributed to the various production departments, where they are processed and returned to maintenance planning (MP) for checking purposes. In the case of damage, a job order is issued and passed to maintenance planning. Tasks are defined by:

— operating instructions and manufacturers' instructions,

— experience gathered by production and workshop supervisors,

— the accumulated experience of maintenance employees.

It is important that completely uncoded text can be used for the individual job elements, so that the user can see at a glance the type and extent of the work to be carried out. The Alsen-Breitenburg system now schedules approximately 44 000 hours per year for inspection, maintenance and routine work, of which approximately

— 3 000 hours per year are for mechanical inspection,
— 8 000 for fitters and electricians,
— 33 000 for production.

It is important to emphasise that the input required from the production departments (in particular routine checks and cleaning) are part of their job, to be discharged without additional man hours. Two reliable and competent employees are permanently assigned to systematic inspection and checking work, for which they are equipped with modern measuring equipment, e.g.

— shock pulse measurement device with documentation and analysis software for assessing the condition of rolling bearings and lubrication,

— ultrasonic testing device for assessing wear,

— electronic stethoscope with cassette store for measuring noise,

— vibration analyzer for assessing how smoothly the machines are running.

- Ultraschallprüfgerät zur Verschleißmessung,
- elektronisches Stethoskop mit Kassettenspeicher zur Geräuschmessung,
- Schwingungsanalysator zur Beurteilung der Laufruhe von Maschinen.

5. Auftragsabwicklung

In **Bild 3** ist der Durchlauf von Instandhaltungsaufträgen von der Eröffnung über die EDV-Erfassung, Vorbereitung, Planung, Ausführung bis zur Rückmeldung im vereinfachten Schema dargestellt. Aufträge werden grundsätzlich auf einem Formular festgehalten. Sie können aus allen Werksbereichen kommen; meistens werden sie durch die periodischen Inspektionen ausgelöst.

Die Aufträge haben eine aufgedruckte fortlaufende Nummer, gehen zunächst an die IHP, werden dort mit der Maschinennummer verknüpft und zusätzlich u. a. nach den folgenden Kriterien klassifiziert:

- Priorität,
- Erledigung bei Kurzstillstand oder Grundreparatur,
- geschätzter Arbeitsumfang in Mann-Stunden,
- Plandatum für Reparaturbeginn und -ende.

Arbeitsaufträge für Instandhaltungsarbeiten werden grundsätzlich eröffnet, wenn der geschätzte Aufwand mehr als 3 Handwerkerstunden beträgt. Diese Schwelle hat sich als richtig erwiesen, weil dadurch auch mehrfach vorkommende Kleinreparaturen aufgezeigt und durch eine Schwachstellenanalyse abgestellt werden können. Aufträge mit kleinerem Stundenaufwand werden mündlich erteilt und über die Maschinennummer abgerechnet.

6. Auftragsplanung

Eine wesentliche Bedeutung kommt der Auftragsplanung zu. Ihre Durchführung weicht bei Alsen-Breitenburg von einer Auftragsvorbereitung im eigentlichen Sinne mit genauen Vorgaben für Personal- und Materialaufwand ab. Eine so arbeitsintensive Auftragsvorbereitung ist für ein Werk dieser Größenordnung noch nicht zweckmäßig und rechtfertigt nicht diesen besonderen Aufwand. Außerdem stehen gut ausgebildete und motivierte Handwerkermeister zur Verfügung, wobei auch die Handwerker selbst einen hohen fachlichen Standard besitzen.

Die Auftragsplanung hat demnach im wesentlichen folgende Auftragsschwerpunkte:

- Terminliche und sachliche Koordinierung zwischen Instandhaltung, Produktion und Fremdfirmen,
- Bereitstellung von Stücklisten, Hilfs- und Arbeitsmitteln sowie Ersatz- und Reserveteilen,
- Hinweise zur Arbeitssicherheit, zum Umweltschutz und für wiederkehrende Reparaturen,
- Kapazitätsermittlungen für Eigen- und Fremdpersonal,
- Aufstellen von Ablaufplänen für Großreparaturen.

Die bei der IHP eingehenden Aufträge werden bei den täglichen Zusammentreffen des Instandhaltungsleiters mit seinen Meistern vorgelegt und abgesprochen. Zum Wochenende treffen sich überdies die Werkstattmeister und ein Mitarbeiter der IHP und legen gemeinsam das Wochenprogramm fest, d. h. das Arbeitsprogramm für die Folgewoche. Hierfür sind etwa erforderliche Anlagenstillstände bereits vorher durch die IHP mit den Produktionsabteilungen abgeklärt. Diese Besprechungen dauern in der Regel nicht länger als eine halbe Stunde, wobei übersichtliche EDV-Ausdrucke der vorliegenden Aufträge nach verschiedenen Sortierkriterien sehr hilfreich sind. Das gemeinsam erarbeitete Wochenprogramm wird daraufhin am Bildschirm eingegeben und für die einzelnen Werkstattmeister ausgedruckt. Dieser Ausdruck dient den Meistern als Grundlage zur Abarbeitung der Aufträge. Die Produktionsabteilungen erhalten ebenfalls eine Liste über die geplanten Wochenarbeiten in ihrem Bereich, so daß sie über die Arbeitsvorhaben informiert sind. Die Wochenplanung wird nach der Priorität

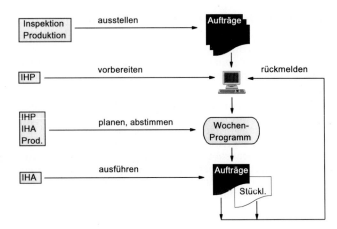

BILD 3: Schema der Auftragsabwicklung
FIGURE 3: Diagram of order processing

Inspektion Produktion	= inspection production
ausstellen	= issue
Aufträge	= job orders
IHP	= MP
vorbereiten	= prepare
rückmelden	= feedback
IHP, IHA, Prod.	= maintenance planning maintenance tasks prod.
planen, abstimmen	= plan, agree
Wochenprogramm	= weekly programme
ausführen	= carry out
Stückl.	= parts list

5. Order processing

Fig. 3 is a simplified diagram of how maintenance jobs are handled, from initiation by EDP data capture, through preparation, planning, implementation and feedback. All jobs are recorded on a form. They can come from any part of the works; they originate mostly from the periodic inspections.

Each job bears a serial number and passes first to MP where they are linked with a machine number and also classified in accordance with the following key criteria:

- priority,
- to be carried out during temporary stoppages or major overhaul,
- estimated labour input in man hours,
- planned repair start and finish dates.

Maintenance job orders are issued when the estimated work content exceeds 3 craftsman hours. This threshold has been borne out in practice, since it enables frequent small repairs to be highlighted by a trouble spot analysis. Orders requiring less labour are issued orally and charged against the machine number.

6. Job order planning

Job order planning has a high priority. The procedure at Alsen-Breitenburg differs from job order preparation proper in the precise standards set for labour and material inputs. Such labour-intensive job order preparation is disproportionate for work of this magnitude, and does not justify this level of effort. There are moreover well trained and well motivated skilled craftsmen available, and the craftsmen themselves have a high standard of technical competence.

Job order planning thus has the following key features:

- coordination of schedules and resources between maintenance, production and outside companies,
- provision of parts lists, process materials and replacement parts and spare parts,
- information on workplace safety, environmental protection and for periodic repairs,
- determining capacity for in-house and contract labour,
- establishing schedules for major repairs.

BILD 4: Informationen einer Stückliste
FIGURE 4: Parts list information

Magazingeführte Teile	= parts held in stock
Qualitätssicherung	= quality assurance
Artikelbezeichnung	= item designation
eingebaute Stückzahl	= quantity installed
Lagerplatz	= stock location
Preis, Bestellmenge	= price, order quantity
Stückliste nach Masch.-Nr. (auch Magazin-Entnahmeschein)	= parts list by machine number (including stores withdrawal slip)
Werkstoffauswahl	= material selection
Bearb.-Vorschriften	= process instructions
Dokumentation	= documentation
Montagehilfe	= assembly aid
Zeichnungen, techn. Bearbeitungen	= drawings, technical processes
Reparatur-Anweisungen	= repair instructions
besondere Werkzeuge	= special tools
Arbeitssicherheit	= safety

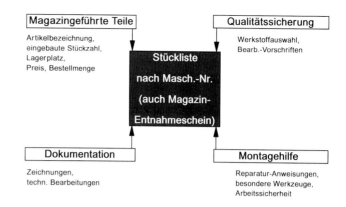

festgelegt, wobei Maßnahmen zur Unfallverhütung vorrangig berücksichtigt werden.

Es hat sich für die Planung auch als vorteilhaft erwiesen, die Kapazitäten der einzelnen Meisterbereiche nur zu 75 bis 80 % auszulasten. Damit bleiben dem Meister Personalreserven für die Beseitigung von nicht vermeidbaren, unvorhersehbaren Schäden. Ist das Wochenprogramm erledigt und sind noch Personalkapazitäten frei, so werden gemeinsam mit der IHP die nächstdringlichen Aufträge festgelegt und dann abgearbeitet.

Grundsätzlich sind in den Meisterbüros nur „aktive", d. h. in Arbeit befindliche Aufträge, während die „ruhenden" Aufträge zunächst bei der IHP verbleiben. Dadurch werden die Meister nicht durch eine Vielzahl von Auftragspapieren erdrückt. Sie behalten einen guten Überblick und haben durch die zügige Abarbeitung des Wochenprogramms ein nicht zu unterschätzendes und motivierendes Erfolgserlebnis. Da alle Aufträge im EDV-System gespeichert sind, können bei unvorhergesehenen Anlagenstillständen rasche Entscheidungen und Maßnahmen getroffen werden, was die Verfügbarkeit der Anlagen nicht unerheblich erhöht.

7. Integrierte Materialwirtschaft

Ein wesentliches Element der geplanten Instandhaltung ist eine gut organisierte Materialwirtschaft. Die Lagerhaltungskosten in einem Zementwerk sind nicht unerheblich. Das früher vorherrschende Sicherheitsdenken für die Bevorratung von Reserveteilen wird durch eine termingerechte Bereitstellung – just in time – abgelöst. Das ist nur möglich mit der geplanten Instandhaltung und mit den Informationen aus einer langjährigen Maschinengeschichte. Das Festlegen der Mindestbestände ist also nicht in erster Linie eine Frage der Wiederbeschaffungszeit, sondern eine aus der Maschinengeschichte ablesbare Erfahrung. Zielgerichtete Inspektionen tragen dazu bei, den Lagerbestand zu minimieren.

Selbstverständlich brauchen die Instandhalter einen vollständigen Einblick in die Magazinbestände und Bestellvorgänge. Überdies müssen sie einen guten Zugriff zu den einzelnen Positionen durch sachgerechte Katalogisierung und Lagerplatzzuordnung haben.

8. Stücklisten

Für die geplante Instandhaltung sind gut aufgebaute Stücklisten dringend erforderlich. Für die Arbeitsvorbereitung und -abwicklung ist es sehr zeitsparend, wenn solche Stücklisten für die Maschinen mit aktuellem Lagerbestand über Bildschirm oder Ausdruck abgerufen werden können. Die Kataloge der Maschinenhersteller sind zwar in vielen Fällen sehr hilfreich, sie stellen jedoch keine Verbindung zu den lagermäßig geführten Ersatzteilen her.

Als sehr praktikabel hat sich ein Stücklisten-Modul erwiesen, in welchem die in **Bild 4** aufgeführten Informationen gespeichert und weitergegeben werden. Neben dem aktuel-

The orders received by MP are discussed at the daily meetings of the maintenance manager with his supervisors. The workshop supervisors also meet with a MP employee at the end of the week to prepare the programme of work for the following week. Any necessary plant shutdowns are agreed in advance by MP with the production departments. These discussions generally take no longer than half an hour, and are greatly facilitated by summarized EDP printouts of jobs on hand, sorted by various criteria. The jointly agreed week's programme of work is then input at the screen, and printed out for the individual workshop supervisors. This printout gives the supervisors a basis for carrying out their jobs. The production departments also receive a list of the planned week's work in their area of responsibility, ensuring that they are informed of the planned schedule of work. The weekly plan is prioritized, with prime consideration given to accident prevention.

It has proved helpful for planning purposes to work on a 75 to 80 % loading of the capacity of the various trades. This leaves the supervisors with spare capacity for dealing with inevitable unforeseen damage. If there is still capacity free when the week's programme has been completed, the next most urgent jobs are identified together with MP and then carried out.

Only "active" jobs (i.e. jobs currently in progress) are located in the supervisors' offices whilst the "pending" jobs are initially held back by MP. This prevents the supervisors being flooded with job paperwork. They still have a good overview, and working briskly through the week's programme provides a success experience which has a significant motivating effect. Since all orders are logged in the EDP system, quick decisions can be made and action taken in the event of unforeseen plant shutdowns, contributing considerably to plant availability.

7. Integrated materials management and control

Well organised materials management and control is a key element in planned maintenance. The costs of holding stock in a cement works are substantial. The previous defensive philosophy of holding stocks of spare parts has been replaced by the just-in-time approach. This is only made possible by planned maintenance, and by the information derived from many years of machine history. The minimum stock level is determined principally not by the reorder time, but by experience derived from machine history. Targeted inspections contribute to minimising stock levels.

Of course the maintenance team needs a complete picture of the stock on hand and ordering processes. They also need good access to the individual items by suitable cataloguing and systematic allocation of storage areas.

8. Parts lists

Planned maintenance is critically dependent on well structured parts lists. It saves a great deal of time in production scheduling and management if such parts lists can be acces-

len Lagerbestand, der Bestellmenge, der eingebauten Stückzahl mit den artikelspezifischen Daten können noch weitere beliebige Informationen gespeichert werden, wie z. B. Qualitätsangaben, Vermerke über Montagehilfen, Hinweise auf Arbeitssicherheit und Dokumentationen usw.

Die Stücklisten können auch als Magazin-Entnahmeschein dienen, so daß zusätzliche Such- und Schreibarbeit entfällt. Da die Stücklisten für einzelne Maschinen z. T. sehr umfangreich sind, sollten am Bildschirm über ein Suchfeld nur bestimmte Maschinenteilbereiche herausgegriffen bzw. in Listenform ausgedruckt werden können. In dieser Form aufgebaute Stücklisten haben sich bewährt. Sie werden von den Handwerkern als Arbeitserleichterung angesehen, weil sie Wege- und Wartezeiten einsparen. Da in den Stücklisten auch die eingebauten Artikelmengen geführt werden, wird eine gezielte Mengenbevorratung möglich.

9. Auftragsbezogene Kostenrechnung

Im Hinblick auf die hohen Kosten der Instandhaltung ist die Instandhaltungskostenrechnung von ausschlaggebender Bedeutung. Kosteninformationen, welche lediglich an einer Maschine oder auf einem Kostenplatz gesammelt werden, sind wenig aussagefähig und, wenn sie mit über vier Wochen Nachlaufzeit ausgewiesen werden, sind sie als Planungs- oder Steuerunginstrument völlig ungeeignet.

Es ist daher vorteilhaft und erforderlich, die Kosten für jeden Auftrag zu erfassen und spätestens in der Folgewoche darzustellen.

Die dafür entwickelten Module erfüllen diese Forderung, sie sind in **Bild 5** aufgezeigt. Der bei der Auftragseröffnung von der IHP eingegebene Datensatz wird hierfür benutzt. Die Kostenrechnung ist in zwei Stufen gegliedert: in die Kostenschnellerfassung und in die Auftragsendabrechnung.

Die Kostenschnellerfassung ist abgekoppelt von der Betriebskostenrechnung und dient ausschließlich der Instandhaltung. Die eigenen und fremden Lohnstunden werden auftragsgerecht von der Lohnbuchhaltung erfaßt und mit Standardverrechnungssätzen in die Schnellabrechnung überspielt. Die Materialentnahmen werden ebenfalls mit der entsprechenden Auftragsnummer verbucht.

Diese Kosteninformation liegt bereits am Mittwoch der Folgewoche vor, sie ist also sehr zeitnah und hat gegenüber der Endabrechnung eine Abweichung von weniger als 5 %. Die Kostenschnellerfassung hat sich besonders bei Grundreparaturen als sehr vorteilhaft erwiesen, weil hierfür — auf wenige Wochen zusammengedrängt — fast die Hälfte des Instandhaltungsbudgets verbraucht wird. Kostenabweichungen sind sofort ersichtlich, ihre Ursachenanalyse bereitet keine Schwierigkeiten, und es kann rechtzeitig eingegriffen werden. Die Auftragsendabrechnung fließt in die Betriebskostenrechnung. Dabei werden die eigenen Lohnkosten mit Zuschlägen aus der Lohnabrechnung übernommen. Die Fremdleistungen werden nach Rechnungseingang verbucht.

Die ermittelten Kosten gehen danach in den Betriebswirtschaftsdienst, dienen folglich auch dem Budgetabgleich und werden in die Maschinengeschichte übernommen. Bei Alsen-Breitenburg liegt seit 1984 eine komplette Übersicht über alle IH-Aufträge, nach Maschinen sortiert, vor.

Es sind folgende Daten gespeichert und abgreifbar:

— Art der IH-Maßnahmen, d. h. vollständiger Auftragstext,

— Kosten für Einzelaufträge mit Eigen- und Fremdlohnkosten, Materialkosten, Gesamtkosten,

— Unterscheidung nach laufenden und Grundreparaturen,

— Kennzeichnung von außergewöhnlichen IH-Maßnahmen.

Diese Informationen dienen als Basis für folgende Aufgaben:

— Die Planung von IH-Maßnahmen,
    die durch Vergleich mit wiederkehrenden Aufträgen erleichtert wird.

sed for the machines with the current stock level on screen or printed out. Although machine manufacturers' catalogues are in many cases very helpful, they do not provide a link to the spare parts held in store.

A parts list module, which stores and distributes the information indicated in **Fig. 4**, has proved very practical. Any other information such as quality data, notes on assembly aids, notes on workplace safety and documentation, etc. required can be stored along with the current stock level, order quantity, the number of items installed with the data for that particular item.

The parts lists can also be used as a stores withdrawal slip, dispensing with further searching and paperwork. Since the parts lists for individual machines are in some cases very extensive, a facility is needed to select only certain categories of machine part via an on-screen search field, and/ or to print them out selectively in list form. Parts lists constructed in this way have proved effective. Shopfloor staff find them helpful because they save transport and waiting time. Since the number of items installed are also incorporated in the parts lists, it is possible to achieve targeting stocking levels.

9. Job-related costing

In view of the high costs of maintenance, maintenance costing is of crucial importance. Cost information collected from one machine or one cost centre gives an incomplete picture, and if it is not available until more than four weeks after the event, it is completely unsuitable for planning or control purposes.

It is therefore essential that the costs for each job are gathered and made available the following week at the latest.

The modules developed for this application meet this requirement, and are shown in **Fig. 5**. The record input by MP when the job is initiated is used for this purpose. There are two stages in the costing process: rapid "spot" acquisition of costing data, and final job costing.

Spot acquisition of costing data is independent of works cost accounting, and is used exclusively for maintenance purposes. In-house and contract labour hours are collected by the payroll department by job, and transferred to spot costing at standard cost rates. Materials issued are also booked against the corresponding job number.

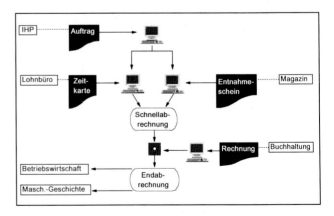

BILD 5: Beispiel für eine auftragsbezogene Kostenabrechnung
FIGURE 5: Example of job-related costing

IHP = MP
Auftrag = job order
Lohnbüro = payroll office
Zeitkarte = time card
Entnahmeschein = withdrawal slip
Magazin = stores
Schnellabrechnung = spot costing
Rechnung = invoice
Buchhaltung = accounts
Betriebswirtschaft = finance and administration
Masch.-Geschichte = machine history
Endabrechnung = final costing

- Die Budgeterstellung,
  welche durch Unterscheidung von laufenden Reparaturen und Sondermaßnahmen einen hohen Genauigkeitsgrad erhält.
- Die Ersatzteilbeschaffung,
  bei der die Entscheidungen durch Standzeitanalysen und die Reparaturhäufigkeit unterstützt wird.
- Die Ersatz- bzw. Neubeschaffung
  durch Offenlegung der Reparaturanfälligkeit der Aggregate, wodurch auch Hinweise auf die Lieferantenauswahl abgeleitet werden können.
- Die Schwachstellenbeseitigung,
  die durch das Aufzeigen von wiederkehrenden Reparaturen ermöglicht wird.

10. Schlußbemerkungen

Bei Alsen-Breitenburg wurde das System der geplanten Instandhaltung konsequent umgesetzt. Nach fast 10jähriger Erfahrung ergibt sich die Frage nach dem Nutzen. Durch die geplante Instandhaltung konnten die ad-hoc-Reparaturen von früher 80 % auf etwa 25 % reduziert werden, ebenso die Überstunden der Handwerker.

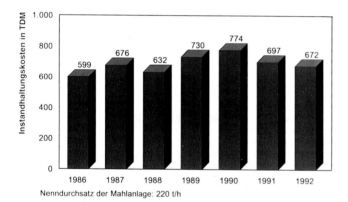

BILD 6: Entwicklung der Instandhaltungskosten für eine Zementmahlanlage (einschl. Abtransport)
FIGURE 6: Maintenance costs of a cement grinding plant (including transport system)

Instandhaltungskosten in TDM = maintenance costs in DM'000
Nenndurchsatz der Mahlanlage = nominal throughput of the grinding plant

Als ein Beispiel für den Erfolg der geplanten Instandhaltung sind in **Bild 6** die Instandhaltungskosten der Zementmahlanlage 1 für die Jahre 1986 bis 1992 dargestellt. Diese Mühle wurde während des Berichtszeitraums mit annähernd gleicher Laufzeit pro Jahr voll ausgelastet. In den nicht inflationsbereinigten Kosten sind auch die Kosten für Mahlkugeln und Mühlenauskleidung enthalten. Aus dem Diagramm ist ersichtlich, daß die Instandhaltungskosten trotz erheblich gestiegener Personal- und Materialkosten auf etwa gleichem Niveau gehalten werden konnten. Dieses Ergebnis ist auf die praktische Umsetzung der geplanten Instandhaltung zurückzuführen.

This costing information is available by the Wednesday of the following week, and is therefore very timely and deviates from the final costing figure by less than 5 %. Spot costing has proved very useful, especially for major overhauls, since these consume almost half the maintenance budget in a period of just a few weeks. Cost variances are immediately apparent, and their cause can be analyzed easily and appropriate measures taken in good time.

Final job costing is transferred to works cost accounting, together with in-house labour costs including bonuses from payroll. Contract services are entered after receipt of invoice.

The costs derived are then passed to Finance and Administration; they are then used for budget reconciliation and are added to the machine history. There is a complete overview of all maintenance jobs carried out at Alsen-Breitenburg since 1984, sorted by machine.

The following data are stored and can be accessed:

- type of maintenance activity, i.e. complete description of job,
- cost of individual jobs with costs of in-house labour, contract labour, materials, and total costs,
- differential listing by running repairs and major overhauls,
- identification of exceptional maintenance work.

This information provides the basis for the following tasks:

- Planning maintenance work
  which is facilitated by comparison with recurring jobs.
- Establishing budgets
  which achieves a high level of accuracy by distinguishing between running repairs and special jobs.
- Procuring spare parts
  supporting decision making with standard time analyses and repair frequency data.
- Procuring replacements and new equipment
  by revealing the reliability of the units, which also helps in the selection of suppliers.
- Eliminating trouble spots
  which is facilitated by highlighting recurring repairs.

10. Concluding remarks

The planned maintenance system at Alsen-Breitenburg has been implemented as a coherent strategy. After almost 10 years' experience there is no longer any question as to its usefulness. Planned maintenance has enabled ad hoc repairs to be reduced from the previous level of 80 % to about 25 %; shopfloor overtime has also been reduced.

The maintenance costs of Cement Mill 1 for the years 1986 to 1992 are given in **Fig. 6** as an example of the success of planned maintenance. This mill was fully loaded throughout the reporting period with roughly the same running time per year. The costs are not adjusted for inflation, and include the cost of grinding balls and mill lining. The diagram shows that maintenance costs have remained at roughly the same level despite significant increases in costs of labour and materials. This is the result of the practical application of planned maintenance.

Mensch im Betrieb*)
People at work*)
L'homme au travail

El hombre y su trabajo

Von **U. Jönsson,** Skövde/Schweden

Fachbericht 6.2 · Zusammenfassung – Die Wechselwirkung zwischen Mensch und Technik im Betrieb ist der Schlüssel für ein erfolgreiches Unternehmen. Arbeitsplatzbedingungen sind gesetzlich geregelt mit der Zielsetzung, Unfälle und zu hohe physische Belastungen zu vermeiden. Ein erweiterter gesetzlicher Rahmen umschließt heutzutage das gesamte Arbeitsmilieu. Das reine Unfallschutzdenken der Mitarbeiter wird erweitert mit gezielter Weiterbildung. In der modernen Zementindustrie ist die Zahl der Unfälle, verglichen mit anderen industriellen Bereichen, verhältnismäßig gering. Für eine erfolgreiche Unfallverhütung sind eine systematische Planung des Arbeitsablaufs und dessen ständige Kontrolle erforderlich. Die Entwicklung der eigenständigen Kompetenz der Mitarbeiter wird gefördert durch ein breites Angebot von theoretischen und praktischen Einsatzmöglichkeiten im Betrieb. Das traditionelle Berufswissen sollte jedoch durch Schulung des abstrakten Denkvermögens und die Fähigkeit, gegebene Information verarbeiten zu können, weiter komplettiert werden. Generell sollte auch das betriebliche Vorschlagswesen verbessert werden. Ein Blick in die Zukunft deutet darauf hin, daß der Weg in eine zweite industrielle Revolution, die sich auf individuelle Kompetenz und moderne Informationstechnologie stützt, bereits beschritten wurde. Diese Entwicklung führt zu einer neuen Denkweise und damit zu einer neuen Betriebskultur. Die Betriebe, denen es gelingt, diese neue Denkweise weiter zu entwickeln, gehören zu den Gewinnern.

Mensch im Betrieb

Special report 6.2 · Summary – The interaction between people and technology at work is the key to a successful business. Workplace conditions are controlled by law with the object of preventing accidents and excessive physical stresses. Nowadays an extended legal framework includes the entire working environment. Specific continued training is being used to extend the employees' ideas beyond pure accident prevention. The number of accidents in the modern cement industry is comparatively low when compared with other industrial sectors. Systematic planning and constant checking of work procedures are necessary for successful accident prevention. Development of the employees' independent capabilities is promoted by a wide range of theoretical and practical opportunities for using them at work. However, the traditional professional knowledge should be perfected by training abstract reasoning power and the ability to process given information; the plant suggestion book system should be generally improved. A glance at the future indicates that we are on the way to a second industrial revolution based on individual competence and modern information technology. This is leading to a new mental attitude and hence to a new working culture. The businesses which are successful in developing this new mental attitude further will be the winners.

People at work

Rapport spécial 6.2 · Résumé – Au sein de l'entreprise, les échanges entre l'homme et la machine sont la clé du succès d'une entreprise. Les conditions des postes de travail sont régis par la loi dans le but d'éviter des accidents et des charges physiques trop grandes. Un cadre légal élargi inclut de nos jours l'environnement du travail considéré comme un tout. La prévention proprement dite des accidents de travail est élargie à une formaton ciblée. Dans l'industrie cimentière moderne le nombre des accidents, comparés à ceux des autres branches industrielles, est relativement bas. Pour que la prévention des accidents soit satisfaisante, il est indispensable de prévoir une programmation systématique des travaux et d'en assurer sans cesse le contrôle. Le développement de la compétence propre du personnel est encouragé, dans l'entreprise, par un vaste éventail de possibilités d'utilisation tant théoriques que pratiques. Les connaissances professionnelles traditionnelles devraient être complétées par des stages de formation développant le raisonnement abstrait et la capacité d'assimiler des informations données, et les propositions d'amélioration interne à l'entreprise devraient être améliorées. Un simple regard vers l'avenir montre que nous sommes sur le chemin de la deuxième révolution industrielle, qui s'appuie sur la compétence individuelle et une technologie d'information moderne. Ceci conduit à une nouvelle forme de réflexion et, par là même, à une nouvelle gestion de l'entreprise. Les entreprises capables de continuer sur cette voie nouvelle, seront du côté des gagnants.

L'homme au travail

---

*) Überarbeitete Fassung eines Vortrages zum VDZ Kongreß '93, Düsseldorf (27. 9.–1. 10. 1993)
Revised text of a lecture to the VDZ Congress '93, Düsseldorf (27.9.–1.10.1993)

Informe de ramo 6.2 · Resumen – La interacción entre el hombre y la técnica es la clave del éxito de una empresa. Las condiciones que deben reinar en los puestos de trabajo estan reguladas por la Ley, con el fin de prevenir accidentes e impedir cargas físicas excesivas. Un amplio marco legal encierra, hoy en día, todo el entorno laboral. La prevención de accidentes propiamente dicha se ve ampliada con una formación continuada de los trabajadores. En la industria del cemento, el número de accidentes es relativamente bajo, en comparación con otros sectores industriales. Para que la prevención de accidentes tenga éxito, se requiere una planificación sistemática de los trabajos y un constante control de los mismos. El desarrollo de la competencia propia de cada trabajador se forma mediante una amplia oferta de puestos de trabajo dentro de la empresa, tanto de orientación teórica como práctica. Los conocimientos profesionales tradicionales deben completarse, sin embargo, mediante cursillos de entrenamiento del razonamiento abstracto y mediante el desarrollo de la facultad de asimilar las informaciones recibidas, a la vez que debe ampliarse el sistema interno de sugerencias de los trabajadores. Mirando hacia el futuro, todo parece indicar que vamos camino de una segunda revolución industrial, basada en la competencia individual y la moderna tecnología de la información. Ello nos llevará a una nueva forma de pensar y, por consiguiente, a una nueva cultura del trabajo. Las empresas que consigan desarrollar este nuevo pensamiento, contarán entre los ganadores.

El hombre y su trabajo

1. Einleitung

Die Konkurrenzstärke und der Erfolg eines Unternehmens sind in hohem Maße von dessen Fähigkeit abhängig, neue Technologien zu entwickeln und anzuwenden. Im Prinzip sind neue Technologien allen Ländern der Welt zugänglich, wenn dazu neben den menschlichen auch die ökonomischen und politischen Voraussetzungen gegeben sind. Die Schwierigkeiten liegen vor allem darin, angebotene technische Möglichkeiten auch zu nutzen. Deutliche Unterschiede sind z. B. zwischen Industrie- und Entwicklungsländern festzustellen. Damit wird auch sichtbar, daß die Wechselwirkung zwischen Mensch und Technik der Schlüssel für eine erfolgreiche Unternehmensführung ist. In der Erweiterung dieser Gedanken ist eine positive persönliche Entwicklung der Mitarbeiter eines Unternehmens auch als gute Entwicklung der Gesellschaft zu sehen.

2. Gesetzgebung

Schon zu Beginn der Industrialisierung wie auch heute geht das Bestreben dahin, die Mitarbeiter der Unternehmen gegen alle Formen von Gesundheitsschäden und zu hohen Belastungen durch die Arbeit zu schützen.

In folgenden Bereichen hat der Gesetzgeber hierzu besondere Bestimmungen erlassen:

— Arbeitsmilieu,

— Arbeit für Minderjährige,

— Arbeitszeit,

— Urlaub.

Der Inhalt der Arbeitsschutzgesetzgebung ist nach und nach ergänzt worden. Es wurde zum Beispiel angestrebt, ein Höchstmaß für die Arbeitszeit und eine Mindestdauer für den Urlaub durch Gesetze zu garantieren.

Die Arbeitsschutzgesetze haben in erster Linie das Ziel, Unfälle und zu hohe physische Belastungen zu verhindern. Nach und nach erhielten sie einen erweiterten Rahmen und umschlossen das gesamte Arbeitsmilieu. Allmählich wurde klar, daß das äußere und innere Milieu fortlaufend verbessert werden muß. Durch Luftverunreinigungen verursachte Krankheiten und durch schlechte ergonomische Verhältnisse entstandene Verschleißschäden führten zu zunehmender Beunruhigung. Auch Krankheiten oder Unbehagen, deren Ursachen in psychischen und sozialen Belastungen zu suchen waren, erweckten zunehmende Aufmerksamkeit. Gleichzeitig wurde dadurch der Begriff des Arbeitsmilieus erweitert. Fragen der Personalorganisation, der Arbeitsplanung sowie deren Einfluß auf die Ausführung der Arbeit gehören ebenfalls dazu. Auch derartige Fragen sind nunmehr in Gesetzen und Richtlinien in größerem Umfang geregelt.

Wird die Entwicklung aus einer mehr auf den Prozeß ausgerichteten Sicht betrachtet, so kann man feststellen, daß

1. Introduction

The competitiveness and success of a company depend largely on its ability to develop and apply new technologies. New technologies are in principle accessible to all countries of the world, provided there are the necessary human, economic and political conditions. The chief obstacle is making effective use of the technical potential available. There are, for example, significant differences between the industrialized countries and the developing countries. This also illustrates the interplay between people and technology as the key to successful organization management. This principle implies that positive personal development of employees is an effective means of organisational development.

2. Legislation

As in the early days of industrialization, the concern is to protect and ensure all aspects of health and safety at work.

The legislation concerns the following specific areas:

— systems of work,

— employment of minors,

— hours of work,

— holidays.

The scope of employment protection legislation has been gradually extended. There has, for example, been an attempt to establish legal maximum working hours and minimum vacation entitlement.

Employment protection legislation is primarily aimed at preventing accidents and excessive physical stress. The legislation has gradually expanded to include the workplace as a whole. By degrees it became clear that continuous improvement was necessary both within the workplace and as regards the environment. There was increasing concern at illness caused by air pollution, and injury resulting from poor ergonomic design. Diseases or discomfort caused by mental and social stress were also coming under increasing scrutiny. The concept of the workplace was also expanding. Questions of task allocation and work planning, and their influence on work performance were also covered by this concept. Such matters are now fairly widely covered by legislation and codes of practice.

If these developments are regarded from a more process-oriented point of view it becomes evident that current thinking has extended beyond mere accident prevention to include the personal development of employees, emphasizing recognition and application of individual talent. The emphasis has shifted from hard physical work to skilled, flexible discharge of tasks. Modern research has revealed that only a small part of the capacity of the human brain is used, and that the brain is capable of expanding its capacity by ongoing training.

heute neben dem reinen Unfallschutzdenken vor allem auch die Weiterbildung des einzelnen Mitarbeiters angestrebt wird. Seine persönlichen Voraussetzungen müssen erkannt und genutzt werden. Anzustreben ist keine harte physische Arbeit, sondern eine gewandte, flexible Durchführung von Aufgaben. Durch die moderne Forschung ist bekannt, daß nur ein kleiner Teil der Gehirnkapazität des Menschen ausgenutzt wird und daß das Gehirn die Fähigkeit besitzt, seine Kapazität durch ständiges Training zu erweitern.

3. Arbeitsschäden

In der modernen Zementindustrie sind erfreulicherweise die Anzahl und das Ausmaß von Körperschäden durch Arbeitsunfälle verhältnismäßig gering. Die Unfallhäufigkeit hat heute einen Stand erreicht, den man vor 25 Jahren kaum für möglich gehalten hätte (**Bild 1**). Eine Erklärung hierfür bieten der hohe technische Stand der Produktionsanlagen, die weit vorangeschrittene Automation und der Einsatz moderner Hilfsmittel bei Reparatur- und Wartungsarbeiten.

Diese Feststellung soll jedoch nicht zu der Annahme verleiten, daß alle Probleme aus der Welt geschafft sind. Die technische Entwicklung führt ständig zu neuen Risiken, die erkannt und in ihrer Wirkung durch Aufklärung entschärft werden müssen. Dabei sollten u. a. bedacht werden:

— Zusatzmittel und die Anwendung verschiedener chemischer Präparate, unter anderem bei Reparaturen,
— neue Brennstoffe,
— Ein-Mann-Arbeit (Einzelarbeitsplätze).

Latent sind ebenfalls die bekannten, lokal konzentrierten Schwierigkeiten mit Staub- und Lärmbelästigung. Wesentlich für eine erfolgreiche Sicherheitsarbeit sind die fortlaufende und systematische Planung von Arbeitsabläufen, die systematische Anleitung zur sicheren Arbeit und nicht zuletzt auch die Kontrolle, um ein hohes Sicherheitsniveau beizubehalten. In Schweden wird zur Zeit eine Arbeitsmethode für die interne Kontrolle eingeführt, die große Ähnlichkeit mit dem Qualitätssicherungssystem ISO 9000 aufweist. Dadurch soll die Sicherheitsarbeit in das Streben nach hoher Produktivität und guter Qualität integriert werden.

4. Entwicklung der Kompetenz

Ein breites Ausbildungsangebot hat zum Ziel, das Wissen und damit die Kompetenz des Arbeitnehmers zu erweitern. Es ist nicht nur die Ausbildung im engeren Sinn, die zum Beispiel durch eine Fortbildung erreicht werden kann. Bei der Ausbildung sollte auch eine deutliche Aufteilung zwischen den Aufgaben des Staates und der Wirtschaft vorgenommen werden. Der Staat soll eine gute Grundausbildung vermitteln, so zum Beispiel in den Grundfächern wie Mathematik, Sprache, Ökonomie und Technik. Die Wirtschaft strebt eine Zusammenarbeit mit den Schulen an, um die Verbindung zur Praxis herzustellen und die Anforderungen an gute Grundkenntnisse — zumindest für die Technik und deren Entwicklung — zu nennen. Anschließend müssen in den Unternehmen diese grundlegenden Kenntnisse mit den vorliegenden Erfahrungen und Fertigkeiten gefestigt und komplettiert werden. Hierbei wird es als nicht ausreichend angesehen, wenn ein Mitarbeiter nur die Bedienung einer neuen Maschine in einem Kursus erlernt. Das Resultat wird verbessert, wenn gleichzeitig die allgemeine Qualifizierung des Arbeitnehmers erhöht wird, so daß er in die Lage versetzt wird, durch selbständige Entscheidungen den Arbeitsablauf zu beeinflussen. Durch eine laufende Erweiterung des Arbeitsbereichs und durch laufende Schulung des abstrakten Denkvermögens auf der Basis des traditionellen Berufswissens werden so die Voraussetzungen für eine positive persönliche Entwicklung und eine effektivere Ausführung der Arbeit geschaffen.

5. Betriebliches Vorschlagswesen

Das gelobte Land der Verbesserungsvorschläge scheint Japan zu sein. Japans Fortschritt als Industrieland wird in

3. Safety at work

The number and extent of injuries sustained in accidents at work are fortunately relatively low in the modern cement industry. The accident rate has reached a level that would hardly have been considered possible 25 years ago (**Fig. 1**). This is because of the technical sophistication of production equipment, the advance of automation, and the use of modern resources for repair and maintenance work.

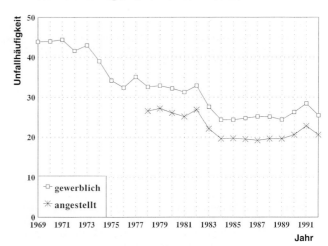

BILD 1: Unfallhäufigkeit in deutschen Zementwerken
FIGURE 1: Accident rates in German cement works

Unfallhäufigkeit	= accident rate
Jahr	= year
gewerblich	= shopfloor
angestellt	= staff

This is not however to suggest that all problems have been eliminated. Technical developments are constantly generating new hazards which have to be identified and countered by appropriate education and training measures. Relevant considerations include:

— reagents and the use of different chemical compounds for repairs and other purposes,
— new fuels,
— single manning (individual workplaces).

There are also latent problems with the known, locally concentrated, dust and noise hazards. Effective safety management requires continuous and systematic planning of work, systematic training in safe working practices and, importantly, monitoring to ensure that a high standard of safety is maintained. In Sweden a system of internal monitoring is currently being introduced which is very similar to the ISO 9000 quality assurance standard. This is intended to ensure that safety is an integral part of the push for increased productivity and high quality standards.

4. Developing competence

The purpose of a broad-based training and development programme is to increase the knowledge and competence of the employee. It is not just skills development in the narrow sense which can, for example, be achieved by training. A clear distinction also has to be made between the roles of government and of industry. The government's role is to provide good basic education, for example in mathematics, languages, economics and technology. The role of industry is to cooperate with the schools to create a link with the practical world, and to define the basic knowledge required — at least in the area of the development and application of technology. This basic knowledge must then be reinforced and complemented at the workplace with appropriate experience and skills. It is not enough for a worker go on a course to learn how to operate a new machine. Better results are achieved if the general level of qualification of the employee is increased, enabling him to make autonomous decisions about the process. Ongoing job enlargement and

hohem Maße dem Engagement der Mitarbeiter bei der Arbeit und dem Willen, Vorschläge zur Verbesserung der Technik und der Arbeitsmethoden zu machen, zugeschrieben.

Die Anzahl der Vorschläge ist in Japan etwa hundertmal höher als in Schweden. In japanischen Unternehmen tragen fast alle Arbeitnehmer mit Vorschlägen zur Verbesserung des Arbeitsablaufs bei, während es sich in Schweden nur um eine kleinere Mitarbeitergruppe handelt, die Verbesserungsvorschläge einreicht.

Es ist jedoch ein steigendes Interesse und eine Neuorientierung im Vorschlagswesen festzustellen, die an eine ausgeprägte Dezentralisierung gebunden sind. Wesentlich gefördert wird das Vorschlagswesen, wenn

— von seiten der Unternehmensführung ein ausgesprochenes Interesse an Vorschlägen zur Verbesserung besteht,
— das Interesse der Abteilungsleiter zunimmt und dieser Personenkreis die Verantwortung für Verbesserungsvorschläge übernimmt,
— die Bearbeitungszeit verkürzt und die Verwaltung vereinfacht werden,
— ein persönlicher Kontakt zwischen dem, der den Vorschlag einreicht, und demjenigen, der ihn beurteilt, aufgebaut wird,
— es zu einer schnellen Einführung eines gutgeheißenen Vorschlags kommt,
— großzügige Belohnungsregeln eingeführt werden, da Quantität immer zu Qualität führt.

6. Firmeninterne Sozialleistung

Es sind derzeitig in steigendem Umfang größere Unterschiede in der Handhabung personeller und sozialer Fragen zwischen der Gesellschaft und den Wirtschaftsunternehmen festzustellen. Die Unternehmen unterstützen in größerem Umfang

— vorbeugende Gesundheitsmaßnahmen,
— Leistungssport mit lokaler Anbindung,
— lokale kulturelle Veranstaltungen und den
— Betriebssport.

Die Entwicklung der Arbeitsplatzbedingungen in den Betrieben wird jedoch regional oder national unterschiedlich weitergeführt.

7. Die Zementfabrik von morgen — eine Vision

7.1 Personal

Ein modernes Unternehmen muß sich einem Gesamtdenken zuwenden und davon ausgehen, daß sowohl auf ökonomische, technische und menschliche Fakten Rücksicht zu nehmen ist. Durch ständige Diskussion und Information sollen die Mitarbeiter die Zukunft nicht als etwas Drohendes erleben, sondern eher als etwas, was durch sie zu beeinflussen ist. Der Wille und die Fähigkeit zu Veränderungen auf technischem und administrativem Gebiet sind dadurch verstärkt vorhanden.

Die Personalorganisation ist so anzulegen, daß Tätigkeit und Verantwortung stark dezentralisiert sind. Die meisten Beschäftigten haben demnach ein relativ breites Arbeitsfeld mit einer Vielzahl von Arbeitsaufgaben. Das Prozeßpersonal zum Beispiel arbeitet oft nicht nur in der Produktion, sondern führt ebenfalls einfachere Wartungsarbeiten aus, wie Schmierung, Reinigung, Administration sowie Umbau- und Montagearbeiten. Reparatur- und Wartungsarbeiter sind in zunehmendem Maße höher qualifiziert, teils durch ein besseres Fachwissen mit erhöhter Flexibilität für unterschiedliche Arbeitsaufgaben in der Nähe der Produktionseinheiten, teils durch eine Spezialisierung, die für eine effektive Suche nach Fehlern erforderlich ist. Der einzelne Mitarbeiter übernimmt somit eine größere Verantwortung bei der Ausführung seiner Tätigkeit, was schließlich zu größerer Zufriedenheit und Freude an der Arbeit führt.

ongoing development of the capacity for abstract thought on the basis of traditional professional knowledge will create the necessary conditions for positive personal development and effective working.

5. Suggestion schemes

Japan appears to be the shining example as far as suggestion schemes are concerned. Japan's advance as an industrialized nation has been largely due to the commitment of its workforce, and the readiness of employees to suggest improvements in technology and working methods.

The number of suggestions is approximately 100 times greater in Japan than in Sweden. In Japanese companies, almost all employees suggest improvements in working methods, whereas in Sweden there is only a small group of employees who make suggestions.

There is however increasing interest and a new attitude to suggestion schemes, linked to a distinct trend towards decentralization. Key factors contributing to improved suggestion schemes are:

— management showing a positive interest in improvement suggestions,
— department managers taking greater interest, and accepting responsibility for improvement suggestions,
— reducing the time taken to process suggestions, and simplifying administration,
— establishing personal contact between the person making the suggestion and the person assessing it,
— implementing approved suggestions rapidly,
— providing a generous reward structure, since quantity always leads to quality.

6. Company welfare

There are increasingly significant differences in the relationship between industry and the community in the field of welfare and social provision. Employers are increasingly supporting:

— preventative health programmes,
— competitive sports with local involvement,
— local cultural events and
— company sports.

However, there are differences in the regional and national trends in working conditions.

7. The cement plant of the future – a vision

7.1 Personnel

A modern company must adopt a comprehensive approach which takes into account economic, technical and human factors. By process of constant discussion and information, employees should come to regard the future not as a threat, but rather as something they can influence. This enhances the will and capability to make technical and administrative changes.

Work must be organised in such a way that both activity and responsibility are decentralized to a large degree, increasing the scope of most jobs accordingly by including a variety of tasks. Process personnel for example often carry out not only production tasks, but also simple maintenance tasks such as lubrication, cleaning, administration and conversion and fitting work. The level of qualification of repair and maintenance employees is increasing, partly as a result of greater specialist knowledge with increased flexibility for different tasks at production unit level, partly because of the specialisation necessary for effective fault finding. In this way the individual employee takes on greater responsibility for carrying out his task, which ultimately leads to greater job satisfaction.

Die Tätigkeit in der Verwaltung sollte aus einer Mischung von Spezialisierung und Generalisierung bestehen. Jeder einzelne hat eine Hauptaufgabe, auf die er spezialisiert ist, ist jedoch gleichzeitig ausgebildet, um bei Spitzenbelastungen und Vakanzen die Arbeit anderer zu übernehmen.

7.2 Management

Der Bedarf an Führungskräften innerhalb der Administration ist weiterhin dadurch zu vermindern, daß die meisten Mitarbeiter eine gute Übersicht über ihre Arbeitsaufgaben erhalten und selbst darüber bestimmen können, welche Vorgänge eine erhöhte Priorität besitzen oder aufgeschoben werden können, um in der Zwischenzeit etwa einem Kollegen zu helfen. In der Zukunft wird es wie heute bestimmte Führungskräfte geben, die für die Koordination der Arbeiten und den Zusammenhalt zwischen den Abteilungen sorgen. Die Abteilungsebene ist dann verantwortlich für übergeordnete Aktivitäten und für die Kontrolle ausgeführter Arbeiten. Sie benutzt einen großen Anteil ihrer Zeit zur Information und übergeordneten Ausbildung von Mitarbeitern. Das führt unter anderem auch dazu, daß den meisten Mitarbeitern die betriebswirtschaftlichen Zusammenhänge verständlich werden und sie großes Interesse an den laufenden Betriebsergebnissen entwickeln.

Diese Entwicklung ist jedoch generationsabhängig. Es müssen deshalb jüngere Leute rekrutiert und gleichzeitig den älteren Mitarbeitern Möglichkeiten zur Weiterbildung gegeben werden, damit sie die neuen Zusammenhänge verstehen lernen. Bei Neueinstellungen ist es wichtiger, die Fähigkeit im Umgang und Kontakt zu unterstelltem Personal und den Willen zum Lernen zu bewerten, als die formellen Kompetenzen in den Vordergrund zu stellen. Das Unternehmen muß ebenfalls großes Gewicht auf die bisherige Entwicklung eines Bewerbers im Arbeitsleben und weniger auf seine Kenntnisse legen, die er für seine Einstellung nachweisen kann.

Eine zukunftsorientierte Personalpolitik baut demnach in hohem Maße auf der internen Aus- und Weiterbildung auf. Die Spezialisierung für eine einzige Arbeitsaufgabe vermindert die Einsatzfähigkeit und Beweglichkeit. Darum muß es der Wunsch des Betriebs sein, daß der Mitarbeiter sich ein breites Wissen aneignet.

Administrative activity should consist of a mixture of specialised and generalised work. Each individual has a specialised core task, but is also trained to take on other people's jobs during peak times and vacations.

7.2 Management

The need for management within the administrative process can be reduced by giving most employees a good overview of their tasks, allowing them to decide for themselves which processes have a higher priority or can be postponed in order to provide assistance to a colleague. In future there will still be some managers to coordinate activities and liaise between departments. Superordinate activities and checking work carried out would then be the responsibility of departmental management, which would spend much of its time on information and overall development of employees. This would also lead to most employees understanding the economic environment, and developing a greater interest in the running of the organisation.

Employee acceptance of this approach will vary by age group. It is therefore necessary to recruit younger people and at the same time to provide older employees with development opportunities so that they can understand the new environment. In the case of new recruits it is more important to assess ability to communicate and make contact with subordinate personnel, and the will to learn, than to place the main emphasis on formal competence. The organisation must also place great emphasis on the applicant's development to date, and less on the knowledge he brings to his task.

A forward-looking personnel policy thus rests to a large extent on internal training and development. Specialisation for a single task reduces flexibility and adaptability. It is therefore in the organisation's interest to expand the knowledge base of all employees.

Verfahren für den sicheren Ausbruch feuerfester Baustoffe aus Drehrohröfen

Safe methods of breaking refractory materials out of rotary tube kilns

Méthodes pour le dégarnissage en toute sécurité des matériaux réfractaires dans les fours rotatifs

Métodos para la demolición segura de materiales refractarios empleados en los hornos rotativos

Von **G. Adam,** Hardegsen/Deutschland

Zusammenfassung – Bei thermischen und mechanischen Schäden am Mauerwerk sowie zu starker Ansatzbildung müssen feuerfeste Baustoffe ersetzt bzw. Ansätze aus den Drehrohröfen entfernt werden. Hierbei sind aus Gründen der Arbeitssicherheit und des Gesundheitsschutzes einige Grundvoraussetzungen zu erfüllen. Beim Ofenfutter-Ausbruch ist die Unfallverhütungsvorschrift der Steinbruch-BG 7.1; VBG 47a Schacht- und Drehrohröfen zu beachten. Die in der Vergangenheit und auch noch heute teilweise praktizierten manuellen Ofenausbrucharbeiten lassen sich durch den Einsatz von Maschinen und anderen Hilfsmitteln sicherer und effektiver durchführen. Zum Einsatz kommt hierbei hydraulisch bzw. pneumatisch betriebenes, fahrbares Ausbruchgerät, das teilweise mit Fernbedienung ausgestattet ist. Im Fall der Ansatzbeseitigung ist nach wie vor der Einsatz von Industriekanonen gebräuchlich. Beim Ausbruch von feuerfesten Baustoffen mit maschinellen Hilfsmitteln nimmt die Unfallgefahr drastisch ab. Die Kosten und der Zeitaufwand können sich bis um 80 % vermindern.

Verfahren für den sicheren Ausbruch feuerfester Baustoffe aus Drehrohröfen

Summary – If there is thermal and mechanical damage to the brickwork or excessively thick coating then the refractory materials have to be replaced or the coating has to be removed from the rotary tube kiln. Some fundamental conditions have to be fulfilled for safe working and health protection. When breaking out kiln linings the quarry regulations for prevention of accidents – BG7.1; VBG 47a shaft kilns and rotary tube kilns, have to be observed. The manual breaking-out methods which were used in the past and are still sometimes practiced today can be carried out more safely and effectively by using machines and other aids. Hydraulically or pneumatically driven mobile breaking equipment is used, sometimes with remote controls. As in the past it is still customary to use industrial guns for removing coating. The danger of accident is decreased drastically when using mechanical aids for breaking out refractory materials. The costs and time expenditure can be reduced by up to 80%.

Safe methods of breaking refractory materials out of rotary tube kilns

Résumé – A l'occurrence de dommages thermique et mécanique de la maçonnerie et en cas de forte formation de croûtages, les briques réfractaires ou les croûtages doivent être enlevés du four rotatif. Dans ce contexte, quelques règles fondamentales sont à observer en raison de la sécurité du travail et de la protection de la santé. Lors du dégarnissage de la maçonnerie de four, il faut remplir les conditions de la directive de prévention d'accidents de la BG 7.1 (carrières); VBG 47a fours droits et rotatifs. Les travaux manuels de dégarnissage du four, pratiqués aussi en partie encore aujoud'hui, peuvent être exécutés plus efficacement et avec plus de sécurité par l'utilisation de machines et d'autres artifices. Y sont mis en oeuvre des équipements de dégarnissage hydrauliques ou pneumatiques, automoteurs, parfois à télécommande. Pour l'élimination des croûtages, les canons pour l'industrie sont encore et toujours usuels. Quand le dégarnissage est effectué à l'aide de machines, le danger d'accident diminue très fortement. Les coûts et le temps dépensé peuvent être réduits de 80% dans les meilleurs cas.

Méthodes pour le dégarnissage en toute sécurité des matériaux réfractaires dans les fours rotatifs

Resumen – Cuando se producen daños térmicos o mecánicos en el revestimiento refractario o cuando se aprecia una formación excesiva de adherencias, hay que desmontar los materiales refractarios o eliminar las adherencias que se hayan formado en los hornos rotativos. A este respecto, hay que cumplir una serie que requisitos básicos, por razones de seguridad en el trabajo y de protección sanitaria. En cuanto a la demolición del forro refractario de los hornos, hay que tener en cuenta las prescripciones de prevención de accidentes de la BG 7.1 (canteras); VBG 47a hornos de cuba y rotativos. Los trabajos de demolición manuales, practicados en el pasado y también algunas veces hoy en día, pueden efectuarse de forma más segura y eficaz mediante el empleo de máquinas y otros artificios. Se emplean para ello aparatos de demolición móviles, de accionamiento hidráulico o neumático, equipados en parte con mando a distancia. Para la eliminación de las adherencias se siguen utilizando los cañones industriales. Llevando a cabo la demolición de materiales refractarios por medio de artefactos mecánicos, se reduce drásticamente el peligro de accidentes. En cuanto a coste y tiempo invertido, se puede conseguir un ahorro de hasta un 80 %.

Métodos para la demolición segura de materiales refractarios empleados en los hornos rotativos

Der Ausbruch von feuerfesten Ausmauerungen und die Entfernung von Ansätzen im Drehrohofen mit maschineller Hilfe entsprechen dem derzeitigen Stand der Technik.

Die manuelle Arbeitsweise ist aus diesem Grund sowie aus Sicht der Arbeitssicherheit und der Wirtschaftlichkeit überholt und nicht mehr zeitgemäß. In fast allen Fällen läßt sich neben einer Verbesserung der Arbeitssicherheit eine Verringerung der Kosten und des Zeitaufwandes bis zu 80 % erreichen, was den Einsatz von maschinellem Ausbruchgerät rechtfertigt.

1. Einleitung

Thermische und mechanische Beschädigungen der feuerfesten Ausmauerung sowie eine zu starke Ansatzbildung zwingen die Betreiber von Drehrohröfen, die Ansätze bzw. das feuerfeste Mauerwerk ganz oder teilweise auszubrechen. Hierbei ist die Begehung des Drehrohres durch Wartungspersonal unumgänglich, so daß Maßnahmen ergriffen werden müssen, um ein effizientes und vor allem sicheres Beräumen des Drehrohofens zu ermöglichen.

2. Sicherheitsmaßnahmen

Die Sicherheitsmaßnahmen, die beim Ausbruch des Ofenfutters zu beachten sind, sind in der Unfallverhütungsvorschrift „7.1 / VGB 47 a, Schacht- und Drehrohröfen" der Steinbruchs-Berufsgenossenschaft zusammengestellt. Demzufolge sind vor der Beräumung der feuerfesten Ausmauerung folgende Vorbereitungen und Sicherheitsmaßnahmen zu treffen:

— Der Ofenantrieb muß gegen unbefugtes Einschalten gesichert werden.

— Vorhandene Luftdruckstoßgeräte müssen drucklos gemacht und gegen Wiederaufladen gesichert werden.

— Es muß eine Sicherheitsunterweisung des Ausbruchpersonals durchgeführt werden.

— Es ist ein Verantwortlicher für die Ausbrucharbeiten zu benennen.

— Eine sichere Arbeitsbühne ist am Ofenauslauf zu errichten.

— Im Drehrohr muß eine ausreichende Beleuchtung (42 Volt) vorgesehen werden.

— Die Ansatzverhältnisse müssen durch den Verantwortlichen mittels Sichtkontrolle und Stocherprüfung kontrolliert werden.

— Der Ansatz muß vor und während der Arbeiten im Drehrohr ständig beobachtet werden.

— Die Arbeitsstelle im Ansatzbereich ist durch ein geeignetes Schutzdach bzw. Schutzvorrichtung zu sichern.

— Vom Personal muß die persönliche Schutzausrüstung getragen werden (Schutzhelm, -brille, -handschuhe, Sicherheitsschuhe, Gehörschutz, Staubschutz „P2").

3. Ausbrucharbeiten

In der Vergangenheit wurde der Ausbruch des feuerfesten Materiales bzw. des Ansatzes manuell durchgeführt. Obwohl dies auch heute noch vielerorts gängige Praxis ist, lassen sich die Ausbrucharbeiten durch den Einsatz von Maschinen und anderen Hilfsmitteln sicherer und effektiver durchführen.

Fahrbare Ausbruchgeräte stellen eine erhebliche Arbeitserleichterung dar. Bei diesen Geräten ist das eigentliche Ausbruchwerkzeug auf einer fahrbaren Basis mit einem schwenk- und drehbaren Arm befestigt, der vom Bedienungspersonal dirigiert werden kann. Die Ausbruchgeräte können sowohl hydraulisch als auch pneumatisch angetrieben werden. Sie sind in der Lage, zentimetergenau auszubrechen bzw. Ansätze zu entfernen. Hierbei ist es erforderlich, daß die Maschine über ein ausreichend dimensioniertes Schutzdach verfügt und das Bedienungspersonal die geforderte Schutzausrüstung trägt. Fahrbare Ausbruchge-

It is now state of the art to use mechanical means for breaking out refractory linings and removing coating from rotary tube kilns.

For this reason manual operation is an outdated method, not in keeping with present-day practice, from the points of view of both industrial safety and profitability. Mechanisation can in nearly all cases reduce costs and the time expended by up to 80% as well as improving industrial safety, so there is good reason to use mechanical breaking-out equipment.

1. Introduction

If there is thermal and mechanical damage to the refractory lining or excessively thick coating in a rotary tube kiln, the operator will be obliged to break out some or all of the coating or lining. It is essential to have maintenance personnel working inside the kiln, so measures must be taken to enable it to be cleared efficiently and above all safely.

2. Safety measures

The safety measures which have to be observed when breaking out the kiln lining are set out in the Quarry-workers Association's regulations for prevention of accidents "7.1 / VGB 47a, Shaft kilns and rotary tube kilns". These state that the following preparations and safety measures must be carried out before clearing a refractory lining:

— Unauthorised switching on of the kiln drive must be prevented.

— Any compressed air blast equipment present must be depressurised and prevented from recharging.

— The personnel involved in the breaking-out work must be given safety instructions.

— A person must be appointed to take responsibility for the breaking out work.

— A safe operating platform must be erected at the kiln outlet.

— Adequate lighting (42 volt) must be provided in the kiln.

— The responsible person must check the condition of the coating by visual inspection and testing with a rod.

— The coating must be under constant observation before and during work in the rotary kiln.

— The workstation in the coating area must be safeguarded by a suitable roof and/or other protective equipment.

— The workers must wear personal protective gear (a hard hat, goggles, safety gloves and shoes, hearing protection and "P2" dust protection).

3. Breaking out work

In the past, refractory material and coating were broken out manually. Although this is still the practice in many places even today, breaking-out work can be carried out more safely and effectively by using machines and other aids.

Mobile breaking-out equipment makes the work considerably easier. The actual breaking-out implement is mounted on a mobile base with a pivotable and rotatable arm which can be directed by the operating personnel. The equipment may have either a hydraulic or a pneumatic drive; it can break out material or remove coating accurately to the nearest centimetre. The machine must have a protective roof of adequate dimensions and the operators must wear the required protective gear. Remote-controlled mobile breaking-out equipment has the further advantage that the operators can carry out and survey the work from a safe distance.

A method which is applied less frequently is to break out brickwork and coating using industrial guns. An adequate slot in an axial direction is cut in the brickwork or ring of coating to be removed, by shooting from a safe distance. This decreases and eliminates the wedging action and ring stress of the brickwork or ring of coating. Slight turning of

räte mit Fernbedienung haben darüber hinaus den Vorteil, daß das Bedienungspersonal aus sicherer Entfernung die Arbeiten ausführen und überwachen kann.

Eine seltener angewandte Variante stellt das Ausbrechen von Mauerwerk und Ansatz mittels Industriekanone dar. Hierbei wird aus sicherer Entfernung ein ausreichender Schlitz in axialer Richtung in das zu entfernende Mauerwerk oder den Ansatzring geschossen. Damit werden die Keiligkeit und die Ringspannung des Mauerwerkes bzw. des Ansatzringes abgebaut bzw. aufgehoben. Durch eine geringe Drehung des Ofenrohres fällt dann der Ring in sich zusammen. Diese Ausbruchvariante ist jedoch nur begrenzt, abhängig von der Lage des Ansatzes bzw. der Ausbruchstelle, durchführbar. Außerdem läßt sie sich erst bei Öfen mit einem Durchmesser größer 4,2 m realisieren.

4. Räumarbeiten

Nach dem Ausbrechen des Ansatzes und des Mauerwerkes ist es aus Gründen der Sicherheit und Effizienz sinnvoll, das angefallene Material mit maschineller Hilfe zu beräumen. Hierzu können entweder spezielle Maschinen wie z. B. kleine Räumbagger oder auch Ausbruchgeräte mit Wechselschaufeln verwendet werden. Auch bei kleineren Ofendurchmessern können spezielle, maschinelle Ausbruch- und Räumhilfen eingesetzt werden. Die Einsatz- und Anwendungsmöglichkeiten maschineller Einrichtungen verbessern sich jedoch mit wachsendem Ofendurchmesser, da der verfügbare Arbeitsraum größer wird.

the kiln tube then makes the ring collapse. However, the use of this method is limited, depending on the positions of the coating or breaking-out location. It is also only feasible for kilns larger than 4.2 m in diameter.

4. Clearing work

When the coating and brickwork have been broken out, the material removed needs to be cleared mechanically for reasons of safety and efficiency. Either special machines such as small clearing excavators or breaking-out machines with interchangeable scoops may be used. For small-diameter kilns special mechanical breaking-out and clearing aids may be employed. But the opportunities for using mechanical equipment increase with the diameter of the kiln, as the available working space becomes larger.

Rationalisierte Instandhaltung von kontinuierlichen Dosiereinrichtungen für Schüttgut unter Berücksichtigung der Qualitätssicherung

Rationalized maintenance of continuous metering equipment for bulk materials with due regard to quality assurance

Maintenance rationalisée d'équipements de dosage en continu pour matières en vrac, en tenant compte de l'assurance de la qualité

Mantenimiento racionalizado de equipos de dosificación continua de materiales a granel, teniendo en cuenta el aseguramiento de la calidad

Von **B. Allenberg**, Darmstadt/Deutschland

Zusammenfassung – Die Forderung nach einer zertifizierten Zementproduktion nach ISO 9000 führt unter anderem auch zu zusätzlichen Aufwendungen in der Instandhaltung und Prüfung der Waagen und Dosierer. Die angewandten Maßnahmen müssen dabei effektiv und wirtschaftlich sein. Der Beitrag zeigt das auf die jeweilige Applikation und den speziellen Waagentyp abgestufte Vorgehen bei der Auswahl und Durchführung der Verfahren zur Geräteüberprüfung. Dabei können auch bereits vorhandene Systeme in ein Qualitätssicherungssystem nach ISO 9000 integriert werden. Ein Vergleich der Anforderungen in anderen Branchen mit denen bei der Zementherstellung kann die Übertragung von Erkenntnissen erleichtern. Das gilt insbesondere für die Anpassung bereits definierter Qualitätsprotokolle. Um die Kosten möglichst gering zu halten, erfordert eine gezielte Instandhaltung häufigere Kontrollen, die jedoch mit wenig Aufwand durchzuführen sind. Die Zentralisierung der Überwachungs- und Wartungsarbeiten auf einen Arbeitsplatz bietet dabei ein großes Einsparungspotential. Insbesondere eine Datenübertragung über Weitverkehrsnetze (z. B. Telefon) erlaubt es, viele Tätigkeiten in lokal weit auseinanderliegenden Fertigungsstätten auf eine Zentrale zu konzentrieren. Das Konzept ermöglicht es zudem, Prüfungen zu einer unabhängigen Stelle zu verlagern und im Notfall schnelle Hilfe durch den Hersteller der Waagen anzufordern.

Summary – The requirement for certified cement production in accordance with ISO 9000 leads, among other things, to additional expenditure on the maintenance and testing of weighers and metering equipment. The measures applied must be effective and economical. This contribution indicates procedures when selecting and carrying out methods for equipment checking which are graded to the particular application and the specific type of weigher. Existing systems can also be integrated into a quality assurance system complying with ISO 9000. Comparison of the requirements in other branches of industry with those in cement production can facilitate the transfer of knowledge. This is particularly true for the adaptation of already defined quality logs. In order to keep the costs as low as possible, a carefully planned maintenance system requires more frequent checks which, however, should be carried out with little expenditure. Centralization of the monitoring and maintenance work in one working position offers great potentials for saving. In particular, data transmission over long-range networks (e.g. telephone) makes it possible to concentrate many activities in production sites located far from one another in one centre. The concept also makes it possible to switch testing to an independent agency and, in an emergency, to call for rapid help from the weigher manufacturers.

Résumé – Le besoin d'une production de ciment certifiée d'après ISO 9000 conduit, entre autre, à des dépenses supplémentaires pour la maintenance et la vérification des bascules et doseurs. Les actions entreprises y doivent être efficaces et économiques. La contribution montre, cas par cas, la procédure taillée sur mesure pour chaque application et type de bascule, lors du choix et de la réalisation des méthodes de vérification des appareils. Y peuvent aussi être intégrés des systèmes déjà existants, dans un système d'assurance de la qualité selon ISO 9000. Une comparaison des exigences dans d'autres branches, avec celles dans la fabrication du ciment, peut faciliter la transposition des connaissances. Cela vaut surtout pour l'adaptation de protocoles de qualité déjà définis. Afin de maintenir les coûts aussi bas que possible, une maintenance ciblée demande des contrôles plus fréquents, réalisables pourtant à peu de frais. La centralisation des tâches de surveillance et de maintenance à un seul poste de travail y offre un fort potentiel d'économie. Surtout la transmission des données par réseaux

à grande distance (p. ex. télécommunication) permet de concentrer une multitude d'activités dans des lieux de production très dispersés, sur une seule centrale. Ce concept permet, en outre, de délocaliser les vérifications en un endroit indépendant et, en cas d'urgence, de demander une intervention rapide du fabricant des bascules.

Resumen – La exigencia de una producción de cemento certificada según ISO 9000 conduce, entre otras cosas, a gastos adicionales por el mantenimiento y control de las básculas y equipos dosificadores. Las medidas tomadas a este respecto deben ser eficaces y económicas. El presente artículo muestra la forma de actuar a la hora de seleccionar y emplear los métodos de control de los aparatos, teniendo en cuenta las distintas aplicaciones y tipos de básculas. Existe la posibilidad de integrar los sistemas existentes en un sistema de aseguramiento de la calidad según ISO 9000. Una comparación de los requerimientos existentes en otros sectores con los usuales en la fabricación del cemento podrá facilitar la transmisión de conocimientos. Esto es el caso, especialmente, de la adaptación de protocolos de calidad ya definidos. Con el fin de reducir los gastos a un mínimo, un mantenimiento bien enfocado requiere controles más frecuentes, pero éstos se pueden llevar a cabo sin mayores gastos. La centralización de los trabajos de control y mantenimiento en un solo puesto de trabajo ofrece, en este sentido, un gran potencial de ahorro. Sobre todo la transmisión de datos a través de redes de comunicación a gran distancia (por ej. teléfono) permite concentrar en una sola central muchas actividades llevadas a cabo en lugares de producción situados a gran distancia los unos de los otros. Este concepto permite, además, encargar los controles a un organismo independiente y solicitar ayuda rápida al fabricante de las básculas, si fuese necesario.

Mantenimiento racionalizado de equipos de dosificación continua de materiales a granel, teniendo en cuenta el aseguramiento de la calidad

1. Einleitung

Für kontinuierliche Waagen ist mit dem Diagnose- und Qualitätssicherungstool nun ein Werkzeug verfügbar, das den Anwender beim Aufbau einer rationalisierten Instandhaltung und kostengünstigen Qualitätssicherung unterstützt. Die Basis des Systems bildet eine zuverlässige Mechanik mit der Wägeelektronik MULTICONT. Prozeßnah werden somit viele eng mit der Wägetechnik verbundenen Aufgaben durch betriebssichere Standardkomponenten erledigt. Das automatische Diagnosesystem erleichtert die Ursachenforschung bei aufgetretenen Störungen. Mittels Datenfernübertragung erhält der Servicetechniker stets zuverlässige Informationen und kann korrigierend eingreifen. Das Qualitätssicherungs-Tool liefert, basierend auf den vom MULTICONT zur Verfügung gestellten Daten, die für den Aufbau einer Qualitätssicherung nach ISO 9000 notwendigen statistischen Auswertungen und Dokumente. On-Stream Materialkontrollen liefern dazu die besten Kontrollwerte.

2. Qualitätssicherung und Instandhaltung

Der Bereich der Anlageninstandhaltung ist in den letzten Jahren vermehrt unter Kostendruck geraten. Gab es früher für die unterschiedlichen Maschinengruppen Spezialisten in ausreichender Zahl, so ist heute eine verkleinerte Mannschaft für alle Anlagenteile, von der Rohstoffgewinnung bis zur Verpackung, zuständig. Diese Konzentration der Aufgaben kann nur mit Hilfe effektiver Werkzeuge bewältigt werden, die teilweise in die elektronischen Steuerungen integriert sind, andererseits aber auch die preiswert verfügbare Rechenleistung von PCs nutzen.

Das Fehlen allgemein anwendbarer Kommunikationswege für die Übertragung von aktuellen Betriebsdaten und der Konfiguration der Teilsysteme wird auch auf mittlere Sicht der gewünschten Einführung des zentralen „Ingenieurplatzes" für alle Ebenen der Anlagenautomatisierung entgegenstehen. Um so größere Bedeutung kommt Werkzeugen zu, die die vorhandenen Ressourcen und bereits akzeptierte Standards, z.B. im Bereich der Bedieneroberflächen von PCs, konsequent nutzen und damit den Einstiegsaufwand in sinnvollen Grenzen halten. Zuverlässige Informationen, die erst die wirtschaftliche Zentralisierung der Wartung ermöglichen, werden damit auch dem Techniker verfügbar, der sich mit den einzelnen Anlagenteilen nur im seltenen Fehlerfall beschäftigen will. Die Qualitätssicherung ist traditionell bereits als zentrale Stelle organisiert, wie dies nun auch durch die ISO 9000 festgelegt wurde. Die Erfassung der notwendigen Daten wird wegen unbefriedigender Kommunikationswege oft noch manuell vorgenommen.

Der vorliegende Bericht zeigt neue Entwicklungen für kontinuierlich arbeitende Schüttgut-Wäge- und -dosierstatio-

1. Introduction

The diagnosis and quality assurance tool for continuous weighers provides the user with an instrument for developing rationalised maintenance and cost-effective quality assurance. The system is based on a reliable mechanism with MULTICONT weighing electronics and is process-oriented. Thus many tasks closely connected with the weighing method are carried out by fail-safe standard components. When trouble arises, the automatic diagnostic system facilitates the search for the cause. By using remote data transmission the service engineer can always obtain reliable information and take corrective action. Based on the data made available by MULTICONT, the quality assurance tool provides the statistical evaluations and documents required to establish a quality assurance system complying with ISO 9000. On-stream material checks provide the best control values for this purpose.

2. Quality assurance and maintenance

Plant maintenance has come under more and more financial pressure of recent years. Whereas in the past there were enough specialists for the different groups of machines, today a smaller team is responsible for all parts of the plant, from quarrying to packaging. This concentration of tasks can only be dealt with by using effective tools, some of them incorporated in the electronic controls but others utilising the cheap computing power of PCs.

Even in the medium term the lack of generally practicable communication routes for transmitting current operating data and the configuration of the subsystems will impede the desired introduction of a central "engineering position" for all levels of plant automation. This makes tools all the more important, since they make consistent use of existing resources and accepted standards, e.g. for the operator surfaces of PCs, and thus keep the cost of access within reasonable limits. By means of these tools reliable information – without which it is impossible to centralise servicing economically – is made available even to technicians who will seldom work on the individual parts of the plant except when dealing with a failure. Quality assurance is already traditionally organised as a central unit as stipulated in ISO 9000. The necessary data is still often recorded manually owing to the unsatisfactory communication routes.

This report describes new developments in continuously operating stations for weighing and metering bulk materials, taking as an example Schenck's MULTICONT electronic weighing system with PC-assisted tools for servicing and quality assurance. These have taken a decisive step towards rationalising maintenance and making quality assurance cost-effective. Particular importance is attached to the design of the weigher and the communication routes

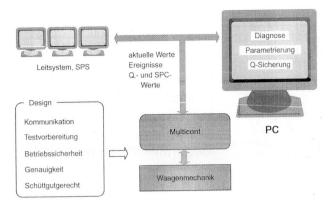
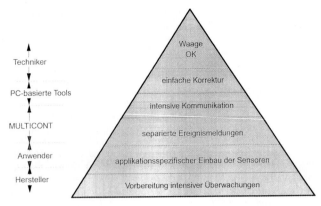

BILD 1: Design und Kommunikationswege für kontinuierliche Waagen
FIGURE 1: Design and communication routes for continuous weighers

Leitsystem, SPS	= control system, PLC system
aktuelle Werte	= current values
Ereignisse	= events
Q.- und SPC-Werte	= Q and SPC values
Diagnose	= diagnosis
Parametrierung	= parameterisation
Q.-Sicherung	= quality assurance
Waagenmechanik	= weigher mechanism
Kommunikation	= communication
Testvorbereitung	= Test preparation
Betriebssicherheit	= Reliability
Genauigkeit	= Accuracy
Schüttgutgerecht	= Suitability for bulk material

BILD 2: Maßnahmen zur effektiven Fehlerdiagnose
FIGURE 2: Measures for effective fault diagnosis

Techniker	= Technician
PC-basierte Tools	= PC-based tools
Anwender	= User
Hersteller	= Manufacturer
Waage	= weigher
einfache Korrektur	= simple correction
intensive Kommunikation	= intensive communication
separierte Ereignismeldung	= separate event messages
applikationsspezifischer Einbau der Sensoren	= application-specific installation of sensors
Vorbereitung intensiver Überwachungen	= preparation of intensive monitoring system

nen am Beispiel der Wägeelektronik MULTICONT der Firma Schenck mit den PC-gestützten Tools für Service und Qualitätssicherung, die einen entscheidenden Schritt in Richtung auf eine rationalisierte Instandhaltung und kostengünstige Qualitätssicherung darstellen. Dem Design der Wägeeinrichtung und den Kommunikationswegen innerhalb der Waage sowie zu ihrem Umfeld kommt nach **Bild 1** dabei entscheidende Bedeutung zu.

3. Perspektiven für die Instandhaltung

Das Rationalisierungspotential im Bereich der Instandhaltung wird durch folgende Maßnahmen weitgehend ausgeschöpft:

- Zuverlässiger Datenaustausch mit entfernt liegenden Anlageteilen.
- Planbarkeit der Servicetätigkeit.
- Automatisierte Erstellung einer Störungsdiagnose.
- Korrektureingriffe ohne lange Einarbeitung.

Bild 2 zeigt die zur effektiven und zentralisierten Durchführung der Instandhaltung kontinuierlicher Waagen notwendigen Schritte. Für die Wägeelektronik MULTICONT ist nun eine Erweiterung, das Diagose-Tool, verfügbar.

3.1 Datenfernübertragung

Für das Personal der Wartungszentralen fallen erhebliche Fahrzeiten zu den verstreut liegenden Anlageteilen an, die dadurch, daß keine gesicherten Informationen über ein Störungsbild vorliegen, notwendig werden. Häufig kommt es auch vor, daß der Spezialist nicht am Ort ist und auch der Anlagenfahrer den Sachverhalt oft nicht darstellen kann.

Eine Datenfernübertragung über weitverbreitete Netze (insbesondere Telefon) schafft hier Abhilfe. Mittels PC und Modem ist es so von fast allen Punkten aus möglich, sich verläßliche Information zu verschaffen und Korrektureingriffe vorzunehmen. Dies erlaubt die Mitnahme der richtigen Werkzeuge und Ersatzteile und erübrigt viele Einsätze vollständig. **Bild 3** zeigt die geeignete Architektur eines Fernwartungssystems für kontinuierliche Waagen. Alle relevanten Daten der angeschlossenen Waagen werden im MULTICONT Master für die Kommunikation zur Verfügung gestellt. Ein Standard-PC oder Lap Top fragt die Daten

within the weigher and between the weigher and its surroundings, as illustrated in **Fig. 1**.

3. Prospects for maintenance

The following measures virtually exhaust the potential for rationalisation in the field of maintenance:

- reliable data exchange with remote parts of the plant,
- the possibility of planning the servicing work,
- automated diagnosis of malfunctions,
- corrective action without a long period of familiarization.

Fig. 2 shows the steps which are necessary for effective and centralised maintenance of continuous weighers. An extension of the MULTICONT electronic weighing system is available in the form of the diagnostic tool.

3.1 Remote data transmission

Engineers at service centres have to spend a considerable time travelling to distant parts of the plant because there is no reliable information about a failure. The relevant specialist is often not there, and even the plant operator is often unable to describe the situation.

The problem can be solved by data transmission over long-range networks (especially the telephone). With PCs and modems reliable information can be obtained almost anywhere and corrective action taken. The engineer knows which tools and spare parts he will have to take, and in many cases he need not even go to the location of the fault. **Fig. 3** shows suitable architecture for a remote servicing system for continuous weighers. All relevant data on the connected weighers are held in the MULTICONT master for communication. A standard or laptop PC extracts the data through a modem-modem connection and the ubiquitous telephone network, puts it into an appropriate form and combines the data to form higher-level conclusions.

Remote data transmission also makes it possible for the manufacturers of metering equipment to be called in for particularly difficult cases, or for servicing capacity to be relocated there with attendant cost benefits.

3.2 Diagnosis

The essential requirement for any fault diagnosis system is the availability of diverse and detailed information from pro-

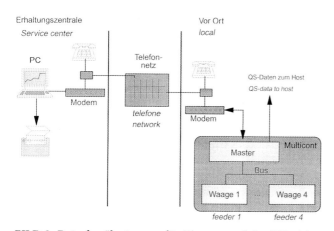

BILD 3: Datenfernübertragung für Diagnose und Qualitätssicherung
FIGURE 3: Remote data transmission for diagnosis and quality assurance

über eine Modem-Modem Verbindung und das überall verfügbare Telefonnetz ab, stellt sie in adäquater Form dar und verknüpft die Daten zu übergeordneten Schlüssen.

Eine solche Fernübertragung ermöglicht weiterhin eine Unterstützung durch den Hersteller einer Dosiereinrichtung in besonders schwierigen Fällen bzw. die kostengünstige Verlagerung von Servicekapazitäten dorthin.

3.2 Diagnose

Grundlage jeder Fehlerdiagnose ist die Verfügbarkeit möglichst vielfältiger und detaillierter Information aus den prozeßnahen Ebenen. Die Wägeelektronik MULTICONT bietet neben zahlreichen automatischen Korrektur-, Kompensations- und Justagealgorithmen eine weite Spanne von Überwachungsmöglichkeiten des Wäge- und Dosierprozesses selbst, aber auch der Waagenperipherie, die sehr häufig Grund für Betriebsstörungen ist.

Die erkannten Ereignisse sind bereits auf der untersten Automatisierungsebene verfügbar und werden dem Leitsystem, einer SPS oder dem Diagnose-Tool gemäß Bild 3 zur Verfügung gestellt. Im Diagnose-Tool wird die Erfahrung aus sehr vielen Applikationen über mehr als 20 Jahre zur Analyse der Ereignisse zur Verfügung gestellt. Ohne langes Nachschlagen in unterschiedlichen Dokumentationen erfährt der Bediener, welche Prozeßvariablen als unregelmäßig erkannt wurden. Mögliche Ursachen werden aufgezeigt und ggf. ergänzende Meßmethoden vorgeschlagen. Ist der Fehler eingegrenzt, so werden die Korrekturmaßnahmen angeboten, die auch den weniger mit der Materie vertrauten Bediener, z.B. das Schichtpersonal, schnell in die Lage versetzen, die Anlage in einen geordneten Betrieb zurückzuführen.

Die Diagnose ist als Hypertext-System ausgeführt, d.h. durch Anfahren von Sprungmarken und Stichworten wird dem Techniker die Funktion einer kontinuierlichen Waage transparent. Spezielle anlagenspezifische Kommentare werden vom Betreiber on line erstellt und sind dann jederzeit kontextsensitiv verfügbar. Das System realisiert damit für kontinuierliche Waagen erstmals einen Bedienkomfort, der aus dem Bereich des Software-Engineering bereits länger bekannt und geschätzt ist.

Der individuellen Programmierung eines so weitgehenden Diagnosesystems in einem Rechner beim Anwender steht entgegen, daß meist das notwendige Expertenwissen fehlt und die notwendigen Programmierzeiten die im Betrieb zu erwartenden Einsparungen bei weitem überschreiten dürften.

3.3 Parametrierung

Es besteht der Wunsch, auch die prozeßnahen Komponenten über den zentralen Engineering-Platz aus zu konfigurieren und zu parametrieren. Dem steht leider die noch unvollständige Normung der Kommunikation entgegen. Daher sind heute nur aufwendige Lösungen für einzelne Leitsy-

cess-oriented levels. The MULTICONT electronic weighing system offers many automatic correction, compensation and adjustment algorithms as well as a wide range of monitoring facilities for the weighing and metering process and also for the weigher peripherals, which frequently cause malfunctioning.

Detected events are already available at the lowest automation level, and are offered to the control system, a PLC system or the diagnostic tool as shown in Fig. 3. Experience from a great many applications over more than 20 years is available in the diagnostic tool to analyze events. The operator can find out which processing variables have been identified as irregular without lengthy consultation of different documents. Possible causes are given and measuring methods may be suggested. Once the fault has been defined corrective measures are offered. They enable even operators who are less familiar with the subject, such as shift workers, to get the plant back into operation in a short time.

Diagnosis is carried out as a hypertext system, i.e. a technician can see how a continuous weigher operates by entering branch labels and key words. Special comments specific to the plant may be recorded on line by the operator and are then available at any time in appropriate situations. An operating aid which has long been known and appreciated in software engineering is thus available for the first time in continuous weighing.

The obstacles to individual programming of such an extensive diagnostic system in a user's computer are that the necessary expert knowledge is not usually available, and the programming times would far exceed the expected savings in operation.

3.3 Setting the parameters

It would be desirable to have process-oriented components configured and parametered from the central engineering position, but this will unfortunately not be possible until communication is fully standardised. At present it is only possible to make expensive arrangements for individual families of control systems and individual process-oriented components. The method adopted in the MULTICONT system is to shift the parameterisation to a PC. By using the standard graphics MS Windows surface all controllers of the system can be shown clearly and at low cost, in a way which would otherwise involve unjustifiable expenditure.

Fig. 4 shows the operator surface, taking a differential metering weigher as an example. New or amended sets of parameters are created en bloc off line and loaded into the electronic weighing system at the desired time. Consistency checks provide an additional safeguard against incorrect setting. The extensive use of symbols gives far greater clarity, and suggested values specific to the application considerably shorten the startup time. Although an electronic weighing system with optimum reliability and full control of peripherals must necessarily be very complex, the method of adjustment remains clear and simple.

The controllers are also available at the serial interface to the control system or the PLC system. In this way individual values can be modified in the course of normal process control, while the service engineer can see the whole setup on the PC. Parameters may of course be set by remote data transmission, so that minor modifications can be made without visiting the location. This means that emergency operation, for example, can be arranged quickly.

4. Quality assurance

The quality of the cement which is the final product depends very much on the accuracy of the continuous metering equipment, as demonstrated by the example of the weighbelt feeder. Quality is made up of two components, accuracy and reliability [1], [2]. The preparation of quality assurance measures at the lowest automation level is of prime importance. All the other measures are founded on this, as illustrated in **Fig. 5**.

Reliability ensures that a plant is stable and continuous in its operation. The prerequisite for this is a perfected weighing

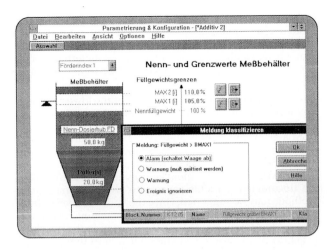

BILD 4: Beispiel einer graphisch unterstützten Parametrierung
FIGURE 4: Example of a graphically supported parameterization system

Parametrierung & Konfiguration — [*Additiv 2]	= Parameterisation and configuration [*additive 2]
Datei, Bearbeiten, Ansicht, Optionen, Hilfe	= File, Process, View, Options, Help
Auswahl	= Selection
Förderindex 1	= Conveying index 1
Nenn- und Grenzwerte Meßbehälter	= Nominal and limit values, metering containers
Meßbehälter Füllgewichtsgrenzen	= Metering containers Filling weight limits
Nennfüllgewicht 100 %	= Nominal filling weight 100 %
Nenn-Dosierhub FD	= Nominal metering range FD
Meldung klassifizieren	= Classify message
Meldung: Füllgewicht > BMAX1, Ok	= Message: Filling weight > BMAX1, OK
Alarm (schaltet Waage ab), Abbrechen	= Alarm (switches off weigher), Stop
Warnung (muß quittiert werden)	= Warning (must be acknowledged)
Warnung, Hilfe	= Warning, Help
Ereignis ignorieren	= Ignore event

stemfamilien und einzelne prozeßnahe Komponenten realisierbar. Im MULTICONT-System wurde daher der Weg beschritten, die Parametrierung auf einen PC zu verlagern. Die Verwendung der standardisierten graphischen Oberfläche MS-Windows erlaubt dabei die kostengünstige und übersichtliche Darstellung aller Einsteller des Systems in einer Art, die anders nur mit unvertretbarem Aufwand geschaffen werden könnte.

Bild 4 zeigt die Bedieneroberfläche am Beispiel einer Differentialdosierwaage. Neue oder geänderte Parametersätze werden en block off line erstellt und zum gewünschten Zeitpunkt in die Wägeelektronik geladen. Konsistenzprüfungen erhöhen dabei die Sicherheit gegen Einstellfehler. Die weitgehende Verwendung von Symbolen erhöht die Übersichtlichkeit entscheidend. Zusätzlich reduzieren applikationsspezifische Vorschlagswerte die Inbetriebnahme erheblich. Obwohl eine auf hohe Betriebssicherheit und vollständige Steuerung der Peripherie optimierte Wägeelektronik eine sehr hohe Komplexität aufweisen muß, bleibt die Einstellung damit übersichtlich und einfach.

Zusätzlich sind die Einsteller auch an der seriellen Schnittstelle zum Leitsystem oder zur SPS verfügbar. Einzelne Werte können so im Zuge der normalen Prozeßsteuerung modifiziert werden, während die Gesamtübersicht für den Servicetechniker auf dem PC entsteht. Selbstverständlich ist die Parametrierung auch zusammen mit der Datenfernübertragung zu nutzen, so daß kleinere Modifikationen ohne Anreise durchgeführt werden können. Damit wird es z.B. möglich, schnell einen Notbetrieb zu realisieren.

4. Qualitätssicherung

Die Qualität des Endprodukts Zement hängt entscheidend von der Genauigkeit der kontinuierlichen Dosiereinrichtungen ab, wie das Beispiel der Dosierbandwaage zeigt. Qualität setzt sich dabei aus den Komponenten „Genauigkeit" und „Betriebssicherheit" zusammen [1], [2]. Entscheidende Bedeutung kommt hier der Vorbereitung qualitätssichernder Maßnahmen auf der untersten Automatisierungsebene zu. Alle weiteren Maßnahmen bauen gemäß **Bild 5** darauf auf.

Hohe Betriebssicherheit stellt einen stabilen und kontinuierlichen Betrieb einer Anlage sicher. Die Basis liefert hier eine ausgereifte Wägemechanik kombiniert mit weitreichenden Überwachungen, die bereits vor einem Störfall über entsprechende Warnungen eine vorbeugende Wartung erlauben. Schenck-Waagen auf der Basis MULTICONT liefern, neben der integrierten Schlupfüberwachung für das Förderband, z.B. eine Bandpositionsüberwachung, die bereits vor einem vollständigen Ablauf des Bandes eine Warnung generiert. Die aktuelle Bandposition kann jederzeit auch per Datenfernübertragung abgerufen werden.

Ein weiterer Faktor für hohe Betriebssicherheit ist die zuverlässige Inbetriebnahme. Neben der graphischen Parametrierung vereinfachen im MULTICONT Einstellprogramme die Justage. Die Verwendung kalibrierter Ele-

mechanism combined with extensive monitoring, with appropriate warnings being issued to allow for preventive servicing before the equipment fails. Schenck weighers based on MULTICONT provide not only integral slip monitoring for the conveyor belt but also, for example, belt position monitoring system which gives a warning signal before the belt has gone completely line. The current position of the belt may be recalled at any time by data transmission.

Another factor involved in high operational reliability is dependable commissioning. Apart from graphic parameterisation, installation programs simplify the setting-up in MULTICONT. The use of calibrated components enables standard weighers to be fully commissioned without any test weights or material checks. The last two processes then provide extra information which can be used for quality assurance. This means that correct installation is also monitored in situ.

The absolute accuracy of a weigher is often of minor importance in cement production, whereas long-term stability is very significant [3]. Ideal monitoring is provided by an on-stream material check, with the amount of bulk material conveyed by the continuous weigher being compared with a quantity determined by an upstream check container. Advances in weighing electronics allow small check quantities to be used when both measuring points are contained in one system. Extremely small check quantities, which can

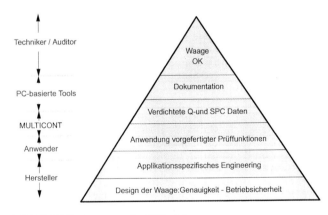

BILD 5: Maßnahmen zur Qualitätssicherung
FIGURE 5: Quality assurance measures

Techniker / Auditor	= Technician / auditor
PC-basierte Tools	= PC-based tools
Anwender	= User
Hersteller	= Manufacturer
Waage	= weigher
Dokumentation	= documentation
Verdichtete Q-und SPC Daten	= compressed quality and SPC data
Anwendung vorgefertigter Prüffunktionen	= application of previously compiled test functions
Applikationsspezifisches Engineering	= application-specific engineering
Design der Waage: Genauigkeit — Betriebssicherheit	= design: accuracy — reliability

mente erlaubt die vollständige Inbetriebnahme der Standardwaage ohne Prüfgewichte oder Materialkontrollen. Die beiden letzten Verfahren liefern nun redundante Informationen, die zur Qualitätssicherung zur Verfügung stehen. Somit wird auch der korrekte Einbau vor Ort zusätzlich überwacht.

Die absolute Genauigkeit einer Waage bei der Zementerzeugung ist oft von untergeordneter Bedeutung, allerdings kommt der Langzeitstabilität entscheidende Bedeutung zu [3]. Die ideale Überwachung leistet eine On-Stream Materialkontrolle, wobei die von der kontinuierlichen Waage geförderte Schüttgutmenge mit der mittels eines vorgeschalteten Kontrollbehälters bestimmten Menge verglichen wird. Die Fortschritte in der Wägeelektronik erlauben dann kleine Kontrollmengen, wenn beide Messungen in einem einzigen System untergebracht werden. Für die Detektion grober Ungenauigkeiten, z.B. durch ein irrtümlich auf die Wägebrücke aufgelegtes Werkzeug oder verklemmtes Schüttgut, eignen sich auch extrem kleine Kontrollmengen, die einen kostengünstigen Aufbau im Gebäude erlauben.

Als Alternative zu Materialkontrollen kann die Prüfgewichtseinrichtung gelten, bei der bei leerem Band ein bekanntes Prüfgewicht auf die Wägebrücke aufgebracht wird. Dieses Verfahren ist jedoch meist mit manuellen Eingriffen verbunden und setzt Betriebsunterbrechungen voraus.

Ereignisse aus Waage und Umfeld, Prüfgewichtskontrollen und Materialkontrollen liefern wichtige Daten, die auf die Qualität der Dosierung zu einem bestimmten Zeitpunkt schließen lassen. Diese Daten sind an der seriellen Schnittstelle zu bauseitigen Rechnern verfügbar. Andererseits werden viele der bauseitigen Systeme mit der Vielzahl der zur Verfügung gestellten Daten überfordert, besonders wenn man bedenkt, daß die Wägetechnik nur einen Teil der gesamten prozeßnahen Komponenten darstellt. Daher werden im MULTICONT zusätzlich zu den aktuellen Daten auch statistische Kenngrößen ermittelt. Ein Beispiel hierfür sind Dosiergenauigkeit und -konstanz als Maß für die Regelgenauigkeit und die dynamische Stabilität der Dosierung.

Alle Daten können mittels der Fernübertragung auch von einem PC aus gelesen werden. Der PC erstellt daraus die für die Qualitätssicherung notwendigen Statistiken und Qualitätsregelkarten. Zyklisch lassen sich so Qualitätsprotokolle erzeugen und einer Produktionscharge zuordnen. Die Protokolle sind eine wichtige Basis für die Planung von Instandhaltungsarbeiten, da durch die Trenddarstellung der Regelkarten zukünftige Betriebsprobleme vorhergesagt werden können. So läßt sich z.B. das Zuwachsen von Abzugsquerschnitten aus einem Behälter anhand der Tendenz der Bandbeladung erkennen, bevor die Wägegenauigkeit beeinträchtigt wird. Korrelationen zu Umfeldbedingungen, z.B. der Verwendung von Rohstoffen mit geänderten Eigenschaften, erlauben es, die Prozesse besser und stabiler zu steuern.

Das Format der Qualitätsprotokolle wird sich stets nach den speziellen Erfordernissen vor Ort richten. Das Qualitätssicherungs-Tool für MULTICONT stellt daher einen Protokollgenerator zur Verfügung, der aus einer Formatdatei und den aus der Waage gelesenen Werten für jede Waage das spezifische Protokoll erzeugt. Tabellarische, graphische Darstellungen und ein Gesamturteil der Waagenqualität können damit vor Ort konfiguriert und bei Bedarf geändert werden.

Das Tool stellt damit eine kostengünstige Alternative zur Programmierung dieser Aufgaben in einem zentralen Qualitätssystem dar oder übernimmt die Aufgaben bis zur Verfügbarkeit eines solchen Systems. Zur Vorverarbeitung der Qualitätsdaten kann es zwischen Waage und zentralem System geschaltet werden. Der Zugriff auf die Qualitätsdaten über das Qualitätsdatennetzwerk ist somit gewährleistet.

be provided at low cost in the building, are suitable for detecting gross inaccuracies, e.g. a tool placed accidentally on the weighbridge or compacted bulk material.

An alternative to material checks is check weight equipment, where a known check weight is put on the weighbridge while the belt is empty. However the process usually involves manual action and production has to be interrupted.

Events from the weigher and its surroundings, check weight controls and material check provide important data indicating the quality of metering at a given time. This data is available at the serial interface with users' computers. On the other hand many of the users' systems are overburdened with all the available data, especially considering that the weighing technique only accounts for one part of the process-oriented components. Statistical quantities are therefore determined in MULTICONT as well as current data. An example of these are metering accuracy and constancy as a measure of the control accuracy and dynamic stability of the metered feeding system.

All the data can also be read by a PC by remote transmission. The PC uses it to draw up the statistics and quality control cards required for quality assurance. Quality reports can thus be drafted cyclically and assigned to a production batch. The reports are an important basis for planning maintenance work, since future operating problems can be predicted from the trends revealed by the control cards. For example, the gradual blockage the of discharge cross-section from a hopper can be recognised from the belt loading tendency before weighing accuracy is affected. Correlation with conditions in the surroundings, e.g. the use of raw materials with different properties, allow better, more stable control of the processes.

The format of the quality report will always depend on special local requirements. The quality assurance tool for MULTICONT therefore provides a report generator which creates the specific report for each weigher from a format file and the values read from the weigher. Tabulated graphic representations and a total assessment of weigher quality can thus be configured in situ and changed as required.

The tool is thus a cost-effective alternative to programming these tasks in a central quality system, or it takes on the tasks until such a system is available. Quality data may be pre-processed by switching between the weigher and the central system, so it can be accessed through the quality data network.

5. Literatur

[1] Allenberg, B., und Jost, G.: Kurzzeitgenauigkeit und Betriebssicherheit bei der Schüttgutdosierung – Fortschritte durch Smart Control Strategies, In: wägen + dosieren, 22. Jg. 3/92, S. 12 17.

[2] Allenberg, B.: Requirements on Continuous Weighers in a Quality Assurance System, In: Bulk Solids Handling, Vol. 13, No. 2, May 1993, pp. 314–318.

[3] Allenberg, B.: BIC – der neue Weg zu höchster Dosiergenauigkeit, In: Zement-Kalk-Gips, 1993.

Computerunterstützte Rohstoffplanung
Computer-aided raw material planning

Planification des matières premières assistée par ordinateur

Planificación de materias primas con ayuda del ordenador

Von **W. Baumgartner,** Holderbank/Schweiz

Zusammenfassung – Die Anforderungen an die Rohstoffplanung nehmen laufend zu. Es ist daher erforderlich, neue Werkzeuge wie Computer für eine wirtschaftliche Planung und Produktion einzusetzen. „Holderbank" hat bereits vor mehr als 15 Jahren begonnen, computerunterstützte Rohstoffbewertungs- und -planungssysteme zu entwickeln und praktisch einzusetzen. Lagerstättenmodellierungssysteme wurden 1977 eingeführt, gefolgt von QSO (Quarrying Scheduling Optimization) im Jahr 1982, um die tägliche Rohstoffgewinnung sowie den gesamten Lebenszyklus eines Steinbruches umfassend zu optimieren. Um die Zielsetzungen Sicherstellung einer stetigen und ständig verfügbaren Rohstoffversorung bei gleichzeitig geringstmöglichen Kosten und einer längstmöglichen Lebensdauer zu erfüllen, besteht das System aus den drei Hauptelementen Inventarisierung der Rohstoffressourcen mittels eines Lagerstättenmodells, strategische, langfrisitige Planung zur optimalen Rohstoffnutzung und tägliche Produktionsplanung und -kontrolle. Die von „Holderbank" durchgeführten Schätzungen ergaben ein jährliches Kosteneinsparungspotential von insgesamt mehr als 100 000 bis über 1 Mio. US$. Das entspricht einer Einsparung von etwa 0,1 bis 0,3 US $/t Rohmehl. Sie wurde durch Verringerung des Abraums, der Korrekturmaterialkosten und durch erhöhte Ofenverfügbarkeit aufgrund einer gleichmäßigeren Rohmischungszusammensetzung erreicht.

Computerunterstützte Rohstoffplanung

Summary – The demands on raw material planning are increasing constantly. It is therefore necessary to use new tools such as computers for cost-effective planning and production. "Holderbank" started more than 15 years ago to develop computer-aided systems for raw material evaluation and planning and put them into practice. Systems for modelling deposits were introduced in 1977, followed in 1982 by QSO (Quarrying Scheduling Optimization) for comprehensive optimization of daily raw material excavation and of the entire life cycle of a quarry. In order to fulfil the objectives of ensuring a constant and continuously available supply of raw material with, at the same time, the lowest possible costs and longest possible life, the system consists of the three main elements of making an inventory of the raw material resources using a model of the deposit, strategic long-term planning for optimum utilization of raw materials, and daily production planning and control. The assessments carried out by "Holderbank" resulted in a yearly savings potential of a total of more than 100 000 to over 1 million US $. This corresponds to a saving of about 0.1 to 0.3 US $/t raw meal. It was achieved by reducing the waste and the costs of correcting material, and by increasing kiln availability through more uniform raw mix composition.

Computer-aided raw material planning

Résumé – Les exigences posées à la planification des matières premières se multiplient toujours. Il est donc nécessaire d'utiliser des outils nouveaux, tels que les ordinateurs, pour une planification et une production économiques. „Holderbank" a commencé il y a déjà plus de 15 ans, à développer des systèmes d'évaluation et de planification des matières premières, assistés par ordinateur et de les mettre en oeuvre dans la pratique. Des systèmes de modelisation de gisements ont été introduits en 1977, suivis de QSO (Quarrying Scheduling Optimization) dans l'année 1982, pour optimiser exhaustivement l'extraction journalière des matières premières ainsi que la totalité du cycle de vie d'une carrière. Afin d'atteindre les objectifs d'assurer un approvisionnement constant et disponible en permanence avec, en même temps, des coûts aussi bas et une durée d'exploitation aussi longue que possibles, le système se compose des trois éléments principaux inventaire des ressources de matières premières au moyen d'un modèle de gisement, planification stratégique à longue date pour l'utilisation optimale des matières premières et planification et contrôle journaliers de la production. Les évaluations conduites par „Holderbank" ont révélé un potentiel annuel d'économie de coûts, d'au total plus de cent mille jusqu'à plus d'un million de $ US. Cela correspond à une économie d'environ 0,1 à 0,3 $ US/t de farine crue. Elle a été obtenue par la diminution des stériles, des coûts de matières de correction et par une plus grande disponibilité du four, grâce à une composition plus régulière du mélange cru.

Planification des matières premières assistée par ordinateur

Resumen – Los requerimientos respecto de la planificación de materias primas aumentan constantemente. Por esta razón, es necesario emplear nuevos útiles, tales como ordenadores, para la planificación económica de la producción. "Holderbank" empezó, hace ya más de 15 años, a desarrollar y aplicar en la práctica sistemas de evaluación y planificación de materias primas, basados en el empleo de ordenadores. En 1977 se introdujeron sistemas de modelación de yacimientos, a los que siguió el QSO (Quarrying Scheduling Optimizacion) en el año 1982, destinado a optimizar, de forma completa, la extracción diaria de materias primas así como el ciclo de vida total de una cantera.

Planificación de materias primas con ayuda del ordenador

STOCKYARD EQUIPMENT

PORTAL SCRAPER RECLAIMERS

PORTAL SCRAPER RECLAIMERS

SEMI PORTAL SCRAPER RECLAIMERS

CANTILEVER SCRAPER RECLAIMERS

BRIDGE SCRAPER RECLAIMERS

PORTAL BRIDGE SCRAPER RECLAIMERS

CIRCULAR STORAGES

STACKER-RECLAIMERS

STACKER-RECLAIMERS

CIRCULAR STORAGES

SLEWING SCRAPER RECLAIMERS

SLEWING STACKERS

SCHADE
DORTMUND

GUSTAV SCHADE MASCHINENFABRIK
Postfach 103 642 · D-44036 Dortmund (Germany) · Phone: (02 31) 45 07-0 · Telex: 822 429 · Telefax: (02 31) 45 23 55

SCHADE PORTAL BRIDGE SCRAPER RECLAIMER

The Portal Bridge Scraper Reclaimer combines the best homogenisation characteristics by reclaiming from the front slope of the stockpile from as many blending beds as required with the facility to reclaim frozen, consolidated or fire hazard stockpiles at the side slope.

GUSTAV SCHADE MASCHINENFABRIK · Postfach 10 36 42 · D 44036 Dortmund · Phone: (02 31) 45 07-0 · Telex: 8 22 429 · Telefax: (02 31) 45 23 55

Para garantizar el objetivo del aseguramiento de un suministro constante y siempre disponible de materias primas, reduciendo al mismo tiempo los gastos a un mínimo y aumentando lo más posible la duración de vida, el sistema se compone de tres elementos principales, que son: inventario de los recursos de materias primas mediante la modelación de los yacimientos, planificación estratégica y a largo plazo de la utilización óptima de las materias primas y planificación y control diario de la producción. Las evaluaciones llevadas a cabo por "Holderbank" han dado como resultado un potencial de ahorro total de gastos de más de 100 000 y hasta más de 1 millón de US$ al año, lo que equivale a un ahorro del orden de 0,1 a 0,3 US$/t de crudo. Este ahorro se ha podido conseguir mediante la reducción de monteras y de material de corrección así como mediante una mayor disponibilidad del horno, debido a una composición más uniforme de la mezcla de crudos.

1. Die Notwendigkeit neuer Werkzeuge

Rohstoffe sind eine der Grundvoraussetzungen zur Zementherstellung. Ohne vorausschauende, sorgfältige Planung ist nicht nur die Rentabilität, sondern auch die mittel- und langfristige Stellung eines Zementherstellers gefährdet.

Die Rohstoffplanung selbst sieht sich laufend steigenden Anforderungen ausgesetzt. In der Zementindustrie wurde nun ein Komplexitätsgrad erreicht, der es nicht mehr erlaubt, ohne neue Werkzeuge – wie den Computer – wirtschaftlich zu planen und zu produzieren.

Auf dem Gebiet der Sicherstellung und Versorgung mit Zementrohstoffen wird die steigende Komplexität im wesentlichen von folgenden Faktoren verursacht:

— Bedarf nach immer größeren Rohstoffmengen je Produktionseinheit, bei

— immer strenger werdenden Qualitätsanforderungen,

— stark zunehmenden behördlichen und umweltbedingten Auflagen und Einschränkungen,

— laufend steigenden Kosten sowie

— wachsender Anzahl von Zementtypen und sich rasch ändernden Produktanforderungen.

Kein Zementhersteller kann es sich daher leisten, die Vorteile der Kosten- und Betriebsoptimierung zu vernachlässigen, falls bestimmte Wirtschaftlichkeitsziele erreicht und eingehalten werden sollen. Daraus folgt die Notwendigkeit einer genauen Lagerstättenbewertung und verläßlichen Abbauplanung unter Einsatz intelligenter und rascher Methoden. Da große Datenmengen aus Exploration und Produktion laufend und unter Berücksichtigung zahlreicher Anforderungen und Einschränkungen verarbeitet werden müssen, ist der Einsatz von Computern unumgänglich.

Aus diesen Gründen hat „Holderbank" bereits vor mehr als 15 Jahren begonnen, computerunterstützte Rohstoffbewertungs- und -planungssysteme zu entwickeln und praktisch einzusetzen. Lagerstättenmodellierungssysteme wurden 1977 eingeführt, gefolgt von QSO (Quarry Scheduling Optimization) im Jahr 1982, um die tägliche Gewinnung sowie den gesamten Lebenszyklus eines Steinbruches vollumfänglich zu optimieren.

2. Rohstoffplanungssystem

Um die Zielsetzungen

— Sicherstellung einer stetigen und ständigen Rohstoffversorgung vom Steinbruch zu den Klinkerproduktionseinheiten bei gleichzeitiger

— wirtschaftlicher Nutzung der verfügbaren Rohstoffe, d. h. zu geringstmöglichen Kosten über die längstmögliche Lebensdauer

bestmöglich zu erfüllen, besteht das „Holderbank"-System aus drei Hauptelementen:

— Genaue Inventarisierung der Rohstoffressourcen mittels eines Lagerstättenmodelles.

— Strategischer mittel- und langfristiger Plan zur optimalen Rohstoffnutzung.

— Tägliches Produktionsplanungs- und -kontrollprogramm.

1. The necessity for new tools

Raw materials are one of the basic requirements for cement manufacture. Without careful, forward-looking, planning not only the profitability, but also the medium- and long-term position of a cement manufacturer are placed at risk.

Raw material planning itself is exposed to continuously rising requirements. The cement industry has now reached a level of complexity where cost-effective planning and production is no longer possible without new tools – like the computer.

The increasing complexity in the field of securing and providing cement raw materials is caused to a great extent by the following factors:

— demand for ever larger quantities of raw materials per production unit, with

— ever stricter quality specifications,

— greatly increasing official and environmental injunctions and restrictions,

— continuously rising costs, and

— growing number of cement types and rapidly changing product specifications.

No cement manufacturer can therefore afford to neglect the advantages of cost and operation optimization if certain economic objectives are to be reached and maintained. From this follows the necessity for accurate evaluation of the deposit and reliable planning of the excavation using intelligent and rapid methods. The large quantities of data from exploration and production must be processed continuously and must take account of large numbers of requirements and constraints, so the use of computers is unavoidable.

For these reasons „Holderbank" started to develop and use computer-aided systems for evaluating and planning raw materials more than 15 years ago. Deposit modelling systems were introduced in 1977, followed by QSO (quarry scheduling optimization) in 1982 to ensure full optimization of the daily excavation work and the total life cycle of a quarry.

2. Raw material planning system

In order to fulfil the objectives

— of securing a constant and continuous supply of raw materials from the quarry to the clinker production units with, at the same time

— cost-effective utilization of the available raw materials, i.e. with the lowest possible costs over the longest possible life

in the best way possible, the „Holderbank" system consists of three main elements:

— taking an accurate inventory of the raw material resources using a model of the deposit,

— strategic medium- and long-term plan for optimum raw material utilization,

— daily production planning and checking programme.

The combined application of the modules developed over the course of time resulted in an array of instruments which

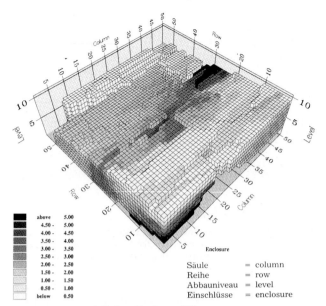

Säule	=	column
Reihe	=	row
Abbauniveau	=	level
Einschlüsse	=	enclosure

BILD 1: Blockmodell einer Tonlagerstätte.
Die verschiedenen Grautöne zeigen die Verteilung des SO_3-Gehaltes
FIGURE 1: Block model of a clay deposit
The different tones of grey show the distribution of the SO_3 content

Durch kombinierte Anwendung der im Laufe der Zeit entwickelten Module entstand ein Instrumentarium, das die oben aufgeführten Aufgaben von der strategischen, langfristigen Planung bis zur täglichen Produktionssteuerung effizient und optimal löst. Die oft gestellte Frage nach Eigenschaften und Nutzen der wichtigsten Module soll in den folgenden Abschnitten behandelt werden.

2.1 Blockmodell

Die Berechnung von Lagerstätten- oder Blockmodellen hat zum Ziel, eine objektive, verläßliche und reproduzierbare Lagerstättenbeschreibung zu erhalten. Ein Blockmodell ist eine vollständige dreidimensionale Beschreibung einer Lagerstätte (**Bild 1**). Zur Berechnung des Modells wird die Lagerstätte in eine große Anzahl kleiner Blöcke unterteilt. Jeder Block repräsentiert eine Abbaumenge, die einer Wochen- oder Monatsproduktion entspricht. Von Bohrungen und anderen Explorationsdaten ausgehend, werden jedem Block mittels Interpolationsverfahren Gehalte und andere Kennwerte zugeordnet. Mit anderen Worten, verschiedene Informationen wie Bohrergebnisse, geologische Untersuchungen, geochemische und geophysikalische Resultate usw. werden in eine bestmögliche und einheitliche Lagerstättenbeschreibung umgewandelt.

Das so entstehende Blockmodell ist ein genaues Lagerstätteninventar, das für jeden Punkt (Block) die zu erwartende Qualität und Menge beschreibt. Es stellt heute eines der wirkungsvollsten Instrumentarien zur Rohstoffbewertung dar.

2.2 Strategische Abbauplanung (QSO)

Um geologisch und chemisch komplexe Lagerstätten bestmöglich zu nutzen, ist die Anwendung von Operations-Research-Methoden erforderlich. Eine erste praktische Anwendung für eine „Holderbank"-Gruppengesellschaft erfolgte bereits 1982. Die damals entwickelten Ideen und die erzielten Resultate waren ermutigend und richtungsweisend. Die Anwendung dieser Methoden war jedoch relativ teuer, da Großcomputer und gut ausgebildetes Personal hierfür erforderlich waren. Der Durchbruch von QSO (Quarry Scheduling Optimization) erfolgte 1985 mit der zunehmenden Verbreitung der Mikrocomputer, da QSO ein Planungswerkzeug ist, das häufig und vor Ort eingesetzt wird.

Während das Blockmodell den Zustand der Lagerstätte und insbesondere der Rohstoffreserven beschreibt, zeigt QSO in Form von Abbauplänen, wie man die Lagerstätte bestmöglich nutzen kann. Technisch durchführbare Abbaupläne, die natürlich immer die Rohmischungsanforderungen in Menge und Qualität erfüllen, werden auf der Grundlage von

can provide efficient and optimum solutions to the problems listed above, ranging from strategic long-term planning to daily production control. The most frequent questions about the characteristics and benefits of the most important modules will be dealt with in the following sections.

2.1 Block model

The object of calculating deposit or block models is to obtain an objective, reliable, and reproducible description of the deposit. A block model is a complete three dimensional description of a deposit (**Fig. 1**). To calculate the model, the deposit is subdivided into a large number of small blocks. Each block represents a quantity of quarried material corresponding to a week's or a month's production. Based on drillings and other exploration data, interpolation methods are used to assign the contents and other parameters to each block. In other words, a variety of information such as drill results, geological investigations, geochemical and geophysical results, etc., are converted into the best possible consistent description of the deposit.

The resulting block model is an accurate inventory of the deposit which describes the expected quality and quantity for each point (block). It now represents one of the most effective instruments for raw material evaluation.

2.2 Quarry scheduling optimization (QSO)

In order to make the best possible use of geologically and chemically complex deposits, it is necessary to apply operations research methods. The first practical application for a „Holderbank" group took place as early as 1982. The ideas developed at that time and the results achieved were encouraging and showed the way forward. The application of these methods was, however, relatively expensive as it required large computers and well-trained personnel. The break-through for QSO (quarry scheduling optimization) took place in 1985 with the increasing spread of microcomputers as QSO is a planning tool which is used frequently and on-site.

The block model describes the state of the deposit and, in particular, the raw material reserves, but the QSO shows in the form of excavation plans how the best possible use can be made of the deposit. Industrially feasible quarrying plans, which naturally have to fulfil the raw mix requirements for quantity and quality, are calculated on the basis of deposit data and raw mix requirements. QSO results can, for example, be shown in the form of perspective representations of the quarry development. From this it can be seen how the quarry will look at a certain point in time if the proposed quarrying plan is followed (**Fig. 2**).

Such detailed representations of the quarry – prepared using the technique of digital terrain modelling – show the quarry development from the operational point of view clearly and in a readily understandable form. The same technique can also be used for planning and evaluating the environmental impact of a proposed quarry in order to meet, as far as is possible, the increasing objective of blending the quarry as inconspicuously as possible into the landscape.

2.3 Production planning and checking

Good production planning – the daily business – requires in many cases the use of optimization methods if a regular raw mix composition is to be achieved; this pays for itself in lower raw mix costs, improved kiln availability and consistent clinker quality.

Short-term excavation planning and production control requires additional data with respect to the raw material composition. Average values which are adequate for medium- and long-term excavation planning are no longer sufficient. The daily fluctuations require special consideration which now necessitates a more detailed databank than for medium- and long-term planning tasks. The use of blasting hole analyses, sampling stations, or other continuous production data are recommended for refining the long-term plans.

Lagerstättendaten und Rohmischungsanforderungen errechnet. QSO-Resultate können beispielsweise in Form von perspektivischen Darstellungen der Steinbruchentwicklung gezeigt werden. Man ersieht daraus, wie der Steinbruch zu einem gewissen Zeitpunkt aussehen wird, falls man dem vorgeschlagenen Abbauplan folgt (**Bild 2**).

Solch detaillierte Steinbruchdarstellungen – erstellt mit der Technik der digitalen Geländemodellierung – zeigen klar und leicht verständlich die Steinbruchentwicklung aus betrieblicher Sicht. Dieselbe Technik kann auch zur Planung und Bewertung der umweltrelevanten Auswirkungen eines vorgeschlagenen Steinbruches eingesetzt werden, um dem immer häufigeren Ziel, den Steinbruch so unauffällig wie möglich in die Landschaft einzubetten, möglichst zu entsprechen.

2.3 Produktionsplanung und Kontrolle

Gute Produktionsplanung, das Tagesgeschäft, benötigt in vielen Fällen den Einsatz von Optimierungsmethoden, will man eine gleichmäßige Rohmischungszusammensetzung erzielen, was sich in geringeren Rohmischungskosten, verbesserter Ofenverfügbarkeit und einheitlicher Klinkerqualität auszahlt.

Die kurzfristige Abbauplanung und Produktionssteuerung erfordern zusätzliche Angaben bezüglich der Rohmaterialzusammensetzung. Durchschnittswerte, wie sie für mittel- und langfristige Abbauplanung genügen, sind nicht mehr ausreichend. Die täglichen Schwankungen verlangen besondere Beachtung, was nun eine detailliertere Datenbasis erfordert als für mittel- und langfristige Planungsaufgaben. Zur Verfeinerung der langfristigen Pläne bietet sich die Verwendung von Sprengbohrlochanalysen, Probenahmestationen oder anderer kontinuierlicher Produktionsdaten an.

Durch die Anwendung der QSO-Technik für diese kurzfristigen Zielsetzungen wird die Produktionssteuerung zu einem maßgebenden Werkzeug für die Qualitätskontrolle. Tatsächlich stellt es den ersten Schritt in einem modernen Qualitätskontrollkonzept dar. Denn Qualität muß geplant und nicht nur kontrolliert werden. Planung und Kontrolle haben eine starke Wechselwirkung, und durch den Einsatz von QSO für die Produktionsplanung ist die Verbindung zwischen Planung und Kontrolle sichergestellt.

Insgesamt gibt es damit ein lückenloses Zusammenspiel von Lagerstättenbewertung über Abbauplanung bis zur Produktionssteuerung und Qualitätskontrolle. Dies ist heute eine Voraussetzung, um die gewünschte Produktqualität und -quantität über eine längere Zeit mit geringstmöglichen täglichen Schwankungen sicherzustellen und dabei die verfügbaren Rohstoffreserven immer noch bestmöglich zu nutzen.

2.4 Anwendbarkeit

Mathematisch gesehen handelt es sich bei der Berechnung optimaler Abbaupläne um eine anspruchsvolle Aufgabe. Die Berechnung von Blockmodellen ist ebenfalls ein nicht einfaches, sehr rechenintensives Unterfangen. Dementsprechend waren am Anfang der Entwicklung große und teure wissenschaftliche Rechner und gut geschultes Personal erforderlich.

Heutzutage können diese Aufgaben auf preisgünstigen Mikrocomputern (Personalcomputer) durchgeführt werden, mit dem Resultat, daß das System rasch und einfach in den Fabriken installiert und betrieben werden kann. Die Bohrlochdatenbank, das Blockmodell und die Abbauplanung werden von Experten erstellt und eingerichtet. Nach einer relativ kurzen Einschulungsphase kann das Personal das System vollumfänglich bedienen und einsetzen. Dies ist eine ganz wesentliche Voraussetzung zur Erzielung bester Ergebnisse sowohl vom Computer-System als auch vom Steinbruchpersonal.

Es ist festzuhalten, daß neben dem Stand der Technik entsprechenden Softwareprodukten einer der Erfolgsfaktoren bei der Implementierung von Rohstoffplanungssystemen

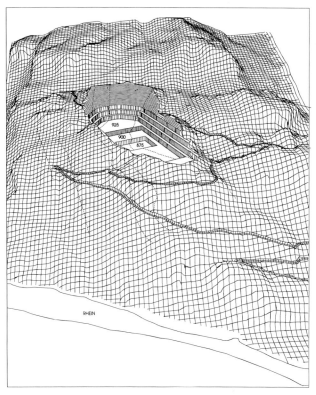

BILD 2: Perspektivische Darstellung einer Steinbruchentwicklung
FIGURE 2: Three-dimensional representation of the development of a quarry

By using the QSO technique for these short-term objectives, the production control system becomes an important tool for quality control. In fact, it represents the first step in a modern quality control scheme as quality must be planned and not just checked. There is a strong interaction between planning and checking, and the use of QSO for production planning ensures the link between planning and checking.

Taking as a whole, it provides uninterrupted coordination from deposit evaluation through excavation planning, right up to production control and quality checking. Nowadays, this is a precondition for ensuring the required quality and quantity of product over a fairly long period with the lowest possible daily fluctuations, and at the same time for continuing to make the best possible use of the available raw material reserves.

2.4 Applicability

From the mathematical point of view, the calculation of optimum excavation plans is a demanding task. Calculation of the block models is also a computationally intensive undertaking which is not at all simple. At the start of the development it was therefore necessary to have large and expensive scientific computers and well-trained personnel.

Nowadays these tasks can be carried out on low-cost microcomputers (personal computers) with the result that the system is rapid and simple to instal and operate in factories. The drill hole databank, the block model, and the excavation planning are prepared and established by experts. The personnel can operate and use the system to its full extent after a relatively short training phase. This is an essential requirement for achieving the best results both from the computer system and from the quarry personnel.

It must be recorded that in addition to the high level of appropriate software products, one of the success factors in implementing raw material planning systems is „experience". Proof of „Holderbank's" expertise in this field is provided by the more than 100 projects throughout the world which have been carried out successfully using computer-aided raw material planning. The present software system can be described as developed „by a cement manufacturer for a cement manufacturer".

die „Erfahrung" ist. Der Nachweis über die Expertise von „Holderbank" auf diesem Gebiet wird durch die mehr als 100 weltweiten Projekte, die erfolgreich mit Hilfe computerunterstützter Rohstoffplanung durchgeführt wurden, erbracht. Das heute vorliegende Softwaresystem kann als ein „von einem Zementhersteller für einen Zementhersteller" entwickeltes bezeichnet werden.

3. Wirtschaftlichkeit

Immer öfter wird, unter dem laufend steigenden Kostendruck, die Frage nach Kosteneinsparungen durch computerunterstützte Rohstoffplanung gestellt. Neben dem indirekten Nutzen – wie beispielsweise besseres und verläßlicheres Wissen um die Lagerstätte, rasche und bestmögliche Resultate in Bewertung und Planung, Quantifizierung von Risiken, Bewertung verschiedener Szenarien – interessieren vor allem die meßbaren Produktionskosteneinsparungen. In der Vergangenheit von „Holderbank" durchgeführte Schätzungen zeigten in verschiedenen Situationen ein jährliches Einsparungspotential von bis zu einigen hunderttausend US$ auf. Auf eine Tonne Rohmischung bezogen, bedeutet dies meist 10–30 Cents Kosteneinsparungen. Diese wurden durch Verringerungen der Abraummenge, der Korrekturmaterialkosten und durch erhöhte Ofenverfügbarkeit aufgrund einer gleichmäßigen Rohmischungszusammensetzung erreicht.

Die gesamten Planungskosten betrugen jeweils 100 000–200 000 US$. Unter Berücksichtigung von Rückzahlung in der oben erwähnten Höhe handelt es sich somit um äußerst lohnende Investitionen, wie Kosteneinsparungen und Rückzahlungsperioden von etwa einem Jahr aufzeigen. Ein anderer Nachweis des Nutzens der computerisierten Rohstoffplanung kann von der bereits erwähnten Projektanzahl abgeleitet werden. Bislang wurden über 100 internationale Projekte durchgeführt. Alleine 1993 waren es 20 Projekte für Zementhersteller und Kalkproduzenten auf allen 5 Kontinenten.

Die Zunahme der Anzahl der Projekte spiegelt die Akzeptanz dieser Produkte durch die Produzenten wider. Diese Akzeptanz ist einerseits in der Notwendigkeit einer zunehmend umsichtiger und umfassender werdenden Rohstoffplanung und andererseits in der Verfügbarkeit von verbesserten und nützlicheren, PC-basierten Programmen begründet. Heute liefert die computerisierte Rohstoffplanung die Führungsinstrumente für eine langfristige, gesellschaftsweite Planung und die operationellen Werkzeuge, um produktionsorientierte Ziele, wie minimale Kosten und optimale Produktqualitäten, zu verwirklichen.

3. Cost-effectiveness

Under the continuously rising pressure of costs the question about cost savings through computer-aided raw material planning appears with increasing regularity. In addition to the indirect benefits – such as, for example, better and more reliable knowledge about the deposit, rapid and optimum results in evaluation and planning, quantification of risks, evaluation of different scenarios – the main interest lies in the measurable savings in production costs. Estimates carried out by „Holderbank" in the past showed an annual savings potential of up to a few hundred thousand US$ in various situations. Relative to one tonne of raw mix this normally means cost savings of 10–30 cents. These were achieved through reductions in the quantity of overburden and in the costs of correction material, and through increased kiln availability because of a more uniform raw mix composition.

At the time the entire planning costs amounted to 100 000–200 000 US$. Bearing in mind the payback at the level mentioned above this is therefore an exceptionally worthwhile investment, as is shown by cost savings and payback periods of about one year. Further evidence of the benefit of computerized raw material planning can be deduced from the number of projects already mentioned. So far over 100 international projects have been carried out. In 1993 alone, there were 20 projects for cement manufacturers and lime producers on all five continents.

The increase in the number of the projects reflects the acceptance of these products by the producers. This acceptance is based on the one hand on the need for an increasingly circumspective and comprehensive raw material planning system and, on the other hand, on the availability of improved and more serviceable PC-based programs. Nowadays the computerized raw material planning system provides a management instrument for long-term company-wide planning, and the operational tools for putting into effect such production-orientated targets as minimum costs and optimum product quality.

Heutiger Stand der vollautomatischen Verpackungstechnik von Baustoffen aller Art

Present state of fully automatic packaging technology for all types of building materials

Etat actuel de la technique d'ensachage totalement automatisée de matériaux de construction de toute sorte

Estado actual de la técnica de envasado totalmente automático de todo tipo de materiales para la construcción

Von **R. Festge,** Oelde/Deutscland

Zusammenfassung – Die Leistung moderner rotierender Zement-Absackmaschinen beträgt heute 5500 Sack/h. Neben der Leistungserhöhung und der Reduzierung des Aufwandes für Bedienungspersonal bestand das wesentliche Ziel der Entwicklung in erster Linie darin, die Betriebskosten weiter zu senken. Hierfür wurden die Bedienung der Maschinen vereinfacht, die Lebensdauer der Verschleißteile erhöht und Einsparungen an Sackmaterial wie auch Verbesserungen in der Wägegenauigkeit vorgenommen. Elektronische Steuerungen unterstützen diese Entwicklung. Wesentliche Voraussetzung für den Erfolg waren jedoch einschneidende Verbesserungen im Bereich der Fülltechnik durch die Sackgewichtskontrolle über eine elektronische Bandkorrektur. Die bei der Sackverpackung anfallenden Daten können elektronisch aufgezeichnet, gespeichert und ausgewertet werden, wodurch die automatische Überwachung des Packbetriebs ermöglicht wird. Bei den Ventilsack-Aufsteckautomaten wurde neben einer stark verbesserten Sicherheit und Trefferquote auch eine deutliche Leistungserhöhung auf ca. 3000 Sack/h erreicht. Zusätzlich wurde ein Depalettiersystem für fabrikneue Leersackbündel entwickelt, die nicht umreift zu sein brauchen. Der Roboter mit seinem Greifersystem ist in der Lage, mehrere Aufsteckautomaten gleichzeitig zu beschicken. Somit ist auch die letzte Lücke im vollautomatischen Packbetrieb – vom Leersacklager bis zur beladenen Palette – geschlossen. Die für Abfüllung von Trockenmörteln, Putzen und Bauchemieprodukten heute gebräuchlichen luftarmen Füllsysteme ermöglichen mit Hilfe der Filterventiltechnik und der Ventilverschließsysteme ein sauberes Füllen nichtperforierter Ventilsäcke für den Verkauf auf Baumärkten.

Summary – Modern rotary cement bagging machines now have capacities of 5500 bags/h. In addition to increasing throughput and reducing the expenditure on operating personnel, the important objective of development has been primarily to lower operating costs still further. To this end the machines were made simpler to operate, the service life of the wearing parts was increased, and savings were made in bag material as well as improvements in the accuracy of weighing. This development was assisted by electronic control systems. However, an essential condition for the success was drastic improvements in the sector of filling technology by checking bag rates over an electronic belt correcting system. The data produced in the bag packing system can be displayed electronically, stored and evaluated, which makes it possible to monitor the packing operation automatically. With the automatic valve bag applicators, not only were sharply improved reliability and percentage success rates achieved, but also a significant increase in output to about 3000 bags/h. A depalletizing system was also developed for empty bag bundles fresh from the factory which do not need to be banded. The robot with its gripper system is capable of feeding several automatic bag applicators simultaneously. This now closes the last gap in fully automatic packing operation – from empty bag store to loaded pallets. The low-air filling systems now normal for bagging dry ready-mix mortars, plasters and building chemical products, aided by filter valve technology and the valve closing system, make it possible to fill unperforated valve bags cleanly for sale in the building material markets.

Résumé – La performance des ensacheuses rotatives modernes pour ciment atteint aujourd'hui 5500 sacs/h. A côté de l'augmentation de la performance et de la diminution des dépenses de main-d'oeuvre de conduite, l'objectif essentiel du développement est, en premier lieu, d'abaisser encore les coûts d'exploitation. Pour ce faire, la conduite des machines a été simplifiée, la durée de service des pièces d'usure a été prolongée et des économies en matière de sacs, comme aussi des améliorations de l'exactitude de pesage, ont été apportées. Des commandes électroniques ont favorisé ce développement. Une condition essentielle du succès étaient néanmoins des améliorations décisives dans le domaine de la technique de remplissage, par le contrôle du poids des sacs au moyen d'une correction électronique de la bande. Les données recueillies lors de l'ensachage peuvent être affichées, stockées et exploitées électroniquement, ce qui rend possible la surveillance automatique du fonctionnement de l'ensachage. Sur les

automates d'enfilage des sacs à valve a pu être atteinte, outre une sécurité et une précision de présentation fortement améliorées, aussi une nette augmentation de la performance, à environ 3000 sacs/h. En complément a été mis au point un système de dépalettisation pour des paquets de sacs vides neufs, qui ne requiert plus le cerclage. Le robot, avec son système de préhension, est à même de servir simultanément plusieurs automates d'enfilage. Ainsi, la dernière brèche dans le fonctionnement automatique de l'emballage – du stock des sacs vides jusqu'à la palette chargée – est maintenant comblée. Les systèmes de remplissage à vide partiel, usuels aujourd'hui pour l'ensachage de mortiers secs, d'enduits et de produits chimiques pour la construction rendent possible, à l'aide de la technique de valve à filtre et des systèmes de fermeture des valves, un remplissage propre de sacs à valve non perforés, pour la vente chez les marchands de matériaux de construction.

Estado actual de la técnica de envasado totalmente automático de todo tipo de materiales para la construcción

Resumen – El rendimiento de las modernas máquinas ensacadoras rotativas asciende, hoy en día, a 5 500 sacos/h. Aparte del aumento del rendimiento y de la reducción de gastos para el personal de servicio, el objetivo principal del desarrollo consistía en reducir aún más los gastos de explotación. Para ello, se ha simplificado el manejo de las máquinas, se ha aumentado la duración de vida de las piezas sometidas a desgaste, se han logrado ahorros en los gastos para sacos y se ha mejorado la precisión de los sistemas de pesado. El empleo de sistemas de mando electrónico ha favorecido este desarrollo. Sin embargo, las condiciones esenciales para tener éxito han sido las mejoras decisivas, logradas en el ámbito de la técnica de llenado, debido al control del peso de los sacos a través de la corrección electrónica de la cinta. Los datos obtenidos durante el ensacado pueden ser registrados, almacenados y evaluados electrónicamente, permitiendo un control automático de la operación de envasado. En las ensacadoras con boquilla de aplicación automática a sacos con válvula se ha conseguido no solamente una mejora notable en cuanto a seguridad y número de aciertos, sino también un notable aumento de la capacidad, que alcanza hasta 3 000 sacos/h aprox. Al mismo tiempo ha sido desarrollado un nuevo sistema de despaletización de fardos de sacos vacíos, nuevos, los cuales no necesitan llevar un cerco. El robot con su sistema de agarre es capaz de alimentar varias boquillas de aplicación automática a la vez. De esta forma queda eliminada la última laguna en el envasado totalmente automático – desde el almacén de sacos vacíos hasta la paleta cargada. Los sistemas casi exentos de aire, usuales hoy en día para el envasado de morteros secos, enlucidos y productos químicos para la construcción permiten, gracias a la técnica de las válvulas de filtro y de los sistemas de cierre de válvulas, un envasado limpio de sacos con válvula, no perforados y destinados a ser vendidos en los comercios de materiales para la construcción.

Durch die Entwicklung des Leersack-Depalettierers konnte für den automatischen Verpackungsbetrieb eine wesentliche Lücke im Packprozeß geschlossen werden (**Bild 1** und **2**). Neben der in Zentraleuropa oft zum Einsatz kommenden Leersackrolle geht die Sackhersteller-Industrie bei Bündelware mehr und mehr dazu über, diese auf Standardpaletten zu liefern, um somit dem Stand der allgemeinen Logistik zu entsprechen. Der Depalettierer kann die Sackbündel in der Packanlage von diesen Paletten entnehmen und dem Aufstecksystem zuführen. Magazine und vor allen Dingen das manuelle Umladen der Leersäcke in die Magazine entfallen.

Basis dieses Depalettierers ist ein Roboter mit hoher Leistung, der in der Lage ist, gleichzeitig mehrere Packlinien auch mit unterschiedlichen Säcken zu beschicken. Eine präzise Sensortechnik sichert hierbei eine hohe Verfügbarkeit auch bei Palettenware, die transportbedingte Verschiebungen aufweist. Dieses System zeichnet sich auch dadurch aus, daß die Sackbündel keine Umreifung benötigen. Dies senkt Kosten in der Sackfabrik wie auch in der Packanlage und erübrigt die Entsorgung dieser Materialien.

Eine weitere Entwicklung stellt der automatische Sackaufstecker Radimat Compact (**Bild 3**) dar. Ziel dieser Entwicklung war es, die nachträgliche Automatisierung bestehender Packanlagen zu erleichtern. Durch die Bauform wird dieser Automat den oft eingeschränkten Platzverhältnissen in alten Packereien gerecht. Umfangreiche Baumaßnahmen können daher entfallen. Die gesamte Maschine wird als komplett installierte Einheit vor die Packmaschine gestellt und ermöglicht dadurch kürzeste Montagezeiten.

Ein wichtiger Baustein für den automatisierten Packbetrieb ist das HAVER-Daten-Prozeß-System DPS. Mit diesem System werden alle Daten der HAVER-Wägeelektronik MEC+N oder der Gewichtskorrekturwaage ausgewertet und über Monitor oder Drucker angezeigt. Schnittstellen ermöglichen den Transfer der Daten auch zur betrieblichen EDV-Einrichtung. Alle Daten, wie Gewichte oder Störmel-

The development of the empty bag depalletizer has closed a serious gap in the automatic packaging process (**Figs. 1** and **2**). Apart from producing rolls of empty bags such as are often used in Central Europe, the bag manufacturing industry is tending to supply more and more bundles of bags on standard pallets in accordance with the general practice. The depalletizer can take the bundles of bags off the pallets in the packaging plant and feed them to the applicator system. Magazines, and above all manual reloading of empty bags into magazines, can be dispensed with.

The basic element of the depalletizer is a high performance robot which can feed several packaging lines simultaneously, even with different bags. Precise sensor technology ensures good availability even when the palletized materials have shifted during transport. A distinctive feature of the system is that the bundles of bags do not need to be banded.

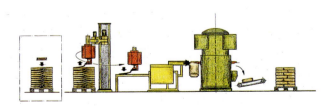

BILD 1: Materialflußschema in einem automatisierten Verpakkungsbetrieb
FIGURE 1: Material flow diagram in an automated packing operation

Sackherstellung = bag production
Abfüllbetrieb = filling operation

BILD 2: Ansicht eines Leersack-Depalettierers
FIGURE 2: View of an empty bag depalletizer

dungen, sind auch für jeden Füllstutzen separat abrufbar. Dies erleichtert die gezielte Wartung wesentlich.

Werden über eine Verpackungslinie verschiedene Produkte, d.h. auch unterschiedliche Gewichte oder Sackabmessungen gefahren, sind alle Anlagenteile mit automatischen Verstelleinrichtungen ausrüstbar. Über den Sortenwahlschalter kann dann mit einem Handgriff die gesamte Linie auf die gespeicherten Parameter umgestellt werden. In Verbindung mit leistungsstarken Packmaschinen und Aufsteckautomaten garantiert die Gewichtskorrekturwaage höchste Gewichtsgenauigkeiten ohne Kapazitätseinbußen.

Die jüngste Entwicklung ist eine neue Sackabwurftechnik. Sie stellt auch bei höchsten Durchsätzen einen störungsfreien und staubarmen Betrieb sicher. Um den Bestrebungen nach dünnwandigerer, d.h. auch kostengünstigerer Sackware zu entsprechen, konnten hier die auf den Sack wirkenden mechanischen Belastungen deutlich reduziert werden.

Ein besonders aktuelles Thema ist die bevorstehende Verminderung der Sackgewichte von 50 auf 25 kg. Daraus ergeben sich Forderungen nach höherer Gewichtsgenauigkeit, gesteigerter Sackzahl und größerer Sacksauberkeit. In Fällen, in denen eine Verminderung der stündlichen Packleistung durch die Umstellung auf 25 kg-Säcke nicht akzeptiert werden kann, ist es jedoch ohne große bauliche Veränderungen möglich, anstelle eines z.B. 8-Stutzen-Packers eine Maschine mit höherer Stutzenzahl zu installieren.

Um die Lagerfähigkeit der abgefüllten Produkte zu verlängern, werden vielfach Säcke mit Kunststoffbeschichtungen oder Kunststoffzwischenlagen als Feuchtigkeitssperrschicht eingesetzt. In der Mörtelindustrie ist diese Sackkonstruktion besonders häufig anzutreffen. Allerdings bedeutet eine solche Sperrschicht nicht nur, daß ein Eindringen von Feuchtigkeit verhindert wird, sondern daß zugleich auch die Entlüftung des Produktes während und nach dem Füllvorgang erschwert wird. Die Folge sind längere Füllzeiten, ein Kapazitätsverlust und größere Verschmutzungen der Säcke durch den hohen Innendruck.

Zur Lösung der technischen Aufgabe wurde die HAVER-Filterventiltechnik entwickelt (**Bild 4**). Hiermit ist es möglich, den Sack während des Füllprozesses überwiegend durch den im Ventil befindlichen Filter zu entlüften. Zur Steigerung der Entlüftungsgeschwindigkeit wird dieser Filter über das Füllrohr mit einem Unterdruck beaufschlagt. Eine besondere Klemmeinrichtung dichtet dabei den Sack am Füllrohr hermetisch ab, so daß der Füllvorgang nahezu rückmehlfrei ist und die Säcke sauber bleiben. Ein weiterer Vorteil dieser Technik sind ein hoher Füllungsgrad der

This lowers costs at both the bag factory and the packaging plant and eliminates the need to dispose of the bands.

Another development is the Radimat Compact automatic bag applicator (**Fig. 3**). Its purpose is to facilitate subsequent automation of existing packaging plants. The machine is shaped so that it can fit into the often limited space in such plant, thus avoiding extensive building work. The whole machine is set up as a fully installed unit in front of the packaging machine, so that erection times can be extremely short.

An important component of automated packaging is the HAVER data processing system DPS. It evaluates all data from the HAVER MEC+N electronic weigher or the weight correcting weigher, displays it on the monitor or prints it out. There are interfaces for transferring the data to the main data processing system. All data, such as weights or fault signals, can be called up separately for each filling spout, a feature which greatly facilitates selective maintenance.

If different products, i.e. products with different weights or bag dimensions, may go down a packaging line, all parts of the plant can be equipped with automatic adjustment devices. The whole line can then be reset virtually instantaneously to stored parameters by operating the type-selection switch. When used in conjunction with high-powered packaging machines and automatic applicators the weight-correcting weigher guarantees extremely accurate weights without any loss of capacity.

The latest development is a new method of discharging bags. It ensures trouble-free and relatively dust-free operation even with a very high throughput. A clear reduction in the mechanical stresses on the bag has been achieved here in order to match the attempts being made by industrialists to use thinner, i.e. cheaper bag material.

BILD 3: Ansicht eines automatischen Sackaufsteckers
FIGURE 3: View of an automatic bag applicator

A particularly topical subject is the planned reduction in bag weights from 50 to 25 kg. This will produce a demand for more accurate weighing, larger numbers of bags and cleaner bags. In cases where the switch to 25 kg bags would produce an unacceptable reduction in the hourly packaging rate it is possible, without any major structural changes, to replace an eight-nozzle packer, for example, by a machine with more nozzles.

Bags with a plastic coating or with intermediate layers of plastic to provide a moisture barrier are commonly used to extend the shelf life of the packed products. This bag construction is encountered especially frequently in the mortar industry. Yet the barrier layer not only prevents moisture from penetrating; it also makes it difficult to deaerate the product during and after filling. Filling times are consequently longer, there is a loss of capacity and the bags become dirtier as a result of the high internal pressure.

HAVER filter valve technology has been developed to solve the technical problem (**Fig. 4**). With this technology the bag can be vented during the filling process, largely through the

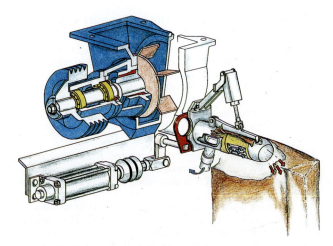

BILD 4: Entlüftung von befüllten Säcken durch ein Filterventil
FIGURE 4: Deaerating of filled bags through a filter valve

Säcke, eine gleichmäßige Befüllung, eine hohe Gewichtsgenauigkeit sowie eine Steigerung der Leistung durch kürzere Füll- und Entlüftungszeiten und der Einsatz kostengünstigerer Sackmaterialien. Mit der HAVER-Ventilverschließtechnik kann der sauber und kompakt gefüllte Sack im Ventil verschlossen werden, um den Anforderungen z. B. von Baumärkten zu genügen.

Im Mörtelbereich weisen die Körnungen der abzufüllenden Produkte meistens sehr große Unterschiede auf. In den meisten Fällen ist eine Abfüllung mit dem HAVER-Turbinensystem möglich. Bei sehr groben Produkten hat die Luftmaschine, die ohne mechanische Elemente arbeitet, Vorteile. Der HAVER-Duopacker vereinigt beide Systeme und ist somit für die gesamte Produktpalette einsetzbar. Eine ebenso große Anwendungsbreite bietet der HAVER-Combimat für das Abfüllen in offene Säcke. Alle Füllsysteme sind darüber hinaus so ausgelegt, daß bei Produktwechsel die Restmaterialien weitgehend selbsttätig ausgetragen werden und eventuelle manuelle Reinigungstätigkeiten auf ein Minimum reduziert sind.

Für die Verpackung von Zement, Kalk, Gips und daraus hergestellter Baustoffe wird derzeit nahezu ausschließlich das Turbinensystem eingesetzt, das bereits vor über 70 Jahren entwickelt wurde. Das Grundsystem hat sich bis heute bewährt. Das Ziel vieler Detailveränderungen bestand im wesentlichen darin, den Massendurchsatz zu steigern und die Verpackungstechnik an sehr unterschiedliche Produkteigenschaften möglichst optimal anzupassen. Die Anpassung des jeweiligen Produkts an das am besten geeignete Füllsystem wird in der HAVER-Versuchsanstalt vorgenommen. Hier werden verschiedene Konstruktionen unter gleichen Rahmenbedingungen einander gegenübergestellt und auf der Grundlage der in **Tabelle 1** zusammengestellten Meßgrößen beurteilt. Diese Versuche lassen eine objektive Beurteilung zu und erleichtern die Entscheidungsfindung.

TABELLE 1: Meßgrößen zur Beurteilung des Betriebsverhaltens von Schüttgütern in „Turbinen"-Abfüllmaschinen

Meßgrößen
– Massenstrom aus Füllgewicht und Füllzeit
– Verdichtungsgrad aus Schüttdichten vor und nach Abfüllung
– Impulskraft des Füllstrahles im Grobstrom
– Statischer Druck des Füllstromes
– Elektrische Leistungsaufnahme
– Verbrauch an elektrischer Energie
– Luftverbrauch zur Fluidisierung je Belüftungsstelle
– Füllzeit
– Abweichung vom Sollgewicht
– Gasdruck im Gebinde
– Antriebsmoment

filter contained in the valve. A negative pressure is applied to the filter through the filling tube to increase the deflating speed. A special clamping device seals the bag hermetically at the filling tube, so that there is hardly any spillage during the filling process and the bags stay clean. Other advantages of this technique are that the bags can be made fuller and filled evenly, that weights can be very accurate and that output can be increased through having shorter filling and venting times and using cheaper bagging materials. The HAVER valve-sealing technology enables the cleanly and compactly filled bag to be sealed at the valve to satisfy the requirements of, for example, the building market.

As far as mortars are concerned the products to be bagged usually have enormous variations in particle sizes. In most cases they can be bagged by the HAVER turbine system. Pneumatic machinery working without any mechanical elements has advantages for very coarse products. The HAVER Duopacker combines both systems and can therefore be used for the whole product range. The HAVER Combimat, designed for filling into open bags, offers an equally wide range of applications. All the filling systems are also designed so that when the product is changed the residual materials are discharged largely automatically, and any manual cleaning work is reduced to a minimum.

The turbine system developed over 70 years ago is still used almost exclusively for packing cement, lime, plaster and

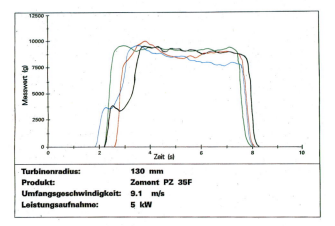

Turbinenradius:	130 mm
Produkt:	Zement PZ 35F
Umfangsgeschwindigkeit:	9.1 m/s
Leistungsaufnahme:	5 kW

BILD 5: Impulskraft des Füllstrahles bei der Verpackung von PZ 35 F, Senkrecht-Maschine
FIGURE 5: Impulsive force of the filling jet when packing PZ 35 F Portland cement, vertical machine

Turbinenradius	= impeller radius
Produkt	= product
Zement	= cement
Umfangsgeschwindigkeit	= circumferential speed
Leistungsaufnahme	= power consumption

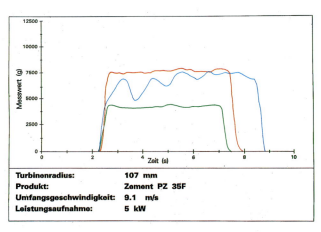

Turbinenradius:	107 mm
Produkt:	Zement PZ 35F
Umfangsgeschwindigkeit:	9.1 m/s
Leistungsaufnahme:	5 kW

BILD 6: Impulskraft des Füllstrahles bei der Verpackung von PZ 35 F, Waagerecht-Maschine
FIGURE 6: Impulsive force of the filling jet when packing PZ 35 F Portland cement, horizontal machine

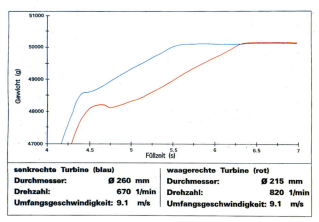

BILD 7: Vergleich der Gewichtseinstellung bei Waagerecht- und Senkrecht-Turbinenpackern
FIGURE 7: Comparison of the weights with horizontal and vertical impeller packers

Senkrechte Turbine (blau) = vertical impeller (blue)
Waagerechte Turbine (rot) = horizontal impeller (red)
Durchmesser = diameter
Drehzahl = rotational speed
Umfangsgeschwindigkeit = circumferential speed

Wie unterschiedlich sich konstruktive Ausführungsformen auswirken, zeigen die Meßkurven über die Impulskraft des Füllstrahles beim Abfüllen von Portlandzement 35 F (**Bild 5 und 6**). Verglichen werden hierbei die waagerechte Turbine mit der Senkrechtturbine bei gleichen Leistungsaufnahmen und gleichen Umfangsgeschwindigkeiten. Entsprechend den Bedingungen in einer Packanlage wurden die Messungen bei unterschiedlicher Belüftung des Produktes durchgeführt. Während die Senkrechtturbine nur unwesentlich auf eine unterschiedliche Belüftung reagierte, sind die Unterschiede bei der Waagerechtturbine deutlich festzustellen. Eine unterschiedliche Impulskraft führt zu unterschiedlich großen Massenströmen, was dann im Feinstrom zu einer Verschlechterung der Gewichtsgenauigkeit führt. Gleichzeitig bedeutet dies auch eine Veränderung der Leistung.

Zum Vergleich des Füllverhaltens der Systeme wurde im **Bild 7** der letzte Abschnitt des Füllvorganges dargestellt. Auch bei diesen Messungen zeigte sich, daß die Senkrechtturbine bessere Ergebnisse aufweist, die entweder zu einer höheren Leistung der Anlage oder aber bei gleicher Leistung zu einer besseren Produktverdichtung führten. Damit können Kosten beim Verpackungsmaterial eingespart werden. Daraus geht hervor, daß das Füllsystem einen spürbaren Einfluß auf die Betriebskosten hat. Der bauliche Mehraufwand, den die Senkrechtturbine mit sich bringt, kann jedoch in kurzer Zeit amortisiert werden.

building materials made from them. The basic system has proved its worth to this day. The purpose of the many changes of details has been to increase the throughput of material and to adapt the packaging technology to very different product properties. Adaptation of any particular product to the most suitable filling system is carried out at the HAVER test institute. Here various designs are compared under the same general conditions and judged on the basis of the measurements listed in **Table 1**. The tests allow an objective assessment to be made and facilitate decision-making.

TABLE 1: Measured variables for judging the operating behaviour of bulk materials in „turbine" bagging machines

Measured variables
— Mass flow from filling weight and filling time
— Degree of compaction from apparent densities before and after filling
— Impulsive force of filling jet in coarse flow
— Static pressure of filling flow
— Electrical power input
— Consumption of electrical power
— Air consumption for fluidisation, per aeration point
— Filling time
— Deviation from desired weight
— Gas pressure in package
— Drive torque

The different effects produced by different designs are illustrated by the graphs showing the impulsive force of the filling jet during the bagging of 35 F Portland cement (**Figs. 5 and 6**). The horizontal turbine is compared with the vertical one for the same power consumption and the same peripheral speed. The measurements were taken with varying aeration of the product, corresponding to the conditions in a packaging plant. Whereas the vertical turbine had an insignificant reaction to differences in aeration, the reaction of the horizontal turbine can be seen clearly. Differences in impulsive force lead to different mass flows, which in turn lead to loss of weighing accuracy during the fine flow. It also signifies a change in output.

The last stage in the filling process is shown in **Fig. 7** to compare the filling actions of the systems. The measurements again show the vertical turbine giving better results, leading either to a higher output from the plant or to better product compaction with the same output so that savings can be made in packaging material. It follows that the filling system has a noticeable effect on operating costs. The additional expenditure on construction which the vertical turbine entails will pay for itself in a short time.

HALBACH & BRAUN

ᛒ Systeme zum Fördern und Zerkleinern

Seit der Gründung im Jahr 1920 hat sich HALBACH & BRAUN der Aufgabe gewidmet, den Bergbaubetrieben in den Bereichen Gewinnung und Förderung sowohl auf der maschinentechnischen als auch auf der konzeptionellen Seite Problemlösungen anzubieten oder diese gemeinsam mit den Betrieben zu erarbeiten. Aus diesem Grund gehört HALBACH & BRAUN zu den erfolgreichsten Unternehmen der deutschen Bergbauzulieferindustrie. So zählen HALBACH & BRAUN-Produkte heute weltweit zu den Standardausrüstungen in vielen Betrieben. Die Aktivitäten erstrecken sich auf den Tiefbau und den Tagebau, aber auch auf Gebiete wie Umschlags- und Lagerplatztechnik, Systemleittechnik sowie Baustoffrecycling.

Schlagkopfbrecher der Baureihe SK
Bisher ausgeführte Baugrößen: SK11/11, SK 11/14, SK 11/18, SK 14/14, SK 14/18 und SK 18/18; wobei die erste Zahl für den Rotordurchmesser (11 : 1100 mm) und die zweite Zahl für die Breite des Einlaufquerschnittes (18 : 1850 mm) steht.

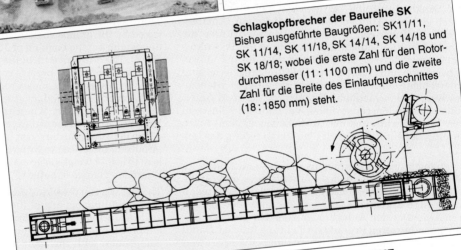

Die HALBACH & BRAUN Kettenförderer gehören nachweislich zu den betriebssichersten und wartungsfreundlichsten Förderausrüstungen, auch unter den schwierigsten Bedingungen.

- Einkettenförderer EKF
- Doppelmittenkettenförderer DMKF
- Doppelaußenkettenförderer DKF
- Bunkerförderer BUKF
- Bunkerabzugförderer
- Ladeförderer
- Schlagkopfbrecher
- Walzenbrecher
- Lamellen-Überlastkupplung
- Hochdruckpumpen
- Elektrische und elektronische Ausrüstungen
- Steuerungs-, Überwachungs- und Automatisierungssysteme
- Getriebe

HALBACH & BRAUN Industries
90 W. Chestnut Street
Millcraft Center
Washington, PA 15301/USA
Phone: ++1 - 412 - 2257840
Telex: 023 - 275444
Telefax: ++1 - 412 - 225 - 3390

HALBACH & BRAUN (PTY) LTD S.A.
50 Gen. Smuts Road, Duncanville
Vereeniging 1930 -Transvaal
South Africa
Phone: ++ 27 -16 - 225735
Telefax: ++ 27 - 16 - 224306

HALBACH & BRAUN Australia Pty. Ltd.
5 Toohey Road
Wetherill Park NSW 2164
P.O. Box 6330
Australia
Phone: ++ 61 - 2 - 7562000
Telefax: ++ 61 - 2 - 7571830

HALACH & BRAUN Katowice
Aleja Korfantego 83
Poland - 40 161 Katowice
Phone: ++48 - 32 - 582832
Telex: 313784
Telefax: ++48 - 32 - 588230

HALBACH & BRAUN Maschinenfabrik GmbH + Co
Postfach 200761, D-42207 Wuppertal, Otto-Hahn-Str. 51, D-42369 Wuppertal-Ronsdorf
Telefon: (0202) 2414-0, Telefax: (0202) 2414-199, Telex: 8591736

Einschichtbetrieb eines Steinbruchs mit einem Durchsatz von 1,8 Mio. t/Jahr

Single-shift operation in a quarry with a throughput of 1.8 million t/year

Exploitation non postée d'une carrière avec un débit de 1,8 Mt/an

Explotación de una cantera en un solo turno, con un caudal de 1,8 millones de t/año

Von **H. Fleck,** Harburg/Deutschland

Zusammenfassung – Zur Versorgung eines Zementwerks, eines Kalkwerks und eines Trockenmörtelwerks wird ein gemeinsamer Steinbruch betrieben. Der Steinbruch liegt in Harburg ca. 2 km vom östlichen Rand des Rieskraters entfernt. Aufgrund der geologischen Entstehungsgeschichte sind Massenkalkschollen in brecciierte Malmkalke eingelagert und häufig von eingequetschten grünlich-braunen Tertiärtonen durchsetzt. Dieser Lagerstättenaufbau behindert eine gezielte Abbauplanung, so daß die Ladestellen an jedem Arbeitstag nach den Anforderungen an die Versorgung der 3 unterschiedlichen Produktionszweige umgestellt werden müssen. Aufgrund dieser Situation erwies sich der Einschichtbetrieb des Steinbruchs als besonders vorteilhaft. Die in den Jahren von 1938 bis 1987 auf 1,1 – 1,2 Mio. t ausgebaute Gesamtförderung wurde 1987/88 durch Erneuerung und Vergrößerung der Brecheranlage, der Ladegeräte und eines Teils der SKWs auf 1,8 Mio. t/a erhöht. 2 E-Bagger und 1 Dieselbagger wurden durch 2 leistungsfähigere größere Dieselbagger ersetzt, die zusätzlich für den Fallkugelbetrieb ausgerüstet sind. 4 SKWs mit 25 t wurden durch 2 SKWs mit 80 t Nutzlast ersetzt. Die Umstellung des Steinbruchs auf größere Einheiten ermöglichte den Einschichtbetrieb und führte zu einer Kosteneinsparung von ca. 20 %. Eine weitere Voraussetzung für die Einsparung ist ein flexibler Einsatz des Personals. Durch regelmäßige Schulung ist jeder Mitarbeiter in der Lage, mehrere Geräte zu bedienen. Spezialreparaturen an den Fahrzeugen werden von Fremdmonteuren, die Wartung und Instandhaltung an den Förderanlagen und Brechern vom Steinbruchpersonal eigenverantwortlich durchgeführt.

Summary – One common quarry is operated to supply a cement works, a lime works, and a dry ready-mix mortar works. The quarry is in Harburg about 2 km from the eastern edge of the "Rieskrater". Due to the geological history blocks of granular limestone are embedded in brecciated Malm limestone and are often permeated by inclusions of green-brown tertiary clays. This structure of the deposit prevents any careful controlled planning of the excavation, with the result that the loading positions have to be changed on each working day to suit the requirements for supplying the 3 different production branches. Because of this situation it proved to be particularly advantageous to work the quarry in a single shift. The 1.1 – 1.2 million t total excavation requirement for the years 1938 to 1987 was increased in 1987/88 to 1.8 million t/a by renewing and expanding the crusher plant, the loading equipment and some of the lorries. Two electric excavators and one diesel excavator were replaced by two larger, more powerful, diesel excavators, which are also equipped for drop-ball work. 4 lorries with payloads of 25 t were replaced by 2 lorries with payloads of 80 t. Changing over the quarry to larger units made the single-shift operation possible, and led to a cost saving of about 20%. Another requirement for the saving is the flexible use of personnel. Through regular training every employee is capable of operating several machines. Special repairs on the vehicles are carried out by outside mechanics but the quarry personnel are responsible for carrying out the maintenance and repairs on the conveying systems and crushers.

Résumé – Pour alimenter une cimenterie, une chaufournerie et une usine à mortier sec, est exploitée une carrière commune. La carrière se situe à Harburg, éloignée d'environ 2 km de la limite du cratère Ries. Suite à la formation géologique, des lentilles de calcaire massif sont incluses dans des calcaires marneux discontinus, fréquement séparés par des argiles tertiaires verdâtremarron emprisonnées. Cette configuration du gisement empêche une planification d'extraction ciblée, obligeant de déplacer les points de chargement au cours de chaque journée de travail, selon les besoins d'alimentation des 3 différentes branches de production. Compte tenu de cette situation, l'exploitation de la carrière en un seul poste s'est révélée particulièrement avantageuse. La masse totale extraite, passée au cours des années 1938 à 1987, à 1,1 – 1,2 Mt, a été augmentée en 1987/88 par la rénovation et l'agrandissement de l'atelier de concassage, des équipements de chargement et d'une partie des dumpers, jusqu'à 1,8 Mt/an. 2 pelles

électriques et 1 pelle diesel ont été remplacées par deux pelles Diesel plus grandes et plus performantes, équipées en outre pour la fragmentation par masse tombante. 4 dumpers de 25 t ont été remplacés par 2 dumpers d'une charge utile de 80 t. La conversion de la carrière à des équipements plus grands a rendu possible le travail en un seul poste et a conduit à une économie de dépenses, d'environ 20%. Une autre condition pour l'économie est un emploi flexible de la main-d'oeuvre. Une formation continue rend chaque membre du personnel apte à conduire plusieurs appareils. Les réparations des véhicules sont exécutées par du personnel extérieur, la maintenance et la remise en état des équipements de transport et des concasseurs sont effectuées sous responsabilité propre, par le personnel de la carrière.

Resumen – Para el abastecimiento de una fábrica de cemento, una fábrica de cal y una fábrica de mortero seco se aprovecha una cantera común. Está situada en Harburg, a unos 2 km de los límites orientales del cráter Ries. Debido a su formación geológica, existen terrones de masas calizas que se encuentran localizados en las brechas de cal del malm y que incluyen muchas veces arcillas terciarias de color verdoso-marrón. Esta formación dificulta una buena planificación de los trabajos de explotación, de modo que cada día hay que cambiar los puntos de carga, de acuerdo con los requerimientos del abastecimiento de 3 ramos distintos de producción. Teniendo en cuenta esta situación, la explotación de la cantera en un solo turno resultó especialmente ventajosa. El volumen de explotación total, que fue aumentado entre 1938 y 1987 a 1,1 – 1,2 millones de toneladas se llevó en 1987/88 a 1,8 millones de t/año, mediante la renovación y ampliación de la instalación de trituración, de la maquinaria de carga y de parte de los camiones pesados. Se sustituyeron 2 excavadoras eléctricas y 1 excavadora diesel por 2 excavadoras diesel mayores y más potentes, que van equipadas adicionalmente con dispositivos de fragmentación por bolas de gran masa. Se sustituyeron 4 camiones pesados de 25 t por 2 camiones pesados de 80 t. La transformación de la cantera, que supone el empleo de unidades mayores, ha permitido trabajar en un solo turno, con lo cual se ha conseguido reducir los gastos en un 20 % aprox. Otro requisito para reducir gastos es el empleo flexible de la mano de obra. Gracias a una formación periódica, cada colaborador es capaz de manejar varios aparatos. Las reparaciones muy complicadas en los vehículos las llevan a cabo mecánicos externos. Los trabajos de reparación y mantenimiento de las cintas transportadoras y de las trituradoras son llevados a cabo bajo la responsabilidad del personal de la cantera.

Explotación de una cantera en un solo turno, con un caudal de 1,8 millones de t/año

Das Familienunternehmen Märker produziert seit mehr als 100 Jahren Baustoffe in Harburg. Zur Versorgung eines Zementwerkes, eines Kalkwerkes und eines Trockenmörtelwerkes wird ein gemeinsamer Steinbruch betrieben, wie auf der Gesamtansicht des Werkes (**Bild 1**) zu sehen ist.

Der Steinbruch befindet sich etwa 2 km vom östlichen Rand des bekannten Rieskraters entfernt, der durch einen Meteoriteneinschlag vor ca. 15 Mio. Jahren entstanden ist. Die Auswirkungen sind daran zu erkennen, daß Massenkalkschollen in breccierte Malmkalke eingelagert sind, oft durch eingequetschte grünlich-braune Tertiärtone abgetrennt. Dieses willkürliche Durcheinander von Tonen, Mergeln und Kalken der verschiedensten Zusammensetzung macht eine gezielte Abbauplanung nicht möglich. Grundsätzlich sind jedoch alle Rohmaterialkomponenten vorhanden, die zum Brennen eines hochwertigen Zementklinkers notwendig sind. Zur Branntkalk- und Kalksteinsanderzeugung sind hochreine Kalke vorhanden.

Eine ständige Umstellung der Ladestellen an jedem Arbeitstag ist unumgänglich, um die Versorgung der 3 Produktionszweige sicherzustellen. Bis 1987 wurde wöchentlich von Montag bis einschließlich Samstag 48 Stunden gearbeitet. Die Jahresmenge lag bei ca. 1 Mio. t. Abgebaut wurde und wird auf 3 Sohlen, wobei das Material über der oberen Sohle durch Reißen und Schieben, die 30 m hohe Wand zwischen Sohle 2 und 3 durch Bohren und Sprengen, hereingewonnen wird.

Beim Herabschieben über die 30 m hohe Wand wird eine Separation des Materials in grob und fein erreicht. Die groben Brocken am Rande des Haufwerkes mit in der Regel höherem $CaCO_3$-Gehalt werden für die Versorgung des Kalkwerkes und die Sandproduktion verwendet.

1985 war, wie aus den Tabellen in **Bild 2** und **3** zu entnehmen ist, eine Schubraupe Cat D 10 mit einer Schildbreite von 6 m und 740 PS im Einsatz. Geladen wurde mit 4 Ladegeräten, 2 Elektrobaggern Demag H 71 mit 220 kW und 3,5 m³-Schau-

The Märker family concern has produced building materials at Harburg for over a century. One common quarry is operated to supply a cement works, a lime works and a dry readymix mortar works. A view of the whole works can be seen in **Fig.1**.

The quarry is about 2 km from the eastern edge of the well-known Ries crater formed by meteorite impact about 15 million years ago. Due to the geological history blocks of granular limestone are embedded in brecciated Malm limestone and are often permeated by inclusions of green-brown tertiary clays. The confusing mixture of clays, marls and limes of widely varying composition prevents any careful controlled planning of the excavation. Yet all the raw material components necessary for burning a high-grade cement clinker are present, as are very pure limes for quick lime and lime-

BILD 1: Gesamtansicht des Märkerwerkes
FIGURE 1: General view of the Märker works

Jahr	Gewinnen			Laden
	Bohren	Schieben	Bagger	Radlader
1985	HBM 60 + Sohlenbohrmaschine	CAT D 10	2 Demag H 71 E 1 Demag H 71 D	1 CAT 988 B
1986				
1987			1 Demag H 71 E 1 Demag H 71 D 1 Demag H 121 D	
1988				
1989			1 Demag H 85 D 1 Demag H 121 D	
1990				
1991				
1992		CAT D 11		
1993				

Jahr	Knäppern	Transportieren	Zerkleinern
		SKW	Brecher
1985	1 CAT 225	4 Euclid R 35	1 Titan D 85
1986		2 Faun K 55	1 Arbed 1500 x 1800
1987	1 CAT 225	2 Faun K 55	1 Arbed 1500 x 1800
1988	1 Fallkugel	2 Faun K 85	1 O & K 1800 x 1800 2400 x 2400
1989	1 CAT 225		
1990	2 Fallkugeln		
1991			
1992		2 Faun K 85	
1993		1 CAT 733 1 CAT 777	

BILD 2 und 3: Tabelle eingesetzter Geräte und Maschinen im Steinbruch
FIGURE 2 and 3: Table of equipment and machines used in the quarry

Jahr	= year	Bohren	= drilling	Radlader	= wheeled loader	Transportieren	= conveying
Gewinnen	= quarrying	Schieben	= pushing	Sohlenbohrmaschine	= level drilling machine	Zerkleinern	= crushing
Laden	= loading	Bagger	= excavator	Knäppern	= boulder work	SKW	= dump truck
						Brecher	= crusher

fel, 1 Dieselbagger mit 220 PS und 4,5 m³-Schaufel und 1 Radlader Cat 988 mit 5 m³-Schaufel. Der Zwischentransport wurde mit 6 SKW's, 4 Euclid mit 25 t und 2 Faun K 55 mit 50 t Nutzlast durchgeführt.

Das Material für die Branntkalk- und Sanderzeugung wird in einem Arbed-Doppelwalzenbrecher mit den Walzenabmessungen 1500 mm × 1800 mm zerkleinert und in der nachgeschalteten Siebanlage in 3 Fraktionen aufgeteilt. Angetrieben werden die Walzen mit 2 × 350 kW. Das Material für die Zementproduktion wurde bei höherem Tonanteil im gleichen Brecher zerkleinert, weniger zum Kleben neigendes Material in einem Doppelwellenhammerbrecher Titan 80 D 75.

1987/88 wurden die Ladegeräte und ein Teil der SKW's erneuert. Die beiden Elektrobagger und der Dieselbagger H 71 wurden durch zwei größere Dieselbagger, einem Demag H 121 mit 8 m³-Schaufel, einem Dienstgewicht von 120 t und einer Antriebsleistung von 750 PS sowie einem Demag H 85 mit 6,5 m³-Schaufel, einem Dienstgewicht von 80 t und einer Antriebsleistung von 640 PS, ersetzt. Die neuen Bagger wurden für den Fallkugelbetrieb ausgerüstet, was zu einer deutlichen Reduzierung der Knäpperarbeiten mit dem Hydraulikhammer führte, da der Bagger in den Beladepausen sofort größere Knäpper zerkleinern kann. Die verwendete Großkugel hat ein Gewicht von ca. 6 t.

4 SKW's mit 25 t wurden durch 2 SKW's, Faun K 85.8 mit 80 t Nutzlast und über 700 PS ersetzt (**Bild 4**).

Für den Doppelwellenhammerbrecher wurde eine zweistufige Brecheranlage mit Walzenbrechern mit einer Leistung von 1100 t/h gebaut (**Bild 5 und 6**). Die Maschinendaten sind in **Tabelle 1** zusammengefaßt. Mit diesen Maßnahmen konnte das Personal um 4 Personen auf 14 Mann plus Meister reduziert werden. Ein Blick auf die beiden nebeneinanderliegenden Brecheranlagen mit Abförderung, Siebturm und Zwischenlager zeigt Bild 5. Auf dem Diagramm in **Bild 7** sind die Kenndaten des Steinbruchs für die Jahre 1985–1993 zusammengestellt.

stone sand production. The loading positions have to be changed on each working day to suit the requirements for supplying the three different production branches. Up to 1987 a 48-hour week was worked, including Saturdays. The annual amount excavated was approx. 1 million tonnes. Excavation is and was carried out on three quarry levels; the material above the upper level is obtained by ripping and pushing up, and the 30 m face between levels 2 and 3 by drilling and blasting. The material separates into coarse and fine components when pushed over the 30 m face. The coarse lumps at the edge of the rock pile generally have a higher $CaCO_3$ content and are used to supply the lime works and for sand production.

As will be seen from the tables in **Figs. 2** and **3**, a Cat D 10 740 HP bulldozer with a 6 m blade was used in 1985. Four

BILD 5: Teilansicht des Steinbruchs mit den beiden Produktionslinien
FIGURE 5: Partial view of the quarry showing the two production lines

BILD 4: Steinbruchgeräte
FIGURE 4: Quarry equipment

BILD 6: Zweistufige Brecheranlage für 1100 t/h Durchsatz
FIGURE 6: Two-stage crusher plant for a throughput of 1100 t/h

TABELLE 1: Technische Daten der Brechanlage IV		TABLE 1: Technical data for the new crushing plant IV	
Anlagenabmessungen	H = 17,20 m, B = 29,00 m, L = 46,6 m	Plant dimensions:	height 17.20 m, width 29.00 m, length 46.6 m
Trichter	3seitig beschickbar, 150 m³ Inhalt, 3 × 7,10 Einschüttbreite	Hopper:	can be fed from 3 sides, 150 m³ capacity, tipping width 3 × 7.10
Plattenband	1800 mm × 11000 mm AA Antrieb: Hägglunds MA 566, 350 Nm/bar, Pumpe 75 kW	Apron conveyor:	1800 mm × 11100 mm between centres Drive: Hägglunds MA 566, 350 Nm/bar, pump 75 kW
Kettenförderer	1800 mm × 9100 mm AA, 2,2 kW Getriebemotor	Chain conveyor:	1800 mm × 9100 mm between centres, 2.2 kW geared motor
Vorbrecher	1800 mm × 1800 mm je 1.250 kW + 360 kW Festwalze: Antrieb Kurzschlußläufer mit Frequenzumformer $n1/n2 = (500-2000\,Upm)/(141-563,4\,Upm)$ v Walze: 4 – 8 m/s Loswalze: Antrieb Schleifringläufer $n1/n2 = 2000\,(563,4\,Upm)$ v Walze: 6 m/s Zahnhöhe 140, Spaltenweite min: 80 mm	Primary crusher:	1800 mm × 1800 mm, 1250 kW each + 360 kW Fixed roll: squirrel cage drive with frequency converter $n1/n2 = (500-2000\,rpm)/(141-563.4\,rpm)$ Roll v: 4–8 m/s Loose roll: slipring drive $n1/n2 = 2000\,(563.4\,rpm)$ Roll v: 6 m/s Tooth height 140, min. gap width 80 mm
Obertrumketten-förderer	2200 × 9780 mm AA, Antrieb: 90 kW, v = 0,23 m/s	Top strand chain conveyor:	2200 × 9780 mm between centres, drive 90 kW v = 0.23 m/s
Nachbrecher	1400 mm × 2400 mm je 1250 kW + 360 kW Festwalze: Antrieb Kurzschlußläufer mit Frequenzumformer $n1/n2 = (500-2000\,Upm)/(80,6-322,6\,Upm)$ v Walze: 8 – 18 m/s Loswalze: Antrieb Schleifringläufer $n1/n2 = 2000\,(322,6\,Upm)$ v Walze: 12 m/s Zahnhöhe 80 mm, Spaltenweite min: 20 mm	Secondary crusher:	1400 mm × 2400 mm, 1250 kW each + 360 kW Fixed roll: squirrel cage drive with frequency converter $n1/n2 = (500-2000\,rpm)/(80.6-322.6\,rpm)$ Roll v: 8–18 m/s Loose roll: slipring drive $n1/n2 = 2000\,(322.6\,rpm)$ Roll v: 12 m/s Tooth height 80 mm, min. gap width 20 mm
Muldengurtförderer	2400 mm × 24800 mm AA, Antrieb: 75 kW, v = 1,1 m/s	Troughed belt conveyor:	2400 mm × 24800 mm between centres, Drive 75 kW, v = 1.1 m/s
Muldengurtförderer	1400 mm × 18000 mm AA, Antrieb: 30 kW, v = 1,92 m/s	Troughed belt conveyor:	1400 mm × 18000 mm between centres, Drive 30 kW, v = 1.92 m/s

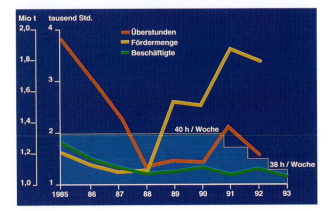

BILD 7: Diagramm der wichtigsten Steinbruch-Kenndaten
FIGURE 7: Diagram of the most important characteristic quarry data

tausend Std.	= thousand hours
Überstunden	= overtime work
Fördermenge	= output
Beschäftigte	= employees
38 h/woche	= working time 38 h/week

machines were used for the loading work, two Demag H 71 220 kW electric excavators with 3.5 m³ buckets, one 220 HP diesel excavator with a 4.5 m³ bucket and one Cat 988 wheeled loader with a 5 m³ bucket. Intermediate transport was provided by six dump trucks – four Euclids with 25 t payloads and two Faun K 55 with 50 t payloads.

The material for quick lime and sand production is crushed in a twin-roll crusher with 1500 mm × 1800 mm diameter rolls, and split into 3 fractions in the downstream screening plant. The drive for the rolls is 2 × 350 kW. Of the material destined for cement production, that with a higher clay content was crushed in the same crusher, and the less sticky material was crushed in a Titan 80 D 75 twin-rotor hammer crusher.

In 1987/88 the loaders and some of the dump trucks were replaced. The two electric excavators and the H 71 diesel excavator were replaced by two larger diesel ones, a Demag H 121 with an 8 m³ bucket, a service weight of 120 t and a 750 HP drive, and a Demag H 85 with a 6.5 m³ bucket, a service weight of 80 t and a 640 HP drive. The new excavators were equipped for drop-ball work; this led to a marked reduction in boulder work with the hydraulic hammer, since the excavators can switch over immediately to crushing large boulders in the intervals between loading. The cast iron ball used weighs about 6 t.

Four 25 t dump trucks were replaced by two Faun K 85.8 dump trucks, each of over 700 HP with a payload of 80 t (**Fig. 4**).

A two-stage crushing plant containing roll crushers with an output of 1 100 t/h was constructed for the twin-rotor hammer crusher (**Figs. 5** and **6**). The machine data is summarised in **Table 1**. This made it possible to reduce the personnel by four, to fourteen men plus a foreman. There is a view of the two adjacent crusher plants with the outgoing conveyors, screening tower and intermediate store in Fig. 5. The quarry data for the years 1985 to 1993 is summarised in the diagram in **Fig. 7**.

Die Jahresleistung hat sich bis heute fast verdoppelt. Die Leistung pro Mann und Jahr stieg von 60000 t auf cirka 130000 t. Die jährliche Zahl der Überstunden ging in dieser Zeit deutlich von 217 h/Mann auf cirka 130 h/Mann zurück, obwohl im selben Zeitraum die wöchentliche Regelarbeitszeit von 40 h auf 38 h im Jahre 1992 verringert wurde. Die Kosten pro Tonne im Steinbruch vom Gewinnen bis zum Zwischenlager haben sich auf Basis von 1987 gleich 100% vor der Umstellung auf Werte zwischen 50 und 70% verringert.

Die Umstellung der Steinbruchorganisation auf größere Einheiten hat sich bewährt und zu einer absoluten Kosten-

einsparung von cirka 20% geführt. Voraussetzung für diese Einsparung ist neben leistungsfähigeren Geräten der flexible Einsatz des Personals, der durch regelmäßige Schulung möglich ist. Jeder Mitarbeiter ist an mehreren Geräten ausgebildet. Damit gibt es auch bei Urlaub oder Krankheit einzelner Mitarbeiter keine Probleme. Anreiz dafür ist eine Zulage zum Stundenlohn.

Nur für Spezialreparaturen an den Fahrzeugen und Geräten werden Fremdmonteure zugezogen. Die Wartung und Instandhaltung wird vom Steinbruchpersonal eigenverantwortlich durchgeführt. Dadurch wird auf die Anlage und die Geräte mehr Rücksicht genommen. Die Ausfälle sind kürzer, und die Ausfallkosten werden verringert.

Zusammenfassend ist festzustellen, daß die mit einem hohen Kapitaleinsatz verbundene Umstellung der Steinbruchorganisation die Erwartungen mehr als erfüllt hat.

Literaturverzeichnis:

Fleck, H.K.: Optimierung der Rohmaterialaufbereitung im Märker Zementwerk Harburg Zement-Kalk-Gips 41, 1988, Nr. 7, S. 317–321.

Annual output has now almost doubled. Output per man per year has risen from 60000 t to about 130000 t. The annual overtime worked has declined sharply from 217 to about 130 h/man in this period, although the normal working week went down from 40 h to 38 h in 1992. If 1987 is taken as the base year before the reorganisation, quarrying costs from extraction to intermediate storage have dropped from 100% in that year to between 50 and 70%.

Changing over the quarry to larger units has been a successful step, leading to an absolute cost saving of about 20%. Another precondition for this saving, in addition to more productive equipment, is the flexible use of personnel. Through regular training every employee is capable of operating several machines and there are no problems when individual workers are on holiday or sick leave. The incentive to train is an increase in the hourly wage.

Outside mechanics are only brought in for special repairs to vehicles and equipment, and the quarry personnel are responsible for normal maintenance and repairs. As a result they take more care of the plant and equipment, shutdown times are shorter and shutdown costs lower.

To summarise, it can be said that the quarry reorganisation entailed a large capital investment but has more than fulfilled expectations.

Transport- und Lagertechnik für Zementklinker – Vorstellung der großen Klinkerlager in Belgien und Thailand

Transport and storage technology for cement clinker – Description of the large clinker stores in Belgium and Thailand

Technique de transport et de stockage du clinker à ciment –
Présentation des grands stockages de clinker en Belgique et en Thaïlande

Técnica de transporte y de almacenamiento de clínker de cemento –
Presentación de los grandes almacenes de clínker existentes en Bélgica y Tailandia

Von **W. Heine**, Rheinberg/Deutschland

Zusammenfassung – Bekannt sind Klinkersilos für eine Lagerkapazität von 30 000 bis 60 000 t. Sie haben einen Durchmesser bis zu 34 m und eine Höhe von bis zu 60 m. Sie werden vorzugsweise mit einem Abzugskanal ausgerüstet, so daß eine gute Restlosentleerung von 80 % erreicht wird. Für große Zementwerke, wie sie in Belgien und Thailand gebaut wurden, waren größere Lagerkapazitäten erforderlich. Es bestand die Möglichkeit, entweder mehrere Klinkersilos in bekannter Bauweise oder aber neue Großraumlager mit einem Fassungsvermögen von etwa 150 000 t zu konzipieren. Aufgrund niedrigerer Baukosten wurde in beiden Werken der Bau eines Großraumlagers mit einer Kapazität von 140 000 bzw. 150 000 t bevorzugt. Die Klinkerlager haben einen Durchmesser von 50 bis 65 m und bestehen aus einem Betonzylinder und einem Stahldach mit Trapezblechverkleidung. Je nach Silodurchmesser werden 2, 3 oder 4 Abzugskanäle angeordnet. Die für die Beschickung und den Siloabzug installierten Plattenbänder ermöglichen bei 1 800 mm Plattenbreite eine Förderleistung von 750 t/h.

Summary – Clinker silos for storage capacities of 30 000 to 60 000 t are familiar. They have diameters up to 34 m and heights up to 60 m. They normally have one extraction duct and achieve good final emptying rates of 80 %. The large cement works, such as have been built in Belgium and Thailand, needed corresponding storage capacities. They had the option either of building several clinker silos using the familiar method of construction, or of designing new large stores with capacities of about 150 000 t. Because of the lower building costs both works preferred to build large stores with capacities of 140 000 and 150 000 t respectively. The clinker stores have diameters of 50 to 65 m and consist of a concrete cylinder and a steel roof with sheet cladding with trapezoidal corrugations. 2, 3 or 4 extractions ducts are provided, depending on the silo diameter. The apron conveyors with widths of 1 800 mm installed for feeding and emptying the silos are capable of conveying capacities of 750 t/h.

Résumé – Des silos à clinker d'une capacité de stockage de 30 000 à 60 000 t sont connus. Ils ont un diamètre allant jusqu'à 34 m et une hauteur jusqu'à 60 m. Ils sont équipés, de préférence, d'une fosse de reprise, permettant d'obtenir une vidange sans perte, de 80 %. Des capacités de stockage conséquentes avaient été nécessaires pour les grandes cimenteries, construites en Belgique et en Thaïlande. Les possibilités offertes étaient la construction de plusieurs silos à clinker de type usuel, ou alors des stockages nouveaux à grande contenance, d'une capacité d'emmagasinage d'à peu près 150 000 t. En raison des coûts plus bas de la réalisation, la construction d'un stockage haute capacité, de respectivement 140 000 et 150 000 t, a été préférée dans les deux usines. Les magasins de clinker ont un diamètre de 50 à 65 m et sont constitués d'un cylindre en béton et d'un toit en acier, à couverture en tôles trapézoidales. 2, 3 ou 4 fosses de reprise sont prévues selon le diamètre du silo. Les convoyeurs à plaques installés pour le remplissage et la vidange du silo autorisent, pour une largeur des plaques de 1 800 mm, une performance de transport de 750 t/h.

Resumen - Como es sabido, los silos para clínker están dimensionados para una capacidad de almacenamiento de 30 000 a 60 000 t. Tienen un diámetro de hasta 34 m y una altura de hasta 60 m. Van equipados preferentemente con un canal de extracción, de modo que se consigue un buen vaciado total de hasta un 80 %. Para grandes fábricas de cemento, tales como las construidas en Bélgica y Tailandia, hacían falta capacidades de almacenamiento apropiadas. Existía la posibilidad de levantar varios silos, de construcción convencional, o nuevos depósitos de gran tamaño, con capacidades

Our Programme for Cement Plants!

MASCHINEN- UND VERFAHRENSTECHNIK
Bernhard Blatton GmbH
P. O. Box 1665 · D- 66 749 Dillingen · Fed. Rep. of Germany
Phone (0 68 31) 7007-0 · Telex 443 107 · Telefax (0 68 31) 7007 58

- Raw Material Transport
- Sampling Plants
- Preblending Plants
- Raw Meal Homogenizing
- Coal Handling
- Clinker Transport and Storage
- Cement Transport and Storage
- Filling and Despatch
- Laboratory Automation

 A FLEXIBLE PROGRAMME ON A SOLID BASIS OF RELIABLE EXPERIENCE.

10.93

Delivery programme

WIR PLANEN UND BAUEN:
WE ARE DESIGNING AND CONSTRUCTING:
NOUS ETUDIONS ET CONSTRUISONS:

- Raw material transport
- Sampling plants
- Pre-blending plants
- Additive plants
- Corrective handling
- Raw meal transport, homogenizing and Kiln / Raw mill dedusting
- Preheater - Kiln - Cooler
- Heat exchanger - Rotary kiln - Cooler
- Coal handling
- Clinker additive and gypsum storage
- Transport and cement storage
- Cement handling
- Filling and despatch
- Mill feeding

PFAFF | AQS
Sampling
Sample conveying
Sample preparation
Laboratory automation

Clinker | Cement | Raw meal

de hasta 150 000 t aprox. Debido a que los gastos de construcción eran más bajos, se prefirió en ambas fábricas el montaje de un depósito de gran capacidad, apto para 140 000 y 150 000 t respectivamente. Los depósitos de clínker tienen un diámetro de 50 a 65 m y consisten en un cilindro de hormigón y un tejado de acero, con revestimiento de chapa trapezoidal. Según el diámetro del silo, se disponen 2, 3 o 4 canales de extracción. Las cintas de placas previstas para la alimentación y el vaciado de los silos permiten un caudal de transporte de 750 t/h, para un ancho de placas de 1 800 mm.

1. Einleitung

Bekannt sind Klinkersilos für eine Lagerkapazität von 30 000 bis 60 000 t. Sie haben einen Durchmesser bis zu 34 m und eine Höhe von 60 m. Sie werden vorzugsweise mit einem Abzugskanal ausgerüstet, so daß eine gute Restlosentleerung von 80 % erreicht wird.

Für die neuen Zementwerke in Thailand, in denen Ofenanlagen mit 9 000 und 10 000 tato installiert wurden, sind Klinkerlager mit einer Kapazität von 150 000 t erforderlich. Die erste Planungsphase berücksichtigte 6 Silos mit je 25 000 t Inhalt. Die Firma AUMUND hat mit den Ingenieuren aus Thailand alternative Klinkerlagerkonzepte entwickelt.

In einer ersten Studie wurde festgestellt, daß die Investitionskosten für Klinkerlager in Europa vergleichbar sind mit denen in Thailand. Weiterhin wurde festgestellt, daß die Investitionskosten deutlich sinken, wenn die Anzahl der Silos reduziert wird. So kosten drei Silos mit je 50 000 t nur noch 83 % und ein Silo mit 150.000 t nur noch 67 % der veranschlagten Bausumme. Praktische Betriebserfahrungen mit solch großen Klinkersilos, die bisher in Thailand nicht üblich waren, konnten insbesondere in den belgischen Zementwerken von CBR/Lixhe und CCB/Gaurain gewonnen werden.

2. Prinzip

Die neuen Klinkerlager für eine Lagerkapazität von 50 000 bis 200 000 t haben einen Durchmesser von 45 bis 65 m. Sie bestehen aus einem Betonzylinder mit einer Höhe von 15 bis 30 m und einem Stahldach, welches auch für die größten geplanten Durchmesser ohne Mittelsäule ausgeführt wird. Diese selbsttragende Dachkonstruktion wird ausgelegt für die Vertikallasten aus den Brückenauflagern, das Gewicht der Filteranlage und den Dachaufbau. Die Vertikallasten können 400 t betragen. Die Dachkonstruktion wird mit Trapezblechen verkleidet, wobei besonders die Nahtstellen staub- und wasserdicht ausgeführt werden.

Um einen guten Entleerungsgrad zu erreichen, werden je nach Lagerdurchmesser 2, 3 oder 4 Kanäle vorgesehen. Jeder Abzugskanal wird mit einem AUMUND-Plattenband ausgerüstet, welches in Verbindung mit mehreren Siloabzügen die staubarme Entleerung durch Schwerkraft, ermöglicht (**Bild 1**). Unter den Rundlagern befinden sich 10–28 Siloabzüge, die durch einen systematischen Wechsel eine kontinuierliche Entleerung über die ganze Grundfläche des Lagers ermöglichen. Diese Abzugsmethode erlaubt die Zusammenführung von unterschiedlichen Kornfraktionen, die sich durch den Entmischungseffekt beim Füllvorgang gebildet haben.

3. Projekte

3.1 Klinkersilo CBR/Lixhe, Belgien

Im Werk Lixhe der CBR befindet sich ein neues Klinkersilo mit einer Lagerkapazität von 80 000 t. Der Betonzylinder hat einen Durchmesser von 45 m bei einer Gesamthöhe von 57,8 m. Aufgrund des großen Durchmessers wurde das Dach als Stahlkonstruktion mit Trapezblechverkleidung ausgeführt. Nur ein Plattenband wurde für die Beschickung und ein weiteres Plattenband für die Entleerung vorgesehen. Um den Abzugsgrad zu verbessern, wurde der Boden trichterförmig durch Auffüllen und Verdichten mit Kalksteinschotter ausgebildet. Für die Siloentstaubung wurde ein Filter mit nur 30 000 m³/h installiert. Über die technischen Besonderheiten wird später berichtet. Um die Wartungsarbeiten an den Transport- und Filteranlagen zu vereinfachen,

1. Introduction

Clinker silos for storage capacities of 30 000 to 60 000 t are well known. They have diameters up to 34 m and heights up to 60 m. They normally have one extraction duct and achieve good final emptying rates of 80 %.

Clinker stores with capacities of 150 000 t are required for the new cement works in Thailand, where kiln plants processing 9 000 and 10 000 tonnes per day have been installed. The first planning phase allowed for six 25 000 t silos. AUMUND has developed alternative clinker storage schemes with the engineers from Thailand.

A first study revealed that investment costs for clinker stores in Europe are comparable with those in Thailand. There was also found to be a marked reduction in these costs when the number of silos is decreased. Thus three 50 000 t silos cost only 83 % of the estimated figure, and one 150 000 t silo only 67 %. Such large clinker silos have not hitherto been used in Thailand, but practical experience in operating them has been gained at CBR's cement works at Lixhe and CCB's works at Gaurain in Belgium.

2. Layout

The new clinker stores have 50 000 to 200 000 t capacities and are 45 to 65 m in diameter. They consist of a concrete cylinder 15 to 30 m high and a steel roof which is also designed for the largest diameters planned, with no central column. The self-supporting roof construction is designed for vertical loads from the bridge supports, the weight of the filter plant and the roof structure, which may amount to 400 t. The roof structure is clad with trapezoidal sheeting and the seams are dust-tight and water-tight.

2, 3 or 4 ducts are provided, depending on the diameter of the store, in order to achieve a good emptying rate. Each extraction duct has an AUMUND apron conveyor which, in conjunction with several silo extraction points, empties by gravity with little dust (**Fig. 1**). Below the circular store there are 10 to 28 silo extraction points which operate systematically in turn, providing continuous emptying over the whole area of the store. This extraction method makes it possible to

BILD 1: CCB Gaurain, Belgien
Siloabzugsbänder 3 × KZB 250–1800 je 750 t/h regelbar
FIGURE 1: CCB Gaurain, Belgium
Silo extraction belts 3 × KZB 250–1 800 each 750 t/h controllable

BILD 2: CBR Lixhe
Klinkersilo 80 000 t; 45 m Durchmesser —
80 000 t; 90 m Durchmesser
FIGURE 2: CBR Lixhe
80 000 t; 45 m diameter – 80 000 t; 90 m diameter

BILD 3: CCB Gaurain, Belgien
Klinkerlager 1 × 150 000 t; 65 m Durchmesser 2 × 75 000 t; 50 m Durchmesser
FIGURE 3: CCB Gaurain, Belgium
Clinker store 1 × 150 000 t; 65 m diameter 2 × 75 000 t; 50 m diameter

wurden zwei Fahrstühle in Betonsäulen installiert. Die Säulen dienen gleichzeitig als Auflager für die Brückenkonstruktion (**Bild 2**).

3.2 Klinkerlager CCB/Gaurain Belgien

Die Zementfabrik CCB in Gaurain besitzt ein Klinkerlager für 140 000 t Lagerkapazität. Der Betonzylinder hat einen Durchmesser von 65 m und das Silo eine Gesamthöhe von 50 m. Das Dach besteht aus einer Stahlkonstruktion, die mit Trapezblechen abgedeckt ist.

Die Entleerung des Silos erfolgt über vier Abzugskanäle, so daß ein guter Entleerungsgrad erreicht wird. Jeder Abzugskanal hat eine Abzugsleistung von 750 t/h. Insgesamt sind 28 Siloabzüge installiert. Durch den geplanten Wechsel der Abzugsöffnung wird eine gleichmäßige Siloentleerung und eine optimale Klinkermischung bei möglichen Qualitätsschwankungen erreicht (**Bild 3**).

Diese Anlage wurde später durch zwei weitere Klinkerlager mit je 70 000 t Lagerkapazität erweitert. Der Lagerdurchmesser ist 55 m. Die Entleerung erfolgt über drei Abzugskanäle. Die Abzugsbänder beschicken zwei Gurtbandanlagen, die die Verbindung zu den Zementmühlen darstellen. Diese Gurtbänder haben eine maximale Leistung von je 1500 t/h. Die Antriebe der Abzugsplattenbänder sind regelbar ausgeführt, so daß jede gewünschte Abzugsleistung eingestellt werden kann.

3.3 Klinkerlager Siam City Cement, Thailand

Aufgrund der praktischen Erfahrungen, die bei den belgischen Zementwerken gesammelt wurden, haben die beiden thailändischen Zementwerke die Firma AUMUND mit der Generalplanung der Klinkertransportwege und Silokonzepte beauftragt. Aufgrund der ungewöhnlich großen Ofenleistungen waren fördertechnische Besonderheiten zu berücksichtigen. Die ungewöhnlichen Regenfälle in Thailand erfordern es, alle Haupttransporteinrichtungen über dem Normalniveau des Geländes anzuordnen. So wurde das erste Plattenband nach dem Kühler so verlängert, daß die Schleppkette des Standardkühlers mit einer zusätzlichen Übergabe Schleppkette-Plattenband entfallen konnte. Eine besondere Staubhaube in Verbindung mit einer guten Taktsteuerung der Kühlerentleerung vereinfachte die Transportaufgabe in diesem Bereich. Praktische Betriebserfahrungen für diese neuartige Anordnung eines Plattenbandes im Kühlerbereich liegen bereits von einigen Zementwerken vor. Der Steigungswinkel dieses Plattenbandes ist auf maximal 28° beschränkt (**Bild 4**). Das Plattenband beschickt einen Eckturm, der mit einem Zwischenbunker zur Aufnahme von Schwachbrand in der Inbetriebnahmephase oder für die Kontrolle der Ofenleistung bestimmt ist. Der Inhalt des Zwischenbunkers kann in dem

recombine different particle size fractions which have been formed by the segregation effect during the filling operation.

3. Project

3.1 CBR's clinker silo at Lixhe, Belgium

At CBR's Lixhe works there is a new clinker silo with a storage capacity of 80 000 t. The concrete cylinder has a diameter of 45 m and a total height of 57.8 m. Owing to the large diameter, the roof has been made as a steel structure clad with trapezoidal sheeting. Only one apron conveyor has been provided for filling and a further one for emptying. To improve the extraction rate the floor has been made funnel-shaped by filling it in and packing it with crushed limestone. A filter with a capacity of only 30 000 m³/h has been installed for dedusting the silo. The special technical features will be reported on later. Two lifts have been installed in concrete columns to simplify maintenance work on the conveying and filtering plants. The columns also act as supports for the bridge structure (**Fig. 2**).

3.2 CCB's clinker store at Gaurain, Belgium

CCB's cement works at Gaurain has a clinker store with a 140 000 t capacity. The concrete cylinder is 65 m in diameter and the silo has a total height of 50 m. The roof is a steel structure clad with trapezoidal sheeting.

The silo is emptied through four extraction ducts, so a good emptying rate is achieved. Each duct extracts 750 t/h and a total of 28 silo extraction points are installed. Uniform emptying of the silo is obtained by planned alternation of the extraction openings, with optimum blending of the clinker where there are possible fluctuations in quality (**Fig. 3**).

The plant was extended later to include two more 70 000 t clinker stores 55 m in diameter, which are emptied through three extraction ducts. The extraction belts feed two belt installations which form the links with the cement mills. Each belt has a maximum output of 1500 t/h. The drives for the discharge apron conveyors are controllable, so any desired extraction rate can be set.

3.3 The Siam City Cement clinker store in Thailand

The two Thai cement works commissioned AUMUND to take over general planning of the clinker transport routes and silo schemes, on the basis of their practical experience at the Belgian cement works. Special conveying arrangements had to be considered owing to the unusually large kiln capacities. Owing to the exceptionally high rainfall in Thailand all the main transporting equipment has to be located above normal ground level. Hence the first apron conveyor was extended towards the cooler, so that it was possible to

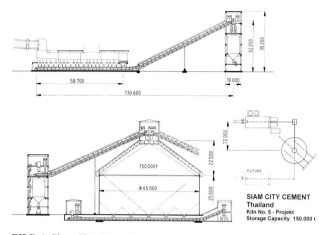

BILD 4: Siam City Cement
Anlagenplan
FIGURE 4: Siam City Cement
Layout plan
Ofen 5-Projekt = kiln No. 5-Project
Lagerkapazität = storage capacity

BILD 5: Siam City Cement
Klinkerlager 1 × 150 000 t; 65 m Durchmesser
KZB 250–1800 mit LKW-Beladung im Zwischenturm
FIGURE 5: Siam City Cement, Thailand
Clinker store 1 × 150 000 t; 65 m diameter
KZB 250–1800 with lorry loading at intermediate tower

Fall auf LKW's verladen oder aber über eine Bypass-Schurre und einen Verbindungsförderer dosiert der Mühlenbeschickung zugeführt werden. Im oberen Teil des Turmes befinden sich die Filteranlagen, die für die Entstaubung der Bandübergabe und der LKW-Beladung ausreichend dimensioniert sind.

Ein zweites Plattenband transportiert im Normalfall den Klinker über den Eckturm zu dem eigentlichen Klinkerlager mit 150 000 t Lagerkapazität. Der Kopf des Klinkerlagers ist so dimensioniert, daß später ein zweites Plattenband für den Anschluß einer neuen Ofenlinie eingebaut werden kann. Für die Entstaubung der beiden Plattenbänder und des Klinkerlagers selbst werden in diesem Dachaufbau Filteranlagen für 40 000 und 15 000 m³/h installiert. Die Dimensionen der Plattenbänder und der Filteranlagen erfordern einen Dachaufbau mit einem Durchmesser von 18,2 m und einer Dachbelastung von 4000 kN. Das Rundlager hat einen Durchmesser von 60 m bei einer Gesamthöhe von 47 m.

Unter dem Klinkerlager sind drei Abzugskanäle angeordnet, die mit Plattenbändern und AUMUND-Schwerkraftabzügen ausgerüstet sind. Die Abzugsleistung je Kanal ist 750 t/h. Die Übergabe des Klinkers erfolgt über Verteilerschurren auf 2 Gurtbandförderer, die eine wahlweise Beschickung der Gurtbandförderer ermöglicht. Alle Übergabestellen werden mit einer Stahlkonstruktion mit Trapezblechverkleidung ausgerüstet und mit Filteranlagen sorgfältig entstaubt. **Bild 5** zeigt eine Übersicht über die gesamte Anlage.

3.4 Siam Cement

In dieser neuen Zementfabrik wird eine Ofenanlage mit einer Tagesproduktion von 10 000 tato installiert. Das Kühlersystem wird konventionell mit einer Schleppkette ausgerüstet, so daß der Klinkertransport in einer Grube mit einer Tiefe von minus 7,5 m beginnt. Aufgrund der Bauhöhe der geplanten Klinkersilos errechnet sich eine Gesamtförderhöhe von 55 m. Für eine geplante Ofenleistung von 700 t/h kann diese Förderhöhe auch bei Einsatz der größten von AUMUND entwickelten Ketten nicht realisiert werden. Es mußten deshalb zwei Plattenbänder gewählt werden, um die Förderhöhe zu halbieren. Der Steigungswinkel der beiden Plattenbänder ist auf 28° begrenzt (**Bild 6**).

Im Bereich der Bandübergabe wurde ein Turm erforderlich, der als Zwischenbunker für Schwachbrand ausgebildet wird. Mit dem Zwischenbunker besteht auch die Möglichkeit, die Ofenleistung zu kontrollieren. Zwischen Silo 1 und Silo 2 wird ein horizontales Plattenband installiert, welches in einer allseitig geschlossenen Bandbrücke arbeitet. Für die wahlweise Beschickung von Silo 1 und Silo 2 sind keine besonderen motorisch betätigten Schieber erforderlich.

dispense with the drag-chain of the standard cooler with an additional drag-chain/conveyor transfer point. A special dust hood together with good timed control of cooler emptying simplifies conveying work in this section.

Practical experience in operating this novel apron conveyor arrangement in the cooler section is already available from some cement works. The angle of ascent of the conveyor is limited to a maximum of 28° (**Fig. 4**). The conveyor feeds a corner tower, which has an intermediate bin for receiving under-burnt material during the start-up phase or for checking kiln performance. The contents of the bin may in that case be loaded onto lorries, or alternatively transported through a bypass chute and on a connecting conveyor to be added in metered quantities to the mill feed. The top of the tower contains the filter plants, which are large enough to dedust the belt transfer point and lorry loading operations.

A second apron conveyor normally transports the clinker through the corner tower to the actual clinker store with a capacity of 150 000 t. The top of the store is dimensioned so that a second apron conveyor can be installed later to connect a new kiln line. Filter plants for treating 40 000 and 15 000 m³/h are being installed in this roof structure to dedust the two apron conveyors and the actual clinker store. The dimensions of the apron conveyors and filter plants require a roof structure with a diameter of 18.2 m and a roof load of 4000 kN. The circular store has a diameter of 60 m and a total height of 47 m.

Below the clinker store there are three extraction ducts, equipped with apron conveyors and AUMUND gravity discharge units. The extraction rate for each duct is 750 t/h. The clinker is transferred to 2 belt-type conveyors through distributor chutes which allow either of the belt conveyors to be fed. All transfer points are fitted with steel structures clad with trapezoidal sheeting, and carefully dedusted by filter plants. **Fig. 5** is a view over the whole installation.

3.4 Siam Cement

A kiln plant producing 10 000 tonnes per day is being installed at this new cement works. The cooler system is equipped with a drag chain in the conventional manner, so transportation of the clinker will start in a pit with a depth of minus 7.5 m. Owing to the height of the planned clinker silo the total conveying height is calculated as 55 m. This height cannot be achieved for a planned kiln output of 700 t/h, even if the largest chains developed by AUMUND are used. Hence two apron conveyors have had to be selected to halve the conveying height. Their angle of ascent is limited to 28° (**Fig. 6**).

It became necessary to have a tower, in the form of an intermediate hopper for under-burnt material, at the belt transfer point. The bin hopper makes it possible to check kiln perfor-

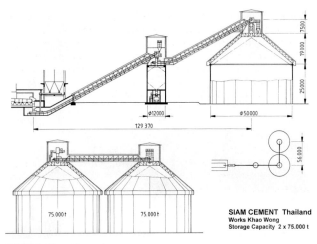

BILD 6: Siam Cement
Anlagenplan
FIGURE 6: Siam Cement
Layout plan
Ofen 5-Projekt = kiln No. 5-Project
Lagerkapazität = storage capacity

BILD 7: Siam Cement, Klinkerlager 2 × 75 000 t; 50 m Durchmesser Silobeschickung KZB 250–1600
FIGURE 7: Siam Cement, Clinker store 2 × 75 000 t; 50 m diameter silo filling KZB 250–1600

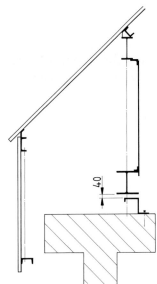

BILD 8: Ringspalt zur Frischluftzuführung
FIGURE 8: Annular gap for fresh air supply

Stoppt das Verbindungsband zum Silo 2, so wird der Klinker durch Überlaufschurren in das Silo 1 geleitet. Dieses simple Konstruktionsprinzip hat sich in vielen Zementwerken bewährt.

Die beiden Klinkersilos haben einen Durchmesser von 50 m. Die Höhe des Betonzylinders ist 25 m bei einer Gesamthöhe des Silos von 44 m. Die Dachkonstruktion besteht aus Stahlträgern mit einer Trapezblechverkleidung. Der Dachaufbau hat einen Durchmesser von 14,8 m und dient der Aufnahme des Plattenbandes und einer Filteranlage. Das Konzept der Siloentstaubung wird in Abschnitt 3.5 behandelt. Unter dem Silo befinden sich drei Abzugskanäle, die aufgrund der zu erwartenden starken Regenfälle über dem Null-Niveau angeordnet sind. Im Rahmen des Bauengineerings hat AUMUND alle Ausführungspläne für die Klinkersilos mit Abzugskanälen, die Dachkonstruktion sowie die Bandbrükken geliefert (**Bild 7**).

3.5 Entstaubung des Klinkerlagers

Bei dem Klinkersilo von CBR, Lixhe/Belgien mit einem Durchmesser von 45 m wurde erstmalig ein anderes Entstaubungsprinzip eingesetzt. Im Bereich des Überganges von dem Betonzylinder zu der Dachkonstruktion wurde ein Luftspalt von 40 mm angeordnet. Dadurch entsteht eine ringförmige Frischluftöffnung, die bei dem gewählten Filter mit 30.000 m^3/h eine kontinuierliche Lufteinzugsgeschwindigkeit von 1,2 m/s gewährleistet. Dieses Entstaubungsprinzip hat sich für dieses Klinkerlager mit 80 000 t Lagerkapazität bestens bewährt (**Bild 8**).

Es wird angenommen, daß die angesaugte Kaltluft die Lufttemperatur im Klinkerlager deutlich reduziert und damit die thermischen Auftriebskräfte günstig beeinflußt. Die Zeichnung zeigt die konstruktive Ausführung des Luftspaltes oberhalb des Betonzylinders als Z-Profil. Diese Ausführung wurde aufgrund der guten Betriebserfahrungen auch für die großen Klinkerlager von Thailand vorgesehen. Der Luftspalt beträgt 4 cm und kann den lokalen Betriebsbedingungen angepaßt werden.

4. Ausblick und Wertung

Klinkerlager mit großen Lagerkapazitäten werden wirtschaftlich als sogenannte Rundlager ohne Mittelsäule gebaut. Das günstige Durchmesser / Höhenverhältnis reduziert die Bauhöhe der Lager und den Investitionsaufwand für die Transporteinrichtungen zur Silobeschickung.

Der Einsatz von mehreren Abzugskanälen ermöglicht die Zusammenführung von verschiedenen Kornfraktionen, die sich durch den Entmischungseffekt beim Füllvorgang gebildet haben.

mance. A horizontal apron conveyor, working in a belt bridge enclosed on all sides, is installed between silo 1 and silo 2. No special motor-operated slides are necessary for selective filling of silo 1 and silo 2. If the belt providing the connection to silo 2 stops, the clinker is passed through overflow chutes into silo 1. This simple design principle has proved successful at many cement works.

The two clinker silos have diameters of 50 m. The height of the concrete cylinder is 25 m and the total height of the silo 44 m. The roof structure consists of steel girders with a cladding of trapezoidal sheeting. The roof structure has a diameter of 14.8 m and is used to contain the apron conveyor and a filter plant. The silo dedusting scheme is dealt with in section 3.5. Below the silo there are three extraction ducts, located above zero level in view of the heavy rainfall expected. On the constructional engineering side AUMUND has supplied all the plans for the clinker silos including the extraction ducts, the roof structure and the belt bridges (**Fig. 7**).

3.5 Dedusting the clinker store

A different dedusting principle has been applied at CBR's clinker silo at Lixhe, Belgium. A 40 mm air gap has been provided at the transition from the concrete cylinder to the roof structure. It forms a circular fresh air aperture, and with the selected filter treating 30 000 m^3/h the aperture ensures a continuous air intake speed of 1.2 m/s. The principle works extremely well at this 80 000 t capacity clinker store (**Fig. 8**).

Die Klinkerlagerbauart kann wirtschaftlich für 50 000 bis 200 000 t Lagerkapazität gebaut werden. Diese Lager werden mit einer selbsttragenden Dachkonstruktion ausgerüstet, die bis heute für Durchmesser von 45–65 m ausgeführt wurde. Im Rahmen eines Entwicklungsprojektes wird derzeit die technische Ausführung für Dachkonstruktionen mit einem Durchmesser von 34–90 m nach Wirtschaftlichkeitsgesichtspunkten geprüft.

It is assumed that the cold air drawn in considerably lowers the air temperature in the clinker store and thus has a favourable effect on thermal lifting forces. The drawing shows the construction of the air gap above the concrete cylinder, in the form of a Z profile. It is also envisaged to use this design for the large clinker stores in Thailand, in view of its satisfactory operation. The air gap is 4 cm wide and may be adapted to local operating conditions.

4. Prospects and assessment

It is cost-effective to build large-capacity clinker stores as so-called circular stores without a central column. The favourable diameter/height ratio reduces the overall height of the store and the capital cost of conveying equipment for filling the silos.

By using several extraction ducts it is possible to recombine the different particle size fractions which have been formed by the segregation effect during filling.

The clinker store design can be built economically for capacities of 50 000 to 200 000 t. These stores have self-supporting roof structures which have so far been made for diameters of 45 to 65 m. The technical design of roof structures with diameters of 34 to 90 m is at present being studied in a development project, from the economic viewpoint.

Bilder mit Genehmigung von

AUMUND Fördertechnik GmbH & Co Rheinberg

Illustrations by courtesy of

AUMUND Fördertechnik GmbH & Co Rheinberg

Einsatz eines Schaufelradbaggers in Kalksteinbrüchen*)

The use of bucket-wheel excavators in limestone quarries*)

Utilisation d'un excavateur à roue-pelle dans des carrières de calcaire

Empleo de una excavadora de rueda de palas en canteras

Von **R. Hoffmann** und **H. Steinberg,** Hannover/Deutschland

Zusammenfassung – Zur Rohmaterialgewinnung wird im Steinbruch der Teutonia Zementwerk AG das Gestein mittels Großbohrlochsprengung gewonnen und mit Hydraulikbaggern geladen. Über ein gleisgebundenes Fördersystem wird das Material zu einem stationären Brecher und anschließend mit Bandanlagen ins Werk gefördert. Mit wachsender Ausdehnung des Steinbruches verlängerten sich die Transportentfernungen und -zeiten von der Bruchwand zum Brecher und vergrößerte sich die Gleisanlage. Um die ständig steigenden Personal- und Betriebskosten zu begrenzen, wurde damit begonnen, das System der Rohmaterialgewinnung umzustellen. Gegenüber der Alternative „Schreitbrecher und Förderband" bot sich das Schaufelradbagger-Fördersystem an. Ausgehend von Versuchen mit einem kleinen Standardgerät und einer vergleichenden Wirtschaftlichkeitsbetrachtung mit den bekannten Gewinnungs- und Fördersystemen wurde im Jahr 1991 einer der beiden Fördergewinnungskreise auf einen Schaufelradbagger-Bänder-Betrieb umgestellt. Diese von O & K (jetzt Krupp Fördertechnik GmbH) gelieferte Anlage besteht aus einem Schaufelradbagger Typ S 400 für das Schälen des Materials von der Wand, aus einem Bandwagen zur Vergrößerung des Aktionsradius sowie einer Bandanlage mit kleinem Walzenbrecher. Der Walzenbrecher dient der Zerkleinerung des Überkorns. Die Bandanlage übernimmt den Transport in das etwa 1,5 km entfernte Werk. Das Investitionsvolumen betrug 6 Mio. DM. Nach einer Betriebszeit von ca. 18 Monaten und einer Förderung von 500000 t wurden pro Tonne bei den Personalkosten etwa 60 %, bei den Reparatur-, Wartungs- und Verschleißkosten etwa 55 % und bei den Energiekosten etwa 50 % gegenüber dem alten Gewinnungssystem eingespart. Eine weitergehende Optimierung der derzeitigen Förderleistung von ca. 600 t/h könnte durch Verbesserung der Zahngeometrie am Schaufelrad und einen anlagengerechten Abbau der Wand erreicht werden.

Summary – In Teutonia Zementwerk AG's quarries the rock is excavated by large-diameter hole blasting and loaded with hydraulic excavators to obtain the raw material. The material is transported on a rail mounted conveying system to a stationary crusher, and then by a belt system to the works. As the quarry extended the transport distances and times from the quarry face to the crusher lengthened and the rail system was extended. A start was therefore made in changing the raw material excavation system in order to limit the continuously increasing personnel and operating costs. There was much to recommend the bucket-wheel excavator conveying system when compared with the alternative of mobile crusher and conveying belt. After trials with a small standard machine and economic comparisons with known excavating and conveying systems, one of the two conveying and excavating areas was changed over to bucket-wheel excavator and belt operation. This plant, supplied by O & K (now Krupp Fördertechnik GmbH), consists of an S 400 bucket-wheel excavator for planing the material from the face, a belt waggon to extend the radius of action, and a belt system with small roller crusher. The roller crusher is used to break down the oversize material. The belt system transports the material to the works about 1.5 km away. The capital costs were about 6 million DM. After an operating period of about 18 months during which 500000 t were transported there was a saving per tonne compared with the old excavation system of about 60% in personnel costs, about 55% in repair, maintenance and wear costs, and about 50% in power costs. Further optimization of the present output capacity of about 600 t/h could be achieved by improving the tooth geometry on the bucket-wheel and by adapting the method of digging the face to suit the equipment.

Résumé – Pour l'obtention de la matière première à la carrière de la Teutonia Zementwerk AG, la roche est abattue par tir à gros trous et chargée au moyen de pelles hydrauliques. Un système de transport par rail achemine la matière vers un concasseur stationnaire, d'où elle est ensuite transportée à l'usine par convoyeurs à courroie. Avec une extension croissante de la carrière, les distances et temps de transport, du mur d'abattage au concasseur, se sont allongés et le système de rails s'est agrandi. Afin de limiter les coûts de main-d'œuvre et d'exploitation toujours croissants, on a commencé

---

*) Überarbeitete Fassung eines Vortrages zum VDZ-Kongreß '93, Düsseldorf (27. 9. – 1. 10. 1993)
   Revised text of a lecture to the VDZ Congress '93, Düsseldorf (27.9.–1.10.1993)

par modifier le système d'extraction de matière première. Contre l'alternative „concasseur marchant et courroie transporteuse", s'est offert le système d'extraction et de transport à roue-pelle. Suite à des essais avec un petit appareil standard et à des évaluations de rentabilité comparatives par rapport aux systèmes connus d'extraction et de transport, un des secteurs d'extraction et de transport a été, en 1991, transformé pour un fonctionnement roue-pelle et courroies. Cette installation, fournie par O & K (maintenant Krupp Fördertechnik GmbH), se compose d'un excavateur à roue-pelle de type S 400, pour extraire la matière au mur, d'un chariot à courroie pour agrandir le rayon d'action et d'une installation à courroie avec un petit concasseur à cylindres. Celui-ci sert à fragmenter les trop gros morceaux. L'installation à courroie se charge de l'acheminement à l'usine, éloignée d'environ 1,5 km. Le montant des investissements s'est élevé à 6 M DM. Après une durée d'exploitation d'environ 18 mois et l'extraction et transport de 500000 t, avaient été économisés, par rapport à l'ancien système d'extraction, environ 60% des coûts de main-d'oeuvre, environ 55% des coûts de réparation, de maintenance et d'usure et environ 50% des coûts d'énergie. Une optimisation ultérieure de l'actuelle performance d'extraction, d'environ 600 t/h, pourrait être obtenue par l'amélioration de la géométrie des dents de la roue-pelle et par une extraction à la paroi, plus adaptée aux particularités de l'installation.

Resumen — Para la extracción de la materia prima, en la cantera de Teutonia Zementwerk AG, se obtiene el material mediante voladura por agujeros grandes y con ayuda de una excavadora hidráulica. Luego el material es transportado, mediante un sistema de transporte sobre carril, hacia una trituradora estacionaria y llevado a continuación a la fábrica por medio de cintas transportadoras. A medida que se extiende la cantera, aumentan los recorridos y la duración del transporte desde el frente de arranque hasta la trituradora, con lo cual aumenta también la longitud de los carriles. Con el fin de limitar el constante aumento de los gastos de personal y de explotación, se ha empezado a cambiar el sistema de extracción de materias primas. Frente a la alternativa de „trituradora caminante y cinta transportadora" se ofrecía el sistema de transporte con excavadora de rueda de palas. Partiendo de unos ensayos realizados con una máquina standard de reducido tamaño y un estudio comparativo de rentabilidad de otros sistemas de extracción y de transporte conocidos, se ha sustituido, en 1991, uno de los dos circuitos de transporte y de extracción por el sistema de excavadora de rueda de palas y de cinta transportadora. Los equipos suministrados por O&K (ahora Krupp Fördertechnik GmbH) consisten en una excavadora de rueda de palas S 400 para la escarificación del material en el frente de arranque, un carro móvil de terminal de cinta, para aumentar el radio de acción, así como una cinta transportadora y una pequeña trituradora de rodillos. Este último sirve para el desmenuzamiento del grano sobre dimensionado. La instalación de cinta transportadora se encarga del transporte del material a la fábrica de cemento situada a 1,5 km de distancia aprox. El volumen de inversiones ha sido de 6 millonas de DM. Después de un tiempo de servicio de unos 18 meses y el transporte de 500000 t de material, se ha podido ahorrar por tonelada un 60% aprox. de los gastos de personal, un 55%, más o menos, de los gastos para reparaciones, mantenimiento y desgaste, y un 50% aprox. de los gastos energéticos, en comparación con el antiguo sistema de explotación. Un perfeccionamiento de la geometria de los dientes, en la rueda de palas, así como un arranque adecuado del material, permiten una optimización de la capacidad de extracción actual.

Empleo de una excavadora de rueda de palas en canteras

1. Ausgangssituation

Das Gebiet östlich von Hannover wird seit mehr als 120 Jahren durch die Zementindustrie geprägt. Durch die Ablagerungen eines Meeres der jüngeren Kreidezeit entstanden hier vor rd. 80 Mio. Jahren ausgedehnte Rohstoffvorkommen zur Herstellung von hochwertigen Zementen.

Die Lagerstätten der Teutonia Zementwerk AG können hinsichtlich ihrer Kalkstein-, Kieselsäure- und Tonkomponenten dem Kalkmergel zugeordnet werden. Im Tagebau wird ein leicht klüftiger Kalkmergel mit einer Grubenfeuchte von 10 – 12% abgebaut. Der Kalkmergel besitzt eine mittlere Druckfestigkeit von 10 – 20 MPa und kann als relativ hart und spröde charakterisiert werden. Wegen seines niedrigen Gehaltes an freiem Siliziumdioxid ist das Gestein nur wenig abrasiv. Über die gesamte Abbaufläche schwankt der Kalkstandard in seiner natürlichen Zusammensetzung zwischen 65 bis 150. Aufgrund der hervorragenden Qualität der Vorkommen kann auf die Zugabe von Tonerde, Kiesabbrand oder hochwertigem Kalkstein zu Korrekturzwecken der Rohstoffmischung verzichtet werden. Die große Schwankungsbreite im Kalkstandard verlangt allerdings den selektiven Abbau von zwei Kalkmergelqualitäten, die auf einen durchschnittlichen Kalkstandard von 98 bis 100 gemischt werden.

Der jährliche Bedarf von rd. 1,2 Mio. Tonnen Kalkmergel wurde bislang (**Bild 1**) ausschließlich mittels Großbohrlochsprengung aus 30 m hohen Gesteinswänden gelöst. Hydraulikbagger luden das Haufwerk in Loren eines gleisgebunde-

1. Original situation

The region to the east of Hannover has been orientated towards the cement industry for more than 120 years. Extensive raw material deposits suitable for producing high grade cements built up here about 80 million years ago by sedimentation from an ocean during the younger Cretaceous period.

The deposits of the Teutonia Zementwerk AG can be classified as lime marl as far as their limestone, silica and clay components are concerned. A slightly fissured lime marl with an as-dug moisture content of 10–12% is extracted in the quarry. The lime marl has an average compressive strength of 10–20 MPa and can be characterized as relatively hard and brittle. The rock is only slightly abrasive because of its low content of free silica. In its natural composition the lime standard fluctuates over the entire quarrying area between 65 and 150. Because of the excellent quality of the deposit it is possible to dispense with the addition of alumina, roasted pyrites or high-grade limestone for the purposes of correcting the raw material mix. However, the wide range of fluctuation in lime standard requires selective quarrying of two grades of lime marl which are then mixed to give an average lime standard of 98 to 100.

The annual requirement of about 1.2 million tonnes of lime marl has so far been obtained entirely by large-diameter bore hole blasting from 30 m high rock faces (**Fig. 1**). Hydraulic excavators have loaded the broken material into trolleys on a rail conveyor system. Trains then transport the

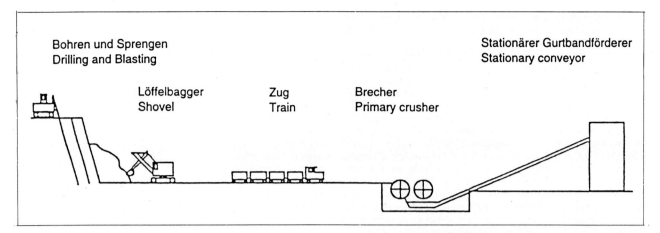

BILD 1: Schematische Darstellung der alten Gewinnungstechnologie mit Sprengbetrieb, gleisgebundenem Materialtransport und stationärer Brecheranlage
FIGURE 1: Diagrammatic representation of the old quarrying technology with blasting, rail transport and stationary crusher plant

nen Fördersystems. Züge transportierten dann das Material zu einem stationären Doppelwellenhammerbrecher, der es auf Korngrößen von 0–80 mm Kantenlänge zerkleinerte. Der so gebrochene Kalkmergel wurde mit einer Gurtförderanlage in das ca. 1,5 km entfernte Zementwerk transportiert und dort in einer Halle zwischengelagert.

Dieses diskontinuierliche Verfahren mit seiner gleisgebundenen Förderung führt zwangsläufig zu steigenden Betriebskosten, da sich mit wachsender Ausdehnung des Steinbruchs die Transportentfernungen vergrößern. Zu den steigenden Betriebskosten sind in den letzten Jahren noch erhebliche Belastungen durch verschärfte Sicherheits- und Umweltauflagen hinzugekommen. Außerdem wurden durch die Behörde zur Genehmigung von Großbohrlochsprengungen immer mehr einschränkende Bestimmungen erlassen. Durch diese Entwicklungen veranlaßt, wurde von der Leitung des Zementwerkes über neue Systeme zur Rohmaterialgewinnung nachgedacht.

Umfangreiche Gesteinsuntersuchungen und ein mehrmonatiger Probebetrieb mit einem kleinen standardmäßigen Schaufelradbagger der Baugrößenbezeichnung S 100 lieferten erfolgversprechende Ergebnisse. Deshalb wurde nach einem Wirtschaftlichkeitsvergleich dieser Abbaumethode mit bekannten Gewinnungs- und Fördersystemen im Jahre 1991 einer der beiden Förderkreise auf den Betrieb eines Schaufelradbaggers in Verbindung mit einem Gurtbandtransport umgestellt.

2. Beschreibung der Anlagen

Bild 2 zeigt schematisch das neue mit Ausrüstungen von O & K (jetzt Krupp Fördertechnik GmbH) kontinuierlich arbeitende Gewinnungs- und Transportsystem. Der Schaufelradbagger mit der Typenbezeichnung S 400/250 und einer Dienstmasse von 186 t fräst im Fallschnitt den Kalkmergel

material to a stationary twin-shaft hammer crusher which breaks it down to a particle size of 0–80 mm edge length. The crushed chalk marl is transported approximately 1.5 km on a belt conveyor system to the cement works where it is placed in intermediate storage in a hall.

This discontinuous process with its rail conveying system inevitably leads to rising operating costs as the transport distances increase with increasing extent of the quarry. In recent years considerable burdens due to more severe safety and environmental regulations have been added to the rising operating costs. The authorities have also been issuing ever more restrictive conditions for the licensing of large-diameter bore hole blasting. These developments caused the management of the cement works to consider new systems for raw material excavation.

Extensive rock investigations and a trial operation lasting several months with a small standard S100 bucket wheel excavator gave promising results. These led, after an economic comparison of this method of extraction with known extracting and conveying systems, to conversion of one of the two conveying circuits in 1991 to operation with a bucket wheel excavator in combination with conveyor belt transport.

2. Description of the plant

The new continuous quarrying and transport system with equipment from O & K (now Krupp Fördertechnik GmbH) is shown diagrammatically in Fig. 2. The S400/250 bucket wheel excavator with a service weight of 186 t cuts the lime marl in the face in drop cut operation into a transportable particle size composition and then feeds it on via a slewing discharge boom. After the bucket wheel excavator there is a crawler-mounted belt wagon which increases the action

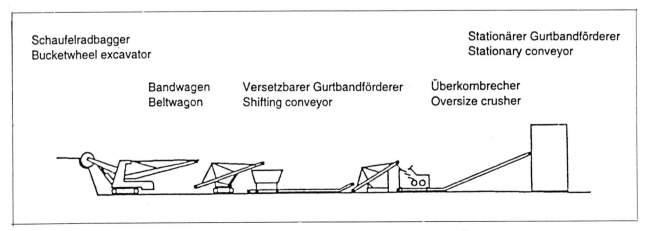

BILD 2: Neue Gewinnungstechnologie mit Schaufelradbagger, versetzbarem Gurtbandförderer und Überkornbrecher
FIGURE 2: New quarrying technology with bucket wheel excavator, movable belt conveyor and oversize crusher

BILD 3: Blick auf das speziell ausgebildete Schaufelrad
FIGURE 3: View of the specially designed bucket wheel

in einer transportfähigen Korngrößenzusammensetzung aus der Wand und gibt ihn dann über einen schwenkbaren Ausleger weiter. Dem Schaufelradbagger ist ein Bandwagen mit Raupenfahrwerk nachgeschaltet, der den Aktionsradius des Gerätes um 30 m auf ca. 60 m vergrößert. Der Bandwagen übernimmt das Material vom Schaufelradbagger und wirft es in den Aufgabebunker eines mit der Kabeltrommel kombinierten Wagens, der über einem rückbaren Gurtförderer angeordnet ist. Zur Anlage gehört auch ein Aufgabeförderband für einen stationären Überkornbrecher mit Vorabsiebung. Eine stationäre Gurtbandanlage verbindet den neuen Förderkreis mit dem vorhandenen Förderweg ins Werk.

Die Förderleistung der Anlage wird entscheidend durch die spezielle Ausbildung des Schaufelrades bestimmt (**Bild 3**). Die optimale Ausbildung konnte erst durch zahlreiche Versuche mit dem Versuchsgerät S 100 ermittelt werden, wobei die verschiedenen Parameter zur konstruktiven Auslegung der Becher und Schneidzähne (Anzahl, Form, Anstellwinkel und Befestigungsart) mehrfach variiert wurden. Zur Ausführung kam schließlich ein Schaufelrad mit 32 Schneidelementen, bestehend aus 16 Eimern und 16 Vorschneidern, die jetzt direkt in das etwas schräg gelagerte Schaufelrad integriert wurden. Bei den sonst üblichen Schaufelrädern sind die einzelnen Eimer und Vorschneider abnehmbar auf dem Radkörper befestigt. Den Boden der Becher bilden Kettenmatten; sie verhindern Anbackungen und sorgen für eine gute Restentleerung. Die Zähne der 1. Generation steckten in Fittings, die an den Bechern angeschweißt waren. Diese Befestigungsart hielt den hohen mechanischen Belastungen nicht stand, die Zähne brachen aus. Um einen ungestörten Kraftfluß von den Zahnspitzen bis hinein in den Radkörper sicherzustellen, wurden die Zähne der 2. Generation direkt an die Becherlippen angeschweißt. Als ein zusätzlicher Vorteil erwies sich bei der werkstoffmäßigen Auslegung der Zähne der Übergang zur Stahlqualität ST 52. Durch diese und andere Maßnahmen konnte die

radius of the machine by 30 m to approximately 60 m. The belt wagon accepts the material from the bucket wheel excavator and throws it into the feed hopper of a wagon which is combined with the cable drum and is located above a movable belt conveyor. The plant also includes a feed conveyor belt for a stationary oversize crusher with prescreening system. The new conveying circuit is linked to the existing conveyor route to the works by a stationary belt conveyor system.

The critical factor determining the conveying capacity of the plant is the special design of the bucket wheel (**Fig. 3**). It was only possible to determine the optimum design through a large number of trials with the S100 test machine in which many variations were made to the different parameters in the design of the bucket and cutting teeth (number, shape, approach angle and method of attachment). The eventual design was a bucket wheel with 32 cutting elements consisting of 16 buckets and 16 pre cutters which were integrated directly into the slightly inclined bucket wheel.

With normal bucket wheels the individual buckets and pre cutters are attached to the body of the wheel in such a way that they can be removed. Chain mats form the base of the bucket; they prevent build-up and ensure complete emptying. The first generation of teeth were inserted into fittings which were welded to the buckets. This type of fitting could not withstand the high mechanical loads and the teeth broke out, so the teeth of the second generation were welded directly to the bucket lips to ensure an uninterrupted force flow from the tip of the tooth into the body of the wheel. In the design of the material for the teeth it also proved beneficial to change to ST 52 grade steel. Through this and other measures it was possible to convert the installed drive power of the bucket wheel of 315 kW into very high specific tooth forces. Of the original tooth length of 110 mm still 65 mm were left after 24 months operating time. It is planned to replace the teeth after they have worn down to about 50 mm. The teeth are shaped so that even when worn they still meet the face with a sharp cutting edge. Attempts are being made at present to increase the service life still further by welding a hard metal layer to the tooth tips.

The bucket wheel has a diameter of 5.6 m and is driven by a 315 kW motor. The circumferential speed of 2.9 m/s is relatively high. The preliminary trials with the S100 standard machine which had a circumferential speed of 1.5 m/s had shown that this is the only way to avoid vibration of the machine and tearing complete blocks of rock out of the face. In spite of this, pieces of rock with edge lengths up to 400 mm become detached from the face in the strongly fissured sections. **Fig. 4** shows the downstream oversize crusher with prescreening system which only allows pieces of rock with edge lengths <120 mm to pass into the transport system to the works. A Mogensen vibrating rod sizer allows 90 to 95% of the incoming material to pass. The oversize material is crushed to the required particle size in a 650 mm diameter × 1200 mm twin-roll crusher with an installed drive power of 2 × 45 kW. The crusher and also the other transfer points are electrically heated to avoid any build-up of the relatively moist marl.

The bucket wheel excavator (**Fig. 5**) works in blocks, i.e. it digs a block approximately 22 m wide from one working position. At each pass the bucket wheel is first placed on the quarry face. The material is then cut in thin shavings from the face in the direction of the strata starting from the top. During this process the bucket wheel penetrates about one half wheel diameter deep into the wall. At the same time the boom slews evenly backwards and forwards by approximately 120°, and is lowered by about 100 mm at each change in direction. From time to time the excavator has to be moved backwards in order to pick up the material spillage, which cannot be completely avoided. The 10.5 m long bucket wheel boom permits a digging height of 9 m and can be slewed through a maximum of 2 × 90° in relation to the 23 m long boom of the discharge boom. Together with the belt wagon (**Fig. 6**) this therefore gives a maximum action radius of 60 m. The conveying capacity, conveyor belt width

BILD 4: Steinbruch mit Überkornbrecher und Vorabsiebung

FIGURE 4: Quarry with oversize crusher and prescreening system

installierte Antriebsleistung des Schaufelrades von 315 kW in sehr hohe spezifische Zahnkräfte umgesetzt werden. Von der ursprünglichen Zahnlänge von 110 mm waren nach 24-monatiger Betriebszeit noch 65 mm vorhanden. Das Auswechseln der Zähne ist bei einer Abnutzung bis auf 50 mm geplant. Die Zähne sind so geformt, daß sie auch bei Verschleiß noch mit einer scharfen Schneide auf die Wand treffen. Durch Aufschweißen einer Hartmetallschicht an den Zahnspitzen wird z. Z. versucht, die Standzeit noch weiter zu verlängern.

Das Schaufelrad hat einen Durchmesser von 5,6 m und wird von einem Motor mit 315 kW angetrieben. Die Umfangsgeschwindigkeit liegt mit 2,9 m/s relativ hoch. Die Vorversuche mit dem Standardgerät S 100, das eine Umfangsgeschwindigkeit von 1,5 m/s besaß, hatten gezeigt, daß nur auf diesem Wege Erschütterungen des Gerätes sowie das Herausreißen von ganzen Gesteinsblöcken aus der Wand zu vermeiden sind. An stark zerklüfteten Wandpartien werden trotzdem Gesteinspartien mit einer Kantenlänge bis zu 400 mm aus der Wand gelöst. **Bild 4** zeigt den nachgeschalteten Überkornbrecher mit Vorabsiebung, der die Aufgabe hat, nur Gesteinsstücke mit einer Kantenlänge < 120 mm für den Transport ins Werk freizugeben. Ein Vibro-Stangensizer von Mogensen läßt bereits 90 bis 95 % des ankommenden Fördergutes passieren. Das Überkorn wird in einem Zweiwalzenbrecher ⌀ 650 mm × 1200 mm mit einer installierten Antriebsleistung von 2 × 45 kW auf die gewünschte Korngröße gebrochen. Um Anbackungen des relativ feuchten Mergels zu vermeiden, wird der Brecher, wie auch die anderen Übergabestellen, elektrisch beheizt.

Der Schaufelradbagger (**Bild 5**) arbeitet im Blockbetrieb, das heißt, er gewinnt aus einer Arbeitsposition einen Block von ca. 22 m Breite. Bei jedem Durchgang wird das Schaufelrad zuerst auf die Bruchwand aufgelegt. Dann wird das Material in dünnen Spänen, von oben beginnend, in Schichtrichtung aus der Wand gefräst. Dabei dringt das Schaufelrad etwa einen halben Raddurchmesser tief in die Wand ein. Der Ausleger schwenkt gleichmäßig um ca. 120 Grad hin und her und wird bei jedem Richtungswechsel ungefähr 100 mm abgesenkt. Von Zeit zu Zeit muß der Bagger zurücksetzen, um den Materialüberwurf aufzunehmen, der sich nicht ganz vermeiden läßt. Die 10,5 m lange Ausladung des Schaufelrades erlaubt eine Abtragshöhe von 9 m und läßt sich gegenüber der 23 m langen Ausladung der Verladebrücke um maximal 2 × 90 Grad schwenken. Zusammen mit dem Bandwagen (**Bild 6**) ergibt sich somit ein Aktionsradius von max. 60 m. Förderleistung, Förderbandbreite und Schaufelraddurchmesser entsprechen den Parametern der Baureihe S 250. Da das Fräsen in relativ hartem Gestein das Gerät sehr stark beansprucht, wurde der Ober-

and bucket wheel diameter correspond to the parameters of the S250 model. Cutting into the relatively hard rock subjects the machine to very severe stresses, so the undercarriage and uppercarriage were designed on the heavier S400 model.

The bucket wheel and hydraulic pumps, and also the slewing and travelling gear, are driven electrically. The power supply is provided from a 6 KV network. A transformer station on the counterweight boom produces an operating voltage of 500 V. The hydraulic units are also located in a weatherproof housing on the counterweight boom.

3. Economic considerations

Approximately 6 million DM were invested for the complete plant. **Table 1** shows the comparison of the specific costs for personnel, energy, repair and maintenance for the period from January 1991 to October 1993. With the conventional rail transport operation up to 8 workers were occupied simultaneously for the drilling, blasting, loading, conveying and crushing work, but for raw material quarrying with the bucket wheel excavator two fairly highly trained plant operators are sufficient. Of the two plant operators, one controls the bucket wheel excavator while the second operates

BILD 5: Schaufelradbagger S 400/250
FIGURE 5: S 400/250 bucket wheel excavator

BILD 6: Fahrbarer Bandwagen mit Raupenfahrwerk

FIGURE 6: Mobile, crawler-mounted, belt wagon

und Unterwagen nach der schwereren S 400 Baureihe ausgeführt.

Schaufelrad und Hydraulikpumpen sowie Schwenk- und Fahrwerk werden elektrisch angetrieben. Die Spannungsversorgung erfolgt aus einem 6 KV-Netz. Eine Trafostation auf dem Gegengewichtausleger erzeugt eine Betriebsspannung von 500 Volt. Auch die Hydraulikaggregate sind auf dem Gegengewichtsausleger wetterfest untergebracht.

3. Wirtschaftliche Betrachtungen

Für die Gesamtanlage wurden ca. 6 Mio. DM investiert. **Tabelle 1** zeigt den Vergleich der spezifischen Kosten für Personal, Energie, Instandhaltung und Wartung für den Zeitraum von Januar 1991 bis Oktober 1993. Während beim konventionellen, gleisgebundenen Förderbetrieb für die Arbeitsschritte Bohren, Sprengen, Laden, Fördern und Brechen bis zu 8 Arbeitskräfte gleichzeitig beschäftigt waren, reichen für die Rohmaterialgewinnung mit dem Schaufelradbagger zwei höher qualifizierte Anlagenfahrer aus. Von den beiden Anlagenfahrern steuert einer den Schaufelradbagger, während der zweite den Bandwagen bedient oder die nachgeschaltete Gurtbandanlage kontrolliert.

Der für die Zementherstellung benötigte Kalkstandard wird durch Mischen von zwei Rohmaterialqualitäten erreicht. Die Laufzeiten der beiden Förder- und Gewinnungssysteme werden durch den qualitativ festgelegten Kalkstandard bestimmt. Insgesamt kann der Steinbruch sehr flexibel betrieben werden, so daß auch bei möglichen Störungen die wöchentlich benötigte Fördermenge ohne zusätzliche Mehrarbeit gewonnen wird. An Stillstandstagen übernimmt die Bedienungsmannschaft die Wartung und Reinigung der Anlage.

Ausgehend von Personalkosten in Höhe von 0,96 DM/t bei der gleisgebundenen Förderung, ergeben sich beim Einsatz des Schaufelradbaggers Kosten von 0,35 DM/t und somit Einsparungen von 0,61 DM/t bzw. 63 %. Steigende Personalkosten durch Lohnerhöhungen und Arbeitszeitverkürzungen in dieser Zeit konnten durch Durchsatzsteigerungen ausgeglichen werden. Die spezifischen Reparatur- und Wartungskosten konnten ebenfalls um 44 %, von 1,12 auf 0,63 DM/t gesenkt werden, wobei beim Schaufelradbagger gerade während der Inbetriebnahmephase nicht zwischen Reparatur- und Optimierungskosten unterschieden wurde. Dieser Kostenposten beinhaltet nicht das Verlegen der Gleise, da dieser Aufwand ungefähr mit der Verlängerung des Strossenförderers vergleichbar ist.

Beim Vergleich der spezifischen Energiekosten, die sich beim herkömmlichen System zunächst aus dem Treibstoff- und Stromverbrauch ergeben, muß auch noch der Sprengstoffverbrauch berücksichtigt werden, da beim Schaufelradbagger ausschließlich elektrische Energie eingesetzt wird, um das Material aus der Wand zu lösen. Tabelle 1 zeigt,

the belt wagon or checks the following conveyor belt system.

The lime standard required for cement production is achieved by mixing two grades of raw material. The running times of the two conveying and quarrying systems are determined by the qualitatively defined lime standard. As a whole the quarry can be operated very flexibly so that the quantities required weekly can be obtained without additional overtime, even if faults occur. The operating team maintains and cleans the plant on days when it is shut down.

The personnel costs for the rail transport system were around 0.96 DM/t, while for the bucket wheel excavator systems the costs are 0.35 DM/t, resulting in savings of 0.61 DM/t or 63 %. Rising personnel costs through increased wages and shortened working hours at that time were offset by increases in throughput. The specific repair and maintenance costs were also lowered by 44 %, from 1.12 to 0.63 DM/t. During the commissioning phase, it was not possible to differentiate between repair and optimization costs with the bucket wheel excavator. These costs do not cover the laying of the rails as this expenditure is roughly comparable with the extension of the face conveyor.

When comparing the specific energy costs which, with the conventional system, arise primarily from the fuel and power consumption, it is also necessary to take the consumption of explosives into account, as the bucket wheel excavator only uses electrical energy to break the material from the face. Table 1 shows that the energy costs were more than halved with a specific value of 0.16 DM/t as opposed to 0.37 DM/t. From these comments it follows that when compared with the rail conveying system, the specific quarrying costs were reduced from 2.45 to 1.14 DM/t, i.e. by 1.31 DM/t. Depreciation and interest as calculated for costing were not taken into account in this comparison.

TABELLE 1: Kostenvergleich zwischen alter und neuer Gewinnungstechnologie
TABLE 1: Cost comparison between the old and new methods of quarrying

Kostenarten Costcategory	Bezogene Kosten DM/t Specific costs	
	Altes System Old system	Neues System New system
Personal Personnel	0,96	0,35
Reparatur/Wartung Repair/Maintenance	1,12	0,63
Energie/Sprengmittel Energy/Exlosive	0,37	0,16
Summe Sum	2,45	1,14

daß bei den Energiekosten mit einem spezifischen Wert von 0,16 DM/t gegenüber 0,37 DM/t mehr als eine Halbierung erzielt werden konnte. Aus den vorangegangenen Ausführungen folgt, daß im Vergleich zur gleisgebundenen Förderung die spezifischen Gewinnungskosten von 2,45 auf 1,14 DM/t, d. h. um 1,31 DM/t gesenkt werden konnten. Bei dieser vergleichenden Gegenüberstellung wurden Abschreibungen und kalkulatorische Zinsen nicht berücksichtigt.

Seit der Inbetriebnahme im Jahre 1991 wurden ca. 1 Mio. t Kalkmergel mit dem neuen Schaufelradbaggersystem gewonnen. Diese Gewinnungsleistung entspricht etwa einem Drittel der gesamten Tagebauförderleistung in dieser Zeit. Der Anteil wird durch das Mischungsverhältnis der beiden Rohmaterialkomponenten zur Einstellung des qualitativ festgelegten Kalkstandards bestimmt. Obwohl die Gewinnungskosten bedeutend gesenkt werden konnten, ist die angestrebte Förderleistung von 800 t/h im Jahresmittel im dritten Jahr noch nicht ganz erreicht worden. **Bild 7** zeigt die kumulierten Förderleistungen des Schaufelradbaggers seit Inbetriebnahme. Aus dem Bild geht hervor, daß die Förderleistung 1992 mit 643 t/h bereits deutlich über der von 1991 mit 578 t/h lag. Ende Oktober 1993 betrug die durchschnittliche Stundenleistung 786 t/h, wobei Einzelwerte schon deutlich über der Zielstellung von 800 t/h lagen.

Bevor der heutige Leistungsstand erzielt wurde, mußten während der Inbetriebnahmephase in den ersten Monaten zahlreiche Schwierigkeiten durch geeignete Maßnahmen beseitigt werden. Starker Wasseraustritt aus der Wand sowie ein hoher Materialüberwurf am Schaufelrad minderten zunächst die Förderleistung des Gerätes ganz erheblich. Außerdem stellte sich heraus, daß der vorhandene Förderweg ins Werk – ein gemuldeter Gurtförderer 1000 mm mit mehreren Übergabestellen – kurzzeitig auftretende Belastungsspitzen bis zu 1500 t/h nicht immer bewältigen konnte. Eine wirksame Maßnahme bildete hier der nachträgliche Einbau einer Bandwaage auf dem Schaufelradbagger, die es dem Anlagenfahrer heute ermöglicht, Förderstromspitzen zu erkennen und zu vermeiden. Auch die zunehmende Erfahrung des Baggerfahrers hat schließlich zur Leistungssteigerung beigetragen.

4. Schlußbetrachtung und Ausblick

Die Senkung der Gewinnungskosten ist neben dem Umweltschutz und der Arbeitssicherheit eine der wichtigsten Aufgaben für jeden Betreiber eines Steinbruchs. Durch die Einführung eines Schaufelradbaggers in Verbindung mit einem Förderbandsystem ist es bei der Teutonia Zementwerk AG gelungen, den steigenden Betriebskosten insbesondere durch den Verbrauch an elektrischer Energie sowie auch den steigenden Personalkosten wirksam entgegenzutreten. Der teilweise Verzicht auf Großbohrlochsprengungen verminderte die Belästigungen durch Sprengerschütterungen und Lärm und hat zu einer höheren Akzeptanz der Anwohner gegenüber dem Zementwerk beigetragen. Andererseits lassen sich auch die Rohstoffvorkommen effektiver nutzen, da der einzuhaltende Sicherheitsabstand gegenüber der nächsten Bebauung deutlich verringert werden kann.

Die vorgestellte Anlage hat sich nach Beseitigung der Anlaufschwierigkeiten als zuverlässig, betriebssicher und wirtschaftlich erwiesen.

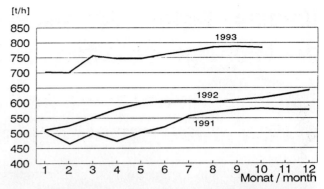

BILD 7: Kumulierte Förderleistung des Schaufelradbaggers
FIGURE 7: Cumulative conveying capacity of the bucket wheel excavator

Since the commissioning in 1991, approximately 1 million t lime marl have been quarried with the new bucket wheel excavator system. This quarrying capacity corresponds to approximately one third of the total quarry output during this period. The proportion is determined by the mixing ratio of the two raw material components to achieve the qualitatively defined lime standard. Although the quarrying costs were significantly lowered the target output for the yearly average of 800 t/h was not fully achieved by the third year. **Fig. 7** shows the cumulative outputs from the bucket wheel excavator since commissioning. From the diagram it can be seen that the 1992 output of 643 t/h was significantly higher than that for 1991 of 578 t/h. At the end of October 1993, the average hourly output was 786 t/h and individual values were significantly higher than the target of 800 t/h.

Large numbers of difficulties had to be eliminated in the first few months by suitable measures during the commission phase before the present output could be achieved. At first the output from the machine was greatly reduced by the heavy flow of water from the face and by the large amount of material spillage from the bucket wheel. It also emerged that the existing conveyor route to the works – a 1000 mm troughed belt conveyor with several transfer points – could not always deal with the short-term peak loads of up to 1500 t/h. An effective measure here was the retro-fitting of a belt weigher on the bucket wheel excavator, so that the plant operator can now recognize and avoid output peaks. The increasing experience of the excavator operator has ultimately also contributed to the increase in output.

4. Final comments and outlook

Lowering the quarrying costs is, in addition to environmental protection and safety at work, one of the most important tasks for any quarry operator. The introduction of a bucket wheel excavator in conjunction with a belt conveyor system has enabled the Teutonia Zementwerk AG to counter the rising operating costs caused mainly by the consumption of electrical power and rising personnel costs. The partial abandonment of large-diameter hole blasting has reduced the nuisance caused by blasting vibration and noise and has contributed to greater acceptance of the cement works by the neighbouring residents. The raw material deposits can also be utilized more effectively because the safety distance which has to be maintained from the nearest buildings can be reduced significantly.

After elimination of the initial difficulties the plant described has proved to be dependable, operationally reliable, and cost effective.

Literature

[1] Schröder, D., und Trümper, R.: Digging rock with bucket wheel excavators. Bulk solids handling, Volume 13, No. 2, May 1993, pp. 265–274.

[2] Schröder, D.: Schaufelradbagger als Alternative zum Sprengbetrieb für semihartes Gestein. Zement-Kalk-Gips 46 (1993) No. 8, pp. 423–429.

Uniaxial-Tester – ein neues Gerät zur Beurteilung der Fließeigenschaften von Zement

The uniaxial tester – a new apparatus for assessment of flow properties of cement

Testeur uniaxial – un appareil nouveau pour tester les propriétés d'écoulement du ciment

Aparato de ensayo uniaxial – un nuevo aparato para el ensayo de las propiedades reológicas de los cementos

Von **L.P. Maltby, G. G. Enstad,** Porsgrunn, und **E. Stoltenberg-Hansson,** Brevik/Norwegen

Zusammenfassung – Die Fließeigenschaften von Zement haben einen großen Einfluß auf seine Handhabungs- und Transportkosten. Da die Mahlhilfsmittel genau dosiert werden müssen, ist eine gute Methode, mit der die Fließeigenschaften des Zements gemessen werden können, besonders wichtig. Zu diesem Zweck wurde ein neues Gerät, der Uniaxial-Tester, entwickelt. Grundsätzlich wird die Zementprobe in eine leicht konische Form gegeben und mittels eines Kolbens verdichtet. Der Kolben kann bewegt werden, was zur Verformung der Probe in vertikaler Richtung führt. Folglich ergibt sich die Festigkeit des Zementbetts als Maß für die uniaxiale Verfestigungslast nach Verdichtung der Zementprobe. Wenn die Prüfungen mit verschiedenen Lasten vorgenommen werden, erhält man eine Funktion, die der bei der Auslegung von Silos für gute Fließfähigkeit ähnlich ist. Dank der guten Reproduzierbarkeit der Prüfergebnisse, kann dieses Prüfgerät zum Nachweis von Abweichungen im Fließverhalten bei der routinemäßigen Qualitätskontrolle, für Zeit-Verfestigungsuntersuchungen oder für verschiedene Experimente mehr wissenschaftlicher Natur eingesetzt werden.

Uniaxial-Tester – ein neues Gerät zur Beurteilung der Fließeigenschaften von Zement

Summary – The flow properties of cement are very important for the costs of handling and transporting the cement. The need for good control of the dosage of grinding aids adds to the importance of a good method for measuring the flow properties of cement. A new apparatus, the Uniaxial Tester, has been developed for this purpose. In principle, the cement sample is confined in a slightly conical die, and consolidated by means of a piston. The piston can be moved, deforming the sample vertically. As a result the compressive failure strength of the cement is given as a function of the uniaxial consolidation stress previously compacting the sample. By undertaking tests at various stress levels, a function similar to the flow function used for designing silos for reliable flow is obtained. Due to good reproducibility of test results, this tester may therefore be used for detecting deviations in flow behaviour as routine quality control, for time consolidation studies or for more scientific experiments of a different kind.

The uniaxial tester – a new apparatus for assessment of flow properties of cement

Résumé – Les propriétés d'écoulement du ciment ont une grande influence sur les frais de manipulation et de transport. Comme les additifs de broyage exigent un dosage précis, il importe de disposer d'une bonne méthode de mesure des propriétés d'écoulement du ciment. Pour ce faire, un nouvel appareil, le testeur uniaxial, a été mis au point. Dans cet appareil l'échantillon de ciment est introduit dans un moule de forme conique puis comprimé à l'aide d'un piston. Le piston se déplaçant entraîne une déformation de l'échantillon dans le sens vertical. Il en résulte, par voie de conséquence, la résistance du lit du ciment comme valeur de durcissement uniaxiale après compression de l'échantilon. Si les tests sont effectués avec application de différentes charges, le fonctionnement obtenu est semblable à celui que l'on obtient lors du dimensionnement des silos pour de bonnes conditions d'écoulement. Etant donné que les résultats des tests sont bien reproductibles, cet appareil de mesure peut être utilisé pour prouver les écarts d'écoulement lors du contrôle de qualité de routine, mais aussi pour des études de durcissement par rapport au temps ou pour diverses expériences à caractère plus scientifique.

Testeur uniaxial – un appareil nouveau pour tester les propriétés d'écoulement du ciment

Resumen – Las propiedades reológicas de los cementos tienen un gran influjo sobre los gastos de manipulación y de transporte de los mismos. Puesto que los coadyuvantes de la molienda tienen que ser dosificados con precisión, un buen método con el que se puedan medir las propiedades reológicas de los cementos resulta particularmente importante. Para ello, se desarrolló el nuevo aparato de ensayo uniaxial. En principio, la muestra de cemento se introduce en un molde cónico, compactándose a continuación mediante un émbolo. Este se puede mover, lo que produce una deformación de la muestra

Aparato de ensayo uniaxial – un nuevo aparato para el ensayo de las propiedades reológicas de los cementos

en sentido vertical. Por consiguiente, la resistencia del lecho de cemento sirve de medida de la carga de compactación uniaxial, después de la compresión de la muestra. Llevando a cabo los ensayos con diferentes cargas, se obtiene una función similar a la del dimensionamiento de los silos de buena fluidificación. Gracias a la buena reproducibilidad de los resultados de los ensayos, se puede emplear este aparato de ensayo para la comprobación de las desviaciones reológicas durante los controles periódicos de calidad, para estudios del endurecimiento en función del tiempo o para diferentes experimentos de carácter más bien científico.

1. Einführung

Während der Herstellung und Handhabung von Zement werden von Zeit zu Zeit eine schlechte Fließfähigkeit und Veränderungen der Fließeigenschaften festgestellt. Die Fließeigenschaften des Zements wirken sich nachhaltig auf die Kosten der Handhabung und des Transports des Materials aus. Angesichts der Forderungen nach besserer Qualität, größerer Feinheit der vermahlenen Zemente und des verbreiteten Einsatzes von Mahlhilfsmitteln kommt der Festlegung von Qualitätskontrollverfahren, die quantitative Angaben über die Veränderung der Fließeigenschaften liefern, entscheidende Bedeutung zu.

Die für die Charakterisierung der Fließeigenschaften von Zement verfügbaren Mittel wurden bisher nur unzureichend untersucht und kranken an ihrer mangelnden Reproduzierbarkeit. Verschiedene verwendete Methoden, wie z.B. der Pack Set Index (PSI), verschiedene Siebverfahren usw. sind unzulänglich, da sie nicht ausreichend reproduzierbar sind und nicht auf einer soliden wissenschaftlichen Grundlage stehen. Darüber hinaus scheint sich keines der aus der Fachliteratur bekannten vorhandenen Schergeräte (siehe die Übersicht in [1]) für Qualitätskontrollverfahren zu eignen. Deshalb wurde für diese Aufgabe ein Uniaxial-Prüfgerät entwickelt.

2. Konstruktion des Uniaxial-Testers

Das von POSTEC-Research entwickelte Uniaxial-Prüfgerät (**Bilder 1 und 2**) wurde schon in früheren Veröffentlichungen beschrieben [2-4]. Die Zementprobe wird in eine leicht konische Form gegeben und mit einem Kolben verdichtet (**Bild 3a**). Der Kolben läßt sich bewegen und verformt dabei die Probe in vertikaler Richtung. Zwischen dem Außenrand des Kolbens und dem Innenrand des unteren Teils der einschließenden Form ist eine flexible Membran gespannt. Eine Schmierstoffschicht sorgt dafür, daß zwischen der flexiblen Membran und der Wand der Form eine möglichst geringe Reibung erfolgt. Da die Membran gespannt ist, zieht sie sich mit der Mehlprobe zusammen und gewährleistet auf diese Weise eine homogene Verdichtung der gesamten Probenmenge.

Vor Beginn des Füllvorgangs wird die Form zusammen mit dem Kolben aus dem Prüfgerät herausgenommen und umgedreht. Die Mehlprobe wird locker in die Form gefüllt, und Luftblasen werden mit leichtem Druck mit Hilfe einer Bürste aus der Probe herausgepreßt, um sicherzustellen, daß die Probe nur marginal vorverfestigt wird. Dadurch soll insbesondere gewährleistet werden, daß die Verdichtung der Probe nur während der Verfestigungsphase des Versuchs innerhalb der einschließenden Form erfolgt. Somit werden die Ergebnisse nicht durch den Packvorgang beeinflußt und hängen weniger stark von dem Bedienungspersonal ab.

Nach Einfüllen der Probe in die Form wird diese mit einer Verschlußkappe abgedichtet, wieder aufrecht gestellt und an den Führungsvorrichtungen des Prüfgeräts befestigt. Die Probe wird durch vertikale Kolbenbewegungen verdichtet, Abbildung 3a. Nach Erreichen einer vorgegebenen Verfestigungslast σ_{1u}, die einer bestimmten Verformung ε_{1u} der Probe entspricht, wird der Kolben angehalten. Nachdem die gewünschte Verfestigungslast erreicht ist, kann sich die Probe unter dieser Verformungshöhe während einer angegebenen Zeitdauer (gewöhnlich 2 min) stabilisieren. Die Last σ_{1u} wird dann auf ein Minimum reduziert, und die Form wirdin den Führungen des Prüfgerätshochgezogen, bis die Probe freisteht, **Bild 3b**.

1. Introduction

Poor flow and variations in flow properties are from time to time experienced during production and handling of cement. The flow properties of the cement have a considerable influence on the costs of handling and transporting the material. Demands for improved quality, increased fineness of the ground cements, and the widespread use of grinding aids, make it of vital importance to establish quality control procedures that give quantitative indications of the variation in flow properties.

The means to characterize the flow properties of cement, have not been satisfactorily investigated and they suffer from the lack of reproducibility. Different methods used, e.g. the Pack Set Index (P.S.I.), various sieve methods etc. have their shortcomings, due to the lack of reproducibility. and they are not based on a sound scientific basis. Furthermore, none of the existing shear devices known from literature (overview given in [1]) seem to be suitable for quality control procedures. So, for this purpose a uniaxial tester has been developed.

2. Design of the uniaxial tester

The uniaxial tester being developed by POSTEC-Research (**Figure 1** and **2**) has been described in earlier publications [2-4]. The cement sample is confined in a slightly conical die, and consolidated by means of a piston, **Figure 3a**. The piston can be moved, deforming the sample vertically. A flexible membrane is stretched between the outer periphery of the piston and the inner periphery of the lower part of the confining die. A layer of lubricant will ensure that there is an absolute minimum of friction between the flexible mem-

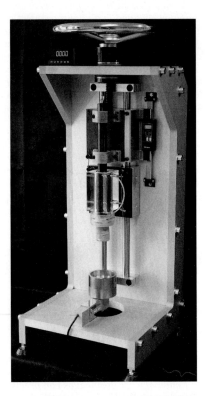

BILD 1: Eine manuelle Version des von POSTEC-Research entwickelten Uniaxial-Prüfgeräts

FIGURE 1: A manual version of the uniaxial tester developed by POSTEC-Research

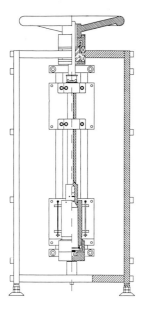

BILD 2: Vorderansicht des Uniaxial-Prüfgeräts
FIGURE 2: Front view of the uniaxial tester

Anschließend wird durch eine erneute Abwärtsbewegung des Kolbens die Festigkeit des Zementbetts gemessen. Die Kolbenbewegungen werden sorgfältig gesteuert, und der Höchstwert f_c wird erreicht, noch bevor die Probe zerfällt. Damit wird eine genaue Untersuchung der Ausbreitung von Scherflächen ermöglicht. Eine Scherfläche sollte mit der Horizontalen einen Winkel bilden; siehe Bild 3b. Entsteht eine komplexere Scherfläche, ist die Probenhöhe aller Wahrscheinlichkeit nach zu niedrig. Der Versuch muß mit einer größeren Probe wiederholt werden. Das Prüfgerät ist jedoch so ausgelegt, daß mit dem Standardpackverfahren normalerweise eine ausreichende Mehlmenge in die Form gefüllt werden kann.

Die Versuchsergebnisse lassen sich durch Auftragen der Bruchstauchungsfestigkeit des Zements in Abhängigkeit von der zuvor die Probe verdichtenden uniaxialen Verfestigungslast darstellen. Die Durchführung von Versuchen mit verschiedenen Belastungsgraden ergibt eine der Flußfunktion, wie sie bei der Konstruktion von auf zuverlässigen Durchfluß ausgelegten Silos verwendet wird, ähnliche Funktion.

3. Verwendung des Prüfgeräts

Detaillierte Untersuchungen haben für den Uniaxial-Tester bei der Verwendung für Standardprüfungen des Fließverhaltens ein optimales Verfahren ergeben. Für das Verfahren gelten ausdrückliche Richtlinien, die eingehalten werden müssen, um reproduzierbare Ergebnisse zu gewährleisten, wie sie in [3] für Kalksteinmehl aufgeführt sind. Die Versuche haben sich als nahezu bedienerunabhängig erwiesen, solange die Leitlinien für das Prüfverfahren beachtet werden. Die Gesamtdauer eines Versuchs liegt bei rund 20 Minuten.

An verschiedenen Zementproben wurden Prüfungen vorgenommen. Wie früher schon beschrieben, wurden die Versuchsergebnisse durch Darstellung der Festigkeit der Probe in Abhängigkeit von der uniaxialen Verfestigungslast angegeben. Die **Bilder 4** und **5** zeigen die bei sieben verschiedenen Zementproben gewonnen Funktionen. Diese Ergebnisse beruhen auf Wiederholungsversuchen mit maximalen Abweichungen der Festigkeitswerte von ± 5 %. Wie klar zu erkennen ist, weichen die Proben in der Festigkeit ab (z.B. beim Fließverhalten), und die Unterschiede werden im hochbelasteten Bereich (~250 kPa) am deutlichsten. Bei diesen hohen Belastungswerten ist eine vage Korrelation mit den Pack Set Index-Werten (auf den Bildern kurz PSI) zu beobachten.

4. Diskussion und Schlußbemerkungen

Unterschiede im Fließverhalten lassen sich durch Festigkeitsmessungen bei hoher wie bei niedriger Belastung erkennen. Allerdings können je nach den in den konkreten

brane and the die wall. Since the membrane is stretched, it will contact with the powder sample, ensuring that the sample is compacted homogeneously throughout the total volume.

Before the filling procedure is initiated, the die together with the piston, is taken out of the tester and turned upside down. The powder sample is filled loosely into the die, and air voids are gently pressed out of the sample by means of a brush, ensuring that the sample is only marginally preconsolidated. The main objective is to ensure that the compaction of the sample only occurs within the confining die under the consolidation phase of the experiment. This will ensure that the results are not influenced by the packing procedure, and hence are less operator dependant.

Once the sample is filled into the die, the die is sealed with a bottom cup, turned back in the upright position, and fixed to the guiding devices of the tester. The sample is compacted by moving the piston vertically, Figure 3a. The piston is stopped when a predetermined consolidation stress, σ_{1u}, is reached, corresponding to a certain deformation, ε_{1u}, of the sample. After the desired level of consolidation stress has been reached, the sample is allowed to stabilize under this deformation height for a specific length of time (usually 2 min.). The stress, σ_{1u}, is then reduced to a minimum value, and the die is pulled up, following the guides of the tester, allowing the sample to stand by itself, **Figure 3b**.

The compressive failure strength is then measured by moving the piston downwards once more. The movements of the piston are well controlled, and the maximum value f_c is reached before the sample falls apart. This allows for close studies of the propagation of shear planes. A shear plane should form an angle α with the horizontal plane, see figure 3b. If a more complex shear plane develops it is most likely that the sample height is too low. The experiment will have to be repeated with a larger sample. However, the tester is designed so that normally a sufficient amount of powder can be filled into the die by applying the standard packing procedure.

Results of the experiments can be presented by plotting the compressive failure strength of the cement as a function of the uniaxial consolidation stress previously compacting the sample. By undertaking tests at various stress levels a function similar to the flow function used for design of silos for reliable flow, is obtained.

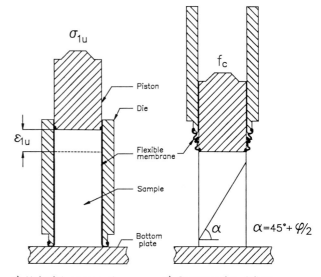

a) Uniaxial compaction b) Compressive failure

BILD 3: Hauptmerkmale des Prüfgeräts
FIGURE 3: Main features of the tester

Kolben	= piston
Form	= die
Flexible Membran	= flexible membrane
Probe	= sample
Bodenplatte	= bottom plate
a) Uniaxiale Verdichtung	= uniaxial compaction
b) Festigkeit des Zementbetts	= compressive failure

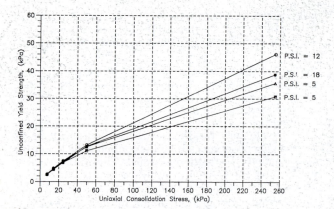

BILD 4: Ergebnisse von vier in dem Prüfgerät untersuchten Zementproben
FIGURE 4: Results of four cement samples investigated in the tester
Druckfestigkeit der Probe (kPa) = Unconfined yield strength, (kPa)
Uniaxiale Verfestigungslast (kPa) = Uniaxial consolidation stress, (kPa)

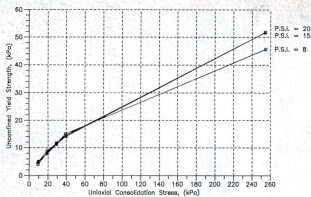

BILD 5: Ergebnisse von drei in dem Prüfgerät untersuchten Zementproben
FIGURE 5: Results of three cement samples investigasted in the tester
Druckfestigkeit der Probe (kPa) = Unconfined yield strength, (kPa)
Uniaxiale Verfestigungslast (kPa) = Uniaxial consolidation stress, (kPa)

Prozessen wirksam werdenden Belastungsniveaus verschiedene Probleme auftreten. Für die in den Bildern 4 und 5 gezeigten Unterschiede im Fließverhalten wurde bisher keine Erklärung gefunden. Hier ist das Fehlen einer eindeutigen Korrelation zwischen der Verfestigungslast und den Pack Set Index-Werten (PSI) hervorzuheben. Die praktischen Erfahrungen liefern Beispiele, bei denen sich die Fließeigenschaften nicht aus den PSI-Werten erklären lassen.

Faktoren wie chemische Zusammensetzung, Korngröße und Formverteilungen, die Menge an Mahlhilfsmitteln und zeitliche Faktoren könne einige der vielen Variablen sein, die das Fließverhalten der Zemente beeinflussen. Für die Untersuchungen dieser und anderer Faktoren bedarf es jedoch eines Geräts, das einen stimmigen Reaktionswert für die Fließeigenschaften liefert.

Die mit Zement und anderen Mehlen [3, 4] durchgeführten Prüfungen zeigen, daß der Uniaxial-Tester Abweichungen im Fließverhalten von Mehlen nachzuweisen vermag. Aufgrund der ausgezeichneten Reproduzierbarkeit der Testergebnisse kann dieses Prüfgerät deshalb für routinemäßige Qualitätskontrollen eingesetzt werden. Die Ergebnisse des Uniaxial-Prüfgeräts liegen gewöhnlich unter denen anderer für die Auslegung von Silos verwendeter Schergeräte [2-4], doch ansonsten besteht eine gute Übereinstimmung. Die niedrigeren Werte lassen sich mit dem gewählten Verfestigungsverfahren erklären.

Darüber hinaus kann der Tester für Zeit-Verfestigungsuntersuchungen und weitere wissenschaftliche Experimente in bezug auf elastische Eigenschaften, Wechselbeanspruchung, Spannungsrelaxation und Kriechvorgänge, wie in [4] kurz beschrieben, eingesetzt werden.

Literatur

[1] Schwedes, J., und Schulze, D.: Proceedings of Second World Congress, Particle Technology, 19.–22. September 1990, Kioto, Japan.

[2] Enstad, G. G., und Maltby, L. P.: Flow Property Testing of Particulate Solids, Bulk Solids Handling, Vol.12, No. 3 (1992), pp. 451–454.

[3] Maltby, L. P., und Enstad, G. G.: Uniaxial Tester for Quality Control and Flow Property Characterization of Powders, Bulk Solids Handling, Vol.13, No.1 (1993), pp. 135–139.

[4] Maltby, L. P., Enstad, G. G., und de Silva, S. R.: Characterization of Flow Properties and Quality Control of Cohesive Particulate Solids by means of a Uniaxial Tester, Proceedings of Powder and Bulk Solids Conference '93, 3.–6.Mai 1993, Rosemont, USA.

3. Use of the tester

Detailed investigations have led to an optimal procedure for the uniaxial tester when used for standardized tests on flow behaviour. The procedure has distinct guidelines that must be followed to guarantee reproducible results as demonstrated for limestone powder in [3]. Experiments have been found to be nearly operator independent, as long as the guidelines of the test procedure are followed. The total time for one experiment is of the order of 20 minutes.

Tests have been undertaken on various cement samples. As described earlier, test results were presented by plotting the strength of the sample as a function the uniaxial consolidation stress. **Figures 4** and **5** show the obtained functions for seven different cement samples. These results are based on replicate tests giving maximum deviations of the strength values of +/− 5 %. As clearly seen, the samples deviate in strength (e. g. in flow behaviour), and the differences are best seen in the high stress region (~ 250 kPa). A vague correlation with the Pack Set Index values (also shown in the figures as P.S.I.) is seen at these high stress levels.

4. Discussion and concluding remarks

Differences in flow behaviour can be detected by strength measurements both at high and low stresses. However, different problems may arise depending on the stress levels acting in the actual process. The differences in flow behaviour shown in Figures 4 and 5 have not been fully explained so far. The lack of a clear correlation between consolidation stress and P.S.I. is noteworthy. Experience from practice gives examples where the flow properties can not be explained by the P.S.I.

Factors like chemical composition, particle size and shape distributions, amount of grinding aid and time factors may be some of the many variables influencing the flow behaviour of the cements. However, in order to start investigating these and other factors, a device is needed to get a consistent value for the flow properties.

It is shown by the tests undertaken with the cement and other powders [3,4], that the uniaxial tester is capable of detecting deviations in flow behaviour of powders. Due to the very good reproducibility of test results, this tester may therefore be used for routine quality control procedures. Results of the uniaxial tester usually fall below other shear testers used for silo design [2-4], but otherwise the agreement is good. The lower values can be explained by the consolidation procedure used.

Furthermore, the tester may be utilized for time consolidation studies and more scientific experiments on elastic properties, repeated cycle consolidations, stress relaxation and creep phenomena as briefly described in [4].

IBAU HAMBURG

Terminals, ship loader, ship unloader, cement tanker

New Technology for the modern Cement-Industry

The IBAU scope of supply for cement handling comprises all kinds of fluidized gravity conveyors, IBAU-Pumps and blow tank conveyors, silo discharge equipment of any size and capacity and IBAU Multi Compartment silos.

IBAU stationary or floating cement terminals as well as ship loading equipment, mobile ship unloaders and equipment for cement tankers.

IBAU HAMBURG · Rödingsmarkt 35 · 20459 Hamburg · Phone (040) 3613090 · Fax (040) 363983 · Telex 2162653 iib d

IBAU HAMBURG

Multi Compartment Silos
Thousands of IBAU Silos are installed

Central cone silo

Thousands of silos

Ciments Luxembourgeois
Ciments d'Origny
Altkirch

Heidelberg Schelklingen
3 x Alamo Cement
Rüdersdorfer Zement
Heidelberg Moldan

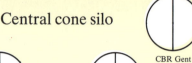
CBR Gent

Ciments Luxembourgeois
Dyckerhoff Mark II
Dyckerhoff
Geseke

2 x Rajashree Cement
Grasim Cement
ENCI Maastricht

Ring silo

Diamond Cement Damoh
RKW Wülfrath Schwelgern
Italcementi Selerno
Polysius AG Cape Portland
Bateman Ltd. Lethabo
Italcementi Vibo Valencia
Schencking KG Lienen
Ciments Francais Bussac
Perlmooser Mannersdorf
Taiwan Cement Nankang
Kedah Cement Langkawi
UNICEM Vernasca

Tasek Cement
Asia Cement Singapore
Kedah Cement Johor Port

Italcementi Vibo Valencia
Cementario Adriatico Pescara
Perlmooser Mannersdorf
Ciments de la Loire Villiers au Bouin
Italcementi Salerno
Ciments Francais Beaucaire
Taiwan Cement Nankang
KRC Umwelttechnik Walheim
Uniland Los Monjos
Ribblesdale Cement Clithero
Kedah Cement Port Prai
UNICEM Vernasca

Dyckerhoff AG
Amöneburg

Ciments Vicat
Montalieu
Deuna Zement

Readymix Dortmund
ENCI Maastricht
CCB Ramecroix
CBR Lixhe

Belapatfalva
Hungaria

Johann Schaefer
Kalkwerke

Dyckerhoff Mark II
Zementwerk Lauffen
Dyckerhoff Neuwied

Heidelberg Schelklingen
Anneliese Zementwerke
Cimento Itambé

TCEC/USCC
Taiwan

Rohrbach
Dotternhausen

Andre Büechl
Regensburg

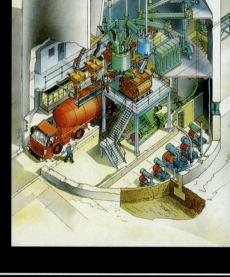

IBAU HAMBURG · Rödingsmarkt 35 · 20459 Hamburg · Phone (040) 36 13 09 0 · Fax (040) 36 39 83 · Telex 2162 653 iib d

Konzept zur Zementwerksmodernisierung und -optimierung

Approach to cement plant upgrading and optimization

Concept de modernisation et d'optimisation d'une cimenterie

Concepto de modernización y optimización de una fábrica de cemento

Von **B. K. Shrikhande, T. N. Tiwari** und **M. A. Purohit**, Thane/Indien

Zusammenfassung – Heute steht die Zementherstellung unter den Zwängen des stetig steigenden Kostendrucks, der Ressourcenknappheit und der Kundenforderungen nach besserer Produktqualität. Um angesichts dieser Situation konkurrenzfähig zu bleiben, müssen die Hersteller Maschinen von hohem Energiewirkungsgrad einsetzen und streng überwachte Kostensenkungsstrategien anwenden. Um mit verschiedenen Problemen und Sachzwängen, wie veraltete Verfahren und Technologien, Einschränkungen bei der Anlagenkonstruktion und Werksgröße, Veränderungen bei den Rohstoffen, niedrige Anlagenverfügbarkeit usw., fertig zu werden, bedarf es einer umfassenden Diagnose der Anlage einschließlich innerbetrieblicher Untersuchungen und Feldstudien. In den internen Untersuchungen sind Werksplanung und Maschinenanordnung, Auslegungsspezifikationen, innerbetriebliches Transportwesen, relevante Betriebskennzahlen einschließlich der Stillstandzeiten, bestehende Reparatur- und Wartungspraxis und der Zeitplan für die vorbeugende Wartung zu analysieren. Bei den Feldstudien geht es um die Messung der Durchflußgeschwindigkeit sämtlicher Eingangs- und Ausgangsmaterialströme aller Werksteile, Verfahrensparameter und Verbrauchszahlen für Brennstoff, Energie und Material. Im allgemeinen erfordert die Optimierung nur wenige große Investitionen. Sie führt zu 10 – 20%igen Produktionssteigerungen, einer Senkung des Brennstoff- und Energieverbrauchs und sehr kurzen Amortisationszeiten.

Summary – Cement manufacturing in today's context is constrained by increasing input costs, resources crunch and customer demand for a better quality of product. To compete in this situation, manufacturers are required to select energy efficient machinery, and closely monitor their operating strategy for cost reduction. To tackle various problems and constraints such as process and technology obsolescence, design and plant size limitation, changes of raw materials, poor plant availability etc., a comprehensive and detailed diagnostic study of the plant including bench and field studies needs to be carried out. In a bench study, the plant and equipment layout, design specifications, material handling procedures, relevant historic operational data including down time reports, existing maintenance practices and preventive maintenance schedule is analyzed. In a field study, the investigations include measurement of flow rates of all input and output streams for all plant subsections, process parameters and consumption factors for fuel, power and materials. In general the optimization exercise now needs substantial investments. It leads to the benefit of a 10 to 20% increased production and a reduction in fuel and energy consumption, with very short payback periods.

Résumé – De nos jours la fabrication du ciment est soumise aux contraintes d'une pression croissante et constante exercée sur les coûts mais aussi à la pénurie des ressources naturelles et aux exigences du client en faveur d'une meilleure qualité du produit. Vu cette situation, pour rester compétitif, il faut que les fabricants utilisent des machines d'un rendement énergétique élevé et mettent en oeuvre des stratégies de réduction des coûts devant faire l'objet d'une surveillance rigoureuse. Pour parvenir à maîtriser ces problèmes et ces sujétions notamment procédés et technologies caducs, limitations imposées à la construction des installations et à la taille des usines, modifications survenues au niveau des matières premières, faible disponibilité des installations, etc., il importe de procéder à un vaste diagnostic de l'installation y compris analyses internes à l'entreprise et études de terrain. Les analyses internes concernent notamment la gestion de l'entreprise et la disposition des machines, les spécifications d'ingénierie, la manutention à l'intérieur de l'entreprise, les données d'exploitation y compris les périodes d'immobilisation, l'application courante de programme de réparation et d'entretien existants et le planning en matière d'entretien préventif. Pour ce qui est des études de terrain, il s'agit de la mesure de la vitesse du débit de tous flux de matière d'entrée et de sortie de tous les secteurs de l'usine, des paramètres d'exploitation et des chiffres de consommation concernant le combustible, l'énergie et la matière. Généralement, l'optimisation n'exige que peu d'investissements importants. Elle entraîne de 10 à 20% d'accroissements de production, une diminution de consommation d'énergie et de combustible et des temps d'amortissements très courts.

Concepto de modernización y optimización de una fábrica de cemento

Resumen – Hoy en día, la fabricación del cemento se tiene que enfrentar al constante aumento de los costes, de la escasez de recursos y de las exigencias de los clientes que piden una mejor calidad del producto. Para poder mantener su competitividad ante esta situación, los fabricantes tienen que emplear máquinas que tengan un alto grado de rendimiento energético y aplicar unas estrategias de reducción de costes rigurosamente controladas. Con el fin de vencer ciertas dificultades y problemas, tales como procedimientos anticuados, limitaciones de tipo constructivo y de tamaño de la instalación, variaciones de las materias primas, reducida disponibilidad de la instalación, etc., se requiere un amplio diagnóstico de la instalación, incluyendo estudios internos y de campo. En cuanto a los estudios internos, cabe analizar la planificación de la fábrica y la disposición de las máquinas, las especificaciones de dimensionamiento, el sistema de transportes dentro de la fábrica, los parámetros relevantes de la instalación, incluyendo los períodos de parada, la forma de llevar a cabo las reparaciones y el mantenimiento así como el horario del mantenimiento preventivo. En los estudios de campo se trata de la medición de la velocidad de paso de todos los flujos de material, de entrada y de salida, de todas las secciones de la fábrica, parámetros tecnológicos y cifras de consumo referentes a combustibles, energía y material. Por regla general, la optimización requiere sólo pocas inversiones importantes. Permite aumentar la producción en un 10 a 20 % y bajar el consumo de combustibles y de energía. Además, los tiempos de amortización resultan muy cortos.

1. Einführung

Der energieintensive Prozeß der Zementherstellung erfordert heute die Wahl einer fortschrittlichen energiesparenden Technologie, wirtschaftliche Produktionskapazitäten und gute Betriebsstrategien, um die Kosten zu reduzieren, die Produktion zu maximieren und die gewünschte Produktqualität aufrechtzuerhalten. Alle Zementhersteller bemühen sich, diese Ziele zu erreichen, und stehen dabei unter folgenden Zwängen:

— veraltete Verfahren, Anlagen und Gerätetechnik,
— Veränderungen oder Schwankungen der eingesetzten Roh- oder Brennstoffe,
— geringe Anlagenverfügbarkeit.

2. Diagnose

Beim Betreiben einer Anlage ist die eindeutige Erkennung der Ursachen für derartige Zwänge von größter Bedeutung. Eine falsche Diagnose der Probleme und daraus resultierende Korrekturmaßnahmen führen sicherlich nicht zur Lösung des Problems, sondern eher zu einer weiteren Verschlechterung günstiger Bedingungen in anderen Teilen der Anlage.

Die einzig klare und logische Lösung in einer solchen Situation besteht in einer umfassenden und gründlichen Diagnose des gesamten Systems unter allen Aspekten, wie Betrieb, Reparatur und Wartung, Verfahren und technische Auslegung der Anlagen.

2.1 Anlagendiagnose

Die Diagnose beinhaltet innerbetriebliche Untersuchungen und Feldstudien. In einer internen Untersuchung sind Werksplanung und Maschinenanordnung, Auslegungsspezifikationen, innerbetriebliches Transportwesen, relevante Betriebskennzahlen einschließlich der Stillstandzeiten, bestehende Reparatur- und Wartungspraxis und der Zeitplan für die vorbeugende Wartung zu analysieren.

Bei den Feldstudien geht es um die Messung der Durchflußgeschwindigkeit sämtlicher Eingangs- und Ausgangsmaterialströme aller Werksteile, Verfahrensparameter und Verbrauchszahlen für Brennstoff, Energie und Material. In der internen Untersuchung sind auch die Erfordernisse eines störungsfreien Betriebs der Anlagen zu beachten. In jedem Prozeßstadium werden Kreislaufproben entnommen und auf ihre physikalischen und chemischen Merkmale hin analysiert. In einer Feldstudie muß auch der mechanische Zustand der technischen Anlagen und Maschinen überprüft werden.

Im allgemeinen ist zu beobachten, daß bei Umsetzung der hieraus abgeleiteten Empfehlungen das Werk eine 10–20%ige Produktionssteigerung und eine entsprechende Senkung des Brennstoff- und Energieverbrauchs ohne

1. Introduction

The energy intensive cement manufacturing process today requires a selection of advanced energy efficient technology, economic plant capacities, and good operating strategies to reduce cost, as well as to maximize production and maintain desired product as well as to maximize production and maintain desired product quality. Every cement manufacturer tries to achieve the above mentioned targets. The constraints they face are:

— Obsolescence of process, equipment and instrumentation,
— Changes or fluctuations in input raw materials or fuel,
— Poor plant availability.

2. Diagnostic study

In operating plants clear identification of the main cause of such undesired occurrences is extremely important. Any wrong diagnosis of problems and corrective actions based thereon can definitely not solve the problem but can cause further deterioration of advantageous conditions in other sections of the plant.

The only clearcut and logical solution in such a situation is to carry out a comprehensive and detailed diagnostic study to look at the entire system from all sides such as operation, maintenance, process and engineering design.

2.1 Plant diagnostic study

The diagnosis comprises bench and field studies. In a bench study, the plant equipment layout, design specifications, material handling procedures, relevant historic operational data including down time reports, existing maintenance practices and preventive maintenance schedules are analyzed.

Field studies include the measurement of flow rates of all input and output streams for all plant subsections, process parameters and consumption factors for fuel, power and material. During the plant study, the operational requirements for smooth plant running are also noted. At each stage of the process, circuit samples are taken and analyzed for their physical and chemical characteristics. In a field study, attention is also given to monitor the mechanical condition of the plant and machinery.

It is generally observed that with the implementation of recommendations of this exercise a plant may derive benefit of 10–20% increased production and corresponding reduction in fuel and energy consumption without substantial investment. The pay back period for such schemes is fairly short.

große Investitionen erreichen kann. Die Amortisationszeit solcher Maßnahmen ist sehr kurz.

3. Fallstudie

Dieses Beispiel beschreibt als typische Fallstudie eine Ofenoptimierung, die in Verbindung mit dem oben ausgeführten diagnostischen Ansatz durchgeführt wurde. Die Studie hatte folgende Ziele:

— Beseitigung von Engpässen im Ofensystem
— Senkung des spezifischen Energieverbrauchs und
— Erarbeitung eines angemessenen Umbauplans bei möglichst geringen Stillstandzeiten der Anlage.

3.1 Zustand vor Optimierung

Die Hauptmerkmale der Anlage vor ihrer Optimierung waren eine Leistung von 850 t/d Klinker und ein spezifischer Brennstoffenergieverbrauch von 955 kcal/kg Klinker. Der hohe Brennstoffverbrauch war im wesentlichen auf die große Abgasmenge sowie Strahlungs- und Kühlerwärmeverluste zurückzuführen.

Auf der Grundlage einer gründlichen Analyse wurde beobachtet, daß die Ofenleistung durch eine zu geringe Kapazität des Rostkühlers eingeschränkt wurde. Ohne große Veränderungen am Rostkühler vorzunehmen wäre es jedoch möglich, die Ofenleistung auf 1220 t/d zu steigern. Die Überprüfung der Vorwärmerauslegung und die Berücksichtigung der technisch-wirtschaftlichen Konsequenzen führten zu dem Entschluß, die Ofenleistung auf ein Niveau von 1000 bis 1050 t/d zu erhöhen; dabei wurde mit einer Senkung des Brennstoffverbrauchs von etwa 8 % gerechnet.

3.2 Umbauplan

Die geplanten und ausgeführten Umbaumaßnahmen umfaßten:

— Vergrößerung des Durchmessers der oberen Vorwärmer-Zyklone um etwa 20 %,
— Vergrößerung der Steigrohr- und Tauchrohrdurchmesser,
— zusätzliche Isolierung der Zyklone und Leitungen,
— Einbau eines Abgasgebläses mit höherer Leistung,
— Vergrößerung der Höhe des Vorcalcinierers und des Durchmessers der Tertiärluftleitung,
— Einbau von stehenden Rostplatten in der Rekuperationszone des Rostkühlers,
— Einbau neuer Druckgebläse und Antriebe in den Rostkühler,
— Erweiterung des ersten Kühlerrosters um zwei Reihen,
— Austausch des vorhandenen konventionellen Brenners gegen einen Hochleistungs-Dreikanalbrenner,
— Optimierung der Rohmischung in Richtung besserer Brennbarkeit.

Die für das Projekt angesetzten Kosten beliefen sich auf US-$ 1 Mio. Die geplante Ofenstillstandzeit betrug 40 Tage.

3.3 Auswertung und Vorteile

Der Ofenbetrieb stabilisierte sich bei einer Leistung von 1040 t/d, was eine Verbesserung von 22 % bedeutete. Der spezifische Brennstoffenergieverbrauch sank auf 865 kcal/kg, d. h. um 9 %, so daß die erreichte Leistung den gesetzten Zielen entsprach. Außerdem ergab sich eine Verminderung des spezifischen Stromverbrauchs um 18 kWh pro Tonne Klinker. Die errechnete einfache Amortisationszeit von weniger als 1,5 Jahre wurde eingehalten.

Abschließend ist festzustellen, daß ein ähnliches Konzept für die Ofenoptimierung in vielen Zementwerken mit großem Nutzen eingesetzt werden kann.

3. Case study

This example presents a typical case study of kiln optimization, especially done in line with the similar diagnostic approach. The objectives of this study were:

— To de-bottleneck the kiln system,
— To reduce specific energy consumption and
— To evolve a suitable modification plan considering minimum shutdown of the plant.

3.1 Pre-optimization status

The main characteristics of the status before optimization consists in an output rate of the plant of 850 tpd of clinker and a specific fuel consumption of 995 kcal/kg clinker. The high fuel consumption was mainly caused by high exhaust gas, radiation and cooler heat losses.

Based on a detailed analysis, it was observed that the kiln production was restricted by the capacity of the grate cooler. However, without initiating a major modification of the grate cooler it was possible to achieve a production of 1220 tpd of clinker from the kiln. Based on a preheater design study and taking into account the techno-economic implications it was decided to upgrade the kiln production to a level of 1000–1050 tpd clinker with an expected reduction in fuel consumption by 8 %.

3.2 Modification plan

The modifications, as planned and carried out, are given below.

— Increase of the diameter of the preheater top cyclones of about 20 %,
— Enlargement of the diameter of the raiser ducts and immersion tubes,
— Provision of backup insulation for the cyclones and ducts,
— Installation of a waste gas fan with higher capacity,
— Increase of the precalciner vessel height and tertiary air duct diameter,
— Installation of dead grate plates in the recuperating zone of the grate cooler,
— Installation of new forced draught fans and drives for the grate cooler,
— Extension of the first grate cooler compartment by two rows,
— Installation of a high efficiency three channel burner in place of the existing conventional burner,
— Optimization of the raw mix with enhanced burnability.

The estimated cost for the project was U.S. $ 1.0 Million. The projected kiln shutdown time was 40 days.

3.3 Evaluation and benefits

The kiln operation was stabilized at a kiln output rate of 1040 tpd, i.e. improved by 22 %. The specific fuel consumption was decreased to 865 kcal/kg, i.e. reduced by 9 %, hence the achieved performance was in line with the set targets. Apart from this reduction in the specific electric energy consumption of 18 kWh/t of clinker was realized. The calculated simple pay back period for the project worked out to be less than 1.5 years.

In conclusion, a similar approach to kiln system upgradation can be adopted in a large number of cement plants for achieving substantial benefits.

Ein rechnergestütztes Planungs- und Informationssystem für Instandhaltung, Materialwirtschaft und Dokumentation

A computer-aided planning and information system for maintenance, material management and documentation

Un système de planification et d'information, assisté par ordinateur, pour la maintenance, la gestion du matériel et la documentation

Un sistema de planificación y de información computerizado para mantenimiento, gestión de materiales y documentación

Von **U. Spielhagen**, Beckum und **H. Karasch**, Hannover/Deutschland

Zusammenfassung – Der Produktionsfaktor „Anlagen" hat in den vergangenen Jahren auch in der Zementindustrie an Bedeutung gewonnen. Deshalb wurde das rechnergestützte Planungs- und Informationssystem INVERS der Firma Data Concept eingeführt. Das System besteht u. a. aus den Funktionsbereichen: Stammdatenpflege für Betriebsmittel und Kapazitäten; Instandhaltungsplanung für periodische Arbeitsgänge, Erstellung von Arbeitsplänen; Auftragssteuerung mit Arbeitsvorbereitung, Ersatzteilreservierung; Rückmeldungen zur Ausführungsbestätigung der IH-Maßnahmen, Buchung von Stunden und verbrauchtem Material sowie Historie als Nachweis der Maßnahmen und Kosten. Die Auswertung nach Kosten weist Daten für die betriebswirtschaftliche Beurteilung von Maschinen und Lagerhaltung aus, wobei die Ausgabe der Ersatzteile gegen Werkstattauftrag oder Kostenstelle über Barcode-Lesestift abgewickelt wird. Relevante Daten werden dabei über Schnittstellen zu übergeordneten Systemen übertragen. Schadensauswertungen zeigen Art und Häufigkeit von Schäden sowie deren Kostenanteile, die der Maschinen-Lebenslaufakte zugeordnet werden. Der Budgetplanung dienen aufgelaufene IH-Kosten als Plankosten für die neue Periode.

Summary – In the past few years the production factor "plant" has gained in importance, including in the cement industry. The computer-aided INVERS planning and information system from the company Data Concept was therefore introduced. The system consists, among other things, of the functional areas: care of master data for resources and capacities; maintenance planning for regular working cycles, drawing up working plans; job management with work preparation, reservation of spare parts; acknowledgement of the confirmation that maintenance measures have been carried out; booking of hours and materials used, and history as evidence of the work and costs. Evaluation based on costs identifies data for industrial management assessment of machines and stocks, in which the issuing of spare parts against workshop orders or cost centres is dealt with using bar-code readers. Relevant data are transmitted via interfaces to higher-order systems. Damage assessments show the type and frequency of the damage and its cost breakdown, which is allocated to the individual machine records. Accumulated maintenance costs serve as planning costs for budget planning for the new period.

Résumé – Le facteur de production „équipements" a aussi gagné en importance, au cours des dernières années, dans l'industrie cimentière. Pour cette raison, a été introduit le système de planification et d'information assisté par ordinateur INVERS de la firme Data Concept. Entre autre, le système se compose des secteurs fonctionnels: mise à jour des données de base pour moyens d'exploitation et capacités; planification de la maintenance pour travaux périodiques, élaboration de plans de travail; gestion des tâches avec préparation du travail et réservation des pièces de rechange; messages pour la confirmation d'exécution des actions de maintenance; enregistrement des heures et du matériel consommé ainsi que historique comme preuve des actions et coûts. L'exploitation d'après les coûts fournit des données pour l'évaluation de la rentabilité de machines et de la tenu des stocks, la sortie des pièces de rechange s'effectuant contre commande de l'atelier ou par le service des coûts, avec crayon lecteur de code à barres. Les données significatives sont transmises à des systèmes supérieurs, au moyen d'interconnexions. Des exploitations des dommages montrent la nature et la fréquence des dommages ainsi que les coûts impliqués, qui sont ajoutés au dossier de la vie de service de la machine. La planification du budget s'appuie sur les coûts de maintenance cumulés, comme coûts de planification pour la nouvelle période.

Resumen – El factor de producción "instalaciones" ha adquirido mayor importancia en los últimos años, también en la industria del cemento. Por esta razón, se ha introducido el sistema de planificación y de información computerizado INVERS de la firma Data Concept. Este sistema se compone, entre otros, de los siguientes sectores funcionales: gestión de datos básicos de insumos y capacidades; planificación del mantenimiento en relación con operaciones periódicas, confección de planes de trabajo; gestión de pedidos, incluyendo preparación de trabajos y reserva de repuestos; avisos de confirmación de la ejecución de las operaciones de mantenimiento, asiento de horas y de materiales empleados e historial de comprobación de las medidas adoptadas y de los costes. La evaluación por costes arroja datos que permiten el enjuiciamiento económico de las máquinas y del almacenamiento. La entrega de los repuestos se efectúa por pedidos del taller o cuentas de costes, mediante lápiz lector de código de barras. A este respecto, los datos relevantes se transmiten a otros sistemas de orden superior, a través de interfaces. La evaluación de averías muestra el tipo y frecuencia de las mismas y los correspondientes gastos, los cuales son asignados al historial de la máquina en cuestión. La planificación presupuestaria se basa en los gastos de mantenimiento acumulados, que sirven de referencia para el nuevo período.

Un sistema de planificación y de información computerizado para mantenimiento, gestión de materiales y documentación

1. Anlageninstandhaltung als Ansatzpunkt zur Rationalisierung

Die Instandhaltungskosten machen bei der Zementherstellung einen hohen Anteil der Gesamtherstellkosten aus. Sie werden sich aufgrund der Werterhaltung alter Anlagen in Zukunft weiter erhöhen. Wachsende Kapitalbindung im Anlagevermögen, hohe Anlagenverfügbarkeit und der Wert der Lagerbestände stehen immer mehr im Blickpunkt des Instandhaltungsmanagements. Ein globalerer Markt, Qualitätsvereinbarungen verlangen ein hohes Maß an gleichmäßiger Produktqualität. Gesetzgeber und Öffentlichkeit erwarten, daß die Instandhaltung die Sicherheit der Anlagen zur Abwehr von Gefahren für Mensch und Umwelt nicht nur aufrecht erhält, sondern auch der Dynamisierung der Regelwerke durch Definition neuer Sollzustände Rechnung trägt. Somit ist die Anlageninstandhaltung besonders in den letzten Jahren immer mehr zu einer zentralen substantiellen Einflußgröße des Unternehmens geworden.

Vor diesem Hintergrund wurde ab 1988 für die Materialwirtschaft und die Instandhaltung das rechnergestützte Planungs- und Informationssystem INVERS bei den Readymix Zementwerken in Beckum mit einer Klinkerproduktion von 3000 t/d mit dem in **Bild 1** dargestellten Zielen eingeführt.

BILD 1: Ziele der rechnergestützten Instandhaltung

- Erhöhung der Verfügbarkeit der Produktionsanlagen
- Senkung von Instandhaltungskosten
- Minimierung der Ersatzteillagerkosten
- Transparenz der Kosten- und Schadenschwerpunkte
- Erfüllung von Forderungen aus Gesetzen und Vorschriften für Mensch und Umwelt

2. Ausgangssituation

2.1 Materialwirtschaft

Die Materialwirtschaft basierte vor Einführung der EDV auf der mangel- und fehlerhaften, aufwendigen Fortschreibung von Ersatzteilen, Werten sowie Einkaufspreisen. Da u.a. nur die Endverbräuche und nicht die Umschlagshäufigkeit von Ersatzteilen betrachtet werden konnten, fehlten wesentliche Voraussetzungen für eine Materialdisposition. Hohe Kapitalbildung, großer zeitlicher Aufwand für Routinearbeiten, fehlende Transparenz waren das Alltagsgeschäft in Lager, Einkauf, Rechnungsprüfung und Werkstätten. Mit Einführung der EDV mußten somit alle Material-, Lager-, Einkaufs- und Lieferanteninformationen grundlegend erarbeitet, strukturiert, klassifiziert und in den kaufmännischen Rahmen eingepaßt werden. Dementsprechend wurden in der EDV bei ca. 2500 Bestellungen pro Jahr etwa 16000 Lagerteile von 340 Lieferanten hinterlegt. Außerdem mußten die techn./kaufm. Betriebsorganisation, das Lager, die Lagerverwaltung- und einrichtung den neuen Bedingungen mit nicht unerheblichen Mitteln angepaßt werden.

1. Plant maintenance as the strating point for improving efficiency

Maintenance costs constitute a major proportion of overall production costs in cement production. They are set to increase further because of the replacement cost of ageing plant. The increasing level of assets tied up in plant, high plant availability and the value of stock are increasingly the focus of maintenance management. Quality agreements and an increasingly global market demand a high level of consistent product quality. Legislators and the public expect that maintenance will not only maintain the safety of systems designed to protect people and the environment, but also take account of the effective implementation of regulations by setting new standards. This has made plant maintenance an ever more central and substantial influencing factor in the company, especially in recent years.

The INVERS computer-aided planning and information system was introduced against this background from 1988 for materials management and maintenance at the Readymix cement works in Beckum, which has a clinker production of 3000 t/d, with the goals set out in **Fig. 1**.

FIGURE 1: Objectives of the compuiter-aided maintenance system

- Increasing production plant availability
- Reducing maintenance costs
- Minimizing spare part stockholding costs
- Highlighting sources of cost and damage
- Meeting statutory requirements and regulations regarding people and environment

2. Strating point

2.1 Materials management

Before the introduction of EDP, materials management was based on the inadequate and error-prone process of forward projection for spare parts, data and purchase prices. Since, for example, data were available only for final consumption and not for the rate of turnover of spare parts, essential inputs for effective inventory control were lacking. High capital formation, large amounts of time-consuming routine work and lack of transparency were the day-to-day reality in the stores, in purchasing, in accounting control and on the shop floor. With the introduction of EDP, a fundamental review of all material, inventory, purchasing and supplier information was necessary to compile, structure, classify and adapt it to the commercial environment. With approximately 2500 purchase orders a year, 16000 stock items from 340 suppliers were stored in the EDP system. Substantial resources had to be devoted to adapting technical/commercial organization, stores, and inventory systems and facilities management to the new conditions.

2.2 Instandhaltung

Ein Werkstattauftragswesen war nicht vorhanden. Die Instandhaltungskosten standen damit nur sehr anonym und global ca. 6 Wochen nach Arbeitsausführung und dann nur auf Kostenstellen- und nicht auf Betriebsmittelebene zur Verfügung. Mit Einführung von INVERS wurde für ca. 12 Anlagen mit ca. 750 Anlagenteilen der gesamte technische Anlagen- und Betriebsmittelstamm größtenteils vor Ort aufgenommen, grundlegend definiert und strukturiert. Außerdem wurde die gesamte Betriebsorganisation neu auf die auftragsorientierte Instandhaltungsabwicklung mit z. Zt. ca. 8500 Instandhaltungsaufträgen pro Jahr eingestellt.

2.3 Datenverarbeitung

Bis zur Systemeinführung beschränkte sich die EDV im wesentlichen auf kaufmännische Bereiche. Mit Einführung von INVERS wurde für Produktionsabteilungen, Instandhaltung, Leitstand, Werkstattbüros, Materialwirtschaft und Rechnungsprüfung ein umfangreiches Datennetz aufgebaut. Auf Basis einer UNIX Hardwareplattform stehen, im gesamten Werk verteilt, an ca. 30 Bildschirmarbeitsplätzen alle systemrelevanten Informationen einschließlich des parallel eingeführten Personalerfassungssystems TIME [1] zur Verfügung. Die Lagerverwaltung setzt für alle Routinevorgänge Barcodetechnologie ein.

3. Systemeinführung

Der Aufbau der EDV-Lösung sollte einem Anforderungsprofil entsprechend schrittweise zunächst für die Materialwirtschaft und anschließend für die Instandhaltung in einem Zeitraster von 18 Monaten eingeführt werden.

Die Systemeinführung wurde nach diesem Plan umgesetzt und zunächst für den relativ kleinen Instandhaltungsbereich Steinbruch/Fuhrpark im Echtbetrieb erprobt. Erst nach dieser Einführungsphase wurde die gesamte Maschinen- und Elektroinstandhaltung in das im **Bild 2** dargestellte Gesamtkonzept einbezogen.

4. Systembausteine

Die Systembausteine der rechnergestützten Instandhaltung sind im **Bild 3** wiedergegeben.

4.1 Stammdatenverwaltung

Grundlage systematischer Auftragsbearbeitung in der rechnergestützten Instandhaltung und der ordnungsgemäßen Abwicklung in der Materialwirtschaft ist eine sorgfältig aufgebaute, übersichtliche, strukturierte und vollständige Datenbasis. Über diese Stammdatenverwaltung stehen im Tagesgeschäft im Dialog alle benötigten Informationen des Systems, u. a. über Lieferanten, Ersatzteile, Anlagen und Betriebsmittel, jedem Anwender zur Verfügung.

4.2 Auftragsbearbeitung

Die Auftragsbearbeitung und Abrechnung von Instandhaltungs-Werkstattaufträgen erfolgt auf der Basis von Betriebsmitteln oder Kostenstellen. Bei Standard-IH-Maßnahmen ist der Werkstattauftrag automatisch mit dem fortgeschriebenen Arbeitsplan verkettet, der die genauen Arbeitsanweisungen, verbale Sicherheitshinweise, Meß- und Schmiervorschriften und den geplanten Werkzeug- und Materialbedarf enthält. Für die Auftragsausführung benötigte Werkzeuge und Materialien können wahlweise automatisch reserviert werden. Werkstattaufträge lassen sich terminieren und bis zur Fertigstellung lückenlos kontrollieren.

4.3 Werkstattberichtswesen

Im Werkstattberichtswesen werden alle Informationen der Auftragsabwicklung über die Auftragsabrechnung zu einem der Werkstattberichte verarbeitet. Diese Berichte enthalten nicht nur eine Übersicht der geplanten und geleisteten Stunden als Soll-/Ist-Vergleich, sondern auch eine Kostenzusammenstellung getrennt nach Löhnen und Materialien sowie nach Eigen- und Fremdleistungen. Weitere

2.2 Maintenance

There was no shop order system in place. Maintenance cost figures were available approximately 6 weeks after the work had been carried out, in a very non-specific and global form, and then only by cost centre, and not broken down by production resources. With the introduction of INVERS, virtually all the equipment and production resources for about 12 plants comprising about 750 units were recorded on the spot, definitively registered and categorized. The entire plant was also reorganized to accommodate the new order-led maintenance process, with currently some 8500 maintenance orders per year.

2.3 Data processing

Until the system was introduced, EDP concentrated chiefly on commercial aspects. With the introduction of INVERS, a comprehensive data network was established for production departments, maintenance, control room, workshop administration, material management and accounting control. All information of relevance to the system is available at approximately 30 workstations distributed throughout the plant, based on a UNIX hardware platform; the TIME [1] personnel data acquisition system was also introduced in parallel. Barcode technology is used for all routine inventory management processes.

3. Implementation of the system

According to the specification, the EDP system was to be introduced in stages over a period of 18 months, first for material management and then for maintenance. Implementation was completed to this schedule, and a pilot scheme implemented in the relatively small quarry/vehicle fleet maintenance area. Only after this pilot phase was the whole mechanical and electrical maintenance system incorporated in the overall system shown in **Fig. 2**.

4. System modules

The computer-aided maintenance system modules are shown in **Fig. 3**

4.1 Database maintenance

Systematic order processing in computer-aided maintenance and proper processing in materials management are based on a carefully constructed, clear, complete and structured database. This database maintenance provides each user with all the system information needed for the day-to-day conduct of business on an interactive basis, such as information on suppliers, spare parts, plant and resources.

4.2 Order processing

Order processing and the settlement of maintenance workshop orders are carried out on the basis of resources or cost centres. For standard maintenance activity the workshop order is automatically linked into the projected schedule of job operations which contains the precise instructions, verbal safety information, measuring and lubrication instructions and the planned tool and material requirements. The tools and materials needed to carry out the order can be automatically reserved if required. Workshop orders can be scheduled and rigorously controlled through to completion.

4.3 Workshop reporting

In workshop reporting, all the order processing information is compiled in one of the workshop reports by Order Processing. These reports contain not only a summary variance report of hours scheduled against hours worked, but also a cost breakdown by labour and materials, and by company and external services. Further analysis can also be carried out to give orders on hand and utilisation for each workshop.

4.4 Maintenance and inspection planning

Maintenance intervals based on operating hours, kilometres or calendar weeks are currently being stored in the system as part of the maintenance and inspection plan which is being

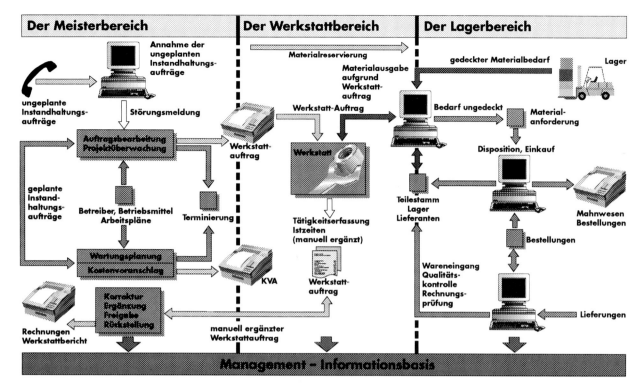

BILD 2: Ablauforganisation der rechnergestützten Instandhaltung
FIGURE 2: Organisational flow chart for the computer-aided planned maintenance system

Der Meisterbereich = Supervision
Annahme der ungeplanten Instandhaltungsaufträge = Accept unplanned maintenance jobs
ungeplante Instandhaltungsaufträge = Unplanned maintenance jobs
Störungsmeldung = Fault indication
Auftragsbearbeitung, Projektüberwachung = Order processing, project monitoring
geplante Instandhaltungsaufträge = Planned maintenance jobs
Betreiber, Betriebsmittel, Arbeitspläne = Operators, resources, work schedules
Terminierung = Scheduling
Wartungsplanung, Kostenvoranschlag = Maintenance planning, cost estimate
Rechnungen, Werkstattbericht = Invoices, workshop report
Korrektur, Ergänzung, Freigabe, Rückstellung = correct, amend, release, reserve

Der Werkstattbereich = Workshop
Materialreservierung = Material reservation
Materialausgabe aufgrund Werkstattauftrag = Material release for workshop order
Werkstatt-Auftrag = Workshop order
Werkstatt = Workshop
Tätigkeitserfassung Istzeiten (manuell ergänzt) = Activity recording actual times (supplemented manually)
Werkstattauftgrag = Workshop order
manuell ergänzter Werkstattauftrag = Manually supplemented works order

Der Lagerbereich = Stores area
gedeckter Materialbedarf = Material requirements covered
Lager = Stores
Bedarf ungedeckt = Requirements not covered
Materialanforderung = Material requirement
Disposition, Einkauf = Reordering, purchasing
Teilestamm, Lager, Lieferanten = Parts database, store, suppliers
Mahnwesen, Bestellungen = Credit control, orders
Bestellungen = Orders
Wareneingang, Qualitätskontrolle, Rechnungsprüfung = Goods inwards, quality control, accounting control
Lieferungen = Deliveries

Management-Informationsbasis = Management Information Base

BILD 3: Bausteine der rechnergestützten Instandhaltung INVERS
FIGURE 3: Modules in the INVERS computer-aided maintenance system

Stammdatenverwaltung = Database administration
Projekt-/abwicklung, -kontrolle = Project management and control
Auftragsbearbeitung = Order processing
Budgetplanung/-kontrolle = Budgetary planning and control
Auftragsabrechnung = Job accounting
Grafik, Statistik = Graphics, statistics
Werkstattberichtswesen = Workshop reporting
Auswertung Kosten/Schäden = Cost/damage analysis
Wartungsplanung = Maintenance planning
Lagerwirtschaft = Stock control
Kapazitäts-/Personalbedarfsplanung = Capacity and manpower planning
Einkauf/Bestellabwicklung = Purchasing/ordering

Auswertungsmöglichkeiten geben u. a. den offenen Auftragsbestand der Werkstätten und deren Kapazitätsauslastung wieder.

4.4 Wartungs- und Inspektionsplanung

Innerhalb der sich im Aufbau befindlichen Wartungs- und Inspektionsplanung werden z. Zt. für alle Maschinen Intervalle wie Betriebsstunden, Kilometer oder Kalenderwochen im System hinterlegt. Hieraus werden über das Programm konkrete Termine für die Ausführung einzelner Arbeiten berechnet. Diese werden dann in Standard-Arbeitsplänen in Planstunden aufgelöst und über die Grundauslastungstabelle der Werkstätten automatisch mit in den Auftragsbestand einbezogen.

4.5 Kapazitäts- und Personalplanung

Mit Hilfe der Kapazitäts- und Personalbedarfsplanung wird über die Kapazitätsmatrix eine langfristige Personalplanung für die Werkstätten erarbeitet. Hierbei werden nicht nur die Plantermine mit vorgegebenen Zeitrahmen, sondern auch die bereits aktiven Aufträge berücksichtigt. Die Gegenüberstellung der Leistungsanforderung der Kapazitätsmatrix mit den verfügbaren Stunden des IH-Personals weisen dann die Kapazitätsauslastung für die einzelnen Werkstätten aus.

4.6 Einkaufs- und Bestellabwicklung

Basis der Einkaufs- und Bestellabwicklung sind unter Berücksichtigung von Reservierungen und offenen Bestellungen vom System statistisch ermittelte Teile-Verbrauchsdaten. Die hieraus ermittelten Bestellvorschläge lösen entweder eine Einkaufsanfrage oder einen Bestellvorgang aus. Der aus der Bestellung resultierende Wareneingang wird im Lager quittiert und gibt die eingegangenen Mengen für die Rechnungsprüfung frei. Die nachgeschaltete Rechnungsprüfung aktualisiert die Preise und stellt gleichzeitig als Schnittstelle Buchungsdaten für die Finanzbuchhaltung zur Verfügung. Sämtliche Vorgänge von Einkaufs- und Bestellabwicklung mit eingebundenem Mahnwesen unterliegen wahlweise einer Terminverfolgung.

4.7 Lagerwirtschaft

In der Lagerwirtschaft übernehmen Barcodelesestifte die Funktionen für die Inventur sowie die automatische Übertragung der Lagerausgabedaten des Werkstattauftrages in den Rechner und ihre direkte Buchung gegen die Verbrauchskostenstelle. Eine Reihe aussagekräftiger Auswertungsmöglichkeiten wie Teileumsatzstatistik, Reichweiten- bzw. Lagerhüterliste und ABC-Analyse erlauben eine permanente Bestandskontrolle des Lagers und stellen der Lagerverwaltung und dem Controlling ein effizientes Werkzeug für die Beurteilung und Optimierung der Ersatzteilevorratung zur Verfügung.

4.8 Analysen

Da das System sämtliche Daten nicht nur zentral sammelt, sondern auch nach differenzierten Kriterien ordnet und in unterschiedlichste Relationen stellt, kann praktisch zu jeder Zeit jede Frage über das Instandhaltungsgeschehen von Anlagen bzw. Maschinen an das System im Dialog gestellt werden. Dabei dient die Kostenzusammenstellung für eine betriebswirtschaftliche und die Forschung nach Schäden oder Ursachen einer technischen Anlagenzustandsbetrachtung. Im Lebenslauf eines technischen Betriebsmittels werden alle seit Inbetriebnahme in aktueller Reihenfolge aufgelaufenen IH-Maßnahmen und deren Kosten ausgewiesen.

4.9 Budgetplanung und Kontrolle

Im Rahmen der eingeführten Budgetplanung und -kontrolle werden auf der Basis der IH-Kosten einer vorangegangenen Instandhaltungsperiode zunächst Plandaten für die neue Periode unter Berücksichtigung von Kostensteigerungen für Lohn und Material erstellt. Nach Festschreibung und Genehmigung der neuen periodischen Instandhaltungsplandaten fließen die täglichen Verbrauchswerte als built up. Based on this data, the program calculates a detailed work schedule for all the individual tasks. These are then broken down into standard work schedules in planned hours, and automatically included in the orders on hand through the basic loading table.

4.5 Capacity and manpower planning

A long-term manpower plan for the workshops was worked out using capacity planning and manpower planning. This takes into account both the schedule deadlines and their specified timescales, and orders currently being processed. Individual workshop utilization is then derived by comparing the workload derived from the capacity matrix with the maintenance man-hours available.

4.6 Purchasing and ordering process

The purchasing and ordering process is based on parts consumption data derived statistically by the system, taking into account reservations and orders pending. This generates order recommendations which in turn generate either a purchase enquiry or a purchase order. The resultant incoming goods are confirmed in the stores, releasing the quantities received for Accounting Control, who now update the prices and provide entry data as an interface for financial accounting. All the purchasing and order processing activities including credit control may be subject to expediting if appropriate.

4.7 Stock control

In stock control, bar code wands take over the functions for stock-taking and automatically transmit the workshop order outgoing stock data to the computer, booking them directly to the consuming cost centre. A number of powerful analytical tools such as parts turnover statistics, coverage and inactive inventory items list and ABC analysis enable the stock level to be permanently monitored, providing inventory management and controlling with an efficient tool for assessing and optimizing the spare parts stock levels.

4.8 Analyses

Since the system not only collects all the data centrally, but also organizes it according to distinct criteria, and analyses them according to the most diverse relations, the system can be interactively interrogated with virtually any question at any time about the current status of maintenance work as regards plant and machinery. Commercial aspects are covered by the cost breakdown, and technical plant status by investigation of damage and its causes. Over the lifespan of an item of plant, all maintenance activity since commissioning is listed in chronological order, with costs.

4.9 Budgetary planning and control

As part of the budgetary planning and control process now in place, the first stage is to generate forecast data for the new period based on the maintenance costs for previous maintenance period, taking into account increased labour and material costs. When the new periodic maintenance schedule data have been recorded and approved, the daily consumption figures are fed in as actual data against the budget forecast. The resultant comparative data provides the plant controller with a constantly updated maintenance cost variance report for the period.

4.10 Project management and control

The project management and control system currently under development supports the creation of project documents for cost approval. Cost estimates are sought for a project. Inputs from skilled company workers are scheduled in and allocated to the project. Implementation of the project is handled by the Order Processing module. Automatic project projection then gives the project manager a constantly updated overview of planned project milestones with associated costs, especially in the case of major planning sequences.

Istdaten gegen die Budgetvorgabe. Mit dem daraus resultierenden Vergleichsergebnis steht dem Anlagencontrolling ein stets zeitaktueller Soll/Ist-Vergleich der Instandhaltungskosten innerhalb der Planperiode zur Verfügung.

4.10 Projektabwicklung und -kontrolle

Die sich im Aufbau befindliche Projektabwicklung und -kontrolle unterstützt die Erstellung von Projektunterlagen für die Kostengenehmigung. Für ein Projekt werden Kostenvoranschläge eingeholt. Die Mitwirkung eigener Handwerker wird vorgeplant und dem Projekt zugeordnet. Die Projektdurchführung selbst wird über den Baustein Auftragsbearbeitung abgewickelt. Durch die automatische Projektfortschreibung steht dann dem Projektverantwortlichen besonders bei großen Planungsabläufen eine jederzeit aktuelle Übersicht über geplante Projekttermine mit den zugehörigen Kosten zur Verfügung.

5. Ergebnis

Die frühe Beteiligung der Mitarbeiter an dem arbeits- und schulungsintensiven Aufbau des Systems hat zu einer schnellen Akzeptanz von INVERS auf breiter Werksebene beigetragen. INVERS ist zu einem selbstverständlichen Hilfsmittel für die Bewältigung der täglichen Routinearbeiten in allen betrieblichen Abteilungen geworden. Die im Instandhaltungsgeschehen bisher insgesamt erzielte Transparenz hat wesentliche Voraussetzungen für das zukünftige Anlagenmanagement geschaffen, so daß auf Dauer die Anlagenverfügbarkeit auf einem hohen Niveau gehalten wird und die Instandhaltungs- und Lagerbestandskosten ständig im Blickpunkt bleiben. Die mit dem System angestrebten Ziele sind somit zunächst erfüllt, wenn auch die erreichten Einsparungen sich nur schwer in konkreten Zahlen ausdrücken lassen. Es bedarf jetzt der menschlichen Intelligenz zur Sichtung und Analyse des stets aktuell zur Verfügung stehenden Datenmaterials, um damit die hieraus gewonnenen Erkenntnisse noch stärker in das tägliche Betriebsgeschehen einfließen zu lassen.

Literaturverzeichnis

[1] Schramm, H. F. W.: Zukunftssicher und flexibel, Automatisierte Zeitwirtschaft für die Lohn- und Gehaltsabrechnung. Erfolg, Büromagazin für den Chef und Leitende in der Industrie (1993), H. 3/4, 42. Jahrgang.

5. Result

A major commitment in terms of labour and training was required to construct the system, and involvement of employees at an early stage helped to secure rapid acceptance of INVERS throughout the works. INVERS has established itself as a natural means of tackling daily routine tasks in all departments. The transparency achieved overall so far has laid the foundation for future plant management, so that in the long run plant availability is kept at a high level, and maintenance and stock-holding costs are constantly monitored. This means the system's targets are achieved, even if the savings are hard to quantify in concrete terms. Human intelligence is now needed to sift and analyse the up-to-date data material always available, to enhance still further the benefits provided by the information for the day-to-day running of the plant.

Strömungsverhältnisse in Turbinenpackern *)

Flow conditions in rotary impeller packers *)

Conditions d'écoulement dans les ensacheuses à turbine

Condiciones de flujo dentro de las turboensacadoras

Von **J. Spiess,** Beckum/Deutschland

Zusammenfassung — *Für die Abfüllung von Zement wird seit Anfang des 20. Jahrhunderts das Turbinen-Füllsystem verwendet. Mitte der 80er Jahre wurden Alternativkonstruktionen entwickelt und ebenfalls praktisch erprobt. In wissenschaftlichen Versuchen wurden die Strömungsverhältnisse in den Laufrädern der Turbinen untersucht und die Meßdaten mit Hilfe bekannter Grundgleichungen ausgewertet. Die dadurch mögliche Optimierung konnte für eine deutliche Verbesserung der Abfülleistung bei Produkten mit Feinheiten von über 10 000 Blaine und für eine erhebliche Verminderung des Verschleißes genutzt werden. Zwischenzeitlich erfolgten praktische Erprobungen, welche die numerischen Berechnungen und die experimentell ermittelten Daten voll bestätigen. Darüber hinaus wurde eine neue Wägeelektronik entwickelt, die den Massestrom kontinuierlich mißt und alle Parameter für die Grob- und Feinstromumschaltung sowie Nachlaufkorrektur in einem Zyklus von 23 ms berechnet und optimiert. Jede Veränderung im Fließverhalten des Produktes wird innerhalb dieser Zykluszeit festgestellt. Sie führt dazu, daß alle Parameter selbständig darauf eingestellt werden, so daß eine außergewöhnliche Gewichtsgenauigkeit bei voller Leistung der Abfüllmaschine garantiert wird. Um die volle Leistungsfähigkeit der neuen Rotopacker kontinuierlich auszunutzen, stehen solche Automatisierungsbausteine wie Leersackaufstecker und Palettierautomaten zur Verfügung, die Umstellungen beim Produkt, beim Sackgewicht oder beim Packmuster vollautomatisch vornehmen.*

Strömungsverhältnisse in Turbinenpackern

Summary — *The rotary impeller filling system has been used for bagging cement since the start of the 20th century. Alternative designs were developed in the mid 80s, and were also put to the test under practical conditions. The flow conditions in the turbine impellers were investigated in scientific trials and the measurements were evaluated with the aid of well-known fundamental equations. The optimization which this made possible was utilized to make a significant improvement in the bagging capacity with products with finenesses of over 10 000 Blaine and to achieve a considerable decrease in wear. In the meantime, practical tests were carried out which fully confirmed the numerical calculations and the experimentally determined data. A new electronic weighing system was also developed which, in a cycle of 23 ms, measures the mass flow continuously and calculates and optimizes all parameters for the changeover between coarse and fine flow as well as the run-on correction. Any change in the flow behaviour of the product is determined within this cycle time. All the parameters are then adjusted independently to the situation so that exceptional weight accuracy is guaranteed when the bagging machine is operating at full capacity. Automation modules such as empty bag applicators and automatic palletizers are available to make continuous use of the full capabilities of the new packer; these automatically undertake any changes in product, bag weight or stacking pattern.*

Flow conditions in rotary impeller packers

Résumé — *Pour l'ensachage du ciment est utilisé, depuis le début du 20ème siècle, le système de remplissage à turbine. Vers le milieu des années 80 ont été conçues des variantes, essayées aussi en service. Au cours d'expériences scientifiques, les conditions d'écoulement dans les rotors des turbines ont été étudiées et les données de mesure ont été exploitées à l'aide des équations fondamentales connues. L'optimisation rendue ainsi possible a pu être mise à profit pour une nette amélioration des performances de remplissage avec des produits de finesses de plus de 10 000 Blaine et pour une réduction non négligeable de l'usure. En même temps, ont eu lieu des expériences pratiques, qui ont confirmé pleinement les calculs numériques et les données obtenues expérimentalement. En complément, a été mise au point une nouvelle électronique du pesage, qui mesure en continu le flux massique et calcule tous les paramètres pour la variation produit gros ou produit fin et qui calcule et optimise la correction à posteriori selon un cycle de 23 ms. Chaque changement du comportement d'écoulement du produit est constaté pendant le temps de ce cycle. Ceci conduit à un ajustement automatique de tous les paramètres, garantissant ainsi une exactitude extraordinaire du poids à pleine capacité de l'ensacheuse. Afin d'exploiter pleinement le rendement en continu des nouvelles ensacheuses rotatives, la gamme est complétée par des auxiliaires d'automatisation tels que enfileuses de sacs vides et palettiseuses automatiques, capables d'opérer la sélection adéquate lors d'un changement de produit, d'un poids par sac et d'un schéma de palettisation.*

Conditions d'écoulement dans les ensacheuses à turbine

---

*) Überarbeitete Fassung eines Vortrages zum VDZ-Kongreß '93, Düsseldorf (27. 9.–1. 10. 1993)
Revised text of a lecture to the VDZ Congress '93 Düsseldorf (27.9.–1.10.1993)

Condiciones de flujo dentro de las turboensacadoras

Resumen — Para el envasado del cemento se viene empleando, desde comienzos del siglo XX, el sistema de turboensacadoras. A mediados de los años 80, fueron desarrolladas soluciones alternativas y sometidas a ensayos prácticos. Se llevaron a cabo investigaciones científicas para estudiar las condiciones de flujo dentro de los rodetes de las turbines, evaluándose los datos de medición con ayuda de ecuaciones básicas ya conocidas. La optimización conseguida con ello ha podido aprovecharse para obtener una notable mejora en el rendimiento de envasado de los productos, cuya finura era superior a 10 000 Blaine, y una considerable disminución del desgaste. Entretanto, se han llevado a cabo ensayos prácticos, que han confirmado plenamente los cálculos numéricos y los datos obtenidos por vía experimental. Aparte de ello, se ha desarrollado un nueva sistema electrónico de pesado, el cual mide el flujo de masas de forma continua, calculando y optimizando todos los parámetros necesarios para la regulación de los flujos grueso y fino y la corrección del chorro, todo ello en un ciclo de 23 ms. Cualquier cambio producido en la reología del producto se detecta dentro del ciclo mencionado, lo que hace que todos los parámetros se regulen automáticamente y que quede garantizada una precisión extraordinaria de pesado, con pleno rendimiento de la máquina ensacadora. Con el fin de aprovechar, de forma continua, la capacidad máxima de las nuevas ensacadoras rotativas, se dispone de elementos de automatización, tales como aplicadores de sacos vacíos y paletizadores automáticos, los cuales proceden automáticamente a los necesarios cambios de producto, peso de los sacos o capas de sacos.

1. Theoretische Grundlagen

In einem Turbinenpacker wird dem Fluid im rotierenden Schaufelrad Energie zugeführt (**Bild 1**). Um zu verstehen, wie es dabei zu einer Druck- und Geschwindigkeitserhöhung kommt, müssen die Strömungsverhältnisse in einem einzelnen Schaufelkanal des Laufrads betrachtet werden (**Bild 2**). Ein auf dem Laufrad „mitfahrender Beobachter" würde das Fluid mit der Relativgeschwindigkeit w_1 und dem statischen Druck p_1 durch den Eintrittsquerschnitt fließen sehen.

Nach der Kontinuitätsgleichung verzögert sich die Relativgeschwindigkeit in dem immer breiter werdenden Schaufelkanal, so daß nach Bernoulli der statische Druck p_2 größer als p_1 werden muß. Für den außenstehenden Betrachter hat das Fluid dagegen die absolute Geschwindigkeit c, die sich vektoriell als Summe aus Relativ- und Umfangsgeschwindigkeit u ergibt. Mit der Zunahme der Umfangsgeschwindigkeit vom Eintritts- zum Austrittsradius wird die Geschwindigkeit des Fluids von c_1 auf c_2 erhöht. Bei der Förderung von Feststoffen muß außerdem der Verschleiß beachtet werden, der proportional zur dritten Potenz der

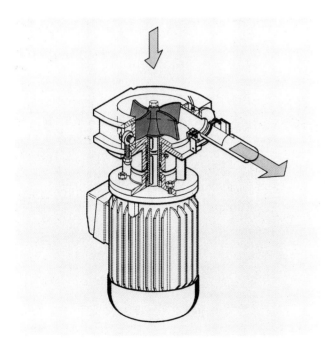

BILD 1: Materialfluß in einem Turbinenpacker mit horizontaler Turbine
FIGURE 1: Material flow in a rotary impeller packer with horizontal impeller

1. Theoretical principles

In a rotary impeller packer, energy is introduced into the fluid in the rotating impeller (**Fig. 1**). In order to understand how this leads to an increase in pressure and velocity it is necessary to examine the flow conditions in a single channel between the blades of the impeller (**Fig. 2**). An observer travelling on the impeller would see the fluid flowing through the inlet cross-section with a relative velocity w_1 and a static pressure p_1.

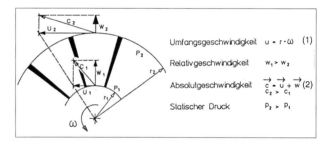

BILD 2: Druck- und Geschwindigkeitsverlauf in einem Schaufelkanal der Füllturbine
FIGURE 2: Pressure and flow in a channel between the blades of the filling impeller

Umfangsgeschwindigkeit	= circumferential velocity
Relativgeschwindigkeit	= relative velocity
Absolutgeschwindigkeit	= absolute velocity
Statischer Druck	= static pressure

The relative velocity slows down in the widening channel between the blades in accordance with the continuity equation, so that according to Bernoulli the static pressure p_2 must be larger than p_1. For the external observer, however, the fluid has the absolute velocity c, given vectorially by the sum of the relative velocity and the circumferential velocity u. With the increase in circumferential velocity from the inlet radius to the outlet radius the velocity of the fluid is raised from c_1 to c_2. In the transport of solids it is also necessary to pay attention to the wear, which increases in proportion to the cube of the relative velocity between the solids and impeller blade. In the vertical rotary impeller the inlet cross-section has approximately the same width as the impeller itself and the average approach flow velocity of, for example, a cement is 0.92 m/s (**Fig. 3**); this then impinges on the outermost radius of the impeller rotating at high speed. The blade ends have a circumferential velocity of approximately 10 m/s and meet the incoming cement almost at right angles with this high velocity. A calculated relative velocity of 11.2 m/s is reached which must be used as the basis when considering wear. Completely different flow conditions are found in the horizontal rotary impeller in **Fig. 4**. The inlet cross-section in this case corresponds approximately to the

Relativgeschwindigkeit zwischen Feststoff und Laufradschaufel ansteigt. Bei der Vertikalturbine ist der Eintrittsquerschnitt etwa genauso breit wie die Turbine selbst, und die mittlere Zuströmgeschwindigkeit z.B. eines Zements beträgt 0,92 m/s (**Bild 3**), der dann auf den äußersten Radius des mit hoher Drehzahl umlaufenden Schaufelrades auftrifft. Die Schaufelenden haben eine Umfangsgeschwindigkeit von ca. 10 m/s und treffen mit dieser hohen Geschwindigkeit fast rechtwinklig auf den zulaufenden Zement. Dabei wird eine rechnerische Relativgeschwindigkeit von 11,2 m/s erreicht, die für Verschleißbetrachtungen zugrunde gelegt werden muß. Völlig andere Strömungsverhältnisse findet man bei der liegenden Horizontalturbine im **Bild 4**. Der Eintrittsquerschnitt entspricht hier etwa dem Durchmesser des Laufrades. Bei diesem großen Strömungsquerschnitt beträgt die mittlere Zuströmgeschwindigkeit nur noch 0,4 m/s. Die Wandreibung des zulaufenden Zements besitzt den Vorteil, daß auf die mit hoher Umfangsgeschwindigkeit umlaufenden Schaufelenden kaum Zement trifft. Bedingt durch die Kernströmung, beträgt die maximale Relativgeschwindigkeit nur noch 5,6 m/s. Neben dem Verschleiß bestimmen diese Strömungsverhältnisse auch den Wirkungsgrad einer Turbine. Eine liegende Turbine benötigt nur eine Antriebsleistung von 4,0 kW gegenüber von 5,5 kW bei einer stehenden Turbine. Ein weiterer Unterschied zwischen beiden Systemen ergibt sich durch das Förderverhalten bei sehr feinen Produkten. Während die Schaufelenden der Vertikalturbine teilweise das Produkt in den Zulauf zurückschleudern, bildet sich bei der Horizontalturbine im Zentrum ein Sog aus, mit dem die Abfüllung von Produkten mit Feinheiten bis zu 20 000 Blaine realisiert werden kann.

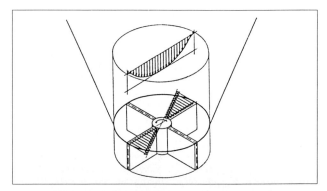

BILD 4: Strömungsverläufe und Geschwindigkeiten in einer Horizontalturbine
FIGURE 4: Flow behaviour and velocities in a horizontal rotary impeller

diameter of the impeller. With this large flow cross-section the average approach velocity is only 0.4 m/s. The wall friction of the incoming cement is an advantage in that the blade ends rotating at high circumferential velocity hardly meet any cement. Because of the core flow the maximum relative velocity is still only 5.6 m/s. In addition to the wear these flow conditions also determine the efficiency of a rotary impeller. A horizontal impeller only requires a drive rating of 4.0 kW as opposed to 5.5 kW for a vertical impeller. Another difference between the two systems arises through the conveying behaviour with very fine products. While to some extent the blade ends of the vertical impeller throw the product back into the inlet the horizontal impeller forms a suction in the centre which means that it can bag products with finenesses of up to 20 000 Blaine.

2. Structure and functioning of modern rotary packers

The basic requirements for the design of the new rotary packer (**Fig. 5**) were a high bagging capacity, simple opera-

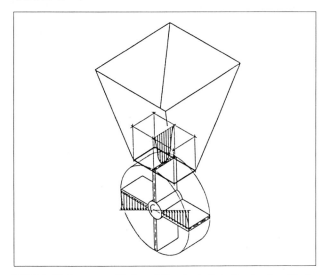

BILD 3: Strömungsverläufe und Geschwindigkeiten in einer Vertikalturbine
FIGURE 3: Flow behaviour and velocities in a vertical rotary impeller

2. Aufbau und Funktion moderner Rotopacker

Für die Konstruktion der neuen Rotopacker (**Bild 5**) waren eine hohe Abfülleistung, einfache Bedienung, große Betriebssicherheit, niedriger Energiebedarf, minimale Staubemissionen und geringste Wartungsansprüche grundlegende Forderungen. Die direkte Verbindung des rotierenden Laufrads mit der Motorwelle hat den Vorteil, daß auf Kraftübertragungselemente verzichtet werden konnte. Darüber hinaus kann der Austausch eines Laufrades in Minutenschnelle (**Bild 6**) vorgenommen werden. Eine Schieberplatte mit zwei Stellungszylindern, die zwischen hartverchromten Flanschen läuft, öffnet und schließt den Füllkanal und wird zur Feindosierung in eine Zwischenposition gestellt. Da sich an dieser Stelle bereits der Übergang zwischen feststehenden Maschinenelementen und den vertikal beweglichen Waageteilen befindet, wird zur Entkopplung ein flexibles Schlauchstück eingesetzt, so daß Staubemissionen völlig vermieden werden können. Der vordere

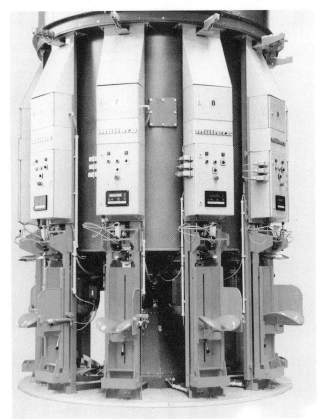

BILD 5: 10-Stutzen-Rotopacker mit horizontalen Turbinen und ROTRONIC®-Wägeelektronik
FIGURE 5: 10-spout rotary packer with horizontal rotary impellers and ROTRONIC® electronic weighing system

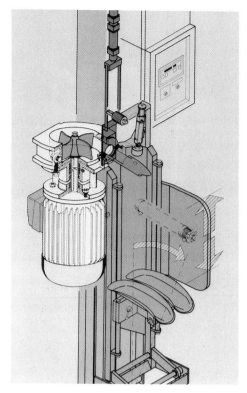

BILD 6: Isometrische Darstellung eines einzelnen Füllmoduls
FIGURE 6: Isometric representation of a single filling module

Flansch ist mit einer Ausblaseinheit für die Füllröhre ausgestattet, die nach Beendigung des Füllvorgangs die im Füllrohr liegenden Reste in den Sack fördert. Die Füllröhre und der Füllrohrhalter sind bereits Teile der Waage. Sie sind ebenso wie der Sackstuhl mit dem Sackabwurfzylinder in einem stabilen, verwindungssteifen Rahmen eingebaut, der mit 3 kurzen Federlenkern am Maschinentragrahmen befestigt ist. Die kurzen Federlenker haben eine hohe Eigenfrequenz und erlauben mit den daraus resultierenden kurzen Einschwingzeiten die Verwendung moderner Digitaltechnik. Im Gegensatz zu der bislang verfügbaren Wägeelektronik mißt die sogenannte ROTRONIC® ständig den Massestrom und berechnet daraus die Grob- und Feinstromabschaltung sowie den Nachlauf des Produktes (**Bild 7**). Sie korrigiert und optimiert kontinuierlich alle Reglereinstellungen, die Filterverzögerung und die Einschwingsperrzeiten. Jede Veränderung im Fließverhalten des Produktes wird innerhalb der Zykluszeit von 23 ms festgestellt. Die erste von drei Filterstufen berechnet dann aus dem in Echtzeit gemessenen realen Massestrom die Steigung der Gewichtsfunktion. Zur Unterdrückung von Fehlmessungen beginnt die reguläre Auswertung nur, wenn das Differential der Gewichtsfunktion gleich Null ist, d.h. bei konstanter Steigung bzw. bei eingeschwungenem Zustand der Waage. Im regulären Zyklus werden die Grob- und Feinstromabschaltungen und der Nachlaufvorhalt errechnet, wobei alle Regler und die Filterverzögerung automatisch mit korrigiert werden. Falls die Hochrechnung ergibt, daß ein Umschalt- oder Abschaltpunkt innerhalb des nächsten Zyklus liegt, werden die fortlaufenden Messungen und internen Prüfroutinen abgebrochen und der exakte Schaltpunkt mit einer Toleranz von weniger als einer Millisekunde berechnet (**Bild 8**). Dieses neue Wägesystem garantiert eine außergewöhnliche Gewichtsgenauigkeit bei voller Leistung der Abfüllmaschine. Es erleichtert darüber hinaus die Bedienung der Packmaschine, da bei einem Produktwechsel nur das Vollsackgewicht und die Sacklänge eingestellt werden müssen, während alle anderen Parameter von der Rotronic selbständig ermittelt, überwacht und optimiert werden.

Um die volle Leistungsfähigkeit moderner Rotopacker kontinuierlich ausnutzen zu können, stehen als Automatisie-

tion, high operational reliability, low power consumption, minimum dust emission and very low maintenance requirements. Direct connection of the rotating impeller to the motor shaft had the advantage that it was possible to dispense with any power transmission elements. It is also possible to replace an impeller in minutes (**Fig. 6**). A sliding plate with two positioning cylinders, which runs between hard-chrome plated flanges, opens and closes the filling duct and is set in an intermediate position for fine feeding. The transition from fixed mechanical elements to the weighing parts with their vertical movements occurs at this point, so they are separated by a flexible section of tube and any dust emission is completely avoided. The front flange has a blow-down unit for the filling tube which, after the filling process is complete, conveys the residue lying in the filling tube into the bag. The filling tube and filling tube holder form part of the weigher. Like the bag support and the bag throw-off cylinder, they are built into a stable, torsionally rigid, frame which is attached to the machine support frame by three short spring links. The short spring links have high natural frequencies, and the resulting short damping times allow modern digital technology to be used. Unlike the electronic weighing systems previously available the ROTRONIC® measures the mass flow continuously, from which it calculates the coarse and fine flow cutoff and the product after-flow (**Fig. 7**). It continuously corrects and optimizes all controller settings, the filter delay and the vibration damping times. Any change in the flow behaviour of the product is detected within the cycle time of 23 ms. The first of the three filter stages then calculates the gradient of the weight function from the true mass flow measured in real time. To suppress false measurements the regular evaluation only begins when the differential of the weight function is equal to zero, i.e. at constant gradient or when the weigher is in a steady state. The coarse and fine flow cutoffs and the after-flow allowance are calculated during the regular cycle, including automatic corrections for all the controllers and the filter delay. If the extrapolation shows that a changeover or cutoff point lies within the next cycle the continuous measure-

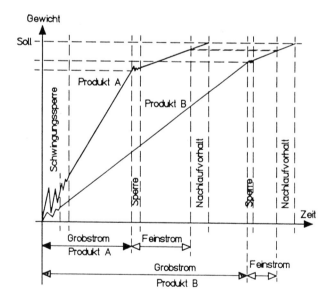

BILD 7: Prinzipieller Verlauf des Füllvorgangs bei Produkten mit unterschiedlichem Fließverhalten und automatische Berechnung der Parameter durch die ROTRONIC®-Waagenelektronik
FIGURE 7: Basic sequence of the filling process for products with different flow characteristics, and automatic calculation of the parameters by the ROTRONIC® electronic weighing system

Gewicht	= weight
Soll	= target
Produkt A	= product A
Produkt B	= product B
Schwingungssperre	= vibration damping
Sperre	= damping
Nachlaufvorhalt	= after-flow allowance
Zeit	= time
Grobstrom	= coarse flow
Feinstrom	= fine flow

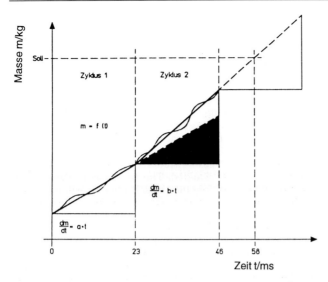

BILD 8: Messung des realen Massestroms, Linearisierung und Vorausberechnung des Folgezyklus durch die ROTRONIC®-Waagenelektronik
FIGURE 8: Measurement of the true mass flow, linearization and advance calculation of the next cycle by the ROTRONIC® electronic weighing system

Masse m/kg	= mass m [kg]
Soll	= target
Zyklus	= cycle
Zeit t/ms	= time t [ms]

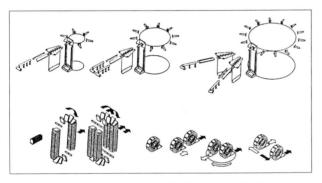

BILD 9: Mögliche Kombinationen der Leersackaufsteckautomaten in den drei Leistungsklassen 2 600, 3 600 und 4 500 Sack/h mit Rotopackern bis zu 8, 10 und 14 Stutzen und verschiedenen Leersackmagazinen
FIGURE 9: Possible combinations of the automatic empty bag applicators in the three output classes of 2 600, 3 600 and 4 500 bags/h with rotary packers up to 8, 10 and 14 spouts and various empty bag magazines

BILD 10: Palettierautomat in einem Zementwerk
FIGURE 10: Automatic palletizer in a cement works

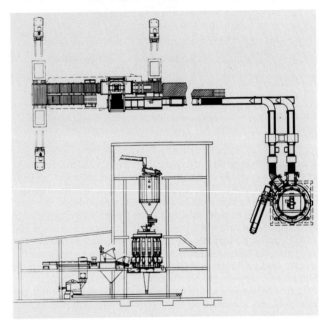

BILD 11: Aufstellungsplan einer kompletten Absack- und Palettieranlage für eine Packleistung von 130 t/h in einem Zementwerk
FIGURE 11: Layout of a complete bagging and palletizing plant for a packing capacity of 130 t/h in a cement works

rungsbausteine Leersackaufstecker für alle gängigen Ventilsackbauarten in drei Leistungsklassen bis zu 4 500 Sack/h zur Verfügung (**Bild 9**). Die gebündelten oder aufgerollten Leersäcke werden einzeln aus dem Magazin entnommen, zentriert und in eine Formhülse eingeführt. Während zwei Treibriemen unterhalb des Ventilbodens den Sack auf eine Geschwindigkeit von 4 m/s beschleunigen, wird das Sackventil beim Lauf durch die Formhülse geöffnet und präzise auf den vorbeilaufenden Füllstutzen aufgesteckt. Ein besonderes konstruktives Merkmal ist die Führung und Ausrichtung der Leersäcke einzig an der Ventilbodenseite mit der Ventilöffnung. Hierdurch können unterschiedliche Sackformate ohne jede Verstellung aufgesteckt werden, und insbesondere für die bevorstehende Umstellung des Sackgewichtes von 50 auf 25 kg bietet dieses System große Vorteile.

3. Stand und Möglichkeiten der Palettiertechnik

Die wirtschaftliche Handhabung der gefüllten Säcke setzt die maschinelle Palettierung auf ausgereiften Anlagen mit hoher Zuverlässigkeit voraus (**Bild 10**). Die Anforderungen der Zement- und Baustoffindustrie an die Stapelqualität

ments and internal checking routines are interrupted and the exact switching point is calculated with a tolerance of less than one millisecond (**Fig. 8**). This new weighing system guarantees an exceptional weight accuracy at full machine bagging capacity. It also makes the packing machine easier to operate as only the full bag weight and the bag length have to be set when the product is changed; all the other parameters are determined, monitored and optimized by the Rotronic itself.

In order to make best use of the full capabilities of modern rotary packers there are automation modules available in the form of empty bag applicators for all the current types of valve bag in three throughput classes up to 4 500 bags/h (**Fig. 9**). The empty bags in bundles or reels are taken out of the magazine individually, centred and inserted into a forming sleeve. Two drive belts under the valve base accelerate the bag to a speed of 4 m/s and the bag valve is opened as it passes through the forming sleeve; it is then applied accurately to the filling spout as it travels past. One particular design feature is that the empty bags are only guided and aligned at the valve base end with the valve opening. This means that the applicator can handle different bag formats

werden auch bei hohen Packleistungen mit weiterentwikkelten Palettierautomaten erfüllt. Die für die zunehmend realisierte Direktanbindung an vollautomatische Hochregallager erforderlichen Toleranzen an die Palettenladung werden mit der neuesten Ausrüstungsgeneration ebenso erfüllt wie die Forderung nach vollautomatischer Umstellung beim Produkt-, Sackgewichts- und Packmusterwechsel. Eine Vielzahl konstruktiver Details, wie das staudrucklose Auftakten und Überheben einzelner Sackzeilen, die Lagenkalibrierung von allen vier Seiten, die Reduzierung der Ablagehöhe zwischen den geteilten Schiebeblechen und der Palette, die wirkungsvolle Egalisierung der Lagen mit dem hohen Preßdruck der Lagenandruckplatte und ein optimal gestalteter Zulauf der Säcke haben zu einer bedeutenden Verbesserung der Stapelqualität geführt. Besonders interessant ist die Flexibilität der Palettierautomaten bei der Bildung unterschiedlicher Packmuster für verschiedene Sackgewichte. So können beispielsweise bei einem PLS-Palettierer, der für die Verbandsstapelung von 50-kg-Säcken mit einer Leistung von 2600 Sack/h ausgelegt ist, durch eine einfache Programmwahl 4000 Sack/h mit 25 kg Gewicht palettiert werden, so daß der Absackbetrieb eine annähernd gleiche Leistung beibehält. **Bild 11** zeigt den Aufstellungsplan einer kompletten, in jedem Detail optimal aufeinander abgestimmten Absack- und Palettieranlage in der beschriebenen Technologie in einem Zementwerk.

without any adjustment; this system has great advantages, especially for the impending change of bag weight from 50 to 25 kg.

3. Current state and potential of palletizing technology

Mechanical palletizing on sophisticated plants of high reliability is required to achieve cost-effective handling of the filled bags (**Fig. 10**). Advanced automatic palletizers can meet the demands made by the cement and construction materials industries on the quality of stacking, even at high packing rates. The increasing use of direct links to fully automatic high bay warehouses requires the palletized load to have tolerances which are met by the latest generation of equipment, as is the demand for fully automatic changeover when changing product, bag weight and stacking pattern. A large number of design details, such as the timed stopping and lifting of single lines of bags without any impact pressure, the sizing of the layer on all four sides, the reduction in the drop height between the divided slide plates and the pallet, the effective flattening of the layers by the high pressure applied with the layer pressure plate, and the optimum configuration of the bag feed have led to a significant improvement in the stacking quality. Of particular interest is the flexibility of the automatic palletizers when forming different stacking patterns for different bag weights. For example, a PLS palletizer which is designed for a capacity of 2600 bags/h with 50 kg bags in an interlocked stacking pattern is able, by a simple program selection, to palletize 4000 bags/h of 25 kg weight, so the bagging operation remains at almost the same capacity. **Fig. 11** shows the layout of a complete bagging and palletizing plant which is optimally coordinated in every detail and uses the technology described in a cement works.

Mischen und Dosieren von Klinker – Segregationseffekt und Gegenmaßnahmen

Blending and dosage of clinker – resulting effect of segregation and ways to counteract it

Mélange et dosage de clinker – effet de ségrégation et remèdes

Mezcla y dosificación de clínker – efecto de segregación y remedios

Von **E. Stoltenberg-Hansson**, Brevik, und **G. Enstad**, Porsg/Norwegen

Zusammenfassung – Die kugelförmigen Klinkerpartikel unterschiedlicher Größenverteilung haben in Silos und Bunkern eine starke Segregationsneigung, die zu Qualitätsschwankungen führen kann. Das gilt auch für Klinkermischungen verschiedener Qualität und Granulometrie. Mit einem besonders konstruierten Entmischungs-Prüfgerät in kleinem Maßstab, das von POSTEC Research entwickelt wurde, wurden vorläufige Versuche durchgeführt. Anschließend wurden Versuche in einer Zementmühle im großen Maßstab mit einem traditionellen zylinderförmigen Klinkerbunker gemacht. Die Ergebnisse zeigten insgesamt deutlich, daß die Klinkersegregation in einer Weise auftritt, die aufgrund des hohen Alkaligehalts in der grobkörnigen Fraktion und des niedrigeren Alkaligehalts in der feinkörnigen Fraktion zu Qualitätsverlusten führen kann. Das wurde offenbar, wenn der Klinkerbunker nach Art eines Trichters entleert wurde und ein stetig zunehmender Alkaligehalt festzustellen war. Selbst bei fast völlig gefülltem Klinkerbunker waren die Schwankungen im Alkaligehalt zu hoch. Deshalb wurde festgelegt, einen zusätzlichen Bunker zur Zudosierung eines zweiten Klinkers zu installieren, um das Mischungsverhältnis der verschiedenen Klinker konstant zu halten. Wenn darüber hinaus die Bunker einen hohen Füllstand hatten, konnten die Schwankungen im Alkaligehalt innerhalb akzeptabler Grenzen gehalten werden. Versuche mit dem Segregationsprüfgerät im kleinen Maßstab haben sich als nützliches Instrument für den Umbau von Silos und Aufgabegeräten erwiesen, um gleichbleibenden Massedurchsatz zu erreichen und die Betriebsweise entsprechend zu ändern.

Summary – The spherical clinker particles of different size distribution have a great tendency for segregation in silos and hoppers which may induce quality variations. This also applies to blends of clinkers of different quality and granulometry. Preliminary tests were performed in a specially designed small scale segregation tester developed by POSTEC Research. Subsequent tests were done in a full scale cement mill with traditional cylindrical clinker hopper. The results altogether clearly showed that the clinker segregates in a way that could affect the quality adversely due to high alkali contents in the coarse sized fraction and lower contents in the fine sized fraction. This became obvious when the clinker hopper was emptied in the funnel flow mode leading to a steadily increasing alkali content. Even when operating a nearly full clinker hopper, the variations in the alkali content were considered too high to be acceptable. It was therefore decided to install an extra hopper for dosage of a second clinker in order to keep the mix ratio of the different clinker constant. By an additional high filling degree of the hoppers the variations in the alkali content were kept within satisfactory limits. Small scale experiments with the segregation tester proved to be a helpful tool for redesigning silos and feeding equipment to give mass flow and change its operational modes.

Résumé – Les particules de clinker en forme de boule de différente répartition dimensionnelle ont, dans les silos, une forte propension à la ségrégation, qui peut conduire à des fluctuations de qualité. Ceci concerne aussi les mélanges de clinker de diverses qualités et la granulométrie. Un appareil de mesure de ségrégation de conception spéciale à petite échelle, mis au point par POSTEC Research, a servi à effectuer des tests préliminaires. Après quoi, des essais ont été faits dans un broyeur à ciment à pleine échelle avec une trémie à clinker classique de forme cylindrique. Globalement, les résultats ont prouvé clairement que la ségrégation du clinker se présente de manière telle qu'elle peut, en raison de la teneur élevée en composés alcalins dans la fraction à gros grains, d'une part, et de la basse teneur en composés alcalins dans la fraction à grains fins, de l'autre, conduire à des pertes de qualité. Ceci est devenu évident quand la trémie à clinker a été vidée à la manière d'un entonnoir, et qu'une teneur croissante en composés alcalins a été enregistrée. Même lorsque la trémie à clinker était presque entièrement pleine, les fluctuations de la teneur en composés alcalins étaient encore trop élevées, Voilà pourquoi il a été fixé d'installer une trémie d'appoint pour le dosage du deuxième clinker, afin de maintenir constant le rapport de mélange des différents clinkers. Si, en plus de cela, les trémis avaient un niveau de charge élevée, les fluctuations de teneur en composés alcalins pouvaient être maintenues

Professional components and systems for bulk technology

INDUSTRIETECHNIK GMBH

The company offers a broad range of special components for the silo and conveying technique. Together with the respective Know-how and engineering, safe and cost-efficient solutions are laid out for the customer.

Most of the products can be delivered on short term from stock and are completed by a detailed documentation.

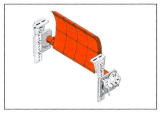 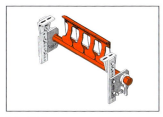
 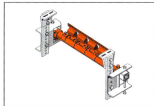

VIBREX Conveyor Belt Scrapers
for keeping conveyor belt installations clean

CAREX® SEGMENTBLOCK SYSTEM
for sealing of transfer chutes

LINEX® Conveyor Belt Centralizing Idlers
for the straight run of conveyor belts

SEALEX® Inspection Windows
for housings in production and transport plants

FRICLESS® Slide Cushions
for the careful support of conveyor belts at transfers

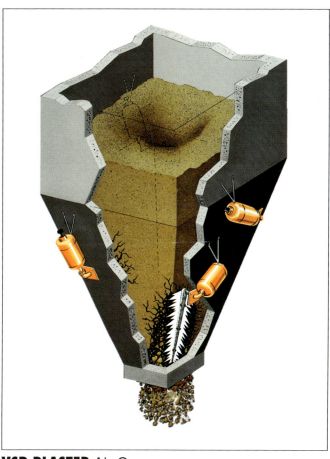

VSR BLASTER Air Cannons
pneumatical removing of cloggings and cakings

VSR BLASTER sword nozzle, operation example

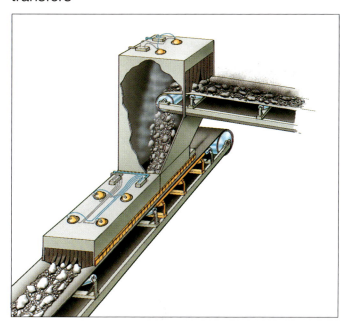

DUSTEX® Water Dispersion System
for suppressing of industrial dust and gas cooling

 **INDUSTRIETECHNIK GMBH**

D-45422 Postfach 10 22 22
D-45472 Hingbergstr. 319
Mülheim an der Ruhr
Tel. 02 08 - 43 46 2 - 22
Fax. 02 08 - 43 46 2 - 62

UMWELT-ENGINEERING GMBH

Professional components and systems for bulk technology

The company supplies special equipment for the storing and conveying technique, either as key components together with the necessary engineering or as complete units. They are increasingly applied in the environmental and recycling technique for waste paper, sewage, sludge, filter cake, building rumble, FGD gypsum, garbage, compost, old tires, fly ash, wooden products, refuse glass.

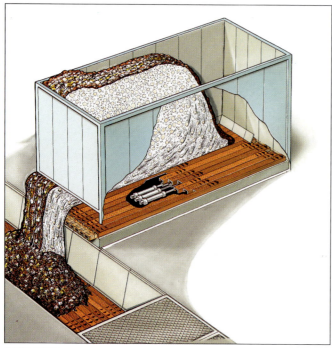

HOREX® Shuttle Floor
for the stationary and mobile, large area conveying of bulk goods

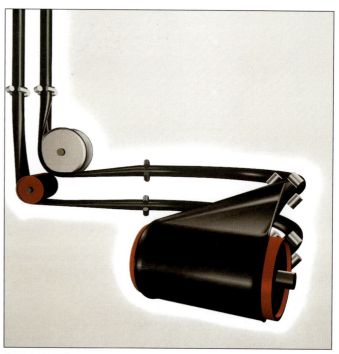

RONDEX® Hose Belt Conveyor
for encapsulated, curved and steep inclining conveying

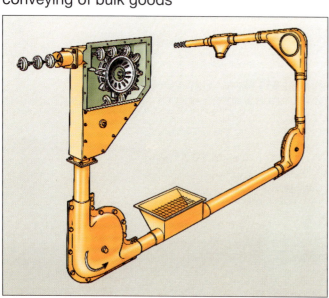

COBREX® Disk Chain Conveyor
for the bulk good transport in complex works

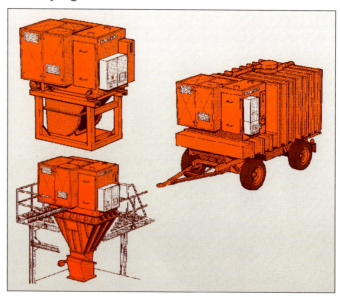

TRANSVAC® Suction Units
for powerful industrial cleaning and bulk good conveying

UMWELT-ENGINEERING GMBH

D-45422 Postfach 10 22 22
D-45472 Hingbergstr. 319
Mülheim an der Ruhr
Tel.   02 08 - 43 46 24 - 0
Fax.   02 08 -   43 60 77

dans des limites acceptables. Les essais réalisés avec l'appareil de mesure de ségrégation à faible échelle ont prouvé que l'on disposait d'un outil utile pour la transformation des silos et appareils d'alimentation afin d'obtenir un débit massique constant, et modifier en conséquence le mode de fonctionnement.

Mezcla y dosificación de clínker – efecto de segregación y remedios

Resumen – *Las partículas esféricas de clínker, de diferente distribución granulométrica, presentan una fuerte tendencia a la segregación en los silos y tolvas, la cual puede conducir a fluctuaciones de calidad. Se puede decir lo mismo de las mezclas de clínker de diferente calidad y granulometría. Mediante un aparato de control de la segregación, de construcción especial, que funciona en pequeña escala y que fue desarrollado por POSTEC Research, se llevaron a cabo ensayos provisionales. A continuación, se hicieron ensayos en gran escala, en un molino de cemento, con una tolva de clínker tradicional, de forma cilíndrica. Los resultados de los ensayos mostraron, en su conjunto, con bastante claridad que la segregación del clínker se presenta de tal forma que puede conducir a una merma de calidad, debido al elevado contenido de álcalis en las fracciones gruesas y al reducido contenido de álcalis en las fracciones finas. Esto se hizo evidente al vaciarse la tolva a modo de embudo, comprobándose un constante aumento del contenido de álcalis. Incluso con una tolva casi llena de clínker, las variaciones del contenido de álcalis eran demasiado elevadas. Por esta razón, se decidió instalar una tolva adicional para la dosificación de un segundo clínker, con el fin de mantener constante la relación de mezcla de los diferentes clínkeres. Si, además, las tolvas tenían un alto nivel de llenado, se podían mantener las variaciones del contenido de álcalis dentro de unos límites aceptables. Los ensayos llevados a cabo mediante el aparato de control de segregación, en pequeña escala, resultaron ser un instrumento útil para la transformación de silos y aparatos de alimentación, con el fin de conseguir un caudal de masas constante, introduciendo las modificaciones adecuadas.*

1. Einführung

Die Zementhersteller unternehmen bei der Kontrolle der Qualität (z.B. der Zusammensetzung) des Rohmehls und des Klinkers große Anstrengungen. Größere Qualitätsschwankungen können jedoch in Klinkerlagern/-silos oder Klinkerbunkern durch Entmischung (Segregation) auftreten. **Bild 1** zeigt das Prinzip der Haufensegregation von körnigen Materialien, wie z.B. Klinker, während der mittigen Füllung eines Silos sowie seiner Entleerung im Kernfluß. Das Feinmaterial verläßt das Silo als erstes, wodurch es im Hinblick auf die grobkörnige Fraktion zu einer Entmischung kommt [1]. Bei einem Silo mit gleichbleibendem Massedurchsatz werden die entmischten Fraktionen demgegenüber während der Entleerung erneut vermischt, da es über den gesamten Siloquerschnitt zu einem gleichmäßigen Durchsatz kommt.

Bei unterschiedlicher Mahlbarkeit und chemischer Zusammensetzung der feinkörnigen und der grobkörnigen Fraktion, die nicht ungewöhnlich sind, können aufgrund einer Entmischung in den Kernflußsilos Störungen des Verfahrensablaufs und eine Beeinträchtigung der Qualität auftreten.

1. Introduction

Cement producers put much effort into controlling the quality (e.g. the composition) of the raw meal and the clinker. Increased quality variations may, however, be introduced in the clinker stocks/silos or clinker hoppers by segregation.

Fig. 1 shows the principles of heap segregation of granulous materials like clinker during central filling of a silo and emptying it in a funnel flow mode. The fine material will leave the silo first causing a separation from the coarse material [1]. In a mass flow silo, however, the segregation fractions are remixed during emptying since even flow occurs in the whole cross section of the silo.

When the fine and coarse material have different grindability and chemical composition, which is not uncommon, disturbances in process operation and quality may be introduced by segregation in funnel flow silos.

This study was performed for the purpose of producing a new cement quality by mixing two types of clinker, standard clinker and reduced alkali clinker.

2. Clinker Composition

	LSF	C_3A	Alkali Na_2O-eq.	Sieve analysis	
				–1 mm	–25 mm
1. Standard clinker	96.2	7.9	1.18	3%	78%
2. Reduced alkali clinker	89.2	5.9	0.65	51%	90%

Chemical analyses of different size fractions of the two types of clinker showed that

— the standard clinker had a homogeneous composition regardless of size.

— In the low alkali clinker the alkali content in the coarsest particles was twice as high as in the particles less than 1 mm.

The alkali content was therefore chosen as the key parameter for monitoring the tendency towards segregation.

3. Segregation tests

Preliminary tests were performed in a specially designed small scale segregation tester developed by POSTEC Research [2]. Subsequent tests were done in a full scale cement mill with a traditional cylindrical clinker hopper.

 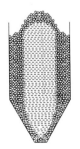 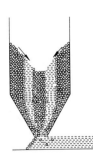

a. Filling of silo.   b. Full silo.   c. Emptying in funnel flow mode.

BILD 1: Grundsätze der Haufenentmischung und der Entleerung im Kernfluß
FIGURE 1: Principles for heap seggregation and emptying in funnel flow mode

a. Silofüllung = filling of silo
b. Gefülltes Silo = full silo
c. Entleerung im Kernfluß = emptying in funnel flow mode

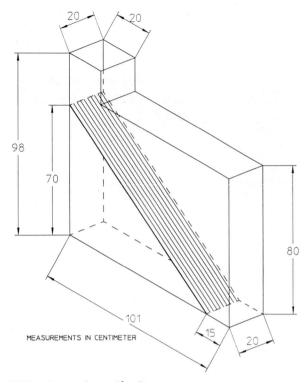

BILD 2: Segregationsprüfgerät
FIGURE 2: Segregation tester
Maße in cm   = Measurements in centimeter

Die vorliegende Studie wurde durchgeführt, um durch Mischen von zwei Klinkertypen — Normalklinker und alkaliarmer Klinker — eine neue Zementqualität herzustellen.

2. Klinkerzusammensetzung

	KSG	C_3A	Alkali Na_2O-Äq.	Siebanalyse -1 mm	-25 mm
1. Normalklinker	96,2	7,9	1,18	3 %	78 %
2. Alkaliarmer Klinker	89,2	5,9	0,65	51 %	90 %

Chemische Analysen verschiedener Korngrößenfraktionen der beiden Klinkertypen ergaben, daß

— der Normalklinker korngrößenunabhängig homogen zusammengesetzt war;

— bei dem alkaliarmen Klinker der Alkaligehalt der gröbsten Teilchen doppelt so hoch war wie bei den Teilchen < 1 mm.

Darum wurde der Alkaligehalt als Hauptparameter für die Überwachung der Segregationsneigung gewählt.

3. Segregationsprüfungen

Vorversuche wurden mit einem von POSTEC Research eigens entwickelten kleinen Segregationsprüfgerät vorgenommen [2]. Die späteren Prüfungen fanden in einer normal großen Zementmühle mit einem herkömmlichen zylindrischen Klinkerbunker statt.

3.1 Kleines Segregationsprüfgerät für Versuche im Labormaßstab

Das in **Bild 2** gezeigte Gerät ist eine modifizierte Ausführung eines Prüfgeräts, das zur Ermittlung der Haufenentmischung wie auch der luftinduzierten Entmischung von Pulvern und körnigen Materialien verwendet wird, die etwas feiner als Klinker sind [1].

Zur Verringerung des Materialbedarfs wurde das Prüfgerät mit einer Platte ausgerüstet, die einen angenommenen Böschungswinkel von ca. 40° aufwies. Rund 90 kg Klinker eines gründlich durchgearbeiteten Gemischs von 50 % alkaliarmem Klinker und 50 % Normalklinker wurden langsam

3.1 Small scale segregation tester

The tester which is shown in **Fig. 2** is a modification of a tester used for determining both heap segregation of powders and granulous materials somewhat finer than clinker [1]. To reduce the quality of material needed, the tester was equipped with a plate having the supposed angle of repose of about 40°. About 90 kg clinker consisting of a through mix of 50 % low alkali and 50 % standard clinker was loaded slowly through a moveable tube at the top of the segregation tester.

Fig. 3 shows the situation when all the material has come to rest. The clinker was fixed in position, then the tester was laid flat, opened, and the content cut in six parts by forcing in thin steel plates. Sieve analyses were performed (**Fig. 4**) and all the material in the six fractions was crushed and analysed. The alkali content is shown in **Fig. 5**.

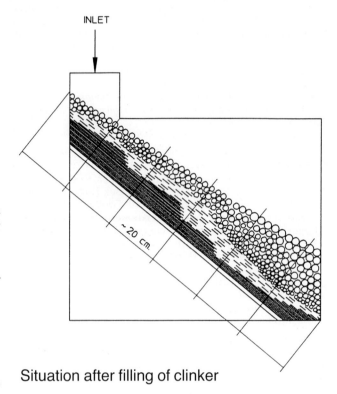

Situation after filling of clinker

BILD 3: Vermischen und Dosieren von Klinker
FIGURE 3: Blending and dosage of clinker
Einlaßöffnung   = inlet
Lage nach dem Einfüllen von Klinker = Situation after filling of clinker

3.2 Full scale production

Cement mill no. 7 of Norcem Dalen Works consisting of a roller press and a ball mill in series was used for the tests. (F. L. Smidth)

Roller press:	1 200 kW
Ball mill:	4 000 kW
Mill capacity at Blaine 330 m²/kg:	200 tons/h
Clinker hopper capacity (only one at the time):	600 tons

The clinker hopper, which is of conventional cylindrical design, was filled from the top with 50/50 low alkali/standard clinker by simultaneous weighing from the respective clinker storage silos.

A. The mill was run emptying the clinker hopper completely without refilling.

Cement was sampled every 10 minutes. These samples were tested for chemical composition and some were tested for standard strength.

The alkali content is given in **Fig. 6**, curve A.

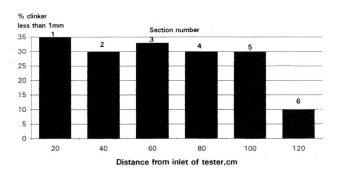

BILD 4: Siebanalyse des Segregationsprüfgeräts
FIGURE 4: Sieve analysis from segregation tester

% Klinker unter 1 mm	= % clinker less than 1 mm
Abschnittsnummer	= section numer
Entfernung zur Einlaßöffnung des Prüfgeräts in cm	= Distance from inlet of tester, cm

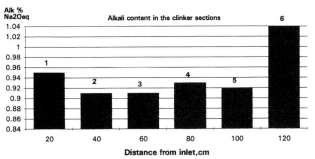

BILD 5: Ergebnisse der Segregationsprüfung
FIGURE 5: Results from segregation tester

% Alkali	= Alk %
Na₂O-Äq.	= Na₂ O eq
Alkaligehalt der Klinkerabschnitte	= alkali content in the clinker sections
Entfernung zur Einlaßöffnung in cm	= distance from inlet, cm

durch ein bewegliches Rohr oben auf dem Segregationsprüfgerät eingefüllt. **Bild 3** zeigt die Verhältnisse, nachdem das gesamte Material zur Ruhe gekommen ist. Der Klinker wurde in seiner Lage fixiert. Dann wurde das Prüfgerät flach hingelegt, geöffnet und der Inhalt durch Hineindrükken dünner Stahlplatten in sechs Teile zerschnitten.

Siebanalysen wurden vorgenommen (**Bild 4**), und das gesamte Material der sechs Fraktionen wurde zerkleinert und analysiert. Der Alkaligehalt ist dem **Bild 5** zu entnehmen.

3.2 Einsatz im industriellen Maßstab

Für die Prüfungen wurde die Zementmühle Nr. 7 der Norcem Dalen-Zementwerke, die aus einer Rollenpresse und einer Kugelmühle (F.L. Smidth) bestand, die in Reihe angeordnet sind, verwendet.

Rollenpresse:	1 200 kW
Kugelmühle:	4 000 kW
Mühlenkapazität bei einem Blaine-Wert von 330 m²/kg:	200 t/h
Klinkersilokapazität:	600 t

Das Klinkersilo in herkömmlicher zylindrischer Bauart wurde von oben durch gleichzeitiges Zuwiegen aus den entsprechenden Klinkersilos jeweils zur Hälfte mit alkaliarmem/normalem Klinker gefüllt.

A. Die Mühle wurde mit völliger Entleerung des Klinkersilos ohne Nachfüllen betrieben.

Alle 10 Minuten wurden Zementproben genommen.

Diese Proben wurden auf ihre chemische Zusammensetzung untersucht und zum Teil auch auf Festigkeit geprüft.

Der Alkaligehalt ist **Bild 6**, Kurve A, zu entnehmen.

B. Die Mühle lief rund 11 Stunden, wobei das Klinkersilo die ganze Zeit über fast ganz gefüllt blieb (60–90 %).

Zementproben wurden alle 10 Minuten genommen.

Der Alkaligehalt dieser Proben ergibt sich aus Bild 6, Kurve B.

4. Diskussion und Schlußbemerkungen

Die Ergebnisse der Segregationsprüfung zeigen eindeutig, daß sich der Klinker auf eine Art und Weise entmischt, die die Qualität beeinträchtigen könnte.

Die großtechnische Prüfung ergab, daß das Klinkersilo eindeutig im Kernfluß entleert wurde, woraus ein stetig zunehmender Alkaligehalt resultiert. Selbst bei Betrieb mit einem fast ganz gefüllten Klinkersilo wurden die Schwankungen des Alkaligehalts als unannehmbar hoch bewertet.

Deshalb wurde beschlossen, ein zusätzliches Silo für die Dosierung eines zweiten Klinkers zu installieren, um das Mischungsverhältnis der verschiedenen Klinker konstant

B. The mill was run for a period of about 11 hours, all the time keeping the clinker hopper almost full (60–90 %). Sampling of cement was carried out every 10 minutes.

The alkali content of these samples is shown in Fig. 6, curve B.

4. Discussion and concluding remarks

The results from the segregation tester clearly show that the clinker segregates in a way that could effect the quality adversely.

The production test showed that it is obvious that the clinker hopper is emptied in the funnel flow mode, giving a steadily increasing alkali content. Even when operating with a nearly full clinker hopper, the variations in the alkali content is considered too high to be acceptable.

BILD 6: Prüfvermahlung
FIGURE 6: Production test grinding

Na₂O-Äq.	= Na₂O eq
Zementmühle Nr.7	= Cement Mill No. 7
Zeit (Stunden)	= Time, hours
Prüfung A: Sich entleerendes Silo	= Test A: emptying hopper
Prüfung B: Volles Silo	= Test B: full hopper

zu halten. Außerdem müssen die Silos während des Betriebs gut gefüllt bleiben. Dann bleiben die Schwankungen des Alkaligehalts innerhalb vertretbarer Grenzen.

Die durchgeführten Prüfungen zeigen, daß die Segregationsneigung und die entsprechenden Auswirkungen auf die Qualität bei Versuchen in kleinem Maßstab mit einfachen Geräten belegt werden können. An diese Versuche können sich Prüfungen und Messungen im Rahmen des vollen Herstellungsbetriebs anschließen. Sollte sich ein Beitrag der Segregation zu den Qualitätsschwankungen ergeben, sind als Gegenmaßnahme die Silos und die Aufgabegeräte so neuzugestalten, daß sie einen gleichbleibenden Massendurchsatz ermöglichen, und/oder die Betriebsweise ist entsprechend zu ändern.

Es wird auf [3], [4] und [5] verwiesen.

It was therefore decided to install an extra hopper for the dosage of a second clinker in order to keep the mix ratio of the different clinker constant. Also the hoppers are to be kept well filled during operation. The variations in the alkali content are then kept within satisfactory limits.

The tests performed show that the tendency to segregation and the resulting effect on quality may be demonstrated in small scale experiments using simple equipment. Such experiments may be followed by tests and measurements in full scale production. If segregation is found to contribute to quality variations one should counter it by redesigning silos and feeding equipment to give a constant mass flow and/or change operational modes.

Reference is given to [3], [4] and [5].

Literature

[1] Enstad, G.G., Shinohara, K., and Johanson, S.T.: Prediction of Segregation During Filling of Silos – An Outline of Previous Work and a Programme for Future Work, POSTEC-Report No. 911301-1, November 1991.

[2] Mosby, J.: Segregation – POSTEC-Newsletter No. 10, April 1992, pp. 23–25.

[3] Carson, J.W., Royal, T.A., and Goodwill, D.A.: Understanding and Eliminating Particle Segregation Problems, Bulk Solids Handling, Vol. 6, No. 1, February 1986, pp. 139-144.

[4] Johanson, J.R.: Solids Segregation – Case Histories and Solutions, Bulk Solids Handling, Vol. 7, No. 2, April 1987, pp. 205–208.

[5] Johanson, J.R.: Causes and Solutions, Powder and Bulk Eng., August 1988, pp. 13–19.

Mischtechnik im Werk II der Anneliese Zementwerke AG, Ennigerloh

Blending technology in Works II of Anneliese Zementwerke AG, Ennigerloh

Technique de mélange à l'usine II de la Anneliese Zementwerke AG, Ennigerloh

Técnica de mezcla en la factoría II de Anneliese Zementwerke AG, de Ennigerloh

Von **B. Thier**, Ennigerloh/Deutschland

Zusammenfassung – Seit 1972/73 wurde das Werk II der Anneliese Zementwerke AG in Ennigerloh systematisch von einem Zementwerk mit Klinkerproduktion auf ein reines Mahl- und Mischwerk mit modernster Technologie umgestellt. Dementsprechend wurden als neue Anlagenteile ein Durchlaufmischer (Lödige), eine Chargenmischanlage (Eirich) und ein Mehrkammersilo mit integrierter Mischanlage in Betrieb genommen. Der Durchlaufmischer ist auf eine Mischleistung von 240 t/h, direkt beschickt von einer Rohmühle und angeschlossen über Dosieranlagen an Zement- und Flugaschesilos, ausgelegt. Die Chargenmischeranlage für körnige und pulverige Sonderprodukte besteht aus 10 Großkomponenten- und 4 Kleinkomponentensilos mit je 2 Behälterwaagen. Die Mischerleistung beträgt abhängig von der Schüttdichte des Fertigproduktes 75 – 120 t/h. Die Komponentensilos sind an Zement-, Flugasche- und Schottersiloanlagen angeschlossen und können aus Silozügen beschickt werden. Die Fertigprodukte werden direkt vom Mischer sowie über 7 Vorratssilos lose verladen oder über 3 Packmaschinen abgesackt und palettiert zum Versand gebracht. Das Mehrkammersilo mit integrierter Mischanlage hat einen Durchmesser von 20 m und eine Gesamthöhe von 61 m. Es besteht aus einem Innensilo und einem Ringsilo mit 8 Kammern und verfügt über 3 Verladestraßen. Über die außenliegenden Verladestraßen kann aus jeder Kammer verladen werden. Die mittlere Verladestraße wird direkt vom Mischer (160 t/h) bedient. Jede Kammer kann als Vorratssilo oder/und als Komponentensilo benutzt werden. Die Beschickung erfolgt von Zement- und Kalksteinmahlanlagen, vom Mischer und aus Silofahrzeugen. Als Siloabzug dient das bewährte IBAU-System mit Zentralkegel. Freiprogrammierbare Steuerungen in Verbindung mit dem Leitstandsystem COROS sowie EDV-Anlagen für Dosierung und Versand ermöglichen die Bedienung und Disposition dieser Anlage von nur einem Mitarbeiter.

Summary – Works II of the Anneliese Zementwerke AG in Ennigerloh has been systematically changed since 1972/73 from a cement works with clinker production to a pure grinding and mixing works using very modern technology. A continuous mixer (Lödige), a batch mixing system (Eirich) and a multi-chamber silo with integral blending plant were brought into operation as new parts of the plant. The continuous mixer designed for a mixing capacity of 240 t/h is fed directly from a tube mill, and connected via metering systems to cement and fly ash silos. The batch mixing system for special granular and powdered products consists of 10 silos for main components and 4 silos for minor components, each with 2 hopper weighers. The mixing capacity is 75–120 t/h depending on the bulk density of the finished product. The component silos are connected to cement, fly ash, and crushed stone silo systems and can be fed from bulk container trains. The finished products are loaded in bulk either directly from the mixer or via 7 storage silos, or are packed in 3 bagging machines and palletized for despatch. The multi-chamber silo with integral mixing plant has a diameter of 20 m and a total height of 61 m. It consists of one inner silo and a ring silo with 8 chambers, and has three loading lanes. Loading in the outer loading lanes can be carried out from any chamber. The central loading lane is served directly from the mixer (160 t/h). Each chamber can be used as a storage silo or/and as a component silo. They are fed from the cement and limestone grinding plants, from the mixer, and from bulk tanker vehicles. The well-tried IBAU system with central cone is used for the silo discharge. Programmable control systems used in conjunction with the COROS control room system and EDP systems for metering and despatch, mean that this plant can be operated and managed by only one employee.

Résumé – Depuis 1972/73, l'usine II de la Anneliese Zementwerke AG à Ennigerloh a été transformée systématiquement, d'une cimenterie avec production de clinker, en une usine purement de broyage et de mélange de la technologie la plus moderne. Par conséquent, ont été mis en service, comme nouveaux équipements de l'installation, un mélangeur en continu (Lödige), un mélangeur par charges (Eirich) et un silo à compartiments multiples avec installation de mélange intégrée. Le mélangeur en continu, directement alimenté par un tube broyeur et connecté par des dispositifs de dosage à des silos à ciment et cendres volantes, est proportionné pour une capacité de mélange

de 240 t/h. L'atelier de mélange par charges, pour les produits spéciaux granulaires et pulvérulents, se compose de 10 silos à composants principaux et 4 silos à composants secondaires, avec respectivement 2 bascules à trémie. En fonction du poids en vrac du produit fini, la capacité du mélangeur est de 75 à 120 t/h. Les silos à composants sont connectés avec des installations de silos à ciment, à cendres volantes et à granulats. Les produits finis sont, soit directement du mélangeur, soit au moyen de 7 silos de réserve, expédiés en vrac ou ensachés à l'aide de 3 machines d'emballage et expédiés sur palette. Le silo à plusieurs compartiments avec atelier de mélange intégré a un diamètre de 20 m et une hauteur totale de 61 m. Il se compose d'un silo central et d'un silo annulaire à 8 chambres et dispose de 3 voies de chargement. Les voies de chargement extérieures permettent de charger à partir de chaque chambre. La voie de chargement au milieu est servie directement par le mélangeur (160 t/h). Chaque chambre peut servir, soit de silo de réserve, soit de silo à composants. L'alimentation s'effectue à partir d'ateliers de broyage de ciment et de farine calcaire, à partir du mélangeur et par camion-silo. Comme reprise au silo sert le système confirmé IBAU avec cône central. Des commandes librement programmables, en liaison avec le système de poste de conduite COROS ainsi qu'avec des équipements d'informatique pour le dosage et l'expédition, permettent la conduite et le service de cette installation, par un seul collaborateur.*

Resumen – *Desde 1972/73, la factoría II de Anneliese Zementwerke AG, de Ennigerloh, que era una fábrica de cemento con producción de clínker, se ha venido transformando sistemáticamente en una fábrica de molienda y mezclado, de modernísima tecnología. Para ello, se pusieron en servicio, como elementos nuevos de la instalación, un mezclador de paso continuo (Lödige), una instalación de mezcla por lotes sucesivos (Eirich) y un silo de varias cámaras, con instalación de mezcla integrada. El mezclador de paso está dimensionado para un rendimiento de mezcla de 240/h, y es alimentado directamente desde un molino de crudo y conectado con silos de cemento y de cenizas volantes, a través de instalaciones de dosificación. La instalación de mezcla por lotes sucesivos, prevista para productos especiales en forma de granos y de polvo, se compone de 10 silos para componentes principales y 4 silos para componentes secundarios, cada uno con 2 básculas de tolva. La capacidad de mezcla es de 75 – 120 t/h, dependiente de la masa volúmica aparente del producto acabado. Los silos para componentes están conectados con silos para cemento, cenizas volantes y grancillas, pudiendo ser alimentados mediante trenes de carga de silos. Los productos acabados se cargan a granel, directamente desde el mezclador y 7 silos de almacenamiento o bien son envasados por medio de 3 máquinas ensacadoras y paletizados para su expedición. El silo de varias cámaras, con instalación de mezcla integrada, tiene un diámetro de 20 m y una altura total de 61 m. Está integrado por un silo interno y otro silo anular, de 8 cámaras, y dispone de 3 vías de carga. Las vías de carga exteriores permiten la carga a partir de cualquier cámara. La vía de carga central recibe el material directamente del mezclador (160 t/h). Cada cámara puede utilizarse como silo de reserva o/y como silo de componentes. La alimentación tiene lugar a partir de instalaciones de molienda de cemento y de caliza, a partir del mezclador así como de camiones-cisterna. Como dispositivo de extracción de silos se emplea el experimentado sistema IBAU, de cono central. Los controladores libremente programables, junto con el sistema COROS, para centro de control, y con equipos informáticos para dosificación y expedición, permiten que una sola persona se encargue de la operación y manejo de esta instalación.*

Técnica de mezcla en la factoría II de Anneliese Zementwerke AG, de Ennigerloh

1. Einleitung

Seit 1972/73 wurde das Werk II der ANNELIESE Zementwerke AG in Ennigerloh systematisch von einem Zementwerk mit Klinkerproduktion auf ein reines Mahl- und Mischwerk mit modernster Technologie umgestellt. **Bild 1** zeigt den Lageplan des Werks.

2. Mischanlage 1

1974 wurde ein Durchlaufmischer der Firma Lödige, Paderborn, mit einer Durchsatzleistung von 240 t/h und einer Antriebsleistung von 75 kW für pulverige Produkte in Betrieb genommen. Der Mischer, in der Zementsiloanlage aufgebaut, ist mit drei Vorratssilos für Fertigprodukte mit einer Lagerkapazität von je 2000 t verbunden. Von einer Mahltrocknungsanlage wird er direkt mit Kalksteinmehlen beschickt. Über eine Airliftanlage, angeschlossen an zwei Zementvorratssilos mit je 2000 t Inhalt und einer nachgeschalteten Dosierbandwaage, wird die Zementkomponente aufgegeben. Flugasche wird aus vier Vorratssilos mit einem Gesamtinhalt von 11000 m³ über eine Großbehälterdosieranlage, System Simplex der Firma Schenck, zudosiert. Die Dosierbandwaagen der Mahltrocknungsanlage für Kalksteinmehl und für Zement sowie die Dosieranlage für Flugasche werden über eine Gemengeregelung mit Rezeptvorwahl gesteuert.

1. Introduction

Since 1972–73 Works II of the Anneliese Zementwerke AG in Ennigerloh has been systematically converted from a cement works with clinker production to a pure grinding and mixing works using the most modern technology. The works layout plan is shown in **Fig. 1**.

2. Mixing plant 1

A continuous mixer from the firm of Lödige, Paderborn, with a throughput capacity of 240 t/h and a drive rating of 75 kW for powder products was brought into operation in 1974. The mixer is installed in the cement silo plant, and is connected to 3 storage silos for finished products each with a storage capacity of 2000 t. It is fed directly with limestone meal from a drying and grinding plant. The cement components are supplied by an airlift plant connected to two cement storage silos, each with a capacity of 2000 t and a downstream weighbelt feeder. Fly ash is metered from four storage silos with a total capacity of 11 000 m³ through a large Simplex hopper metering system from the firm of Schenck. The weighbelt feeders for the drying and grinding plant for limestone meal and for cement, and the metering system for fly ash, are controlled by a mix control system with preselection of the mix formulation.

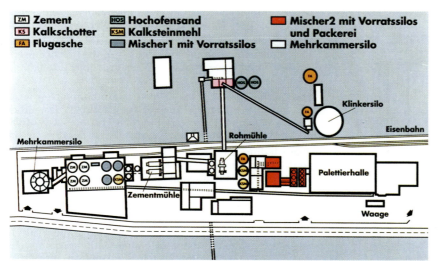

BILD 1: Lageplan
FIGURE 1: Layout plan

Zement	= cement
Kalkschotter	= crushed limestone
Flugasche	= fly ash
Hochofensand	= granulated blast furnace slag
Kalksteinmehl	= limestone meal
Mischer 1 mit Vorratssilos	= mixer 1 with storage silos
Mischer 2 mit Vorratssilos und	= mixer 2 with storage silos and packing plant
Mehrkammersilo	= multi-compartment silo
Zementmühle	= cement mill
Rohmühle	= raw mill
Klinkersilo	= clinker silo
Eisenbahn	= railway
Palettierhalle	= palletizing hall
Waage	= weighbridge

Mit dem in **Bild 2** dargestellten Durchlaufmischer werden Massenprodukte äußerst wirtschaftlich hergestellt. Die variable Anpassung der Produktzusammensetzung und die Erweiterung der Rezepturen sind nur begrenzt möglich.

3. Mischanlage 2

Im Jahr 1982 mußten die stillgelegten Klinkerproduktionsanlagen des Werkes II abgebrochen werden, um Platz zu schaffen für eine Misch- und Palettieranlage.

1983 wurde auf diesem Gelände eine Chargenmischanlage der Firma Eirich für körnige und pulverige Fertigprodukte aufgebaut und in Betrieb genommen. Die Anlage besteht aus zehn Großkomponentensilos mit je 150 m^3 Inhalt, die aus Silokesselfahrzeugen direkt beschickt werden. Sie sind außerdem mit vorhandenen Vorratssiloanlagen für Zement, Flugasche, Kalksteinmehl und Schotter über pneumatische oder mechanische Transporteinrichtungen verbunden.

Die Großkomponenten werden über zwei Behälterwaagen mit je 5 000 kg Inhalt zum Mischer eingewogen.

Vier Kleinkomponentensilos mit je 30 m^3 Inhalt, ebenfalls aus Silokesselfahrzeugen beschickt, dosieren die Zuschlagstoffe über eine 100 kg und eine 500 kg Behälterwaage zum Mischer. Weitere Zugaben erfolgen aus zwei Containern mit je 1 m^3 Inhalt über eine 50 kg Behälterwaage.

Die Leistung des Eirich-Intensiv-Mischers (Bild 2) beträgt, abhängig vom Schüttgewicht des Fertigproduktes, 75 t/h bis 120 t/h. Über den Mischteller, zwei Mischsterne und den Wirbler werden 142 kW als Mischenergie eingeleitet.

Die Fertigprodukte werden über mechanische Zwischentransporte entweder direkt oder über sieben Vorratssilos mit je 350 m^3 Inhalt und zwei zugehörigen Loseverladeeinrichtungen in Silokesselfahrzeugen versandt.

Drei Reihenpacker, je einer für körnige, sandige und pulverige Produkte, sind an dieser Mischanlage 2 angeschlossen. Die Sackware wird über die Palettierhalle mit zwei Palettierern verladen.

Bei dieser Mischanlage traten anfangs Vermischungen körniger und pulveriger Produkte auf, da auf der Fertiggutseite nur ein mechanischer Transportweg vorgesehen war. Für pulverige Fertigprodukte wurde ein separater pneumatischer Transport und vor allen Verladeeinrichtungen eine Absiebung auf eine Körnung < 2 mm nachgerüstet. Damit war das Vermischungsproblem gelöst.

4. Mischanlage 3

4.1 Allgemein

Zur Entlastung des Mischers 2 und in Hinsicht auf die Euro-Norm 197, Teil 1, wurde Anfang des Jahres 1990 beschlossen, im Werk II eine weitere Mischanlage für pulverige Produkte aufzubauen.

Bulk products are produced exceptionally economically with the continuous mixer shown in **Fig. 2**. Variations to the product composition and extensions to the mix formulations are only possible within limits.

3. Mixing plant 2

The clinker production plants of Works II, which had been closed down, had to be demolished in 1982 to make way for a mixing and palletizing plant.

A batch mixing plant from the firm of Eirich for granular and powdered finished products was built on this site and brought into operation in 1983. This plant consists of 10 silos for the main components, each with a capacity of 150 m^3, which are fed directly from bulk tanker vehicles. They are also connected by pneumatic or mechanical transport equipment to the existing storage silo plants for cement, fly ash, limestone meal and crushed stone.

The main components are weighed into the mixer via two hopper weighers, each with a capacity of 5 000 kg.

Four silos for minor components, each with a capacity of 30 m^3 and also filled from bulk tanker vehicles, meter the aggregates to the mixer via a 100 kg and a 500 kg hopper weigher.

BILD 2: Mischer 2
Eirich-Intensiv-Mischer 100 t/h
FIGURE 2: Mixer 2
Eirich intensive mixer 100 t/h

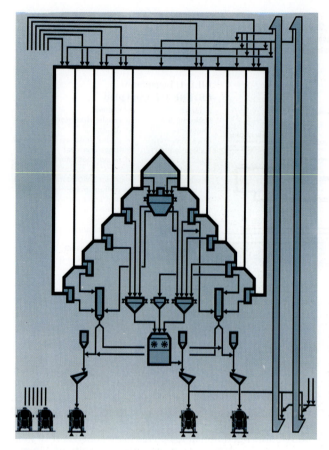

BILD 3: Fließschema Mehrkammersilo
FIGURE 3: Flow sheet – multi-compartment silo

Der Bau eines Mehrkammersilos mit integrierter Mischanlage und zugehörigen Loseversandanlagen (siehe **Bild 3**) wurde beschlossen.

4.2 Anlagenbeschreibung

Das Fließschema (Bild 3) zeigt vereinfacht die Misch-, Versand- und Beschickungsanlagen des IBAU-Mehrkammersilos.

Für die Bevorratung von mindestens fünf Großkomponenten wie z. B. Kalksteinmehl, Flugaschen, Hochofensand und Zemente sowie von drei bis vier Fertigprodukten wurde ein Mehrkammersilo mit neun Zellen gewählt.

Das ausgeführte neunzellige Mehrkammersilo hat einen Durchmesser von 20,6 Metern und eine Gesamthöhe von 61 Metern. Es besteht aus einem Innensilo mit einem lichten Durchmesser von 9,5 Metern und einem nutzbaren Inhalt von 1600 m³ sowie einem, in acht Kammern aufgeteilten, Ringsilo mit einem Außendurchmesser von 20 Metern und einem Innendurchmesser von 10,1 Metern. Jede Kammer hat ein nutzbares Volumen von 800 m³.

Der gesamte Siloboden des Ring- und Innensilos ist nach dem bekannten IBAU-Zentralkegelsystem ausgebildet.

Die Rezepturen für die Mischanlage werden in drei Behälterwaagen zusammengestellt. Zwei Behälterwaagen haben ein Fassungsvermögen von je 6 m³. Aus sechs Kammern werden beide Behälterwaagen beschickt. Die dritte Behälterwaage mit einer Kapazität von 1 m³ ist an einer Kammer angeschlossen.

Der IBAU-Doppelwellenzwangsmischer, System BHS Sonthofen (**Bild 4**), mit einem Chargenvolumen von 5,6 m³ erreicht bei 27 Mischungen pro Stunde eine Leistung von 140 m³/h.

Die über beide Wellen eingeleitete Mischenergie beträgt 100 kW. Über ein hydraulisches, wasserdicht schließendes Klappensystem kann der Mischer restlos entleert werden.

Other additions are made from two containers, each with a capacity of 1 m³, via a 50 kg hopper weigher.

The capacity of the Eirich intensive mixer (Fig. 2) is 75 t/h to 120 t/h depending on the bulk density of the finished product. The mixing power of 142 kW is introduced through the mixing pan, two mixing stars, and the swirler.

The finished products are despatched in bulk tanker vehicles via mechanical intermediate transport systems, either directly or via seven storage silos, each with a capacity of 350 m³, and two associated bulk loading units.

Three in-line packers, one each for granular, sandy, and powdered products, are connected to this mixing plant 2. The bagged products are loaded via the palletizing hall with two palletizers.

Intermixing of granular and powdered products occurred initially with this mixing plant because only one mechanical transport route was provided on the finished product side. A separate pneumatic transport system with screening to a particle size < 2 mm before each loading unit was retrofitted for powdered finished products. This solved the intermixing problem.

4. Mixing plant 3

4.1 General

To relieve the load on mixer 2 and in the light of Euro Standard 197, Part I, the decision was made at the start of 1990 to build a further mixing plant for powdered products in Works II.

It was decided to build a multi-compartment silo with integral mixing plant and associated bulk loading systems (see **Fig. 3**).

4.2 Description of the plant

The mixing, despatch, and feed systems of the IBAU multi-compartment silo are shown in simplified form in the flow diagram (Fig. 3).

A multi-compartment silo with 9 cells was chosen for storing at least 5 main components, such as limestone meal, fly ash, granulated blast furnace slag, and cements, as well as three to four finished products.

The completed nine-cell multi-compartment silo has a diameter of 20.6 metres and a total height of 61 metres. It consists of one inner silo with a clear diameter of 9.5 metres and an effective capacity of 1600 m³ and a ring silo with an external diameter of 20 metres and an internal diameter of 10.1 metres subdivided into 8 compartments. Each compartment has an effective volume of 800 m³.

The entire silo base of the ring and inner silos is designed on the well known IBAU central cone system.

The mix formulations for the mixing plant are assembled in three hopper weighers. Two of the hopper weighers have capacities of 6 m³, and both hopper weighers are fed from 6 compartments. The third hopper weigher with a capacity of 1 m³ is connected to one compartment.

The IBAU BHS Sonthofen twin-shaft positive mixer (**Fig. 4**) with a batch volume of 5.6 m³ achieves a rate of 140 m³/h with 27 mixes per hour.

The mixing power introduced through the two shafts is 100 kW. The mixer can be fully emptied using a hydraulic, watertight, flap closing system.

One special feature of this plant is the loading of the finished product directly into the bulk tanker vehicle from the container under the mixer. This takes place on a road vehicle weighbridge in the central drive-through lane. Direct loading from the mixer avoids high cleaning costs and the risk of intermixing when changing over to sensitive products.

The two north and south loading lanes, also equipped with road vehicle weighbridges, are connected to 7 compartments via bulk loading units and separate pneumatic trough conveyors. The north and south loading units each have a capacity of 200 t/h to 250 t/h.

BILD 4: Mischer 3
Doppelwellenzwangsmischer 150 t/h
FIGURE 4: Mixer 3
twin-shaft positive mixer 150 t/h

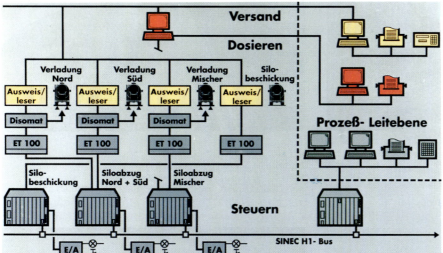

BILD 5: Automation Mehrkammersilo
FIGURE 5: Automation system –
multi-compartment silo

Versand	= despatch
Dosieren	= metered feeding
Verladung Nord	= loading system north
Verladung Süd	= loading system south
Verladung Mischer	= loading system mixer
Silobeschickung	= silo filling
Ausweisleser	= badge reader
Disomat	= Disomat
Silobeschickung	= silo filling
Siloabzug Nord + Süd	= silo extraction north + south
Siloabzug Mischer	= silo extraction mixer
E/A	= I/O
Steuern	= controlling
Prozeß-Leitebene	= process control level

Eine Besonderheit dieser Anlage ist die Verladung des Fertigproduktes von dem Mischernachbehälter direkt in das Silokesselfahrzeug. Dies erfolgt in der mittleren Durchfahrt auf einer Straßenfahrzeugwaage. Mit der Direktverladung aus dem Mischer wird bei der Umstellung auf sensible Produkte hoher Reinigungsaufwand und Vermischungsgefahr vermieden.

Die beiden Verladestraßen Nord und Süd, ebenfalls mit Straßenfahrzeugwaagen ausgerüstet, sind über die Loseverladeeinrichtungen und separate Luftförderrinnenwege mit sieben Kammern verbunden. Die Leistung der Verladeeinrichtungen Nord und Süd beträgt je 200 t/h bis 250 t/h.

Die Materialdosierung zu den Behälterwaagen und zu den Verladeanlagen erfolgt über regelbare Dosierschieber. Die erforderliche hohe Dosiergenauigkeit wird erreicht durch:

– Konstante Vordruckverhältnisse der Belüftungseinrichtungen mittels Überströmventilen,

– kurze Rinnentransportwege zu den Behälterwaagen und den Verladeeinrichtungen,

– Überwachung des Förderstroms über die Meßdosen der Behälter- und Fahrzeugwaagen,

– Grob- und Feinstromdosierung,

– Schnellschlußklappen vor den Behälterwaagen und den Verladeeinrichtungen.

Alle Silokammern sind über zwei Senkrechtbecherwerke mit vorgeschalteten separaten Luftförderrinnentransporten mit dem Mischer, mit zwei vorhandenen Zementmühlen und mit einer Mahltrocknungsanlage für Kalksteinmehl verbunden. Die Materialverteilung in die einzelnen Kam-

The material is metered to the hopper weighers and to the loading plants via controllable flow control valves. The high metering accuracy required is achieved through:

– constant admission pressure conditions from the aeration equipment by means of overflow valves,

– short trough transport routes to the hopper weighers and to the loading units,

– monitoring of the transport flow by the load cells of the hopper weighers and vehicle weighbridges,

– coarse- and fine-flow metered feeding,

– rapid-action, shut-off flaps before the hopper weighers and the loading units.

Two vertical bucket elevators with separate upstream pneumatic trough conveyors connect all the silo compartments to the mixer, to two existing cement mills and to one drying and grinding plant for limestone meal. The material is distributed to the individual compartments by pneumatic trough conveyors with integral cylindrical side emptying.

Six compartments of the multi-compartment silo can also be filled from bulk tanker vehicles.

4.3 Electrical equipment

The mechanical configuration of the plant includes a total of 235 electrical consumers. A high level of automation is demanded of the open- and closed-loop control systems and the monitoring systems, so that only one employee is required per shift for controlling the plant and managing the despatch.

This aim is achieved by the use of automation systems on the despatch, metered feeding, and control levels (**Fig. 5**).

BILD 6: Mehrkammersilo Rohbau
FIGURE 6: Multi-compartment silo carcass

The metered feeding system consists of a control and metering computer. The plant is operated through the control computer.

The metering computer looks after the control and monitoring of the metered feeding equipment and the hopper weighers. It is subordinate to the control computer.

The metering computer is linked to the despatch computer so that mix formulations can be passed from the metering to the despatch computer and orders from the despatch to the metering computer.

Despatch and metering systems control the loading and metered feeding plants through S5 155 U programable control systems from Siemens. The central items of equipment are linked to the COROS process control level by a databus. This level is used for displays as well as for signalling and logging faults.

BILD 7: Ende 1. Gleitabschnitt
Einbringen der Kegelfertigteile
FIGURE 7: End of 1st slip-form lift
installing the prefabricated cone sections

mern erfolgt über Lufttrinnentransporte mit integrierten zylindrischen Seitenentleerern.

Sechs Zellen des Mehrkammersilos können zudem aus Silokesselfahrzeugen beschickt werden.

4.3 Elektrische Ausrüstung

Der maschinelle Aufbau der Anlage enthält insgesamt 235 elektrische Verbraucher. An die Steuerungs-, Regelungs- und Überwachungssysteme wurde ein hoher Automatisierungsanspruch gestellt, um für die Anlagenführung und die Versandabwicklung nur einen Mitarbeiter pro Schicht einzusetzen.

Dieses Ziel wird erreicht durch den Einsatz von Automatisierungssystemen in den Ebenen Versand, Dosieren und Steuern (**Bild 5**).

Das Dosiersystem besteht aus einem Leit- und Dosierrechner. Über den Leitrechner erfolgt die Bedienung der Anlage.

Der Dosierrechner übernimmt die Steuerung und Überwachung der Dosiereinrichtungen und der Behälterwaagen. Er ist dem Leitrechner untergeordnet.

Der Dosierrechner ist mit dem Versandrechner verbunden, so daß Rezepte vom Dosier- zum Versandrechner und Aufträge vom Versand- zum Dosierrechner ausgetauscht werden.

Versand- und Dosiersysteme steuern die Verlade- und Dosieranlagen über frei programmierbare Steuerungen S 5 155 U der Firma Siemens. Die Zentralgeräte sind über einen Datenbus mit der Prozeßleitebene COROS verbunden. Diese Ebene dient der Visualisierung sowie der Meldung und Protokollierung von Störungen.

Alle Silokesselfahrzeugbeladungen und -entladungen im Bereich des Mehrkammersilos werden vom Fahrer selbst vorgenommen. Über Ausweisleser werden im Versandrechner die erforderlichen Daten für die ordnungsgemäße Be- oder Entladung zusammengestellt und die Signale an die Steuerungen weitergegeben.

BILD 8: Mehrkammersilo in Betrieb
FIGURE 8: Multi-compartment silo in operation

4.4 Bautechnik

Statische Probleme traten bei der Herstellung der sechs Öffnungen für die drei Durchfahrten im unteren Bereich des Silomantels auf. Die lichten Durchfahrten der Verladestraßen (**Bild 6**) durften wegen der erforderlichen Sicherheitsabstände nicht weiter eingeengt werden. Der erste Bauabschnitt wurde deshalb in B 55 und Ortsschalung ausgeführt.

Der erste Gleitabschnitt erfolgte dann in 60 cm Wandstärke bis zur halben Kegelhöhe (**Bild 7**). Danach wurde der Kegel in Stahlbetonfertigteilen eingebracht und mit Ortsbeton vergossen. Die Kammerwände des Außenringsilos wurden anschließend bis zur erstellten Siloschafthöhe in Ortsbeton hergestellt.

Im zweiten Gleitabschnitt wurde das Außensilo mit den Kammerwänden und dem Innensilo bis zur Silodecke gezogen. Hier wurde die Schalung für das Innensilo und die Kammerwände abgekoppelt und der Außenschaft bis zum Silodach weitergeglitten.

Der Aufzugs- und Versorgungsschacht wurde, höhenmäßig versetzt, mit dem Silo ebenfalls in Gleitschalung hergestellt.

4.5 Betriebserfahrungen

Nach einer Bau- und Montagezeit von 18 Monaten wurde das Mehrkammersilo mit der Mischanlage (**Bild 8**) am 1. 7. 1992 problemlos in Betrieb genommen. Folgende Garantien wurden im Normalbetrieb erreicht und unterschritten:

+/− 0,1 % Dosiergenauigkeit der Komponenten bezogen auf die maximale Einwaage von 5 600 kg im Dosierbereich 1 zu 10,

Kesselfahrzeugentleerung von 60 t/h.

Die Mischerleistung beträgt 130 t/h bis 140 t/h.

Für das Mischen von Zementen nach der zukünftigen Euronorm ist die neue Mehrkammersiloanlage bestens gerüstet.

All the loading and unloading of bulk tanker vehicles at the multi-compartment silo is carried out by the drivers themselves. The required data for authorized loading and unloading are compiled in the despatch computer via badge readers and the signals are relayed to the control systems.

4.4 Structural engineering

Static problems occurred when making the 6 openings for the 3 drive-through lanes in the lower part of the silo shell. Because of the required safety margins, the openings for the loading lanes could not be narrowed any further (**Fig. 6**). The first stage of the construction was therefore carried out using B 55 and in-situ formwork.

The first slip-form lift was then carried out with a wall thickness of 60 cm up to half the cone height (**Fig. 7**). After which the cone was positioned in precast, reinforced, concrete sections and sealed with in-situ cast concrete. The compartment walls of the outer ring silo were then fabricated in in-situ cast concrete up to the prepared silo shaft level.

The outer silo with the chamber walls and the inner silos were then slip-formed to the silo top in the second lift. At this point the formwork for the inner silo and the compartment walls was disconnected and the slip-forming of the outer shaft was continued to the silo roof.

The lift and utilities shaft was also fabricated by slip-forming with the silo with staggered levels.

4.5 Operating experience

After a construction and installation period of 18 months, the multi-compartment silo with the mixing plant (**Fig. 8**) was brought into operation without any difficulties on 01.07.1992. The following guarantee values were achieved and bettered in normal operation:

± 0.1 % metering accuracy of the components relative to the maximum weighed quantity of 5 600 kg in the metering range 1 to 10,

bulk container vehicle emptying of 60 t/h.

The mixer output is 130 t/h to 140 t/h.

The new multi-compartment silo plant is extremely well equipped for mixing cements to comply with the future Euro Standard.

Verbesserung des Verschleißschutzes der Mahlwalzen von Walzenschüsselmühlen

Improving the wear protection on the grinding rollers in roller grinding mills

Amélioration de la protection contre l'usure des galets de broyeurs à piste

Mejorando la protección contra el desgaste de los rodillos de molienda de los molinos de cubeta y rodillos

Von **W. Wahl,** Ostfildern/Deutschland

Zusammenfassung – Beim Mahlen in Walzenschüsselmühlen tritt an den Auskleidungen und den Mahlwalzen beträchtlicher Verschleiß auf. Früher wurden die Mahlteile bevorzugt aus Stahlguß gefertigt. Die Entwicklung in den letzten 15 Jahren ging dahin, die Verschleißteile der Mahlwerkzeuge in zunehmendem Maß aus Hartguß herzustellen. Abgenutzte Verschleißteile mußten verschrottet werden. Die neue Regenerierungstechnologie, nämlich das Auftragsschweißen abgenutzter Verschleißteile in Walzenschüsselmühlen, kam aus Amerika. Dabei dienen im wesentlichen hochchrom- und -kohlenstoffhaltige Fülldrähte als Auftragsschweißmaterial. Obwohl die abgenutzten Verschleißteile als nicht schweißbar bezeichnet werden, ist es gelungen, durch Auftragsschweißen Panzerungen von bis zu 150 mm Dicke herzustellen. Wegen der werkstoffbedingten Schwierigkeiten bei der Durchführung der Auftragsschweißung kommt der Qualitätssicherung besondere Bedeutung zu. Mit sorgfältig aufgepanzerten Verschleißteilen in Walzenschüsselmühlen können heute Standzeiten erreicht werden, die zwei- bis dreimal höher als die nicht gepanzerter Mahlteile sind.

Summary – Considerable wear takes place at the lining and grinding elements during grinding in roller grinding mills. In the past the grinding parts were normally made from cast steel. Developments in the last 15 years have moved increasingly towards producing the wearing parts for the grinding elements from chilled cast iron. Worn-out wearing parts have had to be scrapped. The new regeneration technology, i.e. build-up welding of worn wearing parts in the roller grinding mills, came from America. Core wires containing high levels of chromium and carbon form the main build-up welding material. Build-up welding has been used successfully to produce armouring up to 150 mm thick although the worn wearing parts are made of materials designated in the literature as not weldable. Quality assurance is particularly important because of the difficulties caused by the materials when carrying out the build-up welding. Wearing parts with carefully applied armouring can now achieve service lives in roller grinding mills which are two to three times higher than those of unarmoured grinding parts.

Résumé – Lors du broyage dans les broyeurs à galets se manifeste, sur les blindages et sur les outils de broyage, une usure considérable. Avant, les outils de broyage avaient été fabriqués de préférence en fonte d'acier. L'évolution au cours des dernières 15 années a conduit à fabriquer, de plus en plus, les outils de broyage en fonte dure. Les pièces d'usure hors service devaient être mises au rebut. La nouvelle technologie de régénération, c'est à dire la reconstitution, par apport de soudure, des pièces d'usure hors service dans les broyeurs à galets, nous vient d'Amérique. Comme matériau de soudure d'apport servent essentiellement des baguettes à haute teneur en chrome et en carbone. Bien que, d'après la littérature, les pièces d'usure hors service soient en matériaux considérés non soudables, on a réussi à constituer, par soudure d'apport, des blindages d'une épaisseur jusqu'à 150 mm. A cause des difficultés inhérentes au matériau, lors de l'exécution de la soudure d'apport, l'assurance de la qualité a une importance particulière. Avec des pièces d'usure de broyeurs à galets, reconstituées avec soin par soudure d'apport, peuvent être atteintes aujourd'hui des durées de service deux à trois fois supérieures à celles d'outils de broyage non blindés.

Resumen – Durante la operación de molienda se produce en los molinos de cubeta y rodillos un desgaste enorme en los revestimientos y los elementos moledores. Antes, los elementos moledores se fabricaban preferentemente de fundición de acero. Durante los últimos 15 años, ha aumentado la tendencia de fabricar las piezas de desgaste de fundición dura. Las piezas muy desgastadas solían ir a la chatarra. La nueva tecnología de regeneración, o sea el aporte de soldadura en las piezas desgastadas de los molinos de cubeta y rodillos vino de los Estados Unidos. Como material de aporte se utilizan, sobre todo, varillas de elevado contenido de cromo y de carbono. Aunque las piezas desgastadas se consideraban como no soldables, se ha conseguido fabricar, por

aporte de soldadura, blindajes de hasta 150 mm de espesor. Puesto que los materiales empleados presentan dificultades para llevar a cabo el aporte de soldadura, es muy importante el aseguramiento de la calidad. Con las piezas de desgaste de los molinos de cubeta y rodillos, cuidadosamente regeneradas mediante aporte de soldadura, se puede alcanzar, hoy en día, una duración de vida dos o tres veces superior a la de los elementos moledores sin blindaje.

1. General

In the cement industry there are a great many individual mechanical elements from a wide variety of units which are subject to wear (**Table 1**). The particular problem of lowering wear costs in the cement industry lies in the fact that once solutions have been found they can rarely be applied with success to other mechanical parts.

As in other branches of industry, the wear costs in the cement industry are hard to estimate because the individual companies use very different accounting systems for dealing with the costs. **Table 2** contains an estimate, carried out by Uetz, of the direct wear costs in the cement industry and, for comparison, in other sectors of the non-metallic minerals industries.

From this it can be seen that costs in the cement industry of 296 million DM are very high in comparison with other branches of industry, and the cement industry is therefore one of the most wear-intensive sectors of the business. The total costs can be estimated from the wear when grinding clinker by assuming similar wear costs per tonne in the rest of the cement works. Based on a cement production of 30 million t/a it can be assumed that the specific wear costs are approximately 2DM/t.

TABLE 1: Wear in the cement industry

Process step	Examples of wear
– raw material excavation and raw meal production	drill crowns, excavator teeth, dump truck bodies, recirculation hoppers, hammer mills, blending reclaimers, mills, conveying equipment
– clinker burning and clinker transport	burner inlet nozzles, transport equipment, kiln tyres, clinker chutes, grinding elements, linings and conveying equipment for coal
– cement grinding	mill linings, grinding balls, classifiers, cyclones, fans
– cement despatch	chutes, pneumatic conveying equipment, bagging machines

TABLE 2: Estimate of the wear costs in the non-metallic minerals industries (acc. to Uetz)

Branch of industry	Wear costs in million DM
– sand and gravel works	18
– cement industry	295
– quarrying and processing of natural stone	323
– gypsum industry	17
– lime industry	88
– concrete industry	41
Intermediate total	782
– brick manufacture – rough stoneware production – asbestos production – production of abrasives	3.9% of the gross output value — 418
Total	1200

BILD 1: Abgenutzte Bandagen einer Walzenschüsselmühle
FIGURE 1: Worn tyres from a roller grinding mill

BILD 2: Aufspannen der abgenutzten Bandage auf einen Drehtisch zum Auftragschweißen
FIGURE 2: Clamping the worn tyre on a turntable for build-up welding

2. Verschleißformen an Mahlwalzen

Zu den stark verschleißgefährdeten Maschinen zählen die Walzenschüsselmühlen. Das Mahlgut wird durch die umlaufenden Mahlwalzen zerkleinert. Dabei kommt es zu Verschleiß durch Abrieb. Bei diesem Vorgang nutzen sich die Mahlwalzen auf der Mantelfläche allerdings ungleichmäßig ab. Es kommt zur Ausbildung von zerklüfteten Verschleißbildern, deren Entstehung noch nicht geklärt ist. Hierdurch wird im Laufe der Zeit die Mahlleistung der Mühle so herabgesetzt, daß es ökonomisch ist, neue Bandagen aufzuziehen.

Die bisher eingesetzten Mahlwalzen waren in der Regel aus Hartguß hergestellt. Statt eines Austauschs könnten sie nach der Abnutzung oder bereits im Neuzustand in geeigneter Weise auftraggeschweißt werden, wodurch sich nicht nur die Mantelfläche sehr viel gleichmäßiger abnutzt, sondern insbesondere die Standzeit wesentlich verlängert wird. Typische Verschleißbilder abgenutzter Mahlwalzen sind im **Bild 1** dargestellt.

3. Regenerieren durch Auftragschweißen

Abgenutzte Teile wie im Bild 1 können durch besondere Formen des Auftragschweißens regeneriert werden, wobei sich diese Technologie in gleicher Weise zum Panzern von Neuteilen wie zum Regenerieren und Panzern abgenutzter Teile eignet. Hierzu müssen die Teile auf eine geeignete Schweißvorrichtung aufgespannt werden, wie das Beispiel im **Bild 2** zeigt. Richtig durchgeführte Panzerungen führen im Vergleich zur Anwendung von ungepanzerten Bandagen zu wesentlichen Standzeitverlängerungen. Aus **Bild 3** geht sehr deutlich hervor, daß eine Hartgußbandage eine kürzere Lebensdauer als eine auftraggeschweißte Bandage aufweist mit der Folge, daß sich bei Anwendung der auftraggeschweißten Lösung die Mahlleistung durch den Abrieb nur langsam und in unbedeutendem Umfang reduziert.

Die Durchführung solcher Auftragschweißarbeiten ist jedoch nur mit beträchtlichem technischen Aufwand und Know-how möglich, da die Werkstoffe der verwendeten Grundkörper in der Literatur als nicht schweißbar gelten. Klassische Werkstoffvarianten, die in der Zementindustrie zum Verschleißschutz eingesetzt werden, sind in **Tabelle 3** zusammengefaßt. Aus der Tabelle kann für die heute üblichen Chromgußlegierungen entnommen werden, daß sich diese im wesentlichen im Kohlenstoff- und Chromgehalt unterscheiden. Der Gehalt an Carbiden beeinflußt das Gefüge, der wie die Ausbildung einer zähen Matrix für den Verschleißschutz entscheidend ist.

Jüngste Untersuchungen haben ergeben, daß es Hartgußlegierungen mit einer kritischen Zusammensetzung gibt, die man möglichst meiden, zumindest aber nicht zum Auftragschweißen vorsehen sollte. Im **Bild 4** ist das Gefügebild einer Legierung unkritischer Zusammensetzung mit ca.

2. Forms of wear on grinding rollers

The machines which are heavily at risk from wear include the roller grinding mills. The material to be ground is comminuted by the revolving grinding rollers, and wear is caused by abrasion. During this process the wear on the grinding rollers on the outer surface is, however, very uneven. Fissured wear profiles are formed, the causes of which have not yet been entirely explained. During the course of time the grinding capacity of the mill is so sharply reduced that it becomes economic to fit new tyres.

Grinding rollers were normally made of chilled cast iron. Instead of replacing them they can be suitably hard-faced by build-up welding, either after they have become worn or even when new. The effect is that not only does the outer surface wear much more evenly, but also the service lives are substantially prolonged. Typical wear patterns of worn grinding rollers are shown in **Fig. 1**.

3. Regeneration by build-up welding

Worn parts like those in Fig.1 can be regenerated by special types of build-up welding. This technology is equally suitable for armouring new parts and for regenerating and armouring worn parts. For this purpose the parts must be clamped in suitable welding equipment as shown by the example in **Fig.2**. Correctly executed armouring results in substantial extension of service life when compared with unarmoured tyres. It can be seen very clearly from **Fig. 3** that a tyre made of chilled cast iron has a shorter service life than

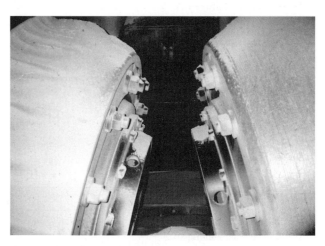

BILD 3: Standzeitvergleich einer Hartgußbandage mit einer auftraggeschweißten Bandage an einer Walzenschüsselmühle zur Rohmehlmahlung
FIGURE 3: Comparison of the service life of a chilled cast iron tyre with a hardfaced tyre in a roller grinding mill for grinding raw meal

TABELLE 3: Hartgußlegierungstypen
TABLE 3: Types of chilled cast iron alloys

Legierungen	Perlitisches weißes Gußeisen	Martensitisches weißes Gußeisen		Chromgußeisen				
		Typ 1	Typ 2	Ni-Cr-Guß	12 Cr-Guß	15 CrMo Guß	20 CrMo Guß	27 Cr-Guß
Kohlenstoff	2,5–3,6	3,0–3,6	<2,9	2,6–3,2	2,0–3,5	2,3–3,6	2,3–2,9	2,3–2,9
Mangan	0,4–1,0	0,3–0,7	0,3–0,7	0,4–0,6	0,5–1,0	0,5–0,9	0,5–0,9	0,5–1,5
Silicium	0,3–1,5	0,3–0,5	0,3–0,5	1,8–2,0	<0,8	<0,8	<0,8	<1,5
Chrom	0–2,0	1,5–2,6	1,4–2,4	8,0–9,0	11,0–14,0	14,0–16,0	18–21	24–28
Nickel	0–1,5	3,3–4,8	3,3–5,0	5,0–6,5	<2,0	<1,2	<1,2	<1,2
Molybdän	0–1,5	0–1,0	0–1,0	0–0,5	<1,5	2,0–3,0	1,4–2,0	<0,6
Kupfer	0–1,5	–	–	–	<1,2	<1,2	<1,2	–
Härte NV$_{30}$								
Guß	350	550	520	550	450	450	450	450
gehärtet				600	600	600	600	550
weichgeglüht					<400	<400	<420	<400

Legierungen	= alloys	Mangan	= manganese
Perlitisches weißes Gußeisen	= perlitic chilled cast iron	Silicium	= silicon
Martensitisches weißes Gußeisen	= martensitic chilled cast iron	Chrom	= chromium
Chromgußeisen	= chromium alloy cast iron	Nickel	= nickel
Ni-Cr-Guß	= Ni-Cr-cast iron	Molybdän	= molybdenum
12 Cr-Guß	= 12 Cr-cast iron	Kupfer	= copper
15 CrMo-Guß	= 15 CrMo cast iron	Härte NV$_{30}$	= NV$_{30}$ hardness
20 CrMo-Guß	= 20 CrMo cast iron	Guß	= cast iron
27 Cr-Guß	= 27 Cr-cast iron	gehärtet	= hardened
Kohlenstoff	= carbon	weichgeglüht	= soft annealed

2,8 % C und 15 % Cr wiedergegeben. Diese Legierung ist gut zum Schweißen geeignet, wie z. B. auch eine Legierung mit 3 % C und 25 % Cr. Beide haben ein Chrom- zu Kohlenstoffverhältnis von deutlich unter 10. Dagegen zeigt **Bild 5** das Gefüge einer Legierung, deren Zusammensetzung mit ca. 2,5 % C und 25 % Cr den kritischen Wert für das Cr/C-Verhältnis von 10 überschreitet. Solche Legierungen müssen als wenig geeignet für das Auftragschweißen angesehen werden.

Zwar bildet sich im Gußzustand erwartungsgemäß ein Gefüge, bestehend aus zähem Austenit und dem verschleißmindernden Carbid mit der Zusammensetzung Cr7%/C3% aus, bei der anschließenden Wärmebehandlung entsteht jedoch zusätzlich ein Carbid der Zusammensetzung Cr23%/C6%. Hierdurch nimmt der Carbidanteil so stark zu, daß die einzelnen Carbide nicht mehr in einer zähen Matrix „schwimmen", sondern ganze Ketten geschlossener Carbidanhäufungen entstehen. Dementsprechend nimmt die Zähigkeit solcher Legierungen beträchtlich ab, wie die in **Tabelle 4** wiedergegebenen Kerbschlagfestigkeiten zeigen. Die Legierung mit 25 % Cr und 2,6 % C hat zwar eine genügend hohe Biegebruchfestigkeit, aber praktisch keine Kerbschlagzähigkeit. Dagegen ist die Legierung mit 25 % Cr und

a tyre with build-up welding. The result of this is that when using the build-up welding solution the grinding capacity is only reduced slowly and to an insignificant extent by the abrasion.

Nevertheless, the execution of such build-up welding work is only possible with considerable technical expenditure and know-how, because the base material used is designated in the literature as not weldable. The classical types of material used in the cement industry for wear protection are listed in **Table 3**. The table shows that the normal types of chromium alloy cast iron differ essentially in their levels of carbon and chromium. The carbide content influences the microstructure which, like the formation of a ductile matrix, is critical for the wear protection.

Very recent investigations have shown that there are chill cast alloys with critical compositions which where possible should be avoided, and at the least should not be designated for build-up welding. **Fig.4** shows a photomicrograph of an alloy with a non-critical composition with approximately 2.8 % C and 15 %Cr. This alloy is highly suitable for welding, as is an alloy with, for example, 3 % C and 25 % Cr. Both have a chromium to carbon ratio of significantly less than 10. **Fig.5**, on the other hand, shows the microstructure of an

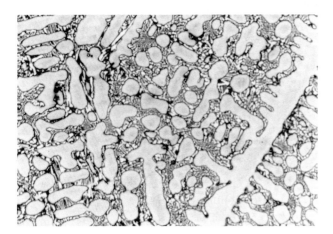

BILD 4: Gefügebild einer Legierung mit ca. 2,8 % Kohlenstoff, 15 % Chrom
FIGURE 4: Micrograph of an alloy with approximately 2.8 % carbon and 15 % chromium

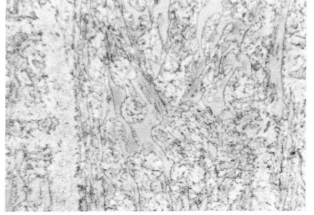

BILD 5: Gefügebild einer für das Auftragschweißen ungeeigneten Legierung mit ca. 2,5 % Kohlenstoff und 25 % Chrom (Cr/C ≥ 10)
FIGURE 5: Micrograph of an alloy which is unsuitable for build-up welding and contains approximately 2.5 % carbon and 25 % chromium (Cr/C ≥ 10)

3% C zum Auftragschweißen als unproblematisch anzusehen, weil sie entsprechend günstige Kerbschlagfestigkeiten aufweist. Dieser Unterschied in der Kerbschlagfestigkeit bedeutet beim Auftragschweißen eine dramatische Zunahme der Gefahr des Rißwachstums im Grundkörper, falls eine ungeeignete Legierung mit zu hohem Chrom-/Kohlenstoffgehalt verwendet wird. Befindet man sich dagegen in einem Bereich, in dem das fragliche Chromcarbid weder im Gußzustand noch während der Wärmebehandlung ausgeschieden wird, kann das Panzern von neuen und abgenutzten Bandagen heute als technisch und ökonomisch sinnvolle Maßnahme angesehen werden, wenn dabei die Verfahrensparameter im Rahmen der technischen Vorgaben liegen. Verwendet werden sollten zum Auftragschweißen demnach Legierungen, die in DIN 8555 in der Legierungsgruppe 10 enthalten sind.

Bei der Verwendung solcher Legierungen entstehen Gefüge, die deutlich mehr und ineinander verzahnte Chromcarbide enthalten, als dies bei der Herstellung durch Guß möglich ist. Legierungen mit 4,5% C und ca. 30% Cr weisen sehr viel höhere Carbidanteile auf, so daß die Standzeit in aller Regel gegenüber konventionellen Hartgußlegierungen mehr als doppelt so hoch ist. Je abrasiver das zu zerkleinernde Gut ist, desto günstiger wirkt sich die Verlängerung der Standzeit auf das Betriebsergebnis aus.

Beim Auftragschweißen kommt es darauf an, die Betriebsparameter innerhalb sehr enger Grenzen zu kontrollieren. Dies gilt für die Temperatur des Grundkörpers, den Schweißstrom, die Schweißspannung, den Drahtvorschub, das freie Drahtende sowie die verwendeten Schweißzusätze. Bei der Erwärmung des Grundkörpers darf eine maximale Temperatur von ca. 100 °C nicht überschritten werden. Geschweißt wird im Gegensatz zur konventionellen Hartauftragschweißung in Strichraupentechnik, um den Energieübergang in den Grundkörper pro Zentimeter Schweißnaht zu vermindern.

alloy with a composition of approximately 2.5% C and 25% Cr which exceeds the critical value of 10 for the Cr/C ratio. Such alloys have to be regarded as rather unsuitable for build-up welding.

As expected, a microstructure is in fact formed in the "as cast" state which consists of ductile austenite and wear-resistant carbide with a composition of Cr 7%/C 3%, but during the subsequent heat treatment a carbide with the composition Cr 23%/C 6% is also formed. This increases the proportion of carbide so sharply that the individual carbides no longer "float" in a ductile matrix, but complete closed carbide chains are formed. This considerably reduces the ductility of such alloys as is shown by the impact strengths given in Table 4. An alloy composed of 25% Cr and 2.6% C does in fact have a sufficiently high ultimate bending strength, but practically no impact strength. On the other hand, an alloy composed of 25% Cr and 3%C is regarded as suitable because it has correspondingly favourable impact strengths. With build-up welding this difference in impact strength signifies a dramatic increase in the risk of crack development in the base material in cases where an unsuitable alloy with excessive levels of chromium and carbon is used. On the other hand, when working in areas in which the chromium carbide in question is not precipitated either as cast or during heat treatment the armouring of new and worn tyres can now be regarded as a technically and economically appropriate procedure provided the process parameters lie within the framework of the technical guidelines.

When such alloys are used microstructures are produced which contain more chromium carbides which are interlocked with one another than is possible during manufacture by casting. Alloys with 4.5% C and approximately 30% Cr contain very much higher proportions of carbide so as a rule the service lives are more than twice as high as those of conventional chill cast alloys. The more abrasive is the material to be ground, the more favourable is the effect of the extension of the service life on the operating results.

During build-up welding it is important to control the operating parameters within very tight limits. This applies to the temperature of the base material, the weld current and weld voltage, the wire feed, the wire projection length and the weld filler materials used. A maximum temperature of approximately 100 °C must not be exceeded when the base material is being heated. Unlike conventional hardfacing the welding is carried out using the so-called stringer bead technique in order to reduce the energy transfer to the base material per centimetre of weld.

4. Build-up welding on dismantled or in situ tyres

A question which is particularly important for the operator is whether the parts can be regenerated in situ or whether they have to be dismantled. At present both technical solutions are possible and are advocated. It is very much more difficult to maintain tight manufacturing tolerances in situ than in a well equipped, quality orientated, welding plant so, in spite of tempting advantages, it is recommended that the armouring should not be carried out in situ. The above-mentioned parameters are only allowed to fluctuate within narrow limits, which often cannot be achieved in multi-shift operation and when welding on site with less competent employees. It is therefore understandable that grinding rollers welded in situ do not always prove successful in practical operation. Only if the same care can be given when build-up welding in situ as for parts which are armoured in a welding shop designed specifically for the purpose, can equally good results be expected.

5. Future developments

Hardface build-up welding of grinding rollers is now one of the most important applications of hardfacing, and is carried out in many countries in the world with the greatest success. In recent years efforts have been directed towards achieving extended service lives through modified alloys for the basic materials. Armouring with thicknesses between 30-50mm

TABELLE 4: Biegebruchfestigkeiten ungeeigneter (25% Chrom, 2,6% Kohlenstoff) und geeigneter (25% Chrom, 3% Kohlenstoff) Legierungen
TABLE 4: Impact strength of unsuitable (25% chromium, 2.6% carbon), and suitable (25% chromium, 3% carbon) alloys

Charge batch	Biegebruch-festigkeit [N/mm³] impact strength	Kerbschlag-arbeit [J] notched bar impact work
G–20–D–3–4λ	1095	0
	1025	0
	595	0,1
	960	0
	1060	0
	1160	
	660	
	1095	
	1125	
	760	
Mittelwert mean value	954	< 0,1
G–20–D–3–5λ	960	2,2
	1290	0,2
	1160	0
	1256	1,5
	1025	0,1
	1255	4,0
	1095	2,9
	1225	
	1225	
	860	
Mittelwert mean value	1135	1,6

4. Auftragschweißen im ausgebauten oder eingebauten Zustand

Eine für den Betreiber besonders wichtige Frage ist, ob die Teile im eingebauten Zustand regeneriert werden können, oder ob sie hierzu ausgebaut werden müssen. Beide technischen Lösungen sind derzeit möglich und werden angeboten. Da im eingebauten Zustand das Einhalten enger Fertigungstoleranzen sehr viel schwieriger als in einem gut eingerichteten und auf Qualität ausgerichteten Schweißbetrieb ist, wird empfohlen, trotz der bestechenden Vorteile für den Vertreiber nicht im eingebauten Zustand zu panzern. Denn dabei dürfen die oben genannten Betriebsparameter nur in engen Grenzen schwanken, was bei einem Mehrschichtbetrieb bzw. beim Schweißen vor Ort mit wenig qualifizierten Mitarbeitern oftmals nicht zu verwirklichen ist. Es ist deshalb verständlich, daß sich im eingebauten Zustand geschweißte Mahlwalzen nicht immer im praktischen Einsatz bewähren. Nur wenn beim Auftragschweißen im eingebauten Zustand die gleiche Sorgfalt aufgewandt werden kann wie bei Teilen, die in einer dafür bestens eingerichteten Schweißerei gepanzert werden, ist mit gleich guten Ergebnissen zu rechnen.

5. Künftige Entwicklungen

Das Auftragschweißen von Mahlwalzen ist heute eine der wichtigsten Anwendungen des Hartauftragschweißens und wird in vielen Ländern der Welt mit allergrößtem Erfolg durchgeführt. In den letzten Jahren ging das Bestreben dahin, durch modifizierte Legierungen für die Grundwerkstoffe Standzeitverlängerungen zu erreichen. Da derzeit bereits 30 bis 50 mm dicke Aufpanzerungen technisch beherrschbar sind, ist die Beanspruchungsfläche völlig unabhängig von den Materialeigenschaften des Grundkörpers. Demnach ist nur das Verschleißverhalten des reinen Schweißguts von Bedeutung. Die Entwicklung hat dazu geführt, daß mit neuen Werkstoffvariationen für den Grundkörper der Mahlwalzen, die über die Chromcarbide hinaus noch weitere hochharte Carbide enthalten, kaum mit einer weitergehenden Standzeitverlängerung zu rechnen ist. Zudem wird das Auftragschweißen durch die Verwendung dickerer Fülldrähte mit höherer Abschmelzleistung bei steigender Strombelastung zunehmend preiswerter (**Bild 6**). Durch die beschriebenen Maßnahmen lassen sich zwar nicht die Materialkosten, wohl aber die spezifischen Lohnkosten pro Kilogramm Auftragsgut senken, die 2/3 des Marktpreises ausmachen. Auftraggeschweißte Teile, bei deren Herstellung die notwendigen qualitätssichernden Maßnahmen beachtet werden, bewähren sich im praktischen Einsatz außerordentlich gut. Hierdurch konnten 2- bis 4fach höhere Standzeiten erreicht werden, abhängig von der Abrasivität des Mahlguts.

Die im Vergleich zu herkömmlichen Mahlwalzen erzielbare Standzeitverlängerung durch Auftragschweißen erwies sich als besonders vorteilhaft bei stark abrasivem Mahlgut.

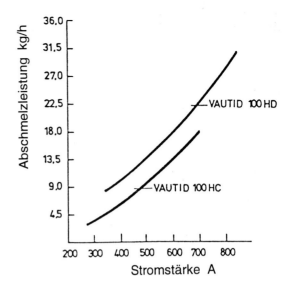

BILD 6: Vergleich der Abschmelzleistung von **VAUTID 100 HD** und **100 HC** in Abhängigkeit von der Stromstärke
FIGURE 6: Comparison of the deposition efficiencies of **VAUTID 100 HD and VAUTID 100 HC** as a function of current strength

Abschmelzleistung kg/h = deposition efficiency kg/h
Stromstärke A = current strength A

can now be managed technically, so the working surface is completely independent of the material properties of the base material. It is therefore only the wear behaviour of the pure weld material which is of interest. The development has led to the situation where new types of material for the base material of the grinding rollers containing further ultra-hard carbides in addition to the chromium carbides are hardly expected to produce any more extensive increases in service life. Build-up welding is also becoming increasingly less expensive through the use of thicker core wires with higher deposition efficiency and increasing current load (**Fig.6**). The measures described do not in fact reduce the cost of materials, but they lower the specific labour costs per kilogramme of deposited material, and these labour costs make up at least 2/3 of the market price. Parts with build-up welding where the necessary quality assurance measures were observed during production have proved exceptionally successful in practical use. Service lives which are 2 to 4 times longer can be achieved, depending on the abrasiveness of the material being ground.

When compared with conventional grinding rollers the extensions to the service life which can be achieved by build-up welding have proved to be particularly advantageous when grinding strongly abrasive material.

FLUITEX®

Pneumatikgewebe maßgeschneidert

Aeroslide Fabrics cut to measure

Lieferbar in Polyester, Baumwolle, Aramid Kevlar / Nomex

Hitzebeständig bis 300 °C

Lieferbar in Breiten bis zu 2400 mm oder konfektioniert

Available in polyester, cotton, aramid kevlar / nomex

Heat resistant up to 300 °C

Available widths up to 2400 mm or according to specification

Staubgut fördern

Conveying powdered materials

Staubgut transportieren, auflockern und entladen

Aerating and discharging of powdered materials from vehicles and trucks

Staubgut lagern und auflockern

Aerating stored powdered materials

MÜHLEN SOHN
P.O. Box 11 65
D - 89130 Blaustein
Tel. +49 - 7304 - 8010
Fax +49 - 7304 - 80123
Telex 07 12 381 museo d

FLUITEX®

Pneumatikgewebe maßgeschneidert

Aeroslide Fabrics cut to measure

Lieferbar in Polyester, Baumwolle, Aramid Kevlar / Nomex

Hitzebeständig bis 300 °C

Lieferbar in Breiten bis zu 2400 mm oder konfektioniert

Available in polyester, cotton, aramid kevlar / nomex

Heat resistant up to 300 °C

Available widths up to 2400 mm or according to specification

Staubgut fördern

Conveying powdered materials

Staubgut transportieren, auflockern und entladen

Aerating and discharging of powdered materials from vehicles and trucks

Staubgut lagern und auflockern

Aerating stored powdered materials

MÜHLEN SOHN
P.O. Box 11 65
D - 89130 Blaustein
Tel. +49 - 7304 - 8010
Fax +49 - 7304 - 80123
Telex 07 12 381 museo d

4. Auftragschweißen im ausgebauten oder eingebauten Zustand

Eine für den Betreiber besonders wichtige Frage ist, ob die Teile im eingebauten Zustand regeneriert werden können, oder ob sie hierzu ausgebaut werden müssen. Beide technischen Lösungen sind derzeit möglich und werden angeboten. Da im eingebauten Zustand das Einhalten enger Fertigungstoleranzen sehr viel schwieriger als in einem gut eingerichteten und auf Qualität ausgerichteten Schweißbetrieb ist, wird empfohlen, trotz der bestechenden Vorteile für den Vertreiber nicht im eingebauten Zustand zu panzern. Denn dabei dürfen die oben genannten Betriebsparameter nur in engen Grenzen schwanken, was bei einem Mehrschichtbetrieb bzw. beim Schweißen vor Ort mit wenig qualifizierten Mitarbeitern oftmals nicht zu verwirklichen ist. Es ist deshalb verständlich, daß sich im eingebauten Zustand geschweißte Mahlwalzen nicht immer im praktischen Einsatz bewähren. Nur wenn beim Auftragschweißen im eingebauten Zustand die gleiche Sorgfalt aufgewandt werden kann wie bei Teilen, die in einer dafür bestens eingerichteten Schweißerei gepanzert werden, ist mit gleich guten Ergebnissen zu rechnen.

5. Künftige Entwicklungen

Das Auftragschweißen von Mahlwalzen ist heute eine der wichtigsten Anwendungen des Hartauftragschweißens und wird in vielen Ländern der Welt mit allergrößtem Erfolg durchgeführt. In den letzten Jahren ging das Bestreben dahin, durch modifizierte Legierungen für die Grundwerkstoffe Standzeitverlängerungen zu erreichen. Da derzeit bereits 30 bis 50 mm dicke Aufpanzerungen technisch beherrschbar sind, ist die Beanspruchungsfläche völlig unabhängig von den Materialeigenschaften des Grundkörpers. Demnach ist nur das Verschleißverhalten des reinen Schweißguts von Bedeutung. Die Entwicklung hat dazu geführt, daß mit neuen Werkstoffvariationen für den Grundkörper der Mahlwalzen, die über die Chromcarbide hinaus noch weitere hochharte Carbide enthalten, kaum mit einer weitergehenden Standzeitverlängerung zu rechnen ist. Zudem wird das Auftragschweißen durch die Verwendung dickerer Fülldrähte mit höherer Abschmelzleistung bei steigender Strombelastung zunehmend preiswerter (**Bild 6**). Durch die beschriebenen Maßnahmen lassen sich zwar nicht die Materialkosten, wohl aber die spezifischen Lohnkosten pro Kilogramm Auftragsgut senken, die 2/3 des Marktpreises ausmachen. Auftraggeschweißte Teile, bei deren Herstellung die notwendigen qualitätssichernden Maßnahmen beachtet werden, bewähren sich im praktischen Einsatz außerordentlich gut. Hierdurch konnten 2- bis 4fach höhere Standzeiten erreicht werden, abhängig von der Abrasivität des Mahlguts.

Die im Vergleich zu herkömmlichen Mahlwalzen erzielbare Standzeitverlängerung durch Auftragschweißen erwies sich als besonders vorteilhaft bei stark abrasivem Mahlgut.

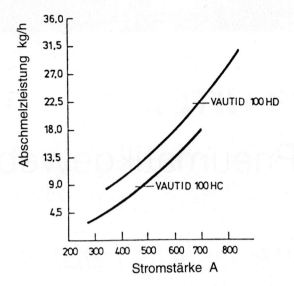

BILD 6: Vergleich der Abschmelzleistung von **VAUTID 100 HD** und **100 HC** in Abhängigkeit von der Stromstärke
FIGURE 6: Comparison of the deposition efficiencies of **VAUTID 100 HD** and **VAUTID 100 HC** as a function of current strength

Abschmelzleistung kg/h = deposition efficiency kg/h
Stromstärke A = current strength A

can now be managed technically, so the working surface is completely independent of the material properties of the base material. It is therefore only the wear behaviour of the pure weld material which is of interest. The development has led to the situation where new types of material for the base material of the grinding rollers containing further ultra-hard carbides in addition to the chromium carbides are hardly expected to produce any more extensive increases in service life. Build-up welding is also becoming increasingly less expensive through the use of thicker core wires with higher deposition efficiency and increasing current load (**Fig.6**). The measures described do not in fact reduce the cost of materials, but they lower the specific labour costs per kilogramme of deposited material, and these labour costs make up at least 2/3 of the market price. Parts with build-up welding where the necessary quality assurance measures were observed during production have proved exceptionally successful in practical use. Service lives which are 2 to 4 times longer can be achieved, depending on the abrasiveness of the material being ground.

When compared with conventional grinding rollers the extensions to the service life which can be achieved by build-up welding have proved to be particularly advantageous when grinding strongly abrasive material.

Wir montieren Industrieanlagen in aller Welt.

Aus unserem Leistungsprogramm: Montagen von schlüsselfertigen Gesamtanlagen, Durchführung und Überwachung von Einzelmontagen aller Art sowie Engineering, Lieferung und Montage von fördertechnischen Anlagen für: Kohle-, Erz- und Salzbergbau, Steine und Erden, Hütten-, Zement- und Papierindustrie, Chemie und Petro-Chemie sowie für die Müllverwertung und Kraftwerke; Projektierung, Lieferung, Vorfertigung und Montage von Rohrleitungen in Stahl und Kunststoff, Behälterbau und Stahlbaumontagen, Umbau-, Reparatur- und Wartungsarbeiten, Demontagen, Krangestellungen; Anlagen-Betriebsführungs- und Managementaufgaben, Entsorgungsdienste für Schlammentwässerung und Abwasserreinigung, Betreibung von Energieversorgungsanlagen, Werkinstandhaltung und Werkstruktur-Planung.

Standort Bochum
Herner Str. 299, D-44809 Bochum
Telefon 0234 / 539-0, Fax 0234 / 539-130
Standort Köln
Deutz-Mülheimer Str. 216, D-51063 Köln
Telefon 0221/822-3501, Fax 0221/822-3503

EIMA
Maschinenbau- und Förderanlagen GmbH

Schneckenflügel
Förderschnecken
Becherwerke
Förderbänder
Behälter
Stahlkonstruktionen

Edewechter Str. 15 – 26160 Bad Zwischenahn-Ekern
Fernruf: 0 44 03/10 62 – Telefax: 0 44 03/5 85 59

Tribol
A BURMAH CASTROL COMPANY

Tribol GmbH
P.O. Box 500210
D-41172 Mönchengladbach
Tel.: 02161-909-30
Fax: 02161-909-400

Partner of the Industry

Whenever you are looking for economical lubrication of your equipment – whether it be crushers or mobile plant equipment – your partner is Tribol. We fight friction losses ... worldwide – reducing costs with Tribol and Molub-Alloy High Performance Lubricants backed up by professional lubriccation management services. You can count on and with us!

Molub-Alloy®
Tribol®

Central-Chain
for high output bucket elevators

- new technical science
- mounting without tools
- increased wear volume
- longer lifetime
- robust and multi-functional

RUD-Kettenfabrik
Rieger & Dietz GmbH u. Co.
D-73428 Aalen
Phone 0 73 61 / 50 40
Facsimile 0 73 61 / 504-450
Germany

RUD CHAINS PTY. LTD.
P.O. Box 536
Archerfield, Queensland 4108
Phone (07) 2 74 36 66
Facsimile (07) 2 74 37 77
Australia

RUD Correntes Industriais Ltda.
Caixa Postal 2666
08780-990 Mogi das Cruzes SP
Phone (011) 4 61 29 44
Facsimile (011) 3 37 70
Brazil

RUD CHAINS LTD.
John Wilson Business Park
Units 10–12, Thanet Way
Whitstable, Kent CT5 3QT
Phone (02 27) 27 66 11
Facsimile (02 27) 27 65 86
Great Britain

RUD CHAIN INC.
1300 Stoney Point Road S.W.
P.O. Box 8145
Cedar Rapids, Iowa 52408
Phone (319) 3 90 40 40
Facsimile (319) 3 90 33 42
USA

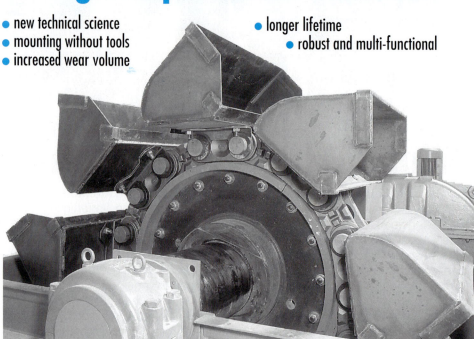

INDUMONT
INDUSTRIE-MONTAGE GMBH
Ein KHD-Unternehmen

Wir montieren Industrieanlagen in aller Welt.

Aus unserem Leistungsprogramm: Montagen von schlüsselfertigen Gesamtanlagen, Durchführung und Überwachung von Einzelmontagen aller Art sowie Engineering, Lieferung und Montage von fördertechnischen Anlagen für: Kohle-, Erz- und Salzbergbau, Steine und Erden, Hütten-, Zement- und Papierindustrie, Chemie und Petro-Chemie sowie für die Müllverwertung und Kraftwerke; Projektierung, Lieferung, Vorfertigung und Montage von Rohrleitungen in Stahl und Kunststoff, Behälterbau und Stahlbaumontagen, Umbau-, Reparatur- und Wartungsarbeiten, Demontagen, Krangestellungen; Anlagen-Betriebsführungs- und Managementaufgaben, Entsorgungsdienste für Schlammentwässerung und Abwasserreinigung, Betreibung von Energieversorgungsanlagen, Werkinstandhaltung und Werkstruktur-Planung.

Standort Bochum
Herner Str. 299, D-44809 Bochum
Telefon 0234 / 539-0, Fax 0234 / 539-130
Standort Köln
Deutz-Mülheimer Str. 216, D-51063 Köln
Telefon 0221/822-3501, Fax 0221/822-3503

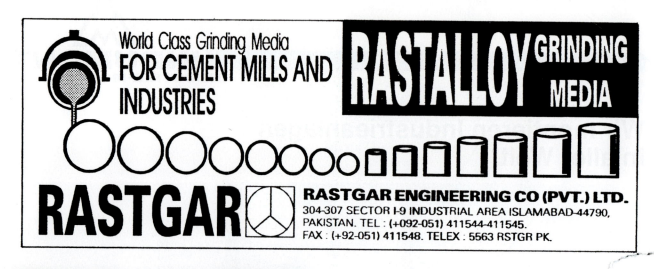

RASTGAR ENGINEERING CO (PVT.) LTD.
304-307 SECTOR I-9 INDUSTRIAL AREA ISLAMABAD-44790,
PAKISTAN. TEL : (+092-051) 411544-411545.
FAX : (+92-051) 411548. TELEX : 5563 RSTGR PK.

EIMA
Maschinenbau- und Förderanlagen GmbH

- Schneckenflügel
- Förderschnecken
- Becherwerke
- Förderbänder
- Behälter
- Stahlkonstruktionen

Edewechter Str. 15 – 26160 Bad Zwischenahn-Ekern
Fernruf: 0 44 03/10 62 – Telefax: 0 44 03/5 85 59

Tribol
A BURMAH CASTROL COMPANY

Tribol GmbH
P.O. Box 500210
D-41172 Mönchengladbach
Tel.: 02161-909-30
Fax: 02161-909-400

Partner of the Industry

Whenever you are looking for economical lubrication of your equipment – whether it be crushers or mobile plant equipment – your partner is Tribol. We fight friction losses ... worldwide – reducing costs with Tribol and Molub-Alloy High Performance Lubricants backed up by professional lubriccation management services. You can count on and with us!

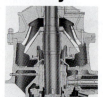

Molub-Alloy®
Tribol®

RUD

Central-Chain
for high output bucket elevators

- new technical science
- mounting without tools
- increased wear volume
- longer lifetime
- robust and multi-functional

Germany
RUD-Kettenfabrik
Rieger & Dietz GmbH u. Co.
D-73428 Aalen
Phone 0 73 61 / 50 40
Facsimile 0 73 61 / 504-450

Australia
RUD CHAINS PTY. LTD.
P.O. Box 536
Archerfield, Queensland 4108
Phone (07) 2 74 36 66
Facsimile (07) 2 74 37 77

Brazil
RUD Correntes Industriais Ltda.
Caixa Postal 2666
08780-990 Mogi das Cruzes SP
Phone (011) 4 61 29 44
Facsimile (011) 3 37 70

Great Britain
RUD CHAINS LTD.
John Wilson Business Park
Units 10–12, Thanet Way
Whitstable, Kent CT5 3QT
Phone (02 27) 27 66 11
Facsimile (02 27) 27 65 86

USA
RUD CHAIN INC.
1300 Stoney Point Road S.W.
P.O. Box 8145
Cedar Rapids, Iowa 52408
Phone (319) 3 90 40 40
Facsimile (319) 3 90 33 42

Complete storage and despatch systems for conveying, storing, filling, distribution, packaging and loading

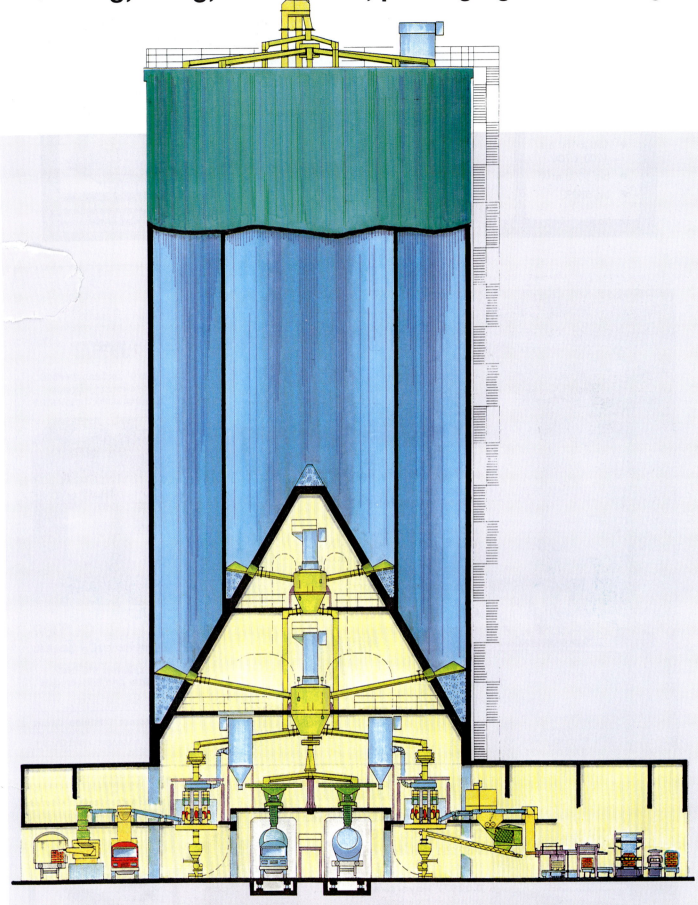

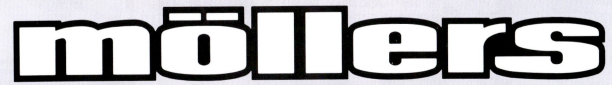

Maschinenfabrik Möllers GmbH u. Co. · 59247 Beckum/Germany · P.O. Box 17 64
Phone (0 25 21) 88-0 · Fax (0 25 21) 88-100 · Telex 89 423

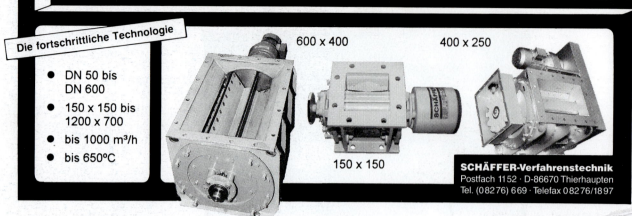

Pneumatische Förderung: Hoch- und Niederdruck aus einem Verdichter

Rotationsverdichter von Mannesmann Demag Verdichter Wittig machen es möglich: sie lassen sich auf verschiedene Betriebsdrücke umschalten. So brauchen Sie nur noch einen Reserveverdichter, um sowohl Ihr Niederdrucknetz für die pneumatische Förderung als auch Ihr Normaldrucknetz für die Werksluft jederzeit zuverlässig versorgen zu können.

Das vollautomatische Regelsystem sorgt für sparsamen Umgang mit Energie, und der integrierte Nachkühler bietet die Gewähr für trockene, technisch ölfreie Qualitätsluft.

Gerne informieren wir Sie ausführlich.

mannesmann *technologie*

Mannesmann Demag Verdichter
Wittig
Johann-Sutter-Straße 6-8, 79650 Schopfheim
Tel. (0 76 22) 3 94-130, Fax (0 76 22) 39 42 00

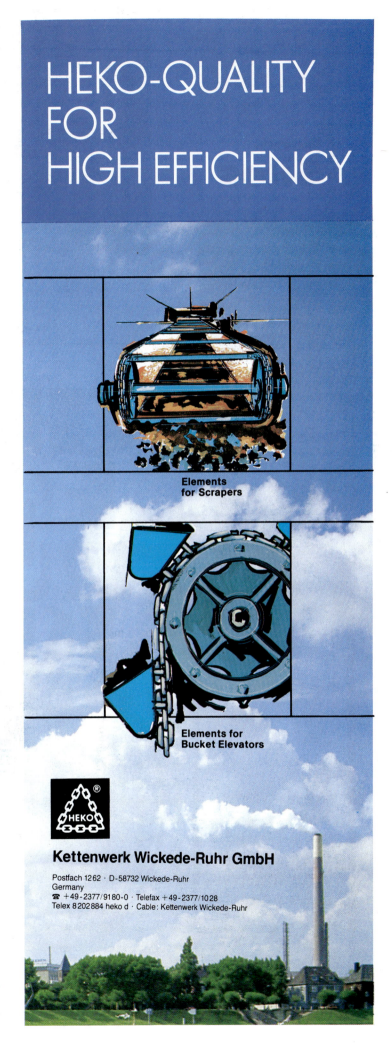

Norm-Silos

für staubförmige und körnige Schüttgüter aller Art

STAHLBLECH-SILOS · STAHLBLECH-SILOS

Gebr. Uhr
Apparate- und Behälterbau
GmbH & Co. · Postfach 3069
D-59 313 Ennigerloh
Telefax 02524/9302-30
Telefon 02524/9302-0

Becherwerke
Schneckenförderer
Doppelwellenmischer
Zellenradschleusen
Flachschieber
Losebelade-Garnituren

Erhard Russig
Fördertechnik
ERU GmbH & Co.

D-59269 Beckum · Siemensstraße 32
Fax 02521/13621 · ☎ 02521/14091

DOSIERROTORWAAGE URW
Unser technischer Beitrag zum Umweltschutz
ROTOR SCALE URW
Our technical contribution to environmental protection

Additive, Kohlenstaub, Flugaschen, Klärschlämme u. a. werden kontinuierlich gravimetrisch mit der Dosierrotorwaage URW direkt pneumatisch in den Prozeß dosiert.

Additives, coal dust, flying ashes, sewage sludges and others will be transported continuously and gravimetrically with the rotor scale URW and then directly fed pneumatically into the process.

PFISTER
WÄGEN · DOSIEREN · STEUERN

PFISTER GmbH · Postf. 410120, D-86068 Augsburg · Stätzlinger Str. 70, D-86165 Augsburg · Tel. (0821) 7949-0 · Fax (0821) 7949-270

CHAINS FOR ROTARY KILNS

ELEVATOR CHAINS

Heinrich Prünte
Fröndenberger Kettenfabrik
GmbH & Co KG
P.O. BOX 1462, D-58720 FRONDENBERG/GERMANY

telex: 8 202 626 fkhp
phone: (0 23 73) 7 00 48
fax: (0 23 73) 7 00 50

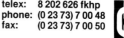

CHAINS FOR THE WORLD

Estb. 1887